The Molecular Pathology of Autoimmune Diseases
Second Edition

The Molecular Pathology ofAutoimmune Diseases

Second Edition

Edited by

Argyrios N. Theofilopoulos
The Scripps Research Institute
La Jolla, California, USA

Constantin A. Bona
Mount Sinai School of Medicine
New York, USA

USA	Publishing Office:	TAYLOR & FRANCIS
		29 West 35th Street
		New York, NY 10001
		Tel: (212) 216-7800
		Fax: (212) 564-7854
	Distribution Centre:	TAYLOR & FRANCIS
		7625 Empire Drive
		Florence, KY 41042
		Tel: 1-800-634-7064
		Fax: 1-800-248-4724
UK		TAYLOR & FRANCIS
		11 New Fetter Lane
		London
		EC4P 4EE
		Tel: +44 (0) 20 7583 9855
		Fax: +44 (0) 20 7842 2298

THE MOLECULAR PATHOLOGY OF AUTOIMMUNE DISEASES: 2nd Edition

1 2 3 4 5 6 7 8 9 0

Printed by Sheridan Books, Ann Arbor, MI.

The paper in this publication meets the requirements of the ANSI Standard Z39.48-1984 (Permanence of Paper).

British Library of Congress Cataloguing-in-Publication Data

A CIP catalog record for this book is available from the British Library.

ISBN 90-5702-645-7

COVER ART: Crystal structure of a T cell receptor in complex with peptide/MHC (Courtesy of Luc Teyton and Ian Wilson, The Scripps Research Institute)

CONTENTS

SECTION II. SYSTEMIC AUTOIMMUNE DISEASES

SECTION III: ORGAN-SPECIFIC AUTOIMMUNE DISEASES

Hematologic Autoimmune Diseases

Endocrine Autoimmune Diseases

CNS Autoimmune Diseases

Autoimmune Diseases of the Muscle

Autoimmune Diseases of the Skin

Other Organ-Specific Autoimmune Diseases

SECTION IV: ASSOCIATED AUTOIMMUNITY

SECTION V: IMMUNOTHERAPEUTIC APPROACHES

CONTRIBUTORS

Yashushi Adachi, MD, PhD
First Department of Pathology, Transplantation Center
Kansai Medical University, 10–15 Fumizono-cho
Moriguchi City, Osaka 570-8506, Japan

David Adams, MD
Liver Research Laboratories
The University of Birmingham Clinical Research Block
Queen Elizabeth Hospital, Edgbaston
Birmingham B15 2TH, UK

Luciano Adorini, MD
Roche Milano Ricerche
Via Olgettina 58, I-20132
Milano, Italy

Marina Afanasyeva, MD, MPH
Department of Molecular Microbiology and Immunology
Department of Pathology
The Johns Hopkins University
School of Medicine
615 Wolfe St.
Baltimore, MD 21205, USA

Iannis Aifantes, PhD
Department of Cancer Immunology and AIDS
Harvard Medical School
Dana-Farber Cancer Institute
44 Binney St.
Boston, Massachusetts 02115, USA

Frank Alderuccio, PhD
Department of Pathology and Immunology
Monash University Medical School
Commercial Rd., Prahran
Victoria 3181, Australia

Amnon Altman, PhD
Division of Cell Biology
La Jolla Institute for Allergy and Immunology
10355 Science Center Dr.
San Diego, California 92121, USA

Alberto Amadori, MD
Institute of Oncology, University of Padua
Department of Oncology and Surgical Sciences
Via Gattemelata 64, I-35128
Padua, Italy

Ana C. Anderson, PhD
Department of Molecular and Cellular Biology
University of California at Berkeley
471 Life Science Addition
Berkeley, California 94720, USA

David E. Anderson, PhD
Department of Medical Microbiology and Immunology
University of California at Davis School of Medicine
One Shields Ave.
Davis, California 95616-8645, USA

Frank C. Arnett, MD
Division of Rheumatology and Clinical Immunogenetics
NIAMS Specialized Center of Research in Scleroderma
Department of Internal Medicine
The University of Texas Medical School at Houston
6431 Fannin, Room 5.270
Houston, Texas 77030, USA

Luis A. Arteaga, MD
Department of Dermatology
University of North Carolina at Chapel Hill
3100 Thurston Bldg., CB#7287
Chapel Hill, North Carolina 27599, USA

Roberto Baccala, PhD
Division d'Hématologie
Centre Hospitalier Universitaire Vaudois (CHUV)
1011 Lausanne, Switzerland

Jean-François Bach, MD
Unite de Recherches de L'INSERM U 25
Maladies Auto-Immunes: Genetique Mechanismes et
Traitements Centre de L'Association
Claude Bernard Immunologie Clinique
Hopital Necker, 161, Rue de Sevres, 75743
Paris, France

Amit Bar-Or, MD
Center for Neurologic Diseases
Brigham and Women's Hospital and
Harvard Medical School
77 Avenue Louis Pasteur, HIM 780
Boston, Massachusetts 02115, USA

Raymond E. Boissy, PhD
Department of Dermatology
University of Cincinnati College of Medicine
231 Bethesda Ave., ML 0592
Cincinnati, Ohio 45267-0592, USA

Dorin-Bogdan Borza, PhD
Department of Biochemistry and Molecular Biology
The University of Kansas Medical Center
3901 Rainbow Blvd.
Kansas City, Kansas 66160, USA

Constantin A. Bona, PhD
Mount Sinai School of Medicine
Department of Microbiology
One Gustav L. Levy Place
Annenberg Building
New York, New York 10029, USA

Dimitrious Boumpas, MD
Division of Rheumatology
Clinical Immunology and Allergy
University of Crete Medical School
715 00 Hereklion
Crete, Greece

Christopher D. Buckley, MRCP, PhD
Division of Immunity and Infection (Rheumatology)
Birmingham Centre for Immune Regulation
University of Birmingham
Edgbastion, Birmingham B152TH, UK

Jan Buer, MD
Mucosal Immunity Group, GBF
National Research Center for Biotechnology
Braunschweig, Germany

Michael C. Carroll, PhD
Department of Pediatrics, The Center for Blood Research
Harvard Medical Center, 200 Longwood Ave.
LMRC, Rm. 502
Boston, Massachusetts 02115, USA

Rachel R. Caspi, PhD
National Eye Institute
National Institutes of Health
10 Center Dr., Bldg. 10, Rm. 10N222
Bethesda, Maryland 20892, USA

Chi-Chao Chan, PhD
National Eye Institute
National Institutes of Health
10 Center Dr., Bldg. 10, Rm. 10N103
Bethesda, Maryland 20892, USA

Jeannine Charreire, Dr Sc
INSERM U477
Unite de Recherches sur La Pathologie Auto-Immune
Université René Descartes
27 Rue du Faubourg Saint-Jacques
75014 Paris, Cedex 14, France

Qiao-Yi Chen, MD, PhD
Research Institute for Children
Department of Pediatrics
Louisiana State University Health Science Center
1542 Tulane Ave.
New Orleans, LA 70112, USA

Fulvia Chieco-Bianchi, MD
Division of Rheumatology
Department of Medical and Surgical Sciences
University of Padua, Via Giustinian 2, I-35128
Padua, Italy

Premkumar Christadoss, MD
Department of Microbiology and Immunology
University of Texas Medical Branch
301 University Boulevard
3-142A Medical Research Bldg.
Galveston, Texas 77555-1070, USA

Irun R. Cohen, MD
Department of Immunology
The Weizmann Institute of Science
Rehovot, 76100, Israel

Madeleine W. Cunningham, PhD
Department of Microbiology and Immunology
University of Oklahoma Health Sciences Center
P.O. Box 26901
Oklahoma City, Oklahoma 73104, USA

Gareth Davies, BDS, MScD
School of Oral Health Sciences
University of Western Australia
Nedlands, WA 6009, Australia

Terry F. Davies, MD
Mount Sinai School of Medicine
One Gustave L. Levy Pl.
Box 1055
New York, New York 10029, USA

Caishu Deng, MD
Department of Microbiology and Immunology
University of Texas Medical Branch
MRB 3.124
Galveston, Texas 77555-1070, USA

Gregory A. Denomme, PhD
Department of Laboratory Medicine and Pathobiology
Faculty of Medicine
University of Toronto Mount Sinai Hospital
University of Toronto
600 University Avenue, Suite 600
Toronto, Ontario M5G 1X5, Canada

Luis A. Diaz, MD
Department of Dermatology
University of North Carolina at Chapel Hill
3100 Thurston Bldg., CB#7287
Chapel Hill, North Carolina 27599, USA

Henrik J. Ditzel, MD, PhD
Department of Immunology
The Scripps Research Institute
10550 North Torrey Pines Rd.
La Jolla, California 92037, USA

Thomas Dörner, MD
Department of Rheumatology
University Hospital Charite
Berlin, Germany

George S. Eisenbarth, MD, PhD
Barbara Davis Center for Childhood Diabetes
Department of Pediatrics, Medicine & Immunology
University of Colorado Health Sciences Center
4200 E. 9th Ave., Box B140
Denver, Colorado 80262, USA

Keith B. Elkon, MD
Hospital for Special Surgery
Weill Medical College of Cornell University
535 E. 70th St.
New York, New York 10021-4872, USA

Sidonia Fagarasan, MD
Department of Medical Chemistry and Molecular
Biology
Graduate School of Medicine
Kyoto University Faculty of Medicine
Yoshida, Sakyo-Ku
Kyoto 606-8501, Japan

Robert I. Fox, MD, PhD, FACP
Allergy and Immunology Clinic
Scripps Memorial Hospital
9850 Genesee Ave., Suite 860
La Jolla, California 92037, USA

Anke Franzke, MD
Medinischke Hochschule Hannover
Dept. of Hematology and Oncology
Hannover, Germany

Francesco Fusi, MD
Department of Obstetrics and Gynecology
Institute Scientific San Raffaele
Via Olgettina 60
Milano, Italy

Renate E. Gay, MD
WHO Collaborating Center for
Molecular Biology and Novel Therapeutic Strategies
for Rheumatic Diseases
University Hospital, Rheumatology Clinic
Gloriastr. 25, CH-8091
Zurich, Switzerland

Steffen Gay, MD
WHO Collaborating Center for
Molecular Biology and Novel Therapeutic Strategies
for Rheumatic Diseases
University Hospital, Rheumatology Clinic
Gloriastr. 25, CH-8091
Zurich, Switzerland

John A. Gebe, PhD
Virginia Mason Research Center
1201 Ninth Ave.
Seattle, Washington 98101-2795, USA

M. Eric Gershwin, MD
Division of Rheumatology, Allergy and Clinical
Immunology
University of California at Davis
School of Medicine
One Shields Ave., TB 192
Davis, California 95616-8660, USA

Igal Gery, PhD
National Eye Institute
National Institutes of Health
10 Center Dr., Bldg. 10, Rm. 10N208
Bethesda, Maryland 20892-0001, USA

Rosana Gonzalez-Quintial, PhD
Division d'Hématologie
Centre Hospitalier Universitaire
Vaudois (CHUV)
1011 Lausanne, Switzerland

Peter A. Gottlieb, PhD
Departments of Pediatrics and Medicine
University of Colorado Health Sciences Center
4200 E. 9th Ave.
Denver, Colorado 80262, USA

Amrie C. Grammer, PhD
Harold C. Simmons Arthritis Research Center and the
Department of Internal Medicine
University of Texas Southwestern Medical Center
Dallas, Texas, USA

David A. Hafler, MD
Center for Neurologic Diseases
Brigham and Women's Hospital
Harvard Medical School
77 Louis Pasteur Ave., HIM785
Boston, Massachusetts 02115, USA

Joichiro Hayashi, DDS, PhD
Department of Immunology
The Scripps Research Institute
10550 North Torrey Pines Rd.
La Jolla, California 92037, USA

Catherine P.M. Hayward, MD, PhD
Departments of Pathology and Molecular Medicine
Faculty of Health Sciences
McMaster University Medical Center, Rm. 2N32
1200 Main St. West Hamilton
Ontario L8N 3Z5, Canada

Susan L. Hill, DVM
Department of Pathology
Department of Molecular Microbiology and Immunology
The Johns Hopkins Medical Institutions
615 N. Wolfe St.
Baltimore, Maryland 21205, USA

Rikard Holmdahl, MD, PhD
Medical Inflammation Research
Lund University
Sölvegatan 19, CMB, S-223 62
Lund, Sweden

Tasuku Honjo, MD
Department of Medical Chemistry and Molecular
Biology
Graduate School of Medicine
Kyoto University, Faculty of Medicine
Yoshida Sakyo-ku, Kyoto, 606-8501, Japan

Billy G. Hudson, PhD
Department of Biochemistry and Molecular Biology
University of Kansas Medical Center School of Medicine
3901 Rainbow Blvd.
Kansas City, Kansas 66160-7421, USA

Susumu Ikehara, MD, PhD
First Department of Pathology, Transplantation Center
Kansai Medical University
10-15 Fumizono-cho
Moriguchi City, Osaka 570-8506, Japan

Fang Jin, MD
Department of Medicine, Rm. C-801
Cornell University Medical College
1300 York Ave.
New York, New York 10021, USA

Roland Jonsson, DMD, PhD
Broegelmann Research Laboratory
University of Bergen
Armauer Hansen Bldg., N-5021
Bergen, Norway

Cees G.M. Kallenberg, MD, PhD
Department of Clinical Immunology
Groningen University Hospital,
Hanzeplein 1, P.O. Box. 30.001
9713 GZ Groningen, Netherlands

Kuppuswamy N. Kasturi, PhD
Department of Microbiology
Mount Sinai Medical Center
Box 1124, One Gustav L. Levy Pl.
New York, New York 10029-6504, USA

Tomoya Katakai, PhD
Center for Molecular Biology and Genetics
Kyoto University, Sakyo-ku
Kyoto 606-8501, Japan

Azad K. Kaushik, DVM, DSc
Department of Pathobiology
University of Guelph
Guelph, Ontario N1G 2W1, Canada

John G. Kelton, MD
Department of Medicine, Faculty of Health Sciences
McMaster University Medical Center, Rm 3W10
1200 Main St. West
Hamilton, Ontario L8N 3Z5, Canada

David L. Kendler, MD
Department of Medicine
University of British Columbia
Vancouver General Hospital – Bone & Mineral Center
Vancouver, Canada

Dwight H. Kono, MD
Department of Immunology
The Scripps Research Institute
10550 North Torrey Pines Rd./IMM3
La Jolla, California 92037, USA

Valerie Kouskoff, PhD
Institute for Gene Therapy and Medicine
Mount Sinai School of Medicine
One Gustav L. Levy Pl., Box 1496
New York, New York 10029-6574, USA

Steven A. Krilis, PhD
Department of Immunology, Allergy & Infectious Diseases
The St. George Hospital, University of NSW
1st Floor, 2 South Street
Kogarah NSW 2217, Australia

Vijay K. Kuchroo, DVM, PhD
Center for Neurologic Diseases
Harvard Medical School
Brigham and Women's Hospital
77 Louis Pasteur Avenue, HIM, Rm. 706
Boston, Massachusetts 02115-5817, USA

Thomas J. Kunicki, PhD
Department of Molecular and Experimental Medicine
Division of Hemostasis and Thrombosis
The Scripps Research Institute
10550 N. Torrey Pines Rd./MEM-150
La Jolla, California 92037, USA

Yoshiyuki Kurata, MD, PhD
Department of Blood Transfusion
Osaka University Hospital
2-15 Yamadaoka
Suita City, Osaka 565-0871, Japan

Patricia Lalor, PhD
The MRC Birmingham Centre for Immune Regulation
The Liver Research Laboratories
The University of Birmingham Clinical Research Block
Queen Elizabeth Hospital
Edgbaston, Birmingham B15 2TH, UK

Astrid Lanoue, PhD
Medical Research Cancer Laboratory of Biology
Cambridge, UK

Antonio Lanzavecchia, MD
Institute for Research in Biomedicine
Via Vela 6.,
6500-Bellinzona, Switzerland

Michael W. Leach, DVM, PhD, DACVP
Schering-Plough Research Institute
P.O. Box 32, 144 Route 94
Lafayette, New Jersey 07848, USA

Oskar Lechner, PhD
INSERM U373, Faculte de Medecine Necker,
156 rue de Vaugirard,
75015 Paris, France

Joong-Won Lee, MD
Department of Pathology, Rm. C352
Cornell University Medical College
1300 York Ave.
New York, New York 10021, USA

Mong-Shang Lin, PhD
Department of Dermatology
University of North Carolina at Chapel Hill
3100 Thurston Bldg., CD#7287
Chapel Hill, North Carolina 27599, USA

Peter E. Lipsky, MD
National Institute of Arthritis and
Muskuloskeletal Diseases
National Institutes of Health
9000 Rockville Pike, Bldg., 10, Rm. 9N228
Bethesda, Maryland 20892-1820, USA

Noel Maclaren, MD
Department of Pediatrics
Weill Medical College at Cornell University
525 E. 68th St., Rm. LC-604
New York, New York 10021, USA

Michael P. Manns, MD
Department of Gastroenterology and Hepatology
Medical School of Hannover
Carl-Neuburg Str. 1, D-30625
Hannover, Germany

Andreas Martin, MD
Department of Medicine
Mount Sinai School of Medicine
One Gustav L. Levy Pl.
New York, New York 10029, USA

Alexander Marx, MD
University of Wurzburg
Pathology Institute
97080 Wurzburg Luitpoldkrankenhaus
Josef-Schneider-Strasse 2, Germany

Tohru Masuda, MD
Kyoto Business Center of Postal Life Insurance
Ministry of Postal Service, Matsugasaki
Sakyo-ku, Kyoto 606-8792, Japan

Lloyd Mayer, MD
Immunobiology Center
Mount Sinai Medical Center
1425 Madison Av., Rm. 11-20C
New York, New York 10029-6564, USA

Sandra M. McLachlan, MD
Cedars-Sinai Medical Center
8700 Beverly Blvd., B-131
Los Angeles, California 90048-1865, USA

Doron Melamed, PhD
Department of Immunology
Bruce Rappaport School of Medicine
Bat-Galim, Haifa 31096, Israel

Paul Michelson, MD
Department of Ophthalmology
Scripps Health Clinical and Research Foundation
9850 Genesee Ave.
La Jolla, California 92037, USA

Hajime Mizutani, MD, PhD
Internal Medicine
Jawasaki Hospital
3-3-1 Higashiyama-cho
Hoyogo-ku, Kobe City
Hyogo-ken 652-0052, Japan

Tomas Mustelin, MD, PhD
The Burnham Institute
La Jolla Cancer Research Center
10901 North Torrey Pines Rd.
La Jolla, California 92037, USA

Gerard B. Nash, PhD
Department of Physiology
University of Birmingham
Edgbaston, Birmingham B1552TH, UK

David Nemazee, PhD
Department of Immunology
The Scripps Research Institute
10550 N. Torrey Pines Rdr./IMM29
La Jolla, California 92037, USA

Gerald T. Nepom, MD, PhD
Virginia Mason Research Center,
1201 Ninth Ave.
Seattle, Washington 98101-2795, USA

James Neuberger, DM
Liver Unit
Queen Elizabeth Hospital
Edgbaston, Birmingham, UK

Lindsay B. Nicholson, MB, BCh, PhD
Center for Neurologic Diseases
Harvard Medical School
Brigham and Women's Hospital
77 Louis Pasteur Avenue
HIM, Rm. 706
Boston, Massachusetts 02115-5817, USA

Akiyoshi Nishio, MD
Department of Gastroenterology
Tenri Hospital
Nara, Japan

Diane J. Nugent, MD
Division of Hematology Oncology
Children's Hospital of Orange County
455 So. Main St.
Orange, California 92868, USA

Robert B. Nussenblatt, MD
National Eye Institute
National Institutes of Health
10 Center Dr., Bldg. 10, Rm. 10N202
Bethesda, Maryland 20892-0001, USA

Petra Obermayer-Straub, PhD
Department of Gastroenterology and Hepatology
Medical School of Hannover
Carl-Neubergstr.1
D-30625 Hannover, Germany

Thomas Pap, MD
WHO Collaborating Center for
Molecular Biology and Novel Therapeutic Strategies
for Rheumatic Diseases
University Hospital, Rheumatology Clinic
Gloriastr. 25
CH-8091 Zurich, Switzerland

Elahna Paul, MD, PhD
Division of Pediatric Nephrology
Children's Hospital and Center for Blood Research
Harvard Medical Center
200 Longwood Ave.
Boston, Massachusetts 02115, USA

Kathleen N. Potter, PhD
Molecular Immunology Group, Tenovus Laboratory
Southampton University Hospitals Trust
Southampton S016 6YD, UK

Mathilde A. Poussin, PhD
Department of Microbiology and Immunology
University of Texas Medical Branch
MRB 3.124
Galveston, Texas 77555-1070, USA

Fiona Powrie, DPhil
University of Oxford
John Radcliffe Hospital
Nuffield Department of Surgery
Headington, Oxford, OX3 9DU, UK

Patricia Price, MSc, PhD
Department of Pathology
University of Western Australia, and
Department of Clinical Immunology
Royal Perth Hospital
WA 6000, Australia

Basil Rapoport, MB, ChB
Cedars-Sinai Medical Center
Beverly Blvd. Suite B-131
Los Angeles, California 90048-1865, USA

Stephen W. Reddel, PhD
Research Lab, Kensington St.
The St. George Hospital
Kogarah Sydney, 2217, Australia

Noel R. Rose, MD, PhD
Department of Immunology & Infectious Diseases
The Johns Hopkins University
School of Hygiene and Public Health
615 N. Wolfe St.
Baltimore, Maryland 21205, USA

Tsuyoshi Sakane, MD, PhD
Departments of Immunology and Medicine
St. Marianna University School of Medicine
2-16-1 Sugao Miyamae-ku
Kawasaki, 216-8511, Japan

Federica Sallusto, PhD
Basel Institute for Immunology
Grenzacherstrasse 487
4005-Basel, Switzerland

Mike Salmon, PhD
The Division of Immunity and Infection (Rheumatology)
The MRC Birmingham Centre for Immune Regulation
University of Birmingham
Edgbaston, Birmingham B152TH, UK

Adelaida Sarukhan, PhD
INSERM U373, Faculte de Medecine Necker
156 rue de Vaugirard,
75015 Paris, France

Stephen P. Schoenberger, PhD
Division of Immune Regulation
La Jolla Institute for Allergy and Immunology
10355 Science Center Dr.
San Diego, California 92121, USA

John Sentry, PhD
Department of Pathology and Immunology
Monash University Medical School
Commercial Rd.
Prahran, Victoria 3818, Australia

Eli E. Sercarz, PhD
Division of Immune Regulation
La Jolla Institute for Allergy and Immunology
10355 Science Center Dr.
San Diego, California 92121, USA

Akira Shimizu, MD
Center for Molecular Biology and Genetics
Kyoto University
Sakyo-ku, Kyoto 606-8507, Japan

Andrei Shustov, MD
Department of Medicine
Division of Rheumatology
University of Maryland at Baltimore
School of Medicine
10 So. Pine St., MSTF Bldg., Rm. 813
Baltimore, Maryland 21201, USA

Francesco Sinigaglia, MD
Roche Milano Richerche
Via Olgettina 58, I-20132
Milano, Italy

Kathrine Skarstein, DDS, PhD
Broegelmann Research Laboratory
University of Bergen
Armauer Hansen Bldg.
N-5021, Bergen, Norway

Freda K. Stevenson, DPhil
Molecular Immunology Group, Tenovus Laboratory
Southampton University Hospitals Trust
Southampton S016 6YD, UK

Christian P. Strassburg, MD
Department of Gastroenterology and Hepatology
Medical School of Hannover
Carl-Neubergstr. 1
D-30625 Hannover, Germany

Thomas Stratmann, PhD
Department of Immunology
The Scripps Research Institute
10550 N. Torrey Pines Rd.
La Jolla, California 92037, USA

Noboru Suzuki, MD, PhD
Departments of Immunology and Medicine
St. Marianna University School of Medicine
2-16-1 Sugao, Miyamae-ku
Kawasaki, 216-8511, Japan

J.W. Cohen Tervaert, MD, PhD
Department of Clinical Immunology
University Hospital Maastricht
P.O. Box 616
6200 MD Maastricht, Netherlands

Cory Teuscher, PhD
Department of Veterinary Pathobiology
University of Illinois at Urbana-Champaign
2001 So. Lincoln Ave.
Urbana, Illinois 61801, USA

Luc Teyton, MD, PhD
Department of Immunology
The Scripps Research Institute
10550 N. Torrey Pines Rd.
La Jolla, California 92037, USA

Argyrios N. Theofilopoulos, MD
Department of Immunology
The Scripps Research Institute
10550 North Torrey Pines Rd./IMM3
La Jolla, California 92037, USA

Ban-Hock Toh, MBBS, DSC, FRACP, FRCPA
Department of Pathology and Immunology
Monash University Medical School
Alfred Hospital Campus
Commercial Rd. Prahran
Victoria 3181, Australia

George C. Tsokos, MD
Department of Medicine
Division of Rheumatology and Immunology
Uniformed Services University of the Health Sciences
Department of Cellular Injury
Walter Reed Army Institute of Research
Bldg. 503, Rm. 1A32
Silver Spring, Maryland 20910-7500, USA

Kenneth S.K. Tung, MD
Department of Pathology and Microbiology
University of Virginia Health Science Center
Box 800214
Charlottesville, Virginia 22908-0001, USA

Charles S. Via, MD
Department of Medicine
Division of Rheumatology
University of Maryland at Baltimore
School of Medicine
10 South Pine St., MSTF. 8-34
Baltimore, Maryland 21201-1116, USA

Harald Von Boehmer, MD, PhD
Department of Cancer Immunology and AIDS
Harvard Medical School
Dana-Farber Cancer Institute
44 Binney St.
Boston, Massachusetts 02115, USA

Ulrich Walter, PhD
INSERM U373
Faculte de Medecine Decker
156 rue de Vaugirard
75015 Paris, France

Simon J.P. Warren, MD
Department of Dermatology
Medical College of Wisconsin
8701 Watertown Plank Rd.
Milwaukee, Wisconsin 53226-3548, USA

Nirohiko Watanabe, MD
Department of Medical Chemistry and
Molecular Biology
Graduate School of Medicine
Kyoto University Faculty of Medicine
Yoshida, Sakyo-ku
Kyoto 606-8501, Japan

Mark E. Weksler, MD
Department of Medicine
Division of Geriatrics and Gerontology
Cornell University Medical College
New York, New York 10021, USA

Georg S. Wick, MD
Institute for Biomedical Aging Research
Austrian Academy of Sciences
Rennweg 10, 6020
Innsbruck, Austria
and
Institute for General and Experimental Pathology
University of Innsbruck
Fritz-Pregl-Strasse 3
6020 Innsbruck, Austria

Curtis B. Wilson, PhD
Department of Immunology
The Scripps Research Institute
10550 North Torrey Pines Rd./1mm5
La Jolla, California 92037, USA

Xu, Qingbo, MD PhD
Department of Experimental Pathology
Austrian Academy of Sciences
Institute for Biomedical Aging Research
Rennweg 10, A-6020, Austria

Carole Zober
INSERM U373, Faculte de Medecine Necker
156 rue de Vaugirard
75015 Paris, France

PREFACE

Since the first edition of this book, research on the pathogenesis of autoimmunity has continued unabated, with striking progress. Every year there are numerous conferences devoted to the topic, and almost every issue of the major immunologic journals contains several articles that attempt to provide another piece to this continuing puzzle. As a consequence, significant advances have been made in both basic and clinical aspects, particularly in elucidating the molecular and genetic basis of autoimmune diseases. The purpose of this book remains the consolidation of this information into a concise and centralized resource.

The second edition includes several thoroughly re-written or updated chapters, as well as many new chapters. The overall organization has been maintained, with basic aspects preceding the chapters on clinical entities, the latter addressed separately from the human and, whenever possible, the corresponding animal model perspective. In this way, the reader can juxtapose the available information and reciprocally address unanswered questions.

The success of a book such as this is almost exclusively dependent on the caliber of its contributors, and we are very fortunate to have world-recognized experts who set aside other matters and enthusiastically took up this demanding task. We would like to thank all of the authors for their truly outstanding contributions in creating a repository of timely and deeply relevant data.

We hope that this text will provide a foundation from which a new generation of scientists will build upon the accomplishments of the clinical and basic scientists before them who have devoted themselves to unraveling the mysteries of autoimmune diseases.

The editors wish to acknowledge the indispensable contributions of M. Kat Occhipinti–Bender, assistant to Dr. Theofilopoulos, who played a central role in the preparation of this text, including chapter editing, project organization and author liaison. We also thank our publisher, Taylor & Francis, as well as Ed Cilurso, Production Director. Argyrios Theofilopoulos also wishes to express his appreciation for the loving support and patience of his wife Ellie and his children, Aliki, Dimitri and Andreas.

Argyrios N. Theofilopoulos
Constantin A. Bona

1 | Molecular Pathology of Autoimmune Diseases: An Overview

Argyrios N. Theofilopoulos

1. INTRODUCTION

The function of the normal immune system is to combat foreign invaders but, under certain circumstances, this protective system attacks the very entity it is supposed to be protecting. The consequence of such an aberrant response, which can be prolonged because the stimulus is continuous, is the development of chronic inflammatory responses and a wide range of diseases (Table 1). In humans, these autoimmune disorders have a high prevalence, generally afflict women more often than men, often strike at the prime of life, and can be crippling or even fatal.

Based on the extent of the tissues involved, these diseases can be divided into two broad categories: Organ-specific and systemic. A further division is based on the immunopathologic mechanisms that mediate tissue damage, i.e., autoantibodies (type II, autoantibodies against cell surface or matrix antigens, and type III, immune complex-mediated) or cytotoxic T cells (type IV). These categories are not absolute, however, since organ-specific autoimmune diseases can be associated with milder manifestations in other tissues and, in some instances, the immunopathogenesis is mixed with autoantibodies and cytotoxic T cells acting in concert either at the initiation or later stages of the disease. Of course, helper T cells appear to be obligatory participants in both humoral- and cell-mediated autoimmune diseases. Although autoantibody responses usually lead to tissue destruction, in some instances autoantibodies directed against certain cell surface receptors stimulate or block specialized cellular functions without destruction of the cell, as in the case of anti-TSH receptor antibodies in Graves' disease and anti-acetylcholine receptor antibodies in myasthenia gravis. Moreover, in some presumed autoimmune diseases, such as

polymyositis, scleroderma and Wegener's granulomatosis, high levels of autoantibodies against intracellular antigens are found, but the means by which they may act to promote inflammation is not clear. Evidence has, however, been presented that autoantibodies may penetrate living cells (1,2) and intracellular antigens may translocate to the cell surface under certain circumstances (3).

An important characteristic of many autoimmune diseases is their heterogeneity in clinical presentation, which gives rise to considerable difficulties in diagnosis, prognosis and treatment. This heterogeneity may be a reflection of diverse genetic contributions that, nevertheless, lead to a more or less similar overall phenotype. Moreover, several of these diseases are characterized by spontaneous remissions and exacerbations, again for reasons that are poorly understood. These fluctuations in the course may be the result of environmental factors and/or exhaustion and re-emergence of the multiple factors that participate in an autoimmune response. Although in most instances these diseases acquire a chronic, if inconsistent, course, in some conditions the autoimmune response subsides and does not reappear, presumably due to the inhibitory effects mediated by regulatory T cells.

The growing list of diseases considered to be autoimmune in nature has often raised questions about the very definition of these entities. The mere detection of autoantibodies or the lack of a clear-cut pathogenesis are obviously insufficient criteria for determining that a disease has an autoimmune basis. This point is well-illustrated by the presence in the serum of so-called "natural" autoantibodies among otherwise normal individuals (4). The exact function of these predominantly IgM class, polyreactive, germline-encoded, low-affinity, non-pathogenic autoantibodies remains unknown. Nevertheless, it has been

Table 1   Diseases with established or postulated autoimmune origin

Disease	Proven or potential effectors	Model
A) Organ-Specific Diseases		
Myasthenia Gravis	Ab against acetylcholine receptor (AchR)	Mice immunized with AchR
Graves' Disease	Ab against thyrotrophin receptor (TSHR)	Mice immunized with fibroblasts or embryonic kidney cells transduced with TSHR and MHC class II, adoptive transfer of TSHR-primed T cells
Hashimoto's Disease	Ab against thyroglobulin (TG) and thyroid peroxidase	Spontaneous thyroiditis in BB/W rats and obese chickens, TG-immunized mice transgenic for human DR3, mice transgenic for IFN-γ under the rat TG promoter
Type 1 (Insulin-Dependent) diabetes mellitus	T cells (CD4+ and CD8+), Abs against islet cells and associated molecules (GAD, insulin, HSP, carboxypeptidase H)	Spontaneous diabetes in the NOD mouse and BB/W rat, streptozotocin-induced diabetes
Multiple Sclerosis (MS)	T cells against myelin proteins	Mice immunized with CNS myelin proteins (MBP, PLP)
Pemphigus (vulgaris, foliaceus)	Ab against desmosomal cadherins (desmoglein 3 in PV, desmoglein 1 in PF)	Passive transfer of patient sera into mice, mice immunized with desmoglein, SCID mice reconstituted with PBL of PV patients
Bullous pemphigoid/ pemphigoid gestationis	Ab against components of the hemidesmosome (BP180, BP230)	Passive serum transfer
Cicatricial pemphigoid	Ab against BP180, laminins (epiligrin)	
Dermatitis herpetiformis	Ab against endomysial antigens/ tissue transglutaminase	
Epidermolysis bullosa aquisita	Ab against type VII collagen	
Erythema multiforme	Ab against desmoplakin I and II	
Primary biliary cirrhosis	Ab against pyruvate dehydrogenase complex (PDC) and other mitochondrial antigens	Mice immunized with intact PDC, GvH
Autoimmune hepatitis	ANA, Ab against smooth muscle, liver and kidney microsomes, soluble liver antigens, and asialoglycoprotein receptor, ANCA, T cells	(NZWxBXSB)F1 (?)
Autoimmune hemolytic anemia	Ab against red blood cells	NZB, IgM or IgG anti-red blood cell transgenic mice
Idiopathic thrombocytopenic purpura	Ab against platelets	male (NZWxBXSB)F1
Primary autoimmune neutopenia	Ab against neutrophils	
Autoimmune gastritis and pernicious anemia	Ab against gastric parietal cells, intrinsic factor and H/K-ATpase	C3H/He, neonatally-thymectomized mice
Addison's Disease	Ab against adrenal cells and steroid 21-hydroxylase	Spontaneous disease in dogs, mice immunized with adrenal extracts
Idiopathic hypoparathyroidism	Ab against parathyroid cells	
Autoimmune premature ovarian failure	Ab against ovarian interstitial cells, corpus luteum, zona pellucida (ZP), 3 beta-hydroxysteroid dehydrogenase, ooplasm specific protein 1	Neonatally thymectomized mice, mice immunized with ZP
Azoospermia	Ab against sperm	
Autoimmune polyendo-crinopathy, candidiasis-ectodermal dystrophy	Mutations in the AutoImmune REgulator gene (AIRE)	

Table 1   (Continued)

Disease	Proven or potential Effectors	Model
Uveoretinitis	Ab against intracellular photoreceptor protein S-Ag (arrestin) and extracellular photoreceptor matrix protein, T cells	Rats and mice immunized with photoreceptor antigens, sensitized T cells
Vitiligo	Ab against melanocytes and tyrosine-related protein 1, T cells	Smyth line chicken, UCD-200 chicken
Autoimmune myocarditis	Ab against cardiac myosin, adenine nucleotide translocator, and branch chained ketoacid dehydrogenase, T cells	Mice immunized with cardiac myosin, Coxsackie virus B3 model, transfer of PBL from patients with myocarditis into SCID mice
Coeliac Disease	Ab against endomysial antigens, tissue transglutaminase	
Bronchial asthma	Ab against beta2-adrenergic receptors	
Stiff-man syndrome	Ab against glutamic decarboxylase (GAD), Ab against amphiphysin	
Tropical spastic paraparesis	A variety of autoantibodies together with HTLV-1 infection	
Rasmussen's encephalitis	Ab against glutamate receptors	
B) Systemic Diseases		
Systemic lupus erythematosus (SLE; also derivative anti-phospholipid syndrome, congenital heart block)	Ab against a plethora of self-antigens, especially nuclear antigens (DNA, histones, RNPs), anti-dsDNA and anti-Sm are pathognomonic	(NZBxW)F1, BXSB, MRL-lpr, several lpr or gld homozygous mice; several induced, transgenic, and knockout mouse models, the (NZBxBXSB)F1 model for anti-phospholipid syndrome
Rheumatoid arthritis and Felty's syndrome	Ab against IgG (rheumatoid factors) and nuclear antigens, synovial antigens, T cells (in Felty's syndrome, also Ab against neurophils)	Collagen and adjuvant-induced arthritis, spontaneous in MRL-lpr and NZB/KN, HTLV-1 tax-transgenic mice, KRN mice, IL-1R antagonist-knockout mice
Sjogren's Syndrome	Ab against Ro/SS-A and La/SS-B (but also in SLE, RA and other conditions)	MRL-lpr
Systemic sclerosis (scleroderma)	Ab against topoisomerase I	TSK mouse, L200 chicken line, GvH
CREST Syndrome	Ab against kinetochore proteins	
Mixed connective tissue disease (MCTD)	Ab against U1RNP	
Polymyositis/Dermatomyositis	Ab against t-RNA synthetase	
Goodpasture's Disease	Ab against the non-collegenous domain of the alpha 3 chain of basement membrane type IV collagen	Masugi nephritis
Wegener's granulomatosis, microscopic polyangiatis, polyarteritis nodosa, and other vasculitides	Ab against proteinase 3 and myeloperoxidase (C-ANCA, P-ANCA) of polymorphonuclear cells, T cells	
Behcet's Syndrome	T cells, cytokines	
Atherosclerosis	Ab against HSP, T cells	Rabbits immunized with atherosclerotic plaque proteins, HSP immunized rabbits

hypothesized that they may be part of innate immunity, acting as a first line of defense against microbial agents and/or as opsonizing "transporteurs" of cellular breakdown products for disposal. Indeed, it has recently been shown that natural antibodies play a critical role in preventing pathogen dissemination to vital organs through improved immunogenicity by enhanced antigen trapping in secondary lymphoid organs (5,6). What remains controversial is whether these natural autoantibodies, under certain circumstances, can give rise to pathogenic autoantibodies, which are usually mutated IgG subclasses (4,7,8). In a similar vein, self-antigen-reactive T cells have been identified in blood and secondary lymphoid organs of normal mice and humans (9,10). Therefore, as stated above, detection of autoantibodies or self-reactive T cells cannot be considered absolute evidence for the autoimmune basis of a given disease–unless it is clearly shown that: a) the autoantibody has been deposited in the afflicted tissue(s), b) the antibodies or T cells react *in vitro* with the disease-relevant autoantigen and, more importantly, c) autoantibodies or T cells can adoptively transfer disease to normal recipients. Obviously, with the exception of disease in fetuses due to transfer of maternal autoantibodies (myasthenia, hyperthyroidism, lupus), and the few reported cases of damage in histocompatible β-cell transplants among diabetic patients (11), this latter requirement is not possible in the clinical situation. If human autoantibodies cross-react, however, they may induce comparable lesions in experimental animals (12). Furthermore, several spontaneous or experimentally-induced animal diseases exist that have been shown to be transferable by autoantibodies or T cells to syngeneic recipients, and to be affected by immunomodulatory procedures. In fact, a large body of knowledge with regard to these diseases has been derived from the study of animal models, and findings from such models have frequently been the template for identification of analogous abnormalities in humans. Descriptions of these models figure prominently in this book.

2. NORMAL AUTORECOGNITION AND SELF-TOLERANCE

The essential task of an adaptive immune system is to maintain the structural integrity of the organism by mobilizing lymphocytes that recognize any foreign substance in order to eliminate or destroy the offending molecule or microorganism. Implicit in this process is the requirement for the immune system to respond to every potential antigenic structure. Indeed, the cardinal characteristic of the system is an enormous diversity, i.e., the presence of a large set of genes from which an almost infinite number of lymphocyte antigen receptors can be created through additional unique and ingenious somatic processes. Although this enormous

diversity is of obvious benefit, there are also inherent risks that some of these receptors will display dangerous anti-self specificities, because the repertoire is initially generated without regard for which antigens are part of self. Despite this potential for adverse responses, the immune system manages exceedingly well in distinguishing self from foreign structures. How this fundamental distinction, and thus tolerance for self, is developed has been the dominant subject of immunologic research, and remarkable advances in the past few years have now put our understanding of this phenomenon on a firm footing. The major developments in this area are outlined below.

Approximately 100 years ago, Ehrlich (recipient, with Mechnikov, of the 1908 Nobel Prize in Physiology or Medicine), clearly expressed the logical necessity that the vertebrate body must, in some way, be inhibited from developing reactivity against its own components, and coined the term "horror autotoxicus" to dramatically depict the dire consequences of failure in this process (13). This constraint was accepted as an evident truth, but without much comprehension of how it was accomplished or maintained until 50 years later, when Burnet (recipient, with Medawar, of the 1960 Nobel Prize in Physiology or Medicine), as a corollary to his visionary "clonal selection" theory of adaptive immunity, formulated the concept of "clonal deletion", in which he proposed that clones expressing receptors for self-antigens are arrested in their development and eliminated, thereby conferring tolerance (14). It was further postulated that emergence of forbidden self-reactive clones later in life results from somatic mutations in the antigen receptor genes that lead to conversion of non-self to self-reactive receptors. Subsequent findings indicating that self-reactive lymphocytes could, under certain circumstances, be identified in the normal mature immune system, raised some doubts about the fidelity of clonal deletions, and led Nossal (15) to formulate the "clonal anergy" theory. According to this concept, emerging self-reactive lymphocytes pass through a tolerance-sensitive phase in which they become functionally inactivated or silenced, but not deleted. Following these concepts, Zinkernagel and Doherty (recipients of the 1996 Nobel Prize in Physiology or Medicine) discovered a completely new and unexpected feature of the immune system: That recognition of self-MHC by T cells was essential to elicit a cytotoxic response (16). This cornerstone observation, defined as "the MHC restriction phenonemon", was subsequently shown to be applicable to helper T cells as well. An additional fundamental finding, reported independently by Bevan (17) and Zinkernagel and associates (18), was that the self-MHC specificity of T cells is acquired during the process of intra-thymic differentiation. The subsequent work of Unanue and associates (19), Townsend et al. (20), and Grey et al. (21) were of great relevance to these findings, showing that the recognition of self-MHC plus antigen was not mediated by dual receptors, but by the interaction of the T cell receptor with a peptide bound on MHC molecules. It was also found that MHC class I molecules expressed on most cell types normally present peptides largely derived from proteins produced endogenously. In contrast, MHC class II molecules (expressed on B cells, macrophages and dendritic cells) present peptides of extracellular proteins that have been internalized. Immature dendritic cells may also secrete proteases that degrade proteins into peptides that can be loaded on empty MHC class II molecules directly, thereby obviating intracellular uptake and presentation (22,23). The sophisticated molecular processes by which peptides are created, bind to MHC molecules and are transported to the cell surfaces have also been defined (24–27). Furthermore, Bjorkman and colleagues (28) obtained the crystal structure of MHC class I molecules, and subsequently these and other investigators also determined the crystal structure of MHC class I and class II/ peptide complexes, as well as of the entire trimolecular TCR/MHC/peptide complex (29–34). Another class of molecules, the CD1 family, was found to mediate an MHC-independent presentation of lipid and gly-colipid antigens, as well as glycosylphosphatidyl- inositol (GPI)-anchored proteins (35,36). In addition, methods were devised by which self and foreign peptides bound to MHC class I and class II molecules could be isolated and charac-terized biochemically and functionally (37–39).

Despite these startling achievements, precise definition of self-tolerance mechanisms could not be accomplished without the parallel identification of the genes encoding B and T cell antigen receptors. The initial landmark studies of Tonegawa (recipient of the 1987 Nobel Prize in Physiology or Medicine) documented that immunoglobulin heavy (H) and light (L, κ and λ) chain gene loci encoding the B cell antigen receptors are made of several gene segments (V, variable; D, diversity for heavy chain only; J, joining; C, constant), each containing multiple discreet units (40). Through a stochastic DNA rearrangement process, one unit from each of these segments is transposed during B cell maturation to create the gene encoding the unique receptor of a B lymphocyte clone, a process termed somatic recom-bination. This process, together with nucleotide additions at the joining points (junctional diversity), random H- and L-chain pairings, and somatic hypermutations during the maturation of an immune response, create an enormously large set of unique B cell receptors. Subsequent studies during the mid-1980's culminated in the discovery of the genes and proteins of the T cell receptor (TCR), with many investigators contributing to this advance (41–52). These receptors consisted of a heterodimeric structure composed of either one α with one β or one γ with one δ polypeptide chain, each encoded by corresponding genes that exhibited an overall organization, structure and assembly similar to immunoglobulin genes.

Based on some of these discoveries, two main experi-mental approaches were developed to accurately define the mechanisms by which self-tolerance is established. The first, pioneered by Kappler and Marrack (53,54), was based on the observation that certain self-molecules—initially defined as Mls and subsequently shown to be encoded by mouse mammary tumor proviral (Mtv) genes integrated on various chromosomes—behave as "superantigens". These molecules bind to MHC class II molecules outside the con-ventional peptide binding groove and interact with all TCRs utilizing particular Vβ segments, almost regardless of the other components of the heterodimeric receptor. It was subsequently documented in the mouse that T cell clones bearing specific Vβs with high affinity for self-superantigens were deleted from the thymic repertoire prior to the exportation of T cells to the periphery. Such endoge-nous superantigens have not yet been clearly identified in other species (55,56). Exonogenous superantigens are, however, also encoded by a variety of bacteria and viruses (57), and are thought by some to play a role in autoimmune disease induction (58).

The second approach to address mechanisms of T cell tolerance was that pioneered by von Boehmer and associ-ates (59,60), who employed mice expressing a transgenic TCR specific for the male H-Y antigen. As in the case of endogenous superantigens, intrathymic deletion of the transgenic T cells in male, but not female, mice was observed. These and other investigators, using variations of this and other transgenic systems, further documented that two self-recognition- based processes take place within the thymus. On one hand, T cells expressing TCR with sufficient affinity/avidity for MHC/self-peptide ligands on thymic epithelial and possibly bone marrow-derived cells are selected/expanded and progress from the immature double-positive (DP, CD4+8+) to the mature single-positive (SP, CD4+8− or CD8+4−) stage, a process termed "positive selection" (61,62). On the other hand, in a separate or at continuum step, cells with high affinity/avidity to self-anti-gens are deleted from the repertoire by apoptotic cell death, a process termed "negative selection", and only those with low affinity/avidity are subsequently allowed to be exported to the periphery (63). Cells that express TCR without sufficient affinity/avidity for self also die intrathymically by apoptosis ("death by neglect"). Ultimately, it appears that only about 1–5% of thymocytes are allowed to emigrate to the periphery. A variety of factors, such as the affinity/avidity that defines positive or negative selection outcomes, whether a T cell recognizes a single peptide or a collection of related peptides, and the role of peptide agonist and antagonists, have also been addressed (63–67).

It was originally assumed that the low-grade anti-self-reactivity of the positively-selected T cells was lost after exportation to the periphery. Recent studies by several groups have, however, shown that low-avidity self-recognition and "stimulation" of naive T cells is maintained and is necessary for their survival in the periphery (68–71). It has further been documented that, in cases of marked loss in T cell numbers, it is this anti-self reactivity that leads to subsequent T cell proliferation and re-establishment of homeostasis (71–75). Most of the above studies suggest that the self-peptides recognized in the periphery are the same as those mediating intrathymic positive selection. It is not clear, however, why the self-recognizing T cells do not proliferate under normal circumstances, except when there is a loss in their numbers. It is possible that, in a normal homeostatic setting, T cells compete for ligands on antigen-presenting cells or inhibit each other by various means, whereas in a lymphopenic state, this competition or inhibition is suppressed. These matters are addressed in detail in Chapter 5.

Because is it expected that not all self-antigens will be present in sufficient quantities, if at all, in the thymus, it has been hypothesized that mechanisms must be in place to ensure tolerance development in the periphery. Extrathymic T cell tolerance, as experimentally assessed in mice expressing transgenic self-antigens under the influence of tissue-specific promoters, was indeed reported and found to be mediated by clonal deletion or anergy (76–80). There is considerable debate whether these experiments unequivocally established the active induction of peripheral T cell tolerance, however, since expression of the transgenic self-antigen was sometimes found in the thymus (81–83). In addition, other studies (discussed in the next section) have shown that transgenic expression of self-antigens in peripheral tissues, together with a specific TCR, does not lead to disease simply because the antigen is inaccessible to the T cells and/or antigen presentation and co-stimulation are inadequate.

It has also long been hypothesized that autoimmunity may be averted by the action of "suppressor" T cells, but this has never been satisfactorily proven. Lately, the term "suppressor" has been replaced by the term "immunoregulatory" cells, and proof of their existence and capacity to inhibit autoimmune phenomena has been documented by a variety of approaches (84–89). Evidence has also been presented that these cells acquire immunoregulatory properties during their intrathymic differentiation (90,91), and that their activity in the periphery can be affected by cytokines such as IL-4, IL-10, TGFβ, and TNF (87,92–97).

Studies have also been conducted in Ig transgenic mice to address the molecular mechanisms of B cell tolerance. Initial studies demonstrated that B cells, early in their ontogeny, are exquisitely sensitive to tolerance induction and show accelerated apoptotic death upon treatment with anti-IgM as surrogate Ag. More recent studies with mice transgenic for Ig recognizing surface-bound or soluble self-antigens have clearly shown the induction of clonal deletions or anergy (98,99 and Chapter 2). Gene expression analysis by micro-

array technology indicated that anergy is associated with induction of a set of genes that included negative regulators of signaling and transcription (100). An additional mechanism of central B cell tolerance involves receptor editing, wherein self- receptor-expressing immature B cells do not die but, via the upregulation of recombinases, rearrange an L-chain different from that initially expressed and create a new receptor without self-reactivity (101). Peripheral deletion of self-reactive B cells has also been reported to occur in double-transgenic mice expressing an anti-self Ig and the self-antigen on the surfaces of hepatocytes and keratinocytes (102,103), but not on the surfaces of thyroid epithelium (104). As is the case with thymocytes, these results provide evidence that B cell tolerance is not actively acquired for most organ-specific antigens that are sequestered behind endothelial barriers. Moreover, in the absence of sufficient T cell help, apoptotic death of self-reactive B cells in germinal centers has been shown to occur, and some of these cells may also die as a consequence of their inability to compete with non-self B cells for follicular niches (105,106). Additionally, killing of Fas-expressing self-reactive anergic B cells by CD4+ T cells has also been described (107,108). Receptor editing has been observed by some investigators in the periphery following immunization of Ig transgenic mice with low avidity antigens (109), but others have failed to observe re-induction of the recombination-activating genes (RAG1, 2) during an immune response (110). Finally, positive selection of autoreactive B cells has recently been documented through the use of transgenic mice expressing a germline gene-encoded specificity for the Thy-1 (CD90) glycoprotein (111).

3. MECHANISMS OF PATHOGENIC AUTOIMMUNITY

Autoimmune diseases are classically throught to be the result of circumvention of the normal self-tolerant state, but the precise mechanisms by which this can occur are mostly unknown. Nevertheless, several non-mutually exclusive theories have been proposed (112) that, in a reductionist and disease- relevant approach, can be encapsulated into two main concepts: a) defects in the induction of central or peripheral tolerance, and b) autoimmunity as a result of an inappropriate, yet conventional, immune response against self-antigens for which tolerance has never been established.

3.1. Tolerance defects Abnormalities in tolerance induction are unlikely to be antigen-selective and, therefore, most likely are of relevance only for systemic autoimmune diseases such as lupus, in which a wide spectrum of autoantigens are involved. Lupus mice, however, even those with genetic defects in the Fas/FasL apoptosis pair of genes (*lpr* and *gld*) delete thymocytes that recognize endogenous superantigens (113–115), as well as TCR-transgenic thymocytes that recognize conventional peptides presented in the context of MHC class I (116) or class II (117,118). Evidence, however, has been presented indicating that, under certain circumstances, Fas may participate in intrathymic negative selection (119), and that some defects in central T cell deletions might exist in lpr homozygous mice (120,121). Nevertheless, the lpr and gld mutations are thought to primarily exhibit considerable defects in peripheral mature T cell activation-induced cell death (AICD) (117,122–125).

B cell tolerance defects have also been found in these mice. Adoptively transferred Ig transgenic B cells carrying desensitized receptors due to chronic exposure to a transgenic self-antigen have been shown, upon antigen challenge, not to produce antibody, but instead to be eliminated by antigen-specific CD4+ T cells. This tolerance-related elimination is apparently mediated by the Fas/FasL pathway, since it does not occur with anergic B cells of *lpr* and *gld* mice (107,108). The potential importance of this mechanism in the elimination of anti-self B cells has been suggested by the demonstration that MRL-Faslpr mice with transgenic T cell-specific expression of normal Fas under the *lck* promoter, although lacking lymphoproliferation, still develop autoantibodies and severe glomerulonephritis (126). It appears, therefore, that autoreactive B cells must be killed through Fas expression and engagement of Fas L-expressing activated T cells in order for lupus disease to be averted. Indeed, as expected, transgenic expression of normal Fas on B cells of MRL-Faslpr mice leads to the absence of autoantibodies (127). Studies in *lpr* mice transgenic for Ig anti-DNA, anti- snRNP or anti-IgG (IgMRF), in contrast to similarly transgenic normal mice, have also shown inefficient silencing, editing, deletion or ignorance of these autoreactive B cells (8,128–132). In addition, developmentally-arrested transgenic Ig anti-dsDNA expressing B cells in normal mice are excluded from the follicles, whereas in similarly transgenic MRL-Faslpr mice, the non- developmentally-arrested B cells enter the follicles (133). It should be noted, however, that central or peripheral B cell tolerance defects have not been observed in Ig transgenic lpr mice recognizing a constitutive (H–2k) or transgenic (hen egg lysozyme) self-antigen (134–136). Therefore, it appears that Fas/FasL participation and defective B cell tolerance may be dependent on the nature of the antigen the B cells recognize.

Overall, it is clear that lupus disease in the Fas and FasL mutant mice is associated with defects in tolerance induction at both the T and B cell levels. It is also evident, however, that several strains of mice spontaneously develop lupus in the absence of defects in Fas/FasL (this, of course, does not exclude other potential defects in apoptosis), that the Fas/FasL defects in themselves are insufficient to induce lupus immunopathology, and that the vast majority of

humans with SLE are devoid of any defects in the expression and function of the Fas/FasL system (137,138, and Chapter 11). There is, however, a small group of patients that exhibit lymphoid cell apoptosis defects and develop the so-called Canale-Smith syndrome or autoimmune lymphoproliferative syndrome (ALPS). These patients carry heterozygous dominant mutations in either Fas (type Ia), FasL (Type Ib), caspase 10 (Type II), or other, as yet undefined, apoptosis genes (Type III) (139–144), and exhibit various manifestations of systemic autoimmunity, but rarely develop the full characteristics of lupus.

In addition to spontaneous lupus-associated Fas/FasL mutations, several gene knock-out and transgenic mice have been shown to exhibit characteristis of systemic autoimmunity (Chapter 21). With few exceptions (complement gene- deleted mice), the relevance of these experimental models to spontaneous lupus has not been established. Nevertheless, they indicate that a variety of disturbances in the immune system can lead to such an outcome, including, other apoptosis defects (Bcl2 or BAFF transgenic mice; Bim, Il-2 or IL 2R knockout mice; increased expression of c-FLIP after retroviral infection); inefficient clearance or degradation of apoptotic bodies and chromatin (Clq or serum amyloid protein knockout mice), defective inhibition of signal transduction (Lyn, CD22, SHP-1 or CTLA-4 knockout mice), defective B cell anergy induction (CD21/CD35 knockout mice, C4-deficient mice), increased MHC expression and antigen presentation (IFNγ transgenic mice), and inadequate cell cycle inhibition (p21 cyclin-dependent kinase inhibitor knockout mice).

3.2. Engagement of ignorant, non-tolerant, T cells

The second hypothesis, more applicable to organ-specific autoimmune diseases, postulates that not all self-antigens are present or sufficiently expressed in the thymus. Therefore, only a fraction of particularly hazardous self-reactive clones are physically eliminated from the repertoire, while the remaining potentially self-reactive T cells are exported to the periphery. These cells, however, remain ignorant and harmless until peripherally-expressed antigens, cryptic self-determinants, neoantigens or foreign mimics become available to them, in which case an immune response and tissue damage might develop.

This postulate is well supported by a variety of experimental and clinical findings. Apart from so-called immunologically "privileged" tissues, such as the eye, the brain and the testis, the sequestration of most peripheral tissue antigens behind anatomic barriers has been strongly suggested by a variety of findings. Thus, intravenous injection of CD8+ cytotoxic T cell lines and clones specific for the hepatitis surface antigen (HBsAg) into mice with transgenic widespread tissue expression of this antigen failed to cause infiltration of any organ but the liver (145). This indicates that the vascular endothelium and basement membranes constitute an extremely effective barrier that normally precludes access of T cells to the parenchymal tissues. Additional findings have clearly established that experimentally-induced thymic availability of peripheral tissue-associated antigens leads to tolerance and loss of susceptibility to corresponding autoimmune diseases. For example, streptozotocin-induced diabetes in mice is inhibited by prior elimination of mature T cells with anti-CD3 mAb, and tolerization of the newly emerging T cells by intrathymic injection of streptozotocin-treated islet cells (146). Similarly, intrathymic injection of islet cells prevents autoimmune disease in the diabetes-prone BB rat (147) and the NOD mouse (148). Diabetes in NOD mice has also been reported to be inhibited by the intrathymic or intra-venous administration of glutamic acid decarboxylase GAD) (149,150), apparently a major autoantigen in this disease (151,152). Finally, prevention of experimental encephalomyelitis by intrathymic injection of myelin basic protein (MBP) or its major encephalitogenic epitope has also been reported (153). The combined evidence indicates that anatomically-sequestered parenchymal antigens that are not usually expressed in the thymus, do not induce tolerance, and their availability in the periphery may lead to engagements of the corresponding mature T cells and disease.

Obviously, for the induction of a sustained autoimmune response and disease, these previously unavailable antigens should be adequately presented by the appropriate MHC and co-stimulation. This latter requirement has clearly been documented in mice double-transgenic for lymphocytic choriomeningitis virus (LCMV) glycoprotein (GP) expressed in the pancreas under the influence of a rat insulin promoter and a GP-specific TCR (154,155). It was observed that the GP-specific T cells were not deleted or anergized in such mice, yet these cells did not attack the GP-expressing pancreas. However, upon infection with the GP-expressing LCMV, T cells were attracted to the pancreas and damage ensued, leading to hyperglycemia. These findings showed that non-tolerant and GP-specific T cells were present, but these cells were not engaged because the transgenic self-antigen (GP) expressing β cells were either not accessible, and/or did not display MHC class II molecules or costimula-tory molecules. However, infection with virus-bearing the self-mimicking epitope, in conjunction with effective presen-tation and co-stimulation by the infected professional APC (156) in secondary lymphoid organs leads to activation of the previously quiescent helper and cytotoxic T cells. How these cells then enter into the parenchyma of the target tissue is not clear, but they may be aided by their activated status and a possible direct viral-induced tissue damage. The role of a preceeding tissue inflammation in permitting accessibility by the activated non-tolerant T cells has been documented in other transgenic models of organ-specific autoimmunity (157,158). It should also be noted that several additional factors determine the outcome, including the nature of the

antigen, dose, number of exposures, precursor frequency and quantity of activated autoreactive T cells, affinity of receptors, expression levels of co-stimulatory and MHC molecules in the tissue, and the type and quantity of cytokines produced (159–163). Of interest, with regard to co-stimulatory molecules, it has been reported that while expression of B7.1 (CD80) alone on pancreatic β cells was insufficient to overcome T cell ignorance, this molecule and LCMV-GP on these cells was sufficient to induce diabetes without superimposed LCMV infection (164). Similarly, in another study, mere expression of B7.1 on pancreatic β cells was also insufficient for diabetes development unless coupled with inflammation caused by transgenic local expression of TNFα (165). Thus, tissue expression of co-stimulatory molecules is not enough to break ignorance but, as expected, it is necessary for an efficient autoimmune attack.

These experiments also bring into the forefront the role of "molecular mimicry" in the pathogenesis of organ-specific autoimmune disorders. Because of the constraints imposed upon construction of coding sequences, closely related or identical polypeptides are often found in unrelated proteins, and many peptide fragments of infectious agents are homologous with self-proteins (166, and Chapter 14). Recent experiments, however, have shown that mimicry does not necessarily require linear concordance in sequence between a foreign mimic and a self-molecule because of degeneracy in TCR specificity, limited promiscuity of MHC peptide presentation, and the overall dependence of TCR-MHC-peptide interactions on peptide conformation rather than linear characteristics. This has clearly been illustrated by the experiments of Wucherpfennig and associates (167,168), who documented that MBP peptide-reactive T cell clones derived from patients with multiple sclerosis could be stimulated by a variety of viral/bacterial peptides, as long as those mimics retained the required conformation for binding and presentation by the predisposing MHC class II haplotype. Additional studies have shown that the complementarity-determining region 3 loops of the TCR α and β chains contribute significantly to the degree of TCR degeneracy in peptide recognition and interaction with foreign mimics (169). These and other findings are important because: a) they indicate that mimeotopes cannot be predicted on simple alignments; b) T cells recognize not only a single peptide, but a limited repertoire of structurally-related peptides derived from different antigens, and c) they suggest that it is unlikely that only a single pathogen with an appropriate mimeotope is responsible for the initiation of a given autoimmune disease. Furthermore, these findings may explain, at least in part, why it has been so difficult to link development of autoimmune diseases with particular pathogens. Additional complications in this regard is that the induction of cross-reactivity does not require a replicating agent, immune-mediated injury can occur after the offending microbe has been removed and, in some instances, different mimics may be involved in priming and disease induction.

An additional consideration about the potential role of viruses in the pathogenesis of organ-specific autoimmune diseases is that they may promote such processes not by mimicry, but by a bystander activation mechanism following tissue damage. The importance of this latter mechanism was recently shown in mice infected with Coxsackie B4 virus, which is strongly associated with the development of IDDM in humans and shares sequence similarity with the islet autoantigen glutamic acid decarboxylase (170). When different strains of mice were infected with Coxsackie B4 virus, it was observed that those with susceptible MHC alleles had no diabetes, but those with a transgenic TCR specific for a different islet autoantigen rapidly developed the disease. It was therefore concluded that diabetes induced by Coxsackie B4 virus infection in this model is a direct result of tissue damage and release of sequestered islet antigens, resulting in the stimulation of resting auto-reactive T cells in the draining lymph nodes. Thus, at least in this case, the viral effect in promoting autoimmunity appeared to be an epiphenomenon resulting from tissue damage and subsequent cross-priming of cytotoxic T cells, and not mediated by mimicry.

Within the context of autoimmunity resulting from the engagement of non- tolerant T cells, some investigators have also suggested that such responses may be initiated by so-called "cryptic" self-antigens (171–174). These investigators have postulated that: a) each self-protein presents a small minority of dominant determinants that are well displayed, are involved in thymic negative selection, and the individual is tolerant to these determinants, and b) the poorly displayed majority of subdominant or cryptic determinants do not induce tolerance and, therefore, a large cohort of potentially self-reactive T cells exist. Thus, notwithstanding anatomically-sequestered antigens, the theory proposes that the available T cell repertoire is directed against cryptic determinants and that some pathogenic autoimmune responses may be directed against such determinants. Several interesting experimental examples in support of this possibility have been presented, and several mechanisms by which such cryptic determinants may be revealed to the immune system have been proposed (175–182) but their relevance to autoimmune disease pathogenesis remains to be established.

Other potentially interesting examples of how neoantigens may engage non- tolerant T cells and induce an autoimmune response are: a) mutated or overexpressed genes in malignant cells, such as p53 (183) and HER-2/neu (184), respectively; b) redistribution of intracellular molecules to the cell surface, such as SS-B expression following viral infection (185); c) binding of drugs, haptenic groups or viral proteins to self-antigens (186–189); d) complexing of antigen (particularly those with repetitively arranged

epitopes) with antibodies, and induction of rheumatoid factors (anti-IgG Fc) (190–192), and others.

In summary, to avoid excessive purging of T cell repertoire during ontogeny, and because a large number of non-hematopoietic cell-associated antigens are not expressed in the thymus, the mature immune system contains a large cohort of T cells with receptors that have the potential to interact with self. Engagement of such cells can be induced by a variety of means, but an extensive set of factors need to be met prior to commencement of deleterious responses. Aside from the nature of the antigen, dose, length and number of exposures, as well as local factors, additional controls (primarily genetic in nature) intervene and, in the majority of cases, are not permissive to the induction or perpetuation of such responses. In a few individuals, however, the genetic make-up is conducive and disease ensues.

4. GENES AND GENETICS OF AUTOIMMUNE DISEASES

Population, family and twin studies have clearly shown that the propensity for autoimmunity is highly dependent on genetic factors. In fact, in some instances, genetic factors appear to be sufficient. For example, several lupus- predisposed strains of mice develop disease in a predictable manner regardless of colony location, or rearing in a germ-free vs. conventional environment (193). Nevertheless, environmental factors can precipitate or accelerate disease development even in these cases (194).

Genes that contribute to the development of autoimmune syndromes can be classified into susceptibility genes that play an etiologic role and effector genes that, although involved in disease immunopathogenesis, do not affect predilection. The former consist of mutations or allelic variants, whereas the latter are normal genes.

A large number of transgenic and gene knockout mouse models have been created, and have provided significant insights into the pathogenesis of autoimmunity, including various immune cell types, autoantigens, MHC, co-stimulatory molecules, signal transduction molecules, chemokines, cytokines, apoptosis-related molecules, and others (rev. in 195–199). These models are thoroughly addressed throughout this book.

A significant advance in the actual identification of predisposing genes for these polygenic disorders has been the development of microsatellite-based dense chromosomal maps, by which the regions of the genome likely to contain susceptibility genes can be identified (200–206). Genome-wide studies have been applied in several spontaneous and induced animal models of autoimmunity, as well as humans with such disorders, with particular emphasis on rheumatoid arthritis, IDDM, multiple sclerosis, inflammatory bowl disease and lupus. The results of these studies are described in the corresponding Chapters. Such studies are at their early phases, and are frequently plagued by incomplete congruency of results, particularly in humans. This, of course, has been anticipated considering genetic heterogeneity, incomplete penetrance of traits, high frequencies of alleles, differences in sample size, judgments in patient stratification, and ethnic or environmental differences (204). Despite these shortcomings, however, improvements in genotyping through the use of denser chromosomal maps based on single nucleotide polymorphisms (SNP) (207–210), the forthcoming complete characterization of the entire mouse and human genomes, and the availability of microarray technologies for efficient expression and functional genomics studies (211–217) will undoubtedly permit the identification of autoimmunity-predisposing genes. Such a development will have profound implications in the understanding of disease pathogenesis, as well as subgenotype classifications, and result in improvements in prognosis and therapy of these complex disorders.

5. CONCLUSIONS

In this golden age of biomedical research, advances in immunology and, by extension, the study of autoimmunity, are particularly impressive. We now have very good understandings of the cells, molecules and pathways that govern the function of the immune system, and have identified several genes and mechanisms that promote or inhibit disease in both organ-specific and systemic models of autoimmunity. This has led to the development of new treatment methodologies that are already in use, or are nearly ready for use, in clinical practice. Concurrent efforts have been made to identify the multiple genes that predispose to autoimmunity in both animals and humans. These forward strides have been encapsulated in this volume. Obviously, there is a great deal in the field of autoimmunity that remains to be defined and explored, but the breakthroughs outlined in this textbook put these goals at last within our grasp.

References

1. Alarcon-Segovia, D., L. Llorente and A. Ruiz-Arguelles. 1996. The penetration of autoantibodies into cells may induce tolerance to self by apoptosis of autoreactive lymphocytes and cause autoimmune disease by dysregulation and/or cell damage. *J. Autoimmun.* 2:295–300.
2. Avrameas, A., T. Ternynck, F. Nato, G. Buttin and S. Avrameas. 1998. Polyreactive anti-DNA monoclonal antibodies and a derived peptide as vectors for the intracytoplasmic and intranuclear translocation of macromolecules. *Proc. Natl. Acad. Sci. USA* 95:5601–5606.
3. Hewins, P., J.W. Tervaert, C.O. Savage, and C.G. Kallenberg. 2000. Is Wegener's granulomatosis an autoimmune disease? *Curr. Opin. Rheumatol.* 12:3–10.
4. Avrameas, S. and T. Ternynck. 1993. The natural autoantibodies system: Between hypotheses and facts. *Mol. Immunol.* 30:1133–1142.
5. Boes, M., A.P. Prodeus, T. Schmidt, M.C. Carroll and J. Chen. 1998. A critical role of natural immunoglobulin M in immediate defense against systemic bacterial infection. *J. Exp. Med.* 188:2381–2386.
6. Ochsenbein, A.F., T. Fehr, C. Lutz, M. Suter, F. Brombacher, H. Hengartner and R.M. Zinkernagel. 1999. Control of early viral and bacterial distribution and disease by natural antibodies. *Science* 286:2156–2159.
7. Dighiero, G. and N.R. Rose. 1999. Critical self-epitopes are key to the understanding of self-tolerance and autoimmunity. *Immunol. Today* 20:423–428.
8. Wang, H. and M.J. Shlomchik. 1999. Autoantigen-specific B cell activation in Fas-deficient rheumatoid factor immunoglobulin transgenic mice. *J. Exp. Med.* 190:639–649.
9. Lohse, A.W. 1989. Autoreactive T cells. *Curr. Opin. Immunol.* 1:718–726.
10. Zauderer, M. 1989. Origin and significance of autoreactive T cells. *Adv. Immunol.* 45:417–437.
11. Sutherland, D.E. 1994. Pancreas transplantation as a treatment for diabetes: Indications and outcome. *Curr. Ther. Endocrinol. Metab.* 5:457–460.
12. Ding, X., L.A. Diaz, J.A. Fairley, G.J. Giudice and Z. Liu. 1999. The anti-desmoglein 1 autoantibodies in pemphigus vulgaris sera are pathogenic. *J. Invest. Dermatol.* 112:739–743.
13. Ehrlich, P. 1900. The Croonian lecture: On immunity. *Proc. R. Soc. London* 66:424–428.
14. Burnet, F.M. The clonal selection theory of acquired immunity. Cambridge: Cambridge Univ. Press, 1959.
15. Nossal, G.J.V. 1983. Cellular mechanisms of immunologic tolerance. *Ann. Rev. Immunol.* 1:33–62.
16. Zinkernagel, R.M. and P.C. Doherty. 1974. Restriction of *in vivo* T cell-mediated cytotoxicity in lymphocytic choriomeningitis within a syngeneic or semiallogeneic system. *Nature* 248:701–703.
17. Bevan, M.J. 1977. In a radiation chimera host H-2 antigens determine immune responsiveness of donor cytotoxic cells. *Nature* 269:417–418.
18. Zinkernagel, R.M., G.N. Callahan, A. Althage, S. Cooper, P.A. Klein and J. Klein. 1978. On the thymus in the differentiation of "H-2 self recognition by T cells": Evidence for dual recognition? *J. Exp. Med.* 147:882–896.
19. Babbitt, B., P.M. Allen, G. Matsueda, E. Haber and E.R. Unanue. 1985. Binding of immunogenic peptides to Ia histocompatibility molecules. *Nature* 317:359–361.
20. Townsend, A.R., J. Rothbard, F.M. Gotch, G. Bahadur, D. Wraith and A.J. McMichael. 1986. The epitopes of influenza nucleoprotein recognized by cytotoxic T lymphocytes can be defined with short synthetic peptides. *Cell* 44:959–968.
21. Buus, S., A. Sette, S.M. Colon, C. Miles and H.M. Grey. 1987. The relationship between major histocompatibility complex (MHC) restriction and the capacity of Ia to bind immunogenic peptides. *Science* 235:1353–1358.
22. Santambrogio, L., A.K. Sato, G.J. Carven, S.L. Belyanskaya, J.L. Strominger and L.J. Stern. 1999. Extracellular antigen processing and presentation by immature dendritic cells. *Proc. Natl. Acad. Sci. USA* 96:15056–15061.

23. Santambrogio, L., A.K. Sato, R.F. Fischer, M.E. Dorf and L.J. Stern. 1999. Abundant empty class II MHC molecules on the surface of immature dendritic cells. *Proc. Natl. Acad. Sci. USA* 96:15050–15055.

24. Cresswell, P., N. Bangia, T. Dick and G. Diedrich. 1999. The nature of the MHC class I peptide loading complex. *Immunol. Rev.* 172:21–28.

25. Wolf, P.R. and H.L. Ploegh. 1995. How MHC class II molecules acquire peptide cargo: Biosynthesis and trafficking through the endocytic pathway. *Annu. Rev. Cell. Dev. Biol.* 11:267–306.

26. Roche, P.A. 1999. Intracellular protein traffic in lymphocytes: "How do I get THERE from HERE"? *Immunity* 11:391–398.

27. Nandi, D., K. Marusina and J.J. Monaco. 1998. How do endogenous proteins become peptides and reach the endoplasmic reticulum. *Curr. Top. Microbiol. Immunol.* 232:15–47.

28. Bjorkman, P.J., M.A. Saper, B. Samraoui, W.S. Bennett, J.L. Strominger and D.C. Wiley. 1987. Structure of the human class I histocompatibility antigen, HLA-A2. *Nature* 329:506–512.

29. Garcia, K.C., L. Teyton and I.A. Wilson. 1999. Structural basis of T cell recognition *Annu. Rev. Immunol.* 17:369–197.

30. Smith, K.J., J. Pyrdol, L. Gauthier, D.C. Wiley and K.W. Wucherpfennig. 1998. Crystal structure of HLA-DR2 (DRA*0101, DRB*1501) complexed with a peptide from human myelin basic protein. *J. Exp. Med.* 188:1511–1520.

31. Ding, Y.H., K.J. Smith, D.N. Garboczi, U. Utz, W.E. Biddison and D.C. Wiley. 1998. Two human T cell receptors bind in a similar diagonal mode to the HLA-A2/Tax peptide complex using different TCR amino acids. *Immunity* 8:403–411.

32. Dessen, A., C.M. Lawrence, S. Cupo, D.M. Zaller and D.C. Wiley. 1997. X ray crystal structure of HLA-DR4 (DRA*0101, DRB1*0401) complexes with a peptide from human collagen II. *Immunity* 7:473–481.

33. Reinherz, E.L., K. Tan, L. Tang, P. Kern, J. Liu, Y. Xiong, R.E. Hussey, A. Smolyar, b. Hare, R. Zhang, A. Joachimiak, H.C. Chang, G. Wagner and J. Wang. 1999. The crystal structure of a T cell receptor in complex with peptide and MHC class II. *Science* 286:1913–1921.

34. Garcia, K.C., M. Degano, L.R. Pease, M. Huang, P.A. Peterson, L. Teyton and I.A. Wilson. 1998. Structural basis of plasticity in T cell receptor recognition of a self peptide-MHC antigen. *Science* 279:1166–1172.

35. Porcelli, S.A. and R.L. Modlin. 1999. The CD1 system: antigen- presenting molecules for T cell recognition of lipids and glycolipids. *Annu. Rev. Immunol.* 17:297–329.

36. Schofield, L., M.J. McConville, D. Hansen, A.S. Campbell, B. Fraser- Reid, M.J. Grusby and S.D. Tachado. 1999. CD1d-restricted immunoglobulin G formation to GPI-anchored antigens mediated by NKT cells. *Science* 283:225–229.

37. Godkin, A., T. Friede, M. Davenport, S. Stevanovic, A. Willis, D. Jewell, A. Hill and H.G. Rammensee. 1997. Use of eluted peptide sequence data to identify the binding characteristics of peptides to the insulin-dependent diabetes susceptibility allele HLA-DQ8 (DQ 3.2). *Int. Immunol.* 9:905–911.

38. Falk, K., O. Rotzschke, S. Stevanovic, G. Jung and H-G Rammensee. 1991. Allele-specific motifs revealed by sequencing of self-peptides eluted from MHC molecules. *Nature* 351:290–296.

39. Pierce, R.A., E.D. field, J.M. denHaan, J.A. Caldwell, F.M. White, J.A. Marto, W. Wang, L.M. Frost, E. Blokland, C. Reinhardus, J. Shabanowitz, D.F. Hunt, E. Goulmy and V.H. Engelhard. 1999. Cutting edge: The HLA-*0101-restricted HY minor histocompatibility antigen originates from DFFRY and contains a cysteinylated cysteine residue as identified by a novel mass spectrometric technique. *J. Immunol.* 163:6360–6364.

40. Tonegawa, S. 1983. Somatic generation of antibody diversity. *Nature* 302:575–581.

41. Arden, B., S.P. Clark, D. Kabelitz and T.W. Mak. 1995. Human T-cell receptor variable gene segment families. *Immunogenet.* 42:455–500.

42. Rowen, L., B.F. Koop and L. Hood. 1996. The complete 685-kilobase DNA sequence of the human beta T cell receptor locus. *Science* 272:1755–1762.

43. Hood, L., L. Rowen and B.F. Koop. 1995. Human and mouse T-cell receptor loci: Genomics, evolution, diversity, and serendipity. *Ann. NY Acad. Sci.* 758:390–412.

44. Allison, J.P., B.W. McIntyre and D. Block. 1982. Tumor-specific antigen of murine T lymphoma defined with a monoclonal antibody. *J. Immunol.* 129:2293–2300.

45. Meuer, S.C., K.A. Fitzgerald, R.E. Hussey, J.C. Hodgdon, S.F. Schlossman and E.L. Reinherz. 1983. Clonotypic structures involved in antigen- specific human T cell function. Relationship to the T3 molecular complex. *J. Exp. Med.* 157:705–719.

46. Tonegawa, S. 1988. Somatic generation of immune diversity. *Bioscience Rep.* 8:3–26.

47. Hedrick, S.M., D.I. Cohen, E.A. Nielson and M.M. Davis. 1984. Isolation of cDNA clones encoding T cell-specific membrane-associated proteins. *Nature* 308:149–153.

48. Chien, Y.H., D.M. Becker, T. Lindsten, M. Okamura, D.I. Cohen and M.M. Davis. 1984. A third type of murine T-cell receptor gene. *Nature* 312:31–35.

49. Chien, Y.H., M. Iwashirua, K.B. Kaplan, J.F. Elliott and M.M. Davis. 1987. A new T-cell receptor gene located within the alpha locus and expressed early in T-cell differentiation. *Nature* 327:677–682.

50. Yanagi, Y., Y. Yoshikai, K. Legget, S.P. Clark, I. Aleksander and T.W. Mak. 1984. A human T cell-specific cDNA clone encodes a protein having extensive homology to immunoglobulin chains. *Nature* 308:145–149.

51. Saito, H., D. Kranz, Y. Takagaki, A. Hayday, H. Eisen and S. Tonegawa. 1984. Complete primary structure of a heterodimeric T-cell receptor deduced from cDNA sequences. *Nature* 309:757–762.

52. Brenner, M.B., J. McLean, D.P. Dialynas, J.L. Strominger, J.A. Smith, F.L. Owen, J.G. Seidman, S. Ip, F. Rosen and M.S. Krangel. 1986. Identification of a putative second T-cell receptor. *Nature* 322:145–149.

53. Kappler, J., N. Roehm and P. Marrack. 1987. T cell tolerance by clonal elimination in the thymus. *Cell* 49:273–280.

54. Kotzin, B.L., D.Y. Leung, J. Kappler and P. Marrack. 1993. Superantigens and their potential role in human disease. *Adv. Immunol.* 54:99–166.

55. Baccala, R., D.H. Kono, S.M. Walker, R.S. Balderas and A.N. Theofilopoulos. 1991. Genomically imposed and somatically modified human thymocyte V-beta gene repertoires. *Proc. Natl. Acad. Sci. USA* 88:2908–2912.

56. Smith, L.R., D.H. Kono, M. Kammuller, R.S. Balderas and A.N. Theofilopoulos. 1992. V-beta repertoire in rats and implications for endogenous superantigens. *Eur. J. Immunol.* 22:641–645.

57. Huber, B.T., P.N. Hsu and N. Sutkowski. 1996. Virus-encoded superantigens. *Microbiol. Rev.* 60:473–482.
58. Schiffenbauer, J., J. Soos and H. Johnson. 1998. The possible role of bacterial superantigens in the pathogenesis of autoimmune disorders. *Immunol. Today* 19:117–120.
59. Kisielow, P., H. Bluthmann, U.D. Staerz, M. Steinmetz and H. von Boehmer. 1988. Tolerance in T cell-receptor transgenic mice involves deletion of nonmature CD4+8+ thymocytes. *Nature* 333:742–746.
60. von Boehmer, H. and P. Kisielow. 1990. Self-nonself discrimination by T cells. *Science* 248:1369–1373.
61. von Boehmer, H. 1994. Positive selection of lymphocytes. *Cell* 76:219–228.
62. Zinkernagel, R. and Al Althage. 1999. On the role of thymic epithelium vs. bone marrow-derived cells in repertoire selection of T cells. *Proc. Natl. Acad. Sci. USA* 96:8092–8097.
63. Nossal, G.J.V. 1994. Negative selection of lymphocytes. *Cell* 76:229–240.
64. Ashton-Rickardt, P.G. and S. Tonegawa. 1994. A differential-avidity model for T-cell selection. *Immunol. Today* 15:362–366.
65. Sebzda, E., V.A. Wallace, J. Mayer, R.S.M. Yeung, T.W. Mak and P.S. Ohashi. 1994. Positive and negative thymocyte selection induced by different concentrations of a single peptide. *Science* 263:1615–1618.
66. Peterson, D.A., R.J. DiPaolo, O. Kanagawa and E.R. Unanue. 1999. Quantitative analysis of the T cell repertoire that escapes negative selection. *Immunity* 11:453–462.
67. Jameson, S.C. and M.J. Bevan. 1995. T cell receptor antagonists and partial agonists. *Immunity* 2:1–11.
68. Freitas, A.A. and B. Rocha. 1999. Peripheral T cell survival. *Curr. Opin. Immunol.* 11:152–156.
69. Kirberg, J., A. Berns and H. von Boehmer. 1997. Peripheral T cell survival requires continual ligation of the T cell receptor to major histocompatibility complex-encoded molecules. *J. Exp. Med.* 186:1269–1275.
70. Viret, C., F.S. Wong and C.A.Jr. Janeway. 1999. Designing and maintaining the mature TCR repertoire: The continuum of self-peptide:self-MHC complex recognition. *Immunity* 10:559–568.
71. Goldrath, A.W. and M.J. Bevan. 1999. Selecting and maintaining a diverse T-cell repertoire. *Nature* 402:255–262.
72. Goldrath, A.W. and M.J. Bevan. 1999. Low-affinity ligands for the TCR drive proliferation of mature CD8+ T cells in lymphopenic hosts. *Immunity* 11:183–190.
73. Kieper, W.C. and S.C. Jameson. 1999. Homeostatis expansion and phenotypic conversion of naive T cells in response to self peptide/MHC ligands. *Proc. Natl. Acad. Sci. USA* 96:13306–13311.
74. Bender, J., T. Mitchell, J. Kappler and P. Marrack. 1999. CD4+ T cell division in irradiated mice requires peptides distinct from those responsible for thymic selection. *J. Exp. Med.* 190:367–174.
75. Ernst, B., D.S. Lee, J.M. Chang, J. Sprent and C.D. Surh. 1999. The peptide ligands mediating positive selection in the thymus control T cell survival and homeostatic proliferation in the periphery. *Immunity* 11:173–181.
76. Lo, D. 1992. T cell tolerance. *Curr. Opin. Immunol.* 4:711–715.
77. Miller, J.F., C. Kurts, J. Allison, H. Kosaka, F. Carbone and W.R. Heath. 1998. Induction of peripheral CD8+ T-cell tolerance by cross-presentation of self-antigens. *Immunol. Rev.* 165:267–277.
78. Miller, J.F. and A. Basten. 1996. Mechanisms of tolerance to self. *Curr. Opin. Immunol.* 8:815–821.

79. Hammerling, G.J., G. Schonrich, I. Ferber and B. Arnold. 1993. Peripheral tolerance as a multi-step mechanism. *Immunol. Rev.* 133:93–104.
80. Heath, W.R., C. Kurts, J.F. Miller and F.R. Carbone. 1998. Cross-tolerance: a pathway for inducing tolerance to peripheral tissue antigens. *J. Exp. Med.* 187:1549–1553.
81. Jolicoeur, C., D. Hanahan and K.M. Smith. 1994. T-cell tolerance toward a transgenic beta-cell antigen and transcription of endogenous pancreatic genes in thymus. *Proc. Natl. Acad. Sci. USA* 91:6707–6711.
82. Antonia, S.J., T. Geiger, J. Miller and R.A. Flavell. 1995. Mechanisms of immune tolerance induction through the thymic expression of a peripheral tissue-specific protein. *Int. Immunol.* 7:715–725.
83. von Herrath, M.G., J. Dockter and M.B. Oldstone. 1994. How virus induces a rapid ro slow onset insulin-dependent diabetes mellitus in a transgenic model. *Immunity* 1:231–242.
84. Moalem, G., R. Leibowtiz-Amit, E. Yoles, F. Mor, I.R. Cohen and M. Schwartz. 1999. Autoimmune T cells protect neurons from secondary degeneration after central nervous system axotomy. *Nature Med.* 5:49–55.
85. Seddon, B. and D. Mason. 1999. Peripheral autoantigen induces regulatory T cells that prevent autoimmunity. *J. Exp. Med.* 189:877–882.
86. Mason, D. and F. Powrie. 1998. Control of immune pathology by regulatory T cells. *Curr. Opin. Immunol.* 10:649–655.
87. Cope, A.P. 1998. Regulation of autoimmunity by pro-inflammatory cytokines. *Curr. Opin. Immunol.* 10:669–670.
88. Kumar, V. and E. Sercarz. 1998. Induction or protection from experimental autoimmune encephalomyelitis depends on the cytokine secretion profile of TCR peptide-specific regulatory CD4 T cells. *J. Immunol.* 161:6585–6591.
89. Groux, H. and F. Powrie. 1999. Regulatory T cells and inflammatory bowel disease. *Immunol. Today* 20:442–445.
90. Saoudi, A., B. Seddon, V. Heath, D. Fowell and D. Mason. 1996. The physiological role of regulatory T cells in the prevention of autoimmunity: The function of the thymus in the generation of the regulatory T cell subset. *Immunol. Rev.* 149:195–216.
91. Heath, V.L., A. Saoudi, B.P. Seddon, N.C. Moore, D.J. Fowell and D.W. Mason. 1996. The role of the thymus in the control of autoimmunity. *J. Autoimmunity* 9:241–246.
92. Falcone, M. and N. Sarvetnick. 1999. Cytokines that regulate autoimmune responses. *Curr. Opin. Immunol.* 11:670–676.
93. Powrie, F., J. Carlino, M.W. Leach, S. Mauze and R.L. Coffman. 1996. A critical role for transforming growth factor-beta, but not interleukin 4, in the suppression of T helper type 1-mediated colitis by CD45RB (low) CD4+ T cells. *J. Exp. Med.* 183:2669–1674.
94. Asseman, C., S. Mauze, M.W. Leach, R.L. Coffman and F. Powrie. 1999. An essential role for interleukin 10 in the function of regulatory T cells that inhibit intestinal inflammation. *J. Exp. Med.* 190:995–1004.
95. Seddon, B. and D. Mason. 1999. Regulatory T cells in the growth of autoimmunity: The essential role of transforming growth factor beta and interleukin 4 in the prevention of autoimmune thyroiditis in rats by peripheral CD4(+)CD45RC(–) cells and CD4(+)CD8(–) thymocytes. *J. Exp. Med.* 189:279–288.
96. Homann, D., A. Holz, A. Bot, B. Coon, T. Wolfe, J. Peterson, T.P. Dyrberg, M.J. Grusby and M.G. von Herrath. 1999. Autoreactive CD4+ T cells protect from autoimmune diabetes via bystander suppression using the IL-4/Stat5 pathway. *Immunity* 11:463–472.

97. Gallichan, W.S., B. Balasa, J.D. Davies and N. Sarvetnick. 1999. Pancreatic IL-4 expression results in islet-reactive Th2 cells that inhibit diabetogenic lymphocytes in the nonobese diabetic mouse. *J. Immunol.* 163:1696–1703.

98. Melamed, D. and D. Nemazee. 1999. Immunoglobulin transgenes in B lymphocyte development, tolerance, and auto-immunity. In: *Curr. Dir. Autoimmunity*, A.N. Theofilopoulos, Ed., Karger, Basel, Switzerland, 1:1–30.

99. Goodnow, C.C., J.G. Cyster, S.B. Hartley, S.E. Bell, M.P. Cooke, J.I. Healy, S. Akkaraju, J.C. Rathmell, S.L. Pogue and K.P. Shokat. 1995. Self- tolerance checkpoints in B lymphocyte development. *Adv. Immunol.* 59:279–368.

100. Glynne, R., S. Akkaraju, J.I. Healy, J. Rayner, C.C. Goodnow and D.H. Mack. 2000. How self-tolerance and the immuno-suppressive drug FK506 prevent B- cell mitogenesis. *Nature* 403:671–676.

101. Nemazee, D. 2000. Receptor editing in B cells. *Adv. Immunol.* 74:89–126.

102. Russell, D.M., Z. Dembic, G. Morahan, J.F. Miller, K. Burki and D. Nemazee. 1991. Peripheral deletion of self-reactive B cells. *Nature* 354:308–311.

103. Lang, J., B. Arnold, G. Hammerling, A.W. Harris, S. Korsmeyer, D. Russell, A. Strasser and D. Nemazee. 1997. Enforced Bcl-2 expression inhibits antigen-mediated clonal elimination of peripheral B cells in an antigen- dependent manner and promotes receptor editing in autoreactive, immature B cells. *J. Exp. Med.* 186:1513–1522.

104. Akkaraju, S., K. Canaan and C.C. Goodnow. 1997. Self-reactive B cells are not eliminated or inactivated by auto-antigen expressed on thyroid epithelial cells. *J. Exp. Med.* 186:2005–2012.

105. Schmidt, K.N. and J.G. Cyster. 1999. Follicular exclusion and rapid elimination of hen egg lysozyme autoantigen-binding B cells are dependent on competitor B cells, but not on T cells. *J. Immunol.* 162:284–291.

106. Cyster, J.G. and C.C. Goodnow. 1995. Antigen-induced exclusion from follicles and anergy are separate and complementary processes that influence peripheral B cell fate. *Immunity* 3:691–701.

107. Rathmell, J.C., M.P. Cooke, W.Y. Ho, J. Grein, S.E. Townsend, M.M. Davis and C.C. Goodnow. 1995. CD95 (Fas)-dependent elimination of self-reactive B cells upon interaction with CD4+ T cells. *Nature* 376:181–184.

108. Rathmell, J.C., S.E. Townsend, J.C. Xu, R.A. Flavell and C.C. Goodnow. 1996. Expansion or elimination of B cells in vivo: Dual roles for CD40- and Fas (CD95)-ligands modulated by the B cell antigen receptor. *Cell* 87:319–329.

109. Hertz, M., V. Kouskoff, T. Nakamura and D. Nemazee. 1998. V(D)J recombinase induction in splenic B lymphocytes is inhibited by antigen-receptor signaling. *Nature* 394:292–295.

110. Yu, W., H. Nagaoka, M. Jankovic, Z. Misulovin, H. Suh, A. Rolink, F. Melchers, E. Meffre and M.C. Nussenzweig. 1999. Continued RAG expression in late stages of B cell development and no apparent re-induction after immunization. *Nature* 400:682–687.

111. Hayakawa, K., M. Asano, S.A. Shinton, M. Gui, D. Allman, C.L. Stewart, J. Silver and R.R. Hardy. 1999. Positive selection of natural autoreactive B cells. *Science* 285:113–116.

112. Theofilopoulos, A.N. 1995. The basis of autoimmunity: Part I. The mechanisms of aberrant self-recognition. *Immunol. Today* 16:90–98.

113. Singer, P.A., R.S. Balderas, R.J. McEvilly, M. Bobardt and A.N. Theofilopoulos. 1989. Tolerance-related Vβ clonal deletions in normal CD4–8-, TCR-α/β+ and abnormal lpr and gld cell populations. *J. Exp. Med.* 170:1869–1877.

114. Singer, P.A. and A.N. Theofilopoulos. 1990. T-cell receptor Vβ repertoire expression in murine models of SLE. *Immunol. Rev.* 118:103–127.

115. Kotzin, B.L., S.K. Babcock and L.R. Herron. 1988. Deletion of potentially self-reactive T cell receptor specificities in L3T4, Lyt-2 T cells of lpr mice. *J. Exp. Med.* 168:2221–2229.

116. Sidman, C.L., J.D. Marshall and H. VonBoehmer. 1992. Transgenic T cell receptor interactions in the lymphoprolifera-tive and autoimmune syndromes of lpr and gld mutant mice. *Eur. J. Immunol.* 22:499–504.

117. Singer, G.G. and A.K. Abbas. 1994. The Fas antigen is involved in peripheral but not thymic deletion of T lymphocytes in T cell receptor transgenic mice. *Immunity* 1:365–371.

118. Sytwu, H.-K., R.S. Liblau and H.O. McDevitt. 1996. The roles of Fas/APO-1 (CD95) and TNF in antigen-induced pro-grammed cell death in T cell receptor transgenic mice. *Immunity* 5:17–30.

119. Kishimoto, H., C.D. Surh and J. Sprent. 1998. A role for Fas in negative selection of thymocytes in vivo. *J. Exp. Med.* 187:1427–1438.

120. Castro, J.E., J.A. Listman, B.A. Jacobson, Y. Wang, P.A. Lopez, S. Ju, P.W. Finn and D.L. Perkins. 1996. Fas modula-tion of apoptosis during negative selection of thymocytes. *Immunity* 5:617–627.

121. Kurasawa, K., Y. Hashimoto and I. Iwamoto. 1999. Fas modu-lates both positive and negative selection of thymocytes. *Cell. Immunol.* 194:127–135.

122. Russell, J.H., B. Rush, C. Weaver and R.D. Wang. 1993. Mature T-cells of autoimmune lpr/lpr mice have a defect in antigen-stimulated suicide. *Proc. Natl. Acad. Sci. USA* 90:4409–4413.

123. Bossu, P., G.G. Singer, P. Andres, R. Ettinger, A. Marshak-Rothstein and A.K. Abbas. 1993. Mature CD4+ T lympho-cytes from MRL/lpr mice are resistant to receptor-mediated tolerance and apoptosis. *J. Immunol.* 151:7233–7239.

124. Mixter, P.F., J.Q. Russell and R.C. Budd. 1994. Delayed kinet-ics of T lymphocyte anergy and deletion in Lpr mice. *J. Autoimmunity* 7:697–710.

125. Mogil, R.J., L. Radvanyi, R. Gonzalez-Quintial, R. Miller, G. Mills, A.N. Theofilopoulos and D.R. Green. 1995. Fas (CD95) participates in peripheral T cell deletion and associated apopto-sis in vivo. *Int. Immunol.* 7:1451–1458.

126. Fukuyama, H., M. Adachi, S. Suematsu, K. Miwa, T. Suda, N. Yoshida and S. Nagata. 1998. Transgenic expression of Fas in T cells blocks lymphoproliferation but not autoimmune disease in MRL-lpr mice. *J. Immunol.* 160:3805–3811.

127. Komano, H., Y. Ikegami, M. Yokoyama, R. Suzuki, S. Yonehara, Y. Yamasaki and N. Shinohara. 1999. Severe impairment of B cell function in lpr/lpr mice expressing trans-genic Fas selectivity on B cells. *Int. Immunol.* 11:1035–1042.

128. Erikson, J., L. Mandik, A. Bui, A. Eaton, H. Noorchashm, K.A. Nguyen and J.H. Roark. 1998. Self-reactive B cells in nonautoimmune and autoimmune mice. *Immunol. Res.* 17:49–61.

129. Roark, J.H., C.L. Kuntz, K.-A. Nguyen, A.J. Caton and J. Erikson. 1995. Breakdown of B cell tolerance in a mouse model of systemic lupus erythematosus. *J. Exp. Med.* 181:1157–1167.

130. Brard, F., M. Shannon, E.L. Prak, S. Litwin and M. Weigert. 1999. Somatic mutation and light chain rearrangement generate autoimmunity in anti- single-stranded DNA trans-genic MRL/lpr mice. *J. Exp. Med.* 190:691–704.

131. Shinde, S., R. Gee, S. Santulli-Marotto, L.K. Bockenstedt, S.H. Clarke and M.J. Mamula. 1999. T cell autoimmunity in Ig transgenic mice. *J. Immunol.* 162:7519–7524.

132. Tighe, H., P.P. Chen, R. Tucker, T.J. Kipps, J. Roudier, F.R. Jirik and D.A. Carson. 1993. Function of B cells expressing a human immunoglobulin M rheumatoid factor autoantibody in transgenic mice. *J. Exp. Med.* 177:109–118.

133. Mandik-Nayak, L., S.J. Seo, C. Sokol, K.M. Potts, A. Bui and J. Erikson. 1999. MRL-lpr/lpr mice exhibit a defect in maintaining developmental arrest and follicular exclusion of anti-double-stranded DNA B cells. *J. Exp. Med.* 189:1799–1814.

134. Rubio, C.F., J. Kench, D.M. Russell, R. Yawger and D. Nemazee. 1996. Analysis of central B cell tolerance in autoimmune-prone MRL/lpr mice bearing autoantibody transgenes. *J. Immunol.* 157:65–71.

135. Kench, J.A., D.M. Russell and D. Nemazee. 1998. Efficient peripheral clonal elimination of B lymphocytes in MRL/lpr mice bearing autoantibody transgenes. *J. Exp. Med.* 188:909–917.

136. Rathmell, J.C. and C.C. Goodnow. 1994. Effects of the lpr mutation on elimination and inactivation of self-reactive B cells. *J. Immunol.* 153:2831–2842.

137. Mysler, E., P. Bini, J. Drappa, P. Ramos, S.M. Friedman, P.H. Krammer and K.B. Elkon. 1994. The apoptosis-1/Fas protein in human systemic lupus erythematosus. *J. Clin. Invest.* 93:1029–1034.

138. Kojima, T., T. Horiuchi, H. Nishizaka, T. Sawabe, M. Higuchi, S.I. Harashime, S. Yoshizawa, H. Tsukoamoto, K. Nagasawa and Y. Niho. 2000. Analysis of fas ligand gene mutation in patients with systemic lupus erythematosus. *Arth. & Rheum.* 43:135–139.

139. Jackson, C.E. and J.M. Puck. 1999. Autoimmune lympho-proliferative syndrome, a disorder of apoptosis. *Curr. Opin. Pediatr.* 11:521–527.

140. Vaishnaw, A.K., J.R. Orlinick, J.L. Chu, P.H. Krammer, M.V. Chao and K.B. Elkon. 1999. The molecular basis for apoptotic defects in patients with CD95 (Fas/Apo-1) mutations. *J. Clin. Invest.* 103:355–363.

141. Straus, S.E., M. Lenardo and J.M. Puck. 1997. The Canale-Smith syndrome. *New Eng. J. Med.* 336:1457–1458.

142. Wang, J., L. Zheng, A. Lobito, F.K. Chan, J. Dale, M. Sneller, X. Yao, J.M. Puck, S.E. Straus and M.J. Lenardo. 1999. Inherited human Caspase 10 mutations underlie defective lymphocyte and dendritic cell apoptosis in auto-immune lymphoproliferative syndrome type II. *Cell* 98:47–58.

143. Straus, S.E., M. Sneller, M.J. Lenardo, J.M. Puck and W. Strober. 1999. An inherited disorder of lymphocyte apopto-sis: The autoimmune lymphoproliferative syndrome. *Ann. Int. Med.* 130:591–601.

144. Fischer, A., F. Rieux-Laucat and F. LeDeist. 1999. A new peak in the ALPS. *Nature Med.* 5:876–877.

145. Ando, K., L.G. Guidotti, A. Cerny, T. Ishikawa and F.V. Chisari. 1994. CTL access to tissue antigen is restricted in vivo. *J. Immunol.* 153:482–488.

146. Herold, K.C., A.G. Montag and F. Buckingham. 1992. Induction of tolerance to autoimmune diabetes with islet antigens. *J. Exp. Med.* 176:1107–1114.

147. Posselt, A.M., C.F. Barker, A.L. Friedman and A. Naji. 1992. Prevention of autoimmune diabetes in the BB rat by intrathymic islet transplantation at birth. *Science* 256:1321–1324.

148. Gerling, I., D. Serreze, S. Christianson and E. Leiter. 1992. Intrathymic islet cell transplantation reduces (beta-cell

149. Kaugman, D.L., M. Clare-Salzler, J. Tian, T. Forsthuber, G.S.P. Ting, P. Robinson, M.A. Atkinson, E.E. Sercarz, A.J. Tobin and P.V. Lehmann. 1993. Spontaneous loss of T-cell tolerance to glutamic acid decarboxylase in murine insulin-dependent diabetes. *Nature* 366:69–72.

150. Tisch, R., X.-D. Yang, S.M. Singer, L.S. Liblau, L. Fugger and H.O. McDevitt. 1993. Immune response to glutamic acid decarboxylase correlates with insulitis in non-obese dia-betic mice. *Nature* 366:72–75.

151. von Boehmer, H. and A. Sarukhan. 1999. GAD, a single autoantigen for diabetes. *Science* 284:1135–1137.

152. Yoon, J.W., C.S. Yoon, H.W. Lim, Q.Q. Huang, Y. Kang, K.H. Pyun, K. Huirasawa, R.S. Sherwin and H.S. Jun. 1999. Control of autoimmune diabetes in NOD mice by GAD expression or suppression in beta cells. *Science* 284:1183–1187.

153. Khoury, S.J., M.H. Sayegh, W.W. Hancock, L. Gallon, C.B. Carpenter and H.L. Weiner. 1993. Acquired tolerance to experimental autoimmune encephalomyelitis by intrathymic injection of myelin basic protein or its major encephalito-genic peptide. *J. Exp. Med.* 178:559–566.

154. Ohashi, P.S., S. Oehen, K. Buerki, H. Pircher, C.T. Ohashi, B. Odermatt, B. Malissen, R.M. Zinkernagel and H. Hengartner. 1991. Ablation of "tolerance" and induction of diabetes by virus infection in viral antigen transgenic mice. *Cell* 65:305–317.

155. Oldstone, M.B.A., M. Nerenberg, P. Southern, J. Price and H. Lewicki. 1991. Virus infection triggers insulin-dependent diabetes mellitus in a transgenic model: Role of anti-self (virus) immune response. *Cell* 65:319–332.

156. Ludewig, B., B. Odermatt, A.F. Ochsenbein, R.M. Zinkernagel and H. Hengartner. 1999. Role of dendritic cells in the induction and maintenance of autoimmune diseases. *Immunol. Rev.* 169:45–54.

157. Limmer, A., T. Sacher, J. Alferink, T. Nichterlein, B. Arnold and G.J. Hammerling. 1998. A two-step model for the induction of organ-specific autoimmunity. *Novartis Found. Symp.* 215:159–167.

158. Limmer, A., T. Sacher, J. Alferink, M. Kretschmar, G. Schonrich, T. Nichterlein, B. Arnold and G.J. Hammerling. 1998. Failure to induce organ- specific autoimmunity by breaking of tolerance: Importance to the microenvironment. *Eur. J. Immunol.* 28:2395–2406.

159. Falcone, M. and N. Sarvetnick. 1999. The effect of local production of cytokines in the pathogenesis of insulin-dependent diabetes mellitus. *Clin. Immunol.* 90:2–9.

160. Bradley, L.M., V.C. Asensio, L.K. Schioetz, J. Harbertson, T. Krahl, G. Patstone, N. Woolf, I.L. Campbell and N. Sarvetnick. 1999. Islet-specific Th1, but not Th2, cells secrete multiple chemokines and promote rapid induction of autoimmune diabetes. *J. Immunol.* 162:2511–2520.

161. von Herrath, M.G. and M.B.A. Oldstone. 1997. Interferon-gamma is essential for destruction of β cells and develop-ment of insulin-dependent diabetes mellitus. *J. Exp. Med.* 185:531–539.

162. von Herrath, M.G., C.F. Evans, M.S. Horwitz and M.B. Oldstone. 1996. Using transgenic mouse models to dissect the pathogenesis of virus-induced autoimmune disorders of the islets of Langerhans and the central nervous system. *Immunol. Rev.* 152:111–143.

163. Ohashi, P., S. Oehen, P. Aichele, H. Pircher, B. Obermatt, P. Herrer, Y. Higuchi, K. Buerki, H. Hengartner and R.M. Zinkernagel. 1993. Induction of diabetes is influenced by the

infectious virus and local expression of MHC class I and tumor necrosis factor-α. *J. Immunol.* 150:5185–5194.

164. von Herrath, M.G., S. Guerder, H. Lewicki, R.A. Flavell and M.B. Oldstone. 1995. Coexpression of B7–1 and viral ("self") transgenes in pancreatic beta cells can break peripheral ignorance and lead to spontaneous autoimmune diabetes. *Immunity* 3:727–738.

165. Guerder, S., D.E. Picarella, P.S. Linsley and R.A. Flavell. 1994. Costimulator B7–1 confers antigen-presenting-cell function to parenchymal tissue and, in conjunction with tumor necrosis factor alpha, leads to autoimmunity in transgenic mice. *Proc. Natl. Acad. Sci. USA* 91:5138–5142.

166. Oldstone, M.B.A. 1987. Molecular mimicry and autoimmune disease. *Cell* 50:819–820.

167. Wucherpfennig, K.W. and J.L. Srtominger. 1995. Molecular mimicry in T cell-mediated autoimmunity: Viral peptides activate human T cell clones specific for myelin basic protein. *Cell* 80:695–705.

168. Hausmann, S. and K.W. Wucherpfennig. 1997. Activation of autoreactive T cells by peptides from human pathogens. *Curr. Opin. Immunol.* 9:831–838.

169. Hausmann, S., M. Martin, L. Gauthier and K.W. Wucherpfennig. 1999. Structural features of autoreactive TCR that determine the degree of degeneracy in peptide recognition. *J. Immunol.* 162:338–344.

170. Horwitz, M.S., L.M. Bradley, J. Harbertson, T. Krahl, J. Lee and N. Sarvetnick. 1998. Diabetes induced by Coxsackie virus: Initiation by bystander damage and not molecular mimicry. *Nature Med.* 4:781–785.

171. Gammon, G., E. Sercarz and G. Benichou. 1991. The dominant self and the cryptic self: Shaping the autoreactive T-cell repertoire. *Immunol. Today* 12:193–195.

172. Moudgil, K.D. and E.E. Sercarz,. 1994. The T cell repertoire against cryptic self determinants and its involvement in autoimmunity and cancer. *Clin. Immunol. Immunopathol.* 73:283–289.

173. Benichou, G., E. Fedoseyeva, C.A. Olson, H.M. Geysen, M. McMillan and E.E. Sercarz. 1994. Disruption of the determinant hierarchy on a self-MHC peptide: Concomitant tolerance induction to the dominant determinant and priming to the cryptic self-determinant. *Int. Immunol.* 6:131–138.

174. Moudgil, K.D., E.E. Sercarz and I.S. Grewal. 1998. Modulation of the immunogenicity of antigenic determinants by their flanking residues. *Immunol. Today* 19:217–220.

175. Warnock, M.G. and J.A. Goodacre. 1997. Cryptic T-cell epitopes and their role in the pathogenesis of autoimmune diseases. *Brit. J. Rheumatol.* 36:1144–1150.

176. Milich, D.R., F. Schodel, D.L. Peterson, J.E. Jones and J.L. Hughes. 1995. Characterization of self-reactive T cells that evade tolerance in hepatitis B e antigen transgenic mice. *Eur. J. Immunol.* 25:1663–1672.

177. Quaratino, S., M. Feldmann, C.M. Dayan, O. Acuto and M. Londei. 1996. Human self-reactive T cell clones expressing identical T cell receptor beta chains differ in their ability to recognize a cryptic self-epitope. *J. Exp. Med.* 183:349–358.

178. Lanzavecchia, A. 1995. How can cryptic epitopes trigger autoimmunity? *J. Exp. Med.* 181:1945–1948.

179. Simitsek, P.D., D.G. Campbell, A. Lanzavecchia, N. Fairweather and C. Watts. 1995. Modulation of antigen processing by bound antibodies can boost or suppress class II major histocompatibility complex presentation of different T cell determinants. *J. Exp. Med.* 181:1957–1963.

180. Salemi, S., A.P. Caporossi, L. Boffa, M.G. Longobardi and V. Barnaba. 1995. HIVgp120 activates autoreactive CD4-specific T cell responses by unveiling of hidden CD4 peptides during processing. *J. Exp. Med.* 181:2253–2257.

181. Wood, P. and T. Elliott. 1998. Glycan-regulated antigen processing of a protein in the endoplasmic reticulum can uncover cryptic cytotoxic T cell epitopes. *J. Exp. Med.* 188:773–778.

182. Djaballah, H. 1997. Antigen processing by proteasomes: Insights into the molecular basis of crypticity. *Mol. Biol. Rep.* 24:63–67.

183. Coomber, D.W., N.J. Hawkins, M.A. clark and R. Ward,L.. 1999. Generation of anti-p53 Fab fragments from individuals with colorectal cancer using phage display. *J. Immunol.* 163:2276–2283.

184. Ward, R.L., N.J. Hawkins, D. Coomber and M.L. Disis. 1999. Antibody immunity to the HER-2/neu oncogenic protein in patients with colorectal cancer. *Human Immunol.* 60:510–515.

185. Baboonian, C., P.J.W. Venables, J. Booth, D.G. Williams, L.M. Roffe and R.N. Maini. 1989. Virus infection induces redistribution and membrane localization of the nuclear antigen La (SS-B): A possible mechanism for autoantibody. *Clin. Exp. Immunol.* 74:454–459.

186. Andreassen, K., U. Moens, H. Nossent, T.N. Marion and O.P. Rekvig. 1999. Termination of human T cell tolerance to histones by presentation of histones and polyomavirus T antigen provided that T antigen is complexes with nucleosomes. *Arth. & Rheum.* 42:2449–2460.

187. Rekvig, O.P., U. Moens, A. Sundsfjord, G. Bredholt, A. Osei, H. Haaheim, T. Traavik, E. Arnesen and H.-J. Haga. 1997. Experimental expression in mice and spontaneous expression in human SLE of polyomavirus T-antigen. A molecular basis for induction of antibodies to DNA and eukaryotic transcription factors. *J. Clin. Invest.* 99:2045 2054.

188. Moens, U., O.-M. Seternes, A.W. Hey, Y. Silsand, T. Traavik, B. Johansen and O.P. Rekvig. 1995. *in vivo* expression of a single viral DNA-binding protein generates systemic lupus erythematosus-related autoimmunity to double- stranded DNA and histones. *Proc. Natl. Acad. Sci. USA* 92:12393–12397.

189. Reeves, W.H., X. Dong, J. Wang and K. Hamilton. 1997. Initiation of autoimmunity to self-proteins complexes with viral antigens. *Ann. NY Acad. Sci.* 815:139–154.

190. Nemazee, D. 1985. Immune complexes can trigger specific, T cell- dependent, autoanti-IgG antibody production in mice. *J. Exp. Med.* 161:242–256.

191. Fehr, T., M.F. Bachmann, E. Bucher, U. Kalinke, F.E. Di Padova, A.B. Lang, H. Hengartner and R.M. Zinkernagel. 1997. Role of repetitive antigen patterns for induction of antibodies against antibodies. *J. Exp. Med.* 185:1785–1792.

192. Posnett, D.N. and J. Edinger. 1997. When do microbes stimulate rheumatoid factor? *J. Exp. Med.* 185:1721–1723.

193. Maldonado, M.A., V. Kakkanaiah, G.C. MacDonald, F. Chen, E.A. Reap, E. Balish, W.R. Farkas, J.C. Jennette, M.P. Madaio, B.L. Kotzin, P.L. Cohen and R.A. Eisenberg. 1999. The role of environmental antigens in the spontaneous development of autoimmunity in MRL-lpr mice. *J. Immunol.* 162:6322–6330.

194. Hang, L.-M., J.H. Slack, C. Amundson, S. Izui, A.N. Theofilopoulos and F.J. Dixon. 1983. Induction of murine autoimmune disease by chronic polyclonal B cell activation. *J. Exp. Med.* 157:874–883.

195. Wong, F.S., B.N. Dittel and C.A.Jr. Janeway. 1999. Transgenes and knockout mutations in animal models of type 1 diabetes and multiple sclerosis. *Immunol. Rev.* 169:93–104.

196. Theofilopoulos, A.N. and D.H. Kono. 1999. The genes of systemic autoimmunity. *Proc. Assoc. Am. Physicians.* 111:228–240.

197. Heward, J. and S.C. Gough. 1997. Genetic susceptibility to the development of autoimmune diseases. *Clin. Sci.* (Colch) 93:479–491.

198. Cooper, G.S., F.W. Miller and J.P. Pandey. 1999. The role of genetic factors in autoimmune disease: Implications for environmental research. *Environ. Health Perspect.* 107:693–700.

199. Garchon, H.-J. 1993. Non-MHC-linked genes in autoimmune diseases. *Curr. Opin. Immunol.* 5:894–899.

200. Lander, E. and L. Kruglyak. 1995. Genetic dissection of complex traits: Guidelines for interpreting and reporting linkage results. *Nature Genet.* 11:241–247.

201. Vyse, T.J. and J.A. Todd. 1996. Genetic analysis of autoimmune disease. *Cell* 85:311–318.

202. Lander, E.S. and N.J. Schork. 1994. Genetic dissection of complex traits. *Science* 265:2037–2048.

203. Dietrich, W., H. Katz, S.E. Lincoln, H-S. Shin, J. Friedman, N.C. Dracopoli and E.S. Lander. 1992. A genetic map of the mouse suitable for typing intraspecific crosses. *Genetics* 131:423–447.

204. Weissman, S.M. 1995. Genetic bases for common polygenic diseases. *Proc. Natl. Acad. Sci. USA* 92:8543–8544.

205. Dib, C., S. Faure, C. Fizames, D. Samson, N. Drouot, A. Vignal, P. Millasseau, S. Marc, J. Hazan, E. Seboun, M. Lathrop, G. Gyapay, J. Morissette and J. Weissenbach. 1996. A comprehensive genetic map of the human genome based on 5,264 microsatellites. *Nature* 380:152–154.

206. Dietrich, W.F., J.C. Miller, R.G. Steen, M. Merchant, D. Damron, R. Nahf, A. Gross, D.C. Joyce, M. Wessel and R.D. Dredge. 1994. A genetic map of the mouse with 4,006 simple sequence length polymorphisms. *Nature Genet.* 7:220–245.

207. Wang, D.G., J.B. Fan, C.J. Sio, A. Berno, P. Young, R. Sapolsky, G. Ghandour, N. Perkins, E. Winchester, J. Spencer, L. Kruglyak, L. stein, L. Hsie, T. Topaloglou, E. Hubbell, E. Robinson, M. Mittmann, M.S. Morris, N. Shen, D. Kilburn, J. Rioux, C. Nusbaum, S. Rozen, T.J. Hudson and E.S. Lander. 1998. Large-scale identification, mapping, and genotyping of single-nucleotide polymorphisms in the human genome. *Science* 280:1077–1082.

208. Cargill, M., D. Altshuler, J. Ireland, P. Sklar, K. Ardlie, N. Patil, C.R. Lane, E.P. Lim, N. Kalayanaraman, J. Nenesh, L. Ziaugra, L. Friedland, A. Rolfe, J. Warrington, R. Lipshutz, G.Q. Daley and E.S. Lander. 1999. Characterization of single-nucleotide polymorphisms in coding regions of human genes. *Nature Genet.* 22:231–238.

209. Kruglyak, L. 1999. Prospects for whole-genome linkage disequilibrium mapping of common disease genes. *Nature Genet.* 22:139!44.

210. Altshuler, D., L. Kruglyak and E. Lander. 1998. Genetic polymorphisms and disease. *New Eng. J. Med.* 338:1626.

211. Cheung, V.G., M. Morley, F. Aguilar, A. Massimi, R. Kucherlapati and G. Childs. 1999. Making and reading microarrays. *Nature Genet.* 21:15–19.

212. Ermolaeva, O., M. Rastogi, K.D. Pruitt, G.D. Schuler, M.L. Bittner, Y. Chen, R. Simon, P. Meltzer, J.M. Trent and M.S. Boguski. 1998. Data management and analysis for gene expression arrays. *Nature Genet.* 20:19–23.

213. Khan, J., M.L. Bittner, Y. Chen, P.S. Meltzer and J.M. Trent. 1999. DNA microarray technology: The anticipated impact on the study of human disease. *Biochim. Biophys. Acta* 1423:17–28.

214. Bowtell, D.D. 1999. Options available—from start to finish—for obtaining expression data by microarray. *Nature Genet.* 21:25–32.

215. Ransay, G. 1998. DNA chips: State-of-the art. *Nature Biotech.* 16:40–44.

216. Heller, R.A., M. Schena, A. Chai, D. Shalon, T. Bedilion, J. Gilmore, D.E. Woolley and R.W. Davis. 1997. Discovery and analysis of inflammatory disease-related genes using cDNA microarrays. *Proc. Natl. Acad. Sci. USA* 94:2150–2155.

217. Lipshutz, R.J., S.P. Fodor, T.R. Gingeras and D.J. Lockhart. 1999. High density synthetic oligonucleotide arrays. *Nature Genet.* 21:20–24.

BASIC ASPECTS OF AUTOIMMUNE PROCESSES

2 | B Cell Tolerance and Autoimmunity

Valerie Kouskoff, Doron Melamed, and David Nemazee

1. INTRODUCTION

B lymphocytes originate in the primary lymphoid organs, which in mice and humans include primarily the fetal liver and the bone marrow, where they assemble antibody genes unique to each cell. The cells undergo a rather complex set of ordered maturation steps guided by, and in turn controlling, the expression of the cell's immunoglobulin chains. The cell surface-bound IgM antibody molecules that are ultimately formed are specialized to act as cell surface antigen receptors and are not secreted, linking the specificity of the antigen receptor to the antibody that may eventually be produced.

Lymphocytes of different specificities are generated randomly, and include autoreactive B lymphocytes, which pose a threat because of their remarkable potential to proliferate and to differentiate to high-rate antibody secretion. In this chapter, we review how such cells are generated and controlled. In particular, we review how various mechanisms reduce the precursor frequency and functionality of self-reactive cells as well as our imperfect knowledge about how tolerance may be broken during memory-type responses.

2. B CELL DEVELOPMENT

Immunoglobulin is made up of light (L) and heavy (H) chains encoded by separate gene loci. The developing B cells first assemble H-chain genes by assembling dispersed variable (V), diversity (D), and joining (J) minigene elements together through specific gene recombination. These rearrangements are regulated both by the expression of the V(D)J recombinase (1) and by receptor gene accessibility to recombination (2). To be recognized by recombinase, the two DNA elements to be joined must each be flanked by a recombination signal sequence (RSS) composed of a conserved heptamer motif that is followed by a spacer, usually of 12 or 23 bases, and conserved nonamer motif (3,4) Recombination occurs by double strand DNA cleavage at the RSSs followed by joining of the four free DNA ends. Depending on the relative transcriptional orientations of gene elements, rearrangement results in excision of intervening DNA or in DNA retention by inversion. Elements flanked by a 12 base pair spacer recombine with those bearing a 23 base pair spacer in *cis* (5,6), i.e., by intrachromosomal recombination.

2.1. V(D)J Recombination

V(D)J recombination is initiated by the lymphocyte-restricted RAG-1 and RAG-2 proteins (7,8). RAG1/2 complexes can carry out much of the reaction *in vitro* (9–12). Terminal deoxynucleotidyltransferase (TdT), which is expressed in some lymphoid cell stages, adds untemplated nucleotides to protein coding segment ends exposed by RAG activity (13–15). A set of ubiquitously expressed proteins that repair DNA damage are required to resolve the coding join reaction (16–20). Because the V(D)J recombination machinery randomly adds or deletes bases during joining, most V(D)J joins are out of frame and therefore fail to encode functional proteins. However, in developing fetal lymphocytes TdT is not expressed (21,22), joins lack N-regions (21–25), and the recombination joins can deviate significantly from random because small sequence identities at the junctions favor precise joins in particular frames (24–27).

2.2. Pro-B Cell Development

The variable portions of the immunoglobulin heavy (H) chain genes in a pro-B cell are assembled in a quasi-random V(D)J recombination process in which D-to-J joining is followed by V-to-DJ joins. The error-prone nature of this gene assembly necessitates monitoring of the production of functional IgM H chain protein (μ-chain), which is subsequently selected on the basis of its ability to assemble with the "surrogate light chain" components λ5 and Vpre-B (28,29). The pre-B receptor is formed when the μ-chain/surrogate light chain complex associates with the B cell-specific signal transducers Ig-α and Ig-β through the μ-chain's transmembrane tail (28,30). The assembled pre-BCR transmits a specific differentiation signal with four important and interrelated consequences: i) B cell developmental progression; ii) downregulation of V(D)J recombination; iii) continued cell survival, and iv) changes in growth factor sensitivity. Support for this model comes from data indicating that a complete membrane form of μ-H chain prevents assembly of a second H chain gene (31,32) and is required for B cell development (33–36). A number of well characterized surface markers and characteristic RNA expression patterns mark this differentiation, as do changes in the ability of cells to proliferate to bone marrow stromal cell-derived cytokines (1,22,37,38). Evidence has been obtained that the signals involved in these steps may be separable. An incompletely formed DJ-Cμ gene can make a protein product that associates with the surrogate L-chains (39) and turns off V(D)J recombination, but cannot foster continued development (32,40). Similarly, in certain Ig transgenic mice, H-chain proteins have been identified that mediate allelic exclusion, but not full developmental progression (41). The efficiency of the downregulation of V(D)J recombination stimulated by the pre-BCR is such that cells with one productive H allele almost always downregulate recombination before a second such rearrangement occurs on the other allele, yielding a significant fraction of B cells with VDJ assembly on only one allele (42). Consistent with the notion that an assembled pre-BCR is necessary for allelic exclusion signals, in the absence of the surrogate L chain component λ5, pre-B cells bearing two functional μ-chains can be found at early developmental stages (40,43). Curiously, allelic exclusion is restored at later stages by unknown mechanisms (43). Conversely, cells with two non-productive H alleles or cells that cannot rearrange Ig genes due to a defect in recombinase do not mature and die rapidly (34,36,40,44,45). This developmental checkpoint is also associated with low level expression of survival genes (22,378). Enforced expression of Bcl-xl can rescue B cells with aberrant H chain rearrangements from apoptosis, but do not allow their developmental progression (46).

2.3.   The Pre-BCR

Although it is widely assumed that pre-BCR signaling is similar to BCR signaling in more mature B cells, there are probably important differences. A clear difference exists in pre-BCR vs. BCR expression levels. There is biochemical evidence for the presence of cell surface-expressed pre-BCR on primary bone marrow pre-B cells (47) at very low levels, but transformed pre-B cell lines express higher levels (48–52). Moreover, the stages of development at which the pre-BCR appears on the cell surface is controversial, and may differ between mouse and human (51–53). In any case, the evidence for the current model for the role of the pre-B receptor is almost exclusively based on genetic rather than biochemical data.

That normal B cell development requires signaling molecules important in BCR-mediated activation of mature B cells, such as syk, btk, and Ig-α/β, has been shown by gene ablation experiments (54–56). Experiments with chimeric μ-chain/Ig-α/β constructs bearing intact or tyrosine-to-phenylalanine-mutated ITAMs revealed the critical nature of intact Ig-α or Ig-β ITAMs, and their sufficiency for signaling developmental progression and downregulation of H loci rearrangements in pre-B cells (57,58). This pre-B cell receptor signal also upregulates L chain gene rearrangements (36,37,49,57–59). Mice deficient in V(D)J recombination, Ig-β, or the membrane form of μ manifest a complete block in B cell development, whereas mice deficient in syk, btk, or λ5 have impaired B cell development. In humans, unlike mice, natural mutations in btk completely ablate B cell production at the pre-B stage (60). Mice deficient in V(D)J recombination or the membrane form of μ manifest a block in B cell development that can be relieved by the introduction of Ig transgenes (61–63), demonstrating that pre-BCR signaling is limited by μ protein expression.

Whether the pre-BCR signals upon interaction with an endogenous or exogenous ligand or has no ligand at all remains to be clarified. λ5-deficient mice are severely impaired in their ability to promote B cell development (43,50), but are somewhat "leaky" due to the rescue afforded by the production of authentic Ig light (L) chains (63,64). These data suggest that surrogate L-chain serves as a μ-chain chaperone. This hypothesis is bolstered by the recent demonstration that a transgene encoding a truncated μ-chain lacking the V and CH1 domains, which is unable to pair with either surrogate or authentic L chains, nevertheless capable of promoting B cell development (65). Similar results were obtained in mice expressing a human heavy chain disease protein transgene (66). In addition, certain μ-chains that fail to foster developmental progression (67,68) cannot associate with surrogate L-chain because of their CDR3 structure (69,70). Taking these data together, it now appears that the ability of μ-chain to assemble into a pre-BCR complex is a major developmental signal, driving pre-B cell proliferation and subsequent differentiation, at least in conventional, B-2 lineage cells. Recent data indicate that pre-B cells of the B-1 lineage may respond to pre-BCR signaling differently than do B-2 pre-B cells (71).

In discussing the early stages in mouse B cell development, it is convenient to refer to the nomenclature of Richard Hardy (37), who extensively characterized cell surface markers defining several intermediate stages in the transition from B-lineage-committed precursors (fraction A) to mature, recirculating B cells (fraction F). In fraction C (CD45R+/CD43+/BP-1+/HAS^{intermediate}), cells undergo V-to-(D)J assembly on the heavy chain loci generating some cells with functional μ-chains. Through the putative signaling mechanisms mentioned above, cells with functional μ-chains are rapidly recruited into the fraction C' compartment, which expresses a higher surface level of HSA (heat stable antigen, CD24) than fraction C and undergoes very rapid proliferation. This expansion is limited by the availability of pre-BCR signaling and lymphokines, such as IL-7 (72,73). At this time cells also downregulate V(D)J recombinase (1,37). After as many as 6 cell divisions, fraction C' cells stop proliferating and differentiate to a small pre-B cell phenotype with distinctive cell surface markers ("fraction D": CD43-/sIg-/CD45R^{intermediate}). The cells now re-express and redirect V(D)J recombinase activity to the Ig L-chain gene loci (59), allowing formation of L-chains and assembly of complete antigen receptors. Little is known about the mechanisms controlling DNA accessibility to recombinase (2,74), but inhibition of V(D)J recombination on the remaining non-productive H chain allele, which often has a D-to-J join, must be uncoupled from H-chain gene expression to maintain H-chain allelic exclusion while allowing light chain gene rearrangement.

2.4. Late B Cell Development

Developmental arrest and subsequent death by neglect follows failure to complete functional L-chain gene rearrangements at the late pre-B stage (35,36). Successful L-chain gene expression expression permits surface IgM expression at the earliest stage in development at which complete sIgM is expressed; such cells are, by definition, immature B cells (Hardy fraction E). This stage is subject to both positive and negative selection, the former promoting further differentiation and the latter imposed by immune tolerance. B cells that meet the selection criteria undergo further differentiation and subsequent maturation to the periphery (75). Some of these cells, by this time at a mature state, will be selected into the long-lived B cell pool, while the majority will be eliminated by peripheral mechanisms (29). In addition to feedback suppression of further V(D)J recombination (1,37,77), this maturation process involves acquisition of maturation markers such as IgD, CD21, CD23 and L-selectin (77,78).

3. THERE ARE MANY KINDS OF B LYMPHOCYTE

B lymphocytes go through a number of developmental stages after acquiring their surface antigen receptors, and it is likely that B-cells are screened for tolerance at all of these stages. Moreover, there are probably developmentally-distinct lineages of B-cells, such as B-1 cells (rev. in 79), and possibly marginal zone B cells that may have unique responses to self- and foreign antigens. For example, the self-renewing sIgM B-1 cell population may only need to be screened for tolerance early in development, whereas the so-called conventional B lymphocyte lineage is continually renewed from sIg bone marrow precursors, and tolerance induction must be ongoing throughout the life of the individual to be effective. Three distinct developmental stages during which B-cells are susceptible to self-tolerance can also be defined: i) immature, sIgM cells in the primary lymphoid organs (e.g., adult bone marrow, fetal liver); ii) small, recirculating sIgM/sIgD cells that are normally the majority of B-cells in the peripheral lymphoid organs; and iii) germinal center B-cells that have recently been activated and have altered their antigen specificity through somatic V-region hypermutation. Although these three types of B cell are phenotypically, functionally, and anatomically distinct, the expression of surface immunoglobulin, which can interact with the extracellular environment, is common to all. Current data now suggest that self-tolerance can be induced during all of these developmental stages, although the precise mechanisms and differences in tolerance-susceptibility remain to be elucidated. Different self-antigens probably induce tolerance primarily at one or another of these developmental stages (or not at all), depending on the pattern of tissue-specific expression, the developmental timing of expression, or the accessibility of the antigen in question.

Finally, that B cells from neonatal and adult mice can differ greatly in tolerance susceptibility has long been known (rev. in 80), and investigations into a number of the key parameters that play a role in this difference are ongoing (e.g., 81–83).

4. CLASSICAL STUDIES ON THE SPECIFICITY OF B CELL TOLERANCE

There is a large body of literature analyzing the specificities of individual B cells or their clonal progeny by scoring antibody secretion. These methods have often detected very high frequencies of autoreactive B cells in normal mice (rev. 84), and these autoreactive B cells demonstrate antibody reactivity to fixed cells, serum proteins, denatured DNA, and cytoskeletal proteins. While these results have often been interpreted as disproving the existence of B-cell tolerance, a different picture has emerged from detailed quantitative studies examining tolerance to experimental tolerogens or cell surface proteins, in which tolerance could greatly diminish the frequency and affinity of reactive B cells that could be induced to secrete antibody. These studies assessed the frequency of clones reactive to foreign antigen detected with and without pretreatments to induce immunological tolerance (85–88).

This experimentally-induced tolerance can be highly specific: For example, tolerance induced to the hapten dinitrophenol (provided in tissue culture as a multimeric conjugate with carrier protein) is profound and fails to tolerize clones reactive to the structurally-similar hapten trinitrophenol (89), and similar results have been obtained in comparisons of human antibody repertoires of different ABO blood groups (90,91). Additional support for the notion of B cell self tolerance is that the absence of B cell tolerance to self makes it difficult to envisage how the science of serology, with its routine demonstration of non-self specificity, could have emerged.

Attempts to resolve the apparent contradictions between the results with experimental tolerogens or immunogens and the studies indicating high frequencies of self-reactive B-cells are perhaps best represented by the work McHeyzer-Williams and Nossal (92), who devised a culture system capable of randomly activating and inducing high-rate antibody production in a high proportion of B cells. These cultures produced IgM in the absence, and IgG1 in the presence, of added IL-4, allowing comparison between the binding of bivalent and decavalent antibody. Culture supernatants were tested for antibody binding to methanol-fixed or -unfixed thymoma cells: The striking finding was that B cells producing IgM antibodies to intracellular antigens were extremely frequent (1 in 37 B cells), whereas the IgG1-form of the same antibodies rarely bound (one B cell in 3×10^6). Neither IgM nor IgG1 antibodies that bound unfixed cells, and were therefore specific for the cell surface, were detectable (less than one B cell in 3×10^6). These results indicate that there is a very stringent tolerance to cell-surface antigens, whereas tolerance to intracellular antigens may spare only low-affinity self-reactive B cells, whose binding requires multivalent contacts to be detectable. These differences probably reflect the relative inaccessibility of intracellular antigens compared to surface antigens.

5. THE MECHANISMS OF B CELL TOLERANCE

5.1. Central Tolerance (Tolerance in Immature B Cells)

The acquisition of sIgM renders developing B cells in the bone marrow susceptible to immune tolerance (29,75,93–95). Two competing concepts have emerged about immune tolerance in B lymphocytes. One postulates that developing lymphocytes pass through a tolerance-susceptible stage in which Ag encounter leads to deletion or functional inactivation. Cells progressing through this stage in the absence of Ag then lose tolerance sensitivity and acquire Ag-inducible effector function (96). The second notion is that B-cells require two inductive signals to become activated, one provided by Ag binding to the BCR and a second provided by a helper T-cell (97); however, cells receiving a BCR signal alone are tolerized. Accumulated experimental data over the

years suggests that both of these models are partly valid, since immature B cells are particularly tolerance-susceptible, but can be stimulated to effector function with strong costimulus (80,87,94,98,99), whereas mature B-cells can be eliminated by a strong BCR signal occurring in the absence of an appropriate co-stimulus (100–103). Early studies in mice demonstrating that chronic treatment from birth with anti-IgM Ab eliminated IgM+ cells in the BM and the periphery provided additional support for these theories, although the possibility of Ab-mediated cytotoxicity could not be ruled out (104). Early *in vitro* studies analyzing the mechanisms for the increased tolerance sensitivity of immature B cells found that BM or fetal B-cells treated with anti-IgM did not die rapidly, but lost sIgM expression, which could not be completely restored upon reculture without anti-IgM, whereas similarly-treated splenic B cells completely recovered sIgM expression (105,106). More recent studies, in contrast, have shown that purified BM immature B cells manifest accelerated death when treated *in vitro* with anti-IgM Abs as surrogate Ag (94,95,107,108). However, a large fraction of the purified B cells in these studies were apoptosis-resistant, suggesting the existence of another mechanism for tolerance induction. Importantly, these studies utilized anti-IgM Abs as surrogate Ag, which might poorly mimic Ag binding to the BCR (94,109). Since different anti-IgM Abs differ in their ability to stimulate B cells, the choice of anti-BCR antibody might also explain the contradictiory results regarding anti-IgM treated immature B cells (rev. in 109).

5.2. Antibody Transgenic Mice

Because classical experiments revealing self-tolerance in the B cell compartment could not directly show how this comes about, elucidating these mechanisms has been the goal of a host of recent experiments in mice that take advantage of transgenic and gene ablation technology. In antibody gene transgenics (Ig Tg), the mouse germline DNA contains pre-rearranged, functional antibody genes encoding an antibody of known specificity. Because antibody chain expression can inhibit further endogenous gene rearrangements, mice with a large clone of B cells of defined specificity can be generated and studied for self-tolerance. Such mice have been generated with specificities to MHC class I molecule K^k or K^b (the 3–83 mouse) (110,111), HEL (112), DNA (112–116), RBC antigen (117), TNP (118), and rheumatoid factor (119,120). Studies in central tolerance in these Tg mice were generally accomplished by breeding the Ig Tg mice with those expressing the appropriate natural or Tg-encoded Ag in the BM or, alternatively, by acute injection of the natural Ag. When 3–83 Tg mice were bred on an H-2^k or H-2^b genetic background, 3–83 Tg B cells were detected in large numbers in the bone marrow, but not in the periphery, indicating efficient deletion of the autoreactive B cells in the bone marrow prior to their complete maturation (99,110). Similarly, anti-HEL Tg B cells were deleted in mice co-expressing a membrane-bound form of HEL (mHEL) (112). Interestingly, immature anti-HEL B cells cultured in the presence of mHEL were developmentally-arrested, but failed to manifest accelerated death (77). In 3H9 H+L Tg mice, which encode anti-double strand DNA specificity, autoreactive B cells were deleted in the bone marrow, and those that matured to the periphery lost their anti-dsDNA specificity through L-chain gene receptor editing (see below) (113). High affinity rheumatoid factor (RF) Tg B cells were also deleted in the bone marrow when Tg mice were bred on a genetic background that generated the cognate Ag (120). In anti-TNP Tg mice, acute administration of Ag by i.v. injection caused deletion of mature and transitional B cells in the bone marrow, but not deletion of the immature IgM^{low} cells (118). In another system, anti-RBC Tg B cells were deleted in the bone marrow, but significant numbers of anti-RBC B1 cells were found in the peritoneal cavity secreting anti-RBC Abs) (117). How these cells managed to escape deletion is not clear, but their survival probably reflected the relative absence of RBCs in the peritoneal cavity, since intraperitoneal injection of RBC caused apoptosis of these cells and cured the hemolytic anemia of these mice (117,121). Despite the fact that these experiments were consistent with the notion of central elimination of self-reactive B cell specificities, significant numbers of apoptotic B cells in the BM were not detected (122).

Cognate Ag in the BM did not result in autoreactive B cell deletion in all cases. In mice transgenic for genes encoding a low affinity RF Ab, B cells expressing the Tg specificity were present in the peripheral lymphoid organs and were fully functional (119); whether these mice develop arthritis remains to be elucidated. Another well described system showed that, in mice double Tg for anti-HEL and a soluble form of HEL, anti-HEL B cells were present in large numbers in the periphery, but appeared to be functional inactivated (anergic) since they failed to signal through the sIg (123). However, this form of tolerance could be reversed by LPS stimulation, but not by T cell help (124). In addition, some groups have been able to demonstrate B cell anergy induction in normal (125–127) and anti-DNA Tg mice (114,115). Reversal of anergy could play a major role in the development of autoimmune diseases, but apparently anergic B cells have a very short lifespan *in vivo* (128), particularly when competing with non-anergic B cells (129), which greatly diminishes the distinction between anergy and deletion.

5.3. Receptor Editing

In two well-studied transgenic mouse models, central tolerance does not eliminate all the B cells in Ig Tg mice

expressing the cognate Ag in the BM (113,130). Instead, peripheral B cells in these mice lose the specificity encoded by the Tg and no longer recognize the original cognate Ag. Most of these B cells retained the Tg H-chain, but expressed a new endogenous κ or λ L-chain that could not be ascribed to a general lack of allelic exclusion, since these Tg demonstrated an excellent exclusion in the absence of Ag (in the 3–83 mouse) (130) or in the H or L-chain only 3H9 Tg mice (113). These findings led to the "receptor editing" hypothesis, which suggests a new mechanism for central tolerance, i.e., that upon encounter with self-Ag, immature sIg$^+$ B cells continue to express or upregulate V(D)J recombinase and undergo secondary rearrangements at the L-chain locus (93,130). Secondary rearrangements readily occur at the L-chain locus, since it is rearranged after H-chain genes and is still accessible in newly generated IgM$^+$ cells, whereas H-chain rearrangements occur at the pro-B/pre-B stage when the H loci are not accessible for further rearrangements (1,2). Furthermore, as shown in Figure 1 and discussed in Radic et al. and Tiegs et al. (93,130), the lack of retention of D elements renders secondary rearrangement at the H-chain loci impossible by conventional V(D)J recombination. [Non-conventional VH gene replacement events can occur by VH to VDJ recombination utilizing a conserved heptamer site within VH gene framework 3, but these events mainly occur in pro-B cells (131)]. In contrast, as shown in Figure 3, secondary rearrangements at the κL-chain loci can occur by deletion or inversion, or can take place in the λ locus (93,130). Moreover, the recombining sequence element (RS) found in the κ locus can rearrange into the Jκ intron or to Vκ genes, effectively silencing the non-productive or self-reactive VJ by deleting Cκ (Figure 3). The first demonstration of the ability of BCR autoreactivity to induce secondary rearrangements was obtained in the 3–83 mouse bred on the H-2^b genetic background (130), and in 3H9 anti-DNA Ig Tg mice (113). In the former study, an increased number of B cells bearing λ-chain and the Tg H chain was found in the periphery accompanied by significantly increased expression of RAG genes and λ rearrangement excision products in the BM. Protection of the B cells in the BM was afforded by enforced expression of bcl-2 Tg, extending the time allowed to undergo receptor editing and resulting in significantly more peripheral B cells that lost their Tg specificity and acquired an endogenous λ L-chain (132). As shown by gene analysis of hybridomas isolated from the 3H9 Tg spleen, in most cases the transgenic L-chain was deleted and replaced by endogenous κ or λ, resulting in loss of anti-dsDNA specificity (113). Since the κ Tg in these mice was not inserted into the normal κ locus, the κ Tg could rarely be extinguished by nested rearrangements, making the secondary rearrangements less effective in rescuing these B cells and resulting in very small numbers of B cells in the periphery

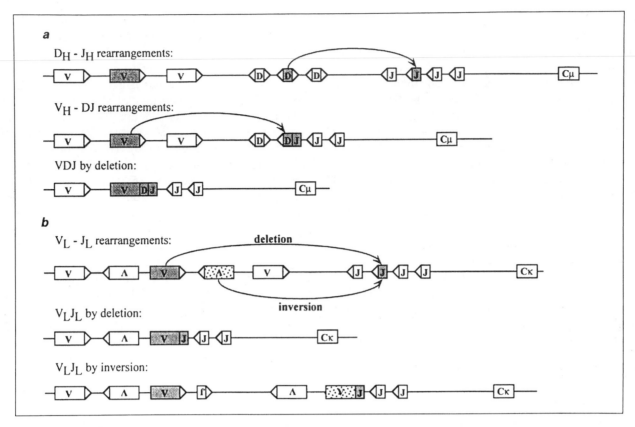

Figure 1.   An ordered pattern of Ig gene rearrangements guides the development of B cells in the BM. *a* Ig gene rearrangement at the pre-B stage starts at the H-chain locus with ligation of DH elements to JH elements (top), followed at the large pre-B II stage by V to DJ rearrangements (middle) to form a complete VDJ (bottom). V to DJ rearrangements occur primarily by deletion of the intervening DNA. *b* Rearrangement in the small pre-B II stage at the L-chain loci occurs by ligation of VL genes to JL. Rearrangements can occur by deletion of the intervening DNA or by inversion, depending on the transcriptional orientation of the VL gene. The structure of the IgL (kappa) locus permits "Leapfrog" secondary rearrangements between V's and J's on either side of an initial VJ join (see Figure 3).

(113.130,133). Although it was still possible that these B cells represented peripheral selection processes of rare variants rather than successfully edited cells, this possibility was excluded by studies generating mice with germline-targeted anti-DNA or anti-MHC class I autoAb genes in their natural genomic context (134,135) wherein receptor editing was extensive and allowed many B cells to mature to the periphery.

Receptor editing has been studied in bone marrow cultures. Treatments of 3–83 Tg bone marrow cells with anti-idiotypic antibody caused significant induction of secondary rearrangements in the relative absence of cell death (136). The occurrence of receptor editing was also demonstrated in IL-7-driven bone marrow cultures from 3–83 Tg and normal mice (133), which produced large numbers of immature B cells with high purity. Grown in the presence of stromal cells bearing the cognate Ag (in the case of 3–83 B cells) or when anti-κ was used as surrogate Ag (in the case of non-Tg B cells), B cells were developmentally-arrested at the immature B cell (fraction E) stage and failed to acquire maturation markers. RAG gene expression, DNA rearrangements, and L-

chain isotype surface expression in the majority of the cells at this stage exhibited intensive attempts to undergo receptor editing. It is noteworthy that no accelerated death was detected in these cultures at this stage of development as a result of the presence of Ag, making it possible that receptor editing is a major mechanism in the development of self-tolerance, although this contradicts studies showing increasing death of immature B cells after binding self-Ag or treatment with anti-BCR Abs (94,95,107,108). In our laboratory, utilizing IL-7-driven BM cultures, we were able to show that both apoptosis and receptor editing contribute to central B cell tolerance (137). Our experiments suggested that the central tolerance mechanism is regulated during B cell development, and receptor selection is compartmentalized from clonal selection. We found that IgM+ cells develop from a receptor editing-competent, apoptosis-resistant stage (immature B cell) to a receptor editing-incompetent, apoptosis-sensitive stage (committed B cell) (137). A schematic diagram demonstrating this model is shown in Figure 4. This finding implies that most B cell central tolerance occurs through receptor editing, since the presence of antigen arrests

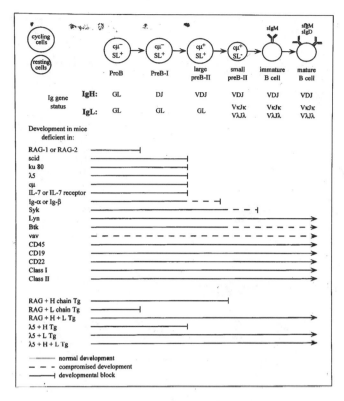

Figure 2. B cell developmental progression requires appropriate gene expression. Progression along the pathway from a pro-B stage to a mature B cell is shown in the upper panel. Disruption of different genes may or may not effect the development of B cells. The effect of specific gene disruptions on B cell development, and the ability of Ig-Tgs to compensate are shown in the bottom panel.

the development in the receptor editing-competent, apoptosis-resistant stage, and does not allow further development to the apoptosis-resistant stage. It is worth noting that these findings are based on a specific *in vitro* antigen:BCR interaction; possibly the nature of the Ag, Ag:BCR affinity, different growth factors, or microenvironmental substances present in the BM may also contribute to the development of B cell tolerance (94,95,138).

5.4. Caveats Regarding the Use of Antibody-Gene Transgenic Mice

The development of Ig Tg B cells in the bone marrow differs from that of non-Tg bone marrow B cells in that they do not need to rearrange Ig genes. For example, Spanopoulou et al. found that bone marrow cells in 3-83 H+L Tg mice lack cells at the intermediate stage between early pre-B and immature B cells, suggesting that the Ig transgenes accelerate development (36). A similar acceleration of B cell development could be detected in mice expressing anti-HEL H+L-chain genes (123). *In vitro* experiments utilizing normal and 3–83 Tg B cell precursors showed that the presence of Ig transgenes

suppresses RAG gene expression, and B cell maturation is not inhibited by the presence of IL-7, as occurs in normal precursors (139). Thus, the presence of an Ig Tg allows B cells to efficiently circumvent developmental checkpoints and promotes their maturation. In addition, the insertion site of the Tg and its copy number are variable and usually non-physiological. To overcome this problem and generate more normal mice, rearranged Ig genes were inserted into the natural loci through homologous recombination. Developing B cells in these "knock-in" mice were predicted to express the inserted genes at the correct developmental stage. It is hoped that the model provided by these mice will allow a better analysis of Ig gene rearrangements and understanding of central and peripheral tolerance mechanisms. It is noteworthy that despite being targeted to the appropriate locus, B cell development in Ig knock-in mice still differs from that in non-manipulated mice, possibly because of premature expression (140), and in the case of IgH loci, because of the artefactual presence of D elements upstream of joined VDJ elements (141–143).

5.5. Peripheral B Cell Tolerance

Not all self-Ags are present in the BM during B cell development. Newly generated B cells migrating from the BM to the periphery are highly sensitive to tolerance (95,111,144). Immature B cells from neonatal spleen (82), immature splenic B cells from irradiated and autoreconstituted mice (144), or certain immature phenotype sIgM+ B cell lymphomas (145) undergo programmed cell death *in vitro* following BCR crosslinking by anti IgM Abs. Acquisition of sIgD, a maturation marker for B cells, generally failed to protect from apoptosis (144), while stimulation of immature B cells with LPS and anti-IgM suppressed their response in one study (82), and protected from apoptosis and induced responsiveness in another study (146). In contrast, mature splenic B cells treated with soluble anti-IgM are activated (147–149), but undergo apoptosis when cultured with plastic-immobilized or hyper crosslinked anti-IgM (101,150). Moreover, Finkelman et al. (160) showed that *in vivo* infusion of anti-IgD monoclonal antibodies could cause B cell deletion in non-Tg mice. The profound differences between immature and mature B cell responses to peripheral antigenic stimulation are beginning to be defined experimentally (94,95).

5.6. The Relevance of Anergy in B Cells

Goodnow, Basten and colleagues carried out the earliest experiments using immunoglobulin transgenic mice to study B-cell tolerance. These investigators bred anti-hen egg lysozyme-transgenic mice with mice expressing a transgene-encoded, soluble form of lysozyme, and found that the resultant mice generated substantial numbers of

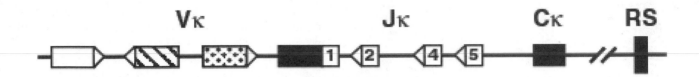

A. secondary VκJκ5 rearrangement by deletion:

B. secondary VκJκ5 rearrangement by inversion:

C. deletion of Cκ by V-RS rearrangement:

Figure 3. Secondary rearrangements and receptor editing can occur in the L-chain locus. A primary VκJκ1 rearrangement is shown in the upper panel. Shown below are three ways that secondary rearrangements can occur. Secondary, nested rearrangements of an upstream Vκ to Jκ5 can delete (line A) or invert (line B) the intervening DNA, depending on the transcriptional orientation of the Vκ gene. Such rearrangements can preserve the functionality of the locus and encode new κ-chains. Alternatively, RS rearrangements can delete the Cκ (line C) and allow further rearrangements at the other κ-allele or at the λ-locus.

splenic B-cells that, upon analysis in a number of functional assays, were "anergic" (i.e., hyporesponsive) and manifested an unusual depletion of B-cells from the marginal zone of the splenic lymphoid follicle (123,151,152). Furthermore, the anergic phenotype was characterized by an inability to signal through sIg, and functional hyporesponsiveness reversible upon lipopolysaccharide stimulation, but not by T cell help (124). Until recently, these results have presented an enigma due to the assumption that the anergic B-cells, being present in the spleen in numbers comparable to those of the antigen-free control mice, were similarly long-lived. The implication was that these B cells might either have an affirmative function in the immune system, or might pose a threat of autoimmunity should their functional inactivation be reversible. It was later shown that the anergic B-cells have an extraordinarily short lifespan *in vivo*, thus greatly diminishing the effective distinction between anergy and deletion (128). This shortened lifespan is thought to depend on competition among B-cells for microenvironmental niches in lymphoid follicles (129), although other models to

explain these results have not been ruled out. Recent experiments in the anti-H-2K^b transgenic model system have demonstrated that central deletion is not dependent on such competition (163), which also suggests that one need not invoke a function for the anergic B-cells that, had they been longer-lived, might have played a suppressive role as toleragenic, self antigen-specific presenting cells (discussed in 153). Other groups have provided evidence for anergy *in vivo* in tolerized normal (125,126,127) and immunoglobulin transgenic mice (114). Anergy in these systems is also associated with short cell lifespan *in vivo*. A B-cell:B-cell competition-dependence of anergic B-cell lifespan implies that a lymphopenic environment could support greater numbers of autoreactive, anergic cells, thus increasing the potential risk of the reversal of tolerance.

Transgenic mice have also provided insights into peripheral tolerance. Again, the use of monoclonal B cell populations with known specificity allowed investigators to follow the B cells *in vivo* and study their response to tissue-specific Ags. Russell et al. and Lang et al. showed that hepa-

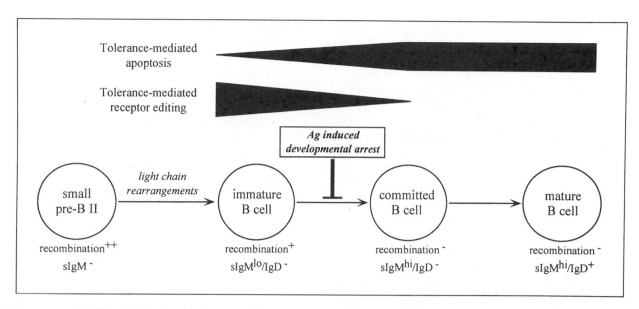

Figure 4. A speculative model for the development of B cell tolerance. It is postulated that IgM+/IgD– B cells progress from a receptor editing-competent, apoptosis-resistant phenotype [immature B cell, IgM(low)/IgD–] to a receptor editing-incompetent, apoptosis-sensitive phenotype [committed B cell, IgM(high)/IgD(low)]. In the absence of Ag, cells can then progress into a mature stage, whereas the presence of Ag in the BM arrests development at the immature stage and induces secondary rearrangements leading to receptor editing.

tocyte or keratinocyte expression of class I molecule K^b resulted in elimination of reactive 3–83 Tg B cells and a profound reduction in anti-K^b serum Ab in genetic crosses (111,132). These mice, lacking B cells in the peripheral lymphoid organs, had normal Tg B cell development in the bone marrow. Co-expression of a bcl-2 Tg completely rescued the Tg B cells when Ag was expressed on keratinocytes, but only partially when Ag was expressed on hepatocytes (132). Similarly, anti-HEL Tg B cells were not deleted when membrane-bound HEL was expressed on thyroid epithelium (154). Thus, it appears that the determination of the fate of the B cell may be influenced by factors such as Ag density, concentration, tissue location, and developmental stage of the B cells at time of antigen encounter. Honjo and colleagues, in studies using anti-RBC Tg mice, showed that peritoneal injection of Ag or forms of anti-Ig Abs resulted in deletion of the B1 (peritoneal) as well as the B2 (conventional) B cells from the peritoneal cavity (103,117). Expression of the anti-apoptotic bcl-2 Tg protected peritoneal B cells from deletion in both 3–83 Tg and anti-RBC Tg mice, although this protection could be overcome by increasing Ag dose (103,121,132). Tolerance-induced peripheral B cell death is the most likely explanation for these results.

5.7. B Cell Tolerance During Germinal Center Responses

T-dependent antigen stimulation through immunoglobulin that induces variable region somatic hypermutation in germinal centers can allow B cells to change their antigen specificity, a process with the potential to generate autoreac-

tive B-cells whose reactivity to self must be controlled. Evidence has been presented for the existence of a second tolerance window, and possibly a unique B-cell sublineage, in memory B cells during or after somatic V-gene hypermutation (155–157). These experiments attempted to tolerize hapten-carrier immunized cells by treatment with the initial hapten on a different carrier, thereby providing sIg crosslinking signals, but no T cell help. Tolerance was assessed by antibody secretion either in whole animals or in splenic fragment culture. Studies by Nossal's group (158,159) showed that the disruption of ongoing T-cell help by high dose, deaggregated carrier protein is a far more important means of memory B-cell tolerance induction than direct sIg crosslinking in these B-cells. These results indicated that antigen-activated B cells that lose affinity or specificity for the immunizing antigen are at a competitive disadvantage in obtaining T-cell help compared with B cells that maintain specificity. This idea might be expanded to suggest that hypermutating germinal center B cells that acquire self-specificity might be eliminated because of restriction in T cell help due to the change in specificity. More recent studies, however, showed that injection of soluble Ag induces apoptosis of high affinity germinal B cells (160–163). The fact that some of these studies were performed in normal mice and used natural Ag rather than anti-IgM Ab strongly support this idea in general (160–162), but it is curious that short-term treatment with soluble immunogen can also induce death in this context. In anti-HEL Tg mice, soluble HEL injection at the peak of the GC response rapidly eliminated anti-HEL-specific B cells in two apoptotic cycles, one within the GC and the second in cells that migrated to lymphoid zones rich in T cells (163). Further studies showed that anti-HEL B cells are excluded from the follicles in the presence of soluble HEL because they fail to compete with non-self B cells for follicular niches and undergo apoptosis (129,164). It has recently become clear that the degree of BCR signaling by Ag and the extent of T cell help determines whether these B cells are activated or undergo apoptosis (165). T cell help in the form of pre-activated CD4+ T cells allowed HEL-binding B cells to migrate into the follicle and form germinal centers (165), and other B cell molecules such as CD45, CD19, and CD22 have been shown to tune the interaction between the BCR and cognate Ag up or down, effectively altering the tolerance process (rev. in 138). Many years ago, Teale and Klinman were able to show in a clonal, splenic focus assay that peripheral tolerance in B cells is an active process requiring synthesis of new RNA and proteins, and the presence of T cell help could reverse this tolerance process (166,167). In the same manner, many studies have shown that anti-CD40 Ab and IL-4 can block apoptosis of resting B cells mediated by multimeric anti-IgM Abs (168–171). Apparently, in the absence of T cell help, self-reactive B cells undergo programmed cell death upon binding self Ags in the periphery.

Several groups have recently conducted investigations into the role of receptor editing in the germinal center response of B cells. The possibility of mature B cells being able to alter their specificity strongly contradicts the strict idea of allelic exclusion. Nevertheless, it has been shown that high levels of RAG-1 and RAG-2 proteins, V(D)J recombination intermediates and increased $V\lambda 1$-$J\lambda 1$ DNA excision products could be found in activated B cells in germinal centers or in resting splenic B cells stimulated with IL-4 in association with several different co-stimuli (172–174). DNA breaks were shown at recombination signal sequences of the κ and H-chain loci in a subset of GC B cells (175), or in resting B cells from H and L knock-in mice stimulated with IL-4 and LPS (176), indicating DNA rearrangements. The biological role, if any, of receptor editing in the periphery in rescuing B cell apoptosis is still unclear. It has been suggested that such secondary rearrangements are induced in B cells with diminished Ag binding as a result of somatic mutation, which are usually eliminated from the GC response (175). However, the biological reason for saving these B cells is still speculative, and it is not clear to what extent receptor editing occurs in the periphery under physiological conditions. Further information about receptor editing in the periphery can be found in reference 177.

6. B CELLS IN AUTOIMMUNITY

6.1. Implications for Autoimmunity

Whether a defect in B-cell tolerance plays a role in autoimmunity remains a key unanswered question. B-cells are obviously critical for many aspects of autoimmunity

(e.g., 178). Theoretically, such a defect alone could cause systemic autoimmune disease. The increasing evidence for an intrinsic B-cell defect in murine autoimmunity is reviewed below.

6.2. Evidence for an Intrinsic B Cell Defect in Murine Autoimmunity

A number of well-known mouse strains spontaneously develop lupus-like autoimmunity, such as MRL-*lpr*, (NZB X NZW)F$_1$, NZM and substrains, and BXSB male mice. As they age, these mice produce large amounts of many kinds of autoantibodies, such as anti-DNA, and suffer from a variety of manifestations of immune complex formation, including kidney failure. Various studies over the years have shown that the autoimmune phenotype can be transferred by bone marrow in a T cell-dependent manner (rev. in 179,180), but there is a considerable body of evidence that these mice have intrinsic B-cell defects as well (181–185). These mice have been analyzed to determine if their intrinsic defect is in self-tolerance in the T helper cells alone, the assumption being that an autoreactive T-cell can recruit normal B-cells into the response against self-tissue. These experiments produced mixed bone marrow chimeras or mixed embryo aggregation chimeras in which T cells and allotype-marked B-cells of the autoimmune-prone and normal genotypes developed together. In this situation, recruitment of normal B cells into the autoimmune response would leave the fraction of auto-antibody produced from B-cells of the autoimmune-prone and normal genotypes the same, or at least proportional to the relative amounts of total serum Ig of the parental allotypes. The striking and consistent finding in all of these studies has been that B-cells of the autoimmune-prone genotype secrete virtually *all* of the autoantibody in these mixed chimeras (185–187). This would indicate an intrinsic B-cell defect in the autoimmune-prone mouse strains. It should be emphasized that there is little evidence for a defect in T-cell thymic deletion by self-superantigen in autoimmune-prone mice (180,188).

Defects in programmed cell death in B cells may play a role in autoimmunity. As a result of chromosomal translocations that bring the Bcl-2 oncogene into proximity with the Ig heavy chain enhancer, the Bcl-2 oncogene is frequently activated in human B-cell follicular lymphomas. Bcl-2 overexpression been shown to inhibit programmed cell death in IL-3-dependent B cell lines deprived of IL-3 (189). Two groups have described transgenic mice expressing Bcl-2 under the control of the immunoglobulin heavy chain enhancer (190,191) wherein these mice manifest B-cell hyperplasia and prolonged B-cell survival *in vivo* and *in vitro*. Interestingly, one of these transgenic mouse lines also manifests early mortality as a result not of B-cell tumors, but of lupus-like glomerulonephritis (191). Since this transgenic mouse line expresses Bcl-2 in the B cells, but not detectably in T cells, a defect in B-cells alone might be responsible for the autoimmunity.

Extensive investigations of a transgenic mouse line expressing heavy and light chain genes encoding an erythrocyte autoantibody have been performed by Honjo and colleagues; surprisingly, some of these transgenic mice develop hemolytic anemia even on a non-autoimmune-prone background (C57BL6). This group has established that B cells of these mice are tolerized to self-red blood cells (RBCs) by clonal deletion, but some B-cells manage to make their way to the peritoneum, where low levels of antigen apparently allow their persistence, expansion, and progression to antibody secretion (192). Intraperitoneal injection of RBCs leads to rapid apoptosis of these B-cells and regression of clinical signs of hemolytic anemia (117). Interestingly, this inducible peripheral deletion could be blocked by Bcl-2 transgene overexpression. These results show that, in this case, induction of antigen-mediated B cell tolerance is sufficient to prevent autoimmune disease, which implies that defects in B cell tolerance alone may sometimes induce autoimmune disease. Additional support for this view has been provided by experiments measuring the *in vivo* tolerance susceptibility of B-cells in autoimmune-prone mice (103).

That pre-B cell clones derived from lupus-prone NZB/NZW F_1 (B/W) mice, but not normal mice, can differentiate in SCID mice and secrete high titer anti-nuclear antibodies, including anti-DNA, was shown by Reininger et al. (193). Most important, these cells gave rise to lupus-like autoimmune disease in the apparent absence of T cells in a subset of recipients, suggesting again that intrinsic B-cell defects can induce autoimmune disease. Taken together, these studies provide strong evidence that intrinsic B-cell defects, possibly including self-tolerance defects, may be at the root of some autoimmune diseases.

6.3. Lack of Allelic Exclusion as a Potential Cause of Autoreactivity

It is accepted that B cells play an important role in the development of autoimmune diseases, but it is not clear whether this reflects a defect in B cell tolerance. Clearly, B cell activation in an autoimmune disease requires the existence of T cells (179,180). Self-reactive B cells can be generated in the BM, but should be deleted there or in the periphery by receptor editing or apoptosis (93–95). In some cases, however, autoreactive B cells can escape central tolerance and mature to the periphery (114,115,119,194–196). A lack of allelic exclusion, i.e., that some cells express two different receptors or continue to undergo Ig rearramgements after emerging from BM is one important mechanism suggested to explain the persistence of autoreactive B cells (115,194). A growing number of recent studies have shown that L-chain allelic exclusion is not complete: For

example, in κ Tg mice, a significant number of B cells apparently ignored the expression of the Tg and rearranged endogenous L-chain genes (197,198). In IL-7-driven B cell cultures, it was also demonstrated that κ-expressing cells could continue to rearrange L-chain and express λ L-chain, or both λ and κ (199,200). The lack of allelic exclusion could also explain the existence of detectable numbers of $\kappa^+\lambda^+$ dual receptor B cells in human peripheral blood (201). Similarly, double recombination Vκ-Jκ products were found in many splenic B cells, indicating that a functional κ expression does not necessarily terminate L-chain rearrangements (202). Although these ongoing rearrangements, especially in normal B cells, can be explained by the acquisition of autoreactivity and consequent receptor editing (93), editing in *trans*, i.e., on the non-autoreactive allele of L-chain genes, might explain how self-reactive B cells can escape deletion. Similarly, in mice transgenic for a self-reactive TCR, T cell expression of a second non self-reactive receptor was able to rescue T cells from thymic deletion (203). This was a result of expression of a second α-chain, since allelic exclusion in the α-chain locus is incomplete, frequently resulting in the generation of T cells that express two receptors (204).

Maturation of autoreactive B cells could be permitted by incomplete allelic exclusion at the H-chain. For example, in R4A-γ2b Tg mice that express the H-chain of an anti-dsDNA Ab, stimulation of spleen cells with LPS resulted in secretion of anti-DNA Abs (194). Gene analysis of hybridomas revealed that many of the anti-DNA producing cells co-expressed an endogenous μ-chain, which probably allowed their development and maturation. However, in long-term BM cultures of H+L Tg B cells (anti HEL) an increased number of cells expressing an endogenous H-chain was detected (205). Further evidence that allelic inclusion can be stably expressed in B cells comes from mice whose IgH alleles were engineered to encode two distinct H-chains in which peripheral B cells express both H-chains (205). Thus, neither allelic exclusion nor B cell tolerance are 100% complete, allowing for the existence of mature B cells in the periphery that express more than one H or L-chain and that might have self-specificity.

6.4. Studies in Autoimmune Experimental Models

B cells are activated to secrete Abs specific to Ags expressed in damaged tissue in many autoimmune diseases. For example, in myasthenia gravis, B cells secrete Abs to the acetylcholine receptor found at the neuromuscular junctions, causing progressive weakness and death. Production of Abs to RBC triggers RBC destruction and the development of autoimmune hemolytic anemia, while autoantibodies to basement membrane collagen cause severe damage to the renal glomeruli in Goodpasture's syndrome. In contrast, there are other, non-organ-specific autoimmune diseases in which B cells secrete a wide range of autoantibody specificities. Systemic lupus erythematosus (SLE) is a well-studied autoimmune disease partially characterized by the production of high titers of antibodies to nuclear antigens such as DNA and Sm proteins (180,206–208). Rheumatoid arthritis (RA) is characterized by the production of high titers of anti-IgG rheumatoid factors (RF), antibodies that generate immune complexes in the joints and cause an inflammatory process contributing to joint destruction (209).

Experimental mice have provided models for many autoimmune diseases. One of the most intensively studied is SLE, where interest has centered on the generation of anti-nuclear Abs. Two genetic murine models for SLE have been intensively studied: (NZBxNZW)F₁ hybrid mice (208) and MRL-*lpr* mice (179,206), spontaneously develop anti-nuclear Abs and other characteristics of SLE, such as glomerulonephritis. The MRL-*lpr* model however, is also characterized by a significant lymphoproliferation as a result of a mutation in the Fas molecule that mediates death signals to cells (210). Studies have shown that lack of Fas is not necessarily the direct reason for the break in B cell self-tolerance in these mice since, in Ig Tg mice bred to the *lpr* background, tolerance was maintained (211–213).

Many hybridomas that secrete anti-nuclear Abs have been generated from these mouse models of SLE over the years, and the expressed Ig genes in these hybridomas have been sequenced and analyzed for possible selection of particular V gene families participating in anti-nuclear Ab production. Generally, the data generated suggested that particular VH and VL genes recurrently participate in the anti-nuclear antibody response (214). Sequence analysis of anti-DNA Abs indicated that arginine residues are important in the CDR3 of the H-chain for dsDNA binding (detailed in ref. 214). Substitution of arginine with glycine could abrogate DNA binding, while additional arginine residues in CDR2 raised the anti-dsDNA affinity by 10-fold (215), indicating that arginine influences the ability of an antibody to bind dsDNA. Another target of autoantibodies in SLE patients, as well as in the experimental murine models, are nuclear proteins (180,206–208). Anti-Sm Abs are produced in high titers and are often able to bind ssDNA and dsDNA. Many anti-Sm hybridomas that bind DNA use similar Ig V genes common in anti-DNA hybridomas and characterized by high content of arginine residues in CDR3 of the H-chain (216). Many studies suggested that the anti-nuclear Ab response undergoes selection, and that both DNA and Sm are selecting Ags in this response (217). This process selects B cells that undergo somatic mutations to increase their affinity to DNA or Sm (217). In the case of the anti-DNA response, acquisition of arginine residues at the binding site increases affinity (215).

RFs are diagnostic autoantibodies in RA that bind the Fc region of IgG and are also found in healthy people (209).

Although experimental models have shown that immune complexes can stimulate RF production (209), the mechanism triggering the activation and secretion of RF Abs (IgM and IgG) in the development of RA remains to be defined. RF Abs are also produced in experimental models of SLE (218).

6.5. Studies in Anti-Self Ig Tg Mice

Many H and L-chain genes used by anti-DNA, anti-Sm, or anti-RF antibodies were cloned from hybridomas isolated from the SLE experimental models, some of which have been used to generate H+L Tg mice or mice with site-specific insertions of the Tg (113–115,219). Unlike the anti-HEL and the anti-class I Tg systems, these mice are Tg for disease-associated Ab, and their experimental use thus provides additional important information regarding the tolerance mechanism.

Erikson et al. (114) described the first disease-associated autoantibody Tg mouse in a study that crossed mice with a non-autoimmune background carrying the anti-DNA specific 3H9 H-chain Tg with mice carrying the Vκ8 L-chain Tg (both strains had good allelic exclusion). Since the combination of VH3H9/Vκ8 was found in many anti-DNA hybridomas, and this particular combination generated an anti-ssDNA antibody, the F1 progeny were analyzed for B cell tolerance. Many B cells in these mice were not deleted in the BM, and DNA-binding B cells could be detected in the periphery. However, no serum anti-DNA Abs could be found, suggesting a state of developmental arrest or anergy in these B cells (114). Similar results were also reported by Tsao and colleagues in an anti-DNA Tg mouse in which the Tg construct encoded an IgM form of the antibody (116). However, in mice in which the construct had an IgG2a form, allelic exclusion was not efficient (described in ref. 194), and many B cells expressed an endogenous IgM resulting in detection of anti-DNA Abs in the serum and development of mild nephritis (115). In both systems, and reminiscent of anti-HEL/sHEL Tg B cells, the unresponsive state could be reversed *in vitro* by LPS stimulation (114,116). In contrast, when 3H9 H-chain Tg mice were bred with Vκ4 L-chain Tg mice generating an anti-dsDNA specificity, B cells were deleted in the BM (113). The remaining B cells in the periphery of these mice showed displacement of the L-chain Tg and the use of endogenous L-chain, suggesting they undergo receptor editing (113). Additional studies by Radic et al. (220) used the 3H9 H-chain Tg, which imposes dsDNA binding on many heterologous L-chains, to further demonstrate receptor editing during B cell development. In 3H9 H-chain Tg mice, B cells with anti-dsDNA specificity were not found in the periphery (114). Gene analysis of hybridomas isolated from these mice revealed a reduced repertoire of the Vκs used by peripheral B cells, and a Jκ usage bias toward the last available segment, Jκ5. Further, no λ1 or λ2 L-chains were found in combination with 3H9, apparently because they confer dsDNA specificity (221). Moreover, incorporation of additional arginine residues in CDR2 of 3H9 (which increase its affinity to dsDNA (215) further reduced the Vκ repertoire and increased the bias towards Jκ5 usage in Tg mice (222). These findings suggested that B cells with anti-dsDNA specificity generated in the BM undergo receptor editing, and only VL genes that do not produce anti-dsDNA specificity in combination with the 3H9 H-chain Tg are selected for maturation (93). More recent studies used mice with site-specific targeting of the H and L-chain of the anti-DNA Ab ("knock in") (135). These mice have the anti-DNA H and L-chain in the natural context and showed that extinguishing the anti-DNA specificity occurs at high frequency by receptor editing through nested rearrangements. Gene analysis of hybridomas revealed that receptor editing could also occur by H-chain replacement (131) but it occurred much more frequently by revision of the knock-in L-chain in H + L Tg knock in mice (135).

Heavy and L-chain genes from hybridomas encoding RF specificity were also cloned and used to generate Tg mice. Shlomchik and colleagues demonstrated that affinity of the Tg BCR to IgG might dictate the state of tolerance of the RF B cells (119,120). One RF Tg mouse was generated from the low affinity hybridoma AM14 (119), and another from the high affinity hybridoma (called 20.8.3) (120); both hybridomas recognized IgG2a of the "a" allotype. A significant tolerance difference was found by Shlomchik and colleagues between the Tg strains in genetic crosses to mice from the "a" allotype or the "b" allotype: In low affinity RF Tg mice, Tg B cells were not deleted, but could be found in the periphery and were able to participate in a primary immune response (119). In contrast, in high affinity RF Tg mice, B cells were deleted in the BM, and those B cells found in the periphery appeared to have undergone receptor editing (120). Thus, studies with Tg mice demonstrated that B cells bearing disease-related specificities could sometimes avoid central tolerance and mature to the periphery where they primarily remain in an unactivated state by peripheral mechanisms. However, because high affinity anti-self (disease-related or non-disease-related) B cells are deleted in the BM, whereas low affinity specificities can mature, the escape from central tolerance appears to be affinity-dependent.

7. CONCLUDING REMARKS

The process of B lymphocyte development in the bone marrow has been well characterized, and shown to be a mechanism that both guides, and is guided by, Ig gene rearrangements. The process is mediated by the expression of other genes, cell contacts, and growth factors, and B cells developing along this pathway are subject to several checkpoints that ensure elimination of self-reactive specificities and perhaps prevent the onset of autoimmune disease. Allelic exclusion is the primary step in this selection process, guiding the B cell to express only one H-chain and one L-chain, promoting its monospecificity. During development, sIgM+ B cells are then subject to central tolerance, where self-reactive specificities are eliminated by receptor editing and apoptosis. In the periphery, autoreactive B cells then undergo apoptosis or become anergic when they bind self-Ag in the absence of T cell help. Despite the efficiency of these checkpoints, however, autoimmune diseases can develop, sometimes with the active participation of B cells. What triggers the development of autoimmune disease and to what extent it represents a defect in the tolerance process remains to be elucidated. The availability of Ig Tg mice has made it possible to show that not all autoreactive B cells are eliminated by central or peripheral tolerance, and the fate of the autoreactive B cell might be determined by factors such as affinity, tissue specificity, and amount of Ag.

References

1. Grawunder, U., T.M. Leu, D.G. Schatz, A. Werner, A.G. Rolink, F. Melchers, and T.H. Winkler. 1995. Down regulation of RAG1 and RAG2 gene expression in preB cells after functional immunoglobulin heavy chain rearrangement. *Immunity* 3:601–608.
2. Sleckman, B.P., J.R. Gorman, and F.W. Alt. 1996. Accessibility control of antigen-receptor variable-region gene assembly: role of cis-acting elements. *Ann. Rev. Immunol.* 14:459–481.
3. Schatz, D.G., M.A. Oettinger, and M.S. Schlissel. 1992. V(D)J recombination: molecular biology and regulation. *Ann. Rev. Immunol.* 10:359–383.
4. Tonegawa, S. 1983. Somatic generation of antibody diversity. *Nature* 302:575–581.
5. Lewis, S.M. 1994. The mechanism of V(D)J joining: lessons from molecular, immunological, and comparative analyses. *Adv. Immunol.* 56:27–150.
6. Lewis, S.M. and G.E. Wu. 1997. The origins of V(D)J recombination. *Cell* 88:159–162.
7. Oettinger, M.A., D.G. Schatz, C. Gorka, and D. Baltimore. 1990. RAG-1 and RAG-2, adjacent genes that synergistically activate V(D)J recombination. *Science* 248:1517–1523.
8. Schatz, D.G., M.A. Oettinger, and D. Baltimore. 1989. The V(D)J recombination activating gene, RAG-1. *Cell* 59:1035–1048.
9. Agrawal, A., Q.M. Eastman, and D.G. Schatz. 1998. Transposition mediated by RAG1 and RAG2 and its implications for the evolution of the immune system. *Nature* 394:744–751.
10. Grawunder, U., R.B. West, and M.R. Lieber. 1998. Antigen receptor gene rearrangement. *Curr. Opin. Immunol.* 10:172–180.
11. Ramsden, D.A., T.T. Paull, and M. Gellert. 1997. Cell-free V(D)J recombination. *Nature* 388:488–491.

12. Schatz, D.G. 1997. V(D)J recombination moves in vitro. *Semin. Immunol.* 9:149–159.

13. Gilfillan, S., A. Dierich, M. Lemeur, C. Benoist, and D. Mathis. 1993. Mice lacking TdT: mature animals with an immature lymphocyte repertoire. *Science* 261:1175–1178.

14. Komori, T., A. Okada, V. Stewart, and F.W. Alt. 1993. Lack of N regions in antigen receptor variable region genes of TdT-deficient lymphocytes. *Science* 261:1171–1175.

15. Landau, N.R., D.G. Schatz, M. Rosa, and D. Baltimore. 1987. Increased frequency of N-region insertion in a murine pre-B-cell line infected with a terminal deoxynucleotidyl transferase retroviral expression vector. *Molec. & Cell. Biol.* 7:3237–3243.

16. Bogue, M.A., C. Wang, C. Zhu, and D.B. Roth. 1997. V(D)J recombination in Ku86-deficient mice: distinct effects on coding, signal, and hybrid joint formation. *Immunity* 7:37–47.

17. Chu, G. 1997. Double strand break repair. *J. Biol. Chem.* 272:24097–24100.

18. Gao, Y., Y. Sun, K.M. Frank, P. Dikkes, Y. Fujiwara, K.J. Seidi, J.M. Sekiguchi, G.A. Rathbun, W. Swat, J. Wang, R.T. Bronson, B.A. Malynn, M. Bryans, C. Zhu, J. Chaudhuri, L. Davidson, R. Ferrini, T. Stamato, S.H. Orkin, M.E. Greenberg, and F.W. Alt. 1998. F. W. A critical role for DNA end-joining proteins in both lymphogenesis and neurogenesis. *Cell* 95:891–902.

19. Gu, Y., K.J. Seidl, G.A. Rathbun, C. Zhu, J.P. Manis, N. van der Stoep, Davidson, H.L. Cheng, J.M. Sekiguchi, K. Frank, P. Stanhope-Baker, M.S. Schlissel, D.B. Roth, and F.W. Alt. 1997. Growth retardation and leaky SCID phenotype of Ku70-deficient mice. *Immunity* 7:653–665.

20. Nussenzweig, A., C. Chen, C.S. da, V, M. Sanchez, K. Sokol, M.C. Nussenzweig, and G.C. Li. 1996. Requirement for Ku80 in growth and immunoglobulin V(D)J recombination. *Nature* 382:551–555.

21. Bogue, M., S. Gilfillan, C. Benoist, and D. Mathis. 1992. Regulation of N-region diversity in antigen receptors through thymocyte differentiation and thymus ontogeny. *Proc. Natl. Acad. Sci. USA* 89:11011–11015.

22. Li, Y.S., K. Hayakawa, and R.R. Hardy. 1993. The regulated expression of B lineage associated genes during B cell differentiation in bone marrow and fetal liver. *J. Exp. Med.* 178:951–960.

23. Feeney, A.J. 1990. Lack of N regions in fetal and neonatal mouse immunoglobulin V-D-J junctional sequences. *J. Exp. Med.* 172:1377–1390.

24. Gu, H., I. Forster, and K. Rajewsky. 1990. Sequence homologies, N sequence insertion and JH gene utilization in VHDJH joining: implications for the joining mechanism and the ontogenetic timing of Ly1 B cell and B-CLL progenitor generation. *EMBO J.* 9:2133–2140.

25. Lafaille, J.J., A. DeCloux, M. Bonneville, Y. Takagaki, and S. Tonegawa. 1989. Junctional sequences of T cell receptor gamma delta genes: implications for gamma delta T cell lineages and for a novel intermediate of V-(D)-J joining. *Cell* 59:859–870.

26. Born, W., G. Rathbun, P. Tucker, P. Marrack, and J. Kappler. 1986. Synchronized rearrangement of T-cell gamma and beta chain genes in fetal thymocyte development. *Science* 234:479–482.

27. Komori, T., L. Pricop, A. Hatakeyama, C.A. Bona, and F.W. Alt. 1996. Repertoires of antigen receptors in Tdt congenitally deficient mice. *Int. Rev. Immunol.* 13:317–325.

28. Karasuyama, H., A. Rolink, and F. Melchers. 1996. Surrogate light chain in B cell development. *Adv. Immunol.* 63:1–41.

29. Rajewsky, K. 1996. Clonal selection and learning in the antibody system. *Nature* 381:751–758.

30. Reth, M. and J. Wienands. 1997. Initiation and processing of signals from the B cell antigen receptor. *Annu. Rev. Immunol.* 15:453–479.

31. Nussenzweig, M.C., A.C. Shaw, E. Sinn, D.B. Danner, K.L. Holmes, H.C. Morse, and P. Leder. 1987. Allelic exclusion in transgenic mice that express the membrane form of immunoglobulin mu. *Science* 236:816–819.

32. Gu, H., D. Kitamura, and K. Rajewsky. 1991. B cell development regulated by gene rearrangement: arrest of maturation by membrane-bound D mu protein and selection of DH element reading frames. *Cell* 65:47–54.

33. Kitamura, D., J. Roes, R. Kuhn, and K. Rajewsky. 1991. A B cell-deficient mouse by targeted disruption of the membrane exon of the immunoglobulin mu chain gene. *Nature* 350:423–426.

34. Reichman-Fried, M., R.R. Hardy, and M.J. Bosma. 1990. Development of B- lineage cells in the bone marrow of scid/scid mice following the introduction of functionally rearranged immunoglobulin transgenes. *Proc. Natl. Acad. Sci. USA* 87:2730–2734.

35. Young, F., B. Ardman, Y. Shinkai, R. Lansford, T.K. Blackwell, M. Mendelsohn, A. Rolink, F. Melchers, and F.W. Alt. 1994. Influence of immunoglobulin heavy- and light-chain expression on B-cell differentiation. *Genes & Development* 8:1043–1057.

36. Spanopoulou, E., C.A. Roman, L.M. Corcoran, M.S. Schlissel, D.P. Silver, D. Nemazee, M.C. Nussenzweig, S.A. Shinton, R.R. Hardy, and D. Baltimore. 1994. Functional immunoglobulin transgenes guide ordered B-cell differentiation in Rag-1-deficient mice. *Genes & Development* 8:1030–1042.

37. Hardy, R.R., C.E. Carmack, S.A. Shinton, J.D. Kemp, and K. Hayakawa. 1991. Resolution and characterization of pro-B and pre-pro-B cell stages in normal mouse bone marrow. *J. Exp. Med.* 173:1213–1225.

38. Rolink, A., U. Grawunder, T.H. Winkler, H. Karasuyama, and F. Melchers. 1994. IL-2 receptor alpha chain (CD25, TAC) expression defines a crucial stage in pre-B cell development. *Int. Immunol.* 6:1257–1264.

39. Tsubata, T., R. Tsubata, and M. Reth. 1991. Cell surface expression of the short immunoglobulin mu chain (D mu protein) in murine pre-B cells is differently regulated from that of the intact mu chain. *Eur. J. Immunol.* 21:1359–1363.

40. Ehlich, A., V. Martin, W. Muller, and K. Rajewsky. 1994. Analysis of the B-cell progenitor compartment at the level of single cells. *Current Biology* 4:573–583.

41. Roth, P.E., L. Doglio, J.T. Manz, J.Y. Kim, D. Lo, and U. Storb. 1993. Immunoglobulin gamma 2b transgenes inhibit heavy chain gene rearrangement, but cannot promote B cell development. *J. Exp. Med.* 178:2007–2021.

42. Alt, F.W., G.D. Yancopoulos, T.K. Blackwell, C. Wood, E. Thomas, M. Boss, R. Coffman, N. Rosenberg, S. Tonegawa, and D. Baltimore. 1984. Ordered rearrangement of immunoglobulin heavy chain variable region segments. *EMBO J* 3:1209–1219.

43. Loffert, D., A. Ehlich, W. Muller, and K. Rajewsky. 1996. Surrogate light chain expression is required to establish immunoglobulin heavy chain allelic exclusion during early B cell development. *Immunity* 4:133–144.

44. Kitamura, D., A. Kudo, S. Schaal, W. Muller, F. Melchers, and K. Rajewsky. 1992. A critical role of lambda 5 protein in B cell development. *Cell* 69:823–831.

45. Bosma, G.C., R.P. Custer, and M.J. Bosma. 1983. A severe combined immunodeficiency mutation in the mouse. *Nature* 301:527–530.

46. Fang, W., D.L. Mueller, C.A. Pennell, J.J. Rivard, Y.S. Li, Hardy, R.R., M.S. Schlissel, and T.W. Behrens. 1996. Frequent aberrant immunoglobulin gene rearrangements in pro-B cells revealed by a bcl-xL transgene. *Immunity* 4:291–299.

47. Karasuyama, H., A. Rolink, Y. Shinkai, F. Young, F.W. Alt, and F. Melchers. 1994. The expression of Vpre-B/lambda 5 surrogate light chain in early bone marrow precursor B cells of normal and B cell-deficient mutant mice. *Cell* 77:133–143.

48. Misener, V., G.P. Downey, and J. Jongstra. 1991. The immunoglobulin light chain related protein lambda 5 is expressed on the surface of mouse pre-B cell lines and can function as a signal transducing molecule. *Int. Immunol.* 3:1129–1136.

49. Tsubata, T., R. Tsubata, and M. Reth. 1992. Crosslinking of the cell surface immunoglobulin (mu-surrogate light chains complex) on pre-B cells induces activation of V gene rearrangements at the immunoglobulin kappa locus. *Int. Immunol.* 4:637–641.

50. Pillai, S. and D. Baltimore. 1988. The omega and iota surrogate immunoglobulin light chains. *Current Topics in Microbiology & Immunology* 137:136–139.

51. Rolink, A., H. Karasuyama, U. Grawunder, D. Haasner, A. Kudo, and F. Melchers. 1993. B cell development in mice with a defective lambda 5 gene. *Eur. J. Immunol.* 23:1284–1288.

52. Lassoued, K., C.A. Nunez, L. Billips, H. Kubagawa, Monteiro, R.C., T.W. LeBlen, and M.D. Cooper. 1993. Expression of surrogate light chain receptors is restricted to a late stage in pre-B cell differentiation. *Cell* 73:73–86.

53. Cherayil, B.J. and S. Pillai. 1991. The omega/lambda 5 surrogate immunoglobulin light chain is expressed on the surface of transitional B lymphocytes in murine bone marrow. *J. Exp. Med.* 173:111–116.

54. Tybulewicz, V.L. 1998. Analysis of antigen receptor signalling using mouse gene targeting. *Curr. Opin. Cell Biol.* 10:195–204.

55. Khan, W.N., F.W. Alt, R.M. Gerstein, B.A. Malynn, I. Larsson, G. Rathbun, L. Davidson, S. Muller, A.B. Kantor, and L.A. Herzenberg. 1995. Defective B cell development and function in Btk-deficient mice. *Immunity* 3:283–299.

56. Gong, S. and M.C. Nussenzweig. 1996. Regulation of an early developmental checkpoint in the B cell pathway by Ig beta. *Science* 272:411–414.

57. Papavasiliou, F., M. Jankovic, H. Suh, and M.C. Nussenzweig. 1995. The cytoplasmic domains of immunoglobulin (Ig) alpha and Ig beta can independently induce the precursor B cell transition and allelic exclusion. *J. Exp. Med.* 182:1389–1394.

58. Papavasiliou, F., Z. Misulovin, H. Suh, and M.C. Nussenzweig. 1995. The role of Ig beta in precursor B cell transition and allelic exclusion. *Science* 268:408–411.

59. Constantinescu, A. and M.S. Schlissel. 1997. Changes in locus-specific V(D)J recombinase activity induced by immunoglobulin gene products during B cell development. *J. Exp. Med.* 185:609–620.

60. Tarakhovsky, A. 1997. Xid and Xid-like immunodeficiencies from a signaling point of view. *Curr. Opin. Immunol.* 9:319–323.

61. Reichman-Fried, M., R.R. Hardy, and M.J. Bosma. 1990. Development of B- lineage cells in the bone marrow of scid/scid mice following the introduction of functionally rearranged immunoglobulin transgenes. *Proc. Natl. Acad. Sci. USA* 87:2730–2734.

62. Young, F., B. Ardman, Y. Shinkai, R. Lansford, T.K. Blackwell, M. Mendelsohn, A. Rolink, F. Melchers, and F.W. Alt. 1994. Influence of immunoglobulin heavy- and light-chain expression on B-cell differentiation. *Genes & Development* 8:1043–1057.

63. Spanopoulou, E., C.A. Roman, L.M. Corcoran, M.S. Schlissel, D.P. Silver, D. Nemazee, M.C. Nussenzweig, S.A. Shinton, R.R. Hardy, and D. Baltimore. 1994. functional immunoglobulin transgenes guide ordered B-cell differentiation in Rag-1-deficient mice. *Genes & Development* 8:1030–1042.

64. Papavasiliou, F., M. Jankovic, and M.C. Nussenzweig. 1996. Surrogate or conventional light chains are required for membrane immunoglobulin mu to activate the precursor B cell transition. *J. Exp. Med.* 184:2025–2030.

65. Shaffer, A.L. and M.S. Schlissel. 1997. A truncated heavy chain protein relieves the requirement for surrogate light chains in early B cell development. *J. Immunol.* 159:1265–1275.

66. Corcos, D., O. Dunda, C. Butor, J.Y. Cesbron, P. Lores, D. Bucchini, and J. Jami. 1995. Pre-B-cell development in the absence of lambda 5 in transgenic mice expressing a heavy-chain disease protein. *Current Biology* 5:1140–1148.

67. Decker, D.J., N.E. Boyle, and N.R. Klinman. 1991. Predominance of nonproductive rearrangements of VH81X gene segments evidences a dependence of B cell clonal maturation on the structure of nascent H chains. *J. Immunol.* 147:1406–1411.

68. Ye, J., S.K. McCray, and S.H. Clarke. 1995. The majority of murine VH12-expressing B cells are excluded from the peripheral repertoire in adults. *Eur. J. Immunol.* 25:2511–2521.

69. Keyna, U., S.E. Applequist, J. Jongstra, G.B. Beck-Engeser, and II.M. Jack. 1995. Ig mu heavy chains with VH81X variable regions do not associate with lambda 5. *Ann. NY Acad. Sci.* 764:39–42.

70. Ye, J., S.K. McCray, and S.H. Clarke. 1996. The transition of pre-BI to pre-BII cells is dependent on the VH structure of the mu/surrogate L chain receptor. *EMBO J.* 15:1524–1533.

71. Wasserman, R., Y.S. Li, S.A. Shinton, C.E. Carmack, T. Manser, D.L. Wiest, K. Hayakawa, and R.R. Hardy. 1998. A novel mechanism for B cell repertoire maturation based on response by B cell precursors to pre-B receptor assembly. *J. Exp. Med.* 187:259–264.

72. Sudo, T., M. Ito, Y. Ogawa, M. Iizuka, H. Kodama, T. Kunisada, S. Hayashi, M. Ogawa, K. Sakai, and S. Nishikawa. 1989. Interleukin 7 production and function in stromal cell-dependent B cell development. *J. Exp. Med.* 170:333–338.

73. Grabstein, K.H., T.J. Waldschmidt, F.D. Finkelman, B.W. Hess, A.R. Alpert, N.E. Boiani, A.E. Namen, and P.J. Morrissey. 1993. Inhibition of murine B and T lymphopoiesis in vivo by an anti-interleukin 7 monoclonal antibody. *J. Exp. Med.* 178:257–264.

74. Gorman, J.R. and F.W. Alt. 1998. Regulation of immunoglobulin light chain isotype expression. *Adv. Immunol.* 69:113–181.

75. Pillai, S. 1999. The chosen few? Positive selection in the generation of naïve B lymphocytes. *Immunity* 10:493–502.

76. Storb, U., P. Roth, and B.K. Kurtz. 1994. Gamma 2b transgenic mice as a model for the role of immunoglobulins in B cell development. *Immunol. Res.* 13:291–8.

77. Hartley, S.B., M.P. Cooke, D.A. Fulcher, A.W. Harris, S. Cory, A. Basten, and C.C. Goodnow. 1993. Elimination of self-reactive B lymphocytes proceeds in two stages: arrested development and cell death. *Cell* 72:325–335.

78. Hardy, R.R. and K. Hayakawa. 1995. B-lineage differentiation stages resolved by multiparameter flow cytometry. *Ann. NY Acad. Sci.* 764:19–24.

79. Hardy, R.R. and K. Hayakawa. 1994. CD5 B cells, a fetal B cell lineage. *Adv. Immunol.* 55:297–339.

80. Nossal, G.J. 1983. Cellular mechanisms of immunologic tolerance. *Ann. Rev. Immunol.* 1:33–62.

81. Yellen-Shaw, A. and Monr. 1992. Developmentally regulated association of a 56-kD member of the surface immunoglobulin M receptor complex. *J. Exp. Med.* 176:129–137.

82. Brines, R. and G.G. Klaus. 1992. Inhibition of lipopolysaccharaide-induced activation of immature B cells by anti-μ and anti-δ and 'yntibodies and its modulation by interleukin-4. *Int. Immmunol.* 4:765–771.

83. Igarashi, H., K. Kuwahara, J. Nomura, A. Matsuda, K. Kikuchi, S. Inui, and N. Sakaguchi. 1994. B cell Ag receptor mediates different types of signals in the protein kinase activity between immature B cell and mature B cell. *J. Immunol.* 153:2381–2393.

84. Nossal, G.J. 1989. Immunological tolerance then and now: was the Medawar school right?. *Immunology Supplement.* 2:2–5.

85. Chiller, J.M., G.S. Habicht, and W.O. Weigle. 1971. Kinetic differences in unresponsiveness of thymus and bone marrow cells. *Science* 171:813–815.

86. Nossal, G.J. and B.L. Pike. 1975. Evidence for the clonal abortion theory of B-lymphocyte tolerance. *J. Exp. Med.* 141:904–917.

87. Metcalf, E.S. and N.R. Klinman. 1977. In vitro tolerance induction of bone marrow cells: a marker for B cell maturation. *J. Immunol.* 118:2111–2116.

88. Etlinger, H.M. and J.M. Chiller. 1979. Maturation of the lymphoid system. I. Induction of tolerance in neonates with a T-dependent antigen that is an obligate immunogen in adults. *J. Immunol.* 122:2558–2563.

89. Teale, J.M., and N.R. Klinman. 1980. Tolerance as an active process. *Nature* 288:385–387.

90. Rieben, R., A. Tucci, U.E. Nydegger, and R.H. Zubler. 1992. Self tolerance to human A and B histo-blood group antigens exists at the B cell level and cannot be broken by potent polyclonal B cell activation in vitro. *Eur. J. Immunol.* 22:2713–2717.

91. Galili, U., J. Buehler, S.B. Shohet, and B.A. Macher. 1987. The human natural anti-Gal IgG. III. The subtlety of immune tolerance in man as demonstrated by crossreactivity between natural anti-Gal and anti-B antibodies. *J. Exp. Med.* 165:693–704.

92. McHeyser-Williams, M.G. and G.J. Nossal. 1988. Clonal analysis of autoantibody-producing cell precursors in the preimmune B cell repertoire. *J. Immunol.* 141:4118–4123.

93. Radic, M.Z. and M. Zouali. 1996. Receptor editing, immune diversification, and self-tolerance. *Immunity* 5:505–511.

94. Klinman, N.R. 1996. The "clonal selection hypothesis" and current concepts of B cell tolerance. *Immunity* 5:189–195.

95. Monroe, J.G. 1996. Tolerance sensitivity of immature-stage B cells: can developmentally regulated B cell antigen receptor (BCR) signal transduction play a role? *J. Immunol.* 156:2657–2660.

96. Lederberg, B. 1959. Genes and antibodies: do antigens bear instructions for antibody specificity or do they select cell lines that arise by mutation. *Science* 129:1653.

97. Bretscher, P. and M. Cohn. 1970. A theory of self-nonself discrimination. *Science* 169:1042–1049.

98. Pike, B.L., T.W. Kay, and G.J. Nossal. 1980. Relative sensitivity of fetal and newborn mice to induction of hapten-specific B cell tolerance. *J. Exp. Med.* 152:1407–1412.

99. Lang, J., M. Jackson, L. Teyton, A. Brunmark, K. Kane, and D. Nemazee. 1996. B cells are exquisitely sensitive to central tolerance and receptor editing induced by ultralow affinity, membrane-bound antigen. *J. Exp. Med.* 184:1685–1697.

100. Finkelman, F.D., J.M. Holmes, O.I. Dukhanina, and S.C. Morris. 1995. Cross-linking of membrane immunoglobulin D, in the absence of T cell help, kills mature B cells in vivo. *J. Exp. Med.* 181:515–525.

101. Parry, S.L., J. Hasbold, M. Holman, and G.G. Klaus. 1994. Hypercross-linking surface IgM or IgD receptors on mature B cells induces apoptosis that is reversed by costimulation with IL-4 and anti-CD40. *J. Immunol.* 152:2821–2829.

102. Tsubata, T., M. Murakami, S. Nisitani, and T. Honjo. 1994. Molecular mechanisms for B lymphocyte selection: induction and regulation of antigen-receptor-mediated apoptosis of mature B cells in normal mice and their defect in autoimmunity-prone mice. *Philosophical Transactions of the Royal Society of London—Series B: Biological Sciences* 345:297–301.

103. Tsubata, T., M. Murakami, and T. Honjo. 1994. Antigen-receptor cross-linking induces peritoneal B-cell apoptosis in normal but not autoimmunity-prone mice. *Current Biology* 4:8–17.

104. Lawton, A.D. 1972. Suppression of immunoglobulin class synthesis in mice. I. Effects of treatment with antibody to μ-chain. *J. Exp. Med.* 277–297.

105. Raff, M.C., J.J. Owen, M.D. Cooper, A.R. Lawton, M. Megson, and W.E. Gathings. 1975. Differences in susceptibility of mature and immature mouse B lymphocytes to anti-immunoglobulin-induced immunoglobulin suppression in vitro. Possible implications for B-cell tolerance to self. *J. Exp. Med.* 142:1052–1064.

106. Sidman, C.L. and E.R. Unanue. 1975. Receptor-mediated inactivation of early B lymphocytes. *Nature* 257:149–151.

107. Nossal, G.J. 1994. Negative selection of lymphocytes. *Cell* 76:229–239.

108. Norvell, A., L. Mandik, and J.G. Monroe. 1995. Engagement of the antigen-receptor on immature murine B lymphocytes results in death by apoptosis. *J. Immunol.* 154:4404–4413.

109. Parry, S.L., M.J. Holman, J. Hasbold, and G.G. Klaus. 1994. Plastic-immobilized anti-mu or anti-delta antibodies induce apoptosis in mature murine B lymphocytes. *Eur. J. Immunol.* 24:974–979.

110. Nemazee, D.A. and K. Burki. 1989. Clonal deletion of B lymphocytes in a transgenic mouse bearing anti-MHC class I antibody genes. *Nature* 337:562–566.

111. Russell, D.M., Z. Dembic, G. Morahan, J.F. Miller, K. Burki, and D. Nemazee. 1991. Peripheral deletion of self-reactive B cells. *Nature* 354:308–311.

112. Hartley, S.B., J. Crosbie, R. Brink, A.B. Kantor, A. Basten, and C.C. Goodnow. 1991. Elimination from peripheral lymphoid tissues of self-reactive B lymphocytes recognizing membrane-bound antigens. *Nature* 353:765–769.
113. Gay, D., T. Saunders, S. Camper, and M. Weigert. 1993. Receptor editing: an approach by autoreactive B cells to escape tolerance. *J. Exp. Med.* 177:999–1008.
114. Erikson, J., M.Z. Radic, S.A. Camper, R.R. Hardy, C. Carmack, and M. Weigert. 1991. Expression of anti-DNA immunoglobulin transgenes in non-autoimmune mice. *Nature* 349:331–334.
115. Tsao, B.P., K. Ohnishi, H. Cheroutre, B. Mitchell, M. Teitell, P. Mixter, M. Kronenberg, and B.H. Hahn. 1992. Failed self-tolerance and autoimmunity in IgG anti-DNA transgenic mice. *J. Immunol.* 149:350–358.
116. Tsao, B.P., A. Chow, H. Cheroutre, Y.W. Song, M.E. McGrath, and M. Kronenberg. 1993. B cells are anergic in transgenic mice that express IgM anti-DNA antibodies. *Eur. J. Immunol.* 23:2332–2339.
117. Murakami, M., T. Tsubata, M. Okamoto, A. Shimizu, S. Kumagai, H. Imura, and T. Honjo. 1992. Antigen-induced apoptotic death of Ly-1 B cells responsible for autoimmune disease in transgenic mice. *Nature* 357:77–80.
118. Carsetti, R., G. Kohler, and M.C. Lamers. 1995. Transitional B cells are the target of negative selection in the B cell compartment. *J. Exp. Med.* 181:2129–2140.
119. Hannum, L.G., D. Ni, A.M. Haberman, M.G. Weigert, and M.J. Shlomchik. 1996. A disease-related rheumatoid factor autoantibody is not tolerized in a normal mouse: implications for the origins of autoantibodies in autoimmune disease. *J. Exp. Med.* 184:1269–1278.
120. Wang, H. and M.J. Shlomchik. 1997. High affinity rheumatoid factor transgenic B cells are eliminated in normal mice. *J. Immunol.* 159:1125–1134.
121. Nisitani, S., T. Tsubata, M. Murakami, M. Okamoto, and T. Honjo. 1993. The bcl-2 gene product inhibits clonal deletion of self-reactive B lymphocytes in the periphery but not in the bone marrow. *J. Exp. Med.* 178:1247–1254.
122. Lu, L. and D.G. Osmond. 1997. Apoptosis during B lymphopoiesis in mouse bone marrow. *J. Immunol.* 158:5136–5145.
123. Goodnow, C.C., J. Crosbie, S. Adelstein, T.B. Lavoie, S.J. Smith-Gill, R.A. Brink, H. Pritchard-Briscoe, J.S. Wotherspoon, R.H. Loblay, K. Raphael, et al. 1988. Altered immunoglobulin expression and functional silencing of self-reactive B lymphocytes in transgenic mice. *Nature* 334:676–682.
124. Cooke, M.P., A.W. Heath, K.M. Shokat, Y. Zeng, F.D. Finkelman, P.S. Linsley, M. Howard, and C.C. Goodnow. 1994. Immunoglobulin signal transduction guides the specificity of B cell-T cell interactions and is blocked in tolerant self-reactive B cells. *J. Exp. Med.* 179:425–38.
125. Nossal, G.J. and B.L. Pike. 1980. Clonal anergy: persistence in tolerant mice of antigen-binding B lymphocytes incapable of responding to antigen or mitogen. *Proc. Natl. Acad. Sci. USA* 77:1602–1606.
126. Pike, B.L., A.W. Boyd, and G.J. Nossal. 1982. Clonal anergy: the universally anergic B lymphocyte. *Proc. Natl. Acad. Sci. USA* 79:2013–2017.
127. Gause, A., N. Yoshida, C. Kappen, and K. Rajewsky. 1987. In vivo generation and function of B cells in the presence of a monoclonal anti-IgM antibody: implications for B cell tolerance. *Eur. J. Immunol.* 17:981–990.
128. Fulcher, D.A. and A. Basten. 1994. Reduced life span of anergic self-reactive B cells in a double-transgenic model. *J. Exp. Med.* 179:125–134.
129. Cyster, J.G., S.B. Hartley, and C.C. Goodnow. 1994. Competition for follicular niches excludes self-reactive cells from the recirculating B-cell repertoire. *Nature* 371:389–395.
130. Tiegs, S.L., D.M. Russell, and D. Nemazee. 1993. Receptor editing in self-reactive bone marrow B cells. *J. Exp. Med.* 177:1009–1020.
131. Chen, C., Z. Nagy, E.L. Prak, and M. Weigert. 1995. Immunoglobulin heavy chain gene replacement: a mechanism of receptor editing. *Immunity* 3:747–755.
132. Lang, J., B. Arnold, G. Hammerling, A.W. Harris, S. Korsmeyer, D. Russell, A. Strasser, and D. Nemazee. 1997. Enforced Bcl-2 expression inhibits antigen-mediated clonal elimination of peripheral B cells in an antigen dose-dependent manner and promotes receptor editing in autoreactive, immature B cells. *J. Exp. Med.* 186:1513–1522.
133. Melamed, D. and D. Nemazee. 1997. Self-antigen does not accelerate immature B cell apoptosis, but stimulates receptor editing as a consequence of developmental arrest. *Proc. Natl. Acad. Sci. USA* 94:9267–9272.
134. Pelanda, R., S. Schwers, E. Sonoda, R.M. Torres, D. Nemazee, and K. Rajewsky. 1997. Receptor editing in a transgenic mouse model: site, efficiency, and role in B cell tolerance and antibody diversification. *Immunity* 7:765–775.
135. Chen, C., E.L. Prak, and M. Weigert. 1997. Editing disease-associated autoantibodies. *Immunity* 6:97–105.
136. Hertz, M. and D. Nemazee. 1997. BCR ligation induces receptor editing in IgM + IgD- bone marrow B cells in vitro. *Immunity* 6:429–436.
137. Melamed, D., R.J. Benschop, J.C. Cambier, and D. Nemazee. 1998. Developmental regulation of B lymphocyte immune tolerance compartmentalizes clonal selection from receptor selection. *Cell* 92:173–182.
138. Goodnow, C.C. 1996. Balancing immunity and tolerance: deleting and tuning lymphocyte repertoires. *Proc. Natl. Acad. Sci. USA* 93:2264–2271.
139. Melamed, D., J.a. Kench, K. Grabstein, A. Rolink, and D. Nemazee. 1997. A functional B-cell receptor transgene allows IL-7-independent maturation of B-cell precursors. *J. Immunol.* 159:1233–1239.
140. Pelanda, R., S. Schaal, R.M. Torres, and K. Rajewsky. 1996. A prematurely expressed Ig(kappa) transgene, but not V(kappa)J(kappa) gene segment targeted into the Ig(kappa) locus, can rescue BG cell defelopment in lambda5-deficient mice. *Immunity* 5:229–239.
141. Cascalho, M., A. Ma, S. Lee, L. Masat, and M. Wabl. 1996. A quasi-monoclonal mouse. *Science* 272:1649–1652.
142. Chen, C., Z. Nagy, E.L. Prak, and M. Weigert. 1995. Immunoglobulin heavy chain gene replacement: a mechanism of receptor editing. *Immunity* 3:747–755.
143. Taki, S., M. Meiering, and K. Rajewsky. 1993. Targeted insertion of a variable region gene into the immunoglobulin heavy chain locus. *Science* 262:1268–1271.
144. Norvell, A. and J.G. Monroe. 1996. Acquisition of surface IgD fails to protect from tolerance-induction. Both surface IgM- and surface IgD-mediated signals induce apoptosis of immature murine B lymphocytes. *J. Immunol.* 156:1328–32.
145. Hasbold, J. and G.G. Klaus. 1990. Anti-immunoglobulin antibodies induce apoptosis in immature B cell lymphomas. *Eur. J. Immunol.* 20:1685–1690.

146. Wechlser, R. and J.G. Monroe. 1996. Lipopolysaccharide prevents apoptosis and induces responsiveness to antigen receptor cross-linking in immature B cells. *Immunology* 89:362.

147. Soares, M., X. Havaux, F. Nisol, H. Bazin, and D. Latinne. 1996. Modulation of rat B cell differentiation in vivo by the administration of an anti-mu monoclonal antibody. *J. Immunol.* 156:108–118.

148. Klaus, G.G., M. Bijsterbosch, A. O'Garra, M.M. Harnett, and K. Rigley. 1987. Receptor signalling and crosstalk in B lymphocytes. *Immunol. Rev.* 99:19–38.

149. Cambier, J.C. and J. Ransom. 1987. Molecular mechanisms of transmembrane signaling in B lymphocytes. *Ann. Rev. Immunol.* 5:175–199.

150. Gaur, A., X.R. Yao, and D.W. Scott. 1993. B cell tolerance induction by cross-linking of membrane IgM, but not IgD, and synergy by cross-linking of both isotypes. *J. Immunol.* 150:1663–1669.

151. Adams, E., A. Basten, and C.C. Goodnow. 1990. Intrinsic B-cell hyporesponsiveness accounts for self-tolerance in lysozyme/anti-lysozyme double-transgenic mice. *Proc. Natl. Acad. Sci. USA* 87:5687–91.

152. Mason, D.Y., M. Jones, and C.C. Goodnow. 1992. Development and follicular localization of tolerant B lymphocytes in lysozyme/anti-lysozyme IgM/IgD transgenic mice. *Int. Immunol.* 4:163–75.

153. Nemazee, D. 1992. Mechanisms and meaning of B-lymphocyte tolerance. *Res. Immunol.* 143:272–275.

154. Akkaraju, S., K. Canaan, and C.C. Goodnow. 1997. Self-reactive B cells are not eliminated or inactivated by autoantigen expressed on thyroid epithelial cells. *J. Exp. Med.* 186:2005–2012.

155. Walker, S.M. and W.O. Weigle. 1985. Primed lymphoid cell tolerance. II. In vivo tolerization of highly tolerogensensitive hapten-primed, potentially IgG-producing B cells. *Cell. Immunol.* 90:331–338.

156. Linton, P.-J., D. Decker, and N.R. Klinman. 1989. Pimary antibody-forming cells and secondary B cells are generated from separate precursor populations. *Cell* 59:1049–1059.

157. Linton, P.J., A. Rudie, and N.R. Klinman. 1991. Tolerance susceptibility of newly generating memory B cells. *J. Immunol.* 146:4099–4104.

158. Nossal, G.J., M. Karvelas, and B. Pulendran. 1993. Soluble antigen profoundly reduces memory B-cell numbers even when given after challenge immunization. *Proc. Natl. Acad. Sci. USA* 90:3088–3092.

159. Pulendran, B., M. Karvelas, and G.J. Nossal. 1994. A form of immunologic tolerance through impairment of germinal center development. *Proc. Natl. Acad. Sci. USA* 91:2639–2643.

160. Smith, K.G., U. Weiss, K. Rajewsky, G.J. Nossal, and D.M. Tarlinton. 1994. Bcl-2 increases memory B cell recruitment but does not perturb selection in germinal centers. *Immunity* 1:803–13.

161. Pulendran, B., K.G. Smith, and G.J. Nossal. 1995. Soluble antigen can impede affinity maturation and the germinal center reaction but enhance extrafollicular immunoglobulin production. *J. Immunol.* 155:1141–1150.

162. Han, S., B. Zheng, J. Dal Porto, and G. Kelsoe. 1995. In situ studies of the primary immune response to (4-hydroxy-3-nitrophenyl)acetyl. IV. Affinity-dependent, antigen-driven B cell apoptosis in germinal centers as a mechanism for maintaining self-tolerance. *J. Exp. Med.* 182:1635–1644.

163. Shokat, K.M. and C.C. Goodnow. 1995. Antigen-induced B-cell death and elimination during germinal-centre immune responses. *Nature* 375:334–338.

164. Cyster, J.G. and C.C. Goodnow. 1995. Antigen-induced exclusion from follicles and anergy are separate and complementary processes that influence peripheral B cell fate. *Immunity* 3:691–701.

165. Fulcher, D.A., A.B. Lyons, S.L. Korn, M.C. Cook, C. Koleda, C. Parish, d. Fazekas, and A. Basten. 1996. The fate of self-reactive B cells depends primarily on the degree of antigen receptor engagement and availability of T cell help. *J. Exp. Med.* 183:2313–2328.

166. Teale, J.M. and N.R. Klinman. 1980. Tolerance as an active process. *Nature* 288:385–387.

167. Teale, J.M. and N.R. Klinman. 1984. Membrane and metabolic requirements for tolerance induction of neonatal B cells. *J. Immunol.* 133:1811–1817.

168. Mayumi, M., Y. Ohshima, D. Hata, K. Kim, T. Heike, K. Katamura, and K. Furusho. 1995. IgM-mediated B cell apoptosis. *Crit. Rev. Immunol.* 15:255–269.

169. Nakanishi, K., K. Matsui, S. Kashiwamura, Y. Nishioka, J. Nomura, Y. Nishimura, N. Sakaguchi, S. Yonehara, K. Higashino, and S. Shinka. 1996. IL-4 and anti-DC40 protect against Fas-mediated B cell apoptosis and induce B cell growth and differentiation. *Int. Immunol.* 8:791–798.

170. Wang, J., I. Taniuchi, Y. Maekawa, M. Howard, M. Cooper, and T. Watanabe. 1996. Expression and function of Fas antigen on activated murine B cells. *Eur. J. Immunol.* 26:92–96.

171. Choi, Y. 1997. Differentiation and apoptosis of human germinal center B-lymphocytes. *Immunol. Res.* 16:161–174.

172. Hikida, M. and H. Ohmori. 1998. Rearrangement of lambda light chain genes in mature B cells in vitro and in vivo. Function of reexpressed recombination-activating gene (RAG) products. *J. Exp. Med.* 187:795–799.

173. Han, S., B. Zheng, D.G. Schatz, E. Spanopoulou, and G. Kelsoe. 1996. Neoteny in lymphocytes: Rag1 and Rag2 expression in germinal center B cells. *Science* 274:2094–2097.

174. Hikida, M., M. Mori, T. Takai, K. Tomochika, K. Hamatani, and H. Ohmori. 1996. Reexpression of RAG-1 and RAG-2 genes in activated mature mouse B cells. *Science* 274:2092–2094.

175. Han, S., S.R. Dillon, B. Zheng, M. Shimoda, M.S. Schlissel, and G. Kelsoe. 1997. V(D)J recombinase activity in a subset of germinal center B lymphocytes. *Science* 278:301–305.

176. Papavasiliou, F., R. Casellas, H. Suh, X.F. Qin, E. Besmer, R. Pelanda, D. Nemazee, K. Rajewsky, and M.C. Nussenzweig. 1997. V(D)J recombination in mature B cells: a mechanism for altering antibody responses. *Science* 278:298–301.

177. Hertz, M. and D. Nemazee. 1998. Receptor editing and commitment in B lymphocytes. *Curr. Opin. Immunol.* 10:208–213.

178. Shlomchik, M.J., M.P. Madaio, D. Ni, M. Trounstein, and D. Huszar. 1994. The role of B cells in lpr/lpr-induced autoimmunity. *J. Exp. Med.* 180:1295–1306.

179. Theofilopoulos, A.N. and F. Dixon. 1985. Murine models of systemic lupus erythematosus. *Adv. Immunol.* 37:269–390.

180. Theofilopoulos, A.N., R. Kofler, P. Singer, and F. Dixon. 1989. Molecular genetics of murine lupus models. *Adv. Immunol.* 46:61–109.

181. Jyonouchi, H., W. Kincade, R. Good, and M. Gershwin. 1983. B lymphocyte lineage cells in newborn and very

young NZB mice: evidence for regulatory disorders affecting B-cell formation. *J. Immunol.* 131:2219–2225.

182. Jyonouchi, H., P.W. Kincade, and R.A. Good. 1985. Changes in B-lymphocyte lineage cell populations of autoimmune-prone BXSB mice. *J. Immunol.* 134:858–864.

183. Gershwin, M., J. Castles, K. Erickson, and A. Ahmed. 1979. Studies of congenitally immunologic mutant New Zealand mice. II. Absence of T cell progenitor populations and B cell defects of congenitally athymic (nude) NZB mice. *J. Immunol.* 122:2020–2025.

184. Mihara, M., Y. Ohsugi, K. Saito, T. Miyai, M. Togashi, S. Ono, S. Murakami, K. Dobashi, F. Hirayama, and T. Hamaoka. 1988. Thymus-independent occurrence of B cell abnormality and requirement for T cells in the development of autoimmune disease, as evidenced by an analysis of the athymic nude individuals. *J. Immunol.* 141:85–90.

185. Sobel, E.S., C. Mohan, L. Morel, J. Schiffenbauer, and E.K. Wakeland.1999. Genetic dissection of SLE pathogenesis: Adoptive transfer of SLE1 mediates the loss of tolerance by bone marrow-derived B cells. *J. Immunol.* 162:2415–2421.

186. Sobel, E., T. Katagiri, K. Katagiri, S. Morris, P. Cohen, and R.A. Eisenberg. 1991. An intrinsic B cell defect is required for the production of autoantibodies in the lpr model of murine systemic autoimmunity. *J. Exp. Med.* 173:1441–1449.

187. Merino, R., L. Fossati, M. Lacour, and S. Izui. 1991. Selective autoantibody production by Yaa+ B cells in autoimmune Yaa+_Yaa- bone marrow chimeric mice. *J. Exp. Med.* 174:1023–1029.

188. Kotzin, B., S. Babcock, and L. Herron. 1988. Deletion of potentially self-reactive T cell receptor specificities in L3T4-, Lyt-2- cells of lpr mice. *J. Exp. Med.* 168:2221–2229.

189. Vaux, D.L., S. Cory, and J.M. Adams. 1988. Bcl-2 gene promotes haemopoietic cell survival and cooperates with c-myc to immortalize pre-B cells. *Nature* 335:440–442.

190. McDonnell, T.J., N. Deane, F.M. Platt, G. Nunez, U. Jaeger, J.P. McKearn, and S.J. Korsmeyer. 1989. bcl-2-immunoglobulin transgenic mice demonstrate extended B cell survival and follicular lymphoproliferation. *Cell* 57:79–88.

191. Strasser, A., A.W. Harris, D.L. Vaux, E. Webb, M.L. Bath, J.M. Adams, and S. Cory. 1990. Abnormalities of the immune system induced by dysregulated bcl-2 expression in transgenic mice. *Current Topics in Microbiology & Immunology* 166:175–81.

192. Okamoto, M., M. Murakami, A. Shimizu, S. Ozaki, T. Tsubata, S. Kumagai, and T. Honjo. 1992. A transgenic model of autoimmune hemolytic anemia. *J. Exp. Med.* 175:71–79.

193. Reininger, L., T. Radaszkiewicz, M. Kosco, F. Melchers, and A.G. Rolink. 1992. Development of autoimmune disease in SCID mice populated with long-term "in vitro" proliferating (NZBxNZW)F1 pre-B cells. *J. Exp. Med.* 176:1343–1353.

194. Iliev, A., L. Spatz, S. Ray, and B. Diamond. 1994. Lack of allelic exclusion permits autoreactive B cells to escape deletion. *J. Immunol.* 153:3551–3556.

195. Nguyen, K.-A., L. Mandik, A. Bui, H. Noorchashm, A. Eaton, and J. Erikson. 1997. Characterization of anti-single-stranded DNA B cells ina non-autoimmune background. *J. Immunol.* 159:2633–2644.

196. Mandik-Nayak, L., A. Bui, H. Noorchashm, A. Eaton, and J. Erikson. 1997. Regulation of anti-double-stranded DNA B cells in nonautoimmune mice: localization to the T-B interface of the splenic follicle. *J. Exp. Med.* 186:1257–1267.

197. Manz, J., K. Denis, O. Witte, R. Brinster, and U. Storb. 1988. Feedback inhibition of immunoglobulin gene rearrangement by membrane mu, but not by secreted mu heavy chains. *J. Exp. Med.* 168:1363–1381.

198. Carmack, C.E., S.A. Camper, J.J. Mackle, W.U. Gerhard, Weigert, and MG. 1991. Influence of a V kappa 8 L chain transgene on endogenous rearrangements and the immune response to the HA(Sb) determinant on influenza virus. *J. Immunol.* 147:2024–2033.

199. Rolink, A., U. Grawunder, D. Haasner, A. Strasser, and F. Melchers. 1993. Immature surface Ig+ B cells can continue to rearrange kappa and lambda L chain gene loci. *J. Exp. Med.* 178:1263–1270.

200. Ghia, P., A. Gratwohl, E. Signer, T.H. Winkler, F. Melchers, and A.G. Rolink. 1995. Immature B cells from human and mouse bone marrow can change their surface light chain expression. *Eur. J. Immunol.* 25:3108–3114.

201. Giachino, C., E. Padovan, and A. Lanzavecchia. 1995. kappa+lambda+ dual receptor B cells are present in the human peripheral repertoire. *J. Exp. Med.* 181:1245–1250.

202. Harada, K. and H. Yamagishi. 1991. Lack of feedback inhibition of V kappa gene rearrangement by productively rearranged alleles. *J. Exp. Med.* 173:409–415.

203. Zal, T., S. Weiss, A. Mellor, and B. Stockinger. 1996. Expression of a second receptor rescues self-specific T cells from thymic deletion and allows activation of autoreactive effector function. *Proc. Natl. Acad. Sci. USA* 93:9102–9107.

204. Borgulya, P., H. Kishi, Y. Uematsu, and H. von Boehmer. 1992. Exclusion and inclusion of alpha and beta T cell receptor alleles. *Cell* 69:529–537.

205. Sonoda, E., Y. Pewzner-Jung, S. Schwers, S. Taki, S. Jung, D. Eilat, and K. Rajewsky. 1997. B cell development under the condition of allelic inclusion. *Immunity* 6:225–233.

206. Cohen, P. and R.A. Eisenberg. 1991. Lpr and gld: single gene models of systemic autoimmunity and lymphoproliferative disease. *Annu. Rev. Immunol.* 9:243–269.

207. Tan, E. 1989. Antinuclear antibodies: diagnostic markers for autoimmune diseases and probes for cell biology. *Adv. Immunol.* 44:93–151.

208. Alarcon, S. and A. Cabral. 1996. Autoantibodies in systemic lupus erythematosus. *Curr. Opin. Rheumatol.* 8:403–407.

209. Mageed, R., M. Borretzen, S. Moyes, K. Thompson, M.C. Nawijn, and J. Natvig. 1997. Rheumatoid factor autoantibodies in health and disease. *Ann. NY Acad. Sci.* 815:296–311.

210. Nagata, S. and T. Suda. 1995. Fas and Fas ligand: lpr and gld mutations. *Immunol. Today* 16:39–43.

211. Rathmell, J.C. and C.C. Goodnow. 1994. Effects of the lpr mutation on elimination and inactivation of self-reactive B cells. *J. Immunol.* 153:2831–2842.

212. Rubio, C.F., J. Kench, D.M. Russell, R. Yawger, and D. Nemazee. 1996. Analysis of central B cell tolerance in autoimmune-prone MRL/lpr mice bearing autoantibody transgenes. *J. Immunol.* 157:65–71.

213. Kench, J.A., D.M. Russell, and D. Nemazee. 1998. Efficient peripheral clonal elimination of B lymphocytes in MRL-lpr mice bearing autoantibody transgenes. *J. Exp. Med.* 188:909–917.

214. Radic, M.Z. and M. Weigert. 1994. Genetic and structural evidence for antigen selection of anti-DNA antibodies. *Annu. Rev. Immunol.* 12:487–520.

215. Radic, M.Z., J. Mackle, J. Erikson, C. Mol, W.F. Anderson, and M. Weigert. 1993. Residues that mediate DNA binding of autoimmune antibodies. *J. Immunol.* 150:4966–4977.

216. Bloom, D.D., J.L. Davignon, P. Cohen, R.A. Eisenberg, and S.H. Clarke. 1993. Overlap of the anti-Sm and anti-DNA responses of MRL/Mp-lpr/lpr mice. *J. Immunol.* 150:1579–1590.

217. Retter, M.W., P.L. Cohen, R.A. Eisenberg, and S.H. Clarke. 1996. Both Sm and DNA are selecting antigens in the anti-Sm B cell response in autoimmune MRL/lpr mice. *J. Immunol.* 156:1296–1306.

218. Howard, T., M. Iannini, J. Burge, and J. Davis. 1991. Rheumatoid factor, cryoglobulinemia, anti-DNA, and renal disease in patients with systemic lupus erythematosus. *J. Rheumatol.* 18:826–830.

219. Prak, E.L. and M. Weigert. 1995. Light chain replacement: a new model for antibody gene rearrangement. *J. Exp. Med.* 182:541–548.

220. Radic, M.Z., M.A. Mascelli, J. Erikson, H. Shan, and M. Weigert. 1991. Ig H and L chain contributions to autoimmune specificities. *J. Immunol.* 146:176–182.

221. Radic, M.Z., J. Erikson, S. Litwin, and M. Weigert. 1993. B lymphocytes may escape tolerance by revising their antigen receptors. *J. Exp. Med.* 177:1165–1173.

222. Chen, C., M.Z. Radic, J. Erikson, S.A. Camper, S. Litwin, R.R. Hardy, and M. Weigert. 1994. Deletion and editing of B cells that express antibodies to DNA. *J. Immunol.* 152:1970–1982.

3 | Genetic Origin of Human Autoantibodies

Kathleen N. Potter and Freda K. Stevenson

1. INTRODUCTION

The study of human autoimmune disease has been greatly facilitated by clear parallels between the murine and human immune systems, which allow us to draw relevant conclusions concerning human immunity and tolerance from murine knockout and transgenic models (1,2). However, in spite of similarities between the two immune systems, there are also differences that need to be recognized. The first is that the human V_H germline repertoire is more limited than the murine, with only 39–51 expressed V_H gene segments (3,4). Perhaps to compensate for this, there appears to be greater complexity in CDR_H3, with wide variation in length and composition (5,6). Maturation of the humoral response by somatic mutation, antigen selection and isotype switch leads to high affinity antibodies, some of which are likely able to mediate autoimmune disease. In terms of following these processes at the V-gene level, studies in humans have the advantage in that the V_H, D, J_H, V_L and J_L gene segments have been fully documented (3,4,7–13), making it possible to investigate bias in variable domain gene segment usage in health and autoimmune disease. We are also able to assess the nature and level of somatic mutations. The combination of observations in mouse and human are generating new concepts that should help our understanding of autoimmune disease and perhaps lead to new approaches to treatment.

2. THE HUMAN GENE SEGMENT REPERTOIRE

Immunoglobulins are glycoproteins in which the heavy (H) and light (L) chain variable domains are products of genes assembled in an ordered fashion from three unlinked multigene families referred to as variable (V), diversity (D), and joining (J) genes (14). The H chain variable domain is assembled by recombination of V_H, D and J_H gene segments, and the L chain variable domain by V_L and J_L gene segment recombination. The majority of human Ig H chain gene segments are located on the long arm of chromosome 14. A minor locus of V_H and D segments lies on chromosomes 15 and 16, but these gene segments do not contribute to the expressed antibody repertoire (8). Gene segments encoding light chains are on chromosome 2 for κ chains, and chromosome 22 for λ chains.

The 957 kb region of DNA on chromosome 14q32.3, which encompasses the human V_H locus, has been completely sequenced (3,4,7,9). Cook and Tomlinson, (15) defined a V_H segment as functional if it contains an open reading frame (ORF) and is rearranged as an in-frame VDJ gene. Based on this definition, they have identified 95 V_H gene segments, 51 with functionally rearranged ORF (Table 1), 9 with non-rearranged ORF (ORF not found rearranged in vivo), 30 pseudogenes and 5 segments that were unsequenced. There is some discrepancy regarding the available repertoire since Matsuda et al., (4) included the stipulation that expression should be confirmed by identification of the V_H sequence in databases of full-length V_H cDNA. This reduces the available repertoire from 51 to 39 (Table 1). Irrespective of definition of a functional V_H segment, the exact number of V_H segments depends on the haplotype, as a 50 kb insertion polymorphism between segments 3–30 and 4–31 (resulting in the duplication of 3–30 and 4–31) is present in approximately one quarter of the Caucasian population (16), and a second insertion occurs in 50% of donors near segment 1–69 that results in duplication of segment 1–69.

The V_H gene segments are grouped into seven families, V_H1-V_H7, based on > 80% homology at the DNA level (4).

Table 1 Human gene segment repertoires

Repertoire	Gene segments						
	V_H	D	J_H	$V\kappa$	$J\kappa$	$V\lambda$	$J\lambda$
Potential repertoire	95 (123)*	27	9	76	5	70	7
Functional repertoire	51 (39)*	25	6	32	5	30	4
Expressed adult repertoire (predominantly used)	9	25	6	6	5	3	4

*Matsuda et al., 1998.

Based on Cook and Tomlinson (15), potentially functional gene segments per family are as follows: 11 V_H1, 3 V_H2, 22 V_H3 gene segments, 11 V_H4, 2 V_H5, 1 V_H6 and 1 V_H7 family gene segment. Table 2 includes both analyses. Members of the different families are interspersed, not clustered, on the chromosome. A nomenclature has been proposed for V_H gene segments that contains the family number, a dash and a number indicating relative proximity to the J_H cluster (7).

The D region cluster consists of four 9.5 kb tandem repeat units containing a set of six D family segments in the order 5'-D_M-D_{LR}-D_{XP}-D_A-D_K-D_N-3'. Twenty-six D segments were sequenced between 53 and 14 kb upstream of the J_H segments, plus the single member of the seventh family, the D_{Q52} (D7–27) gene segment, located in the J_H gene cluster. There are 25 functional D gene segments and 2 pseudogenes (4,9,17) (Table 1). The D segments have also been renamed using D followed by the family number, a dash and position 1–27 in the direction 5' to 3' on the DNA between the V_H6 segment and the J_H locus (9). There are nine J_H segments, of which six are functional and three are pseudogenes (18) (Table 1), with polymorphic nucleotides in J_H 3–6.

In human light chains, approximately 60% are κ and 40% λ. The ~76 $V\kappa$ gene segments are organized in two cassettes, with 40 in the $J\kappa$ proximal region and 36 in the distal inverted region, with segments in the proximal region expressed more frequently (10,19,20). The distal region is a partial duplicate of the proximal region. The $V\kappa$ includes 7 $V\kappa$ subgroups ($V\kappa$I-$V\kappa$VII), with ~32 functional $V\kappa$ gene segments, 5 functional $J\kappa$ gene segments (Table 1) and a single $C\kappa$ gene (11,20,21).

The $V\lambda$ gene segments are located upstream of seven $J\lambda$-$C\lambda$ pairs, and are grouped into ten $V\lambda$ families (22,23). They are arranged in three distinct clusters, each containing members of the 10 different $V\lambda$ families (12). Cluster A is closest to the $J\lambda$-$C\lambda$ region and contains the $V\lambda$2 and $V\lambda$3 families and one member of the $V\lambda$4 family. Cluster B contains the $V\lambda$1, $V\lambda$5, $V\lambda$7 and $V\lambda$9 families, while cluster C, the most 5' cluster, contains two members of the $V\lambda$4 family and the $V\lambda$6, $V\lambda$8 and $V\lambda$10 families (12,22). There are 70 $V\lambda$ segments, of which ~30 are functional, depending on haplotype (Table 1). These include 5 $V\lambda$1, 5 $V\lambda$2, 8 $V\lambda$3, 3 $V\lambda$4, 3 $V\lambda$5, 1 $V\lambda$6, 2 $V\lambda$7, 1 $V\lambda$8, 1 $V\lambda$9, 1 $V\lambda$10. There are 4 functional $J\lambda$ gene segments (Table 1).

The functional repertoire is therefore considerably more limited than the germ line repertoire, and it is from these potentially functional genes that the actual expressed Ig found in B cells is drawn (Table 1).

3. V(D)J RECOMBINATION

Variable genes of H and L chains undergo sequential combination of the five sets of V(D)J gene segments, under the influence of the recombination activating genes RAG-1 and RAG-2 (24). The processes occur at different developmental stages of the B cell (Figure 1), beginning at the pro-B cell stage, where a single D gene segment recombines with a single J_H gene segment. This is followed by the joining of a V_H gene segment to DJ_H (rev. in 25), and pre-B cells can

Table 2 Summary of the human functional V_H segments

V_H family	Functional gene segments (Cook and Tomlinson et al., 1995)	Functional gene segments (Matsuda et al., 1998)
V_H1	11	9
V_H2	3	3
V_H3	22	19
V_H4	11	6
V_H5	2	1
V_H6	1	1
V_H7	1	0
Total	51	39

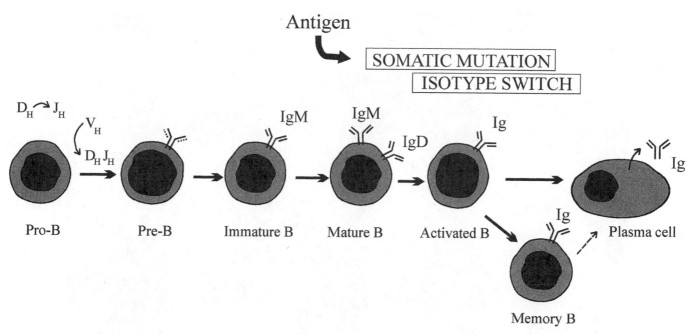

Figure 1. Changes in Ig genes occurring during normal B-cell development. Following recombination of $V_H DJ_H$ genes, pre-B cells express μ chains with a surrogate light chain. Subsequently, V_L-J_L recombination occurs leading to IgM/IgD expression. Binding of antigen then initiates further processes of somatic mutation and isotype switching with generation of memory cells and plasma cells.

express surface Ig with μ chains associated with surrogate L chains (Figure 1). Recombination between V_L and J_L subsequently occurs to generate functional light chain genes, leading to expression of IgM. Imprecision can occur during these V(D)J recombinations since non-templated nucleotides can be inserted at the joints of both H and L chains by the action of the enzyme terminal deoxynucleotidyl transferase (TdT), or deleted by the action of exonuclease. The 5′ ends of certain J_H segments, like J_H 4 and J_H 6, are particularly prone to truncation during the recombination process. These mechanisms generate unique sequences, particularly at the third complementarity determining region (CDR3), which can act as a clonal signature of a B cell. The primary B cell repertoire develops in the bone marrow, with diversity created by choice of gene segments, the recombination processes, and the association of different H and L chains.

Since J_H segments have only one correct reading frame, an in-frame or productive joining occurs in only one out of every three joining events. D segments, however, can be translated in all three reading frames. However, one reading frame tends to encode a stop codon, one encodes glycine residues in conjunction with polar/hydrophilic residues, and a third is hydrophobic in character. The reading frame with the stop codon can still be used if deleted by exonuclease activity during VDJ recombination. Human D segments have a family-specific distribution of these properties over the three reading frames (rev. in 26). In general, short D segments are seen more often than long ones. It is clear that

CDR3 with hydrophilic sequences are dominant, being used by 60% of sequenced Igs (9).

If VDJ recombination at the H locus on the first allele is non-productive (out of frame), an attempt is made at the second allele. If this is also non-productive, the cell will die. A study of the genomic DNA of single B cells indicated that 62/75 had single productive heavy chain VDJ rearrangements, while only 9/75 contained both functional and non-functional rearrangements. These data indicate that most B cells generate a productive VDJ rearrangement on the first attempt. Similar events occur for the light chain where the κ chain rearranges first at one allele and, if non-productive, will attempt a recombination at the second allele. If this second attempt at the κ locus fails, the cells rearrange DNA at the λ locus, first on one allele, then the second.

The surface expression of IgM is required before an immature B cell can leave the bone marrow. Once in the periphery, mature naïve B cells express surface IgM and IgD. When a naive B cell interacts with antigen, it becomes activated and can enter into a germinal center reaction during which somatic mutation occurs either with or without isotype switching (27). These B cells further differentiate either into memory cells, or into plasma cells that secrete high affinity antibody of either of the IgG, IgA or IgE class, depending on which factors are present in the microenvironment (Figure 1).

The V_H domain is composed of seven regions in which FR and CDRs alternate in the order: FR1, CDR1, FR2, CDR2, FR3, CDR3, and FR4. The FRs, while primarily

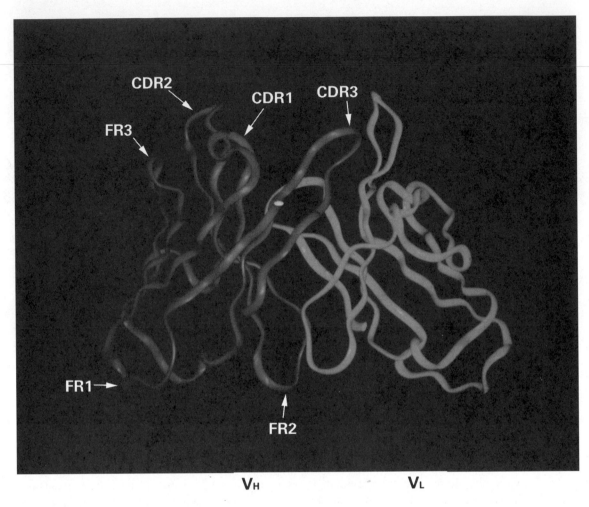

Figure 2.   Ribbon diagram of a molecular model of an Ig Fv region showing V_H (darker shading) and V_L (paler shading). The three framework regions (FRI, FR2 and FR3) and three complementarity-determining regions (CDR1, CDR2 and CDR3) of VH are indicated.

responsible for maintaining the structure of the immuno-globulin fold, also influence antigen specificity. In fact, it is clear from the ribbon diagram (Figure 2) that FR1 and FR3 are exposed to the external environment and have the potential to interact directly with antigen. This may be important in considering how superantigens, or "unconventional" antigens, can bind to FR regions of certain V regions (see below). Figure 2 also illustrates the disposition of the CDRs of the H chain, with CDR3 sited at the center of the combining site. The CDRs in V_H and V_L are critical contact points for interaction with conventional antigen. V_L has been shown to modulate antibody binding, even when the V_H dominates the binding interaction (28). It appears from analyses of IgM[+] (29) and IgG[+] (30) B cells from peripheral blood that H and L chain pairings are largely random. Also there is no correlation between the CDR3 lengths of Vκ and V_H in individual B cells.

3.1.   V(D)J Gene Segment Utilisation:

The frequency with which a gene segment rearranges is potentially influenced by several intrinsic mechanisms: 1) proximity to J gene segments, 2) accessibility to the recombinase machinery, 3) sequence of the recombination signal sequence and the spacer length, 4) efficiency of promoters, 5) proximity to enhancer sequences and 6) stage of B cell ontogeny. Matsuda et al., (4) found that conservation of the promoter and RSS was higher in functional V_H segments than in pseudogenes and that V_H segments preferentially used in the early stages of ontogeny were not necessarily clustered in the J_H proximal region.

It is clear (Table 1) that V_H gene segments in the potential repertoire are not equally represented in the adult expressed repertoire. To assess early influences prior to antigen selection, Kraj et al., (31) compared the V_H3 and

V_H4 family repertoires of pre-B and immature B cells from bone marrow and mature B cells from peripheral blood. They found that over-representation of certain V genes could arise from mechanisms operating at both early and later stages of B cell development. The V4–34, V4–59 and V3–23 gene segments dominated the repertoires of the surface Ig-negative pre-B cell as well as immature and mature B cells, consistent with intrinsic preferences prior to IgM expression. It also appears that certain segments that are favored during early stages of B cell development can become subsequently disfavored during later stages of development.

Biased gene segment utilization in the adult is obviously influenced by antigen exposure. To gain insight into this influence, IgM^+ cells were single cell sorted and both non-productive and productive V_H rearrangements were sequenced (6). A further point investigated was whether there was a difference in V_H gene usage between $CD5^+$ and $CD5^-$ B cells. $CD5^+$ cells are thought to be the major source of polyreactive serum IgM, and a potential source of auto-antibodies. Over 75% of the expressed repertoire was found to be derived from nine V_H gene segments, V1–18, V3–07, V3–23, V3–30, V3–30.3, V4–34, V4–39, V4–59, -/DP-58 (Table 3). In both subsets, V_H3 was overrepresented, with V3–23 being most frequently used in productive rearrangements. In the $CD5^+$ subset, all other V_H families were found at a frequency expected from random utilization. In the $CD5^-$ population however, V_H4 was highly represented in the non-productive repertoire but not in the productive, suggesting a negative selection pressure. Other interesting features of the productive repertoire included higher usage of V_H1 in $CD5 +$ cells. The frequency of specific V_H segments in the mutated population differed from that in the nonmutated population. These results suggest that there is a preferential expression of particular V_H gene segments during ontogeny based on a recombinational bias, as well as selection after interaction with antigen.

There is evidence that structural or selective pressures acting at the amino acid level affect D segment usage. D3–10 is used twice as often as D3–3 even though both have identical RSS and are the same length; similarly, D2–2 is used five times as often as D2–8 (9). These authors use stringent criteria involving at least ten consecutive nucleotides for assigning D segment usage, making it quite difficult to assign short CDR_H3 sequences, and calling into question the existence of DIR segments, inverted D segments or D-D recombination. In the analysis of 893 rearranged CDR_H3 sequences, it was not possible to confidently assign a D segment for 49.5% of the sequences (9). In single cell analysis, using less stringent criteria, D3 was found at the highest frequency in both productive and non-productive rearrangements, in agreement with Corbett et al., (9). Similarly, J_H4 genes were shown to be preferentially used in both productive and non-productive sequences (6).

There appear to be both genetic and structural length constraints on CDR_H3 until two months after birth (32). In the fetus, there is a predilection for short D gene segments such as D7–27 (DQ52) and against long ones such as D3 (D_{XP}). Length is also controlled by limited N addition in the fetal bone marrow, with D7–27 DJ joins having extensive N additions, while D3-containing DJ joins show few. In addition to genetic mechanisms, somatic selection appears to result in restriction of CDR_H3 length in both the fetus and adult.

Single cell PCR was used to examine both productive and nonproductive $V\kappa J\kappa$ rearrangements (33). Nearly every functional $V\kappa$ gene segment was used, although six gene segments were used preferentially (A27, L2, L6, L12a and A17, all from the proximal locus, and O12/O2). A17 was over-utilized because of a rearrangement bias. $J\kappa1$ and $J\kappa2$ were overrepresented and $J\kappa3$ and $J\kappa5$ were underrepresented in the nonproductive repertoire, indicating a molecular basis for the bias. Junctional diversity due to N nucleotide addition was found in 60% of the rearrangements, and by exonuclease trimming in 75% of rearrangements. A strict length of CDR3 was maintained in both repertoires.

Analysis of a library of 7600 λ cDNA clones from peripheral blood lymphocytes indicated that only a few $V\lambda$ gene segments are frequently expressed, with 3 of the 30 functional $V\lambda$ segments (2a2, 1e and 2c) encoding half of the expressed $V\lambda$ repertoire (13). Most of the expressed λ-chains are derived from the A cluster. The $V\lambda$-$J\lambda$ repertoires in normal adults, both productive and nonproductive, was also analyzed by single-cell PCR (34). All 10 $V\lambda$ families were detected in both, with the $V\lambda1$ and $V\lambda2$ found most frequently. $V\lambda4$ was underrepresented and $V\lambda3$ over-represented in the productive repertoire. Evidence for TdT and exonuclease activity was detected in approximately 80% of nonproductive rearrangements. The length of the CDR3 was 30 bp in both repertoires, and all 4 functional $J\lambda$ segments were represented with $J\lambda7$ being observed most often, followed by $J\lambda2/3$, and $J\lambda1$ used least often.

Table 3 Summary of preferentially used gene segments in the adult human

V_H	V_H1	1–18			
	V_H3	3–07	3–23	3–30	3–30.3
	V_H4	4–34	4–39	4–59	
D	D_{XP} (D3)				
J_H	J_H4				
$V\kappa$	A27	L2	L6	L12a	A17	O12/O2
$J\kappa$	$J\kappa1$	$J\kappa2$			
$V\lambda$	2a2	1e	2c		
$J\lambda$	$J\lambda7$				

4. SUPERANTIGENS

An increased understanding of the relationship between the potential and expressed repertoires reveals that much of the previously suspected bias in V_H gene segment usage by autoantibodies is likely to reflect the fact that the normal repertoire itself is biased. However, there are examples of genuine bias in V-gene usage, both in antibody responses to pathogens, and in some autoantibodies. These biases appear to arise from the fact that certain antigens, including autoantigens, bind to the FR of the V domain, and can be considered as B-cell superantigens (SAgs).

SAgs originally referred to intact proteins, primarily of bacterial and viral origin, capable of activating large numbers of T cells, irrespective of the antigen specificity, by binding through the variable region of a specific family of $V\beta$ chains of the T cell receptor outside of the normal binding groove (35). It was first proposed by Pascual and Capra (36) that there could be B cell SAgs that would inter-act with soluble and membrane-bound immunoglobulins outside of the conventional binding-sites. Examples of these include Staphylococcal protein A (SPA), which binds to VH3-encoded antibodies (37), HIV gp120 interacting with V_H3-encoded antibodies (38), and protein L from *Peptococcus magnus*, which binds to a subset of $V\kappa$-encoded light chains (39). Since B-cell superantigens bind outside the conventional binding site, they can stimulate large numbers of B cells via FR of the expressed Ig, resulting in polyclonal responses.

A mature naïve B cell in the periphery displaying unmutated surface IgM, therefore, has the ability to interact with antigens in its environment either through the conventional combining site involving CDRs, or through the external surfaces involving FRs (Figure 2). A structural investigation of the interaction between SPA and a V_H3-encoded antibody indicated that the solvent-exposed surface of the antibody comprising residue 57 of CDR2, and sequences in FR1 and FR3 are all simultaneously required for SPA binding (40,41). It is apparent from sequence analysis that the Ig shape required for SPA binding can be formed from different amino acid combinations, as there is no common sequence motif among human SPA binders or between human and mouse SPA binders. Many, but not all, somatically-mutated V_H3-encoded Ig lose their ability to bind SPA. It appears that both the position and nature of the mutations affect SPA binding. B cell SAgs have the potential to activate B cells and/or induce their deletion. The interaction of gp120 with V_H3-expressing B cells is thought to account for their deletion (42). In a study *in vitro*, Staphylococcal enterotoxin D (SED) induced the survival of B cells uniquely expressing V_H4 encoded IgM Ig. The interaction was not gene segment-specific, and involved a variety of different D and J_H gene segments (43).

The best example of true bias in autoimmunity is the dra-matic finding that all cold agglutinins, with specificity for the I/i antigen of red cells, are derived from one V_H gene segment, V4–34 (44). The superautoantigen of the red cell may reflect molecular mimicry between the I/i carbo-hydrate and a pathogen, but this remains unproven. The I/i antigens bind specifically to FR1 of V4–34-encoded Ig (45, see below). Information on expression of V4–34-encoded Ig has been facilitated by the availability of our MoAb, 9G4, which is specific for this gene product (46). It was found to be quite commonly expressed by normal fetal B cells (~6% of B cells at 20 weeks gestation) and is found in various normal adult lymphoid tissues; 10.8% of bone marrow B cells, 6.9% of peripheral blood B cells, 2.9% of total tonsil B cells, 5.0% of spleen cells and 3.2% of lymph node B cells (46). In contrast, < 1% of circulating IgM and IgG is V4–34-encoded. However, following infection with Epstein-Barr virus (EBV) in infectious mononucleosis, serum levels of V4–34-encoded antibodies are selectively raised (47). Similarly, V4–34 Ig is found following infections with cytomegalovirus, and *Mycoplasma pneumonia*, and, although it is unlikely that all the V4–34 Ig has anti-red cell activity, CA can develop (48).

Interestingly, raised serum levels of V4–34-encoded IgM and IgG, some of which have anti-DNA activity, are found in patients with SLE (49). This rise seems rather specific for SLE, since increases have not been found RA, Sjögren's syndrome, or scleroderma (49,50). It would appear, there-fore, that 9G4-expressing B cells are capable of secreting antibody under certain conditions of stimulation, and that the Ig produced can include various autoantibody specificities.

In the context of SAgs, the fact that a large proportion of B cells express V4–34-encoded Ig creates a situation whereby a potential superantigen, either on the surface of a microorganism, or a self-antigen, could induce polyclonal stimulation. If this occurs frequently, it could set the scene for neoplastic transformation. Such a process may account for the relatively common IgM monoclonal gammopathy giving rise to cold agglutinin disease. Analysis of the struc-ture, specificities and control of expression of V4–34 encoded antibodies, made possible by the availability of the 9G4 MoAb, should reveal information relevant to our understanding of superautoantigens and their role in the immune response.

5. SOMATIC HYPERMUTATION

By comparing the nucleotide sequence of a V-region derived from a B cell with the closest germ line V, D, J sequences in the databases, nucleotide changes arising from somatic mutation can be detected. A few deviations may be due to inherited polymorphic differences in germ line sequences, but these tend to be minor, and generally somatic events are clear. Somatic mutation occurs in the highly organized microenvironment of the germinal center after B cells have become centroblasts (51). In contrast to the mouse, where memory B cells predominantly have class switched, there is a substantial IgM-expressing memory B cells population in the human (52). Nucleotide changes are introduced across the variable domain genes and tend to concentrate in CDR1 and CDR2, known contact points for antigen (53,54). The mechanism of somatic mutation is not fully understood, however, there are intrinsic "hot spots" that mutate in the absence of antigen such as the serine AGY triplets clustered in the CDRs of $V\kappa$ and V_H gene segments (55), and GCY, GTA, TAY and RGYW motifs (rev. in 56). A study of nonproductive V_H DJ_H rearrangements indicated that the mutational machinery targets RGYW sequences on both DNA strands (56). Nucleotide changes in the CDR_H3 sequence are more difficult to locate due to the complexity at this site, and difficulties in identifying the D-segment genes. The environmental factors involved in induction and maintenance of somatic mutation include CD40 ligand, cytokines, follicular dendritic cells, and probably ligation of the surface Ig.

Productively and nonproductively rearranged V genes can apparently be targets for somatic mutation. Clustering of replacement mutations in the CDRs clearly occurs in both and is likely to reflect an intrinsic property of the interaction between variable domain gene sequences and the mutator mechanism (56). The main differences found were that productive rearrangements had a significantly lower mutational frequency and a smaller number of replacement mutations than the nonproductive, particularly in FR1 and FR3. The most obvious feature of antigen-selected sequences, therefore, appears to be conservation of FR sequence It is likely that single amino acid changes have a critical influence on affinity, and these can be discerned only in individual antibodies. These findings have rendered deductions from the overall distribution of replacement or silent mutations about whether an antibody has been selected by antigen rather less secure.

Levels of mutation in V_H are generally higher than in $V\kappa$, but a significant correlation was found between mutational frequencies in heavy and light chain pairs of single B cells, particularly in CDRs (29). Interestingly, > 23% of productive rearrangements, but only 7% of nonproductive rearrangements of $V\kappa$ contained somatic mutations (33).

Even B cells whose receptors are not autospecific have the potential of becoming so following somatic mutation in the germinal center. We have evidence that the mutated monoclonal antibody PD2 derived from a patient with the characteristic anti-mitochondrial antibody of primary biliary cirrhosis that specifically binds lipoylated PDC-E2 (57) no longer binds the antigen when reverted to germline (Potter et al., unpublished observation). This indicates that the high affinity binding by PD2 for its autoantigen is the result of somatic mutation. In order to prevent the proliferation of autoreactive B cells, peripheral processes of negative selection must operate. These could involve counter selection against sequences with mutations in CDRs that result in increased affinity to an autoantigen, a feature described by Borretzen et al., (58) or receptor editing.

5.1. Receptor Editing

Once a B cell has undergone productive a V_H D J_H and a V_L and J_L rearrangement, it may not be irreversibly committed to these choices. It is becoming evident from both mouse and human studies that receptor editing leading to replacement of one V_L gene by another can occur in both the bone marrow (59–63) and in the periphery (30, 64–,66); rev. in 66). The question is whether receptor editing is operative in the periphery in patients with autoimmune disease, and, if so, whether production of autoantibodies arises because of this process, or in spite of it.

Clearly autoreactive IgM + B cells are present in healthy individuals. However, autoimmune disease is averted by several strategies, including cellular deletion, functional silencing (anergy), the absence of T cell help or by editing. Receptor editing apparently plays a role in both central (bone marrow) and peripheral (germinal center) tolerance. The term "receptor editing" includes several concepts: Replacement of a transgenic receptor by an endogenous one (59,60), light chain replacement by a secondary (tertiary or more) $V_L J_L$ recombination (54,64,67,68), heavy chain receptor editing (69), and heavy chain V_H replacement (70,71).

The first reports of receptor editing were in transgenic mice where autoreactive B cells in the bone marrow developed new receptors that lacked autospecificity (59,60). In developing B cells, high affinity receptor binding to self-antigen induces secondary V(D)J recombination resulting in the replacement of autoreactive receptor genes with those encoding non-self reactive receptors (62,63,72). While there is evident screening for autoreactive B cells in the bone marrow, autoreactive IgM^+ B cells with receptors in germline configuration appear to be allowed to exit to the periphery. Ongoing V(D)J recombination in peripheral B cells is indicated by expression of RAG1 and RAG2 in germinal center B cells (73,74). However, in mature human tonsil B cells, cross-linking of Ig receptors turns off RAG gene expression (65). In contrast to bone marrow, receptor editing in the periphery (sometimes called revision) may contribute to the processes of intraclonal competition and diversification by stimulating rearrangement in B cells that do not bind antigen.

There are two reports of light chain receptor editing in systemic lupus erythematosus (SLE) (56,75). Mockridge et al., (56) studied single-cell sorted B cells and found that the IgG^+B cells had mutated V_H paired with unmutated V_L. In contrast, most IgG^+B cells in healthy individuals had mutations in both V_H and V_L (Mockridge and Stevenson, unpublished observations). Dorner et al. (57) studied a patient with SLE and found increased usage of the Jκ5 segment, the most 3' of the 5 Jκ gene segments, and over-representation of the Vκ1 and Vκ4 families especially the L15 (distal region), 014/04 and B3 (proximal region) gene segments. These data indicate that the mechanism of receptor editing is functional and possibly excessive in SLE patients, but that it is not effective in preventing autoimmune disease. The presence of receptor-edited V_L in B cells in SLE is in sharp contrast to the findings in cloned IgG^+ autoantibodies derived from patients. Although only limited data are available, both V_H and V_L tend to have mutations (49). This is also the pattern most often found in antibodies against exogenous infectious agents. Receptor editing also includes V_H gene replacement (71). So far, most of the data on this mechanism, however, are from mice. Using DNA substrates in a pre-B cell line, Covey et al., (76) showed that V_H replacement recombination is mediated by a heptamer-like sequence in the 3' region of the V_H segment in the assembled $V_H DJ_H$ join and the RSS of the unrearranged V_H segment.

5.2. Autoantibodies and Autoimmune Disease

It is against this backdrop of the human repertoire that we analyse the gene segment utilization by autoantibodies. Natural autoantibodies constitute a large proportion of the pre-immune repertoire and are characterized as being polyreactive, IgM isotype, germline encoded and low affinity binders. When the V_H region usage from 28 polyreactive antibodies from fetal and adult splenic B cells were analysed, all six families were represented in proportion to their estimated family size (77). Amino acid substitution experiments demonstrated that junctional sequences in the CDR3 generated somatically during Ig gene rearrangement can result in autospecificity (78).

Hundreds of antibodies from a variety of autoimmune disorders have been sequenced and analysed. In 1993, Capra and Natvig (79) analyzed over 100 V_H genes from a wide variety of diseases and found no evidence of restriction. Evidently, human autoantibodies are derived from the same repertoire as those used by the normal antibodies to exogenous antigens, and there is not a selected set of immunoglobulin genes expressed by individuals with autoimmune disease. The study indicated that almost any V_H and V_L gene segment can make a rheumatoid factor, anti-acetylcholine antibody, anti-striated muscle antibodies and anti-thyroglobulin antibodies. Even with the larger database accumulated since 1993, the results are generally indicative of no restriction in gene segment usage by autoantibodies, with the notable exception of the mandatory use of the V4–34 gene by cold agglutinins (see below).

The roles of autoantibodies in the pathology of autoimmune disease remains a topic of debate. B cells and autoantibodies were previously not regarded as primary initiators of autoimmune disease since it had not been possible to transfer autoimmune disease with either B lymphocytes alone or with autoantibodies (80). However, some autoantibodies, such as IgG anti-dsDNA in lupus, appear to be involved in disease pathology, with a correlation between the serum levels of antibody and the degree of renal damage in patients with nephritis (81). Also, it has been recently demonstrated that a subset of human IgG anti-DNA antibodies can directly precipitate nephritis in SCID mice (82,83). In cold agglutinin disease, and other autoimmune hemolytic anemias, antibodies have a clear direct pathological effect (48). The importance of autoantibody specificity is illustrated by the fact that anti-PDC-E2 antibodies from patients with primary biliary cirrhosis inactivate the enzymatic activity of PDC-E2, while natural autoantibodies from healthy people bind the autoantigen but do not inhibit enzymatic activity (84). In a study of rheumatoid factor, Bonagura et al., (85) mapped IgG epitopes bound by IgM RF from immunized controls compared with RF isolated from patients with rheumatoid arthritis (RA). They found that some RA-RF showed novel specificities for IgG that may identify disease-specific autoantibodies in patients with RA.

Antigen presentation by B cells appears crucial in the development of autoimmune disease, as shown using knock out mice (86). In this model, surface-expressed Ig rather than secreted antibody was essential for the pathogenesis of murine lupus. Presentation by B cells can influence the processing of antigen (reviewed in Mamula, (87). One model predicts that when B cells bind multi-determinant self-proteins and present novel self-peptides, or cryptic peptides, the result is a diversification of autoreactive T cells. B cells, therefore, are considered central to the phenomenon of epitope spreading. The fine specificity of the surface immunoglobulin thus can directly influence the peptide transported to the surface of the B cell (88).

5.3. Usage of V Gene Families in Autoantibodies Specific for Various Autoantigens V4–34 Usage in Cold Agglutinins

A dramatic example of biased use of V_H genes has been revealed in cold agglutinin disease. Cold agglutinins (CA) recognize the I/i carbohydrate antigens on the surface of red blood cells (RBC) and at low temperatures cause hemagglutination and hemolysis (48,89). Pathogenic CA are generally of the IgM isotype, and derive from monoclonal B cell expansions (90). Nucleotide sequence analysis of CA V-genes revealed universal usage of the V4–34 gene segment in association with different D, J_H, V_L and J_L gene segments (44,91) with the result that each $CDR_H 3$ and $CDR_L 3$ has a unique sequence with no evident sequence motifs. Although there is some preference for κ light chains, a wide variety of V_L gene segments can be used.

The anti-idiotopic MoAb 9G4 is specific for V4–34-encoded Ig (see above), and the fact that it also inhibited cold agglutination indicated the closeness of the idiotope to the red cell binding site (46). Mutational analysis using CA expressed in insect cells precisely localized the 9G4 Id to FR1 involving residues at positions 7 (92) and 23–25 (93). Localization of the 9G4 Id to FR1, and the established preference of V4–34 encoded antibodies to bind I/i antigens, suggested that FR1 may be involved in I/i binding, rendering this interaction superantigen-like. Further analysis confirmed that FR1 is critical for the interaction between CA and the I antigen determinants. However, although V4–34-encoded CA have different CDR_H3 sequences, there is an influence of this sequence on the binding site in FR1. It appears that the last six residues of CDR_H3 are particularly influential, and that an overall positive charge on CDR3 was incompatible with anti-I/i activity (49).

We studied the role of the light chain in I binding by producing combinatorial antibodies composed of either CA-derived H chains and ten L chains from various antibodies of different specificities. In general, non-parental L-chains paired with the H-chains, lost I binding. Therefore, even though the H chain FR1 and CDR_H3 are the major mediators, the L-chain can also have a major effect on antigen binding activity. These results support the view that for every V_H gene segment and CDR_H3 in an expressed antibody, there is a L-chain that has been co-selected by antigen.

The V4–34 gene is mandatory for encoding CA autoantibodies, but the gene is also used to encode other antibodies. It is quite frequently used to encode IgM antibodies with specificity for Rh D antigen, and, interestingly, these antibodies also have CA activity, although the reverse is not true (94). As mentioned above, there are increased levels of V4–34-encoded Ig in patients with SLE, and some anti-DNA autoantibodies, of both IgM and IgG class, are derived from V4–34 (49). However, interaction with DNA appears to be mediated by positively charged CDR_H3 sequences, and does not involve the Id-associated sequence of FR1. This was shown by mutating FR1 of an anti-ssDNA antibody to replace Trp by Ser. As expected, the mutant antibody was 9G4-negative, but there was no reduction in binding to ssDNA (92). It would appear that there are antigens, such as the I/i antigens, which interact with the FR as well as the CDR_H3, while others, such as DNA, bind V4–34 encoded Ig irrespective of the presence of the 9G4 Id in FR1. Interestingly, certain V4–34-encoded IgM MoAbs can directly kill normal B cells, apparently by punching holes in the cell membrane (77). Ability to mediate killing is associated with an overall positive charge on the CDR3, and in fact these MoAbs frequently also bind to DNA.

6. RHEUMATOID FACTORS

Rheumatoid factors (RF) are IgM molecules with specificity for the Fc portion of IgG, and are characteristically found in patients with rheumatoid arthritis (RA), although they can be produced in healthy people after infection or immunization. IgMs with RF activity are also secreted by a subset of neoplastic B cells, and these were among the first Igs to show restriction in their V regions (95). Interestingly, the bias towards usage of V_H1 and VκIIIb seen in the neoplastic B cells was not found in RF from patients with RA (96), and there is no evidence of biased usage in the pathogenic RFs in RA (rev. in 97). The first crystal structure of a human IgM RF bound to its autoantigen gave a surprising result (98). RF-AN interacted with the Cγ2/Cγ3 cleft region of Fc using only the edge of the conventional combining site surface, leaving much of the site available for interaction with a different antigen. Since an important contact residue was due to somatic mutation, there is evidence for somatic mutation followed by antigen selection. These data may indicate that Fc binding is an epiphenomenon of RA and the important autoantigen is not yet identified.

When somatic mutation and affinity maturation of IgM RF from immunized healthy individuals was studied, it was found that the V regions underwent extensive mutation, but there was a strong selection against replacement amino acids in the CDRs. In addition, there was no increase in affinity for antigen with increasing mutations (58). These findings suggested that normal individuals selected against production of high affinity RFs. In terms of the light chains, an analysis of κ transcripts from synovium-derived cells indicated that many contained non-templated N additions (99). However, this is not a particular feature of RA since N addition in L chains is a normal mechanism of diversity (100).

7. ANTI-DNA ANTIBODIES

While not all anti-DNA antibodies are pathogenic, IgG antibodies to dsDNA in SLE appear to be involved in the pathogenesis of the disease. Analysis of both IgM and IgG anti-DNA antibodies indicates no evidence of restriction in the use of V_H families (50,101,102), and a wide range of light chains has been found. Therefore, gene segment utilization by anti-DNA antibodies largely reflects the normal B cell-expressed repertoire. Several anti-DNA antibodies have been isolated whose V_H CDR3 is rich in positively-charged amino acids, as noted for V4–34-encoded anti-DNA antibodies (see above). However, this is not an absolute requirement for dsDNA binding, as one third of anti-dsDNA antibodies do not have arginine in CDR_H3. Conversely, antibodies with positively-charged residues in the V_H CDR3 do not always demonstrate DNA binding (103). Mockridge et al., (92) used mutagenesis of a human monoclonal anti-dsDNA antibody encoded by the V4–34 gene to demonstrate that basic residues in the V_H CDR3 were important in binding. Krishnan et al. (104) found a correlation between the residue position of arginine in the CDR_H3 and DNA specificity. Somatic mutations in the L-chain, but not the H-chain, may also be involved in antigen binding (105).

Combinatorial IgG Fab phage display libraries with anti-DNA activity were obtained from a patient with SLE and revealed a predominance of basic residues towards the N-terminus of CDR3 (106). The crucial role of the CDR3 in recognition of DNA was confirmed by transplanting this sequence into an unrelated antibody. Roben et al. (102) were able to create phage display IgG Fab libraries from twins in which only one had SLE. H-L-chain shuffling experiments demonstrated a case in which the *in vitro* creation of anti-dsDNA binding activity required restrictive pairing of a heavy chain with $V\lambda$ light chains, similar to those in circulating anti-dsDNA antibodies. In contrast, IgG anti-dsDNA antibodies could not be recovered from the healthy twin, even from shuffled libraries. These findings indicate how useful information on the development of autoantibodies may be obtained from careful use of phage library display.

8. ANTI-ACETYLCHOLINE RECEPTOR ANTIBODIES IN MYASTHENIA GRAVIS

Myasthenia gravis (MG) is a B cell-mediated autoimmune disease in which autoantibody has a clear pathological role. MG is caused by interference with neuromuscular transmission by autoantibodies against the nicotinic acetylcholine receptor (AchR) on muscle (107). Autoantibodies against the contractile elements of striated muscle (StrAb) are found in approximately 30% of patients with MG. Three monoclonal anti-StrAb were isolated from thymic B lymphocytes from patients with MG and thymoma. The IgM antibody SA-1A was a direct copy of the germline V5–51 gene segment, and used a D3–3 D (D_{XP4}) segment with a germline J_H6 gene segment. The light chain was $V\kappa I$ encoded using a germline $J\kappa4$ gene segment, both in germline configuration. This is an example of gene segments in unmutated form encoding an autoantibody likely to be pathogenic. Two heavily somatically-mutated IgG autoantibodies with anti-titin specificity were also isolated. SA-4A was V4–34 (96.6%)/D1–7 (DM1)/germline J_H4 encoded. The SA-4A light chain was $V\kappa3b$ (A11)/$J\kappa3$ encoded. SA-4B was V_H3/J_H4 encoded, and the light chain was $V\kappa I/J\kappa I$ encoded. In another study, two IgM anti-AchR antibodies were encoded by germline V_H5 gene

segments, while the two IgG antibodies (V_H2 and V_H3 encoded) showed evidence of extensive somatic mutation (108). It is apparent that there is a diversity of gene segments used by anti-AchR autoantibodies.

9. ANTI-PC-E2 ANTIBODIES

Monoclonal antibodies have been isolated from patients with primary biliary cirrhosis that recognize the inner lipoyl domain of the E2 protein of the pyruvate dehydrogenase complex (PDC-E2) (57,109,110). These antibodies may be revealing since they are highly specific for the disease, and patients tend not to have other autoantibodies. Although the panel is quite small, there appears to be a bias towards λ light chains in both IgM and IgG anti-PDC autoantibodies, with V_H gene segments being mostly V_H3. Some of the IgM antibodies were germline encoded, while the IgG antibodies were extensively mutated. Since both the autoantibodies and the autoantigens in this disease are available in recombinant form, it should be possible to assess exactly how these potentially pathological antibodies perceive the autoantigenic determinants.

10. CONCLUSIONS

In the human, the variable domains of H and L chains are assembled from a relatively small number of V,D and J gene segments. It is this same set of gene segments that is used in the production of protective antibodies against infectious pathogens, and by autoantibodies. The central role of autoreactive Ig expressed by B cells in the pathogenesis of autoimmune disease has been underlined by recent data from knock-out mouse models. It is now feasible to reveal the binding sites of this autoreactive Ig at the level of the genes used to encode the variable regions. In humans, analysis of autoantibody V_H and V_L genes is particularly informative, since both the potential and expressed repertoires have been mapped. V-gene sequence data can indicate bias in V-gene usage, a feature consistent with superantigenic stimulation of B cells via framework region sequences. Comparison with germ line gene sequences also delineates somatic mutations, of crucial importance in high affinity antibodies. We have yet to understand how an autoantigen can provide the limited antigen concentration required for affinity maturation. However, the clear importance of amino acid changes in recognition of autoantigen can be investigated using recombinant manipulated antibody fragments. In fact, this technology allows us to re-create the unmutated correlate of high affinity autoantibodies and follow the changing specificity as mutations accumulate.

Receptor editing of normal B cells in the periphery increases the possibility for further generation of auto-reactivity. Evidence for increased editing in patients with lupus raises questions over the role of this phenomenon in the disease process. At present it is unclear if it is a cause or a consequence of autoreactivity. The combination of increased understanding of immune pathways from mouse models together with the gene-based knowledge of anti-body structures in humans will surely solve the puzzle of autoimmune disease, and should lead to a rational basis for new therapeutic strategies for patients.

ACKNOWLEDGEMENTS

The authors would like to thank the Arthritis Research Campaign and Tenovus UK for support, and Dr. Brian Sutton for help with Figure 2.

References

1. Tuaillon, N., P.W. Tucker, and J.D. Capra. 1995. V_HDJ_H recombination in a human immunoglobulin heavy chain tansgenic minilocus. The Immunologist 3/5–6:269–274.

2. Mendez, M.J., L.L. Green, J.R. Corvalan, X.C. Jia, C.E. Maynard- Currie, X.D., Yang, M.L. Gallo, D.M. Louie, D.V. Lee, K.L. Erickson, J. Luna, C.M.-N Roy, H. Abderrahim, F. Kirshenbaum, M. Noguchi D.H. Smith, A. Fukushima, J.F. Hales, M.H. Finer, C.G. Davis, K.M. Zsebo, and A. Jakobovits. 1997. Functional transplant of megabase human immunoglobulin loci recapitulates human antibody response in mice. Nat. Genet. 15: 146–156.

3. Cook, G.P., I.M. Tomlinson, G. Walter, H. Riethman, N.P. Carter, L. Buluwela, G. Winter, and T.H. Rabbitts. 1994. A map of the human immunoglobulin V_H locus completed by analysis of the telomeric region of chromosome 14q. Nat. Genet. 7:162–168.

4. Matsuda, F., K Ishii, P. Bourvagnet, K. I. Kuma, H. Hayashida, T. Miyata, and T. Honjo. 1998. The complete nucleotide sequence of the human immunoglobulin heavy chain variable region locus. J. Exp. Med. 188:2151–262.

5. Wu, T.T., G. Johnson, and E.A. Kabat. 1993. Length distribution of CDRH3 in antibodies. Proteins: Struct. Funct. Genet. 16:1–7.

6. Brezinschek, H.-P., S.J. Foster, R.I. Brezinschek, T. Dorner, R. Domiati-Saad, and P.E. Lipsky. 1997. Analysis of the human V_H gene repertoire. Differential effects of selection and somatic hypermutation on human peripheral CD5+/IgM+ and CD5-/IgM+ B cells. J. Clin. Invest. 99:2488–2501.

7. Matsuda, F., E-K. Shin, H. Nagaoka, R. Matsumura, M. Haino, Y. Fukita, S. Taka-ishi, T. Imai, J.H. Riley, R. Anand, E. Soeda, and T. Honjo. 1993. Structure and physical map of 64 variable segments in the 3'0.8-megabase region of the human immunoglobulin heavy-chain locus. Nature genetics 3:88–94.

8. Tomlinson, I.M., G.P. Cook, N.P. Carter, R. Elaswarapu, S. Smith, G. Walter, L. Buluwela, T.H. Rabbitts, and G. Winter. 1994. Human immunoglobulin V_H and D segments on chromosomes 15q11.2 and 16p11.2. Hum. Mol. Genet. 3:853–860.

9. Corbett, S.J., I.M. Tomlinson, E.L.L. Sonnhammer, S. Buck, and G. Winter. 1997. Sequence of the human immuno-globulin diversity (D) segment locus: a systematic analysis provides no evidence for the use of DIR segments, inverted D segments, "minor" D segments or D-D recombination. J. Mol. Biol. 270:587–597.

10. Zachau, H.G. 1993. The immunoglobulin κ locus-or-what has been learned from looking closely at one-tenth of a percent of the human genome. Gene 135:167–173.

11. Tomlinson, I.M., P.L. Cox, E. Gherardi, A.M. Lesk, and C. Chothia. 1995. The structural repertoire of the human Vκ domain. EMBO J. 14:4628–4638.

12. Williams, S.C., J-P. Frippiat, I.M. Tomlinson, O. Ignatovich, M-P. Lefranc, and G. Winter. 1996. Sequence and evolution of the human germline Vλ repertoire. J. Mol. Biol. 264:220–232.

13. Ignatovich, O., I.M. Tomlinson, P.T. Jones, and G. Winter. 1997. The creation of diversity in the human immuno-glonulin Vλ repertoire. J. Mol. Biol. 268:69–77.

14. Tonegawa, S. 1983. Somatic generation of antibody diversity. Nature 302:575–581.

15. Cook, G.P., and I.M. Tomlinson. 1995. The human immuno-globulin V_{II} repertoire. Immunol. Today 16:237–242.

16. Walter, M.A., U. Surti, M.H. Hofker, and D.W. Cox. 1990. The physical organization of the human immunglobulin heavy chain gene complex. EMBO J. 9:3303 3313.

17. Buluwela, L., D.G. Albertson, P. Sherrington, P.H. Rabbitts, N. Spurr, and T.H. Rabbitts. 1988. The use of chromosomal translocations to study human immunoglobulin gene organi-zation: mapping DH segments within 35 kb of the Cκ gene and identification of a new DH locus. EMBO J. 7:2003–2010.

18. Ravetch, J.V., U. Siebenlist, S. Korsmeyer, T. Waldmann, and P. Leder. 1981. Structure of the human immunoglobulin μ locus: characterization of embryonic and rearranged J and D genes. Cell 27:583–591.

19. Schable, K.F, and H.G. Zachau. 1993. The variable genes of the human immunoglobulin μ locus. Biol. Chem. Hoppe-Seyler 374:1001–1007.

20. Zachau, H. The human immunoglobulin κ genes. In: Immunoglobulin Genes, 2nd ed., T. Honjo, and F.W. Alt, editors. Academic Press, London, pp, 173–192, 1995.

21. Heiter, P.A., J.V. Maizel, and P. Leder. 1982. Evolution of human immunoglobulin k J regions genes. J. Biol. Chem. 257:1516–1522.

22. Frippiat, H.J.P., S.C. Williams, I.M. Tomlinson, G.P. Cook, D. Cherif, D. Le Pasler, J.E. Collins, I. Dunham, G. Winter, and M.P. Lefranc. 1995. Organization of the human immunoglobulin lambda light-chain locus on chromosome 22q11.2. Hum. Mol. Genet. 4:983–991.

23. Kawasaki, K., S. Minoshima, E. Nakato, K. Shibuya, A. Shintani, J.L. Schmeits, J. Wang, and N. Shimizu. 1995. One-megabase sequence analysis of the human immuno-globulin λ gene locus. Genome Res. 7:250–261.

24. Gellert, M. 1992. Molecular analysis of V(D)J recombina-tion. Ann. Rev. Genet. 22:425–446.

25. Okada, A., and F.W. Alt. The variable region gene assembly mechanism. In: Immunoglobulin Genes, 2nd edition, T. Honjo, and F.W. Alt, eds., Academic Press, London, pp. 205–234, 1995.

26. Raaphorst, F.M., C.S. Raman, B.T. Nall, and J.M. Teale. 1997. Molecular mechanisms governing reading frame choice of immunoglobulin diversity genes. Immunol. Today 18:37–43.

27. Berek C., and C. Milstein. 1988. The dynamic nature of the antibody repertoire. *Immunol. Rev.* 105:5–26.

28. Radic, M.Z., M.A. Mascelli, J. Erikson, H. Shan, and M. Weigert. 1991. IgH and L chain contributions to auto-immune specificities. *J. Immunol.* 146:176–182.

29. Brezinschek, H.-P., R.I. Brezinschek, T. Dorner, and P.E. Lipsky. 1998. Similar characteristics of the CDR3 of V(H)1–69/DP10 rearrangements in normal peripheral blood and chronic lymphocytic leukaemia cells. *Brit. J. Haem.* 102:516–521.

30. de Wildt, R.M.T., R.M.A. Hoet, W.J. van Venfooij, I.M, Tomlinson, and G. Winter. 1999. Analysis of heavy and light chain pairings indicates that receptor editing shapes the human antibody repertoire. *J. Mol. Biol.* 285:895–901.

31. Kraj, P., S.P. Rao, A.M. Glas, R.R. Hardy, E.C.B. Milner, and L.E. Silberstein. 1997. The human heavy chain Ig V region gene repetoire is biased at all stages of B cell ontogeny, including early pre-B cells. *J. Immunol.* 158:5824–5832.

32. Shiokawa, S., F. Mortari, J.O. Lima, C. Nunez, F.E. 3rd Bertrand, P.M. Kirkham, S. Zhu, A.P. Dasanayake, and H.W. Schroeder. Jr. 1999. IgM heavy chain complementarity-determining region 3 diversity is constrained by genetic and somatic mechanisms until two months after birth. *J. Immunol.* 162:6060–6070.

33. Foster, S.J., H-P. Brezinschek, R.I. Brezinschek, and P.E. Lipsky. 1997. Molecular mechanisms and selective influences that shape the kappa gene repertoire of IgM + B cells. *J. Clin. Invest.* 99:1614–1620.

34. Farner, N.L., T. Dorner, and P.E. Lispky. Molecular mechanisms and selection influence the generation of the human V lambda, and J lambda repertoire. *J. Immunol.* 15: 162:2137–2145.

35. Herman, A., J.W. Kappler, P. Marrack, and A.M. Pullen. 1991. Superantigens: mechanism of T-cell stimulation and role in immune responses. *Annu. Rev. Immunol.* 9:745–772.

36. Pascual, V., and J.D. Capra. 1991. B cell superantigens? *Current Biol.* 1:315–316.

37. Sasso, E.H., G.J. Silverman, and M. Mannik. 1991. Human IgA and IgG F(ab')₂ that bind to staphylococcal protein A belong to the V_HIII subgroup. *J. Immunol.* 147:1877–1883.

38. Berberian, L., L. Goodglick, T.J. Kipps, J. Braun. 1993. Immunoglonbulin V_H3 gene products: natural ligands for HIV gp120. *Science* (Washington, DC) 261:1588–1591.

39. Nilson, B.H.K., A. Solomon, L. Bjorck, and B. Akerstrom. 1992. Protein L from *Peptococcus magnus* binds to the κ light chain. *J. Biol. Chem.* 267:2234–2239.

40. Randen, I., K.N. Potter, Y.-C. Li, K.M. Thompson, V. Pascual, O. Forre, J.B. Natvig, and J.D. Capra. 1993. *Eur. J. Immunol.* 23:2682–2686.

41. Potter, K.N., Y.-C. Li, and J.D. Capra. 1996. Staphylococcal protein A simultaneously interacts with framework region 1, complementarity determining region 2 and framework region 3 on human V_H3 encoded immunoglobulins. *J. Immunol.* 157:2982–2988.

42. Berberian, L., J. Shukla, R. Jefferis, and J. Braun. 1994. Effects of HIV infection on V_H3 (D12 idiotope) B cells in vivo. *J. Acquired Immune Defic. Syndr.* 7:641–646.

43. Domiati-Saad, R., J.F. Attrep, H.-P. Brezinschek, A.H. Cherrie, D.R. Karp, and P.E. Lipsky. 1996. Staphylococcal enterotoxin D functions as a human B cell superantigen by rescuing V_H4-expressing B cells from apoptosis. *J. Immunol.* 156:3608–3620.

44. Pascual, V., K. Victor, M. Spellberg, T.J. Hamblin, F.K. Stevenson, and J.D. Capra. 1992. V_H restriction among human cold agglutinins. The V_H4–21 gene segment is required to encode anti-I and anti-i specificities. *J. Immunol.* 149:2337–2344.

45. Li, Y.-C., M.B. Spellberg, F.K. Stevenson, J.D. Capra, and K.N. Potter. 1996. The I binding specificity of human V_H4–34 (V_H4–21) encoded antibodies is determined by both V_H framework region 1 and complementarity determining region 3. *J. Mol. Biol.* 256:577–589.

46. Stevenson, F.K., G.J. Smith, J. North, T.J. Hamblin, and M.J. Glennie. 1989. Identification of normal B-cell counter-parts of neoplastic cells which secrete cold agglutinins of anti-I and anti-i specificity. *Brit. J. Haem.* 72:9–15.

47. Chapman, C.J., M.B. Spellberg, G.A. Smith, S.J. Carter, T.J. Hamblin, and F.K. Stevenson. 1993. Auto-red cell antibodies synthesized by patients with infectious mono-nucleosis utilize the V_H4–21 gene segment *J. Immunol.* 151:1051–1061.

48. Roelcke D. 1989. Cold agglutination. *Transfusion Med. Rev.* 2:140–166.

49. Stevenson, F.K., C. Longhurst, C.J. Chapman, M. Ehreustein, M.B. Spellberg, T.J. Hamblin, C.T. Ravirajan, D. Latchman, and D. Isenberg. 1993. Utilization of the V_H4–21 gene segment by anti-DNA antibodies from patients with systemic lupus erythematosus. *J. Autoimm.* 6:809–825.

50. Isenberg, D., M. Spellberg, W. Williams, M. Griffiths, and F. Stevenson. 1993. Identification of the 9G4 idiotope in systemic lupus erythematosus. *Br. J. Rheum.* 32:876–882.

51. Pascual, V., Y.-J. Liu, A. Magalski, O. de Bouteiller, J. Banchereau, and J.D. Capra. 1994. Analysis of somatic mutation in five B cell subsets of human tonsil. *J. Exp. Med.* 180:329–339.

52. Klein, U., K. Rajewsky, and R. Kuppers. 1998. Human immunoglobulin (Ig)M⁺ peripheral blood B cells expressing the CD27 cell surface antigen carry somatically mutated variable region genes: CD27 as a general marker for somatically mutated (memory) B cells. *J. Exp. Med.* 188:1679–1689.

53. Betz, A.G., M.S. Neuberger, and C. Milstein. 1993. Discriminating intrinsic and antigen-selected mutational hotspots in immunoglobulin V genes. *Immunol. Today* 14:405–411.

54. Dorner, T., H.-P. Brezinschek, S.J. Foster, R.I. Brezinschek, N.L. Farner, and P.E. Lipsky. 1998. Comparable impact of mutational and selective influences in shaping the expressed repertoire of peripheral IgM⁺/CD5⁻ and IgM⁺/CD5⁺ B cells. *Eur. J. Immunol.* 28:657–668.

55. Wagner, D.S., C. Milstein, and M.S. Neuberger. 1995. Codon bias targets mutation. *Nature* 376:732–754.

56. Dorner, T., S.J. Foster, N.L. Farner, and P.E. Lipsky. 1998. Somatic hypermutation of human immunoglobulin heavy chain genes: targeting of RGYW motifs on both DNA strands. *Eur. J. Immunol.* 28:3384–3396.

57. Thomson, R.K., Z. Davis, J.M. Palmer, M.J.P. Arthur, S.J. Yeaman, C.J. Chapman, M.B. Spellberg, and F.K. Stevenson. 1998. Immunogenic analysis of a panel of monoclonal IgG and IgM anti-PDC-E2/X antibodies derived from patients with primary biliary cirrhosis. *J. Hepatol.* 28:582–594.

58. Borretzen, M., I. Randen, E. Zdarsky, O. Forre, J.B. Natvig, and K.M. Thompson. 1994. Control of autoantibody

affinity by selection against amino acid replacements in the complementarity-determining regions. *Proc. Natl. Acad. Sci. USA* 91:1217–1221.

59. Tiegs, S.L., D.M. Russell, and D. Nemazee. 1993. Receptor editing in self-reactive bone marrow B cells. *J. Exp. Med.* 177:1009–1020.

60. Gay, D., T. Saunders, S. Camper, and M. Weigert. 1993. Receptor editing: an approach by autoreactive B cells to escape tolerance. *J. Exp. Med.* 177:999–1008.

61. Hertz, M., and D. Nemazee. 1997. BCR ligation induces receptor editing in IgM+IgD-bone marrow B cells in vitro. *Immunity* 6:429–436.

62. Chen, C., E.L. Prak, and M. Weigert. 1997. Editing disease-associated autoantibodies. *Immunity* 6: 97–101.

63. Pelanda, R., S. Schwers, E. Sonoda, R.M. Torres, D. Nemazee, and K. Rajewsky. 1997. Receptor editing in a transgenic mouse model: site, efficiency, and role in B cell tolerance and antibody diversification. *Immunity* 7:765–777.

64. Giachino, C., E. Padovan, and A. Lanzavecchia. 1998. Re-expression of RAG-1 and RAG-2 gene and evidence for secondary rearrangements in human germinal center B lymphocytes. *Eur. J. Immunol.* 28:3506–3513.

65. Meffre, E., F. Papavasiliou, P. Cohen, O. de Bouteiller, D. Bell, H. Karasuyama, C. Schiff, J. Banchereau, Y. Liu, and M.C. Nussenzweig. 1998. Antigen receptor engagement turns off the V(D)J recombination machinery in human tonsil B cells. *J. Exp. Med.* 188:765–772.

66. Nussenzweig, M.C. 1998. Immune receptor editing: revise and select. *Cell* 95:875–878.

67. Hikida, M., and H. Ohmori. 1998. Rearrangement of λ light chain genes in mature B cells in vitro and in vivo. Function of reexpressed recombination-activating (RAG) products. *J. Exp. Med.* 187:795–799.

68. Mockridge, C.I., C.J. Chapman, M.B. Spellerberg, B. Sheth, T.P. Fleming, D.A. Isenberg, and F.K. Stevenson. 1998. Sequence analysis of V4-34 encoded antibodies from single B cells of two patients with SLE. *Clin. Exp. Immunol.* 114:129–136.

69. Qin, X. F., S. Schwers, W. Yu, F. Papavasiliou, H. Suh, A. Nussenzweig, K. Rajewsky, and M.C. Nussenzweig. 1999. Secondary V(D)J recombination in B-1 cells. *Nature* 397:355–359.

70. Stamatopoulos, K., C. Kosmas, N. Stavroyianni, and D. Loukapoulos. 1996. Evidence for immunoglobulin heavy chain variable region gene replacement in a patient with B cell chronic lymphocytic leukemia. *Leukemia* 10:1551–1556.

71. Wilson, P.C., K. Wilson, Y-J. Liu, J. Banchereau, V. Pascual and J.D. Capra. 2000. Receptor revision of immunoglobulin heavy chain variable region genes in normal human B lymphocytes. *J. Exp. Med.* 191: 1881-1894..

72. Melamed, D., R.J. Benschop, J.C. Cambier, and D. Nemazee. 1998. Developmental regulation of B lymphocyte immune tolerance compartmentalized clonal selection from receptor selection. *Cell* 92:173–182.

73. Han, S., B. Zheng, D.G. Schatz, E. Spanopoulou, and G. Kelsoe. 1996. Neoteny in lymphocytes: Rag1 and Rag2 expression in germinal center B cells. *Science* 274:2094–2097.

74. Hikida, M., M. Mori, T. Takai, K. Tomochika, K. Hamatani, H. Ohmori. 1996. Reexpression of RAG-1 and RAG-2 genes in activated mature B cells. *Science* 274:2092–2094.

75. Dorner, T., N.L. Foster, and P.E. Lipsky. 1998. Immuno-globulin kappa chain receptor editing in systemic lupus erythematosus. *J. Clin. Invest.* 102:688–694.

76. Covey, L.R., P. Ferrier, and F.W. Alt. 1990. VH to VHDJH rearrangement is mediated by the internal VH heptamer. *Int. Immunol.* 2:579–583.

77. Bhat, N.M., M.M. Bieber, F.J. Hsu, C.J. Chapman, M. Spellerberg, and F.K. Stevenson. 1997. Rapid cytotoxicity of human B lymphocytes induced by VH4–34 (VH4.21) gene-encoded monoclonal antibodies. II. *Clin. Exp. Immunol.* 108:151–159.

78. Martin, T., R. Crouzier, J.C. Weber, T.J. Kipps, and J.-L. Pasquali. 1994. Structure-function studies on a polyreative (natural) autoantibody. Polyreactivity is dependent on somatically generated sequences in the third complementar-ity-determining region of the antibody heavy chain. *J. Immunol.* 152:5988–5996.

79. Capra, J.D., and J.B. Natvig. 1993. Is there V region restriction in autoimmune disease? *The Immunologist* 1:16–19.

80. Bendelac, A., C. Carnaud, C. Boitard, and J.F. Bach. 1987. Syngeneic transfer of autoimmune diabetes from diabetic NOD mice to healthy neonates. Requirement for both L3T4+ and Lyt-2 + T cells. *J. Exp. Med.* 166:823–832.

81. Okamura, M., Y. Kanayama, K. Amastu, N. Negoro, S. Kohda, T. Takeda, and T. Inoue. 1993. Significance of enzyme linked immunosorbent assay (ELISA) for antibodies to double stranded and single stranded DNA in patients with lupus nephritis:correlation with severity of renal histology. *Ann. Rheum. Dis.* 52:14–20.

82. Ehrenstein, M.R., D.R. Katz, M.H. Griffiths, L. Papadaki, T.H. Winkler, J.R. Kalden, and D.A. Isenberg. 1995. Human IgG anti-DNA antibodies deposit in kidneys and induce pro-teinuria in SCID mice. *Kidney Int.* 48:705–711.

83. Ravirajan, C.T., M.A. Rahman, L. Papadaki, M.H. Griffiths, J. Kalsi, A.C. Martin, M.R. Ehrenstein, S. Latchman, and D.A. Isenberg. 1998. Genetic, structural and functional properties of an IgG DNA-biding monoclonal antibody from a lupus patient with nephritis. *Eur. J. Immunol.* 28:339–350.

84. Chen, Q.-Y., I.R. Mackay, S. Fida, M.A. Myers, and M.J. Rowley. 1998. Natural and disease associated auto-antibodies to the autoantigen, dihydrolipoamide acetyl-transferase, recognize different epitopes. *J. Autoimmunity* 11:151–161.

85. Bonagura, V.R., N. Agostino, M. Borretzen, K.M. Thompson, J.B. Natvig, and S.L. Morrison. 1998. Mapping IgG epitopes bound by rheumatoid factors from immunized controls identifies disease-specific rheumatoid factors pro-duced by patients with rheumatoid arthritis. *J. Immunol.* 160:2496–2505.

86. Chan, O.T., L.G. Hannum, A.M. Haberman, L.M.P. Madaio, and M.J. Shlomichik. 1999. A novel mouse with B cells but lacking serum antibody reveals an antibody-independent role for B cells in murine lupus. *J. Exp. Med.* 189:1639–1648.

87. Mamula, M.J. 1998. Epitope spreading: the role of self pep-tides and autoantigen processing by B lymphocytes. *Immunol. Rev.* 164:231–239.

88. Davidson, H.W., and C. Watts. 1989. Epitope directed pro-cessing of specific antigen by B lymphocytes. *J. Cell Biol.* 109:85–90.

89. Pruzanski, W., and A. Katz. 1984. Cold agglutinins: anti-bodies with biological diversity. *Clin. Immunol. Rev.* 3:131–150.

90. Feizi, T. 1981. The blood group II system: A carbohydrate antigen system defined by naturally monoclonal or oligoclonal autoantibodies of man. *Immunol. Commun.* 10:127–156.

91. Silberstein, L.E., L.C. Jeffries, J. Goldman, D. Friedman, J.S. Moore, P.C. Nowell, D. Roelcke, W. Pruzanski, J. Roudier, and G.J. Silverman. 1991. Variable region gene analysis of pathologic human autoantibodies to the related i and I red blood cell antigens. *Blood* 9:2372–2386.

92. Mockridge, C.I., C.J. Chapman, M.B. Spellerberg, D.A. Isenberg and F.K. Stevenson. 1996. Use of phage surface expression to analyze regions of a human V_H4–34 (V_H4–21)-encoded IgG autoantibody required for recognition of DNA. No involvement of the 9G4 idiotope. *J. Immunol.* 157:2449–2454.

93. Potter, K.N., Y.-C. Li, V. Pascual, R.C. Williams, L.C. Byres, M. Spellerberg, F.K. Stevenson, and J.D. Capra. 1993. Molecular characterization of a cross-reactive idiotope on human immunoglobulins utilizing the V_H4–21 gene segment. *J. Exp. Med.* 178:1419–1428.

94. Thorpe, S.J., C.E. Turner, F.K. Stevenson, M.B. Spellerberg, R. Thorpe, J.B. Natvig, and K.M. Thompson. 1998. Human monoclonal antibodies encoded by the V4–34 gene segment show cold agglutinin activity and variable multireactivity which correlates with the predicted charge of the heavy-chain variable region. *Immunol.* 93:129–136.

95. Kunkel, H.G., V. Agnello, F.G. Joslin, R.J. Winchester, and J.D. Capra. 1973. Cross-idiotypic specificity among mono-clonal IgM proteins with anti-—globulin activity. *J. Exp. Med.* 137:331–342.

96. Victor, K.D., I. Randen, K. Thompson, M. Sioud, O. Forre, J. Natvig, and J.D. Capra. 1991. The complete nucleotide sequences of the heavy chain variable regions of six mono-specific RFs derived from EBV-transformed B cells isolated from the synovial tissue of patients with RA. *J. Clin. Invest.* 86:1320–1328.

97. Borretzen, M., O.J. Mellbye, K.M. Thompson, and J.B. Natvig. 1996. Rheumatoid factors. In: *Autoantibodies*, Elsvier Science B.V., J.B. Peter, and Y. Shoenfeld, eds., pp. 706–715.

98. Corper, A.L., M.K. Sohi, V.R. Bonagura, M. Steinitz, R. Jefferis, A. Feinstein, D. Beals, M.J. Taussig, and B.J. Sutton. 1997. Structure of human IgM rheumatoid factor Fab bound to its autoantigen IgG Fc reveals a novel topology of antibody-antigen interaction. *Nature Structural Biol.* 4:374–381.

99. Lee, S.K., S.L. Bridges, Jr, W.J. Koopman, H.W. Schroeder, and H.W. Jr. 1992. The immunoglobulin kappa light chai repertoire expressed in the synovium of a patient with rheumatoid arthritis. *Arth. Rheum.* 35:905–913.

100. Victor, K.D., and J.D. Capra. 1994. An apparently common mechanism of generating antibody diversity: length varia-tion of the V_L-J_L junction. *Mol. Immunol.* 31:39–46.

101. Isenberg, D.A. 1994. Anti-DNA antibodies-some enigma variations. *The Immunologist* 2/6:190–193.

102. Roben, P., S.M. Barbas, L. Sandoval, J.-M. Lecert, B.D. Stollar, and A. Solomon. Repertoire cloning of lupus anti-DNA autoantibodies. *J. Clin. Invest.* 98: 2827–2837.

103. Rahman, A., S. Menon, D.S. Latchman, and D.A. Isenberg. 1996. Sequences of monoclonal antiphospholipid antibodies: variation on an anti-DNA antibody theme. *Sem. Arthritis Rheum.* 26:515–525.

104. Krishnan, M.R., N.T. Jou, and T.N. Marion. 1996. Correlation between the amino acid position of arginine in V_H-CDR3 and specificity for native DNA among autoim-mune antibodies. *J. Immunol.* 157:2430–2439.

105. Davidson, A., A. Manheimer-Lory, C. Aranow, R. Peterson, N. Hannigan, B. Diamond. 1990. Molecular characterization of a somatically mutated anti-DNA antibody bearing two systemic lupus erythematosus-related idiotypes. *J. Clin. Invest.* 85:1401–1409.

106. Barbas, S.M., H.J. Ditzel, E.M. Salonen, W.P. Yang, G.J. Silverman, D.R. Burton. 1995. Human autoantibody recognition of DNA. *Proc. Natl. Acad. Sci. USA* 92:2529–2533.

107. Patrick, J., and J. Lindstrom. 1973. Autoimmune response to acetylcholine receptor. *Science* 180:871–872.

108. Victor, K.D., V. Pascual, A.K. Lefvert, and J.D. Capra. 1992. Human anti- acetylcholine receptor antibodies use variable gene segments analogous to those used in autoantibodies of various specificities. *Mol. Immunol.* 29:1501–1506.

109. Matsui, M., M. Nakamura, H. Ishibashi, K. Koike, J. Kudo, and Y. Niho. 1993. Human monoclonal antibodies from a patient with primary biliary cirrhosis that recognize two dis-tinct autoepitopes in the E2 component of the pyruvate dehydrogenase complex. *Hepatology* 18:1069–1076.

110. Pascual, V., S. Cha, M.E. Gershwin, J.D. Capra, and P.S.C. Leung. 1994. Nucleotide sequence analysis of natural and combinatorial anti-PDC-E2 antibodies in patients with primary biliary cirhosis. Recapitulating immune selection with molecular biology. *J. Immunol.* 152:2577–2585.

4 | Genetic Origin of Murine Autoantibodies

Azad K. Kaushik and Constantin A. Bona

1. INTRODUCTION

Systemic autoimmune diseases are of multifactorial origin and are the result of complex interactions among immune effector cells, T and B lymphocytes, and the target multiple organs. In human systemic lupus erythematosus (SLE), the direct role of T and B lymphocytes in the initiation, development and progression of autoimmune pathologic lesion is mostly circumstantial due to the presence of T cells or autoantibody in the inflamed organ (1–3). However, a direct role for both T and B lymphocytes is clearly demonstrable in autoimmune-prone mice (4–8). For example, intrinsic T cell defects have been demonstrated in MRL-*lpr/lpr* (9), BXSB and NZB (10) autoimmune mice. In MRL-*lpr/lpr* mice, thymectomy (11) and treatment with anti-Thy1 (12) or anti-CD4 (13) antibodies helps to ameliorate lymphoproliferation and delay progression of the inflammatory lesion. In addition, a role of B cells in initiating T cell autoreactivity has been suggested to be important in developing an autoimmune cascade reaction (14,15). B cells from MRL-*lpr/lpr* and NZB mice secrete autoantibodies (AAb) because of an intrinsic defect (16–19). Indeed, a direct role for B cells in the development of autoimmune disease is demonstrable in the "Jh knockout" mutation crossed onto the autoimmune MRL-*lpr/lpr* background, resulting in a lack of B cells, as these mice do not show glomerulonephritis/vasculitis, in contrast to wild-type mice (20). Recent studies using MRL/MpJ-Fas[lpr] mice, which express a mutant transgene B as cell surface Ig without its secretion, further demonstrate that functional B cells are also essential for disease expression independent of serum autoantibody (21). These observations suggest that, in addition to autoantibody-dependent immunopathology, B cells also directly contribute to local inflammation and/or

serve as antigen presenting cells for antigen-specific T cells. Direct evidence supporting a primary role of AAbs in autoimmune pathogenesis also exists. For example, some anti-DNA AAbs cause nephritis upon injection into normal mice (22,23). Similarly, rheumatoid factor (RF) of the IgG3 isotype has been observed to cause vasculitis and nephritis upon injection (24). However, these observations are complicated by the fact that physiological natural autoantibodies have also been isolated from the autoimmune inflammatory lesion (25). The natural autoantibodies, encoded by germline genes, have been suggested to be associated with the "immunological homunculus" that reflects physiological images of the self-enshrined in a specific network of lymphocytes (26). An evidence for the existence of immunological homunculus is demonstrable in the serum IgM repertoires of several mice strains based on analysis of global natural autoantibody profiles (27).

2. B Lymphocytes and Autoimmunity

Naïve mature B cells, generated in the bone marrow, migrate to secondary lymphoid organs where they either undergo apoptosis or are recruited into circulation. Interestingly, autoreactive B cells constitute a significant proportion of the mature naïve antigen reactive circulating B-cell pool (28,29). These autoreactive B cells have low to moderate affinity for structurally-dissimilar conserved self-antigens. Normally resting autoreactive B cells are harmless unless their immunoglobulin (Ig) receptor mutates during proliferation to provide a better auto-antigen combining site. The role of naïve mature B cells in antigen processing and presentation of self-antigen is not fully understood, but it could be argued that T cell help for critical self-antigens are not available, since they are deleted during negative T cell selection in the thymus. The occurrence of autoreactive B cells in the preimmune repertoire challenges Burnet's concept of deletion of "forbidden" clones, since their occurrence would result in *horror autotoxicus*.

2.1   Positive and Negative Selection of B Lymphocyte: Peripheral Versus Central Tolerance

The murine immune system comprises 500×10^6 mature B cells (30), and 50×10^6 immature B cells are generated daily in the adult mouse bone marrow (31) with a majority (98%) having a half life of 2 to 4 days. Of these, approximately $2–3 \times 10^6$ cells are recruited into the peripheral pool of mature antigen-responsive B cells with a half life lasting 4 to 6 weeks or longer (32). During B cell development, immunoglobulin (Ig) rearrangement (33) begins with DH to JH at both heavy chain alleles, followed by VH to DH-JH rearrangements. Subsequent to VH-DH-JH rearrangements, $V\kappa$ to $J\kappa$ rearrangements begin at the (κ-locus and, if non-productive, $V\lambda$ to $J\lambda$ rearrangement occurs at the λ-light chain locus. A majority of pre-B cells (>95%) with productive VH-DH-JH rearrangements express μ-heavy chains in their cytoplasm and are, thus, positively selected in contrast to B cells with non-productive VH-DH-JH rearrangements. The expression of μ-heavy chain together with surrogate light- chain as pre-B cell receptor permits extensive expansion of B-lineage committed cells such that they outnumber those with non-productive VH-DH-JH rearrangements. Immature B cells that express IgM, but lack IgD on the cell surface, undergo negative (deletion, receptor editing, anergy) or positive (by as yet unidentified self-antigens) selection before being recruited into the peripheral pool of mature B cells each day.

Studies employing various transgenic mice have helped elucidate some of the mechanisms involved in the selection of B lymphocyte during its developmental pathway. Central deletion of autoreactive B cells at pre-B (sIg⁻) to immature (sIg⁺) B cell stage is noted in mice transgenic for an antibody to MHC class I (K^k or K^b) that also express MHC $K^{k/b}$ as ubiquitous self antigen (34). However, complete deletion does not occur at the pre-B to immature B cell stage, since continuous rearrangements at the κ- or λ-light chain loci, because of upregulated RAG1 and RAG2, permit immature B cells to express an Ig receptor with a new light-chain (34). This process is termed receptor editing, and is aimed at eliminating developing autoreactive B cells with potential pathogenic ability. For receptor editing to operate, a signaling threshold via Ig receptor present on immature B cell is required that would retain activated recombination machinery i.e., RAG1 and RAG2. At this stage, the possibility for positive selection of immature B cells by unknown autoantigens exists, as

evidenced by rapid and reversible downregulation of RAG1 and RAG2 in immature B cell lines from N-myc transgenic mice as a result of Ig-receptor cross-linking (35). In mouse transgenic for H- and L-chain genes of an antibody to DNA, similar mechanisms of central tolerance to ubiquitous DNA are evident (36,37). The B cell deletion occurs during the transition from pre-B to immature B cell stage in the bone marrow but receptor editing permits the escape of B cells expressing Ig receptor with the transgenic L-chain replaced by an endogenously rearranged L-chain to the periphery. When the self-antigen MHC-K^b is expressed peripherally, for example, in the liver or kidney, deletion of immature B cells occurs during the transition from immature to mature as well as mature B cell stage without any possibilities for receptor editing. Thus, deletion of autoreactive B cells, in contrast to central tolerance, probably occurs at a late stage when an auto-antigen is expressed peripherally.

Studies with mice transgenic for H- and L-chain genes for antibody (IgM and IgD) to hen egg lysozyme (HEL) have shown that the membrane form of HEL expressed as a self-antigen does not result in deletion of HEL-binding B cells in the bone marrow, but in the peripheral lymphoid organs (38,39). This observation is consistent with studies of B cells against MHC class I ($K^{k/b}$) where in such B cells are deleted in transition from immature to mature B cell stage without any receptor editing. Similar encounter of transgenic HEL-specific B cells with the soluble form of HEL, however, renders them anergic, but continued presence of the autoantigen is essential. Further complexity is evident from experiments aimed at understanding the threshold of receptor occupancy with remarkably different signaling outcomes. Transgenic lysozyme specific splenic B cells bind to HEL with affinities ranging from 10^{-5} to 10^{-9}. When HEL is expressed as a soluble self-molecule at a concentration of 1nM in blood, high affinity, but not low affinity, self-reactive B cells are selectively deleted. Expression of HEL at still higher concentrations as the soluble antigen results in deletion of both high and low affinity B cells. The B cells that bind HEL with still lower affinity escape deletion by soluble HEL but are eliminated by the membrane-bound form of antigen exposed on the cell surface, probably as a result of increased avidity via enhanced cross- linking of HEL-binding immunoglobulin (40). In contrast to deletion of autoreactive B cells triggered by the ubiquitous membrane bound expression of HEL, selective expression of HEL on thyroid cell membrane does not lead to either deletion or inactivation of circulating HEL-reactive B cells (41). Thus, tolerance is not actively acquired to organ-specific antigens in the mature B cell pool and some mechanism other than central tolerance operates to prevent the generation of pathogenic B cells in organ-specific diseases. These experiments provide evidence for the deletion of self-reactive clones, but the threshold for deletion appears to vary considerably depending upon the nature of autoantigen and where it is encountered. Although negative selection of strongly autoreactive B cells is effected via deletion or anergy in the bone marrow, recent experiments suggest that bone marrow also exports immature B cells (HSA+) to outer T cell zones of the white pulp of the spleen (42). These B cells undergo negative selection in the outer T cell zone and the small number of B cells that persist express a skewed V-region repertoire (43,44). Experimental data suggest that self-reactive B cells to circulating HEL accumulate in the T cell zone of the spleen and are excluded from migration into the B cell follicles depending upon the concentration of self ligand, affinity of the receptor, signaling threshold and the presence of competing B cells (45). The B cells that are not allowed to enter the follicle are short-lived in contrast to those entering the B cell follicles. Similarly, immature B cells against double stranded DNA (dsDNA) are known to localize at the interface of T and B cell zones in the spleen (46). These observations suggest that autoreactive B cells undergo negative selection in the periphery as well. It has been suggested that pathogenic autoantibody-secreting B cells in autoimmune mice accumulate in the outer T cell zone probably as a result of failure of negative selection at this site. At the same time, recent observations suggest that receptor revision in the periphery via secondary rearrangements during the development of an immune response may create Ig receptors capable of binding to self-antigens (47).

Overall, the following conclusions can be drawn with regard to positive and negative selection of B cells in the context of autoimmunity and tolerance:

1. Central clonal deletion and receptor editing occurs during pre-B (sIg−) and immature (sIg+) B cell transition as a result of exposure to self antigens. Exposure to self-antigen expressed in the periphery also results in clonal deletion of self-reactive B cells, though at a late stage, during transition from immature to mature B cell stage without any possibility for receptor editing. Stringent deletion of critical self-antigen-reactive B cells e.g., blood group antigens, is essential to survival, whereas tolerance exists for low to moderate avidity self-reactive B cells to a number of conserved self-antigens, including self-Ig (rheumatoid factor) itself.

2. Exposure of a soluble form of self-antigen to B cells, depending upon the threshold of receptor occupancy, renders them anergic, and these anergic B cells die within a few days.

3. Immature self-reactive B cells escaping into the periphery accumulate in the splenic T cell zone, where they further undergo negative selection and are prevented from entering B cell follicles to prevent autoimmunity.

4. While positive selection of B cells operates both centrally and in the periphery, the self-ligands involved in such a selection process remain to be identified.

2.2 Positive and Negative Selection of Autoreactive B Lymphocyte: Contribution of Cell Signalling Molecules

Expression of the signaling competent B cell antigen receptor (BCR) is essential for positive and negative selection of B cells during B cell development as well as during the B cell response to a self- or foreign antigen (rev. in 48). The BCR is characterized by a complex hetero-oligomeric structure with distinct receptor subunits engaged in ligand binding and subsequent signal transduction. The B cell surface immunoglobulin (sIg) functions as a ligand binding receptor, while Igα (CD79α) and Igβ (CD79β) contribute to signaling cascade and, thus, determine the signaling outcome. The ability of Igα and Igβ to engage in cell signaling depends upon the presence of immunoreceptor tyrosine based activation motif (ITAM), composed of two YXXL separated by six to eight amino acid spacers, in the cytoplasmic tail (49,50). Cross-linking of the BCR results in phosphorylation of tyrosine residues of the ITAM mediated by Src (Lyn, Blk, Fyn), Syk and Btk phosphotyrosine kinases (PTKs). The activity of Syk and src-like PTKs is inhibited by the protein tyrosine phosphatase (PTPs), SHP-1, when recruited in close proximity of BCR by CD22 or FcγRIIB, which dephosphorylates this motif. While SHP-1 is a negative regulator of cell signaling, another phosphatase CD45 acts as a positive regulator of cell signaling. As a result of signaling through BCR, immature B cells reactivate and/or sustain recombination machinery and edit the receptors. The B cells unable to eliminate undesirable autoreactivity via receptor editing undergo apoptosis. The responding mature B cells are activated, express CD86, and upregulate MHC class II expression. A number of intrinsic factors, including BCR component expression levels, co-receptors, accessory molecules and signaling molecules, determine the fate of the B cell and its threshold of responsiveness.

The Src-PTK Lyn −/− B cells show hyper-reactivity, increased IgM levels and occurrence of autoantibodies to dsDNA and other nuclear antigens (51,52). The net effect of Lyn PTK is considered to be negative rather than positive B cell activation. The Lyn −/− mice carrying anti-HEL BCR as a transgene show an increased negative selection to HEL expressed as self-antigen, as expected from enhanced signaling properties of Lyn −/− B cells (53). But the production of autoantibodies in Lyn −/− mice is surprising, since autoreactive B cells would be expected to be negatively selected in the absence of Lyn. It has been suggested that Lyn, CD22 and SHP-1 operate in a single pathway and Lyn functions both as negative (by phosphorylating inhibitory receptors such as CD22) and positive regulator in cell signaling. The Syk tyrosine kinase is required for positive selection of immature B cells into the circulating B cell pool as Syk-deficient B cells have the life-span of normal immature B cells. Further, mutation in the Syk gene completely blocks recruitment of immature B cells into the circulating B cell pool as well as their entry in the B cell follicle (54). A comparison of VH gene usage in immature and circulating B cells shows that only a subset of VH genes expressed in immature B cells is evident in circulating B cells. This suggests a selective positive selection of B cells via binding of the BCR by endogenous molecules into the circulating B cell pool. A point mutation in the Btk PTK (55,56) in mice causes X-linked immunodeficiency that results in over- representation of the peripheral immature B cell phenotype (IgM^{high}IgD^{low}) with a striking absence of the peritoneal B-1 cell population. These xid mice express skewed V-gene repertoire (57,58) in defective xid female mice compared to their normal counterparts (Figure 1). The mutant Btk B cells are unable to respond to thymus-independent type II antigens indicating a role for Btk in BCR signaling.

CD45 –/– mice generate mature B cells, but they express IgM ^{high}IgD^{low}, in contrast to normal mature B cells that are mostly IgM ^{low}IgD^{high} (59). This led to the suggestion that CD45 PTP contributes to the positive selection of B cells. When the anti-HEL transgene is bred into CD45 –/– mice, the HEL autoantigen (present in soluble form) binding cells are positively selected in contrast to HEL- binding CD45$^{+/+}$ B cells that are negatively selected (60). It is argued that even in the absence of CD45, a weak signal is induced by the antigen and reduced BCR signal strength in CD45 –/– mice results in autoantibody production because of inhibition of negative selection. Motheaten viable (me^v/me^v) mice express mutant SHP-1 PTP (61) and, thus markedly exhibit decreased SHP-1 phosphatase activity. These mice show hypergammaglobulinemia, autoantibody production and a predominant B-1 cell population (62). The SHP-1, in contrast to CD45 PTP, influences the threshold of responsiveness by which B cells respond to an antigen and, thus, is a negative regulator of B cell signaling. The absence or significant loss of function of SHP-1 is expected to result in increased negative selection of self-reactive B cells due to the lowered threshold of responsiveness to an autoantigen. Thus, the production of autoantibodies in me^v/me^v B cells is paradoxical, similar to Lyn –/– and CD45 –/– B cells. SHP-1 deficiency results in skewed V$_H$ gene, especially Q52 and J558, expression in me^v/me^v mice (Figure 2) compared to background mice, which appear to be related to increased negative B cell

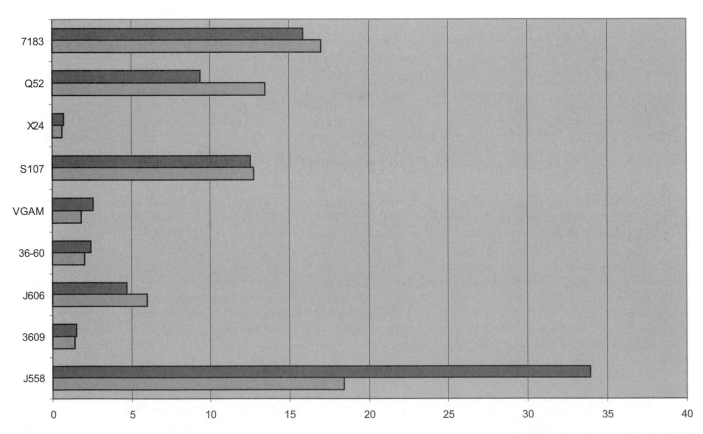

Figure 1.   VH gene expression in immunodeficient xid male as compared to normal xid female mice (Data from Feng and Stein, 1991)

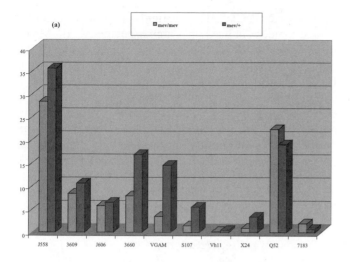

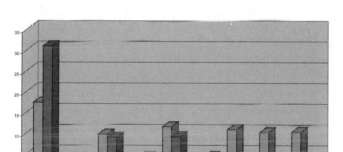

Figure 2. V_H and V_k gene expression in mev/mev and mev/+ mice.

PTPs may be expected to overcome effects of the motheaten mutation associated with SHP-1 deficiency. Indeed, BCR signaling due to loss of CD45 in me^v/me^v mice largely, but not completely, restores the effects SHP-1 deficiency (65).

Overall, experimental data suggest that defects in either positive or negative B cell signaling, as evidenced by analysis of various PTKs and PTPs, paradoxically results in autoantibody production. Thus, it appears that occurrence of autoreactive B cells is intrinsic to B cell development processes since either positive or negative selection fails to eliminate the autoreactive B cell population. Further, impaired positive or negative selection leads to skewed V-gene expression and, in some instances, contributes toward generation of pathogenic autoantibodies, e.g., increased negative selection in me^v/me^v mice. Thus, V-gene expression in autoantibodies is intricately associated with the generation and recruitment of autoreactive B cells into the periphery.

3. V-GENE EXPRESSION

The functional V_H gene is assembled from 100–500 V- (variable), 12 D- (diversity) and 4 functional J- (joining) gene segments upstream of constant region genes (33). V_H genes located on chromosome 12 (Table 1) tend to be clustered with some interspersion (66). About 300 Vκ genes with downstream, 4 functional Jκ genes, located on chromosome 6, contribute to kappa-variable region diversity 67). Unlike Vκ-light chain, the λ-light chain locus (68,69) is organized as four main recombination units (Figure 3) with the restricted ability to generate antibody diversity. Unlike some mammalian species, 95% of murine antibodies bear kappa-light chain.

selection either early in B cell ontogeny or during recruitment of B cells in the periphery (63). Further, analysis of randomly isolated V_HJ558$^+$ IgM antibodies from me^v/me^v mice at the terminal stage of autoimmune disease revealed use of rarely expressed members of the J558$^+$ V_H gene family and targeted selection for restricted CDR3 length (10 amino acids) of the heavy chain with similar hydrophobicity indices (64). These observations suggest that increased negative selection of B cells because of SHP-1 deficiency contributes to the emergence of B cells with particular VDJ rearrangements that secrete autoantibody. It could be argued that B-1 cells secreting autoantibody in me^v/me^v mice escape increased negative selection because of differences in signaling pathways employed by the B-1 cell population. Since SHP-1 and CD45 PTPs play a negative and positive role in signal transduction, respectively, B cells doubly deficient in these

Table 1   Position of variable heavy (a), and kappa (b) and lambda (c) light chain genes of the murine immunoglobulin (*Meek et al., 1990, Tutter et al., 1991; **D'Hoostelaere et al., 1988; ***Carson and Wu, 1989, Storb et al., 1989)

5′-(J558-3609)-V_H10-J606-V_H12-3609N-VGAM-36.60-S107-VGAM-V_H11-X24-SM7-S107-(Q52-7183)-3′
(a) V_H locus*
5′-(Vκ11, Vκ24, Vκ9, Vκ26)-(Vκ1, Vκ9)-(Vκ4, Vκ8, Vκ10, Vκ12, Vκ13, Vκ19)-(Vκ28, Rn&S)-Vκ23-(Vκ21, Jκ,Cκ)-3′
(b) Vκ locus**
5′-L-Vλ2-VλX-Jλ2-Cλ2-(Jλ4)-(C λ4)-L-Vλ1-J λ3-Cλ3-Jλ1-C λ1-3′
(c) Vλ locus***

3.1. V-Gene Expression in the Primary Antibody Repertoire

Four different research groups using three different techniques, Northern hybridization of RNAs from hybridoma or pre-B cell lines, single cell *in situ* hybridization and cDNA library screening, have analyzed V-gene expression in neonatal and adult B cells (rev. in 70). These experiments have shown remarkable differences in V-gene expression in the newborn compared to the adult mice as evidenced by:

1. Preferred V_H gene expression, especially 81X member of 7183 gene family, in B cells early during ontogeny (71). The V_H genes are expressed stochastically in the adult mouse (72). A predominant role for 7183, Q52, Vh11 and J558 V_H gene expression early in ontogeny is evident in the development of primary antibody repertoire (70).
2. Shorter germline D gene segment expression in fetal B cells (73).
3. Few, if any, N-region insertions in the CDR3H length in fetal B cells (74).
4. Preferred $V\kappa1$, $V\kappa22$ and $V\kappa4$ gene expression in neonatal B cells (75).
5. Predominant expression of the small $V\kappa1$ (75) gene family compared to the largest $V\kappa4/5$ gene family in both adult and neonatal B cells.

These observations suggest that restricted genetic elements are expressed in the neonatal B cells as compared to the adult and, thus, support programmed development of the primary antibody repertoire.

3.2. V_H and $V\kappa$ Gene Expression in Normal and Autoimmune Mice

Differences in V-gene expression have been noted in MRL-*lpr/lpr* and me^v/me^v mice that provide clues to the development of skewed B cell repertoire either early during ontogeny or later in the periphery coincident with the manifestation of systemic autoimmune disease. Expression of V_H genes in non-stimulated MRL-*lpr/lpr* B cells decreases by almost 50% for 3′ Q52 and 7183 V_H gene families (76), which suggests skewing in the expressed antibody repertoire. Indeed, J558 gene expression has been positively correlated with disease severity as determined by concentration of serum antibody to DNA and degree of glomerular nephritis in MRL-*lpr/lpr* mice (77). Similarly, altered V_H gene expression is noted in me^v/me^v mice compared to the wild type background mice as evidenced by an increased Q52 and decreased J558 gene expression (Figure 2). Remarkably, randomly isolated J558+ IgM antibodies in me^v/me^v mice at the terminal stage of the disease

show restricted CDR3H size (ten amino acids), similar hydrophobicity indices and monospecificity. In MRL-*lpr/lpr* mice, differences in V-gene expression may also occur because of unusual VDJ rearrangements in neonatal B cells remarkable for atypical junctions and frequent D-J rearrangements (78), probably associated with intrinsic B cell defects. Interestingly, no remarkable differences with regard to $V\kappa$ gene expression, with the possible exception of the $V\kappa8$ gene family, are evident in autoimmune compared to normal mice. Thus, selection of the B cell repertoire in autoimmune prone mice is altered during B cell development and recruitment of newly emerged B cells into the periphery as evidenced by differences in V-gene expression in autoimmune strain as compared to normal mice.

3.3. V_H + $V\kappa$ Family Pairings in Normal and Autoimmune Mice

Data on V_H + $V\kappa$ gene family pairings in me^v/me^v, and background C57BL/6 mice provide a molecular basis of the alteration of B cell repertoire in the autoimmune mice compared to normal background strain (63). In addition to differences in global V_H and $V\kappa$ gene expression in normal and autoimmune mice, altered V_H + $V\kappa$ gene family pairings (Figure 3) in me^v/me^v mice suggest that global skewing in the B cell repertoire is perhaps characteristic of autoimmune-prone mice that may be linked to one of many genetic traits such as SHP-1 deficiency. It may be recalled that such skewed V_H + $V\kappa$ gene family pairings may also reflect an altered "immunological homunculus" associated with the lymphocyte network in autoimmune-prone mice. Several polygeneic traits, e.g., various PTPs and PTKs, may influence positive or negative selection of B cells and, thus, qualitatively affect central and peripheral tolerance.

4. GENETIC ELEMENTS ENCODING AUTOANTIBODY

Extensive data on genetic elements encoding autoantibody rule out the possibility that autoantibody specific genes exist in the individuals or animals prone to develop systemic autoimmune disease. Although some subtle RFLP differences (79–82) are evident at the V_H- and $V\kappa$ loci in autoimmune-prone strains compared to normal mice, these do not seem to reflect existence of autoantibody-specific genes. Similarly, polymorphism observed in humans (83) using oligonucleotide probes specific to the framework and CDR of various V_H genes seems to reflect allelic differences rather than the existence of autoantibody-specific genes.

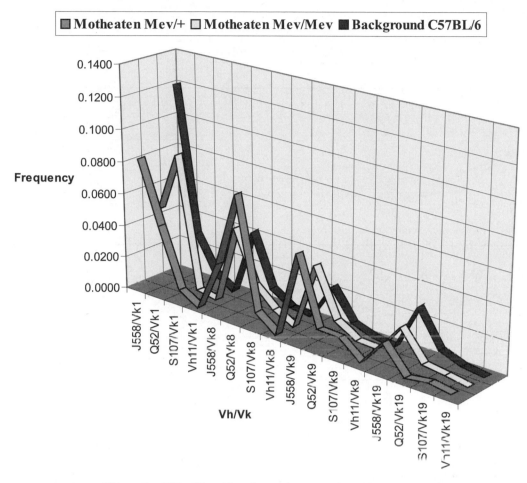

Figure 3. VH + Vk pairings in mev/mev, mev/+ and background C57BL/6 mice.

4.1. V-Gene Expression in Autoantibodies

A comprehensive analysis of V_H gene expression in 211 randomly selected hybridoma-producing autoantibodies (72,84–117) from several mouse strains and sources has led to the conclusion that V_H gene expression in B cells secreting autoantibody does not correspond to genomic complexity ($p > 0.05$). Certain V_H gene families, such as VH11, VH12, S107 and 7183 are over-represented, while others (VGAM3.8, J606 and 36–09) are under-represented in B cells secreting autoantibody (Figure 4). Except for these differences, these data are comparable to random utilization of V_H gene families observed in polyclonally activated B cells. Nevertheless, these differences are significant and reflect skewed B cell repertoire selection either during development or their recruitment in the periphery. As for D-gene usage in autoantibodies, no significant differences ($p < 0.05$) are apparent, since the observed D-gene expression is comparable to the expected frequencies (Figure 5a). Further, recombination analysis of four V_H (J558, 7183, Vh11 and Q52) and three D-gene (SP2, FL16 and Q52) families (Figure

6b) shows that it occurs randomly ($p < 0.05$), and no particular D-gene can be attributed to CDR3H length generation in autoantibodies. In contrast, non-random expression of J_H genes is evident in B cells secreting autoantibody ($p > 0.05$), since J_H1 is over- and J_H3 is under-represented (Figure 6c). Such preferential J_H gene segment usage, similar to V_H gene families, reflects the influence of selection processes in generation of B cells secreting autoantibody.

Analysis of $V\kappa$ gene expression in 211 randomly selected autoantibody-secreting B cells, rescued as hybridomas, revealed that $V\kappa4$, $V\kappa9$, $V\kappa10$ and $V\kappa24$ gene families are overexpressed (Figure 6) compared to empirically determined values for adult B cells. Careful scrutiny of data shows that $V\kappa10$ and $V\kappa24$ genes selectively encode autoantibodies (118), while $V\kappa1$, $V\kappa4$ and $V\kappa9$ gene family expression is similar to mitogen-activated B cells (119). No remarkable differences are seen with regard to other $V\kappa$ gene families. While $V\kappa$ gene families are known to randomly associate with the four $J\kappa$ genes, $V\kappa$- $J\kappa$ recombination among autoantibodies shows a preference for $J\kappa2$ and underexpression of $J\kappa5$ (Figure 7).

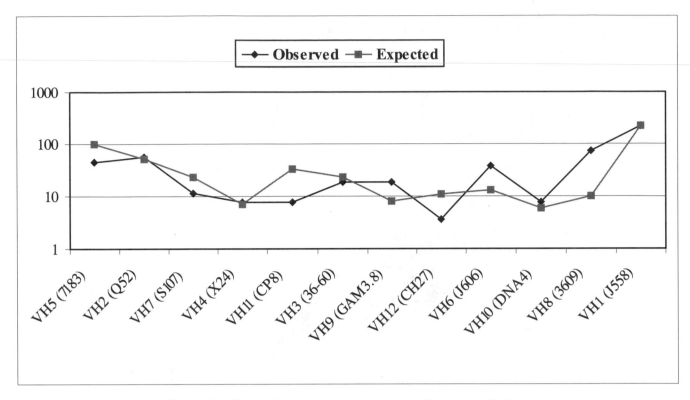

Figure 4.   VH gene family expression in murine self reactive antibodies.

4.2. V$_H$ + Vκ Gene Family Pairings in Autoantibodies

Analysis of V$_H$ + Vκ family pairings in a panel of 447 autoantibody-secreting hybridomas showed that restricted V$_H$ + Vκ family pairings occur in some autoantibodies, most remarkably against histone, bromelain- treated red blood cells (BrMRBC) and red blood cells (Figure 8). However, diverse V$_H$ + Vκ pairings are noted among autoantibodies against many antigens such as DNA,

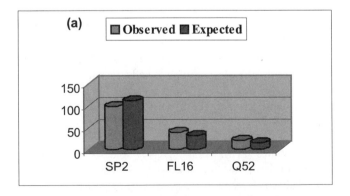

Figure 5.   Expected and observed D (a) gene expression and its recombination frequencies with over- (Vh11, 7183), under- (VHQ52) and normally- (VHJ558) VH genes and expected and observed J-gene (c) expression in self-reactive antibodies.

thyroglobulin, immunoglobulin (rheumatoid factor) and thymocyte. The autoantibodies against BrMRBC represent an example of non-random V$_H$ + Vκ pairings, i.e., Vh11 + Vκ9 and Vh12 + Vκ4, in contrast to multispecific auto-antibodies that essentially show random associations consistent with those noted in mitogen-activated B cells (119). It should be noted that many autoantibodies analyzed may recognize an antigen with varying degrees of affinity and may not yet represent the final clonally selected B cell. In addition, diverse V$_H$ + Vκ pairings indicate plasticity in the ability of V-genes in encoding an autoantibody specificity. Thus, it appears that both restricted and unrestricted V$_H$ + Vκ pairings occur in autoantibodies depending upon the nature of the autoantigen and stage of clonal selection and, therefore, the relevance of such autoantibodies in autoimmune pathology can not be completely excluded.

Overall, expression of various genetic elements encoding autoantibodies differs considerably compared to normal mitogen-activated B cells. The skewed V-gene expression in autoantibodies is reminiscent of skewed V-gene repertoire development in autoimmune-prone mice. Therefore, positive or negative B cell repertoire selection during development and/or recruitment of B cells into the periphery influences the generation of autoantibodies.

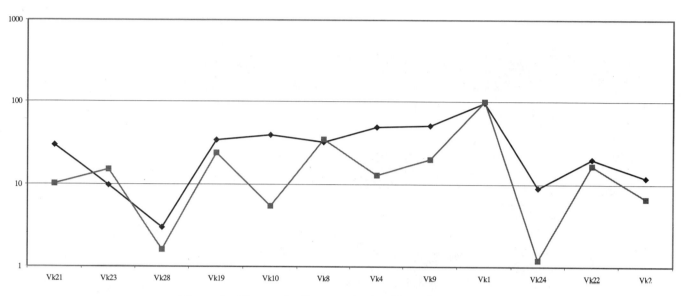

Figure 6.   Vk gene family expression in self-reactive antibodies.

5.   PATHOGENIC AUTOANTIBODIES: POLYCLONAL (NON-SELECTIVE) VERSUS CLONAL (SELECTIVE) ORIGIN

Studies of anti-DNA autoantibodies in MRL-*lpr/lpr* mice (120) have shown that these are of oligoclonal origin based on recurrent genetic elements for both V_H and V_L noted within this mouse. Such clonal relatedness is evident from the use of an identical CDR3 sequence encoded by a set of V_H, D_H and J_H gene segments. Yet, no restriction in V_H, D_H and J_H gene usage exists in autoantibodies to DNA (121–123). In MRL-*lpr/lpr* mice just one or a few clones produce 50–100% of anti-DNA autoantibodies. Many intraclonal differences in B cells secreting anti-DNA autoantibody suggest occurrence of somatic mutations, remarkable for arginine and asparagine substitutions, especially in the CDR3H. Arginine residues may interact with either the phosphate moiety or form stable hydrogen bonds with the heterocyclic bases of DNA. Although anti-DNA V_H segments pair with different Vκ genes, the Vκ1A subgroup appears to be particularly suitable in generation of an anti-DNA binding site (122). In fact, the first codon arginine (121) encoded by Jκ1 instead of germline encoded tryptophan in anti-DNA autoantibodies arises either as a result of somatic mutation or junctional flexibility and further suggests a selective clonal expansion of B cells. The arginine-rich residues in the CDR3, thus, appear to increase affinity of anti-DNA autoantibodies consistent with the development of an anti-DNA autoimmune response similar to foreign antigens. Most of the lupus-pathogenic anti-dsDNA autoantibodies are somatically mutated. Extensive analysis of IgG autoantibodies to a number of other self-antigens, e.g., laminin (123) immunoglobulin (rheumatoid factor; 124,125), histone (120), cardiolipin (126), and red blood cells (109), has shown that these AAbs are generated in a manner similar to those resulting from an exogenous antigen-driven response (124). In rheumatoid arthritis patients, N-region insertions (127) and an unusually long CDR3 (128,129) of the kappa-light chain of rheumatoid factor also suggests antigen-based selection. These observations provide evidence for selective expansion of B cells secreting pathogenic autoantibody similar to those responding to a foreign antigen.

Autoantibodies first appear as IgM and subsequently switch to IgG, similar to the development of an immune response to an antigen. Structural analysis of IgM autoantibodies to DNA, characteristic of systemic lupus erythematosus (SLE), shows that both IgM and IgG autoantibodies have similar variable-region structure and arise from the same clonal B cell precursors in the mouse. Unmutated germline V_H genes also encode the autoantibodies to DNA and unique or rarely-expressed germline VH genes may encode some of these. For example, an inspection of DNA sequences from different strains of mice, such as NZB/NZW F_1, MRL-*lpr/lpr* show that they are closely related (122). The anti-DNA autoantibodies, therefore, are encoded by a restricted set of V_H, e.g., 3H9 (J558), and V_L, e.g., DNA5 (Vκ8), genetic elements. Expression of recurrent germline-and somatically-mutated V_H and V_L

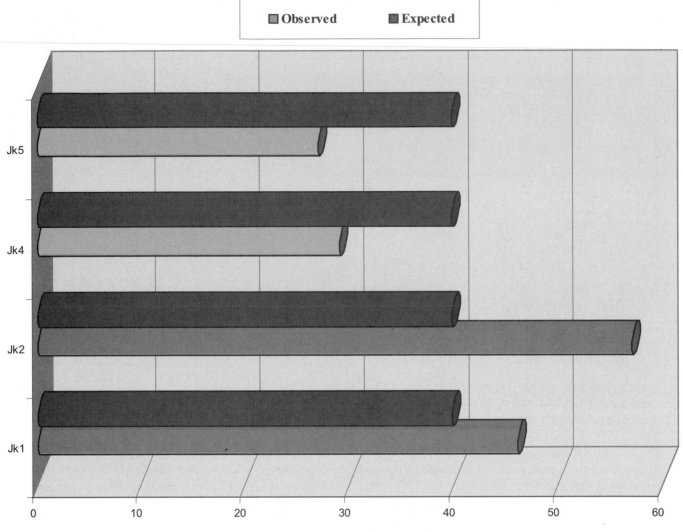

Figure 7.   Jk gene use in autoantibodies.

genetic elements influences the specificity of autoantibodies to native or denatured forms of DNA.

Thus, IgM anti-DNA autoantibodies, although of germline origin and encoded by recurrent V-genes (121), reflect a product of clonal selection. The majority of mono-specific IgM autoantibodies to DNA have at least one argi-nine residue in the CDR3H, probably generated via N-region insertions, as well as a shift in reading frame and D-D fusion. These observations indicate that IgM anti-DNA autoantibodies are also produced by clonally-selected and specifically stimulated B cells and do not arise as a result of non-selective polyclonal stimuli. The anti-DNA autoantibodies from SHP–1 deficient me^v/me^v, similar to other autoimmune strains, are encoded by a restricted set of germline V-genes. The fact that some of the anti-DNA autoantibody-secreting B cells express the same heavy chain in association with different light chains suggests that positive selection by an autoantigen contributes to clonal expansion of the B cells. Interestingly, a restricted set of rarely used V-genes, CDR3H length restriction and mono-specificity of randomly isolated J558$^+$ IgM autoantibodies from me^v/me^v mice at the terminal stage of autoimmune disease (62) suggests that impaired negative or positive selection of autoreactive B cells might also contribute to the origin of autoantibodies via lowered threshold of responsiveness to a self-antigen. A lack of CDR3H length restriction in neonatal me^v/me^v B cells (R. Cassady and A. Kaushik, unpublished) suggests that such increased nega-tive or positive selection for autoantibody-secreting B cells operates in the periphery either during recruitment of newly released B cells from the bone marrow into the circulation or during the development of an immune response to an autoantigen. Indeed, a population of B cells that produce germline-encoded anti-DNA autoantibody has been identified that escapes tolerance induction because of 1 to 4 logs lower affinity compared to the apparent affinities of

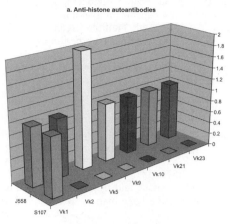

a. Anti-histone autoantibodies

□Vk1 ■Vk2 □Vk5 □Vk9 ■Vk10 ■Vk21 ■Vk23

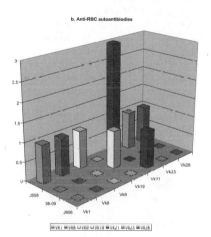

b. Anti-RBC autoantibodies

□Vk1 ■Vk8 □Vk19 □Vk19 ■Vk21 ■Vk23 ■Vk28

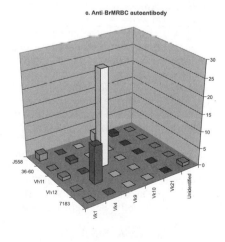

c. Anti-BrMRBC autoantibody

□Vk1 ■Vk4 □Vk9 □Vk10 ■Vk21 ■Unidentified

Figure 8. Restricted VH + Vκ pairings in autoantibodies against histone (a), red blood cells (b) and bromelain treated red blood cells (c).

the autoantibodies (130). These B cells may subsequently be recruited to the development of an immune response and undergo somatic mutation and class switching to produce high-affinity anti-DNA autoantibody. Overall, these observations suggest that both IgM and IgG autoantibodies arise as a result of specifically stimulated and clonally-selected B cell populations, similar to an immune response to a foreign antigen. However, impaired negative or positive selection during B cell development or recruitment of newly generated B cells into the periphery is at the root of subsequent selection and expansion of autoreactive B cells.

ACKNOWLEDGEMENT

The research support from NSERC Canada (A. Kaushik) and NIH, USA (C. Bona) is acknowledged.

References

1. Winfield, J.B., I. Faiferman, and D. Kofler. 1977. Avidity of anti-DNA antibodies in serum and IgG glomerular eluates from patients with systemic lupus erythematosus. Association of high avidity anti-native DNA antibody with glomerulonephritis. *J. Clin. Invest.* 59:90–96.
2. Couser, W.G., D.J. Salant, M.P. Madio, S. Adler, and G.C. Groggel. 1982. Factors influencing glomerular and tubulointerstitial patterns of injury in SLE. *Am. J. Kidney Dis.* 2 (1 Suppl 1): 126–134.
3. Foster, M.H., J. Abbaga, S.R.R. Line, S.K. Thompson, K.J. Barrett, and M. Madio. 1993. Molecular analysis of spontaneous nephrotropic anti-laminin antibodies in an autoimmune MRL-lpr/lpr mouse. *J. Immunol.* 151:814–824.
4. Kelley, V.R., G.C. Diaz, A.M. Jevinkar, and G.G. Singer. 1993. Renal tubular epithelial and T cell interactions in autoimmune renal disease. *Kidney Int. Suppl.* 39:S108–115.
5. Moyer, C.F., J.D. Standberg, and C.L. Reinisch. 1987. Systemic mononuclear-cell vasculitis in MRL/Mp-lpr/lpr mice. A histologic and immunocytochemical analysis. *Am. J. Pathol.* 127:229–242.
6. Dixon, F.J., M.B.A. Oldstone, G. Tonietti. 1971. Pathogenesis of immune complex glomerulonephritis of New Zealand mice. *J. Exp. Med.* 134(Suppl.) 65s–71s.
7. Vlahakos, D.V., M.H. Foster, S. Adams, M. Katz, A.A. Ucci, K.J. Barrett, S.K. Datta, and M.P. Madio. 1992. Anti-DNA antibodies form immune deposits at distinct glomerular and vascular sites. *Kidney Int.* 41:1690–700.
8. Jabs, D.A., and R.A. Prendergast. 1987. Reactive lymphocytes in lacrimal gland and vasculitic renal lesions of autoimmune MRL/lpr mice express L3T4. *J. Exp. Med.* 166:1198–1203.
9. Katagiri, T., P.L. Cohen, and R.A. Eisenberg. 1988. The lpr gene causes an intrinsic T cell abnormality that is required for hyperproliferation. *J. Exp. Med.* 167:741–751.
10. Taurog, J.D., E.S. Raveche, P.A. Smathers, M.H. Glimoner, D.P. Houston, C.T. Hansen, and A.D. Steinberg. 1981. T cell abnormalities in NZB mice occur independently of autoantibody production. *J. Exp. Med.* 153:221–234.
11. Steinberg, A.D., J.B. Roths, E.D. Murphy, R.T. Steinberg, and E.S. Raveche. 1980. Effects of thymectomy or androgen

administration upon the autoimmune disease of MRL/Mp-lpr/lpr mice. *J. Immunol.* 125:871–873.

12. Wofsky, D., J.A. Ledbetter, P.L. Hendler, and W.E. Seaman. 1985. Treatment of murine lupus with monoclonal anti-T cell antibody. *J. Immunol.* 134:852–857.

13. Santoro, T.J., J.P. Portanova, and B.L. Kotzin. 1988. The contribution of L3T4 + T cells to lymphoproliferation and autoantibody production in MRL-lpr/lpr mice. *J. Exp. Med.* 167:1713–1718.

14. Lin, R.H., M.J. Mamula, J.A. Hardin, and C.A. Janeway. 1991. Induction of autoreactive B cells allows priming of autoreactive T cells. *J. Exp. Med.* 173:1433–1439.

15. Mamula, M.J., R.H. Lin, C.A. Janeway, and J.A. Hardin. 1992. Breaking T cell tolerance with foreign and self co-immunogens. A study of autoimmune B and T cell epitopes of cytochrome c. *J. Immunol.* 149:789–795.

16. Klinman, D.M., and A.D. Steinberg. 1986. Proliferation of anti-DNA-producing NZB B cells in a non-autoimmune environment. *J. Immunol.* 137:69–75.

17. Herron, L.R., R.L. Coffman, and B.L. Kotzin. 1988. Enhanced response of autoantibody-secreting B cells from young NZB/NZW mice to T-cell-derived differentiation signals. *Clin. Immunol. Immunopathol.* 46:314–327.

18. Nemazee, D., C. Guiet, K. Buerki, and A. Mark-Rothstein. 1991. B lymphocytes from the autoimmune-prone mouse strain MLR/lpr manifest an intrinsic defect in tetraparental MRL/lpr in equilibrium DBA/2 chimeras. *J. Immunol.* 147:2536–2539.

19. Sobel, E.S., T. Katagiri, S. Morris, P.L. Cohen, and R.A. Eisenberg. 1991. An intrinsic B cell defect is required for the production of autoantibodies in the lpr model of murine systemic autoimmunity. *J. Exp. Med.* 173:1441–1449.

20. Shlomchik, M.J., M.P. Madio, D. Trounstein, M. Ni, and D. Huszar. 1994. The role of B cells in lpr/lpr-induced autoimmunity. *J. Exp. Med.* 180:1295–1306.

21. Chan, O.T.M., L.G. Hannum, A.M. Haberman, M.P. Madio, and A. Shlomchik. 1999. Novel mouse with B cells but lacking serum antibody reveals an antibody-independent role for B cells in murine lupus. *J. Exp. Med.* 17:1639–1648.

22. Dang, H., and R.J. Harbeck. 1984. The in vivo and in vitro glomerular deposition of isolated anti-double-stranded-DNA antibodies in NZB/W mice. *Clin. Immunol. Immunopathol.* 30:265–278.

23. Pankewycz, O.G., P. Migliorini, and M.P. Madio. 1987. Polyreactive autoantibodies are nephritogenic in murine lupus nephritis. *J. Immunol.* 139:3287–3294.

24. Reininger, L., T. Shibata, S. Ozaki, T. Shirai, J.-C. Jaton, and S. Izui. 1990. Variable region sequences of psthogenic anti-mouse red blood cell autoantibodies from autoimmune NZB mice. *Eur. J. Immunol.* 20:771–777.

25. Avrameas, S. 1991. Natural autoantibodies: from "horror autotoxicus" to "gnothi seauton". *Immunol. Today* 12:154–159.

26. Cohen, I.R. 1992. The cognitive paradigm and the immunological homunculus. *Immunol. Today* 13:490–494.

27. Nobrega, A., M. Haury, A. Grandien, E. Malanchere, A. Sundblad, and A. Coutinho. 1993. Global analysis of antibody repertoires. II. Evidence for specificity, self-selection and the immunological "homunculus" of antibodies in normal serum. *Eur. J. Immunol.* 23:2851–2859.

28. Dighiero, G., P. Lymberi, J. Mazie, S. Rouyse, Butler-Browne, R.G. Whalen, and S. Avermas. 1983. Murine hybridomas secreting natural monoclonal antibodies reacting with self antigens. *J. Immunol.* 131:2267–2272.

29. Prbhakar, B.S., J. Saegvza, T. Onodera, and A.L. Notkins. 1984. Lymphocytes capable of making monoclonal autoantibodies that react with multiple organs are a common feature of the normal B cell repertoire. *J. Immunol.* 133:2815–2817.

30. Weissman, I.L. 1994. Developmental switches in the immune system. *Cell* 76:207–218.

31. Osmond, D.G. 1986. Population dynamics of bone marrow B lymphocytes. *Immunol. Rev.* 93:103–124.

32. Schittek, B., and K. Rajewsky. 1990. Maintenance of B-cell memory by long-lived cells generated from proliferating precursors. *Nature* 346:749–751.

33. Tonegawa, S. 1983. Somatic generation of antibody diversity. Nature. 302:575–581.

34. Tiegs, S.L., D.M. Russel, and D. Nemazee. 1993. Receptor editing in self-reactive bone marrow B cells. *J. Exp. Med.* 177:1009–1020.

35. Ma, A., P. Fisher, R. Dildrop, E. Oltz, G. Rathburn, P. Achacoso, A. Stall, and F. Alt. P. Fisher, R. Dildrop, E. Oltz, G. Rathbun, P. Achacoso, A. Stall, and F. Alt. 1992. Surface IgM mediated regulation of RAG gene expression in E mu-N-myc B cell lines. *EMBO J.* 11:2727–2734.

36. Gay, D., T. Saunders, S. Camper, and M. Weigert. 1993. Receptor editing: an approach by autoreactive B cells to escape tolerance. *J. Exp. Med.* 177:999–1008.

37. Radic, M.Z., J. Erikson, S. Litwin, and M. Weigert. 1993. B lymphocytes may escape tolerance by revising their antigen receptors. *J. Exp. Med.* 177:1165–1173.

38. Goodnow, C.C., J. Crosbie, H. Jorgensen, R.A. Brink, and A. Basten. 1989. Induction of self-tolerance in mature peripheral B lymphocytes. *Nature* 342:385–3891.

39. Hartley, S.B., M.P. Cooke, D.A. Fulcher, A.W. Harris, S. Cory, A. Basten, and C.C. Goodnow. 1993. Elimination of self-reactive B lymphocytes proceeds in two stages: arrested development and cell death. *Cell* 72:325–335.

40. Goodnow, C.C. 1997. Balancing immunity, autoimmunity, and self-tolerance. *Ann. NY Acad. Sci.* 815:55–66.

41. Akkaraju, S., K. Cannan, and C.C. Goodnow. 1997. Self-reactive B Cells are not eliminated or inactivated by autoantigen expressed on thyroid epithelial cells. *J. Exp. Med.* 186:2005–2012.

42. Lortan, J.E., C.A. Roobottom-Oldfields, and I.C. MacLennan. 1987. Newly produced virgin B cells migrate to secondary lymphoid organs but their capacity to enter follicles is restricted. *Eur. J. Immunol.* 17:1311–1316.

43. Malynn, B.A., G.D. Yancopoulos, J.E. Barth, C.A. Bona, and F.W. Alt. 1990. Biased expression of JH-proximal VH genes occurs in the newly generated repertorie of neonatal and adult mice. *J. Exp. Med.* 171:843–859.

44. Gu, H., D. Tarlington, W. Muller, K. Rajewsky, and I. Forster. 1991. Most peripheral B cells in mice are ligand selected. *J. Exp. Med.* 173:1357–1371.

45. Townsend, S.E., B.C. Weinturb, and C.C. Goodnow. 1999. Growing Up On the Streets: Why B-cell development differs from T-cells development. *Immunology Today* 20:217–220.

46. Mandik-Nayak, L., A. Bui, H. Noorchashm, A. Eaton, and J. Erikson. 1997. Regulation of anti-double-stranded DNA B cells in nonautoimmune mice: localization to the T-B interface of the splenic follicle. *J. Exp. Med.* 186:1257–1267.

47. Hertz, M., V. Kouskoff, T. Nakamura, and D. Nemazee. 1998. VDJ recombinase induction in splenic B lymphocytes is inhibited by antigen-receptor signaling. *Nature* 394:292–295.

48. Kurosaki, T. 1999. Genetic analysis of B cell antigen receptor signaling. *Annu. Rev. Immunol.* 17:555–592.

49. Law, D.A., V.W.F. Chan, S.K. Datta, and A.L. Defranco. 1993. B-cell antigen receptor motifs have redundant signaling capabilities and bind the trosine kinases PTK72, Lyn, and Fyn. *Curr. Biol.* 3:645–657.

50. Sanchez, M., Z. Misulovin, A.L. Burkhardt, S. Mahajan, T. Costa, R. Franke, J.B. Bolen, and M. Nussenweig. 1993. Signal transduction by Immunoglobulin is mediated through Ig *J. Exp. Med.* 178:1049–1055.

51. Hibbs, M.L., D.M. Tarlinton, J. Armes, D. Rail, G. Hodgson, R. Maglitto, S.A. Stacker, and A.R. Dunn. 1995. Multiple defects in the immune system of Lyn deficient mice culminating in autoimmune disease. *Cell* 83:301–311.

52. Nishizumi, H., I. Taniuchi, Y. Yamanashi, D. Kitamura, D. Ilic, S. Mori, I. Watanabe, and T. Yamamoto. 1995. Impaired proliferation of peripheral B cells and indication of disease in Lyn-deficient mice. *Immunity* 3:549–560.

53. Cornall, R.J., J.G. Cyster, M.L. Hibbs, A.R. Dunn, K.L. Otipoby, E.A. Clark, and C.C. Goodnow. 1998. Polygenic autoimmune traits: Lyn, CD22, and SHP-1 are limiting elements of a biochemical pathway regulating BCR signaling and selection. *Immunity* 8:497–508.

54. Turner, M., A. Gulbranson-Judge, M.E. Quinn, A.E. Walters, I.C.M. Maclennan, and V.L.J. Tybulcwicz. 1997. Syk tyrosine kinase is required for the positive selection of immature B cells into the recirculating B cell pool. *J. Exp. Med.* 186:2013–2021.

55. Tsukada, S., D.C. Saffran, D.J. Rawlings, O. Parolino, R.C. Allen, H. Kubagawa, I. Mohandas, S. Quan, J.W. Belmont, M.D. Cooper, M.E. Conley, and O.N. Witte. 1993. Deficient expression of a B cell cytoplasmic tyrosinc kinase in huma X-linked agammaglobulinemia. *Cell* 72:279–290.

56. Vetrie, D., I. Verochovysky, P. Sideras, J. Holland, A. Davies, F. Flinter, L. Hammarstom, C. Kinnon, R. Levinsky, M. Bobrow, C.I.E. Smith, and D.R. Bently. 1993. The gene involved in X-linkcd agamaglobulinemia is a member of the protein-tyrosine kinase. *Nature* 361:226–233.

57. Dighiero, G., A. Lim, M.P. Lembezat, A. Kaushik, L. Andrade, and A. Freitas. 1988. Comparative study of VH gene family usage by newborn xid and non-xid mice, newborn NZB and adult NZB mice, and splenic and peritoneal cavity B cell compartments. *Eur. J. Immunol.* 18:1979–1983.

58. Feng, S., and K.E. Stein. 1991. VH gene family expression in mice with xid defect. *J. Exp. Med.* 174:45–51.

59. Kishihara, K., J. Penninger, V.A. Wallace, T.M. Kundig, K. Kawai, P.S. Ohashi, M.L. Thomas, C. Furlonger, C.J. Paige, and T.W. Mak. 1993. Normal B lymphocyte development but impaired T cell maturation in CD45-exon6 protein tyrosine phosphatase-deficient mice. *Cell* 74:143–156.

60. Cyster, J.G., J.I. Healy, K. Kishihara, T.W. Mak, M.L. Thomas, and C.C. Goodnow. 1996. Regulation of B lympocyte negative and positive selection by tyrosine phosphatase CD45. *Nature* 381:325–328.

61. Tsui, F.W.L., and H.W. Tsui. 1994. Molecular basis of the motheaten phenotype. *Immunol. Rev.* 138:185–206.

62. Shultz, L.D., D.R. Coman, C.L. Bailey, W.G. Beamer, C.L. Sidman. 1984. Viable motheaten, a new allele at the motheaten locus. I. Pathology. *Am. J. Pathol.* 116:179–192.

63. Saitoh, Y., A. Kaushik, G. Kelose, and C. Bona. 1995. The expression and pairing of Vh and Vk gene families in motheaten mice. *Autoimmunity* 21:185–193.

64. Lipsanen, V., B. Walter, M. Emara, K. Siminovitch, J. Lam, and A. Kaushik. 1997. Restricted CDR3 length of the heavy chain is characteristic of six randomly isolated disease-associated VH J558 + IgM autoantibodies in lupus prone motheaten mice. *Int. Immunol.* 9:655–664.

65. Pani, G., K.A. Siminovitch, and C.J. Paige. 1997. The motheaten mutation rescues B cell signaling and development in CD45-deficient mice. *J. Exp. Med.* 186:581–588.

66. Kofler, R., R. Strohal, S. Balderas, N.E. Johnson, D.J. Noonan, M.A. Duchosal, F.J. Dixon, and A.N. Theofilopoulos. 1988. Immunoglobulin kappa-light chain variable gene complex organization and immunoglobulin gene encoding anti-DNA autoantibodies in lupus mice. *J. Clin. Invest.* 82:852–8666.

67. D'Hoostelaere, L.A., K. Huppi, B. Mock, C. Mallett, and M. Potter. 1988. The Ig kappa L chain allelic groups among the Ig kappa haplotypes and Ig kappa crossover populations suggest a gene order. *J. Immunol.* 141:652–661.

68. Carson, S., and G. Wu. 1989. A linkage map of the immunoglobulin lambda light chain locus. *Immunogenetics* 29:173–179.

69. Storb, U., D. Haasch, B. Arp, P. Sanchez, P.A. Cazenave, and J. Miller. 1989. Physical linkage of mouse lambda genes by pulsed-field gel electrophoresis suggests that the rearrangement process favours proximate target sequences. *Mol. Cell Biol.* 9:711–718.

70. Kaushik, A., and W. Lim. 1996. The primary antibody repertoire of normal, immunodeficient and autoimmune mice is characterized by differences in V gene expression. *Res. Immunol.* 147:9–26.

71. Yancopoulos, G.D., D.V. Desiderio, M. Paskind, J.F. Kearney, D. Baltimore, and F.W. Alt. 1984. Preferential utilization of the most JH proximal VH gene segments in pre-B cell lines. *Nature* 311:727–733.

72. Kelsoe, G. 1992. The primary antibody repertoire: The somatic genotype and paratopic phenotype of B-cell populations. In: *Molecular Immunobiology of Self Reactivity* (C.A. Bona & A. Kaushik, eds.) pp. 81–92, Marcel-Dekker, Inc., New York.

73. Bangs, L.A., I.E. Sanz, and M.J. Teale. 1991. Comparison of D. JH, and Junctional diversity in thc fetal, adult and aged B cell repertoires. *J. Immunol.* 146: 1991–2004.

74. Feeney, A.J. 1990. Lack of N regions in fetal and neonatal mouse immunoglobulin V-D-J junctional sequences. *J. Exp. Med.* 172:1377–1390.

75. Kaushik, A., D.H. Shulze, C.A. Bona, and G. Kelsoe. 1989. Murine VK gene expression violates the VH paradigm. *J. Exp. Med.* 169:1859–1864.

76. Kastner, D.I., T.M. McIntyre, C.P. Mallet, A.B. Hartman, and A.D. Steinberg. 1989. Direct quantitative in situ hybridization studies of IgVH utilization. A comparison between unstimulated B cells autoimmune and normal mice. *J. Immunol.* 143:2761–2767.

77. Komisar, J.L., K.Y. Leung, R.R. Crawley, N. Talal, and J.M. Teale. 1989. Immunoglobulin VH gne family repertoire of plasma cells derived from lupus-prone MRL/1pr and MRL/++ mice. *J. Immunol.* 143:340–347.

78. Klonowski, K.D., L.L. Primiano, and M. Monestier. 1999. Atypical VH-D-JH rearrangements in newborn autoimmune MRL mice. *J. Immunol.* 162(3):1566–1572.

79. Kofler, R., D.J. Noonan, D.E. Levy, M.C. Wilson, N.P.H. Moller, J. Dixon, and A.N. Theophilopoulos. 1985. Genetic elements used for a murine lupus anti-DNA autoantibody are closely related to those for antibodies to exogenous antigens. *J. Exp. Med.* 151:805–815.

80. Painter, C., M. Monestier, B. Bonin, and C.A. Bona. 1986. Functional and molecular studies of V genes expressed in autoantibodies. *Immunol. Rev.* 94:75–83.

81. Kasturi, K., M. Monestier, R. Mayer, and C. Bona. 1988. Biased Usage of certain VK gene families by autoantibodies In their polymorphism in autoimmune strains. *Molec. Immunol.* 25:213–219.

82. Bona, C. 1988. V genes encoding atoantibodies: Molecular and phenotypic characteristics. *Ann. Rev. Immunol.* 6:327–358.

83. Milner, E.C., W.O. Hufnagle, A.M. Glas, and I. Suzuki. 1995. Alexander C Polymorphism and utilization of human VH Genes. *Ann NY Acad. Sci.* 764:50–61.

84. Monestier, M., A. Manheimer-Leroy, B. Bellon, C. Painter, H. Dang, N. Talal, M. Zanetti, R. Schwartz, D. Pisetsky, R. Kuppers, N. Ose, J. Brochier, L. Klareskog, R. Holmdahl, B. Erlanger, F. Alt, and C. Bona. 1986. Shared idiotopes and restricted VH genes characterized murine autoantibodies of various specificities. *J. Clin. Invest.* 78:753–759.

85. Monestier, M. 1991. Variable region genes of Anti-histone autoantibodies from a MRL/mp-lpr/lpr mouse. *Eur. J. Immunol.* 21:1725–1731.

86. Manheimer-Leroy, A.J., M. Monestier, B. Bellon, F.W. Alt, and C.A. Bona. 1986. Fine specificity, idiotype and nature of cloned heavy chain variable region genes of murine monoclonal rheumatoid factor antibodies. *Proc. Natl. Acad. Sci USA* 83:8293–8297.

87. Painter, C.J., M. Monestier, A. Chew, A. Bona-Dimitriu, K. Kasturi, C. Bailey, V.E. Scott, C.L. Sidman C.A. Bona. 1988. Specificities and V-genes encoding monoclonal autoantibodies from viable motheaten mice. *J. Exp. Med.* 167:1137–1153.

88. Bailey, N.C., V. Fidanza, R. Mayer, G. Mazza, M. Fougereeau, and C. Bona. 1989. Activation of clones producing self reactive antibodies by foreign antigens antiidiotype antibody carrying the internal image of the antigen. *J. Clin. Invest.* 84:744–756.

89. Fidanza, V., R. Mayer, H. Zaghouani, M.A. Diliberti, and C.A. Bona. 1990. Autoantibodies Ly-1 and immunoglobulin V-gene expression in hybridomas obtained from young and old New Zealand black mice. *Arth. and Rheum.* 33:711–723.

90. Kaushik, A. 1991. A synopsis of self reactivity. In: *Molecular Biology of Self Reactivity* (C. Bona, and A. Kaushik, eds.) Marcel Dekker Inc., New York, pp. 1–24.

91. Kasturi, N.K., R. Mayer, C.A. Bona, W.T. Scott, and C.L. Sidman. 1990. Structure of V-genes encoding autoantibodies specific for thymocytes and red blood cells in mev mice. *J. Immunol.* 145:2304–2234.

92. Bailey, N.C., K.N. Kasturi, T.K. Blackwell, F.W. Alt, and C.A. Bona. 1991. Complexity of immunoglobulin light chain VK1 gene family in the new zealand black mouse. *Intern. Immunol.* 3:751–760.

93. Zaghouani, H., F.A. Bonilla, K. Meek, and C.A. Bona. 1989. Molecular basis for expression of the A48 regulatory idiotope on antibodies encoded by V genes from various families. *Proc. Natl. Acad. Sci. USA* 86:2341–2345.

94. Shlomchik, M., D.A. Nemazee, V.L. Sato, J. Van Snick, D.A. Carson, and M.G. Weigert. 1986. Variable region sequence of murine IgM anti-IgG monoclonal autoantibodies (rheumatoid factors): a structural explanation for the high frequency of IgM anti-IgG B cells. *J. Exp. Med.* 164:407–427.

95. Shlomchik, M.J., B. Marshak-Rothstein, C.B. Wolfwicz, T.L. Rothestein, and M.G. Weigert. 1987. The role of clonal selection and somatic mutation in autoimmunity. *Nature* 328:805–811.

96. Kofler, R., D.J. Noonan, R. Strohal, R.S. Balderas, N.P.H. Moller, K.J. Dixon, and A.N. Theofilopoulos. 1987. Molecular analysis of the murine lupus associated anti-self response: involvement of a large number of heavy and light chain variable region genes. *Eur. J. Immunol.* 17:91–95.

97. Conger, J.D., H.J. Sage, and R.B. Corely. 1989. Diversity in the available repertoire of murine antibodies reactive with bromlain-treated isologous erythrocytes. *J. Immunol.* 143:4044–4052.

98. Conger, J.D., H.J. Sage, S. Kawaguchi, and R. Corely. 1991. Properties of murine antibodies from different V-region families specific for bromelain-treated mouse erythrocytes. *J. Immunol.* 146:1216–1219.

99. Hardy, R.R., C.E. Carmack, S.A. Shinton, R.J. Riblet, and K. Hayakawa. 1989. A single is utilized predominantly in anti-BrMRBC hybridomas derived from purified Ly1 B cells. *J. Immunol.* 142:3643–3651.

100. Hayakawa, K., C.E. Carmack, R. Hyman, and R.R. Hardy. 1990. Natural autoantibodies to thymocytes: origin, VH genes, fine specificity, and role of Thy-1 glycoproteins. *J. Exp. Med.* 172:869–878.

101. Panosian-Sahakian, N., J.L. Koltz, F. Ebling, M. Kronenberg, and B. Hahn. 1989. Diversity of Ig-V gene segments found in anti-DNA autoantibodies from a single (NZBXNZW) F1 mouse. *J. Immunol.* 142:4500–4506.

102. Pennell, C.A., T.J. Mercolino, T.A. Gardina, L.W. Arnold, G. Haugton, and S.H. Clarke. 1989. Biased immunoglobulin variable region expression by Ly-1 B cells due to clonal selection. *Eur. J. Immunol.* 19:1289–1294.

103. Trepicchio, W., A. Maruya, and K.J. Barret. 1987. The heavy chain genes of lupus anti-DNA autoantibody are encoded in the germline of a non autoimmune strain of mouse and conserved in strain of mice polymorphic for this gene locus. *J. Immunol.* 139:3139–3145.

104. Trepicchio, W., and K.J. Barret. 1987. Eleven MRL-lpr/lpr anti-DNA autoantibodies are encoded by genes from four VH gene families: a potentially biased usage of VH genes. *J. Immunol.* 138:2323–2331.

105. Gleason, S.I., R. Gearhart, N.R. Rose, and R.C. Kuppers. 1990. Autoantibodies to thyroglobulin are encoded by diverse V-gene segments and recognize restricted epitopes. *J. Immunol.* 145:1768–1775.

106. Duchosal, M.A., R. Kofler, R.S. Balderas, T.M. Aguado, F.J. Dixon, and A.N. Theofilopoulos. 1989. Genetic diversity of murine rheumatoid factors. *J. Immunol.* 142:1737–1742.

107. Baccala, R., T.V. Quang, M. Gilbert, T. Ternynck, and S. Avrameas. 1989. Two murine natural polyreactive autoantibodies are encodee by nonmutated germline genes. *Proc. Natl. Acad. Sci. USA* 86:4624–4628.

108. Reninger, L., T. Berney, T. Shibata, F. Spertini, R. Merino, and S. Izui. 1990. Cryoglobulinemia induced by a murine IgG3 rheumatoid factor: skin vasculitis and glomerulonephritis arise from distinct pathogenic mechanisms. *Proc. Natl. Acad. Sci. USA* 87:10038–10042.

109. Reininger, I., P. Ollier, A. Kaushik, and J.C. Jaton. 1987. Novel V genes encode virtually identical and variable regions of six murine monoclonal anti-bromlain-treated red blood cell autoantibodies. *J. Immunol.* 138:316–323.

110. Poncet P., F. Huetz, M.A. Marcos, and L. Andrade. 1990. All VH11 genes expressed in peritoneal lymphocytes encode anti-bromelain-treated mouse red blood cell autoantibodies, but other VH gene families contribute to this specificity. *Eur. J. Immunol.* 20:1583–1589.

111. Smith, R.G., and E.W. Voss, Jr. 1990. Variable region primary structures of monoclonal anti-DNA autoantibodies from NZB F1 mice. *Molec Immunol.* 27:463–470.

112. Marion, N.T., D.M. Tilman, and N.T. Jou. 1990. Interclonal and intraclonal diversity among anti-DNA antibodies from an (NZBXNZW) mouse. *J. Immunol.* 145:2322–2332.

113. Ewulonu, U.K., L.J. Nell, and J.W. Thomas. 1990. VH and VL gene usage by murine IgG antibodies that bind atologous insulin. *J. Immunol.* 144:3091–3098.

114. Okamoto, M., and T. Honjo. 1990. Nucleotide sequences of the gene/cDNA coding for anti-murine erythrocyte autoantibody produced by a hybridoma from NZB mouse. *Nucleic Acid. Res.* 18:1895–1895.

115. Shefner, R., G. Kleiner, A. Turken, L. Papazian, and B. Diamond. 1991. A novel class of anti-DNA antibodies identified in BALAB/c mice. *J. Exp. Med.* 173:287–296.

116. Brigido, M.M., and B.D. Stollar. 1991. Two induced anti-Z-DNA monoclonal antibodies use VH gene segments related to those of anti-DNA autoantibodies. *J. Immunol.* 146:2005–2009.

117. Eilat, D., and R. Fischel. 1991. Recurrent utilization of genetic elements in V regions of antinucleic acid antibodies form autoimmune mice. *J. Immunol.* 147:361–368.

118. Bona, C.A., Y. Saitoh, and G. Kelsoe. 1990. Pairing of VH and VK gene in self reactive antibodies. *J. Clin. Immunol.* 10:223–236.

119. Kaushik, A., D.H. Schultze, F.A. Bonilla, C. Bona, and G. Kelsoe. 1990. Stochastic pairing of VH and VK gene families occurs in polyclonally activated B cells. *Proc. Natl. Acad. Sci. USA* 87:4932–4936.

120. Shlomchik, M., M. Mascelli, H. Shan, Z.M. Radic, D. Pistesky, A. Marshak-Rothstein, and M. Weigert. 1990. Anti-DNA antibodies from autoimmune mice arise by clonal expansion and somatic mutation. *J. Exp. Med.* 171:265–297.

121. Tillman, D.M., N. Jou, R.J. Hill, and T.N. Marion. 1992. Both IgM and IgG anti-DNA antibodies are the products of clonally selective B cell stimulation in (NZB * NZW)F1 Mice. *J. Exp. Med.* 176:761–779.

122. Eilat, D. 1990. The role of germline gene expression and somatic mutation in the generation of autoantibodies to DNA. *Mol. Immunol.* 27:203–210.

123. Foster, M.H., B. Cizman, and M.P. Madio. 1993. Nephritogenic autoantibodies in systemic lupus erythematosus: immunochemical properties, mechanisms of immune deposition, and genetic origins. *Lab. Invest.* 69:494–507.

124. Krishnan, M.R., and T.N. Marion. 1993. Structural similarity of antibody variable regions from immune and autoimmune anti-DNA antibodies. *J. Immunol.* 150:4948–4957.

125. Radic, M.Z., and M. Weigert. 1994. Genetic and structural evidence for antigen selection of anti-DNA antibodies. *Annu. Rev. Immunol.* 12:487–520.

126. Kita, Y., T. Sumida, K. Ichikawa, T. Maeda, F. Yonaha, I. Iwamoto, S. Yosida, and T. Koike. 1993. V gene analysis of anticardiolipin antibodies from MRL-lpr/lpr mice. *J. Immunol.* 151:849–856.

127. Bridges, S.L., Jr. 1998. Frequent N addition and clonal relatedness among immunoglobulin lambda light chains expressed in rheumatoid arthritis synovia and PBL, and the influence of V lambda gene segment utilization on CDR3 length. *Mol. Med.* 4:525–553.

128. Moyes, S.P., R.N. Maini, and R.A. Mageed. 1998. Differential use of immunoglobulin light chain genes and B lymphocyte expansion at sites of disease in rheumathoid arthritis (RA) compared with circulating B lymphocytes. *Clin. Exp. Immunol.* 113:276–288.

129. Lee, S.K., J.B. Kim, Y.J. Chwae, S.L.J.R. Bridges, W.J. Koopman, and H.W. Schroeder, Jr. 1998. Enhanced expression of immunoglobulin kappa light chains with usually long CDR3 regions in patients with reumathoid arthritis. *J. Rheumatol.* 25:1067–1071.

130. Boyne, M.S., L. Spatz, B. Diamond. 1999. Characterization of anti DNA B cells that escape negative selection. *Eur. J. Immunol.* 29:1304–1313.

5 | T Cell Development and Tolerance

Harald von Boehmer and Iannis Aifantis

1. INTRODUCTION

As anticipated by Burnet (1), Lederberg (2) and Jerne (3) many years ago, T cell development is tightly linked to T cell repertoire selection. This is so because T cell development is controlled by T cell receptors for antigen at immature stages (4) preceding the mature stage, at which time engagement of the T cell receptor by antigen can result in the gain of effector function, such as cytokine secretion or cytolytic activity.

Initial hints that T cell receptors for antigen may have something to do with T cell development came from experiments showing that MHC molecules expressed on radioresistant thymic stroma could somehow influence the T cell repertoire (5). Since it was not clear, however, whether they did so by affecting mature T cells through either enhancing or suppressing responsiveness, this issue required a more thorough analysis. The development of transgenic and gene-deficient mice opened new possibilities to address this issue in a more straightforward manner.

The first conclusive evidence that T cell development was tightly controlled by the expression of distinct T cell receptor genes was obtained in rearrangement-deficient, and therefore immuno-deficient [severe combined immuno- deficient (SCID)] mice, that were reconstituted with defined T cell receptor transgenes (4). These experiments linked the known CD4/8 subsets in a developmental order and defined two distinct thymic checkpoints at which the expression of certain T cell receptor genes was essential for developmental progression (6). The first checkpoint affected a thymic subset that underwent efficient TCR β, γ, and δ, but inefficient α, rearrangement (7,8): While productive γ and δ rearrangement was essential for the generation of $\gamma\delta$ T cells (whose significance in the mammalian immune system is still largely obscure), productive TCRβ rearrangement was essential for effective progression along the $\alpha\beta$ pathway. While at this checkpoint, the maturation of CD4-8- cells into CD4+8+ thymocytes was controlled by the TCRβ chain (4, 9), it was found that the specificity of the complete $\alpha\beta$TCR determined the transition of short-lived and non-functional CD4+8+ cells into long-lived and functional CD4+8- and CD4-8+ T cell subsets (10). As a rule, the binding of the $\alpha\beta$TCR to thymic class I or class II MHC molecules in the absence of the agonist peptide resulted in the generation of CD4-8+ and CD4+8- T cells, respectively (11,12). It turned out that the $\alpha\beta$TCR exerted further control on the survival of mature T cells in peripheral lymphoid tissue (13,14), and thus these experiments defined three distinct maturation stages at which T cell receptor genes play an obligatory role in mediating survival (or even cell death). In the following, the essential new features of these checkpoints will be briefly discussed.

2. SURVIVAL AND DEVELOPMENT CONTROLLED BY THE PRE-T CELL RECEPTOR

Data obtained in rearrangement-deficient mice reconstituted with TCRβ transgenes as well as data in TCRβ-deficient mice indicated that productive TCRβ rearrangement was sufficient and essential for developmental progression beyond the CD4-8- stage of maturation (4, 9, 15). Since this occurred in the absence of TCRα chains, and since a single partner chain of disulfide- linked heterodimers can usually not be expressed in a stable manner, the search for a partner chain that could covalently associate with a TCRβ chain at this developmental stage revealed the pre-TCRα (pTα) protein (16, 17). The pTα protein, in fact, exhibits little homology with the TCRα chain in its extracellular Ig-like domain, but the transmembrane portion shares polar residues required for the association with signal- transducing CD3 molecules at the same position. It was subsequently shown that in pTα-/- (18), TCRβ-/- (19) and CD3-/- (20, 21) mice thymic development was arrested at the CD25+44- stage of development, i.e. at a stage where T cell receptor β chains are for the first time abundantly expressed. The arrest was more complete in the CD3-/- compared to pTα-/- or TCRβ-/- mice, indicating that both the γδ TCR and the αβ TCR could rescue, albeit inefficiently, some development of αβ lineage cells (22). The pTα-TCRβ heterodimer is, in fact, expressed on the cellsurface of CD25+44- as well as CD25-44- thymocytes (23) (Fig. 1) and recent unpublished evidence suggests that the initiation of pre-TCR signaling may not require ligation by a ligand on thymic stromal cells: Unlike the TCRα chain, the pTα chain contains a cysteine just below the transmembrane portion (24) that is likely to be palmitoylated, thus allowing spontaneous and cell-autonomous segregation of the pre-TCR into glycolipid-enriched membrane (GEM) domains or rafts that contain other palmitoylated proteins, like p56lck and the adapter-protein LAT, which are involved in pre-TCR signaling (25). In a pre-TCR-competent cell line, but not pre-TCR-incompetent parental line, phosporylation of CD3ε was cell-autonomously initiated and Zap 70 was recruited to the cell membrane and phosphorylated (26). These data are compatible with an earlier notion that a pre-TCR largely devoid of extracellular domains (27) nevertheless appeared competent to signal developmental progression, even though spontaneous aggregation and/or a contribution of the intact pTα-protein was not excluded in the latter experiments. These data suggest that there is no selection of specific TCRβ chains at this stage of development, and that any cell expressing a pre-TCR is selected for survival, followed by seven to nine rounds of division and further maturation that includes expression of CD4 and CD8 molecules as well as effective TCRα rearrangement (28). From data in various genetically-modified mice, it is concluded that p56lck (29), ZAP-70 /syk (30) LAT (31), SLP–76 (32, 33), raf, ras (34, 35) and the MAP kinase pathway (36) are involved in pre-TCR signaling. It is of interest to note that there is a split in signaling pathways that control proliferation and CD4/8

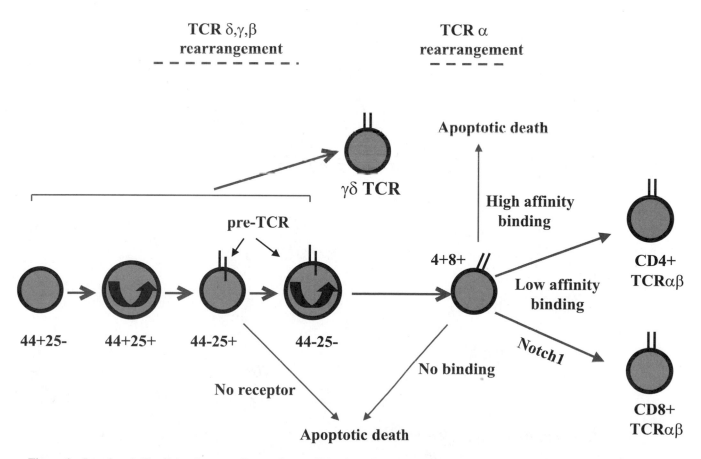

Figure 1   Intrathymic T cell development. Rescue from cell death by the pre-TCR and the αβTCR at checkpoints 1 and 2, respectively, and induction of cell death by high affinity binding of the αβTCR at checkpoint 2 only.

expression on the one hand (raf, ras, MAP-kinase) and TCRβ allelic exclusion through feed-back inhibition by the pre-TCR on further TCRβ rearrangement on the other hand (37–39). These pathways diverge below the SLP76 adapter protein that is required for both developmental progression as well as allelic exclusion (23) (Fig 2).

From these data, it would appear that the pre-TCR acts as a cell-autonomous developmental switch selecting cells with a single productive TCRβ allele for extensive proliferation and differentiation. The latter includes the induction of effective TCRα rearrangement, thus enabling a single TCRβ chain to pair with a great variety of different TCRα chains. A further consequence of pre-TCR signaling may be the diversion of TCRβ expressing cells into the αβ lineage away from the γδ lineage because, contrary to earlier reports, γδ T cells in normal, but not pTα$^{-/-}$, mice are selected against in frame TCRβ rearrangements (40). However, it remains to be

demonstrated that the pre-TCR signals differently from the γδ TCR, and that such signals directly determine commitment to the αβ lineage.

3. SURVIVAL AND DEATH CONTROLLED BY THE αβTCR ON IMMATURE T CELLS

Because of continuing rearrangement of the TCRα loci on both chromosomes, one CD4$^+$8$^+$ T cell can try different TCRα chains in combination with one TCRβ chain in order to generate a receptor that binds to intrathymic ligands (41, 42). Ligation of the αβTCR by different ligands may have drastically different consequences, known as positive or negative selection. In either case, ligation of the αβTCR on immature T cells results in the shut-off of further TCRα rearrangement by downregulation of RAG gene expression

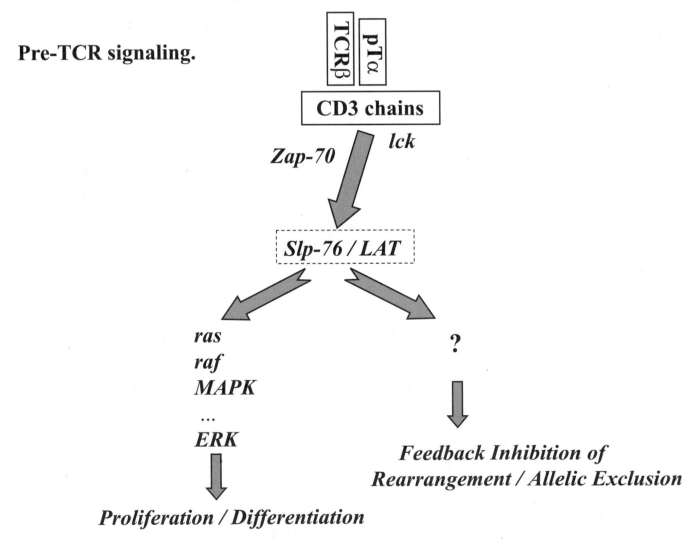

Figure 2   The TCR signal cascades that diverge below the SLP–76 adapter protein to induce expansion and differentiation on the one hand and feedback inhibition resulting in allelic exclusion on the other.

(41, 43, 44). If the $\alpha\beta$TCR does not bind to any ligand, rearrangement continues until the CD4$^+$8$^+$ cell dies after a short lifespan of three to four days ("death by neglect") (45, 46). Low affinity binding of the $\alpha\beta$TCR results in shut-down of rearrangement, downregulation of either CD4 or CD8 surface expression, upregulation of TCR surface levels, and eventually commitment to either the CD4 or CD8 lineage of mature T cells. These events are dependent on signaling pathways very similar to the pre-TCR signaling pathway, i.e. they involve the src kinases and the MAP kinases (47, 48). It has not been established whether CD4 or CD8 lineage commitment depends on different signals generated when the $\alpha\beta$TCR and/or the CD4/8 coreceptors bind to class II and class I MHC molecules (49), respectively, or whether commitment occurs independently of $\alpha\beta$TCR signals (50). It appears that Notch1 is involved in lineage decision, but it is not clear whether Notch expression is regulated by specific signals that depend on the TCR (51). It would appear that cells expressing Notch preferentially enter the CD8 lineage. Some experiments have indicated that low affinity binding resulting in positive selection requires certain specific peptides (52), while other experiments suggest that such positively-selecting peptides do not need to bear any functional or obvious structural relationship to agonist peptides that can induce effector function in immature T cells expressing the same receptor (53). Thus, the structural requirements of the TCR MHC-peptide configuration resulting in positive selection is, despite claims to the contrary, still enigmatic, but it may correlate with the affinity of a particular TCR for a certain MHC-peptide complex (54). As postulated by Burnet and Lederberg, TCR engagement by the MHC-agonist peptide complex on immature CD4$^+$8$^+$ thymocytes results in rapid induction of apoptotic cell-death (4, 55, 56). As anticipated, it is the quality of the ligation rather than the timing of the ligation that determines whether the outcome is positive or negative selection (57, 58). In fact, it appears that there is a relatively large window of time when positive or negative selection may take place: T cells express $\alpha\beta$ TCRs for the first time in the cortex such that selection will initially be guided primarily by MHC-peptide complexes expressed by cortical epithelial cells. It appears that proteins present in the circulation rarely reach the outer cortex (59), and that cortical epithelial cells, because of their poorly-developed lysosomal compartments, mostly present endogenous peptides (60, 61). This would result in positive and negative selection by these peptide-MHC complexes, but not necessarily by peptide-MHC complexes as expressed in hemopoietic cells, including dendritic cells located in the corticomedullary junction or in the medulla. Such a scenario would explain why T cells positively and negatively selected by MHC-peptide complexes expressed on cortical epithelial only would still be reactive to MHC-peptide complexes expressed on hemopoietic cells (62). Thus, positive and/or negative selection may continue in the deep cortex, the corti-

comedullary junction and the medulla (63). This agrees well with the observation that immature medullary thymocytes can still be negatively selected (64), and perhaps explains the notion that MHC molecules expressed on hemopoietic cells may mediate positive selection, albeit inefficiently (65, 66). Thus, the ligand for positive selection may not have to be necessarily expressed on epithelial cells as long as hemopoietic cells and epithelial cells are present in close proximity. In summary, it is likely to be the location of ligands that determines whether positive selection or negative selection of thymocytes with certain receptors occurs early or late in thymocyte maturation.

If negative selection was only dependent on peptides expressed by thymocytes themselves, that would leave a relatively large fraction of cells with receptors for a variety of tissue-specific antigens not expressed in the thymus. Two mechanisms are responsible for negative selection of such T cells within the thymus: One involves the uptake of proteins that enter the thymic medulla in sufficient in sufficient quantity by thymic dendritic cells. The other is represented by the curious finding that many proteins believed to be expressed only in extrathymic tissue are, in fact, also expressed by thymic epithelial cells. Such antigens include insulin and other proteins from exocrine or endocrine pancreatic tissue (67) as well as "liver-specific proteins" (68). In thymic transplantation experiments, it was actually shown that such antigens expressed in the thymus can induce tolerance in CD4 T cells (69). There is a rapidly growing list of such proteins, but it remains to be shown for many of them that they are tolerogenic and if so, by what mechanism. It is sufficient to state here that a recent report indicates that presentation of a certain type of self antigen by thymic epithelial cells is a decisive factor whether or not mice develop autoimmunity (70).

4. SURVIVAL AND CELL DEATH CONTROLLED BY THE $\alpha\beta$TCR ON MATURE T CELLS

The survival of mature T cells appears to depend on the continuous ligation of the $\alpha\beta$TCR by polymorphic MHC molecules. This notion is supported by several studies in MHC-deficient mice (71–73) as well as studies in TCR-transgenic mice in which it was actually shown that the appropriate MHC molecules required for the presentation of the agonist ligand have to be expressed in the absence of the agonist peptide in order to mediate long-term survival of naïve CD4 and CD8 T cells (13, 14). While it was clear that the agonist peptide was not required for this effect, the precise ligand still needs to be defined. Some recent data were interpreted to indicate that the ligand was the same as that responsible for positive selection in the thymus, in one particular case an MHC molecule plus an antagonist peptide. It remains to be explained, however, how an "antagonist" peptide can be responsible for survival of mature T cells without antagonizing them (74–76).

There is also some debate concerning the cell-cycle status in which truly naïve T cells survive for long time periods. Initial experiments have indicated that naïve T cells can survive for months in an intermitotic phase as resting cells even in the absence of other naïve T cells, i.e. in relatively low numbers in T cell-deficient mice (77). More recent studies claim, however, that naïve T cells assume a memory-like phenotype when transferred into T cell-deficient hosts, where they divide extensively (74, 76). These studies need to be interpreted with caution, since the expansion and generation of memory phenotype T cells by the availability of space is not exactly a palatable concept, especially when other data indicate that some naïve T cells remain resting even when surrounded by ample space (76, 77). A potential explanation for the induction of cell division in apparently naïve T cells in T cell-deficient hosts is that these cells, at least in part, are recognizing some endogenous or environmental antigens, perhaps in some specialized tissue, and that this is required for their expansion. In this regard, it is curious to note that among these so-called naïve T cells, there is often a small proportion of cells that express the CD44 activation marker even before transfer into immuno-

deficient hosts (78). An alternative explanation would be that there is a range of affinity of TCRs for self-MHC molecules, relatively low affinities resulting in survival without cell division, while somewhat higher affinities induce some division concomitantly with markers that are usually expressed on memory cells that have been induced with an agonist ligand.

The fact that pre-activated memory-type T cells can survive and expand in antigen-free and T cell-deficient hosts is well established, and underscores the view that memory is long-lived in the absence of antigen (13, 79, 80). This makes some sense, since such findings correspond to the definition of immunologic memory. Long-lived memory in the absence of antigen is an experimental fact irrespective of whether this type of memory affords protection from infection or not in certain experimental systems (79). The memory response is regularly more rapid (sometimes not rapid enough to cope with the spreading of infection) and more effective in terms of cytokine production and cytolytic activity (81–83). The issue of whether the survival of memory cells (and their division) requires some ligation of the $\alpha\beta$ TCR is controversial at present (13, 80, 84). The reasons for these discrepant results are not clear, and further studies in cleaner systems are required.

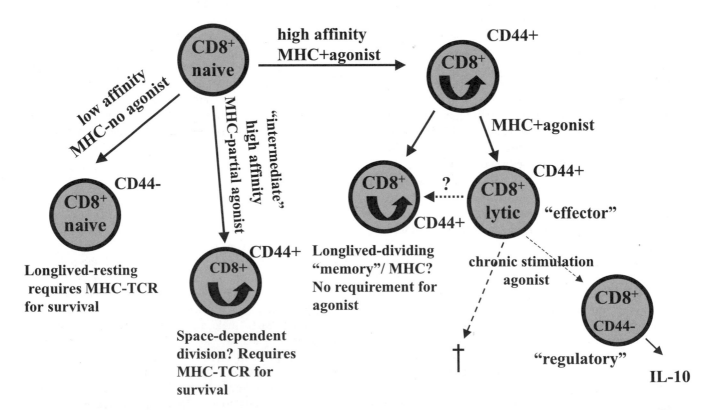

Figure 3 Various options of a naïve CD8⁺ T cell: low affinity binding results in survival as a resting cell, "intermediate" affinity binding in survival associated with some space-dependent division, expression of some activation markers (CD44), but no effector function and high affinity MHC-agonist peptide binding resulting in the generation of effector (cytolytic T cells) as well as CD44⁺ memory cells that continue to cycle in the absence of the agonist peptide. Chronic stimulation by the agonist peptide results in "activation-induced cell death" as well as cells that cease to proliferate but secrete higher levels of IL10 (regulatory T cells).

The fact that naïve cells may not divide while memory T cells do divide in the absence of deliberate antigenic stimulation raises the question of why memory T cells do not generally take over the space of peripheral T cells. Experiments by Rocha and colleagues (85, 86) suggest that this is not so, because the number of naïve and memory T cells is under independent homeostatic control, much like the numbers of granulocytes and monocytes present in peripheral blood are independently-regulated.

It is also clear that the $\alpha\beta$ TCR not only confers survival on mature T cells: Chronic engagement even to ligands on professional APC leads to activation-induced cell death that, in contrast to cell death induced in immature thymocytes, can be preceded by a brief phase of cell division and effector function (77, 87, 88). Moreover, long-term exposure to high doses of antigen can result in the generation of so-called "anergic" cells that do not mediate typical effector function, but rather produce cytokines that regulate the response of other T cells (77, 89) (Fig. 3). These issues are beyond the scope of this article and are the subject of other reviews included in this volume.

5. CONCLUDING REMARKS

Proteins of the T cell receptor for antigen have a decisive role in controlling T cell survival and differentiation at different checkpoints: The pre-TCR represents a cell-autonomous developmental switch, since pre-TCR expression without binding to any ligand confers survival, allelic exclusion, lineage commitment and differentiation, including effective TCR α rearrangement. It would appear that all those events are initiated by segregation of the pre-TCR into kinase-rich glypolipid-enriched membrane domains. The $\alpha\beta$ TCR is required for the generation of functional T cells from nonfunctional precursors and low affinity binding to intrathymic ligands may be one crucial event. High affinity binding to ligands in the cortex and medulla, including peptides that were previously believed to be expressed only extrathymically, purges the developing T cell repertoire of a large number of autoreactive T cells. Further survival of mature T cells requires ligation of the $\alpha\beta$ TCR to polymorphic MHC molecules in the absence of agonist peptides. Naïve T cells may survive for long time periods in the inter-mitotic phase, whereas memory T cells may continue to divide and survive in the absence of the ligand that induced their function.

References

1. Burnet, F.M. 1959. *The clonal selection theory*. London Cambridge University Press.
2. Lederberg, J. 1959. Genes and antibodies: do antigens bear instructions for antibody specificity or do they select cell lines that arise by mutation? *Science* 129:1649.
3. Jerne, N.K. 1971. The somatic generation of immune recognition. *Eur. J. Immunol.* 1:1.
4. von Boehmer, H. 1990. Developmental biology of T cells in T cell- receptor transgenic mice. *Annu. Rev. Immunol.* 8:531–556.
5. Zinkernagel, R.M., A. Althage, S. Cooper, G. Callahan, and J. Klein. 1978. In irradiation chimeras, K or D regions of the chimeric host, not of the donor lymphocytes, determine immune responsiveness of antiviral cytotoxic T cells. *J. Exp. Med.* 148:805–810.
6. Kisielow, P., and H. von Boehmer. 1995. Development and selection of T cells: facts and puzzles. *Adv. Immunol.* 58:87–209.
7. Godfrey, D.I., J. Kennedy, P. Mombaerts, S. Tonegawa, and A. Zlotnik. 1994. Onset of TCR-beta gene rearrangement and role of TCR-beta expression during CD3-CD4-CD8-thymocyte differentiation. *J. Immunol.* 152:4783–4792.
8. Godfrey, D.I., J. Kennedy, T. Suda, and A. Zlotnik. 1993. A developmental pathway involving four phenotypically and functionally distinct subsets of CD3-CD4-CD8- triple-negative adult mouse thymocytes defined by CD44 and CD25 expression. *J. Immunol.* 150:4244–4252.
9. Kishi, H., P. Borgulya, B. Scott, K. Karjalainen, A. Traunecker, J. Kaufman, and H. von Boehmer. 1991. Surface expression of the beta T cell receptor (TCR) chain in the absence of other TCR or CD3 proteins on immature T cells. *EMBO J.* 10:93–100.
10. Kisielow, P., H.S. Teh, H. Bluthmann, and H. von Boehmer. 1988. Positive selection of antigen-specific T cells in thymus by restricting MHC molecules. *Nature* 335:730–733.
11. Teh, H.S., P. Kisielow, B. Scott, H. Kishi, Y. Uematsu, H. Bluthmann, and H. von Boehmer. 1988. Thymic major histocompatibility complex antigens and the alpha beta T-cell receptor determine the CD4/CD8 phenotype of T cells. *Nature* 335:229–233.
12. Kaye, J., M.L. Hsu, M.E. Sauron, S.C. Jameson, N.R. Gascoigne, and S.M. Hedrick. 1989. Selective development of CD4+ T cells in transgenic mice expressing a class II MHC-restricted antigen receptor. *Nature* 341:746–749.
13. Tanchot, C., F.A. Lemonnier, B. Perarnau, A.A. Freitas, and B. Rocha. 1997. Differential requirements for survival and proliferation of CD8 naive or memory T cells. *Science* 276:2057–2062.
14. Kirberg, J., A. Berns, and H. von Boehmer. 1997. Peripheral T cell survival requires continual ligation of the T cell receptor to major histocompatibility complex-encoded molecules. *J. Exp. Med.* 186:1269–1275.
15. Groettrup, M., A. Baron, G. Griffiths, R. Palacios, and H. von Boehmer. 1992. T cell receptor (TCR) beta chain homodimers on the surface of immature but not mature alpha, gamma, delta chain deficient T cell lines. *EMBO J.* 11:2735–2745.
16. Groettrup, M., K. Ungewiss, O. Azogui, R. Palacios, M.J. Owen, A.C. Hayday, and H. von Boehmer. 1993. A novel disulfide-linked heterodimer on pre-T cells consists of the T cell receptor beta chain and a 33 kd glycoprotein. *Cell* 75:283–294.
17. Saint-Ruf, C., K. Ungewiss, M. Groettrup, L. Bruno, H.J. Fehling, and H. von Boehmer. 1994. Analysis and expression of a cloned pre-T cell receptor gene. *Science* 266:1208–1212.
18. Fehling, H.J., A. Krotkova, C. Saint-Ruf, and H. von Boehmer. 1995. Crucial role of the pre-T-cell receptor alpha

gene in development of alpha beta but not gamma delta T cells. *Nature* 375:795–798.

19. Mombaerts, P., A.R. Clarke, M.A. Rudnicki, J. Iacomini, S. Itohara, J.J. Lafaille, L. Wang, Y. Ichikawa, R. Jaenisch, M.L. Hooper, and et al. 1992. Mutations in T-cell antigen receptor genes alpha and beta block thymocyte development at different stages. *Nature* 360:225–231.

20. Malissen, M., A. Gillet, B. Rocha, J. Trucy, E. Vivier, C. Boyer, F. Kontgen, N. Brun, G. Mazza, E. Spanopoulou, and et al. 1993. T cell development in mice lacking the CD3-zeta/eta gene. *EMBO J.* 12:4347–4355.

21. Malissen, M., A. Gillet, L. Ardouin, G. Bouvier, J. Trucy, P. Ferrier, E. Vivier, and B. Malissen. 1995. Altered T cell development in mice with a targeted mutation of the CD3-epsilon gene. *EMBO J.* 14:4641–4653.

22. Buer, J., I. Aifantis, J.P. DiSanto, H.J. Fehling, and H. von Boehmer. 1997. Role of different T cell receptors in the development of pre-T cells. *J. Exp. Med.* 185:1541–1547.

23. Aifantis, I., V.I. Pivniouk, F. Gartner, J. Feinberg, W. Swat, F.W. Alt, H. von Boehmer, and R.S. Geha. 1999. Allelic exclusion of the T cell receptor beta locus requires the SH2 domain-containing leukocyte protein (SLP)- 76 adaptor protein. *J. Exp. Med.* 190:1093–1102.

24. Del Porto, P., L. Bruno, M.G. Mattei, H. von Boehmer, and C. Saint- Ruf. 1995. Cloning and comparative analysis of the human pre-T-cell receptor alpha- chain gene. *Proc. Natl. Acad. Sci. USA* 92:12105–12109.

25. Zhang, W., R.P. Trible, and L.E. Samelson. 1998. LAT palmitoylation: its essential role in membrane microdomain targeting and tyrosine phosphorylation during T cell activation. *Immunity* 9:239–246.

26. Saint-Ruf, C. Pamigada, M., Azogui, O., Debby, P., von Boehmer, A. and Grassi, F. 2000. Different initiation of pre-TCR and γδ (gamma/delta) TCR signaling *Nature* 406:524–527.

27. Irving, B.A., F.W. Alt, and N. Killeen. 1998. Thymocyte development in the absence of pre-T cell receptor extracellular immunoglobulin domains. *Science* 280:905–908.

28. von Boehmer, H., I. Aifantis, J. Feinberg, O. Lechner, C. SaintRuf, U. Walter, J. Buer, and O. Azogui. 1999. Pleiotropic changes controlled by the pre-T cell receptor. *Curr. Opin. Immunol.* 11:135–142.

29. Anderson, S.J., K.M. Abraham, T. Nakayama, A. Singer, and R.M. Perlmutter. 1992. Inhibition of T-cell receptor beta-chain gene rearrangement by overexpression of the non-receptor protein tyrosine kinase p56lck. *EMBO J.* 11:4877–4884.

30. Cheng, A.M., I. Negishi, S.J. Anderson, A.C. Chan, J. Bolen, D.Y. Loh, and T. Pawson. 1997. The Syk and ZAP–70 SH2-containing tyrosine kinases are implicated in pre-T cell receptor signaling. *Proc. Natl. Acad. Sci. USA* 94:9797–9801.

31. Zhang, W., C.L. Sommers, D.N. Burshtyn, C.C. Stebbins, J.B. DeJarnette, R.P. Trible, A. Grinberg, H.C. Tsay, H.M. Jacobs, C.M. Kessler, E.O. Long, P.E. Love, and L.E. Samelson. 1999. Essential role of LAT in T cell develop-ment [In Process Citation]. *Immunity* 10:323–332.

32. Clements, J.L., B. Yang, S.E. Ross-Barta, S.L. Eliason, R.F. Hrstka, R.A. Williamson, and G.A. Koretzky. 1998. Requirement for the leukocyte-specific adapter protein SLP–76 for normal T cell development. *Science* 281:416–419.

33. Pivniouk, V., E. Tsitsikov, P. Swinton, G. Rathbun, F.W. Alt, and R.S. Geha. 1998. Impaired viability and profound

34. Chu, D.H., C.T. Morita, and A. Weiss. 1998. The Syk family of protein tyrosine kinases in T-cell activation and development. *Immunol. Rev.* 165:167–180.

35. Swat, W., Y. Shinkai, H.L. Cheng, L. Davidson, and F.W. Alt. 1996. Activated Ras signals differentiation and expansion of CD4+8+ thymocytes. *Proc. Natl. Acad. Sci. USA* 93:4683–4687.

36. Crompton, T., K.C. Gilmour, and M.J. Owen. 1996. The MAP kinase pathway controls differentiation from double-negative to double-positive thymocyte. *Cell* 86:243–251.

37. Iritani, B.M., J. Alberola-Ila, K.A. Forbush, and R.M. Perimutter. 1999. Distinct signals mediate maturation and allelic exclusion in lymphocyte progenitors. *Immunity* 10:713–722.

38. Gartner, F., F.W. Alt, R. Monroe, M. Chu, B.P. Sleckman, L. Davidson, and W. Swat. 1999. Immature thymocytes employ distinct signaling pathways for allelic exclusion versus differentiation and expansion. *Immunity* 10:537–546.

39. Henning, S.W., and D.A. Cantrell. 1998. GTPases in antigen receptor signalling. *Curr. Opin. Immunol.* 10:322–329.

40. Aifantis, I., O. Azogui, J. Feinberg, C. Saint-Ruf, J. Buer, and H. von Boehmer. 1998. On the role of the pre-T cell receptor in alphabeta versus gammadelta T lineage commitment. *Immunity* 9:649–655.

41. Borgulya, P., H. Kishi, Y. Uematsu, and H. von Boehmer. 1992. Exclusion and inclusion of alpha and beta T cell receptor alleles. *Cell* 69:529–537.

42. Casanova, J.L., P. Romero, C. Widmann, P. Kourilsky, and J.L. Maryanski. 1991. T cell receptor genes in a series of class I major histocompatibility complex-restricted cytotoxic T lymphocyte clones specific for a Plasmodium berghei nonapeptide: implications for T cell allelic exclusion and antigen-specific repertoire. *J. Exp. Med.* 174:1371–1383.

43. Kouskoff, V., J.L. Vonesch, C. Benoist, and D. Mathis. 1995. The influence of positive selection on RAG expression in thymocytes. *Eur. J. Immunol.* 25:54–58.

44. Turka, L.A., D.G. Schatz, M.A. Oettinger, J.J. Chun, C. Gorka, K. Lee, W.T. McCormack, and C.B. Thompson. 1991. Thymocyte expression of RAG-1 and RAG-2: termination by T cell receptor cross-linking. *Science* 253:778–781.

45. Scott, B., H. Bluthmann, H.S. Teh, and H. von Boehmer. 1989. The generation of mature T cells requires interaction of the alpha beta T-cell receptor with major histocompatibility antigens. *Nature* 338:591–593.

46. Huesmann, M., B. Scott, P. Kisielow, and H. von Boehmer. 1991. Kinetics and efficacy of positive selection in the thymus of normal and T cell receptor transgenic mice. *Cell* 66:533–540.

47. Peterson, E.J., J.L. Clements, N. Fang, and G.A. Koretzky. 1998. Adaptor proteins in lymphocyte antigen-receptor signaling. *Curr. Opin. Immunol.* 10:337–344.

48. van Leeuwen, J.E., and L.E. Samelson. 1999. T cell antigen-receptor signal transduction. *Curr. Opin. Immunol.* 11:242–248.

49. von Boehmer, H. 1986. Selection of the alpha, beta heterodimeric T cell receptor for antigen. *Immunol. Today* 7:333.

50. Robey, E.A., B.J. Fowlkes, J.W. Gordon, D. Kioussis, H. von Boehmer, F. Ramsdell, and R. Axel. 1991. Thymic selection in CD8 transgenic mice supports an instructive

block in thymocyte development in mice lacking the adaptor protein SLP-76. *Cell* 94:229–238.

model for commitment to a CD4 or CD8 lineage. *Cell* 64:99–107.

51. Robey, E., D. Chang, A. Itano, D. Cado, H. Alexander, D. Lans, G. Weinmaster, and P. Salmon. 1996. An activated form of Notch influences the choice between CD4 and CD8 T cell lineages. *Cell* 87:483–492.

52. Hogquist, K.A., S.C. Jameson, W.R. Heath, J.L. Howard, M.J. Bevan, and F.R. Carbone. 1994. T cell receptor antagonist peptides induce positive selection. *Cell* 76:17–27.

53. Pawlowski, T.J., M.D. Singleton, D.Y. Loh, R. Berg, and U.D. Staerz. 1996. Permissive recognition during positive selection. *Eur. J. Immunol.* 26:851–857.

54. Matsui, K., J.J. Boniface, P. Steffner, P.A. Reay, and M.M. Davis. 1994. Kinetics of T-cell receptor binding to peptide/I-Ek complexes: correlation of the dissociation rate with T-cell responsiveness. *Proc. Natl. Acad. Sci. USA* 91:12862–12866.

55. Swat, W., L. Ignatowicz, H. von Boehmer, and P. Kisielow. 1991. Clonal deletion of immature CD4 + 8 + thymocytes in suspension culture by extrathymic antigen-presenting cells. *Nature* 351:150–153.

56. Martin, S., and M.J. Bevan. 1997. Antigen-specific and non-specific deletion of immature cortical thymocytes caused by antigen injection. *Eur. J. Immunol.* 27:2726–2736.

57. von Boehmer, H., P. Kisielow, H. Kishi, B. Scott, P. Borgulya, and H.S. Teh. 1989. The expression of CD4 and CD8 accessory molecules on mature T cells is not random but correlates with the specificity of the alpha beta receptor for antigen. *Immunol. Rev.* 109:143–151.

58. Savage, P.A., J.J. Boniface, and M.M. Davis. 1999. A kinetic basis for T cell receptor repertoire selection during an immune response. *Immunity* 10:485–492.

59. Nieuwenhuis, P., R.J. Stet, J.P. Wagenaar, A.S. Wubbena, J. Kampinga, and A. Karrenbeld. 1988. The transcapsular route: a new way for (self-) antigens to by-pass the blood-thymus barrier? *Immunol. Today* 9:372–375.

60. Kasai, M., E. Kominami, and T. Mizuochi. 1998. The antigen presentation pathway in medullary thymic epithelial cells, but not that in cortical thymic epithelial cells, conforms to the endocytic pathway. *Eur. J. Immunol.* 28:1867–1876.

61. Oukka, M., P. Andre, P. Turmel, N. Besnard, V. Angevin, L. Karlsson, P.L. Trans, D. Charron, B. Bihain, K. Kosmatopoulos, and V. Lotteau. 1997. Selectivity of the major histocompatibility complex class II presentation pathway of cortical thymic epithelial cell lines. *Eur. J. Immunol.* 27:855–859.

62. Laufer, T.M., J. DeKoning, J.S. Markowitz, D. Lo, and L.H. Glimcher. 1996. Unopposed positive selection and autoreactivity in mice expressing class II MHC only on thymic cortex. *Nature* 383:81–85.

63. Swat, W., H. von Boehmer, and P. Kisielow. 1994. Central tolerance: clonal deletion or clonal arrest? *Eur. J. Immunol.* 24:485–487.

64. Swat, W., Dessing, M., Bavon, A., Kisielow, P. and von Boehmer, H. 1992 Phenotypic changes accompanying positive selection of CD4+8+ thymocytes *Eur. J. Immunol.* 22:2367–2372.

65. Zinkernagel, R.M., and A. Althage. 1999. On the role of thymic epithelium vs. bone marrow-derived cells in repertoire selection of T cells. *Proc. Natl. Acad. Sci. USA* 96:8092–8097.

66. Bix, M., and D. Raulet. 1992. Inefficient positive selection of T cells directed by haematopoietic cells. *Nature* 359:330.

67. Jolicoeur, C., D. Hanahan, and K.M. Smith. 1994. T-cell tolerance toward a transgenic beta-cell antigen and transcription of endogenous pancreatic genes in thymus. *Proc. Natl. Acad. Sci. USA* 91:6707–6711.

68. Klein, L., T. Klein, U. Ruther, and B. Kyewski. 1998. CD4 T cell tolerance to human C-reactive protein, an inducible serum protein, is mediated by medullary thymic epithelium. *J. Exp. Med.* 188:5–16.

69. Klein, L., M. Klugmann, K.A. Nave, and B. Kyewski. 2000. Shaping of the autoreactive T-cell repertoire by a splice variant of self protein expressed in thymic epithelial cells. *Nat. Med.* 6:56–61

70. Klein, B., B. Kyewski. 2000. Self-antigen presentation by thymic stromal cells: a subtle division of labor. *Curr. Opin. Immun.* 12. 179–186.

71. Takeda, S., H.R. Rodewald, H. Arakawa, H. Bluethmann, and T. Shimizu. 1996. MHC class II molecules are not required for survival of newly generated CD4 + T cells, but affect their long-term life span. *Immunity* 5:217–228.

72. Brocker, T. 1997. Survival of mature CD4 T lymphocytes is dependent on major histocompatibility complex class II-expressing dendritic cells. *J. Exp. Med.* 186:1223–1232.

73. Rooke, R., C. Waltzinger, C. Benoist, and D. Mathis. 1997. Targeted complementation of MHC class II deficiency by intrathymic delivery of recombinant adenoviruses. *Immunity* 7:123–134.

74. Goldrath, A.W., and M.J. Bevan. 1999. Low-affinity ligands for the TCR drive proliferation of mature CD8 + T cells in lymphopenic hosts. *Immunity* 11:183–190.

75. Viret, C., F.S. Wong, and C.A. Jr. Janeway. 1999. Designing and maintaining the mature TCR repertoire: the continuum of self-peptide:self-MHC complex recognition. *Immunity* 10:559–568.

76. Ernst, B., D.S. Lee, J.M. Chang, J. Sprent, and C.D. Surh. 1999. The peptide ligands mediating positive selection in the thymus control T cell survival and homeostatic proliferation in the periphery. *Immunity* 11:173–181.

77. Rocha, B., and H. von Boehmer. 1991. Peripheral selection of the T cell repertoire. *Science* 251:1225–1228.

78. Oehen, S., and K. Brduscha-Riem. 1999. Naive cytotoxic T lymphocytes spontaneously acquire effector function in lymphocytopenic recipients: A pitfall for T cell memory studies? *Eur. J. Immunol.* 29:608–614.

79. Bruno, L., J. Kirberg, and H. von Boehmer. 1995. On the cellular basis of immunological T cell memory. *Immunity* 2:37–43.

80. Murali-Krishna, K., L.L. Lau, S. Sambhara, F. Lemonnier, J. Altman, and R. Ahmed. 1999. Persistence of memory CD8 T cells in MHC class I-deficient mice. *Science* 286:1377–1381.

81. Bachmann, M.F., M. Barner, A. Viola, and M. Kopf. 1999. Distinct kinetics of cytokine production and cytolysis in effector and memory T cells after viral infection. *Eur. J. Immunol.* 29:291–299.

82. Kedl, R.M., and M.F. Mescher. 1998. Qualitative differences between naive and memory T cells make a major contribution to the more rapid and efficient memory CD8 + T cell response. *J. Immunol.* 161:674–683.

83. Zimmermann, C., A. Prevost-Blondel, C. Blaser, and H. Pircher. 1999. Kinetics of the response of naive and memory CD8 T cells to antigen: similarities and differences. *Eur. J. Immunol.* 29:284–290.

84. Swain, S.L., H. Hu, and G. Huston. 1999. Class II-independent generation of CD4 memory T cells from effectors. *Science* 286:1381–1383.

85. Tanchot, C., and B. Rocha. 1995. The peripheral T cell repertoire: independent homeostatic regulation of virgin and activated CD8 + T cell pools. *Eur. J. Immunol.* 25:2127–2136.

86. Tanchot, C., and B. Rocha. 1998. The organization of mature T-cell pools. *Immunol. Today* 19:575–579.

87. Moskophidis, D., F. Lechner, H. Pircher, and R.M. Zinkernagel. 1993. Virus persistence in acutely infected immunocompetent mice by exhaustion of antiviral cytotoxic effector T cells. *Nature* 362:758–761.

88. Gallimore, A., A. Glithero, A. Godkin, A.C. Tissot, A. Pluckthun, T. Elliott, H. Hengartner, and R. Zinkernagel. 1998. Induction and exhaustion of lymphocytic chori- omeningitis virus-specific cytotoxic T lymphocytes visual- ized using soluble tetrameric major histocompatibility complex class I-peptide complexes. *J. Exp. Med.* 187:1383–1393.

89. Buer, J., A. Lanoue, A. Franzke, C. Garcia, H. von Boehmer, and A. Sarukhan. 1998. Interleukin 10 secretion and impaired effector function of major histocompatibility complex class II-restricted T cells anergized in vivo. *J. Exp. Med.* 187:177–183.

6 | T Cell Antigen Receptors and Autoimmune Repertoires

Roberto Baccala, Rosana Gonzalez-Quintial, and Argyrios N. Theofilopoulos

1. INTRODUCTION

The immune system of higher vertebrates is composed of a large complement of cellular and molecular tools that, while respecting self-constituents, detects and eliminates invading microorganisms, and sets up an immunological memory that decreases severity of recurrent infections. A prominent role in orchestrating such protective responses is played by specific regulatory or *effector* T cells, which, upon recognition of antigenic determinants at the surface of target cells, may promote B cell differentiation and antibody production, or directly destroy virally-infected, stressed and malignant cells. Paradoxically, however, T cell activity may, under certain circumstances, lead to very deleterious pathological consequences associated, for example, with inflammation, allergy or autoimmunity. In particular, as indicated by cell transfer experiments, antibody treatments and genetic manipulations in appropriate animal models, T cells acting as helper cells for autoantibody production or as cytotoxic cells appear to be an absolute requirement for the expression of both organ-specific and systemic autoimmune diseases. For these reasons, a growing interest in autoimmune pathology focuses on the characterization of specific T cell clonotypes that accumulate in association with disease progression, particularly in regard to the molecular features of their antigen T cell receptors (TCR), the identity of the inciting antigenic epitopes and the development of inhibitory therapeutic approaches. This chapter will review recent reports on the TCR as they relate to structure, gene diversity, antigen recognition and repertoire selection, and will summarize studies attempting to correlate autoimmune disease with TCR genomic composition, polymorphism and expression.

2. GENERAL STRUCTURE OF T CELL RECEPTORS

TCR are polymorphic glycoproteins constituted by two chains non covalently associated to the signal transducing CD3 complex at the T cell surface. For most mature T cells in blood and peripheral lymphoid organs of primates and rodents, TCR are composed of α and β chains. Expressed by CD8 positive cytotoxic T cells, by CD4 positive helper T cells, or by natural killer T cells, α/β TCR mainly interact with peptide antigens presented by class I or class II major histocompatibility complex (MHC) molecules (1), with superantigens (SAg) associated to class II MHC molecules (2), or with lipids and glycolipids bound to CD1 molecules (3). A second type of TCR is composed of γ and δ chains. This TCR is used by a minority (1–5%) of T cells in blood and peripheral lymphoid organs of primates and rodents, although it predominates in epithelia and mucosa of various tissues, including gut and reproductive tracts (4,5). Ligands for γ/δ TCR are less well characterized than those of α/β TCR, but it has been proposed this TCR might recognize antigen in a antibody-like manner, without need of processing (6–8). Mixed α/γ or β/δ TCR do not arise normally, but may occur in some tumor cells (9,10).

Initial biochemical studies on the TCR were performed using antibodies able to discriminate specific determinants on isolated T cell clones. Immunoprecipitation experiments identified a disulfide-bonded heterodimer of 85–90 kDa, while two-dimensional electrophoresis designated a more acidic α chain of 49–52 kDa and a more basic β chain of 41–43 kDa (11). Similar studies allowed the characterization of γ and δ polypeptides (12–17).

The complete primary structure of the TCR β chain was the first to be identified, deduced from the nucleotide sequence of human cDNA clones from a leukaemic T cell line that differentially hybridized to T cell RNA, but not to B cell RNA (18), and mouse cDNA clones that hybridized T cell-derived cDNA probes from which B cell-specific sequences had been selectively subtracted (19). Analogous approaches were used to identify cDNA clones for the TCR α chain (20–22), as well as for γ (23) and δ chains (24).

From these and other sequencing studies, the primary structure of TCR chains appeared similar to that of antibodies, with an amino-terminal leader peptide, followed by a variable (V) region and a carboxy-terminal constant (C) region. The leader peptide is a short hydrophobic sequence, composed of approximately 15 amino acids, that targets nascent polypeptides to the endoplasmic reticulum and is lost in mature TCR chains. The V region of TCR β and δ chains results from the rearrangement of variable (V), diversity (D) and joining (J) genetic elements, while that of TCR α and γ chains from the rearrangement of V and J gene segments. TCR V regions consist of 100 to 120 amino acids and contain N-linked glycosilation sites and two cysteine residues, involved in an intrachain disulfide loop spanning approximately 65 amino acids, as generally observed in immunoglobulins (Ig) (25). Like antibodies, three hypervariable regions were identified through the comparison of multiple TCR V sequences (26,27). Predicted to contribute most of the contacts with ligands, they are referred to as complementarity determining regions (CDR), and are separated by alternating framework regions (FR). A fourth hypervariable region (HV4) can be identified within the FR3, particularly in the Vβ segment, and is implicated in the interaction with superantigens (2). As is the case for Ig, CDR1 and CDR2 (but also HV4) are encoded within the germline V gene segment. In contrast, CDR3 are formed by the fusion of α and γ V-J segments, or β and δ V-D-J segments, which explains their higher level of variability. The C regions of TCR are composed of four domains, the largest of which is an Ig-like constant domain of 90 to 100 amino acids that includes the two characteristic cysteines (25). In α and δ chains, however, only 50–51 amino acids separate these two cysteines, and it was therefore proposed that both Cα and Cδ could display altered Ig folds (28). The C region continues with an hinge or connecting domain, generally presenting an extra cystein involved in interchain sulphydral bonds. In the case of the γ-chain, however, two types of polypeptides have been described, one of which, Cγ2 (GC2) lacks this extra cystein (29). The C region is then completed by a mainly hydrophobic membrane spanning region of 20 to 24 residues, and a short cytoplasmic region.

Significant advances towards the understanding of TCR structure and antigen recognition have been made in the last few years through crystallographic three-dimensional analysis of TCR fragments (30–33), intact α/β TCR (34–36), and their complexes with peptide-MHC (34,37–41) or SAg-MHC

(32,42,43). As predicted (28,44,45), the tertiary and quaternary structures of TCR are similar to those of Ig Fab fragments. The overall dimensions of TCR are 61 by 56 by 33 Å, compared to 65 by 46 by 30 Å for a representative Fab (34). The Vα and Vβ domains are related to v-type nine-stranded Ig folds typical of antibody V domains, whereas Cβ more distantly resembles a c-type seven-stranded Ig fold. In contrast, and as anticipated, the Cα domain deviates from canonical structures, particularly due to a deletion of 12 to 15 amino acids that precludes the formation of the typical top (g, f and c strands) β-sheet, normally connected by a cystine bond to the bottom (a, b, c and d strands) β-sheet of c-type Ig folds (25). The antigen-combining site of the TCR structures reported so far are relatively flat, similar to anti-protein antibodies. CDR1 and CDR2 are similarly positioned in the various TCR, while considerable differences were observed in the conformations of the centrally-disposed CDR3, a reflection of their varied lengths and sequences. Most TCRs have a cleft between the two CDR3, presumably to accommodate a central upfacing peptide side chain from the peptide-MHC complex.

3.   T CELL RECEPTOR GENES AND LOCI

3.1.   Overall Characteristics

The chromosomal organization of genes encoding TCR α, β, γ and δ chains is, in most instances, similar in mice and humans. The various loci span regions varying from 160 kb to 1,000 kb, each composed of a number of V gene segments followed, in the case of β and δ chains, by D gene segments, and then by J and C gene segments. Only the mouse TCR γ locus differs from other loci, since it is composed of four V-J-C clusters, one of which is in inverted transcriptional orientation compared to the other elements of the locus (46). In the case of the human and mouse TCR β loci, the D-J-C cluster is duplicated, and the same occurs for the J-C cluster of the human TCR γ locus. One Vβ (BV) gene segment, i.e. mouse BV14s1 and human BV20s1, and one Vδ (DV) gene segment, i.e. mouse DV105s1 (Vδ5) and human DV103s1, were found to be placed 3' to the last corresponding C gene segment in an inverted transcriptional orientation. For both mice and humans, the TCR α and δ loci are located on the same region of chromosome 14, with Vα (AV) and DV gene segments interdispersed or even coinciding, and with the δ D-J-C cluster located between V gene segments and the α J-C cluster. It follows that a rearrangement of α V-J gene segments deletes the δ D-J-C cluster, thereby precluding expression of a δ chain from the same locus.

Each germline V gene segment is composed of two exons, separated by an intron 95 to 305 bp long (47). The first exon includes 5' untranslated sequences and most of the leader peptide, while the second exon encodes the last three amino acids of the leader sequence along with the entire *V* gene region. *D* gene segments are encoded by a unique exon that can be used in all three reading frames, whereas *J* gene segments correspond to single exons with unique reading frame. *Cα* (*AC*), *Cβ* (*BC*) and *Cδ* (*DC*) gene segments consist of four exons each. The first two exons encode most of the extracellular C domain, the third encodes a major portion of the transmembrane segment, and the last exon includes the cytoplasmic coding sequence as well as 3′ untranslated nucleotides. In contrast, *Cγ* (*GC*) gene segments exist in different forms constituted by three, four or five exons (46,48).

Functional *V* and *D* gene segments display, at their 3′ ends, conserved heptamer and nonamer sequences, separated by a 23 bp spacer. Almost complementary to these sequences, heptamers and nonamers separated by a 12 bp spacer are found at the 5′ boundary of *D* and *J* gene segments. These are characteristic recombination signal sequences, similar to those identified adjacent to Ig *V*, *D* and *J* gene segments (49,50), and recognized by enzyme complexes of the recombinase activating gene (RAG) family involved in *V-D*, *V-J*, *D-D* and *D-J* somatic gene rearrangements. In addition, *J* gene segments have at their 3′ ends an RNA splice site.

Based on their nucleotide sequences, *V* gene segments displaying more than 75% homology are grouped into subfamilies, a classification criteria related to the sensitivity of the Southern blot (originally used in genomic studies) in differentiating nucleotide sequences. In humans, the number of TCR *V* gene subfamilies varies from 6 in the *TCR γ* locus to 41 in the *TCR α* locus, while the largest subfamilies were found in the *TCR β* locus (*BV6* and *BV13* with 9 members each). The complexity of multimember subfamilies has created some difficulty in the assignment of certain *V* gene segments to one or another subfamily. For example, in humans, the single member of the *BV1* subfamily and two of the eight members of the *BV5* subfamily (*5s3* and *5ts2*) share more then 75% homology, as is observed for one of the three *BV12* members (*12s3*) and one of the nine *BV13* genes (*13s4*) (51). On the other hand, within the *BV13* subfamily, *13s5* shares lesser than 75% homology with *13s7* and *13s8P*.

Not all *V* gene segments identified are functional. Certain genes can display an open reading frame (ORF) but cannot produce a functional TCR chain, either because of altered recombination or splice sequences that preclude their usage in rearrangement or splicing reactions, or because they lack amino acid residues that are essential for appropriate protein folding. Other *V* gene segments correspond to pseudogenes, either due to nucleotide changes creating in-frame stop codons, or to frameshift nucleotide insertions or deletions that alter the normal downstream reading frame. Non functional ORF, or pseudo, *V* genes can occur within both multimember or single member subfamilies.

Two types of polymorphisms have been found for TCR *V* genes. The first corresponds to the insertion or deletion of chromosomal sections that can alter the germline potential in affected individuals. In humans, this type of polymorphism has been found in the *TCR β* locus (one section including two functional genes and a pseudogene) and in the *TCR γ* locus (one section including a pseudogene, and another section including two functional genes). The second type of polymorphism corresponds to allelic variations in the TCR *V* gene sequences found among different individuals. Such sequence variations can occur at the nucleotide level without consequences for the expressed polypeptides, or may introduce relevant amino acid changes, for example in the CDRs, that could significantly alter TCR antigen specificity. Alternatively, sequence polymorphisms can create non functional ORF or pseudo *V* genes.

3.2. Nomenclature of T Cell Receptor Gene Segments

The large number of *V* gene segments, the complexity of certain subfamilies and the existence of polymorphic allelic variants has created considerable confusion in the literaure, as different groups have often given the same gene different names, or different genes the same name. To clarify this issue and unify the nomenclature, an international group of experts has been constituted (52), and the resulting alignments and official designation have been published for both human (51) and mouse (53) TCR *V* gene segments. According to this nomenclature, TCR genes are named with letters and numbers disposed in the following order: a *letter* indicating the TCR chain (A for *α*, B for *β*, G for *γ* and D for *δ*); a *letter* indicating the type of gene segment (*V* for variable, *D* for diversity, *J* for joining, *C* for constant); a *number* indicating the subfamily (*1* for subfamily 1, etc.); a *number* preceded by the letter *s* indicating the subfamily member (*s1* for segment 1, etc.); when known, the allelic variant *number* preceded by the letter *A* or* (*A1* or *01*, for allele 1, etc.); a last *letter* giving additional information on the gene (*P* for pseudogene, *O* for orphon gene, *T* for tentative if the sequence must be confirmed, etc.). For example, the third member of the subfamily 8 of *β chain* V genes, previously named *Vβ8.3*, is named *BV8s3* based on this nomenclature.

More recently, an alternative nomenclature has been proposed (54) based on the complete sequence analysis of the human *TCR β* locus, which considers the actual position of each gene in the locus. Such numbering procedure assigns the *V* subfamilies consecutive numbers, starting at the 5′ end of the locus. The individual subfamily members are then numbered sequentially after the subfamily designation. For example, since *BV8s3* is the 5th member of the 12th subfamily in the human *TCR β* locus (counting 5′ to 3′), this gene segments is named *BV12–5* based on this nomenclature.

In the following figures and tables, both nomenclatures are given (when known) to facilitate comparisons, while in the text, we will refer primarily to the nomenclature reported for human (51) and mouse (53) V gene segments. Additional information regarding TCR gene nomenclature, sequence alignments and allele designations, can be obtained through the internet at the "IMGT, the international ImMunoGeneTics database" (http://imgt.cnusc.fr: 8104), a site initiated and coordinated by Lefranc et al. (55).

3.3. Human T Cell Receptor α and δ Chain Genes

The locus for human TCR α and δ genes has been mapped on chromosome 14 at 14q11–14q12 (56). A complete sequence analysis of this locus has recently been accomplished by Boysen et al. and the corresponding data

reported in Genbank (http://www2.ncbi.nlm.nih.gov/genbank/query_form.html) under the accession numbers AE000658, AE000660 and AE000661. These data complete previous genomic and cDNA sequencing reports on α and/or δ V (51), D (57,58), J (58–61) and C (57,62), gene segments. A graphic representation of the locus is displayed in *Figure 1*. The locus spans a chromosom fragment of approximately 1,000 kb. It includes an initial 5′ region of 820 kb that contains 56 V gene segments, followed by a region of 50 kb that includes the δ D-J-C cluster (3 DD, 4 DJ, 1 DC) as well as an additional V gene segment, DV103s1 (DV3), placed in an inverted transcriptional orientation. The locus in completed at the 3′ end by a region of 71 kb that includes 61 AJ gene segments, and by the AC gene segment. No insertion/deletion polymorphisms have been described for this locus.

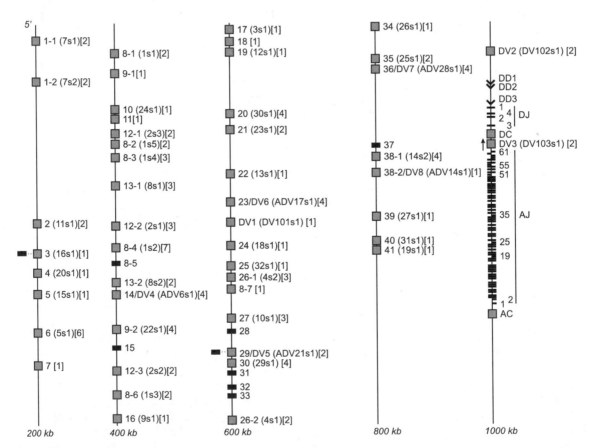

Figure 1.   Locus for human α and δ TCR genes on chromosome 14 (14q11–14q12). Approximate position of functional V gene segments (shadowed boxes) and pseudogenes (filled boxes) according to reported sequencing data (GeneBank, http://www2.ncbi.nlm.nih.gov/genbank/query_form.html, accession numbers AE000658, AE000660 and AE000661). The locus representation was adapted from "IMGT, the international ImMunoGeneTics database" (http://imgt.cnusc.fr:8104). Each V gene segment is named based on two different nomenclatures, the first according to the gene position in the locus, and the second (in parenthesis) corresponding to published alignments (51). The number of functional alleles for each V gene segment is given in brackets. V gene segments used only for δ chains, or for both α and δ chains, are indicated as DV or ADV, respectively. For AJ gene segments, lines ponting to the right indicate pseudogenes and lines pointing to the left indicate possibly non-functional genes with open reading frame (ORF). The arrow close to DV3 (DV103s1) indicate the opposite transcriptional orientation of this gene.

Of the 57 *V* gene segments included in this locus, 49 are *AV*, 3 are *DV*, while 5 are *ADV*, *i.e.*, gene segments that were found to be used for the expression of both α and β TCR chains (63,64).

For the α chain, the 54 V gene segments identified (49 *AV* plus 5 *ADV*) can be grouped into 41 subfamilies, of which 34 are composed of a single member, while 7 subfamilies include 2 to 7 members each. Of these 54 *AV/ADV* gene segments, 45 appear generally functional, 7 are pseudogenes, while the remaining 2 are, due to allelic polymorphisms, either functional or pseudogenes. Consequently, the functional repertoire for the human TCR α chain ranges from 45 *V* gene segments (from 33 subfamilies) to 47 *V* gene segments (from 35 subfamilies).

The potential repertoire of TCR α V gene segments, however, is considerably larger, due to the existence of allelic variants. In fact, several methods and criteria have been developed to distinguish products of different loci from alleles at the same locus. Southern blot analysis of genomic DNA revealed no deletion, limited RFLP, but extensive heterogeneity in regard to the combinations of haplotypic markers, suggesting that frequent recombination events may occur in this locus (65–67). Comparison of *AV* genes from different individuals by sequence analysis (68), single strand conformation polymorphism (SSCP) (69,70), denaturing gradient gel electrophoresis (DGGE) (71), or restriction enzyme analysis (72), identified allelic variants for several *AV* genes that typically differ by one or two, and generally not more than five, nucleotides. Furthermore, comparison of GenEMBL database entries identified a number of *AV* genes differing in few nucleotides that were tentatively classified as alleles (51). The availability of the sequence of the entire locus obviously facilitates such assignments. By combining these data, Lefranc et al. (55) composed a list of alleles that can be consulted at the "IMGT, the international ImMunoGeneTics data base" (http://imgt.cnusc.fr:8104). As summarized in Figure 1 (see numbers in brackets), most *AV1* gene segments are represented in the human population by different allelic variants. For example, up to 7 functional alleles have been reported for *AV1s2* (*AV8–4*), while up to 100 functional variants have been described so far for human *AV* genes.

The 61 *AJ* gene segments include at least 3 pseudogenes (*AJ51, AJ55, AJ60*) displaying stop codons or frame-shift insertions, while several others might represent non functional genes (nevertheless displaying an ORF), either due to altered recombination or splice signals, lack of relevant amino acids, or combinations of these defects (59).

For the δ chain, 8 *V* gene segments (3 *DV* plus 5 *ADV*) have been found in the locus. These are distributed into 8 different subfamilies, 7 of which contain a single member. All are functional and, with the exception of *DV101s1* (*DV1*) and *ADV14s1* (*DV8*), most are represented by 2 to 4

functional alleles. A total of 20 *DV/ADV* allelic variants have been reported to date.

3.4. Mouse T Cell Receptor α *and* δ Chain Genes

Southern blot analysis, deletional mapping, cosmid mapping and analysis of large restriction fragments by field-inversion electrophoresis, indicated that the mouse *TCR* α/δ locus spans a region of 1,000 kb on chromosome 14 (73–76). The locus organization is similar to that in humans, with 75 to 100 *V* gene segments followed by the δ *D-J-C* cluster (with 2 *DD*, 2 *DJ* and 1 *DC*), *DV5* in inverted orientation, and finally by the α J-C cluster (with 50 *AJ* and 1 *AC*). No insertions or deletions have been described for this locus.

Initial estimates suggested that the mouse *TCR* α/δ locus included approximately 100 *V* gene segments (75). Accordingly, alignment of *V* gene nucleotide sequences reported in the GenEMBL database and/or in the literature identified 100 unique mouse *AV* and/or *DV* sequences (53). As illustrated in Table 1, of these 100 gene segments, 84 are *AV*, 11 are *DV* while 4 are *ADV*, *i.e.* gene segments that can be used in both α and δ chain transcripts (77,78).

Based on 75% homology at the nucleotide level, the 101 *AV/DV* gene segments can be grouped into 24 subfamilies, of which 2 (*V11* and *V17*) include *AV* and *ADV* gene segments, 4 (*V2, V4, V6* and *V10*) include *AV* and *DV* gene segments, and 1 (*V7*) include *DV* and *ADV* gene segments. Interestingly, and in contrast to humans, mouse *AV/DV* subfamilies appear mostly composed by multiple members, subfamily *AV8* being the largest with 15 members. On the basis of this analysis, several *V* gene segments were reclassified in a different subfamily and, accordingly, renamed (53). These include *Vδ8* (now *DV2s8*), *Vδ3* (now *DV6s2*), members of the subfamily *Vδ6* (now *DV7*), *Vδ7* (now *DV10s7*), *Vδ9* (now *ADV17s2*), *Vδ1* (now *DV101s1*), *Vδ2* (now *DV102s1*), *Vδ4* (now *DV104s1*) and *Vδ5* (now *DV105s1*).

Only 2 gene segments (*AV16s1* and *AV16s2*) have been described as pseudogenes, suggesting that mice could dispose of 98 functional *V* gene segments. Some of these, however, could correspond to allelic variants of the same gene. In fact, only in a few instances were allelic polymorphisms addressed (73,78,79). For example, *AV6s1* (previously *V* α6) and *DV6s2* (previously *Vδ3*) probably represent two different alleles of the same locus, since Southern blots indicated that *ADV6* is a single-member subfamily (73,78). In addition, seven *Vα* haplotypes were described, according to the presence or absence of particular bands in restriction fragment length polymorphism (RFLP) analysis (80–83).

The 50 *AJ* gene segments are located in a 60 kb region between the inverted *DV5* and the *AC*. They have been named according to their position on the locus, *AJ1* to

Table 1   Mouse T cell receptor AV and DV gene segments

(a)	(b)	(a)	(b)	(a)	(b)	(a)	(b)
1s1		4s1	V α 4	8s3		11s2	V α 11.2
1s2	V α 1	4s2	V α 4	8s4	V α 8.4	11s3	V α 11.3
1s3		4s3	V α 4.3	8s5		11s4	
1s4		4s4	V α 4.4	8s6		ADV11s5	V α 11.1a, V δ
1s5		4s5		8s7		11s6	V α 11.1d
1s6		4s6		8s8		11s7	V α 11.3a
1s7		4s7		8s9		12s1	
1s8		DV4s8		8s10		13s1	V α 13.1
2s1	V α 2	4s9		8s11		13s2	V α 13.2
2s2	V α 2.2	4s10		8s12		14s1	V α 14.1
2s3	V α 2.3	4s11		8s13		14s2	V α 14.2
2s4		4s12		8s14		15s1	
2s5		5s1	V α 5	8s15		16s1P	
2s6		5s2	V α 5	9s1		16s2P	
2s7		5s3		9s2		17s1	
DV2s8	V δ 8	6s1	V α 6	10s1	V α 10.1	ADV17s2	V δ 9
2s9		DV6s2	V δ 3	10s2	V α 10.2	17s3	
3s1		ADV7s1	V α 7, V δ 7.1	10s3	V α 10.3	18s1	
3s2		ADV7s2	V α 7.2, V δ 7.2	10s4		18s2	
3s3		DV7s3	V δ 6	10s5		19s1	
3s4		DV7s4	V δ 6	10s6		20s1	
3s5		DV7s5	V δ 6	DV10s7	V δ 7	DV101s1	V δ 1
3s6		DV7s6	V δ 6.2	10s8		DV102s1	V δ 2
3s7		8s1	V α 8	10s9		DV104s1	V δ 4
3s8		8s2		11s1	V α 11.1	DV105s1	V δ 5

(a) Nomenclature according to Arden et al. (53). V gene segments encoding delta chains are designated as DV, those encoding both alpha and delta chains are indicated as ADV, whereas all others encode primarily alpha chains.
(b) Alternative names. See reference 53 for additional information.

AJ50, beginning at the 3′ end (84). At least 10 of these *AJ* gene segments were found to be pseudogenes due to stop codons, altered recombination or splice signals, lack of relevant amino acids, frameshift nucleotide deletion, or combinations of these defects (84).

3.5.   Human T Cell Receptor β Chain Genes

The human *TCR β* locus has been mapped to the long arm of chromosome 7, at 7q32–7q35 (85,86). Genomic studies of cosmid and phage clones (87–89), Southern blots of conventional and pulsed field gels (89–91), and sequence analysis of *V* genes (51,92), *D-J-C* clusters (93) and the entire locus (54), indicated that TCR β chain genes are located on a region that spans approximately 600 to 650 kb. As illustrated in Figure 2, the locus is characterized by a centromeric section of approximately 500 kb that include 64 of the 65 *V* gene segments, followed by two tandemly arrayed *D-J-C* clusters, located within 15 kb and separated by 2.5 kb, each consisting of 1 *BD* gene segment, 6 (cluster 1) or 7 (cluster 2) *BJ* gene segments and 1 *BC* gene segment. Located 3′ to *BC2*, the last *V* gene segment,

BV20s1 (*BV30*), is placed in inverted transcriptional orientation compared to all the other TCR genes in the locus. In addition to TCR β chain encoding genes, eight trypsinogen genes (three at the 5′ end of the locus, and five just 5′ to the first *D-J-C* cluster) have been identified (54). Two insertion/deletion polymorphisms have been described for this locus. One includes three *BV* genes, *i.e. BV9s2* (*BV3–2*), *BV7s2* (*BV4–3*) and *BV13s2b* (*BV6–3*), while the second includes the trypsinogen genes *T6* and *T7*.

A second group of *BV* gene segments has been localized to the short arm of the chromosome 9 (9p13) near the centromer (94). However, since these gene segments are not linked to any *D-J-C* cluster, they cannot be expressed and, accordingly, have been designated orphon (*O*) *BV* genes. This is probably the result of a translocation from chromosome 7 (7q35) of a 75 kb fragment. The tanslocated region seems to contain at least the orphon genes *BV2s2O* (*BV20–2*), *BV10s2O* (*BV21–2*), *BV29s2O* (*BV22–2*), *BV19s2O* (*BV23–2*), *BV15s2O* (*BV24–2*), *BV11s2O* (*BV25–2*), *BV4s2O* (*BV29–2*), as well as the functional trypsinogen gene *T9* (54).

The 62 (in the deleted haplotypes) to 65 *BV* gene segments identified in the *TCR β* locus of chromosome 7, can

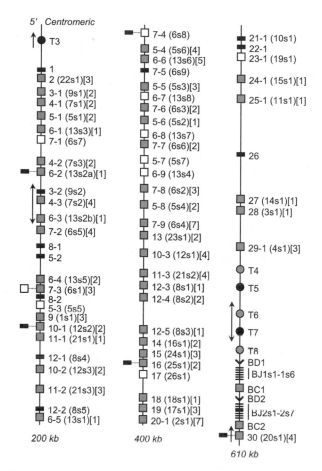

5' *Centromeric*

T3

1
2 (22s1)[3]
3-1 (9s1)[2]
4-1 (7s1)[2]
5-1 (5s1)[2]
6-1 (13s3)[1]
7-1 (6s7)

4-2 (7s3)[2]
6-2 (13s2a)[1]

3-2 (9s2)
4-3 (7s2)[4]
6-3 (13s2b)[1]
7-2 (6s5)[4]
8-1
5-2

6-4 (13s5)[2]
7-3 (6s1)[3]
8-2
5-3 (5s5)
9 (1s1)[3]
10-1 (12s2)[2]
11-1 (21s1)[1]

12-1 (8s4)
10-2 (12s3)[2]

11-2 (21s3)[3]

12-2 (8s5)
6-5 (13s1)[1]

200 kb

7-4 (6s8)
5-4 (5s6)[4]
6-6 (13s6)[5]
7-5 (6s9)

5-5 (5s3)[3]
6-7 (13s8)
7-6 (6s3)[2]
5-6 (5s2)[1]
6-8 (13s7)
7-7 (6s6)[2]

5-7 (5s7)
6-9 (13s4)

7-8 (6s2)[3]

5-8 (5s4)[2]

7-9 (6s4)[7]
13 (23s1)[2]

10-3 (12s1)[4]

11-3 (21s2)[4]
12-3 (8s1)[1]
12-4 (8s2)[2]

12-5 (8s3)[1]
14 (16s1)[2]
15 (24s1)[3]
16 (25s1)[2]
17 (26s1)

18 (18s1)[1]
19 (17s1)[3]
20-1 (2s1)[7]

400 kb

21-1 (10s1)
22-1
23-1 (19s1)

24-1 (15s1)[1]

25-1 (11s1)[1]

26

27 (14s1)[1]
28 (3s1)[1]

29-1 (4s1)[3]

T4
T5

T6
T7

T8
BD1
BJ1s1-1s6

BC1
BD2
BJ2s1-2s7
BC2
30 (20s1)[4]

610 kb

Figure 2. Locus for human β TCR genes on chromosome 7 (7q32–7q35). Approximate position of functional V gene segments (shadowed boxes), non-functional V gene segments with ORF (open boxes) and pseudogenes (filled boxes) is according to reported sequencing data (GeneBank, http://www2.ncbi.nlm.nih.gov/genbank/query_form.html, accession numbers U66059, U66060 and U66061). The locus representation was adapted from "IMGT, the international ImMunoGeneTics database" (http://imgt.cnusc.fr:8104). Each V gene segment is named based on two different nomenclatures, the first according to the gene position in the locus (54), and the second (in parenthesis) corresponding to published alignments (51). The number of functional alleles for each V gene segment is given in brackets. Trypsinogen genes (T) located in the locus are also displayed. Single arrows indicate opposite transcriptional orientation, whereas double arrows correspond to insertion/deletion polymorphisms.

be grouped into 30 subfamilies, 19 of which are composed of a single member, while 11 include 2 to 9 members each. As indicated in Figure 2, members of the various subfamilies are not clustered, but rather interdispersed and distributed over the entire locus. Of these 62 to 65 BV gene segments, 38 to 40 are functional, 10 to 11 are pseudogenes, 8 are non-functional V genes with ORF, 4 are (due to polymorphisms) either functional or pseudogenes, 1 either functional or ORF, and 1 either ORF or pseudogene. Based on these data, the potential repertoire for the human

TCR β chain ranges from 38 to 45 functional BV gene segments organized into 24 subfamilies.

As discussed above for the TCR α locus, the total complement of available BV genes in humans is, in fact, considerably larger due to allelic polymorphisms. These significantly contribute to increase TCR diversity in each individual and, in general, in the human population. Studies aimed at defining the number and the distribution of allelic BV variants among individuals have been performed by several methods, including single-stranded conformational polymorphism (SSCP) analysis (69,95–98), RNase protection assay (99), combinations of PCR and restriction enzyme digestion (96,100–103) or sequencing (104–106), the use of allele-specific antibodies (107) and oligonucleotides (106,108), or the analysis of cosmid clones and genomic DNA (92). Data from these studies, from the comparative analysis of sequences reported in databases (51), and from the complete sequence of the TCR β locus (54), are summarized in Table 2, which highlights the potential extent of allelic polymorphisms associated to human BV genes. Only 13 of the 45 functional BV genes appear to be monoallelic. Most BV genes display 2 or more functional alleles, with a maximum of 7 reported for BV7–9 (BV6s4), and a total of 111 BV gene variants described up to date. Interestingly, most of these variants display coding region polymorphisms, and in several instances the associated amino acid changes affect the CDRs and could, therefore, alter the antigen and superantigen specificity of the corresponding TCR, as demonstrated for allelic variants of BV1s1 (BV9) (109) and BV6s5 (BV7–2, also named BV6s7) (110).

3.6. Mouse T Cell Receptor β Chain Genes

The mouse TCR β locus maps to chromosome 6, as determined by Southern blot analysis of somatic cell hybrid lines (86). In most mice, this locus spans 700 to 800 kb of chromosomal DNA (111), and is organized essentially as in humans, with all but one V gene segment located at the 5' end of the locus, followed by two D-J-C clusters, each including 1 BD, 6 or 7 BJ, and 1 BC (112,113), and finally an additional V gene segment, BV14s1 (BV31), placed 3' to BC2 in inverted transcriptional orientation (111).

Insertion/deletion polymorphisms within the TCR β locus have been observed in several mouse strains. Southern blot analysis of genomic DNA identified 5 Vβ haplotypes. The Vβ haplotype b, exhibited by most mice, disposes of all known Vβ genes. In contrast, large deletions of chromosomal sections are associated to the Vβ haplotypes a, c, d and e (114–119).

Alignment of V gene nucleotide sequences reported in the GenEMBL database and/or in the literature identified 29 unique mouse BV sequences that could be assigned to

Table 2 Human BV genes and alleles

TCRBV gene nomenclature						
(a)	(b)	(c)	(d)	(e)	Accession Number	Sequence differences from allele *01
1s1	9	*01	1s1	9	U66059	
		*02			AF009660	g165c, Q55H;
		*03		HBVT96	M27380	g165a;
2s1	20-1	*01	2s1	MT1-1G	M11955	
		*02		2s1*02	X72719	t28a, W10R;
		*03		Vβ 2.2/MT1.1	M11954	c179g, S60C;
		*04		Vβ 2.3/ph34	M14263	t28a, W10R; c76t; c142a, Q48K;
		*05		Vβ 2.1b/HT120	X57604	t28a, W10R; c142a, Q48K;
		*06		BV2.3a/WBDM11D	D13088	c27t; t28a, W10R; c142a, Q48K;
		*07		Vβ 2.1/4.49	X74852	t28a, W10R; c142a, Q48K; t155a, L52Q; g159c, M53I;
3s1	28	*01	3s1	3s1	U08314	
4s1	29-1	*01	4s1	29-1	U66061	
		*02		PL5.7	M13847	c263g, T88S;
		*03		Vβ 4.3/HBP48	X04926	g117a, M39I;
5s1	5-1	*01	5s1	5-1	U66059	
		*02		Vβ 5.4/ph24	M14271	a2g,K1R; a9g; t28c, Y10H; a64g, S22G; c137t, P46L; c215t, P72L;
5s2	5-6	*01	5s2	5-6	U66060	
5s3	5-5	*01	5s3	5-5	U66060	
		*02		Vβ 5.3a/HT415.9	X57611	a54c, Q18H;
		*03		IGRb08	X58801	g83a, G28E;
5s4	5-8	*01	5s8	5-8	U66060	
		*02		IGRb06	X58803	t154c, F52L;
5s6	5-4	*01	5s6	5-4	U66060	
		*02		5s6	AF009662	g257a, S86N;
		*03		Vβ 5.5	S50547	t60a;
		*04		IGRb07	X58804	t212c, F71S;
6s1	7-3	*01	6s1	Vβ 6.1/9	X61440	
		*04		Vβ 6.1/1.26	X74843	a297t;
		*05		4D1	M13550	t154c, F52L;
6s2	7-8	*01	6s3	ATL12-2G	M11953	
		*02		Vβ 6.3/4-1	X61441	c286a, Q96K;
		*03		Vβ 6.3/HBVT10	M27384	g138c
6s3	7-6	*01	6s4	7-6	U66060	
		*02		Vβ 6/IGRb11	X58806	t78c;
6s4	7-9	*01	6s5	7-9	U66060	
		*02		L17β	M15564	g33c, K11N;
		*03			AF009663	a19g, N7D;
		*04		Vβ 6.8/ph22	M14261	g1a,a2t,t3a,D1l; a4t,T2S; g18c,Q6H; c134a, T45N; t137c, g138t,L46P;a198g; a208g,R70G; t226a,F76l;
		*05		Vβ 6.5/HBVT116	M27385	t256c,F86L;

(a) Nomenclature according to Arden et al. (51).
(b) Nomenclature according to Rowen et al. (54), which considers the position of the gene segment in the locus.
(c) Functional allelic variants, according to the ImmunoGene Tics database, IMGT (http://imgt.cnusc.fr:8104)
(d) Nomenclature according to Wei et al. (72).
(e) Alternative names (see reference 51 or GeneBank for additional information)

Table 2 Human BV genes and alleles (continued)

(a)	(b)	(c)	(d)	(e)	Accession Number	Sequence differences from allele *01

TCRBV gene nomenclature

(a)	(b)	(c)	(d)	(e)	Accession Number	Sequence differences from allele *01
6s4	7-9	*06		Vβ 6.5/1.40	X74844	t256c,c258t,F86L;
		*07			L14854	c286g,Q96E;
6s5	7-2	*01	6s7	Vβ 6.7a/5–2	X61442	
		*02		Vβ 6.7b/GL–PA	X61443	c135g,S45R; g251a,G84E;
		*03		6s7	U07975	c135g,S45R; g251a,G84E; g313a,A105T;
		*04		Vβ 6.6/HBVT45	M27387	c12t
6s6	7-7	*01	6s14	7–7	U66060	
		*02		Vβ 6.7/HT147	X57607	c89t,A30V;
7s1	4-1	*01	7s1	7s1	U07977	
		*02		Vβ 7.1/PL4.9	M13855	t93a;
7s2	4-3	*01	7s2	7s3	U07978	
		*02		Vβ 7/IGRb18	X58812	t263c,F88S;
		*03		Vβ 7.n2	L06888	g183t;
		*04		Vβ 7.2b/HT267.1	X57616	t84g;
7s3	4-2	*01	7s3	7s2	U07975	
		*02		Vβ 7/IGRb17	X58811	t263g,F88C;
8s1	12-3	*01	8s1	Vβ 8.1/M18H7.1B7	X07192	
8s2	12-4	*01	8s2	M3–2	K02546	
		*02		Vβ 8.4/ph8	M14264	c87t; a269g,K90R;
8s3	12-5	*01	8s3	Vβ 8.3/λ VB8.3	X07223	
9s1	3-1	*01	9s1	9s1	U07977	
		*02		Vβ 9.n	L06889	t174c; c181a,L61I; c225a; c256a,c258a,H86K;
11s1	25-1	*01	11s1	BV11	L27610	
12s1	10-3	*01	12s2	12s2/HT7.1	U03115	
		*02		12s2	U17047	c72t;
		*03		Vβ 12.2 Allele 3	L33101	t93c; g156a;
		*04		Vβ 12.2 Allele 4	L33102	t93c;
12s2	10-1	*01	12s4	12s4	U17050	
		*02			AF009660	a180c,Q60H;
12s3	10-2	*01	12s3	12s3	U17049	
		*02		12s3	U17048	t228c;
13s1	6-5	*01	13s1	6–5	U66060	
13s2a	6-2	*01	13s2a	Vβ 13.2/5–2	X61445	
13s2b	6-3	*01	13s2b	13s2	U07978	
13s3	6-1	*01	13s3	Vβ 13.3/11	X61446	
13s5	6-4	*01	13s5	Vβ 13.5/9	X61653	
		*02		13s5	AF009660	t2c,I1T; c52a,R18S;
13s6	6-6	*01	13s6	6–6	U66060	
		*02		13s6	AF009662	a70g,T24A; t201c;
		*03		Vβ 13/IGRb16	X58815	a70g,T24A;
		*04		Vβ 13.4/3.1	X74848	a88g,c90a,N30E; c231t;
		*05		Vβ 13.n3	L06892	a70g,T24A; t201c; c286g,P96A;
14s1	27	*01	14s1	27	U66061	
15s1	24-1	*01	15s1	Vβ 15.1/ATL2-1G	M11951	

Table 2   Human BV genes and alleles (continued)

					Accession Number	Sequence differences from allele *01
TCRBV gene nomenclature						
(a)	(b)	(c)	(d)	(e)		
16s1	14	*01	16s1	Vβ 16	X06154	
		*02		Vβ 16b/HT219	X57722	g204a;
17s1	19	*01	17s1	17s1	U48260	
		*02		17s1	U48259	a134t,D45V; g171c,Q57H;
		*03		Vβ 17.1-V/S30.10	M97725	g171c,Q57H;
18s1	18	*01	18s1	18	U66060	
20s1	30	*01	20s1	30	U66061	
		*02		Vβ 18A/H29	Z13967	t285c
		*04		Vβ 20.1/HUT102β	M13554	g169a,V57I; g179a,G60D; c264t;
		*05		Vβ 20.n	L06893	t60c; c141a; a142c; t285c;
21s1	11-1	*01	21s1	BV21.1/H18.1	M33233	
21s2	11-3	*01	21s4	V21.2	M33234	
		*02		Vβw21/IGRb02	X58797	a48g; t128g,L43R;
		*03		Vβ 21	M62377	a17g; a45g; t75a; t279a; g301a,V101M;
		*04		Vβ 21.4a		t128g,L43R;
21s3	11-2	*01	21s3	11–2	U66059	
		*02		BV21.3	M33235	g292a,D98N;
		*03		Vβw21/IGRb01	X58796	g276a;
22s1	2	*01	22s1	2	U66059	
		*02		Vβ 23	M62379	g65a,R22H;
		*03		Vβ 22	M64351	a237g;
23s1	13	*01	23s1	23s1/H7.1	U03115	
		*02		Vβ 22	M62378	a38g,K13R; t135c; c157t,L53F;
24s1	15	*01	24s1	24s1	U03115	
		*02		Vβw24/IGRb05	X58800	a295g,T99A;
		*03		Vβ 24	M62376	a32g,Q11R; g172a,D58N; t267a; a295g, T99A; t308a,L103Q;
25s2	16	*01	25s2	25s1/HVB30.A	L26231	
		*03		Vβ 25/HsVB25	L26054	a160g,I54V

25 distinct subfamilies (53). More recently, the entire locus, corresponding to haplotype *b*, has been sequenced and the corresponding data deposited in Genbank (http://www2.ncbi.nlm.nih.gov/genbank/query-form.html), under the accession numbers AE000663, AE000664 and AE000665. This study identified 6 additional *BV* genes and subfamilies. As summarized in Table 3, the combined data indicate that the mouse *TCR β* locus includes 35 *BV* gene segments belonging to 31 subfamilies, most of which include a single member. In fact, *BV5* (2 genes and 1 pseudogene) and *BV8* (3 genes) are the only multimember mouse TCR β V gene subfamilies identified. Of the 35 *BV* segments, 21 are functional, 12 are pseudogenes, and 2 (*BV17s1* and *BV19s1*) are due to polymorphisms either functional or pseudogenes. Thus, the potential germline TCR repertoire of mouse *Vβ* haplotype *b* appears to include 21 to 23 functional *BV* gene segments, belonging to 18 to 20 subfamilies.

3.7. Human T Cell Receptor γ Chain Genes

The human *TCR γ* locus has been mapped to chromosome 7 (120), at band 7p15 (121). As indicated in Figure 3, the locus spans approximately 160 kb and includes 15 *V* gene segments (122–124), located within 100 kb centromeric to the locus, followed by 2 *J-C* clusters covering 37 to 40 kb, one consisting of 3 *GJ* and 1 *GC*, and the second including 2 *GJ* and 1 *GC* (29,29,48,86,125,126). Only 16 kb separates the most 3′ *V* gene, *GV11* (*GV4s1*), from the most 5′ *J* gene, *JP1*. Two insertion/deletion polymorphisms have been described, one including the pseudogene *GV3P*, and the second including *GV4* (*GV1s4*) and *GV5* (*GV1s5*) (127,128).

The 12 (in the deleted haplotypes) to 15 *GV* genes identified in this locus can be grouped into 6 subfamilies, of which 5 (*GV2*, *GV3*, *GV4*, *GV5* and *GV6*) are composed of a single member each. Subfamily *GV1* includes 7 to 10 members, 5 of them functional. Of these 12 to 15 *GV* genes, 4 to 6 are functional, 6 to 7 are pseudogenes and 2 are non-functional *V* genes with ORF. Thus, the potential repertoire for the human TCR γ chain ranges from 4 to 6 functional *GV* genes, forming 2 subfamilies.

Polymorphisms associated with *GV* genes are mainly due to the mentioned insertions/deletions or to sequence alterations affecting restrictions enzyme sites. Of the 7 *GV* gene haplotypes identified in such analysis, those with 14 *GV*, *i.e.* with the insertion of *GV4* (*GV1s4*) and *GV5* (*GV1s5*), are the most frequent (127, 128). In addition, analysis of database sequences has suggested the presence of several *GV* alleles in the human population, with at least 8 functional *GV* variants reported. Finally, the *GC2* gene displays polymorphisms resulting from the presence of either 4 or 5 distinct exons due to duplication or triplication of exon 2, as well as from nucleotide sequence variations (29,127).

3.8. Mouse T Cell Receptor γ Chain Genes

The mouse *TCR* γ locus maps to chromosome 13 (76) and spans 205 kb (46,129–131). The nucleotide sequence of the locus has recently been completed and the results are available through Genbank (http://www2.ncbi.nlm.nih.gov/genbank/query_form.html) under accession numbers AF021335 and AF037352. This locus consists of 4 *V-J-C* clusters (named, from 5′ to 3′, γ1, γ3, γ2 and γ4) and display an organization clearly different from that of TCR α/δ, β, and human TCR γ loci. Cluster γ1 includes 4 *GV* genes, 1 *GJ* gene and 1 *GC* gene. Clusters γ3, γ2 and γ4, in contrast, include 1 *GV*, 1 *GJ* and 1 *GC* gene each. Cluster γ3 is not expressed, while the orientation of cluster γ2 is opposed to that of other γ gene segments. Table 4 summarizes information regarding the various nomenclatures of *GV* genes, their position on the chromosome and the cluster to which they are associated.

Table 3   Mouse T Cell Receptor BV gene segments

(a)	(b)	(c)	(d)	(a)	(b)	(c)	(d)	(a)	(b)	(c)	(d)
1s1	5	Vβ1, Vβ11	2	9s1	17	Vβ2	1	21s1P	25		0
2s1	1	Vβ2, Vβ6	1	10s1	4	Vβ3	2	22s1P	22	Vβ3.3P	0
3s1	26		2	11s1	16	Vβ5	1	23s1P	18		0
4s1	2	Vβ4, Vβ9	1	12s1	15	Vβ7	1	24s1P	9		0
5s1	12–2	Vβ5.1, Vβ8	1	13s1	14	Vβ10	1	25s1P	10		0
5s2	12–1	Vβ5.2	2	14s1	31	Vβ14	1	26s1	6		0
5s3P	12–3	Vβ5.3P	0	15s1	20	Vβ15	2	27s1	7		0
6s1	19	Vβ1	2	16s1	3	Vβ16	2	28s1	8		0
7s1	29		1	17s1(P)	24	Vβ17a, Vβ17b	2	29s1	11		0
8s1	13–3	Vβ8.1	*1*	18s1	30	Vβ18	1	30s1	27		0
8s2	13–2	Vβ8.2, Vβ4	3	19s1(P)	21	Vβ19a, Vβ19b	1	31s1	28		0
8s3	13–1	Vβ8.3	1	20s1	23	Vβ20	1				

(a) Nomenclature according to Arden et al. (53)
(b) Nomenclature according to the V gene position on the locus, as reported by Rowen et al. (GenBank, accession numbers AE000663, AE000664, AE000665)
(c) Additional names (see reference 53 for additional information)
(d) Number of functional allelic variants.

Table 4   Mouse T cell receptor GV gene segments

(a)	(b)	(c)	(d)	(e)	(f)	(g)
1s1	4	1	V γ 5	V γ 3	V γ 4.1	1
2s1	3	1	V γ 6	V γ 4	V γ 4.2	2
3s1	2	1	V γ 4	V γ 2	V γ 4.3	3
4s1	1	1	V γ 7	V γ 5	V γ 4.4	2
5s1	5–3	4	V γ 1	V γ 1.1	V γ 2	4
5s2	5–2	2	V γ 2	V γ 1.2	V γ 1	2
5s3	5–1	3	V γ 3	V γ 1.3	V γ 3	2

(a) Nomenclature according to Arden et al. (53).
(b) Position in the locus, 5′ to 3′.
(c) Cluster
(d) Hayday et al. (131)
(e) Garman et al. (130)
(f) Pelkonen et al. (132)
(g) Number of functional allelic variants.

The clustered organization of the mouse *TCR* γ locus correlates with the rearrangement patterns of the corresponding *GV* genes. In fact, rearrangements take place mainly within the clusters, although occasional rearrangements of *GV* and *GJ* genes from different clusters have been observed (132). In addition, the order of the *GV* genes in the *γl* cluster reflects their order of activation in ontogeny and, in part, their tissue-restricted distribution (133,134). The 7 mouse *GV* genes can be organized into 5 subfamilies, one of which, *GV5*, is composed of 3 members that belong to different clusters (*γ3*, *γ2* and *γ4*).

Comparisons of reported *GV* gene sequences identified possible allelic polymorphisms for 6 of these genes (Table 4) (53). In addition, *GC4*, in contrast to the other 3 *GC* genes, display 4 exons instead of 3 (46). This is similar to what observed for human *GC2*, also encoded by 4 (or 5) exons (see above).

4.   GENERATION OF DIVERSITY

As discussed above, V, D and J gene segments are scattered over large fragments of genomic DNA, forming the various loci. Their number and complexity constitute the germline repertoire of TCR diversity. This is significantly increased by allelic polymorphisms, particularly those encoding amino acid replacements within the complementarity determining regions (CDR) 1 and CDR2 of V gene segments. However, since CDR1 and CDR2 are predicted to contribute most contacts with MHC residues during antigen recognition (see below), it is reasonable to think that genes encoding V regions and MHC molecules have been submitted to common evolutionary pressures, which has presumably limited their respective germline diversity.

In contrast, CDR3 loops, predicted to interact primarily with antigenic peptides bound to MHC molecules (see below), display most of the diversity associated to TCR polypeptides. As is the case for Ig, TCR CDR3 results from the combinatorial association of the relatively small number of V, D and J gene segments available in the genome (*Table 5*). Through chromosomal rearrangements, specific *V-J* or *V-D-J* segments are fused together during T cell ontogeny to create unique functional transcription units for α and γ, or β and δ, TCR chains. Particularly significant is the imprecision of the *V(D)J* joining process, with the possibility of deleting a few nucleotides at the 5′ and/or 3′ boundaries of the gene segments, and introducing variable numbers of nucleotides, which can be either germline

Table 5   Sequence diversity for human T cell receptor chains

	αβ TCR		γδ TCR	
	α	β	γ	δ
V segments	47	45	6	8
D segments	–	2	–	3
D in all frames	–	yes	–	yes
J segments	50	13	5	4
N addition	V-J	V-D, D-J	V-J	V-D, D-D, D-J
V-J or V-D-J combinations	2′350	1′170	30	96
V region combinations	2′749′500		2′880	
Junctional combinations	~10^{15}		~10^{18}	

Numbers correspond to functional gene segments only.
Sequence diversity due to allelic polymorphisms is not considered.
Junctional combinations are estimated according to Elliott et al. (78)

encoded and palyndromic (P), or non-templated (N). Further diversity is due to the use of *D* segments in all three reading frames, the possibility of incorporating two *D* segments in δ gene rearrangements, and finally to the combinatorial association of the completed TCR chains forming α/β or γ/δ heterodimers. In addition, although most T cells display a unique clonotypic TCR, dual receptor T cells have also been described (135–137).

Based on the common transcriptional orientation of *V*, *D*, *J* and *C* gene segments, most rearrangements appear to occur via looping-out and deletion of the intervening regions, reactions mediated by conserved heptamer and nonamer sequences spaced by 12 or 23 nucleotides, and resulting in the formation of circular episomes as byproducts. Such extrachromosomal excision circles have, in fact, been observed, and can be used to identify recent thymic emigrants in the peripheral T cell pool (138). In contrast, chromosomal inversion has been proposed for mouse *BV14s1* (*BV31*) and *DV105s1* (*Vδ5*), and for human *BV20s1* (*BV30*) and *DV103s1* (*DV3*), since these genes are located downstream of their respective *C* genes (Figures 1 and 2).

An important diversification mechanisms for Ig is the introduction of somatic hypermutations in a section of DNA spanning the *V(D)J* rearrangement (139). In most instances, such a mechanisms is not used by T cells, although mutations have been observed in T cell subpopulations within germinal centers of lymphoid organs (140,141) and in hybridomas (142).

5. ANTIGEN RECOGNITION

5.1. Peptide Antigens and MHC Restriction for α/β T Cell Receptors

Early studies on antigen recognition indicated a number of differences between T cells and B cells. One of these concerned the conformation of the recognized antigens. While Ig, the B cell receptor, were often specific for determinants dependent on the native conformation of the protein antigen, T cells could be stimulated by both denatured antigens (143) and synthetic peptides (144). It was also observed that T cells recognize antigens that are presented by a second cell type, the antigen presenting cell (APC), upon processing. Experiments showed that antigen presentation requires cellular metabolism, and can be inhibited if, early during presentation, APC are treated with agents that raise lysosomal pH (145–147) or are fixed with glutaradehyde (148). Yet, fixed APC could still present peptidic fragments of specific antigens (149). An additional difference regarded the simultaneous recognition by T cells of antigens and MHC determinants. Demonstration of this phenomenon, termed MHC restriction, was first obtained in experiments using cytotoxic T cells specific for viral (150), minor histocompatibility (151,152) and haptenic (153) determinants. From these and other data, it appeared that T cells mostly recognize processed peptide fragments in association with MHC molecules. It also appeared that cytotoxic T cells were mainly class I MHC restricted, whereas helper T cells were primarily class II MHC restricted.

Class I MHC molecules are heterodimers composed of a polymorphic 46 kDa heavy chain non-covalently associated with the 12 kDa β2-microglobulin (154). The heavy chain is encoded within the *MHC* locus on mouse chromosome 17, and human chromosome 6 (155). Three genes encodes different class I isotypes, *i.e.*, human A, B and C, and mouse K, D and L. Because of the extensive genetic polymorphism, most individuals in outbread populations express two allelic forms of each isotype. Crystallographic structural analysis of class I molecules revealed a typical peptide binding cleft composed of a floor of eight β strands flanked by helical walls, all encoded by the heavy chain (154). The peptide bound in this cleft was 8–9 amino acid long and appeared in an extended conformation due to conserved pockets fixing the peptide amino and carboxyl termini (156). Most contacts were with the peptide main-chain atoms, which explains the ability of MHC molecules to present diverse peptide antigens. Nonetheless, a deep pocket in the class I binding groove accommodated one of the peptide side chains, termed anchor residue (156). In different class I molecules, these pockets are variable in number, position, shape and chemical characteristics, which reflects the ability of each allelic form of MHC molecules to bind a different set of peptide antigens. These data were in agreement with binding studies (157), and with sequencing data of natural peptides, directly acid-eluted from affinity purified class I molecules (158). These experiments showed 8–9 amino acid peptides with diverse residues in almost any position, but with recurrent anchor residues, the nature and the position of which correlated with the class I allotype analyzed. Class I molecules are expressed by most cells, and bind and present mainly peptides derived from cytosolic proteins, degraded through the ubiquitin-proteosome pathway (159), and transported into the endoplasmic reticulum by ATP-dependent transporters of antigen presentation (TAP) (160). Here, the peptides bind to, and allow proper folding of, nascent class I molecules, and the resulting complexes are transported to the cell surface for T cell inspection. Because class I molecules display a binding site for CD8 (161,162), cytotoxic T cells expressing this coreceptor are primarily recruited. This mechanis ensures that each cell presents to cytotoxic T cells a representative sample of all expressed cellular proteins, including viral and other unwanted gene products.

Class II MHC molecules, on the other hand, are composed of α (33 kDa) and β (29 kDa) chains, both encoded in the *MHC* locus (155). Like class I molecules, class II molecules exist as different isotypes (human DR, DP and DQ, and mouse IE and IA), all highly polymorphic. Three-dimensional analysis indicated that class II molecules also present a typical peptide binding cleft (although each polypeptide chain contributes one of the helices and four of the β strands) and display specific pockets to accommodate corresponding peptide anchor residues (163,164), in agreement with peptide-binding (165) and sequencing (158) studies. In contrast to class I molecules, not all cells express class II molecules, but primarily B cells, macrophages, monocytes, dendritic cells and endothelium. Class II are dedicated to the presentation of antigens that enter the endocytic pathway, and therefore their binding in the endoplasmic reticulum of cytosolic peptides is prevented by the invariant chain (166) and its derived peptide, CLIP (167). Transported to the endosomes, class II molecules bind to HLA-DM (H2-M in the mouse), release CLIP and bind peptides from endocytosed proteins (168). Since class II molecules interact with CD4 (161), the bound peptides from extracellular antigens are mainly presented to helper T cells, normally expressing this coreceptor.

5.2. Altered Peptide Ligands

Peptides presented by MHC molecules bind specific TCR and activate a number of T cell functions, including expression of cytokines and proliferation. Peptide variants have been found, however, that induce only part of these effects, and can even inhibit stimulation by the original peptide (169,170). These findings suggested that peptide antigens can be classified according to the set of activation events they evoke (171). Full agonists are peptides that induce full T cell activation, while weak agonists elicit all T cell functions but need higher concentration. Partial agonists, in contrast, mediate only a limited number of functions; for example, only cytokine production but not proliferation (169). Antagonists are ligands that inhibit the effect of agonist peptides when presented on the same APC (170). Finally, null ligands induce no response from the T cell. Compared to full agonist, altered peptides bind similarly to MHC (172), but induce faster TCR dissociation kinetics (173), incomplete phosphorylation-associated TCR signaling (174), and lower level of TCR internalization (175).

5.3. Superantigens and other Antigens for α/β T Cell Receptors

Peptide antigens presented by class I or class II MHC molecules constitute the most common type of ligand for α/β TCR. A second class of antigens, however, has been described. These antigens behave as potent mitogens, stimulate proliferation of large populations of T cells (5–20%), mostly expressing particular $V\beta$ gene products, and were therefore termed superantigens (SAg) (2). This class of molecules include pyrogenic toxins like staphylococcal enterotoxins (SEA through E, except F), staphylococcal toxic shock syndrome toxin-1 (TSST-1), streptococcal superantigen (SSA), and streptococcal pyrogenic exotoxins (SPE-A through C and F). Other SAg that are not pyrogenic toxins include staphylococcal exfoliative toxins (ET-A and B) (176), Mycoplasma arthritis mitogen (MAM) (177), Yersinia pseudotoberculosis mitogen (178), and urtica dioica (179). SAg of viral origin have also been described, particularly those encoded by ORF sequences in the 3′ long terminal repeats of mouse mammary tumor virus (MMTV) (180). Unlike conventional peptide antigens, SAg are normally not processed and bind to MHC class II molecules (2), although binding to class I molecules has also been observed (179). Crystallographic data confirmed earlier biochemical and functional studies indicating that binding of SAgs to MHC molecules is different compared to peptides, i.e., it occurs outside the peptide-binding groove and is dependent on the native conformation of the ligands. Thus, SEB binds mainly to the $\alpha 1$ domain of class II molecules contacting residues from both β-sheets and α-helix, whereas TSST-1 contacts both helices as well as the bound peptide (181).

In addition to peptides and SAg, α/β TCR have been found to react with other ligands, including modified peptides (182), glycopeptides (183), or glycolipids (184) presented by class I molecules, or with lipids and glycolipids presented by CD1 molecules (3).

5.4. Ligands for γ/δ T Cell Receptors

Ligands for γ/δ TCR are less well characterized than those of α/β TCR. It has been reported that clones expressing γ/δ TCR also responded in a self-MHC-restricted fashion, for example to tetanus toxoid (185) or to synthetic Glu-Tyr copolymer (186). Other studies, however, suggest that antigen recognition of proteins like MHC class II (6), non classical MHC molecules (8) and a glycoprotein from herpes-simplex virus (7), occurs without antigen processing or involvement of bound peptides. This suggests that γ/δ TCR might behave more like antibodies during antigen recognition. In addition to protein antigens, γ/δ TCR seem to interact with phosphate-containing products, including isopentenyl pyrophosphate derivatives (187) and nucleotide derivatives (188), as well as with alkylamines (189).

5.5. T Cell Receptor—Ligand Interaction

Several studies have been performed in the attempt to understand and, eventually, predict specific TCR-antigen interactions. Based on their vast diversity, it was suggested that TCR CDR3, like the corresponding regions in antibodies, would contribute most contacts with peptide-MHC complexes (28). Initial TCR mutagenesis studies have, in fact, confirmed the importance of CDR3, showing that single amino acid replacements could alter or event abolish antigen recognition (190). Similar studies further indicated the relevance of CDR3 (191–195), but also of CDR1 (191,192,196–198) and CDR2 (191,192,197,199) in both TCR α and β chains, and both class I-(194,195,198) and class II-(191,192,196,197,199) restricted interactions. Furthermore, alanine scanning mutagenesis revealed that CDR1 and CDR2 contribute most (up to 63%) binding energy (200). Interestingly, transplantation of CDR loops onto another Vβ framework failed to transfer specificity, highlighting the importance of structural context and fine positioning of CDR residues (201). Additional information was obtained with mutant peptides used to immunize mice transgenic for a TCR chain (α or β) specific for the original class I-presented peptide (202). In these experiments, relevant changes (e.g., Lys to Glu) in the peptide induced selection of TCR with compensating amino acid changes (e.g., Glu to Lys) in the CDR3 of the non-transgenic chain. In particular, it was deduced that the peptide N-terminus interacted mainly with Vα, while the peptide C-terminus interacted with Vβ. Based on similar experiments, it was proposed that TCR binds perpendicularly over the face of peptide-MHC class II (203). Finally, studies with class I MHC mutants indicated that changes in the C-terminus (but not in the N-terminus) of the α-helices abolished TCR binding, suggesting a diagonal orientation of the interaction (204).

Such a diagonal orientation has been confirmed by the resolution of the four crystal structures of TCR-peptide-MHC class I complexes available up to date (34,37–40), while a more perpendicular orientation has been observed in the only complex involving class II MHC (41). In these complexes, CDR1 and CDR2 of the α chain are positioned above the N-terminus of the peptide, contacting both peptide and MHC α-helices, particularly the C-terminal residues of the class 1 α2, and class II β1, domains. On the other hand, CDR1 and CDR2 of the TCR β chain are oriented over the C-terminus of the peptide, although contacts from these regions vary among the reported complexes. Finally, both CDR3 lie over the central surface of the complex, providing most contacts with protruding peptide side chains, but also with α-helix residues. Interestingly, two of these complexes involved the binding of different TCR (displaying different α-chain and different CDR3 β) to the same peptide-MHC class I (37,40). The diversity of the contacts that nonetheless give rise to interactions of similar stability documents the plasticity of the immune system, with the ability of even single peptides to engage broad TCR repertoires, or of single TCR to crossreact with large numbers of different ligands (205).

TCR interaction with SAg presented by class II MHC molecules have also been studied. The most obvious characteristic of T cell responses to SAg is the limited number of Vβ regions engaged (2). Such Vβ specificity

suggested that most contacts with SAg are contributed by germline encoded Vβ residues outside the regions involved in peptide recognition. Indeed, experiments with mutagenized or naturally polymorphic TCR variants indicated the importance of Vβ HV4 in the interaction with bacterial and viral SAg (206–210). However, evidences indicating the possible involvement of Vβ CDR1 and CDR2 have also been reported (198,201,207,209,211). Additionally, a number of studies have indicated that β-chain CDR3 (212,213) and Vα (2,214,215) can also contribute to stabilize the complex.

More recently, analysis of crystal structures of TCR β-chain bound to the SAg SEC2, SEC3 and SEB has confirmed the primary importance of germline encoded Vβ residues (32,42). Most contacts result from CDR2 (50 to 63%), with smaller contributions from FR3 (32 to 32%) and HV4 (7 to 9%), suggesting that the binding sites for SAg and peptide-MHC are at least partially overlapping (181). Although no structure of trimolecular complex has been reported until now, a model of interaction between TCR, SAg and MHC class II has been proposed (181). In this model, Vβ interacts with SAg with no contacts with the MHC, while Vα interacts with the β1 domain of MHC. This model also predicts that the binding of TCR to SAg-MHC class II is similar to that with peptide-MHC class II, i.e., it occurs in a more perpendicular orientation as compared to the diagonal binding to peptide-MHC class I.

The development of surface plasmon resonance techniques for detecting macromolecular interactions has allowed precise measurements of kinetics and dissociation constants (Kd) for TCR binding to both peptide-MHC and SAg-MHC (173). The results showed that TCR binding is characterized by low affinities (Kd ranging from 0.1 to 90 μM) compared to antibodies (generally 10 nM to 10 pM), and fast on- and off-rates (half-life of 12 to 30 sec). Interestingly, comparison of altered peptides ligands showed a correlation between TCR binding and biological outcome, with weak agonists, antagonists and null-ligands displaying increasing dissociation-rates (half-life 0.14 to 12 sec).

5.6. Consequences of T Cell Receptor Engagement

As discussed above, TCR interact with peptides or SAg bound to MHC molecules. Engaged TCR transduce intracellular signals resulting in activation of various T cell effector functions and proliferation (174,216,217). Initial steps include the recruitment of the src kinase Lck, which is coupled to CD4 or CD8 and phosphorylate the TCR-associated zeta- and CD3-subunits, thereby mediating recruitment and activation of the syk kinase ZAP-70. This complex phosphorylates several adaptor proteins, leading to activation of Ser/Thr kinase (MAPK) cascades, calcium mobilization and, finally, to the nuclear accumulation of a number of active transcription factors. Specific changes in gene regulation result in expression of cytokine receptors, secretion of cytokines, DNA replication, cell division and cell differentiation. Oligomerization of TCR, CD3, the Lck-associated CD4 or CD8 and adhesion molecules appears to be important for the initial phosphorylation events, in part because it brings into contact the elements involved in kinase reactions (218,219). In addition, it has been reported that, after binding, engaged TCR are downregulated, which allows a single peptide-MHC to trigger several TCR (up to 200), thereby amplifying and sustaining the signaling process (175). This could be one of the reasons for the short life of TCR-ligand interactions (173).

An important question is how a TCR distinguishes different antigens, particularly in view of the fact that peptide-MHC, the most common ligands, appear as flat surfaces, with only few protruding residues coming from the peptide. For example, a peptide derived from a foreign antigen should act as an agonist, and its recognition by the TCR should trigger full T cell activation. In contrast, a peptide from a self antigen, even if homologous to the exogenous peptide, should act as null-agonist, partial-agonist or antagonist, and provide, if any, only negative signals (to inactivate or eliminate strongly autoreactive T cells) or survival signals (in the thymus to positively select self-MHC restricted TCR, or at the periphery to maintain alive a diverse repertoire of naïve T cells). Based on the crystallographic studies summarized above, sufficient TCR contacts are made with the few specific peptide residues protruding from the peptide-MHC complex, to create various levels of binding stability for different peptides. This is also supported by kinetic and affinity studies, showing that binding of TCR to partial agonist and antagonist is characterized by lower stability and faster off-rate (173). It appears, hence, that the shorter time of interaction with altered peptides prevents the TCR to activate a sufficient number of downstream signaling events, leading for example to partial activation (173–175), as suggested by the kinetic proofreading model (220).

6. METHODS FOR T CELL RECEPTOR REPERTOIRE ANALYSIS

Several approaches have been developed to analyze T cell repertoires. Basically, three main groups can be defined, that include: a) methods utilizing *V* gene products as markers to broadly analyze large T cell populations and evaluate their diversity; b) methods focusing on highly specific, junctional *V(D)J* encoded, CDR3 sequences to identify and monitor relevant T cell clonotypes; c) methods based on antigen binding specificity to quantify and isolate T cell subpopulations and clones. This section summarizes the characteristics of the most commonly used approaches.

6.1. *V* Gene Expression Analysis

TCR *V* gene products are markers that can be used to dissect T cell populations into a number of subsets and evaluate repertoire changes during thymic selection, antigen responses and disease. In the case of the human TCR *β* chain, for example, 38 to 45 functional BV gene segments, organized into 24 subfamilies, can be defined. Accordingly, human T cell repertoires can be divided into 24 to 45 groups, depending on whether analysis is performed with subfamily- or gene-specific reagents. Obviously, each of these groups will include millions of different TCR, displaying the same (or homologous) V*β*, but diverse CDR3 and paired *α* chains (see above). The sensitivity of *V* gene analysis is therefore limited, and only changes involving large T cell populations, generally more then 1%, are detectable. Nonetheless, in several instances, similar approaches can give invaluable information, most notably in regard to responses to SAg (2), but also in the case of peptide antigens that preferentially bind particular *V* gene products (221–223).

Because of the large number and close homology of *V* genes, the development of methodologies allowing both discrimination and quantification of their expression has constituted a considerably challenging task. Among the several approaches used, sequence analysis of conventional (224), or anchored-PCR amplified (100,225), cDNA libraries has the advantage of providing, in addition to an evaluation of *V* gene expression frequencies, detailed information on the contribution of allelic polymorphisms and junctional variability. However, this approach is very laborious, implies the analysis of several hundreds of molecular clones for each sample and, therefore, cannot be easily implemented for population studies.

A second method is based on flow cytometry using monoclonal antibodies recognizing specific *V* gene products. This approach is efficient and reproducible, is not influenced by possible differences in transcription activity among genes, gives direct information on the number of cells expressing a specific *V* segment, and can be used in multiparameter analysis in conjunction with antibodies recognizing other relevant T cell markers. Its application to the analysis of mouse TCR V*β* repertoires led to very important advances in understanding basic immunological mechanisms, such as thymic selection and tolerance (226,227). However, the production of such reagents is particularly time-consuming and the fact that, for example, only a portion of the human T cell repertoire can be covered by the limited set of available anti-V*β* antibodies, reflects the complexity of human BV gene families. In addition, although some anti-V*β* reagents display high specificity and can distinguish alleles differing in only two amino acids (107), others are less specific and crossreact with gene products differing in as much as nine residues (228). Finally, the number of cells available from certain tissues are not always sufficient for analysis by antibody staining.

An additional method is based on the quantification of TCR cDNA, reverse-transcribed from rearranged transcripts and amplified by polymerase chain reaction (PCR) through the use of large panels of V and C specific oligonucleotide primers (229–232). V-specific primers can also be degenerate (233), or used in multiplex reactions (234). Products can be quantified either during PCR using fluorescent probes (TaqMan) (235), or after PCR by ELISA (236), but generally products are electrophoresed and then quantified by various means, including radioactivity, fluorescence, and Southern blot. Although in most instances PCR primers were designed to amplify all members of a given V gene family, gene-specific oligonucleotides have also been used. This technology is very sensitive, flexible in the sense that new sets of primers can be quickly redesigned as needed, and allows rapid analysis of V gene repertoires expressed in various tissues starting from very limited numbers of cells. This method is also very specific, although examples of unexpected crossreactivities were observed even with primers meticulously selected on the basis of appropriate thermodynamic parameters (230). An additional concern is associated with the small, but exponentially amplifiable, differences in hybridization efficiencies among primers, which can introduce significant biases in the calculated BV gene expression levels. For example, comparisons with anti-Vβ antibodies indicated that quantitative PCR, in some instances, can give over- or underestimated values (228). Coamplification (instead of the commonly used C gene sequences) of reference templates displaying different sizes but flanked by the same sequences recognized by the primers, can help in normalizing the results (237). Alternatively, anchored- (236) or inverse-PCR (238), in which all cDNA sequences are simultaneously amplified in the same reaction tube by a single pair of primers, were also successfully used.

Multiprobe RNase protection assay (RPA) has also been applied to assess BV gene expression profiles in mice (239–242), rats (243) and humans (99). This method utilizes radiolabeled riboprobes to hybridize cellular RNA. When insufficient, cellular RNA can be enzymatically amplified before analysis by combined anchored-PCR and T7-RNA polymerase-mediated transcription (244). After hybridization, unpaired sequences are digested by RNase A and T1, a very sensitive step in which even single mismatches are often recognized and cleaved. Double-stranded hybrids, in contrast, remain protected, and can be purified and separated by electrophoresis. The autoradiographic signals obtained are proportional to the amount of corresponding BV mRNA in the analyzed sample. Obviously, by using a set of riboprobes of different sizes and specific to each gene, expression of several BV genes can be simultaneously determined. The main advantage of this approach is the high specificity, allowing differentiation of individual subfamily members and even allelic variants of V gene segments (99,245). Like other methodologies, RPA also raises a number of possible problems, e.g., riboprobes cannot distinguish functional from non-functional transcripts, hybridization efficiency could be influenced by the topology of target sequences, and specificity of RNase digestion might not be always absolute, with certain mismatches only partially cleaved and, depending on the sequence context, certain homoduplexes may also be attacked.

6.2. CDR3 Analysis

V gene expression analysis is useful, but not very sensitive, and other methods are necessary to identify small, but relevant, T cell clonal expansions. Obviously, VDJ encoded CDR3 sequences constitute the most appropriate markers to identify specific clonotypes. As mentioned above, cloning and sequencing allow a detailed description of VDJ junctions, but identification of TCR sequences from clonally-expanded T cells requires analysis of large numbers of molecular clones to reach statistical significance. Other approaches, however, have been developed to preselect slightly over-represented VDJ sequences before analysis. These methods rely on unique molecular characteristics of diverse CDR3 sequences, in terms of electrophoretic mobility, hybridization kinetics and duplex stability.

One of these approaches focuses on the heterogeneity of VDJ-lengths, and was termed CDR3-size analysis, CDR3-spectratyping or Immunoscope (234,246–248). VDJC cDNA sequences are amplified by PCR using V and C primers, and the obtained products are separated by electrophoresis on denaturing sequencing gels. Several bands (typically 8 to 12), differing in 3 nucleotides (i.e. 1 codon) and showing a characteristic gaussian distribution of relative intensities, are observed when polyclonal cells are analyzed. In contrast, one or a few bands of increased intensity are obtained when clonal T cells are present in the sample. The sensitivity of the method can be improved by further dissecting the repertoire using J-specific primers, instead of the unique C primer.

An additional method is the single stranded conformational polymorphism (SSCP) analysis, based on the fact that electrophoretic mobility of single stranded DNA in nondenaturing gels is influenced by its nucleotide sequence (249). PCR amplified VDJC sequences are denatured and separated by non-denaturing gel electrophoresis. Based on their CDR3 sequence, different molecules will migrate at different rate, whereas dominant rearrangements from clonal T cells will resolve as unique bands that can be recovered and sequenced.

A variation of SSCP involves a step of renaturation before electrophoresis to create homo-and heteroduplexes. This method, referred to as hetero-duplex analysis (250) or double-stranded conformational polymorphism (DSCP) analysis

(251), appears to partially obviate comigration of different sequences, frequently observed in SSCP studies (251). Separation of diverse hetero and homoduplexes can also be achieved based on different melting properties in denaturing conditions, as provided by denaturing gradient gel electrophoresis (DGGE) and temperature gradient gel electrophoresis (252).

An alternative approach is the sequence enrichment nuclease assay (SENA), based on the reassociation of heat-separated complementary cDNA strands (244). This process follows second order kinetics, *i.e.*, its efficiency increases exponentially with the concentration of the reassociating DNA species. SENA utilizes this kinetic step to further enrich double stranded *VDJ* sequences from clonal T cells. Non-hybridized and heteroduplex sequences (from polyclonal T cells) are then eliminated by a single strand specific nuclease, whereas reassociated homoduplexes (mostly from dominant clonotypes) are selectively amplified by PCR and characterized by sequencing.

6.3. Analysis Based on Antigen Specificity

When the antigen is known or suspected, the presence and size of specific T cell populations can be evaluated by functional *in vitro* assays, using relevant peptides. Such approaches include analysis of T cell proliferation, cytokine production and cytotoxicity, and generally require expansion *in vitro* and limiting dilution to determine the frequency of specific cells. *In vitro* culture can however bias the T cell repertoire. Preactivated T cells often die due to activation induced cell death (AICD), when restimulated *in vitro*, leading to underevaluation of the corresponding frequency (253). Other assays, requiring no expansion but short-time culture *in vitro*, include ELISPOT and detection of intracellular cytokines (254).

An alternative method has been developed in the recent years, that allows direct *ex vivo* quantification and isolation of specific T cell clonotypes. This assay is based on flow cytometry using multimers (*e.g.*, tetramers) of soluble peptide-MHC molecules as staining reagents (253). Although most studies involved class I molecules, class II-based reagents have also been reported (255,256). Such products are obtained by recombinant technology, mostly in prokaryotic expression systems. For example, class I heavy chain is modified by substitution of the transmembrane and cytosolic domains with a biotinylation signal sequence, refolded in the presence of β2-microglobulin and a specific peptide, enzymatically biotinylated, and induced to form tetramers by addition of fluorescently-labeled streptavidin. This staining approach has been used to study T cell responses in several systems, including HIV (253), EBV (257), hepatitis C virus (258) and melanoma-associated antigens (259) in humans, and LCMV (254), influenza pneumonia virus (260) and herpes virus (261) in mice.

7. T CELL RECEPTOR REPERTOIRES IN NORMAL STATES

The efficiency of immune responses depends, in part, on the diversity of T cell populations available at the time of immunization. T cell repertoires are selected primarily in the thymus, where most T cells differentiate from bone marrow, or fetal liver-derived precursors (262,263). Thymocyte differentiation is characterized by sequential expression of various surface molecules, including CD4 and CD8. Initially, double negative (DN) for these coreceptors, thymocyte precursors upregulate CD4 and CD8 after a phase of proliferation. The resulting double positive (DP) cells need to be engaged through their TCR in low affinity interactions with peptide-MHC complexes on thymic epithelial cells to survive and further differentiate. This process, referred to as positive selection, ensures preferential maturation of T cells displaying self-MHC restricted TCR, and determines commitment to the CD4 or CD8 lineages resulting from preferential binding to class II or class I molecules, respectively (264). In contrast, too strong TCR interactions with peptides or SAg presented by MHC molecules on bone marrow-derived cells, induce negative selection or deletion of maturing thymocytes, a mechanism that contributes to maintain tolerance to self constituents (2,265,266). In addition, recent evidence indicated that low affinity recognition of self peptide-MHC complexes is also important at the periphery for the survival and homeostatic expansion of naïve mature T cells (267).

The analysis of thymic and peripheral TCR repertoires in humans and animal models has been a subject of intensive investigation over the past several years. As summarized in the preceding section, several methodologies have been developed for this purpose. The obtained data are particularly useful to interpret repertoire alterations and clonal accumulations observed during responses to pathogens, but also during the development of autoimmune diseases. Baseline levels of TCR *V* gene expression have been estimated in immature DP thymocytes (before thymic selection) from normal mice (241,242), rats (243) and humans (99). For example, marked differences were observed within each thymus, even among homologous TCR *BV* genes classified in the same family. In humans, *BV2s1, 3s1, 5s1, 5s2, 8s1, 13s1* and *17s1* were the most abundant (>2%) and *BV8s3, 11s1, 13s2* and *20s1* the least abundant (<0.5%), while intermediate levels (0.5–2%) were exhibited by the remaining *BV* (99). The fact that as few as 6 to 7 genes of the 45 available encompass up to 50% of the functional β chain transcripts is intriguing and might be indicative of an extensive redundancy in the *BV* genetic reservoir. The molecular mechanisms leading to this unequal expression are, however, still unknown. As is the case for Ig *VH* genes (268), a correlation between expression frequency and chromosomal location of mouse TCR *GV* genes has been observed during ontogeny. The first wave of virgin T cells express TCR *Vγ3* (*GV1s1*), followed by a second wave of TCR *Vγ4* (*GV2s1*) cells (269,270). In late fetal life, development of these two subpopulations ends or slows considerably, and is followed by an adult program with successive waves of appearance (without disappearance) of *Vγ2* (*GV3s1*) and *Vγ5* (*GV4s1*) expressing cells (270). These *γ/δ* cells are exported to the periphery and exhibit tissue-restricted distribution, *i.e.* *Vγ3* in skin, *Vγ4* in the female reproductive tract and in the tongue, *Vγ2* in spleen and lymph nodes, *Vγ5* in intestinal epithelium (271). As discussed above (see Table 4), such order of rearrangement and expression of *GV* genes is colinear with their physical map on cluster *γ1* of chromosome 13 (76). Similar studies in fetal mice (272) and humans (245,273) failed to reveal analogous ontogeny programs for *BV* genes, which are generally used independently of their chromosomal position. Thus, other factors than the distance between *BV* genes and *D-J-C* clusters in the locus should be evoked to explain expression dominance, including differences in transcriptional rates, Vβ/Vα pairing [as suggested by studies with cell lines or transgenic mice (274,275)], chromatin conformation and accessibility, as well as recombinase signal sequences. For example, the extensive variations in the expression of *BV3s1* observed among individuals (99,276) appear to correlate with a single nucleotide difference within the 23 bp spacer that separates heptamer and nonamer recombination sequences (103). It is possible that similar polymorphisms could affect the rearrangement efficiency of different *BV* genes, explaining the observed variations in expression levels.

Overall, the TCR *V* gene repertoires established in the thymus are conserved by mature T cells in peripheral blood and lymphoid organs, and *BV* genes that dominate thymic β chain transcripts remain dominant at the periphery (99,243,276–280). However, thymic selection results in dramatic reductions in the expression of several *BV* gene segments at the periphery of various mouse strains (226,227). Such Vβ deletions correlated with the expression of minor lymphocyte stimulating (Mls) determinants, later found to be encoded by mouse mammary tumor virus (MMTV) sequences integrated in the genomes of these mouse strains (2,180). In particular, it was found that ORF sequences in the 3' long terminal repeats of *MMTV* genes encode endogenous SAg that, presented in the thymus, induce negative selection of specific thymocyte populations expressing the engaged Vβ products. Thus, due to the expression of various MMTV combinations, T cells expressing *BV6s1, 7s1, 8s1* and *9s1* are deleted in MLs[a] mice, T cells expressing *BV3s1* are deleted in MLs[c] mice, and T cells expressing *BV5s1, 11s1, 12s1, 16s1* and *17s1* are deleted, either partially or completely, in I-E[+] mice. In the mouse, Vβ deletions were also observed when bacterial SAg were given intrathymically, neonatally, or peripherally

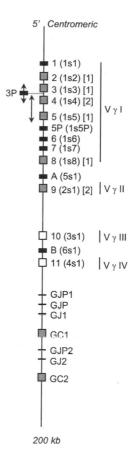

Figure 3. Locus for human γ TCR genes on chromosome 7 (7p15). Approximate position of functional *V* gene segments (shadowed boxes), non-functional *V* gene segments with ORF (open boxes) and pseudogenes (filled boxes) is according to reported sequencing data. The locus representation was adapted from "IMGT, the international ImMunoGeneTics database" (http://imgt.cnusc.fr:8104). Each *V* gene segment is named based on two different nomenclatures, the first according to the gene position in the locus, and the second (in parenthesis) corresponding to published alignments (51). The number of functional alleles for each *V* gene segment is given in brackets. Single arrows indicate opposite transcriptional orientation, whereas double arrows correspond to insertion/deletion polymorphisms.

in adult individuals (2), and similar observations were made in the rat (243). In humans, the possibility of intrathymic deletions could be demonstrated using homozygous *C.B-17* scid/scid (SCID) mice implanted with fragments of human fetal thymus and liver, and injected with the bacterial SAg SEB (273). In contrast to mice, however, no complete deletions were observed in untreated rats and humans. Nonetheless, differences between immature DP thymocytes and syngeneic mature SP thymocytes (99), or between peripheral T cells from different donors (243,276–280), were identified for several *BV* genes in both rats and humans. These differences could be due to thymic selection by weak SAg or by peptide-MHC complexes displaying preferential affinity for specific Vβ products, as well as to peripheral repertoire selection associated to homeostasis, tolerance or infection.

Theoretical calculations estimated the potential T cell repertoire in mice and humans as composed of approximately 10^{15} to 10^{18} clonotypes (Table 5). The effective number of different clonotypes in normal individuals is, however, much smaller. Indeed, the total number of T cells is limited to approximately 10^8 in mice and 10^{12} in humans. Furthermore, and as exemplified for instance by the unequal V gene expression by immature DP thymocytes (99), preferential VDJ rearrangements along with thymic and peripheral selection events significantly shape the repertoire, only retaining the most useful clonotypes. Accordingly, TCR gene amplification and sequencing studies indicated that humans have about 10^6 different β chains in the blood, each pairing with at least 25 different α chains (281). In contrast, in the memory subset, the diversity appeared limited to 10^5 different β chains, each pairing with only a single α chain (281).

Preferential expression in either the CD4 or CD8 T cell subsets was observed for several Vβ (99,240,243,282–284) and Vα (285–287) gene products. As observed in genetically identical mice (240), the extent of such biased representation appeared more conserved in identical twins or in siblings sharing HLA haplotypes than in unrelated individuals, suggesting the role of antigen presenting MHC class I and class II molecules (276–280). Studies with SCID mice reconstituted with human fetal liver and thymus from syngeneic or allogeneic donors gave similar conclusions. Indeed, when identical stem cells developed in different thymic environments, significant differences in the BV gene usage by mature SP CD4 and CD8 cells were observed, indicating that thymocyte selection is influenced by polymorphic determinants on thymic epithelial cells (245). Whether an increased representation of T cells expressing a defined Vβ, for example, in the CD8 subset is due to preferential positive selection by class I molecules or to negative selection by class II molecules, remains unknown. Polymorphic residues in Vα CDR1 and CDR2 that are associated to CD4/CD8 skewing (286,287) were found in crystal structures to correspond to positions implicated in the interaction with MHC molecules (288). In addition, Vα mutagenesis studies indicated that amino acids determining preferential expression in the CD8 subset actually reduce affinity to class I molecules, an observation compatible with a quantitative-instructional model of thymocyte maturation or with decreased negative selection (289).

With regard to the effects of thymic involution and aging, Vβ repertoires remain relatively stable in both total thymocytes and splenocytes throughout the lifespan of the mouse, and without leakage of endogenous SAg-reactive cells (242). Moreover, syngeneic bone marrow transplantation experiments showed that an involuted thymus is perfectly capable of supporting T cell subset development and full Vβ repertoire acquisition (242). Recent studies in humans addressing thymic repertoire diversity and presence of recent thymic emigrants in blood also indicated that thympoiesis remains active in aging individuals (290). Nevertheless, age-associated changes could be observed, particularly in the peripheral CD8 subset of older mice. Thus, some of the endogenous SAg-reactive clonotypes that, early in life, showed incomplete deletions in the CD8 subset (Vβ7 in BALB/c and Vβ5.1, 7 and 12 in DBA/2 mice) declined further with age, whereas other Vβ clonotypes without such reactivity (Vβ2, 14, 15 in DBA/2) significantly expanded with age (242). Subsequent studies noted even more pronounced, stochastically-imposed, Vβ expansions in the CD8 subset (291). Interestingly, similar expansions have also been observed in normal humans and appear accentuated in elderly individuals, sometimes giving the false impression of a T cell malignancy (248,292). Further studies indicated that oligoclonally expanded CD8 T cells may express CD45RO (248) and CD57 (293), lack CD28, and display shorter telomeric lengths (294), consistent with a state of replicative senescence. Also, Vβ repertoire differences in these subsets were identified even in monozygotic twins (280). Although the reasons for such expansions remain unknown, their frequent presence in normal individuals should be taken into consideration while interpreting the significance of expanded clonotypes in immune responses and disease.

The extent of TCR V gene diversity in antigen specific responses has been analyzed with various foreign and self-antigens involving MHC class I and class II restriction. Overall, the complexity of V gene usage reflects the complexity of the antigen, with broad V gene and junctional diversity in response to allogeneic stimuli (295,296), heterogeneous (297–299) or more restricted (222,300–303) TCR in response to whole protein antigens, and rather limited V gene usage in response to individual peptides (300,304–312). Several studies, however, indicated that even single peptides can engage diverse TCR (205), or TCR displaying restricted V regions, but heterogeneous CDR3 (221,313). More recently, experiments with tetramers of peptide-MHC complexes showed that antigen-specific T cells, that can contribute up to 70% of the CD8 population at the pick of the response (253,254,257–261) are often polyclonal, although they generally present preferential V gene expression and CDR3 motifs (314,315). Studies with in vivo- and in vitro-stimulated T cells suggested, that the Vβ restriction observed in some TCR-peptide-MHC interactions may be due to CDR1 and CDR2 residues providing optimal interaction with defined peptide side chains, as single amino acid replacements in the peptide C-terminus can induce a change in the dominant Vβ used by the responding T cells (223). Of note, the restricted V gene usage in TCR specific for certain self-antigens has raised the possibility of TCR-targeting immune interventions in autoimmune diseases (300).

8. T CELL RECEPTOR REPERTOIRES IN AUTOIMMUNE DISEASES

8.1 Rheumatoid Arthritis

Rheumatoid arthritis (RA) is a chronic joint disease of unknown etiology, characterized by a progressive inflammatory response involving synovia and, less frequently, extra-articular organs, finally leading to erosion and destruction of cartilage and bone (316). The autoimmune basis of RA is suggested by the infiltration of affected joints by macrophages, B cells and T cells. The role of T cells is also supported by the disease association with certain MHC variants that presumably present pathogenic self-antigens to specific T cell clones. Most RA patients express, in fact, products of alleles *HLA-DRB1 *0401* (DR4Dw4), *0404/*0408* (DR4Dw14), *0405* (DR4Dw15) and *0101* (DR1) (317), all sharing a stretch of polymorphic residues (codons 67–74) on the β chain α-helix, that profoundly influences peptide binding and T cell recognition (318).

To investigate whether TCR gene loci exhibit linkage with RA, mapping studies using microsatellite markers and TCR *V* gene polymorphisms were performed on large groups of RA patients and controls. Overall, the results showed no association with RA (319), although a possible role of polymorphic *AV8s1* (320), *AV5s1* (321), *BV6s7, BV13s5* and *BV12s4* (322), has been suggested.

Considerable effort has been made to examine TCR diversity in RA synovial fluid, synovial tissue, blood and, in fewer cases, T cell subsets identified with markers like CD4, CD8, CD45RO, CD28 and CD57. Various methods of *V* gene expression analysis have been employed, including flow cytometry with specific antibodies (323,324), quantitative PCR (325–330), and RNase protection assay (244). In general, the results have favored the view of a very diverse *V* gene usage, suggesting that polyclonal T cells are present at the site of inflammation and that no specific TCR *V* gene can be identified as a disease marker for all RA patients. However, depending on the study and the patient analyzed, enhanced expression has been reported for particular *AV* (323,325 327,331), *BV* (244,324–326, 328–330, 332–335), *GV* (336–339) and *DV* (339,340) gene segments.

Dominant *V* gene expression has often been interpreted as suggesting antigen-driven T cell expansions. Based on homologies in the HV4 of *BV3s1, BV14s1* and *BV17s1*, preferentially used in some patients (328), and the dominant expression of *BV14s1* without *VDJ* restriction in

another group of patients (329), it was suggested that SAg might induce proliferation of autoreactive T cell clones, although additional groups of affected individuals did not show the same *BV* restriction. On the other hand, characterization of TCR *VDJ* junctions by CDR3-size analysis, SSCP, SENA and sequencing, clearly indicated that clonally-expanded T cells are present in RA synovia and blood, suggesting the role of unknown conventional self peptides (244,323,323,326,328,329,332,334–336,340–346). Although clonotypes are generally different among individuals, examples of shared sequences have been reported (244,335,346). Within individual patients, T cell clones can be repeatedly found at different sites of the same lesion (343), in different joints (344,345,347), in both synovia and blood (244,345), and some of them persisted for over one year (332,344,346). In addition, expanded T cell clonotypes can be identified in both CD4 and CD8 subsets (244,323,332,342,347). Particularly relevant are CD4 T cells lacking CD28, which are found to display killer activity (348) and to be protected from apoptosis induction, in part through elevated expression of Bcl-2 (349). The origin and precise antigen specificity of clonally expanded T cells in RA remains unknown, although isolated synovial T cell clones and cell lines were found to react *in vitro* with several antigens, including EBV transactivators (350), mycobacterial heat-shock protein (351), cartilage proteoglycans and mycobacterium tuberculosis (352), determinants presented by autologous APC (353), or with various collagen types (352,354). Nevertheless, recent studies using tetramers of peptide-HLA-DR4 complexes failed to identify significant numbers of synovial CD4 T cells specific for type II collagen and cartilage gp39 in RA patients (355).

Animal models of RA include collagen-induced arthritis (CIA), observed in predisposed rodents upon intradermal injection of native type II collagen (356). In the mouse, susceptibility is linked to MHC genes of the H2^q and H2^r haplotypes, but RA-associated HLA transgenes were also found to predispose to CIA (357,358). Restricted TCR repertoires were observed in CIA mice (359). In particular, CD4 T cells expressing Vα11.1 (*AV11s1*) and Vβ8.2 (*BV8s2*) appear to play a key role in DBA/1 mice (H2^q) (360). Accordingly, reduction of CIA was observed in mice carrying genomic *BV* gene deletions (116), or treated with either anti-Vβ8.2 antibodies (360) or peptides derived from Vα11.1 and Vβ8.2 (361,362). Interestingly, studies with TCR transgenic mice indicated a role of *AV11s1* gene polymorphisms in CIA (363).

Another recently reported model of RA involves C57BL/6xNOD mice transgenic for the TCR KRN (364). It was shown that this TCR that reacts with a peptide derived from glucose-6-phosphate isomerase presented by the NOD-derived MHC class II A^{g7} (365), and confers broad autoreactivity, leading to B cell activation and production of arthritogenic Ig, that can induce disease even in the absence of lymphocytes (366).

8.2   Insulin-Dependent Diabetes Mellitus

Insulin-dependent diabetes mellitus (IDDM) results from a progressive autoimmune reaction that ultimately causes the selective elimination of insulin-secreting pancreatic β cells. A number of studies documented the MHC association of IDDM, with a predisposition conferred by HLA-DQ8 (*DQA1 *0301/DQB1 *0302*) and the protective effect of HLA-DQ6 (*DQA1 *0102/DQB1 *0602*) (367). Several candidate antigens have been described as antibody and/or T cell targets, including insulin, glutamic acid decarboxilase (GAD) and the tyrosine phosphatase ICA512 (IA-2) (368).

Initial studies have reported an association between IDDM and a *TCRBC* defined RFLP (369). Additional experiments, however, were unable to confirm these results, suggesting that *TCR* α, β and γ loci are not likely to include genes conferring susceptibility to IDDM (370–372).

TCR repertoire analysis of infiltrated pancreatic islets indicated the presence of diverse TCR, although increased expression was reported, especially in early disease, for various *AV* and *BV* genes (373,374), including *AV14* and *BV3* (375), *BV8* (376), or *BV7* (377). In the latter study, the diverse CDR3 and *AV* gene products associated with *BV7* evoked the possible involvement of a SAg (377). Indeed, a novel endogenous retrovirus related to mouse mammary tumor virus and possessing an *env* gene that encodes a *BV7*-specific SAg, was found to be preferentially expressed in patients (378). However, additional studies have failed to confirm the association between expression of this retrovirus and IDDM (379–381).

The model constituted by nonobese diabetic (NOD) mice has been extensively studied. NOD mice develop a pancreatic disease dependent on CD4 and CD8 T cells. The main susceptibility gene locus has been mapped to the MHC region of chromosome 17. In particular, NOD mice were found to express a unique class II molecule, IA^{g7}, that like DQ8 in humans display a non-Asp amino acid in position 57 of the β chain, and display reduced stability and peptide binding (382).

Initial TCR analysis indicated heterogeneous *V* gene usage by infiltrating T cells of NOD mice (383). However, biased expression was often observed in early disease (384–386), including biases for *BV1* (387) and *BV8s2* (388) sequences displaying limited CDR3 diversity. In addition, islet T cells expressing a transgenic β chain obtained from a pathogenic clonotype, were found to preferentially use an endogenous α-chain identical to that identified in the same clonotype (389). Diverse *BV* gene usage was also found in several T cell clones and lines (390,391), including those displaying defined anti-insulin (392) and anti-GAD65 (393) specificities, while reduced complexity was reported upon

analysis of α chain CDR3 in anti-insulin (394), and in CD8 positive cytotoxic (395, 396), T cell clones.

The pathogenic role of subsets displaying specific *V* gene products was studied in experiments aimed at modifying the complexity of TCR repertoires. No effect on IDDM was found in NOD mice carrying a genomic deletion affecting half of the *BV* genes (397), in NOD mice expressing, on most T cells, a transgenic TCR unrelated to the disease (398), and in NOD mice treated with antibodies to several *BV* gene products, including *BV3*, *BV5*, *BV8*, *BV11* and *BV14* (399). In contrast, treatment with anti-Vβ8 prevented cyclophosphamide-induced disease in NOD/Wehi mice (400), and depletion with anti-Vβ6 reduced T cell pathogenicity in transfer experiments (401). The importance of TCR repertoire diversity was also addressed in experiments evaluating the protective effect of class II molecules of other haplotypes in trasgenic and congenic NOD mice. These studies suggested that, due to its instability and inefficient peptide binding, IA^{g7} may select T cells displaying increased affinity for self-peptides (402,403). Moreover, and in contrast to the protecting transgenic IA^{b}, IA^{g7} may be defective in selecting a larger repertoire of T cell specificities (404).

8.3   Multiple Sclerosis

Considered as the most common autoimmune neurological disease, multiple sclerosis (MS) is charaterarized by an inflammatory infiltrate of the central nervous system leading to demyelination of myelin sheaths, inhibition of nerve conduction, and impared visual, motor and sensory functions. Many genetic loci are involved in determining susceptibility, including the *MHC* locus. Target antigens include myelin basic protein (MBP), proteolipid protein (PLP) and myelin oligodendrocyte glycoprotein. Association of MS with RFLP identified with TCR *BV* and *BC* probes has been reported in several studies (405–409), but was not confirmed by other reports (410–413).

TCR *V* gene analysis in MS lesions and plaques indicated, depending on the study, restricted *AV* gene expression with limited *AJ* usage in Vα12.1 (*AV12s1*) associated CDR3 sequences (414), polyclonal TCR in active, but restrictions in chronic, lesions (415), and increased usage of Vβ5.2 (*BV5s2*) with CDR3 motifs similar to anti-MBP clones (416), or Vβ5.3 (*BV5s3*) rearranged to *BJ1s4* (417). Similarly, analysis of γ/δ T cells indicated increased expression for Vδ1, Vδ2 and Vγ2 genes (418), for Vγ2 and Vδ2 sequences displaying patient-specific CDR3 (419), or for Vδ2–*DJ3* rearrangements (420).

Cerebrospinal fluid (CSF) was also examined using various approaches and heterogeneous results were obtained. Comparison of different patients for TCR usage in samples analyzed either *ex vivo* or after culture in the presence of IL2 and IL4, revealed identical clones in CSF and blood and common usage of *Vβ12* (421), predomince of *Vβ2* (422), overrepresentation of *Vβ6.5* and *Vβ6.7* (423), or increased usage of *Vβ17–BJ1s6* rearrangements of identical size (424). Moreover, analysis of γ/δ T cells in CSF indicated expansion of oligoclonal T cells expressing *Vδ1* and *Vδ2* genes (425), biased usage of *Vγ1* and *Vδ1* sequences bearing diverse CDR3 in recent onset disease (426), or clonal populations expressing *Vγ9-Jγ1*, *Vγ10-JγP1* and *Vγ9-JγP* (427).

Characterization of MBP-specific T cell lines and clones derived from CSF or blood of MS patients revealed diverse TCR *AV* and *BV* usage in some studies (428–430), although restricted, mostly patient-specific, *V* gene expression was observed in several other studies (431), including for *Vβ17* and *Vβ12* (432), *Vβ5.2* and *Vβ6.1* (433), and *Vβ7*, *Vβ17* and *Vα2* (434). Interestingly, *in vitro* stimulation with MBP or tetanus toxoid resulted in the selection of different TCR *AV* gene repertoires in discordant monozygotic twins, but not in concordant and control twin sets (435), suggesting that factors possibly associated to disease shape the T cell repertoire in MS patients. However, the *V* gene usage was not conserved between sets of twins.

The restricted *V* gene repertoires observed in certain patients as well as in animal models (see below) suggested the possibility of using TCR vaccines to modulate the presence and activity of presumably pathogenic T cell populations. TCR peptides corresponding to the CDR2 of V*β5.2* and V*β6.1* chains were found to induce T cell immunity and antibodies in several patients (436,437). Moreover, results of a phase I trial of a Vβ6 CDR2 peptide vaccine in 5 patients indicated a slight decreased CSF cellularity and a marked diminution of Vβ6 mRNA in cytokine supplemented CSF expansion cultures (438).

Experimental autoimmune encephalomyelitis (EAE) is a rodent model of MS that can be induced by immunization with spinal cord homogenate, with components of the myelin sheath like MBP and PLP and with peptides thereof. Several studies have examined the TCR repertoire in EAE-susceptible PL/J and B10.PL mice, and reported restricted usage of *Vβ8.2*, *Vβ13*, *Vα2* and *Vα4* genes displaying conserved, junctional sequences (439). Similar restrictions were observed in Lewis rats, even though these responses were directed to different peptide portions of MBP. Importantly, antibodies to such predominant TCR were found to prevent and even cure EAE (440,441), as did vaccination with inactivated encephalitogenic T cell clones (442,443) with synthetic TCR peptides (444,445), or with naked DNA encoding Vβ8.2 (446), although a more severe disease was observed in a group of *Vβ*-peptide immunized rats (447).

8.4   Autoimmune Thyroid Diseases

Endocrine disorders are frequently due to autoimmunity to thyroid antigens. Graves' disease and Hashimoto's thyroiditis, the most common of these pathologies, are characterized by the production of antibodies to thyroglobulin, thyroid peroxidase or thyroid-stimulating hormone receptor. These pathologies also present a typical lymphocyte infiltration of the thyroid and other extrathyroidal sites, including eye orbital tissue and pretibial tissue.

No linkage with Graves' disease was found using polymorphic microsatellite markers for immunoregulatory genes, including *TCRAV*, *TCRBV*, *IgH* and *CTLA-4* (448). On the other hand, TCR repertoire analysis revealed restricted *AV* (449) and, to a lesser extent, *BV* (450) gene usage in early Graves' disease, suggesting intrathyroidal T cell clonal expansions, although broader repertoires were detected in later stages of both Graves' and Hashimoto diseases (451–453). Biased *V* gene expression appeared different in unrelated patients. Interestingly, similar TCR *V(D)J* sequences were reported in thyroid and extrathyroid tissues of a few patients with Graves' disease (454).

8.5   Myasthenia Gravis

Myasthenia Gravis (MG) is an autoimmune disorder characterized by muscle weakness caused by antibodies to nicotinic acetylcholine receptors (AChR) at the post-synaptic membrane of neuromuscular junctions. Several studies imply AChR-specific helper T cells in the production of pathogenic autoantibodies. On this basis, thymectomy is a common therapeutic option for MG.

TCR *V* gene usage was studied in thymus, blood and isolated T cell clones and lines from patients and controls. Depending on the study, increased Vβ expression was observed in the thymus of patients compared to healthy controls, for *BV5s1* and *BV8* (455), or *BV2* (456). Similarly, in patient serum, increased expression was detected for *BV12* (457), *BV1*, *BV13s2*, *BV17* and *BV20* (458), or *BV4*, *BV6*, *BV15*, *BV16* and *BV24* (456). In particular, *BV4* and *BV6* also appeared increased in patient serum *versus* thymuses (456), and were frequently used by AChR-specific T cell clones, although several other *BV* genes were also observed in clones from different patients (459).

Experimental autoimmune myasthenia gravis (EAMG) can be induced in predisposed mice by immunization with AChR. TCR repertoire analysis in C57BL/6 mice (H2^b) showed restricted usage of *BV8* sequences (460), or *BV6* sequences displaying homologous CDR3 with a conserved glutamic acid (461). Also, preferential *AV1s8* expression was observed in mice expressing a transgenic *BV8s2* (462). In contrast, the relatively resistant bm12 mice, showing only 3 amino acid differences from H2^b class II molecules, presented broader TCR repertoires and AChR peptide specificities (463). Treatments *in vivo* with anti-Vβ8 antibodies inhibited proliferation of specific T cells in C57BL/6 mice (460), and EAMG in *BV*8s2-transgenic mice (462), while anti-Vβ6 antibodies did not have effect on EAMG

(464). Reduced progression of EAMG was also observed using soluble peptide-MHC complexes (465), with peptide analogs acting as partial agonists (466), and after induction of neonatal tolerance to an immunodominant AChR peptide (467).

8.6   Systemic Lupus Erythematosus

Systemic lupus erythematosus (SLE) is the prototype of systemic autoimmune diseases. Involving multiple organs, SLE is associated with diverse clinical manifestations that can vary from transient forms of skin rash and joint pain to a severe kidney disease. The hallmark of SLE is increased production of autoantibodies to various self-constituents, particularly of nuclear origin, including chromatin components like nucleosomes, histones and DNA.

TCR repertoire studies by CDR3 size and SSCP analysis indicated clonal expansions in blood of SLE patients for both α/β (468–470) and γ/δ (471,472) T cells, although no preference for particular V gene segments has been suggested. In addition, the antigen specificity of such *in vivo* accumulating clonotypes remains to be clarified.

T cell clones, acting *in vitro* as helper cells for anti-nuclear antibody production, have been obtained from several SLE patients. Some of these clones appeared to recognize non-histone chromosomal proteins, other were specific for histone epitopes, and most used TCR chains displaying charged amino acids in their CDR3 (473). However, autoantibody-inducing T cell clones expressing diverse TCR BV genes without preference for charged amino acid containing junctional sequences have also been reported (474).

Inbred animal models that consistently develop a disorder closely matching human SLE include New Zealand black (NZB) mice, NZB crossed with New Zealand white (NZB × NZW) mice, NZB × SWR mice, MRL mice homozygous for the lymphoproliferation (*lpr*) gene (MRL-*lpr*), and BXSB mice carying the disease-accelarating, Y chromosome-associated, *Yaa* gene (475). These mice, and their genetic variants obtained using transgenic and gene-knockout technologies, have significantly contributed to the clarification of the role of T cells in SLE.

Genomic studies indicated no association with particular $V\alpha$ or $V\beta$ haplotypes (82), while the chromosomal *BC1-BD2-BJ2* deletion in the *TCR* β locus of NZW mice (476,477) appeared to play no role in the NZB × NZW disease (478). In addition, more recent mapping studies using microsatellite markers indicated that TCR loci are not significantly associated with SLE (479).

TCR repertoire analysis indicated that lupus developing mice, at both young and advanced ages, exhibit deletion of $V\beta$ clonotypes similar to those of MHC and MMTV-matched normal controls (480,481). Overall, animals with disease displayed heterogenous BV gene repertoires,

although perturbations were observed, particularly in aged mice (482,483). Evidences that some of these alterations are due to peripheral clonal expansions has been obtained for MRL-*lpr* mice (484).

Anti-DNA inducing T cell clones and lines have been obtained from NZB × SWR lupus mice and found to exhibit diverse BV repertoires (485). As observed for similar clones derived from human patients, some of these T cells appear to recognize nucleosomal histone epitopes and display a significant frequency of negatively charged amino acids in their CDR3, particularly in the α chain, which seems to confer MHC-unrestricted antigen recognition even when paired with TCR β chains of irrelevant specificity (486).

9.   CONCLUDING REMARKS

Considerable advances have been made during the past few years towards better comprehension of TCR biology. Since most human and mouse TCR gene loci have been entirely sequenced, it is reasonable to assume that most of the corresponding V, D and J gene segments have been identified. This information will greatly facilitate the characterization of the full reservoir of allelic variants, the definition of their involvement in various immune responses and diseases, and the design of more efficient methods of TCR repertoire analysis.

Recent crystallographic studies have established that TCR are structurally similar to antibody Fab, and bind peptide-MHC in various orientations that can be more diagonal in the case of MHC class I, and more orthogonal in the case of MHC class II. Future reports will undoubtedly provide more information necessary to understand and eventually predict the details of TCR ligand interactions, and explain the plasticity and degeneracy of antigen recognition often observed *in vitro* and *in vivo*.

Current methods of TCR repertoire analysis, generally focusing on the usage of V genes and hypervariable CDR3 sequences, are imperfect. A number of studies, however, have indicated that combinations of these approaches can give a fairly precise picture of the complexity of T cell repertoires in various physiological and pathological situations. For instance, it has been established that the functional TCR repertoire is considerably more limited than the potential repertoire predicted by theoretical calculations. Preferential rearrangements and positive/negative selection in the thymus, as well as homeostatic regulation and antigen selection at the periphery, obviously contribute to reduce repertoire diversity, although further experiments are needed to fully clarify the exact mechanisms governing each of these steps.

With regard to the contribution of specific T cell sub-populations to the development of autoimmune diseases, several questions remain to be answered. Although TCR restrictions and clonally-expanded T cells have been repeatedly observed in most autoimmune diseases, the results in different studies and/or in different patients are, overall, not congruent. This is likely caused by the potential degeneracy of antigen recognition, allowing even a single ligand to select several different TCR. It is still not clear, however, whether the observed restrictions are due to the proliferation of relevant T cells actually involved in the autoimmune response, or to the secondary attraction and sequestration in inflammation sites of previously expanded T cells of diverse specificities. In addition, important information is still lacking concerning the antigen recognized by the dominant clonotypes identified in patients as well as in apparently normal individuals. Studies addressing specificity and frequency of similar T cell clones will be necessary to evaluate the relevance of disease-associated vs. naturally-occurring clonal expansions, and to design more efficient therapies.

References

1. Germain, R.N. 1994. MHC-dependent antigen processing and peptide presentation: providing ligands for lymphocyte activation. *Cell* 76:287–299.
2. Kotzin, B.L., D.Y. Leung, J. Kappler, and P. Marrack. 1993. Superantigens and their potential role in human disease. *Adv. Immunol.* 54:99–166.
3. Porcelli, S.A. and R.L. Modlin. 1999. The CD1 system: antigen-presenting molecules for T cell recognition of lipids and glycolipids. *Annu. Rev. Immunol.* 17:297–329.
4. Haas, W., P. Pereira, and S. Tonegawa. 1993. Gamma/delta T cells. *Annu. Rev. Immunol.* 11:637–685.
5. Born, W., C. Cady, J. Jones-Carson, A. Mukasa, M. Lahn, and R. O'Brien. 1999. Immunoregulatory functions of gamma delta T cells. *Adv. Immunol.* 71:77–144.
6. Schild, H., N. Mavaddat, C. Litzenberger, E.W. Ehrich, M.M. Davis, J.A. Bluestone, L. Matis, R.K. Draper, and Y.H. Chien. 1994. The nature of major histocompatibility complex recognition by gamma delta T cells. *Cell* 76:29–37.
7. Sciammas, R., R.M. Johnson, A.I. Sperling, W. Brady, P.S. Linsley, P.G. Spear, F.W. Fitch, and J.A. Bluestone. 1994. Unique antigen recognition by a herpesvirus-specific TCR-gamma delta cell. *J. Immunol.* 152:5392–5397.
8. Weintraub, B.C., M.R. Jackson, and S.M. Hedrick. 1994. Gamma delta T cells can recognize nonclassical MHC in the absence of conventional antigenic peptides. *J. Immunol.* 153:3051–3058.
9. Davodeau, F., M.A. Peyrat, I. Houde, M.M. Hallet, G. De Libero, H. Vie, and Bonneville. 1993. Surface expression of two distinct functional antigen receptors on human gamma delta T cells. *Science* 260:1800–1802.
10. Hochstenbach, F. and M.B. Brenner. 1989. T-cell receptor delta-chain can substitute for alpha to form a beta delta heterodimer. *Nature* 340:562–565.
11. Acuto, O., R.E. Hussey, K.A. Fitzgerald, J.P. Protentis, S.C. Meuer, S.F. Schlossman, and E.L. Reinherz. 1983. The human T cell receptor: appearance in ontogeny and biochemical relationship of alpha and beta subunits on IL-2 dependent clones and T cell tumors. *Cell* 34:717–726.
12. Brenner, M.B., J. McLean, D.P. Dialynas, J.L. Strominger, J.A. Smith, F.L. Owen, J.G. Seidman, S. Ip, F. Rosen, and M.S. Krangel. 1986. Identification of a putative second T-cell receptor. *Nature* 322:145–149.
13. Bank, I., R.A. DePinho, M.B. Brenner, J. Cassimeris, F.W. Alt, and L. Chess. 1986. A functional T3 molecule associated with a novel heterodimer on the surface of immature human thymocytes. *Nature* 322:179–181.
14. Lew, A.M., D.M. Pardoll, W.L. Maloy, B.J. Fowlkes, A. Kruisbeek, S.F. Cheng, R.N. Germain, J.A. Bluestone, R.H. Schwartz, and J.E. Coligan. 1986. Characterization of T cell receptor gamma chain expression in a subset of murine thymocytes. *Science* 234:1401–1405.
15. Moingeon, P., S. Jitsukawa, F. Faure, F. Troalen, F. Triebel, M. Graziani, F. Forestier, D. Bellet, C. Bohuon, and T. Hercend. 1987. A gamma-chain complex forms a functional receptor on cloned human lymphocytes with natural killer-like activity. *Nature* 325:723–726.
16. Nakanishi, N., K. Maeda, K. Ito, M. Heller, and S. Tonegawa. 1987. T gamma protein is expressed on murine fetal thymocytes as a disulphide-linked heterodimer. *Nature* 325:720–723.
17. Littman, D.R., M. Newton, D. Crommie, S.L. Ang, J.G. Seidman, S.N. Gettner, and A. Weiss. 1987. Characterization of an expressed CD3-associated Ti gamma-chain reveals C gamma domain polymorphism. *Nature* 326:85–88.
18. Yanagi, Y., Y. Yoshikai, K. Leggett, S.P. Clark, I. Aleksander, and T.W. Mak. 1984. A human T cell-specific cDNA clone encodes a protein having extensive homology to immunoglobulin chains. *Nature* 308:145–149.
19. Hedrick, S.M., D.I. Cohen, E.A. Nielsen, and M.M. Davis. 1984. Isolation of cDNA clones encoding T cell-specific membrane-associated proteins. *Nature* 308:149–153.
20. Chien, Y., D.M. Becker, T. Lindsten, M. Okamura, D.I. Cohen, and M.M. Davis. 1984. A third type of murine T-cell receptor gene. *Nature* 312:31–35.
21. Saito, H., D.M. Kranz, Y. Takagaki, A.C. Hayday, H.N. Eisen, and S. Tonegawa. 1984. A third rearranged and expressed gene in a clone of cytotoxic T lymphocytes. *Nature* 312:36–40.
22. Sim, G.K., J. Yague, J. Nelson, P. Marrack, E. Palmer, A. Augustin, and J. Kappler. 1984. Primary structure of human T-cell receptor alpha-chain. *Nature* 312:771–775.
23. Saito, H., D.M. Kranz, Y. Takagaki, A.C. Hayday, H.N. Eisen, and S. Tonegawa. 1984. Complete primary structure of a heterodimeric T-cell receptor deduced from cDNA sequences. *Nature* 309:757–762.
24. Chien, Y.H., M. Iwashima, K.B. Kaplan, J.F. Elliott, and M.M. Davis. 1987. A new T cell receptor gene located within the alpha locus and expressed early in T cell differentiation. *Nature* 327:677–682.
25. Bork, P., L. Holm, and C. Sander. 1994. The immunoglobulin fold. Structural classification, sequence patterns and common core. *J. Mol. Biol.* 242:309–320.
26. Kabat, E.A., T.T. Wu, M. Reid-Miller, H.M. Perry, and K.S. Gottesman. 1987. Sequences of Proteins of Immunological Interest. US Dept. of Health and Human Services.
27. Jores, R., P.M. Alzari, and T. Meo. 1990. Resolution of hypervariable regions in T-cell receptor beta chains by a modified Wu-Kabat index of amino acid diversity. *Proc. Natl. Acad. Sci. USA* 87:9138–9142.
28. Davis, M.M. and P.J. Bjorkman. 1988. T-cell antigen receptor genes and T-cell recognition. *Nature* 334:395–402.
29. Lefranc, M.P., A. Forster, and T.H. Rabbitts. 1986. Genetic polymorphism and exon changes of the constant regions of

the human T-cell rearranging gene gamma. *Proc. Natl. Acad. Sci. USA* 83:9596–9600.

30. Bentley, G.A., G. Boulot, K. Karjalainen, and R.A. Mariuzza. 1995. Crystal structure of the β chain of a T cell antigen receptor. *Science* 267:1984–1987.

31. Fields, B.A., B. Ober, E.L. Malchiodi, M.I. Lebedeva, B.C. Braden, X. Ysern, J.K. Kim, X. Shao, E.S. Ward, and R.A. Mariuzza. 1995. Crystal structure of the V alpha domain of a T cell antigen receptor. *Science* 270:1821–1824.

32. Fields, B.A., E.L. Malchiodi, H. Li, X. Ysern, C.V. Stauffacher, P.M. Schlievert, K. Karjalainen, and R.A. Mariuzza. 1996. Crystal structure of a T-cell receptor beta-chain complexed with a superantigen. *Nature* 384:188–192.

33. Li, H., M.I. Lebedeva, A.S. Llera, B.A. Fields, M.B. Brenner, and R.A. Mariuzza. 1998. Structure of the Vdelta domain of a human gammadelta T-cell antigen receptor. *Nature* 391:502–506.

34. Garcia, K.C., M. Degano, R.L. Stanfield, A. Brunmark, M.R. Jackson, P.A. Peterson, L. Teyton, and I.A. Wilson. 1996. An alpha/beta T cell receptor structure at 2.5 Å and its orientation in the TCR-MHC complex. *Science* 274:209–219.

35. Housset, D., G. Mazza, C. Grégoire, C. Piras, B. Malissen, and J.C. Fontecilla-Camps. 1997. The three-dimensional structure of a T-cell antigen receptor Valpha Vbeta heterodimer reveals a novel arrangement of the Vbeta domain. *EMBO J.* 16:4205–4216.

36. Wang, J.H., K. Lim, A. Smolyar, M.K. Teng, J.H. Liu, A.G. Tse, J. Liu, R.E. Hussey, Y. Chishti, C.T. Thomson, R.M. Sweet, S.G. Nathenson, H.C. Chang, J.C. Sacchettini, and E.L. Reinherz. 1998. Atomic structure of an alpha/beta T cell receptor (TCR) heterodimer in complex with an anti-TCR Fab fragment derived from a mitogenic antibody. *EMBO J.* 17:10–26.

37. Garboczi, D.N., P. Ghosh, U. Utz, Q.R. Fan, W.E. Biddison, and D.C. Wiley. 1996. Structure of the complex between human T-cell receptor, viral peptide and HLA-A2. *Nature* 384:134–141.

38. Teng, M.K., A. Smolyar, A.G. Tse, J.H. Liu, J. Liu, R.E. Hussey, S.G. Nathenson, H.C. Chang, E.L. Reinherz, and J.H. Wang. 1998. Identification of a common docking topology with substantial variation among different TCR-peptide-MHC complexes. *Curr. Biol.* 8:409–412.

39. Garcia, K.C., M. Degano, L.R. Pease, M.D. Huang, P.A. Peterson, L. Teyton, and I.A. Wilson. 1998. Structural basis of plasticity in T cell receptor recognition of a self peptide MHC antigen. *Science* 279:1166–1172.

40. Ding, Y.H., K.J. Smith, D.N. Garboczi, U. Utz, W.E. Biddison, and D.C. Wiley. 1998. Two human T cell receptors bind in a similar diagonal mode to the HLA-A2/Tax peptide complex using different TCR amino acids. *Immunity* 8:403–411.

41. Reinherz, E.L., K.M. Tan, L. Tang, P. Kern, J.H. Liu, Y. Xiong, R.E. Hussey, A. Smolyar, B. Hare, R.G. Zhang, A. Joachimiak, H.C. Chang, G. Wagner, and J.H. Wang. 1999. The crystal structure of a T cell receptor in complex with peptide and MHC class II. *Science* 286:1913–1921.

42. Li, H., A. Llera, D. Tsuchiya, L. Leder, X. Ysern, P.M. Schlievert, K. Karjalainen, and R.A. Mariuzza. 1998. Three-dimensional structure of the complex between a T cell receptor beta chain and the superantigen staphylococcal enterotoxin B. *Immunity* 9:807–816.

43. Andersen, P.S., P.M. Lavoie, R.P. Sekaly, H. Churchill, D.M. Kranz, P.M. Schlievert, K. Karjalainen, and R.A. Mariuzza. 1999. Role of the T cell receptor alpha chain in stabilizing TCR-superantigen-MHC class II complexes. *Immunity* 10:473–483.

44. Hedrick, S.M., E.A. Nielsen, J. Kavaler, D.I. Cohen, and M.M. Davis. 1984. Sequence relationships between putative T-cell receptor polypeptides and immunoglobulins. *Nature* 308:153–158.

45. Chothia, C., D.R. Boswell, and A.M. Lesk. 1988. The outline structure of the T-cell alpha beta receptor. *EMBO J.* 7:3745–3755.

46. Vernooij, B.T., J.A. Lenstra, K. Wang, and L. Hood. 1993. Organization of the murine T-cell receptor gamma locus. *Genomics* 17:566–574.

47. Siu, G., S.P. Clark, Y. Yoshikai, M. Malissen, Y. Yanagi, E. Strauss, T.W. Mak, and L. Hood. 1984. The human T cell antigen receptor is encoded by variable, diversity, and joining gene segments that rearrange to generate a complete V gene. *Cell* 37:393–401.

48. Buresi, C., N. Ghanem, S. Huck, G. Lefranc, and M.P. Lefranc. 1989. Exon duplication and triplication in the human T-cell receptor gamma constant region genes and RFLP in French, Lebanese, Tunisian, and black African populations. *Immunogenetics* 29:161–172.

49. Tonegawa, S. 1983. Somatic generation of antibody diversity. *Nature* 302:575–581.

50. Honjo, T. 1983. Immunoglobulin genes. *Annu. Rev. Immunol.* 1:499–528.

51. Arden, B., S.P. Clark, D. Kabelitz, and T.W. Mak. 1995. Human T-cell receptor variable gene segment families. *Immunogenetics* 42:455–500.

52. Williams, A.F., J.L. Strominger, J. Bell, T.W. Mak, J. Kappler, P. Marrack, B. Arden, M.P. Lefranc, L. Hood, S. Tonegawa, and M.M. Davis. 1993. Nomenclature for T-cell receptor (TCR) gene segments of the immune system. *The Immunologist* 1:94–96.

53. Arden, B., S.P. Clark, D. Kabelitz, and T.W. Mak. 1995. Mouse T-cell receptor variable gene segment families. *Immunogenetics* 42:501–530.

54. Rowen, L., B.F. Koop, and L. Hood. 1996. The complete 685-kilobase DNA sequence of the human β T cell receptor locus. *Science* 272:1755–1762.

55. Lefranc, M.P., V. Giudicelli, C. Ginestoux, J. Bodmer, W. Muller, R. Bontrop, M. Lemaitre, A. Malik, V. Barbie, and D. Chaume. 1999. IMGT, the international ImMunoGeneTics database. *Nucl. Acids Res.* 27:209–212.

56. Caccia, N., G.A. Bruns, I.R. Kirsch, G.F. Hollis, V. Bertness, and T.W. Mak. 1985. T cell receptor alpha chain genes are located on chromosome 14 at 14q11–14q12 in humans. *J. Exp. Med.* 161:1255–1260.

57. Takihara, Y., D. Tkachuk, E. Michalopoulos, E. Champagne, J. Reimann, M. Minden, and T.W. Mak. 1988. Sequence and organization of the diversity, joining, and constant region genes of the human T-cell delta-chain locus. *Proc. Natl. Acad. Sci. USA* 85:6097–6101.

58. Loh, E.Y., S. Cwirla, A.T. Serafini, J.H. Phillips, and L.L. Lanier. 1988. Human T-cell-receptor delta chain: genomic organization, diversity, and expression in populations of cells. *Proc. Natl. Acad. Sci. USA* 85:9714–9718.

59. Koop, B.F., L. Rowen, K. Wang, C.L. Kuo, D. Seto, J.A. Lenstra, S. Howard, W. Shan, P. Deshpande, and L. Hood. 1994. The human T-cell receptor TCRAC/TCRDC (C alpha/C delta) region: organization, sequence, and evolution of 97.6 kb of DNA. *Genomics* 19:478–493.

60. Isobe, M., G. Russo, F.G. Haluska, and C.M. Croce. 1988. Cloning of the gene encoding the delta subunit of the human T-cell receptor reveals its physical organization within the

alpha-subunit locus and its involvement in chromosome translocations in T-cell malignancy. *Proc. Natl. Acad. Sci. USA* 85:3933–3937.

61. Satyanarayana, K., S. Hata, P. Devlin, M.G. Roncarolo, J.E. de Vries, H. Spits, J.L. Strominger, and M.S. Krangel. 1988. Genomic organization of the human T-cell antigen-receptor alpha/delta locus. *Proc. Natl. Acad. Sci. USA* 85:8166–8170.

62. Yoshikai, Y., S.P. Clark, S. Taylor, U. Sohn, B.I. Wilson, M.D. Minden, and T.W. Mak. 1985. Organization and sequences of the variable, joining and constant region genes of the human T-cell receptor alpha-chain. *Nature* 316:837–840.

63. Takihara, Y., J. Reimann, E. Michalopoulos, E. Ciccone, L. Moretta, and T.W. Mak. 1989. Diversity and structure of human T cell receptor delta chain genes in peripheral blood gamma/delta-bearing T lymphocytes. *J. Exp. Med.* 169:393–405.

64. Migone, N., S. Padovan, C. Zappador, C. Giachino, M. Bottaro, G. Matullo, C. Carbonara, G.D. Libero, and G. Casorati. 1995. Restriction of the T-cell receptor V delta gene repertoire is due to preferential rearrangement and is independent of antigen selection. *Immunogenetics* 42:323–332.

65. Robinson, M.A. and T.J. Kindt. 1987. Genetic recombination within the human T-cell receptor alpha-chain gene complex. *Proc. Natl. Acad. Sci. USA* 84:9089–9093.

66. Chan, A., R.P. Du, M. Reis, E. Baillie, L.M. Meske, M. Sheehy, and T.W. Mak. 1989. Polymorphism of the human T cell receptor alpha chain variable genes: identification of a highly polymorphic V gene probe. *Int. Immunol.* 1:267–272.

67. Grier, A.H., M.P. Mitchell, and M.A. Robinson. 1990. Polymorphism in human T cell receptor alpha chain variable region genes. *Exp. Clin. Immunogenetics* 7:34–42.

68. Wright, J.A., L. Hood, and P. Concannon. 1991. Human T-cell receptor V alpha gene polymorphism. *Hum. Immunol.* 32:277–283.

69. Cornelis, F., K. Pile, J. Loveridge, P. Moss, R. Harding, C. Julier, and J. Bell. 1993. Systematic study of human alpha beta T cell receptor V segments shows allelic variations resulting in a large number of distinct T cell receptor haplotypes. *Eur. J. Immunol.* 23:1277–1283.

70. Ibberson, M.R., J.P. Copier, E. Llop, C. Navarrete, A.V. Hill, J.K. Cruickshank, and A.K. So. 1998. T-cell receptor variable alpha (TCRAV) polymorphisms in European, Chinese, South American, AfroCaribbean, and Gambian populations. *Immunogenetics* 47:124–130.

71. Charmley, P., D. Nickerson, and L. Hood. 1994. Polymorphism detection and sequence analysis of human T-cell receptor V alpha-chain-encoding gene segments. *Immunogenetics* 39:138–145.

72. Wei, S., P. Charmley, and P. Concannon. 1997. Organization, polymorphism, and expression of the human T-cell receptor AV1 subfamily. *Immunogenetics* 45:405–412.

73. Arden, B., J.L. Klotz, G. Siu, and L.E. Hood. 1985. Diversity and structure of genes of the alpha family of mouse T-cell antigen receptor. *Nature* 316:783–787.

74. Becker, D.M., P. Pattern, Y. Chien, T. Yokota, Z. Eshhar, M. Giedlin, N.R. Gascoigne, C. Goodnow, R. Wolf, K. Arai, and M.M. Davis. 1985. Variability and repertoire size of T-cell receptor V alpha gene segments. *Nature* 317:430–434.

75. Wang, K., J.L. Klotz, G. Kiser, G. Bristol, E. Hays, E. Lai, E. Gese, M. Kronenberg, and L. Hood. 1994. Organization

of the V gene segments in mouse T-cell antigen receptor alpha/delta locus. *Genomics* 20:419–428.

76. Kranz, D.M., H. Saito, C.M. Disteche, K. Swisshelm, D. Pravtcheva, F.H. Ruddle, H.N. Eisen, and S. Tonegawa. 1985. Chromosomal locations of the murine T-cell receptor alpha-chain gene and the T-cell gamma gene. *Science* 227:941–945.

77. Bluestone, J.A., R.Q. Cron, M. Cotterman, B.A. Houlden, and L.A. Matis. 1988. Structure and specificity of T cell receptor gamma/delta on major histocompatibility complex antigen-specific CD3+, CD4−, CD8− T lymphocytes. *J. Exp. Med.* 168:1899–1916.

78. Elliott, J.F., E.P. Rock, P.A. Patten, M.M. Davis, and Y.H. Chien. 1988. The adult T-cell receptor delta-chain is diverse and distinct from that of fetal thymocytes. *Nature* 331:627–631.

79. Jameson, S.C., P.B. Nakajima, J.L. Brooks, W. Heath, O. Kanagawa, and N.R. Gascoigne. 1991. The T cell receptor V alpha 11 gene family. Analysis of allelic sequence polymorphism and demonstration of J alpha region-dependent recognition by allele-specific antibodies. *J. Immunol.* 147:3185–3193.

80. Klotz, J.L., R.K. Barth, G.L. Kiser, L.E. Hood, and M. Kronenberg. 1989. Restriction fragment length polymorphisms of the mouse T-cell receptor gene families. *Immunogenetics* 29:191–201.

81. Jouvin-Marche, E., M.G. Morgado, N. Trede, P.N. Marche, D. Couez, I. Hue, C. Gris, M. Malissen, and P.A. Cazenave. 1989. Complexity, polymorphism, and recombination of mouse T-cell receptor alpha gene families. *Immunogenetics* 30:99–104.

82. Singer, P.A., R.J. McEvilly, R.S. Balderas, F.J. Dixon, and A.N. Theofilopoulos. 1988. T-cell receptor alpha-chain variable-region haplotypes of normal and autoimmune laboratory mouse strains. *Proc. Natl. Acad. Sci. USA* 85:7729–7733.

83. Lund, T., S. Shaikh, M. Hattori, and S. Makino. 1992. Analysis of the T cell receptor (TcR) regions in the NOD, NON and CTS mouse strains define new TcR V alpha haplotypes and new deletions in the TcR V beta region. *Eur. J. Immunol.* 22:871–874.

84. Koop, B.F., R.K. Wilson, K. Wang, B. Vernooij, D. Zallwer, C.L. Kuo, D. Seto, M. Toda, and L. Hood. 1992. Organization, structure, and function of 95 kb of DNA spanning the murine T-cell receptor C alpha/C delta region. *Genomics* 13:1209–1230.

85. Caccia, N., M. Kronenberg, D. Saxe, R. Haars, G.A. Bruns, J. Goverman, M. Malissen, H. Willard, Y. Yoshikai, M. Simon, L.E. Hood, and T.W. Mak. 1984. The T cell receptor beta chain genes are located on chromosome 6 in mice and chromosome 7 in humans. *Cell* 37:1091–1099.

86. Lee, N.E., P. D'Eustachio, D. Pravtcheva, F.H. Ruddle, S.M. Hedrick, and M.M. Davis. 1984. Murine T cell receptor beta chain is encoded on chromosome 6. *J. Exp. Med.* 160:905–913.

87. Siu, G., E.G. Strauss, E. Lai, and L. Hood. 1986. Analysis of a human Vβ gene subfamily. *J. Exp. Med.* 164:1600–1614.

88. Lai, E., P. Concannon, and L. Hood. 1988. Conserved organization of the human and murine T cell receptor β gene families. *Nature* 331:543–546.

89. Robinson, M.A. 1991. The human T cell receptor beta-chain gene complex contains at least 57 variable gene segments. Identification of six V beta genes in four new gene families. *J. Immunol.* 146:4392–4397.

90. Kimura, N., B. Toyonaga, Y. Yoshikai, R.P. Du, and T.W. Mak. 1987. Sequences and repertoire of the human T cell receptor α and β chain variable region genes in thymocytes. *Eur. J. Immunol.* 17:375–383.

91. Concannon, P., R.A. Gatti, and L. Hood. 1987. Human T cell receptor Vβ gene polymorphism. J. Exp. Med. 165:1130–1140.

92. Wei, S., P. Charmley, M.A. Robinson, and P. Concannon. 1994. The extent of the human germline T-cell receptor V beta gene segment repertoire. *Immunogenetics* 40:27–36.

93. Toyonaga, B., Y. Yoshikai, V. Vadasz, B. Chin, and T.W. Mak. 1985. Organization and sequences of the diversity, joining, and constant region genes of the human T-cell receptor beta chain. *Proc. Natl. Acad. Sci. USA* 82:8624–8628.

94. Robinson, M.A., M.P. Mitchell, S. Wei, C.E. Day, T.M. Zhao, and P. Concannon. 1993. Organization of human T-cell receptor beta-chain genes: clusters of V beta genes are present on chromosomes 7 and 9. *Proc. Natl. Acad. Sci. USA* 90:2433–2437.

95. Day, C.E., T. Zhao, and M.A. Robinson. 1992. Silent allelic variants of a T-cell receptor V beta 12 gene are present in diverse human populations. *Hum. Immunol.* 34:196–202.

96. Li, Y., G.R. Sun, Q. Zheng, D.H. Yoo, N. Bhardwaj, D.N. Posnett, M.K. Crow, and S.M. Friedman. 1996. Allelic variants of human TCR *BV*17S1 defined by restriction fragment length polymorphism, single strand conformation polymorphism, and amplification refractory mutation system analyses. *Hum. Immunol.* 49:85–95.

97. Barron, K.S. and M.A. Robinson. 1994. The human T-cell receptor variable gene segment TCRBV6S1 has two null alleles. *Hum. Immunol.* 40:17–19.

98. Pile, K., P. Wordsworth, F. Liote, T. Bardin, J. Bell, and F. Cornelis. 1993. Analysis of a T-cell receptor V beta segment implicated in susceptibility to rheumatoid arthritis: V beta 2 germline polymorphism does not encode susceptibility. *Ann. Rheum. Dis.* 52:891–894.

99. Baccalày, R., D.H. Kono, S. Walker, R.S. Balderas, and A.N. Theofilopoulos. 1991. Genomically imposed and somatically modified human thymocyte Vβ gene repertoire. Proc. Natl. Acad. Sci. USA 88:2908–2912.

100. Plaza, A., D.H. Kono, and A.N. Theofilopoulos. 1991. New human Vβ genes and polymorphic variants. J. Immunol. 147:4360–4365.

101. Vissinga, C.S., P. Charmley, and P. Concannon. 1994. Influence of coding region polymorphism on the peripheral expression of a human TCR Vβ gene. J. Immunol. 152:1222–1227.

102. Charmley, P., K. Wang, L. Hood, and D.A. Nickerson. 1993. Identification and physical mapping of a polymorphic human T cell receptor V beta gene with a frequent null allele. *J. Exp. Med.* 177:135–143.

103. Posnett, D.N., C.S. Vissinga, C. Pambuccian, S. Wei, M.A. Robinson, D. Kostyu, and P. Concannon. 1994. Level of human TCRBV3S1 (V beta 3) expression correlates with allelic polymorphism in the spacer region of the recombination signal sequence. *J. Exp. Med.* 179:1707–1711.

104. Gomolka, M., C. Epplen, J. Buitkamp, and J.T. Epplen. 1993. Novel members and germline polymorphisms in the human T-cell receptor Vb6 family. *Immunogenetics* 37:257–265.

105. Robinson, M.A. 1989. Allelic sequence variations in the hypervariable region of a T-cell receptor beta chain: correlation with restriction fragment length polymorphism in human families and populations. *Proc. Natl. Acad. Sci. USA* 86:9422–9426.

106. Luyrink, L., C.A. Gabriel, S.D. Thompson, A.A. Grom, W.P. Maksymowych, E. Choi, and D.N. Glass. 1993. Reduced expression of a human V beta 6.1 T-cell receptor allele. *Proc. Natl. Acad. Sci. USA* 90:4369–4373.

107. Li, Y., P. Szabo, M.A. Robinson, B. Dong, and D.N. Posnett. 1990. Allelic variations in the human T cell receptor V beta 6.7 gene products. *J. Exp. Med.* 171:221–230.

108. Hansen, T., K.S. Ronningen, R. Ploski, A. Kimura, and E. Thorsby. 1992. Coding region polymorphisms of human T-cell receptor V beta 6.9 and V beta 21.4. *Scand. J. Immunol.* 36:285–290.

109. Vessey, S.J., J.I. Bell, and B.K. Jakobsen. 1996. A functionally significant allelic polymorphism in a T cell receptor V beta gene segment. *Eur. J. Immunol.* 26:1660–1663.

110. Liao, L., A. Marinescu, A. Molano, C. Ciurli, R.P. Sekaly, J.D. Fraser, A. Popowicz, and D.N. Posnett. 1996. TCR binding differs for a bacterial superantigen (SEE) and a viral superantigen (Mtv-9). *J. Exp. Med.* 184:1471–1482.

111. Lindsten, T., N.E. Lee, and M.M. Davis. 1987. Organization of the T-cell antigen-receptor beta-chain locus in mice. *Proc. Natl. Acad. Sci. USA* 84:7639–7643.

112. Gascoigne, N.R., Y. Chien, D.M. Becker, J. Kavaler, and M.M. Davis. 1984. Genomic organization and sequence of T-cell receptor beta-chain constant- and joining-region genes. *Nature* 310:387–391.

113. Malissen, M., K. Minard, S. Mjolsness, M. Kronenberg, J. Goverman, T. Hunkapiller, M.B. Prystowsky, Y. Yoshikai, F. Fitch, T.W. Mak, and L.E. Hood. 1984. Mouse T cell antigen receptor: structure and organization of constant and joining gene segments encoding the beta polypeptide. *Cell* 37:1101–1110.

114. Louie, M.C., C.A. Nelson, and D.Y. Loh. 1989. Identification and characterization of new murine T cell receptor beta chain variable region (V beta) genes. *J. Exp. Med.* 170:1987–1998.

115. Huppi, K.E., L.A. D'Hoostelaere, B.A. Mock, E. Jouvin-Marche, M.A. Behlke, H.S. Chou, R.J. Berry, and D.Y. Loh. 1988. T-cell receptor V beta genes in natural populations of mice. *Immunogenetics* 27:51–56.

116. Behlke, M.A., H.S. Chou, K. Huppi, and D.Y. Loh. 1986. Murine T-cell receptor mutants with deletions of beta-chain variable region genes. *Proc. Natl. Acad. Sci. USA* 83:767–771.

117. Haqqi, T.M., S. Banerjee, G.D. Anderson, and C.S. David. 1989. RIII S/J (H-2r). An inbred mouse strain with a massive deletion of T cell receptor V beta genes. *J. Exp. Med.* 169:1903–1909.

118. Jouvin-Marche, E., N.S. Trede, A. Bandeira, A. Tomas, D.Y. Loh, and P.A. Cazenave. 1989. Different large deletions of T cell receptor V beta genes in natural populations of mice. *Eur. J. Immunol.* 19:1921–1926.

119. Pullen, A.M., W. Potts, E.K. Wakeland, J. Kappler, and P. Marrack. 1990. Surprisingly uneven distribution of the T cell receptor V beta repertoire in wild mice. *J. Exp. Med.* 171:49–62.

120. Rabbitts, T.H., M.P. Lefranc, M.A. Stinson, J.E. Sims, J. Schroder, M. Steinmetz, N.L. Spurr, E. Solomon, and P.N. Goodfellow. 1985. The chromosomal location of T-cell receptor genes and a T cell rearranging gene: possible correlation with specific translocations in human T cell leukaemia. *EMBO J.* 4:1461–1465.

121. Murre, C., R.A. Waldmann, C.C. Morton, K.F. Bongiovanni, T.A. Waldmann, T.B. Shows, and J.G. Seidman. 1985.

Human gamma-chain genes are rearranged in leukaemic T cells and map to the short arm of chromosome 7. *Nature* 316:549–552.

122. Lefranc, M.P., A. Forster, R. Baer, M.A. Stinson, and T.H. Rabbitts. 1986. Diversity and rearrangement of the human T cell rearranging gamma genes: nine germ-line variable genes belonging to two subgroups. *Cell* 45:237–246.

123. Huck, S., P. Dariavach, and M.P. Lefranc. 1988. Variable region genes in the human T-cell rearranging gamma (TRG) locus: V-J junction and homology with the mouse genes. *EMBO J.* 7:719–726.

124. Zhang, X.M., G. Cathala, Z. Soua, M.P. Lefranc, and S. Huck. 1996. The human T-cell receptor gamma variable pseudogene V10 is a distinctive marker of human speciation. *Immunogenetics* 43:196–203.

125. Huck, S. and M.P. Lefranc. 1987. Rearrangements to the JP1, JP and JP2 segments in the human T-cell rearranging gamma gene (TRG gamma) locus. *FEBS Lett.* 224:291–296.

126. Lefranc, M.P., P. Chuchana, P. Dariavach, C. Nguyen, S. Huck, F. Brockly, B. Jordan, and G. Lefranc. 1989. Molecular mapping of the human T cell receptor gamma (TRG) genes and linkage of the variable and constant regions. *Eur. J. Immunol.* 19:989–994.

127. Ghanem, N., Z. Soua, X.G. Zhang, M. Zijun, Y. Zhiwei, G. Lefranc, and M.P. Lefranc. 1991. Polymorphism of the T-cell receptor gamma variable and constant region genes in a Chinese population. *Human Genetics* 86:450–456.

128. Font, M.P., Z. Chen, J.C. Bories, N. Duparc, P. Loiseau, L. Degos, H. Cann, D. Cohen, J. Dausset, and F. Sigaux. 1988. The V gamma locus of the human T cell receptor gamma gene. Repertoire polymorphism of the first variable gene segment subgroup. *J. Exp. Med.* 168:1383–1394.

129. Traunecker, A., F. Oliveri, N. Allen, and K. Karjalainen. 1986. Normal T cell development is possible without 'functional' gamma chain genes. *EMBO J.* 5:1589–1593.

130. Garman, R.D., P.C. Doherty, and D.H. Raulet. 1986. Diversity, rearrangement, and expression of murine T cell gamma genes. *Cell* 45:733–742.

131. Hayday, A.C., H. Saito, S.D. Gillies, D.M. Kranz, G. Tanigawa, H.N. Eisen, and S. Tonegawa. 1985. Structure, organization, and somatic rearrangement of T cell gamma genes. *Cell* 40:259–269.

132. Pelkonen, J., A. Traunecker, and K. Karjalainen. 1987. A new mouse TCR V gamma gene that shows remarkable evolutionary conservation. *EMBO J.* 6:1941–1944.

133. Havran, W.L., Y.H. Chien, and J.P. Allison. 1991. Recognition of self antigens by skin-derived T cells with invariant gamma delta antigen receptors. *Science* 252:1430–1432.

134. Boismenu, R. and W.L. Havran. 1997. An innate view of gamma delta T cells. *Curr. Opin. Immunol.* 9:57–63.

135. Balomenos, D., R.S. Balderas, K.P. Mulvany, J. Kaye, D.H. Kono, and A.N. Theofilopoulos. 1995. Incomplete T cell receptor Vβ allelic exclusion and dual Vβ-expressing cells. J. Immunol. 155:3308–3312.

136. Padovan, E., C. Giachino, M. Cella, S. Valitutti, O. Acuto, and A. Lanzavecchia. 1995. Normal T lymphocytes can express two different T cell receptor β chains: Implications for the mechanism of allelic exclusion. J. Exp. Med. 181:1587–1591.

137. Davodeau, F., M.A. Peyrat, F. Romagné, A. Necker, M.M. Hallet, H. Vié, and Bonneville. 1995. Dual T cell receptor β chain expression on human T lymphocytes. J. Exp. Med. 181:1391–1398.

138. Kong, F.K., C.L. Chen, A. Six, R.D. Hockett, and M.D. Cooper. 1999. T cell receptor gene deletion circles identify recent thymic emigrants in the peripheral T cell pool. *Proc. Natl. Acad. Sci. USA* 96:1536–1540.

139. Berek, C. and C. Milstein. 1988. The dynamic nature of the antibody repertoire. *Immunol. Rev.* 105:5–26.

140. Zheng, B., W. Xue, and G. Kelsoe. 1994. Locus-specific somatic hypermutation in germinal centre T cells. *Nature* 372:556–559.

141. Cheynier, R., S. Henrichwark, and S. Wain-Hobson. 1998. Somatic hypermutation of the T cell receptor Vβ gene in microdissected splenic white pulps from HIV-1-positive patients. Eur. J. Immunol. 28:1604–1610.

142. Marshall, B., R. Schulz, M. Zhou, and A. Mellor. 1999. Alternative splicing and hypermutation of a nonproductively rearranged TCR alpha-chain in a T cell hybridoma. *J. Immunol.* 162:871–877.

143. Gell, P.G. and B. Benacerraf. 1959. Studies on hypersensitivity. II. Delayed hypersensitivity to denatured proteins in guinea pigs. *Immunol.* 2:64–69.

144. Sela, M. 1969. Antigenicity: some molecular aspects. *Science* 166:1365–1374.

145. Ziegler, H.K. and E.R. Unanue. 1982. Decrease in macrophage antigen catabolism caused by ammonia and chloroquine is associated with inhibition of antigen presentation to T cells. *Proc. Natl. Acad. Sci. USA* 79:175–178.

146. Grey, H.M., S.M. Colon, and R.W. Chesnut. 1982. Requirements for the processing of antigen by antigen-presenting B cells. II. Biochemical comparison of the fate of antigen in B cell tumors and macrophages. *J. Immunol.* 129:2389–2395.

147. Lee, K.C., M. Wong, and D. Spitzer. 1982. Chloroquine as a probe for antigen processing by accessory cells. *Transplantation* 34:150–153.

148. Shimonkevitz, R., J. Kappler, P. Marrack, and H. Grey. 1983. Antigen recognition by H-2-restricted T cells. I. Cell-free antigen processing. *J. Exp. Med.* 158:303–316.

149. Shimonkevitz, R., S. Colon, J.W. Kappler, P. Marrack, and H.M. Grey. 1984. Antigen recognition by H-2-restricted T cells. II. A tryptic ovalbumin peptide that substitutes for processed antigen. *J. Immunol.* 133:2067–2074.

150. Zinkernagel, R.M. and P.C. Doherty. 1974. Restriction of in vitro T cell-mediated cytotoxicity in lymphocytic choriomeningitis within a syngeneic or semiallogeneic system. *Nature* 248:701–702.

151. Bevan, M.J. 1975. Interaction antigens detected by cytotoxic T cells with the major histocompatibility complex as modifier. *Nature* 256:419–421.

152. Gordon, R.D., E. Simpson, and L.E. Samelson. 1975. *In vitro* cell-mediated immune responses to the male specific(H-Y) antigen in mice. *J. Exp. Med.* 142:1108–1120.

153. Shearer, G.M., T.G. Rehn, and C.A. Garbarino. 1975. Cell-mediated lympholysis of trinitrophenyl-modified autologous lymphocytes. Effector cell specificity to modified cell surface components controlled by H-2K and H-2D serological regions of the murine major histocompatibility complex. *J. Exp. Med.* 141:1384–1364.

154. Bjorkman, P.J. and P. Parham. 1990. Structure, function, and diversity of class I major histocompatibility complex molecules. *Annu. Rev. Biochem.* 59:253–288.

155. The MHC sequencing consortium. 1999. Complete sequence and gene map of a human major histocompatibility complex. *Nature* 401:921–923.

156. Fremont, D.H., M. Matsumura, E.A. Stura, P.A. Peterson, and I.A. Wilson. 1992. Crystal structures of two viral

peptides in complex with murine MHC class I H-2Kb. *Science* 257:919–927.

157. Cerundolo, V., T. Elliott, J. Elvin, J. Bastin, H.G. Rammensee, and A. Townsend. 1991. The binding affinity and dissociation rates of peptides for class I major hostocompatibility complex molecules. *Eur. J. Immunol.* 21:2069–2075.

158. Rammensee, H.G., T. Friede, and S. Stevanovic. 1995. MHC ligands and peptide motifs: first listing. *Immunogenetics* 41:178–228.

159. Rock, K.L. and A.L. Goldberg. 1999. Degradation of cell proteins and the generation of MHC class I-presented peptides. *Annu. Rev. Immunol.* 17:739–779.

160. Momburg, F. and G.J. Hammerling. 1998. Generation and TAP-mediated transport of peptides for major histocompatibility complex class I molecules. *Adv. Immunol.* 68:191–256.

161. Janeway, C.A., Jr. 1992. The T cell receptor as a multi-component signalling machine: CD4/CD8 coreceptors and CD45 in T cell activation. *Annu. Rev. Immunol.* 10:645–674.

162. Luescher, I.F., E. Vivier, A. Layer, J. Mahiou, F. Godeau, B. Malissen, and P. Romero. 1995. CD8 modulation of T-cell antigen receptor-ligand interactions on living cytotoxic T lymphocytes. *Nature* 373:353–356.

163. Wilson, I.A. 1996. Another twist to MHC-peptide recognition. *Science* 272:973–974.

164. Stern, L.J. and D.C. Wiley. 1994. Antigenic peptide binding by class I and class II histocompatibility proteins. *Structure* 2:245–251.

165. Hammer, J., T. Sturniolo, and F. Sinigaglia. 1997. HLA class II peptide binding specificity and autoimmunity. *Adv. Immunol.* 66:67–100.

166. Teyton, L., D. O'Sullivan, P.W. Dickson, V. Lotteau, A. Sette, P. Fink, and P.A. Peterson. 1990. Invariant chain distinguishes between the exogenous and endogenous antigen presentation pathways. *Nature* 348:39–44.

167. Ghosh, P., M. Amaya, E. Mellins, and D.C. Wiley. 1995. The structure of an intermediate in class II MHC maturation: CLIP bound to HLA-DR3. *Nature* 378:457–462.

168. Kropshofer, H., S.O. Arndt, G. Moldenhauer, G.J. Hömmerling, and A.B. Vogt. 1997. HLA-DM acts as a molecular chaperone and rescues empty HLA-DR molecules at lysosomal pH. *Immunity* 6:293–302.

169. Evavold, B.D. and P.M. Allen. 1991. Seperation of IL-4 production fom Th cell proliferation by an altered T cell receptor ligand. *Science* 252:1308–1310.

170. De Magistris, T.M., J. Alexander, M. Coggeshall, A. Altman, F.C. Gaeta, H.M. Grey, and A. Sette. 1992. Antigen analog-major histocompatibility complexes act as antagonists of the T cell receptor. *Cell* 68:625–634.

171. Sloan-Lancaster, J. and P.M. Allen. 1996. Altered peptide ligand-induced partial T cell activation: molecular mechanisms and role in T cell biology. *Annu. Rev. Immunol.* 14:1–27.

172. Ding, Y.H., B.M. Baker, D.N. Garboczi, W.E. Biddison, and D.C. Wiley. 1999. Four A6-TCR/peptide/HLA-A2 structures that generate very different T cell signals are nearly identical. *Immunity.* 11:45–56.

173. Davis, M.M., J.J. Boniface, Z. Reich, D. Lyons, J. Hampl, B. Arden, and Y.H. Chien. 1998. Ligand recognition by alpha beta T cell receptors. *Annu. Rev. Immunol.* 16:523–534.

174. Germain, R.N. and I. Stefanova. 1999. The dynamics of T cell receptor signaling: Complex orchestration and the key roles of tempo and cooperation. *Annu. Rev. Immunol.* 17:467–522.

175. Lanzavecchia, A., G. Iezzi, and A. Viola. 1999. From TCR engagement to T cell activation: A kinetic view of T cell behavior. *Cell* 96:1–4.

176. Abe, J., J. Forrester, T. Nakahara, J.A. Lafferty, B.L. Kotzin, and D.Y. Leung. 1991. Selective stimulation of human T cells with streptococcal erythrogenic toxins A and B. *J. Immunol.* 146:3747–3750.

177. Cole, B.C. and M.M. Griffiths. 1993. Triggering and exacerbation of autoimmune arthritis by the Mycoplasma arthritidis superantigen MAM. *Arthritis Rheum.* 36:994–1002.

178. Ito, Y., J. Abe, K. Yoshino, T. Takeda, and T. Kohsaka. 1995. Sequence analysis of the gene for a novel superantigen produced by Yersinia pseudotuberculosis and expression of the recombinant protein. *J. Immunol.* 154:5896–5906.

179. Rovira, P., M. Buckle, J.P. Abastado, W.J. Peumans, and P. Truffa-Bachi. 1999. Major histocompatibility class I molecules present Urtica dioica agglutinin, a superantigen of vegetal origin, to T lymphocytes. *Eur. J. Immunol.* 29:1571–1580.

180. Acha-Orbea, H. and H.R. MacDonald. 1995. Superantigens of mouse mammary tumor virus. *Annu. Rev. Immunol.* 13:459–486.

181. Li, H.M., A. Llera, E.L. Malchiodi, and R.A. Mariuzza. 1999. The structural basis of T cell activation by superantigens. *Annu. Rev. Immunol.* 17:435–466.

182. Krebs, S., J.R. Lamas, S. Poenaru, G. Folkers, J.A. De Castro, D. Seebach, and D. Rognan. 1998. Substituting non-peptidic spacers for the T cell receptor-binding part of class I major histocompatibility complex-binding peptides. *J. Biol. Chem.* 273:19072–19079.

183. Glithero, A., J. Tormo, J.S. Haurum, G. Arsequell, G. Valencia, J. Edwards, S. Springer, A. Townsend, Y.L. Pao, M. Wormald, R.A. Dwek, E.Y. Jones, and T. Elliott. 1999. Crystal structures of two H-2D(b)/glycopeptide complexes suggest a molecular basis for CTL cross-reactivity. *Immunity* 10:63–74.

184. Zhao, X.J. and N.-K.V. Cheung. 1995. GD2 oligosaccharide: Target for cytotoxic T lymphocytes. *J. Exp. Med.* 182:67–74.

185. Kozbor, D., G. Trinchieri, D.S. Monos, M. Isobe, G. Russo, J.A. Haney, C. Zmijewski, and C.M. Croce. 1989. Human TCR-gamma+/delta+, CD8+ T lymphocytes recognize tetanus toxoid in an MHC-restricted fashion. *J. Exp. Med.* 169:1847–1851.

186. Vidovic, D., M. Roglic, K. McKune, S. Guerder, C. Mackay, and Z. Dembic. 1989. Qa-1 restricted recognition of foreign antigen by a gamma delta T-cell hybridoma. *Nature* 340:646–650.

187. Tanaka, Y., C.T. Morita, E. Nieves, M.B. Brenner, and B.R. Bloom. 1995. Natural and synthetic non-peptide antigens recognized by human gamma delta T cells. *Nature* 375:155–158.

188. Poquet, Y., P. Constant, F. Halary, M.A. Peyrat, M. Gilleron, F. Davodeau, Bonneville, and J.J. Fournié. 1996. A novel nucleotide-containing antigen for human blood gamma delta T lymphocytes. *Eur. J. Immunol.* 26:2344–2349.

189. Bukowski, J.F., C.T. Morita, and M.B. Brenner. 1999. Human gamma delta T cells recognize alkylamines derived from microbes, edible plants, and tea: Implications for innate immunity. *Immunity.* 11:57–65.

190. Engel, I. and S.M. Hedrick. 1988. Site-directed mutations in the VDJ junctional region of a T cell receptor beta chain cause changes in antigenic peptide recognition. *Cell* 54:473–484.

191. Nalefski, E.A., S. Kasibhatla, and A. Rao. 1992. Functional analysis of the antigen binding site on the T cell receptor alpha chain. *J. Exp. Med.* 175:1553–1563.

192. Kasibhatla, S., E.A. Nalefski, and A. Rao. 1993. Simultaneous involvement of all six predicted antigen binding loops of the T cell receptor in recognition of the MHC/antigenic peptide complex. *J. Immunol.* 151:3140–3151.

193. Moreland, L.W., L.W. Heck, Jr., W.J. Koopman, P.A. Saway, T.C. Adamson, Z. Fronek, R.D. O'Connor, E.E. Morgan, J.P. Diveley, and N.M. Chieffo. 1995. V beta 17 T-cell receptor peptide vaccine. Results of a phase I dose-finding study in patients with rheumatoid arthritis. *Ann. NY Acad. Sci.* 756:211–214.

194. Turner, S.J., S.C. Jameson, and F.R. Carbone. 1997. Functional mapping of the orientation for TCR recognition of an H2-Kb-restricted ovalbumin peptide suggests that the β-chain subunit can dominate the determination of peptide side chain specificity. J. Immunol. 159:2312–2317.

195. Bowness, P., R.L. Allen, D.N. Barclay, E.Y. Jones, and A.J. McMichael. 1998. Importance of a conserved TCR J alpha-encoded tyrosine for T cell recognition of an HLA B27/peptide complex. *Eur. J. Immunol.* 28:2704–2713.

196. Nalefski, E.A., J.G. Wong, and A. Rao. 1990. Amino acid substitutions in the first complementarity-determining region of a murine T-cell receptor alpha chain affect antigen-major histocompatibility complex recognition. *J. Biol. Chem.* 265:8842–8846.

197. Hong, S.C., A. Chelouche, R. Lin, D. Shaywitz, N.S. Braunstein, L. Glimcher, and C.A. Janeway, Jr. 1992. An MHC interaction site maps to the amino-terminal half of the T cell receptor alpha chain variable domain. *Cell* 69:999–1009.

198. Bellio, M., Y.C. Lone, O. de la Calle-Martin, B. Malissen, J.P. Abastado, and P. Kourilsky. 1994. The V beta complementarity determining region 1 of a major histocompatibility complex (MHC) class I-restricted T cell receptor is involved in the recognition of peptide/MHC I and superantigen/MHC II complex. *J. Exp. Med.* 179:1087–1097.

199. Brawley, J.V. and P. Concannon. 1996. Modulation of promiscuous T cell receptor recognition by mutagenesis of CDR2 residues. *J. Exp. Med.* 183:2043–2051.

200. Manning, T.C., C.J. Schlueter, T.C. Brodnicki, E.A. Parke, J.A. Speir, K.C. Garcia, L. Teyton, I.A. Wilson, and D.M. Kranz. 1998. Alanine scanning mutagenesis of an alpha/beta T cell receptor: Mapping the energy of antigen recognition. *Immunity* 8:413–425.

201. Patten, P.A., E.P. Rock, T. Sonoda, d.S. Fazekas, J.L. Jorgensen, and M.M. Davis. 1993. Transfer of putative complementarity-determining region loops of T cell receptor V domains confers toxin reactivity but not peptide/MHC specificity. *J. Immunol.* 150:2281–2294.

202. Jorgensen, J.L., U. Esser, d.S. Fazekas, P.A. Reay, and M.M. Davis. 1992. Mapping T cell receptor-peptide contacts by variant peptide immunization of single-chain transgenics. *Nature* 355:224–230.

203. Sant'Angelo, D.B., G. Waterbury, P. Preston-Hurlburt, S.T. Yoon, R. Medzhitov, S.C. Hong, and C.A. Janeway, Jr. 1996. The specificity and orientation of a TCR to its peptide-MHC class II ligands. *Immunity* 4:367–376.

204. Sun, R., S.E. Shepherd, S.S. Geier, C.T. Thomson, J.M. Sheil, and S.G. Nathenson. 1995. Evidence that the antigen receptors of cytotoxic T lymphocytes interact with a common recognition pattern on the H-2Kb molecule. *Immunity* 3:573–582.

205. Mason, D. 1998. A very high level of crossreactivity is an essential feature of the T-cell receptor. *Immunol. Today* 19:395–404.

206. Choi, Y.W., A. Herman, D. DiGiusto, T. Wade, P. Marrack, and J. Kappler. 1990. Residues of the variable region of the T-cell-receptor beta-chain that interact with S. aureus toxin superantigens. *Nature* 346:471–473.

207. Pullen, A.M., T. Wade, P. Marrack, and J.W. Kappler. 1990. Identification of the region of T cell receptor β chain that interacts with the self-superantigen Mls-1ᵃ. *Cell* 61:1365–1374.

208. Cazenave, P.A., P.N. Marche, E. Jouvin-Marche, D. Voegtlé, F. Bonhomme, A. Bandeira, and A. Coutinho. 1990. Vβ17 gene polymorphism in wild-derived mouse strains: two amino acid substitutions in the Vβ17 region greatly alter T cell receptor specificity. *Cell* 63:717–728.

209. Pullen, A.M., J. Bill, R.T. Kubo, P. Marrack, and J.W. Kappler. 1991. Analysis of the interaction site for the super-antigen Mls-1a on T cell receptor Vβ. *J. Exp. Med.* 173:1183–1192.

210. White, J., A. Pullen, K. Choi, P. Marrack, and J.W. Kappler. 1993. Antigen recognition properties of mutant Vβ 3+ T cell receptors are consistent with an immunoglobulin-like structure for the receptor. *J. Exp. Med.* 177:119–125.

211. Baccalà, R., L.R. Smith, M. Vestberg, P.A. Peterson, B.C. Cole, and A.N. Theofilopoulos. 1992. Mycoplasma arthritidis mitogen (MAM): Vβs engaged in mice, rats and humans, and requirement of HLA-DRα for presentation. *Arthritis Rheum.* 35:434–442.

212. Ciurli, C., D.N. Posnett, R.P. Sékaly, and F. Denis. 1998. Highly biased CDR3 usage in restricted sets of β chain variable regions during viral superantigen 9 response. *J. Exp. Med.* 187:253–258.

213. Hodtsev, A.S., Y.W. Choi, E. Spanopoulou, and D.N. Posnett. 1998. Mycoplasma superantigen is a CDR3-dependent ligand for the T cell antigen receptor. *J. Exp. Med.* 187:319–327.

214. Daly, K., P. Nguyen, D. Hankley, W.J. Zhang, D.L. Woodland, and M.A. Blackman. 1995. Contribution of the TCR α-chain to the differential recognition of bacterial and retroviral superantigens. *J. Immunol.* 155:27–34.

215. Donson, D., H. Borrero, M. Rutman, R. Pergolizzi, N. Malhado, and S. Macphail. 1997. Gene transfer directly demonstrates a role for TCR V alpha elements in superantigen recognition. *J. Immunol.* 158:5229–5236.

216. Chan, A.C., D.M. Desai, and A. Weiss. 1994. The role of protein tyrosine kinases and protein tyrosine phosphatases in T cell antigen receptor signal transduction. *Annu. Rev. Immunol.* 12:555–592.

217. Alberola-Ila, J., S. Takaki, J.D. Kerner, and R.M. Perlmutter. 1997. Differential signaling by lymphocyte antigen receptors. *Annu. Rev. Immunol.* 15:125–154.

218. Reich, Z., J.J. Boniface, D.S. Lyons, N. Borochov, E.J. Wachtel, and M.M. Davis. 1997. Ligand-specific oligomerization of T-cell receptor molecules. *Nature* 387:617–620.

219. Grakoui, A., S.K. Bromley, C. Sumen, M.M. Davis, A.S. Shaw, P.M. Allen, and M.L. Dustin. 1999. The immunological synapse: A molecular machine controlling T cell activation. *Science* 285:221–227.

220. McKeithan, T.W. 1995. Kinetic proofreading in T-cell receptor signal transduction. *Proc. Natl. Acad. Sci. USA* 92:5042–5046.

221. Boitel, B., M. Ermonval, P. Panina-Bordignon, R.A. Mariuzza, A. Lanzavecchia, and O. Acuto. 1992. Preferential Vβ gene usage and lack of junctional sequence conservation

among human T cell receptors specific for a tetanus toxin-derived peptide: Evidence for a dominant role of a germline-encoded V region in antigen/major histocompatibility complex recognition. *J. Exp. Med.* 175:765–777.

222. MacDonald, H.R., J.L. Casanova, J.L. Maryanski, and J.C. Cerottini. 1993. Oligoclonal expansion of major histocompatibility complex class I-restricted cytolytic T lymphocytes during a primary immune response in vivo: direct monitoring by flow cytometry and polymerase chain reaction. *J. Exp. Med.* 177:1487–1492.

223. Kalergis, A.M., T. Ono, F.M. Wang, T.P. DiLorenzo, S. Honda, and S.G. Nathenson. 1999. Single amino acid replacements in an antigenic peptide are sufficient to alter the TCR V beta repertoire of the responding CD8(+) cytotoxic lymphocyte population. *J. Immunol.* 162:7263–7270.

224. Jores, R. and T. Meo. 1993. Few V gene segments dominate the T cell receptor beta-chain repertoire of the human thymus. *J. Immunol.* 151:6110–6122.

225. Rosenberg, W.M., P.A. Moss, and J.I. Bell. 1992. Variation in human T cell receptor Vβ and Jβ repertoire: Analysis using anchor polymerase chain reaction. *Eur. J. Immunol.* 22:541–549.

226. Kappler, J.W., T. Wade, J. White, E. Kushnir, M. Blackman, J. Bill, N. Roehm, and P. Marrack. 1987. A T cell receptor Vβ segment that imparts reactivity to a class II major histocompatibility complex product. *Cell* 49:263–271.

227. MacDonald, H.R., R. Schneider, R.K. Lees, R.C. Howe, H. Acha-Orbea, H. Festenstein, R.M. Zinkernagel, and H. Hengartner. 1988. T-cell receptor Vβ use predicts reactivity and tolerance to Mlsa-encoded antigens. *Nature* 332:40–45.

228. Diu, A., F. Romagne, C. Genevee, C. Rocher, J.M. Bruneau, A. David, F. Praz, and T. Hercend. 1993. Fine specificity of monoclonal antibodies directed at human T cell receptor variable regions: comparison with oligonucleotide-driven amplification for evaluation of V beta expression. *Eur. J. Immunol.* 23:1422–1429.

229. Choi, Y.W., B. Kotzin, L.R. Herron, J. Callahan, P. Marrack, and J. Kappler. 1989. Interaction of Staphylococcus aureus toxin superantigens with human T cells. *Proc. Natl. Acad. Sci. USA* 86:8941–8945.

230. Genevee, C., A. Diu, J. Nierat, A. Caignard, P.Y. Dietrich, L. Ferradini, S. Roman–Roman, F. Triebel, and T. Hercend. 1992. An experimentally validated panel of subfamily-specific oligonucleotide primers (V alpha 1-w29/V beta 1-w24) for the study of human T cell receptor variable V gene segment usage by polymerase chain reaction. *Eur. J. Immunol.* 22:1261–1269.

231. Panzara, M.A., E. Gussoni, L. Steinman, and J.R. Oksenberg. 1992. Analysis of the T cell repertoire using the PCR and specific oligonucleotide primers. *BioTechniques* 12:728–735.

232. Hall, B.L. and O.J. Finn. 1992. PCR-based analysis of the T-cell receptor V beta multigene family: experimental parameters affecting its validity. *BioTechniques* 13:248–257.

233. Moonka, D. and E.Y. Loh. 1994. A consensus primer to amplify both alpha and beta chains of the human T cell receptor. *J. Immunol. Methods* 169:41–51.

234. Maslanka, K., T. Piatek, J. Gorski, and M. Yassai. 1995. Molecular analysis of T cell repertoires. Spectratypes generated by multiplex polymerase chain reaction and evaluated by radioactivity or fluorescence. *Hum. Immunol.* 44:28–34.

235. Lang, R., K. Pfeffer, H. Wagner, and K. Heeg. 1997. A rapid method for semiquantitative analysis of the human V beta-repertoire using TaqManR PCR. *J. Immunol. Methods* 203:181–192.

236. Kohsaka, H., A. Taniguchi, P.P. Chen, W.E. Ollier, and D.A. Carson. 1993. The expressed T cell receptor V gene repertoire of rheumatoid arthritis monozygotic twins: rapid analysis by anchored polymerase chain reaction and enzyme-linked immunosorbent assay. *Eur. J. Immunol.* 23:1895–1901.

237. Duchmann, R., W. Strober, and S.P. James. 1993. Quantitative measurement of human T-cell receptor V beta subfamilies by reverse transcription-polymerase chain reaction using synthetic internal mRNA standards. *DNA Cell Biol.* 12:217–225.

238. Hall, M.A. and J.S. Lanchbury. 1995. Healthy human T-cell receptor beta-chain repertoire. Quantitative analysis and evidence for J beta-related effects on CDR3 structure and diversity. *Hum. Immunol.* 43:207–218.

239. Singer, P.A., R.S. Balderas, R.J. McEvilly, M. Bobardt, and A.N. Theofilopoulos. 1989. Tolerance-related Vβ clonal deletions in normal CD4-8-, TCR-$\alpha\beta$ and abnormal lpr and gld cell populations. *J. Exp. Med.* 170:1869–1877.

240. Singer, P.A., R.S. Balderas, and A.N. Theofilopoulos. 1990. Thymic selection defines multiple T cell receptor Vβ "repertoire phenotypes" at the CD4/CD8 subset level. *EMBO J.* 11:3641–3648.

241. Okada, C.Y. and I.L. Weissman. 1989. Relative Vβ transcript levels in thymus and peripheral lymphoid tissues from various mouse strains. *J. Exp. Med.* 169:1703–1719.

242. González-Quintial, R. and A.N. Theofilopoulos. 1992. Vβ gene repertoires in aging mice. *J. Immunol.* 149:230–236.

243. Smith, L.R., D.H. Kono, M.E. Kammuller, R.S. Balderas, and A.N. Theofilopoulos. 1992. Vβ repertoire in rats and implications for endogenous superantigens. *Eur. J. Immunol.* 22:641–645.

244. González-Quintial, R., R. Baccalà, R.M. Pope, and A.N. Theofilopoulos. 1996. Identification of clonally expanded T cells in rheumatoid arthritis using a sequence enrichment nuclease assay. *J. Clin. Invest.* 97:1335–1343.

245. Vandekerckhove, B.A., R. Baccalà, D. Jones, D.H. Kono, A.N. Theofilopoulos, and M.G. Roncarolo. 1992. Thymic selection of the human T cell receptor Vβ repertoire in SCID-hu mice. *J. Exp. Med.* 176:1619–1624.

246. Pannetier, C., M. Cochet, S. Darche, A. Casrouge, M. ZÖller, and P. Kourilsky. 1993. The sizes of the CDR3 hypervariable regions of the murine T-cell receptor β chains vary as a function of the recombined germ-line segments. *Proc. Natl. Acad. Sci.* USA 90:4319–4323.

247. Cochet, M., C. Pannetier, A. Regnault, S. Darche, C. Leclerc, and P. Kourilsky. 1992. Molecular detection and in vivo analysis of the specific T cell response to a protein antigen. *Eur. J. Immunol.* 22:2639–2647.

248. Hingorani, R., I.H. Choi, P. Akolkar, B. Gulwani-Akolkar, R. Pergolizzi, J. Silver, and P.K. Gregersen. 1993. Clonal predominance of T cell receptors within the CD8+ CD45RO+ subset in normal human subjects. *J. Immunol.* 151:5762–5769.

249. Ikeda, Y., K. Masuko, Y. Nakai, T. Kato, T. Hasanuma, S.I. Yoshino, Y. Mizushima, K. Nishioka, and K. Yamamoto. 1996. High frequencies of identical T cell clonotypes in synovial tissues of rheumatoid arthritis patients suggest the occurrence of common antigen-driven immune responses. *Arthritis Rheum.* 39:446–453.

250. Giachino, C., L. Granziero, V. Modena, V. Maiocco, C. Lomater, F. Fantini, A. Lanzavecchia, and N. Migone. 1994. Clonal expansions of V delta 1+ and V delta 2+ cells

increase with age and limit the repertoire of human gamma delta T cells. *Eur. J. Immunol.* 24:1914–1918.

251. Offermans, M.T., L. Struyk, B. de Geus, F.C. Breedveld, P.J. Van den Elsen, and J. Rozing. 1996. Direct assessment of junctional diversity in rearranged T cell receptor beta chain encoding genes by combined heteroduplex and single strand conformation polymorphism (SSCP) analysis. *J. Immunol. Methods* 191:21–31.

252. Bourguin, A., R. Tung, N. Galili, and J. Sklar. 1990. Rapid, nonradioactive detection of clonal T cell receptor gene rearrangements in lymphoid neoplasms. *Proc. Natl. Acad. Sci. USA* 87:8536–8540.

253. Altman, J.D., P.A. Moss, P.J. Goulder, D.H. Barouch, M.G. McHeyzer-Williams, J.I. Bell, A.J. McMichael, and M.M. Davis. 1996. Phenotypic analysis of antigen-specific T lymphocytes. *Science* 274:94–96.

254. Murali-Krishna, K., J.D. Altman, M. Suresh, D.J. Sourdive, A.J. Zajac, J.D. Miller, J. Slansky, and R. Ahmed. 1998. Counting antigen-specific CD8 T cells: a reevaluation of bystander activation during viral infection. *Immunity* 8:177–187.

255. Crawford, F., H. Kozono, J. White, P. Marrack, and J. Kappler. 1998. Detection of antigen-specific T cells with multivalent soluble class II MHC covalent peptide complexes. *Immunity* 8:675–682.

256. Baldwin, K.K., B.P. Trenchak, J.D. Altman, and M.M. Davis. 1999. Negative selection of T cells occurs throughout thymic development. *J. Immunol.* 163:689–698.

257. Callan, M.F., L. Tan, N. Annels, G.S. Ogg, J.D. Wilson, C.A. O'Callaghan, N. Steven, A.J. McMichael, and A.B. Rickinson. 1998. Direct visualization of antigen-specific CD8+ T cells during the primary immune response to Epstein-Barr virus in vivo. *J. Exp. Med.* 187:1395–1402.

258. He, X.S., B. Rehermann, F.X. Lopez-Labrador, J. Boisvert, R. Cheung, J. Mumm, H. Wedemeyer, M. Berenguer, T.L. Wright, M.M. Davis, and H.B. Greenberg. 1999. Quantitative analysis of hepatitis C virus-specific CD8(+) T cells in peripheral blood and liver using peptide-MHC tetramers. *Proc. Natl. Acad. Sci. USA* 96:5692–5697.

259. Romero, P., P.R. Dunbar, D. Valmori, M. Pittet, G.S. Ogg, D. Rimoldi, J.L. Chen, D. Lienard, J.C. Cerottini, and V. Cerundolo. 1998. Ex vivo staining of metastatic lymph nodes by class I major histocompatibility complex tetramers reveals high numbers of antigen-experienced tumor-specific cytolytic T lymphocytes. *J. Exp. Med.* 188:1641–1650.

260. Flynn, K.J., G.T. Belz, J.D. Altman, R. Ahmed, D.L. Woodland, and P.C. Doherty. 1998. Virus-specific CD8+ T cells in primary and secondary influenza pneumonia. *Immunity* 8:683–691.

261. Stevenson, P.G., G.T. Belz, J.D. Altman, and P.C. Doherty. 1998. Virus-specific CD8(+) T cell numbers are maintained during gamma-herpesvirus reactivation in CD4-deficient mice. *Proc. Natl. Acad. Sci. USA* 95:15565–15570.

262. Shortman, K. and L. Wu. 1996. Early T lymphocyte progenitors. *Annu. Rev. Immunol.* 14:29–47.

263. von Bochmer, H., I. Aifantis, J. Feinberg, O. Lechner, C. Saint-Ruf, U. Walter, J. Buer, and O. Azogui. 1999. Pleiotropic changes controlled by the pre-T-cell receptor. *Curr. Opin. Immunol.* 11:135–142.

264. Jameson, S.C., K.A. Hogquist, and M.J. Bevan. 1995. Positive selection of thymocytes. *Annu. Rev. Immunol.* 13:93–126.

265. Kisielow, P. and H. von Boehmer. 1995. Development and selection of T cells: facts and puzzles. *Adv. Immunol.* 58:87–209.

266. Sebzda, E., S. Mariathasan, T. Ohteki, R. Jones, M.F. Bachmann, and P.S. Ohashi. 1999. Selection of the T cell repertoire. *Annu. Rev. Immunol.* 17:829–874.

267. Freitas, A.A. and B. Rocha. 1999. Peripheral T cell survival. *Curr. Opin. Immunol.* 11:152–156.

268. Yancopoulos, G.D., S.V. Desiderio, M. Paskind, J.F. Kearney, D. Baltimore, and F.W. Alt. 1984. Preferential utilization of the most JH-proximal VH gene segments in pre-B-cell lines. *Nature* 311:727–733.

269. Havran, W.L. and J.P. Allison. 1988. **Developmentally** ordered appearance of thymocytes expressing **different T-cell antigen receptors.** *Nature* 335:443–445.

270. Ikuta, K., N. Uchida, J. Friedman, and I.L. Weissman. 1992. Lymphocytes development from stem cells. *Annu. Rev. Immunol.* 10:759–783.

271. Tonegawa, S., A. Berns, Bonneville, A. Farr, I. Ishida, K. Ito, S. Itohara, C.A. Janeway, Jr., O. Kanagawa, M. Katsuki, R. Kubo, J. Lafaille, P. Mombaerts, D. Murphy, N. Nakanishi, Y. Takagi, L. Van Kaer, and U. Verbeek. 1989. Diversity, development, ligands, and probable funtions of gamma delta T cells. *Cold Spring Habor Symp. Quant. Biol.* 54:31–44.

272. González-Quintial, R., R. Baccalà, R.S. Balderas, and A.N. Theofilopoulos. 1995. Vβ gene repertoire in the aging mouse: a developmental perspective. *Int. Rev. Immunol.* 12:27–40.

273. Baccalà, R., B.A. Vandekerckhove, D. Jones, D.H. Kono, M.G. Roncarolo, and A.N. Theofilopoulos. 1993. Bacterial superantigens mediate T cell deletions in the mouse severe combined immunodeficiency-human liver/thymus model. *J. Exp. Med.* 177:1481–1485.

274. Imberti, L., A. Sottini, and D. Primi. 1992. Expression and combinatorial diversity of germ line-encoded T cell receptor V genes in human peripheral blood T cells. *Cell. Immunol.* 141:21–31.

275. Uematsu, Y. 1992. Preferential association of α and β chains of the T cell antigen receptor. *Eur. J. Immunol.* 22:603–606.

276. Malhotra, U., R. Spielman, and P. Concannon. 1992. Variability in T cell receptor V beta gene usage in human peripheral blood lymphocytes. Studies of identical twins, siblings, and insulin-dependent diabetes mellitus patients. *J. Immunol.* 149:1802–1808.

277. Gulwani-Akolkar, B., D.N. Posnett, C.H. Janson, J. Grunewald, H. Wigzell, P. Akolkar, P.K. Gregersen, and J. Silver. 1991. T cell receptor V-segment frequencies in peripheral blood T cells correlate with human leukocyte antigen type. *J. Exp. Med.* 174:1139–1146.

278. Loveridge, J.A., W.M. Rosenberg, T.B. Kirkwood, and J.I. Bell. 1991. The genetic contribution to human T-cell receptor repertoire. *Immunol.* 74:246–250.

279. Akolkar, P.N., B. Gulwani-Akolkar, R. Pergolizzi, R.D. Bigler, and J. Silver. 1993. Influence of HLA genes on T cell receptor V segment frequencies and expression levels in peripheral blood lymphocytes. *J. Immunol.* 150:2761–2773.

280. Davey, M.P., M.M. Meyer, and A.C. Bakke. 1994. T cell receptor V beta gene expression in monozygotic twins. Discordance in CD8 subset and in disease states. *J. Immunol.* 152:315–321.

281. Arstila, T.P., A. Casrouge, V. Baron, J. Even, and P. Kourilsky. 1999. A direct estimate of the human alphabeta T cell receptor diversity. *Science* 286:958–961.

282. Imberti, L., A. Sottini, G. Spagnoli, and D. Primi. 1990. Expression of the human V beta 8 gene product preferentially correlates with class II major histocompatibility complex restriction specificity. *Eur. J. Immunol.* 20:2817–2819.

283. Grunewald, J., N. Shankar, H. Wigzell, and C.H. Janson. 1991. An analysis of alpha/beta TCR V gene expression in the human thymus. *Int. Immunol.* 3:699–702.

284. Cossarizza, A., M. Kahan, C. Ortolani, C. Franceschi, and M. Londei. 1991. Preferential expression of V beta 6.7 domain on human peripheral CD4+ T cells. Implication for positive selection of T cells in man. *Eur. J. Immunol.* 21:1571–1574.

285. DerSimonian, H., H. Band, and M.B. Brenner. 1991. Increased frequency of T cell receptor V alpha 12.1 expression on CD8+ T cells: evidence that V alpha participate in shaping the peripheral repertoire. *J. Exp. Med.* 174:639–648.

286. Sim, B.C., D. Lo, and N.R. Gascoigne. 1998. Preferential expression of TCR Valpha regions in CD4/CD8 subsets: class discrimination or co-receptor recognition. *Immunol. Today* 19:276–282.

287. Sim, B.C. and N.R. Gascoigne. 1999. Reciprocal expression in CD4 or CD8 subsets of different members of the V alpha 11 gene family correlates with sequence polymorphism. *J. Immunol.* 162:3153–3159.

288. Garcia, K.C., L. Teyton, and L.A. Wilson. 1999. Structural basis of T cell recognition. *Annu. Rev. Immunol.* 17:369–397.

289. Manning, T.C., E.A. Parke, L. Teyton, and D.M. Kranz. 1999. Effects of complementarity determining region mutations on the affinity of an alpha/beta T cell receptor: Measuring the energy associated with CD4/CD8 repertoire skewing. *J. Exp. Med.* 189:461–470.

290. Jamieson, B.D., D.C. Douek, S. Killian, L.E. Hultin, D.D. Scripture-Adams, J.V. Giorgi, D. Marelli, R.A. Koup, and J.A. Zack. 1999. Generation of functional thymocytes in the human adult. *Immunity* 10:569–575.

291. Callahan, J.E., J.W. Kappler, and P. Marrack. 1993. Unexpected expansions of CD8-bearing cells in old mice. *J. Immunol.* 151:6657–6669.

292. Posnett, D.N., R. Sinha, S. Kabak, and C. Russo. 1994. Clonal population of T cells in normal elderly humans: the T cell equivalent to "benign monoclonal gammapathy". *J. Exp. Med.* 179:609–618.

293. Morley, J.K., F.M. Batliwalla, R. Hingorani, and P.K. Gregersen. 1995. Oligoclonal CD8+ T cells are preferentially expanded in the CD57+ subset. *J. Immunol.* 154:6182–6190.

294. Monteiro, J., F. Batliwalla, H. Ostrer, and P.K. Gregersen. 1996. Shortened telomeres in clonally expanded CD28-CD8+ T cells imply a replicative history that is distinct from their CD28+CD8+ counterparts. *J. Immunol.* 156:3587–3590.

295. Bill, J., J. Yague, V.B. Appel, J. White, G. Horn, H.A. Erlich, and E. Palmer. 1989. Molecular genetic analysis of 178 I-Abm12-reactive T cells. *J. Exp. Med.* 169:115–133.

296. Garman, R.D., J.L. Ko, C.D. Vulpe, and D.H. Raulet. 1986. T-cell receptor variable region gene usage in T-cell populations. *Proc.Natl.Acad.Sci.USA* 83:3987–3991.

297. Sherman, D.H., P.S. Hochman, R. Dick, R. Tizard, K.L. Ramachandran, R.A. Flavell, and B.T. Huber. 1987. Molecular analysis of antigen recognition by insulin-specific T- cell hybridomas from B6 wild-type and bm12 mutant mice. *Mol.Cell.Biol.* 7:1865–1872.

298. Spinella, D.G., T.H. Hansen, W.D. Walsh, M.A. Behlke, J.P. Tillinghast, H.S. Chou, P.J. Whiteley, J.A. Kapp, C.W. Pierce, and E.M. Shevach. 1987. Receptor diversity of insulin-specific T cell lines from C57BL (H-2b) mice. *J. Immunol.* 138:3991–3995.

299. Johnson, N.A., F. Carland, P.M. Allen, and L.H. Glimcher. 1989. T cell receptor gene segment usage in a panel of hen-egg white lysozyme specific, I-Ak-restricted T helper hybridomas. *J. Immunol.* 142:3298–3304.

300. Acha-Orbea, H., D.J. Mitchell, L. Timmermann, D.C. Wraith, G.S. Tausch, M.K. Waldor, S.S. Zamvil, H.O. McDevitt, and L. Steinman. 1988. Limited heterogeneity of T cell receptors from lymphocytes mediating autoimmune encephalomyelitis allows specific immune intervention. *Cell* 54:263–273.

301. Gold, D.P., M. Vainiene, B. Celnik, S. Wiley, C. Gibbs, G.A. Hashim, A.A. Vandenbark, and H. Offner. 1992. Characterization of the immune response to a secondary encephalitogenic epitope of basic protein in Lewis rats. II. Biased T cell receptor Vβ expression predominates in spinal cord infiltrating T cells. *J. Immunol.* 148:1712–1717.

302. Burns, F.R., X.B. Li, N. Shen, H. Offner, Y.K. Chou, A.A. Vandenbark, and E. Heber-Katz. 1989. Both rat and mouse T cell receptors specific for the encephalitogenic determinant of myelin basic protein use similar V alpha and V beta chain genes even though the major histocompatibility complex and encephalitogenic determinants being recognized are different. *J. Exp. Med.* 169:27–39.

303. McHeyzer-Williams, M.G. and M.M. Davis. 1995. Antigen-specific development of primary and memory T cells in vivo. *Science* 268:106–111.

304. Hedrick, S.M., I. Engel, D.L. McElligott, P.J. Fink, M.L. Hsu, D. Hansburg, and L.A. Matis. 1988. Selection of amino acid sequences in the beta chain of the T cell antigen receptor. *Science* 239:1541–1544.

305. Lai, M.Z., Y.J. Jang, L.K. Chen, and M.L. Gefter. 1990. Restricted V-(D)-J junctional regions in the T cell response to lambda-repressor. Identification of residues critical for antigen recognition. *J. Immunol.* 144:4851–4856.

306. Fink, P.J., L.A. Matis, D.L. McElligott, M. Bookman, and S.M. Hedrick. 1986. Correlations between T cell specificity and structure of the antigen receptor. *Nature* 321:219–226.

307. Winoto, A., J.L. Urban, N.C. Lan, J. Goverman, L. Hood, and D. Hansburg. 1986. Predominant use of a V alpha gene segment in mouse T-cell receptors for cytochrome c. *Nature* 324:679–682.

308. Yanagi, Y., R. Maekawa, T. Cook, O. Kanagawa, and M.B. Oldstone. 1990. Restricted V-segment usage in T-cell receptors from cytotoxic T lymphocytes specific for a major epitope of lymphocytic choriomeningitis virus. *J.Virol.* 64:5919–5926.

309. Danska, J.S., A.M. Livingstone, V. Paragas, T. Ishihara, and C.G. Fathman. 1990. The presumptive CDR3 regions of both T cell receptor alpha and beta chains determine T cell specificity for myoglobin peptides. *J. Exp. Med.* 172:27–33.

310. Lai, M.Z., S.Y. Huang, T.J. Briner, J.G. Guillet, J.A. Smith, and M.L. Gefter. 1988. T cell receptor gene usage in the response to lambda repressor cI protein. An apparent bias in the usage of a V alpha gene element. *J. Exp. Med.* 168:1081–1097.

311. Urban, J.L., V. Kumar, D.H. Kono, C. Gomez, S.J. Horvath, J. Clayton, D.G. Ando, E.E. Sercarz, and L. Hood. 1988. Restricted use of T cell receptor V genes in murine autoimmune encephalomyelitis raises possibilities for antibody therapy. *Cell* 54:577–592.

312. Sakai, K., A.A. Sinha, D.J. Mitchell, S.S. Zamvil, J.B. Rothbard, H.O. McDevitt, and L. Steinman. 1988.

Involvement of distinct murine T-cell receptors in the autoimmune encephalitogenic response to nested epitopes of myelin basic protein. *Proc.Natl.Acad.Sci.USA* 85:8608–8612.

313. Turner, S.J., S.C. Cose, and F.R. Carbone. 1996. TCR alpha-chain usage can determine antigen-selected TCR β-chain repertoire diversity. *J. Immunol.* 157:4979–4985.

314. Bousso, P., A. Casrouge, J.D. Altman, M. Haury, J. Kanellopoulos, J.P. Abastado, and P. Kourilsky. 1998. Individual variations in the murine T cell response to a specific peptide reflect variability in naive repertoires. *Immunity* 9:169–178.

315. Bieganowska, K., P. Hollsberg, G.J. Buckle, D.G. Lim, T.F. Greten, J. Schneck, J.D. Altman, S. Jacobson, S.L. Ledis, B. Hanchard, J. Chin, O. Morgan, P.A. Roth, and D.A. Hafler. 1999. Direct analysis of viral-specific CD8+ T cells with soluble HLA- A2/Tax11–19 tetramer complexes in patients with human T cell lymphotropic virus-associated myelopathy. *J. Immunol.* 162:1765–1771.

316. Harris, E.D., Jr. 1993. Clinical features of rheumatoid arthritis. In Textbook of Rheumatology. W.N. Kelley, E.D. Harris, S. Ruddy, and C.B. Slege, editors. Saunders, Philadelphia. 874–911.

317. Winchester, R. 1994. The molecular basis of susceptibility to rheumatoid arthritis. *Adv.Immunol.* 56:389–466.

318. Hammer, J., F. Gallazzi, E. Bono, R.W. Karr, J. Guenot, P. Valsasnini, Z.A. Nagy, and F. Sinigaglia. 1995. Peptide binding specificity of HLA-DR4 molecules: Correlation with rheumatoid arthritis association. *J. Exp. Med.* 181:1847–1855.

319. Hall, F.C., M.A. Brown, D.E. Weeks, S. Walsh, A. Nicod, S. Butcher, L.J. Andrews, and B.P. Wordsworth. 1997. A linkage study across the T cell receptor A and T cell receptor B loci in families with rheumatoid arthritis. *Arthritis Rheum.* 40:1798–1802.

320. Cornélis, F., L. Hardwick, R.M. Flipo, M. Martinez, S. Lasbleiz, J.F. Prud'homme, T.H. Tran, S. Walsh, A. Delaye, A. Nicod, M.N. Loste, V. Lepage, K. Gibson, K. Pile, S. Djoulah, P.M. Danzé, F. Lioté, D. Charron, J. Weissenbach, D. Kuntz, T. Bardin, and B.P. Wordsworth. 1997. Association of rheumatoid arthritis with an amino acid allelic variation of the T cell receptor. *Arthritis Rheum.* 40:1387–1390.

321. Ibberson, M., V. Peclat, P.A. Guerne, J.M. Tiercy, P. Wordsworth, J. Lanchbury, J. Camilleri, and A.K. So. 1998. Analysis of T cell receptor V alpha polymorphisms in rheumatoid arthritis. *Ann.Rheum.Dis.* 57:49–51.

322. Mu, H., P. Charmley, M.C. King, and L.A. Criswell. 1996. Synergy between T cell receptor beta gene polymorphism and HLA- DR4 in susceptibility to rheumatoid arthritis. *Arthritis Rheum.* 39:931–937.

323. DerSimonian, H., M. Sugita, D.N. Glass, A.L. Maier, M.E. Weinblatt, T. Reme, and M.B. Brenner. 1993. Clonal V alpha 12.1+ T cell expansions in the peripheral blood of rheumatoid arthritis patients. *J. Exp. Med.* 177:1623–1631.

324. Wang, E.C., T.M. Lawson, K. Vedhara, P.A. Moss, P.J. Lehner, and L.K. Borysiewicz. 1997. CD8high+ (CD57+) T cells in patients with rheumatoid arthritis. *Arthritis Rheum.* 40:237–248.

325. Bucht, A., J.R. Oksenberg, S. Lindblad, A. Gronberg, L. Steinman, and L. Klareskog. 1992. Characterization of T-cell receptor alpha beta repertoire in synovial tissue from different temporal phases of rheumatoid arthritis. *Scand.J. Immunol.* 35:159–165.

326. Maruyama, T., I. Saito, S. Miyake, H. Hashimoto, K. Sato, H. Yagita, K. Okumura, and N. Miyasaka. 1993. A possible role of two hydrophobic amino acids in antigen recognition by synovial T cells in rheumatoid arthritis. *Eur.J. Immunol.* 23:2059–2065.

327. Struyk, L., J.T. Kurnick, G.E. Hawes, J.M. Van Laar, R. Schipper, J.R. Oksenberg, L. Steinman, R.R. De Vries, F.C. Breedveld, and P. Van den Elsen. 1993. T-cell receptor V-gene usage in synovial fluid lymphocytes of patients with chronic arthritis. *Hum.Immunol.* 37:237–251.

328. Howell, M.D., J.P. Diveley, K.A. Lundeen, A. Esty, S.T. Winters, D.J. Carlo, and S.W. Brostoff. 1991. Limited T-cell receptor beta-chain heterogeneity among interleukin 2 receptor-positive synovial T cells suggests a role for superantigen in rheumatoid arthritis. *Proc.Natl.Acad.Sci.USA* 88:10921–10925.

329. Paliard, X., S.G. West, J.A. Lafferty, J.R. Clements, J.W. Kappler, P. Marrack, and B.L. Kotzin. 1991. Evidence for the effects of a superantigen in rheumatoid arthritis. *Science* 253:325–329.

330. Jenkins, R.N., A. Nikaein, A. Zimmermann, K. Meek, and P.E. Lipsky. 1993. T cell receptor V beta gene bias in rheumatoid arthritis. *J.Clin.Invest.* 92:2688–2701.

331. Pluschke, G., G. Ricken, H. Taube, S. Kroninger, I. Melchers, H.H. Peter, K. Eichmann, and U. Krawinkel. 1991. Biased T cell receptor V alpha region repertoire in the synovial fluid of rheumatoid arthritis patients. *Eur.J. Immunol.* 21:2749–2754.

332. Goronzy, J.J., P. Bartz-Bazzanella, W. Hu, M.C. Jendro, D.R. Walser-Kuntz, and C.M. Weyand. 1994. Dominant clonotypes in the repertoire of peripheral CD4+ T cells in rheumatoid arthritis. *J.Clin.Invest.* 94:2068–2076.

333. Li, Y., G.R. Sun, J.R. Tumang, M.K. Crow, and S.M. Friedman. 1994. CDR3 sequence motifs shared by oligoclonal rheumatoid arthritis synovial T cells. Evidence for an antigen-driven response. *J.Clin.Invest.* 94:2525–2531.

334. Struyk, L., G.E. Hawes, R.J. Dolhain, A. van Scherpenzeel, B. Godthelp, F.C. Breedveld, and P.J. Van den Elsen. 1994. Evidence for selective in vivo expansion of synovial tissue-infiltrating CD4+ CD45RO+ T lymphocytes on the basis of CDR3 diversity. *Int.Immunol.* 6:897–907.

335. Hingorani, R., J. Monteiro, R. Furie, E. Chartash, C. Navarrete, R. Pergolizzi, and P.K. Gregersen. 1996. Oligoclonality of V beta 3 TCR chains in the CD8+ T cell population of rheumatoid arthritis patients. *J. Immunol.* 156:852–858.

336. Olive, C., P.A. Gatenby, and S.W. Serjeantson. 1992. Molecular characterization of the V gamma 9 T cell receptor repertoire expressed in patients with rheumatoid arthritis. *Eur.J. Immunol.* 22:2901–2906.

337. Kohsaka, H., A. Taniguchi, P.P. Chen, W.E. Ollier, and D.A. Carson. 1993. The expressed T cell receptor V gene repertoire of rheumatoid arthritis monozygotic twins: rapid analysis by anchored polymerase chain reaction and enzyme-linked immunosorbent assay. *Eur.J. Immunol.* 23:1895–1901.

338. Kageyama, Y., Y. Koide, S. Miyamoto, T. Inoue, and T.O. Yoshida. 1994. The biased V gamma gene usage in the synovial fluid of patients with rheumatoid arthritis. *Eur.J. Immunol.* 24:1122–1129.

339. Soderstrom, K., A. Bucht, E. Halapi, C. Lundqvist, A. Gronberg, E. Nilsson, D.L. Orsini, Y. van de Wal, F. Koning, and M.L. Hammarstrom. 1994. High expression of V gamma 8 is a shared feature of human gamma delta T cells

in the epithelium of the gut and in the inflamed synovial tissue. *J. Immunol.* 152:6017–6027.

340. Olive, C., P.A. Gatenby, and S.W. Serjeantson. 1992. Evidence for oligoclonality of T cell receptor delta chain transcripts expressed in rheumatoid arthritis patients. *Eur.J. Immunol.* 22:2587–2593.

341. Lunardi, C., C. Marguerie, and A.K. So. 1992. An altered repertoire of T cell receptor V gene expression by rheumatoid synovial fluid T lymphocytes. *Clin.Exp.Immunol.* 90:440–446.

342. Fitzgerald, J.E., N.S. Ricalton, A.C. Meyer, S.G. West, H. Kaplan, C. Behrendt, and B.L. Kotzin. 1995. Analysis of clonal CD8+ T cell expansions in normal individuals and patients with rheumatoid arthritis. *J. Immunol.* 154:3538–3547.

343. Ikeda, Y., K. Masuko, Y. Nakai, T. Kato, T. Hasanuma, S.I. Yoshino, Y. Mizushima, K. Nishioka, and K. Yamamoto. 1996. High frequencies of identical T cell clonotypes in synovial tissues of rheumatoid arthritis patients suggest the occurrence of common antigen-driven immune responses. *Arthritis Rheum.* 39:446–453.

344. Alam, A., N. Lambert, L. Lulé, H. Coppin, B. Mazières, C. De Préval, and A. Cantagrel. 1996. Persistence of dominant T cell clones in synovial tissues during rheumatoid arthritis. *J. Immunol.* 156:3480–3485.

345. Lim, A., A. Toubert, C. Pannetier, M. Dougados, D. Charron, P. Kourilsky, and J. Even. 1996. Spread of clonal T-cell expansions in rheumatoid arthritis patients. *Hum.Immunol.* 48:77–83.

346. Kato, T., M. Kurokawa, K. Masuko-Hongo, H. Sasakawa, T. Sekine, S. Ueda, K. Yamamoto, and K. Nishioka. 1997. T cell clonality in synovial fluid of a patient with rheumatoid arthritis–Persistent but fluctuant oligoclonal T cell expansions. *J. Immunol.* 159:5143–5149.

347. Striebich, C.C., M.T. Falta, Y. Wang, J. Bill, and B.L. Kotzin. 1998. Selective accumulation of related CD4(+) T cell clones in the synovial fluid of patients with rheumatoid arthritis. *J. Immunol.* 161:4428–4436.

348. Namekawa, T., U.G. Wagner, J.J. Goronzy, and C.M. Weyand. 1998. Functional subsets of CD4 T cells in rheumatoid synovitis. *Arthritis Rheum.* 41:2108–2116.

349. Schirmer, M., A.N. Vallejo, C.M. Weyand, and J.J. Goronzy. 1998. Resistance to apoptosis and elevated expression of Bcl-2 in clonally expanded CD4+CD28– T cells from rheumatoid arthritis patients. *J. Immunol.* 161:1018–1025.

350. Scotet, E., J. David-Ameline, M.A. Peyrat, A. Moreau-Aubry, D. Pinczon, A. Lim, J. Even, G. Semana, J.M. Berthelot, R. Breathnach, Bonneville, and E. Houssaint. 1996. T cell response to Epstein-Barr virus transactivators in chronic rheumatoid arthritis. *J. Exp. Med.* 184:1791–1800.

351. Celis, L., C. Vandevyver, P. Geusens, J. Dequeker, J. Raus, and J.W. Zhang. 1997. Clonal expansion of mycobacterial heat-shock protein-reactive T lymphocytes in the synovial fluid and blood of rheumatoid arthritis patients. *Arthritis Rheum.* 40:510–519.

352. Melchers, I., J. Jooss-Rüdiger, and H.H. Peter. 1997. Reactivity patterns of synovial T-cell lines derived from a patient with rheumatoid arthritis .1. Reactions with defined antigens and auto-antigens suggest the existence of multireactive T-cell clones. *Scand.J. Immunol.* 46:187–194.

353. Behar, S.M., C. Roy, J. Lederer, P. Fraser, and M.B. Brenner. 1998. Clonally expanded Valpha12+ (AV12S1),CD8+ T cells from a patient with rheumatoid arthritis are autoreactive. *Arthritis Rheum.* 41:498–506.

354. Sekine, T., T. Kato, K. Masuko-Hongo, H. Nakamura, S. Yoshino, K. Nishioka, and K. Yamamoto. 1999. Type II collagen is a target antigen of clonally expanded T cells in the

synovium of patients with rheumatoid arthritis. *Ann. Rheum. Dis.* 58:446–450.

355. Kotzin, B.L., M.T. Falta, F. Crawford, E.F. Rosloniec, J. Bill, P. Marrack, and J. Kappler. 2000. Use of soluble peptide-DR4 tetramers to detect synovial T cells specific for cartilage antigens in patients with rheumatoid arthritis. *Proc. Natl. Acad. Sci. U S A.* 97:291–296.

356. Trentham, D.E., A.S. Townes, and A.H. Kang. 1977. Autoimmunity to type II collagen an experimental model of arthritis. *J. Exp. Med.* 146:857–868.

357. Holmdahl, R., E.C. Andersson, C.B. Andersen, A. Svejgaard, and L. Fugger. 1999. Transgenic mouse models of rheumatoid arthritis. *Immunol.Rev.* 169:161–173.

358. Taneja, V. and C.S. David. 1999. HLA class II transgenic mice as models of human diseases. *Immunol.Rev.* 169:67–79.

359. Haqqi, T.M., X.M. Qu, M.S. Sy, and S. Banerjee. 1995. Restricted expression of T cell receptor V beta and lymphokine genes in arthritic joints of a TCR V beta a (H-2q) mouse strain- BUB/BnJ-with collagen-induced arthritis. *Autoimmunity* 20:163–170.

360. Osman, G.E., M. Toda, O. Kanagawa, and L.E. Hood. 1993. Characterization of the T cell receptor repertoire causing collagen arthritis in mice. *J. Exp. Med.* 177:387–395.

361. Rosloniec, E.F., D.D. Brand, K.B. Whittington, J.M. Stuart, and M. Ciubotaru. 1995. Vaccination with a recombinant Valpha domain of a TCR prevents the development of collagen-induced arthritis. *J. Immunol.* 155:4504–4511.

362. Kumar, V., F. Aziz, E. Sercarz, and A. Miller. 1997. Regulatory T cells specific for the same framework 3 region of the Vβ8.2 chain are involved in the control of collagen II-induced arthritis and experimental autoimmune encephalomyelitis. *J. Exp. Med.* 185:1725–1733.

363. Osman, G.E., M.C. Hannibal, J.P. Anderson, S.R. Lasky, W.C. Ladiges, and L. Hood. 1999. FVB/N (H2(q)) mouse is resistant to arthritis induction and exhibits a genomic deletion of T-cell receptor V beta gene segments. *Immunogenetics.* 49:851–859.

364. Kouskoff, V., A.S. Korganow, V. Duchatelle, C. Degott, C. Benoist, and D. Mathis. 1996. Organ-specific disease provoked by systemic autoimmunity. *Cell* 87:811–822.

365. Matsumoto, I., A. Staub, C. Benoist, and D. Mathis. 1999. Arthritis provoked by linked T and B cell recognition of a glycolytic enzyme. *Science.* 286:1732–1735.

366. Korganow, A.S., H. Ji, S. Mangialaio, V. Duchatelle, R. Pelanda, T. Martin, C. Degott, H. Kikutani, K. Rajewsky, J.L. Pasquali, C. Benoist, and D. Mathis. 1999. From systemic T cell self-reactivity to organ-specific autoimmune disease via immunoglobulins. *Immunity* 10:451–461.

367. Nepom, G.T. and W.W. Kwok. 1998. Molecular basis for HLA-DQ associations with IDDM. *Diabetes* 47:1177–1184.

368. Wong, F.S. and C.A. Janeway, Jr. 1999. Insulin-dependent diabetes mellitus and its animal models. *Curr. Opin. Immunol.* 11:643–647.

369. Hoover, M.L., G. Angelini, E. Ball, P. Stastny, J. Marks, J. Rosenstock, P. Raskin, G.B. Ferrara, R. Tosi, and J.D. Capra. 1986. HLA-DQ and T-cell receptor genes in insulin-dependent diabetes mellitus. *Cold Spring Harbor Symp.Quant.Biol.* 51 Pt 2:803–809.

370. Concannon, P., J.A. Wright, L.G. Wright, D.R. Sylvester, and R.S. Spielman. 1990. T-cell receptor genes and insulin-dependent diabetes mellitus (IDDM): no evidence for linkage from affected sib pairs. *Am. J. Hum. Genetics* 47:45–52.

371. Avoustin, P., L. Briant, C. de Preval, and A. Cambon-Thomsen. 1992. Polymorphism study of TCR alpha and gamma genes in insulin dependent diabetes mellitus (IDDM) multiplex families. *Autoimmunity* 14:97–100.

372. McDermott, M.F., G. Schmidt-Wolf, A.A. Sinha, M. Koo, M.A. Porter, L. Briant, A. Cambon-Thomsen, N.K. Maclaren, D. Fiske, S. Bertera, M. Trucco, C.I. Amos, H.O. McDevitt, and D.L. Kastner. 1996. No linkage or association of telomeric and centromeric T-cell receptor beta-chain markers with susceptibility to type 1 insulin- dependent diabetes in HLA-DR4 multiplex families. *Eur. J. Immunogenet.* 23:361–370.

373. Somoza, N., F. Vargas, C. Roura-Mir, M. Vives-Pi, M.T. Fernandez-Figueras, A. Ariza, R. Gomis, R. Bragado, M. Marti, and D. Jaraquemada. 1994. Pancreas in recent onset insulin-dependent diabetes mellitus. Changes in HLA, adhesion molecules and autoantigens, restricted T cell receptor V beta usage, and cytokine profile. *J. Immunol.* 153:1360–1377.

374. Yamagata, K., H. Nakajima, K. Tomita, N. Itoh, J. Miyagawa, T. Hamaguchi, M. Namba, S. Tamura, S. Kawata, N. Kono, M. Kuwajima, T. Noguchi, T. Hanafusa, and Y. Matsuzawa. 1996. Dominant TCR alpha-chain clonotypes and interferon-gamma are expressed in the pancreas of patients with recent-onset insulin- dependent diabetes mellitus. *Diabetes Res. Clin. Pract.* 34:37–46.

375. Santamaria, P., C. Lewis, J. Jessurun, D.E. Sutherland, and J.J. Barbosa. 1994. Skewed T-cell receptor usage and junctional heterogeneity among isletitis alpha beta and gamma delta T-cells in human IDDM. *Diabetes* 43:599–606.

376. Hanninen, A., S. Jalkanen, M. Salmi, S. Toikkanen, G. Nikolakaros, and O. Simell. 1992. Macrophages, T cell receptor usage, and endothelial cell activation in the pancreas at the onset of insulin-dependent diabetes mellitus. *J. Clin. Invest.* 90:1901–1910.

377. Conrad, B., E. Weidmann, G. Trucco, W.A. Rudert, R. Behboo, C. Ricordi, H. Rodriquez-Rilo, D. Finegold, and M. Trucco. 1994. Evidence for superantigen involvement in insulin-dependent diabetes mellitus aetiology. *Nature* 371:351–355.

378. Conrad, B., R.N. Weissmahr, J. Boni, R. Arcari, J. Schupbach, and B. Mach. 1997. A human endogenous retroviral superantigen as candidate autoimmune gene in type I diabetes. *Cell* 90:303–313.

379. Murphy, V.J., L.C. Harrison, W.A. Rudert, P. Luppi, M. Trucco, A. Fierabracci, P.A. Biro, and G.F. Bottazzo. 1998. Retroviral superantigens and type 1 diabetes mellitus. *Cell* 95:9–11.

380. Lower, R., R.R. Tonjes, K. Boller, J. Denner, B. Kaiser, R.C. Phelps, J. Lower, R. Kurth, K. Badenhoop, H. Donner, K.H. Usadel, T. Miethke, M. Lapatschek, and H. Wagner. 1998. Development of insulin-dependent diabetes mellitus does not depend on specific expression of the human endogenous retrovirus HERV-K. *Cell* 95:11–14.

381. Lan, M.S., A. Mason, R. Coutant, Q.Y. Chen, A. Vargas, J. Rao, R. Gomez, S. Chalew, R. Garry, and N.K. Maclaren. 1998. HERV-K10s and immune-mediated (type 1) diabetes. *Cell* 95:14–16.

382. Ridgway, W.M. and C.G. Fathman. 1999. MHC structure and autoimmune T cell repertoire development. *Curr. Opin. Immunol.* 11:638–642.

383. O'Reilly, L.A., P.R. Hutchings, P.R. Crocker, E. Simpson, T. Lund, D. Kioussis, F. Takei, J. Baird, and A. Cooke. 1991. Characterization of pancreatic islet cell infiltrates in NOD mice: effect of cell transfer and transgene expression. *Eur. J. Immunol.* 21:1171–1180.

384. Toyoda, H., A. Redford, D. Magalong, E. Chan, N. Hosszufalusi, B. Formby, M. Teruya, and M.A. Charles. 1992. In situ islet T cell receptor variable region gene usage in the nonobese diabetic mouse. *Immunol. Lett.* 32:241–245.

385. Galley, K.A. and J.S. Danska. 1995. Peri-islet infiltrates of young non-obese diabetic mice display restricted TCR beta-chain diversity. *J. Immunol.* 154:2969–2982.

386. Sarukhan, A., J.M. Gombert, M. Olivi, J.F. Bach, C. Carnaud, and H.J. Garchon. 1994. Anchored polymerase chain reaction based analysis of the V beta repertoire in the non-obese diabetic (NOD) mouse. *Eur. J. Immunol.* 24:1750–1756.

387. Drexler, K., S. Burtles, and U. Hurtenbach. 1993. Limited heterogeneity of T-cell receptor V beta gene expression in the early stage of insulitis in NOD mice. *Immunol. Lett.* 37:187–196.

388. Yang, Y., B. Charlton, A. Shimada, R. Dal Canto, and C.G. Fathman. 1996. Monoclonal T cells identified in early NOD islet infiltrates. *Immunity* 4:189–194.

389. Verdaguer, J., J.W. Yoon, B. Anderson, N. Averill, T. Utsugi, B.J. Park, and P. Santamaria. 1996. Acceleration of spontaneous diabetes in TCR-beta-transgenic nonobese diabetic mice by beta-cell cytotoxic CD8+ T cells expressing identical endogenous TCR-alpha chains. *J. Immunol.* 157:4726–4735.

390. Nakano, N., H. Kikutani, H. Nishimoto, and T. Kishimoto. 1991. T cell receptor V gene usage of islet beta cell-reactive T cells is not restricted in non-obese diabetic mice. *J. Exp. Med.* 173:1091–1097.

391. Prud'homme, G.J., T.Y. Long, D.C. Bocarro, R.S. Balderas, and A.N. Theofilopoulos. 1991. Analysis of pancreas-infiltrating T cells in diabetic NOD mice: fusion with BW5147 yields a high frequency of islet-reactive hybridomas. *Autoimmunity* 10:285–289.

392. Daniel, D., R.G. Gill, N. Schloot, and D. Wegmann. 1995. Epitope specificity, cytokine production profile and diabetogenic activity of insulin-specific T cell clones isolated from NOD mice. *Eur. J. Immunol.* 25:1056–1062.

393. Schloot, N.C., D. Daniel, M. Norbury-Glaser, and D.R. Wegmann. 1996. Peripheral T cell clones from NOD mice specific for GAD65 peptides: lack of islet responsiveness or diabetogenicity. *J. Autoimmun.* 9:357–363.

394. Simone, E., D. Daniel, N. Schloot, P. Gottlieb, S. Babu, E. Kawasaki, D. Wegmann, and G.S. Eisenbarth. 1997. T cell receptor restriction of diabetogenic autoimmune NOD T cells. *Proc. Natl. Acad. Sci. USA* 94:2518–2521.

395. Santamaria, P., T. Utsugi, B.J. Park, N. Averill, S. Kawazu, and J.W. Yoon. 1995. Beta-cell-cytotoxic CD8+ T cells from nonobese diabetic mice use highly homologous T cell receptor alpha-chain CDR3 sequences. *J. Immunol.* 154:2494–2503.

396. DiLorenzo, T.P., R.T. Graser, T. Ono, G.J. Christianson, H.D. Chapman, D.C. Roopenian, S.G. Nathenson, and D.V. Serreze. 1998. Major histocompatibility complex class I-restricted T cells are required for all but the end stages of diabetes development in nonobese diabetic mice and use a prevalent T cell receptor alpha chain gene rearrangement. *Proc. Natl. Acad. Sci. USA* 95:12538–12543.

397. Shizuru, J.A., C. Taylor-Edwards, A. Livingstone, and C.G. Fathman. 1991. Genetic dissection of T cell receptor V beta gene requirements for spontaneous murine diabetes. *J. Exp. Med.* 174:633–638.

398. Lipes, M.A., A. Rosenzweig, K.N. Tan, G. Tanigawa, D. Ladd, J.G. Seidman, and G.S. Eisenbarth. 1993. Progression to diabetes in nonobese diabetic (NOD) mice with transgenic T cell receptors. *Science* 259:1165–1169.

399. Taki, T., K. Yokono, K. Amano, N. Hatamori, Y. Hirao, Y. Tominaga, S. Maeda, and M. Kasuga. 1993. Effect of T-cell receptor V beta-specific monoclonal antibodies on cyclophosphamide-induced diabetes mellitus in non-obese diabetic mice. *Diabetologia* 36:391–396.

400. Bacelj, A., B. Charlton, and T.E. Mandel. 1989. Prevention of cyclophosphamide-induced diabetes by anti-V beta 8 T-lymphocyte-receptor monoclonal antibody therapy in NOD/Wehi mice. *Diabetes* 38:1492–1495.

401. Edouard, P., C. Thivolet, P. Bedossa, M. Olivi, B. Legrand, A. Bendelac, J.F. Bach, and C. Carnaud. 1993. Evidence for a preferential V beta usage by the T cells which adoptively transfer diabetes in NOD mice. *Eur. J. Immunol.* 23:727–733.

402. Ridgway, W.M., H. Ito, M. Fasso, C. Yu, and C.G. Fathman. 1998. Analysis of the role of variation of major histocompatibility complex class II expression on nonobese diabetic (NOD) peripheral T cell response. *J. Exp. Med* 188:2267–2275.

403. Ridgway, W.M., M. Fasso, and C.G. Fathman. 1999. A new look at MHC and autoimmune disease. *Science* 284:749–751.

404. Luhder, F., J. Katz, C. Benoist, and D. Mathis. 1998. Major histocompatibility complex class II molecules can protect from diabetes by positively selecting T cells with additional specificities. *J. Exp. Med.* 187:379–387.

405. Biddison, W.E., S.S. Beall, P. Concannon, P. Charmley, R.A. Gatti, L.E. Hood, H.F. McFarland, and D.E. McFarlin. 1989. The germline repertoire of T-cell receptor beta-chain genes in patients with multiple sclerosis. *Res. Immunol.* 140:212–215.

406. Oksenberg, J.R., M. Sherritt, A.B. Begovich, H.A. Erlich, C.C. Bernard, L.L. Cavalli-Sforza, and L. Steinman. 1989. T-cell receptor V alpha and C alpha alleles associated with multiple and myasthenia gravis. *Proc. Natl. Acad. Sci. USA* 86:988–992.

407. Charmley, P., S.S. Beall, P. Concannon, L. Hood, and R.A. Gatti. 1991. Further localization of a multiple sclerosis susceptibility gene on chromosome 7q using a new T cell receptor beta-chain DNA polymorphism. *J. Neuroimmunol.* 32:231–240.

408. Beall, S.S., W.E. Biddison, D.E. McFarlin, H.F. McFarland, and L.E. Hood. 1993. Susceptibility for multiple sclerosis is determined, in part, by inheritance of a 175-kb region of the TcR V beta chain locus and HLA class II genes. *J. Neuroimmunol.* 45:53–60.

409. Briant, L., P. Avoustin, J. Clayton, M. McDermott, M. Clanet, and A. Cambon-Thomsen. 1993. Multiple sclerosis susceptibility: population and twin study of polymorphisms in the T-cell receptor beta and gamma genes region. French Group on Multiple Sclerosis. *Autoimmunity* 15:67–73.

410. Hashimoto, L.L., T.W. Mak, and G.C. Ebers. 1992. T cell receptor alpha chain polymorphisms in multiple sclerosis. *J. Neuroimmunol.* 40:41–48.

411. Hillert, J., C. Leng, and O. Olerup. 1992. T-cell receptor alpha chain germline gene polymorphisms in multiple sclerosis. *Neurology* 42:80–84.

412. Sherritt, M.A., J. Oksenberg, N.K. de Rosbo, and C.C. Bernard. 1992. Influence of HLA-DR2, HLA-DPw4, and T cell receptor alpha chain genes on the susceptibility to multiple sclerosis. *Int. Immunol.* 4:177–181.

413. Wei, S., P. Charmley, R.I. Birchfield, and P. Concannon. 1995. Human T-cell receptor V beta gene polymorphism and multiple sclerosis. *Am. J. Hum. Genetics* 56:963–969.

414. Oksenberg, J.R., S. Stuart, A.B. Begovich, R.B. Bell, H.A. Erlich, L. Steinman, and C.C. Bernard. 1990. Limited heterogeneity of rearranged T-cell receptor V alpha transcripts in brains of multiple sclerosis patients. *Nature* 345:344–346.

415. Wucherpfennig, K.W., J. Newcombe, H. Li, C. Keddy, M.L. Cuzner, and D.A. Hafler. 1992. T cell receptor V alpha-V beta repertoire and cytokine gene expression in active multiple sclerosis lesions. *J. Exp. Med.* 175:993–1002.

416. Oksenberg, J.R., M.A. Panzara, A.B. Begovich, D. Mitchell, H.A. Erlich, R.S. Murray, R. Shimonkevitz, M. Sherritt, J. Rothbard, and C.C. Bernard. 1993. Selection for T-cell receptor V beta-D beta-J beta gene rearrangements with specificity for a myelin basic protein peptide in brain lesions of multiple sclerosis. *Nature* 362:68–70.

417. Musette, P., D. Bequet, C. Delarbre, G. Gachelin, P. Kourilsky, and D. Dormont. 1996. Expansion of a recurrent V beta 5.3+ T-cell population in newly diagnosed and untreated HLA-DR2 multiple sclerosis patients. *Proc. Natl. Acad. Sci. USA* 93:12461–12466.

418. Wucherpfennig, K.W., J. Newcombe, H. Li, C. Keddy, M.L. Cuzner, and D.A. Hafler. 1992. Gamma delta T-cell receptor repertoire in acute multiple sclerosis lesions. *Proc. Natl. Acad. Sci. USA* 89:4588–4592.

419. Hvas, J., J.R. Oksenberg, R. Fernando, L. Steinman, and C.C. Bernard. 1993. Gamma delta T cell receptor repertoire in brain lesions of patients with multiple sclerosis. *J. Neuroimmunol.* 46:225–234.

420. Battistini, L., K. Selmaj, C. Kowal, J. Ohmen, R.L. Modlin, C.S. Raine, and C.F. Brosnan. 1995. Multiple sclerosis: limited diversity of the V delta 2-J delta 3 T-cell receptor in chronic active lesions. *Ann. Neurol.* 37:198–203.

421. Lee, S.J., K.W. Wucherpfennig, S.A. Brod, D. Benjamin, H.L. Weiner, and D.A. Hafler. 1991. Common T-cell receptor V beta usage in oligoclonal T lymphocytes derived from cerebrospinal fluid and blood of patients with multiple sclerosis. *Ann. Neurol.* 29:33–40.

422. Usuku, K., N. Joshi, C.J. Hatem, Jr., M.A. Wong, M.C. Stein, and S.L. Hauser. 1996. Biased expression of T cell receptor genes characterizes activated T cells in multiple sclerosis cerebrospinal fluid. *J. Neurosci. Res.* 45:829–837.

423. Wilson, D.B., A.B. Golding, R.A. Smith, T. Dafashy, J. Nelson, L. Smith, D.J. Carlo, S.W. Brostoff, and D.P. Gold. 1997. Results of a phase I clinical trial of a T-cell receptor peptide vaccine in patients with multiple sclerosis. I. Analysis of T-cell receptor utilization in CSF cell populations. *J. Neuroimmunol.* 76:15–28.

424. Lozeron, P., D. Chabas, B. Duprey, O. Lyon-Caen, and R. Liblau. 1998. T cell receptor V beta 5 and V beta 17 clonal diversity in cerebrospinal fluid and peripheral blood lymphocytes of multiple sclerosis patients. *Multiple Sclerosis* 4:154–161.

425. Shimonkevitz, R., C. Colburn, J.A. Burnham, R.S. Murray, and B.L. Kotzin. 1993. Clonal expansions of activated gamma/delta T cells in recent-onset multiple sclerosis. *Proc. Natl. Acad. Sci.USA* 90:923–927.

426. Nick, S., P. Pileri, S. Tongiani, Y. Uematsu, L. Kappos, and G. De Libero. 1995. T cell receptor gamma delta repertoire is skewed in cerebrospinal fluid of multiple sclerosis patients: molecular and functional analyses of antigen-reactive gamma delta clones. *Eur. J. Immunol.* 25:355–363.

427. Bieganowski, P., K. Bieganowska, J. Zaborski, and A. Czlonkowska. 1996. Oligoclonal expansion of gamma delta T cells in cerebrospinal fluid of multiple sclerosis patients. *Multiple Sclerosis* 2:78–82.

428. Meinl, E., F. Weber, K. Drexler, C. Morelle, M. Ott, G. Saruhan-Direskeneli, N. Goebels, B. Ertl, G. Jechart, and G. Giegerich. 1993. Myelin basic protein-specific T lymphocyte repertoire in multiple sclerosis. Complexity of the response and dominance of nested epitopes due to recruitment of multiple T cell clones. *J. Clin. Invest.* 92:2633–2643.

429. Richert, J.R., E.D. Robinson, K. Camphausen, R. Martin, R.R. Voskuhl, M.A. Faerber, H.F. McFarland, and C.K. Hurley. 1995. Diversity of T-cell receptor V alpha, V beta, and CDR3 expression by myelin basic protein-specific human T-cell clones. *Neurology* 45:1919–1922.

430. Vandevyver, C., N. Mertens, P. Van den Elsen, R. Medaer, J. Raus, and J. Zhang. 1995. Clonal expansion of myelin basic protein-reactive T cells in patients with multiple sclerosis: restricted T cell receptor V gene rearrangements and CDR3 sequence. *Eur. J. Immunol.* 25:958–968.

431. Hafler, D.A., M.G. Saadeh, V.K. Kuchroo, E. Milford, and L. Steinman. 1996. TCR usage in human and experimental demyelinating disease. *Immunol.Today* 17:152–159.

432. Wucherpfennig, K.W., K. Ota, N. Endo, J.G. Seidman, A. Rosenzweig, H.L. Weiner, and D.A. Hafler. 1990. Shared human T cell receptor V beta usage to immunodominant regions of myelin basic protein. *Science* 248:1016–1019.

433. Kotzin, B.L., S. Karuturi, Y.K. Chou, J. Lafferty, J.M. Forrester, M. Better, G.E. Nedwin, H. Offner, and A.A. Vandenbark. 1991. Preferential T-cell receptor beta-chain variable gene use in myelin basic protein-reactive T-cell clones from patients with multiple sclerosis. *Proc. Natl. Acad. Sci. USA* 88:9161–9165.

434. Satyanarayana, K., Y.K. Chou, D. Bourdette, R. Whitham, G.A. Hashim, H. Offner, and A.A. Vandenbark. 1993. Epitope specificity and V gene expression of cerebrospinal fluid T cells specific for intact versus cryptic epitopes of myelin basic protein. *J. Neuroimmunol.* 44:57–67.

435. Utz, U., W.E. Biddison, H.F. McFarland, D.E. McFarlin, M. Flerlage, and R. Martin. 1993. Skewed T-cell receptor repertoire in genetically identical twins correlates with multiple sclerosis. *Nature* 364:243–247.

436. Bourdette, D.N., R.H. Whitham, Y.K. Chou, W.J. Morrison, J. Atherton, C. Kenny, D. Liefeld, G.A. Hashim, H. Offner, and A.A. Vandenbark. 1994. Immunity to TCR peptides in multiple sclerosis. I. Successful immunization of patients with synthetic V beta 5.2 and V beta 6.1 CDR2 peptides. *J. Immunol.* 152:2510–2519.

437. Chou, Y.K., W.J. Morrison, A.D. Weinberg, R. Dedrick, R. Whitham, D.N. Bourdette, G. Hashim, H. Offner, and A.A. Vandenbark. 1994. Immunity to TCR peptides in multiple sclerosis. II. T cell recognition of V beta 5.2 and V beta 6.1 CDR2 peptides. *J. Immunol.* 152:2520–2529.

438. Gold, D.P., R.A. Smith, A.B. Golding, E.E. Morgan, T. Dafashy, J. Nelson, L. Smith, J. Diveley, J.A. Laxer, S.P. Richieri, D.J. Carlo, S.W. Brostoff, and D.B. Wilson. 1997. Results of a phase I clinical trial of a T-cell receptor vaccine in patients with multiple sclerosis. II. Comparative analysis of TCR utilization in CSF T-cell populations before and after vaccination with a TCRV beta 6 CDR2 peptide. *J. Neuroimmunol.* 76:29–38.

439. Heber-Katz, E. and H. Acha-Orbea. 1989. The V-region disease hypothesis: evidence from autoimmune encephalomyelitis. *Immunol. Today* 10:164–169.

440. Zaller, D.M., G. Osman, O. Kanagawa, and L. Hood. 1990. Prevention and treatment of murine experimental allergic encephalomyelitis with T cell receptor V beta-specific antibodies. *J. Exp. Med.* 171:1943–1955.

441. Steinman, L. 1991. The development of rational strategies for selective immunotherapy against autoimmune demyelinating disease. *Adv. Immunol.* 49:357–379.

442. Ben-Nun, A., H. Wekerle, and I.R. Cohen. 1981. Vaccination against autoimmune encephalomyelitis with T- lymphocyte line cells reactive against myelin basic protein. *Nature* 292:60–61.

443. Lider, O., T. Reshef, E. Beraud, A. Ben-Nun, and I.R. Cohen. 1988. Anti-idiotypic network induced by T cell vaccination against experimental autoimmune encephalomyelitis. *Science* 239:181–183.

444. Offner, H., G.A. Hashim, and A.A. Vandenbark. 1991. T cell receptor peptide therapy triggers autoregulation of experimental encephalomyelitis. *Science* 251:430–432.

445. Howell, M.D., S.T. Winters, T. Olee, H.C. Powell, D.J. Carlo, and S.W. Brostoff. 1989. Vaccination against experimental allergic encephalomyelitis with T cell receptor peptides. *Science* 246:668–670.

446. Waisman, A., P.J. Ruiz, D.L. Hirschberg, A. Gelman, J.R. Oksenberg, S. Brocke, F. Mor, I.R. Cohen, and L. Steinman. 1996. Suppressive vaccination with DNA encoding a variable region gene of the T-cell receptor prevents autoimmune encephalomyelitis and activates Th2 immunity. *Nature Med.* 2:899–905.

447. Desquenne-Clark, L., T.R. Esch, L. Otvos, Jr., and E. Heber-Katz. 1991. T-cell receptor peptide immunization leads to enhanced and chronic experimental allergic encephalomyelitis. *Proc. Natl. Acad. Sci.USA* 88:7219–7223.

448. Barbesino, G., Y. Tomer, E. Concepcion, T.F. Davies, and D.A. Greenberg. 1998. Linkage analysis of candidate genes in autoimmune thyroid disease: 1. Selected immunoregulatory genes. International Consortium for the Genetics of Autoimmune Thyroid Disease. *J. Clin. Endocrinol. Metab.* 83:1580–1584.

449. Davies, T.F., A. Martin, E.S. Concepcion, P. Graves, L. Cohen, and A. Ben-Nun. 1991. Evidence of limited variability of antigen receptors on intrathyroidal T cells in autoimmune thyroid disease. *N. Engl. J. Med.* 325:238–244.

450. Davies, T.F., A. Martin, E.S. Concepcion, P. Graves, N. Lahat, W.L. Cohen, and A. Ben-Nun. 1992. Evidence for selective accumulation of intrathyroidal T lymphocytes in human autoimmune thyroid disease based on T cell receptor V gene usage. *J. Clin. Invest.* 89:157–162.

451. Davies, T.F. 1992. Preferential use of T-cell receptor V genes in human autoimmune thyroid disease. *Autoimmunity* 13:11–16.

452. McIntosh, R.S., N. Tandon, A.P. Pickerill, R. Davies, D. Barnett, and A.P. Weetman. 1993. IL-2 receptor-positive intrathyroidal lymphocytes in Graves' disease. Analysis of V alpha transcript microheterogeneity. *J. Immunol.* 151:3884–3893.

453. Caso-Pelaez, E., A.M. McGregor, and J.P. Banga. 1995. A polyclonal T cell repertoire of V-alpha and V-beta T cell receptor gene families in intrathyroidal T lymphocytes of Graves' disease patients. *Scand. J. Immunol.* 41:141–147.

454. Heufelder, A.E. 1998. T-cell restriction in thyroid eye disease. *Thyroid* 8:419–422.

455. Truffault, F., S. Cohen-Kaminsky, I. Khalil, P. Levasseur, and S. Berrih-Aknin. 1997. Altered intrathymic T-cell repertoire in human myasthenia gravis. *Ann. Neurol.* 41:731–741.

456. Navaneetham, D., A.S. Penn, A.F. Howard, and B.M. Conti-Fine. 1998. TCR-V beta usage in the thymus and blood of myasthenia gravis patients. *J. Autoimmun.* 11:621–633.

457. Grunewald, J., R. Ahlberg, A.K. Lefvert, H. DerSimonian, H. Wigzell, and C.H. Janson. 1991. Abnormal T-cell expansion and V-gene usage in myasthenia gravis patients. *Scand. J. Immunol.* 34:161–168.

458. Gigliotti, D., A.K. Lefvert, M. Jeddi-Tehrani, S. Esin, V. Hodara, R. Pirskanen, H. Wigzell, and R. Andersson. 1996. Overexpression of select T cell receptor V beta gene families within CD4+ and CD8+ T cell subsets of myasthenia

gravis patients: a role for superantigen(s)? *Mol. Med.* 2:452–459.

459. Raju, R., D. Navaneetham, M.P. Protti, R.M. Horton, B.L. Hoppe, J. Howard, Jr., and B.M. Conti-Fine. 1997. TCR V beta usage by acetylcholine receptor-specific CD4+ T cells in myasthenia gravis. *J. Autoimmun.* 10:203–217.

460. Aime-Sempe, C., S. Cohen-Kaminsky, C. Bruand, I. Klingel-Schmitt, F. Truffault, and S. Berrih-Aknin. 1995. In vivo preferential usage of TCR V beta 8 in Torpedo acetylcholine receptor immune response in the murine experimental model of myasthenia gravis. *J. Neuroimmunol.* 58:191–200.

461. Pierce, J.L., K.A. Zborowski, E. Kraig, and A.J. Infante. 1994. Highly conserved TCR beta chain CDR3 sequences among immunodominant acetylcholine receptor-reactive T cells in murine myasthenia gravis. *Int. Immunol.* 6:775–783.

462. Kaul, R., B. Wu, E. Goluszko, C. Deng, V. Dedhia, G.H. Nabozny, C.S. David, I.J. Rimm, M. Shenoy, T.M. Haqqi, and P. Christadoss. 1997. Experimental autoimmune myasthenia gravis in B10.BV8S2 transgenic mice: preferential usage of TCRAV1 gene by lymphocytes responding to acetylcholine receptor. *J. Immunol.* 158:6006–6012.

463. Yang, B., K.R. McIntosh, and D.B. Drachman. 1998. How subtle differences in MHC class II affect the severity of experimental myasthenia gravis. *Clin. Immunol. Immunopathol.* 86:45–58.

464. Kraig, E., J.L. Pierce, K.Z. Clarkin, N.E. Standifer, P. Currier, K.A. Wall, and A.J. Infante. 1996. Restricted T cell receptor repertoire for acetylcholine receptor in murine myasthenia gravis. *J. Neuroimmunol.* 71:87–95.

465. Spack, E.G. 1997. Treatment of autoimmune diseases through manipulation of antigen presentation. *Crit. Rev. Immunol.* 17:529–536.

466. Faber-Elmann, A., M. Paas-Rozner, M. Sela, and E. Mozes. 1998. Altered peptide ligands act as partial agonists by inhibiting phospholipase C activity induced by myasthenogenic T cell epitopes. *Proc. Natl. Acad. Sci. USA* 95:14320–14325.

467. Shenoy, M., M. Oshima, M.Z. Atassi, and P. Christadoss. 1993. Suppression of experimental autoimmune myasthenia gravis by epitope-specific neonatal tolerance to synthetic region alpha 146– 162 of acetylcholine receptor. *Clin. Immunol. Immunopathol.* 66:230–238.

468. Ben-Nun, A., R.S. Liblau, L. Cohen, D. Lehmann, E. Tournier-Lasserve, A. Rosenzweig, J.W. Zhang, J.C. Raus, and M.A. Bach. 1991. Restricted T-cell receptor V beta gene usage by myelin basic protein-specific T-cell clones in multiple sclerosis: predominant genes vary in individuals. *Proc. Natl. Acad. Sci. USA* 88:2466–2470.

469. Kolowos, W., M. Herrmann, B.B. Ponner, R. Voll, P. Kern, C. Frank, and J.R. Kalden. 1997. Detection of restricted junctional diversity of peripheral T cells in SLE patients by spectratyping. *Lupus* 6:701–707.

470. Mato, T., K. Masuko, Y. Misaki, N. Hirose, K. Ito, Y. Takemoto, K. Izawa, S. Yamamori, T. Kato, K. Nishioka, and K. Yamamoto. 1997. Correlation of clonal T cell expansion with disease activity in systemic lupus erythematosus. *Int. Immunol.* 9:547–554.

471. Olive, C., P.A. Gatenby, and S.W. Serjeantson. 1994. Restricted junctional diversity of T cell receptor delta gene rearrangements expressed in systemic lupus erythematosus (SLE) patients. *Clin. Exp. Immunol.* 97:430–438.

472. Rajagopalan, S., C. Mao, and S.K. Datta. 1992. Pathogenic autoantibody-inducing gamma/delta T helper cells from patients with lupus nephritis express unusual T cell receptors. *Clin. Immunol. Immunopathol.* 62:344–350.

473. Desai-Mehta, A., C. Mao, S. Rajagopalan, T. Robinson, and S.K. Datta. 1995. Structure and specificity of T cell receptors expressed by potentially pathogenic anti-DNA autoantibody-inducing T cells in human lupus. *J. Clin. Invest.* 95:531–541.

474. Takeno, M., H. Nagafuchi, S. Kaneko, S. Wakisaka, K. Oneda, Y. Takeba, N. Yamashita, N. Suzuki, H. Kaneoka, and T. Sakane. 1997. Autoreactive T cell clones from patients with systemic lupus erythematosus support polyclonal autoantibody production. *J. Immunol.* 158:3529–3538.

475. Theofilopoulos, A.N., R. Kofler, P.A. Singer, and F.J. Dixon. 1989. Molecular genetics of murine lupus models. *Adv. Immunol.* 46:61–109.

476. Kotzin, B.L., V.L. Barr, and E. Palmer. 1985. A large deletion within the T-cell receptor beta-chain gene complex in New Zealand white mice. *Science* 229:167–171.

477. Noonan, D.J., R. Kofler, P.A. Singer, G. Cardenas, F.J. Dixon, and A.N. Theofilopoulos. 1986. Delineation of a defect in T cell receptor beta genes of NZW mice predisposed to autoimmunity. *J. Exp. Med.* 163:644–653.

478. Kotzin, B.L. and E. Palmer. 1987. The contribution of NZW genes to lupus-like disease in (NZB x NZW)F1 mice. *J. Exp. Med.* 165:1237–1251.

479. Kono, D.H. and A.N. Theofilopoulos. 1996. Genetic contributions to SLE. *J. Autoimmun.* 9:437–452.

480. Kotzin, B.L., J.W. Kappler, P.C. Marrack, and L.R. Herron. 1989. T cell tolerance to self antigens in New Zealand hybrid mice with lupus-like disease. *J. Immunol.* 143:89–94.

481. Baccalà, R., R. González-Quintial, and A.N. Theofilopoulos. 1992. Lack of evidence for central T cell tolerance defects in lupus-mice and for Vβ-deleting endogenous-superantigens in rats and humans. *Res. Immunol.* 143:288–230.

482. Singer, P.A. and A.N. Theofilopoulos. 1990. T-cell receptor V beta repertoire expression in murine models of SLE. *Immunol. Rev.* 118:103–127.

483. Rozzo, S.J., C.G. Drake, B.L. Chiang, M.E. Gershwin, and B.L. Kotzin. 1994. Evidence for polyclonal T cell activation in murine models of systemic lupus erythematosus. *J. Immunol.* 153:1340–1351.

484. Musette, P., C. Pannetier, G. Gachelin, and P. Kourilsky. 1994. The expansion of a CD4+ T cell population bearing a distinctive beta chain in MRL lpr/lpr mice suggests a role for the fas protein in peripheral T cell selection. *Eur. J. Immunol.* 24:2761–2766.

485. Adams, S., P. Leblanc, and S.K. Datta. 1991. Junctional region sequences of T-cell receptor beta-chain genes expressed by pathogenic anti-DNA autoantibody-inducing helper T cells from lupus mice: possible selection by cationic autoantigens. *Proc. Natl. Acad. Sci. USA* 88:11271–11275.

486. Shi, Y., A. Kaliyaperumal, L.J. Lu, S. Southwood, A. Sette, M.A. Michaels, and S.K. Datta. 1998. Promiscuous presentation and recognition of nucleosomal autoepitopes in lupus: Role of autoimmune T cell receptor alpha chain. *J. Exp. Med.* 187:367–378.

7 | Autoimmune T Cell Recognition: Structural Insights

Thomas Stratmann and Luc Teyton

1. INTRODUCTION

Mostly for convenience, a large number of animal and human diseases have been grouped under the category of autoimmune diseases. This group is extremely heterogeneous in nature. First, there are broad differences in the nature of some of these diseases, and large biological discrepancies within each disease. The level of heterogeneity extends from the symptomatology to the etiology, to the T or B cell nature of the autoreactive response, and to the response to immunosuppressive therapies. Nonetheless, it appears that in most of these diseases, the cells of the immune system are involved in tissue damage. Also common to all of these diseases is the absence of a unique causative agent. Viral, bacteria and environmental antigens have been suspected in many cases, but no correlation has been established directly between these factors and a particular disease. In this context, the question is how tolerance to self-antigens is disrupted? Hypotheses are numerous, but again definitive proof is absent. To address some of the most relevant questions, we and others have chosen a reductionist approach by analyzing the T cell receptor-self-peptide MHC complex at the molecular level. This very basic approach answers three questions: 1) Are autoimmune TCRs peculiar in their structure? 2) Are self-peptide MHC complexes structurally unusual? 3) Is the interaction of these molecules different from the interaction of normal TCR/MHC pairs? It is very likely that the answer to all three questions is negative. However, answering these questions allows us to move forward in the understanding of the autoimmune process. In the unexpected case that all autoimmune TCRs display a high affinity for their ligand, further research in thymic negative selection as a cause for autoimmunity would be warranted. Alternatively, very low

affinities would focus our attention on positive selection. In any case, structural information on autoimmune peptide-MHC and TCR molecules could also guide the design of small molecule compounds that can specifically alter this interaction. From a more general perspective, these studies will help in understanding the linkage between human autoimmune diseases, particular major histocompatibility complex (MHC) class I or II haplotypes. This correlation, which was established three decades ago, is still poorly understood. For some diseases, the linkage is established through the nature of the peptide that associates with a given haplotype, while for others, the MHC itself is the culprit in bearing distinctive structural features that could lead to poor peptide binding, high autoreactivity, and instability. These structural studies have been greatly helped by new protein expression systems, but structural studies remain difficult, confidential and are limited in their advances. To date, there is a small set of structural data on MHC molecules associated with autoimmunity, but information on the TCRs that recognize these structures is lacking thus far. We will discuss the conclusions of these initial findings in the following chapter.

2. AUTOIMMUNE DISEASES ASSOCIATE WITH PARTICULAR MHC CLASS I OR II MOLECULES

Autoimmune diseases result from the collision of a complex genetic-predisposing background with unknown environmental factors. Ongoing genetic analyses of human and inbred mice, such as the New Zealand Black (NZB) (1) or the non-obese diabetic (NOD) mouse (2,3), have shown the complexity of non-additive, non-Mendelian-transmitted

Table 1   Autoimmune diseases linked to MHC class I and II alleles

Disease	Positively Associated MHC Alleles	Negatively Associated MHC Alleles	Susceptible/Resistant Polymorphic Residues[a]	Reference
Insulin-Dependent Diabetes Mellitus	DQB1*0201 DQB1*0302 DQB1*0502 DRB1*0405 DRB1*0401	DQB1*0303 DQB1*0301 DQB1*0602 DRB1*0403	Alaβ57, Serβ57 Aspβ57	48
Rheumatoid Arthritis	DRB1*0401 DRB1*0404 DRB1*0101		**Lysβ71**, Argβ71	33, 42
Multiple Sclerosis	DRA*0101 DRB1*1501 DRB1*1502			30, 76
Pemphigus Vulgaris	DRB1*0402 DQB1*05032		**Gluβ71**	45
Spondyloarthropathies (Ankylosing Spondylitis, Reactive Arthritis, Reiter's Disease)	HLA-B27			10, 11
Systemic Lupus Erythematosus	DR2 DR3 (DRB1*0301)			77, 78
Coeliac Disease	DQA1*0501 DQB1*02 DQB1*0302			79
Goodpasture's Disease	DRB1*1501 DRB4*0404 DRB4*040x	DRB1*0101 DRB1*0701		80

[a]Suscebtible residues are printed in bold letters.

genetic traits for the genetic make-up of these diseases. In most cases, the predominant linkage is found with MHC genes (Table 1). Sequencing these polymorphic genes has not revealed additional variations in diseased patients compared to the general population. It is important to notice that a double dose of the susceptible gene is almost always essential to create a significant risk factor.

If newer and better expression systems for recombinant molecules have allowed significant breakthroughs in structural immunology with the determination of structures such as the TCR and TCR/MHC complex (4,5), the exercise of expressing and crystallizing these molecules remains extremely difficult. Only four MHC molecules linked to autoimmune diseases have had their structure resolved by x-ray crystallography. To date, there is no structure of pathogenic TCR alone or in complex with its self- peptide MHC ligand, if we exclude the HLA-A2 tax peptide/A6 complex which is linked to the HTLV–1 neuropathy (6). In this case, the viral or autoimmune nature of the disease is debatable but the peptide bound to the MHC molecule is a viral peptide and not a self-peptide. For two the remaining four structures it can be argued that they are artificial in nature and not really linked to autoimmunity. Indeed, the

HLA-DR-collagen IV peptide and the HLA-DR2-MBP peptide complexes have not proven to be of any importance in rheumatoid arthritis or multiple sclerosis, respectively, but were originally identified in murine models of induced autoimmunity and appear to be of little importance in human diseases (7–9). We need to point out here that the most common animal models for autoimmune diseases are not good mimics for human diseases. The most useful of them are certainly the spontaneous diseases but, unfortunately, they are the exception.

3.   HLA-B27 AND HUMAN SPONDYLOARTHROPATHIES

The first connection associating an HLA antigen with a human disease was established between the class I antigen HLA-B27 and a group of inflammatory spondyloarthropathies comprising ankylosing spondylitis, reactive arthritis and Reiter's disease (10,11). The onset of these diseases in humans has been linked to intestinal infections with gram-negative bacteria, such as *Klebsiella*, but the formal association between these infections, HLA-B27 and the occurrence of arthropathies is yet to be established. The idea that HLA-B27 is directly involved in disease pathogenesis, is supported by transgenic animal models, which have been established by introducing HLA-B27 into rats and mice (12). Disease in HLA-B27-transgenic mice (13) as well as in HLA-B27/human β_2-microglobulin double-transgenic mice (14) starts with discoloration of nails and hypokeratosis, and eventually results in nail loss. Inflammatory arthritis that may lead to ankylosis in the rear paws appears only in males. At a later stage, splenomegaly, skin inflammation and hair loss occur (12). Since HLA-B27 is an MHC class I antigen, it appears logical to associate the disease with the MHC class I presentation pathway and cytotoxic T cells. However, a role for helper cells is not ruled out, especially at initiation phase of the autoimmune response. Indeed, CD4 knockout mice exhibited a reduced form of the disease (12).

The autoantigen or autoantigens responsible for the disease have yet to be identified. Candidate antigens are type II, type IX and type XI collagen as well as proteoglycans (12). T cells specific for collagen-derived peptides have been isolated from ankylosing spondylitis patients, but there is no experimental data that permits any conclusion to be drawn from these findings (15).

The three-dimensional structure of HLA-B27 has been solved at a resolution of 2.1 Å (16,17). In this case, the source of purified material was not recombinant, but was obtained from human lymphoblastoid LG2 cells. The most significant drawback of this approach is that the molecules were loaded with a complex mixture of self-peptides and not with a single peptide. The overall structure of HLA-B27 resembled previously solved MHC class I structures, such as HLA-A2 and HLA-Aw68. The peptide binding site was shaped by the polymorphic $\alpha 1$ and $\beta 2$ domains, whereas the constant $\alpha 3$ domain associated with β_2-microglobulin. Minor packing differences between HLA-B27 and the aforementioned MHC class I molecules resulted in a shift between the $\alpha 3/\beta_2$-microglobulin unit relative to the $\alpha 1 \alpha 2$ unit, but this could not be linked to the pathogenic behavior of HLA-B27. Self-peptides extracted from HLA-B27 had a uniform length of 9 amino acids and without exception contained an arginine residue at position 2 (P2) (18). Arginine anchored into a deep pocket formed within the floor of the peptide binding cleft and the $\alpha 1 \alpha 2$ helices, and the size and nature of this pocket, was mainly determined by four polymorphic residues, His-9, Thr-24, Glu-45, and Cys-67. Importantly, this constellation of side chains was found in all HLA-B27 subtypes and, in addition, is unique among all other class I alleles. Modeling of side chains found in other class I alleles, such as Phe or Try at position 67, would significantly reduce the size of the pocket and create an hydrophobic environment. The replacement of His-9 with a tyrosine side chain, as found in several alleles, would displace a water molecule buried in this pocket, and hydrogen bonds with the P2 side chain. Site-

directed mutagenesis of this side chain in HLA-B27 has been shown to reduce the binding of influenza A nucleoprotein 383–394 peptide (19).

The biological relevance and consequences of this unique feature of the P2 pocket is unknown. Since it allows selection of peptides containing an arginine at P2, it is straightforward to assume that this trait is essential for the pathogenesis of the disease. HLA-B27 could select viral or bacterial peptides with a P2 Arg anchor that other MHC class I molecules would not select. In turn, this peptide could closely mimic a self-antigen present primarily in axial joints and trigger a self-response.

4. HLA-DR2 AND MULTIPLE SCLEROSIS

Multiple sclerosis (MS) is an autoimmune disease involving the central nervous system (7,20). In monozygotic twins, the concordance rate is only 30% (21,22), and the disease is found twice as often in women than in men (7). Given the relatively low concordance rate in monozygotic twins, non-genetic factors are likely to be involved. In Caucasian populations, the global distribution of the disease is predominantly temperate regions (7). The clinical course of the disease usually consists of successive episodes of relapses and remissions. Some relapses follow a viral infection of the gastrointestinal tract or the upper respiratory system (7). Pathologically, MS lesions appear as an inflammation surrounding venules, T cell infiltration and destruction of the myelin sheath. Autoreactive T cells specific for myelin autoantigens and B cells producing autoantibodies may contribute to the disease (23). Myelin basic protein-(MBP) reactive T cells have been isolated from MS patients (24), but are also found in healthy individuals (7). The initial assumption that target antigen(s) such as MBP would not be accessible by the immune system due to the blood-brain barrier was overturned by the discovery that the antigen was present in the thymus (25). Genomic screens to find potential gene candidates have been rather disappointing (26–29). Like many autoimmune diseases, MS is a multigene disease. The number of genes involved is still unknown (27). Although the association with the HLA-DR2 gene seems to be established, linkage to the region is weak.

Determination of the crystal structure of HLA-DR2 (DRA*0101, DRB*1501) complexed with a peptide from myelin basic protein has been achieved at a resolution of 2.6 Å (30). The DR2 molecules were expressed using a baculovirus expression system, and the MBP peptide sequence 85–99 was tethered to the N-terminus of the beta chain. Overall, HLA-DR2 showed no surprises in its structure compared to other MHC class II structures; the overall architecture is identical to all MHC molecules. The $\alpha 1$ and $\beta 1$ domains assemble to form the peptide binding groove. Several residues of the $\beta 1$ domain are distinct from their counterparts in other MHC class II molecules, which affects the specificity of the binding pockets to various degrees. The most prominent feature of the structure is the size and hydrophobicity of the P4 pocket. A lysine residue at position $\beta 71$ of HLA-DR4 is replaced by an alanine in HLA-DR2. This change increases the hydrophobicity of the P4 pocket, and at the same time substantially widens the pocket. In the bottom of the pocket, the aspartate $\beta 28$ creates a negative environment that would allow binding of positively-charged side chains. In the HLA-DR2 structure, a large aromatic side chain (Phe) anchors into the P4 pocket. The DR$\beta 71$ side chain is not solvent-exposed, but its size can influence the position of a solvent-exposed side chain. The relatively small side chain of Ala$\beta 71$ allows Gln$\beta 70$ to assume a horizontal position relative to the peptide binding site in the DR2 structure. As mentioned above, in the DR4 structure, $\beta 71$ is occupied by a lysine residue, that forces Gln$\beta 70$ to point up towards the TCR. This polymorphism has been shown to affect T-cell recognition of bound peptides (30). A second polymorphism at position $\beta 13$ (Arg) limits the size of the P6 pocket, and as a result, the P6 side chain of the peptide (Asn) cannot anchor deeply into the pocket, as seen in other DR alleles. This feature partially explains why the C-terminal portion of the peptide (P5–P10) is positioned higher in the binding groove than other DR alleles.

The lack of knowledge regarding the autoantigens involved in MS has hindered analysis of the autoreactive T cell repertoire. Some efforts have been made to examine the variable gene usage (Vα and Vβ) of T cell repertoires in MS patients compared to healthy individuals. In the case of MBP-reactive T cell clones, a heterogeneous TCR Vα and Vβ usage was found in one study (24). Clonal expansion was observed in some patients, but differed between individuals. A second study revealed the same Vα usage for six MBP specific T cell clones restricted to DR2 (31); however, due to differences in their CDR3 region, none of these clones were identical. Identification of autoreactive T cells in humans has opened the door to establish transgenic animal models. For example, transgenic mice have been created carrying HLA-DR2 (DRA*0101/DRB*1501) and a human TCR specific for MBP peptide 84–102 (32). However, spontaneous disease was observed in only 4% of the animals. Disease prevalance was increased by backcrossing these genes into T cell deficient mice (Rag2-deficient). Whether any conclusion regarding the human disease can be drawn from these models, however, remains an open question.

5. HLA-DR4 AND RHEUMATOID ARTHRITIS

Rheumatoid arthritis (RA) is a disease characterized by the inflammation and ultimate destruction of synovial joints. The disease is predominant in women. Blood-derived cells, such as T cells, macrophages and plasma cells, have been detected in the joints (33). The proliferation of the synovial membrane in combination with the T cell and inflammatory infiltrate forms the pannus. The destruction of cartilage and surrounding bone is believed to be secondary to the secretion of inflammatory lymphokines and proteases by the pannus. Autoantibodies of the IgM class specific for the Fc terminus IgG (so called rheumatoid factors) are often found (33).

The respective roles of B and T cells in this disease are not known. Aside from the Fc of IgG, autoantigens have not been identified for RA, but the long list of candidate antigens can be classified into three groups: 1) xenogenic antigens; 2) autoantigens absent in the joints, joint-specific autoantigens and ubiquitous self-antigens (33,34). For the first group, bacterial heat shock protein 60, DNA J, the bacterial homologue of the mammalian heat shock protein 60, and a glycine-alanine rich protein region of a nuclear antigen encoded by the Epstein-Barr virus (internal repeat region, IR–3) have been proposed. Autoantigens absent from the joint regroup pro-/filaggrin, a protein involved in crosslinking intermediate filament proteins and found in the endothelium, as well as the Sa antigen, a 50 kDa protein found in human spleen and placenta. Joint-specific antigens are the human cartilage protein gp39, the chondrocyte antigen 65, and type II collagen. Type II collagen is an obvious candidate since it is the main component of cartilage. The injection of type II collagen with complete Freund's adjuvants (CFA) can induce a transitory joint disease in many animals, but the pathology of this artificial disease does not resemble RA (35,36). In humans, autoantibodies against type II collagen have been isolated from RA and non-RA patients, and specific T cells could not be found even in joint effusions (37). Finally, the group of ubiquitous autoantigens includes IgG, p68 or BiP, an endoplasmatic reticulum chaperone, and calpastatin, an intracellular inhibitory protein of cysteine proteases. However, no consistent B and T cell response can be found within groups of RA patients (34) with regard to these antigens.

A strong genetic link of RA with the HLA-DRB1 locus was discovered in the late 1970's (38,39). HLA-DR4 (allotype DRB1*0401 and DRB1*0404) and HLA-DR1 (DRB1*0101) are found in more than 80% of caucasian RA patients (33). The highest risk is conferred by DRB1*0401, which is present in approximately 50% of RA patients (39). This number is even higher in patients with Felty's syndrome (90% are HLA- DR4), a particularly severe from of RA (40). DR4 has also been linked to an increased incidence of chronic arthritis following *Borrelia* infections (41). The disease-associated HLA class II antigens share similar amino acids at positions $\beta 67–74$. Of particular interest is $\beta 71$, an arginine residue in DRB1*0101 and DRB1*0404 and a lysine in DRB1*0401, but a negatively-charged residue (glutamic acid) in non-susceptible haplotypes, such as DRB1*0402. The molecular structure of HLA-DR4 (DRA*0101, DRB*0401) has been to 2.5 Å resolution (42). DR4 molecules were expressed as soluble empty molecules in a fly expression system loaded with the human collagen II peptide 1168–1180 after purification. As seen for HLA-DR2, the overall structure of HLA-DR4 was very similar to HLA-DR1. The collagen peptide binds the HLA-DR4 molecule like most MHC class I and class II peptides by anchoring side chains into binding pockets at the bottom of the binding groove. The P1, P4, P6, P7 pocket are filled, but the P9 is unoccupied, given the presence of a glycine residue in the peptide sequence. Experiments using synthetic peptide libraries and phage-displayed peptides suggested preferential binding of selected residues for positions P1, P2, P3, P4, and P7 of HLA-DR4. P1 preferentially selects large, non-polar, and side chains, P2 arginine, P3 small non-polar side chain (glycine or alanine), P4 non-polar, and P7 aliphatic residues. The P1 pocket is deep and non-polar, and that depth is influenced by a glycine/valine HLA-DRβ polymorphism at position β86. The presence of a glycine at this position creates a large pocket, while valine reduces its size. This polymorphism, however, is not related to disease susceptibility, since HLA-DRB1*0401 (susceptible) has a glycine at this position, but DRB1*0404 (susceptible) as well as DRB1*0402 (non- susceptible) both have a valine at that position. Similarly, the P6 pocket is larger in HLA-DR4 than HLA-DR1 due to a polymorphism at position β11: valine in HLA-DR4, leucine in HLA-DR1. This polymorphism, however, like the β86 polymorphism, cannot be associated with disease susceptibility. As with DR2, the most prominent characteristic of this structure is the nature of the P4 pocket. The critical residues forming this pocket are β67, β70, β71, and β74. The lysine β71 of HLA-DR4 forces the glutamine β70 to point outward towards the T cell receptor. The peptide binding experiments using phage libraries and synthetic peptide libraries have suggested that a positive charge at β71 greatly influenced the peptide selectivity of HLA-DR4 (43). Peptides bearing a positive charge at position P4 bind much more strongly to HLA-DR4 alleles that are not associated with the disease and that have a negatively-charged side chain at position β71, such as HLA-DRB1*0402 (glutamic acid). In contrast, peptides with a negative charge at position P4 are preferentially bound by RA-associated alleles compared to non-disease-associated alleles (43). The physiological effects of this peculiar P4 pocket are unclear. Transgenic mice bearing HLA-DRB1*0401 and the human coreceptor CD4 (8,44), as well as transgenics for a modified version of DRB1*0101 bearing the mouse sequence in the CD4-interacting domain (9), mounted a T cell response against collagen II. Interestingly, the same immunodominant core sequence of collagen II, CII

263–270, was identified for both models. Modeling of this peptide into the DR4 structure could satisfactorily position a glutamic acid into the P4 pocket (42). The electrostatic nature of the P4 pocket has been implicated to be important in another autoimmune disease, pemphigus vulgaris (PV). PV has been linked to HLA-DR4 subtype HLA-DRB1*0402 (45), a subtype that differs from DRB1*0401 at only three positions, clustered in the β chain region 67–71. At β71, DRB1*0402 carries, unlike HLA-DRB1*0401, a negatively-charged residue (glutamic acid), demonstrating that a disease cannot simply be correlated to a single charge within one peptide-anchoring pocket.

In the case of RA, spontaneous non-transgenic animal models are absent. Recently, a new animal model was fortuitously created by backcrossing a transgenic mouse bearing a TCR specific for a bovine ribonuclease peptide in the context of the murine MHC class II molecule I-A^k into the non-obese diabetic (NOD) background (46). These mice spontaneously developed symptoms very similar to human RA. In this model, B cells were absolutely required for disease. The self-antigen recognized by the autoreactive B cells and T cells was surprisingly identified as glucose 6-phosphate isomerase (GIP) (47), an ubiquitous enzyme involved in the cellular carbohydrate metabolism. Despite the artificial settings of the model, this example demonstrates that a non-tissue-specific antigen can ultimately lead to a tissue specific autoreactive response.

6. HLA-DQ, I-A^{g7} AND INSULIN-DEPENDENT DIABETES MELLITUS

Insulin-dependent diabetes mellitus (IDDM) is a metabolic disease caused by the destruction of the insulin-producing beta cells of the pancreas. Like all autoimmune diseases, IDDM is multigenic, with a minimum of 16 genes involved in its genetic background (2,3), most of which have not been identified. The strongest genetic link is with the HLA-DQ genes in human and I-A genes in the mouse. Alleles with alanine, valine or serine at position β57 (48) are associated with IDDM, whereas alleles with an aspartic acid at the same position, e.g. HLA-DRB1, are associated with disease protection. Again, unknown environmental factors play a major role in this disease, best illustrated by the fact that monozygotic twins show a concordance rate of only 50% (48). IDDM is unequally geographically distributed in that 1 to 1.5% of the population is affected in the Northern hemisphere, with a declining incidence from north to south (48). Several autoantigens have been proposed as candidates for the onset or the progression of IDDM, including glutamic acid decarboxylase (GAD) 65/67, insulin, peripherine, and heat shock protein 60 (49,50). Recently, attention has been focused on GAD

65, an enzyme required for synthesis of the inhibitory neurotransmitter γ-aminobutyric acid (51) expressed in the brain and the endocrine pancreas (52). Autoantibodies to GAD 65/67 were initially detected in newly-diagnosed diabetic children (53), and anti-GAD T cell responses can also be detected in some patients (54).

IDDM research has greatly benefited from a spontaneous murine counterpart, i.e. the NOD mouse, which develops diabetes at approximately 20 weeks of age (55). The incidence reaches 80–100% in females and 10–20% in males. Depending on the colony, the animal facility, and other unknown environmental factors, this incidence can vary greatly. For instance, disease incidence is higher in animals kept under pathogen-free conditions. Human and NOD IDDMs appear identical. Pathologically, the first step (3–4 weeks) is insulitis induced by the infiltration of T cells, B cells, marcophages and dendritic cells inside the islets of Langerhans (56,57). At that stage, islet cells are intact and can be stained with anti-insulin antibodies. The destruction is progressive over the next 15–20 weeks (55). As in susceptible humans, the NOD mouse expresses an MHC class II molecule that bears a serine at position $\beta57$ (58). This particular MHC molecule, called I-A^{g7}, shares its α-chain with I-A^d, a non- susceptible MHC haplotype (59). The I-A^{g7} β-chain, unlike other murine class II molecules, also carries an unusual histidine residue at position 56 (58). Replacing I-A^{g7} with I-A^k by transgenesis protected mice from diabetes (60). Mutating serine $\beta57$ to aspartic acid reduced incidence of diabetes without preventing insulitis (61). Finally, the replacement of histidine 56 by a proline residue abolished both insulitis and diabetes (62).

Initial biochemical and functional analyses of I-A^{g7} proposed that the molecule was a poor peptide binder and a particularly unstable molecule (63). The argument regarding instability was mostly based on the structure of other class II molecules. Indeed, the aspartic acid usually found at position $\beta57$ forms a salt bridge with the arginine 76 of the α-chain, an interaction thought to be critical for $\alpha\beta$ dimer stabilization. The second potential consequence of the disappearance of the charged residue at 57 could be the compensation provided by peptides carrying a negative charge at position P9. We have produced soluble empty I-A^{g7} and I-A^d molecules in a fly expression system (64,65,81) in which chain correct pairing was forced by the addition of leucine zippers. In our hands, both molecules behaved very similarly, biochemically and functionally. Chain pairing before and after removal of the zippers was stable at room temperature and 37°C. Further analyses of peptide binding using synthetic peptides and random peptide phage libraries revealed similarities as well as some fundamental differences between the two haplotypes. I-A^{g7} and I-A^d recognized some common peptides, and average peptide binding constants (IC50 values) for both molecules

were very similar. However, I-A^{g7} bound significantly more peptides than I-A^d, and screens of random libraries showed an enrichment for peptides with a C- terminal negatively-charged residue by this molecule. To gain further insights, I-A^{g7} was crystallized with single peptides and its structure was solved at a resolution of 2.6 Å (66). The I-A^{g7} structure was compared to the two structures of I-A^d that we had previously determined (67). The structure of I-A^{g7} was determined with two self-peptides derived from GAD65. The first, GAD65 207–220, is the dominant functional epitope of GAD65 in NOD mice (68). The second, GAD65 221–235, is also one of the major epitopes of GAD65 in NOD mice (68). However, the respective role of these two determinants in the pathogeny of IDDM has yet to be determined. As expected, both I-A^{g7} and I-A^d structures were very similar. The binding energy between the MHC binding groove and the bound peptide was mainly provided by interaction of the peptide backbone, with the residues forming the groove in both structures. For both I-A^d and I-A^{g7}, the various pockets were not utilized to accommodate the side chains of anchor residues. Still, affinity and specificity were allele-specific. Pockets P1, P4 and P6 were almost identical for I-A^{g7} and I-A^d, the main difference being the P9 pocket where the replacement of the proline, glutamic acid pair at position 56 and 57 in I-A^d with the doublet histidine, serine, led to the formation of a shallower, but significantly wider, pocket in I-A^{g7} (Figure 1). In addition, this pocket was open to the side, allowing the accommodation of larger side chains, which gave residues at position P9 the alternative between a downward orientation if they were small and a sideways orientation if they were larger. The charged nature of the pocket suggested that either negatively-charged residues or small hydrophobic/ hydrophilic would preferentially bind into the P9 pocket. This change explained the ability of I-A^{g7} molecules to bind even more peptides than I-A^d molecules. These structures were then used to rationalize our binding data. A very promiscuous peptide binding motif for I-A^{g7} emerged from this analysis (66,81). The P1, P4 and P6 pockets of I-A^{g7} and I-A^d molecules can tolerate the same residues. The P1 pocket allows any residue, whereas the P4 and P6 prefer small residues and small hydrophobic residues. The P9 pocket of I-A^d will allow only small residues and exclude larger side chains and negatively-charged residues. To the contrary, the I-A^{g7} P9 pocket will preferentially bind negatively-charged residues and tolerate small residues. In addition, larger amino acids, like methionine, are an additional possibility (69). The different shape of the P9 pocket allows I-A^{g7} to fit twice as many residues (15) at that position compared to I-A^d (7 residues). One possible biological consequence of this higher promiscuity of binding of I-A^{g7} molecules could be a lower average affinity of peptide binding, which could translate into reduced half-lives of I-A^{g7} molecule-peptide complexes at the cell surface. In turn, a possible

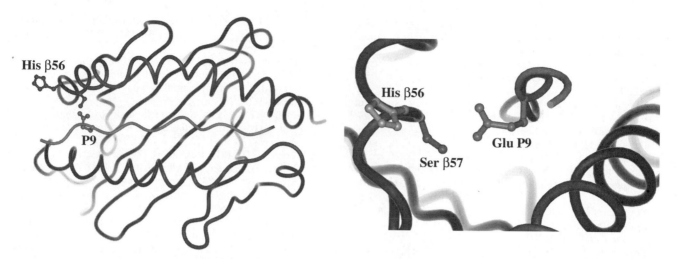

Figure 1. Crystal structure of the I-A^{g7} molecule with the bound dominant epitope of the autoantigen GAD 65 (residues 207–220). Left: Top view of the binding groove. The peptide lays in the middle with its C-terminus to the left. The polymorphic residues His $\beta56$ and Ser $\beta57$ as well as Glu 217 of the peptide are in ball stick representation. His $\beta56$ replaces a conserved proline residue and points out of the binding groove and, thus, possibly interferes with binding of CD4. Right: Perpendicular view of the binding groove. The serine at position $\beta57$ substantially widens the pocket. The side chain of the P9 glutamic acid is filling the side of the pocket and compensates for the lack of the normal aspartic acid found in other MHC class II molecules. Graphic courtesy of A.L. Corper.

consequence could be an altered negative thymic selection and an increased chance for the maturation of autoreactive T cells. The higher promiscuity of binding of I-A^{g7} could also explain the phenotype of the Biozzi HR mouse (70) which, like the NOD mouse, expresses I-A^{g7} molecules. This mouse is characterized by high T and B cell responses to exogenous antigens. The presentation of a larger set of peptides from a single antigen could indeed induce stronger T-cell responses.

7. THE STRUCTURE OF AUTOREACTIVE T CELL RECEPTORS

If data regarding the structures of MHC self-peptide complexes are still scarce and only partially informative, structures of autoreactive T cell receptors are totally absent at the present time for reasons that are multiple and difficult to circumvent. First, the crystallization of TCR remains elusive and largely unsuccessful even though high expression and high degree of purity have been achieved. Second, in most cases, if a pathogenic TCR is isolated, the self-peptide that it recognizes is unknown. The identification of peptides relevant to the pathogeny of any given disease is extremely difficult. Possible antigens are usually initially identified using sera from patients but, for instance, low titers will be ignored or difficult to work with. Subsequently, if the first step is successful, the link between B and T cell responses is also difficult to establish. T cells can be of low frequency, low affinity, or sequestered in tissues. Finally even if T cell reactivity is

demonstrated, the precise peptide has yet to be identified and the pathogenic role of the reactive T cells has to be established. Each potential antigen requires the effort of many to be proven or disproven. Our capacity to use MHC tetramers to track specific T cells and the emergence of DNA arraying and differential display could help us in that search. These great technical difficulties are best illustrated in the case of the murine IDDM. Several autoreactive T cell clones have been isolated from the islets of NOD mice years ago (71–73) and used to produce transgenic T cell receptor mice, such as BDC 2.5 (74,75). The transgenic mouse develops diabetes, but the responsible antigen is still unknown after 10 years of work in many laboratories. This hurdle also limits our ability to measure TCR affinities for self-peptide MHC ligands and establish strong correlations between thymic selection and autoimmunity. However, the effort is warranted for two linked reasons: First, the identification of self-antigens could lead to the development of vaccines for some diseases. Second, and just as important, the description of the structure of pathogenic TCR/self-MHC pairs could help the rational design of small specific compounds able to disrupt T cell activation.

References

1. Morel, L., and E.K. Wakeland. 1998. Susceptibility to lupus nephritis in the NZB/W model system. *Curr. Opin. Immunol.* 10:718–725.
2. Vyse, T.J., and J.A. Todd. 1996. Genetic analysis of autoimmune diabetes. *Cell* 85:311–318.

3. Wicker, L.S., J.A. Todd, and L.B. Peterson. 1995. Genetic control of autoimmune diabetes in the NOD mouse. *Annu. Rev. Immunol.* 13:179–200.

4. Wilson, I.A., and K.C. Garcia. 1997. T cell receptor structure and TCR complexes. *Curr. Opin. Struct. Biol.* 7:839–848.

5. Garcia, K.C., L. Teyton, and I.A. Wilson. 1999. Structural basis of T cell recognition. *Annu. Rev. Immunol.* 17:369–397.

6. Ding, Y.H., K.J. Smith, D.N. Garboczi, U. Utz, W.E. Biddison, and D.C. Wiley. 1998. Two human T cell receptors bind in a similar diagonal mode to the HLA-A2/Tax peptide complex using different TCR amino acids. *Immunity* 8:403–411.

7. Steinman, L. 1996. Multiple sclerosis: A coordinated immunological attack against myelin in the central nervous system. *Cell* 85:299–302.

8. Fugger, L., S.A. Michie, I. Rulifson, C.B. Lock, and G.S. McDevitt. 1994. Expression of HLA-DR4 and human CD4 transgenes in mice determines the variable region beta-chain T-cell repertoire and mediates an HLA-DR-restricted immune response. *Proc. Natl. Acad. Sci. USA* 91:6151–6155.

9. Rosloniec, E.F., D.D. Brand, L.K. Myers, K.B. Whittington, M. Gumanovskaya, D.M. Zaller, A. Woods, D.M. Altmann, J.M. Stuart, and A.H. Kang. 1997. An HLA-DR1 transgene confers susceptibility to collagen-induced arthritis elicited with human type II collagen. *J. Exp. Med.* 185:1113–1122.

10. Brewerton, D.A., F.D. Hart, M. Caffrey, A. Nicholls, D.C.O. James, and R.D. Sturrock. 1973. Ankylosing spondylitis and HL-A 27. *Lancet* i:904–907.

11. Schlosstein, L., P.I. Terasaki, R. Bluestone, and C.M. Pearson. 1973. High association of an HL-A antigen, W27, with ankylosing spondylitis. *N. Engl. J. Med.* 288:704–706.

12. Khare, S.D., H.S. Luthra, and C.S. David. 1998. Unraveling the mystery of HLA-B27 association with human spondyloarthropathies using transgenic and knock out mice. *Seminars Immunol.* 10:15–23.

13. Khare, S.D., H.S. Luthra, and C.S. David. 1995. Spontaneous inflammatory arthritis in HLA-B27 transgenic mice lacking β_2-microglobulin: a model of human spondyloarthropathies. *J. Exp. Med.* 182:1153–1158.

14. Khare, S.D., J. Hansen, H.S. Luthra, and C.S. David. 1996. HLA-B27 heavy chains contribute to spontaneous inflammatory disease in B27/human β_2-microglobulin (beta2m) double transgenic mice with disrupted mouse β_2m. *J. Clin. Invest.* 98:2746–2755.

15. Gao, X.M., P. Wordsworth, and A. McMichael. 1994. Collagen-specific cytotoxic T lymphocyte responses in patients with ankylosing spondylitis and reactive arthritis. *Eur. J. Immunol.* 24:1665–1670.

16. Jardetzky, T.S., W.S. Lane, R.A. Robinson, D.R. Madden, and D.C. Wiley. 1991. The structure of HLA-B27 reveals nonamer self-peptide bound in an extended conformation. *Nature* 353:321–325.

17. Madden, D.R., J.C. Gorga, J.L. Strominger, and D.C. Wiley. 1992. The three-dimensional structure of HLA-B27 at 2.1 Å resolution suggests a general mechanism for tight binding to MHC. *Cell* 70:1035–1048.

18. Jardetzky, T.S., W.S. Lane, R.A. Robinson, D.R. Madden, and D.C. Wiley. 1991. Identification of self peptides bound to purified HLA-B27. *Nature* 353:326–329.

19. Huet, S., D.F. Nixon, J.B. Rothbard, A. Townsend, S.A. Ellis, and A.J. McMichael. 1990. Structural homologies between two HLA B27-restricted peptides suggest residues important for interaction with HLA B27. *Int. Immunol.* 4:311–316.

20. Conlon, P., J.R. Oksenberg, J. Zhang, and L. Steinman. 1999. The immunobiology of multiple sclerosis: An autoimmune disease of the central nervous system. *Neurobiol. Dis.* 6:149–166.

21. Ebers, G.C., A.D. Sadovnick, and N.J. Risch. 1995. A genetic basis for familial aggregation in multiple sclerosis. *Nature* 377:150–151.

22. Ebers, G.C., D.E. Bulman, A.D. Sadovnick, D.W. Paty, S. Warren, W. Hader, T.J. Murray, T.P. Seland, P. Duquette, and T. Grey, et al. 1986. A population-based study of multiple sclerosis in twins. *N. Engl. J. Med.* 315:1638–1642.

23. Wucherpfennig, K.W., J. Zhang, C. Witek, M. Matsui, Y. Modabber, K. Ota and D.A. Hafler. 1994. Clonal expansion and persistence of human T cells specific for an immunodominant myelin basic protein peptide. *J. Immunol.* 152:5581–5592.

24. Vandevyver, C., N. Mertens, P. v. d. Elsen, R. Medaer, J. Raus, and J. Zhang. 1995. Clonal expansion of myelin basic protein-reactive T cells in patients with multiple sclerosis: restricted T cell receptor V gene rearrangements and CDR3 sequence. *Eur. J. Immunol.* 25:958–968.

25. Mathisen, P.M., S. Pease, J. Garvey, L. Hood, and C. Readhead. 1993. Identification of an embryonic isoform of myelin basic protein that is expressed widely in the mouse embryo. *Proc. Natl. Acad. Sci. USA* 90:10125–10129.

26. Kuokkanen, S., M. Sundvall, J.D. Terwilliger, P.J. Tienari, J. Wilkstroem, R. Holmadhal, U. Petterson, and L. Peltonen. 1996. A putative vulnerability locus to multiple sclerosis maps to 5p14-p12 in a region syntenic to the murine locus *Eae2*. *Nature Genet.* 23:477–480.

27. Haines, J.L., M. Ter-Minassian, A. Bazyk, J.F. Gusella, D.J. Kim, H. Terwedow, M.A. Pericak-Vance, J.B. Rimmler, C.S. Haynes, A.D. Roses, A. Lee, B. Shaner, M. Menold, E. Seboun, R.P. Fitoussi, C. Gartioux, C. Reyes, F. Ribierre, G. Gyapay, J. Weissenbach, S.L. Hauser, D.E. Goodkin, R. Lincoln, K. Usuku, J.R. Oksenberg, et al. 1996. A complete genomic screen for multiple sclerosis underscores a role for the major histocompatability complex. *Nature Genet.* 13:469–471.

28. Ebers, G.C., K. Kukay, D.E. Bulman, A.D. Sadovnick, G. Rice, C. Anderson, H. Armstrong, K. Cousin, R.B. Bell, W. Hader, D.W. Paty, S. Hashimoto, J. Oger, P. Duquette, S. Warren, T. Gray, P. O'Connor, A. Nath, A. Auty, L. Metz, G. Francis, J.E. Paulseth, T.J. Murray, W. Pryse-Phillips, and N. Risch, et al. 1996. A full genome search in multiple sclerosis. *Nature Genet.* 13:472–476.

29. Sawcer, S., H.B. Jones, R. Feakes, J. Gray, N. Smaldon, J. Chataway, N. Robertson, D. Clayton, P.N. Goodfellow, and A. Compston. 1996. A genome screen in multiple sclerosis reveals susceptibility loci on chromosome 6p21 and 17q22. *Nature Genet.* 13:464–468.

30. Smith, K.J., J. Pyrdol, L. Gauthier, D.C. Wiley, and K.W. Wucherpfennig. 1998. Crystal structure of HLA-DR2 (DRA*0101, DRB*1501) complexed with a peptide from human myelin basic protein. *J. Exp. Med.* 188:1511–1520.

31. Wucherpfennig, K.W., D.A. Hafler, and J.L. Strominger. 1995. Structure of human T cell receptors specific for an immunodominant myelin basic protein peptide: Positioning of T cell receptors on HLA-DR2/peptide complexes. *Proc. Natl. Acad. Sci. USA* 92:8896–8900.

32. Madsen, L.S., E.C. Andersson, L. Jansson, M. Krogsgaard, C.B. Andersen, J. Enberg, J.L. Strominger, A. Svejgaard,

J.P. Hjorth, R. Holmdahl, K.W. Wucherpfennig, and L. Fugger. 1999. A humanized model for multiple sclerosis using HLA- DR2 and a human T cell receptor. *Nature Genet.* 23:343–347.

33. Feldmann, M., F.M. Brennan, and R.N. Maini. 1996. Rheumatoid Arthritis. *Cell* 85:307–310.

34. Bläβ, S., J.-M. Engel, and G.-R. Burmester. 1999. The immunological homunculus in rheumatoid arthritis. *Arthritis Rheum.* 42:2499–2506.

35. Courtenay, J.S., M.J. Dallman, A.D. Dayan, A. Martin, and B. Mosedale. 1980. Immunisation against heterologous type II collagen induces arthritis in mice. *Nature* 283:666–668.

36. Holmdahl, R. 1998. Genetics of susceptibility to chronic experimental encephalomyelitis and arthritis. *Curr. Opin. Immunol.* 10:710–717.

37. Kotzin, B.L., M.T. Falta, F. Crawford, E.F. Rosloniec, J. Bill, P.Marrack, and J. Kappler. 2000. Use of soluble peptide-DR4 tetramers to detect synovial T cells specific for cartilage antigens in patients with rheumatoid arthritis. *Proc. Natl. Acad. Sci. USA.* 97:291–296.

38. Nepom, G.T., P. Byers, C. Seyfried, L.A. Healey, K.R. Wilske, D. Stage and B.S. Nepom. 1989. HLA genes associated with rheumatoid arthritis. Identification of susceptible alleles using specific oligonucleotide probes. *Arthritis Rheum.* 32:15–21.

39. Nepom, G.T., and H. Erlich. 1991. MHC class-II molecules and autoimmunity. *Ann. Rev. Immunol.* 9:493–525.

40. Lanchbury, J.S., E.E. Jaeger, D.M. Sansom, M.A. Hall, P. Wordsworth, J. Stedeford, J.I. Bell, and G.S. Panayi. 1991. Strong primary selection for the Dw4 subtype of DR4 accounts for the HLA-DQw7 association with Felty's syndrome. *Human Immunol.* 32:56–64.

41. Steere, A.C., E. Dwyer, and R. Winchester. 1990. Association of chronic Lyme arthritis with HLA-DR4 and HLA-DR2 alleles. *N. Engl. J. Med.* 323:219–223.

42. Dessen, A., C.M. Lawrence, S. Cupo, D.M. Zaller, and D.C. Wiley. 1997. X-ray crystal structure of HLA-DR4 (DRA*0101, DRB*0401) complexed with a peptide from human collagen II. *Immunity* 7:473–481.

43. Hammer, J., F. Gallazzi, E. Bono, R.W. Karr, J. Guenot, P. Valasnini, Z. A. Nagy, and F. Sinigaglia. 1995. Peptide binding specify of HLA-DR4 molecules: Correlation with rheumatoid arthritis association. *J. Exp. Med.* 181:1847–1855.

44. Fugger, L., J.B. Rothbard, and G.Sonderstrup-McDevitt. 1996. Specificity of an HLA-DRB*0401-restricted T cell response to type II collagen. *Eur. J. Immunol.* 26:928–933.

45. Wucherpfenning, K.W., and J.L. Strominger. 1995. Selective binding of self peptides to disease-associated major histocompatibility complex (MHC) molecules: A mechanism for MHC-linked susceptibility to human autoimmune disease. *J. Exp. Med.* 181:1597–1601.

46. Kouskoff, V., A.-S. Korgonow, V. Duchatelle, C. Degott, C. Benoist, and D. Mathis. 1996. Organ specific disease provoked by systemic autoreactivity. *Cell* 87:811–822.

47. Matsumoto, I., A. Staub, C. Benoist, and D. Mathis. 1999. Arthritis provoked by linked T and B cell recognition of a glycolytic enzyme. *Science* 286:1732–1735.

48. Tisch, R., and H. McDevitt. 1996. Insulin-dependent diabetes mellitus. *Cell* 85:291–297.

49. Harrison, L.C. 1992. Islet cell antigens in insulin-dependent diabetes: Pandora's box revisited. *Immunol. Today* 13:348–352.

50. Roep, B.O. 1996. T cell responses to autoantigens in IDDM. The search for the holy grail. *Diabetes* 45:1147–1156.

51. Kaufman, D.L., M. Clare-Salzler, J. Tian, T. Frosthuber, G.S.P. Ting, P. Robinson, M.A. Atkinson, E.E. Sercarz, A.J. Tobin, and P.V. Lehmann. 1993. Spontaneous loss of T cell tolerance to glutamic acid decarboxylase in murine insulin-dependent diabetes. *Nature* 366:69–72.

52. Sanjeevi, C.B., A. Falorni, I. Kockum, W. A. Hagopian, and Å. Lernmark. 1996. HLA and glutamic acid decarboxylase in human insulin-dependent diabetes mellitus. *Diabetic Med.* 13:209–217.

53. Bekkeskova, S., J.H. Nielsen, B. Marner, T. Bilde, J. Ludvigsson, and Å. Lernmark. 1982. Autoantibodies in newly diagnosed diabetic childern immunoprecipitate human pancreatic islet cell proteins. *Nature* 298:167–169.

54. Panina-Bordigon, P., R. Lang, P.M. van Ender, E. Benazzi, A.M. Felix, R.M. Pastore, G.A. Spinas, and F. Sinigaglia. 1995. Cytotoxic T cells specific for glutamic acid decarboxylase in autoimmune diabetes. *J. Exp. Med.* 181:1923–1927.

55. Fujita, T., R. Yui, Y. Kusumoto, Y. Serizawa, S. Makino, and Y. Tochino. 1982. Lymphocytic insulitis in a "non-obese diabetic (NOD)" strain of mice: an immunohistochemical and electron microscope investigation. *Biomed. Res.* 3:429–443.

56. Lo, D., C.R. Reilly, B. Scott, R. Liblau, H.O. McDevitt, and L.C. Burkly. 1993. Antigen-presenting cells in adoptively transferred and spontaneous autoimmune diabetes. *Eur. J. Immunol.* 23:1693–1698.

57. Miller, B.J., M.C. Apple, J.J. O'Neil, and L.S. Wicker. 1988. Both the LYT-2+ and L3T4+ T cell subset are required for the transfer of diabetes in nonobese diabetic mice. *J. Immunol.* 140:52–58.

58. Acha-Orbea, H., and H.O. McDevitt. 1987. The first external domain of the nonobese diabetic mouse class II I-A β chain is unique. *Proc. Natl. Acad. Sci. USA* 84:2435–2439.

59. Singer, S. M., R. Tisch, X.-D. Yang, and H.O. McDevitt. 1993. An Aβ^d transgene prevents diabetes in nonobese diabetic mice by inducing regulatory T cells. *Proc. Natl. Acad. Sci. USA* 90:9566–9570.

60. Miyazaki, T., M. Uno, M. Uehira, H. Kikutani, T. Kishimoto, M. Kimoto, H. Nishimoto, J. Miyazaki, and K. Yamamura. 1990. Direct evidence for the contribution of the unique I-A^NOD to the development of insulitis in non-obese diabetic mice. *Nature* 345:722–724.

61. Quartey-Papafio, R., T. Lund, P. Chandler, J. Picard, P. Ozegbe, S. Day, P.R. Hutchings, L. O'Reilly, D. Kioussis, E. Simpson, and A. Cooke. 1995. Aspartate at position 57 of nonobese diabetic I-A^g7 β-chain diminishes the spontaneous incidence of insulin-dependent diabetes mellitus. *J. Immunol.* 154:5567–5575.

62. Lund, T., L. O'Reilly, P. Hutchings, O. Kanagawa, E. Simpson, R. Gravely, P. Chandler, J. Dyson, J.K. Picard, A. Edwards, D. Kioussi, and A. Cooke. 1990. Prevention of insulin diabetes mellitus in non-obese diabetic mice by transgenes encoding modified I-A β-chain or normal I-E α-chain. *Nature* 345:727–730.

63. Carrasco-Martin, E., J. Shimizu, O. Kanagawa, and E.R. Unanue. 1996. The class II MHC I-A^g7 molecules from non-obese diabetic mice are poor peptide binders. *J. Immunol.* 156:450–458.

64. Scott, C.A., K.C. Garcia, F.R. Carbone, I.A. Wilson, and L. Teyton. 1996. Role of chain pairing for the production of soluble IA major histocomapatibility complex class II molecules. *J. Exp. Med.* 183:2087–2095.

65. Scott, C.A., K.C. Garcia, E.A. Stura, P.A. Peterson, I.A. Wilson, and L. Teyton. 1998. Engineering protein for X-ray

chrystallography: the murine major histocompatibility complex class II molecule I-A^d. *Protein Sci.* 7:413–418.

66. Corper, A.L., T. Stratmann, V. Apostolopoulos, C.A. Scott, K.C. Garcia, A.S. Kang, I.A. Wilson, and L. Teyton. 2000. A structural framework for deciphering the link between I-A^{g7} and murine autoimmune diabetes. *Science,* 288:505–511.

67. Scott, C.A., P.A. Peterson, L. Teyton, and I.A. Wilson. 1998. Crystal structure of two I-A^d-peptide complexes reveal that high affinity can be achieved without large anchor residues. *Immunity* 8:319–329.

68. Chao, C.-C., and H.O. McDevitt. 1997. Identification of immunogenic epitopes of GAD 65 presented by I-A^{g7} in non-obese diabetic mice. *Immunogenetics* 46:29–34.

69. Hausmann, D.H.F., B. Yu, S. Hausmann, and K. Wucherpfenning. 1999. pH-dependent peptide binding properties of the type I diabetes-associated I-A^{g7} molecule: rapid release of CLIP at an endosomal pH. *J. Exp. Med.* 189:1723–1733.

70. Liu, G.Y., D. Baker, S. Fairchild, F. Figueroa, R. Quartey-Papfio, M. Tone, D. Healey, A. Cooke, J.L. Turk, and D.C. Wraith. 1993. Complete characterization of the expressed immune response genes in Biozzi AB/H mice: structural and functional identity between AB/H and NOD *A* region molecules. *Immunogenetics* 37:296–300.

71. Haskins, K., M. Portas, B. Bergman, K. Lafferty, and B. Bradley. 1989. Pancreatic islet-specific T cell clones from nonobese diabetic mice. *Proc. Natl. Acad. Sci. USA* 86:8000–8004.

72. Haskins, K., M. Portas, B. Bradley, D. Wegmann, and K. Lafferty. 1998. T-lymphocyte clone specific for pancreatic islet antigen. *Diabetes* 37:1444–1448.

73. Bergman, B., and K. Haskins. 1997. Autoreactive T cell clones from the nonobese diabetic mouse. *Proc. Soc. Exp. Biol. Med.* 214:41–48.

74. Katz, J.D., B. Wang, K. Haskins, C. Benoist, and D. Mathis. 1993. Following a diabetogenic T cell from genesis through pathogenesis. *Cell* 74:1089–1100.

75. Andre, I., A. Gonzalez, B. Wang, J. Katz, C. Benoist, and D. Mathis. 1996. Checkpoints in the progression of autoimmune disease: lessons from diabetes models. *Proc. Natl. Acad. Sci. USA* 93:2260–2263.

76. Sawcer, S., and P.N. Goodfellow. 1998. Inheritance of susceptibility to multiple sclerosis. *Curr. Opin. Immunol.* 10:697–703.

77. Linquist, A.-K.B., and M.E. Alaracón-Riquelme. 1999. The genetics of systemic lupus erythematosus. *Scand. J. Immunol.* 50:562–571.

78. Harley, J.B., K.L. Moser, P.M. Gaffney, and T.W. Behrens. 1998. The genetics of human systemic lupus erythematosus. *Curr. Opin. Immunol.* 10:690–696.

79. Peña, A.S., J.A. Garrote, and J.B. Crusius. 1998. Advances in the immunogenetics of coeliac disease. Clues for understanding the pathogenesis and disease heterogeneity. *Scand. J. Gastro.* (Suppl) 225:56–58.

80. Phelps, R.G., and A.J. Rees. 1999. The HLA complex in Goodpasteure's disease: A model for analyzing susceptibility to autoimmunity. *Kideny Intern.* 56:1638–1653.

81. Stratmann, T., V. Apostolopoulos, V. Mallet-Designe, A.L. Corper, C.A. Scott, I.A. Wilson, A.S. Kang, and L. Teyton, 2000. The I-A^{g7} MHC class II molecule linked to murine diabetes is a promiscous peptide binder. *J. Immunol.* 165:3214–3225.

8 | TCR Signaling Pathways and Their Relevance to Autoimmunity

Tomas Mustelin and Amnon Altman

1. INTRODUCTION

T lymphocytes play a central role in the immune response, both as effectors and as regulatory cells that modulate the function of numerous cell types, primarily those that participate in the body's defense mechanisms. This regulatory function is provided either through direct cell cell contact or via the secretion of various cytokines. Thus, the proper function of T cells is essential for the maintenance of normal homeostasis within and outside the immune system. Conversely, abnormalities in their function can lead to immunological diseases, *e.g.*, autoimmunity or immunodeficiencies. The primary event leading to the activation and differentiation of mature T cells is the triggering of their antigen-specific T cell receptor (TCR) by its specific ligand which consists of a processed antigenic peptide presented in association with major histocompatibility complex (MHC) molecules on the surface of antigen-presenting cells (APCs) or appropriate target cells. This event triggers several signal transduction pathways that involve second messengers, protein kinases, phosphatases and other enzymes and key intermediates. This signaling cascade culminates with the induction of gene transcription according to defined genetic programs that are characteristic of the different T cell subsets, leading to the differentiation and proliferation of these cells. In addition, the proper development and selection of immature T cells in the thymus also depends on incompletely understood signaling events that dictate negative and positive selection processes and determine the specificity repertoire.

Studies on the biology of T lymphocytes have progressed in the past decade from elucidating the structure of the TCR complex (and accessory molecules) and the process of antigen recognition occurring at the cell surface, to analysis of the complex biochemical events that constitute the signal transduction machinery inside the cell (rev. in 1–4). As a result, many new insights were gained, and the process of T cell activation can now be addressed at a more sophisticated mechanistic, rather than phenomenological, level. Nevertheless, many pieces of the puzzle are still missing, and considerable effort is required before the accumulated knowledge can be translated into rational pharmacological approaches that can correct abnormal signaling processes in T cells and modulate immune system diseases in favor of the host.

This chapter provides an overview of TCR-initiated signaling pathways with an emphasis on the aberrant signaling pathways associated with human or experimental autoimmune diseases. It is not meant to be a comprehensive review of the large body of scientific work in this area but, rather, to present the most pertinent observations and open questions. It is for that reason that we chose to refer as much as possible to published review articles, where more complete information regarding aspects covered here is provided.

2. TCR SIGNALING PATHWAYS—AN OVERVIEW

The key recognition and activation element during physiologic T cell responses to antigen is a complex receptor consisting of two clonally distributed, highly polymorphic heterodimeric $\alpha\beta$ or $\gamma\delta$ subunits of the TCR, the three invariant CD3 polypeptides, and two additional homodimeric (ζ-ζ) or heterodimeric (ζ-η) subunits. The function of the $\alpha\beta$ (or $\gamma\delta$) TCR is to recognize and specifically bind nominal antigenic peptides presented by MHC molecules, while the CD3 complex, ζ and η transduce the signal generated by ligand binding to the TCR. The various chains associate noncovalently to form the complete and functional receptor.

Under physiological conditions, T cell activation is induced as the result of a complex and prolonged contact with an APC that presents antigen bound to MHC molecules of the correct type and in the presence of appropriate costimulatory molecules. This cell-cell contact area has recently been found to possess a higher order of structural organization than previously recognized, and is now referred to as the immunological synapse. The recognition of antigen/MHC by the TCR provides the initial event in the formation of this synapse, and is rapidly followed and stabilized by the binding of the CD4 and CD8 coreceptors to non-polymorphic regions of the same MHC molecules. The interaction between the two cells is further strengthened by several additional receptor-counter receptor interactions, most of which play both adhesive and signaling roles. On the T cell, these include CD2, CD28, integrins and glycosylinositolphospholipid-anchored proteins, such as Thy-1 and Ly-6. CD28 is unique among the coreceptors in that it facilitates the movement of cholesterol-rich lipid microdomains, enriched in important signaling molecules, to the T cell-APC contact area (5). Without prior TCR engagement, however, none of these additional interactions become prolonged and they fail to cause any signs of T cell activation.

Among the coreceptors, CD4 and CD8 are particularly important because they associate through their cytoplasmic tails with the lymphoid-specific protein tyrosine kinase (PTK) Lck (6), which is a key mediator for TCR signaling and T cell activation (rev. in 2, 7). One of the earliest detectable responses in activated T cells is the increased phosphorylation of many proteins on tyrosine residues, largely catalyzed by Lck. In fact, inhibition of PTK activity completely prevents T cell activation (8,9), indicating that this phosphorylation is important. Thus, CD4 and CD8 provide a simple mechanism by which an intracellular enzyme can be brought into the vicinity of the TCR (10). Although most mature $\alpha\beta$TCR$^+$ T cells require the presence of the CD4 and CD8 coreceptors for an optimal response to antigen, there are CD4$^-$8$^-$ T cells that respond to antigen or anti-TCR/CD3 antibodies. A prime example is the most widely used model for TCR signaling, the Jurkat T leukemia cell line, which usually is CD8$^-$ and either has a very low level of CD4 or is CD4$^-$. Nevertheless, these cells are among the most easily triggered T cells, perhaps due to their high expression of Lck. It is clear in these cells that Lck actively participates in TCR signaling in a manner that is completely independent of CD4 or CD8 (4).

T cell activation is initiated by increased phosphorylation of the tyrosines in the immunoreceptor tyrosine-based activation motifs (ITAMs) of the signaling subunits of the TCR complex, catalyzed by Lck (and perhaps the related Fyn kinase). This phosphorylation creates docking sites for the tandem Src homology-2 (SH2) domains of a second PTK, called Zap-70 (11). The binding affinity of the two tandem SH2 domains of Zap-70 for the doubly phosphorylated ITAMs is very high, and currently no other signaling molecules are known to bind to doubly phosphorylated ITAMs with affinities even close to those of Zap-70. In contrast, singly phosphorylated ITAMs have a low affinity for Zap-70 and cannot (at least by this mechanism alone) recruit Zap-70. In resting T cells, some of the ITAM tyrosines may already be phosphorylated, but the ITAMs generally do not contain more than one phosphate each. It was recently proposed that the tyrosines of the ζ-chain ITAMs are phosphorylated exclusively in a specific sequence, B1-C2-A2-B2-A1-C1 (where A, B, and C refer to the three ITAMs; 12). If this is found to be a general rule for ITAM phosphorylation in T cells, it implies that efficient Zap-70 recruitment only occurs after the later steps of the sequence. The biological implications of this notion will be discussed below.

Since the most common form of the TCR complex is composed of two dimers of the CD3 subunits (ε-δ and ε-γ) and a ζ-ζ dimer, there are ten ITAMs in each receptor complex. Why does the receptor need so many functional motifs? At least two hypotheses have been proposed to answer this question. First, multiple ITAMs may serve to amplify the signal. In support of this notion, each doubly phosphorylated ζ ITAM readily binds Zap-70 in vitro forming a 1:3 complex. Thus, the fully phosphorylated ζ-ζ dimer in a TCR complex may bind as many as six Zap-70 molecules, and the CD3 subunits four more.

A second possibility for the presence of multiple ITAMs in the TCR complex is that they are functionally different in some way. The amino acid sequences of the six ITAMs are far from identical. In fact, there are some potentially important differences in the nature of the amino acids surrounding the tyrosine in each ITAM. Such differences are likely to result in different affinities of various SH2 domains for these sites (13). In addition, there are differences in the number and placement of acidic residues that have an important impact on the phosphorylation of the tyrosine residues by PTKs, and some ITAMs contain proline residues that may induce a kink in the polypeptide backbone.

Unlike growth factor receptors, the TCR complex is not a simple binary on-off switch. Instead, the recognition of peptide in the groove of an MHC molecule is a delicate balancing act that can result in radically different responses ranging from survival in a resting state to the induction of anergy, partial activation or full activation, which may lead to proliferation or cell suicide. The type of response depends on the avidity of peptide binding, on-off rates and perhaps degrees of receptor aggregation. The conditions that induce anergy generally represent stimulation with peptides that bind to MHC with good affinity, but that are low to intermediate affinity ligands for the TCR.

It has recently been observed that the degree to which ITAMs are phosphorylated correlates with different T cell responses. While unphosphorylated TCR-ζ is approximately 16 kDa, a one-dimensional SDS-PAGE often resolves phospho-ζ into two Mr species of 21 and 23 kDa, the latter presumably corresponding to fully or nearly fully phosphorylated ζ. Under optimal conditions of TCR triggering, the 21-kDa form first increases rapidly and then the 23-kDa form appears. Under conditions that fail to cause T cell activation and instead induce a state of unresponsiveness, referred to as T cell anergy, the 21-kDa form may increase less, equally to, or even more than under activating conditions, but the 23-kDa form can barely be observed (14,15).

Although the model of T cell activation that begins with ζ phosphorylation by Lck (plus Fyn?), followed by Zap-70 recruitment and activation, is well established by multiple studies, reality is probably a bit more complicated. One unsettled question is the possible role of another kinase, Syk (16), which is structurally related to Zap-70. Most mature T cells express Syk although the amount of Syk protein is much lower than the amount of Zap-70. However, since the specific activity of Syk may be as much as 100-fold higher than that of Zap-70, the lower level expression of Syk might not be as significant as it seems at first glance. It has been debated whether Syk is redundant with Zap-70, or acts in a different manner either upstream of Lck, in parallel with Lck or in some combination of these. It is clear that there are many striking differences between the regulation and properties of Syk and Zap-70, but the biological significance of many of these differences remain largely speculative.

In most T cells, very little Syk coimmunoprecipitates with TCR-ζ (our unpublished observation), although Syk readily binds to ITAM-derived phosphopeptides *in vitro*. In intact T cells, however, the much more abundant Zap-70 is likely to effectively compete for a limited number of phosphorylated ITAMs and thereby prevent most of the few Syk molecules from binding. In comparison, the ITAM of the FcεRI γ chain binds Syk much better than it binds Zap-70, and may thus selectively recruit Syk. This may be important in those T cells that express the FcεRI γ chain instead of ζ as part of their TCR, *e.g.*, $\gamma\delta$ T cells. It may also be important that Syk can mediate TCR signals independently of both Lck and Zap-70 (17–19). Since Syk readily autophosphorylates and autoactivates (unlike Zap-70), a relatively small number of Syk molecules could be sufficient to initiate a signaling cascade.

A very important advance for our understanding of the importance of Syk and Zap-70 in lymphocyte activation and development was the recent discovery of patients lacking a functional *zap* gene (20), as well as the generation of mice lacking Zap-70, Syk or both. Lack of Zap-70 in humans leads to a severe immunodeficiency characterized by the absence of CD8$^+$ T cells and TCR-unresponsive mature CD4$^+$ T cells (21). Zap-70-deficient mice lack, in addition, CD4$^+$ T cells while their natural killer (NK) cells are unaffected (22). Mice lacking Syk, on the other hand, die *in utero* from massive hemorrhage (23,24). A lymphocyte-specific lack of Syk (achieved by reconstitution of RAG2$^{-/-}$ mice) resulted in an absence of B cells and many intraepithelial $\gamma\delta$TCR T cells (25). Most other T cells (as well as other leukocyte types) appeared to be normal, although detailed signaling studies have not yet been carried out with them. Compared to the lack of only Zap-70 or Syk, mice lacking both kinases show a much more severe and early arrest in thymic development of T cells (26). At present, it remains unclear whether these largely non-overlapping requirements for Syk and Zap reflect differences in expression pattern or differences in biological function. This question is particularly pertinent in cells expressing both PTKs, such as $\gamma\delta$ T cells and NK cells. Given the crucial role of these kinases, relatively minor disturbances in their expression or regulation may result in significant abnormalities in the immune system. Decreased Syk expression has been reported in cutaneous T leukemia cells (27), which respond poorly to TCR stimulation. Thus, it is conceivable that an increase in Syk expression would lead to hyperreactive T cells and autoimmunity.

3. SIGNALING DURING T CELL DEVELOPMENT

T cells expressing the $\alpha\beta$ TCR develop in the thymus according to a complex genetic program in which CD4$^-$CD8$^-$ (double-negative) thymocytes give rise to CD4$^+$CD8$^+$ (double-positive) cells, which then differentiate into single-positive CD4$^+$ or CD8$^+$ cells. These cells then migrate to the periphery, where they constitute the pool of mature T lymphocytes capable of responding to antigen. The transition from the double-negative to the double-positive stage depends on the assembly and expression of the TCR β chain. This is followed by TCR α chain expression, resulting in a competent TCR capable of transducing signals (rev. in 28). To generate mature T lymphocytes expressssing highly diverse repertoires of antigen receptors and, at the same time, avoid potentially harmful recognition of self antigens, developing T cells are subjected to series of positive and negative selection steps, during which they undergo maturation, programmed cell death and cellular proliferation. These processes are initiated by TCR engagement in combination with costimulatory signals provided by cytokines and adhesion receptors. The ensuing signals are relayed within the cell through the assembly of signaling complexes. Considerable work has recently focused on the signaling pathways that mediate these developmental and selection processes in order to understand the nature of differential signaling, *i.e.*, how signals delivered by the same receptor can lead to diverse biological outcomes depending on the phenotype of the responding cell and its environment (reviewed in 29, 30).

In general, two models have been proposed to explain differential signaling, *i.e.*, the kinetic and conformational models (31–33). Common to the two models is the notion that high affinity TCR-self MHC interactions (which can potentially produce harmful autoreactive T cells) generate complete and strong activation signals resulting in cell death (negative selection). On the other hand, lower affinity interactions that generate weaker (or incomplete) signals cause positive selection, *i.e.*, further differentiation and maturation into functional T cells that can now productively respond to stimulation by foreign antigenic peptides presented by self-MHC molecules on the surface of APCs.

This model implies that abnormalities in the TCR signaling machinery that change the strength of TCR signals in either direction can diminish or increase the threshold for negative and/or positive selection in the thymus. Thus, it can be postulated that diminished signaling will allow self-reactive T cells, which would normally be eliminated by strong TCR signals, to survive, undergo positive selection, mature and emigrate to the periphery, where they could potentially cause autoimmune disease. For example, experimental manipulation of the TCR signal strength by expression of a dominant-negative *Ras* transgene (diminished signaling) or a dominant-negative *SHP1* transgene (enhanced signaling) in the T cell compartment was found to lead to decreased or increased positive selection of CD4⁺ T cells in the thymus, respectively (34,35). A similar situation could account, at least in part, for the autoimmune manifestations in SHP1-deficient motheaten (*me*) mice (see section 7 below), whose thymocytes display hyperresponsiveness to TCR stimulation (36,37).

4.   COSTIMULATION AND AUTOIMMUNITY

In addition to TCR/CD3-derived signals, productive T cell activation requires a second signal that can be provided by a variety of costimulatory receptors expressed on T cells (rev. in 38, 39). Among these, CD28 plays a critical role in IL-2 production, autoimmunity, tumor immunity and anergy (reviewed in 40–43). The CD28 signal cooperates with the TCR/CD3 signal to induce T cell proliferation and high-level production of interleukin-2 (IL-2), other cytokines and chemokines. In addition, CD28 ligation provides an important survival signal that prevents apoptosis (44) or anergy (45), two events that can occur in response to triggering of TCR alone. The B7 family of costimulatory ligands expressed on APCs represents the physiological ligand for CD28 (rev. in 40, 46).

CTLA-4 is the second known receptor for B7. Unlike CD28, which is expressed on resting T cells, CTLA-4 expression is normally very low, and is upregulated following TCR stimulation (reviewed in 47–51). Findings indicating that, in contrast to the positive role of CD28 in enhancing T cell proliferation and IL-2 production, CTLA-4 ligation has an opposite, inhibitory effect (*e.g.*, 52), led to the notion that CTLA-4 engagement delivers inhibitory signals. This suggests that the outcome of TCR stimulation is regulated by CD28 costimulatory signals, as well as by inhibitory signals derived from CTLA-4. This negative immunoregulatory role of CTLA-4 has been firmly established by analyzing CTLA-4-deficient mice. This analysis demonstrated that the loss of CTLA-4 leads to massive lymphoproliferation, massive lymphocytic infiltrates in many organs, eleveted serum immunoglobulin (Ig) levels, and fatal multiorgan tissue destruction (53,54). These studies suggest that negative signaling by CTLA-4 plays an active role in regulating autoreactive T cells.

4.1.   The Role of the B7-CD28/CTLA-4 System in Autoimmunity

The essential regulatory and opposing roles of CD28 and CTLA-4 signals in T cell activation and anergy led to the hypothesis that costimulation may also play a role in autoimmunity. Indeed, numerous studies clearly indicated that manipulation of the B7-CD28/CTLA-4 system can modulate autoimmune diseases, such as experimental allergic encephalytis (EAE), spontaneous diabetes arising in nonobese diabetic (NOD) mice, systemic autoimmunity in lupus-prone (NZBxNZW)F₁ mice, and autoantibody production in hepatitis B antigen-expressing transgenic mice (rev. in 51, 55–57). These studies utilized, for the most part, administration of soluble CTLA-4-Ig fusion proteins or anti B7-1 (CD80)/B7-2 (CD86) antibodies, and this treatments were observed to either protect mice against autoimmune disease or exacerbate it. The different outcomes from manipulating the B7-CD28/CTLA-4 system on autoimmune disease are not surprising given the complexity of the system and the fact that CD28 *vs.* CTLA-4 engagement generally have opposite effects on T cell activation. Furthermore, distinct mechanisms and/or target cells may be involved in mediating the effects of CTLA-4-Ig proteins or anti-B7 antibodies, *e.g.*, anergy induction, skewing of the Th1/Th2 balance, blocking of APC function, and/or direct signaling to T cells.

Recent studies also reported the ectopic expression of the CD28 ligand, B7-1, on T lymphocytes of mice homozygous for the *lpr* or *gld* mutations (see section 5 below) in an age-dependent manner and in all subsets of their T lymphocytes (58). This constitutive expression of B7-1 may contribute to the maintenance and exacerbation of the lymphoproliferation and/or autoimmune manifesttaions in these mice.

The B7-CD28/CTLA-4 system has also been linked to human autoimmune disease. Thus, increased expression of B7-1 has been observed in affected tissues of patients suffering from multiple sclerosis, rheumatoid arthritis, autoimmune thyroiditis, and inflammatory skin diseases (rev. in 51, 56, 59). Taken together, the available evidence suggests that manipulation of the B7-CD28/CTLA-4 system may provide a broadly applicable, antigen-non-specific therapeutic approach for treating autoimmune disease, either by blocking the B7-CD28 costimulatory interaction, or by triggering B7-CTLA-4 interactions that generate negative signals potentially capable of suppressing T cell response to autoantigens.

4.2. Signal Transduction by CD28 and CTLA-4

CD28 signaling. The mechanisms of CD28 signaling are still incompletely understood, but the findings that CD28 becomes phosphorylated on tyrosine, and is associated with three defined signaling proteins, *i.e.*, phosphatidylinositol 3-kinase (PI3-K; 60–62), Grb2 (63–65), and tyrosine kinases of the Tec family (60,66,67), represent important advances in this area. While CD28 and the TCR/CD3 complex share some signaling properties, *e.g.*, the ability to activate some tyrosine kinases and PI3-K, the two receptors clearly differ in several regards. In contrast to the TCR, CD28 crosslinking by its physiological ligand, CD80, fails to stimulate Ras, Raf-1 and extracellular-regulated kinase-2 (ERK2), or induce tyrosine phosphorylation of the adaptor proteins LAT and SLP-76 (however, anti-CD28 monoclonal antibodies do induce these events); under the same conditions, a potent and prolonged phosphorylation of Vav on tyrosine is induced (68). Conversely CD28, but not TCR/CD3, crosslinking induces tyrosine phosphorylation of a 62-kDa protein (69), now known as p62^{dok} (70,71).

The association of CD28 with the Grb2/Sos complex (63–65) provides a potential link to Ras activation. However, it is more likely that ligand binding to CD28 does not activate Ras in itself (68), but rather potentiates Ras activation by TCR/CD3 triggering. This view is consistent with the finding that CD28 costimulation converges with TCR/CD3 signals to activate c-Jun N-terminal kinases (JNKs), AP-1 and the IL-2 promoter in T cells (72,73). The CD28 costimulatory signal is essential since TCR/CD3 ligation alone does not activate JNK (72) or its upstream activating kinase, MEKK-1 (73), in T cells (although it can activate Raf-1 and ERK). The ability of dominant-negative Ras to block JNK activation by TCR plus CD28 crosslinking (73) indicates that Ras regulates JNK activation by these two receptors.

The inducible association of CD28 with PI3-K provides an additional potential mechanism through which CD28 could regulate Ras activation, since PI3-K can activate the Ras signaling pathway in NIH 3T3 fibroblasts and *Xenopus* oocytes (74), or function as an immediate Ras effector (75). However, no definite information exists regarding a function of this putative mechanism in T cells. Although Grb2 and PI3-K (p85) bind to the same site in the phosphorylated cytoplasmic tail of CD28 (and, therefore, should compete for CD28 binding), both signaling proteins coimmunoprecipitate with CD28 from activated T cells. Thus, the relative contribution of PI3-K *vs.* Grb2 association to CD28-mediated costimulatory signals remains to be determined.

Recent studies have also linked CD28 to small GTPases of the Rho family, which play important roles in regulating the actin cytoskeleton and growth signals in different cell types (reviewed in 76, 77), including in T cells (rev. in 78, 79). First, CD28 crosslinking by CD80 induces a rapid and sustained increase in the tyrosine phosphorylation of the hematopoietic cell-specific Rho-family guanine nucleotide exchange factor (GEF), Vav (80), in the absence of TCR ligation (68). Since the exchange activity of Vav is stimulated by tyrosine phosphorylation (81), this suggests that CD28 ligation would stimulate the catalytic activity of Vav toward member(s) of the Rho family. This is consistent with the recent finding that stimulation of T cells with the CD28 ligand B7-2 promoted formation of focal adhesion-like plaques where Rho-family small G proteins accumulated (82). However, it is not known whether CD28-induced tyrosine phosphorylation of Vav enhances its enzymatic activity. Second, as mentioned earlier, CD28 costimulation is required for activation of JNK (72) and MEKK-1 (73) in T cells. Since JNK activation is coupled to Rac and Cdc42 in fibroblasts (83–85) and in T cells (86), this finding suggests that CD28-dependent MEKK-1/JNK activation is mediated by a Vav/Rac (and/or Vav/Cdc42) pathway.

The CD28-induced tyrosine phosphorylation of p62^{dok} (69) provides another potential link to small GTPases, since this pleckstrin homology (PH) domain-containing protein forms a complex with p120 ras-GTPase-activating protein (GAP) and p190 rho-GAP. Tyrosine phosphorylation of p62^{dok} may modulate the cellular distribution of the associated GAPs and sequester them away from their activated Rac and Ras targets, thereby prolonging their activated state (78). Thus, p62^{dok} has the potential to regulate both Ras and Rho GTPases. An additional possibility is that PI3-K could couple immune receptors to Rac/Rho signaling pathways via its coupling to CD28 (or TCR) is based on the finding that PI3-K products can stimulate Rac- and Rho-mediated cytoskeletal responses (87).

The NF-κB cascade represents another important signaling pathway linked to CD28 stimulation. This family of transcription factors plays an important role in regulating inflammatory, immune, and anti-apoptotic responses (rev. in 88–90). Recent studies elucidated a complex cascade of

Ser/Thr kinases that mediate the activation of NF-κB via phosphorylation and concommitant degradation of inhibitory IκB proteins (*e.g.*, 91–95).

The IL-2 gene promoter contains binding sites for several transcription factors, including NFAT, AP-1, NF-κB, and factors interacting with the CD28 response element (CD28RE; 96) (rev. in 97, 98). Although TCR/CD3 ligation is sufficient to stimulate the Ras/Raf/MEK/ERK pathway and the transcription factor NFAT, stimulation of NF-κB (99) and JNK (72) in T cells is strictly dependent on CD28 costimulation. Optimal activation of the CD28RE depends on an adjacent AP-1 binding site (100,101). Recent studies have begun to define the intermediates in the CD28 signaling pathway leading to NF-κB activation in the context of the CD28RE, and established a role for several Ser/Thr protein kinases, such as MEKK, Cot, NIK and IKK in these process (102,103). Thus, the CD28RE element in the IL-2 promoter is an important target for the CD28 costimulatory signals, but considerable effort will be required to define other intermediates and regulatory interactions in this pathway.

Finally, very recent studies called into question the notion that CD28 engagement stimulates signaling pathways distinct from those induced by the TCR/CD3 complex. Rather, these studies indicated that costimualtion by CD28 primarily reflects its ability to enhance TCR/CD3 signaling events by inducing the redistribution and clustering of membrane and intracellular kinase-rich lipid raft microdomains (104) at the contact sites between antigen-specific T cells and APCs, a process that leads to higher and more stable localized tyrosine phosphorylation of several substrates (5), and enhanced Ca²⁺ signaling (105).

CTLA-4 signaling. The finding that CTLA-4 plays a negative regulatory role in TCR-initiated T cell activation (52–54) led to a search of signaling intermediates that mediate this inhibitory function. Consistent with the inhibitory function of CTLA-4, the T cell kinases Lck, Fyn and ZAP-70 were found to be constitutively active in CTLA-4-deficient mice (106). CTLA-4 was found to associate with the TCR-ζ chain in an Lck-dependent manner (107). Furthermore, the cytoplasmic tail of CTLA-4 contains immunoreceptor tyrosine-based inhibitory motifs (ITIMs; rev. in 108), which mediate its association with the SH2 domain of the phosphotyrosine phosphatase (PTPase), SHP2 .(106). Coexpression of SHP-2 resulted in dephosphorylation of the Shc adaptor protein (106) and of CTLA-4-bound TCR-ζ, and abolished the Lck-inducible interaction between TCR-ζ and CTLA-4 (107). In contrast, the SHP1 phosphatase does not appear to function as an intermediate in CTLA-4-mediated negative signaling since CTLA-4 crosslinking inhibits proliferation and IL-2 secretion of T cells from SHP1-deficient motheaten mice

(109). These findings provide a potential mechanistic basis for the negative regulatory function of CTLA-4.

4.3. Other Costimulatory Receptors and Their Role in Autoimmunity

Despite the extensive evidence establishing the critical role of CD28 in costimulation, it is clear that not all T cell responses depend on CD28 (110). This reflects the fact that various other T cell-expressed receptors have been shown to provide costimulatory signals for T cell activation and proliferation (rev. in 111). These include, among others, integrins, CD2, CD5, SLAM, HSA, and members of the TNF receptor (TNFR) family (111). However, many of these do not function in the same way as CD28 in that they do not promote high-level IL-2 production or T cell survival, nor do they prevent anergy induction. Relatively little is known about the association of these costimulatory receptors with autoimmune disease. However, one member of the TNFR family, *i.e.*, OX40 (CD134), deserves special attention since it has been closely linked to experimental autoimmune disease models (rev. in 112). Like other members of the TNFR family, OX40 associates via its cytoplasmic tail with members of the TNFR-associated factor (TRAF) family of adaptor proteins that link TNFRs to downstream signaling pathways (rev. in 113). Stress-activated protein kinases (SAPKs), the transcription factor NF-κB, and the cellular inhibitors of apoptosis (cIAPs) represent the best characterized downstream targets of TNFR/TRAF-initiated signals (114). OX40 associates with TRAF-2, −3 and −5, and was found to activate NF-κB (115,116). Although not demonstrated directly to date, OX40 most likely also activates the JNK family based on its association with TRAF-2.

OX40 expression is upregulated during T cell activation with peak expression at 1–2 days following stimulation either *in vitro* or *in vivo*, followed by its decay after ~4 days. Consequently, OX40 ligation provides a very weak costimulatory signal for the proliferation and IL-2 production of naïve CD4⁺ T cells, but is highly effective in stimulating T cells in conjunction with the CD28 ligand, B7-1 (117). Moreover, the OX40 ligand (OX40L), which is expressed on activated B cells and dendritic cells (112, 118), sustains the production of cytokines by Th cells, and the T-B cell interaction via OX40-OX40L is also necessary for T cell-dependent humoral immune responses (119). OX40 synergizes with B7-1 to activate naïve T cells and, perhaps more importantly, provides an essential costimulatory signal for maintenance of the long-term and repetitive stimulation of activated T cells, once the expression of CD28 has been downregulated. Thus, a primary function of OX40 engagement may be to enhance effector and memmory T cell function by upregulating IL-2 production and increasing the life-span of effector T cells. It has been hypothesized that signals generated upon OX40 ligation inhibit activation-induced T cell death (AICD), thereby increasing the number of activated effector cells differentiating into memory T cells (112).

This implied function of OX40 suggests that it may play a role in the survival, maintenance and expansion of autoreactive T cells and, thus, T cell-dependent autoimmune disease. This notion is supported by several studies that implicated a role for OX40 in experimental autoimmune disease (rev. in 111, 112). Several studies demonstrated an accumulation of OX40+ activated T cells in the inflammatory sites of mice with several autoimmune diseases, e.g., EAE, experimental colitis or spontaneous colitis in IL-2-deficient mice, or $HgCl_2$-induced systemic autoimmunity (rev. in 112). Furthermore, OX40 is selectively expressed on the autoreacive T cells rather than on bystander activated T cells present in the central nervous system of EAE-affected mice (120, 121). Elimination of OX40+ cells by treating mice with an anti-OX40 antibody (122), or blocking of the OX40-OX40L interaction by injecting an OX40-Ig fusion protein (123) was found to ameliorate EAE and ongoing inflammatory bowel disease (124). Thus, the interaction of OX40 with its ligand expressed on APCs may serve as an important therapeutic target in autoimmune disease.

5. SIGNALING DEFECTS IN MURINE MODELS OF AUTOIMMUNITY

5.1. Mutations in the Fas/Fas Ligand (FasL) System

It has long been known that The CD4-CD8-(double-negative) B220+ T cells that accumulate in an age-dependent manner in peripheral lymphoid organs of mice homozygous for the lymphoproliferation (*lpr*) mutation display deficient activation in response to signals initiated by the TCR/CD3 complex (125; rev. in 126). Abnormalities that could contribute to this defect include low expression and rapid TCR/CD3 modulation, loss of coreceptors such as CD4, CD8, CD2 and Ly-6.2, aberrant expression of CD45, and constitutive elevations of inositol phospholipid turnover, tyrosine phosphorylation, and overexpression of the Fyn tyrosine kinase and a few other protooncogenes (126). These properties suggest that the double-negative *lpr* T cells are chronically activated *in vivo*, perhaps by some self antigen(s), rendering them refractory to additional *in vitro* stimulation. It is now well established that *lpr* mutations, as well as *gld* mutations that induce a similar lymphoproliferative and autoimmune phenotype, represent mutations in the genes encoding Fas and its ligand (FasL), respectively (rev. in 127). Thus, the deficient activation and abnormal signaling properties of *lpr* T cells most likely represent a secondary consequence of the Fas/FasL defects whose primary effect is allowing these aberrant cells to escape from the thymus and expand in the periphery. In accor-

dance with this view, it has been suggested that the abnormal T cells in *lpr* mice represent MHC-selected, high-affinity T cells that would normally be deleted in a process of negative selection via a Fas/FasL-mediated apoptotic pathway, but fail to be deleted as a result of the inactivating *lpr/gld* mutations (128,129).

In terms of specific abnormalities of signaling intermediates in the TCR signaling cascade, the most prominent features of the affected T cells in *lpr*- or *gld*-homozygous mice are the constitutive tyrosine phosphorylation of the TCR-associated ζ chain (130), the overexpression of a hyperactive Fyn tyrosine kinase (131), and the deficient TCR-induced inositol phospholipid hydrolysis (132), a process that generates the second messengers essential for T cell activation and proliferation (1). The hyperactivity of Fyn may play a causative role in the systemic autoimmune disease since the lymphadenopathy and other disease manifestations were greatly reduced in Fyn-deficient *lpr* mice (133). This is consistent with an earlier finding that induction of cell cycling in the double-negative *lpr* T cells by combined treatment with IL-2, phorbol ester and Ca^{2+} ionophore reverses the hyporesponsiveness of these cells and normalizes Fyn expression (134). CD4+ T cells are thought to play an important role in the development of SLE-like disease in *lpr* mice, and these cells were found to display lower levels of Lck and decreased CD4-induced cellular tyrosine phosphorylation (135). At the level of the transcription factors known to be involved in IL-2 gene induction, constitutive activation of AP-1, NF-κB and Oct-1, and deficient NFAT activation, were documented in CD4-CD8-*lpr* T cells (136).

5.2. Spontaneous Diabetes in NOD Mice

The spontaneous diabetes occurring in NOD mice has been extensively studied as a model of a T cell-dependent autoimmune disease (rev. in 137–139). The thymocyes of these mice are hyporesponsive to TCR-mediated activation and proliferation, and indirect evidence suggests that this hyporesponsiveness plays a causative role in the autoimmune disease (rev. in 140). Analysis of TCR- and CD4-initiated signaling events in NOD thymocytes revealed several signaling defects, i.e., deficient Ras activation, impaired membrane recruitment of the Grb2/Sos complex known to be important for Ras activation, preferential activation of a non-productive Fyn/Cbl pathway (reviewed in 141), and sequestration of CD4-associated Lck from the TCR complex (142–144).

6.   SIGNALING DEFECTS IN HUMAN SLE

Systemic lupus erythematosus (SLE) is a much studied, but poorly understood, autoimmune condition characterized by multiple symptoms that include inflammation of many organs and circulating autoantibodies. Theoretically, much of the pathology could be explained by an intrinsic defect in the T cells, for example an exaggerated or prolonged response to antigen due to a failure in the negative regulation of the activation machinery (145). To address this possibility, many laboratories have studied the early events that follow TCR ligation using T cells isolated from SLE patients. These experiments have revealed that SLE T cells display an increased tyrosine phosphorylation response *in vitro* (146). An interesting exception in this enhanced response is the markedly decreased phosphorylation of the TCR-ζ, which was found to be the result of drastically reduced expression of this protein (147). It was also reported that SLE T cells are less susceptible to activation-induced cell death, perhaps due to their increased production of soluble Fas protein and enhanced expression of FasL. Finally, abnormalities have also been found in B cells and APCs, and it remains unclear if any of the observed defects are intrinsic or caused by an altered environment in the host. Nevertheless, the notion that SLE T cells may have an etiologically significant alteration in the balance between tyrosine kinases and phosphatases, or some other signaling molecules, is intriguing (rev. in 148).

The human autoimmune lymphoproliferative syndrome (ALPS) recapitulates many of the characteristics of the murine disease occurring in *lpr*- or *gld*-homozygous mice. Thus, children diagnosed with this disease display moderate to massive splenomegaly and/or lymphadenopathy, hypergammaglobulinemia, B cell lymphocytosis, autoimmunity and expansion of a normally rare subset of CD3$^+$ double-negative (CD4$^-$CD8$^-$) T cells (rev. in 149, 150). Similar to the spontaneous murine disease associated with *lpr* or *gld* mutations, ALPS patients display inactivating or dominant-negative mutations in their *FAS* gene (151), and heterozygosity at this locus is sufficient to cause ALPS. More recently, mutations in the enzymatic machinery that drives Fas-mediated apoptosis, *i.e.*, in the gene encoding caspase 10, were also identified in ALPS patients (152). Thus the defective Fas/FasL signal transduction pathway resulting from these mutations is the underlying reason for the expansion of autoreactive T cells in this disease.

Little is known regarding the responsiveness of ALPS T cells to TCR-activating signals. However, it was reported that the normal T cells of these patients are skewed toward the Th2 phenotype, as indicated by the highly elevated circulating IL-10 level, increased *in vitro* production of IL-4 and IL-5, and deficient production of interferon-γ. In contrast, the double-negative ALPS T cells respond to TCR stimulation with minimal lymphokine secretion (153). It is possible, therefore, that the latter cells represent T cells continuously stimulated by self antigens and, as a result,

are refractory to further *in vitro* stimulation, consistent with the hypothesis suggested to account for the deficient TCR signaling observed in *lpr* and *gld* murine T cells (see section 5 above).

7. PHOSPHATASES AND AUTOIMMUNITY

7.1. The Potential Role of Dysregulated PTPase Activity in Autoimmunity

Defects in components of the signal transducing apparatus could have profound effects on the responsiveness of T cells and, thus, on the immune system as a whole. Since the phosphorylation levels of key regulatory enzymes apparently influence signaling processes in an important way, the balance between kinases and phosphatases becomes critical. In general, T cells displaying deficient tyrosine phosphorylation, whether due to elevated kinase activity or dysregulated PTPase activity, would be expected to cause excessive immune responses and, potentially, autoimmunity, while diminished ability to phosphorylate substrates associated with the activating signaling machinery might result in immunodeficiency (145).

Resting T lymphocytes contain very little phosphotyrosine (PTyr), only about 0.01% of acid-stable protein-bound phosphate, compared to phosphorylated serine (95%) and phosphorylated threonine (5%). This indicates that very efficient and specific PTPases must exist in these cells. This notion has been validated by the dramatic increase in intracellular tyrosine phosphorylation that follows a brief treatment of T cells with PTPase inhibitors. The increase in PTyr content after addition of these pharmacological agents is detectable within seconds and quickly reaches levels that far exceed the response to any physiological stimuli. Moreover, these PTPase inhibitors also induce many functional events typical of T cell activation, and prevent reversion of activated T blasts to a resting phenotype (154). Thus, PTPases clearly play a crucial part in the maintenance of a resting T cell phenotype and in limiting the response to antigen. T cell activation has traditionally been assumed to be the result of activation, or at least increased action, of the PTKs, but it has not been formally ruled out that changes in the rate of dephosphorylation may play a role.

Although the PTyr content of cellular proteins is the net result of the opposing effects of PTKs and PTPases, most investigators have in the past concentrated on the PTKs, many of which were cloned and sequenced already in the 1980s. In contrast, most PTPases have only been known for a few years. It seems that the interest in PTPases is now gaining momentum, although little is still known about most of them (rev. in 155, 156). The new excitement was sparked by discoveries such as the finding that SHP1 mediates signaling from inhibitory hematopoietic receptors and is deficient in the autoimmune motheaten mice (157,158), and that PTP1B is involved in insulin receptor signaling (159). It is far from clear which PTPases are involved in lymphocyte activation, how they are regulated or what substrates they act on. The old notion that PTPases are promiscuous and mainly fulfill a house-keeping function has in recent years given way to the realization that they exist as a large family of highly diverse enzymes that are highly specific and tightly regulated.

7.2. PTPases Present in T Lymphocytes

The first PTPase found to be expressed in T cells was the leukocyte common antigen CD45, a transmembrane receptor-like glycoprotein with two tandem catalytic domains in its cytoplasmic tail. It has become clear that CD45 acts as an important positive regulator of Lck and Fyn function by dephosphorylating the C-terminal regulatory tyrosine (160; rev. in 2, 161). Most T cell lines lacking CD45 show a marked absence of basal tyrosine phosphorylation, as would be expected if Lck and Fyn were suppressed. These cells also fail to respond to antigen or antibodies that normally activate the cells.

T cells express at least 20 of the known 35 intracellular PTPases. These enzymes belong to three main classes: 1) the classical PTPases, 2) the low molecular weight PTPases, and 3) the dual-specificity PTPases. The first class contains most of the known PTPases, both the receptor-like transmembrane enzymes such as CD45, as well as several intracellular enzymes containing the same highly conserved catalytic domain. The latter type includes the first cloned PTPase, PTP1B, its T cell homologue, TCPTP, the SH2 domain-containing phosphatases SHP1 and SHP2, and a group of PTPases with elements typically found in cytoskeletal proteins. The dual-specificity phosphatases are in the 20–45 kDa range and are generally nuclear enzymes that dephosphorylate the PTyr and phosphothreonine residues in mitogen-activated protein (MAP) kinases. The low molecular weight PTPases are only distantly related to the other PTPases, and are less than 20 kDa in size.

Currently the best understood intracellular PTPase is SHP1, which is expressed only in cells of hematopoietic lineages. The homozygous loss of SHP1 leads to the motheaten (me) phenotype in mice, which is characterized by a spotty hair loss (hence the name) and a number of abnormalities in the function of the immune system and phagocytic leukocytes. T lymphocytes from these mice are hyperresponsive to TCR stimulation (36,37), suggesting that SHP1 plays an important role as a negative regulator of T cell function. At least part of this function is mediated through a direct dephosphorylation of the Zap-70 and Syk PTKs, resulting in decreased signaling (162). The function of SHP1 is regulated by a class of inhibitory surface receptors that possess an ITIM motif (108) in their intracellular tails. In a manner reminiscent of the ITAM motif, this motif is first phosphorylated by a Src-family PTK, and then recruits the tandem SH2 domains of SHP1. This mechanism also serves to juxtapose SHP1 to its target molecules. The biology of ITIM-containing receptors is a new and promising field of study, that is likely to shed more light on the function of SHP1 and the dampening of immune responses. The autoimmune syndrome resulting from loss of SHP1 in the motheaten mouse sets a strong precedence for a possible role of disturbances in the ITIM-receptor/SHP1 system in human autoimmune conditions.

The hematopoietic protein tyrosine phosphatase, HePTP, is expressed in thymus, spleen, and in most T cell lines examined. The HePTP gene is located on chromosome 1 at q32.1, a site of frequent chromosomal deletions in leukemias and trisomy in myelodysplastic syndrome. HePTP is entirely cytosolic and consists of a single PTPase domain, which occupies the C-terminal 3/4 of the enzyme and is preceded by an ~80-amino acid non-catalytic N-terminus. Recent work has shown that HePTP forms a specific high stoichiometry complex with the MAP kinases ERK and p38 and regulates their catalytic activity (163, 164). These kinases play crucial roles in multiple signaling cascades that regulate gene expression (rev. in 165). In agreement with this notion, HePTP suppresses the transcriptional activation of the IL-2 gene and, presumably, many other genes involved in the immune response (166). Thus, HePTP is a negative regulator of T cell activation, and probably of many other aspects of T cell physiology that depend on MAP kinases, such as development in the thymus, survival and apoptosis.

The proline-, glutamic acid-, serine-, and threonine-enriched PTPase PEP is also expressed primarily in hematopoietic cells and was recently discovered to be physically associated with the 50-kDa cytosolic PTK Csk, an important suppressor of Src family kinases, including Lck and Fyn in T cells. Like HePTP, PEP also has an inhibitory effect on TCR-induced transcriptional activation of immediate early genes and the IL-2 gene (167). Catalytically inactive mutants of PEP had no effects, showing that PEP must dephosphorylate one or several cellular proteins to block signaling. PEP also reduced the activation of MAP kinases, including JNK, which is not regulated by HePTP, and reduced the TCR-induced increase in tyrosine phosphorylation of Lck. Our working hypothesis is that PEP inhibits TCR signaling in cooperation with Csk, and that both enzymes target Src-family kinases that mediate the initial events of TCR signaling. This notion places PEP at the very top of the signaling cascade at a crucial regulatory point, which critically affects the T cell response. Future experiments will reveal whether disturbances in PEP function are associated with human diseases.

Another group, often referred to as the "cytoskeletal" PTPases, are newcomers in the T cell activation field. Four enzymes of this group, PTPH1, PTP-MEG1, PTP36, and PTP-BAS are present in many T cells. All four enzymes are rather large and contain several protein-protein interaction domains and a C-terminal catalytic domain. They are primarily located at the plasma membrane and at least PTPH1 is a powerful negative regulator of TCR signaling (unpublished). In addition, PTP36 may regulate T cell maturation as it is mainly expressed in double-positive thymocytes and less in more mature cells. PTP-BAS again is involved in suppressing apoptosis-inducing signals through the Fas molecule by dephosphorylating some undefined signaling intermediate.

At the amino acid level, the low molecular weight PTPase LMPTP is distantly related to the other PTPases. Nevertheless, the crystallization of LMPTP revealed that the enzyme has a catalytic core that is very similar to that of other PTPases. The primary transcript of the *LMPTP* gene undergoes an alternative splicing event that leads to three productive mRNAs, encoding isoforms A, B, and C. The two former proteins are functional, while the C isoform is a smaller (15 kDa) protein devoid of PTPase activity and of unknown physiological significance (168). In support of a role in T cell physiology, LMPTP contains some phosphate on tyrosine in resting T cells, but is rapidly dephosphorylated (perhaps by itself) upon TCR triggering (169). While this response, as well as the sub-membraneous location of LMPTP are compatible with a role in TCR signaling, the exact function of LMPTP remains unknown.

Finally, the enzyme encoded by the *PTEN* tumor suppressor gene has the PTPase signature sequence, but was recently discovered to mainly dephosphorylate a non-protein substrate: the 3-position of the inositol ring of inositol phospholipids (rev. in 170). In T lymphocytes, PTEN is found in the cytoplasm, but clearly enriched at the plasma membrane where these lipids are located. Expression of PTEN induces rapid apoptosis of T cells (171), an effect that is counteracted by coexpression of constitutively active protein kinase B (also termed c-Akt). This kinase is normally activated by the same phospholipids that PTEN dephosphorylates. This and similar findings suggest that PTEN is an important regulator of inositol phospholipid phosphorylation at the 3-position of the inositol ring, and thereby affects the activity of protein kinase B, a positive regulator of cell survival. Many lymphomas, leukemias and other tumors have deletions or mutations in the *PTEN* gene, and typically display prolonged survival and resistance to receptor-induced cell death. Since antigen-induced cell death is an important mechanism for the downregulation of immune responses, alterations in PTEN expression or function would be expected to cause prolonged T cell activation and, potentially, facilitate the maintenance of autoimmune disease. Indeed mice heterozygous for the PTEN deletion were found to display lymphoimmunity, granulocytosis, autoantibodies and reduced Fas-mediated apoptosis (172)

ACKNOWLEDGMENT

Studies in the authors' laboratories were supported, in part, by U.S. Public Health Service grants CA35299, GM50819, AI-42244 (project 1), AI35603, AI41481 and AI40552, and by Gemini Science, Inc. This is publication number 326 from the La Jolla Institute for Allergy and Immunology.

References

1. Altman, A., K.M. Coggeshall, and T. Mustelin. 1990. Molecular events mediating T cell activation. *Adv. Immunol.* 48:227–360.
2. Mustelin, T. 1994. T cell antigen receptor signaling: three families of tyrosine kinases and a phosphatase. *Immunity* 1:351–356.
3. Weiss, A., and D.R. Littman. 1994. Signal transduction by lymphocyte antigen receptors. *Cell* 76:263–274.
4. Rudd, C.E., and T. Mustelin. 1999. Tyrosine Kinases, phosphatases and lymphoid adapters in T-cell activation. *CRC Crit. Rev. Immunol.* (in press).
5. Viola, A., S. Schroeder, Y. Sakakibara, and A. Lanzavecchia. 1999. T lymphocyte costimulation mediated by reorganization of membrane microdomains. *Science* 283:680–682.
6. Rudd, C.E., J.M. Trevillyan, J.D. Dasgupta, L.L. Wong, and S.F. Schlossman. 1988. The CD4 receptor is complexed in detergent lysates to a protein-tyrosine kinase (pp58) from human T lymphocytes. *Proc. Natl. Acad. Sci. USA* 85:5190–5194.
7. Mustelin, T. 1994. Src family tyrosine kinases in leukocytes. R.G. Landes Co., Austin, TX. 1–155 pp.
8. June, C.H., M.C. Fletcher, J.A. Ledbetter, G.L. Schieven, J.N. Siegel, A.F. Phillips, and L.E. Samelson. 1990. Inhibition of tyrosine phosphorylation prevents T cell receptor-mediated signal transduction. *Proc. Natl. Acad. Sci. USA* 87:7722–7726.
9. Mustelin, T., K.M. Coggeshall, N. Isakov, and A. Altman. 1990. Tyrosine phosphorylation is required for T cell receptor-mediated activation of phospholipase C. *Science* 247:1584–1587.
10. Mustelin, T., and A. Altman. 1989. Do CD4 and CD8 control T-cell activation via a specific tyrosine protein kinase? *Immunol. Today* 10:189–192.
11. Chan, A.C., M. Iwashima, C.W. Turck, and A. Weiss. 1992. ZAP-70: a 70 kd protein-tyrosine kinase that associates with the TCR zeta chain. *Cell* 71:649–662.
12. Neumeister Kersh, E., A.S. Shaw, and P.M. Allen. 1998. Fidelity of T cell activation through multistep T cell receptor ζ phosphorylation. *Science* 281:572–575.

13. Zenner, G., T. Vorherr, T. Mustelin, and P. Burn. 1996. Differential and multiple binding of signal transducing molecules to the ITAMs of the TCR-ζ chain. *J. Cell. Biochem.* 63:94–103.

14. Sloan-Lancaster, J., A.S. Shaw, J.B. Rothbard, and P.M. Allen. 1994. Partial T cell signaling: altered phospho-ζ and lack of Zap70 recruitment in APL-induced T cell anergy. *Cell* 79:913–922.

15. La Face, D.M., C. Couture, K. Anderson, G. Shih, J. Alexander, A. Sette, T. Mustelin, A. Altman, and H.M. Grey. 1997. Differential T cell signaling induced by antagonist peptide-MHC complexes and the associated phenotypic responses. *J. Immunol.* 158:2057–2064.

16. Taniguchi, T., T. Kobayashi, J. Kondo, K. Takahashi, H. Nakamura, J. Suzuki, K. Nagai, T. Yamada, S. Nakamura, and H. Yamamura. 1991. Molecular cloning of a porcine gene *syk* that encodes a 72-kDa protein-tyrosine kinase showing high susceptibility to proteolysis. *J. Biol. Chem.* 266:15790–15796.

17. Kolanus, W., C. Romero, and B. Seed. 1993. T cell activation by clustered tyrosine kinases. *Cell* 74:171–183.

18. Couture, C., G. Baier, A. Altman, and T. Mustelin. 1994. p56^{lck}-independent activation and tyrosine phosphorylation of p72^{syk} by TCR/CD3 stimulation. *Proc. Natl. Acad. Sci. USA* 91:5301–5305.

19. Deckert, M., S. Tartare Deckert, C. Couture, T. Mustelin, and A. Altman. 1996. Functional and physical interactions of Syk family kinases with the Vav proto-oncogene product. *Immunity* 5:591–604.

20. Arpaia, E., S.M., H. Dadi, A. Cohen, and C.M. Roifman. 1994. Defective T cell receptor signaling and CD8$^+$ thymic selection in humans lacking Zap-70 kinase. *Cell* 76:947–958.

21. Elder, M.E., D. Lin, J. Clever, A.C. Chan, T.J. Hope, A. Weiss, and T.G. Parslow. 1994. Human severe combined immunodeficiency due to a defect in ZAP-70, a T cell tyrosine kinase. *Science* 264:1596–1599.

22. Negishi, I., N. Motoyama, K. Nakayama, K. Nakayama, S. Senju, S. Hatakeyama, Q. Zhang, A.C. Chan, and D.Y. Loh. 1995. Essential role for ZAP-70 in both positive and negative selection of thymocytes. *Nature* 376:435–438.

23. Cheng, A.M., B. Rowley, W. Pao, A. Hayday, J.B. Bolen, and T. Pawson. 1995. Syk tyrosine kinase required for mouse viability and B-cell development. *Nature* 378:303–306.

24. Turner, M., P.J. Mee, P.S. Costello, O. Williams, A.A. Price, L.P. Duddy, M.T. Furlong, R.L. Geahlen, and V.L. Tybulewicz. 1995. Perinatal lethality and blocked B-cell development in mice lacking the tyrosine kinase Syk. *Nature* 378:298–302.

25. Mallick-Wood, C.A., W. Pao, A.M. Cheng, J.M. Lewis, S. Kulkarni, J.B. Bolen, B. Rowley, R.E. Tigelaar, T. Pawson, and A.C. Hayday. 1996. Disruption of epithelial gd T cell repertoires by mutation of the Syk tyrosine kinase. *Proc. Natl. Acad. Sci. USA* 93:9704–9709.

26. Cheng, A.M., I. Negishi, S.J. Anderson, A.C. Chan, J. Bolen, D.Y. Loh, and T. Pawson. 1997. The Syk and ZAP-70 SH2-containing tyrosine kinases are implicated in pre- T cell receptor signaling. *Proc. Natl. Acad. Sci. USA* 94:9797–9801.

27. Fargnoli, M.C., R.L. Edelson, C.L. Berger, S. Chimenti, C. Couture, T. Mustelin, and R. Halaban. 1997. TCR diminished signaling in cutaneous T-cell lymphoma is associated with decreased kinase activities of Syk and membrane-associated Csk. *Leukemia* 11:1338–1346.

28. Kisielow, P., and H. von Boehmer. 1995. Development and selection of T cells: facts and puzzles. *Adv. Immunol.* 58:87–209.

29. Alberola-lla, J., S. Takaki, J.D. Kerner, and R.M. Perlmutter. 1997. Differential signaling by lymphocyte antigen receptors. *Annu. Rev. Immunol.* 15:125–154.

30. Jameson, S.C., and M.J. Bevan. 1998. T-cell selection. *Curr. Opin. Immunol.* 10:214–219.

31. Janeway, C.A., Jr. 1995. Ligands for the T-cell receptor: hard times for avidity models. *Immunol. Today* 16:223–225.

32. McKeithan, T.W. 1995. Kinetic proofreading in T-cell receptor signal transduction. *Proc. Natl. Acad. Sci. USA* 92:5042–5046.

33. Rabinowitz, J.D., C. Beeson, D.S. Lyons, M.M. Davis, and H.M. McConnell. 1996. Kinetic discrimination in T-cell activation. *Proc. Natl. Acad. Sci. USA* 93:1401–1405.

34. Swan, K.A., J. Alberola-Ila, J.A. Gross, M.W. Appleby, K.A. Forbush, J.F. Thomas, and R.M. Perlmutter. 1995. Involvement of p21^{ras} distinguishes positive and negative selection in thymocytes. *EMBO J.* 14:276–285.

35. Plas, D.R., C.B. Williams, G.J. Kersh, L.S. White, J.M. White, S. Paust, T. Ulyanova, P.M. Allen, and M.L. Thomas. 1999. The tyrosine phosphatase SHP-1 regulates thymocyte positive selection. *J. Immunol.* 162:5680–5684.

36. Lorenz, U., K.S. Ravichandran, D. Pei, C.T. Walsh, S.J. Burakoff, and B.G. Neel. 1994. Lck-dependent tyrosyl phosphorylation of the phosphotyrosine phosphatase SH-PTP1 in murine T cells. *Mol. Cell. Biol.* 14:1824–1834.

37. Pani, G., K.D. Fischer, I. Mlinaric Rascan, and K.A. Siminovitch. 1996. Signaling capacity of the T cell antigen receptor is negatively regulated by the PTP1C tyrosine phosphatase. *J. Exp. Med.* 184:839–852.

38. Van Seventer, G.A., Y. Shimizu, and S. Shaw. 1991. Roles of multiple accessory molecules in T-cell activation. *Curr. Opin. Immunol.* 3:294–303.

39. Croft, M., and C. Dubey. 1997. Accessory molecule and costimulation requirements for CD4 T cell response. *Crit. Rev. Immunol.* 17:89–118.

40. Lenschow, D.J., T.L. Walunas, and J.A. Bluestone. 1996. CD28/B7 system of T cell costimulation. *Annu. Rev. Immunol.* 14:233–258.

41. Rudd, C.E. 1996. Upstream-downstream: CD28 cosignaling pathways and T cell function. *Immunity* 4:527–534.

42. Chambers, C.A., and J.P. Allison. 1997. Co-stimulation in T cell responses. *Curr. Opin. Immunol.* 9:396–404.

43. Schwartz, R.H. 1997. T cell clonal anergy. *Curr. Opin. Immunol.* 9:351–357.

44. Boise, L.H., A.J. Minn, P.J. Noel, C.H. June, M.A. Accavitti, T. Lindsten, and C.B. Thompson. 1995. CD28 costimulation can promote T cell survival by enhancing the expression of Bcl-XL. *Immunity* 3:87–98.

45. Harding, F.A., J.G. McArthur, J.A. Gross, D.H. Raulet, and J.P. Allison. 1992. CD28-mediated signalling co-stimulates murine T cells and prevents induction of anergy in T-cell clones. *Nature* 356:607–609.

46. Allison, J.P. 1994. CD28-B7 interactions in T-cell activation. *Curr. Opin. Immunol.* 6:414–419.

47. Hutchcroft, J.E., and B.E. Bierer. 1996. Signaling through CD28/CTLA-4 family receptors. Puzzling participation of phosphatidylinositol-3 kinase. *J. Immunol.* 156:4071–4074.

48. Chambers, C.A., M.F. Krummel, B. Boitel, A. Hurwitz, T.J. Sullivan, S. Fournier, D. Cassell, M. Brunner, and J.P. Allison. 1996. The role of CTLA-4 in the regulation and initiation of T-cell responses. *Immunol. Rev.* 153:27–46.

49. Linsley, P.S., and P. Golstein. 1996. Lymphocyte activation: T-cell regulation by CTLA-4. *Curr. Biol.* 6:398–400.

50. Thompson, C.B., and J.P. Allison. 1997. The emerging role of CTLA-4 as an immune attenuator. *Immunity* 7:445–450.

51. Tivol, E.A., A.N. Schweitzer, and A.H. Sharpe. 1996. Costimulation and autoimmunity. *Curr. Opin. Immunol.* 8:822–830.

52. Krummel, M.F., and J.P. Allison. 1995. CD28 and CTLA-4 have opposing effects on the response of T cells to stimulation. *J. Exp. Med.* 182:459–465.

53. Tivol, E.A., F. Borriello, A.N. Schweitzer, W.P. Lynch, J.A. Bluestone, and A.H. Sharpe. 1995. Loss of CTLA-4 leads to massive lymphoproliferation and fatal multiorgan tissue destruction, revealing a critical negative regulatory role of CTLA-4. *Immunity* 3:531–539.

54. Waterhouse, P., J.M. Penninger, E. Timms, A. Wakeham, A. Shahinian, K.P. Lee, C.B. Thompson, H. Griesser, and T.W. Mak. 1995. Lymphoproliferative disorders with early lethality in mice deficient in *Ctla-4. Science* 270:985–988.

55. Guerder, S., and R.A. Flavell. 1995. Costimulation in tolerance and autoimmunity. *Int. Rev. Immunol.* 13:135–146.

56. Harlan, D.M., R. Abe, K.P. Lee, and C.H. June. 1995. Potential roles of the B7 and CD28 receptor families in autoimmunity and immune evasion. *Clin. Immunol. Immunopathol.* 75:99–111.

57. Daikh, D., D. Wofsy, and J.B. Imboden. 1997. The CD28-B7 costimulatory pathway and its role in autoimmune disease. *J. Leuk. Biol.* 62:156–162.

58. Weintraub, J.P., and P.L. Cohen. 1999. Ectopic expression of B7-1 (CD80) on T lymphocytes in autoimmune lpr and gld mice. *Clin. Immunol.* 91:302–309.

59. Perkins, D.L. 1998. T-cell activation in autoimmune and inflammatory diseases. *Curr. Opin. Nephrol. Hypertens.* 7:297–303.

60. August, A., S. Gibson, Y. Kawakami, T. Kawakami, G.B. Mills, and B. Dupont. 1994. CD28 is associated with and induces the immediate tyrosine phosphorylation and activation of the Tec family kinase ITK/EMT in the human Jurkat leukemic T-cell line. *Proc. Natl. Acad. Sci. USA* 91:9347–9351.

61. Prasad, K.V., Y.C. Cai, M. Raab, B. Duckworth, L. Cantley, S.E. Shoelson, and C.E. Rudd. 1994. T-cell antigen CD28 interacts with the lipid kinase phosphatidylinositol 3-kinase by a cytoplasmic Tyr(P)-Met-Xaa-Met motif. *Proc. Natl. Acad. Sci. USA* 91:2834–2838.

62. Truitt, K.E., C.M. Hicks, and J.B. Imboden. 1994. Stimulation of CD28 triggers an association between CD28 and phosphatidylinositol 3-kinase in Jurkat T cells. *J. Exp. Med.* 179:1071–1076.

63. Schneider, H., Y.C. Cai, K.V. Prasad, S.E. Shoelson, and C.E. Rudd. 1995. T cell antigen CD28 binds to the GRB-2/SOS complex, regulators of p21ras. *Eur. J. Immunol.* 25:1044–1050.

64. Kim, H.H., M. Tharayil, and C.E. Rudd. 1998. Growth factor receptor-bound protein 2 SH2/SH3 domain binding to CD28 and its role in co-signaling. *J. Biol. Chem.* 273:296–301.

65. Okkenhaug, K., and R. Rottapel. 1998. Grb2 forms an inducible protein complex with CD28 through a Src homology 3 domain-proline interaction. *J. Biol. Chem.* 273:21194–21202.

66. Marengere, L.E., K. Okkenhaug, A. Clavreul, D. Couez, S. Gibson, G.B. Mills, T.W. Mak, and R. Rottapel. 1997. The SH3 domain of Itk/Emt binds to proline-rich sequences in the cytoplasmic domain of the T cell costimulatory receptor CD28. *J. Immunol.* 159:3220–3229.

67. Yang, W.C., M. Ghiotto, B. Barbarat, and D. Olive. 1999. The role of Tec protein-tyrosine kinase in T cell signaling. *J. Biol. Chem.* 274:607–617.

68. Nunes, J.A., Y. Collette, A. Truneh, D. Olive, and D.A. Cantrell. 1994. The role of p21ras in CD28 signal transduction: triggering of CD28 with antibodies, but not the ligand B7-1, activates p21ras. *J. Exp. Med.* 180:1067–1076.

69. Nunes, J.A., A. Truneh, D. Olive, and D.A. Cantrell. 1996. Signal transduction by CD28 costimulatory receptor on T cells. B7-1 and B7-2 regulation of tyrosine kinase adaptor molecules. *J. Biol. Chem.* 271:1591–1598.

70. Carpino, N., D. Wisniewski, A. Strife, D. Marshak, R. Kobayashi, B. Stillman, and B. Clarkson. 1997. p62dok: a constitutively tyrosine-phosphorylated, GAP-associated protein in chronic myelogenous leukemia progenitor cells. *Cell* 88:197–204.

71. Yamanashi, Y., and D. Baltimore. 1997. Identification of the Abl- and rasGAP-associated 62 kDa protein as a docking protein, Dok. *Cell* 88:205–211.

72. Su, B., E. Jacinto, M. Hibi, T. Kallunki, M. Karin, and Y. Ben Neriah. 1994. JNK is involved in signal integration during costimulation of T lymphocytes. *Cell* 77:727–736.

73. Faris, M., N. Kokot, L. Lee, and A.E. Nel. 1996. Regulation of interleukin-2 transcription by inducible stable expression of dominant negative and dominant active mitogen-activated protein kinase kinase kinase in jurkat T cells. Evidence for the importance of Ras in a pathway that is controlled by dual receptor stimulation. *J. Biol. Chem.* 271:27366–27373.

74. Hu, Q., A. Klippel, A.J. Muslin, W.J. Fantl, and L.T. Williams. 1995. Ras-dependent induction of cellular responses by constitutively active phosphatidylinositol-3 kinase. *Science* 268:100–102.

75. Rodriguez-Viciana, P., P.H. warne, R. Dhand, B. Vanhaesebroeck, I. Gout, M.J. Fry, M.D. Waterfield, and J. Downward. 1994. Phosphatidylinositol-3-OH kinase as a direct target of Ras. *Nature* 370:527–532.

76. Symons, M. 1995. Rho family GTPases: the cytoskeleton and beyond. *Trends. Biochem. Sci.* 21:178–181.

77. Hall, A. 1998. Rho GTPases and the actin cytoskeleton. *Science* 279:509–514.

78. Reif, K., and D.A. Cantrell. 1998. Networking Rho Family GTPases in Lymphocytes. *Immunity* 8:395–401.

79. Altman, A., and M. Deckert. 1998. The functions of small Ras-like GTPases in signaling by hematopoietic cell immune recognition receptors. *Adv. Immunol.* 72:1–101.

80. Collins, T., M. Deckert, and A. Altman. 1997. Views on Vav. *Immunol. Today* 18:221–225.

81. Crespo, P., K.E. Schuebel, A.A. Ostrom, J.S. Gutkind, and X.R. Bustelo. 1997. Phosphotyrosine-dependent activation of Rac-1 GDP/GTP exchange by the *vav* proto-oncogene product. *Nature* 385:169–172.

82. Kaga, S., S. Ragg, K.A. Rogers, and A. Ochi. 1997. Stimulation of CD28 with B7–2 promotes focal adhesion-like contacts where Rho family small G proteins accumulate in T cells. *J. Immunol.* 160:24–27.

83. Coso, O.A., M. Chiariello, J.C. Yu, H. Teramoto, P. Crespo, N. Xu, T. Miki, and J.S. Gutkind. 1995. The small GTP-binding proteins Rac1 and Cdc42 regulate the activity of the JNK/SAPK signaling pathway. *Cell* 81:1137–1146.

84. Minden, A., A. Lin, F.X. Claret, A. Abo, and M. Karin. 1995. Selective activation of the JNK signaling cascade and c-Jun transcriptional activity by the small GTPases Rac and Cdc42Hs. *Cell* 81:1147–1157.

85. Olson, M.F., A. Ashworth, and A. Hall. 1995. An essential role for Rho, Rac, and Cdc42 GTPases in cell cycle progression through G1. *Science* 269:1270–1272.

86. Villalba, M., N. Coudronniere, M. Deckert, E. Texieivo, P. Mas, A. Altman. 2000. A novel functional interaction between Vav and PKCθ is required for TCR-induced T cell activation. *Immunity* 12:151–160.

87. Reif, K., C.D. Nobes, G. Thomas, A. Hall, and D.A. Cantrell. 1996. Phosphatidylinositol 3-kinase signals activate a selective subset of Rac/Rho-dependent effector pathways. *Currr. Biol.* 6:1445–1455.

88. Baeuerle, P.A., and D. Baltimore. 1996. NF-κB: ten years after. *Cell* 87:13–20.

89. Baldwin, A.S., Jr. 1996. The NF-κB and IκB proteins: new discoveries and insights. *Annu. Rev. Immunol.* 14:649–683.

90. Karin, M. 1997. NF-κB: a pivotal transcription factor in chronic inflammatory disease. *N. Engl. J. Med.* 336:1066–1071.

91. DiDonato, J.A., M. Hayakawa, D.M. Rothwarf, E. Zandi, and M. Karin. 1997. A cytokine-responsive IκB kinase that activates the transcription factor NF-κB. *Nature* 388:548–554.

92. Malinin, N.L., M.P. Boldin, A.V. Kovalenko, and D. Wallach. 1997. MAP3K-related kinase involved in NF-κB induction by TNF, CD95 and IL-1. *Nature* 385:540–544.

93. Mercurio, F., H. Zhu, B.W. Murray, A. Shevchenko, B.L. Bennett, J. Li, D.B. Young, M. Barbosa, M. Mann, A. Manning, and A. Rao. 1997. IKK-1 and IKK-2: Cytokine-activated IκB kinases essential for NF-κB activation. *Science* 278:860–866.

94. Woronicz, J.D., X. Gao, M. Rothe, and D.V. Goeddel. 1997. IκB kinase-β: NF-κB activation and complex formation with IκB kinase-α and NIK. *Science* 278:866–869.

95. Zandi, E., D.M. Rothwarf, M. Delhase, M. Hayakawa, and M. Karin. 1997. The IκB kinase complex (IKK) contains two kinase subunits, IKKα and IKKβ, necessary for IκB phosphorylation and NF-κB activation. *Cell* 91:243–252.

96. Fraser, J.D., B.A. Irving, G.R. Crabtree, and A. Weiss. 1991. Regulation of interleukin-2 gene enhancer activity by the T cell accessory molecule CD28. *Science* 251:313–316.

97. Crabtree, G.R., and N.A. Clipstone. 1994. Signal transmission between the plasma membrane and nucleus of T lymphocytes. *Annu. Rev. Biochem.* 63:1045–1083.

98. Jain, J., C. Loh, and A. Rao. 1995. Transcriptional regulation of the IL-2 gene. *Curr. Opin. Immunol.* 7:333–342.

99. Lai, J.H., G. Horvath, Y. Li, and T.H. Tan. 1995. Mechanisms of enhanced nuclear translocation of the transcription factors c-Rel and NF-κB by CD28 costimulation in human T lymphocytes. *Ann. NY. Acad. Sci.* 766:220–223.

100. Shapiro, V.S., K.E. Truitt, J.B. Imboden, and A. Weiss. 1997. CD28 mediates transcriptional upregulation of the interleukin-2 (IL-2) promoter through a composite element containing the CD28RE and NF-IL-2B AP-1 sites. *Mol. Cell. Biol.* 17:4051–4058.

101. McGuire, K.L., and M. Iacobelli. 1997. Involvement of Rel, Fos, and Jun proteins in binding activity to the IL-2 promoter CD28 response element/AP-1 sequence in human T cells. *J. Immunol.* 159:1319–1327.

102. Kempiak, S.J., T.S. Hiura, and A.E. Nel. 1999. The Jun kinase cascade is responsible for activating the CD28RE element of the IL-2 promoter: proof of cross-talk with the IκB kinase cascade. *J. Immunol.* 162:3176–3187.

103. Lin, X., E.T.J. Cunningham, Y. Mu, R. Geleziunas, and W.C. Greene. 1999. The proto-oncogene Cot kinase participates in CD3/CD28 induction of NF-κB acting through the NF-κB-inducing kinase and IκB kinases. *Immunity* 10:271–280.

104. Simons, K., and E. Ikonen. 1997. Functional rafts in cell membranes. *Nature* 387:569–572.

105. Wülfing, C., and M.M. Davis. 1998. A receptor/cytoskeletal movement triggered by costimulation during T cell activation. *Science* 282:2266–2269.

106. Marengere, L.E., P. Waterhouse, G.S. Duncan, H.W. Mittrucker, G.S. Feng, and T.W. Mak. 1996. Regulation of T cell receptor signaling by tyrosine phosphatase SYP association with CTLA-4. *Science* 272:1170–1173.

107. Lee, K.M., E. Chuang, M. Griffin, R. Khattri, D.K. Hong, W. Zhang, D. Straus, L.E. Samelson, C.B. Thompson, and J.A. Bluestone. 1998. Molecular basis of T cell inactivation by CTLA-4. *Science* 282:2263–2266.

108. Cambier, J.C. 1997. Inhibitory receptors abound? *Proc. Natl. Acad. Sci. USA* 94:5993–5995.

109. Chambers, C.A., and J.P. Allison. 1996. The role of tyrosine phosphorylation and PTP-1C in CTLA-4 signal transduction. *Eur. J. Immunol.* 26:3224–3229.

110. Shahinian, A., K. Pfeffer, K.P. Lee, T.M. Kundig, K. Kishihara, K. Wakeham, K. Kawai, P.S. Ohashi, C.B. Thompson, and T.W. Mak. 1993. Differential T cell costimulatory requirements in CD28-deficient mice. *Science* 261:609–612.

111. Watts, T.H., and M.A. DeBenedette. 1999. T cell co-stimulatory molecules other than CD28. *Curr. Opin. Immunol.* 11:286–293.

112. Weinberg, A.D., A.T. Vella, and M. Croft. 1998. OX-40: life beyond the effector T cell stage. *Semin. Immunol.* 10:471–480.

113. Kwon, B., K.Y. Youn, and B.S. Kwon. 1999. Functions of newly identified members of the tumor necrosis factor receptor/ligand superfamilies in lymphocytes. *Curr. Opin. Immunol.* 11:340–345.

114. Gravestein, L.A., and J. Borst. 1998. Tumor necrosis factor receptor family members in the immune system. *Semin. Immunol.* 10:423–434.

115. Arch, R.H., and C.B. Thompson. 1998. 4-1BB and Ox40 are members of a tumor necrosis factor (TNF)-nerve growth factor receptor subfamily that bind TNF receptor-associated factors and activate nuclear factor κB. *Mol. Cell. Biol.* 18:558–565.

116. Kawamata, S., T. Hori, A. Imura, A. Takaori-Kondo, and T. Uchiyama. 1998. Activation of OX40 signal transduction pathways leads to tumor necrosis factor receptor-associated factor (TRAF) 2- and TRAF5-mediated NF-κB activation. *J. Biol. Chem.* 273:5808–5814.

117. Gramaglia, I., A.D. Weinberg, M. Lemon, and M. Croft. 1998. Ox-40 ligand: a potent costimulatory molecule for sustaining primary CD4 T cell responses. *J. . Immunol.* 161:6510–6517.

118. Oshima, Y., Y. Tanaka, H. Tozawa, Y. Takahashi, C. Maliszewski, and G. Delespesse. 1997. Expression and function of OX40 ligand on human dendritic cells. *J. Immunol.* 159:3838–3848.

119. Stuber, E., and W. Strober. 1996. The T cell-B cell interaction via OX40-OX40L is necessary for the T cell-dependent humoral immune response. *J. Exp. Med.* 183:979–989.

120. Weinberg, A.D., M. Lemon, A.J. Jones, M. Vainiene, B. Celnik, A.C. Buenafe, N. Culbertson, A. Bakke, A.A. Vandenbark, and H. Offner. 1996. OX-40 antibody enhances for autoantigen specific Vβ 8.2+ T cells within the

120. spinal cord of Lewis rats with autoimmune encephalo-myelitis. *J. Neurosci. Res.* 43:42–49.

121. Buenafe, A.C., R.C. Tsu, B.J. Bebo, A.C. Bakke, A.A. Vandenbark, and H. Offner. 1997. A TCR V*α* CDR3-specific motif associated with Lewis rat autoimmune encephalomyelitis and basic protein-specific T cell clones. J. Immunol. 158:5472–5483.

122. Weinberg, A.D., D.N. Bourdette, T.J. Sullivan, M. Lemon, J.J. Wallin, R. Maziarz, M. Davey, F. Palida, W. Godfrey, E. Engleman, R.J. Fulton, H. Offner, and A.A. Vandenbark. 1996. Selective depletion of myelin-reactive T cells with the anti-OX-40 antibody ameliorates autoimmune encephalomyelitis. *Nature Med.* 2:183–189.

123. Weinberg, A.D., K.W. Wegmann, C. Funatake, and R.H. Whitham. 1999. Blocking OX-40/OX-40 ligand interaction *in vitro* and *in vivo* leads to decreased T cell function and amelioration of experimental allergic encephalomyelitis. *J. Immunol.* 162:1818–1826.

124. Higgins, L.M., S.A. McDonald, N. Whittle, N. Crockett, J.G. Shields, and T.T. MacDonald. 1999. Regulation of T cell activation *in vitro* and *in vivo* by targeting the OX40-OX40 ligand interaction: amelioration of ongoing inflammatory bowel disease with an OX40-IgG fusion protein, but not with an OX40 ligand-IgG fusion protein. *J. Immunol.* 162:486–493.

125. Altman, A., A.N. Theofilopoulos, R. Weiner, D.H. Katz, and F.J. Dixon. 1981. Analysis of T cell function in autoimmune murine strains. Defects in production and responsiveness to interleukin. *J. Exp. Med.* 154:791–808.

126. Altman, A. 1994. Abnormal antigen receptor-initiated signal transduction in lpr T lymphocytes. *Semin. Immunol.* 6:9–17.

127. Nagata, S. 1997. Apoptosis by death factor. *Cell* 88:355–365.

128. Budd, R.C., and P.F. Mixter. 1995. The origin of CD4⁻CD8⁻ TCR*αβ*⁺ thymocytes: a model based on T-cell receptor avidity. *Immunol. Today* 16:428–431.

129. Suda, T., and S. Nagata. 1997. Why do defects in the Fas-Fas ligand system cause autoimmunity? *J. Allergy Clin. Immunol.* 100, Pt 2:S97–101.

130. Samelson, L.E., W.F. Davidson, H.C.I. Morse, and R.D. Klausner. 1986. Abnormal tyrosine phosphorylation on T-cell receptor in lymphoproliferative disorders. *Nature* 324:674–676.

131. Katagiri, T., K. Urakawa, Y. Yamanashi, K. Semba, T.T., K. Toyoshima, T. Yamamoto, and K. Kano. 1989. Overexpression of *src* family gene for tyrosine-kinase p59ᶠʸⁿ in CD4-CD8- T cells of mice with a lymphoproliferative disorder. *Proc. Natl. Acad. Sci. USA* 86:10064–10068.

132. Scholz, W., N. Isakov, M.I. Mally, A.N. Theofilopoulos, and A. Altman. 1988. Lpr T cell hyporesponsiveness to mitogens linked to deficient receptor-stimulated phosphoinositide hydrolysis. *J. Biol. Chem.* 263:3626–3631.

133. Balomenos, D., R. Rumold, and A.N. Theofilopoulos. 1997. The proliferative in vivo activities of lpr double-negative T cells and the primary role of p59ᶠʸⁿ in their activation and expansion. *J. Immunol.* 159:2265–2273.

134. Clements, J.L., J. Wolfe, S.M. Cooper, and R.C. Budd. 1994. Reversal of hyporesponsiveness in lpr CD4⁻CD8⁻ T cells is achieved by induction of cell cycling and normalization of CD2 and p59ᶠʸⁿ expression. *Eur. J. Immunol.* 24:558–565.

135. Duan, J.M., R. Fagard, and M.P. Madaio. 1996. Abnormal signal transduction through CD4 leads to altered tyrosine phosphorylation in T cells derived from MRL-lpr/lpr mice. *Autoimmunity* 23:231–243.

136. Clements, J.L., S.M. Cooper, and R.C. Budd. 1995. Abnormal regulation of the IL-2 promoter in lpr CD4⁻CD8⁻ T lymphocytes results in constitutive expression of a novel nuclear factor of activated T cells-binding factor. *J. Immunol.* 154:6372–6381.

137. Bach, J.F., and D. Mathis. 1997. The NOD mouse. *Res. Immunol.* 148:285–286.

138. Cohen, I.R. 1997. Questions about NOD mouse diabetes. *Res. Immunol.* 148:286–291.

139. Delovitch, T.L., and B. Singh. 1997. The nonobese diabetic mouse as a model of autoimmune diabetes: immune dysregulation gets the NOD. *Immunity* 7:727–738.

140. Salojin, K.V., J. Zhang, J. Madrenas, and T.L. Delovitch. 1998. T-cell anergy and altered T-cell receptor signaling: effects on autoimmune disease. *Immunol. Today* 19:468–473.

141. Liu, Y.-C., and A. Altman. 1998. Cbl: Complex formation and functional implications. *Cell. Signal.* 10:377–385.

142. Rapoport, M.J., A.H. Lazarus, A. Jaramillo, E. Speck, and T.L. Delovitch. 1993. Thymic T cell anergy in autoimmune nonobese diabetic mice is mediated by deficient T cell receptor regulation of the pathway of p21ʳᵃˢ activation. *J. Exp. Med.* 177:1221–1226.

143. Salojin, K., J. Zhang, M. Cameron, B. Gill, G. Arreaza, A. Ochi, and T.L. Delovitch. 1997. Impaired plasma membrane targeting of Grb2-murine son of sevenless (mSOS) complex and differential activation of the Fyn-T cell receptor (TCR)——Cbl pathway mediate the T cell hyporesponsiveness in autoimmune nonobese diabetic mice. *J. Exp. Med.* 186:887–897.

144. Zhang, J., K. Salojin, and T.L. Delovitch. 1998. Sequestration of CD4-associated Lck from the TCR complex may elicit T cell hyporesponsiveness in nonobese diabetic mice. *J. Immunol.* 160:1148–1157.

145. Mustelin, T.M., and A. Altman. 1993. signal transduction pathways in T lymphocytes and their relevance to autoimmunity. *In* The Molecular Pathology of Autoimmunity. (C.A. Bona, K.A. Siminovitch, M. Zanetti and A.N. Theofilopoulos, editors). Harwood Academic Publishers, New York, NY. pp. 137–147.

146. Vassilopoulos, D., B. Kovacs, and G.C. Tsokos. 1995. TCR/CD3 complex-mediated signal transduction pathway in T cells and T cell lines from patients with systemic lupus erythematosus. *J. Immunol.* 155:2269–2281.

147. Liossis, S.N., X.Z. Ding, G.J. Dennis, and G.C. Tsokos. 1998. Altered pattern of TCR/CD3-mediated protein-tyrosyl phosphorylation in T cells from patients with systemic lupus erythematosus. Deficient expression of the T cell receptor *ζ* chain. *J. Clin. Invest.* 101:1448–1457.

148. Tsokos, G.C., and S.N. Liossis. 1999. Immune cell signaling defects in lupus: activation, anergy and death. *Immunol. Today* 20:119–124.

149. Puck, J.M., and M.C. Sneller. 1997. ALPS: an autoimmune human lymphoproliferative syndrome associated with abnormal lymphocyte apoptosis. *Semin. Immunol.* 9:77–84.

150. Straus, S.E., M. Sneller, M.J. Lenardo, J.M. Puck, and W. Strober. 1999. An inherited disorder of lymphocyte apoptosis: the autoimmune lymphoproliferative syndrome. *Ann. Int. Med.* 130:591–601.

151. Fisher, G.H., F.J. Rosenberg, S.E. Straus, J.K. Dale, L.A. Middleton, A.Y. Lin, W. Strober, M.J. Lenardo, and J.M. Puck. 1995. Dominant interfering Fas gene mutations impair apoptosis in a human autoimmune lymphoproliferative syndrome. *Cell* 81:935–946.

152. Wang, J., L. Zheng, A. Lobito, F.K. Chan, J. Dale, M. Sneller, X. Yao, J.M. Puck, S.E. Straus, and M.J. Lenardo. 1999. Inherited human Caspase 10 mutations underlie defective lymphocyte and dendritic cell apoptosis in autoimmune lymphoproliferative syndrome type II. *Cell* 98:47–58.

153. Fuss, I.J., W. Strober, J.K. Dale, S. Fritz, G.R. Pearlstein, J.M. Puck, M.J. Lenardo, and S.E. Straus. 1997. Characteristic T helper 2 T cell cytokine abnormalities in autoimmune lymphoproliferative syndrome, a syndrome marked by defective apoptosis and humoral autoimmunity. *J. Immunol.* 158:1912–1918.

154. Iivanainen, A.V., C. Lindqvist, T. Mustelin, and L.C. Andersson. 1990. Phosphotyrosine phosphatases are involved in reversion of T lymphoblastic proliferation. *Eur. J. Immunol.* 20:2509–2512.

155. Mustelin, T., J. Brockdorff, A. Gjörloff-Wingren, P. Tailor, S. Han, X. Wang, and M. Saxena. 1998. T cell activation: The coming of the phosphatases. *Frontiers in Bioscience* 3:1060–1096.

156. Mustelin, T., J. Brockdorff, L. Rudbeck, A. Gjörloff-Wingren, S. Han, X. Wang, P. Tailor, and M. Saxena. 1999. The next wave: Protein tyrosine phosphatases enter T cell antigen receptor signaling. *Cell Signaling*, 11:637–650.

157. Kozlowski, M., I. Mlinaric-Rascan, G.-S. Feng, R. Shen, T. Pawson, and K.A. Siminovitch. 1993. Expression and catalytic activity of the tyrosine phosphatase PTP1C is severely impaired in motheaten and viable motheaten mice. *J. Exp. Med.* 178:2157–2163.

158. Tsui, H.W., K.A. Siminovitch, L. de Souza, and F.W.L. Tsui. 1993. Motheaten and viable motheaten mice have mutations in the haematopoietic cell phosphatase gene. *Nature Genetics* 4:124–129.

159. Elchebly, M., P. Payette, E. Michaliszyn, W. Cromlish, S. Collins, A.L. Loy, D. Normandin, A. Cheng, J. Himms-Hagen, C.-C. Chan, C. Ramachandran, M.J. Gresser, M.L. Tremblay, and B.P. Kennedy. 1999. Increased insulin sensitivity and obesity resistance in mice lacking the protein tyrosine phosphatase-1B gene. *Science* 283:1544–1548.

160. Mustelin, T., K.M. Coggeshall, and A. Altman. 1989. Rapid activation of the T-cell tyrosine protein kinase pp56*lck* by the CD45 phosphotyrosine phosphatase. *Proc. Natl. Acad. Sci. USA* 86:6302–6306.

161. Mustelin, T., and P. Burn. 1993. Regulation of src family tyrosine kinases in lymphocytes. *Trends Biochem. Sci.* 18:215–220.

162. Brockdorff, J., S. Williams, C. Couture, and T. Mustelin. 1999. Dephosphorylation of ZAP-70 and inhibition of T cell activation by activated SHP1. *Eur. J. Immunol.*, 29:2539–2550.

163. Saxena, M., S. Williams, J. Brockdorff, J. Gilman, and T. Mustelin. 1999. Inhibition of T cell signaling by MAP kinase-targeted hematopoietic tyrosine phosphatase (HePTP). *Biol. Chem.* 274:11693–11700.

164. Saxena, M., S. Williams, K. Taskén, and T. Mustelin. 1999. Crosstalk between cAMP-dependent kinase and MAP kinase through hematopoietic protein tyrosine phosphatase (HePTP). *Nature Cell. Biol.*, 1:305–311.

165. Treisman, R. 1996. Regulation of transcription by MAP kinase cascades. *Curr. Opin. Cell. Biol.* 8:205–215.

166. Saxena, M., S. Williams, J. Gilman, and T. Mustelin. 1998. Negative regulation of T cell antigen receptor signal transduction by hematopoietic tyrosine phosphatase (HePTP). *J. Biol. Chem.* 273:15340–15344.

167. Gjörloff-Wingren, A., M. Saxena, S. Williams, D. Hammi, and T. Mustelin. 1999. Characterization of TCR-induced receptor-proximal signaling events negatively regulated by the protein tyrosine phosphatase PEP. *Eur. J. Immunol.*, 29:3845–3854.

168. Tailor, P., J. Gilman, S. Williams, and T. Mustelin. 1999. A novel isoform of the low molecular weight protein tyrosine phosphatase, LMPTP-C, arising from alternative mRNA splicing. *Eur. J. Biochem.*, 262:277–282.

169. Tailor, P., J. Gilman, S. Williams, C. Couture, and T. Mustelin. 1997. Regulation of the low molecular weight phosphotyrosine phosphatase by phosphorylation at tyrosines 131 and 132. *J. Biol. Chem.* 272:5371–5374.

170. Maehama, T., and J.E. Dixon. 1999. PTEN: a tumour suppressor that functions as a phospholipid phosphatase. *Trends Cell. Biol.* 9:125–128.

171. Wang, X., A. Gjörloff-Wingren, M. Saxena, N. Pathan, J.C. Reed, and T. Mustelin. 1999. The tumor suppressor PTEN regulates T cell survival and antigen receptor signaling by acting as a phosphatidylinositol 3-phosphatase. *J. Immunol.* 164:1934–1939.

172. Di Cristofano, A., P. Kotsi, Y.F. Peng, C. Cordon-Cardo, K.B. Elkon, and P.P. Pandolfi. 1999. Impaired Fas response autoimmunity in *Ptent*[+/−] mice. *Science* 285:2122–2125.

9 | Cytokines and the Th1/Th2 Paradigm in Autoimmunity

Francesco Sinigaglia

1. INTRODUCTION

The discovery of functionally diverse type 1 and type 2 CD4$^+$ T helper cells has provided a cellular basis to explain the development of polarized versions of the immune response (1–3). An abundance of data now strongly suggests that inflammatory immune responses or DTH reactions are primarily mediated by T helper 1 (Th1) populations that are characterized by secretion of interferon (IFN)-γ and tumor necrosis factor (TNF), but little interleukin (IL)-4. By contrast, Th2 cells produce IL-4, IL-5 and IL-13, and mediate immune responses characterized by high levels of non-complement binding IgG, IgE and eosinophil-mediated cytotoxicity, but no organ-specific tissue destruction. Th2 cells may also exert an anti-inflammatory action by negatively regulating Th1 cell-mediated immune responses. Although individual cells may exhibit a more complex and heterogeneous pattern of cytokine production, many experimentally induced and naturally occurring autoimmune responses show patterns of cytokine production and effector reactions that are clearly indicative of Th1 responses. Thus, although the diversity of T cell-dependent immunity cannot be fit easily into the Th1/Th2 paradigm, understanding the factors that influence the generation and the effector function of Th1 and Th2 cells is important for the development of therapeutic approaches to fighting a variety of human diseases. This chapter focuses on the inflammatory role of Th1 cells and on the role of cytokines in regulating Th1 differentiation and effector function.

2. ROLE OF CYTOKINES IN THE DEVELOPMENT AND REGULATION OF T HELPER SUBSETS

2.1. The Central Role of APC

Th1 and Th2 cells develop from naive CD4$^+$ T cells (Figure 1). The differentiation process is initiated by ligation of the TCR and directed by cytokines present during the initiation of a T cell response. IL-4 promotes Th2 development (4,5), whereas IL-12 plays a central role in controling the development of Th1 cells from naive precursor T cells (6–8). IL-12 is mainly produced by monocytes and dendritic cells (DC) (9,10). Bacteria, viruses and intracellular parasites have been described as inducers of the IL-12 heterodimer both *in vitro* and *in vivo* (6,9,11). DC are the most potent APC for Th cells. After their development in the bone marrow, they transiently reside in nonlynphoid organs. Inflammatory stimuli such as whole bacteria, the microbial cell wall component LPS, and cytokines such as IL-1, GM-CSF and TNF-α all stimulate DC maturation and migration to the T-cell rich areas of lymphoid organs, where DC form clusters with antigen-specific T cells, creating a microenvironment in which Th cells can differentiate (rev. in 12). DC produce IL-12; the level of IL-12 production is controlled to a large extent by the interaction with T cells during the presentation of the antigen. In particular, the interaction between the CD40 molecules on DC and the CD40L present on T cells transmits signals critical to T helper cell development and IL-12 production (13,14), as shown by inhibition with monoclonal antibody (15) and disruption of the CD40L gene (16). IL-10 has the ability to inhibit both DC and macrophage-derived IL-12 production and thus inhibit the development of Th1 cells (17).

2.2. IL-12 Signaling and Th1 Development

Once produced by APC, IL-12 turns on the Th1-specific transcriptional program on T cells. Several pieces of evidence have led to the conclusion that IL-12-driven Th1 differentiation requires the IL-12-responsive transcription factor signal transducer and activator of transcription (STAT) 4 (Figure 2) (18,19), while Th2 differentiation requires the IL-4-responsive transcription factor STAT6 (20,21). Using naive T cells from mice expressing transgenic antigen receptor, Murphy and colleagues have demonstrated that IL-12 selectively activates STAT4 in Th1 but not in Th2 cells (22). These findings have been confirmed in the human system by Rogge et al. and Hilkins

et al. on CD4[+] T cell populations derived from cord blood (23), and on allergen-specific CD4[+] T cell clones generated from the peripheral blood of atopic patients (24), respectively. The cells that exhibited a strongly polarized Th2 cytokine profile were unable to produce IFN-γ and to phosphorylate STAT4 in response to IL-12. In addition to STAT4, other transcription factors may be important in influencing the development of Th1 responses. The interferon regulatory factor-1 (IRF-1) could be a potential candidate since IRF-1 deficient mice have a striking defect in the development of Th1 cells (25,26). However, our analysis of IRF-1 mRNA expression in Th1 and Th2 cells did not reveal significant differences between the two subsets (27). Notably, IL-12 induced a strong up-regulation of IRF-1 transcripts in Th1, but not in Th2 cells, suggesting that some of the IL-12-induced effector functions of Th1 cells may be mediated by IRF-1. Recent studies have

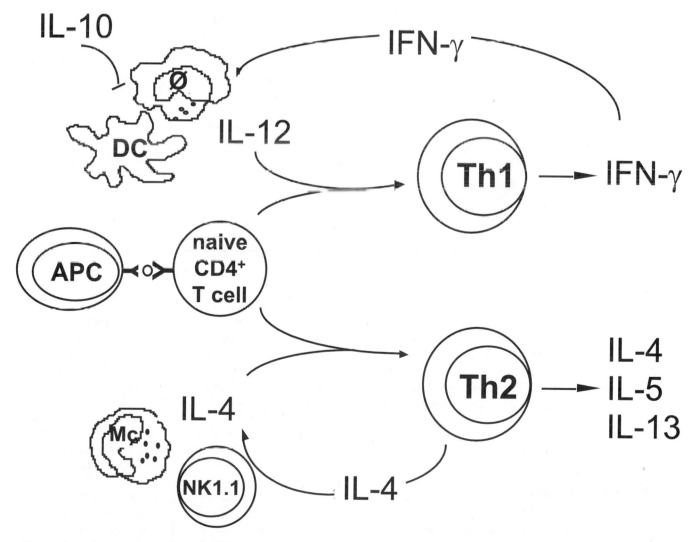

Figure 1.   Cytokine induced T helper cell development. Ø, macrophage; DC, dendritic cell; MC, mast cell; APC, antigen presenting cell.

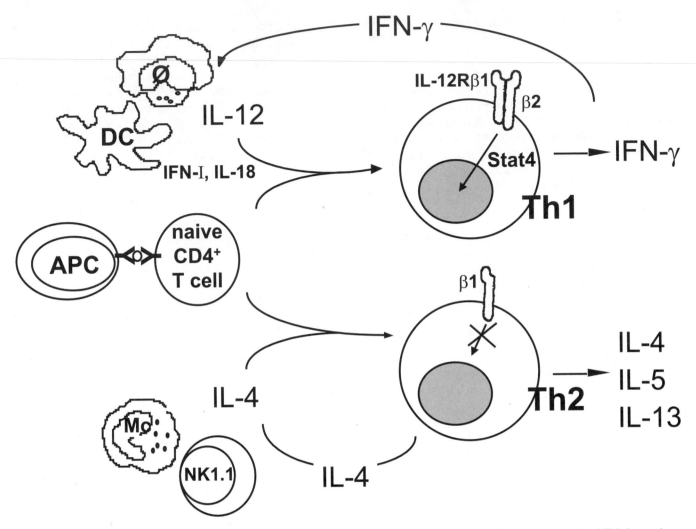

Figure 2. Schematic representation of IL-12R expression during the differentiation of naive T cells into polarized T helper subsets. The IL-12R is not present on naive T cells. Antigen stimulation induces expression of both chains of the receptor. Cells that develop in the presence of IL-12 continue to express both IL-12R subunits. By contrast, differentiation in the absence of IL-12 results in a selective loss of the IL-12Rβ2 chain.

shown that JNK MAP kinase pathway is induced in Th1 but not in Th2 effector cells upon antigen stimulation (28). Furthermore the differentiation of precursor CD4+ T cells into effector Th1 but not Th2 cells is impaired in JNK-deficient mice. The inability of IL-12 to differentiate JNK-deficient CD4+ T cells into effector Th1 cells is caused by a defect in IFN-γ production during the early stage of differentiation. These experiments have thus established that JNK MAP kinase signaling pathway plays a central role in Th1 development.

2.3. Expression and Regulation of IL-12 Receptor in T Helper Cell Differentiation

Lymphocyte response to IL-12 is dependent on the expression of high affinity IL-12R. Gubler and co-workers have identified two IL-12R subunits termed β1 and β2 (29,30).

The IL-12Rβ2 chain in particular is likely to act as a docking site for STAT4 SH2 domains since, by contrast to the IL-12Rβ1, the β2 chain has a cytoplasmic domain that contains three tyrosine residues (30). Studies from Ken Murphy's group in mice and from my group in humans demonstrated that regulation of the IL-12β2 chain is the focal point of early maintenance or loss of IL-12 responsiveness in developing T cells (23,31). The IL-12Rβ2 is not expressed on naïve T cells but is induced at low levels after engagement of the TCR by antigen. This initial expression of functional IL-12Rs is further enhanced when IL-12 is present at the time of priming. In the absence of IL-12, IL-12Rβ2 is expressed at very low levels and becomes undetectable a few days after T cell stimulation (23,32). Thus, while the IL-12Rβ1 subunit is expressed in similar amounts in both Th1 and Th2 cells, the IL-12Rβ2 is selectively expressed by Th1 but not by Th2 cells (Figure 2). The selective expression of the IL-12Rβ2 chain thus accounts for the inability of Th2 cells to phosphorylate Stat4 in response to IL-12.

2.4. Role of Interferons in the Generation of Th1 Cells

Recent reports have shown that, in addition to IL-12, type I interferons (IFN-I, which include IFN-α and IFN-β) can directly induce Th1 cell differentiation. IFN-I has been shown to influence human Th1 cell development most clearly under *in vitro* conditions. When T lymphocytes were cultured in the presence of IFN-α and then cloned (33) or directly stimulated (34) via the CD3/T-cell receptor complex, the proportion of CD4$^+$ T cells secreting IFN-γ was increased. Similar results were obtained with cord blood cells, and with either IFN-α or IFN-β (35). Even when the cloned cells were derived from an atopic donor and were specific for the allergen, circumstances normally favoring a Th2 phenotype, the majority of clones were Th1 when the cloning was done in the presence of IFN-α (36). There is also evidence that *in vivo* administration of IFN-I can favor development of human Th1 cells. IFN-γ-secreting cells temporarily increase in number when patients with multiple sclerosis are treated with IFN-β (37). Furthermore, Maggi et al. cloned CD8+ cells from two HIV-positive individuals before and after treatment with IFN-α for Kaposi's sarcoma. The CD8$^+$ T cells, which were generated from the skin around the tumor, showed a shift towards the Th1 phenotype after treatment (38).

As discussed above, the activation of Stat4 is essential for Th1 cell development. We have recently shown that human IFN-I induces a strong and rapid tyrosine phosphorylation of Stat4 in both human Th1 and Th2 cells (35). These studies extended a previous observation by O'Shea and co-workers, which showed that human IFN-α activates STAT4 in phytoemagglutinin-activated human T cells (39). Thus, IFN-I can induce Th1 development independently of IL-12 (Figure 2). Recently, an important species difference was documented. In contrast to the human case/system, IFN-I was unable to activate STAT4 in mouse T cells and consequently unable to induce *in vitro* development of mouse Th1 cells (31,35).

Th1 development is also dependent on IFN-γ. This effect may be mediated by action on the macrophage to upregulate IL-12 production or by direct effect on T cells. Szabo et al. have shown that in the mouse system IFN-γ can induce expression of the IL-12Rβ2 even when T cells are cultured with IL-4 (40). The data obtained with mouse T cells again diverged from findings in the human system where priming T cells in the presence of IL-4 and IFN-γ resulted in only a marginal IL-12Rβ2 expression (35). Thus, it appears that the effect of IFN-γ on the induction of IL-12 responsiveness is much stronger in mouse T cells than in human T cells. Together these findings indicate that in mice, IFN-γ provides a signal for initiating Th1

development, the full progression of which requires IL-12 signaling. On the contrary, IFN-I is able to induce Th1 development in the absence of IL-12 in human, but not in mouse T cells.

2.5. IL-18 in the Regulation of IFN-γ Production and IL-12R Expression

Although IL-12 and IFNs are undoubtedly key factors in directing the development of Th1 cells from naive precursors, recent studies have suggested a role for IL-18 as a cofactor in Th1 development (41). IL-18 is a cytokine, belonging to the IL-1 cytokine family, which is produced by monocytic cells. It shares some of the biological activities of IL-12 but without significant structure homology. As IL-12, it promotes proliferation and IFN-γ production by Th1, CD8$^+$ and NK cells in mice and humans, but does not appear to drive Th1 development; rather it synergises with IL-12 for IFN-γ production (42). Unlike IL-12, IL-18 does not activate STAT4 in T cells. IL-18 signals via the IL-1 receptor-associated kinase (IRAK) pathway to induce nuclear translocation of the p65/p50 NF-KB complex selectively in Th1 cells (42). Liew and coworkers recently provided a molecular basis for this selectivity by showing that the IL-18R is preferentially expressed on the surface of Th1 cells compared with Th2 cells (43). IL-12 and IL-18 reciprocally upregulated each other's receptor, thus providing a mechanism for the synergistic effect of IL-12 and IL-18 in the development of Th1 cells.

3. ROLE OF TH1 CELLS IN IMMUNOPATHOLOGY

3.1. Experimental Autoimmune Encephalitis (EAE) and Multiple Sclerosis (MS)

EAE is an inflammatory autoimmune disease of the central nervous system and has been extensively used as a model for MS. The disease is characterized clinically by acute onset of paralysis, and histologically by perivascular infiltration of the CNS by mononuclear cells. EAE can be induced in a number of animal species by immunization with myelin basic protein (MBP) or proteolipid protein (PLP), or by adoptive transfer of MBP- or PLP-specific CD4$^+$ T cells. There is now strong evidence that CD4$^+$ Th1 cells are important for the initiation and maintenance of the disease process. The systemic administration of IL-12 to rats immunized with myelin basic protein exacerbated the clinical symptomes of EAE and induced relapse when administered to animals who had recovered from the initial episode of disease (44). Renzetti et al. have found that in a murine model of relapsing remitting EAE, treatment with an antibody to IL-12, starting either at one day after immunization with PLP or during the onset of clinical symptoms prevented the induction or progression of disease (10). These results are consistent with those obtained by Leonard et al. and provide evidence for the involvement of IL-12 in inflammation of the CNS (45). Further evidence for the role of IL-12 in EAE was recently obtained by Shevach and coworkers. The B10.S mice are known to be resistant to EAE. These authors have recently shown that B10.S mice had an antigen-specific defect in their capacity to up-regulate the IL-12Rβ2 subunit (46). This defective expression was due to the failure of the antigen-specific T cells to up-regulate CD40 ligand expression and to induce the production of IL-12. The defective expression of the IL-12Rβ2 chain is likely to be the cause of the inability of the B10.S T cells to differentiate into pathogenic T cells. In addition to the evidence implicating Th1 responses in the pathogenesis of EAE in various animal models, it has been argued that likewise MS is an autoimmune disease in which Th1 cell induction is associated with a worsening of symptoms. The main evidence for this belief is that relapses tend to be preceded by an increase in the number of circulating IFN-γ-secreting T cells (47), and that exogenous IFN-γ worsens symptoms when administered to patients (48). Furthermore, recent studies have suggested the involvement of IL-12 in MS. Using semiquantitative RT-PCR, increased expression of IL-12 p40 mRNA was detected in acute MS plaques, but not in inflammatory lesions to CNS infarcts (49). In addition, increased IL-12 production was observed when antigen-presenting cells were stimulated with T cells isolated from MS patients as compared to control T cells (50).

3.2.   The Puzzle of Type I IFN in MS

On the one hand, IFN-β is licensed for treatment of multiple sclerosis (MS), and decreases the frequency of relapses in about one third of treated patients (51). IFN-β also seems to be effective in reducing symptoms of MS (52). On the other hand, we have argued above that Type I IFN (both α and β), via direct activation of Stat4, are potent inducers of Th1 cells in humans. How can this apparent paradox be resolved, and how do type I IFNs act to alleviate symptoms of the disease?

Recent reviews dispute the view that MS is a purely Th1-mediated disease (53). They point out that, for example, the Th1/Th2 ratio is not markedly shifted in the T cells infiltrating sites of demyelination, and that the effector mechanisms responsible for demyelination are unknown. Furthermore, although some of the symptoms of MS can be reproduced in various EAE models, the extent of the similarity between MS and EAE remains debatable. Thus the Th1/Th2 concept needs to be used cautiously in discussing the pathology of MS. Several possible mechanisms, which are not mutually exclusive and do not depend on the Th1/Th2 paradigm, can be envisaged for the beneficial action of

Type I IFNs in MS (rev. in 54). First, the IFNs might have a direct anti-proliferative effect or an anti-viral effect on a (hypothetical) virus responsible for triggering the disease. A herpes virus (HHV6) DNA has been found in some MS patients (55). Second, Type I IFNs might act to prevent lymphocytic infiltrations into the areas of demyelination that are typical of the disease (rev. in 56). Putative targets for this type of mechanism include integrins and selectins, chemokines, and matrix metalloproteinases (MMPs), all of which are likely to be involved in the migration of leukocytes into the CNS. There is evidence that IFN treatment reduces the adherence of blood leukocytes from MS patients to endothelial cells in vitro, perhaps by affecting VLA-4 expression (57). It has also been shown that IFN-β reduces the production of the T cell protease MMP-9 (58,59), which is important in T cell migratory function in vitro, and may also be so in vivo, since MMP inhibitors reduce symptoms of EAE (60).

3.3.   Role of Th1 Cells in Type I Diabetes (IDDM)

Evidence suggesting a pathogenetic role for Th1 cells in IDDM has been provided by studies of the non-obese diabetic (NOD) mouse, a spontaneous model of human type I diabetes. All NOD mice exhibit lymphocytic infiltration of the islets of Langherans and 60–80% of female NOD mice become hyperglycemic by 30 weeks of age. Although, these infiltrates are comprised of CD4+ and CD8+ T cells, B cells and macrophages, it is the T cells that have been shown in adoptive transfer experiments to play the most prominent role in diabetes induction. T cell clones that are able to accelerate the onset of diabetes in young NOD mice produce Th1 type cytokine when challenged with islets and APC in vitro (61). Similarly, IL-12 administration to NOD mice greatly accelerated the onset of disease, correlating with increased Th1 cytokine production by islet-infiltrating mononuclear cells (62). By contrast, administration of the IL-12 antagonist IL-12 (p40)$_2$ to NOD mice starting at 3 weeks of age, before the onset of insulitis, resulted in deviation of pancreas-infiltrating T cells to the Th2 phenotype and in decreased incidence of both spontaneous and cyclophosphamide-accelerated IDDM (63). These results indicate that Th1 cells play an important role in the pathogenesis of IDDM in NOD mice, and suggest that induction of Th2 cells within the pancreatic infiltrate protects from disease. A role for Th1 cells in the pathogenesis of IDDM in humans has not yet been established.

3.4.   Collagen-induced Arthritis and Rheumatoid Arthritis

Rheumatoid arthritis and its animal model collagen-induced arthritis (CIA) have both been identified as Th1 diseases, as expected of a chronic inflammatory condition (64). IL-12-deficient mice or mice that have been treated with neutralizing anti-IL-12 Abs before onset of disease develop little or no arthritis (65,66). Furthermore, salbutamol, a β2 adrenergic agonist, attenuates ongoing CIA (67) most likely by inhibition of IL-12 production by macrophages and DC (68). In RA detectable levels of IL-12 were found in supernatants of RA joint cells (69) and CD4+ Th1 cells predominate in the synovial fluid of RA patients (70) (Figure 3).

3.5. Inflammatory Bowel Disease

Crohn's disease and ulcerative colitis are the major forms of chronic inflammatory bowel disease (IBD) in developed countries. The cytokine secretion profile of T cells from the lamina propria of patients with Crohn's disease is Th1-like in that elevated IFN-γ secretion was observed, together with decreased secretion of IL-4 and IL-5 (71). In contrast, lamina propria T cells from ulcerative colitis patients were found to secrete increased amounts of IL-5 with decreased IL-4 and normal IFN-γ (71). These data are in agreement with recent findings obtained in the murine model of TNBS-induced colitis, which mimics several important characteristics of CD and can be successfully treated by administration of antibodies to IL-12 (72). IL-12 could be observed in the inflamed bowel and appeared to be produced as a result of CD40-CD40L interaction (73).

Similarly, immunohistochemical analysis of tissue sections from the colons of human patients with Crohn's disease revealed an abundance of IL-12 expressing macrophages and IFN-γ expressing T cells (74). Immunohistochemical analysis on tissue samples from ulcerative colitis patients were not reported.

4. THE ROLE OF CYTOKINES IN THE RECRUITMENT OF TH1 CELLS TO SITES OF INFLAMMATION

Extravasation of effector T cells into tissues is mediated by a series of adhesive interactions between specific ligands expressed on the T cell surface with their respective adhesion molecules expressed on the vascular endothelium (Table 1). As responding cells differentiate into Th1 effectors, they shift their expression of selectins and their receptors, down-regulating L-selectin and up-regulating ligands for E- and P selectins (75). Austrup et al. have shown that Th1, but not Th2, cells express a functional ligand for P- and E-selectin and therefore are selectively recruited to sites where Th1 immune responses occur (76). Expression of P- and E-selectin ligands and cutaneous lymphocyte antigen (CLA) is controlled by the activity of α(1,3)-fucosyltransferases, a family of enzymes which modifies carbohydrate moieties decorating PSGL-1 and other surface receptors (77). Recent studies have reported

Peripheral Blood T cells

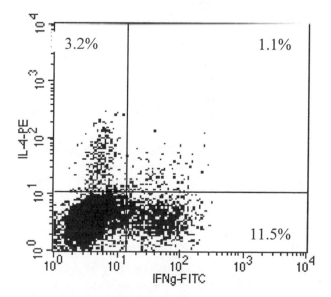

Synovial Fluid T cells

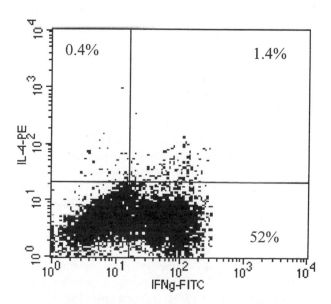

Figure 3.   Intracellular cytokine production of synovial fluid and peripheral blood T cells from a newly diagnosed RA patient. Purified T cells were stimulated with PMA and ionomycin, and stained for intracellular IFN-γ and IL-4.

Table 1 Addressins and chemokine receptors preferentially expressed by Th1 cells

Step 1[a]	Step 2	Step 3–4
CLA	CXCR3	Integrin $\alpha6/\beta1$
P and E selectin ligands	CCR5	

[a] Step 1–4 refer to the multistep model of leukocyte migration. Selectins, chemoattractants, and integrins act in sequence, with some overlap.

that IL-12 up-regulates the expression of CLA on T lymphocytes (78). Consistent with these findings, activation of T cells in the presence of IL-12 induced expression of a particular type of $\alpha(1,3)$-fucosyltransferase (FucT-VII) and enhanced interactions with both E- and P-selectin (79,80). In addition to selectin ligands CD4+, Th1 effectors also express a broad range of chemokine receptors, some of which are preferentially expressed on Th1 subset (i.e. CXCR3, CCR5) compared to the Th2 subset (i.e. CCR3, CCR4 and CCR8) (81–83). Analysis by immunohisto-chemistry of synovial tissue T cells from rheumatoid arthritis patients confirmed that the majority of such cells expressed CXCR3 and CCR5 (84,85). These changes in chemokine receptor expression will affect the ability of the effectors to be recruited to inflammatory sites where other cells are engaged in chemokine synthesis. Thus, chemokine receptor expression could constitute a major regulatory element for the composition of the lymphocytic infiltrates in inflammatory pathologies. In addition to selectins and chemokines, adhesion molecules belonging to the integrin family are also an important component in the tissue-specific recruitment of leukocytes. Recently, we have found differences in the expression and function of these receptors in Th1 and Th2 cells. Among various integrin subunits, Th1 cells were found to express higher levels of integrin $\alpha_6\beta_1$ (laminin receptor) when compared to Th2 cells (86). Notably, surface expression of integrin α_6 in Th1 cells was found to be strongly up-regulated by IL-12, suggesting that the differential expression and regulation of integrins may also contribute to tissue-specific localization of T helper cells. Strong integrin $\alpha_6\beta_1$-mediated adhesion may promote efficient migration of Th1 cells throughout the extracellular matrix (ECM) and/or may help retain extravasated Th1 cells in the tissues. Chronic Th1-dominated immune responses such as the ones discussed above are often characterized by basal membrane thickening and abnormal deposition of ECM (87,88). These conditions may impose a high threshold for leukocyte extravasation. While Th1 cells, by virtue of their high P- and E-selectin ligands and robust adhesion to ECM components (Table 1), may be able to gain access to those inflammatory sites, Th2 cells may not. As discussed above, Th1 cells have been proposed to exert potent pro-inflammatory effects which, if left uncontrolled, may led to tissue-injury and immuno-pathology. Conversely, cytokines produced by Th2 cells such as IL-4, IL-13 and IL-10 have powerful anti-inflammatory effects. This has led to the proposal that one of the major functions of Th2 cells may be to down-regulate Th1 cell-mediated responses and to limit their tissue damaging effects. In this regard the limited ability of Th2 cells, as compared to Th1, to penetrate certain inflammatory sites may be relevant to the maintenance of chronic inflammatory conditions.

5. CONCLUSIONS

The diversity of T-lymphocyte function revealed by studies of T cell subsets is now a well documented feature of the immune system. Learning how to exploit this functional dichotomy for practical applications requires greater knowledge of the molecular events that regulate the development and the effector function of these cell subsets. In this chapter, I have discussed our own and others' studies on the mechanisms that regulate Th cell differentiation. Furthermore, I have discussed recent data indicating that the tissue-specific recruitment of Th1 and Th2 cells is differentially regulated by the intricate interplay of adhesion receptors and chemokines. These novel findings have extended the functional dichotomy of T cell-mediated immune responses at the level of tissue-specific homing. Given the central role of Th1 cells in several chronic inflammatory diseases, manipulating the signals which control differentiation and/or trafficking of these cells may open new therapeutic avenues to the treatment of a number of autoimmune conditions.

References

1. Mosmann, T.R., and R.L. Coffman. 1989. Th1 and Th2 cells: Different patterns of lymphokine secretion lead to different functional properties. *Annu. Rev. Immunol.* 7:145–173.
2. Abbas, A.K., K.M. Murphy, and A. Sher. 1996. Functional diversity of helper T lymphocytes. *Nature* 383:787–793.
3. O'Garra, A. 1998. Cytokines induce the development of functionally heterogeneous T helper cell subsets. *Immunity* 8:275–283.
4. Seder, R.A., W.E. Paul, M.M. Davis, and B. Fazekas de St. Groth. 1992. The presence of interleukin-4 during in vitro priming determines the lymphokine-producing potential of CD4+ T cells from T cell receptor transgenic mice. *J. Exp. Med.* 176:1091–1098.
5. Hsieh, C.S., A.B. Heimberger, J.S. Gold, A. O'Garra, and K.M. Murphy. 1992. Differential regulation of T helper phenotype development by interleukins 4 and 10 in an alpha beta T cell-receptor transgenic system. *Proc. Natl. Acad. Sci. USA* 89:6065–6069.

6. Hsieh, C.-S., S.E. Macatonia, C.S. Tripp, S.F. Wolf, A. O'Garra, and K.M. Murphy. 1993. Development of T$_H$1 CD4$^+$ T cells through IL-12 produced by *Listeria*-induced macrophages. *Science* 260:547–549.

7. Seder, R.A., R. Gazzinelli, A. Sher, and W.E. Paul. 1993. Interleukin 12 acts directly on CD4+ T cells to enhance priming for interferon gamma production and diminishes interleukin 4 inhibition of such priming. *Proc. Natl. Acad. Sci. USA* 90:10188–10192.

8. Manetti, R., P. Parronchi, M.G. Giudizi, M.-P. Piccinni, E. Maggi, G. Trinchieri, and S. Romagnani. 1993. Natural Killer Cell Stimulatory Factor (Interleukin 12 [IL-12]) induces T helper 1 type (Th1)-specific immune responses and inhibits the development of IL-4-producing Th cells. *J. Exp. Med.* 177:1199–1204.

9. Trinchieri, G. 1995. Interleukin-12: A proinflammatory cytokine with immunoregulatory functions that bridge innate resistance and antigen-specific adaptive immunity. *Annu. Rev. Immunol.* 13:251–276.

10. Gately, M.K., L.M. Renzetti, J. Magram, A.S. Stern, L. Adorini, U. Gubler, and D.H. Presky. 1998. The interleukin-12/interleukin-12-receptor system: Role in normal and pathologic immune responses. *Annu. Rev. Immunol.* 16:495–521.

11. Gazzinelli, R.T., S. Hieny, T.A. Wynn, S. Wolf, and A. Sher. 1993. Interleukin 12 is required for the T-lymphocyte-independent induction of interferon of γ by an intracellular parasite and reduces resistance in T-cell deficient hosts. *Proc. Natl. Acad. Sci. USA* 90:6115–6119.

12. Banchereau, J., and R.M. Steinman. 1998. Dendritic cells and the control of immunity. *Nature* 392:245–252.

13. Macatonia, S., N.A. Hosken, M. Litton, P. Vieira, C.-S. Hsieh, I.A. Culpepper, M. Wysocka, G. Trinchieri, K.M. Murphy, and A. O'Garra. 1995. Dendritic cells produce IL-12 and direct the development of Th1 cells from naive CD4+ T cells. *J. Immunol.* 154:5071–5079.

14. Cella, M., D. Scheidegger, K. Palmerlehmann, P. Lane, A. Lanzavecchia, and G. Alber. 1996. Ligation of CD40 on dendritic cells triggers production of high levels of interleukin-12 and enhances T cell stimulatory capacity: T-T help via APC activation. *J. Exp. Med.* 184:747–752.

15. Koch, F., U. Stanzl, P. Jennewein, K. Janke, C. Heufler, E. Kampgen, N. Romani, and G. Schuler. 1996. High level IL-12 production by murine dendritic cells: Upregulation via MHC class II and CD40 molecules and downregulation by IL-4 and IL-10. *J. Exp. Med.* 184:741–746.

16. Grewal, I.S., and R. Flavel. 1998. CD40 and CD154 in Cell-mediated Immunity. *Annu. Rev. Immunol.* 16:111–135.

17. D'Andrea, A., M. Aste-Amezaga, N.M. Valiante, X. Ma, M. Kubin, and G. Trinchieri. 1993. Interleukin 10 (IL-10) inhibits human lymphocyte interferon-γ production by suppressing natural killer cell stimulatory factor/IL-12 synthesis in accessory cells. *J. Exp. Med.* 178:1041–1048.

18. Kaplan, M.H., Y.L. Sun, T. Hoey, and M.J. Grusby. 1996. Impaired IL-12 responses and enhanced development of Th2 cells in Stat4-deficient mice. *Nature* 382:174–177.

19. Thierfelder, W.E., J.M. van Deursen, K. Yamamoto, R.A. Tripp, S.R. Sarawar, R.T. Carson, M.Y. Sangster, D.A.A. Vignali, P.C. Doherty, G.C. Grosveld, and J.N. Ihle. 1996. Requirement for Stat4 in interleukin-12-mediated responses of natural killer, and T cells. *Nature* 382:171–174.

20. Kaplan, M.H., U. Schindler, S.T. Smiley, and M.J. Grusby. 1996. Stat6 is required for mediating responses to IL-4 and for the development of Th2 cells. *Immunity* 4:313–319.

21. Shimoda, K., J. van Deursen, M.Y. Sangster, S.R. Sarawar, R.T. Carson, R.A. Tripp, C. Chu, F.W. Quelle, T. Nosaka, D.A.A. Vignalli, P.C. Doherty, G. Grosveld, W.E. Paul, and J.N. Ihle. 1996. Lack of IL-4-induced Th2 response and IgE class switching in mice with disrupted Stat6 gene. *Nature* 380:630–633.

22. Szabo, S.J., S.J. Szabo, N.G. Jacobson, A.S. Dighe, U. Gubler, and Murphy, K.M. 1995. Developmental commitment to the Th2 lineage by extinction of IL-12 signaling. *Immunity* 2:665–675.

23. Rogge, L., L. Barberis-Maino, M. Biffi, N. Passini, D.H. Presky, U. Gubler, and F. Sinigaglia. 1997. Selective expression of an interleukin-12 receptor component by human T helper 1 cells. *J. Exp. Med.* 185:825–832.

24. Hilkens, C.M.U., G. Messer, K. Tesselaar, A.G.I. Vanrietschoten, M.L. Kapsenberg, and E.A. Wierenga. 1996. Lack of IL-12 signaling in human allergen-specific Th2 cells. *J. Immunol.* 157:4316–4321.

25. Taki, S., T. Sato, K. Ogasawara, T. Fukuda, M. Sato, S. Hida, G. Suzuki, M. Mitsuyama, E.H. Shin, S. Kojima, T. Taniguchi, and Y. Asano. 1997. Multistage regulation of Th1-type immune responses by the transcription factor IRF-1. *Immunity* 6:673–679.

26. Lohoff, M., D. Ferrick, H.-W. Mittrücker, G.-S. Duncan, S. Bischoff, M. Röllinghof, and T.W. Mak. 1997. Interferon regulatory factor-1 is required for a T helper 1 immune response in vivo. *Immunity* 6:681–689.

27. Coccia, E.M., N. Passini, A. Battistini, C. Pini, F. Sinigaglia, and L. Rogge. 1999. Interleukin-12 induces expression of interferon regulatory factor-1 via signal transducer and activator of transcription-4 in human T helper type 1 cells. *J. Biol. Chem.* 274:6698–6703.

28. Yang, D.D., D. Conze, A.J. Whitmarsh, T. Barrett, R.J. Davis, M. Rincon, and R.A. Flavell. 1998. Differentiation of CD4+ T cells to Th1 cells requires MAP kinase JNK2. *Immunity* 9:575–585.

29. Chua, A.O., R. Chizzonite, B.B. Desai, T.P. Truitt, P. Nunes, L.J. Minetti, R.R. Warrier, D.H. Presky, J.F. Levine, M.K. Gately, and U. Gubler. 1994. Expression cloning of a human IL-12 receptor component: A new member of the cytokine receptor superfamily with strong homology to gp130. *J. Immunol.* 153.128–136.

30. Presky, D.H., H. Yang, L.J. Minetti, A.O. Chua, N. Nabavi, C.Y. Wu, M.K. Gately, and U. Gubler. 1996. A functional interleukin 12 receptor complex is composed of two beta-type cytokine receptor subunits. *Proc. Natl. Acad. Sci. USA* 93:14002–14007.

31. Szabo, S.J., N.G. Jacobson, A.S. Dighe, U. Gubler, and K.M. Murphy. 1995. Developmental commitment to the Th2 lineage by extinction of IL-12 signaling. *Immunity* 2:665–675.

32. Rogge, L., A. Papi, D.H. Presky, M. Biffi, L.J. Minetti, D. Miotto, C. Agostini, G. Semenzato, L.M. Fabbri, and F. Sinigaglia. 1999. Antibodies to the IL-12 receptor beta 2 chain mark human Th1 but not Th2 cells in vitro and in vivo. *J. Immunol.* 162:3926–3932.

33. Parronchi, P., M. De Carli, R. Manetti, C. Simonelli, S. Sampognaro, M.P. Piccini, D. Macchia, E. Maggi, G. Del Prete, and S. Romagnani. 1992. Il-4 and IFN (α and γ) Exert Opposite Regulatory Effects on the Development of Cytolytic Potential by Th1 or Th2 Human T Cell Clones. *J. Immunol.* 149:2977–2983.

34. Brinkmann, V., T. Geiger, S. Alkan, and C.H. Heusser. 1993. Interferon α *increases the frequency of interferon γ*-producing CD4$^+$ T cells. *J. Exp. Med.* 178:1655–1663.

35. Rogge, L., D. DAmbrosio, M. Biffi, G. Penna, L.J. Minetti, D.H. Presky, L. Adorini, and F. Sinigaglia. 1998. The role of stat4 in species-specific regulation of Th cell development by type I IFNs. *J. Immunol.* 161:6567–6574.

36. Parronchi, P., S. Mohapatra, S. Sampognaro, L. Giannarini, U. Wahn, P.L. Chong, S. Mohapatra, E. Maggi, H. Ranz, and S. Romagnani. 1996. Effects of interferon-alpha on cytokine profile, T cell receptor repertoire and peptide reactivity of human allergen-specific T cells. *Eur. J. Immunol.* 26:697–703.

37. Dayal, A., M.A. Jensen, A. Lledo, and B.G.W. Arnason. 1995. Interferon-gamma-secreting cells in multiple sclerosis patients treated with interferon beta-1b. *Neurology* 45:2173–2177.

38. Maggi, E., R. Manetti, F. Annunziato, L. Cosmi, M.G. Giudizi, R. Biagiotti, G. Galli, G. Zuccati, and S. Romagnani. 1997. Functional Characterization and Modulation of Cytokine Production by CD8+ T Cells from Human Immunodeficiency Virus-Infected Individuals. *Blood* 89:3672–3681.

39. Cho, S.S., C.M. Bacon, C. Sudarshan, R.C. Rees, D. Finbloom, R. Pine, and J.J. O'Shea. 1996. Activation of STAT4 by IL-12 and IFN-α. *Evidence for the involvement of ligand-induced tyrosine and serine phosphorylation.* J. *Immunol.* 157:4781–4789.

40. Szabo, S.J., A.S. Dighe, U. Gubler, and K.M. Murphy. 1997. Regulation of the interleukin (IL)-12R beta 2 subunit expression in developing T helper 1 (Th1) and Th2 cells. *J. Exp. Med.* 185:817–824.

41. Okamura, H., H. Tsutsui, T. Komatsu, M. Yutsudo, A. Hakura, T. Tanimoto, K. Torigoe, T. Okura, Y. Nukada, K. Hattori, K. Akita, M. Namba, F. Tanabe, K. Konishi, S. Fukada, and M. Kurimoto. 1995. Cloning of a new cytokine that induces IFN-γ production by T cells. *Nature* 378:88–91.

42. Robinson, D., K. Shibuya, A. Mui, F. Zonin, E. Murphy, T. Sana, S.B. Hartley, S. Menon, R. Kastelein, F. Bazan, and A. OGarra. 1997. IGIF does not drive Th1 development but synergizes with IL-12 for interferon-gamma production and activates IRAK and NF kappa B. *Immunity* 7:571–581.

43. Xu, D., W.L. Chan, B.P. Leung, D. Hunter, K. Schulz, R.W. Carter, I.B. McInnes, J.H. Robinson, and F.Y. Liew. 1998. Selective expression and functions of interluekin 18 receptor on T helper (Th) type 1 but not Th2 cells. *J. Exp. Med.* 188:1485–1492.

44. Smith, T., A.K. Hewson, C.I. Kingsley, J.P. Leonard, and M.L. Cuzner. 1997. Interleukin-12 induces relapse in experimental allergic encephalomyelitis in the Lewis rat. *Am. J. Pathol.* 150:1909–1917.

45. Leonard, J.P., K.E. Waldburger, R.G. Schaub, T. Smith, A.K. Hewson, M.L. Cuzner, and S.J. Goldman. 1997. Regulation of the inflammatory response in animal models of multiple sclerosis by interleukin-12. *Crit. Rev. Immunol.* 17:545–553.

46. Chang, J.T., E.M. Shevach, and B.M. Segal. 1999. Regulation of interleukin (IL)-12 receptor beta 2 subunit expression by endogenous IL-12: A critical step in the differentiation of pathogenic autoreactive T cells. *J. Exp. Med.* 189:969–978.

47. Beck, J., P. Rondot, L. Catinot, E. Falcoff, H. Kirchner, and J. Wietzerbin. 1988. Increased production of interferon gamma and tumor necrosis factor precedes clinical manifestations in multiple sclerosis: do cytokines trigger off exacerbations? *Acta Neurol. Scand.* 78:318–323.

48. Panitch, H., R. Hirsch, R. Haley, and K. Johnson. 1987. Exacerbations of multiple sclerosis in patients with gamma interferon. *Lancet* 1:893–895.

49. Windhagen A., J. Newcombe, F. Dangond, C. Strand, M.N. Woodroofe, M.L. Cuzner, and D.A. Hafler. 1995. Expression of costimulatory molecules B7-1 (CD80), B7-2 (CD86), and interleukin 12 cytokine in multiple sclerosis lesions. *J. Exp. Med.* 182:1985–1996.

50. Balashov, K.E., D.R. Smith, S.J. Khoury, D.A. Hafler, and H.L. Weiner. 1997. Increased interleukin 12 production in progressive multiple sclerosis: induction by activated CD4+ T cells via CD40 ligand. *Proc. Natl. Acad. Sci. USA* 94:599–603.

51. The IFNB Multiple Sclerosis Study Group. 1993. Interferon beta-1b is effective in relapsing-remitting multiple sclerosis. Clinical results of a multicenter, randomized, double-blind, placebo-controlled trial. *Neurology* 43:655–661.

52. Rudick, R., D. Goodkin, L. Jacobs, D. Cookfair, R. Herndon, J. Richert, A. Salazar, J. Fischer, C. Granger, J. Simon, J. Alam, N. Simonian, M. Campion, D. Bartoszak, D. Bourdette, J. Braiman, C. Brownscheidle, M. Coats, S. Cohan, D. Dougherty, R. Kinkel, M. Mass, F. Munschauer, R. Priore, R. Whitham, et. al. 1997. Impact of interferon beta-1a on neurologic disability in relapsing multiple sclerosis. The Multiple Sclerosis Collaborative Research Group (MSCRG). *Neurology* 49:358–363.

53. Laman, J. 1998. Balancing the Th1/Th2 concept in multiple sclerosis. *Immunol. Today* 19:489–490.

54. Dhib-Jalbut, S. 1997. Mechanisms of interferon beta action in multiple sclerosis. *Multiple Sclerosis* 3:397–401.

55. Soldan, S.S., R. Berti, N. Salem, P. Secchiero, L. Flamand, P.A. Calabresi, M.B. Brennan, H.W. Maloni, H.F. McFarland, H.-C. Lin, M. Patnaik, and S. Jacobson. 1997. Association of human herpes virus 6 (HHV-6) with multiple sclerosis: increased IgM response to HHV-6 early antigen and detection of serum HHV-6 DNA. *Nat. Med.* 3:1394–1397.

56. Wong, V.W., S. Chabot, O. Stuve, and G. Williams. 1998. Interferon beta in the treatment of multiple sclerosis. *Neurology* 51: 682–689.

57. Calabresi, P.A., C.M. Pelfrey, L.R. Tranquill, H. Maloni, and H.F. McFarland. 1997. VLA-4 expression on peripheral blood lymphocytes is downregulated after treatment of multiple sclerosis with interferon beta. *Neurology* 49:1111–1116.

58. Leppert, D., E. Waubant, M. Burk, J. Oksenberg, and S. Hauser. 1996. Interferon beta-1b inhibits gelatinase secretion and in vitro migration of human T cells: a possible mechanism for treatment efficacy in multiple sclerosis. *Ann. Neurol.* 40:846–852.

59. Stuve, O., N. Dooley, J. Uhm, J. Antel, G. Francis, G. Williams, and V. Yong. 1996. Interferon beta-1b decreases the migration of T lymphocytes in vitro: effects on matrix metalloproteinase-9. *Ann. Neurol.* 40:853–863.

60. Norga, K., L. Paemen, S. Masure, C. Dillen, H. Heremans, A. Billiau, H. Carton, L. Cuzner, T. Olsson, and J. Van Damme. 1995. Prevention of acute autoimmune encephalomyelitis and abrogation of relapses in murine models of multiple sclerosis by the protease inhibitor D-penicillamine. *Inflamm. Res.* 44:529–534.

61. Bergman, B., and K. Haskins. 1994. Islet-specific T-cell clones from the NOD mouse respond to beta-granule antigen. *Diabetes* 43:197–203.

62. Trembleau, S., G. Penna, E. Bosi, A. Mortara, M.K. Gately, and L. Adorini. 1995. Interleukin 12 administration induces

T helper type 1 cells and accelerates autoimmune diabetes in NOD mice. *J. Exp. Med.* 181:817–821.

63. Trembleau, S., G. Penna, S. Gregori, M.K. Gately, and L. Adorini. 1997. Deviation of pancreas-infiltrating cells to Th2 by interleukin-12 antagonist administration inhibits autoimmune diabetes. *Eur. J. Immunol.* 27:2330–2339.

64. Feldmann, M., F.M. Brennan, and R.N. Maini. 1996. Role of cytokines in rheumatoid arthritis. *Annu. Rev. Immunol.* 14:397–440.

65. McIntyre, K.W., D.J. Shuster, K.M. Gillooly, R.R. Warrier, S.E. Connaughton, L.B. Hall, L.H. Arp, M.K. Gately, and J. Magram. 1996. Reduced incidence and severity of collagen-induced arthritis in interleukin-12-deficient mice. *Eur.J.Immunol.* 26:2933–2938.

66. Malfait, A.M., D.M. Butler, D.H. Presky, R.N. Maini, F.M. Brennan, and M. Feldmann. 1998. Blockade of IL-12 during the induction of collagen-induced arthritis (CIA) markedly attenuates the severity of the arthritis. *Clin. Exp. Immunol.* 111:377–383.

67. Malfait, A.-M., A.S. Malik, L. Marinova-Mutachieva, D.M. Butler, R.N. Maini, and F. M. 1999. The *β2-adrenergic agonist salbutamol is a potent suppressor of established collagen-induced arthritis: mechanisms of action.* J. Immunol. 162:6278–6283.

68. Panina Bordignon, P., D. Mazzeo, P. DiLucia, D. DAmbrosio, R. Lang, L. Fabbri, C. Self, and F. Sinigaglia. 1997. beta(2)-agonists prevent Th1 development by selective inhibition of interleukin 12. *J. Clin. Invest.* 100:1513–1519.

69. Bucht, A., P. Larsson, L. Weisbrot, C. Thorne, P. Pisa, G. Smedegard, E.C. Keystone, and A. Gronberg. 1996. Expression of interferon-gamma (IFN-gamma), IL-10, IL-12 and transforming growth factor-beta (TGF-beta) mRNA in synovial fluid cells from patients in the early and late phases of rheumatoid arthritis (RA). *Clin. Exp. Immunol.* 103:357–367.

70. Dolhain, R.J., A.N. van der Heiden, N.T. ter Haar, F.C. Breedveld, and A.M. Miltenburg. 1996. Shift toward T lymphocytes with a T helper 1 cytokine-secretion profile in the joints of patients with rheumatoid arthritis. *Arthritis Rheum.* 39:1961–1969.

71. Fuss, I.J., M. Neurath, M. Boirivant, J.S. Klein, C. de la Motte, S.A. Strong, C. Fiocchi, and W. Strober. 1996. Disparate CD4+ lamina propria (LP) lymphokine secretion profiles in inflammatory bowel disease. Crohn's disease LP cells manifest increased secretion of IFN-gamma, whereas ulcerative colitis LP cells manifest increased secretion of IL-5. *J. Immunol.* 157:1261–1270.

72. Neurath, M.F., I. Fuss, B.L. Kelsall, E. Stuber, and W. Strober. 1995. Antibodies to interleukin 12 abrogate established experimental colitis in mice. *J. Exp. Med.* 182:1281–1290.

73. Stuber, E., W. Strober, and M. Neurath. 1996. Blocking the CD40L-CD40 interaction in vivo specifically prevents the priming of T helper 1 cells through the inhibition of interleukin 12 secretion. *J. Exp. Med.* 183:693–698.

74. Parronchi, P., P. Romagnani, F. Annunziato, S. Sampognaro, A. Becchio, L. Giannarini, E. Maggi, C. Pupilli, F. Tonelli, and S. Romagnani. 1997. Type 1 T-helper cell predominance and interleukin-12 expression in the gut of patients with Crohn's disease. *Am. J. Pathol.* 150:823–832.

75. Bradley, L.M., and S.R. Watson. 1996. Lymphocyte migration into tissue: the paradigm derived from CD4 subsets. *Curr. Opin. Immunol.* 8:312–320.

76. Austrup, F., D. Vestweber, E. Borges, M. Lohning, R. Brauer, U. Herz, H. Renz, R. Hallmann, A. Scheffold, A. Radbruch, and A. Hamann. 1997. P- and E-selectin mediate recruitment of T-helper-1 but not T-helper-2 cells into inflamed tissues. *Nature* 385:81–83.

77. Fuhlbrigge, R.C., R. Alon, K.D. Puri, J.B. Lowe, and T.A. Springer. 1996. Sialylated, fucosylated ligands for L-selectin expressed on leukocytes mediate tethering and rolling adhesions in physiologic flow conditions. *J. Cell. Biol.* 135:837–848.

78. Fuhlbrigge, R.C., J.D. Kieffer, D. Armerding, and T.S. Kupper. 1997. Cutaneous lymphocyte antigen is a specialized form of PSGL-1 expressed on skin-homing T cells. *Nature* 389:978–981.

79. Wagers, A.J., C.M. Waters, L.M. Stoolman, and G.S. Kansas. 1998. Interleukin 12 and interleukin 4 control T cell adhesion to endothelial selectins through opposite effects on α1,3-fucosyltransferase VII gene expression. *J. Exp. Med.* 188:2225–2231.

80. Lichtman, A.H., and A.K. Abbas. 1997. T-cell subsets: recruiting the right kind of help. *Curr. Biol.* 7:242–244.

81. Bonecchi, R., G. Bianchi, P. Panina Bordignon, D. D'Ambrosio, R. Lang, A. Borsatti, S. Sozzani, P. Allavena, P.A. Gray, A. Mantovani, and F. Sinigaglia. 1998. Differential expression of chemokine receptors and chemotactic responsiveness of Th1 and Th2 cells. *J. Exp. Med.* 187:129–134.

82. Sallusto, F., D. Lenig, C.R. Mackay, and A. Lanzavecchia. 1998. Flexible programs of chemokine receptor expression on human polarized T helper 1 and 2 lymphocytes. *J. Exp. Med* 187:875–883.

83. Zingoni, A., H. Soto, J.A. Hedrick, A. Stoppacciaro, C.T. Storlazzi, F. Sinigaglia, D. DAmbrosio, A. OGarra, D. Robinson, M. Rocchi, A. Santoni, A. Zlotnik, and M. Napolitano. 1998. The chemokine receptor CCR8 is preferentially expressed in Th2 but not Th1 cells. *J. Immunol.* 161:547–551.

84. Qin, S.X., J.B. Rottman, P. Myers, N. Kassam, M. Weinblatt, M. Loetscher, A.E. Koch, B. Moser, and C.R. Mackay. 1998. The chemokine receptors CXCR3 and CCR5 mark subsets of T cells associated with certain inflammatory reactions. *J. Clin. Invest.* 101:746–754.

85. Loetscher, P., M. Uguccioni, L. Bordoli, M. Baggiolini, and B. Moser. 1998. CCR5 is characteristic of Th1 lymphocytes. *Nature* 391:344–345.

86. Colantonio, L., A. Iellem, B. Clissi, R. Pardi, L. Rogge, F. Sinigaglia, and D. D'Ambrosio. 1999. Upregulation of integrin α6/β1 and chemokine receptor CCR1 by interleukin-12 promotes the migration of human type 1 helper T cells. *Blood* (in press).

87. Schneider, M., B. Voss, J. Rauterberg, M. Menke, T. Pauly, R.K. Miehlke, J. Friemann, and U. Gerlach. 1994. Basement membrane proteins in synovial membrane: distribution in rheumatoid arthritis and synthesis by fibroblast-like cells. *Clin. Rheumatol.* 13:90–97.

88. Haapasalmi, K., M. Makela, O. Oksala, J. Heino, K.M. Yamada, V.J. Uitto, and H. Larjava. 1995. Expression of epithelial adhesion proteins and integrins in chronic inflammation. *Am. J. Pathol.* 147:193–206.

10 | T Cell Adhesion Molecules in Autoimmunity

Christopher D. Buckley, Patricia Lalor, Mike Salmon, Gerard B. Nash and David H. Adams

1. INTRODUCTION

Adhesive interactions are critical for the effective functioning of normal immune responses. It is, therefore, not surprising that aberrant cell adhesion lies at the heart of many pathological processes, such as inflammation, tumour growth and metastasis, and autoimmunity. Normal immune responses depend on the right cells interacting with each other in the right place at the right time. Most pathological process result, at least in part, because of perturbations in leukocyte activation and adhesion, leading to the wrong cell accumulating in the wrong place at the wrong time. Tremendous progress has been made in the last decade in defining the molecular basis of T cell adhesion and its importance in T cell recognition and migration, and this has led to the development of novel therapeutic targets. In addition, the interdependence of adhesion and apoptosis for the effective resolution of immune responses has been demonstrated at the molecular level by the finding that a number of adhesion molecules play an important role in the recognition and clearance of apoptotic cells by phagocytic cells (1). Ultimately, effective host defense requires effective cross-talk between the regulatory proteins involved in the inflammatory and complement cascades, cells of the immune system and resident tissue stromal cells, such as interstitial macrophages and fibroblasts. To illustrate the critical role of adhesion receptors in the immunopathogenesis of autoimmune diseases, we will highlight their role in leukocyte adhesion, activation and endothelial transmigration. We have emphasized these features of T cell adhesion molecules partly because they are the ones about which most is known, and partly because they are unique to leukocytes.

2. WHAT ARE CELL ADHESION MOLECULES?

To detect changes in their extracellular environment, cells transmit chemical information across their cell membranes. Classically, such chemical signals are mediated through soluble growth factors and cytokines (2). T cell adhesion molecules have, until relatively recently, been thought to function primarily as glue holding cells together or providing traction in order for them to migrate over cellular surfaces or extracellular matrix. However, it is now clear that cell adhesion molecules are able to transmit both mechanical and chemical signals across cell membranes (3). Such adhesion-dependent signals are particularly important for leukocytes where signals from adjacent cells and extracellular matrix proteins provide important cues to allow the cell to position itself appropriately. In addition, these adhesive interactions also regulate many aspects of T cell behavior, such as cell proliferation, differentiation and migration (4–7).

The ability to clone cell surface molecules has led to the isolation and cataloguing of a large number of cell surface proteins that play important roles in cell adhesion (8). Fortunately, many of these proteins fall into a number of distinct families with well-defined structural and functional similarities (Figure 1). The functional analysis of cell adhesion molecules has relied heavily on constructing recombinant chimeric proteins to study how these molecules interact with each other (9). In general, two broad modes of interaction have been defined: Homophilic (binding to self) and heterophilic (binding to non-self ligands). Whilst this first phase in our understanding of how cell adhesion molecules work has been very fruitful, allowing a detailed analysis of binding domains, motifs and key amino acids involved in T cell adhesion events, it

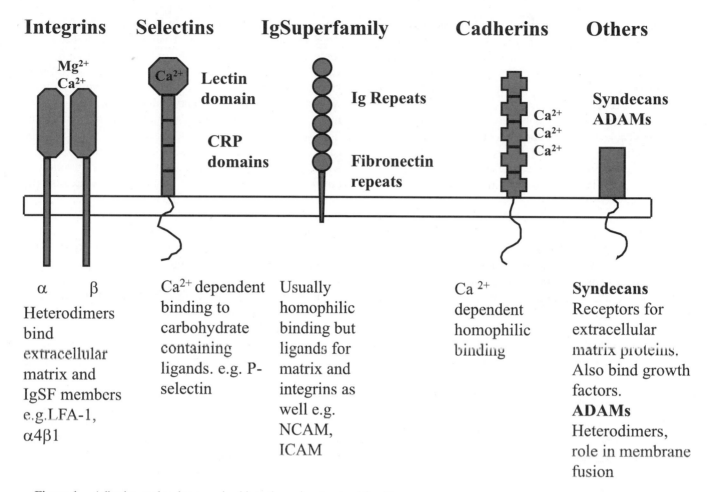

Figure 1.   Adhesion molecules organized into the major structural families.

has become apparent that cell adhesion molecules are more than just structural elements. Often, these molecules are not randomly distributed in cell membranes, but reside in complex structures such as focal adhesion contacts, microvilli, synaptic and adherens junctions, all of which contain specialised cytoskeletal and signalling proteins (10). Some growth factor receptors also appear to reside in adhesive structures, and there is accumulating evidence that cell adhesion molecules not only stick cell together, but can impose order and organization on many other cell surface molecules. In fact these "cis" interactions between cell adhesion molecules and other membrane receptors within T cell membranes appear to be equally as important as "trans" interactions between cells, and provide a wide degree of flexibility to the adhesive repertoire of T cells (10,11).

As the number of cloned T cell surface proteins has increased, the distinction between receptors that are predominantly adhesion receptors or predominantly signalling receptors has become blurred (12). A working definition of a cell adhesion molecule is a cell surface receptor whose interaction with ligand is capable of attaching a cell to either another cell or extracellular matrix. Unlike the recognition and signalling events involved in classical ligand-induced receptor dimerization (13), the linkage of cell adhesion molecules to the cytoskeleton appears to be critical for efficient cell-cell adhesion to take place. In addition, recognition between cell adhesion receptors often leads to receptor multimerization and the redistribution of receptors within the cell membrane. This ability of cell adhesion molecules to cluster within cell membranes, often becoming tethered to components of the cytoskeleton during the process, is an almost universal feature of this family of cell surface proteins and underspins the "spread" phenotype characteristic of adherent leukocytes. This contrasts with "round" phenotype characteristic of leukocytes in flowing blood (14).

The individual affinity constants for the interaction between T cell adhesion molecules is usually low Ka (10^6 to 10^8), unlike growth factors and their receptors where the individual affinity constants are much higher 10^{10} (Fig. 2a). It remains unclear whether affinity constants provide any useful information in the context of leukocyte cell adhesion, particularly since these cellular interactions are rarely irreversible, reflecting the dynamic process in which cell adhesion molecules play key roles, such as cell spreading and migration. In fact, the kinetic parameters of on and off rates are much more useful parameters in the field of cell adhesion, since the time scale for formation and reversal of bonds (characterised by the physical measurements of on and off rates) vary widely between different types of receptors (Table 1). The implication of this dynamic nature of T cell adhesion is that whether a receptor ligand pair acts as an adhesion molecule depends not only on the magnitude of the adhesive force, but the time scale over which cell contact occurs (15).

At one extreme, selectins utilize carbohydrate bonds with fast on and off rates to allow fast-moving leukocytes to be captured from the circulation. In addition to the short lifetime of these selectin-ligand bonds, leukocytes have evolved specialised structural features (microvilli) that facilitate leukocyte-endothelial interactions such that rolling may occur (16,17). In contrast, integrin-mediated adhesion is characterised by receptor-cytoskeletal crosslinking, and is often associated with the formation of stable clusters, leading to strong cell-substrate binding and cell anchorage. Even in this case, however, there is turnover of bonds, since most cells retain the ability to migrate, which requires regulated integrin release during detachment of cells from their substrate (18) (19).

In addition to recognizing and binding their ligands, adhesion molecules must be able to withstand force. How cell adhesion molecules respond to force is an important determinant of their function. The response of a receptor to force can be characterised by its "reactive compliance", which is a measure of how the dissociation rate, or off rate, varies with applied force (20). A low value implies a bond whose tendency to dissociate changes little with applied force, and is therefore likely to maintain adhesion in the face of applied force. Cell surface receptors may be capable of supporting adhesion in experimental models where little force is applied, but may never function as physiologically-relevant adhesion molecules because the environment *in vivo* demands the ability to withstand much greater forces. Therefore because a receptor can support adhesion in static adhesion assays does not mean that it ever does so *in vivo*, especially if it always exists in the presence of other adhesion molecules that are specialised for anchorage functions.

Therefore, the rate at which a receptor makes and breaks its bonds (the on and off rates), and the response of these bonds to applied force define whether it can act as an efficient adhesion molecule in a given environment. In addition to these inherent properties of the receptor-ligand bond, their density, spatial distribution and linkage to microfilaments and the cytoskeleton are also important, because they affect the loading applied to the bonds. In general, the higher the number of bonds at any time, the less the force each needs to support to maintain adhesion. However, the force applied during cell adhesion is unlikely

Table 1   Factors that determine the functional binding of adhesion molecules. See text for details

Parameter	Description	Desirable value for efficient adhesion
On rate	The rate at which bonds form at a given concentration of receptor and ligand. Defines the probability of a bond forming within a given timescale	High on rate for rapid bond formation e.g. selectins
Off rate	The rate at which formed bonds spontaneously disrupt Defines the lifetime of a bond.	Low off rates tend to give inherently stable bonds. High off rates e.g. selectins lead to rolling adhesion.
Affinity	On rate divided by the off rate. Defines the number of bonds formed at equilibrium. Of limited use in cell adhesion because prolonged adhesion itself frequently causes changes in cell behavior affecting receptor density such that equilibrium is rarely reached	Not clear. Successful adhesion molecules may have quite low affinities
Reactive compliance	Rate of change in the off rate with applied force.	Low reactive compliance implies strong adhesion
Distribution (clustering)	Poorly defined parameter often referred to as avidity, but not a property of the individual bond per se.. Represents redistribution of bonds which alters the way in which they bear a load	Clustering (avidity) may represent the means by which adhesion receptors are activated to support adhesion

to be equally spread throughout all bonds at once. Cells may "peel" apart, breaking individual bonds in succession, as occurs during rolling interactions. When bonds are clustered together rather than evenly spread, more bonds share the loading force from the start, and so adhesion strength is significantly increased (21). The biological correlate of such clustering is cytoskeletal crosslinking at sites of focal contacts. Therefore, in molecular terms, the processes required for two cognate cell surface proteins to interact and recognise each other (for example, T cell receptor recognition of peptide and MHC) are very different from those involved in cell adhesion. For cell adhesion to occur, both appropriate recognition and spatial organization of cytoskeletal proteins are required to ensure that strong cell-substrate binding occurs (22) (Fig. 2b). Some of these parameters, which characterise the molecular interactions particularly important for cell adhesion, are summarized in Table 1.

3.   WHY IS T CELL ADHESION IMPORTANT?

3.1.   Gives Flexibility to T Cell Responses

The leukocyte cell surface is a dynamic place. The ability of T cells to change the expression and activity of cell surface receptors with changes in their activation and differentiation state is truly impressive. These changes in "wardrobe" allow for the accurate positioning of T cell subsets within environments as diverse as the blood stream, secondary lymphoid organs and solid tissue. Since leukocytes do not have cilia and cannot swim, they depend on adhesion receptors to provide traction for migration. This requires a cyclical process of adhesive formation and release. T cells migrating through tissue undergo repetitive transient adhesion with surrounding stromal cells, which enables the T cell receptor to scan for foreign antigen present on cells. If foreign antigen is encountered, the T cell receptor triggers the rapid

Cell adhesion is more than just recognition

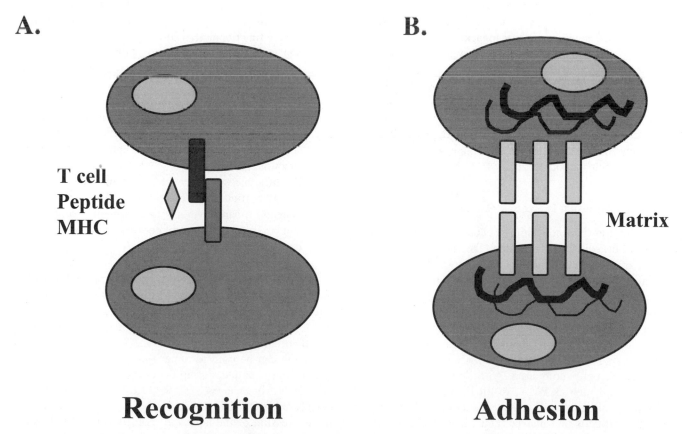

Figure 2.   T cell adhesion involves the transmission of information from the cell to its environment and from the outside into the cell (outside in) and allows the cell to interact actively with its environment.

activation of leukocyte integrins, leading to cell-cell adhesion. In the case of CD8 T cells recognizing antigen in the context of Class I MHC molecules, this leads to rapid killing of the target cell and, in the case of CD4 lymphocytes, to appropriate help for the engaged cell (e.g., T help for B cells). However, if foreign antigen is not recognized, then the T cell rapidly terminates adhesion and begins to migrate in search of other cells. To increase the chances of successful engagement, T cells share chemokine receptors with cells they are destined to meet (such as other T cells, B cells or dendritic cells). This permits the bringing together of multiple cell types, which need each other in order to function efficiently, in the correct microenvironment (23,24). In fact, recent work has pointed to the pivotal role that chemokines and their receptors play in co-ordinating the activation and distribution of T cells subsets within tissue microenvironments during the generation of adaptive immune responses. Perhaps not surprisingly, aberrant expression of chemokines and their receptors leads to pathology. In some cases, chemokines produced by pathogens such as viruses or bacteria can lead to the inappropriate distribution and diversion of T cells away from their intended destination (25). In other cases, the expression of chemokine receptors can lead to their delivery to inappropriate destinations. An example of this occurs in the MRL-*lpr* mouse, which demonstrates many autoimmune features. These mice contain very large numbers of T cells within B cell follicles, a site where their entry is closely guarded. These T cells are unusual in lacking CD4 and CD8, and are thought to contribute to the production of autoantibodies, which characterize these mice. These T cells express high levels of the chemokine receptor CXCR5, which is normally restricted to B cells, allowing them to respond to the B cell chemoattractant chemokine BCA-1. This results in T cell accumulation and retention within B cell follicles, where they provide inappropriate signals to drive the production of autoantibodies (26).

3.2. T Cell Adhesion is Exquisitely Regulated

In order for transient adhesion to occur, T cell adhesiveness must be exquisitely regulated. This is a fundamental property of leukocyte adhesion molecules. The adhesive mechanisms employed by leukocytes appear to be subtly different from those used by "sedentary cells", such as epithelial cells, since leukocytes circulate as non-adherent cells in blood and yet migrate as adherent cells in tissues. There is an intriguing difference between the effects of phorbol esters (which activate protein kinase C pathways) on epithelial cells compared with leukocytes. While phorbol esters, such as PMA, result in the disruption of adhesion in epithelial cells, they enhance the adhesion of leukocyte cell cells and matrix. Thus, it is likely that the signalling connections between leukocyte cell adhesion molecules and the intracellular environment differ when

compared with those in adherent cells. This probably reflects the fact that a crucial function of leukocytes is to make multiple, transient interactions with many cell types rather than maintain a particular position by firm adhesion to extracellular matrix proteins or adjacent cells in a specific tissue (9,17).

Leukocyte integrins are fundamental to efficient T cell function. This has been demonstrated both with the use of blocking anti-integrin antibodies as well as the production of mutant mice in which leukocyte-specific integrins have been knocked out (27–32). In addition, a number of human diseases, the leukocyte adhesion deficiencies (LAD), have been described in which deficiencies of specific adhesion molecules lead to defective adhesion, in effect human "knockouts" (33–36). Allelic variants of adhesion receptors also exist, although at present, their role in disease pathogenesis is unclear.

There are at least five potential ways in which T cell adhesion can be regulated:

1. By increasing the level of expression of adhesion molecules and their receptors in response to activating stimuli, cytokines, or infectious agents, e.g. EBV.
2. By altering the level of functional activation; not all adhesion receptors are constitutively active, and integrins in particular require a triggering signal to induce functional activation.
3. Via cross talk between adhesion receptors, which provides a mechanism for the rapid amplification of adhesive interactions.
4. By the clustering of cell surface receptors. The clustering of adhesion receptors brought about by changes in the cytoskeleton, often in response to signals transduced by adhesion molecules, enhances functional binding.
5. By differences in glycosylation; some adhesion molecules, particularly selectins and selectin ligands, depend on carbohydrates to bind their receptors and can therefore be regulated by post-translational changes in glycosylation.

The best studied and probably most important mechanism by which leukocyte integrins regulate their adhesive interactions is through changes in the level of integrin expression and functional activation. Integrin expression on T cells often goes hand in hand with T cell activation (see later) and can be used as a marker of antigen-experienced T cells. For example, during the progressive differentiation of T cells from naïve (CD45RA+) to memory or effector cells (CD45RO), the level of β1 and β2 integrins as well as the expression of the adhesion molecules CD2 and LFA3 increase dramatically (37). These differences between naïve and memory T cells in terms of adhesion molecule expression not only provide excellent markers of the differentiation state of T cells, but highlight the fact that these two subsets of T cells display different patterns of migration and localization as well as distinct functions (38,39).

T cells undergo multiple adhesive interactions, some of which are relatively short and others more prolonged (Figure 3). For example, an actively-migrating T cell requires multiple short-lived integrin-mediated interactions with its substrate, whereas adhesion between a T cell and dendritic cell needs to be sustained to allow for efficient activation of the T cell. Leukocyte integrins are able to alternate between low avidity and high avidity states to support these functions. Within minutes of cell activation, integrin-mediated adhesion can increase markedly, only to return to a low avidity state within 30–60 minutes. Killing of target cells by cytolytic T cells involves cycles of adhesion mediated by the integrin LFA1 (CD11A, CD18). A single cytolytic T lymphocyte can

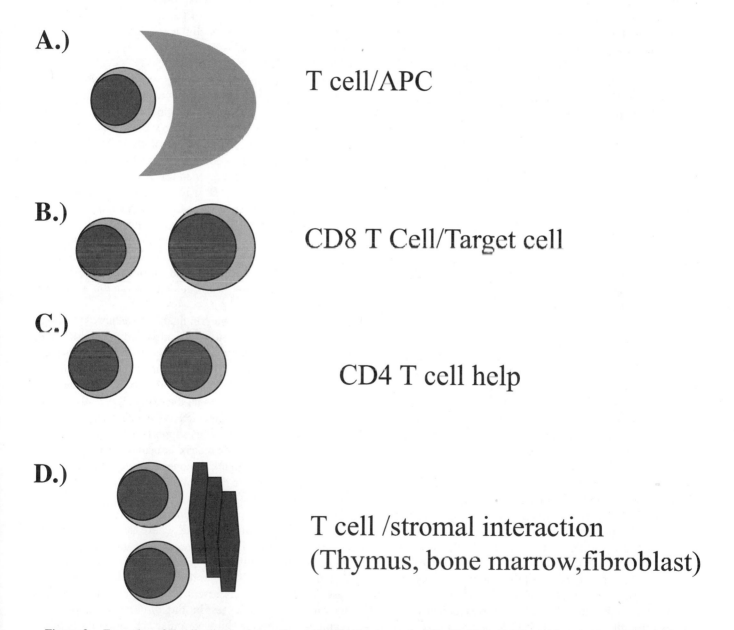

A.)   T cell/APC

B.)   CD8 T Cell/Target cell

C.)   CD4 T cell help

D.)   T cell /stromal interaction (Thymus, bone marrow,fibroblast)

Figure 3.   Examples of T cell adhesive interactions. A) Adhesion is required for T cells to interact with antigen presenting cells and provides important co-stimulatory signals for T cell activation. B) Adhesion allows T cells to talk to each other, for example CD4 interactions with CD8 T cells. C) Adhesion is critical for target cell killing by cytotoxic T cells. D) Adhesion allows T cells to interact with other cell types e.g. stromal cells and with the extracellular matrix.

kill many target cells by migrating from one cell to the next, a process which involves cycles of integrin activation and deactivation. A similar example is found in the scanning of antigen-presenting cells by T cells in the lymph node. The T cell engages and then rapidly disengages from multiple DCs until it finds the cell displaying its unique activating antigen, at which point adhesion is sustained to allow for efficient activation of the T cell. Perhaps the most dramatic cycling of adhesion and de-adhesive events occurs during T cell endothelial interactions (see later).

The signalling events that regulate integrin avidity are poorly understood, although recent work has implicated a large number of intracellular cytoplasmic proteins, which interact with the cytoplasmic domains of integrins (3). During immune surveillance, physiological interactions between T cells and target cells are usually restricted to minutes or one or two hours. During pathological processes, T cells undergo sustained adhesive interactions, often contributing to their accumulation within sites of inflammation. However, it remains unclear how this sustained, and often inappropriate, interaction either with other cells or extracellular matrix is maintained *in vivo*. Changes in deposition of tissue extracellular matrix occur at sites of inflammation, suggesting that signals from this altered tissue microenvironment may regulate lymphocyte response (40). Little is known about how components of the microenvironment affect integrin function, but a variety of inflammatory cytokines, such as IL-1, TNF and interferon-γ, as well as interactions with stromal cells and other leukocytes probably act to regulate T cell adhesion and migration at sites of inflammation.

3.3.   Cell Activation as an Adhesion Response in T Cells

It is clear from the discussion above that T cell adhesion molecules are more than cellular "glue". They are ideally designed to mediate interactions between the T cell and its cellular and extracellular environment and to translate these interactions into information that tells the cell where it is and allows it to respond appropriately. Recent studies have suggested that there is close co-operation between antigen receptors, such as the T cell receptor, and integrins during the activation of T cell adhesion (40). The role of T cell adhesion molecules in costimulation is discussed in detail elsewhere, but is critical for effective T cell activation. Integrins such as LFA-1 and VLA-4 promote adhesion to apposing cells, allowing cognate recognition to occur. In addition, these interactions provide bi-directional co-stimulation. Thus, adhesion through integrins enhances signalling through the TCR, and TCR activation acts as a powerful signal to further activate integrin-mediated adhesion. These interactions occur as part of a cascade of adhesive events that permits adhesion to be rapidly amplified in the presence of cognate recognition of antigen. In addition to integrin-activation, the TCR is a focus for clustering of integrins such that adhesion and co-stimulation are maximized during antigen recognition (41,42). In addition to co-stimulation provided by cell-cell interactions, signals transduced from the ECM by $\beta 1$ integrins, such as VLA-4, VLA-5 and VLA-6, can also deliver co-stimulatory signals to the T cell, allowing a wide range of microenvironmental stimuli to effect T cell activation (43). The molecular mechanisms underpinning this cross-talk between cell activation and integrin function remain largely unknown. Insights from malignant leukemia cells suggest that the altered adhesive phenotype of these cells is a consequence of changes in critical intracellular regulatory proteins, such as phospho-inositide3-kinase or members of the Ras and Rho GTPase family.

Lymphocytes and, to a lesser extent, macrophages, proliferate at sites of inflammation where the shift from a resting to an activated state is a critical determinant during an inflammatory response. It is now clear that changes in adhesion are crucial components of the activated T cell phenotype, and that cellular activation and adhesion are closely linked. Evidence in man for this has come from the finding that lymphocytes from patients with leukocyte adhesion deficiency (LAD), in addition to having defects in migration, are often unable to generate normal immune responses. It remains unclear why integrins are such an important component of leukocyte activation. However, the close integration of adhesive and mitogenic signals is not unique for T cells, but has been demonstrated in many cell types (3).

The requirement for integrin ligation to achieve full leukocyte activation (co-stimulation) also provides a level of control over the inflammatory response. A dynamic aspect of the tissue remodeling that accompanies wound repair, cancer and inflammation is the ability of local proteolytic enzymes to digest and remodel the extracellular matrix (44). This remodeling is accompanied by changes in the ability of the ECM to stimulate and support immune responses. Undigested extracellular matrix is usually monomeric and tends to stimulate immune responses, whereas the fragments released by proteolytic degradation and remodeling are often immunosuppressive (43).

The need for integrin co-stimulation in lymphocyte activation will determine where activation can and cannot take place. Because integrin ligands are primarily found on the endothelium or in the extracellular matrix, cell activation will be confined largely to the extravascular space. This is essential to prevent the premature activation of leukocytes in the circulation. When this regulation is lost and intra-vascular activation takes place, devastating consequences result, including vasculitis and the autoimmune phenomena and multi-organ failure that accompany septic shock.

3.4. Leukocyte Endothelial Interactions

A complex system of leukocyte trafficking has evolved to provide immune surveillance and to allow the immune system to recognise and respond rapidly to foreign antigen wherever it enters the body (45). When the factors that regulate this process are lost or disrupted, leukocyte recruitment continues inappropriately and pathological inflammation ensues (46,47). In many autoimmune diseases, aberrant expression and function of adhesion molecules leads to disordered leukocyte recruitment, retention and activation of lymphocytes within tissue. This inappropriate activation results in tissue destruction and the establishment of chronic inflammation (48). Thus, the expression of adhesion molecules that regulate lymphocyte recruitment is a crucial pathogenetic factor in the development of autoimmune disease.

4. LYMPHOCYTE RECIRCULATION UNDER PHYSIOLOGICAL CONDITIONS

Circulating leukocytes must first recognise and then bind to adhesion molecules on the endothelium before they can be recruited into tissue. This process is regulated by sequential, combinatorial interactions between the leukocyte and the endothelium that lead to the capture, arrest and transmigration of circulating cells. This multi-step model of leukocyte adhesion to vascular endothelium is broadly applicable to most situations, although the details of the signals involved will differ depending on the setting.

In the generally accepted model (45,49,50), the circulating leukocyte is initially captured by receptors on endothelial cells that possess the rapid functional kinetics required to capture a fast-flowing leukocyte and bring it into contact with the vessel wall. Most of these receptors belong either to the selectin (51,52) or the immunoglobulin superfamilies of adhesion proteins (53), and they cause the leukocyte (54) to roll or bump on the endothelium, where it can pick up signals from the endothelial cell surface that activate strong integrin-mediated adhesion and arrest. Most of these activating factors are delivered by members of a family of chemotactic cytokines, "chemokines", which bind to specific G protein-coupled receptors on the leukocyte surface (55–57). Occupancy of these receptors induces a cascade of intracellular signalling events that result in presentation of high-affinity integrin receptors on the leukocyte that binds competently to their immunoglobulin-family counterreceptors on the endothelial surface (58,59). In addition to chemokines, cell surface receptors, including CD31, CD73 and the L-selectin ligand GLYCAM-1 can also, under some circumstances, trigger integrin activation (60,61). If the firmly adherent leukocyte subsequently receives an appropriate migratory signal, it will then undergo transendothelial migration into tissue, where it follows a hierarchy of chemotactic agents towards the focus of inflammation (62).

In order for inflammation to be established, the leukocyte must not only be recruited, but also retained in tissue as a consequence of local activating and survival signals. In the absence of such signals, the lymphocyte will either die by apoptosis or, in the case of lymphocytes, return to the circulation via the lymphatics (63,64).

5. HOMING PATTERNS OF NAÏVE AND MEMORY/EFFECTOR LYMPHOCYTES

Lymphocyte adhesion to endothelial cells appears to follow this paradigm, under physiological and pathological conditions. However, subsets of lymphocytes display different receptors that determine to which anatomical site they are recruited. This is best illustrated by the marked differences between naïve and memory T lymphocytes. Naïve T cells migrate almost exclusively between the circulation and secondary lymphoid tissues, whereas memory cells are largely excluded from lymph nodes and instead migrate into tissue, returning to the circulation via lymphatics (65). The recruitment of T cells to lymph nodes is regulated by the presence of molecules on high endothelial venules (HEV) that are specifically recognised by naive T cells. These molecules include endothelial adhesion molecules, such as the peripheral node addressin, which binds to L-selectin, and the chemokine SLC (secondary lymphoid tissue chemokine), which binds to a specific receptor CCR7 (23). Both L-selectin and CCR7 are preferentially expressed on naive T cells and down-regulated after activation (66–69).

The principle function of secondary lymphoid tissues is to serve as a meeting place for antigen, antigen-presenting cells, and antigen-specific lymphocytes. These different cell types are brought together in the appropriate micro-environments by carefully orchestrated adhesive and migratory signals (23,70). Antigen-bearing dendritic cells exit peripheral sites of infection by altering their adhesive phenotype to allow their release from tissue (71). They then use CCR7 to enter the lymph node and to migrate to the T cell zones, where they present antigen in conjunction with MHC molecules to T cells (24). Circulating naïve T cells use L-selectin and CCR7 to enter lymphoid tissue via high endothelial venules, and then spend several hours migrating through the T cell zone, during which time they make contact with multiple dendritic cells via transient integrin-mediated interactions before reentering the peripheral circulation. Resting B cells travel through lymphoid tissues by the same pathways as T cells, but home to B cell-rich areas, where they reside briefly before returning to the circulation. After binding antigen, B cells relocate to outer T cell zones, promoting an encounter between antigen-specific T and B cells. This highly orchestrated movement of DCs, and T and B lymphocytes into and within secondary tissues depends on the coordinated regulation of adhesion and migration controlled by specific adhesion molecules and chemokine-chemokine receptor interactions (23,72). The activation of T cells occurs as a consequence of interactions with antigen-presenting cells (APCs) that involve integrin-mediated adhesion, and also signals from chemokines, which attract the interacting cells to each other in the appropriate compartment. For example, the chemokine ELC (Ebl1-ligand chemokine), which also binds CCR7, is secreted by a subset of dendritic cells (DCs) in the T cell compartment of the lymph node and acts to bring T cells and DCs together (70,73,74). In addition, DCs that have been activated in tissue also upregulate CCR7, allowing them to respond to SLC and ELC and to co-localize with T cells in the T cell compartment, thereby promoting optimal T-APC interactions (75). Similar mechanisms are also involved in the recruitment and positioning of B cells in lymphoid tissue. Mice lacking CXCR5 (BLR1), a chemokine receptor expressed by B lymphocytes, show defective formation of primary follicles and germinal centres in Peyer's patches and the spleen as well as a loss of inguinal lymph nodes. CXCR5-deficient B cells enter T cell areas within lymphoid tissues, but fail to home to B cell areas (76). The chemokine ligand for CXCR5, BCA-1/BLC, is highly expressed in lymphoid tissues and selectively recruits B lymphocytes (77).

After activation in a lymph node, naïve T cells differentiate into activated effector cells and long-lived memory T cells. These cells exhibit different migratory pathways dictated by changes in their cell surface receptors. For example, memory T cells lose expression of L-selectin, preventing them from binding efficiently to HEV in lymph nodes, but express higher levels of other adhesion molecules, including $\beta 1$ and $\beta 2$ integrins, that promote adhesion to activated endothelium in inflamed tissue (78). Furthermore, effector T cells express high levels of receptors, such as CCR2, CCR3 and CCR5, that interact with inflammatory chemokines and lose expression of CCR7, which is required for entry into lymph nodes (79–82). The result of these changes is that effector cells respond to tissue derived inflammatory signals and not to the physiological signals that drive recruitment to lymph nodes. Recent studies suggest that a subset of Th1 memory T cells maintain CCR7 expression, allowing them to re-circulate through lymph node and spleen during immune surveillance (75,83–87).

6. EVIDENCE FOR TISSUE-SPECIFIC MEMORY T CELL RECIRCULATION

There is accumulating evidence that, during their activation and differentiation in secondary lymphoid tissue, memory T cells are programmed to express specific combinations of homing receptors that bind preferentially to endothelium in the tissue where they were activated. Thus, T cells activated in axillary lymph nodes subsequently recirculate preferentially to the skin, whereas those activated in mesenteric nodes return to the gut (45,65). For this model to work, circulating memory T cells must be able to discriminate between endothelium in different tissues. Several studies report that some tissues display combinations of adhesion molecules on their endothelium that provide a unique molecular address that can be recognised by subsets of circulating leukocytes. These adhesion molecules are called "addressins" and, interestingly, most of them are tethering molecules (65). In addition, some chemokines are expressed in a tissue-specific manner, suggesting that the selection of cells to be recruited occurs at the capture and triggering phase rather than the arrest phase of the cascade. For example, migration of lymphocytes through the gut is mediated by the integrin $\alpha 4\beta 7$, which is expressed by subset of memory T cells that display gut tropism. $\alpha 4\beta 7$ allows these cells to recognise and bind to its ligand MAdCAM-1 (mucosal addressin cell adhesion molecule-1), which is expressed almost exclusively by mucosal vessels (88). In contrast, memory T cells that infiltrate the skin do not express $\alpha 4\beta 7$, but express the cutaneous lymphocyte-associated antigen (CLA) that allows them to bind E-selectin on dermal vessels. CLA is derived by fucosyltransferase VII-mediated glycosylation of the P-selectin glycoprotein ligand-1 (PSGL-1) (89), and is found on memory T cells in the skin, but not on T cells infiltrating other inflammatory sites (90,91). The skin and the gut are the best described examples of tissue-specific memory cell migration, but it seems likely that other tissues and organs are also controlled by distinct combinations of molecules. Preliminary studies suggest this may also apply to the liver, lung and brain (92–96).

There is now evidence that, in addition to classical homing receptors, memory cells that display tissue tropism express particular combinations of chemokine receptors. Thus, skin-homing memory T cells express high levels of the chemokine receptor CCR4 (97), whereas cells that migrate to the liver and gut express high levels of CCR5 and little CCR4 (80,98). Thus, patterns of lymphocyte recirculation will depend upon the combinations of molecules expressed on the lymphocyte and the endothelial chemokines and adhesion molecules that provide tissue with a unique molecular address. The differentiation of particular effector/memory T cells will be determined by the microenvironment in which the T cells was activated by antigen (99). The specific signals responsible are not known, but probably involve a combination of cytokines, stromal factors and the nature of the activating dendritic cells (23,26).

7. LYMPHOCYTE ACTIVATION ALTERS EXPRESSION OF ADHESION MOLECULES AND CHEMOKINE RECEPTORS AND DETERMINES WHERE THE LYMPHOCYTE MIGRATES

As discussed above, lymphocyte activation can alter the adhesive behavior of the cell via several distinct mechanisms. Lymphocyte activation and proliferation is associated with changes in the expression of adhesion molecules and chemokine receptors. The levels of integrins such as LFA-1, VLA-4 and $\alpha4\beta7$ and chemokine receptors such as CXCR3 and CCR5 increase. These changes involve protein synthesis and occur over several hours. A more rapid regulation occurs within seconds of TCR engagement and involves the clustering and conformational activation of cell surface integrins, resulting in increased affinity/avidity for endothelial ligands (100). Thus, TCR activation, particularly when accompanied by co-stimulation through CD28, enhances the ability of $\beta1$ and $\beta2$ integrins to bind to VCAM-1, ICAM-1 and ICAM-2 on endothelium (101,102). A more subtle regulation of adhesion molecule expression has been reported on Th1 and Th2 cells. The preferential accumulation of either Th1 or Th2 cells at a site of inflammation will, in part, determine the outcome of the local inflammatory response and might occur as a consequence either of preferential recruitment, retention or local differentiation of the cell. Differential recruitment appears to be at least partly responsible because of the different requirements these distinct functional subsets have for recruitment to tissue. There are difference in the expression of chemokine receptors and adhesion molecules on Th1 compared with Th2 cells (84). Most lymphocytes express the P-selectin ligand PSGL-1, but the ability to bind E- and P-selectin is restricted by the cell's ability to glycosylate PSGL-1 (103). The expression of the fucosyl transferases (104) responsible for this glycosylation is altered by the local cytokine microenvironment, and IL-4 reduces 3 fucosyltransferase expression in Th2 cells. This means that Th1 cells express more glycosylated PSGL-1 compared with Th2 cells and bind preferentially to P-selectin (104–107). There are also important differences in chemokine receptor expression between Th1 and Th2 cells, such as the fact that Th1 cells express CXCR3, CCR5 and CCR7 whereas Th2 cells express CCR3 and CXCR4. These differences determine where the cell migrates, as elegantly demonstrated for CCR7. The lack of CCR7 expression determines the ability of activated Th2 cells to provide B cell help and to drive antibody production. When murine Th2 cells were transduced with CCR7 and then adoptively transferred, they localized within the spleen in a Th1 distribution and failed to participate in B cell help *in vivo* (87).

8. LYMPHOCYTE ACTIVATION IN TISSUE ALTERS THEIR ADHESIVE/MIGRATORY PHENOTYPE AND PROMOTES RETENTION

A sustained inflammatory response in tissue depends not only on the recruitment of effector cells, but also on their retention and activation at the site of inflammation (63,64). In order for this to happen, the cell needs to adhere to local structures and cease its migratory behavior. This change from a migratory to a resident cells occurs as a consequence of changes in adhesion molecule and chemokine receptor expression in response to local activating signals, such as cytokines, growth factors and, in the case of lymphocytes, engagement of TCR by antigen. As discussed above, activation by antigen and differentiation into effector cells in secondary lymphoid tissue increases expression of adhesion molecules such as LFA-1 and VLA-4, and chemokine receptors such as CXCR3, CCR2 and CCR5 that promote recruitment to tissue. However, subsequent engagement of the TCR at sites of chronic inflammation downregulates many chemokine receptors, including CCR2 and CCR5, while activating integrin-mediated adhesion. The effect of this is to immobilize the lymphocyte at the site of antigen exposure in tissue (108–110). Furthermore, the interaction of lymphocyte integrins with extracellular matrix components in tissue provides co-stimulatory signals that enhance lymphocyte activation and survival/persistence at the site of inflammation (111).

9. TISSUE-SPECIFIC ENDOTHELIAL ADHESION MOLECULES AND THE UNIQUE TISSUE ADDRESS

The endothelial cell is the point of first contact and the site on entry into tissue for a circulating leukocyte. This provides the endothelium with the ability to select how many and which types of leukocytes will be recruited. This is achieved by regulating the expression of combinations of adhesion receptors and chemokines on the endothelial cell surface to provide a molecular address that can recognised by specific subsets of leukocytes. Some molecules, such as ICAM-2 and CD31, are constitutively expressed on most endothelial cells and change little with inflammation. Others, such as ICAM-1, are normally at low levels and increase on activation, and some, such as E-selectin and VCAM-1, only appear following activation (52). Other molecules, such as MAdCAM-1 and VAP-1, define endothelium in a particular microenvironment because they are restricted to particular tissues under basal conditions (88,112,113). The expression of these molecules is presumably determined during endothelial cell differentiation under the influence of specific signals in the local microenvironment. The nature of these stimuli is poorly understood, but likely includes local cytokines, growth factors and signals from surrounding cells and ECM. For example, lymphotoxin and TNF-α both play important roles in inflammation and lymphoid organ development by inducing the expression of endothelial adhesion molecules. However, their effects differ: while both induce endothelium to express VCAM, ICAM and E-selectin, only lymphotoxin induces the expression of MAdCAM-1. Furthermore, neutralization of these cytokines *in vivo* results in distinct anatomical abnormalities (114–116). Another example is the liver, where the unique differentiation of the sinusoidal endothelium appears to be dependent on paracrine growth factors secreted by adjacent cells, including hepatocytes (117,118). Thus, the microenvironment drives endothelial differentiation and the expression of specific adhesion molecules involved in lymphocyte subset recruitment. The nature of these signals is crucial to understanding tissue specificity of homing and disease pathogenesis.

10. ENDOTHELIAL ACTIVATION ENHANCES LEUKOCYTE RECRUITMENT

10.1. Increased Adhesion Molecule Expression with Inflammation

The signals responsible for activating endothelium in the presence of inflammation are better understood, although tissue-specific differences exist in the endothelial response to some cytokines. Broadly, the local presence of proinflammatory cytokines such as IL-1, TNF-α or IFN-γ causes increased expression of a number of endothelial adhesion receptors, thereby extending the range of leukocyte subsets that can be recruited (51,51). The infiltrating leukocytes then amplify the endothelial activation via several mechanisms, including the secretion of cytokines that directly activate endothelium, but also via adhesion-dependent mechanisms. For example, engagement of endothelial ICAM-1 by lymphocyte LFA-1 results in activation of signalling pathways in the endothelium that promote transendothelial migration and increase the expression of other endothelial adhesion molecules, most notably VCAM-1 (119,120). Another mechanism by which lymphocytes have been proposed to activate endothelium is via endothelial CD40. Endothelial cells express this member of the TNF receptor superfamily and its ligand ligand, CD154 is expressed on activated lymphocytes. Upregulation of endothelial CD40 occurs with activation and engagement of CD40 by either antibody or trimeric CD154 increases expression of the adhesion molecules, E-selectin, VACM-1 and ICAM-1 that promote lymphocyte adhesion to endothelium (121). Particular cytokines can regulate the expression of specific adhesion molecules. Thus, a combination of IL-4 and TNF-α selectively upregulates VCAM-1 on endothelial cells in culture (122,123), whereas IFNγ inhibits activation-induced

selectin expression (124). Moreover, temporal expression of endothelial adhesion molecules differs. P-selectin, which is stored as intact protein in endothelial Weibel-Palade bodies, can be rapidly secreted and expressed on the cell surface in response to agonists such as hydrogen peroxide and histamine (125,126), whereas expression of E-selectin requires *de novo* protein synthesis and only appears several hours after endothelial activation.

11. CHEMOKINE SECRETION AND PRESENTATION BY ACTIVATED ENDOTHELIUM

The nature of the inflammatory stimulus will also determine the local secretion and presentation of chemokines by the endothelium. Chemokines are produced by a wide variety of cell types, including immune cells, stromal cells and endothelium. Their expression can be rapidly induced following stimulation by a variety of agents, including bacterial LPS, viruses, proinflammatory cytokines and activation of cell surface receptors (127). The cellular source of chemokines is variable; MCP-1 and IL-8 are almost universally expressed whereas PF4, PBP and CTAP-111 are produced only by platelets. Activated T cells express a range of chemokines at both the mRNA and protein levels, and non-hematopoeitic cells, including endothelial and epithelial cells, are potent sources of many chemokines (80). Infiltrating monocytes and activated lymphocytes are a major source of chemokines, and the chemokines they secrete will determine the subsequent composition and duration of the inflammatory response. For example, antigen-specific CD8+ CTL to myelin proteolipid protein peptides, a putative antigen in multiple sclerosis, secrete the chemokines MIP-1α, MIP-1β, IL-16 and IP-10, which act to recruit CD4+ T cells of the same TCR specificity. Thus, CD8+ cytotoxic T cells can promote and maintain inflammatory responses by recruiting specific CD4 subsets (128). If chemokines are to trigger adhesion and migration effectively, they must be retained at the vessel wall to allow interaction with circulating leukocytes. This is achieved by immobilization on proteoglycans or non-signalling receptors in the endothelial glycocalyx (59,129–131). Chemokines show differential binding to proteoglycans and, since proteoglycans vary from site to site and with activation, this provides a mechanism by which tissues can selectively express a particular cohort of chemokines (132). This process is enhanced because endothelial cells can present chemokines secreted by other underlying cells (52) by a process of transcytosis, and (133) chemokines produced downstream in the circulation can be captured and displayed by endothelial cells (134,135). Thus, the chemokines presented at the endothelium to the circulating leukocyte will reflect not only the activation state of the endothelium, but also the presence of activated stromal and inflammatory cells within the local microenvironment (134,135).

12. ADHESION MOLECULES IN AUTOIMMUNE DISEASE AND POTENTIAL FOR THERAPY

Alterations in adhesion molecule and chemokine/chemokine receptor expression by endothelial cells and lymphocytes will determine when and where leukocytes adhere to the vessel wall and penetrate tissue. In addition, adhesive interactions within tissue at sites of chronic inflammation will

determine the retention, survival and activation of infiltrating leukocytes. It follows that adhesion molecules will be crucial in the pathogenesis of autoimmune disease and is thus unsurprising that aberrant adhesion molecule expression has been described in virtually all the reported autoimmune diseases (rev. in 136–145). There are several ways in which adhesion molecules might affect disease pathogenesis in autoimmune disease:

1) Increased or inappropriate lymphocyte activation via costimulatory signals from adhesion molecules;
2) The enhancement of leukocyte recruitment via activated endothelium;

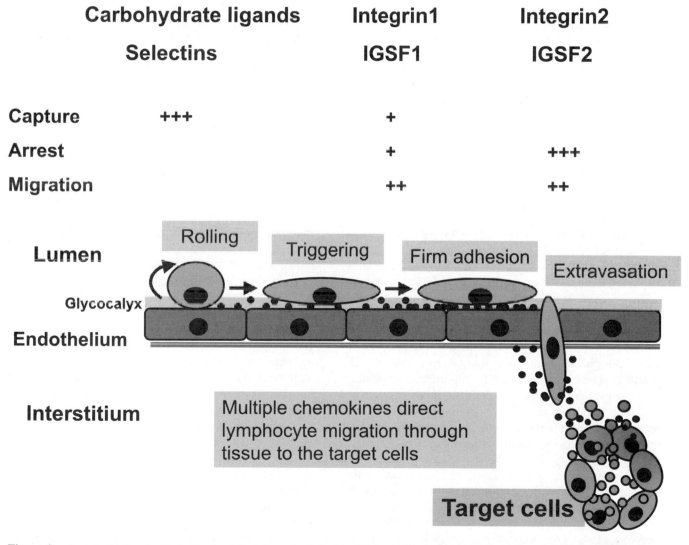

Figure 4. A sequential cascade of adhesive interactions mediates T cell interactions with endothelium. Free-flowing cells in the circulation are captured by tethering receptors (usually carbohydrate dependent) expressed on endothelial cells. This induces the cell to roll on the endothelium, allowing it to pick up activating signals from chemokines localized to the glycocalyx. Chemokines activate G-protein coupled receptors on lymphocytes resulting in the conformational activation of integrins and strong adhesion to endothelial immunoglobulin superfamily members. Chemokine activation also leads to cytoskeletal reorganisation and a migratory phenotype allowing the cell to follow a gradient of chemokines to the site of inflammation.

3) Inappropriate adhesion to the endothelium and perivascular activation resulting in endothelial damage in vasculitis;
4) Retention and survival of activated lymphocytes at sites of chronic inflammation;
5) Activation of effector functions at sites of chronic inflammation by adhesive interactions with activating cells and targets;
6) Dysregulated angiogenesis and neovessel formation at sites of chronic inflammation that promotes continuing cellular recruitment.

Many studies have reported increased expression of endothelial adhesion molecules by activated endothelium in target tissues. A role for these molecules in disease pathogenesis is supported by animal studies in which inflammation can be reduced using either blocking anti-adhesion molecule antibodies or small molecule inhibitors (30,146). Most of these studies have concentrated on the role of the main integrin-mediated pathways and, depending on the disease model, variable anti-inflammatory responses can be induced by inhibiting $\beta1$ or $\beta2$ integrin-mediated adhesion. More interesting is the example provided by inflammatory bowel disease where in both animal models and preliminary patient studies the disease can be significantly ameliorated by blocking the tissue-specific interaction between the $\alpha4\beta7$ integrin on gut-homing T cells and MADCAM-1 on mesenteric vessels (147–150). These studies provide proof that this pathway is of functional importance in lymphocyte homing to the gut. The findings that mucosal lymphoblasts will also bind strongly to synovial endothelium provides a mechanism to explain the link between gut inflammation and arthritis (151,152). Thus, an organ or tissue that demonstrates aberrant expression of homing molecules might become the site for recruitment of lymphocytes activated at a distant site.

The role for selectins in recruitment of lymphocytes to site of chronic inflammation is less clear-cut. Endothelial E- and P-Selectin are particularly important in regulating neutrophil recruitment to sites of tissue injury, as demonstrated by the ability of selectin ligand mimetics to reduce neutrophil accumulation at sites of inflammation (153). Selectins are also important in lymphocyte recruitment, although it appears that selectins play different roles in particular tissues or organs. Thus, Th1 responses in the skin are associated with enhanced lymphocyte binding to endothelial selectins, whereas T cells at Th1-driven sites of inflammation in the gut do not express E or P-selectin receptors (154). This might explain why lymphocytes in the liver of patients with autoimmune liver disease do not bind E-selectin (155). There is now evidence for the involvement of lymphocyte expressed L-selectin in lymphocyte recruitment to chronic inflammatory sites (69,156,157). Lymphocytes fail to migrate into inflamed skin in L-selectin-deficient mice, and functional ligands for L-selectin have been detected on the endothelium of inflamed tissue in animals and in humans (158,159). The ability to inhibit selectin-mediated adhesion with glycans/ oligosaccharides makes them attractive targets for anti-inflammatory therapy (160).

Evidence for activated lymphocyte adhesion molecules in autoimmune diseases comes from studies demonstrating either increased expression or, more impressively, increased functional binding (140,145). Lymphocytes at sites of chronic inflammation in tissue express a range of integrins that allow them to interact with surrounding cells and matrix. The role of these integrins on tissue infiltrating lymphocytes is not fully understood. For instance, many T cells in the inflamed rheumatoid joint and in the liver of patients with autoimmune disease express high levels of the $\alpha E\beta7$ integrin. This molecule is highly expressed by intraepithelial lymphocytes at mucosal sites, where it is believed to regulate survival by interactions with E-cadherin, but what role it plays in the joint or inflamed liver is unclear (161–163).

Potentially, the expression of multiple activated integrins by lymphocytes in vivo could enhance not only their recruitment and retention in tissue, but also their activation. There is evidence of increased expression of molecules such as ICAM-3 and LFA-3 on cells at chronic inflammatory sites in the rheumatoid joint that provide adhesive interactions to trigger the activation of effector functions in infiltrating lymphocytes (164–167). Furthermore, the increased expression of important co-stimulatory molecules such as OX40 and CD40 in the inflamed gut in inflammatory bowel disease might modulate the co-stimulatory signals received by infiltrating T cells and alter their activation status (168,169).

The interplay between adhesion molecules and chemokines will be critical for determining the recruitment and retention of lymphocytes at inflammatory sites. The additional ability of some of the chemokines to provide co-stimulatory signals for T cell activation in association with adhesive interactions suggests they may also modify local lymphocyte activation (170). Many chemokines have been reported at sites of chronic inflammation, and determining which chemokine is critical for pathogenesis can be difficult. The study of the distinct phenotypes associated with various chemokine and chemokine-receptor knockout mice is beginning to elucidate this area. However, several of the chemokine knockouts have provided unexpected results, as illustrated by the recent reports of the CCR1 knockout. In these animals, which lack the CCR1 receptor for MIP and RANTES, Th1 responses in a model of nephritis were enhanced compared with wild-type animals (171).

13. SOLUBLE ADHESION MOLECULES

Adhesion molecule expression is tightly regulated by proteolytic shedding, and the ectodomains of many leukocyte and endothelial adhesion molecules are shed in a soluble form. There is increasing evidence that these soluble molecules are not just waste products, but that the cleavage and shedding of these membrane proteins alters the adhesive properties of the cell and resets the intracellular signalling pathways not only in these cells, but also in neighboring cells (172). Because the levels of many of these molecules is increased in inflammatory disorders, they have generated a great deal of interest as clinical markers of disease activity (173,174). Circulating sCAMS can be measured easily by ELISA kits, and have been shown to be elevated in autoimmune diseases in association with both disease activity and vascular damage. However, most of these studies fail to show that measuring soluble CAMs provides anything more than an expensive marker of inflammation (174,175). In some studies, elevated levels of endothelial CAMs correlate with disease activity and soluble ICAM-3 has been reported to correlate with disease activity in RA and to predict the presence of vasculitis (176,177). Assessing the diagnostic or physiological significance of sCAMs *in vivo* is difficult, since the average level in blood may not be as meaningful as the levels at the site of inflammation. In this context, levels in synovial fluid in RA and bile in inflammatory liver disease have been shown to have more prognostic power than corresponding serum levels (178).

Another diagnostic application of sCAMs is in imaging studies. Radioactively-tagged anti-CAM antibodies have been used for a number of years to image inflamed endothelium in disease models and in clinical studies in RA and IBD. Recently, labelled sCAMs have been used to identify ligands in inflamed tissue (179,180). Fusion proteins that combine the constant region of an immunoglobulin (fc region) with the ligand binding region of an adhesion molecule or cytokine involved in immune reactions have been termed immunoadhesins (181). Immunoadhesions have great potential as therapeutic agents capable of modulating autoimmune and inflammatory diseases because of their exquisite target specificity and pharmacokinetic stability. Both sVCAM-1 and sPECAM-1 exhibit biological activities *in vivo*, in the cases of sVCAM-1, delaying the onset of diabetes in NOD mice in an adoptive transfer model (182,183), and in the case of PECAM-1, inhibiting leukocyte transmigration (184). Despite this great therapeutic potential, it is important to sound a word of caution, because biological agents can have dramatically different results in different autoimmune diseases. For instance, while TNF blockade is very successful in ameliorating inflammation in rheumatoid arthritis, it can be detrimental in other inflammatory diseases including multiple sclerosis. Conversely, anti-interferon-β therapy is showing initial promise in multiple sclerosis but is of no benefit in RA.

14. CONCLUSIONS

The last 10 years have seen major advances in understanding the molecular regulation of lymphocyte adhesion. Understanding the nature and function of these processes has important implications for therapy. The endothelium in the target tissue may be modified by genetic manipulation to block expression of molecules that appear to be critical for lymphocyte entry. The development of drugs, peptides or biologic agents that inhibit adhesion molecule function may add to the armamentarium of immunosuppressive therapy. However, a major problem with such approaches is the unwanted inhibition of lymphocyte recruitment to non-involved tissues, which potentially hinders the protective response to infection. They may be most useful in diseases associated with flares, such as RA, where they can be given in short courses rather than for prolonged periods. One potentially exciting area, with built-in specificity, is the inhibition of molecules that regulate lymphocyte recruitment to specific tissues. The best current example of this is the ability to treat IBD by blocking the $\alpha 4\beta 7$ interaction, but if other tissue specific interactions are uncovered, this raises the possibility of organ-specific inhibition of lymphocyte recruitment. However, perhaps the most enticing prospect is to modulate the nature of the lymphocytes that are activated and recruited. The increasing understanding of the signals, particularly chemokines, that control the recruitment of Th1 versus Th2 lymphocytes to tissue raises the possibility of therapeutic strategies in which the recruitment of tissue-damaging subsets could be selectively inhibited while allowing protective cells to be recruited. However, a recent study utilizing CCR1-deficient mice illustrates the complexity of the mechanisms involved. Despite substantial evidence of a role for CCR1 in driving Th1 nephritis, the disease was more severe in Th1-deficient animals. The authors concluded that rather than simply promoting leukocyte recruitment, CCR1 profoundly alters the effector phase of glomerulonephritis. This example illustrates how disrupting recruitment by therapeutic targeting of adhesion molecules or chemokine receptors may, on occasion, exacerbate the underlying disease (185).

References

1. Savill, J. 1997. Apoptosis in resolution of inflammation. *J. Leukoc. Biol.* 61:375–380.
2. Pawson, T. 1995. Protein modules and signalling networks. *Nature* 373:573–580.

3. Aplin, A.E., A. Howe, S.K. Alahari, and R.L. Juliano. 1998. Signal transduction and signal modulation by cell adhesion receptors: the role of integrins, cadherins, immunoglobulin-cell adhesion molecules, and selectins. *Pharmacol. Rev.* 50:197–263.

4. Gumbiner, B.M. 1996. Cell adhesion: the molecular basis of tissue architecture and morphogenesis. *Cell* 84:345–357.

5. Meredith, J.E.J., S. Winitz, J.M. Lewis, S. Hess, X.D. Ren, M.W. Renshaw, and M.A. Schwartz. 1996. The regulation of growth and intracellular signaling by integrins. *Endocr. Rev.* 17:207–220.

6. Assoian, R.K. 1997. Control of the G1 phase cyclin-dependent kinases by mitogenic growth factors and the extracellular matrix. *Cytokine Growth Factor Rev.* 8:165–170.

7. Assoian, R.K. and X. Zhu. 1997. Cell anchorage and the cytoskeleton as partners in growth factor dependent cell cycle progression. *Curr. Opin. Cell. Biol.* 9:93–98.

8. Hynes, R.O. 1994. The impact of molecular biology on models for cell adhesion. *Bioessays* 16:663–669.

9. Buckley, C.D. and D.L. Simmons. 1997. Cell adhesion: a new target for therapy. *Mol. Med. Today* 3:449–456.

10. Yap, A.S., W.M. Brieher, and B.M. Gumbiner. 1997. Molecular and functional analysis of cadherin-based adherens junctions. *Annu. Rev. Cell. Dev. Biol.* 13:119–146.

11. Porter, J.C. and N. Hogg. 1998. Integrins take partners: cross-talk between integrins and other membrane receptors. *Trends Cell Biol.* 8(10):390–66.

12. Hynes, R.O. 1999. Cell adhesion: Old and new questions. *Trends Cell. Biol.* 9:M33–M37

13. Seed, B. 1994. Making agonists of antagonists. *Chem. Biol.* 1:125–129.

14. Adams, D.H. and G.B. Nash. 1996. Disturbance of leukocyte circulation and adhesion to endothelium as factors in circulatory pathology. *Br. J. Anaesth.* 77:17–31.

15. Bruinsma, R. 1997. Les liaisons dangereuses: adhesion molecules do it statistically. *Proc. Natl. Acad. Sci. USA* 94:375–376.

16. von Andrian, U.H., S.R. Hasslen, R.D. Nelson, S.L. Erlandsen, and E.C. Butcher. 1995. A central role for microvillous receptor presentation in leukocyte adhesion under flow. *Cell* 82:989–999.

17. Shaw, S. and (guest editor of issue). 1993. Introduction: T cell adhesion. *Semin. Immunol.* 5:225–226.

18. Hynes, R.O. and A.D. Lander. 1992. Contact specificities in the associations, migrations and targeting of cells and axons. *Cell* 68:303–322.

19. Huttenlocher, A., R.R. Sandborg, and A.F. Horwitz. 1995. Adhesion in cell migration. *Curr. Opin. Cell Biol.* 7:697–706.

20. Hammer, D.A. and S.M. Apte. 1992. Simulation of cell rolling and adhesion on surfaces in shear flow: General results and analysis of selectin-mediated neutrophil adhesion. *Biophys. J.* 63:35–57.

21. Ward, M.D., M. Dembo, and D.A. Hammer. 1994. Kinetics of cell detachment: peeling of discrete receptor clusters. *Biophys. J.* 67:2522–2534.

22. Shaw, A.S. and M.L. Dustin. 1997. Making the T cell receptor go the distance: a topological view of T cell activation. *Immunity* 6:361–369.

23. Cyster, J.G. 1999. Chemokines and cell migration in secondary lymphoid organs. *Science* 286:2098–2102.

24. Cyster, J.G. 1999. Chemokines and the homing of dendritic cells to the T cell areas of lymphoid organs. *J. Exp. Med.* 189:447–450.

25. Kunkel, S.L. 1999. Promiscuous chemokine receptors and their redundant ligands play an enigmatic role during HIV-1 infection. *Am. J. Respir. Cell Mol. Biol.* 20:859–860.

26. Ansel, K.M., L.J. McHeyzer-Williams, V.N. Ngo, M.G. McHeyzer-Williams, and J.G. Cyster. 1999. In vivo-activated CD4 T cells upregulate CXC chemokine receptor 5 and reprogram their response to lymphoid chemokines. *J. Exp. Med.* 190:1123–1134.

27. Arroyo, A.G., J.T. Yang, H. Rayburn, and R.O. Hynes. 1999. Alpha4 integrins regulate the proliferation/differentiation balance of multilineage hematopoietic progenitors in vivo. *Immunity* 11:555–566.

28. Shier, P., K. Ngo, and W.P. Fung-Leung. 1999. Defective CD8+ T cell activation and cytolytic function in the absence of LFA-1 cannot be restored by increased TCR signaling. *J. Immunol.* 163:4826–4832.

29. Hickey, M.J., D.N. Granger, and P. Kubes. 1999. Molecular mechanisms underlying IL-4-induced leukocyte recruitment in vivo: a critical role for the alpha 4 integrin. *J. Immunol.* 163:3441–3448.

30. Kulidjian, A.A., R. Inman, and T.B. Issekutz. 1999. Rodent models of lymphocyte migration. *Semin. Immunol.* 11:85–93.

31. Wagner, N., J. Lohler, T.F. Tedder, K. Rajewsky, W. Muller, and D.A. Steeber. 1998. L-selectin and beta7 integrin synergistically mediate lymphocyte migration to mesenteric lymph nodes. *Eur. J. Immunol.* 28:3832–3839.

32. Sharpe, A.H. 1995. Analysis of lymphocyte costimulation in vivo using transgenic and 'knockout' mice. *Curr. Opin. Immunol.* 7:389–395.

33. Karsan, A., C.J. Cornejo, R.K. Winn, B.R. Schwartz, W. Way, N. Lannir, R. Gershoni-Baruch, A. Etzioni, H.D. Ochs, and J.M. Harlan. 1998. Leukocyte adhesion deficiency type II is a generalized defect of de novo GDP-fucose biosynthesis. Endothelial cell fucosylation is not required for neutrophil rolling on human nonlymphoid endothelium. *J. Clin. Invest.* 101:2438– 2445.

34. Kuijpers, T.W., R.A. van Lier, D. Hamann, M. de Boer, L.Y. Thung, R.S. Weening, A.J. VerHoeven, and D. Roos. 1997. Leukocyte adhesion deficiency type 1 (LAD-1)/variant. A novel immunodeficiency syndrome characterized by dysfunctional beta2 integrins. *J. Clin. Invest.* 100:1725–1733.

35. Anderson, D.C. and T.A. Springer. 1987. Leukocyte adhesion deficiency: an inherited defect in the Mac-1, LFA-1, and p150,95 glycoproteins. *Annu. Rev. Med.* 38:175–194.

36. Hogg, N., M.P. Stewart, S.L. Scarth, R. Newton, J.M. Shaw, S.K. Law, and N. Klein. 1999. A novel leukocyte adhesion deficiency caused by expressed but nonfunctional beta2 integrins Mac-1 and LFA-1. *J. Clin. Invest.* 103:97–106.

37. Horgan, K.J., Y. Tanaka, and S. Shaw. 1992. Post-thymic differentiation of CD4 T lymphocytes: Naive versus memory subsets and further specialization among memory cells. *Chem. Immunol.* 54:72–102.

38. Akbar, A.N., M. Salmon, and G. Janossy. 1991. The synergy between naive and memory cells during activation. *Immunol. Today* 12:184–188.

39. Mackay, C.R. 1993. Homing of naive, memory and effector lymphocytes. *Curr. Opin. Immunol.* 5:423–427.

40. Stupack, D.G., E. Li, S.A. Silletti, J.A. Kehler, R.L. Geahlen, K. Hahn, G.R. Nemerow, and D.A. Cheresh. 1999. Matrix valency regulates integrin-mediated lymphoid adhesion via Syk kinase. *J. Cell Biol.* 144:777–788.

41. Viola, A., S. Schroeder, Y. Sakakibara, and A. Lanzavecchia. 1999. T lymphocyte costimulation mediated by reorganization of membrane microdomains. *Science* 283:680–682.

42. Ding, Z.M., J.E. Babensee, S.I. Simon, H. Lu, J.L. Perrard, D.C. Bullard, X.Y. Dai, S.K. Bromley, M.L. Dustin, M.L. Entman, C.W. Smith, and C.M. Ballantyne. 1999. Relative contribution of LFA-1 and Mac-1 to neutrophil adhesion and migration. *J. Immunol.* 163:5029–5038.

43. Watts, T.H. and M.A. DeBenedette. 1999. T cell co-stimulatory molecules other than CD28. *Curr. Opin. Immunol.* 11:286–293.

44. Iredale, J.P. and M.J.P. Arthur. 1994. Hepatocyte-matrix interactions. *Gut* 35:729–732.

45. Butcher, E.C. and L.J. Picker. 1996. Lymphocyte homing and homeostasis. *Science* 272:60–66.

46. Adams, D.H. and A.R. Lloyd. 1997. Chemokines: Leucocyte recruitment and activation cytokines. *Lancet* 349:490–495.

47. Jaeschke, H. 1997. Cellular adhesion molecules: regulation and functional significance in the pathogenesis of liver diseases. *Am. J. Physiol.* 273:G602–G611.

48. Kupiec-Weglinski, J.W. and N.L. Tilney. 1989. Lymphocyte migration patterns in organ allograft recipients. *Immunol. Rev.* 108:63–82.

49. Adams, D.H. and S. Shaw. 1994. Leucocyte endothelial interactions and regulation of leucocyte migration. *Lancet* 343:831–836.

50. Springer, T.A. 1994. Traffic signals for lymphocyte recirculation and leukocyte emigration: the multistep paradigm. *Cell* 76:301–314.

51. Bevilacqua, M.P. 1993. Endothelial-leukocyte adhesion molecules. *Annu. Rev. Immunol.* 11:767–784.

52. Ebnet, K., E.P. Kaldjian, A.O. Anderson, and S. Shaw. 1996. Orchestrated information transfer underlying leukocyte endothelial interactions. *Annu. Rev. Immunol.* 14:155–77:155–177.

53. Alon, R., P.D. Kassner, M.W. Carr, E.B. Finger, M.E. Hemler, and T.A. Springer. 1995. The integrin VLA-4 supports tethering and rolling in flow on VCAM-1. *J. Cell Biol.* 128:1243–1253.

54. Lalor, P.E., J.M. Clements, R. Pigott, M.J. Humphries, J.H. Spragg, and G.B. Nash. 1997. Association between receptor density, cellular activation, and transformation of adhesive behavior of flowing lymphocytes binding to vcam-1. *Eur. J. Immunol.* 27:1422–1426.

55. Campbell, J.J., S.X. Qin, K.B. Bacon, C.R. Mackay, and E.C. Butcher. 1996. Biology of chemokine and classical chemoattractant receptors—differential requirements for adhesion-triggering versus chemotactic responses in lymphoid-cells. *J. Cell Biol.* 134:255–266.

56. Luster, A.D. 1998. Chemokines- chemotactic cytokines that mediate inflammation. *New Eng. J. Med.* 338:436–445.

57. Adams, D.H. and A.R. Lloyd. 1997. Chemokines: leucocyte recruitment and activation cytokines. *Lancet* 349:490–495.

58. Carr, M.W., R. Alon, and T.A. Springer. 1996. The c-c chemokine mcp-1 differentially modulates the avidity of beta- 1 and beta-2 integrins on t-lymphocytes. *Immunity* 4:179–187.

59. Tanaka, Y., D.H. Adams, S. Hubscher, H. Hirano, U. Siebenlist, and S. Shaw. 1993. T-cell adhesion induced by proteoglycan-immobilized cytokine MIP-1b. *Nature* 361:79–82.

60. Airas, L., J. Niemela, M. Salmi, T. Puurunen, D.J. Smith, and S. Jalkanen. 1997. Differential regulation and function of CD73, a glycosyl-phosphatidylinositol-linked 70-kd adhesion molecule, on lymphocytes and endothelial cells. *J. Cell Biol.* 136:421–431.

61. Tanaka, Y., S.M. Albelda, K.J. Horgan, G.A. van Seventer, Y. Shimizu, W. Newman, J. Hallam, P.J. Newman, C.A. Buck, and S. Shaw. 1992. CD31 expressed on distinctive T cell subsets is a preferential amplifier of b1 integrin-mediated adhesion. *J. Exp. Med.* 176:245–253.

62. Campbell, J.J., J. Hedrick, A. Zlotnick, M.A. Siani, D.A. Thompson, and E.C. Butcher. 1998. Chemokines and the arrest of lymphocytes rolling under flow. *Science* 279:381–384.

63. Akbar, A.N. and M. Salmon. 1997. Cellular environments and apoptosis: Tissue microenvironments control activated T cell death. *Immunol. Today* 18:72–76.

64. Westermann, J. and U. Bode. 1999. Distribution of activated T cells migrating through the body: a matter of life and death. *Immunol. Today* 20:302–306.

65. Salmi, M. and S. Jalkanen. 1997. How do lymphocytes know where to go: current concepts and enigmas of lymphoctye homing. *Adv. Immunol.* 64:139–202.

66. Bargatze, R.F., M.A. Jutila, and E.C. Butcher. 1995. Distinct roles of l-selectin and integrins $\alpha 4 \beta 7$ and LFA-1 in lymphocyte homing to peyers patch-hev in-situ — the multistep model confirmed and refined. *Immunity* 3:99–108.

67. Girard, J.P. and T.A. Springer. 1995. High endothelial venules (HEVs): specialized endothelium for lymphocyte migration. *Immunol. Today* 16:449–457.

68. Kunkel, E.J., C.L. Ramos, D.A. Steeber, W. Muller, N. Wagner, T.F. Tedder, and K. Ley. 1998. The roles of L-selectin, beta 7 integrins, and P-selectin in leukocyte rolling and adhesion in high endothelial venules of Peyer's patches. *J. Immunol.* 161:2449–2456.

69. Michie, S.A., P.R. Streeter, P.A. Bolt, E.C. Butcher, and L.J. Picker. 1993. The human peripheral lymph node vascular addressin. An inducible endothelial antigen involved in lymphocyte homing. *Am. J. Pathol.* 143:1688–1698.

70. Cyster, J.G. 1999. Chemokines and the homing of dendritic cells to the T cell areas of lymphoid organs. *J. Exp. Med.* 189:447–450.

71. Price, A.A., M. Cumberbatch, I. Kimber, and A. Ager. 1997. Alpha 6 integrins are required for Langerhans cell migration from the epidermis. *J. Exp. Med.* 186:1725–1735.

72. Goodnow, C.C. and J.G. Cyster. 1997. Lymphocyte homing: the scent of a follicle. *Curr. Biol.* 7:R219–R222

73. Gunn, M.D., S. Kyuwa, C. Tam, T. Kakiuchi, A. Matsuzawa, L.T. Williams, and H. Nakano. 1999. Mice lacking expression of secondary lymphoid organ chemokine have defects in lymphocyte homing and dendritic cell localization. *J. Exp. Med.* 189:451–460.

74. Gunn, M.D., K. Tangemann, C. Tam, J.G. Cyster, S.D. Rosen, and L.T. Williams 1998. A chemokine expressed in lymphoid high endothelial venules promotes the adhesion and chemotaxis of naive T lymphocytes. *Proc. Natl. Acad. Sci. USA* 258–633.

75. Forster, R., A. Schubel, D. Breitfeld, E. Kremmer, I. Renner-Muller, E. Wolf, and M. Lipp. 1999. CCR7 coordinates the primary immune response by establishing functional microenvironments in secondary lymphoid organs. *Cell* 99:23–33.

76. Forster, R., A.E. Mattis, E. Kremmer, E. Wolf, G. Brem, and M. Lipp. 1996. A putative chemokine receptor, BLR1, directs B cell migration to defined lymphoid organs and specific anatomic compartments of the spleen. *Cell* 87:1037–1477.

77. Legler, D.F., M. Loetscher, R.S. Roos, I. Clark-Lewis, M. Baggiolini, and B. Moser. 1998 B cell-attracting chemokine 1, a human CXC chemokine expressed in lymphoid tissues, selectively attracts B lymphocytes via BLR1/CXCR5. *J. Exp. Med.* 187:655–600.

78. Shimizu, Y., G.A. van Seventer, K.J. Horgan, and S. Shaw. 1990. Regulated expression and function of three VLA (b1) integrin receptors on T cells. *Nature* 345:250–253.

79. Tang, H.L. and J.G. Cyster. 1999. Chemokine Up-regulation and activated T cell attraction by maturing dendritic cells. *Science* 284:819–822.

80. Shields, P. L., Morland, C. M., Qin, S., Salmon, M., Hubscher, S. G., and Adams, D. H. 1999. Chemokine and chemokine receptor interactions provide a mechanism for selective T cell recruitment to specific liver compartments within hepatitis C infected liver. *J. Immunol.* 163:6236–6243.

81. Bermejo, M., J. Martin-Serrano, E. Oberlin, M.A. Pedraza, A. Serrano, B. Santiago, A. Caruz, P. Loetscher, M. Baggiolini, F. Arenzana-Seisdedos, and J. Alcami. 1998. Activation of blood T lymphocytes down-regulates CXCR4 expression and interferes with propagation of X4 HIV strains *Eur. J. Immunol.* 28:3192–3204.

82. Bleul, C.C., L. Wu, J.A. Hoxie, T.A. Springer, and C.R. Mackay. 1997. The HIV coreceptors CXCR4 and CCR5 are differentially expressed and regulated on human T lymphocytes. *Proc. Natl. Acad. Sci. USA* 94:1925–1930.

83. Sallusto, F., A. Lanzavecchia, and C.R. Mackay. 1998. Chemokines and chemokine receptors in T-cell priming and Th1/Th2- mediated responses. *Immunol. Today* 19:568–574.

84. Sallusto, F., D. Lenig, C.R. Mackay, and A. Lanzavecchia. 1998. Flexible programs of chemokine receptor expression on human polarized T helper 1 and 2 lymphocytes. *J. Exp. Med.* 187:875–883.

85. Sallusto, F., P. Schaerli, P. Loetscher, C. Schaniel, D. Lenig, C.R. Mackay, S. Qin, and A. Lanzavecchia. 1998. Rapid and coordinated switch in chemokine receptor expression during dendritic cell maturation. *Eur. J. Immunol.* 28:2760–2769.

86. Sallusto, F., D. Lenig, R. Forster, M. Lipp, and A. Lanzavecchia. 1999. Two subsets of memory T lymphocytes with distinct homing potentials and effector functions. *Nature* 401:708–712.

87. Randolph, D.A., G. Huang, C.J. Carruthers, L.E. Bromley, and D.D. Chaplin. 1999. The role of CCR7 in TH1 and TH2 cell localization and delivery of B cell help in vivo. *Science* 286:2159–2162.

88. Briskin, M., D. Winsor-Hines, A. Shyjan, N. Cochran, S. Bloom, J. Wilson, L.M. McEvoy, E.C. Butcher, N. Kassam, C.R. Mackay, W. Newman, and D.J. Ringler. 1997. Human mucosal addressin cell adhesion molecule-1 is preferentially expressed in intestinal tract and associated lymphoid tissue. *Am. J. Pathol.* 151:97–110.

89. Fuhlbrigge, R.C., J.D. Kieffer, D. Armerding, and T.S. Kupper. 1997. Cutaneous lymphocyte antigen is a specialized form of PSGL-1 expressed on skin-homing T cells. *Nature* 389:978–981.

90. Picker, L.J., J.R. Treer, B. Ferguson-Darnell, P.A. Collins, P.R. Bergstresser, and L.W. Terstappen. 1993. Control of lymphocyte recirculation in man. II. Differential regulation of the cutaneous lymphocyte-associated antigen, a tissue-selective homing receptor for skin-homing T cells. *J. Immunol.* 1500:1122–1136.

91. Adams, D.H., S.G. Hubscher, N.C. Fisher, A. Williams, and M. Robinson. 1996. Expression of E-selectin (CD62E) and E-selectin ligands in human liver inflammation. *Hepatology* 24:533–538.

92. Salmi, M., D.H. Adams, and S. Jalkanen. 1998. Lymphocyte trafficking in the intestine and liver. *Am. J. Physiol.* 274:G1–G6

93. Yoong, K.F., G. McNab, S.G. Hubscher, and D.H. Adams. 1998. Vascular adhesion protein-1 and ICAM-1 support the adhesion of tumor infiltrating lymphocytes to tumor endothelium in human hepatocellular carcinoma. *J. Immunol.* 160:3978–3988.

94. Abitorabi, M.A., C.R. Mackay, E.H. Jerome, O. Osorio, E.C. Butcher, and D.J. Erle. 1996. Differential expression of homing molecules on recirculating lymphocytes from sheep gut, peripheral, and lung lymph. *J. Immunol.* 156:3111–3117.

95. Erle, D.J., T. Brown, D. Christian, and R. Aris. 1994. Lung epithelial lining fluid t-cell subsets defined by distinct patterns of beta-7 and beta-1 integrin expression. *Am. J. Resp. Cell Mol. Biol.* 10:237–244.

96. Wolber, F.M., J.L. Curtis, A.M. Milik, T. Fields, G.D. Seitzman, K. Kim, S. Kim, J. Sonstein, and L.M. Stoolman. 1997. Lymphocyte recruitment and the kinetics of adhesion receptor expression during the pulmonary immune response to particulate antigen. *Am. J. Pathol.* 151:1715–1727.

97. Campbell, J.J., G. Haraldsen, J. Pan, J. Rottman, S. Qin, P. Ponath, D.P. Andrew, R. Warnke, N. Ruffing, N. Kassam, L. Wu, and E.C. Butcher. 1999. The chemokine receptor CCR4 in vascular recognition by cutaneous but not intestinal memory T cells. *Nature* 400:776–780.

98. Mackay, C.R. 1999. Dual personality of memory T cells. *Nature* 401:659–660.

99. Picker, L.J., J.R. Treer, B. Ferguson-Darnell, P.A. Collins, D. Buck, and L.W. Terstappen. 1993. Control of lymphocyte recirculation in man: I. Differential regulation of the peripheral lymph node homing receptor L-selectin on T cells during the virgin to memory cell transition. *J. Immunol.* 150:1105–1121.

100. Stewart, M. and N. Hogg. 1996. Regulation of leukocyte integrin function: Affinity vs. avidity. *J. Cell Biochem.* 61:554–561.

101. Shimizu, Y., G.A. van Seventer, E. Ennis, W. Newman, K.J. Horgan, and S. Shaw. 1992. Crosslinking of the T cell-specific accessory molecules CD7 and CD28 modulates T cell adhesion. *J. Exp. Med.* 175:577–582.

102. Shimizu, Y., W. Newman, T.V. Gopal, K.J. Horgan, N. Graber, L.D. Beall, G.A. van Seventer, and S. Shaw. 1991. Four molecular pathways of T cell adhesion to endothelial cells: roles of LFA-1, VCAM-1 and ELAM-1 and changes in pathway hierarchy under different activation conditions. *J. Cell Biol.* 113:1203–1212.

103. Austrup, F., D. Vestweber, E. Borges, M. Lohning, R. Brauer, U. Herz, H. Renz, R. Hallmann, A. Scheffold, A. Radbruch, and A. Hamann. 1997. P- and E-selectin mediate recruitment of T-helper-1 but not T-helper-2 cells into inflamed tissues. *Nature* 385:81–83.

104. Borges, E., G. Pendl, R. Eytner, M. Steegmaier, O. Zollner, and D. Vestweber. 1997. The binding of T cell-expressed P-selectin glycoprotein ligand-1 to E- and P-selectin is differentially regulated. *J. Biol. Chem.* 272:28786–28792.

105. Borges, E., W. Tietz, M. Steegmaier, T. Moll, R. Hallmann, A. Hamann, and D. Vestweber. 1997. P-selectin glycoprotein ligand-1 (PSGL-1) on T helper 1 but not on T helper 2 cells binds to P-selectin and supports migration into inflamed skin. *J. Exp. Med.* 185:573–578.

106. van Wely, C.A., A.D. Blanchard, and C.J. Britten. 1998. Differential expression of alpha3 fucosyltransferases in Th1 and Th2 cells correlates with their ability to bind P-selectin. *Biochem. Biophys. Res. Commun.* 247:307–311.

107. Blander, J.M., I. Visintin, C.A.J. Janeway, and R. Medzhitov. 1999. Alpha(1,3)-fucosyltransferase VII and alpha(2,3)-sialyltransferase IV are up-regulated in activated CD4 T cells and maintained after their differentiation into Th1 and migration into inflammatory sites. *J. Immunol.* 163:3746–3752.

108. Sallusto, F., E. Kremmer, B. Palermo, A. Hoy, P. Ponath, S. Qin, R. Forster, M. Lipp, and A. Lanzavecchia. 1999. Switch in chemokine receptor expression upon TCR stimulation reveals novel homing potential for recently activated T cells. *Eur. J. Immunol.* 29:2037–2045.

109. Marelli-Berg, F.M., L. Frasca, L. Weng, G. Lombardi, and R.I. Lechler. 1999. Antigen recognition influences trans-endothelial migration of CD4+ T cells. *J. Immunol.* 162:696–703.

110. van Kooyk, Y., P. van de Wiel-van Kemenade, P. Weder, T.W. Kuijpers, and C.G. Figdor. 1989. Enhancement of LFA-1-mediated cell adhesion by triggering through CD2 or CD3 on T lymphocytes. *Nature* 342:811–813.

111. Shimizu, Y. and S. Shaw. 1991. Lymphocyte interactions with extracellular matrix. *FASEB J.* 5:2292–2299.

112. Salmi, M., K. Kalimo, and S. Jalkanen. 1993. Induction and function of vascular adhesion protein 1 at sites of inflammation. *J. Exp. Med.* 178:2255–2260.

113. McNab, G., J.L. Reeves, M. Salmi, S.G. Hubscher, S. Jalkanen, and D.H. Adams. 1996. Vascular adhesion protein-1 supports adhesion of T lymphocytes to hepatic endothelium. A mechanism for T cell recirculation to the liver? *Gastroenterology* 110:522–528.

114. Cuff, C.A., J. Schwartz, C.M. Bergman, K.S. Russell, J.R. Bender, and N.H. Ruddle. 1998. Lymphotoxin alpha3 induces chemokines and adhesion molecules: insight into the role of LT alpha in inflammation and lymphoid organ development. *J. Immunol.* 161:6853–6860.

115. Ettinger, R., R. Mebius, J.L. Browning, S.A. Michie, S. van Tuijl, G. Kraal, W. van Ewijk, and H.O. McDevitt. 1998. Effects of tumor necrosis factor and lymphotoxin on peripheral lymphoid tissue development. *Int. Immunol.* 10:727–741.

116. Kratz, A., A. Campos-Neto, M.S. Hanson, and N.H. Ruddle. 1996. Chronic inflammation caused by lymphotoxin is lymphoid neogenesis. *J. Exp. Med.* 183:1461–1472.

117. Lalor, P. and D.H. Adams. 1999. The regulation of lymphocyte adhesion to hepatic endothelium. *J. Clin. Mole. Pathol.* 52:214–219.

118. Yamane, A., L. Seetharam, S. Yamaguchi, N. Gotoh, T. Takahashi, G. Neufeld, and M. Shibuya. 1994. A new communication system between hepatocytes and sinusoidal endothelial cells in the liver through VEGF and Fl tyrosine kinase receptor family (Flt-1 and KDR/Flk-1). *Oncogene* 9:2383–2690.

119. Lawson, C., M. Ainsworth, M. Yacoub, and M. Rose. 1999. Ligation of ICAM-1 on endothelial cells leads to expression of VCAM-1 via a nuclear factor-kappaB-independent mechanism. *J. Immunol.* 162:2990–2996.

120. Adamson, P., S. Etienne, P.O. Couraud, V. Calder, and J. Greenwood. 1999. Lymphocyte migration through brain endothelial cell monolayers involves signaling through endothelial ICAM-1 via a rho-dependent pathway. *J. Immunol.* 162:2964–2973.

121. Karmann, K., C.C.W. Hughes, J. Schechner, W.C. Fanslow, and J.S. Pober. 1995. CD40 on human endothelial-cells—inducibility by cytokines and functional regulation of adhesion molecule expression. *Proc. Nat. Acad. Sci. USA* 92:4342–4346.

122. Bergese, S., R. Pelletier, D. Vallera, M. Widmer, and C. Orosz. 1995. Regulation of endothelial VCAM-1 expression in murine cardiac grafts. Roles for TNF and IL-4. *Am. J. Pathol.* 146:989–998.

123. Paleog, E.M., G.R. Aluri, and M. Feldmann. 1992. Contrasting effects of interferon gamma and interleukin 4 on responses of human vascular endothelial cells to tumour necrosis factor alpha. *Cytokine* 4:470–478.

124. Melrose, J., N. Tsurushita, G. Liu, and E.L. Berg. 1998. IFN-gamma inhibits activation-induced expression of E- and P-selectin on endothelial cells. *J. Immunol.* 161:2457–2464.

125. Ley, K. 1994. Histamine can induce leukocyte rolling in rat mesenteric venules. *Am. J. Physiol.* 267:H1010–H1023

126. Jones, D.A., O. Abbassi, L.V. McIntire, R.P. McEver, and C.W. Smith. 1993. P-selectin mediates neutrophil rolling on histamine stimulated endothelial cells. *Biophys. J.* 65:1560–1569.

127. Lukacs, N.W., C. Hogaboam, E. Campbell, and S.L. Kunkel. 1999. Chemokines: function, regulation and alteration of inflammatory responses. *Chem. Immunol.* 72:102–120.

128. Biddison, W.E., W.W. Cruikshank, D.M. Center, C.M. Pelfrey, D.D. Taub, and R.V. Turner. 1998. CD8+ myelin peptide-specific T cells can chemoattract CD4+ myelin peptide-specific T cells: importance of IFN-inducible protein 10. *J. Immunol.* 160:444–448.

129. Rot, A., E. Hub, J. Middleton, F. Pons, C. Rabeck, K. Thierer, J. Wintle, B. Wolff, M. Zsak, and P. Dukor. 1996. Some aspects of IL-8 pathophysiology. III: Chemokine interaction with endothelial cells. *J. Leukoc Biol.* 59:39–44.

130. Lu, Z.H., Z.X. Wang, R. Horuk, J. Hesselgesser, Y.C. Lou, T.J. Hadley, and S.C. Peiper. 1995. The promiscuous chemokine binding profile of the Duffy antigen/receptor for chemokines is primarily localized to sequences in the amino-terminal domain. *J. Biol. Chem.* 270:26239–26245.

131. Hub, E. and A. Rot. 1998. Binding of RANTES, MCP-1, MCP-3, and MIP 1alpha to cells in human skin. *Am. J. Pathol.* 152:749–577.

132. Witt, D.P. and A.D. Lander. 1994. Differential binding of chemokines to glycosaminoglycan subpopulations. *Curr. Biol.* 4:394–400.

133. Middleton, J., S. Neil, J. Wintle, I. Clark-Lewis, H. Moore, C. Lam, M. Auer, E. Hub, and A. Rot. 1997. Transcytosis and surface presentation of IL-8 by venular endothelial cells. *Cell* 91:385–395.

134. Hub, E. and A. Rot. 1998. Binding of RANTES, MCP-1, MCP-3, and MIP-1alpha to cells in human skin. *Am. J. Pathol.* 152:749–757.

135. Wolff, B., A.R. Burns, J. Middleton, and A. Rot. 1998. Endothelial cell "memory" of inflammatory stimulation: human venular endothelial cells store interleukin 8 in Weibel-Palade bodies. *J. Exp. Med.* 188:1757–1762.

136. Magone, M.T. and S.M. Whitcup. 1999. Mechanisms of intraocular inflammation. *Chem. Immunol.* 73:90–119.

137. Sfikakis, P.P. and M. Mavrikakis. 1999. Adhesion and lymphocyte costimulatory molecules in systemic rheumatic diseases. *Clin. Rheumatol.* 18:317–327.

138. Waksman, B.H. 1999. Demyelinating disease: evolution of a paradigm. *Neurochem. Res.* 24:491–495.

139. Kalden, J.R. and B. Manger. 1998. Biologic agents in the treatment of inflammatory rheumatic diseases. *Curr. Opin. Rheumatol.* 10:174–178.

140. Oppenheimer-Marks, N. and P.E. Lipsky. 1998. Adhesion molecules in rheumatoid arthritis. *Springer Semin. Immunopathol.* 20:95–114.

141. Steinman, L. 1995. Escape from "horror autotoxicus": pathogenesis and treatment of autoimmune disease. *Cell* 80:7–10.

142. Pozzilli, P., P. Carotenuto, and G. Delitala. 1994. Lymphocytic traffic and homing into target tissue and the generation of endocrine autoimmunity. *Clin. Endocrinol.* 41:545–554.

143. Krensky, A.M. 1994. T cells in autoimmunity and allograft rejection. *Kidney Int. Suppl.* 44:S50–S56

144. Robertson, C.R. and R.M. McCallum. 1994. Changing concepts in pathophysiology of the vasculitides. *Curr. Opin. Rheumatol.* 6:3–10.

145. Mojcik, C.F. and E.M. Shevach. 1997. Adhesion molecules: a rheumatologic perspective. *Arthritis Rheum.* 40:991–1004.

146. Issekutz, T.B. 1992. Lymphocyte homing to sites of inflammation. *Curr. Opin. Immunol.* 4:287–293.

147. Palecanda, A., J.S. Marshall, X. Li, M.J. Briskin, and T.B. Issekutz. 1999. Selective antibody blockade of lymphocyte migration to mucosal sites and mast cell adhesion. *J. Leukoc. Biol.* 65:649–657.

148. Picarella, D., P. Hurlbut, J. Rottman, X. Shi, E. Butcher, and D.J. Ringler. 1997. Monoclonal antibodies specific for beta 7 integrin and mucosal addressin cell adhesion molecule-1 (MAdCAM-1) reduce inflammation in the colon of scid mice reconstituted with CD45RB^high CD4^+ T cells. *J. Immunol.* 158:2099–2106.

149. Shroff, H.N., C.F. Schwender, A.D. Baxter, F. Brookfield, L.J. Payne, N.A. Cochran, D.L. Gallant, and M.J. Briskin. 1998. Novel modified tripeptide inhibitors of alpha 4 beta 7 mediated lymphoid cell adhesion to MAdCAM-1. *Bioorg. Med. Chem. Lett.* 8:1601–1606.

150. Viney, J.L. and S. Fong. 1998. Beta 7 integrins and their ligands in lymphocyte migration to the gut. *Chem. Immunol.* 71:64–76.

151. Salmi, M., K. Granfors, R. MacDermott, and S. Jalkanen. 1994. Aberrant binding of lamina propria lymphocytes to vascular endothelium in inflammatory bowel diseases. *Gastroenterology* 106:596–605.

152. Brandtzaeg, P. 1997. Review article: Homing of mucosal immune cells—a possible connection between intestinal and articular inflammation. *Aliment Pharmacol. Ther.* 11 Suppl 3:24–37.

153. Mulligan, M.S., J.C. Paulson, S. De Frees, Z.-L. Zheng, J.B. Lowe, and P.A. Ward. 1993. Protective effects of oligosaccharides in P-selectin-dependent lung injury. *Nature* 364:149–151.

154. Chu, A., K. Hong, E.L. Berg, and R.O. Ehrhardt. 1999. Tissue specificity of E- and P-selectin ligands in Th1mediated chronic inflammation. *J. Immunol.* 163:5086–5093.

155. Adams, D.H., J. Fear, S. Shaw, S.G. Hubscher, and S. Afford. 1996. Hepatic expression of macrophage inflammatory protein-1 a and macrophage inflammatory protein-1b after liver transplantation. *Transplantation* 61:817–825.

156. Picker, L.J., S.A. Michie, L.S. Rott, and E.C. Butcher. 1990. A unique phenotype of skin-associated lymphocytes in humans Preferential expression of the HECA-452 epitope by benign and malignant T cells at cutaneous sites. *Am. J. Pathol.* 136:1053–1068.

157. Hanninen, A., C. Taylor, P.R. Streeter, L.S. Stark, J.M. Sarte, J.A. Shizuru, O. Simell, and S.A. Michie. 1993. Vascular addressins are induced on islet vessels during insulitis in nonobese diabetic mice and are involved in lymphoid cell binding to islet endothelium. *J. Clin. Invest.* 92:2509–2515.

158. Rosen, S.D. 1999. Endothelial ligands for L-selectin: from lymphocyte recirculation to allograft rejection [comment]. *Am. J. Pathol.* 155:1013–1020.

159. Toppila, S., T. Paavonen, M.S. Nieminen, P. Hayry, and R. Renkonen. 1999. Endothelial L-selectin ligands are likely to recruit lymphocytes into rejecting human heart transplants. *Am. J. Pathol.* 155:1303–1310.

160. Toppila, S., R. Renkonen, L. Penttila, J. Natunen, H. Salminen, J. Helin, H. Maaheimo, and O. Renkonen. 1999. Enzymatic synthesis of alpha3′sialylated and multiply alpha3fucosylated biantennary polylactosamines. A bivalent [sialyl diLex]-saccharide inhibited lymphocyte-endothelium adhesion organ-selectively. *Eur. J. Biochem.* 261:208–215.

161. Parker, C.M., K.L. Cepek, G.J. Russell, S.K. Shaw, D.N. Posnett, R. Schwarting, and M.B. Brenner. 1992. A family of beta 7 integrins on human mucosal lymphocytes. *Proc. Natl. Acad. Sci. USA* 89:1924–1928.

162. Schon, M.P., A. Arya, E.A. Murphy, C.M. Adams, U.G. Strauch, W.W. Agace, J. Marsal, J.P. Donohue, H. Her, D.R. Beier, S. Olson, L. Lefrancois, M.B. Brenner, M.J. Grusby, and C.M. Parker. 1999. Mucosal T lymphocyte numbers are selectively reduced in integrin alpha E (CD103)-deficient mice. *J. Immunol.* 162:6641–6649.

163. Shaw, S.K. and M.B. Brenner. 1995. The beta 7 integrins in mucosal homing and retention. *Semin. Immunol.* 7:335–342.

164. El-Gabalawy, H., M. Gallatin, R. Vazeux, G. Peterman, and J. Wilkins. 1994. Expression of ICAM-R (ICAM-3), a novel counter-receptor for LFA-1, in rheumatoid and nonrheumatoid synovium. Comparison with other adhesion molecules. *Arthritis Rheum.* 37:846–854.

165. Szekanecz, Z., G.K. Haines, T.R. Lin, L.A. Harlow, S. Goerdt, G. Rayan, and A.E. Koch. 1994. Differential distribution of intercellular adhesion molecules (ICAM-1, ICAM-2, and ICAM-3) and the MS-1 antigen in normal and diseased human synovia. Their possible pathogenetic and clinical significance in rheumatoid arthritis. *Arthritis Rheum.* 37:221–231.

166. De Fougerolles, A.R., X. Qin, and T.A. Springer. 1994. Characterization of the function of intercellular adhesion molecule (ICAM)-3 and comparison with ICAM-1 and ICAM-2 in immune responses. *J. Exp. Med.* 179:619–629.

167. Buckley, C.D., E.D. Ferguson, A.J. Littler, D. Bossy, and D.L. Simmons. 1997. Role of ligands in the activation of LFA-1. *Eur. J. Immunol.* 27:957–962.

168. Souza, H.S., C.C. Elia, J. Spencer, and T.T. MacDonald. 1999. Expression of lymphocyte-endothelial receptor-ligand pairs, alpha4beta7/MAdCAM-1 and OX40/OX40 ligand in the colon and jejunum of patients with inflammatory bowel disease. *Gut* 45:856–863.

169. Higgins, L.M., S.A. McDonald, N. Whittle, N. Crockett, J.G. Shields, and T.T. MacDonald. 1999. Regulation of T cell activation in vitro and in vivo by targeting the OX40–OX40 ligand interaction: amelioration of ongoing inflammatory bowel disease with an OX40-IgG fusion protein, but not with an OX40 ligand-IgG fusion protein. *J. Immunol.* 162:486–493.

170. Ward, S.G., K. Bacon, and J. Westwick. 1998. Chemokines and T lymphocytes: more than an attraction. *Immunity* 9:1–11.

171. Topham, P.S., V. Csizmadia, D. Soler, D. Hines, C.J. Gerard, D.J. Salant, and W.W. Hancock. 1999. Lack of

chemokine receptor CCR1 enhances Th1 responses and glomerular injury during nephrotoxic nephritis. *J. Clin. Invest.* 104:1549–1557.

172. Werb, Z. 1997. ECM and cell surface proteolysis: regulating cellular ecology. *Cell* 91:439–442.

173. Gearing, A.J.H. and W. Newman. 1993. Circulating adhesion molecules in disease. *Immunol. Today* 14:506–512.

174. Buckley, C.D., D.H. Adams, and D. Simmons. 1999. Soluble Leukocyte-endothelial adhesion molecules. In: In Physiology of Inflammation. K. Ley, editor.

175. Littler, A.J., C.D. Buckley, P. Wordsworth, I. Collins, J. Martinson, and D.L. Simmons. 1997. A distinct profile of six soluble adhesion molecules (ICAM-1, ICAM-3, VCAM-1, E-selectin, L-selectin and P-selectin) in rheumatoid arthritis. *Br. J. Rheumatol.* 36:164–169.

176. Denton, C.P., M.C. Bickerstaff, X. Shiwen, M.T. Carulli, D.O. Haskard, R.M. Dubois, and C.M. Black. 1995. Serial circulating adhesion molecule levels reflect disease severity in systemic sclerosis. *Br. J. Rheumatol.* 34:1048–1054.

177. Haskard, D.O. 1995. Cell adhesion molecules in rheumatoid arthritis. *Curr. Opin. Rheumatol.* 7:229–234.

178. Adams, D.H., E. Mainolfi, E. Elias, J.M. Neuberger, and R. Rothlein. 1993. Detection of circulating intercellular adhesion molecule-1 after liver transplantation—evidence of local release within the liver during graft rejection. *Transplantation* 55:83–87.

179. Jamar, F., P.T. Chapman, D.H. Manicourt, D.M. Glass, D.O. Haskard, and A.M. Peters. 1997. A comparison between 111In-anti-E-selectin mAb and 99Tcm-labelled human non-specific immunoglobulin in radionuclide imaging of rheumatoid arthritis. *Br. J. Radiol.* 70:473–481.

180. Chapman, P.T., F. Jamar, E.T. Keelan, A.M. Peters, and D.O. Haskard. 1996. Use of a radiolabeled monoclonal antibody against E-selectin for imaging of endothelial activation in rheumatoid arthritis. *Arthritis Rheum.* 39:1371–1375.

181. Ashkenazi, A. and S.M. Chamow. 1997. Immunoadhesins as research tools and therapeutic agents. *Curr. Opin. Immunol.* 9:195–200.

182. Jakubowski, A., B.N. Ehrenfels, R.B. Pepinsky, and L.C. Burkly. 1995. Vascular cell adhesion molecule-Ig fusion protein selectively targets activated alpha 4-integrin receptors in vivo. Inhibition of autoimmune diabetes in an adoptive transfer model in nonobese diabetic mice. *J. Immunol.* 155:938–946.

183. Jakubowski, A., M.D. Rosa, S. Bixler, R. Lobb, and L.C. Burkly. 1995. Vascular cell adhesion molecule (VCAM)-Ig fusion protein defines distinct affinity states of the very late antigen-4 (VLA-4) receptor. *Cell Adhes. Commun.* 3:131–142.

184. Liao, F., J. Ali, T. Greene, and W.A. Muller. 1997. Soluble domain 1 of platelet-endothelial cell adhesion molecule (PECAM) is sufficient to block transendothelial migration in vitro and in vivo. *J. Exp. Med.* 185:1349–1357.

185. Gao, W., P.S. Topham, J.A. King, S.T. Smiley, V. Csizmadia, B. Lu, C.J. Gerard, and W.W. Hancock. 2000. Targeting of the chemokine receptor CCR1 suppresses development of acute and chronic cardiac allograft rejection. *J. Clin. Invest.* 105:35–44.

11 | Apoptosis and Autoimmunity

Keith B. Elkon

1. INTRODUCTION

By definition, autoimmunity is a consequence of inappropriate survival and activation of self-reactive cells in the immune system. Inappropriate survival may be caused directly by mutations in the molecular pathways that signal cell death, by persistent activation of survival (anti-apoptotic) pathways or by alterations that affect the threshold of signaling through the antigen receptors. In this review, we will briefly discuss why, where and how death decisions are made and the consequences of failure of the normal physiology of apoptosis.

2. WHY DEATH?

Many organ systems in the body such as the gastro-intestinal tract, the skin and the immune system maintain homeostasis by loss of effete cells and replenishment with new cells. By virtue of its function, the immune system has to have three additional skills: A) it has to select cells that have just the right affinity to recognize self but avoid autoimmunity (see below) B) it must be able to selectively induce a single cell with the right specificity to proliferate and generate many thousands of effector cells to attack a foreign organism and then die, and C) it must generate subsets of cells (B lymphocytes) that undergo somatic mutation in order to maximize the efficiency of antibody targeting. As discussed below, these functions are intimately linked to life and death decisions.

3. WHERE DEATH?

In order to achieve selection of lymphocytes with the right avidity for self, the vast majority of potentially self-reactive cells are culled in the thymus (T cells) and bone marrow (B cells). Since B and T cell tolerance are described in detail elsewhere in this volume, this chapter will highlight the apoptotic pathways that achieve deletion of unwanted cells.

3.1. The Thymus

Approximately 95% of all thymocytes undergo apoptosis. There appear to be two fundamentally different pathways to death (1). Thymocytes that respond with high affinity to peptide/MHC complexes are actively induced to die, whereas thymocytes that receive no signal (including those cells with out of frame rearrangements of their T cell receptors) most likely die through a default pathway (lack of rescue). Ultimately, this cell fate decision reflects the integration of signals received from the T cell receptor (TCR), co-receptors (CD4, CD8), co-stimulatory receptors (CD28) and glucocorticoid receptors (2) as well as other unknown receptors on the cell surface. The precise intracellular signal transduction pathways that lead to apoptosis (negative selection) versus survival (positive selection) are not well defined, but appear to diverge after activation of ZAP-70 (Figure 1). Possibly, signaling through the Jun kinase (Jnk) pathway leads to activation of the Nur77 family of transcription factors that appear to be necessary for thymocyte apoptosis [rev. in (1)]. Fas ligand may be partly responsible for death, but other death factors are likely to be additional targets for these transcription factors. In contrast, signaling through the ras-raf-MEK pathway appears

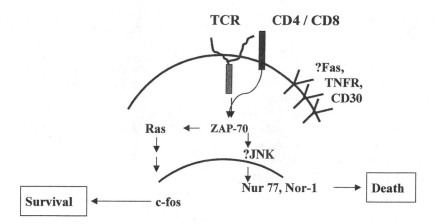

Figure 1. Putative receptors and signal transduction pathways in thymocyte apoptosis. Members of the TNF receptor family have been implicated in thymic selection, but none identified so far are absolutely required. Signals for cell survival versus death appear to diverge downstream from ZAP 70 where Ras activation promotes cell survival and JNK activation has been implicated in promoting cell death.

to be involved in positive selection leading to upregulation of Bcl-2 and thymocyte survival.

Members of the TNFR family play crucial roles in controlling survival of mature lymphocytes (see below). The roles for TNFR, Fas (APO-1/CD95) and CD30 in the thymus, however, are controversial. While Fas deficiency does not appear to affect deletion by Mls superantigens in the thymus, evidence has been presented that Fas-mediated apoptosis is responsible for deletion of "semi-mature", early single-positive thymocytes that encounter high dose antigen in the medulla (3). TNF-α-deficient mice also do not have overt defects in thymic selection, but mice deficient in CD30 were reported to have a decrease in negative selection in the thymus (4). At present, the evidence suggests that members of the TNFR family play ancillary roles in thymic selection, but none has so far been shown to be absolutely required for apoptosis.

3.2. The Bone Marrow

There are many parallels between the maturation of T cells in the thymus and B cells in the bone marrow. Large numbers of pro-B cells die due to faulty rearrangement of the genes encoding components of the B cell receptor (BCR), μ heavy chain and κ or λ light chain. Like CD4$^+$/CD8$^+$ thymocytes, B cells that have reached the maturational stage of "early B cells", sIgM$^+$ IgD-, are very sensitive to apoptosis when engaged by high density cross-linking of their antigen receptor (5,6). Unlike thymocytes, however, early B cells that react with high affinity to self-antigen have a chance to avoid death by expressing a different light chain on the cell surface, "receptor editing". While B cells were thought to lack any element for positive selection, experimental evidence is accumulating that B cells may also be rescued from death by interaction with self-antigen (7).

Dissection of the signaling components of apoptosis in early B cells is even more complex than thymocytes. In view of the capacity for BCR light chain receptor editing on encountering a high avidity interaction with antigen, how does the B cell signal respond differently on a second encounter with a new antigen? In the absence of known death ligands expressed by the B cells themselves, is the death signal intrinsic or does the B cell become highly susceptible to extrinsic death ligands?

3.3. The Peripheral Immune System— Immunoregulation Through Apoptosis

Peripheral lymphocyte tolerance is maintained by a number of mechanisms. These include a possible further round of deletion of self-reactive cells in the early neonatal period (8), induction of anergy that occurs when lymphocytes encounter antigen with low affinity in the absence of co-stimulation (9,10), and apoptosis of activated cells.

Following activation by foreign micro-organisms, lymphocytes become fully armed and express and/or secrete cytotoxic molecules and pro-inflammatory cytokines. It is therefore imperative that, when the initiating organism is destroyed, the activated lymphocytes are also removed. Apoptosis of activated cells [activation induced cell death (AICD)] occurs by two phenomena—cytokine withdrawal, most likely leading to apoptosis by the intrinsic pathway or by active death induction through death receptors (see below). The two death receptors that are implicated in apoptosis of activated peripheral lymphocytes are Fas and TNF-α (11–13), although TRAIL may play a role when induced by cytokines such as type 1 interferons (14). As shown in Figure 2, Fas ligand is expressed on activated CD8, CD4 T cells of the Th1 subset, and NK cells. These cells therefore can kill themselves as well as activated Th2 CD4$^+$ T cells, B cells and macrophages. The role of TNF-α

is less well established, especially since TNF-α-deficient mice do not develop autoimmunity (15,16). Nevertheless, both *in vivo* and *in vitro* experiments suggest that TNF-α plays a role in AICD of subsets of T cells (11–13). Persistence of activated lymphocytes in the presence of highly expressed co-stimulatory molecules on antigen presenting cells allows the activated cells to crossreact with or, in the case of B cells, somatically mutate toward, self specificity. The critical role of apoptosis in the maintenance of peripheral tolerance is illustrated by the autoimmune phenotype of Fas *(lpr)* and Fas ligand *(gld)* mutant mice respectively (see below).

4. HOW DEATH?

A detailed description of the molecular mechanisms of apoptosis and its regulation are beyond the scope of the current chapter and the reader is referred to a series of excellent reviews in *Science*, 28 August 1998. In this chapter, we will highlight the molecular pathways of apoptosis that have potential relevance to autoimmunity.

4.1.   Active Induction of Apoptosis—The TNFR Family

Members of the TNFR family of receptors play pivotal roles in the regulation of cell death in the peripheral immune system. The roles of each receptor, where known, are summarized in Table 1. Surprisingly, many of these receptors are able to signal both apoptosis and activation/survival. The death receptors (DR) identified so far, Fas, TNFR I, DR-3 (TRAMP/wsl/APO-3), DR-4 (TRAIL) and DR-5, share homology in their intra-cellular domains over a 70 amino acid region termed the 'death domain' (17,18). In addition, three

Table 1   The Role of TNFR family members in apoptosis regulation in the immune system

Receptor	Cell type*	Death	Activation/survival
Fas	B and T, Mac	+++	–
TNFR1	All	+	+++
TNFR2	All	+	++
DR3	?	++	?
DR4	?	++	?
DR5	?	++	?
CD30	B and T	+ (inhibition)	?
CD40R	B, Mac	?	+++
TRANCE	DC	–	++

* The predominant cell type in the immune system expressing the receptor. Mac = macrophage, DC = dendritic cell, ? = not fully defined

decoy receptors have been identified, two (DcR1 and DcR2) that bind and inhibit their ligand, TRAIL, and one (DcR3) that binds Fas ligand. These decoy receptors presumably modulate cytotoxic function of the ligands but the biological contexts remain to be defined. Alternative splice forms and shedding of the receptors and ligands also downmodulate their function.

The three-dimensional structure of the death domain has been solved by NMR spectroscopy and has been shown to consist of six amphipathic α-helices that create a unique fold (19). Functionally, the death domain appears to be a novel protein-protein association motif that facilitates homotypic interactions (Fig. 3). Binding of Fas ligand to Fas causes receptor clustering and recruitment of intra-

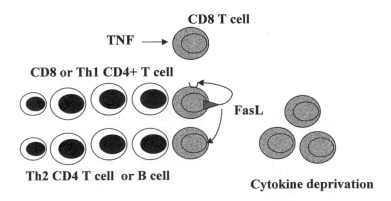

Figure 2. Activation induced cell death of lymphocytes. Following full activation and clonal proliferation of T or B cells, cells may undergo apoptosis through cytokine withdrawal or by active induction of death by Fas ligand (FasL) or TNF-α. FasL is expressed predominantly on activated CD8 and Th1 CD4+ T cells whereas TNF-α is produced both by T cells and activated macrophages. TNF-a may have some selectivity for CD8+ T cells. Since neither B cells nor Th2 CD4+ T cells express significant levels of FasL, their deletion occurs in trans. Apoptotic cells are shaded.

cellular adaptor molecules. Initially, aggregation of Fas induces uptake of the adaptor protein, FADD/MORT1, to the death domain of Fas (20,21). FADD has two structural domains: a C-terminal death domain, which mediates Fas binding, and an N-terminal death effector domain (DED). The FADD DED allows recruitment of pro-caspase 8 (FLICE/MACHα1) (22,23) and pro-caspase 10b (24) via DED–DED interactions. Pro-caspases 8 and 10b have a bipartite structure comprising a DED and an enzymatic caspase domain, the latter linking Fas aggregation with the execution phase of apoptosis. The apposition of pro-caspases 8 and 10b to the activated Fas complex leads to autocatalytic cleavage (22) and conversion of the pro-enzymes to activated proteases, which are released and able to initiate a proteolytic cascade leading to programmed cell death. In some cell lines, caspase 8 cleavage also results in cleavage of the pro-apoptotic molecule Bid that, in turn, activates the mitochondrial amplification cascade [type II pathway (25)].

In addition to interacting molecules promoting cell death, members of the TRAF protein family mediate signaling by TNFR family members directly or indirectly. Six TRAF family members (TRAF 1–6) have been identified and most of these proteins contain a RING-finger motif at their N-termini that is required for their biological activities. TRAF signal transduction is usually linked to the cell survival pathway through activation of NFkB (26).

4.2. Regulation of Apoptosis—The Bcl-2 Family and IAPs

While the upstream molecular pathways through which death signals are transmitted are reasonably well characterized, multiple other signals impinge on downstream events and regulate the the final cell fate decision. Among the most important regulators are proteins belonging to the Bcl family, comprising more than 15 members (rev. in 27.) Bcl-2 homologs such as Bcl-XL (other homologues include mammalian proteins Bcl-w, Mcl-1, A1, and virus-encoded proteins, BHRF1, LMW5-HL, ORF16, KS-Bcl-2, E1B-19K) have at least one of the characteristic Bcl-2 homology (BH) domain motifs and most possess a hydrophobic C-terminal membrane anchor, accounting for their attachment to mitochondrial and nuclear membranes. The pro-apoptotic members of this family such as Bax (others include Bak, Bok, Bik, Blk, Hrk, BNIP3, Bim, Bad, Bid) also contain BH domains.

How do Bcls regulate apoptosis? One level of regulation is conferred by binding interactions (homo- or heterodimerization) between members, effected largely by their BH1, 2 and 3 domains (28). Although the outcomes vary for each specific pair, homodimerization of Bcl-2 or Bax potentiate their anti-or pro-apoptotic function respectively whereas heterodimers may potentiate or abrogate function

of one member of the pair. Bcls such as Bcl-2 and Bcl-XL also bind to Apaf-1 and prevent it from activating caspase 9, analogous to the regulation of CED-4 by CED-9 in *C. elegans* (29).

Bcls regulation of cell death is closely connected to mitochondrial function. The physical association of Bcl-2 family proteins with the outer mitochondrial membrane, as well as the close structural similarity between the BH1 and 2 domains and bacterial pore-forming proteins such as colicins (30), allows them to regulate ion fluxes or the transfer of small molecules from the membrane. *In vitro* models suggest that Bax and Bak promote opening of the VDAC allowing the release of cytochrome c into the cytosol whereas Bcl-2 binds directly to VDAC and closes it (31,32) (Fig. 3). A caspase-independent "Apoptosis Inducing Factor" (AIF) that is also released from the mitochondria under certain circumstances induces cell death by a less well-defined pathway (33).

IAPs (intracellular inhibitors of apoptosis) are a separate family of anti-apoptotic proteins that are highly conserved through evolution. The neuronal IAP (NAIP for neuronal apotosis inhibitory protein) was discovered through the association of NAIP mutations in patients with the severe form of spinal muscular atrophy. Four additional members of the family (c-IAP-1, 2, X-IAP and survivin) that share a baculovirus IAP repeat domain have subsequently been identified. Although IAPs were shown to bind to TRAFs and therefore were thought to function upstream in the apoptotic pathway, more recent studies suggest that IAPs inhibit effector caspases (34). IAPs block apoptosis induced by a variety of stimuli including Fas, TNF-α, UV irradiation and serum withdrawal and survivin is overexpressed in certain cancers as well as in the rheumatoid arthritis synovium (35).

4.3. Execution of Apoptosis—Caspases

Caspases are cysteine-containing proteases that have an unusual substrate specificity for peptidyl sequences with a P1 aspartate residue (rev. in 36,37) These proteases are 30–50 kD in size and comprise a NH2 terminal prodomain, a large subunit domain and a small subunit domain. The active site cysteine residue is contained within the conserved pentapeptide, QACxG, on the large subunit of the enzyme, whereas most of the substrate specificity is determined by the small subunit. The upstream caspases 8, 9, 10, 2 and 4 have large prodomains that interact with regulatory proteins such as FADD for caspases 8 and 10, and Apaf-1 for caspase 9 (Fig. 3). Presumably, clustering of these complexes allows for autocatalytic cleavage of the large and small subdomains to form the active tetramer, as discussed above. Effector caspases such as 3, 6 and 7 have small prodomains and are thought to be cleaved into their active forms by the upstream caspases.

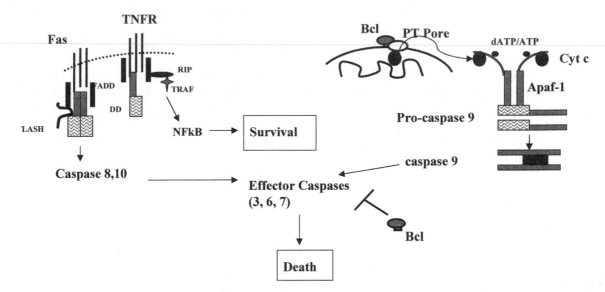

Figure 3. Receptor mediated and intrinsic apoptotic pathways. The extrinsic, ligand induced pathway is shown on the left and the intrinsic pathway induced by growth factor withdrawal or injury to the cell is shown on the right. Both pathways converge on the effector caspases (3, 6 and 7) which lead to cleavage of structural proteins, DNA repair proteins, DNA condensation and release of nucleosomes. Note that TNFR can also signal cell survival by activation of NFkB. See text for a full description of the biochemistry of each pathway.

Members of the caspase family can be divided into 3 functional subgroups based on their substrate specificities (38). Group I (caspases 1, 4 and 5) that are potently inhibited by the serpin, CrmA; Group II (caspases 2, 3 and 7) are specific for DExD; and Group III (caspases 6, 8, 9 and 10) that are specific for I/V/LExD, a sequence that is also contained at the junctions of the caspase subunits themselves. Significantly, granzyme B produced by cytotoxic T cells, has a similar substrate specificity as Group III caspases and is also capable of inducing apoptosis through this pathway. Identification of the substrate specificity of caspases has led to a number of practical applications, including the ability to quantify activity using fluorogenic terapeptide substrates and blockade of proteolytic activity with non-cleavable cell-permeable tetrapeptide analogues.

The effector caspases are necessary for the execution of apoptosis. They cleave specific substrates such as the structural proteins, fodrin, gelsolin and lamins, key intracellular enzymes involved in DNA repair (e.g. poly ADP ribose polymerase, DNA-PK). These changes facilitate inactivation of synthetic functions of the cell, dissolution of the nuclear membrane and the packaging of cellular proteins into apoptotic blebs on the cell surface. In addition, caspases cleave regulatory proteins such as Bcl family members and the Inhibitor of Caspase Activated Dnase (ICAD). Cleavage of ICAD leads to the release of active CAD that, in turn, enters the nucleus and cleaves nucleosomes at the linker region yielding the characteristic "DNA ladder" (39,40). It should be noted that not all caspases are involved in the execution of apoptosis. Caspases 1, 4, 5, 11, 12 and 13 are, most likely, involved in inflammation. The best understood example is caspase 1 (interleukin-1 converting enzyme/ICE), the enzyme that cleaves IL-1 into its active form (41).

5. LIFE AFTER DEATH?

Within the immune system alone, more than 10^8 apoptotic cells are removed from the body each day. These apoptotic cells are generated in vast numbers in the central lymphoid organs such as the thymus and bone marrow but are cleared with remarkable efficiency by phagocytes in these organs. In addition, a significant load of apoptotic cells is also produced in the peripheral immune system due to the relatively short life span of lymphocytes and myeloid cells as well as specialized sites of secondary selection of high affinity B cells in germinal centers. The ligands and receptors responsible for removal of apoptotic cells are incompletely defined but are summarized briefly here.

An early event in apoptotic cells is the appearance of phosphatidylserine (PS) on the cell surface membrane. This membrane asymmetry (PS is usually located on the inner surface of the membrane) is caused by a reduced function of a translocase and possibly, activation of a lipid scramblase (42). PS has been shown to be an important ligand for phagocytosis of apoptotic cells (43), although the receptor remains to be identified. Similarly, certain sugars may be selectively exposed on the membrane triggering their phagocytosis (44).

Diverse receptors have been implicated in the binding and phagocytosis of apoptotic cells by macrophages and/or dendritic cells. Many of the receptors are integrins comprising the vitronectin receptor, $\alpha_v\beta_3$, (45); $\alpha_v\beta_5$ (46); complement receptor 3 (CD11b/CD18) and 4(CD11c/CD18) (47) and class A and B scavenger receptors (48–50). Non-integrin receptors include the ATP-binding cassette transporter, ABC1 (51) and CD14 (52). It is relevant to note that in some cases, phagocytosis of apoptotic cells requires serum factors such thrombospondin to bridge the $\alpha_v\beta_3$ and CD36 receptors (53) or serum complement that is activated by apoptotic cells (47).

In vitro studies suggest that uptake of apoptotic cells induces the expression of immunosuppressive cytokines such as TGF-β1, PGE-2 and possibly IL-10 by macrophages (54,55). These cytokines would tend to dampen an immune response to self-antigens. Since some peptides derived from apoptotic cells can be presented to lymphocytes by dendritic cells and, possibly, macrophages by cross priming (56,57), a question of critical importance to studies of autoimmunity is whether self peptides are presented following phagocytosis of apoptotic cells and whether they induce tolerance or immunity.

5.1.   Apoptotic Cells and Autoimmunity

There is indirect evidence to suggest that the products of apoptotic cells may serve as immunogens in systemic autoimmune disorders such as SLE (Systemic Lupus Erythematosus). This evidence includes the detection of nucleosomes in the circulation of SLE patients with active disease (58); initial production of autoantibodies to nucleosomes (59,60); nucleosomes are more strongly antigenic for T (61) and B cells than DNA or histones alone; nucleosomes, but not isolated DNA or histones, deposit in the glomeruli suggesting that it is *in situ* fixation of nucleosomes, rather than DNA/anti-DNA immune complexes, that causes lupus nephritis (62,63). Patients with SLE also have increased apoptosis of their lymphocytes *in vitro* (64,65), although it is uncertain whether this reflects an intrinsic abnormality or cytokine withdrawal of activated cells.

In addition to these observations in disease, a link between the targets of autoantibodies and caspases has been suggested. Many of the SLE antigens redistribute to apoptotic blebs when cells such as keratinocytes undergo programmed cell death, implying that these antigens may be more accessible to the immune system (66). A particularly relevant antigen in this context is phosphatidylserine that translocates to the cell surface during apoptosis (see above) and is recognized by anti-cardiolipin antibodies (67). A low level of autoantibodies to cardiolipin (but not ds-DNA or protein autoantigens) has been induced by intensive immunization of mice with apoptotic cells (68). Some, *but not all* of the intracellular protein antigens undergo modification, including cleavage and phosphorylation

during apoptosis (69–71). Although it has been suggested that these modified antigens may induce immunity through the production and presentation of cryptic epitopes, it remains to be shown that the immune system is exposed to different epitopes in the periphery compared to the tolerizing compartments in the central immune system.

6.   APOPTOSIS AND AUTOIMMUNE DISEASES

6.1.   Apoptosis is Relevant to Autoimmunity in Several Respects:

A. As mentioned above, a significant body of evidence suggests that autoantibodies target the products of apoptotic cells in some autoimmune disorders.
B. Mutations of the apoptosis receptor, Fas, predisposes to a systemic autoimmune disease.
C. In certain diseases such as rheumatoid arthritis, the rate of cell survival exceeds the rate of cell death leading to excessive growth of tissue.
D. Effectors of apoptosis contribute to tissue injury in organ specific autoimmune diseases such as insulin dependent diabetes mellitus (IDDM), multiple sclerosis (MS) and Hashimoto's thyroiditis.

6.2.   Defects in cell death pathways promote lupus-like autoimmune diseases: Canale Smith Syndrome/ALPS

Mice with mutations of Fas or Fas ligand (FasL) develop a syndrome characterized by lymphoproliferation (*lpr*)/generalized lymphadenopathy (*gld*) together with systemic autoimmunity (72). As might be expected from the key role described for Fas in activation induced cell death (AICD) described above, lymphadenopathy and splenomegaly are the consequence of an absolute increase in the numbers of T and B lymphocytes as well as by the accumulation of an unusual subset of T cells that fail to express either CD4 or CD8 co-receptors ("double negative" T cells). The nature and extent of systemic autoimmunity varies according to the strain into which the Fas or FasL mutation has been bred (72).

A syndrome of massive lymphadenopathy with systemic autoimmunity in children was first recognized by Canale and Smith in 1967 (73). Subsequently, these and other "lpr" patients (74) were found to have mutations in Fas (75–77). The syndrome that now has been identified in more than 50 children [called either Canale Smith Syndrome (CSS), autoimmune lymphoproliferative syndrome (ALPS), or human lymphoproliferative syndrome with autoimmunity (HLSA)] is characterized by lymphadenopathy/ splenomegaly, autoimmune cytopenias (most commonly affecting platelets and red cells) and an increase (>5%) in circulating double-negative T cells. These patients had defective lymphocyte apoptosis in response to anti-Fas antibodies or Fas ligand when tested *in vitro* and most had heterozygous mutations in Fas, affecting the death domain. The majority of these mutations impair Fas mediated apoptosis through a dominant negative effect (78,79). CSS patients without Fas mutations are thought to have alterations in molecules involved in downstream signal transduction (80,81).

Although isolated cases of SLE with either Fas (82) or Fas ligand (83) mutations have been described, most evidence would suggest that Fas and Fas ligand are not mutated in the majority of SLE patients (83–85). Nevertheless, the Canale Smith syndrome is informative since it suggests that defective apoptosis of cells of the immune system can predispose to systemic autoimmune diseases such as SLE, ITP, autoimmune hemolytic anemia and Guillain-Barre Syndrome (82,86). Furthermore, it illustrates (as do the mouse models) that even when a single gene has a powerful effect upon predisposition to systemic autoimmunity, the clinical expression of disease varies depending upon the precise nature of the mutation (78,79) and the interaction of modifiers.

6.3.   Prolonged Exposure to Growth Factors Promotes Anti-apoptotic Gene Expression, Growth and Invasion of Tissue: Rheumatoid Arthritis

Histologically, rheumatoid arthritis (RA) is characterized by an accumulation of inflammatory cells in the synovium leading to pannus formation and destruction of cartilage and bone. Although Fas, FasL and apoptotic cells can be detected in the RA synovium (87,88), the extent of synoviocyte apoptosis is not adequate to counteract ongoing proliferation. This imbalance is explained by a number of factors. First, FasL is expressed at relatively low levels on synovial T cells (97) and soluble Fas or FasL competitors may impede Fas-induced apoptosis. Cytokines such as TNF-α, IL-1β and TGFβ1, which favor synoviocyte proliferation and inhibit susceptibility to apoptosis, are overexpressed in the joints of patients with RA (89–91). These and other signals reduce apoptosis of synoviocytes through increased expression of NFkB, the Bcl-2 family of proteins and the IAP, survivin (88,92–94). Growth of the pannus is compounded by inflammatory changes such as oxidation that result in upregulation (94) and mutations (95) of the tumor suppressor protein, p53.

The observations regarding apoptosis regulators in RA discussed above are significant because they provide an opportunity for therapeutic manipulation (see also below). Local administration of an anti-Fas mAb to HTLV-1 tax transgenic mice or Fas ligand to collgen arthritis (mouse models of RA) led to an improvement in arthritis (96,97). Several strategies to modulate NFkB, attenuate the growth of synovial cells (see below).

6.4.   Effectors of Apoptosis Contribute to Tissue Injury in Organ-Specific Autoimmunity

In contrast to systemic autoimmune diseases that are characterized by B lymphocyte stimulation leading to antibody and immune complex-mediated tissue injury, many organ-specific autoimmune diseases are characterized by a cell-mediated attack leading to the death of specific cell types within the organ. Cell targets are β cells of the islets of Langerhans of the pancreas in insulin-dependent diabetes mellitus (IDDM), oligodendrocytes in the brain in multiple sclerosis (MS), thyrocytes in Hashimoto's thyroiditis [these diseases are reviewed in detail elsewhere (98)], the salivary and lacrimal glands in Sjögren's syndrome and myocytes in polymyositis.

Programmed pathways of cell death (apoptosis) can be implicated in disease pathogenesis as illustrated by the resistance of Fas-deficient (1pr) mice to diseases such as IDDM and EAE (99–101). In many of these diseases, cell death at the site of injury can also be directly demonstrated by DNA fragmentation (TUNEL staining) *in situ*. Apoptosis is usually considered non-inflammatory, but cell fate may be determined as much by energy resources within the cell as by the *effector of death* and several effector pathways may induce features of both apoptosis and necrosis. Although regulation of apoptosis in lymphocytes and antigen-presenting cells is also relevant to the pathogenesis of autoimmunity, as discussed above, here we will only consider how death of target tissues is induced.

In most organ-specific autoimmune diseases, especially those where animal models are available, CD4$^+$ T cells have been shown to be critically involved in the pathogenesis. The disease-promoting CD4 T cells are restricted by MHC class II and therefore unlikely to exert a direct cytotoxic action on the Class I-bearing target cell. CD4 cells may arm other effectors through the production of cytokines, they may induce tissue injury through a "bystander pathway" involving macrophages or induce receptors for cell death on the target cell "assisted suicide pathway". In Sjögren's syndrome in humans, there is controversy as to whether Fas or FasL is constitutively expressed in normal salivary glands, but the co-expression of both molecules in patients with Sjögren's syndrome presumably causes cell death of acinar and ductal cells (102).

A particularly interesting apoptotic mechanism is proposed in NOD.*scid* mice (diabetes-prone mice lacking mature T and B lymphocytes). These mice spontaneously develop loss of submandibular acinar cells associated with increased levels of caspases 1, 2 and 3 in saliva. It has been suggested that intrinsic defects within the submandibular glands promote apoptosis (103). The exocrine glands from both normal and NOD mice constitutively express the ligand for Fas (APO-1/CD95), but only NOD and NOD.scid mice express Fas at 18 weeks, a time at which an inflammatory infiltrate is evident in NOD mice. Expression of Fas appears to be a key factor promoting apoptosis of exocrine glands in NOD mice. This model of "assisted suicide" is similar to that reported in Hashimoto's thyroiditis in humans (104).

Inflammatory myopathies, such as polymyositis (PM) and dermatomyositis (DM), are autoimmune diseases that result in destruction of skeletal muscle fibers. Although Fas is upregulated on the myocytes in these diseases, expression is also increased in non-autoimmune muscle disorders, such as in metabolic myopathies, denervating disorders and muscular dystrophies, but not in normal human muscle tissue (105). Detection of FasL on mononuclear cells invading the muscles in PM and DM patients, with apoptosis of muscle cells implicates Fas/FasL in tissue injury in myositis (106). High levels of expression of the T cell cytotoxic mediator, perforin, in some PM and DM patients (107), indicates that granzyme-mediated myocyte injury is also involved.

7. CONCLUSIONS

The science of apoptosis is in its infancy, some of the biochemical pathways that control it have only been discovered over the last decade. It is highly likely, therefore, that further definition of the regulatory components will lead to insight as to how self-reactive cells in the immune system survive inappropriately in autoimmunity. Most important, these advances will provide new opportunities for therapeutic intervention (Fig. 4). Already, cell permeable caspase inhibitors have been used experimentally to attenuate massive hepatic apoptosis (108) and similar approaches could be applied to human autoimmune diseases characterized by excessive apoptosis. For example, based on the results in animal models (99–101), blockade of Fas selective ligand could possibly attenuate diabetes or multiple sclerosis. For therapeutic efficacy however, two points require attention. It will be important to determine "the point of no return", i.e. the stage at which cells become irreversibly damaged so that apoptosis can be blocked prior to this step. In addition, inhibitors of apoptosis need to be selective for the cell type that is defective. As discussed above, widespread inhibition of apoptosis could potentially predispose the host to cancer and to systemic autoimmune diseases.

In diseases characterized by reduced apoptosis and excessive cell proliferation such as rheumatoid arthritis, pro-apoptotic therapy can be applied. Death-inducing ligands have been successful at treating animal models of rheumatoid arthritis (97) and bisindolylmaleimide VIII has been found to enhance Fas-mediated apoptosis in animals (109). To avoid serious side effects, apoptosis will need to be narrowly targeted to the cell type inducing tissue injury or to the cell type that is proliferating excessively. Given the rate of progress in the field, these therapeutic modalities may not be too far away.

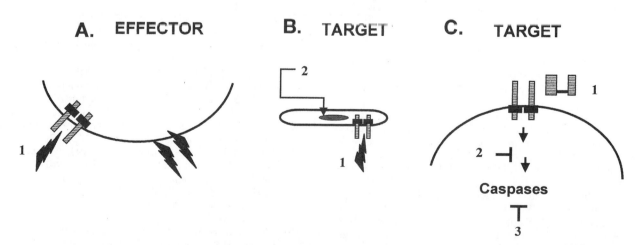

Figure 4. Avenues for therapeutic manipulation of apoptosis. To eliminate effector cells that cause disease in organ specific autoimmunity (e.g. diabetes or multiple sclerosis) (A), or to kill cells that have taken on growth characteristics similar to neoplastic cells in rheumatoid arthritis (B), apoptosis can be directly induced by a death ligand (A1, B1). Pro-apoptotic pathways can also be initiated from within the cell by expression of pro-apoptotic proteins through gene therapy (B2). In diseases where apoptotic cell death leads to loss of organ function, attempts can be made to block death ligands (C1) or interfere with upstream components of apoptosis, for example by introducing anti-apoptotic genes into the cell (C2). Further downstream, cell permeable caspase inhibitors (C3) can block the execution phase of apoptosis. Modified from (110).

ACKNOWLEDGMENTS

The contribution of laboratory members is gratefully acknowledged.

References

1. Amsen, D., and A.M. Kruisbeek. 1998. Thymocyte selection: Not by TCR alone. *Immunol. Rev.* 165:209–229.
2. Vacchio, M.S., J.Y.M. Lee, and J.D. Ashwell. 1999. Thymus-derived glucocorticoids set the thresholds for thymocyte selection by inhibiting TCR-mediated thymocyte activation. *J. Immunol.* 163:1327–1333.
3. Kishimoto, H., C.D. Surh, and J. Sprent. 1998. A role for Fas in negative selection of thymocytes in vivo. *J. Exp. Med.* 187:1427–1438.
4. Amakawa, R., A. Hakem, T.M. Kundig, T. Matsuyama, J.J. Simard, E. Timms, A. Wakeham, H.W. Mittruecker, II. Griesser, H. Takimoto, R. Schmits, A. Shahinian, P. Ohashi, J.M. Penninger, and T.W. Mak. 1996. Impaired negative selection of T cells in Hodgkin's disease antigen CD30-deficient mice. *Cell* 84:551–562.
5. Benhamou, L.E., P.-A. Cazenave, and P. Sarthou. 1990. Anti-immunoglobulins induce death by apoptosis in WEHI-231 B lymphoma cells. *Eur. J. Immunol.* 20:1405–1407.
6. Hasbold, J., and G.G.B. Klaus. 1990. Anti-immunoglobulin antibodies induce apoptosis in immature B cell lymphomas. *Eur. J. Immunol.* 20:1685–1690.
7. Hayakawa, K., M. Asano, S.A. Shinton, M. Gui, D. Allman, C.L. Stewart, J. Silver, and R. R. Hardy. 1999. Positive selection of natural autoreactive B cells. *Science* 285:113–116.
8. Alferink, J., A. Tafuri, D. Vestweber, R. Hallmann, G.J. Hammerling, and B. Arnold. 1998. Control of neonatal tolerance to tissue antigens by peripheral T cell trafficking. *Science* 282:1338–1341.
9. Mueller, D.L., and M.K. Jenkins. 1995. Molecular mechanisms underlying functional T cell unresponsiveness. *Curr. Opin. Immunol.* 7:375–381.
10. Goodnow, C.C. 1992. Transgenic mice and analysis of B cell tolerance. *Ann. Rev. Immunol.* 10:489–518.
11. Singer, G.G. and A.K. Abbas. 1994. The Fas antigen is involved in peripheral but not thymic deletion of T lymphocytes in T cell receptor transgenic mice. *Immunity* 1:365–371.
12. Sytwu, H.-K., R.S. Liblau, and H.O. McDevitt. 1996. The roles of Fas/Apo-1 (CD95) and TNF in antigen-induced programmed cell death in T cell receptor transgenic mice. *Immunity* 5:17–30.
13. Zheng, L., G. Fisher, R.E. Miller, J. Peschon, D.H. Lynch, and M.J. Lenardo. 1995. Induction of apoptosis in mature T cells by tumour necrosis factor. *Nature* 377:348–351.
14. Kayagaki, N., N. Yamaguchi, M. Nakayama, H. Eto, K. Okumura, and H. Yagita. 1999. Type I interferons (IFNs) regulate tumor necrosis factor-related apoptosis-inducing ligand (TRAIL) expression on human T cells: A novel mechanism for the anti-tumor effects of type I IFNs. *J. Exp. Med.* 189:1451–1460.
15. Pasparakis, M., L. Alexopoulou, V. Episkopou, and G. Kollias. 1996. Immune, and inflammatory responses in TNF—deficient mice: A critical requirement for TNF-α in the formation of primary B cell follicles, follicular dendritic cell networks and germinal centers, and in the maturation of the humoral immune response. *J. Exp. Med.* 184:1397–1411.
16. Marino, M.W., A. Dunn, D. Grail, M. Inglese, Y. Noguchi, E. Richards, A. Jungbluth, H. Wada, M. Moore, B. Williamson, S. Basu, and L.J. Old. 1997. Characterization of tumor necrosis factor-deficient mice. *Proc. Natl. Acad. Sci. USA* 94:8093–8098.
17. Itoh, N., and S. Nagata. 1993. A novel protein domain required for apoptosis. Mutational analysis of human Fas antigen. *J. Biol. Chem.* 268:10932–10937.
18. Tartaglia, L.A., T.M. Ayres, G.H.W. Wong, and D.V. Goeddel. 1993. A novel domain within the 55 kd TNF receptor signals cell death. *Cell* 74:845–853.
19. Huang, B., M. Eberstadt, E. Olejniczak, R. Meadows, and S. Fesik. 1996. NMR structure and mutagenesis of the Fas (APO-1/CD95) death domain. *Nature* 384:638–641.
20. Chinnaiyan, A.M., K. O'Rourke, M. Tewari, and V.M. Dixit. 1995. FADD, a novel death domain-containing protein, interacts with the death domain of Fas and initiates apoptosis. *Cell* 81:505–512.
21. Boldin, M.P., E.E. Varfolomeev, Z. Pancer, I.L. Mett, J.H. Camonis, and D. Wallach. 1995. A novel protein that interacts with the death domain of Fas/Apo-1 contains a sequence motif related to the death domain. *J. Biol. Chem.* 270:7795–7798.
22. Muzio, M., A.M. Chinnaiyan, F.C. Kischkel, K. O'Rourke, A. Shevchenko, J. Ni, C. Scaffidi, J.D. Bretz, M. Zhang, R. Gentz, M. Mann, P.H. Krammer, M.E. Peter, and V.M. Dixit. 1996. FLICE, a novel FADD-homologous ICE/CED-3-like protease, is recruited to the CD95 (Fas/APO-1) death-inducing signaling complex. *Cell* 85:817–827.
23. Boldin, M.P., T.M. Goncharov, Y.V. Goltsev, and D. Wallach. 1996. Involvement of MACH, a novel MORT1/FADD-interacting protease, in Fas/Apo-1 and TNF receptor-induced cell death. *Cell* 85:803–815.
24. Vincenz, C., and V.M. Dixit. 1997. Fas-associated death domain protein interleukin-1 beta-converting enzyme 2 (FLICE2), an ICE/Ced-3 homologue, is proximally involved in CD95-and p55-mediated death signaling. *J. Biol. Chem.* 7:6578–6583.
25. Scaffidi, C., S. Fulda, A. Srinivasan, C. Friesen, F. Li, K.J. Tomaselli, K.-M. Debatin, P.H. Krammer, and M.E. Peter. 1998. Two CD95 (APO-1/Fas) signaling pathways. *EMBO J.* 17:1675–1687.
26. Wang, C.-Y., M.W. Mayo, R.G. Korneluk, D.V. Goeddel, and A.S. Baldwin, Jr. 1998. NF-kB anti-apoptosis: Induction of TRAF1 and TRAF2 and c-IAP1 and c-IAP2 to suppress caspase-8 activation. *Science* 281:1680–1684.
27. Adams, J.M., and S. Cory. 1998. The Bcl-s protein family: arbiters of cell survival. *Science* 281:1322–1326.
28. Oltvai, Z.N., and S.J. Korsmeyer. 1994. Checkpoints of dueling dimers foil death wishes. *Cell* 79:189–192.
29. Hu, Y., M.A. Benedict, D. Wu, N. Inohara, and G. Nunez. 1998. Bxl-XL with Apaf-1 and inhibits Apaf-1-dependent caspase-9 activation. *Proc. Natl. Acad. Sci. USA* 95:4386–4391.
30. Muchmore, S.W., M. Sattler, H. Liang, R.P. Meadows, J.E. Harlan, H.S. Yoon, D. Nettesheim, B.S. Chang, C.B. Thompson, S.L. Wong, S.L. Ng, and S.W. Fesik. 1996. X-ray and NMR structure of human Bcl-xL, an inhibitor of programmed cell death. *Nature* 381:335–341.
31. Marzo, I., C. Brenner, N. Zamzami, J.M. Jurgensmeier, S.A. Susin, H.L.A. Vieira, M.-C. Prevost, Z. Xie, S. Matsuyama, J.C. Reed, and G. Kroemer. 1998. Bax and adenine

nucleotide translocator cooperate in the mitochondrial control of apoptosis. *Science* 281:2027–2031.

32. Shimizu, S., M. Narita, and Y. Tsujimoto. 1999. Bcl-2 family proteins regulate the release of apoptogenic cytochrome-c by the mitochondrial channel VDAC. *Nature* 399:483–487.

33. Kroemer, G., N. Zamzami, and S.A. Susin. 1997. Mitochondrial control of apoptosis. [Review] [71 refs.] *Immunology Today* 18:44–51.1

34. Roy, N., Q.L. Deveraux, R. Takahashi, G.S. Salvesen, and J.C. Reed. 1997. The c-IAP-1 and c-IAP-2 proteins are direct inhibitors of specific caspases. *EMBO J* 16:6914–6925.

35. Ambrosini, G., C. Adida, and D.C. Altieri. 1997. A novel anti-apoptosis gene, survivin, expressed in cancer and lymphoma. *Nature Med.* 3:917–921.

36. Thornberry, N.A., and Y. Lazebnik. 1998. Caspases: Enemies within. *Science* 281:1312–1316.

37. Los, M., S. Wesselborg, and K. Schulze-Osthoff. 1999. The role of caspases in development, immunity, and apoptotic signal transduction: lessons from knockout mice. *Immunity* 10:629–639.

38. Garcia-Calvo, M., E.P. Peterson, B. Leiting, R. Ruel, D.W. Nicholson, and N.A. Thornberry. 1998. Inhibition of human caspases by peptide-based and macrophage inhibitors. *J. Biol. Chem.* 273:32608–32613.

39. Enari, M., H. Sakahira, H. Yokoyama, K. Okawa, A. Iwamatsu, and S. Nagata. 1998. A caspase-activated DNase that degrades DNA during apoptosis, and its inhibitor ICAD. *Nature* 391:43–50.

40. Sakahira, H., M. Enari, and S. Nagata. 1998. Cleavage of CAD inhibitor in CAD activation and DNA degradation during apoptosis. *Nature* 391:96–99.

41. Yuan, J., S. Shaham, S. Ledoux, H.M. Ellis, and H.R. Horvitz. 1993. The *C. elegans* cell death gene ced-3 encodes a protein similar to mammalian interleukin-1-converting enzyme. *Cell* 75:641–652.

42. Verhoven, B., R.A. Schlegel, and P. Williamson. 1995. Mechanisms of phosphatidylserine exposure, a phagocyte recognition signal, on apoptotic T lymphocytes. *J. Exp. Med.* 182:1597–1601.

43. Fadok, V.A., D.R. Voelker, P.A. Campbell, J.J. Cohen, D.L. Bratton, and P.M. Henson. 1992. Exposure of phosphatidyl serine on the surface of apoptotic lymphocytes triggers specific recognition and removal by macrophages. *J. Immunol.* 148:2207–2216.

44. Duvall, E., A.H. Wyllie, and R.G. Morris. 1985. Macrophage recognition of cells undergoing programmed cell death. *Immunology* 56:351–358.

45. Savill, J., N. Dransfield Hogg, and C. Haslett. 1990. Vitronectin receptor-mediated phagocytosis of cells undergoing apoptosis. *Nature* 343:170–173.

46. Albert, M.L., S.F.A. Pearce, L.M. Francisco, B. Sauter, P. Roy, R.L. Silverstein, and N. Bhardwaj. 1998. Immature dendritic cells phagocytose apoptotic cells via alpha-v-beta-5 and CD36, and cross-present antigens to cytotoxic T lymphocytes. *J. Exp. Med.* 188:1359–1368.

47. Mevorach, D., J. Mascarenhas, D.A. Gershov, and K.B. Elkon. 1998. Complement-dependent clearance of apoptotic cells. *J. Exp. Med.* 188:2301–2311.

48. Platt, N., H. Suzuki, Y. Kurihara, T. Kodama, and S. Gordon. 1996. Role for the class A macrophage scavenger receptor in the phagocytosis of apoptotic thymocytes *in vitro*. *Proc. Natl. Acad. Sci. USA* 93:12456–12460.

49. Sambrano, G.R., and D. Steinberg. 1995. Recognition of oxidatively damaged and apoptotic cells by an oxidized low density lipoprotein receptor on mouse peritoneal macrophages: Role of membrane phosphatidylserine. *Proc. Natl. Acad. Sci. USA* 92:1396–1400.

50. Fukasawa, M., H. Adachi, K. Hirota, M. Tsujimoto, H. Arai, and K. Inoue. 1996. SRB1, a class B scavenger receptor, recognizes both negatively charged liposomes and apoptotic cells. *Experimental Cell Research* 222:246–250.1

51. Luciani, M.-F., and G. Chimini. 1996. The ATP binding cassette transporter, ABC1, is required for the engulfment of corpses generated by apoptotic cell death. *EMBO J.* 15:226–235.

52. Devitt, A., O.D. Moffatt, C. Raykundalia, J.D. Capra, D.L. Simmons, and C.D. Gregory. 1998. Human CD14 mediates recognition, and phagocytosis of apoptotic cells. *Nature* 392:505–509.

53. Ren, Y., R.L. Silverstein, J. Allen, and J. Savill. 1995. CD36 gene transfer confers capacity for phagocytosis of cells undergoing apoptosis. *J. Exp. Med.* 181:1857–1862.

54. Fadok, V.A., D.L. Bratton, A. Konowal, P.W. Freed, J.Y. Westcott, and P.M. Henson. 1998. Macrophages that have ingested apoptotic cells in vitro inhibit proinflammatory cytokine production through autocrine/paracrine mechanisms involving TGF-beta, PGE2, and PAF. *J. Clin. Invest.* 101:890–898.

55. Voll, R.E., M. Herrmann, E.A. Roth, C. Stach, J.R. Kalden, and I. Girkontaite. 1997. Immunosuppressive effects of apoptotic cells. *Nature* 390:350–351.1

56. Albert, M.L., B. Sauter, and N. Bhardwaj. 1998. Dendritic cells acquire antigen from apoptotic cells and induce class I-restricted CTLs. *Nature* 392:86–89.

57. Bellone, M., G. Iezzi, P. Rovere, G. Galati, A. Ronchetti, M.P. Protti, J. Davoust, C. Rugarli, and A.A. Manfredi. 1997. Processing of engulfed apoptotic bodies yields T cell epitopes. *J. Immunol.* 159:5391–5399.

58. Rumore, P.M., and C.R. Steinman. 1990. Endogenous circulating DNA in systemic lupus erythematosus, occurrence as multimeric complexes bound to histone. *J. Clin. Invest.* 86.69–74.

59. Burlingame, R.W., R.L. Rubin, R.S. Balderas, and A.N. Theofilopoulos. 1993. Genesis and evolution of anti-chromatin autoantibodies in murine lupus implicates T-dependent immunization and self antigen. *J. Clin. Invest.* 91:1687–1696.

60. Amoura, Z., H. Chabre, S. Koutouzov, C. Lotton, A. Cabrespines, J.F. Bach, and L. Jacob. 1994. Nucleosome restricted antibodies are detected before anti-ds DNA and/or anti-histone antibodies in serum of MRL-Mp 1pr/1pr and +/+ mice, and are present in kidney eluates of lupus mice with proteinuria. *Arthritis. Rheum.* 37:1684–1688.

61. Mohan, C., S. Adams, V. Stanik, and S.K. Datta. 1993. Nucleosome: A major immunogen for pathogenic autoantibody-inducing T cells of lupus. *J. Exp. Med.* 177:1367–1381.

62. Termaat, R.M., K.J. Assmann, H.B. Dijkman, F. van Gompel, R.J. Smeenk, and J.H. Berden. 1992. Anti-DNA antibodies can bind to the glomerulus via two distinct mechanisms. *Kidney Int.* 42:1363–1371.

63. Kramers, C., M.N. Hylkema, M.C.J. van Bruggen, R. van de Lagemaat, H.B.P.M. Dijkman, K.J.M. Assmann, R.J.T. Smeenk, and J.H.M. Berden. 1994. Anti-nucleosome antibodies complexed to nucleosomal antigens show anti-DNA

reactivity and bind to rat glomerular basement membrane in vivo. *J. Clin. Invest.* 94:568–577.

64. Emlen, W., J.-A. Niebur, and R. Kadera. 1994. Accelerated in vitro apoptosis of lymphocytes from patients with systemic lupus erythematosus. *J. Immunol.* 152:3685–3692.

65. Perniok, A., F. Wedekind, M. Herrmann, C. Specker, and M. Schneider. 1998. High levels of circulating early apoptotic peripheral blood mononuclear cells in systemic lupus erythematosus. *Lupus* 7:113–118.

66. Casciola-Rosen, L.A., G. Anhalt, and A. Rosen. 1994. Autoantigens targeted in systemic lupus erythematosus are clustered in two populations of surface structures on apoptotic keratinocytes. *J. Exp. Med.* 179:1317–1330.

67. Price, B.E., J. Rauch, M.A. Shia, M.T. Walsh, W. Lieberthal, and H.M. Gilligan. 1996. O'Laughlin T, Koh JS, Levine JS. Antiphospholipid autoantibodies bind to apoptotic, but not viable, thymocytes in a beta2-glycoprotein I-dependent manner. *J. Immunol.* 157:2201–2208.

68. Mevorach, D., J.-L. Zhou, X. Song, and K.B. Elkon. 1998. Systemic exposure to irradiated apoptotic cells induces autoantibody production. *J. Exp. Med.* 188:387–392.

69. Casciola-Rosen, L.A., G.J. Anhalt, and A. Rosen. 1995. DNA-dependent protein kinase is one of a subset of autoantigens specifically cleaved early during apoptosis. *J. Exp. Med.* 182:1625–1634.

70. Utz, P.J., M. Hottelet, P.H. Schur, and P. Anderson. 1997. Proteins phosphorylated during stress-induced apoptosis are common targets for autoantibody production in patients with systemic lupus erythematosus. *J. Exp. Med.* 185:843–854.

71. Casiano, C.A., S.J. Martin, D.R. Green, and E.M. Tan. 1996. Selective cleavage of nuclear autoantigens during CD95 (Fas/APO-1)-mediated T cell apoptosis. *J. Exp. Med.* 184:765–770.

72. Cohen, P.L., and R.A. Eisenberg. 1991. lpr and gld: single gene models of systemic autoimmunity and lymphoproliferative disease. *Annu. Rev. Immunol.* 9:243–269.

73. Canale, V.C., and C.H. Smith. 1967. Chronic lymphadenopathy simulating malignant lymphoma. *J. Pediatr.* 70:891–899.

74. Sneller, M.C., S.E. Straus, E.S. Jaffe, J.S. Jaffe, T.A. Fleisher, M. Stetler-Stevenson, and W. Strober. 1992. A novel lymphoproliferative/autoimmune syndrome resembling murine lpr/gld disease. *J. Clin. Invest.* 90:334–341.

75. Fisher, G.H., F.J. Rosenberg, S.E. Straus, J.K. Dale, L.A. Middelton, A.Y. Lin, W. Strober, M.J. Lenardo, and J.M. Puck. 1995. Dominant interfering Fas gene mutations impair apoptosis in a human lymphoproliferative syndrome. *Cell* 81:935–946.

76. Rieux-Laucat, F., F. Le Deist, C. Hivroz, I.A.G. Roberts, K.M. Debatin, A. Fischer, and J.P. de Villartay. 1995. Mutations in Fas associated with human lymphoproliferative syndrome and autoimmunity. *Science* 268:1347–1349.

77. Drappa, J., A.K. Vaishnaw, K.E. Sullivan, J.L. Chu, and K.B. Elkon. 1996. The Canale Smith syndrome: an inherited autoimmune disorder associated with defective lymphocyte apoptosis and mutations in the Fas gene. *N. Engl. J. Med.* 335:1643–1649.

78. Vaishnaw, A.K., J.R. Orlinick, J.L. Chu, P.H. Krammer, M.V. Chao, and K.B. Elkon. 1999. Molecular basis for the apoptotic defects in patients with CD95 (Fas/Apo-1) mutations. *J. Clin. Invest.* 103:355–363.

79. Martin, D.A., L. Zheng, R.M. Siegel, B. Huang, G.H. Fisher, J. Wang, C.E. Jackson, J.M. Puck, J. Dale, S.E. Straus, M.E. Peter, P.H. Krammer, S. Fesik, and M.J. Lenardo. 1999.

Defective CD95/APO-1/Fas signal complex formation in the human autoimmune lymphoproliferative syndrome, type Ia [In Process Citation]. *Proc. Natl. Acad. Sci. USA* 96:4552–4557.1

80. Dianzani, U., M. Bragardo, D. DiFranco, C. Alliaudi, P. Scagni, D. Buonfiglio, V. Redoglia, S. Bonissoni, A. Correra, I. Dianzani, and U. Ramenghi. 1997. Deficiency of the Fas pathway without Fas gene mutations in pediatric patients with autoimmunity/lymphoproliferation. *Blood* 89:2871–2879.

81. Wang, J., L. Zheng, A. Lobito, F. Chan, J. Dale, M. Sneller, X. Yao, J.M. Puck, S.E. Straus, and M.J. Lenardo. 1999. Inherited human caspase 10 mutations underlie defective lymphocyte and dendritic cell apoptosis in autoimmune lymphoproliferative syndrome type II. *Cell* 98:47–58.

82. Vaishnaw, A.K., E. Toubi, S. Ohsako, J. Drappa, S. Buys, J. Estrada, A. Sitarz, L. Zemel, J.L. Chu, and K.B. Elkon. 1999. Both quantitative and qualitative apoptotic defects are associated with the clinical spectrum of disease, including systemic lupus erythematosus in humans with Fas (APO-1/CD95) mutations. *Arthritis. Rheum..*

83. Wu, J., J. Wilson, J. He, L. Xiang, P.H. Schur, and J.D. Mountz. 1996. Fas ligand mutation in a patient with systemic lupus erythematosus and lymphoproliferative disease. *J. Clin. Invest.* 98:1107–1113.

84. Mysler, E., P. Bini, J. Drappa, P. Ramos, S.M. Friedman, P.H. Krammer, and K.B. Elkon. 1994. The APO-1/Fas protein in human systemic lupus erythematosus. *J. Clin. Invest.* 93:1029–1034.

85. Ohsako, S., M. Hara, M. Harigai, C. Fukusawa, and S. Kashiwazaki. 1994. Expression and function of Fas antigen and Bcl-2 in human systemic lupus erythematosus lymphocytes. *Clin. Immunol. Immunopathol.* 73:109–114.

86. Sneller, M.C., J. Wang, J.K. Dale, W. Strober, L.A. Middelton, Y. Choi, T.A. Fleisher, M. S. Lim, E.S., Jaffe, J.M. Puck, M.J. Lenardo, and S.E. Straus. 1997. Clinical, immunologic, and genetic features of an autoimmune lymphoproliferative syndrome associated with abnormal lymphocyte apoptosis. *Blood* 89:1341–1348.

87. Nakajima, T., H. Aono, T. Hasunuma, K. Yamamoto, T. Shirai, K. Hirohata, and K. Nishioka. 1995. Apoptosis and functional Fas antigen in rheumatoid arthritis synoviocytes. *Arthritis. Rheum.* 38:485–491.

88. Firestein, G.S., M. Yeo, and N.J. Zvaifler. 1995. Apoptosis in rheumatoid arthritis synovium. *J. Clin. Invest.* 96:1631–1638.1

89. Kawakami, A., K. Eguchi, N. Matsuoka, M. Tsubol, Y. Kawabe, T. Aoyagi, and S. Nagataki. 1996. Inhibition of Fas antigen-mediated apoptosis of rheumatoid synovial cells in vitro by transforming growth factor 1. *Arthritis Rheum.* 39:1267–1276.

90. Tsuboi, M., K. Eguchi, A. Kawakami, N. Matsuoka, Y. Kawabe, T. Aoyagi, K. Maeda, and S. Nagataki. 1996. Fas antigen expression on synovial cells was downregulated by interleukin 1. *Biochem. Biophys. Res. Comm.* 218:280–285.

91. Salmon, M., D. Scheel-Toellner, A.P. Huissoon, D. Pilling, N. Shamsadeen, H. Hyde, A. D. D'Angeac, P.A. Bacon, P. Emery, and A.N. Akbar. 1997. Inhibition of T cell apoptosis in the rheumatoid synovium. *J. Clin. Invest.* 99:439–446.

92. Fujisawa, K., H. Aono, T. Hasunuma, K. Yamamoto, S. Mita, and K. Nishioka. 1996. Activation of transcription factor NF-kB in human synovial cells in response to tumor necrosis factor. *Arthritis Rheum.* 39:197–203.

93. Marok, R., P.G. Winyard, A. Coumbe, M.L. Kus, K. Gaffney, S. Blades, P.I. Mapp, C.J. Morris, D.R. Blake, C. Kaltschmidt, and P.A. Baeurle. 1996. Activation of the transcription factor nuclear factor k-B in human inflamed synovial tissue. *Arthritis Rheum.* 39:583–591.

94. Sugiyama, M., T. Tsukazaki, A. Yonekura, S. Matsuzaki, S. Yamashita, and K. Iwasaki. 1996. Localization of apoptosis and expression of apoptosis-related proteins in the synovium of patients with rheumatoid arthritis. *Ann. Rheum. Dis.* 55:442–449.

95. Han, Z., D.L. Boyle, Y. Shi, D.R. Green, and G.S. Firestein. 1999. Dominant-negative p53 mutations in rheumatoid arthritis. *Arthritis Rheum.* 42:1088–1092.

96. Fujisawa, K., H. Asahara, K. Okamoto, H. Aono, T. Hasunuma, T. Kobata, Y. Iwakura, S. Yonehara, T. Sumida, and K. Nishioka. 1996. Therapeutic effect of the anti-Fas antibody on arthritis in HTLV-1 tax transgenic mice. *J. Clin. Invest.* 98:271–278.

97. Zhang, H., Y. Yang, J.L. Horton, E.B. Samoilova, T.A. Judge, L.A. Turka, J.M. Wilson, and Y. Chen. 1997. Amelioration of collagen-induced arthritis by CD95 (Apo-1/Fas)-ligand gene transfer. *J. Clin. Invest.* 100:1951–1957.

98. Ohsako, S., and K.B. Elkon. 1999. Apoptosis in organ specific autoimmunity. *Cell Death and Differentiation* 6:13–21.

99. Itoh, N., A. Imagawa, T. Hanafusa, M. Waguri, K. Yamamoto, H. Iwahashi, M. Moriwaki, H. Nakajima, J. Miyagawa, M. Namba, S. Makino, S. Nagata, N. Kono, and Y. Matsuzawa. 1997. Requirement of Fas for the development of autoimmune diabetes in non-obese diabetic mice. *J. Exp. Med.* 186:613–618.

100. Sabelko, K.A., A.K. Kelly, M.H. Nahm, A.H. Cross, and J.H. Russell. 1997. Fas and Fas ligand enhance the pathogenesis of experimental allergic encephalomyelitis, but are not essential for immune privilege in the central nervous system. *J. Immunol.* 159:3096–3099.

101. Waldner, H., R.A. Sobel, E. Howard, and V.K. Kuchroo. 1997. Fas- and FasL-deficient mice are resistant to induction of autoimmune encephalomyelitis. *J. Immunol.* 159:3100–3103.

102. Kong, L., N. Ogawa, T. Nakabayashi, G.T. Liu, E. D'Souza, H.S. McGuff, D. Guerrero, N. Talal, and H. Dang. 1997. Fas and Fas ligand expression in the salivary glands of patients with primary Sjogren's syndrome. *Archritis. Rheum.* 40:87–97.

103. Robinson, C.P., S. Yamachika, C.E. Alford, C. Cooper, E.L. Pichardo, N. Shah, A.B. Peck, and M.G. Humphreys-Beher. 1997. Elevated levels of cystein protease activity in saliva and salivary glands of the non-obese diabetic (NOD) mouse model for Sjogren syndrome. *Proc. Natl Acad. Sci. USA* 94:5767–5771.

104. Giordano, C., G. Stassi, R. de Maria, M. Todaro, P. Richiusa, G. Papoff, G. Ruberti, M. Bagnasco, R. Testi, and A. Galluzzo. 1997. Potential involvement of Fas and its ligand in the pathogenesis of Hashimoto's thyroiditis. *Science* 275:960–963.

105. Behrens, L., A. Bender, M.A. Johnson, and R. Hohlfeld. 1997. Cytotoxic mechanisms in inflammatory myopathies. Co-expression of Fas and protective Bcl-2 in muscle fibres and inflammatory cells. *Brain* 120:929–938.

106. Sugiura, T., Y. Murakawa, A. Nagai, M. Kondo, and S. Kobayashi. 1999. Fas and Fas ligand interaction induces apoptosis in inflammatory myopathies: CD4+ T cells injury in polymyositis. *Arthritis & Rheumatism* 42:291–298.

107. Goebels, N., D. Michaelis, M. Engelhardt, S. Huber, A. Bender, D. Pongratz, M.A. Johnson, H. Wekerle, J. Tschopp, D. Jenne, and R. Hohlfeld. 1996. Differential expression of perforin in muscle-infiltrating T cells in myositis and dermatomyositis. *J. Clin. Invest.* 97:2905–2910.

108. Rodriguez, I., K. Matsuura, C. Ody, S. Nagata, and P. Vassalli. 1996. Systemic injection of a tripeptide inhibits the intracellular activation of CPP32-like proteases *in vivo* and fully protects mice against Fas-mediated fulminant liver destruction and death. *J. Exp. Med.* 184:2067–2072.

109. Zhou, T., L. Song, P. Yang, Z. Wang, D. Lui, and R.S. Jope. 1999. Bisindolylmaleimide VIII facilitates Fas-mediated apoptosis and inhibits T cell-mediated autoimmune diseases. *Nature. Med.* 5:42–48.

110. Elkon, K.B. 1999. *Kelley's Textbook of Rheumatology, 6th ed.* W.B. Saunders Company, Philadelphia.

12 | Processing and Presentation of Self-Antigens

Constantin Bona

1. INTRODUCTION

While B cells throughout the Ig receptor can recognize conformational or linear epitopes exposed on the surface of native antigens, T cells recognize only peptides derived from the processing of antigens.

CD4 T cells recognize peptides in association with MHC class II molecules. The peptides associated with class II are bound via anchoring amino acid residues to grooves of a given MHC class II allele. This means that the binding of a vast array of peptides is actually allelic-restricted. The class II grooves has pockets with one or two opened end and, therefore, the length of the peptide is not a restrictive element, since it can extend outside the groove. The lengh of class II peptides varies from 10 up to 20 amino acid residues (1).

CD8 T cells recognize peptide in association with MHC class I molecules. The MHC class I molecule is made up of three domains ($\alpha1$, $\alpha2$, $\alpha3$) that in association with $\beta2$-microglobulin, form a closed groove that in turn, can accommodate peptides composed of 8 to 10 amino acid residues. The majority of naturally-occurring peptides extracted from class I molecules are nonapeptides (2). The binding of peptides to class I is also allelic-specific depending on anchoring residues of the peptide and complementarity of MCH class I molecules (3). Changes in amino acids of binding pockets preclude the binding of peptides.

2. EFFECTOR CELLS AND PATHWAYS OF ANTIGEN PRESENTATION

Antigen presentation is the process of degradation of proteins and glycoproteins leading to generation of peptides able to bind to MHC molecules. Since the majority of somatic cells express class I molecules, in principle all of them can present peptides to CD8 T cells.

In contrast, only cells expressing class II molecules can present the peptides to CD4 T cells. The cells able to present peptides to CD4 T cells are divided into two categories: Professional antigen-presenting cells (APC) and facultative or non-professional APC.

Professional APCs are the dendritic cells, B cells and macrophages. However, fibroblasts devoid of class II molecules but transfected with α- and β-chain genes encoding class II molecules can present peptides. This observation indicates that all somatic cells possess the enzymatic machinery for degradation of proteins, and the presence of class II molecule is a requirement for presentation.

The professional APC also express ligands (B7.1, B7.2, LAF-1, LAF-3, ICAM-1) that bind to corresponding receptors on the surface of T cells (CD28, CTLA-4). The interaction of these ligands with molecules expressed by T cells deliver costimulatory signals required for the activation of T cells.

There is a hierarchy in processing and presenting the peptides among APCs. Apparently the dendritic cells are the best, as was clearly demonstrated for a chimeric immunoglobulin (murine IgG2a) bearing influenza virus epitopes in the CDR3 of VH segment. In one molecule, the CDR3 was replaced with HA110–120 peptide of the hemagglutinin (HA) of influenza PR8 virus (4) recognized by CD4 T cells in association with I-E^d, and in another with NP147–161 peptide of nucleoprotein recognized in association with class I K^d by CD8 T cells (5). The maximal activation of a T cell hybridoma specific for HA110–120 peptide was reached with 500 dendritic cells, 8,000 B cells and 10^6 spleen cells (6).

It is noteworthy that, in certain mice prone to autoimmune disease, such as NOD mice, which spontaneously develop IDDM, there is a developmental defect in the differentiation of macrophages. Bone marrow precursors of macrophages exhibit low sensitive to CSF-1 able to induce the differentiation, and NOD macrophages are functionally defective (7).

Non-professional APC can process proteins, but cannot present the peptides in the absence of class II molecules. The expression of class II molecules in these cells is upregulated by INFγ and, in certain cells, eosinophiles by IL-4 and GM-CSF. Several cells are endowed with this property: Eosinophiles, renal proximal tubular cells, myocytes, thyrocytes and intestinal cells.

The differences with regard to which peptide is presented to T cells with MHC class I or II results from different pathways of the processing of antigens (1).

As general rule, the peptides presented by class II are of exogenous origin (exogenous pathway), whereas those presented by class I are of cytoplasmic origin (endogenous pathway).

2.1. Exogenous Pathway

This is the major pathway for processing the proteins engulfed by APCs from the internal millieu by pinocytosis or phagocytosis. Engulfed antigens initially localized in endosomes subsequent to fusion of endocytic vacuoles with lysosomes are degraded in acidic endosomes. Recent studies showed that the generation of peptides takes place within low density endosomes and lysosomal dense compartment (8). The lysosomal enzymes activated in acidic vacuoles play a major role in the degradation of proteins. Bushel et al. (9) brought evidence showing that cathepsin B is responsible for the generation of T cell epitopes from human growth factor self-protein. The peptides generated from degradation within endosomes are sorted by empty class II molecules based on the affinity dictated by the nature of anchoring residues able to interact with complementary residues of class II molecules. The fragmentation of antigen and the presentation of peptides is very fast. Peptide can be detected at the surface of APC within 15 min. when it is internalized via Fc receptor, and within 2 hr. when it is internalized by fluid phase pinocytosis (8).

Class II molecules are synthesized in the ER where the complex composed of α, β, and invariant (Ii) chain, is assembled. The folding and/or oligomerization of class II heterodimer is facilitated in ER by chaperone molecules. Schreiber et al. (10) showed that calnexin is physically associated with αβIi complexes, but not mature αβ dimers. Calnexin may regulate class II intracellular transport by facilitating the egress of competent molecules from the ER. Initially the nascent class II molecules in the ER are empty, but soon the groove is filled with invariant chain-derived

peptides, such as CLIP, which precludes the binding of self peptides to class II molecules in ER (11,12). Invariant chain activity was mapped to exon 3 and particularly to 82–107 amino acid residues (13). CLIP derives from proteolytic digestion of Ii within the ER. CLIP can interact with many class II allelic products, but the affinity of this interaction is controlled by polymorphic residues in class II molecules(13). Invariant chains may also play a role in the transport of class II molecules to endosomal compartment after egress initially from ER and then from Golgi. This was demonstrated by Anderson and Miller (14) who showed that, in the absence of Ii chain, class II could form dimers, but these dimers were misfolded and exhibited altered conformation.

Once class II-CLIP tridimers reach the endosomes, the removal of CLIP is necessary in order to have an empty groove that allows the binding of peptide from exogenous proteins. The non-classical MHC class II molecules HLA-DM has been implicated in the removal of CLIP from class II (15), and cathepsin B in the cleavage of Ii chain (16).

Therefore, two parallel mechanisms occur in the generation of class II-peptide complexes in which the peptide derives from exogenous antigens: One responsible for fragmentation of proteins in endosomes, and another for synthesis of class II molecules in which the groove is filled early with Ii chain-derived peptide, which is released when the complex reaches the endosomes. Fig.1 illustrates the formation steps of class II-peptide complexes in the exogenous pathway.

This paradigm sharply contrasts with the analysis of the origin of peptides eluted from class II molecules of murine and human APC B cells.

Rudensky et al. (17) eluted and sequenced the peptides bound to class II molecules on the surface of a B cell lymphoma line (H-2^{bd}), one of which derived from I-A^b and seven from envelope gp70 protein of murine leukemia virus. The presence of virus in these cells was confirmed by PCR.

Hunt et al. (18) showed that, among twelve peptides isolated from I-A^d expressed on A20 murine B lymphoma, two derived from I-E and one from transferrin receptor.

Chicz et al. (19) carried out a similar study by analyzing the sequences of peptides eluted from human homozygous blastoid lines (HLA-DR2, DR3, DR7, and DR8) and found a predominance of peptides originating from endogenous proteins. For each allele, more than 200 unique peptides were identified by mass spectroscopy. More than 85% of the peptides bound to class II were derived from self-proteins (20).

How can these observations be explained? First, the APC take up the soluble protein molecules from the circulation and internalize them in endosomes, where they are processed in the exogenous pathway. This may be the case with

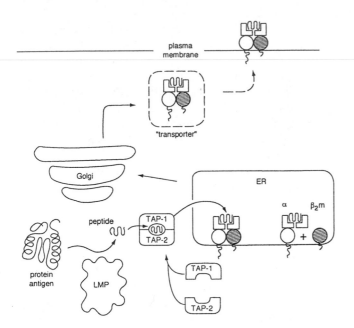

Figure 1. *A model for antigen presentation by MHC class I.* The LMP and/or other proteasome complex may be the site of protein degradation into antigenic fragments. Peptides are then carried into the endoplasm reticulum (ER) by the TAP-1 1/TAP-2 complex. Nascent class 1 molecules (α chains) pair with β_2 microglobun (β_2m). Loading the complex with peptide stabilizes it. Subsequently the complex is carried through the Go and through a transport mechanism to the surface. See text for additional details. (Adapted from Braciale and Braciale, 1991.)

peptides derived from Apoprotein B, serum enzymes, $\alpha 2$ glycoprotein, von Willebrand factor, glucose transporter, coagulation factor V, etc. The peptides from transferrin can derive from the protein internalized via transferrin receptor. However, this mechanism cannot explain the presence of peptides derived from receptors expressed at the surface of B cells (i.e. transferrin, mannose 6-phosphate, IFNγ, IgE Fc, IL-8, serotonin and TCR receptors), peptides derived from class II alloantigens, nor peptides originating from cytoplasmic proteins (i.e. demoglein 3, actin, ras-related RAB7 protein, HSP70, tubulin, 40S ribosomal protein etc.).

The mechanism of generation of peptides from truly intracellular proteins was addressed in an excellent study carried out by Weiss and Bogen (21), who showed that an endogenously produced light chain of MOPC315 meyloma protein was processed and presented to T cells after binding to class II molecules. The peptide derived from $\lambda 2^{315}$ is an idiotype peptide recognized by idiotype-specific T cells (22). Intracellular origin of the idiotype peptide was clearly demonstrated using transfectants induced with engineered vectors and producing $\lambda 2^{315}$ molecules that were neither expressed on the membrane nor secreted. The idiopeptide was generated in the ER, but not within the nucleus. This was demonstrated by the ability of a transfectant expressing

a mutated truncate $\lambda 2$ protein in which the KDEL motif was added to the C-terminus of the light chain to stimulate idiotype-specific T cells. The KDEL motif is responsible for retention of luminal proteins. In contrast, the transfectant in which the NH$_2$-terminus of the $\lambda 2$ chain was replaced with the N-terminus of the large T antigen of the SV40 virus carrying nuclear targeting sequences was unable to stimulate idiotype-specific T cells. Possible generation of idiopeptide in the cytosol was ruled out using a transfectant induced with a vector in which the first 32 N-terminal aminoacids were replaced with NH$_2$-terminal amino acids of rabbit β-globulin. The fusion protein was detected by fluorescence into the cytoplasm, but the transfectant failed to stimulate idiotype-specific T cells.

These interesting findings suggest that the peptides derived from endogenous proteins associated with class II are generated within the ER. Such peptides should have a high affinity for class II molecules, precluding the binding of CLIP peptides. Our interpretation of these findings is supported by an observation of Bikoff (23), who showed that transfectant fibroblast-expressing class II was able to present a peptide derived from endogenous protein (i.e. CH3 domain of IgG2a), but that a transfectant expressing class II, and lacking the invariant chain gene, was clearly defective for the presentation of peptides derived from exogenous proteins. It remains a mystery how class II-self peptide trimers produced within the ER are transferred into the endosomes, since it is generally accepted that invariant chain facilitates the egress from ER into Golgi and then directs heterodimers toward the secretory pathway.

The fact that the majority of MHG associated peptides derive from endogenous proteins is of great interest for understanding tolerance and autoimmunity.

The presence of such peptides on thymic epithelial cells may be responsible for negative selection leading to central tolerance, and those present on APC may be responsible for peripheral tolerance. The generation of high amounts of such peptides subsequent to a primary injury may reach the threshold required for the break-down of self-tolerance, leading to activation of autoreactive lymphocytes that mediate autoimmunity.

Thus, it was of considerable interest to determine whether peptides derived from antigens that are the target of autoantibodies or self-reactive T cells are present on class II molecules. Indeed, such peptides were identified. For example, AchR 289–304 peptide was eluted from DR17 (Ach R is target antigen in Myastenia gravis), and two peptides derived from GAD65 (115–127, 274–286), which are epitopes recognized by diabetogenic T cells were eluted from DR4Dw4. A peptide derived from Histone3 that binds autoantibodies from SLE was eluted from DR4Dw15 molecules and three peptides derived from myelin basic protein (87–99, 131–145 and 139–152) recognized by T cells from patients with multiple sclerosis were eluted from DR40101 molecules.

These few examples clearly demonstrated that peptides derived from self-proteins, which are targets of autoimmune diseases, are associated with class II molecules on the surface of APC.

Since these potentially pathogenic peptides were derived from blastoid lines derived from normal subjects, the ability to generate these peptides may not be pathognomonic for the disease, and other factors may be involved in breaking down peripheral self-tolerance.

It is noteworthy that the presentation of self-peptides is not dependent solely on class II molecules, but may be influenced by nonstructural MHC genes that map to a179kb segment of MHC class II region (24). This suggests that MHC locus encodes factors that prevent or abrogate the presentation of self-peptides (24).

2.2.   Endogenous Pathway

The endogenous pathway is important for the generation of peptides from proteins synthesized within the cytoplasm encoded by host genes or by microbial genes of obligatory intracellular microbes (viruses, rickets and some bacteria and parasites) (25). The peptides generated within endogenous pathway are bound to class I molecules in the ER.

Initially, the heavy chain of class I molecules interacts with chaperon proteins, calnexin and immunoglobulin binding protein (26), which protect the nascent heavy chain molecule from misfolds until it binds the β2-microglobulin, and then the peptide. The loading of class I molecules with peptide is very fast: Within 4 minutes after translation of class I genes. The molecules egress from ER and finally are translocated via cis-Golgi on plasma membrane, where they are recognized by CD8 T cells. Schoel and Kaufmann (27) showed that the β2-microglobulin, considered essential for the function and stability of class I-peptide complex, is not fully required, since the presentation of a peptide can be achieved with macrophages from β2-microglobulin knock out mice. Thus, in the absence of β2-microglobulin, the α-chain can express a low density peptide sufficient for recognition by CD8 CTLs.

The processing of endogenous proteins occurs in the extralysosomal compartment. For proteolysis, mammalian cells use a very conserved apparatus represented by proteasomes (28). Phylogenetic conservation of proteasome function was demonstrated by Niederman et al. (29), who showed that a peptide associated with class I molecules can be generated after incubation *in vitro* of a protein with proteasomes from *D. melanogaster, S. cerevisiaeital* or mammalian cells. There are three major forms of proteasomes: a). 20S, which preferentially degrades misfolded or damaged proteins; b). 26S, and c). 20S proteasome activator

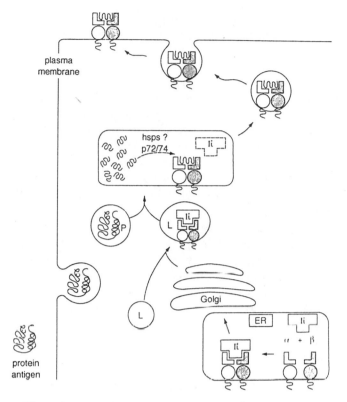

Figure 2. *A model for antigen presentation by MHC class II.* Class II molecules initially form complexes made up of trimers of α + β + Ii (invariant chain). These complexes then enter a distinct intracellular endosomal pathway for loading with peptide. A protein antigen enters the cell by phagocytosis/pinocytosis (left). This may or may not be mediated by various receptors such as FeγR or complement receptors of phagocytes for endocytosis of immune complexes, surface Ig receptors of B cells, etc. (not shown). Eventually, degraded antigen associates with class II molecules which are then transported to the cell surface. The details of antigen degradation are not known, but "classical" lysosomes are probably not involved. The process of removal of Ii and peptide loading onto class II also remains obscure. Members of the hsp 70 family of heat shock proteins may participate (see text). (Adapted from Braciale and Braciale, 1991).

28S complex (30). Generally, the proteasomes have a barrel shape hollow with four layers of rings each composed of seven sub-units located in outer rings. The β sub-unit located in the inner rings is responsible for the degradation of proteins. The protein that penetrates into the proteasomal cavity is associated with ubiquitin, a cytoplasmic chaperone w1 loci (3). The β subunit of three types of proteasomes have IFNγ-inducible homologues and are associated with LMP2 and LMP7 (31). The MHC-linked LMP2 and LMP7 are located in humans between DPβ1 and DQβ1 loci (32), and the genes encoding them exhibit a limited polymorphism. In LMP2, a non-conservative nucleotide base pair change at amino acid position 66 (arginine or histidine) resulted in two alleles, LMP2.1 and 2.2 (33). In contrast, the alleles of LMP7 result from an A-C transversion at nucleotide position 145 in exon2, resulting in glutamine in the case of LMP7.1, or lysine in the case of LMP7.2 allele (34). LMP2 is involved in inhibiting cleavage after acidic residues, whereas LMP7 may enhance cleavage after basic and hydrophobic residues (35).

The role of proteasomes in generation of peptides from endogenous proteins is not generally accepted. However their role is supported by two sets of findings: a). inhibition of assembly of MHC class I-peptide molecules by lactacystin, a proteasome inhibitor (27), and b). in vitro generation of nonapeptides recognized by CD8 T cells subsequent to incubation of native protein with proteasomes (27).

The peptides released from proteasomes are translocated into the ER by protein-specific antigenic peptide (TAP) molecules (36,37). Two types of TAP, TAP-1 and TAP-2, encoded by genes located in MHC class II locus, have been identified in humans and other mammalian species (38). TAPs belong to the ATP-binding cassette family of protein transporters and translocate the peptides to the ER, but also deliver them to class I heterodimers. Kleijmeer et al. (35) provided cytochemical evidence that the TAP-peptide complex is located in the ER membrane and is probably oriented with its ATP-binding domains in the cytosol. In the lumen of ER, nascent α-chain of nascent class I molecule interacts with calnexin and B1P (40) and after formation of heterodimer with β2-microglobulin, the calnexin dissocates and then the heterodimer binds calreticulin (41,42). Class I-calreticulin complex binds the TAP-peptide complex, a step mediated by tapsin, which bridges the TAP with class I-calreticulin complex (43). This step allows for translocation of peptides of a certain size and allelic-specific residues from TAP to class I molecules. Fig. 1 illustrates the major events of class I-peptide trimer formation.

The alterations of TAPs may be important for the initiation or progression of autoimmune diseases. In NOD mice, as in human IDDM, a low expression of class I molecules was observed that correlates with low concentration of peptide transports (44). In B cells from IDDM patients transfected with TAP-1 and TAP-2, Wang et al. (45) showed that the defective expression of class I molecules could be corrected by transfection with TAP genes, which also reduce the number of empty class I molecules. The authors concluded from these observations that TAP gene dysfunction may contribute to the defect in class I phenotype and antigen presentation in IDDM by allowing autoreactive cells to escape negative selection or anergy in the peripheral organs.

The peptides resulting from proteasome digestion are short, 8–10 amino acid long, having an affinity on the order of 10^{-5} to 10^{-7}. The peptides associated with class I molecules have distinct conserved residues called anchoring residues, which are quite restrictive for a given class I allele and permit the insertion of peptide into the pockets of MHC class I groove (46). The peptides originating from endogenous protein express binding motifs specific for a particular MHC class I allele (47,48).

3. EVIDENCE FOR THE PROCESSING AND PRESENTATION OF SELF ANTIGENS

There is no difference between foreign and self antigens with respect to antigen uptake, proteolysis and the presentation of peptides. This was clearly demonstrated in our laboratory through studies of the activation of a T cell clone specific for an immunodominant peptide of influenza virus hemagglutinin. A murine chimeric immunoglobulin expressing HA110–120 peptide in the CDR3 replaced PR8 virus in activating the T cell hybridoma. It is note worthy that the delivery of this neoantigen by a self-immunoglobulin molecule was more efficient than by virus or even synthetic peptide. From 10^9 APC to which 20×10^6 pmoles of peptide was offered, 10^2 pmole were recovered from class II molecules. When 0.4×10^6 pmoles of chimeric IgG molecule was offered, 470 pmole were recovered, and when 0.1×10^6 pmole, PR8 virus was offered, only 10 pmole were recovered from class II . These results suggest that the generation of peptides from self-molecules is more efficient than of foreign molecules. The analysis of the structure of peptides isolated from class II molecules showed that the peptides have an structure identical to the nominal synthetic peptide (49). This observation is in agreement with the results of Winchester et al. (50) showing that APC were equally able to process and present autologous and foreign peptides.

The ability of APCs to process self-molecules and present the peptides was demonstrated in experiments showing that the APCs, directly after removal from the body and without any additional manipulations or incubation with self proteins, stimulated self-peptide-specific T cells. The demonstration of class II-self-peptide complexes on surface of APC was possible after generation of self-reactive T cell clones and T hybridomas.

Lorenz and Paul (51) provided the first evidence of a functional self-peptide-MHC class II complex capable of stimulating autoreactive T cells. Taking advantage of allelic forms of β-chain of hemoglobin, these authors prepared T cell hybridomas specific for the Hb β-chain^d allele. These hybridomas were activated subsequent to in vivo incubation with thymic dendritic cells, spleen, lymph node and small resting B and Kupfer cells (52,53) without addition of exogenous Hb. This observation represents the first demonstration that self-proteins are processed constitutively and can be presented in similar fashion to foreign antigens. This indicates that APC do not distinguish between self and foreign antigens, in agreement with previous observations demonstrating a similar efficiency of guinea pig macrophages to phagocytize foreign and homologous erythrocytes (54).

Stockinger and Lin (55) showed that the pro-C5 precursor is also processed and presented to T cells. While C5 wild mice were tolerant to C5, the C5-deficient mice were not and their macrophages activated class II restricted C5-specific T cell clones without addition of exogenous C5.

Similar results were reported for α1-trypsin in transgenic mice expressing human gene encoding therefore a neoself antigen. Hagerty and Allen (56) studied the presentation of α1-trypsin in transgenic mice expressing a human gene encoding for this neoantigen. T cell-specific probes were used to detect the class II-peptide complex. Interestingly, the authors found that renal proximal tubule epithelial cells, were able to present the antigen a facultative APC which in certain conditions can express class II molecules.

Rider et al. (57) took, a different approach to demonstrate the immunogenicity of peptides derived from the processing of self-antigens by eluting the peptides from murine class II and using them to assay the ability to induce specific T cells. These peptides derived from apolipoprotein E, cystatin-c and transferrin receptor, and were recognized by CD4 T cells in association with I-A^d. It possible that the level of T responses to these self-peptides correlates with the density of peptides expressed on APC, and only those with low density allowed for positive selection of T cells.

Peptides derived from α-and β-chains of class II and from α- and β2-microglobulin chains of class I molecules were eluted from human and murine class II molecules. (21). Allopeptides represent around 10% of eluted peptides, which is very similar to the high frequency (5–10%) of alloreactive T cells. However, it was long unclear whether alloreactive T cells recognized empty class II molecules at the surface of stimulator cells in MLR or, alternatively, they recognize class II-allopeptide complexes. To address this question, Demoz and Lanzavecchia (58) studied the ability of induction of a MLR with a mutant unable to assemble a class II molecule peptide complex or with a soluble class II molecule filled up with a tetanus toxoid derived peptide. Both empty class II molecules and those filled with a foreign-derived peptide class II molecules failed to stimulate alloreactive cells. These results demonstrated that alloreactive T cells stimulated *in vitro* in MLR, or *in vivo* subsequent to allografts recognize class II-allopeptide complexes.

One may ask whether the phenomenon of the processing of self-proteins is relevant for the induction of autoimmune diseases. Unfortunately this question has not been fully investigated, although there are a few examples suggesting that this may be the case.

Myosin is the target antigen of acute myocarditis a T-cell-mediated autoimmune disease. An indirect observation that APC contain class II-myosin-derived peptide *in vivo* was provided by Smith and Allen (58), whom showed that the transfer of T cells from mice with autoimmune myocarditis into SCID mice induced disease, suggesting strongly that SCID APCs expressed class II-myosin peptides capable of activating myosin-specific T cells.

Collagen type II is the target of experimental collagen-induced arthritis and one of the target antigens in human RA disease. Manoury-Schwartz et al. (59) showed that APCs pulsed with fragments of collagen can stimulate collagen-specific T cells. The stimulation capacity was inhibited by leupeptin and chloromethylketone (inhibitors of processing) or by brefaldin, an transport inhibitor of newly-synthesized class II molecules via Golgi. Michaelsson et al. (60) showed cell type differences in the capacity to process collagen type II, since macrophages, but not dendritic cells, were able to activate specific T cells. The same group of investigators (61) showed that collagen-specific T cells actually recognize a glycopeptide in which the carbohydrate was O-linked to the hydroxy-lysines within a portion of peptide corresponding to 256–270 amino acid residues of collagen type II. It is noteworthy that the recognition of glycosylated self peptide is not unique to those derived from the processing of self-molecules, since it was shown that glycopeptides derived from foreign antigens bind MHC molecules and elicit specific T cell responses. Some glycosylated peptides can bind MHC molecules better than non-glycosyalated peptides (62).

4. CONCLUSIONS

From the data reviewed above, several conclusions can be drawn:

a). Peptides from self-proteins represent the vast majority of peptides associated with class II molecules. Among self peptides, 10% derive from MHC antigens.

b). Naturally-occurring peptides from self antigens that are targets of autoimmne B or T cells can be isolated from APC, which can activate auto-reactive T cells without the addition of exogenous antigen. Peptides from myelin basic protein, Ach R, GAD65, myosin, thyroid peroxidase, autoantigens that are targets in multiple sclerosis, myasthenia gravis, autoimmune carditis and Hashimoto's disease, respectively, were eluted from human class II molecules.

c). Self-peptides derived from circulating proteins are generated within exogenous pathways, as are those derived from the processing of foreign antigens.

Self-peptides derived from cytoplasmic proteins are generated within the endogenous pathway, mainly in the ER.

d). There are no differences between the mechanisms of uptake, processing and presentation of peptides derived from self-and foreign-peptides.

e). While the presence of self-peptides on thymic epithelium or peripheral APCs is important in shaping the repertoire by deletion or anergy of autoreactive T cells, they may contribute to breaking down of natural self tolerance, leading to autoimmunity.

References

1. Braciale, T.J., and V.L. Braciale. 1991. Antigen presentation: structural themes and functional variations. *Immunol. Today* 12:124–130.

2. Rotzschike, D., and K. Falk. 1991. Naturally occurring peptide antigens derived from the MHC-class I-restricted processing pathway. *Immunol. Today* 12:447–454.

3. Townsend, A.R.M., C. Ohlen, J. Bastin, H.E. Ljungren, L. Foster, and K. Karre. 1989. Association of class I MHC heavy and light chain induces viral peptides. *Nature* 340:443–448

4. Zaghouani, H., R. Steiman, R. Noncas, H. Shah, W. Gerhard, and C. Bona. 1993. Presentation of a viral T cell epitope expressed in the CDR3 region of a self immunoglobulin molecule. *Science* 259:224–259.

5. Zaghouani, H., M. Krystal, H. Kuzu, T. Moran, H. Shah, Y. Kuzu, J. Schulman, and C. Bona. 1992. Cells expressing a H chain Ig gene carrying a viral T cell epitope are lysed by specific cytolytic T cells . *J. Immunol.* 148:3604–3609.

6. Zaghouani H., Y. Kuzu, H. Kuzu, T.-D. Brumeanu, W.J. Swiggard, R. Steinaman, and C. Bona. 1993. Contrasting efficacy of presentation by MHC class I and II products when peptides are administered within a common carrier, self immunoglobulin. *Eur. J. Immunol.* 23:2746–1750.

7. Serreze, D.V., H. Rex Gaskins, and E.H. Lriter. 1993. Defects in the differentiation and function of antigen presenting cells in NOD/Lt mice. *J. Immunol.* 150;2534–2543.

8. Barnes, K.A., and R.N. Mitchel. 1995. Detection of functional class II-associated antigen: role of low density endosomal compartment in antigen processing. *J. Exp. Med.* 181:1715–1727.

9. Bushell, G., C. Nelson, H. Chiu, C. Grimely, W. Hrnzel, J. Burnier, and S. Fong. 1993. Evidence supporting a role of cathepsin B in the generation of T cell antigenic epitopes of human growth hormone. *Mol. Immunol.* 30:587–591.

10. Schreiber, K.L., M.P. Bell, C.J. Huntoon, S. Rajagopalan, M.B. Brenner, and D.J. McKean. 1993. Class II histocompatibility molecules associated with calnexin during assembly in endoplasmic reticulum. *Int. Immunol.* 6:101–111.

11. Roche, P.A., and P. Cresswell. 1990. Invariant chain asociation with HLA-DR molecules inhibits immunogenic peptide binding. *Nature* 345:615–619.

12. Teyton, L., D. O'Sullivan, P.W. Dickson, V. Lotteau, A. Sette, P. Fink, and P.A. Peterson. 1990. Invariant chain distinguish between the exogenous and endogenous antigen presentation pathways. *Nature* 348:39–43.

13. Sette, A., S. Southwood, J. Miller, and E. Appela. 1995. Binding of MHC class II to the invariant chain-derived peptide, CLIP, is regulated by allelic polymorphism. *J. Exp. Med.* 181:677–683.

14. Anderson, M.S., and J. Miller. 1992. Invariant chain can function as a chaperone protein for class II MHC molecules. *Proc. Natl. Acad. Sci. USA* 89:2282–2286.

15. Kropshofer, H., A.B. Vogt, L.J. Stern, and G. Hammerling. 1995. Self relaese of CLIP in peptide loading HLA-DR molecules. *Science* 270:1357–1359. 16.

16. Reyes, V.E., S. Lu, and R.E. Humphreys. 1991. Cathepsin B cleavege of Ii from class II MHC α *and* β-chains. *J. Immunol.* 146:3877–3880.

17. Rudensky, A.Y., P. Preston-Hurlburt, B.K. Al-Ramdl, J. Rothbard, and C.A. Janeway. 1992. Truncation variants of peptides isolated from class II molecules suggest sequence motifs. *Nature.* 359:429–431.

18. Hunt, D.F., H. Mitchel, T.A. Dickinson, J. Shabanowitz, A.L. Cox, K. Sakaguchi, E. Appela, H.M. Grey, and H. Grey. 1992. Peptides presented to the immune system by murine class II MHC I-A[d]. *Science.* 256:1871–1820.

19. Chicz, R.M., R.G. Urban, W.S. Lane, J.C. Gorga, L.J. Stern, D.A. Vignali, and J.L. Strominger. 1992. Predominant naturally processed peptides bound to HLA-DR1 are derived from MHC-related molecules and are heterogenous in size. *Nature,* 358:764–768.

20. Chicz, R.M., R.G. Uraban, J.C. Gorga, D.A. Vignali, and J. Strominger. 1993. Specificity and promiscuity among naturally processed peptides bound to HLA-DR molecules. *J. Exp. Med.* 178:27–47.

21. Weiss, S., and B. Bogen. 1991. MHC-class II-restricated presentation of intracellular antigen. *Cell* 64:767–776.

22. Hannested, K., G. Kristoffersen, and J.P. Briand. 1986. The T lymphocytes response to syngeneic λ2 chain idiotype. *Eur. J. Immunol.* 16:889–893.

23. Bikoff, E.K. 1992. Formation of complexes between self-peptides and class II molecules in cells defective for presentation of exogenous protein antigens. *J. Immunol.* 148:1–8.

24. Fedosoeyeva, E.V., R.C. Tam, P.L. Orr, M.R. Garovoy, and G. Benichou. 1995. Presentation of a self-peptide for in vivo tolerance induction of CD4 T cells is governed by a processing factor that maps to class II region of MHC locus. *J. Exp. Med.* 182:1481–1491.

25. Townsend, A., and H. Bodmer. 1989. Antigen recognition by class I-restricted T lymphocytes. *Ann. Rev. Immunol.* 7:601–635.

26. Nossner, E., and P. Parham. 1995. Species-specific differences in chaperone interaction of human and mouse MHC complex class I molecules. *J. Exp. Med.* 181:327–337.

27. Schoel, Z.U., and S.H. Kaufmann. 1994. Beta 2-microglobulin independent presentation of exogenously added foreign peptide and endogenous self epitopes by MHC class I alpha-chain to a cross-reactive CD8 CTL clone. *J. Immunol.* 153:4070–4080.

28. Groettrup, M.A., A. Soza, U. Kuckelhorn, and P.-M. Kloetzl. 1996. Peptide antigen production by the proteasomes; complexity provide efficiency. *Immunol. Today* 17:429–435.

29. Niedermann, G., R. Grimm, E. Geoer, M. Maurer, C. Realini, C. Gartman, J. Soll, Omura., M.C. Rechsteiner, W. Baumeister, and K. Eichmann. 1997. Potential immunocompetence of proteolytic fragments produced by proteasomes before evolution of vertebrate immune system. *J. Exp. Med.* 185:209–220.

30. Coux, O., K. Tanaka, and A.L. Goldberg. Structure and function of the 20S and 26S proteasomes. *Annu. Rev. Biochem.* 65:801–847.

31. Driscoll, J., M.G. Brown, D. Finley, and J.J. Monaco. 1993. MHC-linked LMP gene products specifically alter peptidase activity of the proteasomes. *Nature* 365:262–264.

32. Pamer, E., and P. Creswell. 1998. Mechanisms of MHC class I-restricted antigen processing. 16:323–349.

33. Hopkins, L.M., P.J. Bull, J.A. Gerlach, and R.W. Bull. 1997. Further characterization of HLA homozygous typing cell lines at LMP2 polymorphic codon 60 by an ARMS typing method. *Human Immunol.* 53:183–190.

34. Lim, J.K., J. Hunter, M. Fernandez-Vina, and D.L. Mann. 1999. Characterization of LMP polymorphism in homozygous typing cells and a random population. *Human Immunol.* 60:145–151.

35. Gaczynska, M., K.L. Rock, T. Spies, and A.L. Goldberg. 1994. Peptidase activity of proteasomes are differentially regulated by the MHC-encoded genes for LMP2 and LMP7. *Proc. Natl. Acad. Sci. USA* 91:9213–9218.

36. Monaco, J.J., S. Cho, and M. Attaya. 1990. Transport protein genes in the murine MHC: possible implication for antigen presentation. *Science* 250:1723–1730.

37. Spies, T., V. Cerundolo, M. Collona, Cresswell, A. Towsend, and R. DeMars. 1992. Presentation of viral antigen by MHC molecules is dependent on a putative peptide transporter. *Nature* 355:644–646.

38. Kelly, A., S.H. Powis, L.A. Kerr, I. Mockridge, T. Elliot, J. Bastin, B. Ushanska-Ziegler, A. Ziegler, and A. Townsend. 1992. Assembly and function of two ABC transporter proteins encoded in the human MHC. *Nature* 355:641–644.

39. Kleijmeer, M.J., A. Kelly, H.J. Geuze, J.W. Slot, A. Townsend, and J. Trowsdale. 1992. Location of MHC encoded transporters in the endoplasmic reticulum and cis-Golgi. *Nature* 357:342–34.

40. York, I. A., and K.L. Rock. 1996. Antigen presentation by class I MHC. *Annu. rev Immunol.* 14:368–388.

41. Sadasvian, B., P.J. Lhener, B. Ortman, T. Spies, and P. Gresswell. 1996. Role for calreticulin and a novel glyco-protein, tapsin, in the interaction of class I molecules with TAP. *Immunity* 5:103–111.

42. Solheim, J.C., M.R. Harris, C.S. Kindle, and T. H. Hansen. 1997. Prominence of $\beta2$ microglobulin, class I heavy chain conformation, and tapsin in the interaction of class I heavy chain with calreticulin and the transporter associated with antigen processing. *J. Immunol.* 158:2236–2241.

43. Ortman, B., J. Coperman, P.J. Lehner, B. Sadasivan, J.A. Herberg, A.G. Grandea, S.R. Ridddell, R. Tampe, T. Spies, J. Trowsdale, and P. Cresswelll. 1997. A critical role for tapasin in the assembly and function of multimeric MHC class I–TAP complexes. *Science* 277:1306–1310.

44. Faustman, D., Y. Li, H.Y. Lin, R. Huang, and J. Guo. 1992. Expression of intra-MHC transporter genes and class I antigens in diabetes-susceptible NOD mice. *Science* 256:1830–1838.

45. Wang, F., X. Li, and D. Faustman. 1995. Tap-1 and Tap-2 gene therapy selectively restores conformationalley dependent HLA class I expression in type I diabetic cells. *Hum. Gene Ther.* 6:1005–1017.

46. Garrett, T.P.J., M.A. Saper, P.J. Bjorkam, J.L. Strominger, and D.C. Willey. 1989. Specificity pockets for the side chains of peptide antigens in HLA-Aw68. *Nature* 342:692–696.

47. Falk, K., O. Rotzchke, and H.G. Rammensee. 1990. Cellular peptide composition governed by MHC class I molecules. *Nature* 348:248–251.

48. Falk, K., O. Rotzsche, S. Stefanovic, G. Jung, and H. Rammensee. 1991. Allele-specific motifs revealed by sequencing of self-peptides eluted from MHC molecules. *Nature* 351:290–294.

49. Brumeanu, T.-D., W.J. Swiggard, R.M. Steiman, C.A. Bona, and H. Zaghuani. 1993. Efficient loading of identical viral peptides onto class II molecules by antigenized immuno-globulin and influenza virus. *J. Exp. Med.* 178:1795–1799.

50. Winchester, G., G.H. Sunshine, N. Nardi, and A. Mitchison. 1984. Processing of self and foreign antigens. *Immunogenet.* 19:487–491.

51. Lorenz R.G., and P. Allen. 1988. Direct evidence for functional self-peptide/Ia molecules in vivo. *Proc. Natl. Acad. Sci. USA* 85:5220–5233.

52. Hagerty, D.T., B.D. Evavold, and P.M. and Allen. 1991. The processing and presentation of the self-antigen hemoglobin. *J. Immunol.* 147:3282–3288.

53. Bona, C., and Gr. Ghyka. 1968. Abilty of leucocytes to recognize some foreign gamma globulins during pinocytosis. *Nature.* 217:172–173.

54. Stockinger, B., and R.H. Lin. 1989. An intracellular self protein synthesized in macrophages is presented but fails to induce tolerance. *Intern. Immunol.* 1:592–597.

55. Hagerty, D.T., and P.M. Allen. 1993. Tolerance to self and the processing and presentation of antigens. *Intern. Rev. Immunol.* 10:313–321.

56. Rider, B.J., F. Fraga, Q. Yu, and B. Singh. 1996. Immune responses to self peptides naturally presented by murine class II MHC molecules. *Mol. Immunol.* 33:625–633.

57. Demoz, S., and A. Lanzavecchia. 1993. Presentation of self peptides: consequences for self non self discrimination. *Intern. Rev. Immunol.* 10:314–321.

58. Smith, S.C., and P. M. Allen. 1992. Expression of myosin-class II MHC in the normal myocardium occurs before induction of autoimmune myocarditis. *Proc. Natl. Acad. Sci. USA* 89:9131–9135.

59. Manoury-Schwartz, B., G. Chiocchia, and C. Fournier. 1995. Processing and presentation of type II collagen, a fibrillar autoantigen, by H-2q antigen presenting cells. *Eur. J. Immunol.* 25:3235–3242.

60. Michaelsson, E., M. Holmdahl, A. Engstrom, H. Burkhard, A. Scheynius, and R. Holmdahl. 1995. Macrophages, but not dendritic cells, present collagen to T cells. *Eur. J. Immunol.* 25:2234–2241.

61. Michaelsson, E., V. Malmstrom, S. Reis, A. Engstrom, and R. Holmdahl. 1994. T cell recognition of carbohydrates on type II colagen. *J. Exp. Med.* 180:745–449.

62. Harding, C.V., J. Kihlberg, M. Elofsson, G. Magnusson, and E.R. Unanue. 1993. Glycopeptides bind MHC molecules and elicit specific T cell responses. *J. Immunol.* 151:2419–2425.

13 | The Role of HLA in Autoimmune Disease

John A. Gebe and Gerald T. Nepom

1. INTRODUCTION

The major histocompatibility complex (MHC) locus located on the short arm of human chromosome six contains a family of highly polymorphic genes whose predominant role is to present both foreign and self-antigens to the immune system. Within the MHC locus are a set of genes encoding the surface-expressed HLA (human leukocyte antigens) molecules responsible for presenting antigen to T cells. These are divided into Class I molecules (A, B, and C) that present antigen to CD8 T cells and Class II molecules (DR, DQ, and DP) that present antigen to CD4 T cells. In peripheral immunity, T cells recognize foreign proteins as peptides bound to surface-expressed HLA molecules on antigen-presenting cells (APC). Activated T cells then act in a concerted effort with other immune cells to eliminate the source of these non-self antigens. In the context of organ transplantation, incompatibility between donor and recipient is dependent on T cell recognition of differences in the polymorphic MHC regions and is a major factor determining whether the donor organ will be accepted or rejected within a host. In autoimmune disease, self-antigens presented on HLA molecules are a driving force in directing a T cell immune response. A sustained state of this autoreactivity against self-antigens leads to tissue damage; thus, in insulin-dependent diabetes mellitus (IDDM), autoreactivity leads to the destruction of insulin-secreting pancreatic beta cells, autoreactivity in rheumatoid arthritis (RA) results in a loss of joint synovium, and various autoimmune syndromes are associated with multiple tissue damage including skin, nervous system, and kidney.

Over the past 25 years there have been more than 20 autoimmune diseases identified that have been shown to be correlated with specific HLA molecules (1). The results from these population-based studies have also identified HLA alleles that, when present with susceptible HLA alleles, can provide protection from autoimmune disease. A question that has eluded immunologists is the mechanism by which the presence of certain HLA molecules leads to autoimmune states in individuals carrying these susceptible genes and, equally important, how protective alleles negate the effect of diseasesusceptible HLA alleles. Complicating matters further is that even in situations where the correlation between diseased and certain HLA genes is quite high, the absolute predictive value of knowing an individual's HLA and their probability of progression to disease can be quite low. For example, approximately two-thirds of Caucasian individuals with severe erosive forms of RA carry either HLA DRA1*0101/ B1*0401 or DRA1*0101/B1*0404. However, for individuals carrying these haplotypes, the risk of developing RA are only 1 in 20 and 1 in 35 respectively (2). The risk for individuals carrying both DRB1*0401 and DRB1*0404 is 1 in 7, suggesting a synergistic effect among certain HLA molecules. In IDDM 70–95% of Caucasians patients carry the DQB1*0302 haplotype, yet the risk of developing disease is only 1 in 60 (2,3).

The low predictive value of an individual's HLA haplotype for disease progression stems from at least three sources. The first is that most population studies focus on finding correlations between one or two HLA molecules and disease, whereas other linked alleles (Class I A,B, and C and Class II DR,DQ, and DP) can have modulating effects. Indeed, the presence of other HLA molecules can act to either increase or decrease disease susceptibility. While DQB1*0302 is strongly associated with risk for IDDM, the presence of DQB1*0602 in heterozygotes provides a dominant protective effect (4,5). A modulation effect in susceptibility to IDDM in individuals positive for

DQB1*0302 is also observed depending on particular DRB1*04 haplotypes and also class I genes (6,7). This is similar to the mouse model of type I diabetes, NOD (non-obese diabetic), in which mice carry a single class II susceptible H-2 (mouse equivalent of HLA) I-A^{g7} haplotype. The other class II molecule I-E is not expressed due to a mutation in the I-E$_\alpha$ promoter. Diabetes can be prevented in NOD by transgenic expression of the I-E$_\alpha$ (8,9), gene or restricted by expression of protective I-A^k or I-A^d haplotypes (10,11).

The low predictability toward disease progression based solely on HLA typing is further complicated by observations in population studies indicating that non-HLA genes can modify disease susceptibility. In specific ethnic populations, it has been shown that the presence of certain allelic forms of the immune-regulating molecule CTLA4 can alter disease susceptibility to IDDM and Grave's disease (12). A third aspect complicating an understanding of the fundamental mechanisms underlying disease progression with certain HLA genes is that environmental factors most likely play a role in establishing clinical autoimmunity (13). The concordance rate in monozygotic twins for IDDM is somewhere around 50% depending on the study and population (14,15). Low concordance rates have also been observed in other autoimmune diseases, such as RA and MS (Multiple Sclerosis) (16). Several studies have documented correlations between viruses and environmental factors and diabetes (17) and also with other HLA-associated autoimmune diseases such as RA, SLE (systemic lupus erythematosus), MS, and myasthenia (16). The data from twin studies are convincing evidence that HLA molecules and other genes are necessary, but not sufficient for the development of clinical autoimmune disease. It is interesting that not all NOD mice, are clinically diabetic, although, it has been reported that nearly all such mice have autoimmunity (18). Thus, the low concordance rate among monozygotic twins in autoimmune diseases does not necessary imply that the discordant twin is free from ongoing autoimmunity, only that it has not displayed clinical relevance. In light of all the correlation HLA population studies and monozygotic twin studies, it appears that environmental factors are either an initiating or modulating factor in disease progression for an immune system that is already predisposed to autoimmunity due to the presence or absence of specific HLA molecules. Based on the hypothesis that HLA predisposes an immune system towards autoimmunity, environmental factors (pathogens) may skew the autoimmune-prone immune system beyond an activation threshold, resulting in tissue damage. Because HLA molecules are not, in themselves, the causative agent in autoimmune diseases, this perspective has led to extensive investigation into the multiple biological roles HLA plays in shaping and maintaining the immune system and in responding to antigen.

2. BIOLOGICAL ROLE OF HLA MOLECULES

Class I and II HLA molecules exert their functional influence on the immune system as antigen presentation molecules whose primary role is to display both endogenous and exogenous peptides to both developing and mature T cells. Many previous reviews on autoimmune disease have focused on the role of specific peptides and how they bind autoimmune-prone MHC molecules and/or T cells in the pathogenesis of autoimmunity. Here in, we discuss possible mechanisms that may play a role in establishing an immune system prone to autoreactivity, with particular emphasis on the function of autoimmune-associated MHC molecules in selecting a potentially autoreactive T cell repertoire. Inherent in the proposed mechanism for the role of MHC in autoimmunity are the specific peptides and selected T cells that lead to autoimmune damage.

The antigen presentation role of HLA has both activation and regulatory functions. HLA molecules are responsible for the initial generation and maintenance of a repertoire of T cells with minimal reactivity to self-antigens (tolerance), but sufficient variability in their T cell receptors (19,20), such that they are capable of identifying and responding to foreign antigen insults. Tolerance to self-antigens is classified into central tolerance, which resides mainly in the thymus, and peripheral tolerance. Central tolerance to self-antigens in the thymus is mostly through apoptosis of those immature T cells capable of strong interactions with [self-antigens]-[MHC] complexes. Surviving cells represent the TCR repertoire selected for peripheral export, and are a small subset of thymus-derived T cells, each with unique TCR(s) created within the thymus through VDJ beta and VJ alpha recombination events. Peripheral tolerance, on the other hand, maintains these circulating T cells in a regulated state. Until very recently, several autoantigens, including GAD (glutamic acid decarboxylase) and ICA69 (Islet cell antigen 69), which are implicated in IDDM, thyroglobulin, which is implicated in autoimmune thyroid disease, and MBP (myelin basic protein), which is implicated in MS were undetectable in thymic tissues (21). The absence of these antigens in the thymus during thymocyte selection could allow immature T cells to escape central tolerance to these antigens and mount an autoimmune response in peripheral tissues. The detection of mRNA encoding these proteins from thymic tissue has recently been achieved by RT-PCR (22,23). However, a question still remains as to whether these proteins are expressed and, if so, at sufficient quantities available to the MHC compartment to enable central T cell tolerance to them.

Peripheral regulatory mechanisms, sometimes termed anergy, are capable of negative signaling to potentially autoreactive T cells, resulting in a state of immune nonresponsiveness. The phenomenon of T cell proliferation in the *in vitro* autologous mixed lymphocyte reaction

(AMLR) suggests that *in vivo* peripheral mechanisms must exist to regulate T cell responses to peripheral self-antigens (24). Precise molecular mechanisms of peripheral tolerance are complex, and multiple pathways for both anergy and deletion of potentially autoreactive T cells have been demonstrated in TCR tg mice and normal mice.

The multiple functions of a single HLA molecule are, in a sense, a double-edged sword. In the first place, they are responsible for presenting an enormous and diverse set of self-antigens to a large number of immature thymic T cells in establishing a subset of mature naive T cells with a high degree of TCR variability capable of recognizing foreign antigens. In non-thymic tissues, these same self-peptide-bound HLA molecules must maintain the peripheral naive and memory T cell compartment with minimal autoreactivity. On the opposite side, these identical HLA molecules, under pathogen insult, must present foreign antigens to the same T cells in a way that will stimulate an immune response to the foreign antigen and aid in eliminating the pathogen. HLA's role in balancing between self-tolerance and immunity is a delicate one and constantly in flux as a result of the enormous variety of pathogens one is exposed to during life. HLA molecules and their strong correlation with autoimmune diseases present us with the question of whether loss of self tolerance is the result of a deficiency in the T cell generation process consequent to ineffective or incomplete thymic selection (central tolerance) or a breakdown in the maintenance of the peripheral T cell pool. The role of T cell selection in the thymus and its contribution to autoimmunity has lately become a more interesting issue in light of several studies suggesting that the human thymus may be functional throughout life (25–27). This contradicts the common belief that the thymus is relatively non-functional at an early stage in life. If the thymus is, indeed, functional at some level throughout life, then changes in the T cell repertoire may occur that interact with exposure to foreign antigens and also with physiological changes in the aging body, leading to significant alterations in immune function.

3.   HLA ROLE IN CENTRAL TOLERANCE IN THE THYMUS

In the avidity model of thymic T cell selection (28–31), thymocyte TCRs interact with self-peptides bound and presented by Class I-and Class II-expressing thymic epithelial cells. The avidity in this model reflects the overall effective binding constant resulting from the combined interactions of [TCR]-[self-peptide]-[MHC] molecules between any two given cells. Those thymocytes with [TCR]-[self-peptide]-[MHC] avidities beyond a minimum threshold are positively selected (Figure 1.) Those immature T cells that fail this selection event are eliminated through

apoptosis. In the subsequent negative selection stage, immature positively-selected thymocytes interact again with HLA-positive bone marrow derived antigen presenting cells (B cells, dendritic cells, and macrophages) and also medullary epithelial cells. Immature positively-selected thymocytes with [TCR]-[peptide]-[MHC] avidities too high are deleted at this stage through apoptosis to prevent their activation in the periphery. In this avidity model of T cell selection, T cells reactive to self-peptides seen in the thymus are thus eliminated by negative selection prior to peripheral export. In mice, where HLA expression is limited to only cortical thymic epithelial cells (cTEC) using the keratin 14 (K14) promoter driving MHC expression, mature CD4+ T cells are pathogenic when transferred to syngeneic mice, consistent with the critical role of MHC molecules in negative selection (32). Because the thymic APC presents a heterogeneous distribution of self-peptides to the developing thymocyte, the selection outcome is critically dependent on the frequency distribution of specific peptides on the APC; this implies that T cells could escape central tolerance to a self-peptide if it is not presented at sufficient quantities on the selecting APC. The range of moderate avidity to self-peptides presented by MHC molecules in the thymus creates a population of peripheral T cells with sufficient diversity in their T cell receptors to identify the multitude of foreign antigens while minimizing an immune response to self-antigens.

Because of the multiple [TCR]-[peptide]-[MHC] interactions taking place between the developing thymocyte and its selecting APC during central tolerance, establishing tolerance to a particular self-antigen requires antigen expression at sufficient quantities, processed in a form capable of binding a particular MHC molecule, and therefore requires the genetic specificity in which the highly polymorphic MHC molecules determine the specific repertiore of peptides to be seen by developing immature T cells. It has been shown that transgenic targeting of candidate autoantigens to the thymus in mice can modify resistance to autoimmune diseases. EAU (experimental autoimmune uveoretinitis) is an animal model for human intraocular inflammation (uveitis). Targeted expression of ocular autoantigens to the thymus correlates with resistance to EAU (33). In the NOD mouse model of IDDM, targeting proinsulin II to the thymus can prevent development of diabetes (34).

4.   DIVERSITY IN [SELF-PEPTIDE]-[MHC] COMPLEXES IS NEEDED FOR SELECTION AND MAINTENANCE OF A NON-AUTOREACTIVE T CELL REPERTOIRE

An interesting phenomenon of autoreactive T cells arises in mice in which the repertoire of peptides within the class II

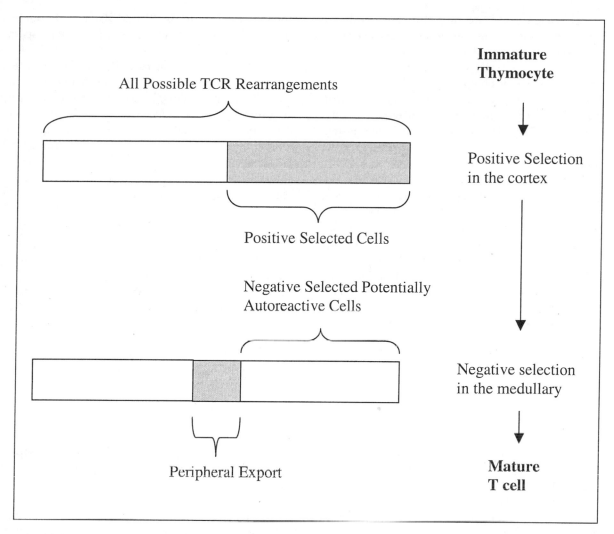

Figure 1. Thymocyte maturation in the thymus. Shaded areas denote thymocytes that exit either cortical or medullar regions during maturation.

binding pocket is severely restricted. Mice expressing the class II I-E$_\alpha$ (52–68aa) peptide covalently linked to the β chain of the I-A^b heterodimer display I-A^b, where nearly all molecules contain the I-E$_\alpha$ (52–68) peptide (35,36). Class II I-E^b expression is absent in these mice due to a mutation in the I-E$_\alpha$ promoter. In these mice, V$_\beta$ usage is similar in diversity to that observed in wild-type I-A^b mice. A similar comparison in TCR V$_\beta$ usage is observed between H-2M$^{-/-}$ and H-2M$^{+/+}$ mice (37–42). H-2M (mouse equivalent of HLA-DM) is responsible for catalyzing the displacement of MHC grove bound CLIP (class II-associated invariant peptide) peptide with other processed peptides. Mice lacking H-2M fail to displace CLIP from the class II peptide grove and, as a result, nearly all surface-expressed class II molecules display only the CLIP peptide. The peripheral CD4$^+$ T cell numbers in H-2M$^{-/-}$ mice has been estimated to be about one-third that of wild-type mice. These mice demonstrate that a diverse peripheral T cell

pool (as measured by V$_\alpha$ and V$_\beta$ usage) can be generated and maintained from a single [peptide]-[MHC] complex. In addition, peripheral T cells generated from a single peptide-MHC complex are capable of responding to alloantigens (38).

More interesting, however, is that a large percentage of CD4$^+$ T cells from I-A^{b}E$_\alpha$ (aa52–68) mice, but not I-A^b mice, will proliferate when presented to syngeneic I-A^b wild-type APCs (43,44). Since the syngeneic APCs present a diverse range of [self-peptide]-[MHC] complexes, the responding T cells are, in this, sense autoreactive. Similar observations are seen in H-2M$^{-/-}$ mice (45). While a large percentage of the APCs in these so-called single [self-peptide]-[MHC] mice express a single dominant peptide, experiments have shown that other self peptides in lower amounts are present and are responsible for the selection of some cells (41).

The loss of tolerance to self-antigens in the T cell repertoire generated in [MHC]-[peptide] restricted mice may be due to the restricted repertoire of [self-peptide]-[MHC] complexes presented in the thymus during T cell selection and/or to a failure in peripheral tolerance mechanisms as a result of a limited complexity of [peptide]-[MHC] complexes. In a recent study comparing the $V\alpha2$-$J\alpha4$ TCR region between H-2M$^{+/-}$ and H-2M$^{-/-}$ TCR$_\beta$ transgenic mice (46), restricted junctional length differences were observed in the H-2M$^{-/-}$ strain. This result is interesting since the $V\alpha$-$J\alpha$ region contains the CDR3α region known to have contact sites to the HLA-bound peptide (47,48), suggesting a possible link between TCR selection and the autoreactive T cells generated and maintained in these so-called single-peptide mice. It is possible that the generation and maintenance of a T cell repertoire with minimal autoreactivity requires MHC molecules to stably present many different self-peptides to developing immature T cells and perhaps also peripheral mature T cells as well. A limitation in the diversity of peptides bound to MHC molecules may itself be an intrinsic defect that results in the development and maintenance of a T cell pool containing potential self-peptide autoreactive T cells.

5. STABILITY OF DIABETES-PRONE MHC MOLECULES

The ability of a peptide-bound MHC molecule to functionally interact with a T cell depends on both the kinetic and thermodynamic stability of the trimolecular [MHC]-[peptide]-[TCR] complex (49–51). Stability comparisons of various human MHC molecules associated with protective and susceptible IDDM alleles have been documented (52–54). Table 1 summarizes the result of an SDS MHC dimer stability study (53) illustrating the strong correlation between MHC dimer stability and susceptibility to IDDM.

In the mouse model of spontaneous diabetes, NOD (non-obese diabetic) mice express the class II I-A^{g7} molecule. The diabetes-prone I-A^{g7} molecule has also been shown to be SDS unstable when compared to protective I-A^d and I-A^b haplotypes (55,56). The source of the instability in diabetes-susceptible MHC molecules is unknown, however,

common to HLA DQB1*0302, DQB1*0201 (human) and H-2 I-A$_\beta$^{g7} (mouse) is the presence of a non-aspartic acid residue at position 57 in the beta chain (57,58). Based on the crystal structure of the HLA-DR1 molecule (59), the aspartic acid at position 57 on the beta chain in non-diabetes-prone haplotypes is thought to form a salt bridge to arginine at position 79 on the alpha chain. It has been demonstrated that some, but not all, class II HLA molecules with non-aspartic acid residues at position 57 in the beta chain have a preference for binding peptides containing a negatively-charged amino acid at peptide position 9 near the beta 57 region (60–62). Additionally, a peptide from mouse hsp60 has been identified that, when bound to I-A^{g7}, stabilizes the MHC to SDS denaturation (63). This same peptide, which has a negatively-charged aspartic acid at position 9, has also been shown to stabilize the human diabetes-susceptible HLA DQA1*0301/B1*0302 molecule (64).

To achieve an avidity threshold suitable for T cell activation, it is likely that unstable MHC molecules can only be biologically active in the presence of strong binding peptides capable of stabilizing the MHC-peptide complex (65). In the case of diabetes-susceptible HLA molecules containing non-aspartic acid residues at beta position 57, the bound peptide repertiore may thus be skewed toward those peptides containing a negative charge at peptide position 9, resulting in a functional restriction in the diversity of HLA-bound peptides. One consequence of this limited MHC-peptide diversity may be the selection of an autoreactive-prone T-cell repertoire similar to the autoreactive T cell repertoire generated in so-called single peptide-HLA mice.

It could be argued that the presence of other more stable MHC molecules in IDDM patients could protect against these selected autoreactive T cells, similar to the diabetes protective effect observed in NOD mice transgenic for non-NOD MHC class II haplotypes (66). However, it has also been observed that for at least one I-A^{g7} restricted diabetogenic T cell that the degree of negative selection of the diabetogenic T cell afforded by protective MHC molecules is variable, depending on specific class II haplotypes (67). Thus, the presence of more stable and diverse peptide-binding MHC molecules in IDDM patients (in this case other DQ, DR and DP molecules) may act to only decrease the precursor frequency of selected autoreactive T cells. Environmental factors may then act to increase the autoreactive precursor pool to a level where autoimmune disease is clinically observed.

6. SHARED EPITOPE AND RHEUMATOID ARTHRITIS

In Rheumatoid arthritis (RA), specific HLA alleles are associated with susceptibility in different ethnic groups, with

Table 1

HLA-DQ	IDDM susceptibility	SDS stability
A1*0102-B1*0602	Dominant Protective	+++
A1*0103-B1*0603	Protective	++
A1*0301-B1*0301	Weakly Protective/Neutral	+
A1*0101-B1*0501	Neutral	+
A1*0102-B1*0602	Neutral/Susceptible	+
A1*0501-B1*0201	Susceptible	0
A1*0301-B1*0302	Susceptible	0/+

Table 2

	67	68	69	70	71	72	73	74	RA Risk
DRB1*0101	L	L	E	Q	R	R	A	A	Susceptible
DRB1*0401	L	L	E	Q	K	R	A	A	Susceptible
DRB1*0404	L	L	E	Q	R	R	A	A	Susceptible
DRB1*0405	L	L	E	Q	R	R	A	A	Susceptible
DRB1*1402	L	L	E	Q	R	R	A	A	Susceptible
DRB1*0402	I	L	E	D	E	R	A	A	Neutral

DRB1*0101 being the susceptible allele in East Asians, DRB1*0401 and DRB1*0404 in Caucasians, DRB1*1402 in Native Indians, and DRB1*0405 in Orientals (68). Common to all these alleles is a stretch of amino acids from residue 67–74 within the β1 domain termed the shared epitope (SE) sequence (Table 2) (69,70).

Population studies have shown that RA susceptibility is increased for those individuals positive for the shared epitope (69,71–73) and is further enhanced synergistically for individuals carrying the shared epitope on both DRB1 alleles (2,74). In a 9-year study on Spanish arthritic patients, the extent of disease correlates well with patients carrying the SE region. In this study, the extent of severity as measured by large joint involvement was 79% in SE (+/+) homozygous patients, 50% in SE (+/–) heterozygous patients, and 32% in patients not carrying the SE region (75). The shared epitope present in nearly all RA-susceptible alleles suggests that this peptide sequence somehow is important in T cell recognition or T cell selection.

Several hypotheses have been put forward to explain the increased risk for RA in individuals carrying the shared epitope on HLA DR molecules. The first hypothesis is that RA-susceptible HLA DR molecules bind specific autoantigens relative to non-RA-associated molecules and result in the presentation of these antigens to peripheral T cells. In support of this hypothesis, crystallography data on peptide-bound DR1, DR3, and DR4 HLA molecules have shown that the shared epitope segment is part of the alpha helical region within the peptide-binding β1 domain (76–78). The DR β1 alpha helical domain interacts both with the bound peptide and the TCR structural elements. The type of amino acid at residue 71 on the HLA-DR β chain (71β) is a major determinant in what type of peptides are capable of binding to HLA-DR molecules (79), and indeed, antigen binding specificity to RA-associated alleles has been demonstrated with a peptide from the candidate autoantigen type II collagen. A peptide comprising residues 261–273 of type II human collagen

has been shown to have a 100–1000 higher binding affinity to the RA-associated alleles DRB1*0401 and DRB1*0404 when compared to the non-RA-associated DRB1*0402 allele. The differences between DRB1*0404 and DRB1*0402 lie within the shared epitope, including residue 71, changing residues in contact with the peptide from lysine (+) to glutamic (–) (Table 2). On the other hand, while this peptide selection model correlates well with the differences between DRB1*0404 and DRB1*0402, it does not provide a satisfactory explanation for the RA association with other "shared epitope" alleles. The class II molecules encoded by DRB1*0101 and DRB1*1402 differ from DRB1*0401, DRB1*0404, and DRB1*0405 in multiple polymorphic sites within the floor of the MHC peptide binding groove. This results in major differences in peptide binding properties between these molecules, making it difficult to envision a unified peptide-binding model accounting for the striking association with the shared epitope. This has led to a search for alternate hypotheses, including the possibility that the shared epitope provides the basis for thymic selection of autoreactive T cells (80). The shared epitope region, which lies in the β1 alpha helical domain, is involved in direct TCR contact (83,84). Based on crystallography, the overall interaction between the T cell TCR and the peptide-MHC has been estimated to involve approximately 41 intermolecular contacts, of which 27 are predicted to be peptide-independent (84). Because of the large number of peptide-independent interactions, it is possible that the strength of interaction (avidity) between TCR and MHC molecules containing the shared epitope are more heavily weighted toward peptide-independent contributions compared to non-shared epitope MHCs. It has been postulated that the thymic-derived T cells selected in this way could be promiscuous in the sense that they may not be so dependent on peptide specificity for activation (68). An MHC-restricted bound peptide with relatively low direct peptide-TCR avidity may be activated due to the stronger avidity contribution of the shared epitope-TCR (Figure 2).

Promiscuous T cells have been found that are capable of stimulation by multiple shared epitope alleles (81 and also unpublished results). In this type of promiscuous T cell recognition, it is not known whether the stimulating peptide is the same in each allele-specific activation. Of interest, however, is the example of a T cell clone (EMO25 from a DRB1*0401 donor) that is capable of being stimulated by both a MT cell line (DRB1*0404) and a DRB1*0404-transfected HLA-DM deficient BLS-1 cell line. The BLS-1 (bare lymphocyte syndrome) cell line lacks endogenous class II expression and is HLA-DM-deficient, and as a result nearly all transfected class II molecules express the CLIP peptide. Stimulation of the EMO25 T cell clone by the MT cell line containing MHC wild-type peptide diversity on the cell surface, and also by the DRB1*0404-transfected BLS-1 where nearly all MHC loaded peptides are CLIP, is consistent with the hypothesis that perhaps some T cells selected on shared epitope MHCs do have TCR avidities directly to the shared epitope.

6. AUTOIMMUNITY AND MHC MOLECULES IN NON-THYMIC ENVIRONMENTS

Outside the thymic environment, HLA molecules on APC exert their dominant role of displaying antigens to circulating T cells. T cells recognizing foreign antigens act in a concerted effort with other immune cells to eliminate the source of the antigen. Key to a successful immune response to foreign antigens is the trimolecular interaction between MHC, foreign-peptide and TCR, which has been shown to involve multiple contacts, including antigen-dependent and antigen-independent regions (82–84). The spatial and temporal interactions between MHC, foreign-peptide, and TCR leading to T cell activation is the sum total of the multiple contacts involved. Changes in specific interactions caused by single amino acids alterations in the MHC or peptide sequence have been shown to be capable of activating the same T cell (85,86). This suggests the possibility that foreign protein antigens with sequences similar to self-peptides could expand a subset of T cells in a normal immune response to

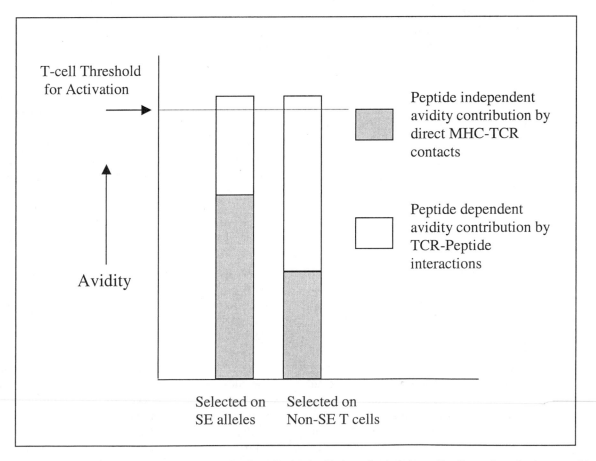

Figure 2.   Model suggesting a possible mechanism for the selection/activation of promiscuous T cells on shared epitope-positive MHC molecules. T cells selected on shared epitope MHC molecules are potentially autoreactive to self antigens due to the relative higher contribution of non-peptide interactions (shaded region) to the overall TCR-Peptide-MHC interaction (avidity) needed for activation.

the foreign antigen and later be activated by the structurally-related self-peptide. In this way foreign peptides may increase the probability of an autoimmune state by expanding a subset of T cells that are capable of responding to an autoantigen, albeit a different peptide. Because the MHC peptide-binding motif is haplotypespecific, the ability of a T cell activating-peptide to be mimicked by a structurally-similar peptide is also MHC haplotype-specific. Thus, this mimicry model as a mode in the pathology of autoimmunity is also MHC haplotypespecific and extends the selection model for T cell autoreactivity discussed above. It seems very likely that the role of HLA molecules in predisposing to autoimmunity initiates with the critical nature of TCR selection, and is then followed by peripheral amplification. A wide variety of mimicry epitopes may suffice for this amplification step, particularly if additional elements of immune triggering, such as the cytokine milieu, expression of co-stimulatory molecules, and specialized APC, are present in the periphery to help the T cell overcome the requisite avidity threshold.

The nature of peripheral antigens that initiate tissue-specific damage in autoimmunity are still unknown. Both specific and non-specific mechanisms are likely: For example, several viruses, including cytomegalo, hepatitis, and coxsackie virus, have a tropism for pancreatic beta cells (87). An association with coxsackie virus antibodies and IDDM has also been observed (88). Coxsackie has held an interest in IDDM progression due to a sequence homology between a region of the protein 2C of Coxsackie and the self-protein glutamic acid decarboxylase (GAD65). It has recently been shown that a peptide derived from 2C is capable of binding the IDDM-susceptible DR3 molecule (89). Coxsackie viral infection in certain mice can also lead to diabetes, however, it has been argued that the cause is a bystander effect and not mimicry to GAD (90). The bystander effect results in the activation of T cells that would normally not be activated to self-antigens, due to the environment (cytokines, chemokines, etc.) setup by the ongoing immune response to the foreign antigen. Whether the virally-induced diabetes in certain mouse haplotypes is bystander effect or mimicry, the model does suggest that potentially autoreactive-specific T cells have escaped MHC-dependent central and peripheral self-tolerance mechanisms. Other studied potential autoimmune-causing proteins that may operate by mimicry to self-peptides and their possible diseases include the mycobacterium tuberculosis heat shock protein (HSP65), which has homologous regions to human HSP65 and has been implicated in IDDM and Arthritis (91), hepatitis B, which has homologous regions to myelin basic protein and can induce experimental allergic encephalomyelitis (EAE) in mice (model for MS) (92), and yeast histone, which has homologous regions to a retinal protein and is capable of causing uveitis in mice (model of human uveitis) (93).

7. CONCLUSIONS

The above discussion on the role of HLA in autoimmunity has placed a major emphasis on a role for MHC molecules in selecting a T cell repertoire that is predisposed toward the potential of autoreactivity as a result of a limited diversity of peptides presented by some MHC molecules to developing thymocytes. Thymocytes matured in an environment where the MHC-presented peptide diversity is limited results in a T cell repertoire in which some of the T cells are auto-reactive to self-antigens when presented on an APC where peptide diversity is not limited. However, because T cells from so-called single peptide mice are not autoreactive within the host animal, one has to postulate that potential autoreactive T cells developed in the thymus in MHC-peptide diversity-limited mice are subsequently autoreactive as a result of a change in diversity of peptides between that seen by the developing thymocyte in the thymus and that seen by mature T cells in the periphery. Naturally-occurring differences in thymic and peripheral APC peptide presentation likely does exist due to differential cathepsin proteases expressed in positive-selecting cortical thymic epithelial cells and peripheral APCs, and also during a normal immune responses to pathogens.

As illustrated in Figure 3, there are parallels between the animal models studied and some human autoimmune

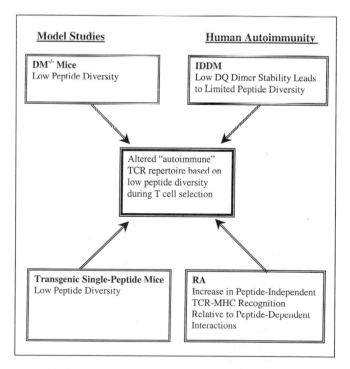

Figure 3. Diagram illustrating the parallels between human autoimmune diseases having autoreactive T cells as a component of the disease and mouse models capable of generating auto-reactive T cells.

diseases. Limited MHC-peptide diversity in DM$^{-/-}$ mice (expressing mostly CLIP-bound MHC molecules) and also I-A^{b}Eα mice (I-Eα peptide covalently linked to I-A^b MHC) select T cells that are autoreactive to self-antigens. The exact mode of selection of these autoreactive T cells is unknown. However, as discussed above, an analogous pathway may contribute to autoreactivity in humans carrying IDDM-and RA-susceptible MHC molecules. The relative instability of the IDDM-associated DQA1*0301/B1*0302 and DQA1*0501/B1*0201 molecules compared to non-IDDM-associated MHC molecules, and the identification of specific peptides that can stabilize IDDM-susceptible MHC molecules, suggests that the selecting structural MHC-peptide variation (and hence the range in [TCR]-[peptide-MHC] interaction energies) seen by developing thymocytes may be limited (as with the mouse models) compared to non-IDDM-associated MHC molecules. In a variation on this theme, RA-associated MHC molecules may focus TCR interactions on peptide-independent binding due to the presence of the shared-epitope, potentially limiting the range of [TCR]-[peptide-MHC] interactions. These parallels between experimental models and human autoimmunity suggest new approaches towards understanding early immune events in the autoimmune-prone genetically-susceptible individual.

References

1. Tiwari J.L., and P.I. Terasaki. 1981. HLA-DR and disease associations. *Prog. Clin. Biol. Res.* 58:151–163.
2. Nepom, G.T. 1995. Class II antigens and disease susceptibility. *Annu. Rev. Med.* 46:17–25.
3. Reijonen, H. and G.T. Nepom. 1997. Molecular pathogenesis of diabetes mellitus. *Front Hormone Res. Basel*, Karger, Vol. 22:46–67.
4. Erlich, H.A, R.L. Griffith, T.L. Bugawan, R. Ziegler, C. Alper, and G. Eisenbarth. 1991. Implication of specific DQB1 alleles in genetic susceptibility and resistance by identification of IDDM siblings with novel HLA-DQB1 allele and unusual DR2 and DR1 haplotypes. *Diabetes* 40:478–481.
5. Kockum, I., C.B. Sanjeevi, S. Eastman, M. Landin-Olsson, G. Dalquist, and A. Lernmark. 1995. Population analysis of protection by HLA-DR and DQ genes from insulin-dependent diabetes mellitus in Swedish children with insulin-dependent diabetes and controls *Eur. J. Immunogenet.* 22:443–465.
6. Nejentsev, S., H. Reijonen, B. Adojaan, L. Kovalchuk, A. Sochnevs, E.I. Schwartz, H.K. Akerblom, and J. Ilonen. 1997. The effect if HLA-B allele on the IDDM risk defined by DRB1*04 subtypes and DQB1*0302. *Diabetes* 46:1888–1892.
7. She, J-X. 1996. Susceptibility to type I diabetes: HLA-DQ and DR revisited. *Immunol. Today* 17:323–329.
8. Nishimoto, H., H. Kakutani, K.I. Yamamura, and T. Kishimoto. 1987. Prevention of autoimmune insulitis by expression of I-E molecules in NOD mice. *Nature* 328:432–434.
9. Uehira, M., M. Uno, T. Kurner, H. Kikutani, K. Mori, T. Inomoto, T. Uede, J. Miyasaki, H. Nishimoto, T. Kishimoto, and K. Yamamura. 1989. Development of autoimmune insulitis is prevented in E$_\alpha$^{d} but not in A$_\beta$^{k} NOD transgenic mice. *Int. Immunol.* 1:209–213.
10. Ridgway, W.M., H. Ito, M. Fasso, C. Yu, and C.G. Fathman. 1998. Analysis of the role if variation of major Histocompatibility complex class II expression on non-obese diabetic (NOD) peripheral T cell response. *J. Exp. Med.* 188:2267–2275.
11. Singer, S.M., R. Tish, X.D. Yang, and H.O. McDevitt. 1993. An A$_\beta$^{d} transgene prevents diabetes in non-obese diabetic mice by inducing regulatory T cells. *Proc. Natl. Acad. Sci. USA* 90:9566–9570.
12. Donner, H., H. Rau, P.G. Walfish, J. Braun, T. Siegmund, R.Finke, J. Herwig, K.H. Usadel, K. Badenhoop. 1997. CTLA4 alanine-17 confers genetic susceptibility to Graves' disease and to type 1 diabetes mellitus. *J. Clin. Endocrinol. Metab.* :143–146.
13. Dahlquist, G. 1994. Non-genetic risk determinants of type I diabetes. *Diabete & Metabolisme* 20:251–257.
14. Tattersall, R.B. and D.A. Pyke. 1972. Diabetes in identical twins. *Lancet* 2:1120–1125.
15. Verge, C.F., R. Gianani, L. Yu, M. Pietropaolo, T. Smith, R.A. Jackson, J.S. Soeldner, and G.S. Eisenbarth. 1995. Late progression to diabetes and evidence for chronic β-cell autoimmunity in identical twins of patients with type I diabetes. Diabetes 44:1176–1179.
16. Leslie, R.D.G. and M. Hawa. 1994. Twin studies in autoimmune disease. *Acta Genet. Med. Gemellol.* 43:71–81.
17. Yoon, J.W. 1990. The role of viruses and environment factors in the induction of diabetes. *Curr. Topics Microbiol. Immunol.* 164:95–123.
18. Gazda, L.S., K.A. Gilchrist, and K.J. Lafferty. 1995. Autoimmune diabetes: caught in the causality trap? *Immunol. Cell Biol.* 73:549–551.
19. Jameson, S.C. and M.J. Bevan. 1998. T cell selection. *Curr. Opin. Immunol.* 10:214–219.
20. Chan, S.H., D. Cosgrove, C. Waltzinger, C. Benoist, and D. Mathis. 1993. Another view of the selective model of thymocyte selection. *Cell* 73:225–236.
21. Heath, V.L., N.C. Moore, S.M. Parnell, and D.W. Mason. 1998. Intrathymic Expression of genes involved in organ specific autoimmune disease. *J. Autoimmunity* 11:309–318.
22. Pugliese, A., M. Zeller, A Jr. Fernandez, L.J. Zalcberg, R.J. Bartlett, C. Ricordi, M. Pietropaola, G.S. Eisenbarth, S.T. Bennett, and D.D. Patel. 1997. The insulin gene is transcribed in the human thymus and transcription levels correlated with allelic variation at the INS VNTR-IDDM2 susceptibility locus for type 1 diabetes. *Nat. Genet* 15:293–297.
23. Sospedra, M., X. Ferrer-Francesch, O. Dominguez, M. Juan, M. Foz-Sala, and R. Pujol-Borrell. 1998. Transcription of a broad range of self-antigens in human thymus suggest a role for central mechanisms in tolerance towards peripheral antigens. *J. Immunol.* 161:5918–5929.
24. Parijs, L.V., and A.K. Abbas. 1998. Homeostasis and self-tolerance in the immune system: turning lymphocytes off. *Science* 280:243–248.
25. McCune, J.M., R. Loftus, D.K. Schmidt, P. Carroll, D. Webster, L.B. Swor-Yim, I.R. Francis, B.H. Gross, and R.M. Grant. 1998. High prevalence of thymic tissue in adults with human immunodeficiency virus-1 infection. *J. Clin. Invest.* 101:2301–2308.

26. Douek, D.C., D.D. McFarland, P.H. Keiser, E.A. Gage, J.M. Massey, B.F. Haynes, M.A. Polis, A.T. Haase, M.B. Feinberg, J.L. Sullivan, B.D. Jamieson, J.A. Zack, L.J. Picker, and R.A. Koup. 1998. Changes in thymic function with age and during the treatment if HIV infection. *Nature* 396:690–695.

27. Jamieson, B.D., D.C. Douek, S. Killian, L.E. Hultin, D.D. Scripture-Adams, J.V. Giorgi, D. Marelli, R.A. Koop, and J.A. Zack. 1999. Generation of functional thymocytes in the human adult. *Immunity* 10:569–575.

28. Jameson, S.C. and M.J. Bevan. 1998. T-cell selection. *Curr. Opinion in Immunol.* 10:214–219.

29. Ashton-Rickardt, P.G. and S. Tonegawa. 1994. A differential model for T-cell selection. *Immunology Today* 15:362–366.

30. Lo, D., C.R. Reilly, L.C. Burkly, J. DeKoning, T.M. Laufer, and L.H. Glimcher. 1997. Thymic stromal cell specialization and the T-cell receptor repertoire. *Immunologic Research* 16:3–14.

31. Bevan, M.J., K.A. Hogquist, and S.C. Jameson. 1994. Selecting the T cell receptor repertoire *Science* 264:796–797.

32. Laufer, T.M. L. Fan, and L.H. Glimcher. 1999. Self-reactive T cells selected on thymic cortical epithelium are polyclonal and are pathogenic in viv. *J. Immunol.* 162:5078–5084.

33. Egwuagu, C.E., P. Charukamnoetkanok, and I. Gery. 1997. Thymic expression of autoantigens correlates with resistance to autoimmune disease. *J. Immunity* 159:3109–3112.

34. French, M.B., J. Allison, D.S. Cram, H.E. Thomas, M. Dempsey-Collier, A. Silvia, H.M. Georgiou, T.W. Kay, L.C. Harrison, and A.M. Lew. 1997. Transgenic expression of mouse proinsulin II prevents diabetes in non-obese diabetic mice [published erratum appears in *Diabetes* 1997 46:924] *Diabetes* 46:34–39.

35. Ignatowicz, L., G. Winslow, J. Kappler, and P. Marrack. 1995. Cell surface expression of class II MHC proteins bound by a single peptide. *J. Immunology* 154:3852–3862.

36. Fukui, Y., T. Ishimoto, M. Utsuyama, T. Gyotoku, T. Koga, K. Nakao, K. Hirokawa, M. Katsuki, and T. Sasazuki. 1997. Positive and negative CD4+ thymocyte selection by a single MHC class II/peptide ligand affected by its expression level in the Thymus. *Immunity* 6:401–410.

37. Miyaszaki, T., P. Wolf, S. Tourne, C. Waltzinger, A. Dierich, N. Barois, H. Ploegh, C. Benoist, and D. Mathis. 1996. Mice Lacking H2-M Complexes, Enigmatic elements of the MHC class II peptide-loading pathway. *Cell* 84:531–541.

38. Surh, C.D., D/Lee, W. Fung-Leung, L. Karlsson, and J. Sprent. 1997. Thymic selection by a single MHC/peptide ligand produces a semidiverse repertoire of CD4+ T cells. *Immunity* 7:209–219.

39. Fung-Leung, W., C.D. Surh, M. Liljedahl, J. Pang, D. Leturcq, P.A. Peterson, S.R. Webb, and L. Karlsson. 1996. Antigen presentation and T cell development in H2-M-deficient mice. *Science* 271:1278–1281.

40. Martin, W.D., G.G. Hicks, S.K. Mendiratta, H.I. Leva, H.E. Ruley, and L. Van Kaer. 1996. H2-M mutant mice are defective in the peptide loading of class II molecules, antigen presentation, and T cell repertoire selection. *Cell* 84:543–550.

41. Grubin, C.E., S. Kovats, P. DeRoos, and A.Y. Rudensky. 1997. Deficient positive selection of CD4 T cells in mice displaying altered repertoires of MHC class II-bound self-peptides. *Immunity* 7:197–208.

42. Tourne, S., T. Miyasaki, A. Oxenius, T. Klein, T. Fehr, B. Kyeski, C. Nenoist, and D.Mathis. 1997. Selection of a broad repertoire of CD4+ T cell in H-2Ma°/° mice. *Immunity* 7:187–196.

43. Ignatowicz, L., W. Rees, R. Pacholczyk, H. Ignatowicz, E. Kushnir, J. Kappler, and P. Marrack. 1997. T cell can be activated by peptides that are unrelated in sequence to their selecting peptide. *Immunity* 7:179.

44. Chmielowski, B., P. Muranski, and L. Ignatowicz. 1999. In the normal repertoire of CD4+ T cells, a class II MHC/peptide complex positively selects TCR with various antigen specificities. *J. Immunol.* 162:95–105.

45. Tourne, S., T. Miyazaki, P. Wolf, H. Ploegh, C. Benoist, and D. Mathis. 1997. Functionality of major histocompatibility complex class II molecules in mice doubly deficient for invariant chain and H-2M complexes. *Proc. Natl. Acad. Sci. USA* 94:9255–9260.

46. Sant'Angelo, D.B., P.G. Waterbury, B.E. Cohen, W.D. Martin, L. Van Kaer, A.C. Hayday, and C.A. Janeway Jr. 1997. The imprint of intrathymic self-peptides on the mature T cell receptor repertoire. *Immunity* 7:517–524.

47. Jorgensen, J.L., U. Esser, B. Fazekas de St. Groth, P.A. Reay, and M.M. Davis. 1992. Mapping T cell receptor-peptide contacts by variant peptide immunization of a single-chain transgenics. *Nature* 355:224–230.

48. Sant' Angelo, D.B., G. Waterbury, P. Preston-Hurlburt, S.T.Yoon, R. Medzhitov, S. Hong, and C.A. Janeway Jr. 1996. The specificity and orientation of a TCR to its peptide-MHC class II ligands. *Immunity* 4:367–376.

49. Williams, C.B., D.L. Engle, G.J. Kersh, J.M. White, and P.M. Allen. 1999. A kinetic threshold between negative and positive selection based on the longevity of the T cell receptor-ligand complex. *J. Exp. Med.* 189:1531–1544.

50. Kwok, W.W., H. Reijonen, B. Falk, D. Koelle, and G.T. Nepom. 1999. Peptide binding affinity and pH variation establish functional thresholds for activation of HLA-DQ-restricted T cell recognition. *J. Immunol.* 60:619–626.

51. Alam, S.M., G.M. Davies, C.M. Lin, T. Zal, W. Nasholds, S.C. Jameson, K.A. Hogquist, N.R.J. Gascoigne, and P. Travers. 1999. Qualitative and quantitative differences in T cell receptor binding of agonist and antagonist ligands. *Immunity* 10:227–237.

52. Buckner, J., W.W. Kwok, B. Nepom, and G.T. Nepom. 1996. Modulation of HLA-DQ binding properties by differences in class II dimer stability and pH-dependent peptide interactions. *Immunology* 157:4940–4945.

53. Ettinger, R.A., A.W. Liu, G.T. Nepom, and W.W. Kwok. 1998. Exceptional stability of the HLA-DQA1*0102/DQB1*0602$\alpha\beta$ protein dimer, the class II MHC molecule associated with protection from insulin-dependent diabetes mellitus. *J. Immunol.* 161:6439–6445.

54. Reizis, B., D.M. Altmann, and I.R. Cohen. 1997. Biochemical characterization of the human diabetes-associated HLA-DQ8 allelic product: similarity to the major Histocompatibility complex class II I-A^{g7} protein of non-obese diabetic mice. *Eur. J. Immunol.* 10:2478–2483.

55. Carrasco-Marin, E., J. Shimizu, O. Kanagawa, and E.R. Unanue. 1996. The class II MHC I-A^{g7} molecules from N-n-obese diabetic mice are poor peptide binders. *J. Immunol.* 156:450–458.

56. Peterson, M. and A.J. Sant. 1998. The inability of the non-obese diabetic class II molecule to form stable peptide complexes does not reflect a failure to interact productively with DM. *J. Immunol.* 161:2961–2967.

57. Chang, Y.W., K.S. Lam, and B.R. Hawkins. 1998. Strong association between DQA1/DQB1 genotype and early-onset

I DDM in Chinese: the association is with alleles rather than specific residues. *Eur. J. Immunogenet.* 25:273–280.

33. Antoniou, A.N., J. Elliott, E. Rosmarakis, and P.J. Dyson. 1998. MHC class II A[b] diabetogenic residue 57 Asp/non-Asp dimorphism influences T-cell recognition and selection. *Immunogenetics* 47:218–225.

59. Stern, L.J., J.H. Brown, T.S. Jardetsky, J.C. Gorga, R.G. Urban, J.L. Strominger, and D.C. Wiley. 1994. Crystal structure of the human class II MHC protein HLA-DR1 complexed with an influenza virus peptide. *Nature* (Lond.) 368:215–221.

60. Quarsten, H., G. Paulsen, B.H. Johansen, C.J. Thorpe, A. Holm, S. Buus, and L.M. Sollid. 1998. The P9 pocket of HLA-DQ2 (non-Aspβ57) *has no particular preference for negatively charged anchor residues found in other type 1 diabetes-predisposing nonAspβ57 MHC class II molecules.* *Int. Immunol.* 10:1229–1236.

61. Nepom, B.S., G.T. Nepom, M. Coleman, and W.W. Kwok. 1996. Critical contribution of *β chain residue 57 in peptide binding ability of both HLA-DR and- DQ molecules.* Proc. Natl. Acad. Sci. USA 93:7202–7206.

62. Kwok, W.W., M.E. Domeier, M.L. Johnson, G.T. Nepom. 1996. HLA-DQB1 codon 57 is critical for peptide binding and recognition. *J. Exp. Med.* 183:1253–1258.

63. Reizis, B., M. Eisenstein, J. Bockova, S. Konen-Waisman, F. Mor, D. Elias, and I.R. Cohen. 1997. Molecular characterization of the diabetes-associated mouse MHC class II protein, I-A[g7]. *Int. Immunology* 9:43–51.

64. Reizis, B., D.M. Altman, and I.R. Cohen. 1997. Biochemical characterization of the human diabetes-associated HLA-D Q8 allelic product: similarity to the major histocompatibility complex class II I-Ag7 protein of non-obese diabetic mice. *Eur. J. Immunol.* 27:2478–2483.

65. Kwok, W.W., H. Reijonen, B.A. Falk, D.M. Koelle, and G.T. Nepom. 1999. Peptide binding affinity and pH variation establish functional thresholds for activation of HLA-DQ-restricted T cell recognition. *Hum Immunol* 60:619–26.

66. Slattery R.M., L. Kjer-Nielsen, J. Allison, B. Charlton, T.E. Mandel, and J.F. Miller. 1990. Prevention of diabetes in non-obese diabetic I-Ak transgenic mice. *Nature* 345:662–663.

67. Schmidt D., J.Verdaguer, N. Averill, and P. Santamaria. 1997. A mechanism for the major histocompatibility complex-linked resistance to autoimmunity *J. Exp. Med.* 186:973–975.

68. Penzotti, J.E., D. Doherty, T.P. Lybrand, and G.T. Nepom. 1996. A structural model for TCR recognition of the HLA class II shared epitope sequence implicated in susceptibility to rheumatoid arthritis. *J. Autoimmunity* 9:287–293.

69. Gregersen, P.K., J. Silver, and R.J. Winchester. 1987. The shared epitope hypothesis. *Arthritis Rheum.* 30:1205–1213.

70. Silver, J., S.M. Goyert. 1985. Epitopes are the functional units of Ia molecules and form the molecular basis for disease susceptibility. *human class II histocompatibility antigens.* Edited by S. Ferrone. B.G. Solheim, and E. Moller. Berlin Springer-Verlag, pp 32–48.

71. Ollier, W., and W. Thomson. Population genetics of rheumatoid arthritis. *In*: Nepom G. (ed) *Rheumatic disease clinics of North America.* 1992. Vol. 18, No.4, p. 741. Saunders, Philadelphia.

72. Winchester, R., E. Dwyer, and S. Rose. The genetic basis of rheumatoid arthritis: the shared epitope hypothesis. *In:*

Nepom G. (ed) *Rheumatic disease clinics of North America.* 1992. Vol. 18, No.4, p. 761. Saunders, Philadelphia.

73. Nepom, G.T. and B.S. Nepom. Prediction of susceptibility to rheumatoid arthritis by human leukocyte antigen genotyping. *In:* Nepom G. (ed) *Rheumatic disease clinics of North America.* 1992. Vol. 18, No.4, p. 785. Saunders, Philadelphia.

74. Weyand C.M. and J.J. Goronzy. 1990. Disease-associated human histocompatibility leukocyte antigen determinants in patients with seropositive rheumatoid arthritis. Functional role in antigen-specific and allogeneic T cell recognition. *J. Clin. Invest.* 85:1051–1057.

75. Valenzuela, A., M.F. Gonzalez-Escribano, R. Rodriguez, I. Moreno, A. Garcia, and A. Nunez-Roldan. 1999. Association of HLA shared epitope with joint damage progression in rheumatoid arthritis. *J. Immunology* 60:250–254.

76. Dessen, A., C.M. Lawrence, S. Cupo, D.M. Zaller, and D.C. Wiley. 1997. X-ray crystal structure of HLA-DR4 (DRA*0101, DRB1*0401) complexed with a peptide from human collagen II. *Immunity* 7:473–481.

77. Ghosh, P., M. Amaya, E. Mellins, and D.C. Wiley. 1995. The structure of an intermediate in class II MHC maturation: CLIP bound to HLA-DR3. *Nature* 378:457–462.

78. Stern, L.J., J.H. Brown, T.S. Jardetzky, J.C. Gorga, R.G. Urban, J.L. Strominger, and D.C. Wiley. 1994. Crystal structure of the human class II MHC protein HLA-DR1 complexed with an influenza virus peptide. *Nature* 368:215–221.

79. Hammer, J., F. Gallazzi, E. Bono, R.W. Karr, J. Guenot, P. Valsasnini, Z.A. Nagy, and F. Sinigaglia. 1995. Peptide binding specificity of HLA-DR4 molecules: correlation with rheumatoid arthritis. *J. Exp. Med.* 181:1847–1855.

80. Nepom, G.T. 1998. Major Histocompatibility complex-directed susceptibility to rheumatoid arthritis. *Adv. Immunol.* 68:315–332.

81. Drover, S., S. Kovats, S. Masewics, J.S. Blum, and G.T. Nepom. 1998. Modulation of peptide-dependent allospecific epitopes on HLA-DR4 molecules by HLA-DM. *Hum. Immunol.* 59:77–86.

82. Chen, Y. and M.M. Davis. 1993. How αβ T-cell receptors "see" peptide/MHC complexes. *Immunol. Today* 14:597–602.

83. Garcia, K.C., M. Degano, R.L. Stanfield, A. Brunmark, M.R. Jackson, P.A. Peterson, L. Teyton, and I.A. Wilson. 1996. An αβ T cell receptor structure at 2.5 A and its orientation in the TCR-MHC complex. *Science* 274:209–219.

84. Garcia, K.C., M. Degano, L.R. Pease, M. Huang, P.A. Peterson, L. Teyton, and I.A. Wilson. 1998. Structural basis of plasticity in T cell receptor recognition of a self peptide-MHC antigen. *Science* 279:1166–1172.

85. Kwok, W.W., M.E. Domeier, M.L. Johnson, G.T. Nepom, and D.M. Koelle. 1996. HLA-DQB1 codon 57 is critical for peptide binding and recognition. *J. Exp. Med.* 183:1253–1258.

86. Ou D. L.A. Mitchell, D. Decarie, A.J. Tingle, and .T. Nepom. 1998. Promiscuous T-cell recognition of a rubella capsid protein epitope restricted by DRB1*0403 and DRB1*0901 molecules sharing an HLQ DR supertype. *Hum. Immunol.* 59:149–157.

87. Thivolet, C. and C. Trepo. 1988. Virus and insulin-dependent diabetes mellitus. Etiopathogenic perspectives. *Pathol. Biol.* 36:182–186.

88. Field, L.L., R.G. McArthur, S.Y. Shin, and J.W. Yoon. 1987. The relationship between Coxsackie-B-virus specific IgG responses and genetic factors (HLA-DR, GM, KM) in insulin-dependent diabetes mellitus. *Diabetes Res.* 6:169–173.

89. Vreugdenhil, G.R., A. Geluk, T.H. Ottenhoff, W.J. Melchers, B.O. Roep, and J.M. Galama. 1998. Molecular mimicry in diabetes mellitus: the homologous domain in coxsackie B virus protein 2C and islet autoantigen GAD65 is highly conserved in the coxsackie B-like enteroviruses and binds to the diabetes associated HLA-DR3 molecule. *Diabetologia* 41:40–46.

90. Horwitz, M.S., L.M. Bradley, J. Harbertson, T Krahl, J. Lee, and N. Sarvetnick. 1998. Diabetes induced by coxsackie virus: initiation by bystander damage and not molecular mimicry. *Nat. Med.* 4:781–785.

91. Elias, D., T. Reshef, O.S. Birk, R.van der Zee, M.D. Walker, and I.R. Cohen. 1991. Vaccination against autoimmune mouse diabetes with a T-cell epitope of the human 65-kDa heat shock protein. *Proc. Natl. Acad. Sci. USA* 88:3088–3091.

92. Fujinami, R.S. and M.B. Oldstone. 1985. Amino acid homology between the encephalitogenic site of myelin basic protein and virus: mechanism for autoimmunity. *Science* (Wash., DC.) 230:1–43.

93. Singh, V.K., H.K. Kalra, K. Yamaki, T. Abe, L.A. Donoso, and T. Shinohara. 1990. Molecular mimicry between a uveitis-pathic site of S-antigen and viral peptides. *J. Immunol.* 144:1282.

14 | Antigenic Mimicry in Autoimmune Diseases

Madeleine Cunningham and Constantin A. Bona

1. INTRODUCTION

While substantial progress has been made in understanding the dysfunction of immune responses during the course of autoimmune diseases, little progress has been made in determining the factors that contribute to the onset of disease. However, several factors have been associated with the development of autoimmune diseases in animals and humans. These include host susceptibility due to genetic factors, dysregulation of the immune responses in the autoimmune disease, and initiation by an environmental insult such as an infectious microorganism (1) as described below. Immune dysregulation by infectious microorganisms may involve polyclonal activation, direct infection of lymphoid cells, alteration of the idiotype-anti-idiotype network or molecular mimicry. Figure 1 illustrates the parameters involved in the development of autoimmune disease.

2. GENETIC FACTORS

Burnet (2) proposed that discrimination between self and nonself is a crucial property of the immune system leading to elimination of self reactive cells (negative selection in modern terms) during the ontogeny. Somatic mutations occurring later in life may lead to conversion of non-self-reactive lymphocytes into autoreactive lymphocytes, which may be involved in the primary event of an autoimmune disease. The analysis of the primary structure of V genes encoding Ig or T cell receptors (TCR) of self-reactive lymphocytes showed that they were encoded by both germline or mutated genes (3,4).

Studies of animal models of autoimmune diseases that contain one gene defect, such as motheaten mice developing multiple autoimmune diseases (5), tight skin mice developing a scleroderma-like syndrome (6), kd/kd mice developing interstitial nephritis (7) and MRL/lpr and gld mice developing systemic lupus erythematosus (SLE)-like syndrome (8), did not identify the exact mechanisms responsible for triggering the onset of autoimmunity. While the genetic defects were present in embryos and manifested at birth, the autoimmunity occurred later in life. Late appearance of autoimmune disease may be related to requirements of environmental factors in addition to expression of genetic defects or alterations during aging. Knockout of some immune genes led to the occurrence and onset of autoimmune diseases (9,10). In these animal models, little is known about why disease is limited to some organs, but it may indicate that some local environmental factors may play a role. In the local environment, cytokines are likely to play a critical role. Animals and humans with enhanced ability to produce certain cytokine reponses that regulate T helper 1 and T helper 2 responses may limit or exacerbate autoimmune disease at different ages and due to responses to infection or other environmental factors (11–14).

In the case of multifactorial diseases, the factors and the mechanisms of activation of susceptibility genes or inactivation of repressor genes are unknown or obscure, respectively. In animal models, some genes (i.e Yaa) are required for the acceleration of the onset of disease (15). Thus, while we have learned a lot about the importance of genetic factors in autoimmunity, little progress has been made concerning its involvement in the onset of autoimmune diseases.

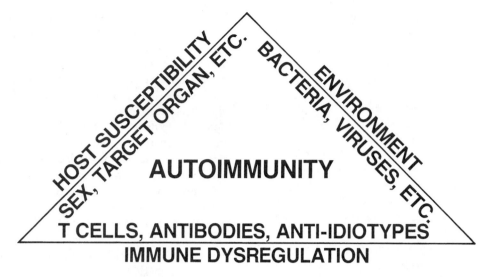

Figure 1. Diagram showing the parameters which affect development of autoimmunity or disease in the host.

There are at least 3 mechanisms by which an infectious microbe may direct the immune system toward production of autoantibodies (Table 1). One way is by activation of host lymphocytes by molecules of the infectious agent that are mitogenic (16–19). Another mechanism is by molecular mimicry of a host antigen by molecules associated with the infectious agent (20–23). Third, the infectious agent may potentially damage host tissues exposing hidden or cryptic determinants that may induce autoantibody production (24,25). These three mechanisms involving bacteria or other infectious agents may participate separately or together to bring about activation of the immune system directed toward host or self-antigens. These mechanisms and their potential to trigger autoimmune disease are discussed below.

3. CRYPTIC EPITOPES

The concept of cryptic epitopes is based on the idea that the antigens, including autoantigens, express immunodominant as well as cryptic less dominant epitopes. The cryptic epitopes can become immunogenic after the digestion of polypeptides. While immunodominant epitopes cause negative selection of autoreactive cells, the cryptic epitopes may perpetuate the autoreactive lymphocytes (26). One may argue that environmental factors such as microbial enzymes/

Table 1 How microbes influence host autoimmunity

Molecular Mimicry
Mitogens and Superantigens
Exposure of Cryptic Epitopes

toxins or viral infection can damage host tissues and generate or expose cryptic epitopes.

While there are numerous data concerning the stimulation of autoreactive T cells by cryptic epitopes, there is little evidence that they are generated in vivo or that their occupancy triggers autoimmune diseases. Generally, cryptic epitopes are linear peptides (26) and, therefore, it is difficult to conceive how cryptic epitopes can stimulate self reactive B cells that are specific for conformational epitopes on the surface of native autoantigens. While there is no doubt that genetic factors and subdominant epitopes may contribute to the pathogenesis of autoimmune disease, external factors may be required to trigger the onset.

4. EXOGENOUS FACTORS

There are multiple theoretical ways by which exogenous factors may contribute to the onset of autoimmune diseases.

4.1. Generation of Neoantigens

Alteration of a self protein by an external agent could create neoantigens that may trigger an autoimmune disease. The best example is drug-induced autoimmune hemolytic anemia in humans. In animal models, it was also shown that hapten coupled-MRDC induced Coombs positivity, and anemia (27).

Expression of a neoantigen, i.e. a viral antigen under the insulin promoter on beta cells, can lead to the occurrence of insulin-dependent diabetes melitis (IDDM) when the animals were infected with a homologous virus (28) or crossed with transgenic mice expressing a TCR specific for the viral protein. However, for a great majority of autoimmune diseases, there is no evidence that exposure or alteration of self proteins precede either organ specific or systemic human or experimental diseases.

4.2. Polyclonal Activators

It is well known that microbes release substances that interact with common receptors expressed on the surface of B or T lymphocytes and activate them in a polyclonal fashion. As classical examples, one may cite lipopolysaccharides, which, subsequent to cross-linking of the LPS receptor, activate resting B cells and induce their differentiation into plasma cells (17) or microbial superantigens, which interact with particular V beta gene products of TCR and activate T cells polyclonally (19). Polyclonal activation involves clones able to recognize both foreign and self antigens. However, polyclonal activation observed in systemic autoimmune diseases may be an epiphenomenon rather than the cause of the disease. B cell polyclonal activators mainly induced synthesis of low affinity IgM autoantibodies, while the majority of pathogenic autoantibodies were high affinity IgG. However, some IgM antibodies have been shown to be cytotoxic and potentially pathogenic in vitro (29,30).

There are a few examples in which polyclonal activators induced autoantibodies (17) and/or accelerated autoimmune diseases. LPS induced the synthesis of rheumatoid factors not associated with occurrence of rheumatoid arthritis (31). LPS also accelerated the onset of disease in animals prone to develop SLE (32,33), and was required to trigger the occurrence of autoimmune hemolytic anemia in transgenic mice expressing V genes encoding for anti-RBC autoantibodies, which otherwise do not develop disease (34). Other observations demonstrated that, in NZB mice, the production of anti-erythrocyte autoantibodies was independent of polyclonal activation and proliferation of cells expressing restricted V beta TCR genes. The activation of T cells in the NZB model led rapidly to a transitory anergy period. Therefore, in the NZB model it is difficult to envision how T cells stimulated by superantigens can trigger autoimmune diseases. Although infectious diseases and their release of cytokines may trigger the onset of autoimmune diseases, there is little epidemiological data demonstrating that infectious agents releasing or secreting polyclonal activators of lymphocytes precede the onset of autoimmune diseases.

5. MOLECULAR MIMICRY

Molecular mimicry has been proposed as a mechanism leading to the breakdown of natural self tolerance and to development of autoimmune diseases. In 1957, Witebsky et al. showed for the first time that the injection of a heterologous thyroglobulin and Freund's complete adjuvant induced an autoimmune thyroiditis in genetically-unrelated animals because of the crossreactivity between host and heterologous thyroglobulins. In 1967, Zabriskie described mimicry between group A streptococci and heart tissues (35). Molecular mimicry has been defined as similar chemical structures shared by molecules encoded by unrelated genes (22,36). Autoimmunity induced by molecular mimicry may occur when microbial and host antigens are similar, but differ enough to break self tolerance. Most studies have identified mimicry between specific antigens by antibody recognition of similar epitopes. B cells can recognize conformational epitopes or linear peptides on the surface of native antigens. In this case, the mimicry antigens inducing antibody synthesis may share not only linear sequences but also conformational fits between self and foreign antigens. On the other hand, T cells recognize peptide associated with MHC gene products. Therefore, the linear microbial peptides mimicking self peptides should be of a certain size to accommodate MHC grooves, to share critical anchoring residues as well as a minimal number of residues recognized by TCR. They also should have flanking regions allowing for correct processing by antigen presenting cells. Mimicry recognized by T cell clones has indicated that mimicking structures may not share amino acid sequence homology (37). Further evidence by Hemmer and colleagues suggest that peptides from randomized peptide libraries could be used to identify multiple crossreactive ligands for a myelin basic protein T cell clone (38,39). Interestingly, some of the crossreactive peptides identified produced proliferative T cell responses at lower concentrations than the original peptide from myelin basic protein. It was concluded that, at least for some autoreactive T cell clones, that antigen recognition was crossreactive and highly degenerate (38,39).

Assessment of the involvement of molecular mimicry in the pathogenesis of autoimmune disease should fill several requirements. First, a significant epidemiological correlation should exist between autoimmune disease and infectious agent bearing epitopes that crossreact with epitopes borne by autoantigens stimulating pathogenic T cells, synthesis of pathogenic autoantibodies, or both. Secondly, infection should precede the occurrence of autoimmune disease. Third, a lag period between the infection and the onset of autoimmune disease should exist in order to break down self tolerance, permitting crossreactive antigens to activate and expand the precursors of self-reactive lymphocytes. How the immune system overlooks self-reactivity or becomes dysregulated to produce damaging crossreactive antibodies or T cells is not completely understood, however, crossreactivity inherently means lower affinity, which may allow crossreactive immune cells to escape deletion.

5.1. Molecular Mimicry by Viruses

The contribution of molecular mimicry to the pathogenesis of autoimmune diseases is based on four categories of observation: Association of viral infections with autoimmune diseases, occurrence of autoantibodies following viral infections, sequence homology between certain viral and host proteins and experimental evidence.

Viral infections associated with autoimmune diseases. There is epidemiological information on viral infections that suggests that the viral infection precedes the onset of the autoimmune disease. Table 2 illustrates those autoimmune diseases preceded by viral infections. For example, Coxsackie B virus was related to the etiology of insulin-dependent diabetes mellitus (IDDM) (40). In certain cases, the virus was isolated from the pancreas of patients with an acute onset.

Isolated virus from the pancreas fills Koch's postulates, since the virus induced the occurrence of IDDM when injected into mice. There is stronger evidence of association of parvovirus B19 and HTLV-1 virus infections with rheumatoid arthritis. T cell leukemia caused by human T cell leukemia virus (HTLV)-1 virus is associated with destructive arthropathies. In Kyusu island, HTLV-1 is highly endemic since 26.7% of inhabitants are seropositive for HTLV-1. Among 7,000 seropositive patients, 1.4% developed RA. Proviral DNA was detected by PCR in synovial tissue of these patients (41). It is also known that acute fibrils diseases caused by parvovirus B19 is associated with prolonged polyarthritis and vasculitis.

Sasaki et al. (42) reported that RA can occur 2 months after infection with parvovirus B19 and the appearance of polyarthritis coincided with the presence of virus in bone marrow cells detected by PCR. Futo et al. (43) detected the

Table 2 Autoimmune diseases associated with virus infections

Virus	Autoimmune disease	Reference
CMV	Guillain-Barre' syndrome	(189)
HSV type 6	Sjogren's syndrome	(190)
Hepatitis A	Cryoglobulinemia	(191)
Hepatitis B	Autoimmune hepatitis	(192)
Hepatitis C	Behcet's syndrome	(193)
HTLV-1	Rheumatoid arthritis	(41)
	Uveitis	(194)
EBV	Idiopathic aplastic anemia	(195)
	Sjogren's syndrome	(196–199)
	Rheumatoid arthritis	(200)
	Systemic lupus erythematosus	(201)
Coxsackie B	IDDM	(40)
	Myocarditis	(202)
Parvovirus B19	Rheumatoid arthritis	(42, 43)
	Periarteritis nodosum	(203)

virus in bone marrow cells of patients with prolonged arthritis. In these studies, it was not shown molecular mimicry of viral antigen and synovial self antigens.

In spite of these elegant observations, there is no compelling evidence of an association of viral infections with the pathogenesis of autoimmune diseases because of a) the inability to isolate viruses consistently from target organs of the autoimmune process and b) a lack of compelling information that one viral infection precedes a specific autoimmune disease. However, it was shown that the same virus can be associated with different autoimmune diseases or alternatively, that multiple viruses can be associated with a single autoimmune disease. For example, Coxsackie B virus has been associated with myocarditis and IDDM, while Sjögren's syndrome has been associated with Epstein Barr virus (EBV), *Herpes simplex* virus (HSV) type 6 and HTLV-1 and systemic lupus erythematosus has been associated with EBV (Table 2). Nevertheless, it is possible that viruses bearing molecular mimicry epitopes are able to breakdown tolerance, trigger the onset of disease, and then are cleared from damaged tissue as a consequence of activation and proliferation of autoreactive lymphocytes.

Occurrence of autoantibodies subsequent to viral infections. Autoantibodies are a hallmark of autoimmune diseases. Some autoantibodies are pathogenic, such in the case of autoimmune thrombocytopenia, myasthenia gravis, autoimmune thyroiditis and potentially in systemic autoimmune diseases, such as systemic lupus erythematosus (SLE). In other diseases they may represent an epiphenomenon resulting from epitope spreading as consequence of primary injury. Numerous viral infections induce the occurrence of autoantibodies exhibiting a variety of specificity for self antigens (Table 3).

The crossreactivity between antibodies induced by viruses with self-antigens was also demonstrated in experimental systems. Thus, Srinvasappa et al. (44) showed that 21 of 635 antiviral mAbs bound to tissue antigens, as assessed by immunofluorescence. Such antibodies were induced in mice subsequent to immunization with Coxsackie, lymphochoriomengitis, measles, rabies, vesicular stomatis, vaccinia and *Herpes simplex* A type 1 viruses.

However, there is no evidence that an anti-viral antibody is pathogenic and causes an organ-specific disease. Such cross-reactive antibodies can be produced subsequent to polyconal activation caused by viral mitogens or can be made by natural antibody forming B cells expanded by viral infections. Presently, there is no strong evidence that anti-viral antibodies which cross react with self antigens play a major role in the pathogenesis of autoimmune diseases.

Sequence homology between viral and self antigens. The concept of molecular mimicry is based on

Table 3 Virus infections associated with occurrence of autoantibodies

Hepatitis A virus
Hepatitis B virus
Mumps
Measles
Rubella
HIV
Infleunza virus
EBV
Coxsackie B virus
Chickenpox
HTLV-1

the idea that homologous sequence between viral proteins and self proteins could play a role in the induction of an autoimmune disease. Accordingly, autoreactive lymphocytes stimulated by viral peptides are activated and then able to launch primary injury leading to an autoimmune phenomenon. In the case of activation of T cells, the viral peptides exhibiting structural homology with self antigens should have a certain size (generally nonamers in the case of CD8$^+$ and a little longer in the case of CD4$^+$ T cells). In addition, they should bear anchor residues to dock into MHC groove and well defined flanking regions permitting a correct processing by APC. Jaknke et al. (45) used computer programs to search for homologous sequences shared by human myelin basic protein (MBP), an autoantigen and target of immune responses in multiple sclerosis and P2 protein target of neuron/brain and viral proteins. Amino acid sequence homology with MBP was found in adenovirus, influenza, measles, canine distemper and EBV viral proteins. The site of homology in MBP was an encephalitogenic site which produced experimental allergic encephalomyelitis in animal models (45).

Ensuing years have seen an increased accumulation of findings of sequence homology between viral proteins and self antigens as a target in autoimmune diseases (46) . We illustrate this point with a few examples of homology of self and viral antigens. In the case of experimental acute encephalitis, Fujinami et al. reported a shared hexapeptide between hepatitis B virus polymerase and MBP responsible for the induction of disease (45) as shown below.

MBP   TTHY<u>GSLP</u>OL
Virus polymerase   IGSY<u>GSLP</u>OE

Wucherpfennig and Strominger (37) searched for peptides that filled the structural requirements for both class II binding and recognition by TCR expressed in myelin basic protein. Human T cell clones recognize MBP 85–99 in association with HLADR2b. This peptide has two hydrophobic residues (Val 89 an Phe-92) that serve as anchors. A set of structural criteria was developed for search of viral peptides exhibiting structural homology with the core of MBP 85–99 peptide, namely anchoring residues mentioned above and TCR contact residues Phe91, Lys93, His-90 and Val 88. A panel of 129 peptides that exhibit molecular motifs were tested in proliferative assay for activation of MBP 85–99 specific T cell clones. Three peptides of different viral origins were found to share anchoring and amino acid residues necessary for recognition by T cells as shown below.

	* * * *
MBP	ENPV<u>VH</u>F<u>F</u>KNIVTPR
	***
Herpes simplex virus	FRQL<u>VH</u>E<u>V</u>RDFAQLL
	* * *
Epstein-Barr virus	TGG<u>V</u>YH<u>F</u>VKKHVHES
	* * *
Influenza A virus	YRNL<u>V</u>W<u>F</u>IKKNTRYP

Stars indicate contact TCR residues and underline anchoring contact residues shared by MBP and viral proteins. All of these peptides were efficiently presented by DR2 molecules to T cells. The observation that certain viral strains were capable of stimulating MBP-specific T cell clones suggested that viruses bearing such peptides may contribute to the pathogenesis of MS. However, this study showed that molecular mimicry occurred by viral peptides having very different amino acid sequences and that their stimulatory activity was related to homology of DR2 anchoring and contact TCR amino acid residues. Furthermore, the diverse viral origin of peptides that stimulate MBP-specific clones makes it unlikely that a single virus is responsible for the induction of MS.

Other examples of molecular mimicry between acetylcholine receptor (AchR) and viral proteins were reported. Autoantibodies against AchR are pathogenic and cause myasthenia gravis. A core aminoacid sequence of AchR peptide shared by VP2 protein of poliovirus with AchR was described by Oldstone (47).

AchR   TVI<u>KES</u>RGTK
VP2 protein TTT<u>KER</u>RGTT

Other sequences of HSVgp D, parvovirus HIW2 and polio middle T antigen exhibit 2–5 amino acid partial homology with PESDQPDL sequence of AchR (48). Sequence homologies were also reported between viral protein and self protein targets of autoantibodies associated with scleroderma such as topoisomerase I fibrillarin (47–51) or CENP-B protein target of antibodies associated with CREST syndrome (52). Pathological implications of these homologous sequences were never proved. In other studies, the sequence PPPGRRP, found in EBV nuclear antigen-1, was reported to be similar to the sequence PPPGMRPP in Sm antigen, an important self antigen recognized by anti-Sm antibodies in approximately 20% patients with lupus (53,54).

In summary, there are countless observations indicating short amino acid sequences shared by virus and self proteins. However, there is little evidence that the molecular mimicry motifs of viruses are indeed responsible for the triggering of autoimmune disease.

Experimental models for molecular mimicry between viral and host antigens. Animal models have significantly contributed to understanding the pathogenesis of disease, characterization of genetic factors, and immune alterations associated with autoimmune disease. Zhao et al. (55) provided strong evidence that an epitope expressed by a coat protein of HSV type 1 was recognized by autoreactive T cells that target corneal antigen in a murine model of autoimmune herpes stromal keratitis.

Zhao et al. (55) have studied the role of molecular mimicry in the herpes stromal keratitis resulting from destruction of corneal tissue by T cells particularly CD4+ T cells. A data search base homology sequence between herpes virus and corneal proteins indicate that the UL-6 HSV-associated protein exhibits high amino acid homology (7 out of 8 amino acid residues) with a corneal protein. A synthetic peptide containing homologous sequence was able to induce the proliferation of keratinogenic T cell clones as well as prevent the disease subsequent to in vivo injection into CAL20 mice, a susceptible strain.

Introduction of a site-directed mutation in UL-6 HSV viral protein prevented the activation of the keratinogenic T cell clone by mutated virus. This observation clearly showed that deletion of UL-6 protein by introduction of a stop codon eliminated the ability of mutant virus to activate the T cells and to cause the disease. Taken together, these findings clearly show that a virus bearing molecular mimicking peptides of self antigens can stimulate the keratinogenic T cells leading to an autoimmune disease.

5.2. Molecular Mimicry and Bacterial Antigens

Bacterial antigen mimicry has been defined as three types of molecular mimicry (Table 4). In the strictest definition, mimicry has been defined as an identical linear amino acid sequence shared between host and bacterial proteins (22). These identical linear amino acid sequences are believed to be the basis of immunological crossreactivity between host and microbial proteins. The advent of monoclonal antibodies simplified the investigation of antibody crossreactivity and molecular mimicry between bacteria and host antigens. Previous studies were dependent on whole serum or affinity purified immunoglobulins from sera as reagents to detect immunological crossreactivities between host and microbial antigens. Monoclonal antibodies are utilized when molecular mimicry or immune crossreactivity are to be investigated. Through the use of

Table 4   Types of bacterial antigen mimicry

Identical or Sequential Epitopes (22)
Similar but Non-Identical Epitopes (29)
Dissimilar Epitopes (68, 101)

monoclonal antibodies, a second and third type of mimicry was identified indicating that not all molecular mimicry or crossreactivity was dependent upon identical amino acid sequences shared between two different proteins (29). Molecular mimicry may occur between homologous sequences sharing 40% amino acid homology or less between the different protein sequences. Some homology may be all that is required to maintain a requisite structure for recognition by antibody molecules. Clearly, T cell crossreactivity requires sequences with contact residues for the TCR and MHC molecules (37–39). Therefore, an alternate form of molecular mimicry may require only structural similarities for immune recognition and crossreaction. One example of the second type of mimicry is the crossreaction observed between the alpha-helical streptococcal M protein and host alpha-helical proteins such as myosin, keratin, tropomyosin, vimentin, and laminin (21,29,56–63). Examples of the third type of mimicry between dissimilar or diverse antigens such as proteins, carbohydrates and deoxyribonucleic acid (DNA) (56,64–71). The basis of these diverse crossreactions is most likely due to aromatic-aromatic and hydrophobic interactions (70), and studies of antibody molecules suggest that the antibody combining site may change conformation in order to bind diverse molecules (72). The plasticity of the T cell receptor has also been reported to be due to large conformational changes associated with the CDR loops upon binding of antigen (73).

Currently, there are several examples of molecular mimicry between bacterial antigens and host tissue antigens, and these examples will be discussed below and are listed in Table 5. Mimicry between the group A streptococcus and host tissues is the most well-studied and evidence supports its role in development of autoimmune sequelae following streptococcal pharyngitis.

Group A streptococcal sequelae: mimicry in acute rheumatic fever and acute glomerulonephritis.

Acute rheumatic fever. The development of acute rheumatic fever following group A streptococcal infection is clearly one of the most well studied examples of molecular mimicry in bacterial autoimmune sequelae. The most serious sequela of rheumatic fever is carditis with the development of chronic rheumatic valvular disease. Autoimmune targets in the host include the heart, joints, brain and skin (74–76). Molecular mimicry was proposed some time ago as a mechanism responsible for the autoimmunity observed in the

Table 5 Autoimmune disease related to bacterial antigen mimicry

Bacteria	Autoimmune disease	Bacterial: host antigens
Group A Streptococci (*Streptococcus pyogenes*)	Rheumatic Fever	Streptococcal M protein N-acetyl-glucosamine: Cardiac myosin Other alpha-helical coiled-coil molecules
Klebsiella, Escherichia Campylobacter, Shigella Salmonella, Yersinia	Ankylosing Spondylitis Reiter's Syndrome	Klebsiella nitrogenase: HLA-B 27
Mycobacteria Streptococci Staphylococci	Arthritis	Heat Shock Protein-65 Peptidoglycan-Polysaccharide Fc receptor protein: Joint/Synovium
Yersinia, Escherichia Proteus, Klebsiella	Myasthenia Gravis	Bacterial Proteins: Acetylcholine Receptor
Chlamydia	Atherosclerosis	Outer Membrane Protein: Cardiac myosin
Borrelia burgdorferi	Lyme Disease	Osp A:LFA-1

disease, because heart reactive antibodies could be absorbed from animal or patient sera by the group A streptococcus (35,77,78) and antibody and complement were found deposited in myocardium (79). The evidence accumulated over the past 15 years has strongly supported the hypothesis that acute rheumatic fever is an autoimmune disease, and epidemiologic evidence strongly linked the disease to preceding streptococcal infection (80,81). Therefore, acute rheumatic fever links autoimmunity and infection.

Studies on the basis of the molecular mimicry between the streptococcus and heart have shown that alpha-helical coiled-coil proteins such as myosin, tropomyosin, keratin, vimentin and laminin in the host were responsible for the crossreactions observed between the host tissues and group A streptococci (60,63,82). The streptococcal M protein and the group A streptococcal carbohydrate epitope, N-acetyl-glucosamine, have been the major streptococcal components shown to be responsible for molecular mimicry between streptococci and host tissues (56,57,59,68–70,83). Synthetic peptides of the streptococcal M protein or human cardiac myosin have been used to map crossreactive B and T cell epitopes (29,59,84). A synthetic peptide containing one of the B cell epitopes of streptococcal M protein (GLN-LYS-SER-LYS-GLN) reacted with myosin-specific antibodies from acute rheumatic fever but not from normal individuals (59). Similar sequences were found in three different rheumatogenic streptococcal M protein serotypes recognized by murine and human monoclonal antibodies crossreactive with M protein and myosin. Although the (GLN-LYS-SER-LYS-GLN) sequence was recognized by all individuals who developed acute rheumatic fever or heart disease, other sites of cross-reactivity have been described (57,83,84). The sites in

human cardiac myosin which crossreacted with strepto-coccal M protein have been mapped to the alpha-helical rod region of the human cardiac myosin molecule (30,85). Several sites have been identified in the light meromyosin tail region. Both streptococcal M protein and myosin amino acid sequences have been shown to possess alpha-helical coiled-coil structures and to have amino acid sequence homology (86,87). Both proteins have a seven-residue periodicity that conforms to the alpha-helical coiled-coil structure. Homology has also been reported to exist between streptococcal M protein and other alpha-helical proteins such as tropomyosin, keratin, and desmin (86,87). The similarity in structure between the streptococcal M protein and family of alpha-helical coiled-coil host proteins is further supported by their immunological crossreactions (56,57, 60,63,83). A summary of the crossreactive anti-streptococcal mAb host tissue targets is found in Table 6. The tissue targets identified by the crossreactive mAbs may be important in the manifestations of group A streptococcal rheumatic fever sequelae of arthritis, carditis, chorea, and erythema marginatum (88,89). A summary of crossreactive human and mouse mAb specificities has been published in previous reviews (90) . The evidence supports the hypothesis that molecular mimicry can occur through structural similarities.

Studies of myosin-specific antibodies in patients with acute rheumatic fever suggest that they contain a cross-reacting idiotype (My1) shared with anti-myosin antibodies present in systemic lupus erythematosus, Sjögren's syndrome and post-streptococcal acute glomerulonephritis (91). Antibodies in sera from normal individuals, acute infection, and other heart, joint and kidney diseases do not have elevated levels of the My1 idiotype. Exactly

Table 6 Crossreactive anti-streptococcal monoclonal antibody host tissue targets

---

Cardiac myosin
Skeletal myosin
Tropomyosin
Keratin
Vimentin
Laminin
Retinal S antigen
DNA

---

how responses to infectious agents might connect these diseases remains to be determined, but certain viruses have been associated with the rheumatic diseases of systemic lupus erythematosus and Sjögren's syndrome. Furthermore, regulation of idiotypes may also play a role in development and prevention of disease, although there is little evidence currently available suggesting that this is the case.

Acute glomerulonephritis.   Sera from patients with post streptococcal acute glomerulonephritis (AGN) were shown by Kefalides and colleagues to contain antibodies against laminin, collagen and other macromolecules found in the glomerular basement membrane (92). The epitope recognized in collagen was identified and shown to be located in the 7-S domain of type IV collagen (93). Streptococcal antigens, immunoglobulins and complement were detected in the kidney glomeruli in AGN (94). Studies by Lange and Markowitz suggested that the glomerular basement membrane shared antigens with streptococcal M12 protein (95) and/or between the group A streptococcus and glomeruli (62,96,97). Kraus and Beachey discovered a renal autoimmune epitope (Ile-Arg-Leu-Arg) in M protein (98). Evidence does suggest that certain M protein serotypes are associated with nephritis, and that molecular mimicry or immunological crossreactivity between glomeruli and M protein could be one mechanism by which anti-glomerular antibodies are produced during infection. In animal models of nephritis induced by nephritogenic streptococci (M type 12), anti-glomerular antibodies eluted from kidney glomeruli reacted with the type 12 streptococcal M protein (99). Furthermore, an anti-glomerular mAb reacted with the M12 protein (97). These studies support immune-mediated mechanisms in post streptococcal AGN. Streptococcal and renal antigens may be an important source of mimicking antigen present in the kidney and play a role in binding Ig and complement and production of nephritis. However, in general, molecular mimicry in acute glomerulonephritis is not as well characterized as that found in rheumatic fever.

Crossreactive subsets of anti-streptococcal/anti-myosin monoclonal antibodies.   Crossreactive anti-streptococcal/ anti-myosin autoantibodies were divided into three major subsets based on their crossreactivity with 1) myosin and other alpha-helical molecules (21,59,60,100,101), 2) DNA (60,101), or 3) N-acetyl-glucosamine (68–70). Figure 2 illustrates the subsets of crossreactive anti-streptococcal/anti-myosin mAbs. All three subsets were identified among mAbs from mice immunized with group A streptococcal components, but in the human, the predominant subset reacted with the N-acetyl-glucosamine epitope and myosin and related molecules. This result is not surprising since rheumatic fever patients do not develop anti-nuclear antibodies during the course of their disease. Elevated levels of polyreactive anti-myosin antibodies found in ARF sera (102) and in animals immunized with streptococcal membranes or walls (78,103–105) most likely account for the reactivity of these sera with myocardium and other tissues. Similar types of crossreactive mouse and human antibodies have been investigated by Lange and colleagues (106) and by Young and colleagues (107), respectively.

Mimicry between the group A streptococcal polysaccharide epitope, N-acetyl-beta-D-glucosamine has proved to be intriguing. The crossreactive human mAbs from rheumatic carditis were highly reactive with myosin and N-acetylglucosamine (30,69). The anti-myosin/anti-N-acetyl-glucosamine antibodies were reactive with specific peptides from cardiac myosin and keratin, both alpha-helical coiled-coil molecules involved in the crossreactivity between group A streptococci and host tissues such as heart and skin (30, 70). A peptide SFGSGFGGGY from keratin was found to mimick N-acetyl-glucosamine by binding to the crossreactive anti-streptococcal mAbs, to N-acetyl glucosamine specific lectins and by inducing an IgG1 response against N-acetyl glucosamine (70). The results

SUBSETS OF CROSSREACTIVE ANTI-MYOSIN/ ANTI-STREPTOCOCCAL ANTIBODIES

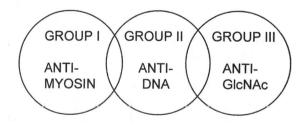

Figure 2.   Human and murine anti-streptococcal/anti-myosin mAbs have been divided into three different subsets based on their reactivity with myosin and N-acetyl-glucosamine, the dominant group A carbohydrate epitope, with DNA and the cell nucleus, a property found among murine mAbs, and reactivity with myosin and a family of alpha-helical coiled-coil molecules. Taken from reference (188) with permission from Indiana University School of Dentistry Press.

have suggested that the binding of some of the crossreactive antistreptococcal mAbs might be "lectin-like". These results clearly indicated that peptide structures could substitute for N-acetylglucosamine and induce an IgG response in animals immunized with the peptide. Mimicry between carbohydrates and peptides may be useful for new directions in vaccine development to avoid the use of poor carbohydrate immunogens from microorganisms (108).

Polyspecific or crossreactive autoantibodies have emerged as a theme in autoimmunity and molecular mimicry (101, 109–111). The V-D-J region genes of the human and mouse crossreactive mAbs have been sequenced (30,63,112), but there is no consensus sequence to explain the molecular basis of polyspecificity and crossreactivity. It is worth noting that the three groups of reactivities in mice do not have restricted antibody V gene families or VH and VL gene combinations that correlate with a specific reactivity. The V genes of crossreactive human mAbs from rheumatic fever were encoded by a heterogeneous group of VH3 family genes (VH-3, VH-8, VH-23, VH-30) and a VH4-59 gene segment (30). Young and colleagues also reported a similar group of V gene sequences for their human anti-streptococcal/anti-myosin antibodies produced from Epstein Barr Virus transformed B cell lines (107). Many sequences were found to be either in germline configuration or were highly homologous with a previously sequenced germline V gene. It has been proposed that a germline antibody may be polyreactive due to conformational rearrangement and configurational change, permitting binding of diverse molecules (72).

Potential role of antibodies in disease: Cytotoxicity linked to anti-myosin/anti-streptococcal antibodies. Although the role of the crossreactive antibodies in disease is not clear, cytotoxic mouse and human mAbs have identified the extracellular matrix protein laminin as a potential tissue target present in basement membrane surrounding myocardium and valve surface endothelium (63,113,114). Crossreactive antibodies may become trapped in extracellular matrix, which may act like a sieve to capture antibody and lead to inflammation in host tissues. Cytotoxic antibodies are rare among the crossreactive antibodies investigated. Therefore, not all molecular mimicry will lead to pathologic consequences. Our hypothesis is that many crossreactive autoantibodies are not capable of damaging tissues because they recognize proteins limited to the inside of the cell.

Animal models of mimicry between streptococcal M protein and cardiac myosin. Although strong evidence existed for immunological and structural mimicry between group A streptococcal M protein and host tissues, demonstration of mimicry causing disease in animal models was lacking. Studies by Huber and Cunningham demonstrated that streptococcal M5 peptide NT4 containing the amino acid sequence GLKTENEGLKTENEGLKTE could produce myocarditis in MRL/++ mice, an autoimmune-prone strain

(23). The studies demonstrated that the myocarditis was mediated by CD4+ T cells and Class II MHC molecules. Antibodies against the IA^k major histocompatibility molecule or antibody against CD4+ T cells abrogated the myocarditis (23). Amino acid sequence homology between the NT4 peptide of M5 and human cardiac myosin demonstrated 80% identity. The homology between the M5 peptide NT4 sequence and cardiac myosin sequence is illustrated below.

NT4 Streptococcal M Peptide   GLKTENEGLKTENEGLKTE

                                : : ::

Human Cardiac Myosin   KLQTENGE

The mimicking sequence in NT4 is repeated four times in the M5 protein, but is not repeated in the cardiac myosins. Repeated regions of the M protein that mimic cardiac myosin may be important in breaking tolerance in the susceptible host and producing autoimmune disease. The data support the hypothesis that epitopes in streptococcal M protein, which mimic cardiac myosin, may be important in breaking tolerance to this potent autoantigen. M protein amino acid sequences observed to induce inflammatory myocardial lesions in mice share sequence homology with cardiac myosins, which may give them the property of producing inflammatory heart lesions (84). Amino acid sequences repeated within the M protein share sequence similarity with cardiac myosins, and these peptides of M protein induced antibody titers ten times higher against cardiac myosin than skeletal myosin (84). Therefore, the streptococcal M protein contains regions of similarity with cardiac myosin, which gives the M protein antigen the ability to break tolerance against cardiac myosin and produce myocardial disease (23,84).

5.3. Mimicry Between Bacteria and Viruses

Streptococcal M proteins not only mimic epitopes in cytoskeletal proteins, they mimic epitopes in other strong bacterial and viral antigens, heat shock protein (Hsp)-65 (115) and coxsackieviral capsid proteins (23,29, 116,117). These immunological crossreactions may be important to host defense mechanisms against pathogens and strongly suggest that some crossreactive antibody molecules recognize and neutralize more than a single infectious agent. Such antibodies may be an important first line of defense and would be highly advantageous for the host. On the other hand, crossreactive antibodies may recognize multiple host tissue antigens in addition to epitopes on bacteria and viruses. Multiple antigen recognition may predispose to the appearance of cytotoxic antibody that recognizes an extracellular antigen and deposits in tissues. One example of multiple recognition and cytotoxicity is an anti-streptococcal mAb, which was shown to neutralize coxsackie virus and to be cytotoxic for heart cells in the presence of complement (29). Multiple pieces of evidence from experiments using synthetic peptides have shown that sites in the streptococcal M protein mimic a site(s) in the VP1 capsid protein of coxsackievirus (29). The homology between coxsackieviral VP1, human cardiac myosin, and streptococcal M protein is shown below.

```
M6 Protein     K E L E E S K K L T E K E K A E L G A K  L E A E A K A L K
                        . :  . .   . : .         : : : :
CVB3 VP1       A K R Y A E W V L T P R Q A A Q L R  R K  L E F F T Y V R F
                        : . . .  . :    : :       . : : :
cardiac myosin A L E E A E A S L E H E E G K I L R  A Q L E F N Q I  K A E
```

Some anti-streptococcal/anti-myosin mAbs have been shown to neutralize enteroviruses and to be cytotoxic for heart cells (29,117). This finding is significant because coxsackieviruses cause autoimmune myocarditis in susceptible hosts (118–120). In fact, one of the anti-streptococcal/anti-myosin mAbs was used to produce viral escape mutants that were shown to cause myocarditis in a different MHC haplotype than the wild-type virus (117).

The Hsp-65 antigen has been shown to play a role in the development of arthritis and diabetes (121,122). Anti-Hsp65 antibodies were shown to react with the streptococcal M protein (115). Crossreactive epitopes shared between streptococcal M proteins and Hsp-65 could play a role in the arthritis sequelae in acute rheumatic fever (115). This finding could explain the relationship between streptococcal arthritis and adjuvant arthritis (123). Antibodies against heat shock protein 60 have also been implicated in cytotoxicity against endothelium (124). Shared epitopes among pathogens may be important in molecular mimicry, may break tolerance to cryptic host epitopes, and may influence the development of autoimmune diseases. Table 7 summarizes bacterial and viral proteins which have immunological similarities and crossreactions with streptococcal M protein.

The evidence suggests that microbial agents such as bacteria are likely to have many crossreactive determinants among themselves as well as with host tissues, however, not all crossreactivity will lead to disease. Whether or not and why a host develops disease from a particular microorganism may depend upon 1) the genetic susceptibility of the host and its ability to respond to disease producing

Table 7 Bacterial and viral antigens with immunological similarities with group A streptococcal

M Proteins
Heat Shock Protein(Hsp)65
Coxsackievirus Capsid Proteins
Group A Streptococcal Carbohydrate*

*N-acetyl-β-D-glucosamine.

epitopes, 2) the location of the infection and cytokine levels produced, and 3) the number of times the host is assaulted by the same or similar crossreactive epitopes.

6. BACTERIAL ANTIGEN MIMICRY ASSOCIATED WITH AUTOIMMUNITY

6.1. Ankylosing Spondylitis and Reiter's Syndrome

Infections of many different etiologies have been reported preceding the onset of ankylosing spondylitis, a major non-rheumatoid arthritis (22). Preceding infections most often reported in ankylosing spondylitis include *Klebsiella, Shigella, Salmonella, Yersinia,* and *Escherichia coli* (125). One of the host susceptibility factors has been identified as the human major histocompatibility complex class I molecule HLA-B27 antigen (125–127). The data reported were striking that >90% of Caucasion patients with ankylosing spondylitis and that >80% with Reiter's Syndrome were HLA-B27+. Only 9% of the normal population were found to be HLA-B27+. It is not certain how the HLA-B27 antigen predisposes the host to disease. One group has suggested that HLA-B27 links abnormal immune responses against infectious agents with the disease, while the other group has proposed that bacterial factors modify the B27 molecule in tissues and damage occurs (128,129). However, molecular mimicry may play an important role in disease. The HLA-B27 molecule has been shown to share amino acid sequence homology with the nitrogenase produced by *Klebsiella pneumoniae* (22). *Klebsiella* has been found present in the intestinal flora of a high percentage of spondylitis patients as compared to healthy controls (130). Three recent reports point to molecular mimicry as a potentially significant phenomenon in the pathogensis of ankylosing spondylitis. A number of reports contribute to the hypothesis that molecular mimicry may play an important role in ankylosing spondylitis. The reports show 1) the sharing of identical amino acid sequences between HLA-B27 and *Klebsiella pneumoniae* (22); 2) the recognition of this epitope by antibodies demonstrating immunologic crossreactivity (131) and 3) the expression of the epitope in affected tissues (132).

When HLA B27 was studied for crossreactive sequences or epitopes shared with infectious agents, a number of bacterial antigens were identified in *Klebsiella pneumoniae, Yersinia enterocolitica, Salmonella, Campylobacter* and *Shigella flexneri* (22,133). The homology shared between the *Klebsiella pneumoniae* nitrogenase and the HLA-B27 molecule has been well-characterized and found to be a 6 amino acid residue identity (QTDRED) beginning at residue 72 of HLA-B27.1 and residue 187 of the nitrogenase (22). The identity of the other *Klebsiella* and gram negative bacterial proteins is not known.

Antibodies that were crossreactive with *K. pneumoniae* antigens and HLA-B27 were then found in the sera of ankylosing spondylitis patients (131). A synthetic peptide of the sequence shared between the *K. pneumoniae* nitrogenase and HLA-B27 molecule was tested against sera of ankylosing spondylitis patients who were HLA-B27+, and the sera were found to react strongly with the peptide. Further studies demonstrated that the shared epitope or sequence was concentrated and expressed in synovial tissues in the ankylosing spondylitis patients (132). The epitope was not expressed in other arthritic diseases. Similar evidence was found in Reiter's syndrome which associated the disease with crossreactivity between HLA-B27 and *K. pneumoniae* nitrogenase (22). The data support the concept that a crossreactive autoimmune response may be a major factor in disease development.

It is established that at least six variants of HLA-B27 exist at the molecular level, with no one variant preferentially associated with disease (134). Studies by Taurog have eluciated the role of HLA-B27 in development of inflammatory disease involving the gastrointestinal tract, peripheral and vertebral joints, male genital tract, skin, nails, and heart. Rats transgenic for HLA-B27 spontaneously developed a pattern of organ system involvement which had a striking resemblance to the HLA-B27-associated human disorders. The transgenic rats have now been investigated for several years as potential models for the human spondyloarthropathies. The results from investigation of the transgenic rat model established that HLA-B27 plays a major role in the pathogenesis of multi-organ system diseases related to spondyloarthropathies (135). Using the transgenic model, disease susceptibility was correlated with gene copy number and level of HLA-B27 expression in lymphoid cells (136). Transgenic rats with a high level of expression of the HLA-B27 develop chronic inflammatory bowel disease (IBD) and arthritis. When the transgenic rats were maintained in a germfree environment, they did not develop disease suggesting that there was a relationship between the presence of the normal flora microorganisms, HLA-B27 and production of disease (137). Assessment of the cecal microflora showed a rise in numbers of *Escherichia coli* and *Enterococcus* spp., which correlated with the presence and severity of inflammatory bowel disease in the HLA-B27 transgenic rats. Although mimicry has not been investigated in the transgenic rat model, the role of bacteria in the development of arthritis and ankylosing spondylitis is very clear. The model is consistent with the hypothesis that bacteria play a major role in the development of disease.

6.2. Mycobacteria and Molecular Mimicry

Mycobacterial antigens have been implicated in crossreactions with host tissues. Recent investigations have shown that mouse and human monoclonal anti-DNA antibodies reacted with mycobacteria and their phospholipids (110,138). The crossreactive anti-DNA antibodies were antinuclear and were inhibited by DNA and polynucleotides (110). Anti-*M. tuberculosis* antibodies crossreacted with brain tissue and were inhibited by mycobacterial glycolipids (139). Some of these anti-DNA antibodies carried an idiotype 16/6 that appeared to be important in disease development (140). Human monoclonal antibodies developed from patients with lepromatous leprosy were shown to crossreact with *Mycobacterium leprae* , DNA, mitochondria, cytoskeletal proteins, vimentin, actin and the acetylcholine receptor (64). Antibodies to host components have often been found in patients with leprosy or its complications.

In another study, a single dose of 2.6×10^7 heat-killed Bacillus Calmette-Guerin (BCG) given i.v. to 8-week-old NOD mice prevented diabetes, but precipitated a syndrome similar to systemic lupus erythematosus (SLE) in which treated mice rapidly developed haemolytic anaemia, high titre anti-DNA and anti-Sm antinuclear autoantibodies, perivascular lymphocytic infiltration in the kidneys and glomerular immune complex deposition (140a).

Adjuvant arthritis can also be induced in rats by immunization with a *Mycobacterium tuberculosis* antigen (141,142). Further studies have shown that *M. tuberculosis* contained an epitope crossreactive with a self antigen expressed in joint cartilage (143). A T cell clone specific for *M. tuberculosis* was strongly arthritogenic (143). The T cell clone also recognized proteoglycans purified from cartilage as well as antigens present in the synovial fluid.

Patients treated with BCG immunotherapy have been shown to develop severe arthritis (144) and peripheral blood lymphocytes from patients with rheumatoid arthritis were highly responsive to the mycobacterial-purified protein derivative (PPD) (145). T cell clones from adjuvant arthritis recognized amino acid residues 180–188 in a 65 kilodalton *Mycobacterium bovis* antigen (146) shown to be a heat shock protein (147). The antigen was found to be related to other bacterial and host heat shock proteins (Hsp). Hsp 65 amino acid residues 180–196 were reported as the mycobacterial epitope recognized by T cell clones or T lymphocytes or hybridomas in adjuvant arthritis (148). The sequence was found in the *M. leprae, M. tuberculosis* and *M. bovis* 65 kilodalton heat shock protein (Hsp 65). The epitope described by van Eden, Born and colleagues shares homology with Hsp 65 of eukaryotic cells, yeast, and *Escherichia coli* (147,148). T cell clones reactive with Hsp 65 were reported to either activate or protect against arthritis. It was also demonstrated that Hsp 65 could induce resistance to arthritis in Lewis rats whereas whole mycobacteria induce disease (147). Arthritogenic or protective clones did not recognize the other recently cloned mycobacterial antigens of 12Kd, 18Kd, 28Kd, and 34Kd (147,149). The data are very strong evidence that Hsp 65 plays a role in crossreactions with joint and synovial tissues and in the development of or protection against arthritis.

6.3. Myasthenia Gravis: Crossreaction Between Bacteria and Acetylcholine Receptor

Myasthenia gravis is a disease that results from defective neuromuscular transmission (150,151). Experimental models of the disease rely on active immunization of animals with the acetylcholine receptor or passive transfer of anti-receptor antibodies (151,152). Furthermore, circulating antibodies against the receptor are present in the majority of patients with myasthenia gravis and their specificities have been well characterized (153). The initiating factors causing the production of antibodies against the acetylcholine receptor are not known, and the disease has not been associated with a particular infectious agent. However, monoclonal antibodies against the alpha-subunit of the acetylcholine receptor were shown to crossreact with proteins in *Escherichia coli*, *Klebsiella pneumoniae*, and *Proteus vulgaris* (154). One of the *E. coli* proteins proposed to crossreact with the acetylcholine receptor was the outer membrane protein, Omp-C, which may form hydrophilic channels across the membrane similar to the acetylcholine receptor .

Other reports have suggested that the antibodies to the acetylcholine receptor may be anti-bacterial antibodies. Infections have been reported to be associated with the onset and trigger of myasthenia gravis (155). In one report, anti-DNA monoclonal antibodies from a patient with lepromatous leprosy crossreacted with *Mycobacterium leprae* and the acetylcholine receptor (64). Further studies have shown that the acetylcholine receptor and alpha-1,3 dextran were linked through an idiotypic network (156). Monoclonal antibodies raised against the alpha-1,3 dextran were shown to react with a monoclonal antibody variable region against the acetylcholine receptor. The alpha-1,3 dextran is present on certain bacteria such as *Enterobacter cloacae* and *Serratia*. The data showed that 12/60 myasthenia gravis patients had antibodies against alpha-1,3 dextran while 40 controls did not. It may be possible that any or all of these bacterial antigens might function by molecular mimicry in genetically predisposed individuals to produce myasthenia gravis.

6.4. Rheumatoid Factors and Bacterial Fc Binding Proteins

Both staphylococci and streptococci possess Fc binding proteins (157–159). It has been suggested that conformational similarities exist between the antigen combining sites of rheumatoid factors and the Fc binding regions of these microbial proteins (160–162). It was shown that affinity-

purified chicken antibodies to staphylococcal protein A also reacted with monoclonal and polyclonal rheumatoid factors (163). The binding was shown to be located in the Fab' region of the crossreactive chicken antibodies.

Staphylococcus aureus strains rich in protein A were shown to induce the production of rheumatoid factor (164, 165) and human rheumatoid factors were shown to possess the internal image of the Fc binding domain of staphylococcal protein A (163). The data would suggest that rheumatoid factors that bind to the immunoglobulin Fc domain may be anti-idiotypes formed against antibodies to protein A or a similar microbial Fc binding protein.

Recent data suggest that staphylococcal protein A binds only to antibodies that use VH3 heavy chains, and thus it has been designated as a B-cell superantigen (166). It has been reported that *Staphylococcus aureus* protein A induced IgM rheumatoid factor (RF) production by human peripheral blood mononuclear cells. Production of IgM rheumatoid factor by staphylococcus protein A was limited to a subset of B cells enriched for VH3 mRNA expression. These results suggested that rheumatoid factors produced in response to *Staphylococcus aureus* strain Cowan were enriched for usage of VH3 heavy chains. The *Staphylococcus aureus*-induced autoantibody response appeared to represent a new B-cell superantigenic property of protein A (166). Whether or not bacterial Fc binding proteins play a role in triggering rheumatoid arthritis is not yet clear.

6.5. Anti-DNA Antibodies Crossreactive with Bacterial Antigens

Serum antibodies against DNA have been detected in a number of diseases, but they are characteristic of systemic lupus erythematosus in humans (167) and mice (168–171). Hybridomas derived from humans and mice without experimental immunization have made it possible to study the crossreactivity of anti-DNA antibodies with other host and foreign antigens. It is now clear that anti-DNA antibodies do react with bacteria or bacterial antigens and Table 8 lists some of the crossreactions reported. Anti-streptococcal monoclonal antibodies that recognized streptococcal M protein and alpha-helical coiled-coil proteins, such as myosin, were

Table 8   Bacterial antigens crossreactive with DNA

Bacteria	Antigen
Group A streptococci	M protein (56)
Group D streptococci	phospholipids (109)
Klebsiella pneumoniae	capsular polysaccharide (67)
Group B streptococci	capsular polysaccharide (66)
Escherichia coli K1	
Salmonella	lipopolysaccharides (33, 204)

anti-nuclear and recognized DNA and polynucleotides poly (dT) and poly (I) (56) in a pattern similar to monoclonal anti-DNA antibodies reported from lupus patients or mice (110, 168, 172). Furthermore, a monoclonal anti-dsDNA antibody R4A, shown to deposit in the glomeruli of mice, and 52b3, a mutant of R4A that demonstrated tubular deposition in the kidney, were both found to be inhibited by specific peptides DWEYSVWLSM and RHEDGDWPRV (173). Immunization of mice with the DWEYSVWLSM peptide resulted in an anti-dsDNA response (71). It was also shown that murine monoclonal anti-DNA antibodies reacted with *Streptococcus faecalis* (109). The data suggest that bacterial phospholipid or protein antigens might be surrogate antigens for DNA and thus be the source of crossreactive anti-DNA antibodies in lupus mice or patients.

Other reports have shown crossreactivity between DNA and polysaccharides of gram negative bacteria. Study of the Waldenstrom macroglobulins demonstrated antibodies crossreactive with *Klebsiella* DNA and nucleic acid antigens (67). Some of these antibodies possessed the lupus anti-DNA autoantibody idiotype 16/6. Furthermore, the alpha(2,8)-linked N-acetyl neuraminic acid polymer, which is the major virulence factor of group B meningococci and *E. coli* K1, reacted with antibody that crossreacted with polynucleotides and denatured DNA(66). The crossreactive antibody was protective against *E. coli* infection of newborn rats. The data support the hypothesis that crossreactivity between dissimilar antigenic structures may be due to charged groups and molecular shape and spacing. Furthermore, it supports the hypothesis that crossreactive antibodies against host antigens may protect against infections. In the study by Kabat and colleagues, the antibody was not injurious to the host even at 23 mg/ml in the serum (66). Whether or not an antibody may become harmful to the host may depend on its properties of specificity and whether or not the antibody recognizes exposed or expressed host antigenic determinants. Many crossreactive antibodies may be protective and noninjurious to the host, while others which recognize exposed or cell surface determinants may cause disease.

Diamond and colleagues have investigated the anti-DNA antibody response in systemic lupus erythematosus and found that individuals immunized with a pneumococcal polysaccharide vaccine demonstrated an increased titer of anti-pneumococcal polysaccharide antibodies that expressed an anti-DNA antibody idiotype found on anti-DNA antibodies in 70–80% systemic lupus erythematosus patients (174). Studies by Adderson have shown that anti-*Haemophilus influenzae* monoclonal antibodies produced from individuals immunized against the type b capsular polysaccharide did give rise to lower affinity anti-type b capsular antibodies that reacted with self-antigens, including DNA (175). In both studies, there was evidence of preferential utilization of the VH3 family genes encoding the antibodies against the polysaccharide antigens (174,175). Other studies of

crossreactive anti-DNA antibodies have shown that anti-DNA/anti-phosphocholine monoclonal antibodies from phosphocholine immunized mice are encoded by V genes not found in the dominant anti-phosphocholine response containing the T15 idiotype (176). An anti-phosphocholine monoclonal antibody bearing the T15 idiotype has been shown to mutate to an anti-DNA antibody by changing one amino acid in CDR1 (177). The T15 idiotype is present on antibodies that protect against infection with *Streptococcus pneumoniae*. Again, the theme emerges that immunologic crossreactivity or mimicry between infectious agents and autoantigens may involve protective antibody against the bacterium and may lead to autoimmunity (178).

6.6. Chlamydia Infections and Inflammatory Heart Disease

Chlamydia, a cause of pneumonia, trachoma, conjunctivitis and sexually-transmitted diseases, have been reported to cause autoimmune sequelae following infection. These sequelae may include arthritis and uveitis, but the latest epidemiological studies suggest that chlamydia may induce inflammatory cardiovascular lesions including blood vessel occlusion in the heart (179,180). Seroepidemiologic studies have shown that *C. pneumoniae* antibody is associated with coronary artery disease, myocardial infarction, carotid artery disease, and cerebrovascular disease. The association of *Chlamydiae pneumoniae* with atherosclerosis is corroborated by evidence of the organism in atherosclerotic lesions throughout the arterial tree and the lack of the organism in healthy arterial tissue. *C. pneumoniae* has been isolated from coronary and carotid atheromatous plaques. To determine whether chronic infection plays a role in initiation or progression of disease, intervention studies of antibiotic regimens in humans have been initiated, and animal models of *C. pneumoniae* infection and atherosclerosis have been developed (181). It is clear that there is current evidence to support the association and potential role of *C. pneumoniae* in cardiovascular disease (179).

Although the evidence has been largely circumstantial, recent investigations have supported the hypothesis that the pathogenesis of cardiovascular lesions could be due to chlamydial infections. Studies by Bachmaier revealed that peptides of the *Chlamydia* 60 KD cysteine-rich outer membrane proteins produce inflammatory heart disease, including myocarditis and perivascular inflammation with fibrinous occlusion of cardiac blood vessels originating from the endothelium, and perivascular fibrosis with thickening of the arterial walls. The chlamydial peptides that caused inflammatory heart disease shared amino acid sequence homology with the alpha isoform of mouse cardiac myosin (180). An example of the sequence homology between cardiac myosin and the chlamydial peptides from the chlamydial 60KD outer membrane proteins (OMP) is shown below.

Alpha cardiac myosin   S L K L M A T L F S T Y A S A D
                              : :   : :
Chlamydial OMP   V L E T S M A E F T S T N V I S

Cardiac myosin is well established as an autoantigen capable of producing myocarditis in susceptible animals (25). Case histories of patients developing myocarditis following documented chlamydial infections have been reported (182–184). Development of inflammatory heart disease, either myocarditis or perivascular inflammation and fibrosis, are expected to be similar to other autoimmune diseases, and occur in genetically susceptible hosts. The data suggest that antigenic mimicry between the chlamydial outer membrane protein and the alpha isoform of mouse cardiac myosin plays a role in development of the disease in BALB/c mice administered the pathogenic peptides from myosin or the chlamydial outer membrane protein.

6.7. Lyme Arthritis: Mimicry Between OspA and Leukocyte Function Antigen (LFA)-1

Infection with *Borrelia burgdorferi* results in Lyme arthritis, which is treatable with antibiotics. However, approximately 10% of patients develop a treatment resistant arthritis that is prolonged and associated with an immune response to the outer surface protein A (OspA). In the group of patents developing the treatment-resistant arthritis, there is an increased frequency of the HLA DR B1*0401 allele. T cells from synovial fluid of the treatment-resistant arthritis respond to Osp A and are the Th1 subset producing gamma-IFN. Twenty-mer synthetic peptides of OspA were tested for binding to DRB1*0401 and peptide spanning residues 154–173 was found to bind DRB1*0401(185). A peptide with highest predicted score for binding to DRB1*0401 was OspA residues 165–173, which supported the hypothesis that the DRB1*0401 could bind a particular OspA peptide (185). To study the relationship of the peptide and the DRB1*0401 molecule, MHC class II deficient mice transgenic for the DRB1*0401 molecule were immunized with the intact OspA molecule and T cells from the lymph nodes were cultured with the OspA peptides and tested for gamma-IFN production (185). The results demonstrated that the mice recognized primarily the 164–183 peptide, while mice with a normal MHC class II expression recognized a group of epitopes. Thus, the immunodominant epitope for the DRB1*0401 allele was identified within the peptide 165–173. Experiments with T cells from the synovial fluid of patients with the treatment resistant arthritis indicated that the 164–173 peptide was the immunodominant Th1 epitope recognized by T cells from the disease (185). Searches of protein databases indicated that the OspA peptide residues 165–173 was homologous to a region in human LFA-1. T cells from patients with the treatment-resistant arthritis responded to the LFA-1 containing the sequence homology with OspA within LFA-1 residues 326–343. *Borrelia burgdorferi sensu stricto* is the only strain that causes treatment-resistant arthritis and also is the only strain with the LFA-1 like sequence in OspA residues 165–173 (185). Once again, host genetic susceptibility and the bacterium play a role in development of an autoimmune sequelae that results from mimicry and sequence homology between OspA, the *Borrelia burgdorferi* protein, and the host protein, LFA-1.

7. CONCLUSIONS

The role of viral and bacterial epitopes in molecular mimicry or immunological crossreactivity is not trival. It is likely that the origins of many autoantibodies are from host responses against infectious agents such as bacteria or viruses. Data suggest that crossreactivity between microbial and host antigens is not unique to the diseased host (186,187). Crossreactive antibodies may be present in normal individuals and may increase during infectious diseases (186). The presence of crossreactive antibodies may support the overall host immune defense system by recognizing multiple infectious agents such as bacteria and viruses. Antibodies that recognize both bacteria and viruses (29) may be advantageous for a host unless the antibody recognizes host cell surface epitopes. In the genetically-susceptible or -predisposed host, crossreactive antibodies that recognize cell surface or exposed tissue determinants may create disease. Immune crossreactivity and molecular mimicry are not limited to antibodies, and manifestations of chronic disease depends on activated and/or crossreactive T lymphocyte responses. In addition to mimicking epitopes, development of autoimmune disease requires a susceptible host with predisposing MHC haplotypes, elevated cytokine production, or target organ sensitivity. As shown in this review, mimicking epitopes may be used to induce tolerance against autoimmune mechanisms of disease. Therefore, mimicry is an important tool for understanding the mechanisms of pathogenesis of autoimmune disease as well as an avenue for development of therapeutic strategies. The reports described in this chapter provide only a birds-eye view of the dysregulation of the immune system by molecular mimicry.

ACKNOWLEDGEMENTS

Gratitude is expressed to all of our students and fellows who have made this work possible. Our research was supported by grants from the National Heart, Lung and Blood Institute HL35280 and HL56267. M.W.C. is the recipient of an NIH Merit Award.

References

1. Shoenfeld, Y., and R.S. Schwartz. 1984. Immunologic and genetic factors in autoimmune diseases. *N. Eng. J. Med.* 311:1019–1029.

2. Burnet, M. 1959. The clonal selection theory of acquired immunity. Vanderbilt University Press, Nashville, TN.

3. Dersimonian, H., A. Long, D. Rubinstein, B.D. Stollar, and R. Schwartz. 1990. VH genes of human autoantibodies. *Int. Rev. Immunol.* 5:253–264.

4. Infante, A.J. 1999. T cell receptor expression in autoimmune diseases. *Int. Rev. Immunol.* 18:1–141.

5. Green, M.C., and L.D. Schultz. 1975. Motheaten, an immunodeficient mutant of the mouse. I. Genetics and pathology. *J. Heredity* 66:250–265.

6. Green, M.C., H.O. Sweey, and L.E. Bunker. 1976. Tightskin, a new mutation of the mouse causing excessive growth of connective tissue and skeleton. *Amer. J. Pathol.* 82:493–512.

7. Smoyer, W.E., and C.J. Kelly. 1994. Inherited interstitial nephritis in kd/kd mice. *Int. Rev. Immunol.* 11:245–253.

8. Bhandoola, A., K. Yui, R.M. Siegel, L. Zerva, and M. Greene. 1994. Gld and lpr mice: single gene mutant models for failed self tolerance. *Int. Rev. Immunol.* 11:231–252.

9. Huang, T.F. 1995. T cell development in CD3-ζ mutant mice. *Int. Rev. Immunol.* 13:29–42.

10. Mombaerts, P. 1995. Lymphocyte development and function in T cell receptor and Tag-1 mutant mice. *Int. Rev. Immunol.* 13:43–64.

11. Mosman, T., and T. Coffman. 1989. Th1 and Th2 cells: different patterns of lymphokine secretion lead to different functional properties. *Ann. Rev. Immunol.* 7:145–173.

12. Modlin, R.L., and T.B. Nutman. 1993. Type of cytokines and negative immune regulation in human infections. *Curr. Opinions Immunol.* 5:511–517.

13. Yamamura, M., X.-H. Wang, O.J.D., K. Uyemura, T.H. Rea, B.R. Bloom, and R.L. Modlin. 1992. Cytokine patterns of immunologically mediated tissue damage. *J. Immunol.* 149:1470–1475.

14. Matyniak, J.E., and S.L. Reiner. 1995. T helper phenotype and genetic susceptibility in experimental lyme disease. *J. Exp. Med.* 181:1251–1254.

15. Izui, S., R. Merino, L. Fossati, and M. Iwamoto. 1994. The role of the Yaa gene in lupus syndrome. *Int. Rev. Immunol.* 11:231–244.

16. Choi, Y., J.W. Kappler, and P. Marrack. 1991. A superantigen encoded in the open reading frame of the 3′ long terminal repeat of mouse mammary tumor virus. *Nature* 350:203–205.

17. Dzarski, R. 1982. Preferential induction of autoantibody secretion in polyclonal activation of peptidoglycan and lipopolysaccharide. II. In vivo studies. *J. Immunol.* 128:1026–1030.

18. Cole, B.C., and M.M. Griffiths. 1993. Triggering and exacerbation of autoimmune arthritis by the Mycoplasma arthritidis superantigen MAM. *Arth. & Rheum.* 36:994–1001.

19. Kappler, J.W., B. Kotzin, L. Herron, E. Gelfand, R. Bigler, A. Boyston, S. Carrell, D. Posnett, and P. Marrack. 1989. Vβ-specific stimulation of human T cells by staphylococcal toxins. Science 244:811–813.

20. Gibofsky, A., S. Kerwar, and J.B. Zabriskie. 1998. Rheumatic fever: the relationships between host, microbe and genetics. In: *Rheumatic disease clinics of North America*, L.R. Espinosa, ed., W.B. Saunders Co., Philadelphia, PA, 24(2), pp. 237–259.

21. Krisher, K., and M.W. Cunningham. 1985. Myosin: a link between streptococci and heart. *Science* 227(4685):413–415.

22. Schwimmbeck, P.L., and M.B.A. Oldstone. 1989. Klebsiella pneumoniae and HLA B27-associated diseases of Reiter's syndrome and ankylosing spondylitis. *Curr. Topics Microbiol. Immunol.* 45:45–56.

23. Huber, S.A., and M.W. Cunningham. 1996. Streptococcal M protein peptide with similarity to myosin induces CD4$^+$ T cell-dependent myocarditis in MRL/++ mice and induces partial tolerance against coxsakieviral myocarditis. *J. Immunol.* 9:3528–34.

24. Neu, N., K.W. Beisel, M.D. Traystman, N.R. Rose, and S.W. Craig. 1987. Autoantibodies specific for the cardiac myosin isoform are found in mice susceptible to coxsackievirus B3 induced myocarditis. *J. Immunol.* 138:2488–2492.

25. Neu, N., N.R. Rose, K.W. Beisel, A. Herskowitz, G. Gurri-Glass, and S.W. Craig. 1987. Cardiac myosin induces myocarditis in genetically predisposed mice. *J. Immunol.* 139:3630–3636.

26. Moudgil, K.D., A. Ametani, I.S. Grewal, V. Kumar, and E.E. Sercarz. 1993. Processing of self-proteins and its impact on shaping the T cell repertoire, autoimmunity and immune regulation. *Int. Rev. Immunol.* 20:365–377.

27. Day, M.J., J. Russel, A.T. Kitwood, M. Pousford, and C.J. Elson. 1989. Expression and regulation of erythrocyte autoantibodies in mice following immunization with rat erythrocytes. *Eur. J. Immunol.* 19:795–801.

28. Oldstone, M. 1998. Molecular mimicry and immune mediated diseases. *FASEB J.* 12:1255–1265.

29. Cunningham, M.W., S.M. Antone, J.M. Gulizia, B.M. McManus, V.A. Fischetti, and C.J. Gauntt. 1992. Cytotoxic and viral neutralizing antibodies crossreact with streptococcal M protein, enteroviruses, and human cardiac myosin. *Proc. Natl. Acad. Sci. USA* 89:1320–1324.

30. Adderson, E.E., A.R. Shikhman, K.E. Ward, and M.W. Cunningham. 1998. Molecular analysis of polyreactive monoclonal antibodies from rheumatic carditis: human anti-N-acetyl-glucosamine/anti-myosin antibody V region genes. *J. Immunol.* 161:2020–2031.

31. Dresser, D.W. 1978. Most IgM-producing cells in mouse secrete autoantibodies (rheumatoid factors). *Nature* 276:480–483.

32. Primi, D., C.I.E. Smith, and G. Moller. 1977. Characterization of self reactive B cells by polyclonal B-cell activators. *J. Exp. Med.* 145:21–32.

33. Izu, S.T., M.J. Kobayakawa, J. Zyrd, J. Louis, and P.H. Lambert. 1977. Mechanism for induction of anti-DNA antibodies by bacterial lipopolysaccharides in mice. II. Correlation between anti-DNA induction and polyclonal antibody formation by various polyclonal B lymphocyte activators. *J. Immunol.* 119:2157–2162.

34. Okamoto, M., M. Murakami, A. Shimizu, S. Ozaki, T. Tsubata, S.-I. Kumagai, and T. Honjo. 1992. A transgenic model of autoimmune hemolytic anemia. *J. Exp. Med.* 175:71–79.

35. Zabriskie, J.B. 1967. Mimetic relationships between group A streptococci and mammalian tissues. *Adv. in Immunol.* 7:147–188.

36. Fujinami, R.S., M.B.A. Oldstone, Z. Wroblewska, M.E. Frankel, and H. Koprowski. 1983. Molecular mimicry in virus infections: cross reaction of measles virus phosphoprotein of herpes simplex virus protein with human intermediate filaments. *Proc. Natl. Acad. Sci. USA* 80:2346–2350.

37. Wucherpfennig, K.W., and J.L. Strominger. 1995. Molecular mimicry in T cell-mediated autoimmunity: viral peptides activate human T cell clones specific for myelin basic protein. *Cell* 80:695–705.

38. Hemmer, B., B.T. Fleckenstein, M. Vergelli, G. Jung, H. McFarland, R. Martin, and K.-H. Weismuller. 1997. Identification of high potency microbial and self ligands for a human autoreactive class II-restricted T cell clone. *J. Exp. Med.* 185:1651–1659.

39. Barnaba, V., and F. Sinigaglia. 1997. Molecular mimicry and T cell-mediated autoimmune disease. *J. Exp. Med.* 185:1529–1531.

40. Oldstone, M.B., M. Nerenberg, and P. Southern. 1991. Virus infection triggers insulin-dependent diabetes mellitus in a transgenic model: role of anti-self (virus) immune response. *Cell* 65:319–331.

41. Hasunuma, T., T. Sumida, and K. Nishioka. 1998. Human T cell leukemia virus type-1 and rheumatoid arthritis. *Int. Rev. Immunol.* 17:291–308.

42. Takahashi, Y., C. Murai, T. Ishii, K. Sugamura, and T. Sasaki. 1998. Human parvovirus B 19 in rheumatoid arthritis. *Int. Rev. Immunol.* 17:232–309.

43. Futo, F., K.G. Sagg, and L.L. Scharosch. 1993. Parvovirus B 19-specific DVH in bone marrow from B19 arthropathy patients: evidence for B19 virus persistence. *J. Infectious Dis.* 167:744–748.

44. Srinivasappa, J., J. Saegusa, B.S. Prabakhar, M.K. Gentry, M.J. Buchmeier, T. Wiktor, H. Koprowski, and M. Oldstone, Notkins, A. 1986. Molecular mimicry: frequency of reactivity of monoclonal antiviral antibodies with normal tissues. *J. Virol.* 57.397–401.

45. Fujinami, R.S., and M.B.A. Oldstone. 1985. Amino acid homology between the encephalitogenic site of myelin basic protein and virus: mechanism of autoimmunity. *Science* 230:1043–1045.

46. Bona, C., C. Murai, and T. Sasaki. 1998. The role of exogenous stimulation in pathogenesis of autoimmune diseases. In: *Contemporary Immunology*: Autoimmune Reactions. Humana Press, Inc., Totowa, NJ. pp. 141–155.

47. Oldstone, M.B.A. 1987. Molecular mimicry and autoimmune disease. *Cell* 50:819–820.

48. Dyrberg, T., and M. Oldstone. 1986. Peptides as probes to study molecular mimicry and virus induced autoimmunity. *Curr. Top. Microbiol.* 130:25–37.

49. Maul, G.G., S.A. Jimenez, E. Riggs, and D. Ziemnica-Kotula. 1989. Determination of an epitope of the diffuse systemic sclerosis marker antigen DNA topoisomerase I: sequence similarity with retroviral p30 gag protein suggest a possible cause for autoimmunity in systemic sclerosis. *Proc. Natl. Acad. Sci. USA* 86:8492–8496.

50. Kasturi, K., A. Hatakeyama, H. Spiera, and C. Bona. 1985. Antifibrillarin autoantibodies present in systemic sclerosis and other connective tissue diseases interact with similar epitopes. *J. Exp. Med.* 181:1027–1036.

51. Muryoi, T., K. Kasturi, M.J. Kafina, D.S. Cram, L.C. Harrison, and T. Sasaki. 1993. Anti-topoisomerase I monoclonal autoantibodies from scleroderma patients and tight skin mouse interact with similar epitopes. *J. Exp. Med.* 175:1103–1109.

52. Douvas, A., and S. Sobelman. 1991. Multiple overlapping homologies between two rheumatoid antigens and immunosuppressive viruses. *Proc. Natl. Acad. Sci. USA* 88:6328–6332.

53. James, J.A., M.J. Mamula, and J.B. Harley. 1994. Sequential autoantigenic determinants of the small nuclear ribonuclear protein SmD are shared by human lupus autoantibodies and MRL *lpr/lpr* antibodies. *Clin. Exp. Immunol.* 98:419–426.

54. Sabbatini, A.S., S. Bombardieri, and P. Migliorini. 1993. Autoantibodies from patients with systemic lupus erythematosus bind a shared sequence of SmD and Epstein-Barr nuclear antigen-1. *Eur. J. Immunol.* 23:1146–1152.

55. Zhao, Z.-S., F. Granucci, I. Yeh, P.A. Schaffer, and H. Cantor. 1998. Molecular mimicry by herpes simplex virus type I: autoimmune disease after viral infection. *Science* 279:1344–1347.

56. Cunningham, M.W., and R.W. Swerlick. 1986. Polyspecificity of anti-streptococcal murine monoclonal antibodies and their implications in autoimmunity. *J. Exp. Med.* 164:998.

57. Dale, J.B., and E.H. Beachey. 1986. Sequence of myosin-crossreactive epitopes of streptococcal M protein. *J. Exp. Med.* 164:1785–90.

58. Bronze, M.S., E.H. Beachey, and J.B. Dale. 1988. Protective and heart-crossreactive epitopes located within the NH2 terminus of type 19 streptococcal M protein. *J. Exp. Med.* 6:1849–1859.

59. Cunningham, M.W., J.M. McCormack, P.G. Fenderson, M.K. Ho, E.H. Beachey, and J.B. Dale. 1989. Human and murine antibodies cross-reactive with streptococcal M protein and myosin recognize the sequence GLN-LYS-SER-LYS-GLN in M protein. *J. Immunol.* 143:2677–2683.

60. Fenderson, P.G., V.A. Fischetti, and M.W. Cunningham. 1989. Tropomyosin shares immunologic epitopes with group A streptococcal M proteins. *J. Immunol.* 1427:2475–2481.

61. Kraus, W., J.M. Seyer, and E.H. Beachey. 1989. Vimentin-cross-reactive epitope of type 12 streptococcal M protein. *Infection and Immunity* 57:2457–61.

62. Kraus, W., J.B. Dale, and E.H. Beachey. 1990. Identification of an epitope of type 1 streptococcal M protein that is shared with a 43-kDa protein of human myocardium and renal glomeruli. *J. Immunol.* 145:4089–4093.

63. Antone, S.M., E.E. Adderson, N.M.J. Mertens, and M.W. Cunningham. 1997. Molecular analysis of V gene sequences encoding cytotoxic anti-streptococcal/ anti-myosin monoclonal antibody 36.2.2 that recognizes the heart cell surface protein laminin. *J. Immunol.* 159:5422–5430.

64. Duggan, D.B., and e. al. 1988. Polyspecificity of human monoclonal antibodies reactive with Mycobacterium leprae, mitochondria, ssDNA, cytoskeletal proteins and the acetylcholine receptor. *Clin. Immunol. Immunopathol.* 49:327–340.

65. Rubin, R.L., and e. al. 1984. Multiple autoantigen binding capabilities of mouse monoclonal antibodies selected for rheumatoid factor activity. *J. Exp. Med.* 159:1429–1440.

66. Kabat, E.A., K.G. Nickerson, J. Liao, L. Grossbard, E.F. Osserman, E. Glickman, L. Chess, J.B. Robbins, R. Schneerson, and Y.H. Yang. 1986. A human monoclonal macroglobulin with specificity for alpha(2–8)-linked poly-N-acetyl neuraminic acid, the capsular polysaccharide of group B meningococci and Escherichia coli K1, which crossreacts with polynucleotides and with denatured DNA. *J. Exp. Med.* 164:642–54.

67. Kabat, E.A., J. Liao, H. Bretting, E.C. Franklin, D. Geltner, B. Frangione, M.E. Koshland, J. Shyong, and E.F. Osserman. 1980. Human monoclonal macroglobulins with specificity for Klebsiella K polysaccharides that contain 3,4-pyruvylated-D-galactose and 4,6-pyruvylated-D-galactose *J. Exp. Med.* 152:979–95.

68. Shikhman, A.R., N.S. Greenspan, and M.W. Cunningham. 1993. A subset of mouse monoclonal antibodies cross-reactive with cytoskeletal proteins and group A streptococcal M proteins recognizes N-acetyl-beta-D-glucosamine. *J. Immunol.* 151:3902–13.

69. Shikhman, A.R., and M.W. Cunningham. 1994. Immunological mimicry between N-acetyl-beta-D-glucosamine and cytokeratin peptides. Evidence for a microbially driven anti-keratin antibody response. *J. Immunol.* 152:4375–4387.

70. Shikhman, A.R., N.S. Greenspan, and M.W. Cunningham. 1994. Cytokeratin peptide SFGSGFGGGY mimics N-acetyl-beta-D-glucosamine in reaction with antibodies and lectins, and induces in vivo anti-carbohydrate antibody response. *J. Immunol.* 153:5593–606.

71. Putterman, C., and B. Diamond. 1998. Immunization with a peptide surrogate for double stranded DNA (dsDNA) induces autoantibody production and renal immunoglobulin deposition. *J. Exp. Med.* 188:29–38.

72. Wedemayer, G.J., P.A. Patten, L.H. Wang, P.G. Schultz, and R.C. Stevens. 1997. Structural insights into the evolution of an antibody combining site. *Science* 276:1665–1669.

73. Garcia, K.C., M. Degano, L. Pease, M. Huang, P.A. Peterson, L. Teyton, and I.A. Wilson. 1998. Structural basis of plasticity in T cell receptor recognition of a self peptide-MHC antigen. *Science* 279:1166–1172.

74. Zabriskie, J.B. 1985. Rheumatic fever: the interplay between host, genetics and microbe. *Circulation* 71:1077–1086.

75. Husby, G., I. van de Rijn, Z.J. B, Z.H. Abdin, and R.C.J. Williams. 1976. Antibodies reacting with cytoplasm of subthalamic and caudate nuclei neurons in chorea and acute rheumatic fever. *J. Exp. Med.* 144:1094–1110.

76. Swerlick, R.A., M.W. Cunningham, and N.K. Hall. 1986. Monoclonal antibodies cross-reactive with group A streptococci and normal and psoriatic human skin. *J. Invest. Dermatol.* 87:367–71.

77. Zabriskie, J.B., K.C. Hsu, and B.C. Seegal. 1970. Heart-reactive antibody associated with rheumatic fever: characterization and diagnostic significance. *Clin. Exp. Immunol.* 7:147–159.

78. Zabriskie, J.B., and E.H. Freimer. 1966. An immunological relationship between the group A streptococcus and mammalian muscle. *J. Exp. Med.* 124:661–678.

79. Kaplan, M.H., R. Bolande, L. Rakita, and J. Blair. 1964. Presence of bound immunoglobulins and complement in the myocardium in acute rheumatic fever. Association with cardiac failure. *New Engl. J. Med.* 271:637–645.

80. Stollerman, G.H. 1997. Rheumatic Fever. *Lancet* 349:935–942.

81. Bisno, A.L. 1995. Non-Suppurative Poststreptococcal Sequelae: Rheumatic Fever and Glomerulonephritis. In: *Principles and practice of infectious diseases*, vol. 2. G.L. Mandell, J. E. Bennett, and R. Dolin, editors. Churchill Livingstone, New York. pp. 1799–1810.

82. Gulizia, J.M., M.W. Cunningham, and B.M. McManus. 1991. Immunoreactivity of anti-streptococcal monoclonal antibodies to human heart valves. Evidence for multiple cross-reactive epitopes. *Amer. J. Pathol.* 138:285–301.

83. Dale, J.B., and E.H. Beachey. 1985. Epitopes of streptococcal M proteins shared with cardiac myosin. *J. Exp. Med.* 162:583–91.

84. Cunningham, M.W., S.M. Antone, M. Smart, R. Liu, and S. Kosanke. 1997. Molecular analysis of human cardiac myosin-cross-reactive B- and T-cell epitopes of the group A streptococcal M5 protein. *Infection and Immunity* 65:3913–3923.

85. Dell, A., S.M. Antone, C.J. Gauntt, C.A. Crossley, W.A. Clark, and M.W. Cunningham. 1991. Autoimmune determinants of rheumatic carditis: localization of epitopes in human cardiac myosin. *European Heart J.* 12(Suppl D):158–62.

86. Manjula, B.N., B.L. Trus, and V.A. Fischetti. 1985. Presence of two distinct regions in the coiled-coil structure of the streptococcal Pep M5 protein: relationship to mammalian coiled-coil proteins and implications to its biological properties. *Proc. Natl. Acad. Sci. USA* 82:1064–68.

87. Manjula, B.N., and V.A. Fischetti. 1986. Sequence homology of group A streptococcal Pep M5 protein with other coiled-coil proteins. *Biochem. Biophys. Res. Commun.* 140:684–90.

88. Jones, T.D. 1944. The diagnosis of rheumatic fever. *J. Am. Med. Assoc.* 126:481–484.

89. Dajani, A.S. 1992. Guidelines for the diagnosis of rheumatic fever (Jones criteria, 1992 update). *J. Am. Med. Assoc.* 268:2069–2073.

90. Cunningham, M.W. 1996. Streptococci and rheumatic fever. In: *Microorganisms and autoimmune disease.* N.R. Rose and H. Friedman, editors. Plenum Publishing Corp., New York. pp. 13–66.

91. McCormack, J.M., C.A. Crossley, E.M. Ayoub, J.B. Harley, and M.W. Cunningham. 1993. Poststreptococcal anti-myosin antibody idiotype associated with systemic lupus erythematosus and Sjogren's syndrome. *J. Infect. Dis.* 168:915–21.

92. Kefalides, N.A., N.T. Pegg, N. Ohno, T. Poon-King, J.B. Zabriskie, and H. Fillit. 1986. Antibodies to basement membrane collagen and to laminin are present in sera from patients with poststreptococcal glomerulonephritis. *J. Exp. Med.* 163:588.

93. Kefalides, N.A., N. Ohno, C.B. Wilson, H. Fillit, J. Zabriskie, and J. Rosenbloom. 1993. Identification of antigenic epitopes in type IV collagen by use of synthetic peptides. *Kidney Internat.* 43:94–.

94. Michael, A.F., Jr., K.N. Drummond, R.A. Good, and R.L. Vernier. 1966. Acute poststreptococcal glomerulonephritis: immune deposit disease. *J. Clin. Invest.* 45:237–48.

95. Markowitz, A.S., and C.F. Lange. 1964. Streptococcal related glomerulonephritis. I. Isolation, immunocytochemistry and comparative chemistry of soluble fractions from type 12 nephritogenic streptococci and human glomeruli. *J. Immunol.* 92:565–575.

96. Lange, C.F. 1969. Chemistry of cross-reactive fragments of streptococcal cell membrane and human glomerular basement membrane. *Trans. Proc.* 1:959–963.

97. Goroncy-Bermes, P., J.B. Dale, E.H. Beachey, and W. Opferkuch. 1987. Monoclonal antibody to human renal glomeruli cross-reacts with streptococcal M protein. *Infect. Immun.* 55:2416–9.

98. Kraus, W., and E.H. Beachey. 1988. Renal autoimmune epitope of group A streptococci specified by M protein tetrapeptide: Ile-Arg-Leu-Arg. *Proc. Nat. Acad. Sci. USA* 85:4516–4520.

99. Lindberg, L.H., and K.L. Vosti. 1969. Elution of glomerular bound antibodies in experimental streptococcal glomerulonephritis. *Science* 166:1032–1033.

100. Cunningham, M.W., N.K. Hall, K.K. Krisher, and A.M. Spanier. 1986. A study of anti-group A streptococcal monoclonal antibodies cross-reactive with myosin. *J. Immunol.* 136:293–298.

101. Cunningham, M.W., and R.A. Swerlick. 1986. Polyspecificity of antistreptococcal murine monoclonal antibodies and their implications in autoimmunity. *J. Exp. Med.* 164:998–1012.

102. Cunningham, M.W., J.M. McCormack, L.R. Talaber, J.B. Harley, E.M. Ayoub, R.S. Muneer, L.T. Chun, and D.V. Reddy. 1988. Human monoclonal antibodies reactive with antigens of the group A Streptococcus and human heart. *J. Immunol.* 141:2760–2766.

103. Kaplan, M.H., and K.H. Svec. 1964. Immunologic relation of streptococcal and tissue antigens. III. Presence in human sera of streptococcal antibody cross reactive with heart tissue. Association with streptococcal infection, rheumatic fever, and glomerulonephritis. *J. Exp. Med.* 119:651–666.

104. Kaplan, M.H., and M.L. Suchy. 1964. Immunologic relation of streptococcal and tissue antigens. II. Cross reactions of antisera to mammalian heart tissue with a cell wall constituent of certain strains of group A streptococci. *J. Exp. Med.* 119:643–650.

105. Kaplan, M.H. 1963. Immunologic relation of streptococcal and tissue antigens. I. Properties of an antigen in certain strains of group A streptococci exhibiting an immunologic cross reaction with human heart tissue. *J. Immunol.* 90:595–606.

106. Lange, C.F. 1994. Localization of [C^{14}] labeled anti-streptococcal cell membrane monoclonal antibodies (anti-SCM mAb) in mice. *Autoimmunity* 19:179–191.

107. Wu, X., B. Liu, P.L. Van der Merwe, N.N. Kalis, S.M. Berney, and D.C. Young. 1998. Myosin-reactive autoantibodies in rheumatic carditis and normal fetus. *Clin. Immunol. Immunopathol.* 87:184–192.

108. Shikhman, A.R., and M.W. Cunningham. 1997. Trick and treat: toward peptide mimic vaccines. *Nature Biotechnol.* 15:512–513.

109. Carroll, P., D. Stafford, R.S. Schwartz, and B.D. Stollar. 1985. Murine monoclonal anti-DNA antibodies bind to endogenous bacteria. *J. Immunol.* 135:1086.

110. Lafer, E.M., J. Rauch, C. Andrezejewski, Jr., D. Mudd, B. Furie, R.S. Schwartz, and B.D. Stollar. 1981. Polyspecific monoclonal lupus autoantibodies reactive with both polynucleotides and phospholipids. *J. Exp. Med.* 153:897–909.

111. Adrezejewski, C., Jr., J. Rauch, B.D. Stollar, and R.S. Schwartz. 1980. Antigen binding diversity and idiotypic cross-reactions among hybridoma autoantibodies to DNA. *J. Immunol.* 126:226–231.

112. Mertens, N.M.J., J.E. Galvin, and M.W. Cunningham. 1999. Anti-streptococcal/anti-myosin mouse monoclonal antibodies: nucleotide sequences for crossreactivity. *Submitted.*

113. Galvin, J.E., M.E. Hemric, K. Ward, and M.W. Cunningham. 1999. Cytotoxic monoclonal antibody from rheumatic carditis reacts with human endothelium: implications in rheumatic heart disease. *Submitted.*

114. Antone, S.M., and M.W. Cunningham. 1992. Cytotoxicity linked to a streptococcal monoclonal antibody which recognizes laminin. In: *New perspectives on streptococci and streptococcal infections. Proceedings of the XI Lancefield International Symposium*, vol. 22. Gustav, Fisher and Verlag, editors. Zbl Bakt Suppl, New York. pp. 189–191.

115. Quinn, A., T.M. Shinnick, and M.W. Cunningham. 1996. Anti-Hsp 65 antibodies recognize M proteins of group A streptococci. *Infection and Immunity* 64:818–824.

116. Huber, S., J. Polgar, A. Moraska, M. Cunningham, P. Schwimmbeck, and P. Schultheiss. 1993. T lymphocyte responses in CVB3-induced murine myocarditis. *Scand. J. Infect. Dis.* 88:67–78.

117. Huber, S.A., A. Moraska, and M. Cunningham. 1994. Alterations in major histocompatibility complex association of myocarditis induced by coxsackievirus B3 mutants selected with monoclonal antibodies to group A streptococci. *Proc. Nat. Acad. Sci. USA* 91:5543–7.

118. Huber, S.A., L.P. Job, and J.F. Woodruff. 1980. Lysis of infected myofibers by coxsackievirus B3 immune lymphocytes. *Am. J. Pathol.* 98:681–694.

119. Huber, S.A., and P.A. Lodge. 1986. Coxsackievirus B-3 myocarditis: identification of different pathogenic mechanisms in DBA/2 and BALB/c mice. *Am. J. Pathol.* 122:284–291.

120. Gauntt, C., S. Tracy, N. Chapman, H. Wood, P. Kolbeck, A. Karaganis, C. Winfrey, and M. Cunningham. 1995. Coxsackievirus-induced chronic myocarditis in murine models. *Eur. Heart J.* 16.56–58.

121. Cohen, I.R. 1991. Autoimmunity to chaperonins in the pathogenesis of arthritis and diabetes. *Ann. Rev. Immunol.* 9:567–589.

122. Cohen, I.R., and D.B. Young. 1991. Autoimmunity, microbial immunity, and the immunological homunculus. *Immunology Today* 12:105–110.

123. Dejoy, S.Q., K.M. Ferguson, T.M. Sapp, J.B. Zabriskie, A.L. Oronsky, and S.S. Kerwar. 1989. Streptococcal cell wall arthritis. Passive transfer of disease with a T cell line with crossreactivity of streptococcal cell wall antigens with *Mycobacterium tuberculosis*. J. Exp. Med. 170:369–382.

124. Schett, G., Q. Xu, A. Amberger, R. Van der Zee, H. Recheis, J. Willeit, and G. Wick. 1995. Autoantibodies against heat shock protein 60 mediate endothelial cytotoxicity. *J. Clin. Invest.* 96:2569–2577.

125. Sheldon, P. 1985. Specific cell-mediated response to bacterial antigens and clinical correlations in reactive arthritis, Reiters syndrome, and ankylosing spondylitis. *Imm. Rev.* 86:5–25.

126. Brewerton, D.A., F.D. Hart, A. Nicholls, M. Caffrey, D.C. James, and R.D. Sturrock. 1973. Ankylosing spondylitis and HL-A 27. *Lancet* 1:7809:904–907.

127. Schlosstein, L., P.I. Terasaki, R. Bluestone, and C.M. Pearson. 1973. High association of an HL-A antigen, W27, with ankylosing spondylitis. *N. Engl. J. Med.* 288:704–706.

128. Keat, A. 1986. Is ankylosing spondylitis caused by Klebsiella? *Immunology Today* 7:144–149.

129. Geczy, A.F., K. Alexander, H.V. Bashir, and J. Edmonds. 1980. A factor(s) in Klebsiella culture filtrates specifically modifies an HLA-B27 associated cell surface component. *Nature* 283:782–784.

130. Ebringer, R., D.R. Cawdell, P. Cowling, and A. Ebringer. 1978. Klebsiella pneumoniae. *Ann. Rheum. Dis.* 6:577.

131. Ewing, C., R. Ebringer, G. Tribbick, and H.M. Geysen. 1990. Antibody activity in ankylosing spondylitis sera to two sites on HLA B27.1 at the MHC groove region (within sequence 65–85), and to a Klebsiella pneumoniae nitrogenase reductase peptide (within sequence 181–199). *J. Exp. Med.* 171:1635–1647.

132. Husby, G., N. Tsuchiya, P.L. Schwimmbeck, A. Keat, J.A. Pahle, M.B. Oldstone, and R.C. Williams, Jr. 1989. Cross-reactive epitope with Klebsiella pneumoniae nitrogenase in articular tissue of HLA-B27+ patients with ankylosing spondylitis. *Arth. & Rheum.* 32:437–445.

133. Lahesmaa, R., M. Skurnik, M. Vaara, M. Leirisalo-Repo, M. Nissila, K. Granfors, and P. Toivanen. 1991. Molecular mimickry between HLA B27 and Yersinia, Salmonella, Shigella and Klebsiella within the same region of HLA alpha 1-helix. *Clin. Exp. Immunol.* 86:399–404.

134. Taurog, J.D., J.P. Durand, F.A. el-Zaatari, and R.E. Hammer. 1988. Studies of HLA-B27-associated disease. *Am. J. Med.* 85:59–60.

135. Hammer, R.E., S.D. Maika, J.A. Richardson, J.P. Tang, and J.D. Taurog. 1990. Spontaneous inflammatory disease in transgenic rats expressing HLA-B27 and human beta 2m: an animal model of HLA-B27-associated human disorders. *Cell* 63:1099–1112.

136. Taurog, J.D., S.D. Maika, W.A. Simmons, M. Breban, and R.E. Hammer. 1993. Susceptibility to inflammatory disease in HLA-B27 transgenic rat lines correlates with the level of B27 expression. *J. Immunol.* 150:4168–78.

137. Taurog, J.D., J.A. Richardson, J.T. Croft, W.A. Simmons, M. Zhou, J.L. Fernandez-Sueiro, E. Balish, and R.E. Hammer. 1994. The germfree state prevents development of gut and joint inflammatory disease in HLA-B27 transgenic rats. *J. Exp. Med.* 180:2359–2364.

138. Shoenfeld, Y., Y. Vilner, A.R. Coates, J. Rauch, G. Lavie, D. Shaul, and J. Pinkhas. 1986. Monoclonal anti-tuberculosis antibodies react with DNA, and monoclonal anti-DNA autoantibodies react with *Mycobacterium tuberculosis*. *Clin. Exp. Immunol.* 66:255–61.

139. Avinoach, I., H. Amital-Teplizki, O. Kuperman, D.A. Isenberg, and Y. Shoenfeld. 1990. Characteristics of anti-neuronal antibodies in systemic lupus erythematosus patients with and without central nervous system involvement: the role of mycobacterial cross-reacting antigens. *Isr. J. Med. Sci.* 26:367–73.

140. Isenberg, D.A., Y. Shoenfeld, M.P. Madaio, J. Rauch, M. Reichlin, B.D. Stollar, and R.S. Schwartz. 1984. Anti-DNA antibody idiotypes in systemic lupus erythematosus. *Lancet* 8400:417–422.

140a. Baxter, A.G., A.C. Horsfall, D. Healey, P. Ozegle, S. Day, D.G. Williams, and A. Cooke. 1994. Mycobacteria precipitate an SLE-like syndrome in diabetes-prone NOD mice. *Immunology* 83:227–231.

141. Pearson, C.M., and e. al. 1961. Studies of arthritis and other lesions induced in rats by injection of mycobacterium adjuvant V. changes affecting the skin and mucous membranes: comparison of the experimental process with human disease. *J. Exp. Med.* 113:485–510.

142. Pearson, C.M., and et. al. 1964. Experimental models in rheumatoid disease. *Arth. & Rheum.* 7:80–86.

143. Holoshitz, J., A. Klajman, I. Druker, Z. Lapidot, A. Yaretzky, A. Frenkel, W. van Eden, and I.R. Cohen. 1986. T lymphocytes of theumatoid arthritis patients show augmented reactivity to a fraction of mycobacteria crossreactive with cartilage. *Lancet* ii:305–309.

144. Torisu, M. 1983. A new side effect of BCG immunotherapy-BCG induced arthritis in man. *Cancer Immunol. Immunother.* 5:77–83.

145. Abrahamsen, T.G., S.S. Froland, and J.B. Natvig. 1978. In vitro mitogen stimulation of synovial fluid lymphocytes from rheumatoid arthritis and juvenile rheumatoid arthritis

patients: dissociation between the response to antigens and polyclonal mitogens. *Scand. J. Immunol.* 7:81–90.

146. Thole, J.E., H.G. Dauwerse, P.K. Das, D.G. Groothuis, L.M. Schouls, and J.D. van Embden. 1985. Cloning of Mycobacterium bovis BCG DNA and expression of antigens in Escherichia coli. *Infect. Immun.* 50:800–6.

147. van Eden, W., et. al. 1988. Cloning of the mycobacterial epitope recognized by T lymphocytes in adjuvant arthritis. *Nature* 331:171–173.

148. Born, W., et. al. 1990. Recognition of peptide antigen by heat shock-reactive $\gamma\delta$ T lymphocytes. *Science* 249:67–69.

149. Young, R.A., V. Mehra, D. Sweetser, T. Buchanan, J. Clark-Curtiss, R.W. Davis, and B.R. Bloom. 1985. Genes for the major protein antigens of the leprosy parasite Mycobacterium leprae. *Nature* 316:450–452.

150. Kao, I., and D.B. Drachman. 1977. Mysasthenic immunoglobulin accelerates acetylcholine receptor degradation. *Science* 196:527–529.

151. Lindstrom, J.M. 1979. Autoimmune response to acetyl choline receptor in myasthenia gravis and its animal model. *Adv. Immunol.* 27:1–50.

152. Lisak, R.P., and R.L. Barchi. 1982. Myasthenia gravis. W.B. Saunders, Philadelphia.

153. Tzartos, S.J., M.E. Seybold, and J.M. Lindstrom. 1982. Specificities of antibodies to acetylcholine receptors in sera from myasthenia gravis patients measured by monoclonal antibodies. *Proc. Natl. Acad. Sci. USA* 79:188–192.

154. Stefansson, K., M.E. Dieperink, D.P. Richman, C.M. Gomez, and L.S. Marton. 1985. Sharing of antigenic determinants between the nicotinic acetylcholine receptor and proteins in Escherichia coli, Proteus vulgaris, and Klebsiella pneumoniae. Possible role in the pathogenesis of myasthenia gravis. *N. Engl. J. Med.* 312:221–225.

155. O'Riordan, J.I., D.H. Miller, J.P. Mottershead, N.P. Hirsch, and R.S. Howard. 1998. The management and outcome of patients with myasthenia gravis treated acutely in a neurological intensive care unit. *Eur. J. Neurol.* 5:137–142.

156. Dwyer, D.S. 1986. Idiotypic network connectivity and apossible cause of myasthenia gravis. *J. Exp. Med.* 164:1310–1318.

157. Guss, B, M. Lindberg and M. Uhlen. 1990. The gene for staphylococcal protein A. In: *Bacterial Immunoglobulin-Binding Proteins*. vol. 1. Boyle, M.D.P., ed. Academic Press, San Diego, CA, pp. 323–327.

158. Raeder, R., and M.D.P. Boyle. 1996. Properties of IgG-binding proteins expressed by Streptococcus pyogenes isolates are predictive of invasive potential. *J. Infect. Dis.* 173:888–895.

159. Myhre, E.B., and G. Kronvall. 1980. Immunochemical aspects of Fc-mediated binding of human IgG subclasses to group A, C and G streptococci. *Mol. Immunol.* 17:1563–73.

160. Nardella, F.A., I.R. Oppliger, G.C. Stone, E.H. Sasso, M. Mannik, J. Sjoquist, A.K. Schroder, P. Christensen, P.J. Johansson, and L. Bjorck. 1988. Fc epitopes for human rheumatoid factors and the relationships of rheumatoid factors to the Fc binding proteins of microorganisms. *Scand. J. Rheumatol. Suppl.* 75:190–198.

161. Nardella, F.A., D.C. Teller, C.V. Barber, and M. Mannik. 1985. IgG rheumatoid factors and staphylococcal protein A bind to a common molecular site on IgG. *J. Exp. Med.* 162:1811–1824.

162. Williams, R.C., Jr. 1988. Hypothesis: rheumatoid factors are antiidiotypes related to bacterial or viral Fc receptors. *Arth. Rheum.* 31:1204–7.

163. Oppliger, I.R., F.A. Nardella, G.C. Stone, and M. Mannik. 1987. Human rheumatoid factors bear the internal image of the Fc binding region of staphylococcal protein A. *J. Exp. Med.* 166:702–10.

164. Levinson, A.I., N.F. Dalal, M. Haidar, L. Tar, and M. Orlow. 1987. Prominent IgM rheumatoid factor production by human cord blood lymphocytes stimulated in vitro with Staphylococcus aureus Cowan I. *J. Immunol.* 139:2237–2241.

165. Levinson, A.I., L. Tar, C. Carafa, and M. Haidar. 1986. Staphylococcus aureus Cowan I. Potent stimulus of immunoglobulin M rheumatoid factor production. *J. Clin. Invest.* 78:612–617.

166. Kozlowski, L.M., S.R. Kunning, Y. Zheng, L.M. Wheatley, and A.I. Levinson. 1995. Staphylococcus aureus Cowan I-induced human immunoglobulin responses: preferential IgM rheumatoid factor production and VH3 mRNA expression by protein A-binding B cells. *J. Clin. Immunol.* 15:145–151.

167. Tan, E.M. 1982. Autoantibodies to nuclear antigens (ANA): their immunobiology in medicine. *Adv. Immunol.* 33:167–240.

168. Stollar, B.D. 1986. Antibodies to DNA. In: *CRC critical reviews in biochemistry*, G.D. Fasman, ed., CRC Press, Inc., New York, NY, pp. 20–21.

169. Tillman, D.M., N.T. Jou, R.J. Hill, and T.N. Marion. 1992. Both IgM and IgG anti-DNA antibodies are the products of clonally selective B cell stimulation in (NZB × NZW)F$_1$ mice. *J. Exp. Med.* 176:761–779.

170. Marion, T.N., M.R. Krishnan, D.D. Desai, N.T. Jou, and D.M. Tillman. 1997. Monoclonal anti-DNA antibodies: structure, specificity, and biology. *Methods* 11:3–11.

171. Datta, S.K., and J. Gavalchin. 1986. Origins of pathogenic anti DNA idiotypes in the NZB X SWR model of lupus nephritis. *Ann. NY Acad. Sci.* 475:47–58.

172. Shoenfeld, Y. 1983. Polyspecificity of monoclonal lupus autoantibodies produced by human–human hybridomas. *N. Eng. J. Med.* 308:414–420.

173. Katz, J.B., W. Limpanasithikul, and B. Diamond. 1994. Mutational analysis of an autoantibody: differential binding and pathogenicity. *J. Exp. Med.* 180:925–932.

174. Grayzel, A., A. Solomon, C. Aranow, and B. Diamond. 1991. Antibodies elicited by pneumococcal antigens bear an anti-DNA-associated idiotype. *J. Clin. Invest.* 87:842–846.

175. Adderson, E.E., P.G. Shackelford, A. Quinn, P.M. Wilson, M.W. Cunningham, R.A. Insel, and W.L. Carroll. 1993. Restricted immunoglobulin VH usage and VDJ combinations in the human response to Haemophilus influenzae type b capsular polysaccharide–nucleotide sequences of monospecific anti-Haemophilus antibodies and polyspecific antibodies cross-reacting with self-antigens. *J. Clin. Invest.* 91:2734–2743.

176. Limpanasithikul, W., S. Ray, and B. Diamond. 1995. Cross-reactive antibodies have both protective and pathogenic potential. *J. Immunol.* 155:967–973.

177. Diamond, B., and M. Scharff. 1984. Somatic mutation of the T15 heavy chain gives rise to an antibody with autoantibody specificity. *Proc. Natl. Acad. Sci. USA* 81:5841–5844.

178. Ray, S.K., C. Putterman, and B. Diamond. 1996. Pathogenic autoantibodies are routinely generated during the response to foreign antigen: a paradign for autoimmune disease. *Proc. Natl. Acad. Sci. USA* 93:2019–2024.

179. Campbell, L.A., C.C. Kuo, and J.T. Grayston. 1998. Chlamydia pneumoniae and cardiovascular disease. *Emerg. Infect. Dis.* 4:571–9.

180. Bachmaier, K., N. Neu, L. de la Maza, S. Pal, A. Hessel, and J.M. Penninger. 1999. Chlamydia infections and heart disease linked through antigenic mimicry. *Science* 283:1335–1339.

181. Grayston, J.T. 1999. Antibiotic treatment trials for secondary prevention of coronary artery disease events [editorial; comment]. *Circulation* 99:1538–1539.

182. Grayston, J.T., C.H. Mordhorst, and S.P. Wang. 1981. Childhood myocarditis associated with Chlamydia trachomatis infection. *J. Am. Med. Assoc.* 246:2823–2827.

183. Ringel, R.E., L.B. Givner, J.I. Brenner, S.W. Huang, S.P. Wang, J.T. Grayston, and M.A. Berman. 1983. Myocarditis as a complication of infantile Chlamydia trachomatis pneumonitis. *Clin. Pediatr. (Phila.)* 22:631–633.

184. Ringel, R.E., J.I. Brenner, M.B. Rennels, S.W. Huang, S.P. Wang, J.T. Grayston, and M.A. Berman. 1982. Serologic evidence for Chlamydia trachomatis myocarditis. *Pediatrics* 70:54–6.

185. Gross, D.M., T. Forsthuber, M. Tary-Lehmann, C. Etling, K. Ito, Z.A. Nagy, J.A. Field, A.C. Steere, and B.T. Huber. 1998. Identification of LFA-1 as a candidate autoantigen in treatment-resistant Lyme disease. *Science* 281:703–706.

186. Cunningham, M.W., and A. Quinn. 1997. Immunological crossreactivity between the class I epitope of streptococcal M protein and myosin. Streptococci and the host. In: T. Horaud, A. Bouvet, R. Leclercq, H. de Montclos, and M. Sicard, eds. Plenum Press, London. pp. 887–892.

187. Tomer, Y., and Y. Shoenfeld. 1988. The significance of natural autoantibodies. *Immunol. Invest.* 17:389–424.

188. Cunningham, M.W. 1998. Molecular mimicry in autoimmunity and infection. In: *Microbial pathogenesis: Current and emerging Issues. Proceedings of the second annual indiana conference, Indiana University School of Dentistry Press, Indianapolis, IN*, pp. 83–99.

189. Hart, I.K., and P.G.E. Kennedy. 1988. Guillain-Barre syndrome associated with cytomegalovirus infection. *Quart. J. Med.* 253:425–430.

190. Baboonian, C., P.J.W. Benables, and R.N. Maini. 1990. Antibodies to human herpesvirus-6 in Sjogren's syndrome. *Arth. & Rheum.* 33:1749–1750.

191. Inman, R.D., M. Hodge, and M.E.A. Johnson. 1986. Arthritis, vasculitis and cryoglobulinemia associated with relapsing hepatitis A virus infection. *Ann. Intern. Med.* 105:700–703.

192. Czaja, A.J., H.A. Carpenter, and P.J. Santrach. 1993. Evidence against hepatitis viruses as important causes of severe autoimmune hepatitis in the United States. *J. Hepatol.* 18:342–252.

193. Munke, H.F., F. Stockmann, and G. Ramdori. 1995. Possible association between Behchet's syndrome and chronic hepatitis C virus infection. *N. Eng. J. Med.* 332:400–401.

194. Sagawa, K., M. Moochizuki, K. Masuoka, T. Katagiri-Katayama, and T. Maida. 1995. Immunopathological mechanisms of HTLV-1 uveitis. *J. Clin. Invest.* 95:852–858.

195. Baranski, B., G. Armstrong, and J.T. Truman. 1988. Epstein-Barr virus in the bone marrow of patients with aplastic anemia. *Ann. Intern. Med.* 109:695–704.

196. Whittingham, S.M., J. McNenage, and I.R. Mackay. 1985. Primary Sjogren's syndrome after infectious mononucleosis. *Ann. Intern. Med.* 102:490–493.

197. Fox, R.I., M. Luppi, H.I. Kang, and P. Pisa. 1991. Reactivation of Epstein-Barr virus in Sjogren's syndrome. *Springer Sem. Immunopathol.* 13:217–231.

198. Deacon, I.M., W.G. Shattles, and J.B. Mathews. 1992. Frequency of EBV DNA detection in Sjogren's syndrome. *Am. J. Med.* 92:453–453.

199. Fox, R.I., M. Luppi, and P. Pisa. 1992. Potential role of EBV in Sjogren's syndrome and rheumatoid arthritis. *J. Rheumatol.* 19:18–24.

200. Wilder, R.L. 1999. Hypothesis for retroviral causation of rheumatoid arthritis. *Curr. Opin. Rheumatol.* 6:295–299.

201. James, J.A., K.M. Kaufman, A.D. Farris, E. Taylor-Albert, T.J.A. Lehman, and J.B. Harley. 1997. An increased prevalence of Epstein-Barr virus infection in young patients suggests a possible etiology for systemic lupus erythematosus. *J. Clin. Invest.* 100:3019–3026.

202. Neumann, D.A., N.R. Rose, A.A. Ansari, and Herkowitz. 1993. Induction of multiple heart autoantibodies in mice with coxsackievirus B3- and cardiac myosin-induced autoimmune myocarditis. *J. Immunol.* 152:343–350.

203. Corman, L.C., and D.J. Dolson. 1992. Polyarteritis nodosa and parvovirus B 19 infection. *Lancet* 339:491.

204. Sumazaki, R., T. Fujita, T. Kabashima, F. Nishikaku, A. Koyama, M. Shibasaki, and H. Takita. 1986. Monoclonal antibody against bacterial lipopolysaccharide cross-reacts with DNA-histone. *Clin. Exp. Immunol.* 66:103–10.

15 | The Idiotypic Network in Autoimmunity

Irun R. Cohen

1. INTRODUCTION

Idiotypy refers to the immune recognition of one lympho-cyte clone (an idiotype; id) by other clones (anti-ids). Idiotypy is a field within immunology with its own history, nomenclature and foundation of knowledge. The chapter written by Maurizio Zanetti in the first edition of *The Molecular Pathology of Autoimmune Diseases* reviewed the classical aspects of idiotypy and how idiotypy is associ-ated with particular autoimmune diseases (1).

A network theory of idiotypy was advanced by Niels Jerne (2). It was well known that anti-ids could be pro-duced by intentional immunization to ids administered in adjuvant (3). There is no surprise here; an id, like most molecules, can act as an antigen. Jerne, however, went a step further and sponsored the idea that every immunization to an antigen involved the automatic activation of anti-ids through a chain reaction; antigens induce ids, and the ids, in their turn, induce anti-ids. This automatic anti-id response, by inhibiting or enhancing the id, was thought to modulate the individual's immune response to the antigen. Jerne sug-gested that network connections between ids and anti-ids were fundamental to immune regulation. However, with few exceptions, researchers have not been able to detect the development of anti-ids during the response to most immu-nizing antigens. Enthusiasm for Jerne's concept of auto-matic, regulatory anti-id networks waned.

Nevertheless, immunization to certain antigens does seem to induce both ids and anti-ids. These exceptional antigens are revealing; they are, for the most part, a set of self-antigens. The observations of natural self-idiotypy suggest that the immune system might have intrinsic prefer-ences for making anti-ids to particular self-antigens. Here, I will briefly review some of the properties of self-idiotypy

and discuss the idea that idiotypy has a natural role in the regulation, not of immune responses in general, but of the autoimmune responses to a selected set of self-antigens. But first let us consider some definitions to help us navigate through the murky waters of idiotypy.

2. DEFINITIONS

2.1. Classical Definition

Ids were defined historically as the unique epitopes of anti-body molecules that could be recognized by other antibod-ies (1). *Idios* in Greek means *one's own*, or *unique*, and *typos* means *blow* or *knock*. The word *type*, by extension, has its origin in the *mark* or *figure* imprinted by a blow. An *idiotype*, therefore, is the unique marker of a particular anti-body, defined (or *struck*) by another antibody. Structurally, ids are associated with the antigen-combining site of an antibody, since that site is the unique structural element of every antibody-producing clone. The particular parts of the antigen receptor bound by an anti-id have a distinct nomen-clature, which can be seen in Zanetti's chapter (1).

2.2. Antigen Mimicry

An important feature of some anti-id antibodies is that they mimic the conformation of the immunizing antigen (4); quite simply, if the antigen combining site of an id antibody can bind specifically both to an antigen and to an anti-id, then the antigen and the anti-id might have some structural similarity, at least in the epitope bound by the id (1). Such anti-ids may act like immune-system images of the antigen that initially activated the id.

2.3.   Current Definition

Today, we extend the term id to include the antigen receptors of B cells and T cells, in addition to those of antibodies. We can also define an id by its interaction with an anti-id T cell and not only by its binding to an anti-id antibody. Note that that these two ways of defining an id rely on different characteristics of the same id molecule. The id structure recognized by antibodies or B cells would include conformational epitopes on the id molecule. The id structure recognized by anti-id T cells, in contrast, is formed by peptides processed from the id molecule and presented on MHC molecules. Different peptides from a single id molecule might be associated with different MHC I or MHC II presenting molecules, and thus the different id peptides might be seen by different anti-id T cells. A T-cell id, for example, might be seen by anti-id B cell (or by anti-id antibody) as a conformation of the T cell's antigen receptor, and the same T-cell receptor id molecule might be detected by anti-id T cells that see id processed peptides. In view of the different ways the immune system can detect and respond to its ids, there is room for much diversity in any anti-id response. There is also room for much confusion among observers.

2.4.   Functional Diversity

Diversity in anti-id reactions is also encoded in the diversity of the biologic effects that can be produced by anti-ids of various kinds. Anti-id antibodies could be of the IgM or of the various IgG isotypes, and so mediate markedly different immune effects (complement fixation, opsonization, blocking, complex formation, etc.). Anti-id T cells, too, could respond to id structures, not only by proliferating, but by secreting different cytokines or by mediating cytotoxicity. For example, anti-id T cells could belong to the Th1 or the Th2 sets of CD4 helper T cells, or to the various sets of CD8 cytotoxic or suppressor T cells. Markedly different outcomes could be produced by these different anti-id reagents. In other words, not only do immunologists detect ids in diverse ways, but the immune system itself can see and respond biologically to ids in diverse ways.

2.5.   DNA Detection

Immunologists now have a new way to detect ids that the immune system does not have; ids may also be detected by their DNA sequences (5). Using the techniques and probes of molecular biology, it is possible to study clones of T cells or B cells that share hyper-variable segments in the sequences of their antigen receptors. Molecular genetics makes it possible to infer family relationships between clones and study the appearance, numbers and distributions of idiotypes. Obviously, one cannot assume that the immune system itself sees and responds to ids that are defined by immunologists using DNA probes. The aptness of the DNA approach to ids obviously depends on the nature of the experimental question.

2.6.   Natural Idiotypy

An additional point to keep in mind is the distinction between natural and contrived idiotypy. Anti-ids that arise in the course of the activation of an immune response to a specific antigen can be called *natural*. Natural anti-ids can be contrasted with the anti-ids that appear only after an id has been used as an immunogenic stimulus; we can call these *contrived* anti-ids. Our discussion here will focus on natural idiotypy.

2.7.   Cross-reactive Idiotypy

Ids and anti-ids that are *common* or *cross-reactive* among the immune systems of different individuals are worth noting (1). The random processes that form lymphocyte antigen receptors cannot explain the prevalence of identical ids in a population of different individuals. Hence, the expression of common idiotypes implies the operation of some selective mechanism. There has to be a biologic reason for common idiotypy.

2.8.   Network

An immune network is formed by a chain of connections between cells and molecules of the immune system. A network emerges when antibodies and lymphocytes recognize each other in addition to their recognition of specific antigens. The connections between elements in immune networks are not exclusively based on ids and anti-ids, but include cytokines and other cell-interaction molecules. The immune system can be viewed as a network of mobile cells and molecules connected functionally by cell migrations, molecular diffusion and ligand-receptor interactions.

With these definitions in mind, we can proceed to outline some of the properties of natural id networks.

3.   NATURAL NETWORK SELF-ANTIGENS

The preference of the immune system for self-idiotypy may be inferred from two classes of observations: The induction of natural anti-ids by immunization to certain self-antigens and the associations of autoimmune diseases with ids and anti-ids.

3.1.   Anti-ids Arising from Immunization

When we examine the situations where anti-ids have been seen to arise naturally in the course of antigen immunization, we find that the antigens tend to be self-antigens. For example, natural anti-ids have been reported following the immunization of standard experimental animals to insulin (6), to agonists of the acetylcholine receptor (7), to thyrotropin (8), and to ligands of the adrenergic receptor (9). The anti-ids in these immunizations actually mimic the structure of the hormone self-antigen, and so the anti-ids activate the hormone receptor to produce the effect of the hormone used for immunization. In other words, an anti-id that mimics the structure of the active site of a hormone antigen can act like an immune image of the hormone itself. Such an anti-id reveals itself by reproducing the biological effects otherwise triggered through the binding of the hormone to the hormone receptor (4); the anti-id here acts like an antibody to the hormone receptor.

3.2.   Autoimmune Disease Idiotypy

Natural anti-ids to autoantibodies and cross-reactive ids have been detected in many autoimmune diseases, such as systemic lupus erythematosus (SLE), myasthenia gravis, thyroiditis, nephritis, scleroderma and others (1).

Although each of these examples is a rich and varied story in its own right, in general, we may conclude that anti-ids may arise naturally following immunization to certain self-antigens, both in experimental animals and spontaneously in diseased humans.

4.   T AND B CELL CONNECTIVITY

As we discussed above, the anti-ids of T cells and B cells see different molecular worlds: The T cells see processed peptide sequences and the B cells see conformations. Nevertheless, T cell and B-cell ids have been observed to be connected; manipulating an autoimmune T cell response can induce autoantibodies, and vice versa (10,11). These observations suggest that specific T cells and B cells, despite their different types of antigen receptors, may share idiotypic connections. It is also conceivable that idiotypic T cells and B cells might be connected by other ligands in the autoimmune network. Natural idiotypic connectivity needs molecular analysis.

5.   DEVELOPMENTAL CONNECTIVITY

The ontogeny of natural ids and anti-ids is another area in need of more detailed study, but is seems that idiotypy may appear during the formative stages of repertoire selection, even before contact with foreign antigens. For example, the early antibodies in newborns were found to interact with a set of anti-id reagents with a frequency ten-fold greater than that characteristic of the antibodies present in

adults (12). In other words, the developing B-cell repertoire begins life with a high degree of idiotypy. It has been suggested that this early B cell connectivity may relate to the idiotypy of the natural autoantibodies positively selected by contact with self-antigens. Later contact with foreign antigens leads to the selection of idiotypically-unconnected antibodies.

Natural idiotypy may also develop during the early selection of T cells in the thymus. My colleagues and I have found that the spontaneous diabetes of NOD strain mice may be inhibited by anti-id T cells that recognize a common T-cell id (13). The id epitope was identified as a peptide sequence in the third complementarity-determining region (CDR3) of the id T-cell antigen receptor. We found that injection of the id peptide into the thymus of very young NOD mice led to a markedly accelerated development of the diabetes, probably through down-regulation of the natural anti-id T-cell repertoire by "thymic tolerance" (13). This observation is compatible with the ideas that the T-cell repertoire generates regulatory T-cell anti-ids in the thymus during T-cell ontogeny, and that negatively affecting the generation of specific anti-id T cells can aggravate spontaneous autoimmune disease. Indeed, thymectomy at a critical developmental stage can, of itself, activate various autoimmune diseases (14). Thus both the B and T cell repertoires would seem to have self-idiotypy built-in, presumably through positive selection for reactivity to certain self-antigens during ontogeny.

6.   NATURAL NETWORK REGULATION OF AUTOIMMUNE DISEASE

The above discussion suggests that natural self-idiotypy functions to control autoimmune disease. We know from several model systems that autoimmunity tends to generate its own regulation. Mice constructed with a transgenic T cell receptor specific for the self-antigen myelin basic protein, for example, do not develop EAE spontaneously if they also express the *rag* gene (15). The product of *rag* is an enzyme essential to the generation of antigen receptors, and the presence of *rag* makes the transgenic T-cell receptor mice "leaky"; a few percent of the animal's T cells manage to circumvent the transgene and generate a number of T cells with their own T cell receptors. Relevant to our discussion is the finding that the *rag*-positive mice are able to generate regulator T cells that control the millions of autoimmune T cells that, otherwise, would cause EAE spontaneously. T cell receptor transgenic mice that lack *rag*, in contrast, do go on to develop overt EAE (15). The nature of the regulatory T cells in the *rag*-positive mice is not yet known, however, so the role of anti-id T cells in this example is still hypothetical.

Anti-id regulatory T cells have been detected in the course of the induction of EAE in standard mice by immunization to myelin antigens (16). Indeed, the administration to rats of doses of anti-myelin T cells too low to cause clinical EAE suffices to activate specific anti-id T cells (17). In other words, the onset of EAE, even sub-clinical EAE, can activate specific anti-id T cells automatically. Analysis of the anti-id network in EAE in mice has uncovered both CD4 and CD8 regulatory T cells and V gene restrictions in these regulatory T cells (18). It is particularly interesting that the epitope in the anti-myelin T cell receptor seen by the regulatory T cells has been mapped to a framework region adjacent to the CDR3 segment of the id molecule (19). This means that the part of the id receptor seen by the regulatory T cells is not strictly an id sequence; different T cell receptors that use the same $V\beta$ 8.2 region gene to recognize different antigens would express this framework sequence in common. This finding shows that anti-receptor regulator T cells may not necessarily be restricted to a truly idiotypic segment of the T cells that they regulate. If this observation can be generalized, it would mean that a single clone of regulatory T cell might be able to control the immune responses to a variety of different antigens, provided that the T cells responding to these antigens use the same $V\beta$ or other framework gene segment (19).

A function for natural anti-ids in regulating a spontaneous autoimmune disease has been observed in the NOD mouse model of type 1 diabetes. The spontaneous onset of insulitis in these mice is accompanied by the spontaneous activation of anti-id T cells. These anti-ids recognize the pathogenic T cells that target the hsp60 self-antigen (13). However, these anti-id T cells spontaneously decline, and the insulitis progresses to beta-cell destruction and overt diabetes. Fortunately, the anti-id T cells can be rescued by vaccination against the id, and the fortified anti-id response down-regulates the id T cells and prevents the diabetes. The natural role of the anti-id T cells was suggested, as I mentioned above, by tolerizing the mice to the id; mice with a down-regulated anti-id response developed markedly accelerated diabetes (13). Thus natural anti-ids do seem to have a role in regulating the spontaneous diabetes of NOD mice.

7.   THERAPEUTIC NETWORK REGULATION OF AUTOIMMUNE DISEASE

7.1.   T-Cell Vaccination

An early hint that anti-id T cells might control autoimmune disease was the effectiveness of T cell vaccination in preventing the active induction of EAE (20). T cell vaccination was found later to be effective in many other autoimmune diseases (21), and is now being used clinically to treat multiple sclerosis (22). The protection induced by T cell vaccination was shown to be mediated by anti-id T cells (17) and by other regulatory T cells, termed anti-ergotypic T cells, that recognize markers on activated T cells (23). The kinetics of the appearance of the regulatory cells suggested that the anti-ids were actually present in the animals even before vaccination; T cell vaccination merely activated natural anti-ids (21).

The activation of anti-id T cells is not due merely to the intrinsic immunogenicity of the id; to respond to T cell vaccination, the subject has to have been primed naturally to the id. T cell vaccination can induce anti-ids only in animals populated with autoimmune ids (24). The T cells of C57BL/6 mice, for example, do not respond to mouse thyroglobulin, and these mice are genetically resistant to experimental autoimmune thyroiditis (EAT; 25). Nevertheless, we succeeded in raising a pathogenic EAT T cell line from these mice by selecting T cells that recognized an epitope of bovine thyroglobulin cross-reactive with mouse thyroglobulin (24). The "artificial" autoimmune T cells could mediate very severe EAT, but these thyroiditogenic T cells could not be made to vaccinate the genetically-resistant mice against the line-mediated disease. Thus, the capacity to respond to T cell vaccination rests on the capacity to generate the specific ids naturally: No natural autoimmunity, no anti-ids. This suggests that natural autoimmune ids are a prerequisite for natural anti-ids, both natural and contrived.

7.2.   Antibody Anti-id Therapy

Natural immunoglobulins (IVIg) have been used to suppress clinical autoimmune disease in many patients, and the effects of IVIg can be explained, at least in part, by the presence of natural anti-id antibodies in the blood of healthy subjects (26). Specific anti-id antibodies have been shown to control various experimental autoimmune diseases (1). The success of anti-id antibodies and of IVIg in treating autoimmune disease suggests that natural anti-ids may function physiologically, as well as therapeutically.

7.3.   Antigen Therapy

Autoimmune diseases in experimental animals, and also recently in humans, can be controlled by the administration of the target self-antigen or its peptides. To suppress disease, the antigen or peptide has to be administered in an immune context or at a body site that fosters a shift in the autoimmune phenotype from a pro-inflammatory Th1 to an anti-inflammatory Th2 response (11). For example, the oral or nasal administration of myelin antigens can downregulate EAE, and some cases of multiple sclerosis (22); the administration of insulin or of a peptide from the hsp60 molecule or from glutamic acid decarboxylase can arrest the development of diabetes in NOD mice (11).

At first glance, one might think that antigen therapy and anti-id therapy are mutually unrelated ways to treat auto-immune diseases. Antigen therapy leads to a cytokine shift in the autoimmune population, while anti-id therapy induces regulatory cells targeted to the pathogenic id. However, recent studies suggest that the two forms of regulation might be mutually connected. Anti-id vaccination both in EAE (using DNA encoding a T cell receptor relevant to EAE; 28) and in NOD diabetes (using a T cell receptor peptide; 13) were found to induce a cytokine shift in the autoimmune effector T cells from a Th1 to a Th2 profile. Similarly, antigen vaccination of NOD mice with a peptide from the hsp60 target antigen, which induces a cytokine shift, also activates the natural anti-id T cells (13). Thus, two different therapeutic modalities, each thought to trigger a unique control mechanism, seem, in the end, to activate a common regulatory mechanism involving both a cytokine shift and an anti-id response. Anti-id therapy and antigen therapy may be seemingly different in concept, but are alike in actual physiology. The immune system, obviously, knows more about immunology than do its student immunologists.

8. THE IMMUNOLOGICAL HOMUNCULUS

The classical clonal selection paradigm asserted that, except for the deletion of self-recognizing clones, the receptor repertoire of the immune system faced the parasitic world as a *tabula rasa* with no intrinsic structure (28). Foreign antigens that happened to enter the body selected for proliferation and differentiation those clones that happened to bear fitting receptors. These chance encounters with foreign antigens, by favoring this or that clone, imposed an evolving structure on the immune system; the immune system was seen to act by principles of Darwinian selection in which environmental antigens were the selecting forces.

The observations of natural idiotypy, in contradiction to classical clonal selection, indicate that the immune system does not register antigenic experience on a clean slate, but rather begins its functional life with a considerable degree of internal self-organization. Clonal selection directed attention to negative selection as the only fate for self-reactive clones, but immunology is now aware that all the T cell clones present in the repertoire had to have been positively selected for a degree of self-reactivity (29).

Natural idiotypy tells us two facts about the outcome of positive self-selection: First, we have seen that certain self-antigens select clones of lymphocytes with genuinely strong self-specific affinity; these dominant self-antigens (see below) are not merely imperfect templates for the altered peptide ligands of the foreign antigens that the immune system will see later. These dominant self-antigens are immunogenic in the periphery. Secondly, the natural id-anti-id structuring of the immune repertoire precedes the

system's contact with the outside world. Unlike clonal selection, idiotypy emphasizes the reflexive strategy of the immune system; the system recognizes its own agents, and not only the external antigens. Idiotypy accepts autoimmunity as a physiological fact of immune life.

I have highlighted the existence of dominant natural autoimmunity and its regulation by natural anti-ids and other mechanisms in the concept of the *immunological homunculus* (30). The central nervous system manages its affairs with a neurological homunculus, an internal image of parts of the body skewed to over-represent those organs that are neurologically predominant, such as the organs of speech, facial expression and the thumbs (in humans). The neurological homunculus is encoded spatially in collections of neural networks. Similarly, the immune system can be seen to manage its affairs with the help of an internal image of the self skewed to those self-antigens that would seem to be immunologically important. This internal image is encoded not spatially, but in the frequency and affinity (avidity) of the clones reactive to these self-antigens and in the networks of anti-ids centered around these clones.

We don't know how the immunological homunculus arises, but it is clear that certain self-antigens are recognized with high affinity. We have found, for example, that transgenic mice made to hyper-express hsp60 in the thymus do not delete their anti-hsp60 T cells (31). Thus, it is conceivable that two pathways of positive self-selection occur during lymphocyte ontogeny: A pathway of low-affinity selection leads to the survival of clones that will later respond with high affinity to foreign antigens in the periphery, and a pathway of high-affinity selection, in contrast, leads to membership in the homunculus set.

Which self-antigens are members of the homuncular set, and what functions does autoimmunity to these antigens serve? Why should the immune system bother building a homunculus? I have discussed these questions at length elsewhere (32), but would like to mention two instances where we have learned something about homunculus function.

Some homuncular antigens are highly conserved throughout evolution, and so are shared by the host and by his or her potential parasites. The stress proteins are such homuncular antigens. Autoimmunity to the hsp60 molecule, for example, can help the host survive microbial infection (33). Indeed, the increased expression of hsp60 in stressed cells can serve as a kind of internal adjuvant and, through natural hsp60 auto-immunity, alert the immune system to critical foreign antigens associated with self-hsp60 (34). As I mentioned above, hyper-expression of hsp60 engineered into the thymus may change the phenotype of hsp60 autoimmunity, but there is no clonal deletion; hsp60 seems resistant to negative selection (31). The molecular basis for the distinction between negative selection and the various forms of positive selection is not known at present.

What about tissue-specific homuncular self-antigens, such as myelin basic protein, thyroglobulin, or insulin? These antigens are not conserved and autoimmunity to them would not be expected to provide any advantage in resisting infection. Why are these antigens included among the dominant self-antigens? I would like to suggest that autoimmunity to tissue antigens can contribute to maintenance of the body. The immune system is not only charged with defending against the dangers of foreign pathogens, but is continuously engaged in maintaining the body. Wound healing, angiogenesis, connective tissue growth, bone remodeling and repair, tissue regeneration, and the disposal of effete cells and molecules all require the actions of immune cells and their molecular products: Cytokines, adhesion molecules, interaction molecules and migration molecules. It is usually thought that only the germ-line, innate arm of the immune system is involved in these maintenance activities. However, my colleagues and I have recently found that autoimmune T cells can markedly contribute to the recovery of the central nervous system from traumatic injury (35). Lines of anti-myelin T cells, which are capable of mediating EAE, provided neuroprotection to the crushed optic nerve (36) or to the spinal cord following contusion (37). These T cells delivered neurotrophic factors to the site of injury and put the damaged nerve to rest by causing a transient paralysis (35).

The discovery of autoimmune tissue maintenance in the central nervous system should stimulate experiments designed to probe the possible role of autoimmune lymphocytes in the physiological maintenance of other body organs (38). Tissue-specific autoimmune lymphocytes could certainly serve as intelligent agents of maintenance, and deliver a dynamically regulated mix of inflammatory signals and growth factors to needy sites. Whether or not tissue maintenance is the key to understanding the evolutionary advantage of homuncular autoimmunity remains to be seen. In any case, natural autoimmunity needs regulation, and the natural anti-id network has good reason to be attached to the homuncular immune system.

References

1. Zanetti, M. 1993. Idiotypy and the idiotype network in autoimmunity. In: The *molecular pathology of autoimmune diseases*, C.A. Bona, K. A. Siminovitch, M. Zanetti, and A.N. Theofilopoulos, editors. Harwood Academic Publishers, Chur, Switzerland. pp. 209–228.
2. Jerne, N.K. 1974. Towards a network theory of the immune system. *Ann. Immunol.* (*Paris*) 125:373–389.
3. Sakato, N., and H.N. Eisen. 1975. Antibodies to idiotypes of isologous immunoglobulins. *J. Exp. Med.*. 141:1411–1426.
4. Sege, K., and P. A. Peterson. 1978. Use of anti-idiotypic antibodies as cell-surface receptor probes. *Proc. Natl. Acad. Sci. USA* 75:2443–2447.
5. Tikochinski, Y., D. Elias, C. Steeg, H. Marcus, M. Kantorowitz, T. Reshef, V. Ablamunits, I.R. Cohen, and A. Friedmann. 1999. A shared TCR CDR3 sequence in NOD mouse autoimmune diabetes. *Int. Immunol.* 11:951–956.
6. Shechter, Y., R. Maron, D. Elias, and I.R. Cohen. 1982. Autoantibodies to insulin receptor spontaneously develop as anti-idiotypes in mice immunized with insulin. *Science* 216:542–545.
7. Wasserman, N.H., A.S. Penn, P.I. Freimuth, N. Treptow, S. Wentzel, W.L. Cleveland, and B.F. Erlanger. 1983. Anti-idiotypic route to anti-acetylcholine receptor antibodies and experimental *myasthenia gravis*. Proc. Natl. Acad. Sci. USA 79:4810–4814
8. Islam, M.N., B. M. Pepper, R. Briones-Urbina, and N.R. Farid. 1983. Biological activity of anti-thyrotropin anti-idiotypic antibody. *Eur. J. Immunol.* 13:57–63.
9. Schreiber, A.B., P.O. Couraud, C. Andre, B. Vray, and A.D. Strosberg. 1980. Anti-aloprenolol anti-idiotypic antibodies bind to -adrenergic receptors and modulate catecholamine-sensitive adeylate cyclase. *Proc. Natl. Acad. Sci. USA* 77:7385–7389.
10. Bedin, C., K. Mignon-Godefroy, M.P. Brazillet, H. Tang, and J. Charreire. 1993. Immunization with thryoglobulin-specific cytotoxic T cell hybridoma induces anti-thyroglobulin antibodies: Characteristics of monoclonal anti-thyroglobulin auto-antibody. *Cell immunol.* 146:227–237.
11. Elias, D., A. Meilin, V. Ablamunits, O.S. Birk, P. Carmi, S. Konen- Weizmann, and I.R. Cohen. 1997. Hsp60 peptide therapy of NOD mouse diabetes induces a Th2 cytokine burst and downregulates autoimmunity to various beta-cell antigens. *Diabetes* 46:758–764
12. Holmberg, D., G. Wennerstorm, L. Andrade, and A. Countinho. 1986 The high idiotypic connectivity of "natural" newborn antibodies is not found in adult mitogen-reactive B cell Repertoires. *Eur. J. Immunol.* 16:82–87
13. Elias, D., Y. Tikochinski, G. Frankel, and I.R. Cohen. 1999. Regulation of NOD mouse autoimmune diabetes by T cells that recognize a TCR CDR3 peptide. *Int. Immunol.* 11:957–66
14. Itoh, M., T. Takahashi, N. Sakaguchi, Y. Kuniyasu, J. Shimizu, F. Otsuka, and S. Sakaguchi. 1999. Thymus and autoimmunity: production of CD25+ CD4+ naturally anergic and suppressive T cells as a key function of the thymus in maintaining immunologic self-tolerance. *J. Immunol.* 162:5317–5326
15. Olivares-Villagommez, D., Y. Wang, and J.J. Lafaille. 1998. Regulatory CD4(+) T cell expressing endogenous T cell receptor chains protect myelin basic protein-specific transgenic mice from spontaneous autoimmune encephalomyelitis. *J. Exp. Med.* 188:1883–1894.
16. Kumar, V., and E. Sercarz. 1996. Dysregulation of potentially pathogenic self reactivity is crucial for the manifestation of clinical autoimmunity. *J. Neurosci. Res.* 45:334–339
17. Lider, O., T. Reshef, E. Beraud, A. Ben-Nun, and I.R. Cohen. 1988. Anti-idiotypic network induced by T cell vaccination against experimental autoimmune encephalomyelitis. Science. 239:181–3.
18. Kumar, V., R. Tabibiazar, H.M. Geysen, and E. Sercarz. 1995. Immunodominant framework region 3 peptide from TCR V beta 8.2 chain controls murine experimental autoimmune encephalomyelitis. *J. Immunol.* 154:1941–1950
19. Kumar, V., F. Aziz, E. Sercarz, and A. Miller. 1997. Regulatory T cells specific for the same framework 3 region of the Vbeta8.2 chain are involved in the control of collagen

II-induced arthritis and experimental autoimmune encephlomyelitis. *J. Exp. Med.* 185:1725–1733.

20. Ben Nun, A., H. Wekerle, and I.R. Cohen. 1981.. Vaccination against autoimmune encephalomyelitis with T-lymphocyte line cells reactive against myelin basic protein. *Nature* 292:60–61.

21. Cohen I.R. 1986.. Regulation of autoimmune disease: physiological and therapeutic. *Immunol. Rev.* 94:5–21.

22. Zhang, J., P. Stinissen, R. Medaer, and J. Raus. 1996. T cell vaccination: clinical application in autoimmune disease. *J. Mol. Med.* 74:653–662.

23. Lohse, A.W., F. Mor, N. Karin, and I.R. Cohen. 1989. Control of experimental autoimmune encephalomyelitis by T cells responding to activated T cells. *Science* 244:820–822.

24. Zerubavel –Weiss, R., D. Markovits, and I. R. Cohen. 1992. Autoimmune thyroiditis (EAT) in genetically resistant mice meditated by a T cell line. *J. Autoimmun.* 5:617–627.

25. Maron R., and I.R. Cohen. 1980. H-2K mutation controls immune response phenotype of autoimmune thyroiditis. Critical expression of mutant gene product in both thymus and thyroid glands. *J. Exp. Med.* 152:1115–1120.

26. Kaveri, S., N. Prasad, T. Vassilev, V. Huretz, A. Pashov, S. Lacroix-Desmazes, and M. Kazatchkine. Modulation of autoimmune responses by intravenous immunoglobulin (IVIg). *Mult. Scler.* 3:121–128.

27. Weiner H.L., and E.P. van Rees. 1999. Mucosal tolerance. *Immunol. Lett.* 68:3–4.

28. Burnet, M. 1959. *Self and not self.* Cambridge University Press, Cambridge.

29. Goldrath A.W., and M.J. Bevan. 1999. Selecting and maintaining a diverse T-cell receptoire. *Nature Med.* 402:255–62.

30. Cohen I.R. 1992. The cognitive paradigm and the immunological homunculus. *Immunol. Today* 13:490–494.

31. Birk, O.S., D.C. Douek, D. Elias, K. Takacs, H. Dewchand, S.L. Gur, M.D. Walker , R. van der Zee, I.R. Cohen, and D.M. Altmann. 1996. The role of hsp60 in autoimmune diabetes: analysis in a transgenic model. *Proc. Natl. Acad. Sci. USA* 93:1032–1037.

32. Cohen, I.R. Tending Adams Garden: Evolving the cognitive immune self. Academic Press, San Diego, 2000.

33. Konen-Waisman, S., A. Cohen, M. Fridkin, and I.R. Cohen. 1999. Self heat shock protein (hsp 60) peptide serves in a conjugate vaccine against a lethal pneumococcal infection. *J. Infec. Dis.* 179:403–413.

34. Birk, O.S., S.L. Gur, D. Elias, R. Margalit, F. Mor, P. Carmi, J. Bockova, D.M. Altmann and I.R. Cohen. 1999. The 60-kDa heat shock protein modulates allograft rejection. *Proc. Natl. Acad. Sci. USA* 96:5159–5163.

35. Schwartz, M., G. Moalem, R. Leibowits-Amit, and I.R. Cohen. 1999. Innate and adaptive immune responses can be beneficial for CNS repair. *Trends Neurosci.* 22:295–299.

36. Moalem G, R. Leibowitz-Amit, E. Yoels, F. Mor, I.R. Cohen, and M. Schwartz. 1999 Autoimmune T calls protect neurons from secondary degeneration after central nervous system axotomy. *Nat. Med.* 5:49–55

37. Hauben, E. U. Nevo, E. Yoels, G. Moalem, E. Agranov, R. Leibowitz-Amit, F. Mor, E. Pevsner, S. Akselrod, M. Neeman, I.R. Cohen, and M. Schwartz. 2000 Autoimmune T cells: a potential neuroprotective therapy for spinal cord injury. *Lancet.* 355:286–287

38. Schwartz, M., and I.R. Cohen. 2000. Autoimmunity can benefit self-maintenance. *Immunol Today.* 21:265–8.

16 | Chemokine Receptors as Regulators of Dendritic Cell and Lymphocyte Traffic: Relevance to Autoimmunity

Federica Sallusto and Antonio Lanzavecchia

1. INTRODUCTION

Protective and pathological immune responses are dependent on the coordinated and timely interaction of various cell types within distinct microenvironments. In the primary response, dendritic cells (DC) migrate from peripheral tissues, where they take up antigen, to the secondary lymphoid organs, where they present the processed antigen to T cells. The helper cells generated during the primary response interact with antigen-specific B lymphocytes to stimulate antibody production, while effector Th1 or Th2 cells enter peripheral inflamed tissues and interact with other leukocytes, such as eosinophils, mast cells and basophils (in the case of Th2 reactions), or macrophages and neutrophils (in the case of Th1 reactions). Tissue-specific pathways for lymphocyte migration to skin, gut and other tissues have been described.

Recent evidence indicates that the selectivity and flexibility necessary to regulate leukocyte traffic under homeostatic and inflammatory conditions, as well as the organization of lymphoid organs, can be provided by a differential tissue distribution of chemokines and a regulated expression of chemokine receptors on different leukocyte subsets. Although there are relatively few studies, it can be anticipated that the basic mechanisms by which chemokines and chemokine receptors control primary and effector immune responses will also apply to autoimmune diseases. Studies in the field of chemokines and their receptors will help not only to shed new light on the pathogenesis, but also will offer therapeutic tools to inhibit autoimmune diseases either at the induction or the effector phase.

2. CHEMOKINES AND CHEMOKINE RECEPTORS IN HOMEOSTASIS AND INFLAMMATION

Chemokines are chemotactic cytokines that signal through seven transmembrane receptors (7TMR) coupled to pertussis toxin-sensitive G_i-proteins (1,2). Chemokines and chemokine receptors are involved in two distinct steps of leukocyte migration (3,4). The first is the extravasation from blood into lymph nodes, Peyer's patches and inflamed tissues. This step is controlled by three sequentially-acting receptor-ligand pairs that determine, in a combinatorial fashion, the specificity of the process. Selectins and their carbohydrate ligands allow rolling of leukocytes on the vascular endothelium, thus exposing chemokine receptors to their ligands displayed on the surface of endothelial cells. The triggering of chemokine receptors results in the rapid activation of integrins, leading to firm adhesion and transmigration.

The second migratory step controlled by chemokines and chemokine receptors is the directional migration and positioning of leukocytes within secondary lymphoid organs or inflamed tissues. It is likely that, in this process, the migrating cell encounters multiple chemokines and chemoattractants in a complex spatial and temporal pattern. It has been shown that neutrophils can use chemokine receptors sequentially to move along different chemotactic gradients (5). This process, defined as "multistep navigation", depends on the property of chemokine receptors to undergo ligand induced desensitization mediated by specific kinases that phosporylate the engaged receptors, resulting in their internalization via clathrin coated pits (6). Thus, it is not surprising that, as a general rule, leukocytes express several types of chemokine receptors to perform distinct steps in the process of extravasation and positioning within the tissue.

There are approximately 40 chemokines identified to date, which are classified according to the configuration of cysteine residues near the N-terminus into four families: CC-, CXC-, C- and CX_3C (Table 1). An operational distinction of

Table 1 Human chemokines and their receptors

Chemokine	Old name	Receptor
CCL1	I-309	CCR8
CCL2	MCP-1	CCR2
CCL3	MIP-1α	CCR1, CCR5
CCL4	MIP-1β	CCR5
CCL5	RANTES	CCR3, CCR5
CCL6	nd	
CCL7	MCP-3	CCR1, CCR2, CCR3
CCL8	MCP-2	CCR2, CCR3
CCL9	nd	
CCL10	nd	
CCL11	eotaxin	CCR3
CCL12	nd	
CCL13	MCP-4	CCR2, CCR3
CCL14	HCC-1	CCR1
CCL15	HCC-2	CCR1, CCR3
CCL16	HCC4	CCR1
CCL17	TARC	CCR4
CCL18	PARC	unknown
CCL19	ELC	CCR7
CCL20	MIP-3α	CCR6
CCL21	SLC	CCR7
CCL22	MDC	CCR4
CCL23	MIP-3	CCR1
CCL24	eotaxin-2	CCR3
CCL25	TECK	CCR9
CCL26	eotaxin-3	CCR3
CCL27	CTACK	unknown
CXCL1	GRO-1	CXCR2
CXCL2	GRO-2	CXCR2
CXCL3	GRO-3	CXCR2
CXCL4	PF4	unknown
CXCL5	ENA-78	CXCR2
CXCL6	GCP-2	CXCR1, CXCR2
CXCL7	NAP-2	CXCR2
CXCL8	IL-8	CXCR1, CXCR2
CXCL9	Mig	CXCR3
CXCL10	IP-10	CXCR3
CXCL11	ITAC	CXCR3
CXCL12	SDF-1	CXCR4
CXCL13	BCA-1	CXCR5
XCL1	lymphotactin	XCR1
XCL2	SCM-1β	XCR1
CX3CL1	fractalkine	CX3CR1

nd = not defined in human. See also reference 94.

chemokines based on the site of production and the nature of eliciting stimuli has been proposed (7,8). Inflammatory chemokines are produced by most cells, including endothelial, epithelial, stromal cells as well as leukocytes. Classical examples include IL-8, RANTES, eotaxin, MIP-1β, MCP-1 and IP-10. These chemokines are rapidly induced to high levels of expression by inflammatory stimuli such as LPS, IL-1 and TNF-α, and their production can be modulated by local cytokines. For instance, IP-10 production is stimulated by IFN-γ at sites of DTH reactions, while eotaxin production is increased by IL-4 in lungs undergoing allergic inflammation (9,10).

Lymphoid chemokines are produced primarily within lymphoid tissues and are involved in maintaining homeostatic leukocyte traffic and compartimentalization within these organs (7,8,11). Examples in this group are the B cell-attracting chemokine BCA-1 (and its mouse homologue BLC), which is produced by cells in B cell follicles, SLC, which is produced by endothelial cells of lymphatics and high endothelial venules (HEV) and by stromal cells in the T cell area of lymph nodes, and ELC, which is produced by interdigitating DC.

While most chemokines fit either the inflammatory or lymphoid type, some may play a dual role depending on the tissue and context in which they are produced. For instance, TARC and MDC can attract effector cells in inflamed liver and lung (12,13), but can also regulate T, B, and DC interactions in secondary lymphoid organs (14). Finally, some chemokines have restricted tissue distribution and may be involved in tissue-specific homing. For instance, TECK is restricted to thymus and intestine, while Lungkine is made exclusively in the lung (15,16).

With the exception of fractalkine (17), which is an integral membrane protein, all chemokines are charged secreted molecules that interact with sulfated proteoglycans present on the cell surface or in the extracellular matrix. Their immobilization is especially important for endothelial cells that are exposed to the blood flow. In this case, chemokines can be either synthesized by the cell itself or transcytosed from the abluminal to the luminal aspect, where they are displayed to rolling leukocytes (18).

3. REGULATION OF PRIMARY IMMUNE RESPONSES BY CHEMOKINES AND CHEMOKINE RECEPTORS

3.1. Dendritic Cell Migration

Primary immune responses take place in the organized structure of secondary lymphoid organs, where encounters between DC, T cells and B cells are orchestrated (Figure 1) (19). Monocytes, the precursors of immature DC, express receptors for inflammatory chemokines (CXCR1, CCR1, CCR2, and CCR5), which account for their capacity to extravasate and migrate into inflamed tissues and migrate toward the source of stimulus (Figure 2) (20,21). The migration up a gradient of inflammatory chemokines allows immature DC to be increasingly exposed to maturation stimuli such as TNF-α, IL-1, LPS or dsRNA (22–24). The maturation process triggered by these stimuli leads to a complete reprogramming of DC function, with down-regulation of endocytosis, upregulation of MHC class I and class II, adhesion and costimulatory molecules, as well as a switch in chemokine receptor expression. Within one to

two hours, maturing DC downregulate receptors for inflammatory chemokines (due to production of the cognate ligands) and therefore become free to sense new chemotactic gradients. Simultaneously, CCR7 is upregulated, allowing maturing DC to be attracted first to the lymphatics by SLC, and subsequently to the T cell areas by SLC and ELC (20,21,25). The role of CCR7 as an essential regulator of DC migration is supported by the observation that in mice lacking CCR7 ligands or CCR7, maturing DC fail to migrate from skin to lymph nodes (26,27).

The migration of DC from the site of antigen capture to the site of antigen presentation represents a general principle, with slight variations in the anatomical details in the case of spleen, Peyer's patches and tonsils. In the latter,

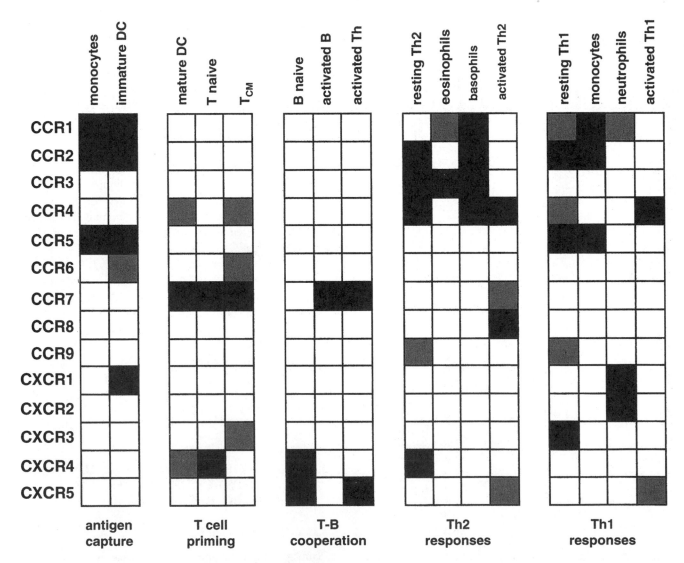

Figure 1. Regulated expression of chemokine receptors on lymphocytes and leukocytes participating in different phases of the immune response. Black and grey filling indicate expression on most, or on a fraction, of cells, respectively. Note the change in chemokine receptor expression following dendritic cell maturation and T cell and B cell activation. Note also the sharing of chemokine receptors among cells that participate in the same phase of the immune response.

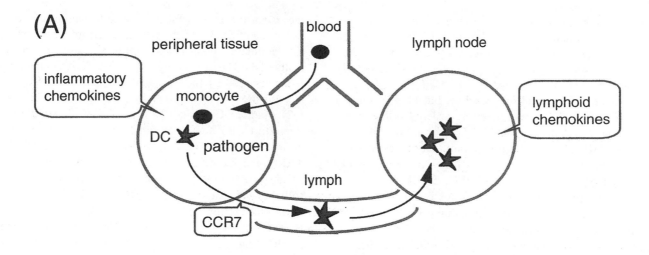

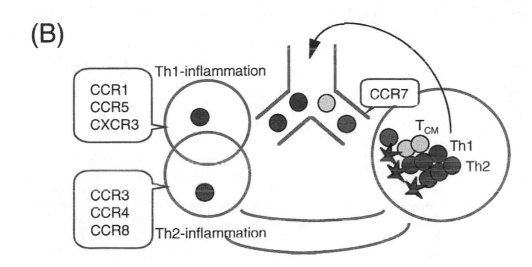

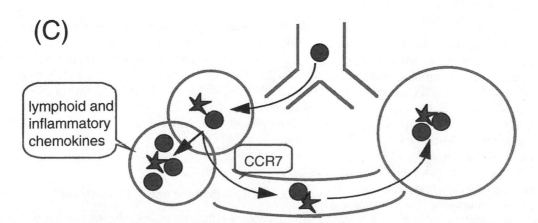

Figure 2. The role of chemokines and chemokine receptors in the orchestration of the immune response. A) Dendritic cell recruitment into inflamed tissues, activation by pathogens and migration to lymph nodes. B) T cell priming and polarization within the specialized lymph node microenvironment, migration of effector T cells into inflamed tissues. C) T cell activation in tissues, migration to lymph nodes or to chronic inflammatory sites. See text for details.

immature Langerhans cells expressing CCR6 are attracted to the crypts where the CCR6 ligand MIP-3α is produced by epithelial cells (25). Therefore, MIP-3α and CCR6 represent a chemokine-chemokine receptor pair particularly relevant in mucosal immunity.

Maturing DC not only respond to, but also produce a variety of chemokines in a precisely ordered fashion (28). Following stimulation with LPS, DC have an initial burst of MIP-1α, MIP-1β and IL-8 production that ceases within a few hours. At later time points, they produce mainly lymphoid chemokines, such as ELC, TARC, and MDC. The ordered production of chemokines is a key aspect of DC biology. On one hand, inflammatory chemokines produced early enhance and sustain the recruitment of monocytes and other inflammatory cells to inflamed tissues, thus sustaining antigen sampling. On the other hand, the late production of ELC, TARC and MDC within the lymphoid microenvironment may facilitate the interaction of mature DC with naive and activated T cells, which express the cognate receptors CCR7 and CCR4, thus sustaining antigenic stimulation.

It has been suggested that DC recruitment and activation may represent a central control mechanism that discriminates between effective T cell priming and tolerance induction (29). In the presence of pathogens, the recruitment of immature DC into tissues and their maturation and migration to draining lymph nodes occurs at high rates, resulting in a strong and sustained stimulation of T cells, which is required to induce their proliferation and polarization (30). In contrast, in the absence of inflammation, the process of DC recruitment and migration occurs at a much lower rate. There is evidence that, in these conditions, antigens carried by the few and poorly stimulatory DC lead to an abortive T cell response and tolerance (31).

3.2. Migration of Naive and Memory Lymphocytes to Secondary Lymphoid Organs

HEV represents ports of entry into the lymph nodes for circulating T and B lymphocytes, while normally excluding other cell types such as neutrophils and monocytes. The critical receptors for lymphocyte arrest and extravasation at this site are L-selectin, CCR7, and LFA-1 that bind to the PNAd, SLC and ICAM-1 expressed on the luminal site of HEV (Figure 1). Mice lacking SLC or CCR7 have a profound defect in T and B cell homing to lymph nodes (26,27). It is possible, however, that under inflammatory conditions, additional chemokines may be displayed on HEV, thus enhancing the recruitment capacity and broadening its specificity.

Having entered the lymph nodes, T and B lymphocytes take different routes marked by specific chemokines. T lymphocytes localize to the surrounding T cell areas, where they scan the DC surface for specific antigen. With the exception of SLC, which is made by stromal cells, most of the landmark chemokines at this site are produced by mature DC. The organizing role of DC-derived chemokines is underscored by the fact that relB-deficient mice, which lack mature DC, have profoundly disorganized T cell areas (32). Mature DC produce ELC and PARC that bind respectively to CCR7 and to an unidentified receptor expressed on naive T cells (33). In addition, they produce MDC and TARC, which bind to CCR4, a receptor that is expressed on a subset of memory T cells and is rapidly upregulated following T cell activation (34,35). The CCR4-MDC/TARC interaction may allow activated T cells to compete more effectively with naive cells for the limiting numbers of DC. Thus, CCR7 and CCR4 may retain stimulated T cells in contact with antigen and in this way promote the polarization process.

Recently, a distinct population of memory T cells that express the lymph node homing receptors L-selectin and CCR7 has been identified in humans (Figure 2) (36). These cells, which have been defined as "central memory" T cells (T_{CM}), lack immediate effector function and are more responsive than naive T cells to antigenic stimulation. A reciprocal subset of memory cells lack CCR7, express receptors for inflammatory chemokines and has immediate effector function. These cells, which have been defined as "effector memory" (T_{EM}), represent classical Th1, Th2 or cytotoxic cells. These findings show that effector function and migratory capacity are coordinately regulated in the T cell differentiation program. It has been suggested that T_{EM} represent a persistent reservoir of effector cells that can be immediately recruited to inflamed tissues to contain invasive pathogens, while T_{CM} represent a pool of clonally expanded T cells that traffic to lymph nodes and, following antigenic challenge, can activate DC, help B cells and generate a new wave of effectors. Helper memory cells capable of stimulating antibody and cytotoxic responses probably belong to this subset.

B lymphocytes that enter into the T cell area through HEV take one further step and migrate to the follicles, where antigen is displayed on follicular dendritic cells (FDC). This step is dependent on CXCR5, which recognizes BLC/BCA-1 produced by FDC themselves (37–39). Interestingly, LTα/β and TNF are required both for FDC development and function as well as for BLC/BCA-1 expression (40).

3.3. T Cell Priming and T Cell—B Cell Cooperation

In secondary lymphoid organs, the segregation of T and B cells in distinct areas allows their separate stimulation by antigen displayed by DC and FDC, respectively. However, once stimulated, the rare antigen-specific cells need to come together to initiate a T-dependent antibody response. Adoptively-transferred T and B cells have been shown to localise initially in their respective areas and, following antigenic challenge, to move synchronously out of these areas towards each other, meeting at the boundary (41). This reciprocal exchange of migratory pathways is regulated by a switch in chemokine receptors (Figure 1). Antigen-stimulated T cells upregulate CXCR5 and CCR4 and thus become sensitive to chemokines present in the B cell areas (42,43). On the other hand, antigen-stimulated B cells acquire responsiveness to ELC, most likely by upregulating CCR7, which drives them to the T cell area (44). This change can occur at any time during B cell migration and, in the case of a soluble self-antigen, may lead to their retention in the T cell area, a phenomenon described as "follicular exclusion" (45). In antigen-stimulated T cells, the balance between CCR7 and CXCR5 responsiveness can be influenced by the strength and nature of the stimulus. For instance, administration of antigen in adjuvant, leading to presentation on activated DC, results in T cell proliferation and migration into B cell areas, while administration of antigen intravenously, resulting in presentation on non activated DC and B cells, leads to an abortive T cell proliferation and failure to migrate to the B cell follicles (46).

Once having interacted with specific T cells, some B cells proliferate and differentiate outside the follicle, while others give rise to the germinal center reaction. It is likely that changes in chemokine receptors are involved in this process as well, but the details are not known. Similarly, there is no information on whether and how chemokines may control homing of memory B cells and plasma cells to medullary cords, splenic red pulp, mucosal-associated tissues and bone marrow.

4. CHEMOKINE RECEPTORS FOR TH1, TH2 AND TISSUE-SPECIFIC EFFECTOR RESPONSES

Th1 and Th2 effector cells mediate different types of protective or pathogenetic responses by secreting distinct sets of cytokines and by interacting with different types of leukocytes. Th1 cells produce IFN-γ and colocalise with macrophages and neutrophils in DTH lesions, while Th2 cells produce IL-4, IL-5 and IL-13 and are present with eosinophils and basophils at sites of allergic inflammation (47). Recent studies have revealed a differential expression of chemokine receptors in Th1 or Th2 cells and a reciprocal regulation of chemokine production in tissues undergoing Th1 or Th2 type immune reactions (Figure 1).

A first example of chemokine receptor and chemokines associated with polarized responses is provided by CCR3

and its ligands eotaxin, eotaxin 2 and MCP-4, which are produced in mucosal tissues undergoing allergic inflammation. CCR3 is expressed on eosinophils and basophils as well as on in vivo and in vitro polarized Th2 cells (48–51). It has been suggested that the sharing of CCR3 may allow these three cell types of colocalise at sites of eotaxin production. Here the IL-4, IL-5 and IL-13 produced by Th2 cells may stimulate the effector function of eosinophils and basophils and boost the production of eotaxin and MDC, thus amplifying and modulating the inflammatory reaction.

Th2 cells also express CCR4 (52,53) and CCR8, the receptor for I-309. It should be noted that while CCR3 is expressed only by Th2 and shows a good correlation with IL-4 producing capacity, CCR4 is also expressed on other cell types, such as skin homing cells (which are mainly Th1), T_{CM} and in vitro activated Th1 cells (34,36,54). The presence of CCR4 on such different subsets may reflect the multiple roles of TARC and MDC, which behave as inflammatory, constitutive or tissue-homing chemokines in different circumstances.

The chemokine receptors expressed preferentially on Th1 cells are CCR5, CXCR3 and CCR1 (52,53,55). In rheumatoid arthritis and multiple sclerosis, thought to be Th1-associated diseases, virtually all T cells in the lesions express CCR5 and CXCR3, although usually only 5–15% of peripheral blood T cells have this phenotype (56). CCR1 and CCR5 are expressed on monocytes and macrophages, consistent with their colocalization with Th1 cells.

The chemokines produced in peripheral tissues during the inflammatory process are expected to determine the extent, quality and duration of the cellular infiltrate. The regulatory network of cytokine-chemokine interactions follows the basic rule of Th1/Th2 regulation. Thus, IL-4 and IL-13 stimulate production of eotaxin and MDC, an effect that is counteracted by IFN-γ (57,58). Conversely, IFN-γ induces IP-10 and Mig and upregulates RANTES, and this effect is antagonised by IL-4 (59). On the other hand, TNF-α, which is associated with both Th1 and type 2 responses, costimulates production of both "type 1" or "type 2" chemokines.

The selective migration of T lymphocytes to skin or gut is also dependent on chemokine receptors. Skin-homing T cells are characterized by the expression of CLA (cutaneous lymphocyte associated antigen) and CCR4 (54). The corresponding ligands, E-selectin and TARC, are expressed on endothelial cells of inflamed skin, but not of inflamed gut. In contrast, gut-homing T cells express high levels of $\alpha4\beta7$ integrin as well as CCR9, while the corresponding ligands, MAdCAM-1 and TECK are expressed on lamina propria venules and Peyer's patch HEV (60). CCR6 is also expressed on a subset of memory T cells and may drive their migration to either skin or gut in response to MIP-3α (61).

4.1.   Migration of Effector T cells Following Antigenic Stimulation

The pattern of chemokine receptor expression described above applies to the resting state of Th1 or Th2 cells, i.e. the cells that are bound to migrate from blood into tissues before encountering antigen. However, antigenic stimulation has been shown to modulate chemokine receptors on polarized cells, changing again their migratory capacity (Figure 1). Within six hours following antigenic stimulation, Th1 and Th2 cells downregulate the receptors for inflammatory chemokines (CCR1, CCR2, CCR3, CCR5, CCR6 and CXCR3) both at the mRNA and protein levels, and concomitantly upregulate CCR7, CCR4, CCR8 and CXCR5 (34,35). This switch in chemokine receptors is transient, since the original pattern is regained within a few days as the cells go back to the resting state. It has been suggested that CCR7 upregulation may allow T cells activated by antigen in peripheral tissues to migrate to the draining lymph nodes following the same pathway of DC (Figure 2) (35). Indeed, activated T cells clustered with DC have been described in the afferent lymph (62). Alternatively, the new set of chemokine receptors expressed on recently activated T cells may relocalize them within the tissues to sites where the corresponding chemokines (ELC, SLC, BCA, I-309 and TARC) are produced in the context of a chronic inflammatory response.

5.   ROLE OF CHEMOKINE RECEPTORS IN AUTOIMMUNE DISEASES

The nature of autoimmune diseases has boosted initial studies on the possible role of inflammatory chemokines in the recruitment of leukocytes that subsequently lead to tissue injury. Several reports have provided evidence of overexpression of inflammatory chemokines and their cognate receptors in tissues undergoing autoimmune reactions. For instance, increased expression of IL-8, GRO-α, MCP-1 and MIP-1α in synovial fluid correlates with infiltration of neutrophils, monocytes and T cells in joints of patients with rheumatoid arthritis (63–65). In addition, CCR5 and CXCR3 are expressed on the infiltrating cells in rheumatoid arthritis (66) or multiple sclerosis (67). Although CCR5 is expressed on most infiltrating cells in rheumatoid arthritis, it remains to be established whether CCR5-deficient individuals are more resistant to arthritis or other autoimmune diseases (68). Similarly, gene targeting experiments have so far failed to provide evidence for a non-redundant role of inflammatory chemokines or their receptors in autoimmune diseases, although mice lacking MIP-1α (69), MCP-1 (70), CCR2 (71), or CCR1 (72) show decreased inflammatory and granulomatous responses. Targeted expression of inflammatory chemokines in peripheral tissues results in development of specific infiltrates that usually do not progress to an overt disease (73–75). Furthermore, disease progression requires complex cellular and molecular changes that have been especially characterized in the NOD mouse model (76). Interestingly, however, chemokine antagonists have been shown to block spontaneous arthritis in MRL-*1pr* mice (77), as well as collagen-induced arthritis (78).

An intriguing observation is that in organs that are targets of autoimmune reactions, most of the infiltrating effector T cells carry specificities for a variety of environmental antigens unrelated to the putative self-antigens. This fact is consistent with a role of inflammatory chemokines in the nonspecific recruitment of resting effector T cells (Th1 or Th2) generated in response to unrelated viral antigens (79,80). In turn, these bystander cells stimulated by the local cytokine milieu may elaborate effector cytokines, thus sustaining the inflammatory process (81).

An hallmark of autoimmune disease is the formation of organized lymphoid structures within target tissues, and it has been suggested that this process, defined as lymphoneogenesis, is key to the development and maintenance of the disease (79,82). It is plausible that lymphoneogenesis may be mediated by the same mechanisms that regulate the normal development of secondary lymphoid organs (40,83, 84). Recent evidence indicates a role for specialized precursor cells as well as lymphoid chemokines in the generation of secondary lymphoid organs. The HEV of developing lymph nodes express MAdCAM-1 and recruit precursor cells that localize to the B cell area (85). These cells express $\alpha4\beta7$ integrin, which binds to MAdCAM-1, and CXCR5, which binds to BLC, and LTβ. The latter is required not only for lymph node development, but also for the expression of lymphoid chemokines produced in the T and B cell areas (86). BCA-1 and its receptor CXCR5 have been found to be highly expressed in newly-formed lymphoid tissue of the gastric mucosa associated with infection with *Helicobacter pylori* (87). It will be interesting to understand whether the expression of lymphoid chemokines such as BCA and SLC in target tissue may play a pathogenetic role in the development of autoimmune diseases by bringing the milieu necessary for T cell priming, i.e. the lymph node, directly in contact with tissue antigens.

6.   CHEMOKINE RECEPTORS AS TARGETS FOR IMMUNOMODULATION

Because of their high degree of specificity and their role in inflammatory responses, chemokine receptors were immediately recognised as attractive targets for drug development (88). We have provided several examples of how chemokine receptors may function in directing cell migration associated with specific immunological and pathological responses. It is conceivable that blocking a single chemokine receptor might inhibit one type of immune response, while leaving others unaffected. The feasibility and the biological consequences of antagonizing chemokine receptors is illustrated by the fact that herpes and pox viruses target chemokines as part of their strategy to deviate the immune response (89,90).

Chemokine receptors can be blocked using antagonists generated by N-terminal truncation or modification of an agonistic chemokine (91). For instance, N-terminal modification of RANTES results in powerful inhibitors of CCR5 and other chemokine receptors (92), and a truncated antagonist of MCP-1 inhibits arthritis in the MRL-*lpr* mouse model (77). The most promising approach to the development of selective chemokine receptor antagonists is represented by small organic molecules, which classically are able to disrupt ligand binding to G-protein coupled receptors (93). To date, a number of small molecule antagonists of chemokine receptors have been identified that are selective for CXCR4, CCR5, CXCR2, CCR1 and others.

7. CONCLUSION

It can be anticipated that a better understanding of the chemokine and chemokine receptor biology combined with the development of selective and powerful inhibitors will offer not only new insights into the pathogenesis of autoimmune diseases, but also identify targets for therapeutic intervention.

References

1. Baggiolini, M., B. Dewald, and B. Moser. 1997. Human chemokines: an update. *Annu. Rev. Immunol.* 15:675–705.
2. Zlotnik, A., J. Morales, and J.A. Hedrick. 1999. Recent advances in chemokines and chemokine receptors. *Crit. Rev. Immunol.* 19:1–47.
3. Springer, T.A. 1994. Traffic signals for lymphocyte recirculation and leukocyte emigration: the multistep paradigm. *Cell* 76:301–314.
4. Butcher, E.C., and L.J. Picker. 1996. Lymphocyte homing and homeostasis. *Science* 272:60–66.
5. Foxman, E.F., J.J. Campbell, and E.C. Butcher. 1997. Multistep navigation and the combinatorial control of leukocyte chemotaxis. *J. Cell Biol.* 139:1349–1360.
6. Premont, R.T., J. Inglese, and R.J. Lefkowitz. 1995. Protein kinases that phosphorylate activated G protein-coupled receptors. *FASEB J.* 9:175–182.
7. Yoshie, O., T. Imai, and H. Nomiyama. 1997. Novel lymphocyte-specific CC chemokines and their receptors. *J. Leukoc. Biol.* 62:634–644.
8. Baggiolini, M. 1998. Chemokines and leukocyte traffic. *Nature* 392:565–568.
9. Jose, P.J., D.A. Griffiths-Johnson, P.D. Collins, D.T. Walsh, R. Moqbel, N.F. Totty, O. Truong, J.J. Hsuan, and T.J. Williams. 1994. Eotaxin: a potent eosinophil chemoattrac-tant cytokine detected in a guinea pig model of allergic airways inflammation. *J. Exp. Med.* 179:881–887.
10. Kaplan, G., A.D. Luster, G. Hancock, and Z.A. Cohn. 1987. The expression of a gamma interferon-induced protein (IP-10) in delayed immune responses in human skin. *J. Exp. Med.* 166:1098–1108.
11. Cyster, J.G. 1999. Chemokines and cell migration in secondary lymphoid organs. *Science* 286:2098–2102.
12. Yoneyama, H., A. Harada, T. Imai, M. Baba, O. Yoshie, Y. Zhang, H. Higashi, M. Murai, H. Asakura, and K. Matsushima. 1998. Pivotal role of TARC, a CC chemokine, in bacteria-induced fulminant hepatic failure in mice. *J. Clin. Invest.* 102:1933–1941.
13. Lloyd, C.M., T. Delaney, T. Nguyen, J. Tian, A.C. Martinez, A.J. Coyle, and J.C. Gutierrez-Ramos. 2000. CC chemokine receptor (CCR)3/eotaxin is followed by CCR4/monocyte-derived chemokine in mediating pulmonary T helper lymphocyte type 2 recruitment after serial antigen challenge in vivo. *J. Exp. Med.* 191:265–274.
14. Tang, H.L., and J.G. Cyster. 1999. Chemokine up-regulation and activated T cell attraction by maturing dendritic cells. *Science* 284:819–822.
15. Vicari, A.P., D.J. Figueroa, J.A. Hedrick, J.S. Foster, K.P. Singh, S. Menon, N.G. Copeland, D.J. Gilbert, N.A. Jenkins, K.B. Bacon, and A. Zlotnik. 1997. TECK: a novel CC chemokine specifically expressed by thymic dendritic cells and potentially involved in T cell development. *Immunity* 7:291–301.
16. Rossi, D.L., S.D. Hurst, Y. Xu, Wang, S. Menon, R.L. Coffman, and A. Zlotnik. 1999. Lungkine, a novel CXC chemokine, specifically expressed by lung bronchoepithelial cells. *J. Immunol.* 162:5490–5497.
17. Imai, T., K. Hieshima, C. Haskell, M. Baba, M. Nagira, M. Nishimura, M. Kakizaki, S. Takagi, H. Nomiyama, T.J. Schall, and O. Yoshie. 1997. Identification and molecular characterization of fractalkine receptor CX3CR1, which mediates both leukocyte migration and adhesion. *Cell* 91:521–530.
18. Middleton, J., S. Neil, J. Wintle, I. Clark-Lewis, H. Moore, C. Lam, M. Auer, E. Hub, and A. Rot. 1997. Transcytosis and surface presentation of IL-8 by venular endothelial cells. *Cell* 91:385–395.
19. Banchereau, J., and R.M. Steinmann. 1998. Dendritic cells and the control of immunity. *Nature* 392:245–252.
20. Sallusto, F., P. Schaerli, P. Loetscher, C. Schaniel, D. Lenig, C.R. Mackay, S. Qin, and A. Lanzavecchia. 1998. Rapid and coordinated switch in chemokine receptor expression during dendritic cell maturation. *Eur. J. Immunol.* 28:2760–2769.
21. Sozzani, S., P. Allavena, G. D'Amico, W. Luini, G. Bianchi, M. Kataura, T. Imai, O. Yoshie, R. Bonecchi, and A. Mantovani. 1998. Differential regulation of chemokine receptors during dendritic cell maturation: a model for their trafficking properties. *J. Immunol.* 161:1083–1086.
22. Sallusto, F., and A. Lanzavecchia. 1994. Efficient presentation of soluble antigen by cultured human dendritic cells is maintained by granulocyte/macrophage colony-stimulating factor plus interleukin 4 and downregulated by tumor necrosis factor alpha. *J. Exp. Med.* 179:1109–1118.
23. Sallusto, F., M. Cella, C. Danieli, and A. Lanzavecchia. 1995. Dendritic cells use macropinocytosis and the mannose receptor to concentrate macromolecules in the major histocompatibility complex class II compartment: downregulation by cytokines and bacterial products. *J. Exp. Med.* 182:389–400.

24. Cella, M., M. Salio, Y. Sakakibara, H. Langen, I. Julkunen, and A. Lanzavecchia. 1999. Maturation, activation, and protection of dendritic cells induced by double-stranded RNA. *J. Exp. Med.* 189:821–829.

25. Dieu, M.-C., B. Vanbervliet, A. Vicari, J.-M. Bridon, E. Oldham, S. Ait-Yahia, F. Briere, A. Zlotnik, S. Lebecque, and C. Caux. 1998. Selective recruitment of immature and mature dendritic cells by distinct chemokines expressed in different anatomic sites. *J. Exp. Med.* 188:373–386.

26. Gunn, M.D., S. Kyuwa, C. Tam, T. Kakiuchi, A. Matsuzawa, L.T. Williams, and H. Nakano. 1999. Mice lacking expression of secondary lymphoid organ chemokine have defects in lymphocyte homing and dendritic cell localization. *J. Exp. Med.* 189:451–460.

27. Forster, R., A. Schubel, D. Breitfeld, E. Kremmer, I. Renner-Muller, E. Wolf, and M. Lipp. 1999. CCR7 coordinates the primary immune response by establishing functional microenvironments in secondary lymphoid organs. *Cell* 99:23–33.

28. Sallusto, F., B. Palermo, D. Lenig, M. Miettinen, S. Matikainen, I. Julkunen, R. Forster, R. Burgstahler, M. Lipp, and A. Lanzavecchia. 1999. Distinct patterns and kinetics of chemokine production regulate dendritic cell function. *Eur. J. Immunol.* 29:1617–1625.

29. Sallusto, F., and A. Lanzavecchia. 1999. Mobilizing dendritic cells for tolerance, priming, and chronic inflammation. *J. Exp. Med.* 189:611–614.

30. Iezzi, G., E. Scotet, D. Scheidegger, and A. Lanzavecchia. 1999. The interplay between the duration of TCR and cytokine signalling determines T cell polarization. *Eur. J. Immunol.* 29:4092–4101.

31. Kurts, C., H. Kosaka, F.R. Carbone, J.F. Miller, and W.R. Heath. 1997. Class I-restricted cross-presentation of exogenous self-antigens leads to deletion of autoreactive CD8(+) T cells. *J. Exp. Med.* 186:239–245.

32. Burkly, L., C. Hession, L. Ogata, C. Reilly, L.A. Marconi, D. Olson, R. Tizard, R. Cate, and D. Lo. 1995. Expression of relB is required for the development of thymic medulla and dendritic cells. *Nature* 373:531–536.

33. Adema, G.J., F. Hartgers, R. Verstraten, E. de Vries, G. Maryland, S. Menon, J. Foster, Y. Xu, P. Nooyen, T. McClanahan, K.B. Bacon, and C.G. Figdor. 1997. A dendritic-cell-derived C-C chemokine that preferentially attracts naive T cells. *Nature* 387:713–717.

34. D'Ambrosio, D., A. Iellem, R. Bonecchi, D. Mazzeo, S. Sozzani, A. Mantovani, and F. Sinigaglia. 1998. Selective up-regulation of chemokine receptors CCR4 and CCR8 upon activation of polarized human type 2 Th cells. *J. Immunol.* 161:5111–5115.

35. Sallusto, F., E. Kremmer, B. Palermo, A. Hoy, P. Ponath, S. Qin, R. Forster, M. Lipp, and A. Lanzavecchia. 1999. Switch in chemokine receptor expression upon TCR stimulation reveals novel homing potential for recently activated T cells. *Eur. J. Immunol.* 29:2037–2045.

36. Sallusto, F., D. Lenig, R. Forster, M. Lipp, and A. Lanzavecchia. 1999. Two subsets of memory T lymphocytes with distinct homing potentials and effector functions. *Nature* 401:708–712.

37. Forster, R., A.E. Mattis, E. Kremmer, E. Wolf, G. Brem, and M. Lipp. 1996. A putative chemokine receptor, BLR1, directs B cell migration to defined lymphoid organs and specific anatomic compartments of the spleen. *Cell* 87:1037–1047.

38. Gunn, M.D., V.N. Ngo, K.M. Ansel, E.H. Ekland, J.G. Cyster, and L.T. Williams. 1998. A B-cell-homing chemokine made in lymphoid follicles activates Burkitt's lymphoma receptor-1. *Nature* 391:799–803.

39. Legler, D.F., M. Loetscher, R.S. Roos, I. Clark-Lewis, M. Baggiolini, and B. Moser. 1998. B cell-attracting chemokine 1, a human CXC chemokine expressed in lymphoid tissues, selectively attracts B lymphocytes via BLR1/CXCR5. *J. Exp. Med.* 187:655–660.

40. Fu, Y.X., and D.D. Chaplin. 1999. Development and maturation of secondary lymphoid tissues. *Annu. Rev. Immunol.* 17:399–433.

41. Garside, P., E. Ingulli, R.R. Merica, J.G. Johnson, R.J. Noelle, and M.K. Jenkins. 1998. Visualization of specific B and T lymphocyte interactions in the lymph node. *Science* 281:96–99.

42. Flynn, S., K.M. Toellner, C. Raykundalia, M. Goodall, and P. Lane. 1998. CD4 T cell cytokine differentiation: the B cell activation molecule, OX40 ligand, instructs CD4 T cells to express interleukin 4 and upregulates expression of the chemokine receptor, Blr-1. *J. Exp. Med.* 188:297–304.

43. Ansel, K.M., L.J. McHeyzer-Williams, V.N. Ngo, M.G. McHeyzer-Williams, and J.G. Cyster. 1999. In vivo-activated CD4 T cells upregulate CXC chemokine receptor 5 and reprogram their response to lymphoid chemokines. *J. Exp. Med.* 190:1123–1134.

44. Ngo, V.N., H.L. Tang, and J.G. Cyster. 1998. Epstein-Barr virus-induced molecule 1 ligand chemokine is expressed by dendritic cells in lymphoid tissues and strongly attracts naive T cells and activated B cells. *J. Exp. Med.* 188:181–191.

45. Cyster, J.G., S.B. Hartley, and C.C. Goodnow. 1994. Competition for follicular niches excludes self-reactive cells from the recirculating B-cell repertoire. *Nature* 371:389–395.

46. Kearney, E.R., K.A. Pape, D.Y. Loh, and M.K. Jenkins. 1994. Visualization of peptide-specific T cell immunity and peripheral tolerance induction *in vivo. Immunity* 1:327–339.

47. Abbas, A.K., K.M. Murphy, and A. Sher. 1996. Functional diversity of helper T lymphocytes. *Nature* 383:787–793.

48. Ponath, P.D., S. Qin, T.W. Post, J. Wang, L. Wu, N.P. Gerard, W. Newman, C. Gerard, and C.R. Mackay. 1996. Molecular cloning and characterization of a human eotaxin receptor expressed selectively on eosinophils. *J. Exp. Med.* 183:2437–2448.

49. Uguccioni, M., C.R. Mackay, B. Ochensberger, P. Loetscher, S. Rhis, G.J. LaRosa, P. Rao, P.D. Ponath, M. Baggiolini, and C.A. Dahinden. 1997. High expression of the chemokine receptor CCR3 in human blood basophils. Role in activation by eotaxin, MCP-4, and other chemokines. *J. Clin. Invest.* 100:1137–1143.

50. Gerber, B., M.P. Zanni, M. Uguccioni, M. Loetscher, C.R. Mackay, W.J. Pichler, N. Yawalkar, M. Baggiolini, and B. Moser. 1997. Functional expression of the eotaxin receptor CCR3 in T lymphocytes co-localising with eosinophils. *Curr. Biol.* 7:836–843.

51. Sallusto, F., C.R. Mackay, and A. Lanzavecchia. 1997. Selective expression of the eotaxin receptor CCR3 by human T helper 2 cells. *Science* 277:2005–2007.

52. Bonecchi, R., G. Bianchi, P.P. Bordignon, D. D'Ambrosio, R. Lang, A. Borsatti, S. Sozzani, P. Allavena, P.A. Gray, A. Mantovani, and F. Sinigaglia. 1998. Differential expression of chemokine receptors and chemotactic responsiveness of type 1 T helper cells (Th1s) and Th2s. *J. Exp. Med.* 187:129–134.

53. Sallusto, F., D. Lenig, C.R. Mackay, and A. Lanzavecchia. 1998. Flexible programs of chemokine receptor expression

on human polarized T helper 1 and 2 lymphocytes. *J. Exp. Med.* 187:875–883.

54. Campbell, J.J., G. Haraldsen, J. Pan, J. Rottman, S. Qin, P. Ponath, D.P. Andrew, R. Warnke, N. Ruffing, N. Kassam, L. Wu, and E.C. Butcher. 1999. The chemokine receptor CCR4 in vascular recognition by cutaneous but not intestinal memory T cells. *Nature* 400:776–780.

55. Loetscher, P., M. Uguccioni, L. Bordoli, M. Baggiolini, B. Moser, C. Chizzolini, and J.M. Dayer. 1998. CCR5 is characteristic of Th1 lymphocytes. *Nature* 391:344–345.

56. Qin, S., J.B. Rottman, P. Myers, N. Kassam, M. Weinblatt, M. Loetscher, A.E. Koch, B. Moser, and C.R. Mackay. 1997. The chemokine receptors CXCR3 and CCR5 mark subsets of T cells associated with certain inflammatory reactions. *J. Clin. Invest.* 101:746–754.

57. Bonecchi, R., S. Sozzani, J.T. Stine, W. Luini, G. D'Amino, P. Allavena, D. Chantry, and A. Mantovani. 1998. Divergent effects of interleukin-4 and interferon-gamma on macrophage-derived chemokine production: an amplification circuit of polarized T helper 2 responses. *Blood* 92:2668–2671.

58. Li, L., Y. Xia, A. Nguyen, Y.H. Lai, L. Feng, T.R. Mosmann, and D. Lo. 1999. Effects of Th2 cytokines on chemokine expression in the lung: IL-13 potently induces eotaxin expression by airway epithelial cells. *J. Immunol.* 162:2477–2487.

59. Luster, A.D., and J.V. Ravetch. 1987. Biochemical characterization of a gamma-interferon-induced protein (IP-10). *J. Exp. Med.* 166:1084–1097.

60. Zabel, B.A., W.W. Agace, J.J. Campbell, H.M. Heath, D. Parent, A.I. Roberts, E.C. Ebert, N. Kassam, S. Qin, M. Zovko, G.J. LaRosa, L.L. Yang, D. Soler, E.C. Butcher, P.D. Ponath, C.M. Parker, and D.P. Andrew. 1999. Human G protein-coupled receptor GPR-9-6/CC chemokine receptor 9 is selectively expressed on intestinal homing T lymphocytes, mucosal lymphocytes, and thymocytes and is required for thymus-expressed chemokine-mediated chemotaxis. *J. Exp. Med.* 190:1241–1256.

61. Liao, F., R.L. Rabin, C.S. Smith, G. Sharma, T.B. Nutman, and J.M. Farber. 1999. CC-chemokine receptor 6 is expressed on diverse memory subsets of T cells and determines responsiveness to macrophage inflammatory protein 3 alpha. *J. Immunol.* 162:186–194.

62. Pope, M., M.G. Betjes, N. Romani, H. Hirmand, P.U. Cameron, L. Hoffman, S. Gezelter, G. Schular, and R.M. Steinman. 1994. Conjugates of dendritic cells and memory T lymphocytes from skin facilitate productive infection with HIV-1. *Cell* 78:389–398.

63. Brennan, F.M., C.O. Zachariae, D. Chantry, C.G. Larsen, M. Turner, R.N. Maini, K. Matsushima, and M. Feldman. 1990. Detection of interleukin 8 biological activity in synovial fluids from patients with rheumatoid arthritis and production of interleukin 8 mRNA by isolated synovial cells. *Eur. J. Immunol.* 20:2141–2144.

64. Koch, A.E., S.L. Kunkel, L.A. Harlow, B. Johnson, H.L. Evanoff, G.K. Haines, M.D. Burdick, R.M. Pope, and R.M. Strieter. 1992. Enhanced production of monocyte chemoattractant protein-1 in rheumatoid arthritis. *J. Clin. Invest.* 90:772–779.

65. Koch, A.E., S.L. Kunkel, L.A. Harlow, D.D. Mazarakis, G.K. Haines, M.D. Burdick, R.M. Pope, and R.M. Strieter. 1994. Macrophage inflammatory protein-1 alpha. A novel chemotactic cytokine for macrophages in rheumatoid arthritis. *J. Clin. Invest.* 93:921–928.

66. Qin, S., J.B. Rottman, P. Myers, N. Kassam, M. Weinblatt, M. Loetscher, A.E. Koch, B. Moser, and C.R. Mackay.

1998. The chemokine receptors CXCR3 and CCR5 mark subsets of T cells associated with certain inflammatory reactions. *J. Clin. Invest.* 101:746–754.

67. Balashov, K.E., J.B. Rottman, H.L. Weiner, and W.W. Hancock. 1999. CCR5(+) and CXCR3(+) T cells are increased in multiple sclerosis and their ligands MIP-1alpha and IP-10 are expressed in demyelinating brain lesions. *Proc. Natl. Acad. Sci. U.S.A.* 96:6873–6878.

68. Garred, P., H.O. Madsen, J. Petersen, H. Marquart, T.M. Hansen, S. Freiesleben Sorensen, B. Volck, A. Svejgaard, and V. Andersen. 1998. CC chemokine receptor 5 polymorphism in rheumatoid arthritis. *J. Rheumatol.* 25:1462–1465.

69. Cook, D.N., M.A. Beck, T.M. Coffman, S.L. Kirby, J.F. Sheridan, I.B. Pragnell, and O. Smithies. 1995. Requirement of MIP-1 alpha for an inflammatory response to viral infection. *Science* 269:1583–1585.

70. Lu, B., B.J. Rutledge, L. Gu, J. Fiorillo, N.W. Lukacs, S.L. Kunkel, R. North, C. Gerard, and B.J. Rollins. 1998. Abnormalities in monocyte recruitment and cytokine expression in monocyte chemoattractant protein 1-deficient mice. *J. Exp. Med.* 187:601–608.

71. Kuziel, W.A., S.J. Morgan, T.C. Dawson, S. Griffin, O. Smithies, K. Ley, and N. Maeda. 1997. Severe reduction in leukocyte adhesion and monocyte extravasation in mice deficient in CC chemokine receptor 2. *Proc. Natl. Acad. Sci. U.S.A.* 94:12053–12058.

72. Gao, J.L., T.A. Wynn, Y. Chang, E.J. Lee, H.E. Broxmeyer, S. Cooper, H.L. Tiffany, H. Westphal, J. Kwon-Chung, and P.M. Murphy. 1997. Impaired host defense, hematopoiesis, granulomatous inflammation and type 1-type 2 cytokine balance in mice lacking CC chemokine receptor 1. *J. Exp. Med.* 185:1959–1968.

73. Fuentes, M.E., S.K. Durham, M.R. Swerdel, A.C. Lewin, D.S. Barton, J.R. Megill, R. Bravo, and S.A. Lira. 1995. Controlled recruitment of monocytes and macrophages to specific organs through transgenic expression of monocyte chemoattractant protein-1. *J. Immunol.* 155:5769–5776.

74. Kolattukudy, P.E., T. Quach, S. Bergese, S. Breckenridge, J. Hensley, R. Altschuld, G. Gordillo, S. Klenotic, C. Orosz, and J. Parker-Thornburg. 1998. Myocarditis induced by targeted expression of the MCP-1 gene in murine cardiac muscle. *Am. J. Pathol.* 152:101–111.

75. Grewal, I.S., B.J. Rutledge, J.A. Fiorillo, L. Gu, R.P. Gladue, R.A. Flavell, and B.J. Rollins. 1997. Transgenic monocyte chemoattractant protein-1 (MCP-1) in pancreatic islets produces monocyte-rich insulitis without diabetes: abrogation by a second transgene expressing systemic MCP-1. *J. Immunol.* 159:401–408.

76. Gonzalez, A., J.D. Katz, M.G. Mattei, H. Kikutani, C. Benoist, and D. Mathis. 1997. Genetic control of diabetes progression. *Immunity* 7:873–883.

77. Gong, J.H., L.G. Ratkay, J.D. Waterfield, and I. Clark-Lewis. 1997. An antagonist of monocyte chemoattractant protein 1 (MCP-1) inhibits arthritis in the MRL-1pr mouse model. *J. Exp. Med.* 186:131–137.

78. Plater-Zyberk, C., A.J. Hoogewerf, A.E. Proudfoot, C.A. Power, and T.N. Wells. 1997. Effect of a CC chemokine receptor antagonist on collagen induced arthritis in DBA/1 mice. *Immunol. Lett.* 57:117–120.

79. Ludewig, B., B. Odermatt, S. Landmann, H. Hengartner, and R.M. Zinkernagel. 1998. Dendritic cells induce autoimmune diabetes and maintain disease via de novo formation of local lymphoid tissue. *J. Exp. Med.* 188:1493–1501.

80. Scotet, E., M.A. Peyrat, X. Saulquin, C. Retiere, C. Couedel, F. Davodeau, N. Dulphy, A. Toubert, J.D. Bignon, A. Lim, H. Vie, M.M. Hallet, R. Liblau, M. Weber, J.M. Berthelot, E. Houssaint, and M. Bonneville. 1999. Frequent enrichment for CD8 T cells reactive against common herpes viruses in chronic inflammatory lesions: towards a reassessment of the physiopathological significance of T cell clonal expansions found in autoimmune inflammatory processes. *Eur. J. Immunol.* 29:973–985.

81. Unutmaz, D., P. Pileri, and S. Abrignani. 1994. Antigen-independent activation of naive and memory resting T cells by a cytokine combination. *J. Exp. Med.* 180:1159–1164.

82. Kratz, A., A. Campos-Neto, M.S. Hanson, and N.H. Ruddle. 1996. Chronic inflammation caused by lymphotoxin is lymphoid neogenesis. *J. Exp. Med.* 183:1461–1472.

83. Kollias, G., E. Douni, G. Kassiotis, and D. Kontoyiannis. 1999. On the role of tumor necrosis factor and receptors in models of multiorgan failure, rheumatoid arthritis, multiple sclerosis and inflammatory bowel disease. *Immunol. Rev.* 169:175–194.

84. Ruddle, N.H. 1999. Lymphoid neo-organogenesis: lymphotoxin's role in inflammation and development. *Immunol. Res.* 19:119–125.

85. Mebius, R.E., P. Rennert, and I.L. Weissman. 1997. Developing lymph nodes collect CD4+CD3- LTbeta+ cells that can differentiate to APC, NK cells, and follicular cells but not T or B cells. *Immunity* 7:493–504.

86. Ngo, V.N., H. Korner, M.D. Gunn, K.D. Schmidt, D.S. Riminton, M.D. Cooper, J.L. Browning, J.D. Sedgwick, and J.G. Cyster. 1999. Lymphotoxin alpha/beta and tumor necrosis factor are required for stromal cell expression of homing chemokines in B and T cell areas of the spleen. *J. Exp. Med.* 189:403–412.

87. Mazzucchelli, L., A. Blaser, A. Kappeler, P. Schaerli, J.A. Laissue, M. Baggiolini, and M. Uguccioni. 1999. BCA-1 is highly expressed in Helicobacter pylori-induced mucosa-associated lymphoid tissue and gastric lymphoma. *J. Clin. Invest.* 104:R49–R54.

88. Baggiolini, M., and B. Moser. 1997. Blocking chemokine receptors. *J. Exp. Med.* 186:1189–1191.

89. Boshoff, C., Y. Endo, P.D. Collins, Y. Takeuchi, J.D. Reeves, V.L. Schweickart, M.A. Siani, T. Sasaki, T.J. Williams, P.W. Gray, P.S. Moore, Y. Chang, and R.A. Weiss. 1997. Angiogenic and HIV-inhibitory functions of KSHV-encoded chemokines. *Science* 278:290–294.

90. Damon, I., P.M. Murphy, and B. Moss. 1998. Broad spectrum chemokine antagonistic activity of a human poxvirus chemokine homolog. *Proc. Natl. Acad. Sci. U.S.A.* 95:6403–6407.

91. Moser, B., B. Dewald, L. Barella, C. Schumacher, M. Baggiolini, and I. Clark-Lewis. 1993. Interleukin-8 antagonists generated by N-terminal modification. *J. Biol. Chem.* 268:7125–7128.

92. Simmons, G., P.R. Clapham, L. Picard, R.E. Offord, M.M. Rosenkilde, T.W. Schwartz, R. Buser, T.N.C. Wells, and A.E. Proudfoot. 1997. Potent inhibition of HIV-1 infectivity in macrophages and lymphocytes by a novel CCR5 antagonist. *Science* 276:276–279.

93. Mackay, C.R., A Lanzavecchia, and F. Sallusto. 1999. Chemoattractant receptors and immune responses. *The Immunologist* 7:112–118.

94. Homey, B., and A. Zlotnik. 1999. Chemokines in allergy. *Curr. Opin. Immunol.* 11:626–634.

17 | Complement and Complement Receptors in Autoimmunity

Elahna Paul and Michael C. Carroll

1. INTRODUCTION

The role of complement in human autoimmune disease is multifaceted, and its involvement in pathogenesis varies from disease to disease. Complement's participation in autoimmunity is most obvious in systemic antibody-mediated diseases complicated by accumulation and deposition of immune complexes (ICs), of which systemic lupus erythematosus (SLE) is the prototype. Organ-specific diseases that seem to have only monoclonal or oligoclonal autoantibody expansion, such as thyroiditis or myasthenia gravis, and T cell-mediated diseases, such as diabetes mellitus and multiple sclerosis, have not yet been causally linked with abnormalities in the complement system.

The ability of complement to amplify inflammation at sites of IC deposition with concomitant depletion of complement components during disease flares suggests that at least some sequelae of autoimmune disease is complement-mediated. However, genetic deficiencies in complement components are, paradoxically, among the strongest of SLE susceptibility genes (1–3), implying that complement's contribution to autoimmune disease is not limited only to those end-stage effector functions that engender progressive organ damage. Indeed, evidence reviewed here suggests that complement participates in immunoregulation of the humoral immune response and may have a role in the induction of normal B cell tolerance.

Following a brief overview of complement in SLE and a synopsis of the complement cascade with its cast of characters, this chapter discusses the housekeeping role of complement in clearance of immune complexes and its stimulatory role in the primary and anamnestic phases of humoral immune responses. Finally, drawing on observations of humans and experimental data from complement gene knockout mice, we propose that tolerance induction is an important mechanism whereby an intact complement system protects against autoimmune disease.

2. COMPLEMENT IN HUMAN SLE

Immune dysregulation in rheumatic diseases, such as SLE, has been investigated from every imaginable vantage point, and clearly there are multiple factors in the immune system that impinge on the autoimmune phenotype. Undoubtedly, without T cells there would be no disease, and yet the putative lupus defect does not reside exclusively within the T cell compartment (4). Similarly, without B cells there would be no disease, yet B cells alone do not cause it (5). Stable foundation has been laid demonstrating that the interactions between various branches of the immune system create autoimmunity (6), and complement is almost certainly one of the culprits. Naturally occurring abnormalities of the complement system therefore provide a springboard to begin dissecting complement's role in immune responses.

One might predict that excess levels of various complement factors could potentiate disease, but such a scenario is not a naturally observed phenomenon in humans. Far more common are deficiencies in various components of the complement system, either primary genetic or secondary acquired (1–3). As in all gene defects, the genetic deficiencies may be absolute, *i.e.* null mutations with no gene product, or functional, *e.g.* point mutations interfering with normal activity of the encoded protein. Deficiencies in many components of the complement system manifest as an increased susceptibility to infection, both acute and chronic, with a decreased ability to clear invading pathogens. In contrast to distal cascade deficiencies, abnormalities of the early classical pathway of complement (components C1, C2, and C4; figure 1) are each associated with an impressive incidence of SLE and other rheumatic diseases (1–3). Although individuals with congenital complement deficiencies constitute only a tiny cohort of all human SLE, this strong association has implicated an important role for complement in the regulation of autoimmunity.

Secondary acquired deficiencies of classical pathway components also correlate with autoreactivity. For example, patients with congenital deficiency of C1-inhibitor have an increased incidence of SLE after unopposed depletion of their C4 and C2 stores, and C4 inactivation has been demonstrated in patients with drug-induced lupus (3). Moreover, phenotypic differences between alternative pathway deficiencies (Factor I, properdin or even C3) in contrast to C1q or C4 deficiency, suggest that the classical complement components participate in immune regulation through mechanisms independent of full cascade activation (1–3).

The role of classical complement in normal immune regulation can be investigated by analysis of humans and animals with spontaneous and/or induced complement deficiencies. Human SLE is primarily an immune complex (IC) disease. Macromolecular complexes of antigen, antibody, complement fragments and other serum proteins, such as serum amyloid P component (SAP), can become insoluble and deposit in a variety of locations, including skin and kidney. C1 or C4 deficiency may lead to an inappropriate accumulation of IC by interrupting normal clearance mechanisms, a situation where poor immune housekeeping leads to buildup of immune debris, followed by IC deposition, and so on (7). Alternatively, or perhaps additionally, C1 or C4 deficiency may predispose to the very generation and overproduction of the autoantibodies found in SLE immune complexes. There is no debate whether IC deposition triggers inflammation and eventual irreversible tissue damage by unleashing the complement cascade. However, while this complement-mediated pathology is clinically devastating, it is only a distal repercussion of more proximal immune dysfunction, particularly as those very individuals lacking early complement function are the ones with increased susceptibility to late IC-mediated damage. Thus in autoimmune prone individuals, the absence of complement is manifest not only in impaired handling and clearance of soluble IC, but also upstream in the very genesis of immune complex disease. If complement deficiencies impair the regulation of autoreactive B cells and thereby predispose to a break in tolerance, then the paradox of complement deficient SLE is explained.

2.1.   Lupus Autoantigens

If we can delineate what breaks tolerance and elicits the pathogenic autoantibodies of SLE then we may be able to design interventions to abrogate the autoimmune response before immune complexes develop, deposit, activate complement, and culminate in irreversible and potentially fatal organ damage. The autoantigens prominent in SLE and SLE-like diseases are themselves one of the intriguing mysteries of disease pathogenesis. Relatively unique to the systemic rheumatic diseases, these autospecificities include nucleic acid molecules (DNA and RNA in various configurations) and the proteins they bind (e.g. histones, splicesomes). Antibodies against double stranded DNA in particular, are hallmarks of SLE.

Apoptosis is a common theme unifying the lupus autoantigen array (8). Many of these antigens are concentrated in the surface blebs and apoptotic bodies of cells undergoing programmed cell death (9). Externalization of these intracellular and nuclear antigens during apoptosis is thought to be a prerequisite for the selection and activation of autoreactive B cells in individuals predisposed to SLE. Furthermore, since defects in the classical pathway predispose to autoreactivity, one infers that complement's normal role in this process, if any, is likely to be protective.

3.   OVERVIEW OF THE COMPLEMENT SYSTEM

While our understanding of the complement system has progressed rapidly over the recent past, it has become clear that this immunomodulatory network has a greater scope than initially appreciated. The interacting molecules include cell surface and soluble serum proteases, regulatory molecules, and receptors (1,3,10), with many bridges linking the innate immune system to adaptive, lymphocyte mediated-responses. In contrast to the adaptive immune system with rearranging, epitope specific T and B cell receptors, the "primitive" system of innate immunity has evolved to recognize conserved antigens shared by a number of non-mammalian pathogens. Complement activation entails serial activation of a team of serum proteases either by antibody bound antigen or by conserved microbial epitopes. This sequential enzymatic cascade culminates either in direct cytotoxic execution of offending pathogens or in recruitment and activation of accessory phagocytes. A basic review of the cascade(s) as presented below is a prerequisite for the discussion that follows, describing complement's role in the regulation of autoimmune disease and the induction of B cell tolerance.

3.1.   The Alternative Pathway

Evolutionarily, the alternative pathway of complement is probably the oldest arm of the innate immune system, constituting an early line of defense against microbial invasion. A rapid, nearly instantaneous response to foreign pathogens is feasible because the system is always "on", with spontaneous, albeit low level C3 activation coupled with constitutive inhibition that can be withdrawn as needed. Activation of the C3 molecule entails a conformational change to expose a highly reactive thioester group that forms covalent bonds indiscriminately with any hydroxyl or amino group in the vicinity. Autologous cells are protected by cell surface regulatory molecules such as MCP (membrane cofactor protein) that promptly inactivate activated C3, whereas C3 bound to non-mammalian MCP-negative cells triggers the rapid fire amplification loop of the alternative pathway cascade. If no covalent bonds form, the fluid phase C3 molecule is inactivated by hydrolysis.

The cascade of the alternative pathway begins gradually when activated C3 binds factor B in the presence of factor D to form C3B. Factor D cleaves C3B to release fragment Ba and generate active C3Bb (figure 1). The cascade accelerates as this C3Bb complex combines with properdin and cleaves fresh C3 molecules (the most abundant complement protein in serum) into C3a and C3b. C3b then recruits more factor B and proteolytic factor D, releases more Ba fragment and generates the C3bBb complex. This complex is called C3 convertase (of the alternative pathway), and will continue generating more C3b until all intact C3 is consumed or until the convertase is inactivated by factor I. The new C3b molecules either remain in the amplification loop forming additional C3 convertase, or progress through the effector limb of the complement pathway (see below). Like C3 itself, if the C3b fragments don't covalently bind to adjacent structures, then they are inactivated by hydrolysis to iC3b.

3.2. The Classical Pathway

Less primitive than the alternative pathway, the classical limb of the complement system probably constitutes the principle site of interaction between innate immunity and the lymphocytic adaptive immune system. This bridge between innate and adaptive immunity is the point of susceptibility where complement deficiencies predispose to autoimmune disease.

The classical pathway begins with C1 (comprised of an antibody and/or antigen binding subunit called C1q and two

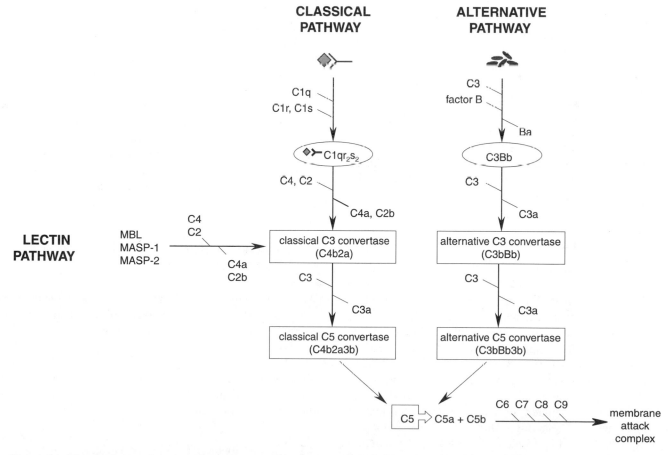

Figure 1. This depiction of the complement cascade shows three triggering pathways (the alternative, classical and lectin pathways) and the direct cytotoxic effector limb that generates the membrane attack complex. See Abbas et al. (10) for a more comprehensive review that includes additional regulatory components of the cascades.

serine proteases C1r and C1s). C1q depends primarily on antibodies for activation, and only then undergoes a conformational change permitting its association with C1r and C1s. There is no spontaneous firing of the classical pathway. C1q is a multivalent molecule with six binding sites for antibody Fc regions and C1q is activated only after two or more of these sites fill. While technically responsive to a single pentameric IgM or a pair of IgGs, the classical cascade is most efficiently triggered when a multivalent antibody/antigen immune complex crosslinks several C1q molecules. After a series of autocatalytic cleavage steps, the fully activated C1 complex then binds and sequentially cleaves C4 and C2 (figure 1). When activated C1 releases C4a from C4 it exposes yet another highly reactive thioester moiety on the remaining C4b fragment. Much like C3b, C4b can form covalent bonds with nearby molecules (see below). By releasing fragments C4a and C2b, activated C1 generates the C4b2a complex, otherwise known as C3 convertase (of the classical pathway). This C3 convertase, like the other, cleaves C3 into C3a and C3b.

3.3.   The Lectin Pathway

The lectin pathway is the most recently recognized branch of the complement system (11). It is triggered when polysaccharides bind to mannan-binding lectin (MBL), a serum protein with structural and functional homology to C1q. MBL associated serine proteases (MASP-1 and MASP-2) are similar to C1r and C1s, and together these three molecules sequentially cleave C4 and C2, feeding into the classical pathway and generating the C3 convertase, C4b2a (figure 1). Defects in MBL may be weak susceptibility genes for human SLE because allelic polymorphisms leading to diminished serum titers of this lectin have been associated with disease (11–14).

3.4.   Effector Functions

All branches of the complement system converge at C3. Continuation of the cascade beyond C3 is one effector function of the innate immune response against invading pathogens. C3b moieties generated by C3 convertase, either by C3bBb (alternative) or by C4b2a (classical), reassociate with these catalytic complexes to form C5 convertase of either the alternative (C3bBb3b) or classical (C4b2a3b) pathways, respectively (figure 1). The C5 convertase, already bound to microbial surfaces via the thioester bond of the first C3b fragment, now cleaves C5, releasing the C5a fragment and retaining C5b. C5b becomes a docking site for C6, the next molecule in the cascade, which in turn sequentially recruits C7, C8 and C9. Assembled C6–C9 constitute the membrane attack complex, a pore forming macromolecular structure that inserts into the plasma membrane of hapless pathogens and thus provokes their ultimate demise.

Proteolysis at various steps of the cascade generates soluble fragments that mediate a second effector function of the complement system, amplifying the immune response beyond a cascade of molecular proteolysis to involve the cell-mediated immune system. The C3a, C4a and C5a fragments are potent vasodilators and leukocyte chemoattractants that recruit and activate both phagocytes and lymphocytes. This amplification of the cell mediated immune response can lead to significant bystander injury to tissues of the surrounding microenvironment, and is likely one molecular mechanism in the pathogenesis of immune complex disease.

These two effector functions, direct cytotoxicity and amplification of the cellular immune response, are the principle ways the innate immune system protects the host against microbial infection. It is thus easy to understand how genetic deficiencies of complement proteins can lead to problems with recurrent and/or chronic bacterial infections. On the other hand, complement deficiency that predisposes to autoimmune disease is harder to understand, but probably arises from those complement effector functions that impinge on the adaptive immune response, primarily via the humoral arm of the immune system. Complement participates in the handling of antigen and trafficking of antibody/antigen immune complexes. As a housekeeping function, C3b and C4b fragments covalently bind IC and transport them via complement receptors to the reticuloendothelial system of the liver and spleen where they are cleared from the circulation (7). Not only does complement mediate IC clearance, but emerging data suggest that complement can specifically deliver antigen to lymphoid organs, optimizing antigen processing and presentation in the follicular sites where B and T cells are concentrated together.

4.   ANTIGEN MODIFICATION AND IMMUNE COMPLEXES

Antigen is recognized and chaperoned through the body by the concerted action of numerous serum proteins. Binding to antigen, either covalently (e.g. C3b and/or C4b fragments) or non-covalently (e.g. antibody), is a method of tagging the antigen. Further processing is then carried out by cells expressing the appropriate cognate receptors, such as Fc receptors for antibody and complement receptors for the complement fragments.

4.1.   Non-Covalent Tagging of Antigen

Immune complexes contain multiple proteins in organized, lattice-like arrays. Antigens within the complexes can be self or foreign, and are not necessarily proteins. Antibodies within the complexes are antigen specific, except perhaps when rheumatoid factor is involved, and the multivalency that permits antigen crosslinking can lead to the formation of rather large and ultimately insoluble multimeric networks. C1q binding to Fc regions within the complexes can limit further growth of the complexes and can sometimes resolubilize those that have precipitated. Complexes that form *in situ* around antigens that are inherently insoluble, such as the glomerular basement membrane, create entirely different problems that are beyond the scope of this discussion (15).

C1q binds to the Fc region of immunoglobulin (Ig) in the complexes in a specific but antigen-independent manner. C1q tagged IC can then propagate the classical cascade, leading to sequential activation and deposition of C4 and then C3 fragments within the array (see below) (7). Alternatively, C1q tagged IC that bind to leukocytes bearing C1q receptors can trigger superoxide production and/or enhance phagocytosis (16).

A variety of non-Ig serum proteins show a propensity for direct binding to those autoantigens that have become known as lupus autoantigens. C1q itself binds to apoptotic blebs independent of Ig (17), and the acute phase reactant serum amyloid P component (SAP) can bind chromatin and activate C1q (18). Mice deficient in SAP develop spontaneous SLE-like autoreactivity, as do mice deficient in C1q, possibly arising from impaired handling of DNA in immune complexes (19,20).

4.2.   Covalent Modification by C3 and C4

Covalent modification of antigen, either alone or in immune complexes, is the basis of complement's ability to mediate both antigen clearance and B cell immunoregulation. As introduced above, cleavage of either C4 or C3 exposes a highly reactive thioester group that is a target of nucleophilic attack by amine and/or hydroxyl groups, thus forming covalent amide and/or ester bonds between C4b or C3b and nearby molecules (*i.e.* antigens). Due to the protective effects of autologous complement regulators, haphazard activation and consumption of complement components is minimized, and C4b and C3b fragments are focussed primarily to foreign antigens or to antigens in immune complexes already tagged with Ig and/or C1q.

The fate of the covalently modified antigens is a subject of much interest and active investigation. Covalent modification with C3b has been long appreciated as a method to opsonize foreign molecules, promoting their uptake and clearance by the very phagocytic cells already recruited by the C4a and C3a chemokines generated early in the cascade. Phagocytosis of immune complexes can be accomplished via macrophage and neutrophil Fc receptors (FcRs) that bind to the Ig constant region within the

antigen/antibody complex. Alternatively, most phagocytes also express cell surface receptors for C3b and thus circumvent the need for antigen-specific Ig in their clearance of complement-opsonized particles, a form of antibody independent humoral immunity.

In contrast to phagocytosis at the single cell level, particles and complexes modified by C4b and C3b fragments can bind to "immune adherence receptors" for macroscopic clearance from the circulation. Probably CR1 on human erythrocytes and as yet unidentified complement receptors on murine neutrophils and platelets render these circulating cells transporters of immune "debris". As these cells percolate through the liver, and possibly the spleen, they are stripped of the complexes through undelineated mechanisms and are returned to the circulation (7,16). It remains unclear how these immune complexes are then handled- *i.e.* are they degraded and recycled within the routine housekeeping function of the reticuloendothelial system? Or are they actually processed and presented to lymphocytes of the adaptive immune system?

In addition to their role in assisting with macroscopic and cell mediated clearance of ICs, covalent modification by complement fragments helps clear complexes at a subcellular level. Similar to C1q tagging, C4b and C3b deposition inhibit precipitation of ICs and help solubilize those already deposited.

5.   COMPLEMENT AND THE ADAPTIVE IMMUNE RESPONSE

Complement receptors on phagocytes, erythrocytes and platelets—immune adherence receptors— are necessary for the sequestration and clearance of immune complexes decorated with complement fragments. A variety of complement receptors have been located on other cell types as well, including B cells and follicular dendritic cells (FDCs; figure 2). In addition to their elevated copy number per cell, the very presence of receptors on these latter cell types raises questions and tantalizing possibilities regarding complement's role in modulation of lymphocyte function, particularly since animals deficient in the receptors or their ligands have impaired humoral responses to T dependent (TD) antigens.

5.1.   Complement Receptors

Complement receptors CR1 and CR2 (CD35 and CD21, respectively) are encoded by two distinct genes in humans and by one alternately spliced gene (*Cr2*) in mice (21,22). CR2 is a homologous, somewhat truncated version of CR1. Both receptors bind an assortment of complement fragments including C3b, C4b, and/or their derivatives (figure 2). Human CR1 also binds C1q (23). Levels of these recep-

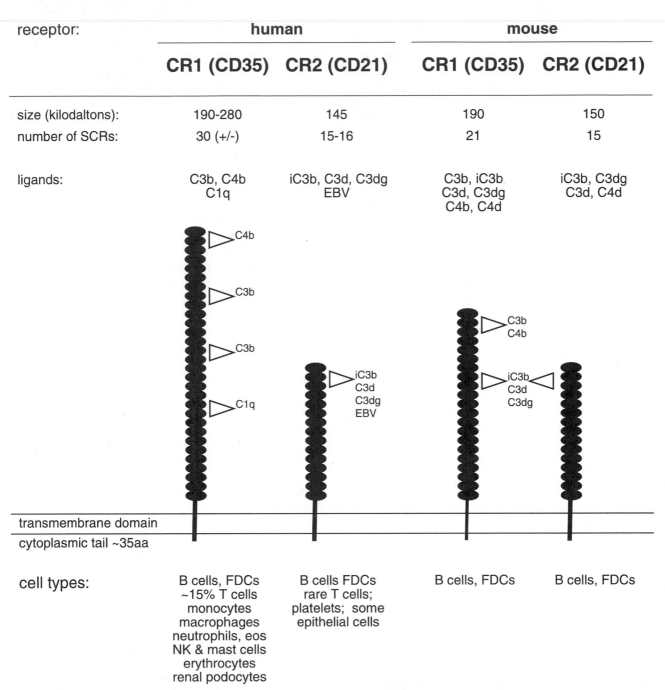

receptor:	human		mouse	
	CR1 (CD35)	**CR2 (CD21)**	**CR1 (CD35)**	**CR2 (CD21)**
size (kilodaltons):	190-280	145	190	150
number of SCRs:	30 (+/-)	15-16	21	15
ligands:	C3b, C4b C1q	iC3b, C3d, C3dg EBV	C3b, iC3b C3d, C3dg C4b, C4d	iC3b, C3dg C3d, C4d
cell types:	B cells, FDCs ~15% T cells monocytes macrophages neutrophils, eos NK & mast cells erythrocytes renal podocytes	B cells FDCs rare T cells; platelets; some epithelial cells	B cells, FDCs	B cells, FDCs

transmembrane domain

cytoplasmic tail ~35aa

Figure 2. CR1 (CD35) and CR2 (CD21) are encoded by two human genes and one alternately spliced mouse gene. Complement fragment ligands are listed in the figure and their known binding sites are depicted schematically. A 60–70 amino acid domain known as a short consensus repeat (SCR) is a tandemly repeated structural motif recurring in many molecules of the complement system. Minor sequence variations within different SCRs translate into altered ligand specificities. Human CR1 is fairly polymorphic; isoforms of different molecular weights (ranging from 190 to 280 kilodaltons) differ primarily in the number of SCRs and hence in the number of C3b fragment binding sites. The most common human CR1 allele is depicted here. In addition to binding complement fragments, CR1 also acts as a cofactor to factor I, an inactivator of C3b and C4b. CR1-bound C3b (or C4b) can be cleaved into CR2 ligands C3dg (or C4d), possibly mediating ligand transfer from one receptor type to another. *(Abbreviations: CR, complement receptor; SCR, short consensus repeat; EBV, Epstein-Barr virus; aa, amino acids; FDC, follicular dendritic cell; NK, natural killer cell; compiled from references 3, 7, 10, 16, 28, 32.*

tors are variably diminished in some patients with SLE (24, 25). While null mutations of human complement receptor genes have not yet been described, polymorphic alleles encode isoforms that differ in both the relative density of cell surface expression as well as the absolute number of complement fragment binding sites per receptor (3). As yet, no firm disease associations have been ascribed to particular complement receptor alleles. Meanwhile, mice deficient in these receptors have been created by targeted deletion of the Cr2 gene locus (26,27). Since Cr2 encodes both CR1 and CR2, these knockout mice lack both receptors, referred to collectively as CR1,2.

A variety of experimental approaches have demonstrated that complement participates in the humoral immune response to TD antigens. Serum can be depleted of complement using cobra venom or via gene knockout technology; complement receptor-ligand interactions can be blocked with anti-receptor antibodies or with soluble receptor analogues (rev. in (28)). In all cases, and most dramatically in CR1,2 deficient animals, immune responses are impaired in a number of ways. Titers of antigen specific IgM and IgG antibodies are diminished, class switching to all IgG isotypes is minimal, and the secondary memory responses lack robust affinity maturation (26,27,29). In some experimental systems, the inadequacy in Ig production is mirrored by alterations in splenic micro-architecture. The B cell follicles in the splenic white pulp remain intact, but there is a reduction in both the size and number of germinal centers (GCs) elicited by immunization with TD antigen (26).

5.2. B Cell Activation

B cells express a plethora of cell surface receptors that transduce combinatorial signals, either positive or negative, impinging on the activation status of the B cell. Negative signal transduction increases the threshold for B cell activation, thus favoring a resting state. The activation threshold is crossed with adequate positive stimulation, pushing the cell to proliferate, class switch, somatically mutate, and/or differentiate further into plasmacytes or memory cells. Cell surface receptors relevant to this discussion include surface IgM [i.e. the B cell receptor (BCR)] that binds antigen, B cell CR1,2 that binds complement fragments, and B cell FcRs that bind Ig. Immune complexes comprised of antigen, antibody and complement fragments contain multiple binding sites for each of these receptors.

CR1,2 is first detectable on the surface of the developing immature B cell and is maximally expressed on the mature resting B cell, co-localized with CD19 (30,31). These complement receptors have very short cytoplasmic tails of approximately 35 amino acids and CD19 is the probable conduit for CR1,2 triggered signal transduction. The two molecules co-localize on the cell surface and co-precipitate

in vitro, and anti-CD19 antibodies can mimic effects of CR1,2 engagement (3,31,32). As positive regulators of B cell activation, co-ligation of CR1,2 and the BCR are additive (33,34). In the presence of the appropriate cognate antigen, engagement of the complement receptor by C3 or C4-derived fragments lowers the threshold for B cell activation (35). In other words, the same amount of antigen is a more potent trigger of B cell activation when it is coated with complement (36). This co-stimulation is one clear example of how complement regulates the adaptive immune response (28).

The impaired reactivity of CR1,2 deficient B cells can be overcome in part by immunization with adjuvant and/or concentrated doses of high affinity antigen (26,27,37). Under these conditions, a stronger signal through the BCR can partially substitute for CR1,2-mediated signal transduction, at least enough to attain the B cell activation threshold in the absence of complement receptor. Most natural immunization, however, is relatively low dose and without adjuvant and so complement receptor signal transduction becomes important for B cell activation outside of the laboratory setting.

It may be that the stoichiometry of antibody, antigen, and complement fragments within the immune complexes, as well as the relative affinities and avidities of each component for its receptor (i.e. FcR, BCR, and CR1,2) provide an internal system of regulatory checks and balances. Antigen specific antibody secretion that is triggered via the BCR and the CR1,2 coreceptor may be ultimately shut down through a combination of negative signaling pathways and withdrawal of positive stimulation. Activated B cells downregulate surface BCR and coreceptor, effectively withdrawing further stimulatory signals through this pathway, and local concentrations of secreted Ig increase enough to preferentially engage the inhibitory FcR, effectively quenching the humoral response (30,38).

5.3. Complement-Deficient Mice

As the ligands for complement receptors, deficiencies in complement components themselves, particularly the early components of the classical pathway, also impinge on the humoral immune response. Similar to CR1,2-deficient mice, mice deficient in C1q, C4 or C3 have impaired primary and memory responses (39–41). Moreover, reminiscent of their human counterparts, these classical complement-deficient mice are predisposed to autoimmune disease (20,42).

Deficiency in C1q diminishes titers of antigen specific IgG2a and IgG3 in both primary and secondary responses to TD antigens. Class switching to these two particular IgG subclasses is driven by interferon gamma and, interestingly, T cells from C1q-deficient mice have diminished secretion of this lymphokine in vitro compared to complement sufficient controls (39). These animals develop a lupus-like disease with increased serum titers of anti-nuclear autoantibodies and a proliferative, crescentic glomerulonephritis with prominent glomerular apoptotic bodies by light microscopy, IgG and C3 immune complexes by immunofluorescence, and electron dense subendothelial and subepithelial deposits (20). As yet there are no published reports of mice uniquely deficient in C1r, C1s, or C2.

C4- and C3-deficient animals are similar to the *Cr2–/–* animals in a variety of ways, including impaired B cell memory (40,41). Their secondary immune response against TD antigens is characterized by reduced antigen-specific IgG titers and decreased size and number of splenic GCs. Nonetheless, there is no evidence of an intrinsic B cell deficit in these mice: polyclonal B cell activation *in vitro* with anti-IgM, CD40-ligand and IL4, as well as mitogens such as PMA and LPS, is indistinguishable from wild type. In addition, CD4 positive T cells from complement-deficient mice are also apparently normal in number and in their ability to support antigen-specific class switching (40). Intriguingly, it seems that local synthesis of C3 by macrophages in the splenic follicle, rather than systemic C3 secreted by hepatocytes, is an important source of ligand for B cell CR1,2. When C3 deficient mice are reconstituted with bone marrow from wild type mice, their C3 serum levels are not restored, but titers of antigen specific antibody and numbers of GCs normalize (43).

5.4.   The Germinal Center Reaction

The very distribution of complement receptors in the splenic white pulp implicates several possible functions for complement in the interactions between B cells, T cells and FDCs. The central T cell-rich zone of the splenic follicle is essentially devoid of CR1 and CR2 receptors, whereas the surrounding B cell zone has copious CR1,2 expression (figure 3). The FDCs that form the infrastructure of the follicle with long thin dendrites intertwined around B and T lymphocytes are also rich in both complement receptors and ligands (*i.e.* C3 fragments). Thus, CR1,2 expression must be explained on both B cells and FDCs. Attracted to FDC rich sub-regions of the follicle, activated B cells undergo clonal expansion to form GC that are dense oligoclonal B cell islands (figure 3). In addition to B cell proliferation, the GC reaction includes B cell differentiation into antibody secreting cells and into memory cells. The process also entails class switching, affinity maturation via Ig gene somatic hypermutation, and possibly light chain receptor editing (44,45).

FDCs, originally thought of as the stromal cells of the lymphoid follicle, seem to have an accessory role in B cell activation and are also long-term reservoirs of antigen, possibly important for maintenance of B cell memory and/or anergy. Purified mouse FDCs amplify B cell responses

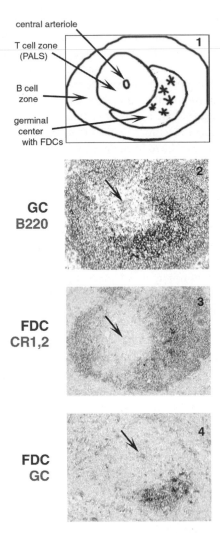

Figure 3.   Panel 1 is a schematic representation of a murine splenic follicle. The peri-arteiolar lymphoid sheath (PALS) is organized around a central arteriole, with a central T cell rich zone surrounded by a circumferential B cell zone. Germinal centers (GCs) develop within B cell regions of the follicle that are particularly rich in follicular dendritic cells (FDCs). Central arterioles are indicated with arrows in each of the three colored panels. **Panel 2** shows B cells in red (stained with rat anti-mouse B220/CD45R clone RA3-6B2) and a large germinal center in blue (stained with peanut agglutinin). **Panel 3** shows B cells in red (stained with rat anti-mouse CR1,2/CD21/CD35 clone 7G6) and FDCs in blue (stained with M1 clone 209). **Panel 4** shows blue FDCs within a red germinal center. (*Photographed at 200x magnification; immunohistochemistry courtesy of A. Shimabukuro-Vornhagen*).

when co-cultured with antigen and T cells, and the effect is C3 and CR1,2 dependent (46). Antigens modified covalently by C3/C4 fragments and/or non-covalently by C1q or Ig tagging can bind to complement receptors or to FcRs on FDCs (figure 4). These complexes are displayed on the FDC surface for retrieval by B cells and amazingly can still be detected *in vivo* months after immunization (47).

Evaluation of mice deficient in FcR or in CR1,2 has demonstrated that the FDC complement receptor is required for FDC-mediated amplification of B cell responses whereas the FDC FcR is expendable. FcR deficient mice actually have increased FDC-mediated antigen trapping (48). Furthermore, FDCs isolated from C3-deficient mice, presumably deficient in C3b coated antigen and thus devoid of FDC cell surface immune complexes, are unable to augment B cell responses (46). Taken together, these observations suggest that FDC complement receptors concentrate complement-decorated antigen within the follicle, offering immunogens to B cells in a setting where T cell help is also available.

Mice expressing transgenic BCRs have been useful tools to study the fate of specific B cell clones during an immune response (49). For example, splenocytes from mice transgenic

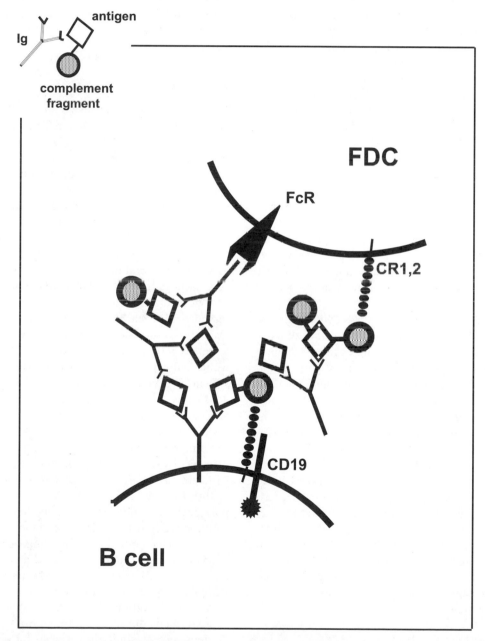

Figure 4.   Immune complexes comprised of antibody (shown as Y-shape) and antigen (shown as diamonds), with and without covalently bound complement fragments (shown as circles) are concentrated in the lymphoid follicle by FDCs and displayed to surrounding B cells. The complexes bind to FDCs via CR1,2 and Fc receptors in a conformational array that leaves some antigenic epitopes, Fc domains and C3b and/or C4d moieties exposed to the surrounding microenvironment. Co-ligation of B cell CR1,2 and BCR by the FDC-anchored complexes triggers signal transduction through CD19, optimizing B cell activation.

for an antibody against hen egg lysozyme (HEL) can be induced to respond to the HEL antigen in the presence of T cell help (50). When the anti-HEL transgenes (heavy and light chain together) are expressed in complement-deficient mice, one can further dissect the role of complement in an antigen driven T cell-dependent humoral response. In one adoptive transfer system, anti-HEL donor B cells are introduced into wild-type mice immunized with low affinity antigen (duck egg lysozyme; DEL). With immunohistochemistry, these transgenic cells can be tracked visually as they traverse the terrain of the recipient splenic follicles and quantitative evaluation is obtained via flow cytometry (37). Over the course of five days, the size of the anti-HEL donor cell population is fairly stable and the cells home to the splenic follicle, where they participate in GC reactions. This paradigm is interrupted when the anti-HEL donor B cells are CR1,2 deficient. The overall survival of the receptor deficient donor population is impaired, follicular localization is poor, and virtually no anti-HEL B cells are detected in GCs. Higher affinity immunogens (turkey egg lysozyme; TEL) tend to improve follicular survival of donor cells, without ameliorating the GC response. Thus, these experiments delineate follicular- and GC-survival as two independent components of B cell positive selection (37).

B cell activation is accompanied by alterations in the composition of the cell surface, including downregulation of the BCR, CD19, and CR1,2 and upregulation of MHC class II and B7.2 (51–53). These latter molecules are vital in cognate B-T cell interactions. The B cell requires T cell help to complete its GC reaction of class switching and affinity maturation. On the other hand, the B cell itself may be the antigen presenting cell, activating cognate T helper cells to begin with (6). B cells are generally poor APCs except when they bear antigen-specific immunoglobulin receptor and interact with antigen-specific T cells. After binding to surface BCR, antigen can be internalized, processed and presented in the context of class II MHC, efficiently activating antigen-specific T cells (51). This process is even more efficient when the BCR bound antigen is opsonized with complement (*i.e.* C3b) so that the CR1,2 coreceptor is also engaged, or possibly even engaged instead of the BCR and thus eliminating the need for the antigen specificity of the surface IgM (52,54). While immune complexes could theoretically be internalized via FcRs as well, *in vitro* data demonstrate that FcR engagement inhibits B cell activation of T cells, suggesting that the complement receptor is more important than the FcR for antigen presentation (34).

6.   B CELL TOLERANCE

As discussed to this point, complement-modified antigen and complement receptors feature prominently in the posi-

tive selection of B cells by TD antigens. However, these vital effects on the humoral immune response fail to suggest how complement deficiencies predispose to autoreactivity. Complement's role in the clearance of immune complexes may contribute to the development of autoimmune disease, where poor immune housekeeping in the setting of complement deficiency permits buildup of stimulatory autoantigens, sufficient to break tolerance. A novel and somewhat controversial model accommodating both complement's role in handling immune complexes as well as its ability to stimulate B cells, proposes that complement also participates in the negative selection of autoreactive B cell clones (55,56).

6.1.   Natural Autoantibodies

The anti-DNA antibodies that are the hallmark of SLE tend to be products of antigen-driven T cell-dependent immune responses, as evidenced by class switching and affinity maturation via somatic hypermutation (57). Although apoptotic material contains multiple autoantigens associated with pathogenicity, the antigen eliciting the canonical anti-DNA response of SLE remains elusive, as do the specificities of the cognate helper T cells (6,15). Not all anti-DNA antibodies are pathogenic, however, and indeed benign autoreactive antibodies are found in healthy individuals. These germline-encoded, moderate to low affinity, polyreactive, IgM anti-DNA antibodies are members of a pool of natural autoantibodies that seem to be highly conserved both in early development and in evolution (58). One infers a survival advantage in skewing the pre-immune B cell repertoire towards autoreactivity, otherwise why would autoantibodies be tolerated in a healthy immune system? We have postulated that natural autoantibodies are not merely tolerated but are in fact required to establish a normal Ig repertoire, and that together with the complement system, establish the foundation for B cell self-tolerance to some highly conserved soluble antigens (55,59). Despite their relatively low affinity for autoantigen, the pentameric structure of the IgM natural autoantibody increases its avidity for both antigen and for C1q. Combined with its polyspecificity, this enhanced avidity enables a single autoreactive IgM pentamer to seed the formation of multimeric immune complexes consisting of antigen, antibody and complement fragments. The C3 and C4 fragments can then mediate efficient delivery and trapping of autoantigen within lymphoid compartments.

A large proportion of the natural autoantibody repertoire is produced by the B cell subset known as B1 cells, and the B1 repertoire is similarly enriched for autoreactivity (60, 61). These relatively primitive B1 cells constitute a large proportion of fetal B cells, but are soon outnumbered by the developing B2 population. In contrast to classical B2 cells that participate in humoral responses to TD antigens, the B1 repertoire is dominated by T cell-independent (TI) antigens. These cells produce germline encoded antibodies of the pre-immune repertoire, rarely undergo class switching or affinity maturation, and are notable for their properties of self-renewal. In the adult, B1 cells are generally sequestered in the peritoneal cavity, whereas B2 cells populate the usual lymphoid organs (60,61).

For reasons yet to be elucidated, animals deficient in either CR1,2 or CD19 maintain a normal size B2 population but are deficient in B1 cells (26,62,63), as are some humans with SLE (64). It may be that B1 self-renewal is dependent on combined stimulation via the BCR and CR1,2. In the absence of CR1,2 signaling, perhaps there is inadequate impetus to maintain proliferation of this lymphocyte population. Whether this translates to a paucity of natural autoantibodies with associated repercussions is currently under investigation. For example, if pre-existing natural autoantibodies mediate delivery of autoantigen to the lymphoid compartment where developing B cells are tolerized, then their absence could interfere with normal tolerance induction.

6.2.   A Unifying Hypothesis

Combining the principles discussed above, we present the following hypothesis. Natural autoantibodies, produced spontaneously and also in response to TI antigens, are part of a housekeeping mechanism that clears away apoptotic debris. These IgM autoantibodies form "natural" immune complexes with DNA, RNA, and associated proteins, and also activate the classical complement pathway via C1q. C1q either binds complement receptors directly or propagates the complement cascade, leading to covalent modification of antigen with C4 and/or C3 fragments. While these immune complexes are by and large disposed of within the hepatosplenic reticuloendothelial system, some are trapped by CR1,2-bearing cells in both primary and secondary lymphoid organs. In this manner, the natural antibody-bound, complement fragment-modified, prototypic antigens of SLE are localized to the bone marrow and presented to developing B cells. In the appropriate microenvironment, an autoreactive immature B cell is susceptible to negative selection (65), and plenty of lupus autoantigens are certainly available in the metabolically active milieu of the bone marrow. Like positive selection in the splenic follicle, negative selection also depends on additive signaling through synergistic cell surface receptors. High affinity autoreactive immature B cells (potentially pathogenic) are deleted even upon antigen encounter alone. Low affinity autoreactive B cells, perhaps including B1 cells, escape clonal deletion because they never attain adequate signaling threshold. Moderate affinity BCRs—the presumed culprits of SLE—reach deletion threshold only in conjunction with concomitant signaling through complement receptor. In the absence of classical complement receptor or ligand, the moderate affinity autoreactive B cells escape negative selection in the bone marrow and graduate to the periphery (figure 5). As systemic "natural" immune complexes accumulate in the background of impaired IC clearance, it is only a matter of time before these moderate affinity autoreactive B cells encounter cognate antigen that they can process and present to neighboring T cells. If autoantigen processing by B cells generates unique epitopes not previously displayed by the non-B cell APCs involved in thymic education, then intrinsic defects in T cell tolerance need not be invoked. Eventually, enough T cells are activated to drive class switching and affinity maturation of these moderate affinity autoreactive B cells, culminating in the pathogenic autoantibodies and immune complexes of SLE.

6.3.   Complement Deficiency Impairs Tolerance Induction

A laboratory model of autoimmunity was created from the anti-HEL transgenic mouse by adding a second transgene encoding a soluble form of hen egg lysozyme (sHEL), thus transforming an avian xeno-antigen into a murine neo-autoantigen. This double transgenic system has revealed several mechanisms of B cell regulation, including central tolerance induction via clonal deletion and maintenance of peripheral tolerance via anergy (66). When this double transgenic set-up is superimposed on a complement-deficient background, there are interesting perturbations in the regulation of the anti-HEL humoral response (42).

Single transgenic mice with the anti-HEL transgene express anti-HEL antibodies in the serum and anti-HEL B cells can be easily identified in spleens and lymph nodes. In contrast, double transgenic mice (both sHEL and anti-HEL) on a normal complement background do not have anti-HEL antibodies in the serum. The anti-HEL B cells from the double transgenic animals have encountered antigen, as evidenced by downregulation of surface IgM, and yet they remain relatively immature, as suggested by their CD23 negative phenotype. Their numbers are markedly diminished and they have a reduced lifespan. When examined *ex vivo*, the anti-HEL B cells are hypo-responsive to activation by antigen or by anti-IgM. These observations indicate that, in the double transgenic sHEL model system, regulation of autoreactive B cells is achieved by a combination of clonal deletion (decreased number of anti-HEL B cells) and induction of anergy (decreased reactivity to antigen) (42,66). Similar results have been obtained in different transgenic model systems, where the autoantibody is transgenic but the antigen itself is a ubiquitous self antigen such as DNA (67–70) or Sm protein (71). In these models that substitute natural autoantigens for the HEL neo-autoantigen, the paradigm of mixing clonal deletion and anergy is preserved.

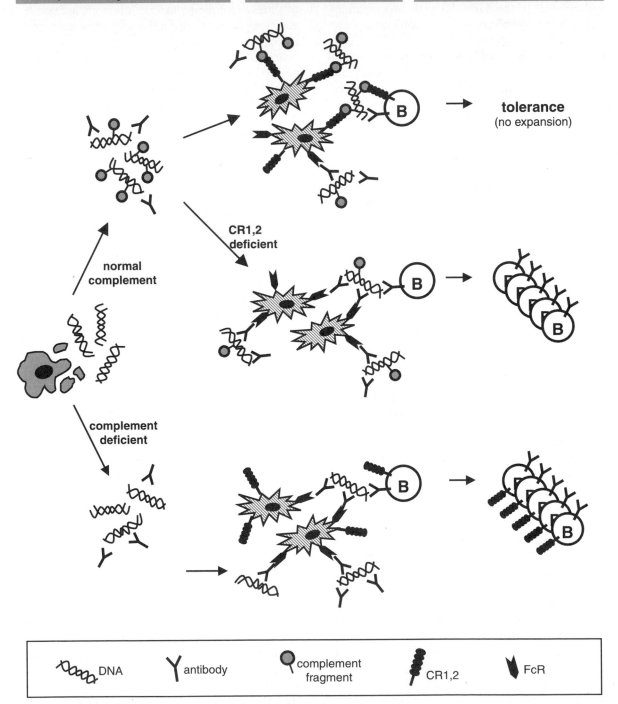

Figure 5. "A unifying hypothesis". Apoptotic debris is regularly released by dying cells during development, aging and life in between. Pre-existing natural IgM autoantibody can bind these prototypical lupus autoantigens and activate the classical pathway of complement, forming immune complexes containing natural autoantibody and covalently modified antigen. Some of the immune complexes are sequestered into lymphoid compartments for trapping and retention by FDCs or FDC-like stromal cells. Effects of BCR and CR1,2 coligation will vary depending on the developmental stage of the B cell; immature B cells are candidates for tolerance induction. In the absence of CR1, 2 or in the absence of classical complement, immune complexes are trapped less efficiently, B cell CR1,2 is not engaged, and autoreactive B cells escape tolerance induction.

When the sHEL/anti-HEL transgenes are bred into mice deficient for CR1,2, regulation of autoreactive B cells is abnormal (42). Although anti-HEL Ig is still undetected in the serum, the number of anti-HEL B cells is increased and their lifespan is longer when compared to complement-sufficient double transgenic animals. Many of the lymph node anti-HEL B cells are CD23-positive, suggesting that maturation of the autoreactive lymphocytes is not arrested. In addition, these B cells respond to HEL antigen *ex vivo* and are therefore not fully anergized. Together these findings implicate a role for complement in normal induction of tolerance to self antigen.

In all likelihood the complement ligand that binds CR1,2 in the induction of anti-HEL tolerance derives from the classical complement pathway because similar autoreactive phenotypes also arise in the absence of C4, but not C3. For example, after bone marrow transplant from anti-HEL donors into sHEL hosts, anti-HEL B cells are tolerized reasonably well even when the recipients are C3-deficient. In contrast, the anti-HEL B cells that develop in the absence of C4 (*i.e.* transplantation into C4-deficient animals) are not properly anergized. Much like the CR1,2-deficient double transgenics, these anti-HEL B cells are present at increased frequency and retain the ability to respond to antigen *ex vivo* (42). These observations suggest that the classical pathway is the principal mediator of CR1,2 based B cell signaling for tolerance induction, possibly in the immature B cell, whereas C3 fragment-mediated complement signaling is reserved for the activation of mature B cells.

7.   COMPLEMENT IN MURINE MODELS OF SLE

Various strains of mice are recognized as model systems of lupus because they develop spontaneous autoimmune disease much like human SLE (72,73). The MRL/*lpr* mouse, for example, features prominently in the murine lupus literature. The MRL genetic background is predisposed to autoimmunity, mechanisms unelucidated, and the *lpr* gene is a recessive, non-productive allele of Fas that, when homozygous on the MRL background, generates a lymphoproliferative phenotype with high serum titers of anti-nuclear antibodies and progressive glomerulonephritis.

Evaluation of these mice has unveiled clues to the role(s) of complement in autoimmunity. For example, the B cells in MRL/*lpr* mice express progressively diminishing levels of surface CR1,2 beginning at approximately 7 weeks of age, fairly early in the course of their disease (30). Since complement receptor is downregulated before serum autoantibody titers rise and before nephritis develops, one might infer that inappropriate modulation of the receptor is a predisposing condition leading to autoimmune disease.

However, impairment of tolerance is likely underway well before manifestations of clinical disease are evident. Consistent with observations of anti-DNA B cells from 3H9 transgenic mice (see below), these data may simply reflect a history of antigen exposure. Emerging evidence suggests that surface CR1,2 is coordinately regulated with the BCR, and that downregulation of surface IgM upon antigen encounter is accompanied by CR1,2 downregulation as well. 3H9 heavy chain mice are transgenic for a rearranged Ig heavy chain gene constructed from an MRL/*lpr* anti-DNA hybridoma called 3H9. Anti-DNA B cells isolated from these mice are anergic and express low levels of surface IgM and complement receptor relative to non-autoreactive controls (67). The transgenic B cells presumably encounter autoantigen fairly frequently *in vivo* and the ensuing co-downregulation of BCR and CR1,2 reflects that exposure. Therefore, it may be that the generalized paucity of surface CR1,2 on B cells from some humans with SLE (24,25) is a marker of high frequency encounters of autoreactive B cells with ubiquitous autoantigen, rather than a causal feature of autoimmune disease.

In general, the lupus-like phenotype of MRL/*lpr* mice varies significantly as the *lpr* mutation is moved to different genetic backgrounds. While the MRL strain is prone to autoimmune disease, many other strains are not. When the *lpr* mutation is homozygous in C57BL/6 (B6) mice or mixed B6 and 129 mice, there is no evident propensity for autoimmunity. By 17 weeks of age B6/*lpr* mice have ANA and anti-dsDNA levels in the normal range whereas titers in MRL/*lpr* mice are quite elevated. When the *lpr* gene is bred into non-MRL complement deficient mice there is a recapitulation of the MRL/*lpr* lupus-like phenotype. Despite the absence of the MRL genetic background, mice homozygous for *lpr* and deficient in either C4 or CR1,2 have increased ANA and dsDNA autoantibody titers and develop severe glomerulonephritis (42).

In contrast to classical pathway deficiencies, deficiency in C3 does not predispose to autoantibody production or to significant renal disease in the *lpr* mice (42). Functional deficits of C3 have been dissociated from lupus-like disease in other model systems as well, even though tissue damage from IC deposition and complement mediated inflammation depends on C3 function. As presented earlier, C1q-deficient mice spontaneously develop anti-nuclear auto-antibodies and renal disease reminiscent of human SLE (20). Somewhat unexpectedly, mice double deficient for classical pathway C1q and alternative pathway factor B, which renders them unable to fire the complement cascade, still develop glomerulonephritis (74). Despite the markedly impaired humoral response of C3-deficient animals, it is becoming quite clear that C3 itself is not protective against autoimmunity. Instead, these results support a critical role for classical complement ligands and receptors in murine models of lupus. It remains to be delineated to what extent complement deficiency intensifies disease by failing to clear complexed autoantigens versus failing to properly target self-antigens to those lymphoid compartments where autoreactive B cells are tolerized.

8. CONCLUSIONS

The molecular and cellular mechanisms whereby deficiencies in components of the classical complement pathway predispose to autoimmune disease are only beginning to be elucidated. In a simplified immune clearance model of disease pathogenesis, complement deficiency predisposes to SLE by impairing clearance of IC from the circulation. The ensuing increased IC load facilitates IC deposition in susceptible tissues, triggering complement-mediated inflammation and propagating disease. Oddly enough, even without C3, classical complement deficiencies are associated with IC deposition that culminates in organ damage, thus dissociating complement-mediated inflammation from autoreactive pathogenesis.

The simplified model proposing that immune complexes cause SLE is unable to account for the presence of pathogenic autoantibodies: What is the mechanism that permits IC containing autoantigens to overcome B cell and presumably T cell tolerance? An enhanced model, acknowledging the contribution of complement dependent IC clearance, proposes that the classical pathway participates in the very induction of self-tolerance, and that without C1q, C4, or their receptors and without natural autoantibodies, potentially pathogenic autoreactive B cells are not properly tolerized.

References

1. Colten, H.R., and F.S. Rosen. 1992. Complement deficiencies. *Annu. Rev. Immunol.* 10:809–34.
2. Lachmann, P.J. 1993. *Complement deficiencies: Genetic and acquired.* 5th ed. In: *Clinical aspects of immunology*, P.J. Lachmann, K. Peters, F.S. Rosen, and M.J. Walport, eds. pp. 1287–1304. Boston: Blackwell Scientific Publications.
3. Moulds, J.M., M. Krych, V.M. Holers, M.K. Liszewski, and J.P. Atkinson. 1992. Genetics of the complement system and rheumatic diseases. *Rheum. Dis. Clin. North Am.* 18:893–914.
4. Seery, J.P., E.C. Wang, V. Cattell, J.M. Carroll, M.J. Owen, and F.M. Watt. 1999. A central role for alpha beta T cells in the pathogenesis of murine lupus. *J. Immunol.* 162:7241–7248.
5. Chan, O., and M.J. Shlomchik. 1998. A new role for B cells in systemic autoimmunity: B cells promote spontaneous T cell activation in MRL-lpr/lpr mice. *J. Immunol.* 160:51–59.
6. Singh, R.R., and B.H. Hahn. 1998. Reciprocal T-B determinant spreading develops spontaneously in murine lupus: implications for pathogenesis. *Immunol. Rev.* 164:201–208.
7. Hebert, L.A. 1991. The clearance of immune complexes from the circulation of man and other primates. *Am. J. Kidney Dis.* 17:352–361.
8. Levine, J.S., and J.S. Koh. 1999. The role of apoptosis in autoimmunity: immunogen, antigen, and accelerant. *Semin. Nephrol.* 19:34–47.
9. Casciola-Rosen, L.A., G. Anhalt, A. Rosen. 1994. Autoantigens targeted in systemic lupus crythematosus are clustered in two populations of surface structures on apoptotic keratinocytes [see comments]. *J. Exp. Med.* 179:1317–1330.
10. Abbas, A.K., A.H. Lichtman, and J.S. Pober. 1997. *The complement system.* Chapter 15, third ed. In: *Cellular and molecular immunology*, 313–338. Philadelphia: W.B. Saunders Company.
11. Turner, M.W. 1996. Mannose-binding lectin: the pluripotent molecule of the innate immune system. *Immunol Today* 17:532–540.
12. Sullivan, K.E., C. Wooten, D. Goldman, and M. Petri. 1996. Mannose-binding protein genetic polymorphisms in black patients with systemic lupus erythematosus. *Arthritis Rheum.* 39:2046–2051.
13. Ip, W.K., S.Y. Chan, C.S. Lau, and Y.L. Lau. 1998. Association of systemic lupus erythematosus with promoter polymorphisms of the mannose-binding lectin gene. *Arthritis Rheum.* 41:1663–1668.
14. Davies, E.J., L.S. Teh, J. Ordi-Ros, N. Snowden, M.C. Hillarby, A. Hajeer, R. Donn, P. Perez-Pemen, M. Vilardell-Tarres, and W.E. Ollier. 1997. A dysfunctional allele of the mannose binding protein gene associates with systemic lupus erythematosus in a Spanish population. *J. Rheumatol.* 24:485–488.
15. Rekvig, O.P., K. Andreassen, and U. Moens. 1998. Antibodies to DNA—towards an understanding of their origin and pathophysiological impact in systemic lupus erythematosus [editorial]. *Scand. J. Rheumatol.* 27:1–6.
16. Nicholson-Weller A., and L.B. Klickstein. 1999. C1q-binding proteins and C1q receptors. *Curr. Opin. Immunol.* 11:42–46.
17. Korb, L.C., J.M. Ahearn. 1997. C1q binds directly and specifically to surface blebs of apoptotic human keratinocytes: complement deficiency and systemic lupus erythematosus revisited. *J. Immunol.* 158:4525–4528.
18. Hicks, P.S., L. Saunero-Nava, T.W. Du Clos, and C. Mold. 1992. Serum amyloid P component binds to histones and activates the classical complement pathway. *J. Immunol.* 149:3689–3694.
19. Bickerstaff, M.C., M. Botto, W.L. Hutchinson, J. Herbert, G.A. Tennent, A. Bybee, D.A. Mitchell, H.T. Cook, P.J. Butler, M.J. Walport, and M.B. Pepys. 1999. Serum amyloid P component controls chromatin degradation and prevents antinuclear autoimmunity. *Nat. Med.* 5:694–697.
20. Botto, M., C. Dell'Agnola, A.E. Bygrave, E.M. Thompson, H.T. Cook, F. Petry, M. Loos, P.P. Pandolfi, and M.J. Walport. 1998. Homozygous C1q deficiency causes glomerulonephritis associated with multiple apoptotic bodies [see comments]. *Nat. Genet.* 19:56–59.
21. Kurtz, C.B., E. O'Toole, S.M. Christensen, and J.H. Weis. 1990. The murine complement receptor gene family. IV. Alternative splicing of Cr2 gene transcripts predicts two distinct gene products that share homologous domains with both human CR2 and CR1. *J. Immunol.* 144:3581–3591.
22. Carroll, M.C., E.M. Alicot, P.J. Katzman, L.B. Klickstein, J.A. Smith, and D.T. Fearon. 1988. Organization of the genes encoding complement receptors type 1 and 2, decay-

accelerating factor, and C4-binding protein in the RCA locus on human chromosome 1. *J. Exp. Med.* 167:1271–1280.

23. Klickstein, L.B., S.F. Barbashov, T. Liu, R.M. Jack, and A. Nicholson-Weller. 1997. Complement receptor type 1 (CR1, CD35) is a receptor for C1q. *Immunity* 7:345–355.

24. Marquart, H.V., A. Svendsen, J.M. Rasmussen, C.H. Nielsen, P. Junker, S.E. Svehag, R.G. Leslie. 1995. Complement receptor expression and activation of the complement cascade on B lymphocytes from patients with systemic lupus erythematosus (SLE). *Clin. Exp. Immunol.* 101:60–65.

25. Wilson, J.G., W.D. Ratnoff, P.H. Schur, and D.T. Fearon. 1986. Decreased expression of the C3b/C4b receptor (CR1) and the C3d receptor (CR2) on B lymphocytes and of CR1 on neutrophils of patients with systemic lupus erythematosus. *Arthritis Rheum.* 29:739–747.

26. Ahearn, J.M., M.B. Fischer, D. Croix, S. Goerg, M. Ma, J. Xia, X. Zhou, R.G. Howard, T.L. Rothstein, and M.C. Carroll. 1996. Disruption of the Cr2 locus results in a reduction in B-la cells and in an impaired B cell response to T-dependent antigen. *Immunity* 4:251–262.

27. Molina, H., V.M. Holers, B. Li, Y. Fung, S. Mariathasan, J. Goellner, J. Strauss-Schoenberger, R.W. Karr, and D.D. Chaplin. 1996. Markedly impaired humoral immune response in mice deficient in complement receptors 1 and 2. *Proc. Natl. Acad. Sci. USA* 93:3357–3361.

28. Carroll, M.C. 1998. The role of complement and complement receptors in induction and regulation of immunity. *Annu. Rev. Immunol.* 16:545–568.

29. Croix, D.A., J.M. Ahearn, A.M. Rosengard, S. Han, G. Kelsoe, M. Ma, and M.C. Carroll. 1996. Antibody response to a T-dependent antigen requires B cell expression of complement receptors. *J. Exp. Med.* 183:1857–1864.

30. Takahashi, K., Y. Kozono, T.J. Waldschmidt, D. Berthiaume, R.J. Quigg, A. Baron, and V.M. Holers. 1997. Mouse complement receptors type 1 (CR1; CD35) and type 2 (CR2; CD21): expression on normal B cell subpopulations and decreased levels during the development of autoimmunity in MRL/lpr mice. *J. Immunol.* 159:1557–1569.

31. Fearon, D.T., and R.H. Carter. 1995. The CD19/CR2/TAPA 1 complex of B lymphocytes: linking natural to acquired immunity. *Annu. Rev. Immunol.* 13:127–132.

32. Tolnay, M., and G.C. Tsokos. 1998. Complement receptor 2 in the regulation of the immune response. *Clin. Immunol. Immunopathol.* 88:123–149.

33. Mongini, P.K., M.A. Vilensky, P.F. Highet, and J.K. Inman. 1997. The affinity threshold for human B cell activation via the antigen receptor complex is reduced upon co-ligation of the antigen receptor with CD21 (CR2). *J. Immunol.* 159:3782–3791.

34. Baiu, D.C., J. Prechl, A. Tchorbanov, H.D. Molina, A. Erdei, A. Sulica, P.J. Capel, and W.L. Hazenbos. 1999. Modulation of the humoral immune response by antibody-mediated antigen targeting to complement receptors and Fc receptors. *J. Immunol.* 162:3125–3130.

35. Carter, R.H., D.T. Fearon. 1992. CD19: lowering the threshold for antigen receptor stimulation of B lymphocytes. *Science.* 256:105–107.

36. Dempsey, P.W., M.E. Allison, S. Akkaraju, C.C. Goodnow, and D.T. Fearon. 1996. C3d of complement as a molecular adjuvant: bridging innate and acquired immunity. *Science* 271:348–350.

37. Fischer, M.B., S. Goerg, L. Shen, A.P. Prodeus, C.C. Goodnow, G. Kelsoe, and M.C. Carroll. 1998. Dependence

of germinal center B cells on expression of CD21/CD35 for survival. *Science* 280:582–585.

38. Coggeshall, K.M. 1998. Inhibitory signaling by B cell Fc gamma RIIb. *Curr. Opin. Immunol.* 10:306–312.

39. Cutler, A.J., M. Botto, D. van Essen, R. Rivi, K.A. Davies, D. Gray, and M.J. Walport. 1998. T cell-dependent immune response in C1q-deficient mice: defective interferon gamma production by antigen-specific T cells. *J. Exp. Med.* 187:1789–1797.

40. Fischer, M.B., M. Ma, S. Goerg, X. Zhou, J. Xia, O. Finco, S. Han, G. Kelsoe, R.G. Howard, T.L. Rothstein, E. Kremmer, F.S. Rosen, and M.C. Carroll. 1996. Regulation of the B cell response to T-dependent antigens by classical pathway complement. *J. Immunol.* 157:549–556.

41. Wessels, M.R., P. Butko, M. Ma, H.B. Warren, A.L. Lage, and M.C. Carroll. 1995. Studies of group B streptococcal infection in mice deficient in complement component C3 or C4 demonstrate an essential role for complement in both innate and acquired immunity. *Proc. Natl. Acad. Sci. USA* 92:11490–11494.

42. Prodeus, A.P., S. Goerg, L.M. Shen, O.O. Pozdnyakova, L. Chu, E.M. Alicot, C.C. Goodnow, and M.C. Carroll. 1998. A critical role for complement in maintenance of self-tolerance. *Immunity* 9:721–731.

43. Fischer, M.B., M. Ma, N.C. Hsu, and M.C. Carroll. 1998. Local synthesis of C3 within the splenic lymphoid compartment can reconstitute the impaired immune response in C3-deficient mice. *J. Immunol.* 160:2619–2625.

44. Tarlinton, D. 1998. Germinal centers: form and function. *Curr. Opin. Immunol.* 10:245–251.

45. Kelsoe, G. 1995. In situ studies of the germinal center reaction. *Adv. Immunol.* 60:267–288.

46. Qin, D., J. Wu, M.C. Carroll, G.F. Burton, A.K. Szakal, and J.G. Tew. 1998. Evidence for an important interaction between a complement-derived CD21 ligand on follicular dendritic cells and CD21 on B cells in the initiation of IgG responses. *J. Immunol.* 161:4549–4554.

47. Tew, J.G., J. Wu, D. Qin, S. Helm, G.F. Burton, and A.K. Szakal. 1997. Follicular dendritic cells and presentation of antigen and costimulatory signals to B cells. *Immunol. Rev.* 156:39–52.

48. Vora, K.A., J.V. Ravetch, T. Manser. 1997. Amplified follicular immune complex deposition in mice lacking the Fc receptor gamma-chain does not alter maturation of the B cell response. *J. Immunol.* 159:2116–2124.

49. Goodnow, C.C. 1992. Transgenic mice and analysis of B-cell tolerance. *Annu. Rev. Immunol.* 10:489–518.

50. Cooke, M.P., A.W. Heath, K.M. Shokat, Y. Zeng, F.D. Finkelman, P.S. Linsley, M. Howard, and C.C. Goodnow. 1994. Immunoglobulin signal transduction guides the specificity of B cell-T cell interactions and is blocked in tolerant self-reactive B cells. *J. Exp. Med.* 179:425–438.

51. Constant, S.L. 1999. B lymphocytes as antigen-presenting cells for CD4+ T cell priming in vivo. *J. Immunol.* 162:5695–5703.

52. Boackle, S.A., M.A. Morris, V.M. Holers, and D.R. Karp. 1998. Complement opsonization is required for presentation of immune complexes by resting peripheral blood B cells. *J. Immunol.* 161:6537–6543.

53. Kozono, Y., R. Abe, H. Kozono, R.G. Kelly, T. Azuma, and V.M. Holers. 1998. Cross-linking CD21/CD35 or CD19 increases both B7-1 and B7-2 expression on murine splenic B cells. *J. Immunol.* 160:1565–1572.

54. Thornton, B.P., V. Vetvicka, and G.D. Ross. 1996. Function of C3 in a humoral response: iC3b/C3dg bound to an

immune complex generated with natural antibody and a primary antigen promotes antigen uptake and the expression of co-stimulatory molecules by all B cells, but only stimulates immunoglobulin synthesis by antigen-specific B cells. *Clin. Exp. Immunol.* 104:531–537.

55. Carroll, M.C. 1998. The lupus paradox [news; comment]. *Nat Genet.* 19:3–4.
56. Paul, E., M.C. Carroll. 1999. SAP-less chromatin triggers systemic lupus erythematosus [news]. *Nat. Med.* 5:607–608.
57. Diamond, B., J.B. Katz, E. Paul, C. Aranow, D. Lustgarten, and M.D. Scharff. 1992. The role of somatic mutation in the pathogenic anti-DNA response. *Annu. Rev. Immunol.* 10:731–757.
58. Dighiero, G. 1997. Natural autoantibodies, tolerance, and autoimmunity. *Ann. NY Acad. Sci.* 815:182–192.
59. Carroll, M.C. 2000. *Adv. Immunol.* (In press.)
60. Hardy, R.R., Y.S. Li, and K. Hayakawa. 1996. Distinctive developmental origins and specificities of the CD5+ B-cell subset. *Semin. Immunol.* 8:37–44.
61. Tarakhovsky, A. 1997. Bar Mitzvah for B-1 cells: how will they grow up? [comment]. *J. Exp. Med.* 185:981–984.
62. Rickert, R.C., K. Rajewsky, and J. Roes. 1995. Impairment of T-cell-dependent B-cell responses and B-1 cell development in CD19-deficient mice. *Nature* 376:352–355.
63. Sato, S., N. Ono, D.A. Steeber, D.S. Pisetsky, and T.F. Tedder. 1996. CD19 regulates B lymphocyte signaling thresholds critical for the development of B-1 lineage cells and autoimmunity. *J. Immunol.* 157:4371–4378.
64. Huck, S., C. Jamin, P. Youinou, and M. Zouali. 1998. High-density expression of CD95 on B cells and underrepresentation of the B-1 cell subset in human lupus. *J. Autoimmun.* 11:449–455.
65. Sandel, P.C., and J.G. Monroe. 1999. Negative selection of immature B cells by receptor editing or deletion is determined by site of antigen encounter. *Immunity.* 10:289–299.

66. Goodnow, C.C., J.G. Cyster, S.B. Hartley, S.E. Bell, M.P. Cooke, J.I. Healy, S. Akkaraju, J.C. Rathmell, S.L. Pogue, and K.P. Shokat. 1995. Self-tolerance checkpoints in B lymphocyte development. *Adv. Immunol.* 59:279–368.
67. Mandik-Nayak, L., A. Bui, H. Noorchashm, A. Eaton, and J. Erikson. 1997. Regulation of anti-double-stranded DNA B cells in nonautoimmune mice: localization to the T-B interface of the splenic follicle. *J. Exp. Med.* 186:1257–1267.
68. Chen, C., Z. Nagy, M.Z. Radic, R.R. Hardy, D. Huszar, S.A. Camper, and M. Weigert. 1995. The site and stage of anti-DNA B-cell deletion. *Nature* 373:252–255.
69. Offen, D., L. Spatz, H. Escowitz, S. Factor, B. Diamond. 1992. Induction of tolerance to an IgG autoantibody. *Proc. Natl. Acad. Sci. USA* 89:8332–8336.
70. Pewzner-Jung, Y., D. Friedmann, E. Sonoda, S. Jung, K. Rajewsky, and D. Eilat. 1998. B cell deletion, anergy, and receptor editing in "knock in" mice targeted with a germline-encoded or somatically mutated anti-DNA heavy chain. *J. Immunol.* 161:4634–4645.
71. Santulli-Marotto, S., M.W. Retter, R. Gee, M.J. Mamula, and S.H. Clarke. 1998. Autoreactive B cell regulation: peripheral induction of developmental arrest by lupus-associated autoantigens. *Immunity* 8:209–219.
72. Theofilopoulos, A.N., and F.J. Dixon. 1985. Murine models of systemic lupus erythematosus. *Adv. Immunol.* 37:269–390.
73. Vyse, T.J., and B.L. Kotzin. 1998. Genetic susceptibility to systemic lupus erythematosus. *Annu. Rev. Immunol.* 16:261–292.
74. Mitchell, D.A., P.R. Taylor, H.T. Cook, J. Moss, A.E. Bygrave, M.J. Walport, and M. Botto. 1999. Cutting edge: C1q protects against the development of glomerulonephritis independently of C3 activation. *J. Immunol.* 162:5676–5679.

SYSTEMIC AUTOIMMUNE DISEASES

18 | Human Systemic Lupus Erythematosus

George C. Tsokos and Dimitrios T. Boumpas

1. INTRODUCTION

Systemic lupus erythematosus (SLE) is a complex autoimmune disease involving several organs (1,2). The main effectors of disease pathology are the diverse autoantibodies, immune complexes and autoreactive cells. Altered biology of immune cells, keratinocytes, endothelial and possibly other cells invariably contributes to disease expression. At the pathogenic level, the disease presents unprecedented complexity, involving an as yet unspecified number of genes, immunoregulatory aberrations, environmental and hormonal factors. Previous and current research efforts have elucidated the complex array of genetic, environmental, hormonal and immunoregulatory factors thought to interplay in the disease pathogenesis. Although certain individuals may have a strong genetic predisposition to develop lupus, it is likely that more than one group of factors need to be present for disease expression. In an occasional patient, certain factors may dominate, e.g., almost all patients with C4 deficiency (3) develop lupus, suggesting that this gene represents a strong single predisposing genetic factor. Theoretically, it can be assumed that if the sum of the effects of different factors exceeds a certain threshold, then clinical manifestation of lupus ensues, while lesser sums may lead to clinical syndromes that do not fulfill the clinical diagnosis of lupus.

2. CLINICAL FEATURES

In the United States, the average incidence of SLE is between 18 and 7.6 cases per 100,000 persons annually and the prevalence from 14.6 to 50.8 cases per 100,000; recent data suggest that the incidence of SLE may be increasing (4,5). Women in their child-bearing years are afflicted 9 times more often than men and African American women seem to be affected more frequently and tend to develop more severe disease (6–8). Diagnosis is generally not difficult when typical signs and symptoms are present, but it may present a differential diagnostic challenge when the disease presents with isolated manifestations, such as seizure disorder in a young woman with anti-nuclear antibodies in the serum. The diagnosis of SLE remains largely clinical with support of laboratory tests (Table 1). The American College of Rheumatology has developed eleven criteria, to classify the disease for research purposes (9): malar or discoid rash, photosensitivity, oral ulcers, arthritis, serositis, renal disorder (active sediment or proteinuria), neurologic disorder (seizures or psychosis), hematologic disorder (hemolytic anemia or leukopenia or lymphopenia or thrombocytopenia) presence of antinuclear antibodies and immunologic disorder (anti-DNA, or anti-Sm or anti-phospholipid antibodies). In general, not all criteria carry the same weight in disease diagnosis. For example, the presence of anti-dsDNA antibodies calls for close observation for the development of additional features, such as nephritis. The presence of 4 or more criteria either serially or simultaneously is sufficient for the diagnosis of SLE.

Constitutional symptoms, malar rash, discoid lesions, photosensitivity and symmetrical, non-erosive arthritis are frequent manifestations (1), whereas inflammatory myositis is uncommon, usually mild and associated with normal creatinine kinase levels but increased serum aldolase levels (10,11).

Antibodies to erythrocytes, neutrophils, lymphocytes, platelet and clotting factors may lead to increased cell destruction, cytopenia and prolonged bleeding. Leukopenia is usually modest and clinically insignificant and leukocyte counts less than 1,000/μl are unusual. Thrombocytopenia

Table 1 Approximate frequency of common clinical and laboratory features of SLE

Feature	Frequency (%)	
	At presentation	Anytime
Clinical		
Arthritis/Arthralgia	50	90
Nephritis	35	70
Fever	10	60
Malar rash	15	50
Discoid rash	10	15
Photosensitivity	15	40
Serositis (pleurisy, pericarditis)	5	40
Oral ulcers	20	50
Raynaud's	30	50
Seizures	5	25
Psychosis	1	10
Lymphadenopathy	15	30
Pneumonitis	1	5
Myositis	1	5
Laboratory		
ANA	80	95
Anti-dsDNA	15	50
Rheumatoid factor	5	20
Thrombocytopenia	5	25
Antiphospholipid antibodies	15	35

occurs in up to 25% of patients but is rarely severe (i.e., platelet count below 20,000/μl). Not uncommonly, lupus presents as idiopathic thrombocytopenic purpura with other manifestations added later in the course of the disease (2).

Pleuritic chest pain occurs in up to 60% of patients and is occasionally associated with pleural effusions. Acute lupus pneumonitis with or without alveolar hemorrhage is rare, but potentially fatal, manifestation of lupus. Chronic interstitial lung disease may lead to pulmonary fibrosis. Pulmonary hypertension is usually mild to moderate and has been associated with Raynaud's phenomenon, anti-ribonucleoprotein antibodies and antiphospholipid antibodies (2). Mild to moderate pericarditis is the most common type of cardiac involvement, whereas pericardial tamponade is rare. Valvular abnormalities, including vegetations, insufficiency and stenosis are increasingly recognized as causes of morbidity and may increase the risk for endocarditis (12). Accelerated premature atherosclerosis of multifactorial etiology has emerged as a significant cause of morbidity and mortality (13–15).

Neuropsychiatric symptoms are common in lupus patients and are either primary, i.e. related to immune-mediated injury, or secondary, resulting from disease in other organs and/or complications of treatment (2). Primary neuropsychiatric events typically occur in a setting of clinical and/or serologic evidence of active SLE and can result from impaired perfusion of the brain due to vasculopathy, vasculitis, leukoaggultination, thrombosis, or to antibody-mediated neuronal cell injury or dysfunction. Antibodies may be produced intrathecally or may access to the central nervous system from a disturbed blood-brain barrier due to the vascular injury. Increased levels of IL-6 or IFN-α in the cerebrospinal fluid (16,17) and increased levels of the excitotoxin quinolinic acid (18), a metabolic product of microglial cells and monocytes stimulated with interferon, may contribute to the pathology. Involvement in the central nervous system can either be diffuse, manifested by encephalopathic brain syndrome, psychosis or major affective disorder, or focal, manifested by seizures, cerebrovascular accidents and transverse myelopathy. Autoantibodies to anti-neuronal antibodies and anti-ribosomal P protein are more likely to occur with diffuse rather focal involvement, suggesting that the former may be autoantibody-mediated (19,20).

The heterogeneity in the expression, course and immunopathogenesis of SLE, even when considered at a single organ level, is best illustrated by renal disease (2). In lupus nephritis, the location of immune complex deposition/formation is linked to the histopathology of disease. Deposition of immune complexes in the mesangium is characteristic of mesangial lupus nephritis and manifests itself with hematuria, mild proteinuria without affecting renal function. Immune complex deposition in the subendothelial area of the capillary loops results in proliferative lupus nephritis (focal or diffuse) with exuberant glomerular hypercellularity that is due to proliferation of mesangial and endothelial cells and leukocytic infiltrates, compromising capillary flow and renal function (Fig. 1). At the clinical level, proliferative nephritis is characterized by hematuria with dysmorphic red blood cells and cellular casts, moderate to severe proteinuria, active serology (anti-DNA antibodies and low C3 levels) and decreased renal function. Membranous nephropathy is defined most rigorously by the presence of epimembranous (subepithelial) deposits along peripheral glomerular capillary loops (Fig. 1) that are diffusely thickened but retain their potency. Classical presentation includes proteinuria, often at nephrotic range with minimal hematuria, no cellular casts, absent or low titer anti-DNA antibodies and normal complement levels. The prognosis is worse when membranous and proliferative lupus nephritis coexist (21).

Antiphospholipid antibody syndrome, defined as a disorder of recurrent venous or arterial thrombosis, pregnancy losses and thrombocytopenia associated with a persistently positive antiphospholipid antibody or lupus anticoagulant test, develops in 5–15% of SLE patients. On the other hand, approximately 30–40% of SLE patients have antiphospholipid antibodies (1), which require β2-

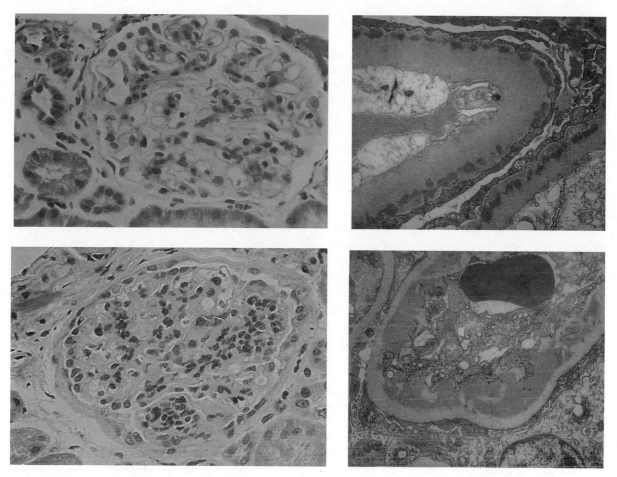

Figure 1.   Lupus nephritis: light(H&E stain × 200) and electron microscopy. (Upper panel). Membranous nephropathy. Capillary loops of the glomerular tuft are prominent and widely patent by light microscopy with subepithelial deposition of darkly staining immune complexes on the basement membrane seen on the electron micrograph. (Lower panel). Proliferative nephritis. Note the subendothelial deposition of immune complexes, which results in a dramatic increase in mesangial and endocapillary cellularity that compromises the patency of capillary loops.

glycoprotein (a naturally-occurring serum anticoagulant, along with protein C, S and antithrombin III) for binding to antigen (see Chapter 20).

ANA screening tests have a high sensitivity (97–100%), but low specificity and predictive value (10–40%). In patients with low-pretest probability, a negative ANA test essentially rules out a diagnosis of SLE. Anti-dsDNA antibodies are found in the sera of 40–75% of lupus patients and are disease-specific. Changes in anti-dsDNA titer in many patients tend to correlate with disease activity and, in some cases, may predict impending recurrences. However, the correlation is poor and has been proven to have no predictive value (22), making preemptive therapeutic strikes inadvisable. Compared to anti-dsDNA titer, C3 level correlates best with disease activity as determined by kidney biopsy (23).

One of the most perplexing aspects of the natural history of lupus is its remitting and relapsing course. Modern treatment neither cures lupus, nor prevents exacerbation in most patients and one third of them relapse after achieving partial or complete remission. Despite advances in diagnosis and management, complications attributable to SLE or its therapy continue to cause substantial morbidity. Current survival rates, in excess of 80%, have been attributed to earlier diagnosis, improved immunosuppressive and general medical treatment, and availability of medical intensive care (1,24). Infections (25,26) and coronary artery disease are the leading causes of death (1). Infections, renal disease requiring renal biopsy and orthopedic management of musculoskeletal manifestations such as osteonecrosis, tendon rupture and fractures are the main reasons for hospitalizations (1,27).

3. AUTONTIBODIES AND IMMUNE COMPLEXES

SLE is characterized by the production of a large list of anti-bodies against an array of non-organ specific self-antigens present in the nucleus, cytoplasm, cell membrane and serum

(Table 2). An immune response against self is commonly a part of the normal immune response and strictly regulated (28). Immune cells with autoreactive potential are present in large numbers in normal subjects and germline genes encoding for antigen-receptors of autoreactive T and B cells are part of the normal gene repertoire. Autoreactive cells are positively selected in an antigen-driven manner (29).

IgM antibodies to DNA are frequently produced in the normal host and bind to ssDNA; they have low affinity for DNA and broad crossreactivity with a variety of other self-antigens. The production of these natural anti-DNA antibodies is tightly regulated and usually, they do not undergo isotype switching and are encoded by germline genes; affinity maturation by the process of somatic mutation does not occur. On the contrary, anti-DNA antibodies in the sera of patients with SLE have quite different features in that they have undergone isotype switching to IgG of various subclasses, and germline genes usually do not encode them because new amino acids are introduced into their variable regions to enhance affinity (somatic mutations and hypermutations) (30–33). Continuous receptor editing in lupus B cells may provide an additional impetus toward production of more "pathogenic" autoantibodies (34,35). Since DNA is a highly anionic macromolecule, positively charged amino acids are introduced into the autoantibody variable regions, particularly arginine, to enhance DNA binding. Lupus anti-DNA antibodies are thus usually charged, IgG, relatively low crossreactive and have high-affinity and antibodies that recognize dsDNA as well; in fact, anti-DNA antibodies that recognize exclusively dsDNA are rather unusual (36–38).

T cells from lupus patients, activated by nucleosomes, provide help to lupus B cells to produce anti-dsDNA of the IgG class (39,40). DNA, nucleosomes and other intracellular autoantigens that are common targets for the lupus immune system are made accessible to immune cells during apoptosis. These autoantigens can be found in surface blebs of dying apoptotic keratinocytes following UV light exposure (41,42). It is possible that exposure of a genetically-susceptible individual to environmental factors causing apoptosis may serve as the trigger to form anti-DNA and other autoantibodies. The ability of an individual to produce anti-DNA antibodies is, to some degree, genetically determined. In the NZM2410 murine lupus model, the ability to respond to nucleosomes and generate anti-histone and anti-DNA antibodies is determined by a defined chromosomal region (43–45).

Anti-DNA and anti-ribosomal P protein antibodies can bind and penetrate live cells and alter their function. Anti-DNA antibodies seem to penetrate the cells after binding to surface antigens via the F(ab')2 and they have been claimed to alter a number of cell functions, including transcription, cell phase transition and proliferation (rev. 46). Anti-ribosomal P antibodies have been associated with neuropsychiatric disease (19,47), unexplained hepatitis (48) and nephritis (49). Anti-P antibodies bind to P protein, which is part of the cell membrane of many cells (50), penetrate live cells and inhibit apolipoprotein B synthesis, cholesterol and neutral lipid accumulation and inhibit overall protein synthesis (51).

Another antibody specificity linked to lupus nephritis is anti-C1q binding. These antibodies are found in the sera of mice and humans with lupus (52–55) and patients with other forms of immune complex glomerulonephritis (56) and fluctuate with disease activity, anti-DNA antibody titer and hypocomplementemia. The antibodies bind to the collagenous portion of the C1q molecule only in the presence of predeposited immune complexes. The clinical

Table 2 Common lupus autoantibodies

Autoantibody	% Positive	Comments
Antinuclear Antibody	>95%	Commonly found in SLE, but also found in other disorders. Many different targets.
Anti-ssDNA	80%	Common, but not specific for SLE.
Anti-dsDNA	50%	Most specific for systemic lupus. Associated with renal disease.
Anti-Ro	40%	Large quantities of antibody. Associated with Photosensitivity, skin rashes, lung disease, low WBC count, anti-La, and rheumatoid factor. Perinatal lupus
Anti-nRNP	40%	Associated with mild disease? myositis.
Anti-Sm	25%	Relatively specific for lupus. Associated with anti-nRNP.
Anti-La	20%	Lack of renal disease? Associated with anti-Ro.
Anti-P	10%	Associated with diffuse CNS manifestations, especially depression and psychosis. Relatively specific for lupus.

association between anti-C1q antibodies and lupus nephritis and their enrichment in diseased glomeruli suggest they become involved in the development of disease pathology by amplifying the effect of deposited immune complexes (57).

The role of certain antibodies in causing disease manifestations is clear. For example, antibodies that bind to antigens on the surface membrane of red cells, neutrophils and platelets and activate complement are obviously causal to the respective cytopenias. Other antibodies, albeit helpful in the differential diagnosis, are of doubtful pathogenic importance. Anti-neuronal antibodies are found in the sera and cerebrospinal fluid of patients with CNS disease and may contribute to disease expression (58,59). The role of anti-phospholipid antibodies in causing thromboses and spontaneous abortions in patients with lupus needs to be determined. Women with antibodies against the Ro/SSA antigen may give birth to children with perinatal lupus, the main feature of which is complete heart block (60) and the risk of similar complications in subsequent pregnancies is very high (20–40%). Anti-Ro antibodies may interfere with the electrophysiology of the cells involved in the conductance within the heart (61).

Several lines of evidence have led to the classification of lupus as an immune complex disease. Immune complexes have been documented in the sera of lupus patients (62–64) and both immunoglobulin and complement factors or split products have been detected in several tissues, including kidneys and skin. In addition, hypocomplementemia and increased levels of complement split factors are found in both serum and urine of lupus patients (65,66).

Cells expressing Fc and complement receptors, such as monocytes, phagocytes, neutrophils, mediate clearance of immune complexes. Immune complex clearance is decreased in lupus patients (67), perhaps due to abnormal Fc receptor function resulting from the expression of alleles that bind IgG with lesser affinity (68) or decreased numbers (69) and/or altered allele expression of complement receptor 1 (CR1) (70–72). Occasionally, patients with lupus may have antibodies against the CR1 receptor that block its function (73). In contrast to what would have been expected from the human studies, Fcγ chain-deficient NZB/NZW mice produce and deposit immune complexes, activate complement, but they are protected from severe nephritis (74).

The number of the involved antigen and antibody molecules defines the lattice of an immune complex. The size and the extent of lattice formation determine the rate of clearance and tissue deposition of immune complexes and the number of the Fc molecules in each immune complex defines its ability to bind to Fc receptors and trigger Fc-mediated cell functions. Large immune complexes (> Ag2Ab2) are quickly cleared by the reticuloendothelial system, and when saturated, they deposit to tissues. Large immune complexes tend to activate complement more efficiently. Small immune complexes have a lesser tendency to deposit to tissues (75). The initial interaction may involve the charge of the antibody, as with deposition to kidneys, binding to an Fc or complement receptor, or by specific or non-specific binding of the F(ab')2 fragment of the involved antibody (76).

Immune complexes in the sera of lupus patients have distinct reasons to deposit to tissues in that those containing cationic anti-DNA antibodies facilitate binding to the kidney basement membrane. These antibodies are of IgG1, IgG2 and IgG3 subclasses and can activate complement effectively (77). In addition to anti-DNA antibodies, other nephrophilic antibodies can bind to the kidneys and instigate the formation of immune complexes and disease (78, 79). Some lupus patients with renal disease and typical renal histopathology and immune deposits do not have serum anti-dsDNA antibodies or antibodies that directly bind to the glomerular basement membrane. The mechanism of renal immune deposition in these patients remains to be determined (78). Histones have high affinity for basement membrane (80) and nucleosomes bind to mesangial cells (81); obviously, immune complexes that contain these components deposit to the kidneys.

Deposited immune complexes initiate complement activation that leads to production of molecules (C3a, C5a) that attract inflammatory cells into the lesion. Immune complexes and complement factor breakdown products (82–84) are also known to regulate lymphocyte function.

4. CELLULAR AND CYTOKINE ABERRATIONS IN LUPUS

A great deal of literature is devoted to the description of the multiple and frequently contradictory immune cell abnormalities encountered in SLE, since aberrations of the immune cells are believed to play a major role in the pathogenesis of lupus. These studies have improved our understanding of the disease, helped us design novel treatments and guided the search for the involved genes (85). While the cellular aberrations in lupus may reflect secondary effects, therapeutic modulation of the cellular abnormalities represents a plausible way to improve the management of lupus patients.

Monocytes from lupus patients display a number of abnormalities including decreases in phagocytic activity (86), interleukin (IL)-1 production (87), expression of surface human leukocyte antigen (HLA)-DR and accessory function for T cell activation (88,89). In addition, lupus monocytes fail to express the costimulatory B7 molecules following activation (90,91). These defects may partially explain the defective responses of lupus T cells to nominal antigens (92). In contrast, peripheral blood monocytes from lupus patients display increased rates of spontaneous apoptosis (93) will be discussed here and elsewhere, which can contribute to the increased levels of circulating autoantigens. Decreased phagocytic function (25) may increase susceptibility of lupus patients to infections.

T cell abnormalities are crucial in the pathogenesis of the disease because they regulate B cell function, and the production of most of the pathogenic autoantibodies is T cell-dependent. T-lymphocyte abnormalities in patients with lupus have been reviewed (85,94). Human studies of lupus lymphocyte biology are limited to peripheral blood lymphocytes and, in general, have established two apparently opposing (T cell enigma; 94) phenomena: The existence of activated T cells that provide excessive help to B cells to produce autoantibodies and yet inability to respond in vitro to nominal antigens and produce IL-2.

The presence of activated T cells in the peripheral blood of lupus patients has been inferred from the presence of increased numbers of cells expressing DR$^+$ antigens (95) and the expression of increased levels of c-myc mRNA (96, 97). Studies of hprt gene-mutated T cells have provided evidence for the existence of rapidly dividing T cells in the peripheral blood of lupus patients. The frequency of hprt-mutated T cells is increased in lupus patients (98) and their numbers correlate with disease activity (99). T cells that carry the hprt gene mutation are phenotypically similar to wild T cells and may, as non-hprt mutated T cells, help autologous B cells produce anti-DNA antibody in a MHC-restricted fashion (100).

The existence of T cells that provide help to B cells to produce autoantibodies has been shown clearly in studies of IL-2-expanded lines and clones. Specifically, IL-2-expanded T cell lines established from lupus nephritis patients helped autologous B cells produce cationic anti-DNA antibodies. The majority of the T-helper cell lines were CD4$^+$, whereas a small fraction were $\alpha\beta$ TCR$^+$ CD4$^-$ CD8$^-$ and $\gamma\delta$ TCR$^+$ (101,102). $\gamma\delta$ TCR$^+$ clones dramatically augmented anti-DNA antibody production in a non-MHC-restricted fashion (101). Sequencing of the α and β TCR chain genes from CD4$^+$ lines from lupus patients revealed recurrent motifs of highly charged residues in their CDR3 loops (103). These results are in keeping with studies in murine lupus (104) and suggest that the autoimmune T helper cells in lupus patients expand in response to charged antigens. Sequencing of γ and δ TCR chains from peripheral blood of lupus patients showed oligoclonal expansion of $\gamma\delta$ T cells in lupus (105) similar to that seen in control disease samples, although another study of patients with lupus nephritis noted a certain bias in the selection of γ and δ chain repertoire (106). Clones reactive with the presumptive autoantigen small nuclear ribonucleoprotein (snRNP) have been established from both lupus patients and normal individuals who carry the HLA-DR2 or HLA-DR4 genotypes following stimulation of peripheral blood MNC with IL2 and snRNP, and they were found to use frequently the Vβ6 TCR gene (107). T cell lines from patients with mixed connective tissue disease that expanded in response to snRNP displayed Vβ1, Vβ3, Vβ5.2, and Vβ14 families at increased frequency (108).

Lupus T cells have been shown to produce decreased amounts of IL-2 (109,110). Both CD4$^+$ and CD8$^+$ cells are responsible for this deficiency (111). Decreased IL-2 secretion in vitro by lupus cells correlates with increased disease activity, lack of previous treatment and increased numbers of spontaneously immunoglobulin-secreting B cells (109,110). Despite the fact that lupus T cells fail to produce IL-2 in vitro, the serum levels of IL-2 increase, indicating the presence of activated cells in vivo (112). The decreased production of IL-2 may represent the result of regulatory factors (113), defective costimulation (90), or the possibility that the T cells may have reached a state of replicative senescence following repeated stimulation with autoantigens, as shown in murine lupus (114). Any of the above conditions should be associated with abnormal activity of factors important in the transcription of the IL-2 gene. Indeed, lupus T cells, but not B cells, have decreased levels of p65/rel A protein that results in decreased NF-κB activity (115).

Peripheral blood mononuclear cells from lupus patients produce less IFNγ both spontaneously (116) and following stimulation with phytohemagglutinin or IL-2 (117) or certain viruses (118,119). Decreased IFNγ may contribute to higher incidence of infections in patients with SLE (120). These findings contrast with reports of patients with rheumatoid arthritis who developed lupus following systemic administration of IFNγ (121,122), suggesting that IFNγ may promote autoimmunity. Lymphocytes from lupus-prone mice express increased levels of IFNγmRNA (123) and IFNγmay promote autoimmunity by enhancing the expression of autoantigens (124). Additional studies have shown clearly that IFNγ promotes autoimmunity. Specifically, NZB/NZW mice treated with an anti-IFNγ antibody (125) and MRL/lpr mice deficient in IFNγ(IFNγ–/–) (126,127) or MRL/lpr (128,129) and NZB/NZW lacking the IFNγ receptor (MRL/lpr γR –/–) (130) have less or no glomerulonephritis or other autoimmune manifestations.

Peripheral blood lymphocytes from HLA-DR2- and DQw1-positive normal individuals produce low amounts of TNFα upon stimulation with antigens. In contrast, lymphocytes from DR3- and DR4-positive individuals produce high levels TNFα. DR2-, DQW1-positive SLE patients are associated with increased frequency of lupus nephritis and produce very low amounts of TNFα under similar conditions. DR2- and DR4-positive SLE patients, who are associated with low frequency of glomerulonephritis, produce high levels of TNFα (131). Decreased TNFα production has been associated with decreased polymorphonuclear cell-mediated phagocytosis(132). Similar to humans, NZB/NZW mice have decreased levels of TNFα, and replacement therapy with recombinant TNFα significant delays the development of nephritis (133). The immuno-modulator AS101 increased levels of TNFα, decreased levels of IL-10 and improved glomerulonephritis in NZB/NZW mice; it also decreased the levels of IL-10 and autoantibody production in SCID mice transplanted with human lupus mononuclear cells (134), suggesting that medicinal modulation of lymphokine production and clinical autoimmune manifestations is plausible.

IL-12, another cytokine in the Th1 group, has also been found to be decreased in the sera of patients with active disease (135); in addition, mononuclear cells from patients with new onset disease (136) and macrophages from NZB/NZW and MRL/*lpr* mice (137) produce decreased amounts of IL-12 (138). Interestingly, IL-12, delivered by means of gene therapy, prevented autoimmune disease development in the graft-versus host (DBA into F_1) model (139), opening the possibility of utilizing IL-12 in the treatment of human disease.

The most important function of IL-6 is the promotion of Ig production by activated B cells and EBV-transformed B cells (140). Lupus mononuclear cells express high levels of mRNA for IL-6 (109). The increased levels of IL-6 that were reported in the cerebrospinal fluid of patients with central nervous system involvement fluctuated with disease activity (141) and may be involved in disease pathology. The possible role of IL-6 in lupus pathogenesis is further supported by the finding that mesangial cells from rats with proliferative glomerulonephritis express increased levels of IL-6 mRNA. Recombinant IL-6 promotes the *in vitro* growth of mesangial cells and may have a role in amplifying the proliferative process in glomerulonephritis (142). IL-6 is also present in involved kidneys of lupus patients (142). Finally, IL-6 has been shown to promote disease expression in NZB/NZW mice (143). The factors leading to constitutive expression of IL-6 in SLE have not been elucidated, but they may involve other regulatory cytokines (144) and genetic factors (145).

Several studies have shown convincingly that IL-10 production is significantly elevated in patients with SLE, and that IL-10 overproduction is implicated in the generation of anti-DNA antibodies (146). Upregulated IL-10 production characterizes not only lupus patients but also healthy members of lupus multiplex families as well as affected first- and even second-degree relatives of SLE patients. The constitutive production of IL-10 in healthy members of lupus families was higher than in healthy unrelated control individuals. Monocytes and a B cell subset are responsible for IL-10 overproduction in both patients and healthy relatives, while IL-10 is absent from normal B cells (147). IL-10 is a potent B cell stimulator and inhibitor of antigen-presenting cell (APC) function, which may explain the defective APC function and B7–1 upregulation previously reported in lupus non-B cells (148). Familial patterns of IL-10 dysregulation point towards a potentially intrinsic

defect (149). Treatment of SCID mice transplanted with human lupus mononuclear cells with an anti-IL-10 antibody caused diminished production of anti-dsDNA antibodies (146). Furthermore, administration of anti-IL-10 antibody to NZB×NZW mice delayed the onset of the lupus. It is possible that the continuously high levels of IL-10 in lupus are responsible for the decreased production of Th1-type cytokines (IL-10 decreases the production of IL-2, TNFα and *IFNγ*) and for the perpetuation of the humoral autoimmune response.

The above cited studies indicate that SLE is characterized by an imbalance in the ratio of type 1 (IFNγ, IL-2): type 2 (IL-4, IL-5, IL-6 and IL-10) cytokine-secreting MNC. In one study, type 1 and type 2 lymphokines were estimated in the same group of patients and confirmed an elevated ratio of IL-10: IFNγ-secreting cells, which correlated with disease activity (116). However, it is simplistic at this point to state that human lupus is a Th2 disease when there is evidence that even in NZB/NZW mice, both type 2 (IL-4) and type 1 (IFNγ and IL-12) lymphokines are needed for the expression of the disease (150).

In contrast to the circulating T cell pool, B cells from lupus patients are not numerically decreased, but display increased proliferation rates, and spontaneously secrete increased amounts of immunoglobulin, including autoantibodies. The amount of spontaneously released immunoglobulin from lupus B cells accurately correlates with disease activity. Polyclonal hypergammaglobulinemia is common in SLE patients. Nevertheless, challenging SLE patients with standard immunizations or stimulating lupus B cells with polyclonal activators *in vitro* results in substantially decreased amounts of specific antibody production compared to the responses of normal B cells (151,152).

A particular B cell subset, the CD5+ cells (B-1 subset), has been reported to be responsible for the production of autoantibodies in NZB/W mice (153,154) and human rheumatoid arthritis (155,156). In contrast, conventional B cells have been found to produce the majority of autontibodies in mice with chronic graft-versus-host disease and MRL-*lpr* mice (157–159). In human lupus, pathogenic autoantibodies are produced by both CD5+ and conventional B cells (160). B-1 cells, a minority among the circulating B cell pool, produce largely IgM crossreactive antibodies of low affinity, uncharacteristic of the autoantibodies detected in SLE sera (156).

Although freshly-isolated B cells and both CD4+ and CD8+ T lymphocyte subpopulations from lupus patients express higher levels of Fas antigen compared to normal cells, mitogenic stimulation of cell subsets causes normal Fas antigen upregulation (161,162). The increased expression of Fas is accompanied by higher rates of spontaneous apoptosis (163). T cells, but not B cells, from lupus patients were also reported to have increased levels of *bcl-2* protein.

The increased apoptotic rate of lupus cells can be reversed following culture *in vitro* with lymphokines, mitogens or superantigens (164). Fas ligand was also found to be expressed at increased amounts in lupus T cells and retained full functional features (165). Increased apoptosis may contribute to the development of cytopenias and the release of nuclear material in the circulation that may, under certain conditions, stimulate the immune system to produce autoantibodies (166). It should be noted that activated T cells from patients with lupus are relatively resistant to a TCR-mediated death stimulus, which may lead to prolonged survival of activated autoreactive cells (167). A Fas protein that lacks the 21-amino acid transmembrane region is secreted in the serum and was reported (168), but not reproduced by others (169,170), to be increased in the sera of lupus patients.

The human "apoptosis deficiency" syndrome, equivalent to the mice carrying the *lpr* and *gld* mutations, was recognized recently as the Canale-Smith syndrome (171) and it is discussed in Chapter 11. Only rare lupus patients have structural defects of the Fas (172) or Fas ligand (173) molecules. It is possible that lupus patients have genetic defects of downstream molecules that participate in the apoptosis pathway.

The CD40–CD40 ligand (CD40L) interaction appears to be aberrant in lupus. Although the number of circulating CD40L+ T cells of lupus patients is not increased compared to normals, its induction in lupus CD4+ and CD8+ T cells is both enhanced and sustained (174). Moreover, it was reported that lupus B cells unexpectedly express CD40L upon stimulation, in an equally intense manner to T cells. This abnormally-regulated molecule in lupus lymphocytes is still functional because anti-CD40L mAb inhibited the production of anti-DNA antibodies (175), and lupus CD40L+ cells induced the expression of CD80 (B7–1) in a B cell line (174). The importance of this interaction in SLE is underscored by experiments showing that administration of a single dose of anti-CD40L monoclonal antibody in mice with lupus caused significant delay in the appearance of nephritis and substantially improved the survival of such animals without compromising the non-autoimmune response (176).

In lupus patients, the CD28+ peripheral blood T cells of both CD4+ and CD8+ subsets are decreased and the circulating CD28− T cell population is expanded. Anti-CD3 antibody-induced apoptosis of CD28+ T cells is significantly accelerated *in vitro* in SLE, providing a possible explanation for the loss of these cells in the peripheral blood *in vivo*, whereas apoptosis of CD28− T cells is barely detected either in lupus patients or in normal persons. The expression and kinetics of upregulation of CTLA-4 in lupus T cells is normal (177).

Abnormalities in the expression of CD80 and CD86 on the cell surface of peripheral blood B cells from patients with SLE have also been reported. Levels of CD86 expression on resting and activated lupus B cells was greater than the levels of normal B cells. CD80 was also significantly overexpressed in activated, but not in resting B cells from lupus patients (91). Therefore, overexpression of costimulatory molecules on circulating B cells in patients with SLE may play a role in the continuous autoreactive T cell help to lupus B cells leading to the production of autoantibodies (178,179).

Non-B APC from lupus patients fail to upregulate *in vitro* surface expression of CD80 following stimulation with IFNγ in a disease-specific fashion (90). Replenishment of functional CD80 molecule in the culture environment significantly increased the responses of SLE T cells to tetanus toxoid and to an anti-CD3 antibody (92). Similarly, the decreased responses of lupus T cells to anti-CD2 antibody are reversed in the presence of CD28-mediated stimulation (180).

T lymphocytes may express abnormally high levels of surface molecules that facilitate attachment to endothelial or other cells. Certain pairs of molecules are involved exclusively in providing the costimulatory signal to T cells, whereas other pairs of molecules are involved both in cell costimulation and adhesion/homing (178,181,182). Distinct sets of surface molecules participate in the cell-cell and cell-matrix adhesion process. Members of the β1-integrin family of adhesion molecules are major adhesive receptors for the extracellular matrix, while members of the β2-integrin family, VLA4 and VLA3 are involved in both cell-cell and cell-extracellular matrix adhesion (183).

The expression of adhesion molecules is upregulated in various disease conditions such as atherosclerosis (184), inflammation (185) and rheumatoid arthritis (186). VLA4 and LFA1 are overexpressed on the surface membrane of SLE lymphocytes. Interestingly, VLA4 is overexpressed only in lymphocytes from patients with vasculitis. Lymphocytes from these patients show increased adhesion to cord vein endothelial cells (187).

Neutrophils from patients with active lupus display increased surface membrane expression of β2-integrin CD1 1b/CD18 (188). Increased expression of the VLA4 partner on the surface of endothelial cells may also be present and further facilitate lymphocyte attachment to endothelial cells and initiation of the inflammatory process. These findings may shed light on the pathogenesis of vasculitis/vasculopathy (189) and other tissue injury in patients with lupus.

5. MEMBRANE-MEDIATED SIGNAL TRANSDUCTION IN LUPUS IMMUNE CELLS

Following the engagement of the antigen-receptor either with a specific antigen or with an anti-receptor antibody, multiple well-regulated intracellular signaling pathways are triggered in the form of biochemical cascades. A critical event in these cascades is the mobilization of Ca^{2+} from intracellular stores, followed by an influx of Ca^{2+} from the extracellular space. The Ca^{2+} response is shared by many cell types, but the presence of specialized Ca^{2+}-sensitive enzymes and transcription factors found in specific tissues dictates the transcription of cell type-specific genes (190,191).

Fresh circulating lupus T cells (whole T cells, CD4$^+$, and CD8$^+$ cells) as well as lupus T cell lines and autoantigen-specific lupus T cell clones displayed significantly increased free intracellular Ca^{2+} [Ca^{2+}]i responses compared to T lymphocytes, T cell lines or autoantigen-specific T cell clones from normal subjects or patients with systemic autoimmune diseases other than SLE. Release of intracellularly stored Ca^{2+} rather than influx from the extracellular space was identified as the major contributor to these abnormally high lupus T-cell responses, and the production of 1,4,5-inositol trisphosphate (IP3, the principal mediator of intracellular Ca^{2+} release) was also enhanced following stimulation with anti-CD3 antibodies (192, 193).

The mechanism for the increased [Ca^{2+}]i responses are not known at this point, although involvement of LAT and overactivation of PLCγ have been suggested. The possibility of defective downregulation of the Ca^{2+} response is attractive and one candidate for such a role is the Ca^{2+}-response-downregulator cAMP-dependent protein kinase A type I (PKA-I). Activation of PKA-I in T cells decreases the efflux of intracellularly-stored free calcium. Kammer et al. has demonstrated that the activity of PKA-I in lupus T cells is significantly impaired (194,195), which may explain the overactive TCR/CD3-initiated lupus calcium response.

The Ca^{2+} pathway critically influences NFAT-mediated transcription of genes such as those of CD40-ligand and Fas-ligand (CD40L and FasL), and therefore, lupus lymphocytes should express more surface membrane CD40L and FasL due to their higher Ca^{2+} responses. Indeed, following activation, lupus T cells overexpress FasL (165) and CD40L (174), which are functional (175).

Stimulation of lupus T cells with anti-CD3 antibodies significantly enhanced production of tyrosyl-phosphorylated cellular proteins, which reached maximal levels earlier than normal cells (196). The reason for this overproduction of phosphotyrosine proteins is currently unknown. Crosslinking of the CD2 T cell surface molecule triggers an intracellular signal thought to be identical to the TCR signal because the CD2-pathway employs TCR to initiate its biochemical pathway. The anti-CD2 antibody-initiated pathway in SLE is defective (197,180). Also, earlier and recent studies have disclosed that following anti-CD2 stimulation, mitogenic responses and the production of TGF-β are decreased in lupus T cells (198).

The TCRζ chain (member of the zeta-family of proteins) is part of the hetero-oligomeric TCR/CD3 complex. While the TCRζ homodimer and the invariant CD3γ, δ, and ε chains do not play a role in antigen-recognition, they are the signal-transducing subunits of the antigen-receptor complex (190,199). Expression of TCRζ chain was found deficient in T cells of two thirds of lupus patients while it was always present in T cells from normal subjects or from non-SLE patients (196,200). In some lupus patients, the TCRζ mRNA was also either absent or decreased correlating with the presence, deficiency or absence of the ζ polypeptide. TCRζ-chain deficiency was first reported in tumor-infiltrating lymphocytes in both humans and experimental models (201–203). Also decreased ζ-chain has been reported in T cells from patients with HIV (204) and rheumatoid arthritis (205,206). In the above conditions, the levels and phosphorylation patterns of TCR-associated protein tyrosine kinases were decreased and the Ca^{2+} responses were also decreased. It has been well established that following activation through the CD3/TCR complex, ζ-chain becomes ubiquitinated and degraded by the proteasome (207). In addition, ζ-chain can be degraded by caspases activated by the Fas pathway (208) and calpain can degrade other TCR-associated molecules, such as ZAP-70 (209). It is possible that one or several of the above mechanisms are responsible for the low levels of ζ-chain in cells from some patients. The absence of both TCRζ and TCRζ mRNA in some patients, however, indicates that additional factors, such as decreased transcription of the ζ-chain gene, are involved. The transcription of the ζ-chain gene depends heavily on the activity of the transcription factor *elf-1* (210). Indeed, in preliminary experiments the levels of *elf-1* were found decreased in lupus T cells (Enyedy and Tsokos unpublished experiments). TCR ζ-chain deficiency can explain a number of lupus T cell defects, such as defective CD2-mediated signaling, defective natural killer activity (86) and activation-induced cell death, which require intact ζ-chain (211).

The study of B cell receptor (BCR)-mediated signaling of SLE B cells provides molecular insights into their disease-characterizing hyperactivity. BCR-mediated signaling abnormalities are remarkably similar to those of lupus T cells. BCR engagement produced significantly increased Ca^{2+} responses (contributed mainly by the intracellular calcium stores), increased production of IP3 and enhanced production of tyrosyl-phosphorylated cellular proteins. It is of interest that proteins with apparent molecular size between 36–64 kDa were hyperphosphorylated in both B and T cells from lupus patients, while the baseline degree of protein tyrosyl phosphorylation did not differ between the two cell types and their normal counterparts (212). The signaling aberrations analyzed above (TCR, BCR, CD2-initiated signaling, and TCRζ deficiency) are independent of disease activity, treatment status, or the presence/absence of specific SLE clinical manifestations, and may thus represent intrinsic abnormalities of the lupus lymphocyte that are also disease-specific, since they were not found in normal lymphocytes or in lymphocytes from patients with other systemic autoimmune rheumatic diseases. The very similar lupus T- and B-cell signaling aberrations mentioned above could substantiate the hypothesis that a common background underlies some heterogeneous lymphocyte functional defects.

6. GENES AND GENETICS

Four major lines of evidence strongly support the genetic basis of human lupus: 1) the familial aggregation of the disease and concordance among twins; 2) the apparent difference in prevalence between ethnic groups; 3) the marked preponderance among females; 4) identified associations between disease manifestations and certain MHC alleles.

The risk for development of lupus in a sibling in families with a member with lupus is 20 times more than in the general population (213). The relatively high concordance rates for SLE in monozygotic twins (25–57%) compared to dizygotic twins (2–9%) supports the importance of the genetic background. It is currently believed that multiple genes confer susceptibility to SLE expression, several of which have been identified to associate with lupus. Most of these associations have been identified while investigating

pathogenic mechanisms in the disease. For example, the conjecture that certain MHC antigens should present autoantigens better than others led to unveiling of associations between the disease or individual manifestations and MHC antigens. On the other hand, the expectation that apoptosis-related genes should be associated with the expression of lupus was not fulfilled.

More recently, genome-wide searches in families with multiple affected members disclosed areas in the genome associated with lupus (Table 3). Obviously, no single gene is sufficient or necessary for disease expression and lupus diathesis is conferred by a yet-unidentified number of genes that differ in individual patients. Still, the fact that almost all of the rare patients with C1r/C1s and complete C4 deficiency develop lupus suggests that the number of the contributing genes may be limited in some patients. The complexity of deciphering genes involved in the expression of lupus is confounded by the number of additional environmental, hormonal and immunoregulatory factors in a given individual or population. Finally, genetic epistasis, the interaction of different genes to produce a disease phenotype, as shown in animal models (45,214–217), may occur in patients with lupus.

The first lupus-family study was conducted by Tsao et al. (218) and supported the linkage between SLE and the genetic locus 1q41–42. Fifty-two multiplex lupus families were analyzed, and the linkage was found to cross ethnic barriers. This study targeted that specific genetic interval because it is syntenic with a known murine lupus-predisposing locus

Table 3   Genome-wide studies in human lupus.

Study	Patients	Race	Association
Tsao et al. (218)	52 sib-pairs	C, As, A	Iq41–Iq42 (true for all 3 ethnic groups)
Tsao et al. (222)	124 families	C, H, As, A	Iq41–q42 (PARP gene) in group C
Delrieu et al. (223)	171 patients	C	No association with PARP alleles
Gaffney et al. (227)	105 families	C, H, As, A	6p11–p21 (HLA locus) 16q13 14q21–23 20p12
Shai et al.	80 families	C, H	1q44, 1p36, 1q24, 1p21, 18q21–22, 16q13, 20p13, 14q23
(228)		H not C	1q44, 1p36
Moser et al. (225)	94 families	A	1q41, 1q23, 11q14–23
		C	14q11, 4p15, 11q25, 2q32, 19q13, 6q26–27, 12p12–11
		All	1q23, 13q32, 20q13, 1q31
Moser et al. (221)	107 sib-pairs	C	1q41(D1S229)
		A	D1S3462 (15cM distal to D1S229)

A, African American; As, Asian; C, Caucasian; H, hispanic.

(219, 220). Study of additional 127 multiplex lupus pedigrees showed linkage with 1q41 (microsatellite marker D1S229) and for African Americans, an even stronger linkage with the marker D1S3462, which is 15 cM distal to D1S229 (221). Of the candidate genes in the 1q41–42 region, poly(ADP-ribose) polymerase (PARP), which encodes a zinc-finger DNA-binding protein involved in DNA repair and apoptosis, alleles were found to be preferentially transmitted in affected off-springs in a large number of families (222). This association was not confirmed in a recent French European PARP allele study of 171 unrelated patients with lupus (223). TGFβ2, which maps within the same region and should have been expected to display strong association with lupus because of its known immunoregulatory effects and abnormal induction in lupus cells (224), was not found to be linked in the same cohort of families.

The cohorts of multiplex lupus families analyzed in two additional studies are larger. Moser et al. (225) studied 94 pedigrees and reported that potential SLE loci (with lod scores >2) were found at chromosomes 1q23, 1q41 and 11q14–23 in African-Americans, and in European-Americans at 14q11, 4p15, 11q25, 2q32, 19q13, 6q26–27, and 12p12–11. In the combined pedigrees, the potential lupus loci were 1q23, 13q32, 20q13, and 1q31, with strongest linkage for locus 1q23 in African-Americans. Candidate genes for this interval are those for FcγRIIA, as well as other nearby genes like FcγRIIIA, FcγRIIIB and the ζ-chain of the T-cell receptor (226).

The study by Gaffney et al. (227) analyzed 105 sib-pair lupus families almost entirely of European-American origin. This genome-wide microsatellite marker screen revealed that the strongest evidence for linkage was found near the MHC locus at 6p11–q21 and at three additional genetic intervals, 16q13, 14q21–23 and at 20p12. The two latter studies did not use the same microsatellite markers for screening, and it is interesting that there was partial agreement on linkage scores for some, but not all, markers that mapped closely. Moreover, the loci with the strongest linkage reported in one study were not found in the other, e.g., linkage with 1q41–42 was strong in two of three studies, but the first reported that it crossed ethnic barriers, while the other showed a predominant effect on African-American families.

The Jacob (228) screen of the human genome in a sample of 188 lupus patients belonging to 80 lupus families with two or more affected relatives per family using 350 polymorphic markers suggested predisposing loci on chromosomes 1 and 18, but no single locus with overwhelming evidence for linkage was found. Regions 1q44 and 1p36 were associated with disease only in Mexican-American families with lupus.

The long arm of human chromosome 1, interestingly, maps several of the potential lupus loci found in these genome-wide screens as well as others implicated in lupus, such as those encoding for FcγRIIA, FcγRIIIA, TCRζ, FasL, IL-10, CR1, CR2 and C1q proteins. These molecules have been implicated in SLE either by small-scale genetic or non-genetic studies (211).

The impact of genetic factors is also underscored by differences in the incidence and prevalence of the disease in various races. SLE has a higher incidence and prevalence in African-Americans, Afro-Caribbeans and East Asians in whom the disease may also have more severe course and prognosis. Certain clinical (e.g. discoid skin lesions, nephritis) and serological (e.g. anti-Sm autoantibodies) manifestations are found more frequently in the African-American lupus population, and the high prevalence of severe nephritis encountered therein is responsible for the severe prognosis. Detailed genetic analysis may reveal the molecular basis for such inter-racial differences (229,230).

Previous studies have focused on the potential association of SLE with MHC alleles or haplotypes. The strongest associations are with DR3, DQ2- and DR2, DQ6-containing MHC Class II haplotypes, but even these confer a relatively low risk for SLE. On the contrary, the association of certain MHC alleles or haplotypes is much stronger with serum autoantibody profiles (rev. in 230).

Nine of ten individuals with complete C1q or C1r/C1s or C4 develop lupus. Thus, genes that encode for early components of the classical complement pathway can single-handedly induce lupus (231). Also, one third of patients with C2 deficiency develop lupus, though the clinical manifestations are milder than those seen with the C1 and C4 deficiency (231). Only 1–2% of lupus patients have a complete deficiency of an early complement pathway component. Serum C4 is composed of C4A and C4B proteins encoded by two >99% identical genes. Of the two proteins, C4A seems more important in mounting a secondary immune response (232). Homozygous C4A deficiency occurs in 10–15% of Caucasian lupus patients compared to less than 2% in controls, and heterozygous C4A deficiency occurs in 50–70% of lupus patients and 20–35% of control subjects (233).

Complement receptor types 1 (CR1) and 2 (CR2) also map on the long arm of chromosome 1, and their expression in SLE is decreased. CR1 binds the C3b and C4b fragments of C3 and exhibits molecular weight polymorphism. CR1 participates in the phagocytic process and provides erythrocytes with the ability to carry and clear immune complexes. CR1 numbers have been reported to be decreased on the surface of SLE erythrocytes, but many studies have discounted the decrease as genetically-determined (234,235).

Mannose binding protein (MBP) is a member of the collectin family and structurally resembles C1q. It is produced in the liver rapidly after challenge and forms trimers that associate with MBP-associated serine proteases and directly mediates cleavage of C3 without activation of the

C1 complex (MBP pathway). Four alleles of MBP have been recognized, each with different biochemical properties: A, wild type; B, G54D, is unstable and the MBP levels are undetectable in homozygotes; C, G57E, intermediate phenotype in terms of serum levels; D, R52C, has a milder phenotype (236,237). The frequency of B and C alleles is increased in Caucasian (238,239) and Chinese (240) lupus populations and the MBP serum levels is decreased. A significant association between MBP deficiency and lupus has been also shown in African-American lupus patients (241).

Lymphokine gene polymorphisms are logically expected to associate with disease manifestations in view of aforementioned multiple abnormalities in lymphokine production. Polymorphisms have been reported for the promoter of TNFα gene, also encoded within the context of MHC-III, in lupus patients in whom its levels are decreased (242). The promoter TNFα-308A (TNF2) polymorphism, is associated with an increased risk of SLE in African-Americans and this relationship was independent of DR alleles (243). TNF2 allele has been associated with lupus in British (244), but not in Chinese, lupus patients (245). Finally, TNF microsatellite markers have been associated with lupus, but not independently of DR alleles (246,247). A TNF receptor 2 polymorphism (196R) was found to be associated with lupus in Japanese patients independent of the DR alleles (248). The TNFβ polymorphism, defined by *Nco I* RFLP, was associated with low production of TNFα, TNFβ and nephritis in Korean lupus patients (249).

A study of a large cohort of Mexican-American lupus patients revealed significantly different allelic distribution for bcl-2, Fas-L, and IL-10 loci compared with controls. More importantly, and highlighting the significance of genetic "build-up", it was noted that individuals carrying specific genotypes of both bcl-2 and IL-10 were at significant risk of developing lupus (250).

IL-1α, IL-1β and IL-1 receptor antagonist genes are located in chromosome 2q and are polymorphic (251). A microsatellite marker in the second intron of the IL-1 receptor antagonist gene is associated with lupus in British (252) and Japanese (253) patients. Also, polymorphisms of the IL-6 gene have been found to be associated with lupus (145).

Receptors for the Fc portion of Ig are involved both in the regulation of the immune response and the clearance of immune complexes. Genes *FcγRIIA* and *FcγRIIIA* encode receptors for the Fc fragment of IgG types IIA (CD32) and IIIA (CD16). A well-described polymorphism of the functional domain of FcγRIIA, which consists of a single amino acid change of an arginine to histidine at position 131, predominates in African-American lupus patients (69) and is associated with defective FcγRIIA function, decreased IgG2 binding, impaired immune-complex handling, and clinically apparent immune-complex deposition in the kidneys. In the African-American SLE patient population, the FcγIIA polymorphism is also a risk factor for invasive pneumococcal infections in SLE (254). Altered distribution of FcγRIIIA alleles was reported in a cohort of Korean patients with lupus nephritis (255). Since these genes alone can not cause disease (74), the epistatic effect of other genes, such as angiotensin I converting enzyme (256), may be needed.

In summary, multiple genetic loci or genes contribute to susceptibility for the development of SLE. Genetic analyses may increase our understanding of the disease heterogeneity and provide a molecular basis for the racial differences in disease prevalence, manifestations, severity and prognosis. The pathogenic contribution and complex interaction of lupus-susceptibility genes needs to be clarified by additional studies and the precise role and contribution of each of the lupus-related genes should be addressed.

7. HORMONES

Although lupus affects prepubertal males and females equally, during puberty it manifests a striking preference for females that is maintained throughout the reproductive years. Thus, female hormonal factors play a permissive role at least, while male hormonal factors play a protective one in the expression of SLE. This has been further supported by studies in murine strains where it has been clearly shown that estrogens have deleterious effects on lupus-prone experimental animals, while androgens are protective (257,258).

Patients with SLE produce increased amounts of the potent estrogens estrone and estriol due to increased 16α hydroxylation of estradiol, and also have decreased levels of androgens (259–262). While estrogens promote (263) and androgens suppress (264) anti-dsDNA antibody production by lupus B cells, the role of sex hormones in the development of the immune system is not clear. While most attention is focused on estrogens, other female hormones (e.g. progesterone, prolactin) may also play a role, since hyperprolactinemia correlates with the appearance of autoantibodies such as anti-dsDNA, anti-Sm and anti-Ro (265). The role of exogenous estrogens was addressed in a long term clinical study where it was reported that long-term estrogen replacement therapy was associated with increased risk for of SLE (266). Past use of oral contraceptives was associated with slightly increased relative risk, but these data refer to a population exposed to high estrogen content oral contraceptives. Case-control studies in populations receiving low-dose oral contraceptives have found no increased risk for lupus (267).

Estrogens act on target cells after binding to their cytoplasmic estrogen receptors, which belong to the group of nuclear receptors. The estrogen-estrogen receptor complex acquires transcription factor activity and, following entrance in the nucleus and binding to specific estrogen-response elements found in the promoters of several genes, it modulates their transcription. A recent report failed to detect major structural defects in the estrogen receptor transcripts from lupus peripheral immune cells (268).

Estrogen receptors are located both in the cytoplasm and on the cell surface membrane. In murine T cells, membrane estrogen receptors, upon binding to estradiol, mediate a rise in the concentration of intracellular calcium ($[Ca^{2+}]_i$), which is a pivotal second messenger. Estrogen response elements are found in the promoters of the protooncogenes *c-fos* and *c-jun* and affect their transcription and, therefore, the levels of the transcription factor AP-1, which is a *fos/jun* heterodimer. Estrogens cause a significant increase in the amounts of calcineurin transcripts and also in calcineurin phosphatase activity in SLE T cells (269). This effect is both lupus-specific, since it was not observed in either normal or disease-control T cells, and estradiol-specific, since other steroid hormones did not share it. Calcineurin activity directly influences the function of transcription factor NFAT, which, along with factor AP-1, determine the transcription rates of early immune response genes (270). In addition, estrogens may contribute to the development of the autoimmune response indirectly by increasing the expression of autoantigens, which define estrogen response elements in the promoter region of the encoding genes. For example, estradiol reportedly increased the expression of 52 and 60-kDa SS-A/Ro antigens in human keratinocytes (271).

8. ENVIRONMENTAL FACTORS

Various environmental factors, such as UV light, heavy metals, organic solvents and infections, influence a genetically-susceptible host in triggering the expression of SLE (Table 4). Exposure to UV light causes photosensitivity (more frequently in the Caucasian lupus population) and is a known disease-exacerbating factor. UV light causes apoptotic cell death of keratinocytes and cell surface expression of autoantigens previously "hidden" in the cytoplasm and/or nucleus. Autoantigens presented on surface membrane blebs of discrete size become accessible for immune recognition and attack. The latter may result in local inflammation and the appearance in the circulation of autoantibodies. UV-light irradiation of cultured human keratinocytes induced changes consistent with apoptosis, and the autoantigens were clustered in two kinds of blebs of the cell surface membrane, the smaller blebs containing endoplasmic reticulum, ribosomes and the (auto) antigen Ro, the larger blebs containing nucleosomal DNA, Ro and La and the small ribonucleoproteins. (41,42). UV-mediated apoptotic cell death may yield increased serum concentrations of autoantigens that activate

Table 4   Environmental factors and lupus

- UV light causes disease flares
- UV-exposed keratinocytes express autoantigen-filled surface blebs
- Cell stressors increase the production of autoantibody-recognizable phosphoproteins
- Phosphoproteins associate with U1 snRNP
- Cell stress leads to activation of mitogen-associated kinases
- Cell stress may facilitate autoantigen expression
- Cell stress may alter gene expression and lead to autoimmunity

immune cells. In addition, the activated apoptotic cascade may allow further degradation of autoantigens, which may lead to the exposure of cryptic and potentially more immunogenic antigens that would permit the expansion of the autoimmune response (272).

UV light and ionizing irradiation, along with other apoptotic stimuli, lead to the generation of new phosphoproteins, which apparently act as autoantigens since they are recognized by lupus sera (273). These phosphoproteins associate with U1-snRNP and may, therefore, alter splicing of various genes (274). In addition, it is known that cell stress induced by irradiation, exposure to heavy metals, toxins and drugs, causes activation of various kinases, including p38 and the Jun N-terminus mitogen-activated protein kinases, which may also contribute to this process (275,276). These studies are important in revealing the biochemistry of the stressed cell and its repercussions in the production of new autoantigens. In addition, the altered cell biochemistry may affect gene transcription in the immune cells, which may render them autoreactive. Such candidate genes may include adhesion and costimulatory molecules (ICAM-1, CD40 ligand, etc.).

Numerous case-reports and uncontrolled studies have suggested that exposure to silica and organic solvents in hair dyes is significant in the development of connective tissue diseases, but controlled studies have put these concerns to rest (277,278). In addition, a prospective study found no increased frequency of immunologic abnormalities in women exposed to silicone breast implants, except for anti-ssDNA antibodies, which do not have known clinical relevance (277).

Smoking also has been reported to correlate with the development of SLE, with increased risk found in two studies, although in the smaller of the two studies the increased risk was not significant. Among others, cigarette smoking affects the activity of enzymes involved in estrogen metabolism, further complicating the already complex interaction between environmental, hormonal and genetic factors for the development of SLE (279).

Clinical experience suggests that SLE may be initiated or recur following an infection (Table 5), but despite

Table 5   Infections in and lupus

- Viruses may break tolerance to autoantigens
- Early scroconversion to EBV antigens is associated with lupus
- Molecular mimicry between viral proteins and self antigens
  Ro/SSA and VSV glycoprotein
  Sm (B/B' subunit) antigen and HIV p24 gag
  Sm (D subunit) and EBNA-1
- Viral proteins may simulate chemokine receptors or ligands
- Viral proteins may interfere with apoptosis-involved proteins
  p53 and bcl-2

repeated efforts, a lupus-causing microorganism has never been identified. It has been hypothesized that infectious agents can disproportionally trigger an endogenously dysregulated immune system for the development or the exacerbation of SLE. Among the common pathogens, the herpesvirus Epstein-Barr virus (EBV) has received most attention. Antibodies against EBV have cross-reactivity with the lupus-specific autoantigen Sm. It was recently reported that newly diagnosed young patients with lupus have a significantly higher percentage of sero-positivity for EBV infection compared to controls. Other tested herpesviruses did not follow this pattern. EBV DNA was found in the lymphocytes of all 32 young lupus patients tested and two thirds of controls (280,281). Whether EBV-infected individuals become more suscept-ible to the development of lupus, or lupus patients are/become more susceptible to EBV infection, or whether a third factor increases susceptibility to both, is currently not known.

Molecular mimicry between autoantigens and antigens expressed by viruses and other pathogens have been extensively considered in the pathogenesis of autoimmune diseases and discussed in a preceding chapter. Epitopes of the SLE- associated 60 kd Ro/SSA autoantigen share sequences with the vesicular stomatitis virus (VSV) nucleocapsid protein, which may explain the presence of anti-Ro/SSA anti-bodies. In addition, immunization of animals with VSV pro-teins causes production of anti-Ro/SSA autoantibodies as well as of anti-VSV antibodies (282). Additional examples of molecular mimicry include that between the B/B' component of the Sm antigen and the HIV-1 p24 gag protein (283,284) and the D component of the Sm antigen and the EBNA-1 protein of EBV (285).

Alternatively, viral infection can break tolerance to self-antigens, as was shown in transgenic mice expressing a VSV glycoprotein. Autoantibodies to VSV glycoprotein cannot be induced by VSV glycoprotein in adjuvant or by recombinant vaccinia virus expressing VSV glycoprotein, but are triggered by infection with wild-type VSV (286). The latter is an attractive mechanism since it can explain disease flares that follow infections.

Viral proteins may interfere with the function of proteins involved in cell death and survival. For example, adenoviral proteins may activate or inhibit p53 and bcl-2 (287). Interestingly, viral proteins may mimic chemokine receptors (288,289) or ligands (288,290). These examples further com-plicate elucidation of the pathogenic involvement of viruses and other infectious agents in the development of the autoim-mune response.

The syndrome of drug-induced lupus has many similari-ties, but also important differences, to the idiopathic SLE syndrome (291). Since it represents a disease entity wherein the inciting factor is known, it is a good model to study certain aspects of SLE pathogenesis. Drugs that cause the SLE-like syndrome have been reported to induce DNA hypomethylation. Lupus T cell DNA is hypomethylated, and the activity of the methylation-inducing enzyme, DNA methyltransferase, is decreased (292). Non-T cells from patients with SLE did not share this abnormality, which affected only half of the lupus patients tested, and this abnormality was not disease-specific. Treatment of T cells with DNA methylation inhibitors induces upregulation of the adhesion/costimulatory molecule lymphocyte function-associated antigen-1 (LFA-1). The significance of this event is underscored by studies in animal models where infusion of T cells overexpressing LFA-1 can mediate the production of anti-dsDNA autoantibodies and the appear-ance of glomerulonephritis. It is thus possible that drugs inducing DNA hypomethylation can initiate an autoim-mune process by upregulating the costimulatory molecule LFA-1 (93,293).

Besides altering the T cell phenotype to facilitate inap-propriate homing and provide help to autoreactive B cells, drugs may interfere with the development of central toler-ance. Mice injected with a hydralazine metabolite produced autoantibodies probably by altering the positive selection of autoreactive T cells (294,295).

9.   FROM PATHOGENESIS TO RATIONAL TREATMENT

Current therapeutic schemes are directed by individual disease manifestations rather than general diagnosis of lupus. Involvement of vital organs (kidney, CNS, lungs, heart, severe hematologic disease) usually calls for intense treatment, including high doses of corticosteroids and cyto-toxic drugs, whereas manifestations from non-vital organs can be treated with moderate or low doses of cortico-steroids, antimalarials, and non-steroidal anti-inflammatory drugs (1,2). Pulse cyclophosphamide has emerged as the treatment of choice for severe visceral involvement (296–300). Corticosteroids, alkylating agents or calcineurin inhibitors (such as and cyclosporine and FK506) may be used in selected cases of patients with membranous lupus nephropathy (301,302).

Mycophenolate mofetil is an inhibitor of purine synthesis that is more effective than azathioprine in reducing the frequency of acute rejection (but not long-term graft survival) in solid organ transplants (303). Preliminary experience with this agent in glomerular diseases, including lupus nephritis, has been encouraging (304). A multicenter controlled trial of mycophenolate in SLE has recently been initiated (Ginzler E, personal communication). Similar to patients receiving organ transplants, combinations of this drug with calcineurin inhibitors may be used in patients with severe disease, refractory to therapy with cyclophosphamide.

Halogenated adenosine analogs, such as cladribine and fludarabine, have been promising in pilot studies (305). These agents become activated by kinases, which are highly expressed in lymphoid cells and therefore display selective toxicity (induction of apoptosis) for both resting and divided lymphocytes (306). Combinations of alkylating agents (cyclophosphamide), which cause DNA strand breaks, and fludarabine, which inhibits DNA repair, have a synergistic effect and are currently being tested in pilot studies in patients with proliferative lupus nephritis. If effective, these agents may decrease the cumulative dose of cyclophosphamide and the resultant amenorrhea, which is almost universal for women over 30 years of age (307).

Understanding the processes involved in the pathogenesis of lupus have made possible the design of rational therapeutic approaches (Table 6). The enormous recent progress in the field of cytokines has raised expectations for their use in the management of autoimmune diseases. Indeed, a fusion molecule between TNF-α receptor and the Fc portion of IgG was shown to significantly improve the clinical course of patients with rheumatoid arthritis (308). It

Table 6   Rationalized therapy for lupus

- Interruption of cognate cell interaction with antibodies or fusion molecules
    comprising a ligand and the Fc portion of human Ig
    - CTLA4-Ig
    - Anti-B7–1 or 2 antibodies
    - Anti-CD40 ligand antibody
- Neutralization of cytokines with antibodies or fusion molecules (cytokine receptor human Ig)
    - Anti-IL-10 antibody
    - Anti-TNFα *antibody*
    - TNFα receptor-FcIgG fusion molecule
- Lymphokines
- Hormones (DHEA)
- Inhibition of complement activation
- Bone marrow transplantation
- Reestablishment of tolerance (T and B cell)
- Modulation of cell signaling

---

should be cautioned that lupus is a pathogenically different disease, and treatment with TNFα-neutralizing biologics may exacerbate it. Patients treated with a humanized anti-TNFα antibody developed autoantibodies, but not clinical disease (309,310), an observation that suggests such reagents should not be used in lupus patients.

In an experimental model of anti-phospholipid syndrome, intraperitoneal injection of IL-3 prevented fetal loss (311). IL-3 has been introduced in clinical trials for treatment of neutropenias in patients with myelodysplastic and aplastic syndromes (312) and has been noted to have minimal toxicity. Although the justification for the use in the prevention of fetal loss originated from the observation that these animals have low serum IL-3 levels, its mechanism of action is not clear.

Neutralization of IL-10 may present another therapeutic option as adjunct therapy for lupus and other systemic autoimmune diseases. Continuous infusion of anti-IL-10 antibody of NZB/NZW mice from birth dramatically improved survival, indicating either that IL-10 is directly involved in promoting disease activity by acting on B lymphocytes and other cells, or that elimination of IL-10 reversed the cytokine ratio in favor of the type 2 group of cytokines (313). A major concern with all candidate lymphokines is the fact that they are pleiotropic. Patients with various forms of malignancies who have been treated with either IL-2 or IFNα have developed a wide range of autoimmune disease including thyroiditis and SLE (120).

Complement activation is abundant in lupus patients, as manifested by the fact that the circulating immune complexes have attached C3 split products (64), and the levels of various complement split products (65) and the membrane attack complex (66) are increased in the sera and urine. The use of biologics that would interfere with complement activation could be of therapeutic value; an anti-C5 antibody was shown to prolong life in NZB/NZW mice (314) and a humanized anti-C5 monoclonal antibody has been considered for use in lupus patients. The use of regulatory surface membrane molecules, such as decay accelerating factor (DAF), in the treatment of lupus should be also considered. Crry-Ig was shown to inhibit nephritis in a nephrotoxic serum animal model (315). Crry is the murine analogue of human DAF and CD69, and inhibits both classical and alternative pathways of complement activation. Similarly, a transgenic mouse secreting Crry was resistant to the effects of nephrotoxic serum (316).

Disruption of the cognate interruption between T and B cell has been predicted and proven to be helpful in lupus. A B7-binding construct (CTLA4 molecule fused with the heavy chain of Ig to provide decreased plasma clearance rates) was given to mice that spontaneously develops lupus and the rate of survival increased along with lower autoantibody titers and reduced kidney pathology (317).

Interruption of the CD40-gp39 (CD-40L) interaction in SNF1 autoimmune mice with anti-gp39 antibody resulted in decreased expansion of autoimmune memory B cells and long term clinical improvement (176). Experiments showing therapeutic effects in autoimmune mice following interruption of the B7-CD28 and CD40-gp39 paths indicate they are involved serially in the production of antibody. The above results have instigated ongoing human trials using humanized anti-CD40 ligand antibodies and combined treatments to achieve better and longer lasting effects (318,319). Indeed, when CTLA41g was combined with anti-gp39, there was sustained inhibition of autoantibody production and renal disease (320).

For patients with severe disease, immunosuppression may be intensified to the point of myelosuppression or hematopoietic ablation. Hematopoiesis and immunity may then be rapidly reconstituted by reinfusion of CD34$^+$ progenitor cells. This approach has been tried in a limited number of patients with lupus and other autoimmune diseases who had failed standard treatment (321,322). Although the number of patients is small and the follow-up period short, it seems that either the severe immunosuppression and the redevelopment of the immune system with possibly less autoimmune features may have beneficial effects on the clinical expression of the disease.

Hormonal treatment of lupus patients to reverse the effect of endogenous androgens with dehydroepiandrosterone has been tested in double-blind controlled studies and found to have a minimal effect, which may justify its use as steroid-sparing agent, or treatment of mild manifestations in selected patients (323–325).

As discussed above, lupus patients have T cells that respond to defined autoantigens and may display certain TCR selection. These studies helped us to understand the pathogenesis of the disease and also have provided rationale for specific immunologic interventions (326). The identification of putative autoantigens may prove useful in reestablishing tolerance. Indeed, myelin basic protein (MBP) disease-inducing T-cells have been successfully deleted by high dose of MBP antigen in murine model of autoimmune encephalomyelitis (327). Additionally, oral administration of chicken collagen II in a double-blinded clinical study to patients with rheumatoid arthritis resulted in significant clinical improvement (328). Recently, histone-defined peptides have been identified in mice (40) and humans (329) that can stimulate T cell-dependent anti-DNA antibody production and provide the basis for the development of peptide-based tolerogenic therapy for humans similar to that demonstrated in mice with lupus (330).

The described cellular biochemical abnormalities that characterize lupus cells may become the target of pharmaceutical intervention. The tyrosine kinase inhibitor, tyrphostin AG490, was shown to prevent binding of freshly isolated mouse lymph node cells and of *in vivo* activated lymphocytes to brain vessel endothelium by inhibiting the expression of adhesion molecules and systemic administration of the drug in a murine model of experimental allergic encephalomyelitis decreased disease activity (331,332). Therefore, drugs that interfere with cell signaling events may prove to be helpful in the treatment of lupus.

10. CONCLUSIONS

In a genetically-susceptible host, exogenous and hormonal factors influence the immune system at multiple levels, leading to numerous immunoregulatory abnormalities. The latter leads to generation of effector mechanisms of autoimmune pathology, including autoantibodies, immune complexes, autoreactive cells and byproducts of immune activation, including lymphokines. T cells interact with B cells by both cognate and non-cognate means to help them produce autoantibodies. Autoantigens are revealed to the immune system because stressful stimuli such as UV-light irradiation of keratinocytes (from exposure to sunlight) induce surface expression of previously hidden nuclear or cytoplasmic constituents. Immune complexes formed in excess amounts are cleared in decreased rates, resulting in increased serum levels and enhanced tissue deposition.

Recent advances in scanning the human genome are expected to identify all genes involved in the expression of the disease. When this is accomplished, we may be able to provide accurate genetic counseling and hopefully identify genes whose products are involved in the pathogenesis of the disease and the expression of pathology.

Although better use of immunosuppressants and improved health care delivery have effectively decreased disease morbidity and mortality, there still is a long way to go before to conquering this disease. Understanding the immunoregulatory abnormalities has helped in designing rational approaches to expand our therapeutic armamentarium. Biologic agents that can specifically reverse certain immune aberrations should further help lupus patients by improving survival, quality of life and limit side effects.

ACKNOWLEDGEMENT

Work was supported by PHS grant RO1-A142269.

References

1. Boumpas, D.T., B.J. Fessler, H.A. Austin, III, J.E. Balow, J.H. Klippel, and M.D. Lockshin. 1995. Systemic lupus erythematosus: emerging concepts. Part 2: Dermatologic and joint disease, the antiphospholipid antibody syndrome,

pregnancy and hormonal therapy, morbidity and mortality, and pathogenesis. *Ann. Intern. Med.* 123:42–53.

2. Boumpas, D.T., H.A. Austin, III, B.J. Fessler, J.E. Balow, J.H. Klippel, and M.D. Lockshin. 1995. Systemic lupus erythematosus: emerging concepts. Part 1: Renal, neuropsychiatric, cardiovascular, pulmonary and hematologic disease. *Ann. Intern. Med.* 122:940–950.

3. Krych, M., J.P. Atkinson, and V.M. Holers. 1992. Complement receptors. *Curr. Opin. Immunol.* 4:8–13.

4. Lawrence, R.C., C.G. Helmick, F.C. Arnett, R.A. Deyo, D.T. Felson, E.H. Giannini, S.P. Heyse, R. Hirsch, M.C. Hochberg, G.G. Hunder, M.H. Liang, S.R. Pillemer, V.D. Steen, and F. Wolfe. 1998. Estimates of the prevalence of arthritis and selected musculoskeletal disorders in the United States. *Arthritis Rheum.* 41:778–799.

5. Uramoto, K.M., C.J. Michet, Jr., J. Thumboo, J. Sunku, W.M. O'Fallon, and S.E. Gabriel. 1999. Trends in the incidence and mortality of systemic lupus erythematosus, 1950–1992. *Arthritis Rheum.* 42:46–50.

6. Austin, H.A., III, D.T. Boumpas, E.M. Vaughan, and J.E. Balow. 1995. High-risk features of lupus nephritis: importance of race and clinical and histological factors in 166 patients. *Nephrol. Dial. Transplant.* 10:1620–1628.

7. Dooley, M.A., S. Hogan, C. Jennette, and R. Falk. 1997. Cyclophosphamide therapy for lupus nephritis: poor renal survival in black Americans. Glomerular Disease Collaborative Network. *Kidney Int.* 51:1188–1195.

8. Bakir, A.A., P.S. Levy, and G. Dunea. 1994. The prognosis of lupus nephritis in African-Americans: a retrospective analysis. *Am. J. Kidney Dis.* 24:159–171.

9. Tan, E.M., A.S. Cohen, J.F. Fries, A.T. Masi, D.J. McShane, N.F. Rothfield, J.G. Schaller, N. Talal, and R.J. Winchester. 1982. The 1982 revised criteria for the classification of systemic lupus erythematosus. *Arthritis Rheum.* 25:1271–1277.

10. Tsokos, G.C., H.M. Moutsopoulos, and A.D. Steinberg. 1981. Muscle involvement in systemic lupus erythematosus. *JAMA* 246:766–768.

11. Wei, N., N. Pavlidis, G. Tsokos, R.J. Elin, and P.H. Plotz. 1981. Clinical significance of low creatine phosphokinase values in patients with connective tissue diseases. *JAMA* 246:1921–1923.

12. Roldan, C.A., B.K. Shively, and M.H. Crawford. 1996. An echocardiographic study of valvular heart disease associated with systemic lupus erythematosus. *N. Engl. J. Med.* 335:1424–1430.

13. Petri, M., D. Spence, L.R. Bone, and M.C. Hochberg. 1992. Coronary artery disease risk factors in the Johns Hopkins Lupus Cohort: prevalence, recognition by patients, and preventive practices. *Medicine* (Baltimore) 71:291–302.

14. Manzi, S., F. Selzer, K. Sutton-Tyrrell, S.G. Fitzgerald, J.E. Rairie, R.P. Tracy, and L.H. Kuller. 1999. Prevalence and risk factors of carotid plaque in women with systemic lupus erythematosus. *Arthritis Rheum.* 42:51–60.

15. Manzi, S., E.N. Meilahn, J.E. Rairie, C.G. Conte, T.A. Medsger, Jr., L. Jansen-McWilliams, R.B. D'Agostino, and L.H. Kuller. 1997. Age-specific incidence rates of myocardial infarction and angina in women with systemic lupus erythematosus: comparison with the Framingham Study. *Am. J. Epidemiol.* 145:408–415.

16. Hirohata, S. and T. Miyamoto. 1990. Elevated levels of interleukin-6 in cerebrospinal fluid from patients with systemic lupus erythematosus and central nervous system involvement. *Arthritis Rheum.* 33:644–649.

17. Tsai, C.Y., T.H. Wu, S.T. Tsai, K.H. Chen, P. Thajeb, W.M. Lin, H.S. Yu, and C.L. Yu. 1994. Cerebrospinal fluid interleukin-6, prostaglandin E2 and autoantibodies in patients with neuropsychiatric systemic lupus erythematosus and central nervous system infections. *Scand. J. Rheumatol.* 23:57–63.

18. Vogelgesang, S.A., M.P. Heyes, S.G. West, A.M. Salazar, P.P. Sfikakis, R.N. Lipnick, G.L. Klipple, and G.C. Tsokos. 1996. Quinolinic acid in patients with systemic lupus erythematosus and neuropsychiatric manifestations. *J. Rheumatol.* 23:850–855.

19. Schneebaum, A.B., J.D. Singleton, S.G. West, J.K. Blodgett, L.G. Allen, J.C. Cheronis, and B.L. Kotzin. 1991. Association of psychiatric manifestations with antibodies to ribosomal P proteins in systemic lupus erythematosus. *Am. J. Med.* 90:54–62.

20. West, S.G., W. Emlen, M.H. Wener, and B.L. Kotzin. 1995. Neuropsychiatric lupus erythematosus: a 10-year prospective study on the value of diagnostic tests. *Am. J. Med.* 99:153–163.

21. Sloan, R.P., M.M. Schwartz, S.M. Korbet, and R.Z. Borok. 1996. Long-term outcome in systemic lupus erythematosus membranous glomerulonephritis. Lupus Nephritis Collaborative Study Group. *J. Am. Soc. Nephrol.* 7:299–305.

22. Esdaile, J.M., L. Joseph, M. Abrahamowicz, Y. Li, D. Danoff, and A.E. Clarke. 1996. Routine immunologic tests in systemic lupus erythematosus: is there a need for more studies? *J. Rheumatol.* 23:1891–1896.

23. Pillemer, S.R., H.A. 3d Austin, G.C. Tsokos, and J.E. Balow. 1988. Lupus nephritis: association between serology and renal biopsy measures. *J. Rheumatol.* 15:284–288.

24. Drenkard, C., A.R. Villa, C. Garcia-Padilla, M.E. Perez-Vazquez, and D. Alarcon-Segovia. 1996. Remission of systematic lupus erythematosus. *Medicine* (Baltimore) 75:88–98.

25. Iliopoulos, A.G. and G.C. Tsokos. 1996. Immunopathogenesis and spectrum of infections in systemic lupus erythematosus. *Semin. Arthritis. Rheum.* 25:318–336.

26. Hellman, D.B., C.M. Kirsch, Q. Whiting-O'Keefe, J. Simonson, N.B. Schiller, M. Petri, G. Gamsu, and W. Gold. 1995. Dyspnea in ambulatory patients with SLE: prevalence, severity, and correlation with incremental exercise testing. *J. Rheumatol.* 22:455–461.

27. Ramsey-Goldman, R., J.E. Dunn, C.F. Huang, D. Dunlop, J.E. Rairie, S. Fitzgerald, and S. Manzi. 1999. Frequency of fractures in women with systemic lupus erythematosus: comparison with United States population data. *Arthritis Rheum.* 42:882–890.

28. Coutinho, A., M.D. Kazatchkine, and S. Avrameas. 1995. Natural autoantibodies. *Curr. Opin. Immunol.* 7:812–818.

29. Hayakawa, K., M. Asano, S.A. Shinton, M. Gui, D. Allman, C.L. Stewart, J. Silver, and R.R. Hardy. 1999. Positive selection of natural autoreactive B cells. *Science* 285:113–116.

30. Davidson, A., A. Manheimer-Lory, C. Aranow, R. Peterson, N. Hannigan, and B. Diamond. 1990. Molecular characterization of a somatically mutated anti-DNA antibody bearing two systemic lupus erythematosus-related idiotypes. *J. Clin. Invest.* 85:1401–1409.

31. Livneh, A., G. Or, A. Many, E. Gazit, and B. Diamond. 1993. Anti-DNA antibodies secreted by peripheral B cells of lupus patients have both normal and lupus-specific features. *Clin. Immunol. Immunopathol* 68:68–73.

32. Livneh, A., E. Behar, A. Many, M. Ehrenfeld, E. Gazit, and B. Diamond. 1992. Lupus anti-DNA antibodies bearing the 8.12 idiotype appear to be somatically mutated. *J. Clin. Immunol.* 12:11–16.

33. Paul, E., A.A. Iliev, A. Livneh, and B. Diamond. 1992. The anti-DNA-associated idiotype 8.12 is encoded by the V lambda II gene family and maps to the vicinity of L chain CDR1. *J. Immunol.* 149:3588–3595.

34. rner, T., N.L. Farner, and P.E. Lipsky. 1999. Ig lambda and heavy chain gene usage in early untreated systemic lupus erythematosus suggests intensive B cell stimulation. *J. Immunol.* 163:1027–1036.

35. Dorner, T., S.J. Foster, N.L. Farner, and P.E. Lipsky. 1998. Immunoglobulin kappa chain receptor editing in systemic lupus erythematosus. *J. Clin. Invest.* 102:688–694.

36. Vlahakos, D.V., M.H. Foster, S. Adams, M. Katz, A.A. Ucci, K.J. Barrett, S.K. Datta, and M.P. Madaio. 1992. Anti-DNA antibodies form immune deposits at distinct glomerular and vascular sites. *Kidney Int.* 41:1690–1700.

37. Vlahakos, D., M.H. Foster, A.A. Ucci, K.J. Barrett, S.K. Datta, and M.P. Madaio. 1992. Murine monoclonal anti-DNA antibodies penetrate cells, bind to nuclei, and induce glomerular proliferation and proteinuria *in vivo*. *J. Am. Soc. Nephrol.* 2:1345–1354.

38. O'Keefe, T.L., S.K. Datta, and T. Imanishi-Kari. 1992. Cationic residues in pathogenic anti-DNA autoantibodies arise by mutations of a germ-line gene that belongs to a large VH gene subfamily. *Eur. J. Immunol.* 22:619–624.

39. Shi, Y., A. Kaliyaperumal, L. Lu, S. Southwood, A. Sette, M.A. Michaels, and S.K. Datta. 1998. Promiscuous presentation and recognition of nucleosomal autoepitopes in lupus: role of autoimmune T cell receptor alpha chain. *J. Exp. Med.* 187:367–378.

40. Kaliyaperumal, A., C. Mohan, W. Wu, and S.K. Datta. 1996. Nucleosomal peptide epitopes for nephritis-inducing T helper cells of murine lupus. *J. Exp. Med.* 183:2459–2469.

41. Casciola-Rosen, L.A., G. Anhalt, and A. Rosen. 1994. Autoantigens targeted in systemic lupus erythematosus are clustered in two populations of surface structures on apoptotic keratinocytes. *J. Exp. Med.* 179:1317–1330.

42. Casciola-Rosen, L., A. Rosen, M. Petri, and M. Schlissel. 1996. Surface blebs on apoptotic cells are sites of enhanced procoagulant activity: Implications for coagualtion events and antigenic spread in systemic lupus erythematosus. *Proc. Natl. Acad. Sci. USA* 93:1624–1629.

43. Mohan, C., L. Morel, P. Yang, and E.K. Wakeland. 1997. Genetic dissection of systemic lupus erythematosus pathogenesis: Sle2 on murine chromosome 4 leads to B cell hyperactivity. *J. Immunol.* 159:454–465.

44. Morel, L., C. Mohan, Y. Yu, B.P. Croker, N. Tian, A. Deng, and E.K. Wakeland. 1997. Functional dissection of systemic lupus erythematosus using congenic mouse strains. *J. Immunol.* 158:6019–6028.

45. Mohan, C., L. Morel, P. Yang, H. Watanabe, B. Croker, G. Gilkeson, and E.K. Wakeland. 1999. Genetic dissection of lupus pathogenesis: a recipe for nephrophilic autoantibodies. *J. Clin. Invest.* 103:1685–1695.

46. Reichlin, M. 1999. Autoantibodies to intracellular antigens in systemic lupus erythematosus patients that bind and penetrate cells. In: Kammer, G.M., and G.C. Tsokos (eds). *Lupus: molecular and cellular pathogenesis.* Humana Press, Totowa, NJ, pp. 389–398.

47. Bonfa, E., S.J. Golombek, L.D. Kaufman, S. Skelly, H. Weissbach, N. Brot, and K.B. Elkon. 1987. Association between lupus psychosis and anti-ribosomal P protein antibodies. *N. Engl. J. Med.* 317:265–271.

48. Hulsey, M., R. Goldstein, L. Scully, W. Surbeck, and M. Reichlin. 1995. Anti-ribosomal P antibodies in systemic lupus erythematosus: a case-control study correlating hepatic and renal disease. *Clin. Immunol. Immunopathol.* 74:252–256.

49. Martin, A.L., and M. Reichlin. 1996. Fluctuations of antibody to ribosomal P proteins correlate with appearance and remission of nephritis in SLE. *Lupus* 5:22–29.

50. Koren, E., M.W. Reichlin, M. Koscec, R.D. Fugate, and M. Reichlin. 1992. Autoantibodies to the ribosomal P proteins react with a plasma membrane- related target on human cells. *J. Clin. Invest.* 89:1236–1241.

51. Koscec, M., E. Koren, M. Wolfson-Reichlin, R.D. Fugate, E. Trieu, I.N. Targoff, and M. Reichlin. 1997. Autoantibodies to ribosomal P proteins penetrate into live hepatocytes and cause cellular dysfunction in culture. *J. Immunol.* 159:2033–2041.

52. Wener, M.H., S. Uwatoko, and M. Mannik. 1989. Antibodies to the collagen-like region of C1q in sera of patients with autoimmune rheumatic diseases. *Arthritis Rheum.* 32:544–551.

53. Coremans, I.E., P.E. Spronk, H. Bootsma, M.R. Daha, E.A. van der Voort, L. Kater, F.C. Breedveld, and C.G. Kallenberg. 1995. Changes in antibodies to C1q predict renal relapses in systemic lupus erythematosus. *Am. J. Kidney Dis.* 26:595–601.

54. Siegert, C.E., M.R. Daha, C.M. Tseng, I.E. Coremans, L.A. van Es, and F.C. Breedveld. 1993. Predictive value of IgG autoantibodies against C1q for nephritis in systemic lupus erythematosus. *Ann. Rheum. Dis.* 52:851–856.

55. Siegert, C., M. Daha, M.L. Westedt, D. van, V, and F. Breedveld. 1991. IgG autoantibodies against C1q are correlated with nephritis, hypocomplementemia, and dsDNA antibodies in systemic lupus erythematosus. *J. Rheumatol.* 18:230–234.

56. Coremans, I.E., M.R. Daha, E.A. van der Voort, Y. Muizert, C. Halma, and F.C. Breedveld. 1992. Antibodies against C1q in anti-glomerular basement membrane nephritis. *Clin. Exp. Immunol.* 87:256–260.

57. Wener, M.H. 1999. Immune complexes and autoantibodies to C1q. In: Kammer, G.M., and G.C. Tsokos (eds). *Lupus: molecular and cellular pathogenesis.* Humana Press, Totowa, NJ, pp. 574–598.

58. Bluestein, H.G., and V.L. Woods, Jr. 1982. Antineuronal antibodies in systemic lupus erythematosus. *Arthritis. Rheum.* 25:773–778.

59. Denburg, J.A., and S.A. Behmann. 1994. Lymphocyte and neuronal antigens in neuropsychiatric lupus: presence of an elutable, immunoprecipitable lymphocyte/neuronal 52 kd reactivity. *Ann. Rheum. Dis.* 53:304–308.

60. Buyon, J.P., S.G. Slade, J.D. Reveille, J.C. Hamel, and E.K. Chan. 1994. Autoantibody responses to the native 52-kDa SS-A/Ro protein in neonatal lupus syndromes, systemic lupus erythematosus, and Sjogren's syndrome. *J. Immunol.* 152: 3675–3684.

61. Buyon, J.P. 1999. Autoimmune associated congenital heart block. Bringing bedside challenges to the bench. In: Kammer, G.M. and G.C. Tsokos (eds). *Lupus: molecular and cellular pathogenesis.* Human Press, Totowa, NJ, pp. 492–513.

62. Theofilopoulos, A.N. and F.J. Dixon. 1979. The biology and detection of immune complexes. *Adv. Immunol.* 28:89–220.

63. Theofilopoulos, A.N. 1981. The Raji, conglutinin, and anti-C3 assays for the detection of complement-fixing immune complexes. *Methods Enzymol.* 74 Pt C:511–530.

64. Aguado, M.T., J.D. Lambris, G.C. Tsokos, R.Z. Burger, D. Bitter-Suermann, J.D. Tamerius, F.J. Dixon, and A.N. Theofilopoulos. 1985. Monoclonal antibodies against com-

plement 3 neoantigens for detection of immune complexes and complement activation. Relationship between immune complex levels, state of C3 and numbers of receptors for C3b. *J. Clin. Invest.* 76:1418–1426.

65. Manzi, S., J.E. Rairie, A.B. Carpenter, R.H. Kelly, S.P. Jagarlapudi, S.M. Sereika, T.A. Medsger, Jr., and R. Ramsey-Goldman. 1996. Sensitivity and specificity of plasma and urine complement split products as indicators of lupus disease activity. *Arthritis Rheum.* 39:1178–1188.

66. Schulze, M., J.V. Donadio, Jr., C.J. Pruchno, P.J. Baker, R.J. Johnson, R.A. Stahl, S. Watkins, D.C. Martin, R. Wurzner, and O. Gotze. 1991. Elevated urinary excretion of the C5b–9 complex in membranous nephropathy. *Kidney Int.* 40:533–538.

67. Schifferli, J.A., and R.P. Taylor. 1989. Physiological and pathological aspects of circulating immune complexes. *Kidney Int.* 35:993–1003.

68. Kimberly, R.P., J.E. Salmon, and J.C. Edberg. 1995. Receptors for immunoglobulin G. Molecular diversity and implications for disease. *Arthritis Rheum.* 38:306–314.

69. Salmon, J.E., S. Millard, L.A. Schachter, F.C. Arnett, E.M. Ginzler, M.F. Gourley, R. Ramsey-Goldman, M.G.E. Peterson, and R.P. Kimberly. 1996. FcgammaRIIA alleles are heritable risk factors for lupus nephritis. *J. Clin. Invest.* 97:1348–1354.

70. Wilson, J.G., W.W. Wong, E.E. 3. Murphy, P.H. Schur, and D.T. Fearon. 1987. Deficiency of the C3b/C4b receptor (CR1) of erythrocytes in systemic lupus erythematosus: analysis of the stability of the defect and of a restriction fragment length polymorphism of the CR1 gene. *J. Immunol.* 138:2708–2710.

71. Ross, G.D., W.J. Yount, M.J. Walport, J.B. Winfield, C.J. Parker, C.R. Fuller, R.P. Taylor, B.L. Myones, and P.J. Lachmann. 1985. Disease-associated loss of erythrocyte complement receptors (CR1, C3b receptors) in patients with systemic lupus erythematosus and other diseases involving autoantibodies and/or complement activation. *J. Immunol.* 135:2005–2014.

72. Moulds, J.M., J.D. Reveille, and F.C. Arnett. 1996. Structural polymorphisms of complement receptor 1 (CR1) in systemic lupus erythematosus (SLE) patients and normal controls of three ethnic groups. *Clin. Exp Immunol.* 105:302–305.

73. Cook, J.M., M.D. Kazatchkine, P. Bourgeois, F. Mignon, J.P. Mery, and M.F. Kahn. 1986. Anti-C3b-receptor (CR1) antibodies in patients with systemic lupus erythematosus. *Clin. Immunol. Immunopathol.* 38:135–138.

74. Clynes, R., C. Dumitru, and J.V. Ravetch. 1998. Uncoupling of immune complex formation and kidney damage in autoimmune glomerulonephritis. *Science* 279:1052–1054.

75. Wener, M.H., and M. Mannik. 1986. Mechanisms of immune deposit formation in renal glomeruli. *Springer Semin. Immunopathol.* 9:219–235.

76. Mannik, M., V.J. Gautheir, S.A. Stapleton, and L.Y.C. Agodoa. 1987. Immune complexes with cationic antibodies deposit in glomeruli more effectively than cationic antibodies alone. *J. Immunol.* 138:4209–4219.

77. Rubin, R.L., F.L. Tang, E.K. Chan, K.M. Pollard, G. Tsay, and E.M. Tan. 1986. IgG subclasses of autoantibodies in systemic lupus erythematosus, Sjogren's syndrome, and drug-induced autoimmunity. *J. Immunol.* 137:2528–2534.

78. Budhai, L., K. Oh, and A. Davidson. 1996. An *in vitro* assay for detection of glomerular binding IgG autoantibodies in patients with systemic lupus erythematosus. *J. Clin. Invest.* 98:1585–1593.

79. Lefkowith, J.B., M. Kiehl, J. Rubenstein, R. DiValerio, K. Bernstein, L. Kahl, R.L. Rubin, and M. Gourley. 1996. Heterogeneity and clinical significance of glomerular-binding antibodies in systemic lupus erythematosus. *J. Clin. Invest.* 98:1373–1380.

80. Schmiedeke, T.M., F.W. Stockl, R. Weber, Y. Sugisaki, S.R. Batsford, and A. Vogt. 1989. Histones have high affinity for the glomerular basement membrane. Relevance for immune complex formation in lupus nephritis. *J. Exp. Med.* 169:1879–1894.

81. Coritsidis, G.N., P.C. Beers, and P.M. Rumore. 1995. Glomerular uptake of nucleosomes: evidence for receptor-mediated mesangial cell binding. *Kidney Int.* 47:1258–1265.

82. Morgan, E.L., and W.O. Weigle. 1983. Polyclonal activation of murine B lymphocytes by immune complexes. *J. Immunol.* 130:1066–1070.

83. Morgan, E.L., and W.O. Weigle. 1980. Polyclonal activation of murine B lymphocytes by Fc fragments. I. The requirement for two signals in the generation of the polyclonal antibody response induced by Fc fragments. *J. Immunol.* 124:1330–1335.

84. Tolnay, M., and G.C. Tsokos. 1998. Complement receptor 2 in the regulation of the immune response. *Clin. Immunol. Immunopathol.* 88:123–132.

85. Tsokos, G.C. 1999. Overview of cellular immune function in systemic lupus erythematosus. In: Lahita, R.G. (ed). *Systemic lupus erythematosus* Academic Press., New York, pp. 17–54.

86. Boswell, J., and P.H. Schur. 1989. Monocyte function in systemic lupus erythematosus. *Cin. Immunol. and Immunopathol.* 52:271–278.

87. Linker-Israeli, M., A.C. Bakke, R.C. Kitridou, S. Gendler, S. Gillis, and D.A. Horwitz. 1983. Defective production of interleukin 1 and interleukin 2 in patients with systemic lupus erythematosus (SLE). *J. Immunol.* 130:2651–2655.

88. Shirakawa, F., V. Vamashito, and H. Susuki. 1985. Decrease in HLA-DR-positive monocyte in patients with SLE. *J. Immunol.* 134:3560–3562.

89. Shirakawa, F., V. Vamashito, and H. Susuki. 1985. Reduced function of the HLA DR positive monocytes in patients with SLE. *J. Clin. Immunol.* 5:396–403.

90. Tsokos, G.C., B. Kovacs, P.P. Sfikakis, S. Theocharis, S. Vogelgesang, and C.S. Via. 1996. Defective antigen presenting cell function in patients with systemic lupus erythematosus. Role of the B7-1 (CD80) costimulatory molecule. *Arthritis Rheum.* 39:600–609.

91. Liossis, S.N., P.P. Sfikakis, and G.C. Tsokos. 1998. Immune cell signaling aberrations in human lupus. *Immunol Res.* 18:27–39.

92. Sfikakis, P.P., R. Oglesby, P. Sfikakis, and G.C. Tsokos. 1996. B7/BB1 provides an important costimulatory signal for CD3- mediated T lymphocyte proliferation in patients with systemic lupus erythematosus (SLE). *Clin. Exp. Immunol.* 1994. Apr. 96:8 14.

93. Richardson, B.C., R.L. Yung, K.J. Johnson, P.E. Rowse, and N.D. Lalwani. 1996. Monocyte apoptosis in patients with acitive lupus. *Arhritis Rheum.* 39:1432–1434.

94. Dayal, A.K., and G.M. Kammer. 1996. The T cell enigma in lupus. *Arthritis Rheum.* 39:23–33.

95. Inghirami, G., J. Simon, J.E. Balow, and G.C. Tsokos. 1988. Activated T lymphocytes in the peripheral blood of patients with systemic lupus erythematosus induce B cells to produce immunoglobulin. *Clin. Exp. Rheumatol.* 6:269–276.

96. Boumpas, D.T., G.C. Tsokos, D.L. Mann, E.G. Eleftheriades, C.C. Harris, and G.E. Mark. 1986. Increased proto-oncogene expression in peripheral blood lymphocytes from patients with systemic lupus erythematosus and other autoimmune diseases. *Arthritis. Rheum.* 29:755–760.

97. Klinman, D.M., J.F. Mushinski, M. Honda, Y. Ishigatsubo, and A.D. Steinberg. 1986. Oncogene expression in autoimmune and normal peripheral blood mononuclear cells. *J. Exp. Med.* 163:1292.

98. Gmelig-Meyling, F., S. Dawisha, and A.D. Steinberg. 1992. Assessment of *in vivo* frequency of mutated T cells in patients with systemic lupus erythematosus. *J. Exp. Med.* 175:297–300.

99. Dawisha, S.M., F. Gmelig-Meylig, and A.D. Steinberg. 1994. Assessment of clinical parameters associated with increased frequency of mutant T cells in patients with systemic lupus erythematosus. *Arthrititis Rheum.* 37:270–277.

100. Theocharis, S., P.P. Sfikakis, R.N. Lipnick, G.L. Klipple, A.D. Steinberg, and G.C. Tsokos. 1995. Characterization of *in vivo* mutated T cell clones from patients with systemic lupus erythematosus. *Clin. Immunol. Immunopathol.* 74:135–142.

101. Rajagopalan, S., T. Zordan, G.C. Tsokos, and S.K. Datta. 1990. Pathogenic anti-DNA autoantibody-inducing T helper cell lines from patients with active lupus nephritis: isolation of CD4–8- T helper cell lines that express the gamma delta T-cell antigen receptor. *Proc. Natl. Acad. Sci. USA* 87:7020–7024.

102. Shivakumar, S., G.C. Tsokos, and S.K. Datta. 1989. T cell receptor alpha/beta expressing double-negative (CD4–/CD8–) and CD4+ T helper cells in humans augment the production of pathogenic anti-DNA autoantibodies associated with lupus nephritis. *J. Immunol.* 143:103–112.

103. Desai-Mehta, A., C. Mao, S. Rajagopalan, T. Robinson, and S.K. Datta. 1995. Structure and specificity of T cell receptors expressed by potentially pathogenic anti-DNA autoantibody-inducing T cells in human lupus. *J. Clin. Invest.* 95:531–541.

104. Mohan, C., and S.K. Datta. 1995. Lupus: key pathogenic mechanisms and contributing factors. *Clin. Immunol. Immunopathol.* 77:209–220.

105. Olive, C., P.A. Gatenby, and S.W. Serjeantson. 1994. Restricted junctional diversity of T cell recepotr delta gene rearrangements expressed in systemic lupus erythematosus (SLE) patients. *Clin. Exp. Immunol.* 97:430–438.

106. Rajagopalan, S., C. Mao, and S.K. Datta. 1992. Pathogenic autoantibody-inducing gamma/delta T helper cells from patients with lupus nephritis express unusual T cell receptors. *Clin. Immunol. Immunopathol.* 62:344–350.

107. Hoffman, R.W., Y. Takeda, G.C. Sharp, D.R. Lee, D.L. Hill, H. Kaneoka, and C.W. Caldwell. 1993. Human T cell clones reactive against U-small nuclear ribonucleoprotein autoantigens from connective tissue disease patients and healthy individuals. *J. Immunol.* 151:6460–6469.

108. Okubo, M., M. Kurokawa, H. Ohto, T. Nishimaki, K. Nishioka, R. Kasukawa, and K. Yamamoto. 1996. Clonotype analysis of peripheral blood T cells and autoantigen- reactive T cells from patients with mixed connective tissue disease. *J. Immunol.* 153:3784–3790.

109. Linker-Israeli, M., and R. Deans. 1991. Dysregulated lymphokine production in systemic lupus erythematosus (SLE). *Ann. NY Acad. Sci.* 567–569.

110. Alcocer-Varela, J., and D. Alarc:on-Segovia. 1982. Decreased production of and response to interleukin-2 by cultured lymphocytes from patients with systemic lupus erythematosus. *J. Clin. Invest.* 69:1388–1392.

111. Murakawa, Y., S. Takada, Y. Ueda, N. Suzuki, T. Hoshino, and T. Sakane. 1985. Characterization of T lymphocyte subpopulations responsible for deficient interleukin 2 activity in patients with systemic lupus erythematosus. *J. Immunol.* 134:187–195.

112. Huang, Y.P., L.H. Perrin, P.A. Miescher, and R.H. Zubler. 1988. Correlation of T and B cell activities *in vitro* and serum IL–2 levels in systemic lupus erythematosus. *J. Immunol.* 141:827–833.

113. Linker-Israeli, M. 1992. Cytokine abnormalities in human lupus. *Clin. Immunol. Immunopathol.* 63:10–12.

114. Sabzevari, H., S. Propp, D.H. Kono, and A.N. Theofilopoulos. 1997. G1 arrest and high expression of cyclin kinase and apoptosis inhibitors in accumulated activated/memory phenotype CD4+ cells of older lupus mice. *Eur. J. Immunol.* 27:1901–1910.

115. Wong, H.K., G.M. Kammer, G. Dennis, and G.C. Tsokos. 1999. Abnormal NF-kappaB activity in T lymphocytes from patients with systemic lupus erythematosus is associated with decreased p65-relA protein expression. *J. Immunol.* 163:1682–1689.

116. Hagiwara, E., M.F. Gourley, S. Lee, and D.K. Klinman. 1996. Disease severity in patients with systemic lupus erythematosus correlates with an increased ratio of interleukin-10:interferon-gamma-secreting cells in the peripheral blood. *Arhtritis Rheum.* 39:379–385.

117. Tsokos, G.C., D.T. Boumpas, P.L. Smith, J.Y. Djeu, and J.E. Balow. 1988. Deficient gamma-IFN production in patients with SLE. *Arthritis Rheum.* 29:1210–1215.

118. Neighbour, P.A., and A.I. Grayzel. 1981. Interferon production *in vitro* by leuocytes from patients with systemic lupus erythematosus and rheumatoid arthritis. *Clin. Exp. Immunol.* 45:576–582.

119. Strannegard, O., S. Hermodsson, and G. Westberg. 1982. Interferon and natural killer cells in systemic lupus erythematosus. *Clin. Exp. Immunol.* 50:246–252.

120. Schattner, A. 1994. Ltmphokines in autoimmunty-a critical review. *Clin. Immmunol. Immunopathol.* 70:177–189.

121. Machold, K.P., and J.S. Smolen. 1990. Interferon-gamma induced exacerbation of systemic lupus erythematosus. *J. Rheumatol.* 17:831–832.

122. Graninger, W.B., W. Hassfeld, B.B. Pesau, K.P. Machold, C.C. Zielinski, and J.S. Smolen. 1991. Induction of systemic lupus erythematosus by interferon-gamma in a patient with rheumatoid arthritis. *J. Rheumatol.* 18:1621–1622.

123. Prud'Homme, G.J., D.H. Kono, and A.N. Theofilopoulos. 1995. Quantitative polymerase chain reaction analysis reveals marked overexpression of interleukin-1 beta, interleukin-1 and interferon- gamma mRNA in the lymph nodes of lupus-prone mice. *Mol. Immunol.* 32:495–503.

124. Seelig, H.P., H. Ehrfeld, and M. Renz. 1994. Interferon-gamma-inducible protein p16. A new target of antinuclear antibodies in patients with systemic lupus erythematosus. *Arthritis Rheum.* 37:1672–1683.

125. Jacob, C.O., P.H. van der Meide, and H.O. McDevitt. 1987. *in vivo* treatment of (NZB X NZW)F1 lupus-like nephritis with monoclonal antibody to gamma interferon. *J. Exp. Med.* 166:798–803.

126. Balomenos, D., R. Rumold, and A.N. Theofilopoulos. 1998. Interferon-gamma is required for lupus-like disease and lymphoaccumulation in MRL-lpr mice. *J. Clin. Invest.* 101:364–371.

127. Peng, S.L., J. Moslehi, and J. Craft. 1997. Roles of interferon-gamma and interleukin-4 in murine lupus. *J. Clin. Invest.* 99:1936–1946.

128. Haas, C., B. Ryffel, and M. Le Hir. 1997. IFN-gamma is essential for the development of autoimmune glomerulonephritis in MRL/lpr mice. *J. Immunol.* 158:5484–5491.

129. Schwarting, A., T. Wada, K. Kinoshita, G. Tesch, and V.R. Kelley. 1998. IFN-gamma receptor signaling is essential for the initiation, acceleration, and destruction of autoimmune kidney disease in MRL- Fas(lpr) mice. *J. Immunol.* 161:494–503.

130. Haas, C., B. Ryffel, and M. Le Hir. 1998. IFN-gamma receptor deletion prevents autoantibody production and glomerulonephritis in lupus-prone (NZB x NZW)F1 mice. *J. Immunol.* 160:3713–3718.

131. Jacob, C.O., and H.O. McDevitt. 1988. Tumour necrosis factor-α *in murine autoimmune "lupus" nephritis.* Nature 331:356–358.

132. Yu, C.L., K.L. Chang, C.C. Chiu, B.N. Chiang, S.H. Han, and S.R. Wang. 1989. Defective phagocytosis, decreased tumour necrosis factor- alpha production, and lymphocyte hyporesponsiveness predispose patients with systemic lupus erythematosus to infections. *Scand. J. Rheumatol.* 18:97–105.

133. Jacob, C.O., and H.O. McDevitt. 1988. Tumour necrosis factor-alpha in murine autoimmune 'lupus' nephritis. *Nature* 331:356–358.

134. Kalechman, Y., U. Gafter, J.P. Da, M. Albeck, D. Alarcon-Segovia, and B. Sredni. 1997. Delay in the onset of systemic lupus erythematosus following treatment with the immunomodulator AS101: association with IL-10 inhibition and increase in TNF-alpha levels. *J. Immunol.* 159:2658–2667.

135. Tokano, Y., S. Morimoto, H. Kaneko, H. Amano, K. Nozawa, Y. Takasaki, and H. Hashimoto. 1999. Levels of IL-12 in the sera of patients with systemic lupus erythematosus (SLE)—relation to Th1- and Th2-derived cytokines. *Clin. Exp. Immunol.* 116:169 173.

136. Horwitz, D.A., J.D. Gray, S.C. Behrendsen, M. Kubin, M. Rengaraju, K. Ohtsuka, and G. Trinchieri. 1998. Decreased production of interleukin-12 and other Th1-type cytokines in patients with recent-onset systemic lupus erythematosus. *Arthritis Rheum.* 41:838–844.

137. Alleva, D.G., S.B. Kaser, and D.I. Beller. 1998. Intrinsic defects in macrophage IL-12 production associated with immune dysfunction in the MRL/++ and New Zealand Black/White F1 lupus-prone mice and the Leishmania major-susceptible BALB/c strain. *J. Immunol.* 161:6878–6884.

138. Liu, T.F. and B.M. Jones. 1998. Impaired production of IL-12 in system lupus erythematosus. II: IL-12 production *in vitro* is correlated negatively with serum IL-10, positively with serum IFN-gamma and negatively with disease activity in SLE. *Cytokine* 10:148–153.

139. Okubo, T., E. Hagiwara, S. Ohno, T. Tsuji, A. Ihata, A. Ueda, A. Shirai, I. Aoki, K. Okuda, J. Miyazaki, and Y. Ishigatsubo. 1999. Administration of an IL-12-encoding DNA plasmid prevents the development of chronic graft-versus-host disease (GVHD). *J. Immunol.* 162:4013–4017.

140. Kishimoto, T., and T. Hirano. 1988. Molecular regulation of B lymphocyte response. *Ann. Rev. Immunol.* 6:485–512.

141. Hirohata, S., and T. Miyamoto. 1990. Elevated levels of interleukin-6 in cerebrospinal fluid from patients with systemic lupus erythematosus and central nervous system. *Arthritis and Rheumatism* 33:644–649.

142. Horii, Y., A. Muraguchi, M. Iwano, T. Matsuda, T. Hirayama, H. Yamada, Y. Fujii, K. Dohi, H. Ishikawa, and Y. Ohmoto. 1989. Involvement of IL-6 in mesangial proliferative glomerulonephritis. *J. Immunol.* 143:3949–3955.

143. Finck, B.K., B. Chan, and D. Wofsy. 1994. Interleukin 6 promotes murine lupus in NZB/NZW F1 mice. *J. Clin. Invest.* 94:585–591.

144. Linker-Israeli, M., M. Honda, R. Nand, R. Mandyam, E. Mengesha, D.J. Wallace, A. Metzger, B. Beharier, and J.R. Klinenberg. 1999. Exogenous IL-10 and IL-4 down-regulate IL-6 production by SLE-derived PBMC. *Clin. Immunol.* 91:6–16.

145. Linker-Israeli, M., D.J. Wallace, J.L. Prehn, R. Nand, L. Li, and J.R. Klinenberg. 1996. A greater variability in the 3′ flanking region of the IL-6 gene in patients with systemic lupus erythematosus (SLE). *Autoimmunity* 23:199–209.

146. Llorente, L., W. Zou, Y. Levy, Y. Richaud-Patin, J. Wijdenes, J. Alcocer-Varela, B. Morel-Fourrier, J.C. Brouet, D. Alarcon-Segovia, and P. Galanaud. 1995. Role of interleukin 10 in the B lymphocyte hyperactivity and autoantibody production of human systemic lupus erythematosus. *J. Exp. Med.* 181:839–844.

147. Llorente, L., Y. Richaud-Patin, R. Fior, J. Alcocer-Varela, J. Wijdenes, B.M. Fourrier, P. Galanaud, and D. Emilie. 1994. *in vivo* production of interleukin-10 by non-T cells in rheumatoid arthritis, Sjogren's syndrome, and systemic lupus erythematosus. A potential mechanism of B lymphocyte hyperactivity and autoimmunity. *Arthritis Rheum.* 37:1647–1655.

148. Sfikakis, P.P., and C.S. Via. 1997. Expression of CD28, CTLA4, CD80, and CD86 molecules in patients with autoimmune rheumatic diseases: implications for immunotherapy. *Clin. Immunol. Immunopathol.* 83:195–198.

149. Llorente, L., Y. Richaud-Patin, J. Couderc, D. Alarcon-Segovia, R. Ruiz-Soto, N. Alcocer-Castillejos, J. Alcocer-Varela, J. Granados, S. Bahena, P. Galanaud, and D. Emilie. 1997. Dysregulation of interleukin-10 production in relatives of patients with systemic lupus erythematosus. *Arthritis Rheum.* 40.1429–1435.

150. Theofilopoulos, A.N., and J.L. Lawson. 1999. Tumor necrosis factor and other cytokines in murine lupus. *Br. J. Rheumatol.*

151. Tsokos, G.C., and J.E. Balow. 1981. Spontaneous and pokeweed mitogen-induced plaque-forming cells in systemic lupus erythematosus. *Clin. Immunol. Immunopathol.* 21:172–183.

152. Blaese, R.M., J. Grayson, and A.D. Steinberg. 1980. Increased immunoglobulin-secreting cells in the blood of patients with active systemic lupus erythematosus. *Amer. J. Med.* 69:345–350.

153. Sidman, C.L., L.D. Shultz, R.R. Hardy, K. Hayakawa, and L.A. Herzenberg. 1986. Production of immunoglobulin isotypes by Ly1+B cells in viable motheaten and normal mice. *Science* 232:1423–1425.

154. Hayakawa, K., R. Hardy, D.R. Parks, L.A. Herzenberg, and A.D. Steinberg. 1984. Ly1 B cells: Functionally distinct lymphocytes that secreta IgM autoantibodies. *Proc. Nat. Acad. Sci. USA* 81:2494–2498.

155. Casali, P., S.E. Burastero, J.E. Balow, and A.L. Notkins. 1989. High-affinity antibodies to ssDNA are produced by CD5- B cells in systemic lupus erythematosus. *J. Immunol.* 143:3476–3483.

156. Casali, P., and A.L. Notkins. 1989. CD5+ B lymphocytes, polyreactive antibodies and the human B-cell repertoire. *Immunol. Today* 10:364–368.

157. Reap, E.A., E.S. Sobel, P.L. Cohen, and R.A. Eisenberg. 1992. The role of CD5+ B cells in the production of autoantibodies in murine systemic lupus erythematosus. *Ann. NY Acad. Sci.* 651:588–590.

158. Reap, E.A., E.S. Sobel, P.L. Cohen, and R.A. Eisenberg. 1993. Conventional B cells, not B-1 cells, are responsible for producing autoantibodies in lpr mice. *J Exp. Med.* 177:69–78.

159. Reap, E.A., E.S. Sobel, J.C. Jennette, P.L. Cohen, and R.A. Eisenberg. 1993. Conventional B cells, not B1 cells, are the source of autoantibodies in chronic graft-versus-host disease. *J. Immunol.* 151:7316–7323.

160. Suzuki, N., T. Sakane, and E.G. Engleman. 1990. Anti-DNA antibodies production by CD5+ and CD5- B cells of patients with systemic lupus erythematosus. *J. Clin. Invest.* 85:238–247.

161. Mysler, E., P. Bini, P. Ramos, S.M. Friedman, P.H. Krammer, and K.B. Elkon. 1994. The apoptosis–1/Fas protein in human systemic lupus erythematosus. *J. Clin. Invest.* 93:1029–1034.

162. Ohsako, S., M. Hara, M. Harigai, C. Fukasawa, and S. Kashiwazaki. 1994. Expression and function of Fas antigen and bcl-2 in human systemic lupus erythematosus lymphocytes. *Clin. Immunol. Immunopathol.* 73:109–114.

163. Emlen, W., J. Niebur, and R. Kadera. 1994. Accelerated *in vitro* apoptosis of lymphocytes from patients with systemic lupus erythematosus. *J. Immunol.* 152:3685–3692.

164. Lorenz, H.M., M. Grunke, T. Hieronymus, M. Herrmann, A. Kuhnel, B. Manger, and J.R. Kalden. 1997. *in vitro* apoptosis and expression of apoptosis-related molecules in lymphocytes from patients with systemic lupus erythematosus and other autoimmune diseases. *Arthritis Rheum.* 40:306–317.

165. Kovacs, B., S.N.C. Liossis, G.J. Dennis, and G.C. Tsokos. 1997. Increased expression of functional Fas-ligand in activated T cells from patients with systemic lupus erythematosus. *Autoimmunity* 25:213–221.

166. Mohan, C., S. Adams, V. Stanik, and S.K. Datta. 1993. Nucleosome: a major immunogen for pathogenic autoantibody- inducing T cells of lupus. *J. Exp. Med.* 177:1367–1381.

167. Kovacs, B., D. Vassilopoulos, S.A. Vogelgesang, and G.C. Tsokos. 1996. Defective CD3-mediated cell death in activated T cells from patients with systemic lupus erythematosus: role of decreased intracellular TNF-alpha. *Clin. Immunol. Immunopathol.* 81:293–302.

168. Cheng, J., T. Zhou, C. Liu, J.P. Shapiro, M.J. Brauer, M.C. Kiefer, P.J. Barr, and J.D. Mountz. 1994. Protection from Fas-mediated apoptosis by a soluble form of the Fas molecule. *Science* 263:1759–1762.

169. Knipping, E., P.H. Krammer, K.B. Onel, T.J. Lehman, E. Myster, and K.B. Elkon. 1995. Levels of soluble Fas/APO-1/CD95 in systemic lupus erythematosus and juvenile rheumatoid arthritis. *Arthritis Rheum.* 38:1735–1737.

170. Goel, N., D.T. Ulrich, E.W. StClair, J.A. Fleming, D.H. Lynch, and M.F. Seldin. 1995. Lack of correlation between serum soluble Fas/APO-1 levels and autoimmune disease. *Arthritis Rheum.* 38:1738–1743.

171. Mysler, E., P. Bini, J. Drappa, P. Ramos, S.M. Friedman, P.H. Krammer, and K.B. Elkon. 1994. The apoptosis-1/Fas protein in human systemic lupus erythematosus. *J. Clin. Invest.* 93:1029–1034.

172. Kovacs, B., T. Szentendrei, J.M. Bednarek, M.C. Pierson, J.D. Mountz, S.A. Vogelgesang, and G.C. Tsokos. 1997. Persistent expression of a soluble form of Fas/APO1 in continuously activated T cells from a patient with SLE. *Clin. Exp. Rheumatol.* 15:19–23.

173. Wu, J., J. Wilson, J. He, L. Xiang, P.H. Schur, and J.D. Mountz. 1996. Fas ligand mutation in a patient with systemic lupus erythematosus and lymphoproliferative disease. *J. Clin. Invest.* 98:1107–1113.

174. Koshy, M., D. Berger, and M.K. Crow. 1996. Increased expression of CD40 ligand on systemic lupus erythematosus lymphocytes. *J. Clin. Invest.* 98:826–837.

175. Desai-Mehta, A., L. Lu, R. Ramsey-Goldman, and S.K. Datta. 1996. Hyperexpression of CD40 ligand by B and T cells in human lupus and its role in pathogenic autoantibody production. *J. Clin. Invest.* 97:2063–2073.

176. Mohan, C., Y. Shi, J.D. Laman, and S.K. Datta. 1995. Interaction between CD40 and its ligand gp39 in the development of murine lupus nephritis. *J. Immunol.* 154:1470–1480.

177. Kaneko, H., K. Saito, H. Hashimoto, H. Yagita, K. Okumura, and M. Azuma. 1996. Preferential elimination of CD28+ T cells in systemic lupus erythematosus (SLE) and the relation with activation-induced apoptosis. *Clin. Exp. Immunol.* 106:218–229.

178. Sfikakis, P.P., and G.C. Tsokos. 1995. Lymphocyte ahesion molecules in autoimmune rheumatic diseases: basic issues and clinical expectations. *Clin. Exper. Rheum.* 13:763–777.

179. Stohl, W. 1995. Impaired polyclonal T cell cytolytic activity. A possible risk factor for systemic lupus erythematosus. *Arhritis Rheum.* 38:506–516.

180. Horwitz, D.A., F.L. Tang, M.M. Stimmler, A. Oki, and J.D. Gray. 1997. Decreased T cell response to anti-CD2 in systemic lupus erythematosus and reversal by anti-CD28: evidence for impaired T cell-accessory cell interaction. *Arthritis Rheum.* 40:822–833.

181. Guinan, E.C., J.G. Gribben, V.A. Boussiotis, G.J. Freeman, and L.M. Nadler. 1994. Pivotal role of the B7:CD28 pathway in transplantation tolerance and tumor immunity. *Blood* 84:3261–3282.

182. June, C.H., J.A. Bluestone, L.M. Nadler, and C.B. Thompson. 1994. The B7 and CD28 receptor families. *Immunol. Today* 15:321–331.

183. Hynes, R.O. 1992. Integrins: versatility, modulation, and signaling in cell adhesion. *Cell* 69:11–25.

184. Cybulsky, M.I., and M.A. Gimbrone. 1990. Endothelial expression of a mononuclear leukocyte adhesion molecule during atherogenesis. *Science* 251:788–791.

185. Argenbright, L.W., and R.W. Barton. 1992. Interactions of leukocyte integrins with intercellular adhesion molecule 1 in the production of inflammatory vascular injury *in vivo*. *J. Clin. Invest.* 89:259–272.

186. Laffon, A., V.R. Garcia, A. Humbria, A.A. Postigo, A.L. Corbi, and M.F. Sanchez. 1991. Upregulated expression and function of VLA4 fibronectin receptors on human activated T cells in rheumatoid arthritis. *J. Clin. Invest.* 88:546–552.

187. Takeuchi, T., K. Amano, H. Sekine, J. Koide, and T. Abe. 1993. Upregulated expression and function of integrin adhesive receptors in systemic lupus erythematosus. *J. Clin. Invest.* 92:3008–3016.

188. Molad, Y., J. Buyon, D.C. Anderson, S.B. Abramson, and B.N. Cronstein. 1994. Intravascular neutrophil activation

in systemic lupus erythematosus (SLE): dissociation between increased expression of CD11b/CD18 and diminished expression of L-selectin on neutrophils from patients with active SLE. *Clin. Immunol. Immunopathol.* 71:281–286.

189. Belmont, H.M., S.B. Abramson, and J.T. Lie. 1996. Pathology and pathogenesis of vascular injury in systemic lupus erythematosus. Interactions of inflammatory cells and activated endothelium. *Arthritis Rheum.* 39:9–22.

190. Musci, M.A., k.M. Latinis, and G.A. Koretzky. 1997. Signaling events in T lymphocytes leading to cellular activation or programmed cell death. *Clin. Immunol. Immunopathol.* 83:205–222.

191. Weiss, A., and D.R. Littman. 1994. Signal transduction by lymphocyte antigen receptors. *Cell* 76:263–274.

192. Vassilopoulos, D., B. Kovacs, and G.C. Tsokos. 1995. TCR/CD3 complex-mediated signal transduction pathway in T cells and cell lines from patients with systemic lupus erythematosus. *J. Immunol.* 155:2269–2281.

193. Liossis, S.N., R.W. Hoffman, and G.C. Tsokos. 1998. Abnormal early TCR/CD3-mediated signaling events of a snRNP- autoreactive lupus T cell clone. *Clin. Immunol. Immunopathol.* 88:305–310.

194. Kammer, G.M., I.U. Khan, and C.J. Malemud. 1994. Deficient type I protein kinase A isozyme activity in systemic lupus erythematosus T lymphocytes. *J. Clin. Invest.* 94:422–430.

195. Laxminarayana, D., I.U. Khan, N. Mishra, I. Olorenshaw, K. Tasken, and G.M. Kammer. 1999. Diminished levels of protein kinase A RI alpha and RI beta transcripts and proteins in systemic lupus erythematosus T lymphocytes. *J. Immunol.* 162:5639–5648.

196. Liossis, S.N., D.Z. Ding, G.J. Dennis, and G.C. Tsokos. 1998. Altered pattern of TCR/CD3-mediated protein-tyrosyl phosphorylation in T cells from patients with systemic lupus erythematosus. Deficient expression of the T cell receptor zeta chain. *J. Clin. Invest.* 101:1448–1457.

197. Fox, D.A., J.A. Millard, J. Treisman, W. Zeldes, A. Bergman, J. Depper, R. Dunne, and W.J. McCune. 1991. Defective CD2 pathway T cell activation in systemic lupus erythematosus. *Arthritis Rheum.* 34:561–571.

198. Ohtsuka, K., J.D. Gray, M.M. Stimmler, B. Toro, and D.A. Horwitz. 1998. Decreased production of TGF-beta by lymphocytes from patients with systemic lupus erythematosus. *J. Immunol.* 160:2539–2545.

199. Shores, E.W., and P.E. Love. 1997. TCR zeta chain in T cell development and selection. *Curr. Opin. Immunol.* 9:380–389.

200. Shores, E.W., T. Tran, A. Grinberg, C.L. Sommers, H. Shen, and P.E. Love. 1997. Role of the multiple T cell receptor (TCR)-zeta chain signaling motifs in selection of the T cell repertoire. *J. Exp. Med.* 185:893–900.

201. Mizoguchi, H., J.J. O'Shea, D.L. Longo, C.M. Loeffler, D.W. McVicar, and A.C. Ochoa. 1992. Alterations in signal transduction molecules in T lymphocytes from tumor-bearing mice. *Science* 258:1795–1798.

202. Reichert, T.E., R. Day, E.M. Wagner, and T.L. Whiteside. 1998. Absent or low expression of the zeta chain in T cells at the tumor site correlates with poor survival in patients with oral carcinoma. *Cancer Res.* 58:5344–5347.

203. Cardi, G., J.A. Heaney, A.R. Schned, D.M. Phillips, M.T. Branda, and M.S. Ernstoff. 1997. T-cell receptor zeta-chain expression on tumor-infiltrating lymphocytes from renal cell carcinoma. *Cancer Res.* 57:3517–3519.

204. Stefanova, I., M.W. Saville, C. Peters, F.R. Cleghorn, D. Schwartz, D.J. Venzon, K.J. Weinhold, N. Jack, C. Bartholomew, W.A. Blattner, R. Yarchoan, J.B. Bolen, and I.D. Horak. 1996. HIV infection—induced posttranslational modification of T cell signaling molecules associated with disease progression. *J. Clin. Invest.* 98:1290–1297.

205. Maurice, M.M., A.C. Lankester, A.C. Bezemer, M.F. Geertsma, P.P. Tak, F.C. Breedveld, R.A. van Lier, and C.L. Verweij. 1997. Defective TCR-mediated signaling in synovial T cells in rheumatoid arthritis. *J. Immunol.* 159:2973–2978.

206. Matsuda, M., A.K. Ulfgren, R. Lenkei, M. Petersson, A.C. Ochoa, S. Lindblad, P. Andersson, L. Klareskog, and R. Kiessling. 1998. Decreased expression of signal-transducing CD3 zeta chains in T cells from the joints and peripheral blood of rheumatoid arthritis patients. *Scand. J. Immunol.* 47:254–262.

207. Yang, M., S. Omura, J.S. Bonifacino, and A.M. Weissman. 1998. Novel aspects of degradation of T cell receptor subunits from the endoplasmic reticulum (ER) in T cells: importance of oligosaccharide processing, ubiquitination, and proteasome-dependent removal from ER membranes. *J. Exp. Med.* 187:835–846.

208. Gastman, B.R., D.E. Johnson, T.L. Whiteside, and H. Rabinowich. 1999. Caspase-mediated degradation of T-cell receptor zeta-chain. *Cancer Res.* 59:1422–1427.

209. Penna, D., S. Muller, F. Martinon, S. Demotz, M. Iwashima, and S. Valitutti. 1999. Degradation of ZAP-70 following antigenic stimulation in human T lymphocytes: role of calpain proteolytic pathway. *J. Immunol.* 163:50–56.

210. Rellahan, B.L., J.P. Jensen, and A.M. Weissman. 1994. Transcriptional regulation of the T cell antigen receptor zeta subunit: identification of a tissue-restricted promoter. *J. Exp. Med.* 180:1529–1534.

211. Tsokos, G.C., and S.N. Liossis. 1999. Immune cell signaling defects in lupus: activation, anergy and death. *Immunol. Today* 20:123–128.

212. Liossis, S.N., B. Kovacs, G. Dennis, G.M. Kammer, and G.C. Tsokos. 1996. B cells from patients with systemic lupus erythematosus display abnormal antigen receptor-mediated early signal transduction events. *J. Clin. Invest.* 98:2549–2557.

213. Deapen, D., A. Escalante, L. Weinrib, D. Horwitz, B. Bachman, P. Roy-Burman, A. Walker, and T.M. Mack. 1992. A revised estimate of twin concordance in systemic lupus erythematosus. *Arthritis Rheum.* 35:311–318.

214. Kono, D.H., and A.N. Theofilopoulos. 1999. Genetic susceptibility to spontaneous lupus in mice. In: Theofilopoulos, A.N. (ed). *Genes and genetics in autoimmunity.* Karger, Basel, pp. 72–98.

215. Theofilopoulos, A.N., and D.H. Kono. 1999. The genes of systemic autoimmunity. *Proc. Assoc. Am. Physicians* 111:228–240.

216. Ibnou-Zekri, N., T.J. Vyse, S.J. Rozzo, M. Iwamoto, T. Kobayakawa, B.L. Kotzin, and S. Izui. 1999. MHC linked control of murine SLE. *Curr. Top. Microbiol. Immunol.* 246:275–280.

217. Vyse, T.J., and B.L. Kotzin. 1998. Genetic susceptibility to systemic lupus erythematosus. *Annu. Rev. Immunol.* 16:261–292.

218. Tsao, B.P., R.M. Cantor, K.C. Kalunian, C.J. Chen, H. Badsha, R. Singh, D.J. Wallace, R.C. Kitridou, S.L. Chen, N. Shen, Y.W. Song, D.A. Isenberg, C.L. Yu, B.H. Hahn, and J.I. Rotter. 1997. Evidence for linkage of a candidate

chromosome 1 region to human systemic lupus erythematosus. *J. Clin. Invest.* 99:725–731.

219. Kono, D.H., R.W. Burlingame, D.G. Owens, A. Kuramochi, R.S. Balderas, D. Balomenos, and A.N. Theofilopoulos. 1994. Lupus susceptibility loci in New Zealand mice. *Proc. Natl. Acad. Sci. USA* 91:10168–10172.

220. Wakeland, E.K., L. Morel, C. Mohan, and M. Yui. 1997. Genetic dissection of lupus nephritis in murine models of SLE. *J. Clin. Immunol.* 17:272–281.

221. Moser, K.L., C. Gray-McGuire, J. Kelly, N. Asundi, H. Yu, G.R. Bruner, M. Mange, R. Hogue, B.R. Neas, and J.B. Harley. 1999. Confirmation of genetic linkage between human systemic lupus erythematosus and chromosome 1q41. *Arthritis Rheum.* 42:1902–1907.

222. Tsao, B.P., R.M. Cantor, J.M. Grossman, N. Shen, N.T. Teophilov, D.J. Wallace, F.C. Arnett, K. Hartung, R. Goldstein, K.C. Kalunian, B.H. Hahn, and J.I. Rotter. 1999. PARP alleles within the linked chromosomal region are associated with systemic lupus erythematosus. *J. Clin. Invest.* 103:1135–1140.

223. Delrieu, O., M. Michel, C. Frances, O. Meyer, C. Michel, F. Wittke, I. Crassard, J.F. Bach, E. Tournier-Lasserve, and J.C. Piette. 1999. Poly(ADP-ribose) polymerase alleles in French Caucasians are associated neither with lupus nor with primary antiphospholipid syndrome. GRAID Research Group. Group for Research on Auto-Immune Disorders. *Arthritis Rheum.* 42:2194–2197.

224. Ohtsuka, K., J.D. Gray, M.M. Stimmler, B. Toro, and D.A. Horwitz. 1998. Decreased production of TGF-beta by lymphocytes from patients with systemic lupus erythematosus. *J. Immunol.* 160:2539–2545.

225. Moser, K.L., B.R. Neas, J.E. Salmon, H. Yu, C. Gray-McGuire, N. Asundi, G.R. Bruner, J. Fox, J. Kelly, S. Henshall, D. Bacino, M. Dietz, R. Hogue, G. Koelsch, L. Nightingale, T. Shaver, N.I. Abdou, D.A. Albert, C. Carson, M. Petri, E.L. Treadwell, J.A. James, and J.B. Harley. 1998. Genome scan of human systemic lupus erythematosus: evidence for linkage on chromosome 1q in African-American pedigrees. *Proc. Natl. Acad. Sci. USA* 95:14869–14874.

226. Harley, J.B., K.L. Moser, P.M. Gaffney, and T.W. Behrens. 1998. The genetics of human systemic lupus erythematosus. *Curr. Opin. Immunol.* 10:690–696.

227. Gaffney, P.M., G.M. Kearns, K.B. Shark, W.A. Ortmann, S.A. Selby, M.L. Malmgren, K.E. Rohlf, T.C. Ockenden, R.P. Messner, R.A. King, S.S. Rich, and T.W. Behrens. 1998. A genome-wide search for susceptibility genes in human systemic lupus erythematosus sib-pair families. *Proc. Natl. Acad. Sci. USA* 95:14875–14879.

228. Shai, R., F.P. Quismorio, Jr., L. Li, O.J. Kwon, J. Morrison, D.J. Wallace, C.M. Neuwelt, C. Brautbar, W.J. Gauderman, and C.O. Jacob. 1999. Genome-wide screen for systemic lupus erythematosus susceptibility genes in multiplex families. *Hum. Mol. Genet.* 8:639–644.

229. Shen, H.H., and R.J. Winchester. 1986. Susceptibility genetics of systemic lupus erythematosus. *Springer Semin. Immunopathol.* 9:143–159.

230. Gulko, P.S., and R.J. Winchester. 1999. Genetics of systemic lupus erythematosus. In: Kammer, G.M. and G.C. Tsokos (eds). *Lupus: molecular and cellular pathogenesis.* Humana Press, Totowa, NJ, pp. 101–123.

231. Agnello, V. 1986. Lupus diseases associated with hereditary and acquired deficiencies of complement. *Springer Semin. Immunopathol.* 9:161–178.

232. Finco, O., S. Li, M. Cuccia, F.S. Rosen, and M.C. Carroll. 1992. Structural differences between the two human complement C4 isotypes affect the humoral immune response. *J. Exp. Med.* 175:537–543.

233. Atkinson, J.P., and P.M. Schneider. 1999. Genetic susceptibility and class III complement genes. In: Lahita, R.G. (ed). *Systemic lupus erythematosus.* Academic Press, New York, pp. 87–102.

234. Birmingham, D.J. 1999. Type one complement receptor and human SLE. In: Kammer, G.M., and G.C. Tsokos (eds). *Lupus: molecular and cellular pathogenesis.* Humana Press, Totowa, NJ, pp. 541–556.

235. Van Dyne, S., V.M. Holers, D.M. Lublin, and J.P. Atkinson. 1987. The polymorphism of the C3b/C4b receptor in the normal population and in patients with systemic lupus erythematosus. *Clin. Exp. Immunol.* 68:570–579.

236. Turner, M.W. 1996. Mannose-binding lectin: the pluripotent molecule of the innate immune system. *Immunol. Today* 17:532–540.

237. Super, M., S.D. Gillies, S. Foley, K. Sastry, J.E. Schweinle, V.J. Silverman, and R.A. Ezekowitz. 1992. Distinct and overlapping functions of allelic forms of human mannose binding protein. *Nat. Genet.* 2:50–55.

238. Davies, E.J., L.S. Teh, J. Ordi-Ros, N. Snowden, M.C. Hillarby, A. Hajeer, R. Donn, P. Perez-Pemen, M. Vilardell-Tarres, and W.E. Ollier. 1997. A dysfunctional allele of the mannose binding protein gene associates with systemic lupus erythematosus in a Spanish population. *J. Rheumatol.* 24:485–488.

239. Davies, E.J., N. Snowden, M.C. Hillarby, D. Carthy, D.M. Grennan, W. Thomson, and W.E. Ollier. 1995. Mannose-binding protein gene polymorphism in systemic lupus erythematosus. *Arthritis Rheum.* 38:110–114.

240. Lau, Y.L., C.S. Lau, S.Y. Chan, J. Karlberg, and M.W. Turner. 1996. Mannose-binding protein in Chinese patients with systemic lupus erythematosus. *Arthritis Rheum.* 39:706–708.

241. Sullivan, K.E., C. Wooten, D. Goldman, and M. Petri. 1996. Mannose-binding protein genetic polymorphisms in black patients with systemic lupus erythematosus. *Arthritis Rheum.* 39:2046–2051.

242. Jacob, C.O. 1992. Studies on the role of tumor necrosis factor in murine and human autoimmunity. *J. Autoimmun.* 5:133–143.

243. Sullivan, K.E., C. Wooten, B.J. Schmeckpeper, D. Goldman, and M.A. Petri. 1997. A promoter polymorphism of tumor necrosis factor alpha associated with systemic lupus erythematosus in African-Americans. *Arthritis Rheum.* 40:2207–2211.

244. Danis, V.A., M. Millington, V. Hyland, R. Lawford, Q. Huang, and D. Grennan. 1995. Increased frequency of the uncommon allele of a tumour necrosis factor alpha gene polymorphism in rheumatoid arthritis and systemic lupus erythematosus. *Dis. Markers* 12:127–133.

245. Fong, K.Y., H.S. Howe, S.K. Tin, M.L. Boey, and P.H. Feng. 1996. Polymorphism of the regulatory region of tumour necrosis factor alpha gene in patients with systemic lupus erythematosus. *Ann. Acad. Med. Singapore* 25:90–93.

246. Hajeer, A.H., J. Worthington, E.J. Davies, M.C. Hillarby, K. Poulton, and W.E. Ollier. 1997. TNF microsatellite a2, b3 and d2 alleles are associated with systemic lupus erythematosus. *Tissue Antigens* 49:222–227.

247. Sturfelt, G., G. Hellmer, and L. Truedsson. 1996. TNF microsatellites in systemic lupus erythematosus-a high frequency of the TNFabc 2–3–1 haplotype in multicase SLE families. *Lupus* 5:618–622.

248. Komata, T., N. Tsuchiya, M. Matsushita, K. Hagiwara, and K. Tokunaga. 1999. Association of tumor necrosis factor receptor 2 (TNFR2) polymorphism with susceptibility to systemic lupus erythematosus. *Tissue Antigens* 53:527–533.

249. Lee, S.H., S.H. Park, J.K. Min, S.I. Kim, W.H. Yoo, Y.S. Hong, J.H. Park, C.S. Cho, T.G. Kim, H. Han, and H.Y. Kim. 1997. Decreased tumour necrosis factor-beta production in TNFB*2 homozygote: an important predisposing factor of lupus nephritis in Koreans. *Lupus* 6:603–609.

250. Mehrian, R., F.P. Quismorio, Jr., G. Strassmann, M.M. Stimmler, D.A. Horwitz, R.C. Kitridou, W.J. Gauderman, J. Morrison, C. Brautbar, and C.O. Jacob. 1998. Synergistic effect between IL-10 and bcl-2 genotypes in determining susceptibility to systemic lupus erythematosus. *Arthritis Rheum.* 41:596–602.

251. Nicklin, M.J., A. Weith, and G.W. Duff. 1994. A physical map of the region encompassing the human interleukin-1 alpha, interleukin-1 beta, and interleukin-1 receptor antagonist genes. *Genomics* 19:382–384.

252. Blakemore, A.I., J.K. Tarlow, M.J. Cork, C. Gordon, P. Emery, and G.W. Duff. 1994. Interleukin-1 receptor antagonist gene polymorphism as a disease severity factor in systemic lupus erythematosus. *Arthritis Rheum.* 37:1380–1385.

253. Suzuki, H., Y. Matsui, and H. Kashiwagi. 1997. Interleukin-1 receptor antagonist gene polymorphism in Japanese patients with systemic lupus erythematosus. *Arthritis Rheum.* 40:389–390.

254. Yee, A.M., S.C. Ng, R.E. Sobel, and J.E. Salmon. 1997. Fc gammaRIIA polymorphism as a risk factor for invasive pneumococcal infections in systemic lupus erythematosus. *Arthritis Rheum.* 40:1180–1182.

255. Salmon, J.E., S. Ng, D.H. Yoo, T.H. Kim, S.Y. Kim, and G.G. Song. 1999. Altered distribution of Fcgamma receptor IIIA alleles in a cohort of Korean patients with lupus nephritis. *Arthritis Rheum.* 42:818–819.

256. Tassiulas, I.O., I. Aksentijevich, J.E. Salmon, Y. Kim, C.H. Yarboro, E.M. Vaughan, J.C. Davis, D.L. Scott, H.A. Austin, J.H. Klippel, J.E. Balow, M.F. Gourley, and D.T. Boumpas. 1998. Angiotensin I converting enzyme gene polymorphisms in systemic lupus erythematosus: decreased prevalence of DD genotype in African American patients. *Clin. Nephrol.* 50:8–13.

257. Theofilopoulos, A.N., and F.J. Dixon. 1985. Murine models of systemic lupus erythematosus. *Adv. Immunol.* 37:269–390.

258. Theofilopoulos, A.N., and F.J. Dixon. 1981. Etiopathogenesis of murine SLE. *Immunol. Rev.* 55:179–216.

259. Sthoeger, Z.M., N. Chiorazzi, and R.G. Lahita. 1988. Regulation of the immune response by sex hormone. I in vitro effects of estradiol and testosterone on pokeweek mitogen-induced human B cell differentiation. *J. Immunol.* 141:91–98.

260. Lahita, R.G., H.L. Bradlow, E. Ginzler, S. Pang, and M. New. 1987. Low plasma androgens in women with systemic lupus erythematosus. *Arthritis Rheum.* 30:241–248.

261. Lahita, R.G., H.G. Kunkel, and H.L. Bradlow. 1983. Increased oxidation of testosterone in systemic lupus erythematosus. *Arthritis Rheum.* 26:1517–1521.

262. Lahita, R.G. 1985. Sex steroids and the rheumatic disease. *Arthritis Rheum.* 28:121–126.

263. Kanda, N., T. Tsuchida, and K. Tamaki. 1999. Estrogen enhancement of anti-double-stranded DNA antibody and immunoglobulin G production in peripheral blood mononuclear cells from patients with systemic lupus erythematosus. *Arthritis Rheum.* 42:328–337.

264. Kanda, N., T. Tsuchida, and K. Tamaki. 1997. Testosterone suppresses anti-DNA antibody production in peripheral blood mononuclear cells from patients with systemic lupus erythematosus. *Arthritis Rheum.* 40:1703–1711.

265. McMurray, R.W., S.H. Allen, A.L. Braun, F. Rodriguez, and S.E. Walker. 1994. Longstanding hyperprolactinemia associated with systemic lupus erythematosus: possible hormonal stimulation of an autoimmune disease. *J. Rheumatol.* 21:843–850.

266. Sanchez-Guerrero, J., M.H. Liang, E.W. Karlson, D.J. Hunter, and G.A. Colditz. 1995. Postmenopausal estrogen therapy and the risk for developing systemic lupus erythematosus. *Ann. Intern. Med.* 122:430–433.

267. Strom, B.L., M.M. Reidenberg, S. West, E.S. Snyder, B. Freundlich, and P.D. Stolley. 1994. Shingles, allergies, family medical history, oral contraceptives, and other potential risk factors for systemic lupus erythematosus. *Am. J. Epidemiol.* 140:632–642.

268. Suenaga, R., M.J. Evans, K. Mitamura, V. Rider, and N.I. Abdou. 1998. Peripheral blood T cells and monocytes and B cell lines derived from patients with lupus express estrogen receptor transcripts similar to those of normal cells. *J. Rheumatol.* 25:1305–1312.

269. Rider, V., R.T. Foster, M. Evans, R. Suenaga, and N.I. Abdou. 1998. Gender differences in autoimmune diseases: estrogen increases calcineurin expression in systemic lupus erythematosus. *Clin. Immunol. Immunopathol.* 89:171–180.

270. Kammer, G.M. and G.C. Tsokos. 1998. Emerging concepts of the molecular basis for estrogen effects on T lymphocytes in systemic lupus erythematosus. *Clin. Immunol. Immunopathol.* 89:192–195.

271. Wang, D. and E.K. Chan. 1996. 17-beta-estradiol increases expression of 52-kDa and 60-kDa SS-A/Ro autoantigens in human keratinocytes and breast cancer cell line MCF-7. *J. Invest. Dermatol* 107:610–614.

272. Rosen, A. and L. Casciola-Rosen. 1999. Autoantigens as substrates for apoptotic proteases: implications for the pathogenesis of systemic autoimmune disease. *Cell Death. Differ.* 6:6–12.

273. Utz, P.J., M. Hottelet, P.H. Schur, and P. Anderson. 1997. Proteins phosphorylated during stress-induced apoptosis are common targets for autoantibody production in patients with systemic lupus erythematosus. *J. Exp. Med.* 185:843–854.

274. Utz, P.J., M. Hottelet, W.J. van Venrooij, and P. Anderson. 1998. Association of phosphorylated serine/arginine (SR) splicing factors with the U1-small ribonucleoprotein (snRNP) autoantigen complex accompanies apoptotic cell death. *J. Exp. Med.* 187:547–560.

275. Gabai, V.L., A.B. Merlin, D.D. Mosser, A.W. Caron, S. Rits, V.I. Shifrin, and M.Y. Sherman, 1997. Hsp70 prevents activation of stress kinases: a novel pathway of cellular thermotolerance. *J. Biol. Chem.* 272:18033–18037.

276. Shifrin, V.I., and P. Anderson. 1999. Trichothecene mycotoxins trigger a ribotoxic stress response that activates c-Jun N-terminal kinase and p38 mitogen-activated protein kinase and induces apoptosis. *J. Biol. Chem.* 274:13985–13992.

277. Karlson, E.W., S.E. Hankinson, M.H. Liang, J. Sanchez-Guerrero, G.A. Colditz, B.J. Rosenau, F.E. Speizer, and P.H. Schur. 1999. Association of silicone breast implants with immunologic abnormalities: a prospective study. *Am. J. Med.* 106:11–19.

278. Sanchez-Guerrero, J., E.W. Karlson, G.A. Colditz, D.J. Hunter, F.E. Speizer, and M.H. Liang. 1996. Hair dye use and the risk of developing systemic lupus erythematosus. *Arthritis Rheum.* 39:657–662.

279. Hardy, C.J., B.P. Palmer, K.R. Muir, A.J. Sutton, and R.J. Powell. 1998. Smoking history, alcohol consumption, and systemic lupus erythematosus: a case-control study. *Ann. Rheum. Dis.* 57:451–455.

280. James, J.A., K.M. Kaufman, A.D. Farris, E. Taylor-Albert, T.J. Lehman, and J.B. Harley. 1997. An increased prevalence of Epstein-Barr virus infection in young patients suggests a possible etiology for systemic lupus erythematosus. *J. Clin. Invest.* 100:3019–3026.

281. Vaughan, J.H. 1997. The Epstein-Barr virus and systemic lupus erythematosus. *J. Clin. Invest.* 100:2939–2940.

282. Huang, S.C., Z. Pan, B.T. Kurien, J.A. James, J.B. Harley, and R.H. Scofield. 1995. Immunization with vesicular stomatitis virus nucleocapsid protein induces autoantibodies to the 60 kD Ro ribonucleoprotein particle. *J. Invest. Med.* 43:151–158.

283. De Keyser, F., S.O. Hoch, M. Takei, H. Dang, H. De Keyser, L.A. Rokeach, and N. Talal. 1992. Cross-reactivity of the B/B' subunit of the Sm ribonucleoprotein autoantigen with proline-rich polypeptides. *Clin. Immunol. Immunopathol.* 62:285–290.

284. Talal, N., R.F. Garry, P.H. Schur, S. Alexander, M.J. Dauphinée, I.H. Livas, A. Ballester, M. Takei, and H. Dang. 1990. A conserved idiotype and antibodies to retroviral proteins in systemic lupus erythematosus. *J. Clin. Invest.* 85:1866–1871.

285. Sabbatini, A., S. Bombardieri, and P. Migliorini. 1993. Autoantibodies from patients with systemic lupus erythematosus bind a shared sequence of SmD and Epstein-Barr virus-encoded nuclear antigen EBNA I. *Eur. J. Immunol.* 23:1146–1152.

286. Zinkernagel, R.M., S. Cooper, J. Chambers, R.A. Lazzarini, H. Hengartner, and H. Arnheiter. 1990. Virus-induced autoantibody response to a transgenic viral antigen. *Nature* 345:68–71.

287. Perl, A., and K. Banki. 1999. Molecular mimicry, altered apoptosis, and immunomodulation as mechanisms of viral pathogenesis in systemic lupus erythematosus. In: Kammer, G.M., and G.C. Tsokos (eds). *Lupus: molecular and cellular pathogenesis.* Humana Press, Totowa, NJ, pp. 43–64.

288. Combadiere, C., S.K. Ahuja, J. Van Damme, H.L. Tiffany, J.L. Gao, and P.M. Murphy. 1995. Monocyte chemoattractant protein-3 is a functional ligand for CC chemokine receptors 1 and 2B. *J. Biol. Chem.* 270:29671–29675.

289. Gao, J.L., and P.M. Murphy. 1994. Human cytomegalovirus open reading frame US28 encodes a functional beta chemokine receptor. *J. Biol. Chem.* 269:28539–28542.

290. Ahuja, S.K., and P.M. Murphy. 1993. Molecular piracy of mammalian interleukin-8 receptor type B by herpesvirus saimiri. *J. Biol. Chem.* 268:20691–20694.

291. Weinstein, A. 1980. Drug-induced systemic lupus erythematosus. *Prog. Clin. Immunol.* 4:1–21.

292. Richardson, B., L. Scheinbart, J. Strahler, L. Gross, S. Hanash, and M. Johnson. 1990. Evidence for impaired T cell DNA methylation in systemic lupus erythematosus and rheumatoid arthritis. *Arthritis and Rheum.* 33:1665–1673.

293. Yung, R., D. Powers, K. Johnson, E. Amento, D. Carr, T. Laing, J. Yang, S. Chang, N. Hemati, and B. Richardson. 1996. Mechanisms of drug-induced lupus. II. T cells overexpressing LFA-1 cause a lupus-like disease in syngeneic mice. *J. Clin. Invest.*

294. Kretz-Rommel, A., and R.L. Rubin. 1997. A metabolite of the lupus-inducing drug procainamide prevents anergy induction in T cell clones. *J. Immunol.* 158:4465–4470.

295. Kretz-Rommel, A., S.R. Duncan, and R.L. Rubin. 1997. Autoimmunity caused by disruption of central T cell tolerance. A murine model of drug-induced lupus. *J. Clin. Invest.* 99:1888–1896.

296. Austin, H.A., III, J.H. Klippel, J.E. Balow, N.G. le Riche, A.D. Steinberg, P.H. Plotz, and J.L. Decker. 1986. Therapy of lupus nephritis. Controlled trial of prednisone and cytotoxic drugs. *N. Engl. J. Med.* 314:614–619.

297. Donadio, J.V., Jr. and R.J. Glassock. 1993. Immunosuppressive drug therapy in lupus nephritis. *Am. J. Kidney Dis.* 21:239–250.

298. Gourley, M.F., H.A. Austin, III, D. Scott, C.H. Yarboro, E.M. Vaughan, J. Muir, D.T. Boumpas, J.H. Klippel, J.E. Balow, and A.D. Steinberg. 1996. Methylprednisolone and cyclophosphamide, alone or in combination, in patients with lupus nephritis. A randomized, controlled trial. *Ann. Intern. Med.* 125:549–557.

299. Boumpas, D.T., S. Barez, J.H. Klippel, and J.E. Balow. 1990. Intermittent cyclophosphamide for the treatment of autoimmune thrombocytopenia in systemic lupus erythematosus. *Ann. Intern. Med.* 112:674–677.

300. Boumpas, D.T., H.A. Austin, E.M. Vaughn, J.H. Klippel, A.D. Steinberg, C.H. Yarboro, and J.E. Balow. 1992. Controlled trial of pulse methylprednisolone versus two regimens of pulse cyclophosphamide in severe lupus nephritis. *Lancet* 340:741–745.

301. Moroni, G., M. Maccario, G. Banfi, S. Quaglini, and C. Ponticelli. 1998. Treatment of membranous lupus nephritis. *Am. J. Kidney Dis.* 31:681–686.

302. Austin, H.A., III, T.T. Antonovych, K. MacKay, D.T. Boumpas, and J.E. Balow. 1992. NIH con-ference. Membranous nephropathy. *Ann. Intern. Med.* 116:672–682.

303. 1995. Placebo-controlled study of mycophenolate mofetil combined with cyclosporin and corticosteroids for prevention of acute rejection. European Mycophenolate Mofetil Cooperative Study Group. *Lancet* 345:1321–1325.

304. Dooley, M.A., F.G. Cosio, P.H. Nachman, M.E. Falkenhain, S.L. Hogan, R.J. Falk, and L.A. Hebert. 1999. Mycophenolate mofetil therapy in lupus nephritis: clinical observations. *J. Am. Soc. Nephrol.* 10:833–839.

305. Davis, J.C., Jr., H. Austin, III, D. Boumpas, T.A. Fleisher, C. Yarboro, A. Larson, J. Balow, J.H. Klippel, and D. Scott. 1998. A pilot study of 2-chloro-2'-deoxyadenosine in the treatment of systemic lupus erythematosus-associated glomerulonephritis. *Arthritis Rheum.* 41:335–343.

306. Frank, D.A., S. Mahajan, and J. Ritz. 1999. Fludarabine-induced immunosuppression is associated with inhibition of STAT1 signaling. *Nat. Med.* 5:444–447.

307. Boumpas, D.T., H.A. Austin, III, E.M. Vaughan, C.H. Yarboro, J.H. Klippel, and J.E. Balow. 1993. Risk for sustained amenorrhea in patients with systemic lupus erythematosus receiving intermittent pulse cyclophosphamide therapy. *Ann. Intern. Med.* 119:366–369.

308. Moreland, L.W., S.W. Baumgartner, M.H. Schiff, E.A. Tindall, R.M. Fleischmann, K. Mohler, M.B. Widmer, and C.M. Blosch. 1997. Treatment of rheumatoid arthritis with a

recombinant human tumor necrosis factor receptor (p75)—Fc fusion protein. *N. Engl. J. Med.* 337:141–147.

309. Present, D.H., P. Rutgeerts, S. Targan, S.B. Hanauer, L. Mayer, R.A. van Hogezand, D.K. Podolsky, B.E. Sands, T. Braakman, K.L. DeWoody, T.F. Schaible, and S.J. van Deventer. 1999. Infliximab for the treatment of fistulas in patients with Crohn's disease. *N. Engl. J. Med.* 340:1398–1405.

310. Baert, F.J., G.R. D'Haens, M. Peeters, M.I. Hiele, T.F. Schaible, D. Shealy, K. Geboes, and P.J. Rutgeerts. 1999. Tumor necrosis factor alpha antibody (infliximab) therapy profoundly down-regulates the inflammation in Crohn's ileocolitis. *Gastroenterology* 116:22–28.

311. Fishman, P., E. Falach-Vaknine, R. Zigelman, R. Bakimer, B. Sredni, M. Djaldetti, and Y. Shoenfeld. 1993. Prevention of fetal loss in experimental antiphospholipid syndrome by *in vivo* administration of recombinant interleukin-3. *J. Clin. Invest.* 91:1834–1837.

312. Ganser, A., G. Seipelt, A. Lindemann, O.G. Ottmann, S. Falk, M. Eder, F. Herrmann, R. Becher, K. Hoffken, and T. Buchner. 1990. Effects of recombinant human interleukin-3 in patients with myelodysplastic syndromes. *Blood* 76:455–462.

313. Ishida, H., T. Muchamuel, S. Sakaguchi, S. Andrade, S. Menon, and M. Howard. 1994. Continuous administration of anti-interleukin 10 antibodies delays onset of autoimmunity in NZB/W F1 mice. *J. Exp. Med.* 179:305–310.

314. Wang, Y., Q. Hu, J.A. Madri, S.A. Rollins, A. Chodera, and L.A. Matis. 1996. Amelioration of lupus-like autoimmune disease in NZB/WF1 mice after treatment with a blocking monoclonal antibody specific for complement component C5. *Proc. Natl. Acad. Sci. USA* 93:8563–8568.

315. Quigg, R.J., Y. Kozono, D. Berthiaume, A. Lim, D.J. Salant, A. Weinfeld, P. Griffin, E. Kremmer, and V.M. Holers. 1998. Blockade of antibody-induced glomerulonephritis with Crry-Ig, a soluble murine complement inhibitor. *J. Immunol.* 160:4553–4560.

316. Quigg, R.J., C. He, A. Lim, D. Berthiaume, J.J. Alexander, D. Kraus, and V.M. Holers. 1998. Transgenic mice overexpressing the complement inhibitor crry as a soluble protein are protected from antibody-induced glomerular injury. *J. Exp. Med.* 188:1321–1331.

317. Singh, R.R., V. Kumar, F.M. Ebling, S. Southwood, A. Sette, E.E. Sercarz, and B.H. Hahn. 1995. T cell determinants from autoantibodies to DNA can upregulate autoimmunity in murine systemic lupus erythematosus. *J. Exp. Med.* 181:2017–2027.

318. Wofsy, D., and D.I. Daikh. 1998. Opportunities for future biological therapy in SLE. *Baillieres Clin. Rheumatol.* 12:529–541.

319. Daikh, D.I., and D. Wofsy. 1998. On the horizon: clinical trials of new immunosuppressive strategies for autoimmune diseases. *Transplant. Proc.* 30:4027–4028.

320. Daikh, D.I., B.K. Finck, P.S. Linsley, D. Hollenbaugh, and D. Wofsy. 1997. Long-term inhibition of murine lupus by brief simultaneous blockade of the B7/CD28 and CD40/gp39 costimulation pathways. *J. Immunol.* 159:3104–3108.

321. Burt, R.K., and A. Traynor. 1998. Hematopoietic stem cell therapy of autoimmune diseases. *Curr. Opin. Hematol.* 5:472–477.

322. Burt, R.K., A.E. Traynor, R. Pope, J. Schroeder, B. Cohen, K.H. Karlin, L. Lobeck, C. Goolsby, P. Rowlings, F.A. Davis, D. Stefoski, C. Terry, C. Keever-Taylor, S. Rosen, D. Vesole, M. Fishman, M. Brush, S. Mujias, M. Villa, and W.H. Burns. 1998. Treatment of autoimmune disease by intense immunosuppressive conditioning and autologous hematopoietic stem cell transplantation. *Blood* 92:3505–3514.

323. van Vollenhoven, R.F., J.L. Park, M.C. Genovese, J.P. West, and J.L. McGuire. 1999. A double-blind, placebo-controlled, clinical trial of dehydroepiandrosterone in severe systemic lupus erythematosus. *Lupus* 8:181–187.

324. Nippoldt, T.B., and K.S. Nair. 1998. Is there a case for DHEA replacement? *Baillieres Clin. Endocrinol. Metab.* 12:507–520.

325. van Vollenhoven, R.F., E.G. Engleman, and J.L. McGuire. 1995. Dehydroepiandrosterone in systemic lupus erythematosus. Results of a double-blind, placebo-controlled, randomized clinical trial. *Arthritis Rheum.* 38:1826–1831.

326. Datta, S.K., A. Kaliyaperumal, and A. Desai-Mehta. 1997. T cells of lupus and molecular targets for immunotherapy. *J. Clin. Immunol.* 17:11–20.

327. Critchfield, J.M., M.K. Racke, J.C. Zuniga-Pfluecker, B. Cannella, C.S. Raine, J. Goverman, and M.J. Lenardo. 1994. T cell deletion in high antigen dose therapy of autoimmune encephalomyelitis. *Science* 263:1139–1143.

328. Trentham, D.E., R.A. Dynesius-Trentham, E.J. Orav, D. Combitchi, C. Lorenzo, K.L. Sewell, D.A. Hafler, and H.L. Weiner. 1993. Effects of oral administration of type II collagen on rheumatoid arthritis. *Science* 261:1669–1670.

329. Lu, L., A. Kaliyaperumal, D.T. Boumpas, and S.K. Datta. 1999. Major peptide autoepitopes for nucleosome-specific T cells of human lupus. *J. Clin. Invest.* 104:345–355.

330. Kaliyaperumal, A., M.A. Michaels, and S.K. Datta. 1999. Antigen-specific therapy of murine lupus nephritis using nucleosomal peptides: tolerance spreading impairs pathogenic function of autoimmune T, and B cells. *J. Immunol.* 162:5775–5783.

331. Constantin, G., C. Laudanna, S. Brocke, and E.C. Butcher. 1999. Inhibition of experimental autoimmune encephalomyelitis by a tyrosine kinase inhibitor. *J. Immunol.* 162:1144–1149.

332. Constantin, G., S. Brocke, A. Izikson, C. Laudanna, and E.C. Butcher. 1998. Tyrphostin AG490, a tyrosine kinase inhibitor, blocks actively induced experimental autoimmune encephalomyelitis. *Eur. J. Immunol.* 28:3523–3529.

19 | Abnormalities in B Cell Activity and the Immunoglobulin Repertoire in Human Systemic Lupus Erythematosus

Amrie C. Grammer, Thomas Dörner, and Peter E. Lipsky

1. INTRODUCTION

Systemic lupus erythematosus (SLE) is a chronic autoimmune disease characterized by the production of multiple autoantibodies. The exact cause of SLE is unclear, but the emergence of disease appears to depend on environmental factors that initiate and/or contribute to pathogenic autoimmunity in genetically-prone individuals. Candidate initiating factors include female sex hormones and infection with viruses and/or bacteria. Characteristics of SLE are a number of profound B cell abnormalities that may either reflect the composite impact of genetic factors or secondary events to other primary immunologic abnormalities. Perturbations of B cell maturation may permit the generation, activation, differentiation and positive selection of B cells that secrete pathogenic autoantibodies. Such autoantibodies may be for intrinsic autoantigens or neoautoantigens that become accessible to the immune system during disease activity. In addition, SLE-prone individuals may have an abnormality in generation and/or maintenance of their functional repertoire of B cells expressing specific Ig genes or may be subjected to conditions that bias specific immunoglobulin variable region (IgV) usage, thereby predisposing to the development of autoantibodies. Furthermore, mechanisms contributing to somatic hypermutation of Ig or subsequent B cell selection may be intrinsically abnormal in patients with SLE and thus permit the preferential emergence of high avidity autoantibodies. Finally, intrinsic B cell overreactivity to normal antigenic stimulation may provide the drive for pathogenic autoantibody formation even though B cell maturation, repertoire generation, maintenance, somatic hypermutation and subsequent selection are not mechanisitically abnormal. For example, intrinsic B cell overactivity could overwhelm the normal mechansims that protect against autoimmunity and thereby permit the escape of pathogenic autoantibody producing B cell progeny.

End-organ damage in SLE has been shown to be initiated by deposition of immune complexes (ICs) in the kidney or by reactivity of the autoantibody with the organ itself or with material bound by the tissue such as dsDNA on the basement membrane of the glomerulus. The most prominent autoantibodies in SLE are specific for components of chromatin and have been shown to induce glomerulonephritis (1). It should be noted that while naturally occurring, polyreactive IgM autoantibodies often have low-affinity specificity for single-stranded (ss) DNA in their germline configuration, pathogenic autoantibodies with specificity for single- and double-stranded DNA are usually of high affinity and have switched to IgG, characteristics of a T cell-dependent (TD) immune response. Injection of anti-DNA Abs into SCID mice induced severe glomerulonephritis with proteinuria, providing strong evidence for pathogenicity of these autoantibodies (2–7). In humans, only a subset of DNA-specific autoantibodies are associated with kidney damage (8); these pathogenic autoantibodies are usually of the IgG subtype and have undergone somatic hypermutation of both V_H and V_L to increase positive charge in the antigen (Ag) combining region, especially the CDR3, and therefore to manifest increased avidity for negatively charged dsDNA. Other pathogenic autoantibodies that emerge in SLE include anti-Ro SSA and -La SSB Abs that are associated with photosensitivity, congenital heart block (9) and glomerulonephritis (10,11), anti-Sm/RNP Abs, whose mechanism of action is unclear (12) and anti-cardiolipin Abs that influence elements of the coagulation system leading to venous or arterial occlusions (13).

2. GENETIC PREDISPOSITION TO HUMAN SLE

Susceptibility to human SLE has been shown to be polygenic and to vary with the ethnic group studied (14–16). In addition, active disease emerges predominantly in females during the second and third decades of life, indicating the importance of a series of initiating events that must occur for active SLE to become apparent. Genetic predisposition to human SLE has been examined extensively by a number of genome scans that have focused on association and/or linkage of chromosmal loci, and in some cases candidate genes in twins and multiplex families of various ethnic origins. The results of these studies emphasize the complexity of genetic and environmental factors that contribute to the appearance of SLE in twenty and thirty-year old women. Moreover, since susceptibility alleles are co-localized to a limited number of chromosomal regions, these studies suggest that autoimmune manifestations of SLE may result from the concerted action of a group of polymorphic alleles that were selectively clustered as a result of evolution to generate an enhanced immune response.

The most consistent data that emerged from studies examining association and/or linkage of human genetic loci with active SLE elucidated the contribution of regions of chromosome 1, q41–42 (17–19) and p13, as well as chromosome 2 (q21–33), 3 (cen–q11), 4 (p15 and q28), 6 (p11–21), 11 (p15), 16 (q13), and 20 (p12 and q13) (18,19). Linkages apparent in one study, but not in others, were also found within regions of chromosomes 1–3, 6, 11–15 and 21.

The importance of association/linkage of chromosome 1 loci with active human SLE is emphasized by previous genetic studies of inbred lupus-prone mouse strains (*sle1* in B6.NZMc1 and *Bxs 1–3* in BXSB) that independently identified multiple, closely-linked loci in the telomeric/distal end of murine chromosome 1 with characteristic manifestations of lupus, such as splenomegaly owing to B cell hyper-proliferation, IgG autoantibodies to nuclear antigens, and glomerulonephritis (16). It should be noted that regions of murine and human chromosome 1 that were found to be associated/linked with SLE are syntenic and contain homologous genes.

The study by Tsao et al. (17) examining SLE affected sib-pairs in multiplex families was the first to demonstrate linkage of chromosome 1q41–42 with IgG anti-chromatin autoantibodies in patients from three different ethnic groups (African-American, Asian and European-American) using D1s229 as a marker. Independent studies of African-Americans (19) and European-Americans (Caucasians; 18) confirmed the importance of this region for SLE susceptibility by using markers D1s3462 and D1s235, respectively. Based on information identifying gene products potentially involved in the immunological dysregulation of B cells in SLE patients, candidate genes in this locus of susceptibility include those encoding the cytokines IL10 (1q32–42; 21) and TGFβ_2 (1q42.2; 21–22), the receptor for IL10 (20), the DNA nucleotide excision repair (NER) enzyme PARP/poly-(ADP-ribose) polymerase (1q41–42; 24), and the CD40- associated adapter protein TRAF5 (1q32.3-1q41.4; 25; A.C. Grammer, personal observation). Of note, although recent studies have elucidated polymorphisms in the promoter of IL10 (26–31) that may contribute to the dysregulated immune responses of SLE patients, these polymorphisms are not unique to patients with SLE (32–40).

Two major genome wide screens identified an association between SLE and chromosomal loci 20q13 and 6p11–21 (18,19). CD40 is a candidate gene with the 20q13 (A.C. Grammer, personal observation), since it has been shown to be essential for B cell responses (41). The linkage between chromosome 6p11–21 and SLE is consistent with previous studies that associated alleles of polymorphic HLA (42–45) and TNFα (46–48), as well as genetic deficiencies of early classical complement components C2, C4 (A and B) and C1q (49–53) that have been shown to be risk factors for SLE.

Numerous studies have examined the association between certain HLA alleles and autoantibody production. For example, HLA-DR3/DR2 has been associated with production of anti-Ro and La and HLA DR4/DQ3 with production of anti-Sm and anti-RNP. Less consistent evidence has associated anti-DNA and anti-cardiolipin and anti phospholipid Ab with DR7, DR4 or DQ7 (45,54–55). Specific HLA alleles have also been shown to be in linkage disequilibrium with other potential SLE susceptibility genes such as TNFα (45), prolactin, C4 and Hsp70-2 (56, 57) that have also been shown to be risk factors for SLE in Northern-Europeans (TNFα, prolactin and C4) and african-americans (Hsp70-2). First, a polymorphism (G to A) in the promoter of the TNFα gene at position -308 to the transcriptional start site has been shown to be associated with SLE (48), although this polymorphism is not unique to patients with SLE (58–61). Linkage to HLA B8-DR3 was apparent in Northern-Europeans (46,47) but not in African-Americans (48). Of note, decreased TNFα production in SLE patients compared to normal controls appears to be associated with disease activity (62), although the association of diminished TNFα production and the G to A polymorphism is uncertain (63,64). In this regard, pharmacological elimination of TNFα in rheumatoid arthritis (RA) has been associated with development of antibodies to dsDNA, but rarely symptoms of SLE (65). Secondly, genes for C2 and C4 are found in the MHC locus and are in linkage disequilibrium with each other (57). Polymorphisms in C4 genes resulting in either partial or total loss of the production of C4 protein (51) are also associated with the development of SLE. Finally, the role of polymorphisms in the prolactin gene in SLE susceptibility is uncertain, especially in view of the recent finding that mice genetically deficient in the receptor for prolactin mount a normal humoral response to TD Ag (66).

Putative linkages between other regions of chromosome 1 and SLE were also elucidated in one study, but not the others. The most well characterized was reported in African-American patients (19) at 1q23. Candidate SLE susceptibility genes in this area include the receptor for IL6 (CD126/CD130; 1q21–23), CD3ζ (1q22–23) and FcγRIIA/CD32 (1q23). Of note, lymphocytes from SLE patients have been observed to express more IL6R (CD126/CD130) (67) and less CD3ζ (68,69) when compared to normal controls. In addition, some SLE patients were found to express a CD3ζ chain that lacked exon 7, and thus a tyrosine residue in the third immunoreceptor tyrosine-based activation motif (ITAM) (69). The importance of this is unclear, however, as recent evidence has shown that the CD3ζ ITAMs are not necesssary for normal T cell development and function (70).

Polymorphisms encoding low affinity versions of FcγRIIA/CD32 (H131 and B176) have been found to be associated with lupus nephritis in Caucasian, Korean and African-American SLE patients (71–75) as well as in murine lupus models (76). FcγRIIA/CD32 is expressed on monocytes, macrophages and neutrophils and is essential for binding and internalization of IgG2 containing immune complexes (ICs), since this isotype does not activate complement dependent mechansims effectively (77). It should be noted that FcγRIIA/CD32 may be in linkage disequilibrium with other candidate genes in this region, including a polymorphism of the FcγRIIIa/CD16 gene (78) that has low affinity for ICs, is predominantly expressed on NK cells, mononuclear phagocytes and renal mesangial cells, and is essential for binding and internalization of IgG1 and IgG3 containing ICs (77).

The other putative linkage that was described between chromosome 1 and SLE in one study, but not the others, was located at 1q31. Candidate genes in this region include complement receptors (79), specifically CD21/CR2 expressed on B cells and follicular dendritic cells (FDCs) and CD35/CR1 expressed on FDCs and RBCs but not B cells (80). Association studies have revealed decreased expression of these receptors on lymphocytes from patients with active SLE when compared to normal controls (81–83).

Recent genetic evidence has cast doubt on several hypotheses regarding SLE susceptibility. Based on data obtained from studies of *lpr/lpr* and *gld/gld* lupus-prone mice, it was suggested that there might be a role for inappropriately expressed CD95/Fas and/or its ligand in the spontaneous apoptosis of lymphocytes observed in human patients with active SLE (84). It should be noted that recent association and linkage studies as well as functional and phenotypic studies examining this hypothesis have come to the conclusion that FasL-CD95/Fas interactions do not play a role in the immunological dysregulation characteristic of human SLE (85–87). Moreover, although it is well established that female sex hormones play a role in the development of SLE and well as in SLE disease activity (88–92), the hypothesis that variations in X chromosome inactivation-induced mosaicism play a role in active SLE was challenged by monozygotic twin studies that found no differences between a healthy twin and the one with active SLE in the methylation status of the CpG region near trinucleotide repeats in exon 1 of the androgen receptor gene on the X chromosome (93).

3. ABNORMALITIES DURING B CELL MATURATION THAT MAY CONTRIBUTE TO INITIATION AND PROGRESSION OF ACTIVE SLE

In SLE-susceptible individuals, it has been hypothesized that genetic abnormalities in components of the T-dependent (TD) and/or T-independent (TI) humoral immune response, coupled with environmental factors such as female sex hormones or infection with bacteria or viruses, leads to the production of pathogenic, IgG autoantibodies that are responsible for end-organ damage. The fact that disease is not usually apparent until the second or third decade of life suggests that many factors over time drive the immune system to generate, activate, differentiate and positively select and/or fail to negatively select B cells that intrinsically, or after mutation and/or receptor editing/replacement, secrete pathogenic autoantibodies with specificities for intrinsic autoantigens or neoautoantigens that become available during disease activity.

Furthermore, SLE-prone individuals may have a genetic abnormality in generation and/or maintenance of their functional repertoire of B cells expressing specific Ig genes or may be subjected to conditions that drive specific immuno-globulin variable region (IgV) usage and/or somatic hypermutation that facilitates reactivity with autoantigens. The mechansims responsible for these phenomena are not yet clear. For example, it is not known whether the abnormal humoral response that emerges in SLE-susceptible individuals is intrinsic or is initiated by environmental events. It is also not clear whether intrinsic or extrinsic stimulation triggers increased survival and/or migration of autoreactive B cells from the bone marrow, B cell hyperreactivity, abnormal T cell help for B cells, and/or aberrant presentation of autoantigenic material by antigen-presenting cells (APCs) via MHC to T cells and/or via FcRs or CRs to B cells. Investigation of these questions is very difficult because of the limited availability of patient material and the ethical constraints involved in invasive procedures in human

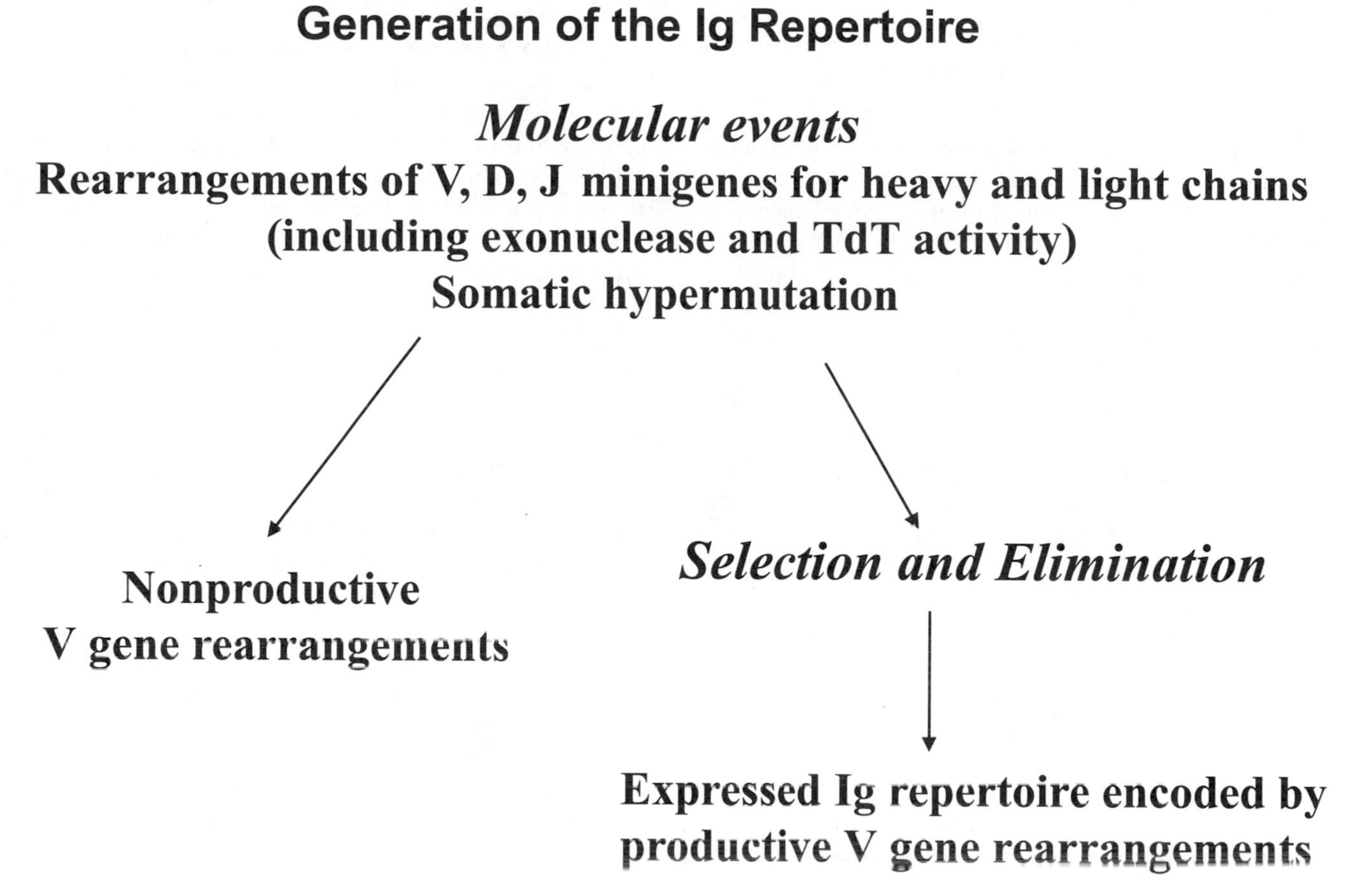

Generation of the Ig Repertoire

Molecular events

Rearrangements of V, D, J minigenes for heavy and light chains (including exonuclease and TdT activity)
Somatic hypermutation

**Nonproductive
V gene rearrangements**

Selection and Elimination

**Expressed Ig repertoire encoded by
productive V gene rearrangements**

Figure 1. Genomic analysis of individual Ig V gene rearrangement allow the differentiation of molecular and selective influences on the V gene repertoire.

subjects. Because of the ease of availability, most of the work has involved an examination of the functional status of circulating mononuclear cells (MCs). It is uncertain whether peripheral blood MCs (PBMCs) are most reflective of the overall immune status of the individual. Moreover, in most studies, the function of these cells has been compared with PBMC from normal individuals. Because the presence of specific MC subpopulations in the blood is governed by systemic effects that influence circulatory patterns as well as a variety of physiologic processes of the MCs themselves such as activation status, apoptosis, adhesion molecule expression and activity, it can be difficult to interpret the results of a comparison between PBMC from normal individuals and SLE patients. A more accurate comparison between SLE and normal lymphoid cells could involve those isolated from secondary lymphoid organs such as lymph node (LN) or tonsil, but samples for such comparisons are difficult to obtain, especially from SLE patients.

Humoral immune responses occur in secondary lymphoid tissues (94–96) that are highly organized structures with distinct T and B cell compartments that collect antigen, APCs expressing Ag in the context of MHC, and recruit lymphocytes. While mature B cells enter LNs and mucosal tissues, such as tonsil and Peyer's patches, in a homing receptor dependent manner through high endothelial venules (HEVs), mature and immature B cells enter the spleen through a specialized area that surrounds the white pulp outside the marginal sinus called the marginal zone (MZ). Recent evidence in the mouse indicates approximately 20×10^6 IgM$^+$ B cells develop in the bone marrow daily (97), of which only 10% reach the spleen and 1–3% become mature B cells (98,99). Only mature B cells are long-lived, can recirculate to LNs and spleen, and be recruited into GCs in normal individuals. Abnormalities in this tightly regulated surface Ig- and adhesion molecule-dependent process could permit access of inappropriate populations to lymphoid organ environments that promote their expansion into the mature B cell pool, rather than deletion, thus potentially predisposing to autoantibody formation. In light of the magnitude of this extensive culling process of immature progenitor cells that routinely occurs, even a subtle abnormality could predispose to the development of autoimmune disease, such as SLE, that take many years to become manifest.

3.1. Abnormalities in Ig Repertoire Generation Can Lead to Production of Pathogenic Autoantibodies

The Ig repertoire is made up of a universe of B cells, each expressing a unique antibody whose specificity for a particular antigen is determined by three complementary-determining regions (CDRs) in the variable (V) region of both heavy (H) and light (L) chains. Each B cell displays a unique set of CDRs that is interposed in the tertiary structure of the protein to form a classical antigenic binding site (100). Amino acids of the CDR contact antigen and composition of the frame work regions (FRs) can indirectly influence the conformation of CDR loops and therefore antigen binding (101). At the molecular level, a high grade of diversity of IgV regions is generated during early B cell development by the somatic recombination process (102) that assembles functional genes by successive rearrangements of one of a number of joining (J), diversity (D) and finally, variable (V) minigene elements of the heavy chain, followed by V-J rearrangement of the light chain.

During recombination, further diversity can be introduced by activity of exonucleases and/or TdT that adds nontemplated N-terminal (N) nucleotides at joining sites (103). A recent study of NZB/NZW F_1 mice that were genetically deficient in TdT has suggested that TdT may drive processes that contribute to pathogenicity of anti-chromatin and anti-histone autoantibodies spontaneously produced by this lupus-prone strain (104). Specifically, knocking out TdT in NZB/NZW F_1 mice decreased the polyreactivity and affinity of anti-chromatin and anti-histone autoantibodies as well as the severity of glomerulonephritis experienced by these mice compared to TdT$^{wild-type}$NZB/NZW F_1 mice. It should be noted that these differences in phenotype were apparent even though the amount of autoantibodies produced in each sub-strain were comparable. Importantly, TdT- mediated complexity induced in the CDR3 during V(D)J recombination in primary lymphoid organs may contribute to the emergence of autoantibodies.

Of importance, the association of heavy and light chain genes is random, generating additional diversity (105,106). In contrast to the T cell receptor repertoire, where all linear epitopes can be recognized by the 10^{14} to 10^{16} diverse receptors that can potentially be generated using the various germline genes, it is estimated that IgV genes are able to generate only 10^8 different B cell receptors that may be insufficient for the recognition of all possible pathogens and conformational epitopes. During TD immune reponses, somatic hypermutation and, occasionally, secondary rearrangement of upstream V gene segments (receptor replacement or editing) and TdT activity (Girshick, H., A.C. Grammer and P.E. Lipsky, in preparation) may further diversify the IgV gene repertoire following antigen exposure (105,107). These mechanisms contribute to the generation of a highly diversified array of IgV gene products that conceivably can recognize a universe of antigens.

B cells develop in primary lymphoid organs such as the BM, liver and peritoneum (108,109). While conventional B cells mature in the BM, B cells that secrete low affinity (10^4–10^7 M^{-1}), polyreactive antibodies have been suggested to develop in the BM, liver, or peritoneum. In the mouse, the former are termed conventional, B-2 cells and the latter are termed CD5$^+$, B-1 cells. Murine and human B-1 B cells can further be identified by their expression of CD11b/CR3/Mac-1 (110). Blood from the adult periphery and the newborn umbilical cord are respective sources for large percentages of the B-2 and B-1 subsets in humans (111). Small numbers of B-1 B cells can be identified in the outer PALS of murine spleens (112) or in the lymphoepithelial regions of human tonsils (113). Moreover, it should be noted that since human B cells can express CD5 as an activation antigen (114), B-1 B cells cannot be conclusively identified in mature humans.

Polyreactive, autospecific B cells. Typically, B-1 B cells have specificity for antigens with repeating epitopes such as exogenous bacterial polysaccharides or phosphorylcholine, termed TI-2 antigens, and endogenous autoantigens such as the Fc region of Ig, ssDNA and membrane proteins expressed by RBCs and thymocytes (108). Antibodies to these antigens are usually of the IgM subtype and have reactivity to more than two structurally unrelated antigens, high idiotypic connectivity and a biased usage of germline genes with little or no somatic mutations (115–117). In this regard, a significantly lower mutational frequency was observed in CD5$^+$ B cells isolated from the peripheral blood of normal humans compared to CD5- B cells (118,119).

It should be noted that recent evidence suggests that another subset of polyreactive B cells in the B-2 lineage may be the "naturally activated", cycling transitional 2 (T2) IgM^{bright}IgD$^+$ CD21^{bright}CD23$^+$ population observed in MZs of spleens of non-autoimmune mice (120). These cells are found in primary follicles and represent a stage of maturation between the IgM^{bright}IgD$^-$ transitional 1 (T1) B cells that migrate to the spleen from the BM and IgM$^+$IgD$^+$ mature B cells capable of recirculating to secondary lymphoid organs. Of note, polyreactive B cells in the BM (121) as well as the MZ of the spleen (120) are frequently found to be in a proliferative state, presumably reflecting constant activation, possibly by self-determinants and/or superantigens.

Naturally occurring, polyreactive, low-affinity IgM autoantibodies must be distinguished from pathogenic IgG autoantibodies such as those that contribute to autoimmune disease activity in SLE patients (122). In normal individuals, polyreactive autoantibodies have been suggested to play important physiological roles in the development of the fetal Ig repertoire and in facilitation of the rapid removal of degraded tissue components such as occurs during fetal remodeling or during apoptosis in later life (115–117; 123). B cells expressing polyreactive Ab have been hypothesized to be essential for the TI humoral response required to combat encapsulated bacteria in adults (124).

Increased and/or aberrant activation of polyreactive, autospecific B cells in SLE-prone individuals may lead to the development of pathogenic autoantibodies that are often isotype switched and/or high affinity. In this regard, in B6.NZM mice, a SLE-susceptibility locus located on the centromeric region of chromosome 4, *sle2*, has been shown to be associated with increased activation of polyreactive B cells (125), suggesting that a similar locus in humans may contribute to the emergence of SLE. Furthermore, recruitment of activated polyreactive B cells into GCs may lead to mutations that allow expressed Ig to react with other Ags. In this circumstance, the Ag driving the immune response may not be the same as the Ag recognized by the resulting pathogenic Ab. For example, individuals often make protective Ig to phosphorylcholine, a dominant Ag on the pneumococcal wall, during the humoral response to this microbe. Purposeful mutation of this protective Ig *in vitro* or mutation that occurs during the course of an *in vivo* humoral reponse in non- autoimmune mice has been shown to lead to loss of binding to phosphorylcholine and the acquisition of binding to DNA (126–129). Binding analysis of a panel of mutants with single amino acid substitutions in the Ig heavy chain documented a change in specificity from phosphorylcholine to DNA that was correlative with the onset of renal damage (129). Futher support for this hypothesis is provided by the findings that non-autoimmune individuals respond to infection with pneumococcus or vaccination with pneumococcal polysaccharides with neutralizing Ig that contains idiotypes associated with pathogenic αDNA Ab that have been described in patients with active SLE (127,130).

Cellular and molecular mechanisms controlling development of the polyreactive, autospecific Ig repertoire. Recent data has suggested that polyreactive autoantibodies are positively selected into the functional human Ig repertoire. First, Ig chain shuffling studies in human transfectomas indicate that polyreactive, autoantibodies are dependent upon a CDR3 generated by somatic recombination (131). As the CDR3 region is intimately involved in Ag binding, these results imply a role for Ag- mediated selection of polyreactive Abs. Of note, since most naturally occurring, polyreactive autoantibodies present in fetal life have been shown to be IgM and to contain few mutations (123), the interactions with Ag involved in their generation may be of lower avidity or intrinsically different than those that stimulate production of pathogenic IgG autoantibodies. Secondly, recent evidence from experiments using transgenic

mice has clearly shown that expression of Ag is required for positive selection of at least one naturally occurring, autoreactive Ab into the murine functional repertoire (132). Since a substantial part of the fetal B cell repertoire expresses IgM with specificity for a variety of structurally unrelated Ags, repertoire selection in the physiologically germ-free intrauterine environment in humans, as well as mice, may be initiated by binding of sIg to developmentally regulated autologous Ags (133). Endogenous superantigens that bind a particular family of Ig receptors in an antigen-independent manner (134,135) may also play a role in the development of the fetal Ig repertoire. This latter hypothesis has not been tested in fetal or neonatal B cells, but is supported by the observation that in mature B cells, expression of one particular V_H3 family member, V_H3–23/DP-47/V_H26 is overrepresented in the productive, but not non-productive, Ig repertoire (118,136,137), in a manner independent of the D_H or J_H segment employed, the length or characteristics of the CDR3, and the pairing of V_k chains. Previous studies had suggested that V_H3–23/DP-47/V_H26 is also overrepresented at the pre- B cell stage of development (138). These findings are consistent with the possibility that selection of V_H3–23/DP-47/V_H26 may have occurred as a result of a B cell superantigen influence at a stage of maturation at which surface μ heavy chain is expressed with surrogate light chain (139,140). Finally, it should be noted that although the number of B cells that enter the mature repertoire in mice raised in a germ-free environment or genetically deficient in T cells is significantly reduced, the number of B cells that enter the polyreactive repertoire is normal (120), implying that neither exogenous microbial Ags nor T cells is essential for the development of these autoreactive B cells.

Limits on the fetal repertoire include molecular processes such as restricted or biased V(D)J recombination (141,142) as well as the requirement for the presence of Ag for selection. In this regard, during murine neonatal development, certain V_H, D and J_H segments are preferentially used (143,144) and diversification mechanisms utilized during the TD immune response, such as N-terminal addition and somatic hypermutation, are not apparent (10). Moreover, the chromosomal order of murine V_H gene segments plays an important role in the usage of V_H gene segments in V_HDJ_H rearrangements (10,143,144). By contrast, germline complexity, but not chromosomal order, of human V_H gene segments appears to play an important role in V_H gene usage in the rearranged Ig heavy chains of mature B cells (118,137,141,142).

The human fetal and neonatal Ig repertoire appear to be restricted in CDR3 size and usage of individual IgV gene elements (141,142,145–149) compared to peripheral B cells from normal adults (150). In contrast to the adult repertoire, restrictions on fetal CDR3 length may not result from selection. Instead, they may relate to a preference for the use of short D_H segments (D7–27 or DQ52) and against the use of long D_H (e.g., D3 or DXP) and J_H6 gene segments by neonatal B cells in fetal liver and BM compared to mature B cells in the adult periphery (150). By contrast, comparison of CDR3 lengths between nonproductive and productive rearrangements in human B cells analyzed by single cell PCR analysis has suggested that Ag contact during adult life positively selects B cells expressing sIg containing V_H rearrangements with shorter CDR3s.

Several observations have suggested that certain IgV genes are associated with reactivity for self-antigens (127, 151,152). Of note, a limited number of individual human IgV gene segments have been frequently observed in the naturally occurring, polyreactive autoantibodies of both the developing/fetal and mature/adult repertoires (106, 147–150,153–160) as well as in pathogenic autoantibodies (126,151,158–163). Finally, the "autoantibody-associated" IgV repertoire is also expressed by CD5$^+$ B cell lines (147) and by certain B cell malignancies such as chronic lymphocytic leukemia (CLL) (151) that are hypothesized to originate from the B-1 lineage. It is important to note, however, that the normal human adult Ig repertoire is also restricted because of utilization of a limited number of IgV genes (118,119,127,136,137,163,164). Moreover, since increased numbers of CD5$^+$ B cells have been detected in patients with SLE and these B cells have been described to produce autoantibodies, it is noteworthy that no difference in V_H gene usage was observed between the CD5$^+$ and CD5$^-$ subsets of human blood B cells (118,119). Furthermore, both molecular and serological findings have suggested a strong overlap between the restricted IgV gene repertoire in normal adults and that used by fetal Ig of the developing repertoire, pathogenic autoantibodies, and B-cell malignancies (118,119,126,127,131,136,137,145–149,151,153–173). For example, it was clearly demonstrated that V_H gene usage by CLL is very similar to that found in the mature, adult B cell repertoire of normal individuals (169). Specific examples of V_H gene segments found to be overexpressed in the repertoires of normal fetuses, neonates, and adults as well as in the polyreactive, autospecific and malignant repertoires include the V_H3 family member V_H3–23/DP-47/V_H26 (136), the V_H4 family member V_H4–34/DP-63/$V_H4.21$ (127,159) and the V_H1 family member V_H1–69/DP-10/51p1 (127). Molecular mechanisms (i.e. enhancer-like sequences, D-proximity, insertion polymorphism, and gene duplication) may account for some of the limited use of V_H genes by B cells in both normal and pathological conditions. However, the details regulating these limitations have not been completely examined. Finally, the possibility that derangements in the molecular, developmental or cellular mechanisms that regulate the B cell repertoire may facilitate the emergence of SLE by increasing the number of B cells expressing autoreactive Ig has not been fully explored.

Cross-reactivity between self and TI-2 antigens A potentially important function of the polyreactive repertoire in normal fetal life may be to opsonize and dispose of apoptotic material and other cellular debris produced during the rapid organ remodeling of the fetus. One result of this process, however, is that a "primitive" broadly specific, polyreactive repertoire may be generated for one purpose, which then may assume different functions in postnatal life (115–117,123). In this regard, a significant portion of polyreactive, autoantibodies from the germ-free fetal environment has been found to react with bacterial components (148), as has been reported for the adult, polyreactive B cell repertoire (115). Cross-reactivities generated *in utero*, encoded by conserved, developmentally- regulated and most likely autoantigen selected IgV genes (148,149) may therefore act as a "first-line defense" of the neonate against bacterial infections when placentally transferred maternal IgG is exhausted (148). Moreover, these cross- reactive B cells may be essential for adult defense against encapsulated bacteria, since individuals lacking these polyreactive B cells as a result of splenectomy are susceptible to acute sepsis because of an inability to mount an effective TI response (124).

Normally, during development of the mature repertoire and exposure to exogenous antigens, the pool of B cells secreting polyreactive autoantibodies diminishes in size, whereas the pool of conventional B cells that does not react with self expands (133,174). In this regard, murine B cells expressing a V_H region transgene expessing the T15 idiotype (175) or a multireactive IgM transgene (176) are selected into the polyreactive splenic B cell subset, but never enter the recirculating mature B pool (120). These findings imply that one checkpoint involved in the normal deletion of polyreactive B cells may occur during differentiation of MZ/T2 immature B cells to recirculating mature B cells. Whether abnormalities in this maturational process occur in SLE remains to be delineated. In contrast to the B cell depletion/anergy induced by binding of conventional B cells to autoantigens (177), it may be possible that polyreactive B cells expressing naturally occurring autoantibodies may be stimulated by diverse, polyvalent, TI antigens to proliferate in a manner that may render these cells more susceptible to abnormal clonal expansion and malignant transformation. This contention is supported by the finding that there is a close relationship of the Ig repertoire expressed by certain B cell malignancies, particularly IgM bearing CLL cells, to the polyreactive repertoire (172, 173). Similarities, strongly indicative of this linkage, include expression of common Ig idiotypes, a similiar pattern of binding specificities (131,151,168–173,178) and low affinity for the recognized antigens (171). It has been hypothesized that pathogenic autoantibodies in SLE may arise by switching and/or somatic hypermutation of polyreactive autoantibodies following abnormal expansion and recruitment of polyreactive B cells into GCs by T cells that have broken tolerance (4,121). One mechanism contributing to this abnormal expansion of polyreactive B cells in SLE-prone individuals may be their exaggerated response to engagement of sIg, as occurs during microbial infection. Whether polyreactive B cells are also involved in processing and presentation of autoantigens has not been sufficiently examined. Of note, B cells secreting pathogenic autoantibodies with germline autoreactivity may have originated from polyreactive B cells that were recruited into GCs to undergo switching to IgG. By contrast, B cells that secrete pathogenic autoantibodies with autoreactivity that can be ablated by eliminating somatic hypermutations must have arisen from a B cell subset that is different from the polyreactive subpopulation.

Pathogenic autospecific B cells. In contrast to the naturally occurring, polyreactive, autospecific antibodies found in healthy humans and normal non-immunized animals, pathogenic autoantibodies are not polyreactive in that they do not react with a wide spectrum of unrelated antigens, are usually IgG and, when sequenced, are often highly mutated and contain increased number of replacement (R) compared to silent (S) mutations in CDR3s, suggesting that they result from a TD immune response (4,7,122,126,179–182). Anti-dsDNA Ab are often cationic because of the presence of arginine (Arg), asparagine (Asp) and lysine (Lys) residues within the CDR3 region that can bind negatively charged conformational determinants in dsDNA. Recent data has suggested that there are at least two different types of autoantibody producing B cell populations in SLE patients. The first is exemplified by anti-dsDNA producing cells. The anti-dsDNA titer exhibits a striking correlation with disease activity, decreasing at the end of a disease flair and in response to conventional anti-proliferative, immunosuppressive therapies such as cyclophosphamide. These results imply that the subpopulation of autoantibody producing cells producing anti-dsDNA Ab may be proliferating lymphoblasts or early plasmablasts that are continuously replenished in a rapid and transient manner that is sensitive to anti-proliferative compounds. The second subpopulation of autoantibody producing cells include cells secreting anti-nuclear Ab and autoantibodies to Ro, La, Sm/RNP and cardiolipin. Titers of these Ab are usually persistent and lack an obvious correlation with disease activity or therapy. These observations suggest that B cells secreting anti-nuclear Ab and autoantibodies to the nucleus, Ro, La, Sm/RNP and cardiolipin may have fully differentiated to long-lived plasma cells that are not dependent upon proliferation for their ability to secrete Ig (183).

Many characteristics of Ig may contribute to the pathogenicity of a given autoantibody. In this regard, the notion that pathogenic anti-DNA antibodies are always high affinity has been challenged recently, since it appears that the charge and affinity for dsDNA are not always predictors of the pathogenic capacity of anti-DNA antibodies (129). Therefore, the fine specificity of the autoantibody rather than avidity alone appears to determine whether it is pathogenic or nonpathogenic (129). In addition, the ability of autoantibodies to enter living cells may contribute to their pathogenic capacity (184). While molecular mechanisms may lead to biased recombination resulting in the preferential usage of certain V, D and J genes, pairing of V_H and V_L, employing a different reading frame in CDR3 or manifesting slightly different exonuclease or TDT activities, cellular mechanisms that have been shown to occur during the TD response to Ag in GCs such as isotype switching, somatic hypermutation, and the re-expression of TdT and/or RAG genes may result in changes in properties of the germline expressed Ig such as the charge and hydrogen bonding capacity of contact residues. Furthermore, the frequency and pattern of somatic hypermutation and/or moderate changes in selective processes during B cell development or activation in GCs may contribute to the production, survival, isotype switching and avidity maturation of autoreactive B cells.

Pathogenic anti-dsDNA antibodies. The most well-characterized pathogenic autoantibodies in SLE are reactive with dsDNA. The majority of human anti-dsDNA Ab are encoded by V_H3 family members, although anti-dsDNA Ab are not restricted to a certain V_H family or an individual IgV gene segment (4,182), in contrast to the V_H4–34 association found in cold agglutinins (161). Of note, anti-dsDNA Ab associated idiotypes encoded by distinct IgV gene segments have been described, although idiotype related gene usage is not absolutely restricted to anti-dsDNA autoantibodies in that the idiotype related IgV gene may be found in Ab that do not react with dsDNA (183–190).

As shown in Table I, the V_H and V_L gene usage of human pathogenic anti-dsDNA Abs has been observed to be similar to that used by mature B cells from normal adults. In contrast, one-third of the approximately 300 murine pathogenic anti-dsDNA Ab that have been sequenced have restricted usage of specific V_H and $V\kappa$ genes (152,191). In this regard, it should be noted that as with other aspects of B cell biology, data from mice are instructive, but not easily applied to the human system. For example, while Abs from normal humans are 60% $V\kappa$ and 40% $V\lambda$, Abs from non-autoimmune mice are 90% $V\kappa$ and 10% $V\lambda$. Conclusive data regarding differences in light chain usage by autoantibodies have not yet been generated. Although there are apparent differences in IgV gene usage, both murine and human pathogenic anti-dsDNA Ab are generally of the IgG isotype and have increased R/S mutational ratios. The enhanced frequency of R mutations in CDRs, especially in the CDR3 region, often results in the accumulation of Arg, Asp and Lys residues that have been shown to bind DNA by charge as well as by the formation of hydrogen bonds (152,181,191–192).

Hypermutation has been observed to introduce Arg in Ig from normal individuals as well as patients with active SLE (4,181,192). Comparison of Arg in productive and nonproductive rearrangements of Ig amplified from B cells isolated from normal donors versus SLE patients elucidated the finding that while Arg substitutions are eliminated from the productive repertoire of normal donors, they are present in the productive repertoire of SLE patients, presumably because of positive selection and/or lack of negative selection in GCs. In the mouse, Arg has been taken as an indication of the acquisition of DNA binding and has been the subject of a variety of investigations (4,181,192), since codons for this amino acid can arise from mutation or from junctional diversity by using an unusual reading frame or an inverted D segment (192). In this regard, the finding that high-titer expression of anti-dsDNA Ab associated idiotypes can be found within SLE kindreds in clinically unaffected family members who do not express anti-dsDNA Abs (185) suggests that SLE-prone individuals may abnormally positively select or fail to negatively select dsDNA binding B cells that are deleted in normal individuals (186,193). Alternatively, genetic factors involved in the generation of an idiotype biased V gene repertoire may predispose to a set of V genes from which somatic hypermutation generates autoantigen binding B cells, although data supporting this possibility are scant since bias in the "autoantibody-associated" Ig repertoire and the normal Ig repertoire appear to be similar.

To define the sequence characteristics of anti-dsDNA Ab that may determine pathogenicity more fully, a broad variety (n = 50) of human B cell hybridomas that secreted anti-dsDNA Ab were analyzed (152). Although the anti-dsDNA Ab analyzed did not preferentially utilize individual IgV genes or IgV gene family members, certain features of the primary sequence of their CDR3s were unusual. Specifically, these anti-dsDNA Ab exhibited D-D fusions, an uncommon reading frame of the D segment, and/or the frequent presence of Arg in their CDR3s (4,126,179–180). Moreover, it was surprising that analyses of both murine and human IgM and IgG anti-dsDNA Ab did not identify a link between Ag driven mutations and the emergence of positively-charged, basic residues, like Arg in the CDRs (152). Furthermore, hybridomas secreting a variety of high avidity, dsDNA-binding Ab were generated from tissues of healthy persons or RA patients who normally do not express these autoantibodies. These results imply that there is unlikely to be a single mechanism by which high avidity autoantibodies are produced. Although in some individuals a genetically-determined bias in the repertoire may increase the tendency for autoantibodies to develop by somatic hypermutation, in others abnormal selection of autoreactive B cells, but not a genetic bias nor a role for somatic hypermutation, appears to play a role.

An analysis of IgV gene usage in autoreactive Ab and IgG F(ab) libraries confirms the conclusion that there is no preferential usage of specifc V_H and V_L genes. In general, the Ig V gene usage of autoantibodies recapitulates the gene usage in B cells from normal individuals with some possible exceptions (Table I). For example, the frequency of usage of V_H3–23, that has been reported to be employed by the 16/6 idiotype, was greater in IgM anti-dsDNA Ab (18.2%) than in the repertoire of normal donors (13%; 118). By contrast, the V_H3–11 gene, which has been shown to be negatively selected in the normal B cell repertoire (137), was found in 4/44 anti-dsDNA clones as well as in a heavily mutated Ig clone derived from an untreated SLE patient that did not express anti-dsDNA Ab (194). Finally, although it is clear that autoantibodies employ V_H genes that are frequent in the normal repertoire, such as V_H3–23, V_H1–69, V_H4–34, V_H4–39 and V_H3–07 (118,137,195), anti-dsDNA antibodies also utilize uncommon V_H genes, such as V_H1–46, V_H3–64, V_H 3–74 and V_H4–61. Similarly, although human anti-dsDNA Ab utilize a broad range of V_L genes, $V\kappa1$–3 family members are slightly more frequent than $V\lambda$ genes in these autoantibodies (Table 1). Thus, the detection of a broad variety of V_H and V_L genes used by anti-dsDNA autoantibodies is consistent with the conclusion that there is no typical pattern of "autoimmune IgV gene" usage although a bias for usage of certain genes over others has been observed in isolated cases.

Further insight into the IgV gene usage by anti-dsDNA Ab was provided by an analysis of 14 anti-dsDNA binding clones from the spleen of a SLE patient (N. Lindsey 1999, personal communication). The observation that all fourteen clones from this SLE patient utilized different V_H and V_L genes is consistent with the conclusion that anti-dsDNA Ab express a variety of V gene segments. Of note and with regard to autoantibody production, usage of V_H4–34 expressing the 9G4 idiotype has been shown to encode cold agglutinins, while this same V_H gene has also been employed for anti-dsDNA binding. As a DNA binder, however, V_H4–34 expresses the T14 idiotype. It is of interest that the frequency of V_H4–34 usage in anti-dsDNA Ab is higher (5/10; 50%; 196) than usage in the normal productive Ig repertoire generated from conventional, CD5- B cells (15.7% of B cells using the V_H4 family; 3.9% overall) as well as from the CD5+ subset (19.2% of B cells employing the V_H4 family; 3.5% overall). Therefore, overrepresentation of V_H4–34 in anti-dsDNA antibodies may reflect an inability of SLE patients to negatively select these B cells as occurs in normal individuals. It should be noted that in normal individuals, V_H4 family members were observed to be negatively selected in CD5$^-$, but not CD5$^+$, B cells by comparing the frequencies of IgV gene usage between the productive and non- productive repertoires (118).

In another study, moderate differences were observed between IgM and IgG human anti-dsDNA Ab with respect

Table I   Human autoantibodies and their gene usage

Name of the sequence or mAb/Isotype	Specificity	V_H Gene Family	V_H Gene	V_H Sequence Homology	V_L Gene Family	V_L Gene	V_L in Sequence Homology	Reference
E-42	anti-60 kD Ro, Fab library	V_H5	V_H5–51/DP73	92.4	Vκ3	L6	94.3	208
E-56	anti-60 kD Ro, Fab library	V_H3	DP31	93.2	Vκ3	L6	94.3	208
E-60	anti-60 kD Ro, Fab library	V_H5	V_H5–51/DP73	92.8	Vλ3	3r/DPL23	95.6	208
E3-MPO	Anti-MPO	V_H4	V_H4–21	90	Vκ1	O2/O12	90	413
AD4–37	DNA, Fab library	V_H3	V_H3–30/DP49	99.3	Vλ3	III.1	99.2	181
AD4–18	DNA, Fab library	V_H5	V_H5–51/DP73	91.9	Vλ2	1a1	85.4	181
SI-32	DNA, Fab library	V_H3	V_H3–23/DP47	97.8	Vλ3	IGLV3S1	99.6	181
SI-13	DNA, Fab library	V_H3	V_H3–23/DP47	91.6	Vλ2	DPL1	96.1	181
SI-1	DNA, Fab library	V_H3	GL-6	97.8	Vκ3	A27/kv325	97.6	181
SI-40	DNA, Fab library	V_H3	Cos3	98.9	Vλ1	Iv1042	95.7	181
SI-22	DNA, Fab library	V_H3	Cos3	97.8	Vλ1	Iv1042	98.8	181
SI-39	DNA, Fab library	V_H1	V_H1–02/DP75	94.9	Vκ3	A27/kv325	97.6	181
A78	DNA	V_H3	V_H3–11/DP35	93.9	Vκ2	A2	87.6	288
B5–3	DNA	V_H3	V_H3–30/DP49	94.2	Vκ3	A27	96.5	288
B12–2	DNA	V_H3	V_H3–48/DP51	90.1	Vκ2	A2	86.3	288
B17–3	DNA	V_H3	V_H3–23/DP47	94.5	Vλ3	IGLV3S2	96.2	288
H17	DNA	V_H3	V_H3–23/DP47	95.9	Vλ3	III.1	95.3	288
M36	DNA	V_H1	hv1263	89.5	Vκ1	O8	96.4	288
MC90	DNA	V_H3	DP58	96.3	Vκ1	O8	98.0	288
O50	DNA	V_H4	V_H4–59/V71–4	96.2	Vλ3	HIv318	96.9	288
I-2a	DNA	V_H3	V_H3–30/hv3005–f3	93.4	Vκ1	L8	95.0	288
H2F	DNA	V_H3	V_H3–23/DP47	89.8	Vκ4	B3	98.0	288
9500	DNA	V_H3	V_H3–33	100	Vκ1	L8/Vd	97.6	414
9604	DNA	V_H3	V_H3–21/DP77	95.6	Vλ3	–	–	414
9702	DNA	V_H1	V_H1–46/DP7	97.6	Vκ3	A27	99.2	414
C119	DNA	V_H3	V_H3–23/DP47	96.3	Vκ3	L6/Vg	100	414
C471	DNA	V_H3	V_H3–64	99.6	Vκ3	A27	99.6	414
III-2R/IgM	DNA mAb IgM	V_H1	V_H1–69/DP-10	96	Vκ1	A20	100	291
III-3R/IgM	DNA mAb IgM	V_H3	V_H3–07	99	Vκ1	O18	99	291

Table I   (Continued)

Name of the sequence or mAb/Isotype	Specificity	V_H Gene Family	V_H Gene	V_H Sequence Homology	V_L Gene Family	V_L Gene	V_L in Sequence Homology	Reference
IC-4/IgM	DNA mAb IgM	V_H4	V_H4–59	97	Vκ1	O18	96	291
II-1/IgM	DNA mAb IgM	V_H5	V_H5–51	98	Vκ3	L16	98	291
NE 1	DNA mAb IgM	V_H4	V_H4–34	100	Vκ1	L5	99.7	415
NE 13	DNA mAb IgM	V_H4	V_H4–34	100	Vκ1	L5	100	415
Kim 11.4	DNA mAb IgM	V_H4	V_H4–39	98.	Vλ1	Vλ1c	98	416
B8807	DNA mAb IgM	V_H3	V_H3–23	100	Vκ1	A30	99.6	414
B8815	DNA mAb IgM	V_H1	V_H1–03	100	Vκ3	L6	99.3	414
B8801	DNA	V_H3	V_H3–11/DP35	95.9	Vκ1	A30	96.8	414
B122	DNA	V_H1	V_H1–18/DP14	96.5	Vκ1	L12a	98.3	414
B6204	DNA	V_H3	V_H3–23/DP47	97.4	Vκ3	A27	99.6	414
32.B9	DNA mAb IgG	V_H3	V_H3–23	97	Vλ8	8a	98	4
33.H11	DNA mAb IgG	V_H3	V_H3–07	95	Vλ2	2a2	99.4	4
33.F12	DNA/cardiolipin, mAb IgG	V_H3	V_H3–11	98	Vκ3	A27	99.7	4
33.C9	DNA mAb IgG	V_H4	V_H4–39	94	Vκ1	L12	98	4
35.12	DNA mAb IgG	V_H3	V_H3–74	95	nd	nd	nd	4
19.E7	DNA mAb IgG	V_H3	V_H3–30	99	Vκ3	L6	99.7	4
T14	DNA mAb IgG	V_H4	V_H4–34	96	Vκ3	A27	99	126
2A4	DNA mAb IgG	V_H4	V_H4–61	92	Vκ1	O2	94	291,417
I–2a	DNA mAb IgG	V_H3	V_H3–30	94	Vκ1	L8	95	291
H2F	DNA mAb IgG	V_H3	V_H3–11	99	Vκ4	B3	99	291
D5	DNA mAb IgG	V_H4	V_H4–34	94	Vκ3	A27	96	418
B3	DNA mAb IgG	V_H3	V_H3–23	94	Vλ2	2a2	93	206
KS3	DNA mAb IgG	V_H4	V_H4–34	94	Vλ2	2a2	96	186
SD6	DNA mAb IgG	V_H3	V_H3–30	97	Vλ2	2a2	96	186
9702	DNA, platelets	V_H1	V_H1–46	98	Vκ3	A27	99	414

Table I   (Continued)

Name of the sequence or mAb/Isotype	Specificity	V_H Gene Family	V_H Gene	V_H Sequence Homology	V_L Gene Family	V_L Gene	V_L in Sequence Homology	Reference
R149	mAb IgG DNA, cardiolipin, mAb IgG	V_H1	V_H1–69	97	Vκ2	A3	99.4	126
RH-14	DNA mAb IgG	V_H3	V_H3–07	96	Vλ2	2a2	99	7

to IgV gene usage (152). For example, anti- dsDNA Ab of the IgG isotype frequently employed the $V_\lambda2a2$ gene (5/6), whereas IgM anti-dsDNA Ab did not utilize the $V_\lambda2a2$ gene at all. Moreover, the V_H3–11 gene was found in the IgG Ab but not in the IgM Ab. This implies that IgV gene usage could play a role in the capacity of B cells to undergo Ig heavy chain isotype switching or to undergo somatic hypermutation and be tested for avidity of Ag binding.

Pathogenic non-dsDNA-reacting autoantibodies (i.e. Ab to the nucleus, Ro, La, Sm/RNP and cardiolipin). Analysis of the productive and non-productive Ig rearrangements from single B cells (Figure 1) isolated from an untreated SLE patient who was positive for anti-nuclear, -Ro, -La, -Sm/RNP and -cardiolipin Ab, but negative for anti-dsDNA Ab (166,197), has provided information regarding potential abnormalities in molecular and/or cellular events that are involved in the generation and/or selection of autoimmune B cells. First, the finding that the distribution of nonproductive V_HDJ_H and V_LJ_L rearrangements in this SLE patient was comparable to a database of normal individuals implied that the V_HDJ_H and V_LJ_L recombination process was unaffected by susceptibility factors for SLE or subsequent disease activity (194,198). Secondly, comparison of V_H gene usage in the productive repertoire of this SLE patient versus a database of normal individuals revealed abnormal representation of the entire V_H4 family as well as certain V_H3 family members (194), indicating the B cells in this patient were exposed to different selection conditions compared to a normal individual. Thirdly, clonal expansion of a B cell expressing IgV genes that are normally negatively selected, V_H3–11 and VIG, was observed in the periphery. Finally, comparison of V_L and J_L usage in the productive versus non-productive Ig repertoires and the pattern of mutations in the productive rearrangements of this SLE patient and a database of normal individuals suggested that editing/ replacement of both V_κ and V_λ had occurred in the SLE patient, but not in normal individuals. Comparision of the frequency of mutations in rearrangements employing V_L distal J_L segments of both SLE

light chains indicated that receptor editing of V_λ occurred before and V_κ began after the initiation of somatic hypermutation in GCs (194,198).

Analysis of productive versus non-productive Ig rearrangements in the entire V_H3 family suggested that this SLE patient (118,136–138,199), like normal individuals, positively selected all V_H3 expressing B cells, regardless of the family member expressed. Importantly, V_H3–23/DP-47/V_H26, the most frequently used V_H3 family member in normal individuals and expressing the 16/6 Id initially identified as associated with anti-DNA Ab (190), was employed to a greater extent in productive Ig rearrangements of this SLE patient when compared with normal individuals. Since this V_H3 family member is overrepresented in the productive, but not non-productive, repertoire (118,136–137) in a manner that is independent of the D_H or J_H segment employed, the hypothesis that further expansion of this V_H3 family member may be related to superantigen stimulation (139–140) will require careful analysis of other SLE patients. Similarly, B cells from this SLE patient preferentially utilized V_H3–11/DP-35 in both non- productive and productive Ig rearrangements significantly more often than normal individuals. Of note, other studies have suggested that V_H3–11 may be negatively selected in some normal individuals (138,199). Overrepresentation of V_H3–11 in this SLE patient was accounted for by the preference in the non-productive repertoire to rearrange this germline gene as well as by the expansion of a V_H3–11/DP-35 expressing clone in this SLE patient. Of importance, non-clonal B cells expressing V_H3–11/DP-35 rearranged to other J_H segments were also overrepresented, suggesting that the negative selection of rearrangements employing V_H3–11/DP-35 previously observed in some normal individuals was disturbed in this SLE patient (138, 199). The issue of whether altered negative selection of certain IgV genes such as V_H3–11/DP-35 is a generalized abnormality in SLE must be examined in other patients with active disease.

The clonal population of B cells expressing V_H3–11 and $V_\lambda1G$ in the periphery of this SLE patient is consistent with other observations that B cells may abnormally clonally expand in the initial stages of SLE (200,201). It should be noted that previous studies did not reveal comparably expanded B cell clones in normal peripheral blood (118; 138, 139,166). The usage of V_H3–11 and $V_\lambda1G$ in this expanded population must be emphasized, since both genes have been reported to be negatively selected in some normal individuals (137,166). These data suggest that an overwhelming Ag stimulus drove the clonal expansion of these potentially autoreactive B cells in a manner similar to that observed lupus-prone mice (201,202). Furthermore, in light of the finding that engagement of sIg induces CD154 expression on B cells (203), the observation that spontaneous proliferation of highly purified B cells from this patient was inhibited by interfering with homotypic CD154-CD40 interactions (204) suggests that interactions between this receptor-ligand pair may have contributed to the expansion and/or maintenance of this clone.

The finding that B cells from this SLE patient had enhanced usage of distal V and J genes for both κ and λ genes in the productive Ig repertoire compared to normal individuals suggested that primary V_L rearrangements were undergoing replacement/editing by subsequent rearrangement more frequently than had been observed in normals (166,167). It is noteworthy that these differences were found despite the fact that the entire B cell population was sampled, consistent with the conclusion that there was a global B cell abnormality in this SLE patient, rather than a defect limited to a subset of B cells or Ig genes. Strikingly, receptor replacement/editing of V_κ and V_λ genes appeared to be occur independently in this SLE patient. Evidence for this conclusion is based on the finding that there was a higher mutational frequency of productive rearrangements employing $J_\kappa1$–4 compared to those using $J_\kappa5$ (198). By contrast, there was an increased usage of 5' V_λ genes represented by genes from cluster C and the 3' $J_\lambda7$ segment, but there was no decrease in the mutational frequency of productive $V\lambda$ rearrangements using these elements nor of the entire productive V_λ repertoire. This comparison of the frequency of mutations in rearrangements using the J distal genes of both SLE light chains compared to rearrangements using J proximal genes indicated that receptor replacement/editing of V_λ occurred before V_κ and before the initiation of somatic hypermutation in GCs. For V_λ, the event may have occurred centrally in the BM before emigration to the periphery. Data from transgenic mice have shown that central receptor editing can operate to replace light chains of B cells expressing autoantibodies (205), but there are no examples of central receptor editing of V_λ chains. One hypothesis generated by these data is that the emergence of V_λ containing autoantibodies during B cell ontogeny may have been the stimulus for central V_λ receptor editing in this SLE patient. In this context, V_λ genes have been shown to be critical parts of a number of human autoantibodies, including those to dsDNA (7,181,186,194,

206), antibodies to La/SS-B and Ro/SS-A, rheumatoid factor (RF) and antibodies to laminin, phospholipids, collagen and histone 2A (4,152,194,206–208). It is noteworthy that certain genes ($V_\lambda4B/J_\lambda2/3$) were found exclusively in the nonproductive repertoires of both the normals and the SLE patient, suggesting that they were eliminated from the productive gene repertoire of each comparably. This implies that some elements of negative selection or receptor editing operated normally in the SLE patient. Similarly, $A30/J_\kappa2$ was exclusively found in the nonproductive repertoire of this SLE patient (19). Productively rearranged $A30/J_\kappa2$ genes have been shown to bind dsDNA in their germline configuration (121,182). Although the binding specificity of $4B/J_\lambda2/3$ gene rearrangements has not been delineated, it was detected only in the nonproductive and not the productive repertoires suggesting the possibility that it might bind an autoantigen. Its elimination from the productive repertoire of normals and the SLE patient might, therefore, result from negative selection and/or receptor editing. Whatever the mechanism of elimination, this process appeared to be intact in this SLE patient and comparable to normals.

3.2. Abberant Selection, Migration and Activation of Immature B Cells

Functional expression of sIg has been shown to be essential for migration of both B-1 and B-2 B cells from primary lymphoid organs (209), development of immature, transitional B cells to mature B cells in the spleen (120) and maintenance of the peripheral B cell repertoire (210). Although sIg engagement on some autoreactive B cells has been shown to be essential for positive selection of these cells into the functional repertoire (132), sIg ligation in primary lymphoid tissues has also been shown to play a role in deleting, anergizing, and inducing receptor editing of autoreactive B cells (109,205). The functional outcome of sIg signaling is hypothesized to depend upon the amount of sIg on the surface of the cell, the amount of Ag present, the affinity of the Ag of sIg and the specific stage of maturation that the B cell is in when sIg is ligated (211). The events that determine whether autoreactive B cells are positively or negatively selected in primary lymphoid tissues are not well characterized, but could be dysregulated in SLE-susceptible individuals who develop differentiated, autoreactive B cells in their periphery.

Anergized, autoreactive B cells that migrate from primary lymphoid tissues are often arrested in development at a transitional, immature stage in normal, nonautoimmune mice (212). Moreover, a higher number of these autoreactive B cells migrate from the BM in *lpr/lpr* autoimmune mice compared to normal, non- autoimmune controls (213). Furthermore, in *lpr/lpr* autoimmune mice, these autoreactive B cells are able to enter the follicular region of the spleen, whereas autoreactive B cells from non-autoimmune mice remain in the splenic outer PALS. Recent studies have demonstrated that signaling through sIgα via syk on transitional B cells is required for maturation to a stage that is able to enter splenic follicles (120). Since SLE B cells have been observed to respond to sIg ligation with higher levels of tyrosine kinases such as syk than non- autoimmune controls (214), more transitional B cells could be stimulated to enter the mature B cell pool than under normal circumstances. This could include B cells that bind autoantigens with low avidity and under normal circumstances may die of neglect. A higher number of mature B cells with specificity for autoantigen in the functional repertoire may over time lead to the development of active SLE. This hypothesis remains to be tested.

Abnormalities in the function of MZ B cells have been suggested to play a role in murine lupus. MZ B cells are a mixed population that includes B1, transitional and mature B cells (215). A subset of MZ B cells expresses CD1, a non-polymorphic, MHC class I-like molecule that presents saccharide and/or lipid antigens to subsets of T cells expressing TCRs that recognize this complex (216). Several studies have suggested a role for this interaction in SLE. For example, transfer of CD4+ or CD8+ T cells from mice transgenic for a TCR that recognizes CD1 into irradiated BALB/c nude hosts induced severe lupus in these normally non-autoimmune mice (217). This finding suggests that T cells recognizing autoantigens in the context of CD1 may provide help for autoreactive B cells in splenic MZs.

Recruitment of polyreactive B cells into a TD humoral response resulting in isotype switching and somatic hypermutation has also been suggested to contribute to the emergence of pathogenic, high affinity, IgG autoantibodies in SLE. Although the elements that may contribute to these events in humans have not been fully characterized, recruitment of autoreactive B cells into TD responses has been examined in lupus-prone *lpr/lpr* mice. Autoreactive B cells were able to form/enter splenic follicles in *lpr/lpr* mice but were retained in the outer PALS in non-autoimmune control mice (213). Furthermore, B cells from these mice producing pathogenic IgG2a rheumatoid factor (RF) with specificity for the Fc portion of Ig were localized in foci in the inner PALS (218). Similar to autoreactive B cells in non-autoimmune mice, anergic B cells generated in the HEL/αHEL double transgenic system were excluded from follicles. Three different mechanisms have been proposed for exclusion of autoreactive B cells from follicles: (1) inability to compete locally with B cells specific for exogenous Ag (219,220), (2) deletion because of strong sIg engagement by high amounts of Ag (221) or (3) exclusion from follicles because of a block in maturation at the T1 stage (120). These mechanisms of exclusion may be dysregulated in an autoimmune mouse.

3.3. Dysregulation of TD Responses to Exogenous or Endogenous Antigen May Contribute to the Emergence of SLE

Abnormal T cell help for B cells may influence differentiation of autoreactive B cells. Stimulation of T cells through the TCR by Ag/MHC complexes has been mimicked *in vitro* by activation with anti-CD3 mAb (222, 223). Many studies have suggested that T cells isolated from SLE patients have decreased proliferative potential as well as abnormalities in signaling following TCR/CD3 ligation such as increased activation of tyrosine kinases and increased Ca^{+2} flux when compared to resting T cells from normal controls (214). It remains uncertain, however, whether these abnormalities reflect an intrinsic defect in T cell activation or are a result of chronic *in vivo* T cell stimulation. Moreover, the findings that SLE T cells have decreased or absent expression of CD3ζ chain compared to normals (68,69) and that the gene for CD3ζ is located within the SLE susceptibility region, 1q22–23 (19), have suggested a potential mechanism for defective T cell activation in SLE. This hypothesis has been called into question recently, however, by the results of a study that examined the role of CD3ζ in TCR/CD3 signaling and found that alterations in the ITAMs in the cytoplasmic domain of CD3ζ that renders them incapable of participating in the transmission of downstream biochemical events did not reduce proximal signaling mediated by Lck, ZAP-70, LAT/SLP-65 and Ca^{+2}, nor affect TCR/CD3 mediated proliferation, cytokine production, upregulation of activation markers such as CD69, CD25, CD95, CD95L or downregulation of TCR/CD3 (70). Most likely the observation that SLE T cells have decreased or absent CD3ζ chain is secondary to degradation following cleavage by caspases as occurs during sustained activation of normal T cells (224), although this hypothesis has not been tested experimentally.

TCR engagement results in a signal that has been shown to induce expression of CD154/CD40-ligand in a manner that is dependent upon the Ca^{+2}/calcineurin signaling pathway leading to NF-AT activation (225–228). CD154-expressing T cells have been localized to the T cell zones of secondary lymphoid tissues where they induce CD40-expressing professional APCs and B cells to express CD80/B7-1 and CD86/B7-2 and then receive further costimulation through the ligand of these molecules, CD28, expressed on their surface (229,230). The importance of this interaction is emphasized by the finding that deficient *in vivo* priming of an allospecific response with CD40−/− APCs could be reversed by the addition of an anti-CD28 mAb (231). Furthermore, blockade of signaling through CD28 with CTLA4.Ig has been shown TD Ab responses (222,233) as well as autoantibody production and glomerulonephritis in SLE- prone NZB/NZW F1 (B/W) mice (233,234) and MRL *lpr/lpr* mice (236). Of importance, various studies have noted aberrant expression of

CD80 and CD86 on PBMC isolated from SLE patients with active disease when compared to normal controls (237–239). It should be noted that the CD28 gene is localized to one of the SLE-susceptibility loci (18–19), 2q33 (240), although preliminary evidence has mitigated against an involvement of this gene in genetically predisposing Mexican-Americans to SLE (241).

CD154-CD40 interactions are absolutely essential for the normal TD humoral immune response (242,243). Of note, blocking these interactions in lupus-prone mice reduces the incidence and severity of disease manifestations such as autoantibody production and glomerulonephritis (235, 244–247). It should be emphasized that several studies have noted that T cells isolated from the periphery of active SLE patients constitutively express CD154 as assayed by flow cytometry (248–250). Elevated Ca^{+2} fluxes observed in SLE T cells may contribute to this phenomenon. This finding is striking, since peripheral blood T cells from normal controls are generally CD154$^-$. However, the question of whether SLE T cells in secondary lymphoid tissues express a higher level of CD154 compared to normal controls remains to be addressed. In addition, it is unknown whether the mechanisms that downregulate CD154 on T cells before they migrate from secondary lymphoid organs are abnormal in SLE. Finally, it should be noted that the genes for CD40 (251,252) and TRAF5 (253), one of the adaptor proteins that mediates CD40-induced signaling events, are located in the SLE- susceptibility loci 20q13 and 1q32.3–41.4 (Grammer, A.C., personal observation). Since CD154-CD40 interactions are crucial for initiation and continuation of humoral responses, it is possible that hyperexpression of CD154 in conjunction with abnormal signaling through its ligand, CD40, over time may lead to aberrant expansion, mutation and/or selection of B cells that secrete autoantibodies.

One caveat to the observation that peripheral blood T cells isolated from SLE patients spontaneously express CD154 on their surface, as detected by flow cytometry, is the finding that soluble CD154 is detected in the sera of patients with active SLE (254). It is possible that soluble CD154 may be associated with SLE T cells in the periphery in a non-specific fashion that is detectable by staining with anti-CD154 mAb. Moreover, it should be noted that serum levels of soluble forms of related TNF-receptor family members, such as CD95 (255), CD27 (256,257), TNFRI (p55), and TNFRII (p75) (258) have been correlated with disease activity in SLE. Matrix metalloproteases (MMPs; 259), such as MMP3/stromelysin-1, that is elevated in the sera of SLE patients (260), may be responsible for some of the cleavage of TNFR family members into soluble forms (Grammer, A.C., C. Pavlovitch and P.E. Lipsky, in preparation).

The issue of whether CD154 is hyperexpressed in SLE secondary lymphoid tissues has been partially addressed by analysis of splenic CD4$^+$ T cells from lupus-prone SNF$_1$ mice *in situ* and immediately *ex vivo*. These experiments revealed that a higher percentage of SNF$_1$ T cells expressed CD154 when compared to T cells from control, non-autoimmune SWR mice. Furthermore, the mean fluorescence intensity (MFI) of both spontaneous and anti-CD3 induced CD154 expression was significantly higher on SNF$_1$ T cells than control mouse T cells (244). Similar experiments have not been performed with human splenic SLE T cells, although CD154 expression was found on T cells in biopsy sections of SLE LNs (250). It should be noted that this study did not compare the percentage or density of expression of CD154 in similar sections of activated LN from non- autoimmune patients.

Downmodulatory signals for CD154 expression include IFNγ and the active form of TGFβ (261,262). In this regard, spontaneous as well as mitogen-induced secretion of these cytokines is decreased in PBMC from patients with active SLE compared to normal controls (263–266). For IFNγ, decreased secretion was correlated with SLE disease activity (264,265). Lupus-prone MRL *lpr/lpr* mice that are genetically deficient in IFNγ do not make IgG2a, the dominant subclass encoding pathogenic anti-dsDNA autoantibodies in these mice, and have greatly reduced glomerulonephritis compared to MRL *lpr/lpr* mice that express functional IFNγ (267). Since IFNγ is required for switching to IgG2a, the lack of this switch factor in these mice most likely contributes to decreased disease activity. The applicability of this finding to human SLE is questionable since the major subclass of anti-dsDNA Ab in SLE patients is IgG1, not IgG2a.

Correlation between secretion of active TGFβ and SLE disease activity has not been examined, although it should be noted that the gene for TGFβ2 is located in the region of the SLE-susceptibility allele, 1q42.2 (17–19). Unexpectedly, when cytokine secretion was analyzed in purified subpopulations, secretion of IFNγ (265) and the active form of TGFβ was comparable between SLE and normal monocytes and T cells, but was decreased in other subsets. For active TGFβ, decreased secretion was observed in NK cells from SLE patients compared to those from normal controls (266). Moreover, it should be noted that mice genetically deficient in TGFβ_1 spontaneously secrete SLE-like pathogenic, IgG autoantibodies (268,269). The contribution of CD154 expression to spontaneous Ig secretion in TGFβ knockout mice is not known.

Potential abnormalities in B cell differentiation during GC reaction formation and perpetuation may lead to the emergence of high affinity, IgG autoantibodies. GCs in secondary lymphoid tissues are thought to be one of the structures in which maturation of the TD response to Ag occurs, fostering expansion, somatic hypermutation, selection, and isotype switching of activated B cells

(94–96). Whereas somatic hypermutation is hypothesized to occur in the dark zone (DZ) of the GC, selection and isotype switching are hypothesized to occur in the light zone (LZ) (270–275). Although the initial phase of TD B cell differentiation described above is restricted by Ag and MHC, later phases are Ag and MHC non-specific and are mediated by contact dependent signals such as CD154-CD40 interactions and cytokines.

Deviant CD154-CD40 interactions during GC reactions may contribute to the differentiation of autoreactive B cells. CD154-expressing T cells initiate the GC reaction by engaging CD40- expressing B cells in the extrafollicular region of secondary lymphoid tissues, thereby inducing them to express CD154 (203) and to proliferate rapidly to form the GC-DZ (94–96). In addition, recent evidence has documented that homotypic B cell interactions are essential for differentiation of $CD38^{++}IgD^-$ GC-DZ $\rightarrow$ $CD38^+IgD^-$ $\rightarrow$ GC-LZ $\rightarrow$ $CD38^-IgD^-$ memory B cell progression as well as for memory B cell progression as well as for differentiation of reactivated memory B cells to form the secondary GC structure that results in Ig secreting cells (203). Studies of SLE patients and their disease-unaffected relatives have found that these individuals, but not normal controls, have peripheral, circulating plasmablasts that spontaneously secrete IgG autoantibodies as well as memory B cells that can be induced to secrete IgG autoantibodies in the presence of the polyclonal memory B cell activator, pokeweed mitogen (276). Since these abnormalities are apparent independent of disease activity, genetically determined factors could contribute to the generation, activation, differentiation and positive selection of memory B cells and plasmablasts that intrinsically, or after mutation, have autoreactive specificity. It should be emphasized that the mechansims that drive these events in SLE are not yet clear. Moreover, mechanisms regulating departure from secondary lymphoid organs may be altered in SLE since activated B cells and plasmablasts are found in increased numbers in the periphery of SLE patients but not in normal controls. Again, mechanisms controlling these potential abnormalities have not been delineated.

CD154-CD40 interactions have been shown to be essential for ongoing GC reactions, since blocking these interactions with anti-CD154 mAb causes GCs to disassemble (277). Hyperexpression of CD154 has been observed by B cells isolated from the periphery of SLE-prone BXSB mice (278) as well as by B cells from human SLE patients (204,248,250). It should be noted that it is not known whether SLE B cells in secondary lymphoid tissues express a higher level of CD154 than their normal counterparts. Despite this caveat, spontaneous proliferation and Ig secretion of highly purified peripheral SLE B cells *in vitro* could be blocked with an anti-CD154 mAb (204). Likewise, spontaneous *in vitro* proliferation of B cells from

BXSB mice was inhibited with an anti-CD154 Ab (278). Furthermore, B cells purified from SLE-prone, SNF_1 mice treated with anti-CD154 mAb spontaneous secrete significantly less total IgM, total IgG and IgG anti-chromatin (ssDNA, dsDNA, histone, histone/DNA) Abs than control mice (244). Finally, the finding that $TCR\alpha\beta^{-/-}$, $CD154^{-/-}$, lupus prone MRL-*lpr/lpr* mice secrete less total IgG1, IgG2a, and IgG2b as well as less IgG anti-nuclear and anti-snRNP Abs than their $CD154^{+/+}$ counterparts (246) is consistent with the possibility that CD154-expressing B cells stimulate autoantibody production in mouse lupus models by homotypic interactions.

A number of observations support the conclusion that spontaneous expression of CD154 by peripheral blood lymphocytes in SLE is unlikely to reflect an intrinsic genetic abnormality, but may be secondary to lymphocyte activation related to disease activity. First, the gene for CD154 is located on Xq26.3–27.1 (279–282) and genetic defects, therefore, are most apparent in males, not females as is the case for SLE. Secondly, CD154 may be considered an early activation marker, since its expression correlates with induction of other early activation markers, such as CD69 by T or B cells (Grammer, A.C., personal observation). Finally, spontaneous CD154 expression on peripheral blood lymphocytes is not specific for SLE, since constitutive CD154 has been observed on highly purified B and T cells from the periphery of patients with other autoimmune diseases, such as RA, Guillain-Barre Syndrome (GBS), chronic inflammatory demyelinating polyneuritis (CIDP) and Myasthenia Gravis (MG) (204). Moreover, homotypic CD154-CD40 interactions between highly purified B cells from these patients mediate spontaneous proliferation and Ig production *in vitro* as has been observed for SLE B cells (204).

Abnormal mutation of Ig genes in GCs may contribute to emergence of autoreactive B cells. The accumulation of mutations in Ig genes by a variety of mechansims during the GC-DZ reaction (270–272, 283–286) may lead to changes in the structure of Ig that allow normal Ig to gain specificity for autoantigen or for autoreactive Ig to increase affinity for autoantigen. There is some evidence that potentially pathogenic autoantibodies may arise by somatic hypermutation during the immune response to foreign antigens from antibodies that have no reactivity to autoantigens in their germline configuration (129,287). Furthermore, it has been suggested that SLE-prone individuals have abnormalies in mechanisms contributing to somatic hypermutation of Ig, since most pathogenic anti-DNA antibodies are heavily mutated (Table I; 288). This remains controversial, however, since it has also been reported that there are no significant differences in the mutational activity directed toward murine antibodies reactive with exogenous compared to endogenous Ags (289).

The importance of somatic mutations and the subsequent introduction of basic residues in the CDRs has been emphasized by several reports (127,206,290). Besides potential differences in the IgV gene usage, somatic hypermutation and the accumulation of basic residues in the CDRs are suspected to be important in the generation of autoantibodies. Five human monoclonal anti-phospholipid antibodies were reported with a high concentration of basic residues in the CDR3 (291) with binding capacities for phospholipids and DNA. A variety of studies (152) have considered enhanced R/S ratios in the CDRs of gene rearrangements encoding autoantibodies as a indication for antigenic selection (292). This assumption has been questioned since nonproductive V_H gene rearrangements without selective influences also exhibited an intrinsically high R/S ratio in the CDRs (197). Thus, increased R/S ratios in the CDRs of V_H genes reflect an intrinsic feature of Ig coding sequences (293) and appear to be a characteristic of the mutator operating on V_H genes independent of antigen.

Most notably, the frequency of somatic mutations was not remarkably enhanced in a battery of human anti-ds/ssDNA antibodies although human IgG monoclonal anti-DNA antibodies were more strikingly mutated. Of importance, however, almost all anti-DNA antibodies carry mutations. As observed in normals, the frequency of somatic hypermutation was higher in V_H compared to V_L rearrangements, indicating the mechanisms driving this outcome are not disturbed in SLE (118,137,294). Although the role of somatic hypermutation in generating autoantibodies remains to be definitively elucidated (4), the finding that six Ab with mutations in the CDRs that introduced basic residues subsequently bound DNA. The importance of introducing these basic residues in CDRs to subsequent DNA reactivity was emphasized by the observation that back mutation to the germline configuration resulted in loss of DNA binding capacity. Similar results were obtained with V_L gene rearrangements (206). These results, as well as the finding that immunization with PC resulted in mutated anti-DNA Ab (128), imply that some autoantibodies are the result of a GC reaction(s) in which Ig genes that had no intrinsic autoantibody reactivity in their germline configuration underwent mutation, gained specificity for autoantigen and were positively selected by autoantigen. Furthermore, consistent with the observation that Ig genes in general are mutated at an increased frequency in SLE patients compared to normal individuals (194,198), a variety of other abnormalities beyond an intrinsic overactivity of the mutational machinery may be involved. In this regard, abnormalities in events that drive GC development and that occur during the GC reaction such as increased antigenic stimulation, a lowered threshold for B cell activation and/or a defect in apoptosis may be involved in generating an increase in mutations in the productive Ig gene repertoire. Of importance, the IgV gene repertoire generated under these pressures seems to be less

influenced by selection than in normal individuals, possibly because of abnormal amounts of autoantigen or abnormalities in apoptosis. Of importance, the increased impact of somatic hypermutation was found to be a generalized abnormality in an SLE patient and not limited to B cells expressing Ig with a unique specificity (194,198).

Detailed analysis of mutations in Ig genes from an untreated SLE patient provided further insights into the generation of diversity during disease (Tables II, III). Markedly increased somatic hypermutation of both nonproductive and productive V_H rearrangements was observed in CD19+ B cells from this SLE patient compared to normal individuals (119,193). Since somatic hypermutation occurs in GCs following challenge with TD antigens (174) and the frequency of mutations in the non-productive repertoire reflects the activity of the mutational machinery without subsequent selection (193), the B cells of this patient appear to have been stimulated in GCs by TD Ag more intensively or more persistently than in normal individuals. It should be emphasized that this untreated SLE patient had a higher percentage of mutated B cells than normal individuals,

Table II Mutational frequency of mutated nonproductively rearranged V genes of normal donors and an untreated SLE patient[+]

	% of B cell mutated	Overall mutational frequency of the V gene repertoire (%)
Nonproductive Ig V gene rearrangements		
Normal		
$V_H DJ_H$	53	3.8 (357/9498)
$V\kappa J\kappa$	24	2.0 (243/11995)
$V\lambda J\lambda$	49	1.2 (76/6263)
SLE		
$V_H DJ_H$	100	6.5 (87/1330)
$V\kappa J\kappa$	82	3.6 (145/4054)
$V\lambda J\lambda$	79	3.1 (412/13194)
Productive Ig V gene rearrangements		
Normal		
$V_H DJ_H$	69	3.3 (1601/47872)
$V\kappa J\kappa$	41	2.6 (754/29184)
$V\lambda J\lambda$	58	2.0 (440/21821)
SLE		
$V_H DJ_H$	90	4.4 (446/10172)
$V\kappa J\kappa$	75	2.8 (379/13569)
$V\lambda J\lambda$	88	3.4 (807/23819)

[+] the mutational frequencies of the particular repertoires of the normals and the SLE patient were significantly different for all rearrangements analyzed (p < 0.001).

Table III Distribution (%) of mutations of individual nucleotides in an untreated SLE patient and normals (NHS).

A. Nonproductive

	V_H NHS	V_H SLE	V_κ NHS	V_κ SLE	V_λ NHS	V_λ SLE
A	23	26	29	12	29	23
G	26	31	28	53	25	41
C	30	27	30	20	16	22
T	21	17	13	15	30	14

B. Productive

	V_H NHS	V_H SLE	V_κ NHS	V_κ SLE	V_λ NHS	V_λ SLE
A	26	17	30	18	29	31
G	26	42	30	50	27	40
C	23	30	25	16	26	19
T	24	10	15	16	17	11

indicating an increased percentage of B cells that had gone through GC reactions to become memory cells. Whether this reflects the intensity or persistence of stimulation or a defect in apoptosis of B cells expressing mutated sIg, as has been suggested (177,287), remains to be determined.

The difference between the frequency of mutations in productive versus the nonproductive Ig rearrangements reflects the influence of selection, with elimination of mutation-generated defective B cell receptors normally more evident than positive selection of those with increased avidity (102,118,119,295). The selection process seems to be generally intact in this SLE patient, even though the overall resulting frequency of mutations in the productive repertoire is much greater than in normal individuals. Further analysis of base pair changes provided evidence that mutations of G were markedly increased in the SLE patient (Table III) compared to normals. These data are consistent with enhanced mutational activity being correlated with enhanced targeted mutations of G/C base pairs as has been previously suggested (119). However, differences in the frequency and the pattern of mutations in SLE need to be evaluated by further studies. Since most of IgV gene rearrangements obtained from SLE patients and anti-DNA antibodies carry mutations and mutations can remarkably influence antigen binding, it seems conceivable that hypermutation and/or subsequent selective processes might be disturbed in SLE.

Taken together, the current available data about the molecular and selective influences of V gene rearrangements in SLE do not allow a final conclusion. However, V_H and V_L gene usage of a variety of autoantibodies seems to reflect the usage found in normals. Strikingly, an enhanced frequency and different pattern of somatic hypermutation was apparent in the untreated SLE patient compared to normal individuals. Although the cause of the disease remains unknown, further studies are needed to evaluate the role of V_H and V_L gene usage, receptor editing of IgV genes, clonal expansion of B cells, and the importance of somatic hypermutation as well as the influence of selection at certain points of B cell development that may all contribute to the emergence of autoimmunity in SLE. Whether the abnormalities in the B cell repertoire reflect a reactive pattern based upon extensive stimulation or represent intrinsic abnormalities in the entire B cell population predisposing to the emergence of autoimmunity remains to be elucidated.

Abnormal selection of autoreactive B cells in GC-LZs may occur in SLE-susceptible individuals. Excess apoptotic material in ICs held on the surface of FDCs in GC-LZs may provide an environment for B cells with intrinsic or acquired specificity for autoantigens to compete for positive selection into the functional repertoire. Accumulation of apoptotic debris in SLE may be related to impaired phagocytic function (296) in conjunction with the increased rate of lymphocyte apoptosis associated with active SLE (297–299). Several factors may contribute to decreased phagocytosis in SLE patients, including abnormalities in the genes expressing $Fc\gamma RIIA/CD32$ and $Fc\gamma RIIIA/CD16$ that result in expression of low affinity versions of these receptors (71–75; 78), as well as decreased levels of components of and receptors for complement (81–83).

B cells expressing sIg with intrinsic specificity for autoantigens may be positively selected in GC-LZs following binding of autoantigen alone or incorporated into an IC. Many factors may contribute to the aberrant recruitment of autoreactive B cells into the functional repertoire of SLE patients. First, decreased phagocytic function in SLE may increase the amount of autoantigen present in GC-LZs, thereby allowing the amount of sIg engagement necessary to positively select autoreactive B cells (211) that would remain anergic or die of neglect in normal individuals. Secondly, SLE B cells may have a lower threshold for positive selection, since ligating sIg on SLE B cells results in enhanced activity of tyrosine kinases and Ca^{+2} release compared to normal controls (214). In addition, the hyperreactive response to sIg engagement may contribute to the enhanced CD154 expression that has been observed on SLE B cells (248–250), since recent evidence has shown that engaging sIg leads to *de novo* synthesis of CD154 in a Ca^{+2}/calcineurin dependent manner (203). Finally, altered signaling through sIg and/or enhanced expression of costimulatory molecules such CD154 may rescue autoreactive SLE B cells from the anergy or deletion that occurs either by neglect or negative signaling following autoantigen binding in normal individuals. In this regard, exposure of anergic B cells generated in a HEL-αHEL double trangenic mouse to exogenous CD154 resulted in proliferation, Ag presentation and Ig secretion at levels comparable to that of normal B cells (300,301).

Following positive selection in GC-LZs, autoreactive B cells may migrate to the apical LZ like normal B cells and present autoantigenic peptides to T cells that have been primed to autoantigen by professional APCs (302). Alternatively, cross-reactive recognition and/or degeneracy of the TCR may allow T cells primed with exogenous Ag to recognize autoepitopes presented in a variety of MHC class II molecules (303,304). Engagement of TCR in this manner may result in rapid expression of CD154 as has been shown for memory T cells isolated from rheumatoid synovium or tonsillar tissue (305,306). Moreover, CD154 engagement of CD40 has been shown to generate signals that are responsible for isotype switching (41) as has been observed in GC-LZs (273–275). Switching of autoantibodies from IgM to IgG has been shown to be crucial for their pathogenicity (8).

In normal individuals, B cells that acquire autoantigen reactivity by somatic mutation in GC-LZ can avoid deletion and be positively selected into the functional repertoire by the process of receptor editing/replacement in GC-LZs (307–310; Girshick, H., A.C. Grammer and P.E. Lipsky, in preparation). In this regard, expression of both of the genes necessary for initiation of this process, RAG1 and RAG2 (311), has been observed in IgG$^+$ tonsillar B cells of the LZ phenotype, CD38$^+$IgD$^-$ (Girshick, H., A.C. Grammer and P.E. Lipsky, in preparation). Although the signals required for RAG expression have not been completely delineated, expression of RAGs in this subset is consistent with previous evidence that RAG2 expression is restricted to cells in G_1/G_0 (312) and that induction of both RAG1 and RAG2 expression in B cells is initiated by engaging sIg (313–315) and maintained by engaging CD40 (308,310), possibly by a BSAP-dependent mechanism (316). By contrast, engaging sIg on B cells that are already positive for the RAGs results in downregulation of RAG expression (310). Moreover, since evidence of receptor editing/replacement has been observed in both V_κ and V_λ chains from a patient with active SLE (194,198), SLE B cells are not deficient and may even be enhanced in their capacity to express RAG genes. Furthermore, comparison of the frequency of mutations in the J distal genes of both SLE light chains compared to the J proximal genes indicated that receptor editing of V_λ occurred before and V_κ began after the initiation of somatic hypermutation in GCs. For V_λ, the event may have occurred centrally in the bone marrow before emigration to the periphery, during a TI response in the extrafollicular region outside GCs, or during expansion of a primary follicle before initiation of mutational mechanisms.

Diversification of the autoimmune response occurs over time by a variety of mechansims, including epitope spreading and antigen spreading (317). Initially, the autoimmune response is restricted to a few determinants. With each subsequent reactivation of memory B cells by Ag, new GCs are formed and B cells undergo more mutation and positive selection. With time, the initial autoimmune response resulting in presentation of a specific dominant peptide of the autoAg diversifies to other epitopes of the same Ag and then to associated Ags by epitope and antigen spreading. For example, mice immunized with the B/B′ peptide of snRNP develop Ab to the immunizing peptide, to subdominant peptides, to the entire snRNP complex and then to DNA (318). These mechansims are hypothesized to contribute to diversification of the autoimmune response in human SLE, although the details have not yet been addressed.

Preliminary evidence suggests that the humoral response may be modulated by anti-idiotypic Abs that recognize the idiotype of the initial Ab. The idiotype of an expressed Ig protein refers to one or more determinants contained within the Ag binding domain. Binding of an anti-idiotypic Ab to sIg on a B cells is hypothesized to downmodulate an ongoing humoral response. While T cells have generally been tolerized to idiotypes of unmutated Ab, T cells can recognize and respond to peptides derived from idiotypes of mutated Ab as foreign Ag in the context of MHC. The principle is that novel peptides derived from Ig may be recognized by T cells if they have not been available during T cell ontogeny and, thus, had not been available to delete or anergize autoreactive T cells. For example, peptides from germline Ig may have been available during T cell development, but peptides from Ig resulting from a TD response, and thus somatic hypermutation, would not have been available to anergize or delete reactive T cells. Recognition of sIg-generated peptides in the context of MHC by T cells in secondary lymphoid organs may be a mechansim that drives anti-idiotypic B cells to expand and be recruited into GCs, where they can receive signals to switch and undergo somatic hypermutation.

The activation and mutational states of a particular B cell may play important roles in determining whether a T cell is able to respond to peptides of Ig presented in the context of MHC class II. For example, a small number of Ig V gene mutations might lead to minimal changes in amino acids of the expressed BCR and the generation of insufficient T cell peptides to stimulate adequete help for their expansion. In contrast, highly mutated BCRs might be the source of sufficient T cell peptides to stimulate help for the expansion of B cells expressing these highly mutated autoimmune Ig gene products. The number of memory B cells and the mutational frequency of B cells increase during life (118,119) and therefore there is the potential for the production of a large number of autoreactive T cell peptides originating from mutated BCRs. Since T cells may not be tolerized to peptides produced from mutated BCRs, this may provide a mechanism for the T cell dependent production of autoantibodies. The increased frequency of many autoantibodies in persons older than 60 years might be explained by this mechanism.

In this context, a study analyzing the mutational frequency and pattern in B cells from a patient with early untreated SLE revealed some remarkable differences in comparison to normals. This appears to reflect an overly activated mutational machinery and is consistent with a hypothesis of others (319) who suggested that "new" V region sequences may be generated as a result of an immune response, that can serve as T cell epitopes. These neopeptides may activate T cell help that in turn further stimulates clonal expansion of B cells expressing the mutated V genes. Whether this circuit contributes to autoimmunity remains to be proven, although evidence of autoimmunity resulting from the expansion of antiidiotypic autoantibodies that cross react with self-tissues is consistent with this idea. There are data suggesting that anti-idiotypes directed against anti-DNA antibodies are deficient in patients with active SLE but emerge during disease quiescence. Moreover, active lupus in mice can be effectively treated by administering anti-idiotypic antibody directed to anti-DNA autoantibodies.

Abnormal cytokine secretion may affect B cell differentiation and survival during GC reactions in SLE-susceptible individuals. CD40 ligation has been shown to be sufficient for limited proliferation, Ig secretion (320,321) and induction of sterile transcripts of all Ig isotypes (322–325), but cytokines amplify and direct the immune response and directly influence heavy chain isotype switching. Engagement of CD40 has been shown to induce B cells to secrete cytokines such as IL6 and IL10 that have been found to be dysregulated in SLE (326). Of note, the genes for the CD126-CD130/IL6R and for IL10 are located at SLE-susceptibility loci, 1q21–23 and 1q32–42, respectively (18, 19). In normals, IL6 and IL10 (327–331) have been shown to promote B cell proliferation and differentiation, whereas IL6 has been shown to promote differentiation with minimal effects on proliferation (331–332). Examination of normal B cells analyzed immediately *ex vivo* from an ongoing humoral immune response in activated secondary lymphoid tissue has revealed that IL6 and IL10 are expressed to varying degrees in all tonsillar B cell subpopulations (Grammer, A.C. and P.E. Lipsky, in preparation).

Purified SLE B cells, but not T cells, spontaneously secrete IL6 and spontaneously express CD126-CD130/ IL6R in a manner that was not correlated with disease activity (67,263, 264). By contrast, other studies observed that levels of IL6 protein in the serum, but not constitutive IL6 mRNA detected in PBMC (333), or mitogen induced IL6 protein from PBMC correlated with SLE disease activity (334–339). The source of elevated IL6 in SLE patients appears to be B cells or monocytes/ macrophages (67,333,340). Of importance, total IgG as well as IgG anti-dsDNA autoantibody production from purified SLE B cells was partially blocked *in vitro* with mAb to CD126/IL6R or IL6 itself (67,333). Moreover, depletion of IL6-secreting macrophages from *in vitro* cultures of (NZB/NZW)F_1 splenocytes decreased production of IgG anti-chromatin Abs (341). Potential factors contributing to elevated levels of IL6 during active SLE include enhanced transcription induced by CpG motifs in bacterial DNA (342), hyperengagement of CD40 by elevated CD154 present in SLE (248–250) and/or elevated levels of ICs engaging FcγRI (340). Since spontaneous expression of CD126/IL6R was not correlated with disease activity and this gene is located at the 1q21–23 SLE susceptibility locus, the B cell response to IL6, in terms of enhanced differentiation to Ig secreting cells, may be genetically influenced.

Elevated levels of spontaneous IL10 secretion have been observed from PBMC isolated from SLE patients or their healthy relatives compared to normal controls. Of note, IL10 levels were not correlated with disease activity (264,343). While the amount of IL10 secreted from monocytes/ macrophages was comparable between SLE patients and normals, elevated IL10 was observed from highly purified peripheral blood B cells (343,344). Moreover, anti-IL10 mAb blocked spontaneous secretion of total IgG as well as IgG anti-chromatin Ab from SLE PBMC cultured *in vitro* or injected into SCID mice (345). Furthermore, IL10 has been shown to inhibit growth (346) and promote apoptosis (347) of SLE T cells, whereas IL10 rescues normal GC B cells from apoptosis (348). However, the effect of IL10 on spontaneous apoptosis of SLE B cells has not been tested. Of importance, levels of IL10 were not correlated with severity of SLE, whereas the ratio of IL10 production compared with secretion of IL2, IFNγ or IL12 has been observed to be an indicator of disease activity (263,265,334,350,351). Decreased spontaneous or mitogen-induced secretion of these cytokines may be caused by direct or indirect down-modulatory influences of excessive IL10. In addition, the finding that there was no correlation between spontaneous or mitogen induced IL4 or GM-CSF production and SLE (265,349,350) indicates that not all cytokines are altered in the course of this disease. Finally, hyperexpression of IL10 during the humoral immune response may be a genetic risk factor for the abnormal B cell function observed in SLE. Evidence for this hypothesis is provided by the findings that the gene for IL10 is located within an SLE-susceptibility locus, 1q32–42 (18, 19), and that polymorphisms in the IL10 promoter correlate with elevated levels of IL10 measured *in vivo* in serum or following *in vitro* culture in the presence or absence of mitogens (26–31). It should be noted that polymorphisms in the promoter region of the IL10 gene are not unique to SLE patients (32–40).

4. FACTORS THAT MAY INITIATE SLE ACTIVITY

4.1.   Female Sex Hormones

Environmental factors contribute to the emergence of SLE in genetically-prone individuals. The principal evidence for this conclusion comes from studies of disease-unaffected relatives of SLE patients (276). For example, B cells in peripheral blood mononuclear cells (PBMC) isolated from female, but not male, blood relatives of SLE patients have been observed to secrete IgG anti-dsDNA autoantibodies spontaneously at levels significantly higher than controls, but less than SLE patients. In contrast, both male and female blood relatives of SLE patients secreted more IgG in response to the polyclonal memory B cell activator, pokeweed mitogen, when compared to controls. Differences between the genders was apparent when Abs to specific Ags were measured. Following stimulation of memory B cells with pokeweed mitogen, male relatives secreted significantly more IgM anti-ssDNA compared to female relatives or male/female controls, whereas female relatives secreted more IgG to the common environmental Ag, influenza hemmaglutinin, compared to male relatives or male/female controls. These data suggest that relatives of SLE patients, like SLE patients themselves, have hyperactive B cells when compared to normal people in the control population. The genetic factors governing this B cell hyperactivity have not been delineated, but candidates include CD126-CD130/IL6R, IL10, CD40 and the CD40-adaptor protein TRAF5 (18,19; Grammer, A.C., personal observation). Finally, the finding that male SLE relatives secrete low affinity, non-switched IgM autoantibodies while female SLE relatives secrete high affinity autoantibodies that have undergone switching to IgG suggests that the female gender may contribute to initiation of SLE disease activity in humans. Additional studies have provided evidence that something expressed in XX females, but not normal XY males, may contribute to SLE-susceptibility. For example, the vast majority of post-pubescent SLE patients are female, whereas the frequency of SLE does not differ between pre-pubescent males and females (352). Moreover, XXY Klinefelter males or the low frequency of post-pubescent XY males with SLE have elevated levels of feminizing metabolites of 17β-estradiol (E_2) compared to normal XY males (353). Finally, estrogen containing oral contraceptives have been found to precipitate SLE or exacerbate existing disease (354). The importance of sex chromosomal effects on the initiation of SLE is emphasized by the finding that mating of NZB or NZW males with BXSB females results in female offspring with severe lupus (355). It should be noted that lupus develops in females, but not males, of the NZB or NZW strains, whereas the Y chromosome in BXSB mice predisposes to

lupus in BXSB males. The mating experiment suggests a role for the recessive X chromosome from the NZB or NZW male parent in the development of lupus. Additional experiments with catrated and noncastrated lupus-prone mice such as NZB/NZW, MRL-*lpr/lpr* or Balb/c immunized with the 16/6 pathogenic idiotype of anti-DNA Ab have demonstrated a disease accelerating effect of estrogens (356–359).

E_2 induces its effects following diffusion across the plasma membrane and binding to cytoplasmic estrogen receptors, hER α and β (360,361), found in human B cells, T cells, and monocytes (362). Nuclear translocation of E_2/ER complexes and association with accessory proteins have been shown to be controlled by serine and tyrosine kinases. Following translocation to the nucleus, E_2/ER complexes bind specific DNA sequences termed estrogen response elements (EREs) in the promoter regions of a number of genes including bcl2 (363), myc (364), fos (365), jun (366) and calcineurin (367). It has also been reported that E_2/ER complexes increase free cytosolic calcium in lymphocytes (368). While bcl2 enhances cell survival and myc influences cellular proliferation, fos, jun and calcium activated calcineurin have been shown to be important for activation of the transcription factors AP-1 and NF-AT, which are important for induction of TNFR family members such as CD154/CD40-ligand (369) as well as cytokines such as IL10 (370) and TNFα (371). In this regard, E_2 has been shown to increase IL10 production from human macrophages (372) and TNFα secretion by human T cells directly (373,374). Secretion of IL1 and IL6 from human PBMC has also been reported following stimulation with E_2 (375,376).

Functionally, E_2 has been reported to rescue autoreactive B cells from tolerization (Bynoe, M.S., C.G. Grimaldi and B. Diamond, submitted), to increase total spontaneous IgG secretion from PBMC isolated from SLE patients and total spontaneous IgG production from PBMC isolated from normal controls (372). Furthermore, E_2 has been shown to increase spontaneous secretion of anti- dsDNA IgG Ab from PBMC isolated from active SLE patients markedly, but not from PBMC isolated from inactive SLE patients or normal controls (372). Spontaneous Ig production from PBMC was partially inhibited by blocking IL10 production from monocytes. When these experiments were repeated with highly purified B cells, E_2-induced Ig was not affected by a blocking mAb to IL10. Together, these results suggest that E_2 may directly costimulate Ig production from B cells that are already producing Ab *in vivo*. Moreover, E_2-induced IL10 from monocytes appears to enhance this response further. In this regard, E_2 induced IL1, IL6 and TNFα act similarly in that all of these cytokines have been shown previously to costimulate Ig production from human B cells (377,378). Finally, female, but not male, relatives of SLE patients secrete autoantibodies that have switched to IgG (276). Since E_2/ER complexes induce signaling cascades leading to activation of NF-AT, the major transcription factor shown to be responsible for induction of the TNF-R family member responsible for B cell expansion as well as Ig class switching and somatic hypermutation, CD154 (369), it is possible that the female sex hormone, E_2, may contribute to initiation of active SLE by polyclonally expanding tolerant B cells and inducing class-switching to high affinity IgG autoantibodies in a CD154-CD40 dependent manner. This hypothetical mechanism has not been directly tested.

An additional mechanism that may contribute to E_2 mediated activation of B cells may be an autocrine feedback loop in which E_2 induced prolactin produced by B cells may bind to prolactin receptors that have been described on human B cells, T cells, monocytes, and neutrophils (379). Of note, prolactin has been detected in human secondary lymphoid tissues such as spleen, lymph node and tonsil (380) that provide the microenvironment necessary for B cell differentiation and selection resulting in secretion of specific, high affinity, class-switched Abs. Furthermore, prolactin is also induced by IL1β, IL2, IL6 and TNFα (381,382), which have all been previously documented to support B cell differentiation into Ig secreting cells (377,378). Of importance, as discussed above, prolactin has been identified as a candidate SLE susceptibility gene in linkage disequilibrium with the HLA locus on chromosome 6p11–21 that was reported to be linked with development of some manifestation of active SLE (56,57). Other studies have associated estrogen-induced hyperprolactinemia, that may occur during pregnancy or postpartum breast-feeding at levels significantly higher than that constitutively present, with glomerulonephritis and the production of a variety of autoantibodies with specificities to ds DNA, cardiolipin and endothelium (383). Engagement of the prolactin receptor induces transcription of a variety of signaling molecules, expression of high affinity IL2Rs on B cells and proliferation of lymphocytes (384–386). It should be noted that the potential role of E_2 and prolactin as initiating factors in the development of SLE has been emphasized by studies with the lupus-prone New Zealand mice (NZB/NZW) (387). Moreover, the possible role of prolactin in ongoing active SLE was supported by the finding that an inhibitor of prolactin, bromocriptine, significantly improved disease activity in a study of Mexican female SLE patients (388). Of note, although prolactin may contribute to hyperreactivity of B cells, this possibility is called into question by the recent observation that mice genetically deficient in prolactin receptors mount a normal humoral response to the TD Ag, NP-CGG (66). Finally, data support the hypothesis that induction of costimulatory molecules and cytokines by female sex hormones in an Ag-independent manner may contribute to initiation of active SLE in susceptible individuals by inducing tolerant B cells to secrete autoantibodies or normal B cells to undergo somatic hypermutation of Ig leading to a gain in reactivity for autoantigens present in secondary lymphoid tissue.

4.2. Viral Infections

A recent comparison of the prevalence of infection with Epstein-Barr Virus (EBV) in young SLE patients compared to normal age-matched controls of European-American, African-American, and hispanic origin suggested that EBV infection may be an etiologic factor in SLE (389). In this study, EBV infection was documented by the presence of IgG anti-VCA (viral capsid antigen) Ab as well as EBV viral DNA in the peripheral blood. Of interest, one of the earliest autoantibodies that emerges in 25–40% of SLE patients is against the B/B' protein of the spliceosome, Sm, and is specific for the sequence PPPGMRPP. Immunization with this peptide or a closely related sequence found in the Epstein-Barr nuclear antigen-1 (EBNA-1), PPPGRRP, has been shown to induce lupus in disease-prone mice by molecular mimicry. It should be noted that the earliest anti-EBV humoral response is specific for the VCA. Fewer people develop an anti-EBNA1 response and fewer still develop an anti-EBNA response to the Sm cross-reactive peptide, PPPGRRP (390).

Many factors may be responsible for the capacity of SLE-susceptible individuals to mount an anti-EBNA1 response cross-reactive to Sm. Following entry into a B cell via CD21/CR2, EBV translocates to the nucleus and transcribes several genes, including EBNA1, EBNA2, LMP1 and LMP2 (391–395). LMP1 and LMP2 are transmembrane proteins that constitutively induce signal transduction cascades similar to CD40 and sIg, respectively. It should be noted that B cells from SLE-susceptible individuals have been shown to mount enhanced responses to sIg engagement (214). Expression of LMP2 may further enhance this dysregulation. Expression of LMP1 leads to proliferation and splenomegaly characteristic of both infectious mononucleosis (EBV-infection) and SLE. In addition, EBV transcribes a viral form of the apoptosis preventing gene, *bcl-2* (BHRF1) (396) as well as a viral form of the cytokine, IL10 (BCRF1) (397). Furthermore, infection of mature human B cells with EBV induces expression of RAG genes (398) that have been shown to be essential for receptor editing/ replacement. In conjunction, these signals could lead to polyclonal B cell activation in a way that might break tolerance of self-reactive B cells and allow their Ig to undergo class switching, somatic hypermutation and possibly receptor editing/replacement. Alternatively, EBV infection may lead to polyclonal activation of B cells in a manner that induces somatic hypermutation and receptor editing so that a B cell gains specificity for autoantigens present in secondary lymphoid tissues. Finally, SLE-prone individuals may permit enhanced EBV control of the humoral arm of the immune system, since SLE T cells have been shown to be deficient in their ability to mount a cytotoxic response against EBV-infected cells (399).

4.3.   Bacterial Infections

Microbial infection of SLE-susceptible individuals has been suggested to play a role in secretion of IgM anti-DNA Ab and to contribute to initiation of active disease. Polyclonal B cell activation leading to secretion of anti-DNA Ab has been observed following stimulation with staphlococcal and streptococcal B cell "superantigens" (sAg), pneumococcal polysaccharides, and DNA in the form of phosphorothioate oligodeoxynucleotides (sODNs) (400–403). Of importance, in light of the finding that individuals susceptible to SLE have an exaggerated response to sIg engagement (214), all of these stimuli utilize sIg ligation in a manner that is independent of the antigen-binding site for all or part of their stimulatory signal.

Staphylococcal superantigens, which have been shown to induce IgM anti-DNA Ab from B cells isolated from normal and SLE donors, include staphylococcal protein A (SPA), staphylococcal enterotoxin (SE) A, and SED. Whereas SPA and SEA bind sIg of B cells in the framework regions of VH_3 (404,405), SED binds sIg of B cells in the framework region of VH_4 (401). All of these sAgs initially induce polyclonal activation of B cells, as assayed by the ability to detect all available VH gene families, that can be sustained in the presence of activated T cells. Of note, the TSST-1 staphylococcal sAg has been shown to stimulate T cells to express CD154 in a manner that was capable of facilitating Ig secretion and Ig heavy chain class switching of bystander B cells (406).

Pneumococcal polysaccharides are classical T-independent (TI) -2 antigens (407) that induce activation of B cells by cross-linking sIg with repeating antigenic epitopes. Recent evidence has shown that ligating sIg induces CD154 expression on the surface of human B cells themselves in a manner that leads to polyclonal expansion (203). In addition, the highest density of CD154 expression in B cell subsets isolated from active human secondary lymphoid tissue undergoing a humoral immune response was observed in the $CD38^{+++}IgD^+$ population hypothesized to be generated by TI Ag stimulation (203). Moreover, CD154-CD40 interactions between human tonsillar B cells have been shown to be essential for differentiation of naïve $CD38^-IgD^+$ B cells to $CD38^{+++}IgD^+$ short-lived Ig-secreting plasmablasts (Grammer, A.C. and P.E. Lipsky, in preparation). Finally, TI-2 antigens induce polyclonal activation of B cells that leads to differentiation into short-lived Ig secreting cells and may activate anergic, autoreactive B cells to secrete autoantibodies. In SLE-susceptible individuals, this response may be enhanced since TI-2 antigens utilize signaling pathways following engagement of sIg that has been shown to be dysregulated in persons prone to active SLE.

DNA in the form of phosphorothioate oligodeoxynucleotides (sODNs) has been shown to activate human B cells polyclonally by a mechanism that utilizes engagement of receptors on the surface of cells (403). More recent data have suggested that sODNs cross-link sIg (Liang, H. and P.E. Lipsky, submitted) in a manner that leads to polyclonal expansion and secretion of Ig, including autoreactive anti-DNA Ab. In addition, immunostimulatory DNA has been shown to induce secretion of IL6 (408), a cytokine that can induce B cells to differentiate into Ig-secreting cells. Furthermore, immunostimulatory bacterial DNA has been described in sera from patients with active SLE (409). In this regard, $IFN\alpha$, a cytokine induced from SLE PBMC by a DNase susceptible factor (410) has been shown to induce autoantibody production and an SLE-like syndrome in non-autoimmune patients and enhance disease in patients with active SLE (411). Recent evidence has demonstrated that $CD4^+MHC$ class $II^+CD3^-CD11c^-$ type 2 dendritic cell precursors (DC2) found in the peripheral blood and in secondary lymphoid tissues, such as tonsil, produce high levels of $IFN\alpha$ in response to infection by viruses or bacteria (412). In conjunction, these data suggest that products of microbial infection such as sAgs, polysaccharides and DNA may contribute to the initiation of active SLE in susceptible individuals by polyclonally activating B cells, leading to expansion of autoreactive clones or induction of somatic hypermutation in normal clones that may change the nature of the Ag that the secreted Ab binds to from foreign to self.

5.   CONCLUSIONS

SLE is a complex, polygenic, chronic autoimmune disease characterized by the production of multiple autoantibodies. Initiating factors, such as female sex hormones or infection with bacteria or viruses, appear to contribute to the emergence of active disease in genetically-prone individuals. Hyperactivity of B lymphocytes is a characteristic of SLE, but which B cell abnormalities are intrinsic and which are acquired remain to be fully elucidated.

ACKNOWLEDGMENT

This work was supported by NIH grant AI-31229. A.C. Grammer was supported by an Arthritis Foundation fellowship and by NIH IRP award RA-AR-0-1001. T. Dorner was supported by Deutsche Forschungsgemeinschaft Grants Dc 491/2–1 and 4–1.

References

1. Burlingame, R.W., and R.L. Rubin. 1996. Autoantibody to the nucleosome subunit (H2A-H2B)-DNA is an early and ubiquitous feature of lupus-like conditions. *Molec. Biol. Reports* 23:159–166.
2. Koffler, D., P.W. Schur, and H.G. Kunkel. 1967. Immunological studies concerning the nephritis of systemic lupus erythematosus. *J. Exp. Med.* 126:607–624.
3. Swanson, P.C., R.L. Yung, N.B. Blatt, E.A. Eagan, J.M. Norris, and B.C. Richardson. 1996. *J. Clin. Invest.* 97:1748–1760.
4. Winkler, T.H., H. Fehr, and J.R. Kalden. 1992. Analysis of immunoglobulin variable region genes from human IgG anti-DNA hybridomas. *Eur. J. Immunol.* 22:1719–1728.
5. Tsao, B.P., F.M. Ebling, C. Roman, N. Panosian-Sahakian, K. Calame, and B.H. Hahn. 1990. Structural characteristics of variable regions of immunoglobulin genes encoding pathogenic autoantibodies in murine lupus. *J. Clin. Invest.* 85:530–540.
6. Ehrenstein, M.R., D.R. Katz, M.H. Griffiths, L. Papadake, T.H. Winkler, J.R. Kalden, and D.A. Isenberg. 1995. Human IgG anti-DNA antibodies deposit in kidneys and induce proteinuria in SCID mice. *Kidney Intl.* 48:705–711.
7. Ravirajan, C.T., M.A. Rahman, L. Papdaki, M.H. Griffiths, J. Kalsi, A.C. Martin, M.R. Ehrenstein, D.S. Latchman, and D.A. Isenberg. 1998. Genetic, structural and functional properties of an IgG DNA-binding monoclonal antibody from a lupus patient with nephritis. *Eur. J. Immunol.* 28:339–350.
8. Okamura, M., Y. Kanayama, K. Amastu, N. Negora, S. Kohda, T. Takeda, and T. Inoue. 1993. Significance of enzyme linked immunosorbent assay (ELISA) for antibodies to double stranded and single stranded DNA in patients with lupus nephritis: correlation with severity of renal histology. *Ann. Rheum. Dis.* 52:14–20.
9. Miranda-Carus, M.E., M. Boutjdir, C.E. Tseng, F. DiDonato, E.K. Chan, and J.P. Buyon. 1998. Induction of antibodies reactive with SSA/Ro-SSB/La and development of congenital heart block in a murine model. *J. Immunol.* 161:5886–5892.
10. Kofler, R., S. Geley, H. Kkofler, and A. Helmberg. 1992. Mouse variable-region gene families: complexity, polymorphism and use in non-autoimmune responses. *Immunol. Rev.* 128:5–21.
11. Tan, E.M. 1991. Autoantibodies in pathology and cell biology. *Cell* 67:841–842.
12. Riemekasten, G., J. Marell, G. Trebeljahr, R. Klein, G. Hausdorf, T. Haupl, J. Schneider-Mergener, G.R. Burmester, and F. Hiepe. 1998. A novel epitope on the C-terminus of SmD1 is recognized by the majority of sera from patients with systemic lupus erythematosus. *J. Clin. Invest.* 102:754–763.
13. Vaarala, O. 1998. Antiphospholipid antibodies and myocardial infarction. *Lupus 7* Suppl. 2:S132–134.
14. Harley, J.B., K.L. Moser, P.M. Gaffney, and T.W. Behrens. 1998. The genetics of human systemic lupus erythematosus. *Curr. Opin. Immunol.* 10:690–696.
15. Tsokos, G.C., and S.N. Liossis. 1999. Immune cell signaling defects in lupus: activation, anergy and death. *Immunol. Today* 20:119–124.
16. Theofilopoulos, A.N., and D.H. Kono. 1999. The genes of systemic autoimmunity. *Proc. Amer. Assoc. Physicians* 111:228–240.
17. Tsao, B.P., R.M. Cantor, K.C. Kalunian, C.-J. Chen, H. Badsha, R. Singh, D.J. Wallace, R.C. Kitridou, S.-l. Chen, N. Shen, Y.W. Song, D.A. Isenberg, C.-L. Yu, B.H. Hahn, and J.I. Rotter. 1997. Evidence for linkage of a candidate chromosome 1 region to human systemic lupus erythematosus. *J. Clin. Invest.* 99:725–731.
18. Moser, K.L., B.R. Neas, J.E. Salmon, H. Yu, C. Gray-McGuire, N. Asundi, G.R. Bruner, J. Fox, J. Kelly, S. Henshall, D. Bacino, M. Dietz, R. Hogue, G. Koelsch, L. Nightingale, T. Shaver, N.I. Abdou, D.A. Albert, C. Carson, M. Petri, E.L. Treadwell, J.A. James, and J.D. Harley. 1998. Genome scan of human systemic lupus erythematosus: evidence for linkage on chromosome 1q in African-American pedigrees. *Proc. Natl. Acad. Sci. USA* 95:14869–14874.
19. Gaffney, P.M., G.M. Kearns, K.B. Shark, W.A. Ortmann, S.A. Selby, M.L. Malmagren, K.E. Rohlf, T.C. Ockenden, R.P. Messner, R.A. King, S.S. Rich, and T.W. Behrens. 1998. A genome-wide search for susceptibility genes in human systemic lupus erythematosus sib-pair families. *Proc. Natl. Acad. Sci. USA* 95:14875–14879.
20. Kim, J.M., C.I. Brannan, N.G. Copeland, N.A. Jenkins, T.A. Khan, and K.W. Moore. 1992. Structure of the mouse IL-10 gene and chromosomal localization of the mouse and human genes. *J. Immunol.* 148:3618–3623.
21. Barton, D.E., B.E. Foellmer, J. Du, J. Tamm, R. Derynck, and U. Francke. 1988. Chromosomal mapping of genes for transforming growth factors beta-2 and beta-3 in man and mouse dipersion of TGF-beta gene family. *Oncogene Res.* 3:323–331.
22. Nishimura, D.Y., A.F. Purchio, and J.C. Murray. 1993. Linkage localization of $TGF\beta_2$ and the human homeobox gene HLX1 to chromosome 1q. *Genomics* 15:357–364.
23. Reboul, J., K. Gardiner, D. Monneron, G. Uze, and G. Lutfalla. 1999. Comparative genomic analysis of the interferon/interleukin-10 receptor gene cluster. *Genome Res.* 9:242–250.
24. Tsao, B.P., R.M. Cantor, J.M. Grossman, N. Shen, N.T. Teophilov, D.J. Wallace, F.C. Arnett, K. Hartung, R. Goldstein, K.C. Kalunian, B.H. Hahn, and J.I. Rotter. 1999. PARP alleles within the linked chromosomal region are associated with systemic lupus erythematosus. *J. Clin. Invest.* 103:1135–1140.
25. Mizushima, S., M. Fujita, T. Ishida, S. Azuma, K. Kato, M. Hirai, M. Otsuka, T. Yamamoto, and J. Inoue. 1998. Cloning and characterization of a cDNA encoding the human homolog of tumor necrosis factor receptor-associated factor 5 (TRAF5). *Gene* 207:135–140.
26. Eskdale, J., P. Wordsworth, S. Bowman, M. Field, and G. Gallagher. 1997. Association between polymorphisms at the human IL-10 locus and systemic lupus erythematosus. *Tissue Antigens* 49:635–639.
27. Lazarus, M., A.H. Hajeer, D. Turner, P. Sinnott, J. Worthington, W.E.R. Ollicr, and I.V. Hutchinson. 1997. Genetic variation in the interleukin 10 gene promoter and systemic lupus erythematosus. *J. Rheum.* 24:2314–2317.
28. Mehrian, R., F.P. Quismorio, G. Strassman, M.M. Stimmler, D.A. Horwitz, R.C. Kitridou, W.J. Gauderman, J. Morrision, C. Brautbar, and C.O. Jacob. 1998. Synergistic effect between IL-10 and bcl-2 geneotypes in determining susceptibility to systemic lupus erythematosus. *Arthritis and Rheum.* 41:596–602.
29. Mok, C.C., J.S. Lanchbury, D.W. Chan, and C.S. Lau. 1998. Interleukin-10 promoter polymorphisms in southern chinese patients with systemic lupus erythematosus. *Arthritis and Rheum.* 41:1090–1095.

30. Rood, M.J., V. Keijsers, M.W. van der Linden, T.Q. Tong, S.W. Borggrave, C.L. Verweij, F.C. Breedveld, and T.W. Huizinga. 1999. Neuropsychiatric systemic lupus erythematosus is associated with imbalance in interleukin 10 promoter haplotypes. *Annals of Rheum. Dis.* 58:85–89.

31. Crawley, E., D. Isenberg, P. Woo, and R. Kay. 1999. Interleukin-10 promoter polymorphism and lupus nephritis: comment on the article by Mok et al. *Arthritis and Rheum.* 42:590–593.

32. Eskdale, J., D. Kube, and G. Gallagher. 1996. A second polymorphic dinucleotide repeat in the 5′ flanking region of the human IL10 gene. *Immunogenet.* 45:82–83.

33. Turner, D.M., D.M. Williams, D. Sankaran, M. Lazarus, P.J. Sinnott, and I.V. Hutchinson. 1997. An investigation of polymorphism in the interleukin-10 gene promoter. *Eur. J. Immunogenet* 24:1–8.

34. Eskdale, J., G. Gallagher, C.L. Verweij, V. Keijsers, R.G.J. Westendorp, and T.W.J. Huizinga. 1998. Interleukin 10 secretion in relation to human IL-10 locus haplotypes. *Proc. Natl. Acad. Sci. USA* 95:9465–9470.

35. Coakely, G., C.C. Mok, A.H. Hajeer, W.E. Ollier, D. Turner, P.J. Sinnott, I.V. Hutchinson, G.S. Panayi, and J.S. Lanchbury. 1998. Interleukin-10 promoter polymorphisms in rheumatoid arthritis and Felty's syndrome. *Brit. J. Rheumatol.* 37:988–991.

36. Eskdale, J., J. McNicholl, P. Wordsworth, B. Jonas, T. Huizinga, M. Field, and G. Gallagher. 1998. Interleukin-10 microsatellite polymorphisms and IL-10 locus alleles in rheumatoid arthritis susceptibility. *Lancet* 352:1282–1283.

37. Cantagrel, A., F. Navaux, P. Loubet-Lescoulie, F. Nourhashemi, G. Enault, M. Abbal, A. Constantin, M. Laroche, and B. Mazieres. 1999. Interleukin-1beta, interleukin-1 receptor antagonist, interleukin-4 and interleukin-10 gene polymorphisms: relationship to occurrence and severity of rheumatoid arthritis. *Arthritis and Rheum.* 42:1093–1100.

38. Huang, D.R., Y.H. Zhou, S.Q. Xia, L. Liu, R. Pirskanen, and A.K. Lefvert. 1999. Markers in the promoter region of interleukin-10 (IL-10) gene in myasthenia gravis: implications of diverse effects of IL-10 in the pathogeneiss of the disease. *J. Neuroimmunol.* 94:82–87.

39. Hobbs, K., J. Negri, M. Klinnert, L.J. Rosenwasser, and L. Borish. Interleukin-10 and transforming growth factor-beta promoter polymorphisms in allergies and asthma. *Am. J. Resp. and Crit. Care Med.* 158:1958–1962.

40. Middleton, P.G., P.R. Taylor, G. Jackson, S.J. Proctor, and A.M. Dickinson. 1998. Cytokine gene polymorphisms associating with severe acute graft-versus- host disease in HLA-identical sibling transplants. *Blood* 92:3943–3948.

41. Grewal, I.S., and R.A. Flavell. 1998. CD40 and CD154 in cell-mediated immunity. *Ann. Rev. Immunol.* 16:111–135.

42. Podrebarac, T.A., D.M. Boisert, and R. Goldstein. 1998. Clinical correlates, serum autoantibodies and the role of the major histocompatability complex in French Canadian and non-French Canadian Caucasians with SLE. *Lupus* 7:183–191.

43. Miyagawa, S., K. Shinohara, M. Nakajima, K. Kidoguchi, T. Fujita, T. Fukumoto, A. Yoshioka, K. Dohi, and T. Shirai. 1998. Polymorphisms of HLA class II genes and autoimmune responses to Ro/SS-A-La/SS-B among Japanese subjects. *Arthritis and Rheum.* 41:927–934.

44. Goldstein, R., J.M. Moulds, C.D. Smith, and D.P. Sengar. 1996. MHC studies of the primary antiphospholipid anti-

body syndrome and of antiphospholipid antibodies in systemic lupus erythematosus. *J. Rheumatol.* 23:1173–1179.

45. Schur, P.H. 1995. Genetics of systemic lupus erythematosus. *Lupus* 4:425–437.

46. Rudwaleit, M., M. Tikly, M. Khamashta, K. Gibson, J. Klinke, G. Hughes, and P. Wordsworth. 1996. Interethnic differences in the asssociation of tumor necrosis factor promoter polymorphisms with systemic lupus erythematosus. *J. Rheumatol.* 23:1725–1728.

47. Hajeer, H.A., J. Worthington, E.J. Davies, M.C. Hillarby, K. Poulton, and W.E.R. Olivier. 1997. TNF microsatellites a2, b3 and d2 alleles are associated with systemic lupus erythematosus. *Tissue Antigens* 49:222–2227.

48. Sullivan, K.E., C. Wooten, B.J. Schmeckpeper, D. Goldman, and M.A. Petri. 1997. A promoter polymorphism of tumor necrosis factor α associated with systemic lupus erythematosus. Arthritis Rheum. 40:2207–2211.

49. Slingsby, J.H., P. Norsworthy, G. Pearce, A.K. Vaishnaw, H. Issler, B.J. Morley, and M.J. Walport. 1996. Homozygous hereditary C1q deficiency and systemic lupus erythematosus. *Arthritis Rheum.* 39:663–670.

50. Reid, K.B.M. 1993. Deficiency of the first component of human complement. In *Immunodeficiencies*, F.S. Rosen and M. Seligman, eds. Harwood Academic Publishers, Philadelphia, PA. pp. 283–293.

51. Fielder, A.H.L., M.J. Walport, J.R. Batchelor, R.I. Rynes, C.M. Black, I.A. Dodi, and G.R.V. Hughes. 1983. Family study of the major histocompatability complex in patients with systemic lupus erythematosus: importance of null alleles of C4A and C4B in determining disease susceptibility. *Brit. Med. J.* 286:425–428.

52. Walport, M.J. 1993. Complement deficiency and diseases. *Brit. J. Rheumatol.* 32:269–273.

53. Provost, T.T., F.C. Arnett, and M. Reichlein. 1983. Homozygous C2 deficiency, lupus erythematosus and anti-Ro(SS-A) antibodies. *Arthritis Rheum.* 26:1279–1282.

54. Granados, J., G. Vargas-Alarcon, C. Drenkard, F. Andrade, H. Melin-Aldana, J. Alcocer-Varela, and D. Alarcon-Segovia. 1997. Relationship of anticardiolipin antibodies and anti-phospholipid syncdrome to HLA-DR7 in Mexican patients with systemic lupus erythematosus. *Lupus* 6:57–62.

55. Arnett, F.C. 1997. The genetics of human lupus. *In* Dubois' Lupus Erythematosus, D.J. Wallace, B.H. Hahn, eds. Wilkins and Wilkins, Baltimore, MD. 77–117.

56. Jarjour, W., A.M. Reed, J. Gauthier, S. Hunt, and J.B. Winfield. 1996. The 8.5 kb *Pst*I allele of the stress protein gene, Hsp70–2. An independent risk factor for systemic lupus erythematosus in African-Americans? *Hum. Immunol.* 45:59–63.

57. Davies, E.J., G. Steers, W.E.R. Ollier, D.M. Grennan, R.G. Cooper, E.M. Hay, and M.C. Hillarby. 1995. Relative contributions of HLA-DQA and complement C4A loci in determining susceptibility to systemic lupus erythematosus. *Brit. J. Rheumatol.* 34:221–225.

58. Skeie, G.O., J.P. Pandey, J.A. Aarli, and N.E. Gilhus. 1999. TNFA and TNFB polymorphisms in myasthenia gravis. *Arch. Neurology* 56:457–461.

59. Albuquerque, R.W., C.M. Hayden, L.J. Palmer, I.A. Liang, R.J. Rye, N.A. Gibson, P.R. Burton, J. Goldblatt, and P.N. Lesouef. 1998. Association of polymorphisms within the tumor necrosis factor (TNF) genes and childhood asthma. *Clin. Exp. Allergy* 28:578–584.

60. Mycko, M., W. Kowalski, M. Kwinkowshi, A.C. Buenafe, B. Szymanska, E. Tronczynska, A. Plucienniczak, and

K. Selmaj. 1998. Multiple sclerosis: the frequency of allelic forms of tumor necrosis factor and lymphotoxin-alpha. *J. Neuroimmunol.* 84:198–206.

61. Fernandez-Real, J.M. C. Gutierrez, W. Ricart, R. Casamitjana, M. Fernandez- Castaner, J. Vendrell, C. Richart, and J. Soler. The TNF-alpha gene NcoI polymorphism influences the relationship among insulin resistance, percent body fat, and increased serum leptin levels. *Diabetes* 46:1468–1472.

62. Horwitz, D.A., J.D. Gray, S.C. Behrendsen, M. Kubin, M. Rengaraju, K. Ohtsuka, and G. Trinchieri. 1998. Decreased production of interleukin-12 and other Th1-type cytokines in patients with recent-onset systemic lupus erythematosus. *Arthritis Rheum.* 41:838–844.

63. Brinkman, B.M., D. Zuijdeest, E.L. Kaijzel, F.C. Breedveld, and C.L. Verweij. 1995. Relevance of the tumor necrosis factor alpha (TNF alpha) -308 promoter polymorphism in TNF alpha gene regulation. *J. Inflamm.* 46:32–41.

64. Kroeger, K.M., K.S. Carville, and L.J. Abraham. 1997. The—308 tumor necrosis factor-alpha promoter polymorphism affects transcription. *Molec. Immunol.* 34:391–399.

65. Maini, R.N., F.C. Breedveld, J.R. Kalden, J.S. Smolen, D. Davis, J.D. Mcfarlane, C. Antoni, B. Leeb, M.J. Elliott, J.N. Woody, T.F. Schaible, and M. Feldmann. 1998. Therapeutic efficacy of multiple intravenous infusions of anti- tumor necrosis factor alpha monoclonal antibody combined with low-dose weekly methotrexate in rheumatoid arthritis. *Arthrits & Rheum.* 41:1552–1563.

66. Bouchard, B., C.J. Ormandy, J.P. DiSanto, and P.A. Kelly. 1999. Immune system development and function in prolactin receptor deficient mice. *J. Immunol.* 163:576–582.

67. Kitani, A., M. Hara, T. Hirose, M. Harugai, K. Suzuki, M. Kawakami, Y. Kawaguchi, T. Hidaka, M. Kawagoe, and H. Nakamura. 1992. Autostimulatory effects of IL-6 on excessive B cell differentiation in patients with systemic lupus erythematosus: analysis of IL-6 production and IL-6R expression. *Clin. Exp. Immunol.* 88:75 83.

68. K. Tsuzaka, T. Takeuchi, N. Onoda, M. Pang, and T. Abe. 1998. Mutations in T cell receptor zeta chain mRNA of peripheral T cells from systemic lupus erythematosus. *J. Autoimm.* 11:381–385.

69. Takeuchi, T., K. Tsuzaka, M. Pang, K. Amano, J. Koide, and T. Abe. 1998. TCR zeta chain lacking exon 7 in two patients with systemic lupus erythematosus. *Int. Immunol.* 10:911–921.

70. Ardouin, L., C. Boyer, A. Gillet, J. Trucy, A.-M. Bernard, J. Nunes, J. Delon, A. Trautmann, H.-T. He, B. Malissen, and M. Malissen. 1999. Crippling of CD3-ζ ITAMs does not affect T cell receptor signaling. *Immunity* 10:409–420.

71. Duits, A.J., H. Bootsma, R.H. Derksen, P.E. Spronk, L. Kater, C.G. Kallenberg, P.J. Capel, N.A. Westerdaal, G.T. Spierenburg, and F.H. Gmelig-Meyling et al. 1995. Skewed distribution of IgG Fc receptor IIa (CD32) polymorphisms is associated with renal disease in systemic lupus erythematosus. *Arthritis & Rheum.* 38:1832–1836.

72. Kimberly, R.P., J.E. Salmon, and J.C. Edberg. 1995. Receptors for immunoglobulin G. *Arthritis & Rheum.* 38:306–314.

73. Salmon, J.E., S. Millard, L.A. Schachter, F.C. Arnett, E.M. Ginzier, M.F. Gourley, R. Ramsey-Goldman, M.G.E. Peterson, and R.P. Kimberly. 1996. FcγRIIA alleles are heritable risk factors for lupus nephritis in african americans. *J. Clin. Invest.* 97:1348–1354.

74. Wu, J., J.C. Edberg, P.B. Redecha, V. Bansal, P.M. Guyre, K. Coleman, J.E. Salmon, and R.P. Kimberly. 1997. A novel polymorphism of FcγRIIA (CD16) alters receptor function and predisposes to autoimmune disease. *J. Clin. Invest.* 100:1059–1070.

75. Clynes, R., C. Dumitru, and J.V. Ravetch. 1998. Uncoupling of immune complex formation and kidney damage in autoimmune glomerulonephritis. *Science* 279:1052–1054.

76. Song, Y.W., C.W. Han, S.W. Kang, H.J. Baek, E.B. Lee, C.H. Shin, B.H. Hahn, and B.P. Tsao. 1998. Abnormal distribution of Fc gamma receptor type IIa polymorphisms in Korean patients with systemic lupus erythematosus. *Arthritis & Rheum.* 41:421–426.

77. Tamm, A., and R.E. Schmidt. 1997. IgG binding sites on human Fc gamma receptors. *Int. Rev. Immunol.* 16:57–85.

78. Koene, H.R., M. Kleijer, A.J. Swaak, K.E. Sullivan, M. Bijl, M.A. Petri, C.G. Kallenberg, D. Roos, A.E. von dem Borne, and M. de Haas. 1998. The Fc gammaRIIIA-158F allele is a risk factor for systemic lupus erythematosus. *Arthritis & Rheum.* 41:1813–1818.

79. Lublin, D.M., M.K. Liszewski, T.W. Post, M.A. Arce, M.M. LeBeau, M.B. Rebentisch, L.S. Lemons, T. Seya, and J.P. Atkinson. 1988. Molecular cloning and chromosomal localization of human membrane cofactor protein (MCP). Evidence for inclusion in the multigene family of complement regulatory proteins. *J. Exp. Med.* 168:181–194.

80. Carroll, M.C. 1998. The role of complement and complement receptors in induction and regulation of immunity. *Ann. Rev. Immunol.* 16:545–568.

81. Kiss, E., I. Csipo, J.H. Cohen, B. Reveil, M. Kavai, and G. Szegedi. 1996. CR1 density polymorphism and expression on erythrocytes of patients with systemic lupus erythematosus. *Autoimm.* 25:53–58.

82. Kumar, A., S. Sinha, P.S., Khandekar, K. Banerjee, and L.M. Srivastava. 1995. Hind III genomic polymorphisms of the C3b receptor (CR1) in patients with SLE: low erythrocyte CR1 expression is an acquired phenomenon. *Immunol. & Cell Biol.* 73:457–462.

83. Marquart, H.V., A. Svendsen, J.M. Rasmussen, C.H. Nielsen, P. Junkder, S.E. Svehag, and R.G. Leslie. 1995. Complement receptor expression and activation of the complement cascade on B lymphocytes from patients with systemic lupus erythematosus. *Clin. & Exp. Immunol.* 101:60–65.

84. Theofilopoulous, A.N., and D.H. Kono. 1999. The genes of systemic autoimmunity. *Proc. Assoc. Am. Physicians* 111:228–240.

85. Mysler, E., B. Paolo, J. Drappa, P. Ramos, S.M. Friedman, P.H. Krammer, and K.B. Elkon. 1994. The apoptosis-1/Fas protein in human systemic lupus erythematosus. *J. Clin. Invest.* 93:1029–1034.

86. McNally, J., D.-H. Yoo, J. Drappa, J.-L. Chu, H. Yagita, S.M. Friedman, and K.B. Elkon. 1997. Fas ligand expression and function in systemic lupus erythematosus. *J. Immunol.* 159:4628–4636.

87. Wu, J., J. Wilson, J. He, L. Xiang, P.H. Schur, and J.D. Mountz. 1996. Fas ligand mutation in a patient with systemic lupus erythematosus and lymphoproliferative disease. *J. Clin. Invest.* 98:1107–1113.

88. Cutolo, M., A. Sulli, B. Villaggio, B. Seriolo, and S. Accardo. 1998. Relations between steroid hormones and cytokines in rheumatoid arthritis and systemic lupus erythematosus. *Ann. Rheum. Dis.* 57:573–577.

89. Kammer, G.M., and G.C. Tsokos. 1998. Emerging concepts of the moleculuar basis for estrogen effects on T lymphocytes in systemic lupus erythematosus. *Clin. Immunol. & Immunopathol.* 89:192–195.

90. Kanda, N., T. Tsuchida, and K. Tamaki. 1999. Estrogen enhancement of anti- double-stranded DNA antibody and immunoglobulin G production in peripheral blood mononuclear cells from patients with systemic lupus erythematosus. *Arthrit. & Rheum.* 42:328–387.

91. Rider, V., R.T. Foster, M. Evans, R. Suenaga, and N.I. Abdou. 1998. Gender differences in autoimmune diseases: estrogen increases calcineurin expression in systemic lupus erythematosus. *Clin. Immunol. & Immunopathol.* 89:171–180.

92. Wilson, K.B., M. Evans, and N.I. Abdou. 1996. Presence of a variant form of the estrogen receptor in peripheral blood mononuclear cells from normal individuals and lupus patients. *J. Reproduct. Biol.* 31:199–208.

93. Huang, Q., A. Parfitt, D.M. Grennan, and N. Manolios. 1997. X-chromosome inactivation in monozygotic twins with systemic lupus erythematosus. *Autoimm.* 26:85–93.

94. MacLennan I.C., A. Gulbranson-Judge, K.M. Toellner, M. Casamayor-Palleja, E. Chan, D.M. Sze, S.A. Luther, and H.A. Orbea. 1997. The changing preference of T and B cells for partners as T-dependent antibody responses develop. *Immunol. Rev.* 156:53–66.

95. Liu, Y.J., and C. Arpin. 1997. Germinal center development. *Immunol. Rev.* 156:111–126.

96. Choi, Y.S. 1997. Differentiation and apoptosis of human germinal center B- lymphocytes. *Immunol. Res.* 16:161–174.

97. Osmond, D.G. 1986. Population dynamics of bone marrow B lymphocytes. *Immunol. Rev.* 93:103–124.

98. Allman, D.M., S.E. Ferguson, V.M. Lentz, and M.P. Cancro. 1993. Peripheral B cell maturation II. Heat stable antigenhi[hi] splenic B cells are an immature developmental intermediate in the production of long lived marrow derived B cells. *J. Immunol.* 151:4431–4444.

99. Melchers, F.A., U. Rolink, U. Grawunder, T.H. Winkler, H. Karasuyama, P, Ghia, and J., Andersson. 1995. Positive and negative selection events during B lymphopoiesis. *Curr. Opin. Immunol.* 7:214–227.

100. Kabat, E.A., T.T. Wu, H.M. Perry, K.S. Gottesmann, and C. Foeller. 1991. *In* Sequences of proteins of immunological interest. NIH press, Bethesda, MD. 91–3242.

101. Pospisil, R, G.O. Young-Cooper, and R.G. Mage. 1995. Preferential expansion and survival of B lymphocytes based on VH framework 1 and framework 3 expression: "positive" selection in appendix of normal and VH-mutant rabbits. *Proc. Natl. Acad. Sci.* 92:6961–6965.

102. Tonegawa, S. 1983. Somatic generation of antibody diversity. *Nature* 302:575–581.

103. Komori, T., L. Pricop, A. Hatakeyama, C.A. Bona, and F.W. Alt. 1996. Repertoires of antigen receptors in Tdt congenitally deficient mice. *Int. Rev. Immunol.* 13:317–325.

104. C. Conde, S. Weller, S. Gilfillan, L. Marcellin, T. Martin, and J.-L. Pasquali. 1998. Terminal deoxynucleotidyl transferase deficiency reduces the incidence of autoimmune nephritis in (New Zealand Black X New Zealand White)F1[1] mice. *J. Immunol.* 161:7023–7030.

105. Rolink, A., and F. Melchers. 1993. Molecular and cellular origins of B lymphocyte diversity. *Curr. Opin. Immunol.* 5:207–217.

106. Brezinschek, H.P., S.J. Foster, T. Dorner, R.I. Brezinschek, and P.E. Lipsky. 1998. Pairing of variable heavy and variable kappa chains in individual naöve and memory B cells. *J. Immunol.* 160:4762–4767.

107. Rajewsky, K. 1998. Burnet's unhappy hybrid. *Nature* 394:624–625.

108. Pillai, S. 1999. The chosen few? Positive selection and the generation of naïve B lymphocytes. *Immunity* 10:493–502.

109. Townsend, S.E., B.C. Weintraub, and C.C. Goodnow. 1999. Growing up on the streets: why B-cell development differs from T-cell development. *Immunol. Today* 5:217–220.

110. Kipps, T.J., and J.H. Vaughan. 1987. Genetic influences on the levels of circulating CD5 B lymphoctyes. *J. Immunol.* 139:1060–1064.

111. Chen, Z.J., C.J. Wheeler, W. Shi, A.J. Wu, C.H. Yarboro, M. Gallagher, and A. L Notkins. 1998. Polyreactive antigen-binding b cells are the predominant cell type in the newborn B cell repertoire. *Eur. J. Immunol.* 28:989–994.

112. Chen, X., F. Martin, K.A. Forbush, R.M. Perlmutter, and J.F. Kearney. 1997. Evidence for selection of a population of multi-reactive B cells into the splenic marginal zone. *Int. Immunol.* 9:27–41.

113. Dono, M., V.L. Burgio, V.L., C. Tacchetti, A. Favre, A. Augliera, S. Zupo, G. Taborelli, N. Chiorazzi, C.E. Grossi, and M. Ferrarini. 1996. Subepithelial B cells in the human palatine tonsil. I. Morphologic, cytochemical, and phenotypic characterization. *Eur. J. Immunol.* 26:2035.

114. Vernino, L.A., D.S. Pisetsky, and P.E. Lipsky. 1992. Analysis of the expression of CD5 by human B cells and correlation with functional activity. *Cellular Immunol.* 139:185–197.

115. Casali, P., and A., Notkins. 1989. Probing the human B cell repertoire with EBV: polyreactive antibodies and CD5[+] B cells. *Ann. Rev. Immunol.* 7:513–535.

116. Guigou, V., B. Guibert, D. Moinier, C. Tonelle, L. Boubli, S. Avrameas, M. Fougereau, and F. Fomoux. 1991. Ig reperotire of human polyspecific antibodies and B cell ontogeny. *J. Immunol.* 146:1368–1374.

117. S. Avrameas. 1991. Natural autoantibodies from "horror autotoxicus" to "gnothi seauton". *Immunol. Today* 12:154–159.

118. Brezinschek, H.P., S.J. Foster, R.I. Brezinschek, T. Dorner, R. Domiati-Saad, and P.E. Lipsky. 1997. Analysis of the human VH gene reperotire. Differential effects of selection and somatic hypermutation on human peripheral CD5(+)/IgM+ and CD5(–)/IgM+ B cells. *J. Clin. Invest.* 99:2488–2501.

119. Dorner, T., H.P. Brezinschek, S.J. Foster, R.I. Brezinschek, N.L. Farner, and P.E. Lipsky. 1998. Comparable impact of mutational and selective influences in shaping the expressed repertoire of peripheral IgM+/CD5– and IgM+/CD5+ B cells. *Eur. J. Immunol.* 28:657–668.

120. Loder, F., B. Mutschler, R.J. Ray, C.J. Paige, P. Sideras, R. Torres, M.C. Lamers, and R. Carsetti. 1999. B cell development in the spleen takes place in discrete steps and is determined by the quality of B cell receptor-derived signals. *J. Exp. Med.* 190:75–89.

121. Fong, S., T.A. Gilbertson, R.J. Hueniken, S.K. Singhal, J.H. Vaughan, and D.A. Carson. 1985. IgM rheumatoid factor autoantibody and immunoglobulin-producing precursor cells in the bone marrow of humans. *Cell. Immunol.* 95:157–172.

122. Tron, F., and J.F. Bach. 1989. Molecular and genetic characteristics of pathogenic autoantibodies. *J. Autoimm.* 2:311–320.

123. Schroeder, H.W., G.C. Ippolito, and S. Shiokawa. 1998. Regulation of the antibody reperotire through control of HCDR3 diversity. *Vaccine* 16:1383–1390.

124. Reid, M.M. 1994. Splenectomy, sepsis, immunization and guidelines. *Lancet* 344:970–971.

125. Morel, L, U.H. Rudofsky, J.A. Longmate, J. Schiffenbauer, and E.K. Wakeland. 1994. Polygenic control of susceptibility to murine systemic lupus erythematosus. *Immunity* 1:219–229.

126. Van Es, J.H., F.H.J. Gmelig, W.R.M. van de Akker, H. Aansfoot, R.H.W.M. Derksen, and T. Logtenberg. 1991. Somatic mutations in the variable regions of a human IgG anti-double-stranded DNA autoantibody suggest a role for antigen induction of systemic lupus erythematosus. *J. Exp. Med.* 173:461–470.

127. Stewart, A.K., C. Huang, A.A. Long, B.D. Stoller, and R.S. Schwartz. 1992. VH-gene representation in autoantibodies reflects the normal human B cell repertoire. *Immunol. Rev.* 128:101–122.

128. Kuo, P., C. Kowal, B. Tadmor, and B. Diamond. 1997. Microbial antigens can elicit autoantibody production: a potential pathway to autoimmune disease. *NY Acad. Sci.* 815:230–236.

129. Putterman, C., W. Limpanasithkul, M. Edelman, and B. Diamond. 1996. The double edge sword of the immune response: mutational analysis of a murine anti-pneumococcal anti-DNA antibody. *J. Clin. Invest.* 97:2251–2259.

130. Paul, E., and B. Diamond. 1993. Characterization of two human anti-DNA antibodies bearing the pathogenic idiotype 8.12. *Autoimm.* 16:13–21.

131. Martin, T., S.F. Duffy, D.A. Carson, and T.J. Kipps. 1992. Evidence for somatic selection of natural autoantibodies. *J. Exp. Med.* 175:983–991.

132. Hayakawa, K., A. Masanao, S.A. Shinton, M. Gui, D. Allman, C.L. Stewart, J. Silver, and R.R. Hardy. 1999. Positive selection of natural autoreactive B cells. *Science* 285:113–116.

133. Couthino, A., A. Grandien, J. Faro-Rivas, and T.A. Mota-Santos. 1988. Idiotypes, tailers and networks. *Ann. Inst. Pasteur Immunol.* 139:599–607.

134. Domiati-Saad, R., and P.E. Lipsky. 1997. B cell superantigens: potential modifiers of the normal human B cell repertoire. *Intl. Rev. Immunol.* 14:309–324.

135. Bernal, A., T. Proft, J.D. Fraser, and D.N. Posnett. 1999. Superantigens in human disease. *J. Clin. Immunol.* 19:949–157

136. Stewart, A.K., C. Huang, B. Stollar, and R.S. Schwartz. 1993. High-frequency representation of a single VH gene in the expressed human B cell repertoire. *J. Exp. Med.* 177:409–418.

137. Brezinschek, H.P., R.I. Brezinschek, and P.E. Lipsky. 1995. Analysis of the heavy chain repertoire of human peripheral blood B cells using single-cell polymerase chain reaction. *J. Immunol.* 155:190–202.

138. Rao, S.P., S.C. Huang, and E.C. Milner. 1996. Analysis of the VH3 repertoire among genetically disparate individuals. *Exp. Clin. Immunogenet.* 13:131–138.

139. Silverman, G.J. 1998. B cell superantigens: possible roles in immunodeficiency and autoimmunity. *Semin. Immunol.* 10:43–55.

140. Domiati-Saad, R., and P.E. Lipsky. 1998. Staphylococcal enterotoxin A induces survival of VH3-expressing human B cells by binding to the VH region with low affinity. *J. Immunol.* 161:1257–1266.

141. Willems van Dijk, K., L.A. Milner, E.H. Sasso, and E.C.B. Milner. 1992. Chromosomal organization of the heavy chain variable region gene segments comprising the human fetal antibody repertoire. *Proc. Natl. Acad. Sci. USA* 89:10430–10434.

142. Pascual, V., L. Verkruyse, M.L. Casey, and J.D. Capra. 1993. Analysis of IgH chain gene segment utilization in human fetal liver. Revisiting the "proximal utilization hypothesis". *J. Immunol.* 151:4164–4172.

143. Yancopoulos, G.D., S.V. Desiderio, M. Paskin, J.F. Kearney, D. Baltimore, and F.W. Alt. 1984. Preferential utilization of the most VH proximal VH gene segments in pre-B cell lines. *Nature* 311:727–733.

144. Perlmutter, R.M., J. Kearney, S.P. Chang, and L.E. Hood. 1985. Developmentally controlled expression of immunoglobulin VH genes. *Science* 227:1597–1601.

145. Schroeder, H.W., and J.Y. Wang. 1987. Early restriction of the human antibody repertoire. *Science* 238:791–793.

146. Schroeder, H.W., and J.Y. Wang. 1990. Preferential utilization of conserved immunoglobulin heavy chain variable gene segments during human fetal life. *Proc. Natl. Acad. Sci. USA* 87:6146–6150.

147. Schutte, M.E.M., S.B. Ebeling, K.E. Akkermans, F.H.J. Gmelig-Meyling, and T. Lottenberg. 1991. Antibody specificity and immunoglobulin VH gene utilization of human monoclonal CD5+ B cell lines. *Eur. J. Immunol.* 21:1151–1121.

148. Hansen, A., S. Jahn, A. Lukowsky, G. Grutz, J. Bohn, R. von Baehr, and U. Settmacher. 1994. VH/VL gene expression in polyreactive-antibody producing human hybridomas from the fetal B cell repertoire. *Exp. Clin. Immunogenet.* 11:1–16.

149. Settmacher, U., S. Jahn, P. Siegel, R. von Baehr, and A. Hansen. 1993. An anti lipid A antibody obtained from the fetal B cell repertoire is encoded by the VH6/Vlambda1 genes. *Mol. Immunol.* 30:953–954.

150. Shiokawa, S., F. Mortari, J.O. Lima, C. Nunez, and F.E. Bertrand. 1999. IgM heavy chain complementarity-determining region 3 diversity is constrained by genetic and somatic mechanisms until two months after birth. *J. Immunol.* 162:6060–6070.

151. Chen, P.P., N.J. Olsen, P.M. Yang, R.W. Soto-Gil, T. Olee, K.A. Siminovitch, and D.A. Carson. 1990. From human autoantibodies to the fetal antibody repertoire to B cell malignancy: it's a small world after all. *Intern. Rev. Immunol.* 5:239–251.

152. Rahman, A., D.S. Latchman, and D.A. Isenberg. 1998. Immunoglobulin variable region sequences of human monoclonal anti-DNA antibodies. *Semin. Arthrit. Rheum.* 28:141–154.

153. Sanz, I., P. Casali, J.W. Thomas, A.L. Notkins, and J.D. Capra. 1989. Nucleotide sequences of eight human natural autoantibody VH regions reveal apparent restricted use of VH families. *J. Immunol.* 142:4054–4061.

154. Hansen, A., D. Roggenbuck, T. Porstmann, R. von Baehr, and L. Franke. 1992. Selection and characterization of a human anti-P24 IgM hybridoma. *Immunobiol.* 189:175. (Abst.)

155. Leibiger, H., A. Hansen, G. Schoenherr, M. Seifert, D. Wustner, R. Stigler, and U. Marx. 1995. Glycosylation analysis of a polyreactive human monoclonal IgG antibody derived from a human-mouse hybridoma. *Mol. Immunol.* 32:595–602.

156. Dersimonian, H., R.S. Schwartz, K.J. Barrett, and B.D. Stollar. 1987. Relationship of human variable region heavy chain germline genes to genes encoding anti-DNA autoantibodies. *J. Immunol.* 139:2496–2501.

157. Cairns, E.P., C. Kwong, V. Misener, P. Ip, D.A. Bell, and L.A. Siminovitch. 1989. Analysis of variable region genes encoding a human anti-DNA antibody of normal origin. *J. Immunol.* 143:685–691.

158. Haindranath, N., I.S. Goldfarb, H. Ikematsu, S.E. Burastero, R.L. Wilder, A.L. Notkins, and P. Casali. 1991. Complete sequence of the genes encoding the VH and VL regions of low- and high-affinity monoclonal IgM and IgA1 rheumatoid factors produced by CD5+ B cells from a rheumatoid arthritis patient. *Int. Immunol.* 3:865–875.

159. Pascual, V., and J.D. Capra. 1992. VH4–21, a human VH gene segment overrepresented in the autoimmune repertoire. *Arthr. Rheum.* 35:11–18.

160. Logtenberg, T., M.E.M. Schutte, S.B. Ebeling, F.H.J. Gmelig-Meyling, and J.H. van Es. 1992. Molecular approaches to the study of human B cell and (auto)antibody repertoire generation and selection. *Immunol. Rev.* 128:23–46.

161. Pascual, V., K. Victor, M. Spellenberg, T.J. Hamblin, F.K Stevenson, and J.D. Capra. 1992. VH restriction among human cold agglutinins. The VH4–21 gene segment is required to encode anti-I and anti-I specificities. *J. Immunol.* 149:2337–2344.

162. Harada, T., N. Suzuki, Y. Mizushima, and T. Sakane. 1994. Usage of a novel class of germline immunoglobulin variable region genes for cationic anti-DNA autoantibodies in human lupus nephritis and its role for the development of the disease. *J. Immunol.* 153:4806–4815.

163. Pascual, V., I. Randen, K. Thompson, S. Mouldy, A. Forre, J. Natvig, and J.D. Capra. 1990. The complete nucleotide sequences of the heavy chain variable regions of six monospecific rheumatoid factors derived from Epstein-Barr-Virus-transformed B cells isolated from the synovial tissue of patients with rheumatoid arthritis. *J. Clin. Invest.* 86:1320–1328.

164. Huang, C., and B.D. Stollar. 1993. A majority of IgH chain cDNA of normal human adult blood lymphocytes resembles cDNA for fetal Ig and natural autoantibodies. *J. Immunol.* 151:5290–5300.

165. Huang, C., A.K. Stewart, R.S. Schwartz, and B.D. Stollar. 1992. Immunoglobulin heavy chain gene expression in peripheral blood B lymphocytes. *J. Clin. Invest.* 89:1331–1343.

166. Farner, N.L., T. Dorner, and P.E. Lipsky. 1999. Molecular mechanisms and selection influence the generation of the human V lambda J lambda repertoire. *J. Immunol.* 162:2137–2145.

167. Foster, S.J., H.P. Brezinschek, R.I. Brezinschek, and P.E. Lipsky. 1997. Molecular mechanisms and selective influences that shape the kappa gene repertoire of IgM+ B cells. *J. Clin. Invest.* 99:1614–1627.

168. Kipps, T.J., B.A. Robbins, and D.A. Carson. 1990. Uniform high frequency of expression of autoantibody-associated crossreactive idiotypes in the primary B cell follicles of human fetal spleen. *J. Exp. Med.* 171:189–196.

169. Fais, F., F. Ghiotto, S. Hashiomoto, B. Sellars, A. Valetto, S.L. Allen, P. Schulman, V.P. Vinciguerra, K. Rai, L.Z. Rassenti, T.J. Kipps, G. Dighiero, H.W. Schroeder, M. Ferrarini, and N. Chiorazzi. 1998. Chronic lymphocytic leukemia B cells express restricted sets of mutated and unmutated antigen receptors. *J. Clin. Invest.* 102:1515–1525.

170. Bahler, D.W., J.A. Miklos, and S.H. Swerdlow. 1997. Ongoing Ig gene hypermutation in salivary gland mucosa-associated lymphoid tissue-type lymphomas. *Blood* 89:3335–3344.

171. Bohme, H., M. Seifert, D. Roggenbuck, W. Docke, R. von Baehrn, and A. Hansen. 1994. Characterization of a B-CLL derived IgM-lambda antibody expressing typical features of a NPAB. *Immunol. Letts.* 41:261–266.

172. Schroeder, H.W., and G. Dighiero. 1994. The pathogenesis of chronic lymphocytic leukemia: analysis of the antibody repertoire. *Immunol. Today* 15:288–294.

173. Brezinschek, H.P., R.I. Brezinschek, T. Dorner, and P.E. Lipsky. 1998. Similar characteristics of the CDR3 of V(H) 1–69/DP-10 rearrangements in normal human peirphral blood and chronic lymphocytic leukaemia B cells. *Br. J. Haematol.* 102:516–521.

174. Kocks, C., and K. Rajewsky. 1989. Stable expression and somatic hypermutation of antibody V regions in B-cell developmental pathways. *Ann. Rev. Immunol.* 7:537–559.

175. Taki, S., M. Meiering, and K. Rajewsky. 1993. Targeted insertion of a variable region gene into the immunoglobulin heavy chain locus. *Science* 262:1268–1271.

176. Chen, X., F. Martin, K.A. Forbush, R.M. Perlmutter, and J.F. Kearney. 1997. Evidence for selection of a population of multi-reactive B cells into the splenic marginal zone. *Int. Immunol.* 9:27–41.

177. Fang, W., B.C. Weintraub, B. Dunlap, P. Garside, K.A. Pape, M.K. Jenkins, C.C. Goodnow, D.L. Mueller, and T.W. Behrens. 1998. Self-reactive B lymphocytes overexpressing Bcl-xL escape negative selection and are tolerized by clonal anergy and receptor editing. *Immunity* 9:35–45.

178. Jahn, S., J. Schwab, A. Hansen, H. Heider, C. Schroeder, A. Lukowsky, M. Achtman, H. Matthes, S.T. Kiessig, H.D. Volk, D.H. Kruger, and R. von Baehr. 1991. Human hybridomas derived from CD5+ B lymphocytes of patients with chronic lymphocytic leukemia (B-CLL) produce multispecific natural IgM (kappa) antibodies. *Clin. & Exp. Immunol.* 83:413–417.

179. Schlomchik, M., M. Mascelli, H. Shan, M.Z. Radic, D. Pisetsky, A. Marshak-Rothstein, and M. Weigert. 1990. Anti-DNA antibodies from autoimmune mice arise by clonal expansion and somatic mutation. *J. Exp. Med.* 171:265–292.

180. Eliat, D., D.M. Webster, and A.R. Rees. 1988. V region sequences of anti-DNA and anti-RNA autoantibodies from NZB/NZW F1 mice. *J. Immunol.* 141:1745–1753.

181. Roben, P., S.M. Barbas, L. Sandoval, J.M. Lecerf, B.D. Stollar, A. Solomon, and G.J. Silverman. 1996. Repertoire cloning of lupus anti-DNA autoantibodies. *J. Clin. Invest.* 98:2827–2837.

182. Suzuki, N., T. Harada, S. Mihara, and T. Sakane. 1996. Characterization of a germline Vk gene encoding cationic anti-DNA antibody and role of receptor editing for development of the autoantibody in patients with systemic lupus erythematosus. *J. Clin. Invest.* 98:1843–1850.

183. Vernino, L., L.M. McAnally, J. Ramberg, and P.E. Lipsky. 1992. Generation of nondividing high rate Ig-secreting plasma cells in cultures of human B cells stimulated with anti-CD3-activated T cells. *J. Immunol.* 148:404–410.

184. Zack, D.J., M. Stempniak, A.L. Wong, C. Taylor, and R.H. Weisbart. 1996. Mechanisms of cellular penetration and nuclear localization of an anti-double strand DNA autoantibody. *J. Immunol.* 157:2082–2088.

185. Livneh, A., A. Halpern, D. Perkins, A. Lazo, R. Halpern, and B. Diamond. 1987. A monoclonal antibody to a cross-reactive idiotype on cationic human anti-DNA antibodies expressing lamda light chains: A new reagent to identify a potentially differential pathogenic subset. *J. Immunol.* 139:123–127.

186. Paul, E., A.A. Iliev, A. Livneh, and B. Diamond. 1992. The anti-DNA-associated idiotype 8.12 is encoded by the V lambda II gene family and maps to the vicinity of L chain CDR1. *J. Immunol.* 149:3588–3595.

187. Hirabayashi, Y., Y. Munakata, T. Sasaki, and H. Sano. 1992. Variable regions of a human anti-DNA antibody 0–81 possessing lupus nephritis-associated idiotype. *Nucleic Acids Res.* 20:2601.

188. Manheimer-Lory A.J., A. Davidson, D. Watkins, N.R. Hannigan, and B.A. Diamond. 1991. Generation and analysis of clonal IgM- and IgG-producing human B cell lines expressing an anti-DNA-associated idiotype. *J. Clin. Invest.* 87:1519–1525.

189. Stevenson, F.K., C. Longhurst, C.J. Chapman, M. Ehrenstein, M.B. Spellerberg, T.J. Hamblin, C.T. Ravirajan, D. Latchman, and D. Isenberg. 1993. *J. Autoimm.* 6:809–825.

190. Buskila, D., and Y. Shoenfeld. 1992. Manipulation of anti-DNA idiotypes: a possible treatment approach to autoimmune diseases. *Concepts in Immunopathol.* 8:114–128.

191. Radic, M.Z., and M. Weigert. 1994. Genetic and structural evidence for antigen selection of anti-DNA antibodies. *Ann. Rev. Immunol.* 12:487–520.

192. Radic, M.Z., and M. Weigert. 1995. Origins of anti-DNA antibodies and their implications for B cell tolerance. *N.Y. Acad. Sci.* 764L384–396.

193. Dorner, T., S.J. Foster, H.P. Brezinschek, and P.E. Lipsky. 1998. Analysis of the targeting of the hypermutations machinery and the impact of subsequent selection on the distribution of nucleotide changes in human VHDJH rearrangements. *Immunol. Rev.* 162:161–171.

194. Dorner, T., N.L. Farner, and P.E. Lipsky. 1999. Ig lambda and heavy chain gene usage in early untreated systemic lupus erythematosus suggest intensive B cell stimulation. *J. Immunol.* 163:1027–1036.

195. Puttemann, C., and B. Diamond. 1998. Immunization with a peptide surrogate for double-stranded DNA (dsDNA) induces autoantibody production and renal immunoglobulin deposition. *J. Exp. Med.* 188:29–38.

196. Mockridge, C.I., C.J. Chapman, M.B. Spellerberg, D.A. Isenberg, and F.K. Stevenson. 1996. Use of phage surface expression to analyze regions of human V4–34 (VH4-21)-encoded IgG autoantibody required for recognition of DNA: no involvement of the 9G4 idiotype. *J. Immunol.* 157.2449–2454.

197. Dorner, T., H.P. Brezinschek, R.I. Brezinschek, S.J. Foster, R. Domiati-Saad, and P.E. Lipsky. 1997. Analysis of the frequency and pattern of somatic mutations within nonproductively rearranged human variable heavy chain genes. *J. Immunol.* 158:2779–2789.

198. Dorner, T., S.J. Foster, N.L. Farner, and P.E. Lipsky. 1998. Immunoglobulin kappa chain receptor editing in systemic lupus erythematosus. *J. Clin. Invest.* 102:688–694.

199. Huang, S.C., R. Jiang, A.M. Glas, and E.C. Milner. 1996. Non-stochastic utilization of Ig V region genes in unselected human peripheral B cells. *Molec. Immunol.* 33:553–560.

200. Ash-Lerner, A., M. Ginsberg-Strauss, Y. Pewzner-Jung, D.D. Desai, T.N. Marion, and D. Eliat. 1997. Expression of an anti-DNA-associated VH gene in immunized and autoimmune mice. *J. Immunol.* 159:1508–1519.

201. Portanova, J.P., G. Creadon, X. Zhang, D.S. Smith, B.L. Kotzin, and L.J. Wysocki. 1995. An early post-mutational selection event directs expansion of autoreactive B cells in murine lupus. *Molec. Immunol.* 32:117–135.

202. Wloch, M.K., A.L. Alexander, A.M. Pippen, D.S. Pisetsky, and G.S. Gilkeson. 1997. Molecular properties of anti-DNA induced in preautoimmune NZB/W mice by immunization with bacterial DNA. *J. Immunol.* 158:4500–4506.

203. Grammer, A.C., R.D. McFarland, J. Heaney, B.F. Darnell, and P.E. Lipsky. 1999. Expression, regulation and function of B cell expressed CD154 in germinal centers. *J. Immunol.* 163:4150–4159.

204. Grammer, A.C., R.D. McFarland, J. Heaney, B.F. Darnell, E. Lightfoot, L. Picker, and P.E. Lipsky. 1999. Regulation and expression of CD154/CD40-ligand on B cells in the normal and autoimmune humoral response. *Scand. J. Immunol.* 50:91 (D13). (Abstr.)

205. M.C. Nussenzweig. 1998. Immune receptor editing: revise and select. *Cell* 95:875–878.

206. Ehrenstein, M.R., B. Hartley, L.S. Wilkinson, and D.A. Isenberg. 1994. Comparison of a monoclonal and polyclonal anti-idiotype against a human IgG anti-DNA antibody. *J. Autoimm.* 7:349–367.

207. Menon, S., M.A. Rahman, C.T. Ravirajan, D. Kandiah, C.M. Longhurst, T. McNally, W.M. Williams, D.S. Latchman, and D.A. Isenberg. 1997. The production, binding characteristics and sequence analysis of four human IgG monoclonal antiphospholipid antibodies. *J. Autoimm.* 10:43–57.

208. Suzuki, H., H. Takemura, M. Suzuki, Y. Sekine, and H. Kashiwagi. 1997. Molecular cloning of anti-SS-A/Ro 60-kDa peptide Fab fragments from infiltrating salivary gland lymphocytes of a patient with Sjogren's syndrome. *Biochem. Biophys. Res. Comm.* 232:101–106.

209. Torres, R.M., H. Flaswinkel, M. Reth, and K. Rajewsky. 1996. Aberrant B cell development and immune response in mice with a comprimised BCR complex. *Science* 272:1804–1808.

210. Lam, K.P., R. Kuhn, and K. Rajewsky. 1997. In vivo ablation of surface immunoglobulin on mature B cells by inducible gene targeting results in rapid cell death. *Cell* 90:1073–1083.

211. Batista, F.D., and M.S. Neuberger. 1998. Affinity dependence of the B cell response to antigen: a threshold, a ceiling, and the importance of off-rate. *Immunity* 8:751–759.

212. Santulli-Marotto, S., M.W. Retter, R. Gee, M.J. Mamula, and S.H. Clarke. 1998. Autoreactive B cell regulation: peripheral induction of developmental arrest by lupus-associated autoantigens. *Immunity* 8:209–219.

213. Mankik-Nayak, L., A. Bui, H. Noorchashm, A. Eaton, and J. Erikson. 1997. Regulation of anti-double-stranded DNA B cells in nonautoimmune mice: localization to the T-B interface of the splenic follicle. *J. Exp. Med.* 8:1257–1267.

214. Stamatis-Nick, C.L., P.P. Sfikakis, and G.C. Tsokos. 1998. Immune cell signaling aberrations in human lupus. *Immunol. Res.* 18:27–39.

215. Spencer, J., M.E. Perry, and D.K. Dunn-Walters. 1998. Human marginal-zone b cells. *Immunol. Today* 19:421–426.

216. Park, S.-H., Y.-H. Chiu, J. Jayawardena, J. Roark, U. Kavita, and A. Bendelac. 1998. Innate and adaptive functions of the CD1 pathway of antigen presentation. *Sem. Immunol.* 10:391–398.

217. Zheng, D., M. Dick, L. Cheng, M. Amano, S. Dejbakhsh-Jones, P. Huie, R. Sibley, and S. Strober. 1998. Subsets of transgenic T cells that recognize CD1 induce or prevent murine lupus: role of cytokines. *J. Exp. Med.* 187:525–536.

218. Jacobson, B.A., T.L. Rothstein, and A. Marshak-Rothstein. 1997. Unique site of IgG2a and rheumatoid factor production in MRL/*lpr* mice. *Immunol. Rev.* 156:103–110.

219. Cyster, J.G., S.B. Hartley, and C.C. Goodnow. 1994. Competition for follicular niches excludes self-reactive cells from the recirculating B-cell repertoire. *Nature* 371:389–395.

220. Cyster, J.G., and C.C. Goodnow. 1995. Antigen-induced exclusion from follicles and anergy are separate and com-

plementary processes that influence peripheral B cell fate. *Immunity* 3:691–701.

221. Fulcher, D.A., A.B. Lyons, S.L. Korn, M.C. Cook, C. Koleda, C. Parish, B. Fazekas de St. Groth, and A. Basten. 1996. The fate of self-reactive B cells depends primarily on the degree of antigen receptor engagement and availability of T cell help. *J. Exp. Med.* 183:1953–1956.

222. Reidel, C., T. Owens, and G.J. Nossal. 1988. A significant proportion of normal resting B cells are induced to secrete immunoglobulin through contact with anti-receptor antibody-activated helper T cells in clonal cultures. *Eur. J. Immunol.* 18:403–408.

223. Hirohata, S., D. Jelinek, and P.E. Lipsky. 1988. T cell-dependent activation of B cell proliferation and differentiation by immobilized monoclonal antibodies to CD3. *J. Immunol.* 140:3736–3744.

224. Gastman, B.R., D.E. Johnson, T.L. Whiteside, and H. Rabinowich. 1999. Caspase-mediated degradation of T-cell receptor zeta-chain. *Cancer Res.* 59:1422–1427.

225. Splawski, J., J. Nishioka, Y. Nishioka, and P.E. Lipsky. 1996. CD40 ligand is expressed and functional on activated neonatal T cells. *J. Immunol.* 156:119–127.

226. Hodge, M.R., A.M. Ranger, F.C. de la Brousse, T. Hoey, M.J. Grusby, and L.H. Glimcher. 1996. Hyperproliferation and dysregulation of IL4 expression in NF-ATp-deficient mice. *Immunity* 4:397–405.

227. Lobo, F.M., R. Zanjani, N. Ho, T.A. Chatila, and R.L. Fuleihan. 1999. Calcium-dependent activation of TNF family gene expression by Ca^{+2}/calmodulin type IV/Gr and calcineurin. *J. Immunol.* 162:2057–2063.

228. Schubert, L.A., G. King, R.Q. Cron, D.B. Lewis, A. Aruffo, and D. Hollenbaugh. 1995. The human gp^{39} promoter: two distinct nuclear factors of activated T cell protein-binding elements contribute indepedently to transcriptional activation. *J. Biol. Chem.* 270:29624–29627.

229. Lederman, S., M.J. Yellin, G. Inghirami, J.J. Lee, D.M. Knowles, and L. Chess. 1992. Molecular interactions mediating T-B lymphocyte collaboration in human lymphoid follicles. Roles of T cell-B cell-activating molecule (5c8 antigen) and CD40 in contact-dependent help. *J. Immunol.* 149:3817–3826.

230. Vyth-Dreese, F.A., T.A. Dellemijn, D. Majoor, and D. de Jong. 1996. Localization in situ of the co-stimulatory molecules B7.1, B7.2, CD40 and their ligands in normal human lymphoid tissue. *Eur. J. Immunol.* 25:3023–3029.

231. Hollander, G.A., E. Castigli, R. Kulbacki, M. Su, S.J. Burakoff, J.C. Gutierrez-Ramos, and R.S. Geha. 1996. Induction of alloantigen-specific tolerance by B cells from CD40-deficient mice. *Proc. Natl. Acad. Sci. USA* 93:4994–4998.

232. M. de Boer, A. Kasran, J. Kwekkeboom, H. Walter, P. Vandenberghe, and J.L. Ceuppens. 1993. Ligation of B7 with CD28/CTLA-4 on T cells results in CD40 ligand expression, interleukin-4 secretion and efficient help for antibody production by B cells. *Eur. J. Immunol.* 23:3120–3125.

233. Linsley, P.S., P.M. Wallace, J. Johnson, M.G. Gibson, J.L. Green, J.A. Ledbetter, C. Singh, and M.A. Tepper. 1992. Immunosuppresion in vivo by a soluble form of the CTLA-4 activation molecule. *Science* 257:792–795.

234. Finck, B.K., P.S. Linsley, and D. Wofsy. 1994. Treatment of murine lupus with CTLA4.Ig. *Science* 265:1225–1227.

235. Daikh, D.I., B.K. Finck, P.S. Linsley, D. Hollenbaugh, and D. Wofsy. 1997. Long-term inhibition of murine lupus by brief simultaneous blockade of the B7/CD28

and CD40/gp^{39} costimulation pathways. *J. Immunol.* 159:3104–3108, 1997.

236. Takiguchi, M., M. Murakami, T. Nakagawa, A. Yamada, S. Chikuma, Y. Kawaguchi, A. Hashimoto, and T. Uede. 1999. Blockade of CD28/CTLA4–B7 pathway prevented autoantibody-related diseases but not lung disease in MRL/lpr mice. *Lab. Invest.* 79:317–326.

237. Sfikakis, P.P., R. Oglesby, P. Sfikakis, and G.C. Tsokos. 1994. B7/BB1 provides an important costimulatory signal for CD3-mediated T lymphocyte proliferation in patients with systemic lupus erythematosus (SLE). *Clin. Exp. Immunol.* 96:8–14.

238. Tsokos, G.C., B. Kovacs, P.P. Sfikakis, S. Theocharis, S. Vogelgesang, and C.S. Via. 1996. Defective antigen-presenting cell function in patients with systemic lupus erythematosus. *Arthrit. & Rheum.* 39:600–609.

239. Abe, K., T. Takasaki, C. Ushiyama, J. Asakawa, T. Fukazawa, M. Seki, M. Hirashima, M. Ogaki, and H. Hashimoto. 1999. Expression of CD80 and CD86 on peripheral blood T lymphocytes in patients with systemic lupus erythematosus. *J. Clin. Immunol.* 19:58–66.

240. Lafagek-Pochitaloff, M., R. Costello, D. Couez, J. Simonetti, P. Mannoni, C. Mawas, and D. Olive. 1990. Human CD28 and CTLA4-Ig superfamily genes are located on chromosome 2 at bands q33–q34. *Immunogenet.* 31:198–201.

241. Mehrian, R., F.P. Quismorio, G. Strassmann, M.M. Stimmler, S.A. Horwitz, R.C. Kitridou, W.J. Gauderman, J. Morrison, C. Brautbar, and C.O. Jacob. 1998. Synergistic effect between IL-10 and bcl-2 genotypes in determining susceptibility to systemic lupus erythematosus. *Arthrit. & Rheum.* 41:596–602.

242. Ochs, H.D., S. Nonoyama, M.L. Farrington, S.H. Fischer, and A. Aruffo. 1993. The role of adhesion molecules in the regulation of antibody responses. *Sem. Hematol.* 30:72–79.

243. Facchetti, F., C. Appiani, L. Salvi, J. Levy, and L.D. Notarangelo. 1995. Immunohistologic analysis of ineffective CD40-CD40 ligand interaction in lymphoid tissues from patients with X-linked immunodeficiency with hyper-IgM. *J. Immunol.* 154:6624–6633.

244. C. Mohan, Y. Shi, J.D. Laman, and S.K. Datta. 1995. Interaction between CD40 and its ligand gp39 in the development of murine lupus nephritis. *J. Immunol.* 154:1470–1480.

245. Early, G.S., W. Zhao, and C.M. Burns. 1996. Anti-CD40 ligand treatment prevents the development of lupus-like nephritis in a subset of New Zealand black X New Zealand white mice. Response correlates with the absence of an anti-antibody response. *J. Immunol.* 157:3159–3164.

246. Peng, S.L., J.M. McNiff, M.P. Madaio, J. Ma, M.J. Owen, R.A. Flavell, A.C. Hayday, and J. Craft. 1997. αβ T cell regulation and CD40 ligand dependence in murine systemic autoimmunity. *J. Immunol.* 158:2464–2470.

247. Kalled, S.L., A.H. Cutler, S.K. Datta, and D.W. Thomas. 1998. Anti-CD40 ligand antibody treatment of SNF1 mice with established nephritis: preservation of kidney function. *J. Immunol.* 160:2158–2165.

248. Desai-Mehta A., L. Lu, R. Ramsey-Goldman, and S.K. Datta. 1996. Hyperexpression of CD40 ligand by B and T cells in human lupus and its role in pathogenic autoantibody production. *J. Clin. Invest.* 97:2063–2073.

249. Koshy, M., D. Berger, and M.K. Crow. 1996. Increased expression of CD40 ligand on systemic lupus erythematosus lymphocytes. *J. Clin. Invest.* 98:826–837.

250. Devi, B.S., S. van Noordin, T. Krausz, and K.A. Davies. 1998. Peripheral blood lymphocytes in SLE – hyperexpression of CD154 on T and B lymphocytes and increased number of double negative T cells. *J. Autoimmunity* 11:471–475.

251. Ramesh, N., V. Ramesh, J.F. Gusella, and R. Geha. 1993. Chromosomal localization of the gene for human B-cell antigen CD40. *Somat. Cell and Molec. Genet.* 19:295–298.

252. Lafage-Pochitaloff, M.P. Hermann, F. Birg, J.-P. Galizzi, J. Simonetti, P. Mannoni, and J. Banchereau. 1993. The human CD40 gene maps to chromosome 20 q12–q13.2. *Leukemia* 8:1172–1175.

253. Nakano, H., M. Shindo, K. Yamada, M.C. Yoshida, S.M. Santee, C.F. Ware, N.A. Jenkins, D.J. Gilbert, H. Yagita, N.C. Copeland, and K. Okumura. 1997. Human TNF receptor-associated factor 5 (TRAF5): cDNA cloning, expression and assignment of the TRAF5 gene to chromosome 1q32. *Genomics* 42:26–32.

254. Vakkalanka, R.K., C. Woo, K.A. Kirou, M. Koshy, D. Berger, and M.K. Crow. 1999. Elevated levels and functional capacity of soluble CD40 ligand in systemic lupus erythematosus. *Arthrit. & Rheum.* 42:871–881.

255. Van Lopik, T., M. Bijl, M. Hart, L. Boeije, T. Gesner, A.A. Creasy, C.G.M. Kallenberg, L.A. Aarden, and R.J.T. Smeenk. 1999. Patients with systemic lupus erythematosus with high plasma levels of sFas risk relapse. *J. Rheum.* 26:60–67.

256. Font, J. L. Pallares, J. Martorell, F. Martinez, A. Gaya, J. Vives, and M. Ingelmo. 1996. Elevated soluble CD27 in serum of patients with systemic lupus erythematosus. *Clin. Immunol. & Immunopathol.* 81:239–243.

257. Swaak, A.J., R.Q. Hintzen, V. Huysen, H.G. van den Brink, and J.T. Smeenk. 1995. Serum levels of soluble forms of T cell activation antigens CD27 and CD25 in systemic lupus erythematosus in relation with lymphocytes count and disease course. 1995. *Clin. Rheum.* 14:293–300.

258. Aderka, D., A. Wysenbeek, H. Engelmann, A.P. Cope, F. Brennan, Y. Molad, V. Hornik, Y. Levo, R.N. Maini, M. Feldmann, and D. Wallach. 1993. Correlation between serum levels of soluble tumor necrosis factor receptor and disease activity in systemic lupus erythematosus. *Arthrit. & Rheum.* 36:1111–1120.

259. Yamamoto S., Y, Higuchi, K. Yoshiyama, E. Shimizu, M. Kataoka, N. Hijiya, and K. Matsuura. 1999. ADAM family proteins in the immune system. *Immunol. Today* 20:278–84.

260. Zucker, S., N. Mian, M. Drews, C. Conner, A. Davidson, F. Miller, P. Birembaut, B. Nawrocki, A.J.P. Docherty, R.A. Greenwald, R. Grimson, and P. Barland. 1999. Increased serum stromelysin-1 levels in systemic lupus erythematosus: lack of correlation with disease activity. *J. Rheum.* 26:78–80.

261. Gauchat, J. -F., J. -P. Aubry, G. Mazzei, P. Life, T. Jomotte, G. Elson, and J. -Y. Bonnefoy. 1993. Human CD40-ligand: molecular cloning, cellular distribution and regulation of expression by factors controlling IgE production. *FEBS Lett.* 315:259–266.

262. Roy, M., T. Waldschmidt, A. Aruffo, J.A. Ledbetter, and R.J. Noelle. 1993. The regulation of the expression of gp39, the CD40 ligand, on normal and cloned CD4+ T cells. *J. Immunol.* 151:2497–2510.

263. Barcellini, W., G.P. Rizzardi, M.O. Borghi, F. Nicoletti, C. Fain, N. del Papa, and P.L. Meroni. 1996. *In vitro* type-1 and type-2 cytokine production in systemic lupus erythematosus: lack of relationship with clinical disease activity. *Lupus* 5:139–145.

264. Hagiwara, E., M.E. Gourley, S. Lee, and D.S. Klinman. 1996. Disease severity in patients with systemic lupus erythematosus correlates with an increased ratio of interleukin-10:interferon-—secreting cells in the peripheral blood. *Arthrit. & Rheum.* 39:379–385.

265. Viallard, J.F., J.L. Pellegrin, V. Ranchin, T. Schaverbeke, J. Dehais, M. Longy-Boursier, J.M. Ragnaud, B. Leng, and J.F. Moreau. 1999. Th1 (IL-2, interferon-gamma (IFN-γ) and Th2 (IL-10, IL4) cytokine production by peripheral blood mononuclear cells (PBMC) from patients with systemic lupus erythematosus (SLE). *Clin. Exp. Immunol.* 115:189–195.

266. Ohtsuka, K., J.D. Gray, M.M. Stimmler, B. Toro, and D.A. Horowitz. 1998. Decreased production of TGF-β by lymphoctyes from patients with systemic lupus erythematosus. *J. Immunol.* 160:2539–2545.

267. Balomenos, D., R. Rumold, and A.N. Theofilopoulos. 1998. Interferon-γ is required for lupus-like disease and lymphoaccumulation in MRL-*lpr* mice. *J. Clin. Invest.* 101:364–371.

268. Dang, H., A.G. Geiser, J.J. Letterio, T. Nakabayashi, L. Kong, G. Fernandez, and N. Talal. 1995. SLE-like autoantibodies and Sjogren's syndrome-like lymphoproliferation in TGF-β knockout mice. *J. Immunol.* 155:3205–3212.

269. Yaswen, L., A.B. Kulkarni, T. Fredrickson, B. Mittleman, R. Schiffman, S. Payne, C. Longenecker, E. Mozes, and S. Karlsson. 1996. Autoimmune manifestations in the transforming factor-beta 1 knockout mouse. *Blood* 87:1439–1445.

270. Apel, M , and C. Berek. 1990. Somatic mutations in antibodies expressed by germinal center B cells early after primary immunization. *Intl. Immunol.* 2:813–819.

271. Jacob, J., and G. Kelsoe. 1992. In situ studies of the primary immune response to (4-hydroxy-3-nitrophenyl)acetyl. II. A common clonal origin for periarteriolar lymphoid sheath-associated foci and germinal centers. *J. Exp. Med.* 176:679–687.

272. Pascual, V., Y.-J. Liu, A. Magalski, O. de Bouteiller, J. Banchereau, and J.D. Capra. 1994. Analysis of somatic mutation in five B cell subsets of human tonsil. *J. Exp. Med.* 180:329–339.

273. Toellner, K.-M., A. Gulbranson-Judge, D.R. Taylor, D.M.-Y. Sze, and I.C.M. MacLennan. 1996. Immunoglobulin switch transcript production *in vivo* related to the site and time of antigen-specific B cell activation. *J. Exp. Med.* 183:2303–2312.

274. Hodgkin, P.D., J.-H. Lee, and A.B. Lyons. 1996. B cell differentiation and isotype switching is related to division cycle number. *J. Exp. Med.* 184:277–281.

275. Liu, Y.-J., F. Malisan, O. de Bouteliller, C. Guret, S. Lebecque, J. Banchereau, F.C. Mills, E.E. Max, and H. Martinez-Valdez. 1996. Within germinal centers, isotype switching of immunoglobulin genes occurs after the onset of somatic hypermutation. *Immunity* 4:241–250.

276. Clark, J., T. Bourne, M.R. Salaman, M.H. Seifert, and D.A. Isenberg. 1996. B lymphocyte hyperactivity in families of patients with systemic lupus erythematosus. *J. Autoimm.* 9:59–65.

277. Han, S., K. Hathcock, B. Zheng, T. Kepler, R. Hodes, and G. Kelsoe. 1995. Cellular interactions in germinal centers: roles of CD40 ligand and B7-2 in established germinal centers. *J. Immunol.* 155:556–567.

278. Blossom, S., E.B. Chu, W.O. Weigle, and K.M. Gilbert. 1997. CD40 ligand expressed on B cells in the BXSB mouse model of systemic lupus erythematosus. *J. Immunol.* 159:4580–4586.

279. Graf, D., U. Korthauer, H. W. Mages, G. Senger, and R. A. Kroczek. 1992. Cloning of TRAP, a ligand for CD40 on human T cells. *Eur. J. Immunol.* 22:3191–3194.

280. Padayachee, M., C. Feighery, A. Finn, C. McKeown, R.J. Levinsky, C. Kinnin, and S. Malcolm. 1992. Mapping the X-linked form of hyper-IgM syndrome (HIGMX1) to Xq26 by close linkage to HPRT. *Genomics* 14:551–553.

281. Aruffo, A., M. Farrington, D. Hollenbaugh, X. Li, A. Milatovich, S. Nonoyama, J. Bajorath, L. S. Grosmaire, R. Stenkamp, M. Neubauer, R. L. Roberts, R. J. Noelle, J. A. Ledbetter, U. Francke, and H. D. Ochs. 1993. The CD40 ligand, gp39, is defective in activated T cells from patients with X-linked hyper-IgM syndrome. *Cell.* 72:291–300.

282. Allen, R. C., R. J. Armitage, M. E. Conley, H. Rosenblatt, N. A. Jenkins, N. G. Copeland, M. A. Bedell, S. Edelhoff, C. M. Disteche, D. K. Simoneaux, W. C. Fanslow, J. Belmont, and M. K. Spriggs. 1993. CD40 ligand gene defects responsible for X-linked hyper-IgM syndrome. *Science* 259:990–993.

283. Kaartinen, M., G.M. Griffiths, A.F. Markham, and C. Milstein. 1983. mRNA sequences define an unusually restricted IgG response to 2-phenyloxazolone and its early diversification. *Nature* 304:320–324.

284. Cumano, A., and K. Rajewsky. 1985. Structure of primary anti-(4-hydroxy-3- nitrophenyl)acetyl (NP) antibodies in normal and idiotypically suppressed C57BL/6 mice. *Eur. J. Immunol.* 15:512–520.

285. Wysocki, L, T. Manser, and M.L. Gefter. 1986. Somatic evolution of variable region structures during an immune response. *Proc. Natl. Acad. Sci.* 83:1847–1851.

286. Klein, U., R. Kuppers, and K. Rajewsky. 1994. Variable region gene analysis of B cell subsets derived from a 4-year-old child: somatically mutated memory B cells accumulate in the peripheral blood already at young age. *J. Exp. Med.* 180:1383–1393.

287. Ray, S.K., C. Putterman, and B. Diamond. 1996. Pathogenic autoantibodies are routinely generated during the response to foreign antigen: a paradigm for autoimmune disease. *Proc. Natl. Acad. Sci. USA* 93:2019–2024.

288. Manheimer-Lory, A.J., G. Zandman-Goddard, A. Davidson, C. Aranow, and B. Diamond. 1997. Lupus-specific antibodies reveal an altered pattern of somatic mutation. *J. Clin. Invest.* 100:2538–2546.

289. D.S. Smith, G. Creadon, P.K. Jena, J.P. Portanova, B.L. Kotzin, and L.J. Wysocki. 1996. Di- and trinucleotide target preferences of somatic mutagenesis in normal and autoreactive B cells. *J. Immunol.* 156:2642–2652.

290. Diamond, B., J.B. Katz, E. Paul, C. Aranow, D. Lustgarten, and M.D. Scharff. 1992. The role of somatic mutation in the pathogenic anti-DNA response. *Ann. Rev. Immunol.* 10:731–757.

291. Manheimer-Lory, A., J.B. Katz, M. Pillinger, C. Ghossein, A. Smith, and B. Diamond. 1992. Molecular characteristics of antibodies bearing an anti-DNA-associated idiotype. *J. Exp. Med.* 176:309.

292. Shlomchik, M., M. Mascelli, H. Shan, M.Z. Radic, D. Pisetsky, A. Marshak-Rothstein, and M. Weigert. 1990. *J. Exp. Med.* 171:265–292.

293. Chang, B., and P. Casali. 1994. The CDR1 sequences of a major proportion of human germline Ig VH genes are inher-ently susceptible to amino acid replacement. *Immunol. Today* 15:367–373.

294. Dorner, T., H.P. Brezinschek, S.J. Foser, R.I. Breqinschek, N.L. Farner, and P.E. Lipsky. 1998. Delineation of selective influences shaping the mutated expressed human Ig heavy chain repertoire. *J. Immunol.* 160:2831–2841.

295. Foster, S.J., T. Dorner, and P.E. Lipsky. Targeting and subsequent selection of somatic hypermutations in the human Vκ repertoire. *Eur. J. Immunol.*, in press.

296. Herrmann, M., R.E. Voll, O.M. Zoller, M. Hagenhofer, B.B. Ponner, and J.R. Kalden. 1998. Impaired phagocytosis of apoptotic cell material by monocyte-derived macrophages from patients with systemic lupus erythematosus. *Arthrit. & Rheum.* 41:1241–1250.

297. Emlen, W., J. Niebur, and R. Kadera. 1994. Accelerated in vitro apoptosis of lymphocytes from patients with systemic lupus erythematosus. *J. Immunol.* 152:3685–3692.

298. Georgescu, L, R.K. Vakkalanka, E.B. Elkon, and M.K. Crow. 1997. Interleukin-10 promoter activation-induced cell death of SLE lymphocytes mediated by Fas ligand. *J. Clin. Invest.* 100:2622–2633.

299. Lorenz, H.M., M. Grunke, T. Hieronymus, M. Herrmann, A. Kuhnel, B. Manger, and J.R. Kalden. 1997. *Arthrit. & Rheum.* 40:1912.

300. Eris, J.M., A. Basten, R. Brink, K. Doherty, M.R. Kehry, and P.D. Hodgkin. 1994. Anergic self-reactive B cells present self antigen and respond normally to CD40-dependent T-cell signals but are defective in antigen-receptor-mediated functions. *Proc. Natl. Acad. Sci. USA* 91:4392–4396.

301. Cooke, M.P., A.W. Heath, K.M. Shokat, Y. Zeng, F.D. Finkelman, P.S. Linsley, M. Howard, and C.C. Goodnow. 1994. Immunoglobulin signal transduction guides the specificity of B cell-T cell interactions and is blocked in tolerant self-reactive B cells. *J. Exp. Med.* 179:425–438.

302. Mamula, M.J., and C.A. Janeway. 1993. Do B cells drive the diversification of immune responses? *Immunol. Today* 14:151–152.

303. Shi, Y., A. Kaliyaperumal, L. Lu, S. Southwood, A. Sette, M.A. Michaels, and S.K. Datta. 1998. Promiscuous presentation and recognition of nucleosomal autoepitopes in lupus: role of autoimmune T cell receptor alpha chain. *J. Exp. Med.* 187:367–378.

304. Singh, R.R., B.H. Hahn, B.P. Tsao, and F.M. Ebling. 1998. Evidence for multiple mechanisms of polyclonal T cell activation in murine lupus. *J. Clin. Invest.* 102:1841–1849.

305. Casamayor-Palleja, M., M. Khan, and I.C.M. MacLennan. 1995. A subset of CD4+ memory T cells contains preformed CD40 ligand that is rapidly but transiently expressed on their surface after activation through the T cell receptor complex. *J. Exp. Med.* 181:1293–1301.

306. MacDonald, K.P., Y. Nishioka, P.E. Lipsky, and R. Thomas. 1997. Functional CD40 ligand is expressed by T cells in rheumatoid arthritis. *J. Clin. Invest.* 100:2404–2414.

307. Han, S., B. Zheng, D.G. Schatz, E. Spanopoulou, and G. Kelsoe. 1996. Neoteny in lymphocytes: Rag1 and Rag2 expression in germinal center B cells. *Science* 274:2094–2097.

308. Hikida, M., M. Mori, T. Takai, K. Tomochika, K. Hamatani, and H. Ohmori. 1996. Reexpression of RAG-1 and RAG-2 genes in activated mature mouse B cells. *Science* 274:2092–2094.

309. Giachino, C., E. Padovan, and A. Lanzavecchia. 1998. Re-expression of RAG-1 and RAG-2 genes and evidence for

secondary rearrangements in human germinal center B lymphocytes. *Eur. J. Immunol.* 28:3506–3513.

310. Meffre, E., F. Papavasiliou, P. Cohen, O. de Bouteiller, D. Bell, H. Karasuyama, C. Schiff, J. Banchereau, Y.-J. Liu, and M.C. Nussenweig. 1998. Antigen receptor engagement turns off the V(D)J recombination machinery in human tonsil B cells. *J. Exp. Med.* 188:765–772.

311. Kelsoe, G. 1999. V(D)J hypermutation and receptor revision: coloring outside the lines. *Curr. Opin. Immunol.* 11:70–75.

312. Lin, W.C., and S. Desiderio. V(D)J recombination and the cell cycle. *Immunol. Today* 16:279–289.

313. Verkoczy, L.K., B.J. Stiernhdm, and N.L. Berinstein. 1995. Up-regulation of recombination activating gene expression by signal transduction through the surface Ig receptor. *J. Immunol.* 154:5136–5143.

314. Ma, A., P. Fisher, R. Dildrop, E. Oltz, G. Rathbun, P. Achacoso, A. Stall, and F.W. Alt. 1992. Surface IgM mediated regulation of RAG gene expression in E mu-N-myc B cell lines. *EMBO J.* 11:2727–2734.

315. Hertz, M., and D. Nemazee. 1997. BCR ligation induces receptor editing in IgM⁺IgD⁻ bone marrow B cells *in vitro*. *Immunity* 6:429–436.

316. Lauring, J., and M.S. Schlissel. 1999. Distinct factors regulate the murine RAG-2 promoter in B- and T- cell lines. *Mol. Cell. Biol.* 19:2601–2612.

317. Singh, R.R., and B.H. Hahn. 1998. Reciprocal T-B determinant spreading develops spontaneously in murine lupus: implications for pathogenesis. *Immunol. Rev.* 164:201–208.

318. James, J.A., T. Gross, R.H. Scofield, and J.B. Harley. 1995. Immunoglobulin epitope spreading and autoimmune disease after peptide immunization. Sm B/B'-derived PPPGMRPP and PPPGIRGP induce spliceosome autoimmunity. *J. Exp. Med.* 181:453–461.

319. Wysocki, L.J., X. Zhang, D.S. Smith, C.M. Snyder, and C. Bonorino. 1998. Somatic origin of T-cell epitopes within antibody variable regions: significance to monoclonal therapy and genesis of systemic autoimmune disease. *Immunol. Rev.* 162:233–246.

320. Nishioka Y., and P.E. Lipsky. 1994. The role of CD40-CD40 ligand interaction in human T cell-B cell collaboration. *J. Immunol.* 153:1027–1036.

321. Tohma, S., and P.E. Lipsky. 1991. Analysis of the mechanisms of T cell-dependent polyclonal activation of human B cells. Induction of human B cell responses by fixed activated T cells. *J. Immunol.* 146:2544–2552.

322. Jumper, M.D., J.B. Splawski, P.E. Lipsky, and K. Meek. 1994. Ligation of CD40 induces sterile transcripts of multiple Ig H chain isotypes in human B cells. *J. Immunol.* 152:438–445.

323. Fujita, K., M.D. Jumper, K. Meek, and P.E. Lipsky. 1995. Evidence for a CD40 response element, distinct from the IL-4 response element, in the germline epsilon promoter. *Int. Immunol.* 7:1529–1533.

324. Warren, W.D., and M.T. Berton. 1995. Induction of germ-line gamma 1 and epsilon Ig gene expression in murine B cells. IL-4 and the CD40 ligand-CD40 interaction provide distinct but synergistic signals. *J. Immunol.* 155:5637–5646.

325. Lin, S.C., H.H. Wortis, and J. Stavnezer. 1998. The ability of CD40L, but not lipopolysaccharide, to initiate immunoglobulin switching to immunoglobulin G1 is explained by differential induction of NF-kappaB/Rel proteins. *Mol. Cell. Biol.* 18:5523–5532.

326. Kirou, K., and M.K. Crow. 1999. New pieces to the SLE cytokine puzzle. *Clin. Immunol.* 91:1–5.

327. Armitage, R.J., B.M. MacDuff, M.K. Spriggs, and W.C. Fanslow. 1993. Human B cell proliferation and Ig secretion induced by recombinant CD40 ligand are modulated by soluble cytokines. *J. Immunol.* 150:3671–3680.

328. Rousset, F., E. Garcia, T. Defrance, C. Peronne, N. Vezzio, D.H. Hsu, R. Kastelein, K.W. Moore, and J. Banchereau. 1992. Interleukin 10 is a potent growth and differentiation factor for activated human B lymphocytes. *Proc. Natl. Acad. Sci. USA* 89:1890–1893.

329. Nonoyama, S., D. Hollenbaugh, A. Aruffo, J.A. Ledbetter, and H.D. Ochs. 1993. B cell activation via CD40 is required for specific antibody production by antigen-stimulated human B cells. *J. Exp. Med.* 178:1097–1102.

330. Inaba, M., K. Inaba, Y. Fukuba, S.-I. Mori, H. Haruna, H. Doi, Y. Adachi, H. Iwai, N. Hosaka, H. Hisha, H. Yagita, and S. Ikehara. 1995. Activation of thymic B cells by signals of CD40 molecules plus interleukin-10. *Eur. J. Immunol.* 25:1244–1248.

331. Burdin, N., C. van Kooten, L. Galibert, J.S. Abrams, J. Wijdenes, J. Banchereau, and F. Rousset. 1995. Endogenous IL-6 and IL-10 contribute to the differentiation of CD40-activated human B lymphocytes. *J. Immunol.* 154:2533–2544.

332. Splawski, J.B., L.M. McAnally, and P.E. Lipsky. 1990. IL-2 dependence of the promotion of human B cell differentiation by IL-6 (BSF-2). *J. Immunol.* 144:562–569.

333. Linker-Israeli, M., R.J. Deans, D.J. Wallace, J. Prehn, T. Ozeri-Chen, and J.R. Klinenberg. 1991. Elevated levels of endogenous IL-6 in systemic lupus erythematosus. *J. Immunol.* 147:117–123.

334. Al-Janadi, M., S. Al-Balla, A. Al-Dalaan, and S. Raziuddin. 1993. Cytokine profile in systemic lupus erythematosus, rheumatoid arthritis, and other rheumatic diseases. *J. Clin. Immunol.* 13:58–67.

335. Teppo, A.M., K. Metsarime, and F. Fyrhquist. 1991. Radioimmunoassay of IL-6 in plasma. *Clin. Chem.* 37:1691–1695.

336. Pelton, B.K., W. Hylton, and A.M. Denman. 1992. Activation of IL6 production by UV-irradiation from patients with SLE. *Clin. Exp. Immunol.* 89:251–254.

337. Alcocer-Varela, J., D. Aleman-Hoey, and D. Alarcon-Segovia. 1992. IL-1 and IL-6 activities are increased in the CSF of patients with CNS lupus erythematosus and correlate with late T cell activation markers. *Lupus* 1:111–117.

338. Nagafuchi, H., N. Suzuki, Y. Mizushima, and T. Sakane. 1993. Constitutive expression of IL-6 receptors and their role in the excessive B cell function in patients with SLE. *J. Immunol.* 151:6525–6534.

339. Stuart, R.A., A.J. Littlewood, A.J. Hall, and P.J. Maddison. 1995. Elevated serum interleukin-6 levels associated with active disease in systemic connective tissue disorders. *Clin. Exp. Rheum.* 13:17–22.

340. Linker-Israeli, M., M. Honda, R. Nand, R. Mandyam, E. Mengesha, D.J. Wallace, A. Metzger, B. Beharier, and J.R. Klinenberg. 1999. Exogenous IL-10 and IL-6 down-regulate IL-6 production by SLE-derived PBMC. *Clin. Immunol.* 91:6–16.

341. Alarcon-Riquelme, M.E., G. Moller, and G. Fernandez. 1993. Macrophage depletion decreases IgG anti-DNA in cultures from (NZB X NZW) F1 spleen cells by eliminating the main source of IL-6. *Clin. Exp. Immunol.* 91:220–225.

342. Klinman, D.M., A.-K. Yi, S.L. Beaucage, J. Conover, and A.M. Krieg. 1996. CpG motifs present in bacterial DNA

rapidly induce lymphocytes to secrete interleukin 6, interleukin 12 and interferon γ. *Proc. Natl. Acad. Sci. USA* 93:2879–2883.

343. Llorente, L., Y. Richaud-Patin, J. Couderc, D. Alarcon-Segovia, R. Ruiz-Soto, N. Alcocer-Castillejos, J. Alcocer-Varela, J. Granados, S. Bahena, P. Galanaud, and D. Emilie. 1997. Dysregulation of interleukin-10 production in relatives of patients with systemic lupus erythematosus. *Arthrit. & Rheum.* 40:1429–1435.

344. Llorente, L., Y. Richaud-Patin, R. Fior, J. Alcocer-Varela, J. Wijdenes, B. Morel-Fourrier, P. Galanaud, and D. Emilie. 1994. In vivo production of interleukin-10 by non-T cells in rheumatoid arthritis, sjogren's syndrome, and systemic lupus erythematosus. *Arthrit. & Rheum.* 11:1647–1655.

345. Llorente, L., W. Zou, Y. Levy, Y. Richaud-Patin, J. Wijdenes, J. Alcocer- Varela, B. Morel-Fourrier, J.-C. Brouet, D. Alascon-Segovia, P. Galanaud, and D. Emilie. 1995. Role of interleukin 10 in the B lymphocyte hyperactivity and autoantibody production of human systemic lupus erythematosus. *J. Exp. Med.* 181:839–844.

346. Taga, K., H. Mostowski, and G. Tosato. 1993. Human interleukin-10 can directly inhibit T-cell growth. *Blood* 81:2964–2971.

347. Georgescu, L., R.K. Vakkalanka, K.B. Elkon, and M.K. Crow. 1997. Interleukin-10 promotes activation-induced cell death of SLE lymphocytes mediated by Fas ligand. *J. Clin. Invest.* 100:2622–2633.

348. Levy, Y., and J.C. Brouet. 1994. Interleukin-10 prevents spontaneous death of germinal center B cells by induction of the bcl-2 protein. *J. Clin. Invest.* 93:424–428.

349. Esdaile, J.M., M. Abrahamowicz, L. Joseph, T. MacKenzie, Y. Li, and D. Danoff. 1996. Laboratory tests as predictors of disease exacerbations in systemic lupus erythematosus. Why some tests fail. *Arthrit. & Rheum.* 39:370–379.

350. Horwitz, D.A., J.D. Gray, S.C. Behrendsen, M. Kubin, M. Rengaraju, K. Ohtsuka, and G. Trinchieri. 1998. Decreased production of interleukin-12 and other Th1-type cytokines in patients with recent-onset systemic lupus erythematosus. *Arthrit. & Rheum.* 41:838–844.

351. Crispin, J.C., and J. Alcocer-Varela. Interleukin-2 and systemic lupus erythematosus: fifteen years later. *Lupus* 7:211–213.

352. Platt, J.L., B.A. Burke, A.J. Fish, Y. Kim, and A.F. Michael. 1982. Systemic lupus erythematosus in the first two decades of life. *Am. J. Kid. Dis.* 2:212–222.

353. Kobayshi, S., T. Shimamoto, O. Taniguchi, H. Hashimoto, and S. Hirose. 1991. Klinefelter's syndrome associated with progressive systemic sclerosis: report of a case and review of the literature. *Clin. Rheumatol.* 10:84–86.

354. Petri, M., and C. Robinson. 1997. Oral contraceptives and systemic lupus erythematosus. *Arthrit. & Rheum.* 40:797–803.

355. Mozes, E., and S. Fuchs. 1974. Linkage between immune response potential to DNA and X chromosome. *Nature* 249:167–168.

356. Carlsten, H., R. Holmdahl, and A. Rarkowski. 1991. Analysis of the genetic encoding of oestradial suppression of delayed-type hypersensitivity in (NZB X NZW) F1 mice. *Immunol.* 73:186–190

357. Brick, J.E., D.A. Wilson, and S.E. Walker. 1985. Hormonal modulation of reponses to thymus-independent and thymus-dependent antigens in autoimmune NZB/W mice. *J. Immunol.* 134:3693–3698.

358. Blank, M., S. Mendlovic, H. Fricke, E. Mozes, N. Talal, and Y. Shoenfeld. 1990. Sex hormone involvement in the induction of experimental systemic lupus erythematosus by a pathogenic anti-DNA idiotype in naïve mice. *J. Rheumatol.* 17:311–317.

359. Carlsten, H., N. Nilsson, R. Jonsson, and A. Tarkowski. 1991. Differential effects of oestrogen in murine lupus: acceleration of glomerulonephritis and amelioration of T cell-mediated lesions. *J. Autoimm.* 4:845–856.

360. Warner, M., S. Nilsson, and J.A. Gustafsson. 1999. The estrogen receptor family. *Curr. Opin. Obstet. & Gynecol.* 11:249–254.

361. Couse, J.F., and K.S. Korach. 1999. Estrogen receptor null mice: what we have learned and where they lead us? *Endocrin. Rev.* 20:358–417.

362. Suenaga, R., M.J. Evans, K. Mitamure, V. Rider, and N.I. Abdou. 1998. Peripheral blood T cells and monocytes and B cell lines derived from patients with lupus express estrogen receptor transcripts similar to those of normal cells. *J. Rheumatol.* 25:1305–1312.

363. Teixeira, C., J.C. Reed, and M.A. Pratt. 1995. Estrogen promotes chemotherapeutic drug resistance by a mechanism involving Bcl-2 proto-oncogene expression in human breast cancer cells. *Cancer Res.* 55:3902–3907.

364. Hurd, C., S. Dinda, N. Khattree, and V.K. Moudgil. 1999. Estrogen-dependent and independent activation of the P1 promoter of the p53 gene in transiently transfected breast cancer cells. *Oncogene* 18:1067–1072.

365. Duan, R., W. Porter, and S. Safe. 1998. Estrogen-induced c-fos protooncogene expression in MCF-7 human breast cancer cells: role of estrogen receptor Sp1 complex formation. *Endocrinol.* 139:1981–190.

366. Hyder, S.M., Z. Nawaz, C. Chiappetta, K. Yokoyama, and G.M. Stancel. 1995. The protooncogene c-jun contains an unusual estrogen-inducible enhancer within the coding sequence. *J. Biol. Chem.* 270:8506–8513.

367. Rider, V., R.T. Foster, M. Evans, R. Suenaga, and N.I. Abdou. 1998. Gender differences in autoimmune diseases: estrogen increases calcineurin expression in systemic lupus erythematosus. *Clin. Immunol. & Immunopathol.* 89:171–180.

368. Benten, W.P., M. Lieberherr, G. Giese, and F. Wunderlich. 1998. Estradiol binding to cell surface raises cytosolic free calcium in T cells. *FEBS Lett.* 422:349–353.

369. Schubert, L.A., G. King, R.Q. Cron, D.B. Lewis, A. Aruffo, and D. Hollenbaugh. 1995. The human gp39 promoter: two distinct nuclear factors of activated T cell protein-binding elements contribute indepedently to transcriptional activation. *J. Biol. Chem.* 270:29624–29627.

370. Eskdale, J., D. Kube, H. Tesch, and G. Gallagher. 1997. Mapping of the human IL10 gene and further characterization of the 5′ flanking sequence. *Immunogenet.* 46:120–128.

371. Tsai, E. Y., J. Yie, D. Thanos, and A. E. Goldfeld. 1996. Cell-type-specific regulation of the human tumor necrosis factor alpha gene in B cells and T cells by NFATp and ATF-2/JUN. *Mol. Cell. Biol.* 16:5232–5244.

372. Kanda, N., T. Tsuchida, and K. Tamaki. 1999. Estrogen enhancement of anti- double-stranded DNA antibody and immunoglobulin G production in peripheral blood mononuclear cells from patients with systemic lupus erythematosus. *Arthrit. & Rheum.* 42:328–337.

373. Piccinini, M.P., M.G. Giudizi, R. Biagiotti, L. Beloni, L. Giannarini, S. Sampognaro, P. Parronchi, R. Manetti, F. Annunziato, and L. Livi. 1996. *J. Immunol.* 155:128–133.

374. Cutolo, M., A. Sulli, B. Seriolo, S. Accardo, and A.T. Masi. 1995. Estrogens, the immune response and autoimmunity. *Clin. & Exp. Rheum.* 13:217–226.

375. Li, Z.G., V.A. Danis, and P.M. Brooks. 1993. Effect of gonadal steroids on the production of IL-1 and IL-6 by blood mononuclear cells in vitro. *Clin. &. Exp. Rheum.* 11:157–162.

376. Hu, S.K., Y.L. Mitcho, and N.C. Rath. 1987. Effect of estradiol on interleukin 1 synthesis by macrophages. *Int. J. Immunopharm.* 10:247–252.

377. Rieckmann, P., J.M. Tuscano, and J.H. Kehrl. 1997. Tumor necrosis factor- alpha (TNF-alpha) and interleukin-6 (IL-6) in B-lymphocyte function. *Methods* 11:128–132.

378. Bertoglio, J.H. 1988. B-cell-derived human interleukin 1. *Crit. Rev. Immunol.* 8:299–313.

379. Clevenger, C.V., D.O. Freier, and J.B. Kline. 1998. Prolactin receptor signal transduction in cells of the immune system. *J. Endocrinol.* 157:187–197.

380. Pellegrini, I., J.J. Lebrun, S. Ali, and P.A. Kelly. 1992. Expression of prolactin and its receptor in human lymphoid cells. *Mol. Endocrinol.* 6:1023–1031.

381. Spangelo, B.L., A.M. Judd, P.C. Isakson, and R.M. MacLeod. 1989. Interleukin-6 stimulates anterior pituitary hormone release in vitro. *Endocrinol.* 125:575–577.

382. Kennedy, R.L., and T.H. Jones. 1991. Cytokines in endocrinology: their roles in health and disease. *Endocrinol.* 129:167–178.

383. Ostensen, M. 1999. Sex hormones and pregnancy in rheumatoid arthritis and systemic lupus erythematosus. *Ann. NY Acad. Sci.* 876:131–143.

384. Lahat, N., A. Miller, R. Shtiller, and E. Touby. 1993. Differential effects of prolactin upon activation and differentiation of human B-lymphocytes. *Neuroimmunol.* 47:35–40.

385. Berczi, I., E. Nagy, S.M. de Toledo, R.J. Matusik, and H.G. Friesen. 1991. Pituitary hormones regulate c-myc and DNA synthesis in lymphoid tissue. *J. Immunol.* 146:2201–2206.

386. Yu-Lee, L, G. Luo, S. Moutoussamy, and J. Finidori. 1999. Prolactin and growth hormone signal transduction in lymphaemopoietic cells. *Cell. & Molec. Sci.* 54:1067–1075.

387. Elbourne, K.B., D. Keisler, and R.W. McMurray. 1998. Differential effects of estrogen and prolactin on autoimmune disease in the NZB/NZW F1 mouse model of systemic lupus erythematosus. *Lupus* 7:420–427.

388. Alvarez-Nemegyei, J., A. Cobarrubias-Cobos, F. Wscalante-Triay, J. Sosa- Munoz, J.M. Miranda, and L.J. Jara. 1998. Bromocriptine in systemic lupus erythematosus: a double-blind, randomized, placebo-controlled study. *Lupus* 7:414–419.

389. James, J.A., K.M. Kaufman, A.D. Farris, E. Taylor-Albert, T.J. Lehman, and J.B. Harley. 1997. An increased prevalence of Epstein-Barr virus infection in young patients suggests a possible etiology for systemic lupus erythematosus. *J. Clin. Invest.* 100:3019–3026.

390. James, J.A., and J.B. Harley. 1998. B-cell epitope spreading in autoimmunity. *Immunol. Rev.* 164:185–200.

391. Rowe, D.T. 1999. Epstein-Barr virus immortalization and latency. *Front. In Biosci.* 4:D346–371.

392. Eliopoulos, A.G., and A.B. Rickinson. 1998. Epstein-Barr virus: LMP1 masquerades as an active receptor. *Curr. Biol.* 8:R196–198.

393. Longnecker, R., and C.L. Miller. 1996. Regulation of Epstein-Barr virus latency by latent membrane protein 2. *Trends in Microbiol.* 4:38–42.

394. Lee, M.A., M.E. Diamond, and J.L. Yates. 1999. Genetic evidence that EBNA-1 is needed for efficient, stable latent infection by Epstein-Barr virus. *J. Virol.* 73:2974–2982.

395. Kaiser, C., G. Laux, D. Eick, N. Jochner, G.W. Bornkamm, and B. Kempkes. 1999. The proto-oncogene c-myc is a direct target gene of Epstein-Barr virus nuclear antigen 2. *J. Virol.* 73:4481–4484.

396. Henderson, S., D. Huen, M. Rowe, C. Dawson, G. Johnson, and A. Rickinson. 1993. Epstein-Barr virus-coded BHRF1 protein, a viral homologue of Bcl-2, protects human B cells from programmed cell death. *Proc. Natl. Acad. Sci.* 90:8479–8483.

397. Touitou, R., C. Cochet, and I. Joab. 1996. Transcriptional analysis of the Epstein-Barr virus interleukin-10 homologue during the lytic cycle. *J. Gen. Virol.* 77:1163–1168.

398. Kuhn-Hallek, I., D.R. Sage, L. Stein, H. Groelle, and J.D. Fingeroth. 1995. Expression of recombination activating genes (RAG-1 and RAG-2) in Epstein-Barr virus-bearing B cells. *Blood* 85:1289–1299.

399. Tsokos, G.C., I.T. Magrath, and J.E. Balow. 1983. Epstein-Barr virus induces normal B cell responses but defective suppressor T cell responses in patients with systemic lupus erythematosus. *J. Immunol.* 131:1797–1801.

400. Kowal, C., A. Weinstein, and B. Diamond. 1999. Molecular mimicry between bacterial and self antigen in a patient with systemic lupus erythematosus. *Eur. J. Immunol.* 29:1901–1911.

401. Domiati Saad, R., J.F. Attrep, H.-P. Brezinschek, A.H. Cherrie, D.R. Karp, and P.E. Lipsky. 1996. Staphylococcal entertoxin D functions as a human B cell superantigen by rescuing VH4-expressing B cells from apoptosis. *J. Immunol.* 156:3608–3620.

402. Rago, J.V., and P.M. Schlievert. 1998. Mechanisms of pathogenesis of staphylococcal and streptococcal superantigens. *Curr. Top. Microbio. & Immunol.* 225:81–97.

403. Liang, H., Y. Nishioka, C.F. Reich, D.S. Pisetsky, and P.E. Lipsky. 1996. Activation of human B cells by phosphorothioate oligodeoxynucleotides. *J. Clin. Invest.* 98:1119–1129.

404. Vasquez-Kristiansen, S., V. Pascual, and P.E. Lipsky. 1994. Staphylococcal protein A induces biased production of Ig by VII3-expressing B lymphocytes. *J. Immunol.* 153:2974–2982.

405. Domiati-Saad, R., and P.E. Lipsky. 1998. Staphylococcal entertoxin A induces survival of VH3-expressing human B cells by binding to the VH region with low affinity. *J. Immunol.* 161:1257–1266.

406. Jabara, H.H., and R.S. Geha. 1996. The superantigen toxic shock syndrome toxin-1 induces CD40 ligand expression and modulates IgE isotype switching. *Int. Immunol.* 8:1503–1510.

407. Snapper, C.M., and J.J. Mond. 1996. A model for induction of T cell- independent humoral immunity in response to polysaccharide antigens. *J. Immunol.* 157:2229–2233.

408. Yi, A.-K., D.M. Klinman, T.L. Martin, S. Matson, and A.M. Krieg. 1996. Rapid immune activation by CpG motifs in bacterial DNA: systemic induction of IL-6 transcription through an antioxidant-sensitive pathway. *J. Immunol.* 157:5394–5402.

409. Krieg, A.M. 1995. CpG DNA: a pathogenic factor in systemic lupus erythematosus? *J. Clin. Immunol.* 15:284–292.

410. Vallin, H., S. Blomberg, C.V. Alm, B. Cederblad, and L. Ronnblom. 1999. Patients with systemic lupus erythematosus (SLE) have a circulating inducer of interferon-alpha (IFN-(α) production acting on leukocytes resembling immature dendritic cells. *Clin. Exp. Immunol.* 115:196–202.

411. Vial, T., and J. Descotes. 1995. Immune-mediated side-effects of cytokines in humans. *Toxicol.* 105:31–57.

412. Siegal, F.P., N. Kadowaki, M. Shodell, P.A. Fitzgerald-Bocarsly, K. Shah, S. Ho, S. Antoneko, and Y.-J. Liu. 1999.

The nature of the principal type I interferon-producing cells in human blood. *Science* 284:1835–1837.

413. Longhurst, C., M.R. Ehrenstein, B. Leaker, F.K. Stevenson, M. Spellerberg, C. Chapman, D. Latchman, D.A. Isenberg, and G. Cambridge. 1996. Analysis of immunoglobulin variable region genes of a human IgM anti-myeloperoxidase antibody derived from a patient with vasculitis. *Immunol.* 87:334–338.

414. Rioux, J.D., E. Zdarsky, M.M. Newkirk, and J. Rauch. 1995. Anti-DNA and anti-platelet specificities of SLE-derived autoantibodies: evidence for CDR2H mutations and CDR3H motifs. *Mol. Immunol.* 32:683–595.

415. Hirabayashi, Y., Y. Munakata, O. Takaai, S. Shibata, T. Sasaki, and H. Sano. 1993. Human B cell clones expressing lupus nephritis-associated anti-DNA idiotypes are preferentially expanded without somatic mutation. *Scand. J. Immunol.* 37:533–540.

416. Daley, M.D., V. Misener, T. Olee, P.P. Chen, and K.A. Siminovitch. 1993. Genetic analysis of the variable region genes encoding a monospecific human natural anti-DNA antibody. *Clin. Exp. Immunol.* 93:1–8.

417. Davidson, A., A. Manheimer-Lory, C. Aranow, R. Peterson, N. Hannigan, and B. Diamond. 1990. Molecular characterization of a somatically mutated anti-DNA antibody bearing two systemic lupus erythematosus-related idiotypes. *J. Clin. Invest.* 85:1401–1409.

418. Kraj, P., D.F. Friedman, F. Stevenson, and L.E. Silberstein. 1995. Evidence for the overexpression of the VH4–34 (VH4–21) Ig gene segment in the normal adult human peripheral blood B cell repertoire. *J. Immunol.* 154:6406–6420.

20 | The Anti-Phospholipid Syndrome

Stephen W. Reddel and Steven A. Krilis

1. INTRODUCTION

The term "antiphospholipid syndrome" (APS) (1) refers to patients with the combination of antiphospholipid antibodies (aPL) and a history of clinical events consistent with the syndrome (2). Most of the clinical events in APS are thrombotic in nature. The term is descriptive rather than pathologically based and there is no gold-standard diagnostic test. The antiphospholipid syndrome can occur independently, thus primary APS (PAPS) (3), or in association with other autoimmune syndromes such as systemic lupus erythematosis, thus secondary APS (SAPS).

Antiphospholipid antibodies can be detected in both solid phase assays and *in vitro* dynamic coagulation tests. The solid phase assay, initially described as a radioimmunoassay (RIA) with the anionic phospholipid cardiolipin (CL) as the antigen (4), now generally uses an enzyme linked immunosorbent assay (ELISA) system to detect anticardiolipin antibodies (aCL). Lupus anticoagulant (LA) is the name given to aPL detected by the prolongation of various coagulation tests that assess the time to clot formation after the addition of an activator and calcium ions.

Early studies noted an association of SLE (5) with a circulating *in vitro* anticoagulant (6) and with a biologic false positive test for syphilis. These assays were later found to correlate with a thrombosis-prone subgroup of SLE patients (7,8). The improvement in sensitivity of the aCL assay compared to the VDRL for detecting this group of patients nevertheless continued to be confounded by positive results for a range of infections and post-infectious disorders, including syphilis and rheumatic fever (9), which are usually not associated with thrombosis (10). An interesting clue to this problem was provided by an investigation of IgG fractions purifed from patients with APS by either phospholipid affinity or by ion exchange chromatography, which found that the IgG fraction failed to bind anionic phospholipid in the absence of human or bovine serum or plasma fractions (11).

The subsequent isolation, purification and amino-terminal sequencing of the essential co-factor for aCL, β_2-glycoprotein 1 (B2GPI) (12), and the confirmation of this by other groups working concurrently (13,14) has lead to a shift in the paradigm of aPL antibodies. Subsequent research has also shown that most cases of clinically significant LA require the presence of prothrombin (15) and/or B2GPI (16,17). In a similar pattern to aCL the binding of LA to prothrombin and/or B2GPI is infrequent in samples from patients with LA secondary to infections (18). "Antiphospholipid antibodies" associated with the clinical manifestations of APS therefore should be considered in that light: they should now be regarded as being directed against the plasma proteins B2GPI and prothrombin, not phospholipids. The requirement for a "cofactor" (in fact, the actual antigen) is not new: In 1959 Loeliger concluded that the presence of prothrombin was essential for the circulating anticoagulant studied (19).

Antibodies directed against B2GPI, although initially thought to be directed against an antigen-phospholipid complex or a cryptic epitope of B2GPI resulting from its interaction with PL (12), can, in fact, bind B2GPI directly when B2GPI is presented on anionic phospholipid, or in the absence of phospholipid on negatively charged irradiated microtitre ELISA plates (20) and, in some circumstances, on standard untreated ELISA plates (21). This reflects heterogeneity in the anti-B2GPI avidities (22). Antibodies detected in the aCL ELISA that bind B2GPI are generally from patients with APS, whereas those detected in the absence of B2GPI are from patients with infections, although antibodies directed against anionic PL itself also can be detected in APS (23). Many antibodies binding anionic PL probably bind by non-specific charge interactions given the ease of dissociation with buffer of increasing ionic strength (24) which may explain the existence of aPL capable of cross reactive binding with the phosphodiester backbone of DNA (25).

More recent research has suggested that antibodies detected in an anti-phosphatidylethanolamine ELISA may require the presence of various kininogens (26). In some instances another anionic phospholipid binding protein, annexin V, may also be bound by aPL (27), although at present this finding has not been demonstrated in a range of populations.

2. ANTI-PHOSPHOLIPID ANTIBODIES

The standard aCL ELISA method is to coat microtitre polystyrene plates with cardiolipin in ethanol, which is then dried by evaporation. Cardiolipin dried in air is rapidly oxidised and this may be important for aPL binding (28). Although the RIA originally described was without the use of serum in the incubation buffer (4), it became apparent that the test is improved by the use of adult bovine serum (ABS) in the buffer (29). In retrospect, this is because B2GPI is provided in the buffer; the original test was, in fact, detecting the associated infectious type antibodies that bind cardiolipin directly.

The provision of the antigen B2GPI in the buffer means that it is admixed with many other plasma proteins, that the quantity and quality of the antigen is not known, and that many different sets of antibodies may be detected in the test. As a result the aCL test, although reasonably sensitive, suffers from very poor specificity with odds ratios for disease after a positive test ranging from only 1.4 to 5 (30). Standardisation of the test across commerically available kits is also extremely poor (31).

Although specific anti-(PL-binding-protein) ELISAs have been developed, the methodology, in particular of the coating concentration and buffer, of an anti-B2GPI ELISA has not been clarified with significant variability between assay methods and sensitivities. However, a uniformly high specificity (0.88–0.99) for thrombosis means that the direct anti-B2GPI ELISA has significant predictive value (30). The methodology of the anti-prothrombin ELISA has also not been clarified and it remains unclear whether anti-prothrombin antibodies increase the risk of thrombosis independently of other aPL (32).

2.1. Phospholipid Specificity of aPL

While antibodies binding plasma proteins such as B2GPI or prothrombin are detected in the aCL or LA assays the degree of binding of these antibodies may reflect changes in the nature and composition of the anionic phospholipid present, be it cardiolipin (CL), phosphatidylserine (PS) or phosphatidylinositol (PI), as this will influence the binding and orientation of the PL binding proteins. The issue of phospholipid specificity was explored in some depth in the 1980s (rev. in 33), but the sometimes contradictory findings of improved binding to specific PL mixtures and ratios of different PL probably related to variations in the binding of B2GPI, something that at the time was not known. The view that PS may in some instances be a better phospholipid than CL for detecting aPL has, however, recently been supported by the slightly higher stoichiometry for B2GPI/PL per mass of PL found for PS than for CL (34).

While antibodies binding to neutral phospholipids such as phosphatidylcholine (PC) are rarely reported, antibodies to the zwitterionic phosphatidylethanolamine (PE) have some interesting associations and a specific subset of antibodies not detected in other assays may bind directly or indirectly to PE. Some groups have found anti-PE in association with APS even in the absence of other aPL (35,36,37) although drug-induced anti-PE unassociated with thrombosis are also common (38). Various other plasma proteins bound to the PE surface may be the antigens detected by "anti-PE" including high and low molecular weight kininogens (26) in addition to B2GPI, which also binds PE well (34). Sugi et al. have also reported that patient antibody binding appears specific for the complex of high molecular weight kininogen and PE even though HMW kininogen also binds well to cardiolipin (39).

It has been recently reported that aPL bind to a unique mitochondrial membrane phospholipid, lysobisphosphatidic acid, which is an anionic di-phospholipid structurally similar to CL (40). The ELISA buffer in this experiment used 10% FCS, therefore the antibodies could in fact be binding a serum protein such as B2GPI.

2.2. Lupus Anticoagulants

The LA is the other major form of aPL, detected by prolongation of *in vitro* clotting assays. The use of limited amounts of phospholipid or other contact surface increases the sensitivity for the detection of LA. As no one test detects all LA (41) typically 2 tests are performed, usually the dilute Russell Viper Venom time (DRVVT) and either an activated partial thromboplastin time (aPTT) or a Kaolin Clotting Time (KCT). The antibodies detected in aCL and in LA are different sets (11) but with overlap in that some CL affinity purified aCL that bind B2GPI will also prolong the DRVVT (16). A consensus statement on the criteria for the diagnosis of lupus anticoagulants has recently been published (42) as has a discussion on recent developments in subtyping lupus anticoagulants (43). There is some evidence in support of the view that the presence of a lupus anticoagulant is the strongest aPL risk factor for thrombosis (44,45).

3. CLINICAL AND PATHOLOGIC MANIFESTATIONS OF APS

The primary clinical manifestations of the antiphospholipid syndrome are mainly the consequence of microangiopathy or thrombosis in the affected organ and effects in almost all organs have been described. There are several clinical manifestations where thrombosis cannot be demonstrated although an effect on the vessel wall may still be implicated. In this review we have stressed the pathological findings, as ultimately the results of *in vitro* research must explain these findings.

The recognised clinical associations of antiphospholipid antibodies and their relative frequencies are listed in Table 1, which is drawn from the series of 667 SLE patients of Alarcón-Segovia et al. (2). Readers are referred to a comprehensive review of the clinical manifestations of the antiphospholipid syndrome by Kandiah et al. (46), as the clinical manifestations here are largely discussed in reference to their histologic and immunologic aspects. Many rare manifestations have been described that, although not found frequently in APS and are therefore difficult to statistically link, nevertheless have a typical thrombotic etiology that suggests their presence in isolated patients is part of the syndrome.

Thrombosis in APS may occur in both the arterial and venous circulations, but tends to be consistent in an individual for one or the other (47). Thrombosis may also affect unusual vascular areas such as the cerebral venous sinuses, hepatic veins, cardiac valves and the fenestrated vasculature of bone. The pathophysiology thus far examined has been limited and is dominated by arterial, cardiac and placental studies with minimal information on the histopathology of venous thrombosis.

3.1. Arterial Pathology

Several reports have dealt with the changes found in arteries and arterioles. Tissues examined have been from meningeal biopsy (48), brain (49) (50) and heart at autopsy (51), renal biopsy (52) (53); and from various other sites (54). The reports are consistent: Characteristic is the finding of involvement of numerous small vessels by a combination of luminal fibrin thrombus (sometimes with recanalisation) and marked myointimal cellular proliferation and fibrous intimal hyperplasia. Elastic laminae are intact and evidence of current or healed inflammation or leukocyte cellular infiltration is rare even in secondary APS with SLE, although occasionally cellular infiltration has

Table 1 Frequency of clinical manifestations associated with aPL applied to 667 SLE patients

Manifestation	n	%
Livido reticularis	202	30
Thrombocytopenia	132	20
Recurrent fetal loss	45*	17
Venous thrombosis	76	11
Hemolytic anemia	59	9
Arterial occlusion	21	3
Leg ulcer	21	3
Pulmonary hypertension	11	2
Transverse myelitis	5	1

*Out of 261 patients at risk
Reproduced with permission from Alarcón-Segovia, D. et al., *Seminars in Arthritis and Rheumatism*, April 1992;21:275–286.

been reported in large leg arteries that may cause occlusion (55,56).

These arteriolar pathologic changes have been decribed as similar to those of accelerated or malignant hypertension. This suggests a role for endothelial injury as the primary event, with *in-situ* thrombosis plausibly secondary, although it has also been interpreted as suggestive of primary *in-situ* thrombosis (49). The recent demonstration on transcranial doppler ultrasound that APS patients have frequent cerebral microemboli does however suggest that their impaction in small arteries and arterioles may trigger a secondary intimal proliferation.

3.2. Neurologic Manifestations

In addition to the cerebral and ophthalmic arterial ischaemic events that are common vascular features of APS (57,50,58), there are several other manifestations worthy of note. Cerebral venous sinuses (59) and jugular veins (60) may become thrombosed and can lead to psuedotumour cerebri. Migraine has often been reported to be associated with APS but this may reflect how common it is in the normal population (61). The association of aPL with epilepsy in SLE (62) or in a more general population (63) appears real, although whether this is secondary to previous cerebral ischaemia or an artifact of anticonvulsant drugs inducing autoantibodies remains to be determined by prospective series. The increased incidence of chorea in APS, possibly modulated by use of the oral contraceptive pill, (64) is not associated with infarcts on cerebral imaging in all cases (65). However, unless MRI is performed soon after onset with fluid attenuation inversion recovery (FLAIR) or other sensitive pulse sequences small basal ganglia lesions may be missed or misinterpreted as perivascular spaces. A similar caveat applies to "transverse myelitis" with APS as MRI in the spinal cord is less sensitive, with some cases possibly being small cord infarcts.

An important but uncommon clinical feature of cerebral APS is a progressive multi-infarct dementia syndrome (66,57,48,67), with the small vessel pathologic findings described above. In one case associated with chorea this was reported to be significantly reversible (68). In APS patients with progressive dementia multiple small infarcts are demonstrable on MRI, which may be mostly in the deep white matter and be misdiagnosed as multiple sclerosis.

Cases of atypical multiple sclerosis, with a normal brain MRI, have been associated with aCL (69). An association with multiple sclerosis and IgM aCL but not IgG aCL (70) was noted although unfortunately this was in a subgroup analysis of the control group in an evaluation of aCL in ischaemic stroke. Some cases diagnosed as multiple sclerosis with brain MRI lesions were found in a large recent study to also have features of APS such as livido reticularis and aPL (71). In neurological practice APS is one of the few disorders (along with vasculitis and Sjogren's syndrome) that can mimic quite typical cases of multiple sclerosis.

Sneddon's syndrome is the association of recurrent stroke and livido reticularis, often leading to dementia. Some cases are clearly APS under an old name and association with aPL is very common (72), with the pathologic changes of myointimal proliferation being essentially the same (73). The reporting of familial Sneddon's syndrome and of cases of Sneddon's without aPL (74,75) may keep this eponymous disease distinct from APS.

3.3. Cardiopulmonary Manifestations

Large vessel arterial occlusions occur commonly in APS although less frequently than small vessel thrombosis. As major proximal artery changes are rarely found in APS on angiography (although this does not show outer wall changes) or pathology, and cardiac valvular abnormalities are frequent most large vessel infarctions would appear to be the result of heart to artery embolism (76) rather than artery to artery embolism or *in situ* thrombosis. Initial reports suggested that Libman-Sacks endocarditis may be a manifestation of aPL in patients with SLE and secondary APS (77). However later echocardiographic studies (78,79) have found that the majority of patients with SLE have heart valve changes including many with masses consistent with Libman-Sacks vegetations and that the incidence is not increased in the presence of aPL. Consistent with that is the lower (but not normal) frequency of heart valve changes in primary APS than in secondary APS (80). It is plausible that aPL increase the risk of thromboembolism from an abnormal valve, although in a report of patients who underwent cardiac valve replacement there was no

increase in the 5 year risk of embolus amongst those who had aPL preoperatively (81).

The primary APS cardiac valvular lesion consists mainly of superficial or intravalvular fibrin deposits and the subsequent changes of organisation: vascular proliferation, fibroblast influx, fibrosis and calcification without prominent inflammation (82). Using immunohistochemistry and anti-idiotype antibody the presence of linear subendothelial deposits of anticardiolipin antibody and complement was described by one group (83). Changes may be confined to the right heart valves and may be associated with pulmonary arterial and deep venous thrombosis (84). Cases of myocardial infarction have been reported in APS (85) but whether aPL have an association with myocardial infarction in a general population remains controversial (86,87).

Pathologic changes of APS in the low pressure pulmonary vasculature follow a similar pattern. Pulmonary emboli from deep venous thrombi occur frequently (88), but as the diagnosis of these is by ventilation/perfusion scanning or angiography, pulmonary or venous tissue is rarely available for study and in some instances when obtained has been considerably after the event. A number of cases have been described where the small pulmonary arteries and arterioles have been occluded with or without recanalisation, with intimal hypertrophy, suggestive of in-situ thrombosis secondary to endothelial/vessel wall changes (89,90). Pulmonary arteritis does not appear to be a feature except in isolated patients with SLE and catastrophic APS (91). A pulmonary capillaritis occurs in isolated cases (89) and may be the explanation for the haemoptysis and pulmonary infiltrates that occurs in some patients with an otherwise thrombotic rather than bleeding tendency and no evidence of hypoprothrombinaemia (see section on prothrombin). Pathologic changes in the deep veins have rarely been described and have been of old lesions where the finding of fibrocalcific bands gives no clue to the etiology of the original thrombosis (84).

3.4. Obstetric Manifestations and Placental Pathology

The association between aPL antibodies and recurrent miscarriage was recognised for lupus anticoagulant (92) and for cardiolipin antibodies (93). The secondary problem of prematurity may be associated either with APS itself or the use of glucocorticoids in some instances (94). Antiphospholipid antibodies only appear to explain about one tenth of recurrent aborters (95,96). Although recurrent miscarriage is often defined as 2 or more events, with a normal incidence of miscarriage at 12–15% this will occur by chance with sufficient frequency to obscure careful evaluation of specific underlying mechanisms of fetal loss (97) and thus 3 or more events is probably more appropriate.

Miscarriage in the second trimester (foetal loss) appears more characteristic for APS but in some studies most miscarriages in APS groups still occur (as in spontaneous cases) in the first trimester (98,99). Although most spontaneously aborted foetuses have chromosomal abnormalities the incidence in APS is reduced to about 20% (100), suggesting that APS has an alternative specific means of inducing miscarriage.

The prevalance of IgG anticardiolipin antibodies remains at normal (101) or near normal (102) levels in normal populations of pregnant women, but the IgM aCL may produce many positive and clinically non-significant results (103). In one study the DRVVT was abnormal in 52/380 women tested (102) with only 6 foetuses being miscarried. The majority of the DRVVT results were in the mean + 2–2.5 SD range suggesting that pregnancy itself may alter the DRVVT test characteristics, whereas the aPTT (102) and the KCT (101) do not appear to be affected. This may influence the interpretation of some papers using the DRVVT, such as the heparin treatment trial of Rai et al. (104).

In placentae, the typical changes are of a uteroplacental vasculopathy with placental infarction sometimes appearing sufficient to lead to foetal death in the first or second trimester (105). There may also be secondary foetal villus damage and chronic villitis in addition to areas of infarction (106) although chronic villitis occurs later in the pregnancy and in association with reduced placental growth may represent a feature of SLE rather than PAPS (105). In the uterus, spiral arteries can be affected by obliterative arteropathy, some with fibrinoid necrosis, some with acute atherosis (107). Intravascular thrombi and excessive fibrin deposition can also be found particuarly in perivillus regions. These changes are not specific and are similar to those of pregnancy induced hypertension. In SLE, inflammatory changes with immunoglobulin and complement deposition has been found in some (108) but not all (109) cases especially when extra-decidual tissue is obtained by retroplacental curettage. Whether these inflammatory changes reflect either primary APS, the underlying SLE component or are normal placental changes postpartum is unclear.

3.5. Dermatologic Manifestations

There are several characteristic lesions of the skin in APS not including Sneddon's syndrome. Common findings in the skin in APS are of livido reticularis and acrocyanosis, which in severe cases, may progress to distal gangrene (110). Livido reticularis is thought to occur because the skin is supplied in 1–4 cm diameter patches by a single ascending arteriole: The livid network represents the watershed regions of impaired perfusion and secondary vascular dilation and stasis (111). Raynaud's phenomenon,

which occurs commonly in SLE, is not a feature of isolated APS.

Another frequent finding is of painful skin nodules or macules (112) that resembles immune complex-mediated vasculitis and are sometimes misleadingly termed livido vasculitis (113); these lesions may progress to extensive ulceration, often on the shins, or heal to leave stellate scars. The histologic finding in this lesion is of non-inflammatory thrombosis of small cutaneous vessels, and these lesions may respond to anticoagulation or if required a low dose infusion of the fibrinolytic tissue plasminogen activator in cases of resistant ulceration (114).

3.6. Haemocytopaenias

It is well recognised that APS is associated with thrombocytopaenia (115) and there is also possibly an association with Coomb's positive haemolytic anaemia (2,116). In one study an aCL IgG titre greater than 7SD above the mean correlated well with thrombocytopaenia, with a weaker correlation with any abnormal titre of aCL and a Coomb's positive state (117). Comparing SLE patients with secondary APS to patients with primary APS a similar frequency of thrombocytopaenia was found but haemolytic anaemia was significantly more frequent in the SLE/APS group, suggesting that while thrombocytopenia is associated with APS the haemolytic anaemia in most cases is a function of SLE (80). Antiphospholipid antibodies are also present in about half the patients with ITP without a diagnosis of APS but the aPL level is not related to the platelet count (118).

3.7. Treatment

The treatment of APS is generally with anticoagulation with warfarin at increased intensity (119,120) or heparin and aspirin in the case of pregnancy and fetal loss (121,104). In some instances empirical plasmapheresis or intravenous immunoglobulin G (IVIG) have been used. Corticosteroids are little used in the treatment of primary APS, although antibody levels may decline with this treatment. The treatment of APS has been the subject of a recent review by Khamashta, M. to which readers are referred (122).

4. PROTEINS BOUND BY ANTIPHOSPHOLIPID ANTIBODIES

4.1. Beta-2 Glycoprotein I

Beta-2 glycoprotein I was first demonstrated as a component of the beta-globulin fraction of human serum (123). The protein has also been termed apolipoprotein H (124) since it circulates in both lipid-bound and free forms and satisfies the criteria for classification as an apolipoprotein (125). However, most B2GPI circulates independently from plasma lipoproteins and the protein structure and surface behavior is quite different from the typical flexible apolipoproteins of class A and C. The purpose of B2GPI is unknown although in *in vitro* systems the protein binds to anionic phospholipid and by occupying these procoagulant sites produces an apparent anticoagulant effect (see proposed pathogenetic mechanisms of APS).

4.2. Beta-2 Glycoprotein I in Plasma

The mean serum level of B2GPI is about 200 mg/l, (SD 35 mg/l),(126), which is more than any of the plasma proteins involved in clotting except fibrinogen. This quantification excludes an outlier group with a phenotypically lower B2GPI level consistently found in approximately 6–10% of English, German and Australian populations amongst others (126,127, 128,129). This suggests an autosomal codominant allelic state with normal and deficient alleles, designated BgN/BgD or alternatively Apo H*N/Apo H*D. The estimated gene frequency for BgD is 3% and subjects with no detectable plasma B2GPI have been found.

The heterozygous state does not appear to be thrombophilic (130) and although subjects have been identified with a complete (i.e., homozygous) deficiency of which 1 had a DVT (130) and 2 had myocardial infarctions in their fifties (131), this may simply reflect the populations screened, which were for unexplained thrombophilia and dyslipidaemia, respectively. An apparently healthy woman identified with homozygous B2GPI deficiency had (bar a minimally shortened DRVVT) normal *in vitro* tests of thrombin generation, protein C activity and fibrinolysis (132). A mouse with a deletion of the B2GPI gene (133) has been developed by Sheng et al. and although this awaits full characterisation the mice appear normal at birth but have a number of abnormal *in vitro* clotting assays (Y. Sheng personal communication).

Beta-2 glycoprotein I is synthesised in the liver (134) and the plasma level is decreased in cirrhosis (135). It is also synthesised in intestinal epithelial cells (136), syncytiotrophoblast and cytotrophoblast (137). The mRNA for B2GPI has been reported in endothelial cells by one group (138) who also found it in astrocytes, neurones and lymphocytes but not in fibroblasts, their negative control. This contradicts a negative study that reported no B2GPI mRNA expression in endothelial cells (139). The level may be increased in chronic infections, but acutely it behaves as a negative acute phase reactant, decreasing with albumin in acute illness without DIC. The level of measured plasma B2GPI dropped more dramatically than albumin in the clinical setting of DIC, suggesting consumption (140), although this was not found in an animal model of DIC (141). Interestingly, the level is increased in patients with aPL (142) but not in the general SLE population (135).

4.3. Beta-2 Glycoprotein I Structure

The human B2GPI protein has 326 amino acids with 4 N-linked glycosylation sites and a predicted amino acid weight on protein sequencing of 36281 (143). The weight average molecular weight measured by sedimentation equilibrium is 43000 (125) and by reduced SDS-PAGE is 54200 (143), although it may appear greater. Both methods are subject to variation in the case of highly glycosylated proteins and, in addition, the oligosaccharides in B2GPI are highly heterogenous in structure (144). On isoelectric focusing there are a number of bands depending upon the number of sialic acid residues (145) and the presence of two isoelectric subspecies of asialo-β_2-glycoprotein I (146). The alteration in glycosylation in recombinant human B2GPI does not appear to affect cardiolipin binding nor the binding of B2GPI dependent anti-cardiolipin antibody (147).

The protein consists of repeating sequences in the form typical of the complement control protein (CCP) module (148), with the first four domains having about 60 amino acid residues and 4 cysteines each, with potential disulphide bridges joining the 1st to 3rd and the 2nd to 4th cysteines (134) to contribute to a "looped-back" structure termed a sushi domain (149). The fifth domain is aberrant with an extra 20 amino acid tail terminating in a cysteine that can form a disulphide link with an extra cysteine found midway between the standard 2nd and 3rd cysteine residue positions (150) (Figure 1). Other members of the CCP superfamily include haptoglobin, complement receptor 1 & 2, C4b binding protein, complement Factor H, clotting Factor XIII (subunit-b) and IL2 receptor (151). Antibodies to B2GPI do not bind haptoglobin,which has a similar MW and is often used as a control protein for B2GPI in binding experiments, nor do they bind Factor H, the domains of which have about 20% amino acid homology to human B2GPI domains 1–4. The medium resolution NMR 3-dimensional structures of modules 15 & 16 of factor H have been used to derive the three dimensional model of the 5th domain of B2GPI (152,46).

In addition to the 326 amino acids of the mature protein, the B2GPI gene encodes a putative leader peptide (134). The gene maps to human chromosome 17q23 → qter, unlike most of the CCP family, the genes for which are found in the regulation of complement activation (RCA) cluster on chromosome 1q32 (153), although other proteins using CCP modules are found elsewhere. On the mouse gene, the different CCP domains are encoded on 1–2 separate exons each (133) with further specific exons for the leader sequence and the aberrant domain V tail (Figure 2). This suggests a

Figure 1.   Amino acid sequence and location of disulphide bonds in bovine β2-glycoprotein I. This demonstrates the 5 "sushi domains" with the aberrant terminal loop of the 5th domain, much of which is predicted to be surface exposed. Unusually the protein ends in a disulphide bond creating a terminal loop rather than in a 'free' tail. Reproduced with permission from Koike, T. *et al*, Lupus 1998;7 (Suppl 2):S14–S17.

process of duplication, dispersion and modification of a founder gene encoding a common CCP domain.

B2GPI is conserved amongst mammals with human, bovine, canine and mouse proteins all having 5 modules except for rat in which the published sequence has only 4 and, on the basis of homology, appears to be missing domain 1. There is 60–80% amino acid sequence identity between species within each module, with module 5 the most conserved (46) (Figure 3).

The folding of the 5th domain of B2GPI is predicted by modelling to consist of 8 strands, organised in 2 distorted β-sheets with long coiled regions connecting the strands, but no helices (46). In domain 5 there is a completely conserved positively charged (multiple lysine) region between a.a. 281–288 that is critical for phospholipid binding (152), although it is not solely responsible for the PL binding site. This was identified initially by three-dimensional modelling of the B2GPI domain V, with an analysis of the likely electrostatic charge (Figure 4). Many of the positively charged side chains are located in a surface region consisting of 3 loops. In

particular, the central loop in this region (284–287 of the B2GPI molecule) contains 3 lysine residues and mutation of any of these to glutamate was predicted to significantly reduce the net positive charge. A subsequent programme of site-directed mutagenesis of this loop confirmed this hypothesis, and also confirmed that disrupting the position of this loop by mutating the flanking cysteines also disrupted phospholipid binding (154).

The aberrant 20 residue tail in the fifth domain (307–326) is identical in human, bovine and canine sequences and differs only in the 3rd last residue in rat and in the 5th to 3rd last residues in mouse. This region of B2GPI is predicted to be largely surface exposed and could well be of functional importance. It is also interesting that this region is susceptible to cleavage by plasmin (155), which clips it at lys 317- thr 318 (156) and that this impairs PL binding. Cleavage at 317–318 has indeed been predicted by modelling to affect the position of the main (a.a.[n] 284–287) PL binding region (152). The presence of this clip has been demonstrated *in vivo* during DIC (157). It has also been demonstrated that domain

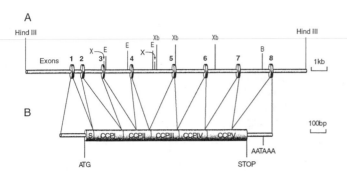

Figure 2. The structure of the mouse B2GPI gene. Each structural region of the protein is encoded separately by 1–2 exons. The more typical complement control protein part of domain V is encoded by exon 7, whereas the aberrant terminal loop of domain V is encoded by exon 8. Reproduced with permission of the authors from Kandiah *et al*, Advances in Immunology 1998;70:507–563

V of B2GPI is heterogenous with polymorphisms in domain V that affect function, specifically 306 (Cys → Gly), which is predicted to disrupt the disulphide bridge forming the terminal tail; and 316 (Trp→Ser), which has been shown to impair binding to PS (158).

Recently, it has been shown that B2GPI is recognised by megalin (159), an endocytic receptor expressed on absorptive endothelia, which has a primary role in the uptake of a number of proteins from the renal proximal tubule. The increase in B2GPI avidity for megalin when complexed to PS containing liposomes suggests that the binding site for megalin might be other than the PL binding site.

The use of bovine serum as a source of B2GPI in the aCL assay relies upon the high homology to human B2GPI (160). It has been established, however, that some antibodies from APS patients will only bind to human B2GPI. Arvieux et al. found that only 7.9% of polyclonal IgG were restricted to human B2GPI, but that 43.7% of IgM fractions were human restricted, many of which were not identified by a standard aCL assay (161).

4.4. Autoantibodies to B2GPI: Binding a Cryptic Epitope or Binding with Low Affinity that Requires Dimerisation for Detection?

The knowledge that patients with APS required B2GPI for binding in the aCL ELISA led to the testing of B2GPI ELISA systems without CL. Although some investigators could demonstrate an ELISA method with some patient samples using B2GPI coated on normal microtitre plates (13,21,162), it was Matsuura E et al. who demonstrated (20) that clinically significant aCL consistently bind B2GPI in a direct ELISA when it is coated to a pre-irradiated polystyrene microtitre plate.

Human anti-B2GPI auto-antibodies generally do not bind well when the B2GPI is coated on standard unirradiated plates. It has been suggested that a comparable amount of B2GPI is bound on the well whether or not the plate is first irradiated (163) when measured with

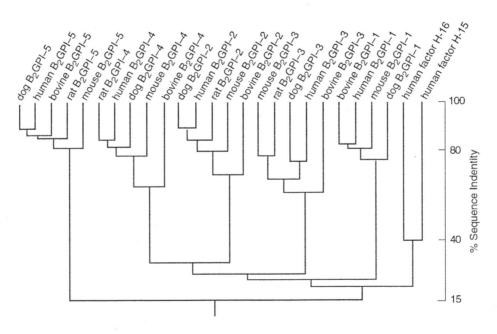

Figure 3. Protein sequence homology tree for the 5 domains of various B2GPIs and the 15th & 16th domains of human complement factor H. Reproduced with permission of the authors from Kandiah *et al*, Advances in Immunology 1998;70:507–563

This may reflect differences in local antigen density or clustering of B2GPI molecules, due to organisation by groups of oxidation radicals induced by irradiation or in the polar head groups of anionic phospholipids (33), which enables dimeric binding by anti-B2GPI antibody (164,165). Alternatively, cryptic epitopes on B2GPI to which human autoantibodies (but not anti-B2GPI induced by immunisation) are directed may be revealed by binding induced conformational changes (163,166).

Fab' fragments of known human anti-B2GPI autoantibodies do not continue to bind B2GPI (164,165), which tends to support the hypothesis that dimeric binding is required and obviates the requirement for invoking cryptic epitopes, although it does not preclude some epitopes from being cryptic. Further work has confirmed that human autoimmune anti-B2GPI can bind native B2GPI but are of low affinity with estimates of Kd ≈ 1 × 10⁻⁵M (164), 3.4–7.2 × 10⁻⁶M (167) and 1.4–4.4 × 10⁻⁶M (165), as opposed to mouse moab antihuman B2GPI with a Kd of 5 × 10⁻⁹M (165). This generally low autoantibody affinity, unlike the mouse antibody, is consistent with the failure of human but not mouse antibodies to bind B2GPI in a standard ELISA (167), which usually requires a Kd lower than 0.5 × 10⁻⁶M (168). A dimerised mutant human B2GPI is bound 3–10 fold more strongly than normal B2GPI, providing supportive evidence for dimeric binding (165). Willems et al. have provided evidence that the initial monomeric binding of aCL to membrane bound B2GPI is rapidly followed by the dimeric binding of an aCL molecule to (B2GPI)₂, with the complex binding the anionic PL 30–40 fold more avidly than B2GPI alone (169).

4.5. Epitope Specificity of Anti-B2GPI Antibodies

Several groups have reported the localisation of epitope(s) on human B2GPI required for antibody binding. A predicted surface exposed loop on domain V of B2GPI with multiple positively charged lysine residues (cys 281—cys 288) was confirmed as the major phospholipid binding site and thus essential for antibody binding in a B2GPI-dependent cardiolipin ELISA (156). Synthetic oligopeptides from domain V including this sequence were found to support the binding to CL of 1/5 human IgM moabs derived from APS patients. In free solution, the direct binding of another one of the moab to B2GPI on an irradiated ELISA plate was inhibited by peptides containing the sequence, suggesting the presence of epitopes on domain V involved in the binding of anti-B2GPI moab from patients with APS (170). More recently, using mutant B2GPI with various domains deleted, the binding of various mouse moabs from mice immunised with human B2GPI did not depend on the presence of domain V in most cases (171). Using the same constructs human polyclonal affinity purified anti-B2GPI antibodies in an ELISA were found

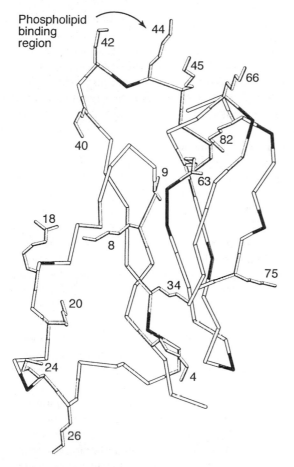

Figure 4.   Main chain trace of the predicted three dimensional model of the 5ᵗʰ domain of human B2GPI.. Residues with positively charged side chains at physiologic pH (lys, his) are represented in grey. The lysines but not the 2 histidines have their side chains shown. Residues with negatively charged side chains (asp, glu) are shown in black. Neutral residues are shown in white. Lysines in positions numbered 42, 44 & 45 of domain V are residues numbered 284, 286 & 287 of the full B2GPI sequence. Reproduced with permission from Sheng, Y et al, The Journal of Immunology Oct 15,1996;157:3744–3751.

high affinity mouse anti-human B2GPI monoclonal antibodies. Roubey et al. differ, finding that as the coating concentration rose, the amount of B2GPI bound to untreated plates reached a plateau, whereas the amount on irradiated plates continued to rise to a saturation level about 1.5 times that of the untreated plates (164). More recently the issue has been clarified by Arvieux et al., who found that anti-B2GPI autoantibodies are heterogenous: Most patients having low affinity antibodies binding to B2GPI only on wells precoated with cardiolipin or on irradiated plates; and a small group of patients with higher affinity autoantibodies that will indeed bind B2GPI coated on standard ELISA plates (22).

to bind the mutants that contained domain IV, but not domains I-III or domain V, suggesting the epitope is on domain IV(172), although a specific domain IV-only construct was not used. Using a similar method with APS patient's anti-B2GPI antibody, but slightly different constructs, another group came to a completely different finding, that the presence of domain I is essential for binding by human autoantibody and that antibody binds domain I alone (173). Whether particular epitopes are crucial for the pathogenesis of APS or only that the epitopes are clustered in a particular region of the protein remains to be seen.

4.6. Prothrombin

Prothrombin is the zymogen precursor of thrombin and, as such, is the key to the coagulation cascade. Thrombin has responsibility for local fibrin clot formation and, upon binding to thrombomodulin, for activating the anticoagulant pathway of protein C to limit distant clot propagation. Thrombin also acts upon the transmembrane G-protein-coupled thrombin receptor to activate platelets (174) and endothelial cells which, by releasing von Willebrand factor (175), enhances the formation of the platelet plug.

Prothrombin is secreted largely by the liver as a single chain 579 amino acid (176) glycoprotein with a MW ~ 72000, of which approximately 8% is carbohydrate at 3 N-linked glycosylation sites (177). It has a phospholipid binding site seperated by 2 kringle domains from the α-thrombin serine protease domain.

The Ca^{2+}-dependent phospholipid binding site has multiple glutamic acid residues and is at the amino-terminal end. The normal post translational processing of this "gla" domain depends upon the vitamin K-dependent carboxylase enzyme. Deficiency or competition with vitamin K results in the reduced secretion of vitamin K dependent proteins (178) and in the appearance in plasma of gla-deficient pro-thrombin with reduced biological activity (179). The α-thrombin serine protease domain is inactive in the prothrombin form.

The prothrombinase complex ($Xa/Va/Ca^{2+}/PL$) modifies prothrombin to thrombin (via several intermediates such as meizothrombin) by cleavage at several sites (180,181). The gla domain/1st kringle domain and the 2nd kringle domain are removed (fragments 1&2, respectively) leaving the conformationally altered and now active serine protease domain of thrombin (182). Thrombin therefore does not have a phospholipid binding site.

4.7. Antibodies to Prothrombin

Antibodies to prothrombin account for many of the LA type of "antiphospholipid antibodies", which, as noted previously, bind phospholipid binding proteins in those samples from patients with clinical APS (183). Most antibodies detected in the aCL ELISA bind B2GPI, and while this also occurs in LA (16,17), many other cases of LA can be absorbed by prothrombin (184), require the presence of prothrombin to be active (15) and can be detected in an anti-prothrombin ELISA (18). In fact, many LA have both anti-prothrombin and anti-B2GPI present (18). These prothrombin antibodies have recently been the subject of a comprehensive review by Galli and Barbui (32). Attention was drawn to the presence of LA associated anti-prothrombin antibodies by the occasional finding of hypoprothrombinaemia and bleeding (185), which was later shown to be the result of high affinity (Kd $\approx$ 10^{-9}–10^{-10}M) anti-prothrombin antibodies (186), presumably resulting in fluid phase binding to form rapidly cleared immune complexes. These antibodies are in fact quite rare and most anti-prothrombin antibodies are not associated with hypoprothrombinaemia (184). It is therefore suggested that in typical LA they bind prothrombin with a lower range of affinities (32). In support of this is the recent demonstration that antibody binding is much better supported by covalently-linked prothrombin dimers and multimers than monomeric prothrombin (187), and that when binding to immobilised prothrombin the higher affinity antibodies (of the LA, not hypoprothrombinaemic group) will bind under less optimal conditions of coating, whereas antibodies in the lower affinity range will only bind under the optimal conditions (187) of prothrombin being preincubated on PS coated wells in the presence of calcium (188). This is similar to the findings with anti-B2GPI. Unlike the anti-B2GPI antibodies, however, is the ability to detect significant anti-prothrombin binding on Western immunoblot, suggesting that some epitopes may be linear peptide sequences or are bound at somewhat higher affinity (183).

4.8. Annexin V

Annexin V is a single chain polypeptide of 319 amino acids to a calculated molecular weight of 35731 Da that is comprised of an amino-terminal tail and 4 linked domains, each with a consensus "endonexin loop sequence" that may potentially act as calcium binding sites. It is unglycosylated and has an isoelectric point of 4.8 (189). It was initially termed placental protein 4, amongst other names, and in the APS literature has also been referred to as placental anticoagulant protein I (190). As it and other members of its family were found to share significant core homology (differing in the amino-terminal tail), they have been retermed the annexins, which also have the characteristic of binding phospholipid, but only in the presence of Ca^{2+} ions (189).

Annexin V was initially isolated from arteries of the umbilical cord (191) and is found throughout many tissues in humans and other animals (192,189), including the non-vascular tissue of the bovine ocular lens (193) and in mononuclear leukocytes, platelets (but not in granules (194)) and red blood cells. It comprises 0.2–0.4% of the protein component of human umbilical vein endothelial cells (HUVEC) in culture (189,195). There is no signal sequence for secretion although annexin I and annexin V have been described as secreted by the prostate gland (196) and annexin V by insect cells but in a baculovirus expression system (197): The evidence for secretion of annexin V as opposed to it entering secretions after cell death is therefore not strong.

Annexin V can be detected at very low levels only in some plasma samples, maximally when collected with EDTA rather than citrate, giving a quantity of 3.1 ± 2.7 ng/ml. However, the majority of normal individuals in one study did not have annexin V in plasma at levels above the limit of detection, which was 0.5 ng/ml (195). The plasma concentrations of prothrombin and B2GPI are therefore about 10^5 fold greater than that of annexin V, if not more. It is unclear if some or all of the plasma component represents annexin V released by cellular blood components or venous endothelial cells during the process of venipuncture and processing. It is possible that even if not present in the plasma *in vivo* significant amounts of annexin V may bind to anionic phospholipid and move to the extracellular surface during membrane lipid redistribution after calcium influx activates the cell, but this has not been demonstrated. Annexin V binds extraordinarily avidly to anionic PL surfaces in the presence of calcium and physiologic [Na] and pH with a Kd outside the limits of determination (i.e., too little remains in solution), thus tighter than 1×10^{-10}M (see table 2)

It is in two quite different respects that annexin V has been examined in APS: As an anticoagulant phospholipid binding protein that might be displaced by the combination of B2GPI bound by anti-B2GPI (*vide infra* in the section on pathogenetic mechanisms), and in quite a different way as another phospholipid binding protein to which so-called antiphospholipid antibodies might be directed.

In the second respect the binding of antibody to annexin V from patients with a history of fetal loss and an LA has been decribed (27), with Nakamura et al. demonstrating that in patients with fetal loss a significant fraction of

purified IgG bound to an annexin V affinity column and that the eluted fraction had LA activity and also caused HUVEC apoptosis in culture (198). These results await repitition in other patient groups, particuarly those with thrombosis.

5. FAMILIAL APS

Cases of APS (or Sneddon's syndrome) in multiple members of families have been documented intermittently (199,200,201) although, as always in a condition that only rarely appears familial, it is difficult to exclude a common environmental influence. Other familial cases of apparent APS have been described in association with inheritance of the Factor V Leiden mutation (APC resistance) (202,203,204), however this common mutation does not appear to be increased in incidence in general populations of patients with APS (205,206). It is important to exclude other inherited disorders of thrombosis, such as the hyper-homocysteinaemia of methylene tetrahydrofolate reductase mutations (207), inherited elevation of plasminogen activator inhibitor (208), the G20210A mutation of prothrombin (209) and the familial haemolytic uraemic syndrome associated with complement factor H deficiency, all of which may resemble APS.

More recently, a number of families have been identified with familial APS with segregation analysis suggesting a susceptibility gene inherited in an autosomal dominant manner (210). Linkage was not, however, demonstrated with a number of candidate genes including HLA, Fas, Fas ligand or B2GPI, with the possibility raised that the genetic influence may be heterogenous. HLA class II associations have been reported for APS (211,212), but only for specific ethnic groups.

6. PROPOSED PATHOGENETIC MECHANISMS OF APL

The mechanism by which aPL cause the thromboses and other clinical manifestations of APS has not yet been eluci-

Table 2 Affinities of anionic PL binding proteins (see also Harper et al. (34))

Substance	Quantity	Molarity	Kd(M)	Mol./Kd	References
β_2-glycoprotein I	200 mg/l	3.7×10^{-6}	6.9×10^{-6}	0.54	(126,143,34)
Prothrombin	100 mg/l	1.4×10^{-6}	5.0×10^{-7}	2.9	(339)
Factor X	5 mg/l	1.7×10^{-7}	1.5×10^{-7}	1.1	(339)
Protein C	4 mg/l	6.5×10^{-8}	2.3×10^{-7}	0.28	(220)
Annexin V	<0.5 ηg/l	$<1.4 \times 10^{-14}$	$<1 \times 10^{-10}$	N/A	(195, 340)

All data given is collated from data given for roughly physiologic conditions: approximately plasma [Na], [Ca] and pH using 20% phosphatidylserine/80% phosphatidylcholine.

dated. Indeed, there are two ways to approach the association between aPL and APS, only one of which is that the antibodies are actually in some way pathogenic. Another view is that an ongoing injury to the vascular/endothelial surface is the pathogenic event by some other means, including a chronic endothelial infection such as EBV (213) or mycoplasma (214), or possibly complement-fixing antibodies directed against endothelial cells. In this case disrupted or apoptotic endothelial cells may express anionic phospholipids and in addition leukocyte adhesion markers, HLA-I and vWF. Anionic PL binding proteins such as B2GPI, prothrombin and annexin V will then be phagocytosed by macrophages or dendritic cells along with the endothelial cells. In individuals with disordered elimination of self-reactive lymphocyte clones such as SLE patients (215) these proteins attached to apoptotic waste may be presented to activatable T_H and B lymphocytes (216), resulting in antibody production to plasma proteins such as B2GPI. In this way the antibodies would be irrelevant bystanders and serve only as markers for another process of endothelial cell injury. This would, of course, be disappointing for many in the field, although the anti-PL binding proteins could also self-induce their production by promoting further FcγR mediated phagocytosis of apoptotic particles (217,218).

The evidence to the contrary, such that aPL are indeed pathogenic, has been presented and various mechanisms have been proposed, with some demonstrations in mouse models including effects on the activated protein C complex, annexin V, apoptosis, atherosclerosis, endothelial cells and platelets. Many of the proposed effects depend upon the binding of aPL and the protein "co-factors" to phospholipid with displacement from and competition for the presumably limited number of anionic PL sites present *in vivo*. An awareness of the relative affinities of the proteins for phospholipid with and without the additional effect of dimeric antibody binding is therefore of relevance (see: Table 2). Given the slight differences in conditions in the different experiments, the 1 log variation in molarity/Kd might be considered to be of questionable significance.

Dimeric binding by antibody/protein complex, or protein/protein complex formation in the case of coagulation factors, has further effects on the measurable affinity to phospholipid surfaces of that complex. The addition of polyclonal aCL to B2GPI (0.1μM) increased binding onto a planar 20% PS membrane at near physiologic ionic conditions some 30–40 fold, with evidence supporting the view that the increased affinity is as a result of dimeric binding (169). Mouse monoclonal antibodies against various domains of human B2GPI (but not against the CL binding domain V) increased B2GPI binding to 20% PS vesicles some 2–3 fold (219), but these antibodies will bind free B2GPI at much high affinity with the Fab' fragment able to

bind alone possibly explaining the lesser augmentation of binding. The binding of prothrombin to immobilised 40%PS in wells (not physiologically equivalent to the above) was increased 2 fold by LA anti-prothrombin antibodies and 4 fold by antiprothrombin antibodies from hypoprothrombinaemic patients (187).

Taken together, the increase in affinity of these proteins for anionic phospholipid upon antibody binding suggests a means by which anti-PL/PL binding protein complexes may compete with and displace the prothrombinase complex in the LA tests, thus prolonging the clotting time. It should however be pointed out that prothrombinase is also a complex of Xa &Va binding to prothrombin and phospholipid, as is APC also in general in a complex form. There is evidence that the PL binding avidity of APC is higher than that of PC, at approximately 15×10^{-8} M compared to 23×10^{-8} M, and that this is increased another log by stoichiometric binding to form a complex with protein S on a phospholipid surface (220). A similar increase in affinity probably occurs with the prothrombinase complex. A direct comparison of the respective molarities and affinities of individual proteins may not reflect the behavior of the *in vivo* binding of protein complexes. The mechanisms by which antibodies increasing the affinity of B2GPI lead to thrombosis are included in the consideration of the various pathogenic mechanisms that follows.

6.1. Effects of aPL on the Activated Protein C Complex (Degradation of Factor V, Va & VIIIa)

An attractive hypothesis for the pathophysiology of thrombosis in APS is interference with the mechanism of degrading factor V by activated protein C (APC) in complex with protein S. Despite the fact that hereditary disorders of this process are rarely associated with arterial or cardiac disease this issue has been extensively explored.

Phospholipid dramatically enhances the activation of protein C by thrombin/thrombomodulin (221) and the proteolytic degradation of factors V, Va and VIIIa. Activated protein C in conjunction with protein S cleaves factor Va at Arg^{506} then Arg^{306} and Arg^{679} (222) (rev. in 223) and, probably of more relevance than thought, the zymogen factor V at Arg^{306} (224), which may significantly limit the quantity of available substrate. Factor Va cleaved at Arg^{506} may also have anticoagulant effects by acting as an additional cofactor for APC (225).

The activation of protein C by thrombin/thrombomodulin takes place on the endothelial cell surface, but there is no evidence for direct anti-thrombomodulin antibody activity in APS (226). A purified immunoglobulin fraction containing an IgM LA had an inhibitory effect on the enhancement by phospholipid of the proteolytic activation of protein C by thrombin with purified thrombomodulin, but with no effect in the absence of phospholipid (227). A similar slowing of the rate of activation of protein C has been reported for endothelial cells previously cultured in the presence of IgG preparations with LA activity (228). A very similar method could not repeat the finding using a number of aPL IgG preparations, although a Methusalian criticism might be that the relative quantity of endothelial cells to thrombin/protein C may have been higher in this study (229).

An interesting hypothesis put forward by de Groot and Derksen (230) is that inhibition of thrombin production (as seen in the LA tests) by anti-prothrombin antibodies may in itself be thrombogenic, citing the finding that a low level intravenous infusion of thrombin (as opposed to the more physiologic high level local production of thrombin) in baboons (231) is primarily anticoagulant and predominantly appears to activate protein C. This assumes first that antiprothrombin antibodies do not act to prolong or potentiate the effect of thrombin and, since most anti-factor II antibodies bind fragment 1 and/or 2 but not thrombin (232), this may be valid, but second, the *in vitro* demonstration of impaired prothrombinase activity in the LA tests is occuring *in vivo*, and since there appears to be increased thrombin generation in APS (233), this assumption may be incorrect.

Many researchers have examined whether aPL can influence the cleavage of Factor Va by the APC complex. After the activation of clotting by the addition of phospholipid, the degradation of Va was found to be markedly slowed in all samples containing LA activity (234). Immunoglobulin fractions with LA activity in a purified system inhibited the cleavage of Va by APC/protein S on a phospholipid (PS/PC) or freeze/thawed platelet surface (235) and also on an endothelial cell surface (236).

Several groups have suggested that this effect may be dependent on the presence of protein S and may inhibit the function of protein S in some instances (237,238), although demonstration of direct anti-protein S antibodies is rare (232). Only about 40% of the protein S in the circulation is free, with the remainder bound to C4b binding protein (C4b-BP). Atsumi et al. suggest that aCL may increase the binding of protein S to C4b-BP and that B2GPI inhibits this binding, although the evidence for this was in a complex ELISA system (239).

Anti-B2GPI antibodies have been assessed by Matsuda et al., who found that rabbit anti-B2GPI antibody inhibited Va degradation and that this effect was B2GPI-dependent (240). A similar finding has been made using monoclonal anti-B2GPI (241). Galli et al. found that in a plasma clotting system, the LA samples with a more abnormal DRVVT than KCT had a delay to peak Va activity, consistent with the LA effect on thrombin generation. In addition, the subsequent degradation of Va was also significantly delayed, and this effect persisted when affinity-purified anti-B2GPI, but not anti-prothrombin antibodies, were used (242).

β_2 glycoprotein I may itself have an effect on the APC complex (241) with both "procoagulant" and "anticoagulant" *in vitro* features. The depletion of B2GPI from plasma appeared to increase the thrombin generation in a Va-dependent system without APC, but also decreased the thrombin generation in the presence of added APC (243). It also, however, prolonged the DRVVT making the interpretation of this experiment assessing thrombin generation somewhat unclear. The 20 amino-terminal residues of a protein reported to bind activated protein C (244) are also identical to B2GPI.

The most likely explanation for the inhibition of APC complex in these various experiments with APS antibodies is competition for anionic phospholipid binding sites akin to the *in vitro* prolongation of clotting times found in the various LA tests. Indeed, increasing the available phospholipid (essentially the platelet neutralisation procedure of LA) diminishes or abolishes the APC inhibition both of aPL (235) and of B2GPI (243). This begs the question of what, if any, of this is significant to the pathological mechanisms of APS *in vivo*.

Smirnov et al. have constructed a formulation of the effects suggesting that in many cases of aPL the APC complex may be more inhibited than the prothrombinase reaction. They firstly demonstrated that the incorporation of phosphatidylethanolamine (PE), and to a lesser extent cardiolipin, into phospholipid vesicles increased factor Va inactivation by APC 10-fold, with very little effect on prothrombin activation (245), and later by creating a protein C chimera with the PL binding (Gla) domain taken from prothrombin that the effect of PE on APC activity is specific to the native protein C PL binding region (246). They have further demonstrated that in the presence of APC at concentrations near that of physiologic protein C, and in the presence of anionic PL + PE, the plasma of some patients with LA actually clots faster than controls, and that very potent inhibition of APC activity could be observed in plasma with relatively modest LA activity (247). This was not, however, a clinically correlated study, nor is it necessarily the same as the finding of anti-PE antibodies (see phospholipid specificity of aPL). Additionally, another group has suggested that the incorporation of PE into PL vesicles does actually significantly increase the activity of the prothrombinase complex (248).

6.2. Anti-Endothelial Cell Antibodies (AECA)

The endothelial proliferation in vessels involved in APS and the occurrence of thrombosis in both arteries and veins suggests any common etiology of the syndrome might well involve a common vessel wall feature such as endothelial cells rather than a disorder of the coagulation cascade, the hereditary disorders of which tend to produce only venous thrombosis.

An effect on endothelial cells in culture was first suggested by Careras et al., who reported a finding of a decrease in the production of prostacycline in the presence of serum containing a lupus anticoagulant (249), although this was not confirmed (250). There is frequently endothelial binding activity in the immunoglobulin fraction of patients with antiphospholipid antibodies (251). More recently, the binding of an APS IgG fraction to human umbilical vein endothelial cells (HUVEC) in culture has been shown to activate the cells (rather than bind to activated cells) to upregulate endothelial leukocyte adhesion molecule (ELAM)-1 in a B2GPI-dependent manner (252,253), although not in all cases (254). Human aPL moabs likewise have been reported to result in an increase in leukocyte adherence to EC (255), which in itself may be thrombogenic. Rather than simply activating endothelial cells the binding of some (B2GPI-independent), AECA may directly induce apoptosis (256), which may involve annexin V-mediated binding (198).

AECA affinity purified against B2GPI has been shown to continue to bind to endothelial cells in the presence of B2GPI, suggesting that at least a part of the antiendothelial activity in APS patients is due to anti-B2GPI antibodies. A similar finding in apoptotic, but not resting thymocytes, suggests a role for anionic phospholipid translocation to the outer membrane lamina as the means of B2GPI binding (257). It is unclear in *in vitro* endothelial cells whether the binding of B2GPI is to anionic PL expressed on the cell surface of the fraction of EC beginning to die, or whether there is a specific EC receptor for B2GPI (258).

An increase in specific endothelial proliferative signals has not been specifically demonstrated in APS, although myointimal cell proliferation is apparent in arterial lesions (*vide supra*) and thrombin is known to have such an effect. Interestingly, the Kringle 2 domain of prothrombin (fragment 2) has been recently demonstrated to have an inhibitory effect (not cytotoxic effect) on bovine endothelial cell proliferative activity (259), as have kringle fragments of plasminogen (260). As an increase in prothrombin fragments 1&2 has been cited as suggestive of evidence for a prothrombotic state (presumably also with increased thrombin levels) in SLE with aCL (233), the balance of the relative pro-&-anti-proliferative effects may be of importance.

Antiendothelial cell antibodies are unlikely to explain all the details of the unique pathophysiology of APS on their own, if for no other reason than their ubiquitousness by current methods of assay (261) in a host of autoimmune disorders associated with arteritis, including cytomegalovirus infection (262), SLE without APS, systemic sclerosis (256), Takayasu's (263), Wegener's (264) and rheumatoid arthritis-associated arteritis (265). The HUVEC plasma membrane contains many different potential protein epitopes and there is evidence to support a range of heterogenous MW epitopes

being detected between different anti-endothelial patient sera (266); as they become better characterised, antibody/epitope associations with clinical features may become apparent. A plausible explanation for why some patients with aPL do not develop APS may be a requirement for 2 pathologic events: AECA to induce apoptotic movement of anionic phospholipid to the surface (267) before it can be bound by B2GPI or prothrombin, which is then available in immobilised form on anionic PL to be bound by the relevant antibody.

6.3. Beta₂-Glycoprotein I and Apoptosis

The process of apoptosis in cells is associated with a loss of the normal bilayer membrane lipid asymmetry with movement of PS to the outer surface (268,269). There is evidence that the identification of PS on the outer surface is a factor in the recognition of apoptotic cells for phagocytosis by scavenger cells (270). Anionic PL binding proteins such as B2GPI would be expected to bind to apoptotic cells and indeed fluorescence-labelled annexin V is used as a FACS marker for this purpose. The binding of B2GPI has been demonstrated for PS containing liposomes (271), apoptotic thymocytes (257) and red blood cell ghosts (272).

B2GPI has also been shown to have a role in the recognition of PS expressing cells by macrophages by Balasubramanian et al., who reported an increased uptake of red blood cell ghosts and apoptotic thymocytes by macrophages in the presence of B2GPI. Although the effect reported in this paper was modest (1.5 fold and 1.2 fold increase from control respectively), there was a more significant increase after the addition of anti-B2GPI antibody (272). It was later found that the addition of F(ab')₂ anti-B2GPI antibody fragments inhibited the B2GPI effect on the uptake by macrophages, suggesting the presence of a macrophage receptor for B2GPI, although the binding of B2GPI to macrophages was not demonstrated in the absence of PS-containing liposomes (273).

The opsonisation of apoptotic cells by anti-B2GPI antibodies/B2GPI antigen may result in preferential internalisation of these cells, by antigen presenting cells, such as dendritic cells (274), possibly reinforcing the immune response to B2GPI. Co-immunisation of mice with apoptotic cells and heterologous B2GPI may also lead to the production of aPL (275).

An important question is whether B2GPI is functioning as a marker specifically of apoptotic cells, which are degraded without activation by phagocytosing cells (276), or whether necrotic cells are also labelled by B2GPI, as the phagocytosis of necrotic cells and necrotic debris may lead to an inflammatory response and antigen presentation by the phagocytosing cell. Given that the exposure of PS also occurs during necrosis (277), the ability of B2GPI to bind to anionic PL may not be the semiotic tag that distinguishes apoptotic from necrotic cells.

6.4. Annexin V as an Anticoagulant Protein Affected by aPL

Annexin V binds to phospholipid in the presence of free calcium ions (191) with an affinity that depends upon the the degree of negative charge found in the phospholipid (278,279), thus cardiolipin, phosphatidylglycerol and phosphatidylserine are bound tightly with weak binding to sphingomyelin and questionable if any binding to phosphatidylcholine. Phosphatidylcholine with sphingomyelin are the dominant phospholipids in the outer membrane lamina of undamaged endothelial cells and unactivated platelets, but phosphatidylserine may move from the inner to the outer lamina during a cell injury that leads to apoptosis (280), and during platelet activation (268). Although this is thought to provide the binding sites for the coagulation factors with γ-carboxylated glutamate residues (Factors II, VII, IX, X and Proteins C & S), it is proposed that annexin V competes with some or all of these factors for the anionic phospholipid binding sites that are necessary for the propagation of the coagulation cascade. This inhibitory effect on the tenase and prothrombinase complexes (281,282) can certainly be demonstrated *in vitro* and is consistent with the affinities for anionic PL of the relevant proteins (see table 2).

The crucial and unanswered question is whether this happens *in vivo*, as annexin V appears to be essentially an intracellular protein and may not function as a circulating natural anticoagulant (195). As discussed in the section on the annexin V protein, most normal individuals do not have detectable annexin V in plasma.

This *in vitro* anticoagulant action of annexin V has been studied in the light of aPL and B2GPI binding to the same negatively-charged phospholipid. Rand et al. initially reported a reduction in the amount of annexin V on the placental villi of women with aPL and recurrent abortions (283), but was not confirmed elsewhere (284). Subsequently Rand et al. reported that antiphospholipid IgG immunoglobulins reduced the quantity of annexin V bound to various phospholipid surfaces, and that the reduction was dependent on the presence of B2GPI (285,286), but that antibody did not bind annexin V. They also found that the presence of B2GPI with aPL reduced the prolongation of the aPTT caused by annexin V, although why B2GPI with aPL had sufficient affinity to displace annexin V but seemingly not compete so well with the coagulation factors was not clear. The converse, that annexin V displaces antiphospholipid antibody, has been demonstrated by Pierangeli et al., who also found that the binding of antibody to anionic phospholipid membranes was increased in the presence of B2GPI but somewhat inhibited when annexin V was then added, suggesting that while a component of the CL affinity purified antibodies may require B2GPI as a cofactor, all 3 components appeared to compete for the same anionic phospholipid sites (190). However, in both studies annexin V levels used were in the order of $\mu g/ml$ ie:10^3-fold or more greater than that expected in the vascular compartment. A simple way to view this function is that, like an LA, the annexin V protein *in vitro* produces an LA effect when assayed in a coagulation test with calcium ions and limited phospholipid.

6.5. APS and Links with Atherosclerosis

Although clearly distinct from the thrombotic vista seen in APS, several links have been drawn with atherosclerosis. Phospholipids comprise an integral part of lipoprotein micelles and oxidation of lipoproteins, particulary low density lipoproteins (LDL), may be a major mechanism of atherosclerosis (287). Macrophages appear to have receptors for oxidatively damaged LDL (288) and in atherosclerotic plaques may become foam cells on phagocytosis of excessive quantities of lipoproteins.

Elevated titres of antibodies against such malondialdehyde modified (oxidised) LDL have been demonstrated in 80% of a group of patients with SLE, with a weak correlation with the presence of aCL (289). Monoclonal antibodies developed against oxidised LDL have been reported to also bind oxidised cardiolipin (28). The presence of anti-ox-LDL antibodies did not associate with the presence of arterial thrombosis in patients with SLE (some with SAPS), whereas anti-B2GPI did so associate (290).

This begs the question of whether B2GPI is involved in the binding of antibody to ox-LDL, given the report of immunoreactivity against B2GPI in atherosclerotic plaque (291). In one study, the addition of B2GPI was found to have no effect on the binding of anti-ox-LDL, although methodologic problems included using a kit-based test not specifically set up for the addition of other agent. In addition, the mean anti-ox-LDL titre found in normals was as high or higher than in the various aPL and SLE patient groups (292). The situation then is not diasporetic, with Horkko et al. finding that some aPL bind a B2GPI/ox-LDL complex (293) and Hasunuma et al. finding that the addition of B2GPI itself inhibited the binding of ox-LDL to macrophages, but the addition of anti-B2GPI plus B2GPI increased the binding (294), possibly by providing a means of receptor (Fcγ) mediated binding (295). It has also been shown that B2GPI may bind to lipoprotein(a) (296): Lp(a) has a protein sequence with homology to plasminogen (297) and the level of Lp(a) may be a risk for atherosclerotic disease.

An alternative approach is whether patients with aPL have increased levels of lipid oxidation as part of a prothrombotic state. Elevated urinary levels of F_2-isoprostanes, an oxidation product of arachidonic acid, was found in SLE patients and correlated with both plasma TNF ($r = 0.84$) and aCL titre ($r = 0.70$) (298). It is not clear then if this is a function of SLE disease activity in which TNF may sometimes be elevated, or of secondary APS. The same group has also demonstrated that treatment with anti-oxidant vitamins reduced the levels of these oxidation products concurrent with a reduction in the plasma level of prothrombin fragments 1 + 2 (299), which are released when prothrombin is cleaved to thrombin.

6.6. Platelets in APS

Thrombocytopaenia is an important manifestation of APS and by analogy with Heparin-Induced Thrombotic Thrombocytopaenic Syndrome (HITTS), it is conceivable that aPL might induce thrombosis in a similar fashion (300). Logically, aCL will bind to platelets upon perturbation as this will induce movement of PS to the outer lamina of the bilayer membrane, as has been demonstrated (301).

Although aPL may be present in many cases of immune thrombocytopaenic purpura (ITP), the titres of antibody do not correlate with the platelet count (118) and patient aPL have been reported not to crossreact with antibodies against platelet glycoproteins (302) even if mouse moabs can be identified that do so (303). The report that some APS patients with thrombocytopaenia have specific anti-platelet glycoprotein antibodies (304) is interesting given the comparative scarcity of marked thrombocytopaenia or purpura in APS, although it is conceivable that these may be of lower avidity than those occuring in ITP. It has also been reported that the inhibition of tenase on activated platelets by B2GPI may be reversed by the addition of anti-cardiolipin antibodies (305).

The more important issue is whether aPL can induce activation of platelets rather than just bind to activated platelets. Unfortunately, platelets assessed by flow cytometry are subject to handling artifacts that may activate them to express the activation markers CD62 (p-selectin) and CD63, and the quantification of platelet granule products such as thromboxane B2, a stable product of platelet derived TX A2, can be subject to measurement artifact and may result from any cause of platelet activation. As such, one group found an increase in the expression of CD63 on APS patient platelets but not CD62 (306), another found an increase in CD62 expression but not CD63 (307) and a third (308) found no increase in CD62 (p-selectin) but a decrease in platelet serotonin levels. Anti-B2GPI moabs in the presence of B2GPI were only able to activate platelets provided other activators were added at subthreshold concentrations (309). An increase in the urinary excretion of 11-dehydro-TX B2 has been reported for patients with aPL (310,311), and purified aCL/B2GPI has been reported to increase the *in vitro* production of TX B2 of normal platelets (311). Elsewhere, however, serum from APS patients was reported to bind to platelets but have no effect on the expression of activation markers such as CD62 (312).

7.   ANIMAL MODELS FOR APS

Humans with APS have a rather unpredictable and heterogenous course with clinical events happening infrequently. There is, therefore, a need in APS for a suitable animal model for therapeutic experiments as well as to better characterise the immune response in APS. Unfortunately there is no ideal model yet available. Various mouse models develop diseases with features of APS, but mice are difficult to venesect repetitively. Rabbits, from which more blood can be obtained, have not been reported to develop APS. The four different mouse models described are of inbred spontaneous autoimmune disease, active immunisation with antigen, passive transfer with human APS antibody and active anti-anti-idiotype immunisation with human APS or SLE antibody.

The immune system of mice is well characterised, with known spontaneous mouse models of SLE (313,314), additional features of an APS-like syndrome having been identified in MRL/++ (315,316) and NZW × BXSB F_1 mice (317). These mice, in addition to SLE-type autoantibodies (ANA, anti-ssDNA, anti-dsDNA), develop direct anticardiolipin antibodies, aCL antibodies that are B2GPI-dependent (317) and direct anti-B2GPI antibodies (318). These mice experience features seen in human APS such as relative thrombocytopaenia, but not thrombosis. They also develop skin changes and nephritis similar to that found in human SLE. The NZW × BXSB F_1 mice, if allowed to live long enough, will develop coronary and sometimes cerebral thrombosis, but this may be secondary to renal hypertension (319).

Mice and rabbits immunised with human B2GPI were reported to develop high titre high affinity polyclonal antibodies in both aCL ELISA (using ABS) and then anti-B2GPI ELISA (320), in that order (321). Others have found that the B2GPI-immunised rabbits will develop anti-B2GPI antibodies, some of which can also bind cardiolipin directly, probably as a result of non-specific charge dependent binding (24). Some phospholipids have been reported to be directly antigenic experimentally, but this was not found for purified cardiolipin, which does not tend to assume a hexagonal configuration (321).

In the MRL/++ model (*vide supra*), it has been reported that the acceleration of the inevitable aCL and ANA antibody production follows immunisation with B2GPI. A study examining a range of manifestations found that this increased the rate of fetal resorption (322), although the control group that was immunised with ovalbumin likewise had a degree of fetal resorption. No neurological manifestations developed.

In other strains of mice not predisposed to autoimmune disease, such as BALB/c, Blank et al. found that successful immunisation with B2GPI lead to features of APS, including thrombocytopaenia and fetal resorptions (323). It was also later reported that this could be ameliorated by the prior oral administration (oral tolerance) of low dose B2GPI (324). However, another group in essentially identical conditions found that immunisation with B2GPI producing anti-B2GPI antibody did not lead to clinical disease; in particular, there was no change in fetal loss or resorption and no change in platelet count (325).

Passive transfer experiments injecting purified polyclonal or monoclonal human aCL antibodies into mice have been examined in several ways. Initially, Branch et al. found that each mouse injected with IgG fractions from aPL patients aborted (326). However, more recently, that group have published that the effect is somewhat variable and that, while some fractions do, other IgG fractions from women with recurrent miscarriage and aPL do not cause murine fetal loss (327). Although transferred aCL antibodies have not been observed to cause deep venous thrombosis in mice, they have been reported to increase the size and duration of the thrombus resulting from a standardised crush injury to the mouse femoral vein after the passive transfer of either an IgG fraction (328) or a monoclonal IgG (329) from patients with APS.

The active immunisation of mice using human (idiotype) aCL to produce anti-idiotype and then anti-anti-idiotype (ie:idiotype) aCL has been widely investigated, largely by one group. This methodology was initially applied to SLE, whence it was reported that a human monoclonal anti-DNA (16/6 idiotype) antibody administered to female C3H.SW mice produced an SLE-like disease (330). The same group have since reported that an APS-like disease including neurological disturbance (331) can be induced in BALB/c mice by the administration of mouse monoclonal aCL and human polyclonal aCL (332), another monoclonal (MIV-7) directed against idiotype 16/6 (although a different clone to the one above) (333) and have made similar findings generating ANCA in mice by immunising with human ANCA (334). Furthermore, they have found that the experimental APS was ameliorated by treatment with IL-3 (335), low molecular weight heparin (336), anti CD-4 moab (333), ciprofloxacin (337), and peptides found to bind the inducing monoclonal in a phage display library (255). Unfortunately, the original finding of induction of an SLE-like disease was not able to be reproduced by an independent group (338), who cited possible environmental factors in their inability to generate any disease or autoantibodies in the appropriately immunised group.

8. CONCLUSION

The antiphospholipid syndrome is associated with antibodies directed against plasma proteins, in particular B2GPI and prothrombin. The molecular and physiologic mechanisms of thrombosis and the other clinical manifestations of APS are currently far from clear, with many inconsistencies in the literature that highlight the incomplete under-

standing of the field. Nevertheless, the antiphospholipid syndrome is clinically important and the mystery surrounding its pathogenesis probably contributes to the interest in this syndrome.

ACKNOWLEDGMENTS

The authors would like to thank Drs Allan Sturgess and Michael Barnett for their critical review of the manuscript, Dr Andrej Sali for the B2GPI modelling predictions, Mr. Marcus Cremonese from Medical illustrations and Ms. Karen Andrews and staff from the J.L. Latham Library.

Reference List

1. Harris, E.N., E. Baguley, R.A. Asherson, and G.R.V. Hughes. 1987. Clinical and serologocal features of the "Anti-Phospholipid Syndrome" (APS) (abstract). *Br. J. Rheumatol.* 26:S19.
2. Alarcon-Segovia, D., M.E. Perez-Vazquez, A.R. Villa, C. Drenkard, and J. Cabiedes. 1992. Preliminary classification criteria for the antiphospholipid syndrome within systemic lupus erythematosus. *Semin. Arthritis Rheum.* 21:275–286.
3. Asherson, R.A., M.A. Khamashta, J. Ordi-Ros, R.H.M.W. Derksen, S.J. Machin, J. Barquinero, H.H. Outt, E.N. Harris, M. Vilardell-Torres, and G.R.V. Hughes. 1989. The "primary" antiphospholipid syndrome: major clinical and serological features. *Medicine* 68:366–374.
4. Harris, E.N., A.E. Gharavi, M.L. Boey, B.M. Patel, C.G. Mackworth-Young, S. Loizou, and G.R.V. Hughes. 1983. Anticardiolipin antibodies: detection by radioimmunoassay and association with thrombosis in systemic lupus erythematosus. *Lancet* 2:1211–1214.
5. Moore, J.E. and C.F. Mohr. 1952. Biologically false positive serologic tests for syphilis: type, incidence, and cause. *JAMA* 150:467–473.
6. Laurell, A.-B. and I.M. Nilsson. 1957. Hypergammaglobulinemia, circulating anticoagulant, and biologic false positive wasserman reaction: a study in two cases. *J. Lab. Clin. Med.* 49:694–707.
7. Bowie, E.J.W., J.H. Thompson, C.A. Pascuzzi, and C.A. Owen. 1963. Thrombosis in systemic lupus erythematosis despite circulating anticoagulants. *J. Lab. Clin. Med.* 63:416–430.
8. Mueh, J.R., K.D. Herbst, and S.I. Rapaport. 1980. Thrombosis in patients with the lupus anticoagulant. *Ann. Intern. Med.* 92:156–159.
9. Figueroa, F., X. Berrios, M. Gutierrez, F. Carrion, J.P. Goycolea, I. Riedel, and S. Jacobelli. 1992. Anticardiolipin antibodies in acute rheumatic fever. *J. Rheum.* 19:1175–1180.
10. Love, P.E. and S.A. Santoro. 1990. Antiphospholipid antibodies: anticardiolipin and the lupus anticoagulant in systemic lupus erythematosus (SLE) and in non-SLE disorders. Prevalence and clinical significance. *Ann. Intern. Med.* 112:682–698.
11. McNeil, H.P., C.N. Chesterman, and S.A. Krilis. 1989. Anticardiolipin antibodies and lupus anticoagulants

comprise separate antibody subgroups with different phospholipid binding characteristics. *Br. J. Haematol.* 73:506–513.

12. McNeil, H.P., R.J. Simpson, C.N. Chesterman, and S.A. Krilis. 1990. Anti-phospholipid antibodies are directed against a complex antigen that includes a lipid-binding inhibitor of coagulation: β_2-glycoprotein-I (apolipoprotein H). *Proc. Natl. Acad. Sci. USA* 87:4120–4124.

13. Galli, M., P. Comfurius, C. Maassen, H.C. Hemker, M.H. De Baets, P.J.C. van Breda-Vriesman, T. Barbui, R.F.A. Zwaal, and E.M. Bevers. 1990. Anticardiolipin antibodies (ACA) directed not to cardiolipin but to a plasma protein cofactor. *Lancet* 335:1544–1547.

14. Matsuura, E., Y. Igarashi, M. Fujimoto, K. Ichikawa, and T. Koike. 1990. Anticardiolipin cofactor(s) and differential diagnosis of autoimmune disease (letter). *Lancet* 336:177–178.

15. Bevers, E.M., M. Galli, T. Barbui, P. Comfurius, and R.F.A. Zwaal. 1991. Lupus anticoagulant IgG's (LA) are not directed to phospholipids only, but to a complex of lipid-bound human prothrombin. *Thromb. Haemost.* 66:629–632.

16. Galli, M., P. Comfurius, T. Barbui, R.F.A. Zwaal, and E.M. Bevers. 1992. Anticoagulant activity of β_2-glycoprotein I is potentiated by a distinct subgroup of anticardiolipin antibodies. *Thromb. Haemost.* 68:297–300.

17. Roubey, R.A.S., C.W. Pratt, J.P. Buyon, and J.B. Winfield. 1992. Lupus anticoagulant activity of autoimmune antiphospholipid antibodies is dependent upon β_2-glycoprotein I. *J. Clin. Invest.* 90:1100–1104.

18. Arvieux, J., L. Darnige, C. Caron, G. Reber, J.C. Bensa, and M.G. Colomb. 1995. Development of an ELISA for autoantibodies to prothrombin showing their prevalence in patients with lupus anticoagulants. *Thromb. Haemost.* 74:1120–1125.

19. Loeliger, A. 1959. Prothrombin as a cofactor of the circulating anticoagulant in systemic lupus erythematosus? *Thromb. Diath. Haemorrh.* 3:237–256.

20. Matsuura, E., Y. Igarashi, T. Yasuda, D.A. Triplett, and T. Koike. 1994. Anticardiolipin antibodies recognize β_2-glycoprotein I structure altered by interacting with an oxygen modified solid phase surface. *J. Exp. Med.* 179:457–462.

21. Viard, J.P., Z. Amoura, and J.F. Bach. 1992. Association of anti-beta 2 glycoprotein I antibodies with lupus-type circulating anticoagulant and thrombosis in systemic lupus erythematosus. *Am. J. Med.* 93:181–186.

22. Arvieux, J., V. Regnault, E. Hachulla, L. Darnige, B. Roussel, and J.C. Bensa. 1998. Heterogeneity and immunochemical properties of anti-β_2-glycoprotein I autoantibodies. *Thromb. Haemost.* 80:393–398.

23. Hunt, J.E., H.P. McNeil, G.J. Morgan, R.M. Crameri, and S.A. Krilis. 1992. A phospholipid β_2-glycoprotein I complex is an antigen for anticardiolipin antibodies occurring in autoimmune disease but not with infection. *Lupus* 1:75–81.

24. Kouts, S., M.X. Wang, S. Adelstein, and S.A. Krilis. 1995. Immunization of a rabbit with β_2-glycoprotein I induces charge-dependent crossreactive antibodies that bind anionic phospholipids and have similar reactivity as autoimmune anti-phospholipid antibodies. *J. Immunol.* 155:958–966.

25. Smeenk, R.J.T., W.A.M. Lucassen, and T.J.G. Swaak. 1987. Is anticardiolipin activity a cross-reaction of anti-DNA or a seperate entity? *Arthritis Rheum.* 30:607–617.

26. Sugi, T. and J.A. McIntyre. 1995. Autoantibodies to phosphatidylethanolamine (PE) recognize a kininogen-PE complex. *Blood* 86:3083–3089.

27. Matsuda, J., N. Saitoh, K. Gohchi, M. Gotoh, and M. Tsukamoto. 1994. Anti-annexin V antibody in systemic lupus erythematosus patients with lupus anticoagulant and/or anticardiolipin antibody. *Am. J. Hematol.* 47:56–58.

28. Horkko, S., E. Miller, E. Dudl, P. Reaven, L.K. Curtiss, N.J. Zvaifler, R. Terkeltaub, S.S. Pierangeli, D.W. Branch, W. Palinski, and J.L. Witztum. 1996. Antiphospholipid antibodies are directed against epitopes of oxidized phospholipids. Recognition of cardiolipin by monoclonal antibodies to epitopes of oxidized low density lipoprotein. *J. Clin. Invest.* 98:815–825.

29. Gharavi, A.E., E.N. Harris, R.A. Asherson, and G.R.V. Hughes. 1987. Anticardiolipin antibodies: isotype distribution and phospholipid specificity. *Ann. Rheum. Dis.* 46:1–6.

30. Reddel, S.W., and S.A. Krilis. 1999. The methodology and clinical significance of anticardiolipin antibodies. *Clin. Diagn. Lab. Immunol.* 6:775–782.

31. Reber, G., J. Arvieux, E. Comby, D. Degenne, P. de Moerloose, M. Sanmarco, and G. Potron. 1995. Multicenter evaluation of nine commercial kits for the quantitation of anticardiolipin antibodies. The Working Group on Methodologies in Haemostasis from the GEHT (Groupe d'Etudes sur l'Hemostase et la Thrombose). *Thromb. Haemost.* 73:444–452.

32. Galli, M., and T. Barbui. 1999. Antiprothrombin antibodies: detection and clinical significance in the antiphospholipid syndrome. *Blood* 93:2149–2157.

33. McNeil, H.P., C.N. Chesterman, and S.A. Krilis. 1991. Immunology and clinical importance of antiphospholipid antibodies. *Adv. Immunol.* 49:193–280.

34. Harper, M.F., P.M. Hayes, B.R. Lentz, and R.A.S. Roubey. 1998. Characterization of β_2- glycoprotein I binding to phospholipid membranes. *Thromb. Haemost.* 80:610–614.

35. Karmochkine, M., P. Cacoub, J.C. Piette, P. Godeau, and M.C. Boffa. 1992. Antiphosphatidylethanolamine antibody as the sole antiphospholipid antibody in systemic lupus erythematosus with thrombosis. *Clin. Exp. Rheumatol.* 10:603–605.

36. Vlachoyiannopoulos, P.G., G. Beigbeder, M. Dueymes, P. Youinou, J.E. Hunt, S.A. Krilis, and H.M. Moutsopoulos. 1993. Antibodies to phosphatidylethanolamine in antiphospholipid syndrome and systemic lupus erythematosus: their correlation with anticardiolipin antibodies and beta 2 glycoprotein-I plasma levels. *Autoimmunity* 16:245–249.

37. Berard, M., R. Chantome, A. Marcelli, and M.C. Boffa. 1996. Antiphosphatidylethanolamine antibodies as the only antiphospholipid antibodies. I. Association with thrombosis and vascular cutaneous diseases. *J. Rheum.* 23:1369–1374.

38. Drouvalakis, K.A., and R.R. Buchanan. 1998. Phospholipid specificity of autoimmune and drug induced lupus anticoagulants; association of phosphatidylethanolamine reactivity with thrombosis in autoimmune disease. *J. Rheum.* 25:290–295.

39. Sugi, T. and J.A. McIntyre. 1996. Phosphatidylethanolamine induces specific conformational changes in the kininogens recognizable by antiphosphatidylethanolamine antibodies. *Thromb. Haemost.* 76:354–360.

40. Kobayashi, T., E. Stang, K.S. Fang, P. de Moerloose, R.G. Parton, and J. Gruenberg. 1998. A lipid associated with the antiphospholipid syndrome regulates endosome structure and function. *Nature* 392:193–197.

41. Kandiah, D.A., and S.A. Krillis. 1998. Heterogeneity of lupus anticoagulant (LA) antibodies: LA activity in dilute Russell's Viper Venom Time and dilute Kaolin Clotting Time detect different populations of antibodies in patients

with the "antiphospholipid" syndrome. *Thromb. Haemost.* 80:250–257.

42. Brandt, J.T., D.A. Triplett, B. Alving, and I. Scharrer. 1995. Criteria for the diagnosis of lupus anticoagulants: an update. On behalf of the Subcommittee on Lupus Anticoagulant/Antiphospholipid Antibody of the Scientific and Standardisation Committee of the ISTH. *Thromb. Haemost.* 74:1185–1190.

43. Exner, T. 1998. Methods for subtyping lupus anticoagulants. *Lupus* 7 Suppl 2:S103–S106.

44. Petri, M., M. Rheinschmidt, Q. Whiting-O'Keefe, D. Hellmann, and L. Corash. 1987. The frequency of lupus anticoagulant in systemic lupus erythematosus. A study of sixty consecutive patients by activated partial thromboplastin time, Russell viper venom time, and anticardiolipin antibody level. *Ann. Intern. Med.* 106:524–531.

45. Horbach, D.A., E. van Oort, R.C.J.M. Donders, R.H.M.W. Derksen, and P.G. de Groot. 1996. Lupus anticoagulant is the strongest risk factor for both venous and arterial thrombosis in patients with systemic lupus erythematosus. Comparison between different assays for the detection of antiphospholipid antibodies. *Thromb. Haemost.* 76:916–924.

46. Kandiah, D.A., A. Sali, Y. Sheng, E.J. Victoria, D.M. Marquis, S.M. Coutts, and S.A. Krilis. 1998. Current insights into the "antiphospholipid" syndrome: clinical, immunological, and molecular aspects. *Adv. Immunol.* 70:507–563.

47. Rosove, M.H., and P.M. Brewer. 1992. Antiphospholipid thrombosis: clinical course after the first thrombotic event in 70 patients. *Ann. Intern. Med.* 117:303–308.

48. Westerman, E.M., J.M. Miles, M. Backonja, and W.R. Sundstrom. 1992. Neuropathologic findings in multi-infarct dementia associated with anticardiolipin antibody. Evidence for endothelial injury as the primary event. *Arthritis Rheum.* 35:1038–1041.

49. Hughson, M.D., G.A. McCarty, C.M. Sholer, and R.A. Brumback. 1993. Thrombotic cerebral arteriopathy in patients with the antiphospholipid syndrome. *Modern Pathology* 6:644–653.

50. Levine, S.R., M.J. Deegan, N. Futrell, and K.M.A. Welch. 1990. Cerebrovascular and neurologic disease associated with antiphospholipid antibodies: 48 cases. *Neurology* 40:1181–1189.

51. Miller, D.J., S.A. Maisch, M.D. Perez, D.L. Kearney, and T.F. Feltes. 1995. Fatal myocardial infarction in an 8-year-old girl with systemic lupus erythematosus, Raynaud's phenomenon, and secondary antiphospholipid antibody syndrome. *J. Rheum.* 22:768–773.

52. Kincaid-Smith, P., K.F. Fairley, and M. Kloss. 1988. Lupus anticoagulant associated with renal thrombotic microangiopathy and pregnancy-related renal failure. *Q. J. Med.* 69(new series):795–815.

53. Nochy, D., E. Daugas, D. Droz, H. Beaufils, J.P. Grunfeld, J.C. Piette, J. Bariety, and G. Hill. 1998. Are there intrarenal vascular lesions characteristic of the primary antiphospholipid syndrome (APS)? (abstract). *Lupus* 7:S224.

54. Hughson, M.D., G.A. McCarty, and R.A. Brumback. 1995. Spectrum of vascular pathology affecting patients with the antiphospholipid syndrome. *Hum. Pathol.* 26:716–724.

55. Alarcon-Segovia, D., M.H. Cardiel, and E. Reyes. 1989. Antiphospholipid arterial vasculopathy. *J. Rheum.* 16:762–767.

56. Goldberger, E., R.C. Elder, R.A. Schwartz, and P.E. Phillips. 1992. Vasculitis in the antiphospholipid syndrome. A cause of ischemia responding to corticosteroids. *Arthritis Rheum.* 35:569–572.

57. Asherson, R.A., M.A. Khamashta, A. Gil, J.J. Vazquez, O. Chan, E. Baguley, and G.R.V. Hughes. 1989. Cerebrovascular disease and antiphospholipid antibodies in systemic lupus erythematosus, lupus-like disease, and the primary antiphospholipid syndrome. *Am. J. Med.* 86:391–399.

58. Brey, R.L., R.G. Hart, D.G. Sherman, and C.H. Tegeler. 1990. Antiphospholipid antibodies and cerebral ischemia in young people. *Neurology* 40:1190–1196.

59. Boggild, M.D., R.V. Sedhev, D. Fraser, and J.R. Heron. 1995. Cerebral venous sinus thrombosis and antiphospholipid antibodies. *Postgrad. Med. J.* 71:487–489.

60. Kale, U.S. and R.G. Wight. 1998. Primary presentation of spontaneous jugular vein thrombosis to the otolaryngologist-in three different pathologies. *J. Laryngol. Otol.* 112:888–890.

61. Tietjen, G.E., M. Day, L. Norris, S. Aurora, A. Halvorsen, L.R. Schultz, and S.R. Levine. 1998. Role of anticardiolipin antibodies in young persons with migraine and transient focal neurologic events: a prospective study. *Neurology* 50:1433–1440.

62. Herranz, M.T., G. Rivier, M.A. Khamashta, K.U. Blaser, and G.R.V. Hughes. 1994. Association between antiphospholipid antibodies and epilepsy in patients with systemic lupus erythematosus. *Arthritis Rheum.* 37:568–571.

63. Verrot, D., M. San-Marco, C. Dravet, P. Genton, P. Disdier, G. Bolla, J.R. Harle, L. Reynaud, and P.J. Weiller. 1997. Prevalence and signification of antinuclear and anticardiolipin antibodies in patients with epilepsy. *Am. J. Med.* 103:33–37.

64. Asherson, R.A., N.E. Harris, A.E. Gharavi, and G.R.V. Hughes. 1986. Systemic lupus erythematosus, antiphospholipid antibodies, chorea, and oral contraceptives (letter). *Arthritis Rheum.* 29:1535–1536.

65. Cervera, R., R.A. Asherson, J. Font, M. Tikly, L. Pallares, A. Chamorro, and M. Ingelmo. 1997. Chorea in the antiphospholipid syndrome. Clinical, radiologic, and immunologic characteristics of 50 patients from our clinics and the recent literature. *Medicine* 76:203–212.

66. Coull, B.M., D.N. Bourdette, S.H. Goodnight, D.P. Briley, and R. Hart. 1987. Multiple cerebral infarctions and dementia associated with anticardiolipin antibodies. *Stroke* 18:1107–1112.

67. Levine, S.R. and R.L. Brey. 1996. Neurologic aspects of antiphospholipid syndrome. *Lupus* 5:347–353.

68. Van Horn, G., F.C. Arnett, and M.M. Dimachkie. 1996. Reversible dementia and chorea in a young woman with the lupus anticoagulant. *Neurology* 46:1599–1603.

69. Fukazawa, T., F. Moriwaka, M. Mukai, T. Hamada, T. Koike, and K. Tashiro. 1993. Anticardiolipin antibodies in Japanese patients with multiple sclerosis. *Acta Neurol. Scand.* 88:184–189.

70. D'Olhaberriague, L., S.R. Levine, L. Salowich-Palm, D. Tanne, K.L. Sawaya, T.K. Aurora, M. Perry, M. Day, T. Spencer, and L. Schultz. 1998. Specificity, isotype, and titer distribution of anticardiolipin antibodies in CNS diseases. *Neurology* 51:1376–1380.

71. Ijdo, J.W., A.M. Conti-Kelly, P. Greco, M. Abedi, M. Amos, J.M. Provenzale, and T.P. Greco. 1999. Antiphospholipid antibodies in patients with multiple sclerosis and MS-like illnesses: MS or APS? *Lupus* 8:109–115.

72. Kalashnikova, L.A., E.L. Nasonov, A.E. Kushekbaeva, and L.A. Gracheva. 1990. Anticardiolipin antibodies in Sneddon's syndrome. *Neurology* 40:464–467.

73. Marsch, W.C. and R. Muckelmann. 1985. Generalized racemose livedo with cerebrovascular lesions (Sneddon syndrome): an occlusive arteriolopathy due to proliferation and migration of medial smooth muscle cells. *Br. J. Dermatol.* 112:703–708.

74. Tourbah, A., J.C. Piette, M.T. Iba-Zizen, O. Lyon-Caen, P. Godeau, and C. Frances. 1997. The natural course of cerebral lesions in Sneddon syndrome. *Arch. Neurol.* 54:53–60.

75. Frances, C., T. Papo, B. Wechsler, J.L. Laporte, V. Biousse, and J.C. Piette. 1998. Sneddon's syndrome (SNS) with or without antiphospholipid antibodies (aPL): a comparative study in 46 patients (abstract). *Lupus* 7:S225.

76. Fulham, M.J., P. Gatenby, and R.R. Tuck. 1994. Focal cerebral ischemia and antiphospholipid antibodies: a case for cardiac embolism. *Acta Neurol. Scand.* 90:417–423.

77. Khamashta, M.A., R. Cervera, R.A. Asherson, J. Font, A. Gil, D.J. Coltart, J.J. Vazquez, C. Pare, M. Ingelmo, J. Oliver, and G.R.V. Hughes. 1990. Association of antibodies against phospholipids with heart valve disease in systemic lupus erythematosus. *Lancet* 335:1541–1544.

78. Gleason, C.B., M.F. Stoddard, S.G. Wagner, R.A. Longaker, S. Pierangeli, Harris, and EN. 1993. A comparison of cardiac valvular involvement in the primary antiphospholipid syndrome versus anticardiolipin-negative systemic lupus erythematosus. *Am. Heart J.* 125:1123–1129.

79. Roldan, C.A., B.K. Shively, C.C. Lau, F.T. Gurule, E.A. Smith, and M.H. Crawford. 1992. Systemic lupus erythematosus valve disease by transesophageal echocardiography and the role of antiphospholipid antibodies. *J. Am. Coll. Cardiol.* 20:1127–1134.

80. Vianna, J.L., M.A. Khamashta, J. Ordi-Ros, J. Font, R. Cervera, A. Lopez-Soto, C. Tolosa, J. Franz, A. Selva, M. Ingelmo, M. Viardell, and G.R.V. Hughes. 1994. Comparison of the primary and secondary antiphospholipid syndrome: a European Multicenter Study of 114 patients. *Am. J. Med.* 96:3–9.

81. Towheed, T.E., P.M. Ford, J.M. Pym, and S.E. Ford. 1995. Cardiac valve replacement and antiphospholipid antibodies (letter). *J. Rheum.* 22:802–803.

82. Garcia-Torres, R., M.C. Amigo, A. de la Rosa, A. Moron, and P.A. Reyes. 1996. Valvular heart disease in primary antiphospholipid syndrome (PAPS): clinical and morphological findings. *Lupus* 5:56–61.

83. Ziporen, L., I. Goldberg, M. Arad, M. Hojnik, J. Ordi-Ros, A. Afek, M. Blank, Y. Sandbank, M. Vilardell-Tarres, I. de Torres, A. Weinberger, R.A. Asherson, Y. Kopolovic, and Y. Shoenfeld. 1996. Libman-Sacks endocarditis in the antiphospholipid syndrome: immunopathologic findings in deformed heart valves. *Lupus* 5:196–205.

84. Brucato, A., F. Baudo, M. Barberis, R. Redaelli, G. Casadei, F. Allegri, E. De Juli, and F. De Cataldo. 1994. Pulmonary hypertension secondary to thrombosis of the pulmonary vessels in a patient with the primary antiphospholipid syndrome. *J. Rheum.* 21:942–944.

85. Asherson, R.A., M.A. Khamashta, E. Baguley, C.M. Oakley, N.R. Rowell, and G.R.V. Hughes. 1989. Myocardial infarction and antiphospholipid antibodies in SLE and related disorders. *Q. J. Med.* 73:1103–1115.

86. Phadke, K.V., R.A. Phillips, D.T.R. Clarke, M. Jones, P. Naish, and P. Carson. 1993. Anticardiolipin antibodies in ischaemic heart disease: marker or myth? *Br. Heart J.* 69:391–394.

87. Vaarala, O., M. Manttari, V. Manninen, L. Tenkanen, M. Puurunen, K. Aho, and T. Palosuo. 1995. Anticardiolipin antibodies and risk of myocardial infarction in a prospective cohort of middle-aged men. *Circulation* 91:23–27.

88. Jeffrey, P.J., R.A. Asherson, and P.J. Rees. 1989. Recurrent deep vein thrombosis, thromboembolic pulmonary hypertension and the "primary" antiphospholipid syndrome (letter). *Clin. Exp. Rheumatol.* 7:567–569.

89. Gertner, E. and J.T. Lie. 1993. Pulmonary capillaritis, alveolar hemorrhage, and recurrent microvascular thrombosis in primary antiphospholipid syndrome. *J. Rheum.* 20:1224–1228.

90. Maggiorini, M., A. Knoblauch, J. Schneider, and E.W. Russi. 1997. Diffuse microvascular pulmonary thrombosis associated with primary antiphospholipid antibody syndrome. *Eur. Respir. J.* 10:727–730.

91. Asherson, R.A. 1992. The catastrophic antiphospholipid syndrome. *J. Rheum.* 19:508–512.

92. Nilsson, I.M., B. Astedt, U. Hedner, and D. Berezin. 1975. Intrauterine death and circulating anticoagulant ("antithromboplastin"). *Acta Med. Scand.* 197:153–159.

93. Hughes, G.R.V. 1983. Thrombosis, abortion, cerebral disease, and the lupus anticoagulant. *Br. Med. J.* 287:1088–1089.

94. Cowchock, S. 1996. Prevention of fetal death in the antiphospholipid syndrome. *Lupus* 5:467–472.

95. Petri, M., M. Golbus, R. Anderson, Q. Whiting-O'Keefe, L. Corash, and D. Hellmann. 1987. Antinuclear antibody, lupus anticoagulant, and anticardiolipin antibody in women with idiopathic habitual abortion. A controlled, prospective study of forty-four women. *Arthritis Rheum.* 30:601–606.

96. Rai, R.S., L. Regan, K. Clifford, W. Pickering, M. Dave, I. Mackie, T. McNally, and H. Cohen. 1995. Antiphospholipid antibodies and beta 2-glycoprotein-I in 500 women with recurrent miscarriage: results of a comprehensive screening approach. *Hum. Reprod.* 10:2001–2005.

97. Stirrat, G.M. 1990. Recurrent miscarriage I: definition and epidemiology. *Lancet* 336:673–675.

98. de Costa, H., M.D. de Moura, R.A. Ferriani, M.I. Anceschi, and J.E. Barbosa. 1993. Prevalence of anti-cardiolipin antibody in habitual aborters. *Gynecol. Obstet. Invest.* 36:221–225.

99. Branch, D.W., J.R. Scott, N.K. Kochenour, and E. Hershgold. 1985. Obstetric compliations associated with the lupus anticoagulant. *N. Engl. J. Med.* 313:1322–1326.

100. Takakuwa, K., K. Asano, M. Arakawa, M. Yasuda, I. Hasegawa, and K. Tanaka. 1997. Chromosome analysis of aborted conceptuses of recurrent aborters positive for anti-cardiolipin antibody. *Fertil. Steril.* 68:54–58.

101. Pattison, N.S., L.W. Chamley, E.J. McKay, G.C. Liggins, and W.S. Butler. 1993. Antiphospholipid antibodies in pregnancy: prevalence and clinical associations. *Br. J. Obstet. Gynaecol.* 100:909–913.

102. Lynch, A., R. Marlar, J. Murphy, G. Davila, M. Santos, J. Rutledge, and W. Emlen. 1994. Antiphospholipid antibodies in predicting adverse pregnancy outcome. A prospective study. *Ann. Intern. Med.* 120:470–475.

103. Bagger, P.V., V. Andersen, B. Baslund, B. Beck, H. Hove, M. Hoier-Madsen, J. Petersen, J. Philip, O. Schaadt, S.O. Skouby, J. Starup, S. Thorsen, and A. Wiik. 1993. Anti-cardiolipin antibodies (IgG and IgA) in women with recurrent fetal loss correlate to clinical and serological characteristics of SLE. *Acta Obstet. Gynecol. Scand.* 72:465–469.

104. Rai, R., H. Cohen, M. Dave, and L. Regan. 1997. Randomised controlled trial of aspirin and aspirin plus heparin in pregnant women with recurrent miscarriage asso-

ciated with phospholipid antibodies (or antiphospholipid antibodies). *BMJ* 314:253–257.

105. Magid, M.S., C. Kaplan, L.R. Sammaritano, M. Peterson, M.L. Druzin, and M.D. Lockshin. 1998. Placental pathology in systemic lupus erythematosus: a prospective study. *Am. J. Obstet. Gynecol.* 179:226–234.

106. Salafia, C.M. and A.L. Parke. 1997. Placental pathology in systemic lupus erythematosus and phospholipid antibody syndrome. *Rheum. Dis. Clin. North. Am.* 23:85–97.

107. Levy, R.A., E. Avvad, J. Oliviera, and L.C. Porto. 1998. Placental pathology in antiphospholipid syndrome. *Lupus* 7:s81–s85.

108. Abramowsky, C.R., M.E. Vegas, G. Swinehart, and M.T. Gyves. 1980. Decidual vasculopathy of the placenta in lupus erythematosus. *N. Engl. J. Med.* 303:668–672.

109. Hanly, J.G., D.D. Gladman, T.H. Rose, C.A. Laskin, and M.B. Urowitz. 1988. Lupus pregnancy. A prospective study of placental changes. *Arthritis Rheum.* 31:358–366.

110. Gibson, G.E., W.P. Su, and M.R. Pittelkow. 1997. Antiphospholipid syndrome and the skin. *J. Am. Acad. Dermatol.* 36:970–982.

111. Dowd, P.M. 1998. Reactions to cold. In: *Rook's textbook of dermatology.* R.H. Champion, J.L. Burton, D.A. Burns, and S.M. Breathnach, editors. Blackwell Science, Oxford. 963–964.

112. Asherson, R.A., S. Jacobelli, H. Rosenberg, P. Mckee, and G.R.V. Hughes. 1992. Skin nodules and macules resembling vasculitis in the antiphospholipid syndrome—a report of two cases. *Clin. Exp. Dermatol.* 17:266–269.

113. Grob, J.J. and J.J. Bonerandi. 1989. Thrombotic skin disease as a marker of the anticardiolipin syndrome. Livedo vasculitis and distal gangrene associated with abnormal serum antiphospholipid activity. *J. Am. Acad. Dermatol.* 20:1063–1069.

114. Klein, K.L. and M.R. Pittelkow. 1992. Tissue plasminogen activator for treatment of livedoid vasculitis. *Mayo Clin. Proc.* 67:923–933.

115. Harris, E.N., R.A. Asherson, A.E. Gharavi, S.H. Morgan, G. Derue, and G.R.V. Hughes. 1985. Thrombocytopenia in SLE and related autoimmune disorders: association with anticardiolipin antibody. *Br. J. Haematol.* 59:227–230.

116. Hughes, G.R.V. 1985. The anticardiolipin syndrome. *Clin. Exp. Rheumatol.* 3:285–286.

117. Harris, E.N., J.K.H. Chan, R.A. Asherson, V.R. Aber, A.E. Gharavi, and G.R.V. Hughes. 1986. Thrombosis, recurrent fetal loss, and thrombocytopenia. Predictive value of the anticardiolipin antibody test. *Arch. Intern. Med.* 146:2153–2156.

118. Stasi, R., E. Stipa, M. Masi, F. Oliva, A. Sciarra, A. Perrotti, Olivieri, G. Zaccari, G.M. Gandolfo, M. Galli, T. Barbui, and G. Papa. 1994. Prevalence and clinical significance of elevated antiphospholipid antibodies in patients with idiopathic thrombocytopenic purpura. *Blood* 84:4203–4208.

119. Derksen, R.H.M.W., P.G. de Groot, L. Kater, and H.K. Nieuwenhuis. 1993. Patients with antiphospholipid antibodies and venous thrombosis should receive long term anticoagulant treatment. *Ann. Rheum. Dis.* 52:689–692.

120. Khamashta, M., M.J. Cuadrado, F. Mujic, N.A. Taub, B.J. Hunt, and G.R.V. Hughes. 1995. The management of thrombosis in the antiphospholipid-antibody syndrome. *N. Engl. J. Med.* 332:993–997.

121. Kutteh, W.H. 1996. Antiphospholipid antibody-associated recurrent pregnancy loss: treatment with heparin and low-dose aspirin is superior to low-dose aspirin alone. *Am. J. Obstet. Gynecol.* 174:1584–1589.

122. Khamashta, M.A. 1998. Management of thrombosis and pregnancy loss in the antiphospholipid syndrome. *Lupus* 7:S163–S169.

123. Schultze, H.E., K. Hiede, and H. Haupt. 1961. Uber ein bisher unbekanntes niedermolekulares beta2-Globulin des Humanserums. *Naturwissenschaften* 48:719.

124. Nakaya, Y., E.J. Schaefer, and H.B. Brewer, Jr. 1980. Activation of human post heparin lipoprotein lipase by apolipoprotein II (β_2-glycoprotein I). *Biochem. Biophys. Res. Commun.* 95:1168–1172.

125. Lee, N.S., H.B.J. Brewer, and J.C. Osborne, Jr. 1983. β_2-Glycoprotein I. Molecular properties of an unusual apolipoprotein, apolipoprotein H. *J. Biol. Chem.* 258:4765–4770.

126. McNally, T., I.J. Mackie, D.A. Isenberg, and S.J. Machin. 1993. Immunoelectrophoresis and ELISA techniques for assay of plasma β_2-glycoprotein-1 and the influence of plasma lipids. *Thromb. Res.* 72:275–286.

127. Koppe, A.L., H. Walter, V.P. Chopra, and M. Bajatzadeh. 1970. Investigations on the genetics and population genetics of the β_2-glycoprotein I polymorphism. *Humangenetik* 9:164–171.

128. Propert, D.N. 1983. β_2-Glycoprotein I phenotypes in chronic schizophrenia. *Biol. Psychiatry* 18:727–731.

129. Cleve, H. 1968. Genetic studies on the deficiency of $\beta_2$glycoprotein I of human serum. *Humangenetik* 5:294–304.

130. Bancsi, L.F.J.M.M., I.K. van der Linden, and R.M. Bertina. 1992. β_2-glycoprotein I deficiency and the risk of thrombosis. *Thromb. Haemost.* 67:649–653.

131. Hoeg, J.M., P. Segal, R.E. Gregg, Y.S. Chang, F.T. Lindgren, G.L. Adamson, Frank, C. Brickman, and H.B. Brewer, Jr. 1985. Characterization of plasma lipids and lipoproteins in patients with β_2-glycoprotein I (apolipoprotein H) deficiency. *Atherosclerosis* 55:25–34.

132. Takeuchi, R., S. Yasuda, T. Atsumi, M. Ieko, H. Takeya, T. Horita, H. Kasahara, Y. Miyoshi, K. Ichikawa, A. Tsutsumi, and T. Koike. 1998. Coagulation and fibrinolytic characteristics in a β_2-glycoprotein I deficiency (abstract). *Lupus* 7:S191.

133. Sheng, Y., H. Herzog, and S.A. Krilis. 1997. Cloning and characterization of the gene encoding the mouse β_2-glycoprotein I. *Genomics* 41:128–130.

134. Steinkasserer, A., C. Estaller, E.H. Weiss, R.B. Sim, and A.J. Day. 1991. Complete nucleotide and deduced amino acid sequence of human β_2-glycoprotein I. *Biochem. J.* 277:387–391.

135. Cohnen, G. 1970. Immunochemical quantitation of β_2-glycoprotein I in various diseases. *J. Lab. Clin. Med.* 75:212–216.

136. Averna, M., G. Paravizzini, G. Marino, E. Lanteri, G. Cavera, Barbagallo, CM, S. Petralia, S. Cavallaro, G. Magro, S. Grasso, A. Notarbartolo, and S. Travali. 1997. Liver is not the unique site of synthesis of β_2-glycoprotein I (apolipoprotein H): evidence for an intestinal localization. *Int. J. Clin. Lab. Res.* 27:207–212.

137. Chamley, L.W., J.L. Allen, and P.M. Johnson. 1997. Synthesis of β_2-glycoprotein 1 by the human placenta. *Placenta* 18:403–410.

138. Caronti, B., C. Calderaro, C. Allesandri, F. Conti, R. Tinghino, G. Palladini, and G. Valesini. 1999. β_2-glycoprotein I (B2-GPI) mRNA is expressed by several cell types involved in anti-phospholipid syndrome-related tissue damage. *Clin. Exp. Immunol.* 115:214–219.

139. Alvarado-de la Barrera, C., S. Bahena, L. Llorente, A. Martinez-Castillo, D. Alarcon-Segovia, and A.R. Cabral. 1998. β_2-glycoprotein-I mRNA transcripts are expressed by

hepatocytes but not by resting or activated human endothelial cells. *Thromb. Res.* 90:239–243.

140. Brighton, T.A., P.J. Hogg, Y.P. Dai, B.H. Murray, B.H. Chong, and C.N. Chesterman. 1996. β_2- glycoprotein I in thrombosis: evidence for a role as a natural anticoagulant. *Br. J. Haematol.* 93:185–194.

141. Brighton, T.A., Y.P. Dai, P.J., Hogg, and C.N. Chesterman. 1998. β_2-glycoprotein I (B2GPI) is not consumed in an animal model of DIC (abstract). *Lupus* 7:S173.

142. Vlachoyiannopoulos, P.G., S.A. Krilis, J.E. Hunt, M.N. Manoussakis, and H.M. Moutsopoulos. 1992. Patients with anticardiolipin antibodies with and without antiphospholipid syndrome: their clinical features and β_2-glycoprotein-I plasma levels. *Eur. J. Clin. Invest.* 22:482–487.

143. Lozier, J., N. Takahashi, and F.W. Putnam. 1984. Complete amino acid sequence of human plasma β_2-glycoprotein I. *Proc. Natl. Acad. Sci. USA* 81:3640–3644.

144. Walsh, M.T., H. Watzlawick, F.W. Putnam, K. Schmid, and R. Brossmer. 1990. Effect of the carbohydrate moiety on the secondary structure of β_2-glycoprotein. I. Implications for the biosynthesis and folding of glycoproteins. *Biochemistry* 29:6250–6257.

145. Schousboe, I. 1983. Characterization of subfractions of β_2-glycoprotein I: evidence for sialic acid microheterogeneity. *Int. J. Biochem.* 15:35–44.

146. Gries, A., J. Nimpf, H. Wurm, G.M. Kostner, and T. Kenner. 1989. Characterization of isoelectric subspecies of asialo-β_2-glycoprotein I. *Biochem. J.* 260:531–534.

147. Kouts, S., C.L. Bunn, A. Steinkasserer, and S. Krilis. 1993. Expression of human recombinant β_2-glycoprotein I with anticardiolipin antibody cofactor activity. *FEBS Lett.* 326:105–108.

148. Patthy, L. 1987. Detecting homology of distantly related proteins with consensus sequences. *J. Mol. Biol.* 198:567–577.

149. Kato, H. and K. Enjyoji. 1991. Amino acid sequence and location of the disulfide bonds in bovine beta 2 glycoprotein I: the presence of five Sushi domains. *Biochemistry* 30:11687–11694.

150. Steinkasserer, A., P.N. Barlow, A.C. Willis, Z. Kertesz, I.D. Campbell, R.B. Sim, and D.G. Norman. 1992. Activity, disulphide mapping and structural modelling of the fifth domain of human β_2-glycoprotein I. *FEBS Lett.* 313:193–197.

151. Reid, K.B.M. and A.J. Day. 1989. Structure-function relationships of the complement components. *Immunol. Today* 10:177–180.

152. Sheng, Y., A. Sali, H. Herzog, J. Lahnstein, and S.A. Krilis. 1996. Site-directed mutagenesis of recombinant human β_2-glycoprotein I identifies a cluster of lysine residues that are critical for phospholipid binding and anti-cardiolipin antibody activity. *J. Immunol.* 157:3744–3751.

153. Steinkasserer, A., D.J. Cockburn, D.M. Black, Y. Boyd, E. Solomon, and R.B. Sim. 1992. Assignment of apolipoprotein H (APOH: beta-2-glycoprotein I) to human chromosome 17q23—qter; determination of the major expression site. *Cytogenetics & Cell Genetics* 60:31–33.

154. Sheng, Y., S.A. Krilis, and A. Sali. 1997. Site-directed mutagenesis of recombinant human β_2-glycoprotein I. Effect of phospholipid binding and anticardiolipin antibody activity. *Ann. NY. Acad. Sci.* 815:331–333.

155. Ohkura, N., Y. Hagihara, T. Yoshimura, Y. Goto, and H. Kato. 1998. Plasmin can reduce the function of human β_2-glycoprotein I by cleaving domain V into a nicked form. *Blood* 91:4173–4179.

156. Hunt, J.E., R.J. Simpson, and S.A. Krilis. 1993. Identification of a region of β_2-glycoprotein I critical for lipid binding and anti-cardiolipin antibody cofactor activity. *Proc. Natl. Acad. Sci. USA* 90:2141–2145.

157. Horbach, D.A., E. van Oort, T. Lisman, J.C.M. Meijers, R.H.M.W. Derksen, and P.G. de Groot. 1998. β_2-glycoprotein I is proteolytically cleaved *in vivo* upon activation of fibrinolysis (abstract). *Lupus* 7:S 174, A007–S174.

158. Sanghera, D.K., D.R. Wagenknecht, J.A. McIntyre, and M.I. Kamboh. 1997. Identification of structural mutations in the fifth domain of apolipoprotein H (β_2-glycoprotein I) which affect phospholipid binding. *Hum. Mol. Genet.* 6:311–316.

159. Moestrup, S.K., I. Schousboe, C. Jacobsen, J.R. Leheste, E.I. Christensen, and T.E. Willnow. 1998. β_2-glycoprotein I (apolipoprotein H) and β_2-glycoprotein I-phospholipid complex harbor a recognition site for the endocytic receptor megalin. *J. Clin. Invest.* 102:902–909.

160. McCarthy, J.M., D.R. Wagenknecht, and J.A. McIntyre. 1994. Activity of antiphospholipid antibody ELISA cofactor in different animal sera. *J. Clin. Lab. Anal.* 8:167–171.

161. Arvieux, J., L. Darnige, E. Hachulla, B. Roussel, J.C. Bensa, and M.G. Colomb. 1996. Species specificity of anti-beta 2 glycoprotein I autoantibodies and its relevance to anti-cardiolipin antibody quantitation. *Thromb. Haemost.* 75:725–730.

162. Cabiedes, J., A.R. Cabral, and D. Alarcon-Segovia. 1995. Clinical manifestations of the antiphospholipid syndrome in patients with systemic lupus erythematosus associate more strongly with anti-beta 2-glycoprotein-I than with antiphospholipid antibodies. *J. Rheum.* 22:1899–1906.

163. Koike, T. and E. Matsuura. 1996. Anti-β_2-glycoprotein I antibody: specificity and clinical significance. *Lupus* 5:378–380.

164. Roubey, R.A.S., R.A. Eisenberg, M.F. Harper, and J.B. Winfield. 1995. "Anticardiolipin" autoantibodies recognize β_2-glycoprotein I in the absence of phospholipid. Importance of Ag density and bivalent binding. *J. Immunol.* 154:954–960.

165. Sheng, Y., D.A. Kandiah, and S.A. Krilis. 1998. Anti-β_2-glycoprotein I autoantibodies from patients with the "antiphospholipid" syndrome bind to β_2-glycoprotein I with low affinity: dimerization of β_2-glycoprotein I induces a significant increase in anti-β_2-glycoprotein I antibody affinity. *J. Immunol.* 161:2038–2043.

166. Pengo, V., A. Biasiolo, and M.G. Fior. 1995. Autoimmune antiphospholipid antibodies are directed against a cryptic epitope expressed when β_2-glycoprotein I is bound to a suitable surface. *Thromb. Haemost.* 73:29–34.

167. Tincani, A., L. Spatola, E. Prati, F. Allegri, P. Ferremi, R. Cattaneo, P. Meroni, and G. Balestrieri. 1996. The anti-β_2-glycoprotein I activity in human anti-phospholipid syndrome sera is due to monoreactive low-affinity autoantibodies directed to epitopes located on native β_2-glycoprotein I and preserved during species' evolution. *J. Immunol.* 157:5732–5738.

168. Steward, M.W. and A.M. Lew. 1985. The importance of antibody affinity in the performance of immunoassays for antibody. *J. Immunol. Methods.* 78:173–190.

169. Willems, G.M., M.P. Janssen, M.M.A.L. Pelsers, P. Comfurius, M. Galli R.F.A. Zwaal, and E.M. Bevers. 1996. Role of divalency in the high-affinity binding of anticardiolipin antibody-β_2-glycoprotein I complexes to lipid membranes. *Biochemistry* 35:13833–13842.

170. Wang, M.X., D.A. Kandiah, K. Ichikawa, M. Khamashta, G. Hughes, T. Koike, R. Roubey, and S.A. Krilis. 1995. Epitope specificity of monoclonal anti-β_2-glycoprotein I

antibodies derived from patients with the antiphospholipid syndrome. *J. Immunol.* 155:1629–1636.

171. Igarashi, M., E. Matsuura, Y. Igarashi, H. Nagae, K. Ichikawa, D.A. Triplett, and T. Koike. 1996. Human β_2-glycoprotein I as an anticardiolipin cofactor determined using mutants expressed by a baculovirus system. *Blood* 87:3262–3270.

172. George, J., B. Gilburd, M. Hojnik, Y. Levy, P. Langevitz, E. Matsuura, T. Koike, and Y. Shoenfeld. 1998. Target recognition of β_2-glycoprotein I (beta2GPI)-dependent anti-cardiolipin antibodies: evidence for involvement of the fourth domain of beta2GPI in antibody binding. *J. Immunol.* 160:3917–3923.

173. Iverson, G.M., E.J. Victoria, and D.M. Marquis. 1998. Anti-β_2-glycoprotein I (beta2GPI) autoantibodies recognize an epitope on the first domain of beta2GPI. *Proc. Natl. Acad. Sci. USA* 95:15542–15546.

174. Vu, T.K.H., D.T. Hung, V.I. Wheaton, and S.R. Coughlin. 1991. Molecular cloning of a functional thrombin receptor reveals a novel proteolytic mechanism of receptor activation. *Cell* 64:1057–1068.

175. Storck, J., B. Kusters, and E.R. Zimmermann. 1995. The tethered ligand receptor is the responsible receptor for the thrombin induced release of von Willebrand factor from endothelial cells (HUVEC). *Thromb. Res.* 77:249–258.

176. Degen, S.J.F., R.T.A. MacGillivray, and E.W. Davie. 1983. Characterization of the complementary deoxyribonucleic acid and gene coding for human prothrombin. *Biochemistry* 22:2087–2097.

177. Mizuochi, T., J. Fujii, W. Kisiel, and A. Kobata. 1981. Studies on the structures of the carbohydrate moiety of human prothrombin. *J. Biochem.* 90:1023–1031.

178. Thompson, A.R. 1977. Factor IX antigen by radioimmunoassay. Abnormal factor IX protein in patients on warfarin therapy and with hemophilia B. *J. Clin. Invest.* 59:900–910.

179. Esmon, C.T., J.W. Suttie, and C.M. Jackson. 1975. The functional significance of vitamin K action. Difference in phospholipid binding between normal and abnormal prothrombin. *J. Biol. Chem.* 250:4095–4099.

180. Esmon, C.T., W.G. Owen, and C.M. Jackson. 1974. A plausible mechanism for prothrombin activation by factor Xa, factor Va, phospholipid, and calcium ions. *J. Biol. Chem.* 249:8045–8047.

181. Krishnaswamy, S., W.R. Church, M.E. Nesheim, and K.G. Mann. 1987. Activation of human prothrombin by human prothrombinase. Influence of factor Va on the reaction mechanism. *J. Biol. Chem.* 262:3291–3299.

182. Stevens, W.K. and M.E. Nesheim. 1993. Structural changes in the protease domain of prothrombin upon activation as assessed by N-bromosuccinimide modification of tryptophan residues in prethrombin-2 and thrombin. *Biochemistry* 32:2787–2794.

183. Permpikul, P., L.V. Rao, and S.I. Rapaport. 1994. Functional and binding studies of the roles of prothrombin and β_2-glycoprotein I in the expression of lupus anticoagulant activity. *Blood* 83:2878–2892.

184. Fleck, R.A., S.I. Rapaport, and L.V. Rao. 1988. Anti-prothrombin antibodies and the lupus anticoagulant. *Blood* 83:2878–2892.

185. Rapaport, S.I., S.B. Ames, and B.J. Duval. 1960. A plasma coagulation defect in systemic lupus erythematosus arising from hypoprothrombinemia combined with antiprothrombinase activity. *Blood* 15:212–227.

186. Bajaj, S.P., S.I. Rapaport, D.S. Fierer, K.D. Herbst, and D.B. Schwartz. 1983. A mechanism for the hypoprothrombinemia of the acquired hypoprothrombinemia-lupus anticoagulant syndrome. *Blood* 61:684–692.

187. Cakir, B., A.C. Rivadeneira, P.M. Hayes, T.L. Ortel, M. Petri, and R.A.S. Roubey. 1998. Anti-prothrombin autoantibodies: detection and characterization (abstract). *Lupus* 7:S219,B067.

188. Galli, M., G. Beretta, M. Daldossi, E.M. Bevers, and T. Barbui. 1997. Different anticoagulant and immunological properties of anti-prothrombin antibodies in patients with antiphospholipid antibodies. *Thromb. Haemost.* 77:486–491.

189. van Heerde, W.L., P.G. de Groot, and C.P.M. Reutelingsperger. 1995. The complexity of the phospholipid binding protein Annexin V. *Thromb. Haemost.* 73:172–179.

190. Pierangeli, S.S., J. Dean, G.H. Goldsmith, D.W. Branch, A. Gharavi, and E.N. Harris. 1996. Studies on the interaction of placental anticoagulant protein 1, beta 2 glycoprotein 1, and antiphospholipid antibodies in the prothrombinase reaction and in the solid phase anticardiolipin assays. *J. Lab. Clin. Med.* 128:194–201.

191. Reutelingsperger, C.P.M., G. Hornstra, and H.C. Hemker. 1985. Isolation and partial purification of a novel anticoagulant from arteries of human umbilical cord. *Eur. J. Biochem.* 151:625–629.

192. Reutelingsperger, C.P.M., W.L. van Heerde, R. Hauptmann, C. Maassen, R.G.J. van Gool, P. de Leeuw, and A. Tiebosch. 1994. Differential tissue expression of Annexin VIII in human. *FEBS Lett.* 349:120–124.

193. Kobayashi, R., R. Nakayama, A. Ohta, F. Sakai, S. Sakuragi, and Y. Tashima. 1990. Identification of the 32 kDa components of bovine lens EDTA-extractable protein as endonexins I and II. *Biochem. J.* 266:505–511.

194. Murphy, C.T., S.H. Peers, R.A. Forder, R.J. Flower, F. Carey, and J. Westwick. 1992. Evidence for the presence and location of annexins in human platelets. *Biochem. Biophys. Res. Commun.* 189:1739–1746.

195. Flaherty, M.J., S. West, R.L. Heimark, K. Fujikawa, and J.F. Tait. 1990. Placental anticoagulant protein-I: measurement in extracellular fluids and cells of the hemostatic system. *J. Lab. Clin. Med.* 115:174–181.

196. Christmas, P., J. Callaway, J. Fallon, J. Jones, and H.T. Haigler. 1991. Selective secretion of annexin I, a protein without a signal sequence, by the human prostate gland. *J. Biol. Chem.* 266:2499–2507.

197. Takehara, K., S. Uchida, N. Marumoto, T. Asawa, S. Osugi, S. Kurusu, I. Hashimoto, and M. Kawaminami. 1994. Secretion of recombinant rat annexin 5 by insect cells in a baculovirus expression system. *Biochem. Biophys. Res. Commun.* 200:1421–1427.

198. Nakamura, N., T. Ban, K. Yamaji, Y. Yoneda, and Y. Wada. 1998. Localization of the apoptosis-inducing activity of lupus anticoagulant in an annexin V-binding antibody subset. *J. Clin. Invest.* 101:1951–1959.

199. Matthey, F., K. Walshe, I.J. Mackie, and S.J. Machin. 1989. Familial occurrence of the antiphospholipid syndrome. *J. Clin. Pathol.* 42:495–497.

200. Ford, P.M., D. Brunet, D.P. Lillicrap, and S.E. Ford. 1990. Premature stroke in a family with lupus anticoagulant and antiphospholipid antibodies. *Stroke* 21:66–71.

201. Pettee, A.D., B.A. Wasserman, N.L. Adams, W. McMullen, H.R. Smith, S.L. Woods, and O.D. Ratnoff. 1994. Familial Sneddon's syndrome: clinical, hematologic, and radiographic findings in two brothers. *Neurology* 44:399–405.

202. Brenner, B., S.L. Vulfsons, N. Lanir, and M. Nahir. 1996. Coexistence of familial antiphospholipid syndrome and factor V Leiden: impact on thrombotic diathesis. *Br. J. Haematol.* 94:166–167.

203. Alarcon-Segovia, D., G.J. Ruiz-Arguelles, J. Garces-Eisele, and A. Ruiz. 1996. Inherited activated protein C resistance in a patient with familial primary antiphospholipid syndrome. *J. Rheum.* 23:2162–2165.

204. Picillo, U., D. De Lucia, E. Palatiello, A. Scuotto, M.R. Marcialis, S. Pezzella, and G. Tirri. 1998. Association of primary antiphospholipid syndrome with inherited activated protein C resistance. *J. Rheum.* 25:1232–1234.

205. Dizon-Townson, D., C. Hutchison, R. Silver, D.W. Branch, and K. Ward. 1995. The factor V Leiden mutation which predisposes to thrombosis is not common in patients with antiphospholipid syndrome. *Thromb. Haemost.* 74:1029–1031.

206. Sasso, E.H., L.A. Suzuki, A.R. Thompson, and M.A. Petri. 1997. Hereditary resistance to activated protein C: an uncommon risk factor for thromboembolic disease in lupus patients with antiphospholipid antibodies. *Arthritis Rheum.* 40:1720–1721.

207. Grandone, E., M., Margaglione, D. Colaizzo, S. Montanaro, G. Pavone, and G. Di Minno. 1997. Presence of FV Leiden and MTHFR mutation in a patient with complicated pregnancies (letter). *Thromb. Haemost.* 77:1036–1037.

208. Joseph, J. and E. Scopelitis. 1994. Seronegative antiphospholipid syndrome associated with plasminogen activator inhibitor. *Lupus* 3:201–203.

209. Grandone, E., M. Margaglione, D. Colaizzo, G. D'Andrea, G. Cappucci, V. Brancaccio, and G. Di Minno. 1998. Genetic susceptibility to pregnancy-related venous thromboembolism: roles of factor V Leiden, prothrombin G20210A, and methylenetetrahydrofolate reductase C677T mutations. *Am. J. Obstet. Gynecol.* 179:1324–1328.

210. Goel, N., T.L. Ortel, D. Bali, J.P. Anderson, I.S. Gourley, H. Smith, C.A. Morris, M. DeSimone, D.W. Branch, P. Ford, D. Berdeaux, R.A.S. Roubey, D.D. Kostyu, S.F. Kingsmore, T. Thiel, C. Amos, and M.F. Seldin. 1999. Familial antiphospholipid antibody syndrome: criteria for disease and evidence for autosomal dominant inheritance. *Arthritis Rheum.* 42:318–327.

211. Arnett, F.C., P. Thiagarajan, C. Ahn, and J.D. Reveille. 1999. Associations of anti-β_2-glycoprotein I autoantibodies with HLA class II alleles in three ethnic groups. *Arthritis Rheum.* 42:268–274.

212. Hashimoto, H., K. Yamanaka, Y. Tokano, N. Lida, Y. Takasaki, K. Kabasawa, and Y. Nishimura. 1998. HLA-DRB1 alleles and beta 2 glycoprotein I-dependent anticardiolipin antibodies in Japanese patients with systemic lupus erythematosus. *Clin. Exp. Rheumatol.* 16:423–427.

213. James, J.A., K.M. Kaufman, A.D. Farris, E. Taylor-Albert, T.J.A. Lehman, and J.B. Harley. 1997. An increased prevalence of Epstein-Barr virus infection in young patients suggests a possible etiology for systemic lupus erythematosus. *J. Clin. Invest.* 100:3019–3026.

214. Blasi, F., F. Denti, M. Erba, R. Cosentini, R. Raccanelli, A. Rinaldi, L. Fagetti, G. Esposito, U. Ruberti, and L. Allegra. 1996. Detection of Chlamydia pneumoniae but not Helicobacter pylori in atherosclerotic plaques of aortic aneurysms. *J. Clin. Microbiol.* 34:2766–2769.

215. Suzuki, N., M. Ichino, S. Mihara, S. Kaneko, and T. Sakane. 1998. Inhibition of Fas/Fas ligand-mediated apoptotic cell death of lymphocytes *in vitro* by circulating anti-Fas ligand autoantibodies in patients with systemic lupus erythematosus. *Arthritis Rheum.* 41:344–353.

216. Herrmann, M., R.E. Voll, O.M. Zoller, M. Hagenhofer, B.B. Ponner, and J.R. Kalden. 1998. Impaired phagocytosis of apoptotic cell material by monocyte-derived macrophages from patients with systemic lupus erythematosus. *Arthritis Rheum.* 41:1241–1250.

217. Manfredi, A.A., P. Rovere, G. Galati, S. Heltai, E. Bozzolo, L. Soldini, J. Davoust, G. Balestrieri, A. Tincani, and M.G. Sabbadini. 1998. Apoptotic cell clearance in systemic lupus erythematosus. I. Opsonization by antiphospholipid antibodies. *Arthritis Rheum.* 41:205–214.

218. Manfredi, A.A., P. Rovere, S. Heltai, G. Galati, G. Nebbia, A. Tincani, G. Balestrieri, and M.G. Sabbadini. 1998. Apoptotic cell clearance in systemic lupus erythematosus. II. Role of β_2-glycoprotein I. *Arthritis Rheum.* 41:215–223.

219. Takeya, H., T. Mori, E.C. Gabazza, K. Kuroda, H. Deguchi, E. Matsuura, K. Ichikawa, T. Koike, and K. Suzuki. 1997. Anti-β_2-glycoprotein I (beta2GPI) monoclonal antibodies with lupus anticoagulant-like activity enhance the beta2GPI binding to phospholipids. *J. Clin. Invest.* 99:2260–2268.

220. Walker, F.J. 1981. Regulation of activated protein C by protein S. The role of phospholipid in factor Va inactivation. *J. Biol. Chem.* 256:11128–11131.

221. Freyssinet, J.M., J. Gauchy, and J.P. Cazenave. 1986. The effect of phospholipids on the activation of protein C by the human thrombin-thrombomodulin complex. *Biochem. J.* 238:151–157.

222. Walker, F.J. and P.J. Fay. 1992. Regulation of blood coagulation by the protein C system. *FASEB J.* 6:2561–2567.

223. Kalafatis, M., P.E. Haley, D. Lu, R.M. Bertina, G.L. Long, and K.G. Mann. 1996. Proteolytic events that regulate factor V activity in whole plasma from normal and activated protein C (APC)-resistant individuals during clotting: an insight into the APC-resistance assay. *Blood* 87:4695–4707.

224. Petaja, J., J.A. Fermandez, A. Gruber, and J.H. Griffin. 1997. Anticoagulant synergism of heparin and activated protein C *in vitro*. Role of a novel anticoagulant mechanism of heparin, enhancement of inactivation of factor V by activated protein C. *J. Clin. Invest.* 99:2655–2663.

225. Thorelli, E., R.J. Kaufman, and B. Dahlback. 1999. Cleavage of factor V at Arg 506 by activated protein C and the expression of anticoagulant activity of factor V. *Blood* 93:2552–2558.

226. Gibson, J., M. Nelson, R. Brown, H. Salem, and H. Kronenberg. 1992. Autoantibodies to thrombomodulin: development of an enzyme immunoassay and a survey of their frequency in patients with the lupus anticoagulant. *Thromb. Haemost.* 67:507–509.

227. Freyssinet, J.M., M.L. Wiesel, J. Gauchy, B. Boneu, and J.P. Cazenave. 1986. An IgM lupus anticoagulant that neutralizes the enhancing effect of phospholipid on purified endothelial thrombomodulin activity—a mechanism for thrombosis. *Thromb. Haemost.* 55:309–313.

228. Cariou, R., G. Tobelem, C. Soria, and J. Caen. 1986. Inhibition of protein C activation by endothelial cells in the presence of lupus anticoagulant (letter). *N. Engl. J. Med.* 314:1193–1194.

229. Oosting, J.D., R.H.M.W. Derksen, T.M. Hackeng, M. van Vliet, K.T. Preissner, B.N. Bouma, and P.G. de Groot. 1991. *In vitro* studies of antiphospholipid antibodies and its cofactor, β_2-glycoprotein I, show negligible effects on endothelial cell mediated protein C activation. *Thromb. Haemost.* 66:666–671.

230. de Groot, P.G. and R.H.M.W. Derksen. 1994. Protein C pathway, antiphospholipid antibodies and thrombosis. *Lupus* 3:229–233.

231. Hanson, S.R., J.H. Griffin, L.A. Harker, A.B. Kelly, C.T. Esmon, and A. Gruber. 1993. Antithrombotic effects of thrombin-induced activation of endogenous protein C in primates. *J. Clin. Invest.* 92:2003–2012.

232. Rao, L.V.M., A.D. Hoang, and S.I. Rapaport. 1996. Mechanism and effects of the binding of lupus anticoagulant IgG and prothrombin to surface phospholipid. *Blood* 88:4173–4182.

233. Ginsberg, J.S., C. Demers, P. Brill-Edwards, M. Johnston, R. Bona, R.F. Burrows, J. Weitz, and J.A. Denburg. 1993. Increased thrombin generation and activity in patients with systemic lupus erythematosus and anticardiolipin antibodies: evidence for a prothrombotic state. *Blood* 81:2958–2963.

234. Marciniak, E. and E.H. Romond. 1989. Impaired catalytic function of activated protein C: a new *in vitro* manifestation of lupus anticoagulant. *Blood* 74:2426–2432.

235. Malia, R.G., S. Kitchen, M. Greaves, and F.E. Preston. 1990. Inhibition of activated protein C and its cofactor protein S by antiphospholipid antibodies. *Br. J. Haematol.* 76:101–107.

236. Borrell, M., N. Sala, C. de Castellarnau, S. Lopez, M. Gari, and J. Fontcuberta. 1992. Immunoglobulin fractions isolated from patients with antiphospholipid antibodies prevent the inactivation of factor Va by activated protein C on human endothelial cells. *Thromb. Haemost.* 68:268–272.

237. Lo, S.C.L., H.H. Salem, M.A. Howard, M.J. Oldmeadow, and B.G. Firkin. 1990. Studies of natural anticoagulant proteins and anticardiolipin antibodies in patients with the lupus anticoagulant. *Br. J. Haematol.* 76:380–386.

238. Oosting, J.D., R.H.M.W. Derksen, I.W.G. Bobbink, T.M. Hackeng, B.N. Bouma, and P.G. de Groot. 1993. Anti phospholipid antibodies directed against a combination of phospholipids with prothrombin, protein C, or protein S: an explanation for their pathogenic mechanism? *Blood* 81:2618–2625.

239. Atsumi, T., M.A. Khamashta, P.R. Ames, K. Ichikawa, T. Koike, and G.R.V. Hughes. 1997. Effect of beta 2glycoprotein I and human monoclonal anticardiolipin antibody on the protein S/C4b-binding protein system. *Lupus* 6:358–364.

240. Matsuda, J., K. Gohchi, K. Kawasugi, M. Gotoh, N. Saitoh, and M. Tsukamoto. 1995. Inhibitory activity of anti β_2-glycoprotein I antibody on factor Va degradation by activated-protein C and its cofactor protein S. *Am. J. Hematol.* 49:89–91.

241. Icko, M., K. Ichikawa, D.A. Triplett, E. Matsuura, T. Atsumi, K. Sawada, and T. Koike. 1999. β_2-glycoprotein I is necessary to inhibit protein C activity by monoclonal anticardiolipin antibodies. *Arthritis Rheum.* 42:167–174.

242. Galli, M., L. Ruggeri, and T. Barbui. 1998. Differential effects of anti-β_2-glycoprotein I and antiprothrombin antibodies on the anticoagulant activity of activated protein C. *Blood* 91:1999–2004.

243. Mori, T., H. Takeya, J. Nishioka, E.C. Gabazza, and K. Suzuki. 1996. β_2-glycoprotein I modulates the anticoagulant activity of activated protein C on the phospholipid surface. *Thromb. Haemost.* 75:49–55.

244. Canfield, W.M. and W. Kisiel. 1982. Evidence of normal functional levels of activated protein C inhibitor in combined Factor V/VIII deficiency disease. *J. Clin. Invest.* 70:1260–1272.

245. Smirnov, M.D. and C.T. Esmon. 1994. Phosphatidylethanolamine incorporation into vesicles selectively enhances factor Va inactivation by activated protein C. *J. Biol. Chem.* 269:816–819.

246. Smirnov, M.D., O. Safa, L. Regan, T. Mather, D.J. Stearns-Kurosawa, Kurosawa, A.R. Rezaie, N.L. Esmon, and C.T. Esmon. 1998. A chimeric protein C containing the prothrombin Gla domain exhibits increased anticoagulant activity and altered phospholipid specificity. *J. Biol. Chem.* 273:9031–9040.

247. Smirnov, M.D., D.T. Triplett, P.C. Comp, N.L. Esmon, and C.T. Esmon. 1995. On the role of phosphatidylethanolamine in the inhibition of activated protein C activity by antiphospholipid antibodies. *J. Clin. Invest.* 95:309–316.

248. Smeets, E.F., P. Comfurius, E.M. Bevers, and R.F.A. Zwaal. 1996. Contribution of different phospholipid classes to the prothrombin converting capacity of sonicated lipid vesicles. *Thromb. Res.* 81:419–426.

249. Carreras, L.O., G. Defreyn, S.J. Machin, J. Vermylen, R. Deman, B. Spitz, and A. van Assche. 1981. Arterial thrombosis, intrauterine death and "lupus" anticoagulant: detection of immunoglobulin interfering with prostacyclin formation. *Lancet* 1:244–246.

250. Hasselaar, P., R.H.M.W. Derksen, L. Blokzijl, and P.G. de Groot. 1988. Thrombosis associated with antiphospholipid antibodies cannot be explained by effects on endothelial and platelet prostanoid synthesis. *Thromb. Haemost.* 59:80–85.

251. Hasselaar, P., R.H.M.W. Derksen, L. Blokzijl, and P.G. de Groot. 1990. Crossreactivity of antibodies directed against cardiolipin, DNA, endothelial cells and blood platelets. *Thromb. Haemost.* 63:169–173.

252. Del Papa, N., L. Guidali, L. Spatola, P. Bonara, M.O. Borghi, A. Tincani, G. Balestrieri, and P.L. Meroni. 1995. Relationship between anti-phospholipid and anti-endothelial cell antibodies III: beta 2 glycoprotein I mediates the antibody binding to endothelial membranes and induces the expression of adhesion molecules. *Clin. Exp. Rheumatol.* 13:179–185.

253. Simantov, R., J.M. LaSala, S.K. Lo, A.E. Gharavi, L.R. Sammaritano, J.E. Salmon, and R.L. Silverstein. 1995. Activation of cultured vascular endothelial cells by antiphospholipid antibodies. *J. Clin. Invest.* 96:2211–2219.

254. Hanly, J.G., C. Hong, and A. Issekutz. 1996. β_2-glycoprotein I and anticardiolipin antibody binding to resting and activated cultured human endothelial cells. *J. Rheum.* 23:1543–1549.

255. Blank, M., Y. Shoenfeld, S. Cabilly, Y. Heldman, M. Fridkin, and E. Katchalski-Katzir. 1999. Prevention of experimental antiphospholipid syndrome and endothelial cell activation by synthetic peptides. *Proc. Natl. Acad. Sci. USA* 96:5164–5168.

256. Bordron, A., M. Dueymes, Y. Levy, C. Jamin, J.P. Leroy, J.C. Piette, Y. Shoenfeld, and P.Y. Youinou. 1998. The binding of some human antiendothelial cell antibodies induces endothelial cell apoptosis. *J. Clin. Invest.* 101:2029–2035.

257. Price, B.E., J. Rauch, M.A. Shia, M.T. Walsh, W. Lieberthal, H.M. Gilligan, T. O'Laughlin, J.S. Koh, and J.S. Levine. 1996. Anti-phospholipid autoantibodies bind to apoptotic, but not viable, thymocytes in a β_2-glycoprotein I—dependent manner. *J. Immunol.* 157:2201–2208.

258. Yan, W.Y., and K.R. McCrae. 1996. β2-glycoprotein-I (B2GP1) binds specifically to human endothelial cells (abstract). *Lupus* 5:504.

259. Lee, T.H., T. Rhim, and S.S. Kim. 1998. Prothrombin kringle-2 domain has a growth inhibitory activity against basic fibroblast growth factor-stimulated capillary endothelial cells. *J. Biol. Chem.* 273:28805–28812.

260. Cao, Y., A. Chen, S.S.A. An, R.W. Ji, D. Davidson, Y. Cao, and M. Linas. 1997. Kringle 5 of plasminogen is a novel inhibitor of endothelial cell growth. *J. Biol. Chem.* 272:22924–22928.

261. Navarro, M., R. Cervera, J. Font, J.C. Reverter, J. Monteagudo, G. Escolar, A. Lopez-Soto, A. Ordinas, and M. Ingelmo. 1997. Anti-endothelial cell antibodies in systemic autoimmune diseases: prevalence and clinical significance. *Lupus* 6:521–526.

262. Toyoda, M., K. Galfayan, O.A. Galera, A. Petrosian, L.S. Czer, and S.C. Jordan. 1997. Cytomegalovirus infection induces anti-endothelial cell antibodies in cardiac and renal allograft recipients. *Transpl. Immunol.* 5:104–111.

263. Eichhorn, J., D. Sima, B. Thiele, C. Lindschau, A. Turowski, H. Schmidt, W. Schneider, H. Haller, and F.C. Luft. 1996. Anti-endothelial cell antibodies in Takayasu arteritis. *Circulation* 94:2396–2401.

264. Del Papa, N., L. Guidali, M. Sironi, Y. Shoenfeld, A. Mantovani, A. Tincani, G. Balestrieri, A. Radice, R.A. Sinico, and P.L. Meroni. 1996. Anti-endothelial cell IgG antibodies from patients with Wegener's granulomatosis bind to human endothelial cells *in vitro* and induce adhesion molecule expression and cytokine secretion. *Arthritis Rheum.* 39:758–766.

265. van der Zee, J.M., A.H.M. Heurkens, E.A.M. van der Voort, M.R. Daha, and F.C. Breedveld. 1991. Characterization of anti-endothelial antibodies in patients with rheumatoid arthritis complicated by vasculitis. *Clin. Exp. Rheumatol.* 9:589–594.

266. Hill, M.B., J.L. Phipps, A. Milford-Ward, M. Greaves, and P. Hughes. 1996. Further characterization of anti endothelial cell antibodies in systemic lupus erythematosus by controlled immunoblotting. *Br. J. Rheum.* 35:1231–1238.

267. Bordron, A., M. Dueymes, Y. Levy, C. Jamin, L. Ziporen, J.C. Piette, Y. Shoenfeld, and P. Youinou. 1998. Anti-endothelial cell antibody binding makes negatively charged phospholipids accessible to antiphospholipid antibodies. *Arthritis Rheum.* 41:1738–1747.

268. Bevers, E.M., P. Comfurius, and R.F.A. Zwaal. 1983. Changes in membrane phospholipid distribution during platelet activation. *Biochim. Biophys. Acta* 57–66.

269. Bevers, E.M., E.F. Smeets, P. Comfurius, and R.F.A. Zwaal. 1994. Physiology of membrane lipid asymmetry. *Lupus* 3:235–240.

270. Fadok, V.A., D.R. Voelker, P.A. Campbell, J.J. Cohen, D.L. Bratton, and P.M. Henson. 1992. Exposure of phosphatidylserine on the surface of apoptotic lymphocytes triggers specific recognition and removal by macrophages. *J. Immunol.* 148:2207–2216.

271. Chonn, A., S.C. Semple, and P.R. Cullis. 1995. Beta 2 glycoprotein 1 is a major protein associated with very rapidly cleared liposomes *in vivo*, suggesting a significant role in the immune clearance of "non-self" particles. *J. Biol. Chem.* 270:25845–25849.

272. Balasubramanian, K., J. Chandra, and A.J. Schroit. 1997. Immune clearance of phosphatidylserine-expressing cells by phagocytes. The role of β_2-glycoprotein 1 in macrophage recognition. *J. Biol. Chem.* 273:31113–31117.

273. Balasubramanian, K. and A.J. Schroit. 1998. Characterization of phosphatidylserine-dependent β_2-glycoprotein 1 macrophage interactions. Implications for apoptotic cell clearance by phagocytes. *J. Biol. Chem.* 273:29272–29277.

274. Rovere, P., A.A. Manfredi, C. Vallinoto, V.S. Zimmermann, U. Fascio, G. Balestrieri, P. Ricciardi-Castagnoli, C. Rugarli, A. Tincani, and M.G. Sabbadini. 1998. Dendritic cells preferentially internalize apoptotic cells opsonized by anti-β_2-glycoprotein I antibodies. *J. Autoimmun.* 11:403–411.

275. Levine, J.S., R. Subang, J.S. Koh, and J. Rauch. 1998. Induction of anti-phospholipid autoantibodies by β_2-glycoprotein I bound to apoptotic thymocytes. *J. Autoimmun.* 11:413–424.

276. Hale, A.J., C.A. Smith, L.C. Sutherland, V.E. Stoneman, V. Longthorne, Culhane, AC, and G.T. Williams. 1996. Apoptosis: molecular regulation of cell death. *Eur. J. Biochem.* 236:1–26 (Erratum Vol. 237:884).

277. Vermes, I., C. Haanen, H. Steffens-Nakken, and C. Reutelingsperger. 1995. A novel assay for apoptosis. Flow cytometric detection of phosphatidylserine expression on early apoptotic cells using fluorescein labelled Annexin V. *J. Immunol. Methods* 184:39–51.

278. Meers, P., D. Daleke, K. Hong, and D. Papahadjopoulos. 1991. Interactions of annexins with membrane phospholipids. *Biochemistry* 30:2903–2908.

279. Meers, P. and T. Mealy. 1993. Calcium-dependent annexin V binding to phospholipids: stoichiometry, specificity, and the role of negative charge. *Biochemistry* 32:11711–11721.

280. Martin, S.J., C.P.M. Reutelingsperger, A.J. McGahon, J.A. Rader, R.C.A.A. van Schie, D.M. LaFace, and D.R. Green. 1995. Early redistribution of plasma membrane phosphatidylserine is a general feature of apoptosis regardless of the initiating stimulus: inhibition by overexpression of Bcl-2 and Abl. *J. Exp. Med.* 182:1545–1556.

281. Reutelingsperger, C.P.M., J.M.M. Kop, G. Hornstra, and H.C. Hemker. 1988. Purification and characterization of a novel protein from bovine aorta that inhibits coagulation. Inhibition of the phospholipid-dependent factor-Xa-catalyzed prothrombin activation, through a high-affinity binding of the anticoagulant to the phospholipids. *Eur. J. Biochem.* 173:171–178.

282. van Heerde, W.L., S. Poort, C.P. Reutelingsperger, and P.G. de Groot. 1994. Binding of recombinant annexin V to endothelial cells: effect of annexin V binding on endothelial-cell-mediated thrombin formation. *Biochem. J.* 302:305–312.

283. Rand, J.H., X.X. Wu, S. Guller, J. Gil, A. Guha, J. Scher, and C.J. Lockwood. 1994. Reduction of annexin-V (placental anticoagulant protein-I) on placental villi of women with antiphospholipid antibodies and recurrent spontaneous abortion. *Am. J. Obstet. Gynecol.* 171:1566–1572.

284. La Rosa, L., P.L. Meroni, A. Tincani, G. Balestrieri, D. Faden, A. Lojacono, L. Morassi, E. Brocchi, N. Del Papa, and A. Gharavi. 1994. Beta 2 glycoprotein I and placental anticoagulant protein I in placentae from patients with antiphospholipid syndrome. *J. Rheum.* 21:1684–1693.

285. Rand, J.H., X.X. Wu, H.A.M. Andree, C.J. Lockwood, S. Guller, J. Scher, and P.C. Harpel. 1997. Pregnancy loss in the antiphospholipid-antibody syndrome-a possible thrombogenic mechanism. *N. Engl. J. Med.* 337:154–160.

286. Rand, J.H., X.X. Wu, H.A.M. Andree, J.B.A. Ross, E. Rusinova, M.G. Gascon-Lema, C. Calandri, and P.C. Harpel. 1998. Antiphospholipid antibodies accelerate plasma coagulation by inhibiting annexin-V binding to phospholipids: a "lupus procoagulant" phenomenon. *Blood* 92:1652–1660.

287. Witztum, J.L. 1994. The oxidation hypothesis of atherosclerosis. *Lancet* 344:793–795.

288. Sambrano, G.R., S. Parthasarathy, and D. Steinberg. 1994. Recognition of oxidatively damaged erythrocytes by a macrophage receptor with specificity for oxidized low density lipoprotein. *Proc. Natl. Acad. Sci. USA* 91:3265–3269.

289. Vaarala, O., G. Alfthan, M. Jauhiainen, M. Leirisalo-Repo, K. Aho, and Palosuo. 1993. Crossrection between antibodies to oxidised low-density lipoprotein and to cardiolipin in systemic lupus erythematosus. *Lancet* 341:923–925.

290. Romero, F.I., O. Amengual, T. Atsumi, M.A. Khamashta, F.J. Tinahones, and G.R.V. Hughes. 1998. Arterial disease in lupus and secondary antiphospholipid syndrome: association with anti-β_2-glycoprotein I antibodies but not with antibodies against oxidized low-density lipoprotein. *Br. J. Rheum.* 37:883–888.

291. George, J., D. Harats, B. Gilburd, A. Afek, Y. Levy, J. Schneiderman, I. Barshack, J. Kopolovic, and Y. Shoenfeld. 1999. Immunolocalization of β_2-glycoprotein I (apolipoprotein H) to human atherosclerotic plaques: potential implications for lesion progression. *Circulation* 99:2227–2230.

292. Matsuda, J., M. Gotoh, K. Kawasugi, K. Gohchi, M. Tsukamoto, and N. Saitoh. 1996. Negligible synergistic effect of β_2-glycoprotein I on the reactivity of antioxidized low-density lipoprotein antibody to oxidized low-density lipoprotein. *Am. J. Hematol.* 52:114–116.

293. Horkko, S., E. Miller, D.W. Branch, W. Palinski, and J.L. Witztum 1997. The epitopes for some antiphospholipid antibodies are adducts of oxidized phospholipid and beta2 glycoprotein 1 (and other proteins). *Proc. Natl. Acad. Sci. USA* 94:10356–10361.

294. Hasunuma, Y., E. Matsuura, Z. Makita, T. Katahira, S. Nishi, and T. Koike. 1997. Involvement of β_2-glycoprotein I and anticardiolipin antibodies in oxidatively modified low-density lipoprotein uptake by macrophages. *Clin. Exp. Immunol.* 107:569–573.

295. Matsuura, E., K. Kobayashi, T. Yasuda, and T. Koike. 1998. Antiphospholipid antibodies and atherosclerosis. *Lupus* 7:S135–S139

296. Kochl, S., F. Fresser, E. Lobentanz, G. Baier, and G. Utermann. 1997. Novel interaction of apolipoprotein(a) with beta-2 glycoprotein I mediated by the kringle IV domain. *Blood* 90:1482–1489.

297. McLean, J.W., J.E. Tomlinson, W.J. Kuang, D.L. Eaton, E.Y. Chen, G.M. Fless, A.M. Scanu, and R.M. Lawn. 1987. cDNA sequence of human apolipoprotein(a) is homologous to plasminogen. *Nature* 330:132–137.

298. Iuliano, L., D. Pratico, D. Ferro, V. Pittoni, G. Valesini, J. Lawson, G.A. FitzGerald, and F. Violi. 1997. Enhanced lipid peroxidation in patients positive for antiphospholipid antibodies. *Blood* 90:3931–3935.

299. Pratico, D., D. Ferro, L. Iuliano, J. Rokach, F. Conti, G. Valesini, G.A. FitzGerald, and F. Violi. 1999. Ongoing prothrombotic state in patients with antiphospholipid antibodies: a role for increased lipid peroxidation. *Blood* 93:3401–3407.

300. Vermylen, J., C. Van Geet, and J. Arnout. 1998. Antibody-mediated thrombosis: relation to the antiphospholipid syndrome. *Lupus* 7:S63–S66

301. Khamashta, M.A., E.N. Harris, A.E. Gharavi, G. Derue, A. Gil, J.J. Vazquez, and G.R.V. Hughes. 1988. Immune mediated mechanism for thrombosis: antiphospholipid antibody binding to platelet membranes. *Ann. Rheum. Dis.* 47:849–854.

302. Lipp, E., A. von Felten, H. Sax, D. Muller, and P. Berchtold. 1998. Antibodies against platelet glycoproteins and antiphospholipid antibodies in autoimmune thrombocytopenia. *Eur. J. Haematol.* 60:283–288.

303. Tokita, S., M. Arai, N. Yamamoto, Y. Katagiri, K. Tanoue, K. Igarashi, M. Umeda, and K. Inoue. 1996. Specific cross-reaction of IgG anti-phospholipid antibody with platelet glycoprotein IIIa. *Thromb. Haemost.* 75:168–174.

304. Godeau, B., J.C. Piette, P. Fromont, L. Intrator , A. Schaeffer, and P. Bierling. 1997. Specific antiplatelet glycoprotein autoantibodies are associated with the thrombocytopenia of primary antiphospholipid syndrome. *Br. J. Haematol.* 98:873–879.

305. Shi, W., B.H. Chong, P.J. Hogg, and C.N. Chesterman. 1993. Anticardiolipin antibodies block the inhibition by β_2-glycoprotein I of the factor Xa generating activity of platelets. *Thromb. Haemost.* 70:342–345.

306. Joseph, J.E., S. Donohoe, P. Harrison, I.J. Mackie, and S.J. Machin. 1998. Platelet activation and turnover in the primary antiphospholipid syndrome. *Lupus* 7:333–340.

307. Fanelli, A., C. Bergamini, S. Rapi, A. Caldini, A. Spinelli, A. Buggiani, and L. Emmi. 1997. Flow cytometric detection of circulating activated platelets in primary antiphospholipid syndrome. Correlation with thrombocytopenia and anticardiolipin antibodies. *Lupus* 6:261–267.

308. Shechter, Y., Y. Tal, A. Greenberg, and B. Brenner. 1998. Platelet activation in patients with antiphospholipid syndrome. *Blood Coagul. Fibrinolysis* 9:653–657.

309. Arvieux, J., B. Roussel, P. Pouzol, and M.G. Colomb. 1993. Platelet activating properties of murine monoclonal antibodies to β_2-glycoprotein I. *Thromb. Haemost.* 70:336–341.

310. Forastiero, R., M. Martinuzzo, L.O. Carreras, and J. Maclouf. 1998. Anti-β_2-glycoprotein I antibodies and platelet activation in patients with antiphospholipid antibodies: association with increased excretion of platelet-derived thromboxane urinary metabolites. *Thromb. Haemost.* 79:42–45.

311. Robbins, D.L., S. Leung, D.J. Miller-Blair, and V. Ziboh. 1998. Effect of anticardiolipin/β_2-glycoprotein I complexes on production of thromboxane A2 by platelets from patients with the antiphospholipid syndrome. *J. Rheum.* 25:51–56.

312. Ford, I., S. Urbaniak, and M. Greaves. 1998. IgG from patients with antiphospholipid syndrome binds to platelets without induction of platelet activation. *Br. J. Haematol.* 102:841–849.

313. Dixon, F.J., B.S. Andrews, R.A. Eisenberg, P.J. McConahey, A.N. Theofilopoulos, and C.B. Wilson. 1978. Etiology and pathogenesis of a spontaneous lupus-like syndrome in mice. *Arthritis Rheum.* 21:S64–S67

314. Theofilopoulos, A.N. and F.J. Dixon. 1985. Murine models of systemic lupus erythematosus. *Adv. Immunol.* 37:269–390.

315. Cohen, M.G., K.M. Pollard, and L. Schrieber. 1988. Relationship of age and sex to autoantibody expression in MRL-+/+ and MRL-lpr/lpr mice: demonstration of an association between the expression of antibodies to histones, denatured DNA and Sm in MRL-+/+ mice. *Clin. Exp. Immunol.* 72:50–54.

316. Gharavi, A.E., R.C. Mellors, and K.B. Elkon. 1989. IgG anticardiolipin antibodies in murine lupus. *Clin. Exp. Immunol.* 78:233–238.

317. Koike, T. 1994. Anticardiolipin antibodies in NZW x BXSB F1 mice. *Lupus* 3:241–246.

318. Monestier, M., D.A. Kandiah, S. Kouts, K.E. Novick, G.L. Ong, M.Z. Radic, and S.A. Krilis. 1996. Monoclonal antibodies from NZW x BXSB F1 mice to β_2-glycoprotein I and cardiolipin. Species specificity and charge-dependent binding. *J. Immunol.* 156:2631–2641.

319. Hang, L., P. Stephens-Larson, J.P. Henry, and F.J. Dixon. 1983. The role of hypertension in the vascular disease and myocardial infarcts associated with murine systemic lupus erythematosus. *Arthritis Rheum.* 26:1340–1345.

320. Gharavi, A.E., L.R. Sammaritano, J. Wen, and K.B. Elkon. 1992. Induction of antiphospholipid autoantibodies by immunization with beta 2 glycoprotein I (apolipoprotein H). *J. Clin. Invest.* 90:1105–1109.

321. Pierangeli, S.S. and E.N. Harris. 1993. Induction of phospholipid-binding antibodies in mice and rabbits by immunization with human beta 2 glycoprotein 1 or anticardiolipin antibodies alone. *Clin. Exp. Immunol.* 93:269–272.

322. Aron, A.L., M.L. Cuellar, R.L. Brey, S. Mckeown, L.R. Espinoza, Y. Shoenfeld, and A.E. Gharavi. 1995. Early onset of autoimmunity in MRL/++ mice following immunization with beta 2 glycoprotein I. *Clin. Exp. Immunol.* 101:78–81.

323. Blank, M., D. Faden, A. Tincani, J. Kopolovic, I. Goldberg, B. Gilburd, F. Allegri, G. Balestrieri, G. Valesini, and Y. Shoenfeld. 1994. Immunization with anticardiolipin cofactor (beta-2-glycoprotein I) induces experimental antiphospholipid syndrome in naive mice. *J. Autoimmun.* 7:441–455.

324. Blank, M., J. George, V. Barak, A. Tincani, T. Koike, and Y. Shoenfeld. 1998. Oral tolerance to low dose *gb$_2$ glycoprotein I: immunomodulation of experimental antiphospholipid syndrome. *J. Immunol.* 161:5303–5312.

325. Silver, R.M., S.S. Pierangeli, A.E. Gharavi, E.N. Harris, S.S. Edwin, C.M. Salafia, and D.W. Branch. 1995. Induction of high levels of anticardiolipin antibodies in mice by immunization with *gb$_2$ glycoprotein I does not cause fetal death. *Am. J. Obstet. Gynecol.* 173:1410–1415.

326. Branch, D.W., D.J. Dudley, M.D. Mitchell, K.A. Creighton, T.M. Abbott, E.H. Hammond, and R.A. aynes. 1990. Immunoglobulin G fractions from patients with antiphospholipid antibodies cause fetal death in BALB/c mice: a model for autoimmune fetal loss. *Am. J. Obstet. Gynecol.* 163:210–216.

327. Silver, R.M., L.A. Smith, S.S. Edwin, B.T. Oshiro, J.R. Scott, and D.W. Branch. 1997. Variable effects on murine pregnancy of immunoglobulin G fractions from women with antiphospholipid antibodies. *Am. J. Obstet. Gynecol.* 177:229–233.

328. Pierangeli, S.S. and E.N. Harris. 1994. Antiphospholipid antibodies in an in vivo thrombosis model in mice. *Lupus* 3:247–251.

329. Olee, T., S.S. Pierangeli, H.H. Handley, D.T. Le , X. Wei, C.J. Lai, J. En, W. Novotny, E.N. Harris , V.L.J. Woods, and P.P. Chen. 1996. A monoclonal IgG anticardiolipin antibody from a patient with the antiphospholipid syndrome is thrombogenic in mice. *Proc. Natl. Acad. Sci. U.S.A.* 93:8606–8611.

330. Mendlovic, S., S. Brocke, Y. Shoenfeld, M. Ben-Bassat, A. Meshorer, Bakimer, and E. Mozes. 1988. Induction of a systemic lupus erythematosus-like disease in mice by a common human anti-DNA idiotype. *Proc. Natl. Acad. Sci. U.S.A.* 85:2260–2264.

331. Ziporen, L., Y. Shoenfeld, Y. Levy, and A.D. Korczyn. 1997. Neurological dysfunction and hyperactive behavior associated with antiphospholipid antibodies. A mouse model. *J. Clin. Invest.* 100:613–619.

332. Blank, M., J. Cohen, V. Toder, and Y. Shoenfeld . 1991. Induction of anti-phospholipid syndrome in naive mice with mouse lupus monoclonal and human polyclonal anticardiolipin antibodies. *Proc. Natl. Acad. Sci. U.S.A.* 88:3069–3073.

333. Tomer, Y., M. Blank, and Y. Shoenfeld. 1994. Suppression of experimental antiphospholipid syndrome and systemic lupus erythematosus in mice by anti-CD4 monoclonal antibodies. *Arthritis Rheum.* 37:1236–1244.

334. Damianovich, M., B. Gilburd, J. George, N. Del Papa, A. Afek, I. Goldberg, Y. Kopolovic, D. Roth, G. Barkai, P.L. Meroni, and Y. Shoenfeld. 1996. Pathogenic role of antiendothelial cell antibodies in vasculitis. An idiotypic experimental model. *J. Immunol.* 156:4946–4951.

335. Fishman, P., E. Falach-Vaknine, R. Zigelman, R. Bakimer, B. Sredni, M. Djaldetti, and Y. Shoenfeld. 1993. Prevention of fetal loss in experimental antiphospholipid syndrome by in vivo administration of recombinant interleukin-3. *J. Clin. Invest.* 91:1834–1837.

336. Inbar, O., M. Blank, D. Faden, A. Tincani, M. Lorber, and Y. Shoenfeld. 1993. Prevention of fetal loss in experimental antiphospholipid syndrome by low-molecular-weight heparin. *Am. J. Obstet. Gynecol.* 169:423–426.

337. Blank, M., J. George, P. Fishman, Y. Levy, V. Toder, S. Savion, V. Barak, T. Koike, and Y. Shoenfeld. 1998. Ciprofloxacin immunomodulation of experimental antiphospholipid syndrome associated with elevation of interleukin-3 and granulocyte-macrophage colony-stimulating factor expression. *Arthritis Rheum.* 41:224–232.

338. Isenberg, D.A., D. Katz, S. Le Page, B. Knight, L. Tucker, P. Maddison, P. Hutchings, R. Watts, J. Andre-Schwartz, R.S. Schwartz, and A. Cooke. 1991. Independent analysis of the 16/6 idiotype lupus model. A role for an environmental factor?. *J. Immunol.* 147:4172–4177.

339. Cutsforth, G.A., R.N. Whitaker, J. Hermans, and B.R. Lentz. 1999. A new model to describe extrinsic protein binding to phospholipid membranes of varying composition: application to human coagulation proteins. *Biochemistry* 28:7453–7461.

340. Tait, J.F., D. Gibson, and K. Fujikawa. 1989. Phospholipid binding properties of human placental anticoagulant protein-I, a member of the lipocortin family. *J. Biol. Chem.* 264:7944–7949.

21 | Genes and Genetics of Murine Lupus

Dwight H. Kono and Argyrios N. Theofilopoulos

1. INTRODUCTION

The etiology of systemic lupus erythematosus is still unknown, but genetic predilection appears to be a dominant factor. Therefore, elucidation of the genes and their contibutions to the development of lupus should reveal many of the fundamental events underlying the initiation and maintenance of systemic autoimmunity. Identification of the genetic alterations that enhance susceptibility will directly define the causative mechanisms that lead to immune system-mediated self-destruction. Identification of the normal genes involved in the immunological and inflammatory processes that promote end-organ damage will reveal the components crucial for disease manifestations, which can provide targets for therapeutic intervention. Progress in delineating specific genes and pathways participating in immune system activation and homeostasis, along with gene manipulation approaches, have identified many new potential avenues for loss of tolerance and development of systemic autoimmunity. Never has the genetic basis for systemic autoimmunity been documented more clearly and yet, ironically, this progress has also revealed the enormous gaps in our understanding of this complex process. This chapter will focus primarily on new findings since the last edition and will summarize our current view of this rapidly evolving field.

2. SPONTANEOUS AND INDUCED MOUSE MODELS OF LUPUS

Research on SLE has certainly benefited from the many spontaneous and induced mouse models that have provided insights into the genetics and immunopathology of this

Table 1 Spontaneous and induced mouse models of lupus

Spontaneous disease models
  NZ and related strains
    NZB
    NZW
    (NZBxNZW)F1
    (NZBxSWR)F1
    (NZBxNZW)RI lines "NZM/Aeg" lines (192)
    (NZBxSM)RI lines "(NXSM)RI"
    (NZBxC58)RI lines "(NX8)RI"
  MRL (*Fas^lpr* and wild-type) and related strains
    MRL-*Fas^lpr*.*ℓℓ* (long-lived substrain) (193)
    MRL-*Fas^lpr*, *Yaa* (194)
    SCG/Kj-*Fas^lpr* (BXSBxMRL-*lpr*)RI (195)
  BXSB and related strains
    BXSB-*ℓℓ* (long-lived substrain) (196)
    (NZWxBXSB)F1
    (NZBxBXSB)F1
  (SJLxSWR)F1 (197)
  Palmerston North (198)
  Motheaten strains (19, 20, 199)
Induced disease models
  Heavy Metal-Induced Autoimmunity (200)
  Drug-induced lupus (201)
  Pristane-Induced (202)
  Anti-idiotypic (203)
  Graft-versus-host disease

disease (Table 1). Among the spontaneous models, the (NZB x NZW)F$_1$ (BWF$_1$) hybrid, MRL-*Fas^lpr* and BXSB mice are the most commonly studied. These strains share characteristics, such as hypergammaglobulinemia, antinuclear antibodies and glomerulonephritis (GN), but have unique features as well, such as arthritis and expanded

CD4⁻CD8⁻ (double-negative, DN) T cells in MRL-*Fas^{lpr}* mice and hemolytic anemia in NZB mice. Details on the clinical manifestations and immunopathology for these and the other spontaneous and induced models have been previously reviewed (1,2).

3. GENES PREDISPOSING TO SPONTANEOUS MOUSE MODELS OF LUPUS

Only a few genes predisposing to spontaneous lupus have been identified, among them the MHC genes, Fas receptor and ligand, and Hcph (hemopoietic cell phosphatase, SHP-1) (Table 2). The Y chromosome accelerator of autoautoimmunity and lymphoproliferation (*Yaa*) gene, responsible for the male predisposition to lupus-like disease in the BXSB strain, has also been implicated using consomic mice (transfer of the Y chromosome), but the gene is not known. Details of Fas and FasL are covered in chapter 11.

The MHC complex is strongly linked to the development of autoimmunity in most lupus-prone strains, particularly the BWF₁ (3,4) and BXSB (5). However, its contribution is highly dependent on other background genes, e.g. the specific haplotype that confers susceptibility is dependent on the lupus background, and autoimmunity does not develop when the predisposing MHC haplotypes are on normal backgrounds (2). The actual gene(s) within the MHC complex that promotes lupus susceptibility still remains to be resolved. Class II molecules are strong candidates based on H-2 congenic studies (5–7) and their central roles in both repertoire shaping and antigen recognition. Nevertheless, other closely-linked immunologically-relevant genes, which are numerous in this region, have not been excluded. One of the strongest association with lupus disease is the heterozygous H-2^z haplotype in combination with one of several different H-2 haplotypes, including d, b, and v (3,4,8–10). Recent studies attempted to directly implicate class II molecules by expressing I-A^z or I-E^z transgenes in the NZB x C57BL/6 background, however, no increased susceptibility was found in mice expressing either of these transgenes (11,12). Although the inference is that the class II molecules are not responsible, it is also possible that the transgene did not adequately recapitulate the expression patterns required to promote autoimmunity, since it is known that slight changes in class II expression result in substantial effects on disease susceptibility, e.g. homozygous versus heterozygous expression (13) or the presence or absence of I-E (14). In this regard, transgenic mice would not have the same levels of class II molecules as wild-type mice, since they would have normal levels of class II in addition to the transgene.

The role of I-E in lupus-prone mice has been investigated by expressing the Eα transgene in H-2^b BXSB mice (7,14, 15). Interestingly, high levels and, to a lesser extent, lower levels of I-Eα transgene expression suppressed the development of lupus. This was apparently due to competitive inhibition of autoantigen presentation by peptide fragments from processed I-Eα that bound effectively to H-2^b class II molecules.

Other potential genes within the H-2 complex include a polymorphic NZW TNF-α gene that appears to promote lupus in the BWF₁ hybrid (16,17), and a recessive NZW locus (*Sles1*) that appears to suppress the development of autoimmunity in NZW mice (18). However, the relationship of *Sles1* to the known class II and TNF-α polymorphisms in the NZW strain has yet to be determined.

SHP-1 is a protein tyrosine phosphatase ubiquitously expressed in hematopoietic lineage cells. Two spontaneous recessive mutations, the motheaten (*me*) and motheaten viable (*me^v*), both lead to similar early lethal phenotypes, which differ slightly in severity because of more complete gene deletion in the *me* variant. Although increased immunoglobulin levels and autoantibodies are detected, the major disease manifestations, which result from severe abnormalities in virtually all hematopoietic cell lineages, are dissimilar to spontaneous lupus, and are not mediated by autoantibodies or T and B lymphocytes (19–23). The basis for the motheaten phenotype is largely related to the requirement for SHP-1 in inhibiting cell activation, following its recruitment by negative regulatory molecules containing immunoreceptor tyrosine-based inhibitory motifs (ITIM) (24). Autoantibody production is likely due to the inability of CD22 to negatively regulate the B cell antigen receptor in the absence of SHP-1 (25).

4. LOCI PREDISPOSING TO SPONTANEOUS MOUSE MODELS OF LUPUS

Current approaches to identify genes predisposing to quantitative traits generally entail four main steps. First, mapping of traits is performed by genome-wide scans using evenly distributed markers spanning the chromosomes. Next, interval-specific congenic strains are generated that contain a generous portion of an introgressed genomic fragment to confirm the mapping studies and to identify the major intermediate phenotypes. A relatively large region is generally taken to assure that the specific gene or genes are indeed present within the interval. Third, panels of congenics with crossovers or smaller intervals are generated to finely map the location of the susceptibility gene or genes. This may be performed in one or two stages, i.e., localization first to ~5 cM-size fragments and then to <0.1 cM size fragments (26). The final step requires cloning, sequencing and identifying of the genes within the fragment. Candidate genes are then selected based on expression, structure, function or other characteristics for screening the parental

Table 2 Susceptibility genes predisposing to lupus-related traits

Name	Chr	cM	Best assoc. Marker	Cross	Phenotype	Parental Allele	Ref.
Bxs4	1	7.7	D1Mit3	B10x(B10xBXSB)F$_1$	LN	BXSB	(47)
Bxs1	1	32.8	D1Mit5	BXSBx(B10xBXSB)F$_1$	GN/ANA/spleen	BXSB	(46)
—	1	54.0	D1Mit48	(WxBa)F$_1$xW	IgM ssDNA/IgM histone	Balb/c	(204)
Bxs2	1	63.1	D1Mit12	BXSBx(B10xBXSB)F$_1$	GN/ANA/spleen	BXSB	(46)
—	1	65.0	D1Mit494	MRL-*lpr*x(MRL-*lpr*xC3H-*lpr*)F$_1$	sialadenitis	MRL	(42)
Sle1	1	87.9	D1Mit15	(NZMxB6)xNZM	GN	NZM (NZW)	(9)
Sle1	1	87.9	D1Mit15	(NZMxB6)F$_2$	dsDNA/GN/spleen	NZM (NZW)	(38)
Lbw7	1	92.3	D1Mit36	BWF$_2$	chr/spleen	NZB	(27)
Nba2	1	92.3	D1Mit111	(BxSM)xW	GN	NZB	(10)
Nba2	1	92.3	D1Mit148	(BxSM)xW/(B6.H2^{z}xB)xB	ANA/gp70/GN	NZB	(205)
Nba2	1	94.2	Crp/Sap	((B6.H2^z & Ba.II2^r)xB)F$_1$xB	GN	NZB	(206)
Bxs3	1	100.0	D1Mit403	BXSBx(B10xBXSB)F$_1$	dsDNA	BXSB	(46)
—	1	106.3	D1Mit17	(WxBa)F$_1$xW	ssDNA	NZW	(204)
	2		D2Mit12	(MRL-*lpr*xBa)F2	ssDNA/dsDNA	MRL +/+,*lpr*/+	(207)
Sles2	3	35.2	D3Mit137	(B6.NZMc1xNZW)F$_1$xNZW	dsDNA/GN (resistance)	NZW	(18)
Bxs5	3	39.7	D3Mit40	B10x(B10xBXSB)F$_1$	ANA/IgG3	BXSB	(47)
Lprm2	3	66.2	D3Mit16	MRL-*lpr*x(MRL-*lpr*xC3H-*lpr*)F$_1$	vasculitis (resistance)	MRL	(40)
Lprm1	4	32.5	D4Mit82	MRL-*lpr*x(MRL-*lpr*xC3H-*lpr*)F$_1$	vasculitis	MRL	(40)
Acla2	4	40.0	D4Mit79	NZWx(NZWxBXSB)F$_1$	CL	BXSB	(48)
Sle2	4	44.5	D4Mit9	(NZMxB6)xNZM	GN	NZM (NZW)	(9)
Lbw2	4	55.6	D4Nds2	BWF$_2$	mortality/GN/spleen	NZB	(27)
Sles2	4	57.6	D4Mit12	(B6.NZMc1xNZW)F$_1$xNZW	dsDNA/GN (resistance)	NZW	(18)
—	4	62.3	D4Mit70	(BxSM)xW	GN	NZB	(10)
Asm2	4	65.0	D4Mit199	MRL-*lpr*x(MRL-*lpr*xC3H-*lpr*)F$_1$	sialadenitis	MRL female	(42)
nba1	4	65.7	Epb4.1(elp-1)	BWF$_1$xW	GN	NZB	(208)
Lmb1	4	69.8	D4Mit12	(B6-*lpr*xMRL-*lpr*)F$_2$	Lprn/dsDNA	B6	(39)
Imh1/Mott	4	69/69.8	D4Mit66/48	BWF$_1$xW	hyper IgM/GN/dsDNA	NZB	(209, 210)
Sle6	5	20.0	D5Mit4	(B6.NZMc1xNZW)F$_1$xNZW	GN	NZW	(18)
Lmb2	5	41.0	D5Mit356	(B6-*lpr*xMRL-*lpr*)F$_2$	Lprn/dsDNA	MRL	(39)
Lprm4	5	54.0	D5Mit23	MRL-*lpr*x(MRL-*lpr*xC3H-*lpr*)F$_1$	spleen	MRL	(40)
Lbw3	5	84.0	D5Mit101	BWF$_2$	mortality	NZW	(27)
—	6	35.0	D6Mit8	MRL-*lpr*x(MRL-*lpr*xC3H-*lpr*)F$_1$	GN (resistance)	MRL	(211)
Lbw4	6	64.0	D6Mit25	BWF$_2$	mortality	NZB	(27)
—	6	74.0	D6Mit374	(NZMxB6)F$_2$	dsDNA	B6	(38)
Sle5	7	0.5	D7Mit178	(NZMxB6)F$_2$	dsDNA	NZM(NZW)	(38)
Lrdm1	7	6.0	Pou2f2(Otf-2)	(MRL-*lpr*xCAST)F$_1$xMRL-*lpr*	GN	MRL	(41)
Sle3	7	16.0	D7Mit25	(NZMxB6)F$_2$	GN	NZM(NZW)	(38)
Lbw5	7	23.0	D7Nds5(Ngfg)	BWF$_2$	mortality	NZW	(27)
Lmb3	7	27.0	D7Mit211	(B6-*lpr*xMRL-*lpr*)F$_2$	Lprn/dsDNA	MRL	(39)
Sle3	7	28.0	p	(NZMxB6)xNZM	GN	NZM (NZW)	(9)
—	7	51.5	D7Mit17	(BxSM)xW	GN	NZB	(10)
—	7	56.5	D7Mit7	(BxW)F$_1$xW	dsDNA	NZB	(212)
Myo1	7	69.0	D7Mit14	NZWx(NZWxBXSB)F$_1$	MI	BXSB	(48)
Pbat2	8	11	D8Mit96	NZWx(NZWxBXSB)F$_1$	platelet	BXSB	(48)
baa1	9	28.0	D9Mit22	(WxBa)F$_1$xW	IgM ssDNA/IgM histone	Balb/c	(204)
Asm1	10	38/40	D10Mit115/259	MRL-*lpr*x(MRL-*lpr*xC3H-*lpr*)F$_1$	sialadenitis	MRL	(42)
Lmb4	10	51.0	D10Mit11	(B6-*lpr*xMRL-*lpr*)F$_2$	Lprn/GN	MRL	(39)
—	10	69.0	D10Mit35	(NZMxB6)F$_2$	GN	NZM & B6#	(38)
—	11	2.0/17.0	D11Mit2/84	(Ba.H2^{z}xB)F$_1$xB	GN	NZB	(206)
—	11	20.0	D11Mit20	(NZMxB6)F$_2$	GN/dsDNA	NZM	(38)
Lbw8	11	28.0	IL4	BWF$_2$	chr	NZB	(27)
—	11	28.5	D11Mit207	(WxBa)F$_1$xW	ssDNA	NZW	(204)

Table 2   (Continued)

Name	Chr	cM	Best assoc. Marker	Cross	Phenotype	Parental Allele	Ref.
—	11	54.0	D11Mit70	(MRL-*lpr*xBa)F$_2$	dsDNA/ssDNA/CL	MRL	(207)
Lrdm2	12	27.0	D12Nyu3	(MRL-*lpr*xCAST)F$_1$xMRL-*lpr*	GN	MRL	(41)
Yaa1	13	35.0 + 6	D13Mit250	B6x(WxB6-Yaa)F$_1$	gp70	NZW	(213)
—	13	71.0	D13Mit31	(NZMxB6)xNZM	dsDNA	NZM	(9)
—	13	71.0	D13Mit150	(BxSM)xW	GN	NZB	(10)
—	14	19.5	D14Nds4	(WxBa)F$_1$xW	histone	NZW	(204)
Myo2	14	39.0	D14Mit68	NZWx(NZWxBXSB)F$_1$	MI	BXSB	(48)
	14	42.5/40.0	D14Nki41/Mit34	((B6.H2^z & Ba.H2^z)xB)F$_1$xB	GN	NZB	(206)
Lprm3	14	44.0	D14Mit195	MRL-*lpr*x(MRL-*lpr*xC3H-*lpr*)F$_1$	GN (resistance)	MRL	(40)
Lprm5	16	21.0	D16Mit3	MRL-*lpr*x(MRL-*lpr*xC3H-*lpr*)F$_1$	dsDNA	MRL	(40)
nwa1	16	38.0	D16Mit5	(WxBa)F$_1$xW	histone	NZW	(204)
nwa1	16	38.0	D16Mit5	(BxW)F$_1$xW	GN/dsDNA	NZW	(212)
Acla1	17	18.2	D17Mit16	NZWx(NZWxBXSB)F$_1$	CL	NZW/BXSB	(48)
Sles1	17	18.8	H2/D17Mit34	(B6.NZMc1xNZW)F$_1$xNZW	GN/dsDNA (resistance)	NZW	(18)
Pbat1	17	18.9	D17Nds2	NZWx(NZWxBXSB)F$_1$	platelet	NZW/BXSB	(48)
—	18	22.0	D18Mit227	MRL-*lpr*x(MRL-*lpr*xC3H-*lpr*)F$_1$	sialadenitis	MRL	(42)
Lbw6	18	47.0	D18Mit8	BWF$_2$	mortality/GN	NZW	(27)
nwa2	19	41.0	D19Mit11	(WxBa)F$_1$xW	ssDNA	NZW	(204)
—	19	50.0	D19Mit3	(NZMxB6)xNZM	dsDNA	NZM	(9)

Loci with linkages p < 0.01 or lod > 1.9 are included and listed by their approximate chromosomal locations (marker with the highest association).
Chr = chromosome.
cM distances are based on the Mouse Genome Database, The Jackson Laboratory (http://www.informatics.jax.org).
Abbreviations for mouse strains (Cross column): B = NZB, B6 = C57BL/6, B10 = C57BL/10, Ba = Balb/c, CAST = CAST/Ei, lpr = *Fas^lpr*,
NZM = NZM/Aeg2410, W = NZW.
Original phenotypes that mapped to loci are shown: chr = anti-chromatin autoantibody, CL = anticardiolipin autoantibody, dsDNA = anti-dsDNA
autoantibody, GN = glomerulonephritis, gp70 = gp70 immune complexes, histone = anti-histone autoantibody, LN = lymphadenopathy,
Lprn = lymphoproliferation, MI = myocardial infarct, platelet = antiplatelet autoantibody and thrombocytopenia, spleen = splenomegaly.
Autoantibodies are IgG unless otherwise noted.
#complex inheritance: either parental strain promotes GN, but heterozygosity protects.

strains for polymorphisms. The bulk of the work in this final step is the physical cloning and sequencing of the entire interval, which will be expedited by the forthcoming complete sequence of the mouse genome.

Over the past few years, several groups have embarked on delineating lupus-related loci using genome-wide scans. Searches involving a variety of crosses have revealed at least 45 named loci linked to one or more lupus traits that are distributed over 17 of the 19 autosomal chromosomes. These and additional unnamed loci are listed by chromosome and chromosomal location (distance in centimorgans from the centromere) in Table 2. Some of these loci, identified by different groups, appear to be identical, whereas others appear to represent unique loci. Overall, the data suggest that susceptibility to spontaneous lupus in these strains is not due to the presence of a large number of common predisposing loci, but rather to different specific sets of a few major loci.

4.1.   Loci Identified in Crosses of NZB and NZW Mice

The NZB and NZW strains have been the most extensively studied of the lupus-prone strains. Genome-wide scans of intercrosses, backcrosses and crosses to normal background strains have resulted in the identification of loci contributing to one or more lupus-related traits on 15 of the 19 autosomal chromosomes (Table 2). Importantly, non-MHC loci that appear confirmed in more than one cross include *Sle1* (NZW derived) and *Lbw7/Nba2* (NZB-derived) on chromosome 1, *Lbw2/Sle2/nba1/Imh1/Mott* (NZB-derrived) on chromosome 4, *Lbw5/Sle3* (NZW-derived) on chromosome 7, *Lbw8* (NZB-derived) on proximal chromosome 8, and *nwa1* (NZW-derived) on chromosome 16. The mapping studies indicate that inheritance of lupus traits is multiplicative, dependent on the number and specific combination of susceptibility loci (epistasis), and suggests that different sets of loci contribute to different traits, i.e., lymphoid hyperplasia, autoantibody production, GN, and mortality (18, 27–29).

The roles of *Sle1*, *Sle2* and *Sle3* in lupus pathogenesis have been further defined using interval-specific congenic C57BL/6 (B6) mice, which contain introgressed genomic fragments of chromosomes 1, 4 or 7, respectively, from the NZB/NZW-derived recombinant inbred strain, NZM/ Aeg2410 (30,31). *Sle1* congenic mice (B6.NZMc1) developed elevated IgG antinuclear antibodies (particularly targeting H2A/H2B/DNA subnucleosomes), but no GN (32). Adoptive co-transfer of bone marrow from B6.NZMc1 and wild-type B6 mice revealed that autoantibodies were mainly produced by B6.NZMc1 B cells, implying that *Sle1* is functionally expressed in B cells (33). B6.NZMc4 congenic mice exhibited generalized B cell hyperactivity, expansions of B1 cells and elevated levels of polyclonal IgM, but no increases in IgG antinuclear antibodies or GN (34). It was hypothesized that the expanded B1 cell population, which expressed higher levels of co-stimulatory molecules, such as B7, might promote autoimmunity by facilitating the presentation of self antigens to T cells (35). B6.NZMc7 congenic mice developed elevated, but low, levels of antinuclear antibodies and a low incidence of GN. A marked increase in activated T cells, elevated CD4:CD8 ratios and resistance to activation-induced cell death were observed, suggesting that *Sle3* may promote generalized T cell activation (36). Bicongenic B6.NZMc1/7 mice developed severe GN associated with elevated IgG antinuclear antibodies to multiple chromatin components, expanded B and T cell populations, and splenomegaly (37). Although the precise mechanisms have yet to be defined, based on these initial findings, a simple additive model was initially proposed wherein lupus develops in NZM/Aeg2410 mice from distinct gene alterations that cause loss of tolerance to nucleosome components (*Sle1*), a B cell defect (*Sle2*) and a T cell defect (*Sle3*) (29). However, more recent studies involving crosses of these congenic mice have demonstrated a more complex inheritance of lupus traits (18,38).

4.2.   Loci Identified in MRL-Fas^lpr Crosses

Although the *Fas^lpr* mutation promotes loss of tolerance and autoimmunity, the type of manifestations and severity of lupus-like disease is highly dependent on background susceptibility genes. This led several groups to define lupus-related quantitative trait loci (QTL) in crosses of MRL-*Fas^lpr* mice, a strain that develops particularly severe spontaneous accelerated systemic autoimmunity (Table 2). QTL for one or more lupus traits have been identified on 13 of the autosomal chromosomes. The large number of loci may be attributable, in part, to the different traits assessed, such as sialadenitis, GN and vasculitis, which are most likely caused by overlapping, but distinct, sets of susceptibility genes, and to the fact that crosses involved

different strains. Nevertheless, several loci on chromosomes 5 (*Lmb2*, *Lprm4*), 7 (*Lmb3*, *Ldrm1*) and 10 (*Lmb4*, *Asm1*) had overlapping intervals and may be identical (39–42). Interestingly, another locus, *Lmb1* (chromosome 4), which mapped to the non-autoimmune B6 background, had an additive effect on lymphoproliferation equal to the other *Lmb* QTL (39). This demonstrates that so-called non-autoimmune mice can harbor *bona fide* susceptibility genes, but presumably the number and combination of such genes are insufficient for disease induction. Such genes undoubtedly account for the background effects observed when using different strain combinations to map QTLs.

4.3.   Loci Identified in BXSB Crosses

BXSB males develop severe accelerated lupus largely due to the presence of the Y chromosome *Yaa* gene (43), however, other background genes are clearly important since significant autoimmune responses are not observed in consomic non-autoimmune background CBA/J.BXSB-Y (44) or B6.BXSB-Y (45). Genome-wide searches have identified ten BXSB-derived loci encompassing seven chromosomes in backcrosses of BXSB to C57BL/10 (B10) or NZW strains (Table 2). In reciprocal BXSBxB10 backcrosses, five QTL were found linked to antinuclear antibodies, lymphoproliferation, and/or GN, of which four were located in different regions of chromosome 1 and one on chromosome 3 (46,47). In contrast, an entirely different set of BXSB-derived loci, linked to other traits (anti-cardiolipin antibodies, anti-platelet antibodies, thrombocytopenia and/or myocardial infarction), were identified on chromosomes 4, 7, 8, 14 and 17 in the BXSBxNZW backcross study (48). These findings, along with studies in other crosses discussed above, clearly demonstrate that clinical heterogeneity of lupus, at least in spontaneous mouse models, is mainly a consequence of specific combinations of susceptibility genes rather than environmental influences.

4.4.   Summary of Lupus Susceptibility Loci from Different Strains

Overall, from these mapping studies involving a variety of crosses of four different lupus-prone background strains, a number of generalizations can be surmised (Table 2). First, each strain appears to contain a few major susceptibility loci, however, there is minimal overlap of susceptibility genes among the various lupus-predisposed strains indicating the potential for a large pool of susceptibility genes. Second, predisposition to different traits within a strain appears to be governed by different sets of genes, some of which are common to several traits. Third, the genetic contributions are generally additive, but can depend substantially on specific combinations (epistasis). This latter finding suggests that certain immunopathologic manifestations may require several concommitant genetic defects possibly because each has an incremental effect on a common pathway or mechanism, and that a threshold must be exceeded before autoimmunity develops. Thus, intervention directed toward only one of these altered alleles might have a stronger therapeutic impact than expected from the total number of susceptibility genes. Fourth, the fact that interval-specific congenic strains manifest highly penetrant component phenotypes clearly opens the way for ultimately cloning the underlying genes.

5. SYSTEMIC AUTOIMMUNITY IN GENE KNOCKOUT AND TRANSGENIC MICE OF NORMAL BACKGROUNDS

Genetic manipulation of non-autoimmune background strains has helped define not only the role of specific genes in the immune system, but also possible mechanisms of systemic autoimmunity. The number and diversity of genes that have resulted in phenotypes resembling lupus is remarkable considering the total number of genes that have been altered. Although this suggests a large pool of abnormalities that might lead to lupus-like disease, the systemic autoimmunity produced by these mutations is sometimes only vaguely similar to spontaneous lupus, and often there are additional unique findings not observed in SLE. Manifestations of lupus have been observed following manipulation of genes affecting B and T lymphocyte activation, complement, cytokines, apoptosis, and the cell cycle (listed in Tables 3 and 4).

5.1. B Cell Activation Genes

The fate of B cells following antigen receptor (BCR) engagement is a complex process that involves direct or indirect interaction of the BCR with numerous molecules that can promote or inhibit cell activation. Among these are several tyrosine kinases (lyn, fyn, Btk, Blk, Syk), phosphatases (CD45, SHP-1, SHP-2, and SHIP) and accessory molecules (CD19, CD22, FcRγIIB). Genetic manipulation of many of these B cell regulatory molecules have resulted in mice with features similar to lupus, e.g. gene knockouts of lyn (49,50) or CD22 (51–53), and spontaneous mutations of SHP-1 in motheaten mice (see Section 3) (19,20). Lyn is a non-receptor scr-related tyrosine kinase required for CD22-mediated negative regulation of BCR signaling. Although lyn also plays a role in positive signaling, this function appears largely redundant since deficiency of lyn leads to hypersensitivity to BCR-mediated triggering. Mice with homozygous deletions of lyn have increased activation and higher turnover rates of B cells, splenomegaly, elevated IgM levels, autoantibodies and GN. Homozygous deficiency of

CD22, a B cell-specific cell surface sialoadhesin that specifically binds to asialoglycoproteins, results in a similar picture with hyperresponsiveness to BCR signaling, expansion of the peritoneal B1 cell population and autoantibodies. The similarity of the lyn and CD22 knockout phenotypes stems from the fact that the inhibitory action of CD22 requires the recruitment and phosphorylation of lyn, which then, in turn, recruits SHP-1, bringing it in proximity to the BCR where it can dephosphorylate the BCR and down-regulate the response. Heterozygous deletions of CD22, lyn and SHP-1 were shown to have additive effects on B cell abnormalities consistent with the contention that they are limiting elements to a common pathway (54).

CD19, along with CD21 and Tapa-1 (CD81) on B cells, form the functional cell surface receptor complex for C3 fragments, which promotes BCR signal transduction and is crucial for B cell development and tolerance. Mice over-expressing a CD19 transgene develop hyperresponsive B cells to BCR cross-linking, marked expansion of the B1 cell population, hypergammaglobulinemia, and autoantibodies (55). Furthermore, expression of the CD19 transgene in anti-hen egg lysozyme Ig (HEL-Ig)/soluble HEL double transgenic mice led to the appearance of anti-HEL antibodies, suggesting that a defect in anergy to certain soluble antigens might be the underlying mechanism (56). Thus, it appears that in both the CD19 transgene and CD22/lyn/SHP-1 knockouts, lowering of the BCR signaling threshold alters the balance between tolerance and immunity, resulting in the development of autoimmunity.

The FcγRIIb on B cells also inhibits the B cell antigen receptor following cross-linking (57), primarily by recruiting, through its intracytoplasmic immunoreceptor tyrosine-based inhibitory motif, the inositol phosphatase SHIP rather than the phosphotyrosine phosphatase SHP-1 (58,59). Gene knockout of FcγRII amplifies humoral and anaphylactic responses (60) and promotes the development of type II collagen-induced arthritis (61) and type IV collagen-induced Goodpasture's syndrome (62a). Recently, autoantibodies and immune complex GN have also been reported in FcγRII knockout mice, in a strain-dependent manner (62b).

Aiolos is a zinc finger DNA-binding protein that shares a common GGGA core sequence binding motif with the closely homologous nuclear factor Ikaros (63). It is highly expressed in mature B cells and, to a lesser extent, in developing bone marrow B cells and thymocytes. In the nucleus of T cells, Aiolos is mainly localized to the 2 MD chromatin remodeling complex that also contains Ikaros, the Mi-2 ATPase, histone deacetylases and other components of the NURD (nucleosome remodeling histone deacetylase) complex (63). NURD complexes are postulated to participate in transcriptional repression by allowing repressors to gain access to chromatin and by promoting DNA methylation (64–66). Recently, Ikaros was also found to interact with the mSin3 family of corepressors (67). These findings have led to

Table 3 Gene knockout and transgenic normal background mice with manifestations of lupus

Name	Gene	Chr*	cM	Autoimmunity	Ref.
knockouts					
CTLA-4	*Cd152*	1	30.1	multiorgan lymphoproliferative dis., myocarditis, pancreatitis	(77–79)
Fasl (spontaneous)	*Fasl*	1	85.0	lymphoproliferation, DN T cells, autoAbs, GN (gld mutation)	(214)
serum amyloid P component	*Sap*	1	94.2	anti-chromatin Ab, GN, female predom.	(123)
IL-2Rα	Il2ra	2	6.4	lymphoproliferation, hyperIgG, autoAb, anti-RBC Ab	(87)
IL-2	Il2	3	19.2	lymphoproliferation, hyperIgG, autoAb, anti-RBC Ab	(86)
lyn	*Lyn*	4	0.0	enhanced B cell activation, splenomegaly, hyperIgM, autoAb, GN	(49, 50)
C1q	*C1qa*	4	66.1	autoAb, GN	(121, 215)
	C1qb	4	66.1		
	C1qc	4	66.1		
SHP-1 (spontaneous)	*Hcph*	6	60.22	autoantibodies (*me* and *me^v* mutations)	(19, 20)
TGFβ1	Tgfb1	7	6.5	multiorgan lymphocytic and monocytic infiltrates	(94)
CD22	*Cd22*	7	9.0	enhanced B cell activation, autoAb	(51, 216)
Zfp-36 (tristetraprolin)	*Zfp36*	7	10.2	complex systemic disease: cachexia, dermatitis, arthritis	(133
IL-2Rβ	Il2rb	15	43.3	lymphoproliferation, hyperIgG, autoAb, anti-RBC Ab	(88)
p21 cyclin-dependent kinase inhibitor 1A	Cdkn1a	17	15.23	anti-chromatin Ab, GN, female predominance	(140)
Fas (spontaneous)	*Fas*	19	23.0	lymphoproliferation, DN T cells, autoAbs, GN, arthritis (lpr and lpr^{cg} mutations)	(218)
Pten (+/− mice)	*Pten*	19	24.5	lymphadenopathy, autoAb, GN, decr. survival, female predom.	(116)
Aiolos (zinc finger protein, subfamily 1A, 3)	*Znfn1a3*	ND	—	activated B cells, incr. IgG, autoAb	(68)
Bim (Bcl2 interacting mediator of cell death)	*Bim*	ND	—	lymphoid/myeloid cell accumulation, autoAb, GN, vasculitis	(109)
Cbl-b	—	ND	—	multiorgan lymphoid infiltrates, anti-dsDNA Ab	(105)
PD-1	—	hu2	q37.3	prolif. arthritis, GN, glomerular IgG3 deposits.	(144)
transgenics					
FLIP (retrovirus-mediated expression)	*Cflar*	1	30.1	hyperIgG, autoAbs, GN	(120)
Bcl-2 (B cell promotor)	*Bcl2*	1	59.8	lymphoid hyperplasia, hyperIgG, autoAb, GN	(110)
CD19	*Cd19*	7	59.0	increased B cell activation, B1 cell population, IgG, autoAb	(55)
Fli-1 (class I promotor)	*Fli1*	9	16.0	lymphoid hyperplasia, autoAb, GN	(76)
IFN-γ (keratin promotor)	*Ifng*	10	67.0	autoAb, GN, female predom.	(125)
IL-4 (class I promotor)	*Il4*	11	29.0	hyperIgG1/E, autoAb, GN	(126)
BAFF (α1 anti-trypsin or β-actin promotor)	Tnfsf13b	ND	—	autoAb (RF, CIC, dsDNA Ab), GN	(74, 75)

Genes are listed by chromosome and approximate chromosomal locations. Gene names are from the Mouse Genome Database (Jackson Laboratories).
*ND = not determined.

the hypothesis that Ikaros family members regulate gene expression during lymphocyte development by recruiting certain histone deacetylase complexes to specific promotors (67). Mice with homozygous deletions of the Aiolos gene develop defects primarily in the B cell compartment, with hyperreponsiveness to BCR and CD40 stimulation, increased number of conventional B cells, but a marked reduction in B1 cells, increased proportion of B cells with activated phenotype, hypergammaglobulinemia (particularly of IgE and IgG1), a 3-fold reduction in IgM and positive ANAs in about half of animals by 16 weeks of age (68). Detected T cell abnormalities were limited to only a slight increase in

proliferative capacity of thymocytes and mature T cells. Thus, deficiency of Aiolos appears to facilitate B cell entry into cell cycle, maturation to germinal center lymphocytes, and a breakdown of B cell tolerance. Long-term studies will determine if overt autoimmune disease develops in these mice (68).

Tnfsf13b (also called BAFF, BlyS, TALL-1, THANK or zTNF4) is a newly identified member of the TNF ligand superfamily expressed primarily on cells of myeloid origin, such as monocytes and dendritic cells (69–72). Both the transmembrane protein and a secreted homotrimeric form, released by cleavage at a furin canonical motif in the stalk

Table 4 Mechanisms for induction of systemic autoimmunity

Enhanced B Cell Activation*
  Lyn, CD22 or SHP-1 knockout
  Aiolos knockout
  CD19 transgenic
  Tnfsf13b (BAFF, BlyS, TALL-1, or THANK) transgenic
  Fli-1 transgenic
T Cell Activation
  CTLA-4 knockout
  IL-2 or IL-2R knockout
  TGF-β deficiency (knockout/dominant negative)
  Cb1-b knockout (increased activation of T cells, possibly B
    cells)
Defective Apoptosis
  Fas or FasL mutations (*lpr* and *gld* mice, ALPS in humans;
    also caspace 10)
  Bim knockout
  Bcl-2 transgenic
  Pten$^{+/-}$ knockout
  Flip transgenic
Complement and Related Genes
  C1q knockout
  SAP knockout
Cytokine-Mediated Activation
  IL-4 transgenic
  IFN-γ transgenic
  TTP (Zfp-36) deficiency (excessive TNF-α)
  TNFα transgenic
Unopposed cell cycling
  p21$^{cip1/waf1}$ (cyclin kinase inhibitor) (primarily T cells)
Miscellaneous
  PD-1 knockout

*Genes are categorized according to the most likely or predominant mechanism.

region, are active in the co-stimulation of B cells, the only cell type known to express its receptor. Two TNF family members, TACI and BCMA, have been identified as the receptors for Tnfsf13b (73). Overexpression of *Tnfsf13b* in the liver (apha-1 antitrypsin promotor with the APO E enhancer) (74) or in multiple tissues (β-actin promotor) (75) resulted in a similar picture, consisting of elevated numbers of B cells and, to a lesser extent, T cells, as well as increases in activated bcl-2-expressing mature B cells, memory/effector phenotype T cells and syndecan-1-positive plasma cells. B cells from transgenic mice survived longer in culture than those from wild-type controls. Furthermore, transgenic mice developed a lupus-like disease characterized by elevated levels of all immunoglobulin subclasses, rheumatoid factor, anti-DNA antibodies, circulating immune complexes and kidney immunoglobulin deposits with proteinuria and elevated blood urea nitrogen. Importantly, elevations in Tnfsf13b was also recently shown in both BWF$_1$ and MRL-*Fas^{lpr}* mice, and treatment of BWF$_1$ mice with a soluble TACI-IgGFc fusion protein, which blocks

Tnfsf13b function, inhibited proteinuria and prolonged survival (73).

In another study, aberrant expression of a transgenic *ets* family proto-oncogene, Fli-1, under the control of the MHC class I promotor, resulted in a constellation of lupus-like manifestations that included lymphoid hyperplasia, hypergammaglobulinemia, elevated antinuclear antibodies, and severe immune complex GN (76). Fli-1$^{-/-}$ B cells were hyperresponsive to a variety of stimuli, showed resistance to activation-induced cell death and had prolonged survival compared to nontransgenic B cells. Although these findings suggest an important regulatory function for Fli-1 in B cell response and homeostasis, a more limited role was indicated by the finding that immunological defects in Fli-1 knockout mice were confined to a mild generalized thymic hypocellularity

5.2. Genes Related to T Cell Activation

Systemic autoimmunity also develops after knockout deletion of certain genes that primarily effect T cell function (Tables 3 and 4). CTLA-4, a surface glycoprotein expressed exclusively on T cells, acts as an inhibitor of the CD28-B7.1/B7.2 costimulatory pathway in part by binding with higher affinity to B7.1 and B7.2. Consequently, mice with homozygous deletion of CTLA-4 develop a multiorgan lymphoproliferative disease associated with increased frequency of activated B and T cells, hypergammaglobulinemia and early mortality at 3–4 weeks of age with severe myocarditis and pancreatitis (77–79). The abnormal T cell expansion and disease manifestations are not due to alteration in thymocyte development (80), but to a failure to maintain homeostasis of activated peripheral T cells, primarily the CD4$^+$ subset (81, 82). The precise mechanism through which this occurs has yet to be determined (83).

Gene knockouts of IL-2 (84–86) or either of its high (IL-2Rα) (87) or low (IL-2Rβ) (88) affinity receptors result in a similar syndrome consisting of late immunosuppression with defective antibody and CTL responses, but also lymphoproliferation, expansion of memory/effector phenotype T cells, polyclonal hypergammaglobulinemia, autoantibodies and immune-mediated hemolytic anemia. Inflammatory bowel disease similar to human ulcerative colitis occurs in mice lacking IL-2 (89) or IL-2Rα (87), but not IL-2β (88). Autoantibody production depends on the expanded CD4$^+$ T cells (88) and CD40/CD40L interaction (84). The accumulation of T lymphocytes appears to be due to resistance of IL-2-deficient T cells to activation-induced cell death, at least partly from decreased sensitivity to Fas-mediated apoptosis (90). Other studies of mice maintained under germfree conditions demonstrated that colitis, but not other autoimmune manifestations, requires exposure to environmental pathogens (91).

TGF-β1 gene knockout mice rapidly develop massive necrotizing lymphocytic and monocytic infiltrates in multiple organs soon after birth, and succumb by around three weeks of age (92,93). Serum IgG autoantibodies to nuclear antigens as well as Ig glomerular deposits are detected, but appear to play a minor role in overall disease severity (94). Although deficiencies of either class II or class I (β2m) molecules combined with TGF-β1$^{-/-}$ reduced both tissue inflammation and autoimmunity, implicating both CD4$^+$ and CD8$^+$ T cells in these processes, in both instances there was only partial improvement in survival because of the remaining lethal myeoloproliferative abnormalities (95, 96). Direct inhibition of TGF-β1 in T cells by a dominant negative TGFβ receptor type II transgene under the control of a modified CD4 promotor specific for CD4$^+$ and CD8$^+$ T cells resulted in sickness, wasting and diarrhea around 3–4 months of age, monocytic infiltrates in multiple tissues, enlarged peripheral lymphoid organs, increased percentage of memory/effector phenotype T cells, hypergammaglobulinemia, autoantibodies and glomerular immune complex deposits. (97). Thus, TGF-β is crucial for maintenance of T cell homeostasis and suppression of autoimmunity.

Cbl-b is member of the cbl family of adaptor proteins that predominantly function to inhibit receptor and non-receptor tyrosine kinases (98,99). Two members, cbl-b and cbl, share a complex structure consisting of an amino-terminal phosphotyrosine-binding (PTB) domain, a C3HC4 RING finger, a proline-rich region capable of binding proteins with SH3 domains, several phosphotyrosine residues for binding SH2 domains, a ubiquitin recognition sequence and, at the carboxy-terminal, a putative leucine zipper. Recently, a smaller third member (cbl c or cbl 3) has been identified that contains the PTB, RING finger and a truncated proline-rich region, but is missing the rest of the carboxy portion (100,101). Although the precise mechanisms are not known, it is thought that cbl proteins may suppress signaling by directly inhibiting PTKs (through binding to negative regulatory sequences or induction of conformational changes), by sequestering molecules critical to cell activation, by recruiting distinct negative regulators, or by ubiquitination, internalization and degradation of receptor PTKs (98,102,103). Cbl-b is expressed in normal and malignant mammary epithelial cells, a variety of normal tissues, and in hematopoietic tissues and cell lines. In accordance with their negative regulatory role, T cells from mice homozygous for the cbl-b gene knockout exhibit enhanced proliferation to antigen receptor signaling and do not require CD28 co-stimulation for IL-2 production or generation of T-dependent antibodies (104, 105). Significantly, enhanced basal and activated levels of Vav, a guanine exchange factor for Rac-1/Rho/ CDC42, was the only alteration in TCR signaling identified in these knockout mice. This was consistent with the previous findings that cbl-b binds to Vav and, when overexpressed, inhibits Vav stimulation of the c-Jun terminal kinase (106). Cbl-b$^{-/-}$ mice exhibit increased susceptibility to autoimmunity, both to experimental autoimmune encephalomyelitis (104) and to a spontanous generalized auto- immune disease (105). The latter consisted of multiorgan lymphoaccumulation of polyclonal activated B and T cells with parenchymal damage, increased plasma cells, and antibodies to dsDNA by 6 months of age. Curiously, spontaneous autoimmunity occurred in only one (105) of the two Cbl-b knockout studies, suggesting that background strain, environment or other factors are important.

5.3. Apoptosis Genes

Bim is a proapoptotic ligand of the Bcl-2 family that shares homology with other members in only the short (nine amino acid) BH3 motif (107). Through this domain, Bim binds to anti-apoptotic Bcl-2 molecules and blocks their function. Bim is largely bound to the cytoplasmic dynein light chain LC8 that, under normal circumstances, is sequestered in the microtubule-associated dynein motor complex (108). Certain apoptotic signals release LC8, allowing LC8, together with Bim to translocate to Bcl 2 and inhibit its function. Homozygous knockout of Bim, resulted in an incompletely penetrant embryonic lethal phenotype, apparently for nonimmunologic reasons. In the surviving offspring, however, alterations in the homeostasis of multiple hematopoietic cell lineages developed (109). As anticipated, Bim-deficient B and T lymphocytes were resistant to certain apoptosis-promoting signals, but not to FasL. The knockout mice were found to have lymphoid hyperplasia with increases in naive T and B cells, altered thymocyte subset composition, and increases in granulocytes and monocytes in the peripheral blood. With age, these mice developed systemic autoimmunity manifested by progressive lymphadenopathy and splenomegaly, dramatic expansion of plasma cells, hyper IgM, IgG and IgA, antinuclear antibodies, immune complex GN and vasculitis with a 55% survival at 1 year.

Similarly, transgenic expression of the bcl-2 gene in B cells under the immunoglobulin enhancer resulted in lymphoid hyperplasia, hypergammaglobulinemia, high titers of antinuclear antibodies, and immune complex GN (110). Studies thus far have suggested that the constitutive bcl-2 expression may promote autoimmunity by blocking apoptosis of autoantibody producing B cells that normally arise spontaneously in germinal centers during the primary response to foreign antigens (111–113).

PTEN is a protein/lipid phosphatase initially identified as a tumor suppressor gene on chromosome 10q23, which is associated with a wide range of human malignancies (114, 115). Germline mutations of PTEN also cause three autosomal dominant disorders: Cowden disease, Lhermitte-Duclos syndrome, and Bannayan-Zonana syndrome. Homozygous knockouts of PTEN are embryonic lethal, but heterozygous mice develop an autoimmune disorder characterized by severe polyclonal lymphadenopathy, diffuse inflammatory cell infiltrates of most organs, hypergammaglobulinemia, anti-DNA antibodies, immune complex GN and decreased survival (116). Females were more severely affected, with survivals less than 12 months of age compared with over 15 months in males. Defective Fas-mediated activation-induced cell death of T and B lymphocytes was observed in PTEN-deficient mice due to impaired Fas signaling associated with increases in the survival factor, Akt. It was, therefore, postulated that uninhibited increases in phosphatidylinositol (3,4,5)-triphosphate (PIP-3), the major substrate for PTEN (117), leads to the recruitment and activation of Akt and possibly other factors, which then inhibits Fas-mediated killing (116). PTEN-deficient mice, however, in contrast to Fas^{lpr} mice, did not have increases in either B220$^+$ or CD4-CD8-DN T cell populations, and, furthermore, PTEN heterozygous knockouts had more severe disease than non-autoimmune background Fas-deficient mice. Thus, the development of autoimmunity in PTEN-deficient mice cannot be completely accounted for by Fas deficiency.

Finally, FLIP (gene name *Cflar*, for Caspase 8 and FADD-like apoptosis regulator, and also called I-FLICE, CASH, Casper, CLARP, FLAME, MRIT) is a death-effector domain-containing protein similar in structure to the apoptosis-promoting CASP8 (FLICE), but devoid of a caspase domain. In contrast to FLICE, FLIP inhibits death receptor-induced apoptosis by blocking the recruitment and activation of CASP8 (118,119). Transplantation of bone marrow cells retrovirally transfected with FLIP resulted in the resistance of B and T cells to activation-induced cell death, expansion of these cell populations and, in 4–6 month-old animals, the development of hypergammaglobulinemia, anti-dsDNA autoantibodies, and glomerular immunoglobulin deposits with histologic evidence of GN characterized by glomerular sclerosis and thickening of mesangium and basement membranes (120). Despite the fact that FLIP blocks Fas signaling of activation-induced cell death, there were no accumulations of B220$^+$ or CD4$^-$ CD8$^-$ DN T cells, or activated T cells. This suggests that Fas and/or FLIP may affect other non overlapping pathways.

5.4. Complement Genes

It has long been known that deficiencies of early complement components (C1q-s, C2, or C4) in humans predispose to SLE, indicating that they have an important regulatory role in suppressing autoimmunity. Although inadequate clearance of immune complexes is a likely mechanism, this does not fully explain the loss of tolerance to nuclear antigens. Gene knockout mice for C1q and C4 were therefore generated to address this issue. Homozygous C1q-deficient mice recapitulated the human disorder with the development of mild, but typical, features of lupus, including autoantibodies and a 25% incidence of immune complex GN (121). A large number of apoptotic bodies were discovered in the glomeruli of these mice, suggesting that C1q plays an essential role in the clearance of apoptotic bodies. In the case of C4 deficiency, the contribution to lupus susceptibility was evident only on the Fas^{lpr} background (122), indicating that lack of C4 in itself is not sufficient to induce autoimmunity. Using an HEL Ig transgenic model, B cells from C4-deficient mice were shown to have a tolerance defect to soluble HEL. Thus, it was hypothesized that early complement components may be vital for maintaining self-tolerance by virtue of their role in presenting tolerizing antigens to B cells.

A similar mechanism was also proposed to explain the unexpected development of lupus-like autoimmunity in mice homozygous for deletion of serum amyloid P component (SAP), a highly conserved plasma protein originally named for its presence in amyloid deposits (123,124). Lupus manifestations included a female predominance, autoantibodies to chromatin and its components, but not to other nuclear, tissue or organ antigens, immune complex GN and low incidence of mortality. In contrast to C1q deficiency, no accumulation of apoptotic bodies was detected in glomeruli. SAP binds to DNA and chromatin, and can displace H1-histones, thereby increasing solubility and reducing the rate of degradation and clearance of chromatin (123). It was hypothesized that SAP functions to promote self-tolerance to chromatin and its subunits by preventing immunogenic antigen-processing and/or by tolerizing chromatin reactive lymphocytes.

5.5. Cytokine Ligand and Receptor Genes

Under certain circumstances, systemic autoimmunity also develops in mice transgenic for the major Th1 and Th2 cytokines, IFN-γ and IL-4, respectively. Expression of INF-γ in the suprabasal layer of the epidermis under the control of the involucrin promotor resulted in not only a severe inflammatory skin disorder, but also the generation of autoantibodies to dsDNA and histone, and an immune complex proliferative GN, particularly in females (125). This suggests that presentation of nuclear antigen by skin Langerhans cells, and perhaps keratinocytes, may be sufficient for the production of antinuclear autoantibodies, and provides a possible explanation for the UV sensitivity of SLE patients. Similarly, C3H mice transgenic for the IL-4 gene under the control of the MHC class I promotor also developed systemic autoimmunity characterized by elevated MHC class II molecules and CD23, enhanced responses to polyclonal stimuli *in vitro*, increased levels of IgG1 and IgE, anemia, antinuclear antibodies, and immune complex GN (126). These manifestations were likely due to direct IL-4-induced polyclonal activation of B cells, since CD4+ T cells were not required and there was no evidence for inefficient negative selection of B cells. The role of IL-4 in promoting lupus, however, is more complicated, since autoimmunity was not observed in other transgenics expressing IL-4 on B cells or T cells (127–130) and, in a spontaneous model of lupus expression of an IL-4 transgene, did not exacerbate disease, but was protective (131). This would imply that a number of factors, such as the level and site of IL-4 production, and background genetic susceptibility, can modify the effects of IL-4 on systemic autoimmunity.

Tristetraprolin (TTP or Zfp-36) is a widely expressed zinc-binding protein initially throught to function as a transcription factor, particularly in lymphoid tissues, where high levels are found (132). Mice with homozygous knockout of TTP develop a complex syndrome consisting of cachexia, patchy alopecia, dermatitis, conjunctivitis, erosive arthritis, myeloid hyperplasia, glomerular mesangial thickening and anti-nuclear antibodies (133). These manifestations are mainly caused by excessive TNFα production by macrophages and are reversed almost entirely by treatment with anti-TNFα antibody (133,134). Thus, TTP must function as a nonredundant negative regulator of TNF-α. The mechanism was later discovered to be due to the binding of TTP to an AU-rich element contained in the TNF-α mRNA, which destabilizes the mRNA (135). In contrast to the autoimmune-promoting effects of elevated levels of TNF-α, physiological amounts may, under certain circumstances, play a role in suppressing systemic autoimmunity. In studies of TNFR1 (p55) knockout mice, non-autoimmune mice did not develop defects in apoptosis or autoimmunity (136,137), yet the same knockout in the lupus-prone C57BL/6-*Fas^lpr* mice resulted in accelerated lymphoproliferation and autoimmune disease (138). Since TNF can induce death of activated peripheral T cells (139), these findings might be attributable to TNF compensating for the lack of Fas in *Fas^lpr* mice.

5.6. Cell Cycle-Related Genes

Gene knockout of the cyclin inhibitor p21^cip1/waf1 in mixed background C57BL/6 x 129/Sv also resulted in the development of systemic autoimmunity characterized by lymphoid hyperplasia, elevated IgG1, IgG antinuclear antibodies, GN and early mortality (140). *In vitro* T cell proliferation was enhanced in these mice and there was an accumulation of effector/memory phenotype (CD44^high) CD4+ T cells, although levels of other activation/effector/memory cell markers, such CD25, CD62L, CD69, and CD45RB, were similar to wild-type mice. An increased proportion of splenic B cells also expressed an activated (HSA^low, IgG^low) phenotype. Interestingly, females had more severe disease than males, similar to human SLE. Based on these findings, it was suggested that p21 negatively regulates T cell proliferation following long-term stimulation, as is presumably the situation for autoantigen-reactive CD4+ T cells. In sharp contrast, other studies in lupus-prone BXSB mice found an increase in p21 and other cyclin inhibitors in the expanded memory/ effector CD4+ T cells that are predominantly arrested in G1 (141). This led to just the opposite hypothesis, that the accumulating CD4+ T cells, following successive rounds of division, become incapable of entering into cell cycle and are resistant to apoptosis because of the build-up of cyclin inhibitors, a state similar to replicative senescence. Although no longer cycling, such cells may nonetheless secrete cytokines and activate B cells. Overall, it may be that both insufficient and excessive p21 may predispose to lupus by two independent mechanisms.

5.7. Miscellaneous

PD-1 is a 55 kD ITIM-containing transmembrane cell surface glycoprotein expressed on activated T and B lymphocyes and monocytic cells (142). Little is known about its signaling pathways or ligand, however, clues to its function have come from gene knockouts. Mice deficient for PD-1 develop moderate hyperplasia of lymphoid and myeloid cells, increases in several Ig isotypes (particularly IgG3), enhanced responses to IgM cross-linking, and alterations in peritoneal B1 cells (143). Older (14 month-old) C57BL/6-PD-1^−/− mice spontaneously develop GN and proliferative arthritis, but not elevated anti-dsDNA antibodies or rheumatoid factor (144). Further acceleration of GN and arthritis occurs when the PD-1 deletion is combined with the *Fas^lpr* mutation. Interestingly, manifestations appear highly dependent on background genes. For example, in contrast to the findings in the C57BL/6 strain, BALB/c-PD-1^−/− mice reportedly develop lethal pancreatitis with massive thrombosis. Thus PD-1 deficiency may accelerate background predisposition to autoimmunity similar to the *Fas^lpr* and Yaa mutations. Although the cellular and molecular details are not yet defined, the data suggest that PD-1 plays an important nonredundant role in maintaining the homeostasis of lymphocytes and myeloid cells following their activation.

5.8. Summary of Lupus-Like Disease in Genetically-Manipulated Normal Backgrounds

Through the deletion or overexpression of single genes, fascinating examples of loss of tolerance and systemic autoimmunity have been generated that provide new models for dissecting the immunopathogenesis of SLE. Although the relevance of many of these models to spontaneous disease is uncertain and, in some cases, even questionable, they have been particularly important in identifying potential mechanisms of autoimmunity at the molecular level. As outlined in the preceeding section, these can be broadly categorized as gene defects that enhance B or T cell activation, that inhibit certain apoptotic pathways, that reduce complement clearance of apoptotic bodies or soluble self-antigen, that modify certain cytokine milieus or that alter cell cycling (Table 4). In some instances, common mechanisms derived independently from different molecular defects have emerged, e.g. CD22, lyn, SHP-1 and CD19, which all modify the BCR signaling threshold. Overall, these studies, which probably involve only a fraction of the total number of immunologically-relevant genes, suggest that a large number of genetic defects might potentially result in systemic autoimmunity and that many of these may involve common pathways or mechanisms. The important issue that remains, however, is what the relevance of these gene defects, induced in mice by genetic manipulation, is to spontaneous SLE.

6. GENE KNOCKOUT AND TRANSGENIC LUPUS BACKGROUND MICE

Other studies have used congenic lupus background strains to directly examine the role of deleted or overexpressed immune-related genes in systemic autoimmunity (Table 5, listed by chromosome and chromosomal location). In the context of lupus pathogenesis, the effects of gene manipulation on different stages of the disease process has provided important information on the crucial molecules and pathways, as well as a deeper understanding of the molecular basis for the diverse manifestations.

6.1. B Cell Related Genes

The crucial role for B cells in lupus disease manifestations was shown in MRL-Fas^{lpr} mice with deletion of the Jh locus (no B cells) (145). In this instance, despite the presence of lymphoproliferation, no signs of nephritis or vasculitis were evident, confirming the notion that autoantibodies play a central role in these processes. Similar findings were also observed in Fas-intact MRL-+/+ mice deficient for the Jh locus (146). Subsequently, B cells were shown to be important not only for autoantibody production, but also for the spontaneous activation of CD4$^+$ and CD8$^+$ T cells (147,148). Furthermore, genetic manipulation of MRL-Fas^{lpr} mice that resulted in B cells expressing surface, but not secreted, immunoglobulin, demonstrated that B cells themselves, without circulating autoantibodies, could promote local cellular infiltration and inflammation, but not GN (149).

6.2. T Cell-Related Genes

The role of helper and cytotoxic T cells, as well as $\alpha\beta$- and $\gamma\delta$-T cell subsets, in lupus have been assessed with a variety of intercross or congenic lupus mice rendered defective for MHC, CD4, CD8 or T cell receptor genes. Deletion of MHC class II (150) or CD4 (151) in MRL-Fas^{lpr} mice reduced autoantibodies and GN, but did not affect lymphadenopathy, indicating that the class II-selected CD4 T cells are important for autoimmunity, whereas DN B220$^+$ T cells are not selected on class II molecules nor are they necessary for development of auto-immunity. In contrast, β2m- (class I) (152–155) or CD8-deficient (151) MRL-Fas^{lpr} or C3H-Fas^{lpr} mice showed reductions in both lymphoproliferation and expansion of DN B220$^+$ cells, but only partial diminution at most in autoantibody levels and GN. Combined with previous studies on TCR Vβ repertoires and anti-CD8 antibody treatment (rev. in 2), these findings indicate a CD8 origin for the DN B220$^+$ T cells in Fas-deficient mice. NZB mice deficient for β2m knockout were also generated and, in this case, a delayed onset and reduced incidence of anti-RBC antibodies were found despite no change in the characteristic NZB thymic abnormalities nor in IgM or IgG autoantibody levels to ss or dsDNA (156).

Although this and other evidence (rev. in 2) clearly demonstrated the importance of TCR$\alpha\beta$ cells in spontaneous lupus, the deletion of the TCRα gene in MRL-Fas^{lpr} mice only partially inhibited disease (157,158). Since TCR$\alpha\beta$/TCR$\gamma\delta$ double-knockout MRL-Fas^{lpr} mice fail to generate class switched autoantibodies and immune complex GN (159,160), this suggests that TCR$\gamma\delta$ cells can also help drive the autoimmune B cells. Paradoxically, however, MRL-Fas^{lpr} mice deficient for TCR$\gamma\delta$ cells showed disease exacerbation. Taken together, these results suggest that TCR$\alpha\beta$ cells are important for cognate MHC-restricted autoantibody production and are the major provider of B cell help in intact MRL-Fas^{lpr} mice, whereas TCR$\gamma\delta$ cells suppress this process, but simultaneously can provide lesser degrees of non MHC-restricted polyclonal B cell help. In addition, knockout of the perforin gene in MRL-Fas^{lpr} mice resulted in disease exacerbation, suggesting that cytolytic cells may be involved in suppressing autoreactivity (161).

6.3. Costimulatory Molecules

The important role of costimulation for the development of lupus was demonstrated in MRL-Fas^{lpr} mice deficient for CD40L (162,163) and CD28 (164). These results were similar to the reductions in disease progression obtained with BXSB and BWF$_1$ mice treated with anti-CD40L or

Table 5 Effects of genetic manipulation of lupus-prone mice

Name	Gene	Chr#	cM	Strain*	Result	Ref.
CD28	Cd28	1	30.1	MRL-lpr	reduced GN	(164)
FcR γ-chain	Fcer1g	1	93.3	BWF$_1$	same autoAb, glom. deposits reduced GN, mortality	(178)
CD21/CD35	Cr2	1	106.6	lpr	accelerated disease	(122, 219)
MHC class I	β2m	2	69.0	MRL-lpr	reduced lymphoproliferation	(153, 154)
				C3H-lpr & -gld	reduced lymphoproliferation	(152)
				NZB	reduced anti-RBC Ab	(156)
CD8	Cd8a	6	30.5	MRL-lpr	reduced lymphoprolif.	(151)
	Cd8b	6	30.5			
5-lipoxygenase	Alox5	6	53.2	MRL-lpr	slight accelerated mortality (males only)	(220)
CD4	Cd4	6	60.18	MRL-lpr	incr. lymphoproliferation; decr. autoAb, GN, mortality	(151, 221)
ICAM-1	Icam1	9	7.0	MRL-lpr	incr. survival, reduced autoAb, GN, vasculitis	(176)
				MRL-lpr	no pulm. inflam., incr. survival. same lymphoprolif., autoAb, GN.	(175)
IFN-γR	Ifngr	10	15.0	MRL-lpr	reduced GN	(168, 172)
	Ifngr2	16	65.0	BWF$_1$	reduced GN, mortality	(170)
fyn	Fyn	10	25.0	MRL-lpr	reduced DN T cells, autoAb, GN	(174)
perforin	Pfp	10	36.0	MRL-lpr	accelerated disease	(161)
IFN-γ	Ifng	10	67.0	(MRL-lprF$_2$	reduced GN, mortality, autoAb	(167)
				MRL-lpr	reduced GN, mortality, autoAb	(169)
				HgIA	reduced autoAb, GN	(171)
IL-4	Il4	11	29.0	(MRL-lprF$_2$	reduced GN	(167)
				BXSB	no effect	(173)
NOS2	Nos2	11	45.6	MRL-lpr (N4)	same autoAb, GN, arthritis reduced vasculitis	(177) (222)
Jh, mIg Tg (B cells no Abs)		12		MRL-lpr	Infiltrate, no GN	(149)
Jh (no B cells)	Igh-J	12	58.0	MRL-lpr	No disease	(145, 146)
TCRαδ (no αβ+ or γδ+ cells)		14			no IgG autoAb, GN	(223)
TCRδ (no γδ+ cells)	Tcrd	14	19.5		disease acceleration	(224)
TCRα (no αβ+ cells)	Tcra	14	19.7		major disease reduction	(158, 225)
B7.2	Cd86	16	26.9	MRL-lpr	reduced GN	(165)
B7.1	Cd80	16	28.0	MRL-lpr	more severe, distinct GN	(165)
MHC class II	H2-Aa	17	18.65	MRL-lpr	reduced autoAb, GN, same lymphoproliferation	(150)
	H2-Ea					
complement C4	C4	17	18.8	MRL-lpr	accelerated disease	(122)
complement factor B	H2-Bf	17	18.85	MRL-lpr	reduced autoAb, GN, vasculitis	(186)
TdT	Tdt	19	39.5	BWF$_1$	same autoAb, reduced GN and mortality	(226)
CD40L	Tnfsf5	X	18.0	MRL-lpr	reduced autoAb, GN	(163)
protein kinase CK2 alpha	Csnk2a1	ND	–	MRL-lpr	accelerated lymphoproliferation, autoAb,	(227)
(casein kinase II)	Csnk2a2	8	50.0	GN		
(transgenic)	Csnk2b	17	19.02			
MCP-1	ND	ND	–	MRL-lpr	reduced GN, mortality; same autoAb and glom. deposits	(180)

Genes are listed by their approximate chromosomal locations. Genes are deficiencies by homologous recombinant knockout unless otherwise stated.
*MRL-lprF$_2$ = mixed background derived from (MRL-Fas^{lpr}×(B6x129) F$_1$)F$_2$, lpr = Fas^{lpr}, gld = Fasl^{gld}.
#ND = not determined.

CTLA-4Ig (rev. in 2). Further dissection of the CD28-B7 axis has been performed in MRL-Fas^{lpr} mice deficient for either B7.1 or B7.2 (165). Interestingly, although neither of the deletions affected autoantibody levels, when compared to wild-type MRL-Fas^{lpr} mice, GN was substantially worse in B7.1-deleted mice, but less severe in B7.2 knockouts. This is consistent with findings in BWF$_1$ mice showing antibodies to B7.2, but not B7.1, suppressed disease (166), and suggests a more critical role for B7.2 in lupus pathogenesis.

6.4. Cytokine Genes

The effect of the Th1 and Th2 cytokines, IFN-γ and IL-4, respectively, on the development of lupus has been investigated using congenic cytokine knockout or transgenic lupus strains. The results in IFN-γ knockout mice are particularly striking since they clearly contradict the earlier notion that Th1-type responses primarily promote cellular immunity, while Th2-type responses promote mainly humoral immunity. Deficiency of IFN-γ or IFN-γR in MRL-*Fas*^{lpr} or BWF$_1$ mice (167–170) uniformly led to a marked reduction in lupus disease, even in the lupus-like mercury-induced autoimmunity model (171), which was previously considered to be a prototypic Th2- mediated disease. This outcome was, in fact, consistent with previous observations of elevated IFN-γ levels in all lupus-prone strains, accelerated disease with IFN-γ treatment, and reduced severity following anti-IFN-γ treatment (rev. in 2). The data thus far indicates that IFN-γ is important in at least two steps of lupus pathogenesis: First, by promoting the response to self-antigens, perhaps because they are of low antigenicity (169, 171). Second, by preventing accelerated mortality and GN even in IFN-$\gamma^{+/-}$ MRL-*Fas*^{lpr} mice despite levels of auto-antibodies and glomerular immunoglobulin deposits similar to wild-type animals. The implication of a requirement for IFN-γ for the progression of local inflammatory responses (169) is supported by the recent findings that IFN-γ signaling is critical for local production of the nephritis-promoting cytokines, CSF-1 and TNFα (172). Furthermore, IFN-γ may be directly or indirectly responsible for the apoptosis of kidney tubular epithelial cells and mesangial cells, respectively. This uncoupling of immune and inflammatory processes is reminiscent of FcR γ-chain and MCP-1 deletions, as noted below. Clearly, the central role of IFN-γ in the induction and effector phases of systemic autoimmunity provides a strong impetus for developing intervention based on blockade of IFN-γ action.

Deletion of IL-4, in contrast, results in only partial or no reduction in disease depending on the lupus-prone strain. IL-4$^{-/-}$ MRL-*Fas*^{lpr} mice produced less serum IgG1 and IgE (Th2-dependent subclasses), but maintained comparable levels of IgG2a and IgG2b, as well as autoantibodies (167). In spite of this, a reduction in lymphoadenopathy and end-organ disease was observed. In contrast, IL-4-deficient male BXSB mice had similar autoantibody levels, GN severity and mortality as their wild-type counterparts (173), suggesting that IL-4 plays little role in the immuno-pathogenesis of disease in this strain.

6.5. Cell Signaling Molecules

A major pathognomonic finding in Fas deficiency is the expansion of DN B220$^+$ T cells that are predominantly derived from CD8$^+$ precursors. Although early *in vitro* functional studies suggested that these DN cells were inert,

other characteristics suggested an activated state, including constitutive phosphorylation of the CD3ζ chain, as well as increases in p59^{fyn}, certain oncogenes and cell surface activation markers (TCR^{low}, CD44^{high}). This was confirmed *in vivo* using BrdU labeling, wherein DN B220$^+$ T cells were shown to be highly cycling early in life, with the frequency of cycling cells declining with age (174). Furthermore, MRL-*Fas*^{lpr} mice deficient for fyn showed reduction in the frequency of DN B220$^+$ T cells along with decreased autoantibody levels and immunopathology, indicating an essential role for fyn in signal transduction and expansion of DN B220$^+$ T cells (174). Thus, DN B220$^+$ T cells are functional, but lose the capacity to proliferate as they accumulate, presumably due to replicative senescence (a loss of entry into cell cycle after a finite number of divisions). Replicative senescence has also been suggested as an explanation for the rapid accumulation of apoptosis-resistant memory/effector (CD44^{high}) CD4 cells in older BXSB male mice, but, in this case, senescence is thought to be a consequence of increases in cyclin inhibitors (141). Such senescent cells may secrete proinflammatory cytokines that contribute to autoimmunity.

6.6. Regulators of Local Immune and Inflammatory Responses

Local upregulation of a variety of molecules, including the MHC, ICAM-1, nitric oxide and MCP-1, which may contribute to the immune and inflammatory responses essential for end-organ damage, have been reported for lupus strains (rev. in 2). The role of such molecules in disease pathogenesis has been addressed by examining lupus-prone mice with homozygous deletions of these genes. MRL-*Fas*^{lpr} mice deficient for ICAM-1 showed improved survival, primarily due to a reduction in vasculitis rather than changes in autoantibody levels and GN (175,176). Similarly, deletion of the NOS2 gene in the MRL-*Fas*^{lpr} mice resulted in a partial improvement, with significant reductions in vasculitis and IgG rheumatoid factor, but equivalent severity of anti-DNA antibody levels and GN (177). In contrast, deletion of the FcR γ-chain in (NZB x NZW)F$_1$ mice resulted in decoupling of the immune and inflammatory processes-with mice having similar auto-antibody levels and glomerular deposits of immunoglobulin or complement, but siginificantly less glomerular destruction and mortality (178). A similar result was also demonstrated in the anti-GBM antibody model (179), wherein administration of anti-GBM antibody bypassed the requirement for generating pathogenic autoantibodies. Thus, FcRγ-chain appears to play an important nonredundant role in the induction of local inflammatory glomerular injury. A similar decoupling was observed in MRL-*Fas*^{lpr} mice with deletion of MCP-1 (macrophage chemoattractant protein-1), a chemokine that recruits macrophages and T cells to tissues (180). MCP-1 appears to be another factor critical for the progression from immune deposits to inflammatory tissue damage.

6.7. Miscellaneous Genes

Terminal deoxynucleotidyl transferase (TdT) is essential for adding nucleotides to the N-regions at the V-(D)-J junctions during rearrangement of B and T cell antigen receptors, a process that enhances B and T cell repertoire diversity both by the random addition of nucleotides and by disrupting the formation of repetitive homology-directed junctions (181–183). Knockout of the *Tdt* gene in normal background mice resulted in a reduction of diversity of the CDR3 regions with essentially fetal antigen receptor repertoires. Nevertheless, such mice were viable and effectively responded to a wide range of immune challenges. Of possible relevance to lupus, frequencies of anti-DNA expressing B cells following LPS stimulation (natural antibody repertoire) were lower in TdT$^{-/-}$ mice, mainly due to a lower incidence of polyreactivity; furthermore, the lack of N-region diversity appeared to reduce affinity of the anti-DNA antibodies (184). Accordingly, TdT-deficient BWF$_1$ background mice had significant reductions in GN and mortality but, curiously, no substantial differences in immunoglobulin levels or antinuclear antibody profiles. In contrast, TdT$^{-/-}$ C57BL/6-*Fas^{lpr}* mice had substantially lower anti-DNA and rheumatoid factor (RF) levels than their TdT$^{+/+}$ littermates (185). Although the mechanism(s) by which TdT enhances lupus remains to be determined, it appears that it promotes autoreactivity, possibly by directly altering the fine specificies and affinities of autoantibodies, T cell receptor repertoire, or both.

The role of the alternative pathway of complement activation in lupus pathogenesis was investigated in MRL-*Fas^{lpr}* mice deficient for the complement factor B (Bf) (186). Bf, produced by a variety of cells types, including hepatocytes, phagocytes, fibroblasts and endothelial cells, is an acute phase reactant necessary for activation of the alternative pathway. Local production of Bf can be detected in the kidneys of lupus-prone mice (187). Homozygous deletion of Bf resulted in significant reduction in the severity of GN and incidence of vasculitis, along with lower levels of anti-dsDNA, IgG3 anti-IgG2a RF, and IgG3 (186). Serum C3 also remained at normal levels in Bf$^{-/-}$ mice. These findings support a significant role for Bf and the alternative pathway in the immune complex autoimmune pathology of these mice.

In some instances, overexpression or deficiency of genes can lead to various degrees of disease acceleration in lupus-prone mice, yet no reported autoimmunity when imposed on normal background strains. These include overexpression of CK2 transgenes and knockouts of TNFR1 (noted previously) or 5-lipoxygenase. Protein kinase CK2 (casein kinase II) is a ubiquitous heterotetrameric serine-theonine kinase composed of two α or α' catalytic and two β regulatory subunits (188). It can phosphorylate a large range of protein substrates, including those involved in nucleic acid synthesis, nuclear transcription, signal transduction, protein synthesis and the cytoskeleton. CK2 is active in proliferating cells, and high levels are found in certain human cancers. Mice transgenic for the α catalytic subunit of CK2 (CKα) show enhanced susceptibility to T cell lymphomas in a stochastic manner (189), and overexpression of CK2 accelerates the development of lymphomas in myc or tal-1-transgenic or p53-knockout mice (190). Crossing the CK2 transgene onto the MRL-*Fas^{lpr}* background resulted in no increase in lymphomas, but marked acceleration of lymphoproliferation and autoimmunity, with higher levels of serum IgG2a and earlier development of ANA positivity and proliferative GN (190). Accelerated disease was postulated to be due to CK2-mediated enhancement of T cell proliferative responses.

Leukotrienes are potent proinflammatory lipid mediators produced through the 5-lipoxygenase pathway of arachidonic acid metabolism. Studies have suggested that arachidonic acid metabolites, including the leukotrienes, may promote the autoimmune pathogenesis of MRL-*Fas^{lpr}* mice (191). In contrast, however, deficiency of 5-lipoxygenase in MRL-*Fas^{lpr}* mice did not reduce disease manifestations, but resulted in a modest acceleration of mortality and a slight increase in the prevalence, but not severity, of arthritis in males. Females were not affected, nor were there differences in lymphoproliferation, autoantibody levels or histologic severity of GN. These findings suggest that 5-lipoxygenase and/or leukotrienes may normally have some modest disease-suppressing activity in males.

7. CONCLUSIONS

Definition of the genes and genetics of murine models of spontaneous lupus has greatly profitted from advances in transgenic, gene knockout and quantitative trait mapping approaches. Genome-wide QTL scans of several lupus strains have resulted not only in the identification of multiple predisposing loci, but have advanced our understanding of the nature and complexity of genetic susceptibility. Interval-specific congenic lines are being used to define the relationship of these loci to component phenotypes and to precisely mapping the location of specific traits, which will eventually lead to the definitive characterization of the predisposing genetic alterations. Results from genetically-manipulated normal background mice with altered expression of specific immune-related genes have revealed a diversity of perturbations that can lead to defects in tolerance and to manifestations of systemic autoimmunity.

Importantly, these findings suggest that spontaneous systemic autoimmunity can be induced by independent genetic abnormalities acting at the various checkpoints controlling immune responses. Genes affecting B and T cell activation, apoptosis, complement clearance of self-antigens, certain cytokines and the cell cycle have thus far been implicated. Similar studies assessing germ-line alterations in lupus-predisposing mice have also identified a variety of genes that in most instances are inhibitory, although a few can enhance disease severity. The former are likely genes involved in downstream effector mechanisms, whereas the latter might play minor roles in lupus predisposition. From the combined mapping and genetic-manipulation studies, it is seems apparent that susceptibility to murine lupus could involve considerable genetic heterogeneity. Such heterogeneity will undoubtedly make identification of the predisposing genes more difficult and labor-intensive, however, once accomplished, delineation of the specific genotypes with their associated traits should provide a much more rational basis for assessing susceptibility and for devising specific therapeutic interventions.

ACKNOWLEGEMENTS

This is Publication Number 13592-IMM from the Department of Immunology, The Scripps Research Institute, 10550 North Torrey Pines Road, La Jolla, CA 92037. The work of the authors reported herein was supported by National Institutes of Health grants AR31203, AR39555, AG15061 and ES08666. The authors thank M. Kat Occhipinti for editorial assistance.

References

1. Theofilopoulos, A.N., and F.J. Dixon. 1985. Murine models of systemic lupus erythematosus. *Adv. Immunol.* 37:269–390.
2. Theofilopoulos, A.N., and D.H. Kono. 1999. Murine lupus models: gene-specific and genome-wide studies. In: *Systemic lupus erythematosus (3)*, R.G. Lahita, ed., Academic Press, San Diego. pp. 145–181.
3. Hirose, S., R. Nagasawa, I. Sekikawa, M. Hamaoka, Y. Ishida, H. Sata, and T. Shirai. 1983. Enhancing effect of H-2 linked NZW genes on the autoimmune traits of (NZB x NZW)F1 mice. *J. Exp. Med.* 158:228–233.
4. Kotzin, B.L., and E. Palmer. 1987. The contribution of NZW genes to lupus-like disease in (NZB x NZW)F1 mice. *J. Exp. Med.* 165:1237–1251.
5. Merino, R., L. Fossati, M. Lacour, R. Lemoine, M. Higaki, and S. Izui. 1992. H-2- linked control of the Yaa gene-induced acceleration of lupus-like autoimmune disease in BXSB mice. *Eur. J. Immunol.* 22:295–299.
6. Chiang, B., E. Bearer, A. Ansari, K. Dorshkind, and M.E. Gershwin. 1990. The BM12 mutation and autoantibodies to dsDNA in NZB.H-2bm12 mice. *J. Immunol.* 145:94–101.
7. Iwamoto, M., N. Ibnou-Zekri, K. Araki, and S. Izui. 1996. Prevention of murine lupus by an I-E alpha chain transgene:
protective role of I-E alpha chain-derived peptides with a high affinity to I-A^b molecules. *Eur. J. Immunol.* 26:307–314.
8. Kawano, H., M. Abe, D. Zhang, T. Saikawa, M. Fujimori, S. Hirose, and T. Shirai. 1992. Heterozygosity of the major histocompatibility complex controls the autoimmune disease in (NZW x BXSB)F1 mice. *Clin. Immunol. Immunopathol.* 65:308–314.
9. Morel, L., U.H. Rudofsky, J.A. Longmate, J. Schiffenbauer, and E.K. Wakeland. 1994. Polygenic control of susceptibility to murine systemic lupus erythematosus. *Immunity* 1:219–229.
10. Drake, C.G., S.J. Rozzo, H.F. Hirschfeld, N.P. Smarnworawong, E. Palmer, and B.L. Kotzin. 1995. Analysis of the New Zealand Black contribution to lupus-like renal disease. Multiple genes that operate in a threshold manner. *J. Immunol.* 154:2441–2447.
11. Vyse, T.J., S.J. Rozzo, C.G. Drake, V.B. Appel, M. Lemeur, S. Izui, E. Palmer, and B.L. Kotzin. 1998. Contributions of Ea(z) and Eb(z) MHC genes to lupus susceptibility in New Zealand mice. *J. Immunol.* 160:2757–2766.
12. Rozzo, S.J., T.J. Vyse, C.S. David, E. Palmer, S. Izui, and B.L. Kotzin. 1999. Analysis of MHC class II genes in the susceptibility to lupus in New Zealand mice. *J Immunol* 162:2623–30.
13. Ibnou-Zekri, N., T.J. Vyse, S.J. Rozzo, M. Iwamoto, T. Kobayakawa, B.L. Kotzin, and S. Izui. 1999. MHC-linked control of murine SLE. *Curr. Top. Microbiol. Immunol.* 246:275–80.
14. Merino, R., M. Iwamoto, L. Fossati, P. Muniesa, K. Araki, S. Takahashi, J. Huarte, K. Yamamura, J.D. Vassalli, and S. Izui. 1993. Prevention of systemic lupus erythematosus in autoimmune BXSB mice by a transgene encoding I-E alpha chain. *J. Exp. Med.* 178:1189–1197.
15. Ibnou-Zekri, N., M. Iwamoto, L. Fossati, P. McConahey, and S. Izui. 1997. Role of the major histocompatibility complex class II Eα gene in lupus susceptibility in mice. *Proc. Natl. Acad. Sci. USA* 94:14654–14659.
16. Fujimura, T., S. Hirose, Y. Jiang, S. Kodera, H. Ohmuro, D. Zhang, Y. Hamano, H. Ishida, S. Furukawa, and T. Shirai. 1998. Dissection of the effects of tumor necrosis factor-alpha and class II gene polymorphisms within the MHC on murine systemic lupus erythematosus (SLE). *Int. Immunol.* 10:1467–1472.
17. Jacob, C.O., S.K. Lee, and G. Strassmann. 1996. Mutational analysis of TNF-alpha gene reveals a regulatory role for the 3′-untranslated region in the genetic predisposition to lupus-like autoimmune disease. *J. Immunol.* 156:3043–50.
18. Morel, L., X.H. Tian, B.P. Croker, and E.K. Wakeland. 1999. Epistatic modifiers of autoimmunity in a murine model of lupus nephritis. *Immunity* 11:131–9.
19. Shultz, L.D., P.A. Schweitzer, T.V. Rajan, T. Yi, J.N. Ihle, R.J. Matthews, M.L. Thomas, and D.R. Beier. 1993. Mutations at the murine motheaten locus are within the hematopoietic cell protein-tyrosine phosphatase (Hcph) gene. *Cell* 73:1445–1454.
20. Tsui, H.W., K.A. Siminovitch, L. deSouza, and F.W.L. Tsui. 1993. *Motheaten* and *viable motheaten* mice have mutations in the haematopoietic cell phosphatase gene. *Nature Genet.* 4:124–129.
21. Van Zant, G., and L.D. Shultz. 1989. Hematologic abnormalities of the immunodeficient mouse mutant, viable motheaten (me^v). *Exp. Hematol.* 17:81–87.
22. Scribner, C.L., C.T. Hansen, D.M. Klinman, and A.D. Steinberg. 1987. The interaction of the xid and me genes. *J. Immunol.* 138:3611–3617.

23. Yu, C.C., H.W. Tsui, B.Y. Ngan, M.J. Shulman, G.E. Wu, and F.W. Tsui. 1996. B and T cells are not required for the viable motheaten phenotype. *J. Exp. Med.* 183:371–80.

24. Thomas, M.L. 1995. Of ITAMs, and ITIMs:turning on and off the B cell antigen receptor. *J. Exp. Med.* 181:1953–1956.

25. Doody, G.M., L.B. Justement, C.C. Delibrias, R.J. Matthews, J. Lin, M.L. Thomas, and D.T. Fearon. 1995. A role in B cell activation for CD22 and the protein tyrosine phosphatase SHP. *Science* 269:242–244.

26. Darvasi, A. 1998. Experimental strategies for the genetic dissection of complex traits in animal models. *Nat. Genet.* 18:19–24.

27. Kono, D.H., R.W. Burlingame, D.G. Owens, A. Kuramochi, R.S. Balderas, D. Balomenos, and A.N. Theofilopoulos. 1994. Lupus susceptibility loci in New Zealand mice. *Proc. Natl. Acad. Sci. USA* 91:10168–72.

28. Vyse, T.J., and B.L. Kotzin. 1998. Genetic susceptibility to systemic lupus erythematosus. *Annu. Rev. Immunol.* 16:261–92:261–292.

29. Morel, L., and E.K. Wakeland. 1998. Susceptibility to lupus nephritis in the NZB/W model system. *Curr. Opin. Immunol.* 10:718–25.

30. Morel, L., C. Mohan, Y. Yu, B.P. Croker, N. Tian, A. Deng, and E.K. Wakeland. 1997. Functional dissection of systemic lupus erythematosus using congenic mouse strains. *J. Immunol.* 158:6019–6028.

31. Morel, L., Y. Yu, K.R. Blenman, R.A. Caldwell, and E.K. Wakeland. 1996. Production of congenic mouse strains carrying genomic intervals containing SLE-susceptibility genes derived from the SLE-prone NZM2410 strain. *Mamm. Genome* 7:335–9.

32. Mohan, C., E. Alas, L. Morel, P. Yang, and E.K. Wakeland. 1998. Genetic dissection of SLE pathogenesis. Sle1 on murine chromosome 1 leads to a selective loss of tolerance to H2A/H2B/DNA subnucleosomes. *J. Clin. Invest.* 101:1362–72.

33. Sobel, E.S., C. Mohan, L. Morel, J. Schiffenbauer, and E.K. Wakeland. 1999. Genetic dissection of SLE pathogenesis: adoptive transfer of Sle1 mediates the loss of tolerance by bone marrow-derived B cells. *J. Immunol.* 162:2415–2421.

34. Mohan, C., L. Morel, P. Yang, and E.K. Wakeland. 1997. Genetic dissection of systemic lupus erythematosus pathogenesis: Sle2 on murine chromosome 4 leads to B cell hyperactivity. *J. Immunol.* 159:454–465.

35. Mohan, C., L. Morel, P. Yang, and E.K. Wakeland. 1998. Accumulation of splenic B1a cells with potent antigen-presenting capability in NZM2410 lupus-prone mice. *Arthritis. Rheum.* 41:1652–62.

36. Mohan, C., Y. Yu, L. Morel, P. Yang, and E.K. Wakeland. 1999. Genetic dissection of Sle pathogenesis: Sle3 on murine chromosome 7 impacts T cell activation, differentiation, and cell death. *J. Immunol.* 162:6492–502.

37. Mohan, C., L. Morel, P. Yang, H. Watanabe, B. Croker, G. Gilkeson, and E.K. Wakeland. 1999. Genetic dissection of lupus pathogenesis: a recipe for nephrophilic autoantibodies. *J. Clin. Invest.* 103:1685–95.

38. Morel, L., C. Mohan, Y. Yu, J. Schiffenbauer, U.H. Rudofsky, N. Tian, J.A. Longmate, and E.K. Wakeland. 1999. Multiplex inheritance of component phenotypes in a murine model of lupus. *Mamm. Genome.* 10:176–81.

39. Vidal, S., D.H. Kono, and A.N. Theofilopoulos. 1998. Loci predisposing to autoimmunity in MRL-*Fas^lpr* and C57BL/6-*Fas^lpr* mice. *J. Clin. Invest.* 101:696–702.

40. Wang, Y., M. Nose, T. Kamoto, M. Nishimura, and H. Hiai. 1997. Host modifier genes affect mouse autoimmunity induced by the lpr gene. *Am. J. Pathol.* 151:1791–8.

41. Watson, M.L., J.K. Rao, G.S. Gilkeson, P. Ruiz, E.M. Eicher, D.S. Pisetsky, A. Matsuzawa, J.M. Rochelle, and M.F. Seldin. 1992. Genetic analysis of MRL-lpr mice: relationship of the Fas apoptosis gene to disease manifestations and renal disease-modifying loci. *J. Exp. Med.* 176:1645–56.

42. Nishihara, M., M. Terada, J. Kamogawa, Y. Ohashi, S. Mori, S. Nakatsuru, Y. Nakamura, and M. Nose. 1999. Genetic basis of autoimmune sialadenitis in MRL/lpr lupus-prone mice: additive and hierarchical properties of polygenic inheritance. *Arthritis Rheum.* 42:2616–23.

43. Izui, S., R. Merino, L. Fossati, and M. Iwamoto. 1994. The role of the Yaa gene in lupus syndrome. *Intern. Rev. Immunol.* 11:211–230.

44. Hudgins, C.C., R.T. Steinberg, D.M. Klinman, M.J.P. Reeves, A.D. Steinberg, and M.J. Reeves. 1985. Studies of consomic mice bearing the Y chromosome of the BXSB mouse. *J. Immunol.* 134:3849–3854.

45. Izui, S., M. Higaki, D. Morrow, and R. Merino. 1988. The Y chromosome from autoimmune BXSB/MpJ mice induces a lupus-like syndrome in (NZW x C57BL/6)F1 male mice, but not in C57BL/6 male mice. *Eur. J. Immunol.* 18:911–915.

46. Hogarth, M.B., J.H. Slingsby, P.J. Allen, E.M. Thompson, P. Chandler, K.A. Davies, E. Simpson, B.J. Morley, and M.J. Walport. 1998. Multiple lupus susceptibility loci map to chromosome 1 in BXSB mice. *J. Immunol.* 161:2753–2761.

47. Haywood, M.E., M.B. Hogarth, J.H. Slingsby, S.J. Rose, P.J. Allen, E.M. Thompson, M.A. Maibaum, P. Chandler, K.A. Davies, E. Simpson, M.J. Walport, and B.J. Morley. 2000. Identification of intervals on chromosomes 1, 3, and 13 linked to the development of lupus in BXSB mice. *Arthritis. Rheum.* 43:349–55.

48. Ida, A., S. Hirose, Y. Hamano, S. Kodera, Y. Jiang, M. Abe, D. Zhang, H. Nishimura, and T. Shirai. 1998. Multigenic control of lupus-associated antiphospholipid syndrome in a model of (NZW x BXSB) F1 mice. *Eur. J. Immunol.* 28:2694–2703.

49. Nishizumi, H., I. Taniuchi, Y. Yamanashi, D. Kitamura, D. Ilic, S. Mori, T. Watanabe, and T. Yamamoto. 1995. Impaired proliferation of peripheral B cells and indication of autoimmune disease in lyn-deficient mice. *Immunity* 3:549–560.

50. Hibbs, M.L., D.M. Tarlinton, J. Armes, D. Grail, G. Hodgson, R. Maglitto, S.A. Stacker, and A.R. Dunn. 1995. Multiple defects in the immune system of Lyn-deficient mice, culminating in autoimmune disease. *Cell* 83:301–11.

51. O'Keefe T, L., G.T. Williams, S.L. Davies, and M.S. Neuberger. 1996. Hyperresponsive B cells in CD22-deficient mice. *Science* 274:798–801.

52. Otipoby, K.L., K.B. Andersson, K.E. Draves, S.J. Klaus, A.G. Carr, J.D. Kerner, R.M. Perlmutter, C.L. Law, and E.A. Clark. 1996. CD22 regulates thymus-independent responses and the lifespan of B cells. *Nature* 384:634–637.

53. Sato, S., A.S. Miller, M. Inaoki, C.B. Bock, P.J. Jansen, M.L.K. Tang, and T.F. Tedder. 1996. CD22 is both a positive and negative regulator of B lymphocyte antigen receptor signal transduction: altered signaling in CD22 deficient mice. *Immunity* 5:551–562.

54. Cornall, R.J., J.G. Cyster, M.L. Hibbs, A.R. Dunn, K.L. Otipoby, E.A. Clark, and C.C. Goodnow. 1998. Polygenic autoimmune traits: Lyn, CD22, and SHP-1 are limiting elements of a biochemical pathway regulating BCR signaling and selection. *Immunity* 8:497–508.

55. Tedder, T.F., M. Inaoki, and S. Sato. 1997. The CD19-CD21 complex regulates signal transduction thresholds governing humoral immunity and autoimmunity. *Immunity* 6:107–18.

56. Inaoki, M., S. Sato, B.C. Weintraub, C.C. Goodnow, and T.F. Tedder. 1997. CD19-regulated signaling thresholds control peripheral tolerance and autoantibody production in B lymphocytes. *J. Exp. Med.* 186:1923–31.

57. Coggeshall, K.M. 1998. Inhibitory signaling by B cell Fc gamma RIIb. *Curr. Opin. Immunol.* 10:306–12.

58. Ono, M., H. Okada, S. Bolland, S. Yanagi, T. Kurosaki, and J.V. Ravetch. 1997. Deletion of SHIP or SHP-1 reveals two distinct pathways for inhibitory signaling. *Cell* 90:293–301.

59. Nadler, M.J.S., B. Chen, J.S. Anderson, H.H. Wortis, and B.G. Neel. 1997. Protein-tyrosine phosphatase SHP-1 is dispensable for FcgammaRIIB-mediated inhibition of B cell antigen receptor activation. *J. Biol. Chem.* 272:20038–43.

60. Takai, T., M. Ono, M. Hikida, H. Ohmori, and J.V. Ravetch. 1996. Augmented humoral and anaphylactic responses in FcgRII-deficient mice. *Nature* 379:346–349.

61. Yuasa, T., S. Kubo, T. Yoshino, A. Ujike, K. Matsumura, M. Ono, J.V. Ravetch, and T. Takai. 1999. Deletion of fcgamma receptor IIB renders H-2(b) mice susceptible to collagen-induced arthritis. *J. Exp. Med.* 189:187–94.

62. Nakamura, A., T. Yuasa, A. Ujike, M. Ono, T. Nukiwa, J.V. Ravetch, and T. Takai. 2000. Fcγ Receptor IIB-deficient mice develop Goodpasture's Syndrome upon immunization with type IV collagen: a novel murine model for autoimmune glomerular basement membrane disease. *J. Exp. Med.* 191:899–906.

62a. Bolland, S. and J.V. Ravetch. 2000. Spontaneous autoimmune disease in FcγRIIB-deficient mice results from strain-specific epistasis. *Immunity* 13:277–85.

63. Morgan, B., L. Sun, N. Avitahl, K. Andrikopoulos, T. Ikeda, E. Gonzales, P. Wu, S. Neben, and K. Georgopoulos. 1997. Aiolos, a lymphoid restricted transcription factor that interacts with Ikaros to regulate lymphocyte differentiation. *EMBO. J.* 16:2004–13.

64. Xue, Y., J. Wong, G.T. Moreno, M.K. Young, J. Cote, and W. Wang. 1998. NURD, a novel complex with both ATP-dependent chromatin-remodeling and histone deacetylase activities. *Mol. Cell* 2:851–61.

65. Zhang, Y., H.H. Ng, H. Erdjument-Bromage, P. Tempst, A. Bird, and D. Reinberg. 1999. Analysis of the NuRD subunits reveals a histone deacetylase core complex and a connection with DNA methylation. *Genes Dev.* 13:1924–35.

66. Knoepfler, P.S., and R.N. Eisenman. 1999. Sin meets NuRD and other tails of repression. *Cell* 99:447–50.

67. Koipally, J., A. Renold, J. Kim, and K. Georgopoulos. 1999. Repression by Ikaros and Aiolos is mediated through histone deacetylase complexes. *EMBO J.* 18:3090–100.

68. Wang, J.H., N. Avitahl, A. Cariappa, C. Friedrich, T. Ikeda, A. Renold, K. Andrikopoulos, L. Liang, S. Pillai, B.A. Morgan, and K. Georgopoulos. 1998. Aiolos regulates B cell activation and maturation to effector state. *Immunity* 9:543–53.

69. Moore, P.A., O. Belvedere, A. Orr, K. Pieri, D.W. LaFleur, P. Feng, D. Soppet, M. Charters, R. Gentz, D. Parmelee, Y. Li, O. Galperina, J. Giri, V. Roschke, B. Nardelli, J. Carrell, S. Sosnovtseva, W. Greenfield, S.M. Ruben, H.S. Olsen, J. Fikes, and D.M. Hilbert. 1999. BLyS: member of the tumor necrosis factor family and B lymphocyte stimulator. *Science* 285:260–3.

70. Schneider, P., F. MacKay, V. Steiner, K. Hofmann, J.L. Bodmer, N. Holler, C. Ambrose, P. Lawton, S. Bixler, H. Acha-Orbea, D. Valmori, P. Romero, C. Werner-Favre, R.H. Zubler, J.L. Browning, and J. Tschopp. 1999. BAFF, a novel ligand of the tumor necrosis factor family, stimulates B cell growth. *J. Exp. Med.* 189:1747–56.

71. Shu, H.B., W.H. Hu, and H. Johnson. 1999. TALL-1 is a novel member of the TNF family that is down-regulated by mitogens. *J. Leukoc. Biol.* 65:680–3.

72. Mukhopadhyay, A., J. Ni, Y. Zhai, G.L. Yu, and B.B. Aggarwal. 1999. Identification and characterization of a novel cytokine, THANK, a TNF homologue that activates apoptosis, nuclear factor-kappaB, and c-Jun NH2-terminal kinase. *J. Biol. Chem.* 274:15978–81.

73. Gross, J.A., J. Johnston, S. Mudri, R. Enselman, S.R. Dillon, K. Madden, X. Wenfeng, J. Parrish-Novak, D. Foster, C. Lofton-Day, M. Moore, A. Littau, A. Grossman, H. Haugen, K. Foley, H. Blumberg, K. Harrison, W. Kindsvogel, and C.H. Clegg. 2000. TACI and BCMA are receptors for a TNF homologue implicated in B-cell autoimmune disease. *Nature* 404:995–999.

74. Mackay, F., S.A. Woodcock, P. Lawton, C. Ambrose, M. Baetscher, P. Schneider, J. Tschopp, and J.L. Browning. 1999. Mice transgenic for BAFF develop lymphocytic disorders along with autoimmune manifestations. *J. Exp. Med.* 190:1697–710.

75. Khare, S.D., I. Sarosi, X.Z. Xia, S. McCabe, K. Miner, I. Solovyev, N. Hawkins, M. Kelley, D. Chang, G. Van, L. Ross, J. Delaney, L. Wang, D. Lacey, W.J. Boyle, and H. Hsu. 2000. Severe B cell hyperplasia and autoimmune disease in TALL-1 transgenic mice. *Proc. Natl. Acad. Sci. USA* 97:3370–3375.

76. Zhang, L., A. Eddy, Y.T. Teng, M. Fritzler, M. Kluppel, F. Melet, and A. Bernstein. 1995. An immunological renal disease in transgenic mice that overexpress Fli-1, a member of the ets family of transcription factor genes. *Mol. Cell Biol.* 15:6961–6970.

77. Waterhouse, P., J.M. Penninger, E. Timms, A. Wakeham, A. Shahinian, K.P. Lee, C.B. Thompson, H. Griesser, and T.W. Mak. 1995. Lymphoproliferative disorders with early lethality in mice deficient in Ctla-4. *Science* 270:985–988.

78. Tivol, E.A., F. Borriello, A.N. Schweitzer, W.P. Lynch, J.A. Bluestone, and A.H. Sharpe. 1995. Loss of CTLA-4 leads to massive lymphoproliferation and fatal multiorgan tissue destruction, revealing a critical negative regulatory role of CTLA-4. *Immunity* 3:541–547.

79. Tivol, E.A., S.D. Boyd, S. McKeon, F. Borriello, P. Nickerson, T.B. Strom, and A.H. Sharpe. 1997. CTLA4Ig prevents lymphoproliferation and fatal multiorgan tissue destruction in CTLA-4-deficient mice. *J. Immunol.* 158:5091–4.

80. Chambers, C.A., D. Cado, T. Truong, and J.P. Allison. 1997. Thymocyte development is normal in CTLA-4-deficient mice. *Proc. Natl. Acad. Sci. USA* 94:9296–301.

81. Chambers, C.A., T.J. Sullivan, and J.P. Allison. 1997. Lymphoproliferation in CTLA-4-deficient mice is mediated by costimulation-dependent activation of CD4+ T cells. *Immunity* 7:885–95.

82. Chambers, C.A., M.S. Kuhns, and J.P. Allison. 1999. Cytotoxic T lymphocyte antigen-4 (CTLA-4) regulates primary and secondary peptide-specific CD4(+) T cell responses. *Proc. Natl. Acad. Sci. USA* 96:8603–8.

83. Bachmann, M.F., G. Kohler, B. Ecabert, T.W. Mak, and M. Kopf. 1999. Cutting edge: lymphoproliferative disease in the absence of CTLA-4 is not T cell autonomous. *J. Immunol.* 163:1128–31.

84. Schorle, H., T. Holtschke, T. Hunig, A. Schimpl, and I. Horak. 1991. Development and function of T cells in mice

rendered interleukin-2 deficient by gene targeting. *Nature* 352:621–4.

85. Kundig, T.M., H. Schorle, M.F. Bachmann, H. Hengartner, R.M. Zinkernagel, and I. Horak. 1993. Immune responses in interleukin-2-deficient mice. *Science* 262:1059–61.

86. Sadlack, B., J. Lohler, H. Schorle, G. Klebb, H. Haber, E. Sickel, R.J. Noelle, and I. Horak. 1995. Generalized autoimmune disease in interleukin-2-deficient mice is triggered by an uncontrolled activation and proliferation of CD4+ T cells. *Eur. J. Immunol.* 25:3053–3059.

87. Willerford, D.M., J. Chen, J.A. Ferry, L. Davidson, A. Ma, and F.W. Alt. 1995. Interleukin-2 receptor alpha chain regulates the size and content of the peripheral lymphoid compartment. *Immunity* 3:521–530.

88. Suzuki, H., T.M. Kundig, C. Furlonger, A. Wakeham, E. Timms, T. Matsuyama, R. Schmits, J.J.L. Simard, P.S. Ohashi, H. Griesser, T. Taniguchi, C.J. Paige, and T.W. Mak. 1995. Deregulated T cell activation and autoimmunity in mice lacking interleukin-2 receptor. *Science* 268:1472–1476.

89. Sadlack, B., H. Merz, H. Schorle, A. Schimpl, A.C. Feller, and I. Horak. 1993. Ulcerative colitis-like disease in mice with a disrupted interleukin-2 gene. *Cell* 75:253–61.

90. Van Parijs, L., A. Biuckians, A. Ibragimov, F.W. Alt, D.M. Willerford, and A.K. Abbas. 1997. Functional responses and apoptosis of CD25 (IL-2R alpha)-deficient T cells expressing a transgenic antigen receptor. *J. Immunol.* 158:3738–3745.

91. Contractor, N.V., H. Bassiri, T. Reya, A.Y. Park, D.C. Baumgart, M.A. Wasik, S.G. Emerson, and S.R. Carding. 1998. Lymphoid hyperplasia, autoimmunity, and compromised intestinal intraepithelial lymphocyte development in colitis free gnotobiotic IL-2-deficient mice. *J. Immunol.* 160:385–94.

92. Shull, M.M., I. Ormsby, A.B. Kier, S. Pawlowski, R.J. Diebold, M. Yin, R. Allen, C. Sidman, G. Proetzel, D. Calvin, N. Annunziata, and T. Doetschman. 1992. Targeted disruption of the mouse transforming growth factor-1 gene results in multifocal inflammatory disease. *Nature* 359:693–699.

93. Kulkarni, A.B., and S. Karlsson. 1993. Transforming growth factor-beta 1 knockout mice. A mutation in one cytokine gene causes a dramatic inflammatory disease. *Am. J. Pathol.* 143:3–9.

94. Dang, H., A.G. Geiser, J.J. Letterio, T. Nakabayashi, L. Kong, G. Fernandes and N. Talal. 1995. SLE-like autoantibodies and Sjögren's syndrome-like lymphoproliferation in TGF- knockout mice. *J. Immunol.* 155:3205–3212.

95. Letterio, J.J., A.G. Geiser, A.B. Kulkarni, H. Dang, L. Kong, T. Nakabayashi, C.L. Mackall, R.E. Gress, and A.B. Roberts. 1996. Autoimmunity associated with TGF-beta1-deficiency in mice is dependent on MHC class II antigen expression. *J. Clin. Invest.* 98:2109–19.

96. Kobayashi, S., K. Yoshida, J.M. Ward, J.J. Letterio, G. Longenecker, L. Yaswen, B. Mittleman, E. Mozes, A.B. Roberts, S. Karlsson, and A.B. Kulkarni. 1999. Beta 2-microglobulin-deficient background ameliorates lethal phenotype of the TGF-beta 1 null mouse. *J. Immunol.* 163:4013–9.

97. Gorelik, L., and R.A. Flavell. 2000. Abrogation of TGFbeta signaling in T cells leads to spontaneous T cell differentiation and autoimmune disease. *Immunity* 12:171–81.

98. Lupher, M.L., Jr., N. Rao, M.J. Eck, and H. Band. 1999. The Cbl protooncoprotein: a negative regulator of immune receptor signal transduction. *Immunol. Today* 20:375–82.

99. Keane, M.M., O.M. Rivero-Lezcano, J.A. Mitchell, K.C. Robbins, and S. Lipkowitz. 1995. Cloning and characterization of cbl-b: a SH3 binding protein with homology to the c-cbl proto-oncogene. *Oncogene* 10:2367–77.

100. Keane, M.M., S.A. Ettenberg, M.M. Nau, P. Banerjee, M. Cuello, J. Penninger and S. Lipkowitz. 1999. cbl-3: a new mammalian cbl family protein. *Oncogene* 18:3365–75.

101. Kim, M., T. Tezuka, Y. Suziki, S. Sugano, M. Hirai, and T. Yamamoto. 1999. Molecular cloning and characterization of a novel cbl-family gene, cbl- c. *Gene* 239:145–54.

102. Clements, J.L., N.J. Boerth, J.R. Lee, and G.A. Koretzky. 1999. Integration of T cell receptor-dependent signaling pathways by adapter proteins. *Annu. Rev. Immunol.* 17:89–108.

103. Rudd, C.E. 1999. Adaptors and molecular scaffolds in immune cell signaling. *Cell* 96:5–8.

104. Chiang, Y.J., H.K. Kole, K. Brown, M. Naramura, S. Fukuhara, R.J. Hu, I.K. Jang, J.S. Gutkind, E. Shevach, and H. Gu. 2000. Cbl-b regulates the CD28 dependence of T-cell activation. *Nature* 403:216–20.

105. Bachmaier, K., C. Krawczyk, I. Kozieradzki, Y.Y. Kong, T. Sasaki, A. Oliveira-dos-Santos, S. Mariathasan, D. Bouchard, A. Wakeham, A. Itie, J. Le, P.S. Ohashi, I. Sarosi, H. Nishina, S. Lipkowitz, and J.M. Penninger. 2000. Negative regulation of lymphocyte activation and autoimmunity by the molecular adaptor Cbl-b. *Nature* 403:211–6.

106. Bustelo, X.R., P. Crespo, M. Lopez-Barahona, J.S. Gutkind, and M. Barbacid. 1997. Cbl-b, a member of the Sli-1/c-Cbl protein family, inhibits Vav- mediated c- Jun N-terminal kinase activation. *Oncogene* 15:2511–20.

107. O'Connor, L., A. Strasser, L.A. O'Reilly, G. Hausmann, J.M. Adams, S. Cory, and D.C. Huang. 1998. Bim: a novel member of the Bcl-2 family that promotes apoptosis. *EMBO J.* 17:384–95.

108. Puthalakath, H., D.C. Huang, L.A. O'Reilly, S.M. King, and A. Strasser. 1999. The proapoptotic activity of the Bcl-2 family member Bim is regulated by interaction with the dynein motor complex. *Mol. Cell.* 3:287–96.

109. Bouillet, P., D. Metcalf, D.C. Huang, D.M. Tarlinton, T.W. Kay, F. Kontgen, J.M. Adams, and A. Strasser. 1999. Proapoptotic Bcl-2 relative Bim required for certain apoptotic responses, leukocyte homeostasis, and to preclude autoimmunity. *Science* 286:1735–8.

110. Strasser, A., S. Whittingham, D.L. Vaux, M.L. Bath, J.M. Adams, S. Cory, and A.W. Harris. 1991. Enforced bcl-2 expression in B-lymphoid cells prolongs antibody responses and elicits autoimmune disease. *Proc. Natl. Acad. Sci. USA* 88:8661–8665.

111. Ray, S.K., C. Putterman, and B. Diamond. 1996. Pathogenic autoantibodies are routinely generated during the response to foreign antigen: a paradigm for autoimmune disease. *Proc. Natl. Acad. Sci. USA* 93:2019–24.

112. Kuo, P., M. Bynoe, and B. Diamond. 1999. Crossreactive B cells are present during a primary but not secondary response in BALB/c mice expressing a bcl-2 transgene. *Mol. Immunol.* 36:471–9.

113. Mandik-Nayak, L., S. Nayak, C. Sokol, A. Eaton-Bassiri, M.P. Madaio, A.J. Caton, and J. Erikson. 2000. The origin of anti-nuclear antibodies in bcl-2 transgenic mice. *Int. Immunol.* 12:353–364.

114. Cantley, L.C., and B.G. Neel. 1999. New insights into tumor suppression: PTEN suppresses tumor formation by restraining the phosphoinositide 3-kinase/AKT pathway. *Proc. Natl. Acad. Sci. USA* 96:4240–5.

115. Di Cristofano, A., and P.P. Pandolfi. 2000. The multiple roles of PTEN in tumor suppression. *Cell* 100:387–90.

116. Di Cristofano, A., P. Kotsi, Y.F. Peng, C. Cordon-Cardo, K.B. Elkon, and P.P. Pandolfi. 1999. Impaired Fas response and autoimmunity in Pten+/– mice. *Science* 285:2122–5.

117. Maehama, T., and J.E. Dixon. 1998. The tumor suppressor, PTEN/MMAC1, dephosphorylates the lipid second messenger, phosphatidylinositol 3,4,5-trisphosphate. *J. Biol. Chem.* 273:13375–8.

118. Irmler, M., M. Thome, M. Hahne, P. Schneider, K. Hofmann, V. Steiner, J.L. Bodmer, M. Schroter, K. Burns, C. Mattmann, D. Rimoldi, L.E. French, and J. Tschopp. 1997. Inhibition of death receptor signals by cellular FLIP. *Nature* 388:190–5.

119. Tschopp, J., M. Irmler, and M. Thome. 1998. Inhibition of fas death signals by FLIPs. *Curr. Opin. Immunol.* 10:552–8.

120. Van Parijs, L., Y. Refaeli, A.K. Abbas, and D. Baltimore. 1999. Autoimmunity as a consequence of retrovirus-mediated expression of C-FLIP in lymphocytes. *Immunity* 11:763–70.

121. Botto, M., C. Dell'Agnola, A.E. Bygrave, E.M. Thompson, H.T. Cook, F. Petry, M. Loos, P.P. Pandolfi, and M.J. Walport. 1998. Homozygous C1q deficiency causes glomerulonephritis associated with multiple apoptotic bodies. *Nat. Genet.* 19:56–59.

122. Prodeus, A.P., S. Goerg, L.M. Shen, O.O. Pozdnyakova, L. Chu, E.M. Alicot, C.C. Goodnow, and M.C. Carroll. 1998. A critical role for complement in maintenance of self-tolerance. *Immunity* 9:721–31.

123. Bickerstaff, M.C., M. Botto, W.L. Hutchinson, J. Herbert, G.A. Tennent, A. Bybee, D.A. Mitchell, H.T. Cook, P.J. Butler, M.J. Walport, and M.B. Pepys. 1999. Serum amyloid P component controls chromatin degradation and prevents antinuclear autoimmunity. *Nat. Med.* 5:694–7.

124. Paul, E., and M.C. Carroll. 1999. SAP-less chromatin triggers systemic lupus erythematosus. *Nat. Med.* 5:607–8.

125. Seery, J.P., J.M. Carroll, V. Cattell, and F.M. Watt. 1997. Antinuclear autoantibodies and lupus nephritis in transgenic mice expressing interferon gamma in the epidermis. *J. Exp. Med.* 186:1451–1459.

126. Erb, K.J., B. Ruger, M. von Brevern, B. Ryffel, A. Schimpl, and K. Rivett. 1997. Constitutive expression of interleukin (IL)-4 *in vivo* causes autoimmune-type disorders in mice. *J. Exp. Med.* 185:329–339.

127. Muller, W., R. Kuhn, and K. Rajewsky. 1991. Major histocompatibility complex class II hyperexpression on B cells in interleukin 4-transgenic mice does not lead to B cell proliferation and hypergammaglobulinemia. *Eur. J. Immunol.* 21:921–5.

128. Burstein, H.J., R.I. Tepper, P. Leder, and A.K. Abbas. 1991. Humoral immune functions in IL-4 transgenic mice. *J. Immunol.* 147:2950–6.

129. Tepper, R.I., D.A. Levinson, B.Z. Stanger, J. Campos-Torres, A.K. Abbas, and P. Leder. 1990. IL-4 induces allergic-like inflammatory disease and alters T cell development in transgenic mice. *Cell* 62:457–67.

130. Lewis, D.B., C.C. Yu, K.A. Forbush, J. Carpenter, T.A. Sato, A. Grossman, D.H. Liggitt, and R.M. Perlmutter. 1991. Interleukin 4 expressed in situ selectively alters thymocyte development. *J. Exp. Med.* 173:89–100.

131. Santiago, M., L. Fossati, C. Jacquiet, W. Muller, S. Izui, and L. Reininger. 1997. Interleukin-4 protects against a genetically linked lupus-like autoimmune syndrome. *J. Exp. Med.* 185:65–70.

132. Worthington, M.T., B.T. Amann, D. Nathans and J.M. Berg. 1996. Metal binding properties and secondary structure of the zinc-binding domain of Nup475. *Proc. Natl. Acad. Sci. USA* 93:13754–9.

133. Taylor, G.A., E. Carballo, D.M. Lee, W.S. Lai, M.J. Thompson, D.D. Patel, Schenkman, Di, G.S. Gilkeson, H.E. Broxmeyer, B.F. Haynes, and P.J. Blackshear. 1996. A pathogenetic role for TNF alpha in the syndrome of cachexia, arthritis, and autoimmunity resulting from tristetraprolin (TTP) deficiency. *Immunity* 4:445–454.

134. Carballo, E., G.S. Gilkeson, and P.J. Blackshear. 1997. Bone marrow transplantation reproduces the tristetraprolin-deficiency syndrome in recombination activating gene-2 (–/–) mice. Evidence that monocyte/macrophage progenitors may be responsible for TNFalpha overproduction. *J. Clin. Invest.* 100:986–995.

135. Carballo, E., W.S. Lai, and P.J. Blackshear. 1998. Feedback inhibition of macrophage tumor necrosis factor-alpha production by tristetraprolin. *Science* 281:1001–5.

136. Rothe, J., W. Lesslauer, H. Lotscher, Y. Lang, P. Koebel, F. Kontgen, A. Althage, R. Zinkernagel, M. Steinmetz, and H. Bluethmann. 1993. Mice lacking the tumour necrosis factor receptor 1 are resistant to TNF-mediated toxicity but highly susceptible to infection by *Listeria monocytogenes*. *Nature* 364:798–802.

137. Pfeffer, K., T. Matsuyama, T.M. Kundig, A. Wakeham, K. Kishihara, A. Shahinian, K. Wiegmann, P.S. Ohashi, M. Kronke, and T.W. Mak. 1993. Mice deficient for the 55 kd tumor necrosis factor receptor are resistant to endotoxic shock, yet succumb to *L. monocytogenes* infection. *Cell* 73:457–467.

138. Zhou, T., C.K. Edwards, P. Yang, Z. Wang, H. Bluethmann, and J.D. Mountz. 1996. Greatly accelerated lymphadenopathy and autoimmune disease in lpr mice lacking tumor necrosis factor receptor I. *J. Immunol.* 156:2661–2665.

139. Sytwu, H.K., R.S. Liblau, and H.O. McDevitt. 1996. The roles of Fas/APO-1 (CD95) and TNF in antigen-induced programmed cell death in T cell receptor transgenic mice. *Immunity* 5:17–30.

140. Balomenos, D., J. Martín-Caballero, M.I. García, I. Prieto, J.M. Flores, M. Serrano, and A.C. Martínez. 2000. The cell cycle inhibitor p21 controls T-cell proliferation and sex-linked lupus development. *Nat. Med.* 6:171–176.

141. Sabzevari, H., S. Propp, D.H. Kono, and A.N. Theofilopoulos. 1997. G1 arrest and high expression of cyclin kinase and apoptosis inhibitors in accumulated activated/memory phenotype CD4+ cells of older lupus mice. *Eur. J. Immunol.* 27:1901–1910.

142. Ishida, Y., Y. Agata, K. Shibahara, and T. Honjo. 1992. Induced expression of PD-1, a novel member of the immunoglobulin gene superfamily, upon programmed cell death. *EMBO J.* 11:3887–95.

143. Nishimura, H., N. Minato, T. Nakano, and T. Honjo. 1998. Immunological studies on PD-1 deficient mice: implication of PD-1 as a negative regulator for B cell responses. *Int. Immunol.* 10:1563–72.

144. Nishimura, H., M. Nose, H. Hiai, N. Minato, and T. Honjo. 1999. Development of lupus-like autoimmune diseases by disruption of the PD-1 gene encoding an ITIM motif-carrying immunoreceptor. *Immunity* 11:141–51.

145. Shlomchik, M.J., M.P. Madaio, D. Ni, M. Trounstein, and D. Huszar. 1994. The role of B cells in lpr/lpr-induced autoimmunity. *J. Exp. Med.* 180:1295–1306.

146. Chan, O.T., M.P. Madaio, and M.J. Shlomchik. 1999. B cells are required for lupus nephritis in the polygenic, Fas-

intact MRL model of systemic autoimmunity. *J. Immunol.* 163:3592–6.

147. Chan, O., and M.J. Shlomchik. 1998. A new role for B cells in systemic autoimmunity: B cells promote spontaneous T cell activation in MRL-lpr/lpr mice. *J. Immunol.* 160:51–9.

148. Chan, O.T., and M.J. Shlomchik. 2000. B cells promote CD8+ T cell activation in MRL-Fas(lpr) mice independently of MHC class I antigen presentation. *J. Immunol.* 164:1658–62.

149. Chan, O.T., L.G. Hannum, A.M. Haberman, M.P. Madaio, and M.J. Shlomchik. 1999. A novel mouse with B cells but lacking serum antibody reveals an antibody-independent role for B cells in murine lupus. *J. Exp. Med.* 189:1639–48.

150. Jevnikar, A.M., M.J. Grusby, and L.H. Glimcher. 1994. Prevention of nephritis in major histocompatibility complex class II-deficient MRL-lpr mice. *J. Exp. Med.* 179:1137–1143.

151. Koh, D.R., A. Ho, A. Rahemtulla, W.P. Fung-Leung, H. Griesser, and T.W. Mak. 1995. Murine lupus in MRL/lpr mice lacking CD4 or CD8 T cells. *Eur. J. Immunol.* 25:2558–2562.

152. Giese, T., and W.F. Davidson. 1995. In CD8+ T cell-deficient lpr/lpr mice, CD4+B220+ and CD4+B220– T cells replace B220+ double-negative T cells as the predominant populations in enlarged lymph nodes. *J. Immunol.* 154:4986–4995.

153. Ohteki, T., M. Iwamoto, S. Izui, and H.R. MacDonald. 1995. Reduced development of CD4–8–B220+ T cells but normal autoantibody production in lpr/lpr mice lacking major histocompatibility complex class I molecules. *Eur. J. Immunol.* 25:37–41.

154. Maldonado, M.A., R.A. Eisenberg, E. Roper, P.L. Cohen, and B.L. Kotzin. 1995. Greatly reduced lymphoproliferation in lpr mice lacking major histocompatibility complex class I. *J. Exp. Med.* 181:641–648.

155. Christianson, G.J., R.L. Blankenburg, T.M. Duffy, D. Panka, J.B. Roths, A. Marshak-Rothstein, and D.C. Roopenian. 1996. beta2-microglobulin dependence of the lupus-like autoimmune syndrome of MRL-lpr mice. *J. Immunol.* 156:4932–9.

156. Chen, S.Y., Y. Takeoka, L. Pike-Nobile, A.A. Ansari, R. Boyd, and M.E. Gershwin. 1997. Autoantibody production and cytokine profiles of MHC class I (beta2- microglobulin) gene deleted New Zealand black (NZB) mice. *Clin. Immunol. Immunopathol.* 84:318–327.

157. Peng, S.L., J. Cappadona, J.M. McNiff, M.P. Madaio, M.J. Owen, A.C. Hayday, and J. Craft. 1998. Pathogenesis of autoimmunity in alphabeta T cell-deficient lupus-prone mice. *Clin. Exp. Immunol.* 111:107–116.

158. Peng, S.L., M.P. Madaio, D.P. Hughes, I.N. Crispe, M.J. Owen, Wen, A.C. Hayday, and J. Craft. 1996. Murine lupus in the absence of alpha beta T cells. *J. Immunol.* 156:4041–4049.

159. Peng, S.L., and J. Craft. 1997. The regulation of murine lupus. *Ann. NY Acad. Sci.* 815:128–138.

160. Craft, J., S. Peng, T. Fujii, M. Okada, and S. Fatenejad. 1999. Autoreactive T cells in murine lupus: origins and roles in autoantibody production. *Immunol. Res.* 19:245–57.

161. Peng, S.L., J. Moslehi, M.E. Robert, and J. Craft. 1998. Perforin protects against autoimmunity in lupus-prone mice. *J. Immunol.* 160:652–60.

162. Ma, J., J. Xu, M.P. Madaio, Q. Peng, J. Zhang, I.S. Grewal, R.A. Flavell, and J. Craft. 1996. Autoimmune lpr/lpr mice deficient in CD40 ligand: spontaneous Ig class switching with dichotomy of autoantibody responses. *J. Immunol.* 157:417–426.

163. Peng, S.L., J.M. McNiff, M.P. Madaio, J. Ma, M.J. Owen, Flavell, Ra, A.C. Hayday, and J. Craft. 1997. alpha-beta T cell regulation and CD40 ligand dependence in murine systemic autoimmunity. *J. Immunol.* 158:2464–2470.

164. Tada, Y., K. Nagasawa, A. Ho, F. Morito, S. Koarada, O. Ushiyama, N. Suzuki, A. Ohta, and T.W. Mak. 1999. Role of the costimulatory molecule CD28 in the development of lupus in MRL/lpr mice. *J. Immunol.* 163:3153–9.

165. Liang, B., R.J. Gee, M.J. Kashgarian, A.H. Sharpe, and M.J. Mamula. 1999. B7 costimulation in the development of lupus: autoimmunity arises either in the absence of B7.1/B7.2 or in the presence of anti-B7.1/B7.2 blocking antibodies. *J. Immunol.* 163:2322–9.

166. Nakajima, A., M. Azuma, S. Kodera, S. Nuriya, A. Terashi, M. Abe, S. Hirose, T. Shirai, H. Yagita, and K. Okumura. 1995. Preferential dependence of autoantibody production in murine lupus on CD86 costimulatory molecule. *Eur. J. Immunol.* 25:3060–3069.

167. Peng, S.L., J. Moslehi, and J. Craft. 1997. Roles of interferon-gamma and interleukin-4 in murine lupus. *J. Clin. Invest.* 99:1936–1946.

168. Haas, C., B. Ryffel and H.M. Le. 1997. IFN-gamma is essential for the development of autoimmune glomerulonephritis in MRL/Ipr mice. *J. Immunol.* 158:5484–5491.

169. Balomenos, D., R. Rumold, and A.N. Theofilopoulos. 1998. Interferon-gamma is required for lupus-like disease and lymphoaccumulation in MRL-lpr mice. *J. Clin. Invest.* 101:364–371.

170. Haas, C., B. Ryffel, and M. Le Hir. 1998. IFN-gamma receptor deletion prevents autoantibody production and glomerulonephritis in lupus-prone (NZB x NZW)F1 mice. *J Immunol.* 160:3713–3718.

171. Kono, D.H., D. Balomenos, D.L. Pearson, M.S. Park, B. Hildebrandt, P. Hultman, and K.M. Pollard. 1998. The prototypic Th2 autoimmunity induced by mercury is dependent on IFN-gamma and not Th1/Th2 imbalance. *J. Immunol.* 161:234–40.

172. Schwarting, A., T. Wada, K. Kinoshita, G. Tesch, and V.R. Kelley. 1998. IFN-gamma receptor signaling is essential for the initiation, acceleration, and destruction of autoimmune kidney disease in MRL – Fas(lpr) mice. *J. Immunol.* 161:494–503.

173. Kono, D.H., D. Balomenos, M.S. Park, and A.N. Theofilopoulos. 2000. Lupus-like disease in BXSB mice is not dependent on IL-4. *J. Immunol.* 164:38–42.

174. Balomenos, D., R. Rumold, and A.N. Theofilopoulos. 1997. The proliferative in vivo activities of lpr double-negative T cells and the primary role of p59fyn in their activation and expansion. *J. Immunol.* 159:2265–73.

175. Lloyd, C.M., J.A. Gonzalo, D.J. Salant, J. Just, and J.C. Gutierrez-Ramos. 1997. Intercellular adhesion molecule-1 deficiency prolongs survival and protects against the development of pulmonary inflammation during murine lupus. *J. Clin. Invest.* 100:963–71.

176. Bullard, D.C., P.D. King, M.J. Hicks, B. Dupont, A.L. Beaudet, and K.B. Elkon. 1997. Intercellular adhesion molecule-1 deficiency protects MRL/MpJ-Fas(lpr) mice from early lethality. *J. Immunol.* 159:2058–67.

177. Gilkeson, G.S., J.S. Mudgett, M.F. Seldin, P. Ruiz, A.A. Alexander, M.A. Misukonis, D.S. Pisetsky, and J.B. Weinberg. 1997. Clinical and serologic manifestations of autoimmune disease in MRL-lpr/lpr mice lacking nitric oxide synthase type 2. *J. Exp. Med.* 186:365–373.

178. Clynes, R., C. Dumitru, and J.V. Ravetch. 1998. Uncoupling of immune complex formation and kidney damage in autoimmune glomerulonephritis. *Science* 279:1052–1054.

179. Park, S.Y., S. Ueda, H. Ohno, Y. Hamano, M. Tanaka, T. Shiratori, T. Yamazaki, H. Arase, N. Arase, A. Karasawa, S. Sato, B. Ledermann, Y. Kondo, K. Okumura, C. Ra, and T. Saito. 1998. Resistance of Fc receptor- deficient mice to fatal glomerulonephritis. *J. Clin. Invest.* 102:1229–38.

180. Tesch, G.H., S. Maifert, A. Schwarting, B.J. Rollins, and V.R. Kelley. 1999. Monocyte chemoattractant protein 1-dependent leukocytic infiltrates are responsible for autoimmune disease in MRL-Fas(lpr) mice. *J. Exp. Med.* 190:1813–1824.

181. Gilfillan, S., C. Benoist, and D. Mathis. 1995. Mice lacking terminal deoxynucleotidyl transferase: adult mice with a fetal antigen receptor repertoire. *Immunol. Rev.* 148:201–19.

182. Komori, T., A. Okada, V. Stewart, and F.W. Alt. 1993. Lack of N regions in antigen receptor variable region genes of TdT-deficient lymphocytes. *Science* 261:1171–5.

183. Komori, T., L. Pricop, A. Hatakeyama, C.A. Bona, and F.W. Alt. 1996. Repertoires of antigen receptors in Tdt congenitally deficient mice. *Int. Rev. Immunol.* 13:317–25.

184. Weller, S., C. Conde, A.M. Knapp, H. Levallois, S. Gilfillan, J.L. Pasquali, and T. Martin. 1997. Autoantibodies in mice lacking terminal deoxynucleotidyl transferase: evidence for a role of N region addition in the polyreactivity and in the affinities of anti-DNA antibodies. *J. Immunol.* 159:3890–8.

185. Molano, I.D., M.K. Wloch, A.A. Alexander, H. Watanabe, and G.S. Gilkeson. 2000. Effect of a genetic deficiency of terminal deoxynucleotidyl transferase on autoantibody production by C57BL6 fas(lpr) mice. *Clin. Immunol.* 94:24–32.

186. Watanabe, H., G. Garnier, A. Circolo, R.A. Wetsel, P. Ruiz, V.M. Holers, S.A. Boackle, H.R. Colten, and G.S. Gilkeson. 2000. Modulation of renal disease in MRL/lpr mice genetically deficient in the alternative complement pathway factor B. *J. Immunol.* 164:786–94.

187. Passwell, J., G.F. Schreiner, M. Nonaka, H.U. Beuscher, and H.R. Colten. 1988. Local extrahepatic expression of complement genes C3, factor B, C2, and C4 is increased in murine lupus nephritis. *J. Clin. Invest.* 82:1676–84.

188. Allende, C.C., and J.E. Allende. 1998. Promiscuous subunit interactions: a possible mechanism for the regulation of protein kinase CK2. *J. Cell Biochem. Suppl.* 31:129–36.

189. Seldin, D.C., and P. Leder. 1995. Casein kinase II alpha transgene-induced murine lymphoma: relation to theileriosis in cattle. *Science* 267:894–7.

190. Xu, X., E. Landesman-Bollag, P.L. Channavajhala, and D.C. Seldin. 1999. Murine protein kinase CK2: gene and oncogene. *Mol. Cell. Biochem.* 191:65–74.

191. Spurney, R.F., P. Ruiz, D.S. Pisetsky, and T.M. Coffman. 1991. Enhanced renal leukotriene production in murine lupus: role of lipoxygenase metabolites. *Kidney Int.* 39:95–102.

192. Rudofsky, U.H., B.D. Evans, S.L. Balaban, V.D. Mottironi, and A.E. Gabrielsen. 1993. Differences in expression of lupus nephritis in New Zealand mixed H-2z homozygous inbred strains of mice derived from New Zealand black and New Zealand white mice. Origins and initial characterization. *Lab. Invest.* 68:419–426.

193. Fossati, L., S. Takahashi, R. Merino, M. Iwamoto, J.P. Aubry, M. Nose, Spach, R. Motta, and S. Izui. 1993. An MRL/MpJ-lpr/lpr substrain with a limited expansion of lpr double-negative T cells and a reduced autoimmune syndrome. *Int. Immunol.* 5:525–532.

194. Suzuka, H., H. Yoshifusa, Y. Nakamura, S. Miyawaki, and Y. Shibata. 1993. Morphological analysis of autoimmune disease in MRL-lpr,Yaa male mice with rapidly progressive systemic lupus erythematosus. *Autoimmunity* 14:275–282.

195. Kinjoh, K., M. Kyogoku, and R.A. Good. 1993. Genetic selection for crescent formation yields mouse strain with rapidly progressive glomerulonephritis and small vessel vasculitis. *Proc. Natl. Acad. Sci. USA* 90:3413–3417.

196. Kofler, R., P.J. McConahey, M.A. Duchosal, R.S. Balderas, A.N. Theofilopoulos, and F.J. Dixon. 1991. An autosomal recessive gene that delays expression of lupus in BXSB mice. *J. Immunol.* 146:1375–1379.

197. Vidal, S., C. Gelpi, and J.L. Rodriguez-Sanchez. 1994. (SWR x SJL)F1 mice: a new model of lupus-like disease. *J. Exp. Med.* 179:1429–1435.

198. Walker, S.E., R.H. Gray, M. Fulton, R.D. Wigley, and B. Schnitzer. 1978. Palmerston North mice, a new animal model of systemic lupus erythematosus. *J. Lab. Clin. Med.* 92:932–45.

199. Bignon, J.S., and K.A. Siminovitch. 1994. Identification of PTP1C mutation as the genetic defect in motheaten and viable motheaten mice: a step toward defining the roles of protein tyrosine phosphatases in the regulation of hemopoietic cell differentiation and function. *Clin. Immunol. Immunopathol.* 73:168–179.

200. Pollard, K.M., and P. Hultman. 1997. Effects of mercury on the immune system. 14:421–440.

201. Rubin, R.L. 1997. Drug-induced lupus. 47:871–901.

202. Satoh, M., A. Kumar, Y.S. Kanwar, and W.H. Reeves. 1995. Anti-nuclear antibody production and immune-complex glomerulonephritis in BALB/c mice treated with pristane. *Proc. Natl. Acad. Sci. USA.* 92:10934–10938.

203. Mendlovic, S., B.H. Fricke, Y. Shoenfeld, R. Bakimer, and E. Mozes. 1990. The genetic regulation of the induction of experimental SLE. *Immunol.* 69:228–236.

204. Vyse, T.J., L. Morel, F.J. Tanner, E.K. Wakeland, and B.L. Kotzin. 1996. Backcross analysis of genes linked to autoantibody production in New Zealand White mice. *J. Immunol.* 157:2719–27.

205. Vyse, T.J., S.J. Rozzo, C.G. Drake, S. Izui, and B.L. Kotzin. 1997. Control of multiple autoantibodies linked with a lupus nephritis susceptibility locus in New Zealand black mice. *J. Immunol.* 158:5566–5574.

206. Rozzo, S.J., T.J. Vyse, C.G. Drake, and B.L. Kotzin. 1996. Effect of genetic background on the contribution of New Zealand black loci to autoimmune lupus nephritis. *Proc. Natl. Acad. Sci. USA* 93:15164–8.

207. Gu, L., A. Weinreb, X.P. Wang, D.J. Zack, J.H. Qiao, R. Weisbart, and A.J. Lusis. 1998. Genetic determinants of autoimmune disease and coronary vasculitis in the MRL-lpr/lpr mouse model of systemic lupus erythematosus. *J. Immunol.* 161:6999–7006.

208. Drake, C.G., S.K. Babcock, E. Palmer, and B.L. Kotzin. 1994. Genetic analysis of the NZB contribution to lupus-like autoimmune disease in (NZB x NZW)F1 mice. *Proc. Natl. Acad. Sci. USA* 91:4062–4066.

209. Hirose, S., H. Tsurui, H. Nishimura, Y. Jiang, and T. Shirai. 1994. Mapping of a gene for hypergammaglobulinemia to the distal region on chromosome 4 in NZB mice and its contribution to systemic lupus erythematosus in (NZB x NZW)F1 mice. *Int. Immunol.* 6:1857–1864.

210. Jiang, Y., S. Hirose, Y. Hamano, S. Kodera, H. Tsurui, M. Abe, K. Terashima, S. Ishikawa, and T. Shirai. 1997. Mapping of a gene for the increased susceptibility of B1

cells to Mott cell formation in murine autoimmune disease. *J. Immunol.* 158:992–997.

211. Nakatsuru, S., M. Terada, M. Nishihara, J. Kamogawa, T. Miyazaki, W.M. Qu, K. Morimoto, C. Yazawa, H. Ogasawara, Y. Abe, K. Fukui, G. Ichien, M.R. Ito, S. Mori, Y. Nakamura, and M. Nose. 1999. Genetic dissection of the complex pathological manifestations of collagen disease in MRL/lpr mice. *Pathol. Int.* 49:974–82.

212. Vyse, T.J., C.G. Drake, S.J. Rozzo, E. Roper, S. Izui, and B.L. Kotzin. 1996. Genetic linkage of IgG autoantibody production in relation to lupus nephritis in New Zealand hybrid mice. *J. Clin. Invest.* 98:1762–72.

213. Santiago, M.L., C. Mary, D. Parzy, C. Jacquet, X. Montagutelli, R.M. Parkhouse, R. Lemoine, S. Izui, and L. Reininger. 1998. Linkage of a major quantitative trait locus to Yaa gene-induced lupus- like nephritis in (NZW x C57BL/6)F1 mice. *Eur. J. Immunol.* 28:4257–4267.

214. Takahashi, T., M. Tanaka, C.I. Brannan, N.A. Jenkins, N.G. Copeland, T. Suda, and S. Nagata. 1994. Generalized lymphoproliferative disease in mice, caused by a point mutation in the Fas ligand. *Cell* 76:969–976.

215. Walport, M.J., K.A. Davies, and M. Botto. 1998. C1q and systemic lupus erythematosus. *Immunobiology* 199:265–85.

216. O'Keefe, T.L., G.T. Williams, F.D. Batista, and M.S. Neuberger. 1999. Deficiency in CD22, a B cell-specific inhibitory receptor, is sufficient to predispose to development of high affinity autoantibodies. *J. Exp. Med.* 189:1307–13.

217. Lai, W.S., E. Carballo, J.R. Strum, E.A. Kennington, R.S. Phillips, and P.J. Blackshear. 1999. Evidence that tristetraprolin binds to AU-rich elements and promotes the deadenylation and destabilization of tumor necrosis factor alpha mRNA. *Mol. Cell Biol.* 19:4311–23.

218. Watanabe-Fukunaga, R., C.I. Brannan, N.G. Copeland, N.A. Jenkins, and S. Nagata. 1992. Lymphoproliferative disorder in mice explained by defects in Fas antigen that mediates apoptosis. *Nature* 356:314–317.

219. Carroll, M.C. 2000. The role of complement in B cell activation and tolerance. *Adv. Immunol.* 74:61–88.

220. Goulet, J.L., R.C. Griffiths, P. Ruiz, R.F. Spurney, D.S. Pisetsky, B.H. Koller, and T.M. Coffman. 1999. Deficiency of 5-lipoxygenase abolishes sex-related survival differences in MRL-lpr/lpr mice. *J. Immunol.* 163:359–66.

221. Chesnutt, M.S., B.K. Finck, N. Killeen, M.K. Connolly, H. Goodman, and D. Wofsy. 1998. Enhanced lymphoproliferation and diminished autoimmunity in CD4- deficient MRL/lpr mice. *Clin. Immunol. Immunopathol.* 87:23–32.

222. Cattell, V. 1999. Nitric oxide and glomerulonephritis. *Semin. Nephrol.* 19:277–87.

223. Peng, S.L., and J. Craft. 1996. T cells in murine lupus: propagation and regulation of disease. *Mol. Biol. Rep.* 23:247–51.

224. Peng, S.L., M.P. Madaio, A.C. Hayday, and J. Craft. 1996. Propagation and regulation of systemic autoimmunity by gammadelta T cells. *J. Immunol.* 157:5689–98.

225. Wen, L., W. Pao, F.S. Wong, Q. Peng, J. Craft, B. Zheng, G. Kelsoe, L. Dianda, M.J. Owen, and A.C. Hayday. 1996. Germinal center formation, immunoglobulin class switching, and autoantibody production driven by "non alpha/beta" T cells. *J. Exp. Med.* 183:2271–82.

226. Conde, C., S. Weller, S. Gilfillan, L. Marcellin, T. Martin, and J.L. Pasquali. 1998. Terminal deoxynucleotidyl transferase deficiency reduces the incidence of autoimmune nephritis in (New Zealand Black x New Zealand White)F1 mice. *J. Immunol.* 161:7023–30.

227. Rifkin, I.R., P.L. Channavajhala, H.L. Kiefer, A.J. Carmack, E. Landesman-Bollag, B.C. Beaudette, B. Jersky, D.J. Salant, S.T. Ju, A. Marshak-Rothstein and D.C. Seldin. 1998. Acceleration of lpr lymphoproliferative and autoimmune disease by transgenic protein kinase CK2 alpha. *J. Immunol.* 161:5164–70.

22 | Rheumatoid Arthritis

Thomas Pap, Renate E. Gay, and Steffen Gay

1. INTRODUCTION

Rheumatoid arthritis (RA) is a chronic systemic disorder that affects primarily the joints and results in their progressive destruction. Notably, this destruction of articular cartilage and bone represents an unique and most prominent feature of disease and results from the interaction of synovial hyperplasia, chronic inflammation, and autoimmunity (1).

However, progressive joint destruction not only distinguishes RA from other arthritides but also determines its outcome in the majority of patients. Apart from causing significant disability in affected individuals, RA is of considerable burden to the society. As shown by several studies, it is one of the most expensive diseases in terms of total costs for diagnosis, treatment, rehabilitation and income losses (2). In England, the annual economic impact of RA has been estimated as to £1.256 billion, with about half being direct and indirect costs (3). Other data from an analysis of the U.S. National Health Interview Survey revealed total annual costs of $8.74 billion, or about 0.3% of the gross domestic product of the United States (4). These figures underline the importance of current efforts to elucidate key features of disease as well as to develop novel therapeutic strategies for the treatment of RA.

In the last years, there have been considerable advances in understanding the cellular and molecular mechanisms of RA. Specifically, the utilization of molecular biology techniques, the introduction of novel animal models and the observation of early stages of disease together with improved tools for clinical and epidemiological studies have provided exiting insights into disease characteristics. Most intriguingly, we have seen formerly conflicting data linked together by studies demonstrating the interplay of different cells as well as the complexity of intracellular pathways.

2. EPIDEMIOLOGY AND CLINICAL FEATURES

RA is a common disease that affects all ethnic or racial groups in every part of the world. However, prevalence and incidence rates vary considerably between populations. For most western populations, estimates on the prevalence of RA range from 0.6% to about 2%. Interestingly, when other ethnic groups are considered, the prevalence may be as high as 5.3% among Chippewa Indians or as low as just 0.1%–0.3% among rural Chinese communities or certain African groups (5,6). As the latter populations are characterized by a different social structure including high poverty and substantial lack of medical care, it has been suggested that the low prevalence in these populations may be due to less precise diagnosis as well as excess mortality (7). Given the differences between ethnic groups and regions, an overall estimate prevalence of 0.8% to 1% in most western countries is widely accepted today (8). Most recently, Gabriel and coworkers investigated the epidemiology of RA in a population based cohort (9) and demonstrated that the epidemiology of RA is dynamic, with clear secular trends in its incidence. In this study, a steadily increasing incidence was seen from residents at the age of 35–44 years to people between 75 and 84 years of age.

RA affects more than twice as many women as men. The reason for this gender preference is not clear, but several lines of evidence point to a complex interaction of genetical factors as well as hormonal and reproductive exposure (10). Specifically, the premenopausal excess incidence of RA in women (11) together with the increased risk for developing first symptoms of RA after pregnancy (12) strongly suggest an influence of endogenous female hormones linked to reproduction. Also, it has been shown that there is a association between breast feeding and RA (13–16). In the context of these observations, prolactin has been proposed to play a role in modulating disease recently (17–21). A mild hyperprolactinemia has been reported in RA patients (22) and animal studies have demonstrated that prolactin is not only essential for the development of adjuvant arthritis but an increase of prolactin synthesis precedes the occurrence of joint swelling (23). Intriguingly, alterations in the prolactin levels may not only provide a most interesting explanation for the preferential affection of woman. As well be discussed, prolactin is also linked to some aspects of genetic involvement in RA.

2.1. Joint Features

RA is a polyarticular disease that may affect most joints of the body. However, the involvement of the wrist and the small finger joints is the most prominent feature of RA, which has been included into the classification criteria for disease (Table 1). Typically, a morning stiffness in the hands is reported as one of the first symptoms followed by—initially transient—swelling of some finger joints. Local articular hyperthermia, tenderness of the joints upon maximal motion/pressure or swellings in the tendons may constitute early diagnostic signs of RA. The onset of RA is mostly gradual with symmetrical articular symptoms in small peripheral joints. Only in about 7%, RA begins as a monarticular disease involving primarily the knee, hip or shoulder joints. However, 10–25% of patients may also present with an acute onset of disease that is characterized by sudden diffuse swelling of multiple joints.

The swelling of joints is caused by an invasive synovitis that ultimately results in the destruction of articular cartilage and bone (Figure 1). Clinically, these changes in the bone structure are monitored by x-ray examination and present as juxta-articular osteoporosis, early joint space narrowing and later erosive changes. As a consequence of articular destruction and tendon involvement, characteristic changes may occur in the hand (Figure 1). They present as ulnar deviation in the metacarpophalangeal joints as well as swan neck and boutonnière deformities of the proximal interphalangeal joints. Bone destruction in other joints such as the forefoot, knees, elbows and shoulders also result in

Table 1 Diagnostic criteria for rheumatoid arthritis

Criteria
Morning stiffness in and around the joints for at least 1 hour
Involvement of at least 3 joint areas simultaneously
Arthritis of at least one area in the wrist, MCP or PIP joints
Symmetric involvement of at least one pair of joints
Presence of subcutaneous nodules
Demonstration of abnormal amounts of rheumatoid factors
Radiographic changes typical for rheumatoid arthritis

specific deformations that consequently lead to severe disability.

Typically, the course of RA is characterized by the continuous involvement of an increasing number of joints. However, there are major individual differences both in joint patterns and disease progression. Also, the rapidity of changes may vary considerably. Although it has been widely accepted that most severe changes occur in the first 5 to 7 years of disease, and this understanding has resulted in the use of disease modifying antirheumatic drugs (DMARD's) early in the course of disease (24), it is still difficult to predict the outcome of disease in an individual patient.

At late stages of disease, arthritis may also affect the axial skeleton, most prominently the cervical spine. Involvement of the atlanto-axial joint can be seen in about 30% of patients with erosive disease and may result in neurological complications caused by the compression of nerve roots.

2.2. Systemic Involvement

Systemic involvement in RA is seen in changes of laboratory parameters as well as in a variety of symptoms, some of which appear not to be related to the arthritis at first sight. General, rather unspecific signs such as fatigue, weight loss and myalgias are seen in more than 30% of patients at onset of disease and may even predominate early articular symptoms. However, systemic involvement and complications of RA may occur at any stage of disease and have a dramatic impact on its outcome. Most of the systemic findings are related to inflammation and present as hematological abnormalities such as anemia, thrombocytosis and an increase in acute phase proteins. However, most common non arthritic organ manifestations comprise subcutaneous nodules, pericarditis, keratoconjunctivitis and rheumatoid vasculitis.

About 20% to 30% of RA patients present with rheumatic nodules, which are found mostly at the extensor surfaces of the upper extremities and have been described in a great variety. They are characterized histologically by a central fibrinoid necrosis with a surrounding fibroblastic proliferation. Rheumatic nodules are thought to derive from localized small vessel vasculitis. Rarely such nodules may also be present in internal organs such as the central nervous system or the lungs. Small vessel vasculitis also constitutes the basis for several other organ involvement. Extending to the small- and medium size arteries of extremities and nerves, rheumatoid vasculitis may complicate RA and result in sensory or sensorimotory neuropathies as well as nailfold- or digital tip infarction (25). Although systemic vasculitis is rare and mainly seen in long-standing RA, it may also occur at early stages of disease and predispose to a poor outcome (26–28).

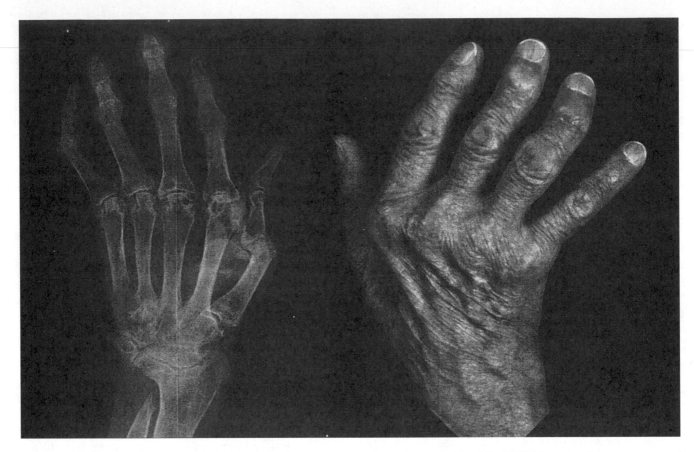

Figure 1. Manifestation of rheumatoid arthritis (RA) in the hand. A mutilating synovitis of the MCP joints constitutes a most prominent feature of disease (swelling and ulnar deviation—right panel). As seen in x-ray picture on the left, severe erosions are present in the MCP joint (Pictures kindly provided by Dr. P. Brühlmann).

3.   ETIOLOGY

Despite the clear description of clinical signs and symptoms, there is still no conclusive concept on the etiology of RA. Although it has been widely accepted that both genetic and environmental factors contribute to the development of disease, their relationship appears complex and is not well understood. Specifically, the question of whether RA is caused by a limited set of genetic determinants in conjunction with a variety of environmental triggers or, on the contrary, a variety of genetic settings together with just one or two different environmental factors is responsible for the development of RA, remains elusive. However, despite a large number of questions that still remain, research of the last years has helped elucidating some key questions of RA etiology.

3.1.   Genetics in RA

Since the initial observation in the 70s that there was an increased risk for developing RA among relatives to RA patients (29), research has undertaken efforts to define the contribution of genetic factors to the etiology of RA. A major tool for determining the genetic component has been the calculation of risk ratios based on epidemiological data. Given a prevalence of RA of about 2–4% among siblings to RA patients (30,31), as well as a background population prevalence of about to 0.8%, a relative risk λ_S between 2.5 and 5 would arise among siblings (32). As for monocygotic twins, early studies suggested a concordance rate between 30 and 50%, although, most recent reports agree that it does not exceed 15% (33–35). From these data, the maximum genetic risk, as represented by the relative risk among identical twins, can be calculated as to a λ_{MZ} of ~ 12 (32).

Major progress in elucidating the contribution of genetical factors to the etiology of RA has come from advances in molecular biology, which showed that twins may differ in important aspects of immune responses, such as the somatic rearrangement of immunoglobulin and T-cell receptor (TCR) genes, and thus perhaps in disease susceptibility, and provided insights into mechanisms that transmit susceptibility into disease. In this context, the analysis of polymorphic markers in affected sibling pairs has been used as the method of choice for investigating complex genetics in RA.

Since first evidence for an association of RA with an antigen on the HLA-D locus was provided by Stastny (36), association of RA with class II HLA molecules has been investigated most intensively (rev. in (37). While initial studies focused primarily on establishing an association between the occurrence of certain class II HLA molecules and disease frequency, molecular biology has provided an understanding of HLA function as well as possible mechanisms by which these molecules could be linked to the development of RA.

Being part of the human major histocompatibility complex (MHC), class II HLA genes encode for a variety of transmembrane heterodimeric glycoproteins that are critically involved in the antigen presentation by macrophages. However, these proteins are also expressed on B-lymphocytes, dendritic cells and also fibroblasts. They are encoded by three distinct, neighboring regions on chromosome 6, termed DP, DQ and DR (Figure 2). For RA, the HLA-DRB genes that encode for the β-chain of HLA-DR specificities have been of particular interest since most of the HLA-DR polymorphism can be found here. While early terminology of these loci was based on serologic tests using various HLA-antisera, current nomenclature includes the specific allelic representation. Thus, the HLA-DR4 locus contains three different alleles, designated HLA-DRB1*0401, *0404, *0405, that have been implicated RA susceptibility to various extents. Specifically, a five-fold risk for RA concordance has been observed in HLA-DRB1*04-positive twins (38), and several studies indicate that HLA-DRB1*04 is found more frequently among RA patients than in the normal population (37). Looking for common features in these allelic variants, it was found that they share a conserved sequence in the third hypervariable region. This conserved DRB1 sequence motif is characterized by the amino acid sequence QK(R)RAA and has been mapped to the positions 70 to 74 of the DR β-chain. Consequently, it has been hypothesized that cross reactivity between this "shared epitope" and one or more arthritogenic antigens may lead to a specific immune response directed against HLA-DR molecules (39–42). Apart from the HLA-DRB1*04 alleles, HLA-DRB1*0101 has been associated with RA (43–47). Interestingly, HLA-DRB1*0101 also possesses a QRRAA motif, supporting a role of the shared epitope in the susceptibility for RA. However, attempts to isolate and characterize unique peptides that are displayed by specific RA-associated DRB1 molecules have thus far failed. In addition, recent studies suggest that a direct interaction of the QK(R)RAA motif and a potentially arthritogenic antigen may not be necessary for regulating the T cell response. Intriguingly, Penzotti and coworkers recently demonstrated direct interactions between the residues 70 (Q) /71(R) and the

TCR CDR3bQ97 of a particular T cell clone (48). This interaction suggests peptide-independent recognition of respective HLA-DRB1 alleles, such as HLA-DRB1*0404. Also, many individuals with the QK(R)RAA motif do not develop RA. Therefore, a more complex relation between the presence of certain HLA-DR alleles and the susceptibility to RA appears likely, and it has been suggested that antigen presentation by shared epitope HLA-DRB1 alleles does not constitute the basis for increases susceptibility of such patients. Rather, the presence of these alleles may result in a specific T-cell/HLA-DR interaction facilitating the development of RA (49).

The genetic involvement, particularly the HLA-DRB1 association, in RA has yet another perspective: Genetic factors may not only contribute to RA as a qualitative trait, namely the question of whether a given individual develops RA, but also appear to contribute to the severity of disease (Figure 2). It has been observed for some time that the presence of HLA-DRB1*04 alleles (HLA-DR4) is associated with more severe disease (50,51). Specifically, a higher likelihood of being rheumatoid factor positive as well as a higher frequency of extraarticular manifestations, such as rheumatoid vasculitis, has been observed in these patients. (52). Moreover, the presence of HLA-DRB1*0401 and HLA-DRB1*0404 has been linked to a higher rate of radiographic erosions and the frequency of joint surgery (Figure 2). Recent data from Weyand and coworkers support the hypothesis that these HLA-DRB1 alleles determine the severity of RA rather than susceptibility to disease (52). They demonstrated that the presence of a second HLA-DR*04 allele in RA patients predisposed to a significantly more severe disease. Moreover, all patients homozygous for HLA-DRB1*0401 developed major organ involvement. Taken together, these data suggest that HLA-DRB1 contributes to the severity of RA in a "dose-dependent" manner, but may not be responsible for the onset of disease.

Apart from the MHC complex, recent interest has focused on other genes that potentially contribute to the development of RA due to evidence that HLA-DR is responsible for little more than 30% of the genetic contribution (53,54) and on our increasing knowledge of molecular genetics. Of the multitude of genes that have been proposed, two, the prolactin and the TNFα genes appear to be of particular interest (37).

The prolactin gene is located ~11 centimorgans (cM) telomeric to the MHC on the chromosome 6p (Figure 2). Interestingly, a recent study by Brennan and coworkers found an association between alleles of two microsatellite markers ~1 and ~3 cM away from the prolactin gene and HLA-DRB1*0401 among women with RA (55) suggesting a possible biologic interaction between the HLA-DRB1 gene and the neighboring prolactin gene. In view of recent data detecting linkage disequilibrium at distances up to

15 cM (56), it can also been hypothesized that this association is due to linkage disequilibrium (55). As mentioned before, such an association could provide an explanation for some aspects of gender preference as well as some specific feature of RA epidemiology.

Interesting data have also come from analyses of the TNFα gene in RA patients. The gene for TNF lies between the loci for HLA class I and class II genes, and two different microsatellites have been described within the locus for TNFα (57). Although one of these has been shown to be in linkage disequilibrium with HLA-DR3 and associated with increased TNFα transcription (58,59), a clear answer as to whether genetic variability in the TNFα locus influences the susceptibility or severity of RA is still missing. While Mulcahy and coworkers (60), and Hajeer and coworkers (55), reported clear associations between the inheritance of TNFα microsatellite alleles and RA, other studies failed to establish such links (61), and no correlation has been found between the levels of TNFα mRNA in peripheral blood or synovial biopsies and the presence of certain TNFα promoter alleles. In view of the aforementioned data on the link between HLA-DRB1 alleles and RA, it is of interest, that the association with a specific TNF microsatellite (a6,

b5, c1,d3, c3) found in the Mulcahy study was independent of shared epitope alleles including HLA-DRB1*0401 (60).

Polymorphisms of T cell receptors (TCR) genes have also been suggested to provide alternative explanations for genetic involvement in the susceptibility to RA. In a recent study, Cornélis and coworkers investigated TCR germline polymorphisms in a case control study of RA (62). Concentrating on the TCRA genes, specifically on the TCRAV6S1, TCRAV7S1, TCRAV8S1, TCRAV10S2 as well as on TCRBV6S1 and TCRBV6S7, they found an association between RA and allele 2 of TCRAV8S1 (62). In another study by Ibberson and So, a significantly increased frequency of the TCRAV5S1 allele was seen in RA patients (63). Other data have suggested TCRVB polymorphisms are prognostic markers for RA (64), but these data also have been questioned (65). However, apart from the conflicting nature of the data (66), there have been methodological concerns with some of these studies and clearly further studies are needed to define the contribution of TCR genes in the etiology of RA.

Taken together, genetic factors, particularly different HLA-DR alleles, appear to contribute to the development of RA. However, the questions of how this association can be linked to molecular events as well as what the specific

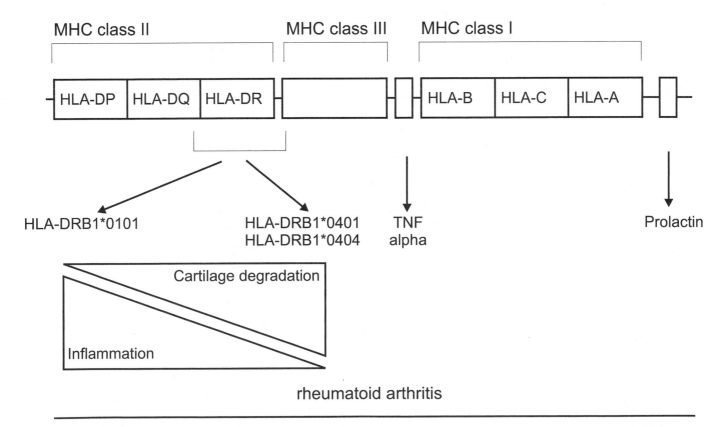

Figure 2. Association of MHC class II with RA. The MHC class II region contains three subregions (HLA-DP, -DQ, and -DR) of which HLA-DR has been associated with susceptibility to and severity of RA. Among the different HLA-DRB1 alleles, HLA-DRB1*0101, HLA-DRB1*0401, and HLA-DRB1*0404 have been associated with RA. While HLA-DRB1*0101 appears to be linked to more severe inflammation, cartilage destruction is more severe in patients with HLA-DRB1*04 alleles. The genes for TNF-alpha and prolactin lie close to the MCH class II.

contribution of other non-MHC genes might be needs further investigation.

3.2. Environmental Factors—The Potential Role of Viruses

The aforementioned epidemiological data, particularly the dynamics in the incidence of RA as well as a genetic contribution of not more than 50%, strongly indicate that environmental factors play a major role in the etiology of RA. Among the factors that may account for the environmental aspect, infectious agents have been implicated most frequently in the etiology of RA. Although a variety of microorganisms have been suggested to cause the disease, no specific agent has been clearly shown to be involved in the development of RA. However, viruses appear to represent the most likely infectious agent responsible for triggering RA (67). At the moment, three hypothetical mechanisms of viral involvement in RA can be envisaged: 1) Possible infection of synovial or peripheral cells with a virus, reinforcing the idea of RA as a primarily infectious disease. 2.) Activation of endogenous retroviral sequences in synovial cells subsequently leading to their activation, and 3.) Viral modulation or triggering in RA either by amplifying cellular activation or contributing to T cell activation on the basis of a certain genetic predisposition.

Viral infection in RA. Infections with several known viruses have been hypothesized as environmental triggers in RA. Among those Epstein-Barr virus (EBV) and parvovirus B19 have been implicated most frequently in the etiology of disease.

Evidence for the involvement of EBV in RA has mainly come from studies demonstrating an altered humoral immune response towards EBV antigens. Specifically, the Epstein-Barr nuclear antigen (EBNA) 1 has been suggested to be involved in inducing a pathological immune response (68–70). Birkenfeld and coworkers showed that p107, the major epitope of EBNA-1, crossreacts with denatured collagen and keratin establishing a potential link between RA, EBV-1, and these potential auto-antigens. Another study by Rumpold and coworkers suggested that antibodies against EBV exhibit crossreactivity with a 62 kD synovial protein (71). However, conflicting data from a more recent study by Buisson and coworkers failed to demonstrate a relationship between the Epstein-Barr virus and the antiperinuclear factor/"perinuclear antigen" system in RA (72). The concept of EBV involvement in RA was reinforced by recent studies that showed homologies in the amino acid sequence of certain EBV-proteins and HLA-DRB1 molecules as well as the occurrence of EBV specific T-cells. Thus, the EBV glycoprotein gp110 has been implicated in molecular mimicry events by containing the shared epitope motif of the HLA-DRB1*0401 molecule (73). Correlating EBV infection and HLA genotypes, a study by Saal and coworkers demonstrated that patients expressing shared epitope HLA-DRB1 alleles and EBV DNA in the synovium were at a more than 40-fold higher risk for RA than patients who were negative for both (74). However, a more recent study by Edinger and coworkers found no abnormal expression of EBV genes in RA synovium despite the presence of EBV antigen-specific T cell clones (75).

Based on similar evidence, human parvovirus B19 (B19) has also been proposed to constitute a causative agent for RA (76–79). A recent study by Takahashi and coworkers suggested that expression of the B19 antigen VP-1 is found specifically in RA synovium, but not in controls. Interestingly, in this study B19-negative macrophages (U-937 cell line) became positive for VP-1 when co-cultured with RA synovial cells (80). In addition, an altered cytokine profile was detected in these cells following the co-culture. From these data, it was concluded that B19 found in RA synovial cells was infective and involved in the initiation and perpetuation of RA. However, numerous other studies have failed to establish a link between B19 infections and RA. (81–83). Investigating serum and tissue samples from 26 patients with RA and 26 patients with OA, Kerr and coworkers found no significant differences. In both groups, about 30% of the patients showed positive PCR results in the tissues as well as serological evidence for a previous B19 infection (83). Taken together, these data provide only circumstantial evidence for the direct involvement of viral infection in the etiology of RA.

Retroviruses and RA. Retroviruses have been frequently implicated in the etiology of RA over the last few years, and it has been hypothesized that a retroviral infection may explain important features of synovial cell activation as well as subsequent inflammation and autoimmunity (67,84). Evidence for the involvement of exogenous retroviruses has been derived from two distinct observations:

First, there are several retrovirus-associated, RA-like arthritides in domestic animals, and retroviruses have also been linked to the development of such diseases in humans. Particularly, human T cell leukemia virus (HTLV)-I infection has been associated with a chronic inflammatory arthropathy in certain populations, such as Japanese (85,86). In these patients, anti-HTLV-1 antibodies, including antibodies against p19, p24, and p28, have been detected in the synovial fluids, and HTLV-I proviral DNA was found to be integrated into the DNA of synovial cells (87). The observation by Iwakura and coworkers (88) that an inflammatory arthropathy resembling human RA is induced in mice transgenic for the HTLV-I env-pX gene strengthened the concept of a potential HTLV-I involvement in RA. Most intriguingly, Fujisawa and coworkers recently showed that, in the arthritic joints of these mice, Env- or Tax-reactive T cells accumulate and are responsible for triggering chronic arthritis (89). In addition, this group demonstrated that extracellular Tax is able to regulate the expression of endogenous cellular genes in synovial cells, potentially contributing to NF-kB-mediated synovial hyperplasia (90). Therefore, a potential role of retroviruses, such as HTLV-I, has been investigated intensively in RA. After Wright and coworkers showed that the zinc finger transcription factor Z-225 (also called Egr-1) is constitutively expressed in HTLV-1-infected cells (91), interest focused on the activation of cellular transcription factors and proto-oncogenes in RA synovial cells. It could be demonstrated that Z-225/Egr-1, which is involved in cell activation and transcribed only transiently in normal cells, is also expressed constitutively in RA synovial fibroblasts over several passages (92). Subsequently, Egr-1 binding sites were found in promoter regions of several genes, and elevated expression of some of these genes have been associated with pathomechanisms of RA (93,94). Other proto-oncogenes activated in the course of a retroviral infection have also been demonstrated in the RA synovium. Among them, c-fos (95), which is co-expressed with egr-1 (96,97) has been found activated specifically in HTLV-I transfected cells (98,99). C-fos encodes for a basic leucin zipper transcription factor and is part of the transcriptional activator AP-1 (jun/fos) that regulates the expression of matrix metalloproteinases (MMPs), such as collagenase (MMP-1). Moreover, as shown in related studies, proto-oncogenes of the egr family are also involved in the activation of the cathepsin L gene (100), a matrix-degrading cysteine proteinase gene that is highly upregulated in RA synovium (95). However, the hypothesis of HTLV-I being directly involved in the pathogenesis of RA has been challenged by several studies. Thus, Bailer and coworkers failed to establish an association between HTLV-I and RA by analyzing peripheral blood mononuclear cells (PBMC's) from patients suffering from RA and systemic lupus erythematosus (101). Therefore, it appears that the abovementioned findings point to possible pathomechanisms associated with cellular activation rather than establish a link between HTLV-I infection and human RA. The data indicate clearly that proto-oncogenes that are constitutively expressed in retrovirally-activated cells as well as upregulated in RA synovial cells play an important role in mediating joint destruction by stimulating the action of matrix degrading enzymes.

The second, even stronger evidence for a potential involvement of exogenous or endogenous retroviruses in the pathogenesis of RA has been derived from the detection of type C virus-like particles in the synovial fluid of RA patients. By analyzing synovial fluid specimens from 16 RA patients, distinct spherical particles resembling type C retroviruses were found in 10 samples (102). In contrast to other known type C retroviruses, such as HTLV-I, however, these particles have a large overall diameter of 200 nm and a 70 nm internal core. They lack homology to any known retrovirus and do not express the epitopes of HTLV-I p19 or HIV p24 (102). However, the exact nature of these type C retroviral particles has not been determined, and the question of whether they constitute replication-competent infectious agents, remains elusive.

As an alternative to the involvement of infectious retroviral agents, endogenous retroviral sequences (ERV's) have been suggested to initiate human RA (1,67,84). It is known that the human genome contains numerous such sequences, most likely derived from exogenous retroviruses that, once integrated into the human genome, lost their ability for full replication during evolution. However, upon stimulation, these sequences can be induced and become active. The question of how the spontaneous activation of ERV is achieved has not been answered, although most recent data point to mechanisms that include the action of retro-transposons. In this context, the MRL-*lpr/lpr* mouse constitutes an interesting animal model in which the effect of retroviral sequences on the initiation of an arthritis can be studied (103–106). In these mice, the homozygous mutation of the *lpr* gene results in the development of an autoimmune disease that includes severe RA-like erosive arthritis. Most intriguingly, the *lpr* mutation represents an insertion of an endogenous retroviral sequence into the *fas* apoptosis gene (103,107), and gene transfer experiments partially correcting this abnormality led to a significant improvement of symptoms (108). Based on these observations, several groups have set out to identify and characterize relevant ERV's in the synovium of RA patients. Takeuchi and coworkers described the expression of ERV-3 and *lambda* 4–1 in synovial tissue specimens from RA patients (109). Isolation of retroviruses or retroviral sequences from human tissue takes advantage of the fact that the retroviral polymerase (*pol*) gene contains sequences that are highly conserved among most retroviruses, and therefore share extensive homology. Using reverse transcriptase polymerase chain reaction (RT-PCR) technique with degenerate primers for this *pol*-region, several homologues to human ERVs could be found. Analyzing a total of 875 different *pol* transcripts from 9 RA patients, as well as 3 osteoarthritis (OA) patients and 4 normal individuals, Nakagawe and coworkers (110) demonstrated the presence of undefined, but primate-type, C retrovirus homologues as well as HERV-9, HERV-K, HERV-L, and retrovirus-like (RTVL) sequences. In addition, there is evidence for the differential expression of specific ERV transcripts in the synovial compartment in RA. A hitherto unknown homologue of the human transducin-like enhancer protein (TLE-2) was identified in synovial fluid samples from 3 of 6 RA patients, but in none of OA specimens (111). Using *in-situ* hybridization, mRNA for this TLE-2 homologue was found abundantly expressed in the lining layer of RA synovial membranes, but was negligible in OA synovial specimens. This is of particular interest because TLE proteins are not only involved in the regulation of nuclear gene transcription, but have also been associated with malignant transformation. Most recently, a retrotransposable L1 element was shown to be expressed in synovial cells at sites of cartilage and bone destruction and to induce human stress-activated protein kinase 2δ/p38δ (112). However, the question of whether ERVs or retroviral sequences are indeed responsible for the initiation and perpetuation of RA as well as mechanisms by which they may act remains to be established.

4. THE RHEUMATOID SYNOVIUM

In RA, hyperplasia of the synovium is a prominent feature, and it has been known for decades that the hyperplastic synovial tissue is responsible for joint destruction (113). Although sometimes still called "pannus", research has demonstrated that the RA synovium is not a homogenous proliferating mass but rather constitutes a highly differentiated tissue displaying distinct changes at different stages of disease as well as in different areas of the synovial membrane (114–117).

Thickening of the RA synovium is largely due to changes in the most superficial cells. These cells form the "lining layer" that, in the course of disease, increases up to 10 cell layers and is characterized by the presence of resident synovial as well as numerous infiltrating cells. Immunohistochemical studies have shown that a considerable proportion of the lining cells are synovial macrophages expressing characteristic surface markers such as CD11b, CD14, CD33, CD68 and MHC class II molecules (118–122). A smaller proportion of lining cells has a fibroblast-like appearance, lack specific macrophage markers and express only low levels of MHC class II (123,124). They are called synovial fibroblasts (SF) and, among others, can be identified by antibodies recognizing prolyl-4-hydroxylase, an enzyme involved in the collagen synthesis. Some of these cells, specifically those in the lining layer, display features that distinguish them from other fibroblasts as well as from normal SFs. This population is referred to as transformed appearing, or activated, SFs (1,114,115).

In RA, not only the lining but also deeper areas of the synovial membrane undergo considerable changes. These are highly variable and characterized by an infiltration with mononuclear cells such as T cells, B cells and macrophages. In contrast to the macrophages, lymphocyte infiltration is mainly perivascular with a clear dominance of T-cells (125). The lymphocytes often organize into aggregates, which resemble those seen in lymph nodes or in extralymphatic sites such as Peyer's patches. However, within the T-cell population the majority of cells are CD4+ and express CD29/CD45R0 antigens that characterizes them as memory T-cells (126,127). This clearly distinguishes RA synovium from lymphatic tissues. Also, while lymph nodes contain a large number of suppressor/cytotoxic lymphocytes, in RA synovium CD8+ cells represent only a minority, and these cells are located at the periphery of the aggregates. The CD4/CD8 ratio in the synovial sublining ranges from 4:1 to 14:1 (128,129), much greater than in the peripheral blood, and also exceeds that in the synovial fluid (123). B lymphocytes make up an only small proportion of synovial cells. However, they have been suggested to contribute to the local inflammation by an increased synthesis of autoantibodies (130–132). Although the specificity of these antibodies has been determined for some B cells, their significance is not clear. There is no evidence that antibodies produced by synovial B lymphocytes initiate disease, but it is obvious that the production of (auto)antibodies perpetuates the inflammation in RA. However, B cells in the synovium also secrete immunoglobulins that bind to the Fc portion of IgG and are called rheumatoid factors (RF).

Vascularization is also altered in the rheumatoid synovium. Specifically, two distinct processes can be distinguished: the formation of high endothelial venules (HEV) that permit the trafficking of lymphocytes into the synovium and the formation of new blood vessels. While at first sight these alterations appear rather unspecific and largely due to the inflammation, research of the last years has demonstrated an interplay between synoviocytes and endothelial cells. As suggested by several studies, the increased blood vessel formation is critical to the hyperplasia of the synovial membrane (133). Recent studies have demonstrated high levels of CD146, which are expressed almost exclusively by vascular endothelium and found in the synovial fluid of patients with early RA (134). These data indicate increased endothelial activity and angiogenesis. Intriguingly, neoangiogenesis is believed to be stimulated mainly by the cells of the activated synovium.

5. CONCEPTS ON THE PATHOGENESIS OF RA

For the pathogenesis of RA it appears well established that both T cell-dependent as well as T cell-independent pathways are involved (Figure 3). Table 2 summarizes the evidence for a pivotal involvement of both T cell-dependent and T cell-independent pathways in the pathogenesis of RA.

5.1. T Cell-Dependent Pathways

TCR receptors have been implicated in the genetics of RA, suggesting that allelic variants of Vα or Vβ elements are associated with the onset of disease. More importantly, determining the expression of TCR specificities has been useful in investigating a potential oligoclonality of infiltrating T cells in the synovium. It has been hypothesized that both thymic selection and post thymic events may result in a specific TCR repertoire in RA. In this context, changes of the TCR repertoire by exposure to superantigens have been studied most intensively. As one of the first, Paliard and coworkers demonstrated a preferential expression of Vβ14 on synovial T cell, and a parallel decrease of Vβ14 T-cells in the periphery (135). Other studies have extended this picture, suggesting an oligoclonal expansion of T cells expressing Vβ3, Vβ14 and Vβ17 (136–139). Interestingly, these Vβ elements show structural homologies that determine susceptibility to superantigen stimulation. Therefore, it was hypothesized that exposure to a common superantigen, such as mycoplasma, may be responsible for selection and oligo-clonal expansion of T cells in RA (140). As selection of T cells in the thymus as well as post-thymic shaping of the T cell repertoire are influenced by MHC molecules, oligoclon-ality of T cells may also provide an explanation for the asso-ciation with HLA-DR alleles. However, this concept has been challenged by several studies that failed to demonstrate a clear clonality of T cells in RA (141,142). In this context, it has also been argued that oligoclonal expansion of T cells may not constitute an early event in RA, but is found predominantly in late, chronic disease (143). This would

imply that the oligoclonality of T cells seen in some studies may be related to cartilage degradation and/or inflammation rather than being an initial, causative event in RA.

Recently, it has been suggested that the clonal expansion of CD4$^+$ T cells that lack the expression of CD28 represents an important step in the pathogenesis of RA (144). Martens and coworkers demonstrated that, in the blood of RA patients, CD4$^+$CD28$^-$ T cells can account for up to 50% of the CD4 subset (145). In addition, CD4$^+$CD28$^-$ T cells were found in the RA synovium (146). Together with the obser-vation that these CD4$^+$CD28$^-$ T cells proliferate in response to immobilized anti-CD3 and are autoreactive to ubiqui-tously distributed autoantigens (146), it has been hypothe-sized that the emergence and survival of CD4$^+$CD28$^-$ T cells constitutes an key event in RA. Indeed, the expansion of CD4$^+$CD28$^-$ T cells has been demonstrated to correlate with extraarticular involvement (145). However, in this study, no correlation with the severity of joint destruction was seen.

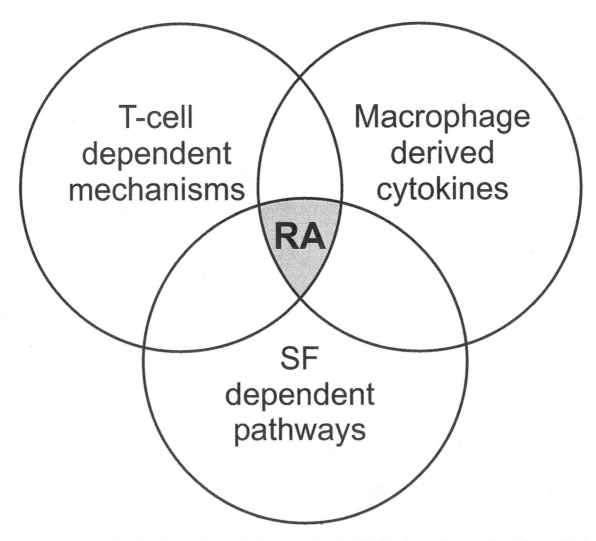

Figure 3. Pathogenesis of RA. T-cell dependent mechanisms as well as T-cell independent pathways and contribute to RA. In addition, an inflammatory pathway (macrophage derived cytokines) enhances the disease process.

Table 2 Pro's and Contra's for the involvement of T-cells in the pathogenesis of RA

Pro	Contra
Presence of T-cells in the RA synovium	Majority of T-cells in the synovium of RA patients are anergic
Association of RA with certain MHC class II alleles	Failure to detect specific etiologic antigens
Expansion of certain TCR Vβ subsets in some patients with RA	*Inconsistent results from different* studies investigating the TCR repertoire
Modest clinical efficacy of some anti-T-cell therapies	Lack of substantial effect of anti-T-cell treatment on the disease outcome
Clinical improvement of RA in AIDS patients	Progression of joint destruction in some RA patients with AIDS
Experimental RA-like arthritis through the injections of antigens	In the SCID mouse, RA-SF destroy cartilage without human T-cells.
	Uncoupling of inflammation and joint destruction in RA,
	Lack of T-cell derived cytokines in RA synovial membranes

Evidence for the involvement of antigen recognition events and T cells in the pathogenesis of RA has also come from the observations that the injection of various antigens can lead to the induction of a relapsing, erosive arthritis in rodents. Thus, immunization of mice and rats with type II collagen results in the development of a RA-like disease (collagen-induced arthritis, CIA) accompanied by high levels of anti-collagen type II antibodies (147–149). Interestingly, depletion of the T helper population not only inhibits the production of anti-collagen antibodies, but also influences the extent of arthritis (150, 151). Recently, other cartilage-derived proteins such as the cartilage oligomeric protein (COMP) or the hyaline cartilage glycoprotein 39 (gp39) have also been associated with antigen induced arthritis (152–154). Consequently, numerous cartilage-derived proteins have been proposed as autoantigens in RA, among them collagen type II (155). Londei and coworkers demonstrated that 12% of T cell clones in RA patients were reactive with collagen type II (156). In addition, collagen type II degradation products have been shown to be present in RA joints. However, collagen type II-reactive T cell clones do not occur in all RA patients, and collagen-specific T and B cells from RA patients as well as those with other specificities are not invariable. In the last years, a multitude of other potential autoantigens have been described, but no clear function in initiating or driving RA could be established (157–162).

The SCW-A model in Lewis rats has been of interest because synovial hyperplasia in this model resembles some important features of human RA: There is a tumor-like proliferation of synovial cells that express high levels of several proto-oncogene products, including *c-fos* and *c-myc*, as well as matrix degrading enzymes such as MMP-1 and MMP-2 (163,164). In addition, like RA synovial fibroblasts, synovial cells in this model do not show contact inhibi-

tion and can be grown under anchorage-independent conditions. The destruction of articular cartilage is clearly IL-1-dependent, while TNFα appears to be involved mainly in the joint swelling. However, arthritis in each of the mentioned animal models is clearly driven by known antigens. Therefore, the question is to what extent their respective pathological processes resemble those of human RA. Most recently, Shiozawa and coworkers provided a potential model for how antigen recognition may enhance arthritis by demonstrating that induction of antigen-induced arthritis in c-fos transgenic mice leads to a severe destruction characterized by the predominant infiltration of fibroblast-like cells and the absence of lymphocytes (165). These data are of interest because overexpression of *c-fos* has been described as one of the intrinsic properties of RA-SF indicating their state of activation (94).

The hypothesis of primary local T cell activation due to (super)antigen exposure has been challenged by the observation that only a minority of T cells are activated within the RA synovium, and that only negligent amounts of T cell-derived cytokines could be detected within the RA synovium (166), specifically, IL-2 and γ-interferon. Although it has been argued that a small and histologically imperceptible number of activated T cells might be sufficient to drive the synovial inflammation (167), this observation has led to the question of the role of T cells in the pathogenesis of RA, and anergy has been proposed to be a dominant feature of synovial T cells. In addition, several studies using anti-CD4 monoclonal antibodies in the treatment of RA have failed to demonstrate a significant and sustained effect (168). In contrast, the paradoxical effects of persistent T cell infiltration, despite peripheral lymphopenia, as well as a shift of peripheral T cell repertoire toward the CD45R0+ memory phenotype was observed (169,170). Although these results were interpreted as being consistent with a critical role for

T cells in the synovial inflammation, alternative explanations include the selective attraction of these cells by fibroblast-and macrophage-derived factors.

The clinical improvement of RA in human immunodeficiency virus (HIV) infected patients has also been used as an argument for the role of T cells in disease. On the other hand, it has also been shown that joint destruction in RA may also proceed in the absence of T cells. In this context, Müller-Ladner et al. reported the case of a 63 year old patient with RA who became infected with HIV. Confirming the aforementioned studies, HIV infection improved the clinical symptoms. However, the HIV infection did not affect the progression of joint destruction in this patient. In addition, histologic evaluation revealed progressive joint destruction with aggressively growing fibroblasts, despite the striking lack of CD4+ T cells (171). Intriguingly, this observation has found indirect confirmation in a recent study showing clinical improvement, but radiological deterioration. Analyzing serial disease activity measures and hand radiographs in 40 patients with RA over 6 years, Mulherin and coworkers observed an significant improvement of both clinical and laboratory parameters, yet at the same time found a significant deterioration in articular erosions (172).

In conclusion, clinical data strongly support the role of T cells in the inflammatory process of RA. However, thus far they provide only circumstantial evidence for the hypothesis that T cell-mediated pathways are primarily involved in driving joint destruction.

5.2.   T Cell-Independent Pathways—The Role of Fibroblasts

Initial observations suggesting a key role for fibroblast-like synoviocytes in the pathogenesis of RA date back to the late seventies. Analyzing large numbers of synovial specimens from RA patients, Fassbender found that RA-SFs exhibit considerable morphological alterations compared with normal SFs (114). He demonstrated that, unlike normal SFs, these cells have an abundant cytoplasm, a dense rough endoplasmatic reticulum and large pale nuclei with several prominent nucleoli. At the same time, Fassbender showed that the invasion of cartilage and subchondral bone by these RA-SF did not require the presence of inflammatory infiltrates (114).

In the last few years, considerable efforts have been made to characterize what Fassbender called the "transformed-appearing" phenotype of synovial fibroblasts, and the interest has focused mainly on the characteristics of this phenotype on a cellular and molecular level as well as on the mechanisms of activation resulting in their destructive behavior.

Two animal models have been of particular importance in establishing the role of SFs in the pathogenesis of RA: The aforementioned MRL-lpr/lpr mouse and the SCID mouse model of RA. The MRL-lpr/lpr mouse model has been of interest because these mice conditionally develop RA-like destructive arthritis resembling that in human RA (105,106, 173). When studied histologically using both light and electron microscopy, it was found that initial joint damage in this model was mediated by proliferating synovial cells (106). Interestingly, cartilage and bone destruction occurs only at sites of synovial attachment (105,106). As demonstrated by Trabandt and coworkers, synovial cells of MRL-lpr/lpr mice constitutively express the MMP-1 gene. Immunohistochemical studies revealed the in-situ expression of MMP-1 in proliferating synovial lining cells as well as in chondrocytes in the first stage of pathological changes in the MRL-lpr/lpr mouse arthropathy (174). Interestingly, MMP-1-expressing synovial lining cells from these mice exhibited markedly elevated RNA levels of the c-fos proto-oncogene in vitro. It is of particular note that inflammatory cells were observed to migrate into the synovium and accelerate the process only after cartilage degradation was initiated. Notably, autoimmunity to collagen type II occurred as a consequence of cartilage damage rather than preceeding it. Most recently, similar observations were obtained by showing that inflammation and articular damage may differ (172), and indicate that critical steps that initiate joint destruction occur very early in the course of disease in the absence of T cells (175).

The SCID-mouse co-implantation model has been widely used to investigate specific aspects of cartilage destruction in human RA based on experiments by Mosier and coworkers demonstrating first that a functionally intact human immune system can survive in SCID recipients (176). Subsequently, Rendt and coworkers (177) set out to study the role of T cells in RA by implanting rheumatoid synovial tissue under the renal capsule of SCID mice. Interestingly, they showed that lymphocyte infiltrates disappear with time, while lining layer synoviocytes survive. Moreover, RA-SFs not only survived in SCID mice but maintained their characteristic biological features. Based on these observations, the SCID mouse co-implantation model for RA was developed as a novel tool for studying the molecular mechanisms of rheumatoid joint destruction in vivo (178). Initially, human RA synovium was co-implanted with normal human cartilage under the renal capsule of SCID mice, imitating the situation in a rheumatoid joint. Both RA synovial tissue and normal human cartilage could be successfully kept in the SCID mice for more than 200 days, and implanted RA synovium showed the same invasive growth and progressive cartilage destruction as in human RA joints (178). Intriguingly, the vast majority of synovial cells found at sites of cartilage invasion resembled activated synovial fibroblasts. To study specifically the molecular properties of these RA-SF, normal human cartilage was implanted together with isolated RA-SFs under the renal capsule of SCID mice (179). This procedure allowed analysis of the matrix-degrading properties of these cells in the absence of human lymphocytes and macrophages. Most intriguingly, RA synovial fibroblasts maintained their aggressive phenotype, particularly at sites of invasion into the co-implanted human cartilage. In contrast, OA synovial fibroblasts did not exhibit this invasive growth (179). Using *in situ* hybridization with RNA probes, a number of cartilage degrading proteases could be demonstrated in the RA-SF actively invading the cartilage. In contrast, few or none of these proteases were found when normal SFs, OA-SFs or dermal fibroblasts were examined.

From these animal models as well as from numerous histological data, it has been concluded that cellular activation, including altered apoptosis, attachment to the cartilage and subsequent degradation of cartilage and bone constitute the main sequelae of RA-SF involvement in RA (Figure 4).

Proto-oncogenes and signaling in RA-SF. The expression of proto-oncogenes and the upregulation of transcriptional factors have been described as a major features of activated RA-SF (rev. in (94). Several observations indicate that both the altered morphology and the aggressive behavior of these cells result from specific changes in intracellular signaling molecules. However, one has to be aware that our understanding of signaling pathways in RA-SFs, particularly the role of individual molecules therein, is still rather vague. Despite major advances in understanding specific pathways, it becomes more and more evident that activation of synovial cells in RA involves the whole complexity of cellular signaling.

The AP-1 transcriptional factor that consists of the c-Fos and c-Jun elements appears to constitute a key element in signaling pathways that ultimately lead to the degradation of cartilage and bone in RA based on the fact that *c-fos* is overexpressed in the RA synovium (96,97,180,181). Importantly, the promoters of several MMP genes contain consensus binding sites for the transcription factor AP-1, and AP-1 sites have been shown to be involved in tissue-specific expression of MMPs (182). Analyzing nuclear extracts by electrophoretic mobility shift assay, Asahara and coworkers found a high DNA binding activity of AP-1 in the synovial tissues of RA patients, but negligible activity in OA samples (183). AP-1 activity was predominately found in adherent cells and correlated with the in situ expression of c-fos and c-jun mRNA as well as with disease activity. However, AP-1 sites do not appear to regulate transcription of MMPs alone. Rather, there are essential interactions with other cis-acting sequences in the promoters and with transcription factors that bind to these sequences (182). Thus, the upregulation of *c-fos* as well as of *fos*-related proto-oncogenes appear to be important for cell activation via AP-1 formation and subsequent matrix degradation.

The transcriptional factor NF-κB is also highly activated in the synovial membrane of RA patients, particularly in RA-SFs (184). NF-κB can be induced not only through pro-inflammatory cytokines but, conversely, NF-κB activation induces the transcription of such cytokines as well as

of adhesion molecules. Therefore, NF-κB activation in synovial cells appears to constitute an important factor in the perpetuation of disease. Specifically, NF-κB appears to play a major role in synovial inflammation (184–187). Overexpression of IκBa, the natural inhibitor of NF-κB, results in the downregulation of pro-inflammatory cytokines, but does not affect anti-inflammatory mediators. In addition, NF-κB has been suggested to contribute to a reduced apoptosis in synovial cells (184) based on observations showing increased apoptosis following suppression of NF-κB. A study by Makarov and coworkers (188) recently demonstrated that inhibition of NF-κB through gene transfer with IκBα may constitute a feasible tool for inhibiting the inflammatory response. Similar to AP-1, NF-κB induces the expression of MMPs, such as MMP-1 (189), but the question of which regulatory pathway contributes most significantly to the altered expression of MMPs in RA remains to be answered.

The proto-oncogenes *ras, raf, sis, myb* and *myc* have also been detected in RA patients to various extents and were predominantly up-regulated in synovial cells attached to cartilage and bone (94). Moreover, binding sites for the aforementioned early response gene *egr-1* could be identified in the promotors of the oncogenes *sis* and *ras*. Trabandt and coworkers demonstrated that the Ras and Myc proteins are co-localized in the synovial tissue in about 70% of the analyzed RA cases (96). Interestingly, the cysteine proteinase, cathepsin L, a major ras-induced protein in *ras*-transformed murine NIH 3T3 cells, was detected in the synovial cells of RA patients (190) together with *ras* and *myc*. Most recent data indicate that some of these proto-oncogenes are also directly involved in the up-regulation of MMPs. Thus, gelatinases (MMP-2 and MMP-9) together with MT1-MMP appear to be regulated by growth factors that mediate their effects through the *ras* proto-oncogene, and c-Ras plays a critical role in the increased expression and proteolytic activation of MMPs in fibroblasts (191,192). Taken together, these data suggest that pathological expression of proto-oncogenes and upregulation of several signaling pathways constitute important steps leading to the overexpression of matrix-degrading enzymes in RA and consecutive destruction of joints.

Recently, alterations in tumor suppressor genes have become of interest in explaining some important features of cell activation and survival in RA. Thus, it was shown that aggressive RA-SFs lack the expression of mRNA for the novel tumor suppressor *PTEN*, which exhibits tyrosine phosphatase activity as well as extensive homology to the cytoskeletal proteins tensin and auxillin (193). *In situ* hybridization on RA synovium revealed a distinct expression pattern of *PTEN* with negligible staining in the lining layer, but abundant expression in the sublining. In contrast, normal synovial tissue exhibited homogenous staining for *PTEN* (194). Interestingly, only 40% of cultured RA-SFs expressed *PTEN*, and SCID-mouse experiments showed no

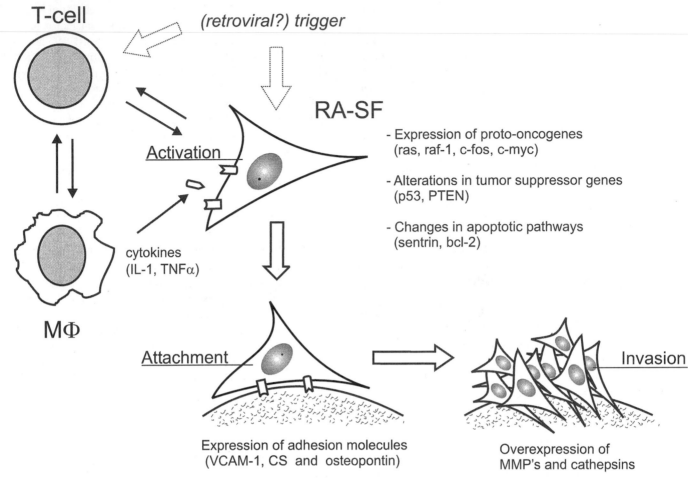

Figure 4.   Activated synovial fibroblasts contribute significantly to the disease process in RA. Early in disease, it comes to the activation by a hitherto unknown (retroviral?) trigger. The activated state of RA-SF can be seen from the upregulation of proto-oncogenes, alterations in apoptotic pathways as well as changes in the expression of tumor suppressors. Subsequently, RA-SF express a variety of adhesion molecules that mediate the attachment to the cartilage. Attachment to the cartilage is an important precondition for the invasion of these cells into the cartilage that is mediated by the concerted action of MMPs and cathepsins.

staining for *PTEN* in those cells aggressively invading the cartilage. Recently, Firestein and coworkers have described somatic mutations of the tumor suppressor gene, p53, in RA synoviocytes (195). Using a mismatch detection system, they demonstrated that tissue from the joints of severe chronic RA patients contained mutant p53 transcripts, which were not found in skin samples from the same patients or in joints of patients with osteoarthritis. In addition, mutant p53 transcripts were identified in synoviocytes cultured from rheumatoid joints. Intriguingly, the predicted amino acid substitutions in p53 were identical or similar to those commonly observed in a variety of tumors. By resulting in a non-functional "cell death suppressor" gene, such mutations, therefore, could contribute to an extended life span of activated cells aggressively destroying cartilage and bone. On the other hand, it is of interest that RA patients from another part of the world did not reveal p53 mutations (196).

Apoptosis in RA-SF.   It is well established that proliferation of synovial cells does not contribute significantly to the hyperplasia of the RA synovium. As shown by Aicher and coworkers, even activated synovial cells from patients with RA do not proliferate faster than synovial cells from osteoarthritis patients (92). Studies using ^{3}H-thymidine incorporation revealed a proliferation rate of only 1–5% (197), and mitosis is also rarely found in RA synovial tissue (198).

On the other hand, less than 1% of lining cells show morphological features of apoptosis (199,200), and recent data have provided evidence for the modulation of apoptotic pathways in the RA synovium. Interestingly, these changes occur predominantly in the lining layer and appear closely linked to the activated phenotype of RA-SF. As a possible mechanism that may extend the life span of RA-SFs, Franz and coworkers have demonstrated that the novel anti-apoptotic molecule, sentrin, (201) is strongly expressed in RA synovium (202). Sentrin has been shown to interact with the signal-competent forms of Fas/Apo-1 and TNF receptor-1 (TNFR-1) and thereby protects cells against anti-Fas-and TNF-induced cell death. Interestingly, sentrin is found mainly in RA synovial lining cells, whereas normal synovium cells express very little sentrin mRNA (202). Quantitative analysis of in cultured SFs revealed a more than 30-fold increase of sentrin mRNA expression compared to OA-SF and normal fibroblasts (202). In related studies, other anti-apoptotic molecules, such as bcl-2, have been found expressed in only few synovial lining cells (203). The aforementioned somatic mutations of p53 tumor suppressor gene in RA-SF, as well as NF-κB-mediated pathways, may also contribute significantly to the inhibition of apoptosis in these cells.

The concept of modulated apoptosis in synovial cells has been confounded by reports on the expression and function of the pro-apoptotic molecule Fas in synovial lining cells (199,204,205). However, some recent findings indicate that intracellular signaling pathways following Fas activation may be modified by additional stimuli that determine whether the affected cell undergoes apoptosis or proliferates (206). These findings could also explain the observation that cultured synovial fibroblasts are resistant to Fas-induced apoptosis despite the surface expression of Fas molecules (207). Taken together, these data suggest that apoptosis-suppressing signals outweigh pro-apoptotic signaling in RA, causing an imbalance of pro- and anti-apoptotic pathways. This imbalance may subsequently lead to an extended life-span of synovial lining cells and result in a prolonged expression of matrix-degrading enzymes at sites of joint destruction.

Attachment.   Attachment of synovial fibroblasts to the joint cartilage appears pivotal for RA compared with other non-destructive forms of arthritides. Not only has it been demonstrated that the adhesion of RA-SF to the cartilage matrix is essential for its subsequent degradation, but recent data indicate that the attachment stimulates the cells as part of the perpetuating principle in RA.

Although integrins have been found predominantly on leukocytes, a number of studies have demonstrated the expression of β1 integrins, such as VLA-3, VLA-4 and VLA-5, on synovial fibroblasts (208,209). Moreover, anti-β1 integrin antibodies inhibited, at least in part, the binding of SFs to extracellular matrix, and the blocking efficacy was significantly higher in RA-SFs compared with normal SFs (208). Since several integrins function as fibronectin receptors, it has been suggested that the fibronectin-rich environment of RA cartilage surface might facilitate adhesion of RA synovial fibroblasts to the cartilage. In this respect, it was shown that CS-1, a spliced isoform of fibronectin, is expressed highly in RA synovium (210). CS-1 appears part of a bi-directional adhesion pathway operative in RA as it binds to the integrin VLA-4 ($\alpha 4\beta 1$; CD49d/ CD29), which also ligates with vascular cell adhesion molecule 1 (VCAM-

1). However, integrins have become of interest for RA not only because of their function as receptor molecules, but also because of their interaction with several signaling pathways and cellular proto-oncogenes (211). Specifically, the expression of early cell cycle genes, such as *c-fos* and *c-myc*, is stimulated by integrin-mediated cell adhesion, and gene expression driven by the *fos* promoter shows strongly synergistic activation by integrin-mediated adhesion (212,213).

Apart from the integrins, VCAM-1 (CD106), a member of the immunoglobulin superfamily, has been characterized as an important adhesion molecule associated with the activated phenotype of RA-SF. Recent studies have also demonstrated increased expression of VCAM-1 in RA synovial fibroblasts particularly its upregulation in the subpopulation of activated lining layer RA-SFs (214,215,216). Studies in the SCID mouse co-implantation model revealed sustained up-regulation of VCAM-1 in RA-SFs even in the absence of human inflammatory cells for at least 60 days (179). Since no specific ligand for VCAM-1 has been found in the extracellular matrix so far, there is only circumstantial evidence for a role of this molecule in the attachment of activated RA-SFs to the cartilage. However, soluble VCAM-1 (sVCAM-1) exhibits chemotactic activity of T cells expressing high affinity VLA-4. Interestingly, the chemotactic activity of sVCAM-1 is inhibited in the presence of anti-VCAM-1 and anti-VLA-4 (217). These data suggest that VCAM-1 produced by RA-SF contributes to the development of the synovial inflammation not only by attracting T cells to the joint, but also by inhibition of the apoptosis of B lymphocytes (218). VCAM-1 may also contribute to T-cell anergy (219) and the induction of angiogenesis (133).

Recently, Petrow and coworkers demonstrated that osteopontin, another extracellular matrix protein that promotes cell attachment, is present in RA-SFs (220). In this study, a stimulatory effect of osteopontin on the secretion of MMP-1 in articular chondrocytes was found. Therefore, osteopontin may not only mediate the attachment of synovial cells to the extracellular matrix, but contribute to the perichondrocytic matrix degradation in RA. Moreover, earlier findings demonstrating that osteopontin stimulates B lymphocytes to produce immunoglobulins (221) as well as chemoattractants for macrophages (222), suggesting that osteopontin produced by SFs might also play an important role in stimulating B lymphocytes to produce rheumatoid factor (RF) in the joint and to attract the influx of macrophages to the rheumatoid synovium. The role of adhesion molecules in RA does not appear restricted to the attachment of synovium to cartilage and bone, but involves the recruitment and survival of T and B lymphocytes as well as the induction of MMPs.

Specific effects of fibroblasts on T-cells.   Recent data suggest that both accumulation and prolonged survival of

T cells within the rheumatoid synovium (223) can be promoted by synovial cells (224), particularly by fibroblast-derived factors (225). Salmon and coworkers showed that isolated synovial T cells rapidly undergo programmed cell death, whereas co-culture with synovial fibroblasts prevented apoptosis (224). McInnes and coworkers demonstrated that interleukin (IL)-15, which attracts CD4+ T cells and may protect T cells from apoptosis, is also produced by RA synovial fibroblasts (225). In addition, RA synovial fibroblasts have been identified as the major source for IL-16 within the synovium (226). IL-16 is not only capable of attracting CD4+ cells, but induces the expression of the IL-2 receptor on resting T cells as well as that of MHC class II molecules. IL-16 also has suppressive properties in that it is capable of inducing T cell anergy (227). Thus, the properties of IL-15 and -16 may help to explain the paradox between the abundance of T cells in the rheumatoid synovium and the

lack of T cell function in the sublining of RA synovial membrane. At this point, it could be speculated that T cell anergy could be a basis for autoimmunity (228), and that the cells attracted to the synovium by RA-SFs and made anergic by these cells have lost their tolerance to "self". Therefore, degraded matrix components of the cartilage might stimulate these T cells to recognize such peptides as autoantigens. As a consequence of these observations, it can be hypothesized that resolution of synovial inflammation may be promoted most effectively by targeting the fibroblast population rather than T cells themselves (224).

5.3. Macrophage-derived Cytokines

Although several lines of evidence indicate that RA-SF maintain their aggressive behavior in the absence of inflammatory cells, it must be stressed that macrophages in

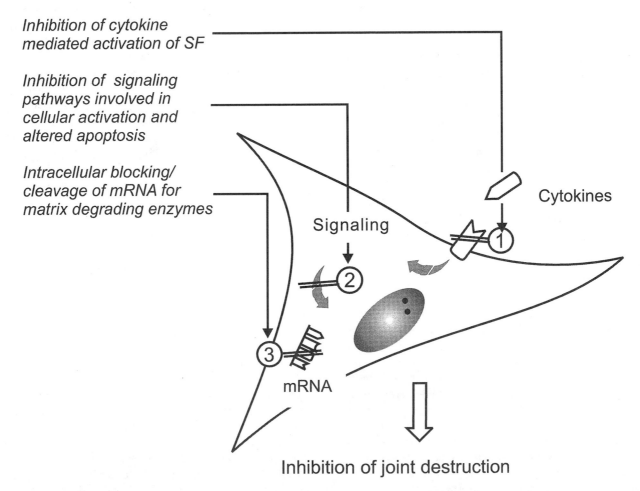

Figure 5. Novel approaches to interfere with the joint destruction in RA. As much as the RA-SFs are concerned, three main approaches are being followed at the moment. Inhibition of pro-inflammatory cytokines (1) may reduce the activation of RA-SFs and thus decrease their invasiveness. Interfering with signaling cascades that are involved in the upregulation of intracellular pathways of activation (2) constitutes a second way to alter the aggressive behavior of these cells. Third, the intracellular inhibition of mRNA for matrix degrading enzymes (3) through the delivery of ribozymes may also reduce joint destruction in RA.

particular are capable of stimulating these cells. Among the cytokines demonstrated to enhance the disease process in RA, IL-1 and TNFα are of particular interest (229). Both cytokines are produced predominantly by macrophage-like synovial lining cells(119), and are capable of inducing a variety of other cytokines, chemokines and prostaglandins. IL-1 and TNFα stimulate the production of matrix degrading enzymes, such as MMPs. In addition, IL-1 also suppresses the synthesis of proteoglycans (PG) (230). In some animal studies, anti-IL-1 treatment was able to normalize chondrocyte synthetic function and reduce the activation of MMPs (231–234). These data correlate with the observations of Müller-Ladner and coworkers, who showed that overexpression of the IL-1 receptor antagonist (IL-1ra) through retroviral gene transfer significantly reduce perichondrocytic matrix degradation in the SCID mouse model (235). Although TNFα has also been demonstrated to enhance the production of MMPs in some experimental systems, recent data point to differences in the importance of TNFα and IL-1. While TNFα appears to be responsible primarily for the extent of the synovitis, IL-1 seems to have a greater impact on the destruction of cartilage. This hypothesis is supported by data from antigen-induced arthritis and zymosan-induced arthritis in mice where the suppression of PG synthesis seen by IL-1 was not detected with TNFα (230). Related studies using anti-TNFα treatment in DBA/1 mice with collagen-induced arthritis (CIA) demonstrated efficacy only shortly after onset of the disease, but had little effect on fully established CIA (231). In contrast, anti-IL-1 α/β treatment ameliorated both early and fully established CIA. This clear suppression of established arthritis was confirmed by administration of high doses of IL-1Ra(236–238). Recent data from the clinical use of specific inhibitors of TNFα and IL-1 appear to be in line with these observations (239).

6. MATRIX DEGRADATION IN RA

Rheumatoid joint destruction is mediated by a concerted action of various proteinases, the most prominent being MMPs (240) and cathepsins (241).

The MMP family contains more than 20 structurally-related members, all of which are characterized by a zinc molecule at the active site. They include collagenase (MMP-1), gelatinases (MMP-2 and MMP-9), and stromelysin (MMP-3). Collagenase 3 (MMP-13) is a recently-described member of the MMP family cloned from mammary carcinoma tissue and, subsequently, from osteoarthritic and rheumatoid synovial tissue (242). Membrane-type MMPs (MT-MMP), characterized by a transmembrane domain and acting on the surface of cells, also belong to the MMP family. MMPs are secreted as inactive pro-enzymes and activated proteolytically by various enzymes such as trypsin,

plasmin, and other proteases. Although there is a broad overlap, MMPs differ with respect to their substrate specificities. While MMP-1 degrades collagen types I, II, III, VII, and X only when they are arranged in a triple helical structure, MMP-2 can also cleave denatured collagen. Interestingly, some MMPs may also cleave other members of this protease family, thus being involved in their activation. Specifically, MMP-3 is able to activate MMP-1 and MMP-9 as well as to degrade proteoglycans (243–245). Several reports have implicated MMPs in rheumatoid joint destruction (246–250). MMP-1 and MMP-3 have been found to be elevated in synovial fluid of patients with RA and are released in large amounts by synovial fibroblast-like cells in culture (251). *In-situ* studies revealed strong MMP-1 and MMP-3 expression within rheumatoid synovium, both at the mRNA and protein level. Using *in-situ* hybridization techniques, other MMPs, such as MMP-2, have been localized to the RA synovial membrane (252). Interestingly, synovial fibroblast-like cells within the lining layer or at the site of cartilage invasion have been identified as the major source of MMPs. This pattern of distribution is similar to that for proto-oncogenes and consistent with the notion that proto-oncogenes are involved in the activation of MMP genes. MMP-13 is also expressed at the mRNA and protein level, especially in the lining layer of rheumatoid synovium (250). Due to this localization, its substrate specificity for collagen type II and its relative resistance to known MMP inhibitors, MMP-13 might play an important role in joint destruction. Most recently, it was also shown that the expression levels of MMP-13 correlated with elevated levels of systemic inflammation markers (253).

MT-MMPs are also abundantly expressed in cells aggressively destroying cartilage and bone in RA (254). This is of particular importance because MT1-MMP degrades extracellular matrix components as well as activates other disease-relevant MMPs, such as MMP-2 and MMP-13 (255). A recent study comparing the expression of MT-MMPs in RA suggested that MT1-MMP is of particular importance in RA. MT1-MMP was shown to be expressed in fibroblasts, macrophages and also in osteoclast-like cells at sites of bone resorption (254). In this analysis, the expression of MT3-MMP mRNA was seen in fibroblasts and some macrophages, particularly in the lining layer. In contrast, expression of MT2- and MT4-MMP was characterized by a scattered staining of only few CD68 negative fibroblasts, and no differences could be found between the lining and the sublining. As concluded from this study, MT1-MMP and potentially MT3-MMP contribute to matrix degradation, while MT2- and MT4-MMP appear not to be involved in rheumatoid joint destruction.

MMP activity is normally balanced by the naturally-occurring tissue inhibitors of metalloproteinase (TIMPs). They interact irreversibly with MMPs such as MMP-1 and MMP-3 and are synthesized and secreted by chondrocytes, synovial fibroblasts and endothelial cells (247,251,256). *In-situ* hybridization studies demonstrated striking amounts of TIMP-1 mRNA in the synovial lining of patients with RA. However, the molar ratio of MMPs to TIMP rather than the absolute levels of TIMP are crucial for joint destruction. In RA, the amount of MMPs produced far outweighs that of the TIMPs, allowing destruction to take place (257).

Cathepsins are the other major group of proteases involved in joint destruction (241). They are classified by their catalytic mechanism and cleave cartilage types II, IX, and XI as well as proteoglycans. The cysteine proteases cathepsin B and L are up-regulated in RA synovium, especially at sites of cartilage invasion (258). Similar to MMPs, cathepsins are activated by proto-oncogenes. Transfection of fibroblasts with the *ras* proto-oncogene leads to cellular transformation and the induction of cathepsin L (259), as corroborated by the finding of combined *ras* and cathepsin L expression (190). Several studies have also shown that proinflammatory cytokines, such as IL-1 and TNFα, can stimulate the production of cathepsins B and L by synovial fibroblast-like cells (260,261). Cathepsin K expression by RA synovial fibroblasts and macrophages has been reported, especially at the site of synovial invasion into articular bone, suggesting that it participates in bone destruction in RA (262).

7.   NOVEL THERAPEUTIC STRATEGIES FOR RA

Thus far, the treatment of RA has been based upon three distinct strategies that comprised the use of non-steroidal anti-inflammatory drugs (NSAIDs), disease-modifying anti-rheumatics (DMARDs) and steroids. Among these, NSAIDS have been used most widely for the treatment of RA, but they provide only symptomatic relief and do not influence the progression of disease. In addition, the long-term use of NSAIDs may cause severe side effects, particularly when high daily doses are needed. Therefore, the majority of patients are treated with a combination of NSAIDs and DMARDs, often requiring the addition of steroids. Recently, a more aggressive use of DMARDs has been favored, including DMARD treatment early in the course of disease as well as the combination of therapies. Using such schemes, a slowdown in disease progression has been observed in some studies (263–265). However, DMARD treatment still fails to achieve satisfactory results in a number of patients. Based on our increasing knowledge about the pathogenesis of RA, several approaches have been developed that specifically target key disease processes.

It is known that the enzyme cyclooxygenase COX-2 is responsible primarily for pain, while COX-1 mediates some important side effects of NSAIDs, such as gastric ulcers. Therefore, the development of novel NSAIDs that specifically inhibit cycloxygenase (COX)-2 has been a major goal of the last years. At the moment, several such novel NSAIDs are being evaluated and the first results appear promising. It has been shown in various studies that COX-2 inhibitors have similar "pain-killing" properties as conventional NSAIDs yet without many of the side effects (266).

Considerable efforts have also been made to develop novel, more specific DMARDs. Particularly promising data have come from leflunomide, a novel DMARD that primarily affects T cells. Recent studies suggest that leflunomide may interfere with the metabolism of pyrimidine nucleotides targeting the enzyme dihydroortate dehydrogenase (DHODH), and it has been proposed that the immunomodulating activity may be related to the inhibition of UMP synthesis in proliferating lymphocytes (267). Clinical studies have demonstrated a significant effect when compared to placebo and other DMARDs (268).

The development of new "biological" agents that interfere with cytokine-mediated inflammatory pathways and activation of synovial cells has been of particular interest over the last years. Specifically, the use of monoclonal antibodies against TNFα and of TNF receptor fusion proteins have provided convincing results. Several studies have demonstrated that inhibition of TNFα inhibits the clinical features of inflammation in the majority of patients (236,269–271). However, the question of whether a significant inhibition of joint destruction can be achieved by these agents has not been established.

Therefore, alternative approaches, including the development of inhibitors for MMPs as well as the use of gene transfer approaches, focus on the direct inhibition of joint destruction in RA (272,273).

8.   CONCLUSIONS

Although substantial progress has been made in elucidating key mechanisms in the pathogenesis of RA, there is still no answer as to what initiates and drives the disease. Specifically, the question of whether autoimmunity or cellular activation represent the earliest pathological events in RA remains elusive. At the moment, it appears that both T cell-dependent and independent pathways contribute significantly to the pathogenesis of RA. While inflammation and inflammation-driven processes appear to be dominated by T cells, non-T cell pathways most likely are of major importance in mediating joint destruction, which represents a unique feature of RA. Intriguingly, there is evidence that inflammation and articular damage are dissociated to a significant extent. In the context of T cell-independent pathways, activated, transformed-appearing RA-SF have been shown to be of particular importance.

Over the last years, these cells have been studied intensively and it has been shown that their specific properties not only cause their aggressive behavior, but may also explain some other aspects of RA pathology. However, no single, specific marker for this phenotype of RA-SF, which is found predominantly in the lining layer of the rheumatoid synovium, has been identified. Rather, it appears that activated RA-SFs are characterized by a set of specific properties that involve alterations in proto-oncogenes, tumor suppressor and adhesion molecules. Therefore, novel therapeutic strategies have to be more specific and also target this particular phenotype of cell.

References

1. Gay, S., R.E. Gay, and W.J. Koopman. 1993. Molecular and cellular mechanisms of joint destruction in rheumatoid arthritis: two cellular mechanisms explain joint destruction? *Ann. Rheum. Dis.* 52 Suppl 1:S39–47.
2. Pincus, T. 1995. The underestimated long term medical and economic consequences of rheumatoid arthritis. *Drugs* 50 Suppl 1:1–14.
3. McIntosh, E. 1996. The cost of rheumatoid arthritis. *Br. J. Rheumatol.* 35:781–790.
4. Yelin, E. 1996. The costs of rheumatoid arthritis: absolute, incremental, and marginal estimates. *J. Rheumatol. Suppl.* 44:47–51.
5. Spector, T.D. 1990. Rheumatoid arthritis. *Rheum. Dis. Clin. North Am.* 16:513–537.
6. Beasley, R.P., P.H. Bennett, and C.C. Lin. 1983. Low prevalence of rheumatoid arthritis in Chinese. Prevalence survey in a rural community. *J. Rheumatol. Suppl.* 10:11–15.
7. Moolenburgh, J.D., S. Moore, H.A. Valkenburg, and M.G. Erasmus. 1984. Rheumatoid arthritis in Lesotho. *Ann. Rheum. Dis.* 43:40–43.
8. Silman, A. 1993. Rheumatoid arthritis. In: *Epidemiology of the rheumatic diseases.* A. Silman and M.C. Hochberg, editors. Oxford University Press, Oxford. pp. 7–68.
9. Gabriel, S.E., C.S. Crowson, and W.M. O'Fallon. 1999. The epidemiology of rheumatoid arthritis in Rochester, Minnesota, 1955–1985. *Arthritis Rheum.* 42:415–420.
10. Brennan, P., and A. Silman. 1995. Why the gender difference in susceptibility to rheumatoid arthritis? *Ann. Rheum. Dis.* 54:694–695.
11. Goemaere, S., C. Ackerman, K. Goethals, F. de Keyser, C. Van der Straeten, G. Verbruggen, H. Mielants, and E.M. Veys. 1990. Onset of symptoms of rheumatoid arthritis in relation to age, sex and menopausal transition. *J. Rheumatol.* 17:1620–1622.
12. Wilder, R.L. 1998. Hormones, pregnancy, and autoimmune diseases. *Ann. NY Acad. Sci.* 840:45–50.
13. Brun, J.G., S. Nilssen, and G. Kvale. 1995. Breast feeding, other reproductive factors and rheumatoid arthritis. A prospective study. *Br. J. Rheumatol.* 34:542–546.
14. Rosenberg, A.M. 1996. Evaluation of associations between breast feeding and subsequent development of juvenile rheumatoid arthritis. *J. Rheumatol.* 23:1080–1082.
15. Jorgensen, C., M.C. Picot, C. Bologna, and J. Sany. 1996. Oral contraception, parity, breast feeding, and severity of rheumatoid arthritis. *Ann. Rheum. Dis.* 55:94–98.
16. Brennan, P., and A. Silman. 1994. Breast-feeding and the onset of rheumatoid arthritis. *Arthritis Rheum.* 37:808–813.
17. Neidhart, M. 1998. Prolactin in autoimmune diseases. *Proc. Soc. Exp. Biol. Med.* 217:408–419.
18. Jorgensen, C., N. Bressot, C. Bologna, and J. Sany. 1995. Dysregulation of the hypothalamo-pituitary axis in rheumatoid arthritis. *J. Rheumatol.* 22:1829–1833.
19. Mateo, L., J.M. Nolla, M.R. Bonnin, M.A. Navarro, and D. Roig-Escofet. 1998. High serum prolactin levels in men with rheumatoid arthritis. *J. Rheumatol.* 25:2077–2082.
20. Chikanza, I.C., P. Petrou, G. Chrousos, G. Kingsley, and G.S. Panayi. 1993. Excessive and dysregulated secretion of prolactin in rheumatoid arthritis: immunopathogenetic and therapeutic implications. *Br. J. Rheumatol.* 32:445–448.
21. Brennan, P., B. Ollier, J. Worthington, A. Hajeer, and A. Silman. 1996. Are both genetic and reproductive associations with rheumatoid arthritis linked to prolactin? *Lancet* 348:106–109.
22. Neidhart, M., R.E. Gay, and S. Gay. 1999. Prolactin and prolactin-like polypeptides in rheumatoid arthritis. *Biomed. Pharmacother.* 53:218–222.
23. Berczi, I., E. Nagy, S.L. Asa, and K. Kovacs. 1984. The influence of pituitary hormones on adjuvant arthritis. *Arthritis Rheum.* 27:682–688.
24. Bensen, W.G., W. Bensen, J.D. Adachi, and P.X. Tugwell. 1990. Remodelling the pyramid: the therapeutic target of rheumatoid arthritis. *J. Rheumatol.* 17:987–989.
25. Watts, R.A., D.M. Carruthers, and D.G. Scott. 1995. Isolated nail fold vasculitis in rheumatoid arthritis. *Ann. Rheum. Dis.* 54:927–929.
26. Vollertsen, R.S., and D.L. Conn. 1990. Vasculitis associated with rheumatoid arthritis. *Rheum. Dis. Clin. North Am.* 16:445–461.
27. Geirsson, A.J., G. Sturfelt, and L. Truedsson. 1987. Clinical and serological features of severe vasculitis in rheumatoid arthritis: prognostic implications. *Ann. Rheum. Dis.* 46:727–733.
28. Panush, R.S., P. Katz, S. Longley, R. Carter, J. Love, and H. Stanley. 1983. Rheumatoid vasculitis: diagnostic and therapeutic decisions. *Clin. Rheumatol.* 2:321–330.
29. Silman, A.J., E. Hennessy, and B. Ollier. 1992. Incidence of rheumatoid arthritis in a genetically predisposed population. *Br. J. Rheumatol.* 31:365–368.
30. Wolfe, F., S.M. Kleinheksel, and M.A. Khan. 1988. Familial vs sporadic rheumatoid arthritis: a comparison of the demographic and clinical characteristics of 956 patients. *J. Rheumatol.* 15:400–404.
31. Deighton, C.M., D.J. Walker, I.D. Griffiths, and D.F. Roberts. 1989. The contribution of HLA to rheumatoid arthritis. *Clin. Genet.* 36:178–182.
32. Seldin, M.F., C.I. Amos, R. Ward, and P.K. Gregersen. 1999. The genetics revolution and the assault on rheumatoid arthritis. *Arthritis Rheum.* 42:1071–1079.
33. Aho, K., M. Koskenvuo, J. Tuominen, and J. Kaprio. 1986. Occurrence of rheumatoid arthritis in a nationwide series of twins. *J. Rheumatol.* 13:899–902.
34. Gregersen, P.K. 1993. Discordance for autoimmunity in monozygotic twins. Are identical twins really identical? *Arthritis Rheum.* 36:1185–1192.
35. Järvinen, P., and K. Aho. 1994. Twin studies in rheumatic diseases. *Semin. Arthritis Rheum.* 24:19–28.
36. Stastny, P. 1978. Association of the B-cell alloantigen DRw4 with rheumatoid arthritis. *N. Engl. J. Med.* 298:869–871.

37. Reveille, J.D. 1998. The genetic contribution to the pathogenesis of rheumatoid arthritis. *Curr. Opin. Rheumatol.* 10:187–200.

38. Jawaheer, D., W. Thomson, A.J. MacGregor, D. Carthy, J. Davidson, P.A. Dyer, A.J. Silman, and W.E. Ollier. 1994. Homozygosity for the HLA-DR shared epitope contributes the highest risk for rheumatoid arthritis concordance in identical twins. *Arthritis Rheum.* 37:681–686.

39. Gregersen, P.K., J. Silver, and R.J. Winchester. 1987. The shared epitope hypothesis. An approach to understanding the molecular genetics of susceptibility to rheumatoid arthritis. *Arthritis Rheum.* 30:1205–1213.

40. Merryman, P.F., R.M. Crapper, S. Lee, P.K. Gregersen, and R.J. Winchester. 1989. Class II major histocompatibility complex gene sequences in rheumatoid arthritis. The third diversity regions of both DR beta 1 genes in two DR1, DRw10-positive individuals specify the same inferred amino acid sequence as the DR beta 1 and DR beta 2 genes of a DR4 (Dw14) haplotype. *Arthritis Rheum.* 32:251–258.

41. Albani, S., and J. Roudier. 1992. Molecular basis for the association between HLA DR4 and rheumatoid arthritis. From the shared epitope hypothesis to a peptidic model of rheumatoid arthritis. *Clin. Biochem.* 25:209–212.

42. Auger, I., E. Toussirot, and J. Roudier. 1997. Molecular mechanisms involved in the association of HLA-DR4 and rheumatoid arthritis. *Immunol. Res.* 16:121–126.

43. Weyand, C.M., K.C. Hicok, D.L. Conn, and J.J. Goronzy. 1992. The influence of HLA-DRB1 genes on disease severity in rheumatoid arthritis. *Ann. Intern. Med.* 117:801–806.

44. Benazet, J.F., D. Reviron, P. Mercier, H. Roux, and J. Roudier. 1995. HLA-DRB1 alleles associated with rheumatoid arthritis in southern France. Absence of extra-articular disease despite expression of the shared epitope. *J. Rheumatol.* 22:607–610.

45. Ploski, R., O. Vinje, K.S. Ronningen, A. Spurkland, D. Sorskaar, F. Vartdal, and O. Forre. 1993. HLA class II alleles and heterogeneity of juvenile rheumatoid arthritis. DRB1*0101 may define a novel subset of the disease. *Arthritis Rheum.* 36:465–472.

46. Toussirot, E., P. Tiberghien, J.C. Balblanc, P. Kremer, J. Despaux, J.L. Dupond, and D. Wendling. 1998. HLA DRB1* alleles in rheumatoid nodulosis: a comparative study with rheumatoid arthritis with and without nodules. *Rheumatol. Int.* 17:233–236.

47. Meyer, J.M., T.I. Evans, R.E. Small, T.W. Redford, J. Han, R. Singh, and G. Moxley. 1999. HLA-DRB1 genotype influences risk for and severity of rheumatoid arthritis. *J. Rheumatol.* 26:1024–1034.

48. Penzotti, J.E., D. Doherty, T.P. Lybrand, and G.T. Nepom. 1996. A structural model for TCR recognition of the HLA class II shared epitope sequence implicated in susceptibility to rheumatoid arthritis. *J. Autoimmun.* 9:287–293.

49. Penzotti, J.E., G.T. Nepom, and T.P. Lybrand. 1997. Use of T cell receptor/HLA-DRB1*04 molecular modeling to predict site-specific interactions for the DR shared epitope associated with rheumatoid arthritis. *Arthritis Rheum.* 40:1316–1326.

50. Roitt, I.M., M. Corbett, H. Festenstein, D. Jaraquemada, C. Papasteriadis, F.C. Hay, and L.J. Nineham. 1978. HLA-DRW4 and prognosis in rheumatoid arthritis. *Lancet* 1:990–990.

51. van Zeben, D., J.M. Hazes, A.H. Zwinderman, A. Cats, G.M. Schreuder, J. D'Amaro, and F.C. Breedveld. 1991. Association of HLA-DR4 with a more progressive disease course in patients with rheumatoid arthritis. Results of a followup study. *Arthritis Rheum.* 34:822–830.

52. Weyand, C.M., C. Xie, and J.J. Goronzy. 1992. Homozygosity for the HLA-DRB1 allele selects for extraarticular manifestations in rheumatoid arthritis. *J. Clin. Invest.* 89:2033–2039.

53. Marlow, A., S. John, A. Hajeer, W.E. Ollier, A.J. Silman, and J. Worthington. 1997. The sensitivity of different analytical methods to detect disease susceptibility genes in rheumatoid arthritis sibling pair families. *J. Rheumatol.* 24:208–211.

54. Leslie, R.D., and M. Hawa. 1994. Twin studies in autoimmune disease. *Acta. Genet. Med. Gemellol. (Roma))* 43:71–81.

55. Brennan, P., A. Hajeer, K.R. Ong, J. Worthington, S. John, W. Thomson, A. Silman, and B. Ollier. 1997. Allelic markers close to prolactin are associated with HLA-DRB1 susceptibility alleles among women with rheumatoid arthritis and systemic lupus erythematosus. *Arthritis Rheum.* 40:1383–1386.

56. Peterson, A.C., A. Di Rienzo, A.E. Lehesjoki, A. de la Chapelle, M. Slatkin, and N.B. Freimer. 1995. The distribution of linkage disequilibrium over anonymous genome regions. *Hum. Mol. Genet.* 4:887–894.

57. Udalova, I.A., S.A. Nedospasov, G.C. Webb, D.D. Chaplin, and R.L. Turetskaya. 1993. Highly informative typing of the human TNF locus using six adjacent polymorphic markers. *Genomics* 16:180–186.

58. Wilson, A.G., N. de Vries, F. Pociot, F.S. di Giovine, L.B. van der Putte, and G.W. Duff. 1993. An allelic polymorphism within the human tumor necrosis factor alpha promoter region is strongly associated with HLA A1, B8, and DR3 alleles. *J. Exp. Med.* 177:557–560.

59. Messer, G., U. Spengler, M.C. Jung, G. Honold, K. Blömer, G.R. Pape, G. Riethmüller, and E.H. Weiss. 1991. Polymorphic structure of the tumor necrosis factor (TNF) locus: an NcoI polymorphism in the first intron of the human TNF-beta gene correlates with a variant amino acid in position 26 and a reduced level of TNF-beta production. *J. Exp. Med.* 173:209–219.

60. Mulcahy, B., F. Waldron-Lynch, M.F. McDermott, C. Adams, C.I. Amos, D.K. Zhu, R.H. Ward, D.O. Clegg, F. Shanahan, M.G. Molloy, and F. O'Gara. 1996. Genetic variability in the tumor necrosis factor-lymphotoxin region influences susceptibility to rheumatoid arthritis. *Am. J. Hum. Genet.* 59:676–683.

61. Brinkman, B.M., T.W. Huizinga, F.C. Breedveld, and C.L. Verweij. 1996. Allele-specific quantification of TNFA transcripts in rheumatoid arthritis. *Hum. Genet.* 97:813–818.

62. Cornélis, F., L. Hardwick, R.M. Flipo, M. Martinez, S. Lasbleiz, J.F. Prud'Homme, T.H. Tran, S. Walsh, A. Delaye, A. Nicod, M.N. Loste, V. Lepage, K. Gibson, K. Pile, S. Djoulah, P.M. Danzé, F. Lioté, D. Charron, J. Weissenbach, D. Kuntz, T. Bardin, and B.P. Wordsworth. 1997. Association of rheumatoid arthritis with an amino acid allelic variation of the T cell receptor. *Arthritis Rheum.* 40:1387–1390.

63. Ibberson, M., and A. So. 1997. Interaction between genes in the T cell receptor-alpha locus and HLA-DR4 in rheumatoid arthritis susceptibility. *J. Rheumatol.* 24:223–223.

64. de Vries, N., C.F. Prinsen, E.B. Mensink, P.L. van Riel, M.A. van't Hof, and L.B. van de Putte. 1993. A T cell receptor beta chain variable region polymorphism associated with radiographic progression in rheumatoid arthritis. *Ann. Rheum. Dis.* 52:327–331.

65. Malhotra, U., and P. Concannon. 1992. T cell receptor beta gene polymorphism and rheumatoid arthritis. *Autoimmunity* 12:75–77.

66. Ibberson, M., V. Peclat, P.A. Guerne, J.M. Tiercy, P. Wordsworth, J. Lanchbury, J. Camilleri, and A.K. So. 1998. Analysis of T cell receptor V alpha polymorphisms in rheumatoid arthritis. *Ann. Rheum. Dis.* 57:49–51.

67. Pap, T., R.E. Gay, and S. Gay. 1998. Do antivirals have any utility in the treatment of arthritis? *Exp. Opin. Invest. Drugs* 7:

68. Billings, P.B., S.O. Hoch, P.J. White, D.A. Carson, and J.H. Vaughan. 1983. Antibodies to the Epstein-Barr virus nuclear antigen and to rheumatoid arthritis nuclear antigen identify the same polypeptide. *Proc. Natl. Acad. Sci. USA* 80:7104–7108.

69. Youinou, P., M. Buisson, J.M. Berthelot, C. Jamin, P. Le Goff, O. Genoulaz, A. Lamour, P.M. Lydyard, and J.M. Seigneurin. 1992. Anti-Epstein-Barr virus-nuclear antigen-1, -2A and -2B antibodies in rheumatoid arthritis patients and their relatives. *Autoimmunity* 13:225–231.

70. Kouri, T., J. Petersen, G. Rhodes, K. Aho, T. Palosuo, M. Heliövaara, H. Isomäki, H. von Essen, and J.H. Vaughan. 1990. Antibodies to synthetic peptides from Epstein-Barr nuclear antigen-1 in sera of patients with early rheumatoid arthritis and in preillness sera. *J. Rheumatol.* 17:1442–1449.

71. Rumpold, H., G.H. Rhodes, P.L. Bloch, D.A. Carson, and J.H. Vaughan. 1987. The glycine-alanine repeating region is the major epitope of the Epstein-Barr nuclear antigen 1 (EBNA-1). *J. Immunol.* 138:593–599.

72. Buisson, M., J.M. Berthelot, P. Le Goff, C. Chastel, A. Lamour, J.M. Seigneurin, and P. Youinou 1994. Lack of relationship between the Epstein-Barr virus and the anti-perinuclear factor/'perinuclear antigen' system in rheumatoid arthritis. *J. Autoimmun.* 7:485–495.

73. Roudier, J., J. Petersen, G.H. Rhodes, J. Luka, and D.A. Carson. 1989. Susceptibility to rheumatoid arthritis maps to a T-cell epitope shared by the HLA-Dw4 DR beta-1 chain and the Epstein-Barr virus glycoprotein gp110. *Proc. Natl. Acad. Sci. USA* 86:5104–5108.

74. Saal, J.G., M. Krimmel, M. Steidle, F. Gerneth, S. Wagner, P. Fritz, S. Koch, J. Zacher, S. Sell, H. Einsele, and C.A. Muller. 1999. Synovial Epstein-Barr virus infection increases the risk of rheumatoid arthritis in individuals with the shared HLA-DR4 epitope. *Arthritis Rheum.* 42:1485–1496.

75. Edinger, J.W., M. Bonneville, E. Scotet, E. Houssaint, H.R. Schumacher, and D.N. Posnett. 1999. EBV gene expression not altered in rheumatoid synovia despite the presence of EBV antigen-specific T cell clones. *J. Immunol.* 162:3694–3701.

76. Saal, J.G., M. Steidle, H. Einsele, C.A. Mller, P. Fritz, and J. Zacher. 1992. Persistence of B19 parvovirus in synovial membranes of patients with rheumatoid arthritis. *Rheumatol. Int.* 12:147–151.

77. Mimori, A., Y. Misaki, T. Hachiya, K. Ito, and S. Kano. 1994. Prevalence of antihuman parvovirus B19 IgG antibodies in patients with refractory rheumatoid arthritis and polyarticular juvenile rheumatoid arthritis. *Rheumatol. Int.* 14:87–90.

78. Takahashi, Y., C. Murai, T. Ishii, K. Sugamura, and T. Sasaki. 1998. Human parvovirus B19 in rheumatoid arthritis. *Int. Rev. Immunol.* 17:309–321.

79. Lunardi, C., M. Tiso, L. Borgato, L. Nanni, R. Millo, G. De Sandre, A.B. Severi, and A. Puccetti. 1998. Chronic par-

vovirus B19 infection induces the production of anti-virus antibodies with autoantigen binding properties. *Eur. J. Immunol.* 28:936–948.

80. Takahashi, Y., C. Murai, S. Shibata, Y. Munakata, T. Ishii, K. Ishii, T. Saitoh, T. Sawai, K. Sugamura, and T. Sasaki. 1998. Human parvovirus B19 as a causative agent for rheumatoid arthritis. *Proc. Natl. Acad. Sci. USA* 95:8227–8232.

81. Nikkari, S., A. Roivainen, P. Hannonen, T. Möttönen, R. Luukkainen, T. Yli-Jama, and P. Toivanen. 1995. Persistence of parvovirus B19 in synovial fluid and bone marrow. *Ann. Rheum. Dis.* 54:597–600.

82. Nikkari, S., R. Luukkainen, T. Möttönen, O. Meurman, P. Hannonen, M. Skurnik, and P. Toivanen. 1994. Does parvovirus B19 have a role in rheumatoid arthritis? *Ann. Rheum. Dis.* 53:106–111.

83. Kerr, J.R., J.P. Cartron, M.D. Curran, J.E. Moore, J.R. Elliott, and R.A. Mollan. 1995. A study of the role of parvovirus B19 in rheumatoid arthritis. *Br. J. Rheumatol.* 34:809–813.

84. Kalden, J.R., and S. Gay. 1994. Retroviruses and autoimmune rheumatic diseases. *Clin. Exp. Immunol.* 98:1–5.

85. Motokawa, S., T. Hasunuma, K. Tajima, A.M. Krieg, S. Ito, K. Iwasaki, and K. Nishioka. 1996. High prevalence of arthropathy in HTLV-I carriers on a Japanese island. *Ann. Rheum. Dis.* 55:193–195.

86. Nishioka, K., I. Maruyama, K. Sato, I. Kitajima, Y. Nakajima, and M. Osame. 1989. Chronic inflammatory arthropathy associated with HTLV-I. *Lancet* 1:441–441.

87. Sato, K., I. Maruyama, Y. Maruyama, I. Kitajima, Y. Nakajima, M. Higaki, K. Yamamoto, N. Miyasaka, M. Osame, and K. Nishioka. 1991. Arthritis in patients infected with human T lymphotropic virus type I. Clinical and immunopathologic features. *Arthritis Rheum.* 34:714–721.

88. Iwakura, Y., M. Tosu, E. Yoshida, M. Takiguchi, K. Sato, I. Kitajima, K. Nishioka, K. Yamamoto, T. Takeda, and M. Hatanaka. 1991. Induction of inflammatory arthropathy resembling rheumatoid arthritis in mice transgenic for HTLV-I. *Science* 253:1026–1028.

89. Fujisawa, K., K. Okamoto, H. Asahara, T. Hasunuma, T. Kobata, T. Kato, T. Sumida, and K. Nishioka. 1998. Evidence for autoantigens of Env/Tax proteins in human T cell leukemia virus type I Env-pX transgenic mice. *Arthritis Rheum.* 41:101–109.

90. Aono, H., K. Fujisawa, T. Hasunuma, S.J. Marriott, and K. Nishioka. 1998. Extracellular human T cell leukemia virus type I tax protein stimulates the proliferation of human synovial cells. *Arthritis Rheum.* 41:1995–2003.

91. Wright, J.J., K.C. Gunter, H. Mitsuya, S.G. Irving, K. Kelly, and U. Siebenlist. 1990. Expression of a zinc finger gene in HTLV-I- and HTLV-II-transformed cells. *Science* 248:588–591.

92. Aicher, W.K., A.H. Heer, A. Trabandt, S.L. Bridges, Jr., H.W. Schroeder, Jr., G. Stransky, R.E. Gay, H. Eibel, H.H. Peter, U. Siebenlist, and et al. 1994. Overexpression of zinc-finger transcription factor Z-225/Egr-1 in synoviocytes from rheumatoid arthritis patients. *J. Immunol.* 152:5940–5948.

93. Gay, S., and R.E. Gay. 1989. Cellular basis and oncogene expression of rheumatoid joint destruction. *Rheumatol. Int.* 9:105–113.

94. Müller-Ladner, U., J. Kriegsmann, R.E. Gay, and S. Gay. 1995. Oncoges in rheumatoid synovium. *Rheum. Dis. Clin. North Am.* 21:675–690.

95. Trabandt, A., R.E. Gay, and S. Gay. 1992. Oncogene activation in rheumatoid synovium. *APMIS* 100:861–875.

96. Trabandt, A., W.K. Aicher, R.E. Gay, V.P. Sukhatme, H.G. Fassbender, and S. Gay. 1992. Spontaneous expression of immediately-early response genes c-fos and egr-1 in collagenase-producing rheumatoid synovial fibroblasts. *Rheumatol. Int.* 12:53–59.

97. Dooley, S., I. Herlitzka, R. Hanselmann, A. Ermis, W. Henn, K. Remberger, T. Hopf, and C. Welter. 1996. Constitutive expression of c-fos and c-jun, overexpression of ets-2, and reduced expression of metastasis suppressor gene nm23- H1 in rheumatoid arthritis. *Ann. Rheum. Dis.* 55:298–304.

98. Fujii, M., P. Sassone-Corsi, and I.M. Verma. 1988. c-fos promoter trans-activation by the tax1 protein of human T cell leukemia virus type I. *Proc. Natl. Acad. Sci. USA* 85:8526–8530.

99. Fujii, M., T. Niki, T. Mori, T. Matsuda, M. Matsui, N. Nomura, and M. Seiki. 1991. HTLV-1 Tax induces expression of various immediate early serum responsive genes. *Oncogene* 6:1023–1029.

100. Ishidoh, K., S. Taniguchi, and E. Kominami. 1997. Egr family member proteins are involved in the activation of the cathepsin L gene in v-src-transformed cells. *Biochem. Biophys. Res. Commun.* 238:665–669.

101. Bailer, R.T., A. Lazo, V. Harisdangkul, G.D. Ehrlich, L.S. Gray, R.L. Whisler, and J.R. Blakeslee. 1994. Lack of evidence for human T cell lymphotrophic virus type I or II infection in patients with systemic lupus erythematosus or rheumatoid arthritis. *J. Rheumatol.* 21:2217–2224.

102. Stransky, G., J. Vernon, W.K. Aicher, L.W. Moreland, R.E. Gay, and S. Gay. 1993. Virus-like particles in synovial fluids from patients with rheumatoid arthritis. *Br. J. Rheumatol.* 32:1044–1048.

103. Chu, J.L., J. Drappa, A. Parnassa, and K.B. Elkon. 1993. The defect in Fas mRNA expression in MRL/lpr mice is associated with insertion of the retrotransponson, ETn. *J. Exp. Med.* 178:723–730.

104. Hang, L., A.N. Theofilopoulos, and F.J. Dixon. 1982. A spontaneous rheumatoid arthritis-like disease in MRL/l mice. *J. Exp. Med.* 155:1690–1701.

105. O'Sullivan, F.X., H.G. Fassbender, S. Gay, and W.J. Koopman. 1985. Etiopathogenesis of the rheumatoid arthritis-like disease in MRL/l mice. I. The histomorphologic basis of joint destruction. *Arthritis Rheum.* 28:529–536.

106. Tanaka, A., F.X. O'Sullivan, W.J. Koopman, and S. Gay. 1988. Etiopathogenesis of rheumatoid arthritis-like disease in MRL/1 mice: II. Ultrastructural basis of joint destruction. *J. Rheumatol.* 15:10–16.

107. Wu, J., T. Zhou, J. He, and J.D. Mountz. 1993. Autoimmune disease in mice due to integration of an endogenous retrovirus in an apoptosis gene. *J. Exp. Med.* 178:461–468.

108. Wu, J., T. Zhou, J. Zhang, J. He, W.C. Gause, and J.D. Mountz. 1994. Correction of accelerated autoimmune disease by early replacement of the mutated lpr gene with the normal Fas apoptosis gene in the T cells of transgenic MRL-lpr/lpr mice. *Proc. Natl. Acad. Sci. USA* 91:2344–2348.

109. Takeuchi, K., K. Katsumata, H. Ikeda, M. Minami, A. Wakisaka, and T. Yoshiki. 1995. Expression of endogenous retroviruses, ERV3 and lambda 4–1, in synovial tissues from patients with rheumatoid arthritis. *Clin. Exp. Immunol.* 99:338–344.

110. Nakagawa, K., V. Brusic, G. McColl, and L.C. Harrison. 1997. Direct evidence for the expression of multiple endogenous retroviruses in the synovial compartment in rheumatoid arthritis. *Arthritis Rheum.* 40:627–638.

111. Müller-Ladner, U., R.E. Gay, and S. Gay. 1998. Retroviral sequences in rheumatoid arthritis synovium. *Intern. Rev. Immunol.* 17:273–290.

112. Neidhart, M., J. Rethage, S. Kuchen, P. Kunzler, R.B. Crowl, M.E. Billingham, R.E. Gay, and S. Gay. 2000. Retrotrausposable L1 elements expressed in rheumatoid arthritis synovial tissue. *Arthritis Rheum.* 43: in press.

113. Harris-ED, J., D.R. DiBona, and S.M. Krane. 1970. A mechanism for cartilage destruction in rheumatoid arthritis. *Trans. Assoc. Am. Physicians.* 83:267–276.

114. Fassbender, H.G. 1983. Histomorphological basis of articular cartilage destruction in rheumatoid arthritis. *Coll. Relat. Res.* 3:141–155.

115. Firestein, G.S. 1996. Invasive fibroblast-like synoviocytes in rheumatoid arthritis. Passive responders or transformed aggressors? *Arthritis Rheum.* 39:1781–1790.

116. Müller-Ladner, U., R.E. Gay, and S. Gay. 1997. Cellular pathways of joint destruction. *Curr. Opin. Rheumatol.* 9:213–220.

117. Hamilton, J.A. 1983. Hypothesis: *in vitro* evidence for the invasive and tumor-like properties of the rheumatoid pannus. *J. Rheumatol.* 10:845–851.

118. Kelly, P.M., E. Bliss, J.A. Morton, J. Burns, and J.O. McGee. 1988. Monoclonal antibody EBM/11: high cellular specificity for human macrophages. *J. Clin. Pathol.* 41:510–515.

119. Burmester, G.R., B. Stuhlmuller, G. Keyszer, and R.W. Kinne. 1997. Mononuclear phagocytes and rheumatoid synovitis. Mastermind or workhorse in arthritis? *Arthritis Rheum.* 40:5–18.

120. Cutolo, M., A. Sulli, A. Barone, B. Seriolo, and S. Accardo. 1993. Macrophages, synovial tissue and rheumatoid arthritis. *Clin. Exp. Rheumatol.* 11:331–339.

121. Helbig, B., W.L. Gross, B. Borisch, H. Starz, and H.K. Muller-Hermelink. 1988. Characterization of synovial macrophages by monoclonal antibodies in rheumatoid arthritis and osteoarthritis. *Scand. J. Rheumatol. Suppl.* 76:61–66.

122. Salisbury, A.K., O. Duke, and L.W. Poulter. 1987. Macrophage-like cells of the pannus area in rheumatoid arthritic joints. *Scand. J. Rheumatol.* 16:263–272.

123. Firestein, G.S. 1998. Rheumatoid synovitis and pannus. In: Rheumatology. J.H. Klippel and P.A. Dieppe, editors. Mosby, London. pp. 5.13.1–5.13.24

124. Hoyhtya, M., R. Myllyla, J. Piuva, K.I. Kivirikko, and K. Tryggvason. 1984. Monoclonal antibodies to human prolyl 4-hydroxylase. *Eur. J. Biochem.* 141:472–482.

125. Ziff, M. 1989. Pathways of mononuclear cell infiltration in rheumatoid synovitis. *Rheumatol. Int.* 9:97–103.

126. Ezawa, K., M. Yamamura, H. Matsui, Z. Ota, and H. Makino. 1997. Comparative analysis of CD45RA- and CD45RO-positive CD4+T cells in peripheral blood, synovial fluid, and synovial tissue in patients with rheumatoid arthritis and osteoarthritis. *Acta Med. Okayama* 51:25–31.

127. Matsuoka, N., K. Eguchi, A. Kawakami, H. Ida, M. Nakashima, M. Sakai, K. Terada, S. Inoue, Y. Kawabe, and A. Kurata. 1991. Phenotypic characteristics of T cells interacted with synovial cells. *J. Rheumatol.* 18:1137–1142.

128. Meijer, C.J., C.B. de Graaff-Reitsma, G.J. Lafeber, and A. Cats. 1982. In situ localization of lymphocyte subsets in synovial membranes of patients with rheumatoid arthritis with monoclonal antibodies. *J. Rheumatol.* 9:359–365.

129. Young, C.L., T.C. Adamson, J.H. Vaughan, and R.I. Fox. 1984. Immunohistologic characterization of synovial mem-

brane lymphocytes in rheumatoid arthritis. *Arthritis Rheum.* 27:32–39.

130. Berek, C., and H.J. Kim. 1997. B-cell activation and development within chronically inflamed synovium in rheumatoid and reactive arthritis. *Semin. Immunol.* 9:261–268.

131. De Clerck, L.S. 1995. B lymphocytes and humoral immune responses in rheumatoid arthritis. *Clin. Rheumatol.* 14 Suppl 2:14–18.

132. Youinou, P.Y., W.L. Irving, M. Shipley, J. Hayes, and P.M. Lydyard. 1984. Evidence for B cell activation in patients with active rheumatoid arthritis. *Clin. Exp. Immunol.* 55:91–98.

133. Koch, A.E. 1998. Review: angiogenesis: implications for rheumatoid arthritis. *Arthritis Rheum.* 41:951–962.

134. Neidhart, M., R. Wehrli, P. Brühlmann, B.A. Michel, R.E. Gay, and S. Gay. 1999. Synovial fluid CD146 (MUC18), a marker for synovial membrane angiogenesis in rheumatoid arthritis. *Arthritis Rheum.* 42:622–630.

135. Paliard, X., S.G. West, J.A. Lafferty, J.R. Clements, J.W. Kappler, P. Marrack, and B.L. Kotzin. 1991. Evidence for the effects of a superantigen in rheumatoid arthritis. *Science* 253:325–329.

136. Howell, M.D., J.P. Diveley, K.A. Lundeen, A. Esty, S.T. Winters, D.J. Carlo, and S.W. Brostoff. 1991. Limited T cell receptor beta-chain heterogeneity among interleukin 2 receptor-positive synovial T cells suggests a role for superantigen in rheumatoid arthritis. *Proc. Natl. Acad. Sci. USA* 88:10921–10925.

137. Goronzy, J.J., P. Bartz-Bazzanella, W. Hu, M.C. Jendro, D.R. Walser-Kuntz, and C.M. Weyand. 1994. Dominant clonotypes in the repertoire of peripheral CD4+ T cells in rheumatoid arthritis. *J. Clin. Invest.* 94:2068–2076.

138. Davey, M.P., and D.D. Munkirs. 1993. Patterns of T-cell receptor variable beta gene expression by synovial fluid and peripheral blood T-cells in rheumatoid arthritis. *Clin. Immunol. Immunopathol.* 68:79–87.

139. Bridges, S.L.J., and L.W. Moreland. 1998. T-cell receptor peptide vaccination in the treatment of rheumatoid arthritis. *Rheum. Dis. Clin. North Am.* 24:641–650.

140. Cole, B.C., and M.M. Griffiths. 1993. Triggering and exacerbation of autoimmune arthritis by the Mycoplasma arthritidis superantigen MAM. *Arthritis Rheum.* 36:994–1002.

141. Brennan, F.M., S. Allard, M. Londei, C. Savill, A. Boylston, S. Carrel, R.N. Maini, and M. Feldmann. 1988. Heterogeneity of T cell receptor idiotypes in rheumatoid arthritis. *Clin. Exp. Immunol.* 73:417–423.

142. Olive, C., P.A. Gatenby, and S.W. Serjeantson. 1991. Analysis of T cell receptor V alpha and V beta gene usage in synovia of patients with rheumatoid arthritis. *Immunol. Cell. Biol.* 69(Pt, 5):349–354.

143. Duby, A.D., A.K. Sinclair, S.L. Osborne-Lawrence, W. Zeldes, L. Kan, and D.A. Fox. 1989. Clonal heterogeneity of synovial fluid T lymphocytes from patients with rheumatoid arthritis. *Proc. Natl. Acad. Sci. USA* 86:6206–6210.

144. Weyand, C.M., and J.J. Goronzy. 1999. T-cell responses in rheumatoid arthritis: systemic abnormalities-local disease. *Curr. Opin. Rheumatol.* 11:210–217.

145. Martens, P.B., J.J. Goronzy, D. Schaid, and C.M. Weyand. 1997. Expansion of unusual CD4+ T cells in severe rheumatoid arthritis. *Arthritis Rheum.* 40:1106–1114.

146. Schmidt, D., J.J. Goronzy, and C.M. Weyand. 1996. CD4+ CD7-CD28- T cells are expanded in rheumatoid arthritis and are characterized by autoreactivity. *J. Clin. Invest.* 97:2027–2037.

147. Trentham, D.E., A.S. Townes, and A.H. Kang. 1977. Autoimmunity to type II collagen an experimental model of arthritis. *J. Exp. Med.* 146:857–868.

148. Holmdahl, R., M.E. Andersson, T.J. Goldschmidt, L. Jansson, M. Karlsson, V. Malmstrom, and J. Mo. 1989. Collagen induced arthritis as an experimental model for rheumatoid arthritis. Immunogenetics, pathogenesis and autoimmunity. *APMIS* 97:575–584.

149. Wilder, R.L., J.P. Case, L.J. Crofford, G.K. Kumkumian, R. Lafyatis, E.F. Remmers, H. Sano, E.M. Sternberg, and D.E. Yocum. 1991. Endothelial cells and the pathogenesis of rheumatoid arthritis in humans and streptococcal cell wall arthritis in Lewis rats. *J. Cell Biochem.* 45:162–166.

150. Ranges, G.E., S. Sriram, and S.M. Cooper. 1985. Prevention of type II collagen-induced arthritis by *in vivo* treatment with anti-L3T4. *J. Exp. Med.* 162:1105–1110.

151. Griffiths, M.M. 1988. Immunogenetics of collagen-induced arthritis in rats. *Int. Rev. Immunol.* 4:1–15.

152. Verheijden, G.F., A.W. Rijnders, E. Bos, C.J. Coenen-de Roo, C.J. van Staveren, A.M. Miltenburg, J.H. Meijerink, D. Elewaut, F. de Keyser, E. Veys, and A.M. Boots. 1997. Human cartilage glycoprotein-39 as a candidate auto-antigen in rheumatoid arthritis. *Arthritis Rheum.* 40:1115–1125.

153. Duric, F.H., R.A. Fava, T.M. Foy, A. Aruffo, J.A. Ledbetter, and R.J. Noelle. 1993. Prevention of collagen-induced arthritis with an antibody to gp39, the ligand for CD40. *Science* 261:1328–1330.

154. Carlson, S., A.S. Hansson, H. Olsson, D. Heinegard, and R. Holmdahl. 1998. Cartilage oligomeric matrix protein (COMP)-induced arthritis in rats. *Clin. Exp. Immunol.* 114:477–484.

155. Tarkowski, A., L. Klareskog, H. Carlsten, P. Herberts, and W.J. Koopman. 1989. Secretion of antibodies to types I and II collagen by synovial tissue cells in patients with rheumatoid arthritis. *Arthritis Rheum.* 32:1087–1092.

156. Londei, M., C.M. Savill, A. Verhoef, F. Brennan, Z.A. Leech, V. Duance, R.N. Maini, and M. Feldmann. 1989. Persistence of collagen type II-specific T-cell clones in the synovial membrane of a patient with rheumatoid arthritis. *Proc. Natl. Acad. Sci. USA* 86:636–640.

157. Cope, A.P., S.D. Patel, F. Hall, M. Congia, H.A. Hubers, G.F. Verheijden, A.M. Boots, R. Menon, M. Trucco, A.W. Rijnders, and G. Sonderstrup. 1999. T cell responses to a human cartilage autoantigen in the context of rheumatoid arthritis-associated and nonassociated HLA-DR4 alleles. *Arthritis Rheum.* 42:1497–1507.

158. Bläss, S., F. Schumann, N.A. Hain, J.M. Engel, B. Stuhlmüller, and G.R. Burmester. 1999. p205 is a major target of autoreactive T cells in rheumatoid arthritis. *Arthritis Rheum.* 42:971–980.

159. Tanaka, M., S. Ozaki, F. Osakada, K. Mori, M. Okubo, and K. Nakao. 1998. Cloning of follistatin-related protein as a novel autoantigen in systemic rheumatic diseases. *Int. Immunol.* 10:1305–1314.

160. Guerassimov, A., Y. Zhang, S. Banerjee, A. Cartman, C. Webber, J. Esdaile, M.A. Fitzcharles, and A.R. Poole. 1998. Autoimmunity to cartilage link protein in patients with rheumatoid arthritis and ankylosing spondylitis. *J. Rheumatol.* 25:1480–1484.

161. Sedlacek, R., S. Mauch, B. Kolb, C. Schätzlein, H. Eibel, H.H. Peter, J. Schmitt, and U. Krawinkel. 1998. Matrix metalloproteinase MMP-19 (RASI-1) is expressed on the surface of activated peripheral blood mononuclear cells and is detected as an autoantigen in rheumatoid arthritis. *Immunobiology* 198:408–423.

162. Bläss, S., C. Haferkamp, C. Specker, M. Schwochau, M. Schneider, and E.M. Schneider. 1997. Rheumatoid arthritis: autoreactive T cells recognising a novel 68k autoantigen. *Ann. Rheum. Dis.* 56:317–322.

163. Case, J.P., H. Sano, R. Lafyatis, E.F. Remmers, G.K. Kumkumian, and R.L. Wilder. 1989. Transin/stromelysin expression in the synovium of rats with experimental erosive arthritis. In situ localization and kinetics of expression of the transformation-associated metalloproteinase in euthymic and athymic Lewis rats. *J. Clin. Invest.* 84:1731–1740.

164. Yocum, D.E., R. Lafyatis, E.F. Remmers, H.R. Schumacher, and R.L. Wilder. 1988. Hyperplastic synoviocytes from rats with streptococcal cell wall-induced arthritis exhibit a transformed phenotype that is thymic-dependent and retinoid inhibitable. *Am. J. Pathol.* 132:38–48.

165. Shiozawa, S., Y. Tanaka, T. Fujita, and T. Tokuhisa. 1992. Destructive arthritis without lymphocyte infiltration in H2-c-fos transgenic mice. *J. Immunol.* 148:3100–3104.

166. Firestein, G.S., W.D. Xu, K. Townsend, D. Broide, J. Alvaro-Gracia, A. Glasebrook, and N.J. Zvaifler. 1988. Cytokines in chronic inflammatory arthritis. I. Failure to detect T cell lymphokines (interleukin 2 and interleukin 3) and presence of macrophage colony-stimulating factor (CSF-1) and a novel mast cell growth factor in rheumatoid synovitis. *J. Exp. Med.* 168:1573–1586.

167. Falta, M.T., and B.L. Kotzin. 1999. T cells as primary players in rheumatoid arthritis. In: *T cells in arthritis.* P. Miossec, B.W. van-den Berg, and G.S. Firestein, editors. Birkenhäuser Verlag, Basel. pp. 201–231.

168. Moreland, L.W., P.W. Pratt, M.D. Mayes, A. Postlethwaite, M.H. Weisman, T. Schnitzer, R. Lightfoot, L. Calabrese, D.J. Zelinger, J.N. Woody, et al. 1995. Double-blind, placebo-controlled multicenter trial using chimeric monoclonal anti-CD4 antibody, cM-T412, in rheumatoid arthritis patients receiving concomitant methotrexate. *Arthritis Rheum.* 38:1581–1588.

169. Jendro, M.C., T. Ganten, E.L. Matteson, C.M. Weyand, and J.J. Goronzy. 1995. Emergence of oligoclonal T cell populations following therapeutic T cell depletion in rheumatoid arthritis. *Arthritis Rheum.* 38:1242–1251.

170. Weyand, C.M., and J.J. Goronzy. 1997. Pathogenesis of rheumatoid arthritis. *Med. Clin. North Am.* 81:29–55.

171. Müller-Ladner, U., J. Kriegsmann, R.E. Gay, W.J. Koopman, S. Gay, and W.W. Chatham. 1995. Progressive joint destruction in a human immunodeficiency virus-infected patient with rheumatoid arthritis. *Arthritis Rheum.* 38:1328–1332.

172. Mulherin, D., O. Fitzgerald, and B. Bresnihan. 1996. Clinical improvement and radiological deterioration in rheumatoid arthritis: evidence that the pathogenesis of synovial inflammation and articular erosion may differ. *Br. J. Rheumatol.* 35:1263–1268.

173. O'Sullivan, F.X., R.E. Gay, and S. Gay. 1995. Spontaneous Arthritis Models. In: *Mechanisms and models in rheumatoid arthritis.* AnonymousAcademic Press Ltd., pp. 471–483.

174. Trabandt, A., R.E. Gay, H. Birkedal Hansen, and S. Gay. 1992. Expression of collagenase and potential transcriptional factors in the MRL/l mouse arthropathy. *Semin. Arthritis Rheum.* 21:246–251.

175. Kraan, M.C., H. Versendaal, M. Jonker, B. Bresnihan, W.J. Post, H.A. ët Hard, F.C. Breedveld, and P.P. Tak. 1998. Asymptomatic synovitis precedes clinically manifest arthritis. *Arthritis Rheum.* 41:1481–1488.

176. Mosier, D.E., R.J. Gulizia, S.M. Baird, and D.B. Wilson. 1988. Transfer of a functional human immune system to mice with severe combined immunodeficiency. *Nature* 335:256–259.

177. Rendt, K.E., T.S. Barry, D.M. Jones, C.B. Richter, S.S. McCachren, and B.F. Haynes. 1993. Engraftment of human synovium into severe combined immune deficient mice. Migration of human peripheral blood T cells to engrafted human synovium and to mouse lymph nodes. *J. Immunol.* 151:7324–7336.

178. Geiler, T., J. Kriegsmann, G.M. Keyszer, R.E. Gay, and S. Gay. 1994. A new model for rheumatoid arthritis generated by engraftment of rheumatoid synovial tissue and normal human cartilage into SCID mice. *Arthritis Rheum.* 37:1664–1671.

179. Müller-Ladner, U., J. Kriegsmann, B.N. Franklin, S. Matsumoto, T. Geiler, R.E. Gay, and S. Gay. 1996. Synovial fibroblasts of patients with rheumatoid arthritis attach to and invade normal human cartilage when engrafted into SCID mice. *Am. J. Pathol.* 149:1607–1615.

180. Grimbacher, B., W.K. Aicher, H.H. Peter, and H. Eibel. 1997. Measurement of transcription factor c-fos and EGR-1 mRNA transcription levels in synovial tissue by quantitative RT-PCR. *Rheumatol. Int.* 17:109–112.

181. Knotny, E., M. Ziolkowska, E. Dudzinka, A. Filipowicz-Sosnowska, and Ry. 1995. Modified expression of c-Fos and c-Jun proteins and production of interleukin-1 beta in patients with rheumatoid arthritis. *Clin. Exp. Rheumatol.* 13:51–57.

182. Benbow, U., and C.E. Brinckerhoff. 1997. The AP-1 site and MMP gene regulation: what is all the fuss about? *Matrix Biol.* 15:519–526.

183. Asahara, H., K. Fujisawa, T. Kobata, T. Hasunuma, T. Maeda, M. Asanuma, N. Ogawa, H. Inoue, T. Sumida, and K. Nishioka. 1997. Direct evidence of high DNA binding activity of transcription factor AP-1 in rheumatoid arthritis synovium. *Arthritis Rheum.* 40:912–918.

184. Miagkov, A.V., D.V. Kovalenko, C.E. Brown, J.R. Didsbury, J.P. Cogswell, S.A. Stimpson, A.S. Baldwin, and S.S. Makarov. 1998. NF-kappaB activation provides the potential link between inflammation and hyperplasia in the arthritic joint. *Proc. Natl. Acad. Sci. USA* 95:13859–13864.

185. Han, Z., D.L. Boyle, A.M. Manning, and G.S. Firestein. 1998. AP-1 and NF-kappaB regulation in rheumatoid arthritis and murine collagen-induced arthritis. *Autoimmunity* 28:197–208.

186. Marok, R., P.G. Winyard, A. Coumbe, M.L. Kus, K. Gaffney, S. Blades, P.I. Mapp, C.J. Morris, D.R. Blake, C. Kaltschmidt, and P.A. Baeuerle. 1996. Activation of the transcription factor nuclear factor-kappaB in human inflamed synovial tissue. *Arthritis Rheum.* 39:583–591.

187. Palombella, V.J., E.M. Conner, J.W. Fuseler, A. Destree, J.M. Davis, F.S. Laroux, R.E. Wolf, J. Huang, S. Brand, P.J. Elliott, D. Lazarus, T. McCormack, L. Parent, R. Stein, J. Adams, and M.B. Grisham. 1998. Role of the proteasome and NF-kappaB in streptococcal cell wall-induced polyarthritis. *Proc. Natl. Acad. Sci. USA* 95:15671–15676.

188. Makarov, S.S., W.N. Johnston, J.C. Olsen, J.M. Watson, K. Mondal, C. Rinehart, and J.S. Haskill. 1997. NF-kappa B as a target for anti-inflammatory gene therapy: suppression of inflammatory responses in monocytic and stromal cells by stable gene transfer of I kappa B alpha cDNA. *Gene Ther.* 4:846–852.

189. Vincenti, M.P., C.I. Coon, and C.E. Brinckerhoff. 1998. Nuclear factor kappaB/p50 activates an element in the distal matrix metalloproteinase 1 promoter in interleukin-1beta-stimulated synovial fibroblasts. *Arthritis Rheum.* 41:1987–1994.

190. Trabandt, A., W.K. Aicher, R.E. Gay, V.P. Sukhatme, H.M. Nilson, R.T. Hamilton, J.R. McGhee, H.G. Fassbender, and S. Gay. 1990. Expression of the collagenolytic and Ras-induced cysteine proteinase cathepsin L and proliferation-associated oncogenes in synovial cells of MRL/I mice and patients with rheumatoid arthritis. *Matrix* 10:349–361.

191. Gum, R., H. Wang, E. Lengyel, J. Juarez, and D. Boyd. 1997. Regulation of 92 kDa type IV collagenase expression by the jun aminoterminal kinase- and the extracellular signal-regulated kinase-dependent signaling cascades. *Oncogene* 14:1481–1493.

192. Korzus, E., H. Nagase, R. Rydell, and J. Travis. 1997. The mitogen-activated protein kinase and JAK-STAT signaling pathways are required for an oncostatin M-responsive element-mediated activation of matrix metalloproteinase 1 gene expression. *J. Biol. Chem.* 272:1188–1196.

193. Li, J., C. Yen, D. Liaw, K. Podsypanina, S. Bose, S.I. Wang, J. Puc, C. Miliaresis, L. Rodgers, R. McCombie, S.H. Bigner, B.C. Giovanella, M. Ittmann, B. Tycko, H. Hibshoosh, M.H. Wigler, and R. Parsons. 1997. PTEN, a putative protein tyrosine phosphatase gene mutated in human brain, breast, and prostate cancer. *Science* 275:1943–1947.

194. Pap, T., K.M. Hummel, J.K. Franz, U. Müller-Ladner, R.E. Gay, and S. Gay. 1998. Downregulation but no mutation of novel tumor suppressor PTEN in aggressively invading rheumatoid arthritis fibroblasts (RA-SF). *Arthritis Rheum.* 41:S239–S239.

195. Firestein, G.S., F. Echeverri, M. Yeo, N.J. Zvaifler, and D.R. Green. 1997. Somatic mutations in the p53 tumor suppressor gene in rheumatoid arthritis synovium. *Proc. Natl. Acad. Sci. USA* 94:10895–10900.

196. Kullmann, F., M. Judex, I. Neudecker, S. Lechner, H.P. Jüsten, D.R. Green, D. Wessinghage, G.S. Firestein, S. Gay, J. Schöllmerich, and U. Müller-Ladner. 1999. Analysis of the p53 tumor suppressor gene in rheumatoid arthritis synovial fibroblasts. *Arthritis Rheum.* 42:1594–1600.

197. Nykanen, P., V. Bergroth, P. Raunio, D. Nordstrom, and Y.T. Konttinen. 1986. Phenotypic characterization of 3H-thymidine incorporating cells in rheumatoid arthritis synovial membrane. *Rheumatol. Int.* 6:269–271.

198. Mohr, W., N. Hummler, B. Peister, and D. Wessinghage. 1986. Proliferation of pannus tissue cells in rheumatoid arthritis. *Rheumatol. Int.* 6:127–132.

199. Matsumoto, S., U. Müller-Ladner, R.E. Gay, K. Nishioka, and S. Gay. 1996. Multistage apoptosis and Fas antigen expression of synovial fibroblasts derived from patients with rheumatoid arthritis. *J. Rheumatol.* 23:1345–1352.

200. Nakajima, T., H. Aono, T. Hasunuma, K. Yamamoto, T. Shirai, K. Hirohata, and K. Nishioka. 1995. Apoptosis and functional Fas antigen in rheumatoid arthritis synoviocytes. *Arthritis Rheum.* 38:485–491.

201. Okura, T., L. Gong, T. Kamitani, T. Wada, I. Okura, C.F. Wei, H.M. Chang, and E.T. Yeh. 1996. Protection against Fas/APO-1- and tumor necrosis factor-mediated cell death by a novel protein, sentrin. *J. Immunol.* 157:4277–4281.

202. Franz, J.K., K.M. Hummel, W.K. Aicher, T. Pap, U. Müller-Ladner, R.E. Gay, and S. Gay. 1998. Invasive synovial fibroblasts express the novel anti-apoptotic molecule sentrin in the SCID mouse model of rheumatoid arthritis. *Arthritis Rheum.* 41:S238.

203. Matsumoto, S., U. Müller-Ladner, R.E. Gay, K. Nishioka, and S. Gay. 1996. Ultrastructural demonstration of apoptosis, Fas and Bcl-2 expression of rheumatoid synovial fibroblasts. *J. Rheumatol.* 23:1345–1352.

204. Asahara, H., T. Hasunuma, T. Kobata, H. Inoue, U. Müller-Ladner, S. Gay, T. Sumida, and K. Nishioka. 1997. In situ expression of protooncogenes and Fas/Fas ligand in rheumatoid arthritis synovium. *J. Rheumatol.* 24:430–435.

205. Firestein, G.S., M. Yeo, and N.J. Zvaifler. 1995. Apoptosis in rheumatoid arthritis synovium. *J. Clin. Invest.* 96:1631–1638.

206. Okamoto, K., T. Kobayashi, T. Kobata, T. Hasunuma, T. Kato, T. Sumida, and K. Nishioka. 2000. Fas-associated death domain protein is a Fas-mediated apoptosis modulator in synoriocytes. *Rheumatology* (Oxford) 39:471–480.

207. Aicher, W.K., H.H. Peter, and H. Eibel. 1996. Human synovial fibroblasts are resistant to Fas induced apoptosis. *Arthritis Rheum.* 39:S75 (Abstract).

208. Rinaldi, N., E.M. Schwarz, D. Weis, J.P. Leppelmann, M. Lukoschek, U. Keilholz, and T.F. Barth. 1997. Increased expression of integrins on fibroblast-like synoviocytes from rheumatoid arthritis *in vitro* correlates with enhanced binding to extracellular matrix proteins. *Ann. Rheum. Dis.* 56:45–51.

209. Ishikawa, H., S. Hirata, Y. Andoh, H. Kubo, N. Nakagawa, Y. Nishibayashi, and K. Mizuno. 1996. An immunohistochemical and immunoelectron microscopic study of adhesion molecules in synovial pannus formation in rheumatoid arthritis. *Rheumatol. Int.* 16:53–60.

210. Müller-Ladner, U., M.J. Elices, J. Kriegsmann, D. Strahl, R.E. Gay, G.S. Firestein, and S. Gay. 1997. Alternatively spliced CS-1 fibronectin isoform and its receptor VLA-4 in rheumatoid synovium demonstrated by in situ hybridization and immunohistochemistry. *J. Rheumatol.* 24:1873–1880.

211. Schwartz, M.A. 1997. Integrins, oncogenes, and anchorage independence. *J. Cell Biology* 139:575–578.

212. Dike, L.E., and S.R. Farmer. 1988. Cell adhesion induces expression of growth-associated genes in suspension-arrested fibroblasts. *Proc. Natl. Acad. Sci. USA* 85:6792–6796.

213. Dike, L.E., and D.E. Ingber. 1996. Integrin-dependent induction of early growth response genes in capillary endothelial cells. *J. Cell Sci.* 109:2855–2863.

214. Kriegsmann, J., G.M. Keyszer, T. Geiler, R. Brauer, R.E. Gay, and S. Gay. 1995. Expression of vascular cell adhesion molecule-1 mRNA and protein in rheumatoid synovium demonstrated by in situ hybridization and immunohistochemistry. *Lab. Invest.* 72:209–214.

215. Morales, D.J., E. Wayner, M.J. Elices, G.J. Alvaro, N.J. Zvaifler, and G.S. Firestein. 1992. Alpha 4/beta 1 integrin (VLA-4) ligands in arthritis. Vascular cell adhesion molecule-1 expression in synovium and on fibroblast-like synoviocytes. *J. Immunol.* 149:1424–1431.

216. Matsuyama, T., and A. Kitani. 1996. The role of VCAM-1 molecule in the pathogenesis of rheumatoid synovitis. *Hum. Cell* 9:187–192.

217. Kitani, A., N. Nakashima, T. Izumihara, M. Inagaki, X. Baoui, S. Yu, T. Matsuda, and T. Matsuyama. 1998. Soluble VCAM-1 induces chemotaxis of Jurkat and synovial fluid T cells bearing high affinity very late antigen-4. *J. Immunol.* 161:4931–4938.

218. Lindhout, E., M. van Eijk, M. van Pel, J. Lindeman, H.J. Dinant, and C. de Groot. 1999. Fibroblast-like

synoviocytes from rheumatoid arthritis patients have intrinsic properties of follicular dendritic cells. *J. Immunol.* 162:5949–5956.

219. Kitani, A., N. Nakashima, T. Matsuda, B. Xu, S. Yu, T. Nakamura, and T. Matsuyama. 1996. T cells bound by vascular cell adhesion molecule-1/CD106 in synovial fluid in rheumatoid arthritis: inhibitory role of soluble vascular cell adhesion molecule-1 in T cell activation. *J. Immunol.* 156:2300–2308.

220. Petrow, P., J.K. Franz, U. Müller-Ladner, K.M. Hummel, R.E. Gay, CW. Prince, and S. Gay. 1997. Expression of osteopontin mRNA in synovial tissue of patients with rheumatoid arthritis(RA) and osteoarthritis (OA). *Arthritis Rheum.* 39:S36 (Abstract).

221. Lampe, M.A., R. Patarca, M.V. Iregui, and H. Cantor. 1991. Polyclonal B cell activation by Eta-1 cytokine and the development of systemic autoimmune disease. *J. Immunol.* 147:2902–2906.

222. Pichler, R., C.M. Giachelli, D. Lombardi, J. Pippin, K. Gordon, C.E. Alpers, S.M. Schwartz, and R.J. Johnson. 1994. Tubulointestinal disease in glomerulonephritis. Potential role of osteopontin (uropontin). *Am. J. Pathol.* 144:915–926.

223. Pap, T., J.K. Franz, R.E. Gay, and S. Gay. 1999. Has research on lymphocytes hindered progress in rheumatoid arthritis? In: *Challenges in rheumatoid arthritis.* H. Bird and M. Snaith, editors. pp.61–77.

224. Salmon, M., D. Scheel Toellner, A.P. Huissoon, D. Pilling, N. Shamsadeen, H. Hyde, A.D. D'Angeac, P.A. Bacon, P. Emery, and A.N. Akbar. 1997. Inhibition of T cell apoptosis in the rheumatoid synovium. *J. Clin. Invest.* 99:439–446.

225. McInnes, I.B., J. al Mughales, M. Field, B.P. Leung, F.P. Huang, R. Dixon, R.D. Sturrock, P.C. Wilkinson, and F.Y. Liew. 1996. The role of interleukin-15 in T-cell migration and activation in rheumatoid arthritis. *Nat. Med.* 2:175–182.

226. Franz, J.K., S. Kolb, K.M. Hummel, F. Lahrtz, M. Neidhart, W.K. Aicher, T. Pap, R.E. Gay, A. Fontana, and S. Gay. 1998. Interleukin-16 produced by synovial fibroblasts mediates chemoattraction to CD4+ T-cells in rheumatoid arthritis. *Eur. J. Immunol.* 28:2661–2671.

227. Ogasawara, H., N. Takeda-Hirokawa, I. Sekigawa, H. Hashimoto, Y. Kaneko, and S. Hirose. 1999. Inhibitory effect of interleukin-16 on interleukin-2 production by CD4+ T cells. *Immunology* 96:215–219.

228. Salojin, K.V., J. Zhang, J. Madrenas, and T.L. Delovitch. 1998. T-cell anergy and altered T-cell receptor signaling: effects on autoimmune disease. *Immunol. Today* 19:468–473.

229. Miossec, P. 1995. Pro- and antiinflammatory cytokine balance in rheumatoid arthritis. *Clin. Exp. Rheumatol.* 13 (suppl. 12):S13–16.

230. van-de Loof, FA., L.A. Joosten, P. van Lent, O.J. Arntz, and B.W. van-den Berg. 1995. Role of interleukin-1, tumor necrosis factor alpha, and interleukin-6 in cartilage proteoglycan metabolism and destruction. Effect of in situ blocking in murine antigen- and zymosan-induced arthritis. *Arthritis Rheum.* 38:164–172.

231. Joosten, L.A., M.M. Helsen, F.A. van-de Loof, and B.W. van-den Berg. 1996. Anticytokine treatment of established type II collagen-induced arthritis in DBA/1 mice. A comparative study using anti-TNF alpha, anti-IL-1 alpha/beta, and IL-1Ra. *Arthritis Rheum.* 39:797–809.

232. van-de Loof, FA, O.J. Arntz, I.G. Otterness, and B.W. van-den Berg. 1992. Protection against cartilage proteoglycan synthesis inhibition by antiinterleukin 1 antibodies in experimental arthritis. *J. Rheumatol.* 19:348–356.

233. van-de Loof, FA, O.J. Arntz, I.G. Otterness, and B.W. van-den Berg. 1993. Modulation of cartilage destruction in murine arthritis with anti-IL-1 antibodies. *Agents Actions* 39 Spec No:C211–C214.

234. van-den Berg, BW, L.A. Joosten, M. Helsen, and F.A. van-de Loof. 1994. Amelioration of established murine collagen-induced arthritis with anti-IL-1 treatment. *Clin. Exp. Immunol.* 95:237–243.

235. Müller-Ladner, U., C.R. Roberts, B.N. Franklin, R.E. Gay, P.D. Robbins, C.H. Evans, and S. Gay. 1997. Human IL-1Ra gene transfer into human synovial fibroblasts is chondroprotective. *J. Immunol.* 158:3492–3498.

236. Maini, R.N., M. Elliott, F.M. Brennan, R.O. Williams, and M. Feldmann. 1997. TNF blockade in rheumatoid arthritis: implications for therapy and pathogenesis. *APMIS* 105:257–263.

237. Evans, C.H., and P.D. Robbins. 1996. The promise of a new clinical trial—intra-articular IL-1 receptor antagonist. *Proc. Assoc. Am. Physicians.* 108:1–5.

238. Arend, W.P., and J.M. Dayer. 1995. Inhibition of the production and effects of interleukin-1 and tumor necrosis factor alpha in rheumatoid arthritis. *Arthritis Rheum.* 38:151–160.

239. Gabay, C., and W.P. Arend. 1998. Treatment of rheumatoid arthritis with IL-1 inhibitors. *Springer Semin. Immunopathol.* 20:229–246.

240. Matrix metalloproteinases. W.C. Parks and R.P. Mecham, editors. Academic Press, San Diego.

241. Müller-Ladner, U., R.E. Gay, and S. Gay. 1996. Cysteine proteinases in arthritis and inflammation. *Perspectives in Drug Discovery and Design* 6:87–98.

242. Wernicke, D., C. Seyfert, B. Hinzmann, and E. Gromnica-Ihle. 1996. Cloning of collagenase 3 from the synovial membrane and its expression in rheumatoid arthritis and osteoarthritis. *J. Rheumatol.* 23:590–595.

243. Ogata, Y., J.J. Enghild, and H. Nagase. 1992. Matrix metalloproteinase 3 (stromelysin) activates the precursor for the human matrix metalloproteinase 9. *J. Biol. Chem.* 267:3581–3584.

244. Ito, A., and H. Nagase. 1988. Evidence that human rheumatoid synovial matrix metalloproteinase 3 is an endogenous activator of procollagenase. *Arch. Biochem. Biophys.* 267:211–216.

245. Ramos-DeSimone, N., E. Hahn-Dantona, J. Sipley, H. Nagase, D.L. French, and J.P. Quigley. 1999. Activation of matrix metalloproteinase-9 (MMP-9) via a converging plasmin/stromelysin-1 cascade enhances tumor cell invasion. *J. Biol. Chem.* 274:13066–13076.

246. Gravallese, E.M., J.M. Darling, A.L. Ladd, J.N. Katz, and L. Glimcher. 1991. In situ hybridization studies on stromelysin and collagenase mRNA expression in rheumatoid synovium. *Arthritis Rheum.* 34:1071–1084.

247. Firestein, G.S., and M. Paine. 1992. Expression of stromelysin and TIMP in rheumatoid arthritis synovium. *Am. J. Pathol* 140:1309–1314.

248. Konttinen, Y.T., A. Ceponis, M. Takagi, M. Ainola, T. Sorsa, M. Sutinen, T. Salo, J. Ma, S. Santavirta, and M. Seiki. 1998. New collagenolytic enzymes/cascade identified at the pannus-hard tissue junction in rheumatoid arthritis: destruction from above. *Matrix Biol.* 17:585–601.

249. Kikuchi, H., W. Shimada, T. Nonaka, S. Ueshima, and S. Tanaka. 1996. Significance of serine proteinase and matrix metalloproteinase systems in the destruction of human articular cartilage. *Clin. Exp. Pharmacol. Physiol.* 23:885–889.

250. Lindy, O., Y.T. Konttinen, T. Sorsa, Y. Ding, S. Santavirta, A. Ceponis, and C. López-Otín. 1997. Matrix-metalloproteinase 13 (collagenase 3) in human rheumatoid synovium. *Arthritis Rheum.* 40:1391–1399.

251. Clark, I.M., L.K. Powell, S. Ramsey, B.L. Hazelman, and T.E. Cawston. 1993. The measurement of collagenase, TIMP and collagenase-TIMP complex in synovial fluids from patients with osteoarthritis and rheumatoid arthritis. *Arthritis Rheum.* 36:372–380.

252. Okada, Y., N. Takeuchi, K. Tomita, I. Nakanishi, and H. Nagase. 1989. Immunolocalization of matrix metalloproteinase 3 (stromelysin) in rheumatoid synovioblasts (B cells): correlation with rheumatoid arthritis. *Ann. Rheum. Dis.* 48:645–653.

253. Westhoff, C.S., D. Freudiger, P. Petrow, C. Seyfert, J. Zacher, J. Kriegsmann, T. Pap, S. Gay, P. Stiehl, E. Gromnica-Ihle, and D. Wernicke. 1999. Characterization of collagenase 3 (matrix metalloproteinase 13) messenger RNA expression in the synovial membrane and synovial fibroblasts of patients with rheumatoid arthritis. *Arthritis Rheum.* 42:1517–1527.

254. Pap, T., S. Kuchen, K.M. Hummel, J.K. Franz, R.E. Gay, and S. Gay. 1998. In-situ expression of membrane-type matrix metalloproteinase 1 (MT1-MMP) and MT3-MMP mRNA in rheumatoid arthritis (RA). *Arthritis Rheum.* 41:S317–S317

255. Sato, H., T. Takino, Y. Okada, J. Cao, A. Shinagawa, E. Yamamoto, and M. Seiki. 1994. A matrix metalloproteinase expressed on the surface of invasive tumour cells. *Nature* 370:61–65.

256. DiBattista, J.A., J.P. Pelletier, M. Zafarullah, N. Fujimoto, K. Obata, and P.J. Martel. 1995. Coordinate regulation of matrix metalloproteases and tissue inhibitor of metalloproteinase expression in human synovial fibroblasts. *J. Rheumatol. Suppl.* 43:123–128.

257. Firestein, G.S., and M.M. Paine. 1992. Stromelysin and tissue inhibitor of metalloproteinases gene expression in rheumatoid arthritis synovium. *Am. J. Pathol.* 140:1309–1314.

258. Cunnane, G., O. Fitzgerald, K.M. Hummel, R.E. Gay, S. Gay, and B. Bresnihan. 1999. Collagenase, cathepsin B and cathepsin L gene expression in the synovial membrane of patients with early inflammatory arthritis. *Rheumatology (Oxford)* 38:34–42.

259. Lemaire, R., R.M. Flipo, D. Monte, T. Dupressoir, B. Duquesnoy, J.Y. Cesbron, A. Janin, A. Capron, and R. Lafyatis. 1994. Synovial fibroblast-like cell transfection with the SV40 large T antigen induces a transformed phenotype and permits transient tumor formation in immunodeficient mice. *J. Rheumatol.* 21:1409–1419.

260. Lemaire, R., G. Huet, F. Zerimech, G. Grard, C. Fontaine, B. Duquesnoy, and R.M. Flipo. 1997. Selective induction of the secretion of cathepsins B and L by cytokines in synovial fibroblast-like cells. *Br. J. Rheumatol.* 36:735–743.

261. Huet, G., R.M. Flipo, C. Colin, A. Janin, B. Hemon, D.M. Collyn, R. Lafyatis, B. Duquesnoy, and P. Degand. 1993. Stimulation of the secretion of latent cysteine proteinase activity by tumor necrosis factor alpha and interleukin-1. *Arthritis Rheum.* 36:772–780.

262. Hummel, K.M., P.K. Petrow, J.K. Franz, U. Müller-Ladner, W.K. Aicher, R.E. Gay, D. Brömme, and S. Gay. 1998. Cysteine proteinase cathepsin K mRNA is expressed in synovium of patients with rheumatoid arthritis and is detected at sites of synovial bone destruction. *J. Rheumatol.* 25:1887–1894.

263. Rich, E., L.W. Moreland, and G.S. Alarcon. 1999. Paucity of radiographic progression in rheumatoid arthritis treated with methotrexate as the first disease modifying antirheumatic drug. *J. Rheumatol.* 26:259–261.

264. Rau, R., B. Schleusser, G. Herborn, and T. Karger. 1997. Long-term treatment of destructive rheumatoid arthritis with methotrexate. *J. Rheumatol.* 24:1881–1889.

265. Pascro, G., F. Priolo, E. Marubini, F. Fantini, G. Ferraccioli, M. Magaro, R. Marcolongo, P. Oriente, V. Pipitone, I. Portioli, G. Tirri, F. Trotta, and C. Della. 1996. Slow progression of joint damage in early rheumatoid arthritis treated with cyclosporin A. *Arthritis Rheum.* 39:1006–1015.

266. Crofford, L.J., R.E. Lipsky, P. Brooks, S.B. Abrahamsou, L.S. Simon, and L.B. van der Putte. 2000. Basic biology and clinical application of specific cyclooxygenase-z inhibitors. *Arthritis Rheum.* 43:4–13.

267. Greene, S., K. Watanabe, J. Braatz-Trulson, and L. Lou. 1995. Inhibition of dihydroorotate dehydrogenase by the immunosuppressive agent leflunomide. *Biochem. Pharmacol.* 50:861–867.

268. Smolen, J.S., J.R. Kalden, D.L. Scott, B. Rozman, T.K. Kvien, A. Larsen, I. Loew-Friedrich, C. Oed, and R. Rosenburg. 1999. Efficacy and safety of leflunomide compared with placebo and sulphasalazine in active rheumatoid arthritis: a double-blind, randomised, multicentre trial. European Leflunomide Study Group. *Lancet* 353:259–266.

269. Moreland, L.W., M.H. Schiff, S.W. Baumgartner, E.A. Tindall, R.M. Fleischmann, K.J. Bulpitt, A.L. Weaver, E.C. Keystone, D.E. Furst, P.J. Mease, E.M. Ruderman, D.A. Horwitz, D.G. Arkfeld, L. Garrison, D.J. Burge, C.M. Blosch, M.L. Lange, N.D. McDonnell, and M.E. Weinblatt. 1999. Etanercept therapy in rheumatoid arthritis. A randomized, controlled trial. *Ann. Intern. Med.* 130:478–486.

270. Moreland, L.W., S.W. Baumgartner, M.H. Schiff, E.A. Tindall, R.M. Fleischmann, A.L. Weaver, R.E. Ettlinger, S. Cohen, W.J. Koopman, K. Mohler, M.B. Widmer, and C.M. Blosch. 1997. Treatment of rheumatoid arthritis with a recombinant human tumor necrosis factor receptor (p75)-Fc fusion protein. *N. Engl. J. Med.* 337:141–147.

271. Moreland, L.W. 1999. Inhibitors of tumor necrosis factor: new treatment options for rheumatoid arthritis. *Cleve. Clin. J. Med.* 66:367–374.

272. Jorgensen, C., and S. Gay. 1998. Gene therapy in osteoarticular diseases: where are we? *Immunol. Today* 19:387–391.

273. Pap, T., R.E. Gay, and S. Gay. 2000. Gene transfer: from concept to therapy. *Curr. Opin. Rheum.* 12:205–210.

23 | Experimental Animal Models of Rheumatoid Arthritis

Rikard Holmdahl

1. INTRODUCTION

Rheumatoid arthritis (RA) is a disease defined by a number of clinically based criteria. It is likely that the described disease is phenotypically and genetically heterogeneous, indicating that RA is a syndrome comprised of several diseases with different pathogenesis. Thus, there is no optimal model for RA, although various animal models can be used to address different parts of the pathogenesis leading to arthritis.

There are, however, some characteristics of RA that lead to understanding regarding the fundamental events in the pathogenesis.

- The disease primarily affects diarthrodial, cartilaginous peripheral joints, although typical systemic manifestations may also occur. Thus, the inflammatory attack is tissue-specific.
- Bone and cartilage are eroded by the inflammatory tissue
- No persisting infectious pathogens have yet been found that could explain inflammation in RA.
- The disease is chronic.
- The disease severity and chronicity is associated with certain MHC class II haplotypes.
- The disease does not develop spontaneously, i.e. it is not a monogenic disease with high penetrance, but is induced by unknown environmental conditions in susceptible individuals that harbor a set of disease-associated genes.

The selection of relevant animal models should be based on the possibilities to address these characteristics. Among the commonly used models that fulfill such criteria are the various cartilage protein-induced models, such as collagen-induced arthritis (CIA), and some of the adjuvant-induced arthritis models, such as pristane-induced arthritis (PIA). These models will be discussed in more detail. Other recently-described models that will most likely turn out to be highly useful are various transgenic mouse strains that spontaneously develop arthritis. Among these are mice over-expressing TNFa, mice expressing HTLV tat and a mouse strain expressing a self-reactive T cell receptor (1–7). A major advantage of these models are the high penetrance of disease, which facilitates the set-up of well-defined experimental approaches. In addition, there are a large number of other models for RA that could reflect specific parts of the different pathogenic mechanisms leading to RA, but that will not be described in detail here.

2. ANIMAL MODEL HISTORY, THE ADJUVANT (MYCOBACTERIUM CELL WALL)-INDUCED ARTHRITIS

An interesting development in identifying useful adjuvants for effective vaccination was the unexpected discovery that complete Freund's adjuvant induced severe arthritis in rats (8). Complete Freund's adjuvant (CFA) is a suspension of freeze-dried tubercle mycobacterium cell wall fragments in mineral oil. It was soon found, however, that the adjuvant induced not only severe arthritis, but also widespread inflammatory infiltrates and granuloma formation in many organs, such as the spleen, liver, bone marrow, meninges, skin and eyes (8,9). The adjuvant-induced disease is severe, but self-limited, and the rat recovers within a few months. The pathology is believed to be caused by spreading of peptidoglycan fragments from the mycobacterium

Table 1 Characteristics of RA and some selected animal models[1]

	RA	OIA in rats	PIA in rats	Heterologous CIA in mice and rats	Homologous CIA in mice and rats	"Spontaneous" arthritis in DBA/1 mice
Bone and cartilage erosions	+++	+/– (20, 22)	+++ (23)	+++ (116)	+++ (55)	–
Enthesopathy and new bone formation	+/–	–	++/–	++/–	++/– (55)	+++ (113)
T cell infiltration in the joints	+++ (137)	+ (22)	+ (23)	+ (86, 138)	+	–
Circulating COMP	+++ (139)	nd	+++ (39)	++	+++ (140)	nd
Symmetric involvement of peripheral joints	+	+ (141)	+	–	–	–
Chronicity (sustained or relapsing joint inflammation and erosion)	+++	–(21)	+++ (23)	+/–	+++ (55, 142, 143)	–
Rheumatoid factors	+++	nd	++	+ (selfassociating RF)(144, 145)	nd	–
Immune response to cartilage	++ (B cell response to CII)(146)	–	–	+++(147–149)	+++ (55, 56)	–
MHC association	DR4, DR1 shared epitope alleles) (150)	Not n in the rat (151)	f>>a,u,c,d,n (23)	u>a>l>c>b>n in rat CIA (117, 152) q and r in mouse CIA (116, 119)	a>u>f>>c,d,n in rat CIA (153) q in mouse CIA (154)	–
Gender preponderance	Females	Females	Females (23)	Females (rat)(155) Males (mouse)(156)	Females (rat) (157) Males (mouse) (55, 158)	Males (112)
Pregnancy mediated suppression and exacerbation postpartum	+	nr	nr	+ (159, 160)	nr	Not applicable
Estrogen mediated suppression	+?	nr	nr	+ (161, 162)	+ (163)	nr

[1] The selected models are described in the main text. The various traits are quantiatively graded using a scvale from +++ (pronounced), ++ (cleali present), + (barely significant), to – (not present). A slash (/) indicate variability. nr = not reported. References are lacking if the description is based on unpublished observation or is common text book knowledge.

cell walls that are difficult for macrophages to digest and thereby are activated (10,11). The disease is T cell-dependent, but there is no consistent evidence for a role of specific T cells crossreactive to joints or other self-tissues or to ubiquitously expressed proteins, such as heat shock proteins (12–14). Interestingly, the defined peptidoglycan component identified to be the arthritogenic principle is not recognized by T cells but has potent adjuvant capacity, indicating that it stimulates antigen-presenting cells or innate immune recognition. Surprisingly, so far only rats (and not mice or primates) have been shown to develop arthritis after mycobacterium challenge (15), although it has been reported that granuloma formation in joints has occurred in humans treated with BCG (16).

Whatever the causes of the inflammatory disease triggered by mycobacterium in oil, it is not an optimal model for RA since it does not primarily affect joints and is not chronic. At the same time, however, two discoveries led to the establishment of more appropriate animal models.

First, it was found that the oil component in which the mycobacterium was suspended enhanced arthritis induction (17), and later it was shown that some of these adjuvants induced arthritis by themselves (18–24). These components are not immunogenic; they are not necessarily derived from infectious pathogens and are not known to trigger innate receptors. Rather, their arthritogenicity seem to be related to their physical structures, which allowed solubilisation in cell membranes. One of the most interesting components is pristane (2,6,10,14-tetramethylpentadecane), which induces disease in both mice and rats and which, in the rat, shows many similarities with RA.

Second, it was found that immunization with cartilage-derived type II collagen induced severe arthritis in rats. The immune-triggered inflammation targeted the same tissues as in RA, the peripheral joints, and could develop chronically. The induction required mineral oil as an adjuvant but not complete Freund's adjuvant. It is likely that most of the cartilage-derived proteins have the capacity to induce arthritis similar to the capacity of most central nervous proteins to induce encephalomyelitis. However, pathogenesis of adjuvant-induced arthritis (such as PIA) is different from that of cartilage protein-induced arthritis (such as CIA).

3. DEFINED, NONBACTERIAL, ADJUVANTS: PRISTANE-INDUCED

4. ARTHRITIS (PIA) IN RATS

Mineral oil-induced arthritis (OIA)(22), avridine-induced arthritis (18) and pristane-induced arthritis (PIA) in the rat (23) share many common features and differ mainly in the severity and chronicity of the disease. They are induced with adjuvant compounds lacking immunogenic capacity; i.e., no specific immune responses are elicited towards them after injection. Instead they are rapidly and widely spread throughout the body after a single subcutaneous injection, and penetrate through cell membranes into cells. After a delay of at least one or two weeks, arthritis

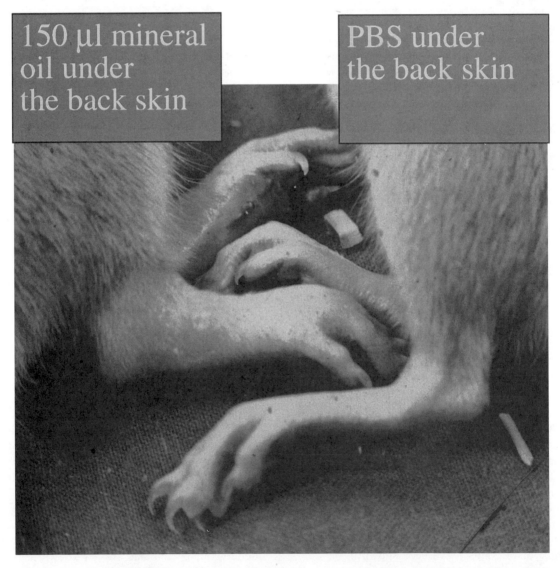

Figure 1.   Mineral oil-induced arthritis in the hind paw of a DA rat as compared with a control DA rat. The arthritic rat had been injected with 150 µl mineral oil in the back skin. The first appearance of PIA and CIA are indistinguishable, but will later develop chronically

suddenly ensues in the peripheral joints, with a similar distribution as seen in RA, and is mainly symmetrical. Occasionally other joints are involved, but systemic manifestations in other tissues have not been reported. In certain rat strains, especially the avridine-induced arthritis and PIA models, arthritis develops as a chronic relapsing disease. Surprisingly, no immune response to cartilage proteins has been observed, although rheumatoid factors are present in serum. A role for cartilage proteins in regulating disease activity is possible since the disease can be prevented and, in fact, therapeutically ameliorated by nasal vaccination with various cartilage proteins (25). Another way to both prevent and therapeutically decrease disease severity in an already established phase is by blocking $\alpha\beta$ T cells (23,26). Together with the observation that the chronic disease course is associated with the MHC region (23,26), this could implicate the activation of T cells recognizing cartilage proteins. However, such T cells have not been observed and T cell transfer of the disease has so far failed to identify antigen-specific T cells (27,28). The inducing agents are all small molecular structures unable to bind to MHC class II molecules and to be recognized by class II-restricted T cells. A role for environmental infectious agents is not likely, since no difference in disease susceptibility could be seen in germ free rats (29). An interesting observation was that the germfree rats did not mount a response mediated by lymphocyte recognition or a specific interaction with receptors involved in the innate immune system. Surprisingly, some of the arthritogenic adjuvants are, in fact, components already present in the body before injection. For example, squalen, a component of cell membranes, is highly arthritogenic in DA rats (24). Furthermore, pristane is also present in normal tissue, since it is a component of chlorophyll and is normally ingested by all mammals, including laboratory rats and humans (30,31). Pristane is taken up through the intestine and spread throughout the body. However, all arthritogenic adjuvants share the capacity to penetrate into cells where they could change membrane fluidity and modulate transcriptional regulation (32,33). The injection route and doses are critical; i.e. it is most likely of critical importance which cell is first activated and to what extent. It is also possible that the arthritogenic adjuvants may have capacity as reactive haptens, or that they are bound to CD1 and recognized by CD1-restricted T cells, although there is no evidence to support these possibilities. The mechanisms of adjuvant actions leading to the development of arthritis are, in addition, likely to be stage-specific. It takes minutes for the adjuvant to spread in the body after injection, but it takes weeks for it to trigger the development of arthritis and months for the arthritis to develop into chronicity. These stages are most likely different processes and consequently should be controlled by distinct sets of genes.

4.1. Genetic Control of PIA

There is a genetic predisposition to develop PIA, as well as for other adjuvant arthritis models. The role of the major histocompatibility region has been addressed for most models using established congenic strains. Thus, studies using MHC congenic LEW rat strains have shown that the MHC locus (designated *Pia1*) contributes to the genetic predisposition (23). The most susceptible haplotype was the RT1-f, which is also associated with the genetic control of avridine-induced arthritis (26) and type XI collagen-induced arthritis (Lu et al., to be published). Interestingly, the effect was seen only on the chronic development of the disease, whereas the day of arthritis onset or incidence was not significantly associated. Analyses of the genetic control outside the already identified MHC region has been initiated not only on PIA, but also other adjuvant arthritis models and collagen-induced arthritis in mice and rats using microsatellite mapping of genetically segregating crosses. Most of the efforts to analyze the genetic control in the rat has taken the advantage of the finding that the DA rat is highly susceptible to several arthritis models, such as mycobacterium-induced arthritis (34), oil-induced arthritis (35), CIA (36,37) and PIA (38). However, only PIA had the advantage of developing a chronic disease course. Thus, after injection of pristane, the DA rat develops disease with high incidence, pronounced joint erosions, high clinical severity and with a chronic relapsing disease course. In contrast, the E3 rats are resistant, and analysis of recombinant inbred strains between DA and E3 has shown that there is a polygenic control of various subtraits of the disease such as onset, erosions and chronicity (23). To clarify the complex genetics of arthritis, the (E3xDA)F2 rats were subjected to a total genome scan using microsatellite markers covering the genome and specific loci linked to different subtraits of the development of arthritis, such as day of onset, erosive arthritis, clinical severity and chronicity could be identified. Arthritis usually develops with a sudden onset in a peripheral joint, and the severity rapidly progresses within the next days. Two loci (*Pia2* on chromosome 4 and *Pia3* on chromosome 6) were found to be associated with onset. Interestingly, these had no effect on the severity or chronicity of the disease. A major locus controlling arthritis severity at the early period between day 20 to day 45, *Pia4*, was found on chromosome 12. DA alleles promoted disease in an additive fashion. The strongest linkage was with clinical scores at day 35, when the erosions of the joints are most pronounced. Consequently, the serum levels of circulating cartilage oligomeric matrix protein (COMP) (39) was raised and linked to the same locus. It was found that loci on chromosome 4 (*Pia5*) and 14 (*Pia6*) were associated with active arthritis at the chronic stage. Interestingly, these loci were not associated with the severity or onset of arthritis phenotypes. These findings suggest that different stages of the disease are controlled by different sets of genes and consequently are dependent on different mechanisms. Interestingly, some, but not all, of the genetic loci are shared between the different arthritis models and also some other autoimmune disease, suggesting that the pathogenesis is partly shared (34,40). Identification of the underlying genes should be a possible and relevant way to dissect the mechanisms in these complex diseases.

4.2. Pristane-Induced Arthritis in Mice

Surprisingly, adjuvant arthritis is not easily inducible in species other than rats. Of the above-mentioned adjuvants, only pristane-induced arthritis (PIA) has been described in the mouse (19,41). The induction of PIA in the mouse requires repeated intraperitoneal injections of pristane that triggers an inflammatory disease with a late and insidious onset. The disease is clearly different from PIA in the rat; the same inducing protocol does not induce disease in the other species and the disease course and characteristics are different. In the mouse, a role for an induced recognition of heat shock proteins has been observed, although such experiments have not been performed using germfree animals (42,43).

However, there are also similarities, such as the chronicity of arthritis, the T cell dependency and maybe also the MHC association (41). The difference between rats and mice is surprising, but may be related to the specific effects of pristane in the mouse with its capacity to induce tumours after intraperitoneal injections (44).

5. CARTILAGE PROTEIN-INDUCED ARTHRITIS

Collagen-induced arthritis (CIA) was first demonstrated in 1977 in the rat (45) and was later reported using other species such as mouse (46) and primates (47). Precaution should be taken using cartilage protein models such as CIA, since it is most likely easily inducible in humans, particularly in individuals with the RA-susceptible HLA haplotypes, DR4 and DR1.

Induction of arthritis with cartilage proteins such as CII requires immunization in an animal strain that is able to mount an autoimmune response. The immunization requires emulsification of CII in adjuvant such as mineral oil or complete Freund's adjuvant, but the disease is different from the various forms of adjuvant-induced arthritides (45) even when a rat strain is used that can develop arthritis with only mineral oil (21).

Not surprisingly, several other cartilage-proteins, such as aggrecan (48), cartilage link protein (49), type XI collagen (CXI) (50), gp39 (51) and cartilage oligomeric matrix protein (COMP) (52) have subsequently been shown to be arthritogenic in different animal strains. All of these models

have been shown to, or are likely to be, associated with immune response genes, i.e., MHC class II genes. It is a surprisingly strict control in which certain haplotypes confer protection whereas others allow the immune response and the disease to develop. Thus, a limited set of peptides is probably recognized, a limitation that may depend on selection of the immune system in response to the presence of these proteins in the cartilage. Thus, there may be more or less pronounced tolerance depending on the location and role of the various proteins in the cartilage. It has, for example, been difficult to show that type IX collagen (CIX) induces arthritis (53), an observation that could depend on the accessibility of CIX on the outer core of the collagen fibrils. A comparison of the various susceptible strains also indicates that genes outside MHC influence disease susceptibility, strongly indicating that the various forms of models differ in pathogenic mechanisms. A common feature, however, is that once arthritis develops, the lesions seem to primarily affect the peripheral cartilaginous joints, thus at least partly reflecting the distribution of the target protein. It could also be that peripheral joints are more sensitive to arthritis due to some other factor such as vulnerability to trauma or the presence of a synovial tissue. This would explain why arthritis does not affect other tissues containing cartilage but lacking synovial tissue or joint functions. Support for this possibility is the development of peripheral arthritis in the adjuvant arthritis models and the spontaneous arthritis developing in T cell receptor transgenic mice producing autoantibodies to a ubiquitous mitochondrial antigen (3,4). Support for a contributing role of immune-directed tissue specificity is the recent demonstration that immunization with matriline 1, a protein that is restricted to tracheal and nasal cartilage, induced a disease affecting these tissues (54) that was inducible in both mice and rats and resembled relapsing polychondritis rather than RA.

Although these various models have different characteristics and genetics, there are also similarities, and the CIA model induced with the major cartilage protein CII will be discussed as a prototype of cartilage protein induced diseases.

5.1. Collagen-Induced Arthritis (CIA)

The CIA model is perhaps the most commonly used model for RA today. However, the model varies considerably depending on the experimental animal species and on whether the type II collagen (CII) used is of self or non-self origin. In both rats and mice immunised with heterologous (non-self) CII, a severe, erosive polyarthritis suddenly develops 2–3 weeks after immunisation. The inflammation usually subsides within 3–4 weeks, although in certain strains a few animals may develop a chronic relapsing disease. Interestingly, arthritis is not as easily inducible with homologous (self) CII in rats and mice, but, once started, it is as severe, and tends to be more chronic than the disease induced with heterologous CII (55,56). The disease is critically dependent on a strong T and B cell immune responses to CII (57,58). Both T and B cells recognise selected epitopes on CII (59–63). The T cells respond to dominant peptides bound to MHC class II molecules and the B cells recognise conformational epitopes on the triple helical part of the molecule. Both T and B cells are selected not only by the CII used for immunisation, but also by the presence of CII in cartilage.

The pathogenic events in the disease induced with homologous as well as heterologous CII are largely unknown, but are likely to develop in different stages, as described above for the PIA model in the rat. The best known events are the priming and the effector phase of the acute and self-limited arthritis induced with heterologous CII. An unresolved issue, however, is the relative importance of T versus B cells in the effector phase. The development of arthritis is a very complex process that can occur through different inflammatory mechanisms and, which varies between strains and species. It is, therefore, sometimes difficult to interpret the large numbers of

profound and often diverging effects on CIA using mice deleted for various genes of putative importance for arthritis. It is clear, however, that a significant part of the inflammatory attack on the joints is mediated by pathogenic antibodies (64–66). There is still a controversy as to whether such antibodies bind to normal cartilage (67–69), but once bound in high enough density, they are likely to fix complement, attract neutrophilic granulocytes and activate macrophages through binding their Fc receptors (70–72). Mice deficient in complement C5 or depleted of complement are resistant to CIA, as are mice lacking FcRI and III, showing that both of these pathways are of importance in the disease. A role for T cell-mediated activation of macrophages and other synovial cells is also likely (73). CIA in both rats and mice has been ameliorated by treatment with antibodies to the T cells after the establishment of the antibody response (74,75). Passive transfer of the disease using CII-reactive T cells is, however, not as readily effective as in other autoimmune models, such as experimental allergic encephalomyelitis (76). Thus, only microscopic arthritis or low incidence of clinical arthritis could be induced by CII-reactive CD4+ T cells (77,78) whereas a somewhat more severe arthritis by CII-reactive

Acute with erosions

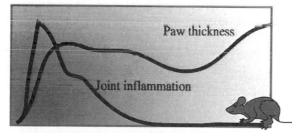

- Mycobacteria induced adjuvant arthritis (AA)
- Collagen induced arthritis (CIA) induced with heterologous cartilage collagen

Acute without erosions

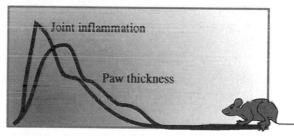

- Mineral oil induced arthritis (OIA) in DA rats

Chronic relapsing with erosions

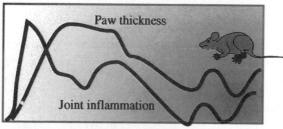

- Avridine induced arthritis (AvIA)
- Pristane induced arthritis (PIA)
- CIA induced with homologous cartilage collagen

Chronic relapsing without erosions

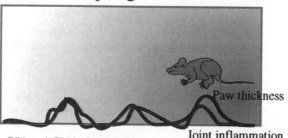

- PIA and CIA in DXEC rats

Figure 2.   Illustration of the disease course of different arthritis models in the rat.

T cells could be induced in immunodeficient mice (SCID mice) (79,80). The relative inefficiency of T cell transfer of arthritis also seems to be the case in other cartilage protein-induced arthritis, models. In proteoglycan(aggrecan)-induced arthritis it is possible to transfer disease with proteoglycan-reactive T cells, but is more effective if the (heterologous) proteoglycan antigen is added in the transfer (81). A cooperative role of T and B cells in the effector phase was suggested by combined transfer experiments (82–84). Histopathological analysis of the inflamed joints supports the notion that the arthritis is mediated by different effector arms and develops through different stages (85–88). Thus, binding of antibodies to the cartilage surface will trigger Fc-receptors on synovial cells and complement activation, leading to infiltration of neutrophilic granulocytes and severe edematous arthritis. A T cell-dependent activation of macrophages, osteoclasts and fibroblasts at the marginal zone and the development of a highly vascularized pannus tissue is, on the other hand, associated with a subchondral erosive process. A healing response with new bone and cartilage formation is often seen a few weeks after the onset of arthritis, leading to restructuring and ankylosis of the joints. The complexity of the inflammatory and healing responses in CIA is also reflected in studies of the involvement of cytokines and trafficking inflammatory cells by therapeutic or genetically manipulative experiments. In general, however, lymphocyte-enhancing costimulatory molecules (89,90) and macrophage-derived cytokines such as IL-1, TNFa and IL-12, have disease-promoting effects (91–96), whereas regulatory cytokines, such as IL-10, and cytokines participating in tissue healing, such as TGFb, ameliorate disease (97–101). These effects are, however, complex and can give widely variable results depending on genetic background, doses, location and the stage of the disease (102–108). The relative importance of the TH1 versus TH2 type of inflammatory responses in CIA has not been clarified and may also vary both on genetics and disease stage(109–111). Different strains develop different types of disease. The DBA/1 strain, which is the most common for induction of CIA, tends to develop severe arthritis that is also self-limited and includes development of new bone formation and ankylosis. In fact, male DBA/1 may spontaneously develop enthesiopathy with a clinical appearance that could resemble CIA, but with a completely different pathogenesis (112,113).

It is clear that, just as discussed with PIA, the disease is very complex and varies depending on animal strain, whether on the induction is with heterologous or homologous proteins and environmental influences. In addition, the disease will occur in stages depending on different mechanisms. The problem will be to identify the critical molecular interactions that lead to the critical pathogenic events in the disease. A comparative analysis of susceptible and resistant strains to find the critical genetic polymorphism controlling such events is one possible approach and several gene regions have been identified controlling CIA in both mice (72,114,115) and rats (36,37). These studies show that the disease is complex and polygenically controlled. In each of these studies, crosses between inbred strains for a selected set of genes seems to operate, and it is likely that there are also specific sets of genes that control specific events in disease development. In the future, the underlying genes and how they interact to control the specific pathogenic events in CIA will be identified. Work on the role of major histocompatibility complex has demonstrated that this is a feasible endeavour and our present knowledge on MHC is useful for further dissection of the disease mechanisms.

5.2. The Role of the Major Histocompatibility Complex

Early observations using the CIA model in both mice and rats induced with heterologous CII indicated a role for the major histocompatibility complex region (116,117). It was later found that the MHC association of CIA induced with homologous CII was even more limited to certain haplotypes, Immunisation of rats with homologous (rat) CII leading to arthritis development is associated with MHC class II genes, with the a haplotype as the most permissible, u, f and l intermediate and n resistant (56). In the mouse, CIA induced with both heterologous and homologous CII is most strongly associated with the H-2q and H-2r haplotypes, although most other haplotypes such as b, s, d and p are not totally resistant to disease induced with heterologous CII (46,116,118). Of interest, it has been possible to further map the MHC association of CIA to the genes coding for the MHC class II A molecule (119). Moreover, the immunodominant peptide derived from the CII molecule bound to the arthritis-associated q variant of the A (A^q) molecule has been found to be located between positions 256 and 270 of CII (59,60). This is a glycopeptide with an oligosaccharide pointing towards the T cell receptor and is recognised by many of the CII-reactive T cells (120). Interestingly, the peptide binding pocket of the A^q molecule is very similar to that of the DR4 (DRB1*0401/DRA) and DR1 molecules, which are associated with RA (121). Furthermore, mice transgenically expressing DR4 or DR1 are susceptible to CIA and respond to a peptide from the same CII region (122–124) and recently reviewed in ((125)). The CII260–273 peptide bind to the DR4*0401 molecule with the major MHC anchors shifted 3 amino acids compared with the binding of the same peptide to the A^q molecule, but the T cell receptor recognition sites are partly shared. Thus, these humanised mouse strains provide us with a CIA model, that shares molecular structures of possible critical importance for the human disease.

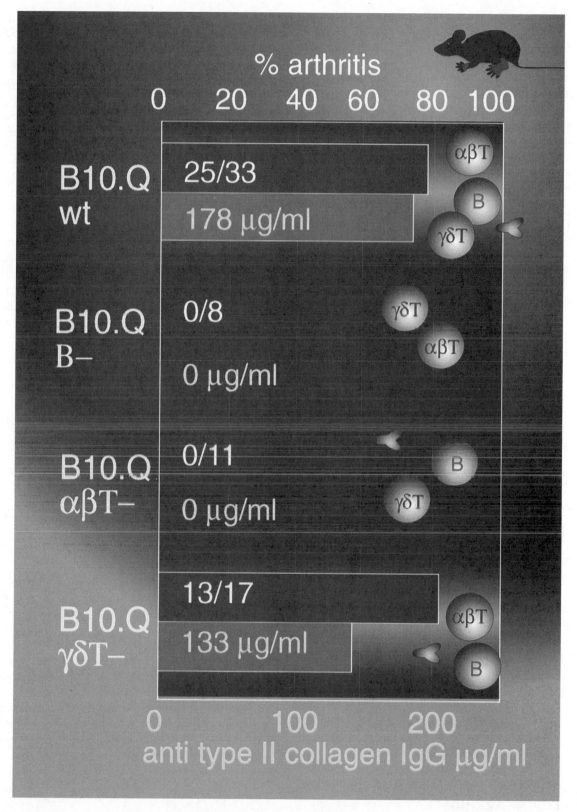

Figure 3. Illustration of the dependency of functional αβ T cells and B cells for development of CIA. The experiments are done with mice genetically deleted for αβ TCR, γδ TCR and μ BCR, respectively, and backcrossed to the B10.Q background (57,58).

5.3. Autoimmune Recognition of Cartilage

It is important to emphasise that the identified structural interaction between MHC class II+peptide complexes and T cells does not give us the answer to the pathogenesis of CIA (or RA), but rather a better tool for further analysis. An important question is how the immune system interacts with the peripheral joints, i.e., how autoreactive T and B cells are normally tolerized and what happens pathologically after their activation by CII immunisation. Most of the T cells reactive with the CII256–270 peptide do not cross-react with the corresponding peptide from mouse CII. The difference between the heterologous and the homologous peptide is position 266, in which the rat has a glutamic acid (E) and the mouse an aspartic acid (D). The importance of this minor difference was demonstrated in transgenic mice that express mutated CII with a glutamic acid at this position. When mutated CII was expressed in cartilage, the T cell response to CII was partially, but not completely tolerized. The mice were susceptible to arthritis, but the incidence was lower, similar to that seen in mice immunised with homologous CII. This finding shows that a normal interaction between cartilage and T cells leads to tolerance manifested by an impaired capacity to give rise to a recall proliferative response to CII. However, the tolerance is not complete, since the T cells can still produce effector cytokines such as gamma interferon and give help to B cells. The observed resistance to arthritis could not be mediated by regulatory effects mediated by the autoreactive T cells since the expression of the mutated CII in mice expressing both H-2r and H-2q haplotypes were highly susceptible to CIA with bovine CII (harbouring a bovine specific Ar-restricted peptide), but showed a lower incidence if immunised with rat CII (126). Moreover, a role for newly exported T cells, which had not yet been subjected to tolerization, was excluded by thymectomy experiments (127). Thus, partially-tolerized T cells may, under extreme circumstances (such as CII immunisation), mediate arthritis. In contrast, B cells reactive with CII are not tolerized and, as soon as the T cells are activated even in a partially-tolerized state, they help B cells produce autoreactive and most likely pathogenic antibodies. It is possible that a similar situation exists in humans, which could explain the difficulties in isolating CII reactive T cells compared with the relative ease with which CII reactive B cells can be detected in the joints.

6.   ANIMAL MODELS WITH "SPONTANEOUSLY" DEVELOPING ARTHRITIS

It is clear from studies of the various induced arthritis models, such as PIA and CIA, that the genetic constitutions of these animals dramatically influences the penetrance of disease susceptibility. This raises the possibilities that arthritis-prone strains may spontaneously develop arthritis. In fact, many inbred mouse and rat strains do occasionally develop low frequency of arthritis (112,128,129). In some strains, such as DBA/1, high frequency of arthritis appears in certain cages with male mice. The grouping of males induces inter-male aggressiveness and such stress seems to be associated with arthritis development (112). Closer analysis of the disease shows that the pathogenesis is very different from collagen-induced arthritis, with no or very minimal erosions and inflammation of the joints. Instead, the tendon insertions are transformed as in enthesopathy, leading to cartilage formation in the tendons, outgrowth of osteophytes and eventually ankylosis. Interestingly, a similar form of enthesopathy can be induced in mice given an overdose of vitamin A (130). The stress-induced arthritis in DBA/1 mice is most likely not immune-mediated since it also develops in mice lacking $\alpha\beta$ T cells (113). It is, perhaps, more useful as a model for psoriasis arthritis or ankylosing spondylitis than RA, but the possible occurrence of spontaneous enthesopathy is important to take into consideration when evaluating animal models for RA.

There are, however, also a number of genetically-defective mouse and rat strains, that spontaneously develop arthritis. One such strain is the MRL/lpr mouse, defective in the Fas gene, which develops microscopic arthritis at older ages (over 5 moths of age) (131). This is characterised by expansion of the synovial tissue with a large number of relatively uniform mesenchymal cells, but scarce in inflammatory cells such as lymphocytes and neutrophilic granulocytes (88,132,133). The development of the joint pathology occurs subsequent to development of severe lupus disease and is, as lupus is, associated with the Fas defect, although other not yet identified genes give important contributions (134).

Spontaneous arthritis has also been observed in a number of different transgenic mouse strains. Examples are mouse strains in which TNFa is overexpressed leading to inflammation in tissues with elevated TNFa (1). Another is a T cell receptor (TCR) transgene backcrossed NOD mouse in which the introduced recombined T cell receptor expressed on peripheral T cells seems to recognise an unknown self-peptide bound to the H-2g7 class II molecule. Surprisingly, this mouse spontaneously developed severe arthritis (4) with the inflammatory attack restricted to the joints in which arthritis develops before puberty (3 weeks of age) with high penetrance. Immune complex precipitations were also found on viscera, and the disease was dependent on B cell activity. Most recently, it was shown that the arthritis in this model is mediated by antigen-specific IgG (3). Transfer of serum or purified serum IgG antibodies from the T cell receptor transgenic mice into wild type mice produce arthritis as early as 24 hours after transfer. The IgG-induced disease had some characteristics similar to the spontaneously-occurring arthritis in T cell receptor transgenic mice, but differed in that it was transient and did not heal with deformed joints. A comparison with earlier established models for rheumatoid arthritis reveals a number of important similarities and differences. The IgG-induced disease has several similarities with various rheumatoid arthritis models induced by cartilage proteins, for example collagen induced arthritis, because the arthritis is dependent on both T and B cells and transient arthritis can be induced with IgG self-reactive antibodies or autoantibodies (rev. in (135). In fact, the kinetics and pathology of the described antibody-mediated arthritis is indistinguishable from anti-CII antibody-mediated arthritis. In addition, dissimilar to rheumatoid arthritis but similar to CIA, there is a paucity of lymphocytes in joints and the healing response may include extensive new bone formation leading to remodelling and deformation of the joints. The main difference between the IgG-induced disease and collagen-induced arthritis, however, is the apparent lack of cartilage-specific immunity in the IgG-induced arthritis. The arthritogenic IgG antibodies recognise an antigen that is ubiquitously-expressed and not specific to cartilage. It remains an important possibility that the ubiquitously expressed protein is exposed in the joints as a neo-epitope, i.e., it is expressed by traumatised, stressed or apoptotic cells in peripheral joints and then released to form precipitating immune complexes. A similar mechanism may occur in the classical antigen-induced arthritis model in which arthritis occurs following the injection of bovine serum albumin in the joints of mice previously immunised with the same foreign protein. Comparison with other spontaneous arthritis models, such as TNFa-overexpressing mice (2) or the spontaneous arthritis in stressed mice (112) reveals several histopathologic similarities, such as the extensive remodelling of the joints. In both of these models, however, the immune system seems to play no role. Nevertheless, the latter model is of general importance because it seems to reflect a common pathway in response to trauma or stress that, in most mouse strains, and most likely also humans, leads to enthesopathy. Comparison with various adjuvant-induced arthritis models, such as pristane-induced arthritis in rats (38), is of interest because they share the lack of arthritogenic autoimmune response to cartilage proteins. There is, therefore, no straightforward explanation for the joint predilection in either model. It is not yet possible to transfer adjuvant-induced arthritis with serum and, in this case, T cells seem to represent the only critical cell type that contains immune specificity.

Another transgenic model in which spontaneous arthritis has been observed is in mice and rats transgenic for the envelope protein of human T cell leukaemia virus 1 (6,7). In this case, there is not only joint inflammation but also widespread inflammatory infiltrates in skin, salivary glands and vessels. In these transgenic models, the spontaneous penetrance of disease is high and is therefore reminiscent of monogenic and highly penetrant diseases in humans, such as the APECED syndrome (136). There are many possibilities about how arthritis may develop in the HTLV transgenic models. Most recently, it was suggested that tolerance to CII is broken down and the arthritis caused by T cells crossreactive with cartilage-derived CII (5).

These more or less "spontaneous" models most likely represent various aspects of the processes leading to arthritis, which will be determined by the transgene or defective gene and strongly influenced by the genetic constitution of the strain. They are useful and efficient to work with and will give answers to specific questions, but are not likely to be optimal models for RA, which is not a spontaneous disease with high penetrance and is not believed to be dependent on a dominant genetic defect.

7. CONCLUSIONS

Undoubtedly, RA is a complex disease. It is both phenotypically and genetically markedly heterogeneous. Thus, there is not one model for RA but several. The advantage with the animal models is that the various mechanisms leading to arthritis can be phenotypically and genetically defined and tested in a suitable biological context. Such models have been available for many years and are now quite useful not only for drug screening, but also for studying the pathogenesis in more detail. In combination with recently developed genetic techniques, such as gene mapping, gene identification and transgenic manipulations, it will now be possible to study the various biological components in a proper biological context.

References

1. Keffer, J., L. Probert, H. Cazlaris, S. Georgopoulos, E. Kaslaris, D. Kioussis, and G. Kollias. 1991. Transgenic mice expressing human tumour necrosis factor: a predicitive genetic model of arthritis. *EMBO J.* 10:4025–4031.
2. Kontoyiannis, D., M. Pasparakis, T.T. Pizarro, F. Cominelli, and G. Kollias. 1999. Impaired on/off regulation of TNF biosynthesis in mice lacking TNF AU- rich elements: implications for joint and gut-associated immunopathologies. *Immunity* 10:387–398.
3. Korganow, A.S., H. Ji, S. Mangialaio, V. Duchatelle, R. Pelanda, T. Martin, C. Degott, H. Kikutani, K. Rajewsky, L. Pasquali, C. Benoist, and D. Mathis. 1999. From systemic T cell self reactivity to organ-specific autoimmune disease via immunoglobulins. Immunity (in press).
4. Kouskoff, V., A.S. Korganow, V. Duchatelle, C. Degott, C. Benoist, and D. Mathis. 1996. Organ-specific disease provoked by systemic autoimmunity. *Cell* 87:811–822.
5. Kotani, M., Y. Tagawa, and Y. Iwakura. 1999. Involvement of autoimmunity against type II collagen in the development of arthritis in mice transgenic for the human T cell leukemia virus type I tax gene [In Process Citation]. *Eur. J. Immunol.* 29:54–64.

6. Iwakura, Y., M. Tosu, E. Yoshida, M. Takiguchi, K. Sato, I. Kitajima, K. Nishioka, K. Yamamoto, T. Takeda, M. Hatanaka, H. Yamamoto, and T. Sekiguchi. 1991. Induction of inflammatory arthropathy resembling rheumatoid arthritis in mice transgenic for HTLV-I. *Science* 253:1026–1028.

7. Yamazaki, H., H. Ikeda, A. Ishizu, Y. Nakamaru, T. Sugaya, K. Kikuchi, S. Yamada, A. Wakisaka, N. Kasai, T. Koike, M. Hatanaka, and T. Yoshiki. 1997. A wide spectrum of collagen vascular and autoimmune diseases in transgenic rats carrying the env-pX gene of human T lymphocyte virus type I. *Int. Immunol.* 9:339–346.

8. Pearson, C.M., and F.D. Wood. 1959. Studies of polyarthritis and other lesions induced in rats by injection of mycobacterial adjuvant. I. General clinical and pathologic characteristics and some modifying factors. *Arthritis Rheum.* 2:440–459.

9. Ward, J.R., and R.S. Jones. 1962. Studies on adjuvant-induced polyarthritis in rats. I. Adjuvant composition, route of injection, and removal of depot site. *Arthritis Rheum.* 5:557–564.

10. Kohashi, O., C.M. Pearson, Y. Watanabe, and S. Kotani. 1977. Preparation of arthritogenic hydrosoluble peptidoglycans from both arthritogenic and non-arthritogenic bacterial cell walls. *Inf. Immun.* 16:861–866.

11. Chang, Y.H., C.M. Pearson, and L. Chedid. 1981. Adjuvant polyarthritis. V. Induction by N-acetylmuramyl-L-alanyl-D-isoglutamine, the smallest peptide subunit of bacterial peptidoglycan. *J. Exp. Med.* 153:1021–1026.

12. Pearson, C.M., and F.D. Wood. 1964. Passive transfer of adjuvant arthritis by lymph node or spleen cells. *J. Exp. Med.* 120:547–573.

13. Yoshino, S., E. Schlipkoter, R. Kinne, T. Hunig, and F. Emmrich. 1990. Suppression and prevention of adjuvant arthritis in rats by a monoclonal antibody to the alpha/beta T cell receptor. *Eur. J. Immunol.* 20:2805–2808.

14. Anderton, S.M., R. van der Zee, B. Prakken, A. Noordzij, and W. van Eden. 1995. Activation of T cells recognizing self 60-kD heat shock protein can protect against experimental arthritis. *J. Exp. Med.* 181:943–952.

15. Bakker, N.P.M., M.G. Van Erck, C. Zurcher, P. Faaber, A. Lemmens, M. Hazenberg, R.E. Bontrop, and M. Jonker. 1990. Experimental immune mediated arthritis in rhesus monkeys. A model for human rheumatoid arthritis? *Rheumatol. Int.* 10:21–29.

16. Torisu, M., T. Miyahara, N. Shinohara, K. Ohsato, and H. Sonozaki. 1978. A new side effect of BCG immunotherapy – BCG-induced arthritis in man. *Cancer Immunol. Immunother.* 5:77–83.

17. Whitehouse, M.W., K.J. Orr, F.W.J. Beck, and C.M. Pearson. 1974. Freund's adjuvants: Relationship to arthritogenicity and adjuvanticity in rats to vehicle composition. *Immunology* 27:311–330.

18. Chang, Y.H., C.M. Pearson, and C. Abe. 1980. Adjuvant polyarthritis. IV. Induction by a synthetic adjuvant: Immunologic, histopathologic, and other studies. *Arthritis Rheum.* 23:62–71.

19. Hopkins, S.J., A.J. Freemont, and M.I.V. Jayson. 1984. Pristane-induced arthritis in Balb/c mice. I. Clinical and histological features of the arthropathy. *Int. Rheumatol.* 5:21–28.

20. Kleinau, S., H. Erlandsson, R. Holmdahl, and L. Klareskog. 1991. Adjuvant oils induce arthritis in the DA rat. I. Characterization of the disease and evidence for an immunological involvement. *J. Autoimmun.* 4:871–880.

21. Holmdahl, R., and C. Kvick. 1992. Vaccination and genetic experiments demonstrate that adjuvant oil induced arthritis and homologous type II collagen induced arthritis in the same rat strain are different diseases. *Clin. Exp. Immunol.* 88:96–100.

22. Holmdahl, R., T.J. Goldschmidt, S. Kleinau, C. Kvick, and R. Jonsson. 1992. Arthritis induced in rats with adjuvant oil is a genetically restricted, alpha beta T-cell dependent autoimmune disease. *Immunology* 76:197–202.

23. Vingsbo, C., P. Sahlstrand, J.G. Brun, R. Jonsson, T. Saxne, and R. Holmdahl. 1996. Pristane-induced arthritis in rats: a new model for rheumatoid arthritis with a chronic disease course influenced by both major histocompatibility complex and non-major histocompatibility complex genes. *Am. J. Pathol.* 149:1675–1683.

24. Lorentzen, J.C. 1999. Identification of arthritogenic adjuvants of self and foreign origin. *Scand. J. Immunol.* 49:45–50.

25. Lu, S., and R. Holmdahl. 1999. Different therapeutic and bystander effects by intranasal administration of homologous type II and type IX collagens on the collagen-induced arthritis and pristane-induced arthritis in rats. *Clin. Immunol.* 90:119–127.

26. Vingsbo, C., R. Jonsson, and R. Holmdahl. 1995. Avridine-induced arthritis in rats; a T cell-dependent chronic disease influenced both by MHC genes and by non-MHC genes. *Clin. Exp. Immunol.* 99:359–363.

27. Taurog, J.D., G.P. Sandberg, and M.L. Mahowald. 1983. The cellular basis of adjuvant arthritis. II. Characterization of the cells mediating passive transfer. *Cell Immunol.* 80:198–204.

28. Svelander, L., A. Mussener, H. Erlandsson-Harris, and S. Kleinau. 1997. Polyclonal Th1 cells transfer oil-induced arthritis. *Immunology* 91:260–5.

29. Björk, J., S. Kleinau, T. Midtvedt, L. Klareskog, and G. Smedegård. 1994. Role of the bowel flora for development of immunity to hsp 65 and arthritis in three experimental models. *Scand. J. Immunol.* 40:648–652.

30. Try, K. 1967. The presence of the hydrocarbons pristane and phytane in human adipose tissue and the occurrence of normal amounts in patients with Refsum's disease. *Scand. J. Clin. Lab. Invest.* 19:385–7.

31. Garrett, L.R., J.G. Chung, P.E. Byers, and M.A. Cuchens. 1989. Dietary effects of pristane on rat lymphoid tissues. *Agents Actions* 28:272–278.

32. Bly, J.E., L.R. Garrett, and M.A. Cuchens. 1990. Pristane induced changes in rat lymphocyte membrane fluidity. *Cancer Biochem. Biophys.* 11:145–54.

33. Lee, S.H., B.C. Ackland, and C.J. Jones. 1992. The tumor promoter pristane activates transcription by a cAMP dependent mechanism. *Mol. Cell Biochem.* 110:75–81.

34. Kawahito, Y., G.W. Cannon, P.S. Gulko, E.F. Remmers, R.E. Longman, V.R. Reese, J. Wang, M.M. Griffiths, and R.L. Wilder. 1998. Localization of quantitative trait loci regulating adjuvant-induced arthritis in rats: evidence for genetic factors common to multiple autoimmune diseases. *J. Immunol.* 161:4411–4419.

35. Lorentzen, J.C., A. Glaser, L. Jacobsson, J. Galli, H. Fakhrai-Rad, L. Klareskog, and H. Luthman. 1998. Identification of rat susceptibility loci for adjuvant-oil induced arthritis. *Proc. Natl. Acad. Sci. USA* 95:6383–6387.

36. Remmers, E.F., R.E. Longman, Y. Du, A. O'Hare, G.W. Cannon, M.M. Griffiths, and R.L. Wilder. 1996. A genome scan localizes five non-MHC loci controlling collagen-induced arthritis in rats. *Nature Genet.* 14:82–85.

37. Gulko, P.S., Y. Kawahito, E.F. Remmers, V.R. Reese, J. Wang, S.V. Dracheva, L. Ge, R.E. Longman, J.S. Shepard, G.W. Cannon, A.D. Sawitzke, R.L. Wilder, and M.M. Griffiths. 1998. Identification of a new non-major histocompatibility complex genetic locus on chromosome 2 that controls disease severity in collagen- induced arthritis in rats. *Arthritis Rheum.* 41:2122–2131.

38. Vingsbo-Lundberg, C., N. Nordquist, P. Olofsson, M. Sundvall, T. Saxne, U. Pettersson, and R. Holmdahl. 1998. Genetic control of arthritis onset, severity and chronicity in a model for rheumatoid arthritis in rats. *Nature Genet.* 20:401–404.

39. Vingsbo-Lundberg, C., T. Saxne, H. Olsson, and R. Holmdahl. 1998. Increased serum levels of cartilage oligomeric matrix protein in chronic erosive arthritis in rats. *Arthritis Rheum.* 41:544–550.

40. Holmdahl, R. 1998. Genetics of susceptibility to chronic experimental encephalomyelitis and arthritis. *Curr. Opin. Immunol.* 10:710 717.

41. Wooley, P.H., J.R. Seibold, J.D. Whalen, and J.M. Chapdelaine. 1989. Pristane-induced arthritis. The immunologic and genetic features of an experimental murine model of autoimmune disease. *Arthritis Rheum.* 32:1022–1030.

42. Thompson, S.J., J.N. Francis, L.K. Siew, G.R. Webb, P.J. Jenner, M.J. Colston, and C.J. Elson. 1998. An immunodominant epitope from mycobacterial 65-kDa heat shock protein protects against pristane-induced arthritis. *J. Immunol.* 160:4628–34.

43. Thompson, S.J., G.A.W. Rook, R.J. Brealey, R. VanderZee, and C.J. Elson. 1990. Autoimmune reactions to heat shock proteins in pristane-induced arthritis. *Eur. J. Immunol.* 20:2479–2484.

44. Potter, M., and J.S. Wax. 1981. Genetics of susceptibility to pristane-induced plasmacytomas in BALB/cAn: reduced susceptibility in BALB/cJ with a brief description of pristane-induced arthritis. *J. Immunol.* 127:1591–1595.

45. Trentham, D.E., A.S. Townes, and A.H. Kang. 1977. Autoimmunity to type II collagen: an experimental model of arthritis. *J. Exp. Med.* 146:857–868.

46. Courtenay, J.S., M.J. Dallman, A.D. Dayan, A. Martin, and B. Mosedal. 1980. Immunization against heterologous type II collagen induces arthritis in mice. *Nature* 283:666–667.

47. Yoo, T.J., S.Y. Kim, J.M. Stuart, R.A. Floyd, G.A. Olson, M.A. Cremer, and A.H. Kang. 1988. Induction of arthritis in monkeys by immunization with type II collagen. *J. Exp. Med.* 168:777–782.

48. Glant, T.T., K. Mikecz, A. Arzoumanian, and A.R. Poole. 1987. Proteoglycan-induced arthritis in Balb/c mice. *Arthritis Rheum.* 30:201–212.

49. Zhang, Y., A. Guerassimov, J.Y. Leroux, A. Cartman, C. Webber, R. Lalic, E. de Miguel, L.C. Rosenberg, and A.R. Poole. 1998. Induction of arthritis in BALB/c mice by cartilage link protein: involvement of distinct regions recognized by T and B lymphocytes. *Am. J. Pathol.* 153:1283–1291.

50. Cremer, M.A., X.J. Ye, K. Terato, S.W. Owens, J.M. Seyer, and A.H. Kang. 1994. Type XI collagen-induced arthritis in the Lewis rat. Characterization of cellular and humoral immune responses to native types XI, V, and II collagen and constituent alpha-chains. *J. Immunol.* 153:824–32.

51. Verheijden, G.F., A.W. Rijnders, E. Bos, C.J. Coenen-de Roo, C.J. van Staveren, A.M. Miltenburg, J.H. Meijerink, D. Elewaut, F. de Keyser, E. Veys, and A.M. Boots. 1997. Human cartilage glycoprotein-39 as a candidate autoantigen in rheumatoid arthritis. *Arthritis Rheum.* 40:1115–1125.

52. Carlsén, S., A.S. Hansson, H. Olsson, D. Heinegård, and R. Holmdahl. 1998. Cartilage oligomeric matrix protein (COMP)-induced arthritis in rats. *Clin. Exp. Immunol.* 114:477–484.

53. Cremer, M.A., X.J. Ye, K. Terato, M.M. Griffiths, W.C. Watson, and A.H. Kang. 1998. Immunity to type IX collagen in rodents: a study of type IX collagen for autoimmune and arthritogenic activities. *Clin. Exp. Immunol.* 112:375–382.

54. Hansson, A.-S., D. Heinegård, and R. Holmdahl. 1999. A new model of polychondritis. Induction with cartilage matrix protein. *J. Clin. Invest.* 104:589–598.

55. Holmdahl, R., L. Jansson, E. Larsson, K. Rubin, and L. Klareskog. 1986. Homologous type II collagen induces chronic and progressive arthritis in mice. *Arthritis Rheum.* 29:106–113.

56. Holmdahl, R., C. Vingsbo, H. Hedrich, M. Karlsson, C. Kvick, T.J. Goldschmidt, and K. Gustafsson. 1992. Homologous collagen-induced arthritis in rats and mice are associated with structurally different major histocompatibility complex DQ-like molecules. *Eur. J. Immunol.* 22:419–424.

57. Corthay, A., Å. Johansson, M. Vestberg, and R. Holmdahl. 1999. Collagen-induced arthritis development requires alphabeta T cells but not gammadelta T cells: studies with T cell-deficient (TCR mutant) mice. *Int. Immunol.* 11:1065–1073.

58. Svensson, L., J. Jirholt, R. Holmdahl, and L. Jansson. 1998. B cell-deficient mice do not develop type II collagen-induced arthritis (CIA). *Clin. Exp. Immunol.* 111:521–526.

59. Michaëlsson, E., M. Andersson, A. Engström, and R. Holmdahl. 1992. Identification of an immunodominant type-II collagen peptide recognized by T cells in H-2q mice: self tolerance at the level of determinant selection. *Eur. . Immunol.* 22:1819–25.

60. Brand, D.D., L.K. Myers, K. Terato, K.B. Whittington, J.M. Stuart, A.H. Kang, and E.F. Rosloniec. 1994. Characterization of the T cell determinants in the induction of autoimmune arthritis by bovine alpha 1(II)-CB11 in H-2q mice. *J. Immunol.* 152:3088–97.

61. Brand, D.D., T.N. Marion, L.K. Myers, E.F. Rosloniec, W.C. Watson, J.M. Stuart, and A.H. Kang. 1996. Autoantibodies to murine type II collagen in collagen induced arthritis. *J. Immunol.* 157:5178–5184.

62. Mo, J.A., and R. Holmdahl. 1996. The B cell response to autologous type II collagen: biased V gene repertoire with V gene sharing and epitope shift. *J. Immunol.* 157:2440–2448.

63. Schulte, S., C. Unger, J.A. Mo, O. Wendler, E. Bauer, S. Frischholz, K. von der Mark, J.R. Kalden, R. Holmdahl, and H. Burkhardt. 1998. Arthritis-related B cell epitopes in collagen II are conformation- dependent and sterically privileged in accessible sites of cartilage collagen fibrils. *J. Biol. Chem.* 273:1551–1561.

64. Stuart, J.M., M.A. Cremer, A.S. Townes, and A.H. Kang. 1982. Type II collagen induced arthritis in rats. Passive transfer with serum and evidence that IgG anticollagen antibodies can cause arthritis. *J. Exp. Med.* 155:1–16.

65. Stuart, J.M., and F.J. Dixon. 1983. Serum transfer of collagen induced arthritis in mice. *J. Exp. Med.* 158:378–392.

66. Takagishi, K., N. Kaibara, T. Hotokebuchi, C. Arita, M. Morinaga, and K. Arai. 1985. Serum transfer of collagen arthritis in congenitally athymic nude rats. *J. Immunol.* 134:3864–3867.

67. Jasin, H.E., and J.D. Taurog. 1991. Mechanisms of disruption of the articular cartilage surface in inflammation. Neutrophil elastase increases availability of collagen type II epitopes for binding with antibody on the surface of articular cartilage. *J. Clin. Invest.* 87:1531–1536.

68. Holmdahl, R., J.A. Mo, R. Jonsson, K. Karlström, and A. Scheynius. 1991. Multiple epitopes on cartilage type II collagen are accessible for antibody binding in vivo. *Autoimmunity* 10:27–34.

69. Mo, J.A., A. Scheynius, S. Nilsson, and R. Holmdahl. 1994. Germline-encoded IgG antibodies bind mouse cartilage in vivo: epitope- and idiotype-specific binding and inhibition. *Scand. J. Immunol.* 39:122–130.

70. Watson, W.C., and A.S. Townes. 1985. Genetic susceptibility to murine collagen II autoimmune arthritis. Proposed relationship to the IgG2 autoantibody subclass response, complement C5, major histocompatibility complex (MHC) and non-MHC loci. *J. Exp. Med.* 162:1878–1891.

71. Wang, Y., S.A. Rollins, J.A. Madri, and L.A. Matis. 1995. Anti-C5 monoclonal antibody therapy prevents collagen-induced arthritis and ameliorates established disease. *Proc. Natl. Acad. Sci. USA* 92: 8955–8959.

72. Mori, L., and G. de Libero. 1998. Genetic control of susceptibility to collagen-induced arthritis in T cell receptor β-chain transgenic mice. *Arthritis Rheum.* 41:256–262.

73. Taylor, P.C., C.Q. Chu, C. Plater Zyberk, and R.N. Maini. 1996. Transfer of type II collagen-induced arthritis from DBA/1 to severe combined immunodeficiency mice can be prevented by blockade of Mac-1. *Immunology* 88:315–321.

74. Goldschmidt, T.J., and R. Holmdahl. 1991. Anti-T cell receptor antibody treatment of rats with established autologous collagen-induced arthritis. Suppression of arthritis without reduction of anti-type II collagen autoantibody levels. *Eur. J. Immunol.* 21:1327–1330.

75. Plater-Zyberk, C., P.C. Taylor, M.G. Blaylock, and R.N. Maini. 1994. Anti-CD5 therapy decreases severity of established disease in collagen type II-induced arthritis in DBA/1 mice. *Clin. Exp. Immunol.* 98:442–447.

76. Zamvil, S., P. Nelson, J. Trotter, D. Mitchell, R. Knobler, R. Fritz, and L. Steinman. 1985. T-cell clones specific for myelin basic protein induce chronic relapsing paralysis and demyelination. *Nature* 317:355–358.

77. Holmdahl, R., L. Klareskog, K. Rubin, E. Larsson, and H. Wigzell. 1985. T lymphocytes in collagen II-induced arthritis in mice. Characterization of arthritogenic collagen II-specific T cell lines and clones. *Scand. J. Immunol.* 22:295–306.

78. Kakimoto, K., M. Katsuki, T. Hirofuji, H. Iwata, and T. Koga. 1988. Isolation of T cell line capable of protecting mice against collagen-induced arthritis. *J. Immunol.* 140:78–83.

79. Williams, R., C. Plater-Zyberk, D. Williams, and R. Maini. 1992. Successful transfer of collagen-induced arthritis to severe combined immunodeficient (SCID) mice. *Clin. Exp. Imunol.* 88:455–460.

80. Kadowaki, K.M., H. Matsuno, H. Tsuji, and I. Tunru. 1994. CD4+ T cells from collagen-induced arthritic mice are essential to transfer arthritis into severe combined immunodeficient mice. *Clin. Exp. Imunol.* 97:212–218.

81. Mikecz, K., T.T. Glant, E. Buzas, and A.R. Poole. 1990. Proteoglycan-induced polyarthritis and spondylitis adoptively transferred to naive (nonimmunized) BALB/c mice. *Arthritis Rheum.* 33:866–876.

82. Seki, N., Y. Sudo, T. Yoshioka, S. Sugihara, T. Fujitsu, S. Sakuma, T. Ogawa, T. Hamaoka, H. Senoh, and H. Fujiwara. 1988. Type II collagen-induced murine arthritis. I. Induction and perpetuation of arthritis require synergy between humoral and cell-mediated immunity. *J. Immunol.* 140: 1477–1484.

83. Taylor, P.C., C. Plater Zyberk, and R.N. Maini. 1995. The role of the B cells in the adoptive transfer of collagen-induced arthritis from DBA/1 (H-2q) to SCID (H-2d) mice. *Eur. J. Immunol.* 25:763–769.

84. Taurog, J.D., S.S. Kerwar, R.A. McReynolds, G.P. Sandberg, S.L. Leary, and M.L. Mahowald. 1985. Synergy between adjuvant arthritis and collagen-induced arthritis in rats. *J. Exp. Med.* 162:962–978.

85. Caulfield, J.P., A. Hein, R. Dynesius-Trentham, and D.E. Trentham. 1982. Morphologic demonstration of two stages in the development of type II collagen-induced arthritis. *Lab. Invest.* 46: 321–343.

86. Holmdahl, R., K. Rubin, L. Klareskog, L. Dencker, G. Gustafsson, and E. Larsson. 1985. Appearence of different lymphoid cells in synovial tissue and in peripheral blood during the course of collagen II-induced arthritis in rats. *Scand. J. Immunol.* 21:197–204.

87. Holmdahl, R., L. Jansson, A. Larsson, and R. Jonsson. 1990. Arthritis in DBA/1 mice induced with passively transferred type II collagen immune serum. Immunohistopathology and serum levels of anti-type II collagen auto-antibodies. *Scand. J. Immunol.* 31:147–157.

88. Holmdahl, R., A. Tarkowski, and R. Jonsson. 1991. Involvement of macrophages and dendritic cells in synovial inflammation of collagen induced arthritis in DBA/1 mice and spontaneous arthritis in MRL/lpr mice. *Autoimmunity* 8:271–280.

89. Tada, Y., K. Nagasawa, A. Ho, F. Morito, O. Ushiyama, N. Suzuki, H. Ohta, and T.W. Mak. 1999. CD28-deficient mice are highly resistant to collagen-induced arthritis. *J. Immunol.* 162:203–208.

90. Webb, L.M., M.J. Walmsley, and M. Feldmann. 1996. Prevention and amelioration of collagen-induced arthritis by blockade of the CD28 co-stimulatory pathway: requirement for both B7-1 and B7-2. *Eur. J. Immunol.* 26:2320–2328.

91. McIntyre, K.W., D.J. Shuster, K.M. Gillooly, R.R. Warrier, S.E. Connaughton, L.B. Hall, L.H. Arp, M.K. Gately, and J. Magram. 1996. Reduced incidence and severity of collagen-induced arthritis in interleukin-12-deficient mice. *Eur. J. Immunol.* 26:2933–8.

92. Hom, J.T., A.M. Bendele, and D.G. Carlson. 1988. *in vivo* administration with IL-1 accelerates the development of collagen-induced arthritis in mice. *J. Immunol.* 141:834–841.

93. Wooley, P.H., J.D. Whalen, D.L. Chapman, A.E. Berger, K.E. Richard, D.G. Aspar, and N.D. Staite. 1993. The efect of an interleukin-1 receptor antagonist protein on type II collagen-induced arthritis in mice. *Arthritis Rheum.* 36:1305–1314.

94. van den Berg, W.B., L.A. Joosten, M. Helsen, and F. van de Loo. 1994. Amelioration of established murine collagen-induced arthritis with anti-IL-1 treatment. *Clin. Exp. Immunol.* 95:237–243.

95. Germann, T., J. Szeliga, H. Hess, S. Storkel, F.J. Podlaski, M.K. Gately, E. Schmitt, and E. Rude. 1995. Administration of interleukin 12 in combination with type II collagen induces severe arthritis in DBA/1 mice. *Proc. Natl. Acad. Sci. USA* 92:4823–4827.

96. Joosten, L.A., M.M. Helsen, F.A. van de Loo, and W.B. van den Berg. 1996. Anticytokine treatment of established type II collagen-induced arthritis in DBA/1 mice. A comparative study using anti-TNF alpha, anti-IL-1 alpha/beta, and IL-1Ra. *Arthritis Rheum.* 39:797–809.

97. Kasama, T., R.M. Strieter, N.W. Lukacs, P.M. Lincoln, M.D. Burdick, and S.L. Kunkel. 1995. Interleukin-10 expression and chemokine regulation during the evolution of murine type II collagen-induced arthritis. *J. Clin. Invest.* 95:2868–76.

98. Kuruvilla, A.P., R. Shah, G.M. Hochwald, H.D. Liggitt, M.A. Palladino, and G.J. Thorbecke. 1991. Protective effect of transforming growth factor beta1 on experimental auto-immune diseases in mice. *Proc. Natl. Acad. Sci. USA* 88:2918–2921.

99. Hess, H., M.K. Gately, E. Rude, E. Schmitt, J. Szeliga, and T. Germann. 1996. High doses of interleukin-12 inhibit the development of joint disease in DBA/1 mice immunized with type II collagen in complete Freund's adjuvant. *Eur. J. Immunol.* 26:187–191.

100. Walmsley, M., P.D. Katsikis, E. Abney, S. Parry, R.O. Williams, R.N. Maini, and M. Feldmann. 1996. Interleukin-10 inhibition of the progression of established collagen-induced arthritis. *Arthritis Rheum.* 39:495–503.

101. Persson, S., A. Mikulowska, S. Narula, A. O'Garra, and R. Holmdahl. 1996. Interleukin-10 suppresses the development of collagen type II-induced arthritis and ameliorates sustained arthritis in rats. *Scand. J. Immunol.* 44:607–614.

102. Tada, Y., A. Ho, D.R. Koh, and T.W. Mak. 1996. Collagen-induced arthritis in CD4- or CD8-deficient mice. CD8+ T cells play a role in initiation and regulate recovery phase of collagen-induced arthritis. *J. Immunol.* 156:4520–4526.

103. Mauritz, N.J., R. Holmdahl, R. Jonsson, P. Van Der Meide, A. Scheynius, and L. Klareskog. 1988. Treatment with inter-feron gamma triggers onset of collagen arthritis in mice. *Arthritis Rheum.* 31:1297–1304.

104. Tada, Y., A. Ho, T. Matsuyama, and T.W. Mak. 1997. Reduced incidence and severity of antigen-induced autoim-mune diseases in mice lacking interferon regulatory factor-1. *J. Exp. Med.* 185:231–238.

105. Kageyama, Y., Y. Koide, A. Yoshida, M. Uchijima, T. Arai, S. Miyamoto, T. Ozeki, M. Hiyoshi, K. Kushida, and T. Inoue. 1998. Reduced susceptibility to collagen-induced arthritis in mice deficient in IFN-gamma receptor. *J. Immunol.* 161:1542–1548.

106. Boissier, M.C., G. Chiocchia, N. Bessis, J. Hajnal, G. Garotta, F. Nicoletti, and C. Fournier. 1995. Biphasic effect of interferon-gamma in murine collagen-induced arthritis. *Eur. J. Immunol.* 25:1184–1190.

107. Joosten, L.A., E. Lubberts, M.M. Helsen, and W.B. van den Berg. 1997. Dual role of IL-12 in early and late stages of murine collagen type II arthritis. *J. Immunol.* 159:4094–4102.

108. Kasama, T., J. Yamazaki, R. Hanaoka, Y. Miwa, Y. Hatano, K. Kobayashi, M. Negishi, H. Ide, and M. Adachi. 1999. Biphasic regulation of the development of murine type II collagen- induced arthritis by interleukin-12: possible involvement of endogenous interleukin-10 and tumor necro-sis factor alpha. *Arthritis Rheum.* 42:100–109.

109. Joosten, L.A., E. Lubberts, P. Durez, M.M. Helsen, M.J. Jacobs, M. Goldman, and W.B. van den Berg. 1997. Role of interleukin-4 and interleukin-10 in murine collagen-induced arthritis. Protective effect of interleukin-4 and

interleukin-10 treatment on cartilage destruction. *Arthritis Rheum.* 40:249–260.

110. Doncarli, A., L.M. Stasiuk, C. Fournier, and O. Abehsira-Amar. 1997. Conversion in vivo from an early dominant Th0/Th1 response to a Th2 phenotype during the develop-ment of collagen-induced arthritis. *Eur. J. Immunol.* 27:1451–1458.

111. Horsfall, A.C., D.M. Butler, L. Marinova, P.J. Warden, R.O. Williams, R.N. Maini, and M. Feldmann. 1997. Suppression of collagen-induced arthritis by continuous administration of IL-4. *J. Immunol.* 159:5687–96.

112. Holmdahl, R., L. Jansson, M. Andersson, and R. Jonsson. 1992. Genetic, hormonal and behavioral influence on spontaneously developing arthritis in normal mice. *Clin. Exp. Immunol.* 88: 467–472.

113. Corthay, A., A. Hansson, and R. Holmdahl. 2000. T lymphocytes are not required for the spontaneous develop-ment of entheseal ossification leading to marginal ankylosis in the DBA/1 mouse. *Arth & Rheum* 43:844–851.

114. Jirholt, J., A. Cook, T. Emahazion, M. Sundvall, L. Jansson, N. Nordquist, U. Pettersson, and R. Holmdahl. 1998. Genetic linkage analysis of collagen-induced arthritis in the mouse. *Eur. J. Immunol.* 28:3321–3328.

115. Yang, H.-T., J. Jirholt, M. Sundvall, L. Jansson, U. Pettersson, and R. Holmdahl. 1999. Chromosome mapping of the suscep-tibility genes of collagen induced arthritis in crosses between DBA/1 and B10.Q mice. *J. Immunol.* 163:2916–2921.

116. Wooley, P.H., H.S. Luthra, J.M. Stuart, and C.S. David. 1981. Type II collagen induced arthritis in mice. I. Major histocompatibility complex (I-region) linkage and antibody correlates. *J. Exp. Med.* 154:688–700.

117. Griffiths, M. 1988. Immunogenetics of collagen-induced arthritis in rats. *Intern. Rev. Immunol.* 4:1–15.

118. Holmdahl, R., L. Jansson, M. Andersson, and E. Larsson. 1988. Immunogenetics of type II collagen autoimmunity land susceptibility to collagen arthritis. *Immunology* 65:305–310.

119. Brunsberg, U., K. Gustafsson, L. Jansson, E. Michaëlsson, L. Ährlund-Richter, S. Pettersson, R. Mattsson, and R. Holmdahl. 1994. Expression of a transgenic class II Ab gene confers susceptibility to collagen-induced arthritis. *Eur. J. Immunol.* 24:1698–1702.

120. Corthay, A., J. Bäcklund, J. Broddefalk, E. Michaëlsson, T.J. Goldschmidt, J. Kihlberg, and R. Holmdahl. 1998. Epitope glycosylation plays a critical role for T cell recognition of type II collagen in collagen-induced arthritis. *Eur. J. Immunol.* 28:2580–2590.

121. Fugger, L., J.B. Rothbard, and G. Sonderstrup-McDevitt. 1996. Specificity of an HLA-DRB1*0401-restricted T cell response to type II collagen. *Eur. J. Immunol.* 26:928–933.

122. Rosloniec, E.F., D.D. Brand, L.K. Myers, K.B. Whittington, M. Gumanovskaya, D.M. Zaller, A. Woods, D.M. Altmann, J.M. Stuart, and A.H. Kang. 1997. An HLA-DR1 trans-gene confers susceptibility to collagen-induced arthritis elicited with human type II collagen. *J. Exp. Med.* 185:1113–1122.

123. Andersson, E.C., B.E. Hansen, H. Jacobsen, L.S. Madsen, C.B. Andersen, J. Engberg, J.B. Rothbard, G. Sönderstrup-McDevitt, V. Malmström, R. Holmdahl, A. Svejgaard, and L. Fugger. 1998. Definition of MHC and T cell receptor contacts in the HLA-DR4 restricted immunodominant epitope in type II collagen and characterization of collagen-induced arthritis in HLA-DR4 and human CD4 transgenic mice. *Proc. Natl. Acad. Sci. USA* 95:7574–7569.

124. Rosloniec, E.F., D.D. Brand, L.K. Myers, Y. Esaki, K.B. Whittington, D.M. Zaller, A. Woods, J.M. Stuart, and A.H. Kang. 1998. Induction of autoimmune arthritis in HLA-DR4 (DRB1*0401) transgenic mice by immunization with human and bovine type II collagen. *J. Immunol.* 160:2573–8.

125. Holmdahl, R., E.C. Andersson, C.B. Andersen, A. Svejgaard, and L. Fugger. 1999. Transgenic mouse models for rheumatoid arthritis. *Immunol. Rev.* 169:161–173.

126. Malmström, V., P. Kjéllen, and R. Holmdahl. 1998. Type II collagen in cartilage evokes peptide-specific tolerance and skews the immune response. *J. Autoimmun.* 11:213–221.

127. Malmström, V., J. Bäcklund, L. Jansson, J. Kihlberg, and R. Holmdahl. 1999. T cells naturally "tolerant" to cartilage-derived type II collagen are important for development of collagen-induced arthritis. 2:315–326.

128. Bouvet, J.P., J. Couderc, Y. Bouthillier, B. Franc, A. Ducailar, and D. Mouton. 1990. Spontaneous rheumatoid-like arthritis in a line of mice sensitive to collagen-induced arthritis. *Arthritis. Rheum.* 33:1716–1722.

129. Nakamura, K., S. Kashiwasaki, K. Takagishi, Y. Tsukamoto, Y. Morohoshi, T. Nakano, and M. Kimura. 1991. Spontaneous degenerative polyarthritis in male New Zealand Black/KN mice. *Arthritis Rheum.* 34:171–179.

130. Boden, S.D., P.A. Labropoulos, B.D. Ragsdale, P.M. Gullino, and L.H. Gerber. 1989. Retinyl acetate-induced arthritis in C3H-Avy mice. *Arthritis Rheum.* 32:625–633.

131. Hang, L., A.N. Theofilopoulos, and F.J. Dixon. 1982. A spontaneous rheumatoid arthritis-like disease in MRL/l mice. *J. Exp. Med.* 155:1690–1701.

132. O'Sullivan, F.X., G. Fassbender H, S. Gay, and W.J. Koopman. 1985. Etiopathogenesis of the rheumatoid arthritis-like disease in MRL/l mice. *Arthritis Rheum.* 28:529–536.

133. Tarkowski, A., R. Jonsson, R. Holmdahl, and L. Klareskog. 1987. Immunohistochemical characterization of synovial cells in arthritic MRL-lpr/lpr mice. *Arthritis Rheum.* 30:75–82.

134. Vidal, S., D.H. Kono, and A.N. Theofilopoulos. 1998. Loci predisposing to autoimmunity in MRL-Fas lpr and C57BL/6-Faslpr mice. *J. Clin. Invest.* 101:696–702.

135. Holmdahl, R., M. Andersson, T.J. Goldschmidt, K. Gustafsson, L. Jansson, and J.A. Mo. 1990. Type II collagen autoimmunity in animals and provocations leading to arthritis. *Immunol. Rev.* 118:193–232.

136. Aaltonen, J., P. Björses, J. Perheentupa, N. Horelli-Kuitunen, A. Palotie, L. Peltonen, Y.S. Lee, F. Francis, S. Hennig, C. Thiel, H. Lehrach, and M.-L. Yaspo. 1997. An autoimmune disease, APECED, caused by mutations in a novel gene featuring two PHD-type zinc-finger domains. *Nature Genet.* 17:399–403.

137. Klareskog, L., U. Forsum, A. Scheynius, D. Kabelitz, and H. Wigzell. 1982. Evidence in support of a self-perpetuating HLA-DR dependent delayed-type hypersensitivity reaction in rheumatoid arthritis. *Proc. Natl. Acad. Sci. USA* 79:3632–3636.

138. Holmdahl, R., R. Jonsson, P. Larsson, and L. Klareskog. 1988. Early appearance of activated CD4 positive T lymphocytes and Ia-expressing cells in joints of DBA/1 mice immunized with type II collagen. *Lab. Invest.* 58:53–60.

139. Forslind, K., K. Eberhardt, A. Jonsson, and T. Saxne. 1992. Increased serum concentrations of cartilage oligomeric matrix protein. A prognostic marker in early rheumatoid arthritis. *Br. J. Rheumatol.* 31:593–598.

140. Larsson, E., A. Mussener, D. Heinegard, L. Klareskog, and T. Saxne. 1997. Increased serum levels of cartilage oligomeric matrix protein and bone sialoprotein in rats with collagen arthritis. *Br. J. Rheumatol.* 36:1258–1261.

141. Cannon, G.W., M.L. Woods, F. Clayton, and M.M. Griffiths. 1993. Induction of arthritis in DA rats by incomplete Freund's adjuvant. *J. Rheumatol.* 20:7–11.

142. Larsson, P., S. Kleinau, R. Holmdahl, and L. Klareskog. 1990. Homologous type II collagen-induced arthritis in rats. Characterization of the disease and demonstration of clinically distinct forms of arthritis in two strains of rats after immunization with the same collagen preparation. *Arthritis Rheum.* 33:693–701.

143. Boissier, M., X. Feng, A. Carlioz, R. Roudier, and C. Fournier. 1987. Experimental autoimmune arthritis in mice. I. Homologous type II collagen is responsible for self-perpetuating chronic polyarthritis. *Annals Rheum. Dis.* 46:691–700.

144. Holmdahl, R., C. Nordling, K. Rubin, A. Tarkowski, and L. Klareskog. 1986. Generation of monoclonal rheumatoid factors after immunization with collagen II – anti collagen II immune complexes. *Scand. J. Immunol.* 24:197–203.

145. Tarkowski, A., R. Holmdahl, and L. Klareskog. 1989. Rheumatoid factors in mice. *Monographs in Allergy* 26:214–229.

146. Tarkowski, A., L. Klareskog, H. Carlsten, P. Herberts, and W.J. Koopman. 1989. Secretion of antibodies to types I and II collagen by synovial tissue cells in patients with rheumatoid arthritis. *Arthritis Rheum.* 32:1087–1092.

147. Trentham, D.E., A.S. Townes, A.H. Kang, and J.R. David. 1978. Humoral and cellular sensitivity to collagen in type II collagen induced arthritis in rats. *J. Clin. Invest.* 61:89–96.

148. Stuart, J.M., M.A. Cremer, A.H. Kang, and A.S. Townes. 1979. Collagen-induced arthritis in rats. Evaluation of early immunologic events. *Arthritis Rheum.* 22:1344–1351.

149. Stuart, J.M., A.S. Townes, and A.H. Kang. 1982. Nature and specificity of the immune response to collagen in type II collagen induced arthritis in mice. *J. Clin. Invest.* 69:673–683.

150. Gregersen, P.K., J. Silver, and R.J. Winchester. 1987. The shared epitope hypothesis. An approach to understanding the molecular genetics of susceptibility to rheumatoid arthritis. *Arthritis Rheum.* 30:1205–1213.

151. Lorentzen, J.C. and L. Klareskog. 1996. Susceptibility of DA rats to arthritis induced with adjuvant oil or rat collagen is determined by genes both within and outside the major histocompatibility complex. *Scand. J. Immunol.* 44:592–598.

152. Griffiths, M.M., and C.W. DeWitt. 1981. Immunogenetic control of experimental collagen-induced arthritis in rats. II: ECIA susceptibility and immune response to type II collagen (calf) are linked to RT1. *J. Immunogenet.* 8:463–470.

153. Holmdahl, R., C. Vingsbo, V. Malmström, L. Jansson, and M. Holmdahl. 1994. Chronicity of arthritis induced with homologous type II collagen (CII) in rats is dependent on anti-CII B-cell activation. *J. Autoimmunity* 7:739–752.

154. Holmdahl, R., L. Klareskog, M. Andersson, and C. Hansen. 1986. High antibody response to autologous type II collagen is restricted to H-2q. *Immunogenetics* 24:84–89.

155. Griffiths, M.M., and C.W. DeWitt. 1984. Modulation of collagen-induced arthritis in rats by non-RT1-linked genes. *J. Immunol.* 133:3043–3046.

156. Holmdahl, R., L. Jansson, and M. Andersson. 1986. Female sex hormones suppress development of collagen-induced arthritis in mice. *Arthritis Rheum.* 29:1501–1509.

157. Holmdahl, R. 1995. Female preponderance for development of arthritis in rats is influenced by both sex chromosomes and sex steroids. *Scand. J. Immunol.* 42:104–109.

158. Boissier, M.C., A. Carlioz, and C. Fournier. 1988. Experimental autoimmune arthritis in mice. II. Early events in the elicitation of the autoimmune phenomenon induced by homologous type II collagen. *Clin. Immunol. Immunopathol.* 48:225–237.

159. Hirahara, F., P.H. Wooley, H.S. Luthra, C.B. Coulam, M.M. Griffiths, and C.S. David. 1986. Collagen-induced arthritis and pregnancy in mice: the effects of pregnancy on collagen-induced arthritis and the high incidence of infertility in arthritic female mice. *Am. J. Reprod. Immunol.* 11:44–54.

160. Mattsson, R., A. Mattsson, R. Holmdahl, A. Whyte, and G.A.W. Rook. 1991. Maintained pregnancy levels of oestrogen afford complete protection from post-partum exacerbation of collagen-induced arthritis. *Clin. Exp. Immunol.* 85:41–47.

161. Jansson, L., A. Mattsson, R. Mattsson, and R. Holmdahl. 1990. Estrogen induced suppression of collagen arthritis. V: Physiological level of estrogen in DBA/1 mice is therapeutic on established arthritis, suppresses anti-type II collagen T-cell dependent immunity and stimulates polyclonal B-cell activity. *J. Autoimmunity* 3:257–270.

162. Larsson, P., and R. Holmdahl. 1987. Oestrogen-induced suppression of collagen arthritis. II. Treatment of rats suppresses development of arthritis but does not affect the anti-type II collagen humoral response. *Scand. J. Immunol.* 27:579–583.

163. Larsson, P., T.J. Goldschmidt, L. Klareskog, and R. Holmdahl. 1989. Oestrogen-mediated suppression of collagen-induced arthritis in rats. Studies on the role of the thymus and of peripheral CD8+ T lymphocytes. *Scand. J. Immunol.* 30:741–747.

24 | Sjögren's Syndrome: An Autoimmune Exocrinopathy

Robert I. Fox, Joichiro Hayashi, and Paul Michelson

1. INTRODUCTION

Sjögren's syndrome (SS) is a chronic disease characterized by severe dryness of the eyes (keratoconjunctivitis sicca), dryness of the mouth (xerostomia) and evidence of systemic autoimmune disease. SS may exist as a primary condition (primary SS, 1° SS) or as a secondary condition (2° SS) in association with rheumatoid arthritis (RA), systemic lupus erythematosus (SLE), or progressive systemic sclerosis (PSS). In some 1° SS patients, there may be involvement of the extraglandular organs, including skin, kidney, liver, lung and nervous system. Furthermore, these patients may develop a lymphoproliferative syndrome that includes increased risk of lymphoma. Thus, SS provides an opportunity to look at the molecular biology of the interactions of the immune, neurological and exoocrine systems.

A particular feature of SS is the ability to look at the actual site of end-organ damage, namely the salivary or lacrimal glands. Due to ease and safety of obtaining minor salivary gland biopsy and of obtaining the efferent secretions (ie. saliva), SS serves as a "model" disease to study the cell-cell and cell matrix interactions that occur in a human organ specific autoimmune disease.

Critical features of pathogenesis include: a) association with particular class II histocompatibility (HLA-DR/DQ) antigens; b) a pattern of particular autoantibodies, including a newly-described antibody against muscarinic M3 receptor for acetylcholine; c) a high proportion of "high endothelial venules" in the lacrimal and salivary gland biopsies, which express increased levels of cell adhesive molecules; d) upregulation of major histocompatibility antigens and adhesive molecules on epithelial cells in salivary and lacrimal glands, probably in response to local production of interferon gamma; e) expression of cell surface (Fas, Fas ligand) and nuclear molecules (bcl-2, bcl-x) on glandular epithelial cells and on lymphocytes infiltrating the glands; f) secretion of pro-inflammatory cytokines (such as IL-1, TNF-α and IFN-γ) by lymphocytes and/or epithelial cells to perpetuate the inflammatory response; g) decreased secretion by the residual glandular acini as a result of their decreased neural innervation and defects in their post-signal transduction.

2. THE PROBLEM OF DEFINITION FOR SJÖGREN'S SYNDROME

In 1932, Henrik Sjögren first reported the triad of keratoconjunctivitis sicca (KCS), xerostomia and rheumatoid arthritis. This syndrome was largely ignored until "rediscovered" by Block et. al in 1956, who presented a detailed description of the clinical spectrum of SS as a systemic autoimmune disease (1) in 1956. SS occurred predominantly in women (ratio 9:1), with peak ages of onset at child-bearing age (35–45 years) and post menopause (55–65 years). It occurs in virtually all ethnic populations, although with different frequencies. In patients with severe keratoconjunctivitis sicca (KCS), xerostomia and autoantibodies against nuclear antigens, there is good agreement on diagnosis among clinicians. However, the diagnosis in patients with milder sicca symptoms and absence of antinuclear antibodies has remained controversial (discussed below), due to the absence of good non-invasive methods for documenting xerostomia.

The gold standard for diagnosis of SS has been the minor salivary gland biopsy, which demonstrates focal lymphocytic infiltrates. This is shown in Figure 1, frame A, which is a minor salivary gland from a SS patient. In comparison,

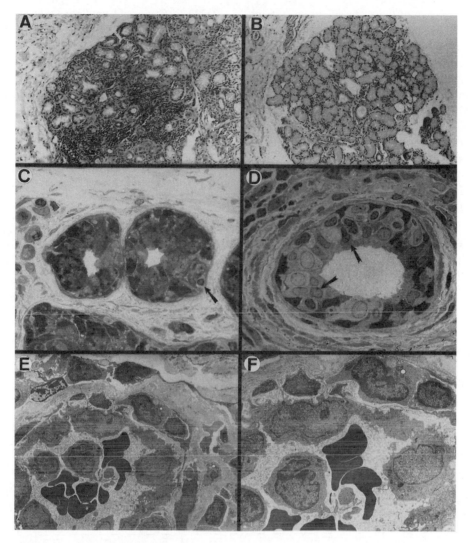

Figure 1. Sjögren's syndrome salivary gland biopsies. Panels A and B show biopsies from an SS patient and from a normal individual, respectively. The initial appearance of lymphocytic infiltrates in the central portion of the salivary gland (frame B) is similar to that noted in the minor salivary gland biopsies of SS patients described above. Under higher magnification, the location of lymphocytes adjacent to salivary gland epithelial cells (i.e., beneath the basement membrane enclosing acini and ducts) can be seen. Under electron microscopy (frames C and D), the appearance of high endothelial venules (containing RBD) and lymphocytes adherent to the vascular endothelium can be noted as well as the absence of electron-dense immune complexes near the basement membrane (frames E and F).

frame B shows a biopsy from a normal individual. Frames C and D show higher power views from frame A; the arrows indicate lymphocytes that are in direct contact with glandular epithelial cells. Frames E and F show low power electron microscropic views of high endothelial venules (contained erythrocytes) and lymphocytes adherent to the vessel wall. Of note, no electron dense clusters are noted along the blood vessel basement membrane.

A cluster of 50 or more lymphocytes is called a "focus", and an average focus score of 2 or more per 4 mm² fulfills the diagnosis of SS (1). Multiple studies have shown that a positive SG biopsy is closely correlated with KCS and anti-nuclear antibodies directed against SS-A (Ro) and SS-B (La) antigens (summarized in ref. 2). However, patients and

physicians have been reluctant to obtain biopsies except in the setting of research protocols. Thus, another classification system is the San Diego criteria, where in patients are diagnosed as SS based on a) objective KCS and xerostomia, or b) a characteristic minor SG biopsy *or* evidence of a systemic autoimmune disease, as manifested by characteristic autoantibodies (3). In 1993, a much less stringent European Economic Community (EEC) criteria was proposed and can be fulfilled without requirement for biopsy or serologic abnormality (4). Only about 10% of patients who fulfill the EEC criteria would fulfill the San Diego or San Francisco criteria. As a result of these different criteria, there is confusion in both the clinical and research literature about the disease associations and prognosis of SS. Features of each of these classification systems are summarized in Table 1. Fortunately, the EEC recently has suggested revision of their criteria to make them more similar to the San Diego criteria (5). However, in the revised EEC criteria, exclusions to diagnosis of SS will still differ from the San Diego and San Francisco criteria. For example, patients with sicca symptoms associated with hepatitis C infection (6,7) are excluded from diagnosis as SS in the San Diego criteria, but included in the EEC criteria where up to 20% of SS patients may have hepatitis C (8). Therefore, it is difficult to compare studies published from Europe (where EEC criteria are frequently used) to studies published in the U.S. where other diagnostic criteria are often used. Also, basic research on pathogenetic mechanisms of SS relies on samples that clinicians provide to researchers. Thus, attempts to find better animal models or to test a particular molecular hypothesis for SS are difficult when there is no "gold" standard for clinical definition of the disease. In this chapter, we will use the San Diego criteria, since these patients share common features of minor salivary gland biopsy, autoantibodies and particular HLA-DR/DQ alleles. Other causes of keratitis and salivary gland enlargement that need to be considered in the differential diagnosis of SS are listed in Table 2.

Table 1   Criteria for diagnosis of primary and secondary SS

I.   **Primary SS**
    A.   Symptoms and objective signs of ocular dryness
        1.   Schirmer test less than 8 mm wetting per 5 minutes, and
        2.   Positive Rose Bengal or fluorescein staining of cornea and conjunctiva to demonstrate keratoconjunctivitis sicca
    B.   Symptoms and objective signs of dry mouth
        1.   Decreased parotid flow rate using Lashley cups or other methods, or
        2.   Abnormal biopsy of minor salivary gland (focus score of ≥ based on average of 4 evaluable lobules)
    C.   Evidence of a systemic autoimmune disorder
        1.   Elevated rheumatoid factor ≥1:320 or
        2.   Elevated antinuclear antibody ≥ 1:320 or
        3.   Presence of anti-SS-A (Ro) or anti-SS-B (La) antibodies
II.   **Secondary SS**
    Characteristic signs and symptoms of SS (described above) plus clinical features sufficient to allow a diagnosis of RA, SLE, polymyositis or scleroderma.
III.   **Exclusions:**
    Hepatitis B or C infection, sarcoidosis, pre-existent lymphoma, acquired immunodeficiency disease and other known causes of keratitis sicca or salivary gland enlargement.

Table 2   Causes of keratitis and salivary gland enlargement other than SS.

Keratitis
    1.   Mucus membrane pemphigoid
    2.   Sarcoidosis
    3.   Infections: virus (adenovirus, herpes, vaccinia), bacteria, or chlymidia (i.e., trachoma)
    4.   Trauma (i.e., from contact lens) and environmental irritant, including chemical burns, exposure to ultraviolet lights or roentgenograms
    5.   Neuropathy, including neurotropic keratitis (i.e., damage to 5th cranial nerve and familial dysautonomia (Reily-Day syndrome)
    6.   Hypovitaminosis A
    7.   Erythema multiforme (Steven-Johnson syndrome)

Salivary gland enlargement
    1.   Sarcoidosis, amyloidosis
    2.   Bacterial (including gonococci and syphilis) and viral infections (i.e., infectious mononucleosis, mumps)
    3.   Tuberculosis, actinomycosis, histoplasmosis, trachoma, leprosy
    4.   Iodide, lead, or copper hypersensitivity
    5.   Hyperlipemic states, especially types IV and V
    6.   Tumors (usually unilateral), including cysts (Warthin tumor), epithelial (adenoma, adenocarcinoma), lymphoma, and mixed salivary gland tumors
    7.   Excessive alcohol consumption
    8.   Human Immunodeficiency Virus (HIV)

3.   CLINICAL MANIFESTATIONS

3.1   Ocular Manifestations

Patients complain of dry eyes and an irritation that is partly relieved by the use of artificial tears. These complaints should be confirmed by objective decrease in tear flow (Schirmer's test done by inserting paper strips under the lower eyelid), where normal flow rate is greater than 8 mm in 5 minutes. It is also useful to determine the stimulated lacrimal flow rate using the Schirmer's II test, in which a Q-tip is gently inserted into the nose to stimulate the naso-lacrimal gland reflex. This serves as a "treadmill" stress test for the lacrimal glands and usefully predicts the response to therapy with pilocarpine.

3.2   Oral symptoms

SS patients describe a dry, painful mouth. A common complaint is difficulty when swallowing food without water. They may have a rapid increase in dental caries and loss of dental enamel. Although periodontal disease is increased, it is difficult to compare this to the increasing rate of periodontal disease in the general population. Patients may have swelling of the parotid and submandibular glands. Most common are intermittent infections, probably related to inspissation of mucus that blocks the draining ducts, increases in pressure in the ductal tree (causing rupture and inflammation), and predisposes to infection by oral organisms.

3.3. Dermatologic Manifestations

Dryness of the skin is a common manifestation of SS. In some patients, dryness has been associated with lymphocytic infiltrates in the eccrine glands (9), but in most patients the histology and immunohistology of clinically "normal" skin is unremarkable. Antibodies against endothelial cells have been found in a subset of SS patients, but are also detected in many other autoimmune disorders and are not closely associated with skin vasculitis. In contrast to SLE patients, immunohistologic examination of uninvolved skin (especially from photo- nonexposed areas) is generally negative for deposition of immunoglobulin or complement (10,11). Upon exposure to TNFα, skin of SS patients may upregulate expression of SS-A and SS-B antigens (12).

SS patients may develop a cutaneous vasculitis with a spectrum of lesions similar to those seen in SLE patients. The skin lesions are usually divided into papable and nonpalpable purpura, with the palpable purpura corresponding to leukocytoclastic vasculitis on biopsy. The accumulation of neutrophils at these sites suggests the antecedent deposition of complement and chemoattractant molecules such as C3a or C5a (13). These skin lesions are usually multiple and symmetrical. When asymmetric skin lesions are present, the possibility of an embolic source (such as a heart valve) should be considered.

Other skin lesions include non-palpable purpura. The biopsy of these lesions does not usually include leukocytoclastic vasculitis (ie, remnants of nuclear debris from granulocytes). These lesions are generally found symmetrically on the lower extremities. The purpuric (reddish) color is derived from red cells that are extruded from the ruptured small dermal vessels. These skin lesions are also called hypergammaglobulemic purpura due to the more frequent association with polyclonal IgG increase. In comparison, among a large cohort of patients with hyperglobulinemic purpura, about 50% have SS (14). The nonpalpable purpura is often associated with rheumatoid factor (esp. IgM- kappa monoclonal rheumatoid factor) containing VKIIIb subclass of light chains (15). Biopsies generally show ruptured blood vessels and deposition of complement. It has been assumed that immune complexes become trapped at the bifurcation of small blood vessels, leading to complement activation by the immune complex. Treatment of the skin lesions is important since they frequently lead to significant sensory peripheral neuropathy, probably as a consequence of inadequate to blood flow to the small cutaneous nerves of the feet and lower legs.

In Japanese SS patients, skin lesions of erythema annulare have been reported to occur relatively frequently (16), including those patients with juvenile onset SS. These skin lesions may have subtle histologic features and distribution on the trunk, which are distinct from SLE patients. Different skin lesions of erythema annulare in the same patient have T-cells that share a preferential expression of particular T-cell antigen receptor variable region sequence (17).

Peripheral swelling of extremity suggests the possibility of venous thrombosis in association with anti-phospholipid antibodies. The usual triad of false positive RPR, prolonged PTT, and anticardiolipin antibodies may be present. Recent studies have also suggested the importance of elevated homocysteine as a co-factor (including elevations induced by use of methotrexate in the absence of folic acid) and the role of autoantibodies against beta2-glycoprotein (18). Anticardiolipin antibodies are found in a subset of SS patients and are generally IgA isotype (20), with lower incidence of thrombosis than found in SLE patients (21).

Antibodies against other matrix elements including fibrillin-1 and golgi components may be found in SS patients with skin lesions, but the association with SS is rather weak and these autoantibodies are found in many different autoimmune disorders (19).

Neonatal lupus may be found in the children of mothers with SS, SLE (22), and in a proportion of mothers bearing antibodies against SS-A and SS-B antigens but lacking clinical SS (23). Mixed cryoglobulinemia also may be associated with leukocytoclastic vasculitis and should initiate a search for occult hepatitis C infection (24).

3.4. Neurologic

Neurologic manifestation in SS patients is generally divided into 3 components, which often show considerable overlap. Peripheral neuropathy is more common sensory than motor, especially in patients with hypergammaglobulinemic purpura. SS patients may develop mononeuritis multiplex, which is more common in patients with leukocytoclastic vasculitis, and motor neuropathy.

Central nervous system involvement may include the spectrum of disorders found in SLE patients, ranging from intracerebral vasculitis (25) to neuropsychiatric manifestations (26). The incidence of CNS lesions in SS patients has remained controversial, with the incidence at our center and other centers being similar to that found in SLE patients (27). The incidence of multiple sclerosis in SS patients has been a particular area of controversy; although rare patients will have both conditions, the general incidence of multiple sclerosis does not appear significantly increased over the general age and sex related population.

Finally, symptoms of chronic fatigue and fibromyalgia are very common in SS patients (28,29). However, they are also common in the general population. A common diagnostic dilemma is the patient with low titer ANA and vague CNS symptoms of chronic fatigue or memory loss. The association with SS have been particularly controversial when the 1993 EEC criteria are used, since chronic fatigue, low titer positive and mild sicca symptoms are common in the generally population (30,31). In our clinic, we attribute the chronic fatigue to an autoimmune process when objective laboratory tests such as ESR or CRP show elevation commensurate with the symptoms. Otherwise, we attribute the fatigue to non-restorative sleep patterns, which appear more common in SS patients. However, we describe below the potential interference of autonomic function (ie. lacrimal and salivary function) due to local production of autoantibodies in cytokines. It is possible that similar mechanisms may play a role in the "fibromyalgia" by influencing function at the hypothalamic level (32). Since fibromyalgia is an enormous "health" problem in terms of dollars spent on patient care and days off work due to disability, the insights about neuro-immune interactions in SS patients may provide insights into better diagnostic and therapeutic approaches, for fibromyalgia.

3.5. Hematologic Abnormalities and Increased Risk of Lymphoma

Neutropenia is also common in SS (33). In general, bone marrow aspirates show adequate myelopoesis and suggest peripheral sequestration or destruction as the mechanism responsible for neutropenia. Antibodies against Fc receptors on neutrophils have been proposed as a target for increased destruction of neutrophils (34). However, these studies have been difficult to reproduce in many patients with neutropenia.

SS patients have increased frequency of serum and urinary paraproteins (35). They exhibit increased levels of cryoglobulins, particularly in association with hypergammaglobulinemic purpura. The cryoglobulin is frequently a type II mixed cryoglobulin containing an IgM-kappa monoclonal rheumatoid factor similar to that found in Waldenstrom's macroglobulinemia (36). In comparison to the type II mixed cryoglobulin in SS patients, the cryoglobulins in SLE and RA patients were polyclonal (type III) (37). Among Japanese SS patients, an increased incidence of non-IgM paraproteins has been reported (38).

The relative risk for primary SS patients of developing lymphoma has been estimated to be approximately 40-fold higher than age- and sex-matched control subjects by investigators at the NIH (39). However, Whaley et al. (40) in Glasgow did not find such an increased prevalence. These discrepancies may be attributed to several factors, including the relatively small number of patients reported with lymphoma and ascertainment bias in patient referral patterns. These differences may also reflect the difficulty in distinguishing lymphoma from extensive polyclonal infitrates in glands or lymph node biopsies (40).

The lymphomas are predominantly non-Hodgkin's B-cell (IgM-K) tumors that arise in the salivary gland and cervical lymph nodes. They have been reported as marginal zone (41) or MALT types (42) in different series. Although the finding of "myoepithelial" islands (i.e., a degenerating tubule surrounded by lymphocytes) is often interpreted as an indication that the tumor is "benign", malignant lymphomas can be found in the same biopsy specimen that contains myoepithelial islands (43,44). T-cell lymphomas also may occur, but are much less common (45). The distinction between malignant lymphoma and "pseudolymphoma" in SS patients is often quite difficult, even when recombinant DNA methods are utilized (46,47). The occurrence of a t (14,18) translocation is closely associated with malignant transformation in SS biopsies and suggests pathogenetic similarity to follicular lymphomas (48). Other forms of non-malignant lymphoid proliferation in primary SS patients include myeloma (49), thymoma (50) and angioimmunoblastic lymphadenopathy (51). In both pseudolymphoma and angioblastic lymphadenopathy, there appears to be a high frequency of progression to frank lymphoma.

4. EPIDEMIOLOGY

SS is subdivided into primary SS (1o SS) and secondary SS (2o SS) where the sicca symptoms are associated with another well-defined autoimmune disorder such as rheumatoid arthritis, systemic lupus erythematosus, polymyositis, and progressive systemic sclerosis (scleroderma). Primary SS (1° SS will be referred to as SS in this chapter for simplicity) is predominant in females (9:1), with peak age incidences in the child bearing age (25–35 yrs) and in the early post-menopausal age (50–55 years). The precise frequency of SS depends on the criteria system used (discussed below). However, the frequency ranges from approximately 0.5% (San Diego Criteria) to 3% (1993 EEC criteria) for primary SS. The frequency of sicca symptoms increase in the elderly, as does the frequency of positive anti-nuclear antibodies (52).

SS has a worldwide distribution. In the United States, it is found in the Caucasoid, Negroid and Mexican populations. SS is predominantly a disease of middle-aged females. Among 200 consecutive SS patients evaluated at Scripps Clinic, 180 were females with a median age of 54 years at the time of onset. In 10 patients, age of onset was 12 years or younger. A positive family history of systemic autoimmune disease involving parent (mother) or sister was confirmed in 14 families. The most commonly reported disease in the family of the SS patient was SLE, occurring in 10/14 families. 1°SS in a sibling was found in only 2/14 families and RA in 2/14 families. These results suggest that an increased risk towards autoimmune disease may exist in families with a 1°SS patient. However, the disease most commonly detected in 1°SS families was SLE and not 1°SS. This suggests that a similar genetic factor(s) predisposed to both 1°SS and SLE. Additional factors, such as environmental agents, may be necessary to lead to the development of a specific disease such as SLE or 1°SS.

Certain HLA-D-associated antigens have increased frequency in SS patients. Initial studies demonstrated the different genetic predisposition between 1°SS patients (i.e., HLA-DR3) and 2°SS patients with associated RA (i.e., HLA-DR4) (53–57). The influence of additional HLA-encoded genes such as HLA-DQ has also been demonstrated (58). In particular, Caucasian patients with the extended haplotype HLA-DR3, DR52a, DQA4 and DQB2 have an increased frequency of antibodies against SS-A and SS-B antigens (58). According to the recently-adopted nomenclature for HLA, this extended haplotype is named HLA-DRB1*0301, -DRB3*0101, -DQA1*0501, -DQB*0201 (59). The extended haplotype containing HLA-DR3 has been associated with other autoimmune disorders, such as thyroiditis and myasthenia gravis, as well as with increased immune responses in normal. In addition, the other chromosome frequently contributes the HLA-DQA1 allele. The presence of heterozygosity at the DQ genetic locus permits the possibility of gene product complementation at the cell surface (58).

In Caucasians, the extended haplotype containing HLA-DR3 and B8 may contain a deletion of the complement protein C4A and other proteins important in the immune response. In Japanese SS patients, a deletion of the complement C4A gene was also noted with increased frequency and complement C2 deficiency is associated with increased frequency of anti-SS A antibody (59). The association of partial complement deficiency and autoimmune diseases may reflect the importance of complement in the normal solubilization and removal of immune complexes. Alteration of this function may lead to excessive stimulation of other cell types.

In other ethnic populations, susceptibility to SS may not be associated with DR3. Increased risk for 1°SS may be associated with HLA-DR5 in Negroids (60), with HLA-DR4 in Japanese, and with HLA-DR8 in Chinese (61,62). The specific mechanism by which HLA confers increased risk of developing SS remains unknown. It has been postulated that specific conformations of HLA-D-associated molecules and of peptides derived from autoantigens may cause CD4+ T-cells to generate an autoimmune response against the salivary gland epithelial cells.

In secondary SS associated with RA, the onset of sicca symptoms is usually many years after the onset of joint symptoms, with ocular symptoms more frequent than oral symptoms. Also, these patients generally lack antibodies against SS-A or SS-B and have associated HLA-DR4, in comparison to the predominance of HLA-DR3 in primary SS. In secondary SS associated with SLE, the clinical distinction is often very difficult, since both groups of patients frequently have a positive ANA, arthralgias, and leukopenia. Indeed, it is useful to consider SLE as a heterogeneous group of disorders in which each subgroup has its own characteristic clinical features and associated pattern of autoantibodies and HLA-DR associations. In this regard, the patients with SS secondary to SLE share the presence of HLA-DR3, positive anti-SS A antibody and generally lack renal involvement or anti-DNA antibodies (more commonly found in a distinct subset of SLE patients that are HLA-DR2). Further, genetic studies on families with SS (as well as identical twin studies) indicate that family members (ie. mothers, sibs, children) usually do not simply inherit SS but may develop either SS or SLE (the SS like subset). This leads to the hypothesis that similar genetic factors predispose to either SS or SLE (SS like subset), and that an additional environment or non-genetic factor (perhaps recombinatorial events involving T-cells receptor or thymic selection) may lead to the emergence of either clinical phenotype. Finally, sicca symptoms may be associated with scleroderma. However, the minor salivary gland biopsies from these patients show a more predominant pattern of sclerosis, similar to that found in other glands (such as those lining the gastrointestinal tract) than in primary SS or secondary SS associated with SLE. These patients have a different pattern of autoantibodies and associated HLA-DR. This suggests that sicca symptoms secondary to scleroderma may have a different pathogenesis and recent suggestions have indicated that the sicca symptoms in scleroderma have more similarity to graft versus host disease (where sicca symptoms have also been reported).

5. OCULAR AND ORAL FEATURES OF SJÖGREN'S SYNDROME: PHYSIOLOGY AND NEUROENDOCRINE ASPECTS

In understanding the clinical manifestations of oral and ocular symptoms, it is important to review the pathophysiology. A common source of confusion is that patients complain about ocular/oral symptoms, while rheumatologists talk about salivary gland biopsies and autoantibodies. As a result, patients with sicca symptoms who lack evidence of systemic autoimmunity (such as fibromyalgia or patients receiving medications with anti-cholinergic side effects) do not receive adequate treatment for their local symptoms.

The normal tear film is composed of three components: Lipid derived from the Meibomian glands, mucus derived from conjunctival goblet cells, and aqueous tears produced by the main and accessory lacrimal glands (54). SS results from decreased production of the aqueous component, which contributes to ocular pathology in several ways. An adequate volume of tear film is necessary for the eyelids to clear debris and bacteria from the conjunctival surface. The proteinaceous fluid contains lysozyme, peroxidase, immunoglobulins (esp. IgA), complement C3 and C4, properdin, lactoferrin, transferrin, epidermal and nerve growth factors, and a group of low molecular weight peptides (termed defensins) that have antibacterial properties. The decreased delivery of these substances to the conjunctiva as a result of diminished tear flow in SS patients contributes to qualitative changes in the epithelial cells measured by exfoliative cytologic methods.

Thus, when a patient talks about ocular symptoms, they are referring to increased viscosity as the upper lid moves over the ocular globe during a blink reflex. This smooth movement is facilitated by the tear films, which is an aqueous "gel" containing mucins (56) as well as water-soluble components (growth factors, nutrients, anti-bacterial proteins, hormones) (63,64). Further, the mucosal cells that line the lid as well as the ocular surface have a transmembrane mucin (muc-1) (65) that further facilitates the "low friction" movement of the lid over the globe. Mucins in the tear film are produced by stratified corneal epithelial cells as well as by goblet cells (66). In the mouth, mucins facilitate the movement of the tongue, protection against bacterial infections (67) and also provide important function in maintenance of taste by sensory nerves (68). The recent cloning of mucins from the upper airways and salivary glands (69–71) will provide new tools to examine their production and regulation in SS patients and perhaps lead to a new generation of therapies.

Aqueous tear production by the lacrimal glands is strongly influenced by neural stimulation. The stromal, periductal, perivascular and acinar areas of the glands are innervated by many parasympathetic, sympathetic and pep-tidergic fibers that harbor immunoactive transmitters, including VIP, substance P, methionine enkephalin, leucine enkephalin, calcitonin gene-related peptide as well as adrenergic and cholinergic agents (103). In general, inner-vation by parasympathetic nerves release acetylcholine, which stimulates the flow of watery secretions and decreases the level of cAMP. Sympathetic stimulation leads to release of norepinephrine, secretion of mucin-rich, pro-teinaceous tears and increased levels of cAMP. In addition, neurotransmitters, such as substance P and vasoactive intestinal peptide (VIP), are produced by autonomic nerves and play a role in tear and saliva flow. Mast cells are present in lacrimal and labial gland biopsies, but their number is not increased in SS patients and electron microscopy did not indicate degranulation of histamine from these cells.

Normal lacrimal and salivary flow is regulated through feedback mechanisms shown schematically in Figure 2, frame A. The mucosal surfaces of the eye or mouth are heavily innervated by unmyelinated fibers that carry afferent signals to the lacrimatory or salvatory nuclei located in the medulla. These medullary nuclei, which are part of the auto-nomic nervous system, are influenced by higher cortical inputs, including taste, smell, anxiety, and depression. The efferent neurons innervate both glandular cells and local blood vessels. The blood vessels provide not only water for tears and saliva, but also growth factors, including hormones (eg. insulin) and matrix proteins (eg. fibronectin and vitronectin) in the perivascular space of the lacrimal and salivary glands. In response to neural stimulation through muscarinic M3 receptors and vasoactive intestinal peptide (VIP) receptors, glandular acinar and ductal cells secrete water, proteins, and mucopolysaccharides (mucins). This complex mixture forms a hydrated gel that lubricates the ocular surface (ie. tears) and the oral mucosa (ie. saliva), containing a variety of anti-bacterial, anti-fungal and wound-healing substances.

Lymphocytes normally transit through the ocular and oral tissues, but their activation is inhibited by the con-stitutive production of Fas ligand by epithelial cells (72). The level of Fas ligand found in the salivary and lacrimal glands of normal individuals and other primates are among the highest levels found in any tissue (72). This is proposed to be the result of the body's inability to tolerate an immune response in such "sensitive" tissues as the eye, and thus a site of "immune privilege" is created. Thus, it appears that the normal lacrimal and salivary glands are structured to normally suppress immune responses by cellular mechanisms and control "infections" by local mucosal barriers augmented by secretory antibodies and anti-infective substances. It may be the failure of these local suppressive mechanisms that leads to emergence of autoimmune reactions that characterize SS.

In the simplest model of SS (Figure 2), the lacrimal or salivary gland is incapable of adequate response to neural signals as a consequence of local immune infiltrates and their derived cytokines. The actual processes in SS or autonomic neuropathy are more complicated than indicated in these schematic diagrams, which are primarily designed to emphasize that salivation or lacrimation are part of a regulatory circuit involving the central nervous system (73). The presence of lymphocytic foci in the lacrimal and salivary glands indicates that these cells have overcome the effects of Fas ligand, which continues to be present in the glands of SS patients (74), perhaps due to increased levels of bcl-2 and bcl-x (discussed below).

The function of epithelial cells is greatly dependent on their interaction with their underlying matrix (75). For example, specific interactions, probably mediated by cell surface integrins, are necessary for the epithelial cell to

Proposed model for Functional Units Involved in Saliva and Tear Flow

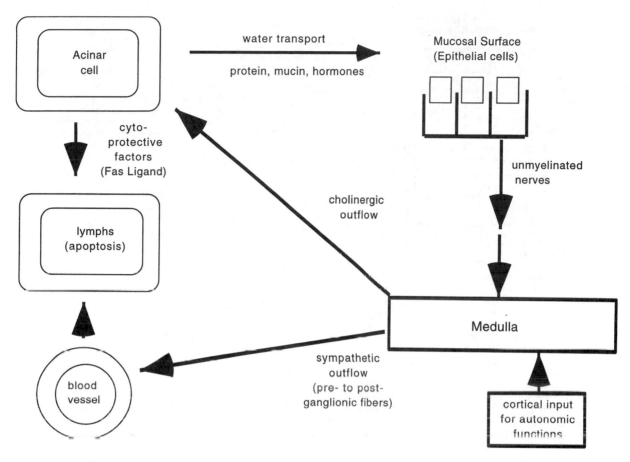

Figure 2. A schematic representation of the neuro-endocrine-immune interactions in saliva or tear production.
Normal efferent sensory neural signals originate from the corneal or oral surface and lead to a specific section of the brain (either the lacrimatory or salivatory nuclei). At this site, higher cortical signals are also integrated and a series of net efferent signals are sent back to blood vessels (the source of water for saliva or tears) and to the glands (that must "pump" the water. The signals to blood vessels are predominantly adrenergic and those to glands are cholinergic. Receptors on the glandular cells are predominantly cholinergic muscarinic type 3 (M3) receptors.
In Sjögren's syndrome, infiltrating lymphocytes release cytokines and autoantibodies that may interrupt the neurosecretory function and response to neurotransmitters.

orient itself in a "frontal" orientation, a process that must occur before upregulation of ATPases and receptors for neural stimulation that are necessary for secretory function (75,76). The production of cytokines may influence this process of epithelial cell differentiation by promotion of metalloproteinase secretion and by directly affecting epithelial cell differentiation (77,78). These interactions are schematically summarized in Figure 3. The epithelial cell sits on a matrix containing collagen (particularly types XIV), laminen, and vitronectin. The characterization of the matrix components is an area for future research. The epithelial cells contain structural proteins (such as fodrin) and nuclear proteins (such as SS-A) that may be targets for the immune response, as well as expression of cell surface receptors for Fas, TGF and EGF (Figure 3). In addition to receptors for EGF and TNFα, these cells may also secrete

growth factors and cytokines that participate in autocrine and paracrine interactions. The epithelial cells may also upregulate HLA-DR, DQ and interferon receptors (77,79–81).

6. PATHOGENESIS: AUTOANTIBODIES AND AUTOANTIGENS

Organ-specific autoantibodies, such as anti-salivary gland, are infrequent in primary SS patients (82). Although anti-salivary gland antibodies have been reported in some RA patients with secondary SS, their absence in most SS patients suggests that anti-duct antibodies are a secondary event rather than playing a primary role in pathogenesis. In this regard, primary SS differs from other autoimmune dis-

Epithelial   Cells   in   Sjogren's   Syndrome   Glands

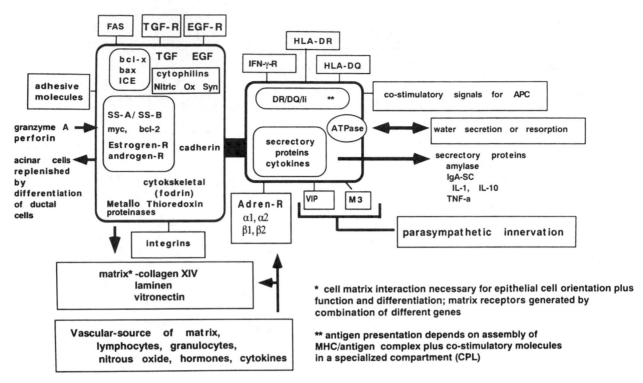

Figure 3.   Schematic Interactions of glandular epithelial cells. The integrity and differentiated function of the epithelial cells depends on its interaction with the underlying matrix. Also, the continued differentiation of the cell depends on input from neural stimulation and a variety of growth factors. Also shown are structural proteins (including fodrin), Sjögren's-associated antigens SS-A, and muscarinic M3 receptor that have been proposed as targets for the immune response in SS.

eases, such as thyroiditis, myasthenia gravis and diabetes mellitus, in which antibodies to target organs are consistently found early in the disease process.

However, there has been recent interest in the demonstration of antibodies in SS patients that react with rodent muscarinic M3 receptor (83). When sera from some SS patients is transferred to animals, an inhibition of secretory response occurs (84). These observations raise the possibility that part of the pathogenesis of decreased secretory function in SS patients is a result of diminished signal transduction from cholinergic nerves to acinar or ductal cells. In this regard, SS might be considered analogous to myasthenia gravis, where in the functional deficit is the result of antibodies directed against a different acetylcholine receptor. However antibodies against human M3 receptor have not been responsibly demonstrated.

Most of the research on autoantigens in sera of SS patients has been directed towards the cloning and isolation of proteins identified by the anti-nuclear antibody reactivity of SS patients. Their positive anti-nuclear antibody (ANA) is most commonly a result of their reactivity with extractable nuclear antigens, such as Sjogren's syndrome A

(SS-A, 60 and 52 kd, also called Ro) and Sjogren's antigen B (SS-B 48 kd, also called La) antigens. They frequently also have rheumatoid factor (IgM anti-IgG Fc). ANA titers are generally measured using a human liver cell line grown in vitro (Hep-2 cells) or a frozen tissue section derived from normal mouse kidney. These sources of ANA substrate differ in both tissue and species origin and occasionally give different results with the same patient's serum. Furthermore, the Hep-2 cells are rapidly dividing and thus express antigens associated with cell proliferation and division.

Antibodies against "SS-A" react with a ribonucleoprotein complex that contains a novel class of small RNAs designated hY1-hY5 (h for human and Y for cytoplasmic) RNAs (85,86), although there is now evidence that such RNAs are also intranuclear (87). The hY RNAs share many similarities, including sequence, size (83–112 bp), secondary sequence and abundance (about 10^5 copies/cell) (85,88).

Initial studies indicated that SS-A was a single 60 kd protein, but subsequent reports demonstrated additional proteins reactive with anti-SS A antibody, including a 52 kd protein in all nucleated cells (89) and possibly with a

54 kd protein in non-nucleated red blood cells (90,91). Some patients produce antibody to the 60 kd only, the 52 kd only, or to both the 52 and 60 kd antigens (92). The SS-A proteins share long consecutive sequences of charged amino acids, a feature in common with other autoantigens recognized by SLE and SS sera (86). It is not known whether all SS-A proteins bind to all hY RNA. Each of the genes encoding SS-A proteins appears to be single-copy and non-polymorphic, although alternative promoters for initiating transcription and alternative splicing of the precursor protein may occur in a tissue-specific manner (56).

B- and T-lymphocytes expressing reactivity with SS-A have been demonstrated in the peripheral blood and salivary gland of SS patients (93). Lymphocytes with anti-SS A reactivity are more frequent in patients with certain HLA-DR/DQ genotypes and particular alleles of the T cell receptor beta variable region (94,93). In Japanese patients, different HLA-DR/DQ are found, but these SS patients also exhibit particular V β gene utilization in the peripheral blood (95) and in the salivary glands (17). The production of autoantibodies to SS-A has been associated with glutamine in position 34 of the DQA1 and leucine in position 26 of DQB1 (96).

The SS-B (La) antigen is a 48 kd protein (89). Ribonucleoprotein SS-B/La belongs to a protein family with consensus sequence for RNA binding (97). Although initially the SS-B was felt to be associated with the same ribonuclear protein complex described above, the 52 kd molecule may not always be associated with the same ribonuclear particle (88), 17-beta-estradiol increases expression of 52-kDa and 60-kDa SS-A/Ro autoantigens in human keratinocytes and breast cancer cell line MCF-7 (98). Also, the SS-A and SS-B antigens may distribute into the cell surface membrane, particularly into the blebs associated with apoptosis (99,100).

Antibodies against SS-B are much more closely associated with clinical symptoms of SS than are anti-SSA antibodies and are present in about 40–50% of SS patients (89). The SS-B protein associates with a class of RNA transcripts derived from RNA polymerase III, including U6 cellular RNA, adenovirus RNAs (VA I and VA II), Epstein-Barr virus RNAs (EBERs) and leader RNAs of some negative strand viruses, such as vesicular stomatitis virus. Recent studies have shown that SS-B protein is essential in the biogenesis of these RNAs. The fact that SS-B is associated with precursors of SS ribosomal and tRNA, but not with their corresponding mature species, has suggested that SS-B may play a role in the maturation of these polymerase III transcripts. In addition, SS-B may bind to U1 RNA, an RNA polymerase II transcript. The 3′ oligouridylate tail of these small RNAs is required for interaction with the SS-B protein.

The SS-B molecule can be split into at least two protease-resistant domains, including a 28 kd domain that binds RNA and a 23 kd domain that is phosphorylated (101). The use of fusion proteins derived from the cloned SS-B gene has allowed recognition of at least 3 immunodominant domains; a the region from amino acids 1 to 107 has strongest reactivity, followed by the region including amino acids 111 to 242, and to a lesser extent with the domain from amino acids 346 to 408 (102). One initial study suggested that the SS-B protein bears sequence homology to a retroviral gag polyprotein) (103). However, subsequent studies indicated that the cross-reactivity was due to polyproline rich sequences and not a specific cross-reactivity with viral proteins (104). It has been proposed that viral (i.e., EBV) transcripts attached to SS-B may lead to immune responses against the ribonuclear protein complex in genetically-predisposed individuals. An alternative hypothesis is that SS-B proteins become antigenic due to molecular mimicry after exposure to a virus such as a retrovirus (105). However, patients with an SS-like syndrome in association with HIV do not develop anti-SS B antibodies (106).

A second 60 kd protein has been cloned using serum from an SS patient and found to have a 94% homology to calreticulin, a high affinity calcium binding protein that resides in the endoplasmic reticulum and to proteins (107). However, subsequent studies have not indicated that this molecule is part of the ribonuclear protein complex formed by the SS-A and SS-B molecules described above.

In the past, a syndrome called "subacute SLE" was defined by the presence of antibody to SS-A in the absence of ANA. In most cases, the finding of an anti-SS A antibody in the absence of ANA suggests a technical error in the measurement of the ANA. SS-A is highly soluble in methanol, and the negative ANA generally represents an "over-fixed" Hep2 cell used in the ANA assay. Also, it is extraordinarily rare for an antibody to SS-B to be present in the absence of anti-SS-A antibody. Again, the finding of positive anti-SS B in the absence of anti-SS A generally indicates an artefactual error in one of these two measurements.

Rheumatoid factor (anti-Fc region of IgG antibody, RF) measurements generally include IgM and IgA isotypes; IgG RF is not measured due to the formation of self-aggregating complex (i.e., IgG anti-IgG Fc). RF are found in low amounts in normal individuals after viral infections, such as EBV (108). RF performs several normal functions, including the ability to facilitate clearance of immune complexes and activate complement (109). In addition, B-cells bearing cell surface RF can bind to immune complexes, internalize these antibody-antigen complexes, and represent antigenic fragments to immune T (109). A similar mechanism might be proposed for the production of anti-nuclear antibodies that bind to viral antigens, leading to re-expression of such antigenic peptides on the cell surface of B cells producing anti-nuclear antibodies.

A cross-reactive idiotype (CRI) in RF of SS patients was first noted using a monoclonal anti-idiotype antibody (MoAb17–109) (110) and the results were expanded using antibodies against synthetic peptides defining framework and hypervariable regions (111). The structural basis for the cross-reactive idiotype was the common use of a particular sub-subgroup of kappa light chains termed the Vkappa IIIb subgroup (15). This kappa light chain is encoded by the germline variable segment gene VK325 (15). To extend these studies, a cDNA library from a SS salivary gland was constructed, and sequence analysis of kappa chain cDNA confirmed the high frequency of VK325 variable region segments (112). Furthermore, B cells expressing VK325 light chains within the salivary gland appeared polyclonal in origin, based on the different N-terminal diversification sequences found in the VK325-containing transcripts (112). In addition to the light chain-associated CRI, a heavy chain idiotype associated with VH1 subgroup had increased frequency of expression in SS biopsies (113).

7. PATHOGENESIS: CELLULAR INFILTRATION OF THE SALIVARY GLANDS

The characteristic finding in SS is lymphocytic infiltration of the salivary gland (Figure 1, frame A). The majority of salivary and lacrimal gland lobules are involved and infiltrates are more prominent in the central portion of the lobule than in the periphery. Well-defined germinal centers are infrequent in minor salivery gland biopsy specimens, but occasionally present in parotid lacrimal, and submandibular infiltrates. Normal parenchymal structures such as acini and ducts may be extensively replaced by these infiltrates. "Activated" lymphocytes with large, vesicular nuclei and prominent mucleoli are frequently present and may be difficult to distinguish from the lymphocytes in lymphoma. At higher power, frame C shows some acini that are relatively intact, but lymphocytes are present within the acinar structure (arrow). Frame D shows a duct with an increased number of infiltrating lymphocytes (arrows) and degenerating epithelial cells. Frames E and F demonstrate the high endothelial venules with RBC in the lumen and lymphocytes adherent to the vessel wall. At higher power electron microscopic level, most acinar cells are collapsed, and the amount of secretory granules are markedly decreased in glands from patients with primary SS. Mitochondria are decreased in number and the remaining ones are swollen. Nuclei of the acinar cells appeared relatively unremarkable except for prominent nucleoli. Tubuloreticular structures may be present and may represent a nonspecific response to epithelial cell injury (114). Of importance, electron-dense immune complex deposits are not present at the basement membrane surrounding the blood vessel or epithelial cells. Taken together, the light

Table 3 Extraglandular manifestations in patients with primary SS

Respiratory
  Chronic bronchitis secondary to dryness of upper and lower airway with mucus plugging
Lymphocytic interstitial pneumonitis
  Pseudolymphoma with nodular infiltrates
  Lymphoma
  Pleural effusions
  Pulmonary hypertension especially with associated scleroderma
Gastrointestinal
  Dysphagia associated with xerostomia
  Atrophic gastritis
  Liver disease, including biliary cirrhosis and sclerosing cholangitis
Skin
  Vaginal dryness
  Hyperglobuliemic purpura-nonthrombocytopenic
Raynaud's phenomena
  Vasculitis
Endocrine, neurologic, and muscular
  Thyroiditis
  Peripheral neuropathy-symmetric involvement of hands and/or feet
Mononeuritis multiplex
  Myalgias
Hematologic
  Neutropenia, anemia, thrombocytopenia
  Pseudolymphoma
  Angioblastic lympadenopathy
  Lymphoma and myeloma
Renal
  Tubulo-interstitial nephritis (TIN)
  Glomerulonephritis-in absence of antibodies to DNA
  Mixed cryoglobulinemia
  Amyloidosis
  Obstructive Nephropathy due to enlarged periaortic lymph nodes
  Lymphoma
  Renal artery vasculitis

and electron microscopie appearance of salivary gland infiltrates in SS suggests a cell-mediated destruction rather than an immune complex deposition mechanism.

To characterize the cell-cell interactions in SS, immunohistologic methods have been used to identify lymphocyte subsets in salivary gland biopsies (115). The majority of lymphocytes are mature T cells (CD3$^+$) of the T helper (CD4$^+$) subset. Some T-suppressor cells (CD8$^+$) were also present. These T cells express the $\alpha\beta$ cell surface receptor, although a small proportion of Tγ cells can be detected (116). Of particular interest, certain lymphocyte subsets were present in the salivary gland infiltrates, but were absent from the blood. Conversely, other lymphocyte subsets (such as natural killer cells) were present in blood, but absent from the salivary gland infiltrates (117). These results emphasize the need to study the target organ in order to elucidate pathogenesis, since alterations in lymphocytes occurring in blood may not be present in the target organ. In lacrimal gland biopsies, a higher proportion of B cells and tendency towards germinal center formation is frequent (118).

Salivary gland epithelial cells in biopsies from SS patients react with anti-HLA DR antibodies, in contrast to normal salivary gland biopsies that lack anti-HLA DR reactivity (119,120). The induction of HLA-DR, DQ and invariant chain on the epithelial cells may play an important role in pathogenesis, since CD4+ T cells (the major lymphocyte subset in the SS salivary gland) interact with peptide antigen presented by HLA-DR molecules. The induction of epithelial cell HLA-DR molecules is probably due to local production of interferon-γ by T cells, since monoclonal antibody against this cytokine eliminates the HLA induction (120).

Other co-stimulatory molecules such as B7 may be expressed on salivary gland epithelial cells (121), although another study did not find expression on salivary gland epithelial cells from biopsies of SS patients (57). In this regard, other tissues, such as keratinocytes and pancreatic islet cells can be induced to express co-stimulatory molecules such as B7 *in vitro*, but they are difficult to detect *in situ* on biopsies from autoimmune tissue.

Integrins including ICAM-1 (122,123), which may serve as co-stimulatory molecules *in vitro*, do appear to be uprog ulated on vascular endothelial cells and acinar cells in biopsics from SS patients. Another upregulated integrin is alpha beta 7, which is also found in intestinal-activated mucosal tissues (124,125).

7.1. Characterization of Immunoglobulin DNA Rearrangements in SS Biopsies

At the DNA level, expansion of one or more B cell clones within salivary gland biopsies of SS patients has been demonstrated by Southern blot methods showing detectable rearrangements of both kappa and heavy chains (47,126). Of particular interest, these biopsies did not fulfill requirements for diagnosis of lymphoma. Further, different rearrangements were seen in serial biopsies from the same patient (48). These studies have emphasized the risk of using DNA rearrangements for the diagnosis of lymphoma in patients with autoimmune disorders associated with lymphoproliferative features (127). Detectable clonal rearrangements of the T cell antigen receptor β-chain gene were much less frequent (116).

Patients with SS are at increased risk for developing lymphoma (39). Prior to overt lymphoma, some patients go through a stage called "pseudolymphoma", characterized

by lymphadenopathy, circulating paraproteins, fever and night sweats (128). However, their biopsies show a predominance of CD4+ T cells, in comparison to the monoclonal B cells that emerge in the same patient's subsequent lymphoma (128). During the pre-lymphoma stage, multiple small B-cell clones undergo transient expansion, perhaps in response to specific antigens or cytokines. A second event, perhaps a karyotypic alteration, subsequently occurs and leads to a neoplastic expansion of a particular B cell clone. A high frequency of translocation (14,18) involving the protooncogene bcl-2 have been noted in these lymphomas (48).

Studies of salivary gland biopsies from Japanese SS patients have indicated a predominance of specific T cell receptor (TCR) utilization in labial salivary glands (129, 130). However, the T cell repertoire expressed in each patient is different, although subsequent biopsies in the same patient show similar T cell variable region utilization. In patients with multiple sites that have been biopsied, it appears that similar TCAR may be found at different sites (130). It has been postulated that T cell receptors from salivary gland biospies may be predicted to show reactivity with SS-A and alpha amylase (131).

7.2   Mechanisms of Glandular Destruction

Glandular destruction in SS may be mediated by either perforin/granzyme (132) pathway. Granzyme is a member of the serine proteinase family and cooperates with the pore-forming protein perforin that forms cylindrical pores in the membraes of target cells (133). It appears that (CD4+/CD28−) as well as CD8+ T cells may perform granzyme/perforin mediately cytolysis.

Recent studies in SS biopsies have also indicated increased fas ligand expression in glandular epithelial cells (74). Among infiltrating lymphocytes, increased expression of bcl-2 and bcl-x may help prevent apoptosis (134,135). In the murine NOD model of SS, increased rates of apoptosis are found even in the NOD.SCID mouse (136), indicating that this process is independent of lymphocytic infiltration (137). Of interest, decreased salivary gland secretion is only noted when lymphocytic infiltrates are present (138).

7.3. *In Vitro* Studies of Lymphocyte Function

Salivary gland lymphocytes (SGL) eluted from SS biopsies also have the ability to produce interleukin-2 (139) and interferon-γ (120). Of particular importance, peripheral blood lymphocytes (PBL) were available from the same SS patients and permitted comparison of PBL and SGL production of IL-2. PBL were often deficient in IL-2 production in the same patient, while SGL production remained intact. Mixing experiments of PBL and SGL from the same patient indicated the "dominant" suppressive effect of the PBL on IL-2 production by SGL. These results demonstrate important functional differences between lymphocytes in the target organ and the peripheral blood of the same patient.

Using antibodies against specific synthetic peptides derived from IL-2, it was possible to show that approximately 2–3% of the CD4$^+$ T cells were producing IL-2 *in situ* (139), and that a similar proportion of T cells expressed high affinity IL-2 receptors. The relatively low percentage of IL-2-producing T cells within the SS salivary gland, as well as the very low percentage of lymphoid cells undergoing DNA replication (S phase) or mitosis suggests that most salivary gland T cells are not generated within the salivary gland by cell division *in situ*. It is more likely that such T cells are generated elsewhere, migrate through the bloodstream and enter the salivary gland through high endothelial venules (140) via specific cell-surface receptors (141) such as ICAM and VLA antigens (125,142).

7.4. Potential Role of Viruses as Co-Factors in Pathogenesis

The initial inciting lesion in SS remains unknown. One candidate is Epstein-Barr virus (118,143–145). In normal individuals, primary infection with EBV (infectious mononucleosis) involves the salivary glands. Although lymphocytes are known to harbor EBV, epithelial cells express receptors for EBV (146,147) and certain tumors (ie. nasopharyngeal carcinoma in China) are closely associated with EBV (148). Infectious EBV can be cultured from the opening of the Stensen ducts in 20% of normal subjects and in >50% of patients receiving corticosteroids or immunosuppressives (149). Thus, these viruses are already present at the relevant target organ, which serves as a site of latency and reactivation (149). Furthermore, EBV can stimulate production of polyclonal antibodies and autoantibodies such as RF (37). Yang et al. (150) reported that anti-SS-B found in many SS patients was directed against a ribonuclear protein that selective associates with EBV-encoded RNA. In nasopharyngeal carcinoma, a disease that is known to be associated with EBV transformation of epithelial cells (151) an intense lymphocytic response occurs that has been termed a "lymphoepithelioma" (152) and these lesions resemble the changes found in primary SS.

In biopsies from SS patients, an increased frequency of salivary gland epithelial cells expressing EBV-associated antigens and EBV DNA can be detected (153,154). Increased EBV DNA can also be detected in SS salivary gland biopsies by polymerase chain reaction (154) and by *in situ* hybridization methods (143). Increased frequency of EBV-associated antigens has also been shown in lacrimal gland biopsies (143). The finding of EBV reactivation is not specific for SS, as indicated by the finding of increased EBV shedding in organ transplant patients receiving immune suppressive medications and increased viral DNA in chronic sialidenitis not associated with SS (155). Using synthetic peptides derived from the early antigen genes of EBV, increased IgA and IgG antibody responses can be detected in SS patients (156,157). Taken together, these findings suggest a potential role for EBV in the pathogenesis of SS. In rare cases of SS, the onset of symptoms occurs at the time of acute EBV infection (i.e., infectious mononucleosis) (158). However, in most cases, the time between initial EBV infection (which occurs before age 20 yrs) and onset of clinical symptoms is greater than 20 years. Further, the antibody titers at disease onset are comparable to patients with other autoimmune diseases and become progressively elevated only later in the course of SS (159). These findings suggested that viral reactivation could occur as a consequence of immune dysregulation in the salivary gland (160). Since EBV is a strong stimulator of immune T cell response (82), reactivation of EBV within the SS salivary gland may serve as a perpetuating factor in salivary gland destruction.

Retroviruses have been suggested as a candidate in some SS patients. Talal et al. (161) found increased reactivity with retroviral protein p24 in a majority of SS sera, and a type A intra-cisternal particle was subsequently isolated from two salivary gland biopsies by Garry et al. (162). Further studies will be required to extend and confirm these interesting results. However, the antibody initially felt to be anti-p24 was subsequently shown to be a cross-reactivity with polyprotein containing proteins including other cellular proteins (104,163).

8. SUMMARY

Sjögren's syndrome (SS) is a systemic autoimmune disease characterized by lymphocytic infiltration of lacrimal and salivary glands. SS can exist as a primary condition (1°SS) or in association with other autoimmune diseases, such as RA or SLE (termed 2°SS). In 1°SS, there may be involvement of extraglandular organs, including skin, nerve, lung and kidney, SS patients produce a variety of autoantibodies, including rheumatoid factor and anti-nuclear autoantibodies. In particular, 1°SS patients produce autoantibodies against ribonuclear protein SS-A (Ro) and SS-B (La), which are involved in the transport and post-transcriptional modification of mRNA. Although the SS-A and SS-B proteins have been cloned, their specific role in pathogenesis remains unclear. Genetic factors, including HLA-DR3 and HLA-DQ, predispose to 1°SS syndrome. In 2°SS associated with RA, the genetic predisposition is HLA-DR and antibodies against SS-B are rarely present.

The precipitating cause of SS remains unknown, but exogenous agents such as Epstein-Barr virus have been proposed as co-factors in perpetuating the immune response against the salivary gland epithelial cells. In contrast to normal salivary glands, the SS gland contains increased proportions of high endothelial venules and the glandular epithelial cells express high levels of HLA-DR antigens. The glands are infiltrated with CD4[+] T cells that can produce cytokines, including IL-2 and interferon-γ. B cells within the salivary gland (SG) produce autoantibodies, including rheumatoid factor. These SG B cells frequently use the VKIIIb sub-subgroup of kappa light chain, a feature that SS patients share with Waldenstrom's macroglobulinemia patients. B cells undergo small clonal expansions that can be detected on Southern blot of immunoglobulin gene rearrangement and SS patients have a markedly increased risk of developing non-Hodgkin's B cell lymphoma involving the salivary glands and cervical lymph nodes. Due to accessibility of the salivary gland for biopsy and the characteristic patterns of autoantibody production, SS provides an opportunity to study the target organ for autoimmune destruction and the transition from autoimmunity to lymphoma.

References

1. Daniels, T.E. 1984. Labial salivary gland biopsy in Sjögren's syndrome. *Arthritis Rheum.* 27:147–56.
2. Daniels, T.E., and J.P. Whitcher. 1994. Association of patterns of labial salivary gland inflammation with keratoconjunctivitis sicca. Analysis of 618 patients with suspected Sjögren's syndrome. *Arthritis Rheum.* 37:869–77.
3. Fox, R.I., C. Robinson, J. Curd, P. Michelson, R. Bone, and F.V. Howell. 1986. First international symposium on Sjögren's syndrome: Suggested criteria for classification. *Scand. J. Rheumatol.* 562:28–30.
4. Vitali, C., S. Bombardieri, H.M. Moutsopoulos, G. Balestrieri, W. Bencivelli, R.M. Bernstein, K.B. Bjerrum, S. Braga, J. Coll, S. De Vita, A.A. Drosos, M. Ehrenfeld, P.Y. Ilatron, E.M. Hay, D.A. Isenberg, A. Janin, J.R. Kalden, L. Kater, Y.T. Konttinen, P.J. Maddison, R.N. Maini, R. Manthorpe, O. Meyer, P. Ostuni, Y. Pennec, J.U. Prause, A. Richards, B. Sauvezie, M. Schiødt, M. Sciuto, C. Scully, Y. Shoenfeld, F.N. Skopouli, J.S. Smolen, M.L. Snaith, M. Tishler, S. Todesco, G. Valesini, P.J.W. Venables, M.J. Wattizux, and P. Youinou. 1993. Preliminary criteria for the classification of Sjögren's syndrome. *Arthritis & Rheum* 36:340–7.
5. Gerli, R., C. Muscat, M. Giansanti, M.G. Danieli, M. Scuito, A. Gabrielli, E. Fiandra and C. Vitale. 1997. Quantitative assessment of salivary gland inflammatory infiltration in primary Sjögren's syndrome, its relationship to different demographic clinical and serologic features of this disorders. *Brit. J. Rheumatol* 36:969–975.
6. Jorgensen, C., M.C. Legouffe, P. Perney, J. Coste, B. Tissot, C. Segarra, C. Bologna, L. Bourrat, B. Combe, F. Blanc, and J. Sany. 1996. Sicca syndrome associated with hepatitis C virus infections *Arthritis & Rheum* 39:1166–1171.
7. Verbaan, H., J. Carlson, S. Eriksson, A. Larsson, R. Liedholm, R. Manthorpe, H. Tabery, A. Widell, and S. Lindgren. 1999. Extrahepatic manifestations of chronic hepatitis C infection and the interrelationship between primary Sjogren's syndrome and hepatitis C in Swedish patients *J. Intern. Med.* 245:127–32.
8. Haddad, J., P. Deny, C. Munz-Gotheil, and J.C. Ambrosini. 1992. Lymphocytic sialadenitis of Sjögren's syndrome associated with chronic hepatitis C virus liver disease. *Lancet* 8:321–323.
9. Navarro, M., R. Cervera, J. Font, J.C. Reverter, J. Monteagudo, G. Escolar, A. Lopez-Soto, A. Ordinas, and M. Ingelmo. 1997. Anti-endothelial cell antibodies in systemic autoimmune diseases: prevalence and clinical significance. *Lupus* 6:521–526.
10. Provost, T.T., R. Watson, and O.B.E. Simmons. 1997. Anti-Ro(SS-A) antibody positive Sjogren's/lupus erythematosus overlap syndrome. *Lupus* 6:105–11.
11. Simmonsobrien, E., S. Chen, A. Watson, R.C.M. Petri, M. Hochberg, M. Stevens, and T. Provost. 1995. One hundred anti-Ro (SS-A) antibody-positive patients: A 10-year follow-up. *Medicine* 74:109–30.
12. Tsubota, K., K. Fukagawa, T. Fujihara, S. Shimmura, I. Saito, K. Saito, and T. Takeuchi. 1999. Regulation of human leukocyte antigen expression in human conjunctival epithelium. *Invest. Ophthalmol. Vis. Sci.* 40:28–34.
13. Alexander, E.L., and T.T. Provost. 1983. Cutaneous manifestations of primary Sjögren's syndrome: a reflection of vasculitis and association with anti-Ro (SSA) antibodies. *J. Invest. Dermatol.* 80:386–391.
14. Kyle, R., G. Gleich, E. Baynd et al. 1971. Benign hyperglobulicmic purpura of Waldenstrom. *Medicine* (Baltimore) 50:113–123.
15. Fox, R.I., D.A. Carson, P. Chen, and S. Fong. 1986. Characterization of a cross reactive idiotype in Sjögren's syndrome. *Scand. J. Rheumatol.* 561:83–88.
16. Ruzicka, T., J. Faes, T. Bergner, R.U. Peter, and O. Braun-Falco. 1991. Annular erythema associated with Sjögren's syndrome: a variant of systemic lupus erythematosus. *J. Am. Acad. Dermatol.* 25:557–560.
17. Sumida, T., Y. Kita, F. Yonaha, T. Maeda, I. Iwamoto, and S. Yoshida. 1994. T cell receptor V alpha repertoire of infiltrating T cells in labial salivary glands from patients with Sjogren's syndrome. *J. Rheumatol.* 21:1655–1661.
18. Merkel, P.A., Y. Chang, S.S. Pierangeli, E.N. Harris, and R.P. Polisson. 1999. Comparison between the standard anti-cardiolipin antibody test and a new phospholipid test in patients with connective tissue diseases. *J. Rheumatol.* 26:591–596.
19. Pollard, K.M., D.K. Lee, C.A. Casiano, M. Bluthner, M.M. Johnston, and E.M. Tan. 1997. The autoimmunity-inducing xenobiotic mercury interacts with the autoantigen fibrillarin and modifies its molecular and antigenic properties. *J. Immunol.* 158:3521–3528.
20. Asherson, R.A., H.M. Fei, H.L. Staub, M.A. Khamashta, G.R.V. Hughes, and R.I. Fox. 1992. Antiphospholipid antibodies and HLA associations in primary Sjögren's syndrome. *Ann. Rheum. Dis.* 51:495–498.
21. Cervera, R., M. Garcia-Carrasco, J. Font, M. Ramos, J.C. Reverter, F.J. Munoz, C. Miret, G. Espinosa, and M. Ingelmo. 1997. Antiphospholipid antibodies in primary Sjögren's syndrome: prevalence and clinical significance in a series of 80 patients. *Clin. Exp. Rheumatol.* 15:361–365.
22. Buyon, J.P. 1996. Neonatal lupus. *Curr. Opin. Rheumatol.* 8:485–90.
23. Buyon, J.P. 1994. Neonatal lupus syndromes. *Curr. Opin. Rheumatol.* 6:523–529.

24. Buezo, G.F., M. Garcia-Buey, L. Rios-Buceta, M.J. Borque, M. Aragues, and E. Dauden. 1996. Cryoglobulinemia and cutaneous leukocytoclastic vasculitis with hepatitis C virus infection. *Int. J. Dermatol.* 35:112–115.

25. Alexander, E.L., K. Malinow, J.E. Lejewski, M.S. Jerdan, T.T. Provost, and G.E. Alexander. 1986. Primary Sjögren's syndrome with central nervous system disease mimicking multiple sclerosis. *Ann. Intern. Med.* 104:323–330.

26. Spezialetti, R., H.G. Bluestein, J.B. Peter, and E.L. Alexander. 1993. Neuropsychiatric disease in Sjogren's syndrome: anti-ribosomal P and anti-neuronal antibodies. *Am. J. Med.* 95:153–160.

27. Fox, R. 1997. Sjogren's syndrome: Progress and controversies. *Med. Clin. NA* 17:441–434.

28. Calabrese, L.H., M.E. Davis, and W.S. Wilke. 1994. Chronic fatigue syndrome and a disorder resembling Sjogren's syndrome: preliminary report. *Clin. Infect Dis.* 18 Suppl 1:S28–31.

29. Fox, R.I., J. Tornwall, T. Maruyama, and M. Stern. 1998. Evolving concepts of diagnosis, pathogenesis, and therapy of Sjögren's syndrome. *Curr. Opin. Rheumatol.* 10:446–456.

30. Tan, E.M., T.E. Feltkamp, J.S. Smolen, B. Butcher, R. Dawkins, M.J. Fritzler, T. Gordon, J.A. Hardin, J.R. Kalden, R.G. Lahita, R.N. Maini, J.S. McDougal, N.F. Rothfield, R.J. Smeenk, Y. Takasaki, A. Wiik, M.R. Wilson, and J.A. Koziol. 1997. Range of antinuclear antibodies in "healthy" individuals. *Arthritis Rheum.* 40:1601–1611.

31. Lightfoot, R. 1997. Cost Effective use of Laboratory Tests in Rheumatology. *Bul. Rheum. Dis.* 46:1–3.

32. Bennett, R.M. 1993. Disabling fibromyalgia: appearance versus reality. *J. Rheumatol.* 20:1821–1824.

33. Bloch, K.J., W.W. Buchanan, M.J. Wohl, and J.J. Bunim. 1956. Sjögren's syndrome: A clinical, pathological and serological study of 62 cases. *Medicine* (Baltimore) 44:187–231.

34. Bux, J., G. Behrens, M. Leist, and C. Mueller-Eckhardt. 1995. Evidence that the granulocyte-specific antigen NC1 is identical with NA2. *Vox Sang.* 68:46–49.

35. Moutsopoulos, H.M., and P. Youinou. 1991. New developments in Sjögren's syndrome. *Curr. Opin. Rheumatol.* 3:815–822.

36. Fox, R.I., P.P. Chen, D.A. Carson, and S. Fong. 1986. Expression of a cross reactive idiotype on rheumatoid factor in patients with Sjögren's syndrome. *J. Immunol.* 136:477–483.

37. Fong, S., P.P. Chen, R.I. Fox, R.D. Goldfien, G.J. Silverman, V. Radoux, F. Jirik, J.H. Vaughan, and D.A. Carson. 1986. Rheumatoid factors in human autoimmune disease: Their origin, development and function. *Path. Immunopath. Res.* 5:305–316.

38. Sugai, T., T. Konda, T. Shirasaka, K. Muragama, and K. Nishikawa. 1980. Non IgM monoclonal gammopathy in patients with Sjögren's syndrome. *Am. J. Med.* 68:861–866.

39. Kassan, S.S., T.L. Thomas, H.M. Moutsopoulos, R. Hoover, R.P. Kimberly, D.R. Budman, J. Costa, J.L. Decker, and T.M. Chused. 1978. Increased risk of lymphoma in sicca syndrome. *Ann. Intern. Med.* 89:888–892.

40. Whaley, K., J. Webb, B. McAvoy, G.R. Hughes, P. Lee, R.N. MacSween, and W.W. Buchanan. 1973. Sjögren's syndrome. 2. Clinical associations and immunological phenomena. *Quart. J. Med.* 66:513–548.

41. Royer, B., D. Cazals-Hatem, J. Sibilia, F. Agbalika, J.M. Cayuela, T. Soussi, F. Maloisel, J.P. Clauvel, J.C. Brouet, and X. Mariette. 1997. Lymphomas in patients with Sjogren's syndrome are marginal zone B-cell neoplasms, arise in diverse extranodal and nodal sites, and are not associated with viruses. *Blood* 90:766–775.

42. Thieblemont, C., F. Berger, and B. Coiffier. 1995. Mucosa-associated lymphoid tissue lymphomas. *Curr. Opin. Oncol* 7:415–420.

43. Lasota, J., and M.M. Miettinen. 1997. Coexistence of different B-cell clones in consecutive lesions of low-grade MALT lymphoma of the salivary gland in Sjogren's disease. *Mod. Pathol.* 10:872–878.

44. Zufferey, P., O.C. Meyer, M. Grossin, and M.F. Kahn. 1995. Primary Sjogren's syndrome (SS) and malignant lymphoma. A retrospective cohort study of 55 patients with SS [see comments]. *Scand. J. Rheumatol.* 24:342–5.

45. Fox, R.I. 1997. Sjogren's syndrome. Controversies and progress. *Clin. Lab. Med.* 17:431–444.

46. Fishleder, A., R. Tubbs, B. Hesse, and H. Levine. 1987. Uniform detection of immunoglobulin-gene rearrangement in benign lymphoepithelial lesions. *New Eng. J. Med.* 3:1118–1121.

47. Freimark, B., and R.I. Fox. 1987. Detection of Clonally Detected T and B Cells in Salivary Gland Biopsies of Sjogren's sjogren lacking myoepithelial islands. *New Eng. J. Med.* 317:1158.

48. Pisa, E.K., P. Pisa, H.I. Kang, and R.I. Fox. 1991. High frequency of t(14;18) translocation in salivary gland lymphomas from Sjögren's syndrome patients. *J. Exp. Med.* 174:1245–1250.

49. Ota, T., A. Wake, S. Eto, and T. Kobayashi. 1995. Sjogren's syndrome terminating with multiple myeloma. *Scand. J. Rheumatol.* 24:316–8.

50. Jordan, R.C., and P.M. Speight. 1996. Lymphoma in Sjogren's syndrome. From histopathology to molecular pathology. *Oral Surg. Oral Med. Oral Pathol. Oral Radiol. Endod.* 81:308–320.

51. Chakravarty, K., M. Goyal, D.G. Scott, and B.G. McCann. 1993. Malignant "angioendotheliomatosis"-(intravascular lymphomatosis) an unusual cutaneous lymphoma in rheumatoid arthritis. *Br. J. Rheumatol.* 32:932–934.

52. Drosos, A., A. Andonopoulos, J. Costopoulos, C. Papadimitriou, and H.M. Moutsopoulos. 1988. Prevalence of primary Sjögren's syndrome in an elderly population. *Br. J. Rheumatol.* 27:123–127.

53. Roitberg-Tambur, A., C.S. Witt, A. Friedmann, C. Safirman, L. Sherman, S. Battat, D. Nelken, and C. Brautbar. 1995. Comparative analysis of HLA polymorphism at the serologic and molecular level in Moroccan and Ashkenazi Jews. *Tissue Antigens* 46:104–110.

54. Miyagawa, S., K. Shinohara, M. Nakajima, K. Kidoguchi, T. Fujita, T. Fukumoto, A. Yoshioka, K. Dohi, and T. Shirai. 1998. Polymorphisms of HLA class II genes and autoimmune responses to Ro/SS-A- La/SS-B among Japanese subjects. *Arthritis Rheum.* 41:927–934.

55. Guggenbuhl, P., S. Jean, P. Jego, B. Grosbois, G. Chales, G. Semana, G. Lancien, E. Veillard, Y. Pawlotsky, and A. Perdriger. 1998. Primary Sjogren's syndrome: role of the HLA-DRB1*0301-*1501 heterozygotes. *J. Rheumatol.* 25:900–905.

56. Scofield, R.H., A.D. Farris, A.C. Horsfall, and J.B. Harley. 1999. Fine specificity of the autoimmune response to the

Ro/SSA and La/SSB ribonucleoproteins. *Arthritis Rheum.* 42:199–209.

57. Fox, R.I., J. Tornwall, and P. Michelson. 1999. Current issues in the diagnosis and treatment of Sjogren's syndrome. *Curr. Opin. Rheumatol.* 11:364–371.

58. Harley, J., M. Reichlin, F. Arnett, E.L. Alexander, W. Bias, and T. Provost. 1986. Gene interaction at HLA-DQ enhances autoantibody production in primary Sjögren's syndrome. *Science* 232:1145–1147.

59. Kang, H.-I., H.M. Fei, I. Saito, S. Sawada, S.-L. Chen, D. Yi, E. Chan, C. Peebles, T.L. Bugawan, H.A. Erlich, and R.I. Fox. 1993. Comparison of HLA class II genes in Caucasoid, Chinese, and Japanese patients with primary Sjögren's syndrome. *J. Immunol.* 150:3615–3623.

60. Reveille, J.D. 1990. Molecular genetics of systemic lupus erythematosus and Sjögren's syndrome. *Curr. Opin. Rheumatol.* 2:733–739.

61. Kang, H.I., H. Fei, and R.I. Fox. 1991. Comparison of genetic factors in Chinese, Japanese and Caucasoid patients with Sjöjren's syndrome. *Arthritis & Rheum.* 34.S41 (supplement).

62. Fei, H.M., H.-I. Kang, S. Scharf, H. Erlich, C. Peebles, and R.I. Fox. 1991. Specific HLA-DQA and HLA-DRB1 alleles confer susceptibility to Sjögren's syndrome and autoantibody SS-B production. *J. Clin. Lab. Analysis* 5:382–391.

63. Bron, A.J. 1997. The Doyne Lecture. Reflections on the tears. *Eye* 11:583–602.

64. Saari, H., S. Halinen, K. Ganlov, T. Sorsa, and Y.T. Konttinen. 1997. Salivary mucous glycoprotein MG1 in Sjogren's syndrome. *Clin. Chim. Acta* 259:83–96.

65. Inatomi, T., S. Spurr-Michaud, A.S. Tisdale, and I.K. Gipson. 1995. Human corneal and conjunctival epithelia express MUC1 mucin. *Invest Ophthalmol. Vis. Sci.* 36:1818–1827.

66. Hicks, S.J., S.D. Carrington, R.L. Kaswan, M.E. Stern, and A.P. Corfield. 1995. Secreted and membrane bound ocular mucins from normal and dry eye dogs. *Biochem. Soc. Trans.* 23:537S.

67. Iontcheva, I., F.G. Oppenheim, and R.F. Troxler. 1997. Human salivary mucin MG1 selectively forms heterotypic complexes with amylase, proline-rich proteins, statherin, and histatins. *J. Dent. Res.* 76:734–743.

68. Matsuo, R., Y. Yamauchi, and T. Morimoto. 1997. Role of submandibular and sublingual saliva in maintenance of taste sensitivity recorded in the chorda tympani of rats. *J. Physiol.* (Lond.) 498:797–807.

69. Bobek, L.A., J. Liu, S.N. Sait, T.B. Shows, Y.A. Bobek, and M.J. Levine. 1996. Structure and chromosomal localization of the human salivary mucin gene, MUC7. *Genomics* 31:277–282.

70. Albone, E.F., F.K. Hagen, C. Szpirer, and L.A. Tabak. 1996. Molecular cloning and characterization of the gene encoding rat submandibular gland apomucin, Mucsmg. *Glycoconj J.* 13:709–716.

71. Zara, J., F.K. Hagen, K.G. Ten Hagen, B.C. Van Wuyckhuyse, and L.A. Tabak. 1996. Cloning and expression of mouse UDP-GaINAc:polypeptide N-acetylgalactosaminyltransferase-T3. *Biochem. Biophys. Res. Commun.* 228:38–44.

72. Griffith, T., T. Brunner, S. Fletcher, D. Green, and T. Ferguson. 1995. Fas ligand-induced apoptosis as a mechanism of immune privilege. *Science* 270:1189–92.

73. Stern, M.E., R.W. Beuerman, R.I. Fox, J. Gao, A.K. Mircheff, and S.C. Pflugfelder. 1998. A unified theory of the role of the ocular surface in dry eye. *Adv. Exp. Med. Biol.* 438:643–651.

74. Kong, L., N. Ogawa, T. Nakabayashi, G.T. Liu, E. D'Souza, H.S. McGuff, D. Guerrero, N. Talal, and H. Dang. 1997. Fas and Fas ligand expression in the salivary glands of patients with primary Sjogren's syndrome. *Arthritis Rheum.* 40:87–97.

75. Baum, B. 1987. Neurotransmitter Control of Secretion. *J. Dent. Res.* 66:628–632.

76. Baum, B.J. 1993. Principles of saliva secretion. *Ann NY Acad. Sci.* 694:17–23.

77. Wu, A., R. Kurrasch, J. Katz, P. Fox, B. Baum and J. Atkinson. 1994. Effect of tumor necrosis factor alpha and interferon gamma on the growth of a human salivary gland cell line. *J. Cell Physiol.* 161:217–226.

78. Wu, A.J., R.M. Lafrenie, C. Park, W. Apinhasmit, Z.J. Chen, H. Birkedal-Hansen, K.M. Yamada, W.G. Stetler-Stevenson, and B.J. Baum. 1997. Modulation of MMP-2 (gelatinase A) and MMP-9 (gelatinase B) by interferon-gamma in a human salivary gland cell line. *J. Cell Physiol.* 171:117–124.

79. Purushotham, K.R., P.L. Wang, and M.G. Humphreys-Beher. 1994. Limited modulation of the mitogen-activated protein kinase pathway by cyclic AMP in rat parotid acinar cells. *Biochem. Biophys. Res. Commun.* 202:743–748.

80. Purushotham, K.R., Y. Nakagawa, M.G. Humphreys Beher, N. Maeda, and C.A. Schneyer. 1993. Rat parotid gland acinar cell proliferation: signal transduction at the plasma membrane. *Crit. Rev. Oral Biol. Med.* 4:537–543.

81. Jones, D.T., D. Monroy, Z. Ji, and S.C. Pflugfelder. 1998. Alterations of ocular surface gene expression in Sjogren's syndrome. *Adv. Exp. Med. Biol.* 438:533–536.

82. MacSween, R.N.M., R.B. Goudie, J.R. Anderson, E. Armstrong, M.A. Murray, D.K. Mason, M.K. Jasani, J.A. Boyle, W.W. Buchanan, and J. Williamson. 1967. Occurrence of antibody to salivary duct epithelium in Sjögren's disease, rheumatoid arthritis, and other arthritides. *Ann. Rheum. Dis.* 26:402–410.

83. Bacman, S., L. Sterin-Borda, J.J. Camusso, R. Arana, O. Hubscher, and E. Borda. 1996. Circulating antibodies against rat parotid gland M3 muscarinic receptors in primary Sjogren's syndrome. *Clin. Exp. Immunol.* 104:454–459.

84. Robinson, C.P., J. Brayer, S. Yamachika, T.R. Esch, A.B. Peck, C.A. Stewart, E. Peen, R. Jonsson, and M.G. Humphreys-Beher. 1998. Transfer of human serum IgG to nonobese diabetic Igmu null mice reveals a role for autoantibodies in the loss of secretory function of exocrine tissues in Sjogren's syndrome. *Proc. Natl. Acad. Sci. USA* 95:7538–7543.

85. Rinke, J., and J. Steitz. 1982. Percursor molecules of both human 5s ribosomal RNA and tRNA are bound by a cellular protein reactive with anti-La lupus antibodies. *Cell* 29:149–159.

86. Lee, L.A., K. Alvarez, T. Gross and J.B. Harley. 1996. The recognition of human 60-kDa Ro ribonucleoprotein particles by antibodies associated with cutaneous lupus and neonatal lupus. *J. Invest. Dermatol.* 107:225–228.

87. Hamilton, R.G., J.B. Harley, W.B. Bias et al. 1988. Two Ro (SS-A) autoantibody responses in systemic lupus erythematous. *Arthritis Rheum.* 31:446–505.

88. Chan, E., and L. Andrade. 1992. Antinuclear Antibodies in Sjogren's Syndrome. *Rheum. Clin. NA* 18:561–567.

89. Ben-Chetrit, E., E.K.L. Chan, K.F. Sullivan, and E.M. Tan. 1988. A 52 KD protein is a novel component of the SS-A-Ro antigenic particle. *J. Exp. Med.* 167:1560–1571.

90. Rader, M.D., C. O'Brien, Y. Liu, J.B. Harley, and M. Reichlin. 1989. Heterogeneity of the Ro/SSA antigen. *J. Clin. Invest.* 83:1293–1298.

91. Chan, E.K., E.M. Tan, D.C. Ward, and A.G. Matera. 1994. Human 60-kDa SS-A/Ro ribonucleoprotein autoantigen gene (SSA2) localized to 1q31 by fluorescence *in situ* hybridization. *Genomics* 23:298–300.

92. Ben-Chetrit, E., R.I. Fox, and E.M. Tan. 1990. Dissociation of immune responses to the SS-A (Ro) 52-kd and 60-kd polypeptides in systemic lupus erythematosus and Sjögren's syndrome. *Arthritis Rheum.* 33:349–355.

93. Halse, A., J.B. Harley, U. Kroneld, and R. Jonsson. 1999. Ro/SS-A-reactive B lymphocytes in salivary glands and peripheral blood of patients with Sjogren's syndrome. *Clin. Exp. Immunol.* 115:203–207.

94. Frank, M.B., R. McArthur, J.B. Harley, and A. Fujisaku. 1990. Anti-Ro(SSA) autoantibodies are associated with T cell receptor beta genes in systemic lupus erythematosus. *J. Clin. Invest.* 85:33–39.

95. Yonaha, F., T. Sumida, T. Maeda, and H. Tomioka. 1992. Restricted junctional usage of T cell receptor V beta 2 and B beta 13 genes, which are overrepresented on infiltrating T cells in the lips of patients with Sjögren's syndrome. *Arthritis Rheum.* 35:1362–1367.

96. Scofield, R., and J. Harley. 1994. Association of anti-SS-A autoantibodies with glutamine in position 34 of the DQA1 and leucine in position 26 of D1B1. *Arth. Rheum.* 37:961–962.

97. Chan, E.K., K.F. Sullivan, and E.M. Tan. 1989. Ribonucleoprotein SS-B/La belongs to a protein family with consensus sequence for RNA-binding. *Nucleic Acids Res.* 17:2233–2244.

98. Wang, D., and E.K. Chan. 1996. 17-beta-estradiol increases expression of 52-kDa and 60-kDa SS-A/Ro autoantigens in human keratinocytes and breast cancer cell line MCF-7. *J. Invest. Dermatol.* 107:610–614.

99. Sontheimer, R.D. 1996. Photoimmunology of lupus erythematosus and dermatomyositis: a speculative review. *Photochem. Photobiol.* 63:583–594.

100. Casiano, C.A., S.I. Martin, D.R. Green, and E.M. Tan. 1996. Selective Cleavage of nuclear autoantigens during CD95 (Fas/Apo-1) mediated T-cell apoptosis. *J. Exp. Med.* 184:765–770.

101. Chan, E.K.L., and E.M. Tan. 1987. Human autoantibody-reactive epitopes of SS-B/La are highly conserved in comparison with epitopes recognized by murine monoclonal antibodies. *J. Exp. Med.* 166:1627–1640.

102. Tseng, C.E., E.K. Chan, E. Miranda, M. Gross, F. Di Donato, and J.P. Buyon. 1997. The 52-kd protein as a target of intermolecular spreading of the immune response to components of the SS-A/Ro-SS-B/La complex. *Arthritis Rheum.* 40:936–944.

103. Talal, N., M.J. Dauphinee, H. Dang, S.S. Alexander, D.J. Hart, and R.F. Garry. 1990. Detection of serum antibodies to retroviral proteins in patients with primary Sjogren's syndrome (autoimmune exocrinopathy). *Arthritis Rheum.* 33:774–781.

104. De Keyser, F., S.O. Hoch, M. Takei, H. Dang, H. De Keyser, L.A. Rokeach, and N. Talal. 1992. Cross-reactivity of the B/B′ subunit of the Sm ribonucleoprotein autoantigen with proline-rich polypeptides. *Clin. Immunol. Immunopathol.* 62:285–290.

105. Talal, N. 1990. Immunologic and viral factors in Sjögren's syndrome. *Clin. Exp. Rheumatol.* 8:23–26.

106. Itescu, S. 1991. Diffuse infiltrative lymphocytosis syndrome in human immunodeficiency virus infection—a Sjögren's-like disease. Rheum Dis *Clin. North Am.* 17:99–115.

107. Rokeach, L., J. Haselby, and J. Meilof. 1991. Characterization of the autoantigen calreticulin. *J. Imm.* 147:3031–3045.

108. Fong, S., T.A. Gilbertson, R.J. Hueniken, S.K. Singhal, J.H. Vaughan, and D.A. Carson. 1985. IgM rheumatoid factor autoantibody and immuhnoglobulin- producing precursor cells in the bone marrow of humans. *Cell. Immunol.* 95:157–172.

109. Carson, D.A., P.P. Chen, R.I. Fox, T.J. Kipps, F. Jirik, R.D. Goldfien, G. Silverman, V. Radoux, and S. Fong 1987. Rheumatoid factors and immune networks. *Ann. Rev. Immunol.* 5:109–126.

110. Fong, S., P.P. Chen, T.A. Gilbertson, J.R. Weber, R.I. Fox, and D.A. Carson. 1986. Expression of three cross-reactive idiotypes on rheumatoid factor autoantibodies from patients with autoimmune diseases and seropositive adults. *J. Immunol.* 137:122–128.

111. Carson, D.A., P.P. Chen, R.I. Fox, T.J. Kipps, I. Jirik, R.D. Goldfein, G. Silverman, V. Radoux and S. Fong. 1988. Rheumatoid factors and immune networks. *Ann. Rev. Immunol.* 5:109–126.

112. Kipps, T.J., E. Tomhave, P.P. Chen, and R.I. Fox. 1989. Molecular characterization of a major autoantibody-associated cross-reactive idiotype in Sjogren's syndrome. *J. Immunol.* 142:4261–4268.

113. Carson, D.A., P.P. Chen, and T.J. Kipps. 1991. New roles for rheumatoid factor. *J. Clin. Invest.* 87:379–383.

114. Daniels, T., R. Sylvester, et al. 1974. Tuboreticular structures within the labial salivary glands in Sjögren's syndrome. *Arthritis Rheum.* 17:593–597.

115. Adamson, T.C., III, R.I. Fox, D.M. Frisman, and F.V. Howell. 1983. Immunohistologic analysis of lymphoid infiltrates in primary Sjögren's syndrome using monoclonal antibodies. *J. Immunol.* 130:203–208.

116. Freimark, B., L. Lanier, J. Phillips, T. Quertermous, and R.I. Fox. 1987. Comparison of T-cell receptor gene rearrangements in patients with large granular T-cell leukemia and Felty's syndrome. *J. Immunol.* 138:1724–1729.

117. Fox, R.I., H.I. Kang, D. Ando, J. Abrams, and E. Pisa. 1994. Cytokine mRNA expression in salivary gland biopsies of Sjogren's syndrome. *J. Immunol.* 152:5532–5539.

118. Jones, D.T., D. Monroy, Z. Ji, S.S. Atherton, and S.C. Pflugfelder. 1994. Sjogren's syndrome: cytokine and Epstein-Barr viral gene expression within the conjunctival epithelium. *Invest. Ophthalmol. Vis. Sci.* 35:3493–3504.

119. Lindahl, G., E. Hedfors, L. Kloreskog, and U. Forsum. 1985. Epithelial HLA-DR expression and T-cell subsets in salivary glands in Sjögren's syndrome. *Clinical Exp. Immunol.* 61:475–482.

120. Fox, R.I., T. Bumol, R. Fantozzi, R. Bone, and R. Schreiber. 1986. Expression of histocompatibility antigen HLA-DR by salivary gland epithelial cells in Sjögren's syndrome. *Arthritis Rheum.* 29:1105–1111.

121. Manoussakis, M.N., I.D. Dimitriou, E.K. Kapsogeorgou, G. Xanthou, S. Paikos, M. Polihronis, and H.M. Moutsopoulos. 1999. Expression of B7 costimulatory mole-

cules by salivary gland epithelial cells in patients with Sjogren's syndrome. *Arthritis Rheum.* 42:229–239.

122. Aziz, K.E., P.J. McCluskey, and D. Wakefield. 1996. Expression of selectins (CD62 E,L,P) and cellular adhesion molecules in primary Sjogren's syndrome: questions to immunoregulation. *Clin. Immunol. Immunopathol.* 80:55–66.

123. Flipo, R.M., T. Cardon, M.C. Copin, M. Vandecandelaere, B. Duquesnoy, and A. Janin. 1997. ICAM-1, E-selectin, and TNF alpha expression in labial salivary glands of patients with rheumatoid vasculitis. *Ann. Rheum. Dis.* 56:41–44.

124. Pang, M., T. Abe, T. Fujihara, S. Mori, K. Tsuzaka, K. Amano, J. Koide, and T. Takeuchi. 1998. Up-regulation of alphaEbeta7, a novel integrin adhesion molecule, on T cells from systemic lupus erythematosus patients with specific epithelial involvement. *Arthritis Rheum.* 41:1456–1463.

125. McMurray, R.W. 1996. Adhesion molecules in autoimmune disease. *Semin. Arthritis Rheum.* 25:215–233.

126. Fishleder, A., R. Tubbs, B. Hesse, and H. Levin. 1987. Immunoglobulin-gene rearrangement in benign lymphoepithelial lesions. *New Eng. J. Med.* 316:1118–1121.

127. Fishleder, A., R. Tubbs, H. Levine, and B. Hesse. 1987. Letter. *New Eng. J. Med.* 317:1158–1159.

128. Fox, R.I., T.C. Adamson III, S. Fong, C.A. Robinson, E.L. Morgan, J.A. Robb, and F.V. Howell. 1983. Lymphocyte phenotype and function of pseudolymphomas associated with Sjögren's syndrome. *J. Clin. Invest.* 72:52–62.

129. Sumida, T., F. Yonaha, T. Maeda, E. Tanabe, T. Koike, H. Tomioka, and S. Yoshida. 1992. T cell receptor repertoire of infiltrating T cells in lips of Sjögren's syndrome patiehts. *J. Clin. Invest.* 89:681–685.

130. Sumida, T., I. Matsumoto, T. Maeda, and K. Nishioka. 1997. T-cell receptor in Sjögren's syndrome. *Br. J. Rheumatol.* 36:622–629.

131. Matsumoto, I., T. Maeda, K. Nishioka, and T. Sumida. 1997. a-amylase function as salivary gland specific T-cell epitope in patients with Sjögren's syndrome (abstract). *Arth. Rheum.* S1044.

132. Alpert, S., H.I. Kang, I. Weissman, and R.I. Fox. 1994. Expression of granzyme A in salivary gland biopsies from patients with primary Sjögren's syndrome. *Arthritis Rheum.* 37:1046–1054.

133. Griffiths, G.M., S. Alpert, E. Lambert, J. McGuire, and I.L. Weissman. 1991. Perforin and granzyme A expression identifying cytolytic lymphocytes in rheumatoid arthritis. *Proc. Natl. Acad. Sci. USA* 89:549–553.

134. Ichikawa, Y., K. Arimori, M. Yoshida, T. Horiki, Y. Hoshina, K. Morita, M. Uchiyama, H. Shimizu, J. Moriuchi, and M. Takaya. 1995. Abnormal expression of apoptosis-related antigens, Fas and bcl-2, on circulating T-lymphocyte subsets in primary Sjögren's syndrome. *Clin. Exp. Rheumatol.* 13:307–313.

135. Nakamura, H., A. Kawakami, M. Tominaga, K. Migita, Y. Kawabe, T. Nakamura, and K. Eguchi. 1999. Expression of CD40/CD40 ligand and Bcl-2 family proteins in labial salivary glands of patients with Sjögren's syndrome [In Process Citation]. *Lab. Invest.* 79:261–269.

136. Robinson, C., S. Yamciuka, C. Alford, C. Cooper, E. Pichardo, N. Shah, M. Peck, and M. Humphrey-Beher. 1997. Elevated levels of cysteine protease activity in saliva and salivary glands of NOD mouse model for Sjögren's syndrome. *Proc. Natl. Acad. Sci. USA* 94:5767–5771.

137. Humphreys-Beher, M.G., S. Yamachika, H. Yamamoto, N. Maeda, Y. Nakagawa, A.B. Peck, and C.P. Robinson. 1998. Salivary gland changes in the NOD mouse model for Sjögren's syndrome: is there a non-immune genetic trigger? *Eur. J. Morphol.* 36:247–251.

138. Robinson, C., J. Cornelius, D. Bounous, H. Yamatomo, M. Humphrey-Beher, and A. Peck. 1998. Characterization of the changing lymphocyte population and cytokine expression in the exocrine tissues of autoimmune NOD mice. *Autoimmunity* 27:29–44.

139. Fox, R.I., A.N. Theofilopoulos, and A. Altman. 1985. Production of interleukin 2 (IL 2) by salivary gland lymphocytes in Sjögren's syndrome. Detection of reactive cells by using antibody directed to synthetic peptides of IL 2. *J. Immunol.* 135:3109–3115.

140. Freemont, A., C. Jones, P. Bromley, and P. Andrews. 1983. Changes in vascular endothelium related to lymphocyte collections in diseased synovium. *Arth. Rheum.* 26:1427–1433.

141. Fox, R.I., and T. Maruyama. 1997. Pathogenesis and treatment of Sjögren's syndrome. *Curr. Opin. Rheumatol.* 9:393–399.

142. Aziz, K.E., and D. Wakefield. 1995. In vivo and in vitro expression of adhesion molcules by peripheral blood lymphocytes from patients with primary Sjogren's syndrome: culture-associated enhancement of LECAM-1 and CD44. *Rheumatol. Int.* 15:69–74.

143. Pflugfelder, S., C. Crouse, D. Monroy, M. Yen, M. Rowe, and S. Atherton. 1993. EBV and the lacrimal gland pathology of Sjogrens syndrome. *Am. J. Path.* 143:49–64.

144. Merne, M.E., and S.M. Syrjanen. 1996. Detection of Epstein-Barr virus in salivary gland specimens from Sjogren's syndrome patients. *Laryngoscope* 106:1534–1539.

145. Saito, I., M. Shimuta, K. Terauchi, K. Tsubota, J. Yodoi, and N. Miyasaka. 1996. Increased expression of human thioredoxin/adult T cell leukemia-derived factor in Sjögren's syndrome. *Arthritis Rheum.* 39:773–782.

146. Sixbey, J., D. Davis, L. Young, L. Hutt-Fletcher, T. Tedder, and A. Rickinson. 1987. Human epithelial cell expression of an EBV receptor. *J. Gen. Virol.* 68:805–811.

147. Birkenbach, M., X. Tong, T. Tedder, and E. Kieff. 1992. Characterization of an EBV receptor on human epithelial cells. *J. Expt. Med.* 176:1405–1414.

148. Wolf, H., H. Zur Hausen, and B. Volker. 1973. EBV genomes in epithelial cells of nasopharyngeal carcinoma. *Nature* 244:245–246.

149. Morgan, D.G., J.C. Niederman, G. Miller, H.W. Smith, and J.M. Dowaliby. 1979. Site of Epstein-Barr virus replication in the oropharynx. *Lancet* 2:1154–1157.

150. Yang, J., N. Zhang, Y. Zeng, and Y. Dong. 1991. Possible etiological associations between Sjögren's syndrome (SS) and Epstein-Barr virus (EBV). *Clin. Exp. Rheu.* 9:338.

151. Busson, P., K. Braham, G. Ganem, F. Thomas, D. Grausz, M. Lipinski, H. Wakasugi, and T. Tursz. 1987. Epstein-Barr virus-containing epithelial cells from nasopharyngeal carcinoma produce interleukin 1α. *Proc. Natl. Acad. Sci. USA* 84:6262–6266.

152. Margan D.G., J.C. Niederman, G. Miller, H.W. Smith, and J.M. Dowaliby. 1979. Site of Epstein-Barr virus replication in the oropharynx. *Lancet* 2:1154–1157.

153. Maitland, N., S. Flint, C. Scully, and S.J. Crean. 1995. Detection of cytomegalovirus and Epstein-Barr virus in labial salivary glands in Sjögren's syndrome and non-specific sialadenitis. *J. Oral Pathol. Med.* 24:293–298.

154. Saito, I., B. Servenius, T. Compton, and R.I. Fox. 1989. Detection of Epstein-Barr virus DNA by polymerase chain reaction in blood and tissue biopsies from patients with Sjögren's syndrome. *J. Exp. Med.* 169:2191–2198.

155. Toda, I., M. Ono, H. Fujishima, and K. Tsubota 1994. Sjogren's syndrome (SS) and Epstein-Barr virus (EBV) reactivation. *Adv. Exp. Med. Biol.* 350:647–650.

156. Fox, R.I., G. Pearson, and J.H. Vaughan. 1986. Detection of Epstein-Barr virus associated antigens and DNA in salivary gland biopsies from patients with Sjögren's syndrome. *J. Immunol.* 137:3162–3168.

157. Fox, R.I., S. Scott, R. Houghton, A. Whalley, J. Geltofsky, J.H. Vaughan, and R. Smith. 1987. Synthetic peptide derived from the Epstein-Barr virus encoded early diffuse antigen (EA-D) reactive with human antibodies. *J. Clin. Lab. Anal.* 1:140–145.

158. Vaughan, J.H., M.D. Nguyen, J.R. Valbracht, K. Patrick, and G.H. Rhodes. 1995. Epstein-Barr virus-induced autoimmune responses. II. Immunoglobulin G autoantibodies to mimicking and nonmimicking epitopes. Presence in autoimmune disease. *J. Clin. Invest.* 95:1316–1327.

159. Fox, R.I. 1994. Epidemiology, pathogenesis, animal models, and treatment of Sjogren's syndrome. *Curr. Opin. Rheumatol.* 6:501–508.

160. Fox, R.I. 1995. Sjogren's syndrome. *Curr. Opin. Rheumatol.* 7:409–416.

161. Talal, N., E. Flescher, and H. Dang. 1992. Evidence for possible retroviral involvement in autoimmune disease. *Ann. Allergy* 69:221–224.

162. Garry, R.F., C.D. Fermin, D.J. Hart, S.S. Alexander, L.A. Donehower, and L.Z. Hong. 1990. Detection of a human intracisternal A-type retroviral particle antigenically related to HIV. *Science* 250:1127–1129.

163. Dang, H., M.J. Dauphinee, N. Talal, R.F. Garry, J.R. Seibold, T.A. Medsger, Jr., S. Alexander, and C.A. Feghali. 1991. Serum antibody to retroviral gag proteins in systemic sclerosis. *Arth. Rheum.* 34:1336–1337.

25 | Experimental Models of Sjögren's Syndrome

Roland Jonsson and Kathrine Skarstein

1. INTRODUCTION

Sjögren's syndrome is the second most common connective tissue disease overall and affects virtually every population or racial group studied (1). It is characterized by progressive destruction and/or a functional defect (2) of exocrine glands by infiltration of inflammatory mononuclear cells in to the salivary and lacrimal glands leading to the classic symptoms of dry mouth and eyes. The triad of dry mouth (xerostomia), dry eyes (keratoconjunctivitis sicca), and a connective tissue disease, usually rheumatoid arthritis (RA) or systemic lupus erythematosus (SLE), is termed secondary Sjögren's syndrome. In the absence of a connective tissue disease the designation primary Sjögren's syndrome (or sicca syndrome) is used. Furthermore, patients frequently have autoantibodies to intracellular ribonucleoproteins, La/SS-B and/or Ro/SS-A and rheumatoid factor (RF) (1).

Although the etiology of Sjögren's syndrome is unknown, there is substantial evidence that immunologically-mediated inflammation may be the most important factor leading to the characteristic chronic salivary and lachrymal gland lesions. Nevertheless, there is considerable evidence that an as yet unknown initiating factor set against the appropriate genetic background may invoke immunologically-mediated inflammatory mechanisms. It has been proposed that T cell-mediated autoimmune responses and apoptosis (3) in the glandular tissue are of central importance in the pathogenesis.

2. MAJOR FEATURES OF EXPERIMENTAL MODEL

To identify potentially important immune reactions leading to inflammatory exocrinopathy, studies of a model resembling the human disease is desirable. Early events in human Sjögren's syndrome are difficult to define due to the relatively late appearance of clinical signs of disease. For prospective analysis of mechanisms underlying the progression of disease, we are also limited in human studies due to the fact that repeated sampling of glandular tissues is hard to justify.

The ideal animal model should have the clinical characteristics of dry eyes and dry mouth. Lacrimal and salivary glands should have mononuclear inflammatory cell infiltration with destruction of normal acinar and ductal epithelium and eventual epimyoepithelial island formation. Characteristic serologic abnormalities such as hypergammaglobulinemia, antinuclear antibodies, antibodies to extractable nuclear antigens Ro/SS-A and La/SS-B, and RF should ideally be present (Table 1) (1).

An appropriate animal model of Sjögren's syndrome could greatly advance our possibilities to identify the target antigens, define the immune mechanisms, and characterize the evolution of tissue pathology. Several experimentally-induced and spontaneous inflammatory reactions with features of Sjögren's syndrome in animals have been reported and previously reviewed (4–9). Experimental models of sialadenitis/lacrimitis have been induced by administration of salivary gland antigens, adjuvants and allogeneic antisera in a variety of animals. However, basic autoimmune

Table 1 Features of Sjögren's syndrome which should be fulfilled by animal models

	Feature
Clinical	Dry eyes, dry mouth
Histological	Inflammatory mononuclear cells infiltrate in lacrimal & salivary glands
	progressive destruction of acinar & ductal epithelium
Serological	Hypergammaglobulinaemia
	ANA: anti-Ro/SS-A, anti-La/SS-B
	Rheumatoid factor
Chronicity	Lesions are destructive and persistent
Immunobiology	
Cellular infiltrate	Focal (>50 mononuclear cells) CD4+ T cells > CD8+ T cells
Abnormal MHC expression	*H-2+* ductal epithelium
Apoptosis	Blocked lymphocytic apoptosis

Table 2 Spontaneous experimental murine models of Sjögren's syndrome

Strain	Lacrimal gland inflammation	Salivary gland inflammation	Anti-Ro/SSA antibodies
NZB	+	+	ND
NZB/NZW	+	+	ND
MRL/lpr	+	+	+
MRL+/lpr	ND	+	ND
MRL/+	+	+	ND
NOD	+	+	+
NFS/sld	+	+	ND
IQI/Jic	+	+	ND

ND = not done.

mechanisms were most often not explored in these studies. Moreover, many models failed to produce destructive and long-lasting lesions in salivary glands, in contrast to Sjögren's syndrome in humans. This review will summarize the knowledge acquired to date, emphasize genetic and molecular aspects and define important topics for future investigations.

3. SPONTANEOUS MODELS OF SJÖGREN'S SYNDROME

A number of studies have confirmed that inflammation of lachrymal and salivary glands is common in not only the traditional autoimmune mice with lupus, including NZB, NZB/NZW (10), the MRL substrains (11), but also in NOD mice (12), which is a model of insulin-dependent diabetes mellitus. A summary of strains with spontaneous inflammation in lachrymal and salivary glands is presented in Table 2. These abnormalities, however, are uncommon in normal mice.

Kessler (10,13) was the first investigator to describe spontaneous changes of Sjögren's syndrome in NZB and NZB/NZW mice. Modified Schirmer test and sialochemical studies were conducted indicating occasional salivary gland pathology. When glands were histologically examined, periductal and perivascular mononuclear cell infiltrates were observed in virtually every animal after four months of age. The cellular infiltrate consisted mainly of small lymphocytes, plasma cells and reticulum cells and epimyoepithelial islands were present. Inflammation was

most severe in lacrimal glands with less extensive involvement of salivary glands. Although the basis for this dichotomy is not clear, it suggests that the autoimmune response is skewed for one organ or another, and suggests that distinct autoantigens might be involved. Moreover, the glandular involvement was more pronounced in females compared to males. This parallels previously described features of autoimmune disease in female NZB/W mice, but the effect of sex is generally less apparent in NZB mice. These reports established in particular the NZB/NZW mouse as a model of Sjögren's syndrome (10,13). Progression of focal sialadenitis in NZB/NZW mice has been studied (14,15). At the ultrastructural level, it was found that mononuclear cells in salivary glands were invading acinar and intercalated duct cells and that glandular epithelial cells appeared damaged (16,17). Histological evidence of conjunctivitis was later shown in NZB and NZB/NZW mice but not in nonautoimmune control strains (13).

MRL mice were first reported to have periductal mononuclear inflammatory cell infiltrates in salivary glands in 1982 (18). Features of Sjögren's syndrome in MRL/*lpr* and MRL/+ mice and normal control strains have subsequently been studied more in detail (11,19–21). Ocular pathology (band keratopathy and posterior uveitis) was common in the MRL substrains (19). Conjunctivitis was associated with inflammation of the adjacent sebaceous glands (11). Sialadenitis in MRL mice was similar to the inflammatory changes in New Zealand mice, but the MRL substrains had earlier onset of inflammation (15,20,21). Ultrastructural study of sialadenitis has shown degenerative changes of acinar epithelium (22).

Interesting observations were subsequently made in the MRL/*lpr* mouse where the *lpr* genotype has been identified as a mutation in the gene encoding Fas, which is a cell surface receptor that mediates apoptosis. Apoptotic cells were absent or appeared at very low frequency among the infiltrating mononuclear cells in salivary glands. Based on the analysis of the apoptotic activity, the T cells seemed to be rescued from apoptosis due to a failure in signalling (23).

A grading system of the salivary gland inflammation was introduced by Jonsson (20), making longitudinal studies feasible and utilizing similar quantitation of sialadenitis as in human Sjögren's syndrome (24). Progressive sialadenitis in MRL/*lpr* mice was documented with the most extensive infiltrates occurring in the submandibular glands starting at two months of age. Early inflammation in salivary glands of MRL/*lpr* mice was characterized by widely scattered focal infiltrates of mainly lymphocytes, which were confined to the perivascular region (Figure 1). With advancing age, the infiltration became more multifocal and the perivascular and periductal aggregates of mononuclear cells became confluent and led to local destruction of secretory acini. Epimyoepithelial island formation, consisting of degenerating salivary gland tissue, was occasionally present.

In NZB/W F$_1$ as well as in MRL/*lpr* mice, CD4$^+$T cells infiltrate the salivary glands (15,20) and the expression of MHC class II products on the salivary gland ductal epithelium in the proximity of lymphoid infiltrates is thought to perpetuate activation of CD4$^+$ T cells.

Lacrimal gland inflammation in NZB/W and MRL mice has also been studied. In both substrains of MRL mice (+/+ and *lpr*), there was focal inflammation detectable at three months of age in the +/+ mice in the absence of any ocular lesions, whereas *lpr* mice had inflammatory lesions by four weeks of age. Significant abnormalities were present at six months of age and persisted and increased throughout life, with all mice having extensive lesions by 18 months (25). In NZB/W mice, no lesions were detected until six months of age and were fully developed by nine months. All three strains of mice showed CD4$^+$ T cells as the main lymphocytic component, but variation in the amount of infiltration by CD8$^+$ T cells and by B cells.

Spontaneous focal inflammatory infiltrates involving the choroid and sclera have also been described in MRL/*lpr* mice (26). Most of the infiltrating cells were CD4$^+$ T cells. No eye lesions were found in the congenic MRL +/+ strain or in NZB/W mice.

NOD mice, in addition to insulin-dependent diabetes mellitus, have been reported to show lymphocytic infiltration in extra-pancreatic organs such as submandibular and lacrimal glands (12). At the age of four weeks insulitis was apparent, but sialadenitis appeared later (8–12 weeks in females and >12 weeks in males) and was characterized by small focal infiltrates of inflammatory cells (<50 cells) prominently located around blood vessels in both sexes. In older animals, inflammatory infiltrates were larger (>200 cells), more frequent and surrounded both blood vessels and ducts. The majority of infiltrating cells were CD4$^+$ and CD8$^+$ cells were comparatively few. Manipulation of MHC class II products in NOD mice by transgenes encoding modified I-Aβ or I-Eα chain prevents insulitis but has no effect on sialadenitis, suggesting that the development of insulin-dependent diabetes mellitus and sialadenitis are two separate events (27).

Apart from lymphocytic infiltrates in exocrine gland tissue, it has been observed that NOD mice also have a corresponding loss of parotid and submandibular gland secretion (9) and thus represent an advance in the identification of an animal model for autoimmune disease-associated sialadenitis. Hyposalivation, as seen in human Sjögren's syndrome, has not been verified in other murine models.

The development of the NOD-*scid* congenic strain has provided a model to investigate the role of immune response in the pathogenesis of Sjögren's syndrome. The NOD *scid* strain lacks functional lymphoid cells and shows little or no serum Ig with age (28). To differentiate autoimmune and non-autoimmune components in the decline of saliva production in NOD mice, the glandular function in NOD-*scid* mice was evaluated (29). As expected, histological analyzes of salivary glands from NOD-*scid* mice did not reveal the presence of lymphocytic infiltrates. Furthermore, total salivary flow and protein concentrations appeared normal. However, the composition of saliva proteins showed significant changes over the months with additional cysteine protease activity (30), and the authors suggest an underlying defect in salivary gland homeostasis in NOD mice.

The appearance of autoimmune diabetes prior to autoimmune exocrinopathy in the NOD mouse suggests that it is a model of secondary, but not primary, autoimmune sicca complications. Since the unique MHC I-A(g7) expression in NOD mice is essential for the development of insulitis and diabetes in these animals, the exocrine gland function in NOD.B10.H2b mice was investigated as a potential model for primary Sjögren's syndrome (31). NOD.B10.H2b mice

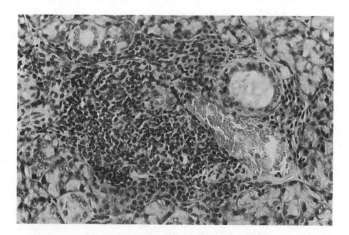

Figure 1. Perivascular and periductal focal accumulation of mononuclear cells in submandibular glands from female MRL/lpr mice (hematoxylin and eosin, original magnification × 250).

exhibited the exocrine gland lymphocytic infiltration typical of Sjögren's syndrome-like disease and dysfunction observed in NOD mice, but without the insulitis and diabetes. This suggests that the unique MHC I-A(g7) is not essential for exocrine tissue autoimmunity. Furthermore, these findings indicate that murine sicca syndrome occurs independently of autoimmune diabetes and that the congenic NOD.B10.H2b mouse represents a novel murine model of primary Sjögren's syndrome.

To clarify the importance of MHC class II region, a series of MHC congenic strains on the NOD background were designed (32). H-2q (NOD.Q) and H-2p (NOD.P) haplotypes were backcrossed into the NOD (H-2g7) and made homozygous at the 10th generation. Histopathological analysis of submandibular glands revealed sialadenitis with a predominance of CD4 positive cells in most of the female NOD congenic mice. None of the NOD congenic mice in the homozygous state developed insulitis or spontaneous diabetes. Our preliminary results implicate and extend previous findings (31) that the unique MHC II H-2g7 is not elementary for the development of sialadenitis in NOD mice.

The NFS/sld mouse also provides an interesting model for primary Sjögren's syndrome, where immune responses against alpha-fodrin may play a role (33). This mouse probably has a defect in development of salivary glands that leads to the cleavage of an important structural protein (fodrin) by caspase, leading to production of an autoantigen (fodrin 125 kd). After neonatal thymectomy, the NSF/sld mouse develops T cell infiltrates in salivary and lacrimal glands that are reactive to the fodrin 125 kd. Sera from Sjögren's syndrome patients also recognized fodrin 125 kd (34). Thymectomy in NFS/sld mouse is one way to induce autoreactive T cells in an animal model. Some effector mechanisms have also been studied in this model (35). Moreover, IQI/Jic has been suggested to be a new animal model for Sjögren's syndrome (36). Focal infiltration of lymphocytes with parenchymal destruction was noted in both lacrimal and salivary glands. The sialadenitis progressed as time went on and was more prominent in females than in males. Interestingly, the lymphocytes in small foci were CD4+ cells, but the majority of infiltrating cells were B cells (B220+), followed by CD4+ T cells in larger lesions.

An additional spontaneous model has been studied where autoimmune disease was found in exocrine organs in immunodeficient alymphoplasia mice (37). This deficiency is due to an autosomal recessive mutation.

4.   TRANSGENIC AND KNOCK-OUT MODELS OF SJÖGREN'S SYNDROME

Transgenic mice containing the HTLV-1 *tax* gene under the control of the viral long terminal repeat have been found to develop an exocrinopathy involving the salivary and lacrimal glands (38). The pathology, also reported in HTLV-1 associated conditions, resembled that of Sjögren's. Proliferation of ductal epithelial cells in the submandibular, parotid and minor salivary glands occurs within several weeks of birth. Over the course of several weeks, ductal proliferation dramatically increases and the structure becomes distorted accompanied by lymphocytic infiltration surrounding the nest of proliferating epithelial cells. Mice surviving 6–8 months exhibit extensive epithelial island enlargement with lymphocytic infiltration leading to destruction of acinar architecture. Salivary gland pathology in these animals correlates with the level of *tax* gene expression in ductal epithelial cells, predominantly found in the nucleus of ductal cells. This model suggests that HTLV-1 is tropic for ductal epithelium of salivary and lacrimal glands and may represent a primary event in the development of exocrinopathy by virally-induced proliferation and perturbation of the function of ductal epithelium followed by a lymphocytic response. This sequence of events may differ slightly from Sjögren's syndrome pathogenesis where lymphocytic infiltration, although not clearly demonstrated, is thought to precede ductal cell proliferation.

Sialadenitis histologically resembling Sjögren's syndrome has also been found in mice transgenic for hepatitis C virus envelope genes (39).

Interleukin (IL)-6-transgenic mice have been studied with the purpose of clarifying the role of this cytokine on the development of autoimmunity (41). A murine graft-versus-host reaction model with MHC class II disparity was used and the autoimmune-like lesions in the transgenic mice were weakened despite graft-versus- host disease and increased levels of IL-6. This indicates a regulating function of this cytokine in the progression of disease.

In order to better understand the role of IL-10 in Sjögren's syndrome, transgenic mice were constructed (41). The transgenic expression of IL-10 induced apoptosis of glandular tissue and infiltration of lymphocytes consisting of primarily FasL+, CD4+ T cells as well as *in vitro* upregulation of FasL expression on T cells. Altogether, this suggested that glandular overexpression of IL-10 and the subsequent Fas/FasL-mediated bystander tissue destruction is a causal factor in the development of Sjögren's syndrome.

Transforming growth factor-β1 (TGF-β1) is a multi-functional growth factor that has profound regulatory effects on many developmental and physiological processes. Disruption of the TGF-β1 gene by homologous recombination in murine embryonic stem cells enables generation of mice that carry the disrupted allele (42,43). Animals homozygous for the mutated TGF-β1 allele will, within 20 days, suffer from a wasting syndrome accompanied by a multifocal, mixed inflammatory cell response and tissue necrosis leading to organ failure and death. This includes exocrine glands with periductal, primarily lymphocytic inflammation in about 50% of the animals. It is hypothesized that absence of TGF-β1 could lead to presentation of self-antigens by inappropriate cells, thereby eliciting an autoimmune response. However, the leucocytic infiltration could be prevented by systemic injections of synthetic fibronectin peptides (43). Further, the acinar and ductal derangements were reversed, suggesting that salivary gland development is not jeopardized in the absence of TGF-β1.

5. TRANSPLANTATION CHIMERAS AS MODELS OF SJÖGREN'S SYNDROME

The graft-versus-host (GVH) mouse, as a model for autoimmunity, was shown to have lymphocytic infiltrations in the salivary glands resembling the exocrinopathy of Sjögren's syndrome (44,45). The model of chronic GVH reaction has been described as an SLE-like autoimmune syndrome early in the response with a Sjögren's syndrome-like glandular effect occurring later (46). In this model, donor spleen cells from DBA/2J/Ssc mice were injected into the tail vein of F_1 hybrids of non-irradiated C3H/Ssc and DBA/2J/Ssc mice of the same sex. Twenty weeks after cell transfer, most GVH animals had affected glands, with the typical morphology of an autoimmune exocrinopathy affecting both male and female animals. Focal lymphocytic infiltration, often periductal, some fibrosis but rare myoepithelial proliferation and acinar atrophy. The exocrinopathy increased in severity from week 8 to week 23 and was concluded to represent the early stages of secondary Sjögren's syndrome with SLE (46).

In a subsequent paper, the same group used donor cells from Balb/c grafted into Balb/c x CBA/H-T6 F_1 hybrids to produce a model of primary Sjögren's syndrome with lupus nephritis (47). In both models, the donor strain has H-2 haplotype identity (H-2^d) and the recipients (CH3 and T6) were both H-2^k. Nine weeks after grafting, sera were positive for ANA and anti-dsDNA and there was no evidence of proteinuria. Twenty weeks after transfer, no autoantibodies were detected but all chimeras had affected glands with the characteristics of exocrinopathy described for the previous model. In neither model there was any attempt made to study the functional aspects of saliva or tear fluid. This latter model is being actively pursued as a model of primary Sjögren's syndrome and represents a significant improvement. This model also allows the study of a normal control group consisting of litter mates of the experimental mice.

6. MODELS OF EXPERIMENTALLY-INDUCED SJÖGREN'S SYNDROME

Several approaches to induce and develop experimental Sjögren's syndrome have been taken. Animals have been injected with salivary gland extracts in complete Freund's adjuvants (CFA) in an attempt to induce experimental sialadenitis, sensitized with foreign protein and challenged with antigens instilled into salivary gland ducts, and been injected with antisera from animals immunized with salivary gland extracts or saliva. Moreover, in mice with chronic graft-versus host disease, it has been suggested that salivary gland inflammation could be induced as part of a generalized autoimmune disorder. Early attempts were done with rabbits, cats and guinea pigs and the interested reader is referred to previous reviews about these potential models (4,5). Most of these studies, however, were performed prior to the recent advances in cellular and molecular biology, and consequently did not explore basic mechanisms of autoimmune glandular inflammation. A summary of murine adjuvant models of Sjögren's syndrome is found in Table 3.

Experimental sialadenitis could be induced by immunization of PL/J mice with carbonic anhydrase (48). In view of the previous observation that a significant percentage of patients with SS had autoantibodies to carbonic anhydrase (CAII) (49), PL/J mice with different H-2 haplotypes were immunized with purified CAII. Compared with control and untreated mice, CAII-immunized mice showed a significant increase in the number and size of mononuclear cell foci in the salivary glands. Among several mouse strains with different H-2 haplotypes, strains bearing H-2^s and H-2^u were susceptible to CAII-induced sialadenitis. Further studies ought to include more detailed characterization of the cellular infiltrates to determine whether the pathogenesis of this animal model is similar to Sjögren's syndrome or not.

Adjuvant models did not always reproduce the histological characteristics of Sjögren's syndrome, but they offer the opportunity to examine factors involved in the induction or protection of sialadenitis. In particular, this includes potential autoantigens and its relation to the genetic background.

7. AUTOANTIGENS AND AUTOANTIBODIES

It is not known why certain self-antigens are selected as targets for the autoimmune response. For many diseases, including spontaneous models of disease in animals, the target antigens have consequently not been identified. The availability of clonal populations of T cells that cause disease, along with recent molecular advances, should facilitate the isolation of T cell target antigens in many of these diseases. Also, T cell clones might be used to analyze tissue-specific expression libraries to identify self-epitopes.

The initiating events in the disease and the reasons why sialadenitis in Sjögren's syndrome persists/progresses over

Table 3   Induced experimental murine models of Sjögren's syndrome (modified from ref nr 5).

Animal	Substance injected	Result	Reference
rats	submand + CFA + Bortadella pertussis vaccine	sialadenitis severe in mature female donated submand; T cell depletion decreased sialadenitis	Sharawy and White 1978; White and Casarett 1974; White 1976
mice	salivary gland + CFA	Autoimmune SL/Ni strain developed severe sialadenitis ASDA[a]	Takeda and Ishikawa 1983
rats	submand + CFA	sialadenitis; ASDA	Dean and Hiramoto 1984
mice (thymectomized)	submand + CFA cells infiltrated	Thy 1.2+, Lyt 2+ submand and parotid; ASDA	Hayashi et al. 1985
mice (thymectomized)	submand + mumps virus	Thy 1.2+, Lyt 2+ cells infiltrated submand; ASDA	Hayashi et al. 1986
mice C3H/He(H-2k) (thymectomized)	submand + CFA submand	L3T4+, Lyt 2+ cells infiltrated ASDA	Hayashi and Hirokawa 1989
Mice PL/J (H-2k)	human carbonic anhydrace II + CFA	sialadenitis	Nishimori et al. 1995

[a] ASDA = anti-salivary duct antibodies.

many years in most patients are unknown. One possible explanation is that the antigens responsible for the immune reaction are persistent. Another source of antigens that could continuously fuel an immune reaction is the normal connective tissue and/or glandular components. These could become targets of a local immune response either as a result of impaired regulation of the immune response, so that self recognition becomes aberrant, or by cross-reactivity with an exogenous agent (i.e., "molecular mimicry").

Interestingly, both anti-Ro (50) and anti-La (51) have been detected in spontaneous murine models. Since these autoantibodies are the dominant serological marker in patients with primary Sjögren's syndrome, this finding is an important take-off for future work. The fine specificity of the immune response to La in MRL/*lpr* mice has been tested for antibody binding to synthetic peptides produced from the cloned cDNA encoding the major antigenic peptide of the La complex (51). It was found that patterns of La antigen recognition displayed by MRL/*lpr* antibodies

differ from those of human autoantibodies, possibly reflecting differences between mouse and man in the induction of these responses.

Recent studies in a model of murine experimental autoimmunity have shown that reciprocal spreading to the Ro52, Ro60 and La polypeptides occurs following immunization with a single component. This intermolecular epitope spreading suggests that there is little or no tolerance to Ro and La in the B cell compartment and limited tolerance in the T cell compartment (52–55). The fine specificity of the autoimmune response to Ro/SSA and La/SSB ribonucleoproteins has recently been reviewed (56).

Anti-salivary duct antibodies were described over 30 years ago as an organ- specific antibody in Sjögren's syndrome, but they have remained poorly characterized and their clinical significance is uncertain. It appears that these antibodies are relatively uncommon in primary Sjögren's syndrome and seen more often in rheumatoid arthritis with secondary Sjögren's syndrome; however, there is considerable variability in the reported frequency of anti-salivary duct anti-bodies in Sjögren's syndrome, possibly reflecting technical problems with the indirect immunofluorescence assay on salivary gland substrate (57,58). Thus, the role of these autoantibodies in models remains uncertain.

Systemic autoimmune disease in humans and mice is characterized by hypergammaglobulinemia and high levels of serum autoantibodies, including RF. We investigated the functional properties of infiltrating mononuclear cells in MRL/*lpr* sialadenitis (59). Spontaneous local immuno-globulin (IgA, IgG, IgM) as well as IgA- and IgM RF pro-duction in salivary glands, lymph nodes, and spleen was analyzed at various ages in autoimmune MRL/*lpr* mice using ·enzymatically-eluted cells in an ELISPOT assay. Local pro-duction of immunoglobulins in salivary glands and lymph nodes occurred with a pattern of IgG >> IgM > IgA. Rheumatoid factors constituted a significant fraction of local IgA and IgM synthesized in salivary glands. The results, however, suggest that local RF production is more likely to be a secondary event in salivary gland inflammation in MRL/*lpr* mice rather than an initiating factor in this process.

The finding of serum autoantibodies directed against the muscarinic M3 receptor (expressed in salivary and lacrimal glands) in the majority of patients is an important advance in understanding the pathogenesis of impaired glandular function in Sjögren's syndrome (60,61). Recent studies in the NOD mouse have indicated that muscarinic receptor autoantibodies are directed against the agonist binding site of the molecule on the cell surface and interfere with secre-tory function of exocrine tissues in Sjögren's syndrome (62). Additional studies have verified a role for such autoantibodies in the NOD mouse (63). However, the clini-cal significance of all these autoantibodies has to be evalu-ated with caution, and their role in Sjögren's syndrome remains to be elucidated.

8. CELLULAR BASIS OF AUTOIMMUNE SIALADENITIS

A fundamental technical problem impeding the dissection of the cellular basis of autoimmune reactions is the isola-tion of important cell populations that are both homoge-neous and representative of an abnormal population that exists *in vivo*. Two complementary approaches are recom-mended to investigate the cellular basis of localized autoimmune inflammation: 1) *in situ* studies directed toward defining the phenotypic characteristics of cells located at sites of parenchymal destruction in inflamed glands and determining whether B cells, T cells, or T cell-derived products, are involved in the initiation of sial-adenitis, and 2) *in vitro* studies directed toward the phenotypic and functional characterization of salivary gland populations present in inflamed glands and analyses

of the effects of T cells (and their products) on glandular cells. Results of *in situ* experiments are essential for proper interpretation of *in vitro* studies.

Populations of cells infiltrating the salivary glands in NZB/NZW mice were first studied by Greenspan (64). Immunoglobulin-positive lymphocytes and plasma cells appeared around blood vessels and ducts. The majority of cells in larger infiltrates stained with anti-mouse thymocyte serum.

With the introduction of monoclonal antibodies (MoAb) defining cell phenotypes it became possible to further characterize the infiltrating cell population. Immunohisto-logical analyses of infiltrating cells in salivary glands in NZB/NZW and MRL/*lpr* mice revealed a high frequency of CD5-(Ly-1) and CD4-(L3T4) expressing lymphocytes at all stages of sialadenitis (Figure 2) (15,20). The majority of the cells in this T cell-dominated infiltrate expressed Ia/H-2 antigen. CD8-(Lyt-2) expressing cells were found in lower frequency and scattered throughout the lymphoid infiltrates. The B220 molecule, recognized by clone 14.8 or 6B2, is normally considered a B cell marker, but is aberrantly found on T cells in *lpr* mice, were seen in considerable numbers (>50%). Ia/H-2 expression was also seen on ductal and glandular epithelial cells in the vicinity of inflammatory lesions (Figure 3A). The presence of Ia antigens and the high frequency of lymphocytes with "helper" phenotype may indicate a perpetuated immunological activation within salivary glands. Similar studies have also been conducted in the lachrymal gland of these mice (65,66).

The normal CD4+ T cell of MRL mice might play a critical role in the autoimmune process since studies have demonstrated that injection of anti-CD4 antisera into MRL/*lpr* mice abolishes the development of autoimmune disease and eliminates production of IgG autoantibodies

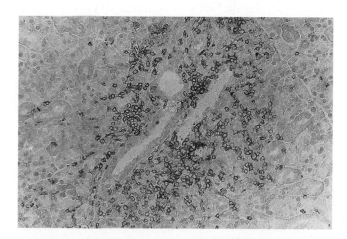

Figure 2. Immunoperoxidase staining of focal sialadenitis in submandibular glands of female MRL/lpr mice with monoclonal antibodies defining CD4 of lymphoid cells (original magnification ×250).

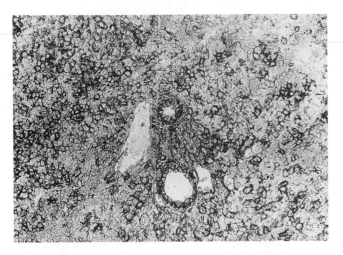

Figure 3. Immunoperoxidase staining of Ia/H-2 expression in MRL/*lpr* sialadenitis. Ia/H-2 is expressed not only by mononuclear and dentritic cells but also by ductal epithelium. (Original magnification × 280)

(67). It should be emphasized that a discrepancy exists between salivary gland and lymph node phenotypes in MRL/*lpr*; higher frequency of CD4+ cells among the infiltrating cells in salivary gland infiltrates compared to the lymph node (20).

Since many T cells at the site of inflammation may have arisen by secondary recruitment and may not be involved in pathology, it is important to determine the role of the T cells under investigation. One of the best ways to do this in the murine system is to establish whether the T cells can transfer disease passively. We examined the possibility of transferring sialadenitis to young "healthy" (prediseased) MRL/*lpr* animals from old MRL/*lpr* mice with established sialadenitis (68,69) using spleen cells and mononuclear cells eluted from salivary glands. The results showed that sialadenitis could be transferred *in vivo* to young MRL/*lpr* mice by splenic and salivary gland mononuclear cells. Furthermore, it was possible to accelerate the inflammatory exocrinopathy with a differentiated pattern depending on the source of cells. Similar studies were also conducted in SCID mice (70). In conclusion, these studies indicate that sialadenitis in MRL/*lpr* mice is mediated by cellular mechanisms and suggest that CD4+ splenic and infiltrating salivary gland mononuclear cells have the ability to accelerate autoimmune disease.

9. EVALUATION OF LOCAL T CELL ACTIVITY IN AUTOIMMUNE SIALADENITIS

This section will provide information about the cellular basis of the infiltrating T cells, particularly the presence of

autoreactive T cells and their ability to secrete factors contributing to autoimmune disease.

The inflammatory lesions of salivary gland and renal vasculitic lesions of autoimmune MRL/*lpr* mice were analyzed for the presence of activated lymphocytes (71). Immunohistological analysis revealed presence of IL-2R-expressing cells in the infiltrates indicating that activated T cells were present.

9.1. T Cell Receptor Usage and Clonal Deletion

In organ-specific autoimmune diseases, autoreactive T cell clones react with specific tissue antigens in an MHC-restricted fashion. Recognition by T cells is carried out by heterodimeric T cell receptor (TCR) generated in a manner analogous to that of immunoglobulins. The random somatic generation of TCR also permits the generation of receptors capable of recognizing self peptides.

Studies of genetic linkage in different autoimmune animal populations implicate the variable region genes of the TCR in the etiology of autoimmune disease. Previous attempts at characterizing T cells in MRL/*lpr* mice have paid little attention towards detailed characterization of T lymphocytes located at sites of tissue injury, undoubtedly due to the difficulty of extracting viable cells from involved tissue. Previous immunohistological findings have shown local activation of T cells in salivary glands (IL-2R expression, IFNγ production) *in situ* (20,71). Moreover, by using MoAbs defining TCR Vβ loci, it has now become possible to characterize *in situ* usage of T cells capable of destroying salivary gland cells.

Interestingly, an expansion of TCRVβ4, Vβ8.1,2 and Vβ10b has been found in salivary glands compared with lymph nodes at 4–5 months of age in MRL/*lpr* (72). This indicates that the response of T cells within the site of inflammation is not completely polyclonal, but that there are sets of T cells that might be responsible for the pathogenesis. The significance of the preferential accumulation of specific Vβ T cells in the salivary glands is for the present unclear.

Our findings are not entirely consistent with results from another study (73) in which MoAbs were used to determine the prevalence of selected V gene elements in salivary glands of MRL/*lpr* mice. Immunohistochemistry and FACS analysis revealed a predominant use of Vβ8 and a small portion of Vβ6 but Vβ4 and Vβ10b were not inlcuded in their analysis. However, in the RT-PCR and Southern blot analysis, the TCR Vβ1–19 were examined and a significant expansion of Vβ8 was found at the ages of two and three months.

In contrast to the sialadenitis in MRL/*lpr* mice, no expansion with age of certain Vβ was found in the NOD mice (74). However, a predominance of TCR Vβ8.1,2 was noted in the NOD animals and this Vβ was also selected in the salivary gland tissue of MRL/*lpr* mice. This common Vβ dominance in both strains could suggest an essential role of these cells in the pathogenesis of sialadenitis.

In view of our previous observations of an oligoclonal expansion of certain Vβ (4,8.1,2, 10b) in salivary glands of MRL/*lpr* mice, a selective antibody therapy was initiated (75). A cocktail containing MoAb against TCRVβ4, Vβ8.1,2, and Vβ10b in one single dose was designed and injected intra-peritoneally into the mice. This *in vivo* did not prevent the development of sialadenitis. However, in the animals with established sialadenitis, treatment with the MoAb significantly decreased the inflammation compared with the control groups.

9.2. Cytokines

The features of the T cells requires further analysis with regard to the capacity to produce cytokines of potentially pathogenic significance. Activated T cells can produce a number of cytokines with each cytokine generally possessing multiple biological activities.

Salivary gland inflammatory lesions and renal vasculitic lesions of autoimmune MRL/*lpr* mice were analyzed for local *in situ* secretion of IFN-γ (72). IFN-γ-producing cells were detected in the periphery of large inflammatory infiltrates (Figure 3B) in comparable numbers to IL-2R-expressing cells, although without the same localization. That subsets among these infiltrating cells produced IFN-γ a further indication of the presence of activated T cells. Local production of IFN-γ by the infiltrating mononuclear cells may induce H-2 antigen expression on epithelial cells (Figure 3A) (as has been shown *in vitro*) and, as a consequence, may enhance presentation of self antigens.

The T cells accumulating in lymph nodes and other organs of MRL/*lpr* mice were initially thought to be immunologically inert due to their inability to proliferate in response to a variety of T cell mitogens. In one study (76), we investigated spontaneous and Con-A stimulated cytokine (IL-2, IL-3, IL-6, TNF-α and IFN-γ) production in purified thymus, lymph node, spleen and salivary gland mononuclear cell preparations from female MRL/*lpr* mice at three weeks, two months and five months of age High levels of IL-6 were spontaneously produced by salivary gland infiltrating cells at the ages investigated, while Con-A stimulated secretion of cytokines was greatly increased in most organs.

A differentiated and age-related pattern of cytokine expression has been suggested to influence the development and progression of the autoimmune sialadenitis in MRL/*lpr* mice (77). In this study, the expression of cytokine genes was assessed by the reverse-transcriptase polymerase chain reaction and by immunohistochemical analysis. The genes for the inflammatory cytokines IL-1β and tumor necrosis factor (TNF) were expressed in the salivary glands before the onset of sialadenitis. IL-6 mRNA expression was clearly detected at the time of onset of salivary gland inflammation and was up-regulated with advancing age.

Another study examined the profiles of cells expressing mRNA for pro-and anti- inflammatory cytokines in submandibular glands of two, three, four and five month- old female MRL/*lpr* mice, using *in situ* hybridization (78). These results suggest a major role of the proinflammatory cytokines TNF-α, IL-1β, IL-6, IL- 12, IFN-γ in initiation and perpetuation of autoimmune sialadenitis in MRL/*lpr* mice in conjunction with an insufficiency of the anti-inflammatory and immunomodulatory cytokines TGF-β and IL-10.

Recently, the cytokine profile in NOD sialadenitis was analyzed (79,80). The particular cytokine pattern in NOD sialadenitis was the local expression of IL-10 and IL-12 mRNA during the course of autoimmune disease but also IL-1β, IL-2, IFN-γ and TNF-α. IL-10 mRNA expression was associated with detectable expression of IL-2 and IFN-γ.

9.3. Chemokines

Chemokines are a superfamily of small proteins that are involved in the migration and activation of cells. They may play a crucial role in immune and inflammatory responses. The involvement of various chemokines in the pathogenesis of diseases has become the subject of intensive research. Members of the β-chemokine sub family include machrophage inflammatory protein (MIP 1α), MIP-1β-monocyte chemoattractant protein (MCP-1) and regulated upon activation, normal T cell expressed and secreted (RANTES). The β-chemokines in general attract monocytes, eosinophils, basophils and lymphocytes with different selectivity (81,82).

Involvement of chemokines in inflammation of submandibular glands of MRL/*lpr* mice was established in a study by using *in situ* hybridisation and immunohistochemistry (83). Young mice with moderate sialadenitis showed an early up-regulation of mRNA expression for MCP-1, MIP-β and RANTES, while MIP-1α mRNA was not affected. Adult mice with more advanced sialadenitis showed a further upregulation of mRNA expression for MCP-1, MIP-1β and RANTES and a remarkably strong up-regulation of MIP-1. These results indicate an involvement of β-chemokine family in autoimmune sialadenitis of MRL/*lpr* mice.

9.4. Adhesion Molecules

A small number of studies have investigated adhesion molecules, which mediate cell to cell and cell to matrix interac-

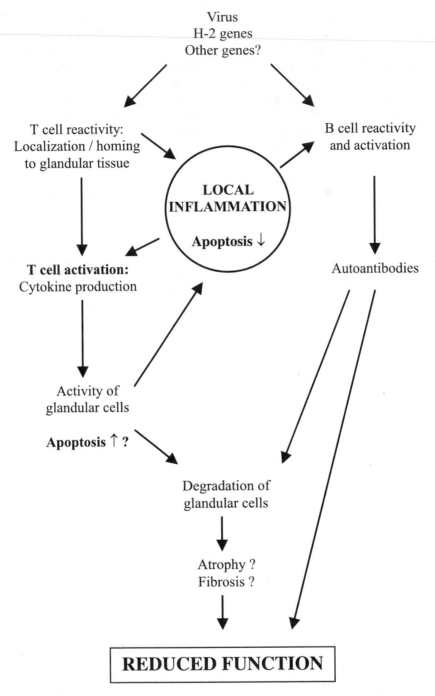

Figure 4.   Proposed scheme of events in murine autoimmune sialadenitis.

tions. The demonstration that antibodies to intracellular adhesion molecule-1 (ICAM-1), in combination with its ligand, lymphocyte function-associated antigen-1 (LFA-1), prevented the adoptive transfer of Sjögre's syndrome from MRL/*lpr* mice to SCID mice, suggested a central role for these molecules in induction and progression of the autoimmune process in submandibular glands of MRL/*lpr* mice. (84). Another study indicates that the ICAM-1/LFA pathway plays a crucial role in the development and subsequent pro-

gression of inflammation in the lachrymal glands in the NFS/sld mouse model for primary Sjögren's syndrome (85).

10. MOLECULAR BASIS AND MECHANISMS OF IMMUNOLOGIC ALTERATIONS

In organ-specific autoimmune diseases, autoreactive Th clones are activated and react with specific tissue antigens in an MHC-restricted manner. In this context, target tissues often aberrantly express H-2 antigens. A theory that aberrant H-2 expression renders these cells immunogenic and may initiate autoimmune responses has been formulated. During antigen recognition, the T cell binds to the MHC+ processed antigen complex via the TCR. Basically, "nonprofessional" antigen presenting cells expressing class I or class II antigens are thought to result in induction of tolerance of T cells. However, under certain inflammatory conditions, local production of cytokines and action by other helper and accessory cells lead to immunostimulation of self-reactive T cells. The activated Th cells are then able to interact with B cells to induce production of autoantibodies against target tissues. Autoantibodies may or may not contribute to the development of tissue injury. Presumably, specifically activated Th cells secrete lymphokines, e.g. IFN-γ, which act on macrophages and/or effector cells to produce tissue injury. Hypothetically, the pathogenesis of murine sialadenitis could follow a proposed schema presented in Figure 4.

There is a significant amount of cell turnover in the immune system, and apoptosis has been found to play a crucial role in the process of negative and positive selection as well as regulation of the immune responses. APO-1/Fas (CD95) belongs to the tumor necrosis factor/nerve growth factor receptor super-family and can directly transduce an apoptotic death signal into cells. The importance of this cell surface molecule in maintaining tolerance is highlighted by the development of lupus- like disease in MRL/lpr mouse, which has a mutation in the gene encoding the Fas receptor (86). The current consensus is that negative selection in thymus is not impaired in MRL/lpr mice (87). It was illustrated that MRL/lpr mice despite a defective genetic background express a detectable amount of apoptosis-related Fas protein on lymphoid cells and in their vicinity in thymus, lymph nodes and in salivary gland infiltrates (23). The observation of quite a few apoptotic cells in the salivary gland infiltrates in MRL+/+ mice and the lack of apoptotic cells in MRL/lpr mice can also reflect an impaired apoptosis in the target organs. Whether this is due to a non-functional FasR/FasL interaction in MRL/lpr mice or other mechanisms in the MRL/+/+ mice is an open question. Gender-associated differences exist in the expression of apoptotic factor mRNA in exocrine tissues of autoimmune mice and some of these differences appear to be due to the influence of androgens (88).

In another recent approach, Fas and TNF receptor I apoptosis pathways have been analyzed in inflammatory salivary gland disease induced by murine cytomegalovirus infection after different strains of mice were infected (89). Although acute salivary gland inflammation developed in all cytomegalovirus-infected mice, only one strain, the B6-lpr/lpr, showed chronic inflammation. Apoptotic cells were detected during the acute, but not the chronic, phase of inflammation. Both Fas- and TNF receptor I-mediated apoptosis was found to contribute to the clearance of murine cytomegalovirus-infected cells in salivary glands. However, because Fas-mediated apoptosis is necessary for the down-modulation of the immune response, a defect in this process can lead to a post-infectious, chronic inflammatory response that resembles Sjögren's syndrome.

Also, excessive synthesis of matrix metalloproteinases in exocrine tissues of NOD mice have been described (90).

An important goal for the future will be to arrange the contributing factors in autoimmune sialadenitis in a temporal sequence (Figure 4). Furthermore, the identification of molecular features that appear to be an intrinsic abnormality of the immune system in autoimmune mice with a defined genetic background promises to offer a more fundamental understanding of the disease process.

Although the mechanisms underlying glandular destruction in SS have not been fully elucidated, available evidence suggests that invading mononuclear cells play an important local role in affecting tissue injury. Delineation of the cellular and molecular mechanisms underlying the parenchymal destruction should provide important insights concerning pathways involved in glandular injury also in human Sjögren's syndrome.

11. IMMUNOMODULATORY AND THERAPEUTIC STRATEGIES

The possibility to utilize animal models for primary therapeutical trials and immunomodulation is advantageous. Immunological and pathological analysis of sialadenitis/lacrimitis in these models, together with clinical studies, are starting to provide an insight as to which of these experimental therapies may be applicable to human disease.

Hormones have long been known to exercise immunoregulatory properties and may be partly responsible for the strong female preponderance. The manifestations of SLE- and Sjögren's syndrome-like disorders in NZB/W mice are sensitive to such manipulations (91); male hormones suppress the manifestations of the autoimmune disease in NZB/W mice, whereas female sex hormones have the opposite effect. Studies by a Dutch group showed that nandrolone decanoate (92) and tibolone (93), compounds with weak androgenic properties, had beneficial effects on NZB/W sialadenitis.

To pursue on the observations of differential expression of murine autoimmune disease, we studied the effect of physiological doses of oestradiol on autoimmune sialadenitis in MRL/lpr mice (94). Treatment with oestradiol significantly ameliorated the focal sialadenitis and also the periarticular inflammation in MRL/lpr mice. However, no alteration of histologically evaluated arthritis could be detected. This differential outcome on the complex autoimmune disease of MRL/*lpr* mice was proposed to represent a dualistic effect of oestradiol on B and T cell-mediated immune responses.

However, there are important bi-directional interactions between the immune and the endocrine systems. Sex hormones influence the immune system throughout life, including postnatal and prenatal stages. In one study, it was shown that mice prenatally exposed to estrogens had accelerated development of autoimmune sialadenitis (95). Moreover, testosterone exposure to female MRL/*lpr* mice dramatically reduced mononuclear infiltration in lacrimal and submandibular glands (96).

Recently, the effects of estrogen deficiency in a murine model for autoimmune exocrinopathy of Sjögren's syndrome was evaluated (97). Severe destructive autoimmune lesions developed in the salivary and lachrymal glands in estrogen- deficient mice, and these lesions were recovered by estrogen administration. Moreover, an increased apoptotic activity of epithelial duct cells was detected. It was also shown that Fas-mediated apoptosis in cultured salivary gland cells was clearly inhibited by estrogens *in vitro*. These results suggest that estrogen deficiency may play an important role on acceleration of organ-specific autoimmune lesions.

By histological and immunohistochemical techniques, we analyzed the immunomodulatory effect of treatment with LS2616, a synthesized oxokinolinamide derivative previously found to ameliorate organ pathology, on sialadenitis in submandibular glands of MRL/*lpr* mice (98). The results were compared with effects obtained after treatment with cyclophosphamide and physiologic saline. However, only cyclophosphamide treatment reduced sialadenitis, while LS2616 increased the semiquantitatively assessed focal inflammation of salivary glands in six month-old mice. This implicated different immunopathogenic mechanisms for the development of sialadenitis versus other organ lesions in the autoimmune disease of MRL/*lpr* mice.

A comparative influence of steroid hormones and immunosuppressive agents on autoimmune expression in lacrimal glands of a female mouse model have been described (99).

The NFS/sld mutant mouse model was used for primary Sjögren's syndrome to investigate the efficacy of topical cyclosporin A in reducing exocrine gland inflammation (100). Topical use of 0.1% cyclosporin A for ten weeks significantly reduced the inflammation in the lachrymal and submandibular glands, but not in the parotid gland. In contrast, systemic cyclosporin A accelerated the infiltration in the exocrine glands.

The new understanding of T cell mediated autoimmunity has led to development of novel therapeutic strategies that initially are being assessed in well-defined animal models. Among the approaches being taken are tissue-specific inhibition of T cell migration. Previous studies have shown that anti-CD4 therapy ameliorate antinuclear antibody production, lymphadenopathy and glomerulonephritis in MRL/*lpr* mice (67,101–103). On the other hand, we and others have shown a significantly increased severity of sialadenitis after anti-CD4 therapy in MRL/*lpr* mice (75,104). Collectively, this suggests that CD4+ cells do not play a similar pathogenic role in each target organ involved in the autoimmune disease of MRL/*lpr* mice. Concerning the lacrimal gland disease in MRL/*lpr* mice, it has been shown that suppression of both CD4+ and CD8+ cells is required to suppress lachrymal gland inflammation in MRL/*lpr* mice (105).

For the future, we also have to count on the possibility of performing local gene therapy in exocrine glands. Such an approach has recently been introduced by applying adenoviral vectors transferred through the ducts of murine glands (106).

ACKNOWLEDGEMENTS

This work was supported by grants from the European BIOMED program (BMH4-CT96–0595, BMH4-CT98–3489), the Research Council of Norway (115563/ 320), The Foundation Health and Rehabilitation and the Broegelmann Foundation. Ms Kate Frøland is acknowledged for editorial assistance and Ms Randi Löfgren for linguistic corrections.

References

1. Jonsson, R., H.-J. Haga, and T. Gordon. 2000. Sjögren's syndrome. In *Arthritis and allied conditions—A textbook of rheumatology*, W.J. Koopman, editor. 14th Edit. Williams & Wilkins, Philadelphia. 1826–1849.
2. Humphreys-Beher, M.G., J. Brayer, S. Yamachika, A.B. Peck, and R. Jonsson. 1999. An alternative perspective to the immune response in autoimmune exocrinopathy: induction of functional quiescence rather than destructive autoaggression. *Scand. J. Immunol.* 49:7–10.
3. Bolstad, A.I., and R. Jonsson. 1998. The role of apoptosis in Sjögren's syndrome. *Ann. Med. Interne.* 149:25–29.
4. Tarkowski, A., R. Jonsson, and L. Klareskog. 1986. Experimental Sjögren's syndrome: a review. *Scand. J. Rheumatol.* Suppl. 61:274–279.
5. Hoffman, R.W., and S.E. Walker. 1987. Animal models of Sjögren's syndrome. In: *Sjögren's syndrome—clinical and immunological aspects*. N. Talal, H.M. Moutsopoulos, S.S. Kassan, editors. Springer-Verlag, Berlin. pp. 266–288.
6. Hayashi, Y., H. Deguchi, A. Nakahata, C. Kurashima, M. Utsuyama, and K. Hirokawa. 1990. Autoimmune sialadenitis: Possible models for Sjögren's syndrome and a common aging phenomenon. *Autoimmunity* 5:215–228.
7. Hayashi, Y., N. Haneji, and H. Hamano. 1994. Pathogenesis of Sjögren's syndrome-like autoimmune lesions in MRL/lpr mice. *Pathol. Int.* 44:559–568.

8. Horsfall, A., K. Skarstein, and R. Jonsson. 1994. Experimental models of Sjögren's syndrome. In: *Autoimmune diseases: Focus on Sjögren's syndrome*. D.A. Isenberg, A. Horsfall, editors. BIOS Scientific Publishers, London. pp. 67–88.

9. Humphreys-Beher, M.G. 1996. Animal models for autoimmune disease-associated xerostomia and xerophthalmia. *Adv. Dent. Res.* 10:73–75.

10. Kessler, H.S. 1968. A laboratory model for Sjögren's syndrome. *Am. J. Pathol.* 52:671–678.

11. Hoffman, R.W., M.A. Alspaugh, K.S. Waggie, J.B. Durham, and S.E. Walker. 1984. Sjögren's syndrome in MRL/1 and MRL/n mice. *Arthritis Rheum.* 27:157–165.

12. Miyagawa, J., T. Hanafusa, A. Miyazaki, K. Yamada, H. Fujino-Kurihara, H. Nakajima, N. Kono, K. Nonaka, Y. Tochino, and S. Tarui. 1986. Ultrastructural and immunocytochemical submandibulitis in the non-obese diabetic (NOD) mouse. *Virchows. Arch. (Cell. Pathol.)* 51:215–225.

13 Kessler, H.S., M. Cubberly, and W. Manski. 1971. Eye changes in autoimmune NZB and NZB x NZW mice. *Arch. Ophthal.* 85:211–219.

14. Keyes, G.G., R.A. Vickers, and J.H. Kersey. 1977. Immunopathology of Sjögren-like disease in NZB/NZW mice. *J. Oral. Pathol.* 6:288–295.

15. Jonsson, R., A. Tarkowski, and K. Backman. 1987. Immunohistochemical characterization of sialadenitis in NZB x NZW F₁ mice. *Clin. Immunol. Immunopathol.* 42:93–101.

16. Carlsöö, B., Y. Östberg. 1978. Ultrastructural observations on the parotitis autoimmunica in the NZB/NZW hybrid mice. *Acta Otolaryngol.* 85:298–306.

17. Carlsöö, B., Y. Östberg. 1979. The autoimmune submandibular sialoadenitis of the NZB/NZW hybrid mice. A light and electron microscopical investigation. *Arch. Otorhinolaryn.* 225:57–65.

18. Hang, L., A.N. Theofilopoulos, and F.J. Dixon. 1982. A spontaneous rheumatoid arthritis-like disease in MRL/1 mice. *J. Exp. Med.* 155:1690–1701.

19. Hoffman, R.W., H.K. Yang, K.S. Waggie, J.B. Durham, J.R. Burge, and S.E. Walker. 1983. Band keratopathy in MRL/1 and MRL/n mice. *Arthritis. Rheum.* 26:645–652.

20. Jonsson, R., A. Tarkowski, K. Bäckman, R. Holmdahl, and L. Klareskog. 1987. Sialadenitis in the MRL-1 mouse: morphological and immunohistochemical characterization of resident and infiltrating cells. *Immunology* 60:611–616.

21. Scott, J., A. Wolff, and P.C. Fox. 1990. Histologic assessment of the submandibular glands in autoimmune-disease-prone mice. *J. Oral Pathol. Med.* 19:131–135.

22. Tanaka, A., F.X. O'Sullivan, W.J. Koopman, and S. Gay. 1988. Ultrastructural study of Sjögren's syndrome-like disease in MRL/1 mice. *J. Oral Pathol.* 17:460–465.

23. Skarstein, K., A.H. Nerland, M. Eidsheim, J.D. Mountz, and R. Jonsson. 1997. Lymphoid cell accumulation in salivary glands of autoimmune MRL mice can be due to impaired apoptosis. *Scand. J. Immunol.* 46:373–378.

24. Daniels, T.E. 1984. Labial salivary gland biopsy in Sjögren's syndrome: assessment as a criterion in 362 suspected cases. *Arhtritis. Rheum.* 27:147–156.

25. Jabs, D., C. Enger, and R. Prendergast. 1991. Murine models of Sjögren's syndrome. Evolution of the lacrimal gland inflammatory lesions. *Invest. Ophthalmol. Vis. Sci.* 32:371–380.

26. Jabs, D., and R. Prendergast. 1991. Ocular inflammation in MRL/Mp-*lpr/lpr* mice. *Invest. Ophthalmol. Vis. Sci.* 32:1944–1947.

27. Lund, T., L. O'Reilly, P. Hutchings, O. Kanagara, E. Simpson, R. Gravely, P. Chandler, J. Dyson, J.K. Picard and A. Edwards 1990. Prevention of insulin-dependent diabetes mellitus in non-obese diabetic mice by transgenes encoding modified I-Aβ chain or normal I-Eα chain. *Nature* 345:727–729.

28. Shultz, L.D., P.A. Schweitzer, S.W. Christianson, B. Gott, I.B. Schweitzer, B. Tennent, S. McKenna, L. Mobraaten, T.V. Rajan, D.L. Greiner, and E.H. Leiter. 1995. Multiple defects in innate and adaptive immunologic function in NOD/LtSz-scid mice. *J. Immunol.* 154:180–191.

29. Robinson, C.P., H. Yamamoto, A.B. Peck, and M.G. Humphreys-Beher. 1996. Genetically programmed development of salivary gland abnormalities in the NOD (Nonobese Diabetic)-*scid* mouse in the absence of detectable lymphocytic infiltration: A potential trigger for sialoadenitis of NOD mice. *Clin. Immunol. Immuopathol.* 79:50–59.

30. Robinson, C.P., S. Yamachika, C.E. Alford, C. Cooper, E.L. Pichardo, N. Shah, A.B. Pcck, and M.G. Humphreys-Beher. 1997. Elevated levels of cysteine protease activity in saliva and salivary glands of the nonobese diabetic (NOD) mouse model for Sjögren's syndrome. *Proc. Natl. Acad. Sci. USA* 94:5767–5771.

31. Robinson, C.P., S. Yamachika, D.I. Bounous, J. Brayer, R. Jonsson, R. Holmdahl, A.B. Peck, and M.G. Humphreys-Beher. 1998. A novel NOD-derived murine model for primary Sjögren's syndrome. *Arthritis. Rheum.* 41:150–156.

32. Sundler, M., K. Skarstein, P. Kjellén, R. Jonsson, and R. Holmdahl. 1998. The influence of different genes on the development of murine sialadenitis. *Arthritis Rheum.* 41:9(suppl)S326 (Abstr.)

33. Haneji, N., H. Hamano, K. Yanagi, and Y. Hayashi. 1994. A new animal model for primary Sjögren's syndrome in NFS/*sld* mutant mice. *J. Immunol.* 153:2769–2777.

34. Haneji, N., T. Nakamura, K. Takio, K. Yanagi, H. Higashiyama, I. Saito, S. Noji, H. Sugino, and Y. Hayashi. 1997. Identification of alpha-fodrin as a candidate autoantigen in primary Sjögren's syndrome. *Science* 276:604–607.

35. Hayashi, Y., N. Haneji, H. Hamano, K Yanagi, M. Takahashi, and N. Ishimaru. 1996. Effector mechanism of experimental autoimmune sialadenitis in the mouse model for primary Sjögren's syndrome. *Cell. Immunol.* 171:217–225.

36. Saegusa, J., and H. Kubota. 1997. Sialadenitis in IQI/Jic mice: A new animal model of Sjögren's syndrome. *J. Vet. Med. Sci.* 59:897–903.

37. Tsubata, R., T. Tsubata, H. Hiai, R. Shinkura, R. Matsumura, T. Sumida, S. Miyawaki, H. Ishida, S. Kumagai, K. Nakao, and T. Honjo. 1996. Autoimmune disease of exocrine organs in immunodeficient alymphoplasia mice: a spontaneous model for Sjögren's syndrome. *Eur. J. Immunol.* 26:2742–2748.

38. Green, J., S.H. Hinrichs, J. Vogel, and G. Jay. 1989. Exocrinopathy resembling Sjögren's syndrome in HTLV-1 *tax* transgenic mice. *Nature* 341:72–74.

39. Koike, K., K. Moriya, K. Ishibashi, H. Yotsuyanagi, Y. Shintani, H. Fujie, K. Kurokawa, Y. Matsuura, and T. Miyamura. 1997. Sialadenitis histologically resembling Sjögren syndrome in mice transgenic for hepatitis C virus envelope genes. *Proc. Natl. Acad. Sci. USA* 94:233–236.

40. Kimura T., K. Suzuki, S. Inada, A. Hayashi, H. Saito, T. Miyai, Y. Ohsugi, Y. Matsuzaki, N. Tanaka, T. Osuga, and M. Fujiwara. 1994. Induction of autoimmune disease by graft-*versus*-host reaction across MHC class II difference:

modification of the lesions in IL-6 transgenic mice. *Clin. Exp. Immunol.* 95:525–529.

41. Saito, I., K. Haruta, M. Shimuta, H. Inoue, H. Sakurai, K. Yamada, N. Ishimaru, H. Higashiyama, T. Sumida, H. Ishida, T. Suda, T. Noda, Y. Hayashi, and K. Tsubota. 1999. Fas ligand-mediated exocrinopathy resembling Sjögren's syndrome in mice transgenic for IL-10. *J. Immunol.* 162:2488–2494.

42. Shull, M., I. Ormsby, A.B. Kier, S. Pawlowski, R.J. Diebold, M. Yin, R. Allen, C. Sidman, G. Proetzel, D. Calvin et al. 1992. Targeted disruption of the mouse transforming growth factor-beta 1 gene results in multifocal inflammatory disease. *Nature* 359:693–699.

43. McCartny-Francis, N.L., D.E. Mizel, R.S. Redman, M. Frazier-Jessen, R. B. Panek, A.B. Julkarni, J.M. Ward, J.B. McCarthy, and S.M. Wahl. 1996. Autoimmune Sjögren's like lesions in salivary glands of TGF-β1-deficient mice are inhibited by adhesion-blocking peptides. *J. Immunol.* 157:1306–1312.

44. Pals, S.T., T. Radaszkiewicz, L. Roozendaal, and E. Gleichmann. 1985. Chronic progressive polyarteritis and other symptoms of collagen vascular disease induced by graft-vs-host reaction. *J. Immunol.* 134:1475–1482.

45. Fujiwara, K., N. Sakaguchi, and T. Watanabe. 1991. Sialoadentis in experimental graft-versus-host disease. An animal model of Sjögren's syndrome. *Lab. Invest.* 65:710–718.

46. Sorensen, I., A.P. Ussing, J.U. Prause, J. Blom, S. Larsen, and J.V. Sparck. 1992. Histopathological changes in exocrine glands of murine transplantation chimeras. I: The development of Sjögren's syndrome-like changes secondary to GVH induced lupus syndrome. *Autoimmunity* 11:261–271.

47. Ussing, A., J.U. Prause, I. Sorensen, S. Larsen, and J.V. Sparck. 1992. Histopathological changes in exocrine glands of murine transplantation chimeras. II: Sjögren's syndrome-like exocrinopathy in mice without lupus nephritis. A model of primary Sjögren's syndrome. *Autoimmunity* 11:273–280.

48. Nishimori, I., T. Bratanova, I. Toshkov, T. Caffrey, M. Mogaki, Y. Shibata, and M. A. Hollingsworth. 1995. Induction of experimental autoimmune sialoadenitis by immunization of PL/J mice with carbonic anhydrase II. *J. Immunol.* 154:4865–4873.

49. Inagaki, Y., Y. Jinno-Yoshida, Y. Hamasaki, and H. Ueki. 1991. A novel autoantibody reactive with carbonic anhydrase in sera from patients with systemic lupus erythematosus and Sjögren's syndrome. *J. Dermatol. Sci.* 2:147–154.

50. Wahren, M., K. Skarstein, I. Blange, I. Pettersson, and R. Jonsson. 1994. MRL/*lpr* mice produce anti-Ro 52000 MW antibodies: detection, analysis of specificity and site of production. *Immunology* 83:9–15.

51. St. Clair, E.W., D. Kenan, J.A. Burch, J.D. Keene, and D.S. Pisetsky. 1991. Anti-La antibody production by MRL-*lpr/lpr* mice: analysis of fine specificity. *J. Immunol.* 146:1885–1892.

52. Topfer, F., T. Gordon, and J. McCluskey. 1995. Intra- and intermolecular spreading of autoimmunity involving the nuclear self-antigens La(SS-B) and Ro(SS-A). *Proc. Natl. Acad. Sci. USA* 92:875–879.

53. Keech, C.L., T.P. Gordon, and J. McCluskey. 1996. The immune response to 52-kDa Ro and 60kDa Ro is linked in experimental autoimmunity. *J. Immunol.* 157:3694–3699.

54. Tseng, C.-E., E.K.L. Chan, E. Miranda, M. Gross, F. Di Donato, and J.P. Buyon. 1997. The 52-kd protein as a target of intermolecular spreading of the immune response to components of the SS-A/Ro-SS-B/La complex. *Arthritis Rheum.* 49:936–944.

55. Reynolds, P., T.P. Gordon, A.W. Purcell, D.C. Jackson, and J. McCluskey. 1996. Hierarchical self-tolerance to T cell determinants within the ubiquitous nuclear self-antigen La (SS-B) permits induction of systemic autoimmunity in normal mice. *J. Exp. Med.* 184:1857–1870.

56. Scofield, R.H., A.P. Farris, A.C. Horsfall, and J.B. Harley. 1999. Fine specificity of the autoimmune response to the Ro/SSA and La/SSB ribonucleoproteins. *Arthritis Rheum.* 42:199–209.

57. MacSween, R.N.M., R.B. Goudie, J.R. Anderson, E. Armstrong, M.A. Murray, D.K. Mason, M.K. Jasani, J.A. Boyle, W.W. Buchanan, and J. Williamson. 1967. Occurrence of antibody to salivary duct epithelium in Sjögren's disease, rheumatoid arthritis and other arthritides—a clinical and laboratory study. *Ann. Rheum. Dis.* 26:402–411.

58. Feltkamp, T.E.W., and A.L. Van Rossum. 1968. Antibodies to salivary duct cells and other autoantibodies in patients with Sjögren's syndrome and other idiopathic autoimmune diseases. *Clin. Exp. Immunol.* 3:1–16.

59. Jonsson, R., A. Pitts, J. Mestecky, and W. Koopman. 1991. Local IgA and IgM rheumatoid factor production in autoimmune MRL/lpr mice. *Autoimmunity* 10:7–14.

60. Bacman, S., L. Sterin-Borda, J.J. Camusso, R. Arana, O. Hubscher, and E. Borda. 1996. Circulating antibodies against rat parotid gland M3 muscarinic receptors in primary Sjögren's syndrome. *Clin. Exp. Immunol.* 104:454–459.

61. Bacman, S., C. Perez Leiros, L. Sterin-Borda, O. Hubscher, R. Arana, and E. Borda. 1998. Autoantibodies against lacrimal gland M3 muscarinic acetylcholine receptors in patients with primary Sjögren's syndrome. *Invest. Ophthalmol. Vis. Sci.* 39:151–156.

62. Robinson, C.P., J. Brayer, S. Yamachika S, T.R. Esch, A.B. Peck, C.A. Stewart, E. Peen, R. Jonsson, and M.G. Humphreys-Beher. 1998. Transfer of human serum IgG to NOD.Igμ^null mice reveals a role for autoantibodies in the loss of secretory function of exocrine tissues in Sjögren's syndrome. *Proc. Natl. Acad. Sci. USA* 95:7538–7543.

63. Esch, T.R., and M.A. Taubman. 1998. Autoantibodies in salivary hypofunction in the NOD mouse. *Ann. NY Acad. Sci.* 842:221–228.

64. Greenspan, J.S., G.A. Gutman, I.L. Weissman, and N. Talal. 1974. Thymus-antigen- and immunoglobulin-positive lymphocytes in tissue infiltrates of NZB/NZW mice. *Clin. Immunol. Immunopathol.* 3:16–31.

65. Jabs, D.A., and R.A. Prendergast. 1987. Reactive lymphocytes in lacrimal glands and renal vasculitic lesions of autoimmune MRL/lpr mice express L3T4. *J. Exp. Med.* 166:1198–1203.

66. Jabs, D.A., and R.A. Prendergast. 1988. Murine models of Sjögren's syndrome: immunohistologic analysis of different strains. *Invest. Ophthalmol. Vis. Sci.* 29:1437–1443.

67. Santoro, T.J., J.P. Portanova, and B.L. Kotzin. 1988. The contribution of L3T4+ T cells to lymphoproliferation and autoantibody production in MRL- *lpr/lpr* mice. *J. Exp. Med.* 167:1713–1718.

68. Skarstein, K., A.C. Johannessen, R. Holmdahl, and R. Jonsson. 1997. Effects on sialadenitis after cellular transfer in autoimmune MRL/*lpr* mice. *Clin. Immunol. Immunopathol.* 84:177–184.

69. Skarstein, K. 1996. Autoimmunity and pathogenesis of murine sialadenitis. Thesis. University of Bergen. ISBN 82-7249-170-2.

70. Hayashi, Y., N. Haneji, H. Hamano, and K. Yanagi. 1994. Transfer of Sjögren's syndrome-like autoimmune lesions into SCID mice and prevention of lesions by anti-CD4 and anti-T cell receptor antibody treatment. *Eur. J. Immunol.* 24:2826–2831.

71. Jonsson, R., and R. Holmdahl, R. 1990. Infiltrating mononuclear cells in salivary glands and kidneys in autoimmune MRL/Mp-lpr mice express IL-2 receptor and produce interferon-γ. *J. Oral Pathol. Med.* 19:330–334.

72. Skarstein, K., R. Holmdahl, A.C. Johannessen, and R. Jonsson. 1994. Oligoclonality of T cells in salivary glands of autoimmune MRL/*lpr* mice. *Immunology.* 81:497–501.

73. Hayashi, Y., H. Hamano, N. Haneji, N. Ishimaru, and K. Yanagi. 1995. Biased T cell receptor Vβ gene usage during specific stages of the development of autoimmune sialadenitis in the MRL/*lpr* mouse model of Sjögren's syndrome. *Arthritis Rheum.* 38:1077–1084.

74. Skarstein, K., M. Wahren, E. Zaura, M. Hattori, and R. Jonsson. 1995. Characterization of T cell receptor repertoire and anti-Ro/SSA autoantibodies in relation to sialadenitis of NOD mice. *Autoimmunity* 22:9–16.

75. Skarstein, K., R. Holmdahl, A.C. Johannessen, T. Goldschmidt, and R. Jonsson. 1995. Short-term administration of selected anti-T-cell receptor Vβ chain specific MoAb reduces sialadenitis in MRL/*lpr* mice. *Scand. J. Immunol.* 42:529–534.

76. Jonsson, R., K. Beagley, J. Mestecky, J. Mountz, and W. Koopman. 1991. Production of IL 2, IL-3, IL-6, TNF-α and IFN-γ in salivary glands and lymphoid organs of autoimmune MRL/lpr mice. *Scand. J. Rheumatol.* 20:216 (Abstr.)

77. Hamano, H., I. Saito, N. Haneji, Y. Mitsuhashi, N. Miyasaka, and Y. Hayashi. 1993. Expression of cytokine genes during development of autoimmune sialadenitis in MRL/lpr mice. *Eur. J. Immunol.* 23:2387–2391.

78. Mustafa, W., J. Zhug, G. Deng, A. Diab, H. Link, L. Frithiof, and B. Klinge. 1998. Augmented levels of macrophage and Th1 cell-related cytokine mRNA in submandibular glands of MRL/lpr mice with autoimmune sialoadenitis. *Clin. Exp. Immunol.* 112:389–396.

79. Yanagi, K., N. Ishimaru, N. Haneji, K. Saegusa, I. Saito, and Y. Hayashi. 1998. Anti-120kDa α-fodrin immune response with Th1-cytokine profile in the NOD mouse model of Sjögren's syndrome. *Eur. J. Immunol.* 2:3336–3345.

80. Yamano, S., J.C. Atkinson, B.J. Baum, and P.C. Fox. 1999. Salivary gland cytokine expression in NOD and normal BALB/c Mice. *Clin. Immunol.* 92:265–275.

81. Schall, T.G., K. Bacon, R.D. Camp, J.W. Kaspari, and D.V. Goeddel. 1993. Human macrophage inflammatory protein alpha (MIP-1 alpha) and MIP-1 beta chemokines attract distinct populations of lymphocytes. *J. Exp. Med.* 177:1821–1826.

82. Baggiolini, M., and Dahinden, C.A. 1994. CC chemokines in allergic inflammation. *Immunol. Today* 15:127–133.

83. Mustafa, W., A. Sharafeldin, A. Diab, M. Huang, H. Bing, J. Zhu, H. Link, L. Frithiof, and B. Klinge. 1998. Coordinated up-regulation of the β-chemokine subfamily in autoimmune sialoadenitis of MRL/lpr mice. *Scand. J. Immunol.* 48:623–628.

84. Hayashi, Y., N. Haneji, K. Yanagi, H. Higashiyama, H. Yagita, and H. Hamano. 1995. Prevention of adoptive transfer of murine Sjögren's syndrome into severe combined immunodeficient (SCID) mice by antibodies against inter-

85. Takahashi, M., Y. Mimura, and Y. Hayashi. 1996. Role of the ICAM-1/LFA-1 pathway during the development of autoimmune dacryoadenitis in an animal model for Sjögren's syndrome. *Pathobiol.* 64:269–274.

86. Watanabe-Fukunaga, R., C.I. Brannan, N.G. Copeland, N.A. Jenkins, and S. Nagata. 1992. Lymphoproliferation disorder in mice explained by defects in Fas antigen that mediates apoptosis. *Nature* 356:314–317.

87. Singer, G.G., and A.K. Abbas. 1994. The Fas antigen is involved in peripheral but not thymic deletion of T lymphocytes in T cell receptor transgenic mice. *Immunity* 1:365–371.

88. Toda, I., L.A. Wickham, and D.A. Sullivan. 1998. Gender and androgen treatment influence the expression of proto-oncognes and apoptotic factors in lacrimal and salivary tissues of MRL/lpr mice. *Clin. Immunol. Immunopathol.* 86:59–71.

89. Fleck, M., E.R. Kern, T. Zhou, B. Lang, and J.D. Mountz. 1998. Murine cytomegalovirus induces a Sjögren's syndrome-like disease in C57Bl/6-lpr/lpr mice. *Arthritis Rheum.* 41:2175–2184.

90. Yamachika, S., J.M. Nanni, K.H. Nguyen, L. Garces, J.M. Lowry, C.P. Robinson, J. Brayer, G.E. Oxford, A. da Silveira, M. Kerr, A.B. Peck, and M.G. Humphreys-Beher. 1998. Excessive synthesis of matrix metalloproteinases in exocrine tissues of NOD mouse models for Sjögren's syndrome. *J. Rheumatol.* 25:2371–2380.

91. Theofilopoulos, A., and Dixon, F. 1981. Etiopathogenesis of murine SLE. *Immunol. Rev.* 55:179–216.

92. Schot, L.P.C., H.A. Verhaul, and A.H. Schuurs. 1984. Effect of nandrolone decanoate on Sjögren's syndrome like disorders in NZB/NZW mice. *Clin. Exp. Immunol.* 57:571–574.

93. Verheul, H.A.M., L.P. Schot, and H.W. Schuurs. 1986. Effects of tibolone, lynestrenol, ethylestrenol and desogestrel on autoimmune disorders in NZB/W mice. *Clin. Immunol. Immunopathol.* 38:198–208.

94. Carlsten, H., N. Nilsson, R. Jonsson, K. Backman, R. Holmdahl, and A. Tarkowski. 1992. Estrogen accelerates immune complex glomerulonephritis but ameliorates T cell-mediated vasculitis and sialadenitis in autoimmune MRL lpr/lpr mice. *Cell. Immunol.* 144:190–202.

95. Ahmed, S.A., T.B. Aufdemorte, J.R. Chen, A.I. Montoya, D. Olive, and N. Talal. 1989. Estrogen induces the development of autoantibodies and promotes salivary gland lymphoid infiltrates in normal mice. *J. Autoimmun.* 2:543–552.

96. Ariga, H., J. Edwards, and D.A. Sullivan. 1989. Androgen control of autoimmune expression in lacrimal glands of MRL/Mp-*lpr/lpr* mice. *Clin. Immunol. Immunopathol.* 53:499–508.

97. Ishimaru, N., K. Saegusa, K. Yanagi, N. Haneji, I. Saito, and Y. Hayashi. 1999. Estrogen deficiency accelarates autoimmune exocrinopathy in murine Sjögren's syndrome through fas-mediated apoptosis. *Am. J. Pathol.* 155:173–181.

98. Jonsson, R., A. Tarkowski, and K. Bäckman. 1988. Effects of immunomodulating treatment on autoimmune sialadenitis in MRL/Mp-lpr mice. *Agents Actions* 25:368–374.

99. Sato, E.H., and D.A. Sullivan. 1994. Comparative influence of steroid hormones and immunosuppressive agents on autoimmune expression in lacrimal glands of a female mouse model of Sjögren's syndrome. *Invest. Ophthalmol. Vis. Sci.* 35:2632–2642.

100. Tsubota, K., I. Saito, N. Ishimaru, and Y. Hayashi. 1998. Use of topical cyclosporin A in a primary Sjögren's syndrome mouse model. *Invest. Ophthalmol. Vis. Sci.* 39:1551–1559.

101. Jabs, D.A., and R.A. Prendergast. 1991. Autoimmune ocular disease in MRL/Mp-lpr/lpr mice is suppresed by anti-CD4 antibody. *Invest. Ophthalmol. Vis. Sci.* 32:2718–2722.

102. Jabs, D.A., C. Lynne Burek, Q. Hu, R.C. Kuppers, B. Lee, and R.A. Prendergast. 1992. Anti-CD4 monoclonal antibody therapy suppresses autoimmune disease in MRL/Mp-*lpr/lpr* mice. *Cell. Immunol.* 141:496–507.

103. Jabs, D.A., R.C. Kuppers, A.M. Saboori, C. Lynne Burek, C. Enger, B. Lee, and R.A. Prendergast. 1994. Effects of early and late treatment with anti-CD4 monoclonal antibody on autoimmune disease in MRL/Mp-*lpr/lpr* mice. *Cell. Immunol.* 154:66–76.

104. O'Sullivan F.X., C.M. Vogelweid, C.L. Besch-Williford, and S.E. Walker. 1995. Differential effects of CD4+ T cell depletion on inflammatory central nervous system disease, arthritis and sialadenitis in MRL/lpr mice. *J. Autoimmun.* 8:163–175.

105. Jabs, D.A., B. Lee, and R.A. Prendergast. 1997. Role of T cells in the pathogenesis of autoimmune lacrimal gland disease in MRL/Mp-*lpr/lpr* mice. *Curr. Eye Res.* 16:909–916.

106. Wang, S., B.J. Baum, S. Yamano, M.H. Mankani, D. Sun, M. Jonsson, C. Davis, F.L. Graham, J. Gauldie, and J.C. Atkinson. 2000. Adenoviral-mediated gene transfer to mouse salivary glands. *J. Dent. Res.* 79:701–708.

26 | Systemic Sclerosis (Scleroderma)

Frank C. Arnett

1. CLASSIFICATION AND DEFINITIONS

Systemic sclerosis (SSc), or systemic scleroderma, is a chronic multisystem connective tissue disease characterized by cutaneous and visceral fibrosis and a widespread obliterative vasculopathy affecting small arteries and capillaries. SSc is the most serious form of scleroderma (thickening of the skin), a heterogeneous group of idiopathic fibrosing skin disorders that also includes localized scleroderma variants in which systemic involvement rarely occurs (1) (Table 1). Moreover, a number of other diseases, some caused by known environmental exposures, resemble scleroderma clinically (pseudoscleroderma syndromes) and may provide clues to pathogenesis (2) (Table 2).

SSc itself is subclassified based on several clinical presentations, most importantly extent of skin involvement, which correlate with prognosis (1) (Table 1). Diffuse SSc refers to skin thickening extending proximal to elbows or knees on the extremities, or present on the trunk, while limited SSc is defined as scleroderma involving only areas distal to elbows and knees and sparing the trunk. Facial changes may be present in either form. Diffuse SSc tends to predict severe visceral involvement and a poor prognosis, while the opposite is true of limited SSc. A variant of limited SSc is the CREST (*c*alcinosis, *R*aynaud's phenomenon, *e*sophageal dysmotility, *s*clerodactyly and *t*elangiectasia) syndrome which also carries a relatively good prognosis. Uncommonly, patients may present with visceral complications of SSc, such as renal crisis, pulmonary fibrosis or hypertension, or gastrointestinal dysmotility, and have no

Table 1 Classification of scleroderma

Systemic sclerosis (SSc) (systemic scleroderma)
    Diffuse cutaneous SSc
    Limited cutaneous SSc
    CREST syndrome*
    SSc *sine* scleroderma
    SSc overlap syndromes (including MCTD)†
Localized scleroderma
    Morphea
    Generalized morphea
    Linear scleroderma
        Facial *en coup de sabre*
        Parry-Romberg syndrome

* CREST = *c*alcinosis, *R*aynaud phenomenon, *e*sophageal dysmotility, *s*clerodactyly, *t*elangiectasia.
† MCTD = mixed connective tissue disease.

Table 2 Pseudoscleroderma syndromes

Diffuse fasciitis with eosinophilia (eosinophilic fasciitis, Shulman syndrome)
Chronic graft-versus-host disease
Scleromyxedema
Digital sclerosis of diabetes mellitus
Scleredema (adultorum of Buschke or diabeticorum)
Exposure-associated syndromes
    Vinyl chloride disease
    Toxic oil syndrome secondary to adulterated rapeseed cooking oil
    Eosinophilia myalgia syndrome secondary to contaminated L-tryptophan
    Bleomycin-associated fibrosis
    Silica-associated scleroderma?
    Others proposed (multiple)

signs of cutaneous scleroderma (SSc *sine* scleroderma). Finally, SSc may occur in overlap with clinical and laboratory features of other connective tissue diseases, including systemic lupus erythematosus (SLE), myositis, or Sjogren's syndrome. When such patients possess autoantibodies to U1-ribonucleoprotein (RNP), they are often classified as mixed connective tissue disease (MCTD).

The American College of Rheumatology (ACR) (formerly American Rheumatism Association) has established classification criteria for SSc (3) (Table 3) which are useful in both diagnosis and selection of cases for studies. Although these criteria show high specificity for SSc (97%–100%), some cases of limited SSc and early diffuse SSc may not meet the requirements for inclusion.

2.   EPIDEMIOLOGY

SSc affects women three times more often than men, and disease-onset typically begins between ages 30–50 years. Children are rarely affected.

The prevalence of SSc in the US has been estimated in several recent studies to be 240–286 per million population, while rates in Britain and Iceland (31 cases/million), Australia (45–86 cases/million), and Japan (38 cases/million) appear to be lower (4,5). Incidence rates in the US indicate approximately 20 new cases per million occur annually, and black women may be 2.5 times more likely than white women to develop the disease (4,6). The highest occurrence of SSc has been reported in Choctaw Native Americans residing in southeastern Oklahoma (4690 cases per million in full-blooded individuals) where common ancestry and several genetic associations rather than an obvious environmental exposure appear to best explain the high disease prevalence (7–9). Smaller clusters of increased SSc prevalence around London's two airports (10), in an Italian village (11), in miners exposed to silica dust (12), and in persons having contact with organic solvents (4,13) suggest the importance of environmental triggers. A greater than expected contemporaneous onset of SSc and breast cancer, as well as other malignancies (14,15), suggests a link to neoplasia, but in the minority of cases.

Mortality from SSc is improving, but is still high. Overall 7 year survival is estimated at 76.5%, with an 81% survival rate for limited disease and 72% for diffuse disease. Pulmonary involvement and renal disease are the leading causes of death (4,6,16)

3.   CLINICAL FEATURES

Two pathological processes characterize SSc and give rise to the clinical manifestations (1). The *first* is widespread endothelial damage to small arteries and capillaries, leading to intimal proliferation and luminal obliteration with tissue infarction, and capillary dropout (17). The arterial disease gives rise to Raynaud's phenomenon—an accentuated vasospastic response to cold and catecholamine stimulation (18–20). Similar arterial lesions to those in the digits have been found spread widely throughout the body, with the curious exception of the brain (21) (Figure 1). This obliterative vasculopathy appears to be directly responsible for SSc renal crisis, pulmonary hypertension, and myocardial damage, and "visceral" Raynaud-like vasospasm is hypothesized and supported by some evidence to contribute to these clinical complications (22–24). Capillary dropout and damaged, enlarged capillary loops can be seen by microscopy *in vivo* in the nailfolds of SSc patients (25,26). Other damaged capillaries give rise to the multiple telangiectasias seen frequently. Finally, a loss of capillaries has been found in the fibrotic skin and in the gut leading to loss of smooth muscle by ischemia.

The *second* pathological process is widespread fibrosis in the skin and other connective tissues which could be a consequence of the ischemia resulting from the vascular lesions (Figure 2). In each of these processes, evidence of inflammation is minimal by usual criteria, although in early cutaneous disease infiltrating mononuclear cells may be found in the dermis (17,26)

4.   CUTANEOUS FEATURES AND TREATMENT

Raynaud phenomenon is usually the first symptom and may precede any other findings by months or even years (19,20). Digital pitted scars or ischemic ulcers may occur on the fingerpads and/or around the nails and tend to predict subsequent progression to SSc. Diffuse swelling of the fingers, hands and/or feet is another early sign that is often misdiagnosed as edema. With time, swelling is replaced with thickened, taut skin over the fingers (sclerodactyly) often causing contractures (Figure 3). In diffuse SSc, skin thickening progresses proximally from distal portions of the extremities (acrosclerosis) to forearms, arms, legs, thighs, face,

Table 3   American College of Rheumatology Preliminary Clinical Criteria for systemic sclerosis*

Major criterion
   Sclerodermatous skin changes proximal to the digits
Minor criteria
   Sclerodactyly
   Digital pitted scars or pulp loss in distal finger pads
   Bibasilar pulmonary fibrosis on chest X-ray

Definite SSc requires the presence of the major criterion or two or more minor criteria.
*reference (3).

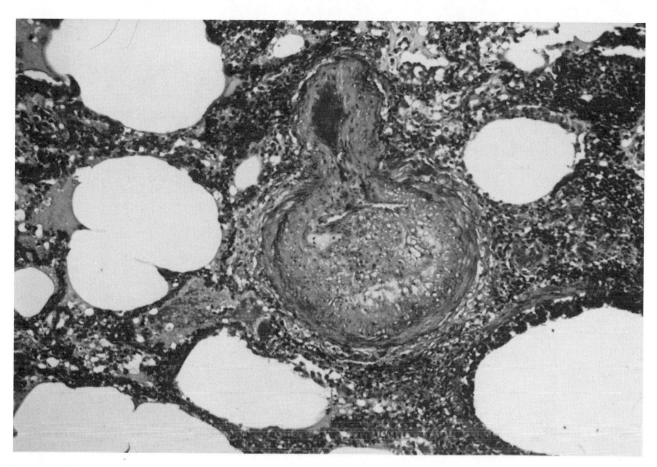

Figure 1. Histopathology of pulmonary parenchyma from a patient with SSc. A small arteriole demonstrates the typical intimal proliferation and obliteration of the lumen. Fibrosis with thickening of interalveolar septae also is apparent.

neck, anterior chest and abdomen. The back is typically spared. Patients with limited SSc often show only sclerodactyly with or without facial changes. Hyperpigmentation and/or hypopigmentation of the affected skin is common but also may occur in areas where thickening is not clinically apparent (27). Thickened, contracted skin may break down over areas of pressure, such as joints, leaving painful ischemic ulcers. Telangiectasias, often multiple, may appear on fingers, hands and face, including the lips and tongue. Vertical furrows appear around the mouth, and vermillion borders of the lips recede, which in combination with facial skin tightening, leads to a typical scleroderma facies.

The frequency and severity of Raynaud phenomenon may be ameliorated by the use of calcium channel blockers, especially long-acting nifedipine, alpha- adrenergic blockers, such as prazosin, the seratonin antagonist, ketanserin, or local application of nitroglycerin (19,20,28). Patients with particularly severe Raynaud phenomenon and digital ulcers may improve with intravenous prostacylin (28). Theoretically, such treatment also may reduce internal organ damage in the heart, kidneys or lungs by reducing the ischemic effects of "visceral" Raynaud phenomenon (22,23). No therapy is

proven to be effective for the skin thickening, although some patients may respond to low dose prednisone in the early "inflammatory" 'or "edematous" phase. D-penicillamine has been used for decades based on retrospective reports of efficacy; however, a recent placebo-controlled trial reported no significant benefit (29). Methotrexate has been reported to be useful in several small uncontrolled series (30). Recombinant human relaxin is currently undergoing clinical trials, and early results suggest that it may improve thickened skin (31).

5. MUSCULOSKELETAL FEATURES

Polyarthralgias and joint stiffness are common symptoms, often early in the disease, and may be due to skin tightness around joints, synovitis, or fibrinous involvement of tendons and tendon sheaths (1). Tendon friction rubs may be felt or ascultated around joints, primarily in diffuse SSc (32). Contractures of joints, including fingers, wrists, elbows, knees or ankles, may occur due to tendinous and periarticular fibrosis and shortening. Ectopic calcification with basic

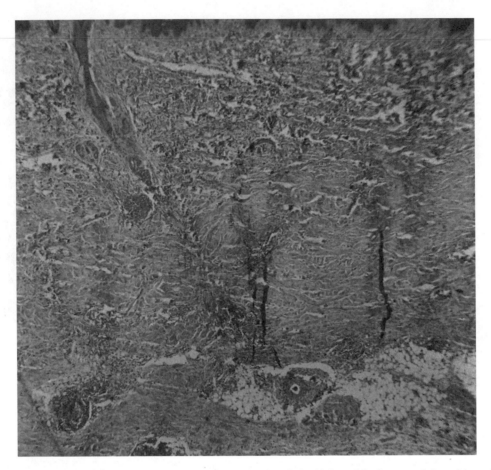

Figure 2. Histopathology of affected skin from a SSc patient. The dermis is thickened by increased deposition of collagen bundles. The epidermis is thinned, and there are diminished numbers of skin appendages.

calcium phosphate (calcinosis) occurs in some patients, especially around joints. Carpal tunnel syndrome occurs commonly. Resorption of bony tufts of fingers and/or toes, probably due to ischemia, may be found radiographically.

Muscle weakness and/or wasting in SSc may occur from disuse atrophy or a non-inflammatory myopathy where muscle fibers are replaced by fibrotic tissue (33). Actual inflammatory myopathy (myositis) with elevated muscle enzymes, abnormal electromyography and typical biopsy picture develops in approximately 20% of SSc patients.

Non-steroidal anti-inflammatory drugs (NSAIDs) and/or low dose prednisone may relieve some musculoskeletal symptoms. Moderate to high doses of corticosteroids are usually required to treat myositis.

6. PULMONARY MANIFESTATIONS

Interstitial fibrosis, typically beginning bilaterally in the lung bases, occurs most commonly and may progress to respiratory failure, pulmonary hypertension, and cor pulmonale (34,35). The earliest manifestation is an alveolitis detectable by reduction in diffusing capacity for carbon monoxide (DLCO), bronchoalveolar lavage and/or high resolution computed tomagraphy (CT) (36–38). Recent clinical trials suggest that early and aggressive treatment of alveolitis with cyclophosphamide, and perhaps corticosteroids, may prevent fibrosis and improve survival (30,35).

Pulmonary hypertension without interstitial fibrosis and due to obliterative pulmonary arteriopathy, indistinguishable histologically from that seen in primary pulmonary hypertension, also may occur. While progressive pulmonary fibrosis may lead to death after several years, pulmonary hypertension usually pursues a relentless downhill course over months. After decades of stable, limited SSc or CREST syndrome, pulmonary hypertension may appear and prove lethal. Prostacyclin analogues have been shown to reduce pulmonary artery pressure, relieve dyspnea, and prolong survival, but the need for constant intra-pulmonary infusion is problematic (28,39,40). Lung transplantation with lasting improvement has been reported in a few patients (41).

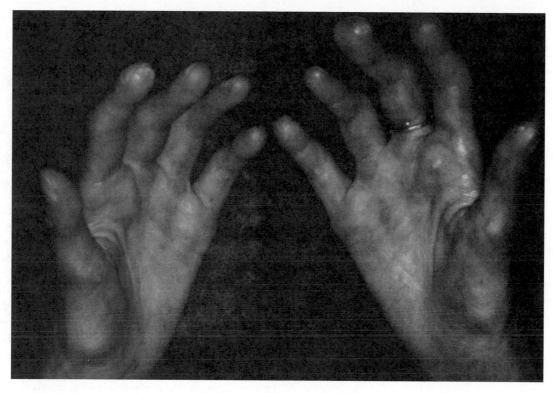

Figure 3. Photograph of the hands of a patient with SSc. The skin is taut and shiny and the interphalangeal joints are contracted. Areas of soft tissue swelling reflect calcium deposition (calcinosis).

7. CARDIAC INVOLVEMENT

Pericardial effusions, usually asymptomatic, are common in patients with SSc. Rarely, pericardial tamponade may prove life-threatening (42,43). Acute pericarditis is rare. Congestive heart failure due to the pulmonary disease (cor pulmonale) is more common than primary myocardial involvement. When the latter occurs due to patchy areas of fibrosis in the myocardium, cardiac arrythmias and sudden death and/or biventricular congestive heart failure may follow.

Since there is no specific treatment for cardiac scleroderma, management depends on pharmacologic suppression of arrythmias and conventional means of treating heart failure. Because of pathological findings in the myocardium of "contraction band necrosis" suggesting reperfusion injury (24) and reversible perfusion abnormalities noted by cardiac scans, suggesting cardiac Raynaud phenomenon (23), the use of vasodilating agents, such as nifedipine or other drugs used for Raynaud phenomenon, should be considered (43).

8. RENAL INVOLVEMENT

Similar arterial obliterative lesions as seen in the digits, pulmonary vasculature and elsewhere causes the so-called "renal crisis" of SSc (44). The typical first symptoms are severe headaches and mental confusion due to malignant systemic hypertension. A microangiopathic hemolytic anemia is usually found due to mechanical destruction of red blood cells in the abnormal renal arteries. The serum creatinine rises rapidly over days to weeks accompanied by oliguria and, ultimately, end stage renal disease ensues. While once nearly invariably fatal, the hypertensive renal crisis can now be treated successfully with ACE- inhibitors, especially if this complication is recognized early. Assiduous control of hypertension is essential and, once accomplished, serum creatinine levels often stabilize or even return to near normal levels.

9. GASTROINTESTINAL (GI) INVOLVEMENT

Dysmotility of the gut is due to loss of smooth muscle, presumably from ischemia (45). Aperistalsis of the distal two-thirds of the esophagus is most common, but loss of motility also may occur in the stomach, small bowel or colon. Gastroesophageal reflux symptoms (GERD) are frequent but are usually responsive to treatment with H-2 blockers or proton pump inhibitors. Lower esophageal stricture and chronic aspiration pneumonitis are now less frequent since the routine use of these drugs. A loss of propulsive ability in the esophagus or other gut segments is often improved by motility promoting agents such as cisapride. Bacterial overgrowth in adynamic small bowel may result in malabsorption which is often amenable to courses of antibiotics. Pseudo-obstruction of the bowel is problematic, and severe motility loss may dictate the chronic use of intravenous parenteral nutrition.

10. OTHER COMPLICATIONS

Hypothyroidism, often occult, from fibrosis of thyroid tissue or, less commonly, Hashimoto's thyroiditis, occurs in approximately 25% of patients and dictates thyroid hormone replacement. Sicca symptoms, dry eyes and/or dry mouth, also are common, and may result from either exocrine glandular fibrosis or cellular infiltration similar to Sjögren's syndrome. Primary biliary cirrhosis occurs more often than expected, usually in patients with the CREST syndrome and Sjogren syndrome (1)

11. LABORATORY ABNORMALITIES

Hematology profiles are usually normal in SSc, although some patients will have a mild anemia of chronic disease, and a brisk microangiopathic hemolytic anemia occurs with renal crisis. Acute phase reactants, such as erythrocyte sedimentation rate (ESR) and C-reactive protein, are typically normal or only modestly elevated. Mild polyclonal hyperglobulinemia is found occasionally. Serum complement levels are normal. Rheumatoid factor activity, usually of low titer, occurs in approximately one-third of patients.

12. ANTINUCLEAR ANTIBODIES

Antinuclear antibodies (ANA) are the serological hallmark of SSc being found in over 90% of patients (1). Immunofluorescent staining patterns of ANA, especially when the HEp-2 cell line is used as antigen substrate, may provide the first clue to more SSc-specific autoantibodies (Table 4). The most common ANA patterns in SSc are speckled, centromere and nucleolar, but rarely mitotic spindle apparatus, centrosome, Golgi bodies or others may be encountered.

A variety of more specific autoantibodies to nuclear antigens, some of which occur only in SSc patients, have been characterized (reviewed in (46)), (47–50) (Table 4). The three most common autoantibody systems accounting for the majority of SSc patients include anti-centromere (ACA), anti-topoisomerase I (or Scl-70), and anti-RNA polymerases (RNAP), most commonly anti-RNAP III.

Anti-U1-RNP antibodies occur in approximately 10% of SSc patients, and a variety of anti-nucleolar specificities account for the rest. It is rare for more than one of these antibodies to be present in any given patient.

There are striking differences in the frequencies of most of these autoantibodies among different ethnic groups (Table 4) (rev. in 46) (51). ACA occurs most commonly in Caucasians of northern and western European descent, and less frequently in southern Europeans, African-Americans, and Asians. On the other hand, anti-topoisomerase I (Scl-70) is much more common in the latter groups, as well as in Native Americans. Anti-PM-Scl is nearly uniquely a specificity of Caucasians, and anti-U3-RNP (fibrillarin) is significantly increased in African- American men compared to other ethnicities and women. Anti-U1-RNP, as in SLE, is increased in African-Americans and Mexican-Americans compared to whites. The reasons for these ethnic differences in autoantibody frequencies probably relate in part to racial differences in MHC allele frequencies (to be discussed); however, this explanation alone is insufficient.

Another remarkable attribute of these autoantibodies, as well as a potential clue to disease pathogenesis, is the specificity of most for SSc and the striking correlations each exhibits for different patterns of clinical expression in terms of patterns of cutaneous and internal organ involvement and long-term prognosis (reviewed in (46)), (47) (Table 4). Patients having ACA typically have limited SSc or CREST syndrome and usually pursue a relatively benign course, although a few develop pulmonary hypertension decades after onset. Similarly, anti-Th/To and anti-U1-RNP positive patients usually have a good prognosis, although myositis requiring corticosteroid therapy often occurs in the latter. On the other hand, anti-topoisomerase 1 is a marker for diffuse skin disease in the majority, as well as pulmonary fibrosis, and anti- RNAP for severe skin and other internal organ involvement with a high likelihood of early mortality. Thus, the concept of autoantibody subsets of SSc has evolved which is clinically useful, although not absolute. There are clearly many patients who demonstrate features and courses distinct from what would be predicted from their autoantibody profile. Moreover, what autoantibody markers tell us about the etiology or pathogenesis of SSc is unclear. Is each autoantibody subgroup a different disease? Do the autoantibodies actually cause some of the clinical manifestations with which they are associated or are they epiphenomena? These questions remain to be answered through further investigations.

13. MOLECULAR PATHOGENESIS

Any concept or hypothesis about the pathogenesis of SSc must take into account and unify three broad areas which

Table 4   Frequent autoantibodies in systemic sclerosis

Autoantibody anti-	ANA pattern	Frequencies in caucasians[+]	SSc specific	Clinical correlations
Centromere (CENP proteins)	Centromere	20%–50%	Yes	Limited SSc CREST Digital ulcers Pulmonary hypertension (late)
Topoisomerase I (Scl-70)	Speckled	20%	Yes	Diffuse SSc Pulmonary fibrosis
RNA polymerase III, II and/or I	Speckled and/or nucleolar*	20%	Yes	Diffuse SSc Renal and cardiac Poor prognosis
U1-RNP	Speckled	10%	No	Limited SSc or MCTD Myositis
U3-RNP (fibrillarin)	Nucleolar	7%	No	Diffuse or limited SSc Pulmonary hypertension Gut involvement Extensive telangiectasia
PM-Scl	Nucleolar	4%	No	Diffuse SSc Myositis, Pulmonary fibrosis Arthritis
Th/To	Nucleolar	4%	Yes	Limited SSc Gut involvement
NOR-90	Nucleolar	<5%	Yes	Unknown
RNA helicase	Nucleolar	Rare	No	Unknown

*RNAP III and II produce speckled and RNAP I nucleolar patterns.
[+] Frequencies may differ significantly in other ethnic groups (see text).

characterize the disease: 1) autoimmunity, 2) endothelial injury and small vessel damage, and 3) widespread fibrosis in connective tissues (17,26,52,53). In addition, etiologic considerations must include either predominantly genetic or environmental contributions, or both (Figure 4) (4,12,13,54). Despite many epidemiological studies, no common environmental exposure leading to SSc has been found. Although there are many instances of toxins which may cause a fibrosing syndrome, none of these examples recapitulates the clinical or autoantibody pattern of SSc (2) (Table 2). Nonetheless, non-apparent and/or ubiquitous exposures to environmental triggers, including known or as yet unidentified microorganisms, remain viable possibilities. Little consideration has been given until recently that genetics might play a major role. Increasingly, evidence is accumulating that SSc, like other diseases considered autoimmune in origin, may represent a complex genetic disease.

13.1. Genetic Factors

SSc appears to cluster in families, albeit infrequently (55–65). Several recent surveys of well-studied cohorts suggest that 1%–2% of cases have one or more affected first-degree relatives with the same disease (66,67). Familial recurrence ratios are 10–20 times what is expected in the population as a whole. Thus, the strongest relative risk factor for SSc identified so far is a positive family history, although the absolute risk for family members is relatively low. The concordance rate for SSc in eight reported pairs of monozygotic twins was 25% (68); however, a formal ongoing survey of disease concordance in both monozygotic and dizygotic twins appears to show no greater disease frequency in the genetically-identical pairs (approximately 6%) (69). The numbers of twins studied, however, remains relatively small, and higher concordance rates for both Raynaud phenomenon and ANA

SCLERODERMA (SYSTEMIC SCLEROSIS)

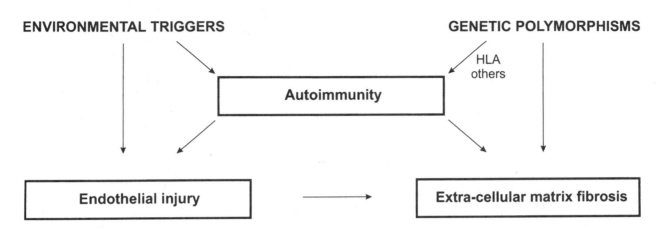

Figure 4.   Schematic representation of potential interacting components in the etiopathogenesis of SSc.

have been found in the monozygotic pairs. Raynaud phenomenon, which occurs in over 90% of SSc cases and is the first symptom in most, shows prevalences of 2%–10% in normal populations (70). Among first-degree relatives of SSc cases, however, 24% have been found to have Raynaud phenomenon compared to only 2% of controls (71). Other autoimmune diseases, including SLE, rheumatoid arthritis, autoimmune thyroid disease, or others occur in approximately 20% of SSc families (67). Positive ANA's have been found in 15–20% of asymptomatic first-degree relatives of SSc patients, although the SSc-specific antibodies, such as ACA or anti- topoisomerase I, have not been detected in relatives (72). Within reported families having two or more affected members with SSc, the same SSc-specific autoantibody tends to recur in the affected members (rev. in 46).

Standard genetic linkage studies in multiplex SSc families have not been reported because of the rarity of the disease and its infrequent familial recurrence. Nonetheless, improvements in genetic mapping strategies, the identification of potential candidate genes from animal models (such as *tsk* mice), molecular studies in SSc tissues, and genetic profiling with cDNA microarrays are now making genetic studies more feasible. There is no question that the major histocompatibility complex (MHC) is one genetic system playing a role in SSc, probably the autoimmune component (discussed below). Other contributing genes are likely, but their definite identification has not been established.

One likely non-MHC candidate genetic region has been localized to human chromosome 15q using linkage disequilibrium mapping and case-control studies in a Native American population with high disease prevalence and a probable founder effect (8). This region contains the fibrillin-1 gene (*FBN1*), mutations of which in humans cause Marfan syndrome (73), and a tandem duplication that is now known to cause the murine *tsk1* fibrotic phenotype (74). It is still unclear whether changes in the *FBN1* gene, or another tightly linked locus, are involved in SSc (8) (discussed later). Preliminary data in this same Native American population suggest possible associations of SSc with the topoisomerase I gene (75), and possibly with several extracellular matrix component genes and/or profibrotic cytokines or their receptors (XD Zhou, FK Tan and FC Arnett: Preliminary data). Polymorphisms in the fibronectin gene recently have been associated with pulmonary fibrosis in SSc patients (76).

13.2.   Associations with Major Histocompatibility Complex (MHC) Alleles

Earlier studies in SSc patients of human leukocyte antigen frequencies (HLA) using serological testing methods showed either no or only weak associations with HLA-class I specificities (46). Stronger, but still relatively modest, correlations were reported for several HLA-class II (DR) types, including DR5 (DR11), DR1, DR3 and DR52 (46). Subsequent investigations using molecular genotyping for specific MHC class II alleles have provided stronger evidence for an effect from MHC genes (77,78). Most strikingly, it is now clear that specific HLA-class II alleles show their strongest associations with specific autoantibodies found in SSc patients (rev. in 46) (Figure 5). Moreover, certain MHC class II amino acid polymorphisms, which are shared among several alleles, appear to be critical in predisposing to each

SCLERODERMA

Genetic - Autoantibody Subsets

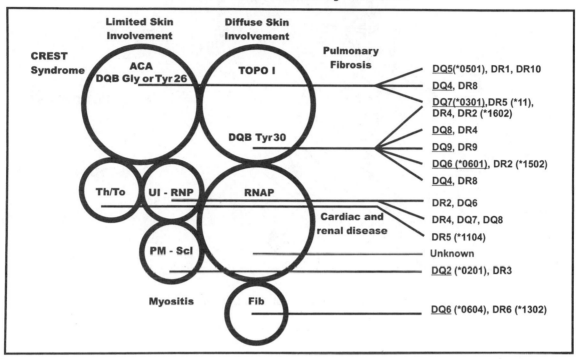

Figure 5. Schematic representation of autoantibody subsets of SSc (circles) showing patterns of clinical expression. Clustered to the right are those autoantibodies associated with diffuse and to the left limited scleroderma. Specific visceral complications are shown adjacent to circles. HLA associations (haplotypes) are shown to the right with lines drawn to each autoantibody subgroup. Revised and reproduced from reference (46), with permission.

specific autoantibody (46). Thus, because normal HLA allelic frequencies may vary considerably among ethnic groups, associations of different HLA alleles with SSc-specific autoantibodies may be found in different populations. The shared class II sequence polymorphism that promotes the autoantibody response, however, is usually found in studies across ethnic lines but expressed in different alleles. In addition, it appears that several different class II loci (DRB1, DQB1, DPB1, etc) may be necessary for mounting a particular autoimmune response. Several examples of these concepts follow below.

Autoantibodies to topoisomerase I in Caucasians and African-Americans are strongly associated with the HLA haplotype *DRB1*11* (DR5 or DR11), *DQB1*0301* (DQ7) (79). Native Americans with this autoantibody instead show an HLA-DR2 haplotype (*DRB1*1602*) that is in linkage disequilibrium with *DQB1*0301* (80). Japanese patients, however, more often have another HLA-DR2 haplotype (*DRB1*1502, DQB1*0601*), but also show a weaker association with HLA-*DQB1*0301* (81) (Figure 5). In studies of the MHC alleles associated with anti-topoisomerase I in various populations, HLA-DQB1 alleles possessing a *tyrosine* residue in position 30 of the

outermost domain appear to be present (such as *DQB1*0301, *0601*, and a few others) (79). This predisposing sequence polymorphism maps to the floor of the peptide binding groove of class II molecules, where it likely determines, by virtue of size and charge, the peptides that can be bound and presented to T cells (? a peptide of topoisomerase I) (Figure 6). In addition, *in vitro* studies of specific T lymphocyte proliferation induced by topoisomerase I in both SSc patients and normal controls is tightly MHC restricted by HLA-DR2, DR5 (DR11) and DR7 alleles, and to a lesser extent by DQ molecules (82). More recently, association studies also have implicated HLA-*DPB1*1301*, and other HLA-DPB1 alleles that possess an *aspartic acid* residue in position 69 as also being important in the anti-topoisomerase I autoimmune response (80,81,83,84). There also is evidence for restricted T cell receptor $\alpha\beta$ usage by autoreactive T cell clones specific for an immunodominant epitope on topoisomerase I (85).

In contrast, anticentromere antibodies (ACA) are strongly associated with HLA-*DQB1*0501*, often in linkage disequilibrium with HLA-DR1 or DR10, as well as several other HLA-DQB1 alleles (*DQB1*0402* linked to

HLA-DQ and Scleroderma Autoantibodies

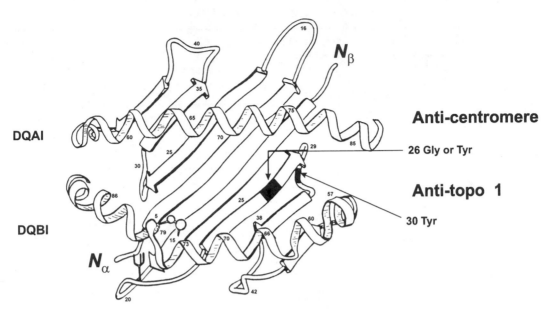

Figure 6. Schematic representation of the outermost domain of an HLA-DQ molecule showing the antigen binding cleft. Numbers represent amino acid positions. Specific amino acids showing the strongest associations with anti-topoisomerase I (Topo I) and anticentromere antibodies are indicated. Reproduced from reference (46), with permission.

DR8, and *DQB1*0301*, linked to DR5 or DR4) (86). All of these ACA-associated DQB1 alleles possess polar uncharged amino acids (*glycine* or *tyrosine*) in position 26 in contradistinction to most other DQB1 alleles that share a *leucine* residue in this position (86) (Figure 6). Whether HLA-DR and/or DP alleles also contribute to ACA is currently unclear.

As shown in Figure 5, most of the SSc-specific autoantibodies have been associated with several different HLA alleles that share certain polymorphisms within the antigen binding groove (46). It should be noted that several of the same HLA alleles are associated with both prognostically "good" and "bad" autoantibodies, i.e. *DQB1*0301* predisposes to anti- topoisomerase I (79–81), a marker for severe SSc, but also to anti-Th/To (87) and weakly to ACA (86), which are markers of more benign SSc. Thus, the autoantibody rather than the HLA allele appears to be more central to how the disease is expressed clinically. Currently, there is no evidence that SSc specific autoantibodies directly cause any of the pathological manifestations of SSc; however, at our current state of knowledge about the pathogenesis of this disease, that possibility requires further exploration. Moreover, the immunologic and immunogenetic data infer that these autoimmune responses are T cell-dependent and autoantigen-driven. Mechanisms underlying the loss of tolerance are not yet understood but several interesting hypothesis, supported by some evidence, are currently being explored.

13.3. Origins of SSc Autoantigens

The specific autoantigens targeted by the immune system in SSc patients provide potentially important clues to pathogenesis. The intracellular functions of autoantigenic molecules, such as topoisomerase I, centromeric proteins (CENPs), RNA polymerases I, II and III, U3-RNP (fibrillarin), and nucleolus organizer regions protein (NOR-90), are quite disparate, as are their structures. Tan has pointed out, however, a unifying attribute of these molecules, namely that they all become enriched in the nucleolus at some time during the cell cycle (88). Casciola-Rosen et al. (89) and Rosen et al. (90), building on observations that metals such as silver, lead, cobalt, zinc, iron, copper and mercury become concentrated in nucleoli, have shown that several of the SSc autoantigens, including topoisomerase I, RNA polymerases and NOR-90, are subject to cleavage and fragmentation at specific sites by metal catalyzed oxidation reactions. Increased production of reactive oxygen species has been previously reported in SSc (90,91). Another autoantigen, the U3-RNP protein, fibrillarin, is not subject to such oxidation reactions but undergoes alteration of disulfide-bonds on exposure to mercury compounds (92). Such changes in autoantigen structure could expose cryptic epitopes leading to T cell activation and an autoimmune response (93). These investigators have proposed that vascular injury leading to perfusion-reperfusion injury and release of reactive oxygen species might be at least one of the critical pathways leading to tissue injury in SSc along with the unique array of autoantibodies seen (89,90). Clearly, this area of investigation needs to be pursued further.

Several additional observations may support a role for metals in the pathogenesis of SSc. Mercuric chloride and silver nitrate injections have been shown to induce anti-U3-RNP (fibrillarin) autoantibodies in certain murine strains bearing a specific H-2 haplotype (94–97). While these animals do not develop a fibrosing syndrome, they do demonstrate abnormal renal and vascular lesions induced by immunological injury.

Urinary mercury levels have been studied in one series of SSc patients (98). Higher mean urinary levels of mercury were found in the SSc patients who were positive for anti-fibrillarin antibodies compared to anti-fibrillarin negative SSc patients and to normal controls, but statistical significance was lost after correction for renal function and body mass. Nonetheless, the results were suggestive of either occult mercury exposure and/or abnormal metabolism of mercury in antifibrillarin positive patients.

13.4. Microchimerism, HLA Compatibility, and Possible Graft-versus Host Mechanisms

SSc affects predominantly women after the childbearing years and in many ways resembles graft-versus-host disease following bone marrow allotransplantation. Recent studies using sensitive PCR techniques have demonstrated retained fetal cells in normal mothers even decades after the birth of a child (microchimerism) (99,100). Moreover, the perinatal transfer of maternal-fetal cellular material appears to be bidirectional, with maternal cells also demonstrable in normal offspring. Thus, the possibility of a graft-vs-host reaction from such retained microchimeric allografts has recently been explored. Nelson et al. reported the PCR detection of male DNA in peripheral blood more frequently and in higher quantities in women with SSc who had borne sons than in normal women having male offspring (101,102). Artlett et al. extended these observations to the skin of women with SSc using a Y-chromosome specific sequence detected by PCR and demonstrated retained male cells using *in situ* hybridization (103). While these studies found significant differences between SSc patients and normal controls, similar investigations of other autoimmune and non-autoimmune diseases to test the specificity of these findings for SSc have not been reported.

The studies by Nelson et al., however, examined HLA relationships between mothers with SSc and their offspring (101). The SSc patients showed significantly more compatibility for paternal HLA-DR alleles than normal mothers and offspring. Although these observations require confirmation, they suggest that HLA class II compatibility between a mother and her offspring predisposes

to later SSc in the mother, possibly because it might allow a higher tolerance to and dosage of retained fetal cells that might later initiate a chronic graft-vs-host reaction. There are, however, significant differences in the pathological findings of SSc and chronic graft-vs-host disease, most notably the absence of Raynaud phenomenon, the typical vascular lesions and SSc specific antibodies in the latter (102). Nonetheless, these novel observations deserve further study.

14. ENDOTHELIAL CELL INJURY

A major pathological hallmark of SSc is progressive and relentless damage to vascular endothelium resulting in widespread obliteration of arterioles and capillaries and intimal proliferation in larger small arteries (17,26,104). These vascular lesions are histopathologically quite different from the inflammatory vasculitis seen in other autoimmune diseases, such as lupus and rheumatoid arthritis. Cellular infiltrates in or around such vessels are only seen occasionally, but activated microvascular pericytes recently have been described (105). Evidence for deposition of typical immunoreactants, such as immunoglobulins and complement, is meager. Nonetheless, it appears likely that endothelial injury occurs as a result of circulating factors, such as autoantibodies, immune complexes, cytokines, proteases or activated immune cells (17). There also is evidence that endothelial cell apoptosis plays an important role (106), perhaps due to anti-endothelial cell binding (107).

For over 20 years it has been documented that SSc sera are selectively cytotoxic to human umbilical vein endothelial cells (26). The best candidate for this cytotoxic factor is granzyme 1, a type IV collagenase, elaborated from cytotoxic T cells (26,108). It has been speculated that granzyme 1 could disrupt the basal lamina and generate fragments eliciting autoimmune responses. Autoantibodies to collagen type IV and the basement membrane adhesion protein, laminin, have been reported in SSc sera but their pathophysiological significance is unclear (26,109). Anti-endothelial cell antibodies could cause vascular damage through antibody-dependent cellular cytotoxicity (109). Cytokines capable of altering endothelial cell function are found in SSc sera and/or tissues, including interleukin-1 (IL 1), IL-2, IL-4, IL-6, IL-8, lymphotoxin, tumor necrosis factor (TNF)α, transforming growth factor (TGF) β, fibroblast growth factor (FGF), and platelet-derived growth factor (PDGF) α and β (17,26,110,111). Platelet activation also is evident with release of mediators such as PDGF, thromboxane A2 and leukotriene B$_4$, which may induce vasoconstriction and stimulate growth of endothelial cells and fibroblasts (26).

That there is perturbation of endothelial cell function in SSc patients is clear (17,26,112,113). Serum levels of tissue plasminogen activator and factor VIII/von Willebrand factor are elevated, and correlate inversely with serum levels of angiotensin-converting enzyme. Endothelin 1, a powerful vasoconstrictor produced by endothelial cells and a fibroblast mitogen, has been reported to be present in high levels in SSc patients (114,115). On the other hand, nitric oxide (NO), a potent endothelial relaxation factor has been found to be low and non-responsive to appropriate physiological stimulation (116–118).

Because of pathological similarities among the vasculopathies seen in SSc, chronic allograft rejection, and coronary artery restenosis after angioplasty, as well as possible links of each to acute or latent infection with cytomegalovirus (CMV) (119), Pandy and LeRoy have recently hypothesized that CMV infection of endothelial cells might initiate or drive the vasculopathy of SSc (120).

15. THE FIBROBLAST AND EXTRA-CELLULAR MATRIX

Fibrosis of the skin, lungs and other tissues in SSc is caused by increased production of extra-cellular matrix (ECM) by activated fibroblasts (26,121–124). A major unanswered question is whether tissue fibrosis is merely a downstream consequence of autoimmune mechanisms and/or ischemia from the vascular lesions, or requires inherent defects in ECM components (17,52,125). As discussed below, recent evidence from both animal models of fibrosis and studies of human SSc fibroblasts suggest that ECM genetic contributions may be necessary for the fibrotic phenotype.

In early SSc cutaneous involvement, mononuclear cells, primarily T lymphocytes, may be found in the deep dermis surrounding small vessels, but disappear with time (26). Mast cells also are present (126). Excessive amounts of collagen, fibrillin-1, fibronectin, osteonectin and glycosaminoglycans are produced by activated fibroblasts and deposited in the deep dermis. The collagen is of several types, including types I, III, V and VI, and is biochemically normal (26). Much of the increased synthesis of ECM proteins, especially collagens, occurs at the level of transcription (127). It appears that SSc fibroblasts are insensitive to feedback inhibition of collagen synthesis, and there is increased intracellular degradation of collagen compared to normal cells. SSc fibroblasts have increased adhesion capacity for collagens I, IV and VI, laminin, and fibronectin and show increased lymphocyte binding due to elevated ICAM-1 levels. Some studies indicate that only a portion of the SSc fibroblast population is activated and responsible for the increased ECM production (128). SSc fibroblasts, however, are not "transformed" since they have the same colony forming efficiencies as control fibroblasts, undergo normal senescence, and lose the activated phenotype after prolonged passage in tissue culture. There also is an increased population of α-actin positive fibroblasts (myofibroblasts) in SSc skin that are important in normal wound healing/contraction and scarring and are induced by the profibrotic cytokine, TGFβ (26).

The mechanism(s) of SSc fibroblast activation is (are) unclear. Autocrine stimulation from cytokines such as TGFβ, PDGF, IL-6 and IL-8 is one proposed mechanism, (110,111,123,129,130), while paracrine activation by endothelial and/or inflammatory cells is another (131–133). TGFβ is definitely a major contributor to tissue fibrosis, and there is abundant evidence that it plays some role in SSc (111,125,134). Immunohistochemical and *in situ* hybridization studies demonstrate increased TGFα1 and TGFβ2 levels in the ECM of affected SSc skin, and upregulation of TGFβ receptors (134). In addition, TGFβ1 has been shown to co-localize to regions of increased collagen expression in the dermis and in the lungs of patients with SSc pulmonary fibrosis, and abnormal TGFβ1 expression may precede the onset of fibrosis (26). Roles for matrix metalloproteinases (MMPs) and tissue inhibitors of metalloproteinases (TIMPs) in this complex and dynamic process of collagen regulation also appear likely (121,124,135)

15.1. ECM Microfibrils and Fibrillin-1

The microfibril is a component of the ECM that, until only recently, has received little attention, despite earlier reports of histopathological abnormalities in microfibrillar structures in SSc skin (136,137). Microfibrils are found alone in connective tissues or are intimately associated with elastin. The most abundant and best characterized microfibrillar protein is fibrillin-1. Mutations in the fibrillin-1 gene (*FBN1*) cause the Marfan syndrome due to disruption of the polymerization of mutant fibrillin-1 into normal microfibrils via a dominant-negative pathogenic effect (73). A possible link between fibrillin-1 and SSc is supported by two recent findings. First, a duplication of the murine *Fbn1* gene causes the tight skin (tsk) 1 phenotype (74), a model for cutaneous fibrosis and possibly for SSc. Second, a genetic haplotype on human chromosome 15 bearing the *FBN1* gene has been linked to SSc in a group of Native Americans with a high disease prevalence (80). Although *FBN1* is not duplicated and a definite genetic abnormality of *FBN1* in humans with SSc has not yet been identified, studies of fibrillin-1 protein in dermal fibrobasts from SSc patients indicate abnormalities in its metabolism. By standard pulse- chase metabolic labeling analyses, fibrillin-1 in synthesized and secreted normally but shows abnormal ECM incorporation in some SSc patients and their unaffected family members (138). Immuno-fluorescence and electron microscopy studies suggest that microfibrils are rapidly degraded in SSc dermal fibroblast cultures (138). The reasons for these *in vitro* findings are as yet unclear. Because fibrillin-1 is known to sequester TGFβ

by virtue of latent transforming growth factor binding proteins (LTBP), it is currently hypothesized that microfibrillar abnormalities, such as those found in *tsk1* mice and human SSc fibroblast cultures, might permit the release of excessive amounts of TGFβ and, thus, promote fibrosis (74,138). It may be notable that the fibrosis which characterizes the *tsk1* mouse still occurs in the absence of T and B lymphocytes (139), thus demonstrating that inherent defects alone, without immune system collaboration, may still lead to a fibrotic phenotype.

Finally, autoantibodies to a recombinant fibrillin-1 protein have recently been detected in the sera of patients with SSc (140), similar to those reported in the tsk1 mouse (141). It is not yet known whether anti-fibrillin-1 autoantibodies play any pathogenetic role in SSc; however, they appear to be relatively specific for scleroderma syndromes, since they were not found often in normal controls or patients with other autoimmune diseases (SLE, rheumatoid arthritis, Sjogren syndrome) (140).

16. CONCLUSIONS

SSc is an extremely complex disease, both clinically and pathogenetically. Recent advances in molecular biology and genetics as applied to SSc are beginning to shed new light on potential mechanisms operating in this disease. Much more research needs to be done in all of the areas discussed above. Only through new knowledge of the pathogenetic pathways leading to the autoimmune, vascular, and ECM abnormalities seen in this disease (Figure 4), are novel therapies and prevention strategies likely to be developed.

References

1. Medsger, Jr., T.A. 1997. Systemic sclerosis (scleroderma): clinical aspects. Arthritis and allied conditions. Edited by W.J. Koopman. Baltimore, Williams and Wilkins, pp. 1433–1464.
2. Silver R.M. 1997. Variant forms of scleroderma. *Arthritis and alliedc conditions*. Edited by W.J. Koopman. Baltimore, Williams and Wilkins, pp. 1465–1480.
3. Masi, A.T., G.P. Rodnan, T.A. Medsger, Jr., R.D. Altman, W.A. D'Angelo, J.F. Fries, E.C. Leroy, A.B. Kirsner, A.H. MacKenzie, D.J. McShane, A.R. Myers, and G.C. Sharp. 1980. Preliminary criteria for the classification of systemic sclerosis (scleroderma). *Arthritis Rheum.* 23:581–590.
4. Mayes, M.D. 1997. Epidemiology of systemic sclerosis and related diseases. *Curr. Opin. Rheumatol.* 9:557–561.
5. Englert, H., J. Small-McMahon, K. Davis, H. O'Connor, P. Chambers, and P. Brooks. 1999. Systemic sclerosis prevalence and mortality in Sydney 1974–88. *Aust. NZ. J. Med.* 29:42–50.
6. Laing, T.J., B.W. Gillespie, M.B. Toth, M.D. Mayes, R.H. Gallavan, Jr., C.J. Burns, J. R. Johanns, B.C. Cooper, B.J. Keroack, M.C. Wasko, J.V. Lacey, Jr. and

D. Schottenfeld. 1997. Racial differences in scleroderma among women in Michigan. *Arthritis Rheum.* 40:734–742.
7. Arnett, F.C., R.F. Howard, F. Tan, J.M. Moulds, W.B. Bias, P.E. Weathers, E. Durban, H.D. Cameron, G. Paxton, T.J. Hodge, and J.D. Reveille. 1996. Increased prevalence of systemic sclerosis in a native American tribe in Oklahoma. *Arthritis Rheum.* 39:1362–1370.
8. Tan, F.K., D.N. Stivers, M.W. Foster, R. Chakraborty, R.F. Howard, D.M. Milewicz, and F.C. Arnett. 1998. Association of microsatellite markers near the fibrillin-1 gene on human chromosome 15q are associated with scleroderma in a Native American population. *Arthritis Rheum.* 41:1729–1737.
9. Harley, J.B., and B.R. Neas. 1998. Oklahoma Choctaw and systemic sclerosis: the founder effect and genetic susceptibility. *Arthritis Rheum.* 41:1725–1728.
10. Silman, A.J., A.J. Hicklin, and C. Black. 1990. Geographical clustering of scleroderma in south and west London. *Brit. J. Rheumatol.* 29:92–96.
11. Valesini, G., A. Litta, M.S. Bonavita, F.L. Luan, M. Purpura, M. Mariani, and F. Balsano. 1993. Geographical clustering of scleroderma in a rural area in the province of Rome. *Clin. Exp. Rheumatol.* 11:41–47.
12. Silman, A.J., and M.C. Hochberg. 1996. Occupational and environmental influences on scleroderma. *Rheum. Dis. Clin. N. Amer.* 22:737–749.
13. Nietert, P.J., S.E. Sutherland, R.M. Silver, J.P. Pandey, R.G. Knapp, D.G. Hoel, and M. Dosemeci. 1998. Is occupational organic solvent exposure a risk factor for scleroderma? *Arthritis Rheum.* 41:1111–1118.
14. Roumm, A.D., and T.A. Medsger. Jr. 1985. Cancer and systemic sclerosis. An epidemiologic study. *Arthritis Rheum.* 28:1336–1340.
15. Lee, P., C. Alderdice, S. Wilkinson, E.C. Keystone, M.B. Urowitz, and D.D. Gladman. 1983. Malignancy in progressive systemic sclerosis—association with breast carcinoma. *J. Rheumatol.* 10:665–666.
16. Bond, C., K.D. Pile, J.D. McNeil, M.J. Ahern, M.D. Smith, L.G. Cleland, and P.J. Roberts-Thompson. 1998. South Australian Scleroderma Register: analysis of deceased patients. *Pathology* 30:386–390.
17. Leroy, E.C. 1996. Systemic sclerosis. A vascular perspective. *Rheum. Dis. Clin. North Am.* 22:675–694.
18. Rodnan, G.P., R.L. Myerowitz, and G.O. Justh. 1980. Morphologic changes in the digital arteries of patients with progressive systemic sclerosis (scleroderma) and Raynaud phenomenon. *Medicine* (Baltimore) 59:393–408.
19. Wigley, F.M. 1996. Raynaud's phenomenon and other features of scleroderma, including pulmonary hypertension. *Curr. Opin. Rheumatol.* 8:561–568.
20. Wigley, F.M., and N.A. Flavahan. 1996. Raynaud's phenomenon. *Rheum. Dis. Clin. North Am.* 22:765–781.
21. D'Angelo, W.A., J.F. Fries, A.T. Masi, and L.E. Shulman. 1969. Pathologic observations in systemic sclerosis (scleroderma). *Am. J. Med.* 46:428–440.
22. Cannon, P.J., M. Hassar, D.B. Case, W.J. Casarella, S.C. Sommers, and E.C. Leroy. 1974. The relationship of hypertension and renal failure in scleroderma (progressive systemic sclerosis) to structural and functional abnormalities of the renal cortical circulation. *Medicine* (Baltimore) 53:1–46.
23. Alexander, E.L., G.S. Firestein, J.L. Weiss, R.R. Heuser, G. Leitl, H.N. Wagner, Jr., J.A. Brinker, A.A. Ciuffo, and L.C. Becker. 1986. Reversible cold-induced abnormalities in myocardial perfusion and function in systemic sclerosis. *Ann. Intern. Med.* 105:661–668.

24. Bulkley, B.H., R.L. Ridolfi, W.R. Salyer, and G.M. Hutchins. 1976. Myocardial lesions of progressive systemic sclerosis. *Circulation* 53:483–490.

25. Maricq, H.R., G. Spencer-Green, and E.C. Leroy. 1976. Skin capillary abnormalities as indicators of organ involvement in scleroderma (systemic sclerosis), Raynaud's syndrome and dermatomyositis. *Am. J. Med.* 61:862–870.

26. Leroy, E.C. 1997. Pathogenesis of systemic sclerosis (scleroderma), Arthritis and allied conditions. Edited by W.J. Koopman Baltimore, Williams and Wilkins, pp. 1481–1490.

27. Pope, J.E., D.T. Shum, R. Gottschalk, A. Stevens, and R. McManus. 1996. Increased pigmentation in scleroderma. *J. Rheumatol.* 23:1912–1916.

28. Kerin, K., and J.H. Yost. 1998. Advances in the diagnosis and management of scleroderma-related vascular complications. *Compr. Ther.* 24:574–581.

29. Clements, P.J., D.E. Furst, W.K. Wong, M. Mayes, B. White, F. Wigley, M.H. Weisman, W. Barr, L.W. Moreland, T.A. Medsger, Jr., V. Steen, R.W. Martin, D. Collier, A. Weinstein, E. Lally, J. Varga, S. Weiner, B. Andrews, M. Abeles, and J.R. Seibold. 1999. High-dose versus low-dose D-penicillamine in early diffuse systemic sclerosis: analysis of a two-year, double-blind, randomized, controlled clinical trial. *Arthritis Rheum.* 42:1194–1203.

30. Pope, J.E. 1996. Treatment of systemic sclerosis. *Rheum. Dis. Clin. North Am.* 22:893–907.

31. Seibold, J.R., P.J. Clements, D.E. Furst, M.D. Mayes, D.A. McCloskey, L.W. Moreland, B. White, F.M. Wigley, S. Rocco, M. Erikson, J.F. Hannigan, M.E. Sanders, and E.P. Amento. 1998. Safety and pharmacokinetics of recombinant human relaxin in systemic sclerosis. *J. Rheumatol.* 25:302–307.

32. Steen, V.D., and T.A. Medsger, Jr. 1997. The palpable tendon friction rub: an important physical examination finding in patients with systemic sclerosis. *Arthritis Rheum.* 40:1146–1151.

33. Olsen, N.J., L.E. King Jr. and J.H. Park. 1996. Muscle abnormalities in scleroderma. *Rheum. Dis. Clin. North Am.* 22:783–796.

34. Minai, O.A., R.A. Dweik, and A.C. Arroliga. 1998. Manifestations of scleroderma pulmonary disease. *Clin. Chest Med.* 19:713–719.

35. Silver, R.M. 1996. Scleroderma. Clinical problems. The lungs. *Rheum. Dis. Clin. North Am.* 22:825–840.

36. Diot, E., E. Boissinot, E. Asquier, J. L. uilmot, E. emarie, C. alat, and P. iot. 1998. Relationship between abnormalities on high-resolution CT and pulmonary function in systemic sclerosis. *Chest* 114:1623–1629.

37. Wells, A.U., D.M. Hansell, M.B. Rubens, A.D. King, D. Cramer, C.M. Black, and R.M. du Bois. 1997. Fibrosing alveolitis in systemic sclerosis: indices of lung function in relation to extent of disease on computed tomography. *Arthritis Rheum.* 40:1229–1236.

38. Domagala-Kulawik, J. 1998. Interstitial lung disease in systemic sclerosis: comparison of BALF lymphocyte phenotype and DLCO impairment. *Respir. Med.* 92:1295–1301.

39. McLaughlin, V.V., and S. Rich. 1998. Pulmonary hypertension—advances in medical and surgical interventions. *J. Heart Lung Transplant.* 17:739–743.

40. Menon, N., L. McAlpine, A.J. Peacock, and R. Madhok. 1998. The acute effects of prostacyclin on pulmonary hemodynamics in patients with pulmonary hypertension secondary to systemic sclerosis. *Arthritis Rheum.* 41:466–469.

41. Pigula, F.A., B.P. Griffith, M.A. Zenati, J.H. Dauber, S.A. Yousem, and R.J. Keenan. 1997. Lung transplantation for respiratory failure resulting from systemic disease. *Ann. Thorac. Surg.* 64:1630–1634.

42. Thompson, A.E., and J.E. Pope. 1998. A study of the frequency of pericardial and pleural effusions in scleroderma. *Br. J. Rheumatol.* 37:1320–1323.

43. Deswal, A., and W.P. Follansbee. 1996. Cardiac involvement in scleroderma. *Rheum. Dis. Clin. North Am.* 22:841–860.

44. Steen, V.D. 1996. Scleroderma renal crisis. *Rheum. Dis. Clin. North Am.* 22:861–878.

45. Rose, S., M.A. Young, and J.C. Reynolds. 1998. Gastrointestinal manifestations of scleroderma. *Gastroenterol. Clin. North Am.* 27:563–594.

46. Arnett, F.C. 1995. HLA and autoimmunity in scleroderma (systemic sclerosis). *Internat. Rev. Immunol.* 12:107–128.

47. Okano, Y. 1996. Antinuclear antibody in systemic sclerosis (scleroderma). *Rheum. Dis. Clin. North Am.* 22:709–735.

48. Amento, E.P. 1998. Immunologic abnormalities in scleroderma. *Semin. Cutan. Med. Surg.* 17:18–21.

49. Kuwana, M., Y. Okano, J. Kaburaki, T.A. Medsger Jr. and T.M. Wright. 1999. Autoantibodies to RNA polymerases recognize multiple subunits and demonstrate cross-reactivity with RNA polymerase complexes. *Arthritis Rheum.* 42:275–284.

50. Arnett, F.C., J.D. Reveille, and B.C. Valdez. 1997. Autoantibodies to a nucleolar RNA helicase protein in patients with connective tissue diseases. *Arthritis Rheum.* 40:1487–1492.

51. Kuwana, M., J. Kaburaki, F.C. Arnett, R.F. Howard, T.A. Medsger, Jr. and T.M. Wright. 1999. Influence of ethnic background on clinical and serologic features in patients with systemic sclerosis and anti-DNA topoisomerase I antibody. *Arthritis Rheum.* 42:465–474.

52. White, B. 1996. Immunopathogenesis of systemic sclerosis. *Rheum. Dis. Clin. North Am.* 22:695–708.

53. Jimenez, S.A. 1996. Pathogenesis of scleroderma. Collagen. *Rheum. Dis. Clin. North Am.* 22:647–674.

54. Burns, C.J., T.J. Laing, B.W. Gillespie, S.G. Heeringa, K.H. Alcser, M.D. Mayes, M.C. Wasko, B. C. Cooper, D.H. Garabrant, and D. Schottenfeld. 1996. The epidemiology of scleroderma among women: assessment of risk from exposure to silicone and silica. *J. Rheumatol.* 23:1904–1911.

55. Sheldon, W.B., D.P. Lurie, H.R. Maricq, M.B. Kahaleh, F.A. DeLustro, A. Gibofsky, and E.C.Leroy. 1981. Three siblings with scleroderma (systemic sclerosis) and two with Raynaud's phenomenon from a single kindred. *Arthritis Rheum.* 24:668–676.

56. McGregor, A.R., A. Watson, E. Yunis, J.P. Pandey, K. Takehara, J.T. Tidwell, A. Ruggieri, R.M. Silver, E.C. Leroy, and H.R. Maricq. 1988. Familial clustering of scleroderma spectrum disease. *Am. J. Med.* 84:1023–1032.

57. Arnett, F.C., W.B. Bias, R.H. McLean, M. Engel, M. Duvic, R. Goldstein, L. Freni-Titulaer, T.W. McKinley, and M.C. Hochberg. 1990. Connective tissue disease in southeast Georgia. A community based study of immunogenetic markers and autoantibodies. *J. Rheumatol.* 17:1029–1035.

58. Whyte, J., C. Artlett, G. Harvey, C.O. Stephens, K. Welsh, C. Black, P.J. Maddison, and N.J. McHugh. 1994. United Kingdom systemic sclerosis study group. HLA-DQB1

associations with anti-topoisomerase-I antibodies in patients with systemic sclerosis and their first degree relatives. *J. Autoimmunity* 7:509–520.

59. McColl, G.J., and R.R.C. Buchanan. 1994. Familial CREST syndrome. *J. Rheumatol.* 21:754–756.

60. Flores, R.H., M.B. Stevens, and F.C. Arnett. 1984. Familial occurrence of progressive systemic sclerosis and systemic lupus erythematosus. *J. Rheumatol.* 11:321–323.

61. Pereira, S., C. Black, K. Welsh, B. Ansell, M. Jayson, P. Maddison, and N. Rowell. 1987. Autoantibodies and immunogenetics in 30 patients with systemic sclerosis and their families. *J. Rheumatol.* 14:760–765.

62. Maddison, P.J., C. Stephens, D. Briggs, K.I. Welsh, G. Harvey, J. Whyte, N. McHugh, and C.M. Black. 1993. The United Kingdom Systemic Sclerosis Study Group. Connective tissue disease and autoantibodies in the kindreds of 63 patients with systemic sclerosis. *Medicine* 72:103–112.

63. Hietarinta, M., S. Koskimites, O. Lassila, E. Soppi, and A. Toivanen. 1993. Familial scleroderma: HLA antigens and autoantibodies. *Brit. J. Rheumatol.* 32:336–338.

64. Manolios, N., H. Dunckley, T. Chivers, P. Brooks, and H. Englert. 1995. Immunogenetic analysis of 5 families with multicase occurrence of scleroderma and/or related variants. *J. Rheumatol.* 22.85–92.

65. Stephens, C.O., D.C. Briggs, J. Whyte, C.M. Artlett, A.B. Scherbakov, N. Olsen, N.G. Gusseva, N.J. McHugh, P.J. Maddison, and K.I. Welsh. 1994. Familial scleroderma—evidence for environmental versus genetic trigger. *Br. J. Rheumatol.* 33:1131–1135.

66. Englert, H., J. Small-McMahon, P. Chambers, H. O'Connor, K. Davis, N. Manolios, R. White, G. Dracos, and P. Brooks. 1999. Familial risk estimation in systemic sclerosis. *Aust. NZ. J. Med.* 29:36–41.

67. Aguilar, M.B., M. Cho, J.D. Reveille, M.D. Mayes, and F.C. Arnett. 1999. Prevalences of familial systemic sclerosis and other autoimmune diseases in three U.S. cohorts. *Arthritis & Rheum.* (Abstract)

68. Cook, N.J., A.J. Silman, J. Propert, and M.D. Cawley. 1993. Features of systemic sclerosis (Scleroderma) in an identical twin pair. *Brit. J. Rheumatol.* 32:926–928.

69. Feghali, C.A., and T.M. Wright. 1995. Epidemiologic and clinical study of twins with scleroderma. *Arthritis Rheum.* 38:S308, (Abstract)

70. Maricq, H.R., P.H. Carpentier, M.C. Weinrich, J.E. Keil, Y. Palesch, C. Biro, M. Vionnet-Fuasset, M. Jiguet, and I. Valter. 1997. Geographic variation in the prevalence of Raynaud's phenomenon: a 5 region comparison. *J. Rheumatol.* 24:879–889.

71. Freedman, R.R., and M.D. Mayes. 1996. Familial aggregation of primary Raynaud's disease. *Arthritis Rheum.* 39:1189–1191.

72. Maddison, P.J., R.P. Skinner, R.S. Periera, C.M. Black, B.M. Ansell, M.I.V. Jayson, N.R. Rowell, and K.I. Welsh. 1986. Antinuclear antibodies in the relatives and spouses of patients with systemic sclerosis. *Annals Rheum. Dis.* 45:793–799.

73. Tsipouras, P., R. DelMastro, M. Sarfarazi, B. Lee, E. Vitale, A.H. Child, M. Godfrey, R.B. Devereux, D. Hewett, B. Steinmann, D. Viljoen, B.C. Sykes, M. Kilpatrick, and F. Ramirez. 1992. Genetic linkage of the Marfan syndrome, ectopia lentis, and congenital contractural arachnodactyly to the fibrillin genes on chromosomes 15 and 5. *N. Engl. J. Med.* 326:905–909.

74. Siracusa, L.D., R. McGrath, Q. Ma, J.J. Moskow, J. Mane, P.J. Christner, A.M. Buchberg, and S.A. Jimenez. 1996. A tandem duplication within the fibrillin 1 gene is associated with the mouse tight skin mutation. *Genome Res.* 6:300–313.

75. Pham, H.T., F.K. Tan, D.N. Stivers, R. Chakraborty, and F.C. Arnett. 1998. Microsatellite polymorphisms near the topoisomerase I gene on chromosome 20 in Choctaw native Americans with scleroderma. *Arthritis & Rheum.* 41:S242(Abstract).

76. Avila, J.J., P.A. Lympany, P. Pantelidis, K.I. Welsh, C.M. Black, and R.M. du Bois. 1999. Fibronectin gene polymorphisms associated with fibrosing alveolitis in systemic sclerosis. *Am. J. Respir. Cell Mol. Biol.* 20:106–112.

77. Briggs, D., C. Stephens, R. Vaughan, K. Welsh, and C. Black. 1993. A molecular and serologic analysis of the major histocompatibility complex and complement component C4 in systemic sclerosis. *Arthritis Rheum.* 36:943–954.

78. Satoh, M., M. Akizuki, M. Kuwana, T. Mimori, H. Yamagata, S. Yoshida, M. Homma, T. Yamamoto, and T. Sasazuki. 1994. Genetic and immunological differences between Japanese patients with diffuse scleroderma and limited scleroderma. *J. Rheumatol.* 21.111–114.

79. Reveille, J.D., E. Durban, M.J. MacLeod, R. Goldstein, R. Moreda, R.D. Altman, and F.C. Arnett. 1992. Association of amino acid sequences in the HLA-DQB1 first domain with the anti-topoisomerase I autoantibody response in scleroderma (progressive systemic sclerosis). *J. Clin. Invest.* 90:973–980.

80. Tan, F.K., D.N. Stivers, F.C. Arnett, R. Chakraborty, R. Howard, and J.D. Reveille. 1999. HLA haplotypes and microsatellite polymorphisms in and around the major histocompatibility complex region in a native American population with a high prevalence of scleroderma (systemic sclerosis). *Tissue Antigens* 53:74–80.

81. Kuwana, M., J. Kaburaki, Y. Okano, H. Inoko, and K. Tsuji. 1993. The HLA-DR and DQ genes control the autoimmune response to DNA topoisomerase I in systemic sclerosis (scleroderma). *J. Clin. Invest.* 92:1296–1301.

82. Kuwana, M., T.A. Medsger, Jr. and T.M. Wright. 1995. T cell proliferative response induced by DNA topoisomerase I in patients with systemic sclerosis and healthy donors. *J. Clin. Invest.* 96:586–596.

83. Reveille, J.D., J. Brady, M. MacLeod-St.Clair, and E. Durban. 1992. HLA-DPB1 alleles and autoantibody subsets in systemic lupus erythematosus, Sjogren's syndrome and progressive systemic sclerosis: A question of disease relevance. *Tissue Antigens* 40:45–48.

84. Rihs, H.P., K. Conrad, J. Mehlhorn, K. May-Taube, B. Welticke, K.H. Frank, and X. Baur. 1996. Molecular analysis of HLA-DPB1 alleles in idiopathic systemic sclerosis patients and uranium miners with systemic sclerosis. *Int. Arch. Allergy Immunol.* 109:216–222.

85. Kuwana, M., T.A. Medsger, Jr. and T.M. Wright. 1997. Highly restricted TCR-alpha beta usage by autoreactive human T cell clones specific for DNA topoisomerase I: recognition of an immunodominant epitope. *J. Immunol.* 158:485–491.

86. Reveille, J.D., D. Owerbach, R. Goldstein, R. Moreda, R.A. Isern, and F.C. Arnett. 1992. Association of polar amino acids at position 26 of the HLA-DQB1 first domain with the anticentromere autoantibody response in systemic sclerosis (scleroderma). *J. Clin. Invest.* 89:1208–1213.

87. Falkner, D., J. Wilson, T.A. Medsger, Jr. and P.A. Morel. 1998. HLA and clinical associations in systemic sclerosis

patients with anti-Th/To antibodies. *Arthritis Rheum.* 41:74–80.

88. Tan, E.M. 1991. Autoantibodies in pathology and cell biology. *Cell* 67:841–842.

89. Casciola-Rosen, L., F. Wigley, and A. Rosen. 1997. Scleroderma autoantigens are uniquely fragmented by metal-catalyzed oxidation reactions: implications for pathogenesis. *J. Exp. Med.* 185:71–79.

90. Rosen, A., L. Casciola-Rosen, and F. Wigley. 1997. Role of metal catalyzed oxidation reactions in the early pathogenesis of scleroderma. *Current. Opin. Rheumatol.* 9:538–543.

91. Stein, C.M., S.B. Tanner, J.A. Awad, L.J. Roberts, and J.D. Morrow. 1996. Evidence of free radical-mediated injury (isoprostane overproduction) in scleroderma. *Arthritis Rheum.* 39:1146–1150.

92. Pollard, K.M., D.K. Lee, C.A.Casiano, M. Bluthner, M.M. Johnston, and E.M. Tan. 1997. The autoimmunity-inducing xenobiotic mercury interacts with the autoantigen fibrillarin and modifies its molecular and antigenic properties. *J. Immunol.* 158:3521–3528.

93. Lanzavecchia, A. 1995. How can cryptic epitopes trigger autoimmunity? *J. Exp. Med.* 181:1945–1948.

94. Reuter, R., G. Tessars, H.W. Vohr, E. Gleichmann, and R. Luhrmann. 1989. Mercuric chloride induces autoantibodies against U3 small nuclear ribonucleoprotein in susceptible mice. *Proc. Natl. Acad. Sci. USA* 86:237–241.

95. Hultman, P., S. Enestrom, K.M. Pollard, and E.M. Tan. 1989. Anti- fibrillarin autoantibodies in mercury-treated mice. *Clin. Exp. Immunol.* 78:470–472.

96. Hultman, P., L.J. Bell, S. Enestrom, and K.M. Pollard. 1992. Murine susceptibility to mercury I. autoantibody profiles and systemic immune deposits in inbred, congenic, and intra-H-2 recombinant strains. *Clin. Immunol. and Immunopath.* 65:98–109.

97. Hultman, P., S. Enestrom, S.J. Turley, and K.M. Pollard. 1994. Selective induction of anti-fibrillarin autoantibodies by silver nitrate in mice. *Clin. Exp. Immunol.* 96:285–291.

98. Arnett, F.C., M.J. Fritzler, C. Ahn, and A. Holian. 1999. Urinary mercury levels in patients with autoantibodies to U3-RNP (fibrillarin). *J. Rheumatol.* (In Press)

99. Nelson, J.L. 1998. Pregnancy immunology and autoimmune disease. *J. Reprod. Med.* 43:335–340.

100. Evans, P.C., N. Lambert, S. Maloney, D.E. Furst, J.M. Moore, and J.L. Nelson. 1999. Long-term fetal microchimerism in peripheral blood mononuclear cell subsets in healthy women and women with scleroderma. *Blood* 93:2033–2037.

101. Nelson, J.L., D.E. Furst, S. Maloney, T. Gooley, P.C. Evans, A. Smith, M.A. Bean, C. Ober, and D.W. Bianchi. 1998. Microchimerism and HLA- compatible relationships of pregnancy in scleroderma. *Lancet* 351:559–562.

102. Nelson, J.L. 1998. Microchimerism and the pathogenesis of systemic sclerosis. *Curr. Opin. Rheumatol.* 10:564–571.

103. Artlett, C.M., J.B. Smith, and S.A. Jimenez. 1998. Identification of fetal DNA and cells in skin lesions from women with systemic sclerosis. *N. Engl. J. Med.* 338:1186–1191.

104. von Bierbrauer, A. 1998. Electron microscopy and capillaroscopically guided nailfold biopsy in connective tissue diseases: detection of ultrastructural changes of the microcirculatory vessels. *Br. J. Rheumatol.* 37:1272–1278.

105. Rajkumar, V.S., C. Sundberg, D.J. Abraham, K. Rubin, and C.M. Black. 1999. Activation of microvascular pericytes in autoimmune Raynaud's phenomenon and systemic sclerosis. *Arthritis Rheum.* 42:930–941.

106. Sgonc, R., M.S. Gruschwitz, H. Dietrich, H. Recheis, M.E. Gershwin, and G. Wick. 1996. Endothelial cell apoptosis is a primary pathogenetic event underlying skin lesions in avian and human scleroderma. *J. Clin. Invest.* 98:785–792.

107. Bordron, A., M.Dueymes, Y. Levy, C. Jamin, J.P. Leroy, J.C. Piette, Y. Shoenfeld, and P.Y. Youinou. 1998. The binding of some human antiendothelial cell antibodies induces endothelial cell apoptosis. *J. Clin. Invest.* 101:2029–2035.

108. Kahaleh, M.B., and Y.F. Fan. 1997. Mechanism of serum-mediated endothelial injury in scleroderma: identification of a granular enzyme in scleroderma skin and sera. *Clin. Immunol. Immunopathol.* 83:32–40.

109. Renaudineau, Y., R. Revelen, Y. Levy, K. Salojin, B. Gilburg, Y. Shoenfeld, and P. Youinou. 1999. Anti-endothelial cell antibodies in systemic sclerosis. *Clin. Diagn. Lab. Immunol.* 6:156–160.

110. Gruschwitz, M.S., M. Albrecht, G. Vieth, and U.F. Haustein. 1997. In situ expression and serum levels of tumor necrosis factor-alpha receptors in patients with early stages of systemic sclerosis. *J. Rheumatol.* 24:1936–1943.

111. Cotton, S.A., A.L. Herrick, M.I. Jayson, and A.J. Freemont. 1998. TGF beta—a role in systemic sclerosis? *J. Pathol.* 184:4–6.

112. Stratton, R.J., J.G. Coghlan, J.D. Pearson, A. Burns, P. Sweny, D.J. Abraham, and C.M. Black. 1998. Different patterns of endothelial cell activation in renal and pulmonary vascular disease in scleroderma. *QJM* 91:561–566.

113. Cailes, J., S. Winter, R.M. du Bois, and T.W. Evans. 1998. Defective endothelially mediated pulmonary vasodilation in systemic sclerosis. *Chest* 114:178–184.

114. Abraham, D.J., R. Vancheeswaran, M.R. Dashwood, V.S. Rajkumar, P. Pantelidis, S.W. Xu, R.M. du Bois, and C.M. Black 1997. Increased levels of endothelin-1 and differential endothelin type A and B receptor expression in scleroderma-associated fibrotic lung disease. *Am. J. Pathol.* 151:831–841.

115. Kazzam, E., A. Waldenstrom, T. Hedner, J. Hedner, and K. Caidahl. 1997. Endothelin may be pathogenic in systemic sclerosis of the heart. *Int. J. Cardiol.* 60:31–39.

116. Yamamoto, T., I. Katayama, and K. Nishioka. 1998. Nitric oxide production and inducible nitric oxide synthase expression in systemic sclerosis. *J. Rheumatol.* 25:314–317.

117. Kharitonov, S.A., J.B. Cailes, C.M. Black, R.M. du Bois, and P.J. Barnes. 1997. Decreased nitric oxide in the exhaled air of patients with systemic sclerosis with pulmonary hypertension. *Thorax* 52:1051–1055.

118. Rolla, G., and A. Calin. 1998. Nitric oxide in systemic sclerosis lung: controversies and expectations. *Clin. Exp. Rheumatol.* 16:522–524.

119. Neidhart, M., S. Kuchen, O. Distler, P. Bruhlmann, B.A. Michel, R.E. Gay, and S. Gay. 1999. Increased serum levels of antibodies against human cytomegalovirus and prevalence of autoantibodies in systemic sclerosis. *Arthritis Rheum.* 42:389–392.

120. Pandey, J.P., and E.C. Leroy. 1998. Human cytomegalovirus and the vasculopathies of autoimmune diseases (especially scleroderma), allograft rejection, and coronary restenosis. *Arthritis Rheum.* 41:10–15.

121. Kuroda, K., and H. Shinkai. 1997. Gene expression of types I and III collagen, decorin, matrix metalloproteinases and tissue inhibitors of metalloproteinases in skin fibroblasts from patients with systemic sclerosis. *Arch. Dermatol. Res.* 289:567–572.

122. Shi-wen, X., C.P. Denton, A. McWhirter, G. Bou-Gharios, D.J. Abraham, R.M. du Bois, and C.M. Black. 1997.

Scleroderma lung fibroblasts exhibit elevated and dysregulated type I collagen biosynthesis. *Arthritis Rheum.* 40:1237–1244.

123. Ichiki, Y., E.A. Smith, E.C. Leroy, and M. Trojanowska. 1997. Basic fibroblast growth factor inhibits basal and transforming growth factor-beta induced collagen alpha 2(I) gene expression in scleroderma and normal fibroblasts. *J. Rheumatol.* 24:90–95.

124. Kikuchi, K., T. Kadono, M. Fusconi, and K. Tamaki. 1997. Tissue inhibitor of metalloproteinase 1 (TIMP-1) may be an autocrine growth factor in scleroderma fibroblasts. *J. Invest. Dermatol.* 108:281–284.

125. Strehlow, D., and J.H. Korn. 1998. Biology of the scleroderma fibroblast. *Curr. Opin. Rheumatol.* 10:572–578.

126. Akimoto, S., O. Ishikawa, Y. Igarashi, M. Kurosawa, and Y. Miyachi. 1998. Dermal mast cells in scleroderma: their skin density, tryptase/chymase phenotypes and degranulation. *Br. J. Dermatol.* 138:399–406.

127. Hitraya, E.G., J. Varga, C.M. Artlett, and S.A. Jimenez. 1998. Identification of elements in the promoter region of the alpha1(I) procollagen gene involved in its up-regulated expression in systemic sclerosis. *Arthritis Rheum.* 41:2048–2058.

128. Jelaska, A., M. Arakawa, G. Broketa, and J.H. Korn. 1996. Heterogeneity of collagen synthesis in normal and systemic sclerosis skin fibroblasts. Increased proportion of high collagen-producing cells in systemic sclerosis fibroblasts. *Arthritis Rheum.* 39:1338–1346.

129. Kadono, T., K. Kikuchi, H. Ihn, K. Takehara, and K. Tamaki. 1998. Increased production of interleukin 6 and interleukin 8 in scleroderma fibroblasts. *J. Rheumatol.* 25:296–301.

130. Zheng, X.Y., J.Z. Zhang, P. Tu, and S.Q. Ma. 1998. Expression of platelet-derived growth factor B-chain and platelet-derived growth factor beta- receptor in fibroblasts of scleroderma. *J Dermatol. Sci.* 18:90–97.

131. Denton, C.P., S. Xu, C.M. Black, and J.D. Pearson. 1997. Scleroderma fibroblasts show increased responsiveness to endothelial cell-derived IL-1 and bFGF. *J. Invest. Dermatol.* 108:269–274.

132. Xu, S., C.P. Denton, A. Holmes, M.R. Dashwood, D.J. Abraham, and C.M. Black. 1998. Endothelins: effect on matrix biosynthesis and proliferation in normal and scleroderma fibroblasts. *J. Cardiovasc. Pharmacol.* 31 Suppl 1:S360–S363.

133. Xu, S.W., C.P. Denton, M.R. Dashwood, D.J. Abraham, and C.M. Black. 1998. Endothelin-1 regulation of intercellular adhesion molecule-1 expression in normal and sclerodermal fibroblasts. *J. Cardiovasc. Pharmacol.* 31 Suppl 1:S545–S547.

134. Kawakami, T., H. Ihn, W. Xu, E. Smith, C. LeRoy, and M. Trojanowska. 1998. Increased expression of TGF-beta receptors by scleroderma fibroblasts: evidence for contribution of autocrine TGF-beta signaling to scleroderma phenotype. *J. Invest. Dermatol.* 110:47–51.

135. Mattila, L., K. Airola, M. Ahonen, M. Hietarinta, C. Black, U. Saarialho-Kere, and V.M. Kahari. 1998. Activation of tissue inhibitor of metalloproteinases-3 (TIMP-3) mRNA expression in scleroderma skin fibroblasts. *J. Invest. Dermatol.* 110:416–421.

136. Fleischmajer, R. 1964. The collagen in scleroderma. *Arch. Dermatol.* 89:437–441.

137. Fleischmajer, R., L. Jacobs, E. Schwartz, and L.Y. Sakai. 1991. Extracellular microfibrils are increased in localized and systemic scleroderma skin. *Lab. Invest.* 64:791–798.

138. Wallis, D.D., F.K. Tan, C.M. Kielty, F.C. Arnett, and D.M. Milewicz. 1999. Abnormalities in fibrillin-1 containing microfibrils in fibroblast cultures from systemic sclerosis (scleroderma) patients. *Arthritis & Rheum.* (In Press).

139. Siracusa, L.D., R. McGrath, J.K. Fisher, and S.A. Jiminez. 1998. The mouse tight skin (Tsk) phenotype is not dependent on the presence of mature T and B lymphocytes. *Mammalian Genome.* 9:907–909.

140. Tan, F.K., F.C. Arnett, S. Stephan, S. Sinichiro, A. Mirachi, S. Takeshi, O. Shoichi, C. LeRoy, A.E. Postlewaite, and C. Bona. 1999. Autoantibodies to the extra-cellular matrix microfibrillar protein, fibrillin-1 in patients with scleroderma and other connective tissue diseases. *J. Immunol.* 163:1066–1072.

141. Murai, C., S. Saito, K.N. Kasturi, and C.A. Bona. 1998. Spontaneous occurrence of anti-fibrillin-1 autoantibodies in tight-skin mice. *Autoimmunity* 28:151–155.

27 | Experimental Models for Scleroderma

Kuppuswamy N. Kasturi and Constantin A. Bona

1. INTRODUCTION

Scleroderma is a polymorphic connective tissue disease comprising diffuse systemic sclerosis, limited scleroderma, linear scleroderma, morphea and CREST syndrome. These clinically well defined but different entities share three common features such as vascular abnormalities, tissue fibrosis and autoimmunity. Though the etiology is not known, it is believed that genetic and environmental factors play a role in the pathogenesis of scleroderma. In humans, it is hard to unravel the underlying cause of scleroderma except in the cases following treatment with bleomycin. Therefore, studies in animal models are important for better understanding of the role of genetic and environmental factors in the development of this disease. Currently available animal models can be grouped into two major categories: 1) inducible and, 2) genetic.

2. INDUCIBLE ANIMAL MODELS OF SCLERODERMA

2.1. Glycosaminoglycan-Induced Scleroderma

Ishikawa et al. (1) reported that glycosaminoglycans (GAG), purified from urine of scleroderma patients and injected into mice, induced scleroderma-like histopathology. Treatment with hyaluronidase or chondroitin sulfatase ABC did not abrogate the property of GAG to induce histopathological changes. It appears that heparin sulfate moiety of it may be responsible for this activity. Mice injected with GAG develop: a) Significant fibrosis, fragmentation of elastic fibers and irregularity in the arrangement of collagen fibers without affecting its binding pattern. Some collagen fibers are of 300–400 A° diameter, intermingled with 650–1000 A° collagen fibers characteristic of the fibrillogenesis process. These collagen fibers are associated with ruthenium red-stained filamentous particles composed of acidic mucopolysaccharide; b) alterations in lungs characterized by perivascular fibrosis, interstitial edema and cellular infiltration, and c) in the esophagus, edematous changes in the mucosal layer, muscular atrophy and fibrosis are characteristic.

It is not known whether these mice develop autoimmunity or not. However, Fox et al. (2) couldn't reproduce these results in mice injected with GAG. Therefore, this is not a well-established reproducible model.

2.2. Bleomycin-Induced Sclerodema

When bleomycin, obtained from Streptomyces venticillus, was given as anti- tumor drug for the treatment of malignant lymphomas and squamous cell carcinomas, some patients developed scleroderma (3,4) with no associated Raynaud phenomenon or internal organ involvement. These observations lead investigators to find a suitable experimental animal that develops scleroderma following injection of bleomycin. Intraperitoneal injection of bleomycin into mice (5) or intratracheal administration into rabbits (6), rats (7), hamsters (8), or baboons (9) induced severe inflammatory and fibrotic alterations in lungs, but not in the skin. The initial site of injury appears to be intima of arteries and veins. The endothelial cells showed edematous degeneracy and detachment from basal cell membrane. Late events are characterized by necrosis of alveolar epithelial cells, fibrinous exudation and fibrosis. The immune system doesn't seem to be involved since similar results were obtained in athymic mice injected with bleomycin (10). This mouse model may, therefore, be useful for studying pulmonary fibrosis.

2.3.   Graft-versus-host Disease

Graft-versus-host disease (GVH) results from activation of T lymphocytes (of donor origin), subsequent to recognition of host tissue alloantigens. In addition to histopathological and immunological alterations, GVH reaction induces scleroderma. Scleroderma is a major therapeutic complication in bone marrow transplant patients. Thus, animal models of GVH reaction were used for studying the pathogenesis of scleroderma. Statsny et al. (11) showed that rats surviving acute phase of GVH developed dermatitis with pronounced cellular infiltration and subdermal fibrosis. Jaffe and Claman (12) reported similar results in Balb/c mice injected with lymphocytes from B10D2 mice, which differ from Balb/c in minor histocompatibility antigens. The alterations included loss of fat tissue, thickening of dermis with condensation of collagen, mononuclear cell infiltration in deep dermis, loss of hair follicles and sebaceous glands. Occasionally these mice also developed anti-nucleolar antibodies(ANA). Deng et al. (13) found that irradiated F1 mice injected with parental DBA/2 cells manifested marked sclerosis and produced anti-skin autoantibodies. This model could be useful for studying the alterations in the cytokine network that is responsible for skin sclerosis since reactive allogenic T lymphocytes are polyclonally activated in this instance.

2.4.   Mutagen-Induced Scleroderma Mouse Model (Tsk-2 Mouse)

Development of scleroderma-like syndrome was described in the male mice of 101/H strain following injection of ethyl nitrosurea, a mutagenic agent (14). This disease is caused by a mutation on chromosome 1 and is inheritable as an autosomal dominant trait (15). Homozygotes die *in utero* and Tsk-2 mice develop skin tightness 10–14 days after birth. The skin thickness is not so marked (800 μm in Tsk-2 mice compared to 500 μm in a normal mouse) but the skin showed excessive deposition of large collagen fibers extending into subdermal adipose tissue and panniculus carnosus. Increased skin collagen content correlates well with the level of α1 collagen transcripts and the content of hydroxyproline in the affected skin. These findings indicate that Tsk-2 mice develop mild cutaneous hyperplasia. Analysis of the expression of Vβ gene families of skin lymphocytes did not show any significant abnormality except that three of the eight TCR Vβ gene families Vβ8.1, Vβ11 and Vβ18 were predominantly used (16). The skewed usage of Vβ gene families suggests that T cells that

recognized limited number of peptides derived from unknown (auto) antigens have been expanded. No further information is available on the specificity of these lymphocytes.

3.   SPONTANEOUSLY OCCURRING MUTANTS DEVELOPING SCLERODERMA-LIKE SYNDROME

Certain strains of experimental animals develop spontaneously scleroderma- like syndrome due to genetic mutations. Currently, two such models are widely used in experimental studies of scleroderma.

3.1.   Avian Model for Scleroderma

In 1942, Bernier described chickens that exhibited inflammatory reactions and fibrosis similar to those described in scleroderma. These chickens were bred for several generations and a homozygous line UCD 200 and a subline 206 were established (17). These chicken lines were homozygous for MHC—B15 allele of B locus. UCD 200 and 206 sublines developed spontaneously scleroderma-like syndrome one to two weeks after hatching. This syndrome is characterized by varying degrees of lesions in the comb leading to necrosis, followed by polyarthritis and dermal lesions in 4 to 6 weeks-old chicken.

Histopathological analysis showed severe mononuclear cell infiltration, starting perivascularly and reaching the peak in about two weeks. Infiltration of mononuclear cells was found in deep dermis and subcutaneous tissue. The infiltrates contained primarily T lymphocytes of CD4$^+$ TCR γ/δ subset and with fewer CD8$^+$ T lymphocytes or B lymphocytes. Blood T lymphocytes responded poorly to activation by mitogens and anti-CD3 monoclonal antibody. These T cells expressed low levels of IL-2 receptors (18) and addition of exogenous IL-2 had no significant effect on the induction of T cell proliferation by mitogens. T lymphocytes from L200 chicken line showed reduced Ca^{++} influx subsequent to stimulation with mitogens (19). These observations suggested that T cells from L200 chickens exhibited an unique genetic defect. Positive and negative selection of T lymphocytes takes place subsequent to the interaction of T cell precursors with thymic epithelial and stromal cells. Staining L200 chicken thymus with MU170 mAb specific for epithelial cells showed striking alterations, staining only medullary perivascular epithelium compared to normal chicken thymus where all type I epithelial cells are stained (20). In addition, reduced apoptosis was observed in the thymocytes of UCD 200 chicken line (21). The major genetic defect appears to result from the dysfunction of negative selection of T cells. The avian scleroderma-like syndrome is associated with the presence of ANA and anticentromere autoantibodies as well. Though the avian model provides some information on the nature of T cell dysfunction and production of autoantibodies, the chicken disease is more severe than scleroderma as the histopathological changes lead to necrosis of the affected tissue, which is uncommon in human scleroderma. Further, these chickens also showed severe infiltration of heart, kidney, and esophagus by 4–5 weeks of age but no fibrosis of internal organs was found.

3.2. Tight-Skin Mouse Scleroderma-Like Syndrome

The tight skin (Tsk) TSK mutation that occurred spontaneously in B10D2/Sn mouse strain was first discovered by Helen Bunker at Jackson Laboratory. This mutation was later characterized as an autosomal dominant trait and transferred to C57BL/6 pa/pa mice (22). The homozygotes die *in utero*. This mutation is located on chromosome 2, 2 cM distal to pallid locus. The Tsk phenotype is characterized by the tightness of skin, marked hyperplasia of subcutaneous loose connective tissue, increased growth of bones, cartilages, and tendons as early as two weeks after birth. In addition to the above morphological and histological changes electron microscopy revealed large accumulations of microfibrils measuring 10 nm in diameter along with scattered collagen fibrils in the intercellular spaces of loose connective tissue. It was believed that the altered phenotype was caused by an alteration in cell receptor affinity for growth factors. But the exact nature of the molecular defect was not known until recently.

Menton and Hess (23) described the physical properties of tight skin mice as comparable to those of human scleroderma, since both the disorders shared common physical properties of stiffness, rigidity and relative elasticity of skin tissue. The hypodermis in scleroderma has fine structural alterations similar to that found in tight skin mice (Figure 1), including the presence of immature densely packed collagen fibrils and increased fibroblasts with dilated endoplasmic reticulum, indicating increased fibrilogenesis (24). Jimenez et al. (25), by *in vitro* biosynthetic studies in cultured dermal fibroblasts, showed that Tsk mouse fibroblasts synthesized five-fold collagen type I α1, α2 and type III α1 transcripts than control cells. The soluble collagen and hydroxyproline content of Tsk mouse skin was also shown to be increased, as in SSc skin (26,27) (Figure 2). Ross et al. (28) showed that there is a significant increase in glycosaminoglycan content in Tsk mouse skin as in SSc.

Tsk mice exhibit cystic emphysema (Figure 3) that differs from the pulmonary sclerosis associated with human scleroderma (29–31) and shows cardiac hypertrophy with an increase of 41% in type I collagen mRNA, 63% in type III and 33% in type IV mRNA in the ventricular myocardium (32,33). The collagen fibers were thicker and

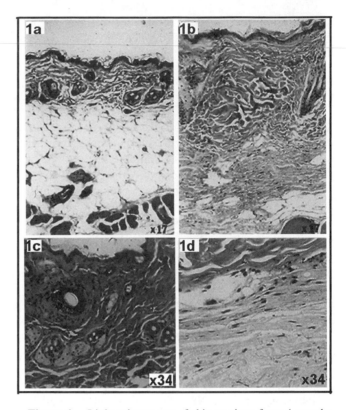

Figure 1. Light microscopy of skin sections from 4 months-old mice stained with hematoxylin-eosin. a,c: C57BL/6 pa/pa mouse; b,d: TSK/+ mouse.

denser in the perivascular areas. The collagen content in urinary bladder is about 70% greater in Tsk mouse than in age matched controls (34).

In 1983, De Lustro et al. (35) demonstrated that Tsk mice developed delayed type hypersensitivity (DTH) response to elastase digested lung peptides, and this response was adoptively transferable with immunocompetent cells to +/+ normal littrermates. However, no antibody activity to these antigens were detected in the sera of Tsk/+ or +/+ mice. Walker et al. (36) found an increase in both the number and proportion of degrnaulated mast cells in the skin of Tsk mice and attributed this to skin fibrosis. However, when experiments showed that fibrosis can be adoptively transferred by bone marrow and spleen cells, they suggested that the Tsk mouse fibrosis is not associated with mast cell proliferation, but with the transferred lymphocytes (37).

Thus, the morphological, phathological and biochemical changes observed in the skin and internal organs of Tsk mice support the view that it represents the best animal model available for studying the pathogenesis of SSc. To further elucidate the molecular and cellular mechanisms that are involved in the pleotrophic effect of the Tsk mutation, a number of groups have started investigations in the past decade.

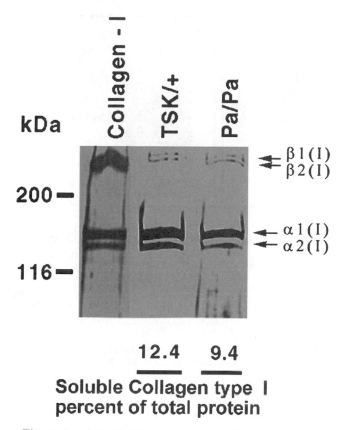

Soluble Collagen type I percent of total protein

Figure 2. SDS-PAGE analysis of pepsin digests of equal amounts of solubilized control and TSK mouse skin samples. Quantitation was performed as described (Hatakeyama et al. 1996).

3.3. Analysis of the Genetic Basis of Tight-Skin Disease

Interspecies backcross studies mapped the Tsk-1 mutation to a region bordered by $\beta2m$ and IL-la close to Fbn-1 gene locus (38,39). Attempts to define the Tsk mutation by RFLP using Bmp-2a, IL-la and $\beta2m$ DNA probes were

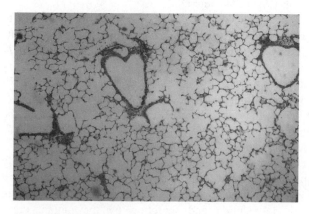

Figure 3. Light microscopic picture of TSK/+ mouse lung tissue showing cystic fibrosis.

however, not successful (40,41). When mutations in human Fbn-1 and Fbn-2 genes were found linked to Marfan syndrome (42) and congenital contractual arachnodactlyly (43), mouse Fbn-1 and 2 genes were mapped to band F, close to the Tsk locus on chromosome 2, and band D-E1 of chromosome 18, respectively (39,44). These findings, together with the observation of the presence of abundant microfibrils in the intercellular matrix of loose connective tissue in Tsk mice skin increased the possibility that Fbn-1 could be a potential candidate gene for Tsk mutation. However, Doute and Clark (45) reported obtaining two recombinants between the Fbn-1 and TSK loci in a study of 130 interspecies backcross progeny.

At about this time, our group and that of Jimenez investigated whether a genetic defect in Fbn-1 gene linked to Tsk phenotype. Towards this goal, we analyzed the restriction fragment length polymorphism (RFLP) of Tsk mouse using a series of Fbn-1-specific gene probes (46) covering the entire length of the mRNA and compared it with parental and unrelated strains (47). While DNA probes corresponding to the nucleotide sequences upstream of 4786 bp and downstream of 5520 bp of Fbn-1 cDNA didn't reveal any RFLP difference between Tsk and C57BL/6 mouse, a DNA probe that contained 4720–5520 bp of the cDNA revealed a distinctive RFLP in the Tsk mouse. The RFLP was characterized by the presence of an additional hybridizing DNA band on digestion with EcoRI and BamH1 (Figure 4, panel A) as well as several other restriction endonucleases (47). This result is typical of an intragenic gene duplication.

To determine whether or not the mutant Fbn-1 allele is expressed in Tsk mice, we examined RNA prepared from Tsk/+ mice and their F_1 progeny. Interestingly, Northern blot analysis of RNA from 8–9 day old embryo fibroblast cells, obtained by breeding heterozygous Tsk/+ mice, showed three distinct hybridization patterns, which is in agreement with the expected number of genotypes in the F_1 generation (Figure 4, Panel B). Of 8 embryos from a single litter, two expressed a Fbn1 mRNA transcript of 11 Kb similar to that of C57BL/6 mice, three expressed a larger Fbn1 RNA transcript of about 14 Kb, and three expressed both the larger and smaller RNA transcripts. RFLP analysis of DNA from these embryos also revealed three distinct patterns: The embryos that expressed the smaller 11 Kb RNA transcript exhibited a wild-type RFLP; those expressing larger and smaller RNA transcripts showed an RFLP pattern identical to Tsk/+ mice, and those expressing only the larger transcripts showed a RFLP pattern that was neither wild-type nor heterozygous Tsk/+ type. The RFLP of the homozygous mutants that expressed only the larger RNA transcript showed higher intensity of the unique DNA band, in EcoRI and BamHI double digests, compared to heterozygotes. These results collectively suggested that the genomic DNA coding for the region of mRNA sequence between 4.72 Kb and 5.52 Kb of Fbn-1 mRNA is altered

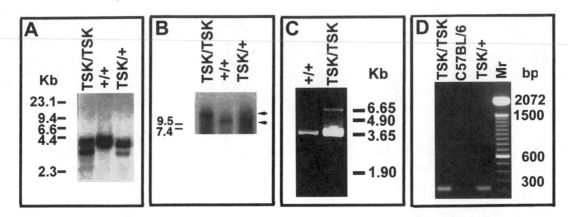

Figure 4.   Analysis of TSK mutation.
(A) autoradiograph of a Southern blot of DNAs digested with BamHI and EcoRI hybridized with Fbn-1 cDNA probe (4720–5446 bp).
(B) Autoradiograph of a Northern blot hybridized with 3′ end Fbn-1 gene probe. The three patterns of expression of mRNA corresponds to the 3 genotypes. Homozygous mutants expressed only the 14 Kb mRNA; Wild type cells expressed the 11 Kb mRNA and TSK/+ cells expressed 11 and 14 Kb mRNAs.
(C) RT-PCR analysis the of mRNAs from wild type and homozygous Fbn-1 mutant embryo fibroblasts. The cDNAs were prepared by reverse transcriptase reaction primed with Fbn-1 specific 3′ end primer. PCR was performed using a forward primer corresponding to 4804–4833 bp of Fbn-1 cDNA and a reverse primer corresponding to 8434–8459 bp. Wild type RNA shows a band of about 3.65 Kb and the mutant shows 3.65 and 6.6 Kb bands.
(D) PCR analysis of the genomic DNA. Forward primer corresponds to 4999–5018 nucleotides of Fbn-1 mRNA (exon 40) and the reverse primer corresponds to a region of intron 17 sequence. The mutants give a PCR product of 245 bp that corresponds to the genomic sequence between the two primers.

and contained an additional sequence that coded for about 3.0 Kb of the mRNA, corresponding to the size difference between smaller and larger mRNA species expressed in heterozygous TSK mice. However, the exact location and nature of the alteration was not known.

To further characterize the mutation, we designed a series of PCR primers that covered upstream and downstream of the hybridizing probe sequence and analyzed the cDNA by PCR. Though only a single RNA band of 14 Kb was detectable in Northerns blots of RNA from homozygous mutants, RT-PCR using FP 4804–4833 bp and RP 8434–8459 bp revealed two PCR products: one similar in size to wild-type, and a second, 3.0 Kb larger, PCR product (Figure 4, Panel C). Reamplification of the larger PCR product gave both the larger and the smaller species, while the smaller PCR product on amplification gave solely the smaller product. Similar results were obtained with three additional sets of oligonucleotide primers. The PCR results suggested that there is duplication of the genome and the insertion of duplicated sequence is somewhere between 4804 and 5222 nucleotides of the cDNA sequence, indicating that the insert contained at least the forward primer sequences that were used in this experiment sequences (between 2500 and 4916 bp of the cDNA). Thus, PCR analysis independently confirmed that a segment of Fbn-1 gene containing at least a portion of the probe sequence has been inserted upstream of 5222 nucleotide of the cDNA. Though the structure of the duplicated sequence or mechanism involved in this process was not known, the inserted gene sequence accounted for the additional length of the mutant mRNA (3.0 Kb).

Sequence analysis of the mutant cDNA downstream of 4720 bp revealed that the insert is 2952 nucleotides long and inserted at base 5065. The sequence, after 5064 base, instead of continuation of Fbn-1 exon 40 sequence as in wild-type, reads the nucleotide 2113 of exon 17 (2952 nucleotides of exon 17–40, i.e. from 2113 to 5065 nucleotides of the mRNA) was found duplicated. The duplicated sequence was inframe. Our results corroborated the finding of Siracusa et al. (48) that a tandem duplication of Fbn-1 gene (from exon 17 to 40) covering genomic sequence of about 30 Kb that coding for the additional 2952 nucleotides of the mutant Fbn-1 mRNA has occurred in the Tsk mutation. Using a forward PCR primer from exon 40 and a reverse primer from intron 17, a unique PCR band is obtained only in Tsk mouse mutants (Figure 4, Panel D).

To determine whether the Fbn-1 mutation is linked to expression of TSK phenotype, we prepared F$_1$ recombinants of Tsk/+ with different strains of mice. Tsk/+ × J$_{II}$D–/– (deficient in mature B lymphocytes), Tsk/+ × RAG2–/– (deficient in both B and T cells) and Tsk/+ × Vit/Vit (develops vitiligo). Analyses of DNA and skin histology from the F$_1$ progeny cleared showed that the inheritance of the Fbn-1 mutation is linked to the development of cutaneous hyperplasia and cellular infiltration (Table 1).

Despite the demonstration of transcription of partially duplicated mutated Fbn-1 genes in Tsk/+ fibroblasts, until recently there was no evidence to show production of mutant Fbn-1 polypeptide and secretion by the mutant cells. Recently Saito et al. (49), have shown by Western blot analysis of culture supernatants that Tsk/+ fibroblasts indeed

Table 1. Correlation between inheritance of defective Fibrillin-1 gene and development of cutaneous hyperplasia in F1 mice

Mouse strain	No. of mice	genotype	Skin thickness μm
TSK	8	Fbn1 +/–	205 ± 15
F1 (TSK/+ × J$_H$D–/–)	4	Fbn1 +/–, J$_H$D+/–	179 ± 33
			$\mid P < 0.05$
	4	Fbn1 +/+, J$_H$D+/–	132 ± 15
F1 (TSK/+ × RAG2–/–)	10	Fbn1 +/–, RAG2+/–	176 ± 59
			$\mid P < 0.05$
	7	Fbn1 +/+, RAG2+/–	119 ± 34
F1 (TSK/+ × vit/vit)	8	Fbn1 +/–, vit/+	206 ± 71
			$\mid P < 0.01$
	8	Fbn1 +/+, vit/+	118 ± 22

produce and secrete mutant Fbn-1. It is intersting to note that the quantity of mutant Fbn-1 detectable is significantly lower than wild type Fbn-1. These results are in accordance with the finding that the amount of mutant Fbn-1 mRNA expressed in Tsk/+ fibroblasts is less than that of wild type mRNA (47). It is conceivable that the mutant protein is stable, but its expression is transcriptionally regulated or the mutant mRNAs are unstable. Following transfection of COS7 cells with mutant Fbn-1 cDNA, (cloned into a vector containing chicken β-actin promoter and flag reporter sequence at the 3′ end), and staining with anti-flag antibodies, Saito et al. (49) showed the presence of mutant Fbn-1 intracytoplasmically. The mutant mRNA codes for a 450 KDa polypeptide that apparently contributes to the formation of beaded microfibrils with altered periodicity (50). Sequencing of partially duplicated mRNA from Fbn-1–/– fibroblasts showed numerous mutations (51) in exons 1–17 and 41–62, reminiscent of the mutations described in Marfan syndrome. Taken together, these results suggest that the Fbn-1 gene displays potential for high frequency of mutations in various species.

3.4.   Epistatic Interaction Between Fbn-1 and Collagen V Genes

Thickening of skin in Tsk mice results from excessive accumulation of extracellular matrix proteins like Col I, Col III and glycoaminoglycans (26,27,52,53). However, the relationship between the presence of defective Fbn-1 gene or expression of mutated Fbn-1 protein and the accumulation of other extracellular proteins that are neither structurally nor functionally associated with it is not well understood. It is conceivable that the other extracellular matrix proteins are expressed more abundantly in a compensatory manner. The type of extracelluar matrix protein that is expressed compensatorily might vary depending upon the tissue involved.

The pN/pN mice, obtained by a targeted deletion in Col Vα2 gene, exhibited skin fragility, skeletal abnormailities, and altered collagen fiber organization (54). Though there are striking differences between Tsk and pN/pN mouse in cellular infiltration of dermis, collagen accumulation, organization of collagen fibres and in the diameter of collagen fibrils, both mutant strains exhibited cutaneous thickening. This observation prompted us to examine whether the phenotypic changes observed in Tsk mouse can be reversed by genetic complementation with a collagen-defective strain, like pN/pN (55). F$_1$ TSK/+, pN/+, or +/+, pN/+ genotype mice showed a normal wild-type histological picture, exhibiting an averge skin thickness of 120–150 nm, which is significantly less than that of TSK/+ or pN/pN mice. Electron microscopic examination of sections showed that collagen bundles are composed of fibrils with normal periodicity, principally 80–110 nm diameter fibrils, similar to control C57/BL/6 mice (Figure 5). Soluble collagen I and V content of F1 mice skin was also significantly lower than that of Tsk/+ mice, thus restoring the normal skin thickness.

The mechanisms that might be involved in this process are complex. It is conceivable that either the heterotypic fibrils formed by Col I and mutant Col V fibrils are more accessible to proteolytic enzymes, or deposition of Col I and Col III fibrils on Col V scaffolding is altered, leading to their easier accessibility to metaloproteinases and degradation.

3.5.   Autoimmunity in Tsk Mice

Individuals affected with systemic sclerosis often develop autoimmunity that is manifested by production of autoantibodies against a variety of nuclear proteins, particularly topoisomerase I, RNA polymerase I, centromere proteins, U3-RNP and in some cases even cell-mediated immunity to collagen I (56). Delustro et al. (35) reported that Tsk mice develop delayed type hypersensitivity (DTH) to elastase

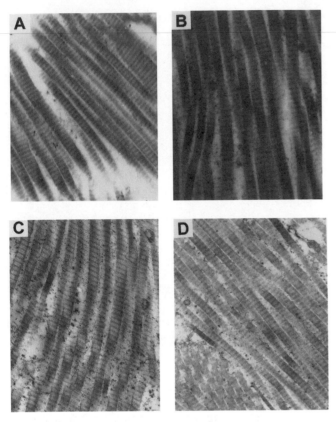

Figure 5. Electron microscic picture of collagen fibrils present in the mid dermis (× 10 K). TSK/+ (A), pN/pN (B), TSK/+ pN/+ (C) and +/+ pN/+ (D).

They speculated that a direct interaction between immune cells and fibroblasts might be responsible for this transferable activity. Studies by Phelps et al. (57) showed that transfer of Tsk mouse bone marrow cells to irradiated control mice induced skin fibrosis, cellular infiltration in skin tissue and production of autoantibodies to topoisomerase I and RNA polymerase I, after a lapse of several months. Infusion of either T or B lymphocytes alone was insufficient to induce skin fibrosis in the recipients, but infusion of B cells was sufficient to augment the titer of autoantibodies.

3.6.   Production of Autoantibodies in Tsk Mice

Analysis of sera from Tsk mice showed that antibodies against topoisomerase I, RNA polymerase I, Fcγ R, Fbn1 and anti-nuclear antibodies of both IgM and IgG isotypes increased with age (58–63). Table 2 illustrates the specificity of autoantibodies present in SSc, CREST syndrome and Tsk mouse. Analysis of the autoantibody repertoire of Tsk mice by producing hybridomas showed that, indeed, these mice had an increased frequncy of autoreactive B cells compared to control mice. Antibodies reactive with topoisomerase I, RNA Polymerase I, Fcγ R, fibrillin-1 are speciifc and are not inhibitable by unrelated antigens in the competitive inhibition RIA. Majority of the anti-nuclear antibodies produced in Tsk mice were anti-nucleolar antibodies (61). The results also demonstrated that Tsk monoclonal anti-topo I antibodies bound to similar epitopes, as do SSc patient antibodies (64). However, only few mAbs displayed inhibitory activity on enzyme function compared to high inhibitory activity of SSc antibodies.

A significant number of Tsk anti-topo I monoclonal antibodies expressed a crossreactive idiotope that was recognized by an anti-idiotype antibody isolated from a two month old Tsk mouse (64). Interestingly, this Tsk mouse

solubilized murine lung peptides. This DTH response was adoptively transferrable to control mice with splenic cells. Anti- Thy-1.2 antibody treatment of spleen cells significantly reduced the effectiveness of the adoptive transfer. Similarly, Walker et al. (37) observed that transfer of immuno competent spleen or bone marrow cells from Tsk mice into control mice resulted in fibrotic skin lesions.

Table 2.   Specificty of autoantibodies present scleroderma and tight skin mouse

Autoantigen specificity	Systemic sclerosis	CREST syndrome	TSK mouse
ANA	+	+	+
topoisomerase I	+	+	+
RNA polymerase I	+	−	−
RNA polymerase III	+	−	ND
U3-RNP	+	±	ND
Centromere	−	+	−
PM-Scl	+	−	ND
NOR 90	+	−	ND
FcγII (CD16)	+	ND	+
Fibrillin-1	+	+	+
FK 506 binding protein 12	+	+	ND
Nucleoli	+	+	+

anti-idiotype antibody also reacted with anti-topo I mAbs isolated from a SSc patient. Thus, both mouse and human antibodies shared an inter-species crossreactive idiotope, suggesting that V genes encoding these autoantibodies are conserved during phylogeny. Similar crossreactive Ids have been described in anti-Sm antibodies produced in MRL mice and human SLE (65). Western blot analysis using a series of topo I fusion proteins showed that both human and mouse monoclonal antibodies reacted more strongly with the NH2 terminal portion of topo I than with C terminal portion (64). A search of the Protein database revealed that a novel pentapeptide sequence present in the NH2 terminal moiety of topo I is shared with the UL40 polypeptide, which is coded by cytomegalovirus genome, suggesting a possible mechanism for activation of autoreactive B cells by molecular mimicry / by infectious agents. This is consistent with the finding by Maul et al. (66) who described sharing of a hexapeptide sequence between topo I and murine retroviral p30 gag protein.

The frequency of anti-nuclear antibody producing cells were also higher in Tsk mice than in controls. Immunofluorescence analysis showed that majority of anti-nuclear antibodies found in Tsk mice were anti-nucleolar (Figure 6),

and a significant number of anti-nuclear antibodies also reacted with RNA polymerase I. These antibodies bound to 190 KD subunit of RNA Pol I, similar to autoantibodies present in SSc sera (62), and recognized a common conserved epitope, since they reacted not only with the mammalian RNA Polymerase I but also with RNA pol I of E. coli and T7 phage. Anti-RNA Polymerase I antibodies present in SSc patients sera exhibited inhibition of catalytic function of the enzyme. By contrast, mouse monoclonal antibodies showed three types of activities on the enzyne function, namely inhibitory, enhancing and neutral. It appears, therefore, mouse mAbs recognized three different epitopes present on native enzymes. Mouse anti-RNA Polymerase I antibodies also showed augmented binding to phosphorylated enzyme than to native enzyme, similar to anti-RNA Polymerase I antibodies found in sera of MRL mouse and SLE patients (67,68).

Tsk mice also produced anti-FcγR antibodies similar to patients with SSc and other autoimmune diseases and the serum antibody titers increased with age. These antibodies might block efficient clearing of immune complexes and may also take part in the inflammatory responses, thus contributing to the pathogenesis of autoimmune diseases (69).

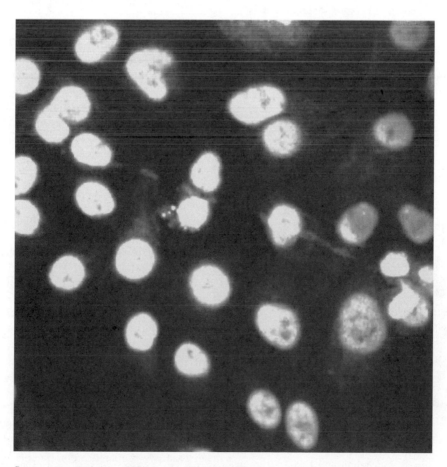

Figure 6.   Immunofluorescence staining of Hep2 cells with tightskin mouse autoantibodies showing strong nucleolar staining.

Further, these antibodies have been shown *in vitro* to trigger the release of hydrolytic enzymes and cytokines from macrophages and natural killer (NK) cells that promote inflammatory reactions. It is particularly interesting that Tsk mice also produce antibodies against Fbn-1, similar to other scleroderma target antigens (Figure 7). Since these mice expressed mutant Fbn-1, production of antibodies specific for this protein is of great interest.

The CD5[+] subset of B cells represents less than 5% of peripheral B lymphocytes, but contributes significantly to the production of autoantibodies in NZB mice and in certain human autoimmune diseases (70). To determine whether or not the CD5[+] subset preferentially contributes to the autoreactive antibody repertoire in the Tsk mouse, we analyzed the autoreactive antibody secreting hybridomas for the expression CD5 molecule by PCR. The results showed that a majority of the B cell clones were derived from B1 cells, the conventional peripheral B lymphocytes (71) suggesting that the expansion of autoreactive B cells may be determined by V genes of Igs rather than by genetic properties linked to a particular B cell subset. Northern hybridization analysis of RNAs showed that over 70% of the clones expressed V genes from J558 V_H family, while genes from V_H J606 and 36–09 families were not represented at all. In contrast, V gene cDNA libraries derived from Tsk mouse splenic RNA showed that the expression of J558 as well as 36–09 gene families were of the expected frequency, corresponding to the genomic complexity. These results suggested that there is a bias toward the use of genes from the J558 family in the autoreactive repertoire, while the usage of V_κ gene families was random. Pairing of V_H: $V\kappa$ genes was stochastic. To understand the molecular and genetic basis of the production of J558 gene-encoded autoantibodies, we analyzed the structure of J558 genes that are expressed in Tsk mouse autoantibodies (72). Sequence homology analysis indicated that these J558 genes were germline-derived with no somatic mutations, and belonged to diverse subfamilies. Autoantibodies with distinctive specificities were generated by pairing of similar or identical V_H genes with different $V\kappa$ genes. Fourteen of 18 genes analyzed shared a unique heptapeptide sequence, YVEKFKG, in the second CDR region of the heavy chain. Use of different germline genes bearing a common peptide motif in the diversity region CDR2 suggests a regulatory role, such as expression of a common idiotope. The data supports the view that the selection and expansion of autoreactive B cells in Tsk mice is V_H gene-mediated.

3.7. B Cells Do Not Play a Role in the Development of Cutaneous Hyperplasia

Studies by Phelps et al. (57) showed that transfer of Tsk mouse bone marrow cells into irradiated control mice induced skin fibrosis and cellular infiltration in skin tissue as well as an increase in serum anti-topo I and anti-RNA pol I antibody titers in the recipients. Infusion of either T or B cells alone did not facilitate skin fibrosis in the recipient mice, but infusion of B lymphocytes was sufficient to augment serum antibody levels.

To determine whether B lymphocytes play a role in the pathogenesis of Tsk syndrome or whether production of autoantibodies represents an epiphenomenon, we prepared Tsk mice defective in the development of mature B cells (J_H D–/– and carrying the Fbn-1 mutant gene (Fbn1+/–) by selecting from N2 backcrosses (47). In addition to physical examination of the N2 backcross progeny to assess the tightness of the skin, skin samples were also examined histologically to determine dermal thickening. Our results clearly demonstrated that mice deficient in mature B cells and inheriting the mutant Fbn1 gene developed cutaneous hyperplasia associated with cell infiltration at levels more evident than in Tsk/+ mice. Thus, production of antibodies or the presence of mature B cells does not play an integral role in the development of cutaneous hyperplasia.

Though these results implied that T cells may play an important role in the pathogenesis of Tsk disease, there was also conclusive evidence in support of this. To determine whether mature T cells play a role in the development of Tsk disease, we prepared recombinants by mating TSK/+ with RAG2–/– (deficient in mature T and B cells). Selected F[1] progeny with the mutant Fbn-1 genotype were backcrossed

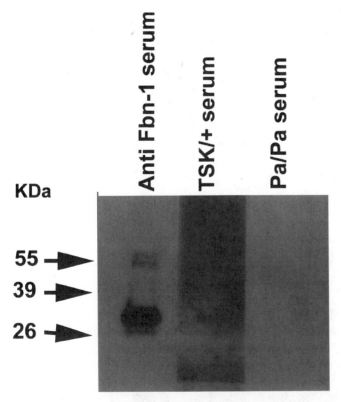

Figure 7. Western blot analysis demonstrating the presence of anti Fbn-1 autoantibodies in TSK mice sera. Western blots of Fbn1 fusion protein incubated with rabbit anti Fbn-1 serum at 1:2000 dilution (lane 1); TSK/+ mouse serum at 1:100 dilution (lane 2):C57BL/6 pa/pa serum at 1:100 dilution (lane 3)

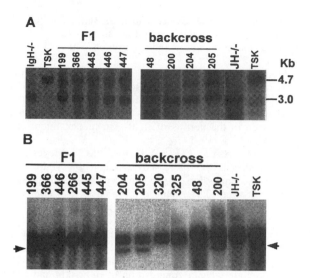

Figure 8. Southern analysis of DNA extracted from TSK, JH$^{+/-}$F$_1$ and backcross progeny obtained from crossing of F$_1$ × JH$^{-/-}$ mice.

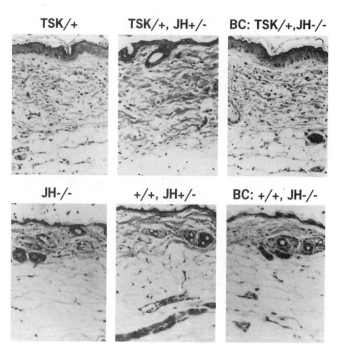

Figure 9. Histology of the skin from TSK/+, JH –/–, and backcross progeny TSK/+, J$_H$H–, +/+, JH+/–, and +/+, JH –/– mice (magnification × 50).

with RAG2–/– mice, and the resulting progeny of the N2 generation were genotyped for inheritence of the Fbn-1 mutation and RAG2 genes, both loci on chromosome 2. Of about one hundred N2 progeny analyzed, we found only one had the desired genotype (RAG2–/– Fbn-1+/–). Unfortunately, this mouse died early and we could not determine the phenotype histologically. Recently, Siracusa et al. (74), after screening 153 offspring of N2 progeny, obtained 10 recombinants that carried the desired recombinant chromosome with the RAG2–/– and Fbn-1 mutation. An analyzing these recombinants, they concluded that the Tsk phenotype is not dependent on the presence of mature T or B lymphocytes. However, other reports strongly suggested that T cells play a role in the development of cutaneous hyperplasia in Tsk mice. Fertin et al. (74) have shown that IL-4 stimulated collagen synthesis in normal, as well as in scleroderma, fibroblasts. They also found that IL-4 was as effective as TGFβ. Ong et al. (75) showed that development of Tsk syndrome can be prevented by injection of anti-IL-4 monoclonal antibody soon after birth. Wallace et al. (76) investigated the role of T lymphocytes subsets in the pathogenesis of skin fibrosis by breeding CD4–/– and CD8–/– mice with Tsk/+ mice and studying the progeny. They found a marked decrease in the thickness and cellularity as well as mild collagen disarrangement in the hypodermis of Tsk/+ CD4–/– mice compared to TSK/+ CD4+/+ CD8+/+ mice, though these animals had serum anti-topo I antibody levels similar to Tsk/+ controls.

Production of increased amounts of col I, col III, glycosaminoglycans and other matrix proteins might compensate for the deficiency of normal Fbn-1 in the ECM of Tsk mice. However, it is not clear how the autoreactive B cells are activated. It is possible that the alteration in the composition of ECM, due to either the lack of functional Fbn-1 or

the presence of mutant Fbn-1 modifies the microenvironment of Langherhan cells which could lead to increased efficiency in presenting autoantigens, thereby activating autoreactive B or T cells.

Taken together, the above observations indicate that: (a) Tsk mice spontaneously produce autoantibodies specific to scleroderma target nuclear antigens. These autoantibodies may result from the ingestion of apoptotic cells or blebs by antigen presenting cells and presenting peptides to lymphocytes. Thus, autoantibody production appears to be an epiphenomenon subsequent to primary injury. However, it is still unclear why some antibodies, like anti-topoisomerase I or anti-centromere proteins antibodies, are specific for SSc and CREST syndrome, respectively. (b) T cells may play a major role in Tsk syndrome. This view is consistent with several observations, such as the presence of T$_H$1 cells specific for elastase-digested lung peptides in Tsk mice, reduction of fibrosis in Tsk+/–, CD4–/– mice and abrogation of development of cutaneous hyperplasia in newborn Tsk mice injected with anti-IL-4 monoclonal antibodies.

References

1. Ishikawa, H., S. Suzuki, and R. Honuchi. 1975. An approach to experimental scleroderma using urinary glycosaminoglycans from patients with systemic sclerosis. *Acta Dermatol. Venerol.* 55:97–107.
2. Fox, P.K., D.D. White, M. Cavenagh, M.E. Davies, and F. Wusterman. 1982. Failure to demonstrate fibrotic changes in the skin of mice injected with glycosaminoglycan factors

from the urine of scleroderma patients. *Dermatologica* 164:80–83.

3. Finch, W.R., G.P. Rodnan, R.B. Buckingham, R.K. Prince, and A. Winkelstein. 1980. Bleomycin induced scleroderma. *J. Rheumatol.* 7:651–658.

4. Kerr, L.D., and H. Spiera. 1992. Scleroderma in association with the use of bleomycin. *J. Rheumatol.* 19:294–296.

5. Adamson, I.R., and D.H. Bowden. 1974. The pathogenesis of bleomycin- induced pulmonary fibrosis in mice. *Am. J. Pathol.* 77:185–189.

6. Laurent, C.J., R.J. McAnulty, B. Corrine, and I.I. Cockeri. 1981. Biochemical and histological changes in pulmonary fibrosis induced in rabbits with intratracheal bleomycin. *Eur. J. Clin. Invest.* 11:441–452.

7. Kelly, J., R.A. Newman, and J.N. Evans.1980. Bleomycin induced fibrosis in the rat. *J. Lab. Clin. Med.* 96:954–966.

8. Clark, J.G., J.E. Overton, B.A. Marieno, J. Uitto, and B.C. Stracher. 1980. Collagen biosynthesis in bleomycin induced pulmonary sclerosis in hamsters. *J. Clin. Lab. Med.* 96:943.

9. Collins, J.F. B. McCullough, J.J. Coalsou, and W.I. Johonson, 1981. Bleomycin induced diffuse interstitial pulmonary fibrosis in babbons. *Am. Rev. Res. Dis.* 123:305–314.

10. Szapiel, S.V., N.A. Elson, J.D. Flumer, G.W. Hunninghake, and R.E. Crystal. 1979. Bleomycin induced interstial pulmonar disease in the nude athymic mice. *Am. Rev. Res. Dis.* 120:893–899.

11. Statsny, P., V.A. Stembridge, and M. Ziff. 1963. Homologous disease in the adult rat, a model for autoimmune disease. *J. Exp. Med.* 118:635–648.

12. Jaffe, B.D., and H.N. Claman. 1985. Chronic graft versus host disease as a model for scleroderma. *Cell. Immunol.* 94:1–12.

13. Deng, J.S., B.K. Nicholes, M.R. Charley R.D. Southeimer, and J.N. Gillien. 1985. Serological studies in mice undergoing chronic graft versus host reaction. *Clin. Res.* 28:163.

14. Peters, J., and S.T. Bell. 1986. Tightskin-2 mouse. *Mouse News Letter* 174:91–91.

15. Christner, P.J., J. Peters, D. Hawkins, L.D. Siracusa, and S.A. Jimenez. 1995. The tight skin 2 mouse. *Arthritis Rheum.* 38:1791–1798.

16. Wooley, P.H., S. Sud, A. Lengendorfer, C. Celkins, P.J. Christiner, J. Peters, and S.A. Jimenez. 1998. T cells infiltrating the skin of tight skin 2 scleroderma like mice exhibit T cell receptor bias. *Autoimmun.* 27:91–98.

17. Gershwin, M.E., H. Abplanalp, J.J. Castles, R.M. Ikeda, J. Van de Water, J. Eklund, and D. Haynes. 1981. Characterization of spontaneous disease of white leghorn chickens resembling progressive systemic sclerosis. *J. Exp. Med.* 1640–1659.

18. Grenschwartz, M.S., S. Moormen, G. Kromer, R. Sgouc, H. Dietrich, G. Boeck, M.E. Gershwin, R. Boyd, and G. Wick. 1991. Phenotypic analysis of skin infiltrates in comparison with peripheral lymphocytes in early avian scleoderma. *J. Autoimmun.* 9:577–593.

19. Wilson, T.J., J. Van de Water, S.C. Mohr, R.L. Boyd, A. Ansari, G. Wick, and M.E. Gershwin. 1992. Avian scleroderma: Evidence for qualitative and quantitative defects. *J. Autoimm.* 5:261–272.

20. Boyd, R.L., T.J. Wilson, J. Van de Water, L.A. Haepanen, and M.E. Gershwin. 1991. Selective abnormalities in the thymic microenvironment associated with avian scleroderma. *J. Autoimmun.* 4:369–380.

21. Sgnoc, R., and G. Wick. 1999. What can we learn from an avian model for scleroderma. In *The decade of autoimmunity*. Elsvier Science, pp. 209–217.

22. Green, M.C., H.O. Sweet, and L.E. Bunker. 1976. Tight-skin,a new mutation of the mouse causing excessive growth of connective tissue and skeleton. *Am. J. Pathol.* 82:493–512.

23. Menton, D.N., R.A. Hess, J.R. Lichenstein, and A.J. Eisen. 1978. The structure and stensile properties of the skin of tight-skin mutant mice. *J. Invest. Dermatol.* 70:4–10.

24. Menton, D.N., and R.A. Hess. 1980. The ultrastructure of collagen in the dermis of tight-skin mutant mice. *Invest. Dermatol.* 74:139–147.

25. Jimenez, S.A., C.J. Williams, J.C. Meyers, and R. Bashey.1986. Increased collagen biosynthesis and increased expression of type I and type III procollagen genes in tight-skin (TSK) mouse fibroblasts. *J. Biol. Chem.* 261:657–662.

26. Osborn, T.G., N.E. Bauer, and S.C. Ross. 1983. The tight-skin mouse: physical and biochemical properties of the skin. *J. Rheumatol.* 10:793–796.

27. Dorner, R.W., T.G. Osborn, and S.C. Ross. 1987. Glycosaminoglycan composition of tight-skin and control mouse skins. *J. Rheumatol.* 14:295–298.

28. Ross, S.C., T.G. Osborn, R.W. Dorner, and J. Juckner. 1983. Glycosaminoglycan content in skin of the tight skin mouse. *Arthritis Rheum.* 26:653–657.

29. Szapiel, S.V., J.D. Fulmer, G.W. Hunninghake, N.A.Elson, O. Kawanami, V.J.Ferrans, and R.G. Crystal. 1981. Hereditary emphysema in the tight-skin mouse. *Amer. Rev. Res. Dis.* 123:680–685.

30. Rossi, G.A., G.W. Hunninghake, J.E. Gadek, S.U. Szapiel, O. Kawanami, V.J. Ferrons, and R.G. Crystal. 1984. Hereditary emphysema in the tight skin mice. *Amer. Rev. Res. Dis.* 129:850–855.

31. Gardi, C., P.A. Martorana, M.M. De Santi, P.V. Even, and G. Lungarella. 1989. A biochemical and morphological investigation of the early development of genetic emphysema in tight-skin mice. *Exp. Mol. Pathol.* 50:398- 410.

32. Osborn, T.G., R.I Bashey, T.L. Moore, and V.W. Fischer. 1987. Collagenous abnormalities in the heart of tight-skin mouse. *J. Mol. Cell. Cardiol.* 19:581–587.

33. Chapman, D., and M. Eghbali. 1990. Expression of fibrillar types I and III and basement membrane collagen type IV genes in myocardium of tight-skin mouse. *Cardio. Res.* 24:578–583.

34. Longhurst, P.A., B. Eika, R.E. Leggett, and R.M. Levin. 1992. Urinary bladder function in the tight-skin mouse. *J. Urol.* 148:1611–1614.

35. Delustro, F.A., A.M. Mackel, and E.C. LeRoy. 1983. Delayed-type hypersensitivity of elastase soluble lung peptides in the tight-skin mouse. *Cell. Immunol.* 81:175–179.

36. Walker, M.A., R.A. Harley, J. Maize, F. Delustro, and C. LeRoy. 1985. Mast cells and their degranulation in the TSK mouse model of scleroderma. *Proc. Soc. Exp. Biol. Med.* 180:323–328.

37. Walker, M.A., R.A. Harley, F.A. Delustro, and E.C. LeRoy. 1989. Adoptive transfer of tight-skin fibrosis to +/+ recipients by tsk bone marrow and spleen cells. *Proc. Soc. Exp. Biol. Med.* 192:196–200.

38. Everett, E.T., J.L. Pablos, S.E. Harris, E.C. Leroy, and J.S. Noris. 1994. The tight-skin mutation is closely linked to B2m on mouse chromosome 2. *Mamm. Genome* 5:55–57.

39. Goldstein, C., P. Liaw, S.A. Jimenez, A.M. Buchberg, and L.D.Siracusa. 1994. Of mice and Marfan: genetic linkage analyses of the fibrillin genes Fbn1 and Fbn2, in mouse genome. *Mamm. Genome* 5:696–700.

40. Siracusa, L.D., A.M. Buchberg, N.G. Copland, and N.A. Jenkins. 1989. Recombinant inbred strain and interspecific backcross analysis of molecular markers flanking the murine agouti coat color locus. *Genetics* 122:669–679.

41. Siracusa, L.D., P. Christner, R. McGrath, S.D. Mowers, K.K. Nelson, and S.A. Jimenez. 1993. The tight skin (Tsk) mutation in the mouse, a model for human fibrotic diseases is tightly linked to the B 2-microglobulin (B2m) gene on chromosome 2. *Genomics* 17:748–751.

42. Dietz, H.C., G.R. Cutting, R.E. Pyeritz, C.L. Maslen, L.Y. Sakai, A. Hamosh, E.J. Nanthakumar, S.M. Curristin, G. Stetten, D.A. Meyers, and C.A. Francomano. 1991. Marfan syndrome caused by a recurrent de novo missense mutation in the fibrillin gene. *Nature* 352:337–339.

43. Putnam, E.A., H. Zhang, F. Ramirez, and D.M. Milewicz. 1995. Fibrillin - 2 mutations result in the Marfan-like disorder, congenital contractural arachnodactyly. *Nat. Genetics* 11:456–458.

44. Li, X., L. Periera, H. Zhang, C. Sanguineti, F. Ramirez, J. Bonadio, and U. Franke. 1993. Fibrillin genes map to regions of conserved mouse/human synteny on mouse chromosome 2 and 18. *Genomics* 18:667–672.

45. Doute, R.C., and S.II. Clark. 1994. Tight-skin maps on mouse chromosome 2within the region of linkage homol ogywith human chromosome 15. *Genomics* 22:223–225.

46. Yin, W., E. Smiley, J. Germiller, C. Sanguinetti, T. Lawton, L. Periera, F. Ramirez, and J. Bonadio. 1995. Primary structure and developmental expression of Fbn-1, the mouse fibrillin gene. *J. Biol. Chem.* 270:1798–1806.

47. Kasturi, K.N., A. Hatakeyama, C. Murai, R. Gordon, R.G. Phelps, and C.A. Bona. 1997. B Cell Deficiency does not Abrogate Development of Cutaneous Hyperplasia in Mice Inheriting the Defective Fibrillin-1 Gene. *J. Autoimmun.* 10:505–517.

48. Siracusa, L.D., R. McGrath, Q. Ma, J.J. Moskow, J. Manne, P.J. Christner, A.M. Buchberg, and S.A. Jimenez. 1996. A tandem duplication within the fibrillin 1 gene is associated with the mouse tight skin mutation. *Genomic Res.* 6:300–313.

49. Saito, S., H. Nishimura, T. Brumeanu, S. Casares, A.C. Stan, T. Honjo, and C.A. Bona. 1999. Characterization of mutated protein encoded by partially duplicated fibrillin-1 gene in tight skin mice. *Mol. Immunol.* 10:1–8.

50. Kielty, J.M., M. Ragunath, L.D. Siracusa, M.J. Sherratt, R. Peters, C.A. Shuttleworth, and S.A. Jimenez. 1998. The tight skin mous: Demonstration of mutant fibrillin-1 Production and Assembly into abnormal Microfibrils. *J. Cell. Biol.* 140:1159–1166.

51. Bona, C.A., C. Murai, S. Casares, K. Kasturi, H. Nishimura, T. Honjo, and F. Matsuda. 1997. Structure of the Mutant Fibrillin-1 Gene in the Tight Skin (TSK) Mouse. *DNA Res.* 4:267–271.

52. Rosenberg, G.T., S.C. Ross, and T.G. Osborn. 1984. Glycosaminoglycan content in the lung of the tight-skin mouse. *J. Rheumatol.* 11:318–320.

53. Hatakeyama, A., K.N. Kasturi, I. Wolf, R.G. Phelps, and C.A. Bona. 1996. Correlation between the concentration of serum antitopoisomerase I autoantibodies and histological and biochemical alterations in the skin of TSK mice. *Cell Immunol.* 167:135–140.

54. Andrikopoulos, K., X. Liu, D.R. Keene, R. Jaenisch, and F. Ramirez. 1995. Targeted mutation in collagen 5 alpha 2 gene reveals a regulatory role for type V collagen during matrix assembly. *Nat. Genetics* 9:31–36.

55. Phelps, R.G., C. Murai, S. Saito, A. Hatakeyama, K. Andrikopoulos, K.N. Kasturi, and C.A. Bona. 1998. Effect of targeted mutation in collagen V alpha2 gene on development of cutaneous hyperplasia in tight skin mice. *Mol. Med.* 4:356–360.

56. Stuart, J.M., A.E. Postlethwaite, and A.H. Kang.1976. Evidence for cell-mediated immunity to collagen in progressive systemic sclerosis. *J. Lab. Clin. Med.* 88:601–607.

57. Phelps, R.G., C. Daian, S. Shinobu, R. Fleischmajer, and C.A. Bona.1993. Induction of skin fibrosis and auto-antibodies by infusion of immunocompetent cells from tight-skin mice into pa/pa mice. *J. Autoimmun.* 6:701–718.

58. Boros, P., J. Chen, C. Bona, and J.C. Unkeless. 1990. Autoimmune mice make anti-FcR antibodies. *J. Exp. Med.* 171:1581–1595.

59. Muryoi, T., K.N. Kasturi, M.J. Kafina, Y. Saito, O. Usuba, J.S. Perlish, R. Fleischmayer, and C.A. Bona. 1991. Self reactive repertoir of tight skin mouse: immunochemical and molecular characterization of anti-topoisomerase I autoantibodies. *Autoimmun.* 9:109–119.

60. Bocchieri, M.H., P.D. Henriksen, K.N. Kasturi, T. Muryoi, C.A. Bona, and S.A. Jimenez. 1991. Evidence for autoimmunity in tight skin mouse model of systemic sclerosis,. *Arthritis Rheum.* 34:599–605.

61. Muryoi, T., J. Andre-Schwartz, Y. Saitoh, A. Dimitriu-Bona, C.A. Bona, and K.N. Kasturi. 1992. Self reactive repertoire of tight-skin mouse: immunochemical and molecular characterization of anticellular auoantibodies. *Cell. Immunol.* 144:43–54.

62. Shibata, S., T. Muryoi, Y. Saitoh, T. Brumeanu, C.A. Bona, and K.N. Kasturi. 1993. Immunochemical and molecular characteristics of anti-RNA pol I autoantibodies produced by TSK mouse. *J. Clin. Invest.* 92:984–992.

63. Murai, C., S. Saito, K.N. Kasturi, and C.A. Bona.1998. Spontaneous occurrence of anti-fibrillin-1 autoantibodies in tight-skin mice. *Autoimmun.* 28:151–155.

64. Muryoi, T., K.N. Kasturi, M.J. Kafina, D.S. Cram, L.C. Harrison, T. Sasaki, and C.A. Bona. 1992. Antitopoisomerase I monoclonal autoantibodies from scleroderma patients and tight skin mous interact with similar epitopes. *J. Exp. Med* 175:1103–1109.

65. Takei, M., H. Dang, R.J. Wang, and N. Jalal. 1988. Characteristrics of a human monoclonal anti-Sm autoantibody expressing an interspecies idiotype. *J. Immunol.* 14:3108–3113.

66. Maul, G.G., S.A. Jimenez; E. Riggs, and D. Ziemnicka-Kotula. 1989. Determination of an epitope of diffuse systemic sclerosis marker antigen, DNA topoisomerase I: sequence similarity with retroviral P30gag protein suggests a possible cause for autoimmunity in systemic sclerosis. *Proc. Natl. Acad. Sci. USA* 86:8492–8496.

67. Stetler, D.A., and S.T. Jacob. 1984. Phosphorylation of RNA polymerase I augments its interaction with autoantibodies of systemic lupus erythematosus patients. *J. Biol. Chem.* 259:13629–13632.

68. Stetler, D.A., D.E. Sipes, and S.T. Jacob. 1985. Anti-RNA polymerase I antibodies in sera of MRL lpr/lpr and MRL +/+ autoimmune mice. *J. Exp. Med.* 162:1760–1770.

69. Boros, P., J.A. Odin, T. Muryoi, S.U. Masur, C. Bona, and J.C. Unkeless. 1991. IgM anti-FcgammaR autoantibodies trigger neutrophil degranulation. *J. Exp. Med.* 175:1473–1482.

70. Hayakawa, K., R.R. Hardyy, D.R. Parks, and L.A. Herzenberg. 1983. The Ly1 B cell subpopulation in normal, immunodefective and autoimmune mice. *J. Exp. Med.* 157:202–218.

71. Kasturi, K.N., C. Diain, Y. Saitoh, T. Muryoi, and C.A. Bona.1993. Autoantibody repertoire of tight-skin mouse: Analysis of VH and VK gene usage. *Mol. Immunol.* 30:969–978.

72. Kasturi, K.N., X.Y. Yio, and C.A. Bona. 1994. Molecular characterization of J558 genes encoding tight-skin mouse autoantibodies: Identical heavy-chain variable genes code for antibodies with different specificities. *Proc. Natl. Acad. Sci. USA* 91:8067–8071.

73. Siracusa, L.D.,R. McGrath, J.K. Fisher, and S.A. Jimenez. 1998. The mouse tight skin (Tsk) phenotype is not dependent on the presence of mature T and B lymphocytes. *Mamm. Genome.* 9:907–909.

74. Fertin, C., J.F. Nicolas, P. Giller, B. Kalis, J. Banchereau, and F.X. Maquart. 1991. Interleukin-4 stimulates collagen synthesis by normal and scleroderma fibroblasts in dermal equivalents. *Cell. Mol. Biol.* 37:823–829.

75. Ong, C., C. Wong, C.R. Roberts, H.S. Teh, and F.R. Jirik.1998. Anti- IL–4 treatment prevents dermal collagen deposition in the tight-skin mouse model of scleroderma. *Eur. J. Immunol.* 28:2619–2629.

76. Wallace, V.A., S. Kondo, T. Kono, Z. Xing, E. Timms, C. Furlonger, E. Keystone, J. Gouldie, D.N. Sander, T.W. Mak, and C.J. Paige. 1994. A role for CD4 T cells in the pathogenesis of skin fibrosis in tight skin mice. *Eur. J. Immunol.* 24:1463–1466.

28 | Systemic Vasculitis

Cees G.M. Kallenberg and J.W. Cohen Tervaert

INTRODUCTION

Vasculitis is a condition characterized by inflammation of blood vessels. Its clinical manifestations are dependent on the localization and size of the involved vessels as well as on the nature of the inflammatory process. Vasculitis can be secondary to other conditions or constitute a primary, in most cases idiopathic, disorder. Underlying conditions in the secondary vasculitides are infectious diseases, connective tissue diseases, and hypersensitivity disorders (Table1). Immune complexes, either deposited from the circulation or formed *in situ*, are in many cases involved, in the pathophysiology of the secondary vasculitides. Those complexes are supposedly composed of microbial antigens in case of underlying infectious diseases, autoantigens in connective tissue diseases, and non-microbial exogenous antigens in hypersensitivity disorders (table 1). Although immune deposits can be demonstrated in the involved vessel wall by direct immunofluorescence of biopsy material, the specificities of the antigens and their corresponding antibodies have not been demonstrated in most cases.

The primary vasculitides (Table 2) are idiopathic systemic diseases with variable clinical menifestations (1). Histopathologically, fibrinoid necrosis of the vessel wall is apparent (2). With the exception of Henoch Schönlein purpura and cryoglobulinemia-associated vasculitis, the lesions are characterized by paucity of immune deposits, and the pathophysiology of vessel wall inflammation in those pauci-immune conditions is far from clear. In the eighties, however, autoantibodies to cytoplasmic constituents of myeloid cells were detected in patients with idiopathic necrotizing small-vessel vasculitis (3), (Table 2). The antineutrophil cytoplasmic autoantibodies (ANCA), particularly those reacting with proteinase 3 and myeloperoxidase, were shown to be sensitive and specific for the mentioned condi-

Table 1 Secondary vasculitides: antigens presumably involved

- Exogenous antigens:
  – microbial antigens
    bacterial
        streptococci
        staphylococci
        mycobacterium leprae
        treponema pallidum
        others
    viral
        hepatitis B/C virus
        human immunodeficiency virus
        cytomegalovirus
        Epstein-Barr virus
        others
  protozoal
        plasmodia
  – non-microbial antigens
        heterologous proteins
        allergens
        drugs
        tumor antigens (?)
- Autologous antigens
        nuclear antigens (antinuclear antibodies)
        immunoglobulin G (rheumatoid factor, cryoglobulins)
        others

tions (4). This suggests that the autoantibodies are involved in the pathogenesis of the associated diseases, and positions those diseases within the spectrum of autoimmune disorders. In addition, autoantibodies to endothelial cells (AECA) have been described in patients with idiopathic systemic vasculitis, notably in Kawasaki disease (5,6). There are, however, many uncertainties regarding their antigenic specificities and their clinical and pathophysiologic significance in systemic vasculitis. Finally, T cell reactivity towards vessel wall components has been suggested to underlie large vessel vasculitis, such as in giant cell arteritis (7).

This chapter will briefly discuss the clinical and histopathological characteristics of the primary systemic vasculitides with the main focus on autoimmune phenomena and their role in disease immunopathogenesis.

2. PRIMAY SYSTEMIC VASCULITIS—CLASSIFICATION AND CLINICAL PRESENTATION

As stated before, a diagnosis of vasculitis is based on the presence of inflammation of the vessel wall. When other underlying conditions (table 1) or direct infection of the vessel wall, either contiguous or via the blood stream, have been excluded (8), the vasculitis is idiopathic or primary. The primary vasculitides are classified based on the size of the vessels involved, the histopathology of the lesions, and the presence of characteristic clinical symptoms. A classification scheme and definitions for the various vasculitic syndromes were proposed in 1993 by an International Consensus Group (1), Table 2. Those definitions were not intended to be used as diagnostic criteria, although many authors currently base their diagnosis on the so-called Chapel Hill Consensus Conference definitions. In clinical practice, however, many patients with histologically-confirmed vasculitis do not fulfill those definitions either because they present with incomplete or overlapping clinical syndromes, or because biopsies fail to demonstrate the pathognomonic lesions as described in the definitions. To classify patients with histologically-or angiographically-proven vasculitis, the American College of Rheumatology (ACR) has developed sets of criteria for the various vasculitides that are based on clinical signs and symptoms (9) although those criteria have their limitations in terms of sensitivity and specificity.

Table 2   Primary vasculitides (ref. 1)

- Large vessel vasculitis
  - Giant cell (temporal) arteritis
  - Takayasu's arteritis
- Medium-sized vessel vasculitis
  - Polyarteritis nodosa
  - Kawasaki disease
- Small vessel vasculitis
  - Wegener's granulomatosis*
  - Churg Strauss syndrome*
  - Microscopic polyangiitis*
  - Henoch-Schönlein purpura
  - Essential cryoglobulinemic vasculitis
  - Cutaneous leukocytoclastic angiitis

*associated with anti-neutrophil cytoplasmic autoantibodies (ANCA)

2.1.   Large Vessel Vasculitides

According to the Chapel Hill definitions (1) the vasculitic process in the large vessel vasculitides is confined to the aorta and its major branches.

The most common form, particularly in the Caucasian population, is *giant cell arteritis*, its name derived from the presence of many Langhans'giant cells in the lesions. Histopathologically, invasion of the vessel wall with macrophages, and, to a lesser extent, lymphocytes and plasma cells, is seen. Giant cells are clearly present and the internal elastic membrane is disrupted. In some areas, calcification of the internal elastic membrane with focal accumulation of giant cells has been observed (10), and it has been suggested that those foreign body giant cells are secondary to degeneration of the internal elastic membrane. Next, an autoimmune response would ensue, since those giant cells present degenerated membrane structures to T lymphocytes (see below). Clinically, the disease frequently presents with headache, tenderness of the scalp, particularly the temporal arteries (which has led to the synonymous designation of temporal arteritis), and claudication of the jaws. In addition, symptoms related to ischaemia of arteries in the upper part of the body may be present, such as loss of vision due to vasculitis of the retinal artery and neurological defects due to vasculitis of the internal carotid artery, the vertebrobasilar artery and intracerebral arteries. Other systems may be involved as well. Moreover, polymyalgia rheumatica syndrome, consisting of pain and stiffness of the proximal extremities, is present in about 50% of patients. Systemic symptoms such as fatique, malaise and fever with highly elevated ESR and anemia are almost invariably present. The disease generally occurs at older age, above 50 years, almost exclusively in Caucasians, with an incidence of 17 per 100,000 persons above 50 years. The disease generally reacts favorably to corticosteroids treatment.

Takayasu arteritis is a second form of large vessel vasculitis, and affects the aorta and its brachiocephalic branches. It may also affect the pulmonary arteries, other visceral arteries, and arteries of the lower extremities. Histopathologically, the disease has various stages. Active lesions are characterized by a granulomatous giant cell arteritis with a lymphoplasmacytic infiltrate with eosinophils, histiocytes and Langhans' cells. As a result of active inflammation, segmental narrowing and dilatation with aneurysm formation may occur, characterized by fibrosis and focal destruction of the musculoelastic layers. At the time of active inflammation, systemic symptoms are present accompanied by an increased acute phase response. Later symptoms are related to the localization and the extent of obstruction of the involved vessels, and may include claudication of upper and lower extremities (so-called "pulseless" disease), cerebral symptoms, ischemic bowel disease, renovascular hypertension, aortic insufficiency, etc. The disease occurs at a younger age than giant cell arteritis, particularly in women between 15 and 45 years of age, is relatively rare (2–20 per million) and is more frequent in Orientals, Africans and Latin Americans. Corticosteroids are the main treatment. In many cases, cytotoxic drugs such as methotrexate must be included to control the disease.

2.2. Medium-Sized Vessel Vasculitis

Classical **polyarteritis nodosa** (PAN), according to the Chapel Hill definitions (1), is a form of vasculitis confined to medium-sized arteries without involvement of smaller vessels. Applying that definition, PAN is now a rare disease (11), and many cases formerly diagnosed as PAN are now classified as microscopic polyangiitis (see below). The disease is histopathologically characterized by fibrinoid necrosis of the vessel wall with frequent microaneurysm formation that can be visualized by visceral angiography and may be a clue to diagnosis.

Clinically, systemic symptoms almost invariably include fever, fatigue, weight loss, arthralgia and myalgia. Other symptoms are related to involvement of specific arteries, such as renal insufficiency and malignant hypertension in renal vasculitis, ischemic bowel disease when the mesenteric arteries are involved, etc. The disease has been associated with Hepatitis B virus infection as well as with HIV-infection, albeit in a minority of the cases.

Another form of vasculitis of medium sized arteries is **Kawasaki syndrome**, also designated as mucocutaneous lymph node syndrome, a disease that particularly occurs in childhood. The disease presents with fever, a polymorphous exanthema especially on the palms and soles, reddening of lips and oral cavity, bilateral conjunctival injection, and cervical lymphadenopathy. These symptoms suggest an infectious etiology but no specific microorganism has been identified. Analysis of the T cell repertoire showed selective expansion of T cells expressing T cell receptor variable regions $V\beta2$ and $V\beta8$ suggesting bacterial superantigen stimulation (12). In 35% of cases, vasculitis of the coronary arteries occurs. The disease is most frequent in Japan, where its incidence is 150 per 100,000 children under the age of five. Aspirin and high-dose intravenous immunoglobulin are standard treatment modalities.

2.3. Small Vessel Vasculitis

Within the spectrum of vasculitis, the idiopathic small vessel vasculitides (Wegener's Granulomatosis, microscopic polyangiitis, and Churg Strauss syndrome), are strongly associated with the presence of anti-neutrophil cytoplasmic antibodies (ANCA), as will be discussed later (4). **Wegener's Granulomatosis** is characterized by the triad of granulomatous inflammation of the respiratory tract, systemic vasculitis, and necrotizing crescentic glomerulonephritis. Limited forms in which the kidney is not involved, may occur. The disease generally follows a biphasic course. Initially, inflammatory

lesions of the upper respiratory tract occur, such as chronic sinusitis, rhinitis, and/or otitis, with ongoing destruction resulting in for example, a saddle nose deformity. Systemic symptoms such as malaise, arthralgias and myalgias, are frequently present. Later, systemic vasculitis with rapidly progressive glomerulonephritis develops in many patients. Other organs, including the lungs (focal granulomatous vasculitic lesions that may cavitate), the eyes, the skin and the peripheral nervous system are frequently involved as well. The disease occurs somewhat more frequently in Caucasian men, with an onset around the fifth decade, and has a higher incidence in northern countries. As in other vasculitides, relapses frequently occur. Without treatment, prognosis is poor, with a mean survival of less than six months. Treatment with cyclophosphamide in combination with corticosteroids has improved 5 years survival to around 70% (13).

Microscopic polyangiitis frequently presents as a renal-pulmonary syndrome with pulmonary capillaritis and necrotizing crescentic glomerulonephritis together with systemic symptoms such as fever, malaise, arthralgia and myalgia. In contrast to Wegener's Granulomatosis, granulomatous inflammation and/or destructive lesions of the upper respiratory tract are absent. Small vessel vasculitis is prominent, whereas vasculitis in classical PAN is confined to medium-sized arteries. In addition to lung- and kidney, other organs such as eyes, skin and peripheral nervous system may be involved as well. A form limited to the renal system has been described as idiopathic necrotizing crescentic glomerulonephritis. As in most idiopathic systemic vasculitides, the lesions are characterized by absence or paucity of immune deposits. Age at onset and sex distribution are comparable to that in Wegener's Granulomatosis, and treatment also consists of cyclophosphamide and corticosteroids.

Churg-Strauss syndrome differs from Wegener's Granulomatosis (WG) and microscopic polyangiitis (MPA) by the presence of asthma and hypereosinophilia. Patients generally have a long history of asthma and hypereosinophilia (>1500/mm^3) before they develop systemic vasculitis. Eosinophil-rich granulomatous inflammation occurs in the respiratory tract clinically manifest as lung infiltrates and rhinitis with polyposis. Necrotizing vasculitis of small and medium-sized vessels with infiltration of eosinophils underlies the clinical manifestations of mononeuritis multiplex, ischemic bowel disease, and purpura or nodules of the skin. Renal involvement occurs less frequently than in WG and MPA. Cardiomyopathy, occurring in about 50% of the patients, is a major cause of mortality. The disease presents at the age of 35–40 years, equally in male and female, and treatment is similar to that for WG and MPA.

Three other disease entities are included in the group of small vessel vasculitides according to the Chapel Hill classification. **Henoch-Schönlein purpura**, predominantly occurring in children, is clinically characterized by attacks of purpura, arthralgia/arthritis, and gastrointestinal symptoms.

Glomerulonephritis occurs less frequently in children, but is frequent when Henoch-Schönlein purpura presents at older age. Skin and intestinal lesions are histopathologically characterized by leucocytoclastic vasculitis with IgA deposits in the vessel wall. Renal pathology varies from mesangial via focal-segmental to diffuse proliferative glomerulonephritis sometimes with extracapillary proliferation as well. IgA deposits are primarily localized in the mesangium, but may also be present in the capillary wall accompanied by C3 (but not C1q and C4) in most cases. Attacks frequently occur in association with infections, particularly of the respiratory tract, but no causative agent or specific antigen has been identified. The disease is selflimiting in most cases but proliferative glomerulonephritis may progress to end-stage renal failure. In the latter cases, the use of corticosteroid treatment has been advocated.

Essential cryoglobulinemic vasculitis is a form of immune complex-mediated vasculitis. Circulating cryoglobulins consist of polyclonal IgG and monoclonal (type II) or polyclonal (type III) IgM with rheumatoid factor activity. IgG- and IgM deposits are detectable in the lesions. The disease is particularly associated with Hepatitis B- and C-virus infection. Clinically, purpura, arthralgia and peripheral neuropathy are most prominent, whereas renal involvement is less common. Depending on the severity of the clinical manifestations, immunosuppressive treatment may be instituted. In the case of underlying viral infection, anti-viral treatment, such as interferon-α, has been suggested.

Cutaneous leukocytoclastic angiitis in most cases is a form of secondary vasculitis. As discussed previously, immune deposits are present within the vessel wall, but this condition may occur as an idiopathic variety with paucity of immune deposits. Since the therapeutic strategy for isolated cutaneous angiitis should be less aggressive than for the systemic vasculitides, cutaneous angiitis has been included in the Chapel Hill classification as a separate category.

3.   AUTOIMMUNITY AND THE SYSTEMIC VASCULITIDES

Are the primary "idiopathic" systemic vasculitides autoimmune diseases? To answer this question, Koch's postulates, in a modified form (14), can be used to define autoimmune etiology. These postulates, briefly, state that an autoimmune response must be present, the target(s) of the response identified, and induction of an analogous autoimmune response in experimental animals should result in the development of a disease similar to that in humans. The primary vasculitides do not fulfill those postulates. Direct and indirect evidence for the autoimmune pathogenesis of the vasculitides is lacking.

There is, however, circumstantial evidence of the autoimmune nature of the systemic vasculitides. First, immunosuppressive treatment is generally effective for those diseases.

Secondly, and more importantly, autoimmune phenomena do occur, and evidence is accumulating that those autoimmune phenomena are involved in disease pathophysiology. We will discuss here the role of ANCA in the small vessel vasculitides, the relevance of anti-endothelial antibodies (AECA) in the vasculitides in general, and Kawasaki's disease in particular, and recent data concerning autoimmune responses to vessel wall components in giant cell arteritis.

3.1.   Anti-Neutrophil Cytoplasmic Antibodies (ANCA)

3.1.1.   Antigenic specificities and disease associations. *c-ANCA and proteinase 3*   ANCA were first described by Davies et al. (15) in 1982 in a few patients with segmental necrotizing glomerulonephritis. In 1985, Van der Woude et al. (3) reported the presence of autoantibodies reacting with cytoplasmic components of neutrophils and monocytes in patients with Wegener's granulomatosis (WG). The antibodies produced a cytoplasmic fluorescence pattern with accentuation of the area within the nuclear lobes when tested by indirect immunofluorescence (IIF) on ethanol-fixed neutrophils (Fig. 1), and were later designated c-ANCA. Goldschmeding et al. (16) demonstrated that the target antigen of c-ANCA was a 29 kD serine protease contained within the azurophilic, or α-granules of neutrophils that was different from elastase and cathepsin G, the other two serine proteaes from these granules (Table 3). Further studies (17–19) confirmed that this 29 kD serine protease was identical to proteinase 3, which had already been described by Kao et al. (20). Proteinase 3 (Pr3) has been cloned and shown to be a 29 kD glycoprotein of 228 aminoacids (21) that is homologous to p29b, an antibiotic protein from human neutrophils (22), and to myeloblastin, a growth-promoting protein from myeloid cells (23). Pr3 is synthesized as a prepro-enzyme. A signal peptide at the N-terminus is first proteolytically cleaved, which results in proteolytic activity of the enzyme. Finally, a peptide of seven residues is cleft from the carboxy terminus, which does not seem to influence enzymatic activity of Pr3 (24).

Pr3 is present both in monocytes and granulocytes and appears early in mono-myeloid differentiation (25). It is thought to be present in myeloid cells only (26) although Mayet et al. described the occurrence of Pr3 at both RNA- and protein levels, in a number of other cell types, including endothelial cells (27–29), renal carcinoma cells (30), alveolar epithelial cells (31) and renal tubular epithelial cells (32). In endothelial cells, cytokine stimulation induced surface expression of Pr3, which would make Pr3 accessible for the autoantibodies and might, thus, be relevant for the pathogenesis of vasculitis, as will be discussed later. Whether or not Pr3 is present in endothelial cells is, however, still under discussion (26–29). Interestingly, Taekema-Roelvink et al. described a specific receptor for Pr3 on endothelial cells (33) that would enable Pr3 released

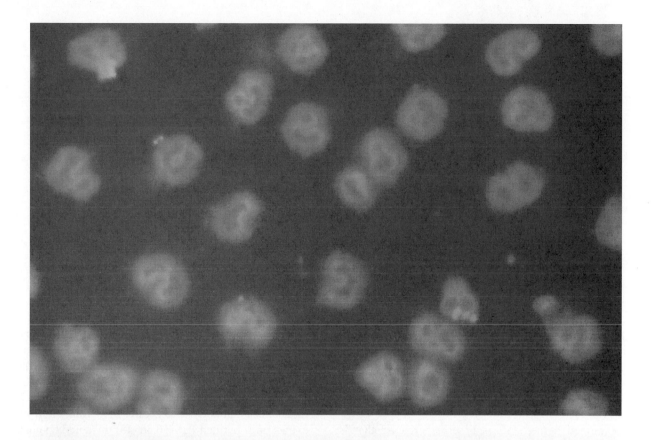

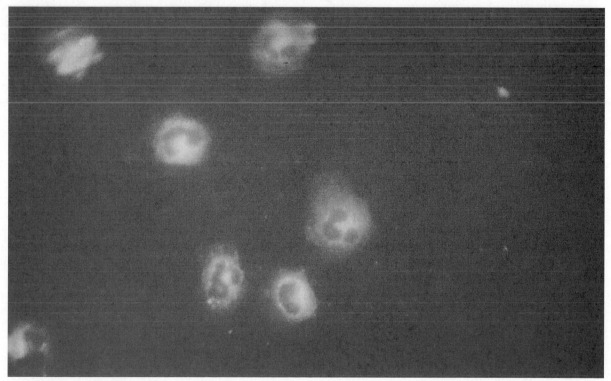

Figure 1.   Staining of ethanol-fixed granulocytes by indirect immunofluorescence. Upper panel: Characteristic cytoplasmic staining pattern with accentuation of the fluorescence intensity in the area within the nuclear lobes (cANCA). Lower panel: Perinuclear staining pattern (pANCA).

Table 3 Cellular localization of ANCA antigens

	Azurophilic granules	Specific granules	Cytosol/ cytoskeleton	Nucleus
Anti-microbial enzymes Serine proteases	myeloperoxidase[2] lysozyme proteinase 3[2] elastase[1,2] cathepsin G[1] azurocidin[2]	lysozyme[1]		
Acid hydrolase Other enzymes/ proteins	β-glucuronidase[1] bactericidal/ permeability- increasing protein[1,2]	lactoferrin[1]	catalase[1] alpha-enolase[1] actin[3]	histone H1[1] HMG1/HMG2[1]

[1]ANCA antigens in inflammatory bowel disease and primary sclerosing cholangitis.
[2]ANCA antigens in systemic vasculitides.
[3]ANCA antigen in autoimmune hepatitis.

from activated leukocytes to bind to endothelial cells and be targeted by the autoantibodies. Pr3, as recognized by human autoantibodies, has been found only in humans and baboons. Recently, a homologue of Pr3 has been described in mice that shows 69% homology with human Pr3 (34). Mouse Pr3 is, however, not recognized by human antibodies to Pr3.

As mentioned before, Pr3 is a serine protease that is slightly cationic (pI 7.6–9.0), has proteolytic activity as measured on several substrates (35), and is physiologically inhibited by α_1-antitrypsin (36). Pr3-α_1-antitrypsin complexes can, indeed, be detected in inflammatory fluids (36) and in sera, as shown in a study of patients with WG (37). Deficiency of α_1-antitrypsin occurs more frequently in patients with anti-Pr3 autoantibodies and WG, but in patients with α_1-antitrypsin deficiency, no increased prevalence of (c-)ANCA has been observed (38). Thus, a causal relationship between α_1-antitrypsin deficiency and the development of c-ANCA/ anti-Pr3 and WG has not been substantiated.

Upon full activation of myeloid cells degranulation occurs and Pr3, together with other granule constituents, is released. *In vitro* stimulation of neutrophils with low doses of pro-inflammatory cytokines, such as tumor necrosis factor α (TNF α), interleukin-1 or interleukin-8, results in membrane expression of Pr3 (and other granule constituents such as MPO and lactoferrin) (39,40). Those so-called primed neutrophils expressing Pr3 and MPO can be further stimulated by ANCA, as will be discussed later. The mechanisms involved in surface expression of Pr3/MPO on neutrophils are not fully understood, but binding due to charge interaction (after minor release of the cationic enzymes Pr3 and MPO during priming) via the serine protease inhibitor enzyme complex (SEC) receptor on neutrophils, or via β_2 integrins, are all likely possibilities (40). Surface expression of Pr3 occurs in some individuals on a low percentage of resting neutrophils (41,42) and, furthermore, on apoptotic neutrophils (43).

Detection of autoantibodies to Pr3 is classically done by IIF on ethanol-fixed neutrophils (44) on which they generally produce a cytoplasmic pattern (Fig. 1). The c-ANCA pattern is, however, not always identical with anti-Pr3, requiring, antigen-specific assays, particularly ELISAs, that either use purified Pr3 (45) in a directly coated system or a Pr3-specific monoclonal antibody to capture Pr3 from an α-granule extract or as purified Pr3 (16,46–48). It should be noted that purification of Pr3 from buffy coats, as done by ion-exchange chromotography, is very laborious and requires large amounts of donor leukocytes. Furthermore, the capture ELISA has the advantage of presenting Pr3 in its native configuration, which is important since the human autoantibodies primarily react with conformational epitopes (see below). This may explain why recombinant proteins are still unsatisfactory for use in ELISA (49), with the exception of Pr3 expressed in a human mast cell line (50,51), and possibly recombinant Pr3 from a baculo-virus expression system (unpublished observation). Standardization of the assays is of utmost importance, as shown in a collaborative European study (45).

p-ANCA and myeloperoxidase. Following the description of c-ANCA as a marker for WG, several groups observed that sera from a number of patients suspected of suffering from vasculitis produced a perinuclear fluorescence pattern on ethanol-fixed neutrophils. Falk and Jennette (52) and our group (53), described myeloperoxidase (MPO) as a major antigen for p-ANCA-positive sera in patients with crescentic glomerulonephritis and/or necrotizing vasculitis. MPO is a highly cationic protein (pI 11.0) with a m.w. of 146 kD, and consists of two chains. By immunoprecipitation, at least 2 bands of approximately 60 and 42 kD, representing a complete and a partly degraded single chain, are recognized by monoclonal and human polyclonal antibodies (46).

MPO is present, like Pr3, in the azurophilic or α-granules of myeloid cells, but not in other cells. Human MPO shows considerable homology with mouse and rat MPO. Immunization of rats with human MPO results in the generation of anti-human MPO antibodies that cross-react with rat MPO (54), thus generating anti-MPO autoantibodies that are relevant for the development of experimental models of anti-MPO-associated vasculitis/glomerulonephritis (55).

MPO plays a critical role in the generation of reactive oxygen species (56), by catalyzing the generation of hypochlorite from hydrogenperoxide and chloride anion.

Ceruloplasmin has been mentioned as its natural inhibitor (57,58). Comparable to Pr3, full activation of neutrophils results in the release of MPO, whereas priming of neutrophils with low concentrations of pro-inflammatory cytokines results in surface expression of MPO relevant for anti-MPO induced neutrophil activation (40) (see below).

Detection of anti-MPO can be done as a screening procedure by IIF on ethanol-fixed neutrophils (44; Fig. 1). While MPO is localized, together with Pr3, in the azurophilic granules, a perinuclear fluorescence pattern is observed when (auto)antibodies to MPO are probed on ethanol-fixed neutrophils. This perinuclear pattern is an artifact of ethanol fixation (59). When neutrophils are fixed with a cross-linking fixative such as paraformaldehyde, anti-MPO-positive sera produce a cytoplasmic staining pattern (59). As MPO is a highly cationic protein (pI 11.0), it moves during ethanol fixation to the negatively-charged nuclear membrane, explaining the perinuclear fluorescence pattern.

As stated before, IIF on ethanol-fixed neutrophils can only be used as a screening procedure for the detection of anti-MPO. It has become clear that ANCA of diverse specificities, such as lactoferrin, β-glucuronidase, elastase, cathepsin G and others, also produce a p-ANCA pattern (60–62). The latter autoantibodies occur in a wide variety of idiopathic inflammatory disorders other than vasculitis (60–62). In addition, many pANCA-positive sera occurring in those conditions have not yet been characterized with respect to their antigenic specificities. Thus, the detection of anti-MPO requires antigen-specific assays.

In contrast to Pr3, purification and isolation of MPO can be performed on a large scale. Commercially-available MPO can be used relatively easily in ELISA, although such preparations may contain impurities, particularly lactoferrin (63). Alternatively, a capture ELISA can be used in which a monoclonal anti-MPO antibody captures MPO from crude α-granule extract (16,46,53).

Other target antigens of ANCA in systemic vasculitis   As will be discussed in the next paragraph, Pr3 and MPO are the major target antigens of ANCA in the systemic vasculitides. Two other antigens are very incidentally recognized by ANCA in those diseases, the first being human leukocyte elastase (HLE), a 30 kD serine protease from azurophilic granules. Autoantibodies to HLE, as tested by capture ELISA, have been detected in a few sera from symptomatic patients, suggesting WG (64). The rare occurrence of anti-HLE does not warrant testing for anti-HLE in routine laboratory work-up for patients with vasculitis.

Bactericidal permeability-increasing protein (BPI) is another antigen incidentally recognized by autoantibodies in systemic vasculitis. BPI, a 54 kD protein localized in the azurophilic granules, is cytotoxic for Gram-negative bacteria by interacting with both the bacterial envelope and cell-free lipopolysaccharide, leading to bacterial killing and neutralization of free endotoxin (65). Zhao et al. (66) reported that ANCA directed to BPI were found in sera from patients with small-vessel vasculitides who were negative for anti-Pr3 or anti-MPO. Further studies have shown that anti-BPI occur in chronic infectious diseases, particularly in patients with cystic fibrosis who are carriers of *Pseudomonas aeruginosa* (67), as well as in idiopathic inflammatory diseases such as primary sclerosing cholangitis (68). The significance of anti-BPI in systemic vasculitis is limited.

Associations of anti-Pr3 and anti-MPO with clinical disease entities in systemic vasculitis. Following the original description of c-ANCA in WG (3), three major studies on more than 200 patients have found a sensitivity of c-ANCA/anti-Pr3 of 90% for extended WG characterized by the triad of granulomatous inflammation of the respiratory tract, systemic vasculitis, and necrotizing crescentic glomerulonephritis (69–71). The sensitivity of anti-Pr3 for limited WG, i.e. disease manifestations without obvious renal involvement, amounted to 75% (70). Specificity of anti-Pr3 for WG or related small-vessel vasculitides was 98% when selected groups of patients with idiopathic inflammatory or infectious diseases were tested (69,70). The aforementioned studies were performed by groups highly experienced in ANCA-testing. The high sensitivity and specificity of cANCA for WG have, however, been questioned (72). Rao et al. (73) did a meta-analysis on the role of cANCA testing in the diagnosis of WG. Summarizing current literature, they found a sensitivity of 66% and a specificity of 98% of c-ANCA for a diagnosis of WG. Sensitivity rose to 91% when only patients with active disease were considered. Anti-Pr3 also occur in primary vasculitides other than WG: 25–40% of patients with microscopic polyangiitis (MPA), 20–30% of patients with idiopathic necrotizing and crescentic glomerulonephritis (NCGN), and a minority of patients with Churg-Strauss syndrome test positive for anti-Pr3 (rev. in 4). Anti-Pr3 has only very incidentally been reported in other disorders (4).

Whereas anti-Pr3 are primarily associated with WG, anti-MPO are found in primary vasculitis patients with a more diverse presentation (53,74). In WG, most of the patients who test negative for anti-Pr3, are positive for anti-MPO. Also, some 60% of patients with MPA, 65% of those with idiopathic NCGN, and 60% of those with Churg-Strauss Syndrome are positive for anti-MPO (53,73,74). Anti-MPO have also been detected in 30–40% of patients with anti-GBM disease (75). Those patients are considerably older than patients with anti-GBM only, may have clinical signs suggesting associated vasculitis, and possibly have a better renal outcome (76). Furthermore, anti-MPO have been reported in connective tissue diseases such as SLE (77) and in various drug-induced disease states, such as hydralazine-induced glomerulonephritis (78), vasculitis-like syndromes associated with propylthiouracil (79), other antithyroid medications, minocycline and penicillamine (80). It is clear from the previous data that there is a clinical overlap between patients with anti-Pr3 and anti-MPO. To further define the clinical characteristics of anti-Pr3 versus anti-MPO associated vasculitis/ glomerulonephritis, Franssen et al. (81,82) analyzed the clinical course of consecutive patients with anti-Pr3- and anti-MPO-associated vasculitis/glomerulonephritis. They observed that patients with anti-Pr3 had a more fulminant course prior to diagnosis, with more severe inflammation in their renal biopsy (81). In addition, anti-Pr3-positive patients had more extra-renal disease manifestations, granuloma formation, and relapsed more frequently than anti-MPO positive patients (82). These differences may, at least in part, be explained by a higher neutrophil activating capacity of anti-Pr3 compared to anti-MPO, as shown by *in vitro* studies (83), see below.

In view of the clinical overlap between anti-Pr3 and anti-MPO, testing for ANCA in patients suspected of vasculitis should include tests for both. A large European collaborative study analyzed the sensitivity and specificity of anti-Pr3/anti-MPO for the idiopathic small vessel vasculitides (84), and found that the combination of cANCA by IIF and anti-Pr3 by ELISA or pANCA by IIF and anti-MPO by ELISA had a specificity of 98.4% in a group of 153 patients with either WG, MPA or idiopathic NCGN compared to 184 disease controls and 740 healthy subjects. In that multi-center study, the sensitivity of cANCA + anti-Pr3 or pANCA + anti-MPO for WG, MPA or idiopathic NCGN was 73%, 67%, and 82%, respectively; these numbers are somewhat lower than in other studies, possibly since patients with minor disease activity were included as well.

In conclusion, anti-Pr3/anti-MPO, but not (p)ANCA alone, are highly specific and reasonably sensitive markers for the idiopathic necrotizing small vessel vasculitides.

3.1.2. Induction of ANCA.   ANCA are primarily of the IgG-class (85), although IgM-class antibodies have also been described (86). Analysis of the IgG-subclass distribution of anti-Pr3 and anti-MPO shows that responses are mostly in the IgG1- and IgG4-subclass (85,87). IgG3-subclasses of ANCA, as studied for anti-Pr3, are present as well, particularly in patients with WG and renal involvement (85). These data are compatible with an antigen-driven, T cell-dependent immune response. In addition, the antibody response to Pr3 and MPO appears polyclonal. Several groups have tried to analyze the epitope specificities of anti-Pr3 and anti-MPO. Using overlapping peptides of 7 amino acids spanning the whole sequence of Pr3, Williams et al. found 11 immune reactive regions (88) which is in contrast with the observation that human anti-Pr3 autoantibodies primarily recognize con-formational structures of Pr3, and consequently, do not react

with recombinant Pr3 expressed in prokaryotic expression systems (49,89). Epitope mapping of Pr3 using a large set of monoclonal antibodies revealed at least 4 major epitope regions on the molecule (90). MPO studies using sets of monoclonal antibodies in inhibition ELISA have shown that different epitopes are recognized on MPO by human anti-MPO autoantibodies (91). Whether epitope spreading occurs during the course of the disease has not been reported, nor has the relation between epitope specificity and functional characteristics of the autoantibodies (described in the next paragraph) been analyzed.

As mentioned before, the Ig-class and -subclass distribution of anti-Pr3/anti-MPO suggests that the antibodies arise during an antigen-driven, T cell-dependent immune response. Several groups therefore, analyzed the *in vitro* proliferative capacity of peripheral blood T cells on stimulation with Pr3 and MPO (92–95). Generally, patients with WG positive for anti-Pr3 responded more frequently and stronger to Pr3 than controls, although a substantial number of controls also showed lymphocyte proliferation in response to Pr3. In contrast, most patients with anti-MPO did not show *in vitro* lymphocyte proliferation to MPO, nor did most of the controls (92–95). No dominant T cell epitopes have yet been identified for Pr3 or MPO, and their determination may give a clue to the inducing agent underlying the development of anti-Pr3 or anti-MPO.

What do we know about the induction or regulation of the autoantibody production in relation to disease expression? Unfortunately, no consistent observations are available showing that the development of anti-Pr3 or anti-MPO pre-ceeds the clinical expression of the associated diseases. Some data, however, suggest that exogenous factors are involved. First, the use of some drugs, particularly propylthiouracil (79) and hydralazine (78), are associated with the development of anti-MPO in conjunction with vasculitis-like diseases or glomerulonephritis or not. Secondly, mercuric chloride-induced autoimmune disease in rats involves the development of anti-MPO and, under certain circumstances, the occurrence of vasculitis, particularly in the gut (96). The mechanisms underlying drug- and mercuric chloride-induced anti-MPO production remain to be clarified.

It has been noted that re-appearance or rises in anti-Pr3 levels are associated with, or followed by, re-activation of disease in many patients (69,97–99), but these findings have been questioned in other studies (100). A meta-analysis of the available literature showed that 94 of 197 rises in ANCA levels were followed by a relapse of WG, whereas 81 of 157 relapses were preceded by a rise in ANCA titer (101). Differences in methodology, however, do not allow firm conclusions as to the diagnostic sensitiv-ity and specificity of a rise in ANCA titer for an ensuing relapse of the associated disease. Although a rise in titer should alert the physician to the possibility of an ensuing relapse, treatment based on rises in ANCA levels cannot be recommended until the association is clearly defined.

Nevertheless, the question of what induces reappearance or rise in titer of anti-Pr3 remains.

In this respect, studies by Stegeman et al. (102,103) may be relevant since they observed that chronic nasal carriage of *Staphylococcus aureus* was a significant risk factor for relapses in patients with WG (102), as was persistence of c-ANCA after induction of remission. Persistence of c-ANCA during remission was not independent of chronic

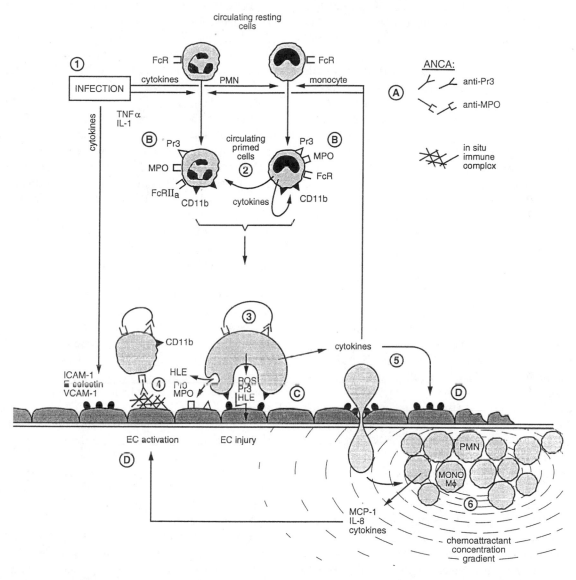

Figure 2. Schematic representation of the immune mechanisms supposedly involved in the pathophysiology of ANCA associated vasculitides (40).

(1) Cytokines released due to (local) infection cause upregulation of adhesion molecules on the endothelium and priming of neutrophils and/or monocytes. (2) Circulating primed neutrophils and/or monocytes express the ANCA antigens on the cell surface. (3) Adherence of primed neutrophils and/or monocytes to the endothelium, followed by activation of these cells by ANCA. Activated neutrophils and/or monocytes release reactive oxygen species (ROS) and lysosomal enzymes, which leads to endothelial cell injury and eventually to necrotizing inflammation. (4) Degranulation of proteinase 3 and myeloperoxidase by these ANCA-activated neutrophils and/or monocytes results in endothelial cell activation, endothelial cell injury or even endothelial cell apoptosis. Furthermore, bound Pr3 and MPO serve as planted antigens, resulting in *in situ* immune complexes, which, in turn attract other neutrophils. (5) ANCA-induced monocyte activation leads to production of monocyte chemoattractant protein-1 (MCP-1) and interleukin 8 (IL-8) production by these cells. The release of chemoattractants by these cells amplifies monocytes and neutrophil recruitment, possibly leading to granuloma formation (6). (A) to (D) represent the four prerequisites for endothelial cell damage by ANCA; (A) the presence of ANCA, (B) expression of the target antigens for ANCA on primed neutrophils and monocytes, (C) the necessity of an interaction between primed neutrophils and endothelium via β_2-integrins, and finally, (D) activation of endothelial cells.

nasal carriage of *S. aureus*, which might suggest that both factors are causally interrelated. Furthermore, Stegeman et al. (103) demonstrated that maintenance treatment with co-trimoxazole in WG patients resulted in a 60% reduction of relapses. How *S. aureus* may induce disease relapses is not clear, but several hypotheses have been formulated, including a role of *S. aureus*-derived superantigens stimulating the (auto)immune response (104).

3.1.3.   Pathogenic potential of ANCA in the systemic vasculitides—*in vitro* experimental findings.   *Interaction of ANCA with polymorphonuclear granulocytes (PMN).* Falk et al. (105) first demonstrated that PMN can be activated by ANCA, resulting in their production of reactive oxygen species and release of lysosomal enzymes. PMN, however, must be primed before they can be activated, which involves a process of pre-activation that can be accomplished *in vitro* with low doses of pro-inflammatory cytokines, such as tumor necrosis factor alpha (TNFα), interleukin-1 or IL-8. Priming results in surface expression of the target antigens of ANCA, such as Pr3 and MPO, as discussed previously. In this way, the antigens are available for interaction with the antibodies. Whether surface expression alone of the ANCA-antigens is sufficient for further activation by the autoantibodies is open to question, since constitutive expression of Pr3, as observed on a percentage of resting PMN in some individuals, does not result in massive activation. The primed state includes other intracellular changes that prepares the PMN for full activation once an appropriate signal is given.

How activation of primed PMN occurs after interaction with ANCA has not been fully analyzed. Kettritz et al. (106) showed that ANCA-induced PMN activation is dependent on cross-linking of surface molecules, since ANCA IgG and F(ab')$_2$ fragments, but not Fab fragments, were capable of stimulating the production of oxygen radicals by primed PMN. Others, however, have not been able to find activation of primed PMN by using F(ab')$_2$ fragment of ANCA (107,108), and suggest that activation occurs via interaction of the autoantibodies with Fcγ-receptors (FcγR), particularly the FcγRIIa receptor. Combined FcγRIIa and FcγRIIIb engagement also has been proposed (109). ANCA-induced PMN activation was strongly inhibited when blocking monoclonal antibodies against the second Fcγ receptor were used. The FcγRIIa is the only Fcγ receptor that interacts with IgG2 and also has a particular affinity for the IgG3 subclass (110). Interestingly, it was observed that increased *in vitro* PMN-activating capacity of serum IgG fractions from remission to relapse in patients with WG and anti-Pr3 correlated with increased levels of the IgG3 subclass of anti-Pr3 in those fractions, and not with that of the other subclasses (111). This also suggests that FcγR-interaction is involved in ANCA-induced PMN activation. Otherwise, as discussed later, Pr3 and MPO may bind to molecules such as the β_2 integrins, and binding of IgG-ANCA or F(ab')$_2$ fragments may result in

signal transduction via those molecules that have a transmembrane domain and a cytoplasmic tail. A detailed discussion on the possible mechanisms and signal transduction routes involved in ANCA-induced PMN activation is given elsewhere (40). Reumaux et al. (112) demonstrated that a third factor, adherence of PMN to a surface, is needed for *in vitro* PMN activation. They also showed that blocking antibodies to β_2-integrins, particularly CD18, inhibited PMN-activation. This suggests that ANCA-induced PMN activation only occurs when primed PMN are bound to a surface, e.g. endothelial cells, a process dependent on adhesion molecules, especially the β_2-integrins.

What are the *in vitro* effects of ANCA-induced PMN activation? As mentioned previously, those activated PMN produce and secrete reactive oxygen species (83,105,106, 113), degranulate lysosomal enzymes, including Pr3, MPO and elastase (83,105,106), and produce inflammatory mediators, such as TNFα, IL-1, IL-8, and leukotriene B4 (113,114). As a consequence, other cells are attracted, primed and activated, resulting in a strong amplification of inflammation. The process is markedly enhanced in the presence of extracellular arachidonic acid (113), also contributing to increased inflammation. Finally, upon activation by ANCA, PMN express increased levels of adhesion molecules, including β_2-integrins, that facilitate binding to the endothelial monolayer (115). *in vitro* studies have shown that interaction between (primed) PMN, ANCA and activated endothelial cells can result in damage to the latter cells by toxic products released from the activated PMN (116,117). In addition, lysosomal constituents, including Pr3 and MPO, may bind to the endothelial cell surface, serving as "planted" antigens and targets for ANCA. Indeed, endothelial cells can be coated *in vitro* with these antigens (118,119). Furthermore, *in vitro* incubation of endothelial cells with Pr3 induces production of IL-8 (120), endothelial cell apoptosis (121), and endothelial cell detachment and lysis (122). These effects were independent of the serine protease activity of Pr3.

Although the exact mechanisms involved are not fully elucidated, there is overwhelming evidence that ANCA *in vitro*, are able to augment and sustain a PMN-mediated inflammatory reaction, particularly affecting endothelial cells.

Interaction of ANCA with monocytes.   Since Pr3 and MPO are also constituents of granules from monocytes, these cells are likely to be targets for ANCA as well. It has been reported that monocytes can be activated by ANCA to produce reactive oxygen species (ROS) (123). This effect was not dependent on priming with TNFα, but priming did enhance ANCA-mediated ROS-production by monocytes. In addition, it has been shown that ANCA stimulate monocytes to produce and release IL-8 (124) and monocyte chemoattractant protein-1 (MCP-1) (125). In vivo, ANCA-induced MCP-1 secretion may play an important role in the formation of granulomas by amplification of local monocyte recruitment. FACS analysis of circulating monocytes in WG patients showed the presence of MPO and Pr3 on the cell surface during active disease and relapses, and monocytes do express CD18 and the FcγRIIa receptor (126). This suggests that the mechanisms underlying ANCA-induced monocyte activation are similar to those observed for PMN.

Interaction of ANCA with endothelial cells. It is generally believed that the ANCA antigens MPO and Pr3 are specific for cells of the monomyeloid lineage. As discussed previously, Mayet et al. also described Pr3 expression by renal carcinoma cells (30), renal tubular epithelial cells (32), pulmonary epithelial cells (31), and endothelial cells (27,28). Although Sibelius et al. confirmed Pr3 expression on endothelial cells (29), others have not been able to confirm this (26). Pr3 expression, as demonstrated by Mayet et al. (27), was increased, and translocation of the enzyme to the cell membrane was observed after stimulation of endothelial cells with either TNFα, IFNγ or IL-1. In this way, Pr3 is available for interaction with circulating ANCA. It should, however, be noted that within vascular lesions deposits of immunoglobulins are absent or scanty in ANCA-associated vasculitis (127). Therefore, direct or indirect binding of ANCA to endothelial cells as a pathophysiological mechanism in the development of ANCA-associated vasculitis has been disputed. However, the absence of immunoglobulin deposits in fully developed vascular lesions does not exclude a role for immune complexes in the initial phase of the disease process. Further *in vitro* studies have demonstrated other potentially pathogenic effects of anti-Pr3 on endothelial cells. De Bandt et al. (128) showed that anti-Pr3 can induce upregulation of adhesion molecules and expression of IL-1 and tissue factor by endothelial cells, although this effect may be due to the presence of anti-endothelial cell antibodies. Sibelius et al. (29) observed that incubation of endothelial cells with anti-Pr3 resulted in phosphoin-ositide hydrolysis-related signal transduction, followed by the synthesis of prostacyclin and platelet-activating factor and by increased endothelial protein leakage. Thus, anti-Pr3 may potentially affect endothelial cells in ANCA-associated vasculitis, but the expression of Pr3 by endothelial cells is still not uniformly accepted.

A schematic representation of the immune mechanisms supposedly involved in the pathophysiology of ANCA-associated vasculitis is given in Figure 2 (40).

Interaction of ANCA with their target antigens. ANCA may also affect the enzymatic properties of their target antigens. Binding of anti-Pr3 to Pr3 can inhibit the irreversible inactivation of Pr3 by α_1-antitrypsin, although most anti-Pr3 positive sera also inhibit the enzymatic activity of Pr3 (129). Binding of anti-Pr3 to Pr3 may interfere with the clearance of Pr3 released from neutrophils. Dissociation of active Pr3 from these complexes may contribute to tissue injury. A longitudinal study of WG patients showed that disease activity correlated better with the amount of inhibitory activity of anti-Pr3 on the inactivation of Pr3 by α_1-antitrypsin than with the anti-Pr3 titer itself (130).

Anti-MPO may likewise interfere with the enzymatic activity of their target enzyme, and additionally interfere with binding of MPO to its natural inhibitor, ceuloplasmin (58). Correlations between clinical disease activity of anti-MPO-associated vasculitis and *in vitro* effects of anti-MPO have not been described.

3.1.4. Pathogenic potential of ANCA in the systemic vasculitides: *in vivo* experimental findings.

Although the aforementioned findings suggest that ANCA are potentially pathogenic in the associated diseases, *in vivo* experimental models are clearly needed to further analyze how those autoantibodies can induce specific lesions in intact animals. To date, several animal models have been described, none of which seems fully equivalent to the human situation.

Development of anti-MPO antibodies have been described in Brown Norway (BN) rats exposed to mercuric chloride (96). Those anti-MPO antibodies develop as part of a polyclonal response, including antinuclear antibodies, antibodies to DNA, collagen, thyreoglobulin and components of the glomerular basement membrane. In those animals, inflammatory lesions developed in many organs, but particularly in the duodenum and cecum, where necrotizing leukocytoclastic vasculitis of submucosal vessels was apparent, accompanied by a neutrophilic and mononuclear infiltrate. Pretreatment of the animals with antibiotics severely reduced the vasculitic lesions, suggesting that microbial factors are involved in the expression of the lesions. Although anti-MPO antibodies were present in HgCl2-treated BN rats, their pathogenic potential is not apparent from this model.

In 1993, Kinjoh et al. described an inbred strain of mice derived from (BXSB x MRL/Mp-*lpr/lpr*)F₁ hybrid mice that spontaneously developed rapidly progressive crescentic glomerulonephritis and necrotizing vasculitis (131). These mice also develop antibodies to MPO, again as part of a polyclonal response involving other autoantibodies, which makes it difficult to define the pathophysiological role of any of these autoantobodies in the development of vasculitis and glomerulonephritis. A novel recombinant congenic strain of mice, designated McH5-*lpr/lpr*, has been established by hybridization of MRL/Mp-*lpr/lpr* mice with C3H/Hej-*lpr/lpr* mice, which are not prone to develop autoimmune diseases (132). McH5-*lpr/lpr* mice spontaneously develop granulomatous arteritis particularly in the kidneys but, in most cases, without glomerulonephritis. Anti-MPO and anti-DNA titers are low in those mice, particularly compared to titers in the "Kinjoh" mice. The impact of these findings for our understanding of human ANCA-associated vasculitis/glomerulonephritis is not clear.

In order to develop a model for anti-MPO-associated glomerulonephritis, Brouwer et al. (133) immunized Brown-Norway (BN) rats with human MPO in complete Freund's adjuvant. MPO-immunized rats developed antibodies to human MPO crossreacting with rat MPO, indicating anti-MPO autoantibodies. Renal arteries of immunized rats were subsequently perfused with products of activated neutrophils (proteolytic enzymes, MPO and its substrate, H_2O_2). In contrast to control immunized rats, MPO-immunized rats developed severe necrotizing crescentic glomerulonephritis (NCGN) with interstitial infiltrates and vasculitis. Immune deposits were present in the initial phase, but disappeared, supposedly due to the neutrophil-activating capacity of the anti-MPO antibodies, which induced necrotizing inflammation with breakdown of immune deposits. In this model, renal clamp ischemia could replace perfusion with H_2O_2, suggesting that renal ischemia contributes to the development of anti-MPO-associated NCGN via induction of the production of reactive oxygen species by endothelial cells (134).

Since ANCA-associated vasculitis is a systemic disease, Heeringa et al. (135) injected MPO-immunized BN-rats systemically in the jugular vein with a neutrophil lysozomal extract and H_2O_2 resulting in development of severe necrotizing vasculitis in the lungs and gut. To further prove the phlogistic potential of anti-MPO antibodies, Heeringa et al. (54) injected MPO-immunized BN rats with a subnephritogenic dose of rabbit anti-glomerular basement membrane antibodies. In control immunized rats, only minor lesions developed, while MPO-immunized rats, which had developed anti-rat MPO antibodies, showed severe renal lesions with fibrinoid necrosis of glomerular capillaries and crescent formation. This demonstrates that the anti-MPO immune response was able to severely aggravate the anti-GBM disease in this model.

The aforementioned models all relate to anti-MPO-associated vasculitides. Induction of autoimmunity to Pr3 in experimental models is hampered by the fact that the equivalent of human Pr3 in rats and mice has only recently been characterized (34), and shows only a restricted homology to human Pr3. Human Pr3-ANCA do not recognize rat-or mouse-Pr3, as tested by IIF, and immunization of rats with human Pr3 does not result in the formation of antibodies that react with cytoplasmic components of rat myeloid cells. Therefore, the role of autoimmunity to Pr3 cannot presently be studied in *in vivo* experimental models.

Induction of autoantibodies to Pr3 has, however, been described in mice by the group of Shoenfeld using idiotypic manipulation (136,137). In this model, affinity-purified human anti-Pr3 are injected into mice, which develop Pr3-specific anti-idiotipic antibodies after two weeks, and anti-anti-idiotypic antibodies that react with Pr3 after 4 months. Surprisingly, sera from these mice also reacted with human MPO and endothelial surface proteins. After 8 months, some mice developed inflammatory lesions in their lungs, but reactivity of the antibodies to mouse Pr3 was not demonstrated, casting doubt on the Pr3-specific auto-immune base of this model.

Taken together, a fully satisfying animal model for anti-Pr3 or anti-MPO-associated vasculitis/glomerulonephritis is presently not available although some data from experimental studies support a pathophysiological role for ANCA.

In summary, Pr3-ANCA and MPO-ANCA, but not ANCA alone as tested by IIF, are important diagnostic markers for the idiopathic small-vessel vasculitides, particularly Wegener's granulomatosis, microscopic polyangiitis, Churg Strauss Syndrome and idiopathic necrotizing crescentic glomerulonephritis. Changes in levels of Pr3-ANCA, and possibly MPO-ANCA, are related to changes in disease activity, although this correlation is far from absolute. *in vitro* studies clearly demonstrate the potential of the antibodies to interact with myeloid cells, resulting in their activation and destructive activity toward endothelial cells. *In vivo* experimental models, although not perfect corrollaries to the human situation, support a pathophysiological role for MPO-ANCA.

3.2.   Anti-endothelial Cell Antibodies (AECA)

Anti-endothelial cell antibodies (AECA) represent a heterogeneous group of antibodies found in various inflammatory diseases. Nearly three decades ago, these antibodies were reported by Lindquist and Osterland, who detected specific staining of the vascular endothelium by several serum samples during testing for antinuclear antibodies on mouse kidney sections as tissue substrate (138). Later, cultured human endothelial cells (EC) were used as a substrate to detect AECA, and it was found that AECA react with different endothelial antigens ranging in molecular weight from 25–200 kDa, as demonstrated by immunoprecipitation using crude extracts of EC (rev. in 139).

In vitro, it has been demonstrated that AECA may directly cause EC injury and/or apoptosis of EC or interfere with several EC functions (140). Therefore, AECA have been postulated to be of pathophysiological relevance, although their pathophysiological role is uncertain. It is not known if these antibodies are generated before or after vascular damage, or if they actually cause vascular dysfunction *in vivo*. Here, we will discuss the possible role of AECA in the pathophysiology of vasculitis and the value of AECA as markers of disease activity.

3.2.1.   Methods of detection. AECA are usually detected by enzyme-linked immunosorbent assays (ELISA) using as a substrate cultured human umbilical vein endothelial cells (HUVEC) that are passaged three to four times (139–144). Confluent cells are usually fixed with gluturaldehyde or paraformaldehyde to avoid non-specific IgG binding. Fixation, however, induces permeabilisation of EC membranes. Reactivity of patient sera in an AECA-ELISA in which fixed EC are used may, therefore, be partially attributed to reaction with intracellular constituents. In addition, fixation induces EC membrane alterations that facilitate binding of antibodies to negatively-charged phospholipids (145). To avoid these artefacts, several groups

have used ELISA assays with unfixed EC to detect AECA (143,144). Unfortunately, very few reports have compared detection of AECA on fixed and unfixed EC. AECA also can be detected by methods such as immunofluorescence assays, radioimmunoassays, FACS analysis, immunoblotting and immunoprecipitation, and by functional assays such as complement-dependent cytotoxicity (CDC) and antibody-dependent cytotoxicity (ADCC) (139). Results obtained in these tests may differ. Substrates other than HUVEC are also sometimes used, for instance, cell membrane extracts, cells from renal or adipose microvasculature, an endothelial-epithelial cell line (ie, EAhy 926), or a spontaneously-transformed EC line, ECV 304 (139). Each of these substrates has its own advantages and disadvantages. Despite these differences, concentrated efforts are currently being made to standardize tests for the detection of AECA.

3.2.2. Endothelial antigens. AECA are a group of auto-antibodies directed against a variety of endothelial autoantigens. The antigens recognized by AECA are, generally, not specific for EC (139). EC-specific AECA are only found in a few instances, e.g., in sera from patients with Kawasaki's disease (KD). In most other cases, AECA react not only with EC, but also with other cells, such as fibroblasts and/or peripheral blood mononuclear cells, as demonstrated by absorption studies (139).

The EC antigens either show a constitutive and stable expression or can be modulated by cytokines (rev. in 139). Cryptic antigens have been described as well (146). In some cases, activation of EC by cytokines may be a pre-requisite for the detection of AECA (139). Finally, antigenic determinants for AECA may, in fact, be molecules that adhere to EC, e.g., deoxyribonucleic acid (DNA), beta 2-glycoprotein I (β_2-GPI) and/or myeloperoxidase (MPO) (139). These so-called "planted" antigens adhere to the endothelium either directly through charge-mediated mechanisms (MPO, β_2-GPI), or indirectly via a DNA/histone bridge (DNA). Serum samples have been shown to contain MPO, DNA, and/or β_2-GPI (139). MPO, β_2-GPI, and/or DNA in these samples may adhere to EC during incubation of EC with those sera. In this way anti-DNA, anti-β_2-GPI and/or anti-MPO, when present in the sera, may falsely be detected as AECA. This phenomenon, which occurs *in vitro*, may also occur *in vivo* although no data are available that prove *in vivo* binding. Many AECA antigens are still not well characterized.

Constitutive endothelial antigens HLA class I antigens, normally present on EC, are not a major antigenic determinant for AECA since, for most AECA-positive sera, no significant loss of AECA activity is observed after absorbtion with other cells expressing HLA class I antigens (139). Furthermore, AECA-positive sera are usually negative for the presence of anti-HLA I antibodies as detected in a cytotoxicity assay against a large panel of human lymphocyte donors (147), and react with EC from

donors with different HLA phenotypes to the same extent. Many constitutive endothelial antigens for AECA are not yet characterized. When proteins are extracted from lysed endothelial cells, transferred to nitrocellulose membranes, and subsequently incubated with serum samples from patients, IgG antibodies to several protein bands can be detected by immunoblotting. When cell-surface membrane proteins are selectively radiolabeled and subsequently used in immunoprecipitation assays, some disaese-specific endothelial antigens can be immunoprecipitated, such as a 125 kDa band in Wegener's granulomatosis (WG) and a 200 kDa band in SLE (142). AECA distinct from those that react with cell surface membrane proteins have also been described, and may react with extracellular matrix (ECM) components. By testing serum samples from patients with various systemic vasculitides, Direskeneli et al. found a 25% reduction of AECA binding after incubation of the sera with a crude extract of ECM components (148). Different extracellular matrix components may be target antigens of AECA, such as collagen type II, IV, VII, vimentin, and/or laminin (139,148,149). Antibodies to collagen II and IV have been described in a large variety of diseases, including vasculitis (148,150), antibodies to laminin in a high proportion of patients with primary Raynaud's phenomenon and/or systemic sclerosis (149), and antivimentin antibodies in patients with SLE (151).

Cryptic antigens. HLA class II determinants are present on activated EC only, and could represent target antigens for AECA. Antibodies to HLA class II antigens are implicated in the pathogenesis of rejection of transplanted organs (139). Proteinase 3 (PR3) may be another cryptic target antigen for AECA, as discussed earlier in this Chapter.

"Planted" antigens. Cationic proteins from leukocytes may be target antigens both for ANCA and, as planted antigens, for AECA. Cationic ANCA antigens, particularly MPO, may bind to EC after being released by activated neutrophils and/or monocytes. Binding to EC probably occurs via a charge-mediated mechanism. β_2-GPI is another example of a "planted" antigen for AECA. Anti-β_2-GPI antibodies have been described in diseases such as SLE, primary antiphospholipid syndrome, and Sneddon's syndrome (139). β_2-GPI adheres to EC membranes (140) and may be present in sera. Furthermore, EC-bound β_2-GPI is recognized by anti-β_2-GPI antibodies (152). DNA is another example of a "planted antigen" (153). During testing for AECA on fixed EC, some groups found a strong association between AECA and anti-DNA antibodies in SLE patients (rev. in 139). In addition, monoclonal anti-DNA antibodies also bind to EC membranes (153) at least partially due to binding of DNA/anti-DNA immune complexes to EC, and such binding was enhanced by the addition of histones. Furthermore, treatment of HUVEC with DNase reduced the binding of monoclonal DNA-antibodies to HUVEC by 20%, suggesting that DNA is, indeed, present on the surface of HUVEC (153).

3.2.3. Disease associations.   AECA are described in a variety of diseases, such as connective tissue diseases, systemic vasculitides, and others (rev. in 140). Their associations with the vasculitides will be discussed here.

Large vessel vasculitis   In 1967, Nakao et al. (154) reported anti-aortic antibodies in patients with Takayasu's arteritis (TA), but these reports could not be reproduced (155). Brasile et al., later described AECA in three patients with temporal arteritis, and one with TA (156). One of the serum samples was also tested for reactivity with frozen cadaveric vessel sections from different anatomic sites and reacted with the iliac artery, the inferior mesenteric artery and veins, but not with the aorta.

High titers of AECA in almost all patients with TA was confirmed by ELISA, FACS analysis, and confocal microscopy (157). Salojin et al. (158), however, found AECA in only 7 of 21 (33%) patients with TA, and 8 of 26 (31%) patients with temporal arteritis.

Medium-sized vessel vasculitis.   Brasile et al. (156) reported AECA in 4 patients tested with polyarteritis nodosa (PAN), whereas only 10 of 32 (31%) patients with PAN reacted positively in the study of Salojin et al. (158). In Kawasaki's disease (KD), IgM AECA were observed in 53% of patients, whereas IgG AECA were infrequent (159). The antigens involved are still not well characterized, but cytokine-inducible epitopes may be important (160). Complement-dependent cytotoxicity has been found in the majority of KD sera (159,160), particularly when cytokine-stimulated EC were used as a substrate (160). By ELISA, however, AECA binding can be detected on both non-stimulated and TNF-stimulated EC (159). AECA may be a marker of disease activity in KD since the antibodies disappear in convalescent patients (160).

Small vessel vasculitis.   In WG, MPA and Churg Strauss Syndrome, AECA are frequently found (prevalence: 40–100%) (141,144,147,161,162). About 40% of AECA in the small vessel vasculitides are, however, only borderline positive (141,144). AECA are more frequently found in patients with MPO-ANCA than in those with PR3-ANCA (141). The endothelial antigens recognized by AECA in those diseases are not well characterized. Some PR3-ANCA may bind to EC (27), but cross-absorption studies show that most serum samples bind to EC independent of their binding to ANCA antigens (163). Protein bands of 25, 68, 125, 155, and 180 kD were detected by immunoblotting. The 68 and 125 kD bands were characteristic of WG patients (142). IgG binding to EC was enhanced by pretreatment of EC with TNF or IL-1 (141). AECA from patients with WG activated EC, resulting in increased expression of adhesion molecules and secretion of IL-6, IL-8, and MCP-1 (144,163). These effects were more prominent in patients with high AECA titers than in borderline positive patients (144). AECA levels correlate with disease activity (147,161), and increasing AECA titers may be used as a marker of relapse, especially in ANCA-negative cases (164). In a prospective study on 10 patients followed during a mean period of 36 weeks, a rise of AECA titers was seen in 8/11 relapses, and patients persistently positive for AECA were at risk of a subsequent relapse (164).

3.2.4. Pathogenetic role.   The correlation between changes in AECA titers and disease activity suggests a role for AECA in the induction of vessel wall damage, although it does not exclude the possibility that AECA are a result of vascular injury. Several mechanisms have been proposed by which AECA may play a role in the pathophysiology of vasculitis. First, binding of AECA to EC may result in activation of EC. Upregulated expression of endothelial adhesion molecules such as E-selectin, ICAM-1, and/or VCAM-1 was found after incubation of EC with IgG AECA (163), and similar findings were reported for anti-PR3-, anti-β_2-GPI-, and anti-HLA class I-antibodies (140,165,166,167). Activation by AECA may also result in the secretion of chemoattractants and/or cytokines (144,163). In addition, binding of AECA to EC may induce inhibition of prostacyclin production by EC (168). More recently, it was shown that AECA from patients with vasculitis and/or systemic sclerosis may induce apoptosis of EC (145,169). During this process, exposure of anionic phospholipids occurs (most notably phosphatidylserine) to the surface of the cells, which become available for binding of β_2-GPI and, subsequently, anti-phospholipid antibodies (170). This may result in proinflammatory clearance of apoptotic cells (171).

An alternative mechanism by which AECA could be a trigger in the pathogenesis of associated diseases is complement-dependent cytotoxicity (CDC) and/or antibody-dependent cellular cytotoxicity (ADCC). CDC towards cytokine-activated EC has been reported in Kawasaki's disease (159,160), but not in other forms of vasculitis, and ADCC has occasionally been demonstrated using serum samples from patients with WG or MPA (141).

An animal model supporting a pathophysiological role of AECA has not yet been developed, but several models are suggestive in this respect. Injection of antibodies to EC antigens, such as angiotensin-converting enzyme (ACE) or factor VIII von Willebrand, induces lung injury and/or glomerulonephritis in rabbits or rats (rev. in 139). Furthermore, AECA can be detected in serum samples from lupus-prone (MRL *lpr/lpr*) mice (140). In an idiotypic manipulation model, AECA can be induced in mice immunized with human IgG fractions with AECA activity, and the appearance of AECA in these animals was associated with glomerular vascular inflammation (172).

AECA are detected in a variety of vasculitic and other inflammatory disorders. Although probably of limited value in the differential diagnosis of these diseases, the detection of these antibodies may be valuable for follow-up studies. Several studies support a role for AECA in the pathophysiology of these diseases. The antibodies may activate EC, induce apoptosis, and induce complement-mediated and/or antibody-dependent vascular damage. Further characterisation of putative target antigens would be helpful in the search for a possible pathophysiological role of these antibodies.

3.3.   Autoimmunity in Giant Cell Arteritis?

Giant cell arteritis (GCA), as discussed previously, is a form of large vessel vasculitis generally occurring at older age. It is characterized by a granulomatous inflammatory reaction within the vessel wall with accumulation of T-cells, macrophages, and multinucleated giant cells. Its etiology is unknown, but variations of incidence with grouping of cases in certain areas have suggested that environmental infections could play a role (173). The predilection of the disease in caucasians from northern European countries and northern areas in the USA suggest that, in addition to environmental factors genetic factors are involved. HLA-typing has shown an association with HLA-DR4 and with HLA-DRB1 alleles, particularly with DRB*04 (174,175), as in rheumatoid arthritis. This association was accompanied by corticosteroid resistance (175), suggesting a more severe variety of the disease, this was not, however, confirmed in another study (174). HLA associations may be mechanistically be explained in different ways, but point to the involvement of antigen-specific T lymphocytes in the pathogenesis of the associated disorder. To further investigate the nature of the antigen-specific T cell response in GCA, Martinez-Taboada et al. (176) analyzed the presence of clonally expanded T cells in biopsies from temporal arteries. Systematic screening of the T cell receptor β-chain repertoire from 8 biopsies revealed clonally expanded T cells in 30% of the V-J combinations. Some of these clones were present at different sites in the biopsy, but not in the peripheral blood. Sequencing showed a diversity of Vβ-chain sequences. One of the T cell clonotypes showed proliferation when incubated with monocytes pulsed with temporal artery extracts from patients, but not with extracts from control temporal arteries. These data point to an (auto)antigen-specific T cell response in which a modified antigen may be involved. In this respect, actinically-degenerated elastic tissue has been suggested as the relevant autoantigen (177), although characterization of the precise antigenic structures and their modification(s) has not been accomplished. Further proof for the role of T cells in the pathogenesis of GCA was obtained by implanting diseased temporal artery specimens from GCA patients into SCID mice to study the T cell-dependency of the lesions (178). The inflammatory infiltrate persisted after implantation, with expansion of certain T cells producing interleukin-2 (IL-2) and interferon γ, inducing the production of IL-1β and IL-6 by monocytes/macrophages. The selective expansion of those T cells in the context of diseased arteries suggests a locally-expressed antigen and an ensuing Th-1 type reaction.

Analysis of cytokine patterns in biopsies from patients with GCA also shows mRNA expression of IL-2 and IFNγ as well as IL-1β, but not of IL-10 (179). Interestingly, the presence of giant cells was associated with local synthesis of IFNγ. Thus, in vivo data from patient material also points to a Th-1 type of response in which macrophages are important effector cells. Those macrophages and multinucleated giant cells produce platelet-derived growth factor (PDGF) A and B (180), which might, at least in part, be responsible for intimal proliferation and luminal narrowing. Such activated macrophages also produce nitric oxide radicals and reactive oxygen species that, together, induce tissue damage by lipid peroxidation and peroxynitrite-induced nitration of tyrosine residues on proteins (181). The dual role of nitric oxide (NO) in vasculitis, i.e. the protective effect of NO generated by endothelium derived nitric oxyde synthase (NOS) with respect to endothelial relaxation and anti-adhesion, versus the harmful effect of NO and NO-radicals induced by inducable NOS from inflammatory cells, is under investigation (182).

In conclusion, T cell-based autoimmunity seems to underlie giant cell arteritis. Characterization of the relevant auto-antigen(s) and epitopes, whether or not they are constitutive or modified proteins, is still a major challenge.

4.   CONCLUSION

The primary systemic vasculitides constitute a group of clinically-heterogeneous disorders characterized by idiopathic inflammation of the vessel wall. Exogenous, genetic and particularly autoimmune factors seem to be involved in their pathogenesis. Autoimmune responses to myeloid lysosomal proteins, particularly proteinase 3 and myeloperoxidase, are involved in the small vessel vasculitides, and a wealth of data now support their pathogenetic role. Anti-endothelial antibodies are present in many cases of systemic vasculitis, but await further characterization. In large vessel vasculitides, specific T cell responses, possibly directed to modified vessel wall proteins, seem to underlie the disease process but, again, characterization of the target antigens remains to be accomplished. Elucidation of the precise pathogenetic mechanisms will allow more specific treatment for this group of diseases, which have a high degree of morbidity and mortality.

References

1. Jennette, J.C., R.J. Falk, K. Andrassy, P.A. Bacon, J. Churg, W.L. Gross, E.C. Hagen, G.S. Hoffman, G.G. Hunder, C.G.M. Kallenberg, R.T. McCluskey, R.A. Sinico, A.J. Rees, L.A. van Es, R. Waldherr, and A. Wiik. 1994. Nomenclature of systemic vasculitides. Proposal of an international consensus conference. *Arthritis Rheum.* 37:187–192.
2. Lie, J.T. 1989. Systemic and isolated vasculitis: a rational approach to classification and pathologic diagnosis. *Pathol. Annu.* 24:25–114.

3. Woude, F.J. van der, N. Rasmussen, S. Lobatto, A. Wiik, H. Permin, L.A. van Es, M. van der Giessen, G.K. van der Hem, and T.H. The. 1985. Autoantibodies to neutrophils and monocytes: a new tool for diagnosis and a marker of disease activity in Wegener's Granulomatosis. *Lancet* ii:425–429.

4. Kallenberg, C.G.M., E. Brouwer, J.J. Weening, and J.W. Cohen Tervaert. 1994. Anti-neutrophil cytoplasmic antibodies: current diagnostic and pathophysiological potential. *Kidney Int.* 46:1–15.

5. Belizna, C., and J.W. Cohen Tervaert. 1997. Specificity, pathogenicity, and clinical value of antiendothelial cell antibodies. *Seminars Arthritis Rheum.* 27:98–109.

6. Leung, D.Y.M., T. Collins, L.A. Lapierre, R.S. Geha, and J.S. Pober. 1986. IgM antibodies present in the acute phase of Kawasaki syndrome lyse cultured vascular EC stimulated by gamma interferon. *J. Clin. Invest.* 77:1428–1435.

7. Weyand, C.M., and J.J. Goronzy. 1999. Arterial wall injury in giant cell arteritis. *Arthritis Rheum.* 42:844–853.

8. Somer, T., and S.M. Finegold. 1995. Vasculitides associated with infections, immunization, and antimicrobial drugs. *Clin. Infect. Dis.* 20:101–136.

9. Hunder, G.G., W.P. Arend, D.A. Bloch, L.H. Calabrese, A.S. Fauci, J.F. Fries, R.Y. Leavitt, J.T. Lie, R.W. Lightfoot, Jr., A.T. Masi, D.J. McShane, B.A. Michel, J.A. Mills, M.B. Stevens, S.L. Wallace, and N.J. Zvaiffler. 1990. The American College of Rheumatology 1990 criteria for the classification of vasculitis. Introduction. *Arthritis Rheum.* 33:1065–1067.

10. Nordberg, E., B.A. Bengtsson, and C. Nordberg. 1991. Temporal artery morphology and morphometry in giant cell arteritis. *Acta Pathol. Microbiol. Immunol. Scand.* 99:1013–1023.

11. Watts, R.A., V.A. Jolliffe, D.M. Carruthers, M. Lockwood, and D.G. Scott. 1996. Effect of classification on the incidence of polyarteritis nodosa and microscopic polyangiitis. *Arthritis Rheum.* 39:1208–1212.

12. Abe, J., B.L. Kotzin, K. Jujo, M.E. Melish, M.P. Glode, T. Kohsaka, and D.Y. Leung. 1992. Selective expansion of T cells expressing T-cell receptor variable regions Vβ2 *and* Vβ8 in Kawasaki disease. *Proc. Natl. Acad. Sci. USA* 89:4066–4070.

13. Hoffman, G.S., G.S. Kerr, R.Y. Leavitt, C.W. Hallahan, R.S. Lebovics, and W.D. Travis. 1992. Wegener Granolomatosis: an analysis of 158 patients. *Ann. Intern. Med.* 116:488–498.

14. Hewins, P., J.W. Cohen Tervaert, C.O.S. Savage, and C.G.M. Kallenberg. 2000. Is Wegener's granulomatosis an autoimmune disease? *Curr. Opin. Rheumatol.*,12:3–10.

15. Davies, D.J., J.E. Moran, J.F. Niall, and G. Ryan. 1982. Segmental necrotizing glomerulonephritis with anti-neutrophil antibody: possible arbovirus aetiology. *Br. Med. J.* 285:606.

16. Goldschmeding, R., C.E. van der Schoot, D. ten Bokkel Huinink, C.E.Hack, M.E. van den Ende, C.G.M. Kallenberg, A.E.G.Kr. von dem Borne. 1989. Wegener's Granulomatosis autoantibodies identify a novel diisopropylfluorophosphate-binding protein in the lysosomes of normal human neutrophils. *J. Clin. Invest.* 84:1577–1587.

17. Niles, J.L., R.T. McCluskey, M.F. Ahmad, and M.A. Arnaout. 1989.: Wegener's Granulomatosis autoantigen is a novel serine proteinase. *Blood* 74:1888–1893.

18. Jennette, J.C., J.R. Hoidal, and R.J. Falk. 1990. Specificity of anti-neutrophil cytoplasmic autoantibodies for proteinase 3. *Blood* 75:2263–2264.

19. Lüdemann, J., B. Utecht, and W.L. Gross. 1990. Anti-neutrophil cytoplasm antibodies in Wegener's Granulo-

matosis recognize an elastinolytic enzyme. *J. Exp. Med.* 171:357–362.

20. Kao, R.C., M.G. Wehner, K.M. Skubitz, B.H. Gray, and J.R. Hoidal. 1988. Proteinase 3. A distinct human polymorphonuclear leucocyte proteinase that produces emphysema in hamsters. *J. Clin. Invest.* 82:1963–1973.

21. Campanelli, D., M. Melchior, Y. Fu, N. Nakata, H. Shuman, C. Nathan, and J.E. Gabay. 1990. Cloning of cDNA for proteinase 3: a serine protease, antibiotic, and autoantigen from human neutrophils. *J. Exp. Med.* 172:1709–1715.

22. Campanelli, D., P.A. Detmers, C.F. Nathan, and J.E. Gabay. 1990. Azurocidin and a homologous serine protease from neutrophils. Differential antimicrobial and proteolytic properties. *J. Clin. Invest.* 85:904–915.

23. Bories, D., M.C. Raynal, D.H. Solomon, Z. Darzynkiewicz, and Y. Cayre. 1990. Down-regulation of a serine protease, myeloblastin, causes growth arrest and differentiation of promyelocytic leukemia cells. *Cell* 59:959–968.

24. Jenne, D.E. 1996. Gene structure of ANCA target antigens: implications for the pathogenesis of vasculitis. *Sarcoidosis Vasculitis and Diffuse Lung Diseases* 13:209–213.

25. Charles, L.A., R.J. Falk, and J.C. Jennette. 1992. Reactivity of neutrophil cytoplasmic autoantibodies with mononuclear phagocytes. *J. Leukocyte Biol.* 51:65–68.

26. King, W.J., D. Adu, M.R. Daha, C.J. Brooks, D.J. Radford, A.A. Pall, and C.O. Savage. 1995. Endothelial cells and renal epithelial cells do not express the Wegener's autoantigen, proteinase 3. *Clin. Exp. Immunol.* 102:98–105.

27. Mayet, W.J., E. Csernok, C. Szymkowiak, W.L. Gross, and K.H. Meyer zum Büschenfelde. 1993. Human endothelial cells express proteinase 3, the target antigen of anticytoplasmic antibodies in Wegener's Granulomatosis. *Blood* 82:1221–1229.

28. Mayet, W.J., A. Schwarting, T. Orth, R. Duchmann, and K.H. Meyer zum Buschenfelde. 1996. Antibodies to proteinase 3 mediate expression of vascular cell adhesion molecule-1 (VCAM-1). *Clin. Exp. Immunol.* 103:259–267.

29. Sibelius, U., K. Hattar, A. Schenkel, T. Noll, E. Csernok, W.L. Gross, W.J. Mayet, H.M. Piper, W. Seeger, and F. Grimminger. 1998. Wegener's granulomatosis: anti-proteinase 3 antibodies are potent inductors of human endothelial cell signaling and leakage response. *J. Exp. Med.* 187:497–503.

30. Mayet, W.J., E. Hermann, E. Csernok, A. Knuth, T. Poralla, W.L. Gross, and K.H. Meyer zum Büschenfelde. 1991. A human renal cancer line as a new antigen source for the detection of antibodies to cytoplasmic and nuclear antigens in sera of patients with Wegener's Granulomatosis. *J. Immunol. Methods* 143:57–64.

31. Brockmann, H., A. Schwarting, J. Kriegsmann, P. Petrow, A. Gaumann, and W.J. Mayet. 1998. Expression of the major target antigen of ANCA (proteinase 3) in Wegener's granulomatosis in lung tissue. *Arthritis Rheum.* 41(S):S347.

32. Schwarting, A., J.F. Schlaak, E. Wandel, K.H. Meyer zum Buschenfelde, and W.J. Mayet. 1997. Human renal tubular epithelial cells as target cells for antibodies to proteinase 3 (c-ANCA). *Nephrol. Dial. Transpl.* 12:916–923.

33. Taekema-Roelvink, M.E.J., C. van Kooten, W. Schroeijers, E. Heemskerk, and M.R. Daha. 1998. Binding of proteinase 3 (Pr3) to human umbilical vein endothelial cells (HUVEC) in vitro. *Clin. Exp. Immunol.* 112 (suppl. 1):42.

34. Jenne, D.E., L. Fr–hlich, A.M. Hummel, and U. Specks. 1997. Cloning and functional expression of the murine homologue of proteinase 3: implications for the design of murine models of vasculitis. *FEBS Lett.* 408:187–190.

35. Kam, C.M., J.E. Kerrigan, K.M. Dolman, R. Goldschmeding, A.E.G.Kr. von dem Borne, and J.C. Powers. 1992. Substrate and inhibitor studies on Proteinase 3. *FEBS Lett.* 297:119–123.

36. Dolman, K.M., B.A. van de Wiel, C.M. Kam, J.J. Abbink, C.E. Hack, A. Sonnenberg, J.C. Powers, A.E.G.Kr. von dem Borne, and R. Goldschmeding. 1992. Determination of proteinase 3/alpha1-antitrypsin complexes in inflammatory fluids. *FEBS Lett.* 314:117–121.

37. Henshaw, T.J., C.C. Malone, J.E. Gabay, R.C. Williams, Jr. 1994. Elevations of neutrophil proteinase 3 in serum of patients with Wegener's granulomatosis and polyarteritis nodosa. *Arthritis Rheum.* 37:104–112.

38. Esnault, V.L.M., A. Testa, M. Audrain, C. Rogé, M. Hamidou, J.H. Barrier, R. Sesboué, J.P. Martin, Ph. Lesavre. 1993. Alpha 1-antitrypsin genetic polymorphism in ANCA-positive systemic vasculitis. *Kidney Int.* 43:1329–1332.

39. Charles, L.A., M.L.R. Caldas, F.J. Falk, R.S. Terrell, and J.C. Jennette. 1991. Antibodies against granule proteins activate neutrophils in vitro. *J. Leukocyte Biol.* 50:539–546.

40. Muller Kobold, A.C., Y.M. van der Geld, P.C. Limburg, J.W. Cohen Tervert, and C.G.M. Kallenberg. 1999. Pathophysiology of ANCA-associated glomerulonephritis. *Nephrol. Dial. Transplant* 14:1366–1375.

41. Halbwachs Mecarelli, L., G. Bessou, P. Lesavre, S. Lopez, and V. Witko Sarsat. 1995. Bimodal distribution of proteinase 3 (Pr3) surface expression reflects a constitutive heterogeneity in the polymorphonuclear neutrophil pool. *FEBS Lett.* 374:29–33.

42. Muller Kobold, A.C., C.G.M. Kallenberg, and J.W. Cohen Tervaert. 1998. Leukocyte membrane expression of proteinase 3 correlates with disease activity in patients with Wegener's granulomatosis. *Brit. J. Rheumatol.* 37:901–907.

43. Gilligan, H.M., B. Bredy, H.R.Brady, M.J. Hebert, H.S. Slayter, Y. Xu, J. Rauch, M.A. Shia, J.S. Koh, and J.S. Levine. 1996. Antineutrophil cytoplasmic autoantibodie interact with primary granule constituents on the surface of apoptotic neutrophils in the absence of neutrophil priming. *J. Exp. Med.* 184:2231–2241.

44. Wiik, A. 1989. Delineation for a standard procedure for indirect immunofluorescence detection of ANCA. *APMIS* 97:S12–S13.

45. Hagen, E.C., K. Andrassy, E. Csernok, M.R. Daha, G. Gaskin, W.L. Gross, B. Hansen, Z. Heigl, J. Hermans, D. Jayne, C.G. Kallenberg, P. Lesavre, C.M. Lockwood, J. Lüdemann, F. Mascart-Lemone, E. Mirapeix, C.D. Pusey, N. Rasmussen, R.A. Sinico, A. Tzioufas, J. Wieslander, A. Wiik, and F.J. van der Woude. 1996. Development and standardization of solid phase assays for the detection of antineutrophil cytoplasmic antibodies (ANCA): a report on the second phase of an international cooperative study on the standardization of ANCA assays. *J. Immunol. Meth.* 196:1–15.

46. Cohen Tervaert, J.W., R. Goldschmeding, J.D. Elema, M. van der Giessen, M.G. Huitema, G.K. van der Hem, T.H. The, A.E.G.Kr. von dem Borne, and C.G.M. Kallenberg. 1990. Autoantibodies against myeloid lysosomal enzymes in crescentic glomerulonephritis. *Kidney Int.* 37:799–806.

47. Baslund, B., M. Segelmark, A. Wiik, W. Sapirt, J. Petersen, and J. Wieslander. 1995. Screening for anti-neutrophil cytoplasmic antibodies (ANCA): is indirect immunofluorescence the method of choice? *Clin. Exp. Immunol.* 99:486–492.

48. Westman, K.W., D. Selga, P. Bygren, M. Segelmark, B. Baslund, A. Wiik, and J. Wieslander. 1998. Clinical evaluation of a capture ELISA for detection of proteinase-3 antineutrophil cytoplasmic antibody. *Kidney Int.* 53:1230–1236.

49. Harmsen, M.C., P. Heeringa, Y.M. van der Geld, M.G. Huitema, A. Klimp, A. Tiran, and C.G.M. Kallenberg. 1997. Recombinant proteinase-3 (Wegener's antigen) expressed in *Pichia pastoris* is functionally active and is recognised by patient sera. *Clin. Exp. Immunol.* 110:257–264.

50. Specks, U., E.M. Wiegert, and H.A. Homburger. 1997. Human mast cells expressing recombinant proteinase 3 (Pr3) as substrate for clinical testing for anti-neutrophil cytoplasmic antibodies (ANCA). *Clin. Exp. Immunol.* 109:286–295.

51. Sun, J., D.N. Fass, J.A. Hudson, M.A. Viss, H.A. Homburger, and U. Specks. 1998. Capture-ELISA based on recombinant proteinase 3 (Pr3) is sensitive for Pr3-ANCA testing and allows detection of Pr3 and Pr3-ANCA/Pr3 immune complexes. *J. Immunol. Meth.* 221:111–123.

52. Falk, R.J., and J.C. Jennette. 1988. Anti-neutrophil cytoplasmic autoantibodies with specificity for myeloperoxidase in patients with systemic vasculitis and idiopathic necrotizing and crescentic glomerulonephritis. *N. Engl. J. Med.* 318:1651–1657.

53. Cohen Tervaert, T.W.R. Goldschmeding, J.D. Elema, P.C. Limburg, M. van der Giessen, M.G. Huitema, M.I. Koolen, R.J. Hené, T.H. The, G.K. van der Hem, A.E.G.Kr. von dem Borne, C.G.M. Kallenberg. 1990. Association of autoantibodies to myeloperoxidase with different forms of vasculitis. *Arthritis Rheum.* 33:1264–1272.

54. Heeringa, P., E. Brouwer, P.A. Klok, M.G. Huitema, J. van den Borne, J.J. Weening, C.G.M. Kallenberg. 1996. Autoantibodies to myeloperoxidase aggravate mild anti glomerular-basement-membrane-mediated glomerular injury in the rat. *Am. J. Pathol.* 149:1695–1706.

55. Heeringa, P., E. Brouwer, J.W. Cohen Tervaert, J.J. Weening, and C.G.M. Kallenberg. 1998. Animal models of anti-neutrophil cytoplasmic antibody associated vasculitis. *Kidney Int.* 53:253–263.

56. Weiss, S.J. 1989. Tissue destruction by neutrophils. *N. Engl. J. Med.* 320:365–376.

57. Segelmark, M., B. Persson, T. Hellmark, and J. Wieslander. 1997. Binding and inhibition of myeloperoxidase (MPO): a major function of ceruloplasmin? *Clin. Exp. Immunol.* 108:167–174.

58. Griffin, S.V., P.T. Chapman, E.A. Lianos, and C.M. Lockwood. 1999. The inhibition of myeloperoxidase by ceruloplasmin can be reversed by anti-myeloperoxidase antibodies. *Kidney Int.* 55:917–925.

59. Charles, L.A., R.J. Falk, and J.C. Jennette. 1989. Reactivity of anti-neutrophil cytoplasmic autoantibodies with HL-60 cells. *Clin. Immunol. Immunopathol.* 53:243–253.

60. Kallenberg, C.G.M., A.H.L. Mulder, and J.W. Cohen Tervaert. 1992. Antineutrophil cytoplasmic antibodies: a still growing class of autoantibodies in inflammatory disorders. *Am. J. Med.* 93:675–682.

61. Hoffman, G.S., and U. Specks. 1998. Antineutrophil cytoplasmic antibodies. *Arthritis Rheum.* 41:1521–1537.

62. Roozendaal, C., and C.G.M. Kallenberg. 1999. Are anti-neutrophil cytoplasmic antibodies (ANCA) clinically useful in inflammatory bowel disease? *Clin. Exp. Immunol.* 116:206–213.

63. Audrain, M.A., T.A. Baranger, C.M. Lockwood, and V.L. Esnault. 1994. High immunoreactivity of lactoferrin contaminating commercially purified myeloperoxidase. *J. Immunol. Methods* 176:23–31.

64. Cohen Tervaert, J.W., A.H.L. Mulder, C.A. Stegeman, J.D. Elema, M.G. Huitema, T.H. The, and C.G.M. Kallenberg. 1993. The occurrence of autoantibodies to human leukocyte elastase in Wegener's Granulomatosis and other inflammatory disorders. *Ann. Rheum. Dis.* 52:115–120.

65. Elsbach, P., and J. Weiss. 1993.The bactericidal/permeability-increasing protein (BPI), a potent element in host-defence against Gram-negative bacteria and LPS. *Immunobiology* 87:417–429.

66. Zhao, M.H., S.J. Jones, and C.M. Lockwood. 1995. Bactericidal/permeability-increasing protein (BPI) is an important antigen for anti-neutrophil cytoplasmic auto-antibodies (ANCA) in vasculitis. *Clin. Exp. Immunol.* 99:49–56.

67. Zhao, M.H., D.R. Jayne, L.G. Ardiles, F. Culley, M.E. Hodson, and C.M. Lockwood. 1996. Autoantibodies against bactericidal/permeability-increasing protein (BPI) in cystic fibrosis. *Q. J. Med.* 89:259–265.

68. Roozendaal, C., A.W.M. van Milligen de Wit, E.B. Haagsma, G. Horst, C. Schwarze, H.H. Peter, J.H. Kleibeuker, J.W. Cohen Tervaert, P.C. Limburg, and C.G.M. Kallenberg.1998. Anti-neutrophil cytoplasmic antibodies (ANCA) in primary sclerosing cholangitis: defined specificities may be associated with distinct clinical features. *Am. J. Med.* 105:393–399.

69. Cohen Tervaert, J.W., F.J. van der Woude, A.S. Fauci, J.L. Ambrus, J. Velosa, W.F. Keane, S. Meijer, M. van der Giessen, T.H. The, G.K. van der Hem, and C.G.M. Kallenberg. 1989. Association between active Wegener's Granulomatosis and anticytoplasmic antibodies. *Arch. Intern. Med.* 149:2461–2465.

70. Nölle, B., V. Specks, J. Lüdemann, M.S. Rohrbach, D.A. De Remee, and W.L. Gross. 1989. Anticytoplasmic autoantibodies: their immunodiagnostic value in Wegener's Granulomatosis. *Ann. Int. Med.* 111:28–40.

71. Weber, M.F.A., K. Andrassy, O. Pullig, J. Koderisch, and K. Netzer. 1992. Antineutrophil cytoplasmic antibodies and antiglomerular basement membrane antibodies in Goodpasture's syndrome and Wegener's granulomatosis. *J. Am. Soc. Nephrol.* 2:1227–1234.

72. Rao, J.K., N.B. Allen, J.R. Feussner, and M. Weinberger. 1995. A prospective study of antineutrophil cytoplasmic antibody (c-ANCA) and clinical criteria in diagnosing Wegener's granulomatosis. *Lancet* 346:926–931.

73. Rao, J.K., M. Weinberger, E.Z. Oddone, N.B. Allen, P. Landsman, and J.R. Feussner. 1995.The role of antineutrophil cytoplasmic antibody testing in the diagnosis of Wegener granulomatosis. *Ann. Intern. Med.* 123:925–932.

74. Cohen Tervaert, J.W., P.C. Limburg, J.D. Elema, M.G. Huitema, G. Horst, T.H. The, and C.G.M. Kallenberg. 1991. Detection of autoantibodies against myeloid lysosomal enzymes: a useful adjunct to classification of patients with biopsy-proven necrotizing arteritis. *Am. J. Med.* 91:59–66.

75. Jayne, D.R.W., P.D. Marshall, S.J. Jones, and C.M. Lockwood. 1990. Autoantibodies to GBM and neutrophil cytoplasm in rapidly progressive glomerulonephritis. *Kidney Int.* 37:965–970.

76. Bosch, X., E. Mirapeix, J. Font, X. Borrellas, R. Rodriguez, A. Lopez-Soto, M. Ingelmo, and L. Revert. 1991. Prognostic implication of anti-neutrophil cytoplasmic autoantibodies with myeloperoxidase specificity in anti-glomerular basement membrane disease. *Clin. Nephrol.* 36:107–113.

77. Merkel, P.A., R.P. Polisson, Y. Chang, S.J. Skates, and J.L.Niles.1997. Prevalence of antineutrophil cytoplasmic antibodies in a large inception cohort of patients with connective tissue disease. *Ann. Intern. Med.* 126:866–873.

78. Nässberger, L., H. Sjöholm, Jonsson, G. Sturfelt, and A. Äkessen. 1990. Autoantibodies against neutrophil cytoplasm components in systemic lupus erythematosus and hydralazine-induced lupus. *Clin. Exp. Immunol.* 81:380–383.

79. Dolman, K.M., R.O.B. Gans, T.H.J. Vervaart, G. Zevenbergen, D. Maingay, R.E. Nikkels, A.J.M. Donker, A.E.G.K.R. Von dem Borne, and R. Goldschmeding. 1993. Vasculitic disorders and anti-neutrophil cytoplasmic autoantibodies associated with propylthiouracil therapy. *Lancet* 342:651–652.

80. Merkel, P.A. 1998. Drugs associated with vasculitis. *Curr. Opinion Rheumatol.* 10:45–50.

81. Franssen, C.F.M., R.O.B. Gans, B. Arends, C. Hageluken, P.M. ter Wee, P.G.G. Gerlag, and S.J. Hoorntje. 1995. Differences between anti-myeloperoxidase and antiproteinase 3 associated renal disease. *Kidney Int.* 47:193–199.

82. Franssen, C.F.M., R. Gans, C.G.M. Kallenberg, C. Hageluken, and S.J. Hoorntje. 1998. Disease spectrum of patients with antineutrophil cytoplasmic autoantibodies of defined specificities: distinct differences between patients with anti-proteinase 3 and anti-myeloperoxidase auto-antibodies. *J. Intern. Med.* 244:209–216.

83. Franssen, C.F.M., M.G. Huitema, A.C. Muller Kobold, W. Oost-Kort, P.C. Limburg, A. Tiebosch, C.A. Stegeman, C.G.M. Kallenberg, and J.W. Cohen Tervaert. 1999. in vitro neutrophil activation by antibodies to proteinase 3 and myeloperoxidase from patients with crescentic glomerulonephritis. *J. Am. Soc. Nephrol.* 10:1506–1515.

84. Hagen, E.C., M.R. Daha, J. Hermans, K. Andrassy, E. Csernok, G. Gaskin, P. Lesavre, J. Lüdemann, N. Rasmussen, R.A. Sinico, A. Wiik, and F.J. van der Woude. 1998. Diagnostic value of standardized assays for anti-neutrophil cytoplasmic antibodies in idiopathic systemic vasculitis. EC/BCR project for ANCA assay standardization. *Kidney Int.* 53:743–753.

85. Brouwer, E., J.W. Cohen Tervaert, G. Horst, M.G. Huitema, M. van der Giessen, P.C. Limburg, and C.G.M. Kallenberg. 1991. Predominance of IgG4 subclass of anti-neutrophil cytoplasmic autoantibodies in patients with Wegener's Granulomatosis and clinically related disorders. *Clin. Exp. Immunol.* 83:379–386.

86. Esnault, V.L., B. Soleimani, M.T. Keogan, A.A. Brownlee, D.R. Jayne, and C.M. Lockwood. 1992. Association of IgM with IgG ANCA in patients presenting with pulmonary hemorrhage. *Kidney Int.* 41:1304–1310.

87. Jayne, D.R., A.P. Weetman, and C.M. Lockwood. 1991. IgG subclass distribution of autoantibodies to neutrophil cytoplasmic antigens in systemic vasculitis. *Clin. Exp. Immunol.* 84:476–481.

88. Williams, R.C., Jr., R. Staud, C.C. Malone, J. Payabyab, L. Byres, and D. Underwood. 1994. Epitopes on proteinase-3 recognized by antibodies from patients with Wegener's granulomatosis. *J. Immunol.* 152:4722–4737.

89. Bini, P., J.E. Gabay, A. Teitel, M. Melchior, J.L. Zhou, and K.B. Elkon. 1992. Antineutrophil cytoplasmic autoantibodies in Wegener's Granulomatosis recognize conformational epitope(s) on proteinase 3. *J. Immunol.* 149:1409–1415.

90. Geld, Y.M., van der Limburg, P.C. and C.G.M. Kallenberg. 1999. Characterization of monoclonal antibodies to Proteinase 3 as candidates tools for epitope mapping of human anti-Proteinase 3 autoantibodies. *Clin. Exp. Immunol.*, 118:487–496.

91. Audrain, M.A.P., T.A.R. Baranger, N. Moguilevski, S.J. Martin, A. Devys, C.M. Lockwood, J.Y. Muller, and V.L.M. Esnault. 1997. Anti-native and recombinant

myeloperoxidase monoclonals and human autoantibodies. *Clin. Exp. Immunol.* 107:127–134.

92. Brouwer. E., C.A. Stegeman, M.G. Huitema, P.C. Limburg, C.G.M. Kallenberg. 1994. T-cell reactivity to proteinase 3 and myeloperoxidase in patients with Wegener's granulomatosis. *Clin. Exp. Immunol.* 98:448–453

93. Griffith, M.E., A. Coulhar, and C.D. Pusey. 1996. T cell responses to myeloperoxidase (MPO) and proteinase 3 (PR3) in patients with systemic vasculitis. *Clin. Exp. Immunol.* 103:253–258.

94. King, W.J., C.J. Brooks, R. Holder, P. Hughes, D. Adu, and C.O.S. Savage. 1998. T lymphocyte response to anti-neutrophil cytoplasmic autoantibody (ANCA) antigens are present in patients with ANCA associated systemic vasculitis and persist during disease remission. *Clin. Exp. Immunol.* 112:539–546.

95. Ballieux, B.E., S.H. van der Berg, E.C. Hagen, F.J. van der Woude, C.J. Melief, and M.R. Daha. 1995. Cell-mediated autoimmunity in patients with Wegener's granulomatosis. *Clin. Exp. Immunol.* 100:186–193.

96. Mathieson, P.W., S. Thiru, and D.B.G. Oliveira. 1992. Mercuric chloride-treated Brown Norway rats develop widespread tissue injury including necrotizing vasculitis. *Lab. Invest.* 67:121–129.

97. Egner, W., and H.M. Chapel. 1990. Titration of antibodies against neutrophil cytoplasmic antigens is useful in monitoring disease activity in systemic vasculitides. *Clin. Exp. Immunol.* 82:244–249.

98. De'Oliviera, J., G. Gaskin, A. Dash, A.J. Rees, and C.D. Pusey. 1995. Relationship between disease activity and anti-neutrophil cytoplasmic antibody concentration in long-term management of systemic vasculitis. *Am. J. Kidney. Dis.* 25:380–389.

99 Jayne, D.R.W., G. Gaskin, C.D. Pusey, and C.M. Lockwood. 1995. ANCA and predicting relapse in systemic vasculitis. *Q. J. Med.* 88:127–133.

100. Kerr, G.R., T.H.A. Fleischer, C.W. Hallahan, R.Y. Leavitt, A.S. Fauci, and G.S. Hoffman. 1993. Limited prognostic value of changes in antineutrophil cytoplasmic antibody titer in patients with Wegener's granulomatosis. *Arthritis Rheum.* 36:365–371.

101. Cohen Tervaert, J.W., C.A. Stegeman, and C.G.M. Kallenberg. 1996. Serial ANCA testing is useful in monitoring disease activity of patients with ANCA-associated vasculitides. *Sarcoidosis* 13:241–245.

102. Stegeman, C.A, J.W. Cohen Tervaert, W.J. Sluiter, W. Manson, P.E. de Jong, and C.G.M. Kallenberg. 1994. Association of chronic nasal carriage of Staphylococcus Aureus and higher relapse rates in Wegener's granulomatosis. *Ann. Intern. Med.* 113:12–17.

103. Stegeman, C.A., J.W. Cohen Tervaert, P.E. de Jong, and C.G.M. Kallenberg. 1996. Trimethoprim-sulfamethoxazole for the prevention of relapses of Wegener's granulomatosis. *N. Engl. J. Med.* 335:16–20.

104. Cohen Tervaert, J.W., E.R. Popa, and N.A. Bos. 1999. The role of superantigens in vasculitis. *Curr. Opinion Rheumatol.* 11:24–33.

105. Falk, R.J., R.S. Terrell, L.A. Charles, and J.C. Jennette. 1990. Anti-neutrophil cytoplasmic autoantibodies induce neutrophils to degranulate and produce oxygen radicals in vitro. *Proc. Natl. Acad. Sci USA* 87:4115–4119.

106. Kettritz, R., J.C. Jennette, and R.J. Falk. 1997. Crosslinking of ANCA antigens stimulates superoxide release by human neutrophils. *J. Am. Soc. Nephrol.* 8:386–394.

107. Porges, A.J., P.B. Redecha, W.T. Kimberly, E. Csernok, W.L. Gross, and R.P. Kimberly. 1994. Anti-neutrophil cytoplasmic antibodies engage and activate human neutrophils via Fc gamma RIIa. *J. Immunol.* 153:1271–1280.

108. Mulder, A.H., P. Heeringa, E. Brouwer, P.C. Limburg, and C.G.M. Kallenberg. 1994. Activation of granulocytes by anti-neutrophil cytoplasmic antibodies (ANCA): a Fc gamma RII-dependent process. *Clin. Exp. Immunol.* 98:270–278.

109. Edberg, J.C., M. Kocher, H. Fleit, and R.P. Kimberly. 1998. Anti-neutrophil cytoplasmic antibodies preferentially engage Fc RIIIb on neutrophils. *Arthritis Rheum.* 41(S):S243.

110. Winkel, J.G.J. van de, and J.A. Capel. 1993. Human IgG Fc receptor heterogeneity: molecular aspects and clinical implications. *Immunology Today* 14:215–221.

111. Mulder, A.H., C.A. Stegeman, and C.G.M. Kallenberg. 1995. Activation of granulocytes by anti-neutrophil cytoplasmic antibodies (ANCA) in Wegener's granulomatosis: a predominant role for the IgG3 subclass of ANCA. *Clin. Exp. Immunol.* 101:227–232.

112. Reumaux, D., P.J. Vossebeld, D. Roos, and A.J. Verhoeven. 1995. Effect of tumor necrosis factor-induced integrin activation on Fc gamma receptor II-mediated signal transduction: relevance for activation of neutrophils by anti-proteinase 3 or anti-myeloperoxidase antibodies. *Blood* 86:3189–3195.

113. Grimminger, F., K. Hattar, C. Papavassilis, B. Temmesfeld, E. Csernok, and W.L. Gross. 1996. Neutrophil activation by anti-proteinase 3 antibodies in Wegener's granulomatosis: Role of arachidonic acid and leukotriene B4 generation. *J. Exp. Med.* 184:1567–1572.

114. Brooks, C.J., W.J. King, D.J. Radford, A. Adu, M. McGrath, and C.O.S. Savage. 1996. IL-1 beta production by human polymorphonuclear leukocytes stimulated by anti neutrophil cytoplasmic autoantibodies: relevance to systemic vasculitis. *Clin. Exp. Immunol.* 106:273–279.

115. Johnson, P.A., H.D. Alexander, S.A. McMillan, and A.P. Maxwell. 1997. Up-regulation of the granulocyte adhesion molecule Mac-1 by autoantibodies in autoimmune vasculitis. *Clin. Exp. Immunol.* 107:513–519.

116. Ewert, B.H., J.C. Jennette, and R.J. Falk. 1992. Anti-myeloperoxidase antibodies stimulate neutrophils to damage human endothelial cells. *Kidney Int.* 41:375–383.

117. Savage, C.O., B.E. Pottinger, G. Gaskin, C.D. Pusey, and J.D. Pearson. 1992. Autoantibodies developing to myeloperoxidase and proteinase 3 in systemic vasculitis stimulate neutrophil cytotoxicity toward cultured endothelial cells. *Am. J. Pathol.* 141:335–342.

118. Savage, C.O., G. Gaskin, C.D. Pusey, and J.D. Pearson. 1993. Anti-neutrophil cytoplasm antibodies can recognize vascular endothelial cell-bound anti-neutrophil cytoplasm antibody-associated autoantigens. *Exp. Nephrol.* 1:190–195.

119. Ballieux, B.E., K.T. Zondervan, P. Kievit, E.C. Hagen, L.A. van Es, F.J. van der Woude, and M.R. Daha. 1994. Binding of proteinase 3 and myeloperoxidase to endothelial cells: ANCA-mediated endothelial damage through ADCC? *Clin. Exp. Immunol.* 97:52–60.

120. Berger, S.P., M.A. Seelen, P.S. Hiemstra, J.S. Gerritsma, E. Heemskerk, F.J. van der Woude, and M.R. Daha. 1996. Proteinase 3, the major autoantigen of Wegener's granulomatosis, enhances IL-8 production by endothelial cells in vitro. *J. Am. Soc. Nephrol.* 7:694–701.

121. Yang, J.J., R. Kettritz, R.J. Falk, J.C. Jennette, and M.L. Gaibo. 1996. Apoptosis of endothelial cells induced by neutrophil serine proteases proteinase 3 and elastase. *Am. J. Pathol.* 149:1617–1626.

122. Ballieux, B.E., P.S. Hiemstra, N. Klar-Mohamad, E.C. Hagen, L.A. van Es, F.J. van der Woude, and M.R. Daha. 1994.

Detachment and cytolysis of human endothelial cells by proteinase 3. *Eur. J. Immunol.* 24:3211–3215.

123. Ewert, B.H., J.C. Jennette, and R.J. Falk. 1991. The pathogenetic role of antineutrophil cytoplasmic autoantibodies. *Am. J. Kidney Dis. 18:188–195.*

124. Ralston, D.R., C.B. Marsh, M.P. Lowe, and M.D. Wewers. 1997. Antineutrophil cytoplasmic antibodies induce monocyte IL-8 release. Role of surface proteinase 3, alpha1-antitrypsin, and Fcgamma receptors. *J. Clin. Invest.* 100:1416–1424.

125. Casselman, B.L., K.S. Kilgore, B.F. Miller, and J.S. Warren. 1995. Antibodies to neutrophil cytoplasmic antigens induce monocyte chemoattractant protein-1 secretion from human monocytes. *J. Lab. Clin. Med.* 126:495–502.

126. Muller Kobold, A.C., C.G.M. Kallenberg, and J.W. Cohen Tervaert. 1999. Monocyte activation in patients with Wegener's granulomatosis. *Ann. Rheum. Dis.* 58:237–245.

127. Horn, R.G., A.S. Fauci, A.S. Rosenthal, and S.M. Wolff. 1974. Renal biopsy pathology in Wegener's granulomatosis. *Am. J. Pathol.* 74:423–440.

128. Brandt, M., V. de Olliviec, O. Meyer, C. Babin-Chevaye, F. Kherhai, D. de Prost, J. Hakim, and C. Pasquier. 1997. Induction of interleukin-1 and subsequent tissue factor expression by anti-proteinase 3 antibodies in human umbilical vein endothelial cells. *Arthritis Rheum.* 40:2030–2038.

129. Wiel, B.A, van de, K.M. Dolman, C.H. van der Meer-Gerritsen, C.E. Hack, A.E.G.Kr. von dem Borne, and R. Goldschmeding. 1992. Interference of Wegener's granulomatosis autoantibodies with neutrophil proteinase 3 activity. *Clin. Exp. Immunol.* 90:409–414.

130. Dolman, K.M., C.A. Stegeman, B.A. van de Wiel, C.E. Hack, A.E.G.Kr. von dem Borne, C.G.M. Kallenberg, and R. Goldschmeding. 1993. Relevance of classic anti-neutrophil cytoplasmic autoantibody (cANCA)-mediated inhibition of proteinase 3α-1-antitrypsin complexation to disease activity in Wegener's granulomatosis. *Clin. Exp. Immunol.* 93:405–410.

131. Kinjoh, K., M. Kyoguko, and R.A. Good. 1993. Genetic selection for crescent formation yields mouse strain with rapidly progressive glomerulonephritis and small vessel vasculitis. *Proc. Natl. Acad. Sci. USA* 90:3413–3417.

132. Nose, M., M. Nishimura, M.R. Ito, J. Itoh, T. Shibata, and T. Sugisaki. 1996. Arteritis in a novel congenic strain of mice derived from MRL/*lpr* lupus mice. *Am. J. Pathol.* 149:1763–1768.

133. Brouwer, E., M.G. Huitema, P.A. Klok, J.W. Cohen Tervaert, J.J. Weening, and C.G.M. Kallenberg. 1993. Anti-myeloperoxidase associated proliferative glomerulonephritis: an animal model. *J. Exp. Med.* 177:905–914.

134. Brouwer, E., P.A. Klok, M.G. Huitema, J.J. Weening, and C.G.M. Kallenberg. 1995. Renal ischemia/reperfusion injury contributes to renal damage in experimental anti-myeloperoxidase-associated proliferative glomerulonephritis. *Kidney Int.* 47:1121–1129.

135. Heeringa, P., P. Foucher, P.A. Klok, M.G. Huitema, J.W. Cohen Tervaert, J.J. Weening, and C.G.M. Kallenberg. 1997. Systemic injection of products of activated neutrophils and H_2O_2 in myeloperoxidase-immunized rats leads to necrotizing vasculitis in the lungs and gut. *Am. J. Pathol.* 151:131–140.

136. Tomer, Y., B. Gilburd, M. Blank, O. Lider, R. Hershkoviz, P. Fishman, R. Zigelman, P.L. Meroni, A. Wiik, and Y. Shoenfeld. 1995. Characterization of biologically active antineutrophil cytoplasmic antibodies induced in mice. *Arthritis Rheum.* 38:1375–1381.

137. Blank, M., Y. Tomer, M. Stein, J. Kopolovic, A. Wiik, P.L. Meroni, G. Conforti, and Y. Shoenfeld. 1995. Immuniza-tion with anti-neutrophil cytoplasmic antibody (ANCA) induces the production of mouse ANCA and perivascular lymphocyte infiltration. *Clin. Exp. Immunol.* 102:120–130.

138. Lindqvist, K.J., and C.K. Osterland. 1971. Human antibodies to vascular endothelium. *Clin. Exp. Immunol.* 9:753–762.

139. Belizna, C., and J.W. Cohen Tervaert. 1997. Specificity, pathogenecity, and clinical value of antiendothelial cell antibodies. *Seminars Arthritis Rheum.* 27:98–109.

140. Meroni, P.L., N. Del Papa, E. Raschi, A. Tincani, G. Balestrieri, and P. Youinou. 1999. Antiendothelial cell antibodies (AECA): from a laboratory curiosity to another useful autoantibody. In: Shoenfeld Y (ed). The decade of autoimmunity. *Elsevier Science BV, Amsterdam.* pp. 285–294.

141. Savage, C.O.S., B.E. Pottinger, G. Gaskin, C.M. Lockwood, C.D. Pusey, and J.D. Pearson. 1991. Vascular damage in Wegener's granulomatosis and microscopic polyarteritis:presence of AECA and their relation to ANCA. *Clin. Exp. Immunol.* 85:14–19.

142. Del Papa, N., G. Conforti, D. Gambini, L. La Rosa, A. Tincani, D. D'Cruz, M. Khamashta, G.V.R. Hughes, G. Balestrieri, and P.L. Meroni. 1994. Characterisation of the endothelial surface proteins recognized by AECA in primary and secondary autoimmune vasculitis. *Clin. Immunol. Immunopathol.* 70:211–216.

143. Hashemi, S., C. Douglas Smith, and C.A. Izaguirre. 1987. AECA: Detection and characterisation using a cellular enzyme-linked immunosorbent assay. *J. Lab. Clin. Med.* 109:434–440.

144. Muller Kobold, A.C., R.T. van Wijk, C.F.M. Franssen, G. Molema, C.G.M. Kallenberg, and J.W. Cohen Tervaert. 1999. In vitro upregulation of E-selectin and induction of interleukin-6 in endothelial cells by autoantibodies in Wegener's granulomatosis and microscopic polyangiitis. *Clin. Exp. Rheumatol.* 17:433–440.

145. Hill, M.B., J.L. Phipps, R.J. Cartwright, A. Milford-Ward, M. Greaves, and P. Hughes. 1996. Antibodies to membranes of endothelial cells and fibroblasts in scleroderma. *Clin. Exp. Immunol.* 106:491–497.

146. Koenig, D.W., L. Barley-Maloney, and T.O. Daniel. 1993. A Western Blot Assay detects autoantibodies to cryptic endothelial antigens in thrombotic microangiopathies. *J. Clin. Immunol.* 13:204–210.

147. Ferraro, G., P.L. Meroni, A. Tincani, R.A. Sinico, W. Barcellini, A. Radice, G. Gregorini, M. Froldi, M.O. Borghi, and G. Balestrieri. 1990. AECA in patients with Wegener granulomatosis and micropolyarteritis. *Clin. Exp. Immunol.* 79:47–53.

148. Direskeneli, H., D. D'Cruz, A.M. Khamashta, and R.V. Hughes. 1994. Autoantibodies against EC,ECM,and human collagen IV in patients with systemic vasculitis. *Clin. Immunol. Immunopathol* 70:206–210.

149. Gabrielli, A., M. Montroni, S. Rupoli, M.L. Caniglia, F. De Lustro, and G. Danieli. 1988. A retrospective study of antibodies against basement membrane antigens (type IV collagen and laminin) in patients with primary and secondary Raynaud's phenomenon. *Arthrios Rheum.* 31:1433–1436.

150. Moreland, L.W., R.E. Gay, and S. Gay. 1991. Collagen antibodies in patients with vasculitis and systemic lupus erythematosus. *Clin. Immunol. Immunopathol.* 60:412–418.

151. Blaschek, M.A., M. Boehmen, J. Jouquan, A.M. Simitizis, S. Fifas, P. Le Goff, and P. Youinou. 1988. Relation of antivimentin antibodies to anticardiolipin antibodies in systemic lupus erythematosus. *Ann. Rheum. Dis.* 47:708–716.

152. Del-Papa, N., L. Guidali, L. Spatola, P. Bonara, M.O. Borghi, A. Tincani, G. Balestrieri, and P.L. Meroni. 1995. Relationship between anti-phospholipid and anti-endothelial cell antibodies III: beta 2 glycoprotein I mediates the antibody binding to endothelial membranes and induces the expression of adhesion molecules. *Clin. Exp. Rheumatol.* 13:179–186

153. Chan, T.M., G. Frampton, N.A. Staines, P. Hobby, G.J. Perry, and J.S. Cameron. 1992. Different mechanisms by which anti-DNA MoAbs bind to human EC and glomerular mesangial cells. *Clin. Exp. Immunol.* 88:68–74.

154. Nakao, K., K. Ikeda, and M. Kimata. 1967. Takayasu's arteritis: clinical report of 84 cases and immunological studies of 7 cases. *Circulation* 53:1141–1144.

155. Chopra, P., R.K. Datta, A. Dasgupta, and S. Bhargava. 1983. Nonspecific aortaarteritis (Takayasu's disease): an immunologic and autopsy study. *Jpn. Heart J.* 24:549–556.

156. Brasile, L., J. Kremer, J. Clarke, and J. Cerrili. 1987. Identification of an autoantibody to vascular endothelial cell-specific antigen in patients with systemic vasculitis. *Am. J. Med.* 87:74–80.

157. Eichhorn, J., D. Sima, B. Thiele, C. Lindschau, A. Turowski, H. Schmidt, W. Schneider, H. Haller, and F.C. Luft. 1996. Anti-endothelial cell antibodies in Takayasu arteritis. *Circulation* 94:2396–2401.

158. Salojin, K.V., M. Le-Tonqueze, E.L. Nassovov, M.T. Blouch, A.A. Baranov, A. Saraux, L. Guillevin, J.N. Fiessinger, J.C. Piette, and P. Youinou. 1996. Anti-endothelial cell antibodies in patients with various forms of vasculitis. *Clin. Exp. Rheumatol.* 14:163–169.

159. Kaneko, K., C.O.S. Savage, B.E. Pottinger, V. Shah, J.D. Pearson, and M.J. Dillon. 1994. AECA can be cytotoxic to EC without cytokine prestimulation and correlate with ELISA antibody measurement in Kawasaki disease. *Clin. Exp. Immunol.* 8:264–269.

160. Leung, D.Y.M., R.S. Geha, J.W. Newburger, J.C. Burns, W. Fiers, L.A. Lapierre, and J.S. Pober. 1986. Two monokines, IL 1 and TNF, render cultured vscular EC susceptible to lysis by antibodies circulating during Kawasaki syndrome. *J. Exp. Med.* 164:1958–1972.

161. Gobel, Y., J. Eichhorn, R. Kottritz, L. Briedigkeit, D. Sima, C. Linschau, H. Haller, and F.C. Luft. 1996. Disease activity and autoantibodies to endothelial cells in patients with Wegener's granulomatosis. *Am. J. Kidney Dis.* 28:186–194.

162. Schmitt, W.H., E. Csernok, S. Kobayashi, A. Klinkenborg, E. Reinhold-Keller, and W.L. Gross. 1998. Churg-Strauss syndrome: serum markers of lymphocyte activation and endothelial damage. *Arthritis Rheum.* 41:445–452.

163. Del-Papa, N., L. Guidali, M. Sironi, Y. Shoenfeld, A. Mantovani, A. Tincani, G. Balestrieri, A. Radice, R.A. Sinico, and P.L. Meroni. 1996. Anti-endothelial cell IgG antibodies from patients with Wegener's granulomatosis bind to human endothelial cells *in vitro* and induce adhesion molecule expression and cytokine secretion. *Arthritis Rheum.* 39:758–766.

164. Chan, T.M., G. Frampton, D.R. Jayne, G.J. Perry, C.M. Lockwood, and J.S. Cameron. 1993. Clinical signifiance of AECA in systemic vasculitis: a longitudinal study comparing AECA and ANCA. *Am. J. Kidney Dis.* 22:387–392.

165. Mayet, W.J., and K.H. Meyer-zum-Buschenfelde. 1993. Antibodies to proteinase 3 increase adhesion of neutrophils to human endothelial cells. *Clin. Exp. Immunol.* 94:440–446.

166. Meroni, P.L., N.Del. Papa, B. Beltrami, A. Tincani, G. Balestrieri, and S.A. Krilis. 1996. Modulation of endo-thelial cell function by antiphospholipid antibodies. *Lupus* 5:448–450.

167. Bian, H., P.E. Harris, A. Mulder, and E.F. Reed. 1997. Anti-HLA antibody ligation to HLA class I molecules expressed by endothelial cells stimulates tyrosine phosphorylation, inositol phosphate generation, and proliferation. *Hum. Immunol.* 53:90–97.

168. Lindsey, N.J., F.I. Henderson, R. Malia, M.A. Milford-Ward, M. Greaves, and P. Hughes. 1994. Inhibition of prostacyclin release by endothelial binding anti-cardiolipin antibodies in thrombosis-prone patients with SLE and the anti-phospholipid syndrome. *Br. J. Rheumatol.* 33:20–26.

169. Bordron, A., M. Dueymes, Y. Levy, C. Jamin, J.P. Leroy, J.C. Piette, Y. Shoenfeld, and P.Y. Youinou. 1998. The binding of some human antiendothelial cell antibodies induces endothelial cell apoptosis. *J. Clin. Invest.* 101:2029–2035.

170. Bordron, A., M. Dueymes, Y. Levy, C. Jamin, L. Ziporen, J.C. Piette, Y. Shoenfeld, and P. Youinou. 1998. Anti-endothelial cell antibody binding makes negatively charged phospholipids accessible to antiphospholipid antibodies. *Arthritis Rheum.* 41:1738–1747.

171. Mantredi, A.A., P. Rovere, S. Heltai, G. Galati, G. Nebbia, A. Tincani, et al. 1998. Apoptotic cell clearance in systemic lupus erythematosus. II. Role of β2-glycoprotein I. *Arthritis Rheum.* 41:215–223.

172. Damianovich, M., B. Gilburd, J. George, N. Del-Papa, A. Afek, I. Goldberg, Y. Kopolovic, D. Roth, G. Barkai, P.L. Meroni, and Y. Shoenfeld. 1996. Pathogenic role of anti-endothelial cell antibodies in vasculitis. An idiotypic experimental model. *J. Immunol.* 156:4946–4951.

173. Hunder, G.G. 1998. Giant cell arteritis. *Lupus* 7:266–269.

174. Combe, B., J. Sany, A. Le Quellec, J. Clot, and J.F. Eliaou. 1998. Distribution of HLA-DRB1 alleles of patients with polymyalgia rheumatica and giant cell arteritis in a Mediterranian population. *J. Rheumatol.* 25:94–98.

175. Rauzy, O., M. Fort, F. Nourhashemi, L. Abric, H. Juchet, M. Ecoiffier, M. Abbal, and D. Adoue. 1999. Relation between JLA DRB1 alleles and corticosteroid resistance in giant cell arteritis. *Ann. Rheum. Dis.* 57:380–382.

176. Martinez-Taboada, V., N.N. Hunder, G.G. Hunder, C.M. Weyand, and J.J.Goronzy. 1996. Recognition of tissue resid-ing antigen by T cells in vasculitic lesions of giant cell arteritis. *J. Mol. Med.* 74:695–703.

177. O'Brien, J.P., Regan, W. 1998. Actinically degenerated elastic tissue: the prime antigen in the giant cell arteritis syndrome? New data from the posterior ciliary arteries. *Clin. Exp. Rheumatol.* 16:39–48.

178. Brack A., Geisler, A., Martinez-Taboada V, Younge B.R., Goronzy, J.J. Weynad, C.M. 1997. Giant cell asculitis is a T-cell dependent disease. *Mol. Med.* 3:530–543.

179. Weyand C.M. Tetzlaff, N., Bjornsson, J., Brack, A., Younge, B., Goronzy, J.J. 1997. Disease patterns and tissue cytokine profiles in giant cell arteritis. *Arthritis Rheum.* 40:19–26.

180. Kaiser M, Weyand, C.M., Bjornsson, J., Goronzy, J.J. 1998. Platelet-derived growth factor, intimal hyperplasia, and ischemic complications in giant cell arteritis. *Arthritis Rheum.* 42:623–633.

181. Weynad, C.M., Goronzy, J.J. 1999. Arterial wall injury in gaint cell arteritis. *Arthritis Rheum.* 42:844–853.

182. Heeringa, P, van Goor H., Huitema, M.G., Klok, P.A., Moshage, H., Kallenberg, C.G.M. 1998. Expression of inducible NOS, endothelial NOD and peroxynitrite modified proteins in experimental anti-myeloperoxidase associated crescentic glomerulonephritis. *Kidneys Int.* 53:382–393.

ORGAN-SPECIFIC AUTOIMMUNE DISEASES

29 | Autoimmune Hemolytic Anemia

Gregory A. Denomme, Catherine P. M. Hayward and John G. Kelton

1. INTRODUCTION

Autoimmune hemolytic anemia (AIHA) is a relatively uncommon autoimmune disease that is unique for two reasons: First, the diagnostic autoantibody for the disorder is the direct mediator of red cell destruction. Second, because both the autoantibody and its target (red cells) can easily be obtained, the antibody specificity and target antigen have been well characterized. While progress has been made in characterizing the target autoantigens and understanding the mechanism of autoantibody-mediated red cell destruction, much less is known about the underlying immune dysfunction that leads to the abnormal generation of the red cell autoantibodies.

AIHA can be classified in several ways: (a) its chronicity (acute or chronic); (b) the mechanism of cell destruction (immunoglobulin or complement); (c) the optimal temperature of autoantibody reactivity (warm or cold); and (d) whether the disorder is idiopathic or secondary to an underlying disorder. Most IgG autoantibodies react at body temperature (warm), while most IgM autoantibodies react at temperatures below 37°C, are termed cold autoantibodies, and activate complement upon warming to room temperature and above. The serological classification of AIHA is summarized in Table 1.

For clinicians, the most useful classification is based on the results obtained through serological investigation by the transfusion service laboratory. Usually, it is in the transfusion setting that the diagnosis of AIHA is first made, because either a positive direct antiglobulin test (DAT), an incompatible immediate spin crossmatch, or a positive antibody screen (preceding crossmatch request) is detected. The results of serological tests, including the immunoglobulin class and thermal range of the antibody, allow the classification of AIHA into IgG-mediated hemolysis versus complement-mediated hemolysis, which is almost always IgM-mediated. Based on thermal reactivity, 59% of 865 patients with AIHA were considered to have warm autoimmune hemolytic anemia, 32% cold autoimmune hemolytic anemia, 7% mixed (warm plus cold), and 2% paroxysmal cold hemoglobinuria (Table 1) (1). However, this study did not show whether hemolysis was due to IgG, complement or both. In another study, Petz and Garrity classified 347 patients with AIHA using monospecific antisera in the DAT to detect the presence of IgG or complement on the patients' red cells (2) Seventy-three percent of patients had IgG on their red cells that was present alone (23%) or with complement (50%). Twenty-six percent had complement only on their red cells. Of the patients with warm (IgG) AIHA, 67% had IgG plus complement, 20% had IgG alone and 13% had complement only. For those patients with cold agglutinin disease, only complement (C3) was detected on the red cells. In this report, a small number of patients had paroxysmal cold hemoglobinuria (IgG-mediated cold hemolysis), and all of these patients had only complement on their red cells.

Our group looked at the presentation of these patients from a different perspective. We analyzed 74 patients who

Table 1 Classification of autoimmune hemolytic anemia

Serological classification	Percentage
Warm	59%
Cold	32%
Paroxysmal Cold Hemoglobinuria	2%
Mixed (Warm and Cold)	7%

had a positive direct antiglobulin test (DAT) identified by routine testing in the transfusion laboratory (3). Forty percent of the positive DAT tests occurred in patients who had warm AIHA that was either primary or secondary (usually to systemic lupus erythematosus). Fourteen percent had cold (IgM) AIHA. However, 27% of patients had a positive DAT due to IgG, but with a negative eluate test (i.e. the red cell-associated IgG when eluted from the patient's red cells did not re-bind to normal red cells). Each of these patients had elevated levels plasma IgG, often to very high levels. Therefore, this work emphasizes that, without supporting clinical and laboratory evidence of hemolysis, the DAT has relatively low specificity for the diagnosis of AIHA. In other words, IgG can bind to the red cell membrane specifically (true autoantibody, as is the case in AIHA), or nonspecifically due to elevated levels of immunoglobulin (hypergammaglobulinemia). The DAT does not discriminate between these two instances.

2. PRESENTATION OF PATIENTS WITH AUTOIMMUNE HEMOLYTIC ANEMIA

Patients with AIHA have an increased rate of red cell destruction that is sufficiently severe to overcome the bone marrow's ability to compensate with increased red cell production. Occasionally, other factors contribute to the anemia including deficiencies of vitamins or trace elements (folate or iron) that are required for red cell production. Some patients with previously stable AIHA can have sudden worsening of anemia caused by suppression of bone marrow production often due to a concomitant bacterial or viral infection. Indeed, parvovirus-associated marrow suppression in patients with AIHA has been reported (4,5). Furthermore, a lag in the bone marrow response to hemolytic stress has been postulated to account for the reticulocytopenia seen in as may as 20% of patients presenting with AIHA (6,7).

Patients with AIHA may look pale because of their anemia, and some are mildly jaundiced as a result of breakdown of the hemoglobin. Other unique symptoms can be observed with certain types of AIHA, and they will be described subsequently. The examination of the blood film often shows spherocytes in both types of AIHA, although spherocytes are more frequently observed with IgG (warm) compared to IgM (cold) AIHA.

Laboratory investigations of patients with both warm and cold AIHA demonstrate evidence of extravascular cell destruction with an increased serum LDH and bilirubin (indirect more elevated than direct), a decreased haptoglobin, and increased reticulocyte count and red cell creatine. In some patients, especially those with cold agglutinin disease, episodes of intravascular hemolysis are associated with free hemoglobin in the serum and urine. Hemoglobin in the urine is followed by hemosiderinuria, which generally indicates chronic intravascular hemolysis. In those disorders of secondary AIHA, the underlying disorder (malignancy, infection, etc.) often produces other symptoms not associated with hemolysis.

3. WARM AUTOIMMUNE HEMOLYTIC ANEMIA

3.1. Overview of the Mechanism of IgG-Mediated Cell Clearance

Warm AIHA has contributed to our understanding of how IgG antibodies cause red cell destruction. Following the initial step of autoantibody binding to the target autoantigen on the red cell, the Fc portion of the autoantibody triggers subsequent events. The circulating opsonized red cell surface is targeted for removal during passage through the spleen. It is splenic macrophages that recognize IgG-sensitized red cells via specific receptors for the Fc portion of IgG molecule that results in complete or partial phagocytosis. Non-primate animal models and *in vitro* models of erythrophagocytosis have helped in our understanding of the principles of IgG-mediated cell clearance (8,9).

3.2 The Functional Regions of the IgG Molecule

IgG, a bifunctional molecule that comprises ~20% of the total plasma, is the most important opsonin for cell clearance. The amino-terminal of the molecule, or antibody binding fragment (Fab) region, has a great degree of variability that is responsible for antigen specificity. The carboxy-terminal, also termed the fragment crystalline (Fc) region, is highly homologous among all IgG subclasses. Indeed, components of the Fc region have a remarkable similarity across many different mammalian species so much so that IgG from one species has effector function in another species. It is the Fc portion of the IgG molecule that interacts with receptors (R) on phagocytic cells. There are four subclasses of human IgG, with subclasses 1 and 3 being the most efficient in activating complement, interacting with Fc gamma (γ) Rs and initiating phagocytosis (10–12). It is not surprising then, given the varying potential for IgG to activate complement, different IgG subclass expression for the autoantibody can impact on overall disease severity. However, polymorphic variants for complement C4 and for the receptors of IgG and complement also exist. The functional differences of these variants also may account for some of the variation seen in the clinical presentation and degree of anemia observed in AIHA. The functional differences for these different variants will be explained below.

3.3. The Receptors for IgG

Fcγ Rs expressed on the surface of monocytes and macrophages are responsible for the biological responses to IgG. The effector cell responses mediated through intracellular tyrosine activation pathways include: phagocytosis of antibody coated cells, clearance of soluble immune complexes, and antibody-dependent cellular cytotoxicity (ADCC). Twelve different FcγR mRNA transcripts have been isolated and are grouped into three different classes: FcγR I, II, and III. The transcripts are derived from eight distinct but highly homologous genes (13). All FcγR are expressed on macrophages with the exception of Fcγ RIIIb. Monocytes must be activated to express Fcγ RIIIa.

Fcγ RI, a 70 kDa molecule expressed on monocytes at 10,000 to 40,000 copies per cell, has the highest affinity for IgG. The affinity of FcγRI is such that it binds monomeric IgG, suggesting that the receptor is occupied at physiological plasma concentrations of IgG (10,14). Moreover, the report of healthy individuals with congenital deficiencies of this receptor raises additional questions concerning its biological importance (10,15).

FcγRIIa is found on many blood cells, including platelets, eosinophils, B-lymphocytes and monocytes. Monocytes express both Fcγ RIIa and RIIb. When engaged by IgG Fc, FcγRIIb, participates in the downregulation of cell activation through the inhibition of tyrosine phosphorylation, whereas Fcγ RIIa results in activation signaling events. How these opposing signaling events are coordinated in the monocyte is not entirely clear, but since monocytes express FcγRI, RIIa, and RIIIa, activation must override Fcγ RIIb down regulatory signals. FcγRII has a much lower affinity for IgG than FcγRI (10,16) and multimeric IgG expressed on the surface of sensitized cells, or in the form of soluble immune complexes, must be present to engage this receptor. However, a di-allelic polymorphic variation of FcγRIIa, found among the healthy population, has shown that the binding affinity varies for the IgG subclasses. The affinity of the FcγRIIa-His[131] variant is significantly higher for human IgG2 than FcγRIIa-Arg[131] (17).

Fcγ RIIIa is found on macrophages, neutrophils, natural killer cells, and eosinophils, but not on resting monocytes (10,18). Like Fcγ RIIa, Fcγ RIIIa has a low affinity for monomeric IgG and requires multimeric IgG expression for efficient interaction. Allelic variants also exist for this receptor with FcγRIIIa-Val[158], binding nearly twice as much IgG1 as Fcγ RIIIa-Phe[158] (19). FcγRIIIa is probably the dominant receptor for the clearance of IgG-sensitized cells since the administration of a blocking monoclonal antibodies to Fcγ RIIIa temporarily prevent immune clearance of sensitized platelets in a patients with autoimmune thrombocytopenia (20–22). Presumably these blocking antibodies would also attenuate the immune destruction of red cells.

For a number of years, our group has been interested in studying the relationship between the plasma concentration of IgG and the *in vivo* function of the Fc receptors (23). Fcγ R function can be studied *in vivo* by measuring the clearance of ^{51}-chromium labeled and anti-D sensitized autologous red cells (Figure 1). Overall, we observed an inverse relationship between the Fcγ R function and the plasma concentration of monomeric IgG (24). In those patients with hypogammaglobulinemia, there is enhanced Fcγ R-mediated cell clearance, which can be normalized by the administration of IgG. Very high concentrations of IgG inhibit Fcγ R-dependent cell clearance (Figure 2) (24). Presumably, this effect of IgG partially explains the efficacy of high dose IVIg in immune cytopenic disorders. Furthermore, in some individuals expressing the HLA-B8/DR3 haplotype, we observed reduced red cell clearance, a finding consistent with other previously published work (25).

3.4. Complement 4 Variants

C4 is a required factor for complement-dependent red cell lysis by the classical pathway. Two highly homologous genes, C4A and C4B, express C4. Reduced levels, or the absence of C4 in rare cases, have been observed and most are due to the inheritance of an inactive "null" gene (25). C4 null genes are present with increased frequency among patients with SLE (26) and their presence inversely correlates with high titered anti-CMV in healthy blood donors (27). Presumably, individuals who inherit at least one C4 null gene must produce increased amounts of IgG in order to activate lower levels of complement C4 in response to certain infections. A corollary is that the injection of IgG antibody-sensitized autologous cells will not activate C4 as readily in these individuals. Furthermore, we now know that the C4A*QO (C4A null) is linked with the HLA-B8/DR3 haplotype (28). Therefore, variability in complement activation by IgG-sensitized autologous red cells that is undetectable using traditional *in vitro* techniques provide a partial explanation why some HLA-8/DR3 normal individuals have reduced red cell clearance. Some of these individuals probably also have lower levels of C4 due to the presence of a C4 null gene.

3.5. Complement Receptor (CR) 1 Variants

CR1 is expressed to a high degree of variability on red cells. Molecular analysis has shown that the extracellular portion is organized as 30 short consensus repeats (SCR) consisting of 60 to 70 amino acids each. The N-terminal 28 SCRs can be grouped into four long homologous repeats (LHR), each with the capacity to bind C3b and with slight variation in their amino acid sequence. Allotypic variants exist that differ in the number of these LHRs (29).

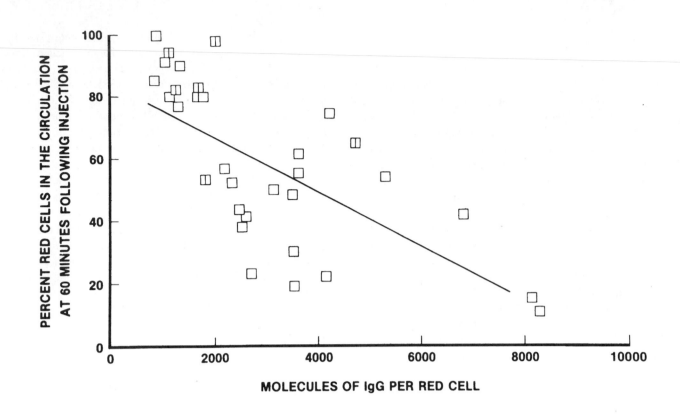

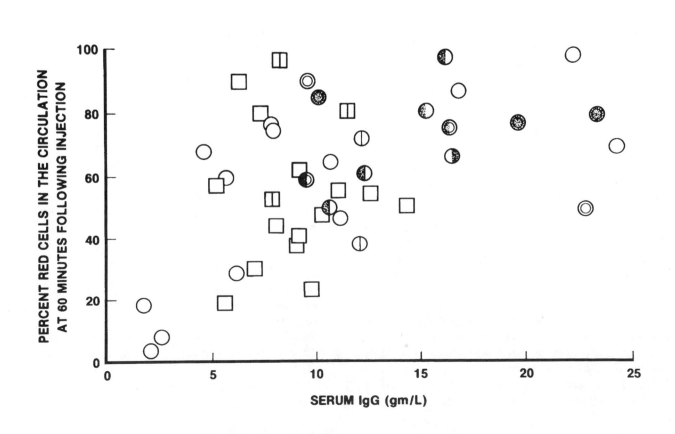

Furthermore, some of these variants differ in the number of copies expressed on the red cell with as much as a 7–10 fold variation in the level of expression (30). It is thought that expression is dependent on the sensitivity to protease digestion rather than on the half life of mRNA transcripts, since not all LHRs are identical (30). Thus, the amount of C3b bound to the red cell is dependent, in part, on the inheritance of receptor alleles that vary in the number of CR1 molecules expressed and the number of sites that can bind C3b.

Phagocytosis of autoantibody-coated red cells not only depends on the quantity of opsonized IgG autoantibody and on the IgG subclass. As few as 230 molecules of IgG3 has been shown to initiate phagocytosis, whereas other subclasses of autoantibodies required more than 2000 IgG molecules to cause phagocytosis (31,32). The inheritance of C4 variants as well as Fcγ and complement receptor variants are also thought to influence the clearance of autoantibody coated red cells. In many patients, *in vitro* phagocytosis and the strength of the DAT correlate with the severity of clinical hemolysis. But given the variability of subclass affinity and the inheritance of complement and receptor variants and other factors that can influence red cell production, variability in the degree of anemia is expected. Patients with DAT negative autoimmune hemolysis are well described.

3.6.   The RE System in Red Cell Destruction

Hemolysis also is dependent on reticuloendothelial (RE) function. Some, but not all, investigators have shown enhanced RE activity in patients with autoimmune hemolysis (31,33,34). Other investigators have described increased expression of FcγRII on circulating monocytes in patients with autoimmune hemolysis (35). Biological mediators such as γ-interferon have been shown to affect FcγR expression and function *in vitro*, but their role in autoimmune red cell clearance is unknown (10).

The RE system functions in two ways (10). In the first and conventional sense, IgG-sensitized red cells bind to the FcγR on phagocytic cells and may be completely or partially phagocytosed. Many of the red cells probably escape complete phagocytosis, since only a small amount of the IgG-sensitized red cell membrane is pinched off the cell. The red cell, now minus a small amount of membrane, causes the red cell to assume a spherical shape. As a second mechanism, the spleen has the ability to clear spherocytic red cells, even when the majority of autoantibody is no longer present on its surface. During each pass through the spleen, red cells must renegotiate the narrow inter-endothelial slits within the splenic sinuses. The sphered red cells have a reduced deformability and are more likely to be retained within the splenic sinus where they are ingested by phagocytic cells (Figure 3).

3.7.   Clinical and Laboratory Findings of Warm Autoantibody-Mediated Hemolytic Anemia

Warm AIHA can be primary or associated to another disease, which is termed secondary. The diagnosis of primary disease, considered the classical form of AIHA, is based on the exclusion of secondary causes. Secondary warm AIHA has been reported in association with drugs, malignancies, collagen vascular disease, among other disorders (Table 2). Chronic lymphocytic leukemia is the malignancy most commonly associated with secondary AIHA (Table 3). Additionally, drugs such as methyldopa and procainamide can induce autoantibodies against red cells and less commonly can cause hemolytic anemia (36,37). The most frequent causes of AIHA are idiopathic and drug induced hemolytic anemias (Table 2) (1). In most patients, a careful history, physical examination and a few serological tests can exclude secondary causes.

Some patients with warm AIHA will present only with symptoms attributable to their anemia. Other patients will have symptoms related to the underlying disorder. The accelerated extravascular red cell destruction is associated with a compensatory increased red cell production evidenced by the appearance of immature red cells (reticulocytes) and less commonly with nucleated red cells.

Confirmation of the diagnosis of AIHA requires demonstration of the autoantibody. Serologically, autoantibody is measured by the direct antiglobulin test (also known as the direct Coomb's test), which detects increased amounts of IgG and/or complement bound to the patient's washed red cells. Serological findings are similar in primary and secondary warm AIHA, although a higher percentage of drug-induced AIHA have IgG only on the red cells (1). Positive direct antiglobulin tests may provide diagnostic evidence in autoimmune hemolysis; however, they are not specific for this disorder. Occasionally, a patient may appear to have warm AIHA, but serological tests are negative (38). This form of hemolytic anemia will be dealt with separately. The clinical significance of the laboratory findings in hemolytic anemia must be evaluated together and in the context of the clinical situation. Because of the panagglutinin, care should be taken to interpret the Rh type. A control that contains the appropriate amount of enhancing protein (e.g. 6% albumin) must be negative (no agglutination) to confirm the Rh status of the test sample. If the control is positive, the test sample is not interpretable. For the most part, the amount of albumin in the ABO typing sera is identical to commericial anti-D and serves as a suitable control for all but group AB patients. For the group AB patient, an autologous control (substituting the patient's own serum for the antisera together with their red cells) or albumin control must be tested. When the control is positive, it may be useful to test the patient's red cells washed in saline. If the control is still positive, saline anti-D (an IgM low protein Rh typing reagent) may be a useful

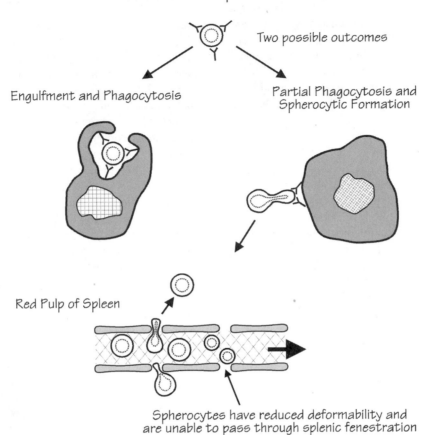

IgG Sensitized Red Cells
interact with the Splenic R.E. Cells

Two possible outcomes

Engulfment and Phagocytosis

Partial Phagocytosis and
Spherocytic Formation

Red Pulp of Spleen

Spherocytes have reduced deformability and
are unable to pass through splenic fenestration

alternative. However, it must be realized that the albumin content in saline anti-D is the same as many of the current routine IgG or monoclonal blended anti-D typing reagents, and a suitable control must be included with the test. If the Rh status is still uncertain, Rh-negative blood should be given until the status can be unequivocally established. Because autoantibodies in the serum will mask a co-existing alloantibody, and because alloantibodies frequently form in transfused patients with autoantibodies, additional testing is required to exclude the presence of alloantibodies. We have found that auto-absorption using the patient's own cells is a useful technique to look for any concomitant alloantibody.

Table 2   Causes of autoimmune hemolytic anemia

	Warm	Cold
Primary (idiopathic)	37%	48%
Secondary	63%	52%
Drugs	31%	
Infection	2%	20%
Malignancy	22%	22%
Collagen Vascular Disease	4%	4%
Pregnancy	1%	
Other	3%	6%

Table 3   Autoimmune hemolytic anemia and malignancy

	Warm autoimmune hemolytic anemia 65%	Cold autoimmune hemolytic anemia 35%
Carcinoma	23%	26%
Lymphoid Malignancies:		
Non-Hodgkin's Lymphoma	18%	26%
Chronic Lymphocytic Leukemia	41%	3%
Multiple Myeloma	4%	5%
Macroglobulinemia	1%	5%
Hodgkin's Disease	8%	11%
Other	5%	11%

3.8. Target Autoantigens in Warm AIHA

For most cases of IgG-mediated AIHA, the specific autoantigen on the red cell is not known. Antigen specificity is determined by the elution of the antibody from the patient's red cells followed by testing against a panel of commercially-available or in-house antigen-typed red cells. Using this technique, some red cell autoantibodies react with the proteins that express the Rh antigens, as evidenced by the lack of reactivity with Rh null cells (39,40). Of these Rh-specific autoantibodies, auto anti-e reactivity is the most common. The antigens in the Rh system are carried on 30–32 kDa non-glycosylated, transmembrane proteins (41). The antigens are encoded by two homologous genes: The *RHD* gene responsible for Rh (D) expression, and the *RHCE* gene that expresses Rh (C), (c), (E), and (c) antigens (42). Often, autoantibodies will react with all red cells, but demonstrate increased reactivity with Rh (e)-positive cells. A variety of other target autoantigens have been reported. Specificities include antibodies directed against Kell, LW, U, Wr^b, Xg^a, Vel, and Jk^a (39,40,43). Occasionally, IgG red cell autoantibodies that are directed against the i^t antigen might be seen in association with Hodgkin's disease (44). The mechanism accounting for the unusual specificity of secondary autoantibodies in Hodgkin's disease is unknown.

Frequently, in drug-induced hemolytic anemia, the red cell-specific autoantibodies are serologically identical to the primary autoantibodies in idiopathic AIHA (1). For example, methyldopa is the agent most often associated with AIHA, and these autoantibodies also target Rh antigens (36). Procainamide is also associated with a high incidence of direct antiglobulin tests due to the production of red cell autoantibodies and, again, the most frequent target of the induced autoantibodies is the Rh antigen system (37).

3.9. Treatment of Warm AIHA

The type of treatment with AIHA depends upon the severity of the hemolysis and resulting anemia. Some patients have only mild anemia because they are able to compensate for the red cell destruction by increased red cell production. These patients should receive supplemental folic acid and possibly iron (depending on iron stores) to ensure that they are able to continue to produce an increased number of red cells. Any secondary cause of AIHA should be considered and excluded by the appropriate investigations.

Patients with more severe AIHA required treatment for the disorder itself. The first line of treatment is corticosteroids in a typical regimen of prednisone given at a dose of 1 mg per kilogram of body weight per day. As soon as the hemoglobin rises to near normal level, the prednisone should be decreased and, if possible, discontinued. Almost every patient who receives long-term corticosteroid therapy

will have adverse effects, and seldom is chronic corticosteroid therapy justified. Some patients may require a more definitive treatment, such as splenectomy. In a series of 52 patients with warm AIHA, splenectomy resulted in long term cure or sustained remission in the two-thirds of patients with IgG-mediated warm AIHA (45). Splenectomy is thought to be effective because it removes a major site of autoantibody production and also the most important organ for the clearance of IgG-sensitized red cells. Splenectomy patients with AIHA should be preceded by a vaccination using a polyvalent pneumococcal vaccine, meningococcal vaccine, and probably H. influenza. Unfortunately, there is no way of predicting which patients will respond to splenectomy, and at least on third of all patients with IgG-mediated AIHA will fail splenectomy and require alternative treatment. The relative uncommonness of the disorder and confounding factors such as an underlying cause for the hemolytic anemia (46) makes it difficult to apply evidence-based approaches to therapy. Most of the reports describing the subsequent management of these patients consist of small series of case reports. Alternative treatments include the use of danazol, immunosuppressive agents such as cyclophosphamide, azathioprine, vincristine, IVIg, and plasmapheresis (47,48).

Some patients have a severe and potentially life-threatening anemia due to the hemolysis, and a blood transfusion should be administered without delay. Because it is essentially impossible to provide completely compatible blood to these patients, some physicians delay the transfusion while trying to identify the most compatible unit. This approach is not acceptable. The additional risk of an alloantibody in a patient who already has a rapid rate of red cell destruction is uncertain, but probably small.

The management of methyldopa-induced autoantibody is much simpler than the management of a patient with a true warm autoantibody. In those patients who are receiving methyldopa and develop a red cell autoantibody, but do not have evidence of increased red cell destruction (evidenced by anemia plus an increased reticulocyte count), it is our practice to allow the patient to continue on the methyldopa. This represents the majority of patients treated with methyldopa in which an autoantibody forms. A small minority of patients who are treated with methyldopa or procainamide will become anemic (36,37). These patients respond to discontinuation of the drug and, although the autoantibody can persist for weeks to months, the hemolysis generally stops within a few weeks.

3.10. Direct Antiglobulin Test-Negative Hemolytic Anemia

Some patients with apparent warm (IgG) AIHA do not have detectable autoantibodies. These patients have a negative direct antiglobulin test and do not have IgG or complement on their red cells. A number of theories have been proposed to explain the mechanism of hemolysis in this subset of patients. First, their AIHA could be caused by potent IgG antibodies present in quantities too low to be detected by the routine DAT. Sensitive assays, capable of detecting low levels of cell bound IgG, have demonstrated that some of these patients do have increased amounts of IgG on their red cells compared to normal red cells (49). Autoantibody subclass may also determine whether hemolysis occurs in individuals with low amounts of red cell bound IgG (50). The direct antiglobulin test measures total IgG, and therefore it is possible for an increase in a specific subclass to occur without elevating the total red cell bound IgG. An alternative theory is that certain autoantibodies are more likely to activate complement (51), and the antigen distribution of certain protein red cells may favor complement activation even when only a few IgG molecules are bound (52). Another theory is that, in some patients, a low serum IgG level or an increased phagocytic activity favors the development of clinically significant red cell destruction at a low level of red cell bound IgG (53).

Autoimmune hemolytic anemia due to IgA autoantibodies has been described in some patients with a negative direct antiglobulin test (54). In some patients, both IgG and IgA are increased, however other patients have only isolated increased red cell IgA (54). In one patient, IgA-sensitized red cells have been demonstrated to undergo antibody (IgA)-dependent cellular cytotoxicity (54). Further evidence supporting the concept of IgA-mediated hemolysis comes from recent identification of an Fcα receptor (55).

Other mechanisms of red cell destruction have been proposed to elucidate unexplained immune-mediated hemolysis. Perforin, a protein released by cytotoxic T-lymphocytes, can cause red cell hemolysis *in vitro*. This lytic factor forms holes in the red cell membrane similar to those produced by the membrane attack complex of the complement system (56). The red cell defense mechanisms that limit perforin-induced hemolysis appear to be distinct from the proteins that inhibit the complement membrane attack complex (57). A potential role for perforin in causing autoimmune hemolysis would involve the attack of red cells by this protein, but although this has been demonstrated *in vitro*, it is uncertain if it occurs *in vivo* (56).

4.   COLD AGGLUTININ DISEASE

Cold agglutinin disease is a relatively uncommon cause of autoimmune hemolytic anemia, but it is important for several reasons. First, it is perhaps the only disorder in the body in which a monoclonal antibody has a readily identifiable target antigen. In contrast to cold agglutinins, malignant (multiple myeloma) and non-malignant mono-clonal antibodies (benign monoclonal gammopathy) have no identifiable target antigens. It remains unknown why the IgM autoantibodies of cold agglutinin disease are unique in this respect. A second reason for the importance of cold agglutinin disease is that these antibodies can be used as probes to study red cell membrane determinates. A third reason for the importance of this syndrome is related to the impetus that cold agglutinin disorders have given to studying pathways of cell destruction. There are no receptors for IgM, yet some patients with cold agglutinin disease have such a rapid rate of red cell destruction that intravascular hemolysis can occur. The mediator of the increased red cell destruction and occasional intravascular hemolysis is complement.

Many aspects involving cold agglutinins remain unexplained. Perhaps one of the most important is the biological function (if any) of cold agglutinins in healthy individuals. For unknown reasons, almost everyone has IgM cold agglutinins that appear not to cause any illness or have any effect on red cell survival. The explanation for this ubiquitous finding and why, if it is biologically irrelevant, it is present in everyone, remains unknown.

IgM cold agglutinins are autoantibodies that bind to red cells, and their greatest reactivity is in the cold, usually at temperatures below normal body temperatures. The reason for their low thermal amplitude of reactivity remains uncertain, but investigators have suggested that it is related to the conformational changes that red cell surface components undergo on warming that render them non-reactive to the autoantibodies at higher temperatures. In cold agglutinin disease, the IgM dissociates from the red cells when warmed, but the cells remain coated with complement. Before discussing several cold agglutinin disorders and the target antigen, we will briefly summarize pathways of complement activation and its control on red cells.

4.1.   Overview of Complement-Mediated Cell Clearance

Traditionally, complement has been considered to mediate red cell destruction by forming holes in the membrane, which results in red cell lysis. In fact, this is an uncommon event and is likely to occur only if there has been a failure in the 'control' mechanisms. The complement system represents a group of plasma proteins that bind to cells after immunoglobulins bind (classical pathway) or without the binding of immunoglobulin (alternative pathway), leading to the destruction of the cells. The complement cascade, like other cascades in the body, is an amplification system in which the initial activation of a small number of complement molecules results in the subsequent generation of many downstream components. Following the binding of several molecules of IgG, or one or more molecules of IgM, to the cell surface, the classical pathway of complement activation is initiated. The initial complement components, the C1qrs complex, activates the next complement components, C4 and C2, by enzymatic cleavage which bind to the membrane, forming a complex call C4b2b. This complex (also known as C3 convertase) can activate many molecules of C3, again by partial cleavage. The largest component of C3, termed C3b, binds to the membrane, and serves as an important opsonin. C3b also can be deposited on the membrane through the alternative pathway of complement activation (10).

The binding of C3b to the membrane represents an important branch point for the complement cascade. C3b can activate subsequent complement components, including C5, 6,7,8. The C5b, 6,7,8 complex activates C9 molecules that enter the membrane, forming a tubule that triggers cell lysis. As one might anticipate, intravascular rupture of a cell can release proteolytic enzymes and other intracellular components that can be dangerous to the host. Not surprisingly, there are numerous mechanisms red cells utilize to minimize complement-mediated hemolysis. Nonetheless, such an event does occur when regulatory proteins are overwhelmed, such as when an individual with cold agglutinin disease is transfused with red cells (10). Intravascular hemolysis also can occur in those disorders associated with abnormal or reduced regulatory protein activity, such as paroxysmal nocturnal hemoglobinuria (PNH) (58).

C3b sensitization leads to the adherence of the sensitized cell to phagocytic cells via specific complement receptors. Although phagocytosis of cells coated only by complement is uncommon, if the cell is also sensitized by another opsonin such as IgG, phagocytosis occurs more readily. As mentioned previously, C3b-sensitized red cells bind to complement receptors type 1 (CR1). Depending upon the type of CR1 variant and presumably the number and density of C3b on the cell surface, the C3b-sensitized cells with either be ingested or the C3b will be partially degraded by Factor I to form iC3b (10). Cells sensitized by iC3b bind strongly to complement receptors type 3 (CR3), which are found on neutrophils, monocytes, and lymphocytes (10).

Several factors limit the destruction of red cells by complement. Those red cells carrying iC3b that escape phagocytosis probably have a normal survival because the iC3b is degraded to C3dg, which, although it can bind weakly to CR3 receptors, probably does not initiate phagocytosis. Additionally, once a red cell has large amounts of C3dg on its surface, it is protected from further complement-mediated lysis because the C3dg rings the IgM binding sites, preventing the deposition of more C3b and limiting both complement-mediated cell lysis and phagocytosis.

CR1 and CR3 receptors, like Fcγ receptors, have a finite capacity for binding and activating complement-sensitized cells, such as red cells. Patients with chronic cold agglu-

tinin disease have an increased rate of red cell destruction without much evidence of intravascular red cell hemolysis as shown by free hemoglobin. This data indicates that most of the red cell destruction in cold agglutinin disease occurs via phagocytosis of the complement-sensitized red cells. However, the transfusion of normal red cells into these patients can result in an episode of intravascular hemolysis. This is thought to occur because the transfused red cells are not 'protected' by C3dg on their surface and, hence, the rapid deposition of C3b by the cold agglutinin often overwhelms the CR1 and CR3 receptors, resulting in complement deposition and intravascular, rather than extravascular, hemolysis.

4.2. Red Cell Defense Mechanisms Against Complement-Mediated Lysis

Inhibitors of complement activation on the red cell membrane include membrane inhibitor of react lysis (MIRL or CD59), decay accelerating factor (DAF), and C8 binding protein (C8bp) (58–60). MIRL inhibits C5b-9 activity by blocking the incorporation C9 (61). DAF accelerates the spontaneous dissociation of C4b and C2a (58). These three proteins share the property of being linked to the red cell membrane by a glycosyl phosphatidylinositol (GPI) anchor (58). These proteins are attached to the extracellular leaflet of the plasma membrane by a carbohydrate-phosphatidyl link that effectively restricts complement activation, since these molecules lack a hydrophobic, transmembrane domain and can move freely about the plane of the membrane (58). Patients with PHN are deficient in these three membrane complement inhibitors (58), and they are at risk of red cell lysis following even minimal complement activation. Due to the clonal nature of the disorder, only the red cells lacking GPI-anchored proteins are at risk.

4.3. Primary and Secondary Cold Agglutinin Disease

Either polyclonal or monoclonal IgM cold agglutinins can mediate hemolysis via complement activation (62–65). Polyclonal cold agglutinins are observed most frequently as acute, transient autoantibodies associated with a viral illness. In contrast, monoclonal antibodies with cold agglutinin activity (monoclonal gammopathies) characteristically cause a chronic hemolytic disorder. Monoclonal cold agglutinins can occur as a primary disorder or secondary to a variety of lymphoproliferative disorders (Tables 2 and 3) (1,64).

4.4. Transient Cold Agglutinins

Several acute infections are associated with the production of transient cold agglutinins and in rare cases these antibodies cause acute hemolytic anemia. The usual cause of post-infectious cold agglutinins are those that form following a *Mycoplasma pneumoniae* infection. At least 50% of patients with pneumonia due to *Mycoplasma* infection produce a cold agglutinin within weeks of the infection (62). A compensated mild hemolysis is common, but severe hemolysis is rare. These cold agglutinins are polyclonal IgM that have specificity against the I-antigen on red cells, and often there is light chain restriction, with kappa being predominant (62).

Infectious mononucleosis is another common cause of cold agglutinins. Again, although the formation of autoantibodies is common in this condition, severe hemolysis is distinctly uncommon. These autoantibodies are usually polyclonal and have anti-i reactivity. Cytomegalovirus, rubella and HIV may also be associated with cold agglutinins (Table 4) (62).

4.5. Primary Cold Agglutinin Disease

Primary cold agglutinin disease is a chronic hemolytic disorder that is almost always caused by a monoclonal IgM cold agglutinin. The monoclonal antibody usually is comprised of kappa light chains and has an anti-I specificity (62). Although this antibody may be produced in high titer and react near 37°C, usually there is not enough monoclonal protein produced to visualize an abnormality of the serum protein electrophoresis, although paraproteins have been observed.

4.6. Clinical and Laboratory Findings of Cold Autoantibody-Mediated Hemolytic Anemia

Cold agglutinins present in normal individuals are not associated with disease. In contrast, symptomatic cold agglutinin disease typically is manifested by Raynaud's phenomenon and acrocyanosis of peripheral tissues following exposure to cold temperature. The tissue cyanosis is thought to be the result of red cell agglutination and complement activation in the peripheral vasculature (64). Paroxysmal cold hemoglobinuria represents a distinct subset of cold-reactive antibodies with different clinical manifestations and will be discussed separately.

Cold agglutinins often cause red cell agglutination in blood samples collected at room temperature. The blood film shows characteristic red cell agglutinates. To correctly evaluate for cold agglutinins bound to the red cells, patient blood pecimens must be maintained at body temperature and washed with warm saline. An EDTA specimen can be taken to evaluate the red cells for *in vivo* complement activation. Anemia and reticulocytosis are frequent findings. Increased unconjugated bilirubin and LDH are markers of the accelerated red cell destruction. Serologically, cold agglutinin disease is associated with complement-activating IgM antibodies in the plasma or serum that have a high thermal amplitude. Complement-activating cold agglutinins are also present in low titers in normal individuals, but are not associated with disease. The major difference between these normal autoantibodies and pathological cold agglutinins is the higher thermal range of the pathological antibodies. Indeed, thermal amplitude rather that antibody titer is the important determinant of pathologic significance (Table 5) (62,66,67). In a subset of patients with cold agglutinin disease, cold-reactive IgG antibodies or, more commonly, a mixture of cold-reactive IgG and IgM antibodies are found (63,65,68). These serological differences are significant, as response to therapy in the latter subset of cases can be quite different. Rarely, cold-reactive monoclonal IgG or IgA antibodies have been implicated as the pathogenic autoantibody in cold agglutinin disease (62,65).

In addition to the increase in red cell-bound C3dg in cold agglutinin disease, there is a reduction in the number of CR1 on red cells (69). Not only does this receptor bind C3b, it also functions as a cofactor with Factor I to degrade C3b to iC3b and C3dg (10). Thus, CR1 functions *in vivo* to clear immune complexes and degrade activated complement. Acquired low levels of CR1 are seen in a number of disorders that involve *in vivo* complement activation, including acute and chronic cold agglutinins, warm AIHA, PNH, and SLE (69). When the different disorders were compared, the most severe reduction in red cell CR1 levels was seen in cold agglutinin disease, and this reduction was associated with highest amount of red cell bound C3dg (69). In patients with *Mycoplasma pneumoniae* infection, the red cell C3dg and CR1 levels return to normal follow-

Table 4 Infections and transient cold agglutinins

	Antibody Class	Incidence	Specificity	Clinical Evidence of Hemolysis
Mycoplasma pneumoniae	IgM	80%	I	80%—rarely significant
Epstein Barr Virus	IgM/G		i	<1%
Cytomegalovirus	IgM		I	rare
HIV	IgM		I, i, Gd	rare
Rubella	IgM		Pr	rare

Table 5 Cold agglutinins

	Antibody class	Titer	Thermal amplitude		
			4°C	24°C	37°C
Nonpathogenic	IgM	Low	++++	–	–
Pathogenic	IgM	Low/High	++++	++++	+/–
	Less often IgG or IgA (anti-Pr)				

ing recovery (69). A similar recovery of red cell CR1 has been observed following treatment of AIHA with splenectomy (69). The deposition of C3dg on red cells treated *in vitro* with either antibody or immune complexes in whole blood do not lead to CR1 loss. In contrast, C3dg deposition on red cells *in vivo* always is correlated with a loss of CR1. Because of these discrepancies, the loss of CR1 has been postulated to occur as the complement-sensitized cells interact with the phagocytic cells in the liver and spleen (69).

As with warm AIHA, red cell transfusions can be life-saving in some patients with cold AIHA. Unfortunately, concern about a transfusion reaction can inappropriately delay this important treatment. Because the cold agglutinin causes agglutination of all target cells, it can be difficult to determine the patient's ABO blood group. To avoid this problem, the blood sample should be collected and maintained at 37°C following the phlebotomy procedure. The patient's red cells should be washed with warm saline to remove the IgM autoantibody. Some blood banks treat the sample with reducing agent to stop IgM-mediated cold agglutinins. IgG anti-A and anti-B in the patient's serum can be detected by testing the serum at 37°C using anti-antiglobulin reagent as the end point. To exclude co-existing alloantibodies, the incubation of the patient's serum with the test panel should be performed at 37°C with the reactivity read using a monospecific anti-IgG. Most secondary causes of cold agglutinin disease can be ruled out on the basis of a careful history, physical examination, abdominal ultrasound and bone marrow examination (Table 3).

4.7. Target Autoantigens

Most cold agglutinins, whether polyclonal or monoclonal, are specific for antigens in the I/i system. These autoantibodies are directed against carbohydrate epitopes on the I and i antigens (62,65), which are located on precursors of the ABH and Lewis blood groups. The expression of red cell I antigens is age-dependent, newborn red cells express the i antigen whereas adult red cells express the I antigen. Serologically, the autoantibodies with the I/i system are distinguished from each other using the selective reactivity of anti-I with adult red cells and anti-i with newborn red cells (63).

Rarely, other target autoantigens may be involved in cold agglutinin disease. Anti-Pr autoantibodies are directed against sialic acid determinants on a proteolytic sensitive red cell glycoprotein (62,63,65). Other cold agglutinin specificities have been reported and include anti-Sa, anti-F1, anti-Lud, anti-Gd, anti-Vo/Li, anti-R^x, anti-P, and anti-M (62,70). A number of cold agglutinins are directed against antigenic determinates whose expression changes during development, evidenced by selective reactivity with either fetal or adult red cells (Table 6).

4.8. Molecular Basis of Autoantibody Synthesis in Cold Autoimmune Hemolytic Anemia

Less is known about the mechanism for generating the red cell autoantibodies in polyclonal cold agglutinin disease; however, important observations have been made in mono

Table 6   Antigen specificity of cold agglutinins [62]

Antibody	Specificity	Protease sensitive	Neuraminidase sensitive	Differentiation antigen
Anti-I	Glycolipid and N-linked glycoproteins	No	No	Yes
Anti-i	Band 3 containing repeating N-acetyllactosamine units I-branched; i-linear Part of the internal structure of ABH antigens			
Anti-Gd	Gangliosides on sialylated I/i glycolipids	No	Yes	No
Anti-F1	Bianntennary glycoproteins and glycolipids on I antigens	No	Yes	Yes
Anti-Lud	Neuraminidase sensitive antigen	Yes	Yes	Yes
Anti-Vo/Li	Sialytated i determinants	No	Yes	Yes
Anti-Pr	O-linked oligosaccharides on glycophorin A and B	Yes	Yes: Pr$_{1,2,3}$,Pr^M No: Pr$_a$	No
Anti-Sa	Glycolipid and glycoprotein (glycophorin A)	Yes	Yes	No

clonal cold agglutinin disease. Several investigators have demonstrated selective gene usage for pathogenic cold agglutinins with I/i blood group specificity.

Early studies of monoclonal, anti-I cold agglutinins demonstrated that the autoantibodies from different patients had common idiotypes (71), suggesting that common genetic elements encoding the variable region were used for these antibodies. Recent molecular studies have confirmed selective immunoglobulin gene utilization (62), which appears to be unique to anti-I/i since cold agglutinins with other specificities do not use common genetic elements (62). Virtually all pathogenic cold agglutinins of I/i blood group specificity used the variable heavy chain $V_H 4–21$ in near germ-line (unmutated) configuration in association with a restricted number of kappa light chains (72). In contrast to these autoantibodies, the VH and kappa light chain genes are expressed in only a small percentage of IgM-κ non-autoantibodies (73) which led to the suggestion that cold agglutinin disease may be due to defects in the regulatory mechanism of these naturally-occurring antibodies (74). Furthermore, the near universal use of a single heavy chain gene suggested that a specific V_H sequence is required for the I/i specificity. Curiously, the other naturally-occurring anti-I cold agglutinins that have been isolated from health individuals do not use the $V_H 4–21$ gene (75). It is not known why pathogenic anti-I/i cold agglutinins are encoded exclusively by the $V_H 4–21$ gene. B cells expressing the anti-I or anti-i $V_H 4–21$ idiotype also have been demonstrated in normal adult lymphoid tissue and in fetal spleen (76). Since the I/i blood group antigens are expressed on B cells and their expression is developmentally regulated, it is possible that these antibodies are necessary during B cell ontogeny or in B cell development and differentiation (77).

Rarely, monoclonal old agglutinins with specificity for other antigens are found in cold agglutinin disease. Unlike pathogenic anti-I cold agglutinins, anti-Pr autoantibodies from different individuals do not share idiotypes (78). Less is known about the cold agglutinins directed at other autoantigens. Dissimilarities between anti-Pr and anti-I cold agglutinins suggest that there may be a number of molecular mechanisms involved in generating autoreactivity in cold agglutinin disease.

4.9.   Treatment of Cold Agglutinin Disease

Transient cold agglutinins rarely cause clinically significant hemolysis, but, in contrast, chronic cold agglutinins can cause a severe and potentially life-threatening anemia. An important component of the treatment for cold agglutinin disease is avoiding exposure to cold, which often precipitates symptoms. Red cell transfusions are often required for patients with severe hemolysis, and these transfusions should be administered slowly. Other treatments for cold agglutinin disease are directed at reducing the autoantibody either by decreasing production or by removing the autoantibody by plasmapheresis. Folate supplements are required to support the increased red cell production.

Traditionally, alkylating agents have been the main treatment for chronic cold agglutinins and are the treatment of choice for patients with lymphoproliferative disorder and symptomatic cold agglutinin disease. Because of the risk of developing a secondary drug-inducted malignancy associated with long-term alkylating agents, plasmapheresis is the preferred treatment for younger patients. Recently, alpha interferon has been used successfully to treat patients with refractory hemolysis (79). Unlike warm AIHA, cold agglutinin disease does not usually respond to prednisone. However, patients who have spherocytes, an acute presentation, or an IgG form of cold agglutinin disease may respond (67,68,80). Splenectomy is seldom curative in these patients and is generally not used.

5.   PAROXYSMAL COLD HEMOGLOBINURIA

5.1.   Etiology and Pathogenesis

This form of immune hemolytic anemia is caused by a unique subset of IgG autoantibodies that are biphasic hemolysins. Serologically, these autoantibodies bind to the red cell in the cold and, following warming, mediate hemolysis by complement activation.

Paroxysmal cold hemoglobinuria (PCH) is a common cause of acute, post-infectious hemolytic anemia (81). Historically, this disorder was first described in association with congenital and chronic syphilis. Currently, viral illnesses such as mumps, measles and chicken pox are the commonest precipitants, but in many cases, the pathogenic agents are not known (81). PCH is one of the most common causes of acute hemolytic anemia in childhood (82).

5.2.   Clinical and Laboratory Findings of Paroxysmal Cold Hemoglobinuria

Most patients with PCH present with anemia, jaundice and hemoglobinuria. Cold exposure can also precipitate hemolysis in PCH associated with syphilis but, in contrast to cold agglutinin disease, most patients do not exhibit acrocyanosis or cold induced symptoms (81). The severity of the anemia can be quite variable, and an inappropriately low reticulocyte count is a frequent finding (81).

The diagnosis of PCH is based on the serological demonstration of the characteristic biphasic hemolysin using the Donath Landsteiner test. The principle of the test is that this autoantibody binds to the red cell optimally in the cold (4°C) and complement-mediated hemolysis follows upon warming to 37°C. Active disease is associated with evidence of *in vivo* complement activation on target red cells, detected by the presence of C3dg on the circulating red cells. Erythrophagocytosis is frequently observed in blood films, a phenomenon that may be triggered by the presence of both C3b and IgG on the red cells of these patients. Surprisingly, IgG is not present in sufficient amounts for detection by standard anti-human globulin tests (81). The unusual biphasic nature of the hemolysin and the clinical presentation distinguish this autoimmune disorder from cold agglutinin disease.

5.3. Target Autoantigens

The Donath Landsteiner antibody in PCH is directed against the glycosphingolipid globoside known as the P antigen and also recognizes Forssman glycolipids. Other autoantibody specificities are extremely rare (81).

5.4. Treatment of PCH

The disorder usually is an acute, self-limiting condition, which is secondary to a viral illness, and treatment is limited to supportive care. Severe anemia may require transfusion and patients should receive folate supplements. In spite of a theoretical antigen incompatibility, P antigen-positive blood is generally used for transfusion with good therapeutic results (81).

ACKNOWLEDGEMENTS

The studies described in this report were supported by a grant from the Canadian Blood Services and by The Medical Research Council of Canada. G.A. Denomme is supported by a Bayer/Canadian Blood Services/MRC Scholarship.

References

1. Sokol, R.J., S. Hewitt, and B.K. Stamps. 1981. Autoimmune haemolysis: an 18-year study of 865 cases referred to a regional transfusion centre. *Br. Med. J.* 282:2023–2027.
2. Petz, L.D., and G. Garratty. 1980. Acquired Immune Hemolytic Anemias. Churchill Livingstone, New York. pp. 190–195.
3. Heddle, N.M., J.G. Kelton, K.L. Turchyn, and M.A.M. Ali. 1988. Hypergammaglobulinemia can be associated with a positive direct antiglobulin test, a nonreactive eluate, and no evidence of hemolysis. *Transfusion* 28:29–33.
4. Bertrand, Y., J.J. Lefrere, G. Leverger, A.M. Courouce, C. Feo, M. Clark, G. Schaison, and J.P. Soulier. 1985. Autoimmune haemolytic anaemia revealed by human parvovirus linked erythroblastopenia. *Lancet* 2:382–383.
5. Lefrere, J.P., A.M. Courouce, Y. Bertrand, R. Girot, and J.P. Soulier, 1986. Human parvovirus and aplastic crisis in chronic hemolytic anemias: a study of 24 observations. *Am. J. Hematol.* 23:271–275.
6. Liesveld, J.L., J.M. Rowe, and M.A. Lichtman. 1987. Variability of the erythropoietic response in autoimmune hemolytic anemia: analysis of 109 cases. *Blood* 69:820–826.
7. Conley, C.L., S.M. Lippman, P.M. Ness, L.D. Petz, D.R. Branch, and M.T. Gallagher. 1982. Autoimmune hemolytic anemia with reticulocytopenia and erythroid marrow. *N. Engl. J. Med.* 306:281–286.
8. Segal, D.M., S.K. Dower, and J.A. Titus. 1983. The role of non-immune IgG in controlling IgG-mediated effector functions. *Mol. Immunol.* 20:1177–1189.
9. Anderson, C.L., L. Shen, D.M. Eicher, M.D. Wewers, and J.K. Gill. 1990. Phagocytosis mediated by three distinct Fc gamma receptor classes on human leukocytes. *J. Exp. Med.* 171:1333–1345.
10. Anderson, D., and J.G. Kelton. Immune Destruction of Red Blood Cells. American Association of Blood Banks, Arlington, VA. pp. 1–52.
11. Abramson, N., and P.H. Schur. 1972. The IgG subclasses of red cell antibodies and relationship to monocyte binding. *Blood* 40:500–508.
12. Wiener, E., A. Atwal, K.M. Thompson, M.D. Melaned, B. Gorick, and N.C. Hugh-Jones. 1987. Differences between the activities of human monoclonal IgG1 and IgG3 subclasses of anti-D(Rh) antibody in their ability to mediate red cell-binding to macrophages. *Immunol.* 62:401–404.
13. Gessner, J.E., H. Heiken, A. Tamm, and R.E. Schmidt. 1998. The IgG Fc receptor family. *Ann. Hematol.* 76:231–248.
14. Vaughn, M., M. Taylor, and T. Mohanakumar. 1985. Characterization of human IgG Fc receptors. *J. Immunol.* 138:4059–4065.
15. Ceuppens, J.L., M.L. Baroja, F. Van Vaeck, and C.L. Anderson. 1988. Defect in the membrane expression of high affinity 72-kD Fc gamma receptors on phagocytic cells in four healthy subjects. *J. Clin. Invest.* 82:571–578.
16. Kelton, J.G., J.W. Smith, A.V. Santos, W.G. Murphy, and P. Horsewood. 1987. Platelet IgG Fc receptor. *Am. J. Hematol.* 25:299–310.
17. Warmerdam, P.A.M., J.G.J. van de Winkel, A. Vlug, N.A.C. Westerdaal, and P.J.A. Capel. 1991. A single amino acid in the second Ig-like domain of the human Fc gamma receptor II is critical for human IgG2 binding. *J. Immunol.* 147:1338–1343.
18. Unkeless, J.C., E. Scigliano, and V.H. Freedman. 1988. Structure and function of human and murine receptors for IgG. *Ann. Rev. Immunol.* 6:251–281.
19. Koene, H.R., M. Kleijer, J. Algra, D. Roos, A.E.G. Kr. von dem Borne, and M. de Haas 1997. Fc gammaRIIIa-158V/F polymorphism influences the binding of IgG by natural killer cell Fc gammaRIIIa, independently of the Fc gammaRIIIa-48L/R/H phenotype. *Blood* 90:1109–1114.
20. Clarkson, S.B., R.P. Kimberly, J.E. Valinsky, M.D. Witmer, J.B. Bussel, R.L. Nachman, and J.C. Unkeless. 1986. Blockade of clearance of immune complexes by an anti-Fc gamma receptor monoclonal antibody. *J. Exp. Med.* 164:474–489.
21. Clarkson, S.B., J.B. Bussel, R.P. Kimberly, J.E. Valinsky, R.L. Nachman, and J.C. Unkeless. 1986. Treatment of refractory immune thrombocytopenic purpura with an anti-Fc gamma-receptor antibody. *N. Eng. J. Med.* 314:1236–1239.
22. Ericson, S.G., K.D. Colemena, K. Wardwell, S. Baker, M.W. Fanger, P.M. Guyre, P. Ely. 1996. Monoclonal antibody 197 (anti-Fc gamma RI) infusion in a patient with immune thrombocytopenia purpura (ITP) results in down-modulation of Fc gamma RI on circulating monocytes. *Br. J. Haematol.* 92:718–724.

23. Kelton, J.G. 1987. Platelet and red cell clearance is determined by the interaction of the IgG and complement on the cells and the activity of the reticuloendothelial system. *Trans. Med. Rev.* 1:75–84.

24. Kelton, J.G., J. Singer, C. Rodger, J. Gauldie, P. Horsewood, and P. Dent. 1985. The concentration of IgG in the serum is a major determinant of Fc-dependent reticuloendothelial function. *Blood* 66:490–495.

25. Kimberly, R.P., A. gibofsky, J.E. Salmon, and M. Fotino. 1983. Impaired FC-mediated mononuclear phagocyte system clearance in HLA-DR2 and MT1-positive healthy young adults. *J. Exp. Med.* 157:1698–1703.

26. Lokki, M.L., A. Circolo, P. Ahokas, K.L. Rupert, C.Y. Yu, and H.R. Colten. 1999. Deficiency of human complement protein C4 due to identical frameshift mutations in the C4A and C4B genes. *J. Immunol.* 162:3687–3693.

27. Reveille, J.D., J.M. Moulds, C. Ahn, A.W. Friedman, B. Baethge, J. Roseman, K.V. Straaton, and G.S. Alarcon. 1998. Systemic lupus erythematosus in three ethnic groups: 1. The effects of HLA class II, C4, and CR1 alleles, socioeconomic factors, and ethnicity at disease onset. LUMINA Study Group. Lupus in minority populations, nature versus nurture. *Arthritis Rheum.* 41:1161–1172.

28. Moulds J.M., and R. DeJongh. 1992. Influence of C4B null genes on cytomegalovirus antibody titers in healthy blood donors. *Transfusion* 32:145–147.

29. Cornillet P., P. Gredy, J.L. Pennaforte, O. Meyer, M.D. Kazatchkine, and J.H. Cohen. 1992. Increased frequency of the long (S) allotype of CR1 (the C3b/C4b receptor, CD35) in patients with systemic lupus erythematosus. *Clin. Exp. Immunol.* 1992 89:22–25.

30. Xiang, L., J.R. Rundles, D.R. Hamilton, and J.G. Wilson. 1999. Quantitative alleles of CR1: coding sequence analysis and comparison of haplotypes in two ethnic groups. *J. Immunol.* 163:4939–4945.

31. Zupanska, B., E. Thompson, E. Brojer, E.E. Thomson, A.H. Merry, and H. Seyfried. 1987. Monocyte-erythrocyte interaction in autoimmune haemolytic anaemia in relation to the number of erythrocyte-bound IgG molecules and subclass specificity of autoantibodies. *Vox. Sang.* 52:212–218.

32. Zupanska, B., E. Thompson, E. Brojer, and A.H. Merry. 1987. Phagocytosis of erythrocytes sensitized with known amounts of IgG1 and IgG3 anti-Rh antibodies Vox Sang. 53:96–101.

33. Kay N.E., and S.D. Douglas. 1977. Monocyte-erythrocyte interaction in vitro in immune hemolytic anemias. *Blood* 50:889–897.

34. Gallagher, M.T., D.R. Branch, A. Mison, and L.D. Petz. 1983. Evaluation of reticuloendothelial function in autoimmune hemolytic anemia using an in vitro assay of monocyte-macrophage interaction with erythrocytes. *Exp. Hematol.* 11:82–89.

35. Fries, L.F., C.M. Brickman, and M.M. Frank. 1983. Monocyte receptors for the Fc portion of IgG increase in number in autoimmune hemolytic anemia and other hemolytic states and are decreased by glucocorticoid therapy. *J. Immunol.* 131:1240–1245.

36. Murphy W.G., and J.G. Kelton. 1988. Methyldopa-induced autoantibodies against red blood cells. *Blood Rev.* 2:36–42.

37. Kleinman, S., R. Nelson, L. Smith, and D. Goldfinger, 1984. Positive direct antiglobulin tests and immune hemolytic anemia in patients receiving procainamide. *N. Eng. J. Med.* 311:809–812.

38. Garratty, G. 1987. The significance of IgG on the red cell surface. *Transfus. Med. Rev.* 1:47–57.

39. Vos, G.H., L.D. Petz, G. Garraty, and H.H. Fudenberg. 1973. Autoantibodies in acquired hemolytic anemia with special reference to the LW system. *Blood* 42:445–453.

40. Issitt, P.D., B.G. Pavone, D. Goldfinger, H. Zwicker, C.H. Issitt, J.A. Tessel, S.W. Kroovand, and C.A. Bell. 1976. Anti-Wrb, and other autoantibodies responsible for positive direct antiglobulin tests in 150 individuals. *Br. J. Haematol.* 34:5–18.

41. Bloy, C., D. Blanchard, P. Lambin, D. Goosens, P. Rouger, C. Salmon, S.P. Masouredis, and J.P. Cartron. 1988. Characterization of the D, C, E and G antigens of the Rh blood group system with human monoclonal antibodies. *Mol. Immunol.* 25:925–930.

42. Cartron, J.P. 1994. Defining the Rh blood group antigens. Biochemistry and molecular genetics. *Blood Rev.* 8:199–212.

43. Marsh, L.W., R. Oyen, E. Alicea, M. Linter, and S. Horton. 1979. Autoimmune hemolytic anemia and the Kell blood groups. *Am. J. Hematol.* 7:155–162.

44. Levine, A.M., P. Thornton, S.J. Forman, P.V. Hale, D. Holdorf, C.L. Roualult, D. Powars, D.I. Feinstein, and R.J. Lukes. 1980. Positive Coomb's test in Hodgkin's disease: significance and implications. *Blood* 55:607–611.

45. Coon, W.W. 1985. Splenectomy in the treatment of hemolytic anemia. *Arch. Surg.* 120:625–628.

46. Akpek, G., D. McAneny, and L. Weintraub. 1999. Comparative response to splenectomy in Coomb's-positive autoimmune hemolytic anemia with or without associated disease. *Am. J. Hematol.* 61:98–102.

47. Ahn, Y.S., W.J. Harrington, R. Mylvaganam, J. Ayug, and L.M. Pall. 1985. Danazol therapy for autoimmune hemolytic anemia. *Ann. Int. Med.* 102:298–301.

48. Mitchell, C.A., M.B. van der Weyden, and B.G. Firkin. 1987. High dose intravenous gammaglobulin in Coomb's-positive hemolytic anemia. *Aust. N. Z. J. Med.* 17:290–294.

49. Gilliland, B.C., E. Baxter, and R.S. Evans. 1971. Red-cell antibodies in acquired hemolytic anemia with negative antiglobulin serum tests. *N. Eng. J. Med.* 285:252–256.

50. Rosse, W.F. 1973. Correlation of in vivo and in vitro measurements of hemolysis in hemolytic anemia due to immune reactions. *Prog. Hematol.* 8:51–75.

51. Englefreit, C.P., A. Borne, D. Beckers, and J.J. Loghem. 1974. Autoimmune haemolytic anaemia: serological and immunochemical characteristics of the autoantibodies; mechanisms of cell destruction. *Ser. Haematol.* 7:328–347.

52. Logue, G., and W. Rosse. 1976. Immunologic mechanisms in autoimmune hemolytic disease. *Semin. Hematol.* 13:277–289.

53. Kelton, J.G. 1985. The concentration of IgG in the serum is a major determinant of Fc-dependent reticuloendothelial function. *Blood* 66:490–495.

54. Clark, D.A., E.N. Dessypris, D.E. Jenkins, and S.B. Krantz. 1984. Acquired immune hemolytic anemia associated with IgA erythrocyte coating: investigation of hemolytic mechanisms. *Blood* 64:1000–1005.

55. Fanger, M.W., L. Shen, J. Pugh, and G.M. Bernier. 1980. Subpopulations of human peripheral granulocytes and monocytes express receptors for IgA. *Proc. Natl. Acad. Sci. USA* 77:3640–3644.

56. Berke, G. 1989. The cytolytic T lymphocyte and its mode of action. *Immunol. Lett.* 20:169–178.

57. Meri, S., B.P. Morgan, M. Wing, J. Jones, A. Davies, E. Podack, and P.J. Lachmann. 1990. Human protectin (CD59), and 18–20-kD homologous complement restriction factor,

does not restrict perforin-mediated lysis. *J. Exp. Med.* 172:367–370.

58. Rosse, W.F. 1990. Phosphatidylinositol-linked proteins and paroxysmal nocturnal hemoglobinuria. *Blood* 75:1595–1601.

59. Holguin, M.H., L.R. Frederick, N.J. Bernshaw, L.A. Wilcox, and C.J. Parker. 1989. Isolation and characterization of a membrane protein from normal human erythrocytes that inhibits reactive lysis of the erythrocytes of paroxysmal nocturnal hemoglobinuria. *J. Clin. Invest.* 84:7–17.

60. Holguin, M.H., L.A. Wilcox, N.J. Bernshaw, W.F. Rosse, and C.J. Parker. 1990. Erythrocyte membrane inhibitor of reactive lysis: effects of phosphatidylinositol-specific phospholipase C on the isolated and cell-associated protein. *Blood* 75:284–289.

61. Rollins, S.A., and P.J. Sims. 1990. The complement-inhibitory activity of CD59 resides in its capacity to block incorporation of C9 into membrane C5b-9. *J. Immunol.* 144:3478–3483.

62. Roelcke, D. 1989. Cold agglutination. *Transf. Med. Rev.* 2:140–166.

63. Pruzanski, W., and K.H. Shumak. 1977. Biologic activity of cold-reacting autoantibodies 1. *N. Engl. J. Med.* 297.538–542.

64. Pruzanskid, W., and K.H. Shumak. 1977. Biologic activity of cold-reacting autoantibodies 2. *N. Engl. J. Med.* 297:538–583.

65. Roelcke, D. 1974. Cold agglutination. Antibodies and antigens. *Clin. Immunol. Immunopath.* 2:266–280.

66. Rosse, W.F., and J.P. Adams. 1980. The variability of hemolysis in the cold agglutinin syndrome. *Blood* 56:409–416.

67. Schreiber, A.D., D.S. Herskovitz, and M. Goldwein. 1977. Low-titer cold-hemagglutinin disease. Mechanism of hemolysis and response to corticosteroids. *N. Engl. J. Med.* 296:14901494.

68. Silberstein, L.E., E.M. Berkman, and A.D. Schreiber. 1987. Cold hemagglutinin disease associated with IgG cold-reactive antibody. *Ann. Int. Med.* 106:238–242.

69. Ross, G.D., W.J. Yount, M.J. Walport, J.B. Winfield, C.J. Parker, C.R. Fuller, R.P. Taylor, B.L. Myones, and P.J. Lachman. 1985. Disease-associated loss of erythrocyte complement receptors (CR1, C3b receptors) in patients with systemic lupus erythematosus and other diseases involving autoantibodies and/or complement activation. *J. Immunol.* 135:2005–2014.

70. Mollison, P.L., C.P. Engelfriet, and M. Contreras. 1987. Blood Transfusion in Clinical Medicine. Blackwell Scientific Publications, Oxford. pp. 417–419.

71. Capra J.D., J.M. Kehoe, R.C. Williams, Jr., T. Feizi, and H.G. Kunkel. 1971. Light chain sequences of human IgM cold agglutinins (variable-region subgroups amino-acid sequence-kappa light chain-N-terminal). *Proc. Natl. Acad. Sci. USA* 1972 69:40–43.

72. Silberstein, L.E., L.C. Jefferies, J. Goldman, D.F. Friedman, J.S. Morre, P.C. Nowell, D. Roelcke, W. Pruzanski, J. Roudier, and G.J. Silverman. 1991. Variable region gene analysis of pathologic human autoantibodies to the related i and I red blood cell antigens. *Blood* 73:2372–2386.

73. Goni, F.R., P.P. Chen, D. McGinnis, M.L. Arjonilla, J. Fernandez, D. Carson, A. Solomon, E. Mendez, and B. Frangione. 1989. Structural and idiotypic characterization of the L chains of human IgM autoantibodies with different specificities. *J. Immunol.* 1452:3158–3163.

74. Pascual, V., K. Victor, M. Spellerberg, T.J. Hamblin, F.K. Stevenson, and J.D. Capra. 1992. VH restriction among human cold agglutinins. The VH4-21 gene segment is required to encode anti-I and anti-i specificities. *J. Immunol.* 149:2337–2344.

75. Jefferies, L.C., C.M. Carchidi, and L.E. Silberstein. 1993. Naturally occurring anti-i/I cold agglutinins may be encoded by different VH3 genes as well as the VH4.21 gene segment. *J. Clin. Invest.* 92:2821–2833.

76. Stevenson, F.K., G.J. Smith, J. North, T.J. Hamblin, and M.J. Glennie. 1989. Identification of normal B-cell counterparts of neoplastic cells which secrete cold agglutinins of anti-I and anti-i specificity. *Br. J. Haemat.* 72:15.

77. Silberstein, L.E. 1993. Natural and pathologic human autoimmune responses to carbohydrate antigens on red blood cells. *Springer Semin. Immunopathol.* 15:139–153.

78. Jeffries, L.C., F.K. Stevenson, J. Goldman, I.M. Bennett, S.L. Spitalnik, and L.E. Silberstein. 1990. Anti-idiotypic antibodies specific for a pathologic anti-Pr2 cold agglutinin. *Transfusion* 30:495–502.

79. O'Connor, B.M., and J.S. Clifford, W.D. Lawrence, and G.L. Logue. 1989. Alpha-interferon for severe cold agglutinin disease. *Ann. Int. Med.* 111:255–256.

80. Meytes, D., M. Adler, I. Virag, D. Feigl, and C. Levene. 1985. High-dose methylprednisolone in acute immune cold hemolysis. *N. Engl. J. Med.* 312:318.

81. Heddle, N.M. 1989. Acute paroxysmal cold hemoglobinuria. *Transf. Med. Rev.* 3:219–229.

82. Gottsche, B., A. Salama, and C. Mueller-Eckhardt. 1990. Donath-Landsteiner autoimmune hemolytic anemia in children. A study of 22 cases. *Vox. Sang.* 58:281–286.

30 | Experimental Models of Autoimmune Hemolytic Anemia

Sidonia Fagarasan, Norihiko Watanabe, and Tasuku Honjo

1. INTRODUCTION

Autoimmune hemolytic anemia (AIHA), characterized by the destruction of red blood cells (RBC) as a result of the production of Coombs antibody (Ab), is one of the oldest autoimmune diseases recognized in humans.

Since New Zealand Black (NZB) mice spontaneously develop AIHA, this murine model has been intensively studied to define the genetic mechanism, etiology and pathogenesis of this disease (1–11). Although the cellular and molecular mechanisms of anti-RBC autoantibody (autoAb) production remain unclear, it has been shown that there are two major groups of anti-mouse RBC autoAbs: One, predominantly of the IgG class, recognizes surface determinants on intact RBCs and the other, predominantly of the IgM class, reacts with antigenic determinants exposed only after treatment of mouse RBC with proteolytic enzymes such as bromelain (2,3). The former autoantigen was initially designated as X antigen and identified later as the erythrocyte anion channel band 3 (9,12), and the latter was named HB antigen, which is revealed only after treatment of RBCs with proteolytic enzymes (2,3).

While autoAbs against intact erythrocytes are encoded by different V_H, V_K, D, J_H and J_K genes (13), those for anti-bromelain-treated RBCs use mainly the V_H11-V_K9 gene combination (14,15). The V_H11-V_K9 family, associated with Ab against bromelain-treated RBC specificity, has been shown to be predominantly expressed by B1 cells (16) and the autoAbs are selectively produced by this subpopulation of B cells (17). The anti-bromelain treated RBC responses are likely to be independent of the presence of T cells, since IgG responses and somatic mutations are absent. On the other hand, the development of IgG and the presence of somatic mutations, even for IgM autoAb, suggest the involvement of T cells for anti-intact RBC autoAb production (10,18). It is likely that two distinct B cell populations are involved in anti-RBC autoAb responses.

Although the antigenic specificity plays a significant role in the development of AIHA, the immunoglobulin (Ig) heavy chain (H chain) of anti-RBC autoAbs appears to be important in the pathogenesis of anemia by determining effector functions such as complement-dependent hemolysis, Fc receptor-mediated phagocytosis and multivalency-induced hemagglutination (8,18). *In vivo* studies revealed that sequestration of agglutinated RBCs in spleens and livers and Fc receptor-mediated phagocytosis, but not complement-mediated hemolysis, are the major mechanisms for the development of AIHA (8).

Transgenic (Tg) mice expressing a pathogenic IgM anti-RBC autoAb-4C8mAb (8,19) that reacts with the 4.1 band from the internal erythrocyte membrane skeleton (12) have been generated in an attempt to understand the biology of anti-RBC autoreactive B cells and factors responsible for triggering and progression of AIHA (20,21). In these mice, almost all conventional B cells are deleted from the periphery due to their reactivity with RBCs. B1 cells expressing the transgenes escape from clonal deletion and expand in the sequestered compartment of the peritoneal cavity. Approximately half the Tg mice suffer from AIHA when bred in the conventional facility (20). The presence of autoreactive B1 cells is required for the development of AIHA in this murine model, since the disease has been cured after elimination of B1 cells by intraperitoneal injection of RBCs, and Tg mice crossed with *xid* mice, in which B1 cells are congenitally absent, do not suffer from anemia (22). However, the simple presence of autoreactive B cells is not sufficient for induction of anemia. The development of disease is most probably due to the activation of autoreactive B cells by environmental factors such as bacterial, viral or parasitic infection, because germfree or pathogen specific-free Tg mice have never suffered from spontaneous AIHA, while a half of the mice in the conventional conditions developed the disease (23). Moreover, the administration of lipopolysaccharides (LPS) of gram negative bacteria or Th2 cytokines such as IL-5 and IL-10 activate autoreactive cells and induce anemia in asymptomatic Tg mice (24–27).

This Tg mouse model strongly supports the idea that B1 cells may play an important role in the development of B-cell dominant autoimmune diseases such as AIHA. In addition, a longterm clinical observation that environmental factors are required for the onset of autoimmune diseases has gained direct evidence in this animal model. The major question to be answered is how and where the autoreactive B1 cells in the peritoneal cavity are activated and induced to secrete a large amount of the autoAbs.

2. CHARACTERIZATION OF TRANSGENIC MICE THAT EXPRESS ANTI-ERYTHROCYTE AUTOANTIBODY GENES

In an attempt to understand the biology of autoreactive B cells and factors responsible for the onset and evolution of AIHA, we generated Tg mouse lines in which all B cells express the anti-RBC autoAb on the surface by introducing the Ig genes derived from the anti-RBC autoAb-producing hybridoma (4C8 mAb)(20,21).

2.1. DNA Construction and Generation of Transgenic Mice

The H chain transgene was constructed using DNA fragments described as follows: The 1.3-kb XhoI-NcoI fragment containing the promoter region and the initiation site was isolated from pSV-Vμ1 (21). The 0.4-kb NcoI-EcoT141 fragment was isolated from cDNA encoding the H chain of the 4C8 mAb (19). The 1.8-kb EcoT141-EcoRI fragment containing the J_H3, J_H4 and enhancer sequences, and the 12-kb EcoRI-XhoI fragment containing the major intron and the Cμ region, were derived from the genomic clones Ch.M.Ig.μ-18 (29) and λ gtWES.IgH714, respectively (30). The pSP72/73-based plasmid containing the complete 4C8 H chain gene was designated pMO-μ4C8 (Fig. 1A). For construction of the L chain transgene, the 5.6-kb BamHI fragment extending from the promoter to the Cκ exon was recloned into Bluescript 13) (-) and the plasmid was designated pMO-κ4C8 (Fig. 1B).

To generate H-chain Tg (H Tg) mice, the 15.5-kb XhoI fragment of pMO-μ4C8 was injected into fertilized eggs of C57BL/6 mice, followed by transfer of viable eggs into the oviducts of the pseudopregnant ICR mice.

L-chain Tg (L Tg) mice were generated similarly using the 5.6-kb BamHI fragment from pMO-κ4C8 plasmid. Double Tg (HLTg) mice were obtained by mating H- and L-chain Tg mice (20).

New lines of 4C8 mAb Tg mice, which carry the tandem joined H- and L-chain transgene, were also generated (21). To construct the tandem-joined transgene, a 5.6-kb BamHI fragment of pMO-κ4C8 plasmid was first subcloned in the BamHI site of pSP73 vector. A 15.5 kb XhoI fragment of pMO-μ4C8 was subcloned between the SalI and XhoI sites. The pSP73 plasmid containing both H and L chain genes was designated as pSP73-$\kappa\mu$4C8. Tandem-joined H + L mice were generated by injecting a 22.0-kb PvuI-XhoI fragment of pSP73-$\kappa\mu$4C8. The presence of the transgenes and the homozygosity of the Tg loci were screened by polymerase chain reaction (PCR) and Southern blot analysis.

2.2. Allelic Exclusion in Transgenic Mice that Express IgM Anti-RBC Autoantibody

In the clonal selection theory, Burnet envisaged that each Ab-forming cell is genetically committed to express a distinct Ab specificity (31). In all mammals, the genes encoding Ab variable (V) regions are generated during B cell development from gene segments V,D, and J for the IgH locus, V and J for the IgL locus through a process of joining or site-specific recombination. For cellular selection, according with Burnet theory, each B lymphocyte should ideally express only one V_H and one V_L region gene. This principle is known as allelic exclusion, or isotype exclusion in the case of the κ and λ L chains (32).

Studies on Tg (33–35) and knockout mice (36) have demonstrated the essential role of surface expression of the Tg H and L chains in supression of endogenous H and L gene rearrangement. Mice expressing the membrane-form μ-chain transgene inhibit expression of the endogenous μ-chain, whereas Tg mice with the secreted form of μ-chain do not show such inhibition (33–35,37). Expression of L-chain transgenes also induces suppression of rearrangement and expression of the endogeneous L-chain locus (35).

In HL Tg mice that carry the Ig genes encoding the 4C8 autoAb against murine RBC, both H- and L-chain genes of the endogenous loci are excluded efficiently (20).

In single H-chain Tg (H Tg) mice, less than 2.5% of the splenic B cells express endogeneous H-chain genes (IgM^{b+}), indicating that these genes are almost completely excluded by the expression of the transgene (IgM^{a+}). Although the allotypic specificity for the Tg L-chain is not known, allelic exclusion appears to be accomplished for

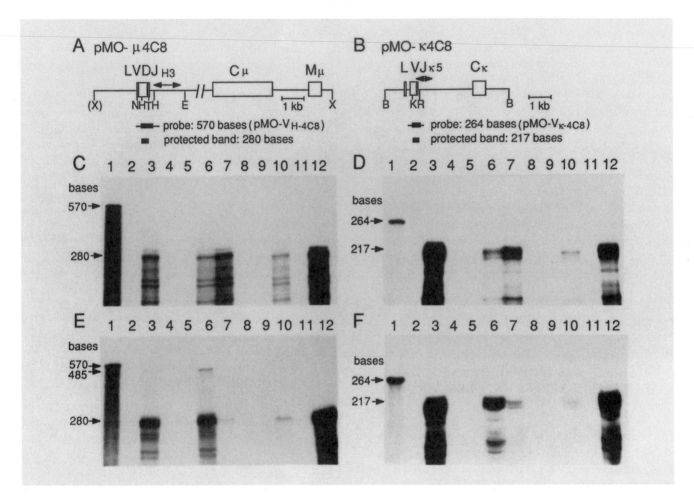

Figure 1. Structures of microinjected Ig genes and their expression. (A) 4C8 H chain gene. (B) 4C8 L chain gene. Coding regions and untranslated regions are represented by open boxes. Probes for RNase protection assay are shown below with expected protection bands. Arrows indicate regions of probes used for Southern blotting analyses. Relevant restriction sites are shown: B, BamHI; N, Nco I; E, EcoR I; T, EcoT14I; X, Xho I; K, Kpn I; H, Hind III; R, Rsa I. A site in a linker is shown with a bracket. RNase protection assays for RNA isolated from (C) H Tg mouse; (D) L Tg mouse; and (E and F) HL Tg mouse. Each line contained 2.5 μg of total RNA. The 485 base-band is due to unspliced mRNA. Origin of RNA: lane 1, RNA without digestion; lane 2, brain; lane 3, lymph node; lane 4, heart; lane 5, lung, lane 6, thymus; lane 7, spleen; lane 8, liver, lane 9, kidney; lane 10, bone marrow; lane 11, muscle; lane 12, 4C8 hybridoma. Data are reproduced from Okamoto et al. (20).

both H-and L-chain genes. In double-Tg (HL Tg) mice, almost all surface IgM[+] splenic cells express the transgene idiotype (Id), which is formed by the combination of the H- and L-chains of the 4C8 mAb and recognized by the S54mAb (Fig. 2A,B). Very few spleen B cells express endogenous allotype-specific IgM, which does not carry the Id of the transgenes.

However, the supression of endogenous H- and L-chain gene expression in Ig Tg mice is not always complete and the efficiencies of allelic exclusion are variable among different lines of Tg mice (31,39–46). Variable integration sites of transgenes may influence the onset and level of the transgene expression, which can affect and explain the differences in the efficiency of the allelic exclusion. To avoid

the difference due to the integration site of the transgenes, we studied the allelic exclusion in Tg mice that carry tandem joined H- and L-chain transgene (H + L Tg mice) (21). In these Tg mice, homozygous mice expressed more Tg H- and L-chains on the surface of B cells than heterozygous mice.

In all H + L Tg lines, the total numbers of endogenous H chain expressing IgM[b+] spleen cells were lower in homozygous than heterozygous mice, indicating that inhibition of endogenous H-chain expression (allelic exclusion) is stronger in homozygous than in heterozygous mice (Fig. 2C). Inhibition of endogenous L-chain expression is also stronger in homozygous than in heterozygous mice as estimated by difference in expression of TgH chain IgM[a] allotype and

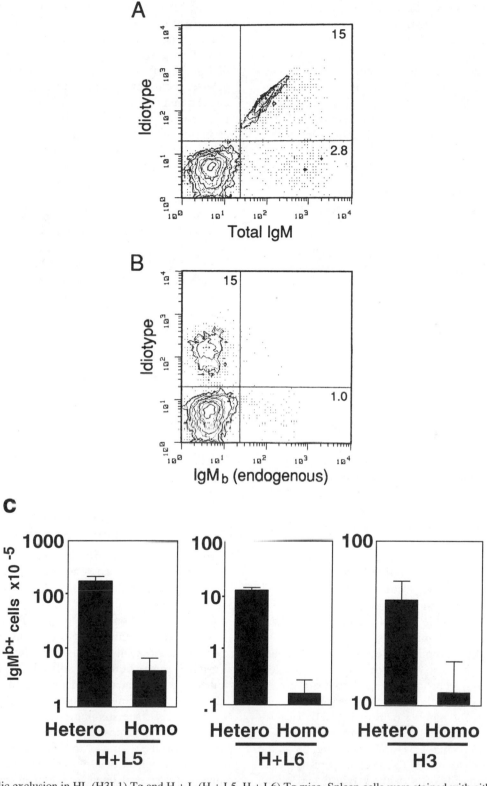

Figure 2. Allelic exclusion in HL (H3L1) Tg and H + L (H + L5, H + L6) Tg mice. Spleen cells were stained with either (A) anti-IgM or (B) anti-IgM[b] mAb and anti-Id mAb (S54). Contour maps show correlated fluorescence in arbitrary relative fluorescence units. Percentages of positively stained cells are indicated in each compartment. Reproduced from Okamoto et al. (20). (C) Total numbers of the cells with endogenous H chains in heterozygous or homozygous H + L5, H + L6 and H3 Tg mice. Spleen cells were stained with anti-IgM[a] (transgenic) and anti-IgM[b] (endogenous) Ab and analyzed from at least three mice of each Tg line. Numbers of indicated cells were calculated by (percentage of IgM[b+] cells) × (number of viable cells). Results are expressed as mean ± SD. Data are taken from Watanabe et al. (21).

S54Id. The total number of cells expressing endogenous L-chain (IgM^{a+} Id^{low}) in bone marrow and spleen were less in homozygous than in heterozygous mice.

Therefore, homozygosity of the transgene Ig loci increases the surface Ig expression in bone marrow B cells and causes stronger allelic exclusion compared with heterozygosity. Since VDJ recombination is all or none in each B cell, it appears that there is a threshold of the B cell receptor(BCR) expression and/or signaling level that induces allelic exclusion.

2.3. Two Distinct Mechanisms of Tolerance: Deletion and Anergy of Autoreactive B cells in HL Tg Mice.

During their development, B lymphocytes are subjected to both positive and negative selection events depending on the expression and antigen specificity of the BCR. Studies on Ig Tg mice have established that interactions between immature B cells and endogeneous antigens through BCR lead to tolerance induction (47–50). Negative selection of autoreactive B cells occurs by at least three distinct mechanisms: Deletion, anergy and receptor editing. Membrane-bound self-antigens eliminate self-reactive B cells by strong cross-linking of surface Ig receptor (clonal deletion) (48,50). In contrast, immature B cells bearing the transgenic anti-self Ig receptor that have encountered soluble self-antigens enter in an inactive state in which the cells are refractory to further stimulation (clonal anergy) (47,49,51,52). Accumulating evidence suggests that self-reactive immature B cells can be rescued from deletion by upregulation of RAG-1 and RAG-2, and replacement of the autoreactive BCR by secondary, Ig gene rearrangement (receptor editing) (53,54).

How are autoreactive B cells regulated in HL Tg mice? The transgene expression analyzed using riboprobes complementary to the V$_H$ and V$_L$ sequences have revealed different patterns of expression in single Tg (H Tg and L Tg) mice compared with double Tg (HL Tg) mice (20). In H Tg and L Tg mice, the transgenes are expressed in all lymphoid tissues/lymph nodes, spleen, thymus and bone marrow. In HL Tg mice, however, expression of the transgenes in the spleen and bone marrow are greatly reduced, probably due to clonal deletion of autoreactive B cells (Fig. 1 C-F). The number of B cells in HL Tg mice was estimated by FACS analysis and compared with those in H Tg, L Tg and normal C57BL/6 mice (Fig. 3). In the spleen and lymph nodes of HL Tg mice, the percentage and the number of mature B cells (namely B220$^+$ IgM$^+$ cells) were drastically reduced compared with those in control mice. The staining patterns with anti-IgM Ab were almost identical to those with the anti-Id Ab. This indicates that autoreactive B cells are deleted, although the degree of deletion is very different among individuals. The ratios of the number of B cells of HL Tg mice to that of H Tg or L Tg mice vary between 0.01 and 1.0 (Fig. 3). The B cells in HL Tg mice are confined to the cells expressing low levels of surface IgM, suggesting that autoreactive B cells with high levels of surface antigen receptors are preferentially deleted. In the bone marrow of HL Tg mice, the percentage and the number of B cells are similar to the other control mice. As in the spleen or lymph nodes, these B cells express low levels of antigen receptors on their surface. B cells with low density of surface receptors are often observed in an anergic state (47,55,56) as they cannot produce the autoAb after in vitro stimulation with LPS (57–59). When cultivated in vitro and stimulated with LPS, splenic B cells of HL Tg mice cannot be activated to produce the autoAb. By ELISA, the total IgM and Id IgM were measured in the supernatant. The IgM production value per B cell was calculated as an index of the reactivity to LPS (Fig. 4A) and these values of HL Tg mice were reduced to less than one-tenth of those of control mice. This indicates that splenic B cells of HL Tg mice that have survived clonal deletion are anergized, although we cannot exclude the possibility that in vivo, where the environmental stimuli are more complex these cells might be activated to produce autoAb. There is no correlation between the extent of clonal deletion and the degree of B cell anergy in each HL Tg individual (Fig. 4B). This suggests that although both deletion and anergy are invoked for the mechanism of B cell tolerance in HL Tg mice, their relative contribution differs greatly in each individual.

2.4. Autoreactive B1 Cells Escape from Clonal Deletion and Expand in the Peritoneal Cavity of anti-RBC Tg mice

Two subsets of B cells designated B1 and B2 cells have been identified based on their anatomical location, developmental origin, cell surface markers, Ab repertoire and self-renewal capacity (60–63). B1 cells form a dominant population in the peritoneal cavity, but are rare in the spleen or lymph nodes of adult mice (62–65). As mentioned above, in the peripheral lymphoid organs of HL Tg mice, conventional B2 cells (B220$^+$ IgM$^+$ Mac1$^-$) are either deleted or anergized. However, the peritoneal cavity of HL Tg mice contains normal numbers of B1 cells as defined by surface staining with anti-IgM and anti-Macl Abs (Table 1). Almost all B1 cells express the transgene idiotype on their surface and most likely are autoreactive cells. As in spleen or lymph nodes, B2 cells in the peritoneal cavity are deleted, resulting in relative enrichment of B1 cells.

In another anti-RBC Tg line (H + L Tg mice), B1 cells in the peritoneal cavity also escape from clonal deletion. Peritoneal cavity cells of H + L Tg mice contained a limited number of B220$^+$Id$^+$ cells compared with HL Tg mice and

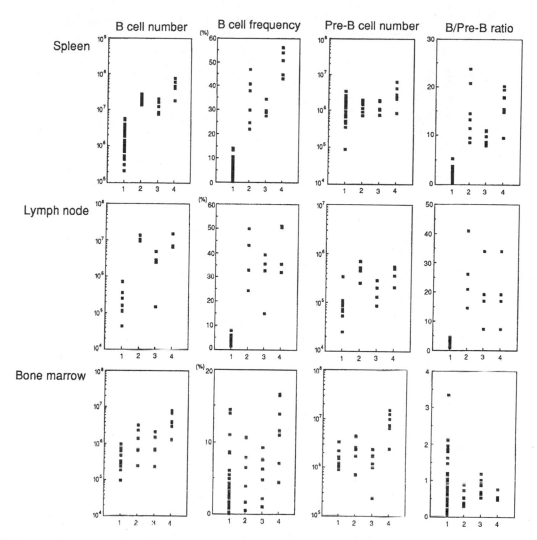

Figure 3.   Deletion of autoreactive B cells in HL Tg mice. Spleen, lymph node and bone marrow cell suspensions of HL Tg, H Tg, L Tg and normal C57BL/6 mice were stained with anti-B220 (RA3 6B2) and either one of S54 and anti-IgM Abs and analyzed by flow cytometry. Numbers of B220⁺ IgM⁺ or B220⁺ IgM⁻ were calculated by (% positive cells) × (cells/organ). Id⁺ B cells number were shown in HL Tg mice. B220⁺ cells without surface IgM and IgG were considered pre-B cells. Total bone marrow cell suspensions were obtained from two femurs and total lymph node cell suspensions were obtained from the inguinal, axillary and submandibular regions for all mice. Each square represents one individual. Column 1, H3L1 Tg mice; column 2, H3 Tg mice; column 3, L1Tg mice; column 4, normal C57BL/6 mice. Reproduced from Okamoto et al. (20).

all of these cells belong to the B1 subset as they express Mac1 on the surface. Moreover, the size of autoreactive B1 cell compartment is larger in homozygous than heterozygous mice (Fig. 5A, B), suggesting that a higher expression level of the autoAb facilitates the increase in autoreactive B1 cells in the peritoneal cavity. That the autoreactivity of Ig expressed on B cells is very important for enlargement of B1 cells subpopulation is demonstrated by analysing H only Tg mice (H3). In H3 mice, although the majority of peritoneal B cells express the Tg H chain (IgMᵃ⁺), they are not autoreactive because of the absence of the Tg L chain. In these mice, the percentage of IgM⁺ Mac1⁺ cells does not

increase in homozygous mice compared with heterozygous mice (Fig. 5C), indicating that increased surface expression of non-autoreactive B cells does not facilitate the enlargement of the B1 cell population in the peritoneal cavity. This observation in anti-RBC Tg mice is consistent with previous observations that defective BCR signaling causes reduction of B1 cell numbers (66–68), and that loss of a BCR inhibitory molecule, SHP-1, increases the B1 cell number (69–73). In summary, increased levels of autoreactive BCR augment clonal deletion of B2 cells in the bone marrow and spleen, but expansion of B1 cells in the peritoneal cavity.

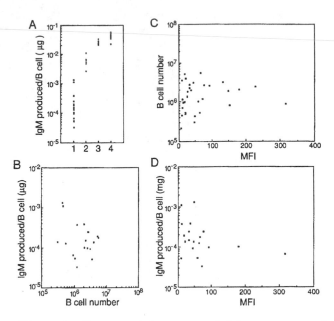

Table 1 Ly-1 B cells in the peritoneal cavity escape from deletion

Mice	Percent Ly-1B	Percent conventional B	Ly-1B/ conventional B
H3L1	33 ± 4.8 (5.3 ± 1.3)	6 ± 2.3 (1.2 ± 0.6)	5.5
H3	7 ± 1.3 (1.6 ± 0.5)	12 ± 0.8 (2.9 ± 0.7)	0.6
L1	21 ± 4.5 (7.3 ± 0.2)	15 ± 3.1 (5.2 ± 0.1)	1.4
Normal C57BL/6	24 ± 4.3 (3.9 ± 0.5)	23 ± 4.1 (3.7 ± 0.2)	1.1

Peritoneal cells were stained and analyzed by two-color FACS® with anti-Id (S54) or anti-IgM and anti-MAC-1 antibodies to discriminate Ly-1 B cells from conventional B cells. Actual cell numbers × 10^{-5} of each category are also shown in parentheses. Cell number was calculated as described in the legend to Fig. 5. Id-5⁺ B cell numbers are shown in H3L1 mice. Results are expressed as the mean ± SD of eight mice.

Figure 4. Poor correlation among deletion, anergy and autoAb amounts. Each square represents ane individual. (A) Reduced LPS-response of autoreactive B cells of HL Tg mice. Total spleen cells were cultured in vitro in the presence of 50 μg/ml LPS at three different cell densities: 10^5/ml, 10^6/ml and 10^7/ml for 3 or 7 days. The highest Ig production by spleen cells of normal H Tg, L Tg and HL Tg mice were found in cultures containing 10^5, 10^5, 10^5, and 10^6 cells/ml, respectively, for 7 days. Amount of secreted IgM in the supernatants of LPS-induced *in vitro* cultures was divided by the number of B cells in the culture which gave the most efficient Ig production. In the case of HL Tg mice Id⁺ IgM/Id⁺ B cells and in other cases total IgM⁺/total B cells are shown. Column 1, HL Tg mice; column 2, H Tg mice; column 3, L Tg mice; column 4, C57BL/6 mice. (B) No correlation between the extent of deletion and anergy of autoreactive B cells in HL Tg mice. LPS-induced Id⁺ IgM per Id⁺ B cells was plotted against remaining Id⁺ B cells number in the spleen. (C and D) Amount of the autoAbs in HL Tg mice is not directly related with either B cell number, LPS-responsive capacity of spleen cells. Id⁺ B cell number(C) and LPS-induced Id⁺ IgM per Id⁺ B cells (D) were plotted against MFI of FITC-anti-IgM bound to autoAbs associated with erythrocytes. Data are taken from Okamoto et al. (20).

3.   BREAKAGE OF TOLERANCE: AUTOIMMUNE HEMOLYTIC ANEMIA IN HL TG MICE

3.1.   Autoimmune Symptoms in Transgenic Mice Carrying Anti-RBC Autoantibody Genes

To examine whether HL Tg mice suffer from AIHA, two parameters were measured: Hematocrit values and the amount of autoAb bound to RBCs, by surface staining with the anti-IgM Ab. The amounts of autoAb associated with erythrocytes, expressed by mean fluorescence intensity (MFI), show wide variations among HL individuals spreading in almost two-log range (Fig. 6A). A reverse correlation of the hematocrit values with MFI indicates: *a.* the total amount of the autoAb produced directly correlates with anemia; *b.* most of the autoAbs produced are pathogenic and not mutated to escape from tolerance.

Based on the hematocrit values, HL Tg mice were classified into three groups: The anemic type with hematocrit <30%, the tolerant type with hematocrit >40% and the intermediate type with the hematocrit values between 30 and 40% (Fig. 6C). About 50% of the HL Tg mice bred in conventional facilities suffered from anemia, indicating a loss of self-tolerance. The fact that even a single mouse often changes from the state of complete tolerance to an acute onset of AIHA indicates that the efficiency of tolerance is variable among individuals in spite of the same genetic background.

Histological examinations have revealed that the anemic phenotype is associated with accumulation of agglutinated RBC in the spleen. The half-life of erythrocytes from the anemic type is half those from control mice, while the half-life of erythrocytes from the intermediate type mice is slightly shortened (Fig. 6B). These results suggest that the anemic phenotype is due to the reduction of the half-life of erythrocytes, which is caused by destruction of erythrocytes associated with increased anti-RBC autoAb in serum.

3.2.   Environmental Stimuli are Responsible for Induction of AIHA

The total numbers of B cells in the spleen (Fig. 4C,D) or peritoneal cavity of HL Tg mice are not related with the amounts of autoAb associated with erythrocytes. Therefore, it is more likely that only a small fraction of B cells escaping from the tolerance mechanisms are activated and

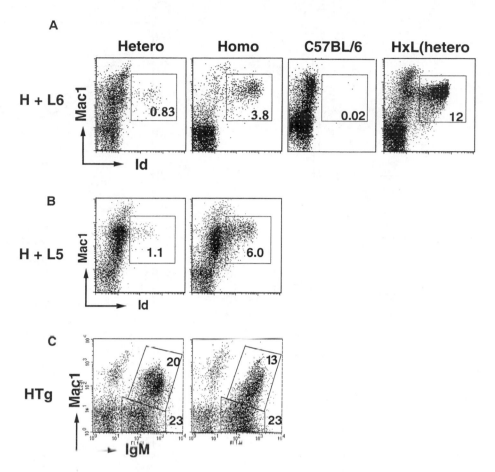

Figure 5. Flow cytometric analysis of peritoneal cells of heterozygous or homozygous H + L Tg(H + L5 and H + L6), HL Tg, H Tg and C57BL/6 mice. Cells were isolated from peritoneal cavity of mice at 10–14 weeks and stained with (A,B) biotin-anti-Id Ab followed by FITC-streptavidin and PE anti-Mac-1, (C) FITC-anti-IgM and PE-anti-Mac-1 Abs. The percentages of each gated region cells in total viable cells are indicated. Total peritoneal cell numbers were constant between heterozygous and homozygous Tg mice. Data are reproduced from Watanabe et al. (21).

differentiate into plasma cells that produced relatively large amount of the autoAb in the anemic phenotype mice.

The wide variation in the severity of anemia in HL Tg mice suggest that the environmental factors may be involved in activation of autoreactive B cells. To demonstrate the involvement of environmental stimuli in the onset of the autoimmune disease, HL Tg mice were bred under different conditions: Germ-free, specific pathogen-free and conventional (23). Germ-free mice are not contaminated with any bacteria or other pathogens. Specific pathogen-free mice were not infected with the following pathogens examined: Sendai virus, mouse hepatitis virus, *Corynebacterium kutcheri*, *Pasteurella pneumotropica*, *Salmonella* species, *Giardia muris*, *Tritrichomonas* species, *Syphacia* species or *Spironucleus muris*. While in conventional conditions about half the mice develop autoimmune disease, in germ-free and specific pathogen-free conditions, no mice suffer from anemia. Although in mice housed in conventional conditions, four microorganisms were detected (Sendai virus, mouse

hepatitis virus, Tyzzer's organism and *Syphacia* species), this result does not imply that any of these pathogens are responsible for the induction of anemia. Another evidence that infection can trigger the onset of autoimmune anemia comes from the observation that approximately one third of the germ-free or pathogen-free mice suffer from anemia when transferred into conventional conditions.

These results clearly demonstrate the relationship between the occurrence of AIHA and infection, although the pathogens or the mechanisms that activate autoreactive B cells in HL Tg mice remain to be elucidate.

3.3. LPS can Activate Autoreactive B Cells in HL Tg Mice

LPS, the invariant molecular structure of Gram-negative bacteria is a well known T-cell independent B lymphocyte mitogen and macrophage/monocyte activator (74,75). The effect of breeding environments on HL Tg mice suggests

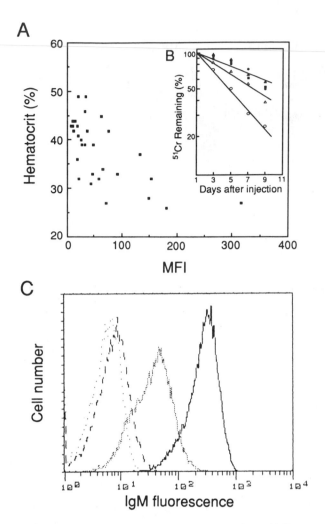

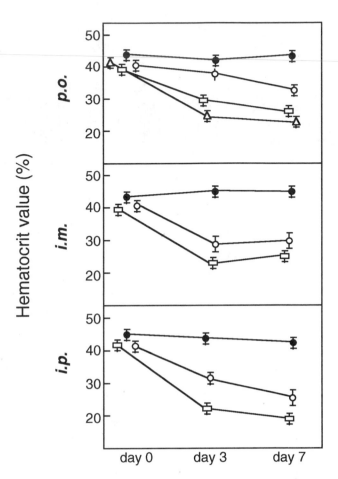

Figure 7. Injection of LPS aggravates hemolytic anemia dose-dependently. *Salmonella minnesota* LPS, suspended in PBS, was administrated orally (p.o.), intramuscularly (i.m.) or intraperitoneally (i.p.) to non-anemic HL Tg mice. Hematocrit value of the peripheral blood from each mouse was measured at days 0,3, and 7. The bars indicate the standard deviation of the values of the three to five mice. Amount of LPS used per mouse are: closed circles, 0 μg; open circles, 10 μg; open squares, 100 μg; open triangles, 200 μg. Data are reproduced from Murakami et al. (24).

Figure 6. Various phenotypes of HL Tg mice. (A) Reverse correlation of hematocrit values with amounts of autoAbs on erythrocytes. 32 HL Tg mice were analyzed by quantitating hematocrit values (vertical axis) and MFI of FITC-anti-IgM Ab bound to erythrocytes (horizontal axis). Each square representss one mouse. (B) Survival of ^{51}Cr-labeled syngeneic erythrocytes in HL Tg mice. The first sample was collected 24h after injection and this value was taken as 100%. Subsequently, samples were collected at a 2-days interval. Closed circles, triangles and squares represent normal C57BL/6, H Tg and L Tg mice, respectively. Open circles and triangles show the anemic and the intermediate types mice, respectively. (C) Erythrocytes were stained with anti-IgM Ab and analyzed on FACScan®: solid line, the anemic type; fine stippled line, the intermediate type; broken line, the tolerant type; rough stippled line, normal C57BL/6 mice. Data are taken from Okamoto et al. (20).

that induction of AIHA could be triggered by exogeneous antigens. To assess this possibility, the effects of LPS injection into HL Tg mice have been examined. LPS derived from *Salmonella minnesota* or *Escherichia coli* was administrated orally, intramuscularly or intraperitoneally into nonanemic HL Tg mice whose hematocrit values were

>40%. Regardless of the route of administration, anemia was induced in HL Tg mice, suggesting that LPS can activate autoreactive B cells to produce autoAb. When LPS is injected intramuscularly or intraperitoneally, lower doses of endotoxin (<100 μg/mouse) induced hemolytic anemia, while higher doses of LPS (>100 μg/mouse) are lethal, probably because of endotoxic shock. On the other hand, oral administration of LPS induced anemia in a dose-dependent manner up to 200 μg/mouse (Fig. 7). The different dose effects of LPS may depend, at least in part, on different rates of LPS absorbtion after oral or systemic injection.

The mucosal immune system has developed a state of tolerance that is an important component of the homeostatic response to comensal bacteria (76–78). In spite of this sophisticated and highly regulated mechanism, oral administration of LPS induced hemolytic anemia in anti-RBC Tg mice. The fact that LPS derived from enteric bacteria can induce symptoms similar to spontaneously-occuring AIHA in HL Tg mice suggests that bacterial infection may trigger the onset of the disease. This correlates well with the observation that the incidence of anemia increases when HL Tg mice are transferred from specific pathogen-free to conventional conditions (23). Therefore, it appears that the anergy of autoreactive B cells is a reversible process (79) and that the silenced self-reactive B cells are reactivated under particular conditions to give rise to autoAb production and anemia in HL Tg mice.

4. PERITONEAL B1 CELLS ARE RESPONSIBLE FOR AUTOANTIBODY PRODUCTION IN HL TG MICE

4.1. Antigen-Induced Apoptotic Death of B1 Cells in HL Tg Mice

Since autoreactive B cells are eliminated at the immature stage in bone marrow, the number of self-reactive B cells is markedly reduced in the peripheral blood, spleen and lymph nodes of HL Tg mice. In contrast, the peritoneal cavity of these mice contains a normal number of auto-reactive B1 cells that express surface Id (Table 1).

How can peritoneal B1 cells escape from clonal deletion? The idea that HL Tg mice allows survival of autoreactive B1 cells in the peritoneal cavity in the absence of the self-antigen has gained support from experiments injecting RBC into the peritoneal cavity (22). After a single injection of erythrocytes, peritoneal B1 cells in HL Tg mice were extinguished as early as 12 hours and did not recover for one week, whereas peritoneal cells of control mice were not affected by the same treatment. Moreover, peritoneal cells from RBC-injected Tg mice showed DNA fragmentation, indicating that apoptosis is involved in clonal deletion of B1 cells *in vivo* after encountering the self-antigen. Elimination of peritoneal B1 cells by self-antigen suggests that B1 cells that have received strong BCR-signaling might be eliminated before homing to the peritoneal cavity in HL Tg mice. The self-reactive B1 cells that have received a lower level of BCR-signaling and escaped from clonal deletion migrate into the peritoneal cavity, where they proliferate due to their self-renewal capacity. The fact that in homozygous anti-RBC Tg mice, the size of autoreactive B1 cells is larger as compared with heterozygous mice (subchapter 2.4.), imply that the level of surface Ig expression influences the expansion of B1 cells in the peritoneal cavity (21).

These results suggest that there are several levels of BCR signaling that regulates self-reactive B1 and B2 differentiation. At a lower level, self-reactive B2 cells can be stimulated to become anergic or to induce receptor editing (47,53,80,81). At an intermediate signaling level, B2 cells are clonally deleted and B-0 (82,83) and/or B-2 are induced to differentiate into B1 cells, which migrate into the peritoneal cavity. At a strong signalling level, B1 cells are also clonally deleted, as in the case of RBC injection into the peritoneal cavity of anti-RBC Tg mice (22). It might be possible that at least a sizable fraction of peritoneal B1 cells originate and expand from auto-reactive B cells stimulated by self-antigens to a level strong enough to be activated, but weak enough to avoid apoptosis.

It is generally accepted that immature B cells and peripheral mature B cells are not only phenotypically, but also functionally, distinct. Immature B cells are refractory to BCR-induced positive signals and are rather induced to undergo programmed cell death, whereas the same stimulation induces DNA synthesis in mature B cells (84–88). In small pre-B cells and immature B cells, the survival gene *bcl-2* is down-regulated, since these cells are programmed to die unless they are rescued into mature, long-lived, peripheral B cells by a BCR-mediated signal (89,90). In mature B cells, *bcl-2* is upregulated again (90) and the *bcl-2* gene product has been shown to inhibit apoptosis of normal lymphocytes (91,92).

To test the effect of the bcl-2 gene product on clonal deletion of autoreactive B cells at the level of bone marrow and periphery, HL Tg mice were crossed with Tg mice carrying the genes for the bcl-2 in the B-lineage cells (93). In *bcl-2* Tg mice, the numbers of both bone marrow and spleen B cells increased in the presence of *bcl-2* (94,95). In *bcl-2* × HL Tg mice, the numbers of B cells in spleen and bone marrow are markedly decreased (5–10 fold) compared with those in mice carrying *bcl-2* and

Table 2   Percentage of B lymphocytes in various organs of transgenic mice carrying either Ig chain or bcl-2 gene

	Nontransgenic littermate	L chain	H chain	H × L	bcl-2	bcl-2 × L	bcl-2 × H	bcl-2 × H × L
Spleen	51.2 ± 5.5	30.2 ± 3.2	36.0 ± 9.2	3.4 ± 1.9	83.0	40.9	72.6	3.4 ± 1.7
Bone marrow	14.4 ± 2.8	6.6 ± 1.9	10.1 ± 2.1	3.1 ± 2.1	75.4	35.9	63.3	6.2 ± 2.1
Peritoneal cavity	42.5 ± 5.3	40.9	41.8	36.7 ± 2.8	79.7	76.8	75.8	35.3 ± 7.4

Numbers with ± SD are from three mice and the other numbers are one representative value.

either IgH or IgL chain, or with those in mice carrying *bcl-2* and IgM incapable of reacting with erythocytes (Table 2). This indicates that bone marrow B cells are eliminated by clonal deletion in spite of the presence of the *bcl-2* transgene. As in the case of HL Tg mice, the peritoneal cavity of *bcl-2* × HL Tg mice contains a normal number of B1 cells, but only few conventional B cells. In HL Tg mice, intraperitoneal injection of the self-antigen almost completely eliminates B1 cells by inducing apoptotic death. In contrast, peritoneal B1 cells of *bcl-2* × HL Tg mice are no longer eliminated upon intraperitoneal injection of RBC. In *bcl-2* × HL Tg mice, injection of 10^9 RBC for 7 consecutive days fails to induce apoptosis, whereas a single injection of 10^7 RBC eliminates B1 cells in HL Tg mice. Therefore, *bcl-2* blocks antigen-induced apoptosis of peritoneal B1 cells but not of bone marrow B cells. Although it was reported that persistance of peritoneal B cells (CD5$^+$ and CD5$^-$) may not be controlled by the bcl-2 gene expression because low levels of bcl-2 mRNA were found in these cell populations (96), introduction of *bcl-2* prevented apoptosis induced by strong BCR-stimulation in anti-RBC Tg mice. This suggests that *bcl-2* has different effects on antigen-induced elimination of immature B cells in bone marrow and mature B cells in periphery.

4.2.   Involvment of B1 Cells in Autoimmune Hemolytic Anemia in HL Tg Mice

The pathogenicity of peritoneal B1 cells in AIHA in the HL Tg mice is suggested by several observations. First, peritoneal B1 cells represent the major B cell population in HL Tg mice, since almost all conventional B cells are either deleted or anergized (20,21,93). Second, elimination of B1 cells by intraperitoneal injection of antigens is followed by a decrease of autoAb bound to RBC and the hematocrit value normalization (Fig. 8) (22). Two other pieces of evidence support the involvement of B1 cells in AIHA: *a*. HL Tg mice bearing the X-linked immune deficiency mutation *xid*, in which B1 cells are congenitally absent (60,97), did not suffer from anemia (24). *b*. Hemolytic anemia is more frequent and severe in *bcl-2* × HL Tg mice that contained a normal number of apoptosis-resistant B1 cells in the peritoneal cavity (93).

Based on these observations, it is more likely that a small number of self-reactive B1 cells that have escaped negative selection and migrated into the peritoneal cavity are able to survive and proliferate in anti-RBC Tg mice. The expansion of the self-reactive B1 cells is possible because of the absence of self-antigen in the peritoneal cavity and the self-renewal capacity of B1 cells. However, activation of B1 cells by environmental factors is probably necessary for the development of AIHA, since the presence of a normal number of B1 cells in the peritoneal cavity is not sufficient

to trigger the onset of the anemia in mice kept in specific pathogen-free conditions.

4.3.   Involvment of Interleukins in Development of AIHA in HL Tg Mice

Several interleukins (IL) have been shown to be involved in proliferation and differentiation of B1 cells. *In vitro* and *in vivo* studies using IL-Tg mice (98) or long-term anti-IL treatment (99) have revealed that proliferation of B1 cells requires different ILs from conventional B2 cells. While IL-4 induces stimulation only of conventional B2 cells, IL-5 acts on progenitor of B1 cells to induce proliferation (100), and on mature B1 cells and conventional B2 cells to induce differentiation into plasma cells (101–104). IL-5 Tg mice (98) show a remarkable increase of B1 cells in spleen and production of polyreactive IgM autoAbs, while IL-5- receptor α-deficient mice have a decreased number of B1 cells in the peritoneal cavity (105). IL-10 appears to be involved in regulation of B1 cells, as continuous administration of the anti-IL-10 Ab depleted peritoneal B1 cells (99). Moreover, LPS-stimulated peritoneal B1 cells produce significant levels of IL-10, raising the possibility that this cytokine is also important for B1 cell function (106).

Are these cytokines important for activation and terminal differentiation of B1 cells in HL Tg mice? To adress this issue, two strategies are envisaged: *a*. Administration of ILs into nonanemic HL Tg mice and monitoring the hematocrit values and the numbers of autoAb-producing cells (25). *b*. Analysis of IL-receptor deficient-HL Tg mice, generated by crossing of HL Tg mice with IL-5Rα$^{-/-}$ mice (26). Intramuscular or intraperitoneal injection of IL-5 and IL-10 induces anemia in HL Tg mice (Fig. 9) whereas the hematocrit values and the amounts of the autoAb do not change after IL-4 or PBS injection. The development of anemia after IL-10 and IL-5 injection suggests that these Th2 cytokines induce differentiation of peritoneal B1 cells into Ab-producing cells in HL Tg mice. The number of peritoneal B1 cells is not affected by IL-4 or IL-10 injection and increases only by 20% after IL-5 administration. It is likely that IL-10 or IL-5 induces not a marked proliferation, but rather differentiation of B1 cells into Ab-producing cells in HL Tg mice. That activation of B1 cells requires IL-5 is further demonstrated in IL5Rα$^{-/-}$ XHL Tg mice. Although these mice contain as much as 30% of the peritoneal cells as compared to HL Tg mice, the serum anti-RBC Ab is undetectable.

IL-10 appears to play essential roles in proliferation and activation of B1 cells because injection of anti-IL-10 mAb drastically reduces the number of peritoneal B1 cells and prevents the development of anemia in HL Tg mice (27). That IL-10 is involved in LPS-induced hemolytic anemia is also suggested by two observations: The serum level of IL-10 markedly increases after LPS administration in HL Tg

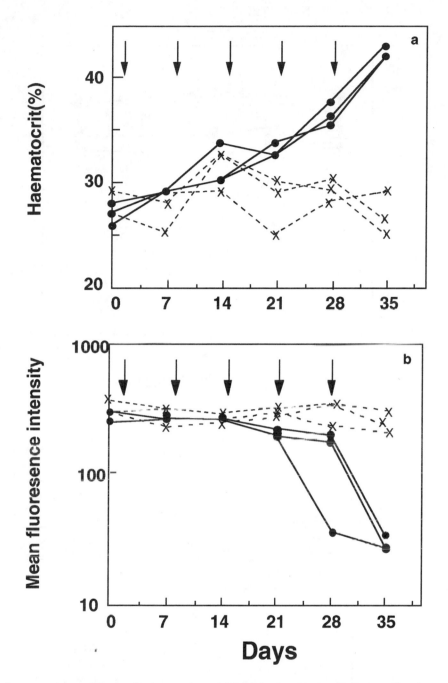

Figure 8. HL Tg mice recover from anemia after intraperitoneal RBC injection. Mouse RBC (1×10^9 cells) were injected every 7 days into the peritoneum of three anemic HL Tg mice (6–8 weeks old) whose hematocrit values were less than 30% (solid lines with filled circles). As controls, similar anemic HL Tg mice were injected with PBS (broken lines with crosses). *a*. Time course of hematocrit values of HL Tg mice injected with RBC or PBS. *b*. Time course of amount of autoAb bound to circulating erythrocytes of the HL Tg mice. Arrows indicate points of injection of RBC. For detecting Ab bound to RBC, circulating RBC were stained with FITC-anti-IgM Ab and analyzed by flow cytometry. The MFI of each sample was calculated. Hematocrits and MFI in normal littermates are 45–50% and 1–10, respectively. Data are reproduced after Murakami et al. (22).

mice (27) and IL-10 neutralization prevents the development of disease by reduction of the amount of anti-RBC Ig probably due to the extinction of autoreactive B1 cells (27).

Since RAG2$^{-/-}$ × HL Tg mice that lack T cells because of the RAG2$^{-/-}$ background also respond by increasing the IL-10 level after LPS injection, it appears that endotoxin stimulates non-T cells to secrete IL-10 in these mice. It is more likely that macrophages are responsible for elevated levels of IL-10 in RAG2$^{-/-}$ × HL Tg mice since it is known that sorted peritoneal macrophages secrete IL-10

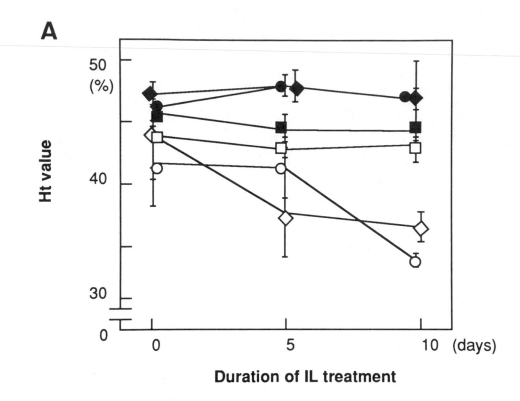

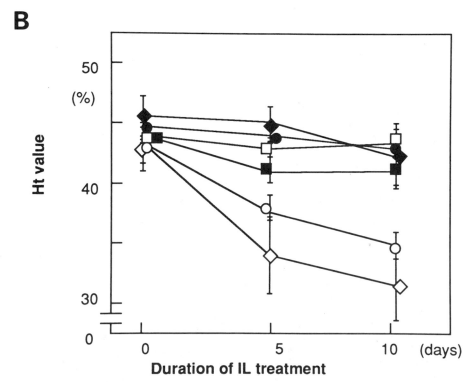

Figure 9.   Effects of injections of PBS, IL-4, IL-5 and IL-10 on HL Tg mice and C57BL/6 mice. PBS (▽ ▼), 25000 U IL-4 (□, ■), 25000 U IL-5 (◇ ◆) or 10μg IL-10 (○ ●) were injected intraperitoneally (A,C) or intramuscularly (B,D) into HL Tg mice (open symbols) or C57BL/6 (closed symbols) for 10 consecutive days. Hematocrit (Ht) values (A,B) of peripheral blood from IL-injected mice were measured at day 5 and 10. Amounts of anti-RBC autoAb bound to circulating RBC (C,D) were measured in IL-injected mice at day 5 and 10. Peripheral blood was incubated with FITC-labeled anti-IgM Ab and analyzed by a FACScan®. Bars indicate SD of values of three to five mice. Data are modified from Nisitani et al. (25).

C

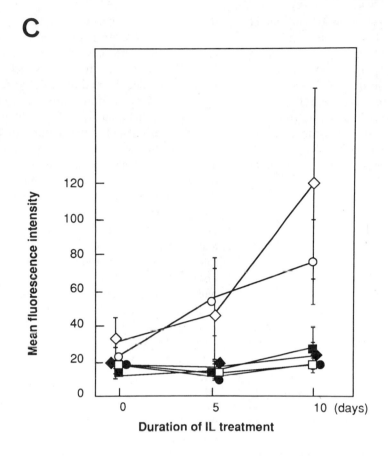

D

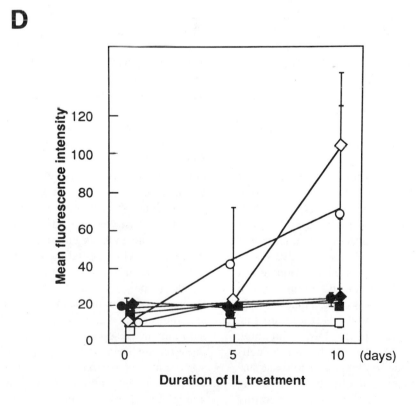

Figure 9. (*Continued*)

after LPS stimulation (107). IL-10 effects on B1 cells might be indirect. In anti-IL-10 Ab-treated normal mice, the effect of IL-10 depletion on peritoneal B1 cells is suggested to be mediated by elevated levels of IFNγ (99). However, this does not appear to be the case in RAG2$^{-/-}$ × HL Tg mice because these mice lack IFNγ-secreting Th1 cytokines. Therefore, it might be possible that IL-10 stimulates non-B non-T cells to secrete IL-5 and other factors that activate B1 cells. It is also possible that IL-10 is required to increase the responsiveness of B1 cells to other cytokines (108,109).

5. CONCLUSIONS

Autoimmune diseases are believed to be caused by the appearance of autoreactive lymphocytes as a result of breakage of self-tolerance (110). Depending on the predominant type of autoreactive lymphocytes autoimmune diseases are generally divided into three types: B-cell dominant autoimmune diseases caused by circulating autoAbs produced by autoreactive B cells, such as AIHA caused by anti-RBC autoAbs (111) and myasthenia gravis (MG) (112) caused by anti-nicotinic acetyl-choline receptor (nAChR) Abs; T-cell dominant autoimmune diseases caused by autoreactive T cells, such as experimental autoimmune encephalomyelitis (EAE), insulin-dependent diabetes mellitus (IDDM) in non-obese diabetic (NOD) mice and collagen-induced arthritis (CIA) (113); and Combinatorial type due to the appearance of autoreactive B and T cells, such as systemic lupus erythematosus (SLE) (114).

One important question regarding autoimmune diseases is how autoreactive lymphocytes are allowed to appear. To address this question, several Tg mice that express pathogenic autoreactive antigen receptors have been established (47–49). These Tg mice models have shown that immunological tolerance of autoreactive lymphocytes is regulated mainly by two mechanisms (110) i.e., deletion and anergy. In the anti-MHC class I Ab Tg model of Nemazee and Burki (48), high concentrations of allotypic H-2 class I antigen expressed on the cell surface as a multivalent antigen lead to clonal deletion, whereas relatively low concentrations of hen egg lysozyme (HEL) in Goodnow's model (47) and single-stranded DNA in soluble form in the Erikson and Weigert model (49) cause clonal anergy. In another Tg line established by Weigert's group that expresses anti-double-stranded DNA IgM with high affinity on the surface (115), the number of B cells decreases in the periphery due to clonal deletion, but surviving B cells appear to change the affinity of surface Ig against the antigen, probably due to the receptor-editing mechanism (53). However, in HL Tg mice, both of these mechanisms are invoked in the same individual, suggesting

that deletion and anergy are not mutually exclusive (20). Moreover, their relative contributions vary among individuals even though they have the same genetic background.

Studies on Tg mice models have led to the proposal that B-cell tolerance is maintained by several checkpoints (116); elimination of immature B cells, follicular competition, T-cell help, anergy, Fas-mediated elimination of anergic cells, and elimination in germinal center. According to this theory, there are several opportunities for autoreactive B cells to leak through each checkpoint and many studies indicate that autoreactive B cells escaping from immunologic surveillance actually exist *in vivo*, although the number and affinity of their autoAbs are low (117–119). Although many autoAb Tg mouse models have a significant number of autoreactive B cells escaping from tolerance, the occurrence of autoimmune diseases is observed in only two (20,120), one being HL Tg mice, which produce the anti- RBC autoAb.

The prevalence of anemia in our HL Tg mice is dependent on environmental factors, suggesting that the simple appearance of autoreactive B cells is not sufficient and that activation of these cells, probably due to infections, is very important for the onset of the autoimmune disease. Approximately 50% of Tg mice suffer from anemia when bred in conventional conditions. In contrast, no mice housed in specific pathogen-free conditions suffer from spontaneous hemolytic anemia unless injected with LPS or/and Th-2 cytokines or transferred to the conventional facility. It appears that autoreactive B1 cells escaping from clonal deletion and expanding in the peritoneal cavity play a major role in the induction of AIHA, since the disease was cured after elimination of B1 cells by intraperitoneal injection of antigens. These observations support a two-step model for development of AIHA in anti-RBCs autoAb Tg mice: Autoreactive B1 cells escape clonal deletion and expand in the compartment free of self-antigen, and activation of these autoreactive cells by additional environmental factors leads to the production of autoAbs and finally to the onset of the disease. That environmental factors can initiate an autoimmune response has been suggested by clinical and experimental observations, and the HL Tg mouse model clearly demonstrates that infection breaks down the B cell tolerance and induces autoimmune symptoms. Although the role of environmental factors in pathogenesis of AIHA is not fully understood, we can speculate that infectious agents can provoke changes in responsiveness to cytokines, cytokine levels, expression of homing receptors or adhesion molecules on autoreactive B1 cells that will induce an autoimmune response. Obviously, the most important questions to be answered in HL Tg mice are how and where autoreactive B1 cells are activated and induced to differentiate into autoAb-producing cells. Because oral administration of LPS-induced anemia in asymptomatic mice, we previously proposed that migration and communication between gut-associated lymphoid tissues (GALT) and peritoneal cavity might be involved in the activation of B1 cells (24). Kroese at al. (122) have shown that many intestinal IgA plasma cells are derived from B1 cells residing in the peritoneal cavity, suggesting that frequent migration of these cells may take place between the peritoneal cavity and lamina propria of the gut. Other studies (123,124, our unpublished data) confirmed that B1 cells from the peritoneal cavity have the homing capacity not only to lamina propria (123), where they can be identified as IgA plasma cells, but also to other lymphoid tissues, giving rise to IgG, IgM and IgA plasma cells in spleen, lymph nodes, and bone marrow. However, the typical surface phenotype of B1 cells, namely B220^{low} IgM^{hi} IgD^{low} Mac1^{low} CD5^{low} /CD5$^-$ are no longer identified after the cells leave the peritoneal cavity, suggesting that after their migration to other lymphoid tissues, they change (downregulate) surface antigens as they begin to differentiate into plasma cells (our unpublished data). In fact, we are unable to reproduce the data to indicate the presence of B1 cells in the lamina propria of small intestine (24).

Based on the observation that LPS administrated orally activates peritoneal B1 cells and induces anemia (24), we suggested a scenario in which B1 cells are first activated in lamina propria and then move to the peritoneal cavity, where they produce autoAb. We previously thought that activated peritoneal B1 cells produced and secreted the anti-RBC Ab *in situ* (24–26) based on the detection of Ig secreting cells by the ELISPOT assay. However, we have been unable to identify typical plasma cells in the peritoneal cavity of anemic HL Tg mice. Another possible site of Ig secretion is mesenteric lymph nodes, where peritoneal B1 cells can home. In fact our original study (20) clearly shows that large amounts of H and L mRNA are expressed in lymph nodes, although the number and the frequency of mature B cells are similar to the levels of spleen and bone marrow. These results suggest that lymph nodes of HL Tg mice may contain a larger number of plasma cells rather than surface IgM$^+$ B cells. Further support for the proposition that the peritoneal cavity is not the major site of autoAb production is the observation that 2 to 3 weeks are

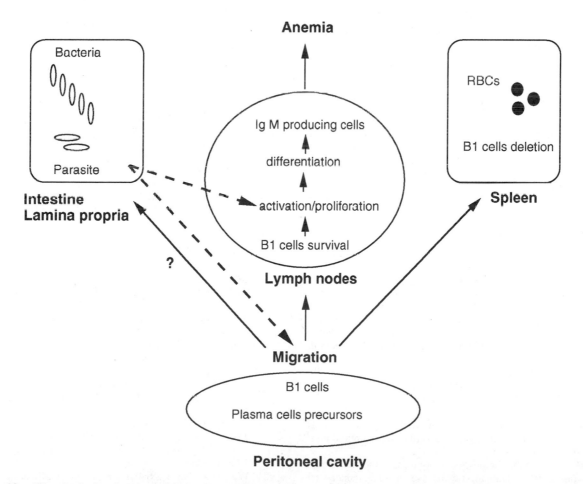

Figure 10. Hypothetical pathway of peritoneal B1 cells migration, activation and differentiation into IgM-producing cells in HL Tg mice. B1 cells leave the peritoneal cavity through draining lymph nodes, become activated in their migration process, proliferate and differentiate into plasma cells in the absence of self-antigens, probably in lymph nodes. Infection and particularly intestinal pathogens induce AIHA by enhancing the homing receptors, adhesion molecules on B1 cells, cytokine levels and responsiveness to cytokines which will influence the migration to GALT, activation and differentiation of B1 cells into Ab-producing cells.

required for the recovery from anemia after clonal deletion of peritoneal B1 cells (Fig. 8). Since the half-life of Ig is almost 7 days, this delay of recovery suggests the presence of activated Ig-secreting cells in places other than the peritoneal cavity. In our latest experiments using a cross of HL Tg mice(HLx KN6xRAG2 –/– mice), we found autoAb-producing cells in mesenteric lymph nodes, but not in the peritoneal cavity, raising the possibility that activation of B1 cells and autoAb secretion could take place in lymph nodes.

As summarized in Fig. 10, we believe that the migration process of peritoneal B1 cells is closely associated with their activation and differentiation into Ig-producing cells. Perhaps B1 cells leave the peritoneal cavity through draining lymph nodes, become activated during or after their migration to other lymphoid organs, proliferate and differentiate into plasma cells in the absence of RBC, most probably in lymph nodes. We know that infection, LPS administration or Th2 cytokine injection can induce differentiation of autoreactive B cells into autoAb-producing cells. Therefore, it is important to know the B1 cell traffic to GALT in order to understand the mechanism of activation of autoreactive B cells by infection, and particularly by intestinal pathogens.

ACKNOWLEDGEMENTS

This work was supported by the C. O. E. Grant from the Ministry of Education, Science, Sports, and Culture of Japan. We thank Ms. Minako Yamaguchi for secretarial assistance.

References

1. Linder, E., and T.S. Edgington. 1973. Immunobiology of the autoantibody response. I. Circulating analogues of erythrocyte autoantigens and heterogeneity of the autoimmune response of NZB mice. *Clin. Exp. Immunol.* 13:279–292.
2. Linder, E., and T.S. Edgington. 1973. Immunobiology of the autoantibody response. II. The lipoprotein-associated soluble HB erythrocyte autoantigen of NZB mice. *J. Immunol.* 110:53–62.
3. DeHeer, D.H., E.J. Linder, and T.S. Edgington. 1978. Delineation of spontaneous erythrocyte autoantibody responses of NZB and other strain of mice. *J. Immunol.* 120:825–830.
4. Lewis, D.E., S. Griswold, and N.L. Warner. 1981. Monoclonal anti-erythrocyte auto-antibodies from unimmunized NZB mice. *Hybridoma* 1:71–76.
5. Cooke, L.A., N.A. Staines, A. Morgan, C. Moorhouse, and G. Harris. 1982. Haemolytic disease in mice induced by transplantation of hybridoma cells secreting monoclonal anti-erythrocyte autoantibodies. *Immunology* 47:569–572.
6. Ozaki, S., R. Nagasawa, H. Saito, and T. Shirai. 1984. Hybridoma autoantibodies to erythrocytes from NZB mice and induction of hemolytic anemia. *Immunol. Lett.* 8:115–119.
7. Caulfield, M.J., D. Stanko, and C. Calkins. 1989. Characterization of the spontaneous autoimmune (anti-erythrocyte) response in NZB mice using a pathogenic monoclonal antibody and its anti-idiotype. *Immunology* 66:233–237.
8. Shibata, T., T. Berney, L. Reininger, Y. Chicheportiche, S. Ozaki, T. Shirai, and S. Izui. 1990. Monoclonal anti-erythrocyte antibodies derived from NZB mice cause autoimmune hemolytic anemia by two distinct pathogenic mechanisms. *Int. Immunol.* 2:1133–1141.
9. Barker, R.N., G.G. De Sa Oliveira, C.J. Elson, and P.M. Lydyard. 1993. Pathogenic autoantibodies in the NZB mouse are specific for erythrocyte Band 3 protein. *Eur. J. Immunol.* 23:1723–1726.
10. Scott, B.B., S. Sadigh, M. Stow, R.A.K. Mageed, E.M. Andrew, and R.N. Maini. 1993. Molecular mechanisms resulting in pathogenic anti-mouse erythrocyte antibodies in New Zealand black mice. *Clin. Exp. Immunol.* 93:26–33.
11. Oliveira, G.G.S., P.R. Hutchings, I.M. Roitt, and P.M. Lydyard. 1994. Production of erythrocyte autoantibodies in NZB mice is inhibited by CD4 antibodies. *Clin. Exp. Immunol.* 96:297–302.
12. De Sa Oliveira G.G., S. Izui, C.T. Ravirajan, R.A. Mageed, P.M. Lydyard, C.J. Elson, and R.N. Barker. 1996. Diverse antigen specificity of erythrocyte-reactive monoclonal autoantibodies from NZB mice. *Clin. Exp. Immunol.* 105:313–320.
13. Reininger, L., T. Shibata, S. Ozaki, T. Shirai, J.-C. Jaton, and S. Izui. 1990. Variable refion sequences of pathogenic anti-mouse red blood cell autoantibodies from autoimmune NZB mice. *Eur. J. Immunol.* 20:771–777.
14. Reininger, L., P. Ollier, P. Poncet, A. Kaushik, and J.-C. Jaton. 1987. Novel V genes encode virtually identical variable regions of six murine monoclonal anti-bromelain-treated red blood cell autoantibodies. *J. Immunol.* 138:316–323.
15. Hardy, R.R., C.E. Cormack, S.A. Shinton, R.J. Riblet, and K. Hayakawa. 1989. A single VH gene is utilized predominantly in anti-BrMRBC hybridomas derived from purified Ly-1 B cells. Definition of VH11 family. *J. Immunol.* 142:3643–3651.
16. Andrade, L., A.A. Freitas, F. Huentz, P. Poncet, and A. Countiho. 1989. Immunoglobulin Vh gene expression in Ly-1+ and conventional B lymphocytes. *Eur. J. Immunol.* 19:1117–1122.
17. Hayakawa, K., R.R. Hardy, M. Honda, L.A. Herzenberg, A.D. Steinberg, and L.A. Herzenberg. 1984. Ly-1 B cells: functionally distinct lymphocytes that secrete IgM autoantibodies. *Proc. Natl. Acad. Sci. USA* 81:2494–2498.
18. Izui, S. 1994. Autoimmune hemolytic anemia. *Current Opinion in Immunology* 6:926–930.
19. Okamoto, M., and T. Honjo. 1990. Nucleotide sequences of the gene/cDNA coding for anti-murine erythrocyte autoantibody produced by a hybridoma from NZB mouse. *Nucl. Acids Res.* 18:1895.
20. Okamoto, M., M. Murakami, A. Shimizu, S. Ozaki, T. Tsubata, S. Kumagai, and T. Honjo. 1992. A transgenic model of autoimmune hemolytic anemia. *J. Exp. Med.* 175:71–79.
21. Watanabe, N., S. Nisitani, K. Ikuta, M. Suzuki, T. Chiba, and T. Honjo. 1999. Expression levels of B cell surface immunoglobulin regulate efficiency of allelic exclusion and size of auto-reactive B1 cell compartment. *J. Exp. Med.* 190:461–469.

22. Murakami, M., T. Tsubata, M. Okamoto, A., Shimizu, S. Kumagai, H. Imura, and T. Honjo. 1992. Antigen-induced apoptotic death of Ly-1 B cells responsible for autoimmune disease in transgenic mice. *Nature* 357:77–80.

23. Murakami, M., K. Nakajima, K. Yamazaki, T. Muraguchi, T. Serikawa, and T. Honjo. 1997. Effects of breeding environments on generation and activation of autoreactive B-1 cells in anti-red blood cell autoantibody transgenic mice. *J. Exp. Med.* 185:791–794.

24. Murakami, M., T. Tsubata, R. Shinkura, S. Nisitani, M. Okamoto, H. Yoshioka, T. Usui, S. Miyawaki, and T. Honjo. 1994. Oral administration of lipopolysaccharides activates B-1 cells in the peritoneal cavity and lamina propria of the gut and induces autoimmune symptoms in an autoantibody transgenic mouse. *J. Exp. Med.* 180:111–121.

25. Nisitani, S., T. Tsubata, M. Murakami, and T. Honjo. 1995. Administration of interleukin-5 or -10 activates peritoneal B1 cells and induces autoimmune hemolytic anemia in anti-erythrocyte autoantibody transgenic mice. *Eur. J. Immunol.* 25:3047–3052.

26. Sakiyama, T., K. Ikuta, S. Nisitani, K. Takatsu, and T. Honjo. 1999. Requirement of IL-5 for induction of autoimmune hemolytic anemia in anti-red blood cell autoantibody transgenic mice. *Int. Immunol.* 11:995–1000.

27. Nisitani, S., T. Sakiyama, and T. Honjo. 1998. Involvement of IL-10 in induction of autoimmune hemolytic anemia in anti-erythrocyte Ig transgenic mice. *Int. Immunol.* 10:1039–1047.

28. Neuberger, M.S. 1983. Expression and regulation of immunoglobulin heavy chain gene transfected into lymphoid cells. *EMBO J.* 2:1373.

29. Shimizu, A., N. Takahashi, Y. Yaoita, and T. Honjo. 1982. Organization of the constant-region gene family of the mouse immunoglobulin heavy chain. *Cell* 28:499–506.

30. Kataoka, T., T. Kawakami, N. Takahashi, and T. Honjo. 1980. Rearrangement of immunoglobulin γl-chain gene and mechanism for heavy-chain class switch. *Proc. Natl. Acad. Sci. USA* 77:919.

31. Burnet, F.M. 1959. The Clonal Selection Theory of Acquired Immunity. The University Press. Cambridge.

32. Rajewsky, K. 1996. Clonal selection and learning in the antibody system. *Nature* 381:751–758.

33. Nussenzweig, M.C., A.C. Shaw, E. Sinn, D.B. Danner, K.L. Holmes, H. C. Morse, and P. Leder. 1987. Allelic exclusion in transgenic mice that express the membrane form of immunoglobulin μ. *Science* 236:816–819.

34. Nussenzweig, M.C., A.C. Shaw, E. Sin, J. Campos-Torres, and P. Leader. 1988. Allelic exclusion in transgenic mice carrying mutant human IgM genes. *J. Exp. Med.* 167:1969–1974.

35. Ritchie, K.A., R.L. Brinster, and U. Storb. 1984. Allelic exclusion and control of endogenous immunoglobulin gene rearrangement in κ transgenic mice. *Nature* 321:517–520.

36. Kitamura, D., and K. Rajewsky. 1992. Targeted disruption of μ chain membrane exon causes loss of heavy-chain allelic exclusion. *Nature* 356:154–156.

37. Manz, J., K. Denis, O. Witte, R. Brister, and U. Storb. 1988. Feedback inhibition of immunoglobulin gene rearrangement by membrane μ, but not by secreted μ heavy chains. *J. Exp. Med.* 168:1363–1381.

38. Stall, A.M., F.G.M. Kroese, F.T. Gadus, D.G. Sieckmann, L.A. Herzenberg, and L.A. Herzenberg. 1988. Rearrangement and expression of endogenous immunoglobulin genes occur in many murine B cells expressing transgenic membrane IgM. *Proc. Natl. Acad. Sci. USA* 85:3546–3550.

39. Muller, W., U. Ruther, P. Vieira, J. Hombach, M. Reth, and K. Rajewsky. 1989. Membrane-bound IgM obstructs B cell development in transgenic mice. *Eur. J. Immunol.* 19:923–928.

40. Kenny, J.J., F. Finkelman, F. Macchiarini, W.C. Kopp, U. Storb, and D.L. Longo. 1989. Alteration of the B cell surface phenotype, immune response to phosphocholine and the B cell repertoire in M169 μ plus κ transgenic mice. *J. Immunol.* 142:4466–4474.

41. Rath, S., J. Durdik, R.M. Gerstein, E. Selsing, and A. Nisonoff. 1989. Quantitative analysis of idiotypic mimicry and allelic exclusion in mice with a μ Ig transgene. *J. Immunol.* 143:2074–2080.

42. Forni, L. 1990. Extensive splenic B cell activation in IgM-transgenic mice. *Eur. J. Immunol.* 20:983–989.

43. Grandien, A., A. Countiho, and J. Andersson. 1990. Selective peripheral expansion and activation of B cells expressing endogenous immunoglobulin in μ-transgenic mice. *Eur. J. Immunol.* 20:991 998.

44. Iacomini, J., N. Yannoutsos, S. Bandyopadhay, and T. Imanishi-Kari. 1991. Endogenous immunoglobulin expression in μ transgenic mice. *Int. Immunol.* 3:185–196.

45. Lam, K.P., L.A. Herzenberg, and A.M. Stall. 1993. A high frequency of hybridomas from M54 μ heavy chain transgenic mice initially co-express transgenic and rearranged endogenous μ genes. *Int. Immunol.* 5:1011–1022.

46. Rusconi, S., and G. Kohler. 1985. Transmition and expression of a specific pair of rearranged immunoglobulin μ and κ genes in a transgenic mouse line. *Nature* 314:330–334.

47. Goodnow, C.C., J. Crosbie, S. Adelstein, T.B. Lavoie, S.J. Smith-Gill, R.A. Brink, H. Pritchart-Briscoe, J.S. Wotherspoon, R.H. Loblay, K. Raphael, K. Raphael, R.J. Trent, and A. Basten. 1988. Altered immunoglobulin expression and functional silencing of self-reactive B lymphocytes in transgenic mice. *Nature* 334:676 682.

48. Nemazee, D.A., and K. Burki. 1989. Clonal deletion of B lymphocytes in a transgenic mouse bearing anti-MHC class I antibody genes. *Nature* 337:562–566.

49. Erikson, J., M.Z. Radic, S.A. Camper, R.R. Hardy, C. Carmack, and M. Weigert. 1991. Expression of anti-DNA immunoglobulin transgenes in non-autoimmune mice. *Nature* 349:331–334.

50. Hartley, S.S., J. Crosbie, R. Brink, A.B. Kantor, A. Basten, and C.C. Goodnow. 1991. Elimination from peripheral lymphoid tissues of self-reactive B lymphocytes recognizing membrane-bound antigens. *Nature* 353:765–769.

51. Fulcher, D.A., and A. Basten. 1994. Reduced life span of anergic self-reactive B cells in a double-transgenic model. *J. Exp. Med.* 179:125–134.

52. Fulcher, D.A., A.B. Lyons, S.L. Korn, M.C. Cook, C. Koleda, C. Parish, B. Fazekas de St Groth, and A. Basten. 1996. The fate of self-reactive B cells depends primarily on the degree of antigen receptor engagement and availability of T cell help. *J. Exp. Med.* 183:2313–2328.

53. Gay, D., T. Saunders, S. Camper, and M. Weigert. 1993. Receptor editing: an approach by autoreactive B cells to escape tolerance. *J. Exp. Med.* 177:999–1008.

54. Tiegs, S.L., D.M. Russell, and D. Nemazee. 1993. Receptor editing in self-reactive bone marrow B cells. *J. Exp. Med.* 177:1009–1020.

55. Nossal, G.V.J. 1983. Cellular mechanisms of immunologic tolerance. *Annu. Rev. Immunol.* 1:33–62.

56. Gause, A., N. Yoshida, C. Kappen, and K. Rajewsky. 1987. *In vivo* generation and function of B cells in the presence of

a monoclonal anti-IgM antibody: Implication for B cell tolerance. *Eur. J. Immunol.* 17:981–990.

57. Pike, B.L., T.W. Kay, and G.J.V. Nossal. 1980. Relative sensitivity of fetal and newborn mice to induction of hapten-specific B cell tolerance. *J. Exp. Med.* 152:1407–1412.

58. Pike, B.L., A.W. Boyd, and G.J.V. Nossal. 1982. Clonal anergy. The universally anergic B lymphocyte. *Proc. Natl. Acad. Sci. USA* 79:2013–2017.

59. Adams, E., A. Basten, and C.C. Goodnow. 1990. Intrinsic B-cell hyporesponsiveness accounts for self-tolerance in lysozyme/antilysozyme double-transgenic mice. *Proc. Natl. Acad Sci. USA* 87:5687–5691.

60. Hayakawa, K., R.R. Hardy, D.R. Parks, and L.A. Herzenberg. 1983. The "Ly-1 B" cell population in normal immunodefective and autoimmune mice. *J. Exp. Med.* 157:202–218.

61. Herzenberg, L.A., A.M. Stall, P.A. Lalor, C. Sidman, W.A. Moore, D.R. Parks, and L.A. Herzenberg. 1986. The Ly-1 B cell lineage. *Immunol. Rev.* 93:81–102.

62. Hayakawa, K., R.R. Hardy, L.A. Herzenberg, and L.A. Herzenberg. 1985. Progenitors for Ly-1 B cells are distinct from progenitors for other B cells. *J. Exp. Med.* 161:1554–1568.

63. Markos, M.A., F. Huetz, P. Pereira, J.L. Andreu, C. Martinez-A., and A. Countiho. 1989. Further evidence for coelomic-associated B lymphocytes. *Eur. J. Immunol.* 11:2031–2035.

64. Kipps, T.J. 1898. The CD5 B cell. *Adv. Immunol.* 47:117–185.

65. Forster, I., H. Gu, W. Muller, M. Schmitt, D. Tarlinton, and K. Rajewsky. 1991. CD5 B cells in the mouse. *Curr. Top. Microbiol. Immunol.* 173:247–251.

66. Rickert, R.C., K. Rajewsky, and J. Roes. 1995. Impairment of T-cell-dependent B-cell responses and B-1 cell development in CD19-deficient mice. *Nature* 376:352–355.

67. Zhang, R.F., F.W. Alt, L. Davisdon, S.H. Orkin, and W. Swat. 1995. Defective signaling through the T- and B-cell antigen receptors in lymphoid cells lacking the vav proto-oncogene. *Nature* 374:470–473.

68. Leitges, M., C. Schmedt, R. Guinamard, J. Davoust, S. Schaal. S. Stabel, and A. Tarakhovsky. 1996. Immuno-deficiency in protein kinase Cβ-deficient mice. *Science* 273:788–791.

69. Sidman, C.L., L.D. Shultz, R.R. Hardy, K. Hayakawa, and L.A. Herzenberg. 1986. Production of immunoglobulin isotypes by Ly-1+ B cells in viable motheathen and normal mice. *Science* 232:1423–1425.

70. Scultz, L.D., P.A. Schweitzer, T.V. Rajan, T. Yi, J.N. Ihle, R.J. Matthews, M.L. Thomas, and D.R. Beier. 1993. Mutation at the murine motheathen locus are within the hematopoietic cell protein-tyrosin phosphatase (Hcph) gene. *Cell* 73:1445–1454.

71. Tsui, H.W., K.A. Siminovitch, L. De Sousa, and F.W.L. Tsui. 1993. Motheathen and viable motheathen mice have mutations in the hematopoietic cell phosphatase gene. *Nature Genet.* 4:124–129.

72. Cyster, J.G., and C.C. Goodnow. 1995. Protein tyrosine phosphatase 1C negatively regulates antigen receptor signaling in B lymphocytes and determine the thresholds for negative selection. *Immunity* 2:13–24.

73. Pani, G., M. Kozlowski, J.C. Cambier, G.B. Mills, and K.A. Siminovitch. 1995. Identification of the tyrosine phosphatase PTP1C as a B cell antigen receptor-associated protein involved in the regulation of B cell signaling. *J. Exp. Med.* 181:2077–2084.

74. Melchers, F., V. Braun, and C. Galanos. 1975. The lipoprotein of the other membrane of *Escherichia coli*: a B-lymphocyte mitogen. *J. Exp. Med.* 142:473–482.

75. Ulevitch, R.J., and P.S. Tobias. 1995. Receptor-dependent mechanisms of cell stimulation by bacterial endotoxin. *Annu. Rev. Immunol.* 13:437–475.

76. Mowat, A.M. 1987. The regulation of immune responses to dietary protein antigens. *Immunology Today* 8:93–98.

77. Mowat, A.M. 1994. Oral tolerance and regulation of immunity to dietary antigens. In: *Handbook of Mucosal Immunology*. Ogra, P.L., J. Mestecky, M.E. Lamm, W. Strober, J.R. McGhee, and J. Bienenstock, eds. San Diego: Academic Press Inc., pp. 185–201.

78. Mowat, A.M., and J.L. Viney. 1997. The anatomical basis of intestinal immunity. *Immunol. Rev.* 156:145–166.

79. Goodnow C.C., R. Brink, and E. Adams. 1991. Breakdown of self-tolerance in anergic B lymphocytes. *Nature* 352:532–536.

80. Melamed, D., and D. Nemazee. 1997. Self-antigen does not accelerate immature B cell apoptosis, but stimulates receptor editing as a consequence of developmental arrest. *Proc. Natl. Acad. Sci. USA* 94:9267–9272.

81. Hertz, M., and D. Nemazee. 1997. BCR ligation induces receptor editing in IgM + IgD-bone marrow B cells *in vitro*. *Immunity* 6:429–436.

82. Arnold, L.W., C.A. Pennell, S.K. McCray, and S.H. Clarke. 1994. Development of B-1 cells: segregation of phosphatidyl choline-specific B cells to B-1 population occurs after immunoglobulin gene expression. *J. Exp. Med.* 179:1585–1595.

83. Clarke, S.H., and L.W. Arnold. 1998. B-1 cell development: evidence for an uncommitted immunoglobulin (IgM+) B cell precursor in B-1 cell differentiation. *J. Exp. Med.* 187:1325–1334.

84. Yellen, A.J., W. Gleen, V.P. Sukhatme, X. Cao, and J.G. Monroe. 1991. Signaling through surface IgM in tolerance-susceptible immature murine B lymphocytes. *J. Immunol.* 146:1446–1454.

85. Yellen-Shaw, A., and J.G. Monroe. 1992. Differential responsiveness of immature-and mature-stage murine B cells to anti-IgM reflects both FcR-dependent and -independent mechanisms. *Cell. Immunol.* 145:339–350.

86. Norvell, A., L. Mandik, and J.G. Monroe. 1995. Engagement of the antigen-receptor on immature murine B lymphocytes results in death by apoptosis. *J. Immunol.* 154:4404–4413.

87. Monroe, J.G. 1996. Tolerance sensitivity of immature stage B cells: can developmentally regulated B cell antigen receptor(BCR) signal transduction play a role? *J. Immunol.* 156:2657–2660.

88. Norvell, A., and J.G. Monroe. 1996. Aquisition of surface IgD fails to protect from tolerance-induction. Both surface IgM-and surface IgD-mediated signals induce apoptosis of immature murine B lymphocytes. *J. Immunol.* 156:1328–1332.

89. Merino, R., L. Ding, D.J. Veis, S.J. Korsmeyer, and G. Nunez. 1994. Developmental regulation of the Bcl-2 protein and susceptibility to cell death in B lymphocytes. *EMBO J.* 13:683–691.

90. Li, Y.S., K. Hayakawa, and R.R. Hardy. 1993. The regulated expression of B lineage associated genes during B cell dif-

ferentiation in bone marrow and fetal liver. *J. Exp. Med.* 178:951–960.

91. Strasser, A., A.W. Harris, and S. Cory. 1991. bcl-2 transgene inhibits T cell death and perturbs thymic self-censorship. *Cell* 67:889–899.

92. Sentman, C.L., J.R. Shutter, D. Hockenbery, O. Kanagawa, and S.J. Korsmeyer. 1991. bcl-2 inhibits multiple forms of apoptosis but not negative selection in thymocytes. *Cell* 67:879–888.

93. Nisitani, S., T. Tsubata, M. Murakami, M. Okamoto, and T. Honjo. 1993. The bcl-2 gene product inhibits clonal deletion of self-reactive B lymphocytes in the periphery but not in the bone marrow. *J. Exp. Med.* 178:1247–1254.

94. McDonnell, T.J., N. Deane, F.M. Platt, G. Nunez, U. Jaeger, J.P. McKearn, and S.J. Korsmeyer. 1989. bcl-2-immunoglobulin transgenic mice demonstrate extended B cell survival and folicular lymphoproliferation. *Cell* 57:79–88.

95. Strasser, A., S. Whittingham, D.L. Vaux, M.L. Bath, J.M. Adams, S. Cory, and A.W. Harris. 1991. Enforced BCL2 expression in B-lymphoid cells prolongs antibody responses and elicits autoimmune disease. *Proc. Natl. Acad. Sci. USA* 88:8661–8665.

96. Haury, M., A. Freitas, V. Hermitte, A. Coutinho, and U. Hibner. 1993. The physiology of bcl-2 expression in murine B lymphocytes. *Oncogene* 8:1257–1262.

97. Sher, I. 1982. CBA/N immune defective mice: evidence for the failure of a B cells subpopulation to be expressed. *Immunol. Rev.* 64:117–136.

98. Tominaga, A., S. Takaki, N. Koyama, S. Katoh, R. Matsumoto, M. Migita, Y. Hitoshi, Y. Hosoya, S. Yamauchi, Y. Kanai, J. Miyazaki, G. Usuku, K. Yamamura, and K. Takatsu. 1991. Transgenic mice expressing a B cell growth and differentiation factor gene (interleukin 5) develop eosinofilia and autoantibody production. *J. Exp. Med.* 173:429–437.

99. Ishida, H., R. Hastings, J. Kearny, and M. Howard. 1992. Continuous anti-IL-10 antibody administration depletes mice CD5 B cells but not conventional B cells. *J. Exp. Med.* 175:1213–1230.

100. Kantor, A.B. 1991. The development and repertoire of B-1 cells (CD5 B cells). *Immunol. Today* 12:389–391.

101. Kinashi, T., N. Harada, E. Severinson, T. Tanabe, P. Sideras, M. Konishi, C. Azuma, A. Tominaga, S. Bergstedt-Lindqvist, M. Takahashi, F. Matsuda, Y. Yaoita, K. Takatsu, and T. Honjo. 1986. Cloning of complementary DNA for T-cell replacing factor and identity with B-cell growth factor II. *Nature* 324:70–73.

102. Hitoshi, Y., N. Yamaguchi, S. Mita, E. Sonoda, S. Takaki, A. Tominaga, and K. Takatsu. 1990. Distribution of IL-5 receptor-positive B cells: expression of IL-5 receptors on Ly-1 (CD5) + B cells. *J. Immunol.* 144:4218–4225.

103. Hitoshi, Y., E. Sonoda, Y. Kikuchi, S. Yonehara, H. Nakauchi, and K. Takatsu. 1993. Interleukin-5 receptor positive B cells, but not eosinophils are functionally and numerically influenced in the mice carried with X-linked immunodeficiency. *Int. Immunol.* 5:1183–1190.

104. Wetzel, G.D. 1989. Interleukin 5 regulation of peritoneal Ly-1 B lymphocyte proliferation, differentiation and autoantibody secretion. *Eur. J. Immunol.* 19:1701–1707.

105. Yoshida, T., K. Ikuta, H. Sugaya, K. Maki, M. Takagi, H. Kanazawa, S. Sunaga, T. Kinashi, K. Yoshimura, J. Miyazaki, S. Takaki, and K. Takatsu. 1996. Defective B-1

cell development and impared immunity against *Angiostrongylus cantonensis* in IL-5Ra-deficient mice. *Immunity* 4:483–494.

106. O'Garra, A.R. Chang, N. Go, R. Hastings, G. Haughton, and M. Howard. 1992. Ly-1 B (B1) cells are the main source of B cell-derived interleukin 10. *Eur. J. Immunol.* 22:711–717.

107. Fiorentino, D.F., A. Zlotnik, T.R. Mosmann, M. Howard, and A. O'Garra. 1991. IL-10 inhibits cytokines production by activated macrophages. *J. Immunol.* 147:3815–3822.

108. Rousset, F., E. Garcia, T. Defrance, C. Peronne, N. Vezzio, D.-H. Hsu, R. Kastelein, K.W. Moore, and J. Banchereau. 1992. Interleukin 10 is a potent growth and differentiation factor for activated human B lymphocytes. *Proc. Natl. Acad. Sci. USA* 89:1890–1893.

109. Fluckiger, A.-C., P. Garrone, I. Durand, J.-P. Galizzi, and J. Banchereau. 1993. Interleukin 10 (IL-10) upregulates functional high affinity IL-2 receptors on normal and leukemia B lymphocytes. *J. Exp. Med.* 178:1473–1481.

110. Goodnow, C.C. 1996. Balancing immunity and tolerance: deleting and tuning lymphocyte repertoires. *Proc. Natl. Acad. Sci. USA* 19:2264–2271.

111. Izui, S. 1994. Autoimmune hemolytic anemia. *Curr. Opin. Immunol.* 6:926–930.

112. Willcox, N. 1993. Myastenia gravis. *Curr. Opin. Immunol.* 5:910–917.

113. Linington, C., and R. Hohfeld. 1993. T cell-mediated autoimmunity: molecular interaction and therapeutic implications. *J. Autoimmun.* 3:501–506.

114. Kotzin, B.L. 1996. Systemic lupus erythematosus. *Cell* 85:303–306.

115. Chen C., M.Z. Radic, J. Erikson, S.A. Camper, S. Litwin, R.R. Hardy, and M. Weigert. 1994. Deletion and editing of B cells that express antibodies to DNA. *J. Immunol.* 152:1970–1982.

116. Cornall R.J., C.C. Goodnow, and J.G. Cyster. 1995. The regulation of self-reactive B cells. *Curr. Opin. Immunol.* 7:804–811.

117. Pisetsky, D.S., D.F. Jelinek, L.M. McAnally, C.F. Reich, and P.E. Lipsky. 1990. In vitro autoantibody production by normal adult and cord blood B cells. *J. Clin. Invest.* 85:899–903.

118. He, X, J.J. Goronzy, and C.M. Weyand. 1993. The repertoire of rheumatoid factor-producing B cells in normal subjects and patients with rheumatoid arthritis. *Arthritis Rheumatol.* 36:1061–1069.

119. Hirohata, S., T. Inoue, and K. Ito. 1991. Frequency analysis of human pheripheral blood B cells producing autoantibodies: differential activation requirements of precursors for B cells producing IgM-RF and anti-DNA antibody. *Cell. Immunol.* 138:445–455.

120. Tsao, B.P., K. Ohnishi, H. Cheroutre, B. Mitchell, M. Teitell, P. Mixter, M. Kronenberg, and B.H. Hahan. 1992. Failed self-tolerance and autoimmunity in IgG anti-DNA transgenic mice. *J. Immunol.* 149:350–358.

121. Offen, D., L. Spatz, H. Escowitz, S. Factor, and B. Diamond. 1992. Induction of tolerance to an IgG autoantibody. *Proc. Natl. Acad. Sci. USA* 89:8332–8336.

122. Kroese, G.G.M., E.C. Butcher, A.M. Stall, P.A. Lalor, S. Adams, and L.A. Herzenberg. 1989. Many of the IgA producing plasma cells in murine gut are derived from self-replenishing precursors in the peritoneal cavity. *Int. Immunol.* 1:75–84.

123. Beagley, K.W., A.M. Murray, J.R. McGhee, and J.H. Eldridge. 1995. Peritoneal cavity CD5(B1a) B cells: Cytokine induced IgA secretion and homing to intestinal lamina propria in SCID mice. *Immunol. Cell. Biol.* 73:425–432.

124. Bos, N.A., J.C.A.M. Bun, S.H. Pompa, E.R. Cebra, G.J. Deenen, M.J.F. van der Cammen, F.G.M. Kroese, and J.J. Cebra. 1996. Monoclonal immunoglobulin A derived from peritoneal B cells is encoded by both germ line and somatically mutated VH genes and is reactive with comensal bacteria. *Infect. Immun.* 64:616–623.

31 | Autoimmune Thrombocytopenic Purpura

Diane J. Nugent and Thomas Kunicki

1. INTRODUCTION

Platelets play a central role in the maintenance of normal hemostasis and vascular repair. A decrease in platelet number, thrombocytopenia, can result in bruising, petechiae or even life-threatening bleeding. Immune-mediated thrombocytopenic purpura (AITP) is one of the most common forms of autoimmune disease affecting both adults and children. Patients with this disease develop autoantibodies that bind to platelet membrane antigens mediating the rapid destruction of antibody-coated cells in the reticuloendothelial system, particularly the spleen and liver. Sensing a decrease in circulating thrombocytes, the bone marrow responds with an increase in platelet production. Specific marrow progenitors, the megakaryocytes, increase in both number and ploidy to compensate for a shortened platelet life span.

Over the years, the triad of thrombocytopenia, increased platelet production and decreased platelet survival has served to define the clinical syndrome of AITP. Like many other syndromes, immune-mediated thrombocytopenia may occur as an isolated phenomenon or in association with other conditions, systemic lupus erythematosus (SLE) or pregnancy, for example. The presentation and prognosis of ITP may vary according to age, sex and familial predisposition to autoimmune disease. In this chapter, the clinical syndromes associated with immune-mediated thrombocytopenia will be reviewed, with special emphasis on the immunogenicity of the autoantigens and the molecular characterization of known human antiplatelet autoantibodies. In addition, this chapter will review the rapidly emerging data regarding the T lymphocyte helper phenotype and cytokine dysregulation that may lead to autoimmune thrombocytopenia, although the true etiology of AITP remains unknown.

2. HISTOPATHOLOGICAL AND CLINICAL FEATURES

In the early 1950's, Harrington and coworkers observed that pregnant women with thrombocytopenia often gave birth to infants with profound, but transient, thrombocytopenia (1). This suggested to them that the same plasma factor causing maternal thrombocytopenia could cross the placenta and mediate thrombocytopenia in the fetus and newborn. To prove that this factor was indeed present in the plasma of thrombocytopenic patients with increased platelet turnover, Harrington infused himself with plasma from a patient with ITP. The results of this infamous experiment left no room for doubt (2). Harrington became acutely thrombocytopenic with bleeding and a brief seizure, necessitating his own admission to the hematology ward for several days until his platelet count recovered. Subsequent studies, performed on volunteers, demonstrated that the antiplatelet activity was isolated in the gamma fraction of ITP plasma (3). Studies undertaken in other laboratories confirmed that IgG antiplatelet autoantibodies were responsible for the thrombocytopenia in ITP (4).

Even in uncomplicated ITP, there is considerable variation in presentation and prognosis. Based on the duration of thrombocytopenia, a distinction is made between acute versus chronic ITP. Thrombocytopenia lasting less than 6 months is termed acute, and greater than 6 months as chronic. Children are more likely to have the acute form of ITP and in 60–75% of the patients, the thrombocytopenia resolves within 2–4 months of diagnosis regardless of therapy (5,6). Childhood ITP occurs with equal frequency in boys and girls, is seasonal in nature, and may follow a recent infectious exposure or immunization, suggesting a role for viral or bacterial antigens in triggering antiplatelet autoantibodies (7). There is no familial predisposition in the acute form of childhood ITP.

Chronic ITP is the more common form in the adult (75–85% of adult cases), and occurs more frequently in women, similar to other autoimmune diseases (3:1 ratio female:male). Chronic ITP is not seasonal in nature and may be associated with autoantibodies to non-platelet antigens (i.e. antinuclear antigen, DNA, or cytoskeletal proteins). Furthermore, chronic ITP is more likely to occur in families where there is an increased incidence of other autoimmune diseases (8,9). The latter pattern suggests an underlying immune dysregulation and, in some cases, a familial predisposition toward autoimmune disease. Unlike SLE or diabetes mellitus (10), no one has identified a significant association with HLA class I, or class II DR, DP, or DQ antigens (11–13). The peak age for acute ITP is 2–5 years of age which is also the time when children experience the greatest frequency of viral infections and the peak incidence of the common form of childhood leukemia (ALL). Children with the chronic form of ITP mirror the adult phenotype, in that females predominate and there are no seasonal fluctuations in incidence of disease.

Once Harrington established the immune nature of AITP, physicians began to rely on the demonstration of increased platelet-associated immunoglobulin (PAIG) or the presence of circulating antiplatelet antibodies in patient plasma to confirm the diagnosis. Initially, assays focused on quantifying of platelet-associated IgG, IgM, and complement. Sensitivity of these assays was significantly limited by the fact that the target tissue, platelets, were markedly decreased in number. Moreover, platelets carry an internal pool of serum IgG in their alpha granules as the result of uptake by the megakaryocyte and packaging during platelet production (14,15). The purpose of this IgG in normal individuals is unknown. With both immune and *non-immune* platelet destruction syndromes, platelet activation triggers the release of alpha granules, thus increasing the amount of PAIG non-specifically. Therefore, increase in platelet-associated IgG alone may support the diagnosis of AITP, but does not rule out the possibility of low grade consumption of platelets from other etiologies associated with platelet activation, such as hemolytic uremic syndrome, vascular abnormalities (arterial-venous shunts, hemangiomas), or massive tissue damage as seen with burns or head trauma.

Work in the 1960's and 1970's focused on the development of sensitive assays to detect circulating antiplatelet antibodies in serum, even though autoantibodies are often of low titer and affinity. These new techniques increased the efficiency of antibody screening so that all patients might be evaluated using indirect binding assays to detect and quantitate the amount of antibody bound to the platelet and free in the plasma. The relative frequency of IgG isotypes mirrors that found in other autoimmune diseases

(16,17). The IgG1 and IgG3 isotypes are dominant in 60% of AITP patients, IgG2 and IgG4 are present in 13% and 9% of serum samples, respectively. Antiplatelet antibodies of the IgM class can also mediate platelet destruction, particularly when they fix complement (18–20). The true incidence of IgM antiplatelet autoantibodies varies somewhat due to the reluctance of researchers to look for IgM in screening AITP serum samples. Increased platelet-associated complement (C3) levels are found in roughly 30–40% of chronic ITP patients (21,22). In children, about one third of the patients have only IgM platelet-reactive antibodies, another 20–30% have both IgG and IgM. In adults, the reported frequency of IgM antiplatelet antibodies is 10% or less, and may reflect the relative dominance of chronic AITP in adults. In the sections that follow, the molecular nature of these autoantibodies and their target autoantigens will be discussed in detail.

Histopathological examination of the bone marrow and spleen has been performed in both acute and chronic ITP. In the early stages of ITP, the bone marrow may appear normal without the expected expansion of platelet precursors. As the bone marrow responds to persistent thrombocytopenia, the megakaryocytes begin to increase in number, particularly in the early, less mature cells (ploidy < 4N) and in the very large megakaryocytes (ploidy > 16N). This pattern is more likely to be seen after four weeks of thrombocytopenia and may be seen with immune or non-immune platelet destruction syndromes. Interestingly, in approximately 15% of adult patients with chronic AITP, the increased number of megakaryocytes may not correlate with increased platelet production (23). Platelet survival studies in these patients suggest that antibody binding to platelet antigens expressed on the megakaryocyte may interfere with the mechanics of platelets demargination or "pinching off" from the megakaryocyte membrane. This may be true in a subset of childhood AITP, but these studies, which would require the infusion of radioisotope-labeled platelets, have not been approved for children due to the risk of radiation exposure. The myeloid and erythroid precursors should be normal both in number and appearance in AITP. Children with the acute form of ITP, and less than four years of age, may have an increased number of immature lymphocytes in the bone marrow. Notably, these may account for 20–30% of nucleated cells in the bone marrow and often express early B cell markers such as CD10 (CALLA) and CD19 (early B cell antigen), as seen in the common form of childhood lymphocytic leukemia (24). Eosinophilia may also be associated with childhood ITP, but has no prognostic value in predicting severity of AITP (25,26).

As a therapeutic maneuver, many chronic AITP patients undergo splenectomy with normalization of platelet count in approximately 70–80% of cases. The spleen plays a major role in the removal of antibody-coated cells and is more efficient than the liver in immune destruction of cells due to a relatively slower rate of blood flow. This difference is accentuated by changes in the splenic microcirculatory pathway associated with lymphoid hyperplasia, a common finding in spleens of patients with chronic AITP (27). These changes result in longer transit time of antibody coated-platelets in areas rich with macrophages.

In addition to eliminating the major site of platelet destruction, splenectomy also removes a significant number of antibody-secreting B lymphocytes normally housed in the spleen. Approximately, 40% of the nucleated cells in the spleen are B cells; the remainder are T lymphocytes, monocytes, and macrophages. Over a period of 4–12 weeks, the serum antiplatelet autoantibody decreases markedly, and eventually PAIG also decreases to normal, or near normal, levels. It is unclear how splenectomy results in the resolution of AITP. Clearly, B cells proliferate and respond appropriately to antigenic challenge in sites other than the spleen, because post-splenectomy patients continue to produce an adequate amount of immunoglobulin of all subclasses. There are patients who also continue to make low levels of autoantibody, but in the absence of splenic destruction, the bone marrow is able to compensate for the remaining low-grade consumption in the reticuloendothelial system. In patients where splenectomy is successful, removal of the spleen may have eliminated the autoantibody-secreting B cells or, more importantly, may have disrupted the T and B cell dysregulation responsible for the autoimmune reactivity. In the 15–20% of cases where immune thrombocytopenia is not resolved by splenectomy, antibody-coated platelets are removed by the liver or may even be destroyed by macrophages in the bone marrow itself (23).

3. AUTOANTIGENS: IMMUNOCHEMICAL AND MOLECULAR CHARACTERISTICS

Both human and murine studies suggest that autoantibodies may arise as a consequence of antigen-driven B cell activation and selection following a polyclonal response (28,29). Clinicians began to look for distinct target antigens associated with the acute versus chronic AITP. Unlike experimental allergic encephalitis where a discreet portion of myelin basic protein is known to encode the immunogenic region of the molecule (30,31), many membrane antigens have been proposed as immunogenic in AITP. Platelet glycoprotein, glycolipid, protein and cytoskeletal elements have all been reported as antigenic targets in AITP. Given the diversity in presentation of immune thrombocytopenia, it is not surprising that there would also be numerous epitopes involved. Current research is directed toward the mapping and identification of these antigenic epitopes in the hope of categorizing the subsets of immune thrombocy-

topenia and examining their possible role in triggering the disease. Comparing the sequences of platelet autoantigens to other known protein sequences may provide insight on the origin of autoantibody for the various forms of AITP. The following section is a review of the known autoantigens in AITP and their disease associations.

3.1. Integrin αIIb-βIIIa

Platelet membrane glycoproteins play an important role as receptors to mediate platelet-platelet cohesion, platelet adhesion and spreading. The most abundant of these, integrin $\alpha_{IIb}\beta_3$, initially known as glycoprotein (GP) IIb-IIIa, is a heterodimer and is a member of the greater family of adhesion receptors known as integrins. This glycoprotein is a major antigenic target in the chronic form of ITP in both children and adults.

Having developed sensitive and specific assays for detecting serum autoantibody, researchers began to direct their attention to the identification of platelet target antigens. In 1982, van Leeuwen et al. (34) used IgG eluted from the platelets of 42 chronic AITP patients to assess antiplatelet reactivity using platelets with known deficiencies of certain membrane glycoproteins. Although all 42 IgG eluates bound to normal platelets, 35/42 did not bind to Glanzmann thrombasthenia platelets that specifically lack $\alpha_{IIb}\beta_3$. Since this landmark experiment, many others have confirmed the dominance of $\alpha_{IIb}\beta_3$ as a target antigen in AITP using sensitive direct and indirect immunoassays (35–37).

The amount of autoantibody needed to effectively destroy the platelet or remove it from circulation may not depend so much on the number of antigen binding sites, but on the isotype and subclass of the autoantibody, as well as the mobility of the antigen itself. The $\alpha_{IIb}\beta_3$ integrin is a very mobile molecule (38) and, in some studies, demonstrates patching and capping when anti-$\alpha_{IIb}\beta_3$ autoantibodies are bound to the complex (39,40). This appears to be a unique characteristic of the $\alpha_{IIb}\beta_3$ complex, because neither murine or human antibodies induce patching and capping of other platelet glycoproteins, such as GPIbα.

Antigen-specific assays allow researchers to screen for reactivity to $\alpha_{IIb}\beta_3$ and GPIb directly. The purified antigen is adsorbed to an ELISA microliter plate or the human autoantibody-antigen complex is "captured" by a murine monoclonal antibody linked to an immobilized matrix such as a bead or microliter plate. The AITP plasma is incubated with the antigen and specific binding of the autoantibody is measured with a goat or rabbit antihuman IgG that has a radioisotope or peroxidase conjugate for detection (41,42). Using these techniques, up to 60–70% of chronic ITP patients were found to have anti $\alpha_{IIb}\beta_3$ antibodies.

To assess the diversity of autoantibody binding sites on $\alpha_{IIb}\beta_3$, murine monoclonal antibodies to $\alpha_{IIb}\beta_3$ were used in competitive inhibition assays with serum from patients with chronic AITP. In one study, binding of a murine monoclonal anti-α_{IIb} antibody, 3B2, was decreased on platelets from all 16 patients studied with chronic immune thrombocytopenia (43). This result suggested that the epitope recognized by 3B2 was at least one member of the autoantigen repertoire, as defined by decreased binding of 3B2 on platelets from those AITP patients. An alternate explanation might be that conformational changes in the α_{IIb}-βIIIa complex associated with any autoantibody binding, masks or "unfolds" the 3B2 site, resulting in decreased binding of the murine antibody. In the ensuing 8 years, with more sensitive assays and larger numbers sampled, no one has demonstrated such a high frequency (94–100%) of anti-α_{IIb} autoantibodies in AITP.

To assess the true clonality of the anti-platelet antibodies from patients with chronic AITP, an expanded panel of 4 murine monoclonal antibodies, each recognizing unique and different sites on the $\alpha_{IIb}\beta_3$ complex, were used to block binding of autoantibodies in the IgG fractions of 6 patients with chronic AITP (44). This study suggested that autoantibodies from AITP patients reacted with a variety of epitopes on $\alpha_{IIb}\beta_3$. Indeed, a report using purified glycoprotein adsorbed to beads in an indirect immunoassay noted that 30/47 (67%) serum samples were positive for at least one of the glycoproteins screened in this assay. Moreover, 16/47 (36%) of these AITP patients had autoantibodies directed against multiple glycoproteins. In all cases where multiple specificities were demonstrated, the adsorption of the patient's serum against antigen-coated beads of one glycoprotein did not remove the reactivity to the other glycoprotein targets. In this study, positive reactions to multiple glycoproteins were found in the following combinations: Ib and α_{IIb} (2%), Ib and β_3 (13%), Ib and α_{IIb} β_3 complex (4%), Ib and α_{IIb} and β_3 (17%).

Immunoblot technology also allowed researchers to identify the platelet protein and glycoproteins targeted by autoantibodies from patients with AITP. To screen serum using the immunoblot technique, platelet proteins are separated by size and relative migration using SDS polyacrylamide gel electrophoresis. The separated proteins are then transferred to nitrocellulose paper and incubated with the patient's plasma. Binding of the antiplatelet autoantibody is detected with a rabbit or goat anti-human IgG antibody, which has a radioisotope or peroxidase conjugate attached so that the target antigen is identified on autoradiography or when the enzyme conjugate catalyzes a color reaction (45–47).

Beardsley et al. were the first to use the immunoblot technique to demonstrate binding of autoantibody to platelet proteins. These studies suggest that when it is possible to detect autoantibodies, the major platelet membrane target antigens are $\alpha_{IIb}\beta_3$ or GPIb in chronic ITP. In this assay, one can demonstrate autoantibody binding to specific proteins and distinguish binding to α_{IIb} versus β_3 for example. Although some plasma contained anti-α_{IIb} activity, serum IgG from the majority (9/13) of chronic AITP patients bound β_3 (45). Other glycoproteins, such as GPIb, $\alpha2\beta1$, and a 90 kD protein were also identified, but were much less frequent than β_3. Not all platelet proteins can maintain their immunogenic epitopes following the denaturation step required for SDS-PAGE, but this preliminary data provided the basis for continued interest in β_3 and the mapping studies that followed.

The importance of this platelet glycoprotein as a dominant target antigen has been challenged by a number of studies suggesting that the anti-$\alpha_{IIb}\beta_3$ autoantibodies are part of the normal repertoire of naturally-occurring autoantibodies that remove activated or senescent platelets (47–53). Therefore, it is not surprising to find this specificity in so many patients with ongoing immune-mediated platelet destruction. Peptide mapping confirms that many of these antibodies react with the cytoplasmic tail of this molecule, which would not be readily accessible in the normal circulating platelet. In addition, the persistence of antibodies to $\alpha_{IIb}\beta_3$ in the plasma of patients with *resolved* thrombocytopenia suggests that they may play a more important role in the clearance of damaged cells rather than as the primary autoantibody marking the platelet for destruction in AITP. Unraveling the role of antibody isotype and changes in affinity of the V gene family used for certain idiotypes may help to clarify the difference between a pathologic versus naturally-occurring autoantibody. Until then, the dominance of antibodies to $\alpha_{IIb}\beta_3$ will continue to be based on detection of these autoantibodies in sensitive immunoassays. Much work is still needed to sort out the pathologic autoantibody from those plasma antibodies that are normally present in the human immunoglobulin repertoire.

3.2. Localization of Autoantigens on $\alpha_{IIb}\beta_3$

A number of factors have facilitated the mapping of autoepitopes on $\alpha_{IIb}\beta_3$ and other platelet membrane proteins. Advances in molecular biology have resulted in the sequencing of the $\alpha_{IIb}\beta_3$ gene along with other platelet membrane target proteins. Examination of the derived amino acid sequences of these molecules have allowed researchers to focus on areas that are potentially immunogenic or located near functionally important sites on the $\alpha_{IIb}\beta_3$ complex. Binding studies on proteolytic fragments of $\alpha_{IIb}\beta_3$ or GPIb are more easily interpreted knowing the likely enzyme cleavage sites and the size of the fragments generated. In addition, synthetic peptides or phage-display peptide libraries may be used to pinpoint the immunoreactive site within a certain fragment of the platelet membrane target antigen. The production of human monoclonal autoantibodies or recombinant Fab fragments from patients with immune thrombocytopenia provides researchers with specific antibody probes to map target antigens. Human monoclonal antibodies have an advantage over serum antibodies in that they are monoclonal and circumvent problems associated with using a polyclonal serum where many different, low titer autoantibodies may be found with reactivity against a variety of epitopes.

Fujisawa generated synthetic peptides representing various regions of the β_3 molecule that were used as antigenic targets to screen sera from patients with chronic AITP (48). He found that autoantibodies from 5 of 13 samples reacted with the carboxyl-terminal region of β_3 represented by peptides spanning residues 721–744 or 742–762 (Figure 1). These autoantibodies are clearly elevated in the sera of the AITP patients and not present in normal or control sera. Their role in the pathogenesis of immune thrombocytopenia is clouded by the fact that this portion of the GPIIIa molecule is located within the cytoplasm, not on the surface, of the resting platelet. However, internal autoantigens are the hallmark of many immune-mediated diseases, such as polymyositis and systemic lupus erythematosus (49,50). As in these diseases, one can postulate that the autoantigen is primarily internal, but at some point is expressed on the surface of the platelet surface,

perhaps with activation or senescence, at which time the autoantibody can bind and remove the cell from circulation.

The report by Fujisawa is one of many suggesting that autoantibodies to $\alpha_{IIb}\beta_3$ may bind to cryptic epitopes or neoantigens that are expressed with platelet activation or senescence (51–53). Platelets normally accumulate IgG on the membrane surface as they age or become activated over 7–10 days, and then they are cleared by reticuloendothelial cells in the spleen (54). Using Fab'2 fraction of IgG, it has been shown that normal individuals have autoantibodies in their sera that specifically bind to senescent antigens on red cells and platelets to mediate clearance of spent or dysfunctional cells (55,56). The role of these naturally-occurring autoantibodies in pathological immune destruction of cells is unclear. Animal studies suggest that the initial binding of the pathogenic antiplatelet autoantibody may trigger the expression of senescent or cryptic antigens on the platelet membrane, resulting in a second wave of immunoglobulin binding and clearance by naturally-occurring autoantibodies (59).

Using high titer antisera to $\alpha_{IIb}\beta_3$ from two different patients with AITP, Kekomaki et al. (60) localized the binding of autoantibodies to a 33 kD chymotryptic fragment derived from the cysteine-rich portion of the mole-

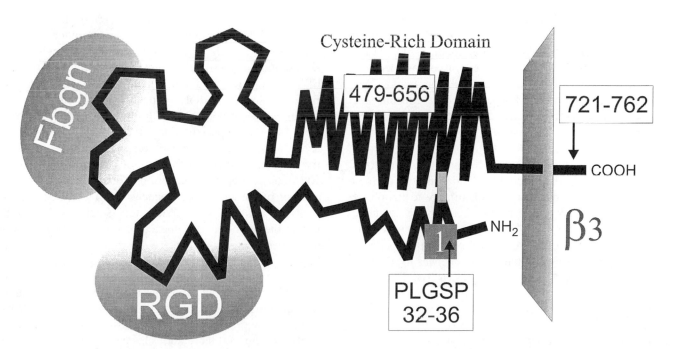

Figure 1.   Schematic representation of the platelet membrane integrin, β_3 (glycoprotein IIIa). The numbers in boxes represent previously described alloantigens on this highly immunogenic molecule. Peptides spanning residues 721–744 or 742–762 in the carboxyl-terminal region of β_3 were used as antigenic targets to screen sera from patients with chronic AITP.[48] Autoantibodies from 5 of 13 samples screened reacted with peptides in this region. The cysteine-rich portion of the molecule which spans residues 479–654 was also found to be a common target for patients with autoantibody in AITP or other "disease-related" thrombocytopenic patients may include patients with lymphoproliferative disease, such as chronic lymphocytic leukemia (CLL), Hodgkin's Disease, or collagen vascular disease patients.[60] Using plasma from patients with AITP to screen a phage display library, a novel peptide did show significant homology with sequence on GPIIIa (32–36) suggesting these site as autoantigen target on this molecule (63).

cule that spans residues 479–654 (Figure 1). Having identified this region, the 33 kD chymotryptic fragment was used in an ELISA to screen sera from patients with thrombocytopenia. Overall, reactivity with the 33 kD fragment was found in 15 of 31 (48%) patients with chronic ITP and 2 of 8 (25%) patients with acute ITP. Unlike many other studies of this nature, these researchers included sera from patients where the thrombocytopenia was not associated with AITP. These "disease-related" thrombocytopenic patients may include patients with lymphoproliferative disease, such as chronic lymphocytic leukemia (CLL), Hodgkin's Disease, or collagen vascular disease patients. One third of these patients also had anti-33 kD fragment reactivity detected in their sera. These data mirror those reported by Berchtold et al., who reported that 60% of all thrombocytopenic patients screened had increased platelet associated immunoglobulin (PAIG), and 30% of these patients reacted with either $\alpha_{IIb}\beta_3$ or GPIb complex. Following extensive adsorption studies, he also demonstrated that the reactivity $\alpha_{IIb}\beta_3$ was distinct from the other primary autoantibodies present in the sera of SLE or myasthenia gravis patients, and thus did not result from crossreactivity of the anti-DNA or anti-acetylcholine receptor antibodies present in these syndromes.

Identification of autoimmune epitopes has been aided by the development of the phage-display peptide library. Bowditch and McMillan used a filamentous phage library that displays random linear hexapeptides to identify peptide sequences recognized by AITP autoantibodies (61). Plasma antibody eluates from patients were used to select for phage-displaying autoantibody-reactive peptides. They identified anti-$\alpha_{IIb}\beta_3$ antibody-specific phage encoding the peptide sequences Arg-Glu-Lys-Ala-Lys-Trp (REKAKW) and Pro-Val-Val-Trp-Lys-Asn (PVVWKN) and the hexapeptide sequence Arg-Glu-Leu-Leu-Lys-Met. Each phage showed saturable dose-dependent binding to immobilized autoantibody, and binding could be blocked with purified $\alpha_{IIb}\beta_3$. The binding of plasma autoantibody to the phage encoding REKAKW could be blocked with a synthetic peptide derived from the β_3 cytoplasmic tail; however, the PVVWKN was not. Using sequential overlapping peptides from the β_3 cytoplasmic region, an epitope was localized to the sequence Arg-Ala-Arg-Ala-Lys-Trp (β_3 734–739) (Figure 1).

Further work by this group made use of a similar technique to map an autoepitope on α_{IIb} (62). They used a filamentous phage library that displays random peptides, 11 amino acids in length flanked on each side by a cysteine, to identify peptide sequences recognized by an antiplatelet autoantibody that blocked fibrinogen binding to $\alpha_{IIb}\beta_3$. Phage from individual colonies, after the fourth purification, were tested for binding to the autoantibody. A phage expressing the sequence CTGRVPLGFEDLC exhibited saturable dose-dependent binding to immobilized

autoantibody. This binding could be blocked with purified $\alpha_{IIb}\beta_3$ and α_{IIb}, but not by β_3. The peptide amino acid sequence has partial identity with amino acids 4–10 and with amino acids 31–35 on α_{IIb} (Figure 2). This work suggests that the autoantibody is binding on the first 35 amino acids of α_{IIb}, and infers that this region of α_{IIb} is critical for ligand binding.

Gevorkian used plasma from patients with AITP to screen their phage display library rather than affinity purified autoantibodies, as described in the previous study (63). They obtained a panel of affinity-selected phage clones that have been shown to react in enzyme-linked immunosorbent assay with autoantibodies from AITP patients. Using whole plasma, one would anticipate finding a wide variety of antigen targets, primarily infectious in origin. None of the peptides obtained has been described previously as possibly being an epitope for antiplatelet antibodies, and the majority of them did not show any homology with known platelet glycoproteins. Even so, three novel peptides did show significant homology with sequences on β_3 and GPIb (Figure 3). They concluded that some of the other peptides identified in this study could represent discontinuous epitopes or mimotopes of natural autoantigens.

Escher's efforts to map autoantigens resulted in the isolation of 2 recombinant autoantibody Fab fragments, A and B (clones A and B), specific for $\alpha_{IIb}\beta_3$ (62). These Fab fragments inhibit the binding of anti-$\alpha_{IIb}\beta_3$ autoantibodies from patients with chronic immune thrombocytopenia. Dissociation of the $\alpha_{IIb}\beta_3$ complex on platelets by 10 mM EDTA completely inhibited binding of both clones A and B, confirming the complex-dependent nature of these Fab recombinant antibodies. In a fibrinogen-binding ELISA, using fibrinogen concentrations 10^{-4}–10^{-6} lower than serum levels, fibrinogen binding was completely inhibited by Fab fragments A and B. The affinity of clones A and B was 10^{-8}M, as defined by Scatchard plot analysis. Comparative analysis of the amino acid sequences of recombinant anti-idiotypic antibodies generated against clones A and B versus the amino acid sequence of $\alpha_{IIb}\beta_3$ showed homologies between complementary determining regions on the Fab fragments and the fibrinogen-binding sites on $\alpha_{IIb}\beta_3$ (Figure 2).

Human monoclonal antibodies have simplified the search for autoantigens in man by circumventing the problems associated with polyreactivity of plasma and crossreactive antibodies. Several laboratories have used this technique to characterize autoepitopes on α_{IIb} β_3 (51,64,65). The best example of how these reagents are utilized to map immunogenic epitopes is exemplified by work with 2E7, a human monoclonal autoantibody that binds to $\alpha_{IIb}\beta_3$ (64,65) (Figures 2 and 4). Using the immunoblot technique, they established that 2E7 only bound to the α_{IIb} molecule, not β_3. A series of 31 synthetic peptides were generated, each

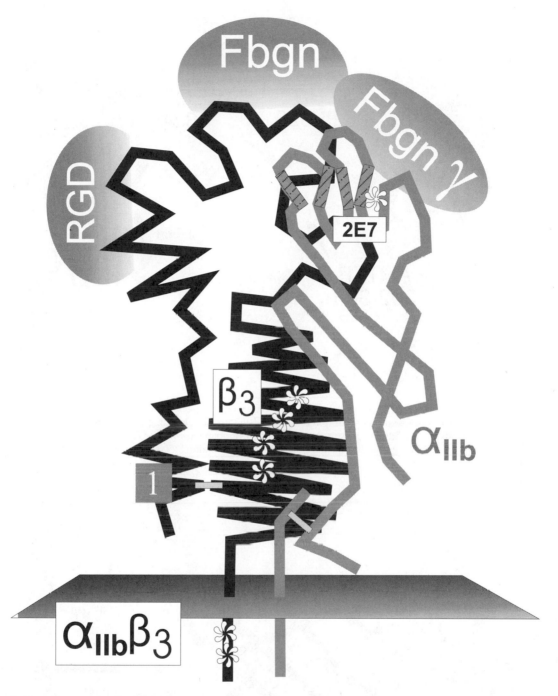

Figure 2. Schematic representation of the platelet membrane integrin α_{IIb} from the complex $\alpha_{IIb}\beta_3$ (glycoprotein IIb–IIIa). The hatched regions indicate sites involved in divalent cation binding to the α_{IIb} molecule. The shaded oval designates the fibrinogen gamma chain binding site. A series of 31 synthetic peptides were generated, each spanning a different portion of the α_{IIb} molecule. The human monoclonal autoantibody, 2E7 was analyzed in a dot blot assay and only reacted with the peptide derived from residues 222–238 of the heavy chain of α_{IIb} (51).

spanning a different portion of the α_{IIb} and β molecules, and reactivity of 2E7 to each peptide was analyzed in a dot blot assay. The monoclonal antibody only reacted with the peptide derived from residues 222–238 of the heavy chain of α_{IIb}. In addition, only the α_{IIb} 222–238 peptide blocked

binding of 2E7 to solid phase $\alpha_{IIb}\beta_3$ in competitive inhibition assays (Figure 5A).

As mentioned previously, the $\alpha_{IIb}\beta_3$ molecule is a member of the greater family of adhesion molecules or integrins. In this system, α_{IIb} has significant homology

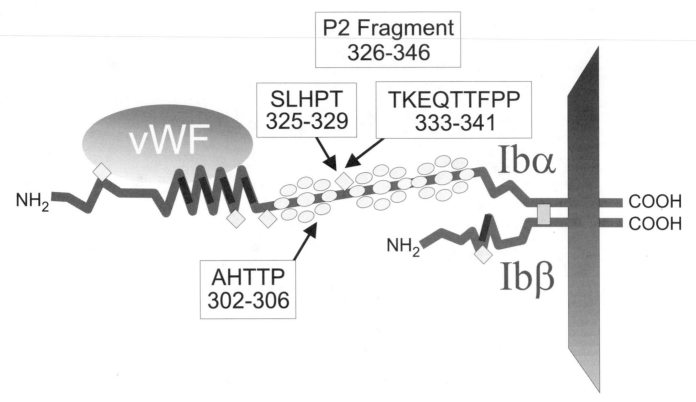

Figure 3. Schematic representation of the platelet membrane glycoprotein Ib α *and* β chains. The GPIbα heavy chain is a large molecule and contains leucine-rich repeats and a region in the middle of the glycoprotein that is heavily glycosylated. The two hatched boxes represent sites of previously described alloantigens. Using plasma from patients with AITP to screen a phage display library two novel peptides did show significant homology with sequences on GPIb (302–306 and 325–329) suggesting these sites as possible novel autoantigen targets on this molecule (63). The P2 fragment and sequence 333–341 were identified as potential target antigens using plasma from chronic ITP patients (140).

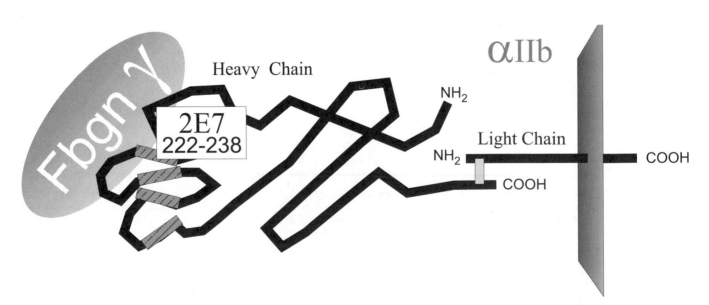

Figure 4. Schematic representation of the platelet membrane integrin complex $\alpha_{IIb}\beta_3$ (glycoprotein IIb–IIIa). This heterodimeric complex is a major glycoprotein receptor and is thought to be the most immunogenic molecules on the platelet surface. The 2E7 binding site overlaps the first calcium binding domain of α_{IIb}. In fact the binding of 2E7 is greatly enhanced in the presence of EDTA, confirming the predicted position of the 2E7 epitope and implying that the binding site is less cryptic when calcium is removed from the complex

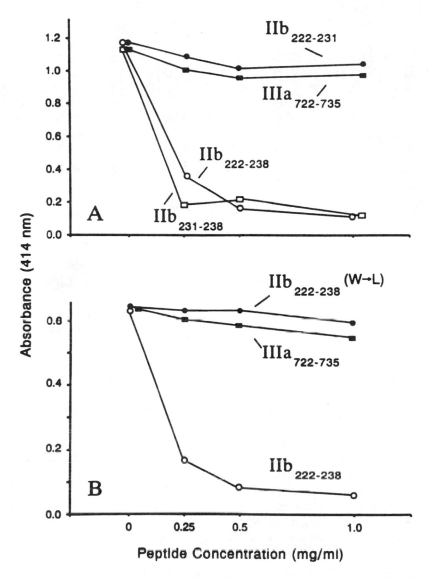

Figure 5. Using immunoblot technique, they established that 2E7 only bound to residues 222–238 of the heavy chain of α_{IIb}. In addition, only the α_{IIb} 222–238 peptide blocked binding of 2E7 to solid phase $\alpha_{IIb}\beta_3$ in competitive inhibition assays (5A). In order to determine the exact location of 2E7 binding in the α_{IIb} heavy chain, a second series of peptides were generated which would span either the heterologous region (222–231) or the homologous region (231–238) of this molecule. In the highly homologous region spanning residues 234–238, α_{IIb} differs from the other integrin α chains only in position 235 (W→L), therefore, an additional peptide, αIIb222–238 (235 W→L), was generated to test the effect of this single amino acid substitution. When competitive inhibition studies were performed only the peptide αIIb222–238 (235 W→L) from the highly homologous region specifically blocked binding of 2E7 to $\alpha_{IIb}\beta_3$ (figure 5B).

with other α-chains. Comparison of the 220–238 sequences of α_{IIb} and seven other integrin alpha subunits is shown in Figure 6. Although there is significant homology between αIIb and other alpha integrin chains, 2E7 did not bind to the other cell types that express these related integrin family members, such as endothelial cells, lymphocytes, monocytes, or neutrophils. To determine the exact location of 2E7 binding in the α_{IIb} heavy chain, a second series of peptides were generated that would span either the heterologous region (222–231) or the homologous region (231–238) of this molecule. In the highly homolo-

gous region spanning residues 234–238, α_{IIb} differs from the other α-chains only in position 235 (W→L), therefore, an additional peptide, αIIb222–238 (235 W→L), was generated to test the effect of this single amino acid substitution. When competitive inhibition studies were performed with these peptides, the peptide from the highly-homologous region specifically blocked binding of 2E7 to $\alpha_{IIb}\beta_3$ (Figure 5B). This inhibitory effect was lost when the substitution was made in position 235, suggesting that the tryptophan plays a critical role in determining the immunogenic epitope recognized by 2E7. As seen in

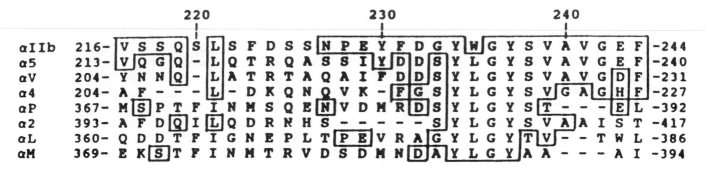

Figure 6. A comparison of the 220–238 sequences of α_{IIb} and seven other integrin alpha subunits. The $\alpha_{IIb}\beta_3$ molecule is a member of the greater family of adhesion molecules or integrins. In the highly homologous region spanning residues 234–238, α_{IIb} differs from the other α chains only in position 235 (W→L).

Figure 4, the 2E7 binding site overlaps the first calcium binding domain of α_{IIb}. In fact, the binding of 2E7 is greatly enhanced in the presence of EDTA, confirming the predicted position of the 2E7 epitope and implying that the binding site is less cryptic when calcium is removed from the complex.

Unfortunately, not all autoepitopes are linear, as described for the 2E7 binding site on α_{IIb}. Other epitopes may be conformational and dependent on the tertiary configuration of the molecule. Given the fluid nature of the platelet membrane and the remarkable metamorphosis that occurs with activation and adhesion, it is not surprising that conformational antigenic sites would come and go. In fact, Shadle et al. describe a neoantigen on α_{IIb}, which is only revealed following the normal binding of fibrinogen to its receptor, the $\alpha_{IIb}\beta_3$ complex (66). Serum autoantibodies have been described that demonstrate increased binding to platelets in the presence of calcium chelators, such as EDTA, heparin, or other drugs (67–69). Some of the autoantibody binding may be directed to neoantigens, and thus is enhanced following changes in membrane glycoproteins induced by agents like EDTA or heparin.

Human monoclonal autoantibodies that bind to conformational epitopes are still very useful in identifying the nature of cryptic or neoantigens on platelet glycoproteins in AITP. 5E5 is a human monoclonal autoantibody that was isolated from a patient with chronic ITP and serum antiplatelet IgG and IgM antibodies reactive with β_3 (51). 5E5 binds to a neoantigen associated with β_3 that is only expressed on platelets following thrombin activation or senescence. Screening of 5E5 against a series of synthetic peptides representing various regions of the β molecule did not prove useful in identifying a specific binding site on this molecule. In collaboration with Wencel-Drake (Chicago, Illinois), immunofluorescence microscopy was used to localize the 5E5 epitope on both resting versus activated platelets, with and without membrane permeabilization. These studies revealed that there is no detectable expression of the 5E5 epitope on the platelet surface, in resting, fixed, non-permeabilized platelets. However, in resting permeabilized platelets, this epitope is found in an internal pool of β_3 that does not co-localize with the α-granule marker, fibrinogen. In contrast, following thrombin activation, the 5E5 epitope is expressed on the platelet surface. This study suggested that the 5E5 epitope in resting platelets is found uniquely in the cytoplasmic pool of β_3, which is expressed following thrombin activation and subsequent mobilization to the platelet surface.

In summary, $\alpha_{IIb}\beta_3$ complex is undoubtedly a frequently recognized target antigen in immunoassays in AITP. However, mapping of the autoepitopes recognized by the pathogenic antibodies in patient serum is complicated by the fact that this glycoprotein complex also appears to express neoantigens or cryptic epitopes that are revealed with normal platelet activation or senescence. Patient serum also contains the naturally-occurring autoantibodies present in all individuals that recognize the $\alpha_{IIb}\beta_3$ neoantigens and trigger immune-mediated clearance of spent or aged platelets. Indeed, the very high frequency of the murine antibody 3B2 platelet epitope described by van Leeuwen initially in AITP, may have been more representative of the naturally-occurring autoantibodies directed against activation or senescence epitopes on α_{IIb}. No dominant autoepitope on $\alpha_{IIb}\beta_3$ has emerged as the key autoantigen for pathologic AITP antibodies.

Current research is focused on this interaction of pathologic and naturally-occurring autoantibodies in the platelet destruction in AITP. The relative amount of the initial pathologic autoantibody reacting with any platelet membrane antigen might be miniscule compared to the secondary wave of naturally-occurring antibody to $\alpha_{IIb}\beta_3$ after the initial destruction begins. Platelet eluates to be used in immunoassays for the identification of target antigens could find a dominance of anti-$\alpha_{IIb}\beta_3$, but one cannot determine whether this is a primary, pathologic antibody or the secondary, naturally-occurring autoantibodies involved in clean-up of activated or damaged cells. The introduction of anti-idiotypic reagents to identify and isolate specific antiplatelet autoantibodies, as described below, has been very useful in linking anti-β_3 antibodies bearing certain idiotypes with known subtypes of AITP. This approach is critical to a clear understanding of whether $\alpha_{IIb}\beta_3$ is a primary or secondary target in AITP.

3.3. Other Platelet Membranes Glycoprotein Targets

Many other glycoproteins have been implicated as target antigens in AITP. These include autoantibodies to another integrin, $\alpha2\beta1$, previously known as glycoprotein Ia-Iia (70), or other proteins of unknown identity that appear on immunoblot when screening sera from patients with AITP (17,45). Rather than create an extensive list of all reported autoantigenic targets in AITP (71–78), it is more interesting to examine the target glycoprotein receptors in relationship to their function and possible association with certain forms of AITP.

Glycoprotein Ib-IX complex. Unlike the $\alpha_{IIb}\beta_3$ complex, autoantibodies directed against glycoprotein Ib (GPIb) are almost universally associated with pathological immune destruction of platelets. Thus far, no one has described naturally-occurring antibodies reacting with GPIb in either resting or activated platelets. Although only 15–30% of AITP patients have autoantibodies that bind to

GPIb, there is no background binding of antibodies in normal plasma, so the characterization of this antigen is much more straightforward. Furthermore, using anti-idiotypic reagents, it appears that the repertoire of idiotypes expressed by anti-GPIb antibodies from patients with AITP is very narrowly defined (79). In this study, affinity-purified antiplatelet GPIb antibody from patient DM, who produced a very high titer anti-GPIb autoantibody, was used to immunize rabbits and mice for the production of rabbit polyclonal and murine monoclonal anti-DM idiotypic antibodies. These reagents were highly specific for the DM idiotype and could be used to screen a large number of serum samples from normal individuals and patients with AITP. Furthermore, the rabbit polyclonal and certain of the murine monoclonal antibodies block binding of the immunizing antibody, DM, to platelets or purified GPIb. Approximately 95% of AITP serum samples from patients positive for the DM idiotype contained anti-GPIb autoantibody ($p < 0.00001$).

The GPIbα heavy chain is a large molecule and contains leucine-rich repeats and a region in the middle of the glycoprotein that is heavily glycosylated (Figure 3). The NH2-terminal region of GPIb is responsible for binding to von Willebrand's antigen, which mediates platelet adhesion. Theoretically, there could be a variety of antigenic sites on this molecule, but in the studies examining frequency of DM

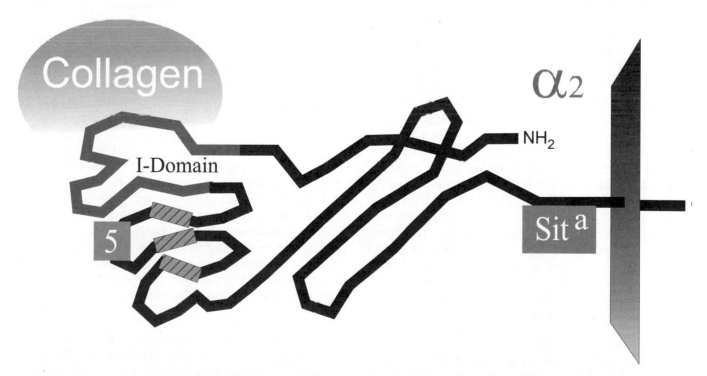

Figure 7.   Schematic representation of the platelet membrane integrin $\alpha2\beta1$ (glycoprotein Ia/IIa). Platelet integrin $\alpha2\beta1$ is one of two important collagen receptors on the platelet membrane. Although present in a much lower concentration than $\alpha_{IIb}\beta_3$ or GP Ib-IX complex, this receptor has also been shown to be a target autoantigen in 10–20% of chronic AITP patients and some SLE patients with an acquired bleeding diathesis.[84,85] The two boxed regions (5, Sit[a]) represent previously described alloantigens. The hatched region designates the divalent cation binding site.

idiotype-positive, anti-GPIb autoantibodies, only 12–15% of the anti-GPIb plasmas were negative for DM idiotypic antibodies. Half of these were from patients who had developed drug-dependent antibodies that bound to the GPIb-IX complex, particularly those associated with the ingestion of quinine or quinidine. These results suggest that the DM idiotype is very tightly associated with autoantibodies against GPIb in patients with AITP, and although DM is not the only autoepitope on GPIb, it is certainly present as one of the anti-GPIb antibodies in the serum of the majority of patients with autoantibodies to this complex.

The DM autoantigen has been mapped to the heavy chain of the GPIb complex. Following trypsin treatment of affinity-purified GPIb complex, the DM autoantibody specifically recognizes a site on the 40–45 kD fragment of the GPIbα molecule, the same region where von Willebrand's protein binds to mediate platelet adhesion. Ongoing research to identify the sequence of the DM autoepitope will provide a critical piece of information that is currently lacking in AITP, namely the role these glycoprotein sequences play in antigen selection of certain pathologic autoantibody idiotypes.

Epitope mapping on GPIb was performed using a series of 22 linear amino acid sequences as target antigens (140). Six of 16 plasma samples from patients with chronic ITP reacted with a peptide generated from amino acid sequence 326–346, suggesting that this may be a dominant epitope for autoantibodies binding to platelet GPIb.

Platelet glycoprotein V. In the mid-1980's, when screening for antigen-specific autoantibodies became available, many researchers hoped that the patterns of target antigen reactivity might be predictive of the clinical course of AITP, i.e. acute versus chronic. In fact, an initial report of 8 children with acute AITP using the immunoblot technique failed to demonstrate anti-β_3 antibody, even though this was the dominant antigenic target recognized by autoantibodies from patients with chronic AITP (45). This was followed by a report from the same laboratory (80) demonstrating serum antibody reactivity to GPV in four pediatric patients with thrombocytopenia associated with varicella (chickenpox) infection. Glycoprotein V (GPV) is a thrombin-sensitive, 85 kD glycoprotein of unknown function on the platelet surface. There was no evidence of immune complexes in these patients or varicella antigens adsorbed to the surface of their platelets. As all of the children with varicella related AITP and acute childhood form of AITP resolved their thrombocytopenia within six months, there was some hope that GPV reactivity would be a marker for mild disease and these patients could escape aggressive immunosuppressive therapy.

Many centers are trying to establish the importance of autoantibody specificity in predicting the severity of ITP and in determining the initial therapeutic modality. If anti-GPV antibodies are more commonly associated with acute ITP and a limited course of thrombocytopenia, then less therapeutic intervention should be required. Alternatively, if GPIb is more likely associated with a more severe and chronic form of the disease, then more aggressive therapy might be justified in hopes of modifying or reversing the likelihood of protracted thrombocytopenia. Large scale cooperative clinical studies with well-entrenched procedures for long-term follow up are necessary to validate these theories. None of the current studies incorporate the relative frequency of autoantibodies to other reported platelet membrane targets, such as proteins, lipids, or glycosphingolipids (81–83). These specificities might also be important in predicting severity of disease or response to certain therapeutic regimens, such as intravenous infusions of immunoglobulin (IVIG).

3.4.   Platelet Integrin a2b1 (Glycoprotein IA/IIA)

Platelet integrin $\alpha2\beta1$ is one of two important collagen receptors on the platelet membrane. Although present in a much lower concentration than $\alpha_{IIb}\beta_3$ or GP Ib-IX complex, this receptor has also been shown to be a target autoantigen in 10–20% of chronic AITP patients and some SLE patients with an acquired bleeding diathesis (84,85). Efforts to map the specific epitope on this molecule have been unsuccessful, even though a number of groups have suggested that the binding of these antibodies impair function, suggesting that it might be near the binding site of its ligand, collagen. (Figure 7). Deckmyn et al. describe a plasma autoantibody directed against a protein comigrating with α_2 and recognized by the patient's antibody when affinity-purified $\alpha_2\beta_1$ was used as antigen (84). The $\alpha_2\beta_1$ complex was immunoprecipitated from a platelet lysate by the patient's plasma, and purified platelet specific IgGs from this patient inhibited aggregation of normal platelets induced by collagen or by wheat germ agglutinin. Dromigny described an increased bleeding time observed in a 48-year-old woman that led to the discovery of SLE confirmed by immunological tests (85). Platelet function and analysis of membrane glycoproteins revealed an isolated impaired collagen-induced platelet aggregation and the presence of autoantibodies directed against $\alpha_2\beta_1$ and GPIb/IX. As detailed in the sections that follow on immunoglobulin gene family VH3, human monoclonal autoantibodies reactive with $\alpha_2\beta_1$ are homologous with those isolated from other SLE patients reacting to single stranded DNA.

4.   IMMUNOGLOBULIN VARIABLE GENE FAMILIES IN IMMUNE-MEDIATED THROMBOCYTOPENIA

In the following sections, the structural and molecular characterization of human autoantibodies in immune-mediated thrombocytopenia will be reviewed.

Researchers have focused their attention on the mechanism controlling selection of the B cell repertoire to identify a predisposition toward the production of pathologic antibodies in a variety of autoimmune disorders. In the previous section, target antigens in AITP were reviewed. Extensive work in animal models suggests that the mechanisms that generate autoantibodies in syngeneic mice may include the preferential use of particular variable (V) heavy (H) or light (L) gene families in conjunction with antigen selection following somatic mutation to generate high affinity IgG autoantibodies. Studies in human autoimmune disease have been limited by genetic diversity from one individual to another, and an inability to study the disease process as it evolves.

VH gene utilization has been studied extensively as a predisposing factor in human autoimmune disease. In light of the genetic diversity in man, it is all that more impressive that patients with AITP or SLE share immunoglobulin idiotypes and demonstrate striking similarities in the preferential use of certain variable gene families. Conversely, deletions of physiologically important VH genes may also increase the risk of autoimmunity through indirect effects on the development and homeostasis of the B cell repertoire (87–90).

As yet, only one human VH gene deletion, hv3005, has been described in association with chronic AITP. This VH gene encodes heavy chains of rheumatoid factors and is located in a complex locus that encompasses a combination of one to four copies of six highly homologous VH3 genes (87–89). Olee et al. hypothesized that a homozygous deletion of this critical autoantibody-associated Ig variable (V) gene altered the immune system and thus predisposed the host to autoimmune disorders (88). Initial experiments with a Humhv3005/P1 probe revealed that one of the four major hybridizing bands was missing in approximately 20% of patients with either rheumatoid arthritis or systemic lupus erythematosus, but only 2% of normal subjects. Mo et al. extended these studies to 44 patients with chronic AITP (89), and found that hv3005, and related genes, were absent in a higher percentage of AITP patients (14/44, 31.8%) in either normal (7/88, 8%, p = 0.002) or thrombocytopenic patients without chronic AITP (6/53, 11.3%, p = 0.042). These data suggest that deletions of hv3005 and/or highly homologous VH genes may predispose individuals to the development of chronic AITP, and may contribute toward the production of pathogenic antiplatelet antibodies. The importance of this deletion relative to other factors predisposing to autoimmune disease was tested by Huang et al. in patients with SLE (90) in a study that suggests that C4A null alleles predispose strongly to development of lupus, whereas the influence of hv3005 deletion is relatively weak. A similar investigation comparing the relative contribution to disease predisposition of various susceptibility genes, such as hv3005, is lacking in AITP.

There is considerably more information regarding the preferential use of selected VH gene families in autoimmune disease. Studying the shifts in utilization of VH gene families may provide insight into the immunoglobulin repertoire from which pathogenic autoantibodies emerge. However, these shifts in circulating B cell populations still may not address the pathologic autoantibodies themselves. There are only three groups of platelet-reactive autoantibodies that have been studied on a molecular level. These include the SLE autoantibody HF2–1/17, which binds to single-stranded DNA, cardiolipin, as well as platelet ceramide (91–93), the anti-$\alpha_{IIb}\beta_3$-specific human monoclonal antibodies 2E7 and 5E5 (65,94), and the DM series of anti-GPIb autoantibodies (79,93). Antibodies HF2–1/17, 2E7, and 5E5 all belong to the VH3 family. The light chains from both 2E7 and HF2–1/17, are members of the Vκ1 subgroup. The DM idiotype-positive antibodies associated with anti-GPIb reactivity are derived primarily from the VH4 family and utilize both kappa and lambda light chains.

Comparison of the monoclonal autoantibodies in this group, which have been isolated from patients with immune thrombocytopenia, demonstrate the challenge faced by researchers attempting to correlate sequence with binding specificities. The HF2–1/17 is a polyreactive autoantibody produced by a human-human hybridoma derived from B lymphocytes of a patient with SLE and thrombocytopenia. Many previously sequenced autoantibodies use the VH3 family (95–98) and exhibit a striking degree of homology with germline sequence. These antibodies, like HF2–1/17, are often polyreactive and seem to arise in response to in vivo or in vitro, polyclonal activation (99–102).

Autoantibodies 2E7 and 5E5 are also derived from the VH3 family and are highly homologous with HF2–1/17 and each other (Figure 8). These autoantibodies were isolated from EBV-transformed B lymphocytes from patients with autoimmune thrombocytopenia and no evidence of SLE (51,56), bind α_{IIb} and β_3, respectively, and show no evidence of binding to either single- or double-stranded DNA. Both the VH genes for HF2–1/17 and 5E5 appear to have arisen from the VH26 germline gene (103) and 2E7 from either VHI.9111 or 56PIVH germline genes, both of which are members of the VH3 family (104,105).

Atkinson et al. suggested that the anti-DNA binding properties of HF2–1/17 might be correlated with the Arg residues at positions 24 and 30 of the VK1 light chains. This hypothesis stems from the fact that HF2–1/17 shares this sequence with other 16/6 idiotype-bearing antibodies, namely, HKIOI, WEA, GAL, 16/6,18/2,21/28, all components of anti-DNA binding antibodies (106). The Arg residues in positions 24 and 30 are preserved in the 2E7 autoantibody, as well. However, the kappa light chain of 2E7, which is virtually identical to that seen in HF2–1/17

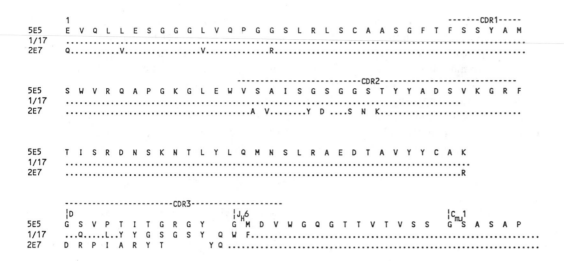

HF2-1/17 : anti-DNA and anti-platelet autoantibody from a patient with systemic lupus erythematosus (SLE)
(Lampmann GW, Furie B, Schwartz RS, Stollar BD, Furie BC: Amino acid sequence of a platelet-binding human anti-DNA monoclonal
autoantibody. Blood 74:262, 1989

2E7 : anti-platelet glycoprotein IIb isolated from a patient with ITP.
(Kunicki TJ, Annis DS, Gorski J, Nugent DJ: Nucleotide sequence of the Human Autoantibody 2E7 specific for the platelet Integrin IIb
Heavy chain, Autoimmunity in press, 1991.)

Figure 8. Comparison of the deduced amino acid sequence of human monoclonal autoantibodies 5E5, HF2–1/17, and 2E7. Autoantibodies 2E7 and 5E5 are also derived from the VH3 family and are highly homologous with HF2–1/17 and each other except for the antigen binding region designated CDR3.

and shares more homology to HF2–1/17 than the other 16/6 idiotype positive antibodies, does not bind to single or double stranded DNA.

Many research centers would like to accumulate sufficient structural information to correlate variable gene region amino acid sequence with autoantibody specificity. The VH3 gene family is used extensively in the formation of antibodies of varied specificities. Unlike the murine gene, the VH family genes on the human chromosome are highly interdigitated, especially VH3, which has the largest number of genes in this region of chromosome 14. Although this family may be used commonly in the naturally-occurring autoantibodies, the more pathologic antibodies IgG anti-DNA autoantibodies also use the same kv325 germline gene commonly associated with polyreactive IgM autoantibodies. However, in the pathologic antibodies, the kv325 gene has undergone somatic mutation, suggesting an antigen-driven selection process in these patients. Furthermore, animal models of autoantibodies suggest that even single amino acid substitutions are enough to influence antigen binding specificity (107,108). Conversely, a single amino acid substitution in the target autoantigen may also result in a complete loss of autoantibody binding, as seen with 2E7 when the immunodominant

trp in position 235 of the α_{IIb} molecule is replaced by a Leu residue (64). Current research to define the structural characteristics of the antibody-binding site and model possible idiotypic determinants will undoubtedly clarify how these subtle changes result in such diverse binding patterns (109).

Recently, another human monoclonal autoantibody has been cloned that also uses the VH3 gene family, but has very clear specificity for platelet GPIa/IIa (110). Human B cell lines were derived by limiting dilutions of Epstein-Barr virus-transformed peripheral B cells from a patient with an autoantibody against glycoprotein Ia/IIa, and who manifested defective collagen-induced platelet aggregation and a bleeding problem. Antibody-producing clones were selected for their reactivity with whole platelets or with affinity-purified GPIa/IIa by enzyme-linked immunosorbent assay (ELISA). One of these cell lines produced an IgM (E3G6) that interfered with platelet aggregation responses. Polymerase chain reaction amplification with two different sets of primers specific for human kappa-chain resulted in the isolation of a unique sequence. Further analysis showed that the kappa variable domain sequence is similar to the germline gene A30, to 2E7, an anti-α_{IIb} human autoantibody, and to HF2–1/17, a systemic lupus erythematosus (SLE)-associated broad-specificity human autoantibody.

The specificity of antibody, E3G6, appears to be determined by the μ-chain. The sequence of this heavy chain is encoded by a VH3 gene segment strongly homologous to the germline gene DP-77 and by JH4 gene segment that is also germline. All four mutations in E3G6, compared to DP-77, are in CDR, and result in amino acid substitutions, which implies that E3G6 may have been derived from an antigen-driven response. The autoantigen epitope for this antibody has not yet been mapped.

4.1. The DM Idiotypic Autoantibodies, G1 and VH4

In AITP patients whose disease is mediated by GPIb-specific autoantibody, a predominant component of the platelet reactive antibody displays a major cross-reactive idiotype termed DMId (79). This idiotype was not detected in 60 age-matched normal individuals nor in over 300 ITP patients who lacked GPIb-specific autoantibodies. The correlation between the presence of DM Id-positive antibodies and anti-GPIb-specific autoantibodies was highly significant ($p < 0.0001$) using a rabbit polyclonal anti-DM reagent. To clarify the molecular genetic origin of these DM Id positive autoantibodies, monoclonal B cell lines producing DM Id-positive IgM antibodies were analyzed (93).

Sequencing of the variable gene regions of 6 human monoclonal antibodies, from a patient with high titer anti-GPIb and DM Id-positive antibodies, demonstrated that the DM-positive IgM monoclonal antibodies were primarily derived from the VH4 family. Two of the 6 monoclonal antibodies from this patient were specific for GPIb, and one of these was also derived from the VH4 family. The other monoclonal anti-GPIb autoantibody from this patient was derived from the VH1 family. These DM idiotype-positive antibodies demonstrated substitutions that were most likely due to somatic mutations, unless they represent currently unknown germline elements. This observation is similar to that made by Logtenberg and coworkers, who found somatically-mutated VH6 elements in anti-DNA antibodies (111). There were no obvious V or J segment differences to account for the DM Id expression among the monoclonal B cell lines. In fact, there were striking differences between the VH4 genes in the length of the D segments, probably due to random nucleotide additions (N regions) to the junctions at the time of VH–D–JH joining (112).

Another human monoclonal IgG antiplatelet antibody has been characterized by Olee et al. (113) that utilizes VH4 and Vλ2 gene families and was isolated from a patient with chronic AITP. Like the DM Id antibodies, this VH region also exhibited very large nucleotide insertion in a CDR of the heavy chain. The antigen specificity of G1 is shared by both $\alpha_{IIb}\beta_3$ complex and tetanus toxoid. The monoclonal antibody-secreting B cell line was selected using an ELISA to detect reactivity to $\alpha_{IIb}\beta_3$, a target antigen in the chronic AITP patient from whom this EBV-transformed cell line was derived. However, G1 also shows a 3–4 fold greater binding to tetanus toxoid compared to $\alpha_{IIb}\beta_3$.

Like so many of the human monoclonal autoantibodies isolated thus far, there appears to be cross-reactivity between autologous antigens and external pathogens. AITP has demonstrated a strong association between infection and the onset of autoimmunity. One hypothesis suggests that infectious agents induce co-stimulatory activity on cells expressing low levels of an autoantigen, in turn activating autoreactive T cells. An alternative therory termed "molecular mimicry" suggests that antibodies or T cells generated in response to an infectious agent cross-react with self-antigens. There is good evidence for the involvement of T cells in autoimmune diseases, but the particular T cells responsible for AITP are much more difficult to isolate than the platelet autoantibodies. The specific T cells that triggered the disease and their initial target antigens are difficult to identify due to a phenomenon called epitope spreading. At the onset, the T cell response might be directed against a specific infectious antigenic epitope, but over time, the T cells expand and diversity. New T cell clones emerge that react with other parts of the same protein or with other molecules in the damaged organ or cell, stimulating the production of autoreactive B cells previously tolerant to these epitopes. Thus, an array of cross-reacting antibodies and T cells are frequently found at diagnosis of autoimmune disease.

Animal studies have demonstrated that both CDR and framework regions of the heavy and light chains play an important role in defining the three-dimensional structure of cross-reactive immunoglobulin idiotypes (114–116). As the EBV transformable population of lymphocytes may not represent the antigen-driven autoantibody-secreting B cell responsible for AITP, it is necessary to isolate and sequence these idiotype-positive autoantibodies using an alternate technology. Currently, heavy and light chain immunoglobulin libraries can be generated in E. coli expression and phage display systems. Using these techniques, idiotype positive clones can be directly sequenced and, in protein expression systems, the relative contribution of heavy and light chain sequences can be analyzed with regard to idiotype expression and antigen binding (117).

5. IMMUNOREGULATORY DYSFUNCTION

The excellent response of AITP patients to the intravenous infusion of high dose immunoglobulin and the demonstration of anti-idiotypic antibodies in these preparations (118,119) provide strong evidence that immunomodulatory therapy is feasible in autoimmune-mediated thrombocytopenia. Most physicians concur that Fc blockade plays a major role in the success of intravenous immunoglobulin infusions, which induces a rise in platelet count in the majority of patients regardless of age or etiology of the autoantibody. Apart from reducing splenic clearance of platelets, there is increasing evidence that treatment also alters T cell subsets and may produce alterations in cytokine production (120). Although articles abound in cytokine response to therapy, there are few that stress the distinction between changes in cytokine mRNA expression or protein production as measured *in vitro* or in the serum of the AITP patient. As many of these immunoregulatory effects may take place in the microenvironment of the spleen or bone marrow, changes in circulating cells or their production of cytokines *in vitro*, should be interpreted with circumspection. The following sections will cover the current data on lymphocyte studies and cytokine production in AITP and changes noted in response to therapy.

5.1. T Lymphocyte and Cytokine Studies in AITP

It is generally held that the production of autoantibodies is driven and controlled by cellular and soluble regulatory mechanisms. Many investigators have examined defects involving T cells or an abnormal dominance of one of the T helper subsets, Th1 or Th2, and their associated cytokines in autoimmune disease. The following sections summarize the observations of many excellent investigators, but as yet no clear T cell pattern has been established in AITP, nor is there any published information on the presence or absence of Th3 subsets in AITP. Perhaps this reflects the heterogeneous nature of AITP and suggests that the T cell dysregulation may evolve in different directions relative to the age, sex, or underlying environmental or viral triggers.

Given the importance of the T cell receptor (TCR) in other autoimmune diseases, it is not surprising that initial studies by Ware et al. focused on characterization of TCR γ/δ utilization (121,122). They studied 11 children with acute ITP and 19 children with chronic ITP and observed elevated numbers of TCR γ/δ^+ T lymphocytes in several patients. In those patients with the highest elevations (TCR $\gamma/\delta^+/CD3^+$ percentage ranging from 37.8 to 48.1% at initial evaluation), the expanded cell population exclusively expressed the surface V δ 2/V γ 9 heterodimer. Analysis of the nucleotide sequences used by these TCR γ/δ^+ cells demonstrated a diverse set of VDJC gene rearrangements, suggesting a superantigen response. There was a close correlation between the number of TCR γ/δ^+ T lymphocytes and the degree of thrombocytopenia in each patient, but no specific platelet-reactive T cell clones were isolated from that group. Continued efforts by Ware et al. resulted

in the isolation of eight T cell clones that showed *in vitro* proliferation against allogeneic platelets from 2 boys with chronic ITP and elevated numbers of Vβ8$^+$ T cells (122). Four of seven positive clones also had measurable interleukin (IL)-2 secretion following platelet stimulation, providing further evidence for T cell reactivity. Their results provided the first evidence that patients with ITP may have platelet-reactive T lymphocytes identifiable at the clonal level, supporting the hypothesis that autoreactive peripheral T lymphocytes in AITP may mediate or participate in the pathogenesis of this disorder.

Using the common AITP target antigen$\alpha_{IIb}\beta_3$, Kuwana et al. extended these T cell studies to examine *in vitro* production of antibody to this platelet glycoproteinby peripheral blood mononuclear cells (PBMC) (123). T cell proliferative responses to platelet membrane $\alpha_{IIb}\beta_3$ were examined in 14 patients with chronic immune thrombocytopenic purpura (ITP), 7 systemic lupus erythematosus (SLE) patients with or without thrombocytopenia, and 10 healthy donors. Although peripheral blood T cells from all subjects failed to respond to the protein complex in its native state, reduced $\alpha_{IIb}\beta_3$ stimulated T cells from three ITP patients and one SLE patient with thrombocytopenia, and tryptic peptides of $\alpha_{IIb}\beta_3$ stimulated T cells from nearly all subject Characterization of T cell response induced by modified $\alpha_{IIb}\beta_3$ showed that the response was restricted by HLA-DR, the responding T cells had a CD4(+) phenotype, and the proliferation was accelerated only in ITP patients, suggesting *in vivo* activation of these T cells. *In vitro* IgG anti-$\alpha_{IIb}\beta_3$ synthesis in PBMC cultures was induced by modified $\alpha_{IIb}\beta_3$ specifically in ITP patients who demonstrated platelet-associated anti-$\alpha_{IIb}\beta_3$ antibody. PBMC cultures without antigenic stimulation did not produce anti-$\alpha_{IIb}\beta_3$ antibody in ITP patients. Anti-$\alpha_{IIb}\beta_3$ antibody produced in supernatants was absorbed by incubation with normal platelets. None of the PBMC culture supernatants from healthy donors contained a significant amount of IgG anti-$\alpha_{IIb}\beta_3$ antibody, but **all** of them showed trypsindigested $\alpha_{IIb}\beta_3$-induced T cell proliferation. They concluded that CD4(+) and HLA-DR-restricted T cells to $\alpha_{IIb}\beta_3$ are involved in production of anti-platelet autoantibody in ITP patients and are related to the pathogenic process in chronic ITP.

Similar T cell proliferation in healthy individuals in response to $\alpha_{IIb}\beta_3$ was described two years previously by Filion (124). They demonstrated the presence of autoreactive T cells to $\alpha_{IIb}\beta_3$, in the periphery of all healthy individuals tested (n = 25). Using an *in vitro* T cell proliferation assay, they showed that activation of these specific $\alpha_{IIb}\beta_3$ autoreactive α/β TCR$^+$ CD4$^+$ CD8$^-$ T cells required internalization and processing of the $\alpha IIb\beta_3$ by antigen-presenting cells, and its presentation by HLA-DR class II molecules, in the presence of exogenous interleukin 2 (IL-2). This implied that some autoreactive T cells directed against membrane antigens present on platelets were not necessarily eliminated by intrathymic deletion.

The pervasive presence of platelet-reactive T cells in the periphery of healthy individuals suggests that there is a tightly regulated network of cells that keeps these autoreactive T cell clones in check. In a related manner, these regulatory cells also participate in the immunoglobulin idiotypic network to limit the expansion of pathologic forms of antiplatelet antibody. Pooled immunoglobulin (IVIg) from healthy donors contains antibodies reactive with idiotypes of natural and disease-related autoantibodies or surface immunoglobulins of B cells. IVIg also contains antibodies reactive with the idiotype, framework and constant regions of the beta chain of the T cell receptor. Infusion of IVIg results in a modulation of synthesis and release of cytokines. Kazatchkine et al. proposed that the immunoregulatory effect of IVIg in autoimmune disease is dependent on the selection of the recipient's immune repertoires by the variable (VII or VL) region reactivities of infused immunoglobulins (125,126). Mehta observed that immunoglobulins (IVIg and anti-D immunoglobulin preparations), and their Fab fragments, inhibited the binding of antiplatelet autoantibodies to normal platelets from 15.8 to 90.7% and 25.6 to 90.08% respectively, whereas, their Fc portion did not show any inhibition (127) The IVIg and Rh immunoglobulin products reacted with the monoclonal antibodies, only through their Fab and not through the Fc portions, thereby confirming its specific anti-idiotype activity. The fact that pooled immunoglobulin is rich in these regulatory elements suggests that normal individuals use these repertoires to maintain self-tolerance and suppress autoreactive T or B cell activity.

There is a high frequency of antiplatelet autoantibodies associated with collagen vascular disease, such as SLE, thyroiditis, Evan's syndrome and primary biliary cirrhosis. They are also present in states of congenital or acquired immune dysfunction, including common variable dysgammaglobulinemia, DiGeorge Syndrome, autoimmune lymphoproliferative syndrome, pregnancy, HIV or EBV for example. Why are platelet antigens such a common target for autoreactive antibodies? Researchers are beginning to examine the role of the platelet itself in the triggering of AITP. It is now clear that the activated platelet shares many of the co-stimulatory receptors commonly found on lymphocytes (CD40L, CD80) and monocytes (CD14) (128,129). In an activated state, the platelet itself might substitute for the accessory cells that trigger autoantibody production, temporarily bypassing the tightly regulated T and B cell idiotypic network. Semple et al. found a significant number of children with acute (80%), chronic (71%), or chronic-complex (55%) had platelets expressing HLA-DR, in contrast to normal controls and patients with non-immune thrombocytopenia (128). HLA-DR was variably co-expressed on distinct smaller and larger-sized platelet populations with CD41, CD45, CD14, CD80, and/or glycophorin molecules. In normal healthy individuals, platelets only express Class I molecules. The authors hypothesize that the HLA-DR expression on these young platelets may play a role in the triggering or perpetuation of AITP itself.

Investigators have focused considerable effort on the measurement of cytokines in AITP in an effort to further characterize the influence of T helper subsets in driving the autoimmune process. Unfortunately, the majority of these studies draw conclusions based on a single determination of a cytokine profile. As cytokine signaling may rapidly fluctuate, it is difficult to draw conclusions from a single assay. In addition, soluble receptor may cloud the results if not measured directly at the same time point the cytokines are assayed. To distinguish between normal and pathological immune responses, multiple samples should be obtained over time, with attention paid to both the cytokine and its soluble receptor, when indicated. Large-scale trials with adults and children with AITP are needed to examine the cytokine profiles over time, and to clarify the differences observed in different populations with either the acute or chronic form of thrombocytopenia. The following studies include multiple samples and controls in their reports, although there is still some disparity between results in children versus adults.

In the pediatric age group, patients appear to have a Th1 type of cytokine response with very low IL-4 and IL-6, and elevated levels of IL-2, IFN-γ, and TNF-β (130,131). Adult chronic AITP, immune thrombocytopenia associated with malignancy, or the Autoimmunity/Lymphoproliferation syndrome (ALPS) with defects in the Fas apoptosis pathway have found elevated levels of IL-10, IL-11, IL-6 and IL-13 (131,132,135–137). Additional work by Zimmerman and Bussel have examined the changes in T cell subsets and cytokine production in response to IVIG versus WinRho (antiD) (133,134). Zimmerman et al. found that intravenous immunoglobulin and dexamethasone induced an alteration of T lymphocyte subsets and suppression of in vitro T lymphocyte proliferation (134). Although equally effective in the treatment of AITP in children, they found that anti-D caused significantly less inhibition than IVIG or dexamethasone in the six children with chronic ITP studied following anti-D administration. Anti-D did not affect T lymphocyte subsets including the T cell receptor variable β repertoire, in vitro T lymphocyte proliferation to mitogens, recall antigens, or interleukin-2, in vitro IgG synthesis induced by pokéweed mitogen, or T lymphocyte cytokine mRNA levels. Bussel et al. compared the changes in cytokine levels following treatment with IVIG and anti-D in thrombocytopenic adults without HIV infection. Patients were treated with either IVIG or anti-D. IL-6, IL-10, MCP-1, and TNFα were measured in duplicate on each sample using an ELISA technique (133).

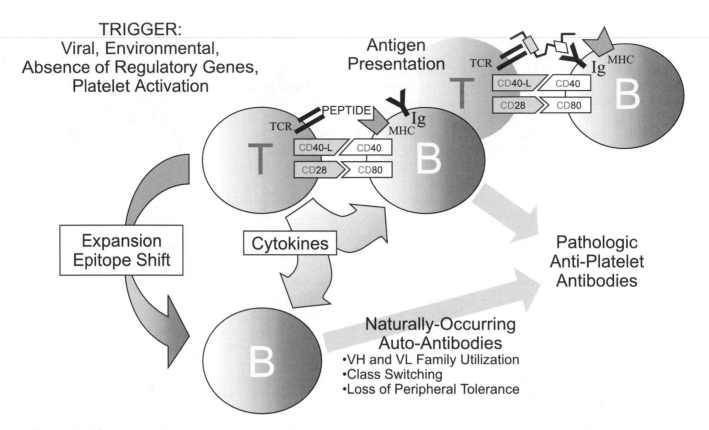

Figure 9. Schematic representation of proposed cellular interactions in the production of pathologic autoantibodies. This simplified model is designed to show sites for possible therapeutic intervention. An example cited in the text is the use of humanized monoclonal antibodies to the CD40L molecule, or other accessory molecules, to disrupt the production of antiplatelet autoantibodies in ITP.[1]

Compared to baseline levels, there was an increase in 3 of 4 cytokines 2 hours following administration of anti-D, with MCP-1 reaching the highest levels; only TNFα failed to change significantly from baseline. After IVIG, there was a significant increase in IL-10 levels at 4 hours, but not 2 hours, post therapy, and in MCP-1 levels at 1 day. No changes were seen in IL-6 or TNFα. They hypothesize that the early increase in these macrophage-synthesized cytokines with anti-D demonstrates the substantial interaction of antibody-coated RBC with macrophages (134). The differences compared to IVIG are presumably due to anti-D creating a large model immune complex of hundreds of antibodies on each RBC. Both IVIG and anti-D result in decreased splenic clearance of antibody coated platelets, but each has additional effects that may be tailored to the AITP population, as Zimmerman and Bussel have done in comparing children with chronic ITP versus HIV-positive adults with immune thrombocytopenia. Immune therapy that may disrupt the pathologic production of autoantibody has also been initiated using humanized monoclonal antibodies to CD40L (Figure 9) or other accessory molecules (138,139). Early trials in refractory adult AITP patients appear promising, but remain to be proven in large scale clinical studies.

6. CONCLUSIONS

Although there have been considerable advances in clarifying the immune nature of AITP, much work remains to be done in identifying the range of immunodominant antigens on the platelet surface and determining how these molecules facilitate the unrestricted expression of antiplatelet autoantibodies in this disease. Such diversity in presentation and prognosis argues for the heterogeneous nature of autoimmune thrombocytopenia. A clearer understanding of the nature of AITP will allow physicians to use less toxic therapy for acute disease and more specific immune modulation for those patients with the more chronic forms of AITP. Current research efforts are directed toward distinguishing the specific etiologies behind the production of antiplatelet autoantibodies in each form of AITP. Emerging data on the autoantigens and immunoglobulin gene sequences provide insight on the pathogenic mechanisms triggering the childhood versus adult forms of AITP. Similarly, new information on the T cell, cytokine and other regulatory elements associated with AITP may explain the production of antiplatelet antibodies in syndromes where changes in self tolerance result in autoimmune disease.

References

1. Harrington, W.J., C.C. Sprague, V. Minnich, M.S. Carl, C.V. Moore, R.C. Aulvin, and R. Dubach. 1953. Immunologic mechanisms in neonatal and thrombocytopenic purpura. *Ann. Intern. Med.* 38:433–469.
2. Altman, L.K. 1987. Who Goes First?: The story of self-experimentation in medicine. New York: Random House, pp. 273–283.
3. Harrington, W.J., V. Minnich, J.W. Hollingsworth, and C.V. Moore. 1951. Demonstration of a circulating factor in the blood of patients with thrombocytopenic purpura. *J. Lab. Clin. Med.* 38:1–10.
4. Shulman, N.R., V.J.K. Marder, and R.S. Weinrach. 1965. Similarities between known antiplatelet antibodies and the factor responsible for thrombocytopenia in idiopathic purpura. *Ann. NY Acad. Sci.* 124:499–504.
5. Lusher, J.M., A. Enami, V. Ravindranath, and A.I. Warrier. 1984. Idiopathic thrombocytopenic purpura in children. *Am. J. Pediatr. Hematol. Oncol.* Summer; 6:149–157.
6. Walker, R.W., and W. Walker. 1984. Idiopathic thrombocytopenic purpura, initial illness and long term follow up. *Arch. Dis. Child.* 59:316–22.
7. Lusher, J.M., and R. Iyer. 1977. Idiopathic thrombocytopenic purpura in children. *Semin. Thromb. Hemostas.* 3:175–199.
8. Lippman, S.M., F.C. Arnett, C.L. Conley, P.M. Ness, D.A. Meyers, and W.B. Bias. 1982. Genetic factors predisposing to autoimmune diseases. *Am. J. Med.* 73:827–840.
9. Panzer, S., E. Penner, W. Graninger, I. Schulz, and J.S. Smolen. 1990. Antinuclear antibodies in patients with chronic idiopathic autoimmune thrombocytopenia followed 2–30 years. *Am. J. Hematol.* 32:100–103.
10. Steinberg, A.D., M.F. Gourley, D.M. Klinman, G.C. Tsokos, D.E. Scott, and A.M. Krieg. 1991. Systemic lupus erythematosus. *Ann. Intern. Med.* 115:548–559.
11. Gratama, J.W., J. D'Amaro, J. de Koning, and G.J. de Otto Lander. 1984. The HLA system in immune thrombocytopenic purpura: its relation to the outcome of therapy. *Br. J. Hematol.* 56:287–293.
12. Mayr, W.R., G. Mueller-Eckhardt, M. Kruger, C. Mueller-Eckhardt, K. Lechner, and H. Niessner. 1981. HLA-DR in chronic ITP. *Tissue Antigens* 18:56–57.
13. Mueller-Eckhardt, G., G. Pawelec, R. Haas, C. Otto, V. Kiefel, N. Odum, and C. Mueller-Eckhardt. 1989. HLA-DP antigens in patients with chronic autoimmune thrombocytopenia (AITP). *Tissue Antigens* 34:121–126.
14. George, J.N., S. Saucerman, S.P. Levine, L.K. Kneriem, and D.F. Bainton. 1985. Immunoglobulin is a platelet alpha granule-secreted protein. *J. Clin. Invest.* 76:2020–2025.
15. Handagama, P.J., J.N. George, M.A. Shuman, R.P., McEver, and D.F. Bainton. 1987. Incorporation of a circulating protein into megakaryocyte and platelet granules. *Proc. Natl. Acad. Sci. USA* 84(3):861–865.
16. Tijhuis, G.J., R.J. Klaassen, P.W. Modderman, W.H. Ouwehand, and A.E. von dem Borne. 1991. Quantification of platelet-bound immunoglobulins of different class and subclass using radiolabeled monoclonal antibodies: assay conditions and clinical application. *Br. J. Haematol.* 77:93–101.
17. Winiarski, J. 1989. IgG and IgM antibodies to platelet membrane glycoprotein antigens in acute childhood idiopathic thrombocytopenic purpura *Br. J. Haematol.* 73:88–92.
18. Kayser, W.C., Mueller-Eckhardt, V. Budde, and R.E. Schmidt. 1981. Complement fixing platelet autoantibodies in autoimmune thrombocytopenia. *Am. J. Haematol.* 11:213–219.
19. Pawha, J., D. Giuliani, and B.S. Morse. 1983. Platelet-associated IgM levels in thrombocytopenia. *Vox. Sang.* 45:97–103.
20. Cines, D.B., S.B. Wilson, A. Thomas, and A.D. Schreiber. 1985. Platelet antibodies of the IgM class in immune thrombocytopenic purpura. *Clin. Invest.* 75:1183–1190.
21. Hauch, T.W., and W.F. Rosse. 1977. Platelet-bound complement (C3) in immune thrombocytopenia. *Blood* 50:1129–1136.
22. Cines, D.B., and A.D. Schreiber. 1979. Immune thrombocytopenia: Use of a Coombs' antiglobulin test to detect IgG and C3 on platelets. *N. Engl. J. Med.* 300:106–111.
23. Ballem, P.J., G.M. Segal, J.R. Stratton, T. Gernsheimer, J.W. Adamson, S.J. Slichter. 1987. Mechanisms of thrombocytopenia in chronic autoimmune thrombocytopenic purpura. *J. Clin. Invest.* 80:33–40.
24. Cornelius, A.S., D. Campbell, E. Schwartz, and M. Poncz. 1991. Elevated common acute lymphoblastic leukemia antigen expression in pediatric immune thrombocytopenic purpura. *Am. J. Pediatr. Hematol. Oncol.* 13:57–61.
25. McClure, P.D. 1975. Idiopathic thrombocytopenic purpura in children: Diagnosis and Management *Pediatrics* 55:68–74.
26. Lusher, J.M., and R. Iyer. 1977. Idiopathic thrombocytopenic purpura in children. *Semin. Thromb. Hemostas.* 3:175–199.
27. Schmidt, E.E., I.C. MacDonald, and A.C. Groom. 1991. Changes in chronic idiopathic thrombocytopenic purpura. *Blood* 78:1485–1489.
28. van Es, J.H., F.H. Gmelig Meyling, W.R. van de Akker, H. Anastoot, R.H. Derksen and T. Logtenberg. 1991. Somatic mutations in the variable regions of a human IgG anti-double stranded DNA autoantibody suggest a role for antigen in the induction of systemic lupus erythematosus. *J. Exp. Med.* 173:461–470.
29. Behar, S.M., D.L. Lustgarten, S. Corbet and M.D. Scharff. 1991. Characterization of somatically mutated S107 VH11-encoded anti-DNA autoantibodies derived from autoimmune (NZB × NZW) Fl mice *J. Exp. Med.* 173:731–741.
30. Davis, C.B., D.J. Mitchell, D.C. Wraith, J.A. Todd, S.S. Zamvil, H.O. McDevitt, L. Steinman, and P.P. Jones. 1989. Polymorphic residues on the I-A beta chain modulate the stimulation of T cell clones specific for the N-terminal peptide of the autoantigen myelin basic protein. *J. Immunol.* 143:2083–2093.
31. Zaller, D.M., G. Osman, O. Kanagawa, and L. Hood. 1990. Prevention and treatment of murine experimental allergic encephalomyelitis with T cell receptor V beta-specific antibodies. *J. Exp. Med.* 171:1943–1955.
32. Kunicki, T.J., D. Pidard, J.P. Rosa, and A.T. Nurden. 1981. The formation of calcium-dependent complexes of platelet membrane glycoproteins IIb and IIIa in solution as determined by crossed immunoelectrophoresis. *Blood* 58:268–278.
33. Pytela R., M. Pierschbacher, M.H. Ginsberg, E.F. Plow, and E. Ruoslanti. 1986. Platelet membrane glycoprotein IIb-IIIa: Member of a family of Arg-Gly-Asp-specific adhesion receptors. *Science* 231:1559–1562.
34. van Leeuwen, E.F., J.T.M. van der Ven, C.P. Engelfriet, and A.E. von dem Borne. 1982. Specificity of autoantibodies in autoimmune thrombocytopenia. *Blood* 59:23–26.

35. Woods, V.L., E.H. Oh, D. Mason, and R. McMillan. 1984. Autoantibodies against the platelet glycoprotein IIb/IIIa complex in patients with chronic ITP. *Blood* 63:368–375.

36. McMillan, R. P. Tani, F. Millard, P. Berchtold, L. Renshaw, and V.L. Woods, Jr. 1987. Platelet-associated and anti-glycoprotein autoantibodies in chronic ITP. *Blood* 70:1040–1045.

37. Kiefel, V., S. Santoso, M. Weisheit, and C. Mueller-Eckhardt. 1987. Monoclonal antibody-specific immobilization of platelet antigens (MAIPA): a new tool for the identification of platelet-reactive antibodies. *Blood* 70:1722–1726.

38. Wencel-Drake, J.D. 1990. Plasma membrane GPIIb/IIIa. Evidence for a cycling receptor pool. *Am. J. Pathol.* 136:61–70.

39. Bourguignon, L.Y. 1984. Receptor capping in platelet membranes. *Cell. Biol. Int. Rep.* 819–26.

40. Santoso, S., V. Kiefel, and C. Mueller-Eckhardt. 1987. Redistribution of platelet glycoproteins induced by allo- and autoantibodies. *Thromb. Haemost.* 58:866–871.

41. Woods, V.L., Y. Kurata R.R. Montogomery, P. Tani, D. Mason, F.H. Oh, and R. McMillan. 1984. Autoantibodies against platelet glycoprotien Ib in patients with chronic immune thrombocytopenic purpura. *Blood* 64:156–160.

42. Furihata, K., D.J. Nugent, A. Bissonette, R.J. Aster, and T.J. Kunicki. 1987. On the association of the platelet-specific alloantigen, Pena, with GPIIB-IIIa. Evidence of Heterogeneity of glycoprotein IIIa. *J. Clin. Invest.* 80:1624–1630.

43. Varon, D., and S. Karpatkin. 1983. A monoclonal anti-platelet antibody with decreased reactivity for autoimmune thrombocytopenic platelets. *Proc. Natl. Acad. Sci. USA* 80:6992–6995.

44. Tsubakio, T., P. Tani, V.L. Woods, Jr., and R. McMillian R. 1987. Autoantibodies against GPIIb-IIIa in chronic ITP react with different epitopes. *Br. J. Haematol.* 67:345–348.

45. Beardsley, D., J. Speigel, M.M. Jacobs, R.I. Handen, and S.E. Lux, 4th. 1984. Platelet membrane glycoprotein IIIa contains target antigens that bind anti-platelet antibodies in immune thrombocytopenias. *J. Clin. Invest.* 74:1701–1707.

46. Szatkowski, N.S., T.J. Kunicki, and R.H. Aster. 1986. Identification of glycoprotein Ib as a target for autoantibody in idiopathic (autoimmune) thrombocytopenic purpura. *Blood* 67:310–315.

46a. Mason, D., and R. McMillan. 1984. Platelet antigens in chronic idiopathic thrombocytopenic purpura. *Br. J. Haematol.* 56:529–534.

47a. Berchtold, P., D. Muller, W.C. Kouns, M.A. Riederer, and B. Steiner. 1998. Plasma autoantibodies against platelet glycoprotein IIb/IIIa from patients with autoimmune thrombocytopenic purpura may recognize different antigenic determinants. *Eur. J. Haematol.* 61:223–228.

47. Fujisawa, K., T.E. O'Toole, D. Tani, J.C. Loftus, E.F. Plow, M.H. Ginsbert, and R. McMillian. 1991. Autoantibodies to the presumptive cytoplasmic domain of platelet glycoprotein IIIa in patients with chronic immune thrombocytopenic purpura. *Blood* 77:2207–2213.

48. Miller, F.W., K.A. White, T. Biswas, and P.H. Plotz. 1990. The role of an autoantigen, histidyl-tRNA synthetase, in the induction and maintenance of autoimmunity. *Proc. Natl. Acad. Sci. USA* 87:9933–9937.

49. Boire, G., and J. Craft. 1990. Human Ro ribonucleoprotein particles: Characterization of native structure and stable association with the La polypeptide. *J. Clin. Invest.* 85:1182–1190.

51. Nugent, D.J., T.J. Kijnicki, C. Berglund, and I.D. Bernstein. 1987. A human monoclonal autoantibody recognizes a neoantigen on glycoprotein nIa expressed on stored and activated platelets. *Blood* 70:16–22.

52. Von dem Borne, A.E. 1986. Antibodies against cryptantigens of platelets. In Platelet Serology, Decary F., and Rock, G. (eds.) S. Karger, Basel, pp. 33–36.

53. Kunicki, T.J., K. Furihata, D. Nugent, R. Kekomaki, and J.P. Scott. 1990. A human monoclonal autoantibody specific for human platelet glycoprotein IIb (integrin αIIb) heavy chain. *Hum. Antibod. Hybridomas* 1:83–95.

54. Kelton, J.G., C.J. Carter, C. Roger, G. Bebenek J. Gauldie, D. Shenidan, Y.B. Kassam, W.F. Kean, W.W. Buchanan, and P.J. Rooney. 1984. The relationship among platelet-associated IgG, platelet lifespan, and reticuloendothelial cell function. *Blood* 63:1434–1444.

55. Kay, M.M. 1983. Appearance of a terminal differentiation antigen on senescent damaged cells and its implications for physiologic autoantibodies. *Biomembranes* 11:119–126.

56. Khansari, N., and H.H. Fudenberg. 1983. Immune elimination of autologous aging platelets by monocytes: Requirement of membrane-specific autoantibody. *Eur. J. Immunol.* 13:990–994.

57. Matthes, T., A. Wolff, P. Soubiran, F. Gres, and G. Dighiero. 1988. Antitubulin antibodies. Natural autoantibodies and induced antibodies recognize different epitopes on tubulin molecule. *J. Immunol.* 141:3135–3145.

58. Dighiero, G., A. Kanshik, P. Poncet, and X.R. Ge. 1987. Origin and significance of autoantibodies. *Concepts Immunopathol.* 4:42–76.

59. Sinha, T.K., P. Horsewood, and J.G. Koltan. 1991. Nonimmune and immune binding of IgG to platelets in an animal model of immune thrombocytopeina. *Blood* 78:344a.

60. Kekomaki, R., B. Dawson, J. McFarland, and T.J. Kunicki. 1991. Localization of human platelet autoantigens to the cysteine-rich region of glycoprotein IIIa. *J. Clin. Invest.* 88:847–854.

61. Bowditch, R.D., P. Tani, K.C. Fong, and R. McMillan. 1996. Characterization of autoantigenic epitopes on platelet glycoprotein IIb/IIIa using random peptide libraries. *Blood* 88:4579–84.

62. Escher, R.J., S. Vogel, S.M. Miescher, M. Stadler, and P. Berchtold. 1999. Recombinant anti-GpIIb/IIIa autoantibodies inhibit binding of fibrinogen to GPIIb/IIIa. *Blood* 94:450a

63. Gevorkian, G., K. Manoutcharian, J.C. Almagro, T. Govezensky, and V. Dominguez. 1998. Identification of autoimmune thrombocytopenic purpura-related epitopes using phage-display peptide library. *Clin. Immunol. Immunopathol.* 86:305–309.

64. Kunicki, T.J., E.F. Plow, and R. Kekomaki. 1991. Human monoclonal antoantibody 2E7 is specific for a peptide sequence of platelet glycoprotein IIb. Localization of the epitope to IIb231–238 with an Immunodominant TrP235. *J. Autoimmun.* 4:415–431.

65. Kunicki, T.J., D.S. Annis, J. Gorski, and D.J. Nugent. 1991. Nucleotide sequence of the human autoantibody 2E7 specific for the platelet integrin IIb heavy chain. *J. Autoimm.* 4:433–446.

66. Shadle, P.J., M.H. Ginsberg, E.F. Plow, and S.H. Barondes. 1984. Platelet-collagen adhesion: Inhibition by a monoclonal antibody that binds glycoprotein IIb. *J. Cell Biol.* 99:2056–2061.

67. Onder, O., A. Weinstein, and L.W. Hoyer. 1980. Pseudothrombocytopenia caused by platelet agglutinins that are reactive in blood anticoagulated with chelating agents. *Blood* 56:177–182.

68. Pegels, J.G., E.C.E. Bruynes, C.P. Engelfriet, and A.E. von dem Borne. 1982. Pseudothrombocytopenia: An immunologic study on platelet antibodies dependent on EDTA. *Blood* 59:157–161.

69. van Vliet, H.H., M.C. Kappers-Klunne, and J. Abels. 1996. Pseudothrombocytopenia: a cold antibody against platelet glycoprotein GPIIb. *Br. J. Haematol.* 62:501–511.

70. Deckmyn, H., S.L. Chew O, and J. Vermulen. 1990. Lack of platelet response to collagen associated with an autoantibody against glycoprotein Ia. *Thrombo. Haemost.* 64:74–79.

71. Lynch, D.M., and S.E. Howe. 1986. Antigenic determinants in idiopathic thrombocytopenic purpura. *Br. J. Haematol.* 63:301–308.

72. Sugiyama, T., M., Okuma, F. Ushikubi, S. Sensaki, K. Kanaji, and H. Uchino. 1987. A novel platelet aggregating factor found in a patient with defective collagen-induced platelet aggregation and autoimmune thrombocytopenia. *Blood* 69:1712–1720.

73. Pfueller, S.L., R. David B.G. Firkin, R.A. Bilston, W.F. Cortizo, and G. Raines. 1990. Platelet aggregating IgG antibody to platelet surface glycoproteins associated with thrombosis and thrombocytopenia. *Br. J. Haematol.* 74:336–341.

74. Honda, S., T. Tsubakio, Y. Tomiyama, H. Take, T. Furubayashi, H. Mitzutami, and Y. Kurata. 1990. Two human monoclonal antiplatelet autoantibodies established from patients with chronic idiopathic thrombocytopenic purpura. *Br. J. Haematol.* 75:245–249.

75. Varon, D., S. Linder, E. Gembom, L. Guedg, H. Langbeheu, A. Berrebi, and Z. Eshhar. 1990. Human monoclonal antibody derived from an autoimmune thrombocytopenic purpura patient, recognizing an intermediate filament's determinant common to vimentin and desmin *Clin. Immunol. Immunopath.* 54:454–468.

76. Barque, J.-P., and A. Kamiguian. 1990. Human autoantibodies identify a nuclear chromatin-associated antigen (PSL or p55) in human platelets. *European J. Cell. Biol.* 51:183–187.

77. Tomiyama, Y., and R. Kekomaki. 1991. Antivinculin antibodies in sera of patients with immune thrombocytopenia and in sera of normal subjects. *Blood* 79:161–168.

78. Pfueller, S.L., D. Logan, T.T. Tran, and R.A. Bilston. 1990. Naturally occurring IgG antibodies to intracellular and cytoskeletal components of human platelets. *Clin. Exp. Immunol.* 79:367–373.

79. Nugent, D.J. 1989. Human monoclonal antibodies in the characterization of platelet antigens. In: Platelet Immunobiology: Molecular and Clinical Aspects. T.J. Kunicki, and J.N. George (eds). Lippincott, Philadelphia, pp. 273–290.

80. Beardsley, D.J.S., and J. Ho. 1989. Varicella-associated thrombocytopenia: Antibodies against an 85-kDa thrombin sensitive protein (GPV). *Blood* 66 (suppl 1):1030.

81. Berchtold, P., and R. McMillan. 1989. Autoantibodies against platelet membrane glycoproteins in children with acute and chronic immune thrombocytopenic purpura. *Blood* 74:1600–1602.

82. Shapiro, S.S., and P. Thiagarajan. 1982. Lupus anticoagulants. *Prog. Hemost. Thromb.* 6:263–269.

83. van Vliet, H.H.D.M., M.C. Kappers-Klunne, J.W.B. van der Hel, and J. Abels. 1987. Antibodies against glycosphingolipids in sera of patients with idiopathic thrombocytopenic purpura. *Br. J. Haematol.* 67:103–108.

84. Deckmyn, H., S.L., Chew, and J. Vermylen. 1990. Lack of platelet response to collagen associated with an autoantibody against glycoprotein Ia: a novel cause of acquired qualitative platelet dysfunction. *Thromb. Haemost.* 64:74–79.

85. Dromigny, A., P. Triadou, P. Lesavre, M.C. Morel-Kopp, and C. Kaplan. 1996. Lack of platelet response to collagen associated with autoantibodies against glycoprotein (GP) Ia/IIa and Ib/IX leading to the discovery of SLE. *Hematol. Cell Ther.* 38:355–357.

86. Murakave, H., Z. Lam, B.C. Furie, V.N. Reinhold, T. Asano, and B. Furie. 1991. Sulfated glycolipids are the platelet autoantigens for human platelet-binding monoclonal anti-DNA autoantibodies. *J. Biol. Chem.* 266:15414–15419.

87. Cho, C.S., X. Wang, Y. Zhao, D.A. Carson, and P.P. Chen. 1997. Genotyping by PCR-ELISA of a complex polymorphic region that contains one to four copies of six highly homologous human VH3 genes. *Proc. Assoc. Am. Physicians* 109:558–564.

88. Olee, T., P.M. Yang, K.A. Siminovitch, N.J. Olsen, J. Hillson, J. Wu, F. Kozin, D.A. Carson, and P.P. Chen. 1991. Molecular basis of an autoantibody-associated restriction fragment length polymorphism that confers susceptibility to autoimmune diseases. *J. Clin. Invest.* 88:193–203.

89. Mo, L., S.J. Leu, C. Berry, F. Liu, T. Olee, X.Y. Yang, D.S. Beardsley, R. McMillan, V.L. Woods, Jr. and P.P. 1996. The frequency of homozygous deletion of a developmentally regulated Vh gene (Humhv3005) is increased in patients with chronic idiopathic thrombocytopenic purpura. *Autoimmunity* 24:257–263.

90. Huang, D.F., K.A. Siminovitch, X.Y. Liu, T. Olee, N.J. Olsen, C. Berry, D.A. Carson, and P.P. Chen. 1995. Population and family studies of three disease-related polymorphic genes in systemic lupus erythematosus. *J. Clin. Invest.* 95:1766–1772.

91. Lampmann, G.W., B.B. Furie, B. Furie, R.C. Schwartz, and B.D. Stollar. 1989. Amino Acid sequence of a platelet-binding human anti-DNA monoclonal autoantibody. *Blood* 74:262–269.

92. Asano, T., B. Furie, and B.B. Furie. 1985. Platelet binding properties of monoclonal lupus autoantibodies produced by human hybridomas. *Blood* 66:1254–1260.

93. Hiraiwa, A., D.J. Nugent, and E. Milner. 1990. Sequence analysis of monoclonal antibodies derived from a patient with idiopathic thrombocytopenic purpura. *Autoimmunity* 8:107–113.

94. Nugent, D., and T. Walker. 1991. Nucleotide sequence of human monoclonal antiplatelet autoantibody 5E5: its relation to antibodies derived from patients with SLE and ITP. *Throm. Hemast.* 65:1062.

95. Hoch, S., and J. Schwaber. 1987. Identification and sequence of the VH gene elements encoding a human anti-DNA antibody. *J. Immunol.* 139:1689–1693.

96. Newkirk, M.M., R.A. Mageed, R. Jefferies, P.P. Chen, and J.A. Capra. 1987. Complete amino acid sequences of variable regions of two human IgM rheumatoid factors, BOR and KAS of the Wa idiotypic family, reveal restricted use of heavy and light chain variable and joining region gene segments. *J. Exp. Med.* 166:550–564.

97. Chen, P.P., M.F. Liu, S. Sinha, and D.A. Carson. 1988. A 16/6 idiotype-positive anti-DNA antibody is encoded by a

conserved VH gene with no somatic mutation. *Arthritis Rheum.* 31:1429–1431.

98. Cairns, E., P.C. Kwong, V. Misener, P. Ip, D.A. Bell, and K.A. Siminovitch. 1989. Analysis of variable region genes encoding a human anti-DNA antibody of normal origin. *J. Immunol.* 143:685–691.

99. Zouali, M., and B.D. Stollar. 1988. Origin and diversification of anti-DNA antibodies. *Immunol. Rev.* 105:137–159.

100. Cairns, E., J. Block, and D.A. Bell. 1984. Anti-DNA autoantibody-producing hybridomas of normal human lymphoid cell origin. *J. Clin. Invest.* 74:880–887.

101. Burton, D.R. 1990. Antibody: the flexible adaptor molecule. *Trends Biochem. Sci.* 15:64–69.

102. Siminovitch, K.A., V. Misener, P.C. Kwong, O.L. Song, P.P. Chen. 1990. A natural autoantibody is encoded by gennline heavy and lambda light chain variable region genes without somatic mutation. *J. Clin. Invest.* 84:1675–1678.

103. Dersimonian, H., R.S. Schwartz, K.J. Barrett, and B.D. Stollar. 1987. Relationship of human variable region heavy chain germline genes to genes encoding anti-DNA autoantibodies. *J. Immunol.* 139:2496–2501.

104. Schroeder, H.W., H.L. Hillson, and R. Perlmutter. 1987. Early restriction of the human antibody repertoire. *Science* 283:791–793.

105. Isenberg, D.A., C. Dudeney, W. Williams, I. Addison, S. Charles, J. Clarke, and A. Rodd-Pokropek. 1987. Measurement of antiDNA antibodies: a reappraisal using five different methods. *Ann. Rheum. Dis.* 46:448–456.

106. Atkinson, P.M., G.W. Lampmann, B.C. Furie, Y. Naparstek, R.S. Schwatz, B.D. Stollar, and B. Furie. 1985. Homology of the NH2-tertninal amino acid sequences of the heavy and light chains of human monoclonal lupus autoantibodies containing the dominant 16/6 idiotype. *J. Clin. Invest.* 75:1138–1143.

107. Eilat, D., M. Hochberg, J. Pumphrey, and S. Rudikoff. 1984. Monoclonal antibodies to DNA and RNA from NZB/NZW F1 mice: Antigenic specificities and NH2 terminal amino acid sequences. *J. Immunol.* 133:489–494.

108. Diamond, B., and M.A. Scharff. 1984. Somatic mutation of the T15 heavy chain gives rise to an antibody with autoantibody specificity. *Proc. Natl. Acad. Sci.* USA 81:5841–5844.

109. Sollazzo, M., D. Castiglia, M. Zanetti, R. Billetta, and A. Tramontano. 1990. Structural definition by antibody engineering of an idiotypic determinant. *Protein. Eng.* 3:531–539.

110. Deckmyn, H., J. Zhang, E. Van Houtte, and J. Vermylen. 1994. Production and nucleotide sequence of an inhibitory human IgM autoantibody directed against platelet glycoprotein Ia/IIa. *Blood* 84:1968–1974.

111. Logtenberg, T., F.M. Young, J.H. Van Es, F.H. Gmelig-Meyling, and F.W. Alt. 1989. Autoantibodies encoded by the most JH-proximal human immunoglobulin heavy chain variable region gene. *J. Exp. Med.* 170:1347–1355.

112. Desiderio, S.V., G.D. Yancopoulos, M. Paskind, E. Thomas, M.A. Boss, and N. Landau. 1988. Insertion of N regions into heavy-chain genes is correlated with expression of terminal deoxytransferase in B cells. *Nature* 311:752–755.

113. Olee, T., J. En, C.J. Lai, L. Mo, C.S. Cho, X. Wei, X.E. Wang, V.L. Woods, Jr., and P. Chen. 1997. Generation and analysis of an IgG anti-platelet autoantibody reveals unusual molecular features. *Br. J. Haematol.* 96:836–845.

114. Haba, S., M.B. Lasconbe, R.J. Poljak, and A. Nisonoff. 1989. Structure of idiotopes associated with antiphenylar-

sonate antibodies expressing an intrastrain crossreactive idiotype. *J. Exp. Med.* 170:1075–1090.

115. Rose, D.R., R.K. Strong, M.N. Margolies, M.I. Gefter, and G.A. Petsko. 1990. The crystal structure of the antigen-binding fragment of the murine anti-arsonate mono-clonal antibody 36–71 at 2.9–A resolution. *Proc. Natl. Acad. Sci. USA* 87:338–342.

116. Victor-Kobrin, C., Z.T. Barak, F.A. Bonilla, B. Kobrin, I. Sanz, D. French, J. Rothe, and C. Bona. 1990. A molecular and structural analysis of the VH and VK regions of monoclonal antibodies bearing the A 48 regulatory idiotype. *J. Immunol.* 144:614–624.

117. Williams, S., G. Creadon, T. Kunicki, and D.J. Nugent. 1999. A protein expression system for evaluating occurring autoantibody fragments. *Molecular Bio. Cell.* 10:147a.

118. Imbach, P., S. Barandun, V. d'Apuzzo, C. Baumgartner, A. Hirt, A. Morell, E. Rossi, M. Schoni, M. Vest, and A.P. Wagner. 1981. High-dose intravemous gammaglobulin for idiopathic thrombocytopenic purpura in childhood. *Lancet* 1:1228–1230.

119. Berchtold, P., G.L. Dale, P. Tani, and R. McMillan. 1989. Inhibition of autoantibody binding to platelet glycoprotein IIb/IIIa by anti-idiotypic antibodies in intravenous gamma-globulin. *Blood* 74:2414–2417.

120. Tsubakio, T., Y. Kurata, S. Katagini, Y. Kanakuna, T. Tamaki, J. Kuyama, Y. Kanayama, T. Yonezawa, and S. Tarni. 1983. Alteration of T cell subsets and immunoglobulin synthesis in vitro during high-dose gammaglobulin therapy in patients with idiopathic thrombocytopenic purpura. *Clin. Exp. Immunol.* 53:697–702.

121. Ware, R.E., and T.A. Howard. 1994. Elevated numbers of gamma-delta (gamma delta+) T lymphocytes in children with immune thrombocytopenic purpura. *J. Clin. Immunol.* 14:237–247.

122. Ware, R.E., and T.A. Howard. 1993. Phenotypic and clonal analysis of T lymphocytes in childhood immune thrombocytopenic purpura. *Blood* 82:2137–2142.

123. Kuwana, M., J. Kaburaki, and Y. Ikeda. 1998. Autoreactive T cells to platelet GPIIb-IIIa in immune thrombocytopenic purpura. Role in production of anti-platelet autoantibody. *J. Clin. Invest.* 102:1393–1402.

124. Filion, M.C., C. Proulx, A.J. Bradley, D.V. Devine, R.P. Sékaly, F. Décary, and P. Chartrand. 1996. Presence in peripheral blood of healthy individuals of autoreactive T cells to a membrane antigen present on bone marrow-derived cells. *Blood* 88:2144–2150.

125. Mouthon, L., S. Kaveri, and M. Kazatchkine. 1994. Immune modulating effects of intravenous immunoglobulin (IVIg) in autoimmune diseases. *Transfus. Sci.* 15:393–408.

126. Lacroix-Desmazes, S., L. Mouthon, S.H. Spalter, S. Kaveri, and M.D. Kazatchkine. 1996. Immunoglobulins and the regulation of autoimmunity through the immune network. *Clin. Exp. Rheumatol. Suppl.* 15:S9–15.

127. Mehta, Y.S., and S.S. Badakere. 1996. In-vitro inhibition of antiplatelet autoantibodies by intravenous immunoglobulins and Rh immunoglobulins. *J. Postgrad. Med.* 42:46–49.

128. Semple, J.W., Y. Milev, D. Cosgrave, M. Mody, A. Hornstein, V. Blanchette, and J. Freedman. 1996. Differences in serum cytokine levels in acute and chronic autoimmune thrombocytopenic purpura: relationship to platelet phenotype and antiplatelet T-cell reactivity. *Blood* 87:4245–4254.

129. Nugent, D., M. Berman, K. Imfeld, V. Dadufalza, and C. Sandborg. 1998. Depressed IL-4 levels in children with

acuate and chronic immune thrombocytopenia(ITP). *FASEB J.* 12:A609.

130. Garcia-Suarez, J., A. Prieto, E. Reyes, L. Manzano, K. Arribalzaga, and M. Alvarez-Mon. 1995. Abnormal γIFN and αTNF secretion in purified CD2+ cells from autoimmune thrombocytopenic purpura (ATP) patients: their implication in the clinical course of the disease. *Am. J. Hematol.* 49:271–276.

131. Andersson, J. 1998. Cytokines in idiopathic thrombocytopenic purpura (ITP). *Acta. Paediatr. Suppl.* 424:61–64.

132. Dianzani, U., M. Bragardo, D. DiFranco, C. Alliaudi, P. Scagui, V. Redonglia, S. Bonissone, A. Correra, I. Dianzani, and U. Ramenghi. 1997. Deficiency of the Fas Apoptosis Pathway without Fas Gene Mutations in Pediatric Patients with Autoimmunity/Lymphoproliferation. *Blood* 89:2871–2879.

133. Bussel, J., N. Heddle, C. Richards, and M. Woloski. 1999. MCP-1, IL-10, IL-6 and TNFα levels in patients with ITP before and after in anti-D and IVIG treatments. 94:15a.

134. Zimmerman, S.A., F.J. Malinoski, and R.E. Ware. 1998. Immunologic effects of anti-D (WinRho-SD) in children with immune thrombocytopenic purpura. *Am. J. Hematol.* 57:131–138.

135. Crossley, A.R., A.M. Dickinson, S.J. Proctor, and J.E. Calvert. 1996. Effects of interferon-alpha therapy on immune parameters in immune thrombocytopenic purpura. *Autoimmunity* 242:81–100.

136. Lazarus, A.H., T. Joy, and A.R. Crow. 1998. Analysis of transmembrane signaling and T cell defects associated with idiopathic thrombocytopenic purpura (ITP) *Acta. Paediatr. Suppl.* 424:21–25.

137. Erduran, E., Y. Aslan, Y. Aliyazicioglu, H. Mocan, and Y. Gedik. 1998. Plasma soluble interleukin-2 receptor levels in patients with idiopathic thrombocytopenic purpura. *Am. J. Hematol.* 57:119–123.

138. George, J., G. Raskob, J. Bussel, E. Cobos, D. Green, J. Tongol, C. Rutherford, J. Wasser, H. Croft, S. Rhinehart, B. Oates, J. Scaramucci, and K. Nadeau. 1999. Safety and effect on platelet count of repeated doses of monoclonal antibody to CD40 ligand in patients with chronic ITP. *Blood* 94:19a.

139. Bussel, J., M. Wissert, B. Oates, J. Scaramucci, K. Nadeau, and B. Adelman. 1999. Humanized monoclonal anti-Cd40 ligand antibody (Hu5c8) rescue therapy of 15 adults with severe chronic refractory ITP. *Blood* 94:646a.

140. He, R., D.M. Reid, C.E. Jones and N.R. Shulman. 1995. Extra-cellular epitopes of platelet glycoprotein Ibα reactive with serum antibodies from patients with chronic idiopathic thrombocytopenic purpura. *Blood* 86:3789–3796.

32 | Experimental Model of Immune Thrombocytopenic Purpura

Susumu Ikehara, Hajime Mizutani, Yasushi Adachi, and Yoshiyuki Kurata

1. INTRODUCTION

Immune (autoimmune) thrombocytopenic purpura (ITP) is a syndrome characterized by persistant thrombocytopenia, which is caused by autoantibodies binding to platelet membrane antigens and mediating the destruction of platelets in the reticuloendothelial system, particularly the spleen and liver. It has been generally believed that the triad of ITP are thrombocytopenia, normal or increased megakariocytosis in the bone marrow, and decreased platelet lifespan (PLS). Although a great deal of progress has been made in understanding the nature of platelet-specific autoantigens, the pathogenicity of antibodies (Abs) and their roles in the etiology of ITP remain unclear. The discovery of a spontaneous animal model for ITP has, therefore, been long awaited. We have found the $(NZW \times BXSB)F_1$ (W/B F_1) mouse to be such a model (1), although the W/B F_1 mouse is not an animal model for pure ITP, since the mice develop SLE and other autoimmune diseases. However, this mouse is a useful animal model for analyzing the etiopathogenesis of ITP.

2. GENERAL FEATURES OF THE $(NZW \times BXSB)F_1$ MOUSE

W/B F_1 mice result from the mating of female NZW mice, which have the latent SLE predisposition, with male BXSB mice, which have the Y-chromosome-linked autoimmune-accelerating factor. Male W/B F_1 mice have been found to develop early onset autoimmune diseases such as SLE and Sjögren's syndrome. This is because the Y-chromosome-linked autoimmune-accelerating factor of the BXSB strain is fully expressed in the male hybrids (2).

The male mice show proteinuria from the age of 8 weeks, and 50% die either of renal failure due to lupus nephritis or of myocardial infarction due to anti-phospholipid Ab syndrome by the age of 5 months (3,4). The mice also show hypertension (> 160 mmHg), autoimmune hepatitis, and Sjögren's syndrome-like lesions in the salivary glands. Serological abnormalities, such as anti-DNA Abs, circulating immune complexes (CICs) and anticardiolipin Abs, are also found (2–4).

3. THE $(NZW \times BXSB)$ F_1 MOUSE MODEL OF ITP

Male W/B F_1 mice develop thrombocytopenia with age, although bone marrow megakaryocyte counts increase (1). A significant decrease in platelet counts is observed in mice with age (Table 1). In contrast, the parental (NZW and BXSB) and female W/B F_1 (3 months) mice show normal platelet counts, although the platelet counts are higher in these mice than in humans. Platelet-associated antibodies (PAAs) are positive in 2-month or older W/B F_1 mice; percent-positive platelets increase with age (Fig. 1 and Table 1). As shown in Table 1, the PLS of BALB/c mice is 2.11 days. Male W/B F_1 mice more than 2 months of age show a significantly shorter PLS than those at the age of 1 month. Platelet-bindable serum antibodies (PBAs) are also found in male W/B F_1 mice; the PBAs belong to IgG- and IgM-classes, but not IgA (5). These findings strongly suggest that antiplatelet Abs are responsible for the development of thrombocytopenia.

To determine whether the reduced PLS is due to abnormalities in the male W/B F_1 platelets themselves or to plasma factors, platelet transfer experiments between male W/B F_1

Table 1 Platelet lifespan (PLS), platelet-associated antibodies (PAAs), and platelet counts in W/B F$_1$ mice

Mice	Age (months)	Sex	Platelet counts ($\times 10^{-3}/\mu$l)	PAAs[a] (%)	PLS[b] (days)
BALB/c	6	M	1,255	1.6	2.11
NZW	6	M	1,005	1.6	1.62
BXSB	4	F	962	1.9	1.57
W/B F$_1$	3	F	1,012	3.0	1.44
W/B F$_1$	1	M	1,063	1.4	1.20
W/B F$_1$	2	M	1,012	10.1	0.56
W/B F$_1$	3	M	842	18.9	0.25
W/B F$_1$	4	M	612	15.9	0.09
W/B F$_1$	5	M	502	25.5	0.12
W/B F$_1$	6	M	491	49.9	0.11

[a]Platelets were labeled with FTTC-conjugated goat-mouse Ig and analyzed using a FACScan.
[b]Platelet lifespan was examined by transferring ^{III}In-labeled platelets from donor mice to age-matched mice of the same strain.

mice and either female W/B F$_1$ mice or normal BALB/c mice were performed. As shown in Table 2, normal BALB/c (6 months) and female W/B F$_1$ (3 months) platelets transferred to male W/B F$_1$ (6 months) mice were immediately cleared from the circulation. Platelets transferred from male W/B F$_1$ mice (6 months) to BALB/c mice (6 months) were also cleared from the circulation immediately, as were platelets transferred from male W/B F$_1$ mice (6 months) to female W/B F$_1$ mice (3 months) (5).

Several causes of thrombocytopenia in patients with ITP have been proposed. One is the presence of antiplatelet Abs, which results in platelet destruction by sequestration by the RES. It has been reported that the Abs in humans belong mainly to the IgG class. In W/B F$_1$ mice, IgG and IgM (but not IgA) Abs have been found in both the plasma and on the platelets. As W/B F$_1$ mice show high levels of CICs from the age of 2.5 months, it is conceivable that CICs are involved in the development of thrombocytopenia, since it is known that they affect platelets by activating complement or platelet-aggregating factor. However, murine platelets have no Fc receptors (6), and sera from BXSB and MRL/lpr mice, which show high CIC levels, did not bind to the platelets of BALB/c mice. Therefore, it seems unlikely that CICs play an important role in the development of thrombocytopenia.

4. EFFECTS OF SPLENECTOMY ON ITP

Splenectomy is probably the most effective therapy for ITP, resulting in remission in up to 80% of patients, the rate being higher than that achieved through corticosteroid therapy (7,8). However, for unknown reasons, in 5 to 20% of patients, splenectomy does not lead to an increase in platelet counts, and other types of therapy need to be developed for such refractory cases (9–11). The effects of splenectomy on platelet kinetics and production of antiplatelet Abs have been studied to address this issues.

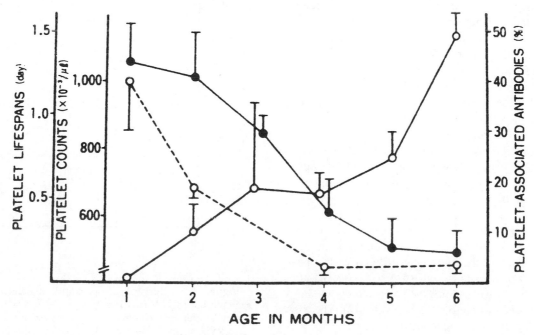

Figure 1. Correlations between increased PAA values (○—-○) and either reduced PLSs (○- - -○) or decreased platelet counts (●——●) in male W/B F1 mice. The data represent mean ± SD of three or four mice (5).

Table 2   Platelet transfer experiments between male W/B F₁, and either female W/B F₁ or normal BALB/c mice

Donor mice				Recipient mice				
Mice	Sex	Age (months)	PAAs (%)	Mice	Sex	Age (months)	PAAs (%)	PLS (days)
BALB/c	M	6	1.6	BALB/c	M	6	1.6	2.11
BALB/c	M	6	1.7	W/B F₁	M	6	44.6	0.15
W/B F₁	M	6	41.2	BALB/c	M	6	1.6	0.11
W/B F₁	F	3	1.9	W/B F₁	M	6	47.0	0.21
W/B F₁	M	6	39.0	W/B F₁	F	3	3.0	0.12

Uptake of ¹¹¹In-labeled platelets by major organs was examined 3 days after injection, by which time labeled platelets had been cleared almost completely from the peripheral blood. The spleen and liver of BALB/c and W/B F₁ mice were the main accumulation sites of labeled platelets. Splenic uptake was significantly greater in male W/B F₁ mice than in BALB/c mice (12). One week after splenectomy, the main accumulation site of labeled platelets was the liver, and no significant uptake of ¹¹¹In by the lungs or bone marrow was observed in either splenectomized or non-splenectomized W/B F₁ mice.

The effects of splenectomy on platelet counts were next examined using 3-month-old W/B F₁ (male) and normal BALB/c mice. As shown in Fig. 2, splenectomy led to a marked increase in platelet counts in both BALB/c mice and W/B F₁ mice 1 week after splenectomy. Two weeks after splenectomy, the increase in the platelet counts of W/B F₁ mice became statistically significant compared to controls. Three weeks after splenectomy, a gradual decrease in platelet counts was observed in both groups. However, high platelet counts were maintained in the splenectomized mice for up to 6 weeks (12).

The PLS of each mouse before and after splenectomy is shown in Table 3. The PLS of male BALB/c (3 months) or female W/B F₁ (3 months) mice is 2.11 and 1.44 days, respectively, whereas the PLS of male W/B F₁ (3 months) mice was significantly shortened (0.56 days). One week after splenectomy, a significant prolongation of the PLS was observed in contrast to the controls; the values were also significantly greater than those before splenectomy, although they did not reach completely normal levels.

One week after splenectomy, a slight decrease in PAA values was observed (in contrast to the controls), although there was no statistically significant difference. The peak channel of fluorescence intensity was reduced after splenectomy, but, 2 weeks after the splenectomy, the PAA values had returned to pre-operative levels (12). The PBA values were not affected by splenectomy.

The limited effect of splenectomy may be explained by the production of autoantibodies in other sites. Slightly decreased PAA values were noted 1 week after splenectomy, but these started to rise again 2 weeks post-operatively. A gradual decrease in platelet counts was also observed 3 weeks post-operatively. These findings suggest that, after splenectomy, Ab production is replaced by other sites (the bone marrow and LNs). Similar findings have been reported in the NZB mouse, which is a well-known model of autoimmune hemolytic anemia (AIHA) (13). Since we have demonstrated that, after splenectomy, the main site of labeled platelet accumulation is the liver, splenic reticulo-endothelial phagocytic functions may be compensated for by systemic RESs (probably by hepatic phagocytic cells) (12).

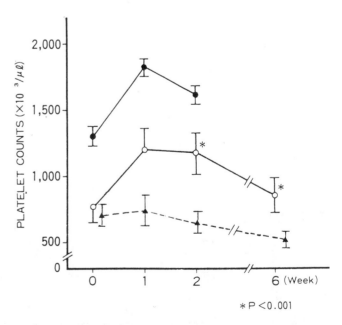

Figure 2. Effect of splenectomy on platelet counts in 3-month-old male W/B F1 mice (○——○, n = 10) 0, 1, 2 and 6 weeks after splenectomy; the platelet counts of non-manipulated W/B F1 (male) mice are shown (▲- - - -▲). Platelet counts of BALB/c mice are also shown (●——●, n = 5). Bars represent mean ± SD (12).

Table 3   Effects of splenectomy on platelet lifespan in W/B F₁ and normal mice

Mice	Age (months)	Sex	Splenectomy	PLS (days)[a]	
				Before splenectomy	After splenectomy[b]
W/B F₁	3	M	+	0.56	1.10
W/B F₁	3	M	−	0.66	0.33
W/B F₁	3	F	−	1.44	ND[c]
BALB/c	3	M	+	2.11	3.02

[a]Platelet lifespan (PLS) was determined by transferring ¹¹¹In-labeled platelets from donor mice to age-matched mice of the same strain.
[b]PLS was determined 1 week after splenectomy.
[c]Not determined.

5.   EFFECTS OF PREDNISOLONE ON ITP

Several mechanisms have been described for the action of corticosteroids in ITP, including reduction of autoantibody production and suppression of reticulo-endothelial phagocytic functions (14–18). More recent kinetic studies suggest that corticosteroids improve platelet counts by increasing bone marrow platelet production (19). In humans, Gernsheimer et al. have reported that the main mechanism of prednisolone in ITP is the correction of impaired platelet production (19). Platelet production is lower in patients with ITP than in normal subjects, but increases after prednisolone treatment. However, in W/B F₁ mice, it has been found that platelet production increases before treatment, indicating that production is not impaired (20). To determine the mechanism by which platelet counts increase after corticosteroid therapy for human ITP, platelet kinetics using prednisolone-treated W/B F₁ mice were studied.

An increase in platelet counts was observed in W/B F₁ mice 4 weeks after treatment with prednisolone (2 mg/kg/d); no increase occurred in nontreated W/B F₁ mice. The prolonged PLS was observed in treated W/B F₁ mice, but not in nontreated controls. No increase in platelet production (platelet turnover) was found in prednisolone-treated W/B F₁ mice, but significant decreases in PAAs and PBAs were noted. Studies on organ localization of radiolabeled platelets showed that hepatic uptake significantly decreased in the treated W/B F₁ mice, but not in nontreated W/B F₁ mice. To elucidate the effect of prednisolone on the reticulo-endothelial phagocytic activity in W/B F₁ mice, in vivo clearance of IgG-sensitized, ⁵¹Cr-labeled autologous erythrocytes was examined. ⁵¹Cr-labeled RBCs coated with IgG antibody were injected into 3-month-old W/B F₁ mice, and the fate of the cells then determined (Fig. 3). Nontreated W/B F₁ mice (n = 4) showed a rapid clearance of these cells, whereas W/B F₁ mice treated with prednisolone

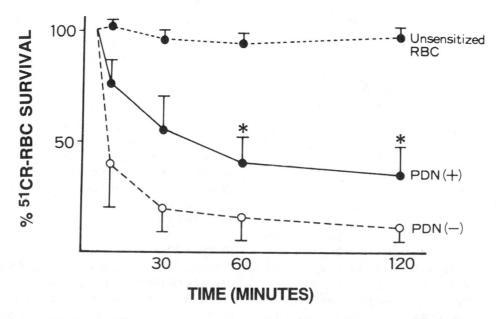

Figure 3.   Clearance of IgG-coated ⁵¹Cr-labeled erythrocytes in W/B F1 mice pretreated with PDN (——, n = 4) and controls (- - -, n = 4). Data points represent the mean values ± SD (20). *$p < 0.05$.

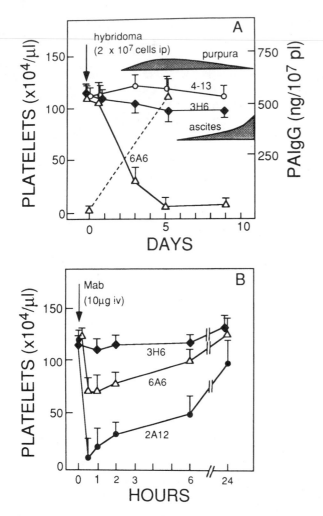

Figure 4. (A) The pathogenicity of hybridomas was evaluated by the IP injection of 2×10^7 hybridoma cells into nu/nu mice (n = 3/clone). Injection of cells of some platelet-reactive clones (4–13 shown; ○) or platelet-unreactive clones (3H6 shown; ◆) did not alter the platelet counts of nu/nu mice (—-). In contrast, reduced platelet counts (△——△), elevated PAIgG values (△- - - -△), and purpura and ascites (shaded areas) developed within 10 days after injecting cells of clone 6A6. (B) The pathogenicity of the mAbs was further evaluated by the IV injection of 10 mg of purified ascites Ig into nu/nu mice (n = 3/clone). Injection of pathogenic mAbs (6A6 [△] or 2A12 [●]) significantly reduced platelet counts, whereas injection of nonpathogenic, platelet-reactive mAbs (data not shown) or platelet-unreactive mAbs (3H6 shown; ◆) did not alter platelet counts (24).

for 4 weeks (n = 4) showed a much lower clearance rate. At 2 hours, 35% of the IgG-coated RBCs remained in the circulation in the treated group compared with 11% of controls. ^{51}Cr-labeled, non-sensitized RBCs were not removed from the circulation (20).

Evaluating the effect of prednisolone on the FcγR number of splenic or hepatic macrophages required analysis of the macrophage FcγR expression using the rat

monoclonal antibody (2.4G2) that specifically binds to FcγR II on mouse macrophages. Although the expression of FcγR II on hepatic or splenic macrophages from prednisolone-treated mice decreased slightly, no statistically significant change was found by FACS analyses, suggesting that prednisolone has little effect on the number of FcγRs (20). These results and the findings of splenectomy suggest that prednisolone improves platelet counts not only by suppressing systemic reticulo-endothelial phagocytic functions but also by reducing Ab production.

6. CHARACTERIZATION OF MONOCLONAL ANTIPLATELET AUTOANTIBODIES

Target antigens in human platelets recognized by antiplatelet Abs obtained from patients with ITP have been shown to be platelet surface membrane glycoprotein GPIb, GPIIb, and GPIIIa (21–23). Using the spleen cells from W/B F_1 mice, hybridomas secreting antiplatelet mAbs were developed. Some recognized epitopes similar to those recognized by platelet-associated IgG, since eluted IgG from platelets of W/B F_1 mice with ITP inhibited platelet binding by mAbs in a competitive microenzyme-linked immunosorbent assay. When hybridoma cells or purified IgG were injected into BALB/c nu/nu mice, they showed purpura, acute thrombocytopenia, and elevated PAIgG levels (Fig 4). Marked megakaryocytosis was evident in the bone marrow and spleen. Immunoblotting analyses revealed that antiplatelet mAbs bound to a 100-kd platelet protein (24). Further studies with these mAbs will contribute to an understanding of the etio-pathogenesis of ITP.

7. EFFECTS OF BONE MARROW TRANSPLANTATION ON ITP

It has previously been demonstrated that BMT can be used to prevent and treat both organ-specific and systemic autoimmune diseases in (NZB × NZW)F1, BXSB, MRL/lpr, NOD, FGS and KK-Ay mice (25–31).

These findings prompted attempts to prevent and treat both autoimmune thrombocytopenic purpura and other autoimmune characteristics in W/B F_1 mice. W/B F_1 mice (3 to 5 months old) were lethally irradiated and reconstituted with T cell-depleted bone marrow cells of BALB/c mice, and the recipients sacrificed 5.5 months later. As shown in Table 4, the levels of both CICs and anti-DNA Abs were reduced to those of young (1.5 months) W/B F_1 mice. Decreased T cell functions were also normalized after BMT. Platelet counts in BALB/c mice were normalized after BMT, and PBAs were reduced (Table 4). Glomerular damage was also repaired and the survival rate

Table 4 Correction of abnormal serological findings by allogeneic bone marrow transplantation (BMT)

Mice (Sex)	Age[a]	CICs (OD$_{492}$)	Anti-DNA Abs (OD$_{492}$)		PBAs
			Anti-ssDNA	Anti-dsDNA	
BALB/c (♀)	1.5	0.019	0.022	0.003	35.3
W/B F$_1$ (♂)	1.5	0.110	0.267	0.053	27.5
	5.5	0.477	1.390	0.400	54.3
(BALB/c → W/BF$_1$(♂))	8.5	0.083	0.484	0.080	35.6
	9.5	0.118	0.147	0.049	43.6
	10.5	0.133	0.084	0.034	34.8

[a]W/B F$_1$ mice were lethally (9.5 Gy) irradiated and then reconstituted with 1×10^7 T cell-depleted bone marrow cells of BALB/c mice. The mice were sacrificed 5 months after BMT.

was greatly improved (3). These findings strongly suggest that BMT is the ultimate tool in the treatment of autoimmune diseases, including ITP.

ACKNOWLEDGMENT

These experiments were carried out in collaboration with researchers who appear in the references of this paper. I would like to express my deep appreciation to them.

We would like to thank Mr. Hilary Eastwick-Field and Ms. K. Ando for their expert help in the preparation of the manuscript.

References

1. Oyaizu, N., R. Yasumizu, M. Miyama-Inaba, S. Nomura, H. Yoshida, S. Miyawaki, Y. Shibata, S. Mitsuoka, K. Yasunaga, S. Morii, R.A. Good, and S. Ikehara. 1988. (NZW × BXSB)F1 mouse. A new animal model of idiopathic thrombocytopenic purpura. *J. Exp. Med.* 167:2017–2022.
2. Hang, L.M., S. Izui, and F.J. Dixon. 1981. (NZW × BXSB)F1 hybrid. A model of acute lupus and coronary vascular disease with myocardial infarction. *J. Exp. Med.* 154:216–221.
3. Adachi, Y., M. Inaba, Y. Amoh, H. Yoshifusa, Y. Nakamura, H. Suzuka, S. Akamatu, S. Nakai, H. Haruna, M. Adachi, H. Genba, and S. Ikehara. 1995. Effect of bone marrow transplantation of anti-phospholipid antibody syndrome in murine lupus mice. *Immunobiol.* 192:218–230.
4. Hang, L.M., P.M. Stephen-Larson, J.P. Henry, and F.J. Dixon. 1984. Transfer of renovascular hypertension and coronary heart disease by lymphoid cells from SLE-prone mice. *Am. J. Pathol.* 115:42–46.
5. Mizutani, H., T. Furubayashi, A. Kuriu, H. Take, Y. Tomiyama, H. Yoshida, Y. Nakamura, M. Inaba, Y. Kurata, T. Yonezawa, S. Tarui, and S. Ikehara. 1990. Analyses of thrombocytopenia in idiopathic thrombocytopenic purpura-prone mice by platelet transfer experiments between (NZW × BXSB)F1 and normal mice. *Blood* 75:1809–1812.
6. Pfueller, S.L. 1985. Immunology of the platelet surface. Platelet Membrane Glycoproteins. New York: Plenum Press, pp. 327–355.
7. DiFino, S.M., N.A. Lachant, J.J. Kirshner, and A.J. Gottlieb. 1980. Adult idiopathic thrombocytopenic purpura. Clinical findings and response to therapy. *Am. J. Med.* 69:430–442.
8. Pizzuto, J., and R. Ambritz. 1984. Therapeutic experience of 934 adults with idiopathic thrombocytopenic purpura: multicentric trial of the cooperative Latin American group on hemostasis and thrombosis. *Blood* 64:1179–1183.
9. Ahn, Y.S., W.J. Harrington, S.R. Simon, R. Mylvaganam, L.M. Pall, and A.G. So. 1983. Danazol for the treatment of idiopathic thrombocytopenic purpura. *N. Engl. J. Med.* 308:1396–1399.
10. Ahn, Y.S., W.J. Harrington, R. Myivaganam, L.M. Allen, and L.M. Pall. 1984. Slow infusion of vinca alkaloids in the treatment of idiopathic thrombocytopenic purpura. *Ann. Intern. Med.* 10:192–196.
11. Kelsey, P.R., K.P. Schofield, and C.G. Geary. 1985. Refractory idiopathic thrombocytopenic purpura (ITP) treated with cyclosporine. *Br. J. Haematol.* 60:197–198.
12. Mizutani, H., T. Furubayashi, H. Kashiwagi, S. Honda, H. Take, Y. Kurata, T. Yonezawa, S. Tarui, and S. Ikehara. 1992. Effects of splenectomy on immune thrombocytopenic purpura in (NZW × BXSB)F$_1$ mice: analyses of platelet kinetics and anti-platelet antibody production. *Thromb. Haemost.* 67:563–566.
13. Holmes, M.C., and M. Burnet. 1963. The influence of splenectomy in NZB mice. *Austral. J. Exp. Biol.* 41:449–455.
14. Kaplan, M.E., and J.H. Jandl. 1961. Inhibition of red cell sequestration by cortisone. *J. Exp. Med.* 114:921–937.
15. Atkinson, J.P., A.D. Schreiber, and M.M. Frank. 1973. Effect of corticosteroids and splenectomy on the immune clearance and destruction of erythrocytes. *J. Clin. Invest.* 52:1509.
16. Friedman, D., F. Nettl, and A.D. Schreiber. 1985. Effect of estradiol and steroid analogues on the clearance of immunoglobulin G-coated erythrocytes. *J. Clin. Invest.* 75:162–167.
17. Rinehart, J.J., A.L. Sagone, S.P. Balcerzak, G.A. Ackerman, and A.F. LoBuglio. 1975. Effect of corticosteroid therapy on human monocyte function. *N. Engl. J. Med.* 292:236–241.
18. Handin, R.I., and T.P. Stossel. 1978. Effect of corticosteroid therapy on the phagocytosis of antibody-coated platelets by human leukocytes. *Blood* 51:771–779.
19. Gernsheimer, T., J. Stratton, P.J. Ballem, and S.J. Slichter. 1989. Mechanisms of response to treatment in autoimmune thrombocytopenic purpura. *N. Engl. J. Med.* 320:974–980.

20. Mizutani, H., T. Furubayashi, Y. Imai, H. Kashiwagi, S. Honda, H. Take, Y. Kurata, T. Yonezawa, S. Tarui, and S. Ikehara. 1992. Mechanism of corticosteroid action in immune thrombocytopenic purpura (ITP): experimental studies using ITP-prone mice. (NZW × BXSB) F1. *Blood* 79:94–947.

21. Kunicki, T.J., and P.J. Newman. 1992. The molecular immunology of human platelet proteins. *Blood* 80:1386–1404.

22. Woods, V.L., Y. Kurata, R.R. Montgomery, P. Tani, D. Mason, E.H. Oh, and R. McMillan. 1984. Autoantibodies against platelet glycoprotein Ib in patients with chronic immune thrombocytopenic purpura. *Blood* 64:156–160.

23. Woods, V.L., E.H. Oh, D. Mason, and R. McMillan. 1984. Autoantibodies against the platelet glycoprotein IIb/IIIa complex in patients with chronic ITP. *Blood* 63:368–375

24. Mizutani, H., R.W. Engelman, Y. Kurata, S. Ikehara, and R.A. Good. 1993. Development and characterization of monoclonal antiplatelet autoantibodies from autoimmune thrombocytopenic purpura-prone (NZW × BXSB)F1 mice. *Blood* 82:837–844, 1993.

25. Ikehara, S., R.A. Good, T. Nakamura, K. Sekita, S. Inoue, Maung Maung Oo, E. Muso, K. Ogawa, and Y. Hamashima. 1985. Rationale for bone marrow transplantation in the treatment of autoimmune diseases. *Proc. Natl. Acad. Sci. U.S.A.* 82:2483–2487.

26. Ikehara, S., H. Ohtuki, R.A. Good, H. Asamoto, T. Nakamura, K. Sekita, E. Muso, Y. Tochino, T. Ida, H. Kuzuya, H. Imura, and Y. Hamashima. 1985. Prevention of type I diabetes in nonobese diabetic mice by allogeneic bone marrow transplantation. *Proc. Natl. Acad. Sci. USA* 82:7743–7747.

27. Yasumizu, R., K. Sugiura, H. Iwai, M. Inaba, S. Makino, T. Ida, H. Imura, Y. Hamashima, R.A. Good, and A. Ikehara. 1987. Treatment of type I diabetes mellitus in non-obese diabetes mice by transplantation of allogeneic bone marrow and pancreatic tissue. *Proc. Natl. Acad. Sci. USA* 84:6555–6557.

28. Ikehara, S., R. Yasumizu, M. Inaba, S. Izui, K. Hayakawa, K. Sekita, J. Toki, K. Sugiura, H. Iwai, T. Nakamura, E. Muso, Y. Hamashima, and R.A. Good. 1989. Long-term observations of autoimmune-prone mice treated for autoimmune disease by allogeneic bone marrow transplantation. *Proc. Natl. Acad. Sci. USA* 86:3306–3310.

29. Ikehara, S., M. Inaba, H. Ishida, H. Ogata, H. Hisha, R. Yasumizu, N. Oyaizu, K. Sugiura, J. Toki, F. Takao, Soe Than, M. Kawamura, N. Nishioka, N. Nagata, and R.A. Good. 1991. Rationale for transplantation of both allogeneic bone marrow and stromal cells in the treatment of autoimmune disease. New Strategies in Bone Marrow Transplantation, UCLA Symposia on Molecular and Cellular Biology, New York: New Series, Wiley-Liss, 137:251–257.

30. Soe Than, H. Ishida, M. Inaba, Y. Fukuba, Y. Seino, M. Adachi, H. Imura, and S. Ikehara. 1992. Bone marrow transplantation as a strategy for treatment of non-insulin-dependent diabetes mellitus in KK-Ay mice. *J. Exp. Med.* 176:1233–1238.

31. Nishimura, M., J. Toki, K. Sugiura, F. Hashimoto, T. Tomita, H. Fujishima, Y. Hiramatsu, N. Nishioka, N. Nagata, Y. Takahashi, and S. Ikehara. 1994. Focal segmental glomerular sclerosis, a type of intractable chronic glomerulonephritis, is a stem cell disorder. *J. Exp. Med.* 179:1053–1058.

32. Mizutani, H., R.W. Engelman, K. Kinjoh, Y. Kurata, S. Ikehara, and R.A. Good. 1993. Prevention and induction of occlusive coronary vascular disease in autoimmune (W/B)F1 mice by haploidentical bone marrow transplantation: possible role for anticardiolipin autoantibodies. *Blood* 82:3091–3097.

33 | Autoimmune Neutropenia

Henrik J. Ditzel, MD, PhD

1. INTRODUCTION

Neutrophilic granulocytes, or simply neutrophils, are phagocytic cells that are an important part of the human body's non-specific defense system. The cells are easily recognized in blood smears by their characteristic multilobed nuclei and the fine stippling throughout the cytoplasmic compartment, representing primary and secondary granules containing distinct sets of proinflammatory mediators. If the number of neutrophils in the body are significantly reduced or their function impaired, the individual becomes susceptible to repeated bacterial or fungal infections. Such neutrophil reduction, resulting in neutropenia, may be caused by an autoimmune mechanism, as detailed in this chapter, but may also be due to a variety of other disease conditions, including infections, malignant disease and congenital disorders. Many aspects of the disease mechanism of autoimmune neutropenia are poorly understood, although it seems clear that the antibodies, either as specific antibodies directed against surface molecules on neutrophils or their precursors, or as part of immune complex, are intimately involved in the peripheral destruction of neutrophils. Currently, a increasing effort is being undertaken to develop more standardized techniques for evaluation of autoimmune neutropenia. In this review the classification of autoimmune neutropenia, the underlying disease autoimmune mechanisms and the current therapeutic possibilities will be considered.

2. CLASSIFICATION OF AUTOIMMUNE NEUTROPENIAS

Autoimmune neutropenia is a relatively rare disease compared to other hematologic autoimmune disorders, such as autoimmune thombocytopenia and autoimmune anemia. The disorder can be divided into primary or secondary autoimmune neutropenia, depending on whether neutropenia is the only disorder or is part of another systemic autoimmune disease. Table 1 lists the different disease conditions wherein autoimmune neutropenia can be observed. Primary and secondary autoimmune neutropenia has been

Table 1 Disorders where autoimmune neutropenia is observed

Primary Autoimmune Neutropenia
Neutropenia as the only condition
   Autoimmune neutropenia of infancy
   Chronic autoimmune neutropenia
   Drug-induced autoimmune neutropenia
Secondary Autoimmune Neutropenia
Neutropenia as part of an autoimmune disease
   Systemic lupus erythematosus (SLE)
   Felty's syndrome
   Sjögren's syndrome
   Scleroderma
   Polymyalgia rhematica
   Mixed connective tissue syndrome
   Evans syndrome
   Autoimmune thrombocytopenia
   Large granular lymphocytosis (LGL)
   Human immunodefiency virus disease
   Hodgkin's disease or Non-Hodgkin's Lymphoma
   Leukemia
Non-Autoimmune Neutropenia
   Alloimmune neutropenia
   Cancers
   Cyclical neutropenia
   Infections: hepatitis virus, HIV-1
   Drug-induced toxic neutropenia

reported to occur with prevalences of 1:100,000 and 1:35,000, respectively (1,2). Large population surveys have shown that primary autoimmune neutropenia is predominantly found in children aged 1/2–4 years, whereas secondary autoimmune neutropenia is more frequent in adults aged 40–60 years (3). While children with primary autoimmune neutropenia have a relatively benign disease course with spontaneous remission, adults with secondary autoimmune neutropenia generally have a more chronic severe disease path. Secondary autoimmune neutropenia is observed in approximately half of all patients with systemic lupus erythematosus (SLE), and in all patients with Felty's syndrome, an extra-articular manifestation of rheumatoid arthritis involving neutropenia with/without splenomegaly (4–6). Autoimmune neutropenia has also been observed associated with other autoimmune diseases including Sjögren's syndrome, Evans syndrome, polymyalgia rheumatica, mixed connective tissue syndrome and scleromyxoedema, although the latter three are extremely rare. Patients with lymphomas and leukemias may also experience neutropenia induced by an autoimmune mechanism, and should be distinguished from those caused by cancer cell expansion in the marrow or by chemotherapy. Autoimmune neutropenia is also observed in some patients with large granular lymphocytosis (LGL), a syndrome recently shown to be closely related to Felty's syndrome. In addition to rheumatoid arthritis and neutropenia, elevated numbers of large granular lymphocytes characterizes LGL. (7,8). LGL may, although rare, evolve into leukemia (8).

Primary autoimmune neutropenia is primarily observed in young children and is, therefore, often referred to as autoimmune neutropenia of infancy. However, primary autoimmune neutropenia may also occur later in life, and in these patients the disease is usually more chronic. Primary autoimmune neutropenia is distinguished from idiopathic neutropenia by the presence of neutrophil-associated antibodies. However, since the sensitivity or specificity of the current assays for detecting anti-neutrophil antibodies are less than optimal, the two types of neutropenias may not always be distinguishable. Drug-induced neutropenias may or may not be autoimmune-related. Alternatively, drug-associated neutropenia may result from a toxic reaction caused by a chemotherapeutic or immunosuppressive agent, or as an idiosyncratic reaction of a otherwise non-toxic agent. Autoimmune neutropenia can also be observed in about half of all patients undergoing allogeneic or autologous bone marrow transplantation. The anti-neutrophil antibodies observed in allografted patients are primarily directed against antigens of the marrow donor type (probably alloantigen) but are, by definition, autoantibodies.

Alloimmune-induced neutropenia in newborns resulting from maternal sensitization by fetal neutrophils during gestation and subsequent transfer of maternal antibodies against the newborn's neutrophils across the placenta should be distinguished from autoimmune neutropenia, although many of the known targets on neutrophils of allo- and auto-antibodies have been found to be the same. A more detailed description of the different conditions resulting in autoimmune neutropenia will follow.

3.   CLINICAL SYMPTOMS AND PATIENT EVALUATION

Human neutrophils have a fast turnover, with a half-life of approximately 8 hours. To maintain the neutrophil pool, 6×10^{10} new neutrophils are produced in the bone marrow every 24 hours. The normal neutrophil count in the circulation of about $1.8–7 \times 10^9$/L, representing less than 5% of neutrophils in the human body. Under normal physiologic conditions, a stable equilibrium exists between production and peripheral utilization. When the production of neutrophils in bone marrow is suppressed or outpaced by utilization in the periphery, the number of neutrophils in the circulation decreases and results in neutropenia. Generally, mild neutropenia is defined as neutrophil count of $1.0–1.5 \times 10^9$/L, moderate neutropenia as $0.5–1.0 \times 10^9$/L, and severe neutropenia as $<0.5 \times 10^9$/L.

Although blood neutrophil counts give no information on neutrophils turnover and deposition in tissue, a reverse correlation between the neutrophil count in patients with severe neutropenia and incidences of serious bacterial infections has been observed. In contrast, the risk of infection in patients with mild-to-moderate neutropenia is difficult to predict. The function of the remaining neutrophils, the duration of the neutropenia and other host defense parameters influence the infection risk. Recent studies have found that the serum concentration of soluble CD16 is directly proportional to the total-body neutrophil mass (9,10), and a new assay based on soluble CD16 has been developed to predict the infection risk (11). Other studies have shown that neutropenic patients with a rapid turnover of neutrophils have a better chance of avoiding serious infection.

Assessment of the susceptibility to infection of patients with moderate to severe autoimmune neutropenia includes a detailed medical history focusing of the frequency of infections. These should include infection of the respiratory and gastrointestinal tracts, oropharyngeal and cutaneous infections, and fevers without identified focus. The medical history should also include drug history and current medications to evaluate possible drug-induced neutropenia. Family history of autoimmune diseases and frequent infections should also be obtained. Laboratory analysis of patients with low neutrophil counts should include full analysis of hematological cell counts, including hematocrit, hemoglobin, reticulocytes, platelet, and white cell differentiation count to assess whether general cytopenia or other hematologic abnormalities exist. Further, bone marrow aspiration and biopsies are indicated to determine precursor frequency. Measurement of the marrow neutrophil reserve by use of endotoxin, etiocholanolone or glucorticosteroids may be included later. The serum should be tested for serum-reactive protein, anti-nuclear antibodies, and rheumatoid factor. Since neutropenia can be observed during viral infections, such as hepatitis, HIV-1, parvovirus and malaria, it would be useful to test for these viruses. The serum should be tested for anti-neutrophil antibodies and the neutrophils for bound immunoglobulin, as detailed below.

The type of bacterial infections observed in autoimmune neutropenia depends on the severity of the neutropenia and the age of the patient. The most common infections in neutropenic patients include mucosal and cutanous inflammations, such as gingivitis, cellulitis, furunculosis, abscesses, pneumonia and septicemia. As consequence of the lack of neutrophils, the inflammation lacks the normal clinical signs of fluctuance, induration and exudate. Severe neutropenia, as sometimes seen in Felty's syndrome, may result in fulminant bacterial infections including deep-tissue infection of the lungs, liver or sinuses, and is the leading cause of death in such patients.

4. MECHANISM OF AUTOIMMUNE NEUTROPHIL DESTRUCTION

Neutropenia may be caused by different autoimmune mechanisms, though peripheral destruction of neutrophils appears to be the dominant one (Fig. 1). Marrow aspirates from patients with accelerated peripheral destruction of neutrophils usually show adequate or accelerated neutrophil maturation. In other instances the bone marrow, although producing sufficient neutrophil precursors, may show failure to develop beyond the myelocyte/metamyelocyte stage, perhaps as a result of antibodies targeting an antigen expressed on more mature neutrophil precursor cells or due to other inhibitors that suppress the myelopoiesis. Indeed, one study reported that phagocytosis of granulocytes by bone marrow macrophages was observed in a few patients, presumably confirming that sensitization of neutrophils occurs in the bone marrow.

Proposed mechanisms for the peripheral destruction of neutrophils includes: 1) antigen-specific antibodies targeting neutrophils or 2) neutrophil precursors, or 3) deposition of immune complexes on neutrophils. T cell-mediated suppression of bone marrow myelopoiesis has also been proposed as a cause of autoimmune neutropenia (12–14).

Although there is strong evidence that antibodies play a major role in the peripheral neutrophils sequestration and reduced survival, the precise mechanism remains unclear, partially because the targets of anti-neutrophil autoantibodies are, in contrast to the red cell and platelet system, not fully defined.

4.1. Methods for Detection of Anti-neutrophil Antibodies

Neutrophils are a difficult cell type to work with in vitro due to problems with clumping, poor viability and rapid development of abnormal morphology. Furthermore, the presence of the low affinity Fc receptors (FcRγII and FcRγRIIIb) and complement receptors on the neutrophil surface can lead to epitope-nonspecific binding of immune complexes that may confuse results interpretation. Numerous assays have been described to detect anti-neutrophil antibodies, but, in contrast to techniques for demonstrating antibodies to red cells, thrombocytes and lymphocytes, the methods for demonstrating neutrophil antibodies are not yet standardized. Methods for demonstration of anti-neutrophil antibodies include the granulocyte agglutination test (GAT) (15), neutrophil opsonization

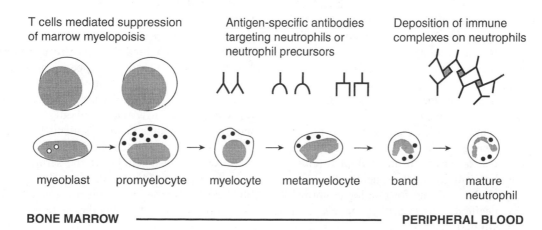

Figure 1.   Immune mechanism involved in autoimmune neutropenia targeting the myelopoisis at different stages.

(15), quantitative anti-globulin consumption (16), radio-labeled anti-globulin binding (17), staphylococcal protein A binding (18), indirect neutrophil immunofluorescence (GIFT) using fluorescence microscopy (19), or flow cytometry (20), and monoclonal antibody immobilization of granulocyte antigens (MAIGA) (21). Most of these methods, however, have limited reproducibility, makes quantification difficult and are time-consuming. Flow cytometry is probably currently the most useful method, offering the potential for rapid evaluation of large numbers of cells with small volumes of serum in a more quantitative and objective manner (20,22). The ability to gate specifically on the granulocyte population makes highly purified preparations of cells unnecessary. Further, from the histogram it may be possible to distinguish anti-neutrophil antibody binding from immune complex binding using this method. Additionally, both IgM and IgG antibodies may be separately, but simultaneously, measured. As flow cytometry becomes more readily available to most hematological laboratories, this seems to be test of choice. GIFT, in combination with the GAT or MAIGA, was found to be the best means of anti-neutrophil antibody detection at the Second International Granulocyte Serology Workshop (23). Most tests, especially the functional tests such as GAT, require freshly-purified (less than 4 hours) neutrophils. Some investigators have found it useful to use paraformaldehyde-fixed neutrophils in the GIFT whereas other obtained better result with the freshly isolated neutrophils. Since neutrophils get very easily activated resulting in release of granule enzyme and oxygen radical, no reliable procedure for freeze storage is available, making the testing for anti-neutrophil antibodies on neutrophils laborious. Recently, CHO cells transfected with the genes encoding for the three allotypic variants (NA1, NA2, SH) of FcγRIIIb were generated and shown by MAIGA to be useful for typing antisera directed against these epitopes (24).

When analyzing for antibodies and immune complexes in autoimmune neutropenia, two reservoirs should be studied, serum and immunoglobulin bound to the few circulating neutrophils (neutrophil-bound immunoglobulin). For many purposes, it is important to distinguish between neutrophil-specific antibodies and immune complex binding to neutrophils. Two tests, the C1q-binding test, based on precipitation of immune complex-fixed C1q by PEG (25), and an acid or ether elution test, have been developed to attempt analysis of the conformation of the neutrophil-bound immunoglobulin.

The acid elution test is based on an attempt to elute bound immunoglobulin from the involved neutrophils. The eluate can subsequently be analyzed for binding to normal neutrophils and for specificity using purified antigens. In a model system, it was found that while neutrophil-specific antibodies could be eluted from neutrophils by low pH,

immune complexes made by heat-aggregation of normal IgG could not (26). The difference was related to the avidity of the molecules interaction with the neutrophils surface; while the anti-neutrophil antibody/antigen interaction was mostly mono or divalent, the immune complexes bound multivalently to a large numbers of Fc and complement receptors. Additionally, it was shown that only larger immune complexes, not smaller bivalent immune complexes (generated by mixing two different anti-tetanus-toxoid human monoclonal antibodies with tetanus toxoid in a molar ratio of 1:1:2), could be recognized in the C1q binding assay, while both were correctly detected in the elution assay. However, the elution process was not very efficient since only a fraction of the antibodies could be eluted and reconstituted in normal state. Further, an antibody to FcRIII was not eluted at all.

4.2.   Neutrophil Cell Surface Binding Antibodies in Different Diseases

The term anti-neutrophil antibodies are sometimes used to describe antibodies binding to either the cell surface, cytoplasm or nucleus of neutrophils. Here, only antibodies binding to the neutrophils surface will be considered. Such antibodies have been detected in the sera of patients with primary autoimmune neutropenia, SLE, Felty's syndrome, Sjögren's syndrome, Evans syndrome, drug reactions and lymphoproliferative disorders, but have also been detected in autoimmune thyroiditis, idiopathic thrombocytopenic purpura, human immunodeficiency virus (HIV) infection, Coomb's positive hemolytic anemia, and primary biliary cirrhosis (3,5,22,27,28). In certain cases, the titers of anti-neutrophil antibodies correlate with the severity of neutropenia, implying an etiologic role for these antibodies (29,30). However, this correlation is lacking in other studies that employed different techniques and patient samples, perhaps reflecting the fact that only a minor part of the anti-neutrophil antibody response is responsible for the neutrophil depletion in these cases.

One problem in detecting antibodies with current flow cytometry methods is that immune sera contain immune complexes that may bind to the Fcγ receptor on the surface of neutrophils (31), and thus appear artefactually to have an antigen specificity usually attributed to the Fab fragment of the autoantibody. By masking the Fc receptor with a monoclonal anti-Fc receptor antibody, reduction of serum binding to neutrophils in some patients has been observed, indicating that both antigen-specific antibodies and immune complexes are responsible for the serum anti-neutrophil antibody response (13). An alternative strategy to differentiate between immune complexes and nomomeric anti-neutrophil specific IgG antibodies is to ultracentrifuge the serum sample before adding it to normal neutrophils for binding studies. By comparing the binding of the supernate and resuspended sediment from the ultracentrifuged serum with non-ultracentrifuged serum, the amount of immune complexes may be determined (32).

The value of testing for anti-neutrophil antibodies in a clinical setting is controversial, partially because the sensitivity and specificity of the current assays are limited, and partially because the therapeutic strategies for patients with autoimmune neutropenia and patients with idiopatic neutropenia is often identical. One area where detection of anti-neutrophil antibodies may influence the therapeutic strategy is in patients with infection and neutropenia. Since neutropenia may occur as a result of infection and vise versa, the finding of anti-neutrophil antibodies may support the therapeutic use of immunosuppressive drugs or corticosteroids which otherwise may be contraindicated.

An additional problem in developing better assays for the detection of anti-neutrophil antibodies, is the lack of information about the precise repertoire of neutrophil antigens that serve as targets for the immune response.

4.3. Targets of Surface Binding Anti-neutrophil Antibodies in Autoimmune Neutropenia

There is currently limited information regarding the antigenic specificities of anti-neutrophil autoantibodies, although a few neutrophil-specific antigens have been identified (Table 2). However, most of these antigens have only been identified with whole sera and neither have purified Ig preparations been evaluated nor human monoclonal antibodies generated to confirm the specificity of the antibody responses observed. No neutrophil antigens solely targeted by autoimmune sera, but not alloimmune sera, have been identified so far.

Neutrophil-reactive antibodies recognizing antigens also present on other cell-types have also been described. The most frequent of which are antibodies to HLA, and, to lesser extent, to blood group I and P antigens. These two blood group antigens are, in contrast to ABO, Rh and Le, present on neutrophils.

The first neutrophil-specific antigens recognized by autoantibodies were the so-called neutrophil-associated antigens (NA, NB, etc.) demonstrated in sera of patients

Table 2 Targets of autoimmune anti-neutrophil antibodies

NA1 and NA2 antigens (Fcγ receptor III, CD16b)
NB1 and NB2 (58–64 kda GP)
CD11b/CD18 (LeuCAMb)
CR$_1$ complement receptor (CD35)
Fcg receptor II (CD32)
HLA
Blood group I and P antigens

with primary autoimmune neutropenia of infancy. The NA1, NA2 and recently-described SH antigens are allotypic forms of the Fc gamma receptor IIIb (FcγRIIIb, CD16b), the major Fcγ-receptor of neutrophils (33,34,34). FcγRIIIb, which is a 233 amino acid N-glycosylated protein anchored to the membrane by phosphatidyl-inositol-glycan, also contain the LAN and SAR antigens. Exclusive cross-linking of the FcγRIIIb receptor may induce Ca^{2+} mobilization and respiratory burst of neutrophils. Of the FcγRIIIb epitopes, autoimmune neutropenia sera most frequently recognize NA1. The precise localization of the NA epitopes on FcγRIIIb is not defined, but NA polymorphism relates to a 4 amino acid difference between the two isoforms, mapped to the first membrane-distal domain (35). Two of the 4 variable amino acids, ser (46) and Asn (65), enable additional glycosylation of the NA2 allele.

Another antigen system containing the epitopes NB1 and NB2 has been found on a 58- to 64-kDa PIG-anchored glycoprotein present on the surface of neutrophils and on intracellular membranes of secondary granules and small vesicles (36,37). The function of NB1 is still unknown. Interestingly, the amount of the NB1 epitope varies considerably between individuals, but seems to be evenly expressed on all neutrophils from the same donor.

Autoantibodies against the other Fc receptor found on neutrophils, FcγRII (CD32), are very rare, but have been found in few patients with autoimmune neutropenia (38). Similarly, autoantibodies against the CR$_1$ complement receptor (CD35) has also been described in a few patients with autoimmune neutropenia (38). CR$_1$ that bind C3b with high affinity is a glycoprotein with a molecular weight that varies between 160 and 250 kDa. The protein consists of a large extracellular portion, a transmembrane-spanning domain of 25 residues and a short cytoplasmic tail. The extracellular portion consist of approximately 30 repeated domains arranged in tandem (39). Each of the short consensus repeats (SCR) contains 60–70 amino acids with four conserved cysteines. In addition, a second conserved larger pattern of 450 amino acid repeats exists in CR1. Each of these long homologous repeats (LHR) contains seven SCR. Four allelic variants containing 4–7 LHRs, respectively, exist, corresponding to the various molecular weights (40).

Autoantibodies against the neutrophil adhesion glycoprotein complex CD11b/CD18 (also referred to as leukocyte adhesion molecule b, LeuCAMb) have been identified. Using an immunobead antigen capture assay, anti-CD11b was detected in 7 of 50 autoimmune neutropenia sera (41), but no common clinical diagnosis correlated with these antibodies. Purified IgG preparations from such patients caused a decrease in the adhesion and/or opsonic receptor-mediated function of freshly-isolated neutrophils in vitro, however no further data on the fine specificity of the antibodies has been reported.

As indicated above, most autoimmune-related neutropenia is associated with adequate or accelerated neutrophil maturation in the bone marrow. However, three reports have described patients with neutropenia and myeloid hypoplasia whose serum deposited IgG on myeloid precursors (42–44). In one such study, sera from three patients with selective myeloid hypoplasia bound significantly more IgG to cells of the myeloid precursor cell line HL-60 than control sera, whereas IgG binding to mature granulocytes was not significantly different (42). Another group used a multilineage cell line, K562 and a primitive myelomonocytic leukemia cell line, KG1a and flow cytometry to evaluate sera from 148 patients with suspected autoimmune neutropenia for the presence of antibodies to myeloid precursor cells. IgG of 28 % and 15% of the sera, respectively, reacted with the two cell lines, however, some sera reacted with both cell lines whereas others with only one. Furthermore, some sera also reacted with normal purified neutrophils or bone marrow myeloid progenitor, whereas others did not. No common clinical diagnosis could be correlated with the reactivity pattern. (43).

In an attempt to identify novel autoantigens involved in autoimmune neutropenia, one group used SDS-PAGE separated neutrophils and Western blotting to analysis serum of patients with Felty's syndrome. A group of antigens with molecular weights in the range of 80–84 kDa were recognized by serum of patients with Felty's syndrome and not by patients with other autoimmune diseases or healthy individuals (45). The authors concluded that the anti-neutrophil antibodies recognized an antigen different from the previously-identified surface structures on neutrophils, however, no further elucidation of the nature of these auto-antibody/antigen pairs has been reported. A strategy for such characterization, which also may be applied to other auto-antibody/autoantigen pairs, could involve the generation of human monoclonal antibodies.

One approach for the production of human monoclonal antibodies is to use lymphocytes isolated from bone marrow or PBLs as the source for Epstein-Barr virus (EBV) transformation or fusion. An alternative approach, based on cloning antibody fragments in phage libraries has been developed (46). Such libraries of phage-displaying antibody fragments may be constructed using bone marrow as immune source, which closely reflects the antibodies in serum. The libraries of 10^7–10^8 members can then be sorted by selection on neutrophils and the cloned antibody Fab fragment analyzed for antigen specificity. Such an approach has been successfully used to generate human monoclonal antibodies to known autoantigens such as thyroid peroxidase (47,48), pANCA (49), U1 RNA-associated A protein (50), DNA (51) and nicotinic acetylcholine receptor (52), but also as a mean to identify new autoantigen/antibody pairs, such as elongation factor-1A in Felty's syndrome (53).

4.4. Effector Mechanisms Involved in the Pathogenesis of Anti-neutrophil Antibodies

As with anemia and thrombocytopenia, different effector mechanisms may be involved in antibody-mediated autoimmune neutropenia. Antigen-specific targeting of neutrophils resulting in neutrophils sensitization, and rapid clearance in the spleen by Fc-mediated phagocytosis is one of the mechanisms (54–56). The opsonic activity of antibodies in serum has been quantitatively studied in vitro by the ability of macrophages to engulf sensitized neutrophils (55,57). Other studies have shown that complement is unable to lyse neutrophils, in contrast to erythrocytes (58). However deposition of complement factor C3 on the neutrophils may add to the opsonic activity of antibodies (59,60). Only one study has been reported wherein sera from autoimmune neutropenic patients was tested in an animal model (61), as described below in the Felty's syndrome section, and no animal study has been reported wherein defined human monoclonal anti-neutrophil antibodies were tested. However, in a guinea pig model, administration of a rabbit anti-guinea pig neutrophil antiserum resulted in rapid disappearance of circulating neutrophils (54). Examination of the bone marrow and lymphoid organs demonstrated that neutrophils were engulfed by macrophages, further indicating that phagocytosis of sensitized neutrophils maybe an important mechanism leading to neutropenia.

5.   SELECTED DISORDERS WITH AUTOIMMUNE NEUTROPENIA

5.1.   Primary Autoimmune Neutropenia

Primary autoimmune neutropenia is also referred to as primary autommune neutropenia of infancy, since children 4 years or younger comprise a large proportion of this population. In contrast to newborn children with alloimmune neonatal neutropenia, children with autoimmune neutropenia have normal neuthropil counts prior to the disease onset. In a larger study of 240 children with primary autoimmune neutropenia, two thirds were diagnosed between 5 and 15 months of age (38), although autoimmune neutropenia have been observed as early as 3 weeks after birth (3). Primary autoimmune neutropenia may also be seen in adults, and in them the disease generally have a more chronic course than in the children. The disease seems to occur with equal frequency in males and females. In young children, neutropenia generally has a relatively benign course, and 95% achieve complete spontaneous remission within 7 to 24 months of onset in one study, and 6 to 48 months in another study (38,62). The median duration of disease is 20 months (62). The condition is caused by anti-neutrophil IgG antibodies binding to neutrophil-surface antigens, which are most effectively detected using the GIFT assay as compared to the other anti-neutrophil antibody assays as described above. In one study 235 of 240 sera from patients with primary autoimmune neutropenia were positive in the GIFT assay. The remaining 5 sera were only positive in the GAT assay (38). However, false positive reactions may be frequently observed with the GIFT assay, due to immune complex-binding to Fc receptors. The process that elicit the anti-neutrophils antibody production is not known, however an association between anti-NA$_1$ autoantibodies and HLA-DR2 has been reported (63). Infection with parvovirus B19 has been suggested to tricker primary autoimmune neutropenia (64). IgM antibodies against parvovirus B19 was observed in a high number of individuals with primary autoimmune neutropenia, however further examination of the sera of an extended panel of sera from autoimmune neutropenia patients did nor confirm the association (38).

While the specificity of the major part of anti-neutrophil antibodies in sera of patients with primary neutropenia has not been identified, reactivity has been described for the following neutrophil antigens: FCγRIIIb (27%), CD11b/CD18 (21%), C3b complement receptor (14%) and FcγRII (2%). In other studies, antibodies to NB1 have been described. The bone marrow of patients with autoimmune neutropenia of infancy is generally normo- or hyper-cellular, with a reduced number of metamyelocytes, band or segmented neutrophils. The "arrest state" varies from patient to patient and may reflect the myeloid antigens (early vs. late) targeted by the autoantibodies.

Autoimmune neutropenia of infancy should be differentiated from other neutropenia in children including alloimmune neonatal neutropenia, cyclic neutropenia, drug-induced neutropenia and neonatal neutropenia, including reticular dysgenesis, Schwachman-Diamond syndrome, and Kostmann syndrome. Alloimmune neonatal neutropenia, which can last up to six months, can be excluded by detection of neutrophil alloantibodies in the maternal serum. Cyclic neutropenia, occurring as a sporadically or dominantly inherited autosomal disorder, is characterized by neutropenic periods of 3 to 6 days occurring approximately every 21 days (65). Kostmann congenital agranulocytosis is an autosomal recessive familial neutropenia caused by a plasma factor deficiency (66).

5.2. Felty's Syndrome

Felty's syndrome is an extra-articular manifestation of rheumatoid arthritis involving neutropenia (4). Until recently Felty's syndrome classically also was described to include spenomegaly, however a recent study has shown that this it is not an essential feature of the disorder (67). Generally, moderate to severe neutropenia is observed in these patients. Most have had rheumatoid arthritis symptoms for several years prior to the occurrence of the

neutropenia, but individuals may also initially present with Felty's syndrome.

The cause of autoimmune neutropenia is controversial, and probably involves several different humoral and cellular immune mechanisms (13,61). Felty's syndrome sera, ultracentrafuged to sediment immune complexes and aggregated IgG, has shown reduced binding to neutrophils compared to non-centrifuged sera (32). Whole sera and sediments of ultracentrafuged sera from Felty's syndrome patients also gave positive signal in an C1q binding assay further indicating the presence of immune complexes. However, in other patients with Felty's syndrome, immune complexes did not seem to be responsible for the neutropenia, since no reduction in IgG binding to neutrophils following ultracentrifugation or C1q binding was observed. In these patients, antibodies specific for neutrophils and not immune complexes are responsible for IgG binding to circulating neutrophils in these patients. These results confirm earlier studies wherein different levels of immune complexes and neutrophil-specific IgG antibodies were observed in larger groups of patients with Felty's Syndrome. In that study, pepsin digestion of the IgG antibody fraction of sera from a subgroup of these patients also exhibited significantly greater binding to neutrophils than normal sera (13). In another study, however, pepsin digestion of the IgG antibody fraction of autoimmune sera eliminated antibody binding to neutrophils (68), but some appropriate controls were lacking in this report. The importance of immune complexes and soluble IgG for development of neutropenia has been the subject of in vivo studies. Injection of the immune complex fraction of sera from Felty's syndrome patients into mice resulted in a reduction in neutrophils, whereas the effect of the IgG fraction was minimal (61). However, it is questionable how well this correlates with the human system, since the specificity of the autoantibodies was not determined, and the corresponding antigen may not exist on mouse neutrophils. Further, the interaction between human immune complexes and aggregated material with mouse neutrophils may be very different from the human counterpart.

Nevertheless, soluble immune complexes can lead to complement activation, adherence of neutrophils to other leukocytes, decreased neutrophil phagocytic ability, or decreased neutrophil chemotaxis. Similar to the monomereic IgG anti-neutrophil antibodies, the specificity and common features of the antibodies involved in these immune complexes has not been determined, and may be valuable for the prevention of immune complex formation, and for understanding of how the antibodies are elicited.

Analysis of sera from Felty's syndrome patients has shown that anti-nuclear antibodies (ANA) of as yet undefined specificities are prevalent in 70% of patients, in contrast to 30% of patients with rheumatoid arthritis without Felty's syndrome (4,69). Recently, elongation factor-1A has been proposed as an important target for such antibodies (53). Further, one third of patients with Felty's syndrome also have anti-neutrophil cytoplasmic antibodies (ANCA), whereas ANCA is not observed in rheumatoid arthritis patients lacking concomitant Felty's syndrome (69–71).

A role for T cell suppression on normal bone marrow granulopoiesis resulting in neutropenia in Felty's syndrome has also been proposed (12,13). T cells from peripheral blood, bone marrow and spleen cells of Felty's syndrome patents, in contrast to rheumatoid arthritis patients and normal individuals, have been shown to suppress the colony forming unit in culture of bone marrow (12,13).

5.3. Large Granular Lymphocytosis

Patients with large granular lymphocytosis can be divided into two groups: Those with rheumatoid arthritis and high prevalence of HLA-DR4 seem to be a variant of Felty's syndrome. In contrast, the frequency of DR4 in patients with LGL without rheumatoid arthritis is the same as in the normal population. In addition to rheumatoid arthritis and neutropenia, patients with LGL have increased numbers of large granular lymphocytes in their blood and bone marrow that represent activated cytotoxic T cells (72,73). The role of these activated T cells is unknown, but one hypothesis has been that the cells reflect chronic antigenic stimulation, perhaps due to viral infection or other agents, leading to low-grade malignant transformation. More detailed analysis of the phenotype of the LGL cells have shown that they are CD45RA+, and CD45RO-, usually associated with naive cells and not memory cells (73). Although the T cells are clonal in origin, leukemia seldom develops. In a recent study, 90% of the LGL patients were alive 3.5 years after diagnosis. In addition to neutropenia, anemia, thrombocytopenia and splenomegaly may also be observed in these patients.

5.4   Evans Syndrome

Evans syndrome, described by Evans and Duane in 1949, includes autoimmune haemolytic anaemia, and idiopatic thrombocytopenia (ITP) and/or autoimmune neutropenia (74,75). Experiments where cell-bound antibodies from patients with Evans syndrome were eluting from PBMC have shown that distinct cell-specific antibodies directed against each type of blood cell (red cells, neutrophils and thrombocytes) are the cause of the anemia, thrombocytopenia and neutropenia (76).

5.5. Systemic Lupus Erythematosus

The nature of anti-neutrophil autoantibodies in systemic lupus erythematosus (SLE) and their importance for neutropenia is controversial. Whereas some investigators claim these antibodies are only **epiphenomenon**, since antineutrophil antibodies are also observed in non-neutropenic SLE patients, others have demonstrated specific immunological functions of the antibodies. In one report, it was shown that IgG purified from 5 of 10 patients with SLE and neutropenia suppressed the colony formation by binding to primitive haematopoietic progenitor cells and suppressing their growth. In contrast, monocytes, lymphocytes and complement had no suppressive effect on colony formation (77). In another study, the contributions of soluble immune complexes or cell-specific antibodies to the levels of IgG neutrophil-binding activity in SLE sera was studied (27). Elevated levels of IgG neutrophil-binding activity were found in gel filtration G-200 excluded pools (immune complexes) in 11 of 18 SLE sera, and in 13 of the G-200 IgG pools. In 9 sera, elevated levels were observed in both pools and F(ab')2 fragments of IgG from the SLE sera bound to normal neutrophils, in contrast to normal sera. No correlation between anti-neutrophil antibody level or immune complex level with the neutrophil count could be made.

5.6. Drug-induced Immune Neutropenia

The mechanisms of drug-induced neutropenia are diverse, and partially depending on the therapeutic drugs involved, may include immune or non-immune mechanisms (78). Table 3 lists the most common drugs inducing immune-mediated neutropenia, including antibiotics, antimalaria

Table 3   Drugs that may cause immune-mediated neutropenia

Antibiotics	*Antiinflammatory and analgesics*
Penicillin	Diclofenac
Oxacillin	Propylphenazone
Ampicillin	Dimethylaminophenazone
Amoxycillin	Aminopyrine
Dicloxacillin	Ibuprofen
Nafcillin	
Ceftriaxone	
Cephradine	*Antiarrythmics*
Cefotaxime	Procainamide
Cefuroxime	Aprindine hydrochloride
Sufphafurazole	Flecanide
Sulfathiazole	Quinidine
Vancomycin	
	Others
Antimalaria agents	Phenytoin
Quinine	Levamisole
Chloroquine	Clozapine
Amodiaquine	Olanzapine
	Mercuhydrin
Antithyroids	
propylthiouracil	
metamizole	

drugs, antiinflammatory agents, analgesics, antiarrhythmics, and antithyroids.

Non-immune neutropenia can be caused by direct toxicity of the drug or their metabolites to the bone marrow, which suppresses neutrophil production. In contrast, drug-induced autoimmune neutropenia may be caused by a immune reaction against the neutrophils or their precursors. This immune reaction involves production of autoantibodies that can be directed against neutrophils themselves without requiring coating with the eliciting drug for neutrophil antibody binding. Such antibodies has been described to be elicited by propylthiouracil, levamisole and clozapine (79,80). Although another report suggest that neutropenia in patients treated with clozapine is caused by an other mechanism than antibodies (81). Antibodies may also be elicited by drug-induced altered neutrophil antigens or by drug inhibition of the T suppressor function that normally controls and inhibits autoantibody formation. The anti-neutrophil autoantibodies elicited by the latter mechanism may persist after the drug is withdrawn. More often, however, the antibodies only react with drug-coated neutrophils, as seen in quinine-, propylphenazone- and flecanide-treated patients (82–84), or drug-metabolite-coated neutrophils, as seen in metamizole-, aminopyrine-, procainamide- and diclofenac-treated patients (81,84). Finally, a third group of antibodies only binds to neutrophils when they are incubated concomitant with both serum and drug, and not when antibodies are incubated with the corresponding drug-coated antibodies. These antibodies seem to be directed against the drug and form immune complexes that then bind to the low affinity Fc receptors, FcγRII and FcγRIIIb on neutrophils, resulting in their depletion. Drugs that reportedly elicit antibodies reacting only with neutrophils when serum and drug are added simultaneously include penicillin and cephalosporin (84,85).

5.7. Autoimmune Neutropenia Following Bone Marrow Transplantation

Donor-recipient mismatches involving neutrophil-specific allogens, may similarly to that observed for the ABO and Rh blood group systems, complicate engraftment following allogeneic bone marrow transplantation and haemapoetic stem cell transplantation, resulting in neutropenia (86). Antibodies to circulating neutrophils of donor origin have been observed in approximately 65% of patients following bone marrow engraftment (87). Anti-neutrophil antibodies have also been observed in patients undergoing autologous bone marrow transplantation, although the mechanism of this phenomena has not yet been elucidated. In one study, it was found that while 76% of patients with anti-neutrophil antibodies developed neutropenia, only 6% of those lacking

anti-neutrophil antibodies did so (88). Mismatches involving the neutrophil alloantigens NA1 and NB1 have been suggested as the cause of late graft failure as a result of autoantibody binding (37,89).

The specificity of the involved anti-neutrophils antibodies seems to be important since, in some patients the antibodies have no apparent effect, while in others, they either cause a transient sharp fall in the neutrophil count that was otherwise recovering or delay the post-graft neutrophil recovery (90). In addition, the ability of the bone marrow to compensate for the antibody-mediated neutrophil destruction may influence the clinical outcome. Finally, the underlying malignancy or the chemotherapeutic agents used before the marrow transplantation may also affect the clinical course.

6. TREATMENT OF AUTOIMMUNE NEUTROPENIA

The lack of a clear understanding of the mechanisms underlying the different types of autoimmune neutropenia has resulted in the widespread use of nonspecific therapies. In severe cases, these include high-dose corticosteroid therapy, splenectomy, plasmapheresis, i.v. immunoglobulin gamma-globulin infusion, immunosuppressive therapy and, more recently, granulocyte macrophage colony-stimulating factor (GM-CSF), or granulocyte colony-stimulating factor (G-CSF) treatment.

The introduction of G-CSF therapy has had a great impact on the treatment of autoimmune neutropenia. G-CSF therapy, which, at the end of the 1980's, was shown to effectively normalize neutrophil counts in cancer patients undergoing chemotherapy, has also been shown to increase neutrophil counts in most other types of neutropenias, including the different subtypes of autoimmune neutropenia (91–95). G-CSF is a naturally-occurring hematopoietic growth factor that regulates the neutrophil normal steady-state levels by stimulating neutrophil production within the bone marrow. Endogenous G-CSF is produced by monocytes, fibroblast and endothelial cells, and increased amount of G-CSF in the blood is observed in response to infection or neutropenia. Mice deficient in endogenous G-CSF exhibit chronic neutropenia and impaired neutrophil mobilization (96). Exogenous G-CSF produced by recombinant technology, like endogenous G-CSF, binds to its cell surface receptor and increases the rate of neutrophil production by stimulating neutrophil growth and differentiation of neutrophil precursors (97–99). Increased G-CSF level also reduces the time of maturation in the circulation from 5 days to 1. In addition, G-CSF is functioning as a chemoattractant and enhances antibody-dependent cellular cytotoxicity, adherence, phagocytosis and killing (100). Compared to GM-CSF, M-CSF and multi-CSF, G-

CSF has been found to have minimal effects on production of other hematopoetic cells and was more potent in inducing sustained elevation of blood neutrophils (101). Currently, there are two recombinant G-CSF products on the market: A nonglycosylated form, Filgrastim, produced in *E. coli*, and a glycosylated form, Lenograstim, produced in CHO cells. Similar biological activities have been observed with the two recombinant forms. The recommended dose is 5 μg/kg/day for the initial induction period until satisfactory normalization of the neutrophil count has been achieved. The frequency of injections can then be lowered to 1 to 3 times a week. Following the first administration of G-CSF, an initial transient drop in neutrophils in the circulation may be seen, presumably due to margination, but as early as 3 or 4 h after administration, the initiation of the neutrophil count normalization may be seen. In some cases, however, normalization is first observed after 10 days. In a large study of patients with chronic idiopathic neutropenia, G-CSF treatment increased the neutrophil count to above 1.5×10^9/L in 93% of 75 patients (75). G-CSF has also been found to effectively maintain normal neutrophil levels during long-term treatment, although a gradual decline in neutrophil counts may occur. However, even a slight increase in neutrophil count compared to disease levels may be sufficient to reduce the frequency and severity of infections. In one study, the mean infection rate decreased from 20.4 events during the year prior to G-CSF treatment to 2.4 events during the first year of treatment (75). Interestingly, the anti-neutrophil antibodies in patients with autoimmune neutropenia disappear following G-CSF treatment (94), and this has been hypothesized to be due to absorption of the anti-neutrophil antibodies by the increased number of neutrophils (101). Others have suggested that G-CSF directly down-regulates the autoantigen in addition to the neutrophil-stimulating effect (102). G-CSF has been shown to be effective in cases where corticosteroids and cyclosporine treatments have failed (94,101). No significant side effects have been observed with G-CSF; medullary bone pain is the only consistent complaint (24% of treated patients), and can be treated with either non-narcotic analgesics or, in more severe cases, narcotic analgesics (75,103,104). Myelodysplasia or leukemia following G-CSF treatment has not been observed. G-CSF treatment of patients with secondary autoimmune neutropenia, particularly Felty's syndrome, has resulted in flares of excising arthritis, bullous gangrenosa, vasculitic skin rashes and hypersensitivity vasculitis although they are rare (105–108). The mechanism for the G-CSF mediated flare of arthritis and induction of vasculitis is unknown, however several mechanisms have been suggested, including increased neutrophil response to TNF-α, upregulation of CD11b and FcγRIII (109,110).

Therapy of autoimmune neutropenia should depend on the severity of the symptoms, i.e. the severity and frequency of infections. The overall treatment strategy aims at preventing infections rather than normalizing the neutrophil count. For children with primary autoimmune neutropenia of infancy, which often manifests only moderate neutropenia, treatment focusing on normalizing the neutrophil count is not needed. Though, it can be difficult to predict the frequency of infection from the neutrophil count, since anti-neutrophil antibodies directed against certain neutrophil antigens may also result in qualitative neutrophil defects. For patients with primary autoimmune neutropenia of infancy, treatment with antibiotics has been shown to be sufficient in greater than 90% of the infections. Most receive prophylactic antibiotic treatment (cotrimoxazole) to prolong infection-free periods. Such treatment should be initiated after recurrent infections, which predominantly manifest as skin infection, otitis media and upper respiratory tract infection (Table 4). Cotrimoxazole is well-tolerated and no significant allergic reaction, delayed marrow recovery or drug resistance has been reported. In other instances, however, the repeated use of broad-spectrum antibiotics has given rise to antibiotic-resistant bacterial variants that are increasingly difficult to treat. In children with more severe infections, treatment with recombinant human G-CSF, as detailed above, has been shown to effectively increase neutrophil levels to above 1.5×10^9/L. Although G-CSF therapy is fairy expensive, it has been shown that the reduction in infections during treatment reduced the yearly estimated cost of medical care to one third (111). Corticosteroids and high-dose i.v. IgG has also be used, but are less effective (62,112). IgG treatment generally induce remission, but the remission only last about a week.

Patients with neutropenia as a result of underlying SLE or Sjögren's syndrome usually have active rheumatic disease. Treatment of the underlying disease also improves

Table 4 Infections (260 events) grouped by type in 240 children with primary autoimmune neutropenia at the time of diagnosis. From Bux J. et al., Blood 91:181–186, 1998

Infection	No. of patient	(%)
Upper respiratory infections	49	19
Otitis media	44	17
Pyodermia	32	12
Fever of unknown origin	31	12
Abscess	26	10
Gastroenteritis	25	10
Pneumonia	19	7
Lymphadenitis colli	11	4
Sepsis	8	3
Urogenital infections	8	3
Conjunctivitis	4	2
Phlegmonia	2	1
Meningitis	1	1

the neutrophil count. In contrast, patients with Felty's syndrome, often do not have active arthritis, and more direct treatment of the neutropenia is required, particularly in patients with severe neutropenia. Patients with Felty's syndrome and neutrophil counts of less than 0.5 10^9/L have a significant increase in mortality compared to rheumatoid arthritis patients without Felty's syndrome. Again, non-specific therapies are used due to the lack of clear understanding of the mechanisms underlying autoimmune neutropenia in Felty's syndrome. Therapies used for treating to rheumatoid arthritis, such as D-penicillamine (113), gold salts (114), corticosteroids, cyclophosphamide, methotrexate (115), azathioprine, and cyclosporine (116) may also be useful in treating patients with Felty's syndrome. Treatment aiming at controlling the neutropenia in Felty's syndrome has historically included splenectomy, high-dose corticosteroid therapy, immunosupressive therapy, G-CSF and GM-CSF therapy (105,117), plasmapheresis (118) and i.v. gammaglobulin infusion. However, many of these therapies often yield only transient benefits at the risk of potentially life-threatening side effects (6,119). Currently, G-CSF treatment, particularly in Felty's syndrome patients presenting with acute infection or severe neutropenia, seems to be the treatment of choice. After an increase in neutrophil count is observed, low-dose methotrexate therapy may be added to maintain stable neutrophil levels without continuous administration of G-CSF. Methotrexate therapy may start at 2.5 to 5 mg/week, slowly increasing the dose by 2.5 mg/week until a level of 10 to 12.5 mg/week is reached. Methotrexate should be avoided in patients with impaired renal or hepatic function. Since methotrexate may cause severe marrow suppression, patients undergoing this treatment should be monitored closely. Successful long-term mono-therapy (>1.5 years) with G-CSF in patients with Felty's syndrome has also been reported (120,121). Other patients with Felty's syndrome, however, may not respond to either G-CSF or methotrexate treatment, or the side effects (although rare for G-CSF, as discussed above) may be too severe, necessitating consideration of spenectomy. In a few patients with Felty's syndrome, combination of G-CSF and prednisolone, cyclophosphamide or methrotrexate have resulted in normalization of neutrophil counts, whereas G-CSF alone had no effect.

In larger studies, splenectomy has been shown to result in increased neutrophil counts in about half of the cases but, in many instances, this increased neutrophil count has not been sustained over years. In a few instances, post-splenectomy sepsis has been seen. Corticosteroid has also been widely used for treatment of Felty's syndrome. In one large study, 56% of Felty's syndrome patients showed positive effects of the treatment. Plasmapheresis and i.v. IgG infusion are not considered suitable therapeutic strategies for Felty's syndrome today.

Patients with LGL, rheumatoid arthritis and neutropenia have also responded with increased neutrophil counts following G-CSF therapy (122,123), while no effect was observed in patients with LGL without rheumatoid arthritis (123). Patients with LGL, both with and without rheumatoid arthritis, have also been shown to respond to methrotrexate treatment (124), with five of 10 showing complete clinical remission, and disappearance of the abnormal T cell clone in three of these patients. These complete and partial responses have been successfully maintained on therapy, with a follow-up period ranging from 1.3 to 9.6 years (108).

G-CSF has also been shown to increase neutrophil counts in patients with SLE and neutropenia, and has been shown to do so in one case of scleromyxoedema and neutropenia (125,126). In the nine SLE patients treated with G-CSF, exacerbation of CNS symptoms in two patients and one case of leukocytoclastic vasculitis were observed. In the patient with scleromyxoedema, G-CSF treatment, in addition to the effect on the neutrophil count, also seemed to have a positive effect on the underlying scleromyxoedema, with reduction in skin thickening and facial sclerosis, as verified by tissue biopsies (126).

Successful treatment of autoimmune neutropenia after allogeneic bone marrow transplantation has included G-CSF and corticosteroid therapy (127). In one bone marrow transplant patient, splenectomy successfully resolved the neutropenia after previous therapeutic attempts with i.v. IgG treatment, plasmapheresis and corcosteroid had failed (128).

In conclusion, although major improvements in the therapy of autoimmune neutropenia have been accomplished with the introduction of G-CSF therapy, further understanding of the underlying mechanisms of the peripheral destruction and involvement of human IgG antibodies in this process may allow the design of more specific therapies.

7. CONCLUSIONS

Autoimmune neutropenia is a disorder observed in a mixed group of diseases that renders the patient susceptible to repeated bacterial or fungal infections. The mechanisms for the development of neutropenia are diverse and include both humoral and cellular mechanisms, although peripheral destruction of neutrophils by either neutrophil-specific antibodies or immune complexes seem to the dominating causes. Although some of the neutrophil antigens recognized by autoantibodies of autoimmune neutropenic patients have begun to be identified, most have only been studied using whole serum, and not confirmed with purified Ig preparations or human monoclonal antibodies. Identification of the target molecules and understanding the antibody/antigen interactions are essential, for development of standardized techniques for clinical diagnostic evaluation of autoimmune neutropenia. The introduction of G-CSF in the therapy of autoimmune neutropenia has had a great impact on patient treatment, but our efforts to clarify the disease mechanisms that lead to autoimmune neutropenia must be pursued.

References

1. Lyall, E.G., G.F. Lucas, and O.B. Eden. 1992. Autoimmune neutropenia of infancy. *J. Clin. Pathol.* 45:431–434.

2. Beatty, P.A. and D.F. Stroncek. 1992. Autoimmune neutropenia in Sheboygan County, Wisconsin. *J. Lab. Clin. Med.* 119:718–723.

3. Bux, J., K. Kissell, K. Nowak, U. Spengel, and C. Mueller-Eckhardt. 1991. Autoimmune neutropenia: clinical and laboratory studies in 143 patients. *Ann. Hematol.* 63:249–252.

4. Goldberg, J. and R.S. Pinals. 1980. Felty's syndrome. *Semin. Arthritis Rheum.* 10:52–65.

5. Shastri, K.A. and G.L. Logue. 1993. Autoimmune neutropenia. *Blood* 81:1984–1995.

6. Bux, J. and C. Mueller-Eckhardt. 1992. Autoimmune neutropenia. *Semin. Hematol.* 29:45–53.

7. Chan, W.C., S. Link, A. Mawle, I. Check, R.K. Brynes, and E.F. Winton. 1986. Heterogeneity of large granular lymphocyte proliferations: delineation of two major subtypes. *Blood* 68:1142–1153.

8. Chan, W.C., E.F. Winton, and T.A. Waldmann. 1986. Lymphocytosis of large granular lymphocytes. *Arch.I ntern. Med.* 146:1201–1203.

9. Huzinga, T.W.J., M. de Haas, J.H. Nuijens, D. Roos, and A.E.G.Kr. von dem Borne. 1990. Soluble Fc gamma receptor III in human plasma originates from release by neutrophils. *J. Clin. Invest.* 86:416–423.

10. Huizinga, T.W., M. de Haas, M.H. van Oers, M. Kleijer, H. Vile, P.A. van der Wouw, A. Moulijn, H. van Weezel, D. Roos, and A.E. von dem Borne. 1994. The plasma concentration of soluble Fc-gamma-RIII is related to production of neutrophils. *Br. J. Haematol.* 87:459–463.

11. Koene, H.R., M. de Haas, M. Leijer, T.W. Huizinga, D. Roos, and A.E. von dem Borne. 1998. Clinical value of soluble IgG Fc receptor type III in plasma from patients with chronic idiopathic neutropenia. *Blood.* 15:3962–3966.

12. Abdou, N.I., C. NaPombejara, L. Balentine, and N.L. Abdou. 1978. Suppressor cell mediated neutropenia in Felty's syndrome. *J. Clin. Invest.* 61:738–743.

13. Starkebaum, G., W.P. Singer, and W.P. Arend. 1980. Humoral and cellular immune mechanisms of neutropenia in patients with Felty's syndrome. *Clin. Exp. Immunol.* 39:307–314.

14. Killick, S.B., J.C.W. Marsh, G. Hale, H. Waldmann, S.J. Kelly, and E.C. Gordon-Smith. 1997. Sustained remission of severe resistant autoimmune neutropenia with Campath-1H. *Br. J. Haematol.* 97:306–308.

15. Hadley, A.G., A.M. Holburn, C. Bunch, and H. Chapel. 1986. Anti-granulocyte opsonic activity and autoimmune neutropenia. *Br. J. Haematol.* 63:581–589.

16. Blumfelder, T.M. and G. Logue. 1981. Human IgG antigranulocyte antibodies: comparison of detection by quantitative antiglobulin consumption and by binding of 125I staph protein A. *Am. J. Hematol.* 11:77–84.

17. Cines, D.B., F. Passero, D. Guerry, M. Bina, B. Dusak, and A. Schreiber. 1982. Granulocyte associated IgG in neutropenic disorders. *Blood* 59:124–132.

18. McAllister, J.A., L.A. Boxer, and R.L. Baehner. 1979. The use and limitation of staphylococcal protein A for study of antineutrophil antibodies. *Blood* 54:1330–1337.

19. Verheugt, F., A. von dem Borne, F. Decary, and C. Engelfreit. 1977. The detection of granulocyte alloantibodies with an indirect immunoflorescence test. *Br. J. Haematol.* 36:533–544.

20. Maher, G.M. and K.R. Hartman. 1993. Detection of antineutrophil autoantibodies by flow cytometry: use of unfixed neutrophils as anitgenic targets. *J. Clin. Lab. Anal.* 7:334–340.

21. Bux, J., B. Kober, and V. Kiefel. 1993. Analysis of granulocyte-reactive antibodies using an immunoassay based upon monoclonal-antibody-specific immobilization of granulocyte antigens. *Transfusion Medicine* 3:157–162.

22. Lamour, A., R.L. Corre, Y.-L. Pennec, J. Carton, and Youinou. 1995. Heterogeneity of neutrophil antibodies in patients with primary Sjögren's syndrome. *Blood* 86:3553–3559.

23. Bux, J. and J. Chapman. 1997. Report on the second international granulocyte serology workshop. *Transfusion* 37:977–983.

24. Bux, J., K. Kissell, C. Hofmann, and S. Santoso. 1999. The user of allele-specific recombinant Fc gamma receptor IIIb antigens for the detection of granulocyte antibodies. *Blood.* 93:357–362.

25. Zubler, R.H., G.L.P.H. Lange, and P.A. Miescher. 1976. Detection of immune complexes in unheated sera by a modified 125I-C1q-binding test. *J. Immunol.* 116:232–235.

26. Klaassen, R.J.L., R. Goldschmeding, A.B.J. Vlekke, R. Rozendaal, and A.E. von dem Borne. 1991. Differentiation between neutrophil-bound antibodies and immune complexes. *Br. J. Hematol.* 77:398–402.

27. Starkebaum, G. and W.P. Arend. 1979. Neutrophil-binding immunoglobulin G in systemic lupus erythematosus. *J. Clin. Invest.* 64:902–912.

28. Murphy, M.F., P. Metcalfe, A.H. Waters, D.C. Linch, R. Cheingsong-Popov, C. Carne, and I.D. Weller. 1985. Immune neutropenia in homosexual men. *Lancet.* 1:217–218.

29. Blumfelder, T.M., G.L. Logue, and D.S. Shimm. 1981. Felty's syndrome: effects of splenectomy upon granulocyte count and granulocyte-associated IgG. *Ann. Intern. Med.* 94:623–628.

30. Logue, G.L. and H.R. Silberman. 1979. Felty's syndrome without splenomegaly. *Am. J. Med.* 66:703–706.

31. van de Winkel, J.G.J. and P.J.A. Capel. 1993. Human IgG Fc receptor heterogeneity: molecular aspects and clinical implications. *Immunol. Today.* 14:215–221.

32. Bux, J., M. Sohn, R. Hachmann, and C. Mueller-Eckhardt. 1992. Quantitation of granulocyte antibodies in sera and determination of their binding sites. *Br. J. Haematol.* 82:20–25.

33. Ory, P.A., M.R. Cark, E.E. Kwoh, S.B. Aarkson, and I.M. Goldstein. 1989. Sequences of complementary DNAs that encode the NA1 and NA2 forms of Fc receptor III on humane neutrophils. *J. Clin. Invest.* 84:1688.

34. Stroncek, D.F., K.M. Skubitz, L.B. Plachta, R.A. Shankar, M.E. Clay, J. Herman, H.B. Fleit, and J. McCullough. 1991. Alloimmune neonatal neutropenia due to an antibody to the neutrophil Fc-gamma with maternal deficiency of CD16 antigen. *Blood* 77:1572–1580.

35. Tamm, A. and R.E. Schmidt. 1996. The binding epitopes of human CD16 (Fc gamma RIII) monoclonal antibodies. Implications for ligand binding. *J. Immunol.* 157:1576–1581.

36. Stroncek, D.F., K.M. Skubitz, and J. McCullough. 1990. Biochemical characterization of the neutrophil-specific antigen NB1. *Blood* 75:744–755.

37. Stroncek, D.F., R.S. Shapiro, A.H. Filipovich, L.B. Plachta, and M.E. Clay. 1993. Prolonged neutropenia resulting from antibodies to neutrophil-specific antigen NB1 following marrow transplantation. *Transfusion* 33:158–163.

38. Bux, J., G. Behrens, G. Jaeger, and K. Welte. 1998. Diagnosis and clinical course of autoimmune neutropenia in infacny: analysis of of 240 cases. *Blood* 91:181–186.

39. Klickstein, L.B., W.W. Wong, J.A. Smith, J.H. Weis, J.G. Wilson, and D.T. Fearon. 1987. Human C3b/C4b receptor (CR1). Demonstration of long homologous repeating domains that are composed of the short consensus repeats characteristics of C3/C4 binding proteins. *J. Exp. Med.* 165:1095–1112.

40. Dykman, T.R., J.A. Hatch, M.S. Aqua, and J.P. Atkinson. 1985. Polymorphism of the C3b/C4b receptor (CR1): characterization of a fourth allele. *J. Immunol.* 134:1787–1789.

41. Hartman, K.R. and D.G. Wright. 1991. Identification of autoantibodies specific for the neutrophil adhesion glycoproteins CD11b/CD18 in patients with autoimmune neutropenia. *Blood* 78:1096–1104.

42. Currie, M.S., J.B. Weinberg, P.K. Rustagi, and G.L. Logue. 1987. Antibodies to granulocyte precursors in selective myeloid hypoplasia and other suspected autoimmune neutropenias: Use of HL-60 cells as targets. *Blood* 69:529–536.

43. Hartman, K.R., V.F. LaRussa, S.W. Rothwell, T.O. Atolagbe, F.T. Ward, and G. Klipple. 1994. Antibodies to myeloid precursor cells in autoimmune neutropenia. *Blood* 84:625–631.

44. Harmon, D.C., S.A. Weitzman, and T.P. Stossel. 1984. The severity of immune neutropenia correlates with the maturational specificity of antineutrophil antibodies. *Br. J. Haematol.* 58:209–215.

45. Rothko, K., T.S. Kickler, M.E. Clay, R.J. Johnson, and D.F. Stroncek. 1989. Immunoblotting characterization of neutrophil antigenic targets in autoimmune neutropenia. *Blood* 74:1698–1703.

46. Burton, D.R. and C.F. Barbas 1994. Human antibodies from combinatorial libraries. *Adv. Immunol.* 57:191–280.

47. Portolano, S., G.D. Chazenbalk, P. Seto, J.S. Hutchison, B. Rapoport, and S.M. McLachlan. 1992. Recognition by recombinant autoimmune thyroid disease-derived Fab fragments of a dominant conformational epitope on thyroid peroxidase. *J. Clin. Invest.* 90:720–726.

48. Hexham, J.M., L.J. Partridge, J. Furmaniak, V.B. Petersen, J.C. Colls, C.A.S. Pegg, B.R. Smith, and D.R. Burton. 1994. Cloning and characterisation of TPO autoantibodies using combinatorial phage display libraries. *Autoimmunity.* 17:167–179.

49. Eggena, M., S.R. Targan, L. Iwanczyk, A. Vidrich, L.K. Gordon, and J. Braun. 1996. Phage display cloning and characterization of an immunogenetic marker (perinuclear anti-neutrophil cytoplasmic antibody) in ulcerative colitis. *J. Immunol.* 156:4005–4011.

50. De Wildt, R.M.T., R. Finnern, W.H. Ouwehand, A.D. Griffiths, W.J. Van Venrooij, and R.M.A. Hoet. 1996. Characterization of human variable domain antibody fragments against the U1 RNA-associated A protein, selected from a synthetic and patient-derived combinatorial V gene library. *Eur. J. Immunol.* 26:629–639.

51. Barbas, S., H.J. Ditzel, E.M. Salonen, W.-P. Yang, G.J. Silverman, and D.R. Burton. 1995. Human auto-antibody recognition of DNA. *Proc. Natl. Acad. Sci. USA* 92:2529–2533.

52. Graus, Y.F., M.H. de Baets, P.W.H.I. Parren, S. Berrih-Aknin, J. Wokke, P.J. van Breda Vriesman, and D.R. Burton. 1997. Human anti-nicotinic acetylcholine receptor recombinant Fab fragments isolated from thymus-derived phage display libraries from myasthenia gravis patients reflect predominant specificities in serum and block the action of pathogenic serum antibodies. *J. Immunol.* 158:1919–1929.

53. Ditzel, H.J., Y. Masaki, H. Nielsen, L. Farnes, and D.B. Burton. 2000. Cloning and expression of a novel human antibody-antigen pair associated with Felty's syndrome. *Proc. Natl. Acad. Sci. USA.* 97:9234–9239.

54. Simpson, D.M. and R. Ross. 1971. Effects of heterologous anti-neutrophil serum in guinea pigs. Hematologic and ultra-structural observations. *Am. J. Pathol.* 65:79–102.

55. Boxer, L.A., M.S. Greenberg, G.J. Boxer, and T.P. Stossel. 1975. Autoimmune neutropenia. *N. Eng. J. Med.* 293:748–753.

56. Hadley, A.G., M.A. byron, H.M. Chapel, C. Bunch, and A.M. Holburn. 1987. Anti-granulocyte opsonic activity in sera from patients with systemic lupus erythematosus. *Br. J. Haematol.* 65:61–65.

57. Boxer, L.A. and T.P. Stossel. 1974. Effects of anti-human neutrophil antibodies in vitro. *J. Clin. Invest.* 53:1534–1545.

58. Stern, M. and W.F. Rosse. 1979. Two popluations of granu-locytes in paroxysmal nocturnal hemoglobinuria. *Blood* 53:928–934.

59. Rustagi, P.K., M.S. Currie, and G.L. Logue. 1982. Activation of human complement by immunoglobulin G antigranulocyte antibody. *J. Clin. Invest.* 70:1137–1147

60. Rustagi, P.K., M.S. Currie, and G.L. Logue. 1985. Complement activating antineutrophil antibody in systemic lupus erythematosus. *Am. J. Med.* 78:971–977.

61. Breedveld, F.C., G.J.M. Lafeber, E. De Vries, J.H.J.M. Van Krieken, and A. Cats. 1986. Immune complexes and the pathogenesis of neutropenia in Felty's syndrome. *Ann. Rheum. Dis.* 45:696–702.

62. Lalezari, P., M. Khorshidi, and M. Petrosova. 1986. Auto-immune neutropenia of infancy. *J. Pediatrics* 109:764–769.

63. Bux, J., G. Mueller-Eckhardt, and C. Mueller-Eckhardt. 1991. Autoimmunization against the neutrophil-specific NA1 antigen is associated with HLA-DR2. *Hum. Immunol.* 30:18–21.

64. Cartron, J., B. Bader-Meunier, M. Deplanche, F. Morinet, E. Vilmer, F. Freycon, and G. Tchernia. 1995. Human par-vovirus B19–associated childhood autoimmune neutropenia. *Int. J. Pediatr. Hematol. Oncol.* 2:471. (Abstr.)

65. Haurie, C., D.C. Dale, and M.C. Mackey. 1998. Cyclical neutropenia and other periodic hematological disorders: a review of mechanisms and mathematical models. *Blood* 92:2629–2640.

66. Bjure, J., L.R. Nilsson, and C.M. Plum. 1962. Familial neutropenia caused by deficiency of a plasma factor. *Acta Paediatr.* 51:497–508.

67. Campion, G., P.J. Maddison, N. Goulding, I. James, M.J. Ahern, I. Watt, and D. Sansom. 1990. The Felty' syndrome: a case-matched study of clinical manifestations and outcome, serologic features, and immuongenetic associations. *Medicine (Baltimore).* 69:69–80.

68. Petersen, J. and A. Wiik. 1983. Lack of evidence for granulocyte specific membrane-directed autoantibodies in neutropenic cases of rheumatoid arthritis and autoimmune neutropenia. *Acta Pathol. Microbiol. Scand.* 91:15–22.

69. Juby, A., C. Johnston, P. Davis, and A.S. Russell. 1992. Antinuclear and antineutrophil cytoplasmic antibodies (ANCA) in the sera of patients with Felty's syndrome. *Br. J. Rheumatol.* 31:185–188.

70. van der Woude, F.J., N.L.S. Rasmussen, H. Permin, M. van der Giessen, N. Rasmussen, A. Wiik, L.A. van Es, and G.K. van der Hem. 1985. Autoantibodies against neutrophils and monocytes: tool for diagnosis and marker of disease acitivty in Wegener's granulomatosis. *Lancet* 1:425–429.

71. Coremans, I.E., E.C. Hagen, E.A. van der Voort, F.J. van der Woude, M.R. Daha, and F.C. Breedveld. 1993. Autoantibodies to neutrophil cytoplasmic enzymes in Felty's syndrome. *Clin. Exp. Rheumatol.* 11:255–262.

72. Starkebaum, G., T.P.Jr. Loughran, L.K. Gaur, P. Davis, and B.S. Nepom. 1997. Immunogenetic similarities between patients with Felty's syndrome and those with clonal expansions of large granular lymphocytes in rheumatoid arthritis. *Arthritis. Rheum.* 40:624–626.

73. Bowman, S.J., G.C. Geddes, V. Corrigal, G.S. Panayi, and J.S. Lanchbury. 1996. Large granular lymphocyte expansions in Felty's syndrome have an unusual phenotype of acitvated CD45RA+ cells. *Br. J. Rheumatol.* 35:1252–1255.

74. Evans, R.S. and R.T. Duane. 1949. Acquired hemolytic anemia: relation of erythrocyte antibody to activity of the disease, significance of thrombocytopenia and leukopenia. *Blood* 4:1196–1213.

75. Dale, D.C., T. Cottle, and A.A. Bolyard. 1997. Long term treatment of chronic idiopathic neutropenia with G-CSF. *Blood* 90:173a. (Abstr.)

76. Pegels, J.G., F.M. Helmerhorst, E.F. van Leeuwen, C. van de Plas-van Dalen, C.P. Engelfriet, and A.E. von dem Borne. 1982. The Evans syndrome: characterization of the responsible autoantibodies. *Br. J. Haematol.* 51:445–450.

77. Liu, H., K. Ozaki, Y. Matsuzaki, M. Abe, M. Kosaka, and S. Saito. 1995. Suppression of haematopoiesis by IgG autoantibodies from patients with systemic lupus erythematosus (SLE). *Clin. Exp. Immunol.* 100:480–485.

78. Stroncek, D.F. 1993. Drug-induced immune neutropenia. *Trans. Med. Rev.* 7:268–274.

79. Pisciotta, A.V., S.A. Konings, L.L. Ciespmier, C.E. Cronkite, and J.A. Lieberman. 1992. Cytotoxic activity in serum of patients with clozapine-induced agranulocytosis. *J. Lab. Clin. Med.* 119:254–266.

80. Thompson, J.S., J.M. Herbick, L.W. Klassen, C.D. Severson, V.L. Overlin, J.W. Blaschke, M.A. Silverman, and C.L. Vogel. 1980. Studies on levamisole-induced agranulocytosis. *Blood* 56:388–396.

81. Uetrecht, J.P. 1996. Reactive metabolites and agranulocytosis. *Eur. J. Haematol.* 57 Suppl:83–88.

82. Stroncek, D.F., G.M. Vercellotti, D.E. Hammerschmidt, D.J. Christie, R.A. Shankar, and H.S. Jacob. 1992. Characterization of multiple quinine-dependent antibodies in a patient with episodic hemolytic uremic syndrome and immune agranulocytosis. *Blood* 80:241–248.

83. Stroncek, D.F., R.A. Shankar, and G.P. Herr. 1993. Quinine-dependent antibodies to neutrophils react with a 60 kD glycoprotein on which NB1 antigen is located and a n 85 kD glycosylphosphati-dylinositol linked N-glycosylated plasma membrane glycoprotein. *Blood* 81:2758–2766.

84. Salama, A., B. Schutz, V. Kiefel, H. Breithau[t, and C. Mueller-Eckhardt. 1989. Immune-mediated agranulocytosis related to drugs and their metabolites: mode of sensitization and heterogeneity of antibodies. *Br. J. Haematol.* 72:127–132.

85. Murphy, M.F., P.C. Metcalfe, P.C. Grint, A.R. Green, S. Knowles, J.A. Amess, and A.H. Waters. 1985. Cephalosporin-induced immune neutropenia. *Br. J. Haematol.* 59:9–14.

86. Klumpp, T.R. and J.H. Herman. 1993. Autoimmune neutropenia after bone marrow transplantation. *Blood* 82:1035. (Abstr.)

87. Minchinton, R.M. and A.H. Waters. 1985. Autoimmune thrombocytopenia and neutropenia after bone marrow transplantation. *Blood* 66:752. (Abstr.)

88. Holland, H.K., T.S. Kickler, M. Zahurak, and H. Braine. 1991. Association of anti-neutrophil antibodies with post-engraftment neutropenia following allogeneic bone marrow transplantation. *Blood* 78S:193a. (Abstr.)

89. Klumpp, T.R., J.H. Herman, K.F. Mangan, and J.S. Macdonald. 1992. The effect of neutrophil alloantigen mismatches on engraftment following allogeneic bone marow transplantation. *Blood* 80S:340a. (Abstr.)

90. Minchinton, R.M. and A.H. Waters. 1984. The occurence and significance of neutrophil antibodies. *Br. J. Haematol.* 56:521–528.

91. McCawley, L.J., H.M. Korchak, S.D. Douglas, D.E. Campbell, P.S. Thornton, C.A. Stanley, L. Baker, and L. Kilpatrick. 1994. In vitro and in vivo effects of granulocyte colony-stimulating factor on neutrophils in glycogen storage disease type 1B: granulocyte colony-stimulating factor therapy corrects the neutropenia and the defects in respiratory burst activity and Ca2+ mobilization . *Pediatr. Res.* 35:84–90.

92. Wang, W.C., Cordoba, A.J. Infante, and M.E. Conley. 1994. Successful treatment of neutropenia in the hyper-immunoglobulin M syndrome with granulocyte colony-stimulating factor. *Am. J. Ped. Hem.-Onc.* 16:160–163.

93. Welte, K., J. Gabrilove, M.H. Bronchud, E. Platzer, and G. Morstyn. 1996. Filgrastim (r-metHuG-CSF): The first 10 years. *Blood* 88:1907–1929.

94. Kuijpers, T.W., M. de Haas, C.J. de Groot, A.E.G.Kr. Von dem Borne, and R.S. Weening.1996. The use of rh-G-CSF in chronic autoimmune neutropenia reversal of autoimmune phenomena, a case history. *Br. J. Haematol.* 94:464–469.

95. Vlasveld, L.T., M. de Haas, A.A.M. Ermens, L. Porcelijn, J.A. van Marion-Kievit, and A.E.G.K. von dem Borne. 1997. G-CSF-induced decrease of the anti-granulocyte autoantibody levels in a patient with autoimmune granulocytopenia. *Ann. Hematol.* 75:58–64.

96. Lieschke, G.J., D. Grail, and G. Hodgson. 1994. Mice lacking granulocyte colony-stimulating factor have chronic neutropenia, granulocyte and macrophage progenitor cell deficiency, and impaired neutrophil mobilization. *Blood* 84:1737–1746.

97. Welte, K., M.A. Bonilla, and A.P. Gillio. 1987. Recombinant human G-CSF: effects on hematopoiesis in normal and cyclophosphamide treated primates. *J. Exp. Med.* 165:941–948.

98. Duhrsen, U., J.L. Villeval, and J. Boyd. 1988. Effects of recombinant human granulocyte colony-stimulating factor on hematopoietic progenitor cells in cancer patients. *Blood* 72:2074–2081.

99. Souza, L.M., T.C. Boone, and J. Gabrilove. 1986. Recombinant human granulocyte colony-stimulating factor: effects on normal and leukemic myeloid cells. *Science.* 232:61–65.

100. Yuo, A., S. Kitagawa, and A. Ohsaka. 1989. Recombinant human granulocyte colony-stimulating factor as an activator of human granulocytes: potentiation of responses triggered by receptor-mediated agonists and stimulation of C3bi receptor expression and adherance. *Blood* 74:2144–2149.

101. Takahashi, K., S. Taniguchi, K. Akashi, K. Fujimoto, T. Sibuya, H. Ishibashi, M. Harada, and Y. Niho. 1991. Human recombinant granulocyte colony-stimulating factor for the treatment of autoimmune neutropenia. *Acta Haematol.* 86:95–98.

102. Rodwell, R.L., P.H. Gray, K.M. Taylor, and R. Minchinton. 1995. Filgrastim for the treatment of immune neonatal neutropenia. *Blood* 86 Suppl:508a. (Abstr.)

103. Bauduer, F. 1998. G-CSF: a very efficient therapy in chronic autoimmune neutropenia. A brief review of the literature. *Hematol. Cell Ther.* 40:189–191.

104. Bonilla, M.A., D. Dale, and C. Zeidler. 1997. Long-term safety of treatment with recombinant human granulocyte colony-stimulating factor (r-metHuG-CSF) in patients with severe congenital neutropenias. *Br. J. Haematol.* 88:723–730.

105. McMullin, M.F. and M.B. Finch. 1995. Felty's syndrome treated with rhG-CSF associated with flare of arthritis and skin rash. *Clin. Rheumatol.* 14:204–208.

106. Ross, H.J., L.A. Moy, R. Kaplan, and R.A. Figlin. 1991. Bullous pyoderma gangrenosum after granulocyte colony stimulating factor treatment. *Cancer* 68:441–443.

107. Wodzinkski, M.A., K.K. Hampton, and J.T. Reilly. 1991. Differential effect of G-CSF and GM-CSF in acquired chronic neutropenia. *Br. J. Haematol.* 77:249–250.

108. Starkebaum, G. 1997. Use of colony-stimulating factors in the treatment of neutropenia associated with collagen vascular disease. *Curr. Opin. Hematol.* 4:196–199.

109. Xu, S., M. Hoglund, and P. Venge. 1996. The effect of granulocyte colony-stimulating factor (G-CSF) on the degranulation of secondary granule proteins from human neutrophils in vivo may be indirect. *Haematol.* 93:558–568.

110. de Haas, M., J.M. Kerst, C.E. van der Schoot, J. Calafat, C.E.N.J.H. Hack, D. Roos, R.H. van Oers, and A.E. von dem Borne. 1994. Granulocyte colony-stimulating factor administration to healthy volunteers: analysis of the immediate activating effects on circulating neutrophils. *Blood* 84:3885–3894.

111. Bernini, J.C., R. Wooley, and G.R. Buchanan. 1996. Low-dose recombinant human granulocyte colony-stimulating factor therapy in children with symptomatic chronic idiopathic neutropenia. *J. Pediatrics.* 129:551–558.

112. Bussel, J., P. Lalezari, M. Hilgartner, J. Partin, S. Fikrig, J. O'malley, and S. Barandun. 1983. Reversal of neutropenia with intravenous gammaglobulin in autoimmune neutropenia of infancy. *Blood* 62:398–400.

113. Lakhanpal, S. and H.S. Luthra. 1985. D-penicillamine in Felty's syndrome. *J. Rheumatol.* 12:703–706.

114. Dillon, A.M., H.S. Luthra, D.L. Conn, and R.H. Ferguson. 1986. Parenteral gold therapy in Felty's syndrome. *Medicine (Baltimore)* 65:107–112.

115. Fiechtner, J.J., D.R. Miller, and G. Starkebaum. 1989. Reversal of neutropenia with methotrexate treatment in patients with Felty's syndrome. *Arthritis Rheum.* 32:194–201.

116. Canvin, J.M.G., B.I. Dalal, F. Baragar, and J.B. Johnston. 1991. Cyclosporin for the treatment of granulotopenia in Felty's syndrome. *Am. J. Hematol.* 36:219. (Abstr.)

117. Joseph, G., D.H. Neustadt, J. Hamm, M. Kellihan, and T. Hadley 1991. GM-CSF in the treatment of Felty syndrome. *Am. J. Hematol.* 37:55–56.

118. Clotet, B., E. Argelagues, J. Junca, M. Grifol, J. Sanz, A. Ribera, E. Lience, and M. Foz. 1985. Plasmapheresis in Felty's syndrome. *Scand. J. Rheumatol.* 14:438.

119. Logue, G.L., A.T. Huang, and D.S. Shimm. 1981. Failure of splenectomy in Felty's syndrome. The role of antibodies supporting granulocyte lysis by lymphocytes. *N. Eng. J. Med.* 304:580–583.

120. Fraser, D.D., G.P. Sartiano, T.W. Butler, and E.L. Treadwell. 1993. Neutropenia of Felty's syndrome successfully treated with granulocyte colony-stimulating factor. *J. Rheumatol.* 20:1447–1448.

121. Graham, K.E. and G.O. Coodley. 1995. A prolonged use of granulocyte colony stimulating factor in Felty's syndrome. *J. Rheumatol.* 22:174–176.

122. Weide, R., J. Heymanns, H. Koppler, M. Tiemann, B. Huss, K.H. Pfluger, and K. Havemann. 1994. Successful treatment of neutropenia in T-LGL leukemia (T gamma-lymphocytosis) with granulocyte colony-stimulating factor. *Ann Hematol.* 69:117–119.

123. Stanworth, S.J., L. Green, R.S. Pumphrey, D.R. Swinson, and M. Bhavnani. 1996. An unusual association of Felty syndrome and TCR gamma delta lymphocytosis. *J. Clin. Pathol.* 49:351–353.

124. Loughran, T.P.Jr., P.G. Kidd, and G. Starkebaum. 1994. Treatment of large granular lymphocyte leukemia with oral low-dose methotrexate. *Blood* 84:2164–2170.

125. Schwab, U.M., P. Harten, R.A. Zeuner, M. Kulper, and H.H. Euler. 1996. G-CSF in patients with lupus-associated neutropenia and infections. *J. Rheumatol.* 55.174–179.

126. Choen, A.M., E. Hodak, M. David, M. Mittleman, R. Gal, and R. Stern. 1996. Beneficial effect of granulocyte-colony-stimulating factor in scleromyxoedema associated with severe idiopathic neutropenia. *Br. J. Dermatol.* 135:626–629.

127. Klumpp, T.R., J.H. Herman, J.S. Macdonald, M.K. Schnell, M. Mullaney, and K.F. Managan. 1992. Autoimmune neutropenia following peripheral blood stem cell transplantation. *Am. J. Hematol.* 41:215–217.

128. Koeppler, H. and J.M. Goldman. 1988. "Auto"-immune neutropenia after allogeneic bone marrow transplantation unresponsive to conventional immunosuppression but resolving promptly after splenectomy. *Eur. J. Haematol.* 41:182–185.

34 | Human Autoimmune Diabetes

Peter A. Gottlieb and George S. Eisenbarth

1. INTRODUCTION

Diabetes mellitus is a complex of syndromes characterized by hyperglycemia and altered glucose metabolism associated with specific microvascular and macrovascular complications as well as neuropathy (1). There are many etiologies of elevated plasma glucose and impaired glucose tolerance (2). This review will concentrate on what has been termed type 1A diabetes, or the form of diabetes resulting from immune-mediated destruction of the cells that produce insulin (3–6). This is the most common form of childhood diabetes in the United States, but type 1A diabetes can present at any age. The term Type 2 diabetes probably encompasses several genetic disorders and is characterized phenotypically by insulin resistance, an association with obesity, an absence of an absolute requirement for insulin and thus a diminished incidence of ketoacidosis. Diabetic syndromes with a known genetic etiology such as several forms of maturity onset diabetes of the young (MODY) are distinct from both type 2 and type 1 diabetes. The clearest distinction between these forms of diabetes relate to the detection of known genetic mutations (e.g. the MODY3 HNF-1α mutation (7) or the presence of anti-islet autoantibodies for type 1A diabetes. Type 1 diabetes or insulin-dependent diabetes mellitus of children is characterized by a more rapid onset of hyperglycemia, a tendency to ketoacidosis, and total dependence on insulin to survive and maintain health. Such a dependence usually only develops several years after diabetes onset. There has been considerable progress in reaching a consensus that the major subset of type 1 diabetes is of immune etiology and thus the designation type 1A diabetes for "immune-mediated" diabetes (2). Type 1B diabetes refers to diabetic syndromes with loss of insulin-producing cells, but where the loss is not immune-mediated.

2. HYPOTHESES REGARDING THE NATURAL HISTORY OF TYPE 1 DIABETES

At present, there are 3 main hypotheses concerning the natural history of type 1A diabetes. A fourth hypothesis originally proposed almost two decades ago arguing for an acute viral etiology has few supporters at this time. The three remaining hypotheses are:

A. Type 1A diabetes is a chronic and progressive disorder resulting from immune-mediated destruction of islet beta cells. An extension of this concept is that as immunologic and immunogenetic assays are refined, one will be able to both predict the risk of diabetes and the approximate rate of progression to diabetes.

B. Type 1A diabetes results from "multiple" hits, some of which may be viral, and thus diabetes develops slowly. Because of the inherent randomness of these "hits" it would be difficult to predict diabetes risk and rate of progression to diabetes.

C. Type 1A diabetes is preceded by a long prodromal phase of autoimmunity, but actual islet beta cell destruction is acute and occurs at the end of the process.

Our group favors the first hypothesis, which is encapsulated in our dividing the development of diabetes into a series of stages (I. Genetic Susceptibility, II. Triggering of Autoimmunity, III. Active Autoimmunity, IV. Loss of Insulin Secretion, V. Overt Diabetes and VI. Loss of all β-cells with insulin dependence (Figure 1)). The stages can occur over months to years and are associated with progressive loss of beta cell function, which usually precedes the

Stages of Autoimmunity Illustrated by Type 1 DM

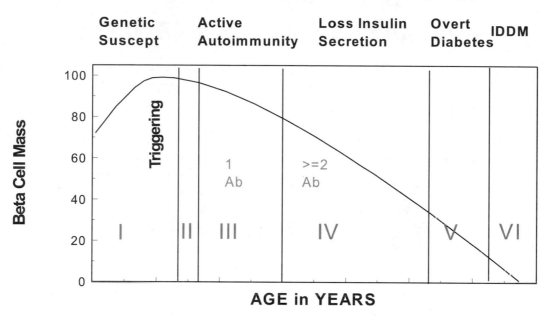

Figure 1.   Hypothetical stages and a hypothesis for loss of beta cell mass prior to the development of type 1 diabetes. Modified from Eisenbarth, G.S. et al., N Engl J Med 314:1360–1368, 1986.

diagnosis of overt diabetes. Palmer and coworkers favor the second hypothesis (8) and Lafferty, Fathman and their coworkers the third (9,10). A difficulty in distinguishing between these hypotheses relates to the current inability to image the islets within the pancreas and a general reluctance to biopsy the pancreas. Recent studies in the NOD mouse provide convincing evidence for progressive loss of islet beta cells prior to development of overt hyperglycemia (Figure 2) (11). In contrast to immunodeficient NOD/SCID mice, β-cell of NOD mice mass is already decreased at 8 weeks of age and declines further prior to the development of diabetes. Even at 8 weeks while the destructive process is taking place, there is increased replication of β-cells of islets of NOD mice noted; an apparently futile attempt to maintain β-cell mass. It is likely that at the end of the process leading to diabetes, there is an acceleration of autoimmunity reflected in immunologic alterations such as a change in cytokines produced by islet-infiltrating T cells (9,10).

In man, the best evidence for a chronic progressive disease course comes from metabolic studies of individuals developing diabetes (in particular insulin secretion following intravenous glucose)(IVGTT) (12) and the observation that at the onset of diabetes significant islet β-cells usually are present, which are lost with time (13). The ability of a model using insulin autoantibodies and first phase insulin release during an intravenous glucose tolerance test to predict the approximate time of onset of diabetes is consistent with the first hypothesis (14). As anti-islet autoanti-

body assays have improved (see below) a number of prospective studies of diabetes development are finding stable expression of anti-islet autoantibodies in the prodromal phase.

Recent studies have filled in an important gap in our knowledge concerning the development of type 1 diabetes, namely the evaluation of children followed from birth until the development of type 1A diabetes (15–18). From these studies it is apparent that anti-islet autoantibodies in highly genetically susceptible individuals (e.g. HLA DR3/4 heterozygous individuals) often develop in the first 6 to 9 months of life (Figure 3). Autoantibodies often develop sequentially over 6 months to one year. Anti-insulin autoantibodies usually develop first, GAD65 autoantibodies occur early, and ICA512 autoantibodies often develop after GAD65 and insulin autoantibodies (16,18,19)

3.   GENETIC SUSCEPTIBILITY

3.1.   Twins

The prevalence of type 1A diabetes in the U.S. population is approximately 1/300. Approximately $\frac{1}{2}$ of monozygotic twins of patients with type 1A diabetes develop diabetes. The concordance of monozygotic (50%) and dizygotic twins (5%) for type 1A diabetes differs dramatically (20). The probability of a discordant monozygotic twin progressing to diabetes decreases with the duration of discordance,

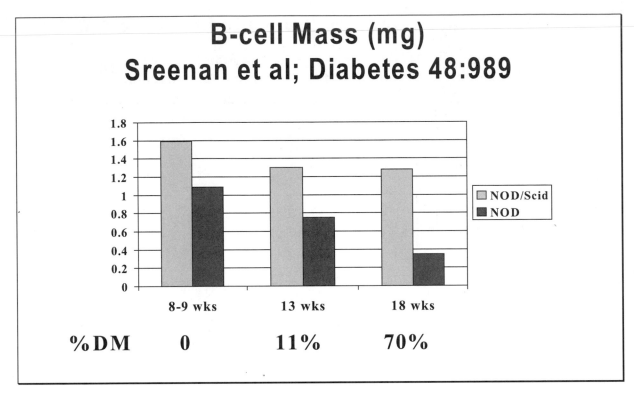

Figure 2.   Beta-cell mass of NOD mice and NOD/SCID mice prior to the development (8 and 13 weeks) of type 1 diabetes. Progressive loss of beta cell mass is evident only in NOD mice and not in the immunodeficient NOD/scid.

but twins can become concordant more than 40 years after the development of diabetes in their twin mate (Figure 4). The risk for diabetes of a dizygotic twin is probably similar to the risk of a sibling of a patient with diabetes. (5%) Thus the shared environment of dizygotic twins does not appear to greatly enhance the development of diabetes. Expression of anti-islet autoantibodies is similarly much greater for monozygotic twins as compared to dizygotic twins. The great majority of monozygotic twins of a patient with type 1 diabetes expressing anti-islet autoantibodies progress to diabetes (21).

Approximately 30 to 50% of monozygotic twins of patients with type 1 diabetes even after long-term follow up do not develop diabetes. This suggests that either stochastic, environmental, or mutational events influence the development of diabetes in twins who are genetically "identical" at birth. One study suggested that either serum IL-4 levels or IL-4 production by NK T cells could aid in the definition of risk of diabetes amongst twins (22). The serum IL-4 assay utilized in this work artifactually measured heterophile antibodies directed against the detecting murine reagents, rather than serum IL-4 itself (23,24). To date, studies of NK T cell IL-4 secretion are not available for twins actually followed to the development of diabetes. The low IL-4 production by NK T cells of diabetic twins described may relate to the presence of diabetes rather than immunologic risk. At present, absent the

expression of anti-islet autoantibodies or impaired glucose tolerance we cannot define risk of diabetes for a discordant monozygotic twin.

3.2.   Autoimmune Polyendocrine Syndrome (APS-I)

Type 1A diabetes is clearly genetically heterogenous. Despite considerable effort in elucidating the genetics of the disorder, the number and specific genes that create susceptibility within most individuals or families have yet to be elucidated. One exception is Autoimmune Polyendocrine Syndrome Type 1 (APS-I) where a single autosomal recessive mutation of a gene termed Autoimmune Regulator (AIRE) is associated with a 15% risk of type 1 diabetes (25). Mutations of this gene lead to mucocutaneous candidiasis, hypoparathyroidism, Addison's disease as well as type 1 diabetes and additional immunologic disorders.

3.3.   MHC Genes

Analysis of the inheritance of diabetes amongst pairs of diabetic siblings have implicated more than 15 loci as contributing to diabetes risk (26). With the exception of the major histocompatibility complex (MHC) and potentially the insulin locus, the familial risk attributed to any single locus is small and has been difficult to replicate in multiple studies. One interpretation of these results is that diabetes

Sequential Development of Autoantibodies Early in Life

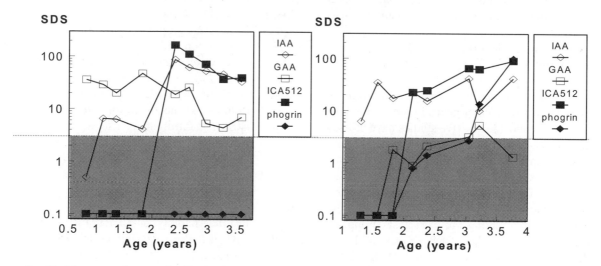

Figure 3. Development of autoantibodies in two children, relatives of patients with type 1 diabetes. The Y-axis, is in standard deviation score (SDS) and plotted on a log scale (IAA = Insulin autoantibodies, GAA = GAD65 autoantibodies, ICA512 = ICAS12/1A-2 autoantibodies, and phogrin = IA-2b autoantibodies).

results from the contribution of many genes, which together create the immunologic milieu for disease pathogenesis. Such a "polygenic" hypothesis would mirror the inheritance of diabetes in the NOD mouse model (27). An alternative hypothesis is that the disease is genetically heterogeneous with major "diabetogenes" and that for any given family, only several of these genes are needed for diabetes susceptibility; however, between individuals, families or countries, those specific "diabetogenes" may differ. We favor the latter hypothesis (see below). For the

Progression to Diabetes in Monozygotic and Dizygotic Twins with Type 1A diabetes

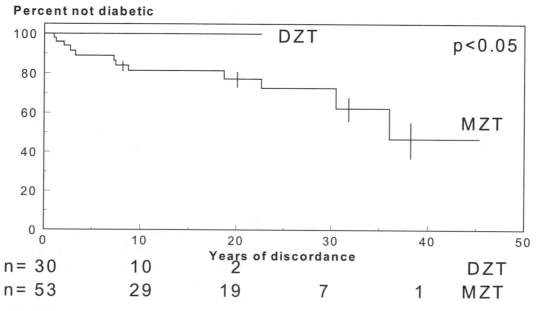

Figure 4. Life Table analysis of progression to diabetes of monozygotic (MZT) and dizygotic (DZT) twins of patients with type 1A diabetes initially studied when non-diabetic (Br Med J 1999, 318:698–702)

great majority of families, genes within the major histocompatibility (MHC) complex appear to be necessary but not sufficient for diabetes risk.

The major genetic locus determining familial aggregation of type 1A diabetes is the MHC, and in man this locus has been designated IDDM1 (28–30). There is almost universal agreement that the class II alleles DQ and DR play a role in determining susceptibility. The establishment of large DNA repositories with complete families and recent developments in genetic screening have led to a resurgence in studies on both HLA and other genes involved in type 1 diabetes. Some of the most instructive studies have utilized the Transmission Dysequilibrium Test (TDT), which examines the transmission of particular alleles to both affected and non-affected siblings from their parents. HLA DR4/DQ8 and HLA DR3/DQ2 are found in the highest frequency among Caucasian patients with diabetes in these families. Interestingly, utilizing the TDT test, it appears that rare HLA haplotypes, such as DR8, DQA1*0401 on a DQB1*0402 also confer risk as high as DR4, DQ8 or DR3, DQ2 (Table 1) (31). An allele closely related to DQB1*0402 is quite common among Korean patients with diabetes, suggesting that HLA alleles and their frequency of distribution among differing populations may partially account for the variability in diabetes incidence seen throughout the world (32). Further analysis indicates that rare alleles such as DRB1*1401 and DQA1*0201, DQB1*0303 may be as protective as the long recognized protective DQ molecule DQA1*0102, DQB1*0602 associated with DR2. At present, we have not seen transmission of either allele from a parent to a patient with anti-islet autoantibody positive diabetes. In the HBDI family dataset, the DQB1*0602 allele was transmitted to only two offspring with diabetes from 149 parents with the allele (1.3% transmission).

It appears that in man, similar to the NOD mouse, both DQ and DR alleles (mouse I-A and I-E respectively) can protect from type 1A diabetes (e.g. DQA1*0102, DQB1*0602 and DRB1*1401). Table 1 summarizes a spectrum of diabetes risk associated with multiple HLA DR and DQ haplotypes. At present, it is not possible to encapsulate the protection and susceptibility associated with the above alleles by reference to specific polymorphic amino acid residues. For example, the hypothesis that aspartic acid at position 57 of DQβ alleles is associated with diabetes risk (33) fails to account for the high risk allele DQB1*0402 or to differentiate a series of alleles with dramatically different "protection" (e.g., DQB1*0602 versus DQB1*0301 both having Asp 57), nor the influence of both HLA DR and DQ alleles (34). Further, the "most" diabetogenic HLA DQ allele, DQ8 (DQA1*0301, DQB1*0302), when associated with DRB1*0403 is not a high risk haplotype. For the haplotype DRB1*1401, DQA1*0101, DQB1*0503, it is the DRB1 allele rather than DQ that is protective (e.g. when DQA1*0101, DQB1*0503 is present without DRB1*1401, it is transmitted to diabetic offspring approximately 50% of the time).

It is likely that additional loci within or close to the major histocompatibility complex will be found that influence the risk of diabetes and the age of diabetes presentation. There are reports that specific class I alleles (e.g. A24) are associated with diabetes onset at a younger age (35,36). Erlich and coworkers are studying DP alleles (37). Recent work in Addison's disease suggests that homozygosity at an allele of the MICA locus is associated with disease risk. An allele (5.1) that creates a premature termination codon of this Class I related molecule is found in nearly 80% of patients with the disorder compared to 40% of unaffected siblings. Finally, a number of groups are evaluating MHC linked microsatellites such as D6S273 and D6S2223 for independent linkage to disease while controlling for DQ and DR alleles (38).

3.4. Insulin Gene (IDDM2)

Upstream of the insulin gene is a Variable Nucleotide Tandem Repeat (VNTR), which is associated with risk of type 1A diabetes (39,40). Variation in the size of this 5′ VNTR appears to correlate with thymic expression of messenger RNA for insulin (41,42). Hanahan and coworkers have developed an elegant hypothesis that Peripheral Antigen Expressing (PAE) cells within the thymus influence the development of "tolerance" to self antigens (43). Lines of transgenic mice, which express the T antigen gene driven off the insulin promoter, with higher level of thymic expression are non-autoimmune prone, while those with decreased or no thymic expression express autoimmunity to varying degrees correlating with the level of thymic expression of antigen. The long form of the insulin VNTR

Table 1   Spectrum of diabetes risk by HLA DR and DQ haplotypes

High Risk	DRB1	DQA1	DQB1
	0301	0501	0201
	0401, 0402	0301	0302
	0405	0301	0401
	0801	0401	0402
	1601	0102	0502
Moderate Risk	0403	0301	0302
	0401, 0402	0301	0301
	1302	0102	0604
	0701	0201	0201
LowRisk	1101	0501	0301
"Protective"	1401	0101	0503
	0701	0201	0303
	1501	0102	0602

associated with decreased risk of type 1A diabetes is associated with greater thymic insulin message. The lack of central tolerance to insulin may be one of the genetic influences on the development of type 1A diabetes in humans.

3.5. Other Non-MHC Genes

It is felt that the combined effects of haplotype sharing for the HLA and insulin (INS) regions account for at most 60–70% of the familial aggregation of type 1A diabetes. In some populations, the combined effects of HLA and INS contribute less than 50% of the familial increased diabetes risk. Therefore, several genome-wide linkage studies have been conducted to identify candidate regions that may contain unidentified susceptibility genes.

About 20 candidate regions for diabetes genes have been reported in linkage studies of affected sib pairs (Table 2) (26,44–53). Follow-up studies have been done in several candidate regions based in part on evidence that these regions show strong linkage to the disease in humans and also because of homology to candidate genes identified in genome-wide linkage studies in the NOD mouse, an animal model of type 1A diabetes. Most of the known or suspected susceptibility loci have been designated IDDMn, where n is a numeric identifier, e.g., IDDM1 refers to genes mapping to the HLA region at 6p21, IDDM2 to the INS region at 11p15. Although the statistical analysis varies across studies, none of the proposed susceptibility loci have an effect as strong as that associated with the HLA region on chromosome 6 (see Table 2). An added complexity has been the inability to confirm certain loci in different population groups. This might occur if the identified regions are "false" positive associations or if different combinations of non-HLA genes contribute to autoimmune diabetes in different populations.

There is considerable controversy regarding the interpretation of linkage studies for complex diseases. The maximum LOD (log10 of the odds favoring linkage over no linkage) score (MLS) is used as a measure of the statistical evidence for linkage between a marker and a gene. The mathematics for determining the frequency of spurious linkages among those meeting the LOD 3.0 criterion are complicated, but the LOD 3.0L, criterion was designed to insure that this frequency (the posterior probability of a type 1 error) is no greater than 5% and applies only to Mendelian disorders. In disorders following a Mendelian pattern of autosomal dominant or recessive transmission, the pattern of inheritance of the disease phenotype is usually obvious. It is much more difficult to confidently define the reported linkages to diabetes susceptibility genes, since the mode of inheritance of the genes causing these complex disorders is unknown. In fact, the primary motivation for genome-wide linkage searches is to provide definitive evidence that one or more of such genes exist. Based on proposed guidelines and evidence from a combination of studies, IDDM4, IDDM5, and IDDM8 meet the criterion for confirmed linkages (MLS exceeding 3.6 with additional confirmation in an independent sample (26,54). Even these loci were however not confirmed in the studies of Concannon and coworkers evaluating a large set of families, including the families in the original reports describing linkage to these loci (53,55). The only other putative gene, rather than genetic locus, identified to date is the CTLA-4 gene (IDDM12). For CTLA-4, there is a polymorphism within the leader sequence with either an alanine (G) or a threonine (A) at position 49. The association with diabetes is weak (e.g. 69% ala for diabetics versus 58% for controls, RR – 1.6 in one study) and not found in all populations (56). Nevertheless, CTLA-4 is an interesting candidate gene in terms of mediating signaling between T cells (CTLA-4) and antigen presenting cells (molecules B7–1 and B7–2). In addition polymorphisms of CTLA-4 are associated with Graves' disease (57).

At this point, IDDM3 and IDDM7 fall short of the above criteria and may represent spurious linkages. Several other reported linkages have not been studied as extensively as 2q, 6q and 11q. These include chromosome 14-linked IDDM11 and chromosome 18-linked IDDM6, both of which await confirmation. Although finding susceptibility genes rather

Table 2 IDDM candidate regions identified by linkage/association studies

Region	Locus	λ_S	Region	Locus	λ_S
ALL		~15	10	IDDM10	1.1–2.2
2q31	IDDM7	1.0–1.6	10q25.1	IDDM17	
2q33	IDDM12		11p15	IDDM2	1.6
2q34	IDDM13		11q13	IDDM4	1.0–1.5
3q	IDDM9	1.0-1.7	14q	IDDM11	
6p21	IDDM1	1.7-4.2	15q	IDDM3	
6q21	IDDM15				
6q25	IDDM5	1.0–3.0	18q	IDDM6	1.0–1.5
6q27	IDDM8	1.0–2.1			

than loci (IDDMn) is critical to gaining greater understanding of the autoimmune process, HLA or IDDM1 remain the strongest factor in explaining the increased risk of diabetes in relatives. All other putative and confirmed susceptibility loci, including INS, have "lamba-s" (λ_s) values (sibling risk of disease/population prevalence) in the range of 1.0 to 1.6 compared to 2.5 to 4.5 for HLA (see Table 2).

3.6.   Oligogenic Inheritance of Type 1A Diabetes (IDDM17)

It is important to note that within a subset of families, a particular locus could be the primary cause of diabetes, especially where high risk HLA alleles are found in high frequency among both affected and non-affected individuals. The current difficulty in defining and confirming non-MHC loci may relate to two hypotheses: Polygenic inheritance vs. genetic heterogeneity. In the polygenic inheritance model, dozens of disease loci exert small effects, which together create susceptibility to disease. In the genetic heterogeneity hypothesis, as few as one locus plus HLA in a given family may underlie susceptibility (MHC + one or two additional loci), but for different families different loci may be important. To evaluate this latter hypothesis, we have performed linkage analysis in a large Bedouin Arab family with 19 members with type 1 diabetes (Figure 5) (58). In this family, homozygosity (suggesting autosomal recessive inheritance) at a locus on chromosome 10 (now termed IDDM17) plus HLA alleles DR3 or DR4 convey a risk of diabetes of approximately 40% (Nonparametric linkage score >5). When the gene underlying this association is identified, based on the genetic heterogeneity model we would predict that this gene contributes to the development of diabetes in a subset of families from other populations.

3.7.   MHC Plus Non-MHC Genes

Studies have reported linkage heterogeneity when a data set is stratified by HLA and/or other factors. For example, the evidence for linkage to IDDM13 is restricted to an affected sibling sharing 0 or 1 HLA haplotypes. This is consistent with genetic heterogeneity in the sibling recurrence risk, with the contribution of non-HLA genes greater among siblings with less risk attributable to HLA (0 or 1 haplotypes) than those with greater risk attributable to HLA

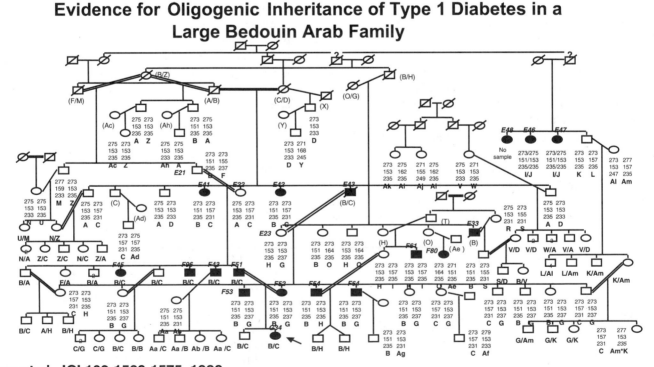

Evidence for Oligogenic Inheritance of Type 1 Diabetes in a Large Bedouin Arab Family

Verge et al. JCI 102:1569-1575, 1998

Figure 5.   Family tree in which 19 members have developed type 1 diabetes and diabetes appears to be inherited as an autosomal recessive mutation at a locus on chromosome 10 plus typical type 1 HLA alleles. The B haplotype on chromosome 10 occurs in the majority of patients with diabetes (Numbers refer to chromosome 10q25.1 microsatellite alleles) and diabetics have homologous B like 10q25.1 regions for their other haplotype (e.g. haplotypes, Ae, o, I, j, c, g, and h).

alleles (2 HLA haplotypes shared). Other studies have subgrouped siblings according to the presence/absence of high risk HLA alleles rather the number of shared haplotypes. Although subdividing the data set in this way may make more biologic sense, it adds to the difficulties of statistical interpretation and analysis.

4. ENVIRONMENTAL FACTORS

The annual incidence of type 1 diabetes amongst children varies dramatically, from more than 40 per 100,000 children in Finland to less than 2/100,000 in Japan (59). In addition, for several countries there is compelling evidence of a temporal increase in the incidence of type 1 diabetes, with countries such as Finland experiencing more than a doubling in incidence over the past three decades (60,61). How much of the variation in incidence between countries is due to environmental factors and how much is due to genetic differences between populations is unknown. For the island of Sardinia, with one of the highest incidences of childhood diabetes in the world, the major factor appears to be genetics, as migrants from Sardinia to the Italian mainland have maintained their high incidence of type 1 diabetes (62).

The temporal increase in the incidence of type 1 diabetes almost certainly is due to environmental factors. One hypothesis is that children with modern sanitation and vaccination for major infectious disorders are living in a "cleaner" environment than in the past and that this "clean" environment favors the development of type 1 diabetes. For example, BB rats and NOD mice raised in more germ-free environments develop diabetes at a higher frequency and at an earlier age (63). The "dirty" environment theoretically may allow for the proper development of the ability of the immune system to discriminate between self and non-self, or indirectly may allow for the development of factors, which can suppress or delay the development of diabetes. It is not yet clear whether type 1 diabetes is actually increasing in incidence or the age of onset of the disorder is decreasing. Given such changes in childhood diabetes incidence and only a 50% concordance of monozygotic twins, investigators have searched for environmental factors influencing the development of diabetes.

The search for viral causes of type 1 diabetes began at a time when it was thought that the disease developed acutely. It was noted that the incidence of diabetes had seasonal variation with an increase in children presenting with the disorder in the fall and winter. With the long prodrome preceding type 1 diabetes, many investigators now believe that this seasonal variation is related to children either coming to the attention of their physician with infection or the added stress associated with infections precipitating metabolic decompensation. Thus, more recent studies have

searched for viruses that might initiate the development of diabetes rather than merely precipitate hyperglycemic symptoms. A series of virus candidates include picornaviruses, rotaviruses, herpesvirses, mumps, rubella, and retroviruses (59,64). A report of the isolation of a retrovirus to which the majority of patients with type 1 diabetes produced antibodies (in contrast to none of the controls) has been withdrawn and not confirmed by multiple additional studies (65,66).

Picornaviruses are small RNA viruses that include enteroviruses, rhinoviruses and the Coxsackie viruses. Previous studies have linked the risk of type 1 diabetes to *congenital* rubella infection, but intriguingly, non-congenital infection does not appear to convey this risk (67,68). Approximately 1/4 of children with congenital rubella will develop diabetes in the future. As many or more of these children also develop autoimmune thyroiditis, suggesting that the virus has altered the immune system thereby increasing risk for a series of autoimmune disorders (67,69). None of the other viral infections are as clearly linked to disease. Coxsackie viruses have been of particular interest because of a homology between the virus and the target antigen glutamic acid decarboxylase (GAD) (70–72). A series of studies have related anti-Coxsackie antibodies to type 1 diabetes, either at onset or during the development of diabetes but both negative and positive studies have been reported (73–78). Viral infection of the mother may also contribute to the development of diabetes (74,79). A possible explanation for the lack of definitive association may derive from the fact that viral factors are ubiquitous and not unique to those with diabetes. Rather, the specific and pathologic immune response they elicit in genetically predisposed individuals may initiate or accelerate the development of beta cell autoimmunity and ultimately type 1 diabetes. Prospective studies of the development of autoimmunity in young children will likely help to elucidate whether any viral infection can be reliably associated with the initiation of autoimmunity (80).

At least one major autoimmune disorder, namely celiac disease, is clearly related to a nutritional environmental factor. Celiac disease is remarkable in that the disorder is dependent upon the ingestion of gliadin, a wheat protein. The major autoantigen of celiac disease has recently been found to be transglutaminase, and the detection of both transglutaminase autoantibodies and abnormal intestinal biopsies are dependent upon ingestion of gliadin (81,82). Within several months of discontinuing the intake of gliadin, the disease remits. Celiac disease is very strongly associated with type 1 diabetes. Approximately 10% of children with type 1 diabetes have transglutaminase autoantibodies, and 5% will have celiac disease diagnosed upon biopsy. We have recently observed that approximately 1/3 of children with type 1 diabetes who are DR3 homozygous have transglutaminase autoantibodies versus approximately

1/10 DR 3 homozygous children from the general population (82) (Figure 6). Thus there is clear evidence that children with diabetes can have autoimmunity which is dependent upon nutritional factors.

The nutritional factor most studied relative to type 1 diabetes is cow's milk (83). Antibodies to milk proteins and T cell responses to these proteins have been reported to be increased amongst children with type 1 diabetes. This includes a molecule termed ICA69 with some homology to bovine albumin (84,85). Elliott and coworkers have suggested that certain forms of milk casein vary in their protein sequence and can be converted to peptides with immunomodulatory properties (86). A number of retrospective studies have implicated early ingestion of milk to the development of diabetes. The increased risk in most of these studies is, however, small and prospective studies following children till the expression of autoimmunity (Baby-Diab study of Munich and DAISY study of Denver) have failed to find an association between milk ingestion and anti-islet autoimmunity (87).

Overall, the search for environmental factors contributing to the development of diabetes have been relatively disappointing. With the exception of congenital rubella infection, none have been confirmed. Prospective studies evaluating the development of autoantibodies in relation to putative environmental factors will hopefully help resolve a number of controversies in this area.

5.   AUTOANTIBODIES

Autoantibodies associated with type 1 diabetes were initially detected using indirect immunofluorescence and sections of normal human pancreas (88). This assay (cytoplasmic islet cell antibodies, or ICA) has utility in the diagnosis and prediction of type 1 diabetes but has proven difficult to standardize (89). The determination of anti-islet autoantibodies has been revolutionized by the cloning of a series of islet autoantigens (90). For autoantibody radioassays, many investigators simply take cloned cDNA of a given target autoantigen and in vitro transcribe and translate the cDNA to produce labeled autoantigen (91). This was first accomplished in the diabetes field for the enzyme glutamic acid decarboxylase (GAD65), which is a cytoplasmic enzyme expressed in all islet cells of man and a series of neuroendocrine tissues. The association of autoantibodies to GAD65 with Stiff Man Syndrome, a rare neuromuscular disease characterized by muscle spasms, contributed to the identification of GAD65 as the elusive islet 64kD autoantigen (92). Autoantibodies to GAD65 are now usually detected in a semi-automated 96-well format.

Rabin and coworkers originally discovered the autoantigen ICA512 by screening an islet expression library with sera from patients with type 1 diabetes (93–96). Prior to the characterization of ICA512, Christie and coworkers had identified autoantibodies reacting with a 40kD and 37kD tryptic fragment of labeled islets (97). It is now understood that the 37kD protein is ICA512 and the 40kD molecule is a fragment of a molecule termed phogrin by Hutton and coworkers, IA-2β by Notkins and coworkers, and IAR by other investigators (98). ICA512 and IA-2β are both molecules associated with neuroendocrine secretory granules (e.g. the islet granules containing insulin). Both molecules have tyrosine phosphatase "like" domains, but functional enzymatic activity has not been demonstrated. The molecules are most homologous in their C-terminal intracytoplasmic domains, which are also the domain to which essentially all of the autoantibodies are directed. By utilizing the cytoplasmic domain, or other N terminal shortened constructs of ICA512 and IA-2β, the specificity of assays for diabetes associated anti-islet autoantibodies can be improved with little if any loss of sensitivity. Using sera from children with new onset diabetes approximately 10% of such children express antibodies which react with ICA512 but fail to react with IA-2β (99). To date, we have only detected 1% of diabetic children with anti-IA-2β autoantibodies, who are negative for ICA512 autoantibodies. In addition the bulk of IA-2β reactivity can be absorbed with synthetic ICA512 antigen altogether this suggests that the majority of autoantibodies directed against IA-2β cross react with ICA512.

Although insulin was the first autoantigen biochemically characterized (100), the assay for anti-insulin autoantibodies has been much less convenient in contrast to assays for GAD65 and ICA512. We currently assay both GAD65 and ICA512 autoantibodies utilizing dual radioactive labeling of these molecules (^{3}H-leucine for GAD65 and ^{35}S methionine for ICA512) and can perform the assay for both autoantibodies in a single well requiring less than 10 μl of sera. Investigators have shared antigen clones with other clinical investigators, and thus the GAD65 and ICA512 assays have been readily adopted in laboratories on four continents. In contrast the standard insulin autoantibody assay utilizes duplicate determinations with and without competition with unlabeled insulin and 600 μl of sera per determination. The insulin assay is performed in centrifuge tubes and is thus labor intensive. Recent workshop comparisons indicated that insulin autoantibody assays that use less than the above 600 μl were significantly less sensitive in detecting relevant anti-insulin autoantibodies (101). Williams and coworkers have also recently described an insulin autoantibody radioassay that utilizes protein A rather than polyethylene glycol (18,102,103). Modifications of this assay now allow insulin autoantibodies to be determined with the same throughput as for GAD and ICA512 autoantibody assays. In addition, it is now apparent that with assays utilizing polyethylene glycol hemolyzed sera creates false positive insulin autoantibody determinations.

TG Ab prevalence in Diabetics, and General Population by HLA

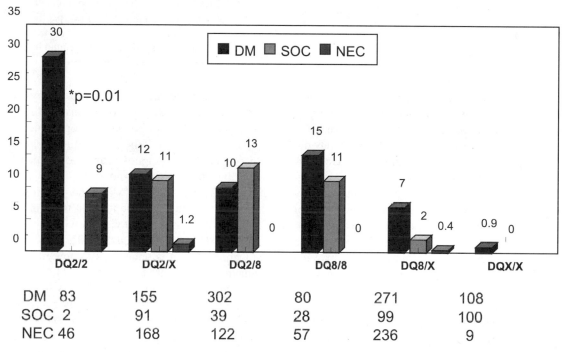

Figure 6. Transglutaminase autoantibody positivity (radioassay) for patients with type 1 diabetes (DM), sibling and offspring of patients with diabetes (SOC) and newborn children from the general population (NEC).

This is abrogated with the protein A microassay. Attempts to measure insulin autoantibodies by standard ELISA techniques led to assays which could detect anti-insulin antibodies following subcutaneous insulin therapy but could not detect the autoantibodies of prediabetic and new onset patients (104). This failure of ELISA assays probably relates to the observation that insulin autoantibodies recognize a conformational determinant and are of very high affinity (10^{10}) and very low capacity (10^{-12}) (105).

Antibodies to carboxypeptidase H are too infrequent to contribute to a standard panel of autoantibodies. In addition to the antigens listed in Table 3, there are a series of additional potential autoantigens with partially characterized molecules, or characterized molecules where the assay format does not allow determination of antibodies for thousands of sera samples (106). It is likely that further characterization of these molecules, as well as potentially unknown molecules, or refinements in assay methodology will provide additional autoantigens of value in the diagnosis and prediction of Type 1 diabetes.

5.1. Combinatorial Autoantibody Determination

With three specific autoantibody assays (GAD65, ICA512 and insulin), the presence of two or more autoantibodies gives a high positive predictive value for progression to diabetes with a sensitivity of approximately 80% (107,108). The association of multiple autoantibodies with high risk for progression to Type 1 diabetes is illustrated in Figure 7. Only one relative of approximately 600 progressed to diabetes without expressing at least one of the autoantibodies at the time of first testing. Less than 20% of relative expressing only a single autoantibody (most often GAD65 or insulin autoantibodies) progressed to diabetes. In contrast expression of two, and more dramatically three, autoantibodies at the time of first testing was associated with an extremely high risk of diabetes with long-term follow up.

An unusual form of cytoplasmic islet cell antibody termed "restricted" or "selective" was found almost a decade ago to be associated with a low risk of progression to diabetes. This

Table 3 Recombinant anti-islet autoantibody assays

Antigen	Sensitivity	Comment
Insulin	40–95%	The titer is inversely correlated with the age of onset
GAD	70%	Prevalence increases with age
IA2	60%	Islet tyrosine phosphatase
Phogrin or lA2b	55%	These autoantibodies are predominantly a subset of IA2
1CA69	?	Radioassay format not diabetes specific
Carboxypeptidase H	10%	Low sensitivity
Casein	?	Probably low specificity since found in controls as well

ICA was composed of antibodies that reacted with human and rat islets but not mouse islets and is now known to represent only GAD65 autoantibodies (109). A significant subset of individuals with this as their only autoantibody did not progress to diabetes and expressed the protective HLA allele, DQB1*0602. We now view restricted ICA as a subset of the "rule" that expression of a single autoantibody (GAD65 in this case) is associated with little evidence of beta cell destruction even with years of follow up and intravenous glucose tolerance testing. One hypothesis is that individuals expressing only a single autoantibody may not be destroying beta cells, but the expression of multiple autoantibodies indicates beta cell destruction with immunogenic presentation of a series of islet antigens. An adult who had an autopsy after detection of expression of only GAD65 autoantibodies was found to have a normal pancreas without evidence of insulitis.

Many of the relatives in the group with a single autoantibody at initial testing in Figure 7 who went on to develop diabetes eventually expressed multiple autoantibodies prior to diabetes onset (16). In young children, autoantibody spreading to multiple autoantigens, to multiple epitopes and within a given autoantigen is readily observed (110). The general rule that expression of multiple autoantibodies is associated with high diabetes risk appears to be accurate for autoantibody testing within the general population. Less than 1/300 individuals in the general population express more than a single autoantibody, which is similar to the risk for type 1 diabetes in the United States, while 80% of children at Type 1 diabetes diagnosis express two or more autoantibodies. The predictive value of a single autoantibody is low. Cytoplasmic ICA positive individuals, (whether relatives or from the general population) who lack all three biochemically-determined autoantibodies, rarely progress to diabetes. Though there are additional islet autoantigens identified with cytoplasmic ICA reactivity not detected by measuring GAD65, ICA512, or insulin autoantibodies, we do not utilize ICA testing. In the absence of any of the biochemical autoantibodies, cytoplasmic ICA is rarely associated with diabetes risk (3445). Individuals positive for cytoplasmic ICA and GAD65$^+$ may express only a single autoantibody, and not two in that antibodies to GAD65 are often (but not always) detected with the ICA test. Though there is no universal agreement on the continued utilization of the cytoplasmic ICA test, we recommend not using the test as it is likely to increase the "noise" of current diagnostic and prognostic tests more than contribute to specific diagnosis.

There is one situation where the prognostic information provided by the expression of multiple anti-islet autoantibodies is unclear. For individuals with protective HLA DQ alleles, DQA1*0102, DQB1*0602, who express multiple autoantibodies, it is unclear as to how many of these individuals will progress to diabetes. (111,112) A small subset will progress, but part of the difficulty in understanding the metabolic progression of such individuals is that they are rare (less than 10% of relatives expressing multiple autoantibodies). The protective HLA DQB1*0602 allele appears to partially protect from the expression of any anti-islet autoantibody, more dramatically protect from the expression of multiple anti-islet autoantibodies, and even more dramatically protect from the development of type 1 diabetes. One situation where DQB1*0602 does not protect from diabetes is the APS-I syndrome. This is probably related to the syndrome's chromosome 21 mutation, which appears to lead to autoimmunity (diabetes and Addison's disease) independent of HLA alleles.

There are reports of the loss of cytoplasmic ICA expression. Transient autoantibodies are less frequently observed with biochemical autoantibody testing but can occur in as many as 5% of young children. A single test of autoantibodies should not be utilized for assessing diabetes risk. With autoantibody tests set at the first percentile, in a certain percentage of tests a given autoantibody will exceed the first percentile in a single test of an individual. For example, utilizing 3 biochemical autoantibodies all set at the 99th percentile of normal, one may expect 3/100 measurements to exceed the 99th percentile for at least one test. This is much less of a problem when expression of multiple autoantibodies (expected by chance < 1/1000) or with confirmation of autoantibody positivity on a subsequent sera sample are also utilized.

Expression of autoantibodies can be useful in determining the type of diabetes among children of differing ethnic

Progression to diabetes in first-degree relatives, by number of autoantibodies

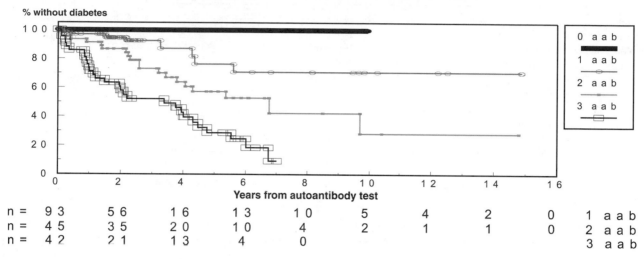

Figure 7. Diabetes progression of first degree relatives divided by number of autoantibodies present at first determination (of GAD65, ICA512 and insulin). Modified from Verge *et al.*, Diabetes.

origin. For example, ¹/₂ of African American and Hispanic American children express no autoantibodies. This is true for less than 10% of Caucasian children in the U.S. HLA analysis of autoantibody negative children reveals that the distribution of their HLA-DQ alleles protects the general population rather than autoantibody positive children with type 1 diabetes.

In contrast to the immunohistochemical cytoplasmic islet cell autoantibody test, autoantibodies reacting with GAD65 or ICA512 often persist for decades after the diagnosis of Type 1 diabetes, and more than 50% of individuals are still positive for one or more of these autoantibodies more than 20 years after diabetes onset. One caveat is that positivity for anti-insulin autoantibodies cannot be relied upon after initiation of insulin therapy. Patients treated for more than several weeks with subcutaneous human insulin uniformly expresses anti-insulin antibodies.

6.   T CELL RESPONSES AND OTHER IMMUNE SYSTEM ABNORMALITIES

6.1.   Is There a Primary Autoantigen?

A large number of islet autoantigens have been identified by virtue of the reactivity of anti-islet autoantibodies including insulin, GAD65, GAD67, IA-2 (ICA512) and IA-2β (phogrin), glima 38, carboxypeptidase H, glycolipids and additional poorly characterized molecules. In addition, analysis of islet infiltrating T cells have revealed that additional antigens appear to be recognized by T cells

which do not react with any of the currently identified autoantigens. With such a long list of antigens, it is almost certain that immune responses to some of these molecules arise secondary to islet destruction. The leading hypothesis is that there is no primary autoantigen but rather recognition of any one of several islet molecules by the immune system leads to type 1 diabetes. Alternatively there may be a primary autoantigen, namely a molecule targeted at the onset of autoimmunity, and a molecule whose absence would preclude diabetes development. The two molecules most studied in this vein are glutamic acid decarboxylase (GAD) and insulin. An unanswered question is what criteria would have to be fulfilled to identify a primary autoantigen, if it existed. The criteria we would propose is that with the engineered absence of the molecule (or alteration of key immunologic epitopes), neither diabetes nor insulitis occurs, and this criteria is satisfied only for one islet molecule. Much less stringent criteria have been evaluated including determination of which molecule is first recognized by autoantibodies or T lymphocytes, whether the molecule can be administered to prevent diabetes and whether genetic engineered widespread expression of the molecule prevents diabetes.

As illustrated in Table 4, both the autoantigens GAD and insulin have been extensively studied in the NOD mouse model of diabetes and man. For both of these autoantigens autoantibodies appear early in man and are highly specific for the development of diabetes. Anti-insulin autoantibodies usually appear first but GAD65 autoantibodies can occur first. For the NOD mouse model of diabetes, most investigators have been unable to detect high affinity GAD autoantibodies

Table 4   Contrasting insulin and GAD as primary autoantigen

	Insulin	GAD
Knockout of gene	N/A[1]	GAD65 no effect DM
Anti-sense constructs	N/A	Decrease DM 2/6 Lines[2]
Antigen on Class/II Promoter	Prevent	No effect
HIgh affinity autoantibodies-NOD	Yes	Not detected
High affinity autoantibodies-MAN	Yes	Yes
Early T-cell responses-NOD	Yes	Yes
Protection with antigen	Yes	Yes
Induction disease immunization	No?	No
T-cell responses MAN	Yes	Yes

1. N/A = Not available
2. Despite a decrease in diabetes in two transgenic GAD anti-sense lines, for both lines *cytoxan* therapy induces diabetes.

with fluid phase radioassays similar to the assays used to predict diabetes in man. In contrast, NOD mice produce anti-insulin autoantibodies beginning as early as 4 weeks of age (113). T cell responses to GAD and insulin are reported to develop as early as 4 weeks of age (114).

More direct information relative to the role of GAD and insulin molecules comes from studies of knockout and transgenic mice. Baekkeskov and coworkers have produced NOD mice with the GAD65 gene deleted. The GAD67 gene was not knocked out. The development of diabetes in these animals was unchanged. This is perhaps consistent with the great difficulty in detecting GAD molecules in mouse islets, in contrast to rat and human islets. In man GAD is expressed in all islet cells, while in the rat it is primarily expressed in β-cells.

Transgenic NOD mice expressing either GAD or insulin in multiple cell types (class II promoters) have been produced (115,116). With the proinsulin gene both insulitis and diabetes was prevented (116). With GAD diabetes was not prevented. This observation has been interpreted that there is an abnormality in developing tolerance to GAD. An alternative hypothesis is that GAD is not an essential autoantigen of NOD mice.

There is one interesting report of transgenic NOD mice with anti-sense to GAD expressed in islets (117). Multiple strains expressing the anti-sense were produced and two were protected from diabetes. With further breeding onto the NOD background, of the two above protected strains, one develops a low incidence of spontaneous diabetes and the other can be induced to develop diabetes following cyclophosphamide administration.

6.2.   What Abnormalities Underlie Loss of Tolerance to Islet Autoantigens?

Approximately $^1/_2$ of identical twins of patients with type 1 diabetes develop the disorder. Patients also frequently develop a series of additional autoimmune diseases. This suggests that there will be generalized immunologic abnormalities that underlie susceptibility to autoimmunity. At present such abnormalities are poorly defined. One could hypothesize that defects which impair proper regulation of the normal immune response could lower the threshold for autoimmunity, while defects which increase general reactivity, either activation of the antigen presenting cell or changes in the responsiveness of T and B cells could increase the number of "autoreactive" cells. Among these hypothesized abnormalities are a deficiency in NK T cells, increased prostaglandin synthetase (COX 2) activity of macrophages, abnormal peptide processing and binding, reduced activation requirements for "autoreactive" T cells and abnormalities in programmed cell death or apoptosis.

NK T cells.   Many studies have implied that normal individuals may harbor T cells directed against self-antigens, termed "autoreactive". Since the vast majority of such individuals do not develop autoimmune disease, it is believed that impaired immune regulation allows for the expansion of the "autoreactive" cell population and for the eventual full manifestation of the immunological response as pathologic islet destruction. One possible cell that might hypothetically fit this regulatory role is the NK+ T cell (118). This particular cell, which recognizes a specific Class I-like allele (119,120), is unusual in that it can produce cytokines such as IL-4 very early in its activation, in contrast to most naïve T cells, which take several days to weeks to develop the same capacity. As noted previously, in human twins discordant for type 1 diabetes, the presence of NK+ T cells, which produced IL-4, appears to be associated with protection from type 1 diabetes even in the presence of autoantibodies (22), while NK+ T cells, which produce only IFN-γ upon stimulation, are associated with type 1 diabetes. Since it is known that naïve ThO cells can be skewed towards either a Th1 or Th2 phenotype depending on the cytokine milieu they are exposed to at the time of initial T cell receptor-MHC-peptide complex engagement, an IL-4 producing cell could theoretically influence the development of a Th2 response to a particular islet antigen if the cell was present and secreting cytokine at this time. Studies in the NOD mouse suggest that these cells can influence the development of type 1 diabetes (121) and further work to better define the nature and control of their particular cytokine responses may shed further light on this novel hypothesis.

Antigen presenting cell abnormalities. Defective antigen presentation might underlie autoimmunity in two major ways. Deletion of self-reactive T cells, which primarily occurs in the thymus but also occurs in the periphery, requires proper signaling through the TCR. If the strength of the interaction of TCR-MHC and peptide is weak, then deletion of self-reactive T cells might not occur and so the pool of autoreactive T cells might be increased in individuals as one step in the process towards autoimmunity. Alternatively and additionally, MHC molecules, which allow for a wider diversity of peptides to be presented, may influence the T cell repertoire during selection or even allow for presentation of self antigens that normally could not be presented (30,122–124). NOD IA^{g7} and human DQ8 have amino acid changes in their structures that can potentially open part of the peptide binding groove and allow for longer and perhaps more diverse peptides to be presented to T cells.

Defects in costimulation could also influence the interaction of antigen presenting cell and T cell and contribute to an autoimmune phenotype. To date, genetic studies have suggested that CTLA4, which receives signals from B7 molecules usually found on antigen presenting cells, may be genetically linked to the development of type 1 diabetes (56). Other abnormalities of IL-2 or its receptor have been implicated in the NOD mouse and suggest interesting loci to examine in human patients. Recent work has also suggested that the processing of antigen via endosomal pathways may be abnormal and so could lead to presentation of usually protected antigenic determinants. Clare-Salzler and colleagues have proposed that elevated levels of protaglandin synthase II, an inducible enzyme that contributes to the production of inflammatory and vasoactive prostaglandins appear to identify new onset diabetic subjects and a high risk population of prediabetic patients who bear islet cell autoantibodies. Whether this defect is primary or secondary to the disease process is currently under evaluation.

6.3. What are the Effecter Mechanisms for β-cell Destruction?

Animal models of diabetes have clearly demonstrated the central role of T lymphocytes in the pathogenesis of type 1 diabetes. The consensus from this work is that both CD4

(4) and CD8 (125,126) T cells normally participate in the destruction of pancreatic islet beta cells. Under unique experimental conditions one or the other cell type may alone be able to cause destruction, but even then there may be help from NK cells and macrophages in the final destructive mechanisms. Although autoantibodies play a crucial role in defining the prediabetic state and diagnosing type 1 diabetes, there appears to be no unique pathologic role that they play in beta cell death. However, B cells themselves appear to be critical to the development of autoimmunity in the NOD mouse, since B cell deficient NOD mice do not become diabetic in spite of the availability of a number of alternative antigen presenting cells: macrophages or dendritic cells (127).

CD4 T cells. Type 1 diabetes in humans has been found to be associated with various abnormalities in peripheral T cell function (128–134). Early studies suggested that differences in the ratios of T cell subsets, CD4/CD8, could be detected between patients and controls. A common abnormality noted across many of these studies is an increase in activated T cells. These cells may express both CD45RA and CD45RO, which represent recently activated T cells or alternatively may express either HLA class II or CD25 (IL-2 Receptor) on their cell surfaces. The apparent activation state may be caused by a maturation defect in individuals with diabetes (135) or alternatively could result from constant priming with islet autoantigens (134).

Autoreactive T cells are felt to be at the heart of the autoimmune process in type 1A diabetes, but these cells have proven elusive to detection and characterization by many investigators. The primary autoantigens for disease prediction and development, insulin, GAD65 and IA-2, have been defined by detection via autoantibody assays (84,136–139), while other candidate antigens have been defined by T-cells (imogen-38, insulin-secretory granule protein of 38kDa) (140,141). Epitope spreading, autoreactivity to multiple autoantigens, whether detected by autoantibody (Figure 7) or T cell reactivity, appears to increase the risk of development of diabetes (108,142).

As noted above, the data reported on T cell reactivity to various candidate autoantigens have been inconsistent. Many explanations have been offered for why this reactivity has been hard to detect reproducibly. Some of these include low precursor frequencies of circulating autoreactive T cells (143), potential peripheral immune regulation of autoreactivity (144), assay systems with low signal to noise ratio (elevated background reactivity) or undue reliance on proliferation of T cells as the sole determinant of reactivity. The importance of obtaining T cells from the site of inflammation have been underscored by Thorsby's work in celiac disease, which demonstrated that lesional T-cells display a different HLA restriction pattern (HLA-DQ) compared to peripheral T-cells (HLA-DP and -DR) (145). In recent years, several reports have been published that show a lack of differential responses between patients and controls (142,146). Similar data has been noted in other T cell-mediated autoimmune diseases (143,147,148). The lack of detectable proliferative responses of autoreactive T-cells may have resulted from anergy or activation-induced cell death through apoptosis (146,149), which can be prevented by addition of low-dose IL-2 (143,146).

The First International Workshop for Standardization of T cell assays sponsored by the Immunology of Diabetes Society has, therefore, set a goal to develop sensitive, specific and reproducible T cell assays that can detect reactivity associated with the pathogenesis of type 1 diabetes. The initial findings suggest that technical issues are critical to detection of reactivity, particularly the use of frozen or fresh cells in the assay and differences in purity and choice of vector (e.g. *E coli* preparations with contaminants) in autoantigen preparations (150). Although this first cooperative study did not indicate a methodology to discriminate type 1 prediabetic or diabetic patients from control subjects, it has set the stage for further work to improve detection of T cell autoreactivity. Ultimately, this would allow for the identification of critical targets for immunointervention therapies, as well as the identification of immunological surrogate markers to monitor disease progression in prediabetic subjects involved in protection studies (151).

CD8 T cells. A number of investigators have now been able to clone CD8 T cells reacting with islet antigens and able to transfer diabetes (125,126). In particular, the T cell clone described by Wong and coworkers has now been identified to react with a B-chain peptide of insulin (126). Santamaria and colleagues have produced a panel of CD8 T cell clones with restricted heterogeneity of their T cell receptor (152,153). A "mimotope" of the unknown antigen for these T cells has been identified. The T cells are restricted by K^d and a Class I tetramer with mimeotope peptide is apparently able to detect in T cells arising in early insulitic lesions.

Cytokines. Pathogenic CD4 or CD8 T cells in type 1 diabetes are felt to express a Th1 phenotype in which IFN-γ, IL-2 and TNFα are produced. In the NOD mouse, from which, much of this data has been derived, protection from diabetes accomplished by various different immune interventions is generally associated with non-destructive inflammation and a shift in the cytokine pattern to one exhibiting Th-2 cytokines, namely IL-4, IL-5, IL-10 and TGFβ (154–159). Several studies investigating either antigen-specific or mitogen-induced cytokine production by T cells from patients with type 1 diabetes support this polarization of the immune response (118–120). Recent innovations in cytokine detection methods, including the use of intracellular staining and ELISPOT assays, may allow us to better determine human reactivity to potential islet autoantigens utilizing a more sensitive readout system than proliferation. Furthermore, if immunotherapies can alter the pattern of cytokine expression, then use of these methods may provide us with important surrogate immunological markers to follow as these clinical trials are being undertaken.

Fas/TNF. Programmed cell death or apoptosis has a role in the normal regulation of the immune response (160,161). Abnormalities in apoptotic pathways appear to predispose to autoimmunity in murine models of lupus. Investigators have been examining these pathways to determine both how beta cells die and whether abnormalities in apoptosis may be found in the pathogenic lymphocytes which cause beta cell destruction (162,163). Several studies in the NOD mouse have demonstrated the presence of Fas on the cell surface of beta cells during the development of insulitis and diabetes. Beta cells appear to die by programmed cell death and engagement of Fas by FasLigand expressed on lymphocytes may play a role in this process. In human type 1 diabetes, there appears to be conflicting findings regarding the beta cell specific nature of Fas expression within a setting of inflammation. One group has noted Fas expression on beta cells only, suggesting a direct role for Fas in beta cell apoptosis, while two groups utilizing confocal microscopy have found Fas expression on all islet cells including α and δ cells, which would argue against this point of view. When intracellular mRNA message for members of the apoptotic pathway was assessed, it appeared that critical caspases needed for signaling via the Fas pathway were reduced in diabetic islets, but not in control islets. If this observation is verified, then it would suggest that other pathways of apoptosis such as TNFR and Bcl2/Bax might be more important for beta cell death than a less than fully competent Fas pathway.

7.   PREDICTION AND PREVENTION

As stated by Yogi Berra, prediction is difficult, especially of the future. Though it is difficult, providing prognostic information is one of the major functions of medicine, historically preceding the ability to influence disease progression. With current autoantibody assays, genetic typing and metabolic testing, it is relatively easy to identify individuals with a risk of type 1 diabetes exceeding 90% in a given time period. For example, relatives of a patient with type 1 diabetes expressing ≥ 2 of GAD65, ICA512, or anti-insulin autoantibodies and with first phase insulin release below the 10th percentile and lacking DQB1*0602 have such a risk (107). It is important to note that by identifying a very high risk group of individuals, (90% risk within five years), it does not in any way imply that the remaining 10%

will never develop diabetes. It is much more likely that such individuals will develop diabetes within ten years. Life Table analysis is the best manner to attempt to assess the future risk of diabetes. Similarly, a moderate risk group (risk approximately 50% within 5 years; ≥ 2 autoantibodies, preserved first phase insulin secretion, lacking DQB1*0602) does not mean that the remaining 50% will not progress to diabetes. It is likely that the 1/2 of this group who do not develop diabetes within five years and whose first phase insulin secretion declines to below the 10th percentile, now have a 90% risk of diabetes within the next five years.

7.1. "Disease" Stage Specific Prevention

The manner by which Type 1 diabetes is likely to be prevented will be dependent upon the stage of the disorder targeted. The earlier the Stage in the natural history of Type 1 diabetes the lower the probability of diabetes (Table 5) and thus the safer the considered therapy should be.

At the Stage of Genetic Risk, in the absence of autoantibodies or decreased insulin secretion, the risk of diabetes is 1/300 in the United States general population without any testing. The risk increases to 6% for DR3/4 (DQB1*0302) individuals from the general population, to 40% for DR3/4 (DQB1*0302) siblings of a patient with type 1 diabetes. Risk approaches 70% in identical twins of patients with type 1 diabetes. The risk of identical twins sets a practical limit for defining inherited genetic risk ("inherited" genes as opposed to hypothesized somatic mutations and T cell receptor rearrangements, which may contribute to genetic risk). Thus, any therapy considered on the basis of inherited

genetic risk will be limited by the lack of diabetes development for many individuals. The highest risk group which might form the basis of a clinical trial are DR 3/4 (DQB1*0302) positive first degree relatives of patients with Type 1 diabetes who appear to have an almost 40% risk of stably expressing anti-islet autoantibodies by age three.

As discussed under "Combinatorial" analysis of anti-islet autoantibodies, individuals with a high risk of progression to Type 1 diabetes can be identified by the expression of multiple biochemically defined autoantibodies, with or without loss of first phase insulin secretion following intravenous glucose. The intravenous glucose tolerance test results would influence primarily the average time to the onset of diabetes. An important caveat is that some individuals expressing even multiple autoantibodies with a protective HLA DQ allele such as DQB1*0602, may never progress to diabetes in their lifetime (164). A number of large national and continental trials are currently based upon the detection of anti-islet autoantibodies, evaluation of insulin secretion and genetic susceptibility. The entry criteria for a preventive trial has, in some cases been less discriminating than would be considered important in providing prognostic information to an individual (e.g. lack of exclusion for DQB1*0602). If large numbers of individuals are recruited to a trial by cytoplasmic ICA testing alone, 7% are ICA positive and have DQB1*0602. This 7% might not materially effect the power calculations of a large trial, though it would be generally recommended that as improved prognostic information becomes available, they be included in the design of future trials out of respect to individuals.

Table 5   Prediction/prevention diabetes by disease "Stage"

Stage	Current best positive predictive value	Comment
Genetic Risk	Twin = 50%, 1° Relative and DR 3/DR4, DQ8/2 = 20–40%, 1° Relative = 5%, DR3/DR4, DQ8/2 and no relative = 6%	Genetic risk is influenced at present by family history and HLA DR and DQ typing. Future consideration is immunologic vaccination.
Autoantibodies present	1 Autoantibody = 10–20% >2 Autoantibodies > 90% in ten years	For prognostic information we only utilize autoantibody radioassays with specificity set at > = 99% for each assay Current trials for prevention target this disease stage.
Autoantibodies + Loss Insulin IVGTT	Any of GAD65, 1CA512 or insulin autoantibodies with IVGTT insulin < first percentile, risk>90% within five years.	With biochemical autoantibody testing IVGTT testing is unlikely to be necessary for the design of the next generation of prevention trials.
Diabetes onset	With autoantibody positivity >90% progression to minimal or no C-peptide secretion within 5 years	Preservation of C-peptide secretion is an important goal in that individuals with retained C-peptide have improved metabolic control.

7.2. Immunosuppression

Trials of generalized immunosuppression have been carried out at the time of onset of diabetes. Prednisone after diabetes onset provides little long term protection. Short courses of anti-T cell antibody therapy with ATGAM (anti-thymocyte globulin), anti-CD5, anti-CD25, and anti-T12 had no permanent effect in terms of maintaining C-peptide. Trials of anti-CD3 were limited by acute inflammatory side effects and it is possible that newer anti-CD3 molecules may be able to avoid this problem. Azathioprine appears to have had relatively little efficacy in preventing progressive loss of C-peptide secretion after diabetes onset (165,166), while cyclosporine A in large randomized blinded trials was very effective in preventing further loss of C-peptide secretion (167–170). Cyclosporine A suppressed anti-insulin autoantibodies, but not cytoplasmic ICA; however, following discontinuation of cyclosporine A, the preservation of C-peptide secretion was rapidly lost. In cyclosporine A-treated patients, hyperglycemia recurred in individuals who could transiently discontinue insulin therapy even though cyclosporine therapy continued. The recurrence of hyperglycemia appeared to be metabolically mediated in a setting where immunosuppression was instituted after the destruction of the majority of insulin secreting islet cells.

The failure of immunosuppression to "cure" diabetes, and to preserve C-peptide secretion after discontinuing the drugs has led to relatively little interest in generalized immunosuppression. The most efficacious drug, cyclosporine A, appears to require relatively large doses, and be associated with both renal toxicity (apparently no permanent toxicity from follow up of patients in these trials) and has a significant risk of malignancy with long term therapy.

With advances in transplantation immunobiology and the development of chimeric molecules for human use, newer agents may soon be available which would meet the criteria for use in new onset and prediabetic subjects. These agents (e.g., anti-CD3 antibodies) appear to be safer in their side effect profiles and more efficacious based on their ability to prevent primary or secondary graft rejection (Mycophenolate mofetil, or Daclizumab, anti-IL2R), or significantly reduce inflammation in rheumatoid arthritis (Enteracept, anti-TNFR). Lastly, they might have the potential to be used either transiently or intermittently rather than continuously which may change the current calculus regarding their use for a primarily immunological disorder, namely type 1A diabetes.

7.3. Immune Deviation and Anti-Inflammatory Therapies

Experiments in the NOD mouse and BB rat models suggested that injection of either Freund's adjuvant or BCG could prevent diabetes (171). Large randomized trials of BCG treatment of new onset patients with diabetes have now been carried out and show that BCG has no beneficial effect in preserving C-peptide secretion or inducing remissions of type 1 diabetes (172,173).

Nicotinamide limits the development of diabetes in NOD mice, and its mechanism of action may relate to reduction of free radical islet destruction, or blockade of lymphokine mediated islet destruction. Several trials of nicotinamide after diabetes onset have been reported. The results from randomized placebo controlled trials failed to detect preservation of C-peptide secretion, and we suspect that nicotinamide is without beneficial effects after diabetes onset in children (174). A meta-analysis reported by Pozzilli and Kolb came to a different conclusion when taking into account less stringent factors than preservation of C-peptide secretion.

Elliott and coworkers in New Zealand, in collaboration with Chase and colleagues from Denver, reported the combined results of a preventative pilot trial of nicotinamide in man (175). This pilot trial was limited in that all of the non-treated control individuals came from Denver, while the majority of the treated individuals were identified by a different cytoplasmic ICA assay utilized in New Zealand. With long-term follow up most of the nicotinamide treated antibody positive individuals have eventually progressed to diabetes, altogether with a significant delay in diabetes progression (~50% compared to controls).

Elliott and coworkers extended these observations into a trial for diabetes prevention in the general population (176). This trial is the first of many other population based prevention trials (after a therapy is shown to be effective and safe in a randomized placebo controlled pilot trial). The trial design screened a large cohort of New Zealand school children for expression of cytoplasmic islet cell antibodies while a matched group of school children were not screened. All ICA positive children were then treated with nicotinamide and compared to the non-screened and non-treated children. Again, a 50% delay in progression to diabetes was claimed. Side effects of nicotinamide therapy have not been reported (a worry was the induction of tumors found in animals treated with nicotinamide and streptozotocin).

These non-randomized trials of nicotinamide have how led to the very large randomized ENDIT trial (European Nicotinamide Diabetes Intervention Trial) and a smaller randomized trial in Germany. Unlike the New Zealand study conducted in the general population, these studies were directed at first degree relatives of patients with type 1 diabetes who expressed cytoplasmic islet cell autoantibodies. The German nicotinamide study has already been discontinued with no difference in progression to diabetes observed (177).

7.4. Antigen-Specific Therapies

Insulin is the only antigen specific therapy to be utilized in man to date. The first non-randomized pilot trial of insulin therapy was based on evidence from the BB rat model that subcutaneous insulin therapy prevented diabetes (178). In this model relatively large doses of insulin were required for protection and it was felt that insulin may affect diabetes development though feedback inhibition of beta cell function (beta cell rest) (179). To be able to effectively suppress insulin secretion in man and also give long term insulin therapy, the pilot trial utilized both intravenous insulin (given at 9 month intervals by continuous infusion) as well as low dose daily subcutaneous insulin (180). In retrospect, with data from the NOD mouse demonstrating that insulin "vaccination" prevents diabetes (Table 6), prevention of diabetes may be related to both beta cell rest and forms of immune regulation induced by insulin administration (181–186). In light of these observations, two extra arms were added to the pilot study after completion of the initial phase, namely subcutaneous insulin alone, and intravenous insulin alone (180). Relatives of patients with type 1 diabetes at very high risk of progression to diabetes were recruited for these studies based upon expression of high titer cytoplasmic ICA, low first phase insulin secretion and/or expression of insulin autoantibodies. All 8 non-treated individuals progressed to diabetes within 3 years of follow-up. In contrast, 3/9 relatives treated with IV and subcutaneous insulin remain non-diabetic, and 6/9 remained non-diabetic for longer than four years, with one relative followed up to 10 years. The subcutaneous group alone, 6/8 individuals remain non-diabetic with the longest follow up, now at 6.6 years, and 6/8 followed for longer than 4 years. This preliminary trial suggests that if there is efficacy in diabetes delay/prevention, subcutaneous insulin alone may be effective. Pilot trials by Ziegler and coworkers in Germany and Vardi and coworkers in Israel have similar results (187).

The above pilot trials, though statistically significant, are too small to conclude that insulin therapy will truly delay or for a subset, prevent the development of diabetes. For example, one of the individuals remaining non-diabetic in the IV/SQ therapy group now followed for 8 years expresses DQB1*0602, and though he expressed all three biochemical autoantibodies one cannot conclude that he would not have remained non-diabetic without any therapy. GAD65 and ICA512 autoantibodies were not altered by insulin therapy while anti-insulin antibodies dramatically increased and then surprisingly decreased after years of therapy.

Given the above pilot data, the National Institutes of Health has instituted a randomized control trial of parenteral insulin therapy and a placebo controlled trial of oral insulin with Jay Skyler at the University of Miami as the principal investigator. Entry criteria for the trial include the expression of ICA in relatives of patients, lack of DQB1 *0602, loss of first phase insulin secretion and/or expression of insulin autoantibodies. More than 60,000 relatives throughout the United States have been screened for participation in the trial, and more than 250 entered into the parenteral portion of the trial and more than 200 in the oral portion of the trial. It will likely require two more years of recruitment to fill the projected trial numbers. The major outcome variable of the trial is the development of diabetes by adult National Diabetes Data Group criteria.

The major concern with insulin therapy, even at low doses in non-diabetic autoantibody positive individuals, is the development of hypoglycemia. With up to ten years of follow up, there have been no severe hypoglycemic episodes. In this trial, individuals eat what they desire, only occasionally monitor blood glucose (less than once per day) and, when asked after three years of therapy whether they wanted to discontinue participation in the trial, all have opted to continue. Families participating in the trials readily recognize that receiving twice a day low dose insulin injections is very different from treating (full blown) insulin-dependent diabetes.

Table 6 Immunologic vaccination

Potential antigens	Studies
Insulin and insulin peptides	Zhou et al: Oral insulin Karounos et al: Mutated insulin Muir et al: Insulin and insulin B chain Wegmann et al: Insulin B:9-23 peptide Gaur et al: Altered peptide ligand of B:9–23
GAD65	Tisch et al: Intrathymic GAD65 Kaufman et al: Intravenous GAD65 Tian et al: Intranasal GAD peptides Tisch et al: GAD65 peptides p217 and p290 late in disease

A study from Japan suggests that insulin therapy may help in preserving C-peptide secretion in autoantibody positive patients presenting with non-insulin dependent diabetes (188). In this small pilot study, adult patients presenting with diabetes and autoantibodies were treated either with subcutaneous insulin or an oral hypoglycemic agent, and those treated with insulin, maintained or improved their C-peptide levels in comparison to oral hypoglycemic-treated patients. Additionally, insulin-treated patients appeared to lose expression of cytoplasmic ICA faster when compared to oral hypoglycemic treated patients. With current autoantibody assays, approximately 5% of patients developing diabetes as adults appear to have slowly progressive autoimmune type 1 diabetes (LADA or latent autoimmune diabetes of adults), and thus, if the results of this pilot study are confirmed in large randomized trials, a large number of patients may benefit from early insulin therapy.

8. CONCLUSIONS

Type 1A diabetes arises from a complex interaction of genes, environment and the immune system to result in a metabolic disorder with multi-system end-organ effects. The rapid advancement in our understanding of the human genome in the next several years will greatly aid in determining the genetic influences of this single or multiple disorder. Therapies based upon this knowledge may lead to new forms of treatment that may prevent the development of immune-mediated diabetes. If specific environmental factors are identified, then reduction in the increasing incidence of type 1A diabetes may be achieved in high-risk regions such as Scandinavia. Current and improved autoantibody assays in combination with HLA determination and other immunological assays will allow for the identification of high risk prediabetic individuals both within the first degree relative pool and the general population. Intervention studies may offer varied forms of therapy based on the stratification of risk for a given individual. Within the next few years, a series of prevention trials using either subcutaneous, oral or nasal insulin, dietary intervention (hydrolyzed casein) and nicotinamide will be completed and will provide useful information upon which to base the next generation of immune intervention trials. Although it is likely that none of these initial approaches will be the final answer to this difficult problem, they hopefully will provide hints at what might be effective alone or in combination in the future.

As results of these trials become known in the new millennium, the data will spur further trials in populations at genetically high risk (e.g., 20%) prior to the development of autoimmunity (autoantibodies). If therapies are found to be safe and effective, then extending these studies to the general population from which more than 80% of the new onset cases will be found, could be considered with the ultimate goal of one day preventing the development of this autoimmune disorder entirely. The design and implementation of trials such as these will require much collaboration between the families involved, practicing physicians and the research community. Based on the organizational success of the DPT-1 and ENDIT multi-center studies, and the rapidity of developments in the field of type 1 diabetes, immunology and genetics, the future looks bright for safe and effective prevention of autoimmune type 1 diabetes.

ACKNOWLEDGMENTS

Supported by grants from the National Institutes of Health DK55364, DK32083, DK55969, DK46639, DK32493, DK50979, AI39213 and grants from the Juvenile Diabetes Foundation, the American Diabetes Association and the Children's Diabetes Foundation. Clinical Research Unit support is essential for patient based studies.

References

1. Krolewski, A.S. and J.H. Warram. 1994. Epidemiology of Late Complications of Diabetes. In: *Joslin's Diabetes Mellitus*. C.R. Kahn and G.C. Weir, editors. Lea and Febiger, Philadelphia, PA. pp. 605–619.
2. American Diabetes Association. 1997. Report of the Expert Committee on the Diagnosis and Classification of Diabetes Mellitus. *Diabetes Care* 20:1183–1197.
3. Pugliese, A. and G.S. Eisenbarth. 1996. Human type I diabetes mellitus: genetic susceptibility and resistance. In Type I Diabetes; Molecular, Cellular, and Clinical Immunology. G.S. Eisenbarth and K.J. Lafferty, editors. Oxford University Press, New York City, New York, pp. 134–152.
4. Haskins, K. 1999. T cell receptor gene usage in autoimmune diabetes. *Intern. Rev. Immunol* 18:61–81.
5. McDevitt, H.O. 1998. The role of MHC class II molecules in susceptibility and resistance to autoimmunity. *Curr. Opin. Immunol.* 10:677–681.
6. Bach, J.-F. 1995. Insulin-dependent diabetes mellitus as a B-cell targeted disease of immunoregulation. *J. Autoimmun.* 8:439–463.
7. Yamagata, K., N. Oda, P.J. Kaisaki, S. Menzel, H. Furtua, M. Vaxillaire, L. Southam, R.D. Cox, G.M. Lathrop, V.V. Boriraj, X. Chen, N.J. Cox, Y. Oda, H. Yano, M.M. Le Beau, S. Yamada, H. Nishigori, J. Takeda, S.S. Fajans, A.T. Hattersley, N. Iwasaki, T. Hansen, O. Pedersen, K.S. Polonsky, R.C. Turner, G. Velho, J.-C. Chevre, P. Froguel, and G.L. Bell. 1996. Mutations in the hepatocyte nuclease factor-1-alpha gene in maturity-onset diabetes of the young (MODY). *Nature* 384:455–457.
8. Greenbaum, C.J., K.L. Sears, S.E. Kahn, and J.P. Palmer. 1999. Relationship of B-cell function and autoantibodies to progression and nonprogression of subclinical type 1 diabetes. *Diabetes* 48:170–175.

9. Gazda, L.S., B. Charlton, and K.J. Lafferty. 1997. Diabetes results from a late change in the autoimmune response of NOD mice. *J. Autoimmun.* 10:261–270.

10. Shimada, A., B. Charlton, C. Taylor-Edwards, and C.G. Fathman. 1996. β-cell destruction may be a late consequence of the autoimmune process in nonobese diabetic mice. *Diabetes* 45:1063–1067.

11. Sreenan, S., A.J. Pick, M. Levisetti, A.C. Baldwin, W. Pugh, and K.S. Polonsky. 1999. Increased β-cell proliferation and reduced mass before diabetes onset in the nonobese diabetic mouse. *Diabetes* 48:989–996.

12. Vardi, P., L. Crisa, R.A. Jackson, R.D. Herskowitz, J.I. Wolfsdorf, D. Einhorn, L. Linarelli, R. Dolinar, S. Wentworth, S.J. Brink, H. Starkman, J.S. Soeldner, and G.S. Eisenbarth. 1991. Predictive value of intravenous glucose tolerance test insulin secretion less than or greater than the first percentile in islet cell antibody positive relatives of type I (insulin-dependent) diabetic patients. *Diabetologia* 34:93 102.

13. Foulis, A.K., C.N. Liddle, M.A. Farquharson, J.A. Richmond, and R.S. Weir. 1986. The histopathology of the pancreas in type I diabetes (insulin dependent) mellitus: a 25-year review of deaths in patients under 20 years of age in the United Kingdom. *Diabetologia* 29:267–274.

14. Eisenbarth, G.S., R. Gianani, L. Yu, M. Pietropaolo, C.F. Verge, H.P. Chase, M.J. Redondo, P. Colman, L. Harrison, and R. Jackson. 1998. Dual parameter model for prediction of type 1 diabetes mellitus. *Proc. Assoc. Am. Physicians* 110:126–135.

15. Rewers, M., T.L. Bugawan, J.M. Norris, A. Blair, B. Beaty, M. Hoffman, R.S. Jr. McDuffie, R.F. Hamman, G. Klingensmith, G.S. Eisenbarth, and H.A. Erlich. 1996. Newborn screening for HLA markers associated with IDDM: diabetes autoimmunity study in the young (DAISY). *Diabetologia* 39:807–812.

16. Yu, L., M. Rewers, R. Gianani, E. Kawasaki, Y. Zhang, C. Verge, P. Chase, G. Klingensmith, H. Erlich, J. Norris, and G.S. Eisenbarth. 1996. Anti-islet autoantibodies develop sequentially rather than simultaneously. *J. Clin. Endocrinol. Metab.* 81:4264–4267.

17. Ziegler, A.-G., M. Hummel, M. Schenker, and E. Bonifacio. 1999. Autoantibody appearance and risk for development of childhood diabetes in offspring of parents with type 1 diabetes. The 2-year analysis of the German BABYDIAB study. *Diabetes* 48:460–468.

18. Naserke, H.E., E. Bonifacio, and A.-G. Ziegler. 1999. Immunoglobulin G insulin autoantibodies in BABYDIAB offspring appear postnatally: sensitive early detection using a protein A/G-based radiobinding assay. *J. Clin. Endocrinol. Metab.* 84:1239–1243.

19. Falorni, A., C.E. Grubin, I. Takei, A. Shimada, A. Kasuga, T. Maruyama, Y. Ozawa, T. Kasatani, T. Saruta, L. Li, and Å. Lernmark. 1994. Radioimmunoassay detects the frequent occurrence of autoantibodies to the Mr 65,000 isoform of glutamic acid decarboxylase in Japanese insulin-dependent diabetes. *Autoimmunity* 19:113–125.

20. Redondo, M.J., M. Rewers, L. Yu, S. Garg, C.C. Pilcher, R.B. Elliott, and G.S. Eisenbarth. 1999. Genetic determination of islet cell autoimmunity in monozygotic twin, dizygotic twin, and non-twin siblings of patients with type 1 diabetes: prospective twin study. *BMJ* 318:698–702.

21. Verge, C.F., R. Gianani, L. Yu, M. Pietropaolo, T. Smith, R.A. Jackson, J.S. Soeldner, and G.S. Eisenbarth. 1995. Late progression to diabetes and evidence for chronic β-cell

22. autoimmunity in identical twins of patients with type I diabetes. *Diabetes* 44:1176–1179.

22. Wilson, S.B., S.C. Kent, K.T. Patton, T. Orban, R.A. Jackson, M. Exley, S. Porcelli, D.A. Schatz, M.A. Atkinson, S.P. Balk, J.L. Strominger, and D.A. Hafler. 1998. Extreme Th1 bias of invariant Vα24JαQ T cells in type 1 diabetes. *Nature* 391:177–181.

23. Ellis, T.M., B.S. Wilson, C. Wasserfall, S. Kent, J. Strominger, D. Hafler, and M.A. Atkinson. 1999. Potential heterophilic antibodies as a confounding variable to measurement of serum interleukin-4 (IL-4). *Diabetes* 48:A210

24. Redondo, M.J., P. Gottlieb, T. Motheral, C.L. Mulgrew, M. Rewers, S. Babu, E. Stephens, D. Wegmann, and G.S. Eisenbarth. 1999. Artifactual serum interleukin 4: heterophile antibodies associated with HLA alleles protective for type 1 diabetes. *Diabetes* 48:A436.

25. Aaltonen, J., P. Björses, J. Perheentupa, N. Horelli-Kuitunen, A. Palotie, L. Peltonen, Y.S. Lee, F. Francis, S. Hennig, C. Thiel, H. Lehrach, and M.-L. Yaspo. 1997. An autoimmune disease, APECED, caused by mutations in a novel gene featuring two PHD-type zinc-finger domains. *Nat. Genet.* 17.399–403.

26. Todd, J.A., and M. Farrall. 1996. Panning for gold: genome-wide scanning for linkage in type I diabetes. *Hum. Mol. Genet.* 5:1443–1448.

27. Wicker, L.S., J.A. Todd, and L.B. Peterson. 1995. Genetic control of autoimmune diabetes in the NOD mouse. *Ann. Rev. Immunol.* 13:179–200.

28. Noble, J.A., A.M. Valdes, M. Cook, W. Klitz, G. Thomson, and H.A. Erlich. 1996. The role of HLA class II genes in insulin-dependent diabetes mellitus: Molecular analysis of 180 Caucasian, multiplex families. *Am. J. Hum. Genet.* 59:1134–1148.

29. Bosi, E., F. Becker, E. Bonifacio, R. Wagner, P. Collins, E.A. Gale, and G.F. Bottazzo. 1991. Progression to type I diabetes in autoimmune endocrine patients with islet cell antibodies. *Diabetes* 40:977–984.

30. Nepom, G.T., and W.W. Kwok. 1998. Perspectives in Diabetes: Molecular basis for HLA-DQ associations with IDDM. *Diabetes* 47:1177–1184.

31. Kawasaki, E., J. Noble, H. Erlich, C.L. Mulgrew, P.R. Fain, and G.S. Eisenbarth. 1998. Transmission of DQ haplotypes to patients with type 1 diabetes. *Diabetes* 47:1971–1973.

32. Park, Y., H. Lee, C.S. Koh, H. Min, M. Rowley, I.R. Mackay, P. Zimmet, B. McCarthy, E. McCanlies, J. Dorman, and M. Trucco. 1996. The low prevalence of immunogenetic markers in Korean adult-onset IDDM patients. *Diabetes Care* 19:241–245.

33. Todd, J.A., J.I. Bell, and H.O. McDevitt. 1987. HLA-DQB gene contributes to susceptibility and resistance to insulin-dependent diabetes mellitus. *Nature* 329:599–604.

34. Harfouch-Hammoud, E., J. Timsit, C. Boitard, J.-F. Bach, and S. Caillat-Zucman. 1996. Contribution of DRB1*04 variants to predisposition to or protection from insulin dependent diabetes mellitus is independent of DQ. *J. Autoimmun.* 9:411–414.

35. Nakanishi, K., T. Kobayashi, T. Murase, T. Nakatsuji, H. Inoko, K. Tsuji, and K. Kosaka. 1993. Association of HLA-A24 with complete β-cell destruction in IDDM. *Diabetes* 42:1086–1098.

36. Yamamoto, A.M., I. Deschamps, H.J. Garchon, H. Roussely, N. Moreau, G. Beaurain, J.J. Robert, and J.F. Bach. 1998. Young age and HLA markers enhance the risk of progres-

sion to type 1 diabetes in antibody-positive siblings of diabetic children. *J. Autoimmun.* 11:643–650.

37. Erlich, H.A., J.I. Rotter, J.D. Chang, S.J. Shaw, L.J. Raffel, W. Klitz, T.L. Bugawan, and A. Zeidler. 1996. Association of HLA-DPB1*0301 with IDDM in Mexican-Americans. *Diabetes* 45:610–614.

38. Moghaddam, P.H., P. de Knijf, B.O. Roep, B. Van der Auwera, A. Naipal, F. Gorus, F. Schuit, M.J. Giphart, and the Belgian Diabetes Registry. 1998. Genetic structure of IDDM1: Two separate regions in the major histocompatibility complex contribute to susceptibility or protection. *Diabetes* 47:263–269.

39. Bell, G.I., S. Horita, and J.H. Karam. 1984. A polymorphic locus near the human insulin gene is associated with insulin-dependent diabetes mellitus. *Diabetes* 33:176–183.

40. Bennett, S.T., A.M. Lucassen, S.C.L. Gough, E.E. Powell, D.E. Undlien, L.E. Pritchard, M.E. Merriman, Y. Kawaguchi, M.J. Dronsfield, F. Pociot, J. Nerup, N. Bouzekri, A. Cambon-Thomsen, K.S. Ronningen, A.H. Barnett, S.C. Bain, and J.A. Todd. 1995. Susceptibility to human type I diabetes at IDDM2 is determined by tandem repeat variation at the insulin gene minisatellite locus. *Nat. Genet.* 9:284–292.

41. Pugliese, A., M. Zeller, A. Fernandez, L.J. Zalcberg, R.J. Bartlett, C. Ricordi, M. Pietropaolo, G.S. Eisenbarth, S.T. Bennett, and D.D. Patel. 1997. The insulin gene is transcribed in the human thymus and transcription levels correlate with allelic variation at the INS VNTR-IDDM2 susceptibility locus for type I diabetes. *Nat. Genet.* 15:293–297.

42. Vafiadis, P., S.T. Bennett, J.A. Todd, J. Nadeau, R. Grabs, C.G. Goodyer, S. Wickramasinghe, E. Colle, and C. Polychronakos. 1997. Insulin expression in human thymus is modulated by INS VNTR alleles at the IDDM2 locus. *Nat. Genet.* 15:289–292.

43. Hanahan, D. 1998. Peripheral-antigen-expressing cells in thymic medulla: factors in self-tolerance and autoimmunity. *Curr. Opin. Immunol.* 10:656–662.

44. Eisenbarth, G.S. 1987. Lilly Lecture: Genes, generator of diversity, glycoconjugates and autoimmune beta cell insufficiency in Type I diabetes. *Diabetes* 36:355–364.

45. Fu, J., H. Ikegami, Y. Kawaguchi, T. Fujisawa, Y. Kawabata, Y. Hamada, H. Ueda, M. Shintani, K. Nojima, N. Babaya, Q.-J. Shen, Y. Uchigata, T. Urakami, Y. Omori, K. Shima, and T. Ogihara. 1998. Association of distal chromosome 2q with IDDM in Japanese subjects. *Diabetologia* 41:228–232.

46. Nakagawa, Y., Y. Kawaguchi, R.C.J. Twells, C. Muxworthy, K.M.D. Hunter, A. Wilson, M.E. Merriman, R.D. Cox, T. Merriman, F. Cucca, P.A. McKinney, J.P.H. Shield, J. Tuomilehto, E. Tuomilehto-Wolf, C. Ionesco-Tirgoviste, L. Nisticó, R. Buzzetti, P. Pozzilli, San-Raffaele Family Study, G. Jones, E. Thorsby, D.E. Undlien, F. Pociot, J. Nerup, K.S. Rönningen, Bart's Oxford Family Study Group, S.C. Bain, and J.A. Todd. 1998. Fine mapping of the diabetes-susceptibility locus, IDDM4, on chromosome 11q13. *Am. J. Hum. Genet.* 63:547–556.

47. Denny, P., C.J. Lord, N.J. Hill, J.V. Goy, E.R. Levy, P.L. Podolin, L.B. Peterson, L.S. Wicker, J.A. Todd, and P.A. Lyons. 1997. Mapping of the IDDM locus *Idd3* to a 0.35-cM interval containing the *Interleukin-2* gene. *Diabetes* 46:695–700.

48. Reed, P., F. Cucca, S. Jenkins, M. Merriman, A. Wilson, P. McKinney, E. Bosi, G. Joner, K. Ronningen, E. Thorsby, D. Undlien, T. Merriman, A. Barnett, S. Bain, and J. Todd.

1997. Evidence for a type 1 diabetes susceptibility locus (IDDM10) on human chromosome 10p11-q11. *Hum. Mol. Genet.* 6:1011–1016.

49. Field, L.L., R. Tobias, and T. Magnus. 1996. A locus on chromosome 15q26(IDDM3) produces susceptibility to insulin-dependent diabetes mellitus. *Nat. Genet.* 8:189–194.

50. Field, L.L., R. Tobias, G. Thomson, and S. Plon. 1996. Susceptibility to insulin-dependent diabetes mellitus maps to a locus (*IDDM11*) on human chromosome 14q24.3-q31. *Genomics* 33:1–8.

51. Luo, D.-F., R. Buzzetti, J.I. Rotter, N.K. Maclaren, L.J. Raffel, L. Nistico, C. Giovannini, P. Pozzilli, G. Thomson, and J.-X. She. 1996. Confirmation of three susceptibility genes to insulin-dependent diabetes mellitus: IDDM4, IDDM5, and IDDM8. *Hum. Mol. Genet.* 5:693–698.

52. Owerbach, D., and K.H. Gabbay. 1996. Perspectives in diabetes: the search for IDDM susceptibility genes the next generation. *Diabetes* 45:544–551.

53. Concannon, P., K.J. Gogolin-Ewens, D.A. Hinds, B. Wapelhorst, V.A. Morrison, B. Stirling, M. Mitra, J. Farmer, S.R. Williams, N.J. Cox, G.I. Bell, N. Risch, and R.S. Spielman. 1998. A second-generation screen of the human genome for susceptibility to insulin-dependent diabetes mellitus. *Nat. Genet.* 19:292–296.

54. She, J.-X. 1996. Susceptibility to type I diabetes: HLA-DQ and DR revisited. *Immunol. Today* 17:323–329.

55. Lernmark, Å., and J. Ott. 1998. Sometimes it's hot, sometimes it's not. *Nat. Genet.* 19:213–214.

56. Nisticò, L., R. Buzzetti, L.E. Pritchard, B. Van der Auwera, C. Giovannini, E. Bosi, M.T.M. Larrad, M.S. Rios, C.C. Chow, C.S. Cockram, K. Jacobs, C. Mikovic, S.C. Bain, A.H. Barnett, C.L. Vandewalle, F. Schuit, F.K. Gorus, Belgian Diabetes Registry, R. Tosi, P. Pozzilli, and J.A. Todd. 1996. The CTLA-4 gene region of chromosome 2q33 is linked to, and associated with, type I diabetes. *Hum. Mol. Genet.* 5:1075–1080.

57. Yanagawa, T., Y. Hidaka, V. Guimaraes, M. Soliman, and L.J. DeGroot. 1995. CTLA-4 gene polymorphism associated with Graves' disease in a Caucasian population. *J. Clin. Endocrinol. Metab.* 80:41–45.

58. Verge, C.F., P. Vardi, S. Babu, F. Bao, H.A. Erlich, T. Bugawan, L. Yu, G.S. Eisenbarth, and P.R. Fain. 1998. Evidence for oligogenic inheritance of type 1A diabetes in a large Bedouin Arab family. *J. Clin. Invest.* 102:1569–1575.

59. Rewers, M., and J.M. Norris. 1996. Epidemiology of type I diabetes. In: Type I Diabetes: Molecular, Cellular, and Clinical Immunology. G.S. Eisenbarth and K.J. Lafferty, editors. Oxford University Press, New York. 172–208.

60. Drykoningen, C.E.M., A.L.M. Mulder, G.J. Vaandrager, R.E. LaPorte, and G.J. Bruining. 1992. The incidence of male childhood type 1 (insulin-dependent) diabetes mellitus is rising rapidly in the Netherlands. *Diabetologia* 35:139–142.

61. Gardner, S.G., P.J. Bingley, P.A. Sawtell, S. Weeks, and E.A.M. Gale. 1997. Rising incidence of insulin dependent diabetes in children aged under 5 years in the Oxford region: time trend analysis. *BMJ* 315:713–717.

62. Muntoni, S., M.T. Fonte, S. Stoduto, G. Marietti, C. Bizzarri, A. Crino, P. Ciampalini, G. Multai, M.A. Suppa, M.C. Matteoli, L. Lucentini, L.M. Sebastiani, N. Visalli, P. Pozzilli, and B. Boscherini. 1997. Incidence of Insulin-dependent diabetes mellitus among Sardinian-

heritage children born in Lazio region, Italy. *Lancet* 349:160–162.

63. Rossini, A.A., R.M. Williams, J.P. Mordes, M.C. Appel, and A.A. Like. 1979. Spontaneous diabetes in gnotobiotic BB/W rat. *Diabetes* 28:1031–1032.

64. Dahlquist, G. 1998. Environmental risk factors in human type 1 diabetes—an epidemiological perspective. *Diabetes Metab. Rev.* 11:37–46.

65. Conrad, B., R.N. Weissmahr, J. Böni, R. Arcari, J. Schüpbach, and B. Mach. 1997. A human endogenous retroviral superantigen as candidate autoimmune gene in type 1 diabetes. *Cell* 90:303–313.

66. Jaeckel, E., S. Heringlake, D. Berger, G. Brabant, G. Hunsmann, and M.P. Manns. 1999. No evidence for association between IDDMK$_{1,2}$22, a novel isolated retrovirus, and IDDM. *Diabetes* 48:209–214.

67. Clarke, W.L., K.A. Shaver, G.M. Bright, A.D. Rogol, and W.E. Nance. 1984. Autoimmunity in congenital rubella syndrome. *J. Pediatr.* 104:370–373.

68. Rubinstein, P., M.E. Walker, B. Fedun, M.E. Witt, L.Z. Cooper, and F. Ginsberg-Fellner. 1982. The HLA system in congenital rubella patients with and without diabetes. *Diabetes* 31:1088–1091.

69. Rabinowe, S.L., K.L. George, R. Loughlin, J.S. Soeldner, and G.S. Eisenbarth. 1986. Congenital rubella. Monoclonal antibody-defined T cell abnormalities in young adults. *Am. J. Med.* 81:779–782.

70. Atkinson, M.A., M.A. Bowman, L. Campbell, B.L. Darrow, D.L. Kaufman, and N.K. Maclaren. 1994. Cellular immunity to a determinant common to glutamate decarboxylase and coxsackie virus in insulin-dependent diabetes. *J. Clin. Invest.* 94:2125–2129.

71. Richter, W., T. Mertens, B. Schoel, P. Muir, A. Ritzkowsky, W.A. Scherbaum, and B.O. Boehm. 1994. Sequence homology of the diabetes-associated autoantigen glutamate decarboxylase with coxsackie B4–2C protein and heat shock protein 60 mediates no molecular mimicry of autoantibodies. *J. Exp. Med.* 180:721–726.

72. Tian, J., P.V. Lehmann, and D.L. Kaufman. 1994. T cell cross-reactivity between coxsackievirus and glutamate decarboxylase is associated with a murine diabetes susceptibility allele. *J. Exp. Med.* 180:1979–1984.

73. Hyöty, H., M. Hiltunen, M. Knip, M. Laakkonen, P. Vähäsalo, J. Karjalainen, P. Koskela, M. Roivainen, P. Leinikki, T. Hovi, H.K. Åkerblom, and The Childhood Diabetes in Finland (DiMe) Study Group. 1995. A prospective study of the role of coxsackie B and other enterovirus infections in the pathogenesis of IDDM. *Diabetes* 44:652–657.

74. Dahlquist, G., G. Frisk, S.A. Ivarsson, L. Svanberg, M. Forsgren, and H. Diderholm. 1995. Indications that maternal coxsackie B virus infection during pregnancy is a risk factor for childhood-onset IDDM. *Diabetologia* 38:1371–1373.

75. Clements, G.B., D.N. Galbraith, and K.W. Taylor. 1995. Coxsackie B virus infection and onset of childhood diabetes. *Lancet* 346:221–223.

76. Schloot, N.C., B.O. Roep, D.R. Wegmann, L. Yu, T.B. Wang, and G.S. Eisenbarth. 1997. T cell reactivity to GAD65 peptide sequences shared with coxsackie virus protein in recent-onset IDDM patients and control subjects. *Diabetologia* 40:332–338.

77. Andreoletti, L., D. Hober, C. Haber-Vandenberghe, S. Belaich, M.C. Vantyghem, J. Lefebvre, and P. Wattre. 1997. Detection of coxsackie B virus RNA sequences in whole blood samples from adult patients at the onset of type I diabetes mellitus. *J. Med. Virol.* 52:121–127.

78. D'Alessio, D.J. 1992. A case-control study of group B Coxsackievirus immunoglobulin M antibody prevalence and HLA-DR antigens in newly diagnosed cases of insulin-dependent diabetes mellitus. *Am. J. Epidemiol.* 135:1331–1338.

79. Dahlquist, G.G., S. Ivarsson, B. Lindberg, and M. Forsgren. 1995. Maternal enteroviral infection during pregnancy as a risk factor for childhood IDDM. A population-based case-control study. *Diabetes* 44:408–413.

80. Graves, P.M., K.J. Barriga, J.M. Norris, M.R. Hoffman, L. Yu, G.S. Eisenbarth, and M. Rewers. 1999. Lack of association between early childhood immunizations and beta-cell autoimmunity. *Diabetes Care.* 22:1694–1697.

81. Dietrich, W., T. Ehnis, M. Bauer, P. Donner, U. Volta, E.O. Riecken, and D. Schuppan. 1997. Identification of tissue transglutaminase as the autoantigen of celiac disease. *Nat. Med.* 3:797–801.

82. Bao, F., L. Yu, S. Babu, T. Wang, E.J. Hoffenberg, M. Rewers, and G.S. Eisenbarth. 1999. One third of HLA DQ2 homozygous patients with type 1 diabetes express celiac disease associated transglutaminase autoantibodies. *J. Autoimmun.* 13:143–148.

83. Martin, J.M., B. Trink, D. Daneman, H.M. Dosch, and B. Robinson. 1998. Milk proteins in the etiology of insulin-dependent diabetes mellitus (IDDM). *Ann. Med.* 23:447–452.

84. Pietropaolo, M., L. Castano, S. Babu, R. Buelow, S. Martin, A. Martin, A. Powers, M. Prochazka, J. Naggert, E.H. Leiter, and G.S. Eisenbarth. 1993. Islet cell autoantigen 69 kDa (ICA69): molecular cloning and characterization of a novel diabetes associated autoantigen. *J. Clin. Invest.* 92:359–371.

85. Stassi, G., N. Schloot, and M. Pietropaolo. 1997. Islet cell autoantigen 69 kDa (ICA69) is preferentially expressed in the human islets of Langerhans than exocrine pancreas. *Diabetologia* 40:120–121.

86. Elliott, R.B., D.P. Harris, J.P. Hill, N.J. Bibby, and H.E. Wasmuth. 1999. Type I (insulin-dependent) diabetes mellitus and cow milk: casein variant consumption. *Diabetologia* 42:292–296.

87. Norris, J.M., B. Beaty, G. Klingensmith, L. Yu, M. Hoffman, H.P. Chase, H.A. Erlich, R.F. Hamman, G.S. Eisenbarth, and M. Rewers. 1996. Lack of association between early exposure to cow's milk protein and β-cell autoimmunity: Diabetes Autoimmunity Study in the Young (DAISY). *JAMA* 276:609–614.

88. Bottazzo, G.F., A. Florin-Christensen, and D. Doniach. 1974. Islet-cell antibodies in diabetes mellitus with autoimmune polyendocrine deficiencies. *Lancet* 2:1279–1283.

89. Greenbaum, C.J., J.P. Palmer, S. Nagataki, Y. Yamaguchi, J.L. Molenaar, W.A.M. Van Beers, N.K. Maclaren, Å. Lernmark, and Participating Laboratories. 1992. Improved specificity of ICA assays in the Fourth International Immunology of Diabetes Exchange Workshop. *Diabetes* 41:1570–1574.

90. Leslie, R.D., M.A. Atkinson and A.L. Notkins. 1999. Autoantigens IA-2 and GAD in Type I (insulin-dependent) diabetes. *Diabetologia* 42:3–14.

91. Falorni, A., E. Örtqvist, B. Persson and Å. Lernmark. 1995. Radioimmunoassays for glutamic acid decarboxylase (GAD65) and GAD65 autoantibodies using ^{35}S or ^{3}H recombinant human ligands. *J. Immunol. Methods* 186:89–99.

92. Solimena M., and P. DeCamilli. 1991. Autoimmunity to glutamic acid decarboxylase (GAD) in Stiff-man syndrome and insulin-dependent diabetes mellitus. *Trends Neurosci.* 14:452–457.

93. Rabin, D.U., S.M. Pleasic, R. Palmer-Crocker, and J.A. Shapiro. 1992. Cloning and expression of IDDM-specific human autoantigens. *Diabetes* 41:183–186.

94. Gianani, R., D.U. Rabin, C.F. Verge, L. Yu, S. Babu, M. Pietropaolo, and G.S. Eisenbarth. 1995. ICA512 autoantibody radioassay. *Diabetes* 44:1340–1344.

95. Lan, M.S., J. LU, Y. Goto, and A.L. Notkins. 1994. Molecular cloning and identification of a receptor-type protein tyrosine phosphatase, IA-2, from human insulinoma. *DNA Cell Biol.* 13:505–514.

96. Zhang, B., M.S. Lan, A.L. Notkins, and M.S. Lan. 1997. Autoantibodies to IA-2 in IDDM: location of major antigenic determinants. *Diabetes* 46:40–43.

97. Christie, M.R. 1996. Islet cell autoantigens in type I diabetes. *Eur. J. Clin. Invest.* 26:827–838.

98. Pietropaolo, M., J.C. Hutton, and G.S. Eisenbarth. 1996. Protein tyrosine phosphatase-like proteins: link with IDDM. *Diabetes Care* 20:208–214.

99. Kawasaki, E., G.S. Eisenbarth, C. Wasmeier, and J.C. Hutton. 1996. Autoantibodies to protein tyrosine phosphatase-like proteins in type I diabetes: overlapping specificities to phogrin and ICA512/IA-2. *Diabetes* 45:1344–1349.

100. Greenbaum, C.J., and J.P. Palmer. 1996. Humoral immune markers: insulin antibodies. In: Prediction, Prevention, and Genetic Counseling in IDDM. J.P. Palmer, editor. Wiley, Chichester, England. pp. 63–76.

101. Verge, C.F., D. Stenger, E. Bonifacio, P.G. Colman, C. Pilcher, P.J. Bingley, G.S. Eisenbarth, and Participating Laboratories. 1998. Combined use of autoantibodies (IA-2ab, Gadab, IAA, ICA) in type 1 diabetes: Combinatorial islet autoantibody workshop. *Diabetes* 47:1857–1866.

102. Williams, A.J.K., P.J. Bingley, E. Bonifacio, J.P. Palmer, and E.A.M. Gale. 1997. A novel micro-assay for insulin autoantibodies. *J. Autoimmun.* 10:473–478.

103. Naserke, H.E., N. Dozio, A.-G. Ziegler, and E. Bonifacio. 1998. Comparison of a novel micro-assay for insulin autoantibodies with the conventional radiobinding assay. *Diabetologia* 41:681–683.

104. Greenbaum, C.J., T.J. Wilkin, and J.P. Palmer. 1992. Fifth international serum exchange workshop for insulin autoantibody (IAA) standardization. *Diabetologia* 35:798–800.

105. Castano, L., A.G. Ziegler, R. Ziegler, S. Shoelson, and G.S. Eisenbarth. 1993. Characterization of insulin autoantibodies in relatives of patients with type 1 diabetes. *Diabetes* 42:1202–1209.

106. Atkinson, M.A., and N.K. Maclaren. 1993. Islet cell autoantigens in insulin-dependent diabetes. *J. Clin. Invest.* 92:1608–1616.

107. Verge, C.F., R. Gianani, E. Kawasaki, L. Yu, M. Pietropaolo, R.A. Jackson, H.P. Chase, and G.S. Eisenbarth. 1996. Prediction of type I diabetes in first-degree relatives using a combination of insulin, GAD, and ICA512bdc/IA-2 autoantibodies. *Diabetes* 45:926–933.

108. Bingley, P.J., M.R. Christie, E. Bonifacio, R. Bonfanti, M. Shattock, M.-T. Fonte, G.-F. Bottazzo, and E.A.M. Gale. 1994. Combined analysis of autoantibodies improves prediction of IDDM in islet cell antibody-positive relatives. *Diabetes* 43:1304–1310.

109. Gianani, R., C.F. Verge, R.I. Moromisato-Gianani, L. Yu, Y.J. Zhang, A. Pugliese, and G.S. Eisenbarth. 1996. Limited loss of tolerance to islet autoantigens in ICA+ first-degree relatives of patients with type I diabetes expressing the HLA DQB1*0602 alle. *J. Autoimmun.* 9:423–425.

110. Kawasaki, E., L. Yu, M.J. Rewers, J.C. Hutton, and G.S. Eisenbarth. 1998. Definition of multiple ICA512/phogrin autoantibody epitopes and detection of intramolecular epitope spreading in relatives of patients with type 1 diabetes. *Diabetes* 47:733–742.

111. Pugliese, A., E. Kawasaki, M. Zeller, L. Yu, S. Babu, M. Solimena, C.T. Moraes, M. Pietropaolo, R.P. Friday, M. Trucco, C. Ricordi, M. Allen, J.A. Noble, H.A. Erlich, and G.S. Eisenbarth. 1999. Sequence analysis of the diabetes-protective human leukocyte antigen-DQB1*0602 allele in unaffected, islet cell antibody-positive first degree relatives and in rare patients with type 1 diabetes. *J. Clin. Endocrinol. Metab.* 84:1722–1728.

112. Huang, W., J.-X. She, A. Muir, D. Laskowsa, B. Zorovich, D. Schatz, and N.K. Maclaren. 1994. High risk HLA-DR/DQ genotypes for IDD confer susceptibility to autoantibodies but DQBI*0602 does not prevent them. *J. Autoimmun.* 7:889–897.

113. Ziegler, A.G., P. Vardi, A.T. Ricker, M. Hattori, J.S. Soeldner, and G.S. Eisenbarth. 1989. Radioassay determination of insulin autoantibodies in NOD mice. Correlation with increased risk of progression to overt diabetes. *Diabetes* 38:358–363.

114. Kaufman, D.L., M. Clare-Salzler, J. Tian, T. Forsthuber, G.S. Ting, P. Robinson, M.A. Atkinson, E.E. Sercarz, A.J. Tobin, and P.V. Lehmann. 1993. Spontaneous loss of T-cell tolerance to glutamic acid decarboxylase in murine insulin dependent diabetes. *Nature* 366:69–72.

115. Geng, L., M. Solimena, R.A. Flavell, R.S. Sherwin, and A.C. Hayday. 1998. Widespread expression of an autoantigen-GAD65 transgene does not tolerize non-obese diabetic mice and can exacerbate disease. *Proc. Natl. Acad. Sci. USA* 95:10055–10060.

116. French, M.B., J. Allison, D.S. Cram, H.E. Thomas, M. Dempsey-Collier, A. Silva, H.M. Georgiou, T.W. Kay, L.C. Harrison, and A.M. Lew. 1996. Transgenic expression of mouse proinsulin II prevents diabetes in nonobese diabetic mice. *Diabetes* 46:34–39.

117. Yoon, J.W., C.S. Yoon, H.W. Lim, Q.Q. Huang, Y. Kang, K.H. Pyun, K. Hirasawa, R.S. Sherwin, and H.S. Jun. 1999. Control of autoimmune diabetes in NOD mice by GAD expression or suppression in beta cells. *Science* 284:1183–1187.

118. Gombert, J.M., A. Herbelin, E. Tancrede-Bohin, M. Dy, C. Carnaud, and J.F. Bach. 1996. Early quantitative and functional deficiency of NK1+-like thymocytes in the NOD mouse. *Eur. J. Immunol.* 26:2989–2998.

119. Park, S.-H., J.H. Roark, and A. Bendelac. 1998. Tissue-specific recognition of mouse CD1 molecules. *J. Immunol.* 160:3128–3134.

120. Chen, Y.H., N.M. Chiu, M. Mandal, N. Wang, and C.R. Wang. 1997. Impaired NK1+ T cell development and early IL-4 production in CD1-deficient mice. *Immunity* 6:459–467.

121. Hammond, K.J.L., L.D. Poulton, L.J. Palmisano, P.A. Silveira, D.I. Godfrey, and A.G. Baxter. 1998. α/β-T cell receptor (TCR)+CD4−CD8− (NKT) thymocytes prevent insulin-dependent diabetes mellitus in nonobese diabetic (NOD)/Lt mice by the influence of interleukin (IL)-4 and/or IL-10. *J. Exp. Med.* 187:1047–1056.

122. Ridgway, W.M., M. Fasso, and C.G. Fathman. 1999. A new look at MHC and autoimmune disease. *Science* 284:749,751.

123. Ridgway, W.M., H. Ito, M. Fasso, C. Yu, and C.G. Fathman. 1998. Analysis of the role of variation of major histocompatibility complex class II expression on nonobese diabetic (NOD) peripheral T cell response. *J. Exp. Med.* 188:2267–2275.

124. Carrasco-Marin, E., J. Shimizu, O. Kanagawa, and E.R. Unanue. 1996. The class II MHC I-A^{g7} molecules from nonobese diabetic mice are poor peptide binders. *J. Immunol.* 156:450–458.

125. Santamaria, P., T. Utsugi, B.-J. Park, N. Averill, S. Kawazu, and J.-W. Yoon. 1995. Beta-cell-cytotoxic CD8+ T cells from nonobese diabetic mice use highly homologous T cell receptor α-chain CDR3 sequences. *J. Immunol.* 154:2494–2503.

126. Wong, F.S., I. Visintin, L. Wen, R.A. Flavell, and C.A. Janeway. 1996. CD8 T cell clones from young nonobese diabetic (NOD) islets can transfer rapid onset of diabetes in NOD mice in the absence of CD4 cells. *J. Exp. Med.* 183:67–76.

127. Serreze, D.V., H.D. Chapman, D.S. Varnum, M.S. Hanson, P.C. Reifsnyder, D.R. Scott, S.A. Fleming, E.H. Leiter, and L.D. Shultz. 1996. B lymphocytes are essential for the initiation of T cell-mediated autoimmune diabetes: analysis of a new "speed-congenic" stock of NOD.Igu^{null} mice. *J. Exp. Med.* 184:2049–2053.

128. Atkinson, M.A., and N.K. Maclaren. 1994. The pathogenesis of insulin-dependent diabetes mellitus. *N. Engl. J. Med.* 331:1428–1436.

129. Roep, B.O. 1996. T-cell responses to autoantigens in IDDM. The search for the Holy Grail. *Diabetes* 45:1147–1156.

130. Peakman, M., J.M. Tredger, E.T. Davies, M. Davenport, J.B. Dunne, R. Williams, and D. Vergani. 1993. Analysis of peripheral blood mononuclear cells in rodents by three-colour flow cytometry using a small-volume lysed whole blood technique. *J. Immunol. Methods* 158:87–94.

131. Smerdon, R.A., M. Peakman, M.J. Hussain, L. Alviggi, P.J. Watkins, R. David, G. Leslie, and D. Vergani. 1993. Increase in simultaneous coexpression of naive and memory lymphocyte markers at diagnosis of IDDM. *Diabetes* 42:127–133.

132. Al-Sakkaf, L., P. Pozzilli, A.C. Tarn, G. Schwarz, E.A. Gale, and G.F. Bottazzo. 1989. Persistent reduction of CD4/CD8 lymphocyte ratio and cell activation before the onset of type 1 (insulin-dependent) diabetes. *Diabetologia* 32:322–325.

133. Buschard, K., P. Damsbo, and C. Ropke. 1990. Activated CD4+ and CD8+ T-lymphocytes in newly diagnosed type 1 diabetes: a prospective study. *Diabet. Med.* 7:132–136.

134. Petersen, L.D., G. Duinkerken, G.J. Bruining, R.A. van Lier, R.R. de Vries, and B.O. Roep. 1996. Increased numbers of in vivo activated T cells in patients with recent onset insulin-dependent diabetes mellitus. *J. Autoimmun.* 9:731–737.

135. Faustman, D., G. Eisenbarth, J. Daley, and J. Breitmeyer. 1989. Abnormal T-lymphocyte subsets in type I diabetes. *Diabetes* 38:1462–1468.

136. Palmer, J.P., C.M. Asplin, P. Clemons, K. Lyen, O. Tatpati, P.K. Raghu, and T.L. Paquette. 1983. Insulin antibodies in insulin-dependent diabetics before insulin treatment. *Science* 222:1337–1339.

137. Schloot, N.C., B.O. Roep, D. Wegmann, L. Yu, H.P. Chase, T. Wang, and G.S. Eisenbarth. 1997. Altered immune response to insulin in newly diagnosed compared to insulin-treated diabetic patients and healthy control subjects. *Diabetologia* 40:564–572.

138. Ellis, T.M., D.A. Schatz, E.W. Ottendorfer, M.S. Lan, C. Wasserfall, P.J. Salisbury, J.-X. She, A.L. Notkins, N.K. Maclaren, and M.A. Atkinson. 1998. The relationship between humoral and cellular immunity to IA-2 in IDDM. *Diabetes* 47:566–569.

139. Roep, B.O., G. Duinkerken, G.M.T. Schreuder, H. Kolb, R.R.P. De Vries, and S. Martin. 1996. HLA-associated inverse correlation between T cell and antibody responsiveness to islet autoantigen in recent-onset insulin-dependent diabetes mellitus. *Eur. J. Immunol.* 26:1285–1289.

140. Arden, S.D., B.O. Roep, P.I. Neophytou, E.F. Usac, G. Duinkerken, R.R.P. De Vries, and J.C. Hutton. 1996. Imogen 38: a novel 38-kD islet mitochondrial auto-antigen recognized by T cells from a newly diagnosed type I diabetic patient. *J. Clin. Invest.* 97:551–561.

141. Roep, B.O., S.D. Arden, R.P. deVries, and J.C. Hutton. 1990. T-cell clones from a type-1 diabetes patient respond to insulin secretory granule proteins. *Nature* 345:632–634.

142. Durinovic-Bello, I., M. Hummel, and A.G. Ziegler. 1996. Cellular immune response to diverse islet cell antigens in IDDM. *Diabetes* 45:795–800.

143. Zhang, J., S. Markovic-Plese, B. Lacet, J. Raus, H.L. Weiner, and D.A. Hafler. 1994. Increased frequency of interleukin 2-responsive T cells specific for myelin basic protein and proteolipid protein in peripheral blood and cerebrospinal fluid of patients with multiple sclerosis. *J. Exp. Med.* 179:973–984.

144. Petersen, L.D., M. van der Keur, R.R.P. De Vries, and B.O. Roep. 1999. Autoreactive and immunoregulatory T-cell subsets in insulin-dependent diabetes mellitus. *Diabetologia* 42:443–449.

145. Lundin, K.E., H. Scott, T. Hansen, G. Paulsen, T.S. Halstensen, O. Fausa, E. Thorsby, and L.M. Sollid. 1993. Gliadin-specific, HLA-DQ(alpha 1*0501, beta 1*0201) restricted T cells isolated from the small intestinal mucosa of celiac disease patients. *J. Exp. Med.* 178:187–196.

146. Miyazaki, I., R.K. Cheung, R. Gaedigk, M.F. Hui, J. Van der Meulen, R.V. Rajotte, and H.M. Dosch. 1995. T cell activation and anergy to islet cell antigen in type I diabetes. *J. Immunol.* 154:1461–1469.

147. Gjertsen, H.A., L.M. Sollid, J. Ek, E. Thorsby, and K.E. Lundin. 1994. T cells from the peripheral blood of coeliac disease patients recognize gluten antigens when presented by HLA-DR, -DQ, or -DP molecules. *Scand. J. Immunol.* 39:567–574.

148. Res, P.C., C.G. Schaar, F.C. Breedveld, W. Van Eden, J.D. van Embden, I.R. Cohen, and R.R. de Vries. 1988. Synovial fluid T cell reactivity against 65 kD heat shock protein of mycobacteria in early chronic arthritis. *Lancet* 2:478–480.

149. LaSalle, J.M., P.J. Tolentino, G.J. Freeman, L.M. Nadler, and D.A. Hafler. 1992. Early signaling defects in human T cells anergized by T cell presentation of autoantigen. *J. Exp. Med.* 176:177–186.

150. Roep, B.O. 1999. Standardization of T-cell assays in Type I diabetes. Immunology of Diabetes Society T-cell Committee. *Diabetologia* 42:636–637.

151. Roep, B.O., and R.R.P. De Vries. 1996. Surrogate end markers during the first year in recent onset IDDM for showing efficacy of treatment and during preventive trials. *Diabetes Prev. Ther.* 10:2–3.

152. Schmidt, D., J. Verdaguer, N. Averill, and P. Santamaria. 1997. A mechanism for the major histocompatibility complex-linked resistance to autoimmunity. *J. Exp. Med.* 186:1059–1075.

153. Verdaguer, J., D. Schmidt, A. Amrani, B. Anderson, N. Averill, and P. Santamaria. 1997. Spontaneous autoimmune diabetes in monoclonal T cell nonobese diabetic mice. *J. Exp. Med.* 186:1663–1676.

154. Cameron, M.J., G.A. Arreaza, P. Zucker, S.W. Chensue, R.M. Strieter, S. Chakrabarti, and T.L. Delovitch. 1997. IL-4 prevents insulitis and insulin-dependent diabetes mellitus in nonobese diabetic mice by potentiation of regulatory T helper-2 cell function. *J. Immunol.* 159:4686–4692.

155. Horwitz, M.S., L.M. Bradley, J. Harbertson, T. Krahl, J. Lee, and N. Sarvetnick. 1998. Diabetes induced by Coxsackie virus: initiation by bystander damage and not molecular mimicry. *Nat. Med.* 4:781–785.

156. Sarvetnick, N. 1997. IFN-γ, IGIF, and IDDM. *J. Clin. Invest.* 99:371–372.

157. Mueller, R., T. Krahl, and N. Sarvetnick. 1996. Pancreatic expression of interleukin-4 abrogates insulitis and autoimmune diabetes in nonobese diabetic (NOD) mice. *J. Exp. Med.* 184:1093–1099.

158. Wogensen, L., M.-S. Lee, and N. Sarvetnick. 1994. Production of interleukin 10 by islet cells accelerates immune-mediated destruction of β cells in nonobese diabetic mice. *J. Exp. Med.* 179:1379–1384.

159. Sarvetnick, N., J. Shizuru, D. Liggitt, L. Martin, B. McIntyre, A. Gregory, T. Parslow, and T. Stewart. 1990. Loss of pancreatic islet tolerance induced by β-cell expression of interferon-γ. *Nature* 346:844–847.

160. Giordano, C., G. Stassi, R. De Maria, M. Todaro, P. Richiusa, G. Papoff, G. Ruberti, M. Bagnasco, R. Testi, and A. Galluzzo. 1997. Potential involvement of Fas and its ligand in the pathogenesis of Hashimoto's thyroiditis. *Science* 275:960–963.

161. Bellgrau, D., D. Gold, H. Selawry, J. Moore, A. Franzusoff, and R. Duke. 1995. A role for CD95 ligand in preventing graft rejection. *Nature* 377:600–602.

162. Chervonsky, A.V., Y. Wang, F.S. Wong, I. Visintin, R.A. Flavell, C.A. Janeway, and L.A. Matis. 1997. The role of Fas in autoimmune diabetes. *Cell* 89:17–24.

163. Signore, A., A. Annovazzi, E. Procaccini, P.E. Beales, J. Spencer, R. Testi, and G. Ruberti. 1997. CD95 and CD95-ligand expression in endocrine pancreas of NOD, NOR and BALB/c mice. *Diabetologia* 40:1476–1479.

164. Pugliese, A., R. Gianani, R. Moromisato, Z.L. Awdeh, C.A. Alper, H.A. Erlich, R.A. Jackson, and G.S. Eisenbarth. 1995. HLA-DQB1 *0602 is associated with dominant protection from diabetes even among islet cell antibody-positive first-degree relatives of patients with IDDM. *Diabetes* 44:608–613.

165. Cook, J.J., I. Hudson, L.C. Harrison, B. Dean, P.G. Colman, G.A. Werther, G.L. Warne, and J.M. Court. 1989. Double-blind controlled trial of azathioprine in children with newly diagnosed type I diabetes. *Diabetes* 38:779–783.

166. Silverstein, J., N. Maclaren, W. Riley, R. Spillar, D. Radjenovic, and S. Johnson. 1988. Immunosuppression with azathioprine and prednisone in recent-onset insulin-dependent diabetes mellitus. *N. Engl. J. Med.* 319:599–604.

167. Feutren, G., G. Assan, G. Karsenty, H. DuRostu, J. Sirmai, L. Papoz, B. Vialettes, P. Vexiau, M. Rodier, A. Lallemand, and J.F. Bach. 1986. Cyclosporin increases the rate and length of remissions in insulin dependent diabetes of recent onset. Results of a multicentre double-blind trial. *Lancet* 2:119–124.

168. Bougneres, P.F., J.C. Carel, L., Castano, C. Boitard, J.P. Gardin, P. Landais, J. Hors, M.J. Mihatsch, M. Paillard, J.L. Chaussain, and J.F. Bach. 1988. Factors associated with early remission of type I diabetes in children treated with cyclosporine. *N. Engl. J. Med.* 318:663–670.

169. Martin, S., G. Schernthaner, J. Nerup, F.A. Gries, V.A. Koivisto, J. Dupre, E. Standl, P. Hamet, R. McArthur, M.H. Tan, K. Dawson, A.E. Mehta, S. Van Vliet, B. Von Graffenried, C. Stiller, and H. Kolb. 1991. Follow-up of cyclosporin A treatment in type I (insulin-dependent) diabetes mellitus: lack of long-term effects. *Diabetologia* 34:429–434.

170. Stiller, C.R., A. Laupacis, J. Dupre, M.R. Jenner, P.A. Keown, W. Rodger, and B.M. Wolfe. 1983. Cyclosporine for treatment of early Type I diabetes: preliminary results. *N. Engl. J. Med.* 308:1226–1227.

171. Shehadeh, N., F. Calcinaro, B.J. Bradley, I. Bruchlim, P. Vardi, and K.J. Lafferty. 1994. Effect of adjuvant therapy on development of diabetes in mouse and man. *Lancet* 343:706–707.

172. Klingensmith, G., H. Allen, A. Hayward, L. Stohry, C. Humber, P. Jensen, E. Simoes, and H.P. Chase. 1997. Vaccination with BCG at diagnosis does not alter the course of IDDM. *Diabetes* 46:193A.

173. Ziegler, A.G., P. Vardi, D.J. Gross, S. Bonner-Weir, L. Villa-Kamaroff, P. Halban, H. Ikegami, J.S. Soeldner, and G.S. Eisenbarth. 1989. Production of insulin antibodies by mice rejecting insulin transfected cells. *J. Autoimmun.* 2:219–227.

174. Chase, H.P., N. Butler-Simon, S. Garg, M. McDuffie, S.L. Hoops, and D. O'Brien. 1990. A trial of nicotinamide in newly diagnosed patients with type 1 (insulin-dependent) diabetes mellitus. *Diabetologia* 33:444–446.

175. Elliott, R.B., H.P. Chase, C.C. Pilcher, and B.W. Edgar. 1990. Effect of nicotinamide in preventing diabetes (IDDM) in children. *J. Autoimmun.* 3:61A.

176. Elliott, R.B., C.C. Pilcher, D.M. Fergusson, and A.W. Stewart. 1996. A population-based strategy to prevent insulin-dependent diabetes using nicotinamide. *J. Pediatr. Endocrinol. Meth.* 9:501–509.

177. Lampeter, E.F., A. Klinghammer, W.A. Scherbaum, E. Heinze, B. Haastert, G. Giani, and H. Kolb. 1998. The Deutsche Nicotinamide Intervention Study: an attempt to prevent type 1 diabetes. DENIS Group. *Diabetes* 47:980–984.

178. Gotfredsen, C.F., K. Buschard, and E.K. Frandsen. 1985. Reduction of diabetes incidence of BB Wistar rats by early prophylactic insulin treatment of diabetes-prone animals. *Diabetologia* 28:933–935.

179. Gottlieb, P.A., E.S. Handler, M.C. Appel, D.L. Greiner, J.P. Mordes, and A.A. Rossini. 1991. Insulin treatment prevents diabetes mellitus but not thyroiditis in RT 6-depleted diabetes resistant BB/Wor rats. *Diabetologia* 34:296–300.

180. Keller, R.J., G.S. Eisenbarth, and R.A. Jackson. 1993. Insulin prophylaxis in individuals at high risk of type 1 diabetes. *Lancet* 341:927–928.

181. Muir, A., R. Luchetta, H.-Y. Song, A. Peck, J. Krischer, and N. Maclaren. 1993. Insulin immunization protects NOD mice from diabetes. *Autoimmunity* 15:58.

182. Muir, A., A. Peck, M. Clare-Salzler, Y.-H. Song, J. Cornelius, R. Luchetta, J. Krischer, and N. Maclaren. 1995.

Insulin immunization of nonobese diabetic mice induces a protective insulitis characterized by diminished intraislet interferon-gamma transcription. *J. Clin. Invest.* 95:628–634.

183. Hancock, W.W., M. Polansky, J. Zhang, N. Blogg, and H.L. Weiner. 1995. Suppression of insulitis in non-obese diabetic (NOD) mice by oral insulin administration is associated with selective expression of interleukin-4 and –10, transforming growth factor-beta, and prostaglandin-E. *Am. J. Pathol.* 147:1193–1199.

184. Zhang, Z.J., L. Davidson, G. Eisenbarth, and H.L. Weiner. 1991. Suppression of diabetes in nonobese diabetic mice by oral administration of porcine insulin. *Proc. Natl. Acad. Sci. USA* 88:10252–10256.

185. Simone, E.A., D.R. Wegmann, and G.S. Eisenbarth. 1999. Immunologic "vaccination" for the prevention of autoimmune diabetes mellitus (type 1A). *Diabetes Care* 22:B7–B15.

186. Daniel, D., and D.R. Wegmann. 1996. Protection of nonobese diabetic mice from diabetes by intranasal or subcutaneous administration of insulin peptide B-(9–23). *Proc. Natl. Acad. Sci. USA* 93:956–960.

187. Füchtenbusch, M., W. Rabl, B. Grassl, W. Bachmann, E. Standl and A.-G. Ziegler. 1998. Delay of type 1 diabetes in high risk, first degree relatives by parenteral antigen administration: the Schwabing Insulin Prophylaxis Pilot Trial. *Diabetologia* 41:536–541.

188. Kobayashi, T., K. Nakanishi, T. Murase, and K. Kosaka. 1996. Small doses of subcutaneous insulin as a strategy for preventing slowly progressive β-cell failure in islet cell antibody-positive patients with clinical features of NIDDM. *Diabetes* 45:622–626.

35 | Animal Models of Type I Diabetes

Jean-François Bach

1. INTRODUCTION

Diabetes mellitus is a chronic disease defined by chronic hyperglycemia. Major progress was achieved in the 70's when the distinction was made between type I and type II diabetes. Type I diabetes is defined as a disease state requiring the daily administration of insulin for metabolic control (insulin dependent diabetes mellitus, IDDM). Type II diabetes, which is also associated with hyperglycemia and glycosuria (but without spontaneous evolution to ketosis), is characterized by usual absence of requirement for insulin therapy (non-insulin dependent diabetes mellitus, NIDDM). In addition to these important clinical differences, type I and type II diabetes differ by their putative underlying mechanisms. The vast majority of cases of type I diabetes relate to an autoimmune pathogenesis according to which the β cells of islets of Langerhans, which produce insulin, are destroyed by autoreactive T cells (1). Type II diabetes, in most cases, is a genetically controlled, purely metabolic disease enhanced by additional factors such as obesity leading to insulin resistance that the patients cannot overcome, probably due to some yet undefined β-cell failure. Recent data have indicated that 5–15% of type II diabetes cases are, in fact, also autoimmune (slow type I or late autoimmune diabetes of the adult, LADA) (2, 3).

The study of the pathogenesis of human IDDM is hampered by the quasi impossible access to the pancreas (only very few biopsy or autopsy specimens have been made available from recently diagnosed subjects). One is thus limited to the study of peripheral blood, which is known in organ-specific autoimmune diseases not to be very reliable, particularly for T cell-mediated diseases where many events take place in the target organ or in its vicinity. Additionally, if one accepts the possible follow-up of diabetic patients' relatives, the disease can only be studied

when it is established and there is only a little time between the onset of clinical disease and the final stage of the immune-mediated β-cell aggression: The totality of β cells is destroyed within 12 to 24 months after appearance of clinical symptoms. Hence the interest in animal models that circumvent these limitations for the study of IDDM.

Several animal models of spontaneous IDDM have been described in various species. We shall only discuss here the two rodent models that have been the subject of extensive studies: The nonobese diabetic (NOD) mouse and the Bio Breeding (BB) rat. We shall also discuss experimental diabetes induced in normal mice by repeated administration of low doses of streptozotocin (a model distinct from that of IDDM induced by the complete destruction of β cells following administration of a high single dose of streptozotocin or alloxan).

We shall finally describe several recent models of diabetes created by transducing various transgenes, either T cell receptor derived from T cells of diabetic mice or genes encoding various proteins whose expression is guided to the β cells when coupling the gene with the rat insulin gene promoter.

The goal of this review is to describe these various models presenting the natural history of the disease, particularly at the immunological level. We shall also discuss the major information gained from these models by the study of genetic factors as well as from therapeutic protocols aiming at preventing or curing the disease.

An effort will be made to correlate each model with human disease keeping in mind in all cases that human IDDM is probably an heterogeneous disease as a result of complex interaction of multiple genetic and environmental factors whereas experimental models represent the reproduction in multiple copies of a single individual.

2. THE NOD MOUSE

2.1. Origin

The NOD mouse was discovered in Japan in the early 80's (4) as a newly acquired phenotype in a colony of mice initially screened for type II diabetes. This mouse was bred to the homozygous state. A control strain (NON) was concomitantly developed, but was not used much thereafter because despite sharing a number of genes with NOD, it differs at the H-2 locus. A large number of NOD mouse colonies have been developed worldwide with extreme variations in the disease incidence. It was initially thought that genetic drifts could explain these variations. It is now understood that most of it is due to differences in pathogen environment: the cleaner the mouse, the higher the incidence of the disease (5).

2.2. Natural History

In colonies with the highest incidence, 80–90% of female mice become diabetic between 3 and 5 months of age. Disease only appears in approximately 50% of male mice. The gender difference is even more striking in low incidence colonies where males are essentially protected from disease.

Insulitis is the hallmark pathologic lesion of the NOD mouse diabetes. It evolves through several stages that successively include peri-insulitis, peripheral insulitis, invading insulitis, and destructive insulitis. Peri-insulitis is first observed at three weeks of age (6). It is controlled by a dominant gene located on Chromosome 1 (close to the bcl2 gene) that also codes for sialitis (i.e., an inflammatory lesion of the salivary glands) (7). Peri-insulitis is also observed in H-2 congenic NOD mice with H-2^k (8). Invading insulitis starts at 8–10 weeks, with destructive insulitis appearing not long before the onset of diabetes (9). Sequential quantitative immunohistochemical studies of the endocrine pancreas have indicated that the number of islets and the volume density of endocrine components were only reduced in mice showing fully invasive insulitis (i.e., heavy mononuclear cell infiltrate).

2.3. Autoimmune Abnormalities

T cells from NOD mice proliferate in the presence of islet extracts or chemically-defined β cell autoantigens, notably insulin and glutamic acid decarboxylase (GAD) (10). These autoreactive cells are found both in the spleen and in the islets.

The presence of pathogenic T cells is confirmed by the capacity of spleen cells from diabetic mice to transfer diabetes in immunoincompetent prediabetic NOD mice [neonates (11), scid NOD (12, 13), irradiated NOD (14)]. This transfer requires the concomitant presence of CD4 and CD8 T cells (11). It can be obtained with T cells from diabetic mice, but only when the spleen is collected shortly before the presumed appearance of diabetes.

Islet-reactive autoantibodies are also found, but at a much lower incidence and titer than in human IDDM (15). The most prominent antibodies are directed against insulin and GAD. Interestingly, the islet cell antibody (ICA) tests using indirect immunofluorescence on pancreas sections, which is widely used for human IDDM, is essentially negative in NOD mice.

NOD mice present manifestations of autoimmunity other than diabetes. They produce anti-nuclear antibodies (16), including anti-Sm antibodies (A.M. Yamamoto, unpublished results), may develop autoimmune hemolytic anemia (17) and often show sialitis and thyroiditis (18).

2.4. Genetic Predisposition

Extensive studies have been devoted to the characterization of the multiple genes predisposing to diabetes in NOD mice (19).

The major genes predisposing to diabetes are MHC genes. The NOD mouse presents a fairly unique class II haplotype (IA^{g7}) (20). It should be mentioned that the β chain of IA^{g7} does not contain an aspartic acid residue at position 57 as most other mouse strains do (20), a property reminiscent of what has been described for human diabetics (position 57 on the DQβ-chain).

NOD mice have been crossed with a number of non-autoimmune strains (F$_2$, backcrosses) and the segregation of the diabetic trait or of partial phenotype has been studied in correlation with polymorphic microsatellite markers. The group of John Todd made a pioneering contribution using this approach. The existence of more than 15 predisposing genes has now been delineated (19). Apart from MHC, which has just been mentioned, only one gene was approached with reasonable certainty: the interleukin (IL-) 2 gene (21). It has not been proven, however, whether the polymorphism found in this gene is at the origin of the modest deficiency of IL-2 secretion found in NOD mice or whether there is any relationship between this abnormality and the pathogenesis of diabetes. The other chromosomal areas are still relatively ill-defined in spite of the generation of very helpful, diabetes-resistant congenic strains that have all the NOD genes except those contained in the chromosomal areas in question. In some instances, these chromosomal areas appear to include two or more closely-linked susceptibility genes, making their fine mapping particularly laborious. The problem is complicated further by the fact that there are also protector genes. In fact, paradoxically, some of these genes might be present in the NOD mouse, as has been demonstrated for the FcgrII gene, which shows an alteration in the NOD mouse that prevents its expression on macrophages (22). More generally, it will be important to analyse the respective roles of these different genes with the plausible assumption that their import-ance is quite variable. In other words, one would like to know what the major predisposing genes are in addition to those contained in the MHC. Furthermore, the fact that some of these loci belong to chromosomal regions that have been associated with susceptibility in other models of autoimmunity, such as EAE or systemic lupus, provides a genetic basis for the existence of two kinds of genes: Genes involved in the non-antigen-specific dysregulation of the immune system (that may be common to different autoimmune diseases), and genes that determine the organ specificity of the autoimmune manifestations.

2.5.   Comparison With Human Disease

There has been some debate on the relevance of the NOD mouse model to human IDDM. The gender difference (high female prevalence) is not found when considering the whole population of diabetic patients. However, it is observed in the subgroup of so-called type Ib patients, who present extrapancreatic autoimmunity in association with diabetes as NOD mice do (23). The absence of ICAs mentioned above, is probably trivial, since other autoantibodies are found in the NOD mouse. The fact that NOD mouse disease can be prevented by a wide spectrum of therapeutic measures is perhaps more worrying because of the large number of efficacious methods reported (1, 24). It should be noted, however, that all these apparently easy successes were obtained when administering these treatments at a young age, before the appearance of a sufficient number of pathogenic T cells. Very few of these methods have been shown to be efficacious when diabetes has appeared. Otherwise, it is generally accepted that NOD mouse diabetes closely resembles human IDDM, making it a very good model for the study of its mechanisms and the evaluation of therapeutic methods.

2.6.   Genetically-Modified NOD Mice

A large number of genetically-modified NOD mouse lines have been raised by many laboratories using either the transgenic or the homologous recombination approaches (Table 1).

Transgenic NOD mice.

Rat insulin promoter (RIP)   Many interesting transgenic mice have been produced after coupling the transgene with the RIP, which allows highly selective expression of the transgene in the β cells (with possible very low level expression in non-pancreatic organs, including the thymus). This approach has been notably used by N. Sarvetnik's group, who obtained expression of several major cytokines in the β cells of NOD mice. The disease was prevented by expression of IL-4 (25), TGFβ (26, 27) and TNF (28). It was accelerated by IL-10 (30, 31) (although systemic IL-10 administration, like that of IL-4, protects NOD mice from diabetes).

Table 1   Genetically-modified NOD mice

			References
Transgenic mice			
RIP		IL-4	25
		TGFβ	26, 27
		TNFα	28
		IL-10	30, 31
		B7.1	32–34
		Fas-L	35
Other promoters		TCR	
		NOD	36
		C57BL/6	37
		scid	38
		Cα KO	39
		Vα14-Jα281	40
		CTLA4-Ig	41
		TNF receptor	29
		pro-insulin	42
		hsp60	43
		GAD	44, 45
		MHC class II	46–49
		MHC class II (IA)	50
		MHC class II (IE)	49, 51–54
Genetically-deficient mice		MHC class I (β2m)	57–59, 62
		MHC class II	60
		CIITA	61
		μ chain	63–67
		IFNγ	68
		IFNγ receptor	69
		TNF receptor	70
		IL-12	71
		CD28	41
		IL-4	72
		Fas (lpr)	73–75
		perforin	76

Interestingly, transgenic expression of the co-stimulatory molecule B7.1 accelerates diabetes onset in NOD mice, which it does not do in non-autoimmune prone C57BL/6 mice (32–34).

It has recently been reported that Fas-Ligand expression in certain privileged sites could protect from immune attack through the destruction of Fas+-activated T cells. Hence, the idea to induce Fas-Ligand expression in β cells. β cell expression of Fas-Ligand has provided variable results, probably linked to the variations in expression levels of the transduced genes. In some cases the disease was accelerated, in others it was slowed down (35, A. Lehuen and N. Glaichenhaus, in preparation).

Other promoters.   Other interesting transgenic NOD mice have been obtained by injection of particular T cell receptor genes in NOD embryos. Transduction of TCR genes derived from a pathogenic T cell clone induced accelerated diabetes on the NOD background (36) and, even more strikingly, on the C57BL/6 background (37). The disease was still more spectacular when transgenic mice were backcrossed to NOD scid (38) or TCR Cα knock-out (39, L. Chatenoud, unpublished observations) mice to exclude the expression of other than the trans-genic TCR. The latter data suggests that immune regulation, which delays the onset of diabetes in the TCR-transgenic NOD mice, is mediated by non-transgenic T cells.

Transgenic mice expressing a Vα14-Jα281 TCR gene which demonstrate overexpression of NK T cells, have been shown to be partially protected from diabetes (40).

Transgenic NOD mice expressing large amounts of CTLA4-Ig, which abrogates the CTLA4-B7 interaction, present acceleration of disease onset, contrasting with the protective effect of late CTLA4-Ig treatment (41).

One should lastly mention transgenic NOD mice overexpressing β cell candidate autoantigens produced with the aim of assessing the capacity of the transgene product to induce tolerance and diabetes protection. Such protection was indeed obtained (although to a variable degree) with pro-insulin (42), hsp60 (43) and in some transgenic lines with GAD (44), but not in some others (44, 45, D. Jeske and A. Lehuen, in preparation), even though GAD expression in the thymus was demonstrated at the protein level (D. Jeske and A. Lehuen, in preparation). Even in protected mice, it appeared for the three antigens that tolerance was probably of the active type rather than deletional, as would have been expected. The fact that GAD transgenic expression can induce tolerance in non-autoimmune strains (45) might suggest that there is indeed an abnormality in tolerization to β cell antigens in NOD mice.

Other interesting types of transgenic NOD mice have been produced that express nonNOD MHC class II molecules. In most models, there was protection from the disease (46–49) and protection was of the dominant type. It could be transferred to non-transgenic mice by T cells from transgenic mice and was abrogated by cyclophosphamide treatment (46,47). The interpretation of these experiments is hampered by the unexpected observation that control mice transduced with the syngenic MHC class II (IA^{g7}) are also protected, indicating that abnormal expression of class II MHC molecules may be sufficient to prevent disease onset independent of MHC polymorphism (50). In the same vein, transgenic expression of IE (which is not expressed in NOD mice) also induces disease protection (49,51–53). The mechanisms of this protection are very elusive but could also relate to immune regulation (54).

Genetically-deficient mice. Diabetes can be prevented in NOD mice by inactivation of a number of genes. This was initially shown by crossing NOD mice with scid (55) or nude mice (56), which do not harbour immunocompetent T cells. This has also been shown by inactivation by homologous recombination of MHC class I (57–59) and MHC class II (60) genes. Interestingly, NOD

mice lacking the CIITA (class II transactivator) molecule, rendering them deficient in MHC class II expression and peripheral CD4 T cells, show insulitis but no diabetes (61). Concerning MHC class I-deficient mice, it has been reported that RIP β2m transgene expression restores insulitis, but not diabetes, in β2m$^{-/-}$ NOD mice (62). The latter data emphasize the role of class I and presumably CD8 T cells in the disease initiation. This conclusion is corroborated by the fact that diabetogenic T cells derived from diabetic mice do not transfer diabetes in class I-deficient mice, but do so when such mice transgenically express β2m in the islets (62).

Several lines of B cell-deficient NOD mice have been developed using inactivation of the μ chain expression. B cell deficiency was reported in four studies (63–66), but not in a fifth, with absence of diabetes prevention (67) through a mechanism yet to be explained. Some doubt persists in some of these cases regarding the direct incrimination of the targeted gene (rather than neighbouring genes).

It should also be noted that the disease course was unchanged in mice deficient in IFNγ (68), a rather unexpected result because of the putative role of IFNγ in the disease pathogenesis as corroborated by the disease prevention observed in IFNγ receptor knockout (KO) mice (69). Diabetes onset was also prevented in mice deficient in TNF receptor (70), but not IL-12 (71). The diabetes course was accelerated in CD28 Kd (41) but not in IL-4 KO mice (72).

Genetically-deficient mice were also used to study the mechanisms of β cell death. Contradictory results were reported on Fas deficiency using NOD mice expressing the lpr gene (Fas mutation). N. Itoh et al. (73) reported that such mice are protected from diabetes. However, other reports indicated that this effect could be related to the immune abnormalities of lpr mice, since islets of NOD lpr mice are normally destroyed in wild-type diabetic NOD mice (74, 75). This latter data would tend to exclude a major role for Fas in β cell death. In this context, the decrease of diabetes development observed in perforin KO mice is interesting to note (76).

3. THE BB RAT

3.1. Origin

The BB rat was discovered in Canada in the late 70's (77). It was principally developed in the 80's by the group of Rossini and Like in Worcester, Mass. (78) who, in fact, were the first to bring direct experimental evidence for the autoimmune origin of diabetes. BB rats can be compared to an interesting control strain sharing the same MHC, but showing diabetes resistance: The BB diabetes-resistant line (DR-BB rat) (78,79).

3.2. Natural History

BB rats first present insulitis and then diabetes, which appears between 60 and 120 days of age. Like NOD mice, BB rats show higher incidence of the disease in pathogen-free conditions. However, there is a virus (Kilham virus) that accelerates the disease onset (80).

A unique peculiarity of BB rats is the existence of a major lymphocytopenia (81) that essentially involves RT6 lymphocytes (82). It is interesting to note that DR-BB rats do not present such lymphocytopenia, and that diabetes may be induced in these rats by depleting RT6+ cells by anti-RT6 antibody treatment (83).

3.3. Genetic Factors

The MHC plays a central role. Four other genes have been described, including a gene located on Chromosome 4, that controls the onset of lymphocytopenia (84).

3.4. Comparison With Human IDDM

The BB rat model is very close to human IDDM in many respects. A highly significant difference is the existence of the lymphocytopenia discussed above. This is an important difference, since such lymphocytopenia may be a major contributing factor to the disease pathogenesis. We shall see below that deliberate induction of a similar lymphocytopenia in certain strains of normal rats is sufficient to induce diabetes.

4.   INDUCED MODELS OF IDDM

4.1. Low-Dose Streptozotocin

Repeated injections of low doses of streptozotocin (STZ) induces the onset of insulitis followed by diabetes (85). Non-autoimmune strains show very variable sensitivity to the induction of such diabetes, with a particular sensitivity of CD-1, C57BL/KS and C57BL/6 mice (86). Interestingly, NOD mice are even more sensitive than C57BL/6 to STZ (87). There has been some argument about the autoimmune origin of this model of diabetes because of the observation that NOD scid mice, which do not harbour immunocompetent lymphocytes, are slightly more sensitive to low doses of streptozotocin than wild type NOD mice (88). However, since scid mice are defective in DNA repair (89), their hypersensitivity to low doses of STZ (which generates DNA alkylation) can be alternatively explained at the level of the β cell itself (88). On the other hand, STZ-induced IDDM is prevented by anti-CD3 immunotoxin (90) and is reduced by TCRα-chain gene inactivation (91). The mechanisms of insulitis and diabetes appear to relate to STZ-induced changes in islet immunogenicity: Insulitis only appears on islets grafted in STZ-treated mice if grafting is performed before STZ administration, or if the islets

are first exposed to STZ in vitro (92). The mechanisms of these changes are not fully understood, but might be related to the induction by STZ of increased expression on β cells of class II molecules of the major histocompatibility complex. This increased expression has been directly visualized, and low dose STZ-induced diabetes is prevented by antiinterferon γ (IFNγ) antibody therapy, which is known to inhibit MHC molecule expression (93).

4.2. Lymphocytopenia-Associated Diabetes

Rats thymectomized as adults and sublethally irradiated may develop diabetes and thyroiditis (94). The disease can be prevented by RT6+ CD62L + T cells (95).

Similarly, mice thymectomized at three days of age may develop insulitis, but not diabetes (96). Recent studies by S. Sakaguchi have allowed the characterization of the regulatory cells depleted by thymectomy, which show a CD4+ CD25+ phenotype (97, 98).

4.3. Autosensitization

Efforts have been made to induce diabetes de novo by immunizing against β cell antigens. In surprising contrast with many other models of experimentally-induced auto-immune diseases, it has not appeared possible to induce IDDM by immunizing mice against islet extracts. Similarly, only very limited results have been obtained with chemically-defined β cell autoantigens. Mice immunized with insulin may show insulitis, but no clinical diabetes (99). C57BL/6 mice immunized with p277 peptide derived from heat shock protein 60 may present diabetes when p277 is coupled to a carrier protein, but the disease is very transient (100). The only convincing, but still indirect, data is the demonstration that GAD-specific T cell clones obtained after immunisation against GAD may transfer diabetes even though the rats from which the clones were obtained did not show diabetes (101). The difficulty in producing such experimental diabetes is not understood. It could be related to the ease, with which one can protect NOD mice from diabetes, i.e. by injecting them with hsp 60, insulin or GAD, whatever the route of administration (sc with adjuvant, iv, oral, nasal or intrathymic). It is possible that autoimmunization against β cell antigen elicits the appearance of regulatory cells at least as easily as effector cells, which results in tolerance rather than autoimmunity.

5.   DIABETES IN GENETICALLY-MODIFIED NON-AUTOIMMUNE STRAINS

5.1. Introduction

IDDM is a T-cell mediated autoimmune disease targeting the β cells of the islets of Langerhans. Arguments to be detailed in the next section indicate that the pathogenic autoimmune response is the result of the development of an immune response against β cell autoantigens and of a deficient immune regulation that renders the β cell-specific response chronic. As first pioneered by D. Hanahan (102), it has appeared relatively easy to express large amounts of the β cell transgenes that had been coupled to the RIP. Such a strategy has been applied to several antigens and cytokines with remarkable success (Table 2).

This approach has been used in a number of different contexts. Schematically, one may distinguish single- from double-transgenic mice. In single-transgenic mice, the protein expression in the β cells is performed in the absence of any other treatment or crossing. Double-transgenic mice are the offspring of the RIP-transgenic mouse (expressing the antigen on the β cells) crossed with another transgenic mouse expressing a given immunological feature, creating the conditions for the development of a diabetogenic autoimmune response.

5.2. β Cell Antigen Overexpression

In his initial experiments, Hanahan used the SV40 oncogene, which triggers oncogenesis, associated with the β-cell-specific expression of the SV40 T antigen (102). Such transgenic mice showed two different patterns of clinical manifestations. When the SV40 T antigen was expressed very early in ontogeny, the mice became tolerant to the antigen and developed insulinoma. Conversely, when the T antigen was expressed late, there was no tolerance and the mice developed insulitis without full IDDM.

Two other viral antigens have also been expressed in β cells: The lymphochoriomeningitis virus (LCMV) glycoprotein (103) and the influenza virus hemagglutinin (HA) (104,105). In the two cases, the transgenic mice became tolerant to these antigens, probably due to minor expression in the thymus as directly demonstrated (105a). Consequently, they did not spontaneously become diabetic. Immunologically-mediated diabetes could, however, be induced when the mice were infected by the corresponding virus (103) or when they were crossed with another transgenic mouse expressing a TCR specific to the antigen expressed in the β cells (104,106–108). This was not sufficient, however, at least when class I restricted TCR were used, since one had to infect the double transgenic mice by the corresponding virus to get diabetes in this case (106). This now classical experiment led to the concept of ignorance (or indifference) to explain the fact that diabetes may not appear in spite of a very high expression of a β cell antigen and a very large number of T cells expressing TCR specific of that antigen. Superadded activation of such T cells may thus be necessary to induce diabetes.

5.3. MHC Molecules

In 1983 Bottazzo and Feldman proposed that aberrant expression of MHC class II molecules could be at the origin of some organ-specific autoimmune diseases, such as

Table 2   Insulitis and immune-mediated diabetes in transgenic non-autoimmune prone strains using the rat insulin promoter (RIP)

	References
Diabetes	
LCMV glycoprotein	103, 105a, 106
Influenza virus hemagglutinin	104, 105, 107, 108
B7.1 + TNF	121
B7.1 + IL-2	122, 127
IFNα	126
IL-10	128
Insulitis only	
SV 40	102
IFNγ	125
TNFα	121, 129, 130
CD86	131
MCP-1	132
IL-6	133

thyroiditis and diabetes (109). This hypothesis was essentially based on the observation that such aberrant expression was observed in the thyroid of patients with thyroiditis (110). Even if this data was not always confirmed, it appeared logical to test the hypothesis by deliberately overexpressing MHC class I (111) or class II (112–114) molecules in β cells. Such induction did not lead to the onset of Type I diabetes except in few cases, but it was then shown that the β cell failure was not autoimmune in nature, but rather due to β cell functional disturbance secondary to overexpression of the transgene (111). More recently, a number of groups have generated transgenic mice expressing HLA molecules selected for their association with human IDDM, notably DQ8 (115–119). These mice do not appear to develop diabetes spontaneously. It would be interesting to determine which added factor is necessary to induce the disease. In any case, these mice have proven to be very useful to study the in vivo immune response of HLA restricted T cells to β cell antigens as GAD and their epitopes.

5.4. B7.1

Overexpression of B7.1 in the β cells is not sufficient to induce diabetes in non-autoimmune prone mice (120). However, when B7.1 transgenic mice are crossed with other transgenic mice expressing various cytokines, such as TNF (121) or IL-2 (122), diabetes develops. Diabetes is also observed when transgenic mice expressing the LCMV-glycoprotein antigen are crossed with a RIP-B7.1 transgenic mouse in the absence of viral infection (123, 124).

5.5. Cytokines

Over the years, Sarvetnik's group has developed a large number of RIP-transgenic mice expressing various cytokines in β cells. The first and remarkable observation was that IFNγ-transgenic mice developed full insulitis that could be transferred to normal islets grafted in transgenic mice (125). The mechanisms of IFNγ insulitis are not totally clear, but probably involve overexpression of MHC molecules in a manner more physiological than the direct transgenic overexpression of MHC molecules discussed above. Interestingly, β cell expression of IFNα has also been shown to induce insulitis and diabetes (126).

Similar data have been reported for IL-2 transgenic mice, but in that case, induction of diabetes sometimes required co-expression of B7.1 (127).

More unexpectedly, IL-10 transgenic mice became diabetic (128), which was surprising since IL-10 is a Th2 cytokine and was not thought to contribute to the expression of a presumably Th1 disease. It should be noted, however, that the pancreatic T cell infiltrate in these mice was not of the usual type seen in NOD mice, since it also involved the exocrine pancreas.

TNF-transgenic mice have been produced by several groups (121, 129, 130). Insulitis was observed, but without diabetes.

5.6. Other Proteins

As just mentioned for TNF, a number of other proteins expressed in the β cells transgenically give rise to insulitis without diabetes. This is the case of CD86 (131), monocyte chemoattractant protein-1 (MCP-1) (132) and IL-6 (133). It is difficult to relate this state of isolated insulitis to real diabetes.

6.   CONCLUSIONS: LESSONS FROM ANIMAL MODELS FOR HUMAN IDDM

The study of these very diverse experimental models of IDDM have made considerable contributions to the understanding of the mechanisms of type 1 diabetes. There is, of course, the limitation of the heterogeneity of human IDDM, which does not have its counterpart in each individual experimental system. Several general lessons can, however, be taken from animal models.

6.1.   Natural History

It is now perfectly clear that clinical diabetes is preceded by a very long phase of non-destructive insulitis that probably starts very early in the life of predisposed individuals. Taken together, the various models described indicate that both CD4 and CD8 T cells can mediate diabetes alone but, at least in the spontaneous models, both CD4 and CD8 cells are implicated.

It remains to be determined what the respective contribution of each cell subset is. The disease is indeed T cell-mediated, since transfer of purified T cells into agammaglobulinemic NOD mice induces diabetes (134). Nevertheless B cell-deficient NOD mice do not develop diabetes, which raises the question of the role of B cells in disease pathogenesis. Autoantibodies are found in experimental diabetes, although less prevalently than in human IDDM. Their role in disease pathogenesis is probably a minor one.

As far as the target antigens and T cell repertoire are concerned, it appears that there is no unique antigen involved, but that many antigens can be targeted to induce diabetes either separately, as exemplified by transgenic mice overexpressing a given antigen in the β cells, or concomitantly, as in the case of NOD mice through a phenomenon of antigen spreading.

6.2.   Genetics

The study of the genetics of diabetes in the various models described, NOD mice and BB rats as well as other models, has been extremely informative. A large number of genes have been mapped and a few identified. One may hope that transgenic models and congenic-resistant NOD mice will provide the means to characterize the genes in question. The task is not easy considering that there are often no mutation in the suspect gene, but only a polymorphism whose fortuitous combination with other gene polymorphisms creates disease predisposition. Hopefully, because of their simplified nature, transgenic models will help to characterize the genes responsible for partial phenotypes, as initiated in the BDC 2.5 mice expressing a diabetogenic T-cell receptor (37).

6.3.   Immune Regulation

The study of NOD mice and also other models has brought direct evidence of the existence of immunoregulatory CD4 cells that control the onset of diabetes. Such regulatory cells may soon be available in a clonal form, which will allow the production of further animal models with transgenic expression of the regulatory T cell TCR.

6.4.   Immune Intervention

Animal models of diabetes have proven to be remarkable models for the study of immunointervention. Results obtained with chemicals and various products of biotechnology such as cytokines, monoclonal antibodies or soluble receptor, open new perspectives in the immunoprevention of diabetes, and can also be used as references for the treatment of many other diseases. As an example, Table 3 presents a list of therapeutic maneuvers that have been shown to prevent diabetes onset or cure established disease in NOD mice.

Table 3   Immunotherapy of diabetes in NOD mice: a non-exhaustive list (adapted from Bach J.F., Endocrine Rev., 1994, 516)

Agent	Prevention treatment started ≤ 3 months of age	Prevention of diabetes transfer (treatment of the recipient)	Prevention of cytoxan-induced IDDM	Treatment of overt diabetes (started after the onset of hyperglycemia)
Immunomanipulation				
Neonatal thymectomy	+			
Allogeneic bone marrow transplantation	+			
Backcross to nude mice	+			
Backcross to scid mice	+			
Intrathymic islet grafting	+			
CD4 T cells		+		
Immunosuppressive agents				
Cyclosporin	+			± −
FK506	+		+	
Deoxyspergualin			+	
Rapamycin	+			−
ALS				+
Monoclonal antibodies				
αCD3	+			
αTCR	+		+	+
αVβ8			+	+
αCD4	+	+		
αCD8	+	+	+	+
CD28, CD40L	+			+
αclass I			+	
αclass II	+	+ (neonates)	+	
αIL-2R	+			
αCD45RA	+			
αγIFN		+	+	
α IL 6				
Cytokines				
IL-1	+	+		
IL-4	+			
IL-10	+			
TNFα	+	+		
IL-2 toxin			+	
Miscellaneous				
CTLA4-Ig	+			
CFA	+			
BCG	+		+	
Antioxidants	+ (with steroids)			
Aminoguanidine (NO inhibition)		+		
Vitamin D3 and analogs	+ (insulitis)			
Gangliosides	+			
Con A	+			
Hsp65/peptide	+			
Insulin (parenteral, nasal, oral)	+			
Insulin choleratoxin conjugate (oral)	+			
Diets	+			
Nicotinamide	+			± −
Immunoglobulins	+			
Silica	+		+	
Viruses	+			

±: suppression of diabetes −: no effect

ACKNOWLEDGEMENTS

The author wishes to thank A. Lehuen, H.J. Garchon and M. Lévi-Strauss for helpful discussion.

References

1. Bach, J.F. 1994. Insulin-dependent diabetes mellitus as an autoimmune disease. *Endocrine Rev.* 15:516–542.
2. Tuomi, T., A. Carlsson, H. Li, B. Isomaa, A. Miettinen, A. Nilsson, M. Nissen, B.O. Ehrnstrom, B. Forsen, B. Snickars, K. Lahti, C. Forsblom, C. Saloranta, M.R. Taskinen, and L.C. Groop. 1999. Clinical and genetic characteristics of type 2 diabetes with and without GAD antibodies. *Diabetes* 48:150–157.
3. Turner, R., I. Stratton, V. Horton, S. Manley, P. Zimmet, I.R. MacKay, M. Shattock, G.F. Bottazzo, and R. Holman. 1997. UKPDS 25: Autoantibodies to islet-cell cytoplasm and glutamic acid decarboxylase for prediction of insulin requirement in type 2 diabetes. *Lancet* 350:1288–1293.
4. Makino, S., K. Kunimoto, Y. Muraoka, Y. Mizushima, K. Katagiri, and Y. Tochino. 1980. Breeding of a non-obese, diabetic strain of mice. *Exp. Anim.* 29:1–13.
5. Ohsugi, T., and T. Kurosawa. 1994. Increased incidence of diabetes mellitus in specific pathogen-eliminated offspring produced by embryo transfer in NOD mice with low incidence of the disease. *Lab. Anim. Sci.* 44:386–388.
6. Fujino-Kurihara, H., H. Fujita, A. Hakura, K. Nonaka, and S. Tarui. 1985. Morphological aspects on pancreatic islets of non-obese diabetic (NOD) mice. *Virchows Arch. B Cell Pathol. Incl. Mol. Pathol.* 49:107–120.
7. Garchon, H.J., P. Bedossa, L. Eloy, and J.F. Bach. 1991. Identification and mapping to chromosome 1 of a susceptibility locus for periinsulitis in non-obese diabetic mice. *Nature* 353:260–262.
8. Wicker, L.S., M.C. Appel, F. Dotta, A. Pressey, B.J. Miller, N.H. Delarato, P.A. Fischer, J.R. Boltz R.C., and L.B. Peterson. 1992. Autoimmune syndromes in major histocompatibility complex (MHC) congenic strains of nonobese diabetic (NOD) mice. The NOD MHC is dominant for insulitis and cyclophosphamide-induced diabetes. *J. Exp. Med.* 176:67–77.
9. Debussche, X., B. Lormeau, C. Boitard, M. Toublanc, and R. Assan. 1994. Course of pancreatic beta cell destruction in prediabetic NOD mice: a histomorphometric evaluation. *Diabetes Metab.* 20:282–290.
10. Tisch, R., X.D. Yang, S.M. Singer, R.S. Liblau, L. Fugger, and H.O. McDevitt. 1993. Immune response to glutamic acid decarboxylase correlates with insulitis in non-obese diabetic mice. *Nature* 366:72–75.
11. Bendelac, A., C. Carnaud, C. Boitard, and J.F. Bach. 1987. Syngeneic transfer of autoimmune diabetes from diabetic NOD mice to healthy neonates. Requirement for both L3T4+ and Lyt-2+ T cells. *J. Exp. Med.* 166:823–832.
12. Rohane, P.W., A. Shimada, D.T. Kim, C.T. Edwards, B. Charlton, L.D. Shultz, and C.G. Fathman. 1995. Islet-infiltrating lymphocytes from prediabetic NOD mice rapidly transfer diabetes to NOD-scid/scid mice. *Diabetes* 44:550–554.
13. Christianson, S.W., L.D. Shultz, and E.H. Leiter. 1993. Adoptive transfer of diabetes into immunodeficient NOD-scid/scid mice. Relative contributions of CD4+ and CD8+ T-cells from diabetic versus prediabetic NOD.NON-Thy-1a donors. *Diabetes* 42:44–55.
14. Wicker, L.S., B.J. Miller, and Y. Mullen. 1986. Transfer of autoimmune diabetes mellitus with splenocytes from nonobese diabetic (NOD) mice. *Diabetes* 35:855–860.
15. Bach, J.F. 1987. The natural history of islet-specific autoimmunity in NOD mice. In: NOD Mice and Related Strains: Research Applications in Diabetes, Aids, Cancer and Other Disease, E. Leiter and M. Atkinson, editors. RG Landes, Austin, pp. 121–144.
16. Humphreys-Beher, M.G., L. Brinkley, K.R. Purushotham, P.L. Wang, Y. Nakagawa, D. Dusek, M. Kerr, N. Chegini, and E.K. Chan. 1993. Characterization of antinuclear auto-antibodies present in the serum from nonobese diabetic (NOD) mice. *Clin. Immunol. Immunopathol.* 68:350–356.
17. Baxter, A.G., and T.E. Mandel. 1991. Hemolytic anemia in non-obese diabetic mice. *Eur. J. Immunol.* 21:2051–2055.
18. Many, M.C., S. Maniratunga, and J.F. Denef. 1996. The non-obese diabetic (NOD) mouse: An animal model for autoimmune thyroiditis. *Exp. Clin. Endocrinol. Diabetes* 104:17–20.
19. Wicker, L.S., J.A. Todd, and L.B. Peterson. 1995. Genetic control of autoimmune diabetes in the NOD mouse. *Annu. Rev. Immunol.* 13:179–200.
20. Acha-Orbea, H., and H.O. McDevitt. 1987. The first external domain of the nonobese diabetic mouse class II I-A beta chain is unique. *Proc. Natl. Acad. Sci. USA* 84:2435–2439.
21. Denny, P., C.J. Lord, N.J. Hill, J.V. Goy, E.R. Levy, P.L. Podolin, L.B. Peterson, L.S. Wicker, J.A. Todd, and P.A. Lyons. 1997. Mapping of the IDDM locus Idd3 to a 0.35-cM interval containing the interleukin-2 gene. *Diabetes* 46:695–700.
22. Luan, J.J., R.C. Monteiro, C. Sautes, G. Fluteau, L. Eloy, W.H. Fridman, J.F. Bach, and H.J. Garchon. 1996. Defective Fc gamma RII gene expression in macrophages of NOD mice—Genetic linkage with up-regulation of IgG1 and IgG2b in serum. *J. Immunol.* 157:4707–4716.
23. Gu, X.F., E. Larger, E. Clauser, and R. Assan. 1992. Similarity of HLA-DQ profiles in adult-onset type 1 insulin-dependent diabetic patients with and without extra-pancreatic auto-immune disease. *Diabete Metab.* 18:306–313.
24. Bowman, M.A., E.H. Leiter, and M.A. Atkinson. 1994. Prevention of diabetes in the NOD mouse: implications for therapeutic intervention in human disease. *Immunol. Today* 15:115–120.
25. Mueller, R., T. Krahl, and N. Sarvetnick. 1996. Pancreatic expression of interleukin-4 abrogates insulitis and autoimmune diabetes in nonobese diabetic (NOD) mice. *J. Exp. Med.* 184:1093–1099.
26. Moritani, M., K. Yoshimoto, S.F. Wong, C. Tanaka, T. Yamaoka, T. Sano, Y. Komagata, J. Miyazaki, H. Kikutani, and M. Itakura. 1998. Abrogation of autoimmune diabetes in nonobese diabetic mice and protection against effector lymphocytes by transgenic paracrine TGF-beta 1. *J. Clin. Invest.* 102:499–506.
27. King, C., J. Davies, R. Mueller, M.S. Lee, T. Krahl, B. Yeung, E. O'Connor, and N. Sarvetnick. 1998. TGF-betal alters APC preference, polarizing islet antigen responses toward a Th2 phenotype. *Immunity* 8:601–613.
28. Grewal, I.S., K.D. Grewal, F.S. Wong, D.E. Picarella, C.A. Janeway, and R.A. Flavell. 1996. Local expression of transgene encoded TNF alpha in islets prevents autoimmune diabetes in nonobese diabetic (NOD) mice by preventing the

development of auto-reactive islet-specific T cells. *J. Exp. Med.* 184:1963–1974.

29. Hunger, R.E., C. Carnaud, I. Garcia, P. Vassalli and C. Mueller. 1997. Prevention of autoimmune diabetes mellitus in NOD mice by transgenic expression of soluble tumor necrosis factor receptor p55. *Eur. J. Immunol.* 27:255–261.

30. Moritani, M., K. Yoshimoto, F. Tashiro, C. Hashimoto, J. Miyazaki, S. Li, E. Kudo, H. Iwahana, Y. Hayashi, T. Sano, and M. Itakura. 1994. Transgenic expression of IL-10 in pancreatic islet A cells accelerates autoimmune insulitis and diabetes in non-obese diabetic mice. *Int. Immunol.* 6:1927–1936.

31. Wogensen, L., M.S. Lee, and N. Sarvetnick. 1994. Production of interleukin 10 by islet cells accelerates immune-mediated destruction of beta cells in nonobese diabetic mice. *J. Exp. Med.* 179:1379–1384.

32. Wong, S.S. Guerder, I. Visintin, E.P. Reich, K.E. Swenson, R.A. Flavell, and C.A. Janeway. 1995. Expression of the co-stimulator molecule B7-1 in pancreatic beta-cells accelerates diabetes in the NOD mouse. *Diabetes* 44:326–329.

33. Wong, F.S., I. Visintin, L. Wen, J. Granata, R. Flavell, and C.A. Janeway. 1998. The role of lymphocyte subsets in accelerated diabetes in nonobese diabetic-rat insulin promoter-B7-1 (NOD-RIP-B7-1) mice. *J. Exp. Med.* 187:1985–1993.

34. Wong, F.S., B.N. Dittel, and C.A. Janeway. 1999. Transgenes and knockout mutations in animal models of type 1 diabetes and multiple sclerosis. *Immunol. Rev.* 169:93–106.

35. Chervonsky, A.V., Y. Wang, F.S. Wong, I. Visintin, R.A. Flavell, J.R. Janeway, Ca and L.A. Matis. 1997. The role of Fas in autoimmune diabetes. *Cell* 89:17–24.

36. Katz, J.D., B. Wang, K. Haskins, C. Benoist, and D. Mathis. 1993. Following a diabetogenic T cell from genesis through pathogenesis. *Cell* 74:1089–1100.

37. Gonzalez, A., J.D. Katz, M.G. Mattei, H. Kikutani, C. Benoist, and D. Mathis. 1997. Genetic control of diabetes progression. *Immunity* 7:873–883.

38. Kurrer, M.O., S.V. Pakala, H.L. Hanson, and J.D. Katz. 1997. Beta cell apoptosis in T cell-mediated autoimmune diabetes. *Proc. Natl. Acad. Sci. USA* 94:213–218.

39. Andre-Schmutz, I., C. Hindelang, C. Benoist, and D. Mathis. 1999. Cellular and molecular changes accompanying the progression from insulitis to diabetes. *Eur. J. Immunol.* 29:245–255.

40. Lehuen, A., O. Lantz, L. Beaudoin, V. Laloux, C. Carnaud, A. Bendelac, J.F. Bach, and R.C. Monteiro. 1998. Overexpression of natural killer T cells protects V alpha 14-J alpha 281 transgenic nonobese diabetic mice against diabetes. *J. Exp. Med.* 188:1831–1839.

41. Lenschow, D.J., K.C. Herold, L. Rhee, B. Patel, A. Koons, H.Y. Qin, E. Fuchs, B. Singh, C.B. Thompson, and J.A. Bluestone. 1996. CD28/B7 regulation of Th1 and Th2 subsets in the development of autoimmune diabetes. *Immunity* 5:285–293.

42. French, M.B., J. Allison, D.S. Cram, H.E. Thomas, M. Dempsey-Collier, A. Silva, H.M. Georgiou, T.W. Kay, L.C. Harrison, and A.M. Lew. 1997. Transgenic expression of mouse proinsulin II prevents diabetes in nonobese diabetic mice. *Diabetes* 46:34–39.

43. Birk, O.S., D.C. Douek, D. Elias, K. Takacs, H. Dewchand, S.L. Gur, M.D. Walker, R. Van Der Zee, I.R. Cohen, and D.M. Altmann. 1996. A role of hsp60 in autoimmune diabetes: analysis in a transgenic model. *Proc. Natl. Acad. Sci. USA* 93:1032–1037.

44. Bridgett, M., M. Cetkovic-Cvrlje, R. O'Rourke, Y.G. Shi, S. Narayanswami, J. Lambert, V. Ramiya, S. Baekkeskov, and E.H. Leiter 1998. Differential protection in two transgenic lines of NOD/Lt mice hyperexpressing the autoantigen GAD65 in pancreatic beta-cells. *Diabetes* 47:1848–1856.

45. Geng, L.P., M. Solimena, R.A. Flavell, R.S. Sherwin, and A.C. Hayday. 1998. Widespread expression of an autoantigen-GAD65 transgene does not tolerize non-obese diabetic mice and can exacerbate disease. *Proc. Natl. Acad. Sci. USA* 95:10055–10060.

46. Allison, J., L. Malcolm, J. Culvenor, R.K. Bartholomeusz, K. Holmberg, and J.F. Miller. 1991. Overexpression of beta 2-microglobulin in transgenic mouse islet beta cells results in defective insulin secretion *Proc. Natl. Acad. Sci. USA* 88:2070–2074.

47. Singer, S.M., R. Tisch, X.D. Yang, and H.O. McDevitt. 1993. An Abd transgene prevents diabetes in nonobese diabetic mice by inducing regulatory T cells. *Proc. Natl. Acad. Sci. USA* 90:9566–9570.

48. Hutchings, P., P. Tonks, and A. Cooke. 1997. Effect of MHC transgene expression on spontaneous insulin autoantibody class switch in nonobese diabetic mice. *Diabetes* 46:779–784.

49. Lund, T., L. O'Reilly, P. Hutchings, O. Kanagawa, E. Simpson, R. Gravely, P. Chandler, J. Dyson, J.K. Picard, A. Edwards, D. Kioussis, and A. Cooke. 1990. Prevention of insulin-dependent diabetes mellitus in non-obese diabetic mice by transgenes encoding modified I-A beta-chain or normal I-E alpha-chain. *Nature* 345:727–729.

50. Wherrett, D.K., S.M. Singer, and H.O. McDevitt. 1997. Reduction in diabetes incidence in an I-A(g7) transgenic nonobese diabetic mouse line. *Diabetes* 46:1970–1974.

51. Uno, M., T. Miyazaki, M. Uehira, H. Nishimoto, M. Kimoto, J. Miyazaki, and K. Yamamura. 1992. Complete prevention of diabetes in transgenic NOD mice expressing I-E molecules. *Immunol. Lett* 31:47–52.

52. Nishimoto, H., H. Kikutani, K. Yamamura, and T. Kishimoto. 1987. Prevention of autoimmune insulitis by expression of I-E molecules in NOD mice. *Nature* 328:432–434.

53. Bohme, J., B. Schuhbaur, O. Kanagawa, C. Benoist, and D. Mathis. 1990. MHC-linked protection from diabetes dissociated from clonal deletion of T cells. *Science* 249:293–295.

54. Brenden, N., and J. Bohme. 1998. Disease-protected major histocompatibility complex Ea- transgenic non-obese diabetic (NOD) mice show interleukin- 4 production not seen in susceptible Ea-transgenic and non- transgenic NOD mice. *Immunology* 95:1–7.

55. Soderstrom, I., M.L. Bergman, F. Colucci, K. Lejon, I. Bergqvist, and D. Holmberg. 1996. Establishment and characterization of RAG-2 deficient non-obese diabetic mice. *Scand. J. Immunol.* 43:525–530.

56. Makino, S., M. Harada, Y. Kishimoto, and Y. Hayashi. 1986. Absence of insulitis and overt diabetes in athymic nude mice with NOD genetic background. *Exp. Anim.* 35:495–498.

57. Sumida, T., M. Furukawa, A. Sakamoto, T. Namekawa, T. Maeda, M. Zijlstra, I. Iwamoto, T. Koike, S. Yoshida, and H. Tomioka. 1994. Prevention of insulitis and diabetes in

beta 2-microglobulin-deficient non-obese diabetic mice. *Int. Immunol.* 6:1445–1449.

58. Serreze, D.V., E.H. Leiter, G.J. Christianson, D. Greiner, and D.C. Roopenian. 1994. Major histocompatibility complex class I-deficient NOD-B2mnull mice are diabetes and insulitis resistant. *Diabetes* 43:505–509.

59. Wicker, L.S., E.H. Leiter, J.A. Todd, R.J. Renjilian, E. Peterson, P.A. Fischer, P.L. Podolin, M. Zijlstra, R. Jaenisch, and L.B. Peterson. 1994. Beta 2-microglobulin-deficient NOD mice do not develop insulitis or diabetes. *Diabetes* 43:500–504.

60. Katz, J., C. Benoist, and D. Mathis. 1993. Major histocompatibility complex class I molecules are required for the development of insulitis in non-obese diabetic mice. *Eur. J. Immunol.* 23:3358–3360.

61. Mora, C., F.S. Wong, C.H. Chang, and R.A. Flavell. 1999. Pancreatic infiltration but not diabetes occurs in the relative absence of MHC class II-restricted CD4 T cells: Studies using NOD/CIITA-deficient mice. *J. Immunol.* 162:4576–4588.

62. Kay, T.W., J.L. Parker, L.A. Stephens, H.E. Thomas, and J. Allison. 1996. RIP-beta2-microglobulin transgene expression restores insulitis, but not diabetes, in beta2-microglobulin-null nonobese diabetic mice. *J. Immunol.* 157:3688–3693.

63. Serreze, D.V., H.D. Chapman, D.S. Varnum, M.S. Hanson, P.C. Reifsyder, S.D. Richard, S.A. Fleming, E.H. Leiter, and L.D. Shultz. 1996. B lymphocytes are essential for the initiation of T cell- mediated autoimmune diabetes: Analysis of a new "speed congenic" stock of NOD.Ig mu(null) mice. *J. Exp. Med.* 184:2049–2053.

64. Akashi, T., S. Nagafuchi, K. Anzai, S. Kondo, D. Kitamura, S. Wakana, J. Ono, M. Kikuchi, Y. Niho, and T. Watanabe. 1997. Direct evidence for the contribution of B cells to the progression of insulitis and the development of diabetes in non-obese diabetic mice. *Int. Immunol.* 9:1159–1164.

65. Serreze, D.V., S.A. Fleming, H.D. Chapman, S.D. Richard, E.H. Leiter, and R.M. Tisch. 1998. B lymphocytes are critical antigen-presenting cells for the initiation of T cell-mediated autoimmune diabetes in nonobese diabetic mice. *J. Immunol.* 161:3912–3918.

66. Falcone, M., J. Lee, G. Patstone, B. Yeung, and N. Sarvetnick. 1998. B lymphocytes are crucial antigen-presenting cells in the pathogenic autoimmune response to GAD65 antigen in nonobese diabetic mice. *J. Immunol.* 161:1163–1168.

67. Yang, M., B. Charlton, and A.M. Gautam. 1997. Development of insulitis and diabetes in B cell-deficient NOD mice. *J. Autoimmun.* 10:257–260.

68. Hultgren, B., X.J. Huang, N. Dybdal, and T.A. Stewart. 1996. Genetic absence of gamma-interferon delays but does not prevent diabetes in NOD mice. *Diabetes* 45:812–817.

69. Wang, B., I. Andre, A. Gonzalez, J.D. Katz, M. Aguet, C. Benoist, and D. Mathis. 1997. Interferon-gamma impacts at multiple points during the progression of autoimmune diabetes. *Proc. Natl. Acad. Sci. USA* 94:13844–13849.

70. Kagi, D., A. Ho, B. Odermatt, A. Zakarian, P.S. Ohashi, and T.W. Mak. 1999. TNF receptor 1-dependent beta cell toxicity as an effector pathway in autoimmune diabetes. *J. Immunol.* 162:4598–4605.

71. Trembleau, S., G. Penna, S. Gregori, H.D. Chapman, D.V. Serreze, J. Magram, and L. Adorini. 1999. Pancreas-infiltrating Th1 cells and diabetes develop in IL-12-deficient nonobese diabetic mice. *J. Immunol.* 163:2960–2968.

72. Wang, B., A. Gonzalez, P. Hoglund, J.D. Katz, C. Benoist, and D. Mathis. 1998. Interleukin-4 deficiency does not exacerbate disease in NOD mice. *Diabetes* 47:1207–1211.

73. Itoh, N., A. Imagawa, T. Hanafusa, M. Waguri, K. Yamamoto, H. Iwahashi, M. Moriwaki, H. Nakajima, J. Miyagawa, M. Namba, S. Makino, S. Nagata, N. Kono, and Y. Matsuzawa. 1997. Requirement of Fas for the development of autoimmune diabetes in nonobese diabetic mice. *J. Exp. Med.* 186:613–618.

74. Allison, J., and A. Strasser. 1998. Mechanisms of beta cell death in diabetes: A minor role for CD95. *Proc. Natl. Acad. Sci. USA* 95:13818–13822.

75. Kim, Y.H., S.S. Kim, K.A. Kim, H. Yagita, N. Kayagaki, K.W. Kim, and M.S. Lee. 1999. Apoptosis of pancreatic beta-cells detected in accelerated diabetes of NOD mice: no role of Fas-Fas ligand interaction in autoimmune diabetes. *Eur. J. Immunol.* 29:455–465.

76. Kagi, D., B. Odermatt, P. Seiler, R.M. Zinkernagel, T.W. Mak, and H. Hengartner. 1997. Reduced incidence and delayed onset of diabetes in perforin-deficient nonobese diabetic mice. *J. Exp. Med.* 186:989–997.

77. Nakhooda, A.F., A.A. Like, C.I. Chappel, F.T. Murray, and E.B. Marliss. 1977. The spontaneously diabetic Wistar rat. Metabolic and morphologic studies. *Diabetes* 26:100–112.

78. Crisa, L., J.P. Mordes, and A.A. Rossini. 1992. Autoimmune diabetes mellitus in the BB rat. *Diabetes Metab. Rev.* 8:4–37.

79. Markholst, H., S. Eastman, D. Wilson, B.E. Andreasen, and A. Lernmark. 1991. Diabetes segregates as a single locus in crosses between inbred BB rats prone or resistant to diabetes. *J. Exp. Med.* 174:297–300.

80. Brown, D.W., R.M. Welsh and A.A. Like. 1993. Infection of peripancreatic lymph nodes but not islets precedes Kilham rat virus-induced diabetes in BB/Wor rats. *J. Virol.* 67:5873–5878.

81. Poussier, P., A.F. Nakhooda, J.A. Falk, C. Lee, and E.B. Marliss. 1982. Lymphopenia and abnormal lymphocyte subsets in the "BB" rat: relationship to the diabetic syndrome. *Endocrinology* 110:1825–1827.

82. Greiner, D.L., E.S. Handler, K. Nakano, J.P. Mordes, and A.A. Rossini. 1986. Absence of the RT-6 T cell subset in diabetes-prone BB/W rats. *J. Immunol.* 136:148–151.

83. Greiner, D.L., J.P. Mordes, E.S. Handler, M. Angelillo, N. Nakamura, and A.A. Rossini. 1987. Depletion of RT6.1+ T lymphocytes induces diabetes in resistant biobreeding/Worcester (BB/W) rats. *J. Exp. Med.* 166:461–475.

84. Hornum, L., M. Jackerott, and H. Markholst. 1995. The rat T-cell lymphopenia resistance gene (Lyp) maps between D4Mit6 and Npy on RN04. *Mamm. Genome* 6:371–372.

85. Kolb, H., and K.D. Kroncke. 1993. Lessons from the low-dose streptozocin model. *Diabetes Rev.* 1:116–126.

86. Rossini, A.A., M.C. Appel, R.M. Williams, and A.A. Like. 1977. Genetic influence of the streptozotocin-induced insulitis and hyperglycemia. *Diabetes* 26:916–920.

87. Orlow, S., R. Yasunami, C. Boitard, and J.F. Bach. 1987. Induction précoce du diabète chez la souris NOD par la streptozotocine. *CR Acad. Sci. III* 304:77–78.

88. Gerling, I.C., H. Friedman, D.L. Greiner, L.D. Shultz, and E.H. Leiter. 1994. Multiple low-dose streptozocin-induced diabetes in NOD-scid/scid mice in the absence of functional lymphocytes. *Diabetes* 43:433–440.

89. Hendrickson, E.A., X.Q. Qin, E.A. Bump, D.G. Schatz, M. Oettinger, and D.T. Weaver. 1991. A link between double-

strand break-related repair and V(D)J recombination: the scid mutation. *Proc. Natl. Acad. Sci. USA* 88:4061–4065.

90. Vallera, D.A., S.F. Carroll, S. Brief, and B.R. Blazar. 1992. Anti-CD3 immunotoxin prevents low-dose STZ/interferon-induced autoimmune diabetes in mouse. *Diabetes* 41:457–464.

91. Elliott, J.I., H. Dewchand, and D.M. Altmann. 1997. Streptozotocin-induced diabetes in mice lacking alpha beta T cells. *Clin. Exp. Immunol.* 109:116–120.

92. Weide, L.G., and P.E. Lacy. 1991. Low-dose streptozocin-induced autoimmune diabetes in islet transplantation model. *Diabetes* 40:1157–1162.

93. Cockfield, S.M., V. Ramassar, J. Urmson, and P.F. Halloran. 1989. Multiple low dose streptozotocin induces systemic MHC expression in mice by triggering T cells to release IFN-gamma. *J. Immunol.* 142:1120–1128.

94. Fowell, D., and D. Mason. 1993. Evidence that the T cell repertoire of normal rats contains cells with the potential to cause diabetes. Characterization of the CD4+ T cell subset that inhibits this autoimmune potential. *J. Exp. Med.* 177:627–636.

95. Seddon, B., A. Saoudi, M. Nicholson, and D. Mason. 1996. CD4+CD8-thymocytes that express L-selectin protect rats from diabetes upon adoptive transfer. *Eur. J. Immunol.* 26:2702–2708.

96. Asano, M., M. Toda, N. Sakaguchi, and S. Sakaguchi. 1996. Autoimmune disease as a consequence of developmental abnormality of a T cell subpopulation. *J. Exp. Med.* 184:387–396.

97. Sakaguchi, S., M. Toda, M. Asano, M. Itoh, S.S. Morse, and N. Sakaguchi. 1996. T cell-mediated maintenance of natural self-tolerance: its breakdown as a possible cause of various autoimmune diseases. *J. Autoimmun.* 9:211–220.

98. Itoh, M., T. Takahashi, N. Sakaguchi, Y. Kuniyasu, J. Shimizu, F. Otsuka, and S. Sakaguchi. 1999. Thymus and autoimmunity: production of CD25+CD4+ naturally anergic and suppressive T cells as a key function of the thymus in maintaining immunologic self-tolerance. *J. Immunol.* 162:5317–5326.

99. Kloppel, G., E. Altenahr, G. Freytag, and F.K. Jansen. 1974. Immune insulitis and manifest diabetes mellitus. Studies on the course of immune insulitis and the induction of diabetes mellitus in rabbits immunized with insulin. *Virchows Arch. A. Pathol. Anat. Histol.* 364:333–346.

100. Elias, D., H. Marcus, T. Reshef, V. Ablamunits, and I.R. Cohen. 1995. Induction of diabetes in standard mice by immunization with the p277 peptide of a 60-kDa heat shock protein. *Eur. J. Immunol.* 25:2851–2857.

101. Zekzer, D., F.S. Wong, O. Ayalon, I. Millet, M. Altieri, S. Shintani, M. Solimena, and R.S. Sherwin. 1998. GAD-reactive CD4+ Th1 cells induce diabetes in NOD/SCID mice. *J. Clin. Invest.* 101:68–73.

102. Adams, T.E., S. Alpert, and D. Hanahan. 1987. Non-tolerance and autoantibodies to a transgenic self antigen expressed in pancreatic beta cells. *Nature* 325:223–228.

103. Oldstone, M.B., M. Nerenberg, P. Southern, J. Price, and H. Lewicki. 1991. Virus infection triggers insulin-dependent diabetes mellitus in a transgenic model: role of anti-self (virus) immune response. *Cell* 65:319–331.

104. Sarukhan, A., A. Lanoue, A. Franzke, N. Brousse, J. Buer, and H. Von Boehmer. 1998. Changes in function of antigen-specific lymphocytes correlating with progression towards diabetes in a transgenic model. *EMBO J* 17:71–80.

105. Roman, L.M., L.F. Simons, R.E. Hammer, J.F. Sambrook, and M.J. Gething. 1990. The expression of influenza virus hemagglutinin in the pancreatic beta cells of transgenic mice results in autoimmune diabetes. *Cell* 61:383–396.

105a. Von Herrath, M.G., J. Dockter, and M.B. Oldstone. 1994. How virus induces a rapid or slow onset insulin-dependent diabetes mellitus in a transgenic model. *Immunity* 1:231–242.

106. Ohashi, P.S., S. Oehen, K. Buerki, H. Pircher, C.T. Ohashi, B. Odermatt, B. Malissen, R.M. Zinkernagel, and H. Hengartner. 1991. Ablation of "tolerance" and induction of diabetes by virus infection in viral antigen transgenic mice. *Cell* 65:305–317.

107. Scott, B., R. Liblau, S. Degermann, L.A. Marconi, L. Ogata, A.J. Caton, H.O. McDevitt, and D. Lo. 1994. A role for non-MHC genetic polymorphism in susceptibility to spontaneous autoimmunity. *Immunity* 1:73–82.

108. Vizler, C., N. Bercovici, A. Cornet, C. Cambouris, and R.S. Libau. 1999. Role of autoreactive CD8(+) T cells in organ-specific autoimmune diseases: insight from transgenic mouse models. *Immunol Rev.* 169:81–92.

109. Bottazzo, G.F., R. Pujol-Borrell, T. Hanafusa, and M. Feldmann. 1983. Role of aberrant HLA-DR expression and antigen presentation in induction of endocrine autoimmunity. *Lancet* 2:1115–1119.

110. Hanafusa, T., R. Pujol-Borrell, L. Chiovato, R.C. Russell, D. Doniach, and G.F. Bottazzo. 1983. Aberrant expression of HLA-DR antigen on thyrocytes in Graves' disease: relevance for autoimmunity. *Lancet* 2:1111–1115.

111. Allison, J., I.L. Campbell, G. Morahan, T.E. Mandel, L.C. Harrison, and J.F. Miller. 1988. Diabetes in transgenic mice resulting from over-expression of class I histocompatibility molecules in pancreatic beta cells. *Nature* 333:529–533.

112. Gotz, J., H. Eibel, and G. Kohler. 1990. Non-tolerance and differential susceptibility to diabetes in transgenic mice expressing major histocompatibility class II genes on pancreatic beta cells. *Eur. J. Immunol.* 20:1677–1683.

113. Bohme, J., K. Haskins, P. Stecha, W. Van Ewijk, M. Lemeur, P. Gerlinger, C. Benoist, and D. Mathis. 1989. Transgenic mice with I-A on islet cells are normoglycemic but immunologically intolerant. *Science* 244:1179–1183.

114. Lo, D., L.C. Burkly, G. Widera, C. Cowing, R.A. Flavell, R.D. Palmiter, and R.L. Brinster. 1988. Diabetes and tolerance in transgenic mice expressing class II MHC molecules in pancreatic beta cells. *Cell* 53:159–168.

115. Liu, J.L., L.E. Purdy, S. Rabinovitch, A.M. Jevnikar, and J.F. Elliott. 1999. Major DQ8-restricted T-cell epitopes for human GAD65 mapped using human CD4, DQA1 *0301, DQB1 *0302 transgenic IA(Null) NOD mice. *Diabetes* 48:469–477.

116. Wen, L., F.S. Wong, L. Burkly, R. Altieri, C. Mamalaki, D. Kioussis, R.A. Flavell, and R.S. Sherwin. 1998. Induction of insulitis by glutamic acid decarboxylase peptide-specific and HLA-DQ8-restricted CD4(+) T cells from human DQ transgenic mice. *J. Clin. Invest.* 102:947–957.

117. Sonderstrup, G., and H. McDevitt. 1998. Identification of autoantigen epitopes in MHC Class II transgenic mice. *Immunol. Rev.* 164:129–138.

118. Raju, R., S.R. Munn and C.S. David. 1997. T cell recognition of human pre-proinsulin peptides depends on the polymorphism at HLA DQ locus: A study using HLA DQ8 and DQG transgenic mice. *Hum. Immunol.* 58:21–29.

119. Wicker, L.S., S.L. Chen, G.T. Nepom, J.F. Elliott, D.C. Freed, A. Bansal, S. Zheng, A. Herman, A. Lernmark,

D.M. Zaller, L.B. Peterson, J.B. Rothbard, R. Cummings, and P.J. Whiteley. 1996. Naturally processed T cell epitopes from human glutamic acid decarboxylase identified using mice transgenic for the type 1 diabetes-associated human MHC class II allele, DRB1 *0401. *J. Clin. Invest.* 98:2597–2603.

120. Green, E.A., and R.A. Flavell. 1999. Tumor necrosis factor-alpha and the progression of diabetes in non-obese diabetic mice. *Immunol. Rev.* 169:11–22.

121. Guerder, S., D.E. Picarella, P.S. Linsley, and R.A. Flavell. 1994. Costimulator B7–1 confers antigen-presenting-cell function to parenchymal tissue and in conjunction with tumor necrosis factor alpha leads to autoimmunity in transgenic mice. *Proc. Natl. Acad. Sci. USA* 91:5138–5142.

122. Elliott, E.A., and R.A. Flavell. 1994. Transgenic mice expressing constitutive levels of IL-2 in islet beta cells develop diabetes. *Int. Immunol.* 6:1629–1637.

123. Von Herrath, M.G., S. Guerder, H. Lewicki, R.A. Flavell, and M.B. Oldstone. 1995. Coexpression of B7–1 and viral ("self") transgenes in pancreatic beta cells can break peripheral ignorance and lead to spontaneous autoimmune diabetes. *Immunity* 3:727–738.

124. Harlan, D.M., H. Hengartner, M.L. Huang, Y.H. Kang, R. Abe, R.W. Moreadith, H. Pircher, G.S. Gray, P.S. Ohashi, G.J. Freeman, and A.L. Et. 1994. Mice expressing both B7–1 and viral glycoprotein on pancreatic beta cells along with glycoprotein-specific transgenic T cells develop diabetes due to a breakdown of T-lymphocyte unresponsiveness. *Proc. Natl. Acad. Sci. USA* 91:3137–3141.

125. Sarvetnick, N., D. Liggitt, S.L. Pitts, S.E. Hansen, and T.A. Stewart. 1988. Insulin-dependent diabetes mellitus induced in transgenic mice by ectopic expression of class II MHC and interferon-gamma. *Cell* 52:773–782.

126. Stewart, T.A., B. Hultgren, X. Huang, S. Pitts-Meek, J. Hully, and N.J. MacLachlan. 1993. Induction of type I diabetes by interferon-alpha in transgenic mice. *Science* 260:1942–1946.

127. Allison, J., L.A. Stephens, T.W. Kay, C. Kurts, W.R. Heath, J.F.A.P. Miller, and M.F. Krummel. 1998. The threshold for autoimmune T cell killing is influenced by B7–1. *Eur. J. Immunol.* 28:949–960.

128. Lee, M.S., R. Mueller, L.S. Wicker, L.B. Peterson, and N. Sarvetnick. 1996. IL-10 is necessary and sufficient for autoimmune diabetes in conjunction with NOD MHC homozygosity. *J. Exp. Med.* 183:2663–2668.

129. Higuchi, Y., P. Herrera, P. Muniesa, J. Huarte, D. Belin, P. Ohashi, P. Aichele, L. Orci, J.D. Vassalli, and P. Vassalli. 1992. Expression of a tumor necrosis factor alpha transgene in murine pancreatic beta cells results in severe and permanent insulitis without evolution towards diabetes. *J. Exp. Med.* 176:1719–1731.

130. Picarella, D.E., A. Kratz, C.B. Li, N.H. Ruddle, and R.A. Flavell. 1993. Transgenic tumor necrosis factor (TNF)-alpha production in pancreatic islets leads to insulitis, not diabetes. Distinct patterns of inflammation in TNF-alpha and TNF-beta transgenic mice. *J. Immunol.* 150:4136–4150.

131. Guerder, S., E.E. Eynon, and R.A. Flavell. 1998. Autoimmunity without diabetes in transgenic mice expressing beta cell-specific CD86, but not CD80: Parameters that trigger progression to diabetes. *J. Immunol.* 161:2128–2140.

132. Grewal, I.S., B.J. Rutledge, J.A. Fiorillo, L. Gu, R.P. Gladue, R.A. Flavell, and B.J. Rollins. 1997. Transgenic monocyte chemoattractant protein-1 (MCP-1) in pancreatic islets produces monocyte-rich insulitis without diabetes: abrogation by a second transgene expressing systemic MCP-1. *J. Immunol.* 159:401–408.

133. Campbell, I.L., M.V. Hobbs, J. Dockter, M.B. Oldstone, and J. Allison. 1994. Islet inflammation and hyperplasia induced by the pancreatic islet-specific overexpression of interleukin-6 in transgenic mice. *Am. J. Pathol.* 145:157–166.

134. Bendelac, A., C. Boitard, P. Bedossa, H. Bazin, J.F. Bach, and C. Carnaud. 1988. Adoptive T cell transfer of autoimmune nonobese diabetic mouse diabetes does not require recruitment of host B lymphocytes. *J. Immunol.* 141:2625–2628.

36 | The Autoimmune Response in a Transgenic Mouse Model of Insulin-dependent Diabetes

Adelaida Sarukhan, Oskar Lechner, Ulrich Walter, Astrid Lanoue, Anke Franzke, Carole Zober, Jan Buer and Harald von Boehmer

1. INTRODUCTION

Thymic deletion ensures T cell tolerance towards ubiquitous antigens expressed in the thymus as well as to blood-borne antigens taken up by thymic and circulating antigen presenting cells (APCs). In addition, it has been shown that low levels of certain peripheral tissue antigens are expressed in the thymus (1,2), thus inducing deletion of high affinity T cells specific for these antigens. Despite these mechanisms, T cells reactive to tissue-specific antigens, such as pancreatic β-cell antigens, can mature and enter the circulation. Activation of such autoreactive T cells can lead to autoimmunity and result in tissue destruction and functional impairment. Although over the recent years knowledge has accumulated concerning possible target antigens and genes involved in several autoimmune diseases, we still remain fairly ignorant of the initial events leading to the triggering of an autoimmune response as well as the precise mechanisms by which T cells induce tissue damage. In the present chapter, we will discuss results obtained with experimental murine models developed in our and other laboratories, which have contributed to a better understanding of autoimmune phenomena.

2. DISSECTION OF THE AUTOIMMUNE RESPONSE IN A TRANSGENIC MODEL OF DIABETES

We used transgenic mice expressing the influenza virus hemagglutinin (HA) under the control of the rat insulin promoter (3) to develop an experimental model of autoimmune diabetes. These mice (INS-HA) were crossed with those (TCR-HA) expressing a transgenic $\alpha\beta$ TCR specific for an HA peptide presented by MHC class II molecules and for which a clonotype-specific antibody exists (4).

These transgenic mice, though representing an artificial model of disease, offer significant advantages over other models of spontaneous diabetes, like the NOD mouse: One can follow in detail the immune response towards a defined pancreatic β-cell-specific antigen and study where and how the autoreactive T cells are activated, which factors control the development of aggressive autoimmunity and how β-cell death is induced.

2.1. Activation of Self-Reactive T Cells by Cross-Presentation

Circulating self-reactive mature T cells exist in all healthy individuals. The crucial question is what triggers the activation of these lymphocytes. Several theories have been proposed to explain the initiation of an autoimmune response.

It has been proposed that viral infections can trigger autoimmunity by two mechanisms; molecular mimicry and/or bystander T-cell activation (5). The molecular mimicry theory proposes that some pathogens express a protein fragment that is structurally related to a particular self-component (for example, the Coxsackie B4 virus is strongly associated with the development of IDDM in humans and shares sequence similarity with the islet antigen glutamic acid decarboxylase). Presentation of the pathogenic epitope by professional APCs could activate self-reactive T cells that, in turn, would attack tissues bearing self-antigens that they previously ignored. This theory has received support from studies in mice where an autoimmune response toward a self-antigen in the pancreas (6,7) or cornea (8) was triggered by infecting mice with virus that share epitopes with such antigens. The other viral-associated theory, that of bystander activation, proposes that pathogens break down self-tolerance in a non-antigen specific way. They can do this in several ways; by provoking cell death and thus the release of intracellular antigens, increasing their abundance, potentiating antigen presentation or locally perturbing the cytokine balance through inflammation. Evidence for this concept came from a study using NOD mice transgenic for an islet-reactive TCR in which diabetes was accelerated by Coxsackie virus infection in an apparently non-antigen-specific manner (9).

A considerable number of studies show that viral infections are, in fact, not necessary to initiate an autoimmune response. These studies show that antigens can be transferred to APCs that can then, under certain conditions, activate self-reactive T cells. In fact, evidence that cellular antigens can be transferred to and presented by APCs came from early experiments showing that in H-2^d × H-2^b mice, injection of H-2^b cells differing in minor histocompatibility antigens resulted in the priming of H-2^d restricted CTLs against such antigens (10). This phenomena was called cross-priming and suggested that capture and presentation of cell-derived antigens by APCs could represent a mechanism by which T cells could be primed to antigens expressed in peripheral tissue. Transgenic mice expressing antigens in an islet-restricted fashion have provided further evidence for cross-presentation. It was shown that both CD8$^+$ (11) and CD4$^+$ (12–14) T cells specific for a β-cell antigen were activated in the lymphnodes draining the pancreas by bone marrow derived APCs. The identity of the cross-presenting APCs has remained elusive. Dendritic cells are good candidates due to their high antigen uptake and mobility capacities, though efforts to isolate them for use as stimulators *in vitro* have not given satisfactory results. We have found evidence that cells obtained from pancreatic, but not axillary or inguinal, lymphnodes of INS-HA mice present the HA antigen since they are capable of inducing a weak but reproducible proliferation of TCR-HA T cells *in vitro* (15). A better response was obtained when using purified dendritic cells from the pancreatic lymphnodes of TCR-HA × INS-HA mice (our unpublished results), which could be due to the fact that the survival of mature antigen-presenting dendritic cells is increased by constant interaction of their TRANCE receptor with the TRANCE ligand expressed on the antigen-specific T cells (16).

It is important to underline that none of the experiments showing cross-presentation of β-cell antigens described above formally excluded the possibility that the antigen could be expressed by, rather than transferred to, the bone marrow-derived APCs or that some T cell-mediated damage to the islet cells had enhanced cross-presentation. We designed an experiment that formally excluded both possibilities using we used an adoptive transfer system of diabetes in which only the bone marrow-derived APC, and not the β-cells, could present antigen to the T cells. For this, RAG$^{-/-}$ INS-HA mice of the H-2^b haplotype were injected with bone marrow from RAG$^{-/-}$ mice of the H-2^d haplotype (source of dendritic cells and macrophages). Successfully reconstituted bone marrow chimeras were transferred with monospecific T cells recognizing an HA peptide presented by I-E^d molecules. The transferred T cells were activated and caused disease, formally demonstrating the transfer of the HA peptide or protein fragments to H-2^d bone marrow derived dendritic cells / macrophages (15) in the pancreatic surrounding in the absence of damage caused by T cells recognizing antigen on islet β-cells (Figure 1).

It is important to note that cross-presentation of self-antigens does not necessarily mean induction of effector T cells and destruction of target cells. In some cases, cross-presentation can lead to tolerance induction of T cells after an activation phase (3,13,17,18). The factors determining tolerance *versus* autoimmunity remain to be determined, but could depend in part on the frequency of the antigen-specific T cells and their activation threshold, as discussed below.

Finally, the fact that no cross-presentation was observed in some other transgenic models expressing a defined antigen under the insulin promoter, even when the frequency of the antigen-specific T cells was high (6,19,20), could be an exception rather than a rule and may depend in part on the cellular localization of the antigen (cytoplasmic *versus* membrane-bound or secreted) and its level of expression.

2.2 Infiltration of Pancreatic Islets with Enrichment of Antigen-Specific T Cells

In single transgenic mice expressing HA under the rat insulin promoter (RIP), no pancreatic infiltration or disease were observed despite the presence of HA-specific T cell precursors and the cross-presentation of the HA antigen by bone marrow-derived APCs. Functional studies showed that, in fact, HA-specific T cells were rendered tolerant (3) and that perhaps this tolerance was not due to expression of HA in the thymus, as was the case for other antigens expressed under the insulin promoter (21). When the frequency of HA-specific T cells was increased by introducing a class II-restricted transgenic $\alpha\beta$TCR specific for the HA peptide 111–119, which is expressed on approximately 5–10% of circulating T cells, the outcome was completely different: Huge pancreatic infiltrations were observed in all TCR-HA × INS-HA double transgenic mice from the age of 2 weeks (22,23) . These data indicate that the frequency of self-reactive cells is an important parameter in deciding whether autoimmunity develops. Accordingly, in mice expressing the simian virus 40 large T antigen (Tag) under control of the insulin promoter, tolerance by deletion and anergy of MHC class II-restricted TCR-Tag cells was observed when their frequency was low (10%), but pancre-

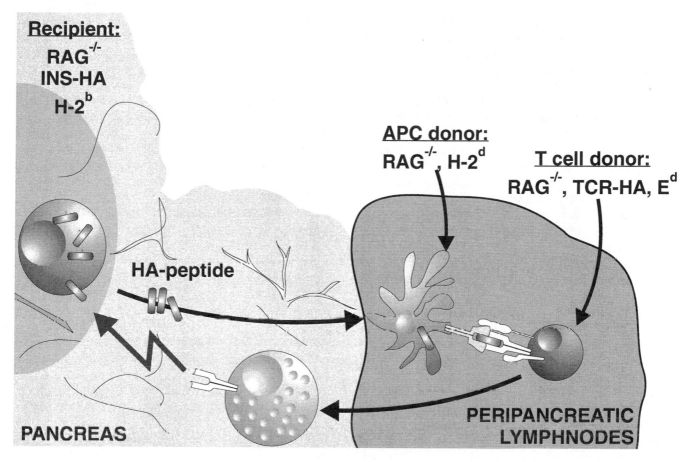

Figure 1. Cross-presentation of antigens expressed by the pancreatic b-cells. Immunodeficient (RAG$^{-/-}$) H-2^b mice expressing the hemagglutinin (HA) exclusively by the pancreatic β-cells were transferred with bone marrow from an H-2^d RAG$^{-/-}$ mouse. After successful reconstitution with H-2^d bone marrow derived APCs (dendritic cells / macrophages), the chimeric mice were transferred with a mono-specific T cell population (obtained from a RAG$^{-/-}$ TCR-HA mouse) recognizing a HA peptide associated to I-E^d molecules. These T cells would only be activated if the HA protein or fragments of it is transferred to local bone-marrow derived APCs. This was indeed the case, since these chimeric mice became diabetic with the same kinetics than the recipient controls which were H-2^d.

atic infiltration was observed when their frequency was high (90%) (21). The affinity of the self-reactive TCR also seems to play an important role since the INS-HA mice develop much milder diabetes when crossed to mice transgenic for another class II-restricted anti-HA TCR (24).

Both immunohistochemistry and isolation of lymphocytes infiltrating the islets revealed that the cellular composition of the islet infiltrates was very similar to that found in the lymph nodes, with a similar distribution and frequency of B-cells, CD4+ and CD8+ T cells, macrophages and dendritic cells. A significant enrichment (4- to 5-fold) for CD4+ T cells expressing the autoreactive TCR was observed in the islets when compared to the axillary or inguinal lymph nodes and a high percentage of these HA-specific T cells had upregulated the surface activation marker CD69. In the lymph nodes draining the pancreas, the percentage of CD4+6.5+ T cells was intermediate between that found in the axillary/inguinal lymphnodes and in the pancreatic islets. Infiltrating T cells that were nega-

tive for the transgenic TCR did not show significant signs of activation (23). These data indicate that entrance to the pancreas is strongly influenced by antigen but is certainly not exclusively restricted to antigen-specific T cells.

2.3. Controlled Progression to Disease

Despite the fact that the pancreata of all TCR-HA × INS-HA double transgenic mice were heavily infiltrated by 4 weeks of age, only 30–40% of these developed overt disease at variable timepoints. No significant difference concerning the degree of infiltration or content of insulin could be observed between diabetic and non-diabetic littermates; both contained "islands" of intact insulin producing β-cells. Thus, surprisingly, heavy infiltration by self-reactive T cells specific for β-cells was not enough to cause immediate overt disease in all mice. Some regulatory mechanisms keeping these cells from attacking β-cells must therefore exist. Similar "check-points" were also observed in NOD mice transgenic for a β-cell specific TCR (BDC2.5): Despite overwhelming insulitis by 3 weeks of age, these mice developed diabetes only several weeks later (25). Such putative regulatory mechanisms cannot be attributed to the presence of other lymphoid cells (i.e. B-cells and non-HA-specific T cells), since a similar uncoupling between insulitis and diabetes was observed in TCR-HA × INS-HA on the RAG background containing a mono-specific TCR-HA population. In fact, in double transgenic mice as well as in the NOD BDC2.5 model, diabetes was induced with a single dose of cyclophosphamide even when mice were on a RAG$^{-/-}$ background (23,25). The mechanism of action of this drug is not known, though it is supposed to interfere with "immunosuppressive" lymphocytes. This would mean that T cells expressing the same TCR can act both as suppressor and effector T cells. Several factors could be regulating the transition to diabetes. One such factor could be the profile of cytokine secretion by the effector T cells. Using a competitive RT-PCR technique, we found that HA-specific T cells infiltrating the islets of diabetic mice secreted significantly more IFN-γ (ten-fold increase) than their counterparts in non-diabetic littermates. Upregulation of this pro-inflammatory cytokine could contribute to β-cell death via mechanisms discussed below. Another factor that could play a role in triggering disease is upregulation of Fas ligand on the antigen-specific cells, though we found no evidence for such an event at the mRNA expression level in the TCR-HA × INS-HA model. Loss of sensitivity to negative signaling through "regulatory" receptors like CTLA-4 and the IL-2R could also play a role. In fact, it was found that in the NOD BDC2.5 model, injection of anti-CTLA4 mAb at the onset of infiltration significantly accelerated the onset of disease (26) and CD25-positive cells have been shown to play an immunoregulatory role (27). The role of CTLA4 and IL-2R on TCR-HA positive cells in the TCR-HA × INS-HA model remains to be addressed. Finally, other extrinsic factors like regulatory cytokines secreted by other cell types such as the antigen presenting cells around and within the islets could also play a role in the transition from "controlled" to "aggressive" autoimmunity. We found a slight downregulation (2-fold) of IL4 but no significant downregulation of TGF-β or IL-10 in the islets of diabetic TCR-HA × INS-HA mice (23 and our unpublished results).

2.4. Insulitis and Diabetes in the Absence of Antigen Recognition on Islet Cells

Diabetes is the result of massive β-cell death provoked by the infiltrating T cells. The role of CD4+ and CD8+ T cells in the development of diabetes has been subject of debate. It has been argued that, in the NOD model, CD8+ T cells may have an early (28–32) or late (33,34) role in the disease process either by destroying some islet cells, resulting in increased release of β-specific antigens, or by destroying islet cells in the diabetic phase. In our and other models of diabetes (23,35), β-cell destruction can occur when the infiltrating T cells are exclusively class II restricted. The mechanism by which class II-restricted T cells can induce β-cell death has remained controversial. The aberrant expression of MHC class II molecules by β-cells and their direct recognition by cytotoxic CD4+ T cells has been proposed but not proven. Destruction of β-cells by soluble factors secreted by infiltrating cells has also been proposed since it has been shown, at least *in vitro*, that pro-inflammatory cytokines like IL-1 and IFN-γ can kill β-cells.

To address this question, we used an adoptive transfer system in which disease was induced in immunodeficient RAG$^{-/-}$ INS-HA mice 8–9 days after transfer of monospecific TCR-HA T cells. We constructed bone marrow chimera mice in which only the bone marrow-derived APC, not the β-cells, could present antigen to the T cells. For this, RAG$^{-/-}$ INS-HA mice of the H-2^b haplotype were injected with bone marrow from RAG$^{-/-}$ mice of the H-2^d haplotype (source of dendritic cells and macrophages). Successfully reconstituted bone marrow chimeras were transferred with monospecific T cells recognizing an HA peptide presented by I-E^d molecules. The only way these T cells could be activated and induce diabetes was by transfer of the HA peptide or protein fragments to H-2^d bone marrow-derived dendritic cells / macrophages. This was indeed the case, since H-2^b RAG$^{-/-}$ INS-HA chimeras having received H-2^d but not H-2^b, bone marrow developed diabetes with a similar kinetics than H-2^d RAG$^{-/-}$ INS-HA controls (15). These results clearly show that insulitis as well as β-cell destruction can proceed in the absence of direct recognition of cells by T cells. Daily insulin treatment of recipients starting at the time of the disease onset protected mice from death which, in the absence of insulin treatment, occurs within 10 days, suggesting that the β-cells and not other components of the pancreas were destroyed (our unpublished results).

2.5. Regulation of β-Cell Death by Intracellular Inhibitors of Death Pathways

The results presented above support the notion that β-cell death in diabetes can be caused by effector mechanisms other than antigen-mediated contact between T cells and β-cells. *In vitro* studies with murine and rat β-cells have implicated several candidate cytokines and factors, such as IL-1, IFN-γ, TNFα or nitric oxide, capable of promoting β-cell death with some synergistic effects (36–38). Nevertheless, it is still not clear which cytokine(s) induces β-cell destruction in autoimmune diabetes. Several receptors belonging to the TNF-receptor (TNFR) superfamily are capable of activating the cascade of proteases involved in apoptosis. In the murine system, these receptors are mainly Fas, TNFR-1 and TNFR-2

(Figure 2). The role of Fas in the induction of β-cell death has been very controversial. Initial studies with Fas-deficient NOD mice showed neither diabetes nor insulitis, suggesting that Fas plays a major role in β-cell destruction (39,40). Nevertheless, further experiments involving transfer of Fas-deficient NOD islets into wild-type NOD mice showed that β-cells could be destroyed even in the absence of Fas (41–44). In the TCR-HA × INS-HA model, we were not able to detect Fas expression on β-cells by immunohistochemistry (23), However, this technique is not very sensitive and cannot clearly discriminate β-cells from surrounding lymphocytes. To circumvent these difficulties and study other death receptors such as TNFR-1 and TNFR-2, we developed a RT-PCR method on single β-cells. We compared death receptor mRNA levels on β-cells from mice at different stages of the disease: non-infiltrated, infiltrated and diabetic. Two diabetes models were compared, the TCR-HA × INS-HA and the NOD models. In both, Fas, TNFR-1 and TNFR-2 were upregulated in infiltrated as compared to non-infiltrated mice. The percentage of receptor positive β-cells did not increase with the onset of diabetes (45).

In a recent paper, islets deficient for Fas, TNFR-1 or TNFR-2 or IFN-γR were grafted into diabetic NOD mice that were subsequently transferred with islet-specific T cells. Only islets deficient for TNFR-1 were not killed unless they were co-grafted with wild-type islets, suggesting that TNFR-1 could be regulating access of antigen to local APCs rather than β-cell death itself (43).

These and our own data suggest that death receptors and their ligands are already expressed by β-cells at the infiltra-

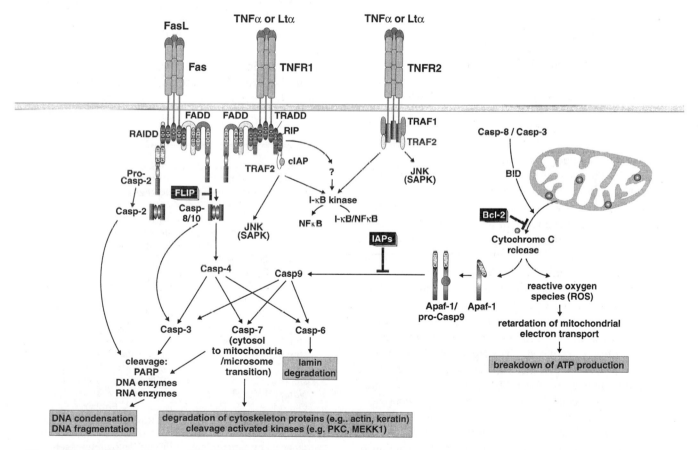

Figure 2.   Principal pathways of Fas, TNFR-1 and -2 signaling.
TNFR-1 and Fas recruit complexes formed by the interactions between RIP (receptor-interacting protein) kinases, TRADD (TNFR associated death domain), FADD (Fas associated death domain) and RAIDD—adapter proteins that contain death domains (DD)—which in turn recruit other proteins to initiate signaling. While Fas activates exclusively pro-apoptotic, caspase (Casp)-activating pathways, TNFR-1 can also induce NF-kB or JNK (Jun NH2-terminal kinase) / SAPK (stress-activated protein kinase) which have been reported to have pro- as well as anti-apoptotic effects, depending on the cell type and the cells metabolic status. TNFR-2 can also induce NF-kB or JNK/SAPK but not caspases. All pro-apoptotic pathways are controlled by specific apoptosis inhibiting proteins, such as FLIP (FADD-like interleukin-1beta-converting enzyme-inhibitory protein), different members of the IAP (inhibitors of apoptosis) family or members of the Bcl-2 family.
DED... death effector domains ; CARD...caspase recruitment domains; TRAF...TNFR associated factor; Apaf-1... apoptosis protease activating factor-1

tion phase. The transition from controlled β-cell death (insulitis) to massive β-cell death (diabetes) could thus be due to downregulation of intracellular inhibitors of apoptosis (i.e. members of the bcl-2 family or death-receptor associated inhibitors such as FLIP or IAPs) or significant changes in the generation of soluble *versus* cell bound TNFR-1, which conceivably could have a major role in the susceptibility to TNFα-mediated autoinflammatory phenotypes (46). We are currently investigating both possibilities, i.e. TNFR-1 shedding by cultured β-cells under the influence of different cytokines as well as expression levels of inhibitors of apoptosis in single β-cells. These experiments should provide further information on the intracellular signaling pathways involved in the regulation of β-cell death and may permit development of therapeutic strategies that specifically interfere with β-cell destruction.

3. CONCLUSIONS

Insulin dependent diabetes (IDDM) is a complex and multigenic disease that results in the destruction of insulin-producing pancreatic β-cells by self-reactive T cells. To better understand certain aspects of the autoimmune response directed against the pancreatic β-cells under controlled conditions, several transgenic murine models of IDDM have been developed wherein a defined antigen is expressed in the pancreas and/or a T cell population specific of islet antigens can be followed.

In our laboratory, we have studied mice expressing the hemagglutinin under the control of the insulin promoter and a transgenic TCR specific for a peptide of the hemagglutin presented by MHC class II molecules. With this model, we have been able to demonstrate that: i) antigens whose expression is restricted to the β-cells can be transferred to local antigen-presenting cells. In some cases, presentation of the islet antigen by these APCs can result in tolerance, in other cases it can result in activation of self-reactive T cells and infiltration of the pancreas. The frequency of self-reactive T cells is a crucial parameter in deciding whether tolerance or autoimmunity ensues. ii) The presence of activated, antigen-specific T cells in the pancreatic islets does not always lead to a massive destruction of the β-cells. There is an uncoupling between the infiltration and the diabetes phase that seems to be controlled by largely unknown regulatory mechanisms; these which may include secretion of certain pro-inflammatory cytokines, surface expression of negative regulatory receptors on the self-reactive T cells, and/or regulation in the sensitivity to apoptosis in the β cells. Direct contact between T cells and β-cells does not seem necessary for destruction of the latter, which may occur through engagement of death receptors belonging to the TNFR superfamily. Overexpression of these receptors by β-cells does not seem to be associated with the progression to diabetes, so variations in the expression of intracellular apoptosis inhibitors could be involved.

The answers obtained with this and other murine models of diabetes should not be extrapolated blindly to human disease, which is much more complex. Nevertheless, they will play an important role in contributing to the development of therapeutic approaches aimed at interfering with disease at different phases; the activation of the self-reactive T cells in subjects at high risk of diabetes, the progression from insulitis to overt disease equally in subjects at high risk of diabetes and the final destruction of the β cells in recently diagnosed patients.

ACKNOWLEDGEMENTS

A.S. is supported by a post-doctoral fellowship from the Juvenile Diabetes Foundation, O.L. is supported by a Marie Curie Training Grant from the European Commission, HvB is supported by the Institut Universitaire de France and the Körber Foundation. This work was supported by the Juvenile Diabetes Foundation, the Institut National de la Santé et Recherche Médicale and the Faculté Necker.

References

1. Jolicoeur,C., D. Hanahan, and K.M. Smith. 1994. T-cell tolerance toward a transgenic beta-cell antigen and transcription of endogenous pancreatic genes in thymus. *Proc. Natl. Acad. Sci. U.S.A* 91:6707–6711.
2. Antonia, S.J., T. Geiger, J. Miller, and R.A. Flavell. 1995. Mechanisms of immune tolerance induction through the thymic expression of a peripheral tissue-specific protein. *Int. Immunol.* 7:715–725.
3. Lo, D., J. Freedman, S. Hesse, R.D. Palmiter, R.L. Brinster, and L.A. Sherman. 1992. Peripheral tolerance to an islet cell-specific hemagglutinin transgene affects CD4+ and CD8+ T cells. *Eur. J. Immunol.* 22:1013–1022.
4. Kirberg, J., A. Baron, S. Jakob, A. Rolink, K. Karjalainen, and H. von Boehmer. 1994. Thymic selection of CD8+ single positive cells with a class II major histocompatibility complex-restricted receptor. *J. Exp. Med.* 180:25–34.
5. Benoist, C. and D. Mathis. 1998. The pathogen connection. *Nature* 394:227–228.
6. Ohashi, P.S., S. Oehen, K. Buerki, H. Pircher, C.T. Ohashi, B. Odermatt, B. Malissen, R.M. Zinkernagel, and H. Hengartner. 1991. Ablation of "tolerance" and induction of diabetes by virus infection in viral antigen transgenic mice. *Cell* 65:305–317.
7. von Herrath, M.G. and M.B. Oldstone. 1996. Virus-induced autoimmune disease. *Curr. Opin. Immunol.* 8:878–885.
8. Zhao, Z.S., F. Granucci, L. Yeh, P.A. Schaffer, and H. Cantor. 1998. Molecular mimicry by herpes simplex virus type-1: autoimmune disease after viral infection. *Science* 279:1344–1347.
9. Horwitz, M.S., L.M. Bradley, J. Harbertson, T. Krahl, J. Lee, and N. Sarvetnick. 1998. Diabetes induced by Coxsackie virus: initiation by bystander damage and not molecular mimicry. *Nat. Med.* 4:781–785.

10. Bevan, M.J. 1976. Cross-priming for a secondary cytotoxic response to minor H antigens with H-2 congenic cells which do not cross-react in the cytotoxic assay. *J. Exp. Med.* 143:1283–1288.

11. Kurts, C., W.R. Heath, F.R. Carbone, J. Allison, J.F. Miller, and H. Kosaka. 1996. Constitutive class I-restricted exogenous presentation of self antigens in vivo. *J. Exp. Med.* 184:923–930.

12. Lo, D., C.R. Reilly, B. Scott, R. Liblau, H.O. McDevitt, and L.C. Burkly. 1993. Antigen-presenting cells in adoptively transferred and spontaneous autoimmune diabetes. *Eur. J. Immunol.* 23:1693–1698.

13. Forster, I. and I. Lieberam. 1996. Peripheral tolerance of CD4 T cells following local activation in adolescent mice. *Eur. J. Immunol.* 26:3194–3202.

14. Hoglund, P., J. Mintern, C. Waltzinger, W. Heath, C. Benoist, and D. Mathis. 1999. Initiation of autoimmune diabetes by developmentally regulated presentation of islet cell antigens in the pancreatic lymph nodes. *J. Exp. Med.* 189:331–339.

15. Sarukhan, A., O. Lechner, and H. von Boehmer. 1999. Autoimmune insulitis and diabetes in the absence of antigen specific contact between T cells and β-islet cells. *Eur. J. Immunol.*

16. Bachmann, M.F., B.R. Wong, R. Josien, R.M. Steinman, A. Oxenius, and Y. Choi. 1999. TRANCE, a tumor necrosis factor family member critical for CD40 ligand- independent T helper cell activation [see comments]. *J. Exp. Med.* 189:1025–1031.

17. Kurts, C., H. Kosaka, F.R. Carbone, J.F.A.P. Miller, and W.R. Heath. 1997. Class I restricted cross presentation of exogenous self-antigens leads to deletion of autoreactive CD8+ T cells. *J. Exp. Med.* 186:239–245.

18. Kirberg, J., W. Swat, B. Rocha, P. Kisielow, and H.VonBoehmer. 1993. Induction of tolerance in immature and mature T cells. *Transplantation* 25:279–280.

19. Soldevila, G., T. Geiger, and R.A. Flavell. 1995. Breaking immunologic ignorance to an antigenic peptide of simian virus 40 large T antigen. *J. Immunol.* 155:5590–5600.

20. Heath, W.R., J. Allison, M.W. Hoffmann, G. Schonrich, G. Hammerling, B. Arnold, and J.F. Miller. 1992. Autoimmune diabetes as a consequence of locally produced interleukin-2. *Nature* 359:547–549.

21. Forster, I., R. Hirose, J.M. Arbeit, B.E. Clausen, and D. Hanahan. 1995. Limited capacity for tolerization of CD4+ T cells specific for a pancreatic beta cell neo-antigen. *Immunity* 2:573–585.

22. Degermann, S., G. Sollami, and K. Karjalainen. 1999. T cell receptor beta chain lacking the large solvent-exposed Cbeta FG loop supports normal alpha/beta T cell development and function in transgenic mice. *J. Exp. Med.* 189:1679–1684.

23. Sarukhan, A., A. Lanoue, A. Franzke, N. Brousse, J. Buer, and H. von Boehmer. 1998. Changes in function of antigen-specific lymphocytes correlating with progression towards diabetes in a transgenic model. *EMBO. J.* 17:71–80.

24. Scott, B., R. Liblau, S. Degermann, L.A. Marconi, L. Ogata, A.J. Caton, H.O. McDevitt, and D. Lo. 1994. A role for non-MHC genetic polymorphism in susceptibility to spontaneous autoimmunity. *Immunity* 1:73–83.

25. Andre, I., A. Gonzalez, B. Wang, J. Katz, C. Benoist, and D. Mathis. 1996. Checkpoints in the progression of autoimmune disease: lessons from diabetes models. *Proc. Natl. Acad. Sci. U.S.A.* 93:2260–2263.

26. Luhder, F., P. Hoglund, J.P. Allison, C. Benoist, and D. Mathis. 1998. Cytotoxic T lymphocyte-associated antigen 4 (CTLA-4) regulates the unfolding of autoimmune diabetes. *J. Exp. Med.* 187:427–432.

27. Papiernik, M., M.L. de Moraes, C. Pontoux, F. Vasseur, and C. Penit. 1998. Regulatory CD4 T cells: expression of IL-2R alpha chain, resistance to clonal deletion and IL-2 dependency. *Int. Immunol.* 10:371–378.

28. Wang, B., A. Gonzalez, C. Benoist, and D. Mathis. 1996. The role of CD8+ T cells in the initiation of insulin-dependent diabetes mellitus. *Eur. J. Immunol.* 26:1762–1769.

29. Sumida, T., M. Furukawa, A. Sakamoto, T.N amekawa, T.Maeda, M.Zijlstra, I.Iwamoto, T.Koike, S.Yoshida, and H.Tomioka. 1994. Prevention of insulitis and diabetes in beta 2-microglobulin-deficient non-obese diabetic mice. *Int. Immunol.* 6:1445–1449.

30. Wicker, L.S., E.H. Leiter, J.A. Todd, R.J. Renjilian, E. Peterson, P.A. Fischer, P.L. Podolin, M. Zijlstra, R. Jaenisch, and L.B. Peterson. 1994. Beta 2-microglobulin-deficient NOD mice do not develop insulitis or diabetes. *Diabetes* 43:500–504.

31. Serreze, D.V., E.H. Leiter, G.J. Christianson, D. Greiner, and D.C. Roopenian. 1994. Major histocompatibility complex class I-deficient NOD-B2mnull mice are diabetes and insulitis resistant. *Diabetes* 43:505–509.

32. Katz, J., C. Benoist, and D. Mathis. 1993. Major histocompatibility complex class I molecules are required for the development of insulitis in non-obese diabetic mice. *Eur. J. Immunol.* 23:3358–3360.

33. Wong, F.S., I. Visintin, L. Wen, R.A. Flavell, and C.A. Janeway, Jr. 1996. CD8 T cell clones from young nonobese diabetic (NOD) islets can transfer rapid onset of diabetes in NOD mice in the absence of CD4 cells. *J. Exp. Med.* 183:67–76.

34. Kagi, D., B. Odermatt, P. Seiler, R.M. Zinkernagel, T.W. Mak, and H. Hengartner. 1997. Reduced incidence and delayed onset of diabetes in perforin-deficient nonobese diabetic mice. *J. Exp. Med.* 186:989–997.

35. Katz, J.D., B. Wang, K. Haskins, C. Benoist, and D. Mathis. 1993. Following a diabetogenic T cell from genesis through pathogenesis. *Cell* 74:1089–1110.

36. Bendtzen, K., K. Buschard, M. Diamant, T. Horn, and M.Svenson. 1989. Possible role of IL-1, TNF-alpha, and IL-6 in insulin-dependent diabetes mellitus and autoimmune thyroid disease. Thyroid Cell Group. *Lymphokine Res.* 8:335–340.

37. Mandrup-Poulsen, T., K. Bendtzen, C.A. Dinarello, and J. Nerup. 1987. Human tumor necrosis factor potentiates human interleukin 1-mediated rat pancreatic beta-cell cytotoxicity. *J. Immunol.* 139:4077–4082.

38. Boehm, U., T. Klamp, M. Groot, and J.C. Howard. 1997. Cellular responses to interferon-gamma. *Annu. Rev. Immunol.* 15:749–795.

39. Chervonsky, A.V., Y. Wang, F.S. Wong, I. Visintin, R.A. Flavell, C.A. Janeway, Jr., and L.A. Matis. 1997. The role of Fas in autoimmune diabetes. *Cell* 89:17–24.

40. Itoh, N., A. Imagawa, T. Hanafusa, M. Waguri, K. Yamamoto, H. Iwahashi, H. Moriwaki, H. Nakajima, J. Miyagawa, M. Namba, S. Makino, S. Nagata, N. Kono, and Y. Matsuzawa. 1997. Requirement of Fas for the development of autoimmune diabetes in nonobese diabetic mice. *J. Exp. Med.* 186:613–618.

41. Thomas, H.E., R. Darwiche, J.A. Corbett, and T.W. Kay. 1999. Evidence that beta cell death in the nonobese diabetic mouse is Fas independent. *J. Immunol.* 163:1562–1569.

42. Kim, Y.H., S. Kim, K.A. Kim, H. Yagita, N. Kayagaki, K.W. Kim, and M.S. Lee. 1999. Apoptosis of pancreatic

beta-cells detected in accelerated diabetes of NOD mice: no role of Fas-Fas ligand interaction in autoimmune diabetes. 29:455–465.

43. Pakala, S.V., M. Chivetta, C.B. Kelly, and J.D. Katz. 1999. In autoimmune diabetes the transition from benign to pernicious insulitis requires an islet cell response to tumor necrosis factor alpha. *J. Exp. Med.* 189:1053–1062.

44. Allison, J. and A. Strasser. 1998. Mechanisms of beta cell death in diabetes: a minor role for CD95. *Proc. Natl. Acad. Sci. U.S.A.* 95:13818–13822.

45. Walter, U., A. Franzke, A. Sarukhan, C. Zober, H. von Boehmer, J. Buer, and O. Lechner. 1999. Monitoring gene expression of TNFR family members by β-cells during development of autoimmune diabetes. *submitted.*

46. McDermott, M.F., I. Aksentijevich, J. Galon, E.M. McDermott, B.W. Ogunkolade, M. Centola, E. Mansfield, M. Gadina, L. Karenko, T. Pettersson, J. McCarthy, D.M. Frucht, M. Aringer, Y. Torosyan, A.M. Teppo, M. Wilson, H.M. Karaarslan, Y. Wan, I. Todd, G. Wood, R. Schlimgen, T.R. Kumarajeewa, S.M. Cooper, J.P. Vella, and D.L. Kastner. 1999. Germline mutations in the extracellular domains of the 55 kDa TNF receptor, TNFR1, define a family of dominantly inherited autoinflammatory syndromes. *Cell* 97:133–144.

37 | Hashimoto's Disease

Sandra M. McLachlan and Basil Rapoport

1. INTRODUCTION

1.1. Spectrum of Autoimmune Thyroid Disease and Associated Autoantibodies

Autoimmune thyroid disease encompasses a spectrum from overt hypothyroidism and/or thyroid inflammation in Hashimoto's disease to hyperthyroidism in Graves disease. The breakdown in tolerance in these diseases is characterized by the production of autoantibodies, predominantly of IgG class, to thyroid specific proteins: Thyroid peroxidase (TPO, formerly known as the "microsomal antigen"), thyroglobulin (Tg), the thyrotropin receptor (TSHR) and the sodium/iodide symporter (Figure 1). Some individuals have autoantibodies to all four autoantigens. However, thyroid infiltration and thyroid damage in Hashimoto's thyroiditis is mainly associated with autoantibodies to TPO and Tg (rev. in 1) On the other hand. TSHR autoantibodies that stimulate the thyroid gland are responsible for Graves' hyperthyroidism (rev. in 2). A small proportion of patients are hypothyroid due to TSHR autoantibodies that block the stimulatory effects of TSH (2).

1.2. Gender Bias, Age of Onset, Prevalence and Incidence

Like many autoimmune diseases, autoimmune thyroid disease clusters in families and occurs predominantly in women. Overt hypothyroidism is typically diagnosed in middle-aged individuals. However, Hashimoto's thyroiditis occurs in some juveniles, usually in families in which both parents have autoantibodies to TPO and/or Tg (3). Among women in a Caucasian community, the prevalence of spontaneous hypothyroidism (Hashimoto's disease) was

7.7% and the incidence rate was 3.5/1000/year; the corresponding values for men were 1.3% and 0.6/1000/year (4).

The prevalence of thyroid autoantibodies in the general population is high, about 25% of women (over 35 years) and 10% of men having autoantibodies to TPO or Tg (5). Although most of these individuals are euthyroid, autopsy studies show a close association between TPO autoantibodies and lymphocytic infiltration of the thyroid (6). Moreover, TPO autoantibodies detected in early pregnancy in euthyroid women are closely associated with the development of thyroid dysfunction in the postpartum period (7). It should be realized that, unlike most other organs subjected to autoimmune attack, the thyroid gland has a trophic hormone (TSH) that stimulates thyroid cell growth. Only when the process of thyroid damage overwhelms the regenerative capacity of the gland will the levels of thyroid hormone decrease sufficiently for the patient to become hypothyroid. The autoimmune response to TPO is, therefore, an invaluable marker of subclinical hypothyroidism.

1.3. Thyroid Autoantigens in Hashimoto's Disease

TPO is the primary enzyme responsible for synthesis of thyroid hormone (thyroxine) from the prohormone thyroglobulin (Tg). TPO is a membrane-bound glycoprotein located primarily at the apical surface of the thyrocyte. It comprises two identical ~ 100 kDa subunits and a heme prosthetic group (Figure 1). Tg is a soluble globular glycoprotein with two ~300 kDa homodimers and constitutes the colloid component of thyroid follicles (Figure 1). Unlike Tg, which is abundant, only limited amounts of TPO can be purified from human thyroid tissue. However, recombinant TPO expressed in eukaryotic cells (not in bacteria) is recognized by human TPO autoantibodies to the same extent as thyroid-derived TPO (reviewed in 1). A stop codon introduced at the TPO ectodomain: Plasma membrane junction converts the membrane-associated protein of 933 amino acids into a soluble 848 residue molecule (8) that facilitates studies of human TPO autoantibodies

2. THE THYROID GLAND—FOCUS OF THE AUTOIMMUNE RESPONSE

2.1. Immunological Changes in the Thyroid Gland

Lymphoid cells, including T and B cells, plasma cells, macrophages and dendritic cells infiltrate the thyroid gland in Hashimoto's thyroiditis and, to a lesser extent, in Graves disease. These populations occur as dense aggregates, often with the architectural features of lymph nodes including germinal centers (9), and/or as diffusely distributed lymphoid cells between thyrocytes (Figure 2). CD4+ T cells predominate in the aggregates whereas CD8+ T cells are often scattered between thyrocytes (10).

Unlike normal thyroid tissue, in autoimmune thyroid disease the thyroid epithelial cells are MHC class II positive (11). These observations formed the basis of the hypothesis that "aberrant" MHC class II expression permits thyrocytes to function as antigen presenting cells and give rise to the autoimmune response (12). Support for this hypothesis is the ability of cytokines, in particular IFN-γ, to induce class II expression on normal thyrocytes (12–14). Although thyrocytes are B7-1 negative (15,16), they express CD40 (17) and B7-2 (18).

2.2. Thyroid Autoantibodies are Synthesized in the Thyroid Gland

Plasma cells specific for Tg were detected by histochemistry in Hashimoto glands (19). Subsequent studies showed that, unlike peripheral blood lymphocytes (which require stimulation), Hashimoto's or Graves' thyroid infiltrating lymphocytes *spontaneously* secrete autoantibodies to TPO

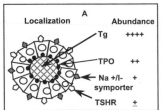
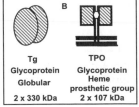

Figure 1.   A. Localization and relative abundance of thyroid autoantigens in a thyroid follicle. Tg, thyroglobulin; TPO, thyroid peroxidase, TSHR, thyrotropin receptor; Na+/I– (sodium iodide) symporter. B, Scheme to illustrate molecular structures of the two major thyroid autoantigens in Hashimoto's disease, Tg and TPO.

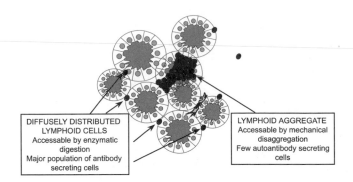

Figure 2.   Lymphoid cells in autoimmune thyroid are present as aggregates and as diffusely distributed cells between thyroid cells. Lymphocytes within aggregates, accessible by mechanical disaggregation, contain few thyroid autoantibody secreting B cells. Diffusely distributed lymphocytes, accessible by digestion, contain the major population of thyroid autoantibody secreting cells.

and/or Tg (20). Similarly, intrathyroidal T cells are activated as reflected in expression of MHC Class II expression (21,22) and the interleukin 2 receptor (23) as well as the ability to "help" peripheral blood B cells secrete thyroid autoantibodies (24,25).

The bone marrow and lymph nodes draining the thyroid gland from some patients also secrete thyroid autoantibodies spontaneously (26,27). However, the magnitude of thyroid autoantibody synthesized by thyroid infiltrating lymphocytes, together with the changes in serum autoantibodies in patients undergoing therapy for autoimmune thyroid disease, point to the thyroid gland as a major site (rev. in 28). Of the two lymphoid populations in diseased thyroid tissue, lymphocytes in aggregates can be isolated by mechanical disaggregation, while scattered lymphocytes are accessible only by digestion. The greatly enhanced autoantibody production by lymphocytes isolated by digestion than by disaggregation indicates that thyroid-autoantibody secreting B cells are located in close proximity to their targets the thyroid cells (10).

2.3. Apoptosis, Cytotoxic T cells and Autoantibody-Mediated Damage

Thyroid damage in Hashimoto's disease, which in some patients leads to end-stage fibrosis, is reflected by increased apoptosis (more APO-1/Fas and decreased BCL2) relative to Graves' or non-autoimmune multinodular thyroid glands (29). Several mechanisms are likely to be involved in thyroid damage. A novel hypothesis proposes that thyroid cells constitutively express Fas-ligand (Fas-L) and undergo apoptosis following induction of Fas by cytokines such as IL-1 β (30). This proposal is controversial because other studies observed constitutive expression of Fas, but not Fas-L and one monoclonal for Fas-L appears to be non-specific (rev. in 31).

Moreover, there is evidence for a labile inhibitor of the Fas death pathway in the thyroid (31).

Conventional mechanisms likely to produce thyroid cell death include cytotoxic T cells and antibody-mediated damage. CD8+ T cells are present in Hashimoto's thyroids (see above) and *in vivo* activated CD8+ cells with NK activity have been cloned from Hashimoto glands (32,33). Moreover, a CD8+ T cell clone lacking NK cell activity and specifically cytotoxic for an undefined antigen on autologous thyroid cells was obtained from intrathyroidal lymphocytes (34). However, other evidence suggests that intrathyroidal T cells are not directly involved in thyrocyte death (35).

Autoantibodies may compound damage initially inflicted by one or both the above mechanisms. TPO antibodies (but not Tg autoantibodies) fix complement (36) and by this means are able to damage thyroid cells *in vitro* (37). However, complement inhibitors are expressed on human thyroid cells *in vivo* (38). Alternatively, TPO autoantibodies, in particular those of IgG1 subclass, can damage thyroid cells (at least *in vitro*) by antibody-dependent cell cytotoxicity (ADCC) (39–41). Although no conclusive data are available, thyroid damage in Hashimoto's disease may involve a combination of thyroid cell "suicide", cytotoxic T cells and ADCC mediated by TPO autoantibodies.

3.   T CELL RECOGNITION OF A MAJOR THYROID AUTOANTIGEN (TPO)

Analysis of T cell recognition (and later B cell recognition) in Hashimoto's disease will focus on TPO because the autoimmune response to this autoantigen is a marker of thyroid inflammation. TPO-specific T cells constitute a minute fraction of the total lymphocyte population. However, an advantage in studying human autoimmune thyroid disease is the occasional availability of thyroid tissue, which contains a population enriched (relative to peripheral blood) in activated, thyroid-specific T cells (for example, 42–44). T cell responses to TPO have been examined in cultures of lymphocytes from thyroid tissue, draining lymph nodes and peripheral blood (45–48). In addition, T cell lines and clones have been generated from peripheral blood lymphocytes using antigen and IL-2 (49) as well as by expanding *in vivo* activated intrathyroidal T cells with IL-2 and anti-CD3 (50).

3.1.   T Cell Epitopes

Synthetic peptides based on the predicted amino acid sequence of TPO and, less commonly, intact protein have been used to study TPO recognition by patients' T cells. Epitopes defined by these approaches encompass most of the TPO molecule, including part of the transmembrane domain

(46–50). Despite the relatively small magnitude of many primary responses, epitopes defined by different laboratories cluster in four regions of TPO (Figure 3A). Moreover, T cell clones derived independently by two groups of investigators recognize similar peptides, namely amino acid residues 632–645 (50) and 625–644 (51).

3.2.   Endogenously Processed TPO

Of potential pathophysiological relevance, *endogenously* processed TPO stimulates T proliferation in T cells from patients with autoimmune thyroid disease (Figure 3B). Cells capable of presenting intact TPO include MHC Class II-positive thyroid cells (50); EBV-transformed B cells (EBVL) transfected with the cDNA for TPO (52,53); and uptake via the B cell Fcε receptor of recombinant IgE-class TPO antibody complexed to TPO (54) TPO-expressing EBVL have been used to derive T cell lines (52) as well as to test the responses of previously isolated T cell clones (53).

3.3.   T Cell Receptor V Genes

TPO-specific T cell lines from the peripheral blood of one patient suggested preferential use of Vα1 and/or Vα3, but diverse Vβ T cell receptor (TCR) genes (52). These findings are consistent with restriction of TCR Vα (but not Vβ) in some (55,56), but not all (57), thyroids from patients with thyroid autoimmunity. However, analysis of four thyroid-derived T cell clones specific for TPO indicated that restricted TCR V region usage does not always correspond to identical epitopic recognition. Thus, clones using identical TCR Vβ chains (1.1), identical Vα chains (15.1), but different Jα regions, recognized, different, non-overlapping epitopes in amino acid-residues 535–551 of TPO (58). Incidentally, there

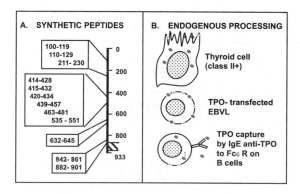

Figure 3.   T cell recognition of TPO. A. Synthetic peptides recognized by lymphocytes from patients with autoimmune thyroid disease based on the linear sequence of TPO (933 amino acids; ectodomain-membrane junction at residue 848). B. Endogenously processed TPO is recognized on MHC class II+ thyroid cells; human EBV L transfected with the cDNA for TPO; and uptake via the B cell Fcε receptor of recombinant IgE-class TPO antibody complexed to TPO.

is no evidence of a role for $\gamma\delta$ T cells in autoimmune thyroid disease (59). Future investigations of clones from multiple patients will be required to characterize the V gene usage of TPO-specific T cells as well as the MHC molecules responsible for presentation of this thyroid autoantigen.

4.   CYTOKINES

Organ-specific autoimmunity is generally considered to be a Th1-mediated condition and "immune deviation" towards Th2-type responses is thought to be important therapeutically (60). Prior to the emergence of the Th1:Th2 paradigm (61), exogenous IFN-γ, IL-2 or TNF-α was observed to inhibit *in vitro* synthesis of TPO (or Tg) autoantibodies by thyroid-infiltrating lymphocytes (62). In contrast, secretion of IFN-γ by T cell clones, derived using IL-2 alone from Hashimoto's thyroid glands, suggested skewing to the Th1 subset (63). These conflicting observations can now be reconciled. Thus, RT-PCR analysis demonstrates mRNA for both Th1- and Th2- cytokines, IFN-γ and IL-4, in Hashimoto's thyroid tissue or infiltrating lymphocytes (64–66). In addition, not only IL-2 but also IL-4 is required to clone thyroid-infiltrating T cells representative of both Th1 and Th2 responses (67). Other cytokine mRNA's detectable in Hashimoto's thyroids include IL-1α, IL-1β, IL-2, IL-6, IL-8, IL-10 and TNF-α (64–66).

Overall, RT-PCR analysis of diseased thyroid tissues (64–66) and flow cytometry of *in vivo* activated thyroid-infiltrating T cells (68) indicate that IFN-γ tends to predominate in thyroid autoimmunity associated with cell-mediated damage (Hashimoto's thyroiditis) (66), while IL-4 is more evident in autoantibody-mediated disease (Graves' hyperthyroidism). However, this Th1/Th2 distinction between Hashimoto's and Graves' diseases is not clear cut. Thus, IL-4 is a switch factor for IgG4 and IgE, and many Hashimoto's patients have TPO and/or Tg autoantibodies of subclass IgG4 and even IgE (see below). In addition, production of TPO autoantibodies is associated with a skewing towards Th2 responses (69). Consequently, immune deviation towards Th2 responses as a form of therapy may simply project the patient from the frying pan (Hashimoto's disease) into the fire (Graves' hyperthyroidism).

5.   B CELL RECOGNITION OF A MAJOR THYROID AUTOANTIGEN (TPO)

Antibodies secreted by plasma cells bear the same antigen-binding sites as the corresponding membrane-bound immunoglobulins on B cells. Consequently, insight into B cell recognition of TPO is provided by patients' serum autoantibodies. However, *human* monoclonal TPO autoantibodies, derived from patients' B cells, are the ultimate tools needed to define the B cells epitopes on TPO in thyroid autoimmunity. Moreover, the genes that encode TPO autoantibody variable regions are the genes for the TPO-specific B cell receptors.

5.1.   Cloning *Human* TPO Autoantibodies

The majority of human monoclonal autoantibodies to TPO (and Tg) have been isolated by the combinatorial immunoglobulin gene library approach (rev. in 70). A source of plasma cells, such as thyroid tissue from a patient with autoimmune thyroid disease, is used to prepare mRNA for reverse transcription of cDNA. DNA for the variable (V) regions of immunoglobulin heavy (H) and light (L) chains is amplified by the polymerase chain reaction and used to construct a "combinatorial" library containing all permutations of H and L pairs. Libraries expressed in bacteriophage lambda vectors are screened using labeled antigen (such as ^{125}I-labeled TPO) to identify individual antigen binding clones in filter lift assays. With "phage display" vectors, specific antibodies are selected by successive rounds of "panning" and amplification for phagemid that bind to antigen on the cell surface or immobilized on a plastic dish. TPO-specific monoclonal autoantibodies expressed as Fab have been obtained by both approaches (rev. in 71).

5.2.   Classes, Subclasses and Affinities of TPO Autoantibodies

Serum TPO autoantibodies are predominantly IgG with IgA (72,73) and IgE TPO (74) antibodies present at low levels in some individuals. These autoantibodies (like those to Tg) are polyclonal, as reflected in the presence of kappa and lambda light chains and several IgG subclasses. Recombinant human TPO-specific Fab isolated from combinatorial immunoglobulin gene libraries closely resemble serum TPO autoantibodies (Table 1): Kappa chains predominate over lambda chains in serum TPO autoantibodies (75) and 118 of 122 recombinant Fab cloned have kappa chains (76–87). In addition, Fab of subclasses IgG1 and IgG4 (78) have been isolated, reflecting the presence of these subclasses in patients' sera (75,88). Finally, the affinities for TPO of most Fab are high (Kd~10^{-10} M), like those for serum TPO autoantibodies (rev. in 1).

5.3.   Variable Region (V) Genes Encoding TPO Autoantibodies

As for many other expressed antibodies, TPO Fab H chains are encoded by genes from VH families 1, 2 and 4 (Table 2) (76–83,87). The germline origin of many D region genes has not been identified because of their high degree of mutation, although DLR2 and DXP2 elements are utilized by some

Table 1 Immunological properties and epitopes of serum TPO antibodies and recombinant TPO-specific *human* monoclonal antibodies expressed as Fab. TPO-specific Fab were obtained by the combinatorial immunoglobulin gene library approach from lymphocytes infiltrating thyroid glands or lymph nodes draining the thyroid. Numbers of individual TPO-specific Fab are in parentheses.

Immunological Properties	Serum anti-TPO	TPO-specific Fab (n)
Class	IgG; some IgA and IgE	IgG (122)
Light chains	Kappa > Lambda	Kappa (118), Lambda (4)
IgG subclass	IgG1>Ig4>IgG2>>IgG3	IgG1 (97), IgG4 (21)
Affinities for TPO	High (Kd ~ 10^{-10} M)	High (Kd ~ 10^{-10} M)
Epitopes — Conformational	Majority	100% (75/75)
— Linear	Probably rare	None
— Immunodominant region	All patients; >75% within an individual serum	97% (34/35)
— Non-immunodominant region	<25%	3% (1/35)

TPO Fab and D-D fusions are common (89). The frequent use of JH4 and JH6 is in accordance with their usage by other autoantibodies and by adult peripheral blood B lymphocytes. The L chains of TPO Fab are encoded by genes from V kappa families I, II and III and V lambda families II, IIIa and IIIb. Some L chain V regions are in germline conformation, whereas somatic mutation is apparent in most VH genes. Such mutations, particularly in complementarity determining regions, are anticipated for antigen-driven, high affinity antibodies.

There are no unusual features of V genes encoding TPO Fab. Many are derived from germline genes that encode other autoantibodies or antibodies. For example, the H chain gene hv1L1 is used in rheumatoid factors (90), an hv1263-like gene is used in anti-cardiolipin/anti-ss DNA (91) and genes derived from hv1263 and 8–1B encode antibodies to rabies virus (92). The L chain genes 012 and A3 are used in cold agglutinins (93) and anti cardiolipin/anti-ss DNA (91), respectively. The frequent use of 012- and kv325-like L chains by TPO Fab likely reflects over-representation of these genes in the expressed repertoire.

5.4. Are TPO-Fab H And L Chains from Combinatorial Libraries Those Used *in Vivo*?

Because of their source, individual H and L chains of TPO Fab must be those used *in vivo*. However, there is debate as to whether the observed H/L pairing occurs *in vivo*. The strong evidence set forth below supports this concept:

i) The high affinities for TPO of these Fab resembles those of serum autoantibodies (Table 1).
ii) When a TPO-Fab L chain was used in "roulette" studies to search the parent H chain library for new TPO-binding Fab, identical or related H chains were obtained (79). Similar observations were made by roulette with a TPO H chain and the parent L chain library.

iii) "Swapping" H and L chains between TPO Fab with unique H and L chain combinations did not produce TPO binding Fab (94).

5.5. Linear Versus Conformational Epitopes Recognized By TPO Autoantibodies

Serum TPO autoantibodies preferentially recognize native rather than denatured TPO (Figure 4A). Similarly, human monoclonal TPO autoantibodies (expressed as Fab) bind better to native than denatured TPO (77,87,95,96). On the other hand, large recombinant TPO polypeptide fragments are recognized by patients' autoantibodies (rev. in 1) and the epitopes of some autoantibodies are reported to be inaccessible in the native molecule (97). Several linear epitopes have been described, in particular the Mab 47/C21 and C2 epitopes corresponding to amino acids residues 713–721 and 590–622 respectively (98,99).

Overall, however, the TPO autoantibody repertoire is biased towards recognition of conformationally intact TPO (rev. in 1). Unlike a panel of mouse monoclonals that recognize *diverse* epitopes, serum TPO autoantibodies interact with a *restricted* region on TPO (100). Human TPO-specific Fab define an "immunodominant region" that is recognized by sera from all individuals and comprises >80% of TPO autoantibodies in the sera of each individual patient. Moreover, the recombinant TPO Fab bind to overlapping epitopes in two domains A and B (78) (Figure 4B). Of interest, binding to domain A is associated with the presence of 012-like L chains (78,101).

5.6. TPO Autoantibody Epitopic "Fingerprints"

Autoantibody "epitopic fingerprints" are determined by quantifying the ability of four recombinant human Fab to inhibit patients' autoantibody binding to TPO (102). As examples (Figure 5A), TPO binding by sera from patients 1 and 5 is strongly inhibited by B1 and B2 domain Fab,

Table 2 Immunoglobulin V region genes utilized by human TPO autoantibodies. Bold type indicates germline genes used by non-thyroid autoantibodies or genes which are over-represented in the expressed repertoire.

	Family	Closest germline gene	D region	J Region
Heavy chains	VH1	**hv1L1**/V1–2; V1–3B; **hv1263**; DP15; DP10; 1–3	DLR2; D21–10; D5/5a; unknown	JH4; JH6
	VH3	**8-1B**; DP42; VH26; **hv3005**; 3–21; 3–23; DP58	DLR2; DK1; D3; D2; unknown	JH1; JH4; JH5; JH6
	VH4	DP-65; 4.34	DXP2; unknown	JH1; JH6
Kappa L chains	V κ I	**O2/O12**; A1; L12; L8; K9		Jκ 1; Jκ 2; Jκ 3 Jκ 4
	V κ II	A3		Jκ 2
	V κ III	**kv325**; L2/ 328h5; Vg		Jκ 5
	V κ IV	**B3**		
Lambda L chains	V λ II	DPL11		J λ 2
	V λ IIIa	Iv318		J λ 1
	V λ IIIb	I150		J λ 2

indicating preferential recognition of the B domain. In contrast, for patient 4, the major inhibition is evident with A1, A2 and B2 (but not B1) domain Fab. Variations in these inhibition patterns are seen in sera from other individuals.

As already mentioned. TPO autoantibodies are present in some clinically euthyroid individuals. TPO epitopic fingerprints are not related to thyroid status (103). However, fingerprints are conserved over time, regardless of fluctuations in TPO autoantibody levels. As illustrated for two women (from a group of 19), TPO epitopic fingerprints were the

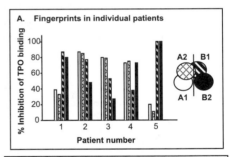

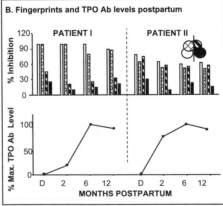

Figure 5. TPO autoantibody epitopic fingerprints. A. Spectrum of fingerprints from 5 patients. Fingerprints are assessed by measuring the inhibition of serum autoantibody binding to ^{125}I-TPO by Fab to the A1, A2, B1 and B2 subdomains in the immunodominant region on TPO (see inset). The % inhibition by each Fab corresponds to the shading for that in the inset.

B. No change in TPO epitopic fingerprints over time (upper panel) despite increases in total TPO antibody levels after delivery (D) and 2, 6 and 12 months postpartum (lower panel). Data for two representative women from a group of 19 individuals studied. Adapted with permission from reference 138.

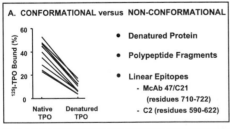

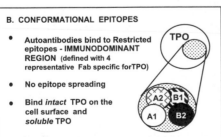

Figure 4. B cell recognition of TPO. A. Evidence for conformational versus non-conformational epitopes. B. Characteristics of conformational epitopes within a restricted (immunodominant) region on TPO.

same at delivery and at 1–2, 5 and 12 months postpartum (Figure 5B)(104). Over the same time intervals, the levels of IgG class TPO autoantibodies rose, in accordance with the rebound observed for many antibodies and autoantibodies after delivery. Conservation of TPO autoantibody fingerprints in the postpartum period and for up to 13 years in juvenile Hashimoto's thyroiditis families (105), suggests that TPO autoantibody epitopic fingerprints may be inherited. Indeed, segregation analysis of epitopic fingerprints in these families and old Amish kindreds revealed genetic transmission of the TPO B domain (106).

5.7. TPO Autoantibody Immunodominant Region

Although an immunodominant region on TPO has been described, the amino acids comprising this region have not been defined. Unlike linear epitopes, the epitopes within this region are conformational and cannot be analyzed using peptides or polypeptide fragments. One approach to circumvent this problem and still retain the native structure is to construct chimeric molecules between TPO and a homologous protein, myeloperoxidase. Studies of TPO/ MPO chimeras, and of monomeric and dimeric TPO, show that the following areas can be excluded from the TPO immunodominant region: i) the contact region between the homodimers; and ii) the amino-terminal 121 amino acids (107). The mMAb47/C21 linear epitope (amino acids 713–721) recognized by some patients' autoantibodies may be on the fringe of the immunodominant region (101). Clearly, localization of the TPO immunodominant region is extremely difficult and will likely require crystallization of a TPO-Fab complex. One TPO-specific human Fab (TR1.9) has been crystallized and its 3-dimensional structure indicates that the immunodominant region, like its antigen binding region, is likely to be relatively flat (108).

6. GENETIC BASIS FOR HASHIMOTO'S DISEASE

Autoantibodies to TPO and/or Tg aggregate in families and simple pedigree analysis suggests either autosomal dominant or multifactorial models of inheritance (rev. in 109). Complex segregation analysis confirmed vertical transmission of TPO or Tg autoantibody production (110) and the data from extended kindreds of Old Order Amish are consistent with a mixed model involving dominant transmission of a major gene (111). Moreover, as already mentioned, recognition of the B domain in the TPO immunodominant region appears to be inherited (106). There is currently no information about the gene loci responsible for transmission of autoantibodies to TPO or Tg.

Turning to clinically manifest disease (as opposed to the presence of autoantibodies alone). HLA DR3, DR4, or DR5

are associated with goitrous Hashimoto's thyroiditis and postpartum thyroiditis in Caucasians (rev. in 109). However, the relative risks are generally low and different associations are seen in other ethnic groups, such as HLA-DR9 in Chinese. Recently, investigators have begun to determine the basis for thyroid autoimmune disease by whole-genome linkage screening (112). Linkage has been excluded between Hashimoto's thyroiditis (and Graves' disease) and the immune response genes HLA, immunoglobulin H chain, T cell receptor and CTLA-4 (113). The power of linkage analysis is in detecting *major* genes, whereas association analysis is more sensitive for detecting genes with *modest* effects (114). Consequently, the data excluding candidate genes from linkage do not negate a role for such genes in autoimmune thyroid disease Instead, these observations imply that a *stronger* genetic susceptibility lies elsewhere in the genome.

7. ENVIRONMENTAL FACTORS

The interaction between environmental factors and genetic factors influences expression of many diseases. In human thyroid autoimmunity, the dominant environmental factor is dietary iodide. Increased iodine intake can cause either hypothyroidism ("iodide myxedema") or iodide-induced thyrotoxicosis (rev. in 115). In iodide-induced hypothyroidism, the individuals most susceptible to iodide are those with antithyroid autoantibodies (116). Variable iodine intake has major effects on animal models of thyroiditis because the antigenicity of Tg is influenced by its iodide content (117,118). Likewise, human T cells recognize iodinated, but not non-iodinated, Tg (119).

8. INFECTIONS AND HASHIMOTO'S THYROIDITIS

Infectious agents have been suggested to be involved in autoimmunity by inducing abnormal antigen presentation, B or T cell activation by superantigens, molecular mimicry, etc. Some evidence points to a role for infectious organisms in thyroid autoimmunity (rev. in 120). Subacute thyroiditis, for example, is likely to have a viral etiology and may be associated with autoantibodies to TPO and/or Tg. On the other hand, thyroid autoantibodies in these patients are usually transient and persistent autoimmune thyroid disease rarely develops. Some Hashimoto's patients have serological evidence for recent bacterial or viral infections. Most attention has been focused on the potential cross-reaction between *Yersinia enterocolitica* proteins and the TSHR. However, there is no conclusive evidence for any candidate infectious organism in the induction of autoimmune thyroid disease (120).

9.   INSIGHT FROM ANIMAL MODELS OF THYROIDITIS

9.1   Spontaneous Thyroiditis

Thyroiditis and Tg autoantibodies develop spontaneously in Obese Strain (OS) chickens and BB/W rats (rev. in 121). As in humans, increased iodine ingestion enhances thyroiditis and autoantibody levels (117). Similarities between the thyroid gland in OS chickens and in Hashimoto's disease include extensive lymphocytic infiltration, MHC II+ thyroid cells and the ability of thyroid-infiltrating lymphocytes to secrete thyroid autoantibodies (122). Although T cells play a major role in thyroid destruction, thyroid histology resembling that in Hashimoto's disease does not develop in OS chickens lacking B cells (122). OS chickens also provide insight into the genetic basis for Hashimoto's disease. Thus, besides MHC genes, the development of thyroiditis is critically dependent on other genes, one set determining abnormal immune responses and the other set influencing the susceptibility of the thyroid gland to damage (123).

9.2.   SCID Mice

Mice with severe combined immunodeficiency (SCID), which lack functional T and B cells, provide a useful system for analyzing human thyroid autoantigen-specific T and B cells. Because thyroid autoantibodies develop in SCIDs engrafted with peripheral or thyroid-derived lymphocytes or with intact thyroid tissue from autoimmune thyroid disease patients (124–127), these mice have been used to develop a model for Graves' disease (rev. in 2). Perhaps the most important insight for Hashimoto's disease provided by these mice is the loss of MHC class II expression on thyroid grafts, confirming that "aberrant" expression requires induction. In accordance with observations of dispersed thyroid cells, IFN-γ administered to SCID mice engrafted with human thyroid and PBMC enhanced thyroid cells expression of HLA-DR (128,129). However, IFN-α (but not IFN-γ) enhanced thyroid autoantibody levels (129).

9.3   Models Induced By Conventional Immunization

Immunization with soluble thyroid antigen (Tg) and adjuvant (Freund's or LPS) induces thyroiditis in rabbits (130) and H2-k strains of mice (rev. in 131). A different MHC susceptibility (H2-b) is required for thyroiditis induction by TPO and CFA (132). Unlike spontaneous thyroiditis in OS chickens (122), none of the induced murine thyroiditis models develops hypothyroidism. MHC antigens and cytotoxic T cells (but not Tg antibodies) play major roles in thyroiditis developing in response to immunization with human or murine Tg (rev. in 131).

9.4.   Cell-Associated (Not Soluble) TPO Induces Hashimoto's-like Antibodies

Patients' autoantibodies to TPO interact with a restricted range of epitopes (see Section 5.7). In contrast, diverse epitopes are recognized by antibodies induced in animals by conventional immunization with purified antigen and adjuvant. Examples include antibodies to Tg (133), TPO (100) and the acetylcholine receptor (134). However, immune responses are influenced by the manner of antigen presentation, as demonstrated by the protocol used to develop the first animal model of Graves' disease, namely injecting mice with fibroblasts co-expressing syngeneic MHC class II and the TSHR (135). This novel approach is based on the "aberrant" MHC class II expression hypothesis (12).

Applying this approach to TPO provides insight into the immunological conundrum described above. High titer antibodies induced by conventional immunization with soluble, eukaryotic TPO and adjuvant have moderate affinities (~Kd ~10^{-9}M) and interact with diverse epitopes. In contrast, antibodies induced with fibroblasts co-expressing MHC class II and TPO resemble patients' autoantibodies in terms of high affinity for TPO (Kd ~ 10^{-10}M) and interaction with restricted epitopes in the immunodominant region (136) (Figure 6A).

Two non-mutually exclusive processes may explain these differences. First it is possible that a different spectrum of peptides is processed and presented by fibroblasts co-expressing TPO and class II than by professional antigen presenting cells following exogenous uptake and processing of purified antigen. Although the fibroblasts usually lack co-stimulatory molecules, there is evidence to suggest that

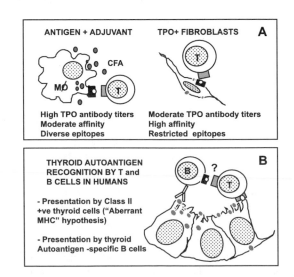

Figure 6.   A. Comparison of TPO antibodies induced in mice by conventional immunization using purified TPO and adjuvant versus injection of TPO+ fibroblasts in the absence of adjuvant. B. Proposed scheme for thyroid autoantigen recognition by T and B cells in human thyroid autoimmunity.

B7-1-independent T cell clones may support autoantibody production in autoimmune thyroid disease (16). Second, unlike the high doses (~100 μg) used in conventional immunization, the very small dose (<0.5 μg) of TPO associated with class II+ cells may lead to selection of B cells with progressively higher affinity receptors. Moreover, antibodies of high (but not low) affinity, either as B cell antigen receptors or complexed with antigen and taken up by professional antigen-presenting cells, can influence the peptides presented to T cells (137). We have hypothesized (54) that the restricted epitopic recognition of human autoantibodies to TPO and Tg results from the influence of autoantibodies on the peptides presented to T cells (54,71).

10. CONCLUSIONS

Although many pieces of the puzzle are missing, the following scenario is proposed for the development of thyroid autoimmunity (Figure 6B). Some form of thyroid damage, possibly a viral infection, induces aberrant MHC class II expression on thyroid cells. In genetically-susceptible individuals, thyrocyte presentation of autoantigens activates T cells which, in turn, provide help for B cells. Recognition of autoantigens on thyrocytes by B cells provides the basis for the extreme bias of the B cell repertoire towards conformational epitopes. Autoantibodies are frequently regarded as useful markers of the autoimmune response. However, B cells are increasingly being recognized as playing a more important role, even in diseases such as type I diabetes and multiple sclerosis which are presumed to be T cell mediated. Indeed, it is likely that the T cells involved in organ-specific tissue damage recognize epitopes on proteins that have been internalized and processed by high affinity, autoantigen-specific B cells.

ACKNOWLEDGEMENTS

Part of the data described in this paper was generated using support from NIH grants DK36182 and DK54684

References

1. McLachlan, S.M., and B. Rapoport. 1992. The molecular biology of thyroid peroxidase cloning, expression and role as autoantigen in autoimmune thyroid disease. *Endocr. Rev.* 13:192–206.
2. Rapoport, B., G.D. Chazenbalk, J.C. Jaume, and S.M. McLachlan. 1998. The thyrotropin receptor Interaction with thyrotropin and autoantibodies. *Endocr. Rev.* 19:673–716.
3. Burek, C.L., W.H. Hoffman, and N.R. Rose. 1982. The presence of thyroid autoantibodies in children and adolescents with autoimmune thyroid disease and in their siblings and parents. *Clin. Immunol. Immunopathol.* 25:395–404.
4. Vanderpump, M.P.J., W.M.G. Tunbridge, J.M. French, D. Appleton, D. Bates, F. Clark, J. Grimley Evans, D.M. Hasan, H. Rodgers, F. Tunbridge, and E.T. Young. 1995. The incidence of thyroid disorders in the community: a twenty-year follow-up of the Whickham survey. *Clin. Endocrinol.* 43:55–68.
5. Prentice, L.M., D.I.W. Phillips, D. Sarsero, K. Beever, S.M. McLachlan, and B. Rees Smith. 1990. Geographical distribution of subclinical autoimmune thyroid disease in Britain: a study using highly sensitive direct assays for autoantibodies to thyroglobulin and thyroid peroxidase. *Acta. Endocrinol. (Copenh).* 123:493–498.
6. Yoshida, H., N. Amino, K. Yagawa, K. Uemura, M. Satoh, K. Miyai, and Y. Kumahara. 1978. Association of serum antithyroid antibodies with lymphocytic infiltration of the thyroid gland Studies of seventy autopsied cases. *J. Clin. Endocrinol. Metab.* 46:859–862.
7. Amino, N., K. Miyai, T. Onishi, T. Hashimoto, K. Arai, K. Ishibashi, and Y. Kumahara. 1976. Transient hypothyroidism after delivery in autoimmune thyroiditis. *J. Clin. Endocrinol. Metab.* 42:296–301.
8. Foti, D., K.D. Kaufman, G.D. Chazenbalk, and B. Rapoport. 1990. Generation of a biologically-active, secreted form of human thyroid peroxidase by site-directed mutagenesis. *Mol. Endocrinol.* 4:786–791.
9. Sodorstrom, N., and A. Biorklund. 1974. Organization of the invading lymphoid tissue in human lymphoid thyroiditis. *Scand. J. Immunol.* 3:295–301.
10. McLachlan, S.M., C.A.S. Pegg, M.C. Atherton, S.L. Middleton, A. Dickinson, F. Clark, S.J. Proctor, G. Proud, and B. Rees Smith. 1986. Subpopulations of thyroid autoantibody secreting lymphocytes in Graves' and Hashimoto thyroid glands. *Clin. Exp. Immunol.* 65:319–328.
11. Hanafusa, T., R. Pujol-Borrell, L. Chiovato, R.C.G. Russell, D. Doniuch, G.F. Bottazzo, and M. Feldmann. 1983. Aberrant expression of HLA-DR antigen on thyrocytes in Graves' disease: Relevance for autoimmunity. *Lancet* ii:1111–1115.
12. Bottazzo, G.F., R. Pujol-Borrell, T. Hanafusa, and M. Feldmann. 1983. Role of aberrant HLA-DR expression and antigen presentation in induction of endocrine autoimmunity. *Lancet* ii:1115–1118.
13. Weetman, A.P., D.J. Volkman, K.D. Burman, T.L. Gerrard, and A.S. Fauci. 1985. The in vitro regulation of human thyrocyte HLA-DR antigen expression. *J. Clin. Endocrinol. Metab.* 61:817–824.
14. Piccinini, L.A., W.A. Mackenzie, M. Platzer, and T.F. Davies. 1987. Lymphokine regulation of HLA-DR gene expression in human thyroid cell monolayers. *J. Clin. Endocrinol. Metab.* 64:543–548.
15. Matsuoka, N., K. Eguchi, A. Kawakami, M. Tsuboi, H. Nakamura, H. Kimura, N. Ishikawa, K. Ito, and S. Nagataki. 1996. Lack of B7-1/BB1 and B7-2/B70 expression on thyrocytes of patients with Graves' disease. Delivery of costimulatory signals from bystander professional antigen-presenting cells. *J. Clin. Endocrinol. Metab.* 81:4137–4143.
16. Lombardi, G., K. Arnold, J. Uren, F. Marelli-Berg, R. Hargreaves, N. Imami, A. Weetman, and R. Lechler. 1997. Antigen presentation by interferon-gamma treated thyroid follicular cells inhibits interleukin (IL-2) and supports IL-4 production by B7-dependent human T cells. *Eur. J. Immunol.* 27:62–71.
17. Faure, G.C., D. Bensoussan-Lejzerowicz, M.C. Bene, V. Aubert, and J. Leclere. 1997. Coexpression of CD40

and class II antigen HLA-DR in Graves' disease thyroid epithelial cells. *Clin. Immunol. Immunopathol.* 84:212–215.

18. Battifora, M., G. Pesce, F. Paolieri, N. Fiorino, C. Giordano, A.M. Riccio, G. Torre, D. Olive, and M. Bagnasco. 1998. B7.1 costimulatory molecule is expressed on thyroid follicular cells in Hashimoto's thyroiditis, but not in Graves' disease. *J. Clin. Endocrinol. Metab.* 83:4130–4139.

19. Mellors, R.C., W.J. Brzosko, and L.S. Sonkin. 1962. Immunopathology of chronic non-specific thyroiditis (autoimmune thyroiditis). *Am. J. Pathol.* 41:425.

20. McLachlan, S.M., A. McGregor, B. Rees Smith, and R. Hall. 1979. Thyroid-autoantibody synthesis by Hashimoto thyroid lymphocytes. *Lancet* i:162–163.

21. Canonica, G.W., M. Bagnasco, M.E. Cosulich, G. Torre, S.M. McLachlan, and B. Rees Smith. 1983. Why thyroid is a major site of thyroid autoantibody synthesis in autoimmune thyroid disease. *Lancet* i:1163–1163.

22. Matsunaga, M., K. Eguchi, T. Fukuda, A. Kurata, H. Tezuka, C. Shimomura, T. Otsubo, N. Ishikawa, K. Ito, and S. Nagataki. 1986. Class II major histocompatibility complex antigen expression and cellular interactions in thyroid glands of Graves' disease. *J. Clin. Endocrinol. Metab.* 62:723–728.

23. Margolick, J.B., A.P. Weetman, and K.D. Burman. 1988. Immunohistochemical analysis of intrathyrodal lymphocytes in Graves' disease: Evidence of activated T cells and production of interferon-gamma. *Clin. Immunol. Immunopathol.* 47:208–218.

24. McLachlan, S.M., G. Proud, C.A.S. Pegg, F. Clark, and B. Rees Smith. 1985. Functional analysis of T and B cells from blood and thyroid tissue in Hashimoto's disease. *Clin. Exp. Immunol.* 59:585–592.

25. Benveniste, P., V.V. Row, and R. Volpe. 1985. Studies on the immunoregulation of thyroid autoantibody production *in vitro*. *Clin. Exp. Immunol.* 61:274–282.

26. Weetman, A.P., A.M. McGregor, M.H. Wheeler, and R. Hall. 1984. Extrathyroidal sites of autoantibody synthesis in Graves' disease. *Clin. Exp. Immunol.* 56:330–336.

27. Atherton, M.C., S.M. McLachlan, C.A.S. Pegg, A. Dickinson, P. Baylis, E.T. Young, S.J. Proctor, and B. Rees Smith. 1985. Thyroid autoantibody synthesis by lymphocytes from different lymphoid organs fractionation of B cells on density gradients. *Immunology* 55:271–279.

28. Rees Smith, B., S.M. McLachlan, and J. Furmaniak. 1988. Autoantibodies to the thyrotropin receptor *Endocr. Rev.* 9:106–121.

29. Hammond, L.J., M.W. Lowdell, P.G. Cerrano, A.W. Goode, G.F. Bottazzo, and R. Mirakian. 1997. Analysis of apoptosis in relation to tissue destruction associated with Hashimoto's autoimmune thyroiditis. *J. Pathol.* 182:138–144.

30. Giordano, C., G. Stassi, R. De Maria, M. Todaro, P. Richiusa, G. Papoff, G. Ruberti, M. Bagnasco, R. Testi, and A. Galluzzo. 1997. Potential involvement of Fas and its ligand in the pathogenesis of Hashimoto's thyroiditis [see comments]. *Science* 275:960–963.

31. Arscott, P.L., and J.R. Baker, Jr. 1998. Apoptosis and thyroiditis. *Clin. Immunol. Immunopathol.* 87:207–217.

32. Del Prete, G., E. Maggi, S. Mariotti, A. Tiri, D. Vercelli, P. Parronchi, D. Macchia, A. Pinchera, M. Ricci, and S. Romagnani. 1986. Cytolytic T lymphocytes with natural killer activity in thyroid infiltrate of patients with Hashimoto's thyroiditis. Analysis at clonal level. *J. Clin. Endocrinol. Metab.* 62:52.

33. Chiovato, L., P. Bassi, F. Santini, C. Mammoli, P. Lapi, P. Carayon, and A. Pinchera. 1993. Antibodies producing complement-mediated thyroid cytotoxicity in patients with atrophic or goitrous autoimmune thyroiditis. *J. Clin. Endocrinol. Metab.* 77:1700–1705.

34. Mackenzie, W.A., and T.F. Davies. 1987. An intrathyroidal T-cell clone specifically cytotoxic for human thyroid cells. *Immunology* 61:101–103.

35. Stassi, G., M. Todaro, F. Bucchieri, A. Stoppacciaro, F. Farina, G. Zummo, R. Testi, and R. De Maria. 1999. Fas/Fas ligand-driven T cell apoptosis as a consequence of ineffective thyroid immunoprivilege in Hashimoto's thyroiditis. *J. Immunol.* 162:263–267.

36. Belyavin, G., and W.R. Trotter. 1959. Investigations of thyroid antigens reacting with Hashimoto sera Evidence for an antigen other than thyroglobulin. *Lancet* i:648–652.

37. Khoury, E.L., L. Hammond, G.F. Bottazzo, and D. Doniach. 1981. Presence of organ specific "microsomal" autoantigen on the surface of human thyroid cells in culture: Its involvement in complement-mediated cytotoxicity. *Clin. Exp. Immunol.* 45:316–328.

38. Tandon, N., S.L. Yan, B.P. Morgan, and A.P. Weetman. 1994. Expression and function of multiple regulators of complement activation in autoimmune thyroid disease. *Immunology* 81:643–647.

39. Bogner, U., H. Schleusener, and J.R. Wall. 1984. Antibody-dependent cell-mediated cytotoxicity against human thyroid cells in Hashimoto's thyroiditis but not Graves' disease. *J. Clin. Endocrinol. Metab.* 59:734–738.

40. Rodien, P., A.-M. Madec, Y. Morel, A. Stefanutti, H. Bornet, and J. Orgiazzi. 1992. Assessment of antibody dependent cell cytotoxicity in autoimmune thyroid disease using porcine thyroid cells. *Autoimmunity* 13:177–185.

41. Guo, J., J.C. Jaume, B. Rapoport, and S.M. McLachlan. 1997. Recombinant thyroid peroxidase-specific Fab converted to immunoglobulin (IgG) molecules: Evidence for thyroid cell damage by IgG1 but not IgG4 autoantibodies. *J. Clin. Endocrinol. Metab.* 82:925–931.

42. Londei, M., G.F. Bottazzo, and M. Feldmann. 1985. Human T cell clones from autoimmune thyroid glands Specific recognition of autologous thyroid cells. *Science* 228:85–89.

43. Weetman, A.P., D.J. Volkman, K.D. Burman, J.B. Margolick, P. Petrick, B.D. Weintraub, and A.S. Fauci. 1986. The production and characterization of thyroid-derived T-cell lines in Graves' disease and Hashimoto's thyroiditis. *Clin. Immunol. Immunopathol.* 39:139–150.

44. Mackenzie, W.A., A.E. Schwartz, E.W. Friedman, and T.F. Davies. 1987. Intrathyroidal T cell clones from patients with autoimmune thyroid disease. *J. Clin. Endocrinol. Metab.* 64:818–824.

45. Fukuma, N., S.M. McLachlan, B. Rapoport, J. Goodacre, S.L. Middleton, D.I.W. Philips C.A.S. Pegg, and B. Rees Smith. 1990. Thyroid autoantigens and human T cell responses. *Clin. Exp. Immunol.* 82:275–283.

46. Tandon, N., M. Freeman, and A.P. Weetman. 1991. T cell responses to synthetic thyroid peroxidase peptides in autoimmune thyroid disease. *Clin. Exp. Immunol.* 86:56–60.

47. Kawakami, Y., M.E. Fisfalen, and L.J. DeGroot. 1992. Proliferative responses of peripheral blood mononuclear cells from patients with autoimmune thyroid disease to synthetic peptide epitopes of human thyroid peroxidase. *Autoimmunity* 13:17–26.

48. Ewins, D.L., P.S. Barnett, S. Ratachaiyavong, C. Sharrock, J. Lanchbury, A.M. McGregor, and J.P. Banga. 1992. Antigen-specific T cell recognition of affinity-purified and recombinant thyroid peroxidase in autoimmune thyroid disease. *Clin. Exp. Immunol.* 90:93–98.

49. Fisfalen, M.E., L.J. DeGroot, W.A. Franklin, and K. Soltani. 1988. Microsomal antigen-reactive lymphocyte lines and clones derived from thyroid tissue of patients with Graves' disease. *J. Clin. Endocrinol. Metab.* 66:776–784.

50. Dayan, C.M., M. Londei, A.E. Corcoran, B. Grubeck-Loebenstein, R.F.L. James, B. Rapoport, and M. Feldmann. 1991. Autoantigen recognition by thyroid-infiltrating T cells in Graves disease *Proc. Natl. Acad. Sci. USA* 88:7415–7419.

51. Fisfalen, M.E., E.M. Palmer, G.A. van Seventer, K. Soltani, Y. Sawai, E. Kaplan, Y. Hidaka, C. Ober, and L.J. DeGroot. 1997. Thyrotropin-receptor and thyroid peroxidase-specific T cell clones and their cytokine profile in autoimmune thyroid disease. *J. Clin. Endocrinol. Metab.* 82:3655–3663.

52. Martin, A., R.P. Magnusson, D.L. Kendler, E. Concepcion, A. Ben-Nun, and T.F. Davies. 1993. Endogenous antigen presentation by autoantigen-transfected Epstein-Barr virus-lymphoblastoid cells. *J. Clin. Invest.* 91:1567–1574.

53. Mullins, R.J., Y. Chernajovsky, C. Dayan, M. Londei, and M. Feldmann. 1994. Transfection of thyroid autoantigens into EBV-transformed B cell lines. *J. Immunol.* 152;5572–5580.

54. Guo, J., S. Quaratino, J.C. Jaume, G. Costante, M. Londei, S.M. McLachlan, and B. Rapoport. 1996. Autoantibody-mediated capture and presentation of autoantigen to T cells via the Fc epsilon receptor by a recombinant human autoantibody Fab converted to IgE. *J. Immunol. Methods* 195:81–92.

55. Davies, T.F., A. Martin, E.S. Concepcion, P. Graves, L. Cohen, and A. Ben-Nun. 1991. Evidence of limited variability of antigen receptors on intrathyroidal T cells in autoimmune thyroid disease. *N. Engl. J. Med.* 325: 238–244.

56. Davies, T.F., A. Martin, E.S. Concepcion, P. Graves, N. Lahat, W.L. Cohen, and A. Ben-Nun. 1992. Evidence for selective accumulation of intrathyroidal T lymphocytes in human autoimmune thyroid disease based on T cell receptor V gene usage. *J. Clin. Invest.* 89:157–162.

57. McIntosh, R.S., P.F. Watson, A.P. Pickerill, R. Davies, and A.P. Weetman. 1993. No restriction of intrathyroidal T cell receptor V alpha families in the thyroid of Graves' disease. *Clin. Exp. Immunol.* 91:147–152.

58. Quaratino, S., M. Feldmann, C.M. Dayan, O. Acuto, and M. Londei. 1996. Human self-reactive T cell clones expressing identical T cell receptor beta chains differ in their ability to recognize a cryptic self-epitope. *J. Exp. Med.* 183:349–358.

59. McIntosh, R.S., N. Tandon, A.P. Pickerill, R. Davies, D. Barnett, and A.P. Weetman. 1993. The gamma delta T cell repertoire in Graves' disease and multinodular goitre. *Clin. Exp. Immunol.* 94:473–477.

60. Liblau, R.S., S.M. Singer, and H.O. McDevitt. 1995. Th1 and Th2 CD4+ T cells in the pathogenesis of organ specific autoimmune diseases. *Immunol. Today* 16:34–38.

61. Mosmann, T.R. and R.L. Coffman. 1989. TH1 and TH2 cells: Different patterns of lymphokine secretion lead to different functional properties. *Ann. Rev. Immunol.* 7:145–173.

62. McLachlan, S.M., J. Taverne, M.C. Atherton, A. Cooke, S.L. Middleton, C. Pegg, F. Clark, and B. Rees Smith. 1990. Cytokines, thyroid autoantibody synthesis and thyroid cell survival in culture *Clin. Exp. Immunol.* 79:175–181.

63. Del Prete, G., A. Tiri, S. Mariotti, A. Pinchera, M. Ricci, and S. Romagnani. 1987. Enhanced production of gamma-interferon by thyroid-derived T cell clones from patients with Hashimoto's thyroiditis. *Clin. Exp. Immunol.* 69:323–331.

64. Paschke, R., F. Schuppert, M. Taton, and T. Velu. 1994. Intrathyroidal cytokine gene expression profiles in auto-immune thyroiditis. *J. Endocrinol.* 141:309–315.

65. Ajjan, R.A., P.F. Watson, R.S. McIntosh, and A.P. Weetman. 1996. Intrathyroidal cytokine gene expression in Hashimoto's thyroiditis. *Clin. Exp. Immunol.* 105:523–528.

66. Heuer, M., G. Aust, S. Ode-Hakim, and W.A. Scherbaum. 1996. Different cytokine mRNA profiles in Graves disease, Hashimoto's thyroiditis, and nonautoimmune thyroid disorders determined by quantitative reverse transcriptase polymerase chain reaction (RT-PCR). *Thyroid.* 6:97–106.

67. Guo, J., B. Rapoport, and S.M. McLachlan. 1997. Cytokine profiles of *in vivo* activated thyroid infiltrating T cells cloned in the presence or absence of interleukin-4. *Autoimmunity* 26:103–110.

68. Roura-Mir, C., M. Catalfamo, M. Sospedra, L. Alcalde, R. Pujol-Borrell, and D. Jaraquemada. 1997. Single-cell analysis of intrathyroidal lymphocytes shows differential cytokine expression in Hashimoto's and Graves disease. *Eur. J. Immunol.* 27:3290–3302.

69. Guo, J., B. Rapoport, and S.M. McLachlan. 1999. Balance of Th1/Th2 cytokines in thyroid autoantibody synthesis in vitro. *Autoimmunity* 30:1–9.

70. Burton, D.R. and C.F. Barbas III. 1994. Human antibodies from combinatorial libraries. *Advances in Immunology* 57:191–281.

71. Rapoport, B., S. Portolano, and S.M. McLachlan. 1995. Combinatorial immunoglobulin gene libraries new insights into human organ-specific autoantibodies. *Immunol. Today* 16:43–49.

72. Weetman, A.P. 1987. IgA class and subclass thyroid auto-antibodies in Graves' disease and Hashimoto's thyroiditis. *Int. Archs. Allergy. Appl. Immun.* 83:432–435.

73. Prummel, M.F., W.M. Wiersinga, B. Rapoport, and S.M. McLachlan. 1993. IgA class thyroid peroxidase and thyroglobulin autoantibodies in Graves' disease: association with the male sex. *Autoimmunity* 16:153–155.

74. Guo, J., B. Rapoport, and S.M. McLachlan. 1997. Thyroid peroxidase autoantibodies of IgE class in thyroid autoimmunity. *Clin. Immunol. Immunopathol.* 82:157–162.

75. Parkes, A.B., S.M. McLachlan, P. Bird, and B. Rees Smith. 1984. The distribution of microsomal and thyroglobulin antibody activity among the IgG subclasses. *Clin. Exp. Immunol.* 57:239–243.

76. Portolano, S., P. Seto, G.D. Chazenbalk, Y. Nagayama, S.M. McLachlan, and B. Rapoport. 1991. A human Fab fragment specific for thyroid peroxidase generated by cloning thyroid lymphocyte-derived immunoglobulin genes in a bacteriophage lambda library. *Biochem. Biophys. Res. Comm.* 179:372–379.

77. Portolano, S., G.D. Chazenbalk, P. Seto, J.S. Hutchison, B. Rapoport, and S.M. McLachlan. 1992. Recognition by recombinant autoimmune thyroid disease-derived Fab fragments of a dominant conformational epitope on human thyroid peroxidase. *J. Clin. Invest.* 90:720–726.

78. Chazenbalk, G.D., S. Portolano, D. Russo, J.S. Hutchison, B. Rapoport, and S.M. McLachlan. 1993. Human organ-specific autoimmune disease: molecular cloning and expression of an autoantibody gene repertoire for a major autoantigen reveals an antigenic dominant region and restricted immunoglobulin gene usage in the target organ. *J. Clin. Invest.* 92:62–74.

79. Portolano, S., G.D. Chazenbalk, J.S. Hutchison, S.M. McLachlan, and B. Rapoport. 1993. Lack of promiscuity in autoantigen-specific H and L chain combinations as revealed by human H and L chain "roulette". *J. Immunol.* 150:880–887.

80. Hexham, J.M., L.J. Partridge, J. Furmaniak, V.B. Petersen, J.C.J. Colls, C.R. Pegg, B. Rees Smith, and D.R. Burton. 1994. Cloning and characterisation of TPO autoantibodies using combinatorial phage display libraries. *Autoimmunity* 17:167–179.

81. Portolano, S., S.M. McLachlan, and B. Rapoport. 1993. High affinity, thyroid-specific human autoantibodies displayed on the surface of filamentous phage use V genes similar to other autoantibodies. *J. Immunol.* 151:2839–2851.

82. McIntosh, R.S., M.S. Asghar, E.H. Kemp, P.F. Watson, A. Gardas, J.P. Banga, and A.P. Weetman. 1997. Analysis of IgG kappa anti-thyroid peroxidase antibodies from different tissues in Hashimoto's thyroiditis *J. Clin. Endocrinol. Metab.* 82:3818–3825.

83. Jaume, J.C., S. Portolano, B. Rapoport, and S.M. McLachlan. 1996. Influence of the light chain repertoire on immunoglobulin genes encoding thyroid autoantibody Fab from combinatorial libraries. *Autoimmunity* 24:11–23.

84. Jaume, J.C., J. Guo, B. Rapoport, and S.M. McLachlan. 1997. The epitopic "fingerprint" of thyroid peroxidase-specific Fab isolated from a patient's thyroid gland by the combinatorial library approach resembles that of autoantibodies in the donor's serum. *Clin. Immunol. Immunopathol.* 84:150–157.

85. Prummel, M.F., S. Portolano, G. Costante, B. Rapoport, and S.M. McLachlan. 1994. Isolation and characterization of a monoclonal human thyroid peroxidase autoantibody of lambda light chain type *Molec. Cell Endocrinol.* 102:161–166.

86. Portolano, S., M.F. Prummel, B. Rapoport, and S.M. McLachlan. 1995. Molecular cloning and characterization of human thyroid peroxidase autoantibodies of lambda light chain type. *Molec. Immunol.* 32: 1157–1169.

87. Guo, J., Y. Wang, J.C. Jaume, B. Rapoport, and S.M. McLachlan. 1999. Rarity of autoantibodies to a major autoantigen, thyroid peroxidase, that interact with denatured antigen or with epitopes outside the immunodominant region. *Clin. Exp. Immunol.* In press.

88. Weetman, A.P., C.M. Black, S.B. Cohen, R. Tomlinson, J.P. Banga, and C.B. Reimer. 1989. Affinity purification of IgG subclasses and the distribution of thyroid auto-antibody reactivity in Hashimoto's thyroiditis. *Scand. J. Immunol.* 30:73–82.

89. McIntosh, R.S. and A.P. Weetman. 1997. Molecular analysis of the antibody response to thyroglobulin and thyroid peroxidase. *Thyroid* 7:471–487.

90. Olee, T., E.W. Lu, D.-F. Huang, R.W. Soto-Gil, M. Deftos, F. Kozin, D.A. Carson, and P.P. Chen. 1992. Genetic analysis of self-associating immunoglobulin G rheumatoid factors from two rheumatoid synovia implicates an antigen-driven response. *J. Exp. Med.* 175:831–842.

91. Van Es, J.H., H. Aanstoot, F.H.J. Gmelig-Meyling, R.H.W.M. Derksen, and T. Logtenberg. 1992. A human systemic lupus erythematosus-related anti-cardiolipin/single-stranded DNA autoantibody is encoded by a somatically mutated variant of the developmentally restricted 51P1 VH gene. *J. Immunol.* 149:2234–2240.

92. Ikematsu, H., N. Harindranath, Y. Ueki, A.L. Notkins, and P. Casali. 1993. Clonal analysis of a human antibody response, II. Sequences of the VH genes of human IgM, IgG, and IgA to rabies virus reveal preferential utilization of VH III segments and somatic hypermutation. *J. Immunol.* 150:1325–1337.

93. Silberstein, L.E., L.C. Jefferies, J. Goldman, D. Friedman, J.S. Moore, P.C. Nowell, N.D. Roelcke, W. Pruzanski, J. Roudier, and G.J. Silverman. 1991. Variable region gene analysis of pathologic human autoantibodies to the related i and l red blood cell antigens. *Blood* 78:2372–2386.

94. Costante, G., S. Portolano, T. Nishikawa, J.C. Jaume, G.D. Chazenbank, B. Rapoport, and S.M. McLachlan. 1994. Recombinant thyroid peroxidase-specific autoantibodies. II. Role of individual heavy and light chains in determining epitope recognition. *Endocrinology* 134:25–30.

95. Chazenbalk, G.D., G. Costante, S. Portolano, S.M. McLachlan, and B. Rapoport. 1993. The immunodominant region on human thyroid peroxidase recognized by autoantibodies does not contain the monoclonal antibody 47/c21 linear epitope. *J. Clin. Endocrinol. Metab.* 77:1715–1718.

96. Czarnocka, B., M. Janota-Bzowski, R.S. McIntosh, M.S. Asghar, P.F. Watson, E.H. Kemp, P. Carayon, and A.P. Weetman. 1997. Immunoglobulin G kappa anti-thyroid peroxidase antibodies in Hashimoto's thyroiditis Epitope mapping analysis. *J. Clin. Endocrinol. Metab.* 82:2639–2644.

97. Arscott, P.L., R.J. Koenig, M.M. Kaplan, G.D. Glick, and J.R. Baker, Jr. 1996. Unique autoantibody epitopes in an immunodominant region of thyroid peroxidase. *J. Biol. Chem.* 271:4966–4973.

98. Finke, R., P. Seto, J. Ruf, P. Carayon, and B. Rapoport. 1991. Determination at the molecular level of a B-cell epitope on thyroid peroxidase likely to be associated with autoimmune thyroid disease. *J. Clin. Endocrinol. Metab.* 73:919–921.

99. Libert, F., M. Ludgate, C. Dinsart, and G. Vassart. 1991. Thyroperoxidase, but not the thyrotropin receptor contains sequential epitopes recognized by autoantibodies in recombinant peptides expressed in the pUEX vector. *J. Clin. Endocrinol. Metab.* 73:857–860.

100. Ruf, J., M. Toubert, B. Czarnocka, J. Durand-Gorde, M. Ferrand, and P. Carayon. 1989. Relationship between immunological structure and biochemical properties of human thyroid peroxidase. *Endocrinology* 125:1211–1218.

101. Guo, J., R.S. McIntosh, B. Czarnocka, A. Weetman, B. Rapoport, and S.M. McLachlan. 1998. Relationship between autoantibody epitopic recognition and immunoglobulin gene usage. *Clin. Exp. Immunol.* 111:408–414.

102. Nishikawa, T., G. Costante, M.F. Prummel, S.M. McLachlan, and B. Rapoport. 1994. Recombinant thyroid peroxidase autoantibodies can be used for epitopic "fingerprinting" of thyroid peroxidase autoantibodies in the sera of individual patients. *J. Clin. Endocrinol. Metab.* 78:944–949.

103. Jaume, J.C., G. Costante, T. Nishikawa, D.I.W. Phillips, B. Rapoport, and S.M. McLachlan. 1995. Thyroid peroxidase autoantibody fingerprints in hypothyroid and euthyroid individuals I Cross-sectional study in elderly women. *J. Clin. Endocrinol. Metab.* 80:994–999.

104. Jaume, J.C., A.B. Parkes, J.H. Lazarus, R. Hall, G. Costante, S.M. McLachlan, and B. Rapoport. 1995. Thyroid peroxidase autoantibody fingerprints. II. A longitudinal study in postpartum thyroiditis *J. Clin. Endocrinol. Metab.* 80:1000–1005.

105. Jaume, J.C., C.L. Burek, W.H. Hoffman, N. Rose, S.M. McLachlan, and B. Rapoport. 1996. Thyroid peroxidase autoantibody epitopic "fingerprints" in juvenile Hashimoto's thyroidits Evidence for conservation over time and in families. *Clin. Exp. Immunol.* 104:115–123.

106. Jaume, J.C., J. Guo, D.L. Pauls, M. Zakarija, J.M. McKenzie, J.A. Egeland, C.L. Burek, N.R. Rose W.H. Hoffman, B. Rapoport, and S.M. McLachlan. 1999. Evidence for genetic transmission of thyroid peroxidase autoantibody epitopic "fingerprints". *J. Clin. Endocrinol. Metab.* 84:1424–1431.

107. Nishikawa, T., B. Rapoport, and S.M. McLachlan. 1994. Exclusion of two major areas on thyroid peroxidase from the immunodominant region containing the conformational epitopes recognized by human autoantibodies. *J. Clin. Endocrinol. Metab.* 79:1648–1654.

108. Chacko, S., E. Padlan, S. Portolano, S.M. McLachlan, and B. Rapoport. 1996. Structural studies of human autoantibodies. Crystal structure of a thyroid peroxidase autoantibody Fab. *J. Biol. Chem.* 271:12191–12198.

109. McLachlan, S.M. and B. Rapoport. 1996. Genetic factors in thyroid disease. *In* Werner and Ingbar's The Thyroid: A Fundamental and Clinical Text, Braverman, L.E. and R.D. Utiger, editors. J.B. Seventh, Lippincott Company Philadelphia PA, 483–496.

110. Phillips, D.I.W., D.C. Shields, J.M. Dugoujon, L. Prentice, P. McGuffin, and B. Rees Smith. 1993. Complex segregation analysis of thyroid autoantibodies: are they inherited as an autosomal dominant trait? *Hum. Hered.* 43:141–146.

111. Pauls, D.L., M. Zakarija, J.M. McKenzie, and J.A. Egeland. 1993. Complex segregation analysis of antibodies to thyroid peroxidase in old order Amish families. *Am. J. Med. Genet.* 47:375–379.

112. Davies, T.F. 1998. Autoimmune thyroid disease genes come in many styles and colors. *J. Clin. Endocrinol. Metab.* 83:3391–3393.

113. Barbesino, G., Y. Tomer, E. Concepcion, T.F. Davies, and D.A. Greenberg. 1998. Linkage analysis of candidate genes in autoimmune thyroid disease. I. Selected immunoregulatory genes *J. Clin. Endocrinol. Metab.* 83:1580–1589.

114. Risch, N. and K. Merikangas. 1996. The future of genetic studies of complex human diseases. *Science* 273:1516–15

115. Ingbar, S.H. 1972. Autoregulation of the thyroid. Response to iodide excess and depletion. *Mayo. Clin. Proc.* 47:814

116. Braverman, L.E., S.H. Ingbar, A.G. Vagenakis, L. Adams, and F. Maloof. 1971. Enhanced susceptibility to iodide myxedema in patients with Hashimoto's thyroidits. *J. Clin. Endocrinol. Metab.* 32:515–521.

117. Bagchi, N., T.R. Brown, E. Urdanivia, and R.S. Sundick. 1985. Induction of autoimmune thyroiditis in chickens by dietary iodine. *Science* 230:325–327.

118. Champion, B.R., D.C. Rayner, P.G. Byfield, K.R. Page, C.T. Chan, and I.M. Roitt. 1987. Critical role of iodination for T cell recognition of thyroglobulin in experimental murine thyroid autoimmunity *J. Immunol.* 139: 3665–3670.

119. Rasooly, L., N.R. Rose, A.M. Saboori, P.W. Ladenson, and C.L. Burek. 1998. Iodine is essential for human T cell recognition of human thyroglobulin. *Autoimmunity* 27:213–219.

120. Tomer, Y. and T.F. Davies. 1993. Infection, thyroid disease, and autoimmunity. *Endocr. Rev.* 14:107–120

121. Weetman, A.P. and A.M. McGregor. 1994. Autoimmune thyroid disease. Further developments in our understanding. *Endocr. Rev.* 15:788–830.

122. Wick, G., J. Most, K. Schauenstein, G. Kromer, H. Dietrich, A. Ziemicki, R. Fassler, S. Schwarz, N. Neu, and K. Hala. 1985. Spontaneous autoimmune thyroiditis—a bird's eye view. *Immunol. Today* 6:359–365.

123. Maczek, C., N. Neu, G. Wick, and K. Hala. 1992. Target organ susceptibility and autoantibody production in an animal model of spontaneous autoimmune thyroiditis. *Autoimmunity* 12:277–284.

124. Macht, L., N. Fukuma, K. Leader, D. Sarsero, C.A.S. Pegg, D.I.W. Philips, P. Yates, S.M. McLachlan, C. Elson, and B. Rees Smith. 1991. Severe combined immunodeficient (SCID) mice: a model for investigating human thyroid autoantibody synthesis. *Clin. Exp. Immunol.* 84:34–42.

125. Davies, T.F., H. Kimura, P. Fong, D. Kendler, L.D. Shultz, S. Thung, and A. Martin. 1991. The SCID-hu mouse and thyroid autoimmunity: characterization of human thyroid autoantibody secretion. *Clin. Immunol. Immunopathol.* 60:319–330.

126. Martin, A., H. Kimura, S. Thung, P. Fong, L.D. Shultz, and T.F. Davies. 1992. Characteristics of long-term human thyroid peroxidase autoantibody secretion in scid mice transplanted with lymphocytes from patients with autoimmune thyroiditis. *Int. Arch. Allergy. Immunol.* 98:317–323.

127. Akasu, F., T. Morita, E. Resetkova, N. Miller, R. Akasu, C. Jamieson, and R. Volpe. 1993. Reconstitution of severe combined immunodeficient mice with intrathyroidal lymphocytes of thyroid xenografts from patients with Hashimoto's thyroiditis. *J. Clin. Endocrinol. Metab.* 76:223–230.

128. Yoshikawa, N., S. Mori, T. Tokoro, S. Ikehara, H. Kumazawa, T. Yamashita, M. Nishikawa and M. Inada. 1996. IFN-g has a protective role against thyroid-specific autoantibody production in severe combined immunodeficient (SCID) mice xenografted with Graves' thyroid tissue. *Thyroid* 6.437–443.

129. Kawai, K., T. Enomoto, V. Fornasier, E. Resetkova, and R. Volpe. 1997. Differential effects of human interferon alpha and interferon gamma on xenografted human thyroid tissue in severe combined immunodeficient and nude mice. *Proc. Assoc. Amer. Physicians* 109:126–135.

130. Rose, N.R. and E. Witebsky. 1956. Studies on organ specificity. V. Changes in the thyroid glands of rabbits following active immunization with rabbit thyroid extracts. *J. Immun.* 76:417–427.

131. Kong, Y.M. 1986. The mouse model of autoimmune thyroid disease. In: Immunology and Medicine, A.M. McGregor, editor. MTP Press Limited Lancaster, 1–24.

132. Kotani, T., K. Umeki, K. Hirai, and S. Ohtaki. 1990. Experimental murine thyroiditis induced by porcine thyroid peroxidase and its transfer by the antigen-specific T cell line. *Clin. Exp. Immunol.* 80:11–18.

133. Nye, L., L.C. Pontes de Carvalho, and I.M. Roitt. 1980. Restrictions in the response to autologous thyroglobulin in the human. *Clin. Exp. Immunol.* 41:252–263.

134. Heidenreich, F., A. Vincent, A. Roberts, and J. Newsom-Davis. 1988. Epitopes on human acetylcholine receptor defined by monoclonal antibodies and myasthenia gravis sera. *Autoimmunity* 1:285–297.

135. Shimojo, N., Y. Kohno, K-I. Yamaguchi, S.-I. Kikuoka, A. Hoshioka, H. Niimi, A. Hirai, Y. Tamura, Y. Saito, L.D.

Kohn, and K. Tahara. 1996. Induction of Graves-like disease in mice by immunization with fibroblasts transfected with the thyrotropin receptor and a class II molecule. *Proc. Natl. Acad. Sci. USA.* 93:11074–11079.

136. Jaume, J.C., J. Guo, Y. Wang, B. Rapoport, and S.M. McLachlan. 1999. Cellular thyroid peroxidase (TPO) unlike purified TPO and adjuvant, induces antibodies in mice that resemble autoantibodies in human autoimmune thyroid disease. *J. Clin. Endocrinol. Metab.* 84:1651–1657.

137. Simitsek, P.D., D.G. Campbell, A. Lanzavecchia, N. Fairweather, and C. Watts. 1995. Modulation of antigen processing by bound antibodies can boost or suppress class II major histocompatibility complex presentation of different T cell determinants. *J. Exp. Med.* 181:1957–1963.

138. McLachlan, S.M. and B. Rapoport, 1995. Genetic and epitopic analysis of thyroid peroxidase (TPO) autoantibodies; markers of the human thyroid autoimmune response. *Clin. Exp. Immunol.* 101:200–206.

38 | Immunological Mechanisms in Graves' Disease

Terry F. Davies, David L. Kendler, and Andreas Martin

1. INTRODUCTION

1.1 The History of Robert Graves' Disease

What we now understand to be autoimmune hyperthyroidism was initially observed by Caleb Parry in 1825, but first described in print by Robert Graves in 1835 (1). These physicians recognized the combination of goiter and prominent staring eyes in patients with heart failure. However, the concept of Graves' disease as a unique organ-specific autoimmune disease was not introduced until 1956, when Adams and Purves discovered TSH receptor autoantibodies [then termed long-acting thyroid stimulator or "LATS" (2)]. Indeed, the discovery of autoantibodies to thyroglobulin in patients with autoimmune thyroiditis as well as patients with Graves' disease (3) captured the imagination of thyroidologists and immunologists and demonstrated that these thyroid diseases were indeed the first confirmed human autoimmune diseases. Since those early days, there has been progress in the understanding of human thyroid autoimmunity, but Graves' disease remains an enigma; a uniquely human disease. In addition, the accessibility of the thyroid gland has caused both autoimmune hyperthyroidism and hypothyroidism to remain models for understanding self tolerance and its breakdown.

1.2 Clinical Features

Autoimmune hyperthyroidism afflicts females more than males (up to a 10:1 ratio) with a 2% prevalence in the population (Table 1). The disorder is usually characterized by a diffuse goiter and the clinical and biochemical features of hyperthyroidism (Figure 1). Patients often demonstrate additional features, including ophthalmopathy (in 30–70%), occasionally dermopathy or "pretibial myxedema" (in less than 1%) and, rarely, acropachy (4). The thyroidal and extrathyroidal manifestations of the disease (typically ophthalmopathy) may be concurrent or may present at different times and run separate courses. While, rarely, some patients with extrathyroidal disease never manifest thyroidal symptoms, most hyperthyroid patients have mild or no clinical eye disease.

1.3. Histopathology

Thyroid gland. There is commonly a diffusely enlarged thyroid gland up to several times its normal size. The follicles are small and lined by both hypertrophied and hyperplastic columnar epithelium with mitoses (Figure 2). There is marked loss of colloid with new small follicle formation. Epithelial cell papillary projections are increased and reach into the colloid, which is decreased in amount and has marginal scalloping and vacuolization. There is a heterogeneous lymphocytic and plasma cell infiltrate which forms aggregates or follicles. This infiltrate has been extensively analyzed and is discussed in Section 6.2.

Retroorbital tissues. The orbit of patients with Graves' infiltrative ophthalmopathy is characterized by an increase in the volume of extraocular muscles and retrobulbar connective tissue. The total orbital tissue volume increases from a normal of about 20cc to over 328cc in the most affected patients (5). There is a mild lymphocyte infiltrate of the extraocular muscle early in the disease, mostly T cells with foci of B cells, and minimal structural muscle damage (6) (Figure 3). The orbital fat and connective tissues are paucicellular. Excess hyaluronic acid and glycosaminoglycan deposition is seen in the

Table 1   Prevalence and incidence of Graves' disease

	Ref.	Prevalence (1/10000)			Incidence (1/10000)		
		Females	Males	All	Females	Males	All
U.K. Doctors	(1)	25					
Wickham, UK	(360)	190–270	16–23	11–16	20–30		
Olmsted Co. USA	(3)				3.7	0.8	2.0
Oakland USA	(361)	54	0	31	7.8	2.1	5.2
Denmark							
old	(1)	9.8					
young		1.4					
South Africa	(1)						
Black							0.4
Colored							1.4
White							2.3
Japan	(1)			8			

(1) (362)
(2) (363;364)
(3) (365)
(4) (366)

tissue interstitium. These substances are hygroscopic and contribute to the increase in the retroorbital tissue volume (7).

Dermal changes.   Similar to the thyroid and orbit, the skin, particularly in the exposed areas such as the pretibia and elbow, may also be infiltrated with lymphocytes although only limited studies on this disorder have been performed (8,9). Glycosaminoglycan deposition and attendant edema fluid cause separation and fragmentation of collagen fibers.

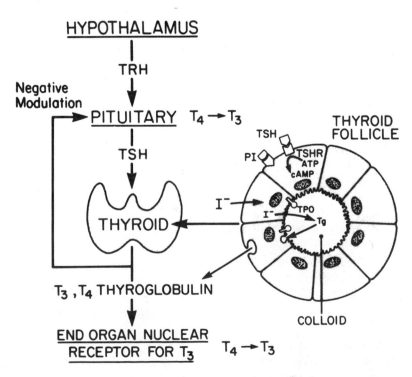

Figure 1.   Physiology of Graves' disease: Normal thyroid hormone homeostasis is controlled by higher cortical centers through the action of TRH on the pituitary gland, causing the release of TSH and thyroid stimulation. In Graves' disease, TSHR-Ab directly stimulates the thyroid gland, leading to thyroid cell growth, excess thyroid hormone and thyroglobulin output and suppression of pituitary TSH release.

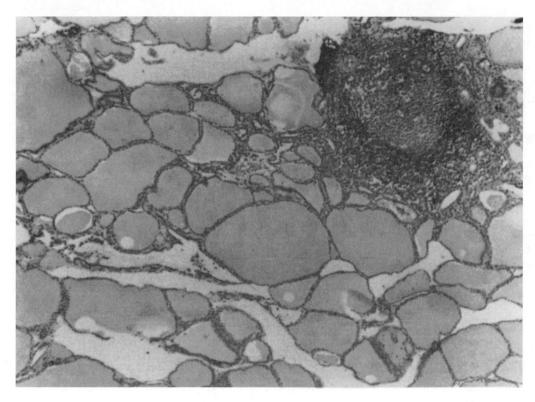

Figure 2.   Histopathology of thyroid gland in Graves' disease: Photomicrograph illustrates the intrathyroidal lymphocytic infiltrate with small irregular thyroid follicles. Occasional germinal centers are present.

1.4. Graves' Disease in the Spectrum of Human Autoimmune Endocrine Disease

Graves' disease should be considered in the context of other autoimmune thyroid diseases: autoimmune thyroiditis with goiter formation (Hashimoto's disease), autoimmune thyroiditis with thyroid atrophy (primary myxedema) and the post-partum and neonatal thyroid syndromes (10). Patients with known autoimmune thyroiditis may present with manifestations of Graves' disease and Graves' patients not uncommonly develop thyroid failure due to autoimmune thyroiditis. Both of these diseases have features of humoral and cellular autoimmunity with high titers of serum autoantibodies to defined thyroid antigens, thyroid antigen-specific T cells, and marked lymphocytic infiltration of the thyroid. Not only do these thyroid autoimmune diseases tend to be familial and multigenerational, but such patients may suffer from, or have a family history of, a variety of other autoimmune endocrine diseases and related autoimmune disorders (11).

2. IMMUNOGENETICS OF GRAVES' DISEASE

The genetic predisposition to Graves' disease, as well as other endocrine autoimmune diseases, has long been suspected from the family history of such patients (9). Children of parents with Graves' disease have a relative risk of only 2.0 for the disease, although in identical twin (monozygotic-MZ) studies, there has been 30–50% disease concordance (12). Such epidemiologic information has suggested a genetic predisposition to disease that may be combined with environmental triggering events.

It must be emphasized that MZ twins are not identical in their immune repertoire due to somatic recombinations of T and B cells as well as individual immune experiences that influence the immune repertoire (13). Therefore, part of the observed discordance between MZ twins may also be due to the discordance in their immune repertoire. Despite such differences, in there remains a substantial inherited susceptibility to Graves' disease, presumably related to both immune and non-immune genes.

2.1. HLA Association, But Not Linkage

Following the original observations of an MHC linkage with thyroiditis in animal models (14), HLA disease associations were sought in patients with Graves' disease using serological typing, restriction fragment length polymorphism (RFLP) analyses and direct sequencing. HLA-B8 was first associated with a 3-fold increased risk of Graves'

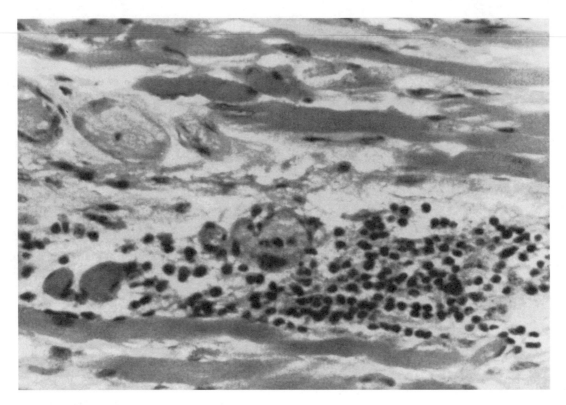

Figure 3.   Histopathology of thyroid opthalmopathy: A patchy paucicellular infiltrate around the muscle fibers is characteristic early in the disease. Later, deposition of glycosaminoglycan and fibrosis occur.

disease in Caucasians, while HLA-B17 and Bw46 conferred additional risks in Blacks and Orientals (15,16). Such observations were subsequently found to be secondary to linkage disequilibrium with HLA-DR3 class II haplotype, which conferred, in some studies up to a 6-fold relative risk of Graves' disease (17,18) (Table 2). RFLP studies also showed similar population associations for HLA-DR3 with the disease (19). Among Caucasians, HLA DQA1*0501 has recently been described as conferring a risk for Graves' disease over and above that of HLA-DR3 (relative risk 3.8) (20,21). However, the risk of developing Graves' disease in HLA-identical siblings was only 7%, not significantly different from the general risk in siblings. Furthermore, this was much lower than the 30% concordance for MZ twins described earlier. In addition, our laboratory has been unable to demonstrate linkage of the HLA gene region to Graves' disease, rather than an association, indicating that HLA genes only contribute a small part of the genetic susceptibility (22–24).

2.2.   Molecular Basis of HLA Associations

Derivation of the typical MHC class II molecular structure has suggested that absence of aspartic acid at position 57 of the DQ β chain may increase the risk of insulin-dependent diabetes mellitus 100-fold (25,26). This molecular change within the peptide-binding groove may influence the type and extent of antigen-peptide binding to the class II molecule and hence antigen presentation itself. In this way, the T cell repertoire would be enhanced or deleted. While it is possible that similar molecular preferences may be associated with Graves' disease susceptibility, a clear pattern has not been demonstrated. One study of RFLP-defined HLA class II heterogeneity associated a DQ β chain polymorphism with Hashimoto's disease. As with the diabetes study, this coded for an amino acid in the binding cleft of the class II molecule (27).

2.3.   Non-HLA Genes and Graves' Disease

There has been much progress in our understanding of the genetics of complex diseases, but most studies in Graves' disease have been association studies. CTLA-4 is an important co-stimulatory molecule necessary for T cell stimulation. There have been several reports demonstrating an association between the CTLA-4 gene and Graves' disease giving a relative risk of 2.1 to 2.8 (28). Analysis of an alanine/threonine (G/A) polymorphism inside the CTLA-4 leader peptide showed a relative risk of approximately 2.0 for the G allele (29). These reports have been consistent in several populations although not all, and there is now evidence that the CTLA-4 gene is also associated

Table 2 Some of the important HLA association studies in Graves' disease performed in Caucasians (for review and list of references see (367;368)(369).

Country	Ethnic group	No. of patients	HLA allele	Relative risk
Canada	Caucasians	175	B8	3.1
			DR3	5.7
UK	Caucasians	127	B8	2.8
			DR3	2.1
France	Caucasians	94	B8	3.4
			DR3	4.2
Hungary	Caucasians	256	B8	3.5
			DR3	4.8
Ireland	Caucasians	86	B8	2.5
			DR3	2.6
Canada	Caucasians	133	DR3	4.6
Sweden	Caucasians	78	B8	4.4
			DR3	3.9
USA	Caucasians	65	DR3	3.4
UK	Caucasians	120	DQA1 *0501	3.8
UK[1]	Caucasians	228	DRB1*0304 DQB1*0301	2.7
			DQA1* 0501	1.9
				3.2
USA	Caucasians	94	DQA1 *0501	3.7

[1]In this study the TDT also showed an association to the extended haplotype DRB1*0304-DQB1*02-DQA1*0501

with autoimmune hypothyroidism and postpartum thyroiditis (30). However, it is unlikely that the CTLA-4 gene itself is a major susceptibility gene for AITD since the relative risks are low, but is likely one of a number of non-specific immune susceptibility genes. The different thyroid autoantigens themselves (the TSHR, Tg, TPO) have also been studied as possible candidates for conferring susceptibility to Graves' disease, but no consistently positive data have been forthcoming.

The first whole genome screening in families with GD and HT was completed in a dataset of 56 multiplex families (354 individuals) (31). This screening revealed 3 new loci linked with Graves' disease (including one on the X chromosome), 2 possible loci that may be linked with Hashimoto's thyroiditis, and one locus linked with both diseases.

2.4. Immunogenetics of Ophthalmopathy

The role of genetics in the development of GO has been less apparent. While there have been a number of case reports of multiple members in families with exophthalmic goiter there have been no family or twin studies performed in GO. In addition, the only candidate genes tested in GO, in the HLA gene region, gave equivocal results (32,33).

3. AUTOANTIGENS IN GRAVES' DISEASE

3.1. The Thyroidal TSH Receptor

Pituitary thyroid stimulating hormone (TSH) is the major trophic hormone for the thyroid gland and influences almost all functions of the thyroid epithelial cell (Figure 1), including expression of the thyroid-specific antigens thyroid peroxidase (TPO) and thyroglobulin (Tg). TSH binds to a defined TSH receptor (TSHR) on the surface of the thyroid epithelial cell, which is G-protein linked and uses cyclic AMP and the phosphoinositol pathways for signal transduction (34). Although the TSHR is present at low density on the thyroid cell surface, with 10^3 to 10^4 sites per cell, it is the primary autoantigen of Graves' disease. Patients with Graves' disease have autoantibodies to the TSHR and T cells that are reactive to the TSHR. Putative extrathyroidal TSH receptors on fibroblasts, adipose tissue, adrenal, gonadal tissue, neutrophils, lymphocytes and muscle cells have also been described, but the physiological role of such expression in most sites remains undefined. However, the retroorbital TSH receptor expression appears to play an important antigenic role in ophthalmic Graves' disease.

Molecular biology. The crystallographic structure of the human TSHR has not yet been determined. The gene has been localized to chromosome 14 and the TSHR-specific mRNA consists of two major transcripts of 4.6 and 1.3 kb secondary to alternate splicing (35,36). The TSHR undergoes extensive post translational modification into a complex subunit structure. Recombinant hTSHR has been studied as a target for human autoantibodies and T cells (see below).

Protein structure and function. The TSHR consists of a 120 kDa, 744 amino acid, sequence cleaved into two subunits, α and β, which are linked by disulfide bonds and have 70% sequence homology with the receptor for luteinizing hormone (LH) (37,38). The 50 kDa α subunit is water-soluble and contains the principal TSH binding sites. The 30 kDa β subunit is insoluble and contains the membrane-spanning domain (Figure 4). Both cleaved and uncleaved receptors can be detected in the cell membrane with an excess of β subunits, suggesting ectodomain shedding or degradation (39).

3.2. Extrathyroidal TSH Receptors

Extraocular muscle. Cross-reacting epitopes between thyroid tissue and retro-orbital tissue have been sought to explain the orbital manifestations of Graves' disease. Sixty-

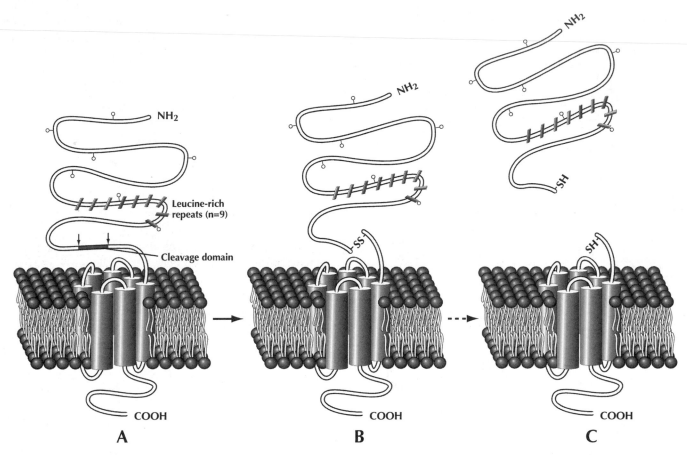

Figure 4.   Three coexisting structural forms of TSH receptors detected in thyroid cell membrane preparations: Current evidence suggests that the receptor first appears on the plasma membrane as intact holoreceptor (a), which is then processed into a two-subunit form (b), comprising separate ectodomain (α subunit) and transmembrane (β subunit) segments linked by disulfide bonds. Subsequently, some ectodomain subunits may be shed, leaving behind membrane-anchored β subunits (c).

four and 73 kDa antigens in porcine extraocular muscle are recognized by sera from patients with Graves' disease, but lack both ophthalmopathy patient specificity and orbital tissue specificity (40). The recognition of succinate dehydrogenase as a secondary muscle antigen has recently been reported and may be a marker of retrobital muscle inflammation rather than a primary event (41).

Retro-orbital fibroblasts and adipocytes.   The retro-orbital fibroblast appears to be the primary target of immune attack in Graves' orbital disease. Retroorbital and dermal fibroblasts are morphologically and functionally distinct *in vitro* (42). For example, dexamethasone and T3 inhibit synthesis of glycosaminoglycans in human dermal, but not retroorbital fibroblast, cultures (43). There is now abundant evidence that the retroorbital fibroblasts and adipocytes express the TSHR antigen (44), which provides the potential specificity cross-over between the retroorbital tissues and the thyroid gland. How this is distinguished from other sites of extrathyroidal TSHR expression remains unclear but may be simply a con-

sequence of the quantity and quality of posttranslational processing.

TSHR expression in other tissues.   In addition to adipocytes and fibroblasts, the mRNA for the TSHR has been found in many tissues, including the thymus, pituitary gland. heart muscle, lymphocytes and brain. The function of such mRNAs is unclear, but many must be considered blind transcripts until TSHR protein expression is demonstrated. Such is the case for thymic epithelium expression, which may be involved in deleting TSHR-specific T and B cells 1 (45,46).

3.3.   Thyroglobulin (Tg)

Patients with Graves' disease frequently have autoantibodies to thyroglobulin, the principal constituents of the follicular colloid, produced by the thyroid epithelial cells and stored in the gland. Thyroid peroxidase (TPO) is responsible for iodination of specific tyrosine residues on thyroglobulin to monoiodotyrosine and diiodotyrosine.

When stimulated by TSH, the thyroid cell phagocytoses thyroglobulin, which is proteolyzed in phagolysosomes yielding thyroxine (T4) and triiodothyronine (T3) that is then released into the extracellular fluid (Figure 1). Tg is detectable in the normal circulation and serum levels are increased in all thyroid diseases, including Graves' disease.

Molecular biology. Genomic DNA encoding Tg is a 200 kb sequence located on chromosome 8, which transcribes both 9.0 kb and 0.9 kb mRNAs (termed Tg-1 and Tg-2) Transcription of Tg mRNAs is increased by TSH, which acts by increasing intracellular cyclic AMP and influencing a series of thyroid-specific transcription factors (TTF1 and TTF2) (47). TSH-stimulated Tg gene transcription is also regulated by a variety of thyroid cell-secreted growth factors (such as IGF-1) and cytokines. Tg mRNAs are enhanced by low-dose IL-1 β and suppressed by high dose IL-1 β (48), while IF-γ inhibits TSH-stimulated Tg-gene transcription (49).

Protein structure and detection. Tg is a large, compact, water-soluble glycoprotein dimer, each subunit being 330 kDa (Figure 5). Although approximately 45 specific tyrosine residues of a total of 72 in the molecule are accessible to iodination by TPO (50), only a few act as primary iodine acceptors and donor sites and these are located in defined polypeptide sequences near the amino and carboxy terminals of the molecule (51). The degree of Tg iodination and post-translational modifications are thought to be important in immunogenicity (see below).

3.4. Thyroid Peroxidase (TPO)

Introduction. Autoantibodies to TPO are common in patients with Graves' disease. TPO is the key enzyme in thyroid physiology, situated on the luminal surface of the microvilli of the thyroid epithelial cell in the appropriate location to catalyze intrafollicular reactions on stored Tg. TPO is the major "microsomal" antigen of autoimmune thyroid disease (52 55).

Molecular biology and regulation. TPO transcription and translation are closely regulated by TSH and are cyclic AMP dependent (56,57). The gene for hTPO is located on the short arm of chromosome 2 (58,59). Two different human TPO cDNA clones, differing by 171 bases, have been isolated and proposed as reason for the 100 and 107 kDa protein doublet often seen on Western blotting. There

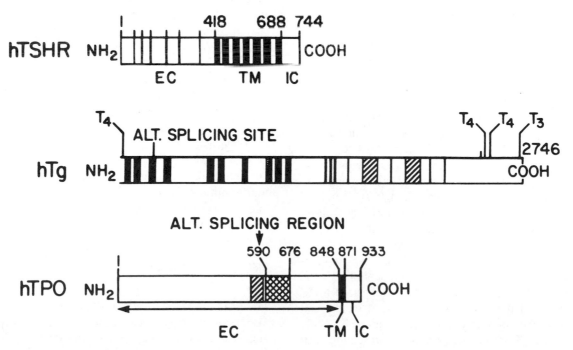

Figure 5. TSHR, TPO, Tg protein structures.
    The three major thyroid autoantigens have been cloned and sequenced. TPO and TSHR have a membrane-spanning region with intra- and extra-cellular domains. Tg is a large dimer which may undergo iodination at several sites.

is also evidence for alternative splicing of TPO message with at least two distinct mRNAs of 3.0 and 3.2 kb. It has been reported that Graves' disease thyroid contains up to 50% of this second form of TPO mRNA (60) raising the question of altered immunogenicity. TPO polymorphisms have also been reported (58,60). TPO induction by TSH is inhibited by IL-1 (61), and IF-γ (62).

Protein structure and function. The deduced amino acid sequence of human TPO is 933 amino acids (63). It is a 107 kDa glycoprotein with 10% glycosylation and a membrane-spanning region close to the carboxy terminus (Figure 5). TPO serves to iodinate tyrosine residues on Tg and to couple iodinated tyrosines to form T3 (monoiodo-tyrosine + di-iodotyrosine) and T4. Thiourylene or anti-thyroid drugs, such as methimazole or propylthiouracil, act by competing for substrate with TPO.

3.5. The Iodide Symporter

Cloning of the sodium-iodide symporter (NIS) (64) has provided valuable information that has led to a number of pathophysiological studies (65,66), including binding studies of Graves' patient IgG to synthetic peptides of the symporter and recombinant NIS, suggesting a possible role of the iodide symporter as autoantigen (66–68). Another group has reported amino acid similarities of NIS with other thyroid antigens (68).

4. ANTIGEN PRESENTATION IN GRAVES' DISEASE

4.1. MHC Expression in the Thyroid Gland

Endogenous antigens are usually processed intracellularly. The resulting peptide fragments bind to newly-synthesized MHC class I molecules to be expressed on the cell surface, where they may be recognized by antigen-specific CD8 T-cells. Normal thyroid epithelial cells express MHC class I antigens, but such expression is enhanced in thyroid tissue from patients with autoimmune thyroid disease and may target CD8 T cells to the thyroid gland (69). Increased MHC class I expression is seen in the majority of Graves' disease thyroids examined at surgery even after extended antithyroid drug treatment.

Exogenous antigens are phagocytosed by antigen-presenting cells (APC) and processed intracellularly. The resulting peptides associate with newly-synthesized MHC class II cells and the complex is expressed on the cell surface, where they may be recognized by antigen-specific CD4+ T cells. This classical antigen-presenting cell function appears to be intact in patients with Graves' disease (70), but, normal human thyroidal epithelial cells are MHC class II-negative. Abnormal MHC class II antigen

expression (HLA-DR, DQ and DP) has been demonstrated in many thyroid epithelial cells from patients with autoimmune thyroid disease in association with lymphocytic infiltration (71,72), and much work has been directed at defining the mechanisms regulating such MHC class II antigen expression in the thyroid gland (Figures 6 and 7). Although SV40 T-antigen immortalized cell lines and malignant thyroid cell lines (73) have been shown to constitutively express HLA class II *in vitro*, and thyroid carcinoma tissue itself may express class II antigens in the absence of significant lymphocytic infiltration (74), most investigations have studied the role of cytokines in inducing class II gene activation secondarily. Little is known about the role of thyroid cell-specific trans-activating factors at the MHC gene 5′ promoter regions in thyroid epithelial cells (75).

4.2. Stimulation of Thyroid Cell HLA Class II Antigen Expression

Cytokines such as IF-γ and TNF-α are capable of influencing HLA class II antigen expression in human thyroid epithelial cells, both *in vitro* and *in vivo*, and may be relevant to the pathophysiology of Graves' disease. Mice treated with IF-γ develop widespread MHC class II-antigen expression and thyroiditis (76), and IF-γ treated thyroid tissue is lysed when returned to syngeneic hosts (77). Human thyroid autoantibodies, thyrocyte HLA-DR expression, and lymphocytic infiltrates have been reported to develop in patients after treatment with α-interferon or IL-2 (78–83). Human IF-γ induces class II antigen expression in cultured human thyrocytes (84,85), an effect potentiated by TSH (86,87) and TSHR-Abs. Induction of HLA-DR antigen was usually greater than HLA-DP which, in turn, was induced to a higher degree than HLA-DQ (71) at both the transcriptional and translational levels (Figure 7). Other known modulators of MHC class II expression include EGF, which may suppress up to 50% of the effects of IF-γ, and TNF-α, which increased IF-γ receptor numbers (87), thereby enhancing the stimulation of HLA-DR antigen expression. Other activators of class II gene transcription have been less well studied. Of particular interest has been the induction of MHC class II antigens by viral infection (88).

4.3. Thyroid Epithelial Cells as Antigen Presenting Cells (APCs)

In situ hybridization studies showed that HLA-DR gene expression was high in Graves' disease thyroids and associated with areas of lymphocytic infiltrate (89) (Figure 6). Such observations that HLA class II antigen expression was localized to areas of the thyroid gland infiltrated with lymphocytes and that thyroid tissue from patients with Graves'

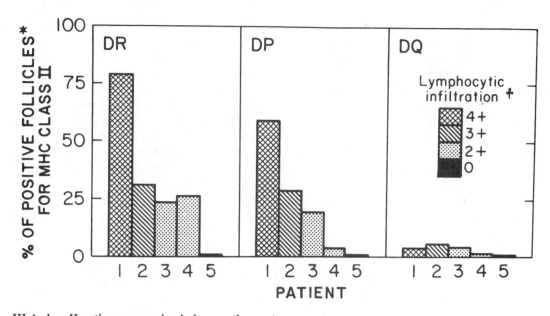

Figure 6.   HLA class II antigen expression in human thyrocytes
Immunocytochemical studies of HLA-DR, -DQ and -DP expression in thyroid follicular epithelium. Frozen sections of thyroid tissue from 4 patients with Graves' disease (1–4) and a normal control thyroid (5) were stained for HLA-DR (left), DP (middle) and DQ (right) antigens. HLA-DR and DP expression correlated with the extent of lymphocytic infiltration. From (357).

disease, but not normal thyroid tissue, expressed MHC class II caused interest in the potential of the thyroid cell to act as a classical antigen-presenting cell. However, much of the evidence for the APC function of human thyroid

cells has been difficult to interpret. This difficulty stems from the inability to obtain purified human thyroid cells, since primary thyroid monolayer cell preparations are contaminated with other potential APCs, such as B-cells and

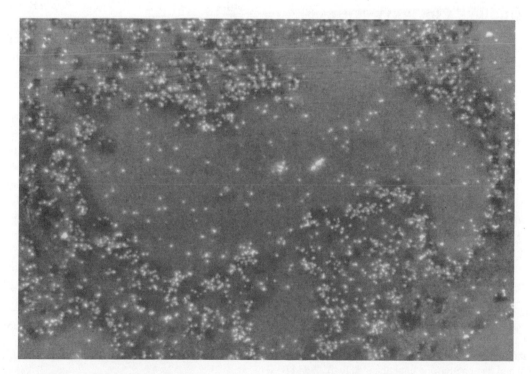

Figure 7.   *In situ* hybridization of HLA-DR α chain RNA in Graves' disease thyroid tissue. Section of Graves' disease thyroid tissue was hybridized with a [³H]-DRα-chain-specific antisense cRNA probe and observed in a dark field. Light grains show HLA-DRα mRNA throughout the epithelial cells of a thyroid follicle (89).

dendritic cells (89). There are, however, some data in support of an APC activity by thyroid cells (Table 3). Hence, once MHC class II antigens are expressed on the surface of thyroid epithelial cells they appear able to act as APCs, initiate or exaggerate thyroid-antigen-specific T-cell expansion, and autoantibody secretion. The factors initiating class II-antigen expression may be primary or secondary, with most of the evidence suggesting that such a reaction is secondary to the presence of activated T cells releasing local cytokines. Although B7-1 and B7-2 antigen expression has not been clearly demonstrated in Graves' thyroid glands, the presence of CD40 on human thyroid cells has added further weight to the APC potential of the human thyroid cell (90).

4.4. Other Intrathyroidal APCs

Dendritic cells are rare in normal thyroid tissue, but have been reported to be increased in Graves' thyroid co-localizing with focal lymphocytic infiltrates (91). Such dendritic cells may also present thyroid antigen to the immune system. Similarly, B cells form a major part of the thyroidal infiltrate and have been shown to be capable of thyroid antigen presentation to T cells (92). Indeed, B cells present autoantigen highly efficiently and such function has been shown to be normal in Graves' disease.

5.   AUTOANTIBODIES AND GRAVES' DISEASE

5.1.   B Cell Function and the Site of Autoantibody Secretion

The B cells that accumulate within the Graves' thyroid gland are large and exhibit reduced proliferation to B cell mitogen and greater basal immunoglobulin secretion than peripheral blood B cells (93). Activated B cell markers are more frequent in intrathyroidal B-cell cultures than in peripheral blood mononuclear cell (PBMC) cultures. IL-6 is produced by activated macrophages and thyroid epithelial cells (94,96) in the autoimmune thyroid gland and may induce differentiation of activated B lymphocytes into

mature immunoglobulin-secreting plasma cells (97), expansion of B cell clones and secretion of autoantibody (98). Lymphocytes from Graves' thyroid tissue may secrete thyroid autoantibodies spontaneously in vitro, again implying pre-activation (99). Hence, the thyroid gland is a primary site of thyroid autoantibody secretion in human autoimmune thyroid disease. Additional evidence comes from animal models of thyroiditis (100) and, indirectly, from the decline in thyroid autoantibodies after antithyroid drug treatment (101), thyroidectomy and radioiodine ablation (102,103) (Figure 8). However, long-term studies on both post-thyroidectomy patients, and those patients receiving radioiodine treatment, show some patients with no decline in autoantibody secretion, implying extrathyroidal sources of continued autoantibody production (104). Here the role of extrathyroidal TSHR expression as an immune stimulant may be of importance.

5.2.   Autoantibodies to the Human TSH Receptor (TSHR-Ab)

Long-acting thyroid stimulator (LATS) was discovered by Adams & Purves (2) over 40 years ago during a search for the thyroid-stimulating activity of Graves' disease using a bioassay for pituitary TSH. Graves' patient serum stimulated radioiodine release from pre-labeled guinea pig thyroid for a much longer time period than a pituitary TSH preparation, and activity was found to reside in the IgG fraction of serum from patients with Graves' disease (2). With the advent of biologically-active radiolabeled TSH, it became possible to probe thyroid membranes for the TSHR: this IgG activity was shown to compete with TSH for receptor occupancy (105) and was indeed a TSHR-Ab acting as a TSH agonist. Hence, in patients with Graves' disease, the thyroid gland is no longer under the control of pituitary TSH but continuously stimulated by the circulating antibody or antibodies with TSH-like activity.

Problems in the measurement of TSHR-stimulating-Ab and TSHR-blocking-Ab.   The development of precise and specific TSHRAb assays using detergent-solubilized porcine thyroidal TSHR's and affinity-purified ^{125}I-labeled TSH allowed the widespread clinical investigation of the

---

Table 3   Evidence for thyrocytes acting as antigen presenting cells

(1) HLA-DR-positive thyroid cells will stimulate an autologous mixed lymphocyte reaction with remarkable proliferation of helper T-cells (370).

(2) Co-culture of Graves' thyroid cells and PBMC leads to IF-γ production and thyroid cell HLA-DR expression (371).

(3) Human thyroid epithelial cells were able to present an influenza-specific peptide to a peptide-specific human T-cell clone which was blocked by HLA-class II antibody (372).

(5) Thyroid epithelial cells were capable of phagocytosis but at a slower rate than macrophages. This function was inhibited by IL-1, methimazole, and dexamethasone but enhanced by IL-2 and IF-β (373).

(6) A cloned line of rat thyroid cells was able to interact directly with cloned thyroid-specific T cells in the absence of APCs (374).

(7) Thyroid epithelial cells express CD40 antigen (90).

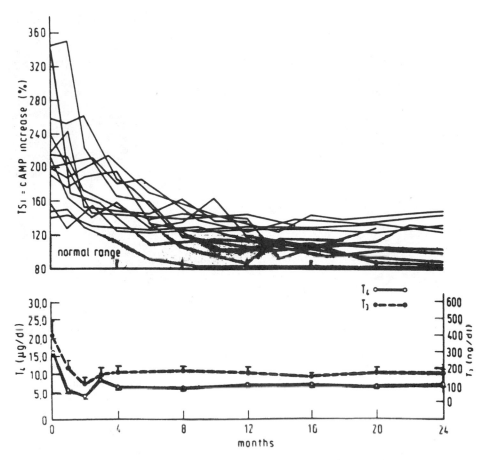

Figure 8. Decline in thyroid autoantibodies with antithyroid drug therapy. Temporal course of TSHR-Ab (top) and thyroid hormone levels (bottom) in 13 individuals with Graves' disease on long-term treatment with methimazole. Antithyroid drug therapy caused a decline in TSHR-Ab and a descent to low levels predicted long-term clinical remission (124). Shaded area indicates normal range.

role of TSHR-Abs in Graves' disease. However, such radioreceptor assays, now possible with recombinant TSH (106) and recombinant TSHR (107), only measure the binding activity of TSHR-Ab and not their degree of biological stimulation of the TSH receptor, two phenomena that have been shown to often diverge. Bioassays for TSHR-Ab bioactivity are now performed with either cultured thyroid cells from established rat cell lines (108), cryopreserved human thyroid cells (109,110) or, most commonly, with hTSHR-transfected Chinese Hamster Ovary (CHO) cells (111) using cyclic AMP generation as the end-point (112). The sensitivities of the radioreceptor and bioassays are similar for TSH (Figure 9). However, there are many reasons for lack of agreement between bioassay and receptor assay results, including the presence of different antibodies with varying degrees of binding and stimulating ability (113). Nevertheless, the commonly used systems do not appear to suffer major interference by species-specific effects (114), and the existence of TSHR-Ab, which act as TSH antagonists rather than agonists, has been clarified (115–117). Such TSHR-Abs bind to the TSH receptor but do not initiate a second signal (Figure 10), and have been termed TSHR-blocking Ab (118), although they may also be neutral in their activity. Further complicating this issue has been the observation of the simultaneous presence of TSHRAb and TSHR-blocking Ab in the same Graves' serum preparations with the effective degree of thyroid stimulation dependent upon the relative concentration and bioactivities of the different autoantibodies (119), a phenomenon reflected in animal models of Graves' disease (120).

Prevalence of TSHR-Ab. TSHR-Abs are detectable only in patients with autoimmune thyroid disease and are, therefore, disease-specific, in contrast to the high prevalence of Tg-Ab and TPO-Ab (up to 20%) in the normal population. Furthermore, TSHR-Ab are uniquely human autoantibodies and are not found constitutively in non-human thyroid disease. Recently, mouse models of Graves' disease have been described following immunization with hTSHR cDNA or native hTSHRs in which stimulating TSHR-Abs can easily be detected (120,121).

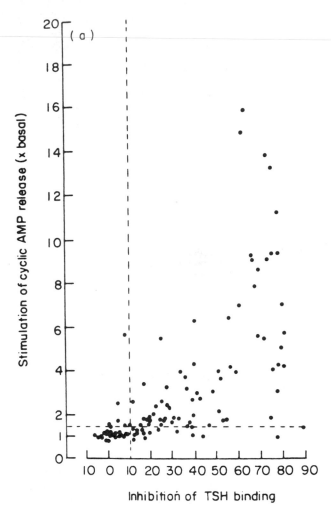

Figure 9. **TSH receptor binding and thyroid stimulation**. Correlation of dialyzed sera from Graves' disease patients in 110 euthyroid and thyrotoxic patients (r = 0.65). Sera may also contain blocking TSHR-Ab which causes poor correlation in some individual samples. From (358).

Eighty to 100% of untreated hyperthyroid patients with Graves' disease have detectable biologically active TSHR-Ab (122,123) that are decreased by treatment of the disease (101;124–126) and, when persistent, are predictive of a response to antithyroid drug treatment (127–130). TSHR-Abs are also detectable in 10–15% of patients with auto-immune thyroiditis (Hashimoto's disease), but are TSHR-blocking Abs on further analysis (131).

Pathophysiological role of TSHR-Ab.
In vivo data.   The original self-infusion of serum from patients with Graves' disease by Adams and colleagues and the resulting thyroid stimulation (132) was the first example of the role of TSHR-Ab in the induction of human hyperthyroidism. Indeed, it was the first example of a pathological human antibody in action. Another early

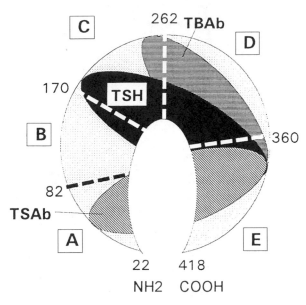

Figure 10. TSHR-stimulating and inhibiting binding domains. The ectodomain of the TSHR, subdivided into 5 arbitrary units, A through E, is depicted in a semicircular form, with the N– and C-termini close to one another. From (34).

demonstration of the *in vivo* effects of TSHR-Ab came from neonatal studies demonstrating the transplacental stimulation of the fetal thyroid in mothers with high titers of TSHR-Ab (133,134). A study of neonates showed that a one week delay in the onset of hyperthyroidism was due to an initial TSHR-blocking Ab effect and a bioactive TSHR-Ab was responsible for persistent hyperthyroidism lasting 6 months (135). The hypothesis was advanced that different functional antibodies had different half-lives in the neonate, perhaps because of different subclasses. However, coexistence in patients with Graves' disease of both thyroid-stimulating and blocking antibodies is indicative of their functional heterogeneity (136), and there is often little relation between the level of TSHR-Ab and the degree of chemical hyperthyroidism (132). The recognition of congenital hyperthyroidism secondary to activating mutations of the TSHR itself also may complicate the clinical picture (137).

In vitro data.   As discussed earlier, a mixture of functional antibodies to the TSHR in serum may determine the overall clinical effect (138), and the presence of both stimulating and blocking TSHR-Ab's has been shown to explain the biphasic dose-response curves seen in TSH bioassays (138). In one investigation, EBV-transformed Graves' disease B cells produced human monoclonal antibodies with stimulating, blocking or mixed activities (139,140). Furthermore, there are indications that TSHR-blocking Ab may become the more prevalent autoantibody after treatment of Graves' disease resulting in hypothyroidism (115,116). These antibodies likely bind to different domains of the TSHR ectodomain and the difference in their functional activity may relate to conformational changes and/or affinity. It has been well shown that stimulating TSHR-Abs are largely dependent on the conformation of the TSHR (34).

TSHR-Ab regulation of TSHR function. TSH and TSHR-Abs affect the function of the thyroid gland causing cyclic AMP-mediated release of thyroid hormone and Tg, stimulation of protein synthesis and iodine uptake. Desensitization of the thyroidal cyclic AMP response by prolonged exposure to high concentrations of either TSH or TSHR-Ab suggested a similar cyclic AMP-mediated mechanism of action (141,142). However, hyperthyroid individuals, by definition, do not have fully desensitized thyroid glands and marked inhibition of endocrine gland function. This differs from what is observed in the gonads following homologous hormonal stimulation (143). Indeed, there is good evidence for positive regulation of the TSH receptor by physiological levels of TSH both *in vivo* and *in vitro* (144,145) and this "relative resistance" to desensitization by lower levels of TSH may allow the hyperthyroid state to persist.

Immunologic Characteristics of TSHR-Ab.

Epitopes used by induced TSHR antibodies. Monoclonal antibodies to the extracellular domain of recombinant hTSHR have been widely employed following immunization with recombinant hTSHR ectodomain, but do not recognize the native conformation of the receptor. The principal B cell epitopes were short linear sequences and the induced antibodies failed to recognize hTSHR expressed in mammalian cells (146,147). To date, no antibodies have been raised to recombinant hTSHR-ectodomain that recognize the native hTSH with high affinity, consistent with the principal epitopes being non-linear, as further demonstrated with human pathological antibodies to the hTSHR (see below).

TSHR-Ab restriction. TSHR-Ab demonstrate light chain restriction in many patients with Graves' disease (148) (Table 4). These data are supportive evidence for oligoclonality and a V gene etiology for the disease on the basis of a restricted clone (148). TSHR-Abs have been shown to exhibit their TSH agonist bioactivity in the IgG1 subclass, again suggesting their oligoclonality (149). The pauciclonal responses of B cells are the consistent with TSHR-Ab being etiologic in disease. However, monoclonality has not been fully established and TSHR-Ab V genes from high affinity TSHR-Abs have not yet been sequenced. One reason for the slow pace of research in this area is the failure to generate immortalized B cells or human hybridomas secreting such high affinity TSHR-Ab

Table 4   Light Chain Restriction of TSHR-Ab

Author	(n)	κ	λ	Both
(375)	11	2	5	4
(376)	11	1	10	0
(377)	3	0	3	0

to analyze their V gene usage (150). Low affinity human monoclonal antibodies to the TSHR have been reported but have not been independently evaluated. IgG with TSHR-stimulating Ab activity can be converted to TSHR-blocking Ab by the addition of anti-IgG or anti-Fab/Fc (151). Also, anti-IgG or anti-Fc but not anti-Fab enhance *in vitro* TSHR-Ab activity (152). Such observations suggest that antibody-induced conformational changes are also important in the structure-function relationship at the TSHR-binding site.

The Immune Etiology of Goiter in Graves' Disease. TSH enhances growth of thyrocytes *in vitro* and causes goiters in humans and animals. The goiter of Graves' disease is secondary to the thyrocyte growth stimulating properties of TSHR-Ab demonstrated in non-human thyroid cells (108,153,154) and in human fetal thyroid cell cultures (155) (Figure 11). Consistent with this observation are the reports of thyroid cancer being more aggressive in Graves' disease patients with high titers of TSHR-Ab (156). Other workers have claimed the presence of TSH receptor-independent thyroid growth-stimulating autoantibodies in a variety of thyroid diseases, but these data have not been confirmed (154,157). TSH induces thyroid cell growth via the cyclic AMP mediated cascade, and/or through activation of phospholipase C (154). However, TSHR-Ab may not activate the phospholipase C pathway in human thyroid (158). Other growth-stimulating and inhibiting factors may also influence thyrocyte growth in Graves' disease by autocrine or paracrine mechanisms. EGF, TGF-α and IGF-1 act in synergy with TSH to promote growth (159). TGF-β is a thyrocyte growth-inhibiting factor (160). Concurrently, thyrocytes *in vitro* produce IGF-1 (161,162) and TGF-β (160). Recently, we have demonstrated the effect of keratinocyte growth factor (KGF) in simulated microgravity (see below) (160a). Hence, keratinocyte growth factor (KGF), which is secreted by fibroblasts and intraepithelial $\gamma\delta$ T cells (163), has a trophic effect on thyroid epithelial cells (164). Intrathyroidal infiltrates of $\gamma\delta$ T cells have been reported (165–167) although, KGF production by intrathyroidal T cells has not yet been clearly demonstrated. Regardless of its site of production, KGF may act in synergy with TSHR Abs and possibly constitute an additional mechanism of thyroid enlargement.

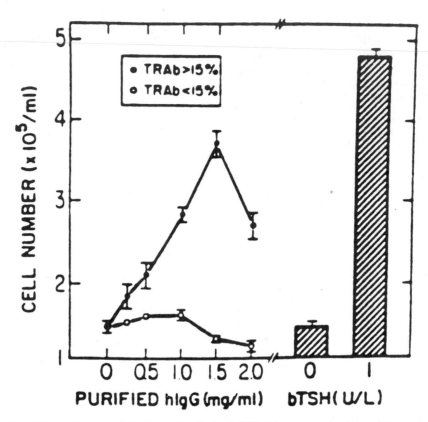

Figure 11.   Growth stimulation of human fetal thyroid cells by TSHR-Ab. Human fetal thyroid monolayers were stimulated with control IgG and Graves' disease IgG (TSHR-Ab >15% binding inhibition). The increase in cell number is shown, with bovine TSH as a control. From (359).

5.3.   Other Human Thyroid-Specific Autoantibodies in Graves' Disease

Graves' patients have a variety of other autoantibodies, including those to thyroid antigens other than the TSHR. These antibodies are considered secondary to disease onset.

Autoantibodies to Thyroglobulin.

Characteristics. Tg-Ab are found in 50–70% of patients with Graves' disease compared to 10–20% of normals (169). Tg-Ab are most commonly IgG and polyclonal; demonstrating only partial restriction to IgG_1 and IgG_4 subclasses (170–172), depending on the monoclonal subclass antibodies used for detection. Such lack of marked restriction in Tg-Ab indicates that these autoantibodies are most likely to be secondary in the immune response of human autoimmune thyroid disease. Although human monoclonal Tg-Ag have been reported (150,173) their V gene characterization is not yet available.

Tg-Ab epitopes. Tg immunization induces anti-thyroglobulin antibodies (Tg-Ab) that recognize conformational epitopes on the large Tg molecule (174,175) Precipitin curves suggested the presence of 4–6 principal epitopes (176) confirmed by reactivity of anti-Tg sera to Tg

fragments prepared by enzymatic digestion of Tg, heat or chemically-denatured Tg, and Tg with differing iodine content (177–179). It has also been shown that these conformational B-cell epitopes depended on the integrity of the disulfide bonds that maintain the structure of Tg (175). The epitopes recognized by such induced anti-sera and monoclonal antibodies have been compared to the epitopes recognized in patients with autoimmune thyroid disease (180,181). Tg autoantibodies were specifically inhibited from binding Tg by some, but not all, murine Tg-mAb and the pattern of inhibition by patient sera was distinct from sera with Tg-Ab, but without disease (182). Using this technique, shared epitopes with TPO have also been described (183). Up to 70% of Tg autoantibodies have been reported to be directed against T4-containing regions of the Tg molecule (184), others have found that the tyrosine residues do not significantly contribute to B cell epitopes (178).

Pathophysiologic role of Tg-Ab. Passive transfer of Tg-Ab in animal models does not cause thyroiditis and human Tg-Ab are not commonly complement fixing (185). However circulating Tg-anti-Tg immune complexes have been demonstrated, as well as thyroid gland basement membrane deposition of Tg immune complexes (186). In rare cases of Graves' disease, there may be an associated glomerulonephritis and nephrotic syndrome (187–189).

Specificity: There is cross-over specificity of Tg with acetylcholinesterase based on sequence analysis and Western blotting, but its significance is uncertain. Anti-human RBC acetylcholinesterase was found in 21% of patients with autoimmune thyroid disease and 4% of normals, but there was no correlation with Tg-Ab titer and antigen inhibition experiments showed limited binding inhibition (190,191).

Autoantibodies to Thyroid Peroxidase (TPO).

Characteristics of TPO-Ab.   TPO autoantibodies are found in up to 20% of normals and 75% of Graves' patients (185). Similar to Tg-Ab, TPO-Ab are also polyclonal with partial restriction to IgG1 and IgG$_4$ subclasses (170,172), indicating their lack of primary involvement in disease etiology (192).

TPO Epitopes.   At least 6 autoantigenic epitopes have been described, including one or both of the enzyme catalytic sites (193,194) Some autoantigenic epitopes have been identified through competition with anti-TPO monoclonal antibodies and recombinant Fab (195,196) and human anti-TPO's of differing subclasses (170). There is evidence that the intrachain loop of amino acids formed by a disulfide bridge is essential to immunogenicity (63), showing that conformational epitopes are important. The poor binding of anti-TPO to reduced and denatured TPO antigen also reflects destruction of such conformational epitopes (197). Studies with tryptic fragments of TPO have suggested that one major epitope is associated with TPO enzyme activity, but such studies were performed with a relatively large fragment (63). Surprisingly, an 85 amino acid fragment of TPO (198) was recognized in a non-glycosylated state by 63% of human sera positive for anti-TPO (199), whereas other non-glycosylated fragments have not been immunoreactive (200). It appears, therefore, that glycosylation may play only a minor role in the B-cell antigenicity of TPO since some sera recognize deglycosylated recombinant human TPO as well as intact TPO (201). Recombinant glycosylated human TPO has been shown to be antigenically similar to natural TPO and can be used as an immune target (202,203).

Pathophysiological Role.   TPO-Ab, unlike Tg-Ab, can mediate complement-dependent cytotoxicity (204,205). There is also a strong correlation between TPO-Ab and the presence of histological thyroiditis (206). Furthermore, TPO-Ab IgG1 subclass has been correlated with higher anti-TPO titers and thyroid failure including in the post-partum period (207). TPO-Ab however, appear not to be highly pathogenic. The fetal human thyroid survives despite transplacental passage of maternal TPO-Ab (208), in contrast with the marked effects of transferred maternal TSHR-Ab in neonates. Though TPO is localized to the thyroid cell follicle membrane, isolated from the circulation, histological studies show that TPO-Ab are able to bind to TPO in vivo (209) and may be capable of *in vitro* inhibition of TPO enzyme activity (193,194,210,211). Because this effect is independent of the titer of anti-TPO, the inhibiting antibody may be directed at specific TPO epitopes (194). Inhibiting antibody is more frequent in Hashimoto's thyroiditis than Graves' disease and TPO- anti-TPO circulating immune complexes cannot be demonstrated in human autoimmune thyroid disease. Crossover specificity with myeloperoxidase and certain epitopes of Tg can also be demonstrated, but are of uncertain significance (212).

Miscellaneous Autoantibodies in Graves' Disease.

TSH and thyroid hormone antibodies.   Anti-TSH has been detected in 0.5% of patients with Graves' disease using bovine TSH as antigen (213) T3-Ab and T4-Ab are also found in Graves' disease (214). All T3-Ab positive patients had Tg-Ab and may represent a subset of antibodies reacting with iodothyronines (215). Such antibodies interfere with hormone assays (especially analog T3 & T4 immunoassays), but not hormone bioactivity (216). A non-TPO cell surface antigen has also been described and proposed as target for antibody-dependent cell-mediated cytotoxicity (ADCC) of thyroid cells and a variety of antigens have been proposed to be involved in extra-ocular muscle cells in thyroid ophthalmopathy (see below).

Autoantibodies to retro-orbital tissues.   Autoantibodies to muscle antigens have long been considered important in the development of Graves' ophthalmopathy. Most data show, however, that such antibodies in Graves' patients are not tissue-specific, but exhibit equal reactivity to both skeletal muscle and eye muscle (41,217,218). However, monoclonal antibody affinity-purified extraocular muscle antigens showed crossreactivity with antibodies to thyroid epithelial cell antigens, suggesting that a common cross-over immune specificity may explain the development of retro-orbital inflammation. Such extraocular muscle antibody titers correlated with the degree of exophthalmos, but not with TSHR-Ab, Tg-Ab, or TPO-Ab (219). The recent cloning of the proposed 64 kDa extraocular muscle antigen as the common enzyme succinate dehydrogenase may shed further light on this important area (220).

Anti-macrophage antibodies.   As demonstrated by complement-dependent cytotoxicity, the presence of anti-macrophage antibodies correlated with TSHR-Ab levels (221) and 28–100% of Graves' sera were positive for anti-ss DNA (222,223), but were usually negative for anti-ds DNA (224). Autoantibodies to tubulin, calmodulin, and dermal antigens have also been reported. IgG from patients with Graves' dermopathy show increased fibroblast DNA, protein, and glycosaminoglycan synthesis (225), but their patho-physiologic relevance and relationship to extrathyroidal TSHR expression have not been elucidated.

6.   T-CELL FUNCTION IN GRAVES' DISEASE

Both helper (CD4) and cytotoxic (CD8) T cells recognize antigens that have been preprocessed by antigen presenting cells (APCs) and expressed in the context of self-MHC molecules. Recognition and quantitation of antigen-specific T cells, therefore, remains more complex than autoantibody determinations. Furthermore, the number and variety of cell surface structures used by T cells for cell-to-cell communication (over 100 to date) has created considerable complexity. Nevertheless, since we know that thyroiditis can be transferred in animal models by thyroid-antigen specific T cells (226–228), but not sustained by thyroid autoantibodies, our understanding of the T cell compartment of the immune system in human autoimmune thyroid disease is critical.

6.1.   T Cell Phenotypes in Graves' Disease

Peripheral blood phenotyping.   The assessment of T-cell surface markers in the peripheral blood of patients with Graves' disease has resulted in widely conflicting data. There are not only differences in the sensitivity and specificity of antibodies and detection techniques used, but also confusion from the differing selection of patients and their clinical status and even the time of day that such testing is performed. A number of general conclusions can, however, be drawn. First, relatives of patients with autoimmune thyroid disease have similar peripheral blood T cell subset phenotypic marker distributions as normal controls (229). Hence, no familial abnormalities have been revealed. Second, in hyperthyroid Graves' disease, there is an increase in activated (for example HLA-DR positive) circulating T cells (118,230–232). The fact that activated T cells can also be induced from Graves' disease PBMC by *in vitro* incubation with thyrocyte antigens (230) suggested that this activation was disease-related. In addition, antigen-specific T cells were raised by using an antibody that stimulates all T cells (CD3). When these T cells were subsequently tested, 10% were reactive against the TSHR (233). Third, the consistent finding of decreased numbers of CD8$^+$ T cells in peripheral blood of hyperthyroid patients appears to be partially secondary to the thyroid hormone excess of hyperthyroidism, since such abnormalities, may correct with antithyroid drug treatment and can be seen in hyperthyroidism not secondary to autoimmune thyroid disease (234). However, some patients with Graves' disease demonstrated not just a decrease in CD8$^+$ T cells, but also a decrease in specific CD4$^+$ subsets (eg. CD4$^+$2H4$^+$) in the peripheral circulation even when rendered euthyroid. This latter observation is reminiscent of studies in multiple sclerosis where a relationship between activity of the disease and a diminution in CD4$^+$ cells was also noted (235). Similar observations have been reported in stratified groups of patients with systemic lupus erythematosus (236). The patients who retain lower circulating numbers of CD8$^+$ T cells were those with the most severe Graves' disease, an observation that originally suggested a "suppressor T-cell defect" in the disease (221,237,238).

Th1 and Th2 phenotypes.   A current concept divides CD4$^+$ T cells into T helper 1 (Th1) and T helper 2 (Th2) based on their cytokine patterns. Several lines of evidence have indicated that Th2 is the predominant T-helper-cell phenotype in Graves' disease:

1. A comparison of T cell phenotypes in Graves' and Hashimoto's thyroiditis.
2. The direct cloning of intrathyroidal T cells and assessment of their cytokine profiles (239),
3. Cytokine characterization of TSHR-responsive T cell clones stimulated with TSHR-transfected EBV lymphoblastoid cells and expanded with anti-CD3 antibody (240)
4. The presence of Th2-typical IgE autoantibodies after removal of other Ig types (241).

While the subdivision into Th1 and Th2 may be an oversimplification (242) that largely applies to mice, its application to thyroid autoimmune disease (Hashimoto's as a Th1 disease and Graves' disease as Th2 dominated) is conceptually helpful. However, this approach has also yielded conflicting evidence since cytokines of both T-helper populations were produced simultaneously when measured at the single-cell level (243). Most studies, however, support the notion that Th2 cells are quantitatively more important than Th1 cells in Graves' disease (see above). Supporting experimental evidence also comes from investigations in the "Shimojo" model of hyperthyroid mice where in 50% of animals became hyperthyroid after administration of the Th2 adjuvant pertussis toxin, whereas a Th1 adjuvant (CFA) led to delayed hyperthyroidism at 12 weeks in 10% of mice after induction (244).

Intrathyroidal T-cell phenotypes.   The intrathyroidal T-cell population shows highly specific changes when compared to the peripheral circulation. As with the PBMC, the data obtained have varied from laboratory to laboratory and appear to be particularly dependent on the method of retrieval from the thyroid tissue (for example, by teasing rather than collagenase digestion). We know now that this can partly be explained by the differences in subset distribution between the interstitium (obtained by teasing) and the follicle-infiltrating T cells (requiring digestion). For example, CD8$^+$ T cells predominantly localize to the follicles (229,245). It is also likely that the influence of disease treatment and disease duration, as well as the technical differences in sample preparation, add to the complexity Nevertheless, a number of clear conclusions have also been possible from analysis of intrathyroidal T-cell populations. First, when Graves' thyroid glands are examined histologically, the intrathyroidal interstitial lymphocytes are more than 80% T cells with CD8+ cells predominating (246). Intraepithelial T cells (in peripolesis) are also predominantly CD8+, with few NK cells and many activated cells (247). Although the CD8+ and regulatory CD4+2H4+ T cell content of the peripheral blood may be depressed in certain Graves' patients, particularly those coming to surgery, there was a relative excess of such cells within the intraepithelial compartment of the Graves' thyroid gland, as evidenced both by flow cytometry and immunocytochemistry (229). This accumulation of CD8+ cells may be secondary to changes in the Th1 and Th2 subsets, which may influence intrathyroidal CD8+ T cell development and function. While MHC class II antigen expression has no direct influence on CD8+ T-cells (248), the presence of the intrathyroidal Th1 cells may enhance local proliferation of CD8+ T cells. Hence, the preservation of antigen-specific intrathyroidal T-cell subsets in Graves' disease, compared to the peripheral circulation, may be an important mechanistic observation. These CD4 subset changes may be reflective of a larger T-cell population in transition. Another attractive explanation for the observed accumulation of CD8+ T cells in Graves' thyroid tissue might be a "homing" mechanism involving undefined receptor molecules on thyroid epithelium. thyroid endothelium or secondary to the presence of thyroid-specific CD8 cells. The mechanisms for such homing are known to involve tissue-specific molecules on lymphocytes (such as LFA-1, and CCR5) (249) as well as molecules expressed by tissues themselves including the expression of ICAM-I by thyroid epithelial cells (250,251).

6.2. Non-antigen-Specific T cell Function in Graves' Disease

Function of peripheral blood T cells. In our experience, non-specific helper and suppressor T-cell function is typically normal in euthyroid patients with Graves' disease (252–254), although some patients, especially those with autoimmune (Hashimoto's) thyroiditis, exhibit a non-specific inability to suppress mitogen-stimulated immuno-globulin secretion *in vitro* (255). Such a phenomenon may be related to the HLA-DR type of the patients examined, since DR3 and DR5 normals have similar defects (256) or to polymorphisms at the T-cell receptor. Much more attention has been directed to thyroid-antigen-specific T-cell function (see below).

Table 5 Influence of IF-γ on thyrocytes

1. Decreased T3 and hTg release (378)
2. Reduced induction of TPO and Tg mRNAs (62;379)
3. Inhibition of amino acid transport (380)
4. Increased basal and stimulated radioiodine uptake (380–382)
5. Decreased insulin- and TSH-stimulated cell growth (87;380;382;383)
6. Decreases susceptibility to cytotoxicity (384)
7. Induces MHC class I and II antigen expression on thyroid cells (385;386)

Function of intrathyoidal T-cells. Autologous mixed lymphocyte reaction studies showed that Graves' disease intrathyroidal CD4+ T cells proliferated, but not CD8+ T cells. Cytokine production is also a useful monitor of T-cell function and responsiveness. It is well known that a non-antigen specific increase in IF-γ production was seen by intrathyroidal T-cell clones in both Hashimoto's disease and Graves' disease (257–259) (*see below*). In addition to its effects on HLA class II expression. IF-γ has a number of direct, principally suppressive, effects on thyrocytes (Table 5). In addition, a variety of other cytokines have been demonstrated to be secreted by intrathyroidal T cells and T cell clones, implicating Graves' disease as a Th2 disease (see above).

6.3. Thyroid Antigen-Specific T-cell Function of Graves' Disease

Thyroid antigenic epitopes.

TSHR. Originally, solubilized TSHR was shown to induce T-cell activation as measured by release of migration inhibition factor (MIF) (260) in patients with Graves' disease. Since then, there have been several studies of TSHR T cell epitopes in human autoimmune thyroid disease (261–270). By combining grading and non-parametric statistics, we identified four important TSHR epitopes (amino acids 247–266,202–221,142–161 and 52–71) (271), that elicited significant PBMC proliferation in Graves' patients more than in normals (p-value for differences in the patient group p < 0.000001, p-value for differences in the control group p = 0.045) (271). One of the epitopes identified was also found in a family by segregation analysis (272). Our data showed that T cell responsiveness to TSHR peptides was more pronounced in patients than in normals, thereby further supporting a critical role for T cells in the autoimmune process in Graves' disease. Anti-TSHR reactivity by intrathyroidal T cells has also been shown using TSHR transfected lymphoblastoid cells as antigen-presenting cells (240,273).

Tg. T cells from animals immunized with Tg do not appear to react preferentially to hormone-containing sites or to Tg with higher iodine content (175). This is because T cell-reactive epitopes are not conformational and are more closely related to the primary structure of Tg. While T-cell reactivity to digested Tg has been reported *in vitro* the epitopes operative in human Tg-antigen-specific T cells have not been fully explored, probably because of the large size of the molecule. In contrast, a number of Tg peptides have been shown able to induce murine thyroiditis (274).

TPO. Although TPO-specific human T cell clones have been reported, data on human TPO T-cell epitopes have mainly been obtained by using PBMC from patients with autoimmune thyroid disease interacting with a series of defined synthetic TPO peptides (275) (Table 7). These studies have defined short linear epitopes within the TPO molecule. There have also been reports of shared T cell epitopes between TPO and Tg (183). Amino acid databank analysis showed greater homology of TPO to Tg than any other human protein; there is a common 8 amino acid sequence with 6 identical and 2 conserved amino acids (276) that may present a linear T cell epitope. In addition, we used an autologous Epstein Barr virus-transformed lymphoblastoid cell line that allowed procurement of TPO-reactive T cell clones from the peripheral blood of a patient with Graves' disease (277). A similar technique was used in another laboratory and showed T-cell reactivity by intrathyroidal T cells to TPO (278). (279)

Peripheral blood T cells. Migration Inhibition Factor (MIF) assays originally provided data on thyroid antigen responsiveness of peripheral blood T cells (168). These early studies, using a crude thyroid extract, suggested in some laboratories that an antigen-specific T-cell defect was present in patients with autoimmune thyroid disease (279–283) while others were unable to show such a defect (284,285). Much of this disparity was possibly secondary to the imprecise nature of the testing procedures used but, applying improved techniques, Volpé et al. continued to generate data in favor of a thyroid-antigen specific defect in T-cell function (286,287).

HLA class II antigen-positive thyroid epithelial cells have also been shown to be effective stimulators of T-cell proliferation in Graves' patients on the basis of a mixed-cell reaction, which may or may not have been thyroid-antigen-dependent (288,289). Such peripheral blood T cells from patients with Graves' disease who had circulating hTPO-Ab and hTg-Ab may also proliferate *in vitro* in response to TPO and/or Tg (290) and to related peptides (see sections on antigen epitopes), but such non-TSHR interactions are unlikely to be etiologic in Graves' disease because of the primary nature of the TSHR in disease pathology. Furthermore, although true frequency data have not been generated, the difficulty in obtaining high stimulation indices in such studies suggests that these autoreactive T cells are of low frequency.

Intrathyroidal T-cells. Intrathyroidal T cell cultures respond more predictably to thyroid antigens and autologous thyroid cells than PBMC. The combination of intrathyroidal T cells with antigen-presenting cells and autologous B cells is known to induce Tg-Ab and/or TPO-Ab secretion (291). Intrathyroidal T cells, in the presence of thyroidal epithelial cells, showed a disparity in T cell subset reactivity as discussed above in antigen non-specific systems (292). When T cell hybridomas were produced from intra-thyroidal lymphoblast cells, none had thyrocyte-specific IL-2 secretion, but many enhanced *in vitro* IgG secretion by B-cells (253).

6.4. Intrathyroidal T Cell Cloning

Cloning T cells from thyroid glands of patients with Graves' disease has provided new insights, although all human studies of T-cell clones have been hampered by the limited longevity of T-cell cultures *in vitro*. In interpreting results, however, it must be understood that the

Table 6   Synthetic peptides or polypeptide fragments observed to stimulate proliferative responses in patients T cells combining a variety of studies, the major T cell epitope regions are shown, as reported in the literature (adapted from reference 387)

Amino acid residue										
0	100	200	300	400	500	600	700	800	900	933
Cloned T cells										
					535–551	632–645				
Primary lymphocyte cultures										
					457–589					
				414–481						
	145–250									
		211–230							882–901	
110–129							842–861			

Table 7 Epitope mapping of murine serum TSHR-Abs. A set of 26 overlapping hTSHR peptides spanning the hTSHR-ecd ectodomain were used as antigens. Positivity was defined as >3 × SDs above control peptide binding. From (307)

	Mouse #	___ Peptide# ___																										EC1	EC2	EC3
		1	2	3	4	5	6	7	8	9	10	11	12	13	14	15	16	17	18	19	20	21	22	23	24	25	26			
Hyper	1																						+	+						
	2	+																												
	3						++				+											+	+	+		+				
	4																					+	+	+						
	5						++			+												+	+	+						
Euthyroid	1						+																+	+	+					
	2						++																+	+						
	3				+		++																							
	4						+			+								+	+						+					
	5																							++						
	6																													
	7																							++						
	8						+																							
Hypo	1																													
	2																													
Controls	1						+																							
	2																													
	3																													
Preimmune	(4)																													

conditions used for cloning may favor the growth of certain subpopulations, giving a biased view of lymphocyte subset frequencies. About 1/10 of activated T cells infiltrating the thyroid gland in patients with autoimmune thyroid disease proliferated in response to autologous thyroid cells or thyroid cell antigens (289,293). Intrathyroidal lymphocytes reacting with self-MHC (AMLR) may comprise up to 50% of T cells, possibly secondary to bias in selection conditions. In Graves' disease, the intrathyroidal lymphocytes from T-cell cloning were 75% CD4$^+$4B4$^+$, of which up to 33% proliferated in response to autologous thyroid epithelial cells, hTg or TPO (294). There have been no antigen-specific cytotoxic T cell clones obtained from the glands of patients with Graves' disease compared to Hashimoto's disease (294,295). Some investigators have, however, found a high prevalence of NK cell clones in Graves' glands when mitogen stimulation was used (296). More than 60% of intrathyroidal T-cell hybridomas were found to secrete helper factors enhancing IgG secretion compared with 10% for peripheral blood T-cell controls (253), providing additional data for a mechanistic basis for TSHR-Ab secretion within the thyroid.

6.5.   T Cell Receptor V Genes

The human T cell receptor (hTcR) V-genes code for the antigen-MHC recognition site on the hTcR, affording antigen specificity. In many human autoimmune diseases and their animal model equivalents, restricted use of hTcR V genes has been described (297). Restricted heterogeneity of hTcR V gene families implicates T cell immunity as possibly etiologic in the disease and identifies potential sites for immunomodulation. In Graves' disease, monoclonal antibodies against the hTcR showed normal TcR $\alpha\beta$ distribution amongst intrathyroidal T cells and an excess of $\gamma\delta$ chain hTcRs intrathyroidally, but not in peripheral blood (298). Others have found many hTcR α/β and γ/δ V and C region rearrangements, suggesting that the T cell response was polyclonal in Graves' and Hashimoto's disease (299). To examine the T cell receptor V α gene usage of intrathyroidal T lymphocytes, we used the polymerase chain reaction (PCR) and employed multiple oligonucleotide amplimers to test initially for V α gene families (#1–18) utilized by intrathyroidal T cells from patients with autoimmune thyroid disease (300). In some patients, our data demonstrated a marked bias in hTcR V α gene utilization by intrathyroidal T cells compared to the TcR peripheral blood from the same individuals (Figure 12). However; the predominant V α genes used differed from patient to patient with no clear disease-related preference. In addition, we also found V β gene family bias, a phenomenon we replicated in humanized *scid* mice (301). Other investigators have reported similar hTcR restrictions in the thyroid gland and extrathyroidal sites, such as the retroorbital

tissues (302,303). Taken together, such information supports the concept of restricted T cell heterogeneity in human autoimmune thyroid disease and points to the primacy of T cells in the etiology of Graves' and Hashimoto's thyroid dysfunction.

7.   ANIMAL MODELS OF GRAVES' DISEASE

A useful model of human Graves' disease should meet several criteria. These include: a) hyperthyroidism, b) goiter formation, c) an intrathyroidal lymphocytic infiltration but an absence of thyroidal destruction, d) retroorbital involvement in the autoimmune process, e) the presence of serum thyroid stimulating antibodies, and f) the ability to transfer the disease. Many attempts have been made to induce Graves' disease in rabbits and mice. Such approaches first used crude thyroid tissue preparations as antigen. We now know that such thyroid tissue preparations contained large quantities of the major thyroid protein product—thyroglobulin (Tg). We also know that immunization with Tg is the most potent inducer of thyroiditis. Hence, all the early studies induced autoimmune thyroiditis rather than Graves' disease. With the cloning of the TSHR, synthetic ectodomain peptides were used to immunize rabbits in attempts to induce Graves' disease (304). Although claims of thyrotoxic rabbits were made by a number of investigators, the data were scanty and have not been confirmed. Subsequent efforts involved immunization of mice with recombinant human or mouse TSHR ectodomain protein made from the cDNA in several expression systems (305). While TSHR-Abs were easily induced, they were all TSHR blocking rather than stimulating.

Recently, both Shimojo et al. (306,168) and Costagliola et al. (121) have reported new approaches to the induction of Graves' disease in mice, the former using fibroblasts expressing the human TSHR and the latter hTSHR cDNA injections. We have confirmed and extended the work of Shimojo et al. using a murine L cell (fibroblast) line spontaneously expressing MHC class II antigen, which was transfected with full length human TSHR cDNA (307). The immunized mice became hyperthyroid and had many, but not all, of the features of Graves' disease, including thyroid-stimulating antibodies. The mouse Graves' models fit many of the criteria with the notable exception of no lymphocytic infiltration in the thyroid. Although a mild lymphocytic infiltration was reported by Shimojo et al. (306), and also by others using recombinant TSHR ectodomain, no infiltrates were seen in our hyperthyroid or hypothyroid mice. Furthermore, administration of an adjuvant favoring a Th1 response where cytotoxic T cells should thrive also failed to induce thyroiditis. Recent observations using recombinant TSHR ectodomain have demonstrated that unique regions of the ectodomain

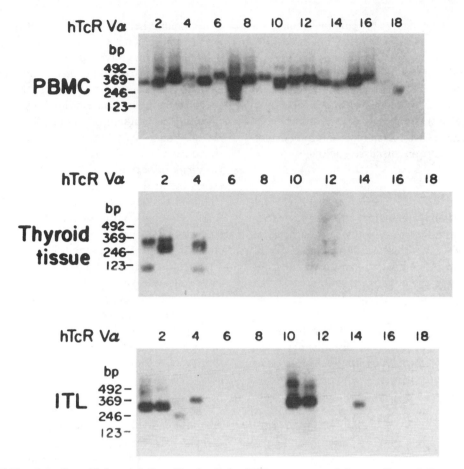

Figure 12. hTcR V α intrathyroidal restriction. Total cellular RNA was prepared from a Graves' disease patient's PBMC (top panels), intrathyroidal lymphocytes (middle panels) and intact thyroid tissue (bottom panels). PCR products from amplimers testing for hTcR V gene families (α 1–18) and (β 1–20) are shown as densitometric contributions to total V gene use. There is marked restriction of intrathyroidal hTcR V α gene, but not V β gene use compared with that seen in peripheral blood.

may be responsible for the development of thyroiditis (274). It is, however, difficult to understand how these epitopes may not have been well represented in the functional, natural receptor used in the recent immunization studies.

Peptide binding data examined the potential epitopes recognized by TSHR antibodies in the Shimojo mouse model (307). Widely recognized was the region incorporating the 50-residue insert in the ectodomain between residues 317–366 (Table 7). However, there was no difference in the epitope mapping using sera from hyperthyroid or hypothyroid mice. This suggested that all the TSHR antibodies recognized the same peptides and that the unique stimulating antibodies were not reflected in the binding to the peptides and required complex conformational epitopes and hence, were negative in Western immunoblots. These results also suggested that the mere presence of stimulating TSHR antibodies was sufficient for hyperthyroidism, but not for full-blown Graves' disease in this model. The studies also fail to explain the role of MHC antigen expression on the immunizing fibroblast. The fibroblast cells employed in this technique were hybrid H-2^k and H-2^d, but the H-2^d antigen sequence was unlikely to have acted as an immune stimulant in disease induction since it does not interact with the T cell receptor (308). The spontaneous expression of the MHC class II antigens may, however, be critical for effective presentation of TSHR antigenic peptides for which there is much evidence. We need to learn a great deal more about this system, as it provides a simple means for developing a true model of Graves' hyperthyroidism in mice.

8. POSSIBLE SEQUENCE OF EVENTS IN GRAVES' DISEASE

Epidemiologic data indicate that Graves' disease occurs in a genetically-susceptible patient after a triggering event. Inherited or environmental factors may come into play perpetuating or ameliorating the disease and its tissue involvement. Although the immunogenetics have been much explored it has been more difficult to quantify the many possible environmental insults that might serve as disease-triggering events. A current concept is that Graves' disease occurs in individuals with T cells and B cells capable of thyroid antigen recognition. Why TSHR-specific T cells and antibody-secreting clones are not deleted, or at least tolerized (anergized) is unknown. Some studies point to several possible trigger factors that require further exploration; some of these factors are discussed below.

8.1. Infection and Graves' Disease

Cross-over specificity/molecular mimicry. Cross-reacting antigens from infective organisms are a potential precipitating cause of Graves' disease. Moreover, they may account for the varied manifestations of the disease, varying with the particular antigens or epitopes to which the aberrant immune response is mounted. The human immune system, depending upon the available repertoire, may see molecular similarities between human thyroid antigens and antigens specific to an organism, perhaps those relatively conserved through evolution. The infective process may also change the context of antigen presentation in cells by the expression of HLA class II antigens on a cell for the first time. The most often quoted example of molecular mimicry in Graves' disease is *Yersinia Enterocolitica* (YE). YE is a gram-negative bacillus that may act as a pathogen, causing chronic enterocolitis in humans. Various autoimmune phenomena have been associated with YE infection, such as skin eruptions and arthralgias (309). There may be an increased incidence of anti-YE antibodies in Graves' disease (310) and Graves' disease patients have positive MIF tests to YE antigen (311). YE antibody binds to a thyroid epithelial cell cytoplasmic antigen (312) and many different YE antigens cross-react with thyroid antigens. Radiolabeled TSH binds to YE with high affinity, suggesting the presence of a TSHR-binding site (313). YE antigenic proteins are encoded by stretches of DNA (42–46 kDa, coding for at least 6 immunogenic proteins essential for virulence) (314). Antibody to YE plasmid has been detected in 80% of recurrent Graves' disease patients, 66% Hashimoto's disease and 35% of normal individuals (314), although others have found no difference from controls (315). However, in a prospective study, all new Graves' disease sera were positive for TSHR-Ab, but negative for YE antibody at the time of disease onset. All patient sera became YE antibody-positive at 6 months, suggesting that the bacterial antibody was not etiologic (316).

Bystander effects. Another important mechanism for the initiation of autoimmune disease by an infective agent is the indirect influence of the resulting inflammatory response on autoantigen-specific lymphocytes. Such an effect has been demonstrated in animal models of autoimmune islet cell destruction (317,318). Furthermore, resident T cells have been recognized in the thyroid glands of patients with Graves' disease (301,319) that could be the target of local cytokine activation initiated by a local immune response to an infectious, or any other toxic, agent. To date there has been no direct experimental evidence of thyroid autoimmunity initiated by such a bystander mechanism. Nevertheless, it remains a very attractive hypothesis for further study.

Virus infection. Although there are animal models of virus-induced thyroiditis, there is no model of virus-induced Graves' disease. However, adenovirus, reovirus and cytomegalovirus infection of rat thyroid cell lines induce MHC Class II antigen expression (320,321) and the resulting antigen presentation may precipitate immune stimulation at the level of the thyroid cell. Viruses may lyse host cells, releasing previously hidden antigens and perhaps virally-altered antigens. In support of this, thyroid autoantibodies may be elevated transiently after subacute thyroiditis. Retrovirus sequences have been postulated in Graves' disease thyroid tissue (322), but these data have not been reproducible (323,324).

8.2. Immune Modulators

Iodine and Graves' disease. Iodine is not a part of the primary Graves' antigen but rather a part of the thyroid hormone molecule on the thyroglobulin backbone. Nevertheless, patients with Graves' disease in remission after anti-thyroid drug therapy, given iodine supplements, are more likely to relapse (325), and relapse after thyroidectomy was also found to be more likely in patients on a high iodine diet. These data suggest that iodine is able to affect immune stimulation in Graves' disease. Autoimmune thyroid disease is rare in iodine-deficient areas (326), but the incidence of AITD and thyroid autoantibodies increases in these areas after the introduction of iodine supplementation (327). Historically, however, this has not always been the case, probably secondary to the degree of iodine deficiency present, the extent of the supplementation and the susceptibility of the population to autoimmune thyroid disease. Further evidence comes from animal studies. The Cornell chicken (CS) has a low incidence of spontaneous thyroiditis, but the incidence of thyroid autoantibodies and lymphocytic thyroid infiltration increases if dietary iodine is supplemented in the first 10 weeks of life (328). The OS chicken has a high incidence of severe spontaneous thyroiditis, but a reduced Tg-Ab titer is seen if the animal is fed a low iodine diet. Similarly, the BB/W rat is autoimmune disease-prone. There is an increased incidence of lymphocyte thyroiditis with dietary iodine supplementation (329) and a decreased incidence of lymphocytic thyroiditis if dietary iodine is reduced (330). The immunogenicity of Tg is increased in OS chickens fed iodine (331) and highly iodinated Tg is better able to stimulate Tg specific T-cell clones (332). This may explain the enhancement of thyroiditis, but does not help explain the enhancement of Graves' disease. However, macrophage phagocytosis of highly iodinated Tg has also been shown to be increased allowing enhanced antigen presentation (333).

Cytokines and Graves' disease. Hyperthyroidism has frequently been reported to occur with cytokine therapy, including interferon-α and interferon-β as well as interleukin 2 (334–336). While the mechanism of activation by interleukin-2 appears straightforward, via T cell activation, the indirect mechanisms for the interferons are less clear.

Stress. Stress has been clinically noted to be a factor associated with the onset of Graves' disease since its first description (337). There are many known effects of stress upon the immune system and a full discussion is outside the scope of this chapter. However, in animal models of stress-induction, there appears to be ample evidence of immune suppression (338). A period of immune hyperreactivity may occur after periods of intense stress, thus precipitating autoimmune thyroid disease in susceptible individuals. Much data have accumulated to show that patients with Graves' disease have an excessive number of stressful life events (339,340).

Pregnancy. The immune suppression of pregnancy has a marked suppressive effect on autoimmune thyroid disease (10). Graves' disease is suppressed during this time along with decreased titers of thyroid autoantibodies. Following such immune suppression, there is a highly predictable form of either post-partum thyroiditis or Graves' disease, which coincides with recovery from suppression and with a short period of immune hyperactivity (341). Indeed, up to 40% of younger women with Graves' disease may give a history of postpartum onset.

Anti-thyroid drugs. Anti-thyroid drug (thionamide) treatment of Graves' disease leads to a fall in thyroid autoantibody titers, including TSHRAb (101,125,342,343), but not a fall in non-thyroid autoantibodies (344). After drug treatment, the presence of TSHR-Ab may be a reliable predictor of relapse, but in its absence a significant number of patients also subsequently relapse (127,130). This immunosuppressive mechanism of action of antithyroid drugs has been widely investigated (345). The normalization of thyroid status and the removal of the effects of hyperthyroidism on the immune system, may be immunomodulatory (346), but methimazole and PTU in physiologic intrathyroidal concentrations (347) inhibit PWM stimulated peripheral lymphocyte IgG, anti-TPO and anti- Tg synthesis. Further evidence of such immuno-suppression comes from the observation that intrathyroidal B-lymphocytes from patients treated with antithyroid drugs had reduced viability and function (as assessed by

antithyroid antibody production) compared with those from patients treated with propranolol alone (348). Methimazole is concentrated in macrophages and may interfere with antigen presentation and hence antibody production via this mechanism (349). Peroxidase inhibition by antithyroid drugs also leads to decreased neutrophil and macrophage free oxygen radical scavenging.

Sex steroids. As in many autoimmune diseases, there is a clear female predisposition to Graves' disease, the basis of which is unknown. Androgens and its analogs prevent, or ameliorate, thyroiditis in animal models, but no data are available for Graves' specific autoimmune phenomena (350). The female predisposition to Graves' disease is not seen in Graves' ophthalmopathy and may be lessened in the elderly (351). Some evidence suggests that the female propensity may be secondary to X chromosome-specific genes or pregnancy-related changes rather than estrogen (10,352).

Hyperthyroidism and the immune system. The hyperthyroidism of Graves' disease may have a profound effect on the immune system. Studies have demonstrated that lymph node size may be increased by paracortical proliferation, medullary plasma cell differentiation, and increased lymph node output of small and large lymphocytes (353). As discussed earlier, Graves' patients may show a relative CD8$^+$ T cell deficiency that normalizes with anti-thyroid drug correction of the hyperthyroidism (346,354,355). Thyroxine has also been shown to directly decrease T cell number and function, and a recent study showed that thyroid hormones had an anti-apoptotic effect in Jurkat cells, a Sézary tumor cell line, as well as in T cells. Furthermore, thyroxine replacement therapy has been suggested to prolong remission in treated Graves' disease even after discontinuation of antithyroid drugs (356), perhaps indicating the ability of thyroxine to suppress immune reactivity. Unfortunately, this observation has not been confirmed.

9. CONCLUSIONS

Graves' disease is the clinical manifestation of a complex series of immunological events. Although the disease is unpredictably "familial" there is no linkage to HLA genes, just a weak association. The presence of an association with HLA in the absence of linkage suggests that other non-HLA genes are more important genetic contributors. Susceptible loci have been located, but the responsible genes remain to be discovered. In addition, the information suggests that a still uncertain precipitating event in a susceptible host may lead to thyroid antigen presentation and initiation of the autoimmune process. We have illustrated how the thyrocyte may function as an antigen-presenting cell under the influence of viruses or cytokines, perpetuating the immune response. The four principal autoantigens of Graves' disease have been sequenced and cloned, leading to major advances in the understanding of their antigenicity to cellular and humoral elements of the immune system. Of paramount significance in the manifestations of clinical disease are TSH receptor autoantibodies, which stimulate growth and function of the thyroid gland, and T cells, which show specificity for thyrocytes. This immune response appears to take place on a background of autoimmune thyroiditis, as evidenced by a variety of thyroid autoantibodies and the presence of a thyroid lymphocytic infiltrate. The emergence of cross-over specificity for TSH receptors expressed in retroobital tissues appears to be the likely explanation for the ophthalmological manifestations of the disease Despite newly described animal models of autoimmune hyperthyroidism, Graves' disease remains a uniquely human autoimmune disease and continues to present a challenge as to its fundamental immunopathological mechanisms.

ACKNOWLEDGEMENT

Supported in part by Public Health Service grants DK45011, and DK35764 and DK52464 from NIDDKD and grant NAG9–816 from NASA.

References

1. Graves, R.J. 1835. Newly observed affection of the thyroid. *London Medical and Surgical Journal* 7:515–523.
2. Adams, D.D., and H.D. Purves. 1956. Abnormal responses in the assay of thyrotropin. *Proceedings of the University of Otago Medical School* 34:11–12.
3. Roitt, I.M., D. Doniach, P.N. Campbell, and R. Vaughan-Hudson. 1956. Autoantibodies in Hashimoto's disease (lymphadenoid goitre). *Lancet* ii:820–822.
4. Gimlette. 1960. Thyroid acropachy. *Lancet* I:22.
5. Feldon, S.E., and J.M. Weiner. 1982. Clinical significance of extraocular muscle volumes in Graves' ophthalmopathy: a quantitative computed tomography study. *Arch. Ophthalmol.* 100:1266–1269.
6. Tallstedt, L., and R. Norberg. 1988. Immunohistochemical staining of normal and Graves' extraocular muscle. *Invest. Ophthalmol. Vis. Sci.* 29:175–184.
7. Weetman, A.P., S. Cohen, K.C. Gatter, P. Fells, and B. Shine. 1989. Immunohistochemical analysis of the retrobulbar tissues in Graves' ophthalmopathy *Clin. Exp. Immunol.* 75:222–227.
8. Fatourechi, V., and A.F. Fransway. 1994. Dermopathy of Graves' disease (pretibial myxedema). *Medicine* 73:1–7.
9. Stadlmayr, W., C. Spitzreg, A.M. Bichlmar, and A.E. Heufelder. 1997. TSH rec eptor transcripts with TSH receptor-like immunoreactivity in orbital and pretibial fibroblasts of patients with Graves' ophthalmopathy and pretibial myxedema. *Thyroid* 7:3–12.
10. Davies, T.F. 1999. The thyroid immunology of the postpartum period. *Thyroid* 9:675–684.
11. Tomer, Y., and T.F. Davies. 1997. The genetic susceptibility to Graves' disease. *Baillière's Clin. Endocrinol. Metab.* 11:431–450.
12. Brix, T.H., K. Christensen, N.V. Holm, B. Harvald, and L. Hegedus. 1998. A population-based study of Graves' disease in Danish twins. *Clinical Endocrinology* 48:397–400.
13. Davies, T.F. 1998. Autoimmune thyroid disease genes come in many styles and colors [editorial; comment]. *J. Clin. Endocrinol. Metab* 83:3391–3393.
14. Vladutiu, A.O., and N.R. Rose. 1975. Cellular basis of the genetic control of immune responsiveness to murine thyroglobulin in mice. *Cell Immunol.* 17:106–113.
15. Ito, M., M. Tanimoto, H. Kamura, M. Yoneda, Y. Morishima, K. Yamauchi, T. Itatsu, K. Takatsuki, and H. Saito. 1989. Association of HLA antigen and restriction fragment length polymorphism of T cell receptor beta-chain gene with Graves' disease and Hashimoto's thyroiditis. *J. Clin. Endocrinol. Metab.* 69:100–104.
16. Yeo, P.P., S.H. Chan, A.C. Thai, W.Y. Ng, K.F. Lui, G.B. Wee, S.H. Tan, B.W. Lee, H.B. Wong, and J.S. Cheah. 1989. HLA Bw46 and DR9 associations in Graves' disease of Chinese patients are age- and sex-related. *Tissue Antigens* 34:179–184.
17. Stenszky, V., L. Kozma, C. Balazs, S. Rochlitz, J.C. Bear, and N.R. Farid. 1985. The genetics of Graves' disease: HLA and disease susceptibility. *J. Clin. Endocrinol. Metab.* 61:735–740.
18. Boehm, B.O., E. Schifferdecker, P. Kuehnl, and K. Schoffling. 1988. Linkage of HLA-DR beta specific restriction fragment length polymorphisms with Graves' disease. *Acta. Endocrinol. Copenh.* 119:251–256.
19. Weetman, A.P., A.K. So, C.A. Warner, L. Foroni, P. Fells, and B. Shine. 1988. Immunogenetics of Graves' ophthalmopathy. *Clin. Endocrinol. Oxf.* 28:619–628.
20. Yanagawa, T., A. Mangklabruks, Y.B. Chang, Y. Okamoto, M.-E. Fisfalen, P.G. Curran, and L.J. DeGroot. 1993. Human histocompatibility leukocyte antigen-DQA1*0501 allele associated with genetic susceptibility to Graves' disease in a caucasian population. *J. Clin. Endocrinol. Metab.* 76:1569–1574.
21. Heward, J.M., A. Allahabadia, J. Daykin, J. Carr-Smith, A. Daly, M. Armitage, P.M. Dodson, M.C. Sheppard, A.H. Barnett, J.A. Franklyn, et al. 1998. Linkage disequilibrium between the human leukocyte antigen class II region of the major histocompatibility complex and Graves' disease: replication using a population case control and family-based study. *J. Clin. Endocrinol. Metab.* 83:3394–3397.
22. Roman, S.H., D. Greenberg, P. Rubinstein, S. Wallenstein, and T.F. Davies. 1992. Genetics of autoimmune thyroid disease: lack of evidence for linkage to HLA within families. *J. Clin. Endocrinol. Metab.* 74:496–503.
23. Barbesino, G., Y. Tomer, E.S. Concepcion, T.F. Davies, and D. Greenberg. 1998. Linkage analysis of candidate genes in autoimmune thyroid disease: 1. Selected immunoregulatory genes. *J. Clin. Endocrinol. Metab.* 83:1580–1584.
24. O'Connor, G., D.S. Neufeld, D.A. Greenberg, L. Concepcion, S.H. Roman, and T.F. Davies. 1993. Lack of disease associated HLA-DQ restriction fragment length polymorphisms in families with autoimmune thyroid disease. *Autoimmunity* 14:237–241.
25. Morel, P.A., J.S. Dorman, J.A. Todd, H.O. McDevitt, and M. Trucco. 1988. Aspartic acid at position 57 of the HLA-DQ beta chain protects against type I diabetes: a family study. *Proc. Natl. Acad. Sci. USA* 85:8111–8115.

26. Demaine, A., K.I. Welsh, B.S. Hawe, and N.R. Farid. 1987. Polymorphism of the T cell receptor beta-chain in Graves' disease. *J. Clin. Endocrinol. Metab.* 65:643–646.

27. Badenhoop, K., G. Schwarz, P.G. Walfish, V. Drummond, K.H. Usadel, and G.F. Bottazzo. 1990. Susceptibility to thyroid autoimmune disease: molecular analysis of HLA-D region genes identifies new markers for goitrous Hashimoto's thyroiditis. *J. Clin. Endocrinol. Metab.* 71:1131–1137.

28. Yanagawa, T., Y. Hidaka, V. Guimaraes, M. Soliman, and L.J. DeGroot. 1995. CTLA-4 gene polymorphism associated with Graves' disease in a Caucasian population. *J. Clin. Endocrinol. Metab.* 80:41–45.

29. Heward, J.M., A. Allahabadia, M. Armitage, A. Hattersley, P.M. Dodson, K. Macleod, J. Carr-Smith, J. Daykin, A. Daly, M.C. Sheppard, R.L. Holder, A.H. Barnett and J.A. Franklyn. 1999. The development of Graves' disease and the CTLA-4 gene on chromosome 2q33. *J. Clin. Endocrinol. Metab.* 84:2398–2401.

30. Waterman, E.A., P.F. Watson, J.H. Lazarus, A.B. Parkes, C. Darke, and A.P. Weetman. 1998. A study of the association between a polymorphism in the CTLA-4 gene and postpartum thyroiditis. *Clin. Endocrinol. (Oxf.)* 49:251–255.

31. Tomer, Y., G. Barbesino, D.A. Greenberg, E.S. Concepcion, and T.F. Davies. 1999. Mapping the major susceptibility loci for familial Graves' and Hashimoto's diseases: evidence for genetic heterogeneity and gene interactions. *J. Clin. Endocrinol. Metab.* 84:4656–4664.

32. Farid, N.R., E. Stone, and G. Johnson. 1980. Graves' disease and HLA: clinical and epidemiologic associations. *Clin. Endocrinol. (Oxf.)* 13.535–544.

33. Kendall-Taylor, P., A. Stephenson, A. Stratton, S.S. Papiha, P. Perros, and D.F. Roberts. 1988. Differentiation of autoimmune ophthalmopathy from Graves' hyperthyroidism by analysis of genetic markers. *Clin. Endocrinol. (Oxf.)* 28:601–610.

34. Rapoport, B., G.D. Chazenbalk, J.C. Jaume, and S.M. McLachlan. 1998. The thyrotropin (TSH) receptor: interaction with TSH and autoantibodies. *Endocr. Rev.* 19:673–716.

35. Nagayama, Y., K.D. Kaufman, P. Seto, and B. Rapoport. 1989. Molecular cloning, sequence and functional expression of the cDNA for the human thyrotropin receptor. *Biochem. Biophys. Res. Commun.* 165:1184–1190.

36. Libert, F., A. Lefort, C. Gerard, M. Parmentier, J. Perret, M. Ludgate, J.E. Dumont, and G. Vassart. 1989. Cloning, sequencing and expression of the human thyrotropin (TSH) receptor: evidence for binding of autoantibodies. *Biochem. Biophys. Res. Commun.* 165:1250–1255.

37. Nagayama, Y., K.D. Kaufman, P. Seto, and B. Rapoport. 1989. Molecular cloning, sequence and functional expression of the cDNA for the human thyrotropin receptor *Biochem. Biophys. Res. Commun.* 165:1184–1190.

38. Sanders, J., Y. Oda, S.A. Roberts, Y. Maruyama, J. Furmaniak, and B. Rees Smith. 1998. Understanding the TSH receptor structure-functional relationship. In: Newer Aspects of Clinical Graves' Disease. T.F. Davies, editor. Balliere, London. pp. 451–480.

39. Couet, J., S. Sar, A. Jolivet, M.-T.V. Hai, E. Milgrom, and M. Misrahi. 1996. Shedding of human TSH receptor ectodomain: involvement of a matrix metalloprotease. *J. Biol. Chem.* 271:4545–4552.

40. Ahmann, A., J. Baker, Jr., A.P. Weetman, L. Wartofsky, T.B. Nutman, and K.D., Burman. 1987. Antibodies to porcine eye muscle in patients with Graves' ophthalmopathy identification of serum immunoglobulins directed against unique determinants by immunoblotting and enzyme-linked immunosorbent assay. *J. Clin. Endocrinol. Metab.* 64:454–460.

41. Kubota, S., K. Gunji, C. Stolarski, J.S. Kennerdell, and J. Wall. 1998. Reevaluation of the prevalences of serum autoantibodies reactive with "64-kd eye muscle proteins" in patients with thyroid-associated ophthalmopathy. *Thyroid* 8:175–179.

42. Smith, T.J., R.S. Bahn, and C.A. Gorman. 1989. Connective tissue, glycosaminoglycans, and diseases of the thyroid. *Endocr. Rev.* 10:366–391.

43. Smith, T.J., R.S. Bahn, and C.A. Gorman. 1989. Hormonal regulation of hyaluronate synthesis in cultured human fibroblasts: evidence for differences between retroocular and dermal fibroblasts. *J. Clin. Endocrinol. Metab.* 69:1019–1023.

44. Bahn, R.S., C.M. Dutton, N. Natt, W. Joba, C. Spitzweg, and A.E. Heufelder. 1998. Thyrotropin receptor expression in Graves' orbital adipose/connective tissues: potential autoantigen in Graves' ophthalmopathy. *J. Clin. Endocrinol. Metab.* 83:998–1002.

45. Murakami, M., Y. Hosoi, T. Negishi, Y. Kamiya, K. Miyashita, M. Yamada, T. Iriuchijima, H. Yokoo, I. Yoshida, Y. Tsushima, and M. Mori. 1996. Thymic hyperplasia in patients with Graves' disease. Identification of thyro-tropin receptors in human thymus. *J. Clin. Invest.* 98:2228–2234.

46. Paschke, R., and V. Geenen. 1995. Messenger RNA expression for a TSH receptor variant in the thymus of a two-year-old child. *J. Mol. Med.* 73:577–580.

47. Civitareale, D., R. Lonigro, A.J. Sinclair, and R. Di-Lauro. 1989. A thyroid-specific nuclear protein essential for tissue-specific expression of the thyroglobulin promoter. *EMBO J.* 8:2537–2542.

48. Kung, A.W., and K.S. Lau. 1990. Interleukin-1 beta modulates thyrotropin-induced thyroglobulin mRNA transcription through 3',5'-cyclic adenosine monophosphate. *Endocrinology* 127:1369–1374.

49. Kung, A.W., and K.S. Lau. 1990. Interferon-gamma inhibits thyrotropin-induced thyroglobulin gene transcription in cultured human thyrocytes. *J. Clin. Endocrinol. Metab.* 70:1512–1517.

50. Turner, C.D., S.B. Chernoff, A. Taurog, and A.B. Rawitch. 1983. Differences in iodinated peptides and thyroid hormone formation after chemical and thyroid peroxidase-catalyzed iodination of human thyroglobulin. *Arch. Biochem. Biophys.* 222:245–258.

51. Dunn, J.T., P.C. Anderson, J.W. Fox, C.A. Fassler, A.D. Dunn, L.A. Hite, and R.C. Moore. 1987. The sites of thyroid hormone formation in rabbit thyroglobulin. *J. Biol. Chem.* 262:16948–16952.

52. Portmann, L., N. Hamada, G. Heinrich, and L.J. DeGroot. 1985. Anti-thyroid peroxidase antibody in patients with autoimmune thyroid disease: possible identity with anti-microsomal antibody. *J. Clin. Endocrinol. Metab.* 61:1001–1003.

53. Mariotti, S., S. Anelli, J. Ruf, R. Bechi, B. Czarnocka, A. Lombardi, P. Carayon, and A. Pinchera. 1987. Comparison of serum thyroid microsomal and thyroid peroxidase autoantibodies in thyroid diseases. *J. Clin. Endocrinol. Metab.* 65:987–993.

54. Libert, F., J. Ruel, M. Ludgate, S. Swillens, N. Alexander, G. Vassart, and C. Dinsart. 1987. Thyroperoxidase, an auto-

antigen with a mosaic structure made of nuclear and mito-chondrial gene modules. *EMBO J.* 6:4193–4196.

55. Seto, P., H. Hirayu, R.P. Magnusson, J. Gestautas, L. Portmann, L.J. DeGroot, and B. Rapoport. 1987. Isolation of a complementary DNA clone for thyroid microsomal antigen. Homology with the gene for thyroid peroxidase. *J. Clin. Invest.* 80:1205–1208.

56. Collison, K.S., J.P. Banga, P.S. Barnett, A.W. Kung, and A.M. McGregor. 1989. Activation of the thyroid peroxidase gene in human thyroid cells: effect of thyrotrophin, forskolin and phorbol ester. *J. Mol. Endocrinol.* 3:1–5.

57. Chiovato, L., P. Vitti, P. Cucchi, C. Mammoli, P. Carajon, and A. Pinchera. 1989. The expression of the microsomal/peroxidase autoantigen in human thyroid cells is thyrotrophin-dependent. *Clin. Exp. Immunol.* 76:47–53.

58. Kimura, S., T. Kotani, O.W. McBride, K. Umeki, K. Hirai, T. Nakayama, and S. Ohtaki. 1987. Human thyroid peroxidase: complete cDNA and protein sequence, chromosome mapping, and identification of two alternately spliced mRNAs. *Proc. Natl. Acad. Sci. USA* 84:5555–5559.

59. McLachlan, S.M., and B. Rapoport. 1992. The molecular biology of thyroid peroxidase cloning expression and role as autoantigen in autoimmune thyroid disease. *Endocrine Reviews* 13:192–206.

60. Zanelli, E., M. Henry, B. Charvet, and Y. Malthiery. 1990. Evidence for an alternate splicing in the thyroperoxidase messenger from patients with Graves' disease. *Biochem. Biophys. Res. Commun.* 170:735–741.

61. Ashizawa, K., S. Yamashita, T. Tobinaga, Y. Nagayama, H. Kimura, H. Hirayu, M. Izumi, and S. Nagataki. 1989. Inhibition of human thyroid peroxidase gene expression by interleukin 1. *Acta Endocrinol. Copenh.* 121:465–469.

62. Ashizawa, K., S. Yamashita, Y. Nagayama, H. Kimura, H. Hirayu, M. Izumi, and S. Nagataki. 1989. Interferon-gamma inhibits thyrotropin-induced thyroidal peroxidase gene expression in cultured human thyrocytes. *J. Clin. Endocrinol. Metab.* 69:475–477.

63. Nakajima, Y., R.D. Howells, C. Pegg, E.D. Jones, and B.R. Smith. 1987. Structure-activity analysis of microsomal antigen/thyroid peroxidase. *Mol. Cell. Endocrinol.* 53:15–23.

64. Dai, G., O. Levy, and N. Carrasco. 1996. Cloning and characterization of the thyroid iodide transporter. *Nature* 379:458–460.

65. Paire, A., V.F. Bernier, R.S. Selmi, and B. Rousset. 1997. Characterization of the rat thyroid iodide transporter using anti-peptide antibodies. Relationship between its expression and activity. *J. Biol. Chem.* 272:18245–18249.

66. Paire, A., V.F. Bernier, R.S. Selmi, and B. Rousset. 1997. Characterization of the rat thyroid iodide transporter using anti-peptide antibodies. Relationship between its expression and activity. *J. Biol. Chem.* 272:18245–18249.

67. Morris, J.C., E.R. Bergert, and W.P. Bryant. 1997. Binding of immunoglobulin G from patients with autoimmune thyroid disease to rat sodium-iodide symporter peptides: evidence for the iodide transporter as an autoantigen. *Thyroid* 7:527–534.

68. Benvenga, S., L. Bartolone, and F. Trimarchi. 1997. Thyroid iodide transporter: local sequence homologies with thyroid autoantigens. *J. Endocrinol. Invest.* 20:508–512.

69. Hanafusa, T., R. Pujol Borrell, L. Chiovato, R.C. Russell, D. Doniach, and G.F. Bottazzo. 1983. Aberrant expression of HLA-DR antigen on thyrocytes in Graves' disease relevance for autoimmunity. *Lancet* 2:1111–1115.

70. De-Bernardo, E., and T.F. Davies. 1986. Antigen presentation in human autoimmune thyroid disease. *Exp. Cell Biol.* 54:155–162.

71. Piccinini, L.A., N.K. Goldsmith, S.H. Roman, and T.F. Davies. 1987. HLA-DP, DQ and DR gene expression in Graves' disease and normal thyroid epithelium. *Tissue Antigens* 30:145–154.

72. Todd, I., B.R. Pujol, L.J. Hammond, G.F. Bottazzo, and M. Feldmann. 1985. Inteferon-gamma induces HLA-DR expression by thyroid epithelium. *Clin. Exp. Immunol.* 61:265–273.

73. Lahat, N., M. Sheinfeld, E. Sobel, A. Kinarty, and Z. Kraiem. 1992. Divergent effects of cytokines on human leukocyte antigen-DR antigen expression of neoplastic and non-neoplastic human thyroid cells. *Cancer* 69:1799–1807.

74. Goldsmith, N.K., S. Dikman, B. Bermas, T.F. Davies, and S.H. Roman. 1988. HLA class II antigen expression and the autoimmune thyroid response in patients with benign and malignant thyroid tumors. *Clin. Immunol. Immunopathol.* 48:161–173.

75. Lahat, N., M.A. Rahat, O. Sadeh, A. Kinarty, and Z. Kraiem. 1998. Regulation of HLA-DR and costimulatory B7 molecules in human thyroid carcinoma cells: differential binding of transcription factors to the HLA-DRalpha promoter. *Thyroid* 8:361–369.

76. Skoskiewicz, M.J., R.B. Colvin, E.E. Schneeberger, and P.S. Russell. 1985. Widespread and selective induction of major histocompatibility complex-determined antigens in vivo by gamma interferon. *J. Exp. Med.* 162:1645–1664.

77. Frohman, M., J.W. Francfort, and C. Cowing. 1991. T-dependent destruction of thyroid isografts exposed to IFN-gamma. *J. Immunol.* 146:2227–2234.

78. Pichert, G., L.M. Jost, L. Zobeli, B. Odermatt, G. Pedia, and R.A. Stahel. 1990. Thyroiditis after treatment with interleukin-2 and interferon alpha-2a. *Br. J. Cancer* 62:100–104.

79. Kauppila, A., K. Cantell, O. Janne, E. Kokko, and R. Vihko. 1982. Serum sex steroid and peptide hormone concentrations, and endometrial estrogen and progestin receptor levels during administration of human leukocyte interferon. *Int. J. Cancer* 29:291–294.

80. Fentiman, I.S., B.S. Thomas, F.R. Balkwill, R.D. Rubens, and J.L. Hayward. 1985. Primary hypothyroidism associated with interferon therapy of breast cancer [letter]. *Lancet* 1:1166.

81. Burman, P., T.H. Totterman, K. Oberg, and F.A. Karlsson. 1986. Thyroid autoimmunity in patients on long term therapy with leukocyte-derived interferon. *J. Clin. Endocrinol. Metab.* 63:1086–1090.

82. Burman, P., T.H. Totterman, K. Oberg, and F.A. Karlsson. 1986. Thyroid autoimmunity in patients on long term therapy with leukocyte-derived interferon. *J. Clin. Endocrinol. Metab.* 63:1086–1090.

83. Fentiman, I.S., F.R. Balkwill, B.S. Thomas, M.J. Russell, I. Todd, and G.F. Bottazzo. 1988. An autoimmune aetiology for hypothyroidism following interferon therapy for breast cancer. *Eur. J. Cancer Clin. Oncol.* 24:1299–1303.

84. Piccinini, L.A., W.A. Mackenzie, M. Platzer, and T.F. Davies. 1987. Lymphokine regulation of HLA-DR gene expression in human thyroid cell monolayers. *J. Clin. Endocrinol. Metab.* 64:543–548.

85. Rayner, D.C., P.M. Lydyard, H.J. de-Assis-Paiva, S. Bidey, P. van-der-Meide, A.M. Varey, and A. Cooke. 1987. Interferon-mediated enhancement of thyroid major histo-

compatibility complex antigen expression. A flow cytometric analysis. *Scand. J. Immunol.* 25:621–628.

86. Todd, I., B.R. Pujol, L.J. Hammond, J.M. McNally, M. Feldmann, and G.F. Bottazzo. 1987. Enhancement of thyrocyte HLA class II expression by thyroid stimulating hormone. *Clin. Exp. Immunol.* 69:524–531.

87. Buscema, M., I. Todd, U. Deuss, L. Hammond, R. Mirakian, B.R. Pujol, and G.F. Bottazzo. 1989. Influence of tumor necrosis factor-alpha on the modulation by interferon-gamma of HLA class II molecules in human thyroid cells and its effect on interferon-gamma binding. *J. Clin. Endocrinol. Metab.* 69:433–439.

88. Tomer, Y., and T.F. Davies. 1993. Infection, Thyroid Disease and Autoimmunity *Endocrine Reviews* 14:107–120.

89. Piccinini, L.A., N.K. Goldsmith, B.S. Schachter, and T.F. Davies. 1988. Localization of HLA-DR alpha-chain messenger ribonucleic acid in normal and autoimmune human thyroid using in situ hybridization. *J. Clin. Endocrinol. Metab.* 66:1307–1315.

90. Metcalfe, R.A., R.S. McIntosh, F. Marelli-Berg, G. Lombardi, R. Lechler, and A.P. Weetman. 1998. Detection of CD40 on human thyroid follicular cells: analysis of expression and function. *J. Clin. Endocrinol. Metab.* 83:1268–1274.

91. Kabel, P.J., H.A. Voorbij, M. De-Haan, R.D. van-der-Gaag, and H.A. Drexhage. 1988. Intrathyroidal dendritic cells. *J. Clin. Endocrinol. Metab.* 66:199–207.

92. Feldmann, M. 1989. Molecular mechanisms involved in human autoimmune diseases: relevance of chronic antigen presentation. Class II expression and cytokine production. *Immunol. Suppl.* 2:66–71.

93. Ueki, Y., K. Eguchi, T. Otsubo, Y. Kawabe, C. Shimomura, H. Tezuka, H. Nakao, A. Kawakami, K. Migita, and T. Ishikawa. 1989. Abnormal B lymphocyte function in thyroid glands from patients with Graves' disease. *J. Clin. Endocrinol. Metab.* 69:939–945.

94. Zheng, R.Q., E. Abney, C.Q. Chu, M. Field, L.B. Grubeck, R.N. Maini, and M. Feldmann. 1991. Detection of interleukin-6 and interleukin-1 production in human thyroid epithelial cells by non-radioactive in situ hybridization and immunohistochemical methods. *Clin. Exp. Immunol.* 83:314–319.

95. Cohen, S.B., and A.P. Weetman. 1988. Activated interstitial and intraepithelial thyroid lymphocytes in autoimmune thyroid disease. *Acta Endocrinol. (Copenh)* 119:161–166.

96. Watson, P.F., A.P. Pickerill, R. Davies, and A.P. Weetman. 1994. Analysis of cytokine gene expression in Graves' disease and multinodular goiter. *J. Clin. Endocinol. Metab.* 79:355–360.

97. Dinarello, C.A., and J.W. Mier. 1987. Lymphokines. *N. Engl. J. Med.* 317:940–945.

98. Sidman, C.L., J.D. Marshall, L.D. Shultz, P.W. Gray, and H.M. Johnson. 1984. Gamma-interferon is one of several direct B cell-maturing lymphokines. *Nature* 309:801–804.

99. Weetman, A.P., A.M. McGregor, J.H. Lazarus, and R. Hall. 1982. Thyroid antibodies are produced by thyroid-derived lymphocytes. *Clin. Exp. Immunol.* 48:196–200.

100. Weetman, A.P., A.M. McGregor, D.P. Rennie, and R. Hall. 1982. Sites of autoantibody production in rats with thyroiditis. *Immunology* 46:465–472.

101. McGregor, A.M., M.M. Petersen, S.M. McLachlan, P. Rooke, B.R. Smith, and R. Hall. 1980. Carbimazole and the autoimmune response in Graves' disease. *N. Engl. J. Med.* 303:302–307.

102. Wilson, R., J.H. McKillop, D.W. Pearson, G.F. Cuthbert, and J.A. Thomson. 1985. Relapse of Graves' disease after medical therapy: predictive value of thyroidal technetium-99m uptake and serum thyroid stimulating hormone receptor antibody levels. *J. Nucl. Med.* 26:1024–1028.

103. McGregor, A.M., S.M. McLachlan, B.R. Smith, and R. Hall. 1979. Effect of irradiation on thyroid-autoantibody production. *Lancet* 2:442–444.

104. De-Bruin, T.W., N.A. Patwardhan, R.S. Brown, and L.E. Braverman. 1988. Graves' disease: changes in TSH receptor and anti-microsomal antibodies after thyroidectomy *Clin. Exp. Immunol.* 72:481–485.

105. Smith, B.R., and R. Hall. 1974. Binding of thyroid stimulators to thyroid membranes. *FEBS Lett.* 42:301–304.

106. Huber, G.K., P. Fong, E.S. Concepcion, and T.F. Davies. 1991. Recombinant human thyroid-stimulating hormone: initial bioactivity assessment using human fetal thyroid cells. *J. Clin. Endocrinol. Metab.* 72:1328–1331.

107. Ludgate, M., J. Perret, M. Parmentier, C. Gerard, F. Libert, J.E. Dumont, and G. Vassart. 1990. Use of the recombinant human thyrotropin receptor (TSH-R) expressed in mammalian cell lines to assay TSH-R autoantibodies. *Mol. Cell Endocrinol.* 73:R13–R18.

108. Zakarija, M., S. Jin, and J.M. McKenzie. 1988. Evidence supporting the identity in Graves' disease of thyroid-stimulating antibody and thyroid growth-promoting immunoglobulin G as assayed in FRTL5 cells. *J. Clin. Invest.* 81:879–884.

109. Hinds, W.E., N. Takai, B. Rapoport, S. Filetti, and O.H. Clark. 1981. Thyroid-stimulating immunoglobulin bioassay using cultured human thyroid cells. *J. Clin. Endocrinol. Metab.* 52:1204–1210.

110. Davies, T.F., M. Platzer, A. Schwartz, and E. Friedman. 1983. Functionality of thyroid-stimulating antibodies assessed by cryopreserved human thyroid cell bioassay. *J. Clin. Endocrinol. Metab.* 57:1021–1027.

111. Costagliola, S., S. Swillens, P. Niccoli, J.E. Dumont, G. Vassart, and M. Ludgate. 1992. Binding assay for thyrotropin receptor autoantibodies using the recombinant receptor protein. *J. Clin. Endocrinol. Metal.* 75:1540–1544.

112. Vitti, P.V., W.A. Valente, F.S. Ambesi-Impiombato, G. Fenzi, A. Pinchera, and L.D. Kohn. 1982. Graves' IgG stimulation of continuously cultured rat thyroid cells: a sensitive and potentially useful clinical assay. *J. Endocrinol. Invest.* 5:179–185.

113. Zakarija, M., and J.M. McKenzie. 1978. Adsorption of thyroid-stimulating antibody (TSAb) of Graves' disease by homologous and heterologous thyroid tissue. *J. Clin. Endocrinol. Metab.* 47:906–908.

114. Vitti, P., W.A. Valente, I.F. Ambesi, G.F. Fenzi, A. Pinchera, and L.D. Kohn. 1982. Graves' IgG stimulation of continuously cultured rat thyroid cells: a sensitive and potentially useful clinical assay. *J. Endocrinol. Invest.* 5:179–182.

115. Hashim, F.A., F.M. Creagh, A.E. Hawrani, A.B. Parkes, P.R. Buckland, and S.B. Rees. 1986. Characterization of TSH antagonist activity in the serum of patients with thyroid disease. *Clin. Endocrinol. Oxf.* 25:275–281.

116. Bech, K., H. Bliddal, N.K. Siersbaek, and T. Friis. 1982. Production of non-stimulatory immunoglobulins that inhibit TSH binding in Graves' disease after radioiodine administration. *Clin. Endocrinol. Oxf.* 17:395–402.

117. Orgiazzi, J., D.E. Williams, I.J. Chopra, and D.H. Solomon. 1976. Human thyroid adenyl cyclase-stimulating activity in immunoglobulin G of patients with Graves' disease *J. Clin. Endocrinol. Metab.* 42:341–354.

118. Kennedy, R., U. Di-Mario, P. Pozzilli, K. Guy, M. Leonardi, M. Sensi, and D. Andreani. 1985. Activated T cells in Graves' disease before treatment. *Clin. Exp. Immunol.* 59:377–382.

119. Zakarija, M., J.M. McKenzie, and W.H. Hoffman. 1986. Prediction and therapy of intrauterine and late-onset neonatal hyperthyroidism. *J. Clin. Endocrinol. Metab.* 62:368–371.

120. Shimojo, N., Y. Kohno, K.-I. Yamaguchi, S.-I. Kikuoka, A. Hoshioka, H. Niimi, A. Hirai, Y. Tamura, Y. Saito, L.D. Kohn, and K. Tahara. 1996. Induction of Graves'-like disease in mice by immunization with fibroblasts transfected with the thyrotropin receptor and a class II molecule. *Proc. Natl. Acad. Sci. USA* 93:11074–11079.

121. Costagliola, S., P. Rodien, M.C. Many, M. Ludgate, and G. Vassart. 1998. Genetic immunization against the TSH receptor causes thyroiditis and allows production of monoclonal antibodies recognizing the native receptor. *J. Immunol.* 160:1458–1465.

122. Reader, S.C., B. Davison, C. Beardwell, J.G. Ratcliffe, and W.R. Robertson. 1986. Protein-A purified human immunoglobulins: a comparison of thyroid stimulating and thyrotrophin receptor binding activities in thyrotoxicosis. *Clin. Endocrinol. Oxf.* 25:441–451.

123. Southgate, K., F. Creagh, M. Teece, C. Kingswood, and S.B. Rees. 1984. A receptor assay for the measurement of TSH receptor antibodies in unextracted serum. *Clin. Endocrinol. Oxf.* 20:539–548.

124. Wenzel, K.W., and J.R. Lente. 1984. Similar effects of thionamide drugs and perchlorate on thyroid-stimulating immunoglobulins in Graves' disease: evidence against an immunosuppressive action of thionamide drugs. *J. Clin. Endocrinol. Metab.* 58:62–69.

125. Pinchera, A., P. Liberti, E. Martino, G.F. Fenzi, L. Grasso, L. Rovis, L. Baschieri, and G. Doria. 1969. Effects of antithyroid therapy on the long-acting thyroid stimulator and the antithyroglobulin antibodies. *J. Clin. Endocrinol. Metab.* 29:231–238.

126. Bech, K., and M.S. Nistrup. 1980. Influence of treatment with radioiodine and propylthiouracil on thyroid stimulating immunoglobulins in Graves' disease. *Clin. Endocrinol. Oxf.* 13:417–424.

127. Davies, T.F., S.M. McLachlan, P.M. Povey, B.R. Smith, and R. Hall. 1977. The influence of propranolol on the thyrotropin receptor. *Endocrinology* 100:974–979.

128. Rapoport, B., F.S. Greenspan, S. Filetti, and M. Pepitone. 1984. Clinical experience with a human thyroid cell bioassay for thyroid-stimulating immunoglobulin. *J. Clin. Endocrinol. Metab.* 58:332–338.

129. Wilson, R., J.H. McKillop, N. Henderson, D.W. Pearson, and J.A. Thomson. 1986. The ability of the serum thyrotrophin receptor antibody (TRAb) index and HLA status to predict long-term remission of thyrotoxicosis following medical therapy for Graves' disease. *Clin. Endocrinol. Oxf.* 25:151–156.

130. Davies, T.F. 1998. Thyroid-stimulating antibodies predict hyperthyroidism. *J. Clin. Endocrinol. Metab.* 83:3777–3781.

131. Endo, K., K. Kasagi, J. Konishi, K. Ikekubo, T. Okuno, Y. Takeda, T. Mori, and K. Torizuka. 1978. Detection and properties of TSH-binding inhibitor immunoglobulins in patients with Graves' disease and Hashimoto's thyroiditis. *J. Clin. Endocrinol. Metab.* 46:734–739.

132. Adams, D.D., F.N. Fastier, J.B. Howie, T.H. Kennedy, J.A. Kilpatrick, and R.D. Stewart. 1974. Stimulation of the human thyroid by infusions of plasma containing LATS protector. *J. Clin. Endocrinol. Metab.* 39:826–832.

133. Sunshine, P., H. Kusumoto, and J.P. Kriss. 1965. Survival time of circulating long-acting thyroid stimulator in neonatal thyrotoxicosis: implications for diagnosis and therapy of the disorder. *Pediatrics* 36:869–876.

134. Zakarija, M., and J.M. McKenzie. 1983. Pregnancy-associated changes in thyroid-stimulating antibody of Graves' disease and the relationship to neonatal hyperthyroidism. *J. Clin. Endocrinol. Metab.* 57:1036–1040.

135. Zakarija, M., A. Garcia, and J.M. McKenzie. 1985. Studies on multiple thyroid cell membrane-directed antibodies in Graves' disease. *J. Clin. Invest.* 76:1885–1891.

136. Beckett, G.J., H.A. Kellett, S.M. Gow, A.J. Hussey, J.D. Hayes, and A.D. Toft. 1985. Raised plasma glutathione S-transferase values in hyperthyroidism and in hypothyroid patients receiving thyroxine replacement: evidence for hepatic damage. *Br. Med. J. Clin. Res. Ed.* 291:427–431.

137. Duprez, L., J. Parma, J. Van Sande, A. Allgeier, J. Leclere, C. Schvartz, M.-J. Delisle, M. Decoulx, J. Orgiazzi, and J. Dumont. 1994. Germline mutations in the thyrotropin receptor gene cause non-autoimmune autosomal dominant hyperthyroidism. *Nature Genetics* 7:396–401.

138. Rees Smith, B., S.M. McLachlan, and J. Furmaniak. 1988. Autoantibodies to the thyrotropin receptor. *Endocr. Rev.* 9:106–121.

139. Yokoyama, N., M. Izumi, S. Katamine, and S. Nagataki. 1987. Heterogeneity of Graves' immunoglobulin G: comparison of thyrotropin receptor antibodies in serum and in culture supernatants of lymphocytes transformed by Epstein-Barr virus infection *J. Clin. Endocrinol. Metab.* 64:215–218.

140. Yoshida, T., Y. Ichikawa, K. Ito, and M. Homma. 1988. Monoclonal antibodies to the thyrotropin receptor bind to a 56-kDa subunit of the thyrotropin receptor and show heterogeneous bioactivities. *J. Biol. Chem.* 263:16341–16347.

141. Vitti, P., L. Chiovato, P. Ceccarelli, A. Lombardi, M. Novaes, G.F. Fenzi, and A. Pinchera. 1986. Thyroid-stimulating antibody mimics thyrotropin in its ability to desensitize the adenosine $3',5'$-monophosphate response to acute stimulation in continuously cultured rat thyroid cells (FRT-L5). *J. Clin. Endocrinol. Metab.* 63:454–458.

142. Damante, G., D. Foti, R. Catalfamo, and S. Filetti. 1987. Desensitization of the thyroid cyclic AMP response to thyroid stimulating immunoglobulin: comparison with TSH. *Metabolism* 36:768–773.

143. Davies, T.F., G.D. Hodgen, M.L. Dufau, and K.J. Catt. 1979. Regulation of primate testicular luteinizing hormone receptors and steroidogenesis. *J. Clin. Invest.* 64:1070–1073.

144. Davies, T.F. 1985. Positive regulation of the guinea pig thyrotropin receptor. *Endocrinology* 117:201–207.

145. Huber, G.K., E.S. Concepcion, P.N. Graves, and T.F. Davies. 1991. Positive regulation of human thyrotropin receptor mRNA by thyrotropin. *J. Clin. Endocrinol. Metab.* 72:1394–1396.

146. Davies, T.F., Y. Bobovnikova, M. Weiss, H. Vlase, T. Moran, and P.N. Graves. 1998. Development and characterization of monoclonal antibodies specific for the murine TSH receptor. *Thyroid* 8:693–701.

147. Nicholson, L.B., H. Vlase, P.N. Graves, M. Nilsson, J. Molne, G.C. Huang, N.G. Morgenthaler, T.F. Davies, A.M. McGregor, and J.P. Banga. 1996. Monoclonal antibodies to the human TSH receptor: epitope mapping and binding to the native receptor on the basolateral plasma membrane of thyroid follicular cells. *J. Mol. Endocrinol* 16:159–170.

148. Zakarija, M. 1980. The thyroid-stimulating antibody of Graves' disease: evidence for restricted heterogeneity. *Horm. Res.* 13:1–15.

149. Weetman, A.P., M.E. Yateman, P.A. Ealey, C.M. Black, C.B. Reimer, R.C.J. Williams, B. Shine, and N.J. Marshall. 1990. Thyroid-stimulating antibody activity between different immunoglobulin G subclasses. *J. Clin. Invest.* 86:723–727.

150. De-Bernardo, E., and T.F. Davies. 1987. A study of human-human hybridomas from patients with autoimmune thyroid disease. *J. Clin. Immunol.* 7:71–77.

151. Amino, N., Y. Watanabe, H. Tamaki, Y. Iwatani, and K. Miyai. 1987. In-vitro conversion of blocking type anti-TSH receptor antibody to the stimulating type by anti-human IgG antibodies. *Clin. Endocrinol. Oxf.* 27:615–624.

152. Saito, T., H. Shimura, T. Endo, and T. Onaya. 1989. Enhancement of the activity of thyroid-stimulating antibodies by anti-human IgG antibodies in vitro. *Clin. Endocrinol. Oxf.* 31:325–334.

153. Marcocci, C., P. Vitti, G. Lopez, F. Santini, C. Mammoli, G.F. Fenzi, and A. Pinchera. 1989. Simultaneous assay of thyroid adenylate cyclase- and growth-stimulating antibodies using FRTL-5 cells. Evidence suggesting their identity in patients with Graves' disease. *Clin. Endocrinol. Oxf.* 30:109–119.

154. Dumont, J.E., P.P. Roger, and M. Ludgate. 1987. Assays for thyroid growth immunoglobulins and their clinical implications: methods, concepts, and misconceptions. *Endocr. Rev.* 8:448–452.

155. Huber, G.K., R. Safirstein, D. Neufeld, and T.F. Davies. 1991. Thyrotropin receptor autoantibodies induce human thyroid cell growth and c-fos activation. *J. Clin. Endocrinol. Metab.* 72:1142–1147.

156. Mazzaferri, E.L. 1990. Thyroid cancer and Graves' disease. *J. Clin Endocrinol. Metab.* 70:826–829.

157. Bliddal, H., L. Hegedus, J.M. Hansen, R. van-der-Gaag, and H.A. Drexhage. 1987. The relationships between serum T3 index, thyroid volume, and thyroid stimulating. TSH receptor binding and thyroid growth stimulating antibodies in untreated Graves' disease. *Clin. Endocrinol. Oxf.* 27:75–84.

158. Laurent, E., J. Van-Sande, M. Ludgate, B. Corvilain, P. Rocmans, J.E. Dumont, and J. Mockel. 1991. Unlike thyrotropin, thyroid-stimulating antibodies do not activate phospholipase C in human thyroid slices. *J. Clin. Invest.* 87:1634–1642.

159. Williams, D.W., E.D. Williams, and T.D. Wynford. 1988. Loss of dependence on IGF-1 for proliferation of human thyroid adenoma cells. *Br. J. Cancer* 57:535–539.

160. Grubeck, L.B., G. Buchan, R. Sadeghi, M. Kissonerghis, M. Londei, M. Turner, K. Pirich, R. Roka, B. Niederle, and H. Kassal. 1989. Transforming growth factor beta regulates thyroid growth. Role in the pathogenesis of nontoxic goiter. *J. Clin. Invest.* 83:764–770.

160a.Martin, A., A. Zhou, R.E. Gordon, S.C. Henderson, A.E. Schwartz, E.W. Friedman, and T.F. Davies. 2000. Thyroid organoid formation in simulated microgravity: Influence of keratinocyte growth factor. *Thyroid* 10:481–487.

161. Tode, B., M. Serio, C.M. Rotella, G. Galli, F. Franceschelli, A. Tanini, and R. Toccafondi. 1989. Insulin-like growth factor-I: autocrine secretion by human thyroid follicular cells in primary culture. *J. Clin. Endocrinol. Metab.* 69:639–647.

162. Williams, D.W., E.D. Williams, and T.D. Wynford. 1989. Evidence for autocrine production of IGF-1 in human thyroid adenomas. *Mol. Cell Endocrinol.* 61:139–143.

163. Boismenu, R., and W.L. Havran. 1994. Modulation of epithelial cell growth by intraepithelial gamma delta T cells. *Science* 266:1253–1255.

164. Finch, P.W., J.S. Rubin, T. Miki, D. Ron, and S.A. Aaronson. 1989. Human KGF is FGF-related with properties of a paracrine effector of epithelial cell growth. *Science* 245:752–755.

165. Paolieri, F., C. Pronzato, M. Battifora, N. Fiorino, G.W. Canonica, and M. Bagnasco. 1995. Infiltrating gamma/delta T-cell receptor-positive lymphocytes in Hashimoto's thyroiditis, Graves' disease and papillary thyroid cancer. *J. Endocrinol. Invest.* 18:295–298.

166. Catalfamo, M., C. Roura Mir, M. Sospedra, P. Aparicio, S. Costagliola, M. Ludgate, R. Pujol Borrell, and D. Jaraquemada. 1996. Self-reactive cytotoxic gamma delta T lymphocytes in Graves' disease specifically recognize thyroid epithelial cells. *J. Immunol.* 156:804–811.

167. Kotani, T., Y. Aratake, K. Hirai, I. Hirai, and S. Ohtaki. 1996. High expression of heat shock protein 60 in follicular cells of Hashimoto's thyroiditis. *Autoimmunity* 25:1–8.

168. Lamki, L., V.V. Row, and R. Volpe. 1973. Cell-mediated immunity in Graves' disease and in Hashimoto's thyroiditis as shown by the demonstration of migration inhibition factor (MIF). *J. Clin. Endocrinol. Metab.* 36:358–364.

169. Kohno, T., Y. Tsunetoshi, and E. Ishikawa. 1988. Existence of anti-thyroglobulin IgG in healthy subjects. *Biochem. Biophys. Res. Commun.* 155:224–229.

170. Davies, T.F., C.M. Weber, P. Wallack, and M. Platzer. 1986. Restricted heterogeneity and T cell dependence of human thyroid autoantibody immunoglobulin G subclasses *J. Clin. Endocrinol. Metab.* 62:945–949.

171. Costin, A., H. Klandorf, J. Persselin, and A.J. Van-Herle. 1987. Characterization of high titer antithyroglobulin antibodies. *J. Endocrinol. Invest.* 10:541–546.

172. McLachlan, S.M., R.U. Feldt, E.T. Young, S.L. Middleton, T.M. Dlichert, N.K. Siersboek, J. Date, D. Carr, F. Clark, and S.B. Rees. 1987. IgG subclass distribution of thyroid autoantibodies: a "fingerprint" of an individual's response to thyroglobulin and thyroid microsomal antigen. *Clin. Endocrinol. Oxf.* 26:335–346.

173. Casali, P., G. Inghirami, M. Nakamura, T.F. Davies, and A.L. Notkins. 1986. Human monoclonals from antigen-specific selection of B lymphocytes and transformation by EBV. *Science* 234:476–479.

174. Dong, Q., M. Ludgate, and G. Vassart. 1989. Towards an antigenic map of human thyroglobulin: identification of ten epitope-bearing sequences within the primary structure of thyroglobulin. *J. Endocrinol.* 122:169–176.

175. Shimojo, N., K. Saito, Y. Kohno, N. Sasaki, O. Tarutani, and H. Nakajima. 1988. Antigenic determinants on thyroglobulin: comparison of the reactivities of different thyroglobulin preparations with serum antibodies and T cells of patients with chronic thyroiditis. *J. Clin. Endocrinol. Metab.* 66:689–695.

176. Roitt, I.M., G. Torrigiani, and D. Doniach. 1968. Immunochemical studies on the thyroglobulin auto-antibody system in human thyroiditis. *Immunology* 15:681–696.

177. Salabe, G.B. 1975. Immunochemistry of the interaction of thyroglobulin (Tg) and its auto- and hetero antibodies. *Acta Endocrinol. Suppl. Copenh.* 196:1–25.

178. Nye, L., C.L. Pontes-de, and I.M. Roitt. 1980. Restrictions in the response to autologous thyroglobulin in the human. *Clin. Exp. Immunol.* 41:252–263.

179. Male, D.K. B.R. Champion, G. Pryce, H. Matthews, and P. Shepherd. 1985. Antigenic determinants of human

thyroglobulin differentiated using antigen fragments *Immunology* 54:419–427.

180. Ruf, J., P. Carayon, P.N. Sarles, F. Kourilsky, and S. Lissitzky. 1983. Specificity of monoclonal antibodies against human thyroglobulin; comparison with autoimmune antibodies. *EMBO J.* 2:1821–1826.

181. Piechaczyk, M., M. Bouanani, S.L. Salhi, L. Baldet, M. Bastide, B. Pau, and J.M. Bastide. 1987. Antigenic domains on the human thyroglobulin molecule recognized by autoantibodies in patients' sera and by natural autoantibodies isolated from the sera of healthy subjects. *Clin. Immunol. Immunopathol.* 45:114–121.

182. Rose, N.R., H.S. Bresler, C.L. Burek, S.L. Gleason, and R.C. Kuppers. 1990. Mapping the autoepitopes of thyroglobulin. *Isr. J. Med. Sci.* 26:666–672.

183. Kohno, Y., N. Naito, Y. Hiyama, N. Shimojo, N. Suzuki, O. Tarutani, H. Niimi, H. Nakajima, and T. Hosoya. 1988. Thyroglobulin and thyroid peroxidase share common epitopes recognized by autoantibodies in patients with chronic autoimmune thyroiditis. *J. Clin. Endocrinol. Metab.* 67:899–907.

184. Pearce, C.J., P.G. Byfield, C.J. Edmonds, M.R. Lalloz, and R.L. Himsworth. 1981. Autoantibodies to thyroglobulin cross reacting with iodothyronines. *Clin. Endocrinol. Oxf.* 15:1–10.

185. Mariotti, S., P. Caturegli, P. Piccolo, G. Barbesino, and A. Pinchera. 1990. Antithyroid peroxidase autoantibodies in thyroid diseases. *J. Clin. Endocrinol. Metab.* 71:661–669.

186. Gruters, A., B. Kohler, A. Wolf, A. Soling, L. de Vijlder, H. Krude, and H. Biebermann. 1996. Screening for mutations of the human thyroid peroxidase gene in patients with congenital hypothyroidism. *Exp. Clin. Endocrinol. Diabetes.* 104:121–123.

187. O'Regan, S., J.S. Fong, B.S. Kaplan, J.P. Chadarevian, N. Lapointe, and K.N. Drummond. 1976. Thyroid antigen-antibody nephritis. *Clin. Immunol. Immunopathol.* 6:341–346.

188. Horvath, F., P. Teague, E.F. Gaffney, D.R. Mars, and T.J. Fuller. 1979. Thyroid antigen associated immune complex glomerulonephritis in Graves' disease *Am. J. Med.* 67:901–904.

189. Jordan, S.C., B. Buckingham, R. Sakai, and D. Olson. 1981. Studies of immune-complex glomerulonephritis mediated by human thyroglobulin. *N. Engl. J. Med.* 304:1212–1215.

190. Ludgate, M., C. Owada, R. Pope, P. Taylor, and G. Vassart. 1986. Cross-reactivity between antibodies to hTG and Torpedo acetylcholinesterase in patients with Graves' ophthalmopathy. *J. Endocrinol. Invest.* 9:Abstract.

191. Weetman, A.P., C.K. Tse, W.R. Randall, K.W. Tsim, and E.A. Barnard. 1988. Acetylcholinesterase antibodies and thyroid autoimmunity. *Clin. Exp. Immunol.* 71:96–99.

192. McLachlan, S.M., and B. Rapoport. 1992. The molecular biology of thyroid peroxidase cloning, expression and role as autoantigen in autoimmune thyroid disease. *Endocr. Rev.* 13:192–206.

193. Kohno, Y., Y. Hiyama, N. Shimojo, H. Niimi, H. Nakajima, and T. Hosoya. 1986. Autoantibodies to thyroid peroxidase in patients with chronic thyroiditis: effect of antibody binding on enzyme activities. *Clin. Exp. Immunol.* 65:534–541.

194. Doble, N.D., J.P. Banga, R. Pope, E. Lalor, P. Kilduff, and A.M. McGregor. 1988. Autoantibodies to the thyroid microsomal/thyroid peroxidase antigen are polyclonal and directed to several distinct antigenic sites. *Immunology* 64:23–29.

195. Ruf, J., M.E. Toubert, B. Czarnocka, G.J. Durand, M. Ferrand, and P. Carayon. 1989. Relationship between immunological structure and biochemical properties of human thyroid peroxidase. *Endocrinology* 125:1211–1218.

196. Chazenbalk, G.D., G. Constante, S. Portolano, S. McLachlan, and B. Rapoport. 1993. The immunodominant region on human thyroid peroxidase recognized by autoantibodies does not contain the monoclonal antibody 47/c21 linear epitope. *J. Clin. Endocrinol. Metab.* 77:1715–1718.

197. Hamada, N., N. Jaeduck, L. Portmann, K. Ito, and L.J. DeGroot. 1987. Antibodies against denatured and reduced thyroid microsomal antigen in autoimmune thyroid disease. *J. Clin. Endocrinol. Metab.* 64:230–238.

198. De Bernardo, E., and T.F. Davies. 1983. Antigen-specific B-cell function in human autoimmune thyroiditis. *J. Clin. Immunol.* 3:392–398.

199. Hamada, N., N. Jaeduck, L. Portmann, K. Ito, and L.J. DeGroot. 1987. Antibodies against denatured and reduced thyroid microsomal antigen in autoimmune thyroid disease. *J. Clin. Endocrinol. Metab.* 64:230–238.

200. Finke, R., P. Seto, and B. Rapoport. 1990. Evidence for the highly conformational nature of the epitope(s) on human thyroid peroxidase that are recognized by sera from patients with Hashimoto's thyroiditis. *J. Clin. Endocrinol. Metab.* 71:53–59.

201. Foti, D., and B. Rapoport. 1990. Carbohydrate moieties in recombinant human thyroid peroxidase: role in recognition by antithyroid peroxidase antibodies in Hashimoto's thyroiditis. *Endocrinology* 126:2983–2988.

202. Kaufman, K.D., S. Filetti, P. Seto, and B. Rapoport. 1990. Recombinant human thyroid peroxidase generated in eukaryotic cells: a source of specific antigen for the immunological assay of antimicrosomal antibodies in the sera of patients with autoimmune thyroid disease. *J. Clin. Endocrinol. Metab.* 70:724–728.

203. Kendler, D.L., A. Martin, R.P. Magnusson, and T.F. Davies. 1990. Detection of autoantibodies to recombinant human thyroid peroxidase by sensitive enzyme immunoassay. *Clin. Endocrinol. Oxf.* 33:751–760.

204. Fenzi, G.F., L. Bartalena, L. Chiovato, C. Marcocci, C.M. Rotella, R. Zonefrati, R. Toccafondi, and A. Pinchera. 1982. Studies on thyroid cell surface antigens using cultured human thyroid cells. *Clin. Exp. Immunol.* 47:336–344.

205. Khoury, E.L., L. Hammond, G.F. Bottazzo, and D. Doniach. 1981. Presence of the organ-specific "microsomal" autoantigen on the surface of human thyroid cells in culture its involvement in complement-mediated cytotoxicity. *Clin. Exp. Immunol.* 45:316–328.

206. Yoshida, H., N. Amino, K. Yagawa, K. Uemura, M. Satoh, K. Miyai, and Y. Kumahara. 1978. Association of serum antithyroid antibodies with lymphocytic infiltration of the thyroid gland: studies of seventy autopsied cases. *J. Clin. Endocrinol. Metab.* 46:859–862.

207. Jansson, R., P.M. Thompson, F. Clark, and S.M. McLachlan. 1986. Association between thyroid microsomal antibodies of subclass IgG-1 and hypothyroidism in autoimmune postpartum thyroiditis. *Clin. Exp. Immunol.* 63:80–86.

208. Dussault, J.H., J. Letarte, H. Guyda, and C. Laberge. 1980. Lack of influence of thyroid antibodies on thyroid function in the newborn infant and on a mass screening program for congenital hypothyroidism. *J. Pediatr.* 96:385–389.

209. Khoury, E.L., G.F. Bottazzo, and I.M. Roitt. 1984. The thyroid "microsomal" antibody revisited. Its paradoxical binding in vivo to the apical surface of the follicular epithelium. *J. Exp. Med.* 159:577–591.

210. Okamoto, Y., N. Hamada, H. Saito, M. Ohno, J. Noh, K. Ito, and H. Morii. 1989. Thyroid peroxidase activity-inhibiting

immunoglobulins in patients with autoimmune thyroid disease. *J. Clin. Endocrinol. Metab.* 68:730–734.

211. Kohno, Y., N. Naito, K. Saito, A. Hoshioka, H. Niimi, H. Nakajima, and T. Hosoya. 1989. Anti-thyroid peroxidase antibody activity in sera of patients with systemic lupus erythematosus. *Clin. Exp. Immunol.* 75:217–221.

212. Banga, J.P., R.W. Tomlinson, N. Doble, E. Odell, and A.M. McGregor. 1989. Thyroid microsomal/thyroid peroxidase autoantibodies show discrete patterns of cross-reactivity to myeloperoxidase, lactoperoxidase and horseradish peroxidase. *Immunology* 67:197–204.

213. Noh, J., N. Hamada, H. Saito, H. Oyanagi, N. Ishikawa, N. Momotani, K. Ito, and H. Morii. 1989. Evidence against the importance in the disease process of antibodies to bovine thyroid-stimulating hormone found in some patients with Graves' disease *J. Clin. Endocrinol. Metab.* 68:107–113.

214. Sakata, S., S. Nakamura, and K. Miura. 1985. Autoantibodies against thyroid hormones or iodothyronine. Implications in diagnosis, thyroid function, treatment, and pathogenesis. *Ann. Intern. Med.* 103:579–589.

215. Wang, P.W., M.J. Huang, R.T. Liu, and C.D. Chen. 1990. Triiodothyronine autoantibodies in Graves' disease: their changes after antithyroid therapy and relationship with the thyroglobulin antibodies. *Acta Endocrinol. Copenh.* 122:22–28.

216. Iitaka, M.I., N. Fukasawa, Y. Hara, M. Yanagisawa, K. Hase, S. Miura, Y. Sakatsume, and J. Ishii. 1990. The mechanism for the discrepancy between serum total and free thyroxine values induced by autoantibodies: report on two patients with Graves' disease. *Acta Endocrinol. Copenh.* 123:123–128.

217. Pope, R.M., and A.M. McGregor. 1989 Clinical studies of the reactivity of serum IgG from patients with Graves' ophthalmopathy with porcine orbital tissue preparations. *Biomed. Pharmacother.* 43:573–579.

218. Wall, J.R., N. Bernard, A. Boucher, M. Salvi, Z.G. Zhang, J. Kennerdell, A. Tyutyunikov, and C. Genovese. 1993. Pathogenesis of thyroid-associated ophthalmopathy an autoimmune disorder of the eye muscle associated with Graves' hyperthyroidism and Hashimoto's thyroiditis. *Clin. Immunol. Immunopathol.* 68:1–8.

219. Chang, T.C., K.M. Huang, T.J. Chang, and S.L. Lin. 1990. Correlation of orbital computed tomography and antibodies in patients with hyperthyroid Graves' disease *Clin. Endocrinol. Oxf.* 32:551–558.

220. Kubota, S., K. Gunji, B.A. Ackrell, B. Cochran, C. Stolarski, S. Wengrowicz, J.S. Kennerdell, Y. Hiromatsu, and J. Wall. 1998. The 64-kilodalton eye muscle protein is the flavoprotein subunit of mitochondrial succinate dehydrogenase: the corresponding serum antibodies are good markers of an immune-mediated damage to the eye muscle in patients with Graves' hyperthyroidism. *J. Clin. Endocrinol. Metab.* 83:443–447.

221. Zeki, K., T. Fujihira, F. Shirakawa, K. Watanabe, K. Yahata, S. Eto, H. Sakamoto, and H. Suzuki. 1987. Circulating monocyte (macrophage)-specific antibodies in patients with autoimmune thyroid diseases. *Nippon. Naibunpi. Gakkai. Zasshi.* 63:773–781.

222. McDermott, M.T., S.G. West, J.M. Emlen, and G.S. Kidd. 1990. Antideoxyribonucleic acid antibodies in Graves' disease. *J. Clin. Endocrinol. Metab.* 71:509–511.

223. McDermott, M.T., S.G. West, J.W. Emlen, and G.S. Kidd. 1990. Antideoxyribonucleic acid antibodies in Graves' disease. *J. Clin. Endocrinol. Metab.* 71:509–511.

224. Katakura, M., T. Yamada, T. Aizawa, K. Hiramatsu, Y. Yukimura, M. Ishihara, N. Takasu, K. Maruyama, M. Kameko, and M. Kanai. 1987. Presence of anti-deoxyribonucleic acid antibody in patients with hyperthyroidism of Graves' disease. *J. Clin. Endocrinol. Metab.* 64:405–408.

225. Tao, T.W., S.L. Leu, and J.P. Kriss. 1989. Biological activity of autoantibodies associated with Graves' dermopathy. *J. Clin. Endocrinol. Metab.* 69:90–99.

226. Maron, R., R. Zerubavel, A. Friedman, and I.R. Cohen. 1983. T lymphocyte line specific for thyroglobulin produces or vaccinates against autoimmune thyroiditis in mice. *J. Immunol.* 131:2316–2322.

227. Romball, C.G., and W.D. Weigle. 1987. Transfer of experimental autoimmune thyroiditis with T cell clones. *J. Immunol.* 138:1092–1098.

228. Kotani, T., K. Umeki, K. Hirai, and S. Ohtaki. 1990. Experimental murine thyroiditis induced by procine thyroid peroxidase and its transfer by the antigen-specific T cell line. *Clin. Exp. Immunol.* 80:11–18.

229. Martin, A., N.K. Goldsmith, E.W. Friedman, A.E. Schwartz, T.F. Davies, and S.H. Roman. 1990. Intrathyroidal accumulation of T cell phenotypes in autoimmune thyroid disease. *Autoimmunity* 6:269–281.

230. Zeki, K., T. Fujihira, F. Shirakawa, K. Watanabe, and S. Eto. 1987. Existence and immunological significance of circulating Ia+ T cells in autoimmune thyroid diseases. *Acta Endocrinol. Copenh.* 115:282–288.

231. Ahmann, A.J., and K.D. Burman. 1987. The role of T lymphocytes in autoimmune thyroid disease. *Endocrinol. Metab. Clin. North Am.* 16:287–326.

232. Jackson, R.A., B.F. Haynes, W.M. Burch, K. Shimizu, M.A. Bowring, and G.S. Eisenbarth. 1984. Ia+ T cells in new onset Graves' disease. *J. Clin. Endocrinol. Metab.* 59:187–190.

233. Mullins, R.J., S.B. Cohen, L.M. Webb, Y. Chernajovsky, C.M. Dayan, M. Londei, and M. Feldmann. 1995. Identification of thyroid stimulating hormone receptor-specific T cells in Graves' disease thyroid using autoantigen-transfected Epstein-Barr virus transformed B cell lines. *J. Clin. Invest.* 96:30–37.

234. Bonnyns, M., J. Bentin, G. Devetter, and J. Duchateau. 1983. Heterogeneity of immunoregulatory T cells in human thyroid autoimmunity: influence of thyroid status. *Clin. Exp. Immunol.* 52:629–634.

235. Morimoto, C., D.A. Hafler, H.L. Weiner, N.L. Letvin, M. Hagan, J. Daley, and S.F. Schlossman. 1987. Selective loss of the suppressor-inducer T-cell subset in progressive multiple sclerosis. Analysis with anti-2H4 monoclonal antibody. *N. Engl. J. Med.* 316:67–72.

236. Morimoto, C., A.D. Steinberg, N.L. Letvin, M. Hagan, T. Takeuchi, J. Daley, H. Levine, and S.F. Schlossman. 1987. A defect of immunoregulatory T cell subsets in systemic lupus erythematosus patients demonstrated with anti-2H4 antibody. *J. Clin. Invest.* 79:762–768.

237. Aoki, N., K.M. Pinnamaneni, and L.J. DeGroot. 1979. Studies on suppressor cell function in thyroid diseases. *J. Clin. Endocrinol. Metab.* 48:803–810.

238. Lamki, L., V.V. Row, and R. Volpe. 1973. Cell-mediated immunity in Graves' disease and in Hashimoto's thyroiditis as shown by the demonstration of migration inhibition factor (MIF). *J. Clin. Endocrinol. Metab.* 36:358–364.

239. Fisfalen, M.E., E.M. Palmer, G.A. van Seventer, K. Soltani, Y. Sawai, E. Kaplan, Y. Hidaka, C. Ober, and L.J. DeGroot. 1997. Thyrotropin-receptor and thyroid peroxidase-specific

T cell clones and their cytokine profile in autoimmune thyroid disease. *J. Clin. Endocrinol. Metab.* 82:3655–3663.

240. Mullins, R.J., S.B. Cohen, L.M. Webb, Y. Chernajovsky, C.M. Dayan, M. Londei, and M. Feldmann. 1995. Identification of thyroid stimulating hormone receptor-specific T cells in Graves' disease thyroid using autoantigen-transfected Epstein-Barr virus-transformed B cell lines. *J. Clin. Invest* 96:30–37.

241. Guo, J., B. Rapoport, and S.M. McLachlan. 1997. Thyroid peroxidase autoantibodies of IgE class in thyroid autoimmunity. *Clin. Immunol. Immunopathol.* 82:157–162.

242. Romagnani, S. 1991. Human TH1 and TH2 subsets: doubt no more. *Immunol. Today* 12:256–257.

243. Roura, M.C., M. Catalfamo, M. Sospedra, L. Alcalde, B.R. Pujol, and D. Jaraquemada. 1997. Single-cell analysis of intrathyroidal lymphocytes shows differential cytokine expression in Hashimoto's and Graves' disease. *Eur. J. Immunol.* 27:3290–3302.

244. Kita, M., L. Ahmad, R.C. Marians, H. Vlase, P. Unger, P.N. Graves, and T.F. Davies. 1999. Regulation and transfer of a murine model of thyrotropin receptor antibody mediated Graves' disease. *Endocrinology* 140:1392–1398.

245. McLachlan, S.M., C.A. Pegg, M.C. Atherton, S.L. Middleton, A. Dickinson, F. Clark, S.J. Proctor, G. Proud, and S.B. Rees. 1986. Subpopulations of thyroid autoantibody secreting lymphocytes in Graves' and Hashimoto thyroid glands. *Clin. Exp. Immunol.* 65:319–328.

246. Margolick, J.B., S.M. Hsu, D.J. Volkman, K.D. Burman, and A.S. Fauci. 1984. Immunohistochemical characterization of intrathyroid lymphocytes in Graves' disease Interstitial and intraepithelial populations. *Am. J. Med.* 76:815–821.

247. Cohen, S.B., and A.P. Weetman. 1988. Activated interstitial and intraepithelial thyroid lymphocytes in autoimmune thyroid disease. *Acta Endocrinol. Copenh.* 119:161–166.

248. Hirose, W., N. Lahat, M. Platzer, S. Schmitt, and T.F. Davies. 1988. Activation of MHC-restricted rat T cells by cloned syngeneic thyrocytes. *J. Immunol.* 141:1098–1102.

249. Campbell, J.J., G. Haraldsen, J. Pan, J. Rottman, S. Qin, P. Ponath, D.P. Andrew, R. Warnke, N. Ruffing, N. Kassam, et al. 1999. The chemokine receptor CCR4 in vascular recognition by cutaneous but not intestinal memory T cells. *Nature* 400:776–780.

250. Martin, A., G.K. Huber, and T.F. Davies. 1990. Induction of human thyroid cell ICAM-1 (CD54) antigen expression and ICAM-1-mediated lymphocyte binding. *Endocrinology* 127:651–657.

251. Weetman, A.P., S. Cohen, M.W. Makgoba, and L.K. Borysiewicz. 1989. Expression of an intercellular adhesion molecule, ICAM-1, by human thyroid cells. *J. Endocrinol.* 122:185–19.

252. Davies, T.F., B. Bermas, M. Platzer, and S.H. Roman. 1985. T-cell sensitization to autologous thyroid cells and normal non-specific suppressor T-cell function inn Graves' disease. *Clin. Endocrinol. Oxf.* 22:155–167.

253. Martin, A., A.E. Schwartz, E.W. Friedman, and T.F. Davies. 1989. Successful production of intrathyroidal human T cell hybridomas: evidence for intact helper T cell function in Graves' disease. *J. Clin. Endocrinol. Metab.* 69:1104–1108.

254. Davies, T.F., and M. Platzer. 1986. The T cell suppressor defect in autoimmune thyroiditis: evidence for a high set "autoimmunostat". *Clin. Exp. Immunol.* 63:73–79.

255. Ueki, Y., K. Eguchi, T. Fukuda, T. Otsubo, Y. Kawabe, C. Shimomura, M. Matsunaga, H. Tezuka, N. Ishikawa, and K.

Ito. 1987. Dysfunction of suppressor T cells in thyroid glands from patients with Graves' disease. *J. Clin. Endocrinol. Metab.* 65:922–928.

256. Okabe, N., R. Mori, S. Miake, and K. Inoue. 1984. Effects of antithyroid drugs on Con A-induced suppressor cell activity. *J. Clin. Lab. Immunol.* 13:167–169.

257. Del Prete, G.F., A., Tiri, S. Mariotti, A. Pinchera, M. Ricci, and S. Romagnani. 1987. Enhanced production of gamma-interferon by thyroid-derived T cell clones from patients with Hashimoto's thyroiditis. *Clin. Exp. Immunol.* 69:323–331.

258. Del Prete, G.F., A. Tiri, M. De Carli, S. Mariotti, A. Pinchera, I. Chretien, S. Romagnani, and M. Ricci. 1989. High potential to tumor necrosis factor alpha (TNF-alpha) production of thyroid infiltrating T lymphocytes in Hashimoto's thyroiditis a peculiar feature of destructive thyroid autoimmunity. *Autoimmunity* 4:267–276.

259. Margolick, J.B., A.P. Weetman, and K.D. Burman. 1988. Immunohistochemical analysis of intrathyroidal lymphocytes in Graves' disease: evidence of activated T cells and production of interferon-gamma. *Clin. Immunol. Immunopathol.* 47:208–218.

260. Mäkinen, T., G. Wägar, L. Apter, E. von Willebrand, and F. Pekonen. 1978. Evidence that the TSH receptor acts as a mitogenic antigen in Graves' disease. *Nature* 275:314–315.

261. Tandon, N., M.A., Freeman, and A.P. Weetman. 1992. T cell responses to synthetic TSH receptor peptides in Graves' disease. *Clin. Exp. Immunol.* 89:468–473.

262. Fan, J.L., R.K. Desai, G.S. Seetharamaiah, J.S. Dallas, N.M. Wagle, and B.S. Prabhakar. 1993. Heterogeneity in cellular and antibody responses against thyrotropin receptor in patients with Graves' disease detected using synthetic peptides. *J. Autoimmun.* 6:799–808.

263. Sakata, S., S. Tanaka, K. Okuda, K. Miura, T. Manshouri, and M.Z. Atassi. 1993. Autoimmune T-cell recognition sites of human thyrotropin receptor in Graves' disease. *Mol. Cell Endocrinol.* 92:77–82.

264. Okamoto, Y., T. Yanagawa, M.E. Fisfalen, and L.J. DeGroot. 1994. Proliferative responses of peripheral blood mononuclear cells from patients with Graves' disease to synthetic peptides epitopes of human thyrotropin receptor. *Thyroid* 4:37–42.

265. Soliman, M., E. Kaplan, T. Yanagawa, Y. Hidaka, M.E. Fisfalen, and L.J. DeGroot. 1995. T-cells recognize multiple epitopes in the human thyrotropin receptor extracellular domain. *J. Clin. Endocrinol. Metab.* 80:905–914.

266. Nagy, E.V., J.C. Morris, H.B. Burch, S. Bhatia, K. Salata, and K.D. Burman. 1995. Thyrotropin receptor T cells epitopes in autoimmune thyroid disease. *Clin. Immunol. Immunopathol.* 75:117–124.

267. Hishinuma, A., K. Kasai, K. Ichimura, T. Emoto, and S. Shimoda. 1992. Effects of epidermal growth factor, phorbol ester, and retinoic acid on hormone synthesis and morphology in porcine thyroid follicles cultured in collagen gel. *Thyroid* 2:351–359.

268. Soliman, M., E. Kaplan, A. Abdel Latif, N. Scherberg, and L.J. DeGroot. 1995. Does thyroidectomy, radioactive iodine therapy, or antithyroid drug treatment alter reactivity of patients' T cells to epitopes of thyrotropin receptor in autoimmune thyroid diseases? *J. Clin. Endocrinol. Metab.* 80:2312–2321.

269. Kellermann, S.-A., D.J. McCormick, S.L. Freeman, J.C. Morris and B.M. Conti-Fine. 1996. TSH receptor sequences recognized by CD4+ T cells in Graves' disease patients and healthy controls. *J. Autoimm.* 8:695–698.

270. Soliman, M., E. Kaplan, V. Guimaraes, T. Yanagawa, and L.J. DeGroot. 1996. T-cell recognition of residue 158–176 in thyrotropin receptor confers risk for development of thyroid autoimmunity in siblings in a family with Graves' disease. *Thyroid* 6:545–551.

271. Martin, A., M. Nakashima, A. Zhou, D. Aronson, A.J. Werner, and T.F. Davies. 1997. Detection of major T-cell epitopes on the human TSH receptor by overriding immune heterogeneity in patients with Graves' disease. *J. Clin. Endocrinol. Metab.* 82(10):3361–3366.

272. Soliman, M., E. Kaplan, V. Guimaraes, T. Yanagawa, and L.J. DeGroot. 1996. T-cell recognition of residue 158–176 in thyrotropin receptor confers risk for development of thyroid autoimmunity in siblings in a family with Graves' disease. *Thyroid* 6:545–551.

273. Martin, A., R.P. Magnusson, D.L. Kendler, E. Concepcion, A. Ben-Nun, and T.F. Davies. 1993. Endogenous antigen presentation by autoantigen-transfected EBV-lymphoblastoid cells: I. Generation of human thyroid peroxidase-reactive T cells and their T-cell receptor repertoire. *J. Clin. Invest.* 91:1567–1574.

274. Wang, S.H., G.C. Caryanniotis, Y. Zhang, M. Gupta, A.M. McGregor, and J.P. Banga. 1998. Induction of thyroiditis in mice after immunization with TSHR lacking serologically dominant regions. *Clin. Exp. Immunol.* 113:119–125.

275. Kawakami, Y., M.E. Fisfalen, and L.J. DeGroot. 1992. Proliferative responses of peripheral blood mononuclear cells from patients with autoimmune thyroid diseases to synthetic peptide epitopes of human thyroid peroxidase. *Autoimmunity* 13:17–26.

276. McLachlan, S.M., and B. Rapoport. 1989. Evidence for a potential common T-cell epitope between human thyroid peroxidase and human thyroglobulin with implications for the pathogenesis of autoimmune thyroid disease. *Autoimmunity* 5:101–106.

277. Martin, A., R.P. Magnusson, D.L. Kendler, E. Concepcion, A. Ben Nun, and T.F. Davies. 1993. Endogenous antigen presentation by autoantigen-transfected Epstein-Barr virus-lymphoblastoid cells. I. Generation of human thyroid peroxidase- reactive T cells and their T cell receptor repertoire. *J. Clin. Invest.* 91:1567–1574.

278. Mullins, R.J., Y. Chernajovsky, C. Dayan, M. Londei, and M. Feldmann. 1994. Transfection of thyroid auto-antigens into EBV-transformed B cell lines. Recognition by Graves' disease thyroid T cells. *J. Immunol.* 152:5572–5580.

279. Topliss, D., J. How, M. Lewis, V. Row, and R. Volpe. 1983. Evidence for cell-mediated immunity and specific suppressor T lymphocyte dysfunction in Graves' disease and diabetes mellitus. *J. Clin. Endocrinol. Metab.* 57:700–705.

280. Okita, N., D. Topliss, M. Lewis, V.V. Row, and R. Volpe. 1981. T-lymphocyte sensitization in Graves' and Hashimoto's diseases confirmed by an indirect migration inhibition factor test. *J. Clin. Endocrinol. Metab.* 52:523–527.

281. Okita, N., V.V. Row, and R. Volpe. 1981. Suppressor T-lymphocyte deficiency in Graves' disease and Hashimoto's thyroiditis. *J. Clin. Endocrinol. Metab.* 52:528–533.

282. Okita, N., V.V. Row, and R. Volpe, 1981. Suppressor T-lymphocyte deficiency in Graves' disease and Hashimoto's thyroiditis. *J. Clin. Endocrinol. Metab.* 52:528–533.

283. Topliss, D.J., N. Okita, M. Lewis, V.V. Row, and R. Volpe. 1981. Allosuppressor T lymphocytes abolish migration inhibition factor production in autoimmune thyroid disease: evidence from radiosensitivity experiments. *Clin. Endocrinol. Oxf.* 15:335–341.

284. Ludgate, M.E., S. Ratanachaiyavong, A.P. Weetman, R. Hall, and A.M. McGregor. 1985. Failure to demonstrate cellmediated immune responses to thyroid antigens in Graves' disease using *in vitro* assays of lymphokine-mediated migration inhibition. *J. Clin. Endocrinol. Metab.* 60:98–102.

285. Okita, N., A. Kidd, V.V. Row, and R. Volpe. 1980. Sensitization of T-lymphocytes in Graves' and Hashimoto's diseases. *J. Clin. Endocrinol. Metab.* 51:316–320.

286. Iitaka, M., J. Bernstein, H.C. Gerstein, Y. Iwatani, V.V. Row, and R. Volpe. 1986. Sensitization of T lymphocytes to thyroid antigen in autoimmune thyroid disease as demonstrated by the monocyte procoagulant activity test. *J. Endocrinol. Invest.* 9:471–478.

287. Iitaka, M., Y. Iwatani, H.C. Gerstein, V.V. Row, and R. Volpe. 1987. Immunomodulatory effect of the treatment of Graves' disease on antigen-specific monocyte procoagulant activity production. *Clin. Endocrinol. (Oxf.).* 27:321–330.

288. Davies, T.F. 1985. Cocultures of human thyroid monolayer cells and autologous T cells: impact of HLA class II antigen expression. *J. Clin. Endocrinol. Metab.* 61:418–422.

289. Londei, M., G.F. Bottazzo, and M. Feldmann. 1985. Human T-cell clones from autoimmune thyroid glands: specific recognition of autologous thyroid cells. *Science* 228:85–89.

290. Aoki, N., and J. DeGroot. 1979. Lymphocyte blastogenic response to human thyroglobulin in Graves' disease, Hashimoto's thyroiditis, and metastatic thyroid cancer. *Clin. Exp. Immunol.* 38:523–530.

291. Fisfalen, M.E., K. Soltani, A.M. Janiga, Y. Kawakami, E. Macchia, J. Quintans, and L.J. DeGroot. 1990. The regulatory role of human helper T-cell clones on antithyroid antibody production by peripheral B-cells. *J. Clin. Endocrinol. Metab.* 71:170–178.

292. Scott, T., and B. Glick. 1987. Organ weights, T-cell proliferation, and graft vs host capabilities of hypothyroidic chickens. *Gen. Comp. Endocrinol.* 67:270–276.

293. Mariotti, S., L. Chiovato, P. Vitti, C. Marcocci, G. F. Fenzi, G.F. Del-Prete, A. Tiri, S. Romagnani, M. Ricci, and A. Pinchera. 1989. Recent advances in the understanding of humoral and cellular mechanisms implicated in thyroid autoimmune disorders. *Clin. Immunol. Immunopathol.* 50:S73–S84.

294. Mackenzie, W.A., and T.F. Davies. 1987. An intrathyroidal T-cell clone specifically cytotoxic for human thyroid cells. *Immunology* 61:101–103.

295. Bagnasco, M., D. Venuti, I. Prigione, G.C. Torre, S. Ferrini, and G.W. Canonica. 1988. Graves' disease: phenotypic and functional analysis at the clonal level of the T-cell repertoire in peripheral blood and in thyroid. *Clin. Immunol. Immunopathol.* 47:230–239.

296. Del Prete, G.F., S. Mariotti, A. Tiri, M. Ricci, A. Pinchera, and S. Romagnani. 1987. Characterization of thyroid infiltrating lymphocytes in Hashimoto's thyroiditis: detection of B and T cells specific for thyroid antigens. Acta Endocrinol. Suppl. (Copenh.) 281:111–114.

297. Davies, T.F. 1995. T cell receptor gene expression in autoimmune thyroid disease. In: T-cell receptor use in human autoimmune diseases. M.M. Davies and J. Buxbaum, editors. New York Academy of Sciences, New York. pp. 331–344.

298. Teng, W.P., S.B. Cohen, D.N. Posnett, and A.P. Weetman. 1990. T cell receptor phenotypes in autoimmune thyroid disease. *J. Endocrinol. Invest.* 13:339–342.

299. Lipoldova, M., M. Londei, B. Grubeck-Loebenstein, M. Feldmann, and M.J. Owen. 1989. Analysis of T-cell receptor usage in activated T-cell clones from Hashimoto's thyroiditis and Graves' disease. *J. Autoimmun.* 2:1–13.

300. Davies, T.F., A. Martin, E.S. Concepcion, P. Graves, L. Cohen, and A. Ben Nun. 1991. Evidence of limited variability of antigen receptors on intrathyroidal T cells in autoimmune thyroid disease. *N. Engl. J. Med.* 325:238–244.

301. Matsuoka, N., A. Martin, E.S. Concepcion, P. Unger, L.D. Shultz, and T.F. Davies. 1993. Preservation of functioning human thyroid organoids in the *scid* mouse: II. Biased use of intrathyroidal T cell receptor V genes. *J. Clin. Endocrinol. Metab.* 77:311–315.

302. Heufelder, A.E., S. Herterich, G. Ernst, R.S. Bahn, and P.C. Scriba. 1995. Analysis of retroorbital T cell antigen receptor variable region gene usage in patients with Graves' ophthalmopathy. *Eur. J. Endocrinol.* 132:266–277.

303. McIntosh, R.S., P.F. Watson, and A.P. Weetman. 1997. Analysis of the T cell receptor V alpha repertoire in Hashimoto's thyroiditis: Evidence for the restricted accumulation of CD8+ T cells in the absence of CD4+ T cell restriction. *J. Clin. Endocrinol. Metab.* 82:1140–1146.

304. Sakata, S., T. Ogawa, I. Matsui, T. Manshouri, and M.Z. Atassi. 1992. Biological activities of rabbit antibodies against synthetic human TSH receptor peptides representing TSH binding regions. *Biochem. Biophys. Res. Commun.* 182:1369–1375.

305. Davies, T.F., H. Vlase, and M. Kita. 1999. The search for an animal model for Graves' disease. In: The decade of autoimmunity. Y. Shoenfeld, editor. Elsevier, Amsterdam. pp. 43–50.

306. Shimojo, N., Y. Kohno, K. Yamaguchi, S. Kikuoka, A. Hoshioka, H. Niimi, A. Hirai, Y. Tamura, Y. Saito, L.D. Kohn, and K. Tahara. 1996. Induction of Graves-like disease in mice by immunization with fibroblasts transfected with the thyrotropin receptor and a class II molecule. *Proc. Natl. Acad. Sci. USA* 93:11074–11079.

307. Kita, M., L. Ahmad, R.C. Marians, H. Vlase, P. Unger, P.N. Graves, and T.F. Davies. 1999. Regulation and transfer of a murine model of thyrotropin receptor antibody mediated Graves' disease. *Endocrinology* 140:1392–1398.

308. Germaine, R.N. 1986. The ins and outs of antigen processing and presentation. *Nature* 322:687–689.

309. Wilkin, T.J. 1990. Receptor autoimmunity in endocrine disorders. *N. Engl. J. Med.* 323:1318–1324.

310. Shenkman, L., and E.J. Bottone. 1976. Antibodies to Yersinia enterocolitica in thyroid disease. *Ann. Intern. Med.* 85:735–739.

311. Bech, K., O. Clemmensen, J.H. Larsen, S. Thyme, and G. Bendixen. 1978. Cell-mediated immunity of Yersinia enterocolitica serotype 3 in patients with thyroid diseases. *Allergy* 33:82–88.

312. Lidman, K., U. Eriksson, R. Norberg, and A. Fagraeus. 1976. Indirect immunofluorescence staining of human thyroid by antibodies occurring in Yersinia enterocolitica infections. *Clin. Exp. Immunol.* 23:429–435.

313. Weiss, M., S.H. Ingbar, S. Winblad, and D.L. Kasper. 1983. Demonstration of a saturable binding site for thyrotropin in Yersinia enterocolitica. *Science* 219:1331–1333.

314. Wenzel, B.E., J. Heesemann, K.W. Wenzel, and P.C. Scriba. 1988. Patients with autoimmune thyroid diseases have antibodies to plasmid encoded proteins of enteropathogenic Yersinia. *J. Endocrinol. Invest.* 11:139–140.

315. Arscott, P., E.D. Rosen, R.J. Koenig, M.M. Kaplan, T. Ellis, N. Thompson, and J.R. Baker, Jr. 1992. Immunoreactivity to Yersinia enterocolitica antigens in patients with autoimmune thyroid disease. *J. Clin. Endocrinol. Metab.* 75:295–300.

316. Wenzel, B.E., and J. Heesemann. 1987. Antigenic homologies between plasmid encoded proteins from enteropathogenic Yersinia and thyroid autoantigen. *Horm. Metab. Res. Suppl* 17:77–78.

317. Ehl, S., J. Hombach, P. Aichele, H. Hengartner, and R.M. Zinkernagel. 1997. Bystander activation of cytotoxic T cells: studies on the mechanism and evaluation of *in vivo* significance in a transgenic mouse model. *J. Exp. Med.* 185:1241–1251.

318. Horwitz, M.S., L.M. Bradley, J. Harbertson, T. Krahl, J. Lee, and N. Sarvetnick. 1998. Diabetes induced by Coxsackie virus: initiation by bystander damage and not molecular mimicry. *Nat. Med.* 4:781–785.

319. Martin, A., M. Valentine, P. Unger, S.W. Yeung, L.D. Shultz, and T.F. Davies. 1994. Engraftment of human lymphocytes and thyroid tissue into Scid and Rag2-deficient mice: absent progression of lymphocytic infiltration. *J. Clin. Endocrinol. Metab.* 79:716–723.

320. Neufeld, D.S., M. Platzer, and T.F. Davies. 1989. Reovirus induction of MHC class II antigen in rat thyroid cells. *Endocrinology* 124:543–545.

321. Khoury, E., L. Pereira, and F. Greenspan. 1991. Induction of HLA-DR expression on thyroid follicular cells by cytomegalovirus infection *in vitro*. *Amer. J. Pathol.* 138:1209–1223.

322. Ciampolillo, A., V. Marini, R. Mirakian, M. Buscema, T. Schulz, R. Pujol Borrell, and G.F. Bottazzo. 1989. Retrovirus-like sequences in Graves' disease: implications for human autoimmunity. *Lancet* 1:1096–1100.

323. Tominaga, T. S. Katamine, H. Namba, N. Yokoyama, S. Nakamura, S. Morita, S. Yamashita, M. Izumi, T. Miyamoto, and S. Nagataki. 1991. Lack of evidence for the presence of human immunodeficiency virus type 1-related sequences in patients with Graves' disease. *Thyroid* 1:307–314.

324. Humphrey, M., J.R. Baker, Jr., F.E. Carr, L. Wartofsky, J. Mosca, J.J. Drabick, D.S. Burke, Y.Y. Djuh, and K.D. Burman. 1991. Absence of retroviral sequences in Graves' disease. *Lancet* 337:17–18.

325. Alexander, W.D., R.M. Harden, D.A. Koutras, and E. Wayne. 1965. Influence of iodine intake after treatment with anti-thyroid drugs. *Lancet* 2:866–868.

326. Headington, J.T., and T. Tantajumroon. 1967. Surgical thyroid disease in northern Thailand. A study in geographic pathology. *Arch. Surg.* 95:157–161.

327. Boukis, M.A., D.A. Koutras, A. Souvatzoglou, A. Evangelopoulou, M. Vrontakis, and S.D. Moulopoulos. 1983. Thyroid hormone and immunological studies in endemic goiter. *J. Clin. Endocrinol. Metab.* 57:859–862.

328. Bagchi, N., T.R. Brown, E. Urdanivia, and R.S. Sundick. 1985. Induction of autoimmune thyroiditis in chickens by dietary iodine. *Science* 230:325–327.

329. Allen, E.M., M.C. Appel, and L.E. Braverman. 1986. The effect of iodide ingestion on the development of spontaneous lymphocytic thyroiditis in the diabetes-prone BB/W rat. *Endocrinology* 118:1977–1981.

330. Safran, M., T.L. Paul, E. Roti, and L.E. Braverman. 1987. Environmental factors affecting autoimmune thyroid disease. *Endocrinol. Metab. Clin. North Am.* 16:327–342.

331. Sundick, R.S., D.M. Herdegen, T.R. Brown, and N. Bagchi. 1987. The incorporation of dietary iodine into thyro-

globulin increases its immunogenicity. *Endocrinology* 120:2078–2084.

332. Champion, B.R., D.C. Rayner, P.G.H. Byfield, K.R. Page, C.T.J. Chan, and I.M. Roitt. 1987. Critical role of iodination for T cell recognition of thyroglobulin in experimental murine thyroid autoimmunity. *J. Immunol.* 139:3665–3670.

333. Weetman, A.P., A.M. McGregor, and R. Hall. 1983. Thyroglobulin uptake and presentation by macrophages in experimental autoimmune thyroiditis. *Immunology* 50:315–318.

334. Schuppert, F., E. Rambusch, H. Kirchner, J. Atzpodien, L.D. Kohn, and M.A. von zur. 1997. Patients treated with interferon-alpha, interferon-beta, and interleukin-2 have a different thyroid autoantibody pattern than patients suffering from endogenous autoimmune thyroid disease. *Thyroid* 7:837–842.

335. Schwid, S.R., A.D. Goodman, and D.H. Mattson. 1997. Autoimmune hyperthyroidism in patients with multiple sclerosis treated with interferon beta-1b. *Arch. Neurol.* 54:1169–1190.

336. Baudin, E., P. Marcellin, M. Pouteau, N. Colas-Linhart, J.P. Le Floch, C. Lemmonier, J.P. Benhamou, and B. Bok. 1993. Reversibility of thyroid dysfunction induced by recombinant alpha interferon in chronic hepatitis C. *Clin. Endocrinol. (Oxf.)* 39:657–661.

337. Parry, C.H. 1825. Disease of the heart. In: Collections form the unpublished writings. Volume 2. Underwoods, London. pp. 111–125.

338. Jain, R., D. Zwickler, C.S. Hollander, H. Brand, A. Saperstein, B. Hutchinson, C. Brown, and T. Audhya. 1991. Corticotropin-releasing factor modulates the immune response to stress in the rat. *Endocrinology* 128:1329–1336.

339. Sonino, N., M. Girelli, M. Boscaro, F. Fallo, B. Busnardo, and G.A. Fava. 1993. Life events in the pathogenesis of Graves' disease. A controlled study. *Acta Endocrinol.* 128:293–296.

340. Winsa, B., H.-O. Adami, R. Bergstrom, A. Gamstedt, P.A. Dahlberg, U. Adamson, R. Jansson, and A. Karlsson. 1991. Stressful life events and Graves' disease. *Lancet* 338:1475–1479.

341. Amino, N., O. Tanizawa, H. Mori, Y. Iwatani, T. Yamada, K. Kurachi, Y. Kumahara, and K. Miyai. 1982. Aggravation of thyrotoxicosis in early pregnancy and after delivery in Graves' disease. *J. Clin. Endocrinol. Metab.* 55:108–112.

342. Marcocci, C., L. Chiovato, S. Mariotti, and A. Pinchera. 1982. Changes of circulating thyroid autoantibody levels during and after the therapy with methimazole in patients with Graves' disease. *J. Endocrinol. Invest.* 5:13–19.

343. Fenzi, G., K. Hashizume, C.P. Roudebush, and L.J. DeGroot. 1979. Changes in thyroid-stimulating immunoglobulins during antithyroid therapy. *J. Clin. Endocrinol. Metab.* 48:572–576.

344. McGregor, A.M., S.B. Rees, R. Hall, P.N. Collins, B.G. Franco, and M.M. Petersen. 1982. Specificity of the immunosuppressive action of carbimazole in Graves' disease. *Br. Med. J. Clin. Res. Ed.* 284:1750–1751.

345. Weetman, A.P. 1994. The immunomodulatory effects of antithyroid drugs. *Thyroid* 4:145–146.

346. Volpe, R., A. Karlsson, R. Jansson, and P.A. Dahlberg. 1986. Evidence that antithyroid drugs induce remissions in Graves' disease by modulating thyroid cellular activity. *Clin. Endocrinol. Oxf.* 25:453–462.

347. Weiss, I., and T.F. Davies. 1981. Inhibition of immunoglobulin-secreting cells by antithyroid drugs. *J. Clin. Endocrinol. Metab.* 53:1223–1228.

348. McLachlan, S.M., C.A. Peggy, M.C. Atherton, S. Middleton, E.T. Young, F. Clark, and B.R. Smith. 1985. The effect of carbimazole on thyroid autoantibody synthesis by thyroid lymphocytes. *J. Clin. Endocrinol. Metab.* 60:1237–1242.

349. Weetman, A.P., A.P. McGregor, and R. Hall. 1983. Methimazole inhibits thyroid autoantibody production by an action on accessory cells. *Clin. Immunol. Immunopathol.* 28:39–45.

350. Ansar, A.S., P.R. Young, and W.J. Penhale. 1986. Beneficial effect of testosterone in the treatment of chronic autoimmune thyroiditis in rats. *J. Immunol.* 136:143–147.

351. Chiovato, L., P. Lapi, E. Fiore, M. Tonacchera, and A. Pinchera. 1993. Thyroid autoimmunity and female gender. *J. Endocrinol. Invest.* 16:384–391.

352. Barbesino, G., Y. Tomer, E.S. Concepcion, T.F. Davies, and D.A. Greenberg. Linkage analysis of candidate genes in autoimmune thyroid disease: 2. Selected gender-related genes and the X-chromosome. *J. Clin. Endocrinol. Metab.*, in press. 1998.

353. Vitti, P.V., W.A. Valente, F.S. Ambesi-Impiombato, G. Fenzi, A. Pinchera, and L.D. Kohn. 1982. Graves' IgG stimulation of continuously cultured rat thyroid cells: a sensitive and potentially useful clinical assay. *J. Endocrinol. Invest.* 5:179–185.

354. Volpe, R. 1988. The immunoregulatory disturbance in autoimmune thyroid disease. *Autoimmunity* 2:55–72.

355. Chan, J.Y. and P.G. Walfish. 1986. Activated (Ia+) T-lymphocytes and their subsets in autoimmune thyroid diseases: analysis by dual laser flow microfluorocytometry. *J. Clin. Endocrinol. Metab.* 62:403–409.

356. Hashizume, K., K. Ichikawa, A. Sakurai, S. Suzuki, T. Takeda, M. Kobayashi, T. Miyamoto, M. Arai, and T. Nagasawa. 1991. Administration of thyroxine in treated Graves' disease. Effects on the level of antibodies to thyroid-stimulating hormone receptors and on the risk of recurrence of hyperthyroidism. *N. Engl. J. Med.* 324:947–953.

357. Piccinini, L.A., N.K. Goldsmith, S.H. Roman, and T.F. Davies. 1987. HLA-DP, DQ and DR gene expression in Graves' disease and normal thyroid epithelium. *Tissue Antigens* 30:145–154.

358. Creach, F., M. Teece, S. Williams, S. Didcote, W. Perkins, F. Hashim, and B. Rees-Smith. 1985. An analysis of thyrotrophin receptor binding and thyroid stimulating activities in a series of Graves sera. *Clin. Endocrinol.* 23:395–404.

359. Huber, G.K., R. Safirstein, D. Neufeld, and T.F. Davies. 1991. Thyrotropin receptor autoantibodies induce human thyroid cell growth and c-fos activation. *J. Clin. Endocrinol. Metab.* 72:1142–1147.

360. Tunbridge, W.M., D.C. Evered, R. Hall, D. Appleton, M. Brewis, F. Clark, J.G. Evans, E. Young, T. Bird, and P.A. Smith. 1977. The spectrum of thyroid disease in a community: the Whickham survey. *Clin. Endocrinol. (Oxf.)* 7:481–493.

361. dos Remedios, L.V., P.M. Weber, R. Feldman, D.A. Schurr, and T.G. Tsoi. 1980. Detecting unsuspected thyroid dysfunction by the free thyroxine index. *Arch. Intern. Med.* 140:1045–1049.

362. Hoffenberg, R. 1974. Aetiology of hyperthyroidism. I. *Br. Med. J.* 3:452–455.

363. Tunbridge, W.M.G. 1983. Prevalence of autoimmune endocrine disease. In: Autoimmune endocrine disease. T.F. Davies, editor. Wiley, New York. pp. 93–100.

364. Vanderpump, M.P.J., W.M.G. Tunbridge, J.M. French, D. Appleton, D. Bates, F. Clark, J. Grimley Evans, D.M.

Hasan, H. Rodgers, F. Tunbridge *et al.* 1995. The incidence of thyroid disorders in the community: a twenty-year follow-up of the Whickham survey. *Clin. Endocrinol.* 43:55–68.

365. Furszyfer, J., L.T. Kurland, W.M. McConahey, and L.R. Elveback. 1970. Graves' disease in Olmsted County, Minnesota 1935 through 1967. *Mayo Clinic Proceedings* 45:636–644.

366. dos Remedios, L.V., P.M. Weber, R. Feldman, D.A. Schurr, and T.G. Tsoi. 1980. Detecting unsuspected thyroid dysfunction by the free thyroxine index. *Arch. Intern. Med.* 140:1045.

367. Tomer, Y., G. Barbesino, D.A. Greenberg, E.S. Concepcion, and T.F. Davies. A new Graves' disease susceptibility locus maps to chromosome 20q11.2. *Amer. J. Human Genetics*, in press.

368. Wall, J.R., J. Henderson, C.R. Strakosch, and D.M. Joyner. 1982. Graves' ophthalmopathy. *Can. Med. Assoc.* 124:856–858.

369. Tomer, Y., and T. Davies. 2000. In: Oxford textbook of endocrinology. J. Wass and Shalet, S., editors. Oxford University Press, Oxford, UK.

370. Eguchi, K., T. Otsubo, Y. Kawabe, Y. Ueki, T. Fukuda, M. Matsunaga, C. Shimomura, N. Ishikawa, H. Tezuka, H. Nakao *et al.* 1987. The remarkable proliferation of helper T cell subset in response to autologous thyrocytes and intrathyroidal T cells from patients with Graves' disease. *Clin. Exp. Immunol.* 70:403–410.

371. Otsubo, T., K. Eguchi, C. Shimomura, Y. Ueki, H. Tezuka, N. Ishikawa, K. Ito, and S. Nagataki. 1988. *In vitro* cellular interactions among thyrocytes, T cells and monocytes from patients with Graves' disease. *Acta Endocrinol. (Copenh.)* 117:282–288.

372. Londei, M., J.R. Lamb, G.F. Bottazzo, and M. Feldmann. 1984. Epithelial cells expressing aberrant MHC class II determinants can present antigen to cloned human T cells. *Nature* 312:639–641.

373. Matsunaga, M., K. Eguchi, T. Fukuda, H. Tezuka, Y. Ueki, Y. Kawabe, C. Shimomura, T. Otsubo, N. Ishikawa, K. Ito *et al.* 1988. The effects of cytokines, antithyroidal drugs and glucocorticoids on phagocytosis by thyroid cells. *Acta Endocrinol. Copenh.* 119:413–419.

374. Kimura, H., and T.F. Davies. 1991. Thyroid-specific T cells in the normal Wistar rat. II. T cell clones interact with cloned wistar rat thyroid cells and provide direct evidence for autoantigen presentation by thyroid epithelial cells. *Clin. Immunol. Immunopathol.* 58:195–206.

375. Williams, R.C.J., N.J. Marshall, K. Kilpatrick, J. Montano, P.M. Brickell, M. Goodall, P.A. Ealey, B. Shine, A.P. Weetman, and R.K. Craig. 1988. Kappa/lambda immunoglobulin distribution in Graves' thyroid-stimulating antibodies. Simultaneous analysis of C lambda gene polymorphisms. *J. Clin. Invest.* 82:1306–1312.

376. Knight, J., P. Laing, A. Knight, D. Adams, and N. Ling. 1986. Thyroid-stimulating autoantibodies usually contain only lambda-light chains: evidence for the "forbidden clone" theory. *J. Clin. Endocrinol. Metab.* 62:342–347.

377. Zakarija, M. 1983. Immunochemical characterization of the thyroid-stimulating antibody (TSAb) of Graves' disease: evidence for restricted heterogeneity. *J. Clin. Lab. Immunol.* 10:77–85.

378. Nagayama, Y., M. Izumi, K. Ashizawa, T. Kiriyama, N. Yokoyama, S. Morita, S. Ohtakara, T. Fukuda, K. Eguchi, I. Morimoto *et al.* 1987. Inhibitory effect of interferon-gamma on the response of human thyrocytes to thyrotropin (TSH) stimulation: relationship between the response to TSH and the expression of DR antigen. *J. Clin. Endocrinol. Metab.* 64:949–953.

379. Graves, P., D.S. Neufeld, and T.F. Davies. 1989. Differential cytokine regulation of MHC class II and thyroglobin mRNAs in rat thyroid cells. *Mol. Endocrinol.* 3:758–762.

380. Misaki, T., D. Tramontano, and S.H. Ingbar. 1988. Effects of rat gamma- and non-gamma-interferons on the expression of Ia antigen, growth, and differentiated functions of FRTL5 cells. *Endocrinology* 123:2849–2857.

381. Weetman, A.P. 1982. Age dependence of spontaneous plaque forming cells in human peripheral blood in Graves' disease, rheumatoid arthritis and normal subjects. *J. Clin. Lab Immunol.* 8:91–93.

382. Zakarija, M., and J.M. McKenzie. 1989. Influence of cytokines on growth and differentiated function of FRTL5 cells. *Endocrinology* 125:1260–1265.

383. Zakarija, M., F.J. Hornicek, S. Levis, and J.M. McKenzie. 1988. Effects of gamma-interferon and tumor necrosis factor alpha on thyroid cells: induction of class II antigen and inhibition of growth stimulation. *Mol. Cell Endocrinol.* 58:129–136.

384. Bogner, U., B. Sigle, and H. Schleusener. 1988. Interferon-gamma protects human thyroid epithelial cells against cell-mediated cytotoxicity. *Immunobiology* 176:423–431.

385. Todd, I., R. Pujol-Borrell, L.J. Hammond, G.F. Bottazzo, and M. Feldmann. 1985. Interferon-gamma induces HLA-DR expression by thyroid epithelium. *Clin. Exp. Immunol.* 61:265–273.

386. Piccinini, L.A., W.A. Mackenzie, M. Platzer, and T.F. Davies. 1987. Lymphokine regulation of HLA-DR gene expression in human thyroid cell monolayers. *J. Clin. Endocrinol. Metab.* 64:543–548.

387. Rapoport, B., and McLachlan. 1994. Thyroid peroxidase as an autoantigen in autoimmune thyroid disease: Update 1994. *Endocrine Reviews* 3:96–102.

39 | Experimental Models of Autoimmune Thyroiditis

Jeannine Charreire

1. INTRODUCTION

The thyroid gland is the target of two main organ-specific autoimmune disorders with opposite clinical outcomes in human: Hashimoto's thyroiditis (HT) can result in hypothyroidism, and Graves' disease in hyperthyroidism. Whereas numerous models of HT have been developed in various species, including mammals and birds, few models of Graves' disease have been reported. Spontaneous as well as experimentally-induced autoimmune thyroiditis have largely contributed to our understanding of the mechanisms involved in HT. One general feature of autoimmune diseases, including thyroid autoimmune disorders, is their Major Histocompatibility Complex (MHC)-linked susceptibility (1). Moreover, autoimmune disorders often exhibit higher incidence in females than in males. Autoimmune processes play an important role in thyroid diseases and during the last ten years, knowledge of their mechanism greatly improved, allowing, at least in experimental models, better targeted and more specific treatments with fewer deleterious secondary effects.

2. CHRONIC SPONTANEOUS AUTOIMMUNE THYROIDITIS (SAT)

2.1. SAT in the Obese Strain (OS) of Chicken

Many species, including mice, rats, guinea-pigs, mastomys, dogs, cows and chickens develop chronic SAT. Among the current laboratory animals, SAT of OS chicken has been the most intensively studied, since it closely resembles human HT in all clinical, endocrinological, histopathological and immunological aspects (2). In the first 2 to 3 weeks post-hatching, the OS chicken develops a severe autoimmune thyroiditis characterized by functional hypothyroidism, complete destruction of the thyroid gland and production of autoantibodies (AAbs) to thyroglobulin (Tg), thyroxine (T4) and tri-iodothyronine. A hallmark of SAT that is also observed in human HT, but not in induced-experimental autoimmune thyroiditis (EAT), is the presence of developed germinal centers inside the thyroid gland. OS chickens also exhibit a general primary lymphocyte hyperreactivity. Activated T cells, the first cells to infiltrate the thyroid gland, can transfer SAT to healthy recipients. The involvement of CD4+ T cells in OS chicken SAT has been demonstrated by the lack of thyroiditis when OS chickens are depleted in CD4+ T cells using mAb at hatching. In contrast, CD8+ T cell depletion induces only partial inhibition of SAT development (3). Moreover, disturbed regulation of glucocorticoide hormones, assessed by the lack of corticosterone response to immune signals and by significantly augmented serum concentrations of corticosteroid binding globulin, is observed (4).

Thyroidal iodine is also involved in the pathogenesis of OS chicken SAT as assessed by the increase in SAT prevalence when chickens are fed with iodine (5), and reciprocally, reduced thyroidal lymphocytic infiltration when iodine depletion is performed (6). OS chickens exhibit an augmented capacity for iodine uptake since early intervention with KClO4, an agent that inhibits uptake and organification of iodine, results in more effective prevention of OS chicken SAT. One regimen initiated *in ovo* and continued until sacrifice is particularly effective and furthermore prevents SAT occurrence during the longest period of time studied, nine weeks. In the same line, it has been demonstrated that transfer of thyroiditis into Cornell Strain of chickens using spleen cells from OS chickens that

Table 1   Characteristics of Chronic SAT

Animal Species	Thyroid gland		Functional hypothyroidism	AAbs to		Associated autoimmune disease	Reference
	Lymphocytic infiltration	Germinal center		Tg	TPO		
OS chicken	+	+	+	+	NT	−	(2)
NOD mouse	+	−	−	+	+	diabetes	(9)
NOD-H-2^{h4}	+	−	−	+	NT	insulitis	(13)
NOD-H-2^{k}	+	+	−	+	NT	−	(15)
MLR-lpr/lpr	+	−	+	+	+	lupus	(23)
Buffalo rat	+	−	(+) increased TSH	+	NT	−	(25)
BB/W rat	+	−	−	+	NT	insulitis, diabetes	(27)

NT = Not tested.

have received an iodine-depleted regimen is markedly reduced (7). Iodine has been shown to induce thyroid cell injury in normal as well as in OS chickens. It can be hypothesized that thyroid cell injury represents an initial event that might be responsible for OS chicken SAT (8). In addition, iodine also increases the immunogenicity of the Tg molecule, thus inducing proliferation of Tg-specific autoreactive cells.

Genetic analyses essentially performed through back-crosses and adoptive transfer experiments to determine MHC susceptibity to SAT provide controversial results that suggest the existence of non-MHC-linked regulation of SAT in OS chicken. Studies aimed at identifying the nature and the number of genes contributing to SAT have led to the hypothesis of one essential recessive autosomal "thyroid susceptibility gene", as yet unidentified, and of a series of modulating genes involved in various reactions (altered glucocorticoide response, altered iodine metabolism, macrophage hyperfunction, MHC-linked immune response genes) (2).

OS chicken SAT is the experimental model of human HT that best mimics the human disease. Furthermore, the bird-specific design of T and B cell immune system allows separate manipulation of humoral and cellular immune responses. For these reasons, OS SAT is one of the most attractive models of experimental autoimune thyroiditis.

2.2. SAT in Rodents

Rodents such as Non Obese Diabetic (NOD) mice, MRL-lpr/lpr mice. Buffalo (BUF) and Biobreeding/Worcester (BB/W) rats spontaneously develop autoimmune thyroiditis, with sometimes functional hypothyroidism, and always thyroid lymphocytic infiltration by T cells, AAbs to thyroid antigens and thyroid follicular cell destruction. In most of those animals, autoimmune thyroiditis is associated with other autoimmune disorders: Spontaneous autoimmune type 1 diabetes in NOD mice and BB/W rats, and systemic

lupus erythematosus in MRL-lpr/lpr mice. The same associations may be observed in patients.

Little is known about the thyroid manifestations in NOD mice except that lymphocytic infiltration of the thyroid gland appears approximately at 15 weeks of age (9) and is less frequent than insulitis or sialitis, even if elevated incidence of thyroiditis has been reported in one particular colony (10). Concordant with HT, AAbs to Tg are correlated with the presence of thyroid lesions. The finding of a unique I-A structure (I–A^{g7}) and the lack of I–E expression in NOD mice (11), has led to the hypothesis that I–A^{g7} plays a major role in the susceptibility of NOD mice to autoimmune manifestations.

MHC-congenic NOD.H–2^{h4} mice derived from a cross between NOD and B10A(4R), and carrying a H-2^{k} haplotype of susceptibility to EAT (12) develop spontaneous thyroiditis, but not diabetes. As reported above in the OS chicken, when NOD.H-2^{h4} mice are given water with 0.05% iodine during 8 weeks, a significant increase in SAT is observed (54% in females, 70% in males), compared to control NOD-H-2^{h4} that have received plain water (5%) (13). In iodine-induced NOD.H-2^{h4} thyroiditis, the levels of thyroid hormones are comparable to those of control NOD.H-2^{h4} mice. Autoantibodies (AAbs) to Tg consist in IgG2a, IgG2b and IgM, while AAbs to thyroid peroxidase (TPO) and IgG1 and IgG3 anti-Tg AAbs are not detected. Interestingly, IgG2b AAbs correlate with the disease. In 100% of NOD mice (14), early administration of iodine at high dose induces cell necrosis and inflammation of the thyroid gland, attested by both the presence of MHC class II^{+} antigen presenting cells (APC) and CD4^{+} and CD8^{+} T cells in the thyroid infiltrate. This iodine-induced NOD thyroiditis, similar to HT after three month iodine treatment, may be considered as a major factor in the induction and persistance of autoimmune thyroiditis in animals carrying a genetic susceptibility to autoimmune diseases.

To characterize the thyroiditis manifestations in NOD mice and evaluate their relevance to HT, we have derived a congenic line of NOD mice with the H-2^{k} haplotype favoring

EAT development (15). We have observed that cumulative incidence of spontaneous thyroiditis is significantly higher in NOD. H-2^k than in NOD mice, but without clinical or biological manifestation of hormonal dysfunction. Furthermore, when NOD or NODH-2^k mice are immunized by murine Tg (mTg) and adjuvants, a severe chronic thyroiditis appears. Tg-immunized NOD.H-2^k also exhibit a marked deficit in Th1-dependent, IgG2a AAbs to mTg that correlates with a peripheral defect in IFN-γ producing CD4$^+$ T cells. In these experiments, elimination of Tg-specific CD4$^+$ T cells might be achieved by syngeneic CD8$^+$ T cells (16), since the idiotypic autoreactivity of CD8$^+$ T cells towards syngeneic CD4$^+$ T cells might be favored by the autoimmune background of the NOD mouse. During Tg-induced chronic EAT of NOD or NOD.H-2^k mice, a five-fold-decrease in CD8$^+$ thyroid-infiltrating lymphocytes (TIL) is observed, compared to levels of TIL from Tg-immunized non-NOD controls. This observation has led us to hypothesize that CD8$^+$ T cells could play a regulatory role in EAT resolution in normal mice, since in NOD mice, CD8$^+$ diminution in TIL leads to chronic EAT. In H-2^k mice, we have previously cloned CD8$^+$ Tg-specific T cells (17) and demonstrated their regulatory role by injecting them into naive syngeneic recipients prior to immunization by Tg plus complete Freund adjuvant (CFA). Such CD8$^+$ T cell line completely inhibited EAT in recipients (18). More recently, CD8$^+$ regulatory T cells have also been reported in a particular form of thyroiditis called granulomatous EAT (gEAT) (19) and in experimental autoimmune encephalomyelitis (20).

To further investigate the role of CD8$^+$ T cells in NOD EAT, we selected NOD mice with a disrupted β-2 microglobulin gene, in which MHC class I antigen and CD8$^+$ T cells are lacking (21). When such mice are immunized with Tg plus CFA, EAT does not develop. However, the percentages of CD4$^+$ T cells that proliferate *in vitro* in response to Tg and that produce IFN-γ are similar to those of wild-type NOD mice. Thus, these experiments demonstrate that CD8$^+$ T cells play a key role in the initiation of EAT independently of CD4$^+$ T cells. Opposite results have been reported in B10BR mice with a disrupted β-2 microglobulin gene (22). However, disease was induced in those mice by immunization with Tg and lipopolysaccharide (LPS) instead of CFA as adjuvant. We hypothesize that LPS might have bypassed the requirement for CD8$^+$ class I-restricted T cells by activating macrophages or dendritic cells (DC) directly and thus providing the right conditions for attracting of CD4$^+$ T cells in the thyroid gland.

Those experiments show that the NOD mouse not only represents a powerful model of diabetes, but also offers a unique opportunity to analyze thyroiditis. For example, the opposing roles of CD8$^+$ T cells, as EAT inducers in NOD β2m$^{-/-}$, or as EAT regulators in chronic NOD EAT can be studied using those animals.

MRL-lpr/lpr mice that are genetically-susceptible to systemic lupus erythematosus-like disease also develop thyroiditis as part of their autoimmune disorders (23). These mice develop a disease very similar to the human HT, including the functional hypothyroidism. During their SAT, MRL-lpr/lpr mice exhibit decreased levels of thyroid hormone T4, increased levels of Thyroid Stimulating Hormone (TSH), extensive lymphocytic infiltration by T lymphocytes of the thyroid tissue and elevated concentrations of circulating AAbs to Tg and TPO. Thyroid cells from diseased MRL-lpr/lpr mice have altered gap junctional communications, assessed by decrease in connexins Cχ43 and Cχ26 (24). This deficiency in connexins initiated by components of the chronic inflammation process is responsible for the impaired cell-cell communication. Because of its recent description, few studies have been devoted to SAT of MRL-lpr/lpr mouse.

BUF rats also develops autoimmune thyroiditis resulting in hypothyroidism with elevated TSH, mononuclear cell infiltration of the thyroid and AAbs to thyroid antigens (25). The disease appears at the age of 4 months and occurs more often in females than in males. The level of AAbs to thyroid antigens parallels the degree of lymphocytic infiltration. Gene products of the rat MHC RT1 affect the severity of clinical symptoms and tissue inflammation. The spontaneous incidence of lymphocytic infiltration, of AAbs to Tg and of augmented TSH, increased after neonatal thymectomy (NTx) are still further augmented when iodine ingestion is associated to NTx (26). As already reported in OS chickens, disease accelerated by iodine could result from a direct cytotoxic effect of iodine on thyroid cell structure, as suggested by a marked accumulation of secondary liposomes and lipid droplets, by swollen and disrupted mitochondria and by extreme dilatation of rough endoplasmic reticulum (27).

The BB/W rat is in addition of thyroiditis a well-known model for the type 1 diabetes and associated autoimmune thyroid disease (28). BB/W rats however, stay euthyroid, and plasma TSH is not significantly raised. AAbs to Tg are detected in the sera from the age of 6 weeks and accumulation of lymphoid cells in the thyroid is seen in 60% of animals at the age of 20 weeks and over. A statistically significant increase in the number of DC has been demonstrated in the thyroid glands of 6–8 week old BB/W rats (29) and remains at high levels from week 8 onward. The incidence of thyroiditis is extremely variable, ranging from 100% in NB line to 4.9% in BE line at 105–110 days of age (30). While lymphopenia is absolutely required for diabetes to occur in BB/W rats, it only confers risk for thyroiditis (31), the MHC conferring dominant susceptibility to BB/W rat thyroiditis. The MHC region to the right of the class I RT1. A strongly correlates with diabetes and insulitis and is also significantly associated with the development of thyroiditis. However, the risk associated with MHC class II alleles u and a for disease development is distinct for insulitis (u) and thyroiditis (a) (32).

Spontaneous models of thyroiditis closely resemble HT. Besides their clinical and endocrinological characteristics, they arise in genetically-susceptible animals without experimental manipulation and never recover. In SAT-prone animals, early iodine diet accelerates the disease, probably through induction of thyroid cell injury, an initial event necessary to SAT development. The role of iodine in human HT has also been suspected in countries in which salt is supplemented in iodine.

3.   EXPERIMENTALLY-INDUCED THYROIDITIS

3.1.   EAT Induction with Tg and Adjuvant

Multiple approaches have been used to induce thyroiditis in mice or rats. One characteristic of induced-EAT is their spontaneous recovery, clearly contrasting with the chronicity of SAT. Immunizations aimed at increasing Tg concentration or immunogenicity are achieved using suitable adjuvant, denaturation or iodination of Tg (33,34). The most frequently used procedure is the subcutaneous injection of 50–100 μg homologous or heterologous Tg into MHC susceptible mice. Tg is emulsified in CFA or LPS, followed 10–15 days later by a booster injection of the same amount of Tg emulsified in incomplete Freund's adjuvant or LPS, respectively. Using this protocol, functional hormonal hypothyroidism is never evidenced and the first signs of thyroiditis (AAbs to Tg and TIL) appear on day 7–9 after the first injection: severe lesions of the thyroid gland are seen between the third and the fifth week and then EAT remits. In situ kinetics analysis of TIL in mice developing EAT (35), shows a shift in the $CD4^+/CD8^+$ ratio directly related to changes in the $CD8^+$ subpopulation which increases and then declines. Anti-Tg AAbs are detectable at their highest levels as soon as the third week post immunization (p.i.), as well as proliferative or cytotoxic T cell responses towards Tg. Tg/CFA-induced EAT is under the control of MHC genes (12): $H-2^k$, and $H-2^s$ are excellent responders, $H-2^q$ good responders, $H-2^a$ fairly good and $H-2^b$ and $H-2^d$ poor responders. Moreover, the use of F_1 and intra-MHC-congenic mice demonstrates that susceptibility to Tg-induced EAT is a dominant trait linked to the 1-A sub-region of the MHC (36). Recently, we have shown that $I-A^{g7}$ confers good responsiveness to thyroiditis (15). The age of mice at time of immunization also interferes with EAT development. After immunization of aged mice by Tg/CFA, thyroid lesions develop either comparable to or only slightly less intense-than those observed in young mice (37). However, drastic reductions in mTg-specific proliferative response, AAbs and delayed type hypersensitivity are observed. Aged thyroids from old mice also exhibit a greater susceptibility to tissue damage. Indeed, transfer of identical numbers of young mTg-sensitized splenocytes into young and aged mice results in a more severe EAT in the aged recipients.

In vitro proliferative responses of spleen or lymph node cells to Tg are detectable as early as day 8 p.i. and are maximum on day 14. These T cell-mediated proliferative responses are abrogated by anti-thy 1–2 or anti-CD4 treatment. Subsequently, it was shown that $CD4^+$ T cells are the helper cells involved in both Tg-specific proliferative responses and AAb production (38). Simultaneously, Tg-specific, MHC class I-restricted cytotoxic T cell response occurs, evidenced either on monolayers of syngeneic thyroid epithelial cells (TEC) (39,40) or on Tg-pulsed syngeneic macrophages. We have cloned such $CD8^+$ MHC class I-restricted cytotoxic T cells specific for Tg from lymph node of Tg/CFA-immunized mice (17) and demonstrated their regulatory role in preventing EAT when inactivated and injected prior to immunization (18).

Various strategies have been designed to determine Tg pathogenic epitopes inducing EAT in the 660 KDa Tg molecule (41). Two types of pathogenic sites, including or not hormonogenic sites are distinguished (42). Among the four hormonogenic sites of the Tg molecule, a 9-mer peptide containing T4 at position 2553 stimulates class II-restricted Tg-specific T cell hybrids that further adoptively transfer EAT when injected into susceptible animals. However, this 9-mer iodinated peptide is not able per se to induce EAT when injected into mice (43). Later a secondary role was attributed to iodination (44). Briefly, through T4 substitution, a non-iodinated Tg peptide at position 2553 has been shown to exhibit a strong immunogenicity in terms of EAT induction and proliferative response to Tg. whereas Tg peptides iodinated in positions 5 and 2567 were poorly immunogenic. It has been concluded that the immunogenicity of a conserved hormonogenic site is more dependent on its amino-acid sequence than on T4 substitution. Recently, it has been demonstrated that iodinated Tg differs from non-iodinated Tg in its reactivity with a panel of mAbs, indicating a loss of some epitopes and the gain of others (45). Iodination of Tg could modify its conformation in a way that affects its uptake by APC.

Pathogenic human or rat Tg peptides not associated with hormonogenic sites are also EAT inducers. We have characterized a 40-mer human Tg peptide F40D (46), located in the middle of the Tg molecule (1672–1711), that induces EAT when injected into $H-2^k$ mice. In addition F40D is recognized by a Tg-specific, class I-restricted cytotoxic $CD8^+$ hybridoma that we have cloned (17). Simultaneously, rat peptide 2495–2511, located in the NH2-end of the Tg, induces EAT and the proliferation of Tg-specific T cell lines in a class II-restricted $H-2^k$ or

H-2^s context (47), whereas rat peptide 2695–2713 exhibits the same properties, but exclusively in the H-2^s context (48) EAT induced by Tg peptides occurs in the quasi absence (46), or with a strongly diminished production of AAbs to Tg (47), further demonstrating that Tg-specific T cells play a major role in EAT induction and development.

Recently, we described a new model of EAT (49) induced by i.v. injection into H-2^k mice of aggregated heat-denatured porcine Tg (hdpTg) without the help of adjuvant. This EAT is mediated by a subset of Tg-specific CD8$^+$ cytotoxic T lymphocytes (CTL) secreting interferon-γ (IFN-γ), thus responding to the definition of Tc1 cells (50). The model is based on the fact that denaturation of exogenous antigen (Ag) induces Ag-specific CTL (51,52), and that CTL play a major role in EAT (17). The kinetics of hdpTg/EAT is comparable to that of EAT induced by Tg/CFA. Whereas T cells from hdpTg-immunized animals are unable to proliferate in response to the immunizing hdpTg. cytotoxic responses to hdpTg are detected. In addition, hdpTg-induced EAT is prevented by in vivo or in vitro treatment by mAb specific for CD8 or IFN-γ. Lastly, titers of antibodies to Tg and the IgG1/IgG2a ratio are twenty times lower in sera from hdpTg-immunized mice than in sera from Tg/CFA immunized mice. This new model of EAT represents a more physiological model of the human HT, since the absence of adjuvant avoids non-specific activation of the immune system and allows thyroiditogenic autoreactive effector cells to be more specifically activated. Furthermore, aggregated Tg has been reported in bovine thyroid glands (49)

3.2. EAT Induction by Lymphoid Cells

In the course of EAT induction by Tg and adjuvant, CD4$^+$ T cells specific for Tg are generated (53). Such thyroid-specific helper T cell lines or clones (54), obtained after in vivo and in vitro stimulation by Tg, transfer EAT when injected into syngeneic naive recipients. The efficiency of EAT transfer is increased by irradiation of the naive recipients (55). In situ analysis (56) of T cell subset composition in adoptively-transferred-EAT shows a higher percentage of CD4$^+$ T cells than CD8$^+$ T cells early after the T cell transfer, followed by an increase in CD8$^+$ T cells, while the percentages of B cells and macrophages remain constant. Passively-transferred-EAT requires CD4$^+$ Tg-specific T cells (38), since EAT does not occur when the inoculum is depleted of CD4$^+$ T cells and the Tg-specific proliferative and AAb responses are abolished. The absence of EAT and of Tg-specific T cell responses are also observed when anti-CD4 mAb is administrated prior to or following transfer of Tg-primed effector T cells. These studies further establish the major role of helper T cells in the pathogenesis of EAT,

and also suggest that therapy with an appropriate mAb can be an effective treatment even when initiated lately in the course of the disease.

Passively-transferred-EAT obtained after in vivo and in vitro stimulations by Tg is generally moderate in severity and characterized by thyroid infiltration consisting primarily of mononuclear cells. It was shown that addition of anti-IL-2R (57) or anti-IFN-γ (58) mAb during the in vitro activation by Tg induced a particular EAT, named granulomatous EAT (gEAT). In this particular form of disease, thyroid lesions show follicular cell proliferation and abundant infiltrates with polymorphonuclear leukocytes, histiocytes and multinucleated giant cells. In contrast, lymphocytes are relatively rare and titers of AAbs to Tg are very high. The mechanisms responsible for gEAT are unclear and further studies are needed to determine pathways used by anti-IL-2R or anti-IFN-γ mAb to induce gEAT.

3.3. Miscellaneous Methods of EAT Induction

The microsomal auto-Ag or its main component the thyroid peroxidase (TPO), an enzyme that catalyzes both the iodination and the coupling of iodinated tyrosines to form thyroid hormones, has also been used to induce EAT (59). Porcine TPO emulsified in CFA and injected into mice using the same protocole as Tg-EAT also induces EAT, with similar kinetics. However, a different genetic restriction from that induced by Tg exists: H-2^b mice are good responders, whereas H-2^a, H-2^d, H-2^k and H-2^s are poor responders. As in Tg-induced EAT, TPO-induced EAT is a transient disease that occurs in the absence of functional hypothyroidism. Four weeks after immunization, mononuclear cell infiltration of the thyroid gland occurs as well as a proliferative response to TPO, but AAbs to TPO are not detected. In TPO-induced EAT, peptide 774–788 from porcine TPO induced EAT as well as TPO-specific proliferative T cell responses and T cell lines (60). TPO-specific T cell lines transfer EAT in syngeneic recipients depleted of T cells or in nude mice, after irradiation or not. Tranferred-TPO EAT occurs three days after T cell line injection and increases in severity until day 50 approximately, whereas the low concentration of anti-TPO AAbs gradually increases. One wonders whether TPO-induced EAT can reflect human HT in which anti-TPO AAbs behave as disease marker.

Another approach to induce thyroid autoimmune reactivity takes advantage of the potential effector or regulatory functions of T lymphocytes. As early as 1973, it was demonstrated (61) that depletion of rat T lymphocytes by (NTx) followed by several low doses of irradiation leads to the spontaneous development of a typical EAT, without Tg injection. Reciprocally, EAT can be prevented by reconstitution of NTx, irradiated rats, by CD4$^+$CD8$^-$ syngeneic T cells. Recently, in this model, the existence of thyroid-specific regulatory T cells evidenced by thyroidectomy *in utero* has been unequivocally demonstrated to be under the control of peripheral thyroid auto-Ags (62). In mice, EAT following NTx and irradiation requires CD5^{dull} CD4^{bright} T cells to develop, but not CD5^{dull} CD8^{bright}. However, CD4^{bright} T cells exhibit a regulatory role, demonstrated in mice by induction of thyroid inflammatory lesions when CD5^{bright} T cells are depleted before transfer of CD5^{dull} CD4^{bright} into syngeneic recipients (63). More precisely, transfer into syngeneic recipients of lymphoid cells depleted in CD5^{bright} T cells or in CD25$^+$ activated T cells spontaneously induces EAT (64). In those experiments only the thyroid gland was the target of damage. In the same line, total lymphoid irradiation of mice caused various organ-specific autoimmune diseases, including thyroiditis (65), that could be transferred by CD4$^+$ T cells. In those experiments, irradiation seems to affect the T cells, rather than the target self-antigens, presumably by altering the regulatory T-cells that control self-reactivity.

More recently, through the use of T cell receptor (TcR) transgenic mice specific for hen egg lysozyme crossed with mice expressing this protein on thyroid epithelium, strong lymphocytic infiltrations have been observed in the thyroid gland from double transgenic mice, but in the absence of clinical symptoms of hypothyroidism (66). In these experiments, T cells that are tolerant to organ-specific antigens become activated when they encounter high localized expression of Ag. However, autoreactive CD4$^+$ T cells that initiate and promote thyroid inflammation are not able to induce thyroid destruction and clinical endocrine insufficiency. One can explain this situation by the lack of hen egg lysozyme specific CD8$^+$ T cells, required for progression from subclinical inflammation to tissue destruction. These animal models of autoimmunity represent useful tools for elucidating the mechanisms of immunological self-tolerance and pathologic mechanisms involved in autoimmune reactivity.

EAT has also been induced in MHC-susceptible mice through i.v. injection of syngeneic DC either pulsed *in vitro* with Tg or obtained from animals immunized with Tg two weeks earlier (67). Interestingly, the histopathological and serological manifestations of EAT are not similar in the two experiments. When animals have received *in vitro* Tg-pulsed DC, EAT reaches its acute phase in less than one-third of mice, two weeks after immunization, while AAbs to Tg are never significantly increased. In contrast, when mice received *in vivo* Tg-pulsed DC, EAT peaks four weeks after the injection in more than 50% of the recipients and AAbs are detectable in 80% of the recipients. Few studies have been devoted to the role of DC in thyroiditis. DC that are significantly increased in BUF rat thyroid when SAT occurs (39), are also found in cultured thyroid cells from normal porcine where they represent 2–3% (68). Do DC inside the thyroid gland behave similarly to the powerful APC from lymphoid organs? Do they present thyroid auto-Ag in an immunogenic or tolerogenic fashion, depending upon external factors such as inflammatory responses or stimulatory signals? Further studies are needed to answer this question.

A different approach has been designed by injecting Reovirus type 1 into Swiss or Balb/c mice. A mild focal thyroiditis, assessed by lymphocytic infiltration of the thyroid glands and AAbs to Tg and second colloid auto-Ag appears three weeks later (69). The role of viruses in thyroid autoimmune reactivity is supported by evidences; in the OS chicken, an endogenous virus, *ev22*, has been detected in Southern blot analysis of OS DNA digests (70), and its presence significantly correlates with the decrease in corticosterone (4). Furthermore, in cultures of human or murine TEC, it has been shown that *in vitro* cytomegalovirus infection (71) as well as Reovirus type 1 infection (72), respectively, induce MHC class II antigen expression on TEC in the absence of pre-existing lymphocyte infiltrates. If aberrant MHC class II expression is now considered as an epiphenomenon that plays a perpetuating rather than a causative role, it cannot be ruled out that virus infection can also induce chemokine or chemokine receptors (73) that would attract lymphocytes into the thyroid. Lastly, it cannot be omitted that viral determinants that mimic host antigens can trigger self-reactive T cells and destroy host tissue (74), a phenomenon demonstrated in murine keratitis induced by herpes simplex virus-type 1 (75). More recently, it was shown that murine adenovirus EB1, which shares >75% homology with the 18-mer Tg peptide 2695–2713 inducing EAT in H-2^s mice, was not able *per se* to induce EAT, but could stimulate T cells previously activated by the 18-mer Tg peptide. In this experiment, the viral peptide behaved as an agonist of pre-activated autoreactive T cells and through this pathway potentiated autoimmunity (76). Finally, viral components such as influenza virus hemaglutinin, can also behave as superantigens that activate B cells in a polyclonal mitogenic way, inducing Ab production that could crossreact with autologous determinants (77).

This brief review shows that induction of thyroiditis is an extremely complex phenomenon. It can be achieved using multiple auto-Ags, various adjuvants, Tg-specific T cells, depletion of regulatory T cells, auto-Ag-pulsed APC, viruses, genetically-defined mice, and probably many other approaches not yet known. This complexity, still amplified in genetically different patients, can explain the great heterogeneity of HT.

4.   T RECEPTOR USAGE IN EXPERIMENTAL MODELS OF THYROIDITIS

Biased or restricted TcR usage has been reported at sites of autoimmune damage in human and murine autoimmune diseases. In TIL from patients with thyroid autoimmune disorders, contrasting results have been obtained using PCR amplification. In a first study (78), restricted Vα, but not Vβ, TcR gene usage has been reported, whereas in a second series of experiments, no restriction has been observed (79). Those discrepant results remain unexplained.

In the spontaneous autoimmune thyroiditis that develops in OS chicken, preferential TcR Vβ1 gene usage has been evidenced in TIL (80). This result obtained after selective depletion of Vβ subset, by injection of mAb specific for Vβ1 or Vβ2 T cells in ovo, is strengthened by the subsequent inhibition of OS chicken SAT development. When only Vβ 1 T cells are depleted, decreases in AAbs to Tg and in thyroid gland lymphocytic infiltrations are observed. Thus, a key role in thyroid autoimmune reactivity of the OS chicken is here attributed to the Vβ 1 T cell subset. TcR Vβ gene usage has also been investigated by PCR amplification of TIL from spontaneous thyroiditis in NOD mice (81). Each NOD mouse shows a clonal expansion in TcR Vβ genes that differs from the others. In addition, CDR3 sequences are heterogenous within expanded Vβ families. The authors conclude that since thyroiditis does not occur uniformly in genetically defined NOD mice reared under similar environmental conditions, it may not be surprising that TCR Vβ genes differ between animals.

TcR Vβ gene usage has been studied in murine EAT induced by Tg/CFA, after transfer of Tg-specific T cells, depletion of particular T cell subsets, in gEAT, and in Tg-specific murine T cell lines. In Tg/CFA-induced EAT, intrathyroidal infiltration is associated with oligoclonally expanded Vβ1 and Vβ13 T cells, as well as in Tg-specific T cell lines (82). In contrast, in CD4$^+$ thyroid specific T cell lines obtained from diseased murine thyroid glands, five Vβ emerged: Vβ 2, 4, 8.3, 14 plus an unidentifed Vβ (83). Furthermore, TIL obtained after transfer of Tg-specific T cell lines exhibit a biased Vβ13 TcR usage bearing two recurrent CDR3 motifs (84). In contrast, multiple TcR Vβ gene usage has been reported in gEAT (85), on day 21 post transfer when maximal severity occurs. A more restricted response, with transcripts expressing Vβ4, Vβ11 and Vβ14, while other genes are undetectable, has been observed on day 11 post transfer when thyroid infiltrates are consistently present. Recently, transgenic mice have been selected in which most of the T cells express an irrelevant Ova-specific Vβ 8.2 TCR, in a k/q or k/k context (86). After immunization by mTg and LPS, significantly reduced AAbs and proliferative responses to mTg have been observed in both strains of transgenic mice. In contrast, histopathological lesions, assessed by the abundance of TIL, are reduced in k/q thyroid glands. Therefore, transgenic k/q mice express a less flexible Tg-specific repertoire than k/k transgenic mice.

Thus, TcR Vβ gene usage is clearly restricted in the OS chickens, and moderately restricted in murine EAT early after immunization. Later on, we can hypothesize that recruitement of bystander cells occurs, that dilute the specific response and results in a greater heterogeneity in TcR Vβ usage.

5. HUMORAL RESPONSE IN EXPERIMENTAL MODELS OF THYROIDITIS

In experimental models of EAT, the pathogenic role of anti-Tg AAbs is not convincingly established. It is not known whether AAbs are pathogenic for the thyroid gland or generated as a secondary event from released Tg after thyroid damage has occurred (87). In the OS chicken, it has been shown that anti-Tg AAbs play a pathogenic role in the spontaneous EAT of the OS B4B4, but do not cause thyroiditis when injected into OS B1B1 chickens, a strain genetically predisposed to EAT, or into normal Cornell strain birds (88). In iodine-induced SAT of the NOD.H-2^{h4} mice, only IgG2b AAbs correlate with the disease (13). In the BB/W rat, titers of thyroid AAbs are closely related to the initial and intermediate stages of EAT (25), but surprisingly animals with high levels of TIL have low titers of AAbs to Tg.

In murine EAT, no correlation has been observed between levels of circulating anti-Tg AAbs and the severity of EAT. Briefly, immunization of H-2^k and H-2^b mice by Tg/CFA induces in both strains of mice comparable levels of anti-Tg AAbs, whereas only H-2^k mice develop thyroid lymphocytic infiltration (12). That immune sera from animals with EAT fail to transfer EAT into syngeneic recipients is now admitted. However, the pathogenic role of anti-Tg AAbs has been revisited through repeated local perfusions of rabbit thyroids with sera from rabbits with thyroiditis containing high titers (> 1/10 000) of AAbs to thyroid AAgs (89). Two to four days following-in situ perfusion of thyroid gland by AAbs to Tg, destructive thyroiditis develops in perfused rabbits with deposits of IgG and C3 in the thyroid follicular basal laminae and infiltration by mononuclear cells and granulocytes. This last experiment demonstrates that the role of anti-Tg AAbs in thyroiditis remains an open issue. It can be hypothesized that the pathogenic AAbs are diluted among the abondant anti-Tg AAbs that recognize numerous non pathogenic epitopes on the large Tg molecule. A few pathogenic AAbs to Tg may netherthess succeed in reaching the thyroid gland and induce injury.

The isotypes of anti-Tg AAbs in sera from mice immunized by Tg/CFA are represented in the following order: IgG1 >IgG2a >IgG2b >IgG3. This hierarchy is also observed when EAT is induced by the 16-mer rat peptide 2495–2511 (90). In contrast, the IgG1/IgG2a ratio is reversed in the Tc1-mediated hdpTg-induced EAT (49), whereas in chronic EAT of NOD. In H-2^k mice, we have observed that anti-Tg AAbs of the IgG2a subclass are decreased in correlation with a detect in IFN-γ producing T cells (10). Using panels of mAbs to Tg produced in mice immunized by porcine or mTg and adjuvant (91,92), it has been shown that most of them recognize conformational epitopes of murine or rat Tg only, whereas a minority bind Tg from other species. Furthermore, mAbs that bind to CNBr-denatured mTg belong to a polyreactive group of mAbs.

The idiotypes borne by anti-Tg AAbs and the determination of their VH family further enable their characterization. In both young and aged BUF rats with thyroiditis, one cross reactive idiotype, id 62, has been evidenced in 20–50% of the AAbs to Tg (93), but also in anti-Tg AAbs from other species including humans. More surprisingly, id 62, which is expressed on both heavy and light chains of anti-Tg mAb, exhibits regulatory functions, since its repeated injection into irradiated BUF rats with EAT significantly reduces levels of AAbs to Tg. One mAb to Tg, 3B8G9 (91), which recognizes the F40D peptide inducing EAT (46), has been used to determine the levels of anti-id AAbs in sera from mice with EAT at various times of the disease. Only the levels of anti-id AAbs that bind to F(ab)$_2$ from 3B8G9, but not to other F(ab)$_2$ from anti-Tg AAbs, parallels EAT activity (94).

Studies of Ig heavy chain variable region genes encoding anti-Tg mAbs have shown that three VH families are primarily utilized during EAT: VH J558, the largest family, VH Q52 and VH 7183, the families the most proximal joining region of the heavy chain genes (95,96). However, in those studies mAbs to Tg have no proved relationship to pathogenicity. The characterization of one EAT-inducing peptide, F40D (46) has allowed us to study the VH usage in splenic B cells after immunization of mice by either F40D, or porcine Tg tryptic fragments inducing EAT or mTg (97). We have observed that immune B cells use VH families related to the size of the immunizing antigen: the larger the antigen, the higher the numbers of VH families used. Moreover, B cell stimulation following immunization with F40D, occurs in VH Q52 family, a VH encoded by D-proximal segment, as previously reported for AAbs in humans (98). Those results have been confirmed by sequences of VH regions of eleven mAbs specific for mTg (96,99). Eight belong to the J558 VH family, two to VH Q52 and one to VH 10. The two VH Q52 mAbs to Tg that have been sequenced separately include 3B8G9, the mAb specific for the F40D peptide inducing EAT (46). They came from two different mice suffering of EAT and show identical amino-acid sequences.

Thus in SAT, AAbs to Tg are mostly related to thyroiditis severity. In contrast, in Tg-induced EAT, AAbs to Tg that mainly recognize conformational epitopes do not correlate to EAT severity. However, some experiments, such as determination of levels of anti-id. AAbs in sera from mice suffering of EAT and sequencing of AAbs to Tg, suggest that pathogenic AAbs to Tg are generated after Tg immunization. Because their small amount in immune sera such AAbs are difficult to detect.

6. AUTOIMMUNE THYROIDITIS AND CYTOKINES

Cytokines are proteins responsible for intercellular communications and cell homeostasis via a complex communication network. It is commonly accepted that Th2 cytokines tend to cure the autoimmune disease and hasten its remission, whereas Th1 cytokines promote disease development (100). In the last few years, the regulatory role of cytokines synthesized by T-lymphocytes and macrophages and the distinction between Th0 (releasing Th1 and Th2 cytokines), Th1 (releasing IL-2, IFN-γ and IL-12), Th2 (releasing IL-4 and IL-10) and Th3 (releasing TGF-β) (50) have improved our understanding of autoimmune responses. Moreover, the development of animals with disrupted genes, the expression of genes in target tissues, the transfection of cells by genes of interest, further contribute to the comprehension of thyroid autoimmunity.

The role of TH1 cytokines has been evidenced in various experimental models of thyroid autoimmunity. In the OS chicken, hyperreactivity of infiltrating T lymphocytes from diseased thyroid gland is attested to by the high amounts of IL-2 receptors on their surfaces and by high levels of IL-2 in sera (101). Immunohistochemical studies, using mAb to the chicken IL-2 receptor reveals that T lymphoblasts are among the first cells invading the thyroid gland. Since IL-2 can disrupt immunological self-tolerance in anergic T cells, this increased level in IL-2 could be responsible for the T cell OS chicken hyperreactivity towards thyroid antigens. In the BB rat, which develops both type 1 diabetes and thyroiditis, TH1 transcripts coding for IFN-γ and IL-12 p40 in the target organs of autoimmune reactivity are increased (102) and probably involved in inflammatory and destructive processes. Altogether, TH1 cytokines favor SAT development in the various models studied.

Another Th1 cytokine, IFN-γ has been extensively studied. IFN-γ induces aberrant expression of MHC class II antigens in cultures of thyrocytes from humans and animals (103,104). In vitro experiments show that TEC expressing MHC class II determinants can present peptidic antigen and stimulate T cells (105,106), but are unable to process antigen. MHC class II antigen expression on TEC is induced by many factors such as IFN-γ (104), PHA (107), TSH, TNF-α (108), and viruses (69,70), substances that synergyze when simultaneously added onto TEC cultures. This Ag-presenting function of class II$^+$ TEC is a peculiarity among the various targets of autoimmune diseases.

In mice, intrahyroidal injections with recombinant IFN-γ induces transient and mild EAT in 30% of animals (109). The role of IFN-γ in the pathogenesis of EAT is also supported by the fact that injection of neutralizing mAb to IFN-γ, from the moment of immunization with Tg/CFA to mouse sacrifice, significantly abrogates thyroid lymphocytic infiltration as well as the levels of circulating AAbs to Tg (110). Moreover, a reduction in the percentage of Tg-specific $CD8^+$ T cells is observed in splenic and lymph node T cells from IFNγ-injected mice. In contrast, in gEAT, administration of mAb to IFN-γ simultaneously to Tg-activated T cells, induces an exacerbated disease (58). A protective, rather than an inducing role for Tg-specific Th1 T is suggested for IFN-γ in gEAT.

Recently, the role of IFN-γ in thyroid autoimmune response has been revisited in EAT induced by Tg/CFA and in gEAT using H-2^k or H-2^q mice with a knock-out IFN-γ receptor (IFN-$\gamma R^{-/-}$) or IFN-γ gene (IFN-$\gamma^{-/-}$), respectively. Compared to immunized control mice, Tg/CFA-induced EAT in IFN-$\gamma R^{-/-}$ H-2^k mice is attenuated, occurs earlier and reveals a Th2-dominated humoral Tg-specific response, whereas the Tg-specific cellular responses are typically Th1 (111). We have explained this complex situation by the unresponsiveness of B cells to IFN-γ in the mutant mice. In IFN-$\gamma^{-/-}$ H-2^q mice, lymphoid cells such as those from control mice induce gEAT when stimulated *in vivo* and *in vitro* by mTg and mAb to IL-2R (112). In this gEAT, thyroid cell infiltrates consist essentially of eosinophils with a decrease in Th1 cytokine transcripts in the thyroid tissue, whereas those coding for Th2 cytokines are not modified. Moreover, IFN-$\gamma^{-/-}$ H-2^q mice produce lower levels of circulating AAbs to Tg after immunization by mTg and LPS. Those results suggest that both IFN-$\gamma R^{-/-}$ or IFN-$\gamma^{-/-}$ mice can develop EAT through the use of different mediators of inflammatory responses.

The role of IL-12 has also been studied in Tg/CFA-induced EAT and transfer to Tg-primed T cells (113). When IL-12 is given to mice simultaneously with Tg/LPS, it increases the severity of EAT and the AAb responses. The same results are obtained when *in vivo* Tg-activated T cells, are cultured *in vitro* with IL-12 before transfer into syngeneic recipients. Moreover, it has been shown that administration of neutralizing anti-IL12 mAb reduces the severity of thyroid lymphocytic infiltration and the production of AAbs to Tg in Tg/CFA immunized mice. In the same line, immunization by Tg/CFA of mice with a disrupted IL-12 gene abolishes EAT occurrence, whereas AAbs to Tg are slightly increased in each isotype. Finally, IL-12 has been injected into normal mice five times a week, from the commencement of Tg/CFA immunization until animal sacrifice Thyroid lymphocytic infiltrations and the productions of AAbs to Tg are abolished. Lastly, in murine gEAT (58), IL-12 further promotes the activation of effector cells responsible for destruction of thyroid tissues (114). Those experiments suggest that IL-12 plays a pivotal role in the severity of EAT, acting both on thyroid lymphocytic infiltration and production of AAbs to Tg. Furthermore, as already observed in SAT, TH1 cytokines actively participate to the development of thyroid autoimmune reactivity, independent of the mode of EAT induction used.

The role of Th2 cytokines has also been studied in EAT, given their potential immunosuppressive activity. We have demonstrated that addition of IL-10 to T cell stimulated *in vivo* and *in vitro* with mTg decreases the Tg-specific proliferative and cytotoxic T cell responses (115). In EAT induced either by mTg/CFA or by adoptive transfer of mTg-specific T lymphocytes, the severity of the diseases are significantly reduced by injection of IL-10, either at the time of priming and challenge with mTg or at the time of T lymphocyte transfer. These effects, dependent upon the amount of IL-10 injected, have been demonstrated to be the consequence of enhanced activation-induced cell death in T lymphocytes.

We have also demonstrated that cultures of *in vivo* primed and *in vitro* re-activated T cells in the presence of mTg and IL-4 over three days induces a decrease in the mTg-specific proliferative response and an increase in the cytotoxic T cell response. Furthermore, EAT induced by transfer of mTg-activated spleen cells cultured in the presence of IL-4 is not significantly decreased (116). We explain this unexpected lack of effect of IL-4 by the too short duration of *in vitro* culture with IL-4 and by the fact that IL-4 is less efficient on activated T cells than on naive T cells. More recently, it has been shown that mice with or without a disrupted IL-4 gene develop a gEAT when immunized by mTg and LPS, or when lymphocytes are cultured with mTg and mAb to IL-2 receptor before transfer into normal or IL-$4^{-/-}$ mice. However, an abolition of the IgG1 AAb response to mTg is observed (117). Productions of IFN-γ, IL-2, IL-4, IL-5, IL-10, IL-13 and TNF-α, in lymphoid cells from spleens or thyroids from IL-$4^{-/-}$ recipients and control animals were comparable, demonstrating that, as in EAT, the presence of IL-4 is not crucial to the induction of gEAT.

Few studies have dealt with the role of inflammatory cytokines in experimental models of thyroiditis. IL-1β has been shown to inhibit iodine uptake when added to cultures of TEC stimulated by TSH (118). Similarly, injection of BB rats with IL-β induces decrease in thyroid hormones, and increase in TSH, accelerating the hypothyroidism of the BB rat (119). Recently, it has been suggested that IL-1β produced by thyroid cells from patients with HT induces Fas expression by normal thyrocytes on which Fas ligand (FasL) is said to be constituvely expressed, and apoptosis of TEC ensues (120). The constitute expression of FasL on normal and pathologic tissues is now clearly requestioned (121). In our laboratory, three lines of mice expressing

FasL in various amounts on their thyroid glands have been produced (122). In those animals, FasL expression prevents the development of EAT in a dose-dependent manner, the best protection being observed in animals with the highest amounts of FasL. Simultaneously, significant decreases in Tg-specific proliferative, cytotoxic and AAb responses are observed. Furthermore, a Th2 immune response is evidenced by the abolition of IgG2a anti-Tg AAb response and IFN-γ production after Tg stimulation. The same inability of mice expressing FasL on their thyrocytes to develop EAT is observed in tansfer-EAT.

The role of another inflammatory cytokine, TNF-α, has also been studied. In thyroids from BB rats treated by iodine for at least two months, transcripts coding for Tg and TPO are decreased whereas those coding for TNF-α are increased (123). These results are in agreement with the *in vitro* studies demonstrating that TNF-α inhibits thyroid cell function. This regulatory function of the thyroid gland by TNF-α, still augmented in the presence of IFN-γ (124), is mediated by receptors located on normal TEC. One can expect that such cytokines could be responsible for the beneficial effect of decreased anti-Tg or anti-TPO AAbs, but also for the adverse effect of induction of MHC class II antigens on TEC, facilitating an efficient immunogenic presentation of AAgs to helper T cells. The final choice resulting of external factors: others cytokines, chemokines or chemokine receptors, and many unknown others.

In the same line, MHC class II antigen expression on TEC and the subsequent T cell stimulation are also involved in cytokine regulation. TEC in which high levels of MHC class II antigens have been induced express low levels of costimulatory molecules B7.1 and B7.2 (125). Consequently, IL-2 production that depends upon B7 for co-stimulation is inhibited, whereas IL-4 release, which is B.7-independent, is not affected. When the same experiments are performed with Th0, B7 independent T cells, antigenic presentation by TEC-expressing MHC class II antigens favors IL-4 production. Thus, antigen presentation by MHC class II antigen expressed by TEC may induce tolerance in autoreactive Th1 cells or favor a Th2 response in naive T lymphocytes. In gEAT, opposing roles for B7.2 have been shown (126). Blocking B7.2 costimulation before effector cells are fully active suppresses the ability of mTg-primed spleen cells to induce gEAT through inhibition of proliferative response to mTg-primed and IL-2 production. However, during the development of thyroid lesions in recipients animals, inhibition of B7.2 increases EAT severity Analysis of cytokine transcripts in the thyroid glands of animals given B7.2 shows decreased expression of IL-2, IL-4, IL-13 and IFN-γ, compared to transcripts in thyroids from control animals. Those results suggests that costimulation is absolutely required during the effector phase of EAT induction. However B7.2 may have opposing roles in the activation versus the effector phase of EAT.

Obviously the role of cytokines on thyroid function is highly complex. The time of their administration, their amounts, the presence of other synergistic or antagonistics cytokines, the state of activation of the target on which they act, and many others unknown factors, render their use extremely delicate. As already suggested (99), TH1 cytokines generally induce or worsen disease In both SAT and EAT, whereas TH2 cytokines ameliorate the disease.

7. PREVENTIVE AND CURATIVE TREATMENTS OF THYROIDITIS

Three main approaches, thyroid-specific or not, have been used to prevent or cure thyroiditis: Suppression, vaccination and tolerance (34). These last years, immunosuppressive drugs such as FK 506 (127), given during three weeks in NTx, irradiated rats significantly reduced CD4$^+$ and CD8$^+$ TIL, inducing an amelioration of EAT. In line with these results, injection of a non-depleting anti-CD4 mAb before immunization of mice by deaggregated mTg and LPS (128), induces a strong inhibition of EAT development. Furthermore, this suppression of EAT by a non-depleting anti-CD4 mAb is transferable by T cells from anti-CD4 injected, Tg/CFA immunized mice (129). Inhibition of EAT development can also be achieved by injection of mTg-specific inactivated T cells (130) or line (131) before immunization of mice. These experiments, that demonstrate the existence of CD4$^+$ regulatory T cells, have also been conducted with CD8$^+$ Tg-specific T cells in EAT (18,130–132) and in gEAT (19). Thus, we took advantage of the cloning of a CD8$^+$, class-I restricted cytotoxic hybridoma, HTC2, (17) specific for the F40D pathogenic peptide of human Tg inducing EAT (46), to demonstrate that injection into normal mice of such inactivated HTC2 cells three weeks before immunization by Tg/CFA, inhibits EAT development (18). Since this inhibition could occur through the production of anti-HTC2 TcR Abs, we produced an anti-clonotypic mAb that recognizes specifically the HTC2 TcR. When this anti-clonotypic mAb was injected one day prior each immunization of mice by Tg/CFA, it fully abolished the thyroid lymphocytic infiltrations of recipients, but was inefficient on the level of circulating AAbs (133). Through those experiments we demonstrated the existence of a regulatory idiotypic network in Tg-induced EAT, mediated by CD8$^+$ T cells and its anti-TcR AAbs (22,32).

Tg administrated orally or i.v. after deaggregation has also been used to induce Ag-specific tolerance and inhibit EAT development (134). When mice are fed orally with Tg, an inhibition of the development of EAT induced by Tg/CFA is observed, accompanied by strong inhibitions of

Tg-specific proliferative response and anti-Tg AAb production. Decreased AAbs to Tg after oral administration of Tg reaches each isotype. Induction of specific tolerance has also been achieved in EAT induced by Tg and LPS and in gEAT (135), using i.v. administration of deaggregated Tg seven and five days before and after immunization, respectively. In both EAT, thyroid infiltrations by lymphocytes are significantly inhibited, as well as the levels of circulating AAbs to Tg and the Tg-specific T cell proliferative response. In spleen cells from mice immunized with deaggregated Tg, IL-2, IFN-γ, IL-4, IL-10, TNF-α and TGF-β transcripts are decreased. This cytokine profile demonstrates that Th1 as well as Th2 cells can be tolerized under particular experimental conditions.

In addition to those successful experiments of EAT prevention, curative attempts have recently been set up. In a first series of experiments, we used the well-known suppressive properties, of IL-10 (136) to inject IL-10 when EAT is already established, i.e. when AAbs to Tg and Tg specific proliferative and cytotoxic responses are detectable. We have observed that EAT is significantly reduced in mice that received high doses of IL-10, whereas levels of anti-Tg AAbs are not affected (115). However, analysis of T cells from spleen and lymph nodes from cured animals show an increase in cells under apoptosis, dependent in the amount of rIL-10 administered. We have confirmed the curative effects of IL-10 in EAT using gene therapy in mTg/CFA-induced EAT (137). IL-10 has been expressed on TEC after DNA deposition on day 21 p.i., when EAT is already established. We have observed a total absence of EAT in treated mice, assessed by a lack of Tg-specific proliferative response and by an absence of lymphocytes infiltrating thyroid tissues. Moreover, a trend towards a TH2 response is evidenced, characterized by decreased production of IFN-γ and an increase of the IgG1/IgG2a ratio.

In a second series of experiments, we deleted Fas-positive (138) Tg-activated lymphocytes in Tg/CFA immunized mice, again when EAT is established. To perform this deletion, expression of FasL was achieved specifically on immune TEC by gene transfer (139). We have observed that in vivo administration of plasmid coding for FasL on thyroid tissue in mice with EAT reduces the lymphocytic infiltration, abrogates the anti-Tg cytotoxic T cell response and induces a selective persistence of IgG1 anti-Tg AAbs.

8. CONCLUSIONS

Experimental models of autoimmune thyroiditis widely contribute to the knowledge of their counterpart human diseases: Hashimoto's thyroiditis and Graves' disease. In recent years, the numerous studies devoted to T and B cell repertoires have clearly demonstrated that each animal suffering spontaneous- or induced-autoimmune thyroiditis develops its own T or B cell repertoire. This individual heterogeneity in the development of the immune response during experimental autoimmune thyroiditis surely reflects the multiple pathways of thyroiditis induction.

During the last decade, invaluable progress has been made in understanding the role of cytokines in thyroiditis induction, progression, perpetuation or remission. While Th1 cytokines favor disease development, Th2 cytokines favor thyroiditis improvement and remission. This general scheme which is also observed in other autoimmune diseases, has allowed strategies of prevention and, more interestingly, the use of Th2 cytokines in curative treatments.

Those experiments represent a real therapeutical progress in thyroiditis. However, their clinical use in human thyroiditis is still hypothetical. A better and more complete knowledge of intimal mechanisms involved in thyroiditis onset and development is necessary to avoid hazardous clinical attempts. Progress in our knowledge of thyroiditis pathophysiology will also benefit to the development of novel therapeutical approaches in other autoimmune diseases.

ACKNOWLEDGMENTS

We thank Dr. Bernard WEILL (Hôpital Cochin, Paris) and Dr. Claude CARNAUD (Hôpital Necker, Paris) for the critical review of this manuscript and for the numerous helpful discussions.

References

1. Nepom, G.T. 1989. Determinants of genetic susceptibility in HLA-associated autoimmune disease. *Clin. Immunol. Immunopathol.* 53:S53–S62.
2. Wick, G., H.P. Brezinschek, K. Hàla, H. Dietrich, H. Wolf, and G. Kroemer. 1990. The obese strain of chickens: an animal model with spontaneous autoimmune thyroiditis. *Adv. Immunol* 47:433–500.
3. Cihak, J., G. Hoffmann-Fezer, M. Wasl, H. Merkle, B. Kaspers, O. Vainio, J. Plachy, K. Hàla, G. Wick, M. Strangassinger, and U. Losch. 1998. Inhibition in the development of spontaneous autoimmune thyroiditis in the obese strain (OS) chickens by *in vivo* treatment with anti-CD4 or anti-CD8 antibodies. *J. Autoimmun.* 11:119–126.
4. Schauenstein, K., R. Fässler, H. Dietrich, S. Schwarz, G. Krömer, and G. Wick. 1987. Disturbed immune-endocrine communication in autoimmune disease. Lack of corticosterone response to immune signals in obese strain chickens with spontaneous autoimmune thyroiditis. *J. Immunol.* 139:1830–1833.
5. Bagchi, N., T.R. Brown, E. Urdanivia, and R.S. Sundick. 1985. Induction of autoimmune thyroiditis in chickens by dietary iodine. *Science* 230:325–327.

6. Brown, T.R., R.S. Sundick, A. Dhar, D. Sheth, and N. Bagchi. 1991. Uptake and metabolism of iodine is crucial for the development of thyroiditis in obese strain chickens. *J. Clin Invest.* 88:106–111.

7. Bagchi, N., R.S. Sundick, L.H. Hu, G.D. Cummings, and T.R. Brown. 1996. Distinct regions of thyroglobulin control the proliferation and suppression of thyroid-specific lymphocytes in obese strain chickens. *Endocrinology* 137:3286–3290.

8. Bagchi, N., T.R. Brown, and R.S. Sundick. 1995. Thyroid cell injury is an initial event in the induction of autoimmune thyroiditis by iodine in obese strain chickens. *Endocrinology* 136:5054–5060.

9. Makino, S., K. Kunimoto, Y. Muraoka, Y. Mizushima, K. Katagiri, and Y. Tochino. 1980. Breeding of a non obese, diabetic strain of mice. *Jikken Dobutsu* 29:1–13.

10. Bernard, N.F., F. Ertug, and H. Margolese. 1992. High incidence of thyroiditis and anti-thyroid autoantibodies in NOD mice. *Diabetes* 41:40–46.

11. Acha-Orbea, H., and H.O. McDevitt. 1987. The first external domain of the non obese diabetic mouse class II I-A beta chain is unique. *Proc. Natl. Acad. Sci. USA* 84:2435–2439.

12. Vladutiu, A.O., and N.R. Rose. 1971. Autoimmune murine thyroiditis. Relation to histocompatibility (H-2) type. *Science* 174:1137–1139.

13. Rasooly, L., C.L. Burek, and N.R. Rose. 1996. Iodine-induced autoimmune thyroiditis in NOD-H-2^{h4} mice. *Clin. Immunol. Immunopathol.* 81:287–292.

14. Many, M.C., S. Maniratunga, I. Varis, M. Dardenne, H.A. Drexhage, and J.F. Denef. 1995. Two-step development of Hashimoto-like thyroiditis in genetically autoimmune prone non-obese diabetic mice: effects of iodine-induced cell necrosis. *J. Endocrinol.* 147:311–320.

15. Damotte D., E. Colomb, C. Cailleau, N. Brousse, J. Charreire, and C. Carnaud. 1997. Analysis of susceptibility of NOD mice to spontaneous and experimentally induced thyroiditis. *Eur. J. Immunol.* 27:2854–2862.

16. Cohen, I.R. 1986. Regulation of autoimmune disease. Physiological and therapeutic. *Immunol. Rev.* 94:5–21.

17. Remy, J.J., B. Texier, G. Chiocchia, and J. Charreire. 1989. Characteristics of cytotoxic thyroglobulin-specific T cell hybridoma. *J. Immunol.* 142:1129–1133.

18. Roubaty, C., C. Bédin, and J. Charreire. 1990. Prevention of experimental autoimmune thyroiditis through the anti-idiotypic network. *J. Immunol.* 144:2167–2172.

19. Braley-Mullen, H., R.W. McMurray, G.C. Sharp, and M. Kyriakos. 1994. Regulation of the induction and resolution of granulomatous experimental autoimmune thyroiditis in mice by CD8+ T cells. *Cell. Immunol.* 153:492–504.

20. Kumar, V., E. Coulsell, B. Ober, G. Hubbard, E. Sercarz, and E.S. Ward. 1997. Recombinant T cell receptor molecules can prevent and reverse experimental autoimmune encephalomyelitis. Dose effect and involvement of both CD4 and CD8 T cells *J. Immunol.* 159:5150–5156.

21. Damotte, D., N. Brousse, J. Charreire, and C. Carnaud. MHC Class I and CD8$^+$ T cells are required for the initiation and resolution of experimental autoimmune thyroiditis in NOD mice (*Submitted for publication*).

22. Lomo, L.C., F. Zhang, D.J. McCormick, A.A. Giraldo, C.S. David, and Y.C. Kong. 1998. Flexibility of the thyroiditogenic T cell repertoire for murine autoimmune thyroiditis in CD8-deficient (beta-2m-/-) and T cell receptor Vβ-congenic mice. *Autoimmunity* 27:127–133.

23. Green, L.M., M. LaBue, J.P. Lazarus, and K.K. Colburn. 1995. Characterization of autoimmune thyroiditis in MRL-lpr/lpr mice. *Lupus* 4:187–196.

24. Green, L.M., J.P. Lazarus, M. LaBue, and M.M. Shah. 1995. Reduced cell-cell communication in a spontaneous murine model of autoimmune thyroid disease. *Endocrinology* 136:3611–3618.

25. Noble, B., T. Yoshida, N.R. Rose, and P.E. Bigazzi. 1976. Thyroid antibodies in spontaneous autoimmune thyroiditis in the Buffalo rat. *J. Immunol.* 117:1447–1455.

26. Allen, E.M., and L.E. Braverman. 1990. The effect of iodine on lymphocytic thyroiditis in the thymectomized Buffalo rat. *Endocrinology* 127:1613–1616.

27. Li, M., and S.C. Boyages. 1994. Iodide induced lymphocytic thyroiditis in the BB/W rat: evidence of direct toxic effects of iodide on thyroid subcellular structure. *Autoimmunity* 18:31–40.

28. Sternthal, E., A.A. Like, K. Sarantis, and L.E. Braverman. 1981. Lymphocytic thyroiditis and diabetes in the BB/W rat. A new model of autoimmune endocrinopathy. *Diabetes* 30:1058–1061.

29. Voorby, H.A.M., J.P. Kabel, M. De Haan, P.H.M. Jeucken, R.D. van der Gaag, M.H. de Baets, and H.A. Drexhage. 1990. Dendritic cells and class II MHC expression on thyrocytes during the autoimmune thyroid disease of the BB rat. *Clin. Immunol. Immunopathol.* 55:9–22.

30. Rajatanavin, R., M.C. Appel, W. Reinhardt, S. Alex, Y.N. Yang, and L.E. Braverman. 1991. Variable prevalence of lymphocytic thyroiditis among diabetes-prone sublines of BB/W or rats. *Endocrinology* 128:153–157.

31. Pettersson, A., D. Wilson, T. Daniels, S. Tobin, H.J. Jacob, E.S. Lander, and A. Lernmark. 1995. Thyroiditis in the BB rat is associated with lymphopenia but occurs independently of diabetes. *J. Autoimmunity* 8:493–505.

32. Awata, T., D.L. Guberski, and A.A. Like. 1995. Genetics of the BB rat: association of autoimmune disorders (diabetes, insulitis and thyroiditis) with lymphopenia and major histocompatibility complex class II. *Endocrinology* 136:5731–5735.

33. Weigle, W.O. 1980. Analysis of autoimmunity through experimental models of thyroiditis and allergic encephalomyelitis. *Adv. Immunol.* 30:159–273.

34. Charreire, J. 1989. Immune mechanisms in autoimmune thyroiditis. *Adv. Immunol.* 46:263–334.

35. Conaway, D.H., A.A. Giraldo, C.S. David, and Y.M. Kong. 1989. *In situ* kinetic analysis of thyroid lymphocyte infiltrate in mice developing experimental autoimmune thyroiditis. *Clin Immunol. Immunopathol.* 53:346–353.

36. Beisel, K.W., C.S. David, A.A. Giraldo, Y.C. Kong, N.R. Rose. 1982. Regulation of experimental autoimmune thyroiditis: Mapping of susceptibility to the 1-A subregion of the mouse H-2. *Immunogenetics* 15:427–430.

37. Romball, C.G., and W.O. Weigle. 1987. The effect of aging on the induction of experimental autoimmune thyroiditis. *J. Immunol.* 139:1490–1495.

38. Stull, S.J., M. Kyriakos, G.C. Sharp, and H. Braley-Mullen. 1988. Prevention and reversal of experimental autoimmune thyroiditis (EAT) in mice by administration of anti-L3T4 monoclonal antibody at different stages of disease development. *Cell. Immunol.* 117:188–198.

39. Creemers, P., N.R. Rose, and Y.C. Kong. 1983. Experimental autoimmune thyroiditis. *In vitro* cytotoxic effects of T lymphocytes on thyroid monolayers. *J. Exp. Med.* 157:559–571.

40. Salamero, J., J. Charreire. 1985. Syngeneic sensitization of mouse lymphocytes on monolayers of thyroid epithelial cells. VII. Generation of thyroid-specific cytotoxic effector cells. *Cell. Immunol.* 91:111–118.

41. Malthiery, Y., and S. Lissitzky. 1987. Primary structure of human thyroglobulin deduced from the sequence of its 8848-base complementary DNA. *Eur. J. Biochem.* 165:491–498.

42. Carayanniotis, G., and V.P. Rao. 1997. Searching for pathogenic epitopes in thyroglobulin: parameters and caveats. *Immunol. Today* 18:83–88.

43. Champion, B.R., K.R. Page, N. Parish, D.C. Rayner, K. Dawe, G. Biswas-Hughes, A. Cooke, M. Geysen, and I.M. Roitt. 1991. Identification of a thyroxine-containing self-epitope of thyroglobulin which triggers thyroid autoreactive T cells. *J. Exp. Med.* 174:363–370.

44. Kong, Y.C., D.J. McCormick, Q. Wan, R.W. Motte, B.E. Fuller, A.A. Giraldo, and C.S. David. 1995. Primary hormonogenic sites as conserved autoepitopes on thyroglobulin in murine autoimmune thyroiditis. Secondary role of iodination. *J. Immunol.* 155:5847–5854.

45. Saboori, A.M., N.R. Rose, H.S. Bresler, M. Vladut-Talor, and C.L. Burek. 1998. Iodination of human thyroglobulin (Tg) alters its immunoreactivity. I. Iodination alters multiple epitopes of human Tg. *Clin. Exp. Immunol.* 113:297–302.

46. Texier, B., C. Bedin, H. Tang, L. Camoin, C. Laurent-Winter, and J. Charreire. 1992. Characterization and sequencing of a 40-amino-acid peptide from human thyroglobulin inducing experimental autoimmune thyroiditis. *J. Immunol.* 148:3405–3411.

47. Chronopoulou, E., and G. Carayanniotis. 1992. Identification of a thyroiditogenic sequence within the thyroglobulin molecule. *J. Immunol.* 149:1039–1044.

48. Carayanniotis, G., E. Chronopoulou, and V.P. Rao. 1994. Distinct genetic pattern of mouse susceptibility to thyroiditis induced by a novel thyroglobulin peptide. *Immunogenetics* 39:21–28.

49. Brazillet, M.P., F. Ratteux, O. Abehsira-Amar, F. Nicoletti, and J. Charreire. 1999. Induction of experimental autoimmune thyroiditis by heat denatured porcine thyroglobulin: a Te1-mediated disease. *Eur. J. Immunol.* 29.1342–1352.

50. Mosmann, T.R., and S. Sad. 1996. The expanding universe of T-cell subsets: Th1, Th2 and more *Immunol. Today* 17:138–146.

51. Schirmbeck, R., W. Bohm, and J. Reimann. 1994. Injection of detergent-denatured ovalbumin primes murine class I-restricted cytotoxic T cells *in vivo*. *Eur. J. Immunol.* 24:2068–2072.

52. Speidel, K., W. Osen, S. Faath, I. Hilgert, R. Obst, J. Braspenning, F. Momburg, G.J. Hämmerling, and H.G. Rammensee. 1997. Priming of cytotoxic T lymphocytes by five heat-aggregated antigens *in vivo*: conditions, efficiency, and relation to antibody responses. *Eur. J. Immunol.* 27:2391–2399.

53. Braley-Mullen, H., M. Johnson, G.C. Sharp, and M. Kyriakos. 1985. Induction of experimental autoimmune thyroiditis in mice with *in vitro* activated splenic T cells. *Cell. Immunol.* 93:132–143.

54. Romball, C.G., and W.O. Weigle. 1987. Transfer of experimental autoimmune thyroiditis with T cell clones. *J. Immunol.* 138:1092–1098.

55. Williams, W.V., M. Kyriakos, G.C. Sharp, and H. Braley-Mullen. 1987. Augmentation of transfer experimental autoimmune thyroiditis (EAT) in mice by irradiation of recipients. *Cell. Immunol.* 109:397–406.

56. Conaway, D.H., A.A. Giraldo, C.S. David, and Y.M. Kong. 1990. *In situ* analysis of T cell subset composition in experimental autoimmune thyroiditis after adoptive transfer of activated spleen cells. *Cell. Immunol.* 125:247–253.

57. Braley-Mullen, H., G.C. Sharp, J.T. Bickel, and M. Kyriakos. 1991. Induction of severe granulomatous experimental autoimmune thyroiditis in mice by effector cells activated in the presence of anti-interleukin 2 receptor antibody. *J. Exp. Med.* 173:899–912.

58. Stull, S.J., G.C. Sharp, M. Kyriakos, J.T. Bickel, and H. Braley-Mullen. 1992. Induction of granulomatous experimental autoimmune thyroiditis in mice with *in vitro* activated effector T cells and anti-IFN-γ antibody. *J. Immunol.* 149:2219–2226.

59. Kotani, T., K. Umeki, K. Hirai, and S. Ohtaki. 1990. Experimental murine thyroiditis induced by porcine thyroid peroxidase and its transfer by the antigen specific T cell line. *Clin. Exp. Immunol.* 80:11–18.

60. Kotani, T., K. Umeki, S. Yagihashi, K. Hirai, and S. Ohtaki. 1992. Identification of thyroiditogenic epitope on porcine peroxidase for C57BL/6 mice. *J. Immunol.* 148:2084–2089.

61. Penhale, W.J., A. Farmer, and W.J. Irvine. 1975. Thyroiditis in T cell-depleted rats: influence of strain, radiation dose, adjuvants, and anti-lymphocyte serum. *Clin. Exp. Immunol.* 21:362–375.

62. Seddon, B., and D. Mason. 1999. Peripheral autoantigen induces regulatory T cells that prevent autoimmunity. *J. Exp. Med.* 189:877–882.

63. Sugihara, S., Y. Izumi, T. Yoshioka, H. Yagi, T. Tsujimura, O. Tarutani, Y. Kohno, S. Murakami, T. Hamaoka, and H. Fujiwara. 1988. Autoimmune thyroiditis in mice depleted of particular T cell subsets. I. Requirement of Lyt-1^{dull} L3T4^{bright} normal T cells for the induction of thyroiditis. *J. Immunol.* 141:105–113.

64. Sugihara, S., H. Fujiwara, H. Niimi, and G.M. Shearer. 1995. Self-thyroid epithelial cell (TEC)-reactive CD8+ T cell lines/clones derived from autoimmune thyroiditis lesions. They recognize self-thyroid antigens directly on TEC to exhibit T helper cell I-type lymphokine production and cytotoxicity against TEC. *J. Immunol.* 155:1619–1628.

65. Sakaguchi, N., K. Miyai, and S. Sakaguchi. 1994. Ionizing radiation and autoimmunity. Induction of autoimmune disease in mice by high dose fractionated total lymphoid irradiation and its prevention by inoculating normal T cells. *J. Immunol.* 152:2586–2595.

66. Akkaraju, S., W.Y. Ho, D. Leong, K. Canaan, M.M. Davis, and C.C. Goodnow. 1997. A range of CD4 T cell tolerance: Partial inactivation to organ-specific antigen allows non destructive thyroiditis or insulitis. *Immunity* 7:255–271.

67. Knight, S.C., J. Farrant, J. Chan, A. Bryant, P.A. Bedford, and C. Bateman. 1988. Induction of autoimmunity with dendritic cells: Studies on Thyroiditis in mice. *Clin. Immunol. Immunopathol.* 48:277–289.

68. Croizet, K., R. Rabilloud, Z. Kostrouch, J.F. Nicolas, and B. Rousset. 2000. Culture of dendritic cells from non-lymphoid organ, the thyroid gland: Evidence for TNFα-dependent phenotypic change of thyroid-derived cells. *Lab. Invest.* 80:215–1225.

69. Srinivasappa, J., C. Garzelli, T. Onodera, U. Ray, and A.L. Notkins. 1988. Virus-induced thyroiditis. *Endocrinology* 122:563–566.

70. Kröemer, G., R. Faessler, K. Hàla, G. Boeck, K. Schauenstein, H.P. Brezinscheck, N. Neu, H. Dietrich,

R. Jakober, and G. Wick. 1988. Genetic analysis of extra-thyroidal features of obese strain (OS) chickens with spontaneous autoimmune thyroiditis. *Eur. J. Immunol.* 18:1499–1505.

71. Khoury, E.L., L. Pereira, and F.S. Greenspan. 1991. Induction of HLA-DR expression on thyroid follicular cells by cytomegalovirus infection *in vitro*. Evidence for a dual mechanism of induction. *Am. J. Pathol.* 138:1209–1223.

72. Gaulton, G.N., M.E. Stein, B. Safko, and M.J. Stadecker. 1989. Direct induction of Ia antigen on murine thyroid-derived epithelial cells by reovirus. *J. Immunol.* 142:3821–3825.

73. Sallusto, F., A. Lanzavecchia, and C.R. Mackay. 1998. Chemokines and chemokine receptors in T-cell priming and Th1/Th2-mediated responses. *Immunol. Today* 19:568–574.

74. Oldstone, M.B.A. 1987. Molecular mimicry and autoimmune disease. *Cell* 50:819–820.

75. Zhao, Z.S., F. Granucci, L. Yeh, P.A. Schaffer, and H. Cantor. 1998. Molecular mimicry by herpes simplex virus-type 1: Autoimmune disease after viral infection. *Science* 279:1344–1347.

76. Rao, V.P., A.E. Kajon, K.R. Spindler, and G. Carayanniotis. 1999. Involvement of epitope mimicry in potentiation but not initiation of autoimmune disease. *J. Immunol.* 162:5888–5893.

77. Cash, E., J. Charreire, and O. Rott. 1996. B-cell activation by superstimulatory influenza virus hemagglutinin: A pathogenesis for autoimmunity? *Immunol. Rev.* 152:67–88.

78. Davies, T.F., E. Martin, E.S. Concepcion, P. Graves, N. Lahat, W.L. Cohen, and A. Ben-Nun. 1992. Evidence for selective accumulation of intrathyroidal T lymphocytes in human autoimmune thyroid disease based on T cell receptor V gene usage. *J. Clin. Invest.* 89:157–162.

79. McIntosh, R.S., P.F. Watson, A.P. Pickerill, R. Davies, and A.P. Weetman. 1993. No restriction of intrathyroidal T cell receptor Vα families in the thyroid of Graves' disease. *Clin. Exp. Immunol.* 91:147–152.

80. Cihak, J., G. Hoffmann-Fezer, A. Koller, B. Kaspers, H. Merkle, K. Hàla, G. Wick, and U. Lösch. 1995. Preferential TCR Vβ1 gene usage by autoreactive T cells in spontaneous autoimmune thyroiditis of the obese strain of chickens. *J. Autoimmun.* 8:507–520.

81. Matsuoka, N., N. Bernard, E.S. Concepcion, P.N. Graves, A. Ben-Nun, and T.F. Davies. 1993. T-cell receptor V region β-chain gene expression in the autoimmune thyroiditis of non-obese diabetic mice. *J. Immunol.* 151:1691–1701.

82. Matsuoka, N., P. Unger, A. Ben-Nun, P.N. Graves, and T. Davies. 1994. Thyroglobulin induced murine thyroiditis assessed by intrathyroidal T cell receptor sequencing. *J. Immunol.* 152:2562–2568.

83. Sugihara, S., H. Fujiwara, and G.M. Shearer. 1993. Autoimmune thyroiditis induced in mice depleted of particular T cell subsets. Characterization of thyroiditis-inducing T cell lines and clones derived from thyroid lesions. *J. Immunol.* 150:683–694.

84. Nakashima, M., Y.M. Kong, and T.F. Davies. 1996. The role of T cells expressing TcR Vβ13 in autoimmune thyroiditis induced by transfer of mouse thyroglobulin-activated lymphocytes: identification of two common CDR3 motifs. *Clin. Immunol. Immunopathol.* 80:204–210.

85. McMurray, R.W., R.W. Hoffman, H. Tang, and H. Braley-Mullen. 1996. T cell receptor Vβ usage in murine experimental autoimmune thyroiditis. *Cell. Immunol.* 172:1–9.

86. Lomo, L.C., R.W. Motte, A.A. Giraldo, G.H. Nabozny, C.S. David, I.J. Rimm, and Y.M. Kong. 1996. Vβ 8.2 transgene expression interferes with development of experimental autoimmune thyroiditis in CBA k/q but not k/k mice. *Cell. Immunol.* 168:297–301.

87. Tomer, Y. 1997. Anti-thyroglobulin autoantibodies in autoimmune thyroid disease: cross-reactive or pathogenic? *Clin. Immunol. Immunopathol.* 82:3–11.

88. Jaroszewski, J., R.S. Sundick, and N.R. Rose. 1978. Effects of antiserum containing thyroglobulin antibody on the chicken thyroid gland. *Clin. Immunol. Immunopathol.* 10:95–103.

89. Inoue, K., N. Niesen, F. Milgrom, and B. Albini. 1993. Transfer of experimental autoimmune thyroiditis by *in situ* perfusion of thyroids with immune sera. *Clin. Immunol. Immunopathol.* 66:11–17.

90. Chronopoulou, E., T.I. Michalak, and G. Carayanniotis. 1994. Autoreactive IgG elicited in mice by the non-dominant but pathogenic thyroglobulin peptide (2495–2511): Implications for thyroid autoimmunity. *Clin. Exp. Immunol.* 98:89–94.

91. Salamero, J., J.J. Remy, and J. Charreire. 1987. Primary syngeneic sensitization on monolayers of thyroid epithelial cells. X. Inhibition of T-cell proliferative response by thyroglobulin-specific monoclonal antibodies. *Clin. Immunol. Immunopathol.* 43:34–47.

92. Kuppers, R.C., H.S. Bresler, C.L. Burek, S.L. Gleason, and N.R. Rose. 1991. Immunodominant determinants of thyroglobulin associated with autoimmune thyroiditis. *Immunol. Series* 55:247–284.

93. Zanetti, M., and J. Rogers. 1987. Independent expression of a regulatory idiotype on heavy and light chains. A further immunochemical analysis with anti-heavy and anti-light chain antibodies. *J. Immunol.* 139:1965–1970.

94. Tang, H., C. Bedin, B. Texier, and J. Charreire. 1990. Autoantibody specific for a thyroglobulin epitope inducing experimental autoimmune thyroiditis or its anti-idiotype correlates with the disease. *Eur. J. Immunol.* 20:1535–1539.

95. Monestier, M., A. Manheimer-Lory, B. Bellon, C. Painter, H. Dang, N. Talal, M. Zanetti, R. Schwartz, D. Pisetsky, R. Kuppers, N. Rose, J. Brochier, L. Klareskog, R. Holmdahl, B. Erlanger, F. Alt, and C. Bona. 1986. Shared idiotopes and restricted immunoglobulin variable region heavy chain genes characterize murine autoantibodies of various specificities. *J. Clin. Invest.* 78:753–759.

96. Gleason, S.L., P. Gearhart, N.R. Rose, and R.C. Kuppers. 1990. Autoantibodies to thyroglobulin are encoded by diverse V-gene segments and recognize restricted epitopes. *J. Immunol.* 145:1768–1775.

97. Mignon-Godefroy, K., A. Ropars, C. Bedin, and J. Charreire. 1993. Ig VH gene family usage in spleen cells of CBA/J mice immunized with experimental autoimmune thyroiditis (EAT) inducer antigens. *Autoimmunity* 14:189–195.

98. Logtenberg, T., F.M. Young, J.H. Van Es, F.H.J. Gmelig-Meyling, and F.W. Alt. 1989. Autoantibodies encoded by the most JH-proximal human immunoglobulin heavy chain variable region gene. *J. Exp. Med.* 170:1347–1355.

99. Bedin, C., A. Ropars, K. Mignon-Godefroy, and J. Charreire. 1995. Molecular heterogeneity of antigen- or idiotype-induced anti-thyroglobulin monoclonal autoantibodies. *Clin. Exp. Immunol.* 100:463–469.

100. Liblau, R.S., S.M. Singer, and H.O. McDevitt. 1995. Th1 and Th2 CD4+ T cells in the pathogenesis of organ-specific autoimmune diseases. *Immunol. Today* 16:34–38.

101. Kromer G., K. Schauenstein, H. Dietrich, R. Fässler, and G. Wick. 1987. Mechanisms of T cell hyperreactivity in obese

strain (OS) chickens with spontaneous autoimmune thyroiditis: lack in nonspecific suppression is due to a primary adherent cell defect. *J. Immunol.* 138:2104–2109.

102. Zipris, D., D.L. Greiner, S. Malkani, B. Whalen, J.P. Mordes, and A.A. Rossini. 1996. Cytokine gene expression in islets and thyroids of BB rats. IFN-γ and IL-12 p40 mRNA increase with age in both diabetic and insulin-treated nondiabetic BB rats. *J. Immunol.* 156:1315–1321.

103. Salamero, J., M. Michel-Béchet, and J. Charreire. 1981. Differences de répartition des antigènes de classe I et II du complexe mejeur d'histocompatibilité (CMH) à la surface des cellules épithéliales de thyroide (CET) murines en culture. *C. R. Acad. Sci. Paris* 293:745–750.

104. Todd, I., R. Pujol-Borrell, L.J. Hammond, G.F. Bottazzo, and M. Feldmann. 1985. Interferon-γ induces HLA-DR expression by thyroid epithelium. *Clin. Exp. Immunol.* 61:265–273.

105. Londei, M., J.R. Lamb, G.F. Bottazzo, and M. Feldmann. 1984. Epithelial cells expressing aberrant MHC class II determinants can present antigen to cloned human T cells. *Nature* 312:639–641.

106. Remy, J.J., J. Salamero, and J. Charreire. 1986. Syngeneic sensitization of mouse lymphocytes on monolayers of thyroid epithelial cells. IX. Thyroid epithelial cells are antigen presenting cells. *Mol. Biol. Med.* 3:167–179.

107. Pujol-Borrell, R., T. Hanafusa, L. Chiovato, and G.F. Bottazzo. 1983. Lectin-induced expression of DR antigen on human cultured follicular thyroid cells. *Nature* 304:71–73.

108. Todd, I., R. Pujol-Borrell, L.J. Hammond, J.M. McNally, M. Feldmann, and G.F. Bottazzo. 1987. Enhancement of thyrocyte HLA class II expression by thyroid stimulating hormone *Clin. Exp. Immunol.* 69:524–531.

109. Remy, J.J., M. Michel-Béchet, and J. Charreire. 1987. Experimental autoimmune thyroiditis induced by recombinant interferon-γ *Immunol. Today* 8:73.

110. Tang, H., K. Mignon-Godefroy, P.L. Meroni, G. Garotta, J. Charreire, and F. Nicoletti. 1993. The effects of a monoclonal antibody to interferon-γ on experimental autoimmune thyroiditis (EAT): prevention of disease and decrease of EAT-specific T cells. *Eur. J. Immunol.* 23:275–278.

111. Alimi, E., S. Huang, M.P. Brazillet, and J. Charreire. 1998. Experimental autoimmune thyroiditis (EAT) in mice lacking the IFN-γ receptor gene. *Eur. J. Immunol.* 28:201–208.

112. Tang, H., G.C. Sharp, K.P. Peterson, and H. Braley-Mullen. 1998. IFN-gamma-deficient mice develop severe granulomatous experimental autoimune thyroiditis with eosinophil infiltration in thyroids. *J. Immunol.* 160:5105–5112.

113. Zaccone, P., P. Hutchings, F. Nicoletti, G. Penna, L. Adorini, and A. Cooke. 1999. The involvement of IL-12 in murine experimentally induced autoimmune thyroid disease. *Eur. J. Immunol.* 29:1933–1942.

114. Braley-Mullen, H., G.C. Sharp, H. Tang, K. Chen, M. Kyriakos, and J.T. Bickel. 1998. Interleukine-12 promotes activation of effector cells that induce a severe destructive granulomatous form of murine experimental autoimmune thyroiditis. *Am. J. Pathol.* 152:1347–1358.

115. Mignon-Godefroy, K., O. Rott, M.P. Brazillet, and J. Charreire. 1995. Curative and protective effects of IL-10 in experimental autoimmune thyroiditis (EAT). Evidence for IL-10-enhanced cell death in EAT. *J. Immunol.* 154:6634–6643.

116. Mignon-Godefroy, K., M.P. Brazillet, O. Rott, and J. Charreire. 1995. Distinctive modulation by IL-4 and IL-10 of the effector function of murine thyroglobulin-primed cells in "transfer-experimental autoimmune thyroiditis". *Cell Immunol.* 162:171–177.

117. Tang, H., G.C. Sharp, K.E. Peterson, and H. Braley-Mullen. 1998. Induction of granulomatous experimental autoimmune thyroiditis in IL-4 gene-disrupted mice. *J. Immunol.* 160:155–162.

118. Nolte, A., G. Bechtner, M. Rafferzeder, and R. Gartner. 1994. Interleukine-1 beta (IL-1β) binds to intact porcine thyroid follicles, decreases iodide uptake but has no effect on cAMP formation or proliferation. *Horm. Metab. Res.* 26:413–418.

119. Reimers, J.I., A.K. Rasmussen, A.E. Karlsen, U. Bjerre, H. Liang, O. Morin, H.U. Andersen, T. Mandrup-Poulsen, A.G. Burger, U. Feldt-Rasmussen, and J. Nerup. 1996. Interleukin-1 beta inhibits rat thyroid cell function *in vivo* and *in vitro* by an NO-independent mechanism and induces hypothyroidism and accelerated thyroiditis in diabetes-prone BB rats. *J. Endocrinol.* 151:147–157.

120. Giordano, C., G. Stassi, R. De Maria, M. Todaro, P. Richiusa, G. Papoff, G. Ruberti, M. Bagnasco, R. Testi, and A. Galluzzo. 1997. Potential involvement of Fas and its ligand in the pathogenesis of Hashimoto's thyroiditis. *Science* 275:960–963.

121. Stokes, T.A, M. Rymaszewski, P.L. Arscott, S.H. Wang, J.D. Bretz, J. Bartron, and J.R. Baker, jr. 1998. Constitutive expression of FasL in thyrocytes. *Science* 279:2015.

122. Batteux, F., B. Malassagne, L. Tourneur, P. Lores, M. Fabre, D. Bucchini, G. Chiocchia. 2000. FasL expression on thyroid follicular cells confers immune privilege to thyroid. *J. Immunol.* 164:1681–1688.

123. Mori, K., M. Mori, S. Stone, L.E. Braverman, and W.J. DeVito. 1998. Increased expression of tumor necrosis factor-alpha and decreased expression of thyroglobulin and thyroid peroxidase mRNA levels in the thyroids of iodide-treated BB/Wor rats. *Eur. J. Endocrinol.* 139:539–545.

124. Tang, K.T., L.E. Braverman, and W.J. DeVito. 1995. Tumor necrosis factor-alpha and interferon-gamma modulate gene expression of type 1 5′-deiodinase, thyroid peroxidase and thyroglobulin in FRTL-5 rat thyroid cells. *Endocrinology* 136:881–888.

125. Lombardi, G., K. Arnold, J. Uren, F. Marelli-Berg, R. Hargreaves, N. Imani, A. Weetman, and R. Lechler. 1997. Antigen presentation by interferon-γ treated thyroid follicular cells inhibits interleukin-2 (IL-2) and supports IL-4 production by B7-dependent human T cells *Eur. J. Immunol.* 27:62–71.

126. Peterson, K.E., G.C. Sharp, H. Tang, and H. Braley-Mullen. 1999. B7-2 has opposing roles during the activation versus effector stages of experimental autoimmune thyroiditis. *J. Immunol.* 162:1859–1867.

127. Tamura, R., N. Woo, G. Murase, G. Carrieri, M.A. Nalesnik, and A.W. Thomson. 1993. Suppression of autoimmune thyroid disease by FK 506: Influence on thyroid infiltrating cells, adhesion molecule expression and anti-thyroglobulin antibody production. *Clin. Exp. Immunol.* 91:368–375.

128. Nabozny, G.H., S.P. Cobbold, H. Waldmann, and Y.M. Kong. 1991. Suppression in murine experimental autoimmune thyroiditis: *In vivo* inhibition of CD4+ T cell-mediated resistance by a non-depleting rat CD4 monoclonal antibody. *Cell. Immunol.* 138:185–196.

129. Hutchings, P.R., A. Cooke, K. Dawe, H. Waldmann, and I.M. Roitt. 1993. Active suppression induced by anti-CD4. *Eur. J. Immunol.* 23:965–968.

130. Flynn, J.C., and Y.M. Kong. 1991. *In vivo* evidence for CD4+ and CD8+ suppressor T cells in vaccination-induced suppression of murine experimental autoimmune thyroiditis. *Clin. Immunol. Immunopathol.* 60:684–694.

131. Maron, R., R. Zerubavel, A. Friedman, and I.R. Cohen. 1983. T lymphocyte line specific for thyroglobulin produces or vaccinates against autoimmune thyroiditis in mice. *J. Immunol.* 131:2316–2322.

132. Nabozny, G.H., J.C. Flynn, and Y.M. Kong. 1991. Synergism between mouse thyroglobulin- and vaccination-induced suppressor mechanisms in murine experimental autoimmune thyroiditis. *Cell Immunol.* 136:340–348.

133. Texier, B., C. Bédin, C. Roubaty, C. Brézin, and J. Charreire. 1992. Protection from experimental autoimmune thyroiditis conferred by a monoclonal antibody to T cell receptor from a cytotoxic hybridoma specific for thyroglobulin. *J. Immunol.* 148:439–444.

134. Guimaraes, V.C., J. Quintans, M.E. Fisfalen, F.H. Straus, K. Wilhelm, G.A. Medeiros-Neto, and L.J. DeGroot. 1995. Suppression of development of experimental autoimmune thyroiditis by oral administration of thyroglobulin. *Endocrinology* 136:3353–3359.

135. Tang, H., and H. Braley-Mullen. 1997. Intravenous administration of deaggregated mouse thyroglobulin suppresses induction of experimental autoimmune thyroiditis and expression of both Th1 and Th2 cytokines. *Int. Immunol.* 9:679–687.

136. Howard, M. and A. O'Garra. 1992. Biological properties of interleukin-10. *Immunol. Today* 13:198–200.

137. Batteux, F., H. Trebeden, J. Charreire, and G. Chiocchia. 1999. Curative treatment of experimental Autoimmune thyroiditis by *in vivo* administration of plasmid DNA coding for interleukin-10. *Eur. J. Immunol.* 29:958–963.

138. Nagata, S. 1997. Apoptosis by death factor. *Cell* 88:355–365.

139. Batteux, F., L. Tourneur, H. Trebeden, J. Charreire, and G. Chiocchia. 1999. Gene therapy of experimental autoimmune thyroiditis by *in vivo* administration of plasmid DNA coding for Fas ligand. *J. Immunol.* 162:603–608.

40 | Autoimmune Polyglandular Syndromes and Addison's Disease

Qiao-Yi Chen and Noel K. Maclaren

1. INTRODUCTION

Individual patients or multiple members of a pedigree may suffer from multiple autoimmune endocrinopathies, or autoimmune polyglandular syndromes (APS). The recognition of APS dates back to the initial description of Addsions' disease, which is a major component disease for patients with APS. Addison's disease was named after Thomas Addison, who initially recognized the co-existence of the clinical manifestations of the disease and adrenalitis and/or adrenal atrophy in 1849 (1). Almost 80 years later, destruction of the adrenal cortex, but not the medulla of adrenal gland, was found to be responsible for the clinical appearance of Addison's disease (2). Adrenal cortical extracts were then demonstrated to be able to maintain the lives of adrenalecotmized animals (3), and be beneficial to patients with Addison's disease (4,5). Adrenal tuberculosis was still the major cause of Addison's disease in the early twentieth century, however, adrenal failure become increasingly recognized in patients without tuberculosis. Hashimoto in 1912 reported lymphocyte infiltration of thyroid gland (6), and Schmidt in 1926 observed simultaneous lymphocytic infiltrations of both adrenal cortices and thyroid gland in two patients (7). Schmidt's syndrome, comprising Addison's disease and Hashimoto's thyroiditis, was expanded to include insulin dependent diabetes by Carpenter et al. in 1964 (8) In 1980, autoimmune polyglandular endocrinopathies were classified into three types, APS-I, II and III, based on clinical features of patients with APS (9,10). Recent studies have provided information on the genetics of Addison's disease and APS, while many autoantigens involved in these autoimmunities have been identified. Autoantibodies to these autoantigens can be used as markers to either facilitate the diagnoses of the involved component

diseases or to predict them. Studies on the subsets of T helper cells (Th1 and Th2) in autoimmune diseases, and the relative balance of polarized Th1/Th2 foundations in the outcome of autoimmunity, have increased our understanding of the pathogenic process of APS. In addition, a gene called autoimmune regulator (AIRE), responsible for APS-I, has recently been identified and multiple mutations have been found in patients with APS-I. Studies of the function of AIRE gene product should promote our understanding of the pathogenic process of APS-I, and perhaps autoimmune diseases in general. Accordingly, this Chapter will summarize the current knowledge on clinical, immunological and genetic underpinnings of Addison's disease and the three types of APS.

2. CLINICAL FEATURES OF ADDISON'S DISEASE AND APS

Autoimmune Addison's disease is a major component disease of APS-I and APS-II, which involve multi-endocrine organs. The diagnosis of APS-I is established when a patient has two of the following three component diseases: Chronic mucocutaneous candidiasis, acquired hypoparathyroidism, and autoimmune Addison's disease. APS-II is diagnosed when a patient has autoimmune Addison's disease together with an autoimmune thyroid disease (Schmidt's syndrome) and/or immune-mediated (type-I) diabetes (IMD) (Carpenters' syndrome). The diagnosis of APS-III is established when a patient has an autoimmune thyroid disease in association with one or more of other organ-specific autoimmune diseases, such as pernicious anemia, vitiligo and/or IMD, but without Addison's disease (9).

2.1. Addison's Disease

Addison's disease due to adrenocortical hormone insufficiency is the consequence of the destruction of steroid-producing cells in the zona glomerulosa, fasciculata and reticularis of adrenal cortex by various causes. These include autoimmune destruction of sterioid producing cells, the most common cause of Addison's disease (11), and other less common causes such as tuberculosis, characterized by calcifications in adrenals and/or other organs, amyloidosis, metastatic neoplasia, lymphoma, non-tuberculous infections, hemorrhage or adrenoleukodystrophy. The autoimmune destruction of adrenal cortex gland is a chronic process and can persist for years before the appearance of overt clinical disease. Pathologically, the adrenal glands in patients with autoimmune Addison's disease are usually atrophic and difficult to be located during autopsy. However, normal-sized adrenal glands detected by CT scan cannot exclude the autoimmune Addison's disease at an early stage. Histologically, the adrenal cortex in autoimmune Addison's disease is infiltrated by lymphocytes, while the adrenal medulla is left intact. In some cases, scattered foci of regeneration with hyperplastic and hypertrophied adreno-cortical cells are present in the adrenal cortex (12). Patients with Addison's disease express the clinical manifestation of hypocortisolism, such as fatigue, nausea, vomiting, weight loss, abdominal pain, and a "muddy" hyperpigmentation, which is often the initial physical sign of Addison's disease. The hyperpigmentation is due to the elevation of melanocyte-stimulating hormones (ACTH/ MSH) caused by the compensatory activation of the hypothalamic-pituitary-adrenocortical axis. However, in rare cases, a defect of melanocyte in response to melanocyte-stimulating hormones could result in the absence of hyperpigmentation in patients with Addison's disease (13). Therefore, patients with clinical symptoms that resemble those of Addison's disease, especially in the presence of hypoglycemia and hyperkalemia, should be examined for Addison's disease even in the absence of hyperpigmentation.

Patients with Addison's disease often develop other autoimmune diseases and thus should be regularly screened for associated endocrine diseases by testing for the corresponding autoantibodies. For example, Addison's disease could be associated with primary hypogonadism, hypoparathyroidism, IMD, chronic active hepatitis, autoimmune thyroid diseases, pernicious anemia and/or vitiligo. The screening for marker autoantibodies for these associated diseases provides for the early diagnosis and treatment of the corresponding diseases. In addition, celiac disease can occur also with Addison's disease or any of the three types of APS. Celiac disease is characterized by the damage of absorptive villi with flattening and hyperplasia of the crypts in the small intestine. Anti-gliadin and anti-endomysial antibodies are often detectable in patients with celiac disease, and the latter appear to be more specific for celiac disease (14). Autoantibodies have been detected in patients with celiac disease to a newly-identified autoantigen, tissue transglutaminase (15), in line with the known evidence for its autoimmune nature (16;17). Here, ingestion of gliadin in wheat appears to provoke a reversible autoimmunity to transglutaminase associated with sprue like symptoms. Thus, the detection of anti-tissue transglutaminase antibodies, anti-gliadin and/or anti-endomysial antibodies in patients with the three types of APS, especially in Addison's disease regardless of its association with APS, would help to identify the existence of celiac disease in the tested patients. An early diagnosis would help to prevent celiac complications by a gliadin-free diet, such as short stature or perhaps, intestinal lymphomas.

2.2. APS-I

APS-I is a rare sporadic childhood disease and the only known systemic autoimmune disease with a Mendelian recessive inheritance mode. APS-I has a relatively high incidence in genetically isolated populations, such as Iranian Jews (1:6000–9000) (18), the Finnish population (1:25,000) (19), and Sardinians (20), affecting males and females equally. The three major component diseases for APS-I are chronic mucocutaneous candidiasis, acquired hypoparathyroidism, and autoimmune Addison's disease. Mucocutaneous candidiasis caused by the infection of *Candida albicans*, usually occurs in early infancy due to potential defect of cellular immunity. The infection of *Candida albicans*, is usually restricted to oral and/or perianal mucosa, the skin and the nails, but oral candidiasis is often the initial manifestation seen in patients with APS-I (10,19,21,22). The candidiasis can also involve gastrointestinal mucosas and result in diarrhea or gastrointestinal hemorrhages, especially when complicated by bacterial overgrowth. Hypoparathyroidism often occurs next, manifested by the symptoms of carpo-pedal spasms, seizures or laryngo-spasm due to severe hypocalcemia (9). Such patients should be tested for adrenal antibodies, since large proportions of such patients are destined to develop Addison's disease within few years. The peak-age-onset for patients with APS-I for Addison's disease is 12 years. The symptoms of hypoparathyroidism can be masked by the development of Addison's disease, and only manifest after the start of steroid replacement therapy (23). Apart from the three core component diseases, patients with APS-I are often afflicted by other autoimmune diseases. These disease include alopecia universalis, hypogonadism, chronic active hepatitis, pernicious anemia, vitiligo, hypophysitis, etc., but seldom IMD. Ectodermal dystrophies of dental enamel and nails, and keratopathy of cornea are also seen in some patients, especially Finns with APS-I. Therefore, APS-I is also known as polyendocrinopathy-candidiasis-ectodermal dystrophy (APECED) in Scandinavian patients (19).

Patients with APS-I are often seen with weight loss, malabsorption and chronic or intermittent diarrhea (19). This may indicate the autoimmune damage of small intestine characterized by the presence of autoantibodies to autoantigens present in intestine cells, such as tryptophan hydroxylase (24). Alternatively, such symptoms could be due to abnormal B-lymphocyte function resulting in IgA deficiency and high levels of IgG and IgE, as observed in some patients with APS-I (25,26). The deficiency of secreted IgA by the intestine may result in small intestinal bacterial overgrowth resulting in diarrhea. Invasive giardiasis occurs in some patients with APS-I.

2.3. APS-II

APS-II is a relatively common disease compared to APS-I. The peak age at onset for APS-II is about 30 years with a distinct female predominance (9). The order of the development of the component disease may vary in individuals. Patients with IMD or thyroid autoimmune disease can be diagnosed concurrently to have Addison's disease, but the disease processes in such cases began many years before the onset of overt clinical symptoms. Thus, the screening of patients with either IMD or autoimmune thyroid disease for marker autoantibodies of Addison's disease is valuable for early identification of those patients with subclinical Addison's disease. Patients with APS-II also frequently present with other diseases, such as pernicious anemia, vitiligo, celiac disease and/or myasthenia gravis. Patients with APS-II are less often seen with hypogonadism, chronic hepatitis, hypophysitis, and alopecia universalis, which are commonly seen in APS-I. However, when such manifestations are present, they are important clinical indicators for the presence of APS-II, especially if they are multiply expressed.

2.4. APS-III

Autoimmune thyroid disease is the commonest autoimmune disease (27), and is often associated with other autoimmune diseases. Accordingly, APS-III appears to be the most common of the three types of APS. If patients with APS-III develop autoimmune Addison's disease, then they will be reclassified as having APS-II. The phenotypic distinction between APS-II and APS-III could be due to effects of environmental factors and/or genetic predispositions, especially HLA-DR/DQ genotypes. The clinical importance of defining the type of APS is to be able to predict the potential occurrence of other associated diseases both in the patients and their immediate family members.

3. HUMORAL AUTOIMMUNITY IN ASSOCIATION WITH APS

The presence of circulating autoantibodies to endocrine autoantigens is a serological characteristic of Addisons' disease and APS. Such autoantibodies occur long before the clinical onset of these diseases. The targeted endocrine autoantigens include intracellular enzymes, surface receptor molecules, and secreted proteins such as hormones produced by the affected organs (Table 1). The expression of some of

Table 1   Major autoantigens in component diseases of APS

	Autoantigen	Tissue/cells	Component disease	Reference
Enzymes	21-OH, 17-OH, P450scc	Adrenal cortex, ovary, testis	Addison disease, hypogonadism	35–37
	GAD$_{65}$	Pancreatic β cells	IMD	139
	Thyroid peroxidase	Thyroid	Hashimoto's thyroiditis	140
	H$^+$, K$^+$ATPase	Gastric parietal cells	Pernicious anemia	75–77
	P450IID6, 2C9, P450 1A2, AADC	Hepatocytes	Chronic active hepatitis	84, 141–143
	Tissue transglutaminase	Small intestine	Celiac disease	15
	Tryptophan hydroxylase	Intestine enterochromaffin cells	Malabsorption	24
	Tyrosinase	Melanocytes	Vitiligo	46
Surface receptors	Calcium sensing receptor	Parathyroid	Hypoparathyroidism	66
	Thyroid stimulating hormone receptor	Thyroid epithelium	Graves' disease	144
	IA2, IA2β	Pancreatic β cells	IMD	145–147
Secreted proteins	Insulin	Pancreatic β cells	IMD	148
	Thyroglobulin	Thyroid	Hashimoto's thyroiditis	149
	Intrinsic factor	Gastric mucosa chief cells	Pernicious anemia	150

21-OH: steroidogenic enzyme 21-hydroxylase. 17-OH: 17 α-hydroxoylase. P450scc: P450 side-chain cleavage enzyme
GAD: glutamic acid decarboxylase. AADC: aromatic L-amino acid decarboxylase.
IA2, IA2b: tyrosine phosphatase-like transmembrane glycoproteins.

the intracellular enzymes are unique to particular endocrine cells, such as thyroid peroxidase, but some enzymes are expressed in more than one tissue types, such as glutamic acid decarboxylase (GAD), which is expressed in pancreatic β cells as well as in central and peripheral neurons (28). Autoantibodies reactive with cell surface receptors may inhibit the action of their respective hormone ligands, such as gonadotropins, insulin, ionized calcium or ACTH (29). While it remains improbable that autoantibodies reactive with intracellular enzymes have any pathogenic significance, it is unclear why enzymes are preferentially targeted in autoimmune diseases. Abnormal metabolic pathways may result in enzyme antigen up-regulation and/or an aberrant presentation to the immune system, and as such could be involved in the initiation of the breakdown of self-immune tolerance against enzyme antigens (30,31). While, the inductive events for the development of the autoantibodies in the autoimmune endocrinopathies remain to be understood, they do serve as diagnostic markers for the corresponding autoimmunity, and thus are useful for the early diagnosis and thus treatment of the corresponding diseases.

3.1. Adrenal Autoantibodies

The majority of patients with new-onset non-tuberculous Addison's disease have adrenal autoantibodies (AA) that react with steroid hormone-secreting cells in zona glomerulosa, fasciculata and reticularis layers of adrenal cortex as detected by indirect immunofluorescence. A proportion of AA-positive patients with Addison's disease also have "distinct steroidal cell" autoantibodies (SCA) that react with all steroid-producing cells, including testicular leydig cells, ovarian granulosa and placental syncytiotrophoblasts depending upon their titers by indirect immunofluorescence. AA and SCA are separate population of autoantibodies, since SCA reactivity can only be removed from serum by preincubation with adrenal, gonadal (ovarian or testicular), or placental homogenates, whereas AA are removed exclusively from positive sera only by prior exposure to adrenal homogenates. SCA indicate a high risk for future Addison's disease and gonadal failure, especially in females with high titers of the latter autoantibodies (32,3), whereas AA confer risk for developing Addison's disease alone (34). The major steroid cell autoantigens involved in the reactions of AA have now been identified as belonging to the p450 family of enzymes, i.e. steroid 21-hydroxylase (21-OH), while the major antigens for SCA including a 55-kDa gonadal and adrenal steroid biosynthetic P_{450} microsomal enzyme have been recognized to be 17α-hydroxylase (17-OH) and P450 side-chain cleavage enzyme (P450scc) (35–37).

Autoantibodies to 21-OH are present in majority of patients with autoimmune Addison's disease, either as an idiopathic disease or in association with APS-I or APS-II,

although the reported frequencies may vary depending with the techniques used (35,36,38–42). The frequencies and titers of autoantibodies to 21-OH are generally higher for patients with Addison's disease in association with an APS than patients with an isolated Addison's disease. Autoantibodies to 21-OH may inhibit the enzyme function of 21-OH *in vitro* (43). However, such effect was not so evident *in vivo* (44), although it is difficult to understand how they could penetrate live cells to do so, rendering them conceptually unlikely to account for the resultant diseases. The dominant epitopes on 21-OH recognized commonly by autoantibodies from patients with Addisons' disease as an isolated disease or in association with APS have been located to in the C-terminal end and in a central region of 21-OH (45–48). The epitopes on 21-OH can either be conformational (49) or linear in nature (46).

Approximately 35% of patients with Addison's diseases are also positive for antibodies to 17-OH and/or P450scc in some series (40,42,50,51). However, almost all patients would have such autoantibodies if patients with both Addison's disease and gonadal failure were studied (42,51). Similar to antibodies to 21-OH, the frequencies of antibodies to 17-OH and/or P450scc are considerably higher in patients with an APS than in patients with isolated Addison's disease. The presence of antibodies to 17-OH or P450scc in patients with isolated Addison's disease like SCA may indicate the progression towards the development of gonadal failure and APS. Four distinct epitope regions were found for 17-OH, and two of them contain an amino acid sequence, which is similar to a partial sequence of 21-OH (52). As discussed later, SCA are a regular feature of APS-I.

Adrenal autoantibodies are useful markers for the prediction of the development of Addison's disease. They are useful markers to screen relatives of patients with APS, especially patients with APS-I or II. Individuals with a positive test for adrenal autoantibodies, such as antibody to 21-OH, often develop overt Addisons' disease over time, particularly children (53). It is thus recommended that patients with features of APS-I or III, but without overt Addison's disease, and relatives of patients with APS-I or II should be tested for adrenal autoantibodies, since classical symptoms and signs of adrenal deficiency may not occur until the majority of the adrenal cortex has been destroyed (54). About 20% of APS-I patients without symptoms of Addison's disease have been found to be positive for antibodies to 21-OH (55). However, 21-OH autoantibody-positive patients with subclinical Addison's disease can be identified most sensitively by increased levels of resting plasma renin activity and/or raised afternoon serum adrenocorticotropin (ACTH) levels tested after patients have been recumbent for 1 hour. As the disease progress, depressed ACTH-stimulated cortisol responses can be seen (54). It is important to diagnose impending

Addison's disease as soon as possible, since stress may precipitate potentially fatal adrenal crisis.

Patients with other autoimmune diseases could have autoantibodies to 21-OH and some of them may eventually develop clinical Addison's disease. In an early study, our group suggested that Addson's disease occurred in about 1:200 (0.5%) patients with IMD (56). Further, up to 2.3% of patients with IMD (57–59) and 3% of patients with Graves' disease (57) were screened to be positive for autoantibodies to 21-OH. Such patients with a positive test for autoantibodies to 21-OH, could be at an early stage of Addison's disease (58). A quarter of patients without clinical symptoms of Addison's disease but with positive test for adrenal cortex autoantibodies, were actually at a stage of subclinical hypoadrenalism (34). Thus, close attention should be paid to patients with a positive test for autoantibodies to 21-OH, especially those with high-risk HLA alleles such as DQB1*0201, which is associated with both IMD and Addison's disease (60,61), or according to one study, the HLA DRB1*0404 allele (62).

3.2. Parathyroid Gland Autoantibodies

Hypoparathyroidism in association with APS-I is regarded as the result of autoimmune destruction of parathyroid gland. Antibodies to parathyroid gland were reported in early studies (63; 64), but are usually not parathyroid-specific (65). The calcium sensing receptor (CaSr) was recently identified as an autoantigen recognized by autoantibodies present in 32% (8/25) of patients with hypoparathyroidism associated with either APS-I or hypothyroidism plus hypoparathyroidism, and the major epitope was located to the external domain of the receptor by Western blotting (66). The potential pathogenic effects of such autoantibodies could either inhibit parathyroid hormone secretion, or damage the gland through an antibody mediated cellular immunity.

3.3. Pancreatic β Cell Autoantibodies

Autoantibodies to pancreatic β cell antigens, such as islet gangliosides, insulin, proinsulin, 65 kDa glutamic acid decarboxylase [GAD_{65}], tyrosine phosphatases (IA–2 and IA2-β), etc. are serological markers for IMD. Autoantibodies to such islet cell antigens have predictive value for developing IMD, especially when they occur as multiple autoantibodies. For example, islet cell antibodies (ICA), when detected by immunofluorescence together with insulin and/or anti-GAD and/or IA-2 antibodies, have high predictive values for developing IMD, but not when they occur alone. Patients with APS especially in association with IMD usually have autoantibodies to these pancreatic β cell antigens. However, autoantibodies to the islet cell antigen, L-dopa decarboxylase, were detected in some

APS-I patients with ICA, but these were not related to diabetes in them (67).

The presence of anti-islet cell autoimmunity in APS-I is not a necessary indication of the destruction of islet cells, since clinically overt diabetes is infrequently seen in patients with APS-I, and if so, it usually occurs late in life, at least in the U.S. (10,68). This could be due to the different reactive antibody characteristics in patients with APS-I and IMD, or IMD in association with APS-II. For example, patients with APS-I may have GAD_{65} autoantibodies readily detectable by Western blotting (68), similar to GAD autoantibodies from patients with Stiffman syndrome, an autoimmune neurological disorder that can be associated with IMD (69). These latter autoantibodies usually recognize linear epitopes on GAD_{65}. However, patients with IMD usually have GAD_{65} autoantibodies reactive with conformational epitopes of undenatured or native GAD proteins (70). This suggests that different immune regulations in APS-I and in IMD could have resulted in the production of these two sets of GAD_{65} autoantibodies. That islet β-cells are not destroyed by the anti-islet cell autoimmune response in APS-I, is presumably due to the lack of other components of the pathogenic factors in the immune system, such as antigen-specific cytotoxic T lymphocytes that are necessary to produce β-cell damage and thus overt hyperglycemia. Alternatively, the islet β-cell non-destructive autoimmune response in patients with APS-I may result from an impaired cellular immune regulatory mechanism, which differs from that in APS-II. These speculations need to be systematically proven through specific studies.

3.4. Autoimmune Thyroid Diseases

Autoantibodies to thyroid gland proteins, such as thyroid peroxidase, thyroglobulin and thyrotropin receptors are characteristics of autoimmune thyroid diseases, which include Graves' disease and Hashmoto's thyroiditis, both of which are usually associated with APS-II and III, but much less so with APS-I. Autoantibodies to the thyrotropin receptor may stimulate or inhibit both thyroid gland activity and growth of thyrocytes, however no consistently discernible effect on thyroid function has been yet attributed to autoantibodies that recognize thyroid peroxidase or thyroglobulin. Experimental allergic thyroiditis (EAT) could be induced by immunization of susceptible strains of mice with thyroglobulin in complete Freund's adjuvant (71) or Graves' disease by immunization with fibroblasts expressing human TSH receptor and MHC class II antigen (7,73).

3.5. Pernicious Anemia

Autoantibodies to gastric parietal cells (PCA) and, less frequently, intrinsic factor (IFA), are the serological markers for patients with achlorhydria and pernicious anemia, which can occur as part of APS. Achlorhydria is seen in some patients with IMD, who have PCA (74). The major autoantigen recognized by PCA is the parietal cell proton pump (H^+, K^+-ATPase) (75–77). PCA appear to be primarily associated with achlorhydria, while IFA may arise secondarily as a consequence of gastric cell damage and are associated with increasing likelihood of vitamin B12 deficiency and clinical pernicious anemia. In APS-I, pernicious anemia is often seen during childhood, whereas in APS-II/III it occurs characteristically in mid to late life, especially in women.

3.6. Vitiligo

Vitiligo due to the depigmentation of skin results from either the destruction or defect of melanocytes by various causes. Vitiligo is often seen in patients with APS, especially APS-I. Autoimmune destruction of melanocytes is one of the major cause of the disease, evidenced by the presence of chronic lymphocyte infiltrations in active vitiligo lesions, and anti-melanocyte autoantibodies (78). Patients with vitiligo have been found to have autoantibodies to tyrosinase, the rate-limiting enzyme for melanin formation (46). The frequencies of antibody to tyrosinase are usually higher for adult patients with diffuse vitiligo or in association with APS, although controversy on the frequencies of autoantibodies to tyrosinase in patients with vitiligo in general have been reported (79–81). Tyrosinase-reactive T cells are present in normal immune system and are responsible for the stimulation of peptides derived from tyrosinase (82). Immunization by tyrosinase-related protein-1 has recently been shown to induce a destruction of melanocytes in mice (83), while they occur spontaneously in the Smyth line chicken model of the disease.

3.7. Autoimmune Liver Diseases

Among those type I APS patients with chronic active hepatitis, autoantibodies against mitochondrial, nuclear, or smooth-muscle antigens are frequently found, although their clinical significances are also unclear. The cytochrome P450 proteins are often the common targets for autoantibodies from patients with APS and patients with autoimmune liver diseases (84). Antibodies to cytochrome p450D6, which is one of the major autoantigen for chronic active hepatitis, were shown to cross-react with 21-OH at their amino acid homologous regions (85), suggesting that such cross-reactive antibody may play roles in patients with APS-I who develop autoimmune hepatitis. In addition, other cytochrome p450 proteins are often targeted by autoantibodies present in patients with APS-I (86). The mechanism behind the wide spectrum of autoreactivity in

patients with APS-I remain to be studied, although it could be due to a defect in the immune regulatory function to maintain immune homeostasis.

4.   CELLULAR IMMUNITY IN APS

Cellular immunity is the major factor involved in pathogenic autoimmune process of the component disease of at least APS-II/III. Although few studies have addressed cellular immune responses in patients with APS as whole, it is believed that the nature of the cellular immune responses in multi-organ autoimmunity is similar to those in isolated component diseases of APS. The cellular defects in patients with APS are likely to affect immune regulation, and may be associated with the defect in respect to their abnormal balance in cytokine production by T cells. T helper cells can be divided into at least two polarized types, T helper-1 (Th1) and T helper-2 (Th2) cells. Polarized Th1 and Th2 cells produce distinct profiles of cytokines (87,88). Th1 cells secrete IFN-γ, interleukin (IL)-2, TNF-α and are involved in cellular immunity defending against intracellular pathogens, and participate in delayed type hypersensitivity reactions, whereas Th2 cells secrete IL-4, 5 and 10 are responsible for phagocyte-independent host defense. An abnormal balance of Th1/Th2 is probably associated with some autoimmune disease, albeit this simple paradigm is undoubtedly going to become more complex with further study. A polarized Th1 response is reported to be associated with IMD and a polarized Th2 response with Graves' disease (89,90), albeit such studies require confirmation. The abnormality in Th1/Th2 balance could be associated with the defect in association with natural killer (NK) T cells. NK-T cells are able to produce large amounts of both IL-4 and IFN-γ rapidly upon activation, and may represent the major source of early IL-4 in certain immune responses driving the Th2 differentiation (91,92). Recent results suggest major roles for NK-T cells in regulating autoimmunity (93,94).

APS-I appears to be possibly related to a defect of Th1 immune responsiveness, with abnormally enhanced Th2 immune responses. Mucocutaneous candidiasis results from a defect of Th1 cell responses and patients with APS-I are often found to have defect T cells (95). Th1 type cellular immunity is the major defending force in response against the yeast *Candida albicans* in healthy individuals (96–98). A defective Th1 response will lead to susceptibility to infection by *Candida*, as demonstrated in animal models (99). Increased Th2 type responses have been observed in patients with chronic mucocutaneous candidiasis (98). These findings suggest that a defective Th1 cell responses may occur in patients with APS-I, since chronic infection by yeast *Candida albicans* is one of the component diseases. Such a defect in Th1 cell responsiveness could lead to the loss of a counter-balancing inhibition of Th2 responsiveness, and result in abnormal, enhanced Th2 responses to self-antigens in patients with APS-I. Indeed, the auto-antibody reactivity to the external domain of the calcium sensing receptor in the hypoparathyroidism of APS-I, raises the possibility that the antibody itself is causing the disease (66). Alternatively, enhanced Th2 responses may occur in patients with APS-I, which could lead in turn to a suppression of protective Th1 responses and ultimately result in defective defense from the invasion of yeast *Candida albicans*.

5. GENETICS

The presence of common clinical features in each type of APS suggests that the underlying inherited susceptibilities might be common to certain of its component diseases. Current genetic findings have clearly distinguished APS-I from the other two types of APS. APS-I has been linked to a non-HLA associated single gene, the autoimmune regulator (AIRE) gene located to chromosome 21q22.3, while APS-II and APS-III have their own distinctive associations with HLA class II genes depending upon their component diseases. However, these two syndromes remain to be further defined genetically, especially with respect to their underlying non-HLA genes.

5.1. GENETIC STUDIES IN APS-I

APS-I is clearly not associated with any class II HLA allele (100). The autosomal recessive pattern of inheritance was observed initially by the analyses of patients with idiopathic Addison's disease and hypoparathyroidism (101) and later associated with APS-I when the three types of APS were so classified (9). The autosomal recessive pattern was also observed in later studies on Finnish patients (19,102) and in Iranian Jewish patients with APS-I (18).

Genetic mapping and identification of APS-I gene, AIRE. A candidate gene, which is potentially responsible for APS-I, was mapped to the long arm of chromosome 21 (21q22.3), using allelic association and linkage analyses initially in 14 Finnish families of patients with APS-I. This candidate gene was located to a genomic interval between two polymorphic microsatellite markers, D21S49 and D21S171. The location of the candidate gene was further narrowed down to a range of less than 500 kb, around a polymorphic microsatellite marker, located near to the gene encoding phospho fructokinase of liver type (PFKL), by linkage analyses (103) and physical mapping (104). These studies involved more Finnish families with APS-I as well as patients from other European countries. We have also observed similar location of the candidate gene by linkage analyses using polymorphic microsatellite markers located at the above genomic interval on families from heterogeneous US patients with APS-I (105). Ultimately, a novel gene named AIRE (autoimmune regulator), located proximal to the gene of PFKL on the long arm of chromosome 21, was identified by two individual groups (106,107). The AIRE gene consists of 14 exons and encodes a protein with an estimated 545 amino acids. It is most likely that this is the disease gene responsible for APS-I, according to the evidence of detected mutations (Figure 1).

Putative role of AIRE protein. An understanding of the biological role of the AIRE protein should provide needed insights into the mechanism of autoimmunity and to APS-I in particular. AIRE gene is expressed as mRNA prevalently in thymus, but is also widely expressed in other tissues such as lymph nodes, pancreas, adrenal cortex and peripheral blood mononuclear cells (106,107). Such tissue distributions were confirmed by anti-serum raised specifically to AIRE protein (108). Attention has already been drawn to its nuclear dot staining pattern and its role in transcriptional regulation of the encoded protein based on the analyses of the predicted amino-acid sequence (106,107), and immunohistochemical staining of either AIRE gene transfected cell lines or human tissues (108–110). The AIRE protein contains two plant homeodomain (PHD) zinc-finger motifs, three LXXLL (L: leucine; X: any amino acid) motifs and a proline-rich region. A conservative domain named SAND was proposed to be a DNA-binding region for AIRE protein (111). The pattern of the two zinc-finger motifs in AIRE-I, has been found in many nuclear proteins, such as chromo-helicase-DNA binding protein (112) and chromatin-modulating proteins (113). These families of proteins containing such motifs are known to play roles in gene expression and regulation (112). In addition, a family of nuclear proteins contains the LXXLL motifs, which are represent the signature sequences that facilitates the interaction of different proteins with nuclear receptors, and related to function as co-activators of transcription (114). Accordingly, AIRE gene is most likely participating in the regulation of the expression of another gene(s).

Characterization of mouse AIRE gene. The counterpart of human AIRE gene in mouse has also been identified with a predicted protein of 552 amino acids (115–117), in a genomic region that contains a cluster of genes corresponding to those in human chromosome 21 (118). The mouse AIRE gene is composed of 14 exons covering a genomic region of approximately 13 kb. There is 75% homology at the nucleotide level over the coding region and a 71% homology at amino acid level between mouse and human AIRE. The motifs present in human AIRE protein also exist in the

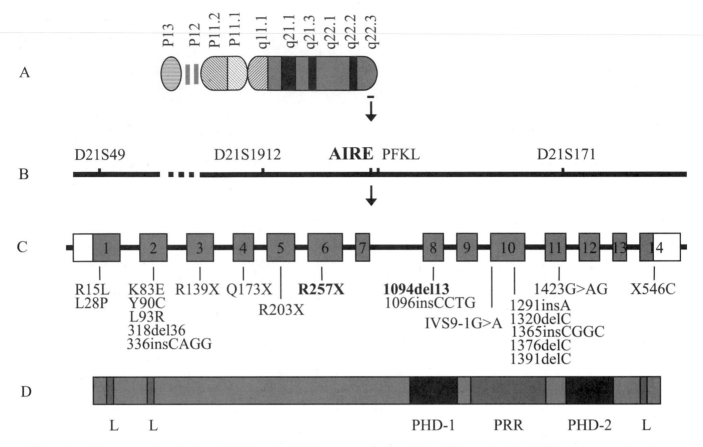

Figure 1. Chromosome localization of AIRE gene, AIRE gene mutations and the AIRE protein. **A**: AIRE gene is mapped to chromosome 21q22.3. **B**: AIRE gene is found to be located at the immediate up-stream of the gene encoding phospho fructokinase of liver type (PFKL). **C**: AIRE gene comprises 14 exons and spans over 11.9 kb genomic DNA. Twenty-one different mutations have been detected in APS-I patients with various ethnic backgrounds. R257X and 1094del13 are the dominant mutations. **D**: AIRE protein contains two PHD zinc-finger motifs (PHD), three LXXLL motifs (L) and a proline-rich region (PRR), typical of a transcription factor.

mouse, which include the highly conserved two PHD motifs, four NR box motifs. The mouse AIRE appears to have similar patterns of nuclear expression in uncharacterized epithelial cells in thymus, lymph node. The identification of mouse AIRE gene will lead to the creation of animal models, such as AIRE gene-knockout mouse, to study the function of AIRE protein in relation to its clear role in the immune system.

AIRE gene mutations. More than twenty different mutations in AIRE gene have been detected in APS-I patients with various ethnic backgrounds. The mutations were present across over most of the exons of AIRE gene (Table 2, and Figure 1), with different polymorphisms detected in coding and non-coding region of AIRE genomic region (119). The nucleotide position of mutations described below will be based on a reported cDNA sequence, (GenBank accession number : AB006682).

There are a total of five missense point mutations, 2 in exon 1 (R15L, L28P) and 3 in exon 2 (K83E, Y90C and L93R), found in AIRE gene of patients with APS-I. R15L

is due to a T to C substitution at nucleotide position of 171 that results in the change of arginine to leucine. R15L was detected in a single allele of a British patient with APS-I (120). L28P is due to a T to C substitution at nucleotide 210, which results in the change of leucine to proline at position 28 of predicted AIRE protein. This mutation was detected in a heterozygous form in a British patient and an American Caucasian patient with APS-I (120,121). K83E and Y90C are due to the substitution of A to G at the nucleotide position of 374 and 396, respectively. K83E was found in a Finnish patient with APS-I (106), and Y90C was detected in a British patient (120). L93R is due to a substitution of T to G at nucleotide position of 407, which results in the change of leucine to arginine. This mutation was detected in one allele of a French-Canadian girl with APS-I (122). It remains to be examined whether these 5 rare mutations that result in the substitution of a single amino acid in exon 1 or 2 could disrupt the function of AIRE protein. Alternatively, some of these mutations could be merely rare polymorphisms of AIRE gene that do not disrupt its function.

Table 2 AIRE gene mutations found in patients with APS-I

Mutation	Exon	Nucleotide change	Nucleotide* position	Effects on AIRE protein	Racial group detected	Allelic presentation	Reference
R15L	1	CGC > CTC	171	Arg → Leu	British	Het**	118
L28P	1	CTG > CCG	210	Leu → Pro	AmeCau, British	Het	118,119
K83E	2	AAG > GAG	374	Lys → Glu	Finnish	Het**	104
Y90C	2	TAT > TGT	396	Tyr → Cys	British	Het**	118
L93R	2	CTG > CGG	405	Leu → Arg	French Canadian	Het**	120
R139X	3	CGA > TGA	542	Stop codon	Sardinian	Hom/Het	21
Q173X	4	CAG > TAG	644	Stop codon	Hispanic	Het**	119
R203X	5	CGA > TGA	734	Stop codon	N. Italian	Het**	117
R257X	6	CGA > TGA	896	Stop codon	***	Hom/Het	21,104,105,117–119
336insCAGG	2	CAGG insertion	336	Frame shift	Arabian	Hom	119
1096insCCTG	8	CCTG insertion	1096	Frame shift	N. Italian	Hom	105,117
1365insCGGC	10	CGGC insertion	1365	Frame shift	Arabian	Hom	*****
1291insA	10	1 bp insertion	1291	Frame shift	Finnish	Het	105
318del36	2	36 bp deletion	318–353	A LXXLL motif deletion	AmeCau	Het**	119
1094del13	8	13 bp deletion	1094–1106	Frame shift	****	Hom/Het	21,105,117–119
1320delC	10	1 bp deletion	1320	Frame shift	French	Hom	105
1376delC	10	1 bp deletion	1376	Frame shift	British	Het**	118
1391delC	10	1 bp deletion	1391	Frame shift	AmeCau, British	Het	118,119
IVS9-1G>A	10	G>A	Acceptor	Exon 10 deletion	Hispanic	Het**	119
1423G>AC	11	G>AC	1423	Frame shift	AmeCau	Het**	119
X546C	14	TGA>TGT	1765	Stop codon destroyed	Finnish	Het	117

Abbreviations: Hom, homozygous allele; Het, heterozygous allele; AmeCau, American Caucasian.

* The nucleotide position of AIRE cDNA refers to GenBank Accession # AB006682. For mutation nomenclature based on the start codon of AIRE subtract 127 from the above number of nucleotides.

** Detected only in one allele of AIRE gene up to now.

*** R257X has been detected in Finnish, Swiss, Italian, German, British, New Zealand, and American Caucasian patients with APS-I.

**** 13 bp deletion has been detected in Finnish, Dutch, Italian, German, British, New Zealander, Asian and American Caucasian patients with APS-I

***** Our unpublished observation.

Four nonsense point mutations due to the introduction of stop codons were detected in 4 separate exons of AIRE gene in patients with APS-I. These mutations include R139X (exon 1), Q173X (exon 4), R203X (exon 5) and R257X (exon 6). R139X is a nonsense mutation due to a C to T transition resulting in the change of an arginine codon to a stop codon. R139X is the dominant mutation for Sardinian patients with APS-I, and has only been found in this ethnic group (20). Q173X is due to a C to T transition that leads to the change of glutamine codon to a stop codon at position of 173 of AIRE protein. Q173X has only been detected in a Hispanic patient with APS-I (121). R203X is due to C to T transition in a CpG dinucleotide resulting in the change of an agrinine codon to a stop codon. This mutation has only been found in one Northern Italian patient with APS-I (119). R257X is a null mutation present in patients with different ethnic backgrounds. R257X is due to transition of C to T at amino-acid position 257 in exon 6. This results in the change of an arginine codon (CCA) to a stop codon (TGA), with the production of a truncated protein with about 256 amino acids (106,107). R257X is a dominant mutation for Finnish patients with APS-I and is also frequently present in patients

with other ethnic backgrounds, such as in north Italians, Swiss, British, Germans, New Zealanders and American Caucasians (119,121).

There are four null mutations caused by the insertion of 1 base pair nucleotide (1291insA, exon 10) or 4 base pair nucleotides (335insCAGG, exon 2; 1096insCCTG, exon 8; 1365insCGGC, exon 10) in the coding region of AIRE gene. These mutations would result in a frame shift during translation and produce premature truncated proteins. 1291insA was detected in 2 Finnish patients with APS-I, and this mutation can result in a frame shift and produce a truncated 422 amino acids (107). 336insCAGG leads to a frame-shift at amino-acid position of 69 of AIRE protein, and produces a premature truncation of a 217 amino acid protein with 148 unrelated C-terminal amino acids (121). The mutation 1096insCCTG can lead to a frame shift at the amino acid codon L323 and produce a premature truncation of a 371-amino acid protein with an unrelated 48 C-terminal amino acids attaching to the truncated AIRE protein (107). This mutation has been found in a North Italian with APS-I (107). 1365insCGGC was recently identified by us in a patient of Arabian descent. This mutation can result in a frame shift and produce a 423 amino acids-truncated protein. Interestingly, all three mutations due to the four base pair nucleotide insertion in the coding region of AIRE gene have the same pattern of two direct repeats of the inserted nucleotides, and were homozygous for the respective mutations in the three patients. Replication errors may have caused such insertion.

Null mutations caused by the deletion of nucleotides in the coding region of AIRE gene were also detected in patients with APS-I. These mutations include three mutations caused by 1 base pair deletion at exon 10 (1320delC, 1376delC and 1391delC), one mutation caused by a 13 base pair deletion at exon 8 (1094del13) and one mutation caused by a 36 base pair deletion at exon 2 (318del36). 1320delC was detected in a French patient (107) and 1376delC was detected in a British patient with APS-I (120). Both of the mutations could result in a frame shift to produce a truncated 478 amino acids. 1391delC is a rare mutation found only in one American patient with APS-I (121), and could result in a frame shift and produce a truncated residue. 1094del13 caused by a 13 base pair deletion at neucleotide position of 1094–1106 is a dominant lesion that occurs to APS-I patients with various ethnic backgrounds (20,107,119–121). The mutation of 318del36 was detected in one allele of an American Caucasian patient with APS-I (121). This mutation is likely caused by mispairing of TCCTGG repeats present at each end of the deletion, and delete the second of the three LXXLL motifs (114).

There are three rare other mutations detected in patients with APS-I, namely IVS9-1G>A (acceptor of exon 10), 1423G>AG (exon 11) and X546C (exon 14). IVS9-1G>A is a rare mutation due to a splice site mutation at the acceptor of exon 10, which leads to the skipping of exon 10 or to a frame shift if the cryptic splice site to exon 10 is used during transcription. Thus, the skipping of exon 10 would result in the deletion of 60 amino acids between the two predicted PHD zinc finger motifs. This mutation was found in an allele of a Hispanic patient with APS-I who has the mutation of Q173X in the other allele (121). 1423G>AC is due to the substitution of G by AC at nucleotide position 1423, and result in a frame shift to produce a premature truncated residue of AIRE protein. This mutation was detected in an American Caucasian patient with APS-I (121). X546C occurs to the stop codon of AIRE gene. It changes the stop codon (TGA) to a cysteine codon (TGT), and results in the AIRE protein attached with a tail of 60 amino acids. X546C was fond in one allele of AIRE gene in two individual Finnish APS-I patients, whose other allele had R257X (119).

In summary, R257X and 1094del13 are the two major mutations found in APS-I patients with different ethnic backgrounds. R257X is the dominant mutation for Finnish patients and North Italian patients with APS-I, while 1094del13 is a dominant mutation for British and American Caucasians, and R139X is the dominant mutation found only in Sardinian patients with APS-I. Other mutations appear to be rare and occur in patients with distinct ethnic backgrounds. The predicted outcomes for most of the AIRE mutations are truncated conceptual protein of AIRE due to either the introduction of a stop codon or frame-shift of the coding gene.

Racial distribution of AIRE gene mutations. Founder effects exist for some genetically-isolated populations according to the analyses on mutations and haplotype of polymorphic markers closely associated with AIRE gene locus. Recombination events become less for two genomic markers that are located more closely than those of remotely located in the same chromosome, i.e. linkage disequilibrium happens more often to closely located genes or polymorphic markers. Thus, individuals are likely to have common ancestors if they share same haplotype for polymorphic markers, which are in linkage disequilibrium, especially for those from genetically isolated populations.

The frequency of each type of the mutations and the haplotypes in relation to the polymorphic markers near AIRE gene differ in the patients from various ethnic backgrounds. Most of the mutations have only been detected in single allele of patients with APS-I, reflecting the sporadic occurrence of the disease in most ethnic groups. That patients with same AIRE gene mutations often had different closely-linked haplotypes suggests that either ancient mutational events or multiple independent events occurred to account for the AIRE mutations and haplotypes observed. For example, R257X is also the major mutation present in patients from European countries other than Finland. Those patients who had R257X tend to have diversified haplotype of D21S1912—PFKL (119). This is also the case for the other major mutation, 1094del13, since different haplotypes were present in patients with different ethnic origins. For example, 9 out of 15 alleles of 1094del13 detected in 13 patients from a group of American Caucasian patients with APS-I had different haplotypes of D21S1912—PFKL (121), suggesting multiple independent events led to the 1094del13 mutation. This should be expected since the patients were of heterogeneous origins typical of the North American population.

However, particular ethnic groups had high frequencies for some mutations accompanied with their distinct haplotypes, suggesting the presence of founder effect in these ethnic groups. Remarkably, R257X is present in up to 82% alleles of Finnish patients with APS-I and accompanied with one haplotype of closely linked polymorphic markers, D21S1912, and PFKL (106,107). This is in concordance with the previous assumption that 85% cases of APS-I in Finnish patients are due to one major mutation commonly present in the ancestors of the Finnish population based on the haplotype analysis on polymorphic markers located closely to AIRE gene (103). Studies of 12 British families with APS-I for AIRE gene mutations found that 17 of the 24 possible mutant AIRE alleles tested had 1094del13 with a common haplotype spanning the AIRE gene locus (120). Similarly, mutation R139X was reported to be present in 90% (18/20) of independent alleles with identical haplotypes for D21S1912—PFKL in Sardinian patients with APS-I (20). The carrier frequencies in general population were 1/250 for R257X in Finns (107), 1/576 for 1094del13 in the British (120), and 5/300 for R139X in Sardinian (20). These data indicate that founder effects of the mutations are present in these three ethnic groups.

Genotypes and phenotype relationship. The correlation between the mutant genotypes of AIRE gene and the expressed clinical phenotypes of APS-I is not so obvious, suggesting that both environmental factors and/or background genes are also involved in the disease process. Patients with same AIRE mutation often present with different component diseases of APS-I or the orders of the appearance of component diseases (20,119,120). Different phenotype expressions are also present in affected siblings. Such variations in the clinical phenotype may be the result of the involved gene interacting with the immune system when exposed to some unknown environment factors. Some patients with a nonsense mutation in one copy and a missense mutation in the other copy of AIRE gene may not develop the full spectrum of clinical manifestations of APS-I, or at least not so by the age of 30 years (120). However, some patients did develop fully the clinical manifestation of APS-I with the presence of a missense mutation of AIRE gene (122).

Background genes within populations could also influence the expressed phenotypes of APS-I. Finnish patients with APS-I often have ectodermal and enamel hypoplasia (19), which appears not to be the result of calcium deficiency due to hypoparathyroidism since ectodermal and enamel hypoplasia can occur in APS-I patients with or without overt hypoparathyroidism (19,123). However, non-Finnish patients with APS-I are seldom seen with enamel hypoplasia (18,105,124). Also, IMD is rarely seen in patients with APS-I in the US, however, it does occur in some Finnish patients with APS-I, especially with increasing age. Since the Finns have the highest incidence rates of IMD in the world, they may coincidentally have a relatively high frequency of IMD genes in their APS-I patients. In addition, mucocutaneous candidiasis is the common component disease in patients with APS-I, but it is a relative rarity among the Iranian Jewish patients with APS-I (18). Presumably, different mutations in the responsible gene could be involved in these various phenotypes as shown in the presence of different haplotypes for genetic markers among patients with different ethnic backgrounds (103). In other words, background genes or genes with

epistatic effects may be responsible for variations in the expressed phenotype. Alternatively, more than one gene may be responsible for the phenotypic development of APS-I, albeit, this is becoming increasingly unlikely. Thus, studies for the function of the identified AIRE gene could shed more light into the pathogenic mechanism of APS-I, especially, autoimmune Addison's disease, which also occurs either as an isolated disease or in association with APS-II.

Role of the AIRE gene in the pathogenesis of APS-I. The biological role of AIRE protein in maintaining self-tolerance remains to be investigated. However, it represents the first clear-cut instance of a Mendelian gene associated with autoimmune disease in man. AIRE protein with a deleted PHD domain showed different nuclear localization compared to wild type (109), suggesting the mutant AIRE molecule would disrupt its function in gene regulation. It has not been established how a defect in the AIRE gene could result in the occurrence of the three core component diseases and the associated autoimmune diseases of APS-I. AIRE protein is prevalently expressed in unknown specific cells in the medulla of thymus, paracortex and medulla of lymph node. This indicates that AIRE gene could be involved in negative selection in the thymus to delete self-reactive T cell clones, and involving in enduring self-reactive T cells to anergy to maintain peripheral tolerance.

5.2. Genetic Studies in APS II/III

APS-II/III express an autosomal dominant pattern with incomplete penetrance, since APS-II often affects individuals in many generations of the same family (26,60,100, 125). Addison's disease, either as an isolated disease or with APS-II is associated with HLA genes, although non-HLA related gene(s) could also be involved in pathogenic process of these two syndromes.

HLA genes and APS II/III. Patients with APS-II often share the same susceptible HLA alleles as patients with only an individual component disease (126,127). For example, DR3-DQB1*0201/ DQB1*0302 are associated with APS-II when there is IMD (128). Such HLA associations suggest that particular molecules of HLA are required in the development of component autoimmune diseases, and the expressions of a particular autoimmune phenotype depends on the involvement of other gene products, especially in a multicomponent autoimmune syndrome like APS-II. For example, DQB1*0302 was no longer in association with APS-II when those patients with APS-II, plus overt clinical IMD or positive for autoantibodies to islet antigens, were excluded from the analyses (60). Also, the development of same disease with different susceptible HLA alleles has been observed in inter-racial studies for HLA susceptibility (129). The frequencies of patients with DQB1*0301 are increased in Hashimoto's thyroiditis, DRB1*03, DRB3 and DQA1*0501 genes are increased in Graves' disease, and perhaps DRB1*13 in vitiligo.

Non-HLA genes and APS II/III. Patients with the appearance of APS-II/III often express different patterns or order of the component diseases, suggesting that environmental factors could be involved in the development of the diseases, as well as the involvement of multiple disease-associated genes. IMD has been linked to more than 10 loci in non-HLA genomic region (130), and susceptible loci have been linked to Graves' disease (131). Recent studies indicate that the gene encoding the cytotoxic T-lymphocyte antigen-4 (CTLA-4), which plays an important role in the down-regulation of T cell activity, has been reported to be associated with the component diseases of APS. For example, CTLA-4 gene polymorphism has been associated with Graves' disease (132–135), IMD (134), autoimmune Addison's disease (136), and vitiligo in associated with Graves' or Addison's disease (137). However, the association of CTLA-4 gene with patients who have isolated Addison's disease or APS-II could differ according to their racial backgrounds (136), or in association with a particular HLA genotype, DQA1*0501 (138), suggesting multigene involvement of the disease.

The AIRE gene appears not to be involved in patients with isolated Addison's disease or APS-II, since we could not find AIRE mutations in more than 20 patients with either isolated Addison's disease or Addison's disease in association with APS-II (personal communication). This suggests the different etiological pathway for APS-I and APS-II, although both diseases are involved with auto-immune destruction adrenal cortex. This again suggests that the cause of pathogenic autoimmunity may involve different pathogenic processes, and that alternative pathways exist for the break down of self-tolerance of immune system. Thus, the genetic studies for autoimmune endocrine syndromes should bring insights to the understanding of the pathogenic process of these diseases, which may be applicable to the component autoimmunities.

6. CONCLUSIONS

The existence of APS continues to fascinate endocrinologists and immunologists alike. The clinical classification of the APS has led to the identification of the AIRE gene, a simple Mendelian recessive gene. The obvious importance of this gene to normal immune functioning, particularly immune regulation and tolerance, is obvious and needs to be actively explored. AIRE gene knockout mice are eagerly awaited for

such studies. However, there are likely to be autoimmune-susceptibility genes underlying APS-II and III also, which need to be actively sought. The products of these susceptibility genes may interact with those of HLA genotypes together with the influence of environmental factors, leading to the development of component diseases. The addition value of the APS to medical science lies in the future discovery of the targeted autoantigens, since some of these may have relevance to their more common simple component diseases. For US patients, we are most interested in their inclusion in a national registry so that investigations can proceed with the numbers required for statistical resolutions. When one considers the progress made in understanding the APS since Addison's descriptions 150 years ago, one can only speculate upon the progress that will be made in the next decades. A new chapter in immunogenetics is anticipated stemming from the clinical observations of these interesting experiments of nature.

References

1. Addison, T. 1849. Anaemia+Disease of the suprarenal capsules. *Lond. Med. Gaz.* 8:517–518.
2. Brenner, O. 1928. Addison's disease with atrophy of the cortex of the suprarenals. *O. J. Med.* 22:121–144.
3. Swingle, W.W., and J.J. Pfiffner. 1931. Studies on adrenal cortex; aqueous extract of adrenal cortex which maintain life of bilaterally adrenal-ectomized cats. *Am. J. Physiol.* 96:164–179.
4. Rowntree, I.G., C.H. Greene, R.G. Ball, W.W. Swingle, and J.J. Pfiffner. 1931. Treatment of Addison's disease with the cortical hormone of the suprarenal gland. *J. Am. Med. Assoc.* 97:1446–1553.
5. Rowntree, L.G., C.H. Greene, W.W. Swingle, and J.J. Pfiffner. 1931. Addison's Disease. *J. Am. Med. Assoc.* 96:231–235.
6. Hashimoto, H. 1912. Zur kenntnis der lymphomatosen veranderung der schilddruse (struma lymphomatosa). *Acta. Klin. Chir.* 97:219–248.
7. Schmidt, M. 1926. Eine biglandulare erkrankung (Nebennieren und Schilddrusse) bei Morbus Addisonii. *Verh. Dtsch. Pathol. Ges.* 21:212–221.
8. Carpenter, C., N. Solomon, S. Silverberg, et al. 1964. Schmidt's syndrome (thyroid and adrenal insufficiency): A review of the literature and a report of fifteen new cases including ten instances of coexistent diabetes mellitus. *Medicine* 43:153–180.
9. Neufeld, M., N. Maclaren, and R. Blizzard. 1980. Autoimmune polyglandular syndromes. *Pediatr. Ann.* 9:154–162.
10. Neufeld, M., N. Maclaren, and R. Blizzard. 1981. Two types of autoimmune Addison's disease associated with different polyglandular autoimmune (PGA) syndromes. *Medicine* 60:355–362.
11. Oelkers, W. 1996. Adrenal insufficiency. *N. Engl. J. Med.* 335:1206–1212.
12. Petri, M., and J. Nerup. 1971. Addison's adrenalitis. Studies on diffuse lymphocytic adrenalitis (idiopathic Addison's disease) and focal lymphocytic infiltration in a control material. *Acta Pathol. Microbiol. Scand.* 79:381–388.

13. Kendereski, A., D. Micic, M. Sumarac, S. Zoric, D. Macut, M. Colic, A. Skaro-Milic, and Z. Bogdanovic. 1999. White Addison's disease: what is the possible cause? *J. Endocrinol. Invest.* 22:395–400.

14. Cuoco, L., M. Certo, R.A. Jorizzo, I. De Vitis, A. Tursi, A. Papa, L. De Marinis, P. Fedeli, G. Fedeli, and G. Gasbarrini. 1999. Prevalence and early diagnosis of coeliac disease in autoimmune thyroid disorders. *Ital. J. Gastroenterol. Hepatol.* 31:283–287.

15. Dieterich, W., T. Ehnis, M. Bauer, P. Donner, U. Volta, E. Riecken, and D. Schuppan. 1997. Identification of tissue transglutaminase as the autoantigen of celiac disease. *Nat. Med.* 3:797–801.

16. Picarelli, A., L. Maiuri, A. Frate, M. Greco, S. Auricchio, and M. Londei. 1996. Production of antiendomysial antibodies after in-vitro gliadin challenge of small intestine biopsy samples from patients with coeliac disease. *Lancet* 348:1065–1067.

17. Maki, M. 1996. Coeliac disease and autoimmunity due to unmasking of cryptic epitopes? *Lancet* 348:1046–1047.

18. Zlotogora, J., and M. Shapiro. 1992. Polyglandular autoimmune syndrome type I among Iranian Jews. *J. Med. Genet.* 29:824 826.

19. Ahonen, P., S. Myllärniemi, J. Perheentupa, and L. Peltonen. 1990. Clinical variation of autoimmune polyendocrinopathy-candidiasis-ectodermal dystrophy (APECED) in a series of 68 patients. *N. Engl. J. Med.* 322:1829–1836.

20. Rosatelli, M., A. Meloni, M. Devoto, et al. 1998. A common mutation in Sardinian autoimmune polyendocrinopathy-candidiasis-ectodermal dystrophy patients. *Hum. Genet.* 103:428–434.

21. Brun, J. 1982. Juvenile autoimmune polyendocrinopathy. *Horm. Res.* 16:308–316.

22. Bottazzo, G., B. Dean, J.M. McNally, E.H. McKay, P.G. Scvift, and D.R. Gamble. 1985. In situ characterization of autoimmune phenomena and expression of HLA molecules in the pancreas in diabetic insulitis. *N. Engl. J. Med.* 313:353–360.

23. Maclaren, N.K., and R.M. Blizzard. 1985. Adrenal autoimmunity and autoimmune polyglandular syndromes. In: *The autoimmune disease.* Rosc N.R. and Mackay I.R., editors. Academic Press, Orlando, pp. 201–225.

24. Ekwall, O., H. Hedstrand, L. Grimelius, J. Haavik, J. Perheentupa, J. Gustafsson, E. Husebye, O. Kampe, and F. Rorsman. 1998. Identification of tryptophan hydroxylase as an intestinal autoantigen. *Lancet* 352:279–283.

25. Arulanantham, K., J. Dwyer, and M. Genel. 1979. Evidence for defective immunoregulation in the syndrome of familial candidiasis endocrinopathy. *N. Engl. J. Med.* 300:164–168.

26. Eisenbarth, G., P. Wilson, F. Ward, C. Buckley, and H. Lebovita. 1979. The polyglandular failure syndrome: Disease inheritance, HLAtype, and immune function studies in patients and families. *Ann. Intern. Med.* 91:528–533.

27. Jacobson, D.L., S.J. Gange, N.R. Rose, and N.M. Graham. 1997. Epidemiology and estimated population burden of selected autoimmune diseases in the United States. *Clin. Immunol. Immunopathol.* 84:223–243.

28. Sorenson, R.L., D.G. Garry, and T.C. Brelje. 1991. Structural and functional considerations of GABA in islets of Langerhans. Beta-cells and nerves. *Diabetes* 40:1365–1374.

29. Wilkin, T. 1990. Receptor autoimmunity in endocrine disorders. *N. Engl. J. Med.* 323:1318–1324.

30. Atkinson, M.A., and N.K. Maclaren. 1993. Islet cell autoantigens in insulin-dependent diabetes. *J. Clin. Invest.* 92:1608–1616.

31. Degli, E.M., and I.R. Mackay. 1997. The GABA network and the pathogenesis of IDDM. *Diabetologia* 40:352–356.

32. Elder, M., N. Maclaren, and W. Riley. 1981. Gonadal autoantibodies in patients with hypogonadism and/or Addison's disease. *J. Clin. Endocrinol. Metab.* 52:1137–1142.

33. Ahonen, P., A. Miettinen, and J. Perheentupa. 1987. Adrenal and steroidal cell antibodies in patients with autoimmune polyglandular disease type I and risk of adrenocortical and ovarian failure. *J. Clin. Endocrinol. Metab.* 64:494–500.

34. Betterle, C., M. Volpato, B. Rees Smith, J. Furmaniak, S. Chen, R. Zanchetta, N.A. Greggio, B. Pedini, M. Boscaro, and F. Presotto. 1997. Adrenal cortex and steroid 21-hyrooxylase autoantibodies in adult patients with organ-specific autoimmune diseases: markers of low progression to clinical Addison's disease. *J. Clin. Endocrinol. Metab.* 82.932–938.

35. Winqvist, O., F. Karlsson, and O. Kampe. 1992. 21Hydroxylase, a major autoantigen in idiopathic Addison's disease. *Lancet* 339:1559–1562.

36. Bednarek, J., J. Furmaniak, N. Wedlock, Y. Kiso, J. Bednarek, J. Baumann-Antczak, C. Morteo, P. Sudbury, A. Hinchcliff, and B. Rees Smith. 1992. Steroid 21-hydroxylase is a major autoantigen involved in adult onset autoimmune Addison's disease. *FEBS Lett.* 309:51–55.

37. Krohn, K., R. Uibo, E. Aavik, P. Peterson, and K. Savilahti. 1992. Identification by molecular cloning of an autoantigen associated with Addison's disease as steroid 17″hydroxylase. *Lancet* 339:770–773.

38. Baumann-Antczak, A., N. Wedlock, J. Bednarek, Y. Kiso, H. Krishnan, S. Fowler, B. Smith, and J. Furmaniak. 1992. Autoimmune Addison's disease and 21-hydroxylase. *Lancet* 340:429–430.

39. Falorni, A., A. Nikoshkov, S. Laureti, E. Grenback, A.L. Hulting, G. Casucci, E. Santeusanio, P. Brunetti, A. Luthman, and A. Lernmark. 1995. High diagnostic accuracy for idiopathic Addison's disease with a sensitive radio-binding assay for autoantibodies against recombinant human 21-hydroxylase. *J. Clin. Endocrinol. Metab.* 80:2752–2755.

40. Chen, S., J. Sawicka, C. Betterle, and M. Powell. 1996. Autoantibodies to steroidogenic enzymes in autoimmune polyglandular syndrome, Addison's disease, and premature ovarian failure. *J. Clin. Endocrinol. Metab.* 81:1871–1876.

41. Tanaka, H., M. Perez, M. Powell, J.F. Sanders, J. Sawicka, S. Chen, L. Prentice, T. Asawa, C. Betterle, M. Volpato, B.R. Smith, and J. Furmaniak. 1997. Steroid 21-hydroxylase autoantibodies: measurements with a new immunoprecipitation assay. *J. Clin. Endocrinol. Metab.* 82:1440–1446.

42. Betterle, C., M. Volpato, B. Pedini, S. Chen, B.R. Smith, and J. Furmaniak. 1999. Adrenal-cortex autoantibodies and steroid-producing cells autoantibodies in patients with Addison's disease: comparison of immunofluorescence and immunoprecipitation assays. *J. Clin. Endocrinol. Metab.* 84:618–622.

43. Furmaniak, J., S. Kominami, T. Asawa, N. Wedlock, J. Colls, and B.R. Smith. 1994. Autoimmune Addison's disease—evidence for a role of steroid 21- hydroxylase autoantibodies in adrenal insufficiency. *J. Clin. Endocrinol. Metab.* 79:1517–1521.

44. Boscaro, M., C. Betterle, M. Volpato, F. Fallo, J. Furmaniak, B. Rees Smith, and N. Sonino. 1996. Hormonal responses during various phases of autoimmune adrenal failure: no evi-

dence for 21-hydroxylase enzyme activity inhibition *in vivo*. *J. Clin. Endocrinol. Metab.* 81:2801–2804.

45. Wedlock, N., T. Asawa, A. Baumann-Antezak, B. Smith, and J. Furmaniak. 1993. Autoimmune Addison's disease. Analysis of autoantibody binding sites on human steroid 21-hydroxylase. *FEBS Lett.* 332:123–126.

46. Song, Y., E. Connor, Y. Li, B. Zorovich, P. Balducci, and N. Maclaren. 1994. The role of tyrosinase in autoimmune vitiligo. *Lancet* 344:1049–1052.

47. Volpato, M., L. Prentice, S. Chen, C. Betterle, B. Rees Smith, and J. Furmaniak. 1998. A study of the epitopes on steroid 21-hydroxylase recognized by autoantibodies in patients with or without Addison's disease. *Clin. Exp. Immunol.* 111:422–428.

48. Chen, S., J. Sawicka, L. Prentice, J.F. Sanders, H. Tanaka, V. Petersen, C. Betterle, M. Volpato, S. Roberts, M. Powell, B.R. Smith, and J. Furmaniak. 1998. Analysis of auto-antibody epitopes on steroid 21-hydroxylase using a panel of monoclonal antibodies. *J. Clin. Endocrinol. Metab.* 83:2977–2986.

49. Asawa, T., N. Wedlock, A. Baumann-Antczak, B. Smith, and J. Furmaniak. 1994. Naturally occurring mutations in human steroid 21-hydroxylase influence adrenal autoantibody binding. *J. Clin. Endocrinol. Metab.* 79:372–376.

50. Winqvist, O., J. Gustafsson, F. Rorsman, F. Karlsson, and O. Kampe. 1993. Two different cytochrome P450 enzymes are the adrenal antigens in autoimmune polyendocrine syndrome type I and Addison's disease. *J. Clin. Invest.* 92:2377–2385.

51. Seissler, J., M. Schott, H. Steinbrenner, P. Peterson, and W. Scherbaum. 1999. Autoantibodies to adrenal cytochrome P450 antigens in isolated Addison's disease and auto-immune polyendocrine syndrome type II. *Exp. Clin. Endocrinol. Diabetes* 107:208–213.

52. Peterson, P., and K. Krohn. 1994. Mapping of B cell epitopes on steroid 17 alpha-hydroxylase, an autoantigen in autoimmune polyglandular syndrome type I. *Clin. Exp. Immunol.* 98:104–109.

53. Betterle, C., M. Volpato, S.B. Rees, J. Furmaniak, S. Chen, R. Zanchetta, N.A. Greggio, B. Pedini, M. Boscaro, and F. Presotto. 1997. II. Adrenal cortex and steroid 21hydroxylase autoantibodies in children with organ-specific autoimmune diseases: markers of high progression to clinical Addison's disease. *J. Clin. Endocrinol. Metab.* 82:939–942.

54. Ketchum, C., W. Riley, and N. Maclaren. 1984. Adrenal dysfunction in asymptomatic patients with adrenocortical autoantibodies. *J. Clin. Endocrinol. Metab.* 58:1166–1170.

55. Uibo, R., E. Aavik, P. Peterson, J. Perheentupa, S., Aranko, R. Pelkonen, and K. Krohn. 1994. Autoantibodies to cytochrome P450scc, P450c17, and P450c21 in autoimmune polyglandular disease types I and II and in isolated Addion's disease. *J. Clin. Endocrinol. Metab.* 78:323–328.

56. Riley, W., N. Maclaren, and M. Neufeld. 1980. Adrenal autoantibodies and Addison disease in insulin-dependent diabetes mellitus. *J. Pediatr.* 97:191–195.

57. Falorni, A., S. Laureti, A. Nikoshkov, M.L. Picchio, B. Hallengren, C.L. Vandewalle, F.K. Gorus, C. Tortoioli, H. Luthman, P. Brunetti, and F. Santeusanio. 1997. 21-hydroxylase autoantibodies in adult patients with endocrine autoimmune diseases are highly specific for Addison's disease. Belgian Diabetes Registry. *Clin. Exp. Immunol.* 107:341–346.

58. Brewer, K., V. Parziale, and G. Eisenbarth. 1997. Screening patients with insulin-dependent diabetes mellitus for adrenal insufficiency. *N. Engl. J. Med.* 337:202.

59. Peterson, P., H. Salmi, H. Hyoty, A. Miettinen, H. Reijonen, M. Knip, H.K. Akerblom, and K. Krohn. 1997. Steroid 21-hydroxylase autoantibodies in insulin-dependent diabetes mellitus. Childhood Diabetes in Finland (DiMe) Study Group. *Clin. Immunol. Immunopathol.* 82:37–42.

60. Huang, W., E. Connor, T.D. Rosa, A. Muir, D. Schatz, J. Silverstein, S. Crockett, J.X. She, and N.K. Maclaren. 1996. Although DR3-DQB1*0201 may be associated with multiple component diseases of the autoimmune polyglandular syndromes, the human leukocyte antigen DR4-DQB1*0302 haplotype is implicated only in beta-cell autoimmunity. *J. Clin. Endocrinol. Metab.* 81:2559–2563.

61. Peterson, P., R. Uibo, J. Peranen, and K. Krohn. 1997. Immunoprecipitation of steroidogenic enzyme autoantigens with autoimmune polyglandular syndrome type I (APS-1) sera; further evidence for independent humoral immunity to P450c17 and p450c21. *Clin. Exp. Immunol.* 107:335–340.

62. Yu, L., K.W. Brewer, S. Gates, A. Wu, T. Wang, S.R. Babu, P.A. Gottlieb, B.M. Freed, J. Noble, H.A. Erlich, M.J. Rewers, and G.S. Eisenbarth. 1999. DRB1*04 and DQ alleles: expression of 21-hydroxylase autoantibodies and risk of progression to Addison's disease. *J. Clin. Endocrinol. Metab.* 84:328–335.

63. Blizzard, R., D. Chee, and W. Davis. 1966. The incidence of parathyroid and other antibodies in the sera of patients with idiopathic hypoparathyroidism. *Clin. Exp. Immunol.* 1:119–128.

64. Chapman, C., A. Bradwell, and P. Dykks. 1986. Do parathyroid and adrenal autoantibodies coexist? *J. Clin. Pathol.* 39:813–814.

65. Betterle, C., A. Caretto, M. Zeviani, B. Pedini and C. Salvate. 1985. Demonstration and characterization of antihuman mitochondria autoantibodies in idiopathic hypoparathyroidism and in other conditions. *Clin. Exp. Immunol.* 62:353–360.

66. Li, Y., Y. Song, N. Rais, E. Connor, D. Schatz, A. Muir, and N. Maclaren. 1996. Autoantibodies to the extracellular domain of the calcium sensing receptor in patients with acquired hypoparathyroidism. *J. Clin. Invest.* 97:910–914.

67. Rorsman, F., E.S. Husebye, O. Winqvist, E. Bjork, F.A. Karlsson, and O. Kampe. 1995. Aromatic-L-amino-acid decarboxylase, a pyridoxal phosphate-dependent enzyme, is a beta-cell autoantigen. *Proc. Natl. Acad. Sci. USA* 92:8626–8629.

68. Velloso, L., O. Winqvist, J. Gustafsson, O. Kampe, and F. Karlsson. 1994. Autoantibodies against a novel 51 kDa islet antigen and glutamate decarboxylase isoforms in autoimmune polyendocrine syndrome type I. *Diabetologia* 37:61–69.

69. Solimena, M., F. Folli, R. Aparisa, G. Pozza, and P. De Camilli. 1990. Autoantibodies to GABA-ergic neurons and pancreatic beta cells in stiff-man syndrome. *N. Engl. J. Med.* 322:1555–1560.

70. Tuomi, T., M.J. Rowley, W.J. Knowles, Q.Y. Chen, T. McAnally, P.Z. Zimmet, and I.R. Mackay. 1994. Autoantigenic properties of native and denatured glutamic acid decarboxylase: evidence for a conformational epitope. *Clin. Immunol. Immunopathol.* 71:53–59.

71. Elrehewy, M., Y. Kong, A. Giraldo, and N. Rose. 1981. Syngeneic thyroglobulin is immunogenic in good responder mice. *Eur. J. Immunol.* 11:146–151.

72. Shimojo, N., Y. Kohno, K. Yamaguchi, S. Kikuoka, A. Hoshioka, H. Niimi, A. Hirai, Y. Tamura, Y. Saito, L.D. Kohn, and K. Tahara. 1996. Induction of Graves-like disease

in mice by immunization with fibroblasts transfected with the thyrotropin receptor and a class II molecule. *Proc. Natl. Acad. Sci. USA* 93:11074–11079.

73. Kikuoka, S., N. Shimojo, K.I. Yamaguchi, Y. Watanabe, A. Hoshioka, A. Hirai, Y. Saito, K. Tahara, L.D. Kohn, N. Maruyama, Y. Kohno, and H. Niimi. 1998. The formation of thyrotropin receptor (TSHR) antibodies in a Graves' animal model requires the N-terminal segment of the TSHR extracellular domain. *Endocrinology* 139:1891–1898.

74. Maclaren, N., and W. Riley. 1985. Thyroid, gastric, and adrenal autoimmunities associated with insulindependent diabetes mellitus. *Diabetes Care 8(Suppl. 1)* 34–38.

75. Karlsson, F.A., P. Burman, L. Loof, and S. Mardh. 1988. Major parietal cell antigen in autoimmune gastritis with pernicious anemia is the acid-producing H+, K+-adenosine triphosphatase of the stomach. *J. Clin. Invest.* 81:475–479.

76. Burman, P., S. Mardh, L. Norberg, and F. Karlsson. 1989. Parietal cell antibodies in pernicious anemia inhibit H+, K+adenosine triphosphatase, the proton pump of the stomach. *Gastroenterology* 96:1434–1438.

77. Toh, B.H., P.A. Gleeson, R.J. Simpson, R.L. Moritz, J.M. Callaghan, I. Goldkorn, C.M. Jones, T.M. Martinelli, F.T. Mu, and D.C. Humphris. 1990. The 60- to 90-kDa parietal cell autoantigen associated with autoimmune gastritis is a beta subunit of the gastric H+/K(+)-ATPase (proton pump). *Proc. Natl. Acad. Sci. USA* 87:6418–6422.

78. Betterle, C., R. Mirakian, D. Doniach, G.I. Bottazo, W. Riley, and N.K. Maclaren. 1984. Antibodies to melanocytes in vitiligo. *Lancet* 1:159.

79. Baharav, E., O. Merimski, Y. Shoenfeld, R. Zigelman, B. Gilbrud, G. Yecheskel, P. Youinou, and P. Fishman. 1996. Tyrosinase as an autoantigen in patients with vitiligo. *Clin. Exp. Immunol.* 105:84–88.

80. Xie, Z., D. Chen, D. Jiao, and J. Bystryn. 1999. Vitiligo antibodies are not directed to tyrosinase. *Arch. Dermatol.* 135:417–422.

81. Fishman, P., O. Merimski, E. Baharav, and Y. Shoenfeld. 1997. Autoantibodies to tyrosinase: the bridge between melanoma and vitiligo. *Cancer* 79:1461–1464.

82. Visseren, M., A. van Elsas, E. van der Voort, M. Ressing, W. Kast, P. Schrier, and C. Melief. 1995. CTL specific for the tyrosinase autoantigen can be induced from healthy donor blood to lyse melanoma cells. *J. Immunol.* 154:3991–3998.

83. Overwijk, W., D. Lee, D. Surman, K. Irvine, C.E. Touloukian, C.C. Chan, M.W. Carroll, B. Moss, S.A. Rosenberg, and N.P. Restifo. 1999. Vaccination with a recombinant vaccinia virus encoding a "self" antigen induces autoimmune vitiligo and tumor cell destruction in mice: requirement for CD4(+) T lymphocytes. *Proc. Natl. Acad. Sci. USA* 96:2982–2987.

84. Manns, M. 1997. Recent developments in autoimmune liver diseases. *J. Gastroenterol. Hepatol.* 12:S256–271.

85. Choudhuri, K., G.V. Gregorio, G. Mieli-Vergani, and D. Vergani. 1998. Immunological cross-reactivity to multiple autoantigens in patients with liver kidney microsomal type 1 autoimmune hepatitis. *Hepatology* 28:1177–1181.

86. Gebre-Medhin, G., E.S. Husebye, J. Gustafsson, O. Winqvist, A. Goksoyr, F. Rorsman, and O. Kampe. 1997. Cytochrome P450IA2 and aromatic L-amino acid decarboxylase are hepatic autoantigens in autoimmune polyendocrine syndrome type I. *FEBS Lett.*, 412:439–445.

87. Mosmann, T.R., and S. Sad. 1996. The expanding universe of T-cell subsets: Th1, Th2 and more. *Immunol. Today* 17:138–146.

88. Del Prete, G.F., M. De Carli, M. Ricci, and S. Romagnani. 1991. Helper activity for immunoglobulin synthesis of T helper type 1 (Th1) and Th2 human T cell clones: the help of Th1 clones is limited by their cytolytic capacity. *J. Exp. Med.* 174:809–813.

89. Berman, M., C. Sandborg, Z. Wang, K. Imfeld, F.J. Zaldivar, V. Dadufalza, and B. Buckingham. 1996. Decreased IL-4 production in new onset type I insulin-dependent diabetes mellitus. *J. Immunol.* 157:4690–4696.

90. Kallmann, B., M. Huther, M. Tubes, J. Feldkamp, J. Bertrams, F. Gries, E. Lampeter, and H. Kolb. 1997. Systemic bias of cytokine production toward cell-mediated immune regulation in IDDM and toward humoral immunity in Graves' disease. *Diabetes* 46:237–243.

91. Exley, M., J. Garcia, S.P. Balk, and S. Porcelli. 1997. Requirements for CD1d recognition by human invariant Valpha24+ CD4-CD8- T cells. *J. Exp. Med.* 186:109–120.

92. Chen, H., and W.E. Paul. 1997. Cultured NK1.1+ CD4+ T cells produce large amounts of IL-4 and IFN-gamma upon activation by anti-CD3 or CD1. *J. Immunol.* 159:2240–2249.

93. Mieza, M.A., T. Itoh, J.Q. Cui, Y. Makino, T. Kawano, K. Tsuchida, T. Koike, T. Shirai, H. Yagita, A. Matsuzawa, H. Koseki, and M. Taniguchi. 1996. Selective reduction of V alpha 14+ NK T cells associated with disease development in autoimmune-prone mice. *J. Immunol.* 156:4035–4040.

94. Wilson, S., S. Kent, K. Patton, T. Orbans, R.A. Jackson, M. Exley, S. Porcelli, D.A. Schatz, M.A. Atkinson, S.P. Balk, J.L. Strominger, and D.A. Hafler. 1998. Extreme Th1 bias of invariant Valpha24JalphaQ T cells in type 1 diabetes. *Nature* 391:177–181.

95. O'Sullivan, D.J., C. Cronin, D. Buckley, T. Mitchell, D. Jenkins, J. Greally, and T. O'Brien. 1997. Unusual manifestations of type 1 autoimmune polyendocrinopathy. *Ir. Med. J.* 90:101–103.

96. Puccetti, P., L. Romani, and F. Bistoni. 1995. A TH1-TH2-like switch in candidiasis: new perspectives for therapy. *Trends. Microbiol.* 3:237–240.

97. Ashman, R.B., and J.M. Papadimitriou. 1995. Production and function of cytokines in natural and acquired immunity to Candida albicans infection. *Microbiol. Rev.* 59:646–672.

98. Lilic, D., A.J. Cant, M. Abinun, J.E. Calvert, and G.P. Spickett. 1996. Chronic mucocutaneous candidiasis. I. Altered antigen-stimulated IL-2, IL-4, IL-6 and interferon-gamma (IFN-gamma) production. *Clin. Exp. Immunol.* 105:205–212.

99. Romani, L., S. Mocci, C. Bietta, L. Lanfaloni, P. Puccetti, and F. Bistoni. 1991. Th1 and Th2 cytokine secretion patterns in murine candidiasis: association of Th1 responses with acquired resistance. *Infect. Immun.* 59:4647–4654.

100. Maclaren, N. and W. Riley. 1986. Inherited susceptibility to autoimmune Addison's disease is linked to human leukocyte antigensDR3 and/or DR4, except when associated with type I autoimmune polyglandular syndrome. *J. Clin. Endocrinol. Metab.* 62:455–459.

101. Spinner, M., R. Blizzard, and B. Childs. 1968. Clinical and genetic heterogeneity in idiopathic Addison's disease and hypoparathyroidism. *J. Clin. Endocrinol. Metab.* 28:795–804.

102. Ahonen, P. 1985. Autoimmune polyendocrinopathy-candidosis-ectodermal dystrophy (APECED): autosomal recessive inheritance. *Clin. Genet.* 27:535–542.

103. Bjorses, P., J. Aaltonen, A. Vikman, J. Perheentupa, G. Ben-Zion, G. Chiamello, N. Dahl, P. Heideman, J.J. Hoorweg-Nijman, L. Mathiron, P.E. Mullis, M. Pohl, M. Ritzen, G. Romeo, M.S. Shapiro, C.S. Smith, J. Solyom, J. Zlotogora, and L. Peltonen. 1996. Genetic homogeneity of autoimmune polyglandular disease type I. *Am. J. Hum. Genet.* 59:879–886.

104. Aaltonen, J., N. Horelli-Kuitunen, J. Fan, P. Bjorses, J. Perheentupa, R. Meyers, A. Palotie, and L. Peltonen. 1997. High-resolution physical and transcriptional mapping of the autoimmune polyendocrinopathy-candidiasis-ectodermal dystrophy locus on chromosome 21q22.3 by FISH. *Genome Research* 7:820–829.

105. Chen, Q.Y., M.S. Lan, J.X. She, and N.K. Maclaren. 1998. The gene responsible for autoimmune polyglandular syndrome type 1 maps to chromosome 21q22.3 in US patients. *J. Autoimmun.* 11:177–183.

106. Nagamine, K., P. Peterson, H. Scott, J. Kudoh, S. Minoshima, M. Heino, K.J. Krohn, M.D. Lalioti, P.E. Mullis, S.E. Antonarakis, K. Kawasaki, S. Asakawa, I. Ito, and N. Shimizu. 1997. Positional cloning of the APECED gene. *Nat. Genet.* 17:393–398.

107. The Finnish-German APECED Consortium. 1997. An autoimmune disease, APECED, caused by mutations in a novel gene featuring two PHD-type zinc-finger domains. *Nat. Genet.* 17:399–403.

108. Bjorses, P., M. Pelto-Huikko, J. Kaukonen, J. Aaltonen, L. Peltonen, and I. Ulmanen. 1999. Localization of the APECED protein in distinct nuclear structures. *Hum. Mol. Genet.* 8:259–266.

109. Rinderle, C., H.M. Christensen, S. Schweiger, H. Lehrach, and M.L. Yaspo. 1999. AIRE encodes a nuclear protein co-localizing with cytoskeletal filaments: altered sub-cellular distribution of mutants lacking the PHD zinc fingers. *Hum. Mol. Genet.* 8:277–290.

110. Heino, M., P. Peterson, J. Kudoh, K. Nagamine, A. Lagerstedt, V. Ovod, A. Ranki, I. Rantala, M. Nieminen, J. Tuukkanen, H.S. Scott, S.E. Antonarakis, N. Shimizu, and K. Krohn. 1999. Autoimmune regulator is expressed in the cells regulating immune tolerance in thymus medulla. *Biochem. Biophys. Res. Commun.* 257:821–825.

111. Gibson, T.J., C. Ramu, C. Gemund, and R. Aasland. 1998. The APECED polyglandular autoimmune syndrome protein, AIRE-1, contains the SAND domain and is probably a transcription factor [letter]. *Trends. Biochem. Sci.* 23:242–244.

112. Woodage, T., M. Basrai, A. Baxevanis, P. Hieter, and F. Collins. 1997. Characterization of the CHD family of proteins. *Proc. Natl. Acad. Sci. USA* 94:11472–11477.

113. Aasland, R., T.J. Gibson, and A.F. Stewart. 1995. The PHD finger: implications for chromatin-mediated transcriptional regulation. *Trends Biochem. Sci.* 20:56–59.

114. Heery, D.M., E. Kalkhoven, S. Hoare, and M.G. Parker. 1997. A signature motif in transcriptional co-activators mediates binding to nuclear receptors. *Nature* 387:733–736.

115. Blechschmidt, K., M. Schweiger, K. Wertz, R. Poulson, H.M. Christensen, A. Rosenthal, H. Lehrach, and M.L. Yaspo. 1999. The mouse Aire gene: comparative genomic sequencing, gene organization, and expression. *Genome. Res.* 9:158–166.

116. Mittaz, L., C. Rossier, M. Heino, P. Peterson, K.J. Krohn, A. Gos, M.A. Morris, J. Kudoh, N. Shimizu, S.E. Antonarakis, and H.S. Scott. 1999. Isolation and characterization of the mouse Aire gene. *Biochem. Biophys. Res. Commun.* 255:483–490.

117. Wang, C.Y., J.D. Shi, A. Davoodi-Semiromi, and J.X. She. 1999. Cloning of Aire, the mouse homologue of the autoimmune regulator (AIRE) gene responsible for autoimmune polyglandular syndrome type 1 (ASP1). *Genomics* 55:322–326.

118. Cole, S.E., and R.H. Reeves. 1998. A cluster of keratin-associated proteins on mouse chromosome 10 in the region of conserved linkage with human chromosome 21. *Genomics* 54:437–442.

119. Scott, H., M. Heino, P. Peterson, L. Mittaz, M.D. Lalioti, C. Betterle, A. Cohen, M. Seri, M. Lerone, G. Romeo, P. Collin, M. Salo, R. Metcalfe, A. Weetman, M.P. Pappasavas, C. Rossier, K. Nagamine, J. Kudoh, N. Shimizu, K.J. Krohn, and S.E. Antonarakis. 1998. Common mutations in autoimmune polyendocrinopathy-candidiasis-ectodermal dystrophy patients of different origins. *Mol. Endocrinol.* 12:1112–1119.

120. Pearce, S., T. Cheetham, H. Imrie, B. Vaidya, N.D. Barnes, R.W. Bilous, D. Carr, K. Meeran, N.J. Shaw, C.S. Smith, A.D. Toft, G. Williams, and P. Kendall-Taylor. 1998. A Common and Recurrent 13-bp Deletion in the Autoimmune Regulator Gene in British Kindreds with Autoimmune Polyendocrinopathy Type 1. *Am. J. Hum. Genet.* 63:1675–1684.

121. Heino, M., H.S. Scott, Q. Chen, P. Peterson, U. Maebpaa, M.P. Papasavvas, L. Mittaz, C. Barras, C. Rossier, G.P. Chrousos, C.A. Stratakis, K. Nagamine, J. Kudoh, N. Shimizu, N. Maclaren, S.E. Antonarakis, and K. Krohn. 1999. Mutation analyses of North American APS-1 patients. *Hum. Mutat.* 13:69–74.

122. Ward, L., J. Paquette, E. Seidman, C. Huot, F. Alvarez, P. Crock, E. Delvin, O. Kampe, and C. Deal. 1999. Severe autoimmune polyendocrinopathy-candidiasis-ectodermal dystrophy in an adolescent girl with a novel AIRE mutation: response to immunosuppressive therapy. *J. Clin. Endocrinol. Metab.* 84:844–852.

123. Lukinmaa, P., J. Waltimo, and S. Pirinen. 1996. Polyendocrinopathy-candidiasis-ectodermal dystrophy (APECED): report of three cases. *J. Craniofac. Genet. Dev. Biol.* 16:174–181.

124. Perniola, R., G. Tamborrino, S. Marsigliante, and C. De Rinaldis. 1998. Assessment of enamel hypoplasia in autoimmune polyendocrinopathy-candidiasis-ectodermal dystrophy (APECED). *J. Oral. Pathol. Med.* 27:278–282.

125. Butler, M., M. Hodes, P. Conneally, A. Biegel, and J. Wright. 1984. Linkage analysis in a large kindred with autosomal dominant transmission of polyglandular autoimmune disease type II (Schmidt syndrome). *Am. J. Med. Genet.* 18:61–65.

126. Santamaria, P., J. Barbosa, A., Lindstrom, T. Lemke, F. Goetz, and S. Rich. 1994. HLA-DQB1-associated susceptibility that distinguishes Hashimoto's thyroiditis from Graves' disease in type I diabetic patients. *J. Clin. Endocrinol. Metab.* 78:878–883.

127. Tamai, H., A. Kimura, R. Dong, S. Matsubayashi, K. Kuma, S. Nagataki, and Sasazuki. 1994. Resistance to autoimmune thyroid disease is associated with HLA-DQ. *J. Clin. Endocrinol. Metab.* 78:94–97.

128. Boehm, B., B. Manfras, S. Seidl, G. Holzberger, P. Kuhnl, C. Rosak, K. Schoffling, and M. Trucco. 1991. The HLA-DQ beta non-Asp-57 allele: a predictor of future insulin-dependent diabetes mellitus in patients with autoimmune Addison's disease. *Tissue Antigens* 37:130–132.

129. She, J. 1996. Susceptibility to type I diabetes: HLA-DQ and DR revisited. *Immunol. Today* 17:323–329.

130. Todd, J. 1997. Genetics of type 1 diabetes. *Pathol. Biol. (Paris)* 45:219–227.

131. Tomer, Y., G. Barbesino, D.A. Greenberg, E. Concepcion, and T.F. Davies. 1998. A new Graves disease-susceptibility locus maps to chromosome 20q11.2. International Consortium for the Genetics of Autoimmune Thyroid Disease. *Am. J. Hum. Genet.* 63:1749–1756.

132. Yanagawa, T., Y. Hidaka, V. Guimaraes, M. Soliman, and L.J. DeGroot. 1995. CTLA-4 gene polymorphism associated with Graves' disease in a Caucasian population. *J. Clin. Endocrinol. Metab.* 80:41–45.

133. Kotsa, K., P.F. Watson, and A.P. Weetman. 1997. A CTLA-4 gene polymorphism is associated with both Graves disease and autoimmune hypothyroidism. *Clin. Endocrinol. (Oxf.)* 46:551–554.

134. Donner, H., H. Rau, P.G. Walfish, J. Braun, T. Siegmund, R. Finke, J. Herwig, K.H. Usadel, and K. Badenhoop. 1997. CTLA4 alanine-17 confers genetic susceptibility to Graves' disease and to type 1 diabetes mellitus. *J. Clin. Endocrinol. Metab.* 82:143–146.

135. Vaidya, B., H. Imrie, P. Perros, E.T. Young, W.F. Kelly, D. Carr, D.M. Large, A.D. Toft, M.I. McCarthy, P. Kendall-Taylor, and S.H. Pearce. 1999. The cytotoxic T lymphocyte antigen-4 is a major Graves' disease locus. *Hum. Mol. Genet.* 8:1195–1199.

136. Kemp, E.H., R.A. Ajjan, E.S. Husebye, P. Peterson, R. Uibo, H. Imrie, S.H. Pearce, P.F. Watson, and A.P. Weetman. 1998. A cytotoxic T lymphocyte antigen-4 (CTLA-4) gene polymorphism is associated with autoimmune Addison's disease in English patients. *Clin. Endocrinol. (Oxf.)* 49:609–613.

137. Kemp, E.H., R.A. Ajjan, E.A. Waterman, D.J. Gawkrodger, M.J. Cork, P.F. Watson, and A.P. Weetman. 1999. Analysis of a microsatellite polymorphism of the cytotoxic T-lymphocyte antigen-4 gene in patients with vitiligo. *Br. J. Dermatol.* 140:73–78.

138. Donner, H., J. Braun, C. Seidl, H. Rau, R. Finke, M. Ventz, P.G. Walfish, K.H. Usadel, and K. Badenhoop. 1997. Codon 17 polymorphism of the cytotoxic T lymphocyte antigen 4 gene in Hashimoto's thyroiditis and Addison's disease. *J. Clin. Endocrinol. Metab.* 82:4130–4132.

139. Baekkeskov, S., H.J. Aanstoot, S. Christgau, A. Reetz, M. Solimena, M. Cascalho, F. Folli, H. Richter-Olesen, P. DeCamilli, and P.D. Camilli. 1990. Identification of the 64K autoantigen in insulin-dependent diabetes as the GABA-synthesizing enzyme glutamic acid decarboxylase. *Nature* 347:151–156.

140. Czarnocka, B., J. Ruf, M. Ferrand, and P. Carayon. 1985. [Antigenic relation between thyroid peroxidase and the microsomal antigen implicated in auto-immune diseases of the thyroid]. *C. R. Acad. Sci. III* 300:577–580.

141. Manns, M.P., K.J. Griffin, L.C. Quattrochi, M. Sacher, H. Thaler, R.H. Tukey, and E.F. Johnson. 1990. Identification of cytochrome P450IA2 as a human autoantigen. *Arch. Biochem. Biophys.* 280:229–232.

142. Manns, M.P., K.J. Griffin, K.F. Sullivan, and E.F. Johnson. 1991. LKM-1 autoantibodies recognize a short linear sequence in P450IID6, a cytochrome P-450 monooxygenase. *J. Clin. Invest.* 88:1370–1378.

143. Gebre-Medhin, G., E.S. Husebye, J. Gustafsson, O. Winqvist, A. Goksoyr, F. Rorsman, and O. Kampe. 1997. Cytochrome P450IA2 and aromatic L-amino acid decarboxylase are hepatic autoantigens in autoimmune polyendocrine syndrome type I. *FEBS Lett.* 412:439–445.

144. Makinen, T., G. Wagar, L. Apter, E. von Willebrand, and F. Pekonen. 1978. Evidence that the TSH receptor acts as a mitogenic antigen in Graves' disease. *Nature* 275:314–315.

145. Rabin, D.U., S.M. Pleasic, J.A. Shapiro, H. Yoo-Warren, J. Oles, J.M. Hicks, D.E. Goldstein, and P.M. Rae. 1994. Islet cell antigen 512 is a diabetes-specific islet autoantigen related to protein tyrosine phosphatases. *J. Immunol.* 152:3183–3188.

146. Lan, M.S., C. Wasserfall, N.K. Maclaren, and A.L. Notkins. 1996. IA-2, a transmembrane protein of the protein tyrosine phosphatase family, is a major autoantigen in insulin-dependent diabetes mellitus. *Proc. Natl. Acad. Sci. USA* 93:6367–6370.

147. Lu, J., Q. Li, H. Xie, Z.J. Chen, A.E. Borovitskaya, N.K. Maclaren, A.L. Notkins, and M.S. Lan. 1996. Identification of a second transmembrane protein tyrosine phosphatase, IA 2beta, as an autoantigen in insulin-dependent diabetes mellitus: precursor of the 37-kDa tryptic fragment. *Proc. Natl. Acad. Sci. USA* 93:2307–2311.

148. Pal, S., U.C. Chaturvedi, R.M. Mehrotra, N.N. Gupta, and A.R. Sircar. 1969. Insulin "auto-antibodies" in diabetes mellitus. *Indian J. Med. Sci.* 23:598–601.

149. Roitt, I., D. Doniach, P. Campbell, and R. Hudson. 1956. Autoantibodies in Hashimoto's disease (lymphadenoid goitre). *Lancet* 2:820–821.

150. Samloff, I.M., and E.V. Barnett. 1965. Identification of intrinsic factor autoantibody and intrinsic factor in man by radioimmunodiffusion and radioimmunoelectrophoresis. *J. Immunol.* 95:536–541.

41 | Molecular Pathology of Multiple Sclerosis

David E. Anderson, Amit Bar-Or, and David A. Hafler

1. INTRODUCTION

Multiple sclerosis (MS) is an inflammatory disease of the white matter within the central nervous system (CNS), and is characterized by demyelination, axonal injury, focal T cell and macrophage infiltration and loss of neurological function (1–3). An estimated 350,000 people in the United States have MS, with 10,000 new cases reported each year. The disease typically manifests between the ages of 20 and 40, and affects women twice as often as it does men. It is the major cause of neurological disability in young people in the Western Hemisphere. MS is generally categorized as being either relapsing-remitting (RR) or primary-progressive (PP) in onset. The course of disease in about 40% of RR patients ultimately changes to a progressive form known as secondary-progressive (SP) multiple sclerosis. The RR form of disease is characterized by a series of attacks that result in varying degrees of disability from which the patients recover partly or completely. This is followed by a remission period of variable duration before another attack. The progressive forms of disease lack the acute attacks and instead typically involve a gradual clinical decline.

Most MS patients experience limb weakness and difficulty with coordination and balance at some time during their disease. Blurred vision, abnormalities of sensation, spasms and fatigue are also common symptoms. The majority of MS patients experience some degree of cognitive impairment such as difficulties with concentration, attention and memory. Because these symptoms are mild when they occur early in disease, they are often overlooked. Depression is another common feature of the illness. As the disease progresses, sexual dysfunction may arise, and bowel and bladder control can become a problem.

Contributing to our understanding of the pathogenesis of MS have been studies of the animal model experimental autoimmune encephalomyelitis (EAE). Accordingly, seminal discoveries relating to the molecular pathogenesis of EAE will be outlined. A major emphasis will be placed on observations that have been substantiated by studies using cells and tissues obtained from MS patients. Four major themes which have been associated with the pathogenesis of EAE and with MS lesions or cells obtained from MS patients will be discussed: (1) The differential activation states of myelin-reactive T cells from MS patients versus normal individuals, (2) the selective expression of chemokines, adhesion molecules and matrix metalloproteinases, (3) the proposed roles of the B7 costimulatory pathway, and (4) the proinflammatory cytokines. A summary of epidemiological data suggesting a role of environmental antigens in MS susceptibility will then be presented, followed by a discussion of the necessary events for the initiation of MS that incorporates these observations.

2. THE DIFFERENTIAL ACTIVATION STATES OF MYELIN-REACTIVE T CELLS FROM MS PATIENTS VERSUS NORMAL INDIVIDUALS

2.1. Elimination of Most, But Not All, Autoreactive T cells by Thymic Selection

T cells are critically dependent upon antigen presenting cells (APCs) such as macrophages, dendritic cells and B cells for their activation. CD4 cells recognize short linear fragments of processed foreign antigens (peptides) presented on the surface of APCs by major histocompatibility complex class II (MHC class II) molecules. While it is relatively easy for us to conceive of a protein fragment, or peptide, from a bacterium or virus as being foreign, it is not as readily apparent to the immune system. There is, after all, no inherent biochemical or structural difference between a peptide derived from a damaged cell within the body and a peptide from the membrane protein of a virus. Yet the immune system must manage to reliably and consistently activate itself only in response to the viral protein fragment and not the fragment from its own tissues. This explains the necessity for immune tolerance, that is, the ability of the immune system to be tolerant of antigens from its own tissues yet respond effectively to antigens from environmental sources, including bacteria, viruses, and parasites.

Ultimately, the T cell repertoire develops to recognize foreign antigens presented in the context of self-MHC molecules. Yet, during T cell selection within the thymus, self-MHC molecules predominantly present self-antigens rather than foreign antigens. Two factors underlie this apparent disparity. First, T cell reactivity to foreign antigens is achieved by recognition of MHC molecules presenting cross-reactive self-antigens (a process termed "positive selection"), giving rise to thymocytes that are inherently autoreactive. Second, "negative selection" mediated by the same MHC complexes presenting self-antigens eliminates all highly autoreactive thymocytes before they emigrate to the periphery as mature T cells. This process of negative selection is arguably the most important mechanism of ensuring immune self-tolerance. Nonetheless, some autoreactive T cells do emigrate from the thymus into the periphery. Peripheral mechanisms of tolerance prevent these cells from becoming pathogenic.

Several ideas have been put forth to explain the presence of autoreactive T cell within the periphery of normal healthy individuals and experimental animals. One explanation has been that many tissue- or organ-specific autoantigens, such as those found in the CNS, are locally sequestered. If this were true, such antigens would not be present within the thymus during T cell development and "negative selection" or deletion of the autoreactive clones would fail to occur. Recently, however, reverse transcriptase PCR analysis was used on normal human thymic tissue to look for a panel of autoantigens implicated in the autoimmune diseases diabetes, thyroiditis and MS (4–6). Among 12 thymi, insulin, glucagon and GAD-67 were detected in six, thyroglobulin was detected in five, myelin basic protein (MBP) and retinal S-antigen were detected in three, and GAD-65 was detected in one. Some of these autoantigens were expressed in an age-dependent fashion. MBP, for example, was detected only in thymi early within development (up to 2-month old thymi). These data suggest that some autoantigens may indeed be present in the thymus at the correct time during T cell repertoire development and should thus be able to delete autoreactive T cells.

Wraith and colleagues have theorized that low affinity interactions between autoantigens and MHC molecules may undermine the efficacy of their presentation in the thymus, thus enabling autoreactive T cells to escape "negative selection". To test this theory, his group used TCR transgenic mice specific for the epitope MBP (p1–11) and altered peptide ligands (APLs) based on this MBP epitope that had greater binding affinity for MHC class II molecules (7). Consistent with their theory, they found that the *in vivo* administration of the higher affinity APLs resulted in deletion of T cells, while administration of the native peptide had no effect. Similar mechanisms have been thought to underlie the tendency of certain strains of mice to be more susceptible to particular autoimmune diseases, such as diabetes in the susceptible NOD mice (8,9).

As described later in this chapter, MHC class II molecules, especially the HLA DR2 alleles, are associated with increased susceptibility to MS (10–15). In most populations studied, approximately 50–70% of all MS patients carry the DR2 gene, a frequency that is 2–3 times higher than in normal matched controls.

This MHC class II molecule could be involved in MS susceptibility for a number of reasons. One possibility, in light of the discussion above, is that this MHC molecule inefficiently binds MBP epitopes and thus fails to negatively select MBP-reactive T cells. However, studies have demonstrated that the binding of the immunodominant MBP epitope p85–99, identified using T cells obtained from the peripheral blood of both MS patients and normal individuals, binds to the MHC class II molecule comprised of the HLA DRB1*1501 allele with very high affinity (Kd approximately 4nM). These data, in addition to experiments demonstrating the presence of MBP protein in the thymus, argue against the notions that MBP-reactive T cells are present within the periphery due to a lack of the autoantigen presence within the thymus or due to low affinity interactions between MBP epitopes and MHC molecules. An alternate hypothesis is that these MBP-reactive T cells express TCRs with relatively low affinity for MBP (p85–99)/DR2 complexes compared to TCRs of T cells reactive against foreign antigens and presented by the same DR2 complexes. Direct binding measurements between the MBP (p85–99)/DR2 complex and MBP-reactive TCRs obtained from T cells of normal and MS patients will soon allow this hypothesis to be tested.

2.2. Autoreactive T cells in Normal Individuals versus MS Patients

EAE is an inflammatory condition that bears similarities to MS and is also characterized by multifocal perivascular CNS inflammatory infiltrates primarily comprised of T cells and monocytes. EAE can be induced in animals by injection of immunodominant MBP or proteolipid protein

(PLP) peptides or by transfer of CD4$^+$ MHC class II-restricted T cells reactive with these peptides (16–19). Among MS patients with the DR2 haplotype, T cells have been identified which react with epitopes from MBP and PLP (20,21). Using T cells and antibodies obtained from the blood and cerebrospinal fluid (CSF) of MS patients, additional autoantigens have also been implicated in the disease (22,23).

It is clear that autoreactive T cells can also be found in the peripheral blood of normal individuals (20). Thus, the mere presence of autoreactive cells in the periphery is an insufficient explanation for the development of autoimmunity. The prediction would be that myelin-specific T cells in MS patients differ in some way from those found in normal individuals. Specifically, MS autoreactive cells may be present in greater numbers, in different functional states, have lower thresholds of activation, or have different effector profiles.

A number of studies have recently demonstrated that MBP-reactive T cells obtained from MS patients are indeed in an enhanced state of activation compared to those isolated from normal individuals. One study found that the IL-2 receptor, a hallmark of activated T cells, was expressed on MBP-reactive T cells from MS patients but not on those obtained from normal individuals (24). Similar frequencies of MBP- and PLP-reactive T cells were obtained from PBMCs of relapsing-remitting MS patients and controls after primary stimulation with antigen. However, if the PBMCs were first expanded for 7 days in rIL-2 and then further expanded with antigen, there was a much higher frequency of MBP- and PLP-reactive T cells obtained from the PBMCs of MS patients. Furthermore, after initial expansion in rIL-2, MBP-reactive T cells were obtainable from the CSF of MS patients but not from the CSF of control individuals. Two recent reports demonstrate that MBP-reactive T cells from the peripheral blood of MS patients, but not normal individuals, are less dependent upon B7 costimulation for their activation. One study used stable cell transfectants expressing either the DR2 molecule alone, or in combination with either the human B7-1 or B7-2 costimulatory molecules, to present the immunodominant peptide MBP (p85–99) to highly purified CD4$^+$ T cells obtained from relapsing-remitting MS patients or controls (25). As expected, transfectants expressing the DR2 molecule alone could not expand MBP-reactive T cells from normal individuals; when either B7–1 or B7–2 were present, MBP-reactive T cell lines could readily be expanded. In marked contrast, transfectants expressing the DR2 molecule alone could expand MBP-reactive T cells obtained from MS patients. In other experiments, the addition of blocking anti-CD28 mAb or of CTLA-4-Ig (to block the engagement of B7 molecules) to PBMC cultures, did not inhibit the expansion of MBP-reactive T cells from MS patients but did inhibit expansion of these T cells when using PBMCs isolated from normal individuals (26). These results confirm the earlier observation that MBP-reactive T cells in MS patients are in a different functional state and further demonstrate that these autoreactive T cells have less stringent requirements for their activation.

3. SELECTIVE EXPRESSION OF CHEMOKINES IN MS

Activated Th1-type, myelin-reactive T cells must migrate from the periphery into the CNS to participate in the demyelination and pathology associated with EAE and MS. Chemokines can enhance T cell and monocyte migration through direct chemoattraction and by activating leukocyte integrins to bind their adhesion receptors on endothelial cells. Certain α and β chemokines have recently been identified that appear to selectively recruit these cells into the CNS and are associated with EAE disease activity (27). Karpus and colleagues directly compared the roles of MIP-1α, MCP-1 and MIP-2 in the induction of EAE (28,29). *In vivo* administration of monoclonal antibodies against MIP-1α inhibited adoptively transferred EAE whereas antibodies directed against MCP-1 inhibited relapses. Consistent with these observations, immunohistochemical analysis of postmortem brain tissue from MS patients has demonstrated that astrocytes, but not perivascular or parenchymal microglia, express MCP-1 in both active demyelinating and chronic active lesions (30). In another study, mRNA levels of RANTES, chemotactic for lymphocytes and monocytes, were examined in brain samples of MS patients (31). RANTES was expressed by activated perivascular T cells that were localized predominantly at the edge of active plaques. A recent comprehensive study by Sorensen and colleagues (32) examined the expression levels of chemokines and chemokine receptors on cells in the CSF of MS patients, neurological control patients and normal individuals, as well as the expression of chemokines and their receptors by brain tissue obtained from MS patients. The authors noted an increase in RANTES levels and a 3-fold increase in IP-10, chemotactic for activated T cells, in the CSF of MS patients compared to controls. There were no differences among the study populations in the levels IL-8 or GROα, which are chemotactic for neutrophils. An increased frequency of CD4$^+$ and CD8$^+$ T cells from the CSF expressed the IP-10 receptor CXCR3. Immunohistochemical staining confirmed these results. There was an increased perivascular expression of CXCR3 in MS lesions, while RANTES staining was also upregulated in MS plaques but was expressed in a more diffuse manner. CCR5, the receptor for RANTES, was also detected on lymphocytes, macrophages and microglia in actively demyelinating lesions. Thus, in addition to MIP-1α and MCP-1, the data clearly suggest that the IP-10/CXCR3 and

RANTES/CCR5 pathways have selective roles in MS pathogenesis.

Recent data indicate that some of the chemokines and chemokine receptors that have been implicated in both EAE and MS are preferentially chemotactic for the proinflammatory Th1 T cells. Th1 cells preferentially migrated in response to the CC chemokines MIP-1α, MIP-1β and RANTES, whereas neither Th1 cells, nor the anti-inflammatory Th2 cells, responded to CXC chemokines (33). The extent to which chemokines and their receptors contribute to MS pathogenesis is unclear. For example, absence of a functional RANTES/CCR5 system alone, is not sufficient to protect against the development of MS, as several MS patients have been identified in whom a homozygous mutation prevents the expression of CCR5 (34).

4. SELECTIVE EXPRESSION OF ADHESION MOLECULES IN MS

Perivascular infiltration of inflammatory cells in the CNS represents one of the pathological hallmarks of MS lesions and requires adhesion and transmigration of these cells across the blood brain barrier (BBB) (35–37). Based on their structure, three families of immunologically active adhesion molecules can be identified: Members of the immunoglobulin (Ig) superfamily, integrins and selectins (38).

4.1. ICAM-1, VCAM-1 and the Integrins

ICAM-1, a 76-114 kDa surface glycoprotein with five extracellular Ig domains, binds to the membrane-bound integrin receptors LFA-1 (CD11a/CD18) and Mac-1 (CD11b/CD18) on the surface of leukocytes (39). VCAM-1, also a member of the Ig superfamily, has seven extracellular Ig domains and binds with $\alpha4$ integrins, including VLA-4 ($\alpha4\beta1$ integrin), which is constitutively expressed on most mononuclear cells (40,41). Both ICAM-1 and VCAM-1 play important roles in endothelial-leukocyte interactions and in leukocyte extravasation (38,42). ICAM-1 and VCAM-1 are also implicated as constimulatory molecules in T cell activation (43). Elevated levels of ICAM-1 have been identified on endothelial cells of both acute and chronic-active MS lesions and these levels were shown to correlate with the extent of leukocyte infiltration (44–46). VCAM-1 was detected in chronic-active MS lesions, on both endothelial cells and on microglia, in contrast to normal brain in which no VCAM-1 was identified (47). The ligands for ICAM-1 and VCAM-1 (LFA-1 and VLA-4, respectively) have been identified on the perivascular inflammatory cells of MS lesions (48). More direct implication in pathophysiology is derived from observations in the EAE model. Myelin autoreactive T cells fail to cross the BBB in VLA-4 deficient animals (49,50). Moreover, treatment of wild type animals with monoclonal antibody directed against VLA-4 results in diminished infiltration of the CNS by inflammatory cells with a concomitant diminution in disease severity (51,52). The specificity of these molecular interactions is underscored by the demonstration that while the interaction between the $\alpha4\beta1$ integrin (VLA-4) and VCAM-1 is required for the development of EAE, blockade of the interaction between the $\alpha4\beta7$ integrin and VCAM-1 does not influence disease development (53).

Several studies have demonstrated adhesion molecules on the surface of CNS glial cells. ICAM-1 positive astrocytes are found both within and around active MS lesions, but not in normal brain (44,47). VCAM-1 and LFA-1 are detectable on microglial cells in chronic active MS lesions (47). In addition to the possible role in inflammatory cell migration, it has been proposed that glial cell expression of adhesion molecules may play roles in antigen presentation and T cell costimulation (43,48,54,55) and in glial-extracellular matrix interactions (56).

The local immune microenvironment may have important ramifications on the expression of adhesion molecules, in turn influencing the recruitment of further inflammatory cells. ICAM-1 expression on cultured human astrocytes is enhanced by proinflammatory cytokines (57). A similar observation is made in murine systems where this upregulatory effect can be countered by anti-inflammatory cytokines (58,59). This cytokine-mediated regulation of adhesion molecule expression on glial cells appears to be cell-type specific (59) and the molecular mechanisms underlying these effects are currently under active study (38).

4.2. Soluble Adhesion Molecules

Elevated levels of the soluble forms of adhesion molecules, including sICAM-1 and sVCAM-1, have been detected in the serum (60–62) and CSF (63–65) of MS patients compared to controls. These levels appear to correlate with clinical and MRI indicators of disease activity and with the pattern of MS (62,66,67). Proteolytic cleavage of membrane associated adhesion molecules appears to be the most likely source of sICAM-1 and sVCAM-1 and it is of interest to note that matrix metalloproteinases (see below) may be involved in this process (68,69). While no consistent association has been found between levels of soluble E-selectin (sE-selectin) and disease activity (62,65,70), there appears to be a selective elevation of sE-selectin in primary progressive MS patients that is not seen in patients with the relapsing remitting pattern (71,72).

Interestingly, during treatment with IFNβ-1b (which has been shown to decrease MS activity), levels of sVCAM and sICAM were increased and these elevations correlated with a decrease in the MRI lesion burden (73–76). In view of these seemingly contradicting observations, it remains difficult to assign a clear pathophysiologic role to soluble adhesion molecules. A possible explanation may stem from the additional observation that VLA-4 expression on peripheral blood lymphocytes is decreased in MS patients during treatment with IFNβ (77). Conceivably, the elevated levels of soluble adhesion molecules (e.g. sVCAM-1) associated with IFNβ treatment may result in quenching or downregulation of their ligands (e.g. VLA-4) on the inflammatory cells and thereby inhibit the immune response (38,77).

5. SELECTIVE EXPRESSION OF MATRIX METALOPROTEINASES

Matrix metalloproteinases (MMPs) comprise a family of tightly regulated proteolytic enzymes that are secreted into the extracellular matrix (ECM). Degradation of the ECM plays an important role in many normal physiological processes, including angiogenesis, endometrial cycling and bone remodeling (78). Under pathologic conditions, ECM degradation may promote tissue invasion by neoplastic or inflammatory cells. MMPs are expressed by activated T cells (79), monocytes (80), astrocytes and microglial cells (81,82) and are typically secreted as proezymes, requiring proteolytic cleavage for their own activation and undergoing down-regulation by tissue inhibitors of metalloproteinases (TIMPS). Potential mechanisms of MMP contribution to MS pathophysiology (recent reviews by *Yong* et al. (83) and *Kieseier* et al. (84) include: (1) Disruption of basement membrane of the BBB (85,86) thereby facilitating trans-migration of inflammatory cells; (2) breakdown of ECM enabling infiltration into the neuropil (79,87); (3) proteolytic cleavage of membrane bound proinflammatory cytokines such as TNFα (88,89) and (4) direct damage to the myelin sheath (90).

In pathological studies of MS tissue, gelatinase B (MMP-9) is expressed in white matter perivascular mononuclear cells and, together with other MMPs, is associated with both monocytes and astrocytes in demyelinating lesions (81,82). Microglia in active MS lesions express a range of inflammatory cytokines that have been shown to induce gelatinase B expression by inflammatory cells *in vitro* (46). Participation of MMPs in MS lesion development may result from aberrant overproduction of the proteases and/or from failure to sufficiently downregulate their actions.

Increased activity of CSF proteolytic enzymes has been reported in MS patients compared to normal individuals (91,92). Gelatinase B levels are increased in both serum and CSF of MS patients during an acute relapse and the elevated levels correlate with the degree of BBB disruption as evidenced by the number of gadolinium enhancing lesions

on MRI (93,94). Furthermore, treatment with corticosteroids which are known to suppress MMP transcription, is associated with a reduction in both CSF gelatinase levels and in the number of enhancing MRI lesions (95).

In vitro studies demonstrate that Interferon β-1b inhibits T cell expression of MMPs and subsequent T cell migration. These observations may explain in part the ability of Interferon β-1b to diminish the gadolinium-enhancing burden and modify the progression of disease in patients with MS (96,97,98). In EAE, the interaction between T cells and endothelial cells, mediated by the adhesion molecule VCAM-1 (see above), induces T cell secretion of the MMP gelatinase (99). MMPs, in turn, are able to participate in the release of adhesion molecules from the cell surface (69). Furthermore, MMPs have been shown to cleave surface bound proinflammatory molecules including TNFα that has been shown to promote BBB breakdown and tissue injury (88). Ongoing studies in EAE allow more direct examination of the complex interactions between chemo-kines, adhesion molecules, MMPs and proinflammatory cytokines and provide the opportunity to study novel therapeutics aimed at these molecular targets.

Several MMP-inhibitors have been shown to ameliorate the clinical course of EAE (100,101) though effects on the degree of inflammation and demyelination have been variable. More recently, MMP inhibition was shown to block and reverse clinical disease in a chronic-relapsing EAE model, in association with downregulation of TNFα mRNA and concomitant upregulation of the Th2 anti-inflammatory cytokine, IL-4 (102). Broad spectrum MMP inhibitors are currently under investigation in early phase clinical trials of rheumatoid arthritis and a range of neoplastic disorders (84).

6. THE ROLE OF B7–1 COSTIMULATION IN MS PATHOGENESIS

6.1. B7:CD28/CTLA-4 Costimulatory Pathway

Activation of autoreactive T cells is required for successful passive transfer of disease in EAE. Similarly, the presence of activated myelin reactive T cells appears to distinguish MS patients from normal subjects, in whom circulating myelin reactive clones are not activated. Efficient activation of naïve T cells is dependent upon two signals: an antigen-specific signal delivered through the TCR, and a second costimula-tory signal that functions to induce the secretion of T cell growth factors such as IL-2. The best-characterized costimu-latory pathway is the B7 pathway (103,104). Two related B7 costimulatory molecules, B7–1 and B7–2, are expressed on APCs, although with different kinetics and expression patterns. B7–2 is found on most APCs at low, but constitu-tive, levels, whereas B7–1 is generally absent until an APC becomes activated, at which time it upregulates the expression of both molecules. These molecules direct signals into T cells through two receptors, CD28 and CTLA4. It has become clear that signals directed through CD28 enhance T cell activation, while signals delivered through CTLA-4 serve to attenuate T cell activation (105,106).

Jenkins and Schwartz first reported the consequence of T cell activation with peptide/MHC class II complexes in the absence of B7 costimulation (107). They found that when splenocytes treated with ECDI (which effectively fixes the cell surface and inactivates many surface molecules), were used *in vitro* as APCs, they failed to stimulate proliferation by antigen-specific normal T cell clones and instead induced a state of long-term unresponsiveness termed anergy. This T cell unresponsiveness was also induced *in vivo* by the I.V. administration of antigen-coupled splenocytes prepared by ECDI treatment. The results were not due to extensive MHC class II complex alteration, as anti-MHC class II mAbs prevented this anergy induction, suggesting that antigen presentation was taking place and was needed for the anergy induction. The authors proposed that the ECDI treatment impaired an additional APC signal necessary to induce IL-2 production and T cell proliferation. The anergic state did not seem to involve inhibition of the IL-2 receptor pathway, however, as T cell clones unable to respond to Ag/MHC restimulation responded normally to exogenous IL-2.

6.2. B7:CD28/CTLA-4 Costimulatory Interactions in EAE

A considerable amount of evidence in the EAE model of MS suggests that the B7:CD28/CTLA-4 costimulatory pathway may well play a role in the pathogenesis and/or regulation of MS (108). Immunohistochemical analysis of the *in vivo* expression of various costimulatory molecules during the course of EAE disease, remission and relapse, demonstrated that members of the B7:CD28 costimulatory pathway were differentially regulated at different times during the course of disease (109). Expression of B7–2 was seen during acute disease and during relapse while B7–1 was expressed by relatively few cells and only during remission. FACS analysis confirmed that astrocytes and infiltrating cells stained positive for B7–2 during relapse, whereas both astrocytes and neurons stained positive for B7–1 during remission only. Splenocytes were positive for B7–1 (by FACS) during acute disease at low levels, suggesting that regulation of B7 molecules may differ between the periphery and the target organ (CNS) at different stages in the disease. Very few cells were positive for CTLA-4 expression, which was the case during all phases of disease. In contrast, CD28 was expressed on a large proportion of cells during acute disease and during relapses but was not expressed at high levels during disease remission.

Several groups have used a system very similar to the one used by Jenkins and Schwartz to investigate the role of B7-mediated T cell costimulation in the activation of autoreactive T cells in EAE. They have demonstrated that chemical cross-linking of APCs pulsed with a variety of autoantigens known to produce EAE can induce tolerance and ameliorate and/or delay EAE (110,111). The ability of CTLA-4-Ig administration *in vivo* to block EAE has also been examined. Using an adoptive transfer model of EAE in which autoreactive T cells from a donor mouse are transferred into syngeneic mice, CTLA-4-Ig present during induction of the autoreactive T cells *in vivo* or present *in vitro* during restimulation of these T cells, ameliorated the disease in the recipient animals. The diminution in disease correlated with reduced *in vitro* production of IL-2 and IL-4, as well as decreased proliferation. However, CTLA-4-Ig treatment of recipient mice after the transfer of autoreactive T cells did not influence either disease course or disease severity. Thus, CTLA-4-Ig was assumed to block the induction phase but not the effector phase of T cells in EAE. Arima and colleagues similarly examined the effects of CTLA-4-Ig administration on induction of EAE in Lewis rats (112). CTLA-4-Ig administration 8 times before/immediately after immunization with spinal cord homogenate was able to prevent the development of EAE. This was reversed by the *in vivo* administration of rIL-2, suggesting that the cells had been rendered anergic *in vivo*. However, the authors noted that administration of CTLA-4-Ig twice after immunization with spinal cord homogenate slightly enhanced the severity of disease, again suggesting that blocking the CD28 pathway is most useful before T cells have been activated and/or clonally expanded. Another explanation for the inability of CTLA-4-Ig to inhibit EAE after induction of the disease is that the CTLA-4-Ig does not efficiently enter the CNS to block the B7:CD28 interactions at the site of effector T cell activation. A recent report suggests that this explanation does have merit, in that local CNS delivery of CTLA-4-Ig using a non-replicating adenoviral vector was able to ameliorate ongoing EAE (113). Collectively, the results indicate that both the timing and route of delivery of CTLA-4-Ig are important factors in treating *in vivo* disease states.

6.3. Implicating B7–1 in Disease Pathogenesis

Early studies investigating the ability of B7-1 and later B7-2 to enhance T cell activation clearly demonstrated that both molecules could costimulate T cells through CD28. When directly compared in their ability to enhance T cell proliferation and secretion of Th1- and Th2-associated cytokines, several studies found that B7-1 and B7-2 costimulation were comparable in their costimulatory abilities, both *in vitro* and *in vivo*.

Nonetheless, a considerable amount of data exists indicating that B7-1 and B7-2 costimulation have different roles in the initiation or regulation of autoimmune diseases. Treatment of mice with anti-B7-1 mAb *in vivo* during the induction of relapsing-remitting EAE protected mice from disease, whereas treatment with anti-B7-2 mAb exacerbated disease severity (114,115). Treatment with anti-B7-1 mAb also skewed the autoantigen-reactive T cells towards a Th2 phenotype, while blockade of B7-2 was correlated with skewing towards a Th1 phenotype. Consistent with these observations, administration of anti-B7-1 Fab, but not anti-B7-2 antibody, after the first remission of EAE, significantly decreased the incidence of relapses in these mice (116). The authors argued that the mechanism was due to a reduction in the number of T cells generated against subsequently exposed myelin epitopes, such that B7-1 blockade prevented epitope spreading. The same group later demonstrated that CNS APCs isolated from the spinal cords of mice with acute EAE preferentially utilized B7-1 costimulation, since anti-B7-1 mAb, but not anti-B7-2 mAb, significantly inhibited T cell activation when the isolated cells were used as a source of APCs (117).

Similar observations have been made in MS suggesting a preferential role for B7-1 in the disease process. Immunohistochemical staining of MS plaques and inflammatory stroke lesions from the same brain demonstrated that while B7-2 expression was expressed in both types of lesions, B7-1 was uniquely associated with the MS plaques (118,119). Another study examined B7-1 and B7-2 expression on B cells and monocytes in PBMCs obtained from active or stable MS patients, and from normal individuals (120). A greater percentage of B7-1-expressing B cells was observed during the active phase of MS, which was not true for B7-2. There was no statistically significant difference in B7-1 vs. B7-2 expression on monocytes between the two groups. Interestingly, after treatment with IFNβ-1b, the frequency of B7-1 expressing B cells decreased, while the frequency of B7-2 expressing B cells remained unchanged. This study provides further support for the notion that B7-1 and B7-2 have different functions and, moreover, that B7-1 is implicated in MS pathogenesis.

7.   PROINFLAMMATORY CYTOKINES AND MS

CD4+ T helper cells can be broadly categorized into one of several subsets based on the cytokines they produce upon activation (121,122). T helper 1 (Th1) cells secrete proinflammatory cytokines such as IFNγ, TNFα and lymphotoxin (LT), which enhance APC activation and the

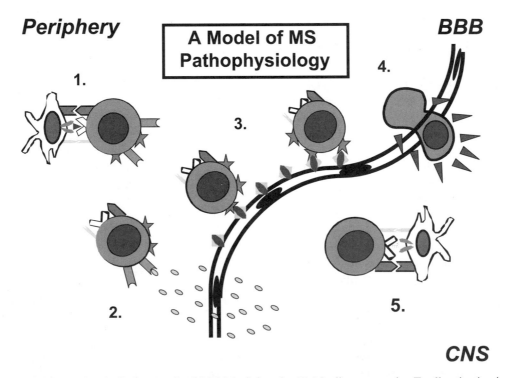

Figure 1.   A model of the Molecular Pathogenesis of Multiple Sclerosis: (1) Myelin autoreactive T cell activation in the periphery (2) Chemoattraction of activated cells to CNS mediated by chemokine/chemokine receptor interactions (3) Adhesion of activated cells to blood-brain-barrier (BBB) endothelium via adhesion molecules and their receptors (4) Matrix metalloproteinase release facilitates infiltration of activated cells across BBB basement membrane and into CNS parenchyma (5) Reactivation of autopathogenic cells in CNS mediates damage to myelin and axonal injury.

clearance of many intracellular pathogens, whereas Th2 cells secrete cytokines such as IL-4, IL-5, and IL-13, which aid in antibody class-switching and elimination of many blood-borne infectious agents (121). The cytokines produced by each T helper cell subset can themselves negatively regulate the differentiation of the other subset. Thus, IFNγ produced by Th1 cells inhibits the differentiation of naïve T cells into Th2 cells and IL-4 secreted by Th2 cells can inhibit the differentiation of Th1 cells. The cytokines IL-10 and TGFβ, secreted by T regulatory 1 (Tr1) and Th3 cells, respectively, are very potent suppressors of T cell activation. The presence of Th1-promoting cytokines, such as IL-12 or IFNγ, or the Th2-promoting cytokine IL-4, are the most potent factors in T cell differentiation and immune deviation. However, antigen dose and B7 costimulation have also been demonstrated to indirectly influence Th differentiation (122).

In studies of EAE, T cells secreting Th1 cytokines have been associated with the pathogenesis of the disease. Several studies have examined brain tissue for cytokine expression in lesions from Lewis rats or mice with EAE (123–124). Th1-associated cytokines such as IFNγ, TNFα, and IL-12 were present during acute disease and during relapses, but not during remissions.

Studies have also confirmed the association of Th1 cytokines with T cells and monocytes from MS patients. Immunohistochemical studies of MS plaques *in situ* have demonstrated the presence of the proinflammatory cytokine TNFα (125,126). TNFα was detected in chronic lesions from MS patients, but not detected in the CNS of neurological controls or in the spleen or PBMCs of MS patients, suggesting that there was a specific association of TNFα with the CNS MS lesions (126). Within the MS plaques, astrocytes and macrophages were found to be the source of the TNFα (125). In another study, IL-12 was found in MS plaques that also expressed B7-1 (119).

Correale and colleagues studied the cytokine profile secreted by PLP-autoreactive T cell clones generated from MS patients at different clinical stages of disease (127). They found that the cytokine profiles of clones changed according to the stage of disease.

During acute attacks, T cells clones had Th1-like phenotypes, with no TGFβ secretion. During remission in the same patients, however, the clones showed Th0, Th1 and Th2 cytokine profiles. Most striking, the levels of IL-10 secreted from clones generated during remission were significantly higher than those measured from clones derived during acute attacks or from normal subjects. When comparing stable versus active MS patients, significant differences in T cell secretion of TGFβ have also been observed (128).

Converging studies of EAE indicate that IL-12 may be a critical factor in the pathogenesis of the disease. Shevach and colleagues have demonstrated that endogenous production of IL-12 is critical for the generation of autoreactive Th1 T cells, as EAE cannot be induced in IL-12-deficient mice (129). Consistent with these results, several groups have demonstrated that the *in vivo* administration of IL-12 to mice or rats could exacerbate EAE or induce clinical relapses, while administration of anti-IL-12 monoclonal antibody ameliorated disease severity and prevented relapses (130–132). Similarly, a novel anti-inflammatory drug that inhibits IL-12 mediated signaling (and blocks IL-12 driven Th1 differentiation *in vitro*), but not IL-12 secretion, was shown to protect against both actively-and passively-induced EAE (133).

Recently, intracellular cytokine staining confirmed that PBMCs from chronic progressive (CP) MS patients express more IL-12 upon activation than those from normal individuals (134). Moreover, treatment of CP MS patients with cyclophosphamide and methylprednisolone reduced the frequency of IL-12-staining monocytes to normal levels. In the same study, a greater frequency of T cells from untreated patients secreted IFNγ and TNFα, compared to T cells from normal controls. Frequencies of T cells expressing Th2 cytokines were comparable among MS patients and normal individuals.

8. EPIDEMIOLOGICAL STUDIES IMPLICATE GENTIC AND ENVIRONMENTAL FACTORS IN MS PATHOGENESIS

In spite of the accumulating insights into the mechanisms of ongoing disease pathophysiology, the inciting cause of MS is difficult to ascertain. Critical analysis of the vast body of epidemiological literature is most consistent with the model of an environmentally triggered disease in a polygenetically susceptible host (135–139). Race and sex are clearly, and independently, associated with MS risk (140–142). In several linkage-analysis studies, the only consistently replicated gene association with disease susceptibility was that of the major histocompatibility complex (MHC) alleles, more specifically, the class II MHC alleles DRB1*1501 or DRB5*0101, DQA1 *0102 and DQB1*0602 (143,144) on chromosome 6p. Weaker associations have been reported, though not always replicated, for other MHC alleles, T-cell receptor and immunoglobulin-chain encoding regions, myelin basic protein and tumor necrosis factor genes (145–147).

Application of microsatelite polymorphism markers to full genome searches in large numbers of affected sibling pairs (148–151) identified chromosomes 2p23, 5q13, 6p21 and 19q13, as susceptibility regions. There is no established evidence for a genetic marker that confers protection from the disease. In combination, these studies support the existing hypothesis that genetic susceptibility to MS is conferred by multiple interacting genes, each with a relatively modest individual contribution (152,153). Results of five population-based twin studies demonstrated a concordance rate of approximately 27% in monozygotic twins and 2.4% concordance in dizygotic, same-sex pairs (139). The lack of 100% concordance underscores the importance of environmental contributors to disease pathogenesis.

In population-based studies, the highest prevalence rates of MS are found in the temperate zones of both the northern and southern hemispheres. This non-random distribution may represent genetic variations in some populations studied (154). In both Europe and the United States, the prevalence of MS may reflect the degree of Scandinavian and northern European heritage in resident populations (155). Nonetheless, the several fold difference in the south-to-north prevalence in Australia, across a genetically homogeneous population, argues for a non-genetic effect (156). The identity of such a latitude-based risk factor remains elusive.

In studies of populations emigrating from areas of high prevalence to areas of low prevalence, those emigrating after the age of 15 retained the higher risk of their country of origin (157,158). These and other studies have suggested that the environmental contribution to MS susceptibility is acquired by about age 15. Reports of disease clusters and epidemics have fueled multiple epidemiological studies addressing a range of putative infectious and other environmental triggers of MS (critically reviewed in 139,159–162). To date, no single infectious agent has been established as the cause of MS. Nonetheless, several well-designed prospective studies have demonstrated a strong association between exacerbations of MS and preceding viral infections (163,164). Such results support a role for infectious agents in the molecular pathogenesis of MS flares.

9. INITIATION OF THE DISEASE PROCESS

Molecular mimicry has been proposed as a mechanism for the induction of, or exacerbation of MS. According to this hypothesis, certain infectious agents are comprised of proteins containing peptide sequences that mimic autoantigen epitopes. Upon infection, presentation of these viral or bacterial peptides in the periphery by infected APCs inadvertently activates autoreactive T cells. In their activated state, these T cells cross the blood-brain-barrier and recognize the autoantigens within the CNS, initiating an inflammatory response that ultimately leads to myelin destruction and axonal damage. The discovery in recent years that there is considerable degeneracy in TCR recognition of peptide/ MHC complexes and that a given TCR can react with multiple peptides, lends support to this theory. In fact, viral and bacterial epitopes have been identified that trigger human MBP-reactive T cells obtained from MS patients (165). Using the immunodominant epitope MBP (p84–102), a database search was performed based on those residues needed for MHC binding

and TCR recognition. More than 600 viral and bacterial epitopes were identified, and narrowed down further to those epitopes from viruses known to cause human disease, those from viruses prevalent in the Northern Hemisphere (where MS is more prevalent), and those from bacterial sequences associated with CNS inflammation. These peptides were then tested to see if they could stimulate proliferation among any of seven different MBP-reactive T cell clones. Ultimately, eight peptides were found to efficiently activate three of the clones.

While the possibility of molecular mimicry as a cause of MS has been supported *in vitro*, few studies have evaluated the ability of molecular mimicry to induce disease *in vivo*. Recently, however, a similar approach was employed to identify viral and bacterial epitopes that could activate a panel of PLP-reactive T cell clones and hybridomas from mice susceptible to the induction of EAE (166). Consistent with previous observations, several bacterial and viral epitopes were identified which could activate PLP-specific T cell clones. However, when genetically susceptible mice were actively immunized with these epitopes, disease was not induced. Of great interest, however, immunization of mice with these epitopes rendered mice susceptible to EAE induction with suboptimal amounts of the autoantigen itself. Thus, expansion of autoreactive T cells with bacterial or viral peptides that mimicked the autoantigen could potentiate, though not induce, disease; activation of autoreactive T cells with the autoantigen itself was necessary for the actual induction of disease. Using a similar approach, another group found that microbial peptides could potentiate, but not induce, autoimmune thyroiditis (167). Based on these results, Carrizosa and colleagues suggested a modified theory of molecular mimicry to explain the induction of MS. They suggested that two different episodes of infection, with either the same or different infectious agents, were necessary for the induction of MS. The first infection serves to expand autoreactive T cells, based on an epitope contained within the infectious agent that cross-reacts with the autoantigen. These cells differentiate into Th1 effector cells due to the proinflammatory conditions created by the viral/bacterial infection. At a later time, the same or different neurotropic infectious agent induces some degree of CNS damage with the release of autoantigens. Small amounts of these autoantigens then manage to activate the previously expanded autoreactive T cells either in the periphery or in the CNS. As these T cells accumulate within the CNS, an autoimmune response is established.

Another *in vivo* model of molecular mimicry was recently described (168). A protein from mouse herpes stromal keratitis (HSK) virus was shown to contain a peptide sequence that activated T cells that cross-reacted with a corneal antigen and induced autoimmune keratitis. T cells from an infected animal could adoptively transfer disease. A mutant form of the virus lacking the cross-reactive peptide sequence generated T cells that were unable to transfer disease, thus demonstrating a direct role of molecular mimicry in this model of autoimmunity. Consistent with the observations of Carrizosa and colleagues, peptide immunization could not induce disease after transfer of the T cells, even when performed in the presence of IL-12. However, adoptive transfer of the cross-reactive T cells into naïve mice in addition to administration of infectious virus at the target site was able to induce the disease. These results again emphasize that molecular mimicry plays a role in the induction of autoimmunity, but cannot itself induce disease without tissue destruction and/or release of autoantigen to further expand or select pathogenic T cells.

Anderson and colleagues have demonstrated that B7-1 sends a quantitatively stronger signal than B7-2 during the activation of T cells with weak peptide agonists and similar results have been obtained by another group (169,170). Based on this data, and the preferential association of B7-1 with MS plaques and EAE induction, they have hypothesized that B7-1 might be associated with MS because it can activate autoreactive T cells recognizing relatively weak mimicry peptides derived from infectious agents. Indeed, most viral and bacterial peptides identified, which serve as molecular mimics to activate autoreactive T cells, behave as weak peptide agonists in that they induce only modest levels of proliferation and cytokine production (165,166).

Recently, the relative roles of B7-1 versus B7-2 costimulation have been investigated in the HSK molecular mimicry model (171). Disease was induced in mice by infection with the virus, resulting in an autoimmune disease as characterized previously by Zhao and colleagues (168). The effects of anti-B7-1 or anti-B7-2 mAb administration on the disease were then investigated. Of great importance, both B7-1 and B7-2 were expressed within the local environment, indicating that both molecules were potentially capable of activating T cells. The authors found that blocking B7-1 prevented viral-induced disease, whereas blocking B7-2 did not reduce the incidence of disease and only ameliorated the severity of the disease. This *in vivo* model of an autoimmune disease induced by a viral epitope that cross-reacts with a self-antigen is consistent with the hypothesis that B7-1 costimulation is uniquely able to activate autoreactive T cells stimulated with cross-reactive peptides derived from infectious agents.

10. CONCLUSIONS

Significant strides have been made towards elucidating the molecular pathophysiology of multiple sclerosis. Carefully designed epidemiological studies in MS patient populations identify putative risk factors and generate hypotheses regarding pathogenesis. The ongoing study of autoimmune processes in animal models such as EAE continues to contribute important insights. Translational studies that combine careful clinical monitoring with modern imaging modalities and state of the art immunological assays are likely to become the mainstay approach to the investigation of the disease mechanisms and therapeutics.

Our current understanding invokes proinflammatory cells and mediators that may be triggered by environmental factors to mediate disease in a genetically susceptible host. A complex interplay exists between activated autoreactive Th-1 cells, the chemokine/chemokine receptors (including IP-10/CXCR3 and RANTES/CCR5) that regulate migration towards the target tissue, and the tightly controlled expression of adhesion molecules (e.g. ICAM-1, VCAM-1) and matrix metalloproteinases (e.g. Gelatinase B) that are required for BBB and tissue infiltration. The resultant effector profiles of autoreactive cells and their interactions with glial cells and infiltrating monocytes, is context dependent and influenced by factors including strength of activating signal, the costimulatory environment and the cytokine milieu (172). Converging animal and human studies implicate the costimulatory molecule B7-1 and the cytokines TNFα and IL-12 in disease pathophysiology. Locally expressed proinflammatory mediators contribute directly to tissue damage and to further disruption of the BBB and chemokine release, which result in subsequent waves of leukocyte infiltration.

The steady dissection of MS pathophysiology continues to implicate novel molecular interactions and, in spite of the apparent complexity, identifies exciting targets for the development of the next generation of therapeutics.

ACKNOWLEDGEMENTS

AB was supported by a Multiple Sclerosis Society of Canada Research Fellowship and by the Clinical Investigator Training Program: Harvard/MIT Health Sciences and Technology -Beth Israel Deaconess Medical Center, in collaboration with Pfizer, Inc.

References

1. Prineas, J.W., and C.S. Raine. 1976. Electron microscopy and immunoperoxidase studies of early multiple sclerosis lesions. *Neurology* 26:29–32.
2. Prineas, J.W., and R.G. Wright. 1978. Macrophages, lymphocytes and plasma cells in the perivascular compartment in chronic multiple sclerosis. *Lab. Invest.* 38:409–421.
3. Adams, C.W., R.N. Poston, and S.J. Buk. 1989. Pathology, histochemistry and immunocytochemistry of lesions in acute multiple sclerosis. *J. Natural Sci.* 92:291–306.

4. Sospedra, M., X. Ferrer-Francesch, O. Dominguez, M. Juan, M. Foz-Sala, and R. Pujol-Borrell. 1998. Transcription of a broad range of self-antigens in human thymus suggests a role for central mechanisms in tolerance toward peripheral antigens. *J. Immunol.* 161:5918–5929.

5. Fritz, R.B., and M.L. Zhao. 1996. Thymic expression of myelin basic protein (MBP). Activation of MBP-specific T cells by thymic cells in the absence of exogenous MBP. *J. Immunol.* 157:5249–5253.

6. Pribyl, T.M., C. Campagnoni, K. Kampf, V.W. Handley, and A.T. Campagnoni. 1996. The major myelin protein genes are expressed in the human thymus. *J. Neuroscience Res.* 45:812–819.

7. Liu, G.Y., P.J. Fairchild, R.M. Smith, J.R. Prowle, D. Kioussis, and D.C. Wraith. 1995. Low avidity recognition of self-antigen by T cells permits escape from central tolerance. *Immunity* 3:407–415.

8. Kanagawa, O., S.M. Martin, B.A. Vaupel, E. Carrasco-Martin, and E.R. Unanue. 1998. Autoreactivity of T cells from nonobese diabetic mice: an I-Ag7-dependent reaction. *Proc. Natl. Acad. Sci.* 95:1721–1724.

9. Ridgway, W.M., M. Fasso, and C.G. Fathman. 1999. A new look at MHC and autoimmune disease. *Science* 284:749–751.

10. Ho, H.Z., J.L. Tiwari, R.W. Haile, P.I. Terasaki, and N.E. Morton. 1982. HLA-linked and unlinked determinants of multiple sclerosis. *Immunogenetics* 15:509–517.

11. Spielman, R.S., and N. Nathanson. 1982. The genetics of susceptibility to multiple sclerosis. *Epidem. Rev.* 4:45–65.

12. Francis, D.A., J.R. Batchelor, W.I. McDonald, J.E. Hern, and A.W. Downie. 1986. Multiple sclerosis and HLA DQw1. *Lancet* 1:211.

13. Olerup, O., J. Hillert, S. Fredrikson, T. Olsson, S. Kam-Hansen, E. Moller, B. Carlsson, and J. Wallin. 1989. Primarily chronic progresive and relapsing/remitting multiple sclerosis: two immunogenetically distinct disease entities. *Proc. Natl. Acad. Sci.* 86:7113–7117.

14. Vartdal, F., L.M. Sollid, B. Vandvik, G. Markussen, and E. Thorsby. 1989. Patients with multiple sclerosis carry DQB1 genes which encode shared polymorphic amino acid sequences. *Hum. Immunol.* 25:103–110.

15. Clerici, N., M. Hernandez, M. Fernandez, J. Rosique, and J. Alvarez-Cermeno. 1989. *Tissue Antigens* 34:309–311.

16. Mokhtarian, F., D.E. McFarlin, and C.S. Raine. 1984. Adoptive transfer of myelin basic protein-sensitized T cells produces chronic relapsing demyelinating disease in mice. *Nature* 309:356–358.

17. Zamvil, S., P. Nelson, J. Trotter, D. Mitchell, R. Knobler, R. Fritz, and L. Steinman. 1985. T-cell clones specific for myelin basic protein induce chronic relapsing paralysis and demyelination. *Nature* 317:355–358.

18. Fritz, R.B., M.J. Skeen, C.H. Chou, M. Garcia, and I.K. Egorov. 1985. Major histocompatibility complex-linked control of the murine immune response to myelin basic protein. *J. Immunol.* 134:2328–2332.

19. Touhy, V.K., Z. Lu, R.A. Sobel, R.A. Lauresen, and M.B. Lees. 1989. Identification of an encephalitogenic determinant of myelin proteolipid protein for SJL mice. *J. Immunol.* 142:1523–1526.

20. Ota, K., M. Matsui, E.L. Milford, G.A. Mackin, H.L. Weiner, and D.A. Hafler. 1990. T-cell recognition of an immunodominant myelin basic protein epitope in multiple sclerosis. *Nature* 346:183–187.

21. Markovic-Plese, S., H. Fukaura, J. Zhang, A. al-Sabbagh, S. Southwood, A. Sette, V.K. Kuchroo, and D.A. Hafler. 1995. T cell recognition of immunodominant and cryptic proteolipid protein epitopes in humans. *J. Immunol.* 155:982–992.

22. Colombo, E., K. Bauki, A.H. Tatum, J. Dancher, P. Ferrante, R.S. Murray, P.E. Philips, and A. Pearl. 1997. Comparative analysis of antibody and cell-mediated autoimmunity to transaldolase and myelin basic protein in patients with multiple sclerosis. *J. Clin. Invest.* 99:1238–1250.

23. Walsh, M.J., J.M. Murray. 1998. Dual implication of 2′, 3′-cyclic nucleotide 3′ phosphodiesterase as major autoantigen and C3 complement-binding protein in the pathogenesis of multiple sclerosis. *J. Clin. Invest.* 101:1923–1931.

24. Zhang, J., S. Markovic-Plese, B. Lacet, J. Raus, H.L. Weiner, and D.A. Hafler. 1994. Increased frequency of inter-leukin 2-responsive T cell specific for myelin basic protein and proteolipid protein in peripheral blood and cerebrospinal fluid of patients with multiple sclerosis. *J. Exp. Med.* 179:973–984.

25. Scholz, C., K.T. Patton, D.E. Anderson, G.J. Freeman, and D.A. Hafler. 1998. Expansion of autoreactive T cells in multiple sclerosis is independent of exogenous B7 costimulation. *J. Immunol.* 160:1532–1538.

26. Lovett-Racke, A.E., J.L. Trotter, J. Lauber, P.J. Perrin, C.H. June, and M.K. Racke. 1998. Decreased dependence of myelin basic protein-reactive T cells on CD28-mediated costimulation in multiple sclerosis patients. A marker of activated/memory T cells. *J. Clin. Invest.* 101:725–730.

27. Karpus, W.J., and R.M. Ransohoff. 1998. Cutting Edge Commentary: Chemokine regulation of experimental autoimmune encephalomyelitis: temporal and spatial expression patterns govern disease pathogenesis. *J. Immunol.* 161:2667–2671.

28. Karpus, W.J., N.W. Lukacs, B.L. McRae, R.M. Streiter, J.L. Kunkel, and S.D. Miller. 1995. An important role for the chemokine macrophage inflammatory protein-1α in the pathogenesis of the T cell-mediated autoimmune disease, experimental autoimmune encephalomyelitis. *J. Immunol.* 155:5003–5010.

29. Karpus, W.J., and K.J. Kennedy. 1997. MIP-1alpha and MCP-1 differentially regulate acute and relapsing autoimmune encephalomyelitis as well as Th1/Th2 lymphocyte differentiation. *J. Leukoc. Biol.* 62:681–687.

30. Van Der Voorn, P., J. Tekstra, R.H. Beelen, C.P. Tensen, P. Van Der Valk, and C.J. De Groot. 1999. Expression of MCP-1 by reactive astrocytes in demyelinating multiple sclerosis lesions. *Am. J. Pathol.* 154:45–51.

31. Hvas, J., C. McLean, J. Justesen, G. Kannourakis, L. Steinman, J.R. Oksenberg, and C.C. Bernard. 1997. Perivascular T cells express the pro-inflammatory chemokine RANTES mRNA in multiple sclerosis lesions. *Scand. J. Immunol.* 46:195–203.

32. Sorensen, T.L., M. Tani, J. Jensen, V. Pierce, C. Lucchinetti, V.A. Folcik, S. Qin, J. Rottman, F. Sellebjerg, R.M. Strieter, J.L. Frederiksen, and R.M. Ransohoff. 1999. Expression of specific chemokines and chemokine receptors in the central nervous system of multiple sclerosis patients. *J. Clin. Invest.* 103:807–815.

33. Siveke, J.T., and A. Hamann. 1998. T helper 1 and T helper 2 cells respond differentially to chemokines. *J. Immunol.* 160:550–554.

34. Bennetts, B.H., S.M. Teutsch, M.M. Buhler, R.N. Heard, and G.J. Stewart. 1997. The CCR5 deletion mutation fails to protect against multiple sclerosis. *Hum. Immunol.* 58:52–59.

35. Raine, C.S., and B. Cannella. 1992. Adhesion molecules and CNS inflammation. *Semin. Neurosci.* 4:201–211.

36. Springer, T.A. 1994. Traffic signals for lymphocyte recirculation and leukocyte emigration: the multistep paradigm. *Cell* 76:301–314.

37. Raine, C.S. 1994. The Dale E. McFarlin memorial lecture: the immunology of the MS lesion. *Ann. Neurol.* 36:S61–S72.

38. Lee, S.J., and E.N. Benveniste. 1999. Adhesion molecule expression and regulation on cells of the CNS. *J. Neuroimmunol.* 98:77–88.

39. Sanchez-Madrid, F., J.A. Nagy, E. Robbins, P. Simon, and T.A. Springer. 1983. A human leukocyte differentiation antigen family with distinct α-subunits and a common β-subunit. *J. Exp. Med.* 158:1785–1803.

40. Elices, M.J., L. Osborn, Y. Takada, C. Crouse, S. Luhowskyj, M.E. Hemler, and R.R. Lobb. 1990. VCAM-1 on activated endothelium interacts with the leukocyte integrin VLA-4 at a site distinct from the VLA-4/fibronectin binding site. *Cell* 60:577–584.

41. Foster, C.A. 1996. VCAM-1/α4-integrin adhesion pathway: therapeutic target for allergic inflammatory disorders. *J. Allerg. Clin. Immunol.* 98:S270–S277.

42. Sligh, J.E., Jr., C.M. Ballantyme, S.S. Rich, H.K. Hawkins, C.W. Smith, A. Bradley, and A.L. Beaudet. 1993. Inflammatory and immune responses are impaired in mice deficient in ICAM-1. *Proc. Natl. Acad. Sci. USA* 90:8529–8533.

43. Damle, N.K., K. Klussman, G. Leytze, A. Aruffo, P.S. Linsley, and J.A. Ledbetter. 1993. Costimulation with intergin ligands ICAM-1 or VCAM-1 augmenst activation-induced death of antigen-specific CD4+ T lymphocytes. *J. Immunol.* 151:2368–2379.

44. Sobel, R.A., M.E. Mitchell, and G. Fondren. 1990. ICAM 1 in cellular immune reactions in the human CNS. *Am. J. Pathol.* 136:1309–1316.

45. Washington R., J. Burton, R.F. Todd. III, W. Newman, L. Dragovic and P. Dore-Duffy. 1994. Expression of immunologically relevant endothelial cell activation antigens on isolated CNS vessels from patients with MS. *Ann. Neurol.* 35:89–97.

46. Cannella, B., and S.C. Raine. 1995. The adhesion molecule and cytokine profile in MS lesions. *Ann. Neurol.* 37:424–435.

47. Brosnan, C.F., B. Cannella, L. Battistini, and C.S. Raine. 1995. Cytokine localization in MS lesions: correlation with adhesion molecule expression and reactive nitrogen species. *Neurology* 45:S16–S21.

48. Bo, L., J.W. Peterson, S. Mork, P.A. Hoffman, W.M. Gallatin, R.M. Ransohoff, and B.D. Trapp. 1996. Distribution of immunoglobulin family members ICAM-1, -2, -3 and the β2 integrin LFA-1 in multiple sclerosis lesions. *J. Neuropathol. Exp. Neurol.* 55:1060–1072.

49. Kuchroo, V.K., C.A. Martin, J.M. Greer, S.-T. Ju, R.A. Sobel, and M.E. Dorf. 1993. Cyokines and adhesion molecules contribute the ability of myelin proteolipid protein-specific T cell clones to mediate EAE. *J. Immunol.* 151:4371–4382.

50. Baron, J.L., J.A. Madri, N.H. Ruddle, G. Hashim, and C.A. Janeway, Jr. 1993. Surface expression of αV4 integrin by CD4 T cells is required for their entry into brain parenchyma. *J. Exp. Med.* 177:57–68.

51. Yednock, T.A., C. Cannon, L.C. Fritz, F. Sanchez-Madrid, L. Steinman, and N. Karin. 1992. Prevention of EAE by antibodies against α4β1 integrin. *Nature* 356:63–66.

52. Soilu-Hanninen, M., M. Royna, A. Salmi, and R. Salonen. 1997. Therapy with antibody against leukocyte integrin VLA-4 (CD49d) is effective and safe in virus-facilitated EAE. *J. Neuroimmunol.* 72:95–105.

53. Engelhardt, B., M. Laschinger, M. Schultz, U. Samulovitz, D. Vebstweber, and G. Hoch. 1998. The development of EAE in the mouse requires α4-integrin but not α4β7-imtegrin. *J. Clin. Invest.* 102:2096–2105.

54. Moingeon, P., H.C. Chang, B.P. Wallner, C. Stebbins, A.Z. Frey and E.L. Reinherz. 1989. CD2-mediated adhesion facilitates T lymphocyte antigen recognition function. *Nature* 339:312–314.

55. Damle, N.K., and A. Aruffo. 1991. VCAM-1 induces T-cell antigen receptor-dependent activation of CD4+ T lymphocytes. *Proc. Natl. Acad. Sci. USA* 88:6403–6407.

56. Aloisi, F., G. Borcellino, P. Samoggia, U. Testa, C. Chelucci, G. Russo, C. Peschle, and G. Levi. 1992. Astrocyte cultures from human embryonic brain: characterization and modulation of surface molecules by inflammatory cytokines. *J. Neurosci. Res.* 32:494–506.

57. Frohman, E.M., T.C. Frohman, M.L. Dustin, B. Vayuvegula, B. Choi, A. Gupta, S. van den Noort, and S. Gupta. 1989. The induction of ICAM-1 expression on human fetal astrocytes by interferon-γ, tumor necrosis factor-α, lymphotoxin and IL-1: relevance to intracerebral antigen presentation. *J. Neuroimmunol.* 23:117–124.

58. Shrikant, P., I.Y. Chung, M. Ballestas, and E.N. Benveniste. 1994. Regulation of ICAM-1 gene expression by TNF-α, IL-1-β and Interferon-γ in astrocytes. *J. Immunol.* 151:209–220.

59. Shrinkat, P., E. Weber, T. Jilling, and E.N. Benveniste. 1995. ICAM-1 gene expression by glial cells: differential mechanisms of inhibition by IL-10 and IL-6. *J. Immunol.* 155:1489–1501.

60. Hartung, H.P., M. Michels, K. Reiners, P. Seeldrayers, J.J. Archelos, and K.V. Toyka. 1993. Soluble ICAM-1 srum levels in MS and viral encephalitis. *Neurology* 43:2331–2335.

61. Rieckman P., S. Martin, I. Weichselbraun, M. Albrecht, B. Kitze, T. Weber, H. Tumani, A. Broocks, W. Luer, A. Helwig, and S. Poser. 1994. Serial analysis of circulating adhesion molecules and TNF receptor in serum from patients with MS. *Neurology* 44:2367–2372.

62. Hartung, H.P., K. Reiners, J.J. Archelos, M. Michels, P. Seeldrayers, F. Heidenreich, K.W. Pfulghaupt, and K.V. Toyka. 1995. Circulating adhesion molecules and TNF-receptor in MS: correlation with MRI. *Ann. Neurol.* 38:186–193.

63. Sharief, M.K., M.A. Noori, M. Ciardi, A. Cirelli, and E.J. Thompson. 1993. Increased levels of circulating ICAM-1 in serum and CSF of patients with active MS: correlation with TNF-α and blood-brain barrier damage. *J. Neuroimmunol.* 43:15–22.

64. Tsukada, N., M. Matsuda, K. Miyagi, and N. Yanagisawa. 1993. Increased levels of ICAM-1 and TNF receptor in the CSF of patients with MS. *Neurology* 43:2679–2682.

65. Droogan A.G., S.A. McMillan, J.P. Douglas, and S.A. Hawkins. 1996. Serum and cerebrospinal fluid levels of soluble adhesion molecules in multiple sclerosis: ant intrathecal release of VCAM-1. *J. Neuroimmunol.* 64:185–191.

66. Rieckman P., B. Altenhofen, A. Riegel, J. Baudewig, and K. Flegenhauer. 1997. Soluble adhesion molecules (sVCAM-1 and sICAM-1) in CSF and serum correlate with MRI activity in MS. *Ann. Neurol.* 41:326–333.

67. Giovannoni, G., M. Lai, J.W. Thorpe, D. Kidd, V. Chamoun, A.J. Thompson, D.H. Miller, M. Feldmann, and E.J. Thompson. 1997. Longitudinal study of soluble adhesion molecules in MS: correlation with gadolinium enhanced MRI. *Neurology* 48:1557–1565.

68. Leca, G., S.E. Mansur, and A. Bensussan. 1995. Expression of VCAM-1 (CD106) by a subset of TCR-γ–bearing lymphocyte clones. *J. Immunol.* 154:1069–1077.

69. Lyons, P.D., and E.N. Benveniste. 1998. Cleavage of membrane-associated ICAM-1 from astrocytes: involvement of a metalloprotease. *GLIA* 22:103–112.

70. Dore-Duffy, P., W. Newman, R. Balabanov, R.P. Lisak, E. Mainolfi, R. Rothlein, and M. Peterson. 1995. Circulating soluble adhesion proteins in cerebral spinal fluid and serum of patients with multiple sclerosis: correlation with clinical activity. *Ann. Neurol.* 37:55–62.

71. Giovannoni, G., J.W. Thorpe, D. Kidd, B.E. Kendall, I.F. Moosley, A.J. Thompson, G. Keir, D.H. Miller, M. Feldmann, and E.J. Thompson. 1996. Soluble E-selectin in MS: Raised concentration in patients with primary progressive disease. *J. Neurol. Neurosurg. Psych.* 60:20–26.

72. McDonnell, G.V., S.A. McMillan, J.P. Douglas, A.G. Droogan, and S.A. Hawkins. 1998. Raised CSF levels of soluble adhesion molecules across the clinical spectrum of MS. *J. Neuroimmunol.* 85:186–192.

73. Calabresi, P.A., L.A. Stone, C.N. Bash, J.A. Frank, and H.F. McFarland. 1997. Interferon beta results in immediate reduction of contrast-enhanced MRI lesions in MS patients followed by weekly MRI. *Neurology* 4:1446–1448.

74. Calabresi, P.A., L.R. Tranquill, J.M. Dambrosia, L.A. Stone H. Maloni, C.N. Bash, J.A. Frank, and H.F. McFarland. 1997. Increases in soluble VCAM-1 correlate with a decrease in MRI lesions on MS treated with interferon β-1b. *Ann. Neurol.* 41:669–674.

75. Trojano, M., C. Avolio, M. Ruggieri, G. Defazio, F. Giuliani, D. Paolicelli, and P. Liverea. 1997. Antiinflammatory effects of rIFNβ-1b in MS: longitudinal changes of the serum sICAM-1, TNFα and T-cell subsets. *Neurology* 48:A244.

76. Trojano, M., C. Avolio, M. Ruggieri, F. De Robertis, F. Giuliani, D. Paolicelli, and P. Livrea. 1998. sICAM-1 in serum and CSF of demyelinating diseases of the central and peripheral nervous system. *Multiple Sclerosis* 4:39–44.

77. Calabresi, P.A., C.M. Pelfrey, L.R. Tranquill, H. Maloni, and H.F. McFarland. 1997. VLA-4 expression on peripheral blood lymphocytes is downregulated after treatment of MS patients with interferon beta. *Neurology* 49:1111–1116.

78. Woessner, J.F., Jr. 1994. The family of matrix metalloproteinases. *Ann. NY Acad. Sci.* 732:11–21.

79. Leppert, D., E. Waubant, R. Galardy, N.W. Bunnett, and S.L. Hauser. 1995. T cell gelatinases mediate basement membrane transmigration *in vitro*. *J. Immunol.* 154:4379–4389.

80. Nielsen, B.S., S. Timshel, L. Kjeldsen, M. Sehested, C. Ryke, N. Borregaard, and K. Dano. 1996. 92 kDa type IV collagenase (MMP-9) is expressed in neutrophils and macrophages but not in malignant epithelial cells in human colon cancer. *Int. J. Cancer* 65:57–62.

81. Cuzner, M.L., D. Gveric, C. Strand, A.J. Laughlin, L. Paemen, G. Opdenakker, and J. Newcombe. 1996. The expression of tissue-type plasminogen activator, matrix metalloproteinases and endogenous inhibitors in the CNS in MS: comparison of stages in lesion evolution. *J. Neuropathol. Exp. Neurol.* 55:1194–1204.

82. Maeda, A., and R.A. Sobel. 1996. Matrix metalloproteinases in the normal human central nervous system, microglial nodules and MS lesions. *J. Neuropathol. Exp. Neurol.* 55:300–309.

83. Yong, V.W., C.A. Krekowski, P.A. Forsyth, R. Bell, and D.R. Edwards. 1998. Matrix metalloproteinases and diseases of the CNS. *Trends Neurosci.* 21:75–80.

84. Kieseier, B.C., T. Seifert, G. Giovannoni, and H.P. Hartung. 1999. Matrix metalloproteinases in inflammatory demyelination: targets for treatment. *Neurology* 53:20–25.

85. Rosenberg, G.A., M. Kornfeld, E. Estrada, R.O. Kelley, L.A. Liotta, and W.G. Stetler-Stevenson. 1992. TIMP-2 reduces proteolytic opening of blood-brain barrier by type IV collagenase. *Brain. Res.* 576;203–207.

86. Rosenberg, G.A., J.E. Dencoff, P.G. McGuire, L.A. Liotta, and W.G. Stetler-Stevenson. 1994. Injury-induced 92-kD gelatinase and urokinase expression in rat brain. *Lab. Invest.* 71:417–422.

87. Anthony, D.C., K.M. Miller, S. Fearn, M.J. Towsend, G. Opdenakker, and G.M. Wills. 1998. Matrix metalloproteinase expression in an experimentally induced DTH model of MS in the rat CNS. *J. Neuroimmunol.* 87:62–72.

88. Hartung, H.P., S. Jung, G. Stoll, J. Zielasck, B. Schmidt, J.J. Archelos, and K.V. Toyka. 1992. Inflammatory mediators in demyelinating disorders of the CNS and PNS. *J. Neuroimmunol.* 40:197–210.

89. Black, R.A., C.T. Rauch, C.J. Kozlosky, J.A. Peschon, J.L. Slack, M.J. Wolfson, B.J. Castner, K.L. Stocking, P. Reddy, S. Srinivasan, N. Nelson, N. Boini, K.A. Schooley, M. Gerhart, R. Davis, J.N. Fitzner, R.S. Johnson, R.J. Paxton, C.J. March, and P. Cerretti. 1997. A metalloproteinase disintegrin that releases TNF-α from cells. *Nature* 385:729–733.

90. Proost, P., J. Van Damme, and G. Opdenakker. 1993. Leukocyte gelatinase B cleavage releases encephalitogens from human myelin basic protein. *Biochem. Biophys. Res. Commun.* 192:1175–1181.

91. Cuzner, M.L., A.N. Davison, and P. Rudge. 1978. Proteolytic enzyme activity of blood leukocytes and CSF in MS. *Ann. Neurol.* 4:337–344.

92. Gijbels, K., S. Masure, H. Carton, and G. Opdenakker. 1992. Gelatinase in the CSF of patients with MS and other inflammatory neurological disorders. *J. Neoroimmunol.* 41:29–34.

93. Leppert, D., J. Ford, G. Stabler, C. Grygar, C. Lienert, S. Huber, K.M. Miller, S.L. Hauser, and L. Kappos. 1998. Matrix metalloproteinase-9 (Gelatinase B) is selectively elevated in CSF during relapses and stable phase of MS. *Brain* 121:2327–2334.

94. Lee, M.A., J. Palace, G. Stabler, J. Ford, A. Gearing, and K. Miller. 1999. Serum gelatinase B, TIMP-1 and TIMP-2 levels in MS: a longitudinal clinical and MRI study. *Brain* 122:191–197.

95. Rosenberg, G.A., J.E. Dencoff, N. Correa, M. Reiners, and C.C. Ford. 1996. Effect of steroids on CSF matrix metalloproteinases in MS: relation to blood-brain barrier injury. *Neurology* 46:1626–1632.

96. Leppert, D., E. Waubant, M.R. Burk, J.R. Oksenberg, and S.L. Hauser. 1996. Interferon beta-1b inhibits gelatinase secretion and *in vitro* migration of human T cells: a possible mechanism for treatment efficacy in MS. *Ann. Neurol.* 40:846–852.

97. Stuve, O., N.P. Dooley, J.H. Uhm, J.P. Antel, G.S. Francis, G. Williams, and V.W. Yong. 1996. Interferon β-1b decreases the migration of T lymphocytes *in vitro*: effects on matrix metalloproteinase-9. *Ann. Neurol.* 40:853–863.

98. Stone, L.A., J.A. Frank, P.S. Albert, C. Bash, M.E. Smith, H. Maloni, and H.F. McFarland. 1995. The effect of inter-

feron-beta on blood-brain barrier disruptions demonstrated by contrast-enhanced MRI in relapsing-remitting MS. *Ann. Neurol.* 37:611–619.

99. Romanic, A.M., and J.A. Madri. 1994. The induction of a 72-kD gelatinase in T cells upon adhesion to endothelial cells is VCAM-1 dependent. *J. Cell. Biol.* 15:1165–1178.

100. Gijbels, K., R.E. Galardy, and L. Steinman. 1994. Reversal of EAE with a hydroxamate inhibitor of matrix metalloproteinases. *J. Clin. Invest.* 94:2177–2182.

101. Hewson, A.K., T. Smith, J.P. Leonard, and M.L. Cuzner. 1995. Suppression of EAE in the Lewis rat by the matrix metalloproteinase inhibitor Ro31–9790. *Inflamm. Res.* 44:345–349.

102. Liedtke, W., B. Cannella, R.J. Mazzaccaro, J.M. Clements, K.M. Miller, K.W. Wucherpfennig, A.J. Gearing, and C.S. Raine. 1998. Effective treatment of models of MS by matrix metalloproteinase inhibitors. *Ann. Neurol.* 44:35–46.

103. Lenschow, D.J., T.L. Walunas, and J.A. Bluestone. 1996. CD28/B7 system of T cell costimulation. *Ann. Rev. Immunol.* 14:233–258.

104. Tivol, E.A., A.N. Schweitzer, and A.H. Sharpe. 1996. Costimulation and autoimmunity. *Curr. Opin. Immunol.* 8(6):822 830.

105. Linsley, P.S. 1995. Distinct roles for CD28 and cytotoxic T lymphocyte-associated molecule-4 receptors during T cell activation? *J. Exp. Med.* 182:289–292.

106. Thompson, C.B., and J.P. Allison. 1997. The emerging role of CTLA-4 as an immune attenuator. *Immunity* 7:445–450.

107. Jenkins, M.K., and R.H. Schwartz. 1987. Antigen presentation by chemically modified splenocytes induces antigen-specific T cell unresponsiveness *in vitro* and *in vivo*. *J. Exp. Med.* 165:302–319.

108. Karandikar, N.J., C.L. Vanderlugt, J.A. Bluestone, and S.D. Miller. 1998. Targeting the B7/CD28:CTLA-4 costimulatory system in CNS autoimmune disease. *J. Neuroimmunol.* 89:10–18.

109. Issazadeh, S., V. Navikas, M. Schaub, M. Sayeg, and S. Khoury. 1998. Kinetics of expression of costimulatory molecules and their ligands in murine relapsing experimental autoimmune encephalomyelitis *in vivo*. *J. Immunol.* 161:1104–1112.

110. Vandenbark, A.A., B. Celnik, M. Vainiene, S.D. Miller, and H. Offner. 1995. Myelin antigen-coupled splenocytes suppress experimental autoimmune encephalomyelitis in Lewis rats through a partially reversible anergy mechanism. *J. Immunol.* 155:5861–5867.

111. Kennedy, K.J., W.S. Smith, S.D. Miller, and W.J. Karpus. 1997. Induction of antigen-specific tolerance for the treatment of ongoing, relapsing autoimmune encephalomyelitis: a comparison between oral and peripheral tolerance. *J. Immunol.* 159:1036–1044.

112. Arima, T., A. Rehman, W.F. Hickey, and M.W. Flye. 1996. Inhibition by CTLA4Ig of experimental allergic encephalomyelitis. *J. Immunol.* 156:4916–4924.

113. Croxford, J.L., J.K. O'Neill, R.R. Ali, K. Browne, A.P. Byrnes, M.J. Dallman M.J. Wood, M. Fedlmann, and D. Baker. 1998. Local gene therapy with CTLA4-immunoglobulin fusion protein in experimental allergic encephalomyelitis. *Eur. J. Immunol.* 28:3904–3916.

114. Kuchroo, V.K., M.P. Das, J.A. Brown, A.M. Ranger, S.S. Zamvil, R.A. Sobel, H.L. Weiner, N. Nabavi, and L.H. Glimcher. 1995. B7-1 and B7-2 costimulatory molecules activate differentially the Th1/Th2 developmental pathways: application to autoimmune disease therapy. *Cell* 80:707–718.

115. Racke, M.K., D.E. Scott, L. Quigley, G.S. Gray, R. Abe, C.H. June, and P.J. Perrin. 1995. Distinct roles for B7–1 (CD-80) and B7–2 (CD-86) in the initiation of experimental allergic encephalomyelitis. *J. Clin. Invest.* 96:2195–2203.

116. Miller, S.D., C.L. Vanderlugt, D.J. Lenschow, J.G. Pope, N.J. Karandikar, M.C. Dal Canto, and J.A. Bluestone. 1996. Blockade of CD28/B7–1 interaction prevents epitope spreading and clinical relapses of murine EAE. *Immunity* 3:739–745.

117. Karandikar, N.J., C.L. Vanderlugt, T. Eagar, L. Tan, J.A. Bluestone, and S.D. Miller. 1998. Tissue-specific up-regulation of B7–1 expression and function during the course of murine relapsing experimental autoimmune encephalomyelitis. *J. Immunol.* 161:192–199.

118. Williams, K., E. Ulvestad, and J.P. Antel. 1994. B7/BB-1 antigen expression on adult human microglia studied *in vitro* and in situ. *Eur. J. Immunol.* 24:3031–3037.

119. Windhagen, A., J. Newcombe, F. Dangond, C. Strand, M.N. Woodroofe, M.L. Cuzner, and D.A. Hafler. 1995. Expression of costimulatory molecules B7–1 (CD80), B7–2 (CD86), and interleukin 12 cytokine in multiple sclerosis lesions. *J. Exp. Med.* 182:1985–1996.

120. Genc, K., D.L. Dona, and A.T. Reder. 1997. Increased CD80+ B cells in active multiple sclerosis and reversal by interferon β-1b therapy. *J. Clin. Invest.* 99:2664–2671.

121. Abbas, A.K., K.M. Murphy, and A. Sher. 1996. Functional diversity of helper T lymphocytes. *Nature* 383: 787–793.

122. O'Garra, A. 1998. Cytokines induce the development of functionally heterogeneous T helper cell subsets. *Immunity* 8:275–283.

123. Khoury, S.J., W.W. Hancock, and H.L. Weiner. 1992. Oral tolerance to myelin basic protein and natural recovery from experimental autoimmune encephalomyelitis are associated with downregulation of inflammatory cytokines and differential upregulation of transforming growth factor beta, interleukin 4, and prostaglandin E expression in the brain. *J. Exp. Med.* 176:1355–1364.

124. Begolka, W.S., C.L. Vanderlugt, S.M. Rahbe, and S.D. Miller. 1998. Differential expression of inflammatory cytokines parallels progression of central nervous system pathology in two clinically distinct models of multiple sclerosis. *J. Immunol.* 161:4437–4446.

125. Hofman, F.M., D.R. Hinton, K. Johnson, and J.E. Merrill. 1989. Tumor necrosis factor identified in multiple sclerosis brain. *J. Exp. Med.* 170:607–612.

126. Selmaj, K., C.S. Raine, B. Cannella, and C.F. Brosnan. 1991. Identification of lymphotoxin and tumor necrosis factor in multiple sclerosis lesions. *J. Clin. Invest.* 87:949–954.

127. Correale, J., W. Gilmore, M. McMillan, S. Li, K. McCarthy, T. Le, and L.P. Weiner 1995. Patterns of cytokine secretion by autoreactive proteolipid protein-specific T cell clones during the course of multiple sclerosis. *J. Immunol.* 154:2959–2968.

128. Mokhtarian, F., Y. Shi, D. Shirazian, L. Morgante, A. Miller, and D. Grob. 1994. Defective production of anti-inflammatory cytokine, TGF-beta by T cell lines of patients with active multiple sclerosis. *J. Immunol.* 152:6003–6010.

129. Segal, B.M., B.K. Dwyer, and E.M. Shevach. 1998. An interleukin (IL)-10/IL-12 immunoregulatory circuit controls susceptibility to autoimmune disease. *J. Exp. Med.* 187:537–546.

130. Leonard, J.P., K.E. Waldburger, R.G. Schaub, T. Smith, A.K. Hewson, M.L. Cuzner, and S.J. Goldman. 1997. Regulation of the inflammatory response in animal models

of multiple sclerosis by interleukin-12. *Crit. Rev. Immunol.* 17:545–553.

131. Bright, J.J., B.F. Musuro, C. Du, and S. Sriram. 1998. Expression of IL-12 in CNS and lymphoid organs of mice with experimental allergic encephalitis. *J. Neuroimmunol.* 82:22–30.

132. Constantinescu, C.S., M. Wysocka, B. Hilliard, E.S. Ventura, E. Lavi, G. Trinchieri, and A. Rostami. 1998. Antibodies against IL-12 prevent superantigen-induced and spontaneous relapses of experimental autoimmune encephalomyelitis. *J. Immunol.* 161:5097–5104.

133. Bright, J.J., C. Du, M. Coon, S. Sriram, and S.J. Klaus. 1998. Prevention of experimental allergic encephalomyelitis via inhibition of IL-12 signaling and IL-12-mediated Th1 differentiation: an effect of the novel anti-inflammatory drug lisofylline. *J. Immunol.* 161:7015–7022.

134. Comabella, M., K. Balashov, S. Issazadeh, D. Smith, H.L. Weiner, and S.J. Khoury. 1998. Elevated interleukin-12 in progressive multiple sclerosis correlates with disease activity and is normalized by pulse cyclophosphamide therapy. *J. Clin. Invest.* 102:671–678.

135. Compston, A. 1994. The epidemiology of multiple sclerosis: principles, achievements, and recommendations. *Ann. Neurol.* 36:S211–S217.

136. Weinshenker, B. 1996. Epidemiology of multiple sclerosis. *Neurol. Clin.* 14:291–308.

137. Lublin, F.D., and S.C. Reingold. 1996. Defining the clinical course of multiple sclerosis: results of an international survey. National Multiple Sclerosis Society (USA) Advisory Committee on Clinical Trials of New Agents in Multiple Sclerosis. *Neurology* 46:907–911.

138. Paty, D.W., and G.C. Ebers. 1998. Multiple sclerosis. In: *Contemporary neurology series.* Vol. 50. D.W. Paty, and G.C. Ebers, editors. F.A. Davis Company, Philadelphia. pp. 150–174.

139. Bar-Or, A., and D. Smith. 1999. Epidemiology of multiple sclerosis. In: *Principles of neuroepidemiology.* T. Batchelor and M. Cudcowicz, editors. Butterworth-Heinemann, Boston. In press.

140. Kurtzke, J.F., G.W. Beebe, and J.E. Norman, Jr. 1979. Epidemiology of multiple sclerosis in U.S. veterans: 1. Race, sex, and geographic distribution. *Neurology* 29:1228–35.

141. Sadovnick A., and G. Ebers. 1993. Epidemiology of multiple sclerosis: a critical overview. *Can. J. Neurol. Sci.* 20:17–29.

142. Irizarry, M.C. 1997. Multiple sclerosis. In: *Neurologic disorders in women.* M.E. Cudkowicz and M.C. Irizarry, editors. Butterworth-Heinemann, Boston. pp. 85–99.

143. Hauser, S., E. Fleischnick, H. Weiner, D. Marcus, Z. Awdeh, E. Yunis, and C. Alper. 1989. Extended major histocompatibility complex haplotypes in patients with multiple sclerosis. *Neurology* 39:275–277.

144. Hillert, J., and O. Olerup. 1993. Multiple sclerosis is associated with genes within or close to the HLA-DR-DQ subregion on a normal DR15 DQ6, Dw2 haplotype. *Neurology* 43:163–168.

145. Olerup, O., and J. Hillert. 1991. HLA class II-associated genetic susceptibility in multiple sclerosis: A critical evaluation. *Tissue Antigens* 38:1–15.

146. Oksenberg, J., E. Seboun, and S. Hauser. 1996. Genetics of demyelinating diseases. *Brain Path.* 6:289–302.

147. Dyment, D., A. Sadnovick, and G. Ebers. 1997. Genetics of multiple sclerosis. *Hum. Mol. Genet.* 6:1693–8.

148. Ebers, G., K. Kukay, D. Bulman, A.D. Sadovnick, G. Rice, and the Multiple Sclerosis Genetics Group. 1996. A full genome search in multiple sclerosis. *Nat. Genet.* 13:472–476.

149. Haines, J., M. Ter-Minassian, A. Bazyk, J.F. Gusella, D.J. Kim, and the Multiple Sclerosis Genetics Group. 1996. Complete genomic screen for multiple sclerosis underscores a role for the major histocompatibility complex. The Multiple Sclerosis Genetics Group. *Nat. Genet.* 13:469–471.

150. Sawcer, S., H. Jones, R. Feakes, J. Gray, N. Smaldon, J. Chataway, N. Robertson, D. Clayton, P.N. Goodfellow, and A. Compston. 1996. A genome screen in multiple sclerosis reveals susceptibility loci on chromosome 6p21 and 17q22. *Nat. Genet.* 13:464–468.

151. Sawcer, S., P. Goodfellow, and A. Compston. 1997. The genetic analysis of multiple sclerosis. *Trend. Genet.* 13:234–239.

152. Ebers, G. 1996. Genetic epidemiology of multiple sclerosis. *Curr. Opin. Neurol.* 9:155–158.

153. Sadovnick, A., D. Dyment, and G. Ebers. 1997. Genetic epidemiology of multiple sclerosis. *Epidem. Rev.* 19:99–106.

154. Bulman, D., and G. Ebers. 1992. The geography of Multiple sclerosis reflects genetic susceptibility. *J. Trop. Geogr. Neurol.* 2:66–72.

155. Hogancamp, W., M. Rodriguez, and B. Weinshenker. 1997. The epidemiology of multiple sclerosis. *Mayo Clin. Proc.* 72:871–878.

156. Hammond, S., J. McLeod, K. Millingen, E. Stewart-Wynne, D. English, J. Holland, and M. McCall. 1988. The epidemiology of multiple sclerosis in three Australian cities: Perth, Newcastle and Hobart. *Brain* 111:1–25.

157. Kurtzke, J., G. Dean, and D. Botha. 1970. A method of estimating age at immigration of white immigrants to South Africa, with an example of its importance. *S. Afr. Med. J.* 44:663–669.

158. Dean, G., and J. Kurtzke. 1971. On the risk of Multiple sclerosis according to age at immigration to South Africa. *Br. Med. J.* 3:725–729.

159. Lauer, L. 1995. Environmental associations with the risk of Multiple sclerosis: The contribution of ecological studies. *Acta. Neurol. Scand.* Suppl. 161:77–88.

160. Cook, S., C. Rohowsky-Kochan, S. Bansil, and P. Dowling. 1995. Evidence for multiple sclerosis as an infectious disease. *Acta Neurol. Scand.* Suppl. 161–169.

161. Granieri, E., and I. Casetta. 1997. Common childhood and adolescent infections and Multiple sclerosis. *Neurology* 49:S42–S54.

162. Hodge, M., and C. Wolfson. 1997. Canine distemper virus and multiple sclerosis. *Neurology* 49:S62–S69.

163. Sibley, W., C. Bamford, and K. Clark. 1985. Clinical viral infections and multiple sclerosis. *Lancet* (i):1313–1315.

164. Panitch, H. 1994. Influence of infections on exacerbations of multiple sclerosis. *Ann. Neurol.* 36:S25–S28.

165. Wucherpfennig, K.W., and J.L. Strominger. 1995. Molecular mimicry in T cell-mediated autoimmunity: viral peptides activate human T cell clones specific for myelin basic protein. *Cell* 80:695–705.

166. Carrizosa, A.M., L.B. Nicholson, M. Farzan, S. Southwood, A. Sette, R.A. Sobel, and V.K. Kuchroo. 1998. Expansion by self antigen is necessary for the induction of experimental autoimmune encephalomyelitis by T cells primed with a

cross-reactive environmental antigen. *J. Immunol.* 161:3307–3314.

167. Rao, V.P., A.E. Kajon, K.R. Spindler, and G. Carayanniotis. 1999. Involvement of epitope mimicry in potentiation but not initiation of autoimmune disease. *J. Immunol.* 162:5888–5893.

168. Zhao, Z.S., F. Gramicci, L. Yeh, P.A. Schaffer, and H. Cantor. 1998. Molecular mimicry by herpes simplex virus-type 1: autoimmune disease after viral infection. *Science* 279:1344–1347.

169. Anderson, D.E., L.J. Ausubel, J. Krieger, P. Höllsberg, G.J. Freeman, and D.A. Hafler. 1997. Weak peptide agonists reveal functional differences in B7-1 and B7-2 costimulation of human T cell clones. *J. Immunol.* 159:1669–1675.

170. Fields, P.E., R.J. Finch, G.S. Gray, R. Zollner, J.L. Thomas, K. Sturmhoefel, K. Lee, S. Wolf, T.F. Gajewski, and F.W. Fitch. 1998. B7.1 is a quantitatively stronger costimulus than B7.2 in the activation of naive CD8+ TCR-transgenic T cells. *J. Immunol.* 161:5268–7525.

171. Chen, H., and R.L. Hendricks. 1998. B7 costimulatory requirements of T cells at an inflammatory site. *J. Immunol.* 160:5045–5052.

172. Bar-Or, A., E.M.L. Oliveria, D.E. Anderson, and D.A. Hafler. 1999. Molecular pathogenesis of multiple sclerosis (review). *J. Neuroimmunol.* 100:252–259.

42 | Experimental Autoimmune Encephalomyelitis

Ana C. Anderson, Lindsay B. Nicholson, and Vijay K. Kuchroo

1. INTRODUCTION

Multiple Sclerosis (MS) is an inflammatory demyelinating disease of the central nervous system (CNS) believed to be autoimmune in origin. MS affects approximately 300,000 people in the United States with a female to male ratio of 2:1. Experimental allergic encephalomyelitis (EAE) is a T cell-mediated autoimmune disease of the CNS inducible in experimental animals by immunization with protein constituents of CNS myelin. EAE is a paralytic disease characterized by the presence of inflammatory infiltrates and demyelination in the CNS. Because of these and other features, EAE has been used as a model of MS. Studies in the EAE model have contributed greatly to our understanding of the mechanisms involved in the pathogenesis of CNS autoimmunity. The EAE model has also been used to evaluate therapeutic approaches for the treatment of MS. This chapter will review the history of the EAE model and its contribution to our understanding of the molecules and genes involved in initiating CNS autoimmunity.

2. THE EAE MODEL

2.1. History

EAE was first discovered at the turn of the century as an "allergic" side effect observed in individuals who had received the Pasteur rabies vaccine. This vaccine consisted of fixed rabies virus grown in rabbit brain tissue. Approximately 0.1% of vaccine recipients developed a monophasic paralytic illness, acute disseminated encephalomyelitis (ADE) (1). Perivascular mononuclear cell infiltrates and focal areas of demyelination in CNS

tissue from these individuals was observed. As ADE was known not to be a sequelae of rabies infection and there were cases of ADE in individuals who received the rabies vaccine but had not been exposed to the rabies virus, it was hypothesized by Remlinger that the CNS tissue in which the vaccine was grown was the cause of illness (2). It was later shown that preparations of normal rabbit or human spinal cord tissue could induce ADE when injected into rabbits (1). Thus, it was concluded that the CNS tissue and not the rabies virus was responsible for the post-vaccinal paralysis. In 1947 the introduction of adjuvants led to the first reproducible induction of "allergic" encephalomyelitis in experimental animals (3). In 1949 Olitsky was able to induce demyelinating disease in mice using adjuvants (4). Thus murine EAE was established as a model for demyelinating disease of the CNS. To date, EAE has been successfully induced in a number of animal species including mice and rats (5).

In 1947 Kabat proposed an autoimmune etiology for EAE (3) suggesting that self-antigens present in the CNS white matter were the cause of EAE since injection of fetal CNS tissue, which lacks myelin, could not induce EAE. In 1962, Einstein identified myelin basic protein (MBP), which comprises 30% of CNS myelin, as an encephalitogenic antigen in CNS tissue (6). In 1951 proteolipid protein (PLP), which comprises 50% of CNS myelin, was identified (7). Initially there was much debate as to whether PLP was itself an encephalitogen or whether PLP-induced autoimmunity was due to contamination by MBP. The demonstration by Tuohy in 1988 that a synthetic peptide of PLP could induce EAE in mice finally resolved the issue (8). More recently peptide epitopes of myelin oligodendrocyte glycoprotein (MOG), which comprises only 0.01–0.05% of CNS myelin, have been shown to induce

EAE (9). To date, several encephalitogenic epitopes of MBP, PLP, and MOG have been identified in different rodent strains (8–21) (Table 1).

EAE is most widely studied in rodents. Rodent EAE has different manifestations clinically depending on the antigen used to induce EAE and the rodent strain. There are three clinical manifestations of EAE: Acute, chronic, and relapsing-remitting. In acute EAE (eg. MBP 1–11 EAE in H-2^u mice or MBP 68–84 EAE in Lewis rats), animals exhibit a single episode of paralytic disease from which they recover. In chronic EAE (eg. MOG 35–55 EAE in H-2^b mice), animals get progressively worse after the initiation of disease and never recover. In relapsing-remitting EAE (eg. PLP 139–151 EAE in H-2^s mice), animals experience a moderate to severe initial acute episode of disease followed by remission and one or more relapses. Although it is not well understood what mechanisms are responsible for remission, there is evidence that suggests that relapses are due to the priming of T cells specific for secondary myelin epitopes that are released as a consequence of breakdown of the myelin sheath (22). There is also evidence to suggest that relapses could be the result of T cell priming by myelin proteins expressed in peripheral lymphoid tissue during disease course (23).

2.2. Generation and Selection of the Autoreactive Repertoire

Autoreactive T cells are generally deleted in the thymus during T cell development. The fact that EAE can be induced in several rodent strains suggests that myelin-reactive cells that escape thymic deletion are responsible for the induction of EAE. One proposed mechanism for the escape of myelin-reactive cells from thymic deletion has been the sequestration of myelin antigens behind the blood-brain barrier. The anatomy of the blood-brain barrier and the lack of lymphatic drainage from the CNS have been cited to support this hypothesis. However, recent data indicates that expression of myelin antigens (MBP and PLP) is not limited to the CNS. Transcripts for MBP have been detected in both the human (24) and mouse thymus (25). Moreover, there is now evidence for expression of MBP protein in the thymus (26) and peripheral lymphoid organs (23) of mice. PLP transcripts and protein also have been reported in human thymus (27), murine thymus (28) and myocardial cells (29). The discovery of myelin protein expression outside the CNS has led to the re-evaluation of immune tolerance to myelin antigens.

A number of recent studies have used MBP-deficient (shiverer) mice to address the issue of central tolerance to MBP. Neither Balb/c nor C3H mice can mount proliferative responses to MBP and are resistant to MBP-induced EAE. However, shiverer mice on the Balb/c and C3H backgrounds respond well to MBP (30,31). In addition, the MBP-reactive T cells from shiverer mice on the Balb/c background are highly encephalitogenic (30). Collectively, these data suggest that the expression of MBP results in tolerance to MBP in wild-type C3H and Balb/c mice. The exact mechanism of tolerance cannot be determined from these studies. Similar studies of MBP-deficient mice on the B10.PL background have shown that the immunodominant epitopes of MBP differ in the shiverer mice when compared to wild-type mice (32). The conclusion was that the self-reactive T cells that escape central tolerance and form the dominant autoreactive T cell repertoire in the peripheral immune compartment recognize epitopes that bind with low affinity (MBP 1–11) and form unstable complexes with the self MHC (IA^u), while T cells that recognize epitopes that bind with high affinity and form stable MHC-peptide complexes (MBP 121–150) are deleted (32).

Table 1 Encephalitogenic myelin epitopes in different rodent strains

Species	Strain	MHC haplotype	Myelin epitope	Reference
Mouse	Biozzi AB/H	H-2^{dg1}	MOG 1–22	(10)
			MOG 43–57	(10)
			MOG 134–148	(10)
			PLP 56–70	(11)
	C3H	H-2^k	PLP 215–232	(12)
	C57Bl/6	H-2^b	MOG 35–55	(9)
	NOD	H-2^{g7}	PLP 56–70	(11)
			MOG 35–55	(13)
	PL/J	H-2^u	MBP 1–11	(14)
			PLP 43–64	(15)
	SJL	H-2^s	MBP 84–104	(16)
			PLP 104–117	(17)
			PLP 139–151	(18)
			PLP 178–191	(19)
			MOG 92–106	(10)
	SWR	H-2^q	PLP 103–116	(8)
Rat	Lewis	RT1^l	MBP 68–84	(20)
			PLP 217–240	(21)

3. EFFECTOR LYMPHOCYTES IN EAE

3.1. T Lymphocytes

There is ample evidence supporting a central role for the T lymphocyte in EAE. In the early 1970's it was demonstrated that rats depleted of T cells by thymectomy and irradiation were resistant to the induction of EAE (33,34). In the early 1980's several groups demonstrated that CD4$^+$ Th cell lines reactive to myelin proteins could adoptively transfer EAE (35,36). With the introduction of techniques for

Pathogenesis of EAE

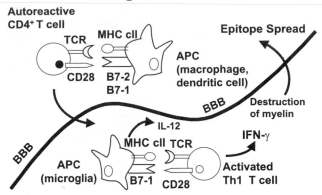

Figure 1. Overview of the pathogenesis of EAE. A myelin reactive CD4+T cell encounters a costimulatory competent APC presenting autoantigen in the periphery. This T cell becomes activated and traffics to the CNS where it differentiates into a Th1 effector cell after it re-encounters autoantigen presented by CNS APCs. Activated Th1 cells mediate destruction of myelin, resulting in the release of new myelin antigens which can in turn activate autoreactive T cells with different myelin specificities.

cloning antigen specific T cells came the observation that reactivity to myelin proteins was necessary, but not sufficient, for encephalitogenicity (37).

In 1986 it was discovered that CD4+ T cells can be subdivided into two distinct subsets (Th1 and Th2) based on their pattern of cytokine secretion (38). Th1 cells secrete IL-2, IFN-γ, and/or LT-α (TNF-β) and mediate delayed-type hypersensitivity (DTH) and class switching to IgG$_{2a}$. Th2 cells secrete IL-4, IL-5 and IL-10 and mediate recruitment of eosinophils and class switching to IgG$_1$ and IgE. These two Th cell populations cross-regulate one another. IL-4 and IFN-γ show reciprocal inhibition and IL-10 indirectly inhibits IFN-γ production by interfering with antigen presentation (39).

Two additional classes of Th cells, Th3 and Tr1, have been described recently. Th3 cells were discovered while studying the mechanism of suppression of EAE induced by oral feeding of MBP (40). Th3 cells produce high levels of TGF-β and promote class switching to IgA. Tr1 cells were first generated *in vitro* by chronic stimulation of CD4+T cells in the presence of IL-10. Tr1 cells produce high levels of IL-10 and low levels of IL-2 and have been shown to prevent colitis (41). Although the suppression of EAE by myelin-reactive Tr1 cells has not yet been described, the importance of IL-10 in suppressing EAE has been demonstrated by studies of EAE in IL-10 deficient and IL-10 transgenic mice (42,43). IL-10 transgenic mice were shown to be completely resistant to EAE, whereas IL-10-deficient mice developed chronic EAE, thereby confirming the role of IL-10 in suppressing disease.

The Th1/Th2 paradigm has proved to be of considerable value in our understanding of EAE. Most encephalitogenic T cell clones isolated from mice immunized with MBP and PLP have a Th1 phenotype (44–46). Moreover, Th1 cytokines as well as other proinflammatory cytokines such as TNF-α are present in inflammatory CNS lesions during EAE, whereas Th2 cytokines are absent (47). In contrast, Th2 cytokines and TGF-β are present in the CNS during disease remission (48). These observations have led to the hypothesis that myelin-reactive T cells that are encephalitogenic will have a Th1 phenotype whereas myelin-reactive T cells that have a Th2 phenotype will be protective.

There is ample evidence to support a protective role for myelin-reactive Th2 cells in EAE. Th2 T cell clones specific for an encephalitogenic epitope of PLP have been shown to be protective if given at the time of immunization and to reverse the course of EAE if administered at the first signs of disease onset (49). In addition, Th2 T cell lines cross-reactive with PLP have been shown to protect against EAE (50).

Despite the observations that support a simple Th1/Th2 paradigm in EAE, there is increasing evidence that the role of Th1/Th2 cytokines in EAE is much more complex than has been proposed. That T cells with a Th2 phenotype are protective for EAE has been challenged recently by at least two studies. In the first study co-transfer of PLP-specific Th2 lines was unable to prevent EAE mediated by Th1 lines (51). Secondly, adoptive transfer of MBP-specific Th2 cells into immunocomprimised animals was shown to result in EAE (52). However, the relevance of this study is called into question by the fact that the disease observed was atypical in that it was characterized by a polymorphonuclear CNS infiltrate as opposed to the mononuclear infiltrate that is typical of EAE.

Furthermore, the surprising and often contradictory results of recent EAE studies in cytokine transgenic and knockout mice have forced the re-evaluation of the role of Th1, Th2, and other inflammatory and suppressive cytokines in the development of EAE (Table 2). For example, EAE in IFN-γ-deficient mice is associated with increased disease severity and mortality rather than protection (53–55). In addition, studies of EAE in IL-4 deficient mice have yielded contradictory results regarding the role of IL-4 in EAE. Although the majority have shown that EAE is not exacerbated by the absence of IL-4 (42) (43,56) one study has shown that EAE is more severe in IL-4-deficient mice (57). There is also considerable controversy surrounding the role of TNF-α in EAE. TNF-α-transgenic mice have been shown to develop spontaneous disease (58). However, studies in TNF-α-deficient mice less clearly support an EAE-promoting role for TNF-α. One study of EAE in TNF-α-deficient mice backcrossed onto the C57BL/6 background found an increase in EAE severity and mortality (59). However, when TNF-α deficiency was directly

Table 2   EAE phenotype of cytokine transgenic and knockout mice

Mutation	Cytokine	EAE phenotype relative to wild-type	Reference
Transgenic	IL-4	reduced incidence	(42)
	IL-10	completely resistant	(42)
	TNFα	spontaneous disease	(58)
Knockout	IL-4	comparable to WT	(42),(43),(56)
	IL-4	increase in severity	(57)
	IL-10	severe chronic disease	(42),(43)
	IL-12	completely resistant	(43)
	IFNγ	increase in mortality	(53)
	IFNγ	increase in severity and mortality	(54)
	IFNγ	susceptible	(55)
	LTα	resistant	(62)
	TNFα	increase in severity and mortality	(59)
	TNFα	delay in onset	(60)
	LTα × TNFα	increase in severity and mortality	(61)

introduced on the C57BL/6 background a delay in onset of EAE was observed (60). Because the TNF-α gene is located within the MHC locus it is possible that carryover of a MHC-linked gene may be confounding the results in the first study. In another study, mice deficient in both TNF-α and LT-α backcrossed onto the SJL background were shown to have increased severity and mortality (61). The problem with this study is that the SJL mice used expressed the H-2^b haplotype rather than the normal H-2^s, raising the possibility that effects of other MHC-linked genes may be contributing to the observed phenotype.

Nevertheless, the roles of some cytokines in EAE have been confirmed by studies using gene-targeted mice. The role of IL-12 in promoting disease has been confirmed by the demonstration that IL-12-deficient mice are completely resistant to EAE (43). The role of LT-α has also been substantiated by the demonstration that LT-α-deficient are resistant to EAE (62). Finally, as mentioned earlier, the role of IL-10 in suppressing disease has been confirmed by the demonstration that IL-10 transgenic mice are completely resistant to EAE and that IL-10-deficient mice succumb to a severe chronic form of EAE (42,43).

One caveat with interpreting the studies in such animals is that the immune system that develops in the absence of a particular cytokine may not be directly comparable to that of a wild-type animal. Moreover, cytokine deficiencies on different genetic backgrounds may give different EAE phenotypes, making it more difficult to distinguish cytokine effects from those of other genes on disease. Thus studies in such artificial systems although valuable should be interpreted with caution.

3.2.   Role of Costimulatory Molecules in EAE

It is well accepted that T cells need to receive two distinct signals to become activated (63). The first signal is ligation of the T cell receptor by peptide/MHC and provides specificity to the interaction. The second signal is provided by the interaction of costimulatory molecules on the surface of antigen presenting cells (APCs) such as B cells, dendritic cells and macrophages with their counter-receptors on the surface of T cells. The most well-characterized family of costimulatory molecules is the B7 family. The B7 family includes at least two molecules, B7-1 and B7-2, which are expressed on the surface of APCs and interact with their counter-receptors, CD28 and CTLA-4, on the surface of T cells. Ligation of the TCR in the absence of B7-1-mediated costimulation results in a state of T cell non-responsiveness or T cell anergy (64). The demonstration of the importance of this family of molecules in T cell activation by this and other studies prompted the examination of the role of B7-mediated costimulation in EAE.

Initial studies using an immunoglobulin fusion protein of CTLA-4, CTLA-4Ig, to block B7-1 and B7-2 molecules suggested that B7 signals were necessary for priming MBP responses *in vivo*. However, administration of CTLA-4Ig *in vivo* did not abrogate adoptive transfer of EAE by MBP-specific T cell lines (65). The inability of CTLA-4Ig to block adoptive transfer of EAE may be due to inability of CTLA-4Ig to enter the CNS, since delivery of CTLA-4Ig directly into the CNS has been shown to inhibit EAE (66). Studies in which specific antibodies against B7-1 and B7-2 were used to address the role of these molecules in EAE demonstrated that blocking B7-1 *in vivo* protected against EAE, whereas blocking B7-2 exacerbated EAE and blocking both B7-1 and B7-2 either had no effect (49) or exacerbated disease (67). The effects of anti-B7-1 and anti-B7-2 antibodies were the result of differential polarization of PLP-specific T cells into Th1 and Th2 subsets. Blocking B7-1 signals *in vivo* was shown to promote differentiation into Th2 cells whereas blocking B7-2 signals promoted differentiation into Th1 cells (49). The exacerbation of disease by anti-B7-1 and anti-B7-2 treatment was correlated with increased TNF-α production by MBP-specific T cells (67). Another study found that administration of anti-B7-1 F(ab) fragments blocked clinical relapses by preventing epitope spreading (68). Taken together, these data suggest that B7-1 signals are critical for the development and progression of EAE.

Some recent studies have specifically addressed the role of CD28 in EAE. One study analyzed the effects of CD28 deficiency on the development of spontaneous and peptide-induced EAE on the PL/J background (69). CD28 deficiency abrogated the spontaneous development of EAE in mice bearing a TCR transgene specific for MBP. Interestingly, transgenic T cells from CD28-deficient mice were able to proliferate to MBP and produce IL-2, excluding anergy as the mechanism of disease regulation. Similarly, MBP-induced EAE was also blocked in CD28-deficient mice; however, this block could be overcome by immunizing with large quantities of MBP peptide. Thus, CD28 appears to be important for the initiation of both forms of EAE. Another study addressed the effects of CD28 deficiency on MOG-induced EAE in the C57BL/6 background (70). Interestingly, CD28-deficient C57BL/6 mice immunized with MOG developed autoimmune meningitis and not the encephalomyelitis commonly observed after immunization with MOG. From the above studies, one can conclude that the presence or absence of CD28 during the initial priming of myelin-reactive T cells can have a profound effect not only on the disease outcome but also on the nature of disease.

Unlike B7 signals through CD28, which are critical for T cell activation, B7 signals through CTLA-4 have been shown to be critical for the regulation of activated T cells and the maintenance of peripheral tolerance. The negative regulatory function of CTLA-4 is best demonstrated by the phenotype of CTLA-4-deficient mice, which succumb to a massive lymphoproliferative disorder characterized by infiltration of several organs (71,72). Studies addressing the role of CTLA-4 in EAE have shown that treatment with anti-CTLA-4 antibody at the time of disease onset (73) or at the time of adoptive transfer (74) markedly exacerbates disease. Exacerbation of disease was associated with increased production of TNF-α, IFN-γ and IL-2 (73). Treatment of mice with anti-CTLA-4 during disease remission resulted in increased severity and a higher incidence of clinical relapses (74).

CD40 is a costimulatory molecule constitutively expressed on the surface of APCs. Its counter-receptor, CD40L, is expressed on activated CD4$^+$ T cells. The importance of the CD40-CD40L interaction was first appreciated in the generation of humoral responses to T-dependent antigens. T cells from mice deficient in CD40 were unable to costimulate B cells to undergo immunoglobulin class switching (75). It was later discovered that these molecules are important for T cell activation in that T cells from CD40L-deficient mice cannot be primed and expanded in vivo (76). Lack of in vivo priming was shown to be responsible for the inability to induce EAE in MBP TCR-transgenic mice deficient in CD40L (77). Other studies addressing the role of CD40-CD40L interactions in EAE demonstrated that treatment with anti-CD40L antibodies could prevent the development of disease even after disease onset (78). The mechanism by which anti-CD40L antibodies could block EAE was later correlated with a selective inhibition of IFNγ production by myelin-reactive T cells (79,80) and an increase in IL-4 production (79).

The effects of the various costimulatory molecules on the development of EAE are underscored by the expression of these molecules in the peripheral immune system and the CNS during the course of disease. B7-1 expression is upregulated relative to B7-2 on CNS-infiltrating mononuclear cells prior to disease onset and during remission (68). In addition, functional studies indicate that B7-1 is the dominant costimulatory molecule present on splenic and CNS APCs from mice with ongoing EAE (81). CD40 expression in the CNS increases significantly during acute disease and relapse and decreases during remission (82). CD40L expression is highest during relapse consistent with the presence of activated T cells in the CNS. The examination of the kinetics of expression of these costimulatory molecules in the peripheral immune system and in the CNS during EAE have largely confirmed the results of the antibody blocking studies and have been critical to our understanding of how autoreactive T cell responses are primed and controlled in vivo.

3.3.   T Cell-mediated Cytotoxicity in EAE

While it is clear that myelin-reactive CD4$^+$ T cells are required for the induction of EAE, the mechanism by which they mediate damage in the CNS is not. As discussed above, the production of proinflammatory cytokines by T cells is believed to be one of the contributing factors in CNS pathology. Another mechanism by which CD4$^+$T cells can mediate tissue destruction in the CNS is through cell-mediated cytotoxicity. The finding that the ability of PLP-specific T cell clones to induce EAE after adoptive transfer correlates with their ability to lyse PLP-pulsed targets in vitro supports a role for cell-mediated cytotoxicity in EAE (45).

One pathway of cell-mediated cytotoxicity is the Fas/FasL pathway. Recent studies have focused on the role of this pathway in EAE. Fas is a member of the TNF/nerve growth factor receptor family and is constitutively expressed on the surface of T cells. FasL is a member of the TNF family and is expressed on the surface of T cells upon activation. Ligation of Fas by FasL induces apoptosis in Fas$^+$ targets. That the proinflammatory cytokines IFN-γ and TNF-α have been shown to upregulate Fas in vitro on microglia, the APCs resident in the CNS, supports the involvement of this pathway in EAE (83).

The roles of Fas and FasL in EAE were initially studied in lpr and gld mice, which harbor mutations that result in non-functional Fas and FasL proteins, respectively. Both lpr and gld mice were found to be resistant to disease induction, thereby supporting a role for the Fas/FasL pathway in EAE. This resistance was not due to a defect in T cell priming or production of proinflammatory cytokines (84–87). Unfortunately, these studies could not distinguish between the roles of apoptotic death of T cells in the CNS or of CNS resident cells in EAE.

In the CNS, Fas is constitutively expressed by oligodendrocytes and microglia. During acute EAE, both oligodendrocytes and microglia upregulate FasL but only microglia appear to undergo apoptosis in situ (88). Infiltrating T cells are positive for both Fas and FasL and undergo apoptosis in situ. T cell apoptosis is most prominent just prior to remission (88,89). Taken together, these observations support a model in which apoptosis of both T cells and microglia is important in down-regulating the immune response in the CNS. Support for this model comes from a recent study in which adoptive transfer of MBP-reactive wild-type T cells into *gld* mice resulted in chronic disease presumably due to the inability of recipient mice to curb T cell expansion and eliminate CD4$^+$ T cells in the CNS (90). Moreover, adoptive transfer of MBP-reactive wild-type T cells into *lpr* mice resulted in both reduced incidence and severity of EAE thereby supporting a requirment for apoptosis of Fas$^+$ cells in the CNS in EAE. Although Fas$^+$ oligodendrocytes do not appear to undergo apoptosis, it has been suggested that oligodendrocyte injury may occur via a lytic mechanism involving Fas (88). In conclusion, the fine balance between Fas-mediated death of effector T cells and Fas-mediated death of CNS resident cells will in part determine the clinical outcome in EAE.

Cell mediated cytotoxicity can also be mediated by CD8$^+$T cells via perforin. Perforin has been shown to be cytotoxic for oligodendrocytes *in vitro* (87). However, EAE in perforin deficient mice is more severe than in wild-type mice suggesting that perforin-mediated killing by CD8$^+$T cells may have a regulatory role in EAE (87). A regulatory role for CD8$^+$T cells in EAE has already been suggested by experiments in which CD8 deficient mice exhibited a higher incidence of relapse after immunization with MBP (91,92).

3.4. B Lymphocytes

Whether or not B cells play a role in EAE either as APCs or through production of anti-myelin antibodies has been controversial. Early studies in rats demonstrated that EAE could not be induced by immunization with MBP in animals depleted of B cells from birth (93). This defect was not due to an inability to generate effector T cells capable of transferring disease and could be reversed by injection of anti-MBP antisera into the B cell depleted rats (94). Moreover, EAE could be induced in B cell depleted rats by adoptive transfer of T cells (94). Taken together, these results suggest that B cells are not necessary for, but can facilitate, EAE in rats. Similar studies in mice gave different results. Mice depleted of B cells from birth could not mount good T cell proliferative responses to MBP and EAE could not be induced in treated animals by co-injecting spinal cord homogenate and anti-MBP antibody, however reconstitution of treated mice with B cells prior to immunization with spinal cord homogenate was able to reverse the defect (95). Moreover, adoptive transfer of EAE into B cell-depleted mice could only be achieved by

injecting anti-MBP antibody into recipient mice or by transferring very large numbers of T cells. These results suggest that the APC function of B cells is critical for the development of murine EAE. The discrepancy between the rat and mouse experiments has been ascribed to the higher susceptibility of rats to EAE (95).

More recently the role of B cells in EAE has been examined in mice genetically engineered to lack B cells. B cell-deficient (H-2^b) mice immunized with MOG developed disease comparable to wild-type mice (96). However, B cell-deficient (H-2^u) mice immunized with MBP had variable disease onset and severity and developed a chronic disease not observed in wild-type mice (97). In contrast with the earlier studies, the results of both of these studies suggest that B cells are not essential for the induction of EAE. Moreover, B cells could have a modulatory role in EAE that may depend on the antigen used for disease induction and/or the genetic background of the animal. Clearly the elucidation of the role of B cells in EAE will require further study.

4. ASSOCIATION OF MHC CLASS II WITH AUTOIMMUNITY

Particular human autoimmune diseases are commonly associated with particular MHC haplotypes. Specifically, the strongest association is with genes encoding MHC Class II molecules. This has raised much interest because Class II MHC molecules present antigen to CD4$^+$ helper T cells that are believed to play a central role in most autoimmune diseases.

The mechanism of the association of MHC Class II molecules with autoimmune disease is not well understood. Initially, it was proposed that disease-associated Class II molecules can efficiently bind self-antigens that activate T cells that in turn mediate autoimmune damage. There are at least two problems with this hypothesis. First, it supposes that the particular self-antigen is not present in the thymus during selection, because if it were it would bind well enough to the MHC to mediate negative selection. Second, we know that all MHC molecules bind self-antigens and yet most individuals do not succumb to autoimmune disease. The second explanation that has been proposed is that disease-associated MHC alleles do not bind self-antigens well and therefore cannot mediate negative selection effectively in the thymus. There is now data that suggests this could be the mechanism underlying the association of IA^{g7} with diabetes in the NOD mouse (98,99). As discussed above, the recent studies in MBP-deficient mice also support this hypothesis. Whether similar mechanisms are operative in the generation of the myelin-reactive repertoire to other myelin antigens in other mouse strains remains to be seen.

5.　GENETICS AND SUSCEPTIBILITY TO EAE

It has been well known for some time that predisposition to autoimmune disease is heritable and therefore has a genetic component. In humans, it is estimated from studies comparing identical twin pairs with non-twin siblings and other relatives that genetics contributes up to 50% of the observable risk of disease. However, defining the mode of inheritance has been difficult. In EAE, genetic studies have identified patterns of inheritance ranging for autosomal dominant to polygenic. The current consensus is that predisposition to disease is polygenic and follows a threshold liability model where effects of alleles at several loci incrementally contribute to susceptibility and define a threshold for disease induction. In this model, any different combination of alleles that contribute to susceptibility (either by increasing or decreasing predisposition) will produce a susceptible phenotype, as long as their cumulative effect surpasses the threshold. Therefore, different strains or individuals may be susceptible to a particular disease due to the inheritance of different sets of disease-associated alleles. Although in EAE an incremental increase in disease severity or incidence according to the number of disease-associated alleles inherited has not yet been shown, an allele that is required or sufficient for disease susceptibility has not been found. Thus disease susceptibility is likely the result of multiple genetic factors acting together. The studies leading to this conclusion are discussed below.

5.1.　Strategies for Genetic Studies of EAE

A genetic analysis of EAE requires that a susceptible phenotype be defined. Paralysis is often chosen, however the definition of less obvious phenotypes, or phenotypes that mark an earlier stage in the progression of disease, has proven to be extremely useful in the identification of disease loci in genetic studies of other autoimmune diseases, such as insulin-dependent diabetes and systemic lupus erythematosus. In EAE, CNS inflammation and demyelination, weight loss, and drop in body temperature have all been used as indicators of disease in addition to paralysis. The use of intermediate disease phenotypes may be important in the future, since such phenotypes depend on a smaller number of genes than the complete disease phenotype, and are therefore easier to analyze.

Classical genetic studies aim to find genetic polymorphisms, i.e. allelic differences in genes, between individuals, that correlate with susceptibility. Early genetic studies of EAE took a candidate gene (or candidate locus) approach and attempted to correlate the inheritance of specific polymorphic genes with susceptibility to disease. In inbred mice, investigators set up a test cross between two strains that show a difference in the phenotypic expression of the trait. The test cross is either a backcross, where the F_1 progeny of a mating between the two strains is crossed with one of the parental strains, or an intercross, where two F_1 animals are mated to produce an F_2. This will generate progeny that, because of the random assortment of chromosomes and the occurrence of crossovers during meiosis, will have some of their genes from the susceptible parent and the rest from the resistant parent. In the case of a backcross, an average of 50% of the genome will be homozygous and identical to the parental strain to which it was backcrossed and 50% will be heterozygous and a hybrid of the two parental strains. In the case of an F_2 intercross, 25% will be homozygous and identical to one parental strain, 25% will be homozygous and identical to the other parental strain, and 50% will be heterozygous and a hybrid of the two parental strains. Backcross or F_2 animals are then phenotyped to determine their expression of the trait and genotyped to determine which parental alleles of the gene in question they carry. If there is a significant correlation between phenotype and genotype, the gene studied is said to be linked with the gene controlling the trait.

More recently, the technology to screen the whole genome at a density of one marker per 20–40 cM, using simple sequence length polymorphisms, has been developed (100). Using the same test crosses as above, the genome can be analyzed in an unbiased fashion and regions that associate with disease can be identified. Because many markers are tested (compared with analyses of candidate genes), the likelihood of associations arising by chance is greater, and statistical tests of significance are necessarily much more stringent than in less complex studies (101). Once an initial genetic analysis has identified important loci, loci believed to contribute to susceptibility can be bred onto a resistant genetic background and those believed to contribute to resistance can be bred onto a susceptible genetic background to produce congenic mice with candidate loci isolated from other confounding genetic effects. This reduces the complexity of the system and makes the fine mapping of each locus possible. With the development of congenic lines, it becomes possible eventually to shorten the susceptibility locus intervals to lengths amenable to positional cloning, or at least to narrow the list of candidate genes contained within the loci to a small enough number to make searching for gene polymorphisms worthwhile. In the NOD mouse model, this approach has been successful for accurately mapping loci which contribute both to diabetes and EAE (102–104).

5.2.　Studies of Candidate Genes

MHC genes have been studied in both rats and mice, because of the clear association between human autoimmune diseases and genes at the HLA locus. In rats early data pointed to an association between MHC and disease resistance (105), but later studies found that susceptibility was likely outside of the MHC locus (106,107) and studies of congenic rats with identical MHC genes also found evidence for disease susceptibility genes outside the MHC locus (108). In mice, tests have been made on inbred lines and lines congenic to them to look at the effect of H-2 (equivalent to the human HLA) haplotype on EAE susceptibility. Replacing H-2 haplotypes that were generally found in resistant strains (a or k) with those that were generally found in susceptible strains (s, b, or q) did not enhance susceptibility (109). In agreement with these findings, a more extensive study of congenic strains found that susceptibility correlated better with background genotype than H-2 haplotype (110). The only H-2 haplotype that appeared to have any effect was H-2^b, since C57BL/10Sn and the H-2 congenic BALB/B were both very resistant compared to others of the same background. Also, crosses between susceptible SJL/J mice and resistant inbred strains sometimes result in an F$_1$ hybrid that is more susceptible to EAE than SJL/J and sometimes result in an F$_1$ hybrid that is less susceptible, indicating that there is not dominant mendelian inheritance of susceptibility to EAE (111). Therefore, while MHC genes are clearly involved, genes outside the MHC are also important determinants of disease susceptibility.

Another gene that has raised interest as a possible modifier of EAE susceptibility in mice is that for the T cell receptor β chain. In many studies, a preferential usage of certain T cell receptor β chain variable regions on encephalitogenic T cells has been seen. When this question was investigated in an F$_2$ intercross between resistant RIIIS/J mice, which have a large deletion in their TCR β-chain V genes, and susceptible B10.RIII, which have no known deletion, there was no association between TCR β haplotype and susceptibility to disease (112). This suggests that if there is a genetic explanation for these observations, it does not reside in the TCR locus.

5.3. Genome-Wide Screening for Susceptibility Loci

Genome wide screening for susceptibility to EAE has been adopted by several groups and applied to both rat and mouse models of EAE (113–115). Taken as a whole, the data strongly supports the polygenic model of disease susceptibility. Several loci have been identified more than once and the simplest interpretation of this is that linkage is with the same gene. However, because many of the intervals are relatively large, it is also likely that some will contain more than one susceptibility or resistance allele. The data from these studies in mice is summarized in Table 3 (116–121).

Several important observations have emerged in recent years from these studies. The first is that susceptibility loci identified in several autoimmune diseases overlap, raising the possibility that "autoimmune gene(s)" common to multiple autoimmune diseases may exist (104,122–124). One example of a locus shared among autoimmune diseases is *Idd3*, located on chromosome 3 and first identified as a suscept-

ibility locus for diabetes in the NOD mouse. *Idd3* has now been identified as a susceptibility gene for EAE in mice (104) and rats (125), and in experimental allergic orchitis (EAO) (122). Although the hypothesis of shared autoimmune gene(s) is attractive, the large confidence intervals of each locus makes it is difficult to determine whether the observed overlap is due to the same locus/gene or a closely linked locus/gene. Ultimately this issue will have to be addressed by physical mapping. Congenic mice in which defined congenic intervals that overlap between different autoimmune diseases have been introduced provide a powerful tool to address this issue, because they allow the susceptibility loci to be physically defined. This strategy has been developed in NOD mice and used to define narrow intervals within which it is possible to identify specific genes. Using this strategy, *Idd3* has been narrowed to a 0.15 cM interval that encompasses two genes, *Fgf2* and *Il2*. Within the *Idd3* locus, the *Il2* gene is the obvious candidate gene and polymorphisms in exon 1 are shared by the susceptible NOD and SJL mouse strains (104). However, the mechanism by which IL-2 polymorphisms affect disease susceptibility is not known and proof that *Idd3* is *Il2* awaits the generation of IL-2 knock-in mice.

The second important observation is that alleles from resistant strains of mice can result in increased susceptibility to disease. Therefore the effect of a gene associated with

Table 3 EAE modifying loci

EAE locus	Phenotype	Chromosome: distance	Reference
eae1	Incidence	17: 18–19cM	(118)
eae2	Incidence	15: 11–20cM	(118)
eae3	Incidence	3: 30–41cM	(118)
	Incidence	3: 45–53cM	(116)
	Incidence/disease subtype	3: 29.5cM	(119), (120)
eae4	Incidence	7: 25–53cM	(117)
	Disease Index	7: 25–51cM	(120)
eae5	Incidence	17: 24–33cM	(121)
	Disease Index	17: 22–23cM	(119), (120)
eae6	Incidence	11: 1–2cM	(121)
	Incidence	11: 2–28cM	(117)
	Severity, duration	11: 20–28cM	(119), (120)
eae7	Severity, duration	11: 48–61cM	(119), (120)
eae8	Incidence	2: 99–107cM	(116)
	severity	2: 105–107cM	(119), (120)
eae9	Duration	9: 28–34cM	(119), (120)
eae10	Onset	3: 64–80cM	(119), (120)
eae11	Incidence/ histopatholgogy	16:41cM	(119)
eae12	Incidence/ disease subtype	7:16cM	(119)
eae13	Incidence/ disease subtype	13:37cM	(119)

disease cannot be predicted on the basis of the disease phenotype of the mouse from which the gene is derived. This apparent paradox arises because the threshold for disease depends on the sum of effects, both positive and negative, of all susceptibility genes. The Idd7 and Idd8 diabetes loci are a good example of this (126). The Idd7 and Idd8 alleles from the NOD mouse contribute to disease resistance whereas the alleles from resistant C57BL/10 contribute to susceptibility. Thus, if the C57BL/10 alleles are introgressed onto the NOD mouse the result is increased susceptibility to diabetes.

Although the power of genome wide screening for disease association is its lack of bias towards known genes in general, and genes implicated in immunoregulation in particular, many other candidate genes have been discussed in the literature on the basis of their location within disease-associated intervals. One region that has been implicated in a number of studies is on chromosome 11, which is syntenic with human 7p11–13, and contains the Th2 cytokine gene cluster (120). A study in rats also identifies a locus associated with EAE which is syntenic with the same region on human chromosome 7 (127). With a combination of genome mapping and candidate gene analysis, it is likely that many autoimmune associated genes will be defined in the next decade.

6. EAE AND MS

6.1. Similarities and Differences

The relevance of the EAE model to the study of MS has been debated. Nevertheless, the numerous similarities between EAE and MS support its value in increasing our understanding of the pathogenesis of MS. These similarities are summarized in Table 4. The greatest difference between EAE and MS lies in our understanding of the events responsible for disease induction. While in EAE it is known that immunization with any of several myelin antigens emulsified in adjuvant is responsible for disease induction, the triggering antigen in MS remains unknown. In fact some still question whether MS is really an autoimmune disease. In addition, evidence from clinical and epidemiologic studies implicating environmental agents in the etiology of MS (128,129) suggests that the etiology of MS is more complex than that of EAE.

6.2. Role of the Environment

Several models have been proposed to explain how environmental agents play a role in the development of CNS autoimmunity. The molecular mimicry model proposes that an immune response generated against environmental agents may crossreact with self-antigens with shared sequence or structural homologies, ultimately resulting in tissue damage and autoimmune disease. It has also been proposed that the *de novo* release of myelin antigens during infection in the CNS could lead to autoimmunity. Lastly, it has been suggested that polyclonal activation of T cells by microbial superantigens could result in autoimmunity.

In support of the molecular mimicry model, peptides derived from measles, influenza, adenovirus, and Epstein-Barr virus that share linear sequence homology to myelin basic protein (MBP) have been identified (130). Furthermore, immunization of rabbits with a peptide derived from the Hepatitis B viral polymerase, shown to have linear sequence homology with MBP resulted in CNS inflammation but not clinical EAE (131). More recently, microbial peptides that share structural homology with an epitope of MBP (MBP 84–102) were identified using motifs to search a protein database. Some of these peptides were then shown to activate human MBP 84–102 specific T cell clones *in vitro* (132). Peptides that activate a human MBP specific T cell clone have also been identified using a random peptide library approach (133).

Several recent studies have tested whether EAE can be induced via molecular mimicry. In one study, a peptide derived from herpesvirus Saimiri with limited homology to MBP 1–11 was shown to crossreact with MBP 1–11. This peptide induced clinical EAE in 40% of the mice immunized. The disease induced with this peptide was similar histologically, although less severe than that observed in mice immunized with the native MBP peptide (134). In the second study, several microbial peptides cross reactive with PLP 139–151 were identified. Despite the cross reactivity observed *in vitro*, none of these peptides were able to induce EAE upon immunization. However, these peptides could be used to prime animals for EAE in that suboptimal amounts of the native PLP peptide could be used to induce disease in treated animals. Furthermore, T cell lines from animals immunized with the microbial peptides and restimulated *in vitro* with PLP 139–151 could transfer EAE into naïve recipients (135). Lastly, another study tested the ability of a human papillomavirus peptide that crossreacts with MBP 87–99 to induce EAE. This peptide could not induce EAE upon immunization unless extremely high quantities were used. However,

Table 4 Similarities between MS and EAE

	EAE	MS
CNS demyelination	+	+
CD4+ T cells in inflammatory lesions	+	+
Relapsing disease	+	+
Chronic disease	+	+
Linkage to MHC	+	+

T cell lines from animals immunized with the papillomavirus peptide and stimulated *in vitro* with either MBP 87–99 or low doses of the viral peptide could adoptively transfer EAE (136).

Despite the variability in results, it is clear from the above studies that crossreactive environmental antigens can be involved in the development of CNS autoimmunity. Moreover, the results of the latter two studies raise an important issue. In both of these studies the crossreactive peptides that were identified could not be used to induce EAE, yet they could prime a T cell population crossreactive with self. EAE could only be induced by adoptive transfer if the fraction of self-reactive T cells within this population was maintained either by stimulation with the self-peptide itself or by stimulating with doses of the crossreactive peptide low enough to preclude selection of cells specific for the cross-reactive peptide alone. These results point to the importance of T cell population dynamics in the development of autoimmunity and further emphasize that the mechanism by which cross reactive antigens may lead to autoimmunity is complex.

There is now also experimental support for the initiation of CNS autoimmunity as a result of viral infection in the CNS. Persistent infection of SJL mice with Theiler's murine encephalomyelitis virus (TMEV) results in chronic demyelinating disease initially mediated by CD4+ T cells targeting virus. It has now been shown that 3–4 weeks after disease onset T cell responses to myelin epitopes appear and may contribute to pathology (137). Extensive analysis demonstrated that myelin-reactive T cells did not arise because of cross reactivity between viral and myelin epitopes. Rather, myelin-reactive T cells arose as a result of the *de novo* release of myelin antigens secondary to tissue destruction mediated by virus-specific T cells. Thus, epitope spreading to myelin antigens is one possible mechanism for viral initiation of CNS autoimmunity.

Microbial superantigens have been implicated in the induction of autoimmune disease. In fact, there is some evidence that supports a role for superantigens in the etiology of autoimmune diabetes (138,139). In EAE, superantigens have been shown to trigger relapses in mice with a normally non-relapsing from of disease (140,141). However, direct triggering of disease in normal animals by superantigens has not been demonstrated.

7. IMMUNE INTERVENTION IN EAE

7.1. Antibody-based Therapy

Antibodies to the MHC have been shown to prevent EAE. Injection of anti-IA antibodies prior to immunization with spinal cord homogenate prevented disease, presumably due to blockade of T cell priming (142). Another study demonstrated that treatment with anti-IA antibodies reduced the number of CNS infiltrating lymphocytes in EAE (143). Moreover, anti-IA treatment also reduced the mortality and relapse rate in animals with chronic relapsing EAE (144).

Because of their central role in EAE, CD4+ T cells have also been the targets of antibody therapy. Anti-CD4 antibodies have been shown to reverse EAE in both rats (145) and mice (146). Later studies using chimeric anti-CD4 antibodies demonstrated that cytotoxic antibodies were most potent in blocking EAE (147).

Lastly, the discovery that the majority of T cell clones reactive to MBP in both mice (148,149) and rats (150,151) express TCR Vβ8.2 paved the way for treating EAE with anti-TCR antibodies. Depletion of Vβ8+ T cells in mice that had been immunized with MBP 1–11 in adjuvant significantly reduced the incidence of EAE and reversed the course of EAE in mice that had been adoptively transferred with an encephalitogenic Vβ8+ MBP 1–9-reactive T cell clone (149).

The success of these antibody-based therapies in preventing EAE confirmed the central role of the T cell in EAE. However, because antibody targeting of the MHC, CD4, or TCRs that express certain Vβ genes will likely suppress immune responses to other antigens, such therapies are not good candidates for the treatment of MS. Therapies that target myelin-reactive T cells specifically will stand the greatest chance of being useful in treating human autoimmune disease.

7.2. T Cell Vaccination

The term T cell vaccination was first coined in 1981 by I.R. Cohen (152) and refers to the phenomenon whereby administration of attenuated MBP-specific T cell lines protects against induction of EAE. Intravenous administration of encephalitogenic MBP-specific T cell lines attenuated by irradiation or chemical fixation was shown to protect rats against active induction, but not adoptive transfer of EAE (152,153). Renewed interest in this therapeutic approach came after the observation that pathogenic T cells in both rat and mouse models of EAE had limited T cell receptor diversity (148–151). Now T cell receptor Vβ 8.2 peptides could be used to vaccinate against EAE (154,155). The protection induced by immunization with TCR Vβ 8.2 peptides is apparently mediated by regulatory T cells that recognize TCR Vβ8.2 peptides (156). In addition, a regulatory T cell population that predominantly expresses TCR Vβ 14 and recognizes TCR Vβ 8.2 peptides has been associated with natural recovery from MBP-induced EAE in mice (157).

In spite of its demonstrated efficacy, T cell vaccination has met with increasing opposition as our understanding of T cell receptor usage in CNS autoimmune disease has grown. It is now clear that in some models of EAE and most individuals with MS T cell receptor usage is diverse. This poses a significant challenge to the widespread use of TCR peptide vaccination in the treatment of MS (158).

7.3. Altered Peptide Ligands

For a long time it was believed that T cell activation occurred in a simple on/off fashion. The demonstration that a CD4+ T cell clone stimulated with its native antigen altered at a single amino acid residue led to secretion of IL-4 in the absence of proliferation challenged this notion and opened a new avenue of investigation into T cell biology (159). Activation of T cells with altered peptide ligands (APLs) has led to the discovery to T cell antagonism (160) and partial agonism (161).

Altered forms of encephalitogenic peptides were shown to protect against EAE even before the term APL was introduced. Several studies suggested that the mechanism of protection by these altered ligands was MHC blockade (162,163). However, data from other studies raised the possibility that other mechanisms were likely (164,165). APLs of the encephalitogenic epitope of PLP in H-2^s mice, PLP 139–151 have been shown to inhibit EAE by T cell receptor antagonism (166,167). These antagonist APLs could inhibit EAE mediated by a diverse T cell repertorie (166) and were 10-fold more effective in inhibiting EAE than the strongest MHC blocker (167). However, the observation that mice with no clincal EAE that had been coimmunized with one of these antagonist APLs and PLP 139–151 exhibited inflammatory foci in the CNS suggested that the APL was not simply antagonizing the PLP 139–151 T cell response. It was later shown that the mechanism of protection of this APL *in vivo* was not TCR antagonism, but stimulation of a regulatory population of T cells that crossreacts with PLP 139–151 and can mediate bystander suppression (168). Another APL of PLP 139–151 has been shown to prevent EAE by inducing Th2 cells crossreactive with PLP 139–151 (50). This subversion of what would normally be a Th1 response has been referred to as immune deviation. An APL of MBP 87–99 has also been shown to operate via a mechanism involving Th2 cytokines (169). Clearly it is important to understand the numerous mechanisms of action of APLs so that they can be used effectively to treat autoimmune disease.

7.4. Copolymer 1

Copolymer 1 (Cop 1) is a random synthetic copolymer of L-alanine, L-glutamic acid, L-lysine, and L-tyrosine in a molar ratio of 6.1 : 1.9 : 4.7 : 1.0. It was first developed in an attempt to simulate the activity of MBP in inducing EAE but was found to suppress EAE in various animal species (170–172). The mechanism of action of Cop 1 is still unclear. It has been suggested that it acts via MHC blockade (173). The fact that Cop 1 suppresses EAE induced by MBP, PLP, MOG, and whole spinal cord homogenate supports this hypothesis (174–176). However, the fact that Cop 1 does not suppress other autoimmune diseases argues against this (177). It has been shown that T cells reactive to Cop 1 have a Th2 phenotype and crossreact with MBP by secreting IL-4, IL-6, and IL-10 (176). Because Cop 1 crossreacts with MBP and not other myelin antigens, it has been proposed that Cop 1 may act via bystander suppression (178). Cop 1 is currently being used in the treatment of MS under the trade name Copaxone. Whether the proposed mechanisms of action for Cop 1 are the same in MS remains to be seen.

7.5. Oral Tolerance

Oral tolerance refers to the state of immune hyporesponsiveness that results after an antigen is fed. Oral tolerance was first described in 1911 when guinea pigs fed hen egg proteins were found to be resistant to anaphylaxis upon subsequent challenge with hen egg proteins (179). Since then, this therapeutic approach has been validated in a number of systems and has been used successfully in the treatment of EAE and other autoimmune diseases (180). As for EAE, feeding of low doses of MBP has been shown to result in active suppression of disease by MBP-reactive regulatory CD4+T cells (Th3 cells) that secrete TGF-β, IL-4 and IL-10 (40). There is also evidence from studies in MBP-fed rats that oral tolerance can operate via clonal anergy (181). Because of its success in treatment of EAE, oral tolerance has been evaluated in the treatment of MS (182).

7.6. Cytokine and Gene Therapy

Direct manipulation of the Th1/Th2 cytokine balance has been the target of numerous therapies for EAE. Administration of IL-4 in mice after adoptive transfer of encephalitogenic MBP-specific T cell lines has been shown to dramatically reduce both clinical and histological disease (183). The protective effect of IL-4 administration was associated with the induction of MBP-specific Th2 cells and inhibition of pro-inflammatory cytokine gene expression (TNFα) in the CNS. Thus, factors that promote Th2 activity can be effective in treating EAE even in the presence of primed Th1 cells. Despite its clinical success, there are two main problems with this type of approach. Systemic administration of Th2 promoting cytokines may have unwanted side-effects and perhaps more importantly may not be as effective as direct delivery into the target organ.

Two recent studies have taken advantage of the homing properties of myelin-reactive T cells to deliver cytokine directly to the CNS. In the first study, a retroviral construct encoding the IL-4 gene was used to transduce an MBP-specific T cell hybridoma (184). These hybridomas could ameliorate EAE when administered to mice 10 days after immunization with MBP. This study provided the first proof of concept for a T cell-based gene therapy approach for the treatment of EAE. In the second study, lymph node T cells from mice immunized with PLP 139–151 were transfected with a construct expressing IL-10 under the control of the IL-2 promoter (185). These cells could ameliorate EAE if given either before or after disease onset. Another gene therapy approach has been the intracerebral injection of cytokine DNA-cationic liposome complexes. When injected on day 12 post immunization, DNA-cationic liposome complexes encoding IL-4, TGF-β, or TNF receptor inhibited the development of clinical EAE (186). Whether such approaches will be of use in the treatment of MS and which cytokines or combinations of cytokines will be most effective clinically remains to be determined.

8. CONCLUSION

The EAE model has proven itself of great value in our understanding of the mechanisms of autoimmune disease. It has clearly demonstrated that CNS autoimmunity is a complex process in which the T lymphocyte has a central role. Furthermore, it has shown that factors (genetic and environmental) that affect selection of the T cell repertoire and the activation/regulation of autoreactive T cells in the periphery are important determinants of CNS autoimmunity. EAE has also been of great use in the evaluation of therapeutic approaches with potential application for MS and other autoimmune diseases. Further study of the EAE model will likely continue to identify factors of relevance to the pathogenesis of MS.

References

1. Stuart, G., and K.S. Krikorian. 1928. The neuro-paralytic accidents of anti-rabies treatment. *Ann. Trop. Med. Parasitol.* 22:327–377.
2. Remlinger, J. 1905. Accidents paralytiques au cours du traitment anti-rabique. *Ann. Inst. Pasteur.* 19:625–646.
3. Kabat, E., A. Wolf, and A.E. Bozer. 1947. The rapid production of acute dessimenated encephalomyelitis in rhesus monkeys by injection of heterologous and homologous brain tissue with adjuvants. *J. Exp. Med.* 85:117–130.
4. Olitsky, P.K., and R.H. Yager. 1949. Experimental disseminated encephalomyelitis in white mice. *J. Exp. Med.* 90:213–223.
5. Paterson, P.Y. 1976. Experimental autoimmune (allergic) encephalomyelitis: induction, pathogenesis and suppression. In: *Textbook of immunopathology.* P.A. Mescher and H.S. Mueller-Eberhard, editors. Grune & Stratton, New York. pp. 179–213.

6. Einstein, E.R., D.M. Robertson, J.M. DiCarpio, and W. Moore. 1962. The isolation from bovine spinal cord of a homogenous protein with encephalitogenic activity. *J. Neurochem.* 9:353–361.
7. Folch, J., and M. Lees. 1951. Proteolipides, a new type of tissue lipoproteins. Their isolation from brain. *J. Biol. Chem.* 191:807–817.
8. Tuohy, V.K., Z. Lu, R.A. Sobel, R.A. Laursen, and M.B. Lees. 1988. A synthetic peptide from myelin proteolipid protein induces experimental allergic encephalomyelitis. *J. Immunol.* 141:1126–1130.
9. Mendel, I., N. Kerlero de Rosobo, and A. Ben-Nun. 1995. A myelin oligodendrocyte glycoprotein peptide induces typical chronic experimental autoimmune encephalomyelitis in H2-^{b} mice: fine specificity and T cell receptor V beta expression of encephalitogenic T cells. *Eur. J. Immunol.* 25:1951–1959.
10. Amor, S., N. Groome, C. Linington, M.M. Morris, K. Dornmair, M.V. Gardinier, J.M. Matthieu, and D. Baker. 1994. Identification of epitopes of myelin oligodendrocyte glycoprotein for the induction of experimental allergic encephalomyelitis in SJL and Biozzi AB/H mice. *J. Immunol.* 153:4349–56.
11. Amor, S., D. Baker, N. Groome, and J.L. Turk. 1993. Identification of a major encephalitogenic epitope of proteolipid protein (residues 56–70) for the induction of experimental allergic encephalomyelitis in Biozzi AB/H and nonobese diabetic mice. *J. Immunol.* 150.5666–5672.
12. Endoh, M., T. Kunishita, J. Nihei, M. Nishizawa, and T. Tabira. 1990. Suceptibility to proteolipid apoprotein and its encephalitogenic determinants in mice. *Int. Arch. Allergy Appl. Immunol.* 93:433–438.
13. Slavin, A., C. Ewing, J. Liu, M. Ichiwaka, J. Slavin, and C.C. Bernard. 1998. Induction of a multiple sclerosis-like disease in mice with an immunodominant epitope of myelin oligodendrocyte glycoprotein. *Autoimmunity.* 28.109–120.
14. Zamvil, S., D.J. Mitchell, A.C. Moore, K. Kitamura, L. Steinman, and J.B. Rothbard. 1986. T cell epitope of the autoantigen myelin basic protein that induces encephalomyelitis. *Nature* 324:258–260.
15. Whitham, R.H., R.E. Jones, G.A. Hashim, C.M. Hoy, R.-Y. Wang, A.A. Vandenbark, and H. Offner. 1991. Location of a new encephalitogenic epitope (residues 43–64) in proteolipid protein that induces relapsing experimental autoimmune encephalomyelitis in PL/J and (SJL X PL)F1 mice. *J. Immunol.* 147:3803–3808.
16. Sakai, K., S.S. Zamvil, D.J. Mitchell, M. Lim, J.B. Rothbard, and L. Steinman. 1988. Characterization of a major encephalitogenic T cell epitope in SJL/J mice with synthetic oligopeptides of myelin basic protein. *J. Neuroimmunol.* 19:21–32.
17. Tuohy, V.K., and D.M. Thomas. 1995. Sequence 104–117 of myelin proteolipid protein is a cryptic encephalitogenic T cell determinant for SJL/J mice. *J. Neuroimmunol.* 56:161–170.
18. Tuohy, V.K., Z. Lu, R.A. Sobel, R.A. Laursen, and M.B. Lees. 1989. Identification of an encephalitogenic determinant of myelin proteolipid protein for SJL mice. *J. Immunol.* 142:1523–1527.
19. Greer, J.M., V.K. Kuchroo, R.A. Sobel, and M.B. Lees. 1992. Identification and characterization of a second encephalitogenic determinant of myelin proteolipid protein (residues 178–191) for SJL mice. *J. Immunol* 149:783–788.

20. Hashim, G.A. 1978. Myelin basic protein: structure, function and antigenic determinants. *Immunol. Rev* 39:60–107.

21. Zhao, W., K.W. Wegmann, J.L. Trotter, K. Ueno, and W.F. Hickey. 1994. Identification of an N-terminally acetylated encephalitogenic epitope in myelin proteolipid apoprotein for the Lewis rat. *J. Immunol* 153:901–909.

22. McRae, B.L., C.L. Vanderlugt, M.C. Dal Canto, and S.D. Miller. 1995. Functional evidence for epitope spreading in the relapsing pathology of EAE in the SJL/J mouse. *J. Exp. Med.* 182:75–85.

23. MacKenzie-Graham, A.J., T.M. Pribyl, S. Kim, V.R. Porter, A.T. Campagnoni, and R.R. Voskuhl. 1997. Myelin protein expression is increased in lymph nodes of mice with relapsing experimental autoimmune encephalomyelitis. *J. Immunol* 159:4602–4610.

24. Pribyl, T.M., C.W. Campagnoni, K. Kampf, T. Kashima, V.W. Handley, J. McMahon, and A.T. Campagnoni. 1996. The human myelin basic protein gene is included within a 179-kilobase transcription unit: expression in the immune and central nervous systems. *Proc. Natl. Acad. Sci. USA* 90:10695–10699.

25. Mathisen, P.M., S. Pease, J. Garvey, L. Hood, and C. Readhead. 1993. Identification of an embryonic isoform of myelin basic protein that is expressed widely in the mouse embryo. *Proc. Natl. Acad. Sci. USA* 90:10125–10129.

26. Fritz, R.B., and M.-L. Zhao. 1996. Thymic expression of myelin basic protein (MBP). *J. Immunol* 157:5429–5253.

27. Pribyl, T.M., C.W. Campagnoni, K. Kampf, T. Kashima, V.W. Handley, J. McMahon, and A.T. Campagnoni. 1996. Expression of the myelin proteolipid protein gene in the human fetal thymus. *J. Neuroimmunol* 67:125–130.

28. Voskuhl, R. 1998. Myelin protein expression in lymphoid tissues: implications for peripheral tolerance. *Immunol. Rev* 164:81–92.

29. Campagnoni, C.W., B. Garbay, P. Micevych, T. Pribyl, K. Kampf, V.W. Handley, and A.T. Campagnoni. 1992. DM20 mRNA splice product of the myelin proteolipid protein gene is expressed in the murine heart. *J. Neurosci. Res* 33:148–155.

30. Yoshizawa, I., R. Bronson, M.E. Dorf, and S. Abromson-Leeman. 1998. T-cell responses to myelin basic protein in normal and MBP-deficient mice. *J. Neuroimmunol* 84:131–138.

31. Targoni, O.S., and P.V. Lehman. 1998. Endogenous myelin basic protein inactivates the high avidity T cell repertorie. *J. Exp. Med* 187:2055–2063.

32. Harrington, C.J., A. Paez, T. Hunkapiller, V. Mannikko, T. Brabb, M. Ahearn, C. Beeson, and J. Goverman. 1998. Differential tolerance is induced in T cells recognizing distinct epitopes of myelin basic protein. *Immunity* 8:571–580.

33. Gonatas, N.K., and J.C. Howard. 1974. Inhibition of experimental allergic encephalomyelitis in rats severely depleted of T cells. *Science* 186:839–841.

34. Ortiz-Ortiz, L., and W.O. Weigle. 1976. Cellular events in the induction of experimental allergic encephalomyelitis in rats. *J. Exp. Med* 144:604–616.

35. Ben-Nun, A., H. Wekerle, and I.R. Cohen. 1981. The rapid isolation of clonable antigen-specific T lymphocyte lines capable of mediating autoimmune encephalomyelitis. *Eur. J. Immunol* 11:195–199.

36. Mokhtarian, F., D.E. McFarlin, and C.S. Raine. 1984. Adoptive transfer of myelin basic protein-sensitized T cells produces chronic relapsing demyelinating disease in mice. *Nature* 309:336–358.

37. Zamvil, S.S., P.A. Nelson, D.J. Mitchell, R.L. Knobler, R.B. Fritz, and L. Steinman. 1985. Encephalitogenic T cell clones specific for myelin basic protein. An usual bias in antigen recognition. *J. Exp. Med.* 162:2107–2124.

38. Mosmann, T.R., H. Cherwinski, M.W. Bond, M.A. Giedlin, and R.L. Coffman. 1986. Two types of murine helper T cell clones. *J. Immunol* 136:2348–2357.

39. Fiorentino, D.F., A. Zlotnik, P. Viera, T.R. Mosmann, M. Howard, and A. O'Garra. 1991. IL-10 acts on the antigen-presenting cell to inhibit cytokine production by Th1 cells. *J. Immunol* 146:3444–3451.

40. Chen, Y., V.K. Kuchroo, J. Inobe, D.A. Hafler, and H.L. Weiner. 1994. Regulatory T cell clones induced by oral tolerance: suppression of autoimmune encephalomyelitis. *Science* 265:1237–1240.

41. Groux, H., A. O'Garra, M. Bigler, M. Rouleau, S. Antonenko, J.E. de Vries, and M. Roncarolo. 1997. A CD4+ T-cell subset inhibits antigen-specific T-cell responses and prevents colitis. *Nature* 389:737–742.

42. Bettelli, E., M. Prabhu Das, E.D. Howard, H.L. Weiner, R.A. Sobel, and V.K. Kuchroo. 1998. IL-10 is a critical in the regulation of autoimmune encephalomyelitis as demonstrated by studies of IL-10- and IL-4-deficient and transgenic mice. *J. Immunol* 161:3299–3306.

43. Segal, B.M., B.K. Dwyer, and E.M. Shevach. 1998. An interleukin (IL)-10/IL-12 immunoregulatory circuit controls susceptibility to autoimmune disease. *J. Exp. Med.* 187:537–546.

44. Baron, J.L., J.A. Madri, N.H. Ruddle, G. Hashim, and J. Janeway, C.A. 1993. Surface expression of $\alpha 4$ integrin by CD4 Tcells is required for their entry into brain parenchyma. *J. Exp. Med.* 177:57–68.

45. Kuchroo, V.K., C.A. Martin, J.M. Greer, S.-T. Ju, R.A. Sobel, and M.E. Dorf. 1993. Cytokines and adhesion molecules contribute to the ability of myelin proteolipid protein-specific T cell clones to mediate experimental allergic encephalomyelitis. *J. Immunol* 151:4371–4382.

46. Miller, S., and W. Karpus. 1994. The immunopathogenesis and regulation of T cell mediated demyelinating diseases. *Immunol. Tod.* 15:356–361.

47. Baker, D., J.K. O'Neill, and J.L. Turk. 1991. Cytokines in the central nervous system of mice during chronic relapsing experimental allergic encephalomyelitis. *Cell. Immunol* 137:505.

48. Khoury, S.J., W.W. Hancock, and H.L. Weiner. 1992. Oral tolerance to myelin basic protein and natural recovery from experimental allergic encephalomyelitis are associated with downregulation of inflammatory cytokines and differential upregulation of transforming growth factor β, interleukin 4, and prostaglandin E expression in the brain. *J. Exp. Med.* 176:1355–1364.

49. Kuchroo, V.K., M. Prabhu Das, J.A. Brown, A.M. Ranger, S.S. Zamvil, R.A. Sobel, H.L. Weiner, N. Nabavi, and L.H. Glimcher. 1995. B7-1 and B7-2 costimulatory molecules activate differentially the Th1/Th2 developmental pathways: application to autoimmune disease therapy. *Cell* 80:707–718.

50. Nicholson, L.B., J.M. Greer, R.A. Sobel, M.B. Lees, and V.K. Kuchroo. 1995. An altered peptide ligand mediates immune deviation and prevents autoimmune encephalomyelitis. *Immunity* 3:397–405.

51. Khoruts, A., S.D. Miller, and M.K. Jenkins. 1995. Neuroantigen-specific Th2 cells are inefficient supressors of

experimental autoimmune encephalomyelitis induced by effector Th1 cells. *J. Immunol* 155:5011–5017.

52. Lafaille, J.L., F. Van de Keere, A.L. Hsu, J.L. Baron, W. Haas, C.S. Raine, and S. Tonegawa. 1997. Myelin basic protein-specific T helper 2 (Th2) cells cause experimental autoimmune encephalomyelitis in immunodeficient hosts rather than protect them from disease. *J. Exp. Med.* 186:307–312.

53. Ferber, I.A., S. Brocke, C. Taylor-Edwards, W. Ridgway, C. Dinisco, L. Steinman, D. Dalton, and C.G. Fathman. 1996. Mice with a disrupted IFN-γ gene are susceptible to the induction of Experimental Autoimmune Encephalomyelitis (EAE). *J. Immunol* 156:5–7.

54. Zhang, B., T. Yamamura, T. Kondo, M. Fujiwara, and T. Tabira. 1997. Regulation of experimental autoimmune encephalomyelitis by natural killer (NK) cells. *J. Exp. Med* 186:1677–1687.

55. Krakowski, M., and T. Owens. 1996. Interferon-γ confers resistance to experimental allergic encephalomyelitis. *Eur. J. Immunol* 26:1641–1646.

56. Liblau, R., L. Steinman, and S. Brocke. 1997. Experimental autoimmune encephalomyelitis in IL-4-deficient mice. *Int. Immunol* 9:799–803.

57. Falcone, M., A.J. Rajan, B.R. Bloom, and C.F. Brosnan. 1998. A critical role for IL-4 in regulating disease severity in experimental allergic encephalomyelitis as demonstrated in IL-4 deficient C57BL/6 mice and BALB/c mice. *J. Immunol* 160:4822–4830.

58. Probert, L., K. Akassoglou, M. Pasparakis, G. Kontogeorgos, and G. Kollias. 1995. Spontaneous inflammatory demyelinating disease in transgenic mice showing central nervous system specific expression of tumor necrosis factor α. *Proc. Natl. Acad. Sci. USA* 92:11294–11298.

59. Liu, J., M.W. Marino, G. Wong, D. Grail, A. Dunn, J. Bettadapura, A.J. Slavin, L. Old, and C.C.A. Bernard. 1998. TNF is a potent anti-inflammatory cytokine in autoimmune mediated demyelination. *Nature Med.* 4:78–83.

60. Korner, H., D.S. Riminton, D.H. Strickland, F.A. Lemckert, J.D. Pollard, and J.D. Sedgwick. 1997. Critical points of tumor necrosis factor action in central nervous system autoimmune inflammation defined by gene targeting. *J. Exp. Med.* 186:1585–1590.

61. Frei, K., H. Eugster, M. Bopst, C.S. Constantinescu, E. Lavi, and A. Fontana. 1997. Tumor necrosis factor a and lymphotoxin are not required for induction of acute experimental autoimmune encephalomyelitis. *J. Exp. Med.* 185:2177–2182.

62. Suen, W.E., C.M. Bergman, P. Hjelmstrom, and N.H. Ruddle. 1997. A critical role for lymphotoxin in experimental allergic encephalomyelitis. *J. Exp. Med.* 186:1233–1240.

63. Bretscher, P., and M. Cohn. 1970. A theory of self-nonself discrimination. *Science* 169:1042–1049.

64. Jenkins, M.K., and R.H. Schwartz. 1987. Antigen presentation by chemically modified splenocytes induces antigen-specific T cell unresponsiveness *in vitro* and *in vivo*. *J. Exp. Med.* 165:302–319.

65. Perrin, P.J., D. Scott, L. Quigley, P.S. Albert, O. Feder, G.S. Gray, R. Abe, C.H. June, and M. Racke. 1995. Role of B7:CD28/CTLA-4 in the induction of chronic relapsing experimental allergic encephalomyelitis. *J. Immunol.* 154:1481–1490.

66. Croxford, J.L., J.K. O'Neill, R.R. Ali, K. Browne, A.P. Byrnes, M.J. Dallman, M.J. Wood, M. Fedlmann, and D. Baker. 1998. Local gene therapy with CTLA4-

67. Perrin, P.J., D. Scott, T.A. Davis, G.S. Gray, M.J. Doggett, R. Abe, C.H. June, and M.K. Racke. 1996. Opposing effects of CTLA4-Ig and anti-CD80 (B7–1) plus anti-CD86 (B7–2) on experimental allergic encephalomyelitis. *J. Neuroimmunol.* 65:31–39.

68. Miller, S.D., C.L. Vanderlugt, D.J. Lenschow, J.G. Pope, N.J. Karandikar, M.C. Dal Canto, and J.A. Bluestone. 1995. Blockade of CD28/B7-1 interaction prevents epitope spreading and clinical relapses of murine EAE. *Immunity* 3:739–745.

69. Oliveira-dos-Santos, A.J., A. Ho, Y. Tada, J.J. Lafaille, S. Tonegawa, T.W. Mak, and J.M. Penninger. 1999. CD28 costimulation is crucial for the development of spontaneous autoimmune encephalomyelitis. *J. Immunol.* 162:4490–4495.

70. Perrin, P.J., E. Lavi, C.A. Rumbley, S.A. Zekavat, and S.M. Philips. 1999. Experimental autoimmune meningitis: a novel neurological disease in CD28-deficient mice. *J. Appl. Biomater.* 91:41–49.

71. Tivol, E.A., F. Borriello, A.N. Schweitzer, W.P. Lynch, J.A. Bluestone, and A.H. Sharpe. 1995. CTLA-4 deficient mice exhibit massive lymphoproliferation and multi-organ lymphatic infiltration: a critical negative immunoregulatory role of CTLA-4. *Immunity* 3:541–547.

72. Waterhouse, P.W., J.M. Penninger, E. Timms, A. Wakeham, A. Shahinian, K.P. Lee, C.B. Thompson, H. Griesser, and T.W. Mak. 1995. Lymphoproliferative disorders with early lethality in mice deficient in CTLA-4. *Science* 270:985–988.

73. Perrin, P.J., J.H. Maldonado, T.A. Davis, C.H. June, and M.K. Racke. 1996. CTLA-4 blockade enhances clinical disease and cytokine production during experimental allergic encephalomyelitis. *J. Immunol.* 157:1333–1336.

74. Karandikar, N.J., C.L. Vanderlugt, T.L. Walunas, S.D. Miller, and J.A. Bluestone. 1996. CTLA-4: a negative regulator of autoimmune disease. *J. Exp. Med.* 184:783–788.

75. van Essen, D., H. Kikutani, and D. Gray. 1995. CD40 ligand-transduced costimulation of T cells in the development of helper function. *Nature* 376:620–623.

76. Grewal, I.S., J. Xu, and R.A. Flavell. 1995. Impairment of antigen-specific T-cell priming in mice lacking CD40 ligand. *Nature* 378:617–620.

77. Grewal, I.S., H.G. Foellmer, K.D. Grewal, J. Xu, F. Hardardottir, J.L. Baron, C.A. Jr. Janeway, and R.A. Flavell. 1996. Requirement for CD40 ligand in costimulation induction, T cell activation, and experimental allergic encephalomyelitis. *Science* 273:1864–1867.

78. Gerritse, K., J.D. Laman, R.J. Noelle, A. Aruffo, J.A. Ledbetter, W.J.A. Boersma, and E. Claassen. 1996. CD40-CD40 ligand interactions in experimental allergic encephalomyelitis and multiple sclerosis. *Proc. Natl. Acad. Sci. USA* 93:2499–2504.

79. Samoilova, E.B., J.L. Horton, H. Zhang, and Y. Chen. 1997. CD40L blockade prevents autoimmune encephalomyelitis and hampers Th1 but not Th2 pathway of T cell differentiation. *J. Mol. Med.* 75:603–608.

80. Howard, L.M., A.J. Miga, C.L. Vanderlugt, M.C. Dal Canto, J.D. Laman, R.J. Noelle, and S.D. Miller. 1999. Mechanisms of immunotherapeutic intervention by anti-CD40L (CD154) antibody in an animal model of multiple sclerosis. *J. Clin. Invest.* 103:281–290.

81. Karandikar, N.J., C.L. Vanderlugt, T. Eager, L. Tan, J.A. Bluestone, and S.D. Miller. 1998. Tissue-specific up-regulation of B7–1 expression and function during the course of murine relapsing experimental autoimmune encephalomyelitis. *J. Immunol.* 161:192–199.

82. Issazadeh, S., V. Navikas, M. Schaub, M. Sayegh, and S. Khoury. 1998. Kinetics of expression of costimulatory molecules and their ligands in murine relapsing experimental autoimmune encephalomyelitis *in vivo*. *J. Immunol.* 161:1104–1112.

83. Spanaus, K.S., R. Schlapbach, and A. Fontana. 1998. TNF-alpha and IFN-gamma render microglia sensitive to Fas ligand-induced apoptosis by induction of Fas expression and down-regulation of Bcl–2 and Bcl-xL. *Eur. J. Immunol* 28:4398–4408.

84. Waldner, H., R.A. Sobel, E. Howard, and V.K. Kuchroo. 1997. Fas- and FasL-deficient mice are resistant to induction of autoimmune encephalomyelitis. *J. Immunol.* 159:3100–3103.

85. Sabelko, K.A., K.A. Kelly, M.H. Nahm, A.H. Cross, and J.H. Russell. 1997. Fas and Fas ligand enhance the pathogenesis of experimental allergic encephalomyelitis, but are not essential for immune privilege in the central nervous system. *J. Immunol.* 159:3096–3099.

86. Okuda, Y., C.C. Bernard, H. Fujimura, T. Yanagihara, and S. Sakoda. 1998. Fas has a crucial role in the progression of experimental autoimmune encephalomyelitis. *Mol. Immunol* 35:317–326.

87. Malipiero, U., K. Frei, K. Spanaus, C. Agresti, H. Lassmann, M. Hahne, J. Tschopp, H. Eugster, and A. Fontana. 1997. Myelin oligodendrocyte glycoprotein-induced autoimmune encephalomyelitis is chronic/relapsing in perforin-knockout mice, but monophasic in Fas- and Fas ligand-deficient *lpr* and *gld* mice. *Eur. J. Immunol.* 27:3151–3160.

88. Bonetti, B., J. Phol, Y. Gao, and C.S. Raine. 1997. Cell death during autoimmune demyelination: effector but not target cells are eliminated by apoptosis. *J. Immunol.* 159:5733–5741.

89. White, C.A., P.A. McCombe, and M.P. Pender. 1998. The roles of Fas, Fas ligand and Bcl-2 in T cell apoptosis in the central nervous system in experimental autoimmune encephalomyelitis. *J. Neuroimmunol.* 82:47–55.

90. Sabelko-Downes, K.A., A.H. Cross, and J.H. Russell. 1999. Dual role for Fas ligand in the initiation of and recovery from experimental allergic encephalomyelitis. *J. Exp. Med.* 189:1195–1205.

91. Koh, D., W. Fung-Leung, A. Ho, D. Gray, H. Acha-Orbea, and T. Mak. 1992. Less mortality but more relapses in experimental allergic encephalomyelitis in CD8 –/– mice. *Science* 256:1210–1213.

92. Jiang, H., S. Zhang, and B. Pernis. 1992. Role of CD8[+] T cells in murine experimental allergic encephalomyelitis. *Science* 256:1213–1215.

93. Willenborg, D.O., and S.J. Prowse. 1983. Immunoglobulin-deficient rats fail to develop experimental allergic encephalomyelitis. *J. Neuroimmunol* 5:99–109.

94. Willenborg, D.O., P. Sjollema, and G. Danta. 1986. Immunoglobulin deficient rats as donors and recipients of effector cells of allergic encephalomyelitis. *J. Neuroimmunol* 11:93–103.

95. Myers, K.J., J. Sprent, J.P. Dougherty, and Y. Ron. 1992. Synergy between encephalitogenic T cells and myelin basic protein-specific antibodies in the induction of experimental autoimmune encephalomyelitis. *J. Neuroimmunol.* 41:1–8.

96. Hjelmstrom, P., A.E. Juedes, J. Fjell, and N.H. Ruddle. 1998. Cutting Edge: B cell-deficient mice develop experimental allergic encephalomyelitis with demyelination after myelin oligodendrocyte glycoprotein sensitization. *J. Immunol.* 161:4480–4483.

97. Wolf, S.D., B.N. Dittel, F. Hardardottir, and J. Janeway, C.A. 1996. Experimental autoimmune encephalomyelitis induction in genetically B cell-deficient mice. *J. Exp. Med.* 184:2271–2278.

98. Ridgway, W.M., M. Fasso, and C.G. Fathman. 1999. A new look at MHC and autoimmune disease. *Science* 284:749–751.

99. Carrasco-Marin, E., J. Shimizu, O. Kanagawa, and E. Unanue. 1996. The class II MHC-IAg[7] molecules from non-obese diabetic mice are poor peptide binders. *J. Immunol.* 156:450–458.

100. Dietrich, W.F., J.C. Miller, R.G. Steen, M. Merchant, D. Damron, R. Nahf, A. Gross, D.C. Joyce, M. Wessel, R.D. Dredge, A. Marquis, L.D. Stein, N. Goodman, D.C. Page, and E.S. Lander. 1994. A genetic map of the mouse with 4,006 simple sequence length polymorphisms. *Nat. Genet.* 7:220–245.

101. Lander, E., and L. Kruglyak. 1995. Genetic dissection of complex traits: Guidelines for interpreting and reporting linkage results. *Nat. Genet.* 11:241–247.

102. McAleer, M.A., P. Reifsnyder, S.M. Palmer, M. Prochazka, J.M. Love, J.B. Copeman, E.E. Powell, N.R. Rodrigues, J. Prins, D.V. Serreze, N.H. DeLarato, L.S. Wicker, L.B. Peterson, N.J. Schork, J.A. Todd, and E.H. Leiter. 1995. Crosses of NOD mice with the related NON strain. A polygenic model for IDDM. *Diabetes* 44:1186–1195.

103. Denny, P., C.J. Lord, N.J. Hill, J.V. Goy, E.R. Levy, P.L. Podolin, L.B. Peterson, L.S. Wicker, J.A. Todd, and P.A. Lyons. 1997. Mapping of the IDDM locus Idd3 to a 0.35-cM interval containing the interleukin-2 gene. *Diabetes* 46:695–700.

104. Encinas, J.A., L.S. Wicker, L.B. Peterson, A. Mukasa, C. Teuscher, R. Sobel, H.L. Weiner, C.E. Seidman, J.G. Seidman, and V.K. Kuchroo. 1999. QTL influencing autoimmune diabetes and encephalomyelitis map to a 0.15-cM region containing *IL2*. *Nat. Genet.* 21:158–160.

105. Gasser, D.L., C.M. Newlin, J. Palm, and N.K. Gonatas. 1973. Genetic control of susceptibility to experimental allergic encephalomyelitis in rats. *Science* 181:872–873.

106. Happ, M.P., P. Wettstein, B. Dietzschold, and E. Heber-Katz. 1988. Genetic control of the development of experimental allergic encephalomyelitis in rats. Separation of MHC and Non-MHC gene effects. *J. Immunol.* 141:1489–1494.

107. Chung, I.Y., J.G. Norris, and E.N. Benveniste. 1991. Differential tumor necrosis factor alpha expression by astrocytes from experimental allergic encephalomyelitis-susceptible and -resistant rat strains. *J. Exp. Med.* 173:801–811.

108. Gasser, D.L., J. Palm, and N.K. Gonatas. 1975. Genetic control of susceptibility to experimental allergic encephalomyelitis and the Ag-B locus of rats. *J. Immunol.* 115:431–433.

109. Levine, S., and R. Sowinski. 1974. Experimental allergic encephalomyelitis in congenic strains of mice. *Immunogenetics* 110:139–143.

110. Montgomery, I.N., and H.C. Rauch. 1982. Experimental allergic encephalomyelitis (EAE) in mice: Primary control of EAE susceptibility is outside the H-2 complex. *J. Immunol.* 128:421–425.

111. Teitelbaum, D., Z. Lando, and R. Arnon. 1978. Genetic control of susceptibility to experimental allergic encephalomyelitis-immunological studies. In: Genetic Control of Autoimmune Disease. N. Rose, M. Bigazzi, and L. Warner, editors. Elsevier North Holland, Amsterdam.

112. Jansson, L., T. Olsson, and R. Holmdahl. 1993. Influence of T-cell receptor genes on chronic experimental autoimmune encephalomyelitis. *Immunogenetics* 37:466–68.

113. Encinas, J.A., H.L. Weiner, and V.K. Kuchroo. 1996. Inheritance of susceptibility to experimental autoimmune encephalomyelitis. *J. Neurosci. Res.* 45:655–669.

114. Holmdahl, R. 1998. Genetics of susceptibility to chronic experimental encephalomyelitis and arthritis. *Curr. Opin. Immunol.* 10:710–717.

115. Encinas. J.A., and V.K. Kuchroo. 1999. Genetics of experimental autoimmune encephalomyelitis. In: Genes and Genetics of Autoimmunity, vol. 1. A.N. Theofilopoulos, editor. Karger, Basel. pp. 247–272.

116. Encinas, J.A., M.B. Lees, R.A. Sobel, C. Symonowicz, J.M. Greer, C.L. Shovlin, H.L. Weiner, C.E. Seidman, J.G. Seidman, and V.K. Kuchroo. 1996. Genetic analysis of susceptibility to experimental autoimmune encephalomyelitis in a cross between SJL/J and B10.S mice. *J. Immunol.* 157:2186–2192.

117. Baker, D., O.A. Rosenwasser, J.K. O'Neill, and J.L. Turk. 1995. Genetic analysis of experimental allergic encephalomyelitis in mice. *J. Immunol.* 155:4046–4051.

118. Sundvall, M., J. Jirholt, H.-T. Yang, L. Jansson, A. Engstrom, U. Petterson, and R. Holmdahl. 1995. Identification of murine loci associated with susceptibility to chronic experimental autoimmune oneophalomyelitis. *Nat. Genet* 10:313–317.

119. Butterfield, R.J., E.P. Blankenhorn, R.J. Roper, J.F. Zachary, R.W. Doerge, J. Sudweeks, J. Rose, and C. Teuscher. 1999. Genetic analysis of disease subtypes and sexual dimorphisms in mouse experimental allergic encephalomyelitis (EAE): relapsing/remitting and monophasic remitting/nonrelapsing EAE are immunogenetically distinct. *J. Immunol.* 162:3096–3102.

120. Butterfield, R.J., J.D. Sudweeks, E.P. Blankenhorn, R. Korngold, J.C. Marini, J.A. Todd, R.J. Roper, and C. Teuscher. 1998. New genetic loci that control susceptibility and symptoms of experimental allergic encephalomyelitis in inbred mice. *J. Immunol.* 161:1860–1867.

121. Croxford, J.L., J.K. O'Neill, and D. Baker. 1997. Polygenic control of experimental allergic encephalomyelitis in Biozzi ABH and Balb/c mice. *J. Neuroimmunol.* 74:205–211.

122. Teuscher, C., B.B. Wardell, J.K. Luncerford, S.D. Michael, and K.S.K. Tung. 1996. Aod2, the locus controlling development of atrophy in neonatal thymectomy-induced autoimmune ovarian dysgenesis, co-localizes with *IL2, Fgfb* and *Idd3 J. Exp. Med.* 183:631–637.

123. Vyse, T.J., and J.A. Todd. 1996. Genetic analysis of autoimmune disease. *Cell* 85:311–318.

124. Becker, K.G., R.M. Simon, J.E. Bailey-Wilson, B. Freidlin, W.E. Biddison, H.F. McFarland, and J.M. Trent. 1998. Clustering of non-major histocompatibility complex susceptibility candidate loci in human autoimmune diseases. *Proc. Natl. Acad. Sci. USA* 95:9979–9984.

125. Dahlman, I., L. Jacobsson, A. Glaser, J.C. Lorentzen, M. Andersson, H. Luthman, and T. Olsson. 1999. Genome-wide linkage analysis of chronic relapsing experimental autoimmune encephalomyelitis in the rat identifies a major

126. Ghosh, S., S.M. Palmer, N.R. Rodrigues, H.G. Cordell, C.M. Hearne, R.J. Cornall, J.-B. Prins, P. McShane, G.M. Lathrop, L.B. Peterson, L.S. Wicker, and J.A. Todd. 1993. Polygenic control of autoimmune diabetes in nonobese diabetic mice. *Nat. Genet* 4:404–409.

127. Roth, M.-P., C. Viratelle, L. Dolbois, M. Delverdier, N. Borot, L. Pelletier, P. Druet, M. Clanet, and H. Coppin. 1999. A genome-wide search identifies two susceptibility loci for experimental autoimmune encephalomyelitis on rat chromosomes 4 and 10. *J. Immunol.* 162:1917–1922.

128. Kurtzke, J. 1985. Epidemiology of multiple sclerosis. In: Handbook of clinical neurology. P.J. Vinken, G.W. Bruyn, H.L. Klawans, and J.C. Koetsier, editors. Elsevier Science Publishing, New York. pp. 259–287.

129. Johnson, R.T., D.E. Griffin, J.S. Hirsch, J.S. Wolinsky, S. Rodenbeck, I. Lindo De Soriano, and A. Vaisberg. 1984. Measles and encephalomyelitis: clinical and immunological studies. *N. Engl. J. Med* 310:137–141.

130. Jahnke, U., E.H. Fischer, and E.C. Alvord, Jr. 1985. Sequence homology between certain viral proteins and proteins related to encephalomyelitis and neuritis. *Science* 229:282–284.

131. Fujinami, R.S., and M.B. Oldstone. 1985. Amino acid homology between the encephalitogenic site of myelin basic protein and virus: mechanism for autoimmunity. *Science* 230:1043–1045.

132. Wucherpfennig, K.W., and J.L. Strominger. 1995. Molecular mimicry in T cell-mediated autoimmunity: viral peptides activate human T cell clones specific for myelin basic protein. *Cell* 80:695–705.

133. Hemmer, B., B.T. Fleckenstein, M. Vergelli, G. Jung, H. McFarland, R. Martin, and K.-H. Wiesmuller. 1997. Identification of high potency microbial and self ligands for a human autoreactive class II restricted T cell clone. *J. Exp. Med.* 185:1651–1659.

134. Gautam, A.M., R. Liblau, G. Chelvanayagam, L. Steinman, and T. Boston. 1998. A viral peptide with limited homology to a self peptide can induce clinical signs of experimental autoimmune encephalomyelitis. *J. Immunol.* 161:60–64

135. Carrizosa, A.M., L.B. Nicholson, M. Farzan, S. Southwood, A. Sette, R.A. Sobel, and V.K. Kuchroo. 1998. Expansion by self antigen is necessary for the induction of experimental autoimmune encephalomyelitis by T cells primed with a cross-reactive environmental antigen. *J. Immunol.* 161:3307–3314.

136. Ufret-Vincenty, R.L., L. Quigley, N. Tresser, S.H. Pak, A. Gado, S. Hausmann, K.W. Wucherpfennig, and S. Brocke. 1998. *In vivo* survival of viral antigen-specific T cells that induce experimental autoimmune encephalomyelitis. *J. Exp. Med.* 188:1725–1738.

137. Miller, S.D., C.L. Vanderlugt, W. Begolka Smith, W. Pao, R.L. Yauch, K.L. Neville, Y. Katz-Levy, A. Carrizosa, and B.S. Kim. 1997. Persistent infection with Theiler's virus leads to CNS autoimmunity via epitope spreading. *Nature Med.* 3:1333–1337.

138. Conrad, B., E. Weidmann, G. Trucco, W.A. Rudert, R. Behboo, C. Ricordi, H. Rodriguez-Rilo, D. Finegold, and M. Trucco. 1994. Evidence for superantigen involvement in insulin-dependent diabetes mellitus aetiology. *Nature* 371:351–355.

139. Conrad, B., R.N. Weissmahr, J. Boni, R. Arcari, J. Schupbach, and B. Mach. 1997. A human endogenous retro-

viral superantigen as candidate autoimmune gene in type I diabetes. *Cell* 90:303–313.

140. Brocke, S., A. Gaur, C. Piercy, A. Gautam, K. Gijbels, C.G. Fathman, and L. Steinman. 1993. Induction of relapsing paralysis in experimental autoimmune encephalomyelitis by bacterial superantigen. *Nature* 365:642–644.

141. Schiffenbauer, J., H.M. Johnson, E.J. Butfiloski, L. Wegrzyn, and J.M. Soos. 1993. Staphylococcal enterotoxins can reactivate experimental allergic encephalomyelitis. *Proc. Natl. Acad. Sci. USA* 90:8543–8546.

142. Steinman, L., J. Rosenbaum, S. Sriram, and H.O. McDevitt. 1981. *In vivo* effects of antibodies to immune response gene products: Prevention of experimental allergic encephalomyelitis. *Proc. Natl. Acad. Sci. USA* 78:7111–7114.

143. Steinman, L., D. Solomon, S. Zamvil, M. Lim, and S. Sriram. 1983. Prevention of EAE with anti I-A antibody: Decreased accumulation of radiolabeled lymphocytes in the central nervous system. *J. Neuroimmunol.* 5:91–97.

144. Sriram, S., and L. Steinman. 1983. Anti-I-A antibody suppresses active encephalomyelitis: Treatment model for disease linked to Ir genes. *J. Exp. Med.* 158:1362–1367.

145. Brostoff, S.W., and D.W. Mason. 1984. Experimental allergic encephalomyelitis: successful treatment *in vivo* with a monoclonal antibody that recognizes T helper cells. *J. Immunol.* 133:1938–1942.

146. Waldor, M.K., S. Sriram, R. Hardy, L.A. Herzenberg, L. Lanier, M. Lim, and L. Steinman. 1985. Reversal of experimental allergic encephalomyelitis with a monoclonal antibody to a T cell subset marker (L3T4). *Science* 227:415–417.

147. Alters, S.E., L. Steinman, and V.T. Oi. 1989. Comparison of rat and rat-mouse chimeric anti-murine CD4 antibodies *in vitro*: Chimeric antibodies lyse low density CD4+ cells. *J. Immunol.* 142:2018–2023.

148. Zamvil, S.S., D.J. Mitchell, N.E. Lee, A.C. Moore, M.K. Waldor, K. Sakai, J.B. Rothbard, H.O. McDevitt, L. Steinman, and H. Acha-Orbea. 1988. Predominant expression of a T cell receptor Vβ gene subfamily in autoimmune encephalomyelitis. *J. Exp. Med.* 167:1586–1596.

149. Acha-Orbea, H., D.J. Mitchell, L. Timmerman, D.C. Wraith, G.S. Tausch, M.K. Waldor, S.S. Zamvil, H.O. McDevitt, and L. Steinman. 1988. Limited heterogeneity of T cell receptors from lymphocytes mediating autoimmune encephalomyelitis allows specific immune intervention. *Cell* 54:263–273.

150. Burns, F.R., X.B. Li, N. Shen, H. Offner, Y.K. Chou, A.A. Vandenbark, and E. Heber-Katz. 1989. Both rat and mouse T cell receptors specific for the encephalitogenic determinant of myelin basic protein use similar V alpha and V beta chain genes even though the major histocompatibility complex and encephalitogenic determinants being recognized are different. *J. Exp. Med.* 169:27–39.

151. Chluba, J., C. Steeg, A. Becker, H. Wekerle, and J.T. Epplen. 1989. T cell receptor beta chain usage in myelin basic protein-specific rat T lymphocytes. *Eur. J. Immunol.* 19:279–284.

152. Ben-Nun, A., and I.R. Cohen. 1981. Vaccination against autoimmune encephalomyelitis (EAE): attenuated autoimmune T lymphocytes confer resistance to induction of active EAE but not to EAE mediated by the intact T lymphocyte line. *Eur. J. Immunol.* 11:949–952.

153. Ben-Nun, A., H. Wekerle, and I.R. Cohen. 1981. Vaccination against autoimmune encephalomyelitis with T-lymphocyte line cells reactive against myelin basic protein. *Nature* 292:60–61.

154. Howell, M.D., S.T. Winters, T. Olee, H.C. Powell, D.J. Carlo, and S.W. Brostoff. 1989. Vaccination against experimental allergic encephalomyelitis with T cell receptor peptides. *Science* 246:688–670.

155. Vandenbark, A.A., G. Hashim, and H. Offner. 1989. Immunization with a synthetic T-cell receptor V-region peptide protects against experimental autoimmune encephalomyelitis. *Nature* 341:541–544.

156. Vandenbark, A.A., G.A. Hashim, and H. Offner. 1996. T cell receptor peptides in treatment of autoimmune disease: rationale and potential. *J. Neurosci. Res.* 43:391–402.

157. Kumar, V., K. Stellrecht, and E. Sercarz. 1996. Inactivation of T cell receptor peptide-specific CD4 regulatory T cells induces chronic experimental autoimmune encephalomyelitis (EAE). *J. Exp. Med.* 184:1609–1617.

158. Hafler, D.A., M.G. Saadeh, V.K. Kuchroo, E. Milford, and L. Steinman. 1996. TCR usage in human and experimental demyelinating disease. *Immunol. Tod.* 17:152–159.

159. Evavold, B.D., and P.M. Allen. 1991. Separation of IL-4 Production from Th Cell Proliferation by an Altered T Cell Receptor Ligand. *Science* 252:1308–1310.

160. De Magistris, M.T., J. Alexander, M. Coggeshall, A. Altmann, F.C.A. Gaeta, H.M. Grey, and A. Sette. 1992. Antigen analog-major histocompatibility complexes act as antagonists of the T cell receptor. *Cell* 66:625–634.

161. Evavold, B.D., J. Sloan-Lancaster, and P.M. Allen. 1993. Tickling the TCR: selective T-cell functions stimulated by altered peptide ligands. *Immunol. Tod.* 14:602–609.

162. Sakai, K., S.S. Zamvil, D.J. Mitchell, S. Hodgkinson, J.B. Rothbard, and L. Steinman. 1989. Prevention of experimental encephalomyelitis with peptides that block interaction of T cells with major histocompatibility complex proteins. *Proc. Natl. Acad. Sci. USA* 86:9470–9474.

163. Lamont, A.G., A. Sette, R. Fujinami, S.M. Colon, C. Miles, and H.M. Grey. 1990. Inhibition of experimental autoimmune encephalomyelitis induction in SJL/J mice by using a peptide with high affinity for IAˢ molecules. *J. Immunol.* 145:1687–1693.

164. Wauben, M.H., C.J. Boog, R. van der Zee, I. Joosten, A. Schlief, and W. van Eden. 1992. Disease inhibition by major histocompatibility complex binding peptide analogues of disease-associated epitopes: more than blocking alone. *J. Exp. Med.* 176:667–677.

165. Smilek, D.E., D.C. Wraith, S. Hodgkinson, S. Dwivedy, L. Steinman, and H.O. McDevitt. 1991. A single amino acid change in a myelin basic protein peptide confers the capacity to prevent rather than induce experimental autoimmune encephalomyelitis. *Proc. Natl. Acad. Sci USA* 88:9633–9637.

166. Kuchroo, V.K., J.M. Greer, D. Kaul, G. Ishioka, A. Franco, A. Sette, R.A. Sobel, and M.B. Lees. 1994. A Single TCR Antagonist Peptide Inhibits Experimental Allergic Encephalomyelitis Mediated by a Diverse T Cell Repertoire. *J. Immunol.* 153:3326–3336.

167. Franco, A., S. Southwood, T. Arrhenius, V.K. Kuchroo, H.M. Grey, A. Sette, and G.Y. Ishioka. 1994. T Cell Receptor Antagonist Peptides are Highly Effective Inhibitors of Experimental Allergic Encephalomyelitis. *Eur. J. Immunol.* 24:940–946.

168. Nicholson, L.B., A. Murtaza, B.P. Hafler, A. Sette, and V.K. Kuchroo. 1997. A T cell receptor antagonist peptide induces T cells that mediate bystander suppression and prevent autoimmune encephalomyelitis induced with multiple myelin antigens. *Proc. Natl. Acad. Sci. USA* 94:9279–9284.

169. Brocke, S., K. Gijbels, M. Allegretta, I. Ferber, C. Piercy, T. Blankenstein, R. Martin, U. Utz, N. Karin, D. Mitchell, T. Veromaa, A. Waisman, A. Gaur, P. Conlon, N. Ling, P.J. Fairchild, D.C. Wraith, A. O'Garra, C.G. Fathman, and L. Steinman. 1996. Treatment of experimental encephalomyelitis with a peptide analogue of myelin basic protein. *Nature* 379:343–346.

170. Teitelbaum, D., A. Meshorer, T. Hirshfeld, R. Arnon, and M. Sela. 1971. Suppression of experimental allergic encephalomyelitis by a synthetic polypeptide. *Eur. J. Immunol.* 1:242–248.

171. Teitelbaum, D., C. Webb, A. Meshorer, R. Arnon, and M. Sela. 1973. Suppression by several synthetic polypeptides of experimental allergic encephalomyelitis induced in guinea pigs and rabbits with bovine and human basic encephalito-gen. *Eur. J. Immunol.* 3:273–279.

172. Teitelbaum, D., C. Webb, M. Bree, A. Meshorer, R. Arnon, and M. Sela. 1974. Suppression of experimental allergic encephalomyelitis in rhesus monkeys by a synthetic basic copolymer. *Clin. Immunol. Immunopathol.* 3:256–262.

173. Racke, M.K., R. Martin, H. McFarland, and R.B. Fritz. 1992. Copolymer-1 induced inhibition of antigen-specific T cell activation: interference with antigen presentation. *J. Neuroimmunol.* 37:75–84.

174. Ben-Nun, A., I. Mendel, R. Bakimer, M. Fridkis-Hareli, D. Teitelbaum, R. Arnon, M. Sela, and N. Kerlero de Rosbo. 1996. The autoimmune reactivity to myelin oligodendrocyte glycoprotein (MOG) in multiple sclerosis is potentially pathogenic: effect of copolymer 1 on MOG-induced disease. *J. Neurol.* 243 Suppl. 1:S14–S22.

175. Teitelbaum, D., M. Fridkis-Hareli, R. Arnon, and M. Sela. 1996. Copolymer 1 inhibits chronic relapsing experimental allergic encephalomyelitis induced by proteolipid protein (PLP) peptides in mice and interferes with PLP-specific T cell responses. *J. Neuroimmunol.* 64:209–217.

176. Aharoni, R., D. Teitelbaum, M. Sela, and R. Arnon. 1997. Copolymer 1 induces T cells of the T helper type 2 that crossreact with myelin basic protein and suppress experimental auotimmune encephalomyelitis. *Proc. Natl. Acad. Sci. USA* 94:10821–10826.

177. Arnon, R. 1996. The development of Cop-1 (Copaxone), an innovative drug for the treatment of multiple sclerosis. *Immunol. Lett.* 50:1–15.

178. Aharoni, R., D. Teitelbaum, M. Sela, and R. Arnon. 1998. Bystander suppression of experimental autoimmune encephalomyelitis by T cell lines and clones of the Th2 type induced by copolymer 1. *J. Neuroimmunol.* 91:135–146.

179. Wells, H.G. 1911. Studies on the chemistry of anaphylaxis (III). Experiments with isolated proteins, especially those of the hen's egg. *J. Infect. Dis.* 8:147–171.

180. Weiner, H.L. 1997. Oral tolerance: immune mechanisms and treatment of autoimmune diseases. *Immunol. Tod.* 18:335–343.

181. Whitacre, C.C., I.E. Gienapp, C.G. Orosz, and D.M. Bitar. 1991. Oral tolerance in experimental autoimmune encephalomyelitis. III. Evidence for clonal anergy. *J. Immunol.* 147:2155–63.

182. Weiner, H.L., G.A. Mackin, M. Matsui, E.J. Orav, S.J. Khoury, D.M. Dawson, and D.A. Hafler. 1993. Double-blind pilot trial of oral tolerization with myelin antigens in multiple sclerosis. *Science* 259:1321–1324.

183. Racke, M.K., A. Bonomo, D.E. Scott, B. Cannella, A. Levine, C.S. Raine, E.M. Shevach, and M. Rocken. 1994. Cytokine-induced immune deviation as a therapy for inflammatory autoimmune disease. *J. Exp. Med.* 180:1961–1966.

184. Shaw, M.K., J.B. Lorens, A. Dhawan, R. DalCanto, H.Y. Tse, A.B. Tran, C. Bonpane, S.L. Eswaran, S. Brocke, N. Sarvetnick, L. Steinman, G.P. Nolan, and C.G. Fathman. 1997. Local delivery of interleukin 4 by retrovirus-transduced T lymphocytes ameliorates experimental autoimmune encephalomyelitis. *J. Exp. Med.* 185:1711–1714.

185. Mathisen, P.M., M. Yu, J.M. Johnson, J.A. Drazba, and V.K. Tuohy. 1997. Treatment of experimental autoimmune encephalomyelitis with genetically modified memory T cells. *J. Exp. Med.* 186:159–164.

186. Croxford, J.L., K. Triantaphyllopoulos, O.I. Podhajcer, M. Feldmann, D. Baker, and Y. Chernajovsky. 1998. Cytokine gene therapy in experimental allergic encephalomyelitis by injection of plasmid DNA-cationic liposome complex into the central nervous system. *J. Immunol.* 160:5181–5187.

43 | Myasthenia Gravis

Alexander Marx

1. INTRODUCTION

Myasthenia gravis (MG) is a prototypic autoimmune disease that fulfills the strict criteria for an autoantibody-mediated disorder against a known target autoantigen, the AChR at the postsynaptic membrane of the neuromuscular junction (NMJ). First of all, immunization of several animal species with the presumed autoantigen, the AChR, can induce a disease called experimental autoimmune MG (EAMG) that is similar to human MG (1). Second, passive transfer of autoantibodies to experimental animals can induce MG symptoms (2). Third, antibodies can be identified at the presumed pathogenetically-relevant site of the disease, the NMJ and AChRs isolated from mysthenic muscle have IgG bound to the receptors (3). Furthermore, the patients often have a family history of autoimmune diseases, they respond to plasmapheresis, and autoantibodies can pass across the placenta from mother to child with the eventual consequence of congenital autoimmune MG or even intrauterine MG resulting in arthrogryposis multiples congenita (4). Knowledge about the molecular structure and structure-function relations of the main target autoantigen of MG, the AChR, is now far advanced (5,6), as the understanding of the structure and function of the NMJ (7). Anti-AChR autoantibodies impair neuromuscular transmission by complement-mediated destruction of the NMJ, by increasing AChR turnover following AChR cross-linking, and by interference with ion channel function (8). Anti-AChR-reactive T-cells of the CD4+ subset are also crucial for the pathogenesis of MG by providing help to anti-AChR antibody-producing B-cells. Given that CD8+ T-cells may be important for the pathogenesis of EAMG (9), a role of CD8+ T-cells with specificity for antigenic structures of the postsynaptic side of the NMJ has recently been considered to play a role in the maintenance of human MG. By direct cytotoxic attack on the NMJ, CD8+ cells may provide continuous release and supply of autoantigen (e.g. AChR) for the ongoing interaction of antigen presenting cells (APCs), autoreactive CD4+ T cells and B cells (10). A genetic predisposition to MG on the MHC level has long been recognized in a majority of patients, but other genetic prerequisites and etiologic mechanisms involved in the activation of anti-AChR T-cells are less well understood, with the exception of certain MG-inducing drugs like D-Penicillamine (11). Likewise, the exact mechanisms by which the heterogeneous thymic pathologies are involved in the pathogenesis of MG are only partially resolved, and what gives rise to the various thymic alterations is unknown (12). In thymic lymphofollicular hyperplasia (thymitis) there is good evidence that the autoimune process occurs completely inside the thymus and might be elicited there: AChRs expressed on thymic myoid cells are presented by APCs to CD4+, AChR-reactive T cells, which, in turn, provide help to B-cells that produce anti-AChR autoantibodies. By contrast, it appears that there is no active autoimmune process going on inside the vast majority of thymomas and in thymic atrophy. In particular, autoantibody production is virtually absent inside these thymic alterations. Instead, there is data from thymoma patients suggesting that abnormal microenvironments in neoplastic thymuses might result in the generation of insufficiently tolerized T cells. After export to the extratumorous immune system, thymoma-derived T cells might gradually replace the normally tolerant by a more autoimmunity-prone T cell repertoire (13,14). The stimuli that might ultimately trigger the activation of AChR-reactive T cells of the abnormal T-cell repertoires in thymoma patients are presently unknown (15). Although MG in patients with thymic atrophy shares some features with thymoma-associated MG (autoantibody profile; poor response to surgery), other features are different (HLA-association, age of onset). Therefore, it is presently unknown whether the generation of abnormal T cells inside atrophic thymuses is also involved in the pathogenesis of atrophy-associated MG or whether its pathogenesis is completely different. Whether still another pathogenesis has to be invoked for some or all cases of so-called seronegative MG is not clear but appears very likely (16,17). In spite of these uncertainties it is now safe to assume that MG is not one disease. Instead MG has to be considered the common symptom shared by a variety of pathogenically different disorders. In support of this notion, surgical as well as immunosuppressive therapies have different effects on the outcome of MG patients as far as MG symptoms are concerned. Antigen-specific immunomodulatory therapies, which are presently being tested in animal models, might overcome the varying degrees of therapeutic success and avoid the side effects of general immunosuppression. For further detailed reviews see also (7,18–20). This review does not deal with congenital MG caused by constitutively abnormal presynaptic ACh or vesicle biology, mutations of the endplate asymmetric form of acetylcholine esterase (AChE) or mutations/ deficiencies of AChR subunits as reviewed recently (21–24). For a detailed review of the Lambert-Eaton myasthenic syndrome, that is only shortly covered here, see also Lang (25,26), Sherer (27) and Dalmau (28).

2. CLINICAL FEATURES

Prevalence, incidence, and clinical findings. Myasthenia gravis is a rare disease with a prevalence of 12–18/100 000 and an annual incidence of 1.1–1.5/100 000 population (29,30). The apparent incidence is slightly rising with the increasing age of population and improved recognition and diagnosis (20). The condition usually presents after puberty and seldom remits spontaneously. In contrast, spontaneous remissions occur in up to 50% of MG cases with prepubertal onset (31). In addition, prepubertal MG is distinguished by a lack of gender preference and associated other autoimmune diseases, and a higher frequency of ocular, seronegative and clinically severe cases (31). In postpubertal cases, life-long therapy is often required, although thymectomy may be curative in a minority of cases (see below).

Myasthenia gravis patients suffer from muscle weakness and fatigue that mainly affect ocular, bulbar and proximal extremity muscles. In less than 5% of patients distal weakness may be more severe than proximal weakness (32). Weakness of respiratory muscles can be life-threatening (33). Weakness usually improves following rest or medical treatment with AChE inhibitors that prolong the action of ACh.

Ocular muscle weakness is the initial manifestation of MG in the majority of patients and ultimately occurs in at least 90% of myasthenics. Ocular symptoms are roughly equally distributed between ptosis and diplopia (34). Within six months of presentation, about 50% of patients develop generalized MG, and 75% of patients will have bulbar or extremity weakness within the first year. After 3 years of onset, only 6% of ocular MG patients develop generalized myasthenic symptoms. Rarely, patients may present with weakness of isolated muscles other than ocular muscles, including pure respiratory muscle failure (7,35).

Myasthenia gravis may concur with other autoimmune diseases at a higher frequency than would be expected by chance. Such significantly more common diseases in non-thymoma MG patient are Graves' disease (36), rheumatoid arthritis, lupus erythematosus and Hashimoto's thyroiditis (37). Other presumed autoimmune diseases occurring in association with MG are polymyositis, autoimmune hemolytic anemia, Wegener's granulomatosis, Crohn's disease, vitiligo, and alopecia areata (38). Paraneoplastic syndromes other than MG in thymoma patients are discussed in the paragraph about thymomas. The clinical findings in MG are reviewed in ref. 8.

Neonatal MG, Arthrogryposis multiplex congenita and MG during pregnancy. Neonatal MG refers to the development of weakness among infants born to myasthenic mothers, in contrast to congenital MG that is usually due to AChR or AChE mutations or presynaptic metabolic deficiencies (24,39). Neonatal weakness is transient in most cases. Neonatal MG does not correlate with the severity of maternal MG symptoms or autoantibody titers in the mother. Indeed, most newborns of MG mothers do not show MG symptoms, although they have maternally-derived antibodies (40,41). Instead, development of neonatal MG correlates with the titer of autoantibodies specific for the fetal AChR (42).

The most dramatic neonatal abnormality due to maternal anti-AChR autoimmunity is arthrogryposis multiplex congenita (AMC) (43). This pathogenetically heterogeneous syndrome is characterized by bizarre and often irreversible contractures of muscles and results from a failure of intrauterine muscle movement, be it due to various defects of the central nervous system, peripheral neuropathies or severe failure of intrauterine neuromuscular transmission. In case of MG-related AMC the mothers usually do not suffer from MG symptoms although the autoantibodies are produced by the maternal immune system. The explanation is that the bulk of maternal anti-AChR autoantibodies are specifically directed towards the fetal AChR. These antibodies pass the placenta and cause severe intrauterine MG while doing no harm to maternal NMJs. Remarkably, the mothers usually do not exhibit ocular MG. For a detailed review of neonatal MG, AMC and the interrelationship of MG and pregnancy see (4,16,41)

2.1. Diagnosis of MG

Although anti-AChR autoantibodies are a hallmark of autoimmune MG, such antibodies are detectable in only about 85% of patients for technical reasons or because of autoantibody affinities that are too low. Therefore, MG diagnosis rests on typical clinical findings, the effect of AChE inhibitors ("edrophonium test"), and electromyography. Nevertheless, serum testing for AChR autoantibodies is obligatory. As to the edrophonium test, improvement of ptosis or the strength of extraocular muscles are most easily evaluated by intravenous edrophonium injection, while improvement of function of other muscles is more difficult to assess (44,45). False-positive edrophonium tests may occur in Lambert-Eaton myasthenic syndrome, motor neuron disease, intracranial mass lesions, and rarely other processes (46). False-negative tests are quite frequent in MG patients, and repeated tests are recommended when MG is clinically considered.

Electromyography (EMG) (47) may show a decrement in the compound muscle action potential during repetitive stimulation at low frequency. Decremetal response of the compound muscle action potential is detectable in 74% of patients with generalized MG (46). Ocular myasthenics, despite a lack of generalized clinical symptoms, may demonstrate a decrement response, though at a lower frequency than generalized myasthenics (7). Single fiber EMG is the most sensitive test to detect neuromuscular transmission abnormalities, particularly in pure ocular MG (48). However, the pitfall of false positive findings has to be considered (47).

Anti-AChR autoantibodies are detected by radioimmunoassays. The standard assay uses human muscle derived from amputation specimens as an antigenic source. Since amputations are usually performed for ischemia resulting in partial denervation, the AChRs obtained are a mixture of adult and fetal isotypes. In addition, the TE671 rhabdomyosarcoma cell line may serve as a more standardized source of mainly fetal AChRs, although the TE671-based assays are usually a bit less sensitive and are not advantageous in seronegative cases (49–51). TE671-based assays may eventually be more sensitive when anti-AChR-γ-subunit antibodies contribute significantly to the autoantibody titer. This might be important in some neonatal MG cases (42) and in mothers of children with arthrogryposis multiplex congenita (43). In other patients, immunoassays that use cell lines engineered to produce high concentrations of adult AChRs may detect anti-AChR autoantibodies at higher concentrations than the above mentioned assays (52). In particular, autoantibodies can be detected with this "epsilon-rich" assay in up to 60% of pure

ocular myasthenic patients, whereas the conventional assay has a sensitivity of only 35% in this MG subgroup (52). This is not entirely surprising, given the observation that extraocular muscles are enriched in ε-subunit mRNA (52).

Recently, a flow-cytometry-based technique has been reported that detects IgG characteristic for "seronegative" myasthenia gravis patients (52a).

3. ANATOMY AND PHYSIOLOGY OF THE NMJ

The neuromuscular junction is the communication site between the nerve and muscle. The action potential that is propagated down the axon is transformed into a chemical signal, the release of acetylcholine (ACh) from nerve terminals of the unmyelinated terminal branches of each nerve fiber. ACh is stored within vesicles which fuse with the nerve terminal membrane upon depolarization-induced calcium influx via voltage-gated $Ca2^+$ channels. Vesicle content (ACh, ATP) is then released into the synaptic cleft at sites that lie directly opposite to the tops of the secondary synaptic folds of the postsynaptic muscle membrane (7). The depolarization response of the postsynaptic membrane to the release of the contents of one vesicle is called the miniature end plate potential (MEPP). The depolarization of the postsynaptic membrane induced at the endplate by all vesicles released following a nerve action potential is the endplate potential. The number of vesicles (about 50–300) triggering the endplate potential is the quantal content. The nerve terminal branches lie in depressions of the postsynaptic muscle called primary synaptic clefts. The width between the nerve terminal and the postsynaptic membrane is about 50 nm. The postsynaptic surface of the muscle membrane is increased by invaginations called secondary synaptic clefts or folds. AChRs are clustered at high concentration at the tops of the secondary synaptic folds, while Na-channels are concentrated at the depths (53) (Fig. 1a). Acetylcholine esterase (AchE) is primarily synthesized by muscle and located in the basal lamina of the secondary synaptic folds. AChE terminates the action of ACh and prevents repeated activation of AChRs (54).

A useful concept to understand neuromuscular transmission is the "safety factor" (7). The safety factor is defined as the ratio of the endplate plate potential to the difference between the membrane potential and the threshold potential for initiating an action potential (55). When the threshold potential is achieved, the action potential will initiate Ca^{2+} release from the sarcoplasmic reliculum and normal contraction will occur. Quantal release, AChR density, AChR conduction properties, and AChE activity influence the safety factor since they contribute to the endplate potential, while Na^+ channel density and the architecture of the

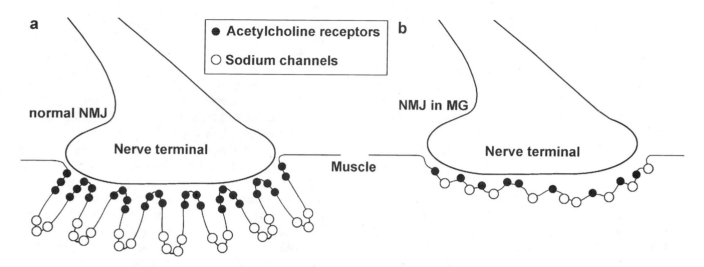

Figure 1. Neuromuscular junction in (a) the normal state and in (b) autoimmune myasthenia gravis (MG). (a) AChRs are concentrated on the tops of the secondary synaptic folds just opposite the nerve terminal. (b) Complement-mediated loss of postsynaptic membrane in MG results in widening of the synaptic cleft, simplification of synaptic folds and loss of AChRs and Na⁺-channels. All these alterations reduce the safety factor.

NMJ affects the action potential threshold (56,57). As described below, the reduction of both AChR and Na⁺ channel densities after complement-mediated destruction of the NMJ contribute to the transmission failure in MG (Fig. 1b). Folding of the postsynaptic membrane is mainly related to the expression of utrophin (58) that is modulated by heregulin/ARIA (59). Clustering of AChRs and AChR gene transcription depends on the nerve-derived 200 kD glycoprotein agrin that triggers of a signaling cascade downstream of the agrin receptor α-dystroglycan and muscle-specific tyrosine kinase receptors (60–62), resulting in localized gene transcription in subsynaptic nuclei (63). Furthermore, rapsyn is indispensible for clustering by anchorage of AChRs to the dystrophin-related complex (DRC) (Fig. 1) (64–66). However, other signaling cascades have yet to be discovered given that agrin knock-out mice have poor clustering of AChRs but not of other synapse-specific proteins (67).

The concentration of AChRs at the NMJ is about 15–20 000 AChRs per μm^2, while it is only 10–20 per μm^2 outside the NMJ (68). AChRs exhibit a continuous turnover with a half-life of 13 to 24 hours at fetal NMJs and of 8–11 days at mature endplates (7). Similar half lifes have been reported for denervated versus innervated muscles (69). Interestingly, the subunit composition by itself is not responsible for the different turnover times of the AChR isoforms (70). Acetylcholine esterase (AChE), which is synthesized by muscle and stored in the basal lamina of the secondary synaptic folds, terminates the action of ACh and prevents the repeated activation of AChRs (54).

4. CLINICAL HETEROGENEITY OF MG, OCULAR MG AND THE MOLECULAR HETEROGENEITY OF NMJS

The molecular basis for the heterogeneity of weakness among muscles is not well understood. Variations in ACh sensitivity have been taken as evidence for differences in the AChR channel properties of NMJs among muscle groups (7). Furthermore, the density of sodium channels and thus the safety factor of NMJs differs among muscles (71) and likely contributes to the heterogeneity of muscle weakness.

Even more enigmatic is the preferential weakness of extraocular muscles (EOM), causing double vision (72). As explanations have been suggested the following (73): (1) Minimal weakness of EOM may become symptomatic earlier than in other muscles.(2) EOM are anatomically and physiologically different from other muscle. 80% are single-innervated twitch fibers with "en grappe" nerve endings; 20% are "multiply innervated fibers" (MIFs) with en grappe endings some of which have a single "en plaque" ending. Since the twitch fibers work at high nerve firing frequency they may be more susceptible to fatigue.(3) The AChRs on EOM may be more accessible to circulating autoantibodies and may have a low safety factor.(4) The antigenic properties of the AChR in EOM may be different from those in other muscles (74), including the occurrence of fetal AChRs at the NMJs of MIFs as detected by immunohistochemistry in rat EOM (75). A role of fetal AChRs among the synaptic AChRs of EOMs has particularly been suggested by the observation that some myasthenic sera specifically target MIF synapses (76) and that sera of pure ocular MG patients react more strongly or become reactive at all when extraocular muscles are used as the source of antigen (34,77). On the other hand, there have been arguments against a role of fetal AChRs in the pathogenesis of ocular MG: (1) It has been suggested that human EOM may not have an increased content of fetal AChR compared to other human muscles but instead exhibit higher levels of ε-subunit message (52). This observation might be pathogenetically relevant given that some sera from ocular MG patients preferentially bind to adult type AChRs (52). (2) About 40% of patients with pure ocular MG have no detectable anti-AChR autoantibodies at all in conventional immunoprecipitation assays (73). (3) Mothers to children with immune-mediated blockade of fetal NMJs in utero (resulting in arthrogryposis multiplex, see above), can be completely asymptomatic although the fetal paralysis is mediated by maternal anti-fetal AChR autoantibodies passing through the placenta (16,43). Thus, despite high levels of anti-γ-subunit autoantibodies, the mothers do not suffer from ocular MG. (4) It has been proposed that the mechanism underlying ptosis in MG may be different from the mechanisms underlying double vision based on the observation that the Levator palpebrae muscle exhibits expression only of adult AChRs, while the other extraocular muscles express the fetal isoform in addition to the predominant adult AChR (34,78).

In summary, the susceptibility of EOMs and the Levator palpebrae muscle to MG cannot easily be related to the expression of fetal AChRs. Therefore, the other possible explanations for the preferential weakness of EOM enumerated above have to be considered. In addition, given that 40% of ocular MG patients remain "seronegative" irrespective of the immunoprecipitation assay applied, it can not be excluded that post-translational AChR modifications or non-AChR epitopes specific to EOM might be important in directing the antibody response preferentially towards ocular muscles.

5. AUTOANTIGENS IN MYASTHENIA GRAVIS

5.1. Acetylcholine Receptor

The essential autoantigen in almost all MG patients is the AChR. Only in some cases of seronegative MG may muscle surface molecules other than the AChR be the primary targets of pathogenic autoantibodies (16). The AChR is a pentameric glycoprotein that is composed of four different subunits and exists in two isoforms (5). The adult AChR is composed of two α-subunits and one copy of each of the β-, δ-, and ε-subunits. The fetal AChR has a γ-subunit in place of the ε-subunit (79). Both isoforms can

be distinguished by their single channel conductance and channel open times that are higher and shorter, respectively, in the adult receptor. Muscle fibers express both fetal and adult AChRs (with a predominance of fetal AChRs) before there is neuromuscular interaction and AChRs form clusters prior to innervation (80). Clustering of fetal and adult AChRs is, however, enhanced by, and in the vicinity of, the contact between a neurite and a muscle fiber, while suppression of γ-subunit expression and enhanced expression of the ε-subunit occurs later during the maturation of the endplate (63,81). After denervation there is overexpression of all AChR subunit genes, the γ-subunit is again expressed and fetal AChRs reappear at the endplate and all along the extrajunctional surface of the muscle (63,82). After long-term denervation (> 4 months), expression of AChR subunit genes (particularly the g-subunit) is partially downregulated again (83).

As an exception to these general rules, the "multiply innervated fibers" (MIFs), which are a minority among the innervated fibers of extraocular muscles, express both adult and fetal AChRs (52,84). The only other site of physiological fetal AChR expression in humans after birth appears to be the thymus, particularly the non-innervated myoid cells located in the thymic medulla (85; Geuder, 1990).

All AChR subunits share substantial homologies at the amino acid level, and appear to have their N-terminal and C-terminal regions in the extracellular space (rev. in 5–7). Each subunit contains four hydrophobic α helices (M1-M4) that span the membrane and the large extracellular loops between M2 and M3 of each subunit form the wall around the channel extracellular orifice. The regions between M1 and M2 and between M3 and M4 are involved in the formation of the vestibules that surround the four intracellular orifices of the ion channel. The ion channel is cation-selective with a low selectivity for cations that can pass due to a relatively wide minimal pore size of 6.5A (86). In the closed state, portions of the M1 and M2 helices form a gate that occludes the channel (87). The amino acid composition of M2 and the amount of positive charge in the segments flanking M2 influence the single channel conductance as far as fetal versus adult channels are concerned (Herlitze, 1996). Replacement within the M2 segment of a nonpolar alanine in the γ-subunit by a polar serine in the ε-subunit appears to decrease the channel open time of the adult AChR (88). In addition, the distinct compositions of the M4 helix and the intracellular linkers between the M3 and M4 helices influence channel open times of the AChR subtypes (89). Recently, high resolution imaging of Torpedo AChR tubular cristals revealed that tunnels within the extracellular wall of channel connect the water-filled vestibule of the receptor with the putative Ach binding sites, while openings between ä-helical segments of the cytoplasmic wall of the channel may allow lateral flow of permeant cations but serve as as filters to exclude anions and impermeant species from the pore (Fig. 1) (6).

Each α-subunit contains one ACh binding site located close to a disulfide bond formed between cystains at position $\alpha192$ and $\alpha193$. One of the binding sites is of low-affinity and located at the interface between the α and the δ-subunit, with a contribution of the δ-residues Trp57, Thr119, Asp180 and Glu189. The other binding site is of high-affinity and located between the α-subunit and either the γ- or ε-subunit. There is positive cooperativity for ACh binding and binding to both sites is required for channel opening to occur (87). α-bungarotoxin (αBTX), which competitively blocks AChR function, binds to a site close to, but distinct from, the ACh binding site on the N-terminal region of the α-subunit. The aBTX binding region includes as a minimum region $\alpha189$–195, but does not depend on the disulfide bond at $\alpha192$–193 (90,91). The binding of ACh to the AChR leads to a change in pore structure that allows ions to traverse the channel (88).

5.2. The Main Immunogenic Region and the VICE-α and VICE-β of the AChR

The main immunogenic region (MIR) of the AChR is the main target region to which autoantibodies of MG patients bind (92). In EAMG, antibodies to the MIR are induced when native AChR is used for immunization (93) and anti-MIR monoclonal antibodies are highly effective in passively transferring EAMG to experimental animals. The MIR is on the extracellular N-terminal segment of the AChR α-subunit and includes the sequence $\alpha67$ to $\alpha76$ in the close vicinity of the ACh binding site (94). The $\alpha67$–76 segments are located on the upper outer portion of the extracellular domain of the AChR at the extreme synaptic end of the α-subunits, exhibiting easy accessibility for autoantibodies (95). Using anti-MIR monoclonal antibodies, competition studies showed that the MIR is mainly restricted to a small region on the AChR α-subunit but clearly not a single epitope (5).

The injection of SDS-denatured AChR or single AChR subunits typically results in the production of antibodies to cytoplasmic sites of the AChR on various subunits (96–99). The main binding site on the α-subunit has been called the "Very Immunogenic Cytoplasmic Epitope alpha" or VICE-α, corresponding to the a373–380 sequence (96). Although autoantibodies and T-cells reactive against VICE-α have not yet been identified by some authors (100), a minority of MG patients appears to have anti-VICE-α antibodies, though at low titers (101) (see below). Nevertheless, VICE-α is potentially interesting because sequences highly homologous with to VICE-α occur repetitively in a 153 kd protein (equivalent to the 160 kD neurofilament) that is a

characteristic, abnormally expressed protein of MG-associated cortical thymomas (see below)(102,103). Cross-reactivity of the anti-VICE-α antibody mAb155 has also been reported for the striational muscle protein fast troponin I (104).

Another major epitope recognized after immunization with denatured AChR is the VICE-β epitope located at $\beta354$–359. This sequence is one of the phosporylation sites on the AChR (97). VICE-β is of potential importance in seronegative MG, since it is assumed, that autoantibodies in this MG subgroup may bind to non-AChR structures on the postsynaptic membrane which may indirectly affect AChR phosphorylation status and channel function (105). A comprehensive review of the AChR and its immunological features is given by Tzartos and coworkers (5).

5.3. Autoantigens other than the AChR in MG Patients

Autoantigens in addition to the AChR are important mainly in thymoma patients. These autoantigens are typically cytoplasmic muscle antigens, called strational antigens, such as α-actinin, actinin, myosin, titin (106–110) Other receptor autoantigens are the ryanodine receptor (important mainly in thymoma patients)(111,112) and the β-2-adrenergic receptor (113). Non-muscle autoantigens are neurofilaments (103) and the cytokines interferon-α and IL-12 (114), to which autoreactive T-cells (in the case of neurofilament) and neutralizing antibodies (in the case of IFN-α and IL-12) have been detected. The specific autoantigens (if any) responsible for the predominant susceptibility of extraocular muscle to develop weakness, have not yet been identified.

6. AUTOANTIBODIES IN MG

6.1. Autoantibodies to the AChR

Autoantibodies in MG are polyclonal, heterogeneous with respect to idiotypes, and recognize different epitopes on the AChR (77,115–121) and striational antigens (107,110,122). With few exceptions, differences in antigen specificity result from somatic hypermutation (123–126). The majority of autoantibodies (about 60–70%) directed to the α-subunit of the AChR recognize a region including the sequence a67–76. This region has been called the main immunogenic region (MIR) "(127) (see above). While this probably reflects an immunodominance of the MIR (11), certain arguments against this notion have been proposed (127a). Indeed, direct proof for the role of anti-MIR antibodies in humans is still awaited. Nevertheless, it is clear that anti-MIR autoantibodies are of utmost pathogenetic importance as far as the functional impairment of the NMJ is concerned. Using an *in vitro* AChR protection assay, Tzartos and coworkers demonstrated that univalent anti-MIR monoclonal antibody fragments efficiently prevented antigenic modulation induced by addition of MG sera to huma TE671 cells (128,129). In contrast, an antibody fragment reacting with the β-subunit provided only weak protection (5). From these findings it has been concluded that the very different pathogenic potential of MG patient sera may in part be related to the different titers of anti-α-subunit (MIR) antibodies (130).

Whether autoantibodies against the cytoplasmic region of the AChR occur in MG sera is controversial: While this has been denied by some authors as far as the a373–380 sequence is concerned (100), others reported on a significant minority of MG patients (13 and 9%, respectively) with autoantibodies inhibiting the binding of monoclonal antibodies to the VICE-α (a374–380) and VICE-β (b354–360) sequence by > 50% (101). The pathogenetic significance of such antibodies has not been resolved so far.

While the α-subunit appears to be the main autoantigen in most MG patients, there are non-thymoma patients with MG in whom the γ-subunit can be the dominant autoantigen (131,132). In contrast, sera in pure ocular MG were found to exhibit a greater reactivity with AChR extracted from innervated muscle compared with that from partially denervated muscle (133), suggesting a role of the epsilon subunit as an autoantibody target (52,73). In addition, the preferred reactivity of autoantibodies from ocular MG patients with extraocular muscle (133) and (ocular) multiple-innervated endplates (74) suggests the occurrence of autoantibodies with specificity for so far uncharacterized antigenic determinants on ocular muscle AChR (15,34,73,74).

To study human anti-AChR antibodies in more detail, hybridoma technology has been applied, to establish stable clones (124). However, the monoclonal human antibodies obtained did not bind to the MIR and were probably not representative of the normal repertoire. Therefore, two groups set out to produce cloned Fabs from thymuses of MG patients with lymphofollicular thymic hyperplasia (i.e., germinal center formation in the thymus; see below) using combinatorial libraries (126,134). The Fabs bound to human AChRs in immunoprecipitation assays with specificity either for the MIR (126,134) or the γ-subunit of fetal AChRs (126). These Fabs probably exhibit features that are more representative of the natural antibody repertoire, although their derivation from random combinations of heavy and light immunoglobulin chains suggests that they may not reflect the patient's *in vivo* antibodies. Nevertheless, these studies revealed the heterogeneity of germ-line origins of human autoantibodies and their usual derivation from an antigen-driven germinal center reactions with hypermutation of the CDR3 regions.

While most autoantibodies against the AChR and striational antigens are IgGs with high affinities, symptoms in seronegative MG" seem to be caused by low-affinity IgM autoantibodies (16,135,136) (see below). At least part of the autoantibody heterogeneity outlined here is thought to result from different pathogenetic mechanisms underlying tolerance breakdown in the various forms of MG (15).

6.2. Mechanisms of NMJ impairment by anti AChR Autoantibodies

Autoantibodies to the AChR produce the impairment of neuromuscular transmission by one of three different mechanisms (8). First, autoantibodies can block AChR function by binding to the ACh binding site, preventing opening of the ion channel. Such autoantibodies are rare in patients and, with few exceptions (137), contribute only marginally to the transmission failure at the NMJ. In animals, blocking antibodies can elicit what has been called hyperacute experimental autoimmune MG (EAMG) (138).

Second, autoantibodies can accelerate the degradation rate of AChR, thereby lowering the concentration of AChR at the postsynaptic membrane. This mechanism has been called antigenic modulation and results from the crosslinking of AChRs and subsequent increased internalization of receptor aggregates, reducing the half-life of the receptors. The presence of such autoantibodies appears to better correlate with the degree of clinical MG symptoms than the autoantibody titer (137).

Third, and most important, antibodies can cause complement-mediated NMJ destruction (3). Deposition of complement at the endplate is associated with a loss and simplification of junctional folds, widening of the synaptic cleft and reduction of both AChR-rich and Na^+ channel-rich areas (139) (Fig. 2). As a consequence, not only is AChR density reduced, but Na^+ channels are lost as well (140). The high concentration of voltage-gated sodium channels in the depth of the postsynaptic folds reduces the action potential threshold at the endplate and thus increases the safety factor for neuromuscular transmission by > 50% (57,140). Conversely, the increased action potential threshold in myasthenic muscle reduces the safety factor by 30% (140). Although destruction of the endplate is followed by increased presynaptic release of ACh (141) and increased transcription of adult-type AChR subunit genes (142), these mechanisms can not compensate for the functional deficit (143).

For a given patient, variations in serum anti-AChR antibody titer correlate well with clinical status. By contrast, studies of populations of MG patients show only a weak correlation between total anti-AChR autoantibody concentrations and disease severity (121,144). The reasons for this phenomenon is not yet clearly. Comparative studies on

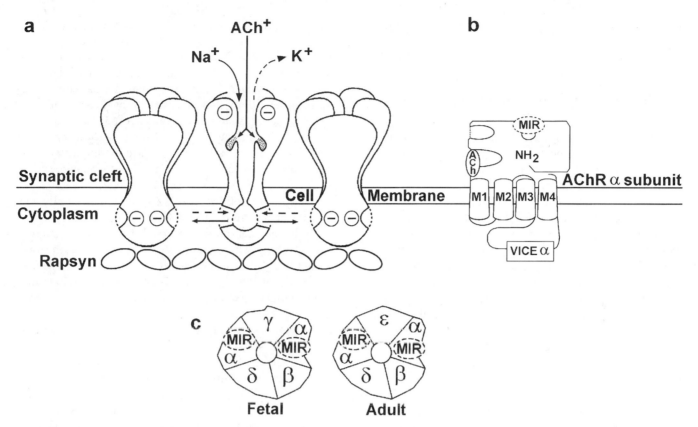

Figure 2. (a) Architecture of the postsynaptic membrane as proposed by Miyazawa et al (6). According to this model, ACh molecules pass through the central vestibule of the AChR channel before entering tunnels to their binding sites. Cations leave or enter the channel on the cytoplasmic side of the membrane through narrow openings in the channel wall. These narrow openings are framed by negatively charged residues excluding anions from the vicinity of the transmembrane pore. Rapsyn is now thought to be attached to the innermost end of the AChR but not to the lateral wall of the cytoplasmic portion of the channel as believed previously. (b) Diagram of the AChR α-subunit (16) depicting the four hydrophobic transmembrane domains (M1–M4), the ACh binding site (ACh), the "very immunogenic cytoplasmic epitope-α" (VICE α) and the "main immunogenic region" (MIR). (c) Diagram of the fetal and the adult AChR subtypes and the location of the MIR. The broken lines in (b) and (c) are to indicate that the MIR is not a single epitope. Instead, for different anti-MIR antibodies, different nearby or remote residues contribute to the whole MIR in addition to the critical α67–76 core epitope.

anti-fetal versus anti-adult AChR autoantibody titers have revealed, that the fetal/adult AChR antibody ratio appears to be quite constant over time in a given patient irrespective of titer variations, suggesting that the intrapersonal auto-antibody repertoire (i.e. B-cell/plasma cell repertoire) is relatively constant (42). In contrast, the remarkable hetero-geneity of autoantibody fine specificities among different patients has long been recognized, suggesting that certain antibody subtypes might be more myathenogenic than others. In support of this hypothesis, titers of autoantibodies with blocking activity or with anti-AChR α-subunit specifi-city have been reported to correlate better with clinical status than total anti-AChR antibody titers (rev. in 2,5,8). However, a contribution of antibody-independent factors, like the heterogeneity of individual safety factors or complement levels, cannot yet be excluded.

6.3. Antibodies to Antigens Other Than the AChR

Autoantibodies to non-AChR proteins are characteristic for MG patients with thymomas. Most important are auto-antibodies to other striational muscle antigens, including the ryanodine receptor (111), titin (106,110), myosin, actin, tropomyosin and α-actinin (107). While detection some of these autoantibodies has been used as a diagnostic tool, their pathogenetic significance is questionable given that their targets are intracellular proteins (20). Thymoma patients have also autoantibodies to non-muscle antigens. High levels of autoantibodies against IL-12 and IFN-α occur in the majority of MG/thymoma patients, but in less than 30% of other MG patients (114). Again, the pathogenetic significance of these autoantibodies has yet to be shown. Anti-axonal autoantibodies are also characteristic for MG/thymoma patients (145). Some of these autoantibodies appear to be directed against neurofilaments (103).

6.4. Autoantibodies in seronegative MG

Sera from seronegative MG patients can impair neuromusclular transmission and reduce AChR density when infused into mice, and the respective patients respond to plasma exchange (146,147). The responsible serum factors copurify with IgM, do not directly bind to AChRs but, nevertheless, reduce ACh-induced currents both in fetal AChR-expressing TE671 (rhabdomyosarcoma) cells and their transfected, ε-subunit rich derivatives (16,136,148). These effects depend on intracellular calcium and can be modulated by agents modifyinfg intracellulat c-AMP levels (16,149). Given that the function of AChRs depends on the phosphorylation status and the action of protein (tyrosine and other) kinases (150,151), it has been concluded that autoantibodies in seronegative patients may indirectly act on intracellular signaling cascades that modify AChR phosporylation (16). However, the extracellular targets of these antibodies at the neuromuscular junction have not yet been identified.

Recently, IgG against a non-AChR antigen expressed by the muscle cell line TE671 has been reported in "seronegative" MG patients (52a).

7. CELLULAR IMMUNITY IN MG

7.1. T-Cells in non-Thymoma MG Patients

The production of the disease-related autoantibodies in MG is dependent on MHC class II-restricted T cells (152–156). In particular, Th1 lymphocytes have been identified by proliferation assays, but Th2 cells thought to be involved in T cell help for antibody production have also been identified (157–159). For the role of cytokines in MG and EAMG see refs 160–162.

Using AChR peptide libraries to study T-cell reactivities, Conti-Fine and coworkers found that most non-thymoma patients with generalized MG recognized all AChR subunits (155,163). Only 6 of their 22 patients did not recognize all AChR subunits and MG in these patients was of recent onset. Only one of the 22 MG patients did not recognize the γ-subunit and another one was unreactive towards the α-subunit. Interestingly, the δ-subunit was recognized by these patients' T-cells significantly less than the α or γ-subunit. Considering that the non-neoplastic thymus is almost certainly of major importance for the triggering and maintainance of most early onset generalized MG cases, the low anti-δ-subunit response may be related to the recent observation that AChR δ-subunit mRNA and protein

are expressed at much lower levels than the α and γ-subunit on thymic myoid cells (164,165). With respect to the anti-γ-subunit T-cell response, it has been reported to correlate with ocular manifestations (163,166,167). Furthermore, CD4$^+$ T-cells of ocular MG patients recognized preferentially and consistently only γ-subunit, but virtually never ε-subunit, peptides (163). This is very surprising considering the observation that, on the B-cell level, the ε-subunit appears to be a preferential target of pure ocular myasthenics (52). In contrast to the high anti-γ-subunit T-cell responses, T-cell reactivities towards α, β, and δ-subunit peptides were variable and generally low 'in pure ocular MG, and CD4$^+$ responses were generally much lower that those observed for generalized MG patients (163). This finding is in aggreement with lower autoantibody titers in ocular MG patients (168). The heterogeneity of the T-cell AChR epitope specificities in a given patient and among different patients has been interpreted to indicate that there is no clearly immunodominant T-cell epitope in humans (163,169) (see below). In any case, the situation is clearly different from the situation in Lewis rats (170).

The relevance of investigations based on T-cell stimulation with small peptides, however, has been questioned (11,171) since it is generally assumed that processed native AChR stimulates the T-cells involved in directing autoantibody synthesis *in vivo* (15). Using recombinant mouse (172) and human AchR subunits, T-cell lines have been raised and the human epitopes have been mapped by synthetic peptides on the α-subunit (α146–160; α59-p3a-65) (2 cases of early onset MG and 1 thymoma) and the ε-subunit (ε201–219) (173–175) (3 cases of early onset MG). All the T-cell clones derived from early onset MG patients exhibited a TH1 cytokine profile. While the α142–160 peptide could efficiently be presented by both MHC class II DR4 and DR52a molecules, the ε201–219 peptide in all three different MG patients were recognized only in conjunction with the DR52a molecule (175). In addition, a clone with anti-ε201–219 reactivity has also been obtained from a patient with D-penicillamine-induced MG (20). This observation suggests that this type of MG might have a specific etiology (D-penicillamine), but a pathogenesis that is otherwise similar to that of early onset MG, involving "determinant spreading" and the processing of native AChR to finally result in a typical anti-AchR antibody response (see below). Importantly, all the clones reported by the Oxford group recognize native human AChR, demonstrating that the relevant peptide can be generated by AChR processing *in vivo*. Furthermore, the restricted epitope specificity of the clones from Oxford (175,176) as opposed to the broad peptide-reactivity of T-cells reported by Conti-Fine and coworkers (163) suggests that there are only few T-cells capable of recognizing epitopes naturally processed from the AChR protein. Therefore, these data have been interpreted by the Oxford group to indicate the existence of an immunodominant T-cell epitope on the AChR ε-subunit in humans with a new candidate susceptibility allele, DR52a (20). Clearly, much more clones need to be studied before definite conclusions can be drawn.

7.2. T-cells in thymoma patients

Data on autoreactive T-cells and their specificity are even more scarce in MG/thymoma patients. Using again the peptide library approach, Conti-Fine and coworkers (163) studied 3 generalized MG patients with thymoma. One of them recognized all AChR subunits strongly (as do the vast majority of patients with early onset MG), while the other two did not recognize the γ-subunit. These authors suggest that this might be related to the unique pathogenesis of paraneoplastic MG.

Using recombinant AChR fragments to raise T-cell clones and check their epitope specificity with synthetic peptides, Willcox and coworkers established two T-cell lines, both of TH0 type, one specific for α146–160 in the context of DR52a (the new susceptibility allee, the other specific for α75–90 in the context of DP14 (175,177). These authors also concluded that their findings support the concept that the pathogenesis of paraneoplastic and early onset MG might be different. Although these findings point to a major role of the AChR α-subunit as the relevant immunogen involved in the early pathogenesis of paraneoplastic MG, the preference of the autoantibodies for the fetal AChR is not only found in most MG patients in general, but also in MG/thymoma patients (176,178).

While the previous studies were based on relatively few patients and clones, our group has investigated a larger series of patients using various recombinant AChR α-subunit fragments and the recombinant 160kd neurofilament (NF) fragment NFM553–737 to study the antigen-specific proliferation of uncloned thymoma-derived thymocytes in short term primary cultures (103,179,180). The use of NFM was motivated by the observations that (1) various NF proteins are expressed in MG-associated thymomas, particularly of cortical type (181); (2) NFM mRNA is detectable in virtually all thymomas (181) and (3) NFM exhibits repetitive AChR-like epitopes (103) recognized by the anti-AChR antibody mAb155 raised against the VICE-α epitope α373–380 (5). We found that MG-associated thymomas were enriched for AChR-reactive T-cells compared with thymomas not associated with MG and with non-myasthenic control thymuses (103,179). These findings are in agreement with a previous study (182). As a new finding, we observed that a large recombinant fragment of the cytoplasmic sequence of the α-subunit (α301–398) containing the VICE-α sequence was also stimulatory for MG/ thymoma-derived T-cells (103), although we and others (100,177) were unable to detect

T-cells that recognize the VICE-α sequence in synthetic peptides. Furthermore, T-cell proliferative responses to the α301–398 sequence were not restricted to MG/thymoma patients but also found in patients with lymphofollicular hyperplasia (thymitis). In contrast, anti-NFM553–537 reactivity of T-cells was significantly restricted to MG/thymoma patients and particularly undetectable in MG patients with thymitis (103). Thus, NFM-reactive T-cells appear to be charcteristic for thymoma patients. Surprisingly, there was no obvious correlation between the intratumorous expression of NFM and the occurrence of NFM-reactive T-cells in thymoma patients and the relevance of NFM-reactive T-cells for the pathogenesis of paraneoplastic MG has yet to be show (103).

AChR-reactive T-cell repertoires in MG patients versus controls. Whether autoreactive T-cell repertoires in the various myasthenia subtypes are different is controversial (11,14,163,182–184). Most importantly, AChR-reactive T cells occur both in the majority of non-myasthenic controls (182,184,185) and a wide range of animal species (rev. in 186). These T cells, belonging to the normal T cell repertoire, are not anergized but are naive ("ignorant") (11), suggesting that MG does not result from a lack of AChR-specific T-cell tolerance. However, future studies have to exclude the existence of "MG-specific" autoaggressive T cells (i.e. of T cells absent from the normal repertoire) in terms of antigen specificity, cytokine profile or ability to provide B cell help.

Although experimental autoimmune MG (EAMG) has been elicited in the absence of a CD8/MHC class I interaction (186–188), the role of CD8+ T cells needs further study considering the strong association of non-thymoma related MG with MHC class I molecules B8 and B7 (Table 1) and the conflicting results in different EAMG models that report on either an aggravating or an attenuating role of CD8+ T-cells (9,188–190). In humans, there is preliminary data suggesting an immunosuppressive role of CD8+ T cells in MG with thymitis/hyperplasia (155,166), which appears not to be the case in thymoma-associated MG (personal observation).

8. GENETIC CONTRIBUTIONS TO THE PATHOGENESIS OF MG

Genetic factors must play an important role in the susceptibility to MG given the 40% concordance rate in pairs of monozygotic twins and the increased risk of MG and anti-AChR autoantibodies in relatives of MG patients (15). As with other autoimmune diseases, the genetic predisposition to MG probably involves multiple genes (191,192). Of these, the contribution of MHC loci is most obvious in non-

Table 1 MG subtypes according to thymus pathology:
Correlation with clinical and epidemiological findings[1]

	Hyperplasia	Thymoma	Atrophy
Onset of symptoms Age (years)	10–39	15–80	>40
Sex m : f	1 : 3	1 : 1	2 : 1
HLA-association	B8; DR3	(DR2)	B7; DR2
Autoantibodies against			
AChR	30–80%	>90%	90%
Striated muscle	10–20%	>90%	30–60%
Titin	<5%	>90%	30–40%

[1] see reference (14).
[2] WDTC = well differentiated thymic carcinoma (215).

thymoma patients (Table 1), but weaker associations of class II genes with MG have also been reported in patients with thymoma (193,194). Surprisingly, for a disease assumed to depend on MHC class II-restricted T cell help, the strongest association is with the class I molecule B8 (15). This may hint at an immunoregulatory role of CD8[+] T cells, as described above. Alternatively, other genes close to the class I locus, such as genes for TNF (195–197), heat shock proteins or TAP transporters, have been discussed with respect to the pathogenesis of MG (15,119,198). Of non-MHC related genes, a polymorphic marker on the switch region of the immunoglobulin heavy chain gene has been reported to be associated with late-onset MG in patients with thymic atrophy (119), suggesting a peculiar humoral immune response as crucial for the pathogenesis of this MG type. Significantly associated polymorphisms of immunoglobulin light chains with MG have been reported (199), as have polymorphisms of the Fc-gammaR IIA (also in thymoma) (200), IL-1 (201), IL-10 (202) and the Ctla-4 gene(particularly in thymoma) (203). How the polymorphism of other non-MHC related genes such as the AChR alpha subunit gene (191) may contribute to MG susceptibility is currently unknown. As to the polymorphism of the IL-4 gene, there appears to be no high-susceptibility group among MG patients (204). In contrast, it has been reported that IL-4 deficiency facilitates EAMG (205), although it is not entirely essential for progression (206).

9. NK-CELLS

Recently, an increased frequency of natural killer T-cells have been reported to occur in the blood of MG patients (207). These cells expressed the nonpolymorphic T-cell receptor α-chain V α24. These cells have been considered to be of potential pathogenetic significance, since they were found to express the proinflammatory cytokines IFN-γ and

IL-4, thought to be mainly essential in the initiation and maintenance of MG, respectively.

10. THE THYMUS IN MG

Thymic alterations are so frequent in MG (90%) that a role of the thymus in the pathogenesis of MG is almost certain. This is supported by the association between pathological changes of the thymus and clinical and epidemiological findings as given in Table 1.

10.1. Histopathology

The diagnosis of *thymic lympho-follicular hyperplasia (Thymitis)* is made in 70% of MG patients (208). Histologically, perivascular spaces (PVS) are extended by B cells forming follicles and germinal centers and the basal membrane become interrupted, which results in a fusion of both compartments (68,209). In the medulla, the numbers of CD11c-, HLA-DR-and CD1-positive dendritic cells are also increased (210). In contrast, myoid cells, a non-inner-vated and usually non-innervated muscle cell of unknown function occur in normal numbers exclusively in the medulla as in the normal thymus and only outside germinal centers (8,211). The only abnormality of myoid cells in MG thymuses is their close apposition to dendritic cells (209), a contact very scarce in normal thymuses. as in the normal thymus, myoid cells in thymitis are MHC class II-negative and express AChR (85,131,209). The thymus cortex in lympho-follicular thymitis shows the normal age-dependent morphology.

Thymitis in seronegative myasthenia gravis has been reported to be a separate entity (107). In this type of thymitis, germinal centers are rare and the number of B cells is almost normal. This type of thymitis results from an increase of mature T cells in extended perivascular spaces (PVS) (107), but whether it is accompanied by a breakdown of the barriers between the PVS and medulla has not been reported.

Thymus atrophy in myasthenia gravis is encountered in 10–20% of MG patients (208). Because of distinct epidemiological and genetic findings (Table 1) and a short course of disease, thymus atrophy is not considered an end stage after thymitis. Morphologically, except from a slight increase in medullary B cells and interdigitating reticulum cells (210), the thymuses in these patients are equivalent to age-matched controls. In particular, the number of myoid cells per thymic tissue area (measured morphometrically in sections) follows the same age-related decline (209).

Thymic epithelial tumors in MG Single cases of neurobalstoma, esophageal, thyroid and breast carcinoma, mature ovarian teratoma, chordoma, pheochromocytoma, lymphoma and Hodgkin's disease associated with MG have been described (14,38). Recently, paraneoplastic MG was also reported in patients with neuroendocrine carcinomas of the lung (atypical carcinoids) (212) and in a patient with a small cell lung cancer (SCLC) whose cells exhibited an exceptional expression of muscle-type AChR in addition to the more common neuronal AChR(213). Whether the association of these tumors and MG is a pathogenetically significant co-occurrence is uncertain, particularly with regard to patients with rhabdomyosarcoma, an immature skeletal muscle tumor that regularly expresses muscular AChRs(214) but is never associated with MG. In contrast, the overwhelming number of tumor-associated MG cases occur in organotypic thymic epithelial tumors (mainly with cortical features) (215) and it is undisputed that this association is of etiologic and pathogenetic significance (11,14,15,105,177,180). The histomorphology of these tumors and clinico-pathological correlations were recently published in detail (12,216a). The features of these tumors in relation to the pathogenesis of MG are discussed in detail below. Interestingly, purely non-organotypic thymic epithelial tumors like squamous cell carcinomas of the thymus resemble their extrathymic counterparts and are virtually never associated with MG (14,217).

11. PATHOGENETIC CONCEPTS IN MG

Pathogenetic models will be discussed here mainly with relation to the thymic alterations encountered in MG patients. However, there has been almost a complete lack of experimental data, and thus of pathogenetic models, for MG in patients with thymus atrophy or seronegative MG, in which the usefulness of thymectomy awaits further statistical support (115,116,176,218,219). Therefore, the main focus here will be on the pathogenesis of MG in lympho-follicular thymitis and thymic epithelial tumors. Since "determinant spreading" is probably an essential common theme in the different pathogenetic models, it will first follow a discussion of experimental autoimmune MG, in which determinant spreading can clearly been observed.

Experimental autoimmune myasthenia gravis (EAMG) and determinant spreading. Experimental autoimmune MG (EAMG) has been induced in rabbits, mice, rats, chicken, monkeys and sheep by injection of xenogeneic or syngeneic AChRs or AChR fragments, or by passive transfer of polyclonal or monoclonal anti-AChR antibodies (220). Immunization with AChR(fragments) induces AChR-specific T-cells and antibodies. Usually, electric fish AChR is applied. However, the inducibilty of EAMG by syngeneic adult AChR or by purified fetal AChR without adjuvant suggests that there is little tolerance towards self-AChR (221). In the acute form of

EAMG, muscle weakness develops around day 10, while in the chronic form it starts after three or more weeks (15). Remarkably, less than 5% of the antibodies against the fish AChR crossreact with the autologous muscle AChR (16). When non-denatured whole AChR of either xenogeneic, syngeneic or fetal origin is used to induce EAMG, a usually predominant fraction of autoantibodies is directed against the conformational-sensitive "main immunogenic region" (MIR) on the α-subunit, to which the sequence a67–76 makes a major contribution (94). Remarkably, competition experiments using monoclonal antibodies with defined fine specificities suggest that the antibody repertoire in rats immunized with whole native AChR is similar to that of MG patients (5). In contrast, when denatured AChR is used for immunization, antibodies react towards the cytoplasmic region of the α-ubunit. The preferential binding site there is called the "very immunogenic cytoplasmic epitope-α" (VICE-α), α373–380 (96). Since VICE-α is an intracellular epitope, antibodies to VICE-α do not induce EAMG. The T-cell response to fish AChR in EAMG is also skewed towards the α-subunit. There is preferred reactivity to a100–116 in Lewis rats (222), and to a146–162 in C57B1/6 mice (223,224). T-cells recognizing these sequences provide help to B-cells primed with the whole AChR and induce the production of anti-AChR antibodies in vitro and in vivo (225–228). Vice versa, induction of tolerance to a146–162 suppresses anti-AChR antibody and Th1 cytokine responses (229). Furthermore, in mice, whose MHC class II molecules do not bind a146–162, antibodies can still be induced, but are less pathogenic (230,231).

The essential role of CD4+ T cells in the pathogenesis of EAMG is clear (186). In contrast, the role of CD8+ cells is less so, since some authors found CD8+ T-cells to be essential for EAMG induction (9), while others found depletion of CD8+ T-cells to aggravate EAMG (188). Determinant spreading refers to the broadening of an immune response against many epitopes of one or more antigen(s) when the triggering immunogen has only a limited number of epitopes. When the immunogen is part of the target antigen, the process is called intramolecular determinant spreading. When the immune reaction spreads to unrelated target antigens (e.g. from an immunizing virus to an autoantigen with shared epitopes), the process is referred to as intermolecular determinant spreading and is now considered one of the pivotal mechanisms of autoimmunization (20,232). In EAMG, there is evidence for both intra- and intramolecular determinant spreading. When rabbits are immunized with peptides covering the human sequence α138–199, most of the animals initially produce non-myasthenogenic antibodies against the human peptide. However, later and after boosting with the human peptides, many animals start producing antibodies to native rabbit AChR epitopes and develop EAMG (20). Importantly, the myasthenogenic autoantibodies react more strongly with rabbit than with human AChR and a majority is directed against the rabbit MIR while they do not (cross-) react with the triggering α138–199. *Vice versa*, the anti-peptide antibodies do not crossreact with the native rabbit AChR and the anti-peptide antibody levels are not correlated with the anti-native AChR autoantibody levels. From these findings, it was concluded that immunization with the human sequence α138–199 led to autoimmunization to self-(rabbit) AChR (20).

The mechanisms of determinant spreading probably involve T-cell epitopes (Hohlfeld, 1990;) that are part of the peptide used for immunization (e.g. the sequence α146–162 within the α138–199 peptide). T-cells activated by a linear, MHC-presented xenogeneic (e.g. Torpedo α-subunit) AChR sequence may subsequently recognize the same or a homologous sequence processed from the autologous AChR, providing help to autoreactive B-cells with a diverse reactivity against various AChR subunits. The source of the autologous AChR for determinant spreading in EAMG is not clear so far. Autologous AChRs are probably not derived from the myoid cells of the thymus (see below), given the observation that immunization of rats with Torpedo AChR does not induce lymphofollicular thymic hyperplasia (thymitis), which is characteristic of early onset myasthenia in man (233). More likely, anti-Torpedo-peptide antibodies may induce an inital release of autologous AChR from the neuro-muscular junction after complement-mediated lysis. After onset of an intramuscular inflammation, cytotoxic T-cells may also be involved in muscular damage and thus in AChR release, given that inflammation inside muscle (as it occurs in EAMG and MG) can induce muscular MHC class I (and II) (234) making muscle cells become targets for CD8+ cytotoxic T-cells (10). Which antigen-presenting cells subsequently present MHC class II-bound autologous AChR peptides to CD4+ helper T-cells is not clear. MCH class II+ muscle cells presenting their endogenous AChR (235), muscle infiltrating macrophages (10,20) and dendritic cells in local lymph nodes (12,14) have all been considered. In spite of these unresolved questions, the finding that immunization with short linear peptides can induce a complex antibody response against conformation-dependent epitopes of the native AChR, clearly raises the possibility that cross-reacting T-cell epitopes could be involved in the pathogenesis of the autoantibody response in MG (12,20,102,152).

11.1.   The thymus and the Various Pathogenetic Pathways of MG

The definition of autoimmune MG is based on common clinical criteria concerning only the terminal stages of this typical antigen driven, T-cell dependent and antibody-mediated autoimmune disease. The defining criteria apply to all patients with autoimmune MG (though not to patients with congenital MG forms due to AChR mutations). However, it is now generally accepted that the heterogeneity that has been recognized for almost 20 years with respect to clinical symptoms, thymus pathology, immunogenetics and immunological findings in MG patients (11,236) is mirrored by a heterogeneity of quite different etiologies and pathogenetic pathways that terminate in the stereotypic MG symptoms. Some autoimmune, but seronegative, MG cases in which AChR function is disturbed, although the AChR is not the target autoantigen, might further augment pathogenetic hereterogeneity (52a,105).

Given the clinical and therapeutic importance of the various MG-associated thymic pathologies and their typical associations with immunogenetic and epidemiological findings (Table. 1), the different pathogenetic models will be discussed in relation to the associated thymic alterations. In addition, pathogenestic hypotheses concerning sero-negative MG will be discussed. For a discussion of drug-associated MG and MG following bone marrow transplantation see (11,237–239).

11.2.   Pathogenesis of MG in Thymic Lymphofollicular Hyperplasia (Thymitis)

Some authors consider lympho-follicular thymitis a secondary phenomenon following the sensitization of T cells in the periphery, recirculation to the thymus and restimulation there (198, 240). However, we and others favor a primary intrathymic pathogenesis of MG, as suggested by Wekerle almost 20 years ago (241). According to this hypothesis, AChR on thymic myoid cells are primarily involved in the triggering of MG in lympho-follicular thymitis. Three findings support this notion: 1) A substantial percentage of autoantibodies in thymitis-associated MG specifically recognize the fetal type of AChR (242); 2) fetal type AChR (i.e. AChR with a gamma instead of an epsilon subunit) are expressed on thymic myoid cells, but not on extrathymic muscle (85,242) except, probably, for multiple-innervated ocular muscles (84); 3) extrathymic immunization with the AChR can induce experimental autoimmune MG (EAMG) in animals, but does not elicit lympho-follicular thymitis (233). In support of this concept, Kirchner et al. (209) reported abnormal clusters of myoid cells and antigen-presenting dendritic cells in thymitis. Since myoid cells remain negative for MHC class II in MG and therefore are unable to present antigen to T cells (85,209), it is thought that the abnormal clustering enables dendritic cells to more efficiently take up AChR released from myoid cells. Processing of engulfed AChR in dendritic cells might result in a quantitatively improved presentation of AChR peptides to potentially AChR-specific T cells that have been found in increased numbers in thymuses with thymitis (182,183). Finally, the thymus with lympho-follicular thymitis is known to be the single most important organ where anti-AChR autoantibodies are produced, both in absolute terms and on a per plasma cell basis (243). Once produced, the autoantibodies may not only react with peripheral muscle AChR but also with AChR on thymic myoid cells. Whether such an antibody-mediated or a cytotoxic mechanism is the basis of the increased apoptosis of thymic myoid cells in MG (211) has yet to be investigated. As elegantly shown by the transplantation of thymitis specimens into SCID mice (resulting in the prolonged production of human anti-AChR autoantibodies in these immunodeficient mice), such thymuses contain all the necessary constituents of a complete and self-sustaining autoimmune reaction (244–246). The mechanisms eliciting lymphofollicular thymitis are not known (see below).

11.3.   Pathogenesis of Paraneoplastic MG

It is generally agreed that the pathogenesis of paraneoplastic MG differs from the pathogenesis of MG in lymphofollicular thymitis (12,14). The absence of an intratumorous autoantibody production is the most striking difference between thymoma and lymphofollicular thymitis (247). Different clinical, epidemiological and genetic features strengthen this statement (Table 1). Furthermore, the pathogenesis of paraneoplastic MG might be heterogeneous considering the heterogeneous morphological and functional findings in the various thymoma subtypes (208). Only about 20% of patients with paraneoplastic MG exhibit lympho-follicular thymitis in the residual thymus, while 80% show thymic atrophy (169). Nevertheless, MG-associated thymic epithelial tumors share common features: MG-associated Thymic Epithelial Tumors (TET) are organotypic, i.e. "thymus-like". Squamous cell carcinomas, carcinoids, and other (non-organotypic) category II malignant TET (12) are never associated with MG. Moreover, it is a striking observation that rhabdomyosarcomas that express large numbers of adult and fetal type AChR are not associated with MG. It can be concluded that the organotypic property of MG-associated TET to provide homing and maturation of immature T cells is an indispensable prerequisit of autoimmunization (14). In fact, we could show that MG-associated TET provide T cell development from the most immature precursors to phenotypically mature thymocytes (179,248).

Potentially myasthenogenic antigens occur in thymic epithelial tumors. Concurrent autoimmunity against three apparently unrelated types of autoantigens is highly characteristic of paraneoplastic MG: a) The AChR, b) the striated muscle antigen titin and c) neuronal antigens. Autoimmunity to the ryanodine receptor is also highly characteristic but less frequent (112,249). With respect to anti-AChR autoimmunity the occurrence of mRNAs of muscular and neuronal AChR has been reported (250–253), although, the respective *proteins* have not been detected in TET

(215,254,255). However, a recent study reports the detection of minute amounts of AChR protein in non-neoplastic thymic epithelial cells (256), suggesting that our inability to detect AChR protein in thymomas might be a problem related to the sensitivity of our method. Furthermore, we recently found a positive correlation between the expression of AChR α-subunit mRNAs and the occurrence of MG (164). On the other hand, Kirchner and coworkers (215) clearly demonstrated the abnormal expression of an AChR-likeepitope related to alpha 373–380 in cortical type TET and a highly significant correlation between the expression of this epitope and the presence of MG. In analogy to our findings, neither the ryanodine receptors nor titin protein have been detected in thymomas, but both titin mRNA (257) and ryanodine receptor mRNA (258) have been found in addition to epitopes of both autoantigens in unrelated proteins (181,249). In contrast, anti-neuronal autoimmunity in paraneoplastic MG (102) led to the detection of hypophosphorylated neurofilaments that are abnormally expressed in MG-associated TET (181). Even more striking was the finding that the medium molecular weight neurofilament expressed in TET contains epitopes equivalent to the AChR epitope alpha 373–380 and epitopes of titin (103). How the abnormal expression of a single molecule with crossreacting epitopes of the two most frequent autoantigens may evoke the simultaneous autoimmunity against the AChR, titin and neuronal antigens will be discussed below.

Autoaggressive T cells occur in MG-associated thymoma. Recently, data has been published on AChR-specific T cells in paraneoplastic MG (103,163,177,179, 185). In our own experiments, we always find better AChR-specific T cell responses in the peripheral blood and residual thymus than in thymomas (unpublished). It is also noteworthy that T cells against intracytoplasmic epitopes of the AChR-alpha subunit occur in MG-associated thymomas (103). However, T cells reactive with the AChR peptide alpha 373–380 (215) have not been detected (100), although the intratumorous expression of this epitope is associated with anti-AChR autoimmunity. Whether this is a paradox is explained in connection with the pathogenetic model given below. Another controversial question is whether intratumorous autoaggressive thymocytes are generated *in situ* or preferably activated there. By three-color-FACS, all MG-associated mixed or cortical-type TET investigated so far were devoid of CD4$^+$ T cells with an activated phenotype (CD25$^+$ CD54$^+$) while a single medullary thymoma contained activated T cells (179). In summary, TET subtypes highly associated with MG (mixed and cortical thymomas and well differentiated thymic carcinomas) seem to lack a substantial number of activated T cells.

A pathogenetic model of paraneoplastic myasthenia gravis. Despite the considerable progress outlined above, there is no unequivocal hypothesis for the pathogenesis of MG (11,12,176). In our opinion, the following findings await explanation by an appropriate pathogenetic model: 1) Paraneoplastic MG occurs only in organotypic TET that contain CD1+ *immature* T cells (208); 2) in cortical type TETs and probably mixed thymomas (unpublished), the occurrence of abnormally expressed proteins with AChR- and titin-like epitopes is associated with the autoimmunity against AChR, titin and neuronal structures (181,215); 3) AChR and neurofilament-reactive T cells occur in mixed and cortical thymomas and well differentiated thymic carcinomas (103,179,185), 4) activated mature T cells are absent in almost all MG-associated TET (248), and there is no intratumorous antibody production (247). Considering experimental evidence from mice for the involvement of endogenous protein in the process of positive selection (259,260), we favor the hypothesis that the aberrant expression of neurofilaments with AChR and titin epitopes in neoplastic epithelial cells might cause false-positive selection of immature T cells. In particular, we suggest that the intratumorous peptide homologues of the MG-associated AChR epitope alpha373–380 could function as one of the selecting peptides for *immature* T cells with prospective AChR specificity (145). The reduced expression of MHC class II expression that is characteristic of most MG-associated thymomas might further skew T-cell selection and disturb intratumorous tolerance induction (13,261). Since selecting peptides in the thymus are either non-stimulatory or antagonistic for *mature* T cells (259,260) this scenario would explain the apparent paradox that we and others (15,100) did not find T cell responses to alpha373–380 *in vitro*.

To become pathogenetically relevant, non-activated autoantigen-specific T cells have to be exported from thymoma to the "periphery," where they could provide help for autoantibody-producing B cells after adequate activation (38). This model implies that there should be a particular population of AChR-specific, autoaggressive T cells in thymoma patients that is essentially absent from the normal T cell repertoire. Thus, in our opinion, thymomas gradually replace a normally tolerant by an autoimmunity-prone T-cell repertoire (12,216a). Indeed, the occurrence of neurofilament-reactive T-cell occurring in MG/thymoma patients, but not controls, is a first hint that this hypothesis could be correct. The „periphery" where emigrant T-cells become activated can clearly be the residual thymus, which we found enriched in autoreactive T cells in most thymoma cases (unpublished). However, other lymphoid organs and probably the bone marrow also have to play a role in this process (14), because complete surgical removal of a thymoma together with the residual thymus is *not* followed by a decline of autoantibody titers (262).

It has not been elucidated which autoantigens maintain this prolonged autoantibody response after thymoma surgery, but the AChR itself is an obvious candidate (10,13,176). In analogy to the postulated release of AChR from thymic myoid cells (see above), the destruction of skeletal muscle endplates by autoantibodies or cytotoxic mechanisms could release AChR and striational antigens which may be processed and presented to autoreactive T cells by the intramuscular inflammatory infiltrate (10,263,264) or by antigen-presenting cells in regional lymph nodes (13,14).

11.4. Pathogenesis of MG in Thymic Atrophy

There have been few experimental data on the pathogenesis of MG associated with thymic atrophy that occurs in the late onset type of myasthenia gravis (265). Data on the heterogeneous occurrence of autoantibodies to striational antigens (107), particularly titin (265), also suggest a heterogeneity among late-onset MG patients. Pathogenetical models have not been suggested so far. Given that thymoma-associated MG and late onset MG associated with thymic atrophy share similarities with respect to striational autoantibodies (Table 1) and the usefulness of thymus surgery (107,266), it is tempting to speculate that some atrophic thymuses, in conjunction with genetic and environmental susceptibility factors, might contribute to the pathogenesis of MG by similar mechanisms as thymomas. Specifically, we speculate that in late onset MG the thymus might still contribute new T-cells to the T-cell repertoire. By this principally physiological mechanism (267), the newly generated but autoimmunity-prone T-cells may gradually replace the more or less tolerant "historic" T-cells (from the pre-atrophic era) (12). According to this model, activation of autoreactive T-cells should occur outside the atrophic thymus, in agreement with a recent histological study (265).

11.5. Pathogenesis of Seronegative MG

About 10–15% of patients with otherwise typical generalized MG do not have detectable antibodies using standard immunoprecipitation assays. Nevertheless, an autoimmune pathogenesis, possibly involving IgG and IgM, is likely, given the observations that (1) patients respond to plasma exchange and immunosuppression and that (2) *in vivo* injection of patient immunoglobulins can evoke MG-like effects at NMJs of experimental animal and human muscle fibers *in vitro* (16,52a,147). Using plasmas of seronegative MG patients, modulation of AChR function could be demonstrated in conventional and ε-transfected TE671 rhabdomyosarcoma cells that express fetal and mainly adult AChR, respectively (16,149). $22Na^+$ influx into TE671 cells is reduced by up to 85% by patient plasma (17), depending on intracellular calcium (149). Based on the earlier observation that AChR phosphorylation by cAMp-dependent protein kinase A (PKA) can reduce AChR function by increased desensitization (150), it was shown that PKA activity in TE671 cells can be stimulated by CGRP, salbutamol, dibutyryl cAMP and cholera toxin, all resulting in reduced AChR function (149). However, plasmas of seronegative patients have no effect on PKA activity in TE671-ε cells. Therefore, the plasma factors of seronegative MG patients probably act on a signalling pathway that does not depend on PKA. From these findings, it has been concluded that autoantibodies not directed against the AChR might act on other constituents of the post-synaptic membrane, which then affect AChR function by increasing its phosphorylation. (16,17,52a). Clearly, the targets at the NMJ have yet to be identified. Furthermore, the etiology and early pathogenesis await elucidation.

12. ETIOLOGY OF MG

While many steps in the pathogenesis of MG have been clarified (see above), the etiology, i.e. the disease trigger, has remained enigmatic except for D-Penicillamine-induced MG (11,237). Infections have long been thought to be such triggers, but how they could break T cell tolerance remains controversial. It has been suggested (268) that superantigens expressed by bacteria or viruses might unspecifically stimulate antigen presenting cells and unprimed AChR-specific T cells that occur in the normal human T cell repertoire. Autoimmunity in this situation may be elicited only in the context of a suitable genetic background, as suggested for multiple sclerosis. More popular has been the view that microbial antigens cross-reacting with self antigens may trigger autoreactivity (8,108). However, experimental evidence for such molecular mimicry is still lacking (see ref. 8). Cross-reactivity could happen both at the B and T cell level. Molecular mimicry on the B cell level concerning epitopes of the AChR or other autoantigens relevant in MG has been reported previously (11,104,109,249,269,270), but whether these epitopes are important *in vivo* has not been elucidated (15,104,198). In particular, it has not been demonstrated that B cells can elicit antigen-specific autoreactive T cell activation by shared B plus T cell epitopes, although B cells in mice can contribute to the diversification of immune responses (271). There is some experimental evidence that molecular mimicry on the T cell level could play a role in initiating autoreactivity even if only one T cell epitope is involved (153). This situation is now described as "determinant spreading", as described above (20,232). In the context of the human disease (MG), endogenous AChR could be released from peripheral skeletal muscle or thymic myoid cells as a consequence of either an inflammatory response in the vicinity of MG end-plates (263,264) or the abnormal attack of interdigitating dendritic cells on myoid cells in thymitis (209,211). The mechanisms of autoantigen release and autopresentation, however, are unknown (10,15). Once initiated, the process may be self-sustaining due to the constant release of endogenous autoantigen. Detecting the (triggering) events among the secondary effects will obviously be the challenge of future experiments and epidemiologic investigations addressing the etiology of MG.

References

1. Patrick, J., and J. Lindstrom. 1972. Autoimmune response to acetylcholine receptor. *Science* 180:187–971.
2. Toyka, K.V., D.B. Drachman, D.E. Griffin, A. Pestronk, J.A. Winkelstein, K.H. Fishbeck, and I. Kao. 1977. Myasthenia gravis. Study of humoral immune mechanisms by passive transfer to mice. *N. Engl. J. Med.* 291:125–131.
3. Engle, A.G., E.H. Lambert, and F.M. Howard. 1997. Immune complexes (IgG and C3) at the motor rastructural and light microscopic localization and *Clin. Proc.* 52:267–280.
4. Riemersma, S., A. Vincent, D. Beeson, C. Newland, S. Hawke, B. Vernet-der, B. Garabedian, B. Eymard, and J. Newsom-Davis. 1996. Association of arthrogryposis multiplex congenita with maternal antibodies inhibiting fetal acetylcholine receptor function. *J. Clin. Invest.* 98.2358–2363.
5. Tzartos, S.J., T. Burkus, M.T. Cung, A. Mamalaki, M. Marraud, P. Orlewski, D. Papanastasiou, C. Sakurellos, M. Sakarellos-Daitsiotis, P. Tsantili, and V. Tsikaris. 1998. Anatomy of the antigenic structure of a large membrane autoantigen, the muscle-type nicotinic acetylcholine receptor *Immunol. Rev.* 163:89–120.
6. Miyazawa, A., Y. Fujiyoshi, M. Stowell, and N. Unwin. 1999. Nicotinic acetylcholine receptor at 4.6 A resolution: transverse tunnels in the channel wall. *J. Mol. Biol.* 288:765–786.
7. Boonyapisit, K., H.J. Kaminski, and R.L. Ruff. 1999. Disorders of neuromuscular junction ion channels. *Am. J. Med.* 106:97–113.
8. Drachman, D.B. 1994. Myasthenia gravis. *N. Engl. J. Med.* 330:1797–1810.
9. Zhang, G.X., B.G. Xiao, M. Bakhiet, P. van der Meide, H. Wigzell, H. Link, and T. Olsson. 1996. Both CD4+ and CD8+ T cells are essential to induce experimental autoimmune myasthenia gravis. *J. Exp. Med.* 184:349–356.
10. Vincent, A., and N. Willcox. 1999. The role of T-cells in the initiation of autoantibody responses in thymoma patients [In Process Citation]. *Pathol. Res. Pract.* 195:535–540.
11. Willcox, N. 1993. Myasthenia gravis. *Curr. Opin. Immunol.* 5:910–917.
12. Marx, A., and H.K. Muller-Hermelink. 1999. From basic immunobiology to the upcoming WHO-classification of tumors of the thymus. The Second Conference on Biological and Clinical Aspects of Thymic Epithelial Tumors and related recent developments. *Pathol. Res. Pract.* 195:515–533.
13. Muller-Hermelink, H.K., A. Wilisch, A. Schultz, and A. Marx. 1997. Characterization of the human thymic microenvironment: Lymphoepithelial interaction in normal thymus and thymoma. *Arch. Histol. Cytol.* 60:9–28.

14. Marx, A., A. Wilisch, A. Schultz, S. Gattenlohner, R. Nenninger, and H.K. Muller-Hermelink. 1997. Pathogenesis of myasthenia gravis. *Virchows Arch.* 430:355–364.

15. Vincent, A. 1994. Aetiological factors in development of myasthenia gravis. *Adv. Neuroimmunol.* 4:355–371.

16. Vincent, A., L. Jacobson, P. Plested, A. Polizzi, T. Tang, S. Riemersma, C. Newland, S. Ghorazian, J. Farrar, C. McLennan, N. Willcox, D. Beeson, and J. Newsom-Davis. 1998. Antibodies affecting ion channel function in acquired neuromyotonia, in seropositive and seronegative myasthenia gravis, and in antibody-mediated arthrogryposis multiplex congenita. *Ann. NY Acad. Sci.* 841:482–496.

17. Plested, C.P., J. Newsom-Davis, and A. Vincent. 1998. Seronegative myasthenia plasmas and non-IgG fractions transiently inhibit nAChR function. *Ann. NY Acad. Sci.* 841:501–504.

18. Lewis R.A., J.F. Selwa, and R.P. Lisak. 1995. Myasthenia gravis: Immunological mechanisms and immunotherapy. *Ann. Neurol.* 37:S51–62.

19. Hoedemaekers, A.C., P.J. van Breda Vriesman, and M.H. De Baets. 1997. Myasthenia gravis as a prototype autoimmune receptor disease. *Immunol. Res.* 16:341–354.

20. Vincent, A., N. Willcox, M. Hill, J. Curnow, C. MacLennan, and D. Beeson. 1998. Determinant spreading and immune responses to acetylcholine receptors in myasthenia gravis. *Immunol. Rev.* 164:157–168.

21. Middleton, L.T. 1996. Congenital myasthenic syndromes. 34th ENMC International Workshop, 10–11 June 1995. *Neuromuscul. Disord.* 6:133–136.

22. Middleton, L.T., K. Christodoulou, F. Deymeer, P. Serdaroglu, C. Ozdemir, A.K. al-Qudah, A. al-Shehab, I. Mavromatis, I. Mylonas, A. Evoli, M. Tsingis, E. Zamba, and K. Kyriallis. 1998. Congenital myasthenic syndrome (CMS) type Ia. Clinical and genetic diversity. *Ann. NY Acad. Sci.* 841:157–166.

23. Engel, A.G. 1999. Congenital myasthenic syndromes. *J. Child. Neurol.* 14:38–41.

24. Beeson, D., C. Newland, R. Croxen, A. Buckel, F.Y. Li, C. Larsson, M. Tariq, A. Vincent, and J. Newsom-Davis. 1998. Congenital myasthenic syndromes. Studies of the AChR and other candidate genes. *Ann. NY Acad. Sci.* 841:181–183.

25. Lang, B., and J. Newsom-Davis. 1995. Immunopathology of the Lambert-Eaton myasthenic syndrome. *Springer Semin. Immunopathol.* 17:3–15.

26. Lang, B., S. Waterman, A. Pinto, D. Jones, F. Moss, J. Boot, P. Brust, M. Williams, K. Stauderman, M. Harpold, M. Motomura, J.W. Moll, A. Vincent, and J. Newsom-Davis. 1998. The role of autoantibodies in Lambert-Eaton myasthenic syndrome. *Ann. NY Acad. Sci.* 841:596–605.

27. Sherer, Y., and Y. Shoenfeld. 1999. A malignancy work-up in patients with cancer-associated (paraneoplastic) autoimmune diseases: pemphigus and myasthenic syndromes as cases in point (review). *Oncol. Rep.* 6:665–668.

28. Dalmau, J. 1999. Carcinoma associated paraneoplastic peripheral neuropathy. *J. Neurol. Neurosurg. Psychiatry* 67:4.

29. Incidence of myasthenia gravis in the Emilia-Romagna region: a prospective multicenter study. Emilia-Romagna Study Group on Clinical and Epidemiological Problems in Neurology. *Neurology* 1998; 51:255–258.

30. Robertson, N.P., J. Deans, and D.A. Compston. 1998. Myasthenia gravis: A population based epidemiological study in Cambridgeshire, England. *J. Neurol. Neurosurg. Psychiatry* 65:492–496.

31. Evoli, A., A.P. Batocchi, E. Bartoccioni, M.M. Lino, C. Minisci, and P. Tonali. 1998. Juvenile myasthenia gravis with prepubertal onset. *Neuromuscul Disord* 8:561–7.

32. Nations, S.P., G.I. Wolfe, A.A. Amato, C.E. Jackson, W.W. Bryan, and R.J. Barohn. 1999. Distal myasthenia gravis. *Neurology* 52:632–634.

33. Thomas, C.E., S.A. Mayer, Y. Gungor, R. Swarup, E.A. Webster, I. Chang, T.H. Brannagan, M.E. Fink, and L.P. Rowland. 1997. Myasthenic crisis: Clinical features, mortality, complications, and risk factors for prolonged intubation. *Neurology* 48:1253–1260.

34. Kaminski, H.J. 1998. Acetylcholine receptor epitopes in ocular myasthenia. *Ann. NY Acad. Sci.* 841:309–319.

35. Grob, D., E.L. Arsura, N.G. Brunner, and T. Namba. 1987. The course of myasthenia gravis and therapies affecting outcome. *Ann. NY Acad. Sci.* 505:472–499.

36. Jacobson, D.M. 1995. Acetylcholine receptor antibodies in patients with Graves' ophthalmopathy. *J. Neuroophthalmol.* 15:166–170.

37. Christensen, P.B., T.S. Jensen, I. Tsiropoulos, T. Sorensen, M. Kjaer, E. Hojer-Pedersen, M.J. Rasmussen, and E. Lehfeldt. 1995. Associated autoimmune diseases in myasthenia gravis. A population-based study [see comments]. *Acta Neurol. Scand.* 91:192–195.

38. Marx, A., A. Schultz, A. Wilisch, R. Nenninger, and H.K. Muller-Hermelink. 1996. Myasthenia gravis. *Verh. Dtsch. Ges. Pathol.* 80:116–126.

39. Engel, A.G., K. Ohno, and S.M. Sine. 1999. Congenital myasthenic syndromes: Recent advances. *Arch. Neurol.* 56:163–167.

40. Morel, E., B. Eymard, B. Vernet-der Garabedian, C. Pannier, O. Dulac, and J.F. Bach. 1988. Neonatal myasthenia gravis: a new clinical and immunologic appraisal on 30 cases. *Neurology* 38:138–142.

41. Batocchi, A.P., L. Majolini, A. Evoli, M.M. Lino, C. Minisci, and P. Tonali. 1999. Course and treatment of myasthenia gravis during pregnancy. *Neurology* 52:447–452.

42. Gardnerova, M., B. Eymard, E. Morel, M. Faltin, J. Zajac, O. Sadovsky, P. Tripon, M. Domergue, B. Vernet-der Garabedian, and J.F. Bach. 1997. The fetal/adult acetylcholine receptor antibody ratio in mothers with myasthenia gravis as a marker for transfer of the disease to the newborn. *Neurology* 48:50–54.

43. Vincent, A., C. Newland, L. Brueton, D. Beeson, S. Riemersma, S.M. Huson, and J. Newsom-Davis. 1995. Arthrogryposis multiplex congenita with maternal autoantibodies specific for a fetal antigen. *Lancet* 346:24–25.

44. Oh, S.J., and H.K. Cho. 1990. Edrophonium responsiveness not necessarily diagnostic of myasthenia gravis [see comments]. *Muscle Nerve* 13:187–191.

45. Newsom-Davis, J. 1990. Edrophonium responsiveness not necessarily diagnostic of myasthenia gravis [letter; comment]. *Muscle Nerve* 13:1186.

46. Oh, S.J., D.E. Kim, R. Kuruoglu, R.J. Bradley, and D. Dwyer. 1992. Diagnostic sensitivity of the laboratory tests in myasthenia gravis. *Muscle Nerve* 15:720–724.

47. Maselli, R.A. 1998. Electrodiagnosis of disorders of neuromuscular transmission. *Ann. NY Acad. Sci.* 841:696–711.

48. Durand, M.C., C. Goulon-Goeau, and P. Gajdos. 1997. Importance of neuromuscular "jitter" under stimulation in the diagnosis of myasthenia gravis. *Neurophysiol. Clin.* 27:471–482.

49. Somnier, F.E. 1994. Anti-acetylcholine receptor (AChR) antibodies measurement in myasthenia gravis: the use of cell

line TE671 as a source of AChR antigen [see comments]. *J. Neuroimmunol.* 51:63–68.

50. Voltz, R., R. Hohlfeld, A. Fateh-Moghadam, T.N. Witt, M. Wick, C. Reimers, B. Siegele, and H. Wekerle. 1991. Myasthenia gravis: Measurement of anti-AChR autoantibodies using cell line TE671. *Neurology* 41:1836–1838.

51. Kennel, P., J.T. Vilquin, P. Fonteneau, C. Tranchant, P. Poindron, and J.M. Warter. 1993. Value of TE671 cells in the detection of anti-acetylcholine receptor antibodies. *Rev. Neurol.* 149:771–775.

52. MacLennan, C., D. Beeson, A.M. Buijs, A. Vincent, and J. Newsom-Davis. 1997. Acetylcholine receptor expression in human extraocular muscles and their susceptibility to myasthenia gravis. *Ann. Neurol.* 41:423–431.

52a. Blaes, F., D. Beeson, P. Plested, B. Lang, and A. Vincent. 2000. IgG from "seronegative" myasthenia gravis patients binds to a muscle cell line, TE671, but not to human acetylcholine receptor. *Ann. Neurol.* 47:54–510.

53. Wood, S.J., and C.R. Slater. 1998. beta-Spectrin is colocalized with both voltage-gated sodium channels and ankyrinG at the adult rat neuromuscular junction. *J. Cell Biol.* 140:675–684.

54. Colquhoun, D., and B. Sakmann. 1985. Fast events in single-channel currents activated by acetylcholine and its analogues at the frog muscle end-plate. *J. Physiol.* (Lond.) 369:501–557.

55. Banker, B.Q., S.S. Kelly, and N. Robbins. 1983. Neuromuscular transmission and correlative morphology in young and old mice. *J. Physiol.* (Lond.) 339:355–377.

56. Slater, C.R., C. Young, S.J. Wood, G.S. Bewick, L.V. Anderson, P. Baxter, P.R. Fawcett, M. Roberts, L. Jacobson, J. Kuks, A. Vincent, and J. Newsom-Davis. 1997. Utrophin abundance is reduced at neuromuscular junctions of patients with both inherited and acquired acetylcholine receptor deficiencies. *Brain* 120:1513–1531.

57. Wood, S.J., and C.R. Slater. 1997. The contribution of postsynaptic folds to the safety factor for neuromuscular transmission in rat fast- and slow-twitch muscles. *J. Physiol.* (Lond.) 500:165–176.

58. Deconinck, A.E., A.C. Potter, J.M. Tinsley, S.J. Wood, R. Vater, C. Young, L. Metzinger, A. Vincent, C.R. Slater, and K.E. Davies. 1997. Postsynaptic abnormalities at the neuromuscular junctions of utrophin-deficient mice. *J. Cell Biol.* 136:883–894.

59. Gramolini A.O., L.M. Angus, L. Schaeffer, E.A. Burton, J.M. Tinsley, K.E. Davies, J.P. Changeux, and B.J. Jasmin. 1999. Induction of utrophin gene expression by heregulin in skeletal muscle cells: Role of the N-box motif and GA binding protein. *Proc. Natl. Acad. Sci. USA* 96:3223–3227.

60. Kleiman, R.J., and L.F. Reichardt. 1996. Testing the agrin hypothesis. *Cell* 85:461–464.

61. Jones, G., T. Meier, M. Lichtsteiner, V. Witzemann, B. Sakmann, and H.R. Brenner. 1997. Induction by agrin of ectopic and functional postsynaptic-like membrane in innervated muscle. *Proc. Natl. Acad. Sci. USA* 94:2654–2659.

62. Hesser, B.A., A. Sander, and V. Witzemann. 1999. Identification and characterization of a novel splice variant of MuSK. *FEBS Lett.* 442:133–137.

63. Brenner, H.R., V. Witzemann, and B. Sakmann. 1990. Imprinting of acetylcholine receptor messenger RNA accumulation in mammalian neuromuscular synapses. *Nature* 344:544–547.

64. Gautam, M., P.G. Noakes, J. Mudd, M. Nichol, G.C. Chu, J.R. Sanes, and J.P. Merlie. 1995. Failure of postsynaptic specialization to develop at neuromuscular junctions of rapsyn-deficient mice. *Nature* 377:232–236.

65. Cornish, T., J. Chi, S. Johnson, Y. Lu, and J.T. Campanelli. 1999. Globular domains of agrin are functional units that collaborate to induce acetylcholine receptor clustering. *J. Cell Sci.* 112:1213–1223.

66. Gautam, M., T.M. DeChiara, D.J. Glass, G.D. Yancopoulos, and J.R. Sanes. 1999. Distinct phenotypes of mutant mice lacking agrin, MuSK, or rapsyn. *Brain Res. Dev. Brain Res.* 114:171–178.

67. Gautam, M., P.G. Noakes, L. Moscoso, F. Rupp, R.H. Scheller, J.P. Merlie, and J.R. Sanes. 1996. Defective neuromuscular synaptogenesis in agrin-deficient mutant mice. *Cell* 85:525–535.

68. Land, B.R., E.E. Salpeter, and M.M. Salpeter. 1981. Kinetic parameters for acetylcholine interaction in intact neuromuscular junction. *Proc. Natl. Acad. Sci. USA* 78:7200–7204.

69. Xu, R., and M.M. Salpeter. 1999. Rate constants of acetylcholine receptor internalization and degradation in mouse muscles. *J. Cell Physiol.* 181:107–112.

70. Sala, C., M. Francolini, D. Di Mauro, and G. Fumagalli. 1998. Role of subunit composition in determining acetylcholine receptor degradation rates in rat myotubes. *Neurosci. Lett.* 256:1–4.

71. Ruff, R.L., and D. Whittlesey. 1993. Na+ currents near and away from endplates on human fast and slow twitch muscle fibers. *Muscle Nerve* 16:922–929.

72. Kaminski, H.J., E. Maas, P. Spiegel, and R.L. Ruff. 1990. Why are eye muscles frequently involved in myasthenia gravis? *Neurology* 40:1663–1669.

73. Newsom-Davis, J. 1997. Myasthenia gravis and the Miller-Fisher variant of Guillain-Barre syndrome. *Curr. Opin. Neurol.* 10:18–21.

74. Oda, K., and H. Shibasaki. 1988. Antigenic difference of acetylcholine receptor between single and multiple form endplates of human extraocular muscle. *Brain Res.* 449:337–340.

75. Kaminski, H.J., L.L. Kusner, and C.H. Block. 1996. Expression of acetylcholine receptor isoforms at extraocular muscle endplates. *Invest. Ophthalmol. Vis Sci* 37:345–351.

76. Oda, K. 1993. Differences in acetylcholine receptor-antibody interactions between extraocular and extremity muscle fibers. *Ann. NY Acad. Sci.* 681:238–255.

77. Vincent, A., and J. Newsom-Davis. 1982. Acetylcholine receptor antibody characteristics in myasthenia gravis. II. Patients with penicillamine-induced myasthenia or idiopathic myasthenia of recent onset. *Clin. Exp. Immunol.* 49:266–272.

78. Kaminski, H.J., L.L. Kusner, K.V. Nash, and R.L. Ruff. 1995. The gamma-subunit of the acetylcholine receptor is not expressed in the levator palpebrae superioris. *Neurology* 45:516–518.

79. Mishina, M., T. Takai, K. Imoto, M. Noda, T. Takahashi, S. Numa, C. Methfessel, and B. Sakmann. 1986. Molecular distinction between fetal and adult forms of muscle acetylcholine receptor. *Nature* 321:406–411.

80. Brehm, P. 1989. Resolving the structural basis for developmental changes in muscle ACh receptor function: it takes nerve. *Trends neurosci.* 12:174–177.

81. Witzemann, V., B. Barg, M. Criado, E. Stein and B. Sakmann. 1989. Developmental regulation of five subunit specific mRNAs encoding acetylcholine receptor subtypes in rat muscle. *FEBS Lett.* 242:419–424.

82. Kues, W.A., H.R. Brenner, B. Sakmann, and V. Witzemann. 1995. Local neurotrophic repression of gene transcripts encoding fetal AChRs at rat neuromuscular synapses. *J. Cell. Biol.* 130:949–957.

83. Adams, L., B.M. Carlson, L. Henderson, and D. Goldman. 1995. Adaptation of nicotinic acetylcholine receptor, myogenin, and MRF4 gene expression to long-term muscle denervation. *J. Cell Biol.* 131:1341–1349.

84. Horton, R.M., A.A. Manfredi, and B.M. Conti-Tronconi. 1993. The "embryonic" gamma subunit of the nicotinic acetylcholine receptor is expressed in adult extraocular muscle. *Neurology* 43:983–986.

85. Schluep, M., N. Willcox, A. Vincent, G.K. Dhoot, and J. Newsom-Davis. 1987. Acetylcholine receptors in human thymic myoid cells in situ: an immunohistological study. *Ann. Neurol.* 22:212–222.

86. Ruff, R.L. 1986. Ionic channels: I. The biophysical basis for ion passage and channel gating. *Muscle Nerve* 9:675–699.

87. Karlin, A., and M.H. Akabas. 1995. Toward a structural basis for the function of nicotinic acetylcholine receptors and their cousins. *Neuron* 15:1231–44.

88. Unwin, N. 1995. Acetylcholine receptor channel imaged in the open state. *Nature* 373:37–43.

89. Bouzat, C., N. Bren, and S.M. Sine. 1994. Structural basis of the different gating kinetics of fetal and adult acetylcholine receptors. *Neuron* 13:1395–1402.

90. McLane, K.E., X.D. Wu, R. Schoepfer, J.M. Lindstrom, and B.M. Conti-Tronconi. 1991. Identification of sequence segments forming the alpha-bungarotoxin binding sites on two nicotinic acetylcholine receptor alpha subunits from the avian brain. *J. Biol. Chem.* 266:15230–15239.

91. McLaughlin, J.T., E. Hawrot, and G. Yellen. 1995. Covalent modification of engineered cysteines in the nicotinic acetylcholine receptor agonist-binding domain inhibits receptor activation. *Biochem. J.* 310:765–769.

92. Tzartos, S., L. Langeberg, S. Hochschwender, and J. Lindstrom. 1983. Demonstration of a main immunogenic region on acetylcholine receptors from human muscle using monoclonal antibodies to human receptor. *FEBS Lett.* 158:116–118.

93. Tzartos, S.J., P. Tsantili, D. Papanastasiou, and A. Mamalaki. 1998. Construction of single-chain Fv fragments of anti-MIR monoclonal antibodies. *Ann. NY Acad. Sci.* 841:475–477.

94. Tzartos, S.J., A. Kokla, S.L. Walgrave, and B.M. Conti-Tronconi. 1988. Localization of the main immunogenic region of human muscle acetylcholine receptor to residues 67–76 of the alpha subunit. *Proc. Natl. Acad. Sci. USA* 85:2899–2903.

95. Beroukhim, R., and N. Unwin. 1995. Three-dimensional location of the main immunogenic region of the acetylcholine receptor. *Neuron* 15:323–331.

96. Tzartos, S.J., and M.S. Remoundos. 1992. Precise epitope mapping of monoclonal antibodies to the cytoplasmic side of the acetylcholine receptor alpha subunit. Dissecting a potentially myasthenogenic epitope. *Eur. J. Biochem.* 207:915–922.

97. Tzartos, S.J., C. Valcana, R. Kouvatsou, and A. Kokla. 1993. The tyrosine phosphorylation site of the acetylcholine receptor beta subunit is located in a highly immunogenic epitope implicated in channel function: antibody probes for beta subunit phosphorylation and function. *Embo. J.* 12:5141–5149.

98. Tzartos, S.J., E. Tzartos, and J.S. Tzartos. 1995. Monoclonal antibodies against the acetylcholine receptor gamma-subunit as site specific probes for receptor tyrosine phosphorylation. *FEBS Lett.* 363:195–198.

99. Tzartos, S.J., R. Kouvatsou, and E. Tzartos. 1995. Monoclonal antibodies as site-specific probes for the acetylcholine-receptor delta-subunit tyrosine and serine phosphorylation sites. *Eur. J. Biochem.* 228:463–472.

100. Nagvekar, N., L.W. Jacobson, N. Willcox, and A. Vincent. 1998. Epitopes expressed in myasthenia gravis (MG) thymomas are not recognized by patients' T cells or autoantibodies. *Clin. Exp. Immunol.* 112:17–20.

101. Tzartos, S.J., and M. Remoundos. 1999. Detection of antibodies directed against the cytoplasmic region of the human acetylcholine receptor in sera from myasthenia gravis patients. *Clin. Exp. Immunol.* 116:146–152.

102. Marx, A., R. O'Connor, K.I. Geuder, F. Hoppe, B. Schalke, S. Tzartos, I. Kalies, T. Kirchner, and H.K. Muller-Hermelink. 1990. Characterization of a protein with an acetylcholine receptor epitope from myasthenia gravis-associated thymomas. *Lab. Invest.* 62:279–286.

103. Schultz, A., V. Hoffacker, A. Wilisch, W. Nix, R. Gold, B. Schalke, S. Tzartos, H.K. Muller-Hermelink, and A. Marx. 1999. Neurofilament is an autoantigenic determinant in myasthenia gravis. *Ann. Neurol.* 46:167–175.

104. Osborn, M., A. Marx, T. Kirchner, S.J. Tzartos, U. Plessman, and K. Weber. 1992. A shared epitope in the acetylcholine receptor-alpha subunit and fast troponin I of skeletal muscle. Is it important for myasthenia gravis? *Am. J. Pathol.* 140:1215–1223.

105. Vincent, A. 1999. Antibodies to ion channels in paraneoplastic disorders. *Brain Pathol.* 9:285–291.

106. Aarli, J.A., K. Stefansson, L.S. Marton, and R.L. Wollmann. 1990. Patients with myasthenia gravis and thymoma have in their sera IgG autoantibodies against titin. *Clin. Exp. Immunol.* 82:284–288.

107. Williams, C.L., J.E. Hay, T.W. Huiatt, and V.A. Lennon. 1992. Paraneoplastic IgG striational autoantibodies produced by clonal thymic B cells and in serum of patients with myasthenia gravis and thymoma react with titin. *Lab. Invest.* 66:331–336.

108. Gautel, M., A. Lakey, D.P. Barlow, Z. Holmes, S. Scales, K. Leonard, S. Labeit, A. Mygland, N.E. Gilhus, and J.A. Aarli. 1993. Titin antibodies in myasthenia gravis: identification of a major immunogenic region of titin. *Neurology* 43:1581–1585.

109. Mohan, S., R.J. Barohn, and K.A. Krolick. 1992. Unexpected cross-reactivity between myosin and a main immunogenic region (MIR) of the acetylcholine receptor by antisera obtained from myasthenia gravis patients. *Clin. Immunol. Immunopathol.* 64:218–226.

110. Aarli, J.A., G.O. Skeie, A. Mygland, and N.E. Gilhus. 1998. Muscle striation antibodies in myasthenia gravis. Diagnostic and functional significance. *Ann. NY Acad. Sci.* 841:505–515.

111. Mygland, A., O.B. Tysnes, R. Matre, P. Volpe, J.A. Aarli, and N.E. Gilhus. 1992. Ryanodine receptor autoantibodies in myasthenia gravis patients with a thymoma. *Ann. Neurol.* 32:589–591.

112. Mygland, A., G. Kuwajima, K. Mikoshiba, O.B. Tysnes, J.A. Aarli, and N.E. Gilhus. 1995. Thymomas express epitopes shared by the ryanodine receptor. *J. Neuroimmunol.* 62:79–83.

113. Yi, Q., W. He, G. Matell, R. Pirskanen, Y. Magnusson, H. Eng, and A.K. Lefvert. 1996. T and B lymphocytes reacting with the extracellular loop of the beta 2-adrenergic receptor (beta 2AR) are present in the peripheral blood of

patients with myasthenia gravis. *Clin. Exp. Immunol.* 103:133–140.

114. Meager, A., A. Vincent, J. Newsom-Davis, and N. Willcox. 1997. Spontaneous neutralising antibodies to interferon-alpha and interleukin-12 in thymoma-associated autoimmune disease. *Lancet* 350:1596–1597.

115. Andrews, P.I., J.M. Massey, and D.B. Sanders. 1993. Acetylcholine receptor antibodies in juvenile myasthenia gravis. *Neurology* 43:977–982.

116. Andrews, P.I., J.M. Massey, J.F. Howard, Jr., and D.B. Sanders. 1994. Race, sex, and puberty influence onset, severity, and outcome in juvenile myasthenia gravis. *Neurology* 44:1208–1214.

117. Balass, M., Y. Heldman, S. Cabilly, D. Givol, E. Katchalski-Katzir, and S. Fuchs. 1993. Identification of a hexapeptide that mimics a conformation-dependent binding site of acetylcholine receptor by use of a phage-epitope library. *Proc. Natl. Acad. Sci. USA* 90:10638–10642.

118. Chiu, H.C., A. Vincent, J. Newsom-Davis, K.H. Hsieh, and T. Hung. 1987. Myasthenia gravis: population differences in disease expression and acetylcholine receptor antibody titers between Chinese and Caucasians. *Neurology* 37:1854–1857

119. Demaine, A., N. Willcox, M. Janer, K. Welsh, and J. Newsom-Davis. 1992. Immunoglobulin heavy chain gene associations in myasthenia gravis: new evidence for disease heterogeneity. *J. Neurol.* 239:53–56.

120. Vincent, A., and J. Newsom-Davis. 1985. Acetylcholine receptor antibody characteristics in myasthenia gravis. III. Patients with low anti-AChR antibody levels. *Clin. Exp. Immunol.* 60:631–636.

121. Vincent, A., and J. Newsom-Davis. 1985. Acetylcholine receptor antibody as a diagnostic test for myasthenia gravis: results in 153 validated cases and 2967 diagnostic assays. *J. Neurol. Neurosurg. Psychiatry* 48:1246–1252.

122. Skeie, G.O., A. Mygland, J.A. Aarli, and N.E. Gilhus. 1995. Titin antibodies in patients with late onset myasthenia gravis: Clinical correlations. *Autoimmunity* 20:99–104.

123. Victor, K.D., V. Pascual, C.L. Williams, V.A. Lennon, and J.D. Capra. 1992. Human monoclonal striational autoantibodies isolated from thymic B lymphocytes of patients with myasthenia gravis use VH and VL gene segments associated with the autoimmune repertoire. *Eur. J. Immunol.* 22:2231–2236.

124. Cardona, A., H.J. Garchon, B. Vernet-der-Garabedian, E. Morel, P. Gajdos, and J.F. Bach. 1994. Human IgG monoclonal autoantibodies against muscle acetylcholine receptor: Direct evidence for clonal heterogeneity of the antiself humoral response in myasthenia gravis. *J. Neuroimmunol.* 53:9–16.

125. Graus, Y., F. Meng, A. Vincent, P. van Breda Vriesman, and M. de Baets. 1995. Sequence analysis of anti-AChR antibodies in experimental autoimmune myasthenia gravis. *J. Immunol.* 154:6382–6396.

126. Farrar, J., S. Portolano, N. Willcox, A. Vincent, L. Jacobson, J. Newsom-Davis, B. Rapoport, and S.M. McLachlan. 1997. Diverse Fab specific for acetylcholine receptor epitopes from a myasthenia gravis thymus combinatorial library. *Int. Immunol.* 9:1311–1318.

127. Mamalaki, A., and S.J. Tzartos. 1994. Nicotinic acetylcholine receptor: structure, function and main immunogenic region. *Adv. Neuroimmunol.* 4:339–354.

127a.Lennon, V.A., and G.E. Griesmann. 1989. Evidence against acetylcholine receptor having a main immunogenic region as target for autoantibodies in myasthenia gravis. *Neurology* 39:1069-1076.

128. Tsantili, P., S.J. Tzartos, and A. Mamalaki. 1999. High affinity single-chain Fv antibody fragments protecting the human nicotinic acetylcholine receptor. *J. Neuroimmunol.* 94:15–27.

129. Papanastasiou, D., A. Mamalaki, E. Eliopoulos, K. Poulas, C. Liolitsas, and S.J. Tzartos. 1999. Construction and characterization of a humanized single chain Fv antibody fragment against the main immunogenic region of the acetylcholine receptor. *J. Neuroimmunol.* 94:182–95.

130. Loutrari, H., A. Kokla, N. Trakas, and S.J. Tzartos. 1997. Expression of human-Torpedo hybrid acetylcholine receptor (AChR) for analysing the subunit specificity of antibodies in sera from patients with myasthenia gravis (MG). *Clin. Exp. Immunol.* 109:538–546.

131. Geuder, K.I., A. Marx, V. Witzemann, B. Schalke, K. Toyka, T. Kirchner, and H.K. Muller-Hermelink. 1992. Pathogenetic significance of fetal-type acetylcholine receptors on thymic myoid cells in myasthenia gravis. *Dev. Immunol.* 2:69 75.

132. Weinberg, C.B., and Z.W. Hall. 1979. Antibodies from patients with myasthenia gravis recognize determinants unique to extrajunctional acetylcholine receptors. *Proc. Natl. Acad. Sci. USA* 76:504–508.

133. Vincent, A., and J. Newsom-Davis. 1982. Acetylcholine receptor antibody characteristics in myasthenia gravis. I. Patients with generalized myasthenia or disease restricted to ocular muscles. *Clin. Exp. Immunol.* 49:257–265.

134. Graus, Y.F., M.H. de Baets, P.W. Parren, S. Berrih-Aknin, J. Wokke, P.J. van Breda Vriesman, and D.R. Burton. 1997. Human anti-nicotinic acetylcholine receptor recombinant Fab fragments isolated from thymus-derived phage display libraries from myasthenia gravis patients reflect predominant specificities in serum and block the action of pathogenic serum antibodies. *J. Immunol.* 158:1919–1929.

135. Vincent, A., Z. Li, A. Hart, R. Barrett-Jolley, T. Yamamoto, J. Burges, D. Wray, N. Byrne, P. Molenaar, and J. Newsom-Davis. 1993. Seronegative myasthenia gravis. Evidence for plasma factor(s) interfering with acetylcholine receptor function. *Ann. NY Acad. Sci.* 681:529–638.

136. Yamamoto, T., A. Vincent, T.A. Ciulla, B. Lang, I. Johnston, and J. Newsom-Davis. 1991. Seronegative myasthenia gravis: A plasma factor inhibiting agonist-induced acetylcholine receptor function copurifies with IgM. *Ann. Neurol.* 30:550–557.

137. Schonbeck, S., S. Chrestel, and R. Hohlfeld. 1990. Myasthenia gravis: prototype of the antireceptor autoimmune diseases. *Int. Rev. Neurobiol.* 32:175–200.

138. Mihovilovic, M., D. Donnelly-Roberts, D.P. Richman, and M. Martinez-Carrion. 1994. Pathogenesis of hyperacute experimental autoimmune myasthenia gravis. Acetylcholine receptor/cholinergic site/receptor function/autoimmunity. *J. Immunol.* 152:5997–6002.

139. Engel, A.G., and G. Fumagalli. 1982. Mechanisms of acetylcholine receptor loss from the neuromuscular junction. *Ciba Found Symp.* 90:197–224.

140. Ruff, R.L., and V.A. Lennon. 1998. End-plate voltage-gated sodium channels are lost in clinical and experimental myasthenia gravis. *Ann. Neurol.* 43:370–379.

141. Molenaar, P.C. 1990. Synaptic adaptation in diseases of the neuromuscular junction. *Prog. Brain Res.* 84:145–149.

142. Guyon, T., A. Wakkach, S. Poea, M. Mouly, I. Klingel-Schmitt, P. Levasseur, D. Beeson, O. Asher, S. Tzartos, and

S. Berrih-Aknin. 1998. Regulation of acetylcholine receptor gene expression in human myasthenia gravis muscles. Evidences for a compensatory mechanism triggered by receptor loss. *J. Clin. Invest.* 102:249–263.

143. Asher, O., D. Neumann, V. Witzemann, and S. Fuchs. 1990. Acetylcholine receptor gene expression in experimental autoimmune myasthenia gravis. *FEBS Lett.* 267:231–235.

144. Lindstrom, J.M., M.E. Seybold, V.A. Lennon, S. Whittingham, and D.D. Duane. 1976. Antibody to acetylcholine receptor in myasthenia gravis. Prevalence, clinical correlates, and diagnostic value. *Neurology* 26:1054–1059.

145. Marx, A., T. Kirchner, A. Greiner, H.K. Muller-Hermelink, B. Schalke, and M. Osborn. 1992. Neurofilament epitopes in thymoma and antiaxonal autoantibodies in myasthenia gravis. *Lancet* 339:707–708.

146. Mossmann, H., U. Bamberger, B.A. Velev, M. Gehrung, and D.K. Hammer. 1986. Effect of platelet-activating factor on human polymorphonuclear leukocyte enhancement of chemiluminescence and antibody-dependent cellular cytotoxicity. *J. Leukoc. Biol.* 39:153–165.

147. Burges, J., D.W. Wray, S. Pizzighella, Z. Hall, and A. Vincent. 1990. A myasthenia gravis plasma immunoglobulin reduces miniature endplate potentials at human endplates in vitro. *Muscle Nerve* 13:407–413.

148. Barrett-Jolley, R., N. Byrne, A. Vincent, and J. Newsom-Davis. 1994. Plasma from patients with seronegative myasthenia gravis inhibit nAChR responses in the TE671/RD cell line. *Pflugers Arch.* 428:492–498.

149. Li, Z., N. Forester, and A. Vincent. 1996. Modulation of acetylcholine receptor function in TE671 (rhabdomyosarcoma) cells by non-AChR ligands: Possible relevance to seronegative myasthenia gravis. *J. Neuroimmunol.* 64:179–183.

150. Huganir, R.L., A.H. Delcour, P. Greengard, and G.P. Hess. 1986. Phosphorylation of the nicotinic acetylcholine receptor regulates its rate of desensitization. *Nature* 321:774–776.

151. Wagner, K., K. Edson, L. Heginbotham, M. Post, R.L. Huganir, and A.J. Czernik. 1991. Determination of the tyrosine phosphorylation sites of the nicotinic acetylcholine receptor. *J. Biol. Chem.* 266:23784–23789.

152. Hohlfeld, R. 1990. Myasthenia gravis and thymoma: Paraneoplastic failure of neuromuscular transmission. *Lab. Invest.* 62:241–243.

153. Hohlfeld, R., and H. Wekerle. 1994. The role of the thymus in myasthenia gravis. *Adv. Neuroimmunol.* 4:373–386.

154. Hohlfeld, R., K.V. Toyka, K. Heininger, H. Grosse-Wilde, and I. Kalies. 1984. Autoimmune human T lymphocytes specific for acetylcholine receptor. *Nature* 310:244–246.

155. Protti, M.P., A.A. Manfredi, R.M. Horton, M. Bellone, and B.M. Conti-Tronconi. 1993. Myasthenia gravis: recognition of a human autoantigen at the molecular level. *Immunol. Today* 14:363–368.

156. Vincent, A., and N. Willcox. 1994. Characterization of specific T cells in myasthenia gravis. *Immunol. Today* 15:41–42.

157. Link, J., M. Soderstrom, A. Ljungdahl, B. Hojeberg, T. Olsson, Z. Xu, S. Fredrikson, Z.Y. Wang, and H. Link. 1994. Organ-specific autoantigens induce interferon-gamma and interleukin-4 mRNA expression in mononuclear cells in multiple sclerosis and myasthenia gravis. *Neurology* 44:728–734.

158. Link, J., B. He, V. Navikas, W. Palasik, S. Fredrikson, M. Soderstrom, and H. Link. 1995. Transforming growth factor-beta 1 suppresses autoantigen-induced expression of pro-inflammatory cytokines but not of interleukin-10 in multiple sclerosis and myasthenia gravis. *J. Neuroimmunol.* 58:21–35.

159. Yi, Q., and A. Osterborg. 1996. Idiotype-specific T cells in multiple myeloma: targets for an immunotherapeutic intervention? *Med. Oncol.* 13:1–7.

160. Zhang, G.X., V. Navikas, and H. Link. 1997. Cytokines and the pathogenesis of myasthenia gravis. *Muscle Nerve* 20:543–551.

161. Bongioanni, P., R. Ricciardi, and M.R. Romano. 1999. T-lymphocyte interferon-gamma receptor binding in patients with myasthenia gravis. *Arch. Neurol.* 56:933–938.

162. Saoudi, A., I. Bernard, A. Hoedemaekers, B. Cautain, K. Martinez, P. Druet, M. De Baets, and J.C. Guery. 1999. Experimental autoimmune myasthenia gravis may occur in the context of a polarized Th1- or Th2-type immune response in rats. *J. Immunol.* 162:7189–7197.

163. Conti-Fine, B.M., D. Navaneetham, P.I. Karachunski, R. Raju, B. Diethelm-Okita, D. Okita, J. Howard, Jr., and Z.Y. Wang. 1998. T cell recognition of the acetylcholine receptor in myasthenia gravis. *Ann. NY Acad. Sci.* 841:283–308.

164. Wilisch, A., S. Gutsche, V. Hoffacker, A. Schultz, S. Tzartos, W. Nix, B. Schalke, C. Schneider, H.K. Muller-Hermelink, and A. Marx. 1999. Association of acetylcholine receptor alpha-subunit gene expression in mixed thymoma with myasthenia gravis. *Neurology* 52:1460–1466.

165. Marx, A., and H.K. Muller-Hermelink. 1999. Thymoma and thymic carcinoma. *Am. J. Surg. Pathol.* 23:739–742.

166. Yuen, M.H., M.P. Protti, B. Diethelm-Okita, L. Moiola, J.F. Howard, Jr., and B.M. Conti-Fine. 1995. Immunoregulatory CD8+ cells recognize antigen-activated CD4+ cells in myasthenia gravis patients and in healthy controls. *J. Immunol.* 154:1508–1520.

167. Yuen, M.H., K.D. Macklin, and B.M. Conti-Fine. 1996. MHC class II presentation of human acetylcholine receptor in Myasthenia gravis: binding of synthetic gamma subunit sequences to DR molecules. *J. Autoimmun.* 9:67–77.

168. Vincent, A., and J. Newsom Davis. 1980. Anti-acetylcholine receptor antibodies. *J. Neurol. Neurosurg. Psychiatry* 43:590–600.

169. Conti-Tronconi, B.M., K.E. McLane, M.A. Raftery, S.A. Grando, and M.P. Protti. 1994. The nicotinic acetylcholine receptor: structure and autoimmune pathology. *Crit. Rev. Biochem. Mol. Biol.* 29:69–123.

170. Zhang, Y., T. Barkas, M. Juillerat, B. Schwendimann, and H. Wekerle. 1988. T cell epitopes in experimental autoimmune myasthenia gravis of the rat: Strain-specific epitopes and cross-reaction between two distinct segments of the alpha chain of the nicotinic acetylcholine receptor (*Torpedo californica*). *Eur. J. Immunol.* 18:551–557.

171. Vincent, A., L. Jacobson, and P. Shillito. 1994. Response to human acetylcholine receptor alpha 138–199: Determinant spreading initiates autoimmunity to self-antigen in rabbits. *Immunol. Lett.* 39:269–275.

172. Melms, A., S. Chrestel, B.C. Schalke, H. Wekerle, A. Mauron, M. Ballivet, and T. Barkas. 1989. Autoimmune T lymphocytes in myasthenia gravis. Determination of target epitopes using T lines and recombinant products of the mouse nicotinic acetylcholine receptor gene. *J. Clin. Invest.* 83:785–790.

173. Ong, B., N. Willcox, P. Wordsworth, D. Beeson, A. Vincent, D. Altmann, J.S. Lanchbury, G.C. Harcourt, J.I. Bell, and J. Newsom-Davis. 1991. Critical role for the Val/

Gly86 HLA-DR beta dimorphism in autoantigen presentation to human T cells. *Proc. Natl. Acad. Sci. USA* 88:7343–7347.

174. Harcourt, G., A.P. Batocchi, S. Hawke, D. Beeson, N. Pantic, L. Jacobson, N. Willcox, A. Vincent, and J. Newsom-Davis. 1993. Detection of alpha-subunit isoforms in human muscle acetylcholine receptor by specific T cells from a myasthenia gravis patient. *Proc. R. Soc. Lond. B. Biol. Sci.* 254:1–6.

175. Hill, M., D. Beeson, P. Moss, L. Jacobson, A. Bond, L. Corlett, J. Newsom-Davis, A. Vincent, and N. Willcox. 1999. Early-onset myasthenia gravis: a recurring T-cell epitope in the adult-specific acetylcholine receptor epsilon subunit presented by the susceptibility allele HLA-DR52a. *Ann. Neurol.* 45:224–231.

176. Beeson, D., A.P. Bond, L. Corlett, S.J. Curnow, M.E. Hill, L.W. Jacobson, C. MacLennan, A. Meager, A.M. Moody, P. Moss, N. Nagvekar, J. Newsom-Davis, N. Pantic, I. Roxanis, E.G. Spack, A. Vincent, and N. Willcox. 1998. Thymus, thymoma, and specific T cells in myasthenia gravis. *Ann. NY Acad. Sci.* 841:371–387.

177. Nagvekar, N., A.M. Moody, P. Moss, I. Roxanis, J. Curnow, D. Beeson, N. Pantic, J. Newsom-Davis, A. Vincent, and N. Willcox. 1998. A pathogenetic role for the thymoma in myasthenia gravis. Autosensitization of IL-4- producing T cell clones recognizing extracellular acetylcholine receptor epitopes presented by minority class II isotypes. *J. Clin. Invest.* 101:2268–2277.

178. Beeson, D., M. Amar, I. Bermudez, A. Vincent, and J. Newsom-Davis. 1996. Stable functional expression of the adult subtype of human muscle acetylcholine receptor following transfection of the human rhabdomyosarcoma cell line TE671 with cDNA encoding the epsilon subunit. *Neurosci. Lett.* 207:57–60.

179. Nenninger, R., A. Schultz, V. Hoffacker, M. Helmreich, A. Wilisch, B. Vandekerokhove, T. Hunig, B. Schalke, C. Schneider, S. Tzartos, H. Kalbacher, H.K. Muller-Hermelink, and A. Marx. 1998. Abnormal thymocyte development and generation of autoreactive T cells in mixed and cortical thymomas. *Lab. Invest.* 78:743–753.

180. Marx, A., A. Schultz, A. Wilisch, M. Helmreich, R. Nenninger, and H.K. Muller-Hermelink. 1998. Paraneoplastic autoimmunity in thymus tumors. *Dev. Immunol.* 6:129–140.

181. Marx, A., A. Wilisch, A. Schultz, A. Greiner, B. Magi, V. Pallini, B. Schalke, K. Toyka, W. Nix, T. Kirchner, and H.K. Muller-Hermelink. 1996. Expression of neurofilaments and of a titin epitope in thymic epithelial tumors. Implications for the pathogenesis of myasthenia gravis. *Am. J. Pathol.* 148:1839–1850.

182. Sommer, N., N. Willcox, G.C. Harcourt, and J. Newsom-Davis. 1990. Myasthenic thymus and thymoma are selectively enriched in acetylcholine receptor-reactive T cells. *Ann. Neurol.* 28:312–319.

183. Melms, A., B.C. Schalke, T. Kirchner, H.K. Muller-Hermelink, E. Albert, and H. Wekerle. 1988. Thymus in myasthenia gravis. Isolation of T-lymphocyte lines specific for the nicotinic acetylcholine receptor from thymuses of myasthenic patients. *J. Clin. Invest.* 81:902–908.

184. Melms, A., G. Malcherek, U. Gern, H. Wietholter, C.A. Muller, R. Schoepfer, and J. Lindstrom. 1992. T cells from normal and myasthenic individuals recognize the human acetylcholine receptor: heterogeneity of antigenic sites on the alpha- subunit. *Ann. Neurol.* 31:311–318.

185. Sommer, N., G.C. Harcourt, N. Willcox, D. Beeson, and J. Newsom-Davis. 1991. Acetylcholine receptor-reactive T lymphocytes from healthy subjects and myasthenia gravis patients. *Neurology* 41:1270–1276.

186. Kaul, R., M. Shenoy, E. Goluszko, and P. Christadoss. 1994. Major histocompatibility complex class II gene disruption prevents experimental autoimmune myasthenia gravis. *J. Immunol.* 152:3152–3157.

187. Asher, O., W.A. Kues, V. Witzemann, S.J. Tzartos, S. Fuchs, and M.C. Souroujon. 1993. Increased gene expression of acetylcholine receptor and myogenic factors in passively transferred experimental autoimmune myasthenia gravis. *J. Immunol.* 151:6442–6450.

188. Shenoy, M., R. Kaul, E. Goluszko, C. David, and P. Christadoss. 1994. Effect of MHC class I and CD8 cell deficiency on experimental autoimmune myasthenia gravis pathogenesis. *J. Immunol.* 153:5330–5335.

189. Zhang, G.X., C.G. Ma, B.G. Xiao, M. Bakhiet, H. Link, and T. Olsson. 1995. Depletion of CD8+ T cells suppresses the development of experimental autoimmune myasthenia gravis in Lewis rats. *Eur. J. Immunol.* 25:1191–1198.

190. Zhang, G.X., C.G. Ma, B.G. Xiao, M. Bakhiet, A. Ljungdahl, T. Olsson, and H. Link. 1995. Suppression of experimental autoimmune myasthenia gravis after CD8 depletion is associated with decreased IFN-gamma and IL 4. *Scand. J. Immunol.* 42:457–465.

191. Garchon, H.J., F. Djabiri, J.P. Viard, P. Gajdos, and J.F. Bach. 1994. Involvement of human muscle acetylcholine receptor alpha-subunit gene (CHRNA) in susceptibility to myasthenia gravis. *Proc. Natl. Acad. Sci. USA* 91:4668–4672.

192. Nicolle, M.W., S. Hawke, N. Willcox, and A. Vincent. 1995. Differences in processing of an autoantigen by DR4:Dw4.2 and DR4:Dw14.2 antigen-presenting cells. *Eur. J. Immunol.* 25:2119–2122.

193. Campbell, R.D., and C.M. Milner. 1993. MHC genes in autoimmunity. *Curr. Opin. Immunol.* 5:887–893.

194. Vieira, M.L., S. Caillat-Zucman, P. Gajdos, S. Cohen-Kaminsky, A. Casteur, and J.F. Bach. 1993. Identification by genomic typing of non-DR3 HLA class II genes associated with myasthenia gravis. *J. Neuroimmunol.* 47:115–122.

195. Hjelmstrom, P., C.S. Peacock, R. Giscombe, R. Pirskanen, A.K. Lefvert, J.M. Blackwell, and C.B. Sanjeevi. 1998. Polymorphism in tumor necrosis factor genes associated with myasthenia gravis. *J. Neuroimmunol.* 88:137–143.

196. Bongioanni, P., R. Ricciardi, D. Pellegrino, and M.R. Romano. 1999. T-cell tumor necrosis factor-alpha receptor binding in myasthenic patients. *J. Neuroimmunol.* 93:203–207.

197. Huang, D.R., R. Pirskanen, G. Matell, and A.K. Lefvert. 1999. Tumour necrosis factor-alpha polymorphism and secretion in myasthenia gravis. *J. Neuroimmunol.* 94:165–171.

198. Willcox, N., M. Schluep, M.A. Ritter, and J. Newsom-Davis. 1991. The thymus in seronegative myasthenia gravis patients. *J. Neurol.* 238:256–261.

199. Dondi, E., P. Gajdos, J.F. Bach, and H.J. Garchon. 1994. Association of Km3 allotype with increased serum levels of autoantibodies against muscle acetylcholine receptor in myasthenia gravis. *J. Neuroimmunol.* 51:221–224.

200. Raknes, G., G.O. Skeie, N.E. Gilhus, S. Aadland, and C. Vedeler. 1998. FcgammaRIIA and FcgammaRIIIB polymorphisms in myasthenia gravis. *J. Neuroimmunol.* 81:173–176.

201. Huang, D., R. Pirskanen, P. Hjelmstrom, and A.K. Lefvert. 1998. Polymorphisms in IL-1beta and IL-1 receptor antagonist genes are associated with myasthenia gravis. *J. Neuroimmunol.* 81:76–81.

202. Huang, D.R., Y.H. Zhou, S.Q. Xia, L. Liu, R. Pirskanen, and A.K. Lefvert. 1999. Markers in the promoter region of interleukin-10 (IL-10) gene in myasthenia gravis: implications of diverse effects of IL-10 in the pathogenesis of the disease. *J. Neuroimmunol.* 94:82–87.

203. Huang, D., L. Liu, K. Noren, S.Q. Xia, J. Trifunovic, R. Pirskanen, and A.K. Lefvert. 1999. Genetic association of Ctla-4 to myasthenia gravis with thymoma. *J. Neuroimmunol.* 88:192–198.

204. Huang, D., S. Xia, Y. Zhou, R. Pirskanen, L. Liu, and A.K. Lefvert. 1998. No evidence for interleukin-4 gene conferring susceptibility to myasthenia gravis. *J. Neuroimmunol.* 92:208–211.

205. Karachunski, P.I., N.S. Ostlie, D.K. Okita, and B.M. Conti-Fine. 1999. Interleukin-4 deficiency facilitates development of experimental myasthenia gravis and precludes its prevention by nasal administration of CD4+ epitope sequences of the acetylcholine receptor. *J. Neuroimmunol.* 95:73–84.

206. Balasa, B., C. Deng, J. Lee, P. Christadoss, and N. Sarvetnick. 1998. The Th2 cytokine IL-4 is not required for the progression of antibody- dependent autoimmune myasthenia gravis. *J. Immunol.* 161:2856–2862.

207. Reinhardt, C., and A. Melms. 1999. Elevated frequencies of natural killer T lymphocytes in myasthenia gravis. *Neurology* 52:1485–1487.

208. Muller-Hermelink, H.K., A. Marx, and T.K. Thymus. 1996. In: *Anderson's Pathology*. Tenth ed. Damjanov ILJ, editor St. Louis: Mosby; pp. 1218–1243.

209. Kirchner, T., B. Schalke, A. Melms, T. von Kugelgen, and H.K. Muller-Hermelink. 1986. Immunohistological patterns of non-neoplastic changes in the thymus in Myasthenia gravis. *Virchows. Arch. B. Cell. Pathol. Incl. Mol. Pathol.* 52:237–257.

210. Kirchner, T., S. Tzartos, F. Hoppe, B. Schalke, H. Wekerle, and H. K. Muller-Hermelink. 1988. Pathogenesis of myasthenia gravis. Acetylcholine receptor-related antigenic determinants in tumor-free thymuses and thymic epithelial tumors. *Am. J. Pathol.* 130:268–280.

211. Bornemann, A., and T. Kirchner. 1996. An immuno-electron-microscopic study of human thymic B cells. *Cell. Tissue. Res.* 284:481–487.

212. Burns, T.M., V.C. Juel, D.B. Sanders, and L.H. Phillips, 2nd. 1999. Neuroendocrine lung tumors and disorders of the neuromuscular junction. *Neurology* 52:1490–1491.

213. Sciamanna, M.A., G.E. Griesmann, C.L. Williams, and V.A. Lennon. 1997. Nicotinic acetylcholine receptors of muscle and neuronal (alpha7) types coexpressed in a small cell lung carcinoma. *J. Neurochem.* 69:2302–2311.

214. Gattenloehner, S., A. Vincent, I. Leuschner, S. Tzartos, H.K. Muller-Hermelink, T. Kirchner, and A. Marx. 1998. The fetal form of the acetylcholine receptor distinguishes rhabdomyosarcomas from other childhood tumors. *Am. J. Pathol.* 152:437–444.

215. Kirchner, T., B. Schalke, J. Buchwald, M. Ritter, A. Marx, and H.K. Muller-Hermelink. 1992. Well- differentiated thymic carcinoma. An organotypical low-grade carcinoma with relationship to cortical thymoma. *Am. J. Surg. Pathol.* 16:1153–1169.

216. Quintanilla-Martinez, L., E.W. Wilkins, Jr., J.A. Ferry, and N.L. Harris. 1993. Thymoma-morphologic subclassification correlates with invasiveness and immunohistologic features: A study of 122 cases. *Hum. Pathol.* 24:958–969.

216a.Müller-Hermelink, H.K., and A. Marx. 2000. Thymoma. *Curr. Opin. Oncol.* 12:426–433.

217. Levy, Y., A. Afek, Y. Sherer, Y. Bar-Dayan, R. Shibi, J. Kopolovic, and Y. Shoenfeld. 1998. Malignant thymoma associated with autoimmune diseases: a retrospective study and review of the literature. *Semin. Arthritis. Rheum.* 28:73–79.

218. Wilkins, K.B., and G.B. Bulkley. 1999. Thymectomy in the integrated management of myasthenia gravis. *Adv. Surg.* 32:105–133.

219. Christensen, P.B., T.S. Jensen, I. Tsiropoulos, T. Sorensen, M. Kjaer, E. Hojer-Pedersen, M.J. Rasmussen, and E. Lehfeldt. 1998. Mortality and survival in myasthenia gravis: A Danish population based study. *J. Neurol. Neurosurg. Psychiatry* 64:78–83.

220. Drachman, D.B., K.R. McIntosh, and B. Yang. 1998. Factors that determine the severity of experimental myasthenia gravis. *Ann. NY Acad. Sci.* 841:262–282.

221. Jermy, A., D. Beeson, and A. Vincent. 1993. Pathogenic autoimmunity to affinity-purified mouse acetylcholine receptor induced without adjuvant in BALB/c mice. *Eur. J. Immunol.* 23:973–976.

222. Fujii, Y., and J. Lindstrom. 1988. Specificity of the T cell immune response to acetylcholine receptor in experimental autoimmune myasthenia gravis. Response to subunits and synthetic peptides. *J. Immunol.* 140:1830–1837.

223. Shenoy, M., E. Goluszko, and P. Christadoss. 1994. The pathogenic role of acetylcholine receptor alpha chain epitope within alpha 146–162 in the development of experimental autoimmune myasthenia gravis in C57BL6 mice. *Clin. Immunol. Immunopathol.* 73:338–343.

224. Oshima, M., A.R. Pachner, and M.Z. Atassi. 1994. Profile of the regions of acetylcholine receptor alpha chain recognized by T-lymphocytes and by antibodies in EAMG-susceptible and non-susceptible mouse strains after different periods of immunization with the receptor. *Mol. Immunol.* 31:833–843.

225. Fujii, Y., and J. Lindstrom. 1988. Regulation of antibody production by helper T cell clones in experimental autoimmune myasthenia gravis. *J. Immunol.* 141:3361–3369.

226. Yeh, T.M., and K.A. Krolick. 1990. T cells reactive with a small synthetic peptide of the acetylcholine receptor can provide help for a clonotypically heterogeneous antibody response and subsequently impaired muscle function. *J. Immunol.* 144:1654–1660.

227. Bellone, M., N. Ostlie, S. Lei, A.A. Manfredi, and B.M. Conti-Tronconi. 1992. T helper function of CD4+ cells specific for defined epitopes on the acetylcholine receptor in congenic mouse strains. *J. Autoimmun.* 5:27–46.

228. Rosenberg, J.S., M. Oshima, and M.Z. Atassi. 1996. B-cell activation in vitro by helper T cells specific to region alpha 146–162 of Torpedo californica nicotinic acetylcholine receptor. *J. Immunol.* 157:3192–3199.

229. Wu, B., C. Deng, E. Goluszko, and P. Christadoss. 1997. Tolerance to a dominant T cell epitope in the acetylcholine receptor molecule induces epitope spread and suppresses murine myasthenia gravis. *J. Immunol.* 159:3016–3023.

230. Bellone, M., P.I. Karachunski, N. Ostlie, S. Lei, and B.M. Conti-Tronconi. 1994. Preferential pairing of T and B cells for production of antibodies without covalent association of T and B epitopes. *Eur. J. Immunol.* 24:799–804.

231. Karachunski, P.I., N. Ostlie, M. Bellone, A.J. Infante, and B.M. Conti-Fine. 1995. Mechanisms by which the I-ABM12 mutation influences susceptibility to experimental myasthenia gravis: A study in homozygous and heterozygous mice. *Scand. J. Immunol.* 42:215–225.

232. Lehmann, P.V., E.E. Sercarz, T. Forşthuber, C.M. Dayan, and G. Gammon. 1993. Determinant spreading and the dynamics of the autoimmune T-cell repertoire. *Immunol. Today* 14:203–208.

233. Meinl, E., W.E. Klinkert, and H. Wekerle. 1991. The thymus in myasthenia gravis. Changes typical for the human disease are absent in experimental autoimmune myasthenia gravis of the Lewis rat. *Am. J. Pathol.* 139:995–1008.

234. Hohlfeld, R., and A.G. Engel. 1994. The immunobiology of muscle. *Immunol. Today* 15:269–274.

235. Baggi, F., M. Nicolle, A. Vincent, H. Matsuo, N. Willcox, and J. Newsom-Davis. 1993. Presentation of endogenous acetylcholine receptor epitope by an MHC class II-transfected human muscle cell line to a specific CD4+ T cell clone from a myasthenia gravis patient. *J. Neuroimmunol.* 46:57–65.

236. Compston, D.A., A. Vincent, J. Newsom-Davis, and J.R. Batchelor. 1980. Clinical, pathological, HLA antigen and immunological evidence for disease heterogeneity in myasthenia gravis. *Brain* 103:579–601.

237. Pascuzzi, R.M. 1998. Drugs and toxins associated with myopathies. *Curr. Opin. Rheumatol.* 10:511–520.

238. Mackey, J.R., S. Desai, L. Larratt, V. Cwik, and J.M. Nabholtz. 1997. Myasthenia gravis in association with allogeneic bone marrow transplantation: Clinical observations, therapeutic implications and review of literature. *Bone Marrow. Transplant.* 19:939–942.

239. Sherer, Y., and Y. Shoenfeld. 1998. Autoimmune diseases and autoimmunity post-bone marrow transplantation. *Bone Marrow Transplant* 22:873–881.

240. Wilisch, A., A. Schultz, A. Greiner, T. Kirchner, H.K. Muller-Hermelink, and A. Marx. 1996. Molecular mimicry between neurofilaments and titin as the basis for autoimmunity towards skeletal muscle in paraneoplastic myasthenia gravis. *Verh. Dtsch. Ges. Pathol.* 80:261–266.

241. Wekerle, H., and U.P. Ketelsen. 1977. Intrathymic pathogenesis and dual genetic control of myasthenia gravis. *Lancet* 1:678–680.

242. Geuder, K.I., A. Marx, V. Witzemann, B. Schalke, T. Kirchner, and H.K. Muller-Hermelink. 1992. Genomic organization and lack of transcription of the nicotinic acetylcholine receptor subunit genes in myasthenia gravis-associated thymoma. *Lab. Invest.* 66:452–458.

243. Scadding, G.K., A. Vincent, J. Newsom-Davis, and K. Henry. 1981. Acetylcholine receptor antibody synthesis by thymic lymphocytes: correlation with thymic histology. *Neurology* 31:935–943.

244. Schonbeck, S., F. Padberg, A. Marx, R. Hohlfeld, and H. Wekerle. 1993. Transplantation of myasthenia gravis thymus to SCID mice. *Ann. NY Acad. Sci.* 681:66–73.

245. Spuler, S., A. Marx, T. Kirchner, R. Hohlfeld, and H. Wekerle. 1994. Myogenesis in thymic transplants in the severe combined immunodeficient mouse model of myasthenia gravis. Differentiation of thymic myoid cells into striated muscle cells. *Am. J. Pathol.* 145:766–770.

246. Spuler, S., A. Sarropoulos, A. Marx, R. Hohlfeld, and H. Wekerle. 1996. Thymoma-associated myasthenia gravis. Transplantation of thymoma and extrathymomal thymic tissue into SCID mice. *Am. J. Pathol.* 148:1359–1365.

247. Fujii, Y., Y. Monden, K. Nakahara, J. Hashimoto, and Y. Kawashima. 1984. Antibody to acetylcholine receptor in myasthenia gravis: production by lymphocytes from thymus or thymoma. *Neurology* 34:1182–1186.

248. Nenninger, R., A. Schultz, H.K. Muller-Hermelink, and A. Marx. 1996. Abnormal T cell maturation in myasthenia gravis associated thymomas. *Verh. Dtsch. Ges. Pathol.* 80:256–260.

249. Mygland, A., J.A. Aarli, R. Matre, and N.E. Gilhus. 1994. Ryanodine receptor antibodies related to severity of thymoma–associated myasthenia gravis. *J. Neurol. Neurosurg. Psychiatry* 57:843–846.

250. Gattenlohner, S., T. Brabletz, A. Schultz, A. Marx, H.K. Muller-Hermelink, and T. Kirchner. 1994. Cloning of a cDNA coding for the acetylcholine receptor alpha-subunit from a thymoma associated with myasthenia gravis. *Thymus* 23:103–113.

251. Hara, H., K. Hayashi, K. Ohta, N. Itoh, and M. Ohta. 1993. Nicotinic acetylcholine receptor mRNAs in myasthenic thymuses: association with intrathymic pathogenesis of myasthenia gravis. *Biochem. Biophys. Res. Commun.* 194:1269–1275.

252. Kornstein, M.J., O. Asher, and S. Fuchs. 1995. Acetylcholine receptor alpha-subunit and myogenin mRNAs in thymus and thymomas. *Am. J. Pathol.* 146:1320–1324.

253. Mihovilovic, M., and A.D. Roses. 1993. Expression of alpha-3, alpha-5, and beta-4 neuronal acetylcholine receptor subunit transcripts in normal and myasthenia gravis thymus. Identification of thymocytes expressing the alpha-3 transcripts. *J. Immunol.* 151:6517–6524.

254. Marx, A., T. Kirchner, A. Greiner, B. Schalke, and H.K. Muller-Hermelink. 1993. Myasthenia gravis-associated thymic epithelial tumors express neurofilaments and are associated with antiaxonal autoimmunity. *Ann. NY Acad. Sci.* 681:107–109.

255. Siara, J., R. Rudel, and A. Marx. 1991. Absence of acetylcholine-induced current in epithelial cells from thymus glands and thymomas of myasthenia gravis patients. *Neurology* 41:128–131.

256. Wakkach, A., T. Guyon, C. Bruand, S. Tzartos, S. Cohen-Kaminsky, and S. Berrih-Aknin. 1996. Expression of acetylcholine receptor genes in human thymic epithelial cells: Implications for myasthenia gravis. *J. Immunol.* 157:3752–3760.

257. Skeie, G.O., A. Freiburg, B. Kolmerer, S. Labeit, J.A. Aarli, and N.E. Gilhus. 1998. Titin transcripts in thymomas. *Ann. NY Acad. Sci.* 841:422–426.

258. Kusner, L.L., A. Mygland, and H.J. Kaminski. 1998. Ryanodine receptor gene expression thymomas. *Muscle Nerve* 21:1299–1303.

259. Ashton-Rickardt, P.G., and S. Tonegawa. 1994. A differential-avidity model for T-cell selection. *Immunol. Today* 15:362–366.

260. Hogquist, K.A., S.C. Jameson, W.R. Heath, J.L. Howard, M.J. Bevan, and F.R. Carbone. 1994. T cell receptor antagonist peptides induce positive selection. *Cell* 76:17–27.

261. Inoue, M., M. Okumura, S. Miyoshi, H. Shiono, K. Fukuhara, Y. Kadota, R. Shirakura, and H. Matsuda. 1999. Impaired expression of MHC class II molecules in response to interferon- gamma (IFN-gamma) on human thymoma neoplastic epithelial cells. *Clin. Exp. Immunol.* 117:1–7.

262. Somnier, F.E. 1994. Exacerbation of myasthenia gravis after removal of thymomas. *Acta. Neurol. Scand.* 90:56–66.

263. Maselli, R.A., D.P. Richman, and R.L. Wollmann. 1991. Inflammation at the neuromuscular junction in myasthenia gravis. *Neurology* 41:1497–1504.

264. Nakano, S., and A.G. Engel. 1993. Myasthenia gravis: Quantitative immunocytochemical analysis of inflammatory cells and detection of complement membrane attack complex at the end-plate in 30 patients. *Neurology* 43:1167–1172.

265. Myking, A.O., G.O. Skeie, J.E. Varhaug, K.S. Andersen, N.E. Gilhus, and J.A. Aarli. 1998. The histomorphology of the thymus in late onset, non-thymoma myasthenia gravis. *Eur. J. Neurol.* 5:401–405.

266. Lubke, E., A. Freiburg, G.O. Skeie, B. Kolmerer, S. Labeit, J.A. Aarli, N.E. Gilhus, R. Wollmann, M. Wussling, J.C. Ruegg, and W.A. Linke. 1998. Striational autoantibodies in myasthenia gravis patients recognize I- band titin epitopes. *J. Neuroimmunol.* 81:98–108.

267. Rodewald, H.R. 1998. The thymus in the age of retirement. *Nature* 396:630–631.

268. Sprent, J. 1993. The thymus and T-cell tolerance. *Ann. NY Acad. Sci.* 681:5–15.

269. Dardenne, M., W. Savino, and J.F. Bach. 1987. Thymomatous epithelial cells and skeletal muscle share a common epitope defined by a monoclonal antibody. *Am. J. Pathol.* 126:194–198.

270. Gilhus, N.E., J.A. Aarli, B. Christensson, and R. Matre. 1984. Rabbit antiserum to a citric acid extract of human skeletal muscle staining thymomas from myasthenia gravis patients. *J. Neuroimmunol.* 7:55–64.

271. Mamula, M.J., R.H. Lin, C.A. Janeway, and J.A. Jr., Hardin. 1992. Breaking T cell tolerance with foreign and self co-immunogens. A study of autoimmune B and T cell epitopes of cytochrome c. *J. Immunol.* 149:789–795.

44 | Experimental Autoimmune Myasthenia Gravis

Premkumar Christadoss, Mathilde A. Poussin, and Caishu Deng

INTRODUCTION

Myasthenia gravis (MG) is an antibody-mediated, autoimmune neuromuscular disease in which autoantibodies directed against the nicotinic acetylcholine receptor (AChR) activate the complement cascade after binding to AChR. Antibody- and complement-mediated destruction and/or cross-linking of AChR by antibodies, which lead to antigenic modulation, cause a reduction in the number of AChR molecules expressed at the neuromuscular junction (NMJ) (1–4). Anti-AChR antibodies can also bind to AChR and interfere with neuromuscular transmission (5). The above pathological events lead to neuromuscular transmission defects and culminate in weakness and fatigue of skeletal muscles in MG patients.

MG is a classic antibody-mediated autoimmune disease because: 1) serum AChR antibodies are present in 90% of MG patients (6–8); 2) IgG, C3 and C9 complement components are localized at the NMJ (9–11); 3) an MG-like disease can be transferred into mice by passive serum transfer from MG patients (12); 4) mice develop MG-like symptoms when monoclonal-anti-AChR antibodies are passively transferred (13); and 5) experimental autoimmune MG (EAMG) can be induced in vertebrates by immunization with AChR proteins in adjuvants (14–18).

The autoantigen AChR has been cloned, sequenced and biochemically characterized. The ligand for AChR is acetylcholine (ACh), which binds to AChR in the muscle channel that allows release of sodium ions into muscle cells. This process leads to local electrical depolarization of the membrane. The AChR is a complex pentameric structure containing 2 α subunits, one β and one δ. The last fetal γ-subunit is replaced by a ϵ-subunit before birth. The molecular weights of the subunits range between 45 and 55 kDa (19). The α-subunit is considered the most highly immunogenic region, because both T and B cells in MG patients and rodents with EAMG are predominantly stimulated by or react to the α-subunit (19,20). Mice are the ideal species to study the immunopathogenesis of MG, because of the availability of congenic, recombinant, mutant, transgenic, and immune response and cytokine gene knockout (KO) mice. Such mouse models allow us to evaluate the immunogenetic and cytokine gene regulation of EAMG. Further, the availability of antibodies to the mouse lymphoid cell surface markers enable us to extrapolate the precise cellular and molecular mechanisms of disease pathogenesis.

In this review, first we will summarize the similarities between MG and EAMG. Next, we will discuss the roles of a) MHC class II molecules and CD4 cells, b) TCR genes and T cell epitopes, c) B cells, d) costimulator molecules, and e) cytokines in EAMG pathogenesis. In the last part of this chapter, we will discuss recent studies related to antigen-specific therapy in EAMG, and its mechanisms, and conclude with the current hypothetical model of EAMG immunopathogenesis.

2. Clinical and Immunopathologic Similarities Between Human MG and Mouse EAMG

EAMG is induced in susceptible strains of mice by subcutaneous immunization with AChR purified from *Torpedo californica* (tAChR) emulsified in complete Freund's adjuvant (CFA). Susceptible mouse strains (e.g., C57BL/6) exhibit muscle weakness, usually after two AChR immunization in CFA (15–17,21,22). Muscle weakness in human MG and EAMG is more prominent in the upper half of the body. Like humans with MG, mice with EAMG demonstrate a significant decremental response in the repetitive nerve stimulation test and a reduction in miniature end plate potential amplitude (16,17,23). Temporary improvement in muscle weakness in cases of both MG and EAMG could be achieved by administering anti-cholinesterase medication (e.g., neostigmine bromide). Around 90% of MG patients and C57BL/6 (B6) mice with EAMG have serum anti-AChR antibodies, while approximately 10–15% of patients with MG and mice with EAMG are seronegative for anti-AChR antibody. IgG and C3 complement component are deposited at the NMJ in MG patients and EAMG (9–11), and C5-deficiency prevented the development of murine EAMG (24). Loss of AChR in muscle is the primary pathology observed in both MG and EAMG (1,18,21,25). Therefore, murine EAMG is an ideal prototype for studying antibody-mediated autoimmune disease pathogenesis and for specific immunointervention.

However, MG and EAMG do differ in a few aspects. The factor(s) initiating MG development is not known, while EAMG is induced by AChR plus CFA immunization. Approximately 10% of MG patients have thymomas and most have thymic hypertrophy. This is important because the human thymus contains germinal centers, while the thymus in EAMG cases does not (26). In human MG, the thymus may be a potential site for autosensitization, because of AChR expression by thymic epithelial and myoid cells and the presence of T and germinal center B cells (27–33). However, in murine EAMG, the AChR draining lymph node cells are the potential site for autosensitization.

3. THE ROLE OF MAJOR HISTOCOMPATIBILITY COMPLEX (MHC) GENES

The MHC class II genes control the lymphocyte proliferative response to AChR (34,35). Even in cases of human MG, MHC class II (HLA-DQ, HLA-DR) genes control the *in vitro* lymphoproliferative response to AChR (20,36,37). A high incidence (>50%) of clinical EAMG can be induced in mice carrying the $H-2^b$ haplotype [e.g., B6 or C57BL/10 (BI0)] (15–17), $H-2^d$ haplotype (B10.D2) (24) and $H-2^v$ (RIIIS/J) mice (38) by 2–3 immunizations with AChR. Strains with $H-2^p$ or $H-2^k$ haplotypes are relatively resistant to EAMG induction (15–17). We demonstrated, for the first time, that genes within MHC (H-2) control EAMG susceptibility (15). B10 recombinant strain analysis mapped the immune response gene controlling cellular and humoral immune responses to AChR, and the development of EAMG to the MHC I-A locus (16). The IgH chain gene also appears to influence EAMG pathogenesis (17).

3.1. The Role of MHC Class II Molecules and CD4 Coreceptor

The MHC molecules are responsible for presenting antigenic peptides processed from endogenous cytosolic proteins or exogenous antigens through the class I and class II pathways, respectively. Antigen processing, which includes degradation of the antigen into peptide fragments, occurs in intracellular organelles. In the Golgi apparatus, MHC class II molecules ($\alpha\beta$) can be found complexed to a polypeptide called the invariant chain (Ii). The $\alpha\beta$-Ii complex is transported through the Golgi complex to an endosomal acidic compartment where Ii is released. This event allows binding on the class II molecule of the peptide, coming from endosomes. MHC class II heterodimers are not expressed on all cell types, but primarily on the specialized antigen-presenting cells (APC) (B-lymphocytes, monocytes/macrophages, Langerhans' cells and dendritic cells). After immunization with AChR in CFA, AChR is processed and dominant peptides are presented in the context of class II molecules for recognition of CD4+ T helper cells (21,22,34,35).

CD4+ T lymphocytes are involved in MG, as well as in EAMG. *In vitro*, anti-CD4 mAb could suppress AChR-specific lymphocyte proliferation (22). *In vivo* administration of anti-CD4 mAb not only prevented clinical EAMG, but also reversed the established disease (22). An MG patient also derived benefit from anti-CD4 antibody treatment (39). MHC class II deficiency and, therefore, CD4+ T cell deficiency in MHC class II gene disrupted mice, rendered them resistant to induction of clinical EAMG (40,41). Virtually no anti-AChR antibody or AChR-specific lymphocytes proliferative responses were observed in MHC class II-deficient mice immunized with AChR (40). Further, EAMG development is suppressed in CD4 gene knockout (KO) mice having intact MHC class II molecules (42). Few CD4 gene KO mice still developed EAMG (Dedhia et al., unpublished), suggesting that non-CD4 cells (e.g., CD4− CD8−, NK, etc) could contribute to EAMG pathogenesis. MHC class I-restricted CD8+ cells do not play a significant role in the genesis of EAMG (43). Although CD8 cell depletion in CD8 gene KO mice has been shown to partially reduce the incidence of EAMG (42), we did not observe any significant reduction in EAMG incidence in CD8 gene KO mice (Dedhia et al., unpublished). Therefore, CD8 cells play a minimal role in EAMG development. Even in human MG, CD4 cells, but not CD8 cells, have been shown to be involved in the autoimmune response to AChR (44). EAMG is suppressed in mice deficient in NK cells (Shi et al., personal communication). Although MHC class II-restricted CD4 cells play a crucial role, NK cells could also contribute to EAMG development. However, in MHC class II-deficient mice with intact NK cells, EAMG was completely prevented (40). The data implicate that NK cells could be involved in the disease only in the presence of MHC class II molecules.

3.2. The Role of MHC Class II Molecules Utilizing MHC Class II Mutant and Transgenic Mice

The murine MHC class II molecules have two isotypes, I-A and I-E molecules. Each isotype is encoded by an α and β gene, and the Aα and Aβ genes encode, respectively, the α and β chains of the I-A molecule. The Eα and Eβ genes encode the α and β chains of the I-A molecule, respectively. The MHC class II molecules present endogenously processed exogenous antigenic peptides to $\alpha\beta$ T cell receptors (TCR) on CD4$^+$ cells. At the same time, the CD4 coreceptor molecule binds to the monomorphic region of MHC class II α and β chains (45). This interaction is crucial for the initiation of MHC class II-restricted, antigen-specific immune responses.

The I-A gene primarily controls the immune responses to AChR *in vitro* and *in vivo*, and the I-A gene dose influences the development of clinical EAMG (16,34,35,41). Treatment of SJL mice (H-2^s) with antibodies to the I-A^s molecule before immunization with AChR in CFA suppressed the antibody response to AChR (46). The first direct genetic evidence for the importance of the MHC class II molecule in EAMG and induced autoimmune diseases came from studies of B6.C-H-2^{bml2} (bml2) and B6 mice (21,47). The bml2 is the first and only I-A locus mutant. The B6 and bml2 strains differ by only 3 amino acids in the β chain of the I-A molecule (Ile^{67}→Phe, Arg^{70}→Gln, Thr^{71}→Lys). The Aβ mutation (gene conversion) in bml2 mice suppressed both the cellular and humoral immune response to AChR, and thus suppressed the development of EAMG (21,35). It is interesting that in human MG, HLA-DQ (equivalent of mouse I-A) polymorphism has been observed (48–51). The bml2 mouse T cells give a reduced response to a few T cell epitopes: peptide α146–162 (52–54), α150–169, α182–198, α181–200, and α360–368 (55,56). These sequences induce good lymphoproliferation in AChR-immunized B6 mice. However, the bml2 mice generate a good lymphoproliferative response after *in vitro* stimulation with the peptide α111–126 (52,53). F$_1$ mice (B6xbml2) develop EAMG with the same frequency as B6 mice, and also recognize the same T cell epitopes (56). The bml2 mouse MHC class II molecule I-A^{bml2} binds peptide α146–162 with a lower affinity than the MHC class II of B6 mice (57). AChR-immune bml2 mice produce less IgG$_{2b}$ anti-AChR antibody compared to B6 mice (54). We hypothesize that the resistance of the bml2 mice is related to a defect in the processing of AChR and/or recognition of pathogenic epitope α146–162, while their APC MHC class II molecules can bind α146–162 and the subdominant epitope

α111–126 with equivalent affinity. This last epitope would induce a lesser pathogenic response when presented to CD4$^+$ cells than would the dominant epitope α146–162 (Poussin et al., unpublished).

Mice from the H-2^b haplotype have a constitutive deletion of the Eα gene, and therefore, fail to express the I-E molecule on their cell surface. Two strains of mice can allow us to determine the relative importance of I-A and I-E molecules in EAMG development. The Eα^k transgene renders mice (B10.Aα^b:β^b. Eα^k:β^b or B10A$^+$ E$^+$) more resistant to EAMG development than B10 mice. The lower disease incidence in B10 A$^+$ E$^+$ transgenic mice correlates with a lower autoantibody level (58), which could explain that in A$^+$ E$^+$ mice either clonal deletion of I-E-reactive cells involved in EAMG or induction of suppressor/regulator cells in the presence of Eα molecule. Another explanation could be that A$^+$ E$^+$ transgenic mice carry two MHC class II isotypes (I-A and I-E) instead of the one (I-A) found in B10 mice. Therefore, the I-A^b molecule of B10 A$^+$ E$^+$ mice can present the pathogenic epitopes in the same manner as B10 mice, but the hybrid class II I-E (Eα^k:β^b) presents other peptides (Poussin et al., unpublished), leading to two possible effects: a) presentation of protective epitopes and generation of regulatory cytokines (e.g., TGF-β), or b) polyclonal activation and dilution of the antigenic response against pathogenic epitopes.

Another strain allowed us to study the importance of I-A$\alpha\beta$ genes in EAMG pathogenesis. This strain is transgenic for I-Aα^k in the B10 background. The I-Aα^k transgene product pairs with the endogenous I-Aβ^b chain to lead to expression of a hybrid MHC class II molecule (I-Aα^k:β^b). In B10.TgAα^k transgenic mice, the disease incidence is reduced and the anti-AChR antibody production impaired after AChR immunization (59). This strain also has two MHC class II isotypes; one hybrid Aα^k:β^b and the other endogenous Aα^b:β^b, but they have no I-E molecule expressed. The explanations for the observed resistance could be: a) The Aα^k:β^b hybrid fails to present the pathogenic T cell epitope (e.g., 146–162); b) Aα^k:β^b presents protective epitope and generates regulatory cytokines (TGF-β); or c) there is polyclonal activation and dilution of antigenic response against pathogenic epitopes. A summary of MHC class II molecules involved in EAMG susceptibility is presented in Table 1.

Recently, two strains have been developed that carry human MHC class II gene (HLA-DQ6 and HLA-DQ8) (60). HLA-DQ6 mice are more resistant to the development of EAMG than HLA-DQ8 mice (61). The HLA-class II transgenic mice will be very useful to study the role of human MHC class II molecules in EAMG pathogenesis, and for specific immunotherapeutic analysis. The association of specific MHC class II molecules with the development of MG is primarily due to the binding and presentation of processed AChR peptides to T cells.

Table 1   Summary of MHC class II molecules involved in EAMG susceptibility*

Strain	Expression of						Clinical EAMG incidence
	Gene				Molecule		
	I-Aα	I-Aβ	I-Eα	I-Eβ	I-A	I-E	
C57BL/10	b	b	☒	b	$\alpha^b\beta^b$	☒	High
B.10.BR	k	k	k	k	$\alpha^k\beta^k$	$\alpha^k\beta^k$	Low
B10.Aβ^0	b	☒	☒	b	☒	☒	No
B10.TgEα^k	b	b	k	b	$\alpha^b\beta^b$	☒	Intermediate
B10.TgAα^k	k,b	b	☒	b	$\alpha^k\beta^b$, $\alpha^b\beta^b$	☒	Low
Bm12	b	bm12	☒	b	$\alpha^b\beta^{bm12}$	☒	Low

☒: no gene or molecule
*Reprinted from MYASTHENIA GRAVIS: Disease Mechanism and Immunointervention, Editor P. Christadoss, ©Narosa Publishing House, New Delhi, 1999.

Therefore, I-A^b, I-E^k, DQ2, and DR3 class II molecules present the AChR-dominant epitope to T cells efficiently, and therefore these class II molecules (alleles) are associated with susceptibility (41,47,48–51). In humans, either a molecular mimic of AChR epitope(s) could be derived from viral or bacterial antigens or from released degraded AChR peptides during normal turnover of AChR in the muscle.

4.   CHARACTERISTICS OF THE PATHOGENIC T CELL EPITOPE

EAMG-susceptible B6 mouse lymphocytes *in vitro* respond well to AChR α-chain peptides 146–162, 182–198, and 111–126. The epitopes within the region of 146–162 in the α-chain of the AChR (α146–162) are among the pathogenic T cell epitopes in susceptible mice (e.g., B6, H-2^b). (i) T cells from AChR-primed, EAMG-susceptible B6 mice respond very well to α146–162, but lymphocytes from EAMG-resistant bml2 mice failed to respond (52–55). (ii) Neonatal injection of soluble α146–162 suppressed EAMG when the mice were immunized with AChR in CFA at adulthood (52). (iii) B6 mice developed EAMG when primed with AChR in CFA and boosted with peptide α146–162, but not with a control peptide (62). Therefore, peptide α146–162-reactive T cells help B cells to produce pathogenic anti-AChR antibodies. Lymphocytes derived from human MG patients respond to numerous T cell epitopes in the AChR molecule (44). In one study, MG patient T cells responded to the human AChR α146–162 and α182–198 sequence (63), suggesting unique autoantigenic determinants within these sequences of *Torpedo* and human AChR α-subunit.

5.   MULTIPLE TCR Vβ GENE USAGE IN EAMG

Several studies have also been performed to determine the TCR genes used in the immune response to AChR. The importance of Vβ6 TCR expression has been demonstrated by the association of reactivity to AChR with expression of the *Mls* gene. Mice homozygous for H-2^b (associated with susceptibility), but lacking expression of Vβ6, respond poorly to AChR (64). The α146–162-reactive T cell lines and clones generated from B6 mice expressed predominantly the Vβ6 gene product (65). *In vivo*, Vβ8-expressing cells were preferentially used in B6 mice with EAMG (66). Vβ8 can also have a pathologic function, since Vβ8 depletion in B10.Vβ8.2 mice, which show preferential usage of the TCRVα1S8 gene, prevented EAMG (67). Studies comparing B6 and bml2 mice showed that, in bml2 mice, Vβ6$^+$ cells do not expand in response to AChR *in vitro*, although non-immune B6 and bml2 mice have similar levels of Vβ6$^+$ cells (53,65). On the other hand, *in vivo* depletion of Vβ6$^+$ cells neither abrogates *in vitro* responsiveness to AChR nor prevents clinical EAMG (68). Moreover, the B10.TCR^c strain and RIIIS/J (with a genomic deletion of ~70% TCR Vβ genes, including Vβ6 and Vβ8) and the B10.Vβ8.2 transgenic strain with a restricted TCRVβ8S2 repertoire responded to AChR and developed EAMG (38,67,68). Our data implicate multiple Vβ genes in EAMG pathogenesis. Even in human MG, there was no restricted AChR-specific TCR-Vα or -Vβ genes used (69). Therefore, highly selective immunotherapy directed at the TCR will be less efficient (69).

6.   THE EFFECT OF B CELL DEFICIENCY

B-lymphocytes have two important functions in immune and autoimmune responses: They differentiate to plasmocytes, the antibody-producing cell, and they are excellent APC. They have long been suspected to be of major importance in EAMG, as anti-AChR antibodies mediate the disease. Mice deficient in B cells failed to generate anti-AChR antibodies and were completely resistant to EAMG induction after immunization with AChR (70,71). In those mice, clinical EAMG could only be induced by passive transfer with antibodies to the main immunogenic region (70). Li et al. have suggested that B cells are not important APC in EAMG (71). However, our studies clearly implicate the importance of B cells during the priming of AChR immune response, but they are not required once the animals are primed (70). In B cell-deficient mice, the early lymphoproliferative response to AChR and the immunodominant peptide α146–162 were impaired, but the late lymphoproliferative response (90 days after the first immunization with AChR) to whole AChR and the dominant peptide α146–162 was not impaired (70). This indicates that in EAMG, B-lymphocytes are essential in disease induction because they are antibody-producing cells, and APC. The APC function can be partially compensated for by other APC (70). Anti-AChR antibody alone are not efficient in causing EAMG. Terminal complement activation is required for full development of clinical EAMG (24). Elimination of the B cells involved in the production of pathogenic anti-AChR antibodies presents a promising strategy for treating MG in an antigen- or clone-specific manner.

7. THE ROLE OF COSTIMULATOR MOLECULES

Activation of T cells requires two signals, one provided by the cognate TCR/MHC-peptide interaction, and the second costimulatory signal mediated through interaction of CD28/CTLA-4 and B7 molecules. In addition to conventional signals 1 and 2, a direct signal delivered to T cells via triggering or cross-linking of the CD40 ligand (CD40L) is also required for full activation of T cells to perform effector functions and to produce cytokines. The cells activate resting B cells through TCR recognition of MHC class II-peptide complexes and costimulation through CD40L/CD40 interaction.

The development of EAMG was investigated in CD28- and CD40L-deficient mice (CD28$^{-/-}$ and CD40L$^{-/-}$). Compared to wild-type mice, the CD28$^{-/-}$ mice became less susceptible, and CD40L$^{-/-}$ mice were completely resistant to EAMG induction (72). Analysis of T helper functions revealed a switch to a Th$_1$ profile in CD28$^{-/-}$ mice. Levels of serum AChR-specific antibodies of the IgG$_1$ isotype were decreased in CD28$^{-/-}$ mice. In the CD40L$^{-/-}$ mice, both Th$_1$ and Th$_2$ cytokine responses were diminished, and T cell-dependent AChR-reactive B cell responses were

more severely impaired than in the CD28$^{-/-}$ mice. Therefore, CD28 and CD40L are differentially required for EAMG induction (72). However, ongoing clinical EAMG could not be reversed when mice were treated with antibody to CD40L (Deng et al., unpublished).

In vivo, CTLA-4Ig can bind to B7.1/B7.2 molecules on APC with high affinity, therefore blocking B7/CD28 interaction. During a primary response, CTLA-4Ig treatment inhibited AChR antibody production profoundly, and induced a shift of AChR antibody isotypes from the normally predominant IgG$_2$ isotype pattern toward an IgG$_1$ response. Challenged rats previously treated with CTLA-4Ig produced markedly lower AChR antibody responses compared to untreated controls, as well as persistent inhibition of the IgG$_{2b}$ isotype, and lack of EAMG development. Treatment of a secondary AChR response with CTLA-4Ig moderately inhibited AChR antibody responses and clinical EAMG (73). On the other hand, CTLA-4 blockade by antibody to CTLA-4 enhanced T cell proliferative responses to AChR and anti-AChR antibody production, while provoking a rapid onset and severe EAMG, which confirmed that CTLA-4 is a negative regulator in antibody-mediated disease as well (Shi and Hans-Gustaf Ljunggren, personal communication). Therapy of MG with CTLA-4Ig or antibody to CD40L are not wise approaches, because blocking the costimulatory function via CD28- or CD40L-CD40 interaction will lead to global immunosuppression, and these approaches are not clone-/antigen-specific.

8. THE ROLE OF CYTOKINES

Cytokines are small proteins (8 to 80 kDa) involved in regulation of immune reactions, and in the communication functions of the immune system. They act by binding onto receptors on cellular membranes, and can have autocrine as well as paracrine action. Production of an AChR-induced soluble "helper" factor by immune cells was described more than a decade ago (23). Since then, many studies have tried to elucidate the roles of Th$_1$ and Th$_2$ cytokines in the pathogenesis of EAMG.

One of the easiest and most direct ways to study their implications in the development of EAMG is through the use of cytokine (or cytokine receptor) gene KO mice, which lack completely the expression of at least one cytokine (or receptor for cytokines). But this approach does not provide all the answers. Often the functions can be shared by several cytokines, and compensating mechanisms exist in the immune system. Also, one cytokine can regulate the function of another cytokine. However, it is possible to determine whether a given cytokine plays a role in EAMG development and one could specifically neutralize or augment the function of a given cytokine and study its role during the afferent and efferent autoimmune responses to AChR.

8.1.   The Role of IFN-α

The influence of IFN-α, which downregulates MHC class II expression on lymphoid cells, has been studied by treating AChR-immunized mice with IFN-α. IFN-α-mediated reduction of MHC class II expression was believed to be due to down-regulation of IFN-γ by IFN-α (74). After immunization with AChR, IFN-α treatment suppressed EAMG development in B6 mice (74). IFN-α treatment, after established disease, induced significant disease remission, with reduction of anti-AChR antibodies belonging to the IgG$_1$ and IgG$_{2b}$ isotypes (75).

8.2.   The Role of IFN-γ and IL-12 (Th$_1$)

Local IFN-γ production at the NMJ in γ-IFN-ε transgenic mice led to the development of an MG-like syndrome (76), which may be caused by the direct destruction of AChR at the postsynaptic membrane by proinflammatory cytokine IFN-γ. However, there is no evidence to suggest IFN-γ can directly mediate damage in the postsynaptic membrane in MG or EAMG.

IFN-γ gene KO mice failed to develop muscle weakness after immunization with AChR (77). The resistance to clinical EAMG is associated with a dramatic reduction in the levels of circulating anti-AChR antibodies, compared to those in wild type mice. Reduction of antibodies affected all the IgG isotypes. AChR-primed IFN-γ KO mice demonstrated a good lymphoproliferative response to AChR and dominant peptide α146–162 (77). Likewise, IFN-γ receptor gene KO mice exhibited a lower incidence of disease compared to wild-type littermates (78), and their production of AChR-specific antibodies was also reduced, an effect mainly due to decreases in specific IgG$_{2a}$ and IgG$_3$ antibodies. One should note that IgG$_{2a}$ gene is deleted in B6 mice. Therefore, the IgG$_{2a}$ anti-AChR antibody detected in AChR-immunized B6 mice could be due to cross reactivity with IgG$_{2b}$ isotype. The lymphoproliferative response to AChR was not affected in IFN-γ receptor gene KO mice (78). The above studies are being repeated in congenic IFN-γ KO mice, housed in a pathogen-free environment, in the B6 background, to study the precise role of IFN-γ in EAMG pathogenesis.

IL-12 treatment of B6 mice increased the lymphoproliferative response and IFN-γ production in cells from AChR-immunized mice stimulated in vitro with AChR. AChR-primed IL-12 gene (p40) KO mice have shown a reduced incidence of EAMG and production of IFN-γ after in vitro stimulation with AChR (79). Even IL-12 p35 gene

KO mice demonstrated a reduced incidence of EAMG (Goluszko et al., manuscript in preparation).

8.3.   The Role of IL-4 and IL-10 (Th$_2$)

IL-4 gene KO mice developed EAMG and had an intact in vitro lymphoproliferative response to AChR and peptide α146–162 (80). They also produced in vitro more IL-2 and IFN-γ than did wild-type B6 mice after priming with AChR. Anti-AChR IgG antibody production by IL-4 KO mice in vivo was equivalent to that of B6 mice, but IgG$_1$ levels were decreased (80). Balaji et al. did not observe any significant increase in the incidence of clinical EAMG in IL-4-deficient mice (80). However, Karachunski et al. observed a slight increase in the incidence of clinical EAMG in IL-4 deficient mice (81), a difference that may be due to the methods the two laboratories used to screen mice for clinical EAMG. Even in human MG, there is no evidence for IL-4 gene polymorphism (82). IL-4 deficiency precludes EAMG prevention by nasal administration of CD4$^+$ epitopes of AChR (81), since IL-4 plays a more crucial role as a mucosal cytokine and could be protective during mucosal tolerance induction.

Data indicate that IL-12 and IFN-γ, but not IL-4, are required for EAMG development. However, IL-4 appears to play a role during mucosal AChR peptide-induced tolerance. Another Th$_2$ cytokine, IL-10, facilitated the development of EAMG (Poussin et al., in J. Neuroimmunol, in press) in a preliminary study. Proinflammatory cytokines, IL-6, TNF-α, and lymphotoxins also appear to facilitate EAMG development (Deng et al.; Goluszko et al., manuscripts submitted). Because of the cross-regulatory and overlapping functions between cytokines, it is crucial to study the effect of individual cytokines during the afferent and efferent limbs of an autoimmune response to a given autoantigen. Our goal is to study the precise cellular and molecular mechanism by which each cytokine contributes to EAMG and MG pathogenesis. These studies could lead to cytokine-directed therapy and/or combination immunotherapy, by blocking the function of facilitative cytokine(s) and maintain remission by T cell and/or B cell AChR epitope(s) tolerance. All of the data obtained with cytokines or their receptor gene KO mice are summarized in Table 2.

9.   ANTIGEN-SPECIFIC IMMUNOINTERVENTION IN EAMG

Because EAMG in mice mimics human MG in clinical and immunopathological manifestation and the cellular mechanism of disease is well understood, the EAMG model serves as a good system for pre-clinical evaluation of specific therapeutic agents. Conventionally, MG has been

Table 2   The effect of cytokine- or cytokine receptor-gene disruption in EAMG*

Gene Knockout	Clinical EAMG (%)	Anti-AChR Antibodies	Anti-AChR Isotypes	Proliferative response to AChr	$\alpha146$–162	Reference
No	50–80	High	$\uparrow$IgG$_1$, and IgG$_{2b}$	High	High	
IFN-γ	0	Low	$\downarrow$IgG$_1$, IgG$_{2a}$, and IgG$_{2b}$	High	High	77
IFN-γ R	19–23	Low	$\downarrow$IgG$_{2a}$, and IgG$_{2b}$	High	ND	78
IL-12	17	Low	$\downarrow$IgG$_{2a}$, and IgG$_3$	ND	ND	79
IL-4	60–80	High	$\downarrow$IgG$_1$, and $\uparrow$IgG$_{2a}$	High	High	80,81

ND: Not done

*Reprinted from MYASTHENIA GRAVIS: Disease Mechanism and Immunointervention, Editor P. Christadoss, ©Narosa Publishing House, New Delhi, 1999.

treated with immunosuppressive drugs, such as cortico-steroids, azathioprine, cyclophosphamide, cyclosporin A, and symptom-relieving drugs. These medications may be effective, but the main drawback of immunosuppressive therapy is its non-specificity, which causes side effects due to global immunosuppression (83–87).

Ideally, treatment of autoimmune diseases, including MG, should specifically delete or suppress the function of the lymphocyte clones involved in disease pathogenesis. High-dose tolerance to a given autoantigen has long been proposed as an effective mean of treating autoimmune diseases. This tolerance approach is antigen-specific, which only suppresses the antigen-specific immune responses and leaves the rest of the immune responses intact. Pathogenesis of EAMG and MG depends on T cell help for B cell production of pathogenic antibodies. It is suggested that $\alpha146$–162 peptide-reactive CD4$^+$ cells interact with AChR-specific B cells to produce pathogenic anti-AChR antibodies. Thus, suppression or deletion of AChR-specific CD4 cells should either prevent and/or induce clinical remission of EAMG. In this section, we will provide an overview of studies performed in different laboratories on antigen-specific interventions of EAMG, and the possible mechanisms which underlie each therapeutic approach.

9.1.   Systemic Tolerance with AChR T Cell Epitope $\alpha146$–162 Peptide

To evaluate the effect of high-dose $\alpha146$–162 peptide tolerance in EAMG, we performed a series of experiments. The $\alpha146$–162 peptide-induced tolerance not only suppressed the response to $\alpha146$–162 peptide itself, but also the responses to the entire auto-antigen AChR and the other AChR subdominant peptide $\alpha182$–198 due to a possible epitope spread (bystander effect), which is caused by reduction of IL-2, IFN-γ, or IL-10 production (88). High doses of peptide $\alpha146$–162 failed to suppress the KLII-primed T cell response, implicating the antigen-specificity of $\alpha146$–162 peptide-induced tolerance.

The incidence of clinical EAMG was significantly lower in the high-dose $\alpha146$–162-treated group compared to that of a control group. The serum anti-AChR antibody IgG$_{2b}$ isotype was significantly suppressed by high-dose $\alpha146$–162 tolerance. Thus, suppressing a fraction of pathogenic anti-AChR IgG$_{2b}$ antibody, rather than the overall antibody response could suppress EAMG (88). The data also implicate, for the first time, the critical role played by the anti-AChR IgG$_{2b}$ isotype in EAMG pathogenesis.

In our system, we could tolerize the T cell response and clinical EAMG after the primary immune response has been established. It remains to be seen as to whether this approach will be effective in reversing the established disease.

9.2.   Oral Tolerance

Oral tolerance with the whole AChR molecule. Wang et al. fed Lewis rats with different doses of AChR prior to immunization, and consequently prevented EAMG development. EAMG was believed to be suppressed through active suppression by endogenous cyokines, such as TGF-β, and by suppression of both Th$_1$ and Th$_2$ cytokine production (89–91). Similar data were obtained by Okumura et al. (92).

Drachman et al. tested the effect of high-dose AChR oral tolerance on the ongoing immune response to AChR (93). Oral high-dose AChR significantly reduced the clinical score. However, the antibody response to either *Torpedo* or rat AChR was enhanced in the AChR-fed rats. The cellular response to AChR and IL-2 production were not inhibited by oral high doses of AChR. The authors concluded that the extra antibodies produced after oral tolerance were not pathogenic. Therefore, oral high doses of AChR in an ongoing immune response resulted in a paradoxical inverse relationship between the increased antibody response and clinical benefit.

Oral tolerance with α146–162 peptide. In a recent study, oral tolerance with α146–162 peptide prevented EAMG and reduced the anti-AChR antibody response (94). Interestingly, oral tolerance to α146–162 peptide suppressed the AChR-specific lymphocyte response to peptides α146–162, α182–198 and whole AChR, as well as both Th_1 (IL-2, IFN-γ) and Th_2 (IL-10) cytokine production, as observed for systemic tolerance with α146–162 peptide (88). Again, studies are required to observe the effect of oral tolerance to AChR or the peptides on established clinical EAMG, in order to apply this approach in the treatment of human MG.

9.3. Nasal Tolerance

Nasal tolerance with the whole AChR molecule. Nasal administration of antigen requires very low doses without peptic proteolytic degradation of the antigen. Ma et al. reported that nasal administration of micrograms of AChR could prevent EAMG, suppress serum anti-AChR antibodies, and decrease the delayed-type hypersensitivity response to AChR (95). Nasal tolerance also suppressed the AChR-specific, IgG-secreting cells, the AChR-reactive, IFN-γ secreting cells, and T cell proliferation to AChR. Nasal tolerance not only prevented EAMG, but also induced protective tolerance in primed animals. Furthermore, nasal administration of higher doses of AChR to Lewis rats two weeks after immunization with AChR in CFA was shown to ameliorate clinical EAMG (96). The numbers of IFN-γ and TNF-α mRNA-expressing LNC cells were suppressed, while IL-4, IL-10 and TGF-β, mRNA-expressing cells were unaffected. These data indicated that prevention of EAMG by nasal administration of high doses of AChR is associated with selective suppression of Th_1 functions, without influencing Th_2 cell functions. The impaired Th_1 functions may result in the production of fewer pathogenic anti-AChR antibodies and contribute to the amelioration of EAMG severity.

Nasal tolerance with AChR peptide. Karachunski et al. observed the effect of nasal delivery of synthetic sequences of *Torpedo* AChR (TAChR) α-chain epitopes recognized by anti-AChR CD4 T cells (α150–169, α181–200, and α360–378), given before and during immunization with TAChR, on clinical EAMG. Nasal administration of α150–169, alone or in combination with the above 3 peptides either before or during immunization, induced lower CD4 cell responsiveness to those epitopes and to AChR, while this method reduced the synthesis of anti-AChR Ab and prevented EAMG (97). Studies on cytokine profiles by splenocytes showed that in sham-tolerized mice only Th_1 (IL-2 only) cells responded to TAChR, while peptide-treated mice had AChR-specific Th_2 (IL-4 and IL-10) responses. AChR peptide treatment induced anergy *in vitro*, and IL-2 could reverse these anergized T cell responses.

However, Zhang et al. showed that synthetic peptides of the AChR α-chain failed to induce nasal tolerance in EAMG (98). EAMG was not prevented by the nasal administration of different doses of AChR α subunit peptides α61–76, α100–116, α146–162, α261–277, or σ subunit σ345–367, or of combinations of these peptides to Lewis rats pre-immunized with AChR (98). This treatment also failed to affect the levels or affinities of anti-AChR antibodies in peptide(s)-treated rats when compared to control rats. Further studies are required to show the effective suppression of ongoing clinical EAMG with nasal tolerance to AChR or its peptides.

9.4. Tolerance With AChR Extracellular Domain

Recombinant human AChR α-chain fragments, Hα1–210 and Hα1–121, contain the main immunogenic region. Nasal administration of these fragments to rats over 10 consecutive days prior to immunization with AChR in CFA markedly prevented EAMG development. T cell responses to AChR and the production of IL-2 and IFN-γ were reduced in comparison to the controls (99).

9.5. Other Approaches toward Antigen-Specific Suppression

Tolerance with AChR analog peptide. Single, substituted human AChR α-chain peptides Hα195–212 and Hα259–271 analogs and dual analog, which contains the two single substituted analogs, may bind to MHC class II molecules on APC as effectively as the original peptides. These analogs were shown to significantly inhibit T cell responses induced by their respective myasthenogenic peptides (97). Hα195–212 and Hα259–271 peptides are immunodominant T cell epitopes in SJL and BALB/c mice, respectively. The single, substituted analogs inhibited the proliferative response of T cell lines specific to relevant peptide and lymph node cells of mice immunized to Hα195–212 and Hα259–271,259–271. The analog inhibited the *in vivo* priming of lymph node cells from both SJL and BALB/c mice, when administered iv, ip, or *per os*. The dual analog also suppressed the manifestation of EAMG induced by inoculation of a pathogenic T cell line. Therefore, a single peptide composed of analogs to two epitopes specificity can be used to regulate T cell responses and the diseases associated with each epitope (100).

With complementary peptide for the AChR. Theoretically, induction of anti-idiotypic (Id) antibodies (Abs) should be a highly specific treatment for the disease by virtue of their potential ability to neutralize Abs to the AChR. Araga et al. (101) have tested this idea by attempting to evoke such anti-Id Abs by immunization with a peptide (termed RhCA 67–16) encoded by RNA complementary to the *Torpedo* AChR main immunogenic region and determining whether such treatment will prevent the development of EAMG. Immunization with RhCA 67–16, but not a control peptide termed PBM 9–1, was found to elicit the production of anti-Id Abs that blocked recognition of native *Torpedo* AChR by its Ab. This anti-Id Ab activity was ablated by incubation of the anti-RhCA 67–16 serum with RhCA 67–16 prior to the assay for Ab binding to AChR. RhCA 67–16 immunization also prevented the development of EAMG in Lewis rats challenged with Torpedo AChR (25% incidence versus 90% in the controls) and diminished the AChR Ab levels in animals injected with low doses of AChR (101).

MHC class II: AChR peptide complex. An MHC class II molecule derived from a rat was coupled with one of the AChR-dominant T cell epitope α100–116 peptides (102). MHC class II: AChR α100–116 could bind to TCR with higher avidity and consequently initiate the early T cell signal events ordinarily induced by APC-associated MHC class II and peptide binding to TCR, but cannot activate T cells, probably due to the lack of a second signal (102). *In vivo* treatment with soluble MHC class II: AChR α100–116 reduced the *in vivo* proliferative response to AChR and peptide α100–116. *In vivo* MHC class II: AChR α100–116 treatment of rats co-immunized with peptides AChR α100–116 and MBP 69–88 inhibited the subsequent *in vitro* proliferative response to AChR α100–116, but not to MBP 69–88 peptide. The EAMG rats were treated once a week for 5 weeks with 25 μg MHC II: AChR α100–116, 25 μg MHC II:HSP 180–188, and 25 μg MHC II alone, respectively. The treatment of individual rats was started when net weight loss exceeded 10 g, and clinical EAMG was reached. Four of six (67%) rats in the MHC class II:AChR α100–116-treated group survived and showed clinical improvement, compared with a 0–20% survival rate in all other treated groups. These data demonstrated that the MHC class II:AChR (α110–116 complex not only specifically suppressed the T cell response to AChR, but also improved established clinical EAMG, which is more representative of the clinical situation encountered in human MG (102).

10. CONCLUSIONS

The afferent and efferent limb of an autoimmune response to AChR in B6 mice is schematically shown in Figure 1. First, the AChR bind to autoreactive B cells via B-cell receptor/surface IgG. AChR is processed and presented by B cells and other APC. Dominant (e.g., α146–162) or sub-dominant (e.g., α111–126 or α182–198) AChR peptides are presented by the I–A^b molecule. Specific CD4$^+$ cells recognize the peptide/MHC class II complex and are activated. APC secrete IL-12, and IL-1 [either IL-12 or IL-1 deficiency suppresses EAMG development (79, Huang and Lefvert, personal communication)] activates the Th cells specific for AChR peptides. Also, NK cells appear to participate in EAMG pathogenesis (Shi and Hans-Gustaf-Ljumggren, personal communication). NK cells might upregulate IL-12 and IFNγ, and therefore could participate in the primary immune response to AChR. Further, not only the above first signal is required, but also the second costimulatory signal via B7/CD28 interaction leads to activation of T cells specific for AChR-dominant peptides. The importance of the CD28 molecule in EAMG has been demonstrated by partial suppression of EAMG in CD28$^{-/-}$ mice. (72). Further, CTLA–4 activation negatively regulates EAMG pathogenesis, since anti-CTLA–4 antibody treatment enhanced EAMG development (73).

The cooperation between T and B cells requires specific cytokines. For example, downregulation of IL-12 and IFN-γ, but not IL-4, prevent EAMG in IFN-γ gene and receptor KO mice, but not in IL-4 gene KO mice (77–80). Further preliminary studies (unpublished) in our laboratory have suggested a facilitative role for IL-10, IL-6, TNF-α, and LT in EAMG pathogenesis. These cytokines appear to activate the B cells specific for production of high affinity anti-AChR antibodies. The interactions between B7/CD28 and CD40/CD40L appear to be crucial in the development of primary T and B cell autoimmune responses to AChR. The CD40/CD40L interaction is absolutely required in EAMG induction, as CD40L gene KO mice are completely resistant to EAMG (72). However, once autoimmunity to AChR is established, antibody to CD40L failed to reverse the established disease (unpublished), suggesting that CD40L interaction is necessary for the primary immune response to AChR, but not absolutely required for the maintenance of an autoimmune response to AChR. Anti-AChR antibodies of high pathogenic potential either destroy AChR by antigen modulation or by antibody and complement mediated destruction of AChR (24). Once memory B cells specific for AChR are established, activation of these memory B cells might require only the cytokines (e.g., IL-10, IL-6, TNF, and LT), which promote activation and differentiation of B cells. Therefore, treatment of established EAMG/MG could be achieved by

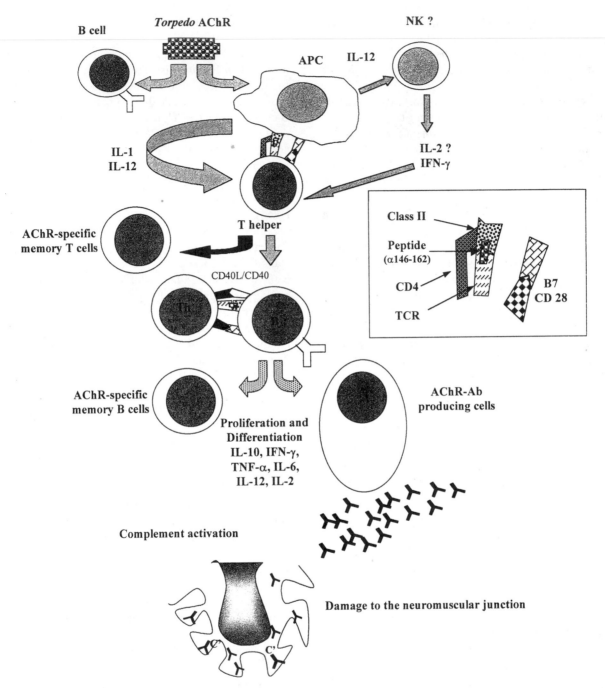

Figure 1.   Hypothetical model of afferent and efferent limb of an autoimmune response to AChR in EAMG.

suppressing the production of cytokines involved in the activation and differentiation of AChR-specific B cells. This form of non-specific therapy (instituted for a temporary period) could be complemented by peptide-specific therapy to maintain remission of clinical EAMG/MG.

ACKNOWLEDGEMENT

Our studies reported in this review were supported by grants from NIH, Muscular Dystrophy Association, Association Française Contre les Myopathies (AFM), Myasthenia Gravis Foundation of America, Sealy Foundation and James W.

McLaughlin Foundation. M. Poussin is an AFM postdoctoral fellow, and C. Deng is a James W. McLaughlin Foundation postdoctoral fellow. Portions of the text appearing in this article have been reproduced from MYASTHENIA GRAVIS: Disease Mechanisms and Immunointervention, Editor P. Christadoss, © Narosa Publishing House, New Delhi, 2000.

References

1. Fambrough, C.M., D.B. Drachman, and S. Satymurti. 1973. Neuromuscular junction in myasthenia gravis: decreased acetylcholine receptors. *Science* 182:293–295.

2. Kao, I., and D.B. Drachman. 1977. Myasthenic immunoglobulin accelerates receptor degradation. *Science* 196:527–529.

3. Heinemann, S., S. Bevan, R. Kullerg, J. Lindstrom, and J. Rice. 1977. Modulation of acetylcholine receptor by antibody against the receptor. *Proc. Natl. Acad. Sci. USA* 7:3090–3094.

4. Drachman, D.B., R.N. Admas, F.F. Stanley, and A. Pestronk. 1980. Mechanisms of acetylcholine receptor loss in myasthenia gravis. *J. Neurol. Neurosurg. Psych.* 43:601–610.

5. Almon, R.R., C.G. Andrew, and S.H. Appel. 1974. Serum globulin in myasthenia gravis: Inhibition of α-bungarotoxin binding to acetylcholine receptors. *Science* 186:55–57.

6. Lindstrom, J.M., M.E. Seybold, V.A. Lennon, S. Whittingham, and D.D. Duane. 1976. Antibody to acetylcholine receptor in myasthenia gravis. *Neurol.* 26:1054–1059.

7. Lefvert, A.K., K. Bengstrom, G. Matell, P.O. Osterman, and R. Pirskanen. 1978. Determination of acetylcholine receptor antibodies in myasthenia gravis clinical usefulness and pathogenic implications. *J. Neurol. Neurosur. Psych.* 4:394–403.

8. Vincent, A., and J. Newsome-Davis. 1985. Acetylcholine receptor antibody as a diagnostic test for myasthenia gravis: results in 153 validated cases and 2967 diagnostic assays. *J. Neurol. Neurosur. Psych.* 48:1246–1252.

9. Engel, A.G., E.H. Lambert, and F.H. Howard. 1977. Immune complexes (IgG and C3) at the motor end-plate in myasthenia gravis: ultrastructural and light microscopic localization and electrophysiologic correlation. *Mayo Clin. Proc.* 52:267–280.

10. Sahashi, K., A.G. Engel, E.H. Lambert, and F.N. Howard, Jr. 1980. Ultrastructural localization of the terminal and lytic ninth complement component (C9) at the motor end-plate in myasthenia gravis. *Neuropathol. Exp. Neurol.* 39:160–172.

11. Sahashi, S., A.G. Engel, J. Lindstrom, E.H. Lambert, and V.A. Lennon. 1978. Ultrastructural localization of immune complexes (IgG and C3) at the end-plate in experimental autoimmune myasthenia gravis. *J. Neuropathol. Exp. Neurol.* 37:212–223.

12. Toyka, K.N., D.B. Drachman, D.E., Griffin, A. Pestronk, J.A. Winkelstein, K.H. Fishbeck, and J. Kao. 1977. Myasthenia gravis: study of humoral immune mechanisms by passive transfer to mice. *N. Engl. J. Med.* 296:125–131.

13. Richman, D.P., C.M. Gomez, P.W. Berman, S.A. Burres, F.W. Fitch, and B.G. Amason. 1980. Monoclonal anti-acetylcholine receptor antibodies can cause experimental myasthenia. *Nature* 286:738–739.

14. Patrick, J., and J.M. Lindstrom. 1973. Autoimmune response to acetylcholine receptor. *Science* 180:871–872.

15. Christadoss P., V.A. Lennon, E.H. Lambert, and C.S. David. 1979. Genetic control of experimental autoimmune myasthenia gravis in mice. In: T and B Lymphocytes: Recognition and Function. *Academic Press*, New York, pp. 249.

16. Christadoss P., V.A. Lennon, C.J. Krco, E.H. Lambert, and C.S. David. 1981. Genetic control of autoimmunity to acetylcholine. Role of Ia molecules. *Ann. NY Acad. Sci.* 377:258–277.

17. Berman P.W., and J. Patrick. 1980. Linkage between the frequency of muscular weakness and loci that regulate immune responsiveness in murine experimental autoimmune myasthenia gravis. *J. Exp. Med.* 152:507–520.

18. Lennon, V., J. Lindstrom, and M.E. Seybold. 1975. Experimental autoimmune myasthenia (EAMG): A model of myasthenia gravis in rats and guinea pigs. *J. Exp. Med.* 141:1365–1375.

19. Tzartos S.J., T. Barkas, M.T. Cung, A. Mamalaki, M. Marraud, P. Orlewski, D. Papanastasiou, C. Sakarellos, M. Sakarellos-Daitsiotis, P. Tsantili, and V. Tsikaris. 1998. Anatomy of the antigenic structure of a large membrane antigen, the muscle-type nicotinic acetylcholine receptor. *Immunol. Rev.* 163:89–120.

20. Vincent A., and N. Willcox. 1994. Characterization of specific T cells in myasthenia gravis. *Immunol. Today* 15:41–42.

21. Christadoss P., J.M. Lindstrom, R.W. Melvold, and N. Talal. 1985. Mutation at I-A beta chain prevents experimental autoimmune myasthenia gravis. *Immunogenetics* 21:33–38.

22. Christadoss P., and M.J. Dauphinee. 1986. Immunotherapy for myasthenia gravis. A murine model. *J. Immunol.* 136:2437–2440.

23. Christadoss P., J.M. Lindstrom, N. Talal, C.R. Duvic, A. Kalantri, and M. Shenoy. 1986. Immune response gene control of lymphocyte proliferation induced by acetylcholine receptor-specific helper factor derived from lymphocytes of myasthenic mice. *J. Immunol.* 137:1845–1849.

24. Christadoss P. 1988. C5 gene influences the development of murine myasthenia gravis. *J. Immunol.* 140:2589–2592.

25. Christadoss, P., J. Lindstrom, S. Munro, and N. Talal. 1985. Muscle acetylcholine receptor loss in murine experimental autoimmune myasthenia gravis: correlated with cellular, humoral and clinical responses. *J. Neuroimmunol.* 8:29–41.

26. Meinl, E., W.E. Klinkert, and H. Wekerle. 1991. The thymus in myasthenia gravis. Changes typical for the human disease are absent in experimental autoimmune myasthenia gravis of the Lewis rat. *Am. J. Pathol.* 139:995–1008.

27. Schluep, M., N. Willcox, A. Vincent, G.K. Dhoot, and J. Newsom-Davis. 1987. Acetylcholine receptors in human thymic myoid cells *in situ*: an immunohistological study. *Ann. Neurol.* 22:212–222.

28. Wekerle, H., and U.P. Ketelsen. 1977. Intrathymic pathogenesis and dual genetic control of myasthenia gravis. *Lancet* 1:678–680.

29. Kornstein, M.J., O. Asher, and S. Fuchs. 1995. Acetylcholine receptor alpha-subunit and myogenin mRNAs in thymus and thymomas. *Am. J. Pathol.* 146:1320–1324.

30. Kirchner, T., S. Tzartos, F. Hoppe, B. Schalke, H. Wekerle, and H.K. Müller-Hermelink. 1988. Pathogenesis of myasthenia gravis. Acetylcholine receptor-related antigenic determinants in tumor-free thymuses and thymic epithelial tumors. *Am. J. Pathol.* 130:268–280.

31. Marx, A., R. O'Connor, K.I. Geuder, F. Hoppe, B. Schalke, S. Tzartos, I. Kalies, T. Kirchner, and H.K. Müller-Hermelink. 1990. Characterization of a protein with an

acetylcholine receptor epitope from myasthenia gravis-associated thymomas. *Lab. Invest.* 62:279–286.

32. Andretta, F., F. Baggi, C. Antozzi, E. Torchiana, P. Bernasconi, O. Simoncini, F. Cornelio, and R. Mantegazza. 1997. Acetylcholine receptor alpha-subunit isoforms are differentially expressed in thymuses from myasthenic patients. *Am. J. Pathol.* 150:341–348.

33. Sommer, N., N. Willcox, G.C. Harcourt, and J. Newsom-Davis. 1990. Myasthenic thymus and thymoma are selectively enriched in acetylcholine receptor-reactive T cells. *Ann. Neurol.* 28:312–319.

34. Christadoss P., V.A. Lennon, and C.S. David. 1979. Genetic control of experimental autoimmune myasthenia gravis in mice. 1. Lymphocyte proliferative response to acetylcholine receptor is under H-2 linked ir gene control. *J. Immunol.* 123:2540–2543.

35. Christadoss P., V.A. Lennon, and C.S. David. 1982. Genetic control of experimental autoimmune myasthenia gravis in mice. III. Ia molecules mediate cellular immune responsiveness to acetylcholine receptors. *J. Immunol.* 128:1141–1144.

36. Hohlfeld R., B. Conti-Tronconi, I. Kalies, J. Bertrams, and K.V. Toyka. 1985. Genetic restriction of autoreactive acetylcholine receptor-specific T lymphocytes in myasthenia gravis. *J. Immunol.* 135:2393–2399.

37. Brocke S., C. Brautbar, L. Steinman, O. Abramsky, J. Rothbard, D. Neumann, S. Fuchs, and E. Mozes. 1988. *In vitro* proliferative response and antibody titers specific to human acetylcholine receptor in patients with myasthenia gravis and relation to HLA class II genes. *J. Clin. Invest.* 82:1894–1900.

38. Shenoy, M., B. Wu, R. Kaul, E. Goluszko, C. David, and P. Christadoss. Possible linkage of myasthenia gravis susceptibility in R111/SJ mice to the genomic deletion of chromosomal segment coding the T cell receptor Vβ genes. In: Fundamental and Experimental Aspect of Autoimmunity, M. Zouali. Springer-Verlag, editor. Series H: Cell Biology, 1994. Vol. 80. pp. 213–220.

39. Ahlberg, R., Q. Yi, R. Matell, G. Swerup, P. Rieber, G. Reithmuller, G. Holm, and A.K. Lafvert. 1993. Clinical improvement of myasthenia gravis by treatment with a chimeric anti-CD4 monoclonal antibody. *NY Acad. Sci.* 681:552–555.

40. Kaul R., M. Shenoy, E. Goluszko, and P. Christadoss. 1994. Major histocompatibility complex class II gene disruption prevents experimental autoimmune myasthenia gravis. *J. Immunol.* 152:3152–3157.

41. Kaul R., M. Shenoy, and P. Christadoss. 1994. The role of major histocompatibility complex genes in myasthenia gravis and experimental autoimmune myasthenia gravis pathogenesis. *Adv. Neuroimmunol.* 4:387–402.

42. Zhang G.-X., B.-G. Xiao, M. Bakhiet, P. van Der Meide, H. Wigzell, H. Link, and T. Olsson. 1996. Both CD4+ and CD8+ T cells are essential to induce experimental autoimmune myasthenia gravis. *J. Exp. Med.* 184:349–356.

43. Shenoy M., R. Kaul, E. Goluszko, C. David, and P. Christadoss. 1994. Effect of MHC class I and CD8 cell deficiency on experimental autoimmune myasthenia gravis pathogenesis. *J. Immunol.* 153:5330–5335.

44. Conti-Fine, B.M., Z.-Y. Wang, R. Raju, J.F. Howard, Jr., and D. Navaneetham. 1999. Anti-acetylcholine receptor CD4+ T cells in myasthenia gravis: epitope repertoire and T cell receptor gene usage. In: Myasthenia Gravis: Disease Mechanisms and Immunointervention. P. Christadoss, editor. Narosa Publishing House, New Delhi, India (in press).

45. König R., S. Fleury, and R.N. Germain. 1996. The structural basis of CD4-MHC interactions: coreceptor contributions to T cell recognition and oligomerization-dependant signal transduction. *Cur. Topics Microbiol. Immunol.* 205:19–46.

46. Waldor, M.K., S. Sriram, H.O. McDewitt, and L. Steinman. 1983. *In vivo* therapy with monoclonal anti-I-A antibody suppresses immune response to acetylcholine receptor. *Proc. Natl. Acad. Sci. USA* 80:2713–2717.

47. Christadoss P. 1989. Immunogenetics of experimental autoimmune myasthenia gravis. *Crit. Rev. Immunol.* 9:247–278.

48. Bell J., L. Rassenti, S. Smoot, K. Smith, C. Newby, R. Hohlfeld, K. Toyka, H. McDevitt, and L. Steinman. 1986. HLA-DQ beta-chain polymorphism linked to myasthenia gravis. *Lancet* 1:1058–1060.

49. Hjelmström P., R. Giscombe, A.K. Lefvert, R. Pirskanen, I. Kockum, M. Landin-Olsson, and C.B. Sanjeevi. 1996. Polymorphic amino acid domains of the HLA-DQ molecule are associated with disease heterogeneity in myasthenia gravis. *J. Neuroimmunol.* 65:125–131.

50. Hjelmström P., C. De Weese-Scott, J.E. Penzotti, T.P. Lybrand, and C.B. Sanjeevi. 1998. Structural differences between HLA-DQ molecules associated with myasthenia gravis characterized by molecular modeling. *J. Neuroimmunol.* 85:102–105.

51. Hjelmström, P., and C.B. Sanjeevi. 1999. Molecular mechanisms of MHC associations with myasthenia gravis. In: Myasthenia Gravis: Disease Mechanisms and Immunointervention. P. Christadoss, editor. Narosa Publishing House, New Delhi, India (in press).

52. Shenoy M., M. Oshima, M.Z. Atassi, and P. Christadoss. 1993. Suppression of experimental autoimmune myasthenia gravis by epitope-specific neonatal tolerance to synthetic region alpha 146–162 of acetylcholine receptor. *Clin. Immunol. Immunopathol.* 66:230–238.

53. Infante A.J., P.A. Thompson, K.A. Krolick, and K.A. Wall. 1991. Determinant selection in murine experimental autoimmune myasthenia gravis. Effect of the bml2 mutation on T cell recognition of acetylcholine receptor epitopes. *J. Immunol.* 146:2977–2982.

54. Yang B., K.R. McIntosh, and D.B. Drachman. 1998. How subtle differences in MHC class II affect the severity of experimental autoimmune myasthenia gravis. *Clin. Immunol. Immunopathol.* 86:45–58.

55. Bellone M., N. Ostlie, S.J. Lei, X.-D. Wu, and B. Conti-Tronconi. 1991. The I-A^{bm12} mutation, which confers resistance to experimental myasthenia gravis, drastically affects the epitope repertoire of murine CD4+ cells sensitized to nicotinic acetylcholine receptor. *J. Immunol.* 147:1484–1491.

56. Karachunski P.I., N. Ostlie, M. Bellone, A.J. Infante, and B.M. Conti-Fine. 1995. Mechanisms by which the I-A^{bm12} mutation influences susceptibility to experimental autoimmune myasthenia gravis: a study in homozygous and heterozygous mice. *Scand. J. Immunol.* 42:215–225.

57. Oshima M., and M.Z. Atassi. 1995. Effect of amino acid substitutions within the region 62–76 of I-Aβ^b on binding with and antigen presentation of *Torpedo* acetylcholine receptor α-chain peptide 146–162. *J. Immunol.* 154:5245–5254.

58. Christadoss P., C.S. David, M. Shenoy, and S. Keve. 1990. Eκα transgene in B10 mice suppresses the development of myasthenia gravis. *Immunogenetics* 31:241–244.

59. Christadoss P., C.S. David, and S. Keve. 1992. I-Aακ transgene pairs with I-Aβ^b gene and protects C57BL10 mice

from developing autoimmune myasthenia gravis. *Clin. Immunol. Immunopathol.* 62:235–239.

60. Cheng S., J. Baisch, C. Krco, S. Savarirayan, J. Hanson, K. Hodgson, M. Smart, and C. David. 1996. Expression and function of HLA-DQ8 (DQA1*0301/DQB1*0302) genes in transgenic mice. *Eur. J. Immunogenetics* 23:15–20.

61. Raju R., W.-Z. Zhan, P. Karachunski, B. Conti-Fine, G.C. Sieck, and C. David. 1998. Polymorphism at the HLA-DQ locus determines susceptibility to experimental autoimmune myasthenia gravis. *J. Immunol.* 160:4169–4174.

62. Shenoy, M., Kaul, R., Goluszko, E., and Christadoss, P. 1994. The pathogenic role of acetylcholine receptor α chain T cell epitope within region 146–162 in the development of experimental autoimmune myasthenia gravis in C57BL6 mice. *Clin. Immunol. Immunopath.* 73:338–343.

63. Oshima, M., T. Ashizawa, M. Pollack, and M.Z. Atassi. 1990. Autoimmune T cell recognition of human acetylcholine receptor: the sites of T cell recognition in myasthenia gravis on the extracellular part of the α subunit. *Eur. J. Immunol.* 20:2563–2569.

64. Krco C.J., C.S. David, and V.A. Lennon. 1991. Mouse T lymphocyte response to acetylcholine receptor determined by T cell receptor for antigen Vβ gene products recognizing MIs-1α. *J. Immunol.* 147:3303–3305.

65. Infante A.J., H. Levcovitz, V. Gordon, K.A. Wall, P.A. Thompson, and K.A. Krolick. 1992. Preferential use of a T cell receptor Vβ gene by acetylcholine receptor reactive T cells from myasthenia gravis-susceptible mice. *J. Immunol.* 148:3385–3390.

66. Aime-Sempe, C., S. Cohen-Kaminsky, C. Bruand, I. Klingel-Schmitt, F. Truffault, and S. Berrih-Aknin. 1995. *In vivo* preferential usage of TCR V beta 8 in *Torpedo* acetylcholine receptor immune response in the murine experimental model of myasthenia gravis. *J. Neuroimmunol.* 58:191–200.

67. Kaul R., D. Wu, E. Goluszko, C. Deng, V. Dedhia, G.H. Nabozny, C.S. David, I.J. Rimm, M. Shenoy, T.M. Haqqi, and P. Christadoss. 1997. Experimental autoimmune myasthenia gravis in B10.BV8S2 transgenic mice. Preferential usage of TCRAV1 gene by lymphocytes responding to acetylcholine receptor. *J. Immunol.* 158:6006–6012.

68. Wu B., M. Shenoy, E. Goluszko, R. Kaul, and P. Christadoss. 1995. TCR gene usage in experimental autoimmune myasthenia gravis pathogenesis. Usage of multiple. TCRBV genes in the H-2^b strains. *J. Immunol.* 154:3603–3610.

69. Melms, A., J.R. Oksenberg, G. Malcherek, R. Schoepfer, C.A. Miiller, J. Lindstrom, and L. Steinman. 1993. T-cell receptor gene usage of acetylcholine receptor-specific T-helper cells. *Ann. NY Acad. Sci.* 681:313–314.

70. Dedhia V., E. Goluszko, B. Wu, C. Deng, and P. Christadoss. 1998. The effect of B cell deficiency on the immune response to acetylcholine receptor and the development of experimental autoimmune myasthenia gravis. *Clin. Immunol. Immunopathol.* 87:266–275.

71. Li, H., S. Fu-Dong, H. Bing, M. Bakheit, B. Wahren, A. Berglöf, K. Sandtedt, and H. Link. 1998. Experimental autoimmune myasthenia gravis induction in B cell-deficient mice. *Internatl. Immunol.* 10:1359–1365.

72. Shi F.-D., B. He, H. Li, D. Matusevicius, H. Link, and H.-G. Ljunggren. 1998. Differential requirements for CD28 and CD40 ligand in the induction of experimental autoimmune myasthenia gravis. *Eur. J. Immunol.* 28:3587–3593.

73. McIntosh, K.R., P.S. Linsley, P.A. Bacha, and D.B. Drachman. 1998. Immunotherapy of experimental autoim-

mune myasthenia gravis: selective effects of CTLA4Ig and synergistic combination with an IL2-diphtheria toxin fusion protein. *J. Neuroimmunol.* 87:136–146.

74. Shenoy M., S. Baron, B. Wu, E. Goluszko, and P. Christadoss. 1995. IFN-α treatment suppresses the development of experimental autoimmune myasthenia gravis. *J. Immunol.* 154:6203–6208.

75. Deng C., E. Goluszko, S. Baron, and Christadoss P. 1996. Interferon α therapy is effective in suppressing the clinical experimental myasthenia gravis. *J. Immunol.* 157:5675–5682.

76. Gu, D., L.L. Wogensen, N.A. Calcutt, C. Xia, X. Zhu, J.P. Merlie, H.S. Fox, J. Lindstrom, H. Powell, and N. Sarvetnick. 1995. Myasthenia gravis-like syndrome induced by expression of γ-IFN in the neuromuscular junction. *J. Exp. Med.* 181:547–557.

77. Balasa B., C. Deng, J. Lee, L.M. Bradley, D.K. Dalton, P. Christadoss, and N. Sarvetnick. 1997. Interferon γ (IFN-γ) is necessary for the genesis of acetylcholine receptor-induced clinical experimental autoimmune myasthenia gravis in mice. *J. Exp. Med.* 186:385–391.

78. Zhang G.-X., B.-G. Xiao, X.-F. Bai, P.H. van Der Meide, A. Om, and H. Link. 1999. Mice with IFN-γ receptor deficiency are less susceptible to experimental autoimmune myasthenia gravis. *J. Immunol.* 162:3775–3781.

79. Moiola L., F. Galbiati F, G. Martino, S. Amadio, E. Brambilla, G. Comi, A. Vincent, L.M. Grimaldi, and L. Adorini. 1998. IL 12 is involved in the induction of experimental autoimmune myasthenia gravis, an antibody-mediated disease. *Eur. J. Immunol.* 28:2487–2497.

80. Balasa B., C. Deng, J. Lee, P. Christadoss, and N. Sarvetnick. 1998. The Th$_2$ cytokine IL 4 is not required for the progression of antibody-dependent autoimmune myasthenia gravis. *J. Immunol.* 161:2856–2862.

81. Karachunski, P.I., N.S. Ostlie, D.K. Okita, and B.M. Fine. 1999. Interleukin-4 deficiency facilitates development of experimental myasthenia gravis and precludes its prevention by nasal administration of CD4$^+$ epitope sequences of the acetylcholine receptor. *J. Neuroimmunol.* 95:73–84.

82. Huang D., S. Xia, Y. Zhou, R. Pirskanen, L. Liu, and A.K. Lefvert. 1998. No evidence for interleukin-4 gene conferring susceptibility to myasthenia gravis. *J. Neuroimmunol.* 92:208–211.

83. Johns, T.R. 1987. Long term corticosteroid treatment of myasthenia gravis. *Ann. NY Acad. Sci.* 505:568–583.

84. Matell, G. 1987. Immunosuppressive drugs: Azathioprine in the treatment of myasthenia gravis. *Ann. NY Acad. Sci.* 505:588–594.

85. Drachman, D.B. 1993. Myasthenia gravis. In: Current Therapy in Neurologic Disease. 4th ed. R.T. Johnson, and J.W. Griffin, editors. St. Louis: Mosby-Year Book, pp. 379–384.

86. Tindall, R.S., J.T. Phillips, J.A. Rollins, L. Wells, and K. Hall. 1993. A clinical therapeutic trial of cyclosporine in myasthenia gravis. *Ann. NY Acad. Sci.* 681:539–551.

87. Wilensky, R., B. Dwyer, and R.F. Mayer. 1993. Relapses in patients with myasthenia gravis treated with azathioprine. *Ann. NY Acad. Sci.* 681:591–593.

88. Wu, B., C. Deng, E. Goluszko, and P. Christadoss. 1997. Tolerance to a dominant T cell epitope in the acetylcholine receptor molecule induces epitope spread and suppresses murine myasthenia gravis. *J. Immunol.* 159:3016–3023.

89. Wang, Z.Y., J. Qiao, and H. Link. 1993. Suppression of experimental autoimmune myasthenia gravis by oral

administration of acetylcholine receptor. *J. Neuroimmunol.* 44:209–214.

90. Ma, C.G., G.X. Zhang, B.G. Xiao, Z.Y. Wang, J. Link, T. Olsson, and H. Link. 1996. Mucosal tolerance to experimental autoimmune myasthenia gravis is associated with down-regulation of AChR-specific IFN-gamma-expressing Th$_1$-like cells and up-regulation of TGF-beta mRNA in mononuclear cells. *Ann. NY Acad. Sci.* 778:273–287.

91. Wang, Z.Y., H. Link, A. Ljungdahl, B. Hojeberg, J. Link, B. He, J. Qiao, A. Melms, and T. Olsson. 1994. Induction of interferon-γ, interleukin-4, and transforming growth factor-β in rats orally tolerized against experimental autoimmune myasthenia gravis. *Cell. Immunol.* 157:353–368.

92. Okumura, S., K. McIntosh, and D.B. Drachman. 1994. Oral administration, of acetylcholine receptor: effects on experimental myasthenia gravis. *Ann. Neurol.* 36:704–713.

93. Drachman, D.B., S. Okumura. R.N. Adams, and K.R. McIntosh. 1996. Oral tolerance in myasthenia gravis. *Ann. NY Acad. Sci.* 778:258–272.

94. Baggi, F., F. Andreetta, E. Caspani, M. Milani, O. Simonicini, R. Longhi, F. Cornelio, R. Mantegazza, and C. Antozzi. In: Myasthenia Gravis: Disease Mechanism and Immunointervention. P. Christadoss, editor. Narosa Publishing House, New Delhi, India (in press).

95. Ma, C.G., G.X. Zhang, B.G. Xiao, J. Link, T. Olsson, and H. Link. 1995. Suppression of experimental autoimmune myasthenia gravis by nasal administration of acetylcholine receptor. *J. Neuroimmunol.* 58:51–60.

96. Shi, F.D., X.F. Bai, H.L. Li, Y.M. Huang, P.H. Van Der Meide, and H. Link. 1998. Nasal tolerance in experimental autoimmune myasthenia gravis (EAMG): induction of protective tolerance in primed animals. *Clin. Exp. Immunol.* 11:506–512.

97. Karachunski, P.I., N.S. Ostlie, D.K. Okita, and B.M. Conti-Fine. 1997. Prevention of experimental myasthenia gravis by nasal administration of synthetic acetylcholine receptor T epitope sequences. *J. Clin. Invest.* 100:3027–3035.

98. Zhang, G.X., F.D. Shi, J. Zhu, B.G. Xiao, M. Levi, B. Wahren, L.Y. Yu, and H. Link. 1998. Synthetic peptides fail to induce nasal tolerance to experimental autoimmune myasthenia gravis. *J. Neuroimmunol.* 85:96–101.

99. Barchan, D., O. Asher, S.J. Tzartos, S. Fuchs, and M.C. Souroujon. 1998. Modulation of the anti-acetylcholine receptor response and experimental autoimmune myasthenia gravis by recombinant fragments of the acetylcholine receptor. *Eur. J. Immunol.* 28:616–624.

100. Katz-Levy, Y., M. Dayan, I. Wirguin, M. Fridkin, M. Sela, and E. Mozes. 1998. Single amino acid analogs of a myasthenogenic peptide modulate specific T cell responses and prevent the induction of experimental autoimmune myasthenia gravis. *J. Neuroimmunol.* 85:78–86.

101. Araga, S., R.D. LeBoeuf, and J.E. Blalock. 1993. Prevention of experimental autoimmune myasthenia gravis by manipulation of the immune network with a complementary peptide for the acetylcholine receptor. *Proc. Natl. Acad. Sci. USA* 90:8747–8751.

102. Spack, E.G., M. McCutcheon, N. Corbelletta, B. Nag, D. Passmore, and S.D. Sharma. 1995. Induction of tolerance in experimental autoimmune myasthenia gravis with solubilized MHC class II:acetylcholine receptor peptide complexes. *J. Autoimmun.* 8:787–807.

45 | Vitiligo

Raymond E. Boissy

1. VITILIGO AS A POSSIBLE AUTOIMMUNE DISEASE

Vitiligo is an acquired cutaneous disease in which the melanocyte component of the skin is destroyed, resulting in amelanotic lesions of variable size and extent (1–3). Generally, vitiligo initially develops on the hands, wrist, body folds, and orifices such as the eyes, mouth, nose, etc. However, vitiligo lesions can develop anywhere on the body. In some cases the disease is confined to those initial sites, however, in most cases it progresses and can affect the entire body surface (Figure 1). In addition, depigmentation of the uveal tract of the eye and putatively of the ear can also occur resulting in minimal night blindness/photophobia (4) and sensorineural hypoacusis/high frequency hearing loss (5,6), respectively. As a health consequence, the white lesions of the skin become immunocompromised, exhibiting a muted response to contact antigens (7,8). In addition, amelanotic vitiligo lesions can become susceptible to the damaging effects of solar irradiation (*i.e.*, premature aging/actinic damage (9) and possibly cancer of the skin). The average age of onset of vitiligo is 22 years with no significant differences in mean age at onset or prevalence between males and females (10,11). From the patient's perspective, the more devastating consequences of this disfiguring loss of skin and eye pigment are the psychological and social problems that result (12,13). This is especially true for African-Americans and American or Asian Indians. Cases of social ostracism and suicide have been occasionally reported.

The etiology of vitiligo is not clearly understood. At present there have been various causative factors implicated in the depigmentary processes of vitiligo. These processes included the cytological (*i.e.*, genetic). environmental (*i.e.*,

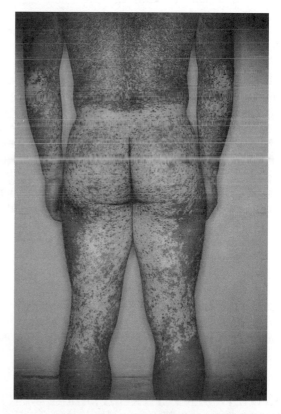

Figure 1. Patient with vitiligo in which the amelanotic lesions are distributed throughout most of the body surface. [Photograph courtesy of James J. Nordlund, Univ. of Cincinnati.]

occupational). immunological (*i.e.*, autoimmune), and neurological destruction of the melanocyte (2,14–16). However, in the last few years, a consensus seems to be forming that vitiligo represents a group of disorders with a variety of causes (2). In epidemiological studies based primarily on clinical records of hospitals and dermatology clinics, the prevalence of vitiligo has been demonstrated to be 6% in West Africa, 2% in Japan, 1% in the United States of America and Egypt, 0.5% in India and Korea, 0.38% in Denmark, 0.33% in Libya, 0.24% in the United Kingdom and 0.14% in Russia (10,17–26). There is clearly a multi-factorial genetic component to vitiligo that appears to predispose individuals to this disease (11,19, 27,28) and that may be responsible for the complex nature of its presentation. Vitiligo is not inherited by a simple Mendelian mechanism. Studies on familial inheritance patterns in India and the United States have demonstrated a complex expression for the inheritance of vitiligo (29,30). In a recent epidemiological study of 15,685 individuals from 298 families, the investigators found a 7-fold increase of vitiligo in primary family members of individuals with vitiligo (11). The investigators developed a genetic model by computer that postulated that a minimum of three diallelic genes might be coordinately involved with the expression of vitiligo i.e., as a polygenic disorder (28).

There are many features of vitiligo suggesting that the development and/or progression of this disease may have immunologically components. In addition, some aspects of vitiligo implicate this disease, or more specifically a subtype of this disease, as an autoimmune disorder. Patients with vitiligo frequently express defined autoimmune syndromes. These include autoimmune based thyroid diseases (i.e., Graves' disease, hyperthyroidism, Hashimoto's disease, hypothyroidism), Addison's disease, pernicious anemia, diabetes mellitus, alopecia areata, and uveitis (31,32). Additional multiple endocrinopathies also are prevalent in patients with vitiligo (32). It has also been demonstrated that patients with vitiligo have a significant increased frequency of organ-specific serum autoantibodies, specifically anti-thyroglobulin, anti-thyroid microsomal, and anti-parietal cell (33–35). Ocular inflammation (i.e., uveitis) also frequently develops in patients with vitiligo (36,37). Because of the association of vitiligo with autoimune disorders it has been proposed that the underlying mechanism of melanocyte destruction in patients with vitiligo was immunologically based.

2. INVOLVEMENT OF THE HUMORAL IMMUNE SYSTEM IN VITILIGO

Since epidermal melanocytes were considered to be immunological targets in an autoimmune response leading to vitiligo, investigators began in the early 1980 to search for melanocyte specific serum autoantibodies (38). Extensive studies by J-C Bystryn have demonstrated that patients with vitiligo can express serum immunoglobulins that recognize melanocyte surface antigens (39,40). Up to

five antigens (35,40–45,75,90, and 150 kDa) were subsequently demonstrated (41). The level of these serum autoantibodies appeared to increase with extent and/or activity of the disease (42,43) and to decrease after theraputic success (44). However, with the exception of a 90 kDa antigen, the expression of these molecules is not restricted to melanocytes but expressed by other cell types in culture. The 40–45 kDa antigen shares a cross-reacting epitope with or is tightly bound to class I HLA. The identity of the other vitiligo antigens is unconfirmed.

More recent reports have demonstrated that vitiligo serum can react with some melanocyte specific proteins (i.e., differentiation antigens). Reactivity to tyrosinase (45–47) and tyrosinase-related protein-1 (TRP-1) (48), both of which are approximately 75 kDa in molecular weight, has been demonstrated. Both of these molecules participate in the catalytic conversion of tyrosine to melanin in the melanocytes (rev. in 49 and 50). However, not all investigators concur that vitiligo patients express serum antibodies that can recognize tyrosinase (51) or TRP-1 (45). In addition, serum antibodies recognizing other gene products involved in the synthesis of melanin, i.e., tyrosinase-related protein-2 (TRP-2) (52,53) and pmel 17 (54), have also been identified in a minority of patients with vitiligo.

Regardless of the molecular nature of the antigen(s), serum from patients with vitiligo exhibits functional destruction of melanocytes in vitro. Specifically, vitiligo serum can cause antibody dependent cellular cytotoxicity as well as complement dependent cytotoxicity of cultured melanocytes (43,55). In addition, injection of the IgG fraction of serum from patients with vitiligo into human skin grafted on a nude mice results in melanocytes destruction in the graft (56). These data suggest that the humoral component of the immune system may be involved in the destruction of the melanocytes. However, histological evidence of B cells or immunoglobulin deposits in the epidermis of advancing lesions of vitiligo has not been demonstrated (38,57).

3. INVOLVEMENT OF THE CELLULAR IMMUNE SYSTEM IN VITILIGO

Pertaining to a cellular immune response associated with vitiligo, inflammatory cells are occasionally observed in vitiligo lesions (58,59). Earlier studies have demonstrated alteration in peripheral T cell subtypes in patients with vitiligo. Specifically, both CD4+ T cells and the CD4+/CD8+ ratio have been reported to either increase (60,61) or decrease (62–64) in the peripheral blood of patients with vitiligo. However, an apparent cutaneous inflammation (65–67) and an abundance of lymphocytes at the inflamed border of an advancing vitiligo lesion rarely occur (68,69). In contrast, these characteristics of inflammation are relatively prevalent in a minor phenotypic subtype of vitiligo termed inflammatory vitiligo (Figure 2) (2). The infiltrating T cells in the perilesional skin exhibit an increase CD8/CD4 ratio plus an increase in the cutaneous lymphocyte antigen and the interleukin-2 receptor (70). In addition, macrophages (71), specifically CD68⁺ OKM5⁻ macrophages (70), are abundant in the dermis at the border of a vitiligo lesion. Also occasionally upregulated in perilesional vitiligo skin is the expression of class II HLA antigens and ICAM-1 (72). Recently, using tetrameric complexes of human histocompatibility leukocyte antigen class I to identify antigen-specific T cells *in vivo*, Ogg et al. (73) demonstrated a high frequency of circulating MelanA-specific, A *0201-restricted cytotoxic T lymphocyte in seven of nine patients with vitiligo. In addition, these CTLs isolated from the vitiligo patients expressed high levels of the skin homing receptor, cutaneous lymphocyte-associated antigen (73). These data suggest that a cellular immune response may be associated with the pathogenesis of the inflammatory form of vitiligo. However, it was recently demonstrated that a CTLA-4

polymorphism associated with various autoimmune disorders such as Graves' disease, autoimmune Addison's disease, and autoimmune hypothyroidism is not associated with the expression of vitiligo (74).

4. VITILIGO AND MELANOMA

There is an interesting connection between the collateral expression of melanoma and vitiligo. Several studies indicate that the prevalence of vitiligo in patients with melanoma is slightly increased over the prevalence of vitiligo in the general population (75–78). Patients with melanoma who develop vitiligo tend to have a good prognosis for post-tumor survival (77–81). However, several other reports failed to demonstrate improved prognosis of a melanoma in the presence of vitiligo (75,78,82). On occasion, the skin surrounding either a primary or a metastatic melanoma may become depigmented resulting in a halo nevus, resembling vitiligo. It has been hypothesized that the immunosurvalence mechanisms mounted in melanoma patients to combat the

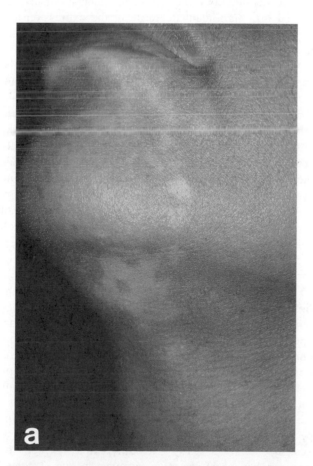

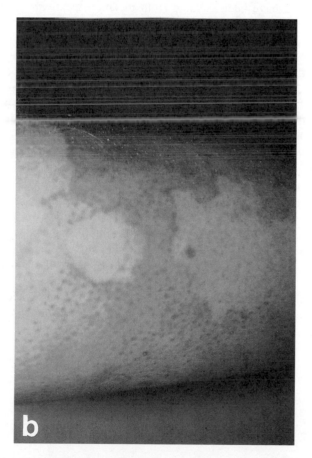

Figure 2.   Vitiligo lesions: (a) Typical vitiligo lesions in which there is a clear demarcation between the amelanotic lesion and the surrounding normally pigmented skin. (b) Atypical vitiligo lesions of a subtype of vitiligo termed "inflammatory vitiligo" in which there is a distinct rim of inflammation along the perimeter of the advancing amelanotic lesion. [Photographs courtesy of James J. Nordlund, Univ. of Cincinnati.]

tumor may spill over and result in the immunological attack of normal epidermal melanocytes. However, this specific hypothesis, as well as the general phenomena that there is an autoimmune mechanism underlying vitiligo that can be harnessed to reject melanoma, has been thoroughly contested (83).

It has been demonstrated that patients with melanoma can produce an immune response to melanocytes differentiation antigens expressed within the melanosomal organelle of normal melanocytes. Some of these antigens have been demonstrate to be normal melanocytes specific proteins. These latter consist of tyrosinase, TRP-1 and pmel-17, which are enzymes and regulatory proteins in the normal biochemical conversion of the amino acid tyrosine to melanin (49,50). Some antigens, like MART-1/melan A, have yet to be molecularly and functionally identified and their role remains unknown (84). The generation of serum antibodies to tyrosinase related protein-1 (85,86) and of CD8$^+$ lymphocytes to various melanocyte differentiation antigens (87–89) in melanoma patients have been described. In addition, tumor-infiltrating lymphocytes targeted to MART-1/melan A, gp100, tyrosinase, TRP-1, and TRP-2 have been isolated from melanoma patients with diverse HLA types (84,90–94).

The documentation of an immune response to normal melanocyte antigens in patient's with melanomas resulted in the development of therapeutic antimelanoma vaccines (88,95–98). Various designed experimental therapies for melanoma treatment can result in the development of vitiligo. A chemotheraputic treatment accompanied with high doses of interferon gamma in patients with melanoma occasionally resulted in the development of vitiligo concurrent with the regression of the melanoma (99,100). In contrast, some melanocyte antigen targeted immunotherapies in treating melanoma have resulted in tumor regression, however the collateral development of vitiligo was not observed (97). In a mouse model, the simultaneous development of vitiligo and tumor regression after immunotherapy has been documented (101). Specifically, mice inoculated with a vaccinia virus encoding murine TRP-1 exhibited pelage depigmentation. In addition, these vaccinated mice readily rejected exogenously administered B16 melanomas by a CD4$^+$ T lymphocyte process.

5. ANIMAL MODELS OF VITILIGO

5.1. The Smyth Chicken

The investigation of the few animal models for vitiligo has provided some insights into the pathophysiology of vitiligo, particularly with respect to immune involvement (102,103). The most well-documented is an avian model, the Smyth

chicken (104), which can develop feather amelanosis generally around the age of sexual maturity. The extent of feather amelanosis that subsequently develops is highly variable, resembling vitiligo in humans. Birds of the Smyth line also express many of the associated autoimmune disorders presented by patients with vitiligo, including ocular depigmentation accompanied by uveitis, an autoimmune-based alopecia areata-like trait, and autoimmune thyroditis. The involvement of the immune system in the etiology of vitiligo in the Smyth chickens was demonstrated by an influx of immunocytes prevalent in the dermal pulp and occasionally in the epidermal layer of the feather during the development of amelanosis (105). In addition, lymphocytes and macrophages become abundant in the uveal tract in concert with ocular melanocyte damage (105–107). Corticosterone therapy of young pre-amelanotic Smyth chicks significantly decreased the incidence of feather amelanosis (108). This suggests that the immune system plays an active role in melanocyte destruction and/or removal in the Smyth line chicken.

The role of antibodies producing B-cells in the Smyth chicken was initially indicated when neonatal bursectomy resulted in a significant decrease in the expression and severity of the feather amelanosis (109). It was subsequently demonstrated that serum autoantibodies, specific for melanocytes, developed immediately prior to or at the time of feather amelanosis (110,111). The predominant serum autoantibody detected in vitiliginous Smyth chickens recognized tyrosinase-related protein-1 (112), analogous to similar reports in human vitiligo (48). Involvement of the cellular component of the immune system was initially revealed by cyclosporin-A treatment of neonatal Smyth line chickens, which effectively reduced the incidence and severity of feather amelanosis and ocular pathology (113,114). In addition, concurrent with the development of the vitiligo, mononuclear cell infiltration and altered T cell profiles were observed in the pulp of developing feathers (115,116) and altered leukocyte profiles existed in the blood (117) of the Smyth chickens.

These observations implicate the immune system in the etiology of vitiligo in the Smyth chicken. However, the melanocyte-specific immune response in this model appears to be secondary. Melanocyte cultures derived from neural crest cells of Smyth line embryos ultimately exhibit altered morphology and premature cell death in isolation from immune cells (118). In addition, neonatal bursectomy that suppresses the development of feather amelanosis (109) did not influence the development of Smyth line specific morphologic abnormalities in the melanocytes of pigmented feathers (119). The Smyth line chickens do appear to have a hyperactive immune system. Affected Smyth birds can produce higher serum antibody titers after administration of exogenous antigens than can their normally pigmented siblings (120). In fact, the extent of this

elevated antibody response correlates positively with the severity of the amelanosis and visual defect, an indication that the development of amelanosis is associated with the general immune competence (*i.e.*, responsiveness) of the animal (119).

5.2.   The Sinclair Pig

A second animal model for vitiligo that exhibits an immune component to melanocyte destruction, as well as an association between melanoma regression and the development of vitiligo, is the Sinclair miniature pig. Sinclair swine exhibit a high incidence of congenital malignant melanoma that spontaneously regress in most affected pigs (121). Tumor regression appears to be mediated by an immune response (122,123), and is usually associated with the development of vitiligo that may progress over most of the body of the pig (123). It has been suggested that the development of vitiligo in the Sinclair miniature pig may be secondary to the immune processes involved in regression of the melanoma (124).

6.   CONCLUSIONS

Vitiligo is a complex disorder that appears to be based on a genetic propensity to develop a pathologic process resulting in the destruction of the melanocyte population of the body. There is clearly an immunological component to the pathophysiology of the disease. Patients developing vitiligo can exhibit serum autoantibodies and autoreactive T cells that may be specific for melanocyte differentiation antigens. Two of the major forms of treatment for vitiligo, application of corticosteroids or psoralen plus ultraviolet light exposure (i.e., PUVA), are both immunosuppressive modalities. The association of other autoimmune disorders with vitiligo and the concurrent expression of vitiligo with melanoma regression also implicate an autoimmune response for the melanocyte destruction in vitiligo. However, it is still unconfirmed whether the immune involvement in melanocyte destruction is the causative factor for vitiligo or alternatively a subsequent recruited process to perpetuate vitiligo in general and inflammatory vitiligo in specific.

References

1. Ortonne, J.-P., and S.K. Bose. 1993. Vitiligo: Where do we stand? *Pigment. Cell. Res.* 6:61–72.
2. Boissy, R.E., and J.J. Nordlund. 1995. Biology of vitiligo. In: Cutaneous Medicine and Surgery: An Integrated Program in Dermatology. K.A. Arndt, P.E. LeBoit, J.K. Robinson, and B.U. Wintroub, editors. W.B. Saunders Company, Philadelphia. pp. 1210–1218.
3. Le Poole, C., and R.E. Boissy. 1997. Vitiligo. In: Seminars Cutaneous Medicine and Surgery, vol. 16. K.A. Arndt, P.E. Le Boit, J.K. Robinson, and B.U. Wintroub, editors. W.B. Saunders Company, Philadelphia. pp. 3–14.
4. Nordlund, J.J., and J.-P. Ortonne. 1998. Vitiligo vulgaris. In: The Pigmentary System. Physiology and Pathophysiology. J.J. Nordlund, R.E. Boissy, V.J. Hearing, R.A. King, and J.-P. Ortonne, editors. Oxford University Press, New York. pp. 513–551.
5. Tosti, A., F. Bardazzi, G. Tosti, and L. Monti. 1987. Audiologic abnormalities in cases of vitiligo. *J. Am. Acad. Dermatol.* 17:230–233.
6. Ardic, F.N., S. Aktan, C.O. Kara, and B. Sanli. 1998. High-frequency hearing and reflex latency in patients with pigment disorder. *Am. J. Otolaryngol.* 19:365–369.
7. Uehara, M., H. Miyauchi, and S. Tanaka. 1984. Diminished contact sensitivity response in vitiliginous skin. *Arch. Dermatol.* 120:195–198.
8. Hatchome, N., S. Aiba, T. Kato, W. Torinuki, and H. Tagami. 1987. Possible functional impairment of Langerhans cells in vitiliginous skin: Reduced ability to elicit dinitrochlorobenzene contact sensitivity reaction and decreased stimulatory effect in the allogeneic mixed skin cell lymphocyte culture reaction. *Arch. Dermatol.* 123:51–54.
9. Calanchini-Postizzi, E., and E. Frenk. 1987. Long-term actinic damage in sun-exposed vitiligo and normally pigmented skin. *Dermatologica* 174:266–271.
10. Nordlund, J.J., and P.P. Majumder. 1997. Recent investigations on vitiligo vulgaris. *Dermatol. Clin.* 15:69–78.
11. Nath, S.K., P.P. Majumder, and J.J. Nordlund. 1994. Genetic epidemiology of vitiligo: multilocus recessivity cross validated. *Am. J. Hum. Genet.* 55:981–990.
12. Porter, J., A. Beuf, J.J. Nordlund, and A.B. Lerner. 1979. Psychological reaction to chronic skin disorders. A study of patients with vitiligo. *Gen. Hosp. Psychiatry* 1:73–77.
13. Porter, J.R., A.H. Beuf, A. Lerner, and J.J. Nordlund. 1986. The psychosocial effect of vitiligo: A comparison of vitiligo patients with "normal" controls, with psoriasis patients, and with patients with other pigmentary disorders. *J. Am. Acad. Dermatol.* 15:220–224.
14. Lerner, A.B. 1971. On the etiology of vitiligo and grey hair. *Am. J. Med.* 51:141–147.
15. Ortonne, J.-P., D.B. Mosher, and T.B. Fitzpatrick. 1983. In: Topics in Dermatology: Vitiligo and Other Hypomelanoses of Hair and Skin. Plenum Medical Book Company, New York.
16. Le Poole, I.C., P.K. Das, R.M. van den Wijngaard, J.D. Bos, and W. Westerhof. 1993. Review of the etiopathomechanism of vitiligo: a convergence theory. *Exp. Dermatol.* 2:145–153.
17. Mehta, N.R., K.C. Shah, C. Theodore, V.P. Vyas, and A.B. Patel. 1973. Epidemiological study of vitiligo in Surat area, South Gujarat. *Indian J. Med. Res.* 61:145–154.
18. Howitz, J., H. Brodthagen, M. Schwartz, and K. Thomsen. 1977. Prevalence of vitiligo: Epidemiological survey on the Isle of Bornholm, Denmark. *Arch. Dermatol.* 113:47–52.
19. Hafez, M., L. Sharaf, and S.M.A. El-Nabi. 1983. The genetics of vitiligo. *Acta Derm. Venereol.* 63:249–251.
20. Das, S.K., P.P. Majumder, R. Chakraborty, T.K. Majumdar, and B. Haldar. 1985. Studies on vitiligo I. Epidemiological profile in Calcutta, India. *Genet. Epidemiol.* 2:71–78.
21. Majumder, P.P. 1999. Genetics and prevalence of vitiligo vulgaris. In: Vitiligo: A Comprehensive Monograph on Basic and Clinical Science. J.J. Nordlund, and S.K. Hann, editors. Blackwell Science, Ltd., Oxford, UK, in press.

22. Kenney, J.A., Jr. 1970. Dermatoses seen in American Negroes. *Int. J. Dermatol.* 9:110–113.

23. Singh, M., G. Singh, A.J. Kanwar, and M.S. Belhaj. 1985. Clinical pattern of vitiligo in Libya. *Int. J. Dermatol.* 24:233–235.

24. George, A.O. 1989. Vitiligo in Ibadan, Nigeria. Incidence, presentation, and problems in management. *Int. J. Dermatol.* 28:385–387.

25. Hann, S.K., H.S. Chung, and Y.K. Park. 1997. Epidemiologic case-control study in patients with vitiligo. *J. Am. Acad. Dermatol.* 36:282–283.

26. Kim, S.M., H.S. Chung, and S.K. Hann. 1998. The genetics of vitiligo in Korean patients. *Int. J. Dermatol.* 37:908–910.

27. Majumder, P.P., D.K. Das, and C.C. Li. 1988. A genetical model for vitiligo. *Am. J. Hum. Genet.* 43:119–125.

28. Majumder, P., J.J. Nordlund, and S.K. Nath. 1993. Pattern of familial aggregation of vitiligo. *Arch. Dermatol.* 129:994–998.

29. Shah, V.C., M.V. Mojamdar, and K.S. Sharma. 1975. Some genetic, biochemical and physiological aspects of leucoderma vitiligo. *J. Cytol. Genet. Congr.* Suppl.:173–178.

30. Shah, V.C., P.B. Haribhakti, M.V. Mojamdar, and K.S. Sharma. 1977. Statistical study of 600 vitiligo cases in the city of Ahmedabad. *Gujarat Med. J.* 42:51–59.

31. Cunliffe, W.J., R. Hall, D.J. Newell, and C.J. Stevenson. 1968. Vitiligo, thyroid diseases and autoimmunity. *Br. J. Dermatol.* 80:135–139.

32. McGregor, B.C., H.I. Katz, and R.P. Doe. 1972. Vitiligo and multiple glandular insufficiencies. *JAMA* 219.724–725.

33. Brostoff, J. 1969. Autoantibodies in patients with vitiligo. *Lancet* 2:177–178.

34. Korkij, W., K. Soltani, S. Simjee, P.G. Marcincin, and T.Y. Chuang. 1984. Tissue-specific autoantibodies and autoimmune disorders in vitiligo and alopecia areata: a retrospective study. *J. Cutan. Pathol.* 11:522–530.

35. Mandry, R.C., L.J. Ortiz, A. Lugo-Somolinos, and J.L. Sanchez. 1996. Organ-specific autoantibodies in vitiligo patients and their relatives. *Int. J. Dermatol.* 35:18–21.

36. Albert, D.M., J.J. Nordlund, and A.B. Lerner. 1979. Ocular abnormalities occurring with vitiligo. *Ophthalmology* 86:1145–1158.

37. Wagoner, M.D., D.M. Albert, A.B. Lerner, J.M. Kirkwood, B.M. Forget, and J.J. Nordlund. 1983. New observations on vitiligo and ocular disease. *Am. J. Ophthalmol.* 96:16–26.

38. Hertz, K., L. Gazze, C. Kirkpatrick, and S. Katz. 1977. Autoimmune vitiligo. Detection of antibodies to melanin-producing cells. *N. Engl. J. Med.* 297:634–637.

39. Naughton, G.K., M. Eisinger, and J.-C. Bystryn. 1983. Antibodies to normal human melanocytes in vitiligo. *J. Exp. Med.* 158:246–251.

40. Bystryn, J.C., and G.K. Naughton. 1985. The significance of vitiligo antibodies. *J. Dermatol.* 12:1–9.

41. Cui, J., R. Harning, M. Henn, and J.-C. Bystryn. 1992. Identification of pigment cell antigens defined by vitiligo antibodies. *J. Invest. Dermatol.* 98:162–165.

42. Harning, R., J. Cui, and J.C. Bystryn. 1991. Relationship between the incidence and level of pigment cell antibodies and disease activity in vitiligo. *J. Invest. Dermatol.* 97:1078–1080.

43. Cui, J., Y. Arita, and J.-C. Bystryn. 1993. Cytolytic antibodies to melanocytes in vitiligo. *J. Invest. Dermatol.* 100:812–815.

44. Hann, S.-K., D.L. Chen, and J.-C. Bystryn. 1997. Systemic steroids suppress anti-melanocyte antibodies in vitiligo. *J. Cutan. Med. Surg.* 1:193–195.

45. Song, Y.H., E. Connor, Y. Li, B. Zorovich, P. Balducci, and N. Maclaren. 1994. The role of tyrosinase in autoimmune vitiligo. *Lancet* 344:1049–1052.

46. Baharav, E., O. Merimski, Y. Shoenfeld, R. Zigelman, B. Gilbrud, G. Yecheskel, P. Youinou, and P. Fishman. 1996. Tyrosinase as an autoantigen in patients with vitiligo. *Clin. Exp. Immunol.* 105:84–88.

47. Kemp, E.H., D.J. Gawkrodger, S. MacNeil, P.F. Watson, and A.P. Weetman. 1997. Detection of tyrosinase autoantibodies in patients with vitiligo using 35S-labeled recombinant human tyrosinase in a radioimunoassay. *J. Invest. Dermatol.* 109:69–73.

48. Kemp, E.H., E.A. Waterman, D.J. Gawkrodger, P.F. Watson, and A.P. Weetman. 1998. Autoantibodies to tyrosinase-related protein-1 detected in the sera of vitiligo patients using a quantitative radiobinding assay. *Br. J. Dermatol.* 139:798–805.

49. Spritz, R.A., and V.J. Hearing. 1994. Genetic orders of pigmentation. In: Advances in Human Genetics, 22nd ed. K. Hirschhorn, and H. Harris, editors. Plenum Publishing, New York. pp. 1–45.

50. Boissy, R.E., and J.J. Nordlund. 1997. Molecular basis of congenital hypopigmentary disorders in humans. A review. *Pigment Cell Res.* 10:12–24.

51. Xie, Z., D. Chen, D. Jiao, and J.C. Bystryn. 1999. Vitiligo antibodies are not directed to tyrosinase. *Arch. Dermatol.* 135:417–422.

52. Kemp, E.H., D.J. Gawkrodger, P.F. Watson, and A.P. Weetman. 1997. Immunoprecipitation of melanogenic enzyme autoantigens with vitiligo sera: evidence for cross-reactive autoantibodies to tyrosinase and tyrosinase-related protein-2 (TRP-2). *Clin. Exp. Immunol.* 109:495–500.

53. Okamoto, T., R.F. Irie, S. Fujii, S.K. Huang, A.J. Nizze, D.L. Morton, and D.S. Hoon. 1998. Anti-tyrosinase-related protein-2 immune response in vitiligo patients and melanoma patients receiving active specific immunotherapy. *J. Invest. Dermatol.* 111:1034–1039.

54. Kemp, E.H., D.J. Gawkrodger, P.F. Watson, and A.P. Weetman. 1998. Autoantibodies to human melanocyte-specific protein pmel17 in the sera of vitiligo patients: a sensitive and quantitative radioimmunoassay (RIA). *Clin. Exp. Immunol.* 114:333–338.

55. Norris, D.A., G.M. Kissinger, G.M. Naughton, and J.-C. Bystryn. 1998. Evidence for immunologic mechanisms in human vitiligo: Patients' sera induce damage to human melanocytes in vitro by complement-mediated damage and antibody-dependent cellular cytotoxicity. *J. Invest. Dermatol.* 90:783–789.

56. Gilhar, A., B. Zelickson, Y. Ulman, and A. Etzioni. 1995. In vivo destruction of melanocytes by the IgG fraction of serum from patients with vitiligo. *J. Invest. Dermatol.* 105:683–686.

57. Bleehen, S.S. 1979. Histology of vitiligo. In: Pigment Cell 5: Part II of Proceedings of the Xth International Pigment Cell Conference, Cambridge, Massachusetts, 1977. S. N. Klaus, editor. S. Karger, Basel/New York. pp. 54–61.

58. Hann, S.K., Y.K. Park, K.G. Lee, E.H. Choi, and S. Im. 1992. Epidermal changes in active vitiligo. *J. Dermatol.* 19:217–222.

59. al-Badri, A.M., P.M. Todd, J.J. Garioch, J.E. Gudgeon, D.G. Stewart, and R.B. Goudie. 1993. An immunohistological study of cutaneous lymphocytes in vitiligo. *J. Pathol.* 170:149–155.

60. Halder, R.M., C.S. Walters, B.A. Johnson, S.G. Chakrabarti, and J.R. Kenney, Jr. T-lymphocytes and interleukin 2 activ-

ity are decreased in vitiligo. Presented at the XIIIth International Pigment Cell Conference, Tucson, AZ, U.S.A.

61. Grimes, P.E., M. Ghoneum, T. Stockton, C. Payne, P. Kelly, and L. Alfred. 1986. T-cell profiles in vitiligo. *J. Am. Acad. Dermatol.* 14:196–201.

62. Soubiran, P., S. Benzaken, C. Bellet, J.P. Lacour, and J.P. Ortonne. 1985. Vitiligo: peripheral T-cell subset imbalance as defined by monoclonal antibodies. *Br. J. Dermatol.* 113:124–127.

63. D'Amelio, R., C. Frati, A. Fattorossi, and F. Aiuti. 1990. Peripheral T-cell subset imbalance in patients with vitiligo and in their apparently healthy first-degree relatives. *Ann. Allerg.* 2:143–145.

64. al-Fouzan, A., M. al-Arbash, F. Fouad, S.A. Kaaba, M.A. Mousa, and S.A. al-Harbi. 1995. Study of HLA class I/IL and T lymphocyte subsets in Kuwaiti vitiligo patients. *Eur. J. Immunogenet.* 22:209–213.

65. Garb, J., and F. Wise. 1948. Vitiligo with raised borders. *Arch. Dermatol. Syphil.* 58:149–153.

66. Buckley, W.R., and W.C. Lobitz, Jr. 1953. Vitiligo with a raised inflammatory border. *Arch. Dermatol. Syphil.* 67:316–320.

67. Michaelsson, G. 1968. Vitiligo with raised borders. Reports of two cases. *Acta Derm. Venereol.* 48:158–161.

68. Gokhale, B.B., and L.N. Mehta. 1983. Histopathology of vitiliginous skin. *Int. J. Dermatol.* 22:477–480.

69. Abdel-Naser, M.B., H. Gollnick, and C.E. Orfanos. 1991. Evidence for primary involvement of keratinocytes in vitiligo. *Arch. Dermatol. Res.* 283.47(Abstr.).

70. Le Poole, I.C., R.M.J.G.J. van den Wijngaard, W. Westerhof, and P.K. Das. 1996. Presence of T cells and macrophages in inflammatory vitiligo skin parallels melanocyte disappearance. *Am. J. Pathol.* 148:1219–1228.

71. Abdel-Naser, M.B., S. Kruger-Krasagakes, K. Krasagakis, H. Gollnick, and C.E. Orfanos. 1994. Further evidence for involvement of both cell mediated and humoral immunity in generalized vitiligo. *Pigment Cell Res.* 7:1–8.

72. al Badri, A.M., A.K. Foulis, P.M. Todd, J.J. Gariouch, J.E. Gudgeon, D.G. Stewart, J.A. Gracie, and R.B. Goudie. 1993. Abnormal expression of MHC class II and ICAM-1 by melanocytes in vitiligo. *J. Pathol.* 169:203–206.

73. Ogg, G.S., P. Rod Dunbar, P. Romero, J.L. Chen, and V. Cerundolo. 1998. High frequency of skin-homing melanocyte-specific cytotoxic T lymphocytes in autoimmune vitiligo. *J. Exp. Med.* 188:1203–1208.

74. Kemp, E.H., R.A. Ajjan, E.A. Waterman, D.J. Gawkrodger, M.J. Cork, P.F. Watson, and A.P. Weetman. 1999. Analysis of a microsatellite polymorphism of the cytotoxic T-lymphocyte antigen-4 gene in patients with vitiligo. *Br. J. Dermatol.* 140:73–78.

75. Laucius, J.F., and M.J. Mastrangelo. 1979. Cutaneous depigmentary phenomena in patients with malignant melanoma. In: Human Malignant Melanoma. W.H. Clark, L.I. Goldman, and M. Mastrangelo, editors. Grune & Stratton, New York. PP. 209–225.

76. Nordlund, J.J., and A.B. Lerner. 1982. Vitiligo: It is important. *Arch. Dermatol.* 118:5–8.

77. Bystryn, J.-C., D. Rigel, R.J. Friedman, and A. Kopf. 1987. Prognostic significance of hypopigmentation in malignant melanoma. *Arch. Dermatol.* 123:1053–1055.

78. Schallreuter, K.U., C. Levenig, and J. Berger. 1991. Vitiligo and cutaneous melanoma: A case study. *Dermatologica* 183:239–245.

79. Nordlund, J.J., J. Kirkwood, B.M. Forget, G. Milton, and A.B. Lerner. 1983. Vitiligo in patients with metastatic melanoma: A good prognostic sign. *J. Am. Acad. Dermatol.* 9:689–695.

80. Koh, H.K., A.J. Sober, H. Nakagawa, D.M. Albert, M.C. Mihm, and T.B. Fitzpatrick. 1983. Malignant melanoma and vitiligo-like leukoderma: An electron microscopic study. *J. Am. Acad. Dermatol.* 9:696–708.

81. Nordlund, J.J., J.M. Kirkwood, B.M. Forget, A. Scheibner, D.M. Albert, E. Lerner, and G.W. Milton. 1985. A demographic study of clinically atypical (dysplastic) nevi in patients with melanoma and comparison subjects. *Cancer Res.* 45:1855–1861.

82. Milton, G.W., W.H. McCarthy, and A. Carlon. 1971. Malignant melanoma and vitiligo. *Australas. J. Dermatol.* 12:131–142.

83. Berd, D., M.J. Mastrangelo, E. Lattime, T. Sato, and H.C. Maguire, Jr. 1996. Melanoma and vitiligo: immunology's Grecian urn. *Cancer Immunol. Immunother.* 42:263–267.

84. Kawakami, Y., and S.A. Rosenberg. 1997. Immunobiology of human melanoma antigens MART-1 and gp 100 and their use for immuno-gene therapy. *Int. Rev. Immunol.* 14:173–192.

85. Mattes, M.J., T.M. Thomson, L.J. Old, and K.O. Lloyd. 1983. A pigmentation-associated differentiation antigen of human melanoma defined by a precipitating antibody in human serum. *Int. J. Cancer* 32:717–721.

86. Vijayasaradhi, S., and A.N. Houghton. 1991. Purification of an autoantigenic 75-kDa human melanosomal glycoprotein. *Int. J. Cancer* 47:298–303.

87. Overwijk, W.W., A. Tsung, K.R. Irvine, M.R. Parkhurst, T.J. Goletz, K. Tsung, M. W. Carroll, C. Liu, B. Moss, S.A. Rosenberg, and N.P. Restifo. 1998. gp100/pmel 17 is a murine tumor rejection antigen: induction of "self"-reactive, tumoricidal T cells using high affinity, altered peptide ligand. *J. Exp. Med.* 188:277–286.

88. Boon, T., P.G. Coulie, and B. Van den Eynde. 1997. Tumor antigens recognized by T cells. *Immunol. Today* 18:267–268.

89. Rosenberg, S.A. 1997. Cancer vaccines based on the identification of genes encoding cancer regression antigens. *Immunol. Today* 18:175–182.

90. Kawakami, Y., S. Eliyahu, C.H. Delgado, P.F. Robbins, K. Sakaguchi, E. Appella, J. R. Yannelli, G.J. Adema, T. Miki, and S.A. Rosenberg. 1994. Identification of a human melanoma antigen recognized by tumor-infiltrating lymphocytes associated with *in vivo* tumor rejection. *Proc. Natl. Acad. Sci. USA* 91:6458–6462.

91. Kawakami, Y., S. Eliyahu, C.H. Delgado, P.F. Robbins, L. Rivoltini, S.L. Topalian, T. Miki, and S.A. Rosenberg. 1994. Cloning of the gene coding for a shared human melanoma antigen recognized by autologous T cells infiltrating into tumor. *Proc. Natl. Acad. Sci. USA* 91:3515–3519.

92. Robbins, P.F., M. El-Gamil, Y. Kawakami, and S.A. Rosenberg. 1994. Recognition of tyrosinase by tumor-infiltrating lymphocytes from a patient responding to immunotherapy. *Cancer Res.* 54:3124–3126.

93. Wang, B.R.-F., P.F. Robbins, Y. Kawakami, X.-Q. Kang, and S.A. Rosenberg. 1995. Identification of a gene encoding a melanoma tumor antigen recognized by HLA-A31-restricted tumor-infiltrating lymphocytes. *J. Exp. Med.* 181:799–804.

94. Wang, R.F., E. Appella, Y. Kawakami, X. Kang, and S.A. Rosenberg. 1996. Identification of TRP-2 as a human tumor antigen recognized by cytotoxic T lymphocytes. *J. Exp. Med.* 184:2207–2216.

95. Barth, R.J., Jr., S.N. Bock, J.J. Mule, and S.A. Rosenberg. 1990. Unique murine tumor-associated antigens identified by tumor infiltrating lymphocytes. *J. Immunol.* 144:1531–1537.

96. Pan, Z.K., G. Ikonomidis, A. Lazenby, D. Pardoll, and Y. Paterson. 1995. A recombinant Listeria monocytogenes vaccine expressing a model tumour antigen protects mice against lethal tumour cell challenge and causes regression of established tumours. *Nat. Med.* 1:471–477.

97. Kawakami, Y., S. Eliyahu, C. Jennings, K. Sakaguchi, X. Kang, S. Southwood, P.F. Robbins, A. Sette, E. Appella, and S.A. Rosenberg. 1995. Recognition of multiple epitopes in the human melanoma antigen gp 100 by tumor-infiltrating T lymphocytes associated with *in vivo* tumor regression. *J. Immunol.* 154:3961–3968.

98. Rosenberg, S.A., J.C. Yang, D.J. Schwartzentruber, P. Hwu, F.M. Marincola, S.L. Topalian, N.P. Restifo, M.E. Dudley, S.L. Schwarz, P.J. Spiess, J.R. Wunderlich, M.R. Parkhurst, Y. Kawakami, C.A. Seipp, J.H. Einhorn, and D.E. White. 1998. Immunologic and therapeutic evaluation of a synthetic peptide vaccine for the treatment of patients with metastatic melanoma. *Nat. Med.* 4:321–327.

99. Richards, J.M., N. Mehta, K. Ramming, and P. Skosey. 1992. Sequential chemoimmunotherapy in the treatment of metastatic melanoma. *J. Clin. Oncol.* 10:1338–1343.

100. Rosenberg, S.A., and D.E. White. 1996. Vitiligo in patients with melanoma: normal tissue antigens can be targets for cancer immunotherapy. *J. Immunother. Emphasis Tumor Immunol.* 19:81–84.

101. Overwijk, W.W., D.S. Lee, D.R. Surman, K.R. Irvine, C.E. Touloukian, C.-C. Chan, M.W. Carroll, B. Moss, S.A. Rosenberg, and N.P. Restifo. 1999. Vaccination with a recombinant vaccinia virus encoding a "self" antigen induces autoimmune vitiligo and tumor cell destruction in mice: Requirement for CD4+ T lymphocytes. *Proc. Natl. Acad. Sci. USA* 96:2982–2987.

102. Boissy, R.E., and M.L. Lamoreux. 1988. Animal models of an acquired pigmentary disorder—vitiligo. In: Proceedings of the XIII International Pigment Cell Conference. Tucson, Arizona, 1986. J. Bagnara, editor. Alan R. Liss, Inc., New York. pp. 207–218.

103. Lamoreux, M.L., and R.E. Boissy. 1999. Animal models. In: Vitiligo: A Comprehensive Monograph on Basic and Clinical Science. S.K. Hann, and J.J. Nordlund, editors. Blackwell Science, Ltd., Oxford, UK, in press.

104. Smyth, J.R., Jr. 1989. The Smyth chicken: A model for autoimmune amelanosis. *CRC Crit. Rev. Poult. Biol.* 2:1–19.

105. Boissy, R.E., J.R. Smyth, Jr., and K.V. Fite. 1983. Progressive cytologic changes during the development of delayed feather amelanosis and associated choroidal defects in the DAM chicken line. A vitiligo model. *Am. J. Pathol.* 111:197–212.

106. Smyth, J.R., Jr., R.E. Boissy, and K.V. Fite. 1981. The DAM chicken: A model for spontaneous postnatal cutaneous and ocular amelanosis. *J. Hered.* 72:150–156.

107. Fite, K.V., N. Montgomery, T. Whitney, R. Boissy, and J.R. Smyth. 1983. Inherited retinal degeneration and ocular amelanosis in the domestic chicken (Ballus domesticus). *Curr. Eye Res.* 2:109–115.

108. Boyle, M.L., III, S.L. Pardue, and J.R. Smyth, Jr. 1987. Effect of corticosterone on the incidence of amelanosis in Smyth delayed amelanotic line chickens. *Poult. Sci.* 66:363–367.

109. Lamont, S.J., and J.R. Smyth, Jr. 1981. Effect of busectomy on development of a spontaneous postnatal amelanosis. *Clin. Immunol. Immunopathol.* 21:407–411.

110. Austin, L.M., R.E. Boissy, B.S. Jacobson, and J.R. Smyth, Jr. 1992. The detection of melanocyte autoantibodies in the Smyth chicken model for vitiligo. *Clin. Immunol. Immunopathol.* 64:112–120.

111. Searle, E.A., L.M. Austin, Y.L. Boissy, H. Zhao, J.J. Nordlund, and R.E. Boissy. 1993. Smyth chicken melanocyte autoantibodies: Cross-species recognition, *in vivo* binding, and plasma membrane reactivity of the antiserum. *Pigment Cell Res.* 6:145–157.

112. Austin, L.M., and R.E. Boissy. 1995. Mammalian tyrosinase related protein-1 is recognized by autoantibodies from vitiliginous Smyth chickens. *Am. J. Pathol.* 146:1529–1541.

113. Pardue, S.L., K.V. Fite, L. Bengston, S.J. Lamont, M.L. Boyle, and J.R. Smyth, Jr. 1987. Enhanced integumental and ocular amelanosis following the termination of cyclosporine administration. *J. Invest. Dermatol.* 88:758–761.

114. Fite, K.V., S. Pardue, L. Bengston, D. Hayden, and J.R. Smyth, Jr. 1986. Effects of cyclosporine in spontaneous, posterior uveitis. *Curr. Eye Res.* 5:787–796.

115. Erf, G.F., A.V. Trejo-Skalli, and J.R. Smyth, Jr. 1995. T cells in regenerating feathers of Smyth line chickens with vitiligo. *Clin. Immunol. Immunopathol.* 76:120–126.

116. Erf, G.F., A.V. Trejo-Skalli, M. Poulin, and J.R. Smyth, Jr. 1997. Lymphocyte populations in dermal lymphoid aggregates of vitiliginous Smyth line and normally pigmented light brown Leghorn chickens. *Vet. Immunol. Immunopathol.* 58:335–343.

117. Erf, G.F., and J.R. Smyth, Jr. 1996. Alterations in blood leukocyte populations in Smyth line chickens with autoimmune vitiligo. *Poult. Sci.* 75:351–356.

118. Boissy, R.E., G.E. Moellmann, A.A. Trainer, J.R. Smyth, Jr., and A.B. Lerner. 1986. Delayed-amelanotic (DAM-Smyth) chicken: Melanocyte dysfunction *in vivo* and *in vitro*. *J. Invest. Dermatol.* 86:149–156.

119. Boissy, R.E., S.J. Lamont, and J.R. Smyth, Jr. 1984. Persistence of abnormal melanocytes in immunosuppressed chickens of the autoimmune "DAM" line. *Cell Tissue Res.* 235:663–668.

120. Lamont, S.J., R.E. Boissy, and J.R. Smyth, Jr. 1982. Humoral immune response and expression of spontaneous postnatal amelanosis in the DAM line chickens. *Immunol. Commun.* 11:121–127.

121. Millikan, L.E., J.L. Boylan, R.R. Hook, and P.J. Manning. 1974. Melanoma in Sinclair swine: A new animal model. *J. Invest. Dermatol.* 62:20–30.

122. Cui, J., D. Chen, M.L. Misfeldt, R.W. Swinfard, and J.C. Bystryn. 1995. Antimelanoma antibodies in swine with spontaneously regressing melanoma. *Pigment Cell Res.* 8:60–63.

123. Morgan, C.D., J.W. Measel, Jr., M.S. Amoss, Jr., A. Rao, and J.F. Greene, Jr. 1996. Immunophenotypic characterization of tumor infiltrating lymphocytes and peripheral blood lymphocytes isolated from melanomatous and non-melanomatous Sinclair miniature swine. *Vet. Immunol. Immunopathol.* 55:189–203.

124. Misfeldt, M.L., and D.R. Grimm. 1994. Sinclair miniature swine: an animal model of human melanoma. *Vet. Immunol. Immunopathol.* 43:167–175.

46 | Pemphigus

Mong-Shang Lin, Luis A. Arteaga, Simon J.P. Warren, and Luis A. Diaz

1. INTRODUCTION

Pemphigus is a group of chronic cutaneous autoimmune diseases that are characterized by blistering of skin and mucous membranes, epidermal cell-cell detachment, also termed acantholysis, and IgG autoantibodies directed against desmosomal antigens (1–4). There are five major forms of pemphigus: 1) Pemphigus vulgaris (PV) and its variant, pemphigus vegetans; 2) pemphigus foliaceus (PF) and its variant, Brazilian pemphigus foliaceus, fogo selvagem (FS); 3) pemphigus erythematosus; 4) drug-induced pemphigus, and 5) paraneoplastic pemphigus. The autoantibodies are important diagnostic and pathological markers of these diseases. Several passive transfer experiments have convincingly demonstrated that the IgG fraction from PV and PF sera reproduce the key clinical, histological and immunological features of humans in neonatal mice (5,6). The autoantigens targeted by PV and PF autoantibodies have been well characterized at the molecular level, however, the etiology of these diseases remains unclear. PV and PF are strongly associated with certain MHC II alleles, hence, it is thought that genetic predisposition plays an important role in the development of these diseases (7–9). Recently, we have demonstrated that T cells from patients with PV and PF react with the same desmosomal antigens that are recognized by the autoantibodies produced by the patients. This observation indicates that self-reactive T lymphocytes are also important elements in the pathogenesis of these diseases (10,11). Since PV and PF account for the majority of patients with pemphigus and most of the current research is devoted to explore the pathogenesis of these diseases, this chapter will focus on recent advances in the understanding of these two autoimmune disorders.

2. PEMPHIGUS VULGARIS

2.1. Clinical and Histological Features of PV

Clinical features. PV is considered the most severe form of pemphigus. Although this disease may occur at any age, the onset is usually seen in the fourth, fifth and sixth decades of life. PV is rare, with an approximate worldwide incidence of 1 to 5 cases per million per year in the general population, but a greater incidence (3.2 cases per 100,000) in certain Jewish populations (12–15).

Clinically, PV is characterized by the presence of flaccid, exceedingly fragile noninflammatory bullae that usually arise on normal appearing skin. These bullae have a tendency to coalesce and rupture easily, resulting in large denuded areas of skin (Figure 1, top). When pressure or friction is applied to lesional skin, sheets of epidermis are easily removed leaving an area of naked skin (Nikolsky's sign) (1).

PV involves both skin and mucous membranes, however, in about half of patients the disease may begin with oral lesions (1). The disease may remain localized to the mucous membranes for months and then spread to involve glabrous skin. The skin lesions are common in the head and neck regions and then progress to involve large areas of the body. In contrast, mucosal involvement is absent in PF (16). Prior to steroid therapy, 50% of PV patients died during the first year of their illness due to dehydration, electrolyte imbalance, malnutrition, and/or sepsis.

Histological features. Acantholysis (epidermal cell detachment) is the histological hallmark of all forms of pemphigus. In PV, the detachment of keratinocytes occurs in the suprabasal layer of the epidermis. Basal cells remain attached to the dermis but detached from each other laterally, a phenomenon that has been compared to a "row of tombstones" (Figure 1, center) (1,2). Although inflammation is generally absent in early lesions, in rare cases, spongiosis associated with an epidermal infiltrate of eosinophils can precede the appearance of blisters (17,18). As bullae develop, a cellular infiltrate made up primarily of eosinophils can be observed around the upper dermal vessels, between basal

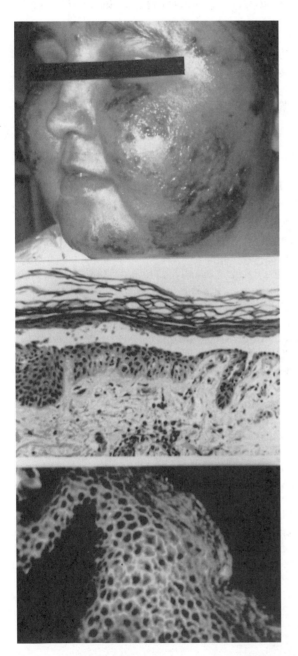

Figure 1. Pemphigus Vulgaris. Top: Picture of PV patients showing vesicles and erosions. Middle: Histological section of lesional skin showing suprabasilar acantholysis. Bottom: PV autoantibodies binding to normal epidermis showing intercellular staining pattern by direct immunofluorescent techniques.

cells, and within blister cavities. Finally, in older lesions, an infiltrate of plasma cells can also be observed. The significance of eosinophils in the pathogenesis of skin lesions in PV remains unknown (19,20).

Immunoelectron microscopy (Immuno-EM) studies using peroxidase-labeled probes have demonstrated that PV autoantibodies are localized to the outer leaflet of the cell membrane of detached keratinocytes with no preferential binding to desmosomes (21). However, using an immuno-gold technique, PV autoantibodies have been detected within the desmosomal cores (22). Skin organ cultures and neonatal mice treated with PV IgG show that the detachment process in the epidermis begins in the inter-desmosomal regions of the intercellular space (ICS). The initial detachment is followed by splitting of the desmosomal units, dissolution of desmosomal plaques, and perinuclear retraction of tonofilaments (23,24).

Diagnosis and treatment. The typical histological features of PV are suprabasilar acantholysis and intradermal vesicles. PV autoantibodies are always detected by direct immunofluorescence (IF) bound to the epidermal ICS of perilesional skin. The sera of over 90% of the patients possess autoantibodies against the epidermal ICS as detected by indirect IF analysis. The titers of these autoantibodies in the sera of patients roughly correlate with the disease activity. Other laboratory techniques, such as ELISA (25), immunoblotting (26), and immuno-precipitation (27) have also been used in research laboratories to make a precise diagnosis of PV.

The treatment of PV is aimed at the elimination of autoantibodies from sera and the epidermis of patients. Corticosteroids are the drugs of choice, with addition of immunosuppressive agents that are used to reduce the side effects of steroids. The immunosuppressive agents found to be beneficial in the therapy of PV are cyclophosphamide, azathioprine, and methotrexate. Other drugs such as gold sodium thiomalate and gold thioglucose have been used as steroid sparing agents, however, their use is limited due to serious side effects. Plasmapheresis has been of short-term benefit in controlling PV (28–30).

2.2. The Autoantigen of Pemphigus Vulgaris

The target antigen of PV autoantibodies is an epidermal 130-kD desmosomal glycoprotein (31) that has been cloned, sequenced, and named desmoglein-3 (Dsg3) (32). Dsg3 has extensive sequence homology to desmoglein-1 (Dsg1), the target antigen of PF autoantibodies (33). Both antigens belong to the cadherin family of calciumdependent cell adhesion molecules, which also include E-cadherin and desmocollins (34). As shown in Figure 2, the extracellular portion of Dsg3 has five major domains and six calcium binding motifs believed to be involved in maintaining the conformation and adhesive function of Dsg3 (35). Recently, the entire extracellular domain of Dsg3 was expressed as a soluble glycopeptide in a baculovirus expression system (27). This recombinant Dsg3 (rDsg3) has been used to map the locations of the epitopes recognized by PV autoantibodies. It was demonstrated that all PV sera are able to immunoprecipitate rDsg3 (Figure 3) (27) and immunoabsorb the reactivity of PV serum by indirect IF (36–38). Moreover, affinity-purified anti-Dsg3 autoantibodies from PV sera induce skin lesions in mice by passive transfer experiments (27,36–38). It can be concluded from these studies that the ectodomain of Dsg3 bears the epitopes that are recognized by pathogenic PV autoantibodies. As discussed in the following sections, we have also demonstrated that the ectodomain of Dsg3 exhibits epitope(s) that are recognized by autoimmune T cells from PV patients. Therefore, the ectodomain of Dsg3 is the target of both pathogenic autoantibodies and self-reactive T cells in PV patients (10).

2.3. The Autoimmune Response in Pemphigus Vulgaris Autoantibodies.

Pemphigus autoantibodies are detected by indirect IF techniques using monkey esophagus or human skin substrates. PV autoantibodies bind the epidermal ICS producing a fishnet smooth staining pattern that reflects the cell surface distribution of Dsg3 epitopes recognized by these autoantibodies (Figure 1, bottom). PV and PF autoantibodies have distinct specificities, but the direct and indirect IF staining patterns are generally indistinguishable. The reactivity of PV autoantibodies is restricted to keratinizing and non-keratinizing squamous epithelia, but is not species-specific (3, 39, 40).

The pathogenic role of PV autoantibodies have been supported by the following clinical and laboratory observations: a) PV autoantibodies are detected bound to lesional epidermis (41–43); b) there is a rough correlation between autoantibody titers and severity of the disease (44); c) the existence of a transient neonatal form of PV in babies born to mothers with active PV as is thought that transfer of maternal autoantibodies via the placenta may cause disease in the neonate (45–47); d) PV and PF IgG fractions induce cell detachment of keratinocytes in cultures, and e) IgG fractions from the serum of PV patients reproduce the key clinical and histological features of the human disease in neonatal mice by passive transfer experiments (5) (Figure 4).

An intriguing feature of the immune response in PV is the mixed autoantibody reactivity with Dsg3 and Dsg1 that is found in approximately 50% of the patients (48). This finding has led investigators to postulate that a certain PV autoantibody profile may be associated with a determined phenotype of PV (49–51). For example, a mucosal subset of PV patients may have an anti-Dsg3 profile alone, whereas a mucocutaneous subset may show both anti-Dsg3 and anti-Dsg1 autoantibody response (49–54). It has been demonstrated that both populations of autoantibodies, anti-Dsg1 and anti-Dsg3, are pathogenic when passively transferred into neonatal mice (27). These observations may explain the progression from PV to PF that has been reported in some of rare cases of PV (52,54,55).

The molecular mechanisms of acantholysis induced by PV autoantibodies is an area of active investigation by several groups. There are three hypothesis that have been tested experimentally in the last several years.

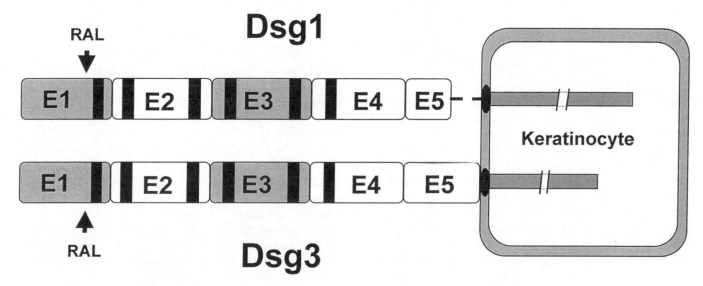

Figure 2.   Structures of Dsg1 and Dsg3. Both Dsg1 and Dsg3 contain five major cadherin-like domains on the extracellular portion. The black strips are the Ca^{2+} binding sites.

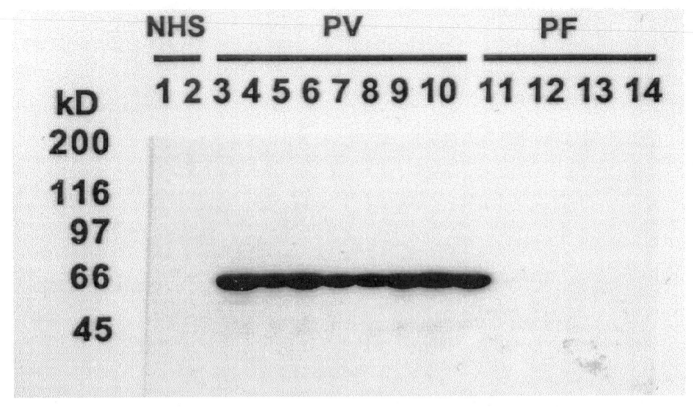

Figure 3. Sera from PV patients recognize the ectodomain of Dsg3 by immunoprecipitation. The presence of anti-Dsg3 auto-antibodies in sera from PV patients was detected by immunoprecipitation. The ectodomain of Dsg3 used in this experiment was prepared using a baculovirus expression system. Recombinant Dsg3 was labeled with ^{125}I following the chloramine T method.

Role of complement: it was suggested that complement activation may be relevant in the acantholytic process seen in the epidermis of PV patients since lesional epidermis of patients always shows strong C3 deposition on the surface of detached cells. In addition, other complement proteins, including the membrane attack complex (C5–C9) (56) are detected bound to the epidermal ICS, where PV auto-antibodies are also located (57, 58). Early studies also demonstrated that blister fluid from skin lesions of PV patients activate complement (56). These observations were reinforced by *in vitro* studies demonstrating that addition of a complement source to epidermal cell cultures previously treated with PV IgG further enhanced the detachment process induced by autoantibodies alone (59–61).

It has been known that the titers of PV autoantibodies are higher in the IgG_4 subclass than in the IgG_1, IgG_2 or IgG_3 (62–64). Since IgG4 antibodies are not efficient at fixing complement, the role of complement activation in PV acantholysis may be marginal (64). Other studies, including more recent findings in our laboratory, have demonstrated that $F(ab)_2$ and F(ab′) from PV IgG were able to induce cutaneous lesions in C5-deficient mice as well as in complement-depleted mice (61,66). These results suggest that

complement activation by PV autoantibodies may not be an essential step in the pathogenesis of the epidermal lesions of PV.

Role of proteases: Others investigators have proposed that PV autoantibodies may activate epidermal proteases, such as plasminogen activator (PA) and plasmin that, in turn, may lead to epidermal cell detachment (67). This hypothesis was supported by studies that showed the epidermal cell detachment induced by PV antibodies in murine epidermal cultures could be blocked by the addition of proteinase inhibitors (68, 69). It was also shown that PA was actively synthesized and released into the supernatant by epidermal cell cultures treated with PV IgG (67). In addition, an increase in urokinase activity was detected in the ICS of lesional epidermis of PV patients (70). More encouraging results implicating PA in the pathogenesis of PV skin lesions were derived from an *in vivo* study using the PV mouse model. In this study, the skin disease induced by passive transfer of PV IgG in neonatal mice could be blocked by pre-treating the animals with the protease inhibitor aprotinin (69).

On the other hand, it has also been reported that dexamethasone, an inhibitor of PA synthesis, does not

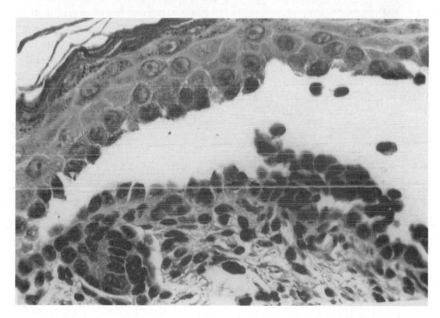

Figure 4.   Animal model of pemphigus vulgaris. Top: A neonatal mouse showing skin lesions induced by transfer of PV antibodies. Bottom: Section of skin from neonatal mouse with induced PV, showing suprabasilar acantholysis.

prevent the acantholysis induced by passive transfer of PV IgG in neonatal mice, despite the fact that PA activity was abolished in the animals undergoing skin disease (71). These studies have been expanded recently to include mice that are genetically ablated of the urokinase-type PA (uPA) or tissue-type PA (tPA) genes. These mice continue to develop acantholysis when passively injected with PV IgG (72).

Impairment of the Dsg3 adhesive function and intracellular signal transduction: Based on *in vitro* studies, we proposed that PV autoantibodies may impair the adhesive function of the epidermal antigen they bound and cause acantholysis (73). This hypothesis is strengthened by the fact

that the epidermal antigens recognized by PV and PF autoantibodies are members of the cadherin family of cell adhesion molecules (32–34). Simple binding of the extracellular domain of Dsg1 or Dsg3 (containing adhesive sites) may be enough to cause acantholysis. In fact, in recent studies univalent Fab derived from both PV and PF IgG was found to be pathogenic by passive transfer experiments in mice (66).

Recently, it has been reported that there is a transient intracellular increase in phospholipase C (PLC), inositol 1,4,5-triphosphate (IP$_3$) production and increased mobilization of intracellular calcium in epidermal cell cultures incubated with PV IgG (74). In addition, there is also a concomitant formation of 1,2 diacylglycerol (DAG), which is an endogenous activator of protein kinase C (PKC). It is known that PKC activation is a major step in the signal transduction pathways of cells undergoing growth and differentiation (75). It was also shown that the intracellular events described above in cells treated with PV IgG are associated with an increase in PA activity (76,77) and translocation of PKC isoforms from the cytosolic pool to the cytoskeleton fraction of keratinocytes (78). Moreover, it was shown recently that PV-IgG causes a rapid breakdown of free Dsg3 that subsequently leads to the formation of Dsg3-depleted desmosomes and phosphorylation of Dsg3 and its dissociation from plakoglobin (79,80). These results may explain the pathogenic mechanisms underlying the development of PV lesions. These encouraging studies need to be confirmed using affinity-purified Dsg1 and Dsg3 autoantibodies from the sera of patients.

T cells. Although the pathogenic role played by autoantibodies is well-characterized in PV, the cellular mechanisms underlying the initiation of the autoimmune response in these patients, that is, the function of T and B lymphocytes, are still not clear. Some preliminary studies demonstrated that peripheral blood lymphocytes from PV patients stimulated *in vitro* with phytohemagglutinin have delayed IL-2 production and decreased IL-2 receptor

expression (81). These characteristics were shown to correlate with autoantibody titers and disease severity in patients, hence, it was concluded that PV patients may have a defect in the secretion of IL-2 and the expression of its receptors. Alternatively, since suppressor T cells (Ts) are thought to play a role in the suppression of autoreactive B cells, a defect in Ts function might account for defective cell-mediated immunity and autoantibody production in PV. On the other hand, an earlier study showed no defects in Ts function in PV patients (82). More recently, peptides containing candidate PV epitopes were synthesized by predicting the possible Dsg3 domains that may interact with the PV HLA-DR alleles (83). It was shown that T cells from PV patients expressing certain HLA-DR alleles responded to these peptides. Specifically, T cells from PV patients expressing HLA-DRB1*0402 and 1401 responded to Dsg3 peptide 190–204 and exhibited a type 2 helper T cell (Th2)-like cytokine profile (83). Similar results have been reported by our group and others (10,84). In our study, it was shown that PVT cells reacted with at least one of the three Dsg3 ectodomain segments (Figure 5). Based on surface marker analysis and cytokine studies, it was concluded that these Dsg3-specific T cells are α/β T cells expressing a CD4-positive memory phenotype and a Th2-like cytokines, a subset of T cells promoting the production of antibodies. Since the development of PV skin lesions is

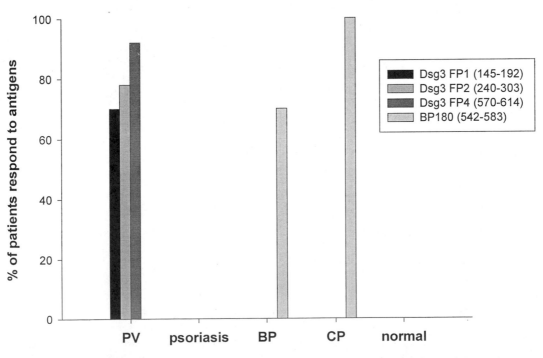

Figure 5.   T lymphocytes from PV patients specifically respond to Dsg3 fusion proteins. The proliferative response of T cells from patients with PV, bullous pemphigoid (BP), cicatricial pemphigoid (CP), psoriasis and healthy controls to Dsg3 and BP180 fusion proteins was examined. A stimulation index greater than or equal to 3 was recognized as a positive response. Numbers of patients in this study: PV, n = 13; BP, n = 3; CP, n = 2; psoriasis, n = 4; and healthy controls, n = 8.

mediated by autoantibodies against Dsg3, we believe that these Dsg3-reactive T lymphocytes may participate in the pathogenesis of this disease at an early stage, leading to the production of autoantibodies. Moreover, our study demonstrated that the response of PV T cell clones to Dsg3 fragments was restricted to the expression of HLA-DRB1*0402 and 1401 alleles. This conclusion was drawn from the observation that T cell clones from PV patients only proliferate to Dsg3-pulsed antigen-presenting cells expressing these two alleles and the fact that anti-HLA-DR antibodies inhibited the cell proliferation to Dsg3 peptides.

It has been well documented that a subset of PV patients express IgG activity to both Dsg1 and Dsg3. Some of these patients may eventually develop PF (52,54,55). Our recent findings clearly indicate that sera from these patients are capable of immunoprecipitating Dsg1 and Dsg3, and both the purified anti-Dsg1 and anti-Dsg3 autoantibodies from these patients are shown to be capable of inducing skin lesions in host animals in a passive transfer model (27). Hence, this study provides the first direct evidence that both populations of antibodies exhibit pathogenic activity and that they account for the clinical progression from PV to PF in these patients. We also found that T cells from this subset of PV patients responded to both Dsg1 and Dsg3 antigens (85). Because both Dsg1 and Dsg3 co-localize to the epidermal desmosomal core, it raises the possibility that T cells from this subset of PV patients may recognize Dsg3 and Dsg1 in a stepwise manner, as stated by the determinant spreading theory (86,87). This study also demonstrates the complexity of autoimmune responses in a particular subset of PV patients. Further characterization of these T cells may help to elucidate the autoimmune mechanism(s) underlying the transition of PV to PF.

2.4. Immunogenetic Features

Genetic factors play an important role in the development and progression of PV. PV occurs in all ethnic and racial groups, but the frequency is higher in those patients with a Jewish background. HLA-A26, HLA-B38, SC21, and HLA-DR4 have been detected in Ashkenazi Jews with PV, whereas HLA-DRw6 was found in non-Ashkenazi Jews. These markers were present in greater than 95% of PV patients (88–90). It has been recently reported that 92% of Ashkenazi Jewish PV patients expressed the HLA-DR4 and HLA-DQw3 (the majority were DR4, DQW8-positive) (7,8). Seventy-five percent of these patients, expressed the extended haplotypes HLA-B38, SC21, DRB1*0402, DQB1*0302 or HLA-B35, SC31, DRB1*0402, DQB1*0302. Interestingly, Iranian PV patients also display similar percentages of these haplotypes (91), whereas PV patients from India and Pakistan show a non-Jewish PV haplotype, HLA-B55, SB45, DRB1*1401, DQB1*0503 (92,93). In Japanese PV patients, both DRB1 alleles, 04

and 14, are present (94), whereas in Korean PV patients, the DRB1*01 allele is more frequent (95). In European PV patients, the Jewish PV haplotype is more frequently observed (96,97). It was suggested that the third hypervariable region of the beta-1 chain of DRB1 allele may constitute a susceptibility locus in PV patients (83,85). Other alleles present in PV patients are DR11 and DQB1*0301 (84).

3. PEMPHIGUS FOLIACEUS

3.1. Clinical, Histological and Epidemiological Features

Clinical features. Pemphigus foliaceus (PF) is a milder form of pemphigus. Clinically, this disease is differentiated from PV by the lack of mucosal involvement and the location of the vesicles, i.e. subcorneal area of the epidermis (98–101). Two forms of PF are described in the literature, one affecting individuals of all ethnic groups and the other, which is known as Fogo Selvagem (FS), is endemic to certain regions of Brazil. These two forms of PF share the same clinical, histological, and immunological features.

PF produces fragile superficial blisters that easily rupture, leaving areas of denuded skin (Figure 6). Initial involvement of the scalp, face, back, and chest are common, but mucous membranes are spared. The disease may spread to involve the entire body, producing an exfoliative erythroderma. The Nikolsky's sign is always positive. In contrast to PV, neonatal PF or FS has been rarely observed in babies born to mothers with active disease (102–104).

Histological features. The chief histological feature of both forms of PF is the presence of acantholytic intraepidermal vesicles located above or below the granular layer of the epidermis. Some of these vesicles may be filled with neutrophils and eosinophilic spongiosis may be present in some biopsies (105,106). Older lesions of PF may show hyperkeratosis and thickening of the epidermis. Bullous impetigo and subcorneal pustular dermatosis may share the same histological features as those of PF. In these cases, direct and indirect IF studies are required to differentiate these conditions (107). Patients with subcorneal pustular dermatosis have been shown to possess IgA anti-desmocollin autoantibodies (108). The relevance of these IgA autoantibodies in the pathogenesis of blister formation is unknown.

Light microscopic examination of skin biopsies reveal subcorneal acantholysis, where the stratum corneum forms the roof of the vesicle and the stratum spinosum forms the floor. EM studies, however, show epidermal cell detachment affecting keratinocytes in all layers of the epidermis

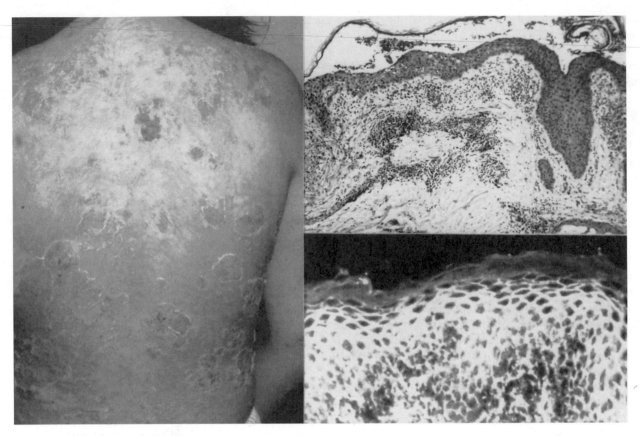

Figure 6.   Brazilian pemphigus foliaceus or fogo selvagem. Left: FS patients showing typical epidermal lesions. Right, top: lesional biopsy showing subcorneal vesiculation and acantholysis. Right, bottom: FS autoantibodies binding to normal epidermis showing intercellular staining by indirect immunofluorescence techniques.

(109,110). The cell detachment is visible on the lateral surfaces of cells of the basal cell layer and extends upwards toward the stratum spinosum and granulosum. The detachment is complete at the level of the granular cell layer. A progressive loss of desmosomes and perinuclear retraction of tonofilments are also features of this disease. Immuno-EM studies, using immunoperoxidase labeled secondary antibodies, demonstrate PF autoantibodies bound diffusely on the surface of keratinocytes; this binding is not restricted to desmosomes (110,111). In contrast, Immuno-EM studies using gold particles, demonstrated that PF autoantibodies bound to the desmosomal core (112,113).

Diagnosis and treatment.   The diagnosis of nonendemic PF and FS is based on clinical, histological, and the immunological features described above. The absence of oral lesions distinguishes both forms of PF from PV. The early and invasive stages of PF should be distinguished from other bullous diseases producing intraepidermal and subepidermal blisters. Histological examination of skin lesions showing subcorneal vesicles favors the diagnosis of PF. Skin biopsies from PF patients show deposits of IgG (and sometimes C3) in the epidermal ICS by direct IF. The titers of anti-Dsg1

autoantibodies in patients' sera roughly correlate with disease extent and activity (114).

Western blot analysis of PF serum is not a viable diagnostic tool since Dsg1 is denatured during the procedure and conformational epitopes are destroyed. Thus, only 30% of PF or FS sera can be detected by this technique. The immunoprecipitation assay, however, is a sensitive and accurate tool to detect PF autoantibodies in the sera of these patients (115,116). This technique requires radiolabeled human or bovine epidermal Dsg1 or recombinant Dsg1 that preserves conformational sensitive epitopes that would otherwise destroyed in the immunoblotting assays. Recently, ELISA kits, using recombinant Dsg1 and Dsg3, have become commercially available, providing information similar to that obtained by Immunoprecipitation tests, but that require less time and laboratory resources (25). The presence of autoantibodies to Dsg3 detected by ELISA and immunoprecipitation is supportive of a diagnosis of PV, however, 40% of PV patients possess anti-Dsg1 autoantibodies in their sera. Interestingly enough, 40% of these PV patients may have anti-Dsg1 autoantibodies. The presence of autoantibodies to Dsg1 alone is supportive of a diagnosis of PF or FS.

Prior to corticosteroids, no effective treatment existed for both FS and PF. Fewer than 10% of FS patients underwent spontaneous remission, while 40% died within the first two years of the disease. Approximately 50% of the patients have a chronic disease characterized by periodic exacerbations that, on occasion, may lead to death. The introduction of steroids in the treatment of PF and FS has radically changed the course and prognosis of the disease. At present, the mortality rates for PF and FS patients is approximately 10%. Compared to PV patients, PF patients respond more rapidly and to smaller doses of corticosteroids. Due to the prolonged use of this medication to control the disease, side effects such as obesity, hirsutism, acne, striae and osteoporosis may develop. In these patients, immunosuppressive drugs such as azathioprine and cyclophosphamide have been used in association with steroids, with excellent results. Other, less frequently used, modalities of treatment include plasmapheresis. Death may occur during the invasive stage of the disease or as a result of complications, such as bacterial or viral infections.

Epidemiological features. Non-endemic pemphigus foliaceus occurs sporadically throughout the world. By contrast, FS shows geographic, temporal and familial clustering in certain endemic regions of Brazil (117). Although FS occurs mainly in Brazil, other minor foci of endemic PF have been reported in Colombia, and recently in Tunisia (118–120). The remarkable epidemiologic features of FS (117) unique to this disease are: a) the disease is common in poor farmers and equally distributed among males and females; b) FS tends to affect several genetically-related members of a family; c) FS was commonly seen in the early half of the century in regions of Brazil being colonized, and disappeared following urbanization; this feature has been clearly documented in the state of Sao Paulo where, in the 30's and 40's, there were hundreds of cases, but currently only a few; d) there is a high frequency of cases in Brazilian Indians that have settled in endemic areas of FS (121,122); one of these Indian tribes, the Terenas, who live in the Limao Verde reservation in the state of Mato Grosso do Sul, exhibits a high prevalence of FS (122), and e) FS is anecdotally reported to undergo spontaneous remission when patients move from an endemic to an urban area; relapses may occur when the patients return to the endemic area. Importantly, since the earliest descriptions of this disease, there have been no reports of transmission of the disease to hospital personnel. These findings strongly suggest that FS is precipitated by exposure to an environmental factor(s) in genetically- predisposed individuals (123–125).

Several independent lines of evidence, including a case-control epidemiological study, point to an insect vector ("black fly", family *Simuliidae*) as an etiological agent of FS (124,125). This study showed that farmers exposed to black fly bites were, as estimated by the odds ratio, 4.7 times more likely to develop FS than those not exposed (124). Recently, we have identified a particular species of simuliidae, *S. nigrimanum*, as the dominant anthropophilic blood feeding black fly in the Terena reservation of Limao Verde, where FS commonly seen (125). In other neighboring disease-free Indian reservations, however, this black fly is absent (125). These findings suggest that *S. nigrimanum* may play a role in the pathogenesis of FS, however, there is no laboratory evidence to support this hypothesis. Based on an antigenic mimicry model, the environmental antigen(s) responsible for triggering FS may exhibit homology to epidermal Dsg1. Antibodies directed against this foreign antigen may cross-react with epidermal dsg1 and precipitate FS.

3.2. The Autoantigen of Pemphigus Foliaceus and Fogo Selvagem

The target antigen of PF and FS has been identified as Dsg1, a 160-kDa desmosomal glycoprotein that shares a similar molecular structure with Dsg3. As shown in Figure 2, the extracellular domain of Dsg1 also exhibits five major regions and six calcium-binding sites (34). Like Dsg3, the intracellular domain of Dsg1 is associated with plakoglobin (126) When immunoblotting assays are used to detect anti-Dsg1, only one-third of FS sera recognized Dsg1 (127–129), however, if recombinant Dsg1 is utilized to perform immunoprecipitation and ELISA techniques, almost all FS sera bind this antigen (130,131). Prior to the construction and expression of Dsg1 in the baculoviral system (132), radiolabeled Dsg1 proteolytic fragments from human or bovine epidermal epidermis were used to test FS sera (115,116). All FS sera immunoprecipitated a 45-kD peptide that by amino acid sequence analysis showed complete homology with the ectodomain of Dsg1 (133).

3.3. The Autoimmune Response in Pemphigus Foliaceus and Fogo Selvagem

Autoantibodies. The pioneering studies of Beutner et al. (134) showed that PF and FS sera exhibit autoantibodies against the epidermal ICS by indirect immunofluorescence (IF). The fluorescent staining pattern of the ICS was similar to that produced by PV sera. Recent studies using affinity-purified anti-Dsg3 antibodies (PV autoantibodies) and anti-Dsg1 antibodies (PF autoantibodies) from the sera of patients, however, demonstrated that the staining patterns produced by these two populations of autoantibodies are different. It appears that anti-Dsg1 autoantibodies bind the surface of suprabasilar keratinocytes, whereas anti-Dsg1 autoantibodies bind the upper layers of the epidermis (135). It has been reported that the majority of FS patients show autoantibodies bound to the ICS of oral and esophageal mucosae by indirect IF despite the fact that these patients do not display mucosal lesions (16,136). This finding leads to a hypothesis that the expression of Dsg3 and Dsg1 may modulate the level of acantholysis seen in PV and PF and the presence or absence of disease in mucosal surfaces (50).

Compelling proof that PF autoantibodies are pathogenic was derived from the observation that intraperitoneal injections of PF IgG into neonatal mice induced cutaneous blisters and erosions in less than 12 hours post-injection (Figure 7). In these experiments, the animals showed a good correlation between the extent of the skin disease and the indirect IF titers of human PF autoantibodies detected in the mouse serum. Light and EM studies of the skin lesions seen in these animals revealed typical subcorneal blisters. Importantly, suprabasilar acantholysis, which is characteristic of PV lesions, was not present in these animals. This study clearly demonstrated that FS (or PF) autoantibodies induce specific epidermal lesions that are histologically distinct from those caused by PV autoantibodies.

The IgG autoantibody response in FS is predominantly of the IgG4 subclass according to indirect IF analysis (136). In fact, IgG4 is the only IgG anti-Dsg1 antibody subclass detected in some patients (136). The IgG4 anti-Dsg1 autoantibodies, however, are pathogenic on the passive transfer animal model (137). Similar to the PV, complement activation is not essential for the induction of FS lesions, as shown by recent studies demonstrating that both F(ab')$_2$ and Fab fragments of FS IgG are pathogenic when tested on passive transfer mouse model (138). This conclusion has been further strengthened by the recent demonstration that PF IgG can also induce skin disease in C5-deficient or complement-depleted mice (139). Hence, it appears that FS autoantibodies may induce acantholysis by impairing the adhesive function of Dsg1 rather than by activation of the complement cascade.

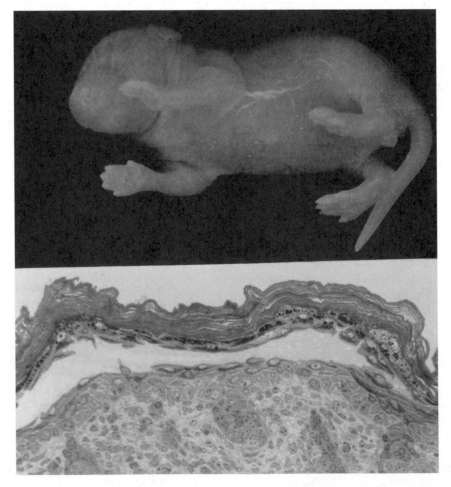

Figure 7.   Animal model of fogo selvagem. Top: Mouse with skin lesions after receiving autoantibodies from patients with FS. Bottom: lesion in neonatal mouse showing typical subcorneal vesiculation and cell detachment of the granular cell layer of the mouse epidermis.

An early attempt to map the Dsg1 epitopes recognized by pathogenic FS autoantibodies had limited success (140). In these studies, segments of human Dsg1 were expressed as bacterial fusion proteins and used in immunoblotting assays to test the reactivity of FS sera (140). It was shown that only 30 to 40% of PF sera reacted with one or more of these Dsg1 fusion proteins. These Dsg1 fusion proteins, however, failed to adsorb pathogenic autoantibodies from FS sera. It is likely that disease-relevant Dsg1 epitopes were not expressed in these fusion proteins or they were denatured during immunoblotting procedures. Recently, Dsg1 constructs have been expressed in mammalian cells or in the baculovirus system (48,130) and utilized to identify the reactivity of FS (or PF) sera by immunoprecipitation and ELISA assays. The results of these studies convincingly demonstrate that Dsg1 epitopes recognized by pathogenic FS (or PF) autoantibodies are conformational and calcium-dependent (48,130,141). Furthermore, these epitopes are located on the ectodomain of the molecule.

T cells. At present, the mechanisms of immuno-regulation mediated by T lymphocytes in FS have not yet been elucidated. In 1976, Guerra et al. (142) found that FS patients have a marked decrease in the total T cell count, as well as a depletion of T cells in the paracortical areas of lymph nodes, suggesting a possible defect in cellular immunity in these patients (142). Other studies have shown elevated thymosin a_1 levels in these patients. Alpha$_1$ thymosin is known to increase the number of helper T cells and enhance resistance to infectious agents (143). Accumulation of T cells within lesional sites of PF patients was also reported, indicating a possible role for T cells in the autoimmune response in PF (144). Ongoing studies in our laboratory show that FS T cells from 13 of 15 patients specifically responded to the ectodomain of Dsg1, but not to other epidermal antigens, such as Dsg3 and BP180 (a glycoprotein recognized by bullous pemphigoid auto-antibodies) (Figure 8) (11). It is worth noting that in T cells from two patients that were unresponsive to Dsg1, FS was inactive at the time of the studies. Furthermore, Dsg1-specific FS T cell clones were generated from five patients (n = 72) and shown to produce a cytokine profile of IL-4, IL-5, and IL-6, but not interferon-γ, suggesting that they are Th2 cells (Figure 9). Phenotypic analysis of these T cell clones demonstrated that they express a CD 4 memory T cell phenotype. It is known that the CD4$^+$ T cell subset is involved in antibody production, therefore, it is believed that these Dsg1-responsive T cell clones may be relevant in the generation of pathogenic autoantibodies in FS. Interestingly, preliminary studies in our laboratory show that T cells from FS patients expressing HLA-DRB1*0404 (or other FS-associated alleles) proliferate in response to

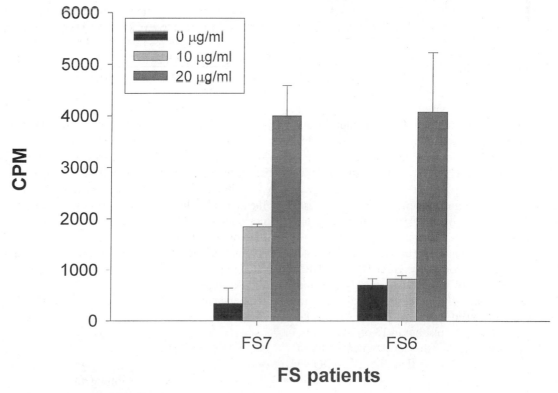

Figure 8.   T lymphocytes of FS patients respond to rDsg1. The proliferative responses of FS T cells to rDsg1 were examined using proliferation assays. Data are expressed as average cpm±SD.

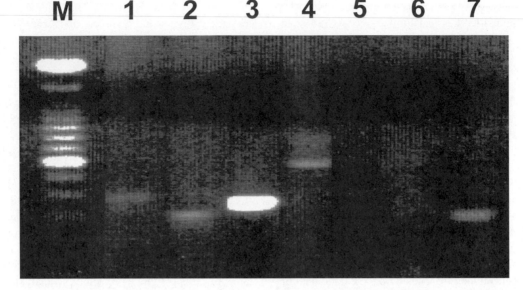

Figure 9.   FS T cells express a Th2-like cytokine profile. The cytokine mRNA messages produced by FS T cells were examined by reverse-transcription-polymerase chain reaction analysis. Lane M, molecular weight marker; lane 1, IL-2; lane 2, IL-4; lane 3, IL-5; lane 4, IL-6; lane 5, γ-IFN; lane 6, TGF-β; and lane 7, β-actin.

two Dsg1 15 amino acid peptides (Lin et al., unreported observation). Thus, it may be concluded from these studies that the extracellular domain of Dsg1 bears epitopes that are recognized not only by T cells, but also by pathogenic autoantibodies. The precise epitope(s) recognized by both T cells and autoantibodies, however, have not yet been characterized.

3.4   Immunogenetic Features

It has been observed that in endemic areas of FS, a large segment of a population is exposed to the environmental antigen(s), but only a small fraction of these individuals develop FS. The increased frequency of familial cases of FS strongly suggests a genetic influence in the onset and progression of FS. For example, in a series of 2,800 FS patients, 18% were familial cases, and 93% of these familial cases represented genetically-related family members (109). These findings indicate that individuals at risk to develop FS in an endemic community are those sharing certain genetic predisposition. HLA studies on FS patients from the Brazilian state of Parana showed that one or both of the HLA-DR1 and DR4 genes were present in 88% of the patients, but 34% of the controls (146). An additional HLA study of 37 patients from Sao Paulo and Brasilia showed that HLA-DRB1*0102 gene was present in only 4 of 47 controls, but in 15 of 37 patients, conferring a relative risk of 7.3 for the development of FS in this area. This study also showed that the HLA-DQw2 (DQB1*0201) allele may confer resistance to FS (this allele was detected in 11 of 49 controls) (146). HLA studies have also been carried out in a Brazilian Amerindian population

where FS is endemic. Patients from the Xavante Indian community expressed the HLA-DRB1*0404 and 1402 alleles more frequently than the control population (combined relative risk = 18.3) (147,148). In another Amerindian population, the Terenas, where the prevalence of FS is 2.6%, 19 of 20 unrelated patients were positive for DRB1* 0404, 1402 or 1406 (relative risk of 14). These alleles are present in 17–35% of the healthy controls. Significantly, the amino acid sequence of residues 67 to 74 of the third hypervariable region of the β chain of the DRB1 gene, i.e., LLEQRRAA, is shared by DRB1*0102 (susceptibility gene of non-Indian FS patients) and DRB1*0404, DRB1*1402, and DRB1*1406 (susceptibility genes of Xavante and Terena FS patients) (147,148). When patients and control data from several studies were pooled, the possession of the LLEQRRAA sequence gave a relative risk for disease of 6.4 (149) Accordingly, this "shared epitope", as originally proposed for rheumatoid arthritis (150), may provide susceptibility for the development of FS in individuals exposed to the appropriate environmental antigen(s).

Finally, due to the rarity of cases, few immunogenetic studies have been performed in the non-endemic form of PF. A single Japanese study from 1981 found an increased incidence of HLA-DR4 in these patients (151)

4.   CONCLUSION

The humoral and cellular immune mechanisms involved in the pathogenesis of PV and PF (or FS) are the focus of intense investigation by different research groups. PV and

PF are human organ-specific autoimmune diseases that offer the unique opportunity to explore the mechanisms of autoantibody-induced injury of the skin, and to investigate the cellular mechanisms involved in autoantibody production. It is well documented that PV and PF autoantibodies are pathogenic by passive transfer experiments, and the antigens have been cloned and characterized as Dsg1 (PF antigen) and Dsg3 (PV antigen). At present, strong epidemiological evidence points to an environmental antigen(s) as the precipitating factor of FS. Hence, these diseases provide a unique opportunity to identify the etiologic agents involved in the development of FS.

The humoral response of PV and PF is characterized by autoantibodies that specifically bind to the ectodomain of Dsg3 and Dsg1, respectively, and induce epidermal cell detachment and blister formation in experimental animals. It has been suggested that simple binding of PV or PF autoantibodies to their respective epitopes on Dsg3 or Dsg1 directly disrupt epidermal cell-cell adhesion and lead to acantholysis. This hypothesis is strengthened by the fact that ablation of the Dsg3 gene in experimental animals induces PV-like lesions (152). It is feasible that the epidermal cell detachment induced by these autoantibodies may trigger cellular responses in epidermal cells that lead to cell detachment. Another promising area of investigation is in the development of animal models of PV and PF by active immunization protocols using Dsg1 and Dsg3 epitopes. If successful, these experimental models may offer new approaches to study autoimmunity in general.

Although autoimmune T cells have been implicated in these diseases, the functions of these cells remain elusive at present. It is accepted that the production of antibodies by B cells in the T-dependent immune responses usually requires the participation of T helper cells. During an immune reaction, T cells form a cognate interaction with antigen-specific B cells and promote the production of antibodies. Due to the fact that PV and PF are antibody-mediated diseases, it is expected that autoimmune T lymphocytes stimulate B cells to produce pathogenic autoantibodies. Our preliminary observations demonstrate that T cell clones from PV and PF patients are capable of inducing antibody production when cultured with autologous B cells. We are currently mapping T cell epitopes on Dsg3 and Dsg1 that are responsible for the pathological autoimmune humoral response against these molecules in PV and PF, respectively.

ACKNOWLEDGEMENT

This work was supported in part by U.S. Public Health Service Grants R37-AR30281, RO1-AR32599, and a Merit Award from the Veterans Administration Central Office (L.A.D.) and a Dermatology Foundation Career Development Award and RO1-AI48427 (M.S.L). Dr. Warren is a Fellow supported by the Dermatology Foundation and Novartis Pharmaceuticals Corporation. Dr. Arteaga is a Visiting Scientist from the Facultad de Medicina, Trujillo, Peru.

References

1. Lever, W.F. 1953. Pemphigus. *Medicine* 32:1–123.
2. Civatte, A. 1943. Diagnostic histopathologique de la dermatite polymorphe douloureseou maladie de During-Brocq. *Ann. Dermatol. Syph. (8th series)* 3:1–30.
3. Beutner, E.H., and R.E. Jordon. 1964. Demonstration of skin antibodies in sera of patients with pemphigus vulgaris by indirect immunofluorescent staining. *Proc. Soc. Exp. Biol. Med.* 117:505–510.
4. Lin, M.S., J.M. Mascaro, Jr., Z. Liu, A. Espana, and L.A. Diaz. 1997. The desmosome and hemidesmosome in cutaneous autoimmunity. *Clin. Exp. Immunol.* 107:9–15.
5. Anhalt, G.J., R.S. Labib, J.J. Voorhees, T.F. Beals, and L.A. Diaz. 1982. Induction of pemphigus in neonatal mice by passive transfer of IgG from patients with the disease. *N. Eng. J. Med.* 306:1189–1196.
6. Roscoe, J.T., L.A. Diaz, S.A.P. Sampaio, R.M. Castro, R.S. Labib, Y. Takahashi, H.P. Patel, and G.J. Anhalt. 1985. Brazilian pemphigus foliaceus autoantibodies are pathogenic to BALB/c mice. *J. Clin. Invest.* 85:296–299.
7. Ahmed, A.R., R. Wagner, K. Khatri, G. Notani, Z. Awdeh, C.A. Apler, and E.J. Yunis. 1991. Major histocompatibility complex haplotypes and class II genes in non-Jewish patients with pemphigus vulgaris. *Proc. Natl. Acad. Sci. USA* 88:7658–7662.
8. Ahmed, A.R., E.J. Yunis, K. Khatri, R. Wagner, G. Notani, Z. Awdeh, and C.A. Apler. 1990. Major histocompatibility haplotypes studies in Ashkenazi Jewish patients with pemphigus vulgaris. *Proc. Natl. Acad. Sci. USA* 86:6215–6219.
9. Moraes, J.R., M.E. Moraes, M. Fernandez-Vina, L.A. Diaz, H. Friedman, I.T. Campbell, R.R. Alvarez, S.A.P. Sampaio, E.A. Rivitti, and P. Stastny. 1991. HLA antigens and the risk for development of pemphigus foliaceus ("fogo selvagem") in endemic areas of Brazil. *Immunogenetics* 33:388–391.
10. Lin, M.S., S.J. Swartz, A. Lopez, X. Ding, M.A. Fernandez-Vina, P. Stastny, J.A. Fairley, and L.A. Diaz. 1997. Development and characterization of T cell lines responding to desmoglein-3. *J. Clin. Invest.* 99:31–40.
11. Lin, M.S., C.-L. Fu, V. Aoki, G. Han-Filho, E.A. Rivitti, J.R. Moraes, M.E. Moraes, A. Lazaro, G.J. Giudice, P. Stastny, and L.A. Diaz, and the Cooperative Group on Fogo Selvagem Research. 2000. Development and characterization of desmoglein-1 specific T lymphocytes from patients with endemic pemphigus foliaceus (fogo selvagem). *J. Clin. Invest.* 105:207–213.
12. Pisanti, S., Y. Sharav, E. Kaufman, and L.N. Pasner. 1974. Pemphigus vulgaris: Incidence in Jews of different ethnic groups, according to age, sex and initial lesion. *Oral. Surg.* 38:382–387.
13. Lynch, P., R.E. Gallego, and N.K. Saied. 1976. Pemphigus—a review. *Ariz. Med.* 33:1030–1037.
14. Hietanen, J., and O.P. Salo. 1982. Pemphigus: An epidemiological study of patients treated in Finnish hospitals between 1969 and 1978. *Acta Dermatol. Vebereol.* 62:491–496.

15. Simon, D.G., D. Kutchoff, R.A. Kaslow, and R. Zabro. 1980. Pemphigus in Hartford County, Connecticut, from 1972 to 1977. *Arch. Dermatol.* 116:1035–1037.

16. Rivitti, E.A., J.A. Sanchs, L.M. Miyauchi, S.A.P. Sampaio, V. Aoki, and L.A. Diaz. 1994. Pemphigus foliaceus autoantibodies bind both epidermis ans squamous mucosal epithelium, but tissue injury is detected only in the epidermis. *J. Am. Acad. Dermatol.* 31:954–958.

17. Wilgram, G.F., J.B. Caufield, and W.F. Lever. 1961. An electron microscopic study of acantholysis in pemphigus vulgaris. *J. Invest. Dermatol.* 36:373–382.

18. Santi, C.G., C.W. Maruta, V. Aoki, M.N. Sotto, E.A. Rivitti, L.A. Diaz, and the Cooperative Group on Fogo Selvagem Research. 1996. Pemphigus herpetiform is rare clinical expression of nonendemic pemphigus foliaceus, fogo selvagem, and pemphigus vulgaris. *J. Am. Acad. Dermatol.* 34:40–46.

19. Moqbel, R., S. Ying, J. Barkens, T.M. Newman, P. Kimmitt, M. Wakelin, L. Taborda-Barata, Q. Meng, C.J. Corrigan, S.R. Durham, and A.B. Kay. 1995. Identification of messenger RNA for IL-4 in human eosinophils with granule localization and release of the translated product. *J. Immunol.* 155:4939–4947.

20. Sur, S., G.J. Glerch, M.C. Swanson, K.R. Bartemes, and D.H. Broide. 1995. Eosinophilic inflammation is associated with elevation of interleukin-5 in the airway of patients with spontaneous symptomatic asthma. *J. Allergy. Clin. Immunol.* 96:661–668.

21. Wolff, K., and E. Scheiner. 1971. Ultrastructural localization of pemphigus autoantibodies within the epidermis. *Nature* 229:59–61.

22. Karpati, S., M. Amagai, R. Prussick, K. Cehrs, and J.R. Stanley. 1993. Pemphigus vulgaris antigen, a desmoglein type of cadherin, is localized within keratinocyte desmosomes. *J. Cell. Biol.* 122:409–415.

23. Hu, C.H., B. Michel, and J.R. Schiltz. 1978. Epidermal acantholysis induced *in vitro* by pemphigus autoantibody: An ultrastructural study. *Am. J. Pathol.* 90:345–351.

24. Takahashi, Y., D. Mutasim, H.P. Patel, G.J. Anhalt, R.S. Labib, and L.A. Diaz. 1985. Experimentally induced pemphigus vulgaris in neonatal BALB/c mice: A time-course study of clinical, immunologic, ultrastructural, and cytochemical changes. *J. Invest. Dermatol.* 84:41–46.

25. Ide, A., T. Hashimoto, M. Amagai, M. Tanaka, and T. Nishikawa. 1995. Detection of autoantibodies against bullous pemphigoid and pemphigus antigens by an enzyme-linked immunosorbent assay using the bacterial recombinant proteins. *Exp. Dermatol.* 4:112–116.

26. Cozzani, E., J. Kanitakis, F. Nicolas, D. Schmitt, and J. Thivolet. 1994. Comparative study of indirect immunofluorescence and immunoblotting for the diagnosis of autoimmune pemphigus. *Arch. Dermatol. Res.* 286:295–299.

27. Ding, X., L.A. Diaz, J.A. Fairley, G.J. Giudice, and Z. Liu. 1999. The anti-desmoglein 1 autoantibodies in pemphigus vulgaris sera are pathogenic. *J. Invest. Dermatol.* 112:739–743.

28. Roujeau, J.C. 1983. Plasma exchange in pemphigus. *Arch. Dermatol.* 119:215–221.

29. Auberbach, R., and J. Bystryn. 1979. Plasmapheresis and immunosuppressive therapy. *Arch. Dermatol.* 115:728–730.

30. Swanson, D.L., and M.W. Dahl. 1981. Pemphigus vulgaris and plasma exchange: clinical and serologic studies. *J. Am. Acad. Dermatol.* 4:325–328.

31. Stanely, J.R., L. Koulu, and C. Thivolet. 1984. Distinction between epidermal antigens binding pemphigus vulgaris and pemphigus foliaceus autoantibodies. *J. Clin. Invest.* 74:313–320.

32. Amagai, M., V. Klaus-Kovtun, and J.R. Stanley. 1991. Autoantibodies against a novel epithelial cadherin in pemphigus vulgaris, a disease of cell adhesion. *Cell* 67:869–877.

33. Wheeler, G.N., A.E. Parker, C.L. Thomas, P. Ataliotis, D. Poynter, J. Amemann, A.J., Rutman, S.C. Pidsley, F.M. Watt, and D.A. Rees. 1991. Desmosomal glycoprotein DG1, a component of intercellular junctions, is related to the cadherin family of cell adhesion molecules. *Proc. Natl. Acad. Sci. USA* 88:4796–4800.

34. Buxton, R.S. and A.I. Magee. 1992. Structure and interactions of desmosomal and other cadherins. *Cell Biol.* 3:157–167.

35. Amagai, M., S. Karpati, V. Klaus-Kovtun, M.C. Udey, and J.R. Stanley. 1994. Extracellular domain of pemphigus vulgaris antigen (desmoglein 3) mediates weak homophilic adhesion. *J. Invest. Dermatol.* 102:402–408.

36. Amagai, M., M. Hashimoto, and N. Shimizu. 1994. Absorption of pathogenic autoantibodies by the extracellular domain of pemphigus vulgaris antigen (Dsg3) produced by baculovirus. *J. Clin. Invest.* 94:59–67.

37. Memar, O.M., S. Rajaraman, and R. Thotakura. 1996. Recombinant desmoglein 3 has the necessary epitopes to absorb and induce blister-causing antibodies. *J. Invest. Dermatol.* 106:261–268.

38. Amagai, M., S. Karpati, V. Klaus-Kovtun, M.C. Udey, and J.R. Stanley. 1992. Autoantibodies against the amino-terminal cadherin-like binding domain of pemphigus vulgaris antigen are pathogenic. *J. Clin. Invest.* 90:919–926.

39. Amagai, M., P.J. Koch, T. Nishikawa, and J.R. Stanley. 1996. Pemphigus vulgaris antigen (desmoglein 3) is localized in the lower epidermis, the site of blister formation in patients. *J. Invest. Dermatol.* 106:351–355.

40. Diaz, L.A., H.J. Weiss, and N.J. Calvanico. 1978. Phylogenetic studies with pemphigus and pemphigoid antibodies. *Acta. Dermatol. Venereol.* 58:537–540.

41. Beutner, E.H., et al. 1965. Autoantibodies in pemphigus vulgaris. *JAMA* 192:682–688.

42. Schiltz, J.R., and B. Michel. 1976. Production of epidermal acantholysis in normal human skin *in vitro* by the IgG fraction from pemphigus serum. *J. Invest. Dermatol.* 67:254–260.

43. Schiltz, J.R., B. Michel, and R. Papa. 1979. Appearance of "pemphigus acantholytic factor" in human skin cultured with pemphigus antibody. *J. Invest. Dermatol.* 73:575–581.

44. Fitzpatrick, R.E., and V.D. Newcomer. 1980. The correlation of disease activity and pemphigus antibody titers in pemphigus. *Arch. Dermatol.* 116:282–290.

45. Green, D., and J.C. Maize. 1982. Maternal pemphigus vulgaris with *in vivo* bound antibodies in the stillborn fetus. *J. Am. Acad. Dermatol.* 7:388–392.

46. Moncada, B., S. Kettelsen, J.L. Hernandez-Moctezuma, and F. Ramirez. 1982. Neonatal pemphigus vulgaris: role of passively transferred pemphigus antibodies. *Br. J. Dermatol.* 115:316–319.

47. Goldberg, N.S., C. DeFeo, and N. Kirshenbaum. 1993. Pemphigus vulgaris and pregnancy:risk factors and recommendations. *J. Am. Acad. Dermatol.* 28:877–879.

48. Emery, D.J., L.A. Diaz, J.A. Fairley, A. Lopez, A.F. Taylor, and G.J. Giudice. 1995. Pemphigus foliaceus and pemphigus vulgaris autoantibodies react with the extracellular domain of desmoglein-1. *J. Invest. Dermatol.* 104:323–328.

49. Sams, W.M., and P.H. Schur. 1973. Studies of the antibodies in pemphigoid and pemphigus. *J. Lab. Clin. Med.* 82:249–254.

50. Mahoney, M.G., Z.H. Wang, K. Rothenberger, P.J. Koch, M. Amagai, and J.R. Stanley. 1999. Explanations for the clinical and microscopic localization of lesions in pemphigus foliaceus and vulgaris. *J. Clin. Invest.* 103:461–468.

51. Amagai, M., K. Tsunoda, D. Zillikens, T. Nagai, and T. Nishikawa. 1999. The clinical phenotype of pemphigus is defined by the anti-desmoglein autoantibody profile. *J. Am. Acad. Dermatol.* 40:167–170.

52. Hashimoto, T., M. Amagai, K. Watanabe, M. Dmochowski, M.A. Chidgey, K.K. Yue, D.R. Garrod, and T. Nishikawa. 1995. A case of pemphigus vulgaris showing reactivity with pemphigus antigens (Dsg1 and Dsg3) and desmocollins. *J. Invest. Dermatol.* 104:541–544.

53. Ding, X., V. Aoki, J.M. Mascaro Jr., A. Lopez-Swiderski, L.A. Diaz, and J.A. Fairley 1997. Mucosal and mucocutaneous (generalized) pemphigus vulgaris show distinct autoantibody profiles. *J. Invest. Dermatol.* 109:592–596.

54. Iwatsuki, K., M. Takigawa, T. Hashimoto, T. Nishikawa, and M. Yamada. 1991. Can pemphigus vulgaris become pemphigus foliaceus. *J. Am. Acad. Dermatol.* 25:797–780.

55. Kawana, S., T. Hashimoto, T. Nishikawa, and S. Nishiyama. 1994. Changes in clinical features, histologic findings, and antigen profiles with development of pemphigus foliaceus from pemphigus vulgaris. *Arch. Dermatol.* 130:1534–1538.

56. Jordon, R.E., N.K. Day, J.R. Luckasen, and R.A. Good. 1973. Complement activation in pemphigus vulgaris blister fluid. *Clin. Exp. Immunol.* 15:53–63.

57. Jordon R.E., A.L. Sohroeter, R.S. Rogers, and H.O. Perry. 1974. Classical and alternate pathway activation of complement in pemphigus vulgaris lesions. *J. Invest. Dermatol* 63:256–259.

58. Nishikawa, T., S. Kurihara, and H. Hatano. 1979. Comparison of *in vivo* and *in vitro* capability of complement fixation by pemphigus autoantibodies. *Dermatologica* 159:290–294.

59. Kawana, S., M. Janson, and R.E. Jordon. 1984. Complement fixation by pemphigus antibody. I. *In vitro* fixation to organ culture and tissue culture skin. *J. Invest. Dermatol.* 82:506–510.

60. Kawana, S., W.D. Geoghegan, and R.E. Jordon. 1989. Deposition of membrane attack complex of complement in pemphigus vulgaris and pemphigus foliaceus skin. *J. Invest. Dermatol.* 92:588–592.

61. Anhalt, G.J., G.O. Till, L.A. Diaz, R.S. Labib, H.P. Patel, and N.F. Eaglstein. 1986. Defining the role of complement in experimental pemphigus vulgaris in mice. *J. Immunol.* 137:2835–2840.

62. Jones, C.C., R.G. Hamilton, and R.E. Jordon. 1988. Subclass distribution of human IgG autoantibodies in pemphigus. *J. Clin. Immunol.* 8:43–49.

63. Kim, Y.H., W.D. Geoghegan, and R.E. Jordon. 1990. Pemphigus immunoglobulin G subclass autoantibodies: studies of reactivity with cultured human keratinocytes. *J. Lab. Clin. Med.* 115:324–331.

64. Bhol, K., A. Mohimen, and A.R. Ahmed. 1994. Correlation of subclasses of IgG with disease activity in pemphigus vulgaris. *Dermatol.* 189:85–89.

65. Shakib, F. 1987. Aspects of the pathological significance of IgG4. *Exp. Clin. Immunogenetics* 4:193–200.

66. Mascaro, J.M., Jr., A. Espana, Z. Liu, X. Ding, S.J. Swartz, J.A. Fairley, and L.A. Diaz. 1997. Mechanisms of acantholysis in pemphigus vulgaris: role of IgG valence. *Clin. Immunol. Immunopathol.* 85:90–96.

67. Hashimoto, K., K.H. Shafron, D.S. Webber, G.S. Lazarus, and K.H. Singer. 1983. Anti-cell surface pemphigus auto-antibody stimulates plasminogen activator activity of human epidermal cells. A proposed mechanism for the loss of epidermal cohesion and blister formation. *J. Exp. Med.* 157:259–272.

68. Lotti, T., P. Bonan, G. Cannarozzo, and E. Panconesi. 1988. *In vivo* studies on the involvement of urokinase in pemphigus acantholysis. *J. Invest. Dermatol.* 91:372–373.

69. Spillman, D.H., P.H. Magnin, L. Roquel, and M. Mitsui. 1988. Aprotinin inhibition of experimental pemphigus in Balb/c mice following passive transfer of pemphigus foliaceus serum. *Clin. Exp. Dermatol.* 13:321–327.

70. Pasricha, J.S., and S.S. Das. 1992. Curative effect of dexamethasone-cyclophosphamide pulse therapy for the treatment of pemphigus vulgaris. *Int. J. Dermatol.* 31:875–877.

71. Anhalt, G.J., H.P. Patel, R.S. Labib, L.A. Diaz, and D. Proud. 1986. Dexamethasone inhibits plasminogen activator activity in experimental pemphigus *in vivo* but does not block acantholysis. *J. Immunol.* 136:113–117

72. Wang, Z., M.G. Mahoney, and J.R. Stanley. 1999. Plasminogen activator is not required for blister formation in pemphigus. *J. Invest. Dermatol.* 112:532A.

73. Diaz, L.A., and C.L. Marcelo 1978. Pemphigoid and pemphigus antigens in cultured epidermal cells *Br. J. Dermatol.* 98:631–637.

74. Esaki, C., M. Seishima, T. Yamada, K. Osada, and Y. Kitajima. 1995. Oharmacologic evidence for involvement of phospholipase C in pemphigus IgG-induced inositol 1,4,5-trisphosphate generation, intracellular calcium increase, and plasminogen activator secretion in DJM-1 cells, a squamous cell carcinoma line. *J. Invest. Dermatol.* 105:329–333.

75. Kirajima, Y., S. Inoue, S. Nagao, K. Nagata, H. Yaoita, and Y. Nozawa. 1988. Biphasic effects of 12-o-tetradecanoylphorbol-13-acetate on the cell morphology of low calcium-grown human epidermal carcinoma cells: involvement of translocation and down regulation of protein kinase C. *Cancer Res.* 48:964–970.

76. Hashimoto, K., K.M. Shafan, P.A. Webber, G.S. Lazarus, and K.H. Singer. 1983. Anti-cell surface pemphigus antibody stimulates plasminogen activator activity of human epidermal cells. A mechanism for the loss of epidermal cohesion and blister formation. *J. Exp. Med.* 157:259–272.

77. Hashimoto, K., T.C. Wun, J. Baird, G.S. Lazarus, and P.J. Jensen. 1989. Characterization of keratinocyte plasminogen activator inhibitors and demonstration of the prevention of pemphigus IgG-induced acantholysis by a purified plasminogen activator inhibitor. *J. Invest. Dermatol.* 92:310–315.

78. Osada, K., M. Seishima, and Y. Kitajima. 1997. Pemphigus IgG activates and translocates protein kinase C from the cytosol to the particulate/cytoskeleton fractions in human keratinocytes. *J. Invest. Dermatol.* 108:482–487.

79. Aoyama, Y., and Y. Kitajima. 1999. Pemphigus vulgaris-IgG causes a rapid breakdown of free desmoglein 3 (Dsg3), leading to the formation of Dsg3-depleted desmosomes. *J. Invest. Dermatol.* 112:67–71.

80. Aoyama, Y., M.K. Owada, and Y. Kitajima. 1999. A pathogenic autoantibody, pemphigus vulgaris-IgG, induces phosphorylation of desmoglein 3 and its dissociation from plakoglobin in cultured keratinocytes. *Eur. J. Immunol.* 29:2233–2240.

81. Kermani-Arab, V., K. Hirji, A.R. Ahmed, and J.L. Fahey. 1984. Deficiency of interleukin-2 production and interleukin-2 receptor expression on peripheral blood leukocytes after phytohemagglutinin stimulation in pemphigus. *J. Invest. Dermatol.* 83:101–104.

82. King, A.J., S.A. Schwartz, and D. Lopatin. 1982. Suppressor cell function is preserved in pemphigus and pemphigoid. *J. Invest. Dermatol.* 79:183–185.

83. Wucherpfennig, K.W., B. Yu, K. Bhol, D.S. Monos, E. Argyris, R.W. Karr, A.R. Ahmed, and J.L. Strominger. 1995. Structural basis for major histocompatibility complex (MHC)-linked susceptibility to autoimmunity: Charged residues of a single MHC binding pocket confer selective presentation of self-peptides in pemphigus vulgaris. *Proc. Natl. Acad. Sci. USA* 92:11935–11939.

84. Hertl, M., M. Amagai, H. Sundaram, J. Stanley, K. Ishii K, and S.I. Katz. 1998. Recognition of desmoglein 3 by autoreactive T cells in pemphigus vulgaris patients and normals. *J. Invest. Dermatol.* 110:62–66.

85. Lin, M.S., S.J. Swartz, A. Lopez, X. Ding, J.A. Fairley, and L.A. Diaz. 1997. T lymphocytes from a subset of patients with pemphigus vulgaris respond to both desmogelin-3 and desmoglein-1. *J. Invest. Dermatol.* 109:734–737.

86. Lehmann, P.V., T. Forsthuber, A. Miller, and E.E. Sercarz. 1992. Spreading of T-cell autoimmunity to cryptic determinant of an autoantigen. *Nature* 358:155–157.

87. Kaufman, D.L., M. Clare-Salzler, and J. Tian. 1993. Spontaneous loss of T cell tolerance to glutamic acid decarboxylase in murine insulin-dependent diabetes. *Nature* 366:69–72.

88. Scharf, S.J., A. Freidmann, C. Brautbar, F. Szafer, L. Steinman, G. Horn, U. Gyllensten, and H.A. Erlich. 1988. HLA class II allelic variation and susceptibility to pemphigus vulgaris. *Proc. Natl. Acad. Sci. USA* 85:3504–3508.

89. Sinha, A.A., C. Brautbar, F. Szafer, A. Friedmann, E. Tzfoni, J.A. Todd, L. Steinman, and H.O. McDevitt. 1988. A newly characterized HLA DQB allele associated with pemphigus vulgaris. *Science* 239:1026–1029.

90. Szafer, F., C. Brautbar, E. Tzfoni, G. Frankel, L. Sherman, I. Cohen, S. Hacham-Zadeh, W. Aberer, G. Tappeiner, G. et al. 1988. Detection of disease-specific restriction fragment length polymorphism in pemphigus vulgaris linked to the DQw1 and DQw3 alleles of the HLA-D region. *Proc. Natl. Acad. Sci. USA* 84:6542–6545.

91. Mobini, N., E.J. Yunis, C.A. Alper, J.J. Yunis, J.C. Delgado, D.E. Yunis, A. Firooz, Y. Dowlati, K. Bahar, P.K. Gregersen, and A.R. Ahmed. 1997. Identical MHC markers in non-Jewish Iranian and Ashkenazi Jewish patients with pemphigus vulgaris: possible common central Asian ancestral origin. *Human Immunol.* 57:62–67.

92. Delgado, J.C., D.E. Yunis, M.V. Bozon, M. Salazar, R. Deulofeut, D. Turbay, N.K. Mehra, J.S. Pasricha, R.S. Raval, H. Patel, B.K. Shah, K. Bhol, C.A. Alper, A.R. Ahmed, and E.J. Yunis. 1996. MHC class II alleles and haplotypes in patients with pemphigus vulgaris from India. *Tissue Antigens* 48:668–672.

93. Delgado, J.C., A. Hameed, J.J. Yunis, K. Bhol, A.I. Rojas, S.B. Rehman, A.A. Khan, M. Ahmad, C.A. Alper, A.R. Ahmed, and E.J. Yunis. 1997. Pemphigus vulgaris autoantibody response is linked to HLA-DQB1*0503 in Pakistani patients. *Human Immunology* 57:110–119.

94. Yamashina, Y., S. Miyagawa, T. Kawatsu, T. Iida, I. Higashimine, T. Shirai, and T. Kaneshige. 1998. Polymorphisms of HLA class II genes in Japanese patients with pemphigus vulgaris. *Tissue Antigens* 52:74–77.

95. Lee, C.W., H.Y. Yang, S.C. Kim, J.H. Jung, and J.J. Hwang. 1998. HLA class II allele associations in Korean patients with pemphigus. *Dermatol.* 197:349–352.

96. Carcassi, C., F. Cottoni, L. Floris, A. Vacca, M. Mulargia, M. Arras, R. Boero, G. La, G. Nasa, A. Ledda, A. Pizzati, D. Cerimele, and L. Contu. 1996. HLA haplotypes and class II molecular alleles in Sardinian and Italian patients with pemphigus vulgaris. *Tissue Antigens* 48:662–667.

97. Bhol, K., J. Yunis, and A.R. Ahmed. 1996. Pemphigus vulgaris in distant relatives of two families: association with major histocompatibility complex class II genes. *Clin. Exp. Dermatol.* 21:100–103.

98. Brown, M.V. 1954. Fogo selvagem (pemphigus foliaceus). *AMA Arch. Dermatol. Syphilol.* 69:589–599.

99. Perry, H.O. 1961. Pemphigus foliaceus. *Arch. Dermatol.* 83:52–57.

100. Azulay, R.D. 1982. Brazilian pemphigus foliaceus. *Int. J. Dermatol.* 21:121–124.

101. Diaz, L.A., S.A.P. Sampaio, E.A. Rivitti, C.R. Martins, P.R. Cunha, C. Lombardi, F.A. Almeida, R.M. Castro, M.L. Macca, C. Lavrado, et al. 1989. Endemic pemphigus foliaceus-"fogo selvagem" I. Clinical feature and immunopathology. *J. Am. Acad. Dermatol.* 20:657–669.

102. Walker, D.C., K.A. Kolar, A.A. Hebert, and R.E. Jordon. 1995. Neonatal pemphigus foliaceus. *Arch. Dermatol.* 131:1308–1311.

103. Rocha-Alvarez, R.H. Friedman, I.T. Campbell, L. Souza-Aguiar, R. Martins-Castro, and L.A. Diaz. 1992. Pregnant women with endemic pemphigus foliaceus (Fogo Selvagem) give birth to disease-free babies. *J. Invest. Dermatol.* 99:78–82.

104. Avalos-Diaz, E., M. Olague-Marchan, A. Lopez-Swiderski, R. Herrera-Esparaza, and L.A. Diaz. Transplacental passage of maternal pemphigus foliaceus autoantibodies induce neonatal pemphigus. Submitted.

105. Vieira, J.P. 1948. Consideracoes sobre o penfigo foliaceo no Brazil, Empresa Grafica da Revista dos Tribunais, Sao Paulo, Brazil.

106. Furtado, T.A. 1959. Histopathology of pemphigus foliaceus. *Arch. Dermatol.* 80:66–71.

107. Santi, C.G., C.W. Maruta, V. Aoki, M.N. Sotto, E.A. Rivitti, L.A. Diaz, and the Cooperative Group on Fogo Selvagem Research. 1996. Pemphigus herpetiform is rare clinical expression of nonendemic pemphigus foliaceus, fogo selvagem, and pemphigus vulgaris. *J. Am. Acad. Dermatol.* 34:40–46.

108. Hashimoto, T.C. Kiyokawa, O. Mori, M. Miyasato, M.A. Chidgey, D.R. Garrod, Y. Kobayashi, K. Komori, K. Ishii, M. Amagai and T. Nishikawa. 1997. Human desmocollin 1 (Dsc 1) is an autoantigen for the subcorneal pustular dermatosis type of IgA pemphigus. *J. Invest. Dermatol.* 109:127–131.

109. Auad, A. 1972. Penfigo Foliaceo Sul-Americano no Estado de Goias. *Rev. Patol. Trop.* 1:293–346.

110. Barros, C. 1972. Ultrastructura da lesao bolhosa e do sinal de nikolski no penfigo foliaceo. *Int. J. Dermatol.* 16:799–806.

111. Sotto, M.N., S.H. Shimizu, J.M. Costa, and T. DeBrito. 1980. South American pemphigus foliaceus: electron microscopy and immunoelectron localization of bound immunoglobulin in the skin and oral mucosa. *Br. J. Dermatol.* 102:521–527.

112. Futamura, S., C. Martins, E.A. Rivitti, R.S. Labib, L.A. Diaz, and G. Anhalt. 1989. Ultrastructural studies of acantholysis induced *in vivo* by passive transfer of IgG from endemic pemphigus foliaceus (Fogo selvagem). *J. Invest. Dermatol.* 93:480–485.

113. Iwarsuki, K., M. Takigawa, F. Jin, and M. Yamada. 1991. Ultrastructural binding site of pemphigus foliaceus autoantibodies: comparison with pemphigus vulgaris. *J. Cutaneous Pathol.* 18:160–163.

114. Squiquera, H.L., L.A. Diaz, S.A. Sampaio, E.A. Rivitti, C.R. Martins, P.R. Cunha, C. Lombardi, C. Lavrado, P. Borges, H. Friedman, et al. 1988. Serologic abnormalities in patients with endemic pemphigus foliaceus (Fogo selvagem), their relatives, and normal donors from endemic and non-endemic areas of Brazil. *J. Invest. Dermatol.* 91:189–191.

115. Calvanico, N.J., S.J. Swartz, and L.A. Diaz. 1993. Affinity immunoblotting studies on the restricting autoantibodies from endemic pemphigus foliaceus patients. *J. Autoimm.* 6:145–157.

116. Olague-Alcala, M., G.J. Giudice, and L.A. Diaz. 1994. Pemphigus foliaceus sera recognize an N-terminal fragment of bovine desmoglein-1. *J. Invest. Dermatol.* 102:882–885.

117. Diaz, L.A., S.A.P. Sampaio, E.A. Rivitti, C.R. Martins, P.R. Cunha, C. Lombardi, F.A. Almeida, R.M. Castro, M.L. Macca, and C. Lavrado. 1989. Endemic pemphigus foliaceus-"fogo selvagem." II. Current and historical epidemiological aspects. *J. Invest. Dermatol.* 92:4–12.

118. Robledo, M.A., S. Prada, D. Jaramillo, and W. Leon. 1988. South-American pemphigus foliaceus: study of an epidemic in El Barge and Nechi, Colombia 1982 to 1986. *Br. J. Dermatol.* 118:737–744.

119. Bastuji-Garin, S., R. Souissi, L. Blum, H. Turki, R. Nouira, B. Jomaa, A. Zahaf, A. Ben Osman, I. Mokhtar, B. Fazza, J. Revuz, J.C. Roujeau, and M.R. Kamoun. 1995. Comparative epidemiology of pemphigus in Tunisia and France: Unusual incidence of pemphigus foliaceus in young Tunisia women. *J. Invest. Dermatol.* 104:302–305.

120. Morini, J.P., B. Jomaa, Y. Gorgi, M.H. Saguem, R. Nourina, J.C. Roujeau, and J. Revuz. 1995. Pemphigus foliaceus in young women. An endemic focus in the sousse area of Tunisia. *Arch. Dermatol.* 129:69–73.

121. Friedman, H., I. Campbell, R. Rocha-Alvarez, I. Ferrari, C.E. Coimbra, J.R. Moraes, N.M. Flowers, P. Stastny, M. Fernandez-Vina, and M. Olague-Alcala. 1995. Endemic pemphigus foliaceus (fogoselvagem) in native Americans from Brazil. *J. Am. Acad. Dermatol.* 32:949–56.

122. Hans-Filho, G., V. Santos, J.H. Katayama, V. Aoki, E.A. Rivitti, S.A.P. Sampaio, and H. Friedman. 1996. An active focus of high prevalence of fogo selvagem on an Amerindian reservation in Brazil. *J. Invest. Dermatol.* 107:68–75.

123. Aranha-Campos, J. 1943. Penfigo foliaceos (fogo selvagem). Aspectos clinicos epeidemiologicos. Comp Melhoramentos de Sao Paulo, Industrias de Papel Brazil.

124. Lombardi, C., D.C. Borges, F. Chaul, S.A.P. Sampaio, E.A. Rivitti, E.A., C.R. Martins et al. 1992. Environmental risk factors in endemic pemphigus foliaceus (fogo selvagem). *J. Invest. Dermatol.* 98:847–850.

125. Eaton, D.P., L.A. Diaz, G. Hans-Filho, V. dos Santos, V. Aoki, H. Friedman, E.A. Rivitti, S.A.P. Sampaio, M.S. Gottlieb, G.J. Giudice, A. Lopez, and E.W. Cupp 1998. Characterization of black fly species (Diptera: Simuliidae) on an Amerindian reservation with a high prevalence of fogo selvagem and neighboring disease-free sites in the state of Mato Grosso do Sul Brazil. *J. Med. Entomol.* 35:120–131.

126. Palka, H.L. and K.J. Green. 1997. Roles of plakoglobin end domains in desmosome assembly. *J. Cell. Sci.* 110:2359–2371.

127. Koulu, L., A. Kusumi, M.S. Steinberg, V. Klaus-Kovtun, and Stanley. 1984. Human antibodies against a desmosomal core protein in pemphigus foliaceus. *J. Exp. Med.* 160:1509–1518.

128. Jones, J.C.R., K.M. Yokoo, and R.D. Goldman. 1986. Further analysis of pemphigus autoantibodies and their use in studies on the heterogeneity, structure, and function of desmosome. *J. Cell. Biol.* 102:1109–1117.

129. Stanley, J.R., L. Koulu, V. Klaus-Kovtun, and M.S. Steinberg. 1986. A monoclonal antibody to the desmosomal glycoprotein desmoglein 1 binds the same polypeptide as human autoantibodies in pemphigus foliaceus. *J. Immunol.* 136:1227–1230.

130. Amagai, M., A. Komai, T. Hashimoto, Y. Shirakata, K. Hashimoto, T. Yamada, Y. Kitajima, K. Ohya, H. Iwanami, and T. Nishikawa. 1999. Usefulness of enzyme linked immunoabsorbent assay using recombinant desmogleins 1 and 3 for serodiagnosis of pemphigus. *Br. J. Dermatol.* 140:351–357.

131. Warren, P.J., M.S. Lin, G.J. Giudice, R.G. Hoffmann, G. Hans-Filho, V. Aoki, E.A. Rivitti, V. dos Santos, and L.A. Diaz for the Cooperative Group on Fogo Selvagem. Endemic pemphigus foliaceus (fogo selvagem) as a model of an environmentally-induced human autoimmune disease. Submitted.

132. Ishi, K., M. Amagai, R.S. Hall, T Hashimoto, A. Takayanagi, S. Gamou, N. Shimizu, and T. Nishikawa. 1997. Characterization of autoantibodies in pemphigus using antigen-specific enzyme-linked immunosorbent assays with baculovirus-expressed recombinant desmogleins. *J. Immunol.* 159.2010–2017.

133. Abreu-Velez, A.M., M. Olague-Marchan, A. Lopex-Swiderski, J.M. Mascaro Jr, G.J. Giudice, and L.A. Diaz. 1997. Characterization of a 45 kD epidermal tryptic peptide recognized by pemphigus foliaceus sera. *J. Invest. Dermatol.* 108:541A (Abstr.).

134. Beutner, E.H., L.S. Prigenzi, W. Hale, C.A. Leme, and O.G. Bier. 1968. Immunofluorescent studies of autoantibodies to intercellular areas of epithelia in Brazilian pemphigus foliaceus. *Proc. Soc. Exp. Biol. Med.* 127:81–86.

135. Amagai, M., P.J. Koch, T. Nishikawa, and J.R. Stanley. 1996. Pemphigus vulgaris antigen (desmoglein 3) is localized in the lower epidermis, the site of blister formation in patients. *J. Invest. Dermatol.* 106:351–355.

136. Takahashi, M.D.F. 1981. Imunopatologia do penfigo foliaceo sulamericano. Estudo por imunofluorescencia direta e indireta. Faculdade Medicina da Universidada de Sao Palo, Brazil.

137. Rock, B., C.R. Martins, A.N. Theofilopoulos, R.S. Balderas, G.J. Anhalt, R.S. Labib, S. Futamura, E.A. Rivitti, and L.A. Diaz. 1989. The pathogenic effect of IgG4 autoantibodies in endemic pemphigus foliaceus (endemic pemphigus foliaceus). *N. Eng. J. Med.* 320:1463–1469.

138. Rock, B., R.S. Labib, and L.A. Diaz. 1990. Monovalent Fab' immunoglobulin fragments from endemic pemphigus foliaceus autoantibodies reproduce the human disease in neonatal BALB/c mice. *J. Clin. Invest.* 85:296–299.

139. España A., L.A. Diaz, G.J. Giudice, J.A. Fairley, G.O. Till, and Z. Liu. 1997. The role of complement in experimental

pemphigus foliaceus. *Clin. Immunol. Immunopathol.* 85:83–89.

140. Allen, E.M., G.J. Giudice, and L.A. Diaz. 1993. Subclass reactivity of pemphigus foliaceus autoantibodies with recombinant human desmoglein. *J. Invest. Dermatol.* 100:685–691.

141. Amagai. M., T. Hashimoto, K.J. Green, N. Shimizu, and T. Nishikawa. 1995. Antigen-specific immunoadsorption of pathogenic autoantibodies in pemphigus foliaceus. *J. Invest. Dermatol.* 104:895–901.

142. Guerra, H.A., A.P. Reis, and M.V.N. Guerra. 1976. T and B lymphocytes in South American pemphigus foliaceus. *Clin. Exp. Immunol.* 23:477–480.

143. Baxevanis, C.N., S. Frillingos, K. Seferiadis, G.J. Reclos, P. Arsenis, A. Katsiyiannis, E. Anastasopoulos, O. Tsolas, and M. Papamichail. 1990. Enhancement of human T lymphocyte function by prothymosin alpha: increased production of interleukin-2 and expression of interleukin-2 receptors in normal human peripheral blood T lymphocytes. *Immunopharmacol. Immunotoxicol.* 12:595–617.

144. Zillikens, D., A. Ambach, A. Zenter, R. Dummer, M. Schussler, G. Burg, and E.B. Brocker. 1993. Evidence for cell-mediated immun mechanisms in the pathology of pemphigus. *Br. J. Dermatol.* 128:636–643.

145. Petzl-Erler, M.L. and J. Santamaria. 1989. Are HLA class II genes controlling susceptibility and resistance to Brazilian pemphigus foliaceus (fogo selvagem)? *Tissue Antigens* 33:408–411.

146. Moraes, J.R., M.E. Moraes, M. Fernandez-Vina, L.A. Diaz, H. Friedman, I.T. Campbell, R.R. Alvarez, S.A.P. Sampaio, E.A. Rivitti, and P. Stastny, P. 1991. HLA antigens and the risk for development of pemphigus foliaceus ("fogo selvagem") in endemic areas of Brazil. *Immunogenetics* 33:388–391.

147. Cerna, M., M. Fernandez-Vina, H. Friedman, M.E. Moraes, J.R. Moraes, L.A. Diaz, and P. Stastny. 1993. Genetic markers for susceptibility to endemic Brazilian pemphigus foliaceus (Fogo Selvagem) in Xavante Indians. *Tissue Antigens* 42:138–140.

148. Moraes, M.E., M. Fernandez-Vina, I. Salatiel, S. Tsai, J.R. Moraes, and P. Stastny. 1993. HLA class II DNA typing in two Brazilian populations. *Tissue Antigens* 41:238–242.

149. Moraes, M.E., M. Fernandez-Vina, A. Lazaro, L.A. Diaz, G. Hans-Filho, H. Friedman, E. Rivitti, V. Aoki, P. Stastny, and J.R. Moraes. 1997. An epitope in the third hypervariable region of the DRB1 gene is involved in the susceptibility to endemic pemphigus foliaceus (fogo selvagem) in three different Brazilian populations. *Tissue Antigens* 49:35–40.

150. Gregersen, P.K., J. Silver, and R.J. Winchester. 1987. The shared epitope hypothesis. An approach to understanding the molecular genetics of susceptibility to rheumatoid arthritis. *Arthritis Rheum.* 30:1205–1212.

151. Matsuyama, M., K. Hashimoto, Y. Yamasaki, R. Shirakura, R. Higuchi, T. Miyajima, and H. Amemiya. 1981. HLA-DR antigens in pemphigus among Japanese. *Tissue Antigens* 17:238–239.

152. Koch, P.J., M.G. Mahoney, H. Ishikawa, L. Pulkkinen, J. Uitto, L. Shultz, G.F. Murphy, D. Whitaker-Menezes, and J.R. Stanley. 1997. Targeted disruption of the pemphigus vulgaris antigen (desmoglein 3) gene in mice causes loss of keratinocyte cell adhesion with a phenotype similar to pemphigus vulgaris. *J. Cell. Biol.* 137:1091–102.

47 | Immunopathological Conditions Affecting Oral Tissues

Patricia Price and Gareth Davies

1. INTRODUCTION

This chapter describes conditions that affect the oral tissues specifically (eg: periodontitis), conditions affecting several mucosal sites and often the skin (eg: pemphigoid) and the salivary manifestations of a polysystemic disease (Sjögren's syndrome). The characteristic clinical features of the syndromes are presented to provide a perspective for scientists rather than as a medical text. Our aim is to summarize the current thinking with regards autoantigens and pathogenic mechanisms, in the hope that this may facilitate the development of unifying hypotheses. Conditions were selected for their common occurrence and/or interesting pathogenesis.

2. INFLAMMATION OF PERIODONTAL TISSUES

Gingivitis and periodontitis are initiated by gram-negative bacteria, which induce inflammation in gingival tissues and lead, in some patients, to the destruction of the underlying periodontal ligament and alveolar bone (1). However, the rates of tissue destruction vary between individuals, and neither the bacterial load nor the spectrum of bacterial species present adequately predict disease incidence or severity. Some of the underlying mechanisms can be elucidated by studies of factors that potentiate the disease process. The American Academy of Periodontology has established the following system for the classification of periodontal diseases:

  I. Gingivitis
  II. Early onset periodontitis
  III. Adult onset periodontitis (adult-type, rapidly progressive or refractory)
  IV. Periodontal abscess
  V. Periodontitis associated with systemic disease (HIV, diabetes, neutropenia)

2.1. Gingivitis

There is little information about the pathogenesis of gingivitis, which may be regarded as a non-specific, non-progressing reaction to the accumulation of dental plaque. Tissues revert to their pre-inflamed state when the bacterial plaque is removed. Inflamed tissue contains elevated levels of IL-1β and IL-6, and reduced levels of collagen types I and III (4), though even higher IL-1β levels are reported in inflamed gingival tissue from patients with periodontal attachment loss (5). Genetic factors clearly affect the development of gingivitis, as different frequencies are described in populations from distinct gene pools residing in equivalent geographic areas (2). However, the genes involved have not been identified, and there is potential for numerous behavioural and dietary factors to impact on the findings. A study of complement alleles in acute necrotizing ulcerative gingivitis yielded no associations suggestive of a role for MHC genes (3).

2.2. Periodontitis

This represents a spectrum of conditions which affect most people at some time in their lives. Periodontal attachment loss is potentiated by type 1 diabetes, HIV infection, smoking and old age (2,6), with evidence of a genetic predisposition (7,8).

Pathogenesis. Numerous surveys have sought correlations between the microbial species present and disease progression. However, to date only *Porphyromonas gingivalis, Porphyromonas intermedia* and *Bacteroides forsythus* emerge as significant prognostic indicators. Smoking clearly affects disease, perhaps via changes in the spectrum of bacteria present, poor microbiocidal responses and/or impaired healing of lesions (2). The effects of age have been reproduced *in vitro* with studies of human gingival and periodontal ligament fibroblasts passaged repeatedly. Lipopolysaccharide (LPS)-stimulated production of PGE_2, IL-1β, IL-6 and plasminogen activator was increased in the aged cells (9). High levels of PGE_2, TNFα and IL-1β have been demonstrated in gingival crevicular fluid (GCF) from diabetic patients with periodontitis compared with diabetics without periodontitis or non-diabetics with periodontitis (10). IL-1ra protein (11) and mRNA (Rodanant, Price and Davies, unpublished data) have been also been demonstrated in affected tissues and may dampen inflammation. Interestingly, cells staining with CD4 and IL-4 were more abundant in gingivitis than periodontitis. *In vitro* studies suggested that IL-4 may downregulate gingival inflammation by inducing apoptosis of infiltrating macrophages (12). In addition, IL-8 may contribute to periodontal disease, promoting the accumulation of neutrophils that produce other inflammatory mediators (1). Blood leukocytes from IDDM patients with mild or moderate to severe periodontitis stimulated with LPS also produced more TNFα than cells from non-diabetic patients with gingivitis or adult periodontitis. This effect was proportional to the severity of disease (13) and suggested that the development of periodontitis may be promoted by high cytokine production induced by LPS stimulation from the oral microflora. It remains to be established whether cytokine levels are elevated by diabetes or by a genetic factor that promotes both diabetes and periodontal changes.

Although it is unlikely that periodontal disease is initiated by anti-self reactivity, autoantibodies reactive against collagen type I are produced by mononuclear cells infiltrating the inflamed gingiva in adult periodontitis. IgG predominated over IgA and IgM in chronic disease (14).

As the later stages of periodontal disease involve the destruction of bone, there is potential for factors implicated in osteoclast activity to play a role. For example, IL-6 levels were correlated with indices of gingivitis and attachment loss in healthy adults with destructive periodontal disease (15). Metalloproteinases, which degrade extracellular matrix (collagen, gelatin, elastin, fibronectin, laminin and proteoglycan core protein) are essential for tooth eruption and have increased activity at diseased sites. This has been confirmed by the demonstration of increased levels of mRNA for several known proteases, showing they are produced by epithelial and endothelial cells, keratinocytes, fibroblasts, osteoblasts, macrophages, lymphocytes and (importantly) neutrophils. Their production by osteoblasts is implicated in the degradation of the osteoid layer and the exposure of the underlying mineralising surface to the osteoclast, thus initiating bone destruction. Bacterial enzymes may contribute to the degradation of collagen directly and via the induction of protease production by fibroblasts, neutrophils and perhaps other cell types (Rev. in 1). The ability of tetracycline to inhibit matrix metalloproteinases has lead to investigations of novel modes of therapy to combat bone loss (16). Tetracycline has long been used to treat periodontitis, but it was assumed its efficacy arose from its bacteriocidal activity, which had been demonstrated using organisms implicated in periodontitis.

The role of TNFα and IL-1 in periodontal damage has now been demonstrated directly in monkeys (*Macaca fascicularis*). Osteoclast activation and mandibular bone loss was stimulated by the application of ligatures soaked in *P. gingivalis*, and inhibited by local treatment with soluble TNF and IL-1 receptors (17). Other members of the TNF/TNFR family have been identified *in silico*, cloned and implicated in bone metabolism in the last 3 years. Knockout mice lacking the TNFR analogue, osteoprotegerin (OPG; also known as Osteoclastogenesis-Inhibitory Factor, OCIF) display osteoporosis because OPG is a soluble decoy receptor for Osteoclast Differentiation Factor (ODF) (18). Similarly, overexpression of OPG and underexpression of ODF lead to osteopetrosis and a defect in tooth eruption (19). The bone resorbative properties of TNFα, IL-1, 1α25-OH-vitamin D3 and prostaglandin E2 are associated with increased levels of ODF MRNA mRNA (20). It is now apparent that this family of molecules will link bone metabolism, inflammation and adaptive immunity. The cell-bound receptor for ODF (ODFR) is identical to Receptor Activator of NFκB (RANK), which was previously implicated in the activation of T-cells and differentiation of dendritic cells (21). Moreover OPG is one of five known ligands for TNF-Related Apoptosis-Inducing Ligand (TRAIL) and acts as a decoy, inhibiting apoptosis mediated by related cell-bound molecules, Death Receptors (DR) 2 and 3 (22). If periodontal damage is viewed as bone resorbtion within reach of LPS (from the oral microflora), one can take short odds that a role for OPG and its ligands in oral inflammation and periodontal bone resorbtion will emerge in the next few years.

A corollary of these associations is that osteopenia (low bone mass) and osteoporosis (fractures associated with osteopenia) may be predisposing factors for periodontal attachment loss, particularly in post-menopausal women. Assessment of the jaws of osteopenic women by dual photon absorptiometry has shown a reduction in mandibular (mainly cortical) and maxillar (mainly trabecular) bone density, but the literature is complicated by the variations in the sites and sensitivity of methods used to quantitate systemic loss of bone mineralisation (Rev. in 23). Nonetheless, oestrogen replacement therapy may improve mineralisation of the jaw as effectively as the lumbar spine (24).

Genetics. Clinical sub-types of periodontal disease are clearly more common in some racial groups. For example, juvenile periodontitis has an incidence of 1.5% in Afro-caribbeans and 0.02% in Europeans (25,26). Moreover, there is anecdotal and documented evidence of familial associations. Early studies failed to demonstrate an association between HLA phenotypes and periodontal disease (27,28). However, patients with localized juvenile periodontitis or adult periodontitis appear to have a higher prevalence of HLA-A2, A9 (and its subtype A24), A28 and B15 than the general population, and patients with rapidly progressive periodontitis have a higher frequency of these alleles and DR4 (4,7). DQB1 has also been implicated in periodontal disease in Japanese patients (29). It is notable that HLA-A9 (A24) also associated with rapid progression of IDDM (30), suggesting a common pathogenic mechanism. Lympho-proliferative responses to *Porphyromonas gingivalis* and *Capnocytophaga sputigena* presented by HLA-DR antigens were elevated in diabetic patients and correlated with periodontal attachment loss. In this cohort, periodontitis was associated with carriage of DR4, DR53 or DQ3, but it was not possible to establish that these proteins present critical bacterial antigens and thus mediate the enhanced proliferative responses (31).

Polymorphisms in the Fcγ receptors on neutrophils have been examined as risk factors for periodontal disease, since there is evidence linking periodontitis with bacterial overgrowth secondary to neutrophil dysfunction. The Fcγ-RIIIb-NA2 allotype was an independent risk factor for disease recurrence in non-smoking Japanese patients (32). This allele is associated with impaired phagocytosis of IgG$_1$-opsonised bacteria in the absence of complement (33).

A single report published in 1980 described an association between juvenile periodontitis and an autosomal dominant locus on chromosome 4, close to a locus affecting Dentinogenesis Imperfecta and the locus encoding Vitamin D-binding group-specific component (Gc) (34). However this requires confirmation.

There is now considerable interest in the finding that patients with severe periodontal disease have an increased frequency of a haplotype comprising allele 2 at both the IL-1A-889 and IL-1B+3953 loci (35). These genes lie approximately 50 kb apart within the IL-1 cluster on chromosome 2q12-q21 and are in moderate linkage disequilibrium (36). The results were replicated in a more recent study that showed the clearest link was with IL-1B+3953 and cited some evidence that allele 2 is associated with higher IL-1

production (37). However, we are not aware of definitive studies of the effects of the polymorphisms on IL-1 production using reporter constructs. Moreover we found no correlation between allele 2 and periondontal attachment loss or the failure of dental implants (Rogers et al submitted). Nonetheless, a screening assay for the polymorphisms is now available commercially.

2.3. HIV-associated periodontitis and gingivitis

Periodontal disease is also relatively common and severe in HIV patients (38). This may be associated with the elevated levels of inflammatory cytokines, such as TNFα and IL-1, produced by unstimulated monocytes from HIV patients (39). Severe damage, denoted as necrotising ulcerative gingivitis (NUG) and necrotising ulcerative periodontitis (NUP), is seen in severely immunodeficient patients and is considered to represent a continuum of the same disease process. The conditions are associated with a spectrum of microbial species similar to that seen in adult periodontitis, with the exception of high levels of *Candida sp.* However, oral candidiasis is now controlled in well managed HIV patients, and periodontal problems still remain. Whilst systemic activation of neutrophils may be impaired by HIV disease, most patients with HIV associated oral lesions display elevated neutrophil activation and release of IL-1β and metalloproteinases in the oral cavity (38), so there is no real basis to conclude that HIV-associated periodontitis is potentiated by a failure to control oral pathogens.

We investigated MHC associations in HIV patients who had detectable inflammatory periodontal damage (quantitated as clinical attachment loss; CAL), but were not experiencing symptoms of periodontal disease (40). The study focussed on the 8.1 ancestral haplotype (HLA-A1, B8, DR3), as carriers have higher risks of several immunopathological disorders, including type 1 diabetes, systemic lupus erythematosus (SLE) and myasthenia gravis (Rev. in 41). Although the haplotype has not been associated with periodontal disease in the absence of HIV, HIV infection invokes high levels of inflammatory cytokines, so different genes may limit and/or potentiate inflammation of the periodontium. CAL values were elevated in 13/16 HLA-B8 patients and 0/16 controls matched for other risk factors. Moreover, the effect mapped to the central MHC. This region encompasses about 50 genes, including the TNF cluster and many novel genes encoding proteins with homology to known immunoregulatory molecules. TNFA-308 (allele 2) is characteristic of the 8.1 haplotype and has been associated with increased TNF production by blood leukocytes (42). However, this polymorphism did not explain the periodontal inflammation as it was carried by 4/16 non-B8 controls. Alleles expressed at IL-1A-889 and IL-1B+3953 loci also showed no association with periodontal attachment loss in our HIV cohort (40).

3. INFLAMMATORY DISORDERS AFFECTING THE ORAL MUCOSA

3.1. Oral Lichen Planus (OLP)

Although OLP is primarily a disease of women aged 50–55 years, it can affect adults of both sexes. It is a long-standing condition with short periods of activity and remission, and may be stress-related. OLP is characterized by painful lesions of mucosal and/or gingival tissues with atrophic ulcerated epithelium, a hyperplastic epithelial reaction or a combination of presentations. Treatment usually involves corticosteroids, but cyclosporine can be used.

Whilst most OLP patients have no other immunopathic disorders, type 1 diabetes, coeliac disease and myasthenia gravis can promote OLP (43). SLE may lead to oral lesions resembling OLP, prompting several studies discussing methods of differential diagnosis. Hyperkeratotic LP-like oral plaques are often seen in SLE patients whose disease is otherwise quiescent (44). The histological changes are not sufficiently distinct to be useful (45), but OLP itself is not associated with autoantibodies characteristic of SLE (46). Inflamed areas display intense monocytic infiltrates, with T-cells, monocytes and XIIIa⁺ dendrocytes at the interface between the epithelium and connective tissue. Abnormal expression of the adhesion markers ICAM-1, VCAM-1, PECAM-1 and ELAM-1 on endothelial cells and L-selectin, LFA-1 and VLA-4 on the infiltrating cells probably mediates the accumulation of inflammatory cells. These markers can be induced by TNFα and IL-1 produced by dermal keratinocytes, but the nature of the initiating stimulus is unclear. Dermal keratinocytes undergo death by apoptosis, but a role for TNFα has not been confirmed (47).

The phenotype and distribution of CD4⁺ intra-epithelial lymphocytes and Langerhans cells are modified in OLP. This involves a selective increase in CD45RA⁺ memory T-cells, whilst the proportion of $\gamma\delta$T-cells appears normal. The accumulation of T-cells correlated with expression of ICAM-1 on epithelial cells (48).

Reports of familial clustering suggest a genetic component to susceptibility. OLP has been associated with several HLA-DR alleles; notably DR2 in Jewish patients (49), DR3 in Swedes (50) and DR9 in a Japanese cohort. This variation suggests disease may be promoted by a gene in linkage disequilibrium with the MHC class II region, which has a polymorphism common to haplotypes containing DR2, DR3 and DR9. It is notable that generalized LP is associated with DR1 (51). The association between OLP and HLA-DR3 is particularly interesting as it involved a dramatic association (relative risk = 21) with the 8.1 haplotype (HLA-A1, B8, DR3) (Section 2.3 41, 50).

3.2. Recurrent Aphthous Ulcers (RAU)

The aetiology of recurrent ulceration is unknown, but the condition is relatively common and has many hallmarks of autoimmune disease/immunopathology. Minor RAU affect ~25% of the population, usually in late childhood and adolescence. A family history of similar ulceration is common. A characteristic feature is healing without scarring, so many sufferers have occasional single ulcers for which they do not seek treatment. Others present with up to six ulcers and exhibit cycles with variable periods of remission. The female: male incidence is about 1.5:1. Ulcers are usually < 5 mm diameter, last 10–14 days and are painful if secondarily infected. The aetiology is obscure; food intolerance and infection have been implicated. Iron and vitamin deficiency and inflammatory bowel disease are risk factors. Treatments include protective pastes, antiseptics and topical corticosteroids.

Major RAU are regarded as a distinct disease in the aphthous spectrum. Ulcers are > 0.5 cm in diameter (often 1–2 cm) and occur in crops of one or two. They are deep, last weeks or months, and heal with scarring. This condition is associated with HIV disease and Behcets' syndrome. Treatment involves corticosteroids, colchicine, thalidomide or surgical disruption (52). No useful genetic studies are available.

Mast cells have been demonstrated in the subepithelial lamina propria, with evidence of activation/degranulation especially around the aphthae (53). Factor XIIIa⁺ cells are also present in affected tissues with a perivascular distribution corresponding with infiltration of mononuclear inflammatory cells. This is consistent with Factor XIIIa⁺ cells being potent antigen-presenting cells (54). Expression of MHC class II antigens *in vivo* and by Langerhans and epithelial cells stimulated by γ-interferon produced by infiltrating CD4 T-cells may also initiate ulcers by promoting antigen-presentation, though no reports address the antigen-specificity of these T-cells. However, patients' circulating CD4 T-cells can lyse autologous cultured epithelial cells (55,56), so it is likely that T-cells have a pathogenic role even though no antigens have been defined.

Although the low numbers of activated B-cells evident around the aphthae do not suggest a primary role for antibodies (56), levels of anti-endothelial cell antibody (AECA) were increased in patients with recurrent ulceration whilst anti-neutrophil antibodies (ANCA) were normal. Binding of AECA to an endothelial cell line was increased by TNFα and decreased by γIFN (57). TNFα is clearly implicated in the pathogenesis of ulcers, as levels correlate with disease activity in normal subjects (58) and HIV patients (59). Although thalidomide is an effective treatment, many HIV patients display increased HIV RNA levels and, paradoxically, increased levels of circulating TNFα and soluble TNF receptors (60). It will be interesting to see if treatment with soluble TNF receptor constructs ameliorates RAU with less side effects.

3.3. Autoimmune Blistering Diseases

Pemphigus vulgaris (PV).   This is the most severe of the pemphigus diseases affecting the skin and mucosal tissues, with a 90% fatality rate without treatment and around 10% with treatment (61). Onset is usually between 40 and 60 years of age, and oral lesions often precede skin lesions. Although relatively few inflammatory cells are seen in the lesions, treatment generally involves corticosteroids. Immunohistochemical studies show characteristic autoantibodies (mostly IgG) deposited along the keratinocytes of the lower epidermis. The antibodies can be demonstrated in serum in 90% of PV patients and are directed against a 130 kD desmosomal glycoprotein identical to Desmoglein 2 or 3. Binding causes release of plasminogen activator, which disrupts cell to cell adhesions and hence causes

blisters to form (acantholysis). Intercellular oedema is notable, with characteristic changes to the basal epithelium (62,63). Passive transfer of serum from patients to neonatal mice or skin cell cultures reproduces the lesions, demonstrating a pathological role for the autoantibodies (64). Few genetic studies are available, but the disease may be associated with HLA-DR4 (65).

Pemphigoid diseases: general features.   The following sections describe a group of relatively common blistering conditions of the oral mucosa, classified as pemphigoid. The nomenclature is confusing, since the terms "benign mucous membrane pemphigoid", "mucosal pemphigoid" and "cicatricial pemphigoid" having been used indiscriminately to describe clinical features of this group of diseases. However, immunological differences between the conditions can be utilized to form a more scientific classification (66,67, 68). This is important as the prognoses and management strategies differ (See Table 1, Figure 1)

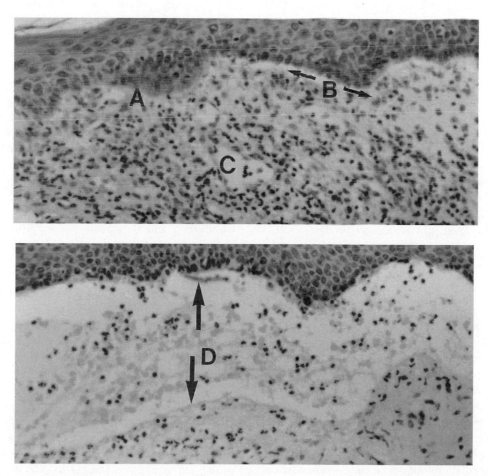

Figure 1.   (i) Haematoxylin and eosin stained section of pemphigoid showing epithelium-connective tissue breakdown [A] and clefting [B]. Note the chronic inflammatory infiltrate [C]. Magnification × 200

(ii) Higher power field showing detail of epithelium-connective tissue interface and blood-filled subepithelial bulla [D]. These features are consistent with pemphigoid but do not distinguish sub-types of the disease. Magnification × 400

Table 1   Oral Blistering Diseases

Condition	Tissues affected	Immunological characteristics
Bullous pemphigoid	Skin and oral mucosa	IgG & IgM deposition at basal membrane zone. Circulating autoantibodies detectable.
Angina bullosa haemorrhagica	Oral tissues involved exclusively. No scarring	Apparently immunologically inert
Mucosal pemphigoid	Oral, ocular and other mucosal sites. No scarring.	IgG, IgM & C3 deposition at basal membrane zone.
Cicatricial pemphigoid	Oral, ocular and other mucosal sites. Ocular and oral scarring	IgG, IgM, C3 & fibrin deposition at basal membrane zone.
Linear IgA bullous dermatosis	Oral, ocular and other mucosal sites. No scarring.	IgA deposition in a linear pattern at basal membrane zone.

Bullous pemphigoid is the classical dermatological blistering disease that affects skin with or without mucosal involvement. It can be distinguished clinically from the mucosal pemphigoides since these do not involve skin and have distinct autoantibody profiles (66). Angina bullosa haemorrhagica has no known immunopathology and hence will not be discussed further. It generally does not require treatment (69,70). The conditions described below are clearly immunopathological disorders as they respond to corticosteroids or dapsone (for IgA-mediated disease).

Oral cicatricial and benign mucous membrane pemphigoid.   Benign mucous membrane pemphigoid is a chronic condition affecting stratified squamous mucous membranes. It is most common in elderly women, predominantly affects the oral and ocular mucosa and heals without scarring (67). In the mouth, gingival tissues are most commonly affected. Linear deposits of IgG and C3 can be demonstrated in the basement membrane zone (62,66). Some authors consider mucous membrane pemphigoid to be a clinical variant of oral cicatricial pemphigoid.

Oral cicatricial pemphigoid is a more severe condition that affects oral and ocular tissues and displays exacerbations and remissions. Lesions heal with mucosal scarring, particularly in the conjunctiva. Oral lesions are usually on the gingiva, but can affect all parts of the mouth. They appear as erosions or ulcers with distinct margins, and arise when the epithelium or basement membrane detaches from the underlying connective tissue or lamina propria. There is considerable inflammatory infiltration, often with abundant neutrophils reflecting secondary infection. Immunofluorescence studies reveal immunoglobulin deposition (usually IgG, C3 and fibrin) in the basement membrane. Many patients have circulating antibodies that recognize epiligrin, a protein similar to laminin V. Others have antibodies reactive with a 230 kD

protein found in the desmosomes and related to desmoplakin, or to type XVII collagen. Variations in the autoantibody profiles may correlate with the clinical features of the disease (62,63). In addition, there may be a purely ocular form of cicatricial pemphigoid, characterized by greater deposition of fibrin and lower levels of autoantibodies (66).

Cicatricial pemphigoid is associated with HLA-DRB1*04 (like PV) and particularly with HLA-DQB1*0301 (71). However the latter may (71) or may not (72) be apparent in patients with oral lesions and no ocular involvement. The HLA associations have been attributed to differential presentation of critical epitopes, as there is evidence of common motifs in the binding grooves of all associated alleles.

Linear IgA bullous dermatosis (LABD).   This is now considered to be distinct from other pemphigoid conditions and dermatitis herpetiformis. Hence LABD lacks the association with HLA-A1, B8, DR3, which is a striking feature of dermatitis herpetiformis and has been described in other childhood bullous disorders (41,73). Oral tissues are involved in 85% of LABD patients and often precede lesions on the skin. The lesions begin with the IgA-mediated accumulation of activated neutrophils in the connective tissue at the basement membrane (74). This leads to necrosis and the separation of the epithelium to form blisters. Immunofluorescence studies show a linear deposition of IgA in the basement membrane zone at the borders of the lesions. More detailed examination reveals two staining patterns involving antibodies reactive with a 97 kD antigen in the lamina lucida or with a larger molecule (290 kD) identified as type VII collagen and a 145 kD dermal antigen. In addition, IgA reactive with basement membrane is often detectable in serum from young patients, but rarely in adults. These differences are consistent with LABD representing a family of related conditions (62,63).

4. CONDITIONS AFFECTING THE SALIVARY TISSUE

Sjögren's Syndrome (SS) is part of the spectrum of chronic systemic rheumatic diseases and affects up to 3% of women by 55 years of age (75). SS has been described as an inflammatory exocrinopathy syndrome, but also affects non-exocrine tissues. Secretions from affected glands are reduced leading to a dry mouth (xerostomia) or dry eyes (xerophthalmia) (76). SS may also involve inflammation of the joints, skin and nervous system. Raynaud's phenomenon and autoimmune thyroiditis are seen in some patients, and many sufferers complain of chronic fatigue. SS is often secondary to rheumatoid arthritis, SLE or scleroderma. Biopsies taken during advanced disease reveal total destruction of the acini and replacement with keratin-containing epithelial cells (Figure 2). These may contribute to disease pathogenesis

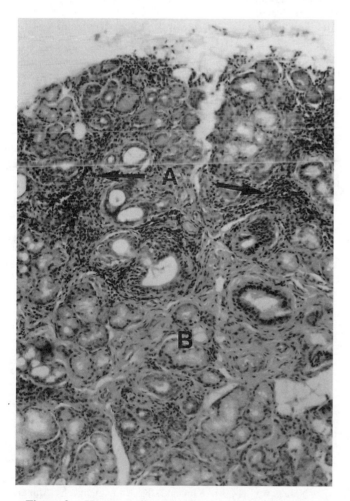

Figure 2.   Haematoxylin and eosin stained section of a minor salivary gland gland from a patient with SS, showing foci of lymphocytes [A] and degeneration of acini [B]. Magnification × 200.

by inappropriately expressing HLA-DR, producing proto-oncogene mRNA and/or pro-inflammatory cytokines. IL-6 mRNA was demonstrated in salivary epithelial cells. This cytokine may be secreted in increased concentrations by parotid glands in SS and can be detected in parotid saliva (77). Mononuclear infiltrates include many CD4+T-cells (75). γδT-cells are not enriched in the infiltrates, but there is some evidence of abnormalities in Vβ usage. Up to 20% of the infiltrate comprises B-cells. The accumulating B-cells may be monoclonal, with lymphomas developing in about 15% of individuals.

Female patients may have infants with neonatal lupus. This may include congenital heart block or other irreversible cardiac defects, but other symptoms resolve by six months of age and hence may be mediated by maternal autoantibodies (78). Anti-Ro and anti-La antibodies reactive with ribonucleo-proteins are present in about of SS patients (75). Anti-La antibodies target a 48 kDa polypeptide now identified as an RNA polymerase III transcription termination factor. The functions of the Ro ribonucleo-protein are less clear. Antibodies bind a complex comprising a 60 kD protein (Ro60), a 52 kD protein (Ro52) and one of four RNA molecules of 83–112 bases (designated hY1, hY3, hY4, hY5). The spectrum of epitopes targeted by autoantibodies varies between studies, as a result of different methodology or variation between individuals. The latter may reflect MHC differences and/or epitope spreading (recognition of additional epitopes in the same protein over time). In addition, different epitopes are recognized by SLE and SS patients (79).

SS has some characteristics of a viral disease with evidence of a pathogenic role for cross-reactions between self and viral antigens. Potential viral triggers include sialotropic viruses (EBV, CMV and HHV-6) and lymphotropic viruses (HIV-1 and HTLV-1) (75). EBV DNA and antigens are found in biopsies of affected glands, whilst normal salivary tissue and other tissue from the patients were negative (80). This suggests that disease may arise from impaired clearance of common viruses, with collateral damage arising from an ineffectual immune response.

SS patients have increased titres of cross-reactive antiretroviral antibodies (75). In addition, a condition with similar clinical manifestations to SS occurs in some HIV patients, although there is a difference in the pattern of the lymphocytic infiltrate with less acinar atrophy (81).

Fas and Fas ligand are expressed on acinar epithelial cells, consistent with evidence that these cells die by apoptosis. In contrast, the infiltrating lymphocytes express Fas and the bcl-2 and are resistant to apoptosis (82), consistent with the histological features of SS.

Numerous studies of cytokine mRNA and protein suggest a pathogenic role for Th1 cytokines particularly γ-interferon in SS, with IL-4 and IL-5 generally undetectable. IL-2 is produced and probably amplifies autoantibody production. Genetic factors are also important. There is a clear association with HLA-DR3 and DQ2 (75), and hence with the 8.1 haplotype (41).

5. CONCLUDING REMARKS

The term "autoimmune disease" is used to describe a wide range of conditions characterised by persistent inflammation that cannot be attributed to a persistent infection. However, it can be argued that its use should be restricted to diseases where autoreactive antibodies or T-cells have demonstrable primary pathogenic effects. Periodontitis and gingivitis do not meet this criteria and are considered here under the broader heading of "immunopathological disorders". Aphthous ulcers are clearly T-cell mediated, but the critical antigens have not been defined. Pemphigus vulgaris, pemphigoid conditions and Sjögren's syndrome are associated with specific autoantibodies in the majority of patients. However, some patients lack the characteristic antibody profiles and the epitopes recognized by individual patients vary, leaving open the question of whether these conditions involve autoimmune recognition as a secondary and variable consequence of inflammation and tissue damage. Thus genes associated with increased risk may be regulatory elements which magnify the inflammatory process (eg: alleles of the TNF and IL-1 genes), rather than genes encoding restriction elements (HLA class II molecules) or proteins that crossreact with viral or bacterial antigens.

ACKNOWLEDGMENTS

The authors thank Dr Peter Hollingsworth and Dr Nicholas Boyd for critical appraisals of this manuscript, and Mr Nick Acquerola and Mr Dennis Barnden for the photography. PP is a Senior Research Officer supported by the National Health and Medical Research Council, Australia. This is manuscript 9912 of the Dept of Clinical Immunology, Royal Perth Hospital.

References

1. Alexander, M.B., and P.D. Damoulis. 1994. The role of cytokines in the pathogenesis of periodontal disease. *Curr. Opin. Periodontol.* 39–53.
2. Genco, R.J. 1996. Current view of risk factors for periodontal diseases. *J. Periodontol.* 67:1041–1049.
3. Melnick, S.L., R.C. Go, R.B. Cogen, and J.M. Roseman. 1988. Allelic variants for complement factors C3, C4, and B in acute necrotizing ulcerative gingivitis. *J. Dental Res.* 67:851–854.
4. Yakovlev, E., I. Kalichman, S. Pisanti, S. Soshan, and V. Barak. 1996. Levels of cytokines and collagen type I and type III as a function of age in human gingivitis. *J. Periodontol.* 67:788–793.
5. Tokoro, Y., T. Yamamoto, and K. Hara. 1996. IL-1β mRNA as the predominant inflammatory cytokine transcript: correlation with inflammatory cell infiltration into human gingiva. *J. Oral Pathol. & Medicine* 25:225–231.
6. Yoshimitus, A., S. Noriyoshi, Y. Masaru, S. Hiroshi, and T. Hisashi. 1998. Effect of aging on functional changes of periodontal tissue cells. *Annal. Periodontol.* 3:350–369.
7. Firatli, E., A. Kantarci, I. Cebeci, H. Tanyeri, G. Sonmez, M. Carin, and O. Tuncer. 1996. Association between HLA antigens and early onset periodontitis. *J. Clin. Periodontol.* 23:563–566.
8. Hart, T.C. 1994. Genetic considerations of risk in human periodontal disease. *Curr. Opin. Periodontol.* pp. 3–11.
9. Abiko, Y., N. Shimizu, M. Yamaguchi, H. Suzuki, and H. Takiguchi. 1998. Effect of aging on functional changes of periodontal tissue cells. *Annal. Periodontol.* 3:350–369.
10. Salvi, G.E., J.D. Beck, and S. Offenbacher. 1998. PGE₂, IL-1β, and TNF-α responses in diabetics as modifiers of periodontal disease expression. *Annal. Periodontol.* 3:40–50.
11. Ishihara, Y., T. Nishihara, T. Kuroyanagi, N. Shirozu, E. Yamagishi, M. Ohguchi, M. Koide, N. Ueda, K. Amano, and T. Noguchi. 1997. Gingival crevicular interleukin-1 and interleukin-1 receptor antagonist levels in periodontally healthy and diseased sites. *J. Periodontal Res.* 32:524–529.
12. Yamamoto, M., K. Kawabata, K. Fujihashi, J.R. McGhee, T.E. van Dyke, T.V. Bamberg, T. Hiroi, and H. Kiyono. 1996. Absence of exogenous interleukin-4-induced apoptosis of gingival macrophages may contribute to chronic inflammation in periodontal diseases. *Am. J. Pathol.* 148:331–339.
13. Collins, S.G.E., Y. Barr, N.P. Lang, and S. Offenbacher. 1997. Monocytic TNFα secretion patterns in IDDM patients with periodontal diseases. *J. Clin. Periodontol.* 24:8–16.
14. Johnson, R., A. Pitts, C. Lue, S. Gay, and J. Mestecky. 1991. Immunoglobulin isotype distribution of locally produced autoantibodies to collagen type I in adult periodontitis. *J. Clin. Periodontol.* 18:703–707.
15. Geivelis, M., D.W. Turner, E.D. Pederson, and B.L. Lamberts. 1993. Measurements of interleukin-6 in a gingival crevicular fluid from adults with destructive periodontal disease. *J. Periodontol.* 64:980–983.
16. Ryan, M.E., S. Ramamurthy, and L.M. Golub. 1996. Matrix metalloproteinases and their inhibition in periodontal treatment. *Curr. Opin. Periodontol.* 3:85–96.
17. Assuma, R., T. Oates, D. Cochran, S. Amar, and D.T. Graves. 1998. IL-1 and TNF antagonists inhibit the inflammatory response and bone loss in experimental periodontitis. *J. Immunol.* 160:403–409.
18. Nakagawa, N., M. Kinosaki, K. Yamaguchi, N. Shima, H. Yasuda, K. Yano, T. Morinaga, and K. Higashio. 1998. RANK is the essential signalling receptor for osteoclast differentiation factor in osteoclastogenesis. *Biochem. Biophys. Res. Commun.* 253:395–400.

19. Kong, Y.Y., H. Yoshida, I. Sarosi, H.L. Tan, E. Timms, C. Capparelli, S. Morony, A.J. Oliveira-dos-Santos, A. Itie, W. Khoo, A. Wakeham, C.R. Dunstan, D.L. Lacey, T.W. Mak, W.J. Boyle, and J.M. Penninger. 1999. OPGL is a key regulator of osteoclastogenesis, lymphocyte development and lymph-node organogenesis. *Nature* 397:315–323.

20. Tsukii, K., N. Shima, S. Mochizuki, K. Yamaguchi, M. Kinosaki, K. Yano, O. Shibata, N. Udagawa, H. Yasuda, T. Suda, and K. Higashio. 1998. Osteoclast differentiation factor mediates an essential signal for bone resorption induced by 1α, 25-dihydroxyvitamin D3, prostaglandin E2, or parathyroid hormone in the micro-environment of bone. *Biochem. Biophys. Res. Commun.* 246:337–341.

21. Galibert, L., M.E. Tometsko, D.M. Anderson, D. Cosman, and W.C. Dougall. 1998. The involvement of multiple tumor necrosis factor receptor (TNFR)-associated factors in the signalling mechanisms of receptor activator of NF-κB, a member of the TNFR superfamily. *J. Biol. Chem.* 273:34120–34127.

22. Emery, J.G., P. McDonnell, M.B. Burke, K.C. Deen, S. Lyn, C. Silverman, E. Dul, E.R. Appelbaum, C. Eichman, R. DiPrinzio, R.A. Dodds, I.E. James, M. Rosenberg, J.C. Lee, and P.R. Young. 1998. Osteoprotegerin is a receptor for the cytotoxic ligand TRAIL. *J. Biol. Chem.* 273:14363–14367.

23. Wactawski-Wende, J., S.G. Grossi, M. Trevisan, R.J. Genco, M. Tezal, R.G. Dunford, A.W. Ho, E. Hausmann, and M.H. Myroslaw. 1996. The role of osteopenia in oral bone loss and periodontal disease. *J. Periodontal.* 67:1076–1084.

24. Jacobs, R., J. Ghyselen, P. Koninckx, and D. van Steenberghe. 1996. Long-term bone mass evaluation of mandible and lumbar spine in a group of women receiving hormone replacement therapy. *Eur. J. Oral Sci.* 104:10–16.

25. Cogen, R.G., J.T. Wright, and A.C. Tate. 1992. Destructive periodontal disease in healthy children. *J. Periodontol.* 63:761–765.

26. Saxby, M.S. 1987. Juvenile periodontitis: an epidemiological study in the West Midlands of the United Kingdom. *J. Clin. Periodontol.* 14:594–598.

27. Saxen, L., and S. Koskimies. 1984. Juvenile periodontitis—no linkage with HLA-antigens. *J. Periodontal Res.* 19:441–444.

28. Cullinan, M.P., J. Sachs, E. Wolf, and G.J. Seymour. 1980. The distribution of HLA-A and -B antigens in patients and their families with periodontosis. *J. Periodontal Res.* 15:177–184.

29. Ohyama, H., S. Takashiba, K. Oyaizu, A. Nagai, T. Naruse, H. Inoko, H. Kurihara, and Y. Murayama. 1996. HLA Class II genotypes associated with early-onset periodontitis: DQB1 molecule primarily confers susceptibility to the disease. *J. Periodontol.* 67:888–894.

30. Nakanishi, K., T. Kobayashi, T. Murase, T. Nakasuji, H. Inoko, K. Tsuji, and K. Kosaka. 1993. Association of HLA-A24 with complete β-cell destruction in IDDM. *Diabetes* 42:1086–93.

31. Seagren Alley, C., R.A. Reinhardt, C.A. Maze, L.M. DuBois, T.O. Wahl, W.C. Duckworth, J.K. Dyer, and T.M. Petro. 1993. HLA-D and T lymphocyte reactivity to specific periodontal pathogens in type 1 diabetic periodontitis. *J. Periodontol.* 64:974–979.

32. Kobayashi, T., N.A.C. Westerdaal, A. Miyazaki, W.L. van der Pol, T. Suzuki, H. Yoshie, J.G.J. van de Winkel, and K. Hara. 1997. Relevance of immunoglobulin G Fc receptor polymorphism to recurrence of adult periodontitis in Japanese patients. *Infection & Immunity* 65:3556–3560.

33. Bredius, R.G.M., C.A.P. Fijen, M. de Haas, E.J. Kuijper, R.S. Weening, J.G.J. van de Winkel, and T.A. Out. 1994. Role of neutrophil FcγRIIa (CD32) and FcγRIIIb (CD16) polymorphic forms in phagocytosis of human IgG1- and IgG3-opsonized bacteria and erythrocytes. *Immunology* 83:624–630.

34. Boughman, J.A., S.L. Halloran, D. Roulston, S. Schwartz, J.B. Suzuki, L.R. Weitkamp, R.E. Wenk, R. Wooten, and M.M. Cohen. 1986. An autosomal-dominant form of juvenile periodontitis: its localization to chromosome 4 and linkage to dewntinogenesis imperfecta and Gc. *J. Craniofacial Genet. & Devel. Biology* 6:341–350.

35. Kornman, K.S., A. Crane, H.Y. Wang, F.S. di Giovine, M.G. Newman, F.W. Pirk, T.G. Wilson Jr, F.L. Higginbottom, and G.W. Duff. 1997. The interleukin-1 genotype as a severity factor in adult periodontal disease. *J. Clin. Periodontol.* 24:72–77.

36. Cox, A., N.J. Camp, M.J. Nicklin, F.S. di Giovine, and G.W. Duff. 1998. An analysis of linkage disequilibrium in the interleukin-1 gene cluster, using a novel grouping method for multiallelic markers. *Am. J. Hum. Genet.* 62:1180–1188.

37. Gore, E.A., J.J. Sanders, J.P. Pandey, Y. Palesch, and G.M.P. Galbraith. 1998. Interleukin-1b^{+3953} allele 2: association with disease status in adult periodontitis. *J. Clin. Periodontol.* 25:781–785.

38. Lamster, I.B., J.T. Grbic, A.M. Dennis, M.D. Begg, and A. Mitchell. 1998. New concepts regarding the pathogenesis of periodontal disease in HIV infection. *Annal. Periodontol.* 3:62–75.

39. Roux-Lombard, P., C. Modoux, A. Cruchaud, and J. Dayer. 1989. Purified blood monocytes from HIV 1-infected patients produce high levels of TNFα and IL-1. *Clin. Immunol. Immunopathol.* 50:374–384.

40. Price, P., D.M. Calder, C.S. Witt, R.J. Allcock, F.T. Christiansen, G.R. Davies, P.U. Cameron, M. Rodgers, K. Baluchova, C.B. Moore, and M. French. 1999. Periodontal attachment loss in HIV-infected patients is associated with the major histocompatibility complex 8.1 haplotype (A1,B8,DR3). *Tissue Antigens* 53:391–399.

41. Price, P., C.S. Witt, R.J. Allcock, D. Sayer, M. Garlepp, C.C. Kok, M. French, S. Mallal, and F. Christiansen. 1999. The genetic basis for the association between the 8.1 ancestral haplotype (A1,B8,DR3) with multiple immunopathological disorders. *Immunological Reviews* 167:257–274.

42. McManus, R., A.G. Wilson, J. Mansfield, D.G. Weir, G.W. Duff, and D. Kelleher. 1996. TNF2, a polymorphism of the TNF-α gene promoter, is a component of the celiac disease major histocompatibility complex haplotype. *Eur. J. Immunol.* 26:2113–2118.

43. Fortune, F., J.A.G. Buchanan. 1993. Oral lichen planus and coeliac disease. *Lancet* 341:1660.

44. Burge, S.M., P.A. Frith, R.P. Juniper, and F. Wojnarowska. 1989. Mucosal involvement in systemic and chronic cutaneous lupus erythematosus. *British J. Dermatol.* 121:727–741.

45. Sanchez, R., R. Jonsson, E. Ahlfors, K. Backman, and C. Czerkinsky. 1988. Oral lesions of lupus erythematosus patients in relation to other chronic inflammatory oral diseases: an immunologic study. *Scand. J. Dent. Res.* 96:569–578.

46. Shuttleworth, D., R.A.C. Graham-Brown, and A.C. Campbell. 1986. The autoimmune background in lichen planus. *British J. Dermatol.* 115:199–203.

47. Lozada-Nur, F., and C. Miranda. 1997. Oral lichen planus: epidemiology, clinical characterisitics and associated diseases. *Sem. Cut. Med. & Surg.* 16:273–277.

48. Walton, L.J., M.G. Macey, M.H. Thornhill, and P.M. Farthing. 1998. Intra-epithelial subpopulations of T-lymphocytes and Langerhans cells in oral lichen planus. *J. Oral Pathol. Med.* 27:116–23.

49. Roitber-Tambur, A., A. Friedmann, S. Korn, A. Markitziu, S. Pisanti, C. Safirman, D. Nelken, and C. Brautbar. 1994. Serologic and molecular analysis of the HLA system in Israeli Jewish patients with oral erosive lichen planus. *Tissue Antigens* 43:219–223.

50. Jontell, M., P.A. Stahlblad, I. Rosdahl, and B. Lindblom. 1987. HLA-DR3 antigens in erosive oral lichen planus, cutaneous lichen planus, and lichenoid reactions. *Acta. Odontol. Scand.* 45:309–312.

51. Powell, F.C., R.S. Rogers, E.R. Dickson, and S.B. Moore. 1986. An association between HLA DRI and lichen planus. *British J. Dermatol.* 114:473–478.

52. Ship, J.A. 1996. Recurrent aphthous stomatitis. An update. *Oral Surg., Oral Med., Oral Pathol., Oral Radiol. & Endodon.* 81:141–147.

53. Natah, S.S., R. Hayrinen-Immonen, J. Hietanen, M. Malmstrom, and Y.T. Konttinen. 1998. Quantitative assessment of mast cells in recurrent aphthous ulcers (RAU). *J. Oral Pathol. & Med.* 27:124–129.

54. Natah, S.S., R. Hayrinen-Immonen, J. Hietanen, M. Malmstrom, and Y.T. Konttinen. 1997. Factor XIIIa-positive dendrocytes are increased in number and size in recurrent aphthous ulcers. *J. Oral Pathol. & Med.* 26:408–413.

55. Savage N.W., and G.J. Seymour. 1994. Specific lymphocytotoxic destruction of autologous epithelial cell targets in recurrent aphthous stomatitis. *Aust. Dent. J.* 39:98–104.

56. Hayrinen-Immonen, R. 1992. Immune activation in recurrent oral ulcers (ROU). *Scand. J. Dent. Res.* 100:222–227.

57. Healy, C.M., D. Carvalho, J.D. Pearson, and M.H. Thornhill. 1996. Raised anti-endothelial cell autoantibodies (AECA), but not anti-neutrophil cytoplasmic autoantibodies (ANCA), in recurrent oral ulceration: Modulation of AECA binding by tumour necrosis factor-alpha (TNF-α) and interferon-gamma (IFN-γ). *Clin. & Exp. Immunol.* 106:523–528.

58. Taylor, L.J., J. Bagg, D.M. Walker, and T.J. Peters. 1992. Increased production of tumour necrosis factor by peripheral blood leukocytes in patients with recurrent oral aphthous ulceration. *J. Oral Pathol. & Med.* 21:21–25.

59. Macphail, L.A., and J.S. Greenspan. 1997. Oral ulceration in HIV infection: investigation and pathogenesis. *Oral Diseases* 3:S190–S193.

60. Jacobson, J.M., J.S. Greenspan, J. Spritzler, N. Ketter, J.L. Fahey, J.B. Jackson, L. Fox, M. Chernoff, A.W. Wu, L.A. MacPhail, G.J. Vasquez, and D.A. Wohl. 1997. Thalidomide for the treatment of oral aphthous ulcers in patients with human immunodeficiency virus infection. *New Engl. J. Med.* 336:1487–1493.

61. Robinson, J.C., F. Lozada-Nur, and I. Frieden. 1997. Oral pemphigus vulgaris: A review of the literature and a report on the management of 12 cases. *Oral Surg., Oral Med., Oral Pathol., Oral Radiol. & Endodon.* 84:349–355.

62. Weinberg, M.A., M.S. Insler, and R.B. Campen. 1997. Mucocutaneous features of autoimmune blistering diseases. *Oral Surg., Oral Med., Oral Pathol., Oral Radiol. & Endodon.* 84:517–534.

63. Eversole, L.R. 1994. Immunopthology of oral mucosal ulcerative, desquamative, and bullous diseases. *Oral Surg., Oral Med., Oral Pathol., Oral Radiol. & Endodon.* 77:555–571.

64. Takahashi, Y., D.F. Mutasim, H.P. Patel, G.J. Anhalt, R.S. Labib, and L.A. Diaz. 1985. Experimentally induced pemphigus vulgaris in neonatal BALB/c mice: A time course study of clinical, immunologic, ultrastructural and cytochemical changes. *J. Invest. Dermatol.* 84:41–46.

65. Williams, D.M. 1989. Vesiculobullous mucocutaneous disease: pemphigus vulgaris. *J. Oral Pathol. Medicine* 18:544–553.

66. Chan L.S., K.B. Yancey, J.A. Regizi, C. Hammerberg, K. Johnson, K.D. Cooper, and H.K. Soong. 1993. Immune mediated subepithelial blistering diseases of mucous membranes. *Arch. Dermatol.* 129:448–455.

67. Williams D.M., G.P. Haffenden, J.N. Leonard, R.M. McMinn, P. Wright, L. Fry, and J.J. Gilkes. 1984. Benign mucous membrane (cicatricial) pemphigoid revisited. A clinical and immunological reappraisal. *Brit. Dent. J.* 157:313–316.

68. Vincent S.D., G.E. Lilly, and K.A. Baker. 1993. Clinical, historic and therapeutic features of cicatriciäl pemphigoid. *Oral Surg., Oral Med., Oral Pathol., Oral Radiol. & Endodon.* 76:453–9.

69. Hopkins R., and D.M. Walker. 1985. Oral blood blisters: angina bullosa haemorrhagica. *Br. J. Oral Maxillofac. Surg.* 23:9–16.

70. Stephenson P., P.J. Lamey, C. Scully, and S.S. Prime. 1987. Angina Bullosa Haemorrhagica: Clinical and laboratory features in 30 patients. *Oral Surg., Oral Med., Oral Pathol., Oral Radiol. & Endodon.* 63:560–5.

71. Delgado, J.C., D. Turbay, E.J. Yunis, J.J. Yunis, E.D. Morton, K. Bhol, R. Norman, C.A. Alper, R.A. Good, and R. Ahmed. 1996. A common major histocompatibility complex class II allele HLA-DQB1* 0301 is present in clinical variants of pemphigoid. *Proc. Natl. Acad. Sci. USA* 93:8569–8571.

72. Chan, L.S., C. Hammerberg, and K.D. Cooper. 1997. Significantly increased occurrence of HLA-DQB1*0301 allele in patients with ocular cicatricial pemphigoid. *J. Invest. Dermatol.* 108:129–132.

73. Lawley, T.J., W. Strober, H. Yaoita, and S.I. Katz. 1980. Small intestinal biopsies and HLA types in dermatitis herpetiformis patients with granular and linear IgA skin deposits. *J. Invest. Dermatol.* 74:9–12.

74. Hendrix, J.D., K.L. Mangum, J.J. Zone, and W.R. Gammon. 1990. Cutaneous IgA deposits in bullous diseases function as ligands to mediate adherence of activated neutrophils. *J. Invest. Dermatol.* 94:667–672.

75. Price, E.J., and P.J.W. Venables. 1995. Etiopathogenesis of Sjögren's Syndrome. *Semin. Arthritis Rheumatol.* 25:117–133.

76. Moutsopolous, H.M. 1994. Sjögren's syndrome; Autoimmune epithelitis. *Clin. Immunol. Immunopathol.* 72:693–696.

77. Grisius, M.M., P.C. Fox, and D.K. Bermudez. 1997. Salivary and serum interleukin 6 in primary Sjögren's Syndrome. *J. Rheumatol.* 24:1089–91.

78. Buyon, J.P. 1997. Autoantibodies reactive with Ro(SSA) and La(SSB) and pregnancy. *J. Rheumatol.* 24:12–16.

79. Wahren-Herlenius M., S. Muller, and D. Isenberg. 1999. Analysis of B-cell epitopes of the Ro/SS-A autoantigen. *Immunol. Today.* 20:234–240.

80. Fox, R.I., G. Pearson, and J.H. Vaughan. 1986. Detection of EBV associated antigens and DNA in salivary gland biop-

sies from patients with Sjögren's Syndrome. *J. Immunol.* 137:3162–3168.

81. Ioachim, H.L., J.R. Ryan, and S.M. Blaugrund. 1988. Salivary gland lymph nodes: The site of lymphadenopathies and lymphomas associated with human immunodeficiency virus infection. *Arch. Pathol. Lab. Med.* 112:1224–1228.

82. Kong, L., N. Ogawa, T. Nakabayashi, G.T. Liu, E. D'Souza, H.S. McGuff, D. Guerrero, N. Talal and H. Dang. 1997. Fas and Fas ligand expression in the salivary glands of patients with primary Sjögren's syndrome. *Arthritis and Rheumatism* 40:87–97.

48 | Behçet's Disease

Tsuyoshi Sakane and Noboru Suzuki

1. INTRODUCTION

Behçet's disease (BD) is recognized as a systemic inflammatory disease of unknown etiology. The disease is not a chronic inflammatory disease, is rather recurrent attacks of acute inflammation and progressively deterioration. Previous reports have shown three major pathophysiologic changes in BD; excessive neutrophil functions, vascular injuries with thrombotic tendency, and autoimmune responses. Association of HLA-B51 with the disease suggests importance of genetic predisposition for the development of this disease. Many reports suggest intimate linkage of immunological abnormalities and neutrophil hyperfunction to the disease manifestations. HLA-B51 molecules alone may be partly responsible for neutrophil hyperfunction in BD. Low levels of prostacycline and increased levels of von Willebrand factor in patients' sera, and increased production of oxygen metabolites by their endothelium, may indicate endothelial cell damage in BD. Recent demonstration of enhanced responsiveness of BD patient T lymphocytes to 60 kDa human-derived heat shock protein (HSP) suggests that this response may reflect cross-reaction between human-derived HSP and HSP from microbial causative agents. We therefore propose a working hypothesis for the etiopathogenesis of BD: In a genetically susceptible individual, an exogenous antigen such as a streptococcal or a viral antigen, of which HSP shares in an epitope with human-derived HSP, may stimulate HSP-specific T lymphocytes in addition to monocytes, and subsequently induce production of Th1 cytokines, IL-8, TNF, and IL-6. These cytokines stimulate monocytes and lymphocytes in an autocrine/paracrine fashion, and stimulate the vascular endothelium as well. Stimulation of polymorphonuclear cells by these cytokines also is of a

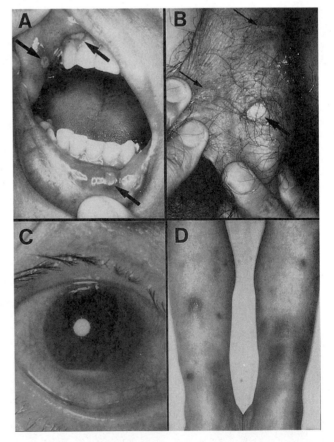

Figure 1. Common symptoms of BD (reproduced from Ref. No. 4)

(A) Multiple oral aphthous ulcers in buccal membrane, gingiva, and labial mucosal membrane, as arrows indicate. (B) An active genital ulcer (thick arrow) and scars in the scrotum (thin arrows). (C) Hypopyon, a visible horizontal layer in the anterior ocular chamber, and deformity of iris. (D) Fresh and old lesions of erythema nodosum in both legs.

particular importance for the pathogenesis of BD because of a genetic hyperresponsiveness of the polymorphonuclear cells in BD. The final outcome will be systemic vascular damage leading to various clinical manifestations of the disease.

2. HLA AND GENETICS

A Mendelian inheritance pattern has not emerged, and no heritable factors are identified, although clustering of BD patients in families is well recognized. Up to 4~5% of BD patients report that some BD-like symptoms develop in a first-degree relative (7). Increased disease risk is conferred to offspring by more severely affected parents. In addition, patients with familial occurrence of BD have a poorer prognosis (8). The data support the hypothesis of a genetic component in the pathogenesis of BD. It is further suggested that inheritance of risk factor(s) in an autosomal dominant manner may be involved in the familial occurrence of BD, but spontaneous development of the disease in offspring of healthy parents also implicates recessively inherited risk factors, predominant environmental influences, or both (9).

2.1. Primary Association of HLA-B51 with BD

BD is associated strongly with the HLA-B51 antigen (5,10,11). The frequency of the B51 antigen is about 60% in patients with BD in Japan and thus is high in the patient population compared with around 10% in the normal

Japanese population, suggesting a strong correlation between BD and B51 (12). This association has been confirmed in BD patients of many other countries (13). Therefore, BD exhibits the same HLA association in different ethnic groups. However, it has remained uncertain whether the HLA-B51 gene is responsible for BD, or alternatively, whether some other nearby genes that are in linkage disequilibrium with B51 are responsible.

As shown in Figure 2, a new family of non-classical MHC molecules, the MHC class I chain-related protein (MIC) including MIC-A and MIC-B, encoded by genes located in the MHC, has recently been identified (14). On the basis of the location of MIC genes and the structure and expression of MIC molecules, it has been postulated that MIC may be a disease-susceptible gene in BD (15). Mizuki et al. found that although the association with MIC-A is highly significant, the association with HLA-B51 independently remains the most significant factor for the development of BD, suggesting that MIC-A gene is unlikely to be the disease-susceptible gene for BD (15). Similarly, the MIC-B gene itself is not responsible for the development of BD (16). The latest genetic linkage studies indicate that the HLA-B 51 gene itself is a most possible candidate gene in the HLA region for the development of BD (17).

Alleles encoding the HLA-B51 antigen are known to include HLA-B*5101–5108 (18). Because of a marked increase in the frequency of the B51 antigen in BD patients, the frequency of individual alleles encoding the antigen is also increased. However, there is no significant difference in the frequency of each allele between B51 antigen-

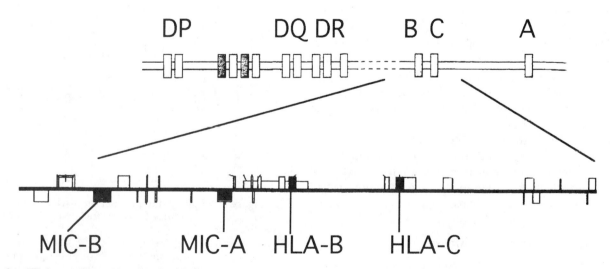

Figure 2.   HLA gene locus associated with BD
Position of MIC-A, MIC-B and HLA-B loci, all of which have been reported to have significant association with development of BD is depicted.

positive patients and healthy controls (14,19). Thus, some amino acids of the HLA-B51 molecule that fulfill the following criteria may be responsible for the development of BD: 1.) They are common to all B51-encoding alleles, but not any particular B51-encoding alleles, and 2.) they do not reside in other HLA-B antigen genes (19). It has previously been reported that the frequency of the B52 antigen, which is a split antigen of the B5 antigen as is B51 and differs from B51 in the amino acid sequence only at two positions, is not increased in BD patients (19). The two amino acids, asparagine at position 63 and phenylalanine at position 67 of HLA-B51, may thus be associated with the development of BD.

Pocket B in HLA grooves of the class I molecule is a major pocket for an antigenic peptide, and is formed by the amino acid residues at positions 63 and 67 (20,21). In most class I alleles, the amino acids in peptides for binding with HLA molecules are dependent on amino acid composition constituting pocket B of each HLA molecule (22). Due to HLA polymorphism, peptides for binding vary largely with different amino acids specific for each HLA allele, particularly amino acids constituting pocket B (22,23). Among HLA alleles there are completely different motifs of the binding peptide in general. The binding peptide motifs differ entirely between B51 and B52, because the amino acids at positions 63 and 67, which constitute pocket B, are different between B51 and B52 (22,23). HLA-B51, but not B52, may have a high affinity for some BD-provoking extrinsic factors such as infectious agents. It is thus possible that B51 molecules are specifically involved in development of BD.

In addition, HLA binds self-peptides during intrathymic T lymphocyte differentiation and plays an important role in the education of T lymphocytes in thymus and thus in the formation of T cell repertoires (24,25). Even in the case of the same exogenous antigen, the peptides derived from the antigen for HLA binding and T cell repertoire for recognizing the antigenic peptides vary with different HLA alleles, causing allele specific immune responses to the antigen (24,25). Collectively, disease susceptibility seems to be dependent on, or at least related to, the selected allele (24,25); development of BD may be at least partly due to the high affinity of B51 molecules for peptide(s) derived from certain infectious agent(s) responsible for BD. It is also possible that B51 is directly involved in the disease onset through aberrant immune responses by skewed T cell repertoires which have been formed under the influence of HLA-B51 antigen in the thymus.

2.2. HLA-B51 Transgenic Mouse

Neutrophil hyperactivity accounts for most of the pathological changes in BD (26). Increased neutrophil function is evident in both B51-positive BD patients and B51-positive healthy individuals (27) (see section 5). We have conducted studies concerning neutrophil functions using B51 transgenic mice produced by human HLA-B51 gene (HLA-B*5101) transfer into mice (28) (Figure 3).

Two distinct stimuli are required for neutrophils to generate maximal amounts of superoxides (29). The first step is priming and the following is triggering. Pure triggering stimuli, such as formyl-methionine leucine phenylalanine (fMLP), induce oxidative burst by primed, but not unprimed, neutrophils. Neutrophils from HLA-B51 transgenic mice show hypersensitivity to fMLP, while HLA-B35 transgenic fail to respond (28). In contrast, superoxide production induced by phorbol myristate acetate (PMA) and opsonized zymosan (OZ), neither of which require the priming process, is comparable between the two transgenic mice (28). Thus, the increased superoxide production by neutrophils in the B51 transgenic mice is not ascribable to the xenogeneic MHC gene transfection, but is considered to be a specific phenomenon resulting from the expression of HLA-B51 molecules (28). It is possible that the B51 antigen expressed in mice presents some exogenous antigenic peptides to T lymphocytes, provokes immune responses and as a result, primes the neutrophils, leading to increased superoxide production in the mice. Thus, the molecular cascade from B51 antigen expression to neutrophil activation may be associated with BD, and B51 antigen is directly involved in the disease onset.

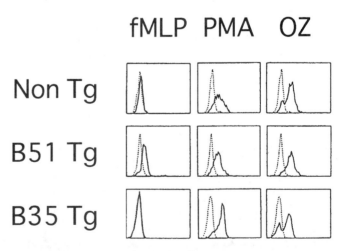

Figure 3. Superoxide production of HLA B51 transgenic (Tg) mice (reproduced from Ref. No. 28 with minor modification)

Neutrophils from non Tg, HLA-B51 Tg, and HLA-B35 Tg mice were stimulated with fMLP, PMA or OZ. Subsequent H_2O_2 production was measured by flow cytometric analysis. Neutrophils from B-51 Tg, but not non Tg nor B35 Tg, respond to fMLP stimulation.

3. IMMUNOLOGICAL FINDINGS RELEVANT TO THE PATHOGENESIS OF BD

Several abnormalities of immune responses have been reported in patients with BD that are quite similar to those observed in the certain autoimmune disorders (30,31). These include induction of lymphocyte transformation by oral mucosa, cytotoxicity lymphocytes on oral mucosa, delayed-type hypersensitivity skin reaction to skin homogenates, histologic features characterized by an early intense lymphocyte infiltration in oral aphthous ulceration, increased numbers of CD4$^+$, CD8$^+$ and $\gamma\delta$ T lymphocytes (32,33), suppressor T cell dysfunction (34), defective IL-2 activity of mitogen-activated T cells (35), increased phenotypically-activated or memory circulating T cells (36), increased serum concentrations of soluble CD8 and CD25 (37), and polyclonal B cell activation (38).

Polarized Th1/Th2 lymphocytes play an important role in the induction and regulation of autoimmunity (39). Recent studies have shown that organ-specific autoimmune diseases involve Th1 lymphocytes, whereas Th2 type immune responses participate in systemic autoimmune diseases with a strong humoral component (40–42). Indeed, Th1 polarization of immune response is found in the experimental autoimmune uveoretinitis (43). Analysis of Th1/Th2 type cytokines in CD3$^+$ lymphocytes of BD patients reveals a polarized Th1 immune responses that parallel disease progression (44). BD patients with active disease show increase in IFN-γ producing CD3$^+$ lymphocytes compared to patients in complete remission and healthy subjects (44). On the contrary, no difference of IL-4 producing CD3$^+$ lymphocytes is observed among active patients, inactive patients and normal individuals (44). Thus, a polarization of Th1/Th2 balance toward the Th1 phenotype is evident in patients with active disease (44). In experimental uveitis rats developed by HSP peptide immunization, administration of IL-4, a Th2-type cytokine that suppresses Th1 responses, significantly decreased the development of uveitis (45), further supporting involvement of Th1 polarization for the pathogenesis of experimental uveitis and possibly BD. Collectively, Th1 cytokine-producing lymphocytes play an important role in the immunopathogenesis of BD.

The coincidence of a unique anatomical localization of $\gamma\delta$ T lymphocytes with the common mucocutaneous lesions in BD (46) stimulated us to investigate the pathogenic role of specific T cell population(s) in the development of BD. We found both phenotypic and functional abnormalities of $\gamma\delta$ T lymphocytes that are characteristics of BD (27,33). Most $\gamma\delta$ T lymphocytes of the BD patients express activation markers such as CD25, CD69, HLA-DR and CD99 (47) indicating that the $\gamma\delta$ T lymphocytes have already been activated *in vivo* (47). Moreover, the BD $\gamma\delta$ T lymphocytes spontaneously produce the proinflammatory

cytokines IFN-γ and TFN-α, and secrete much larger amounts of these cytokines in response to PMA + ionomycin than healthy individuals (47).

It has been generally known that CD45RA and CD45RO are markers for naive and memory cells, respectively. Of interest is that most $\gamma\delta$ T lymphocytes of BD patients consist of CD45RA+ cells and exhibit characteristic features of memory T cells; $\gamma\delta$ T lymphocytes of the patients express a unique CD45RA isoform, 205 kDa (memory cells) but not 220 kDa (naive cells) (27,33). $\gamma\delta$ T lymphocytes in BD strongly express CD16 and CD56, and have perforin granules, indicating phenotypical and functional similarities with NK cells (27,33). The proportion of circulating $\gamma\delta$ T lymphocytes is increased in BD patients compared to controls; remarkable accumulation of $\gamma\delta$ T lymphocytes is frequently observed in their peripheral blood, and $\gamma\delta$ T lymphocytes sometimes constitute more than 60% of the peripheral blood lymphocytes (46). Thus, $\gamma\delta$ T lymphocytes may play a certain role in the tissue injuries observed in BD (27,33).

Immune response to retinal autoantigens may be important for the pathogenesis of uveitis of BD (47). There is a common or highly homologous sequence between several HLA-B molecules associated with uveitis, such as HLA-B27 and B51, and a retinal S antigen (47,48). Retinal-S antigen is shown to be immunogenic in both human and experimental uveitis (49). It has been shown that peripheral blood mononuclear cells (PBMC) from BD patients have higher responses than those from controls for both the HLA-B related peptide and the homologous retinal-S antigen peptide. Thus, cellular immunity to cross reactive retinal-S and HLA-B derived peptides may be involved in the pathogenesis of posterior uveitis in BD (47–49).

4. HUMAN HEAT SHOCK PROTEIN (HSP)-60 AND BD

As summarized above, aberrant immune responses resembling autoimmune diseases are evident in patients with BD. Thus, we have been focusing on analysis of their autoimmune responses. It was reported initially that autoimmune responses to oral epithelial antigen of BD patients turned out to be the identification of circulating anti-65-kDa HSP autoantibody (50). Originally, HSP are reported to be involved in the pathophysiology of autoimmune diseases (51). HSP are unique antigens with a potent immunostimulatory property, and have an extraordinarily high sequence conservation throughout eukaryotic and prokaryotic kingdoms. 65-kDa HSP is an immunodominant antigen that induces bacterial and mammalian 60/65-kDa HSP immune responses and crossreactivity between bacterial and mammalian 60/65-kDa HSP is considered to be a plausible mechanism for the pathogenesis of BD (52,53).

It has been shown that selected peptides derived from the sequences of human 60-kDa HSP induce significant proliferation of T lymphocytes in patients with BD in England. In addition, two peptides of human 60-kDa HSP are most frequently recognized by T lymphocytes from patients with ocular type BD (54). Furthermore, these HSP 60 peptides can induce anterior uveitis in Lewis rats (55). The human HSP peptide 336–351 has high sequence homology with mycobacterial HSP and streptococcal HSP (52,53,55). We also found that the HSP peptide provoked a significant proliferation of T lymphocytes in BD patients in Japan (52) (Figure 4A). That epitope is specific for BD is confirmed by the fact that T lymphocytes from patients with rheumatoid arthritis do not respond to the peptide (52). In addition, we found the importance of the HSP peptide 336–351 for the development of eye symptoms; there exists a significant association between the presence of ocular lesion and the proliferative responses to this peptide by lymphocytes (52) (Figure 4B).

We have examined TCR usage of T lymphocytes responsive to the HSP peptide by means of monoclonal antibody specific for TCR Vβ subfamily. Excessive expansion of T cells with selected TCR Vβ genes is noted in most patients with BD. Thus, it is possible that a self HSP-derived peptide stimulates T cells with restricted TCR repertoires. When we analyzed clonotypes by TCR sequencing of Vβ5.2+T cells that expanded approximately 20 times in response to the HSP peptide stimulation, we found that the T cell clonotypes are quite limited. Four TCR sequences are shared with the 21 clones studied, and the 4 TCR sequences are highly homologous with each other. Thus, it is evident that in BD patients, self HSP peptide-reactive T cells showing quite restricted usage of TCR CDR3 region are accumulating, suggesting involvement of antigen-specific immune responses to HSP in the pathogenesis of this disease.

We further analyzed TCR usage of the T lymphocytes by polymerase chain reaction (PCR) single-strand conformation polymorphism (SSCP)-based technique that enabled us to observe T cell clonotypes qualitatively (52). Using the PCR and subsequent SSCP analysis, PCR products of freshly-isolated peripheral blood lymphocytes from normal

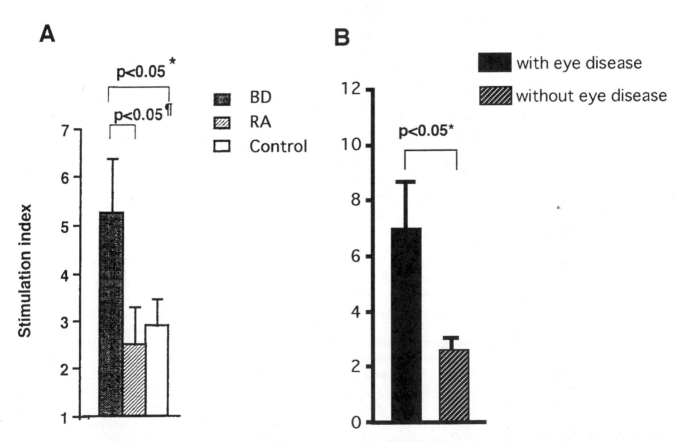

Figure 4.   T lymphocyte responses to the self HSP peptide in patients with BD (reproduced from Ref. No. 52 with major modification)
T cells from patients with BD proliferate in response to human-HSP derived peptide 336–351 (QPHDLGKVGEVIVTKD) stimulation. This T cell response is specific for patients with BD, because T cells from patients with rheumatoid arthritis and those from normal individuals do not respond to this peptide at all.

subjects show a smearing pattern. Interestingly, PCR products of freshly-isolated unstimulated lymphocytes of BD patients result in formation of several distinct bands (Figure 5), some bands becoming prominent after lymphocyte culture with the HSP peptide. T lymphocytes with the same clonotypes proliferate vigorously in response to the same peptide at every attack of their uveitis in the patients for several years (52). These data suggest that a molecular mimicry mechanism induces and exacerbates BD; self-HSP and microbial HSP homologous to the self HSP activate self-reactive T cells specific for the HSP peptide.

Furthermore, activation of PBMC of BD patients by the self-HSP peptide results in the predominant production of IL-12, TNF-α, and IFN-γ in BD patients (52). While B lymphocytes, in comparison with monocytes, produce large amounts of IL-12 in response to HSP peptide stimulation. This suggests that not only T, but also B lymphocytes, are sensitized to the self-HSP peptide. Of note is clear correlation between IL-12 secretion and IFN-γ secretion by BD patients' PBMC in response to the self-HSP peptide stimulation. Thus, IL-12 secreted by HSP-specific B lymphocytes in response to self-HSP peptide stimulation may promote preferential induction of Th1 lymphocytes in BD patients.

In accordance with the above findings, we found that IL-12 receptor β2-chain, a reliable cell surface marker for Th1 lymphocytes, is expressed on self-HSP peptide-reactive T cells, suggesting that self-reactive T cells are polarized in Th1 cells in BD patients. In fact, there is a strong correlation between IL-12 secretion by B cells and TNF-α

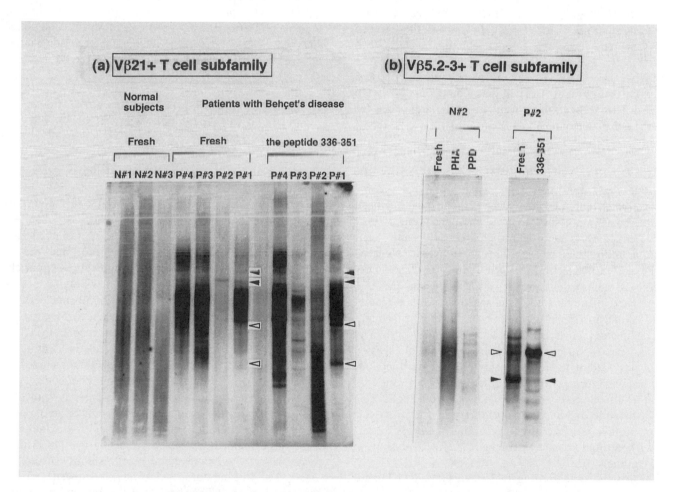

Figure 5. TCR clonotype analysis of T cells that react with the HSP-peptide 336–351 in patients with BD by PCR-single strand conformational polymorphism (SSCP) method (reproduced from Ref. No. 52)

(a) Freshly isolated T cells (Fresh) or T cells that had been stimulated for 9 days with the peptide 336–351 *in vitro* were recovered from normal subjects (N#1, N#2, N#3) and those from patients with BD (P#1, P#2, P#3, P#4). The cDNA was amplified by PCR method using Vβ21 specific primers and visualized by SSCP-Southern blotting analysis. After 9 days culture with the peptide 336–351 lower bands (2 open triangles) became prominent whereas upper bands (2 closed triangles) became faint (P#1). (b) Similar experiments were performed using Vβ5.2–3 specific primers in the normal subject (N#2) and the patient (P#2). These results suggests that self HSP reactive T lymphocytes are clonally expanded in vivo in patients with BD. The clonally expanded T cells of BD patients further proliferated in in vitro culture with the HSP peptide (open triangles).

secretion by T cells in response to the self-HSP peptide in BD patients. The overall results can be summarized that self-antigen first promotes secretion of IL-12 by B lymphocytes, after which IL-12 induces preferential induction of Th1 cells. The polarized Th1 cells stimulate other cell types as well as themselves to secrete various cytokines, including TNF-α. TNF-α may be one of the final afferent arms that provoke tissue inflammation in BD. Indeed, the plasma level of TNF-α has been shown to be elevated in patients with BD (44).

We have next analyzed cytokine production by the PBMC more precisely in association with their clinical pictures. All patients whose PBMC produced low levels of IL-12 and TNF-α can be categorized as having inactive disease. In contrast, the majority of patients in the IL-12 and TNF-α high producer groups suffer from active BD. Thus, self-reactive T cells may skew toward Th1 cells by IL-12 and induce TNF-α secretion. In contrast, those whose PBMC secret considerable amounts of either IL-4, IL-10 or TGF-β have inactive disease. These results suggest that excessive Th1 function may be pathogenic in BD and strategies that promote Th2 cell development and maintain original Th1/Th2 equilibrium may become applicable for treating BD patients.

5.   EXCESSIVE NEUTROPHIL FUNCTIONS IN BD

The activated polymorphonuclear cell function has been considered an important area in solving the pathogenesis of BD. Leukocytosis with neutrophilia and aseptic neutrophilic infiltration into the lesions are frequently observed in the active phase of BD (56). Furthermore, the neutrophils of such patients reveal various abnormalities, e.g. upregulated chemotaxis (57), enhanced superoxide synthesis (58), and increased production of chemical inflammatory mediators such as lysosomal enzymes (59). The patients' neutrophils express an increased level of adhesion molecules, leading to elevated neutrophil effector functions and active interactions with other cell types, such as endothelial cells (60). All these data suggest a central role for neutrophils in the pathophysiology of the disease, a notion further supported by the fact that therapies for targeting neutrophils, such as colchicine, effectively relieve the symptoms (61).

The genetic predisposing factor, an HLA-B51 phenotype, appears to be associated with neutrophil hyperfunction in BD, as discussed in section 2.2. Chajek-Shaul et al. reported that significantly enhanced chemotaxis of neutrophils is observed in HLA-B51-positive patients compared with HLA-B51-negative patients (11). Our studies and those of Sensi et al. demonstrated that superoxide production by neutrophils is elevated in HLA-B51-positive individuals, irrespective of having the disease or not, com-

pared with HLA-B51 negative individuals (28,62). We have also shown the excessive neutrophil function in HLA-B51 (B*5101) transgenic mice (28), findings that further support the association of HLA-B51 molecules with induction of excessive neutrophil functions.

Early studies have indicated that patients' sera possesses stimulatory effects on neutrophils from normal individuals, suggesting that some humoral factors, such as cytokines and immune complexes, are involved in priming of neutrophils. Indeed, recent investigations have disclosed increased serum levels of TNF-α, IL-1β, IL-8 and immune complexes (63–66). Thus, excessive humoral factors in the sera contribute to neutrophil priming, the underlying condition of this disease. More recent studies have confirmed a role for neutrophils in producing proinflammatory cytokines, such as TNF-α, IL-6 and IL-8 (69,70). As expected, neutrophils in BD patients produce considerable amounts of TNF-α and IL-8 spontaneously (71) and therefore, may also contribute to their own activation via an autocrine loop of these cytokines.

6.   CHARACTERIZATION OF CELLS INFILTRATING IN SKIN AND OCULAR LESIONS OF BD

The skin pathergy reaction is a non-specific hyperreactive response that develops at the site of needle-prick or minor trauma in BD patients and is very similar to the erythematosus papules and pustules that spontaneously manifest in these patients (72,73). Ergun et al. studied the histopathology of pathergy extensively and found that at time zero, skin is normal, whereas 4 hours after skin prick, neutrophils are present, usually admixed with lymphocytes (74). Intraepidermal pustules and polymorphonuclear cell aggregates within the needle tract are seen as early as 4 hours, and inflammatory cell density reaches peak by 24 hours ~ 48 hours. Sparse leukocytocasis is identifiable from 4 hours to 48 hours. These investigators suggested that early pathergy is mediated by polymorphonuclear cells and lymphocytes without vasculitis, and excessive function of chemotaxis may explain the rapid accumulation of polymorphonuclear cells along the injection site (74). Similarly, Gul et al. examined the pathergy reaction in BD (75) and found that positive skin pathergy reaction biopsy specimens obtained at 48 hours after skin pricking showed variable degrees of epidermal thickening and cell vacuolization, as well as subcorneal pustule formation. In the *pathergy reaction dermis*, a variable dense focal mononuclear cell infiltrate is seen around vessels and skin appendages, extending into the deep dermis. The mononuclear cell infiltrate is predominantly composed of T lymphocytes and monocytes/macrophages. The majority of the T lymphocytes are CD4+, and almost all the cells express CD45RO.

Approximately half of the infiltrating cells strongly express HLA-DR. Neutrophils constitute less than 5% of the infiltrating cells, but are present as clusters of elastase-positive cells at the needle-prick sites. Vessels within the lesion show marked congestion and endothelial swelling. The endothelial cells express ICAM-1 strongly, and E-selectin moderately. VCAM-1 is not expressed on endothelial cells. The basal and mid-epidermal layers of keratinocytes express HLA-DR and ICAM-1 strongly, particularly in areas close to the dermal mononuclear cell infiltrates (75).

Feron et al. reported CD4+ T lymphocyte involvement in ocular lesions in BD patients by immunohistochemical examination (76). T lymphocytes infiltrating in ocular lesions are activated in vivo and express CD25. There are small numbers of cells in the optic nerve head, retinal vascular endothelium, and retinal pigment epithelium, all of which are HLA-DR positive (77). These investigators pointed out that interaction between infiltrating CD4+ T lymphocytes and HLA-DR positive cells in the eye may induce development of the ocular lesions in this disease.

We found that CD4+ T lymphocytes predominantly infiltrate in BD patient skin lesions (78). As shown in Figure 6A, these lymphocytes mainly produce IFN-γ, but not IL-4. Massive IL-12 is also produced. The data suggest the involvement of Th1 lymphocytes in skin lesions in BD patients as well. TNF-α, IL-6 and IL-1β-producing cells are also accumulating in the lesions. Thus, these pro-inflammatory cytokines may exacerbate acute inflammation in the skin lesions. Furthermore, human HSP 60-expressing cells are prevalent within the lymphocyte-infiltrating area in the skin lesions of BD patients (Figure 6B). Similarly, HSP is spontaneously expressed by PBMC of BD patients. In contrast, skin lesions of patients with atopic dermatitis do not express HSP (Figure 6B). Our results suggest that human HSP 60 expressed in the lesion and HSP-reactive T lymphocytes may be intimately associated with the development of skin lesions in BD and may be the case of other lesions in the disease.

7. HYPERCOAGULABILITY AND VASCULAR LESIONS

Vascular complications in BD can be observed with a prevalence of 5–30% and a clear preponderance of venous lesions compared with arterial involvement (79). Vascular injuries with thrombotic tendencies are another characteristic feature of BD vascular lesions (80).

Vasculitis is observed as the common pathology in all organs involved in this disease, including oral ulcers, skin lesions and posterior uveitis (79). There is a dense perivascular infiltration of lymphocytes and occasionally neu-trophils, primarily around the small venules in BD. The majority of mononuclear cells located at the base of the oral ulcers and at the perivascular infiltrates are CD3+, CD4+ and HLA-DR+ lymphocytes. These findings are suggestive of an "immune cell-mediated vasculitis" due to T lymphocyte-mediated cellular immune response targeting to the components of the vessel wall (49).

Anti-endothelial cell antibodies are present in 20–30% of BD patients (81). Circulating anti-phospholipid and anti-neutrophil cytoplasmic antibodies are not detected in active BD patients. Anti-endothelial cell antibodies of BD patients are shown to enhance expression of ICAM-1, VCAM-1, and E-selectin on the endothelial cell surfaces thus promoting adherence of T lymphocytes to the endothelial cells, leading to their mutual activation (49).

Several molecules are synthesized by vascular endothelial cells that exhibit anti-thrombotic activities, such as prostacycline (PG12), nitric oxide (NO), thrombomodulin (a thrombin receptor on the vascular endothelial cell surface) and tissue plasminogen activator (81,82). The plasma 6-keto prostaglandin F2, a stable metabolite of PG12, is reduced in BD patients. In addition, PG12 biosynthesis is markedly impaired in vitro. Plasma thromboxane B levels in BD patients are increased. These data suggest that the cyclooxygenase pathway in the platelets is intact and the impairment of prostanoid synthesis in BD patients can be confined to the vessel walls or endothelial cells. Elevated plasma thrombomodulin values are reported in active BD (81,82). Endothelin (ET) is a novel vasoconstrictor peptide synthesized by vascular endothelial cells (83,84). The concentrations of ET are also high in patients with active vascular BD. These findings are highly suggestive of endothelial cell damage due to vascular injury accompanying thrombotic events (83,84). Other investigators reported increased plasma levels of von Willebrand factor in BD, patients indicating vascular endothelium damage due to inflammatory processes (85). In addition, circulating proinflammatory cytokines may be involved in activation of not only neutrophils, but also endothelial cells and platelets. Collectively, the mechanism of endothelial cell damage in BD, where vascular thrombosis frequently appears, can be related to direct endothelial cell damage by the cellular infiltrate or indirect cytopathic action of the cytokines. Natural inhibitors of coagulation, including anti-thrombin III, protein C and protein S in circulation, are normal, indicating that there is no primary coagulation inhibitor abnormality in BD (86).

In addition to the overexpression of adhesion molecules on endothelial cells in BD lesions, the sera are shown to enhance the production of soluble ICAM-1 from endothelial cells as well as the expression of Mac-1 and LFA-1 on neutrophils (87). These phenotypical changes facilitate a series of adhesion molecule-mediated interactions between endothelial cells and leucocytes, including neutrophils and

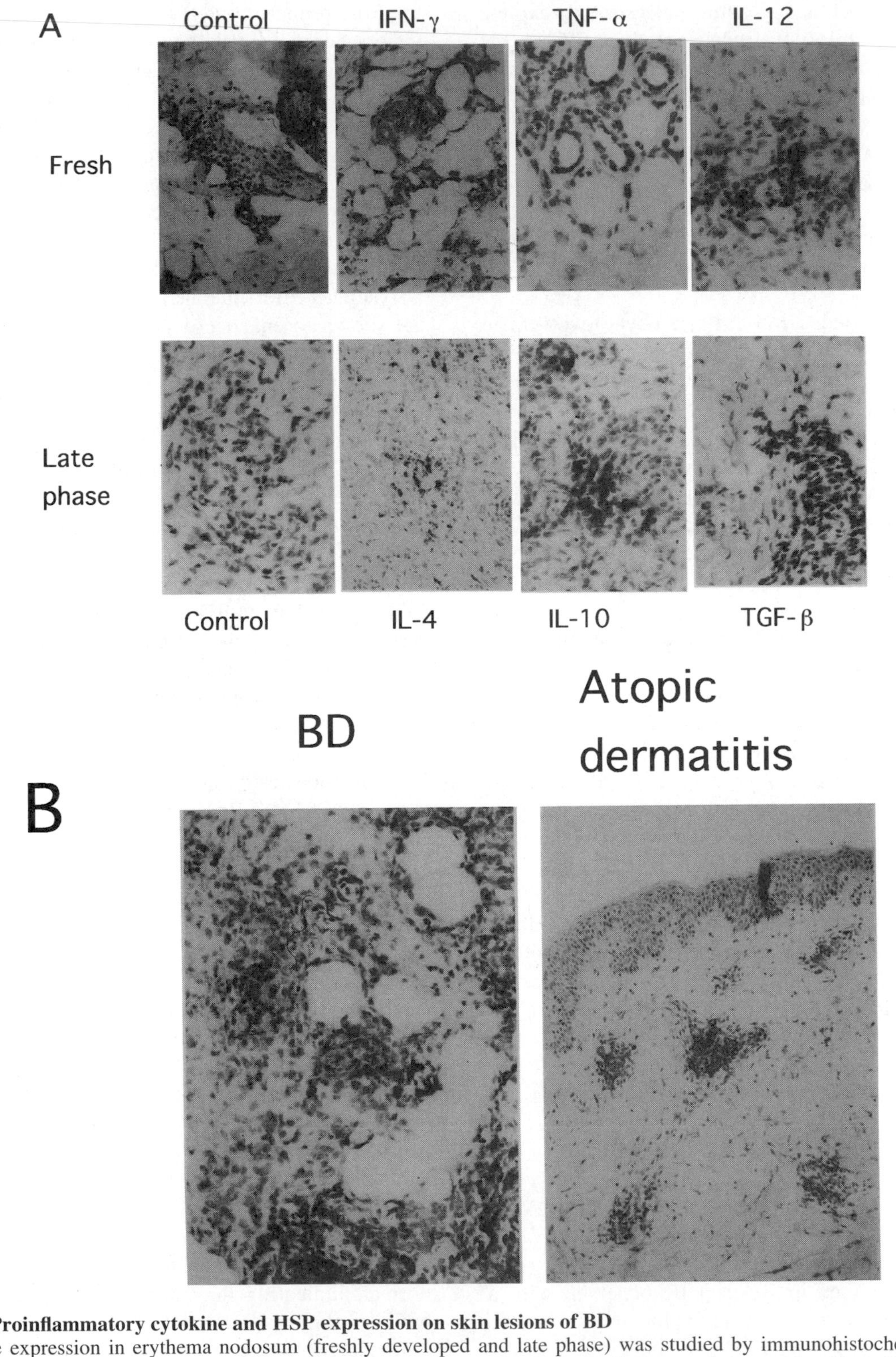

Figure 6.   Proinflammatory cytokine and HSP expression on skin lesions of BD

(A) Cytokine expression in erythema nodosum (freshly developed and late phase) was studied by immunohistochemical staining in patients with BD. We found that IFN-γ, TNF-α and IL-12 are expressed in active lesions, whereas either IL-4, IL-10 or TGF-β is expressed when the eruption is going to subside.

(B) Similarly human HSP-60 expression was studied. Active skin lesions of BD patients but not those of atopic dermatitis expressed human HSP-60.

lymphocytes, leading to acceleration of inflammatory cell activation. In fact, Inaba et al. demonstrated prompt shedding of L-selectin on neutrophils in BD patients suffering active ocular attacks (88), which might reflect the first step of the endothelial cell-leucocyte interaction during the extravasation of leucocytes, as observed in animal models of experimental uveitis. Furthermore, it has been shown that the patient's sera stimulate endothelial cells to secrete IL-8, which also activates neutrophils in concert with adhesion-dependent signaling (60). Thus, there are multiple factors that facilitate excessive leucocyte-endothelial cell interactions in BD. 1.) action of proinflammatory cytokines in paracrine and/or autocrine fashion, 2.) enhanced expression of adhesion molecules on their surfaces, 3.) thrombotic tendencies perhaps secondary to endothelial cell injuries, and 4.) neutrophil hyperactivity based on genetic backgrounds, as shown by B51-transgenic mice. These factors are not mutually exclusive, but rather form a vicious circle to promote the lesions

8. CONCLUSION

BD is a disease caused by multiple etiology, including intrinsic factors and exogenous agents. Recent studies have shown the importance of crossreactive autoimmune responses between bacterial HSP and self HSP for the pathogenesis. The elucidation of molecular pathogenesis would lead to the development of specific intervention of the pathological responses of BD, and to ultimately cure of the disease.

References

1. Kastner, D.L. 1997. Intermittent and periodic arthritic syndromes. In: *Arthritis and allied conditions: a textbook of rheumatology*, 13th edition, W.J. Koopman, editor, Williams & Wilkins, Baltimore, MD. pp. 1279–1306.
2. Kaklamani, V.G., G. Vaiopoulos, and P.G. Kaklamanis. 1998. Behçet's disease. *Semin. Arthritis Rheum.* 27:197–217.
3. Ehrlich, G.E. 1997. Vasculitis in Behçet's disease. *Int. Rev. Immunol.* 14:81–88.
4. Sakane, T., M. Takeno, N. Suzuki, and G. Inaba. 1999. Current concept; Behçet's disease. *New Engl. J. Med.* 341:1284–1291.
5. Ohno, S., M. Ohguchi, S. Hirose, H. Matsuda, A. Wakisaka, and M. Aizawa. 1982. Close association of HLA-Bw51 with Behçet's disease. *Arch. Ophthalmol.* 100:1455–1458.
6. Sakane, T. 1997. New perspective on Behçet's disease. *Int. Rev. Immunol.* 14:89–96.
7. Fresko, I., M. Soy, V. Hamuryudan, S. Yurdakul, S. Yavuz, Z. Tumer, and H. Yazici. 1998. Genetic anticipation in Behçet's syndrome. *Ann. Rheum. Dis.* 57:45–48.
8. Nishiura, K., S. Kotake, A. Ichiishi, and H. Matsuda. 1996. Familial occurrence of Behçet's disease. *Jpn. J. Ophthalmol.* 40:255–259.
9. Stewart, J.A. 1986. Genetic analysis of families of patients with Behçet's syndrome: data incompatible with autosomal recessive inheritance. *Ann. Rheum. Dis.* 45:265–268.
10. Zouboulis, C.C., P. Büttner, D. Djawari, W. Kirch, W. Keitel, H.J. von Keyserlingk-Eberius, and C.E. Orfanos. 1993. HLA-class I antigens in German patients with Adamantiades-Behçet's disease and correlation with clinical manifestations. In: Behçet's Disease, B. Wechsler, and P. Godeau, editors. Elsevier Science Publishers, Amsterdam. pp. 175–180.
11. Chajek-Shaul, T., S. Pisanty, H. Knobler, Y. Matzner, M. Glick, N. Ron, E. Rosenman, and C. Brautbar. 1987. HLA-B51 may serve as an immunogenetic marker for a subgroup of patients with Behçet's syndrome. *Am. J. Med.* 83:666–672.
12. Nakae, K., F. Masaki, T. Hashimoto, G. Inaba, M. Mochizuki, and T. Sakane. 1993. Recent epidemiological features of Behçet's disease in Japan. In: Behçet's Disease, B. Wechsler, P. Godeau, editors. Elsevier Science Publishers, Amsterdam. pp. 145–151.
13. Zouboulis, C.C., I. Kötter, D. Djawari, W. Kirch, P.K. Kohl, F.R. Ochsendorf, W. Keitel, R. Stadler, U. Wollina, F. Proksch, R. Söhnchen, H. Weber, H.P.M. Gollnick, E. Hölzle, K. Fritz, T. Licht, and C.E. Orfanos. 1997. Epidemiological features of Adamantiades-Behçet's disease in Germany and in Europe. *Yonsei Med. J.* 38:411–422.
14. Mizuki, N., H., Inoko, and S. Ohno. 1997. Pathogenic gene responsible for the predisposition to Behçet's disease. *Int. Rev. Immunol.* 14:33–48.
15. Mizuki, N., M. Ota, M. Kimura, S. Ohno, H. Ando, Y. Katsuyama, M. Yamazaki, K. Watanabe, K. Goto, S. Nakamura, S. Bahram, and H. Inoko. 1997. Triplet repeat polymorphism in the transmembrane region of the MICA gene: a strong association of six GCT repetitions with Behçet disease. *Proc. Natl. Acad. Sci. USA* 94:1298–1303.
16. Kimura, T., K. Goto, K. Yabuki, N. Mizuki, G. Tamiya, M. Sato, M. Kimura, H. Inoko, and S. Ohno. 1998. Microsatellite polymorphism within the MICB gene among Japanese patients with Behçet's disease. *Hum. Immunol.* 59:500–502.
17. Wallace, G.R., D.H. Verity, L.J. Delamaine, S. Ohno, H. Inoko, M. Ota, N. Mizuki, K. Yabuki, E. Kondiatis, H.A. Stephens, W. Madanat, C.A. Kanawati, M.R. Stanford, and R.W. Vaughan. 1999. MIC-A allele profiles and HLA class I associations in Behçet's disease. *Immunogenetics* 49:613–617.
18. Vilches, C., M. Bunce, R. de Pablo, A.K. Murray, C.A. McIntyre, and M. Kreisler. 1997. Complete coding regions of two novel HLA-B alleles detected by phototyping (PCR-SSP) in the British caucasoid population: B*5108 and B*5002. *Tissue Antigens* 50:38–41.
19. Mizuki, N., S. Ohno, H. Ando, L. Chen, G.D. Palimeris, E. Stavropoulos-Ghiokas, M. Ishihara, K. Goto, S. Nakamura, Y. Shindo, K. Isobe, N. Ito, and H. Inoko. 1997. A strong association between HLA-B*5101 and Behçet's disease in Greek patients. *Tissue Antigens* 50:57–60.
20. Cano, P., B. Fan, and S. Stass. 1998. A geometric study of the amino acid sequence of class I HLA molecules. *Immunogenetics* 48:324–334.
21. Buxton, S.E., R.J. Benjamin, C. Clayberger, P. Parham, and A.M. Krensky. 1992. Anchoring pockets in human histocompatibility complex leukocyte antigen (HLA) class I molecules: analysis of the conserved B ("45") pocket of HLA-B27. *J. Exp. Med.* 175:809–820.

22. Smith, K.J., S.W. Reid, K. Harlos, A.J. McMichael, D.I. Stuart, J.I. Bell, and E.Y. Jones. 1996. Bound water structure and polymorphic amino acids act together to allow the binding of different peptides to MHC class I HLA-B53. *Immunity* 4:215–228.

23. Sette, A., and J. Sidney. 1998. HLA supertypes and super-motifs: a functional perspective on HLA polymorphism. *Curr. Opin. Immunol.* 10:478–482.

24. Elliott, J.I. 1997. T cell repertoire formation displays characteristics of qualitative models of thymic selection. *Eur. J. Immunol.* 27:1831–1837.

25. Nijman, H.W., J.G. Houbiers, S.H. van der Burg, M.P. Vierboom, P. Kenemans, W.M. Kast, and C.J. Melief. 1993. Characterization of cytotoxic T lymphocyte epitopes of a self-protein, p53, and a non-self-protein, influenza matrix: relationship between major histocompatibility complex peptide binding affinity and immune responsiveness to peptides. *J. Immunother.* 14:121–126.

26. Bacon, P.A. 1991. Vasculitic syndromes associated with other rheumatic conditions and unclassified systemic vasculitis. *Curr. Opin. Rheumatol.* 3:56–61.

27. Yamashita, N. 1997. Hyperreactivity of neutrophils and abnormal T cell homeostasis: a new insight for pathogenesis of Behcet's disease. *Int. Rev. Immunol.* 14:11–19.

28. Takeno, M., A. Kariyone, N. Yamashita, M. Takiguchi, Y. Mizushima, H. Kaneoka, and T. Sakane. 1995. Excessive function of peripheral blood neutrophils from patients with Behcet's disease and from HLA-B51 transgenic mice. *Arthritis Rheum.* 38:426–433.

29. Saeki, K., S. Kitagawa, E. Okuma, S. Hagiwara, M. Yagisawa, and A. Yuo. 1998. Cooperative stimulatory effects of tumor necrosis factor and granulocyte-macrophage colony-stimulating factor on the particular respiratory burst activity in human neutrophils: synergistic priming effect on con-canavalin A-induced response, no interactive priming effect on the chemotactic peptide-induced response and additive triggering effect. *Int. J. Hematol.* 68:269–278.

30. Lehner, T., K.I. Welsh, and J.R. Batchelor. 1982. The relationship of HLA-B and DR phenotypes to Behcet's syndrome, recurrent oral ulceration and the class of immune complexes. *Immunology* 47:581–587.

31. Arbesfeld, S.J., and A.K. Kurban. 1988. Behcet's disease. New perspectives on an enigmatic syndrome. *J. Am. Acad. Dermatol.* 19:767–779.

32. Suzuki, Y., K. Hoshi, T. Matsuda, and Y. Mizushima. 1992. Increased peripheral blood gamma delta+ T cells and natural killer cells in Behcet's disease. *J. Rheumatol.* 19:588–592.

33. Yamashita, N., H. Kaneoka, S. Kaneko, M. Takeno, K. Oneda, H. Koizumi, M. Kogure, G. Inaba, and T. Sakane. 1997. Role of γδ T lymphocytes in the development of Behcet's disease. *Clin. Exp. Immunol.* 107:241–247.

34. Sakane, T., N. Suzuki, Y. Ueda, S. Takada, Y. Murakawa, T. Hoshino, Y. Niwa, and T. Tsunematsu. 1986. Analysis of interleukin-2 activity in patients with Behcet's disease. Ability of T cells to produce and respond to interleukin-2. *Arthritis Rheum.* 29:371–378.

35. Sakane, T., H. Kotani, S. Takada, and T. Tsunematsu. 1982. Functional aberration of T cell subsets in patients with Behcet's disease. *Arthritis Rheum.* 25:1343–1351.

36. Feron, E.J., V.L. Calder, and S.L. Lightman. 1992. Distribution of IL-2R and CD45Ro expression on CD4+ and CD8+ T-lymphocytes in the peripheral blood of patients with posterior uveitis. *Curr. Eye Res.* 11:167–172.

37. Akoglu, T.F., H. Direskeneli, H. Yazici, and R. Lawrence. 1990. TNF, soluble IL-2R and soluble CD-8 in Behcet's disease. *J Rheumatol* 17:1107–1108.

38. Suzuki, N., T. Sakane, Y. Ueda, and T. Tsunematsu. 1986. Abnormal B cell function in patients with Behcet's disease. *Arthritis Rheum* 29:212–219.

39. Del Prete, G. 1998. The concept of type-1 and type-2 helper T cells and their cytokines in humans. *Int. Rev. Immunol.* 16:427–455.

40. O'Garra, A., L. Steinman, and K. Gijbels. 1997. CD4+ T-cell subsets in autoimmunity. *Curr. Opin. Immunol.* 9:872–883.

41. King, C., and N. Sarvetnick. 1997. Organ-specific autoimmunity. *Curr. Opin. Immunol.* 9:863–871.

42. Nicholson, L.B., and V.K. Kuchroo. 1996. Manipulation of the Th1/Th2 balance in autoimmune disease. *Curr. Opin. Immunol.* 8:837–842.

43. Caspi, R.R., P.B. Silver, C.C. Chan, B. Sun, R.K. Agarwal, J. Wells, S. Oddo, Y. Fujino, F. Najafian, and R.L. Wilder. 1996. Genetic susceptibility to experimental autoimmune uveoretinitis in the rat is associated with an elevated Th1 response. *J. Immunol.* 157:2668–2675.

44. Turan, B., H. Gallati, H. Erdi, A. Gurler, B.A. Michel, and P.M. Villiger. 1997. Systemic levels of the T cell regulatory cytokines IL-10 and IL-12 in Behcet's disease; soluble TNFR-75 as a biological marker of disease activity. *J. Rheumatol.* 24:128–132.

45. Hu, W., A. Hasan, A. Wilson, M.R. Stanford, Y. Li-Yang, S. Todryk, R. Whiston, T. Shinnick, Y. Mizushima, R. van der Zee, and T. Lehner. 1998. Experimental mucosal induction of uveitis with the 60-kDa heat shock protein-derived peptide 336–351. *Eur. J. Immunol.* 28:2444–2455.

46. Hamzaoui, K., A. Hamzaoui, F. Hentati, A. Kahan, K. Ayed, A. Chabbou, M.B. Hamida, and M. Hamza. 1994. Phenotype and functional profile of T cells expressing gamma delta receptor from patients with active Behcet's disease. *J. Rheumatol.* 21:2301–2306.

47. Nussenblatt, R.B., S.M. Whitcup, M.D. de Smet, R.R. Caspi, A.T. Kozhich, H.L. Weiner, B. Vistica, and I. Gery. 1996. Intraocular inflammatory disease (uveitis) and the use of oral tolerance: a status report. *Ann. NY Acad. Sci.* 778:325–337.

48. Wildner, G., and S.R. Thurau. 1994. Cross-reactivity between an HLA-B27-derived peptide and a retinal autoantigen peptide: a clue to major histocompatibility complex association with autoimmune disease. *Eur. J. Immunol.* 24:2579–2585.

49. Yamamoto, J.H., Y. Fujino, C. Lin, M. Nieda, T. Juji, and K. Masuda. 1994. S-antigen specific T cell clones from a patient with Behcet's disease. *Br. J. Ophthalmol.* 78:927–932.

50. Lehner, T., E. Lavery, R. Smith, R. van der Zee, Y. Mizushima, and T. Shinnick. 1991. Association between the 65-kilodalton heat shock protein, Streptococcus sanguis, and the corresponding antibodies in Behcet's syndrome. *Infect. Immun.* 59:1434–1441.

51. Kaufmann, S.H. 1994. Heat shock proteins and autoimmunity: a critical appraisal. *Int. Arch. Allergy Immunol.* 103:317–322.

52. Kaneko, S., N. Suzuki, N. Yamashita, H. Nagafuchi, T. Nakajima, S. Wakisaka, S. Yamamoto, and T. Sakane. 1997. Characterization of T cells specific for an epitope of human 60-kD heat shock protein (hsp) in patients with Behcet's disease (BD) in Japan. *Clin. Exp. Immunol.* 108:204–212.

53. Lehner, T. 1997. The role of heat shock protein, microbial and autoimmune agents in the aetiology of Behcet's disease. *Int. Rev. Immunol.* 14:21–32.

54. Pervin, K., A. Childerstone, T. Shinnick, Y. Mizushima, R. van der Zee, A. Hasan, R. Vaughan, and T. Lehner. 1993. T cell epitope expression of mycobacterial and homologous human 65-kilodalton heat shock protein peptides in short term cell lines from patients with Behcet's disease. *J. Immunol.* 151:2273–2282.

55. Stanford, M.R., E. Kasp, R. Whiston, A. Hasan, S. Todryk, T. Shinnick, Y. Mizushima, D.C. Dumonde, R. van der Zee, and T. Lehner. 1994. Heat shock protein peptides reactive in patients with Behcet's disease are uveitogenic in Lewis rats. *Clin. Exp. Immunol.* 97:226–231.

56. Inoue, C., R. Itoh, Y. Kawa, and M. Mizoguchi. 1994. Pathogenesis of mucocutaneous lesions in Behcet's disease. *J. Dermatol.* 21:474–480.

57. Matzner, Y., and V. Leibovici. 1988–89. Increased neutrophil chemotaxis. A secondary phenomenon useful in the diagnosis and follow up of diseases with inflammatory component. *Acta Paediatr. Hung.* 29:191–195.

58. Niwa, Y., S. Miyake, T. Sakane, M. Shingu, and M. Yokoyama. 1982. Auto-oxidative damage in Behcet's disease–endothelial cell damage following the elevated oxygen radicals generated by stimulated neutrophils. *Clin. Exp. Immunol.* 49:247–255.

59. Namba, K., and K. Masuda. 1984. Types of ocular attacks and lysosomal enzymes in Behcet's disease. *Jpn. J. Ophthalmol.* 28:80–88.

60. Sahin, S., T. Akoglu, H. Direskeneli, L.S. Sen, and R. Lawrence. 1996. Neutrophil adhesion to endothelial cells and factors affecting adhesion in patients with Bechet's disease. *Ann. Rheum. Dis.* 55:128–133.

61. Celik, G., O. Kalaycioglu, and G. Durmaz. 1996. Colchicine in Behcet's disease with major vessel thrombosis. *Rheumatol. Int.* 16:43–44.

62. Sensi, A., R. Gavioli, S. Spisani, A. Balboni, L. Melchiorri, A. Menicucci, G. Palumbo, S. Traniello, and O.R. Baricordi. 1991. HLA B51 antigen associated with neutrophil hyperreactivity. *Dis. Markers* 9:327–331.

63. Sayinalp, N., O.I. Ozcebe, O. Ozdemir, I.C. Haznedaroglu, S. Dundar, and S. Kirazli. 1996. Cytokines in Behcet's disease. *J. Rheumatol.* 23:321–322.

64. Yosipovitch, G., B. Shohat, J. Bshara, A. Wysenbeek, and A. Weinberger. 1995. Elevated serum interleukin 1 receptors and interleukin 1B in patients with Behcet's disease: correlations with disease activity and severity. *Isr. J. Med. Sci.* 31:345–348.

65. Wang, L.M., N. Kitteringham, S. Mineshita, J.Z. Wang, Y. Nomura, Y. Koike, and E. Miyashita. 1997. The demonstration of serum interleukin-8 and superoxide dismutase in Adamantiades-Behcet's disease. *Arch. Dermatol. Res.* 289:444–447.

66. Ozoran, K., O. Aydintug, G. Tokgoz, N. Duzgun, H. Tutkak, and A. Gurler. 1995. Serum levels of interleukin-8 in patients with Behcet's disease. *Ann. Rheum. Dis.* 54:610.

67. Mochizuki, M., E. Morita, S. Yamamoto, and S. Yamana. 1997. Characteristics of T cell lines established from skin lesions of Behcet's disease. *J. Dermatol. Sci.* 15:9–13.

68. Mege, J.L., N. Dilsen, V. Sanguedolce, A. Gul, P. Bongrand, H. Roux, L. Ocal, M. Inanc, and C. Capo. 1993. Overproduction of monocyte derived tumor necrosis factor alpha, interleukin (IL) 6, IL-8 and increased neutrophil superoxide generation in Behcet's disease. A comparative study with familial Mediterranean fever and healthy subjects. *J. Rheumatol.* 20:1544–1549.

69. Altstaedt, J., H. Kirchner, and L. Rink. 1996. Cytokine production of neutrophils is limited to interleukin-8. *Immunology* 89:563–568.

70. Takeichi, O., I. Saito, T. Tsurumachi, T. Saito, and I. Moro. 1994. Human polymorphonuclear leukocytes derived from chronically inflamed tissue express inflammatory cytokines in vivo. *Cell. Immunol.* 156:296–309.

71. Shimoyama, Y., M. Takeno, H. Nagafuchi, and T. Sakane. 1997. Prolonged survival of autoprimed neutrophils from patients with Behcet's disease. *Arthritis Rheum.* 40:s66.

72. Gilhar, A., G. Winterstein G, H. Turani H, J. Landau J, and A. Etzioni A. 1989. Skin hyperreactivity response (pathergy) in Behcet's disease. *J. Am. Acad. Dermatol.* 21:547–552.

73. Dilsen, N., M. Konice, O. Aral, L. Ocal, M. Inanc, and A. Gul. 1993. Comparative study of the skin pathergy test with blunt and sharp needles in Behcet's disease: confirmed specificity but decreased sensitivity with sharp needles. *Ann. Rheum. Dis.* 52:823–825.

74. Ergun, T., O. Gurbuz, J. Harvell, J. Jorizzo, and W. White. 1998. The histopathology of pathergy: a chronologic study of skin hyperreactivity in Behcet's disease. *Int. J. Dermatol.* 37:929–933.

75. Gul, A., S. Esin, N. Dilsen, M. Konice, H. Wigzell, and P. Biberfeld. 1995. Immunohistology of skin pathergy reaction in Behcet's disease. *Br. J. Dermatol.* 132:901–907.

76. Feron, E.J., V.L. Calder, and S.L. Lightman. 1995. Oligoclonal activation of CD4+ T lymphocytes in posterior uveitis. *Clin. Exp. Immunol.* 99:412–418.

77. Tugal-Tutkun, I., M. Urgancioglu, and C.S. Foster. 1995. Immunopathologic study of the conjunctiva in patients with Behçet disease. *Ophthalmology* 102:1660–1668.

78. Sakane, T., N. Suzuki, and M. Takeno. 1998. Innate and acquired immunity in Behcet's disease. In: 8th international congress on Behcet's disease: program and abstract. 56.

79. Sagdic, K., Z.G. Ozer, D. Saba, M. Ture, and M. Cengiz. 1996. Venous lesions in Behcet's disease. *Eur. J. Vasc. Endovasc. Surg.* 11:437–440.

80. Allen, N.B. 1993. Miscellaneous vasculitic syndromes including Behcet's disease and central nervous system vasculitis. *Curr. Opin. Rheumatol.* 5:51–56.

81. Triolo, G., A. Accardo-Palumbo, G. Triolo, M.C. Carbone, A. Ferrante, and E. Giardina. 1999. Enhancement of endothelial cell E-selection expression by sera from patients with active Behcet's disease: moderate correlation with anti-endothelial cell antibodies and serum myeloperoxidase levels. *Clin. Immunol.* 91:330–337.

82. Aydintug, A.O., G. Tokgoz, D.P. D'Cruz, A. Gurler, R. Cervera, N. Duzgun, L.S. Atmaca, M.A. Khamashta, and G.R. Hughes. 1993. Antibodies to endothelial cells in patients with Behcet's disease. *Clin. Immunol. Immunopathol.* 67:157–162.

83. Uslu, T., C. Erem, M. Tosun, and O. Deger. 1997. Plasma endothelin-1 levels in Behcet's disease. *Clin. Rheumatol.* 16:59–61.

84. Hamzaoui, A., K. Hamzaoui, A. Chabbou, and K. Ayed. 1996. Endothelin-1 expression in serum and bronchoalveolar lavage from patients with active Behcet's disease. *Br. J. Rheumatol.* 35:357–358.

85. Ozoran, K., N. Dugun, A. Gurler, H. Tutkak, and G. Tokgoz. 1995. Plasma von Willebrand factor, tissue plasminogen activator, plasminogen activator inhibitor, and

antithrombin III levels in Behcet's disease. *Scand. J. Rheumatol.* 24:376–382.

86. Nalcaci, M., and Y. Pekcelen. 1998. Antithrombin III, protein C and protein S plasma levels in patients with Behcet's disease. *J. Int. Med. Res.* 26:206–208.

87. Mine, S., Y. Tanaka, M. Suematu, M. Aso, T. Fujisaki, S. Yamada, and S. Eto. 1998. Hepatocyte growth factor is a potent trigger of neutrophil adhesion through rapid activation of lymphocyte function-associated antigen-1. *Lab. Invest.* 78:1395–1404.

88. Sahin, S., T. Akoglu, H. Direskeneli, L.S. Sen, and R. Lawrence. 1996. Neutrophil adhesion to endothelial cells and factors affecting adhesion in patients with Behcet's disease. *Ann. Rheum. Dis.* 55:128–133.

49 | Autoimmune Gastritis and Pernicious Anemia

Ban-Hock Toh, John Sentry, and Frank Alderuccio

1. INTRODUCTION

We have recently reviewed the historical, clinical, pathologic and immunologic features of autoimmune gastritis and pernicious anemia (1,2). Among the organ-specific autoimmune diseases, autoimmune gastritis par excellence, is the one condition in which the causative target autoantigen has been clearly identified. There is now compelling evidence from mouse models of autoimmune gastritis that the gastric H/K ATPase, the enzyme responsible for acidification of gastric juices, is the causative autoantigen. These mouse models of autoimmune gastritis demonstrate the condition is mediated solely by pathogenic CD4 T cells directed against this enzyme and that the pathogenic CD4 T cells are in turn controlled by a population of regulatory CD4 T cells. In this chapter, we provide a short overview of human autoimmune gastritis and pernicious anemia followed by a review of the molecular and cellular basis of CD4 T cell mediated tolerance and autoimmunity to the gastric H/K ATPase in the mouse.

2. HUMAN AUTOIMMUNE GASTRITIS AND PERNICIOUS ANEMIA

2.1. Clinical spectrum

"Thyrogastric autoimmunity" comprising autoimmune gastritis and pernicious anemia together with autoimmune thyroiditis, are prototypes of organ-specific autoimmunity (3). This organ-specific cluster of autoimmune diseases have since been expanded to include autoimmune diseases affecting primarily other endocrine organs including: the islets of Langerhans (autoimmune diabetes) and the adrenal cortex (Addison's disease). Accordingly, this "thyrogastric cluster" is now also known as the "autoimmune endocrinopathies" to distinguish them from the "autoimmune exocrinopathy" of Sjögren's Syndrome.

Organ-specific autoimmune diseases as a group are characterised by organ-restricted pathology and circulating autoantibodies directed to antigens in the affected organ. In the case of autoimmune gastritis, the condition is associated with circulating autoantibodies to gastric parietal cells and to an intrinsic factor, itself a secretory product of human parietal cells. Parietal cell autoantibody is an excellent marker for the presence of the underlying pathologic lesion of autoimmune gastritis and is present in about 90% of patients with pernicious anemia (1,2). Intrinsic factor autoantibodies, present in about 60% of patients with pernicious anemia, segregate tightly with this condition and are therefore useful for its diagnosis (1,2). The lesion of autoimmune gastritis is associated with the loss of gastric parietal cells as well as of zymogenic cells (see Pathology, below). The loss of gastric parietal cells together with the presence of autoantibodies to intrinsic factor in the gastric juice results in deficiency of intrinsic factor. In turn, this leads to failure of intrinsic factor-dependent absorption of vitamin B12 from the terminal ileum with consequent megaloblastic anaemia (1,2).

Although "pernicious" when first described, pernicious anemia can today be readily treated with replacement vitamin B12 therapy. Nonetheless, the condition is still associated with significant morbidity in the general population given that it remains undiagnosed in almost 2% of persons over the age of 60yr (4) and comprises about a third of patients with neurologic disorders due to cobalamin deficiency without any overt signs of anemia (5).

2.2.   Pathology

Autoimmune gastritis or Chronic Atrophic Gastritis Type A is restricted to the fundus and body of the stomach with preservation of the antrum (1,2). The anatomic localisation of the pathologic lesion distinguishes this condition from Chronic Atrophic Gastritis Type B, which involves the antrum of the stomach and is associated with infection by Helicobacter pylori. The gastritis is characterised by a chronic inflammatory infiltrate in the submucosa that extends into the lamina propria of the mucosa with accompanying loss of gastric parietal cells and zymogenic cells and replacement by cells resembling the intestinal mucosa, a change known as "intestinal metaplasia".

2.3.   Factors contributing to disease susceptibility

While pernicious anemia was initially considered restricted to Caucasians of North European origin, the condition is also now reported in other racial groups, including North American blacks and Latin Americans (1,2). There is some evidence for a genetic predisposition to autoimmune gastritis and pernicious anemia, and there are reports of increased incidence of pernicious anemia and of gastric autoantibodies in first degree relatives of patients with pernicious anemia (1,2). Interestingly, there appears to be a genetic predisposition for this group of organ-specific autoimmune diseases as a whole, given the clustering of related organ-specific autoimmune diseases and organ-specific autoantibodies in first degree relatives (1,2). Anecdotal twin studies also suggest the possibility of a genetic predisposition (1,2). However, no consistent association with particular MHC haplotypes has been demonstrated (6). In line with other autoimmune diseases, there is a predisposition of disease in females suggesting a permissive influence of female sex hormones (4).

There is currently no known environmental trigger for the initiation of autoimmune gastritis. Recently, the gastric H/K ATPase was shown to be the major autoantigen reactive with parietal cell autoantibodies in chronic Helicobacter pylori-associated gastritis with atrophy of the body of the stomach (7). This observation suggests that atrophy of the corpus, associated with infection by Helicobacter pylori, may be a consequence of an autoimmune reaction directed towards the gastric H/K ATPase. This has led some to speculation that Helicobacter pylori may be the environmental factor that triggers the development of autoimmune gastritis (8). However, this tenet is as yet unproven.

2.4.   Identification of the Gastric H/K ATPase as the Molecular Target

The first step in any concerted study of an autoimmune disease is the identification of the autoantigen targeted by the immune system. In the case of organ-specific autoimmune diseases, the presence of organ-specific autoantibodies provides a convenient starting point for the identification of molecular targets reactive with these autoantibodies. For autoimmune gastritis and pernicious anemia, circulating parietal cell autoantibodies are found in >90% of patients with this condition (1,2). The molecular targets recognised by parietal cell autoantibodies have been identified as the two polypeptides—the catalytic α and the glycoprotein β subunits-of the gastric H/K ATPase (Figure 1), the enzyme responsible for the secretion of acid in the stomach (1,2). The H/K ATPase is located in the secretory canaliculi of gastric parietal cells and comprises the major protein of these cells (Figure 1). These secretory canaliculi are vastly expanded in gastric parietal cells that have been stimulated to produce hydrocholoric acid. The gastric H/K ATPase is also known as the gastric "proton pump" as it "pumps" H^+ ions into the lumen of the stomach in exchange for K^+ ions in a reaction driven by ATP in each reaction cycle. The H/K ATPase belongs to the family of "P"-type ATPases, as the enzyme is phosphorylated during each reaction cycle. To date, all sera reactive to gastric parietal cells by immunofluorescence have been found to react with both subunits of the ATPase. There is currently no evidence of reactivity initially with one of the subunits with subsequent "spreading" to the other subunit.

2.5.   Immunopathogenesis

Burnet proposed that "self-nonself discrimination" is mediated by clonal deletion of self-reactive lymphocytes during their development so that the fully developed immune system is only directed against non-self molecules. Since then, compelling experimental evidence has been accrued for the clonal deletion of self-reactive T cells by apoptosis when they encounter self-antigens in the thymus during their development (rev. in 9). However deletion of self-reactive T cells in the thymus can only occur if these antigens are represented in the thymus. There is now considerable evidence that T cells reactive with extrathymic self antigens "escape" clonal deletion in the thymus. The mechanisms by which these self-reactive T cells remain tolerant to their self antigens remain poorly understood.

The identification of the gastric H/K ATPase as the molecular target of parietal cell autoantibodies raises the question of whether this gastric enzyme is also the target of autoreactive T cells. To date, there have been no studies to address this question in humans. However a number of mouse models have been established to address

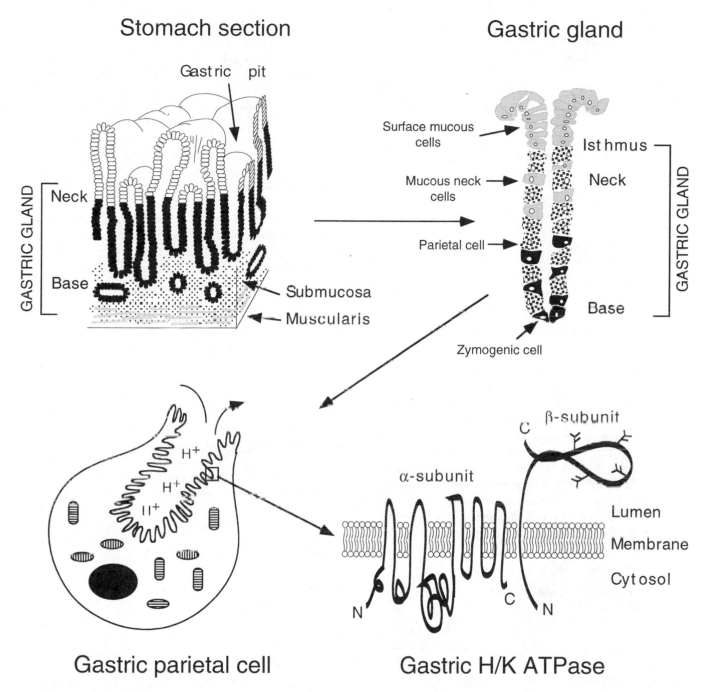

Stomach section

Gastric pit

GASTRIC GLAND

Neck

Base

Submucosa

Muscularis

Gastric gland

Surface mucous cells

Mucous neck cells

Parietal cell

Isthmus

Neck

Base

GASTRIC GLAND

Zymogenic cell

Gastric parietal cell

H⁺

H⁺

H⁺

Gastric H/K ATPase

β-subunit

α-subunit

C

Lumen

Membrane

Cytosol

N

C

N

Figure 1.   Schematic diagram showing the organisation of the gastric mucosa, gastric gland, gastric parietal cell and the gastric H/K ATPase.

the question of the role of the gastric H/K ATPase in the immunopathogenesis of autoimmune gastritis.

3.   MOUSE MODELS OF EXPERIMENTAL AUTOIMMUNE GASTRITIS (EAG)—INSIGHTS INTO IMMUNOPATHOGENESIS

As in man, in the mouse the gastric H/K ATPase is the authentic autoantigen targeted in gastritis (1,2). This

observation has led to the use of transgenic approaches and the H/K ATPase autoantigen in mouse to develop models in which to study experimental autoimmune gastritis (EAG). The mouse gastric lesion is very similar to the human pathology. Indeed, many of the features of human autoimmune gastritis and pernicious anaemia are readily identifiable in EAG (Table 1) In addition, EAG serves as an excellent model for studies of peripheral tolerance and autoimmunity to organ-specific autoantigens.

Table 1 Pathologies common to human autoimmune gastritis and pernicious anemia, and mouse models of experimentally induced autoimmune gastritis

Pathology	human	mouse
Autoantibodies reactive to gastric parietal cell H/K ATPase	+	+
Autoantibodies reactive to intrinsic factor	+	?
Parietal and chief cell loss from the gastric mucosa	+	+
Loss of gastric acid (achlorhydria)	+	+
Mononuclear cell infiltrate in the gastric mucosa	+	+
Megaloblastic anemia	+	?
Hyperplasia of gastric mucosa	–	+
Association with other autoimmune diseases	+	+

3.1. Experimental Autoimmune Gastritis

Experimental autoimmune gastritis can be induced in susceptible strains of mice by a variety of treatments (see Table 2). The initiation of gastritis with circulating autoantibodies to the gastric H/K ATPase in BALBc/CrSIc mice by neonatal thymectomy, adult thymectomy combined with cyclophosphamide treatment (10) or by immunisation with mouse gastric H/K ATPase (11,12) or gastritis that develops in single chain TCR α-transgenic mice (13) indicates that tolerance can be broken by pertubation of T cell homeostasis. In the TCR α-transgenic mice, autoimmunity is mediated by self-reactive T cells expressing endogeneous TCR α-chains, suggesting that transgene expression did not

Table 2 Methods of induction of experimental autoimmune gastritis (EAG) in mouse

Procedures	References
Neonatal thymectomy	17
Adult thymectomy combined with irradiation treatment	1
Adult thymectomy combined cylcophosphamide treatment	10
Neonatal treatment with cyclosporine A	1
High dose fractionated total lymphoid irradiation	1
Transfer of T cells to syngeneic T cell deficient mice	1
Transfer of thymus/thymocyte to syngeneic T cell deficient mice	1
Immunisation with purified gastric H/K ATPase	12
TCR alpha transgenic mice	13
TCR transgenic mice specific for H/K ATPase β subunit (aa 253–277)	Unpubl obs.
Spontaneously in C3H/He mouse strain	14

suppress endogenous α-chain gene rearrangement and may have triggered the expansion/activation of the self-reactive T cells. Immunisation of neonates with the H/K ATPase without adjuvant induced a persistent autoimmune gastritis with H/K ATPase-specific autoantibodies (11). In contrast, adults require repeated immunisations of the H/K ATPase in adjuvant to induce gastritis that is not persistent, but reverses following the cessation of immunisation (12). The recent report of spontaneous autoimmune gastritis associated with circulating parietal cell autoantibodies directed towards the gastric H/K ATPase in C3H/He mice (14) underpins the importance of genetics in the development of this disease. The mouse model, that has given us the most insights into the immunopathogenesis of autoimmune gastritis is that induced by neonatal thymectomy (Figure 2). In the following sections, we summarise the major findings from studies of experimental autoimmune gastritis in mice.

Factors contributing to disease susceptibility. Autoimmune gastritis with circulating autoantibodies to the gastric H/K ATPase only develops in certain susceptible mouse strains while other strains are resistant, implying a strong genetic component. For instance, spontaneous gastritis has been reported only in C3H/He mice, while BALB/cCrS1c mice are highly susceptible to gastritis induced by neonatal thymectomy. Genetic susceptibility of BALB/c mice to thymectomy-induced autoimmune gastritis is not linked to the MHC since H-2 identical DBA mice are resistant to gastritis induced by neonatal thymectomy (1,2). Resistence to the development of autoimmune gastritis can be abolished by crossing gastritis-resistant with gastritis-susceptible mouse strains. In the case of BALB/cCrS1c, a linkage analysis study has been carried out by crossing this mouse strain with gastritis-resistant C57BL/6 mice. These studies led to the identification of two regions on the distal arm of chromosome 4, designated Gasa1 and Gasa2, which are linked to susceptibility to EAG (15). Gasa1 maps within the same chromosomal segment as that conferring susceptibility for Type 1 diabetes mellitus (Idd11) and systemic lupus erythematosus (Nba1). Gasa2 maps to the same chromosomal segment as that conferring susceptibility to the type 1 diabetes mellitus locus, Idd9. While the Gasa2 region contains a number of candidate genes that encode members of the TNF family, susceptibility to EAG may be due to an as yet unidentified gene that maps to this region.

The role of thyroid hormones in the development of gastritis has recently been investigated (16). The time of administration of thyroxine appears to be critical given that gastritis is ameliorated when the hormone is administered to young adult neonatally-thymectomised mice during an active phase of disease development. However, administration of thyroxine before disease development aggravates the disease. The responsible thyroxine-mediated molecular and cellular mechanisms are not known.

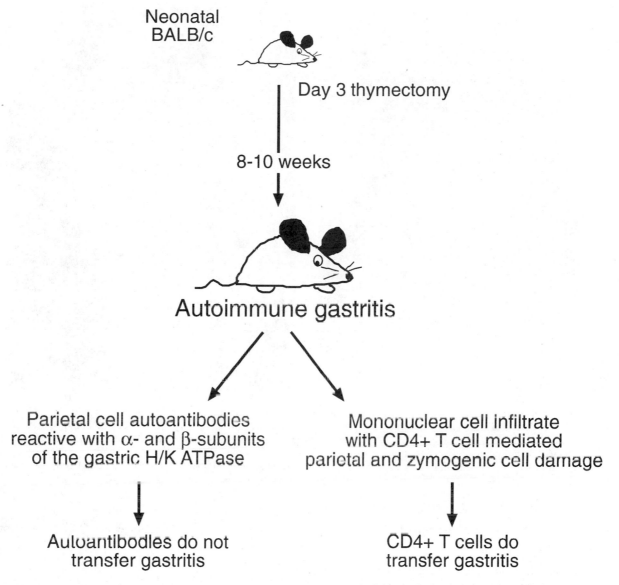

Figure 2. Neonatal thymectomy model of experimental autoimmune gastritis. Experimental autoimmune gastritis is induced by surgically removing the thymus from three day old BALB/c or BALB/cCrSlc mice. Eight to ten weeks following thymectomy, gastritic mice are identified by circulating parietal cell autoantibodies, which are reactive with the the α- and β-subunits of the gastric H/K ATPase. Histologically, gastritic mice display a prominent mononuclear cell infiltrate within the gastric mucosa often accompanied by parietal and zymogenic cell damage. Experimental data indicates that autoimmune gastritis in this model is mediated by CD4[+] T cells and not by autoantibodies.

Pathology. Murine experimental autoimmune gastritis is similar to that of human autoimmune gastritis (see Table 1, above). The chronic inflammatory lesion is associated with degenerative changes in gastric parietal cells and in zymogenic cells. The degeneration of these parenchymal cells is associated with the expansion of a population of smaller cells with morphological features of precursor cells, resulting in "hypertrophy" of the gastric mucosa (Figure 3). The gastritis is also associated with autoantibodies to gastric parietal cells demonstrable by immunofluorescence and reactive to the α and β subunits of the gastric H/K ATPase by ELISA and immunoblotting (17). In the murine model, no investigation for the presence of autoantibodies to intrinsic factor has been undertaken. In the mouse, intrinsic factor is secreted by zymogenic (chief) cells. However, the mononuclear cell infiltrate within the gastric mucosa associated with EAG leads to both parietal cell (containing the gastric H/K ATPase) and zymogenic cell (secreting intrinsic factor) destruction.

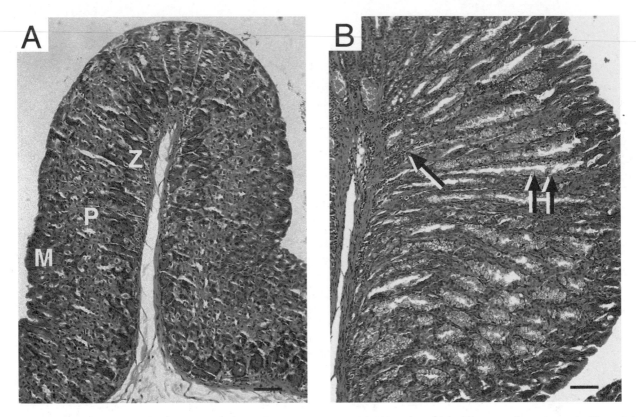

Figure 3.   Haematoxylin and eosin staining of paraffin embedded normal (A) and gastritic (B) stomach sections. A. Normal (non-gastritic) stomach section illustrating typical structure of the gastric gland and the three major cell types including parietal cells (P), zymogenic cells (Z) and surface mucous cells (M). B. Stomach section from mouse with experimental autoimmune gastritis. Destructive gastritis is associated with mononuclear cell infiltrating the gastric mucosa (arrow) and between the gastric glands. Parietal and zymogenic cells are destroyed in the glands resulting in a characteristic morphological change dominated by mucus secreting cells (double arrows) and glandular hypertrophy. Bar: 100 μm

3.2.   Immunopathogenesis of Experimental Autoimmune Gastritis

The development of mouse models of autoimmune gastritis affords the opportunity to investigate the immunopathogenesis of the disease. The identification of the gastric H/K ATPase as the molecular target of parietal cell autoantibodies raises the question of whether this gastric enzyme is also the target of autoreactive T cells.

Mediation by pathogenic Th1 CD4 T cells.   Adoptive transfer studies have shown that gastritis initiated by neonatal thymectomy is mediated by peripheral CD4 T cells and not by CD8 T cells or autoantibodies (1,2). This assertion is further supported by the observation that gastritis is prevented by treatment with depleting antibodies to CD4 T cells, but not with depleting antibodies to CD8 T cells (18). The pathogenic CD4 T cells belong to the CD25$^-$ subset, because depletion of CD25$^+$ CD4 T cells from a total population of CD4 T cells from normal adult BALB/c mice renders the residual population pathogenic when transferred to T cell-deficient hosts. These findings clearly

indicate that potentially pathogenic CD4 T cells are present in normal adult mice. CD4 T cells and macrophages comprise early infiltrates in the gastric mucosa seen at 4 weeks after neonatal thymectomy accompanied by expression of MHC class II molecules in gastric epithelial cells. In contrast, there is only a marginal increase in mucosal CD8 T cells and B cells, with B cells peaking at 8 weeks, largely as follicular aggregates. The CD4 T cells produce a mix of Th1 and Th2 type cytokines, including IFNγ, IL-10, TNFα and GM-CSF, but not IL-4 (19). Splenic T cells from mice with gastritis produced three- to tenfold more IFNγ than T cells from normal animals after stimulation with anti-CD3 antibodies. The gastritis can be prevented by treatment with neutralising antibodies to IFN-γ, indicating that the pathogenic CD4 T cells belong to the Th1 subset (20). These findings are supported by the observation that the infiltrating CD4 T cells express very low levels of IL-4 mRNA, but high levels of INFγ mRNA, and that Th1, but not Th2, CD4 T cells preferentially migrated through an endothelial cell monolayer (21). The vast majority of CD4 T cells infiltrating the gastric mucosa and the draining paragastric lymph node show an activated

activated/memory phenotype, CD45Rb^{low}, CD62L (L selectin)^{low} and CD44^{high}.

Pathogenic CD4 T cells recognise the gastric H/K ATPase subunits. Expression of the gastric H/K ATPase β subunit transgene in the thymus prevented autoimmune gastritis induced by neonatal thymectomy (22), adult thymectomy combined with cyclophosphamide treatment (13) or immunisation with the gastric H/K ATPase (17). The prevention of autoimmunity is specific for gastritis, as transgenic expression of the β subunit did not alter the incidence of associated autoimmune oophoritis, which also develops in about 20% of BALB/c mice after neonatal thymectomy. In contrast, mice harbouring the H/K ATPase α-subunit transgene in the thymus remained susceptible to gastritis (23). Unlike the β-subunit, the α-subunit is expressed in the normal thymus. These observations provide compelling evidence that autoimmune gastritis is initiated by pathogenic CD4 T cells directed towards the β-subunit, thus identifying the β-subunit as the causative autoantigen. This assertion is supported by the failure of β-subunit ^{−/−} mice to develop autoimmune gastritis following neonatal thymectomy (Scarff K, pers comm). However, a pathogenic role for α-subunit-reactive T cells remains likely because pathogenic Th1 and Th2 T cell clones that recognise peptides of the α-subunit of the H/K ATPase has been generated (24) and TCR mRNAs of these T cells have been identified in cells infiltrating the gastric mucosa (25). We suggest that the pathologic lesion of autoimmune gastritis is initiated by β-subunit-reactive CD4 T cells, which may then subsequent recruit α-subunit-reactive T cells to perpetuate the gastric mucosal injury.

Identification of the dominant H/K ATPase target epitope. T cells from mice with gastritis induced by immunisation with gastric H/K ATPase were tested for proliferation to a panel of 21 overlapping peptides spanning the entire sequence of the H/K ATPase β-subunit. A 25-mer at the carboxyl terminus (H/K β ATPase 253–277) stimulated T cell proliferation and induced gastritis in mice without autoantibody production (26) indicating that this peptide is the dominant gastritogenic T cell epitope of the H/K ATPase β-subunit. We have subsequently restricted the gastriogenic epitope to H/Kβ 261–274.

Regulatory CD4 T cells suppress proliferation of pathogenic CD4 T cells. The presence of thymic regulation of pathogenic CD4 T cells is supported by a number of observations. For example, although treatment with cyclophosphamide reduced the numbers of splenic T and B cells by approximately 25-fold, gastritis did not develop unless the adult thymus was also removed (10). Further, the gastritis induced in adult BALB/cCrSlc mice by immunisa-

tions with the gastric H/K ATPase recovers when immunisation ceases (12) Indeed, CD4 T cells contain regulatory as well as pathogenic populations. Pathogenic CD4 T cells are contained in the CD25[−] subset, since complement-mediated depletion of CD25⁺ cells from CD4 T cells from normal mice can render the residual population pathogenic when transfered to T cell-deficient hosts (27). These observations also clearly indicate that potentially pathogenic CD4 T cells are present in normal mice. In contrast, regulatory CD4 T cells reside in the CD25⁺ population, since adoptive transfer of this population prevents autoimmunity induced by neonatal thymectomy or by transfer of CD25[−] CD4 T cells into T cell-deficient mice (28). CD4⁺CD25⁺ regulatory cells can suppress not only disease induced by neonatal thymectomy, but can also efficiently suppress disease induced by cloned autoantigen-specific effector cells. CD25⁺ comprise 5–10% of CD4 T cells. They appear in the periphery 3 days after birth, rapidly increasing to nearly adult levels within 2 weeks. Neonatal thymectomy on day 3 eliminates CD25⁺ T cells from the periphery for several days; inoculation immediately after NTx of CD25⁺ splenic T cells from syngeneic non-Tx adult mice prevents autoimmune disease development, whereas inoculation of CD25[−] T cells even at a larger dose can not. These observations suggest that the regulatory CD4 T cells are programmed to leave the thymus at day 3 after birth and is consistent with the suggestion that autoimmunity induced by neonatal thymectomy is the consequence of removal of this population (29). Indeed, the normal thymus has been shown to produce immunoregulatory CD25⁺ CD4 T cells, which constitute 5% of steroid-resistant mature CD4 T cells (30). Regulatory CD4 T cells are naturally unresponsive (anergic) because, in contrast to the CD25 pathogenic CD4 T cells, they do not proliferate following stimulation with IL-2, ConA, anti-CD3 or anti-CD28 antibody (30,31). Anergic CD25⁺ CD4 T cells can suppress the proliferation of CD4 CD25[−] T cells following stimulation by soluble anti-CD3 antibody or ConA. The antigen concentration required to stimulate CD25⁺ CD4⁺ T cells to exert suppression is much lower than that required to stimulate CD25[−] CD4⁺ T cells to proliferate. The suppression results in inhibition of IL-2 mRNA transcription by the pathogenic CD4 T cells and is not mediated by IL-4, IL-10, TGFβ or any other soluble factor, but requires cell-to-cell contact on the surface of APCs and activation through their TCR. Suppression is not a consequence of Fas-mediated apoptosis and is not MHC-restricted. Inhibition is overcome by breaking anergy by stimulation with IL-2 or anti-CD28 antibody. The anergic/suppressor state of the CD25⁺ CD4 T cells seems to be their basic default state because withdrawal of IL-2 or CD28 antibody resulted in reversion to their original anergic/suppressive state. Abrogation of anergic/suppressor activity of these regulatory cells by treatment with ConA and IL-2 in a mixed population of spleen and lymph node cells from normal mice and adoptive transfer of the mixed population into T cell-deficient mice results in induction of autoimmunity. The demonstration of regulatory CD4 T cells underpins the importance of *dominant* rather than *passive* mechanisms for maintenance of peripheral tolerance to extrathymic antigens.

Mechanism of suppression mediated by regulatory CD4 T cells.   The mechanism responsible for suppression by regulatory CD4 T cells remain largely unknown. The current data suggests that suppression is mediated by competition for antigen or co-stimulatory molecules on the surface of the antigen-presenting cell (Figure 4). However, a role for CD28 and CD40L can be excluded because

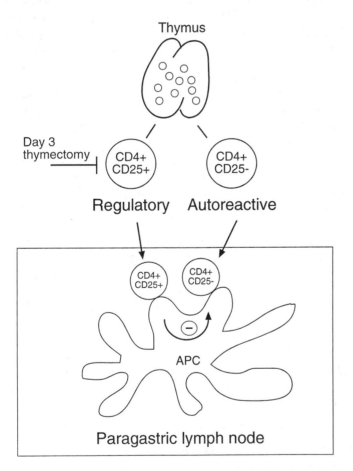

Figure 4.   Regulation by CD4+CD25+ T cells. Autoreactive CD4+CD25− T cells and regulatory CD4+CD25+ regulatory T cells are positively selected in the thymus and exported to the periphery. Neonatal thymectomy prevents emergence of CD4+CD25+ regulatory cells allowing autoreactive CD4+ T cells to expand and initiate disease. Regulation by CD4+CD25+ regulatory cells is not mediated by IL-4, IL-10, TGF-β or other soluble factors and requires that regulatory and effector cells are in contact with the surface of an antigen presenting cell. The signal that induces non-responsiveness in CD4+ effector cells is mediated through the APC.

CD25+CD4 T cells from CD28 and CD40L−/− mice display suppressor activity (31). Induction of CD25 expression on CD25− CD4 T cells *in vivo* or *in vitro* did not result in the generation of suppressor activity. Furthermore, the induction of CD25 expression on a monospecific population of T cells derived from TCR transgenic SCID mice did not result in suppression of post-thymectomy autoimmunity (28). It has been suggested that these observations support the contention that the CD25+ CD4 T cells constitute a distinct lineage of "professional" suppressor cells (30,31).

The suppressor phenomenon observed with the CD25+ CD4 regulatory T cells in autoimmune gastritis is similar to that observed *in vitro* with a human T cell clone and observed *in vivo* as "infectious tolerance". With the *in vitro* observations, cloned human CD4 T cells were first rendered anergic by incubation with soluble peptide in the absence of antigen-presenting cells or by incubation with immobilised antibody to CD3. The anergic T cells caused inhibition of proliferation not only of antigen-specific T cells, but also of T cells with other antigenic specificities, provided the ligands were presented on the surface of the same antigen-presenting cell (33). The mechanism of "linked suppression", otherwise known as "by-stander suppression", also appears to be competition for the surface of the antigen presenting cell and, in this case, for locally-produced interleukin-2 (34). This suppression appeared to be antigen-specific because an anergic clone with a different specificity caused less inhibition of proliferation. Anergic antigen-specific CD4 T cells also inhibited the proliferation of T cells restricted by a different MHC Class II molecule, provided that both restriction elements were presented on the same antigen presenting cell (33). Linked suppression mediated by memory CD4 T cells has also been observed with antigenic determinants presented by MHC class II molecules on the same epithelial cell (34,35).

The *in vivo* observations of suppression induced in CD4 T cells by treatment with non-depleting anti-CD4 antibodies given with antigen or allograft (36,37). Mice made tolerant in this way develop a form of CD4 T cell-mediated immunoregulation that acts to suppress any naïve or primed CD4 or CD8 T cells not only against the same antigen, but also against different ("third party") antigens expressed on the same antigen presenting cell. This form of tolerance was described as "infectious tolerance" presumably because it converts CD4+ T cells that would otherwise have become pathogenic into regulatory T cells. It has been proposed that this form of tolerance develops in the absence of help provided by other pathogenic CD4 T cells ("civil service model"), which then spreads to CD4 T cells reactive with third party antigens linked on the same antigen presenting cell (35). Regulatory CD4 T cells have also been implicated in autoimmune encephalomyelitis (38,39) and in other experimental systems, including oral tolerance, aerosol-induced tolerance and in lymphopenic rats (40).

TCR usage by regulatory CD4 T cells. T cell receptor-(TCR) transgenic mice reactive to MBP develop spontaneous autoimmune encephalomyelitis on a Rag1$^{-/-}$ background (38,39) suggesting that regulatory cells are present in cells with endogenously rearranged TCRs. Observation of differences in disease incidence between TCR transgenic mice on Rag$^{-/-}$ compared to TCR $\alpha^{-/-}$ or TCR β $^{-/-}$ backgrounds indicate that a regulatory CD4 T cell population(s) uses endogenous TCR chains (37). Some preliminary observations suggest the regulatory CD25+ CD4 T cells harbour dual TCRs (30,38). T cells expressing dual receptors have previously been reported (41,42), including self-restricted dual receptor memory T cells (43). Despite advances in our understanding of regulatory CD4 T cells, key questions pertaining to regulatory CD4 T cells remain, including the question of their antigenic specificity, whether they represent a homogeneous or heterogeneous population or distinct lineage(s), whether they express a unique cell surface phenotype and the mechanisms by which they are positively selected in the thymus and by which they mediate suppression.

Pathogenic and regulatory T cells are positively selected in the thymus. The initiation of gastric autoimmunity by neonatal thymectomy and by immunisation with the gastric H/K ATPase indicates that pathogenic T cells have not been deleted in the thymus but have "escaped" to the periphery. It is widely assumed that self-reactive T cells escape deletion because the corresponding self antigens are not represented in the thymus, although there is increasing evidence for the expression of extrathymic antigens in the thymus. In the case of the gastric H/K ATPase, the α-subunit appears to be expressed in the thymus, whereas the β-subunit is not. The presence of pathogenic T cells in the thymus is substantiated by the capacity of thymocytes, particularly CD25$^-$ CD4 thymocytes, to induce gastritis following transfer of these populations to T cell-deficient mice. However, regulatory CD4 T cells are also present in the thymus because the co-transfer of regulatory CD25$^+$ CD4 thymocytes together with pathogenic CD25$^-$ CD4 thymocytes prevents gastritis. These CD25$^+$ CD4 thymocytes comprise 5% of steroid-resistant mature CD4 thymocytes, reside in the thymic medulla and display identical anergic/suppressor activity as the CD25$^+$ CD4 T cells recovered from the periphery (40). These observations suggest that regulatory CD4 T cells are also similarly positively selected in the thymus.

A number of studies suggest these regulatory CD4 T cells are selected on thymic epithelium. For instance, studies with thymic epithelium grafted on to allogeneic athymic mice have shown that the resultant chimeric mice are specifically toler-

ant to grafts of peripheral tissues (e.g., skin and heart) from the thymic epithelium donor strain. Although the chimeric mice harbor peripheral immunocompetent T cells capable of rejecting those grafts, these mice also harbor thymic epithelium-selected CD4 T lymphocytes that inhibited graft rejection by tissue-reactive T cells in immunocompetent recipients. Further the thymic epithelium-selected regulatory T cells were shown to recruit nontolerant tissue-reactive CD4 and CD8 T cells to express similar regulatory functions. Only recent thymic emigrants, but not peripheral resident mature T cells are susceptible to this process of functional education, which also requires exposure to specific antigens and occurs entirely in the periphery (44,45).

4.   MODEL OF AUTOIMMUNE GASTRITIS

Based on the foregoing experimental data derived from mouse models of experimental autoimmune gastritis, we propose the following scenario for autoimmune gastritis (Figure 5).

4.1.   CD4 T Cell-Mediated Tolerance and Autoimmunity to the Gastric H/K ATPase

Our model proposes that autoimmune gastritis is the consequence of pathogenic CD4 T cells activated through their T cell receptor (TCR) by a gastritogenic β subunit peptide presented by MHC Class II molecules of an antigen presenting cell (APC), which may be macrophages and/or dendritic cells. Activation of pathogenic CD4 T cells is controlled by a population(s) of regulatory CD4 T cells. There is some evidence to suggest that control by these regulatory CD4 T is mediated on the surface of the antigen-presenting cell (APC), although the precise mechanism of suppression is not known (Figure 4). We propose these events take place in the draining regional gastric lymph node. This suggestion is supported by the observation that these draining lymph nodes are dramatically enlarged in mice with autoimmune gastritis and contain H/K ATPase reactive T cells as early as 5 weeks after neonatal thymectomy (46). The mechanisms by which the APCs in the stomach are activated is currently unknown. However, we favour the possibility that APCs are activated in the local environment of the stomach by proinflammatory cytokines resulting in the capture of the β-subunit of the H/K ATPase autoantigen released from parietal cells dying as a result of natural cells turnover. Activated APCs then migrate to the regional gastric lymph node, where they present processed β-subunit peptide to CD4 T cells. The activation of the β-subunit-reactive CD4 T cells is then followed by spreading of the response to involve the α-subunit and the recruitment of α-subunit-

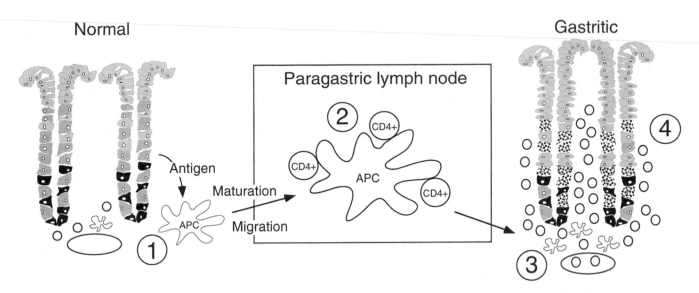

Figure 5. Checkpoints in pathogenesis of experimental autoimmune gastritis. The stages in the initiation and progression of autoimmunity to the parietal and zymogenic cells in the gastric mucosa involve a series of checkpoints that must occur for disease to progress. (1) Local APCs such as dendritic cells, which may constantly be sampling local antigens (including the gastric H/K ATPase) in the gastric glands, are activated through a local inflammatory stimulus. This induces maturation and migration of the dendritic cells to the local draining paragastric lymph node. (2) In the absence of active regulation by CD4+CD25+ regulatory cells, such as in neonatal thymectomy, autoreactive CD4+ T cells are activated and primed to leave the lymph node in search of the target antigen. In the presence of regulatory cells, encounter of autoreactive CD4+ T cells with antigen presented by APCs will be controlled. However, if the stimulus is sufficient to overwhelm regulation, such as by immunisation, then autoimmunity can occur. (3). Following activation, CD4+ effector cells travel through the circulation to the site of inflammation. Controlled by various chemokine and cytokine signals, they migrate across the endothelium into the tissue. (4) Within the gastric gland, the autoreactive CD4+ T cells migrate through the tissue to participate in the cellular destruction associated with autoimmune gastritis. The mechanisms of destruction are not clear, nor is the involvement of other cell types in this destruction.

reactive CD4 T cells. The activated CD4 T cells help H/K ATPase-reactive B cells to produce autoantibodies, and these cells also migrate to the gastric mucosa, resulting in the lesion of autoimmune gastritis (Figure 1). Our model presupposes that gastritis is a consequence of an imbalance favouring pathogenic over regulatory CD4 T cells. The activated CD4 T cells then migrate to the gastric mucosa, traversing the gastric blood vessels using appropriate adhesion molecules and chemokine receptors capable of responding to appropriate chemoattractants (47).

4.2.   Checkpoints in the Progression from Tolerance to Autoimmunity to the Gastric H/K ATPase by CD4 T Cells

Others have proposed that there are likely "checkpoints" (48) in the progression from tolerance to autoimmunity. Based on our model for the genesis of gastric autoimmunity, we suggest that there are minimally at least 4 "checkpoints" present (Figure 5):

Checkpoint 1: Activation of antigen presenting cells
Checkpoint 2: Abrogation of control by regulatory CD4
             T cells

Checkpoint 3: Transmigration of activated CD4 T cells
             across blood vascular barriers
Checkpoint 4: Invasion of activated CD4 T cells into
             connective tissue supporting parenchymal
             cells with the initiation of tissue injury.

The first checkpoint in the initiation of autoimmunity is the activation of antigen-presenting cells. It is currently not known what molecules initiate the activation of these antigen-presenting cells. However, there are a number of candidate cytokines that can fulfill the role of "danger" signals (49) including GM-CSF. GM-CSF is a potent proinflammatory cytokine promoting proliferation, differentiation and activation of macrophages and dendritic cells (50), particularly myeloid-derived dendritic cells (51). Activation of bone marrow-derived macrophages by GM-CSF augments MHC class II expression and antigen presentation (52). Similarly GM-CSF is required for antigen presentation by dendritic cells (53) and augments the primary antibody response by enhancing the function of antigen-presenting cells (54). The second checkpoint requires the pathogenic CD4 T cells to override the control of the regulatory CD4 T cells. This may simply be the result of an excess of pathogenic CD4 T cells over regulatory CD4

T cells or local abrogation of suppression. The third checkpoint is the blood vascular barrier in the stomach, which the activated and pathogenic CD4 T cells have to traverse to reach the gastric mucosa. A recent study has suggested that, during the neonatal period, self-reactive lymphocytes may traffic into non-lymphoid compartments and be rendered tolerant following encounter with "sessile" tissue antigens. With the possible exception of the neonatal period, it is currently widely held that self reactive lymphocytes recirculate only through lymphoid compartments and do not traffic through non-lymphoid tissue (55) in the absence of inflammatory stimuli. The fourth checkpoint is the invasion of the connective tissue which hold the parenchymal cells together followed by the initiation of parenchymal injury. In this context, we have shown increased apoptotic activity in the mucosa of gastritic mice (56).

5. CONCLUSIONS

Studies in man and mice have identified the gastric H/K ATPase as the major molecular target in autoimmune gastritis. Studies in mice indicate that the gastritis is mediated by pathogenic CD4 T cells, which are themselves under the control of regulatory CD4 T cells.

We propose that future directions of research should be directed towards two broad objectives:

1. A better understanding of the immunopathogenesis of autoimmune gastritis and addressing the issue of the 4 checkpoints we have identified above;
2. Development of strategies directed towards the re-establishment of tolerance and the reversal of autoimmunity based on the information obtained from studies carried out to address immunopathogenesis.

The major disadvantage of using non-transgenic animal models for the study of organ-specific autoimmunity is the complexity of the system, especially with respect to the heterogeneity of the TCRs used by pathogenic and regulatory CD4 T cells and the complexity of the cytokines implicated in the initiation of autoimmunity. Therefore, we expect that future studies directed toward unraveling the mechanisms of autoimmunity will increasingly use transgenic approaches.

ACKNOWLEDGEMENTS

Our studies reported in this chapter are supported by grants from the National Health and Medical Research Council of Australia.

References

1. Toh, B.H., J.R. Driel, and P.A. Gleeson. 1997. Pernicious anaemia. *N. Engl. J. Med.* 337:1441–1448.
2. Toh, B.H., P.A. Gleeson, S. Whittingham, and I.R. van Driel. 1998. Autoimmune gastritis and pernicious anemia. N.R. Rose and I.R. Mackay (eds.) Academic Press.
3. Whittingham, S., U. Youngchaiyud, I.R. Mackay, J.D. Buckley, and P.J. Morris. 1975. Thyrogastric autoimmune disease. Studies on the cell-mediated immune system and histocompatibility antigens. *Clin. Exp. Immunol.* 19:289–299.
4. Carmel, R. 1996. Prevalence of undiagnosed pernicious anemia in the elderly. *Arch. Int. Med.* 156:1097–1100.
5. Lindenbaum, J., E.B. Healton, D.G. Savage, J.C. Brust, T.J. Garrett, E.R. Podell, P.D. Marcell, S.P. Stabler, and R.H. Allen. 1988. Neuropsychiatric disorders caused by cobalamin deficiency in the absence of anemia or macrocytosis. *N. Engl. J. Med.* 318:1720–1728.
6. Whittingham, S., I.R. Mackay, and B.D. Tait. 1991. Immunogenetics of pernicious anemia. N.R. Farid (ed.). CRC Press, Inc., Boca Raton, FL.
7. Claeys, D., G. Faller, B.J. Appelmelk, R. Negrini, and T. Kirchner. 1998. The gastric H+,K+-ATPase is a major autoantigen in chronic Helicobacter pylori gastritis with body mucosa atrophy. *Gastroenterology* 115:340–347.
8. Appelmelk, B.J., G. Faller, D. Claeys, T. Kirchner, and G.M.J.E. Vandenbroucke-Grauls. 1998. Bugs on trial: the case of Helicobacter pylori and autoimmunity. *Immunol. Today* 19:296–299.
9. Sprent, J., and H. Kishimoto. 1998. T cell tolerance and the thymus. *Ann. NY Acad. Sci.* 841:236–245.
10. Barrett, S.P., B.H. Toh, F. Alderuccio, I.R. van Driel, and P.A. Gleeson. 1995. Organ-specific autoimmunity induced by adult thymectomy and cyclosphamide-induced lymphopenia *Eur. J. Immunol.* 25:238–244.
11. Claeys, D., E. Saraga, B.C. Rossier, and J.-P. Kraehenbuhl. 1997. Neonatal injection of native proton pump antigens induces autoimmune gastritis in mice. *Gastroenterology* 113:1136–1145.
12. Scarff, K.J., J.M. Pettitt, I.R. van Driel, P.A. Gleeson, and B.H. Toh. 1997. Immunization with gastric H+/K+-ATPase induces a reversible autoimmune gastritis. *Immunology* 92:91–98.
13. Sakaguchi, S., T.H. Ermak, M. Toda, L.J. Berg, W. Ho, B. Fazekas de St. Groth, P.A. Peterson, N. Sakaguchi, and M.M. Davis. 1994. Induction of autoimmune disease in mice by germline alteration of the T cell receptor gene expression. *J. Immunol.* 152:1471–1484.
14. Alderuccio, F., and B.H. Toh. 1998. Spontaneous autoimmune gastritis in C3H/He mice: a new mouse model for gastric autoimmunity. *Am. J. Pathol.* 153:1311–1318.
15. Silvera, P.A., A.G. Baxter, W.E. Cain, and I.R. van Driel. 1999. A major linkage region on distal chromosme 4 confers susceptibility to mouse autoimmune gastritis. *J. Immunol.* 162:5106–5111.
16. Wang, J., N.D. Griggs, K.S. Tung, and J.R. Klein. 1998. Dynamic regulation of gastric autoimmunity by thyroid hormone. *Inter. Immunol.* 10:231–236.
17. Jones, C.M., J.M. Callaghan, P.A. Gleeson, Y. Mori, T. Masuda, and B.H. Toh. 1991. The parietal cell autoantibodies recognised in neonatal thymectomy-induced murine gastritis are the α and β subunits of the gastric proton pump. *Gastroenterology* 101:287–294.

18. de Silva, H.D., I.R. Van Driel, N. La Gruta, B.H. Toh, and P.A. Gleeson. 1998. CD4$^+$ T cells, but not CD8$^+$ T cells, are required for the development of experimental autoimmune gastritis. *Immunology* 93:405–408.

19. Martinelli, T.M., I.R. van Driel, F. Alderuccio, P.A. Gleeson, and B.H. Toh. 1996. Analysis of mononuclear cell infiltrate and cytokine production in murine autoimmune gastritis. *Gastroenterology* 110:1791–1802.

20. Barrett, S.P., P.A. Gleeson, H.D. de Silva, B.H. Toh, and I.R. van Driel. 1996. Interferon-γ is required during the initiation of an orgna-specific autoimmune disease. *Eur. J. Immunol.* 26:1652–1655.

21. Katakai, T., K.J. Mori, T. Masuda, and A. Shimizu. 1998. Differential localization of Th1 and Th2 cells in autoimmune gastritis. *Inter. Immunol.* 10:1325–1334.

22. Alderuccio, F., B.H., Toh, S.S. Tan, P.A. Gleeson, and I.R. van Driel. 1993. An autoimmune disease with multiple molecular targets abrogated by the transgenic expression of a single autoantigen in the thymus. *J. Exp. Med.* 178:419–426.

23. Alderuccio, F., P.A. Gleeson, S.P. Berzins, M. Martin, I.R. van Driel, and B.H. Toh. 1997. Expression of the gastric H/K ATPase α-subunit in the thymus may explain the dominant role of the β-subunit in the pathogenesis of autoimmune gastritis. *Autoimmunity* 25:167–175.

24. Suri-Payer, E., A.Z. Amar, R. Mchugh, K. Natarajan, D.H. Margulies, and E.M. Shevach. 1999. Post-thymectomy auoimmune gastritis: fine specificity and pathogenicity of anti-H/K ATPase-reactive T cells. *Eur. J. Immunol.* 29:669–677.

25. Katakai, T., Y. Agata, A. Shimizu, C. Ohshima, A. Nishio, M. Inaba, S. Kasakura, K.J. Mori, and T. Masuda. 1997. Structure of the TCR expressed on a gastritogenic T cell clone, II-6, and frequent appearance of similar clonotypes in mice bearing autoimmune gastritis. *Inter. Immunol.* 9:1849–1855.

26. de Silva, H.D., P.A. Gleeson, B.H. Toh, I.R. van Driel, and F.R. Carbone. 1999. Identification of a gastritogenic epitope of the H/K ATPase b-subunit. *Immunology* 96:145–151.

27. Sakaguchi, S., M. Sakaguchi, M. Asano, M. Itoh, and M. Toda. 1995. Immunological self-tolerance maintained by activated T cells expressing IL-2 receptor α-chains (CD25)-Breakdown of a single mechanism of self-tolerance causes various autoimmune diseases. *J. Immunol.* 155:1151–1164.

28. Suri-Payer, E., A.Z. Amar, A.M. Thornton, and E.M. Shevach. 1998. CD4+CD25+ T cells inhibit both the induction and effector function of autoreactive T cells and represent a unique lineage of immunoregulatory cells. *J. Immunol.* 160:1212–1218.

29. Asano, M., M. Toda, N. Sakaguchi, and S. Sakaguchi. 1996. Autoimmune disease as a consequence of developmental abnormality of a T cell subpopulation. *J. Exp. Med.* 184:387–396.

30. Itoh, M., T. Takahashi, N. Sakaguchi, Y. Kuniyasu, J. Shimizu, F. Otsuka, and S. Sakaguchi. 1999. Thymus and autoimmunity: Production of CD25$^+$CD4$^+$ naturally anergic and suppressive T cells as a key function of the thymus in maintaining immunological self-tolerance. *J. Immunol.* 162:5317–5326.

31. Thornton, A.M., and E.M. Shevach. 1998. CD4$^+$CD25$^+$ immunoregulatory T cells suppress polyclonal T cell activation *in vitro* by inhibiting interleukin 2 production. *J. Exp. Med.* 188:287–296.

32. Takahashi, T., Y. Kuniyasu, M. Toda, N. Sakaguchi, M. Itoh, M. Iwata, J. Shimizu, and S. Sakaguchi. 1998. Immunologic self-tolerance maintained by CD25$^+$CD4$^+$ naturally anergic and supressive T cells: induction of autoimmune disease by breaking their anergic/suppressive state. *Inter. Immunol.* 10:1969–1980.

33. Lombardi, G., S. Sidhu, P. Batchelor, and R. Lechler. 1994. Anergic T cells as suppressor cells *in vitro*. *Science* 264:1587–1589.

34. Frasca, L., P. Carmichael, R. Lechler, and G. Lombardi. 1997. Anergic T cells effect linked suppression. *Eur. J. Immunol.* 27:3191–3197.

35. Marelli-Berg, F.M., A. Weetman, L. Frasca, S.J. Deacock, N. Imami, G. Lombardi, and R.I. Lechler. 1997. Antigen presentation by epithelial cells induces anergic immunoregulatory CD45RO$^+$ T cells and deletion of CD45RA$^+$ T cells. *J. Immunol.* 159:5853–5861.

36. Cobbold, S., and H. Waldmann. 1998. Infectious tolerance. *Curr. Opin. Immunol.* 10:518–524.

37. Waldmann, H., and S. Cobbold. 1998. How do monoclonal antibodies induce tolerance? A role for infectious tolerance? *Ann. Rev. Immunol.* 16:619–644.

38. Olivares-Villogomez, D., Y. Wang, and J.J. Lafaille. 1998. Regulatory CD4$^+$ T cells expressing endogenous T cell receptor chains protect myelin basic protein-specific transgenic mice from spontaneous autoimmune encephalomyelitis. *J. Exp. Med.* 188:1883–1894.

39. Van de Keere, F., and S. Tonegawa. 1998. CD4$^+$ T cells prevent spontaneous experimental autoimmune encephalomyelitis in anti-myelin basic protein T cell receptor transgenic mice. *J. Exp. Med.* 188:1875–1882.

40. Mason, D., and F. Powrie. 1998. Control of immune pathology by regulatory T cells. *Curr. Opin. Immunol.* 10:649–655.

41. Padovan, E., C. Giachino, M. Cella, S. Valitutti, O. Acuto, and A. Lanzavecchia. 1995. Normal T lymphocytes can express two different T cell receptor beta chains: implications for the mechanism of allelic exclusion. *J. Exp. Med.* 181:1587–1591.

42. Heath, W.R., and J.F.A.P. Miller. 1993. Expression of two α chains on the surface of T cells in T cell receptor transgenic mice. *J. Exp. Med.* 178:1807–1811.

43. Lee, W.T., V. Shiledar-Baxi, G.M. Winslow, D. Mix, and D.B. Murphy. 1998. Self-restricted dual receptor memory T cells. *J. Immunol.* 161:4513–4519.

44. Modigliani, Y., A. Coutinho, P. Pereira, N. Le Douarin, V. Thomas-Vaslin, O. Burlen-Defranoux, J. Salaun, and A. Bandeira. 1996. Establishment of tissue-specific tolerance is driven by regulatory T cells selected by thymic epithelium. *Eur. J. Immunol.* 26:1807–1815.

45. Modigliani, Y., A. Bandeira, and A. Coutinho. 1996. A model for developmentally acquired thymus-dependent tolerance to central and peripheral antigens. *Immunological Reviews* 149:155–120.

46. Suri-Payer, E., P.J. Kehn, A.W. Cheever, and E.M. Shevach. 1996. Pathogenesis of post-thymectomy autoimmune gastritis: Identification of anti-H/K Adenosine triphosphate-reactive T cells. *I. Immunol.* 157:1799–1805.

47. Sallusto, F., A. Lanzavecchia, and C.R. Mackay. 1998. Chemokines and chemokine receptors in T-cell priming and Th1/Th2-mediated responses. *Immunol. Today* 19:568–574.

48. Andre, I., A. Gonzalez, B. Wang, J. Katz, C. Benoist, and D. Mathis. 1996. Checkpoints in the progression of autoimmune disease: Lessons from diabetes models. *Proc. Natl. Acad. Sci. USA* 93:2260–2263.

49. P. Matzinger. 1998. An innate sense of danger. Seminars in Immunology, 10:399–415.

50. Metcalf, D. 1991. The Florey Lecture, 1991. The colony-stimulating factors: discovery to clinical use. Philosophical Transactions of the *Royal Society of London—Series B: Biological Sciences* 333:147–173.

51. Pulendran, B., J.L. Smith, G. Caspary, K. Brasel, D. Pettit, E. Maraskovsky, and C.R. Maliszewski. 1999. Distinct dendritic cell subsets differentially regulate the class of immune response *in vivo. Proc. Natl. Acad. Sci. USA* 96:1036–1041.

52. Fischer, H.G., S. Frosch, K. Reske, and A.B. Reske-Kunz. 1988. Granulocyte-macrophage colony-stimulating factor activates macrophages derived from bone marrow cultures to synthesis of MHC class II molecules and to augmented antigen presentation function. *J. Immunol.* 141:3882–3888.

53. Sallusto, F., and A. Lanzavechia. 1994. Efficient presentation of soluble antigen by cultured human dendritic cells is maintained by granulocyte/macrophage colony-stimulating factor plus interleukin 4 and downregulated by tumor necrosis factor alpha. *J. Exp. Med.* 179:1109–1118.

54. Morrissey, P.J., L. Bressler, L.S. Park, A. Alpert, and S. Gillis. 1987. Granulocyte-macrophage colony-stimulating factor augments the primary antibody response by enhancing the function of antigen-presenting cells. *J. Immunol.* 139:1113–1119.

55. Mondino, A., A. Khoruts, and M.K. Jenkins. 1996. The anatomy of T-cell activation and tolerance. *Proc. Natl. Acad. Sci. USA* 93:2245–2252.

56. Judd, L.M., P.A. Gleeson, B.H. Toh, and I.R. van Driel. 1999. Autoimmune gastritis results in disruption of gastric epithelial cell development. *Am. J. Physiol.* 277:G209–G18.

50 | An Experimental Model of Autoimmune Gastritis

Tomoya Katakai, Tohru Masuda and Akira Shimizu

1. INTRODUCTION

Murine autoimmune gastritis (AIG) is induced experimentally by thymectomy on day 3 after birth (d3-Tx) in BALB/c mice (1,2). It resembles human AIG associated with pernicious anemia, which is a paradigm of organ-specific autoimmune diseases including insulin-dependent diabetes mellitus (IDDM) and Hashimoto's thyroiditis. Histopathologically, there is selective loss of parietal cells from the corpus gland as well as lymphocyte infiltration and the production of autoantibodies to parietal cells (2,3). Although the etiology of autoimmune diseases are largely unclear at present, the studies using this murine AIG model has demonstrated several important findings for elucidating the pathogenesis of organ-specific autoimmune diseases.

The first issue concerns the cellular mechanism. The murine disease is cell-mediated because it can be transferred to syngeneic nude mice by implantation of CD4+ T cells, but not by autoantibodies, of the diseased mouse (4,5). The effecter is parietal cell-specific, belonging to Th1 subset, as in the case of experimental allergic encephalomyelitis (EAE) and IDDM in NOD mice (6). Importantly, the disease onset is inhibited by CD4+ T cells, which exist in the non-diseased mice (7–10). Therefore, cellular interactions within CD4+ T cell subsets seem to maintain peripheral tolerance to organ-specific autoimmune diseases. Indeed, it has been recently reported that the CD4+ regulatory T cells express CD25 (IL-2 receptor α-chain), while pathogenic CD4+ T cells lack it (11,12).

The second concerns autoantigen molecules. There is accumulating evidence from human pernicious anemia that the major major autoantigens in AIG is H^+/K^+-ATPase (proton pump), produced by parietal cells (13). By estab-

lishing hybridomas producing monoclonal anti-parietal cell autoantibodies (14,15) and gastritogenic Th1 clone, IL-6, (6) from AIG BALB/c mice, we determined at least two B cell epitopes on the α- and β-subunits, respectively, and one T cell epitope on the α, all of which are interspecies-specific. It is of note that these epitopes are located in the lumen of the tubulovesicles in parietal cells, where the immune system cannot directly contact (14).

These observations have raised the questions of the molecular mechanism that destroy the target cells, and how autoantigen-specific Th1 and probably Th2 cells infiltrate into the fundic glands and recognize parietal cells. Relative to the former problem, we found that MHC class II together with Fas and ICAM-1 molecules are hyperexpressed on parietal cells at the initial stage of inflammation. In addition, parietal cell-specific Th1 cells, IL-6, express FasL in response to the antigen stimulation, thereby killing antigen-presenting cells tagged with antigen peptide *in vitro*. These findings suggest that cognate interaction between target cells and Th1 killers (16) using abberant class II and adhesion molecules leads target cell lysis using Fas and FasL (17).

This chapter will summarize recent work on the molecular mechanism of Th1 and Th2 responses for the development of AIG in the gastric lesion.

2. CLONALITY ANALYSIS OF GASTRITOGENIC T CELLS IN AIG

2.1. TCR Structure of a Gastritogenic T Cell Clone, IL-6

Complementary DNAs encoding the TCR α- and β-chains expressed by IL-6 cells, which respond to a peptide locating the amino acid position 891 to 905 of the α-subunit of H^+/K^+-ATPase ($^\alpha$891–905) as the antigen, were cloned and sequenced (18) by the 5'-RACE method using C_α-specific primers, and the RT-PCR method using $V_\beta14$- and C_β-specific primers, respectively. After 5'-RACE, we analyzed 12 clones of TCR-α and found 11 of them to be identical, encoding a functional mRNA consisting of $V_\alpha10$-J_αc5a-C_α. The other encodes non-functional $V_\alpha4$-J_αTA28-C_α by a frame-shift at the V-J junction. For TCR β, the sequence of the RT-PCR product was directly determined and confirmed after cloning. Expression of two α and β mRNAs in IL-6 cells was confirmed by RT-PCR using specific V and J region primers synthesized from the determined sequences. Consequently, the TCR clonotype of IL-6 was determined as $V_\alpha10$-J_αc5a-C_α for the α chain and $V_\beta14$-$J_\beta2.3$-$C_\beta2$ for the β chain. V(D)J junctions of the α and β mRNAs cloned from IL-6 cells are shown in Figure 1. It is noted that there are no N insertions in these three genes, suggesting a rather early origin of the IL-6 cells during the development.

Determination of the primary structure of the TCR expressed on the IL-6 clone enabled us to trace this clone specifically with high sensitivity by RT-PCR. We could easily detect IL-6-specific TCR α and β mRNAs in the gastric mucosa (GM) of nude mice transferred IL-6 cells (18). Untransferred nude or normal mice have 10- to 20-fold less amount of TCR β mRNA in total, and have undetectable levels of mRNA specific to the IL-6 TCR. These results imply that, when transferred into nude mice, IL-6 cells home into the GM where few T cells are normally present, resulting in the destruction of parietal cells. These data also indicate that T cells carrying TCR using the same V_α-J_α or V_β-J_β genes to those of IL-6 cells are rare among the T cells in GM of normal mice.

2.2. Frequent Appearance of T Cells Bearing IL-6-like TCR β-Chain in AIG Induced by Neonatal Thymectomy

We next addressed the question of whether T cells expressing the TCR clonotype of IL-6 cells or closely related with it reproducibly appear at the foci of gastritis and are actually involved in the pathogenesis of mouse AIG induced by d3-Tx, and thus whether the antigenic peptide recognized by IL-6 cells acts as a dominant antigen spontaneously inducing AIG in these mice.

To examine the use of IL-6-like clonotypes in d3-Tx AIG mice, cDNA amplified by nested PCR using $V_\beta14$ gene

TCR α (Vα10-Jαc5a)

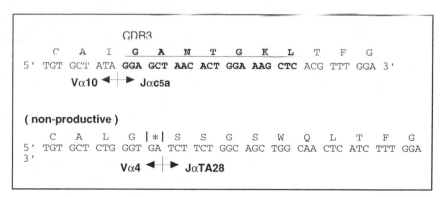

TCR β (Vβ14-Jβ2.3)

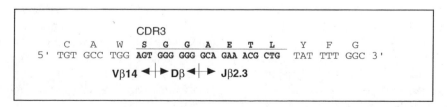

Figure 1. Primary structures of CDR3 regions of TCR α- (upper box) and β- (lower box) chains expressed in IL-6 cells. Nucleotide and deduced amino acid sequences are given in single letter notations with the putative V(D)J junctional points shown.

specific and $J_{\beta}2.3$- or C_{β}-gene specific primers was analyzed. The nested PCR products using $V_{\beta}14$-C_{β} primer combination were further cloned and CDR3 junctional sequences of them were determined (18). Predicted amino acid sequences in junctional regions of TCR β of each clone are shown in Figure 2. From eight of 14 AIG mice analyzed, specific RT-PCR products were detected using $V_{\beta}14$-$J_{\beta}2.3$ primers, suggesting a preferential usage of $J_{\beta}2.3$ for $V_{\beta}14^+$ TCR and IL-6-like TCR β in these animals. As expected, preferential usage of $J_{\beta}2.3$ was seen in all the animals analyzed in which $V_{\beta}14$-$J_{\beta}2.3$ specific RT-PCR products were found. In an AIG mouse at 4 weeks old (Figure 2a), the sequence of six among the eleven clones analyzed completely coincided with that of IL-6 cells. Three of the remaining five clones also gave rise to an almost identical structure with that of IL-6 with only one amino acid (nucleotide) difference of a serine to a threonine (G to C in nucleotide). We designated these two at the amino acid level as the IL-6-like clonotype since they have the identical lengths and sequences at the CDR3 loop except for a homologous amino acid change at the V-D junction. Two other AIG mice at 6 and 12 weeks old shown in Figure 2 (c and d) also demonstrated a similar pattern of CDR3 sequence. Such a preferential use of IL-6-like TCR β genes (more than one-third, up to 100% of analyzed cDNA clones encodes the IL-6-like clonotype) was seen in six of total 14 (43%) AIG mice aged 4–12 weeks old (Table 1). In only two animals, which showed preferential use of $J_{\beta}2.3$ with $V_{\beta}14$, was the IL-6-like clonotype not detected (Figure 2b and Table 1). The other AIG mice analyzed (6 of 14) gave no RT-PCR product using $V_{\beta}14$-$J_{\beta}2.3$-specific primers, and natural preferential usage of $J_{\beta}2.3$ or IL-6-like clonotypes was not seen in the two such animals from which $V_{\beta}14$-C_{β} PCR products were sequenced (Table 1).

These results indicate that cells with the IL-6-like clonotype indeed repeatedly appear in the GM of AIG mice and that this appearance is AIG-specific, since TCR β mRNA encoding even the same V_{β}-J_{β} combination to that of IL-6 was undetectable by RT-PCR in normal GM. They also strongly suggest that the IL-6 cells represent the character of one such rather dominant cell involved in the AIG and are not from the artificially amplified population during the long-term culture with antigen-peptide stimulation *in vitro*. Thus, our finding in this study might support the possibility that cells with the IL-6-like clonotype are actively involved in pathogenesis of AIG, although they are not essential since more than half of AIG mice have no detectable levels of mRNA encoding the IL-6-like TCR.

There was a fairly good coincidence between the preferential expression of IL-6-like TCR β in the GM and proliferative responses to the $^{\alpha}891$–905 peptide of H^+/K^+-ATPase by splenic T cells, although the number of examples is still small. This coincidence might suggest that T cells of the IL-6-like clonotype actually respond to the $^{\alpha}891$–905 peptide in such AIG animals. These findings imply that effecter T cells bearing the same or a similar clonotype to IL-6 cells are actively involved in the pathogenesis of AIG, thus $^{\alpha}891$–905 includes at least one of the dominant gastritogenic epitopes, and that these clones are probably selected when mice underwent thymectomy.

In addition to autoantigens on the α-subunit recognized by the cells such as those characterized in the this study, involvement of the β-subunit antigen in the pathogenesis of AIG has been suggested from the fact that intrathymic expression of the β-subunit of H^+/K^+-ATPase molecules by gene transfection results in tolerance (19). Taking this into consideration, multiple epitopes on both the α- and β-subunits of proton pump seem to be used as gastritogenic antigens in AIG.

3.   Th1 DOMINANT ENVIRONMENT AT THE FOCI OF GASTRITIS

3.1.   Predominant Localization of Th1 cells in Gastric Mucosa of AIG

Due to d3-Tx, a majority of T cells in the whole body in AIG mice have activated/memory phenotype as CD45RB^{low}, L-selectin$^{-/low}$, CD44^{high} and LFA-1^{high}, although the number of T cells are decreased (20). Among the lymphoid organs, gastric lymph nodes(GLN) have an especially high content of such T cells (~70% of CD4$^+$ cells) showing activated/memory phenotype. The foci of gastritis, GM, have even higher content of CD45RB^{low} cells (>90% of CD4$^+$ cells). On the other hand, autoantibodies specific to GM components is produced in a relatively restricted area, GLN and GM, in which higher contents of CD4$^+$ T cells with an activated/memory phenotype were observed. Therefore, further reactions against GM self-antigens seem to occur at the GM and its draining LN (*i.e.*, GLN), and thus these organs are probably the major sites of autoimmune reaction in AIG (20).

Tissue destructive inflammation at sites of GM and anti-GM autoantibody production co-exist in our autoimmune model, and it seems that disease progression correlates to autoantibody titers, although autoantibody *per se* cannot cause the disease (our unpublished data and 21,22). In addition, autoantibodies not only of IgG2a subclass, production of which is augmented by IFN-γ (23), but also of IgG1 subclass, class switching to which IL-4 enhances (23), were detected in the sera of AIG mice. These findings strongly suggested that not only Th1, but also Th2 specific to GM self-antigens are sufficiently activated in the same individuals carrying AIG, but in a relatively restricted area, *i.e.*, GLN and GM.

These observations, however, seem to be rather contradictory in the fact that the functions of two subsets are

Il-6

V	CDR3 loop	J	
C A W TGTGCCTGG	S G G A E T L AGT<u>**GGGGGG**</u>GCAGAAACGCTG	Y F G TATTTTGGC	Jβ2.3

a 4W-2

incidence	V	CDR3 loop	J	
◆ 6/11	C A W TGTGCCTGG	S G G A E T L AGT<u>**GGGGGG**</u>GCAGAAACGCTG	Y F G TATTTTGGC	Jβ2.3
◆ 3/11	C A W TGTGCCTGG	T G G A E T L A<u>**CTGGGGGG**</u>GCAGAAACGCTG	Y F G TATTTTGGC	Jβ2.3
1/11	C A W TGTGCCTGG	D W G E N E Q **<u>GACTGGGGGGAAA</u>**ATGAACAG	Y F G TACTTCGGT	Jβ2.6
1/11	C A W TGTGCCTGG	S L H S G N T L AGTCT**<u>ACA</u>**TTCTGGAAATACGCTC	Y F G TATTTTGGA	Jβ1.3

b 6W-1

incidence	V	CDR3 loop	J	
5/16	C A W TGTGCCTGG	S L G L G G E T L AGTCT**AGGGCTGGGGGGC**GAAACGCTG	Y F G TATTTTGGC	Jβ2.3
2/16	C A W TGTGCCTGG	S L D W G A A E T L AGTCT**AGACTGGGGGGC**TGCAGAAACGCTG	Y F G TATTTTGGC	Jβ2.3
3/16	C A W TGTGCCTGG	S R N R T N S D Y AGTC**GAAACAGGACA**AACTCCGACTAC	T F G ACCTTCGGC	Jβ1.2
3/16	C A W TGTGCCTGG	S R D S Q N T L AG**CCGGGAC**AGTCAAAACACCTTG	Y F G TACTTCGGT	Jβ2.4
1/16	C A W TGTGCCTGG	S S G A N S P L AGT**TCAGGGGC**TAATTCGCCCCTC	Y F A TACTTCGCG	Jβ1.6
1/16	C A W TGTGCCTGG	S L H S G N T L AGTCT**CCAT**TCTGGAAATACGCTC	Y F G TATTTTGGA	Jβ1.3
1/16	C A W TGTGCCTGG	S L D R D N Q D T Q AGTCT**AGACCGGGAT**AACCAAGACACCCAG	Y F G TACTTTGGG	Jβ2.5

c 6W-2

incidence	V	CDR3 loop	J	
◆ 8/13	C A W TGTGCCTGG	T G G A E T L A**<u>CTGGGGG</u>**TGCAGAAACGCTG	Y F G TATTTTGGC	Jβ2.3
2/13	C A W TGTGCCTGG	R G Q G Y E Q AG**<u>GGACAGGGA</u>**TATGAACAG	Y F G TACTTCGGT	Jβ2.6
1/13	C A W TGTGCCTGG	S L G G N Y A E Q AGC**CTTGGGGGG**AACTATGCTGAGCAG	F F G TTCTTCGGA	Jβ2.1
1/13	C A W TGTGCCTGG	S S Q L G D A E Q AGT**TCCCAACTGGGGG**ATGCTGAGCAG	F F G TTCTTCGGA	Jβ2.1
1/13	C A W TGTGCCTGG	S L Y N S P L AGTCT**A**TATAATTCGCCCCTC	Y F A TACTTCGCG	Jβ1.6

d 12W-1

incidence	V	CDR3 loop	J	
◆ 7/12	C A W TGTGCCTGG	T G G A E T L A**<u>CTGGGGG</u>**TGCAGAAACGCTG	Y F G TATTTTGGC	Jβ2.3
◆ 3/12	C A W TGTGCCTGG	T G G A E T L A**<u>CTGGGGGG</u>**GCAGAAACGCTG	Y F G TATTTTGGC	Jβ2.3
1/12	C A W TGTGCCTGG	S P G T G G G E T L AGT**CC<u>GGGGACTGGGGGGGG</u>**AGAAACGCTG	Y F G TATTTTGGC	Jβ2.3
1/12	C A W TGTGCCTGG	S T G G H T G Q L AGT**<u>CTGGGGGGG</u>**CACACCGGGCAGCTC	Y F G TACTTTGGT	Jβ2.2

Figure 2. Structure at the CDR3 region of $V_{\beta}14^{+}$ TCR expressed in the GM of AIG mice between 4 and 12 weeks old age. Complementary DNAs encoding $V_{\beta}14^{+}$ TCR were amplified by nested RT-PCR and cloned, and random clones were sequenced. Nucleotide and deduced amino acid sequences are shown with their incidence (left end) and J_{β} usage (right end). The 3′ and 5′ ends of V and J segments, respectively, (putative junctional points) are indicated by arrowheads. D regions are underlined.

Table 1   Frequent appearance of the IL-6-like clonotype

Animals	Age (weeks after d3-Tx)	Number of sequenced clones of Vβ14-Cβ PCR products			Vβ14-Jβ2.3 PCR product
		Total	Jβ2.3	IL-6-like	
4W-1	4	ND	–	–	–
4W-2	4	11	9	9	+
4W-3	4	ND	–	–	–
6W-1	6	16	7	0	+
6W-2	6	13	8	8	+
8W-1	8	12	5	4	+
8W-2	8	12	6	0	+
8W-3	8	10	0	0	–
12W-1	12	12	11	10	+
12W-2	12	ND	–	–	–
12W-3	12	10	4	4	+
12W-4	12	13	13	13	+
12W-5	12	10	0	0	–
12W-6	12	ND	–	–	–

ND; not done.

mutually exclusive (24–28), and raised the question whether the Th1 and Th2 subsets of CD4+ cells are activated at the same place in AIG mice, especially at GM where sub-organic compartmentation is difficult to imagine. Therefore, we analyzed the location of these subsets in AIG mice using expression of IFN-γ and IL-4 as markers for Th1 and Th2 cells, respectively (20). Both CD4+ CD45RB[high] (naive) and CD4+ CD45RB[low] (activated/memory) T cells were purified from GM or GLN cells by cell sorting, and mRNA expression of IFN-γ and IL-4 in both populations was assessed by RT-PCR followed by Southern hybridization. Little cytokine mRNA was detected in the CD4+ CD45RB[high] populations from every source, in agreement with their "naive" phenotype. In the case of the CD4+ CD45RB[low] population, patterns of mRNA expression were different from each source. The cells from GLN of AIG mice expressed both IFN-γ and a high level of IL-4 mRNAs. In sharp contrast, those from infiltrating CD4+ T cells into the inflamed GM expressed much less IL-4 mRNA, but an increased level of IFN-γ mRNA compared to GLN. CD4+ CD45RB[low] cells sort-purified from LN of normal mice, which are ~20% of total CD4+ cells in LN, also expressed mRNAs of both cytokines in lower amounts than those produced by GLN CD4+ CD45RB[low] cells in AIG.

We further confirmed the expression of IFN-γ or IL-4 in CD4+ T cells as protein at the single cell level by an intracellular staining technique (20). Roughly 20–30% of CD4+ T cells from spleen, GLN and GM of AIG mice expressed detectable levels of either IFN-γ or IL-4 (Figure 3), showing that the majority of CD4+ T cells in AIG mice have an activated/memory phenotype. On the other hand, <6% and ~3% of CD4+ T cells from normal spleen and LN, respectively, expressed such cytokines (Figure 3), also correlating with the finding that CD4+ CD45RB[low]-activated/memory cells were <30% of normal spleen or LN cells. Among cells in GM of AIG mice producing either cytokine, less than one-seventh (3.2% of total CD4+ cells) expressed IL-4 and most of them (20.7% of total CD4+ cells) expressed IFN-γ. In contrast, significant populations, roughly half (14.2% of total CD4+ cells) and one-third (6.6% of total CD4+ cells) of such cells in GLN and spleen of AIG mice, respectively, expressed IL-4 (Figure 3).

These results, both on the expression pattern of the landmark cytokine mRNAs and on the distribution of such cytokine-expressing cells observed by cytoplasmic staining (Figure 3), are quite consistent with each other, and clearly indicated that IFN-γ producers, i.e., Th1 cells, are dominant at GM and IL-4 producers, i.e., Th2 cells are mainly resident in GLN.

Our observations are not from the autoreactive T cells alone, but from the total activated T cells at the inflammation sites. However, autoreactive T cells might not be insignificant among them, since mRNA encoding TCR very similar to that of a T cell clone specific to proton pump (autoantigen) is frequently detectable in GM of mice bearing AIG after d3-Tx, as noted above. Therefore, it is possible to assume that autoreactive Th1 and Th2 cells in this AIG model also show similar differential localization, although formal evidence is lacking. This differential localization of Th1/Th2 subsets even at closely located, but anatomically distinct, sites for autoimmune gastritis, i.e., inflamed lesions of GM and its draining GLN, might explain why tissue destructive inflammation can occur at GM in spite of the fact that the two subsets co-exist in the same animal. Th1 cells predominantly localized in GM could further be activated auto antigen specifically or non-specifically without inhibitory effects of Th2 cells. On the other hand, both Th1 and Th2 cells were observed in GLN, and this apparent co-activation of both subsets might be managed by sub-organic structures such as germinal centers or follicles.

3.2. Selective Transmigration of Th1 Cells from AIG Through Endothelial Cells *In Vitro*

Since the primary cause of infiltration of effector cells into the inflamed tissue is migration of such cells across the blood endothelial wall into the region of inflammation (29), the difference in transmigration ability between two subsets of CD4+ T cells into the inflamed lesion could be one of the reasons for the selective accumulation of Th1, but not Th2, cells in the GM of AIG mice described above. To test the possibility that the two subsets have different trans-endothelial migration (TEM) abilities, we set up an assay

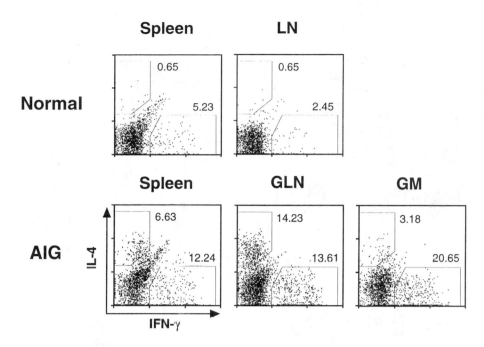

Figure 3. IFN-γ and IL-4-producing CD4+ T cells in lymphoid organs and inflamed GM of AIG mice detected by intracellular staining. Lymphocytes derived from normal or AIG mice were first stimulated by phorbol myristate acetate and ionomycin *in vitro*, followed by the treatment with monensin to stop the secretion of cytokines from the Golgi apparatus, and such cytokines trapped into Golgi were stained together with CD4 antigen on the cell surface. Fluorescence intensities after anti-IFN-γ (horizontal axes) and anti-IL-4 (vertical axes) staining of the cells in CD4+ fraction from the organs indicated are shown as dot plots. Percentages of positively-stained cells (in the enclosed areas) are indicated in the panels.

system using a murine endothelial cell line, F-2, as separator layer cells (Figure 4). Similar systems have previously been shown to be a useful for studying TEM (30,31). In our system, CD4+ T cells prepared from AIG mice were applied onto the confluent monolayer of F-2 cells growing on a pored polycarbonate membrane that is topologically equivalent to inner blood vessels, and after 4 h of incubation, ~5–10% of input T cells reproducibly migrated across the endothelial monolayer and could be collected from the lower compartment equivalent to the tissue fluid of outer blood vessels (Figure 4a). We analyzed the expression levels of IFN-γ and IL-4 mRNAs of these transmigrated cells and compared them to those of input cells or cells remaining in the upper compartment. The transmigrated cells expressed substantial amounts of IFN-γ, but little or no detectable IL-4 mRNA, in contrast to input cells or cells in the upper compartment, which expressed plenty of both types of cytokines. Almost the same results were obtained using cells from spleen and GLN of AIG mice. This ability to support the selective transmigration seems to be specific to some kinds of endothelial cells, since very few cells were recovered from the lower compartment when NIH 3T3 cells were used as the separator layer instead of F2 cells. Furthermore, in such cells (<1% of input cells) recovered from the lower compartment using NIH 3T3, neither IFN-γ nor IL-4 mRNA were detected by RT-PCR even

after normalization of the RNA amount. On the other hand, no bias to the IFN-γ dominancy was seen in the cells that simply passed through the Transwell-pored membrane without a cell monolayer. Intracellular staining of IFN-γ was also performed to analyze the expression at the single cell level, as the marker for activated Th1 cells. As shown in Figure 4b, IFN-γ producing Th1 cells were highly concentrated in the transmigrated compartment of the TEM assay using spleen CD4+ cells from AIG mice.

4. CONCLUSIONS

Murine AIG induced by neonatal thymectomy is one of the best studied experimental systems for autoimmunity. There are great advantages to this model from the abundant basic investigations described in this chapter with regard to the major target antigens, autoantibodies, effector cells and reconstitution of disease by transferring the effector cells to immuno deficient mice (1–4,6,10,14,15,17,21,22,32,33).

We investigated CD4+ T cells the major effector cells of AIG, in AIG mice, from the standpoint of the localization of its subsets, Th1 and Th2, upon which a hypothesis was based that the difference in TEM ability between these two subsets might play an important role in formation of inflammatory autoimmune lesions. In other words, autoreactive (and non-

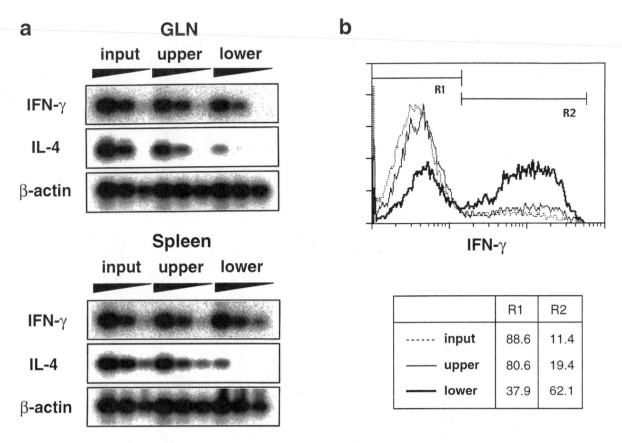

Figure 4. (a) Expression of IFN-γ and IL-4 mRNAs in CD4⁺ T cells purified from GLN or spleen of AIG mice after the TEM assay. Total RNAs were extracted from CD4⁺ T cells recovered from the upper or lower compartment of Transwell plates after the TEM assay, or input cells kept in culture medium and subjected to RT-PCR to quantify mRNAs. cDNAs serially diluted 4 times were used for PCR, and starting amounts of cDNAs for the dilution were normalized by the amounts of β-actin mRNA. (b) Expression of IFN-γ in the individual cells recovered from the TEM assay. Spleen CD4⁺ T cells purified from AIG mice were subjected to the TEM assay, and expression of IFN-γ was detected by intracellular staining and flow cytometry. Fluorescence intensities after anti-IFN-γ staining in CD4⁺ fraction of the cells recovered from the upper and lower compartments or input cells were plotted as histogram. Percentages of IFN-γ negative and positive cells in the region R1 and R2, respectively, are also shown.

specific) Th1 cells might promote self-tissue destruction at the initial phase of disease due to infiltration into the inflamed lesions compared with Th2 cells, which are resident in draining lymph nodes and thus lack the inhibitory effects of Th2 cytokines at the lesions (Figure 3). This model can apply to not only the other autoimmune diseases but immune responses observed in common. Although the molecular mechanisms implicated in this phenomenon remain to be elucidated, several adhesion molecules as well as chemokines might be responsible for differential TEM because the sequential binding and the activation of T cells are required in this process (34). Indeed, we recently found that major part of infiltrating CD4⁺ T cells in GM of AIG strongly expresses either ligand for P-selectin or α4β7-integrin, and that mRNAs encoding chemokines whose receptors are rather specifically expressed on Th1 cells are expressed in GM of AIG mice (Katakai et al. manuscript in preparation).

The cloned T cells with defined TCR clonotype and antigen specificity that represent the major population of cells causing AIG, such as IL-6 (Th1), are very useful tools for analyzing the movement of effector T cells *in vitro* and *in vivo*. Actually, IL-6 cells migrate to GM and induce AIG in nude mice by adoptive transfer. Analyses in TEM system *in vitro* or administration *in vivo* using such clones might help us to understand the molecular mechanisms of the localization and activation of CD4⁺ T cells.

CD4⁺ T cells play a central role in controlling the antigen-specific immune responses by secreting various inter-cellular signaling molecules such as cytokines (24,25). Therefore, it is thought that the major part of autoimmune diseases are caused by the failure of tolerance associated with CD4⁺ T cells. The developing CD4⁺ T cells, together with CD8⁺ T cells, which are reactive to self-components expressed in the thymus, have been shown to be eliminated by 'clonal deletion' (35,36). However, since not all self-antigens are presented there, extrathymic or peripheral mechanisms such as clonal anergy or active suppression are considered to maintain T cell tolerance (37–40), although the fine mechanisms of this peripheral tolerance is not fully understood. Therefore, it is important to determine the fine characters of the effector T cells, including their localization and clonotypes.

It is also not clear why autoimmunity is induced, or why the tolerance associated with CD4[+] T cells is broken by neonatal thymectomy. Since the immune system develops and matures dramatically after the perinatal period, and hence is different in its nature in neonate from the adult life (41–47), it is natural that thymectomy at this period can disturb the subsequent maturation of whole immune system. However, this concept is not of simple as it may first appear. Disease development is controlled by several genetic factors, since there are many murine strains that do not develop AIG and/or other organ-specific autoimmunity even after 3d-Tx (1,2,32). In addition, the observation that a subpopulation of CD4[+] T cells in euthymic normal mice can suppress the development of AIG (4,11,12,21,40,48) indicates that the regulatory cells maintaining some part of peripheral tolerance are also CD4[+] T cells, as are the cells causing the disease. Characterization of these interesting phenomena in 3d-Tx-induced autoimmunity will lead us to essential knowledge on self-tolerance mediated by both the thymus and the periphery.

The extreme complexity of the immune system, composed as it is of various kinds of cells and molecules, makes it difficult to elucidate the etiology of autoimmune diseases, and hence the fundamental methods of treatment and prevention for either organ-specific diseases, such as AIG, or general autoimmune diseases like SLE, is still unclear. Therefore, it is essential to investigate the fine molecular mechanisms that initiate the onset and development of autoimmune diseases using simplified experimental models, like the one shown here, and to extend and to generalize these results to the underlying principles.

ACKNOWLEDGEMENTS

We are grateful to Prof. K.J. Mori for his continuous encouragement.

References

1. Kojima, A., and R.T. Prehn. 1981. Genetic susceptibility to post-thymectomy autoimmune disease in mice. *Immunogenetics* 14:15–27.
2. Mori, Y., M. Hosono, K. Murakami, Y. Yoshikawa, K. Kuribayashi, R. Kannagi, M. Sakai, M. Okuma, and T. Masuda. 1991. Genetic studies on experimental autoimmune gastritis induced by neonatal thymectomy using recombinant inbred strains between a high-incidence strain, BALB/c, and a low-incidence strain, DBA/2. *Clin. Exp. Immunol.* 84:145–152.
3. Fukuma, K., S. Sakaguchi, K. Kuribayashi, W.L. Chen, R. Morishita, K. Sekita, H. Uchino, and T. Masuda. 1988. Immunologic and clinical studies on murine experimental autoimmune gastritis induced by neonatal thymectomy. *Gastroenterology* 94:274–283.
4. Sakaguchi S., K. Fukuma, K. Kuribayashi, and T. Masuda. 1985. Organ-specific autoimmune diseases induced in mice by elimination of T cell subset. I. Evidence for the active participation of T cells in natural self-tolerance; deficit of a T cell subset as a possible cause of autoimmune disease. *J. Exp. Med.* 161:72–87.
5. Murakami, K., M. Hosono, H. Maruyama, Y. Mori, A. Nishio, M. Fukumoto, Y. Watanabe, M. Inaba, K. Kuribayashi, M. Sakai, and T. Masuda. 1993. Concomitant enhancement of the response to Mls-1a antigens and the induction of post-thymectomy autoimmune gastritis in BALB/c mice. *Clin. Exp. Immunol.* 92:500–505.
6. Nishio, A., M. Hosono, Y. Watanabe, M. Sakai, M. Okuma, and T. Masuda. 1994. A conserved epitope on H[+], K[+]-adenosine triphosphatase of parietal cells discerned by a murine gastritogenic T-cell clone. *Gastroenterology* 107:1408–1414.
7. Sakaguchi, S., T. Takahashi, and Y. Nishizuka. 1982. Study on cellular events in post-thymectomy autoimmune oophoritis in mice. II. Requirement of Lyt-1 cells in normal female mice for the prevention of oophoritis. *J. Exp. Med.* 156:1577–1586.
8. Murakami, K., H. Maruyama, A. Nishio, K. Kuribayashi, M. Inaba, K. Inaba, M. Hosono, K. Shinagawa, M. Sakai, and T. Masuda. 1993. Effect of intrathymic injection of organ-specific autoantigens, parietal cells, at the neonatal stage on autoreactive effector and suppressor T cell precursors. *Eur. J. Immunol.* 23:809–814.
9. Nishio, A., T. Katakai, M. Hosono, M. Inaba, M. Sakai, S. Kasakura, T. Masuda. 1995. Breakdown of self-tolerance by intrathymic injection of a T cell line inducing autoimmune gastritis in mice. *Immunology* 85:270–275.
10. Smith, H., Y.H. Lou, P. Lacy, and S.K. Ting. 1992. Tolerance mechanism in experimental ovarian and gastric autoimmune diseases. *J. Immunol.* 149:2212–2218.
11. Sakaguchi, S., N. Sakaguchi, M. Asano, M. Itoh, and M. Toda. 1995. Immunologic self-tolerance maintained by activated T cells expressing IL-2 receptor α-chain (CD25). Breakdown of a single mechanism of self-tolerance causes various autoimmune diseases. *J. Immunol.* 155:1151–1164.
12. Thornton, A.M., and E.M. Shevach 1998. CD4[+] CD25[+] Immunoregulatory T cells suppress polyclonal T cell activation *in vitro* by inhibiting interleukine 2 production. *J. Exp. Med.* 188:287–296.
13. Karlsson, F.A., P. Burman, L. Loof, and S. Mardh. 1988. Major parietal cell antigen in autoimmune gastritis with pernicious anemia is the acid-producing H[+], K[+]-adenosine triphosphatase of the stomach. *J. Clin. Invest.* 81:732–737.
14. Mori, Y., K. Fukuma, Y. Adachi, K. Shigeta, H. Tanaka, M. Sakai, K. Kuribayashi, H. Uchino, and T. Masuda. 1989. Characterisation of parietal cell autoantigens involved in neonatal-thymectomy induced murine autoimmune gastritis using monoclonal autoantibodies. *Gastroenterology* 97:364–375.
15. Jones, C.M., J.M. Callaghan, P. Gleeson, Y. Mori, T. Masuda, and B.H. Toh. 1991. The parietal cell autoantibod-

ies recognize in neonatal thymectomy-induced murine gastritis are the a and b subunits of the gastric proton pump. *Gastroenterology* 101:287–294.

16. Harn, S., R. Gehri, and P. Erb. 1995. Mechanism and biological significance of CD4-mediated cytotoxicity. *Immunol. Rev.* 146:57–79.

17. Nishio, A., T. Katakai, C. Oshima, S. Kasakura, M. Sakai, S. Yonehara, T. Suda, S. Nagata, and T. Masuda. 1996. A possible involvement of Fas-Fas ligand signaling in the pathogenesis of murine autoimmune gastritis. *Gastroenterology* 111:959–967.

18. Katakai, T., Y. Agata, A. Shimizu, C. Ohshima, A. Nishio, M. Inaba, S. Kasakura, K.J. Mori, and T. Masuda. 1997 Structure of the TCR expressed on a gastritogenic T cell clone, IL-6, and frequent appearance of similar clonotypes in mice bearing autoimmune gastritis. *Int. Immunol.* 9:1849–1855.

19. Alderuccio, F., B.H. Toh, S.S. Tan, P.A. Gleeson, and I.R. van Driel. 1993. An autoimmune disease with multiple molecular targets abrogated by the transgenic expression of a single autoantigen in the thymus. *J. Exp. Med.* 178:419–426.

20. Katakai, T., K.J. Mori, T. Masuda, and A. Shimizu. 1998 Differential localization of Th1 and Th2 cells in autoimmune gastritis. *Int. Immunol.* 10:1325–1334.

21. Asano, M., M. Toda, N. Sakaguchi, and S. Sakaguchi. 1996. Autoimmune disease as a consequence of developmental abnormality of a T cell subpopulation. *J. Exp. Med.* 184:387–396.

22. Martinelli, T.M., I.R. van Driel, F. Alderuccio, P.A. Gleeson, and B.H. Toh. 1996. Analysis of mononuclear cell infiltrate and cytokine production in murine autoimmune gastritis. *Gastroenterology* 110:1791.

23. Finkelman, F.D., J. Holmes, I.M. Katona, J.F. Urban, Jr., M.P. Beckmann, L.S. Park, K.A. Schooley, R.L. Coffman, T.R. Mosmann, and W.E. Paul. 1990. Lymphokine control of *in vivo* immunoglobulin isotype selection. *Annu. Rev. Immunol.* 8:303–333.

24. Seder, R.A., and W.E. Paul. 1994. Acquisition of lymphokine-producing phenotype by CD4+ T cells. *Annu. Rev. Immunol.* 12:635–673.

25. Abbas, A.K., K.M. Murphy, and A. Sher. 1996. Functional diversity of helper T lymphocytes. *Nature* 383:787–793.

26. Powrie, F., and R.L. Coffman. 1993. Cytokine regulation of T-cell function: potential for therapeutic intervention. *Immunol. Today* 14:270–274.

27. Fitch, F.W., M.D. McKisic, D.W. Lancki, and T.F. Gajewski. 1993. Differential regulation of murine T lymphocyte subsets. *Annu. Rev. Immunol.* 11:29–48.

28. Reiner, S.L., and R.M. Locksley. 1995. The regulation of immunity to *Leishmania Major. Annu. Rev. Immunol.* 13:151–177.

29. Springer, T.A. 1994. Traffic signals for lymphocyte recirculation and leukocyte emigration: The multistep paradigm. *Cell* 76:301–314.

30. Pietschmann, P., J.J. Cush, P.E. Lipsky, and N. Oppenheimer-Marks. 1992. Identification of subsets of human T cells capable of enhanced tansendothelial migration. *J. Immunol.* 149:1170–1178.

31. Röhnelt, R.K., G. Hoch, Y. Reiß, and B. Engelhardt. 1997. Immunosurveillance modelled *in vitro*: naive and memory 0T cells spontaneously migrate across unstimulated microvascular endothelium. *Int. Immunol.* 9:435–450.

32. Gleeson, P.A., B.-H. Toh, and I.R. van Driel. 1996. Organspecific autoimmunity induced by lymphopenia. *Immunol. Rev.* 149:97125.

33. Suri-Payer, E., P.J. Kehn, A.W. Cheever, and E.M. Shevach. 1996. Pathogenesis of post-thymectomy autoimmune gastritis. Identification of anti-H/K adenosine triphosphatase-reactive T cells. *J. Immunol.* 157:1799–1805.

34. Springer, T.A. 1994. Traffic signals for lymphocyte recirculation and leukocyte emigration: The multistep paradigm. *Cell* 76:301–314.

35. von Boehmer, H. 1994. Positive selection of lymphocytes. *Cell* 76:219–228.

36. Nossal, G.J.V. 1994. Negative selection of lymphocytes. *Cell* 76:229–239.

37. Malvey, E.N., D.G. Telander, T.L. Vanasek, and D.L. Muller. 1998. The role of clonal anergy in the avoidance of autoimmunity: inactivation of autocrine growth without loss of effector function. *Immunol. Rev.* 165:301–318.

38. Salojin, K.V., J. Zhang, J. Madrenas, and T.L. Delovitch. 1998. T-cell anergy and altered T-cell receptor signaling: effects on autoimmune disease. *Immunol. Today* 19:468–473.

39. Hayday, A. 1995. Is antigen-specific suppression now unsupp-ressed? *Curr. Biol.* 5:47–50.

40. Saoudi, A., B. Seddon, V. Heath, D. Fowell, and D. Mason. 1996. The physiological role of regulatory T cells in the prevention of autoimmunity: the function of the thymus in the generation of the regulatory T cell subset. *Immunol. Rev.* 149:195–216.

41. Kovarik, J., and C.A. Siegrist. 1998. Immunity in early life. *Immunol. Today* 19:150–152.

42. Smith, H., I.-M. Chen, R. Kubo, and K.S.K. Tung. 1989. Neonatal thymectomy results in a repertoire enriched in T cells deleted in adult thymus. *Science* 245:749–752.

43. Jones, L.A., L.T. Chin, G.R. Merriam, L.M. Nelson, and A.M. Kruisbeck. 1990. Failure of clonal deletion in neonatally thymectomized mice: tolerance is preserved through clonal anergy. *J. Exp. Med.* 172:1277–1285.

44. Inaba, M., K. Inaba, M. Hosono, T. Kumamoto, T. Ishida, S. Muramatsu, T. Masuda, and S. Ikehara. 1991. Distinct mechanisms of neonatal tolerance induced by dendritic cells and thymic B cells. *J. Exp. Med.* 173:549–559.

45. Bonomo, A., P.J. Kehn, and E.M. Shevach. 1994. Premature escape of double-positive thymocytes to the periphery of young mice. Possible role in autoimmunity. *J. Immunol.* 152:1509–1514.

46. Adkins, B., K. Chun, K. Hamilton, and M. Nassiri. 1996. Naive murine neonatal T cells undergo apoptosis in response to primary stimulation. *J. Immunol.* 157:1343–1349.

47. Adkins, B., and R.-Q. Du. 1998. Newborn mice develop balanced Th1/Th2 primary effector responses *in vivo* but are biased to Th2 secondary responses. *J. Immunol.* 160:4217–4224.

48. Suri-Payer, E., A.Z. Amar, A.M. Thornton, and E.M. Shevach. 1998. CD4+ CD25+ T cells inhibit both the induction and effector function of autoreactive T cells and represent a unique lineage of immunoregulatory cells. *J. Immunol.* 160:1212–1218.

51 | Immunoregulatory Defects in Inflammatory Bowel Disease

Lloyd Mayer

1. INTRODUCTION

The nature of the immune system in the intestinal tract is unique by virtue of its requirement to co-exist intimately with the external environment. The emphasis, therefore, is on rigorous control of immunologic responses in the GI tract, to dietary antigens, luminal bacteria and viruses. When control mechanisms are disrupted, the result is a series of chronic inflammatory diseases of the bowel, specifically ulcerative colitis (UC) and Crohn's disease (CD). These diseases affect over 2 million Americans and are quite prevalent in Europe and Scandinavia. While there is no clear-cut inheritance pattern, the high incidence of disease segregation within families suggests a genetic predisposition (1–7). A number of etiologies have been proposed for these diseases yet, as with many chronic inflammatory diseases, no specific pathogenic mechanism has been clearly defined. In this chapter, we will describe the unique features of each disease and discuss some of the immune and non-immune theories of disease development. The bulk of the chapter will be devoted to the current controversies in the etiopathogenesis of IBD: a directed autoimmune response against elements within the mucosa-associated lymphoid tissue versus immune dysregulation resulting in chronic inflammation and secondary nonspecific tissue injury.

2. Ulcerative Colitis

Ulcerative colitis is a disease defined by specific radiologic and histopathologic features. However, despite the fact that UC was initially described over 100 years ago, the clinical and histologic features overlap with many other inflammatory diseases of the bowel, including Crohn's disease. This underscores one important feature of the intestine: inflammation can only be expressed in a limited fashion in the intestine.

With this important fact in mind, there are still some distinct features ascribed to UC. UC is characterized by inflammation limited to the superficial mucosa. There is a marked inflammatory infiltrate comprised of neutrophils initially, but this is rapidly replaced by a more mononuclear cell infiltrate characteristic of chronic inflammation (3).

Grossly, the inflammation is characterized by erythema, edema, mucosal ulceration, and bleeding that can be monitored by sigmoidoscopy. Since UC is a continuous disease, typically the degree of inflammation in the rectum is reflective of what is occurring higher up. While the mucosa can heal completely after therapy, there is usually evidence of continued inflammation histologically. The repetitive inflammation and healing of ulcers frequently results in the development of mucosal bridges that appear as polyps or pseudopolyps, as they are termed in this disease. These polyps do not have the same malignant potential as adenomatous or villous polyps in non-inflammatory states, and do not require removal prophylactically.

As stated previously, the histologic picture of UC is characterized by its superficial nature. Within the crypts of the epithelial glands, neutrophils transmigrate forming what are called crypt abscesses. While these are not specific for UC, occurring in infectious, ischemic, and Crohn's colitis, they are a more recognized feature of UC. Associated with crypt abscesses is the loss of goblet cells and mucous from the epithelial cell layer. Like crypt abscesses, this is not a specific finding, but is more common in UC.

Despite the active inflammatory process that undergoes remission and exacerbation, fibrosis and scarring are rarely seen. With long term disease, the musculature of the bowel loses its tone, leaving the patient with an ahaustral colon which is a shortened, poorly functioning tube.

Another feature specific for UC is that the disease is typically uniform in its location. If the rectum alone is involved, it is termed proctitis. As the disease progresses up the left colon to the sigmoid and descending colon, it becomes proctosigmoiditis and left-sided colitis, respectively. If the entire colon is involved, it is termed universal colitis. One characteristic of UC is that the disease spreads in a contiguous manner; that is, if the right colon is involved in the process, so is the left colon and rectum, there are no "skip areas" as are seen in Crohn's disease, and the inflammation stays limited to the colon. In some cases, the small bowel is inflamed in a process called "backwash ileitis" involving the distal few centimeters of the terminal ileum. Once the disease is removed, however, (i.e. a total colectomy is performed) the patient is cured. UC does not recur in the small bowel.

One last important feature of UC is its potential to undergo malignant change. The incidence of colon cancer in UC varies in many studies from 10–15%, but it is generally agreed that the duration of disease is an important prognostic indicator (8–11). Disease of greater than 10 years duration is associated with 10% risk of cancer, and the risk roughly increases 10% per decade. Certain histologic features may precede the onset of cancer. Glandular dysplasia in an area without inflammation is suggestive of a premalignant or co-existent malignant process (12–15). Although there is still controversy regarding the meaning of dysplasia, most groups feel that persistent high grade dysplasia is an indication for colectomy. Cancers in UC tend to be more virulent, metastasizing widely before they are recognized, and are of a more poorly differentiated type. Some of this may relate to the fact that symptoms of both cancer and UC are similar, i.e. rectal bleeding, so that early detection is not common. Therefore, programs of surveillance and biopsy are initiated in patients with more than 10 years of disease.

The symptoms of UC reflect the inflammatory process. The ulceration and mucosal friability result in rectal bleeding. The edema and concomitant water loss translate into severe abdominal cramping, which occurs throughout the day and night in severe cases. Stools, which are rarely formed in active disease, may consist of bloody mucosal discharges without any real stool, up to 10–20 times per day. Constitutional symptoms are common and may be insidious early on. Patients "don't feel well" and appear to be more fatigued and irritable. Low grade fevers, headaches, and nausea may accompany flares. The laboratory is rarely helpful in distinguishing disease. Microcytic anemia, hypoalbuminemia, and an elevated sedimentation rate are common features, but certainly not specific. It is critical to rule out infectious etiologies, such as amebiasis, Campylobacter, and Shigella, which can look exactly like UC sigmoidoscopically. Clinical and laboratory findings may be indistinguishable and only stool cultures and/or biopsy allow for definition of the diagnosis. One good feature of UC is that it is usually quite responsive to therapy (i.e. steroids in an acute flare). Lack of response should raise questions as to the validity of the diagnosis.

3. CROHN'S DISEASE

Although technically described in the mid 1800s, Crohn's disease was not recognized as a distinct entity until Crohn, Ginzburg and Oppenheimer reported the gross and pathologic features of several cases in 1932 (16). Crohn's disease encompasses regional enteritis, ileitis, granulomatous colitis, and jejunoileitis under its scope. Unlike UC, this disease can affect any part of the bowel from the oropharynx to the rectum, but most commonly affects the distal ileum and ascending colon (ileocolitis). Pathologically, CD is much different from UC. The inflammatory process initiates as small aphthous ulcers that typically occur over Peyer's patches (17). Ulceration, however, is not a common feature of CD. Rather the inflammation is transmural with marked bowel thickening, fibrosis, and eventual narrowing or stricturing of the bowel. Grossly, the serosa of the bowel is erythematous, and mesenteric fat "creeps" onto the serosal surface (creeping fat) in actively inflamed areas. The disease has a tendency to skip areas of the bowel, resulting in a segmental disease that may be limited in extent at any segment. The fibrosis and scarring results in a classic radiologic feature called the string sign, where a long narrowed segment of bowel fails to open during a barium x-ray. The transmural nature of the inflammation is also responsible for another characteristic feature of CD, intestinal fistulization. With localized obstruction and transmural inflammation, small fistulae develop initially into the mesentery, but eventually into other organs (bladder, skin, vagina) or into other parts of the bowel. These fistulae, which can cause significant symptoms with mesenteric abscesses, pneumaturia, and drainage of stool through the skin or vagina, are easily seen by radiographic studies.

When CD involves the colon, it typically spares the rectum. It may, however, involve the anal verge resulting in perianal CD, a disease that may be independent of other bowel inflammation. Perianal CD presents as fistulae and perirectal abscesses that require frequent surgical intervention (drainage) and antibiotic therapy.

Histologically, the hallmark of CD is the noncaseating granuloma, similar to that seen in sarcoidosis. While not seen in every segment, when present, it is virtually diagnostic of CD. The finding of granulomata in CD is highly suggestive of an exuberant cell-mediated immune response, yet no organism or agent has been reproducibly isolated thus far.

As with UC, the histopathology of CD accounts for the symptoms. The narrowing from fibrosis and segmental inflammation results in the obstructive symptoms of abdominal distention, colicky pain, and obstipation. The inflammatory component of CD results in diarrhea, although usually not as severe as in UC and, unless the colon is involved, is rarely characterized by rectal bleeding. From segmental obstruction, patients develop bacterial overgrowth, secondary lactase deficiency and malabsorption. Thus weight loss and nutritional anemias are common findings. Like UC, low grade fevers and constitutional symptoms (weakness, fatigue, lethargy) are the rule rather than the exception. Fistulae have their own symptom complex as described above.

Surgery for CD is a more common occurrence in the course of this disease. However, unlike UC, the disease can recur anywhere in the bowel, but most often at the anastomosis (18). The incidence of malignancy is lower than that seen in UC, but adenocarcinoma of the small intestine is clearly increased in CD (19,20). Again these cancers are more difficult to detect because the symptoms of the malignancy can mimic disease symptoms and cancers of the small bowel rarely obstruct until very late in the course of their growth since stool in the small bowel is liquid.

3.1. Extraintestinal Manifestations

One feature of IBD that deserves attention is that of extraintestinal manifestations, many of which, such as episcleritis, ankylosing spondylitis, erythema nodosum, are also seen in other autoimmune disorders. The finding of such correlated disorders speaks to the systemic nature of IBD. Some extraintestinal manifestations (episcleritis, aphthous stomatitis, peripheral arthritis, erythema nodosum) parallel disease activity while others (ankylosing spondylitis, pyoderma gangrenosum, sclerosing cholangitis, and uveitis) do not (21–25). There is some distinction between extraintestinal manifestations that occur in UC versus CD. Sclerosing cholangitis, ankylosing spondylitis, and pyoderma are typically associated with UC, while there is a predilection for ocular manifestations in CD. The existence of these systemic manifestations suggests some immunologic mechanism, although no specific defect as such has been defined (e.g. circulating immune complexes) as will be discussed later.

3.2. Genetics of IBD

Like many chronic inflammatory diseases, the development of novel approaches to gene mapping coupled with advances in the human genome project have allowed for the identification of a number of gene associations in IBD. Such studies have largely supplanted prior efforts using a candidate gene approach. Given the focus on immunoregulatory defects, candidate genes encoding immunoregulatory

molecules were used in screens of families and populations. While some weak associations were identified (HLA-DR2 in UC, DR1 DQ5 in CD, ICAM-1 (codon 241 polymorphism) in CD, IL1-RA allele 2 in UC) (26–28), they were not consistently confirmed in larger population studies. The evidence for a genetic basis in CD is stronger than that in UC. Initial studies in monozygotic twins reported a concordance rate of 40% in CD whereas the rate in UC was only 5% (3,29,30). These studies support a role for environmental factors (eg. infections) in these diseases.

Still, the genome-wide screens have provided valuable additional information. Comparable to other chronic inflammatory diseases, IBD appears to be a multigenic disorder. The initial region identified, termed IBD-1, was reported in CD families by the French group GETAID (31–33). This association has been independently confirmed by several other groups. Other regions on chromosomes 1, 3, 7 and 12 have been identified, but confirmation has either been lacking or has not yet been sought (4,34–38). What has been suggested by the group in England is that there may be IBD generic genes (predisposing to either UC or CD) and disease-specific genes (CD- or UC-specific). Furthermore, within a given disease, there may be genes that correlate with a selected subtype of disease (e.g. fistulizing versus stenotic CD). This latter concept is supported by the findings of Bayless et al, who noted that disease subtypes segregate within families and, furthermore, that the concept of genetic anticipation (i.e. diagnosis of the disease at an earlier age in children or siblings of an affected individual) supports the key role that genetics plays in this disease.

4. NON-IMMUNE MODELS OF IBD PATHOGENESIS

As with all chronic inflammatory diseases, a broad search for "the" infectious agent in IBD has been undertaken by many groups with varying results. To this point, however, no single agent has been reproducibly isolated that would fit Koch's postulates for disease induction. This by itself does not exclude an infectious etiology, as the technology for specific virus isolation may progress more slowly than the disease itself. This was certainly true for scrapie until prions were identified (39). However, there is a consistent "inconsistency" in the viruses: Bacteria and, more recently, mycobacteria have been isolated in IBD (39–47).

When UC was first described, it was thought to be some form of dysentery except for the fact that it persisted. After numerous attempts to culture stool and tissues for conventional organisms failed, several groups turned to specialized media to isolate L forms or previously undefined viruses. Despite some initial enthusiasm, these studies could not be reliably reproduced (42,43,48). Chiodini and colleagues identified a mycobacterial species, Mycobacteria linda, a strain of mycobacteria paratuberculosis, in the stools of patients with CD, but not UC or inflammatory controls (41). This organism was an attractive one, since it causes a disease similar to CD in goats (Johne's disease) characterized by wasting, diarrhea, and small bowel inflammation. Several groups reported similar findings, including the presence of antibody to M. Paratuberculosis in the serum (40,49). However, careful analysis utilizing mycobacteria-specific antibodies failed to identify the culprit organism in tissues (50,51). This theory has fallen out of favor, as most groups have now failed to reproduce the original findings (specificity as well as isolation). Based on the initial findings, Thayer et al. initiated a trial of anti-tubercular medications in the treatment of CD (52) that did not result in disease amelioration.

5. IMMUNOLOGIC DEFECTS IN IBD

While no clear-cut immunologic deficit has been described in IBD (as with infectious agents), there are three areas that have developed over the past several years: Autoantibody-mediated tissue damage, T cell-mediated cytotoxicity, and dysregulated T cell activation with nonspecific tissue destruction. The major problem with each of these areas has been the inability to discern whether the phenomena observed are a primary event or secondary to the chronic inflammatory process. At issue is the lack of an identifiable specific antigen that triggers the disease process or is associated with the immunologically-mediated tissue injury. This lack of a known antigen is not unique to IBD, but in the gut, where antigen exposure is overwhelming, it may be impossible to sort out. Furthermore, the aberrant response may be to one of several organisms (see below). The results of the accumulated studies have led to the following hypothesis for the pathogenesis of IBD: In a genetically-predisposed host, there is an antigenic trigger that evokes an inappropriately regulated immune response. In this hypothesis, one has the three components of chronic inflammatory disease; Genetics, environment and immunity. The defect is in the failure to shut down the ongoing immune response that, as will be described below, can be directed against nonpathogenic organisms.

6. EVIDENCE FOR IMMUNOLOGICALLY-MEDIATED TISSUE INJURY IN IBD

As alluded to earlier, a number of plausible pathogenic mechanisms have been proposed for IBD. Largely due to the growth in the area, coupled with negative findings for infectious, metabolic, vascular and neurohormonal etiologies, the immune system had been implicated as a potential culprit in IBD. This is not unreasonable, as the gut is the largest lymphoid organ in the body and possesses an immune system constantly exposed to antigens. Furthermore, the histologic picture of IBD is consistent with immune-mediated tissue injury: The noncaseating granuloma in Crohn's disease (T cell activation) and an Arthus reaction in UC (immune complex-mediated tissue injury). However, despite these findings, the initial studies assessing immunologic mechanisms for IBD were fraught with inconsistency and poor reproducibility. This was, in large part, due to the fact that cells of the peripheral blood, and not gut-associated lymphoid tissue, were studied. Peripheral lymphocytes are not necessarily reflective of events in the gut and may be more susceptible to the modulating effects of therapy (steroids) and nutritional status (malnutrition in CD). Only when techniques were developed to allow for enrichment of gut associated lymphocytes was progress made (53,54).

7. AUTOANTIBODIES IN IBD

Very early on, investigators postulated that antibodies against epithelial cell Ags might be responsible for the complete destruction of epithelial cells in IBD. This was obviously more likely for UC where destruction of the epithelial cell layer is commonplace. Autoantibodies were detected in the serum of patients with UC, binding epithelial cells in fluorescence assays (55–59). These antibodies were thought to arise from cross-reactivity with enterobacterial Ags, specifically E. coli 014, and the common enterobacterial antigen of Kunin. Titers of these antibodies were quite high in UC and, in the initial studies, such Abs were not seen in normal patients. However, several additional findings relegated these autoantibodies to a lesser stature (56). First, when control sera from other non-IBD inflammatory diseases of the intestine (diverticulitis, ischemic colitis, infectious colitis) were obtained, the incidence and titer of anti-colon Abs was comparable to that seen in IBD. Second, the anti-colon Abs were largely noncytotoxic, incapable of focusing ADCC or binding complement (60). This was clearly demonstrated in a rabbit model of colitis, where DNCB sensitization intrarectally resulted in the histologic picture of UC with autoantibodies. In this system, however, intrarectal instillation of the anti-epithelial cell Abs failed to produce intestinal inflammation in naive animals (60,61). Other problems included the lack of correlation of autoantibody titer with disease activity or extent. Thus, patients with pancolitis were found to have similar titers of autoantibody to those with proctosigmoiditis. Lastly, where serum complement levels were measured, consumption of complement components was not evident, although turnover was increased (62–68).

Several autoantibodies have been defined in IBD. Das and co-workers took colonic washings from UC tissues and identified an IgG anti-epithelial cell Ab that could be eluted from tissue (69–80). This colon cell-associated (CCA)-IgG was found in UC, but not in CD or other inflammatory diseases, and detected a 40kd surface epithelial Ag (present on all colonocytes but not small bowel enterocytes). More recently, this Ag has been identified as tropomyosin (72,76,77), allowing for more definitive studies. The distribution of this autoantigen on other epithelial structures revealed a pattern that was reminiscent of organs involved in the extra-intestinal manifestations of IBD (i.e. bile ducts, iris, skin) (73,74). Thus, for the first time, there was an antibody that might not only explain local tissue damage, but also explain pathology outside of the GI tract. Early concerns were raised regarding the isotype of this Ab, IgG, since the predominant Ab in the gut is IgA.

However several studies from MacDermott (81,82) and Brandtzaeg (83) clearly documented a marked increase in IgG-producing cells in the lamina propria of patients with IBD. Furthermore, there is an interesting distribution of IgG subclasses in these tissues. In UC, IgG1 plasma cells predominate while, to a lesser extent, IgG2 plasma cells predominate in CD tissues (82–86). This is not to say that IgA is not present; it is still the predominant immunoglobulin in the GI tract, but the relative increase in IgG in both diseases is remarkable. The mechanism whereby IgG gains access to the lumen is not established. Both IgA and IgM can be transported (transcytosed) via secretory component produced by epithelial cells (83,87). IgG cannot bind to SC and, therefore, must either leak into the lumen between cells or travel through frankly ulcerated areas, although Dickinson et al. described the bi-directional transport of IgG by the class I-like neonatal FcRn in an epithelial cell line (88). Since IgG can bind complement, it is a plausible candidate for a pathogenic autoantibody. Indeed studies by Halstensen et al. have documented complement deposition on epithelial cells in a pattern consistent with CCA-IgG (84,85). Thus, a scenario can be developed where autoantibody-mediated epithelial injury is at the heart of tissue injury in UC. Furthermore, Das has been able to show that CCA-IgG can promote the ADCC-mediated killing of an epithelial cell line (71).

Another autoantibody is that described by Roche et al. who, using sera from patients with UC, was able to detect autoantibodies against epithelial cell-associated components (from cell lysates-ECAC) by immunoblot (89,90). While this Ab also fails to fluctuate with disease activity, it is found at higher incidence in first degree relatives without evidence of disease (89). These findings suggest anti-ECAC Abs may be the result of a dysregulated immune system rendering the individual more susceptible to developing UC. Unlike CCA-IgG, anti-ECAC antibodies recognize multiple cellular components that need to be more clearly defined. However,

antibodies against this group of Ags are cytotoxic to ECAC coated RBC, highly suggestive of a pathogenic autoantibody. Further work is required to define the role(s) of both of these autoantibodies.

Another interesting autoantibody found to be present in UC is the anti-neutrophil cytoplasmic antibody, or ANCA (27,91–94). The titer of this autoantibody, which was initially described in Wegener's granulomatosis, does not correlate with disease activity. In contrast to Wegener's, the fluorescent pattern is perinuclear rather than cytoplasmic (pANCA vs. cANCA), and the antigen against which this Ab is directed is not myeloperoxidase. While this Ab does not appear to provide insight into any pathogenic mechanisms, it may be a useful marker for identifying patients with UC, differentiating them from CD and infectious colitides. pANCA is present in 60–70% of patients with UC and is also present in a small subgroup of first degree relatives consistent with an genetic predisposition to immune dysregulation (27).

8. T CELL ACTIVATION/CYTOTOXICITY IN IBD

One of the first studies assessing cytotoxicity in IBD was that of Perlmann and Broberger, who reported that peripheral WBC from patients with UC could lyse fetal epithelial cells in culture. As work progressed through the years, the responsible cell was found to be a T cell bearing an Fcγ receptor. These data suggested that autoantibodies or cross-reactive Abs could possibly direct ADCC against autologous epithelial cells. While these studies progressed in the early to mid 1970s, once lamina propria lymphocyte isolation was a reproducible technology, it became clear that these FcγR+ T cells were not present in the IBD gut (95), certainly in no greater number than found in normals. Other cytotoxic T cells (NK cells, LAK cells, etc.) were similarly underrepresented in IBD tissues by isolation and immunohistochemical techniques (96). Therefore, it appeared less likely that these cells played any role in the pathogenesis of IBD.

Studies by Shanahan et al. (97) raised this possibility again. Using a system of anti-CD3 mAb-stimulated MHC-unrestricted CTLs (redirected lysis), they found that lamina propria lymphocytes from patients with IBD were enriched for such CTLs. The mechanism of cytotoxicity was thought to be cytokine-mediated.

γ-IFN and TNF production (both T cell products) synergize in mediating epithelial cell line killing. While the level of TNF in the bowel of patients with UC is not particularly elevated, several groups have documented that γ-IFN production is enhanced (98,99). Thus the combination of MHC-unrestricted killing with cytokine-mediated cellular disruption was thought to be a plausible scenario for the development of IBD.

8.1. Immune Dysregulation in Human Disease

As alluded to earlier, it has been known for several years that there is aberrant activation of CD4$^+$ T cells in IBD. This has been noted in the lamina propria as well as in the peripheral blood (DR$^+$ CD4$^+$ T cells) (100–104). The result of this activated state is the secretion of cytokines. Initial studies suggested, similar to the animal models (see below), that there might be a distinct skewing in CD4$^+$ T cell subpopulations activated in either CD or UC. Based upon the histologic picture, it was suggested that CD might represent a state wherein there was aberrant activation of Th1 cells (IL-2, γ-IFN, TNF-α) (105–112) resulting in macrophage activation and granuloma formation, whereas there would be a bias towards Th2 cells in UC promoting Ab production, immune complex formation, complement activation and an Arthus reaction. Indeed, several groups reported a marked increase in γ-IFN in either CD biopsies (by PCR) or from isolated cells. Furthermore, there was some evidence that there was an increase in IL-5 (not IL-4) in UC tissues. While this was a useful paradigm to develop, subsequent studies have questioned the accuracy of these findings. In fact IFN-γ is significantly elevated in UC tissues as well (perhaps not to as great an extent as CD). Interestingly, however, the major regulator of IFN-γ production, IL-12, is increased in CD to a much greater extent than UC (actually IL-12 appears to be absent) (107,110, 113,114), suggesting that other factors (IL-18), (115,116) may account for the γ-IFN production seen in UC and hence the different histologic appearance.

Other evidence supports a dichotomy in T cell subsets activated in the two diseases. Patients with CD enjoy a dramatic clinical response to medications promoting the inhibition of the Th1 and macrophage-derived cytokine TNF-α (117–121), whereas the response to similar therapy in a limited number of UC patients is modest at best. IL-10, an anti-inflammatory Th1 inhibitory cytokine, has some efficiency in the treatment of CD patients, but it does not compare with the response to anti-TNF (infliximab) (122). One interesting finding that adds further support to the concept of dysregulated Th1/Th2 responses in IBD is the epidemiologic observation that CD and UC are diseases of "clean" civilizations. In fact, it has been noted that Westernization of various countries, including changes in sanitation, water supply etc, is associated with the recognition, for the first time, of IBD cases in those countries. One hypothesis put forth by Weinstock and colleagues is that, in a "dirty" environment, intestinal parasites and bacterial infections promote a Th2 dominant environment that inhibits unopposed Th1 (inflammatory) responses (Weinstock, unpublished personal communication). After "sterilization" and removal of these infectious agents, there is a diminution in the Th2 state and unopposed Th1 responses can occur resulting in IBD. This is yet another example of environmental factors influencing immune responses. Whether this hypothesis is validated or not remains to be seen, but it is an intriguing set of observations.

One last series of data supporting the contention of a dysregulated T cell population comes from studies by Blumberg's group using CDR3 display to define clonal expansions within the CD4$^+$ T cell population in the lamina propria (123,124). Previous studies by this group and others had documented oligoclonal expansions of CD8$^+$ T cells in both IEL and LPL populations in normal individuals (125–128). These findings were comparable to those found in peripheral blood of older individuals. Whether these represent expansion of regulatory T cells or cells chronically stimulated by Ag has not been established, however, no studies have reported oligoclonality in the CD4$^+$ T cells in any compartment. When studying IBD patients, such CD4$^+$ T cell clones were noted and these persisted with time, especially at times of flare. Blumberg's group suggested that these might represent the expansion of an Ag-reactive (pathogen or autoantigen) clone, but it could just as easily represent regulatory or inflammatory clones, as well. Interestingly, such expanded CD4$^+$ T cell clones are detected in the TcR$\alpha^{-/-}$ mouse, and it may be that, in this model, the questions relating to their identity can be directly addressed (129).

8.2. Animal Models

Probably the greatest advance in our understanding of IBD came from the fortuitous identification of a number of animal models of colitis. A number of groups seeking to define the role of selected cytokines in systemic immunoregulation utilized targeted gene deletion technology. Interestingly, most of the animals with cytokine gene deletions exhibited normal systemic immune responses (except for the IL-2$^{-/-}$ mouse, which developed autoimmune hemolytic anemia), suggesting that back-up mechanisms controlling systemic immunity were strong. However by 3 months of age the IL-2$^{-/-}$, IL-10$^{-/-}$ and TcR$\alpha^{-/-}$ mice developed a wasting syndrome associated with rectal prolapse and diarrhea (130–133). Histologically, these animals all exhibited some form of chronic colitis without any evidence of infection. At the same time, a C3H/HeJ substrain was identified at Jackson Laboratories that also exhibited prolapse and diarrhea (93,134–138). Those animals had been previously removed from the colony under the assumption that they had an infectious process. When studied more carefully, these animals were noted to have chronic colitis as well. In the past 5 years, a greater number of models have been described (139–143). These seem to segregate into two groups—those with perturbations in T cell regulation and those with defects in mucosal barrier function (144). What these animals have told us is that the mucosal immune system is less tolerant of immunologic imbalance. The default pathway is mucosal inflammation.

These models have taught us a lot more. First, it is clear that the colitis is T cell-mediated. This is most evident in the CD45Rb[hi]/SCID transfer model described by Powrie et al. in which a normal population of phenotypically-naïve (CD45Rb[hi]) CD4[+] T cells are transferred into a SCID recipient (141,145–147). Within 1–2 months, these animals develop active colitis in the absence of B cells. If unseparated CD4[+]T cells or a combination of CD45Rb[hi] and lo cells are used for transfer, no colitis is seen suggesting that pro-inflammatory T cells mediate the disease and that counter-regulatory cells normally inhibit this activity. This regulatory population exists in the CD45Rb[lo] subset. Recent studies by Groux et al. have suggested that this regulatory subset might represent a previously undescribed population of T cells termed TR1 cells (148) that secrete IL-10 upon activation.

The absence of a requirement for B cells has been confirmed in other models where the animals have been crossed with "B-less" mice (IgM [−/−]) in which colitis develops unabated and unaltered. Thus, despite the presence of autoantibodies in some of these mice, antibody appears to play no role in disease pathogenesis.

As the pathogenic T cells were studied further, it became clear that the overwhelming majority of T cells present in the gut of these various models could be categorized as Th1 cells, secreting IFN-γ, and TNF. In fact, anti-IFN-γ or anti-TNF treatment in the IL-10[−/−] and CD45Rb[hi]/SCID transfer model modulated the disease significantly (145), supporting a role for Th1 cytokines in these animals. Indeed the only mouse model of colitis that has not been noted to have a Th2 predominance in the gut has been the TcRα[−/−] mouse (133), which possess T cells in their intestines that produce IL-4; in fact, anti IL-4 mAb therapy abrogates the disease. This is a unique model, as its genetic defect results in the expansion of an unusual population of TcRβ/β cells that may be responsible for the disease. Furthermore, the disease can be "prevented" by early cecectomy (appendectomy), removing a rich source of Peyer's patches (149). It has been suggested that this model may be the most representative of UC.

Two other key pieces of information have been gleaned from these models: The role of enteric flora and the role of genetic background. In every animal model tested to date, the disease is dependent upon the presence of normal flora. In animals derived and maintained in a germ-free environment, no disease occurs (150–152). If these animals are removed from this environment or re-colonized with *normal* bacterial flora from littermates or defined flora, including anaerobic species such as *Bacteroides vulgatus*, disease recurs (142,150,152). Elson's group has defined colonic bacteria-reactive T cell lines in the HeBir mouse that can transfer disease (135). Duchmann and colleagues have data to suggest that, in patients with IBD, there is a loss of tolerance to autologous flora (153). Thus, the

concept of an infectious pathogen initiating (and perpetuating) IBD suffers a setback here. Although some labs have reported that *Helicobacter* species can induce colitis in germ-free IL-10[−/−] mice, the fact that normal flora can do the same renders the notion of specific pathogens less compelling (154). Thus, these mice satisfy the three components of the hypothesis of pathogenesis described above: Genetic predisposition (barrier dysfunction or cytokine gene deletion) resulting in immune dysregulation triggered by an environmental stimulus (normal flora).

The genetic component is further reinforced by the observation that backcrossing IL-2 or IL-10[−/−] mice onto different strains results in disparate expression of disease. Thus, background genes play an important role in the type of disease that develops, echoing the findings described earlier with regard to the results of genome-wide screens.

9. THE EPITHELIAL CELL AS AN ANTIGEN PRESENTING CELL: IMMUNOREGULATORY DEFECT?

The activation of T cells described above requires the interaction of T cells, APCs, and either specific antigen, endogenous mitogens, or superantigens. However, since the T cells lie below the epithelial barrier, Ag entry into the underlying lamina propria is critical. Until recently, only selective pathways for Ag access were reported, i.e. between cells (paracellular), breaching tight junctions, or through specialized epithelium overlying Peyer's patches, M cells (155–166). These latter cells were thought to be the major Ag sampling cells in the gut, transcytosing large particulate Ags from the lumen into the Peyer's patch where a positive immune response could be generated and systemic and local antibody detected. However, in contrast to the peripheral immune system, the immunologic tone of the gut is one of suppression or the active lack of an immune response. This unique feature of the GALT is highlighted by the immunologic phenomenon termed oral tolerance (167–178). Ingestion of certain protein Ags render an animal incapable of mounting an active immune response to that Ag when challenged systemically. Tolerance can be transferred to a naive animal by transferring suppressor cells. The mechanism of generation of those suppressor cells has been undefined to date.

A proposal for an alternate pathway for Ag sampling in the gut was defined: Transcytosis of Ag through epithelial cells. This model developed from the observation that intestinal epithelium express class II Ags constitutively in the small bowel, and expression can be enhanced in inflammatory states (GVHD, IBD, infectious colitis) (179–188). Thus, the epithelium in the gut differs from cells in the thyroid and islets where class II Ag expression is aberrant (and where Ag exposure is limited) (189,190). Interestingly, when normal intestinal epithelial cells from man or rat are used as APCs in Ag-specific responses, the predominant T cells proliferating in these cultures are suppressor cells (CD8+ non-cytotoxic) (191). In man, these are antigen-nonspecific, while in rat, the response is specific (187,188). The reason(s) for this difference is not clear at present, but could clearly shed light on normal immunoregulatory circuits. This model of antigen presentation by epithelial cells has been proposed to explain immunologic suppression in the gut.

The question remains as to whether differences occur in IBD. Several studies have documented that class II Ag expression by enterocytes is enhanced in IBD (183, 192–197), and that an increase in γ-IFN production may be responsible for this finding (99). However, intriguing data relating to the interaction of IBD epithelium with T cells have recently been reported. Here, in contrast to the normal state, where CD8+ T cells proliferate, both UC and CD epithelial cells selectively activate potent CD4+ helper T cells (198). These findings do not relate to the presence of underlying inflammation as epithelial cells from inflammatory controls (ischemic colitis, diverticulitis) stimulate CD8+ T cells preferentially (like normal epithelium) and epithelial cells from uninflamed segments of IBD tissues still stimulate CD4+ T cells. These data suggest that there may be an inherent defect in IBD epithelial cells resulting in the failure to activate normal suppressor mechanisms. A number of studies have documented defective suppressor cell generation in IBD, although the experimental systems have been quite variable (199–201).

More recently, the nature of this defect has been defined by studies attempting to dissect the process of activation of the CD8+ suppressor T cells by IECs. These cells derive from a subpopulation of CD8+ CD28− T cells that utilize class Ib molecules as their restriction element (eg CD1d) and require the presence of a novel co-stimulatory molecules, termed gp180 (202–204). This molecule is a glycoprotein expressed by normal IECs as both an apically-sorted, GPI-anchored, form as well as a transmembrane (basolateral) form. gp180 binds to CD8 and activates the CD8α-associated kinase, p56lck. Inhibition of this kinase blocks IEC individual CD8+ T cell activation. Antibodies to gp180 also inhibit IEC-induced CD8+ T cell proliferation. In IBD, gp180 expression is altered or markedly diminished. In UC, there is a loss of the basolateral form (ie. the form that could interact with T cells in the lamina propria), while in CD, there is a general loss of both forms. The absence of gp180 would result in the failure of these cells to activate CD8+ suppressor T cells (205). The default pathway in the presence of enhanced MHC class II expression would be the activation of CD4+ T cells locally in the lamina propria.

Such a model holds significant promise in the understanding of immunologic defects described in IBD. Uncontrolled CD4+ T cell activation can result in macrophage activation, which has been described in IBD, resulting in nonspecific tissue injury mediated by prostaglandins, leukotrienes, and superoxides (all reportedly elevated in IBD tissues) (206, 207), enhanced secretion of Ab (with potential for autoantibody production), and the differentiation of mature cytolytic T cells (LAK cells, MHC unrestricted killers, etc.) via secretion of cytokines.

The similarities in IBD to a variety of rheumatologic disorders is striking, highlighted by the fact that all therapies used in IBD are currently used or have been tried in rheumatoid arthritis. Inhibition of activated T cells has been approached through the use of cyclosporin with a moderate success rate in CD and UC (117,208–212). Low-dose immunomodulators, such as azathioprine and 6-mercaptopurine, may target T and B cell subpopulations, altering cytokine and autoantibody production (208, 213–218). Whether pathogenetic mechanisms in one disease will aid in the understanding of the other remains an open question.

ACKNOWLEDGEMENT

Supported by PHS grants AI23504 and AI24671

References

1. Louis, E., and J. Belaiche. 1997. Genetics and inflammatory bowel disease: from association studies to wide genome screen. *Acta Gastroenterol. Belg.* 60:201–203.
2. Yang, H., C. McElree, M.P. Roth, F. Shanahan, S.R. Targan, and J.I. Rotter. 1993. Familial empirical risks for inflammatory bowel disease: differences between Jews and non-Jews. *Gut* 34:517–524.
3. Weterman, I.T., and A.S. Pena. 1984. Familial incidence of Crohn's disease in The Netherlands and a review of the literature. *Gastroenterology* 86:449–452.
4. Satsangi, J., M. Parkes, D.P. Jewell, and J.I. Bell. 1998. Genetics of inflammatory bowel disease. *Clin. Sci. (Colch.)* 94:473–478.
5. Roth, M.P., G.M. Petersen, C. McElree, C.M. Vadheim, J.F. Panish, and J.I. Rotter. 1989. Familial empiric risk estimates of inflammatory bowel disease in Ashkenazi Jews *Gastroenterology* 96:1016–1020.
6. Lashner, B.A., A.A. Evans, J.B. Kirsner, and S.B. Hanauer. 1986. Prevalence and incidence of inflammatory bowel disease in family members. *Gastroenterology* 91:1396–1400.
7. Fielding, J.F. 1986. The relative risk of inflammatory bowel disease among parents and siblings of Crohn's disease patients. *J. Clin. Gastroenterol* 8:655–657.
8. Greenstein, A.J., D.B. Sachar, H. Smith, A. Pucillo, A.E. Papatestas, I. Kreel, S.A. Geller, H.D. Janowitz, and A.H. Aufses, Jr. 1979. Cancer in universal and left-sided ulcerative colitis: factors determining risk. *Gastroenterology* 77:290–294.

9. Greenstein, A.J., D.B. Sachar, A. Pucillo, G. Vassiliades, H. Smith, I. Kreel, S.A. Geller, H.D. Janowitz, and A.H. Aufses, Jr. 1979. Cancer in universal and left-sided ulcerative colitis: clinical and pathologic features. *Mt. Sinai J. Med.* 46:25–32.

10. Greenstein, A.J., D.B. Sachar, H. Smith, H.D. Janowitz, and A.H. Aufses, Jr. 1981. A comparison of cancer risk in Crohn's disease and ulcerative colitis. *Cancer* 48:2742–2745.

11. Greenstein, A.J., A. Sugita, and Y. Yamazaki. 1989. Cancer in inflammatory bowel disease. *Jpn. J. Surg.* 19:633–644.

12. Morson, B.C. 1983. Dysplasia in ulcerative colitis. *Scand J. Gastroenterol. Suppl.* 88:36–38.

13. Lennard-Jones, J.E., B.C. Morson, J.K. Ritchie, and C.B. Williams. 1983. Cancer surveillance in ulcerative colitis. Experience over 15 years. *Lancet* 2:149–152.

14. Morson, B.C. 1985. Precancer and cancer in inflammatory bowel disease. *Pathology* 17:173–180.

15. Lennard-Jones, J.E., D.M. Melville, B.C. Morson, J.K. Ritchie, and C.B. Williams. 1990. Precancer and cancer in extensive ulcerative colitis: findings among 401 patients over 22 years. *Gut.* 31:800–806.

16. Crohn, B.B., L. Ginzburg, and G.D. Oppenheimer. 1984. Landmark article Oct 15, 1932. Regional ileitis. A pathological and clinical entity. By Burril B. Crohn, Leon Ginzburg, and Gordon D. Oppenheimer. *Jama.* 251:73–79.

17. Morson, B.C. 1972. The early histological lesion of Crohn's disease. *Proc. R. Soc. Med.* 65:71–72.

18. Lennard-Jones, J.E., and G.A. Stalder. 1967. Prognosis after resection of chronic regional ileitis. *Gut* 8:332–336.

19. Greenstein, A.J., S. Meyers, A. Szporn, G. Slater, H.D. Janowitz, and A.H. Aufses, Jr. 1987. Colorectal cancer in regional ileitis. *Q. J. Med.* 62:33–40.

20. Greenstein, A.J., D.B. Sachar, H. Smith, H.D. Janowitz, and A.H. Aufses, Jr. 1980. Patterns of neoplasia in Crohn's disease and ulcerative colitis. *Cancer* 46:403–407.

21. Apgar, J.T. 1991. Newer aspects of inflammatory bowel disease and its cutaneous manifestations: a selective review. *Semin. Dermatol.* 10:138–147.

22. Das, K.M. 1999. Relationship of extraintestinal involvements in inflammatory bowel disease: new insights into autoimmune pathogenesis. *Dig. Dis. Sci.* 44:1–13.

23. Salmi, M., and S. Jalkanen. 1998. Endothelial ligands and homing of mucosal leukocytes in extraintestinal manifestations of IBD. *Inflamm. Bowel. Dis.* 4:149–156.

24. Levine, J.B., and D. Lukawski-Trubish. 1995. Extraintestinal considerations in inflammatory bowel disease. *Gastroenterol. Clin. North Am.* 24:633–646.

25. Greenstein, A.J., H.D. Janowitz, and D.B. Sachar. 1976. The extra-intestinal complications of Crohn's disease and ulcerative colitis: a study of 700 patients. *Medicine (Baltimore)* 55:401–412.

26. Yang, H., D.K. Vora, S.R. Targan, H. Toyoda, A.L. Beaudet, and J.I. Rotter. 1995. Intercellular adhesion molecule 1 gene associations with immunologic subsets of inflammatory bowel disease. *Gastroenterology* 109:440–448.

27. Yang, H., J.I. Rotter, H. Toyoda, C. Landers, D. Tyran, C.K. McElree, and S.R. Targan. 1993. Ulcerative colitis: a genetically heterogeneous disorder defined by genetic (HLA class II) and subclinical (antineutrophil cytoplasmic antibodies) markers. *J. Clin. Invest.* 92:1080–1084.

28. Toyoda, H., S.J. Wang, H.Y. Yang, A. Redford, D. Magalong, D. Tyan, C.K. McElree, S.R., Pressman, F. Shanahan, S.R. Targan, and et al. 1993. Distinct associations of HLA class II genes with inflammatory bowel disease. *Gastroenterology* 104:741–748.

29. Binder, V., and M. Orholm. 1996. Familial occurrence and inheritance studies in inflammatory bowel disease. *Neth. J. Med.* 48:53–56.

30. Thompson, N.P., R. Driscoll, R.E. Pounder, and A.J. Wakefield. 1996. Genetics versus environment in inflammatory bowel disease: results of a British twin study *Brit. Med. J.* 312:95–96.

31. Hugot, J.P., P. Laurent-Puig, C. Gower-Rousseau, J.M. Olson, J.C. Lee, L. Beaugerie, I. Naom, J.L. Dupas, A. Van Gossum, M. Orholm, C. Bonaiti-Pellie, J. Weissenbach, C.G. Mathew, J.E. Lennard-Jones, A. Cortot, J.F. Colombel, and G. Thomas. 1996. Mapping of a susceptibility locus for Crohn's disease on chromosome 16. *Nature* 379:821–823.

32. Mirza, M.M., J. Lee, D. Teare, J.P. Hugot, P. Laurent-Puig, J.F. Colombel, S.V. Hodgson, G. Thomas, D.F. Easton, J.E. Lennard-Jones, and C.G. Mathew. 1998. Evidence of linkage of the inflammatory bowel disease susceptibility locus on chromosome 16 (IBD1) to ulcerative colitis. *J. Med. Genet.* 35:218–221.

33. Ohmen, J.D., H.Y. Yang, K.K. Yamamoto, H.Y. Zhao, Y. Ma, L.G. Benetley, Z. Huang, S. Gerwehr, S. Pressman, C. McElree, S. Targan, J.I. Rotter, and N. Fischel-Ghodsian. 1996. Susceptibility locus for inflammatory bowel disease on chromosome 16 has a role in Crohn's disease, but not in ulcerative colitis. *Hum. Mol. Genet.* 5:1679–1683.

34. Brant, S.R., Y. Fu, C.T. Fields, R. Baltazar, G. Ravenhill, M.R. Pickles, P.M. Rohal, J. Mann, B.S. Kirschner, E.W. Jabs, T.M. Bayless, S.B. Hanauer, and J.H. Cho. 1998. American families with Crohn's disease have strong evidence for linkage to chromosome 16 but not chromosome 12 [see comments]. *Gastroenterology* 115:1056–1061.

35. Duerr, R.H., M.M. Barmada, L. Zhang, S. Davis, R.A. Preston, L.J. Chensny, J.L. Brown, G.D. Ehrlich, D.E. Weeks, and C.E. Aston. 1998. Linkage and association between inflammatory bowel disease and a locus on chromosome 12. *Am. J. Hum. Genet.* 63:95–100.

36. Cho, J.H., D.L. Nicolae, L.H. Gold, C.T. Fields, M.C.J. aBuda, P.M. Rohal, M.R. Pickles, L. Qin, Y. Fu, J.S. Mann, B.S. Kirschner, E.W. Jabs, J. Weber, S.B. Hanauer, T.M. Bayless, and S.R. Brant. 1998. Identification of novel susceptibility loci for inflammatory bowel disease on chromosomes 1p, 3q, and 4q: evidence for epistasis between 1p and IBD1. *Proc. Natl. Acad. Sci. USA* 95:7502–7507.

37. Satsangi, J., M. Parkes, E. Louis, L. Hashimoto, N. Kato, K. Welsh, J.D. Terwilliger, G.M. Lathrop, J.I. Bell, and D.P. Jewell. 1996. Two stage genome-wide search in inflammatory bowel disease provides evidence for susceptibility loci on chromosomes 3, 7 and 12. *Nat. Genet.* 14:199–202.

38. Parkes, M., J. Satsangi, and D. Jewell. 1997. Mapping susceptibility loci in inflammatory bowel disease: Why and how? *Mol. Med. Today* 3:546–553.

39. Fuzi, M. 1999. Is the pathogen of prion disease a microbial protein? [In Process Citation]. *Med. Hypotheses* 53:91–102.

40. Thayer, W.R., Jr., J.A. Coutu, R.J. Chiodini, H.J. Van Kruiningen, and R.S. Merkal. 1984. Possible role of mycobacteria in inflammatory bowel disease. II. Mycobacterial antibodies in Crohn's disease. *Dig. Dis. Sci.* 29:1080–1085.

41. Chiodini, R.J., H.J. Van Kruiningen, W.R. Thayer, R.S. Merkal, and J.A. Coutu. 1984. Possible role of mycobacteria in inflammatory bowel disease. I. An unclassified Mycobacterium species isolated from patients with Crohn's disease. *Dig. Dis. Sci.* 29:1073–1079.

42. Parent, K., and P. Mitchell. 1978. Cell wall-defective variants of pseudomonas-like (group Va) bacteria in Crohn's disease. *Gastroenterology* 75:368–372.

43. Beeken, W.L. 1980. Transmissible agents in inflammatory bowel disease: 1980. *Med. Clin. North Am.* 64:1021–1035.

44. Whorwell, P.J., W.L. Beeken, and C.A. Phillips. 1977. Absence of reovirus-like agent in Crohn's tissue [letter]. *Lancet* 2:257.

45. Whorwell, P.J., C.A. Phillips, W.L. Beeken, P.K. Little, and K.D. Roessner. 1977. Isolation of reovirus-like agents from patients with Crohn's disease. *Lancet* 1:1169–1171.

46. Beeken, W.L., D.N. Mitchell, and D.R. Cave. 1976. Evidence for a transmissible agent in Crohn's disease. *Clin. Gastroenterol.* 5:289–302.

47. Beeken, W.L., K.K. Goswami, and D.N. Mitchell. 1975. Proceedings: Studies of a viral agent isolated from patients with Crohn's disease and other intestinal disorders. *Gut.* 16:401.

48. Phillpotts, R.J., J. Hermon-Taylor, and B.N. Brooke. 1979. Virus isolation studies in Crohn's disease: a negative report. *Gut.* 20:1057–1062.

49. Cho, S.N., P.J. Brennan, H.H. Yoshimura, B.I. Korelitz, and D.Y. Graham. 1986. Mycobacterial aetiology of Crohn's disease: serologic study using common mycobacterial antigens and a species-specific glycolipid antigen from Mycobacterium paratuberculosis. *Gut* 27:1353–1356.

50. Kobayashi, K., M.J. Blaser, and W.R. Brown. 1989. Immunohistochemical examination for mycobacteria in intestinal tissues from patients with Crohn's disease. *Gastroenterology* 96:1009–1015.

51. Kobayashi, K., W.R. Brown, P.J. Brennan, and M.J. Blaser. 1988. Serum antibodies to mycobacterial antigens in active Crohn's disease. *Gastroenterology* 94:1404–1411.

52. Thayer, W.R. 1992. The use of antimycobacterial agents in Crohn's disease [editorial]. *J. Clin. Gastroenterol.* 15:5–7.

53. Bookman, M.A., and D.M. Bull. 1979. Characteristics of isolated intestinal mucosal lymphoid cells in inflammatory bowel disease. *Gastroenterology* 77:503–510.

54. Bull, D.M., and M.A. Bookman. 1977. Isolation and functional characterization of human intestinal mucosal lymphoid cells. *J. Clin. Invest.* 59:966–974.

55. Bartnik, W., and S. Kaluzewski. 1979. Cellular and humoral response to Kunin antigen (CA) in ulcerative colitis and Crohn's disease. *Arch. Immunol. Ther. Exp. (Warsz.)* 27:531–538.

56. Carlsson, H.E., R. Lagercrantz, and P. Perlmann. 1977. Immunological studies in ulcerative colitis. VIII. Antibodies to colon antigen in patients with ulcerative colitis, Crohn's disease, and other diseases. *Scand. J. Gastroenterol* 12:707–714.

57. Thayer, W.R., Jr., M. Brown, M.H. Sangree, J. Katz, and T. Hersh. 1969. Escherichia Coli O:14 and colon hemagglutinating antibodies in inflammatory bowel disease. *Gastroenterology* 57:311–318.

58. Marcussen, H. 1976. Anti-colon antibodies in ulcerative colitis. A clinical study. *Scand. J. Gastroenterol.* 11:763–767.

59. Marcussen, H., and J. Nerup. 1973. Fluorescent anti-colon and organ-specific antibodies in ulcerative colitis. *Scand. J. Gastroenterol.* 8:9–15.

60. Rabin, B.S., and S.J., Rogers. 1976. Nonpathogenicity of anti-intestinal antibody in the rabbit. *Am. J. Pathol.* 83:269–282.

61. Rabin, B.S., and S.J. Rogers. 1978. A cell-mediated immune model of inflammatory bowel disease in the rabbit. *Gastroenterology* 75:29–33.

62. Ross, I.N., R.A. Thompson, R.D. Montgomery, and P. Asquith. 1979. Significance of serum complement levels in patients with gastrointestinal disease. *J. Clin. Pathol.* 32:798–801.

63. Hodgson, H.J., B.J. Potter, and D.P. Jewell. 1977. Humoral immune system in inflammatory bowel disease: I. Complement levels. *Gut* 18:749–753.

64. Hodgson, H.J., B.J. Potter, and D.P. Jewell. 1977. C3 metabolism in ulcerative colitis and Crohn's disease. *Clin. Exp. Immunol.* 28:490–495.

65. Hodgson, H.J., B.J. Potter, and D.P. Jewell. 1975. Proceedings: Complement in inflammatory bowel disease. *Gut* 16:833–834.

66. Ahrenstedt, O., L. Knutson, B. Nilsson, K. Nilsson-Ekdahl, B. Odlind, and R. Hallgren. 1990. Enhanced local production of complement components in the small intestines of patients with Crohn's disease. *N. Engl. J. Med.* 322:1345–1349.

67. Lake, A.M., A.E. Stitzel, J.R. Urmson, W.A. Walker, and R.E. Spitzer. 1979. Complement alterations in inflammatory bowel disease. *Gastroenterology* 76:1374–1379.

68. Ward, M., and M.A. Eastwood. 1975. Serum C3 and C4 complement components in ulcerative colitis and Crohn's disease. *Digestion* 13:100–103.

69. Das, K.M., M. Vecchi, L. Squillante, A. Dasgupta, M. Henke, and N. Clapp. 1992. Mr 40,000 human colonic epithelial protein expression in colonic mucosa and presence of circulating anti-Mr 40,000 antibodies in cotton top tamarins with spontaneous colitis. *Gut* 33:48–54.

70. Das, K.M., L. Squillante, D. Chitayet, and D.K. Kalousek. 1992. Simultaneous appearance of a unique common epitope in fetal colon, skin, and biliary epithelial cells. A possible link for extracolonic manifestations in ulcerative colitis. *J. Clin. Gastroenterol.* 15:311–316.

71. Biancone, L., K.M. Das, A.I. Roberts, and E.C. Ebert. 1993. Ulcerative colitis serum recognizes the M(r) 40K protein on colonic adenocarcinoma cells for antibody-dependent cellular cytotoxicity. *Digestion* 54:237–242.

72. Das, K.M., A. Dasgupta, A. Mandal, and X. Geng. 1993. Autoimmunity to cytoskeletal protein tropomyosin. A clue to the pathogenetic mechanism for ulcerative colitis. *J. Immunol.* 150:2487–2493.

73. Mandal, A., A. Dasgupta, L. Jeffers, L. Squillante, S. Hyder, R. Reddy, E. Schiff, and K.M. Das. 1994. Autoantibodies in sclerosing cholangitis against a shared peptide in biliary and colon epithelium. *Gastroenterology* 106:185–192.

74. Bhagat, S., and K.M. Das. 1994. A shared and unique peptide in the human colon, eye, and joint detected by a monoclonal antibody. *Gastroenterology* 107:103–108.

75. Dasgupta, A., A. Mandal, and K.M. Das. 1994. Circulating immunoglobulin G1 antibody in patients with ulcerative colitis against the colonic epithelial protein detected by a novel monoclonal antibody. *Gut* 35:1712–1717.

76. Geng, X., L. Biancone, H.H. Dai, J.J., Lin, N. Yoshizaki, A. Dasgupta, F. Pallone, and K.M. Das. 1998. Tropomyosin isoforms in intestinal mucosa: production of autoantibodies to tropomyosin isoforms in ulcerative colitis. *Gastroenterology* 114:912–922.

77. Kesari, K.V., N. Yoshizaki, X. Geng, J.J. Lin, and K.M. Das. 1999. Externalization of tropomyosin isoform 5 in colon epithelial cells [In Process Citation]. *Clin. Exp. Immunol.* 118:219–227.

78. Takahasi, F., H.S. Shah, L.S. Wise, and K.M. Das. 1990. Circulating antibodies against human colonic extract

enriched with a 40 kDa protein in patients with ulcerative colitis. *Gut* 31:1016–1020.

79. Das, K.M., S. Sakamaki, and M. Vecchi. 1989. Ulcerative colitis: specific antibodies against a colonic epithelial Mr 40,000 protein. *Immunol. Invest.* 18:459–472.

80. Takahashi, F., and K.M. Das. 1985. Isolation and characterization of a colonic autoantigen specifically recognized by colon tissue-bound immunoglobulin G from idiopathic ulcerative colitis. *J. Clin. Invest.* 76:311–318.

81. MacDermott, R.P., G.S. Nash, I.O. Auer, R. Shlien, B.S. Lewis, J. Madassery, and M.H. Nahm. 1989. Alterations in serum immunoglobulin G subclasses in patients with ulcerative colitis and Crohn's disease. *Gastroenterology* 96:764–768.

82. Scott, M.G., M.H., Nahm, K. Macke, G.S. Nash, M.J. Bertovich, and R.P. MacDermott. 1986. Spontaneous secretion of IgG subclasses by intestinal mononuclear cells: differences between ulcerative colitis, Crohn's disease, and controls. *Clin. Exp. Immunol.* 66:209–215.

83. Brandtzaeg, P., T.S. Halstensen, K. Kett, P. Krajci, D. Kvale, T.O. Rognum, H. Scott, L.M. Sollid, K. Bjerke, K. Valnes, R. Soderstrom, J. Bjorkander, T. Soderstrom, B. Petrusson, and L.A. Hanson. 1989. Immunobiology and immunopathology of human gut mucosa: humoral immunity and intraepithelial lymphocytes. *Gastroenterology* 97:1562–1584.

84. Halstensen, T.S., K.M. Das, and P. Brandtzaeg. 1995. Epithelial deposits of immunoglobulin G1 and activated complement co-localize with the "M(r) 40kD" putative autoantigen in ulcerative colitis. *Adv. Exp. Med. Biol.* 371B:1273–1276.

85. Halstensen, T.S., K.M. Das, and P. Brandtzaeg. 1993. Epithelial deposits of immunoglobulin G1 and activated complement colocalise with the M(r) 40 kD putative autoantigen in ulcerative colitis. *Gut* 34:650–657.

86. Halstensen, T.S., T.E. Mollnes, P. Garred, O. Fausa, and P. Brandtzaeg. 1990. Epithelial deposition of immunoglobulin G1 and activated complement (C3b and terminal complement complex) in ulcerative colitis. *Gastroenterology* 98:1264–1271.

87. Brandtzaeg, P., K. Valnes, H. Scott, T.O. Rognum, K. Bjerke, and K. Baklien. 1985. The human gastrointestinal secretory immune system in health and disease. *Scand. J. Gastroenterol. Suppl.* 114:17–38.

88. Dickinson, B.L., K. Badizadegan, Z. Wu, J.C. Ahouse, X. Zhu, N.E. Simister, R.S. Blumberg, and W.I. Lencer. 1999. Bidirectional FcRn-dependent IgG transport in polarized human intestinal epithelial cell line. *J. Clin. Invest* 104:903–911.

89. Fiocchi, C., J.K. Roche, and W.M. Michener. 1989. High prevalence of antibodies to intestinal epithelial antigens in patients with inflammatory bowel disease and their relatives [see comments]. *Ann. Intern. Med.* 110:786–794.

90. Roche, J.K., C. Fiocchi, and K. Youngman. 1985. Sensitization to epithelial antigens in chronic mucosal inflammatory disease. Characterization of human intestinal mucosa-derived mononuclear cells reactive with purified epithelial cell-associated components *in vitro*. *J. Clin. Invest.* 75:522–530.

91. Cohavy, O., G. Harth, M. Horwitz, M. Eggena, C. Landers, C. Sutton, S.R. Targan, and J. Braun. 1999. Identification of a novel mycobacterial histone H1 homologue (HupB) as an antigenic target of pANCA monoclonal antibody and serum immunoglobulin A from patients with Crohn's disease. *Infect. Immun.* 67:6510–6517.

92. Saxon, A., F. Shanahan, C. Landers, T. Ganz, and S. Targan. 1990. A distinct subset of antineutrophil cytoplasmic antibodies is associated with inflammatory bowel disease. *J. Allergy Clin. Immunol.* 86:202–210.

93. Seibold, F., S. Brandwein, S. Simpson, C. Terhorst, and C.O. Elson. 1998. pANCA represents a cross-reactivity to enteric bacterial antigens. *J. Clin. Immunol.* 18:153–160.

94. Targan, S.R., C.J. Landers, L. Cobb, R.P. MacDermott, and A. Vidrich. 1995. Perinuclear anti-neutrophil cytoplasmic antibodies are spontaneously produced by mucosal B cells of ulcerative colitis patients. *J. Immunol.* 155:3262–3267.

95. MacDermott, R.P., M.J. Bragdon, I.J. Kodner, and M.J. Bertovich. 1986. Deficient cell-mediated cytotoxicity and hyporesponsiveness to interferon and mitogenic lectin activation by inflammatory bowel disease peripheral blood and intestinal mononuclear cells. *Gastroenterology* 90:6–11.

96. Fiocchi, C., R.R. Tubbs, and K.R. Youngman. 1985. Human intestinal mucosal mononuclear cells exhibit lymphokine-activated killer cell activity. *Gastroenterology* 88:625–637.

97. Shanahan, F., B. Leman, R. Deem, A. Niederlehner, M. Brogan, and S. Targan. 1989. Enhanced peripheral blood T-cell cytotoxicity in inflammatory bowel disease. *J. Clin. Immunol.* 9:55–64.

98. Niessner, M., and B.A. Volk. 1995. Altered Th1/Th2 cytokine profiles in the intestinal mucosa of patients with inflammatory bowel disease as assessed by quantitative reversed transcribed polymerase chain reaction (RT-PCR). *Clin. Exp. Immunol.* 101:428–435.

99. Salomon, P., A. Pizzimenti, A. Panja, A. Reisman, and L. Mayer. 1991. The expression and regulation of class II antigens in normal and inflammatory bowel disease peripheral blood monocytes and intestinal epithelium. *Autoimmunity* 9:141–149.

100. Deem, R.L., F. Shanahan, and S.R. Targan. 1991. Triggered human mucosal T cells release tumour necrosis factor-alpha and interferon-gamma which kill human colonic epithelial cells. *Clin. Exp. Immunol.* 83:79–84.

101. Liu, Z., S. Colpaert, G.R. D'Haens, A. Kasran, M. de Boer, P. Rutgeerts, K. Geboes, and J.L. Ceuppens. 1999. Hyperexpression of CD40 ligand (CD154) in inflammatory bowel disease and its contribution to pathogenic cytokine production. *J. Immunol.* 163:4049–4057.

102. Schreiber, S., R.P. MacDermott, A. Raedler, R. Pinnau, M.J. Bertovich, and G.S. Nash. 1991. Increased activation of isolated intestinal lamina propria mononuclear cells in inflammatory bowel disease [see comments]. *Gastroenterology* 101:1020–1030.

103. Pallone, F., S. Fais, O. Squarcia, L. Biancone, P. Pozzilli, and M. Boirivant. 1987. Activation of peripheral blood and intestinal lamina propria lymphocytes in Crohn's disease. *In vivo* state of activation and *in vitro* response to stimulation as defined by the expression of early activation antigens. *Gut* 28:745–753.

104. Fais, S., F. Pallone, O. Squarcia, M. Boirivant, and P. Pozzilli. 1985. T cell early activation antigens expressed by peripheral lymphocytes in Crohn's disease. *J. Clin. Lab. Immunol.* 16:75–76.

105. James, S.P., G.E. Mullin, M.E. Kanof, and M. Zeitz. 1991. Role of lymphokines in immunoregulatory function of mucosal T cells in humans and nonhuman primates. *Immunol. Res.* 10:230–238.

106. Mullin, G.E., A.J. Lazenby, M.L. Harris, T.M. Bayless, and S.P. James. 1992. Increased interleukin-2 messenger RNA in the intestinal mucosal lesions of Crohn's disease but not ulcerative colitis. *Gastroenterology* 102:1620–1627.

107. Berrebi, D., M. Besnard, G. Fromont-Hankard, R. Paris. J.F. Mougenot, P. De Lagausie, D. Emilie, J.P. Cezard, J. Navarro, and M. Peuchmaur. 1998. Interleukin-12 expression is focally enhanced in the gastric mucosa of pediatric patients with Crohn's disease. *Am. J. Pathol.* 152:667–672.

108. Camoglio, L., A.A. Te Velde, A.J. Tigges, P.K. Das, and S.J. Van Deventer. 1998. Altered expression of interferon-gamma and interleukin-4 in inflammatory bowel disease. *Inflamm. Bowel. Dis.* 4:285–290.

109. Pallone, F., and G. Monteleone. 1998. Interleukin 12 and Th1 responses in inflammatory bowel disease. *Gut* 43:735–736.

110. Parronchi, P., P. Romagnani, F. Annunziato, S. Sampognaro, A. Becchio, L. Giannarini, E. Maggi, C. Pupilli, F. Tonelli, and S. Romagnani. 1997. Type 1 T-helper cell predominance and interleukin-12 expression in the gut of patients with Crohn's disease. *Am. J. Pathol.* 150 3:823–832.

111. Romagnani, P., F. Annunziato, M.C. Baccari, and P. Parronchi. 1997. T cells and cytokines in Crohn's disease. *Curr. Opin. Immunol.* 9:793–799.

112. Romagnani, S. 1999. Th1/Th2 cells. *Inflamm. Bowel. Dis.* 5:285–294.

113. Gately, M.K., L.M. Renzetti, J. Magram, A.S. Stern, L. Adorini, U. Gubler, and D.H. Presky. 1998. The inter-leukin-12/interleukin-12-receptor system: role in normal and pathologic immune responses. *Annu. Rev. Immunol.* 16:495–521.

114. Monteleone, G., L. Biancone, R. Marasco, G. Morrone, O. Marasco, F. Luzza, and F. Pallone. 1997. Interleukin 12 is expressed and actively released by Crohn's disease intestinal lamina propria mononuclear cells. *Gastroenterology* 112:1169–1178.

115. Monteleone, G., F. Trapasso, T. Parrello, L. Biancone, A. Stella, R. Iuliano, F. Luzza, A. Fusco, and F. Pallone. 1999. Bioactive IL-18 expression is up-regulated in Crohn's disease. *J. Immunol.* 163:143–147.

116. Pizarro, T.T., M.H. Michie, M. Bentz, J. Woraratanadharm, M.F. Smith, Jr., E. Foley, C.A. Moskaluk, S.J. Bickston, and F. Cominelli. 1999. IL-18, a novel immunoregulatory cytokine, is up-regulated in Crohn's disease: expression and localization in intestinal mucosal cells. *J. Immunol.* 162:6829–6835.

117. Robinson, M. 1997. Optimizing therapy for inflammatory bowel disease. *Am. J. Gastroenterol.* 92:12S–17S.

118. van Deventer, S.J., P. Rutgeerts, F. Baert, E. Ricart, R. Panaccione, E.V. Loftus, W.J. Tremaine, W.J. Sandborn, S.B. Hanauer, R.A. van Hogezand, and H.W. Verspaget. 1999. Anti-TNF antibody treatment of Crohn's disease: New strategies in the management of inflammatory bowel disease. *Ann. Rheum. Dis.* 5:114–1120.

119. Present, D.H., P. Rutgeerts, S. Targan, S.B. Hanauer, L. Mayer, R.A. van Hogezand, D.K. Podolsky, B.E. Sands, T. Braakman, K.L. DeWoody, T.F. Schaible, and S.J. van Deventer. 1999. Infliximab for the treatment of fistulas in patients with Crohn's disease. *N. Engl. J. Med.* 340:1398–1405.

120. Rutgeerts, P., G. D'Haens, S. Targan, E. Vasiliauskas, S.B. Hanauer, D.H. Present, L. Mayer, R.A. Van Hogezand, T. Braakman, K.L. DeWoody, T.F. Schaible, and S.J. Van Deventer. 1999. Efficacy and safety of retreatment

121. Targan, S.R., S.B. Hanauer, S.J. van Deventer, L. Mayer, D.H. Present, T. Braakman, K.L. DeWoody, T.F. Schaible, and P.J. Rutgeerts. 1997. A short-term study of chimeric monoclonal antibody cA2 to tumor necrosis factor alpha for Crohn's disease. Crohn's Disease. cA2 Study Group. *N. Engl. J. Med.* 337:1029–1035.

122. Narula, S.K., D. Cutler, and P. Grint. 1998. Immunomodulation of Crohn's disease by interleukin-10. *Agents Actions Suppl.* 49:57–65.

123. Chott, A., C.S. Probert, G.G. Gross, R.S. Blumberg, and S.P. Balk. 1996. A common TCR beta-chain expressed by CD8+ intestinal mucosa T cells in ulcerative colitis. *J. Immunol.* 156:3024–3035.

124. Saubermann, L.J., C.S. Probert, A.D. Christ, A. Chott, J.R. Turner, A.C. Stevens, S.P. Balk, and R.S. Blumberg. 1999. Evidence of T cell receptor beta-chain patterns in inflammatory and noninflammatory bowel disease states. *Am. J. Physiol.* 276:613–21.

125. Regnault, A., A. Cumano, P. Vassalli, D. Guy-Grand, and P. Kourilsky. 1994. Oligoclonal repertoire of the CD8 alpha alpha and the CD8 alpha beta TCR-alpha/beta murine intestinal intraepithelial T lymphocytes: evidence for the random emergence of T cells. *J. Exp. Med.* 180:1345–1358.

126. Van Kerckhove, C., G.J. Russell, K. Deusch, K. Reich, A.K. Bhan, H. DerSimonian, and M.B. Brenner. 1992. Oligoclonality of human intestinal intraepithelial T cells. *J. Exp. Med.* 175:57–63.

127. Helgeland, L., F.E. Johansen, J.O. Utgaard, J.T. Vaage, and P. Brandtzaeg. 1999. Oligoclonality of rat intestinal intraepithelial T lymphocytes: overlapping TCR beta-chain repertoires in the CD4 single-positive and CD4/CD8 double-positive subsets. *J. Immunol.* 162:2683–2692.

128. Blumberg, R.S., C.E. Yockey, G.G. Gross, E.C. Ebert, and S.P. Balk. 1993. Human intestinal intraepithelial lymphocytes are derived from a limited number of T cell clones that utilize multiple V beta T cell receptor genes. *J. Immunol.* 150:5144–5153.

129. Mizoguchi, A., E. Mizoguchi, S. Tonegawa, and A.K. Bhan. 1996. Alteration of a polyclonal to an oligoclonal immune response to cecal aerobic bacterial antigens in TCR alpha mutant mice with inflammatory bowel disease. *Int. Immunol.* 8:1387–1394.

130. Kuhn, R., J. Lohler, D. Rennick, K. Rajewsky, and W. Muller. 1993. Interleukin-10-deficient mice develop chronic enterocolitis *Cell* 75:263–274.

131. Davidson, N.J., M.W. Leach, M.M. Fort, L. Thompson-Snipes, R. Kuhn, W. Muller, D.J. Berg, and D.M. Rennick. 1996. T helper cell-1-type CD4+ T cells, but not B cells, mediate colitis in interleukin 10-deficient mice. *J. Exp. Med.* 184:241–251.

132. Sadlack, B., H. Merz, H. Schorle, A. Schimpl, A.C. Feller, and I. Horak. 1993. Ulcerative colitis-like disease in mice with a disrupted interleukin-2 gene *Cell* 75:253–261.

133. Bhan, A.K., E. Mizoguchi, R.N. Smith, and A. Mizoguchi. 1999. Colitis in transgenic and knockout animals as models of human inflammatory bowel disease. *Immunol. Rev.* 169:195–207.

134. Brandwein, S.L., R.P. McCabe, Y. Cong, K.B. Waites, B.U. Ridwan, P.A. Dean, T. Ohkusa, E.H. Birkenmeier, J.P. Sundberg, and C.O. Elson. 1997. Spontaneously colitic

C3H/HeJBir mice demonstrate selective antibody reactivity to antigens of the enteric bacterial flora. *J. Immunol.* 159:44–52.

135. Cong, Y., S.L. Brandwein, R.P. McCabe, A. Lazenby, E.H. Birkenmeier, J.P. Sundberg, and C.O. Elson. 1998. CD4+ T cells reactive to enteric bacterial antigens in spontaneously colitic C3H/HeJBir mice: increased T helper cell type 1 response and ability to transfer disease. *J. Exp. Med.* 187:855–864.

136. Mahler, M., I.J. Bristol, E.H. Leiter, A.E. Workman, E.H. Birkenmeier, C.O. Elson, and J.P. Sundberg. 1998. Differential susceptibility of inbred mouse strains to dextran sulfate sodium-induced colitis. *Am. J. Physiol.* 274:G544–551.

137. Elson, C.O., Y. Cong, S. Brandwein, C.T. Weaver, R.P. McCabe, M. Mahler, J.P. Sundberg, and E.H. Leiter. 1998. Experimental models to study molecular mechanisms underlying intestinal inflammation. *Ann. NY. Acad. Sci.* 859:85–95.

138. Mahler, M., I.J. Bristol, J.P. Sundberg, G.A. Churchill, E.H. Birkenmeier, C.O. Elson, and E.H. Leiter. 1999. Genetic analysis of susceptibility to dextran sulfate sodium-induced colitis in mice. *Genomics* 55:147–156.

139. Louis, E., and J. Belaiche. 1994. Experimental models of inflammatory bowel disease. *Acta. Gastroenterol. Belg.* 57:306–309.

140. Panwala, C.M., J.C. Jones, and J.L. Viney. 1998. A novel model of inflammatory bowel disease: mice deficient for the multiple drug resistance gene, mdr 1a, spontaneously develop colitis. *J. Immunol.* 161.5733–5744.

141. Powrie, F., M.W. Leach, S. Mauze, L.B. Caddle, and R.L. Coffman. 1993. Phenotypically distinct subsets of CD4+ T cells induce or protect from chronic intestinal inflammation in C. B-17 scid mice. *Int. Immunol.* 5:1461–1471.

142. Rath, H.C., K.H. Wilson, and R.B. Sartor. 1999. Differential induction of colitis and gastritis in HLA-B27 transgenic rats selectively colonized with Bacteroides vulgatus or Escherichia coli. *Infect. Immun.* 67:2969–2974.

143. Rudolph, U., M.J. Finegold, S.S. Rich, G.R. Harriman, Y. Srinivasan, P. Brabet, G. Boulay, A. Bradley, and L. Birnbaumer. 1995. Ulcerative colitis and adenocarcinoma of the colon in G alpha i2-deficient mice. *Nat. Genet.* 10:143–150.

144. Hermiston, M.L., and J.I. Gordon. 1995. Inflammatory bowel disease and adenomas in mice expressing a dominant negative N-cadherin. *Science* 270:1203–1207.

145. Powrie, F., M.W. Leach, S. Mauze, S. Menon, L.B. Caddle, and R.L. Coffman. 1994. Inhibition of Th1 responses prevents inflammatory bowel disease in scid mice reconstituted with CD45RB^{hi} CD4+ T cells. *Immunity* 1:553–562.

146. Powrie, F., R. Correa-Olivera, S. Mauze, and R.L. Coffman. 1994. Regulatory interactions between CD45RB^{high} and CD45RB^{low} CD4+ T cells are important for the balance between protective and pathogenic cell-mediated immunity. *J. Exp. Med.* 179:589–600.

147. Powrie, F. 1995. T cells in inflammatory bowel disease: protective and pathogenic roles. *Immunity* 3:171–4.

148. Groux, H., A. O'Garra, M. Bigler, M. Rouleau, S. Antonenko, J.E. de Vries, and M.G. Roncarolo. 1997. A CD4+ T-cell subset inhibits antigen-specific T-cell responses and prevents colitis. *Nature* 389:737–742.

149. Mizoguchi, A., E. Mizoguchi, C. Chiba, and A.K. Bhan. 1996. Role of appendix in the development of inflammatory bowel disease in TCR-alpha mutant mice. *J. Exp. Med.* 184:707–715.

150. Sellon, R.K., S. Tonkonogy, M. Schultz, L.A. Dieleman, W. Grenther, E. Balish, D.M. Rennick, and R.B. Sartor. 1998. Resident enteric bacteria are necessary for development of spontaneous colitis and immune system activation in interleukin-10-deficient mice. *Infect. Immun.* 66:5224–5231.

151. Schultz, M., S.L. Tonkonogy, R.K. Sellon, C. Veltkamp, V.L. Godfrey, J. Kwon, W.B. Grenther, E. Balish, I. Horak, and R.B. Sartor. 1999. IL-2-deficient mice raised under germfree conditions develop delayed mild focal intestinal inflammation. *Am. J. Physiol.* 276:1461–1472.

152. Rath, H.C., H.H. Herfarth, J.S. Ikeda, W.B. Grenther, T.E. Hamm, Jr., E. Balish, J.D. Taurog, R.E. Hammer, K.H. Wilson, and R.B. Sartor. 1996. Normal luminal bacteria, especially Bacteroides species, mediate chronic colitis, gastritis, and arthritis in HLA-B27/human beta2 microglobulin transgenic rats. *J. Clin. Invest.* 98:945–953.

153. Duchmann, R., E. Schmitt, P. Knolle, K.H. Meyer zum Buschenfelde, and M. Neurath. 1996. Tolerance towards resident intestinal flora in mice is abrogated in experimental colitis and restored by treatment with interleukin-10 or antibodies to interleukin-12. *Eur. J. Immunol.* 26:934–938.

154. Kullberg, M.C., J.M. Ward, P.L. Gorelick, P. Caspar, S. Hieny, A. Cheever, D. Jankovic, and A. Sher. 1998. Helicobacter hepaticus triggers colitis in specific pathogen-free interleukin-10 (IL-10)-deficient mice through an IL 12- and gamma interferon-dependent mechanism. *Infect. Immun.* 66:5157–5166.

155. Kabok, Z., T.H. Ermak, and J. Pappo. 1995. Microdissected domes from gut-associated lymphoid tissues: A model of M cell transepithelial transport *in vitro*. *Adv. Exp. Med. Biol.* Vol. 235–238.

156. Kagnoff, M.F. 1993. Immunology of the intestinal tract. *Gastroenterology* 105:1275–1280.

157. Keren, D.F. 1992. Antigen processing in the mucosal immune system. *Semin. Immunol.* 4:211–226.

158. Owen, R.L., N.F. Pierce, R.T. Apple, and W.C. Cray, Jr. 1986. M cell transport of Vibrio cholerae from the intestinal lumen into Peyer's patches: a mechanism for antigen sampling and for microbial transepithelial migration. *J. Infect. Dis.* 153:1108–1118.

159. Paar, M., E.M. Liebler, and J.F. Pohlenz. 1992. Uptake of ferritin by follicle-associated epithelium in the colon of calves. *Vet. Pathol.* 29:120–128.

160. Pabst, R. 1987. The anatomical basis for the immune function of the gut. *Anat. Embryol.* 176:135–144.

161. Samuel, B.U., I. Indrasingh, G. Chandi, and T.J. John. 1992. Ultrastructural specialization of intestinal epithelium over Peyer's patches in the bonnet monkey, Macaca radiata. *Indian. J. Exp. Biol.* 30:1138–1141.

162. Thomas, N.W., P.G. Jenkins, K.A. Howard, M.W. Smith, E.C. Lavelle, J. Holland, and S.S. Davis. 1996. Particle uptake and translocation across epithelial membranes. *J. Anat.* 189:487–490.

163. Gebert, A. 1995. Identification of M-cells in the rabbit tonsil by vimentin immunohistochemistry and *in vivo* protein transport. *Histochem. Cell. Biol.* 104:211–220.

164. Gebert, A., and G. Hach. 1993. Differential binding of lectins to M cells and enterocytes in the rabbit cecum. *Gastroenterology* 105:1350–1361.

165. Ermak, T.H., H.R. Bhagat, and J. Pappo. 1994. Lymphocyte compartments in antigen-sampling regions of rabbit mucosal lymphoid organs. *Am. J. Trop. Hyg.* 50:14–28.

166. Brandtzaeg, P., and K. Bjerke. 1990. Immunomorphological characteristics of human Peyer's patches. *Digestion* 46:262–273.

167. Kagnoff, M.F. 1996. Oral tolerance: mechanisms and possible role in inflammatory joint diseases. *Baillieres Clin. Rheumatol.* 10:41–54.

168. Koh, D.R. 1998. Oral tolerance: mechanisms and therapy of autoimmune diseases. *Ann. Acad. Med. Singapore* 27:47–53.

169. Lundin, B.S., U.I. Dahlgren, L.A. Hanson, and E. Telemo. 1996. Oral tolerization leads to active suppression and bystander tolerance in adult rats while anergy dominates in young rats. *Scand. J. Immunol.* 43:56–63.

170. MacDonald, T.T. 1998. T cell immunity to oral allergens. *Curr. Opin. Immunol.* 10:620–627.

171. Melamed, D., and A. Friedman. 1993. Direct evidence for anergy in T lymphocytes tolerized by oral administration of ovalbumin. *Eur. J. Immunol.* 23:935–942.

172. Mowat, A.M., M. Steel, E.A. Worthey, P.J. Kewin, and P. Garside. 1996. Inactivation of Th1 and Th2 cells by feeding ovalbumin. *Ann. NY Acad. Sci.* 778:122–132.

173. Strober, W., B. Kelsall, and T. Marth. 1998. Oral tolerance. *J. Clin. Immunol.* 18:1–30.

174. Titus, R.G., and J.M. Chiller. 1981. Orally induced tolerance. Definition at the cellular level. *Int. Arch. Allergy Appl. Immunol.* 65:323–338.

175. Tomasi, T.B., W.G. Barr, S.J. Challacombe, and G. Curran. 1983. Oral tolerance and accessory-cell function of Peyer's patches. *Ann. NY Acad. Sci.* 409:145–163.

176. Vistica, B.P., N.P. Chanaud, 3rd, N. Felix, R.R. Caspi, L.V. Rizzo, R.B. Nussenblatt, and I. Gery. 1996. CD8 T-cells are not essential for the induction of "low-dose" oral tolerance. *Clin. Immunol. Immunopathol.* 78:196–202.

177. Weiner, H.L., A. Friedman, A. Miller, S.J. Khoury, A. al-Sabbagh, L. Santos, M. Sayegh, R.B. Nussenblatt, D.E. Trentham, and D.A. Hafler. 1994. Oral tolerance: immunologic mechanisms and treatment of animal and human organ-specific autoimmune diseases by oral administration of autoantigens. *Annu. Rev. Immunol.* 12:809–837.

178. Xiao, B.G., and H. Link. 1997. Mucosal tolerance: a two-edged sword to prevent and treat autoimmune diseases. *Clin. Immunol. Immunopathol.* 85:119–128.

179. Cerf-Bensussan, N., A. Quaroni, J.T. Kurnick, and A.K. Bhan. 1984. Intraepithelial lymphocytes modulate Ia expression by intestinal epithelial cells. *J. Immunol.* 132:2244–2252.

180. Hershberg, R.M., D.H. Cho, A. Youakim, M.B. Bradley, J.S. Lee, P.E. Framson, and G.T. Nepom. 1998. Highly polarized HLA class II antigen processing and presentation by human intestinal epithelial cells. *J. Clin. Invest.* 102:792–803.

181. Kaiserlian, D. 1991. Murine gut epithelial cells express Ia molecules antigenically distinct from those of conventional antigen-presenting cells. *Immunol. Res.* 10:360–364.

182. Kaiserlian, D., K. Vidal, and J.P. Revillard. 1989. Murine enterocytes can present soluble antigen to specific class II-restricted CD4+ T cells. *Eur. J. Immunol.* 19:1513–1516.

183. Mayer, L., D. Eisenhardt, P. Salomon, W. Bauer, R. Plous, and L. Piccinini. 1991. Expression of class II molecules on intestinal epithelial cells in humans. Differences between normal and inflammatory bowel disease [see comments]. *Gastroenterology* 100:3–12.

184. Sanderson, I.R., A.J. Ouellette, E.A. Carter, and P.R. Harmatz. 1993. Ontogeny of Ia messenger RNA in the mouse small intestinal epithelium is modulated by age of weaning and diet. *Gastroenterology* 105:974–980.

185. Sanderson, I.R., A.J. Ouellette, E.A. Carter, and P.R. Harmatz. 1992. Ontogeny of class II MHC mRNA in the mouse small intestinal epithelium. *Mol. Immunol.* 29:1257–1263.

186. Vidal, K., I. Grosjean, and D. Kaiserlian. 1995. Antigen presentation by a mouse duodenal epithelial cell line (MODE-K). *Adv. Exp. Med. Bio.* 225–228.

187. Bland, P.W., and L.G. Warren. 1986. Antigen presentation by epithelial cells of the rat small intestine. I. Kinetics, antigen specificity and blocking by anti-Ia antisera. *Immunology* 58:1–7.

188. Bland, P.W., and L.G. Warren. 1986. Antigen presentation by epithelial cells of the rat small intestine. II. Selective induction of suppressor T cells. *Immunology* 58:9–14.

189. Bottazzo, G.F., I. Todd, R. Mirakian, A. Belfiore, and R. Pujol-Borrell. 1986. Organ-specific autoimmunity: a 1986 overview. *Immunol. Rev.* 94:137–169.

190. Piccinini, L.A., N.K. Goldsmith, S.H. Roman, and T.F. Davies. 1987. HLA-DP, DQ and DR gene expression in Graves' disease and normal thyroid epithelium. *Tissue Antigens* 30:145–154.

191. Mayer, L., and R. Shlien. 1987. Evidence for function of Ia molecules on gut epithelial cells in man. *J. Exp. Med.* 166:1471–1483.

192. Selby, W.S., G. Janossy, D.Y. Mason, and D.P. Jewell. 1983. Expression of HLA-DR antigens by colonic epithelium in inflammatory bowel disease. *Clin. Exp. Immunol.* 53:614–618.

193. Hirata, I., G. Berrebi, L.L. Austin, D.F. Keren, and W.O. Dobbins. 1986. Immunohistological characterization of intraepithelial and lamina propria lymphocytes in control ileum and colon and in inflammatory bowel disease. *Dig. Dis. Sci.* 31:593–603.

194. McDonald, G.B., and D.P. Jewell. 1987. Class II antigen (HLA-DR) expression by intestinal epithelial cells in inflammatory diseases of colon. *J. Clin. Pathol.* 40:312–317.

195. Fais, S., F. Pallone, O. Squarcia, L. Biancone, F. Ricci, P. Paoluzi, and M. Boirivant. 1987. HLA-DR antigens on colonic epithelial cells in inflammatory bowel disease: I. Relation to the state of activation of lamina propria lymphocytes and to the epithelial expression of other surface markers. *Clin. Exp. Immunol.* 68:605–612.

196. Pallone, F., S. Fais, and M.R. Capobianchi. 1988. HLA-D region antigens on isolated human colonic epithelial cells: enhanced expression in inflammatory bowel disease and *in vitro* induction by different stimuli. *Clin. Exp. Immunol.* 74:75–79.

197. Cuvelier, C., H. Mielants, M. De Vos, E. Veys, and H. Roels. 1990. Major histocompatibility complex class II antigen (HLA-DR) expression by ileal epithelial cells in patients with seronegative spondylarthropathy. *Gut* 31:545–549.

198. Mayer, L., and D. Eisenhardt. 1990. Lack of induction of suppressor T cells by intestinal epithelial cells from patients with inflammatory bowel disease. *J. Clin. Invest.* 86:1255–1260.

199. Hodgson, H.J., J.R. Wands, and K.J. Isselbacher. 1978. Decreased suppressor cell activity in inflammatory bowel disease. *Clin. Exp. Immunol.* 32:451–458.

200. Kelleher, D., A. Murphy, C.A. Whelan, C. Feighery, D.G. Weir, and P.W. Keeling. 1989. Defective suppression in the autologous mixed lymphocyte reaction in patients with Crohn's disease. *Gut* 30:839–844.

201. Goodacre, R.L., and J. Bienenstock. 1982. Reduced suppressor cell activity in intestinal lymphocytes from patients with Crohn's disease. *Gastroenterology* 82:653–658.

202. Campbell, N.A., H.S. Kim, R.S. Blumberg, and L. Mayer. 1999. The Nonclassical Class I Molecule CD1d Associates with the Novel CD8 Lignad gp180 on Intestinal Epithelial Cells. *J. Biol. Chem.* 274:26259–26265.

203. Panja, A., R.S. Blumberg, S.P. Balk, and L. Mayer. 1993. CD1d is involved in T cell-intestinal epithelial cell interactions. *J. Exp. Med.* 178:1115–1119.

204. Yio, X.Y., and L. Mayer. 1997. Characterization of a 180-kDa intestinal epithelial cell membrane glycoprotein, gp180. A candidate molecule mediating T cell-epithelial cell interactions. *J. Biol. Chem.* 272:12786–12792.

205. Toy, L.S., X.Y. Yio, A. Lin, S. Honig, and L. Mayer. 1997. Defective expression of gp180, a novel CD8 ligand on intestinal epithelial cells, in inflammatory bowel disease. *J. Clin. Invest.* 100:2062–2071.

206. Sharon, P., and W.F. Stenson. 1984. Enhanced synthesis of leukotriene B4 by colonic mucosa in inflammatory bowel disease. *Gastroenterology* 86:453–460.

207. Donowitz, M. 1985. Arachidonic acid metabolites and their role in inflammatory bowel disease. An update requiring addition of a pathway. *Gastroenterology* 88.580–587.

208. Korelitz, B.I., and D.H. Present. 1996. A history of immunosuppressive drugs in the treatment of inflammatory bowel disease: origins at the Mount Sinai Hospital. *Mt. Sinai. J. Med.* 63:191–201.

209. Kornbluth, A., D.H. Present, S. Lichtiger, and S. Hanauer. 1997. Cyclosporin for severe ulcerative colitis: A user's guide. *Am. J. Gastroenterol.* 92:1424–1428.

210. Lichtiger, S., and D.H. Present. 1990. Preliminary report: cyclosporin in treatment of severe active ulcerative colitis. *Lancet* 336:16–9.

211. Lichtiger, S., D.H. Present, A. Kornbluth, I. Gelernt, J. Bauer, G. Galler, F. Michelassi, and S. Hanauer. 1994. Cyclosporine in severe ulcerative colitis refractory to steroid therapy. *N. Engl. J. Med.* 330:1841–1845.

212. Present, D.H. 1993. Cyclosporine and other immunosuppressive agents: current and future role in the treatment of inflammatory bowel disease. *Am. J. Gastroenterol.* 88:627–630.

213. Present, D.H. 1989. 6-Mercaptopurine and other immunosuppressive agents in the treatment of Crohn's disease and ulcerative colitis. *Gastroenterol. Clin. North. Am.* 18:57–71.

214. Present, D.H., S.J. Meltzer, M.P. Krumholz, A. Wolke, and B.I. Korelitz. 1989. 6-Mercaptopurine in the management of inflammatory bowel disease: short- and long-term toxicity. *Ann. Intern. Med.* 111:641–649.

215. Present, D.H., B.I. Korelitz, N. Wisch, J.L. Glass, D.B. Sachar, and B.S. Pasternack. 1980. Treatment of Crohn's disease with 6-mercaptopurine. A long-term, randomized, double-blind study. *N. Engl. J. Med.* 302:981–987.

216. Korelitz, B.I., and D.H. Present. 1995. Methotrexate for Crohn's disease. *N. Engl. J. Med.* 333:600–601.

217. Klein, M., H.J. Binder, M. Mitchell, R. Aaronson, and H. Spiro. 1974. Treatment of Crohn's disease with azathioprine: a controlled evaluation. *Gastroenterology* 66:916–922.

218. Goenka, M.K., R. Kochhar, B. Tandia, and S.K. Mehta. 1996. Chloroquine for mild to moderately active ulcerative colitis: comparison with sulfasalazine. *Am. J. Gastroenterol.* 91:917–921.

52 | Experimental Models of Inflammatory Bowel Disease

Fiona Powrie and Michael W. Leach

1. INTRODUCTION

Idiopathic inflammatory bowel disease, encompassing Crohn's disease (CD) and ulcerative colitis (UC), is a chronic and relapsing inflammation of the gastrointestinal tract that affects approximately 0.2% of Western populations. The IBD's are particularly insidious diseases that strike young people, often causing severe debilitation and, in some cases, life-threatening sequelae. Characteristic microscopic features of UC include a diffuse infiltration of acute and chronic inflammatory cells limited to the colonic mucosa, as well as crypt abscesses and epithelial cell hyperplasia. In contrast to UC, CD can affect any part of the gastrointestinal tract, with the terminal ileum often being involved. The disease in CD involves inflammatory skip lesions and is characterised histologically by transmural inflammation and submucosal granulomas.

Despite extensive investigation, the aetiology of IBD remains unknown. A current hypothesis that is gaining widespread support is that IBD develops in genetically-susceptible individuals as a result of a dysregulated immune response driven by enteric bacteria (1,2). Indeed, there is substantial evidence to support an immune-mediated pathogenesis in IBD. First, inflammatory lesions involve increases in a number of inflammatory mediators, including IL-1, IL-6, TNF-α and reactive oxygen and nitrogen intermediates. Second, autoantibodies are a common component of UC. Third, IBD patients often exhibit a number of immune-mediated systemic complications, and disease responds to immune suppressive therapy.

Historically, research into the pathogenesis of IBD has been hampered by a lack of good laboratory animal models of chronic intestinal inflammation. However, in the last 7 years, a number of models of IBD have been described in which intestinal lesions resemble those of the human disease. In this chapter, we discuss some of these models, emphasising in particular the immunological features that shed light on primary immune pathogenic mechanisms, and discuss the relevance and utility of these models for the design of more effective therapeutic strategies.

2. MODELS OF IBD

The ideal model of CD or UC would mimic the etiologic, morphologic, immunologic and genetic features of the human diseases, as well as respond to various therapeutic agents in the same way as IBD patients. While some models fulfill these criteria more than others, no model perfectly mimics the human disease. However, study of the now extensive number of models has provided invaluable insights into the development and regulation of intestinal inflammation. Broadly, current models (selected examples listed and referenced in Table 1) can be divided into 5 categories: i) chemically-induced; ii) cell transfer, iii) transgenic; iv) infectious agent; v) spontaneous.

2.1. Chemically-induced

Many models of small and/or large intestinal inflammation have been elicited by administration of exogenous chemicals or bacterial products. Examples include administration of acetic acid into the intestine by enema or injection after laparotomy, administration of carrageenan or dextran sodium sulphate (DSS) in the drinking water, rectal administration of contact sensitizers such as trinitrobenzene sulfonic acid (TNBS) or oxazolone, and intramural injection of peptidoglycan-polysaccharide (PG-PS) polymers into the gut wall.

Table 1 Models of Inflammatory Bowel Disease

	Predominant intestinal immune response	Role of indigenous microbiota in exacerbating disease	Selected reference(s)
Administration of chemicals or bacterial products			
Acetic acid	–	–	(3)
DSS	Th1 and Th2	no	(4,5)
TNBS	Th1 or Th2 (strain-dependent)	yes	(6–8)
Oxazolone	Th2		(9–11)
PG-PS	Th1	–	(12,13)
Caregeenan		yes	(14,15)
Cell Transfer			
CD45RB^{high} or unseparated CD4$^+$ T cell transfer to SCID mice	Th1	yes	(16–20)
T cell-reconstituted Tgϵ26 mice	Th1	–	(21,22)
Hsp60-reactive CD8$^+$ T cells transferred into mice	Th1	no	(23)
Transgenics and gene deletions			
IL-2$^{-/-}$	Th1	yes	(24–26)
IL-2 receptor $\alpha^{-/-}$	–	–	(27)
IL-7 transgenic mouse	Th1	–	(28)
IL 10$^{-/-}$	Th1	yes	(29–32)
LysMcre/Stat3$^{flox/-}$ (cell type specific deletion of Stat 3)	Th1	–	(33)
CRF2–4$^{-/-}$ (lack responsiveness to IL-10)	–	–	(34)
TGF-$\beta 1^{-/-}$	–	–	(35)
TCR$\alpha^{-/-}$	Th2	yes	(36–39)
TCR$\beta^{-/-}$	–		(36)
TNF$^\rho$ARE	–	–	(40)
G$\alpha^{-/-}_{i2}$	Th1	–	(41,42)
STAT-4 transgenic	Th1	yes	(43)
Dominant negative N-cadherin mutant	–	–	(44)
Ganglionic ablation in jejunum and ileum	–	no	(45)
mdrla$^{-/-}$	–	yes	(46)
WASP$^{-/-}$	–	–	(47)
HLA-B27 and human β2-microglobulin transgenic rats	Th1	yes	(48–50)
Infectious agents			
Helicobacter hepaticus-infected immunodeficient mice	Th1	–	(51,52)
Helicobacter bilis-infected immunodeficient rats or mice	–	–	(53)
Citrobacter rodentium-infected mice	Th1	–	(54,55)
Spontaneous Models			
C3H/HeJBir mice	Th1	yes	(56,57)
Senescence accelerated mouse P1/Yit strain	–	yes	(58)
Cotton-top tamarins	–	–	(59)

For chemicals or bacterial products that are administered directly into the intestine, the area affected depends on the administration site. In contrast, chemical/agents that are administered orally, such as carrageenan or DSS, result in large intestinal inflammation. A feature of these models, in contrast to the transfer or transgenic models, is that the inflammation is self-limiting and ultimately resolves. For example, acetic acid results in acute intestinal inflammation related to direct injury of the tissue by the acid. After the initial injury, which is primarily necrosis, acute inflammation occurs, followed by healing. Chronic inflammation is not a major feature of this model and, as such, the model is most useful for the study of non-specific inflammation and repair. Carrageenan and DSS result in acute and chronic intestinal inflammation, as do contact sensitizers and PG-PS administration. Inflammation in carrageenan and DSS models is thought to be related to epithelial damage followed by exposure of the lamina propria to luminal contents. Contact sensitizers are covalently reactive compounds that attach to autologous proteins and cause inflammation by inducing a delayed type hypersensitivity reaction, and in addition by causing direct damage to tissues. The acute and chronic inflammation that develops in the PG-PS model is thought to be related to the reaction against the injected PG-PS.

2.2. Cell Transfer Models

In several models, chronic intestinal inflammation has been shown to develop after transfer of T cell subsets into immune-deficient recipients. These include transfer of small numbers (10^5–10^6) of peripheral CD45RB^{high}CD4$^+$ or unseparated CD4$^+$ T cells to syngeneic severe combined immune deficient (SCID) or recombination activating gene (RAG) 1- or 2-deficient mice, as well as transfer of normal bone marrow into CD3ϵ transgenic mice (Tgϵ26) that have an abnormality in thymic development and are lymphopenic. Colitis in these transfer models was characterised by diffuse chronic large intestinal inflammation with predominant macrophage and CD4$^+$ T cell infiltrates and epithelial hyperplasia (Fig. 1), but without small intestinal involvement. Inflammatory infiltrates included multinucleated giant cells, and crypt abscesses and ulcers were occasionally seen. Intestinal inflammation in these models has features of both CD and UC. In contrast, transfer of an hsp-60-reactive CD8$^+$ T cells clone to TCR$\beta^{-/-}$ mice led to inflammatory lesions predominantly in the small intestine, associated with focal expansion of antigen-specific CD8$^+$ T cells, epithelial degeneration, crypt hyperplasia and villus artophy. In contrast to the CD4$^+$ T cell models of chronic intestinal inflammation, macrophages did not make up a significant component of the inflammatory infiltrate. While the location of the intestinal lesions is similar to CD, this model lacks certain features of CD, such as ulcerations, granulomatous inflammation, and infiltrates with macrophages and CD4$^+$ T cells.

2.3. Transgenic Models

In the past 7 years, numerous transgenic IBD models based on the loss or gain of function of various genes have been described (Table 1). Most of the transgenic models share in common some alteration in the immune system or of epithelial cells that makes the animals susceptible to intestinal inflammation, particularly in the large intestine. These models have various clinical, pathologic, and immunologic features that resemble certain findings in CD and UC. It has been suggested that some models are more like CD or UC, although none has matched exactly the constellation of morphologic and pathologic features seen in CD and UC. In general, most models have Th1-mediated chronic inflammation localised primarily to the large intestine, with epithelial hyperplasia and goblet cell depletion. In most models, the inflammation is diffuse with infiltration of macrophages and T cells. However, different models often have unique changes. For example, the IL-10$^{-/-}$ mouse develops multifocal transmural disease that can also involve the small intestine, and even the stomach, in addition to the large intestine. HLA-B27 and human β2-microglobulin transgenic rats also develop multifocal inflammation throughout the gastrointestinal tract. TNF$^\pi$ARE mice, which overexpress TNF, develop inflammation primarily in the ileum, with only some inflammation reported in the proximal colon. Dominant negative N-cadherin mutant mice developed inflammation only in the small intestine, as did transgenic mice that express herpes simplex virus thymidine kinase in enteric glia and undergo induced ganglionic ablation in the jejunum and ileum.

Several models have been reported to develop invasive or pseudoinvasive epithelium in the large intestine reminiscent of locally-invasive adenocarcinoma, including the G$\alpha_{i2}$$^{-/-}$ and IL-10$^{-/-}$ models. In addition, dominant negative N-cadherin mutant mice develop small intestinal adenomas. Relevant to this, patients with UC have enhanced susceptibility to colon cancer.

2.4. Infectious Models

A growing number of specific infectious causes of IBD in rodents have been identified. Various *Helicobacter* spp have been shown to induce colitis in immune deficient mice and rats. Animals developed multifocal or occasionally diffuse large intestinal inflammation, with prominent epithelial proliferation. Proliferative epithelium extended into the submucosa of rats naturally infected with *Helicobacter bilis. Citrobacter rodentium* infection has also been shown to induce a large intestinal inflammation with epithelial proliferation in some strains of young immune-competent mice. In general, the immune responses (Th1-like) and morphologic findings (proliferative typhlocolitis) in mice and rats with infectious inflammatory bowel disease resemble those observed in transgenic models of colitis, even when known infectious agents are absent from the transgenic models.

2.5. Spontaneous Models

A well-studied spontaneous model is the cotton-top tamarin model, which bears some resemblance to UC, with the for-

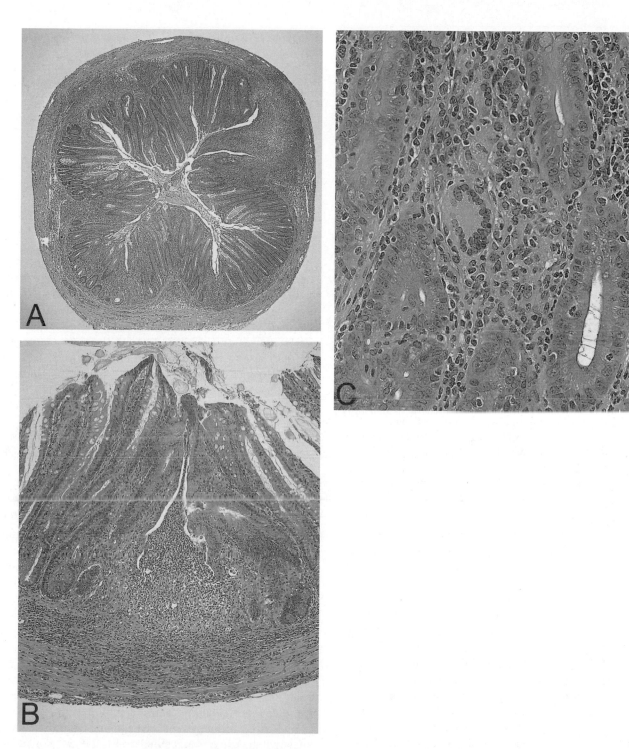

Figure 1.   Colitis induced by transfer of CD45RB[high] CD4[+] T cells into immunodeficient mice. A: Proximal colon from C.B-17 SCID mouse that received 4×10^5 CD45RB[high] CD4[+] T cells. Severe colitis involving the entire cross-section of the colon. Note the marked crypt hypertrophy and inflammation resulting in thickening of the mucosa. There is ulceration and transmural inflammation in the upper right. H&E. ×26. Fig 1A reproduced with permission from (17, 148) B: Distal colon from C.B-17 *scid* mouse that received 4×10^5 CD45RB[high] CD4[+] T cells, demonstrating ulceration, transmural inflammation, and epithelial hyperplasia resulting in crypt hypertrophy, crypt branching, and mucin depletion from goblet cells. H&E ×85. C: Distal colon from C.B-17 *scid* mouse that received 4×10^5 CD45RB[high] CD4[+] T cells. There is chronic inflammation in the mucosa, with numerous mononuclear inflammatory cells and smaller numbers of neutrophils separating intestinal glands. A Langhans'-type multinucleated giant cell is present as well (center). The crypt epithelium is thickened with hyperplastic epithelial cells, and there is mucin depletion from goblet cells. A few neutrophils can be seen within the epithelium. H&E. ×335.

mation of crypt abscesses, neutrophil infiltration and mucin depletion. As with the human disease, the spectrum of colitis is diverse and does not follow Mendelian patterns of inheritance. Colitis has also been described in a substrain of C3H mice termed C3H/HeJBir. By 12 weeks of age, the majority of these mice develop chronic inflammation in the caecum and proximal colon. In contrast to the T cell transfer and genetically-induced models, the inflammation resolves in older C3H/HeJBir mice. Spontaneous inflammation also developed in the ileum and cecum of the senescence-accelerated mouse P1/Yit strain.

3.   PATHOGENESIS OF IBD

3.1.   Cell Types

The normal functioning of the intestine involves a complex interplay between immune cells, epithelial cells and stromal cells as well as the nervous system. The fact that manipulations that affect the immune system, epithelial cells or the enteric nervous system lead to intestinal inflammation highlights the importance of appropriately controlled interactions between these cell types for intestinal homeostasis.

Despite the distinct histological features of many of the models of IBD, there is good evidence that CD4[+] T cells are often involved in the pathogenesis of the diseases. Activated CD4[+] T cells are found in intestinal lesions in a number of models (Table 1), and anti-CD4 mAbs inhibited colitis in the TNBS model (60). Transfer of CD4[+] T cell subsets to SCID or RAG[-/-] mice also led to the induction of colitis, providing direct evidence of the pathological role of these cells (16,20,61). In the majority of cases, TCR $\alpha\beta^+$ CD4[+] cells are involved, however CD4[+] cells also induce colitis in TCR $\alpha^{-/-}$ mice that lack TCR $\alpha\beta^+$ cells. It has been proposed that the pathogenic cells express TCR $\beta\beta$ homodimers, as these cells expand in mice with colitis, however the molecular features of antigen recognition by these T cells has not been studied (62).

While CD8[+] T cells are present in the inflamed colon in many models, they do not seem to be required for development of disease, since IL-2[-/-] mice crossed with β2microglobulin[-/-] mice, which have no CD8[+] cells, developed colitis with somewhat enhanced kinetics (63). However, colitis was reduced in the Tgϵ26 model when donor lymphocytes were deficient in perforin secretion, suggesting cytotoxic T cells contribute to intestinal pathology (64). The finding that hsp-reactive CD8[+] T cells induced inflammatory lesions in the small intestine after transfer to immune-deficient recipients provides direct evidence that these cells can induce immune pathology in the intestine in the absence of CD4[+] cells (23). While T cells clearly play a pivotal role in orchestrating intestinal inflammation, they are not absolutely required, since infections of SCID mice,

which lack T cells, with *Helicobacter* spp led to colitis (51,65). This suggests that, in the presence of the right environmental factors, activation of cells of the innate immune system, such as NK cells and macrophages, can result in intestinal inflammation.

There is no apparent role for B cells in the induction of intestinal inflammation, since IL-10[-/-] mice that lacked B cells developed colitis normally (30), and transfer of CD4[+] T cells, in the absence of B cells, to RAG[-/-] mice was sufficient to induce colitis (66). Colitis in IL-2[-/-] and TCRα[-/-] mice is characterised by the presence of activated B cells as well as autoantibodies (24,36). However, TCRα[-/-] (67) and IL-2[-/-](68) crossed to B cell-deficient mice still developed colitis. Indeed, colitis in TCRα[-/-] mice that lacked B cells was more severe than in TCR α[-/-] mice alone, suggesting that the presence of B cells negatively regulates intestinal inflammation in this model. In the absence of B cells, there was an increase in the number of apoptotic cells in the colon that could be reduced by the administration of Ig isolated from TCR α[-/-] mice, suggesting that the protective effect of B cells may be to facilitate the clearance of apoptotic cells. Recently, impaired clearance of apoptotic cells has been implicated in the development of autoimmune nephritis induced by C1q deficiency (69).

Intraepithelial lymphocytes, together with epithelial cells, constitute a cellular network that controls the passage of substances from the gut lumen to the intestinal mucosa. The importance of the integrity of this barrier for the regulation of intestinal inflammation is illustrated by the finding that mice transgenic for a dominant negative cadherin (N-cadherin) developed IBD (44). Cadherins play an important role in the formation of tight junctions between epithelial cells, and in this model, inflammation develops around epithelial cells containing mutant candherin molecules, suggesting inflammation is a result of a localised disruption of the epithelial cell barrier. Colitis in multidrug-resistance gene (mdr-1)[-/-] mice is also thought to involve alterations in epithelial cell barrier function (46).

Administration of gangcyclovir to transgenic mice that expressed herpes simplex virus thymidine kinase under the glial fibrillic acid promotor led to depletion of glial cells in the jejunal nerves and localised small intestinal pathology (45). Vascular changes in the jejunum led to decreased intestinal motility, suggesting that ischemia can induce intestinal inflammation. However, the link between ischemia and IBD is not well understood.

3.2.   Cytokines

It is now well established that the CD4[+] T cell population in mice and in humans is functionally heterogeneous (70,71). Two major subsets have been identified based on their differential cytokine secretion. Th1 cells secrete IFN-γ and

TNF-α and are involved in macrophage activation and protective cell-mediated immunity. In contrast, Th2 cells do not secrete IFN-γ, but make IL-4, IL-5, IL-10 and IL-13 and stimulate humoral immunity as well as the growth and effector function of mast cells and eosinophils. Analysis of Th subsets in IBD models has revealed differential activation of Th1 cells in the majority of cases (Table 1). This was first demonstrated in the SCID model of colitis, where IFN-γ and TNF-α, but not IL-4 or IL-10 mRNA, levels were found to be increased in mice with colitis (16). Furthermore, comparison of cytokine secretion after polyclonal stimulation of CD4$^+$ T cells isolated from the lamina propria of mice with colitis to that of controls showed significant elevations in INF-γ and IL-3 production, while IL-4 and IL-10 levels were comparable to control colons. INF-γ was shown to be involved in the pathogenesis, since treatment of mice with anti-INF-γ monoclonal antibody prevented the development of colitis (17). Similarly, anti-INF-γ has been shown to inhibit colitis in young IL-10$^{-/-}$ mice (30). There is conflicting data on the absolute requirement for T cell-derived INF-γ in the SCID model, since transfer of CD45RB^{high} CD4$^+$ cells from INF-$\gamma^{-/-}$ mice to SCID mice led to colitis in one study, but not another (72,73). It remains to be established whether NK cells were a source of INF-γ in these studies or whether other cytokines induced the inflammation.

Immune pathology in the absence of INF-γ is most likely the result of TNF-α production, as TNF-α has been shown to be involved in the pathogenesis of a number of models of colitis (17,74,75). Direct demonstration that overproduction of TNF-α can lead to intestinal inflammation comes from the finding that mice in which the AU region of the TNF-α gene was deleted developed granulomatous small intestinal inflammation. TNF-α was over-expressed in these mice, since deletion of regulatory sequences led to enhanced stability of TNF-α mRNA (40). Non-T cells present in the colonic mucosa appear to be the important source of TNF-α, since CD45RB^{high} CD4$^+$ cells from TNF-$\alpha^{-/-}$ mice were able to transfer colitis to immune-deficient recipients capable of producing TNF-α, but failed to induce colitis when transferred to RAG2$^{-/-}$ TNF-$\alpha^{-/-}$ mice. These results suggest that TNF-α production by Th1 cells themselves is not required for induction of intestinal inflammation, but that T cell-induced production of TNF-α by resident host-derived cells is (76).

Development of Th1 responses has been shown to be dependent on IL-12 produced by activated DC and macrophages (77,78). Not too surprisingly, anti-IL-12 mAbs inhibited TNBS colitis (7,25) and CD45RB^{high}CD4$^+$ cells isolated from STAT4$^{-/-}$ mice, which cannot respond to IL-12, were impaired in their ability to transfer colitis (73,79). In addition to its role as a differentiation factor, there is also evidence that IL-12 is involved in the perpetuation of Th1 responses promoting their growth (80) pre-

venting their apoptosis (81). Consistent with this, anti-IL-12, but not anti-INF-γ, reversed established colitis in the TNBS model (7). However, neither anti-IL-12 nor anti-INF-γ could completely reverse established colitis in IL-10$^{-/-}$ mice, suggesting additional factors contribute to disease maintenance (80).

There is also evidence that IL-7 is involved in intestinal inflammation. Development of *Helicobacter hepaticus*-induced colitis in RAG2$^{-/-}$ mice was dependent on IL-7 production (82) and expression of an IL-7 transgene in the colonic mucosa induced colitis (28). Precisely how IL-7 is involved in disease pathogenesis is not known. It has been shown to act as a maintenance factor for lymphocytes and, in the case of T and B cell-independent colitis, is thought to act as a maintenance factor for myeloid cells.

While most models of colitis involve a Th1-mediated pathogenesis, recent evidence suggests that, in some models, differential activation of Th2 cells can induce intestinal inflammation. Elevations in T cells secreting IL-4 were found in TCR$\alpha^{-/-}$ mice as well as in oxazalone-induced colitis, and anti-IL-4 mAbs prevented disease in both of these models (11,38,39). Precisely how IL-4 and the activation of Th2 cells contribute to intestinal immune pathology is not known. Colitis induced by Th2 cells was characterised by a superficial infiltrate containing large numbers of eosinophils, suggesting that activation of other inflammatory cells types may be involved in the immune pathology.

3.3. Costimulatory Molecules

T cell activation involves the integration of two different types of signal (rev. in 83). The first is transmitted via the TCR and associated molecules after recognition of its cognate peptide ligand presented by MHC molecules on antigen presenting cells (APC). The second involves interactions between costimulatory molecules present on T cells with their ligands on APC. Activation of naïve T cells has been shown to be dependent on signalling via CD28 present on the T cell with its ligands CD80 and CD86 on APC. This interaction leads to up regulation of the TNF family molecule CD40L on T cells that, in turn, is able to interact with CD40 present on APC. Cross linking of CD40 on DC and macrophages has been shown to lead to IL-12 secretion and T cell differentiation towards Th1 responses (84). Blockade of CD40-CD40L interactions by administration of an anti-CD40L mAb prevented TNBS-induced colitis (85), indicating that signalling via CD40 is important in the induction of colitis, probably as a result of induced IL-12 secretion and amplification of the Th1 response.

There is evidence that other members of the TNF/TNFR family are involved in IBD. OX-40 is a member of the TNFR family that is expressed primarily on CD4$^+$ T cells after activation (rev. in 86). Its ligand, OX-40L, is expressed on a variety of cell types after activation including B cells, DC and endothelial cells. In contrast to CD28, which is an important costimulatory molecule for primary T cell responses, OX-40 signalling facilitates clonal expansion and the development of secondary T cell responses. Blockade of OX-40 signalling by administration of an OX-40 Fc fusion protein ameliorated established colitis in both the TNBS model and in IL-2$^{-/-}$ mice (87). Recently, DC in the mesenteric lymph nodes (MLN) of mice with colitis were shown to upregulate OX-40L, and administration of an OX-40L mAb inhibited development of colitis induced by transfer of CD45RB^{high} CD4$^+$ cells to SCID mice (V. Malmstrom, N. Barclay and F. Powrie, unpublished results). Both of these studies suggest that signalling via OX-40 plays an important role in the induction and perpetuation of chronic intestinal inflammation, and that activated DC via their expression of OX-40L drive pathogenic T cell responses.

3.4.   Adhesion Molecules

For T cells to induce immune pathology in the intestine, they must first reach mucosal inductive sites, such as the Peyers Patch (PP), where they may encounter antigen and then migrate to the intestinal lamina propria to mediate their effector functions. Study of homing of naive lymphocytes to gut-associated lymphoid tissues (GALT) and subsequent migration to inductive sites has been shown to involve the mucosal homing intergrin $\alpha4\beta7$(88). $\alpha4\beta7$ binds to mucosal addression cell adhesion molecule-1 (mAdCam-1), which is present on high endothelial venuoles in the GALT as well as on ordinary endothelial cells in the intestinal lamina propria. Disruption of lymphocyte homing to GALT by administration of anti- $\alpha4\beta7$ mAb ameliorated colitis in cotton top tamarins (89), and a combination of anti- $\alpha4\beta7$ and anti-mAdCAM-1 mAbs prevented colitis in CD45RB^{high}CD4$^+$ T cell restored-SCID mice (90), indicating the importance of this interaction of influx of pathogenic T cells into the intestine. In addition to homing, adhesion molecules also play a role in retention of leukocytes within tissues. In the intestine, $\alpha\varepsilon\beta7$, via its interaction with e-cadherin on epithelial cells, is thought to mediate retention of lymphocytes within the epithelial layer (91,92). Somewhat surprisingly administration of an anti-$\alpha\varepsilon\beta7$ mAb prevented immunisation-induced colitis in IL-2$^{-/-}$ mice and led to the amelioration of established disease (93). How disruption of interactions between intraepithelial lymphocytes and epithelial cells inhibits accumulation of inflammatory cells in the lamina propria is not clear, but the results suggest that the onset and perpetuation of chronic intestinal inflammation are dependent on retention of T cells in the colon.

Other molecules involved in lymphocyte migration that influence IBD include the CD44 variant isoform, v7. Monoclonal antibodies reactive with this isoform, but not other isoforms of CD44, prevented the development of TNBS colitis and reduced severity of established disease (94). Antibody therapy was accompanied by enhanced production of IL-10, suggesting the mAb may be working to alter T cell activation as opposed to affects on lymphocyte migration.

3.5.   Effector Molecules

While it is clear that elevations in the cytokines IL-12, TNF-α and IFN-γ are involved in the pathogenesis of colitis in many IBD models, precisely how these cytokines induce the characteristic pathological lesions, such as epithelial cell hyperplasia, depletion of mucin-secreting cells and ulceration, is not known. Both TNF-α and INF-γ have been shown to have direct effects on intestinal epithelial cells. TNF-α has been shown to induce epithelial cell apoptosis, and IFN-γ can induce epithelial cell damage and disruption of the epithelial cell barrier (95,96). Both these cytokines also induce No synthetase, leading to enhanced nitric oxide, which has been suggested to be a toxic effector molecule in some models of colitis (97,98).

The maintenance of intestinal architecture is thought be a dynamic process involving a balance between production of epithelial cell growth factors such as keratinocyte growth factor (KGF) with the production of matrix degrading enzymes, the matrix metalloproteinases (MMP's) and their inhibitors, tissue inhibitor of MMP (TIMPS) (reviewed in 99). TNF has been shown to lead to enhanced production of KGF and MMP-3 by lamina propria stromal cells suggesting elevations in TNF-α may alter the equilibrium in the intestine towards epithelial cell growth and ultimately, MMP-induced ulceration.

4.   THE ROLE OF INDIGENOUS MICROBIOTA

The digestive tract, especially the large intestine, normally contains high concentrations of a wide variety of microorganisms, representing the indigenous microbiota. In humans, there are estimated to be 10^{14} bacteria/g of colonic tissue that provide a source of foreign antigen as well as proinflammatory bacterial cell wall products. Despite the presence of this large, potentially proinflammatory stimulus, the mucosal immune system remains tolerant to the indigenous microbiota. This process appears to be highly specific to individuals, since humans and mice were shown to be tolerant of their own autologous microbiota, but not to heterologous microbiota, including that of littermates raised under identical conditions (100,101). It has been suggested that a breakdown in the homeostatic mechanisms that control immune responses to commensal bacteria is involved in the pathogenesis of IBD (1). Indeed, there is abundant evidence that inflammation in the various animals models of IBD is driven by intestinal bacteria.

Evaluation of models under conventional, specific pathogen-free (SPF) or germ-free conditions has provided clear evidence that the indigenous microbiota is involved in the development of intestinal inflammation. Inflammation in many models occurred under conventional housing conditions, but at a lower incidence/severity under reduced flora or SPF conditions, or did not occur at all under germ-free conditions. This has been observed in the carrageenan model (15), in SCID mice restored with CD4$^+$ T cell subsets (19,102), in IL-2$^{-/-}$ (24,26) IL-10$^{-/-}$ (29,32), and TCR$\alpha^{-/-}$ mice (37), and in rats transgenic for HLA-B27 and human β2-microglobulin (49). In the latter case, the simple act of conventionalising germ-free mice resulted in large intestinal inflammation, especially in the cecum (103). Antibiotics have been used to successfully treat intestinal inflammation in a number of models, providing further support that enteric bacteria drive the inflammatory process (18). It should be noted that not all mice remain disease-free when kept under germ-free conditions. For example, older germ-free IL-2$^{-/-}$ mice developed colitis, although in a less severe form than when exposed to microbes (26), and germ-free mice receiving Hsp60-reactive CD8$^+$ T cells still developed intestinal inflammation (23). The latter result demonstrated that activation of autoreactive T cells, which recognise self-antigens present in the intestine, can induce intestinal inflammation in the absence of a bacterial flora.

Exposure of animals to various specific microbes has been used to clarify the role of these micro-organisms in some models. An important role for *Bacteroides* spp. has been demonstrated in HLA-B27/β2-microglobulin transgenic rats. Germ-free rats were colonised with various bacterial cocktails, but only those containing *Bacteroides vulgatus* developed significant gastrointestinal inflammation (50). However, germ-free IL-10$^{-/-}$ mice repopulated with *B. vulgatus* or other selected bacterial species did not have colitis that was as severe as that occurring when mice were exposed to a more diverse microbiota under routine SPF conditions (32). These latter findings highlight differences in the responses to various bacteria between the models and suggest that severe disease requires more than one enteric bacteria.

Precisely how the indigenous microbiota induces intestinal inflammation is not know. Enteric bacteria may stimulate antigen-specific T cells, act as superantigens or mitogens, or lead to induction of non-specific inflammation and polyclonal T cell activation. There is now accumulating evidence that antigen-specific CD4$^+$ T cells, which recognise bacterial antigens, induce colitis. CD4$^+$ T cells from the spontaneously-colitic C3H/HeJBir mouse were found to proliferate and produce IFN-γ in response to antigens from enteric bacteria, but not food or autoantigens (57). Importantly, CD4$^+$ Th1 cell lines reactive with bacter-

ial antigens were able to induce colitis when transferred to immune-deficient recipients. Similarly, CD4$^+$ T cells from STAT-4 transgenic mice reacted strongly to autologous bacterial antigens (43), and lymphocytes from *mdr 1a*$^{-/-}$ mice with active colitis demonstrated increased reactivity to intestinal bacterial antigens compared to non-colitic *mdr 1a*$^{-/-}$ and control mice (46). In TCR$\alpha^{-/-}$ mice, an alteration of the recognition of aerobic bacteria from a polyclonal to an oligoclonal response was shown, associated with development of autoantibodies (104), and lamina propria CD4$^+$ TCR$\alpha^-\beta^+$ T cells in this model reacted to bacterial antigens (105). In the latter case, a monoclonal accumulation of CD4$^+$ TCR$\alpha^-\beta^+$ T cells was found in the colon of mice with colitis, suggesting the response was antigen-driven. Nonviable bacterial products can also elicit inflammation. Immune responses to bacterial cell wall fragments (12) or chemotactic peptides (106) have been used to elicit inflammation in some models. Furthermore, bacterial intimin from killed bacteria resulted in a strong Th1 response in the mouse, with significant epithelial hyperplasia (107).

Although most efforts have focused on the potential detrimental effects of indigenous microbiota, recent data suggests that certain indigenous bacteria can also play a beneficial role by ameliorating intestinal inflammation. Rectal administration of *Lactobacillus reuteri*, or oral lactulose treatment (which increases intestinal *Lactobacillus* spp.) ameliorates colitis in IL-10 KO mice (108). At least part of the beneficial effect appeared to be related to reducing mucosal aerobic adherent and translocated bacteria. Administration of *L. reuteri* also ameliorated colitis in rats receiving intracolonic acetic acid, and reduced intestinal permeability (109). In a more general sense, the indigenous microbiota can prevent inflammation by blocking colonisation by pathogens. Taken together, these results suggest that intestinal inflammation may be related to an imbalance in the microbiota, with insufficient protective species relative to potentially pathogenic species, and further that administration of protective species (probiotics) or agents to support protective species (prebiotics) can be therapeutic (110).

Animal models of IBD have also provided additional insights into the interaction between the host and indigenous microbiota. Immunodeficient mice infected with *Helicobacter hepaticus* (51,111) and immunodeficient mice or rats infected with *H. bilis* (53,65) developed intestinal inflammation. In contrast, these agents did not cause intestinal disease in immunocompetent strains (52,112) nor in immunodeficient mice that lacked the ability to produce IL-7 (Rag-2$^{-/-}$/IL-7$^{-/-}$ mice) (82). These findings clearly illustrate that certain agents may only be pathogenic under selected conditions, highlighting the complexity of the interaction between the host and microbiota. In a more general sense, the emerging role of various *Helicobacter* spp. in the development of gastrointestinal inflammation in humans and other species demonstrates that unrecognised pathogenic agents may still exist, even after extensive searches for an infectious agent in selected diseases.

While the above data clearly indicates a role of indigenous microbiota in animal models of IBD, the precise interrelationship between the microbiota and the development of inflammation remain unclear. It is possible that the inflammation represents a normal response to an as yet unrecognised and persistent or recurrent antigen, or alternatively the inflammation may represent a more general dysregulated response to the microbiota ordinarily present in the intestine, secondary to a failure of normal immunoregulatory mechanisms.

5. REGULATORY T CELLS CONTROL INTESTINAL INFLAMMATION

A characteristic of the normal intestine is its ability to activate repair mechanisms and resolve inflammatory responses once the inciting stimulus, be it an exogenous chemical or an infectious agent, has been removed. This indicates that potent homeostatic mechanisms control intestinal responses, making it possibility that these mechanisms are deficient in individuals with IBD where intestinal inflammation and injury persist.

The precise features of intestinal homeostasis have not been fully elucidated. However, evidence from the SCID model of colitis suggests that a functionally-distinct population of regulatory T cells (Treg) plays a key role in controlling inflammatory responses in the intestine (113). Colitis induced in SCID mice after transfer of CD45RB^{high}CD4$^+$ T cells could be inhibited by co-transfer of the reciprocal CD45RB^{low}CD4$^+$ T cell subset. These results suggest that the pool of antigen-experienced CD4$^+$ T cells in normal mice contain Treg cells capable of inhibiting colitis. Analysis of the mechanism of immune suppression revealed that it was dependent on TGF-β and independent of IL-4, suggesting that these cells were functionally distinct from Th2 cells (66).

There is evidence that IL-10 plays an important role in mucosal immune regulation, since mice with a targeted disruption of the IL-10 gene developed enterocolitis (29). In addition, administration of murine rIL-10 prevented colitis in SCID mice restored with CD45RB^{high}CD4$^+$ T cells (17) and in IL-10$^{-/-}$ mice treated from weaning (30). Furthermore, CD45RB^{high}CD4$^+$ T cells isolated from transgenic mice that expressed IL-10 under the IL-2 promoter failed to transfer colitis, but rather were able to inhibit colitis induced by normal CD45RB^{high}CD4$^+$T cells (114). IL-10 secretion by Treg cells appears to be essential for their function, since CD45RB^{low}CD4$^+$T cells from IL-10$^{-/-}$ mice failed to inhibit colitis induced by transfer of CD45RB^{high}CD4$^+$ T cell (115).

In addition, blockade of IL-10 function by administration of a monoclonal antibody against the IL-10R completely abrogated protection transferred by CD45RB^{low}CD4$^+$ T cells from normal mice. Studies in the SCID model of colitis suggest that IL-10 and TGF-β both play non-redundant roles in the functioning of Treg cells, which control inflammatory responses towards intestinal antigens, since the neutralisation or absence of either one of these cytokines is sufficient to abrogate protection. TGF-β has also been shown to regulate induction of immunisation induced colitis in IL-2$^{-/-}$ mice (116).

The factors driving the induction of IL-10 and/or TGF-β expressing cells in the gut are not yet fully understood. It has been suggested that it is the balance between proinflammatory and anti-inflammatory signals acting on antigen-presenting cells (APCs) that, dictates T cell subset differentiation (117,118). In the TNBS model, mononuclear cells were shown to mount Th1 responses to bacterial antigens, whereas administration of rIL-10 or treatment with anti-IL-12 mAbs inhibited intestinal inflammation and restored tolerance towards bacterial antigens (101). Interestingly, recent studies of *Helicobacter hepaticus* infection in mice showed that normal mice mounted an IL-10-dependent response, whereas IL-10$^{-/-}$ mice developed a pathogenic Th1 response towards the bacterium (52). These studies support the hypothesis that, in immunocompetent hosts, enteric antigens induce IL-10 secreting T cells that are immune suppressive and prevent inflammatory responses towards intestinal antigens.

Precisely how IL-10 and TGF-β induced immune suppression *in vivo* is not known. Both cytokines have well-characterised immune suppressive properties active on both the inductive and effector phases of T cell-mediated inflammatory responses (119,120). IL-10 has been shown to mediate a range of anti-inflammatory activities both on T cells and APCs, including the down-regulation of antigen-induced proliferation, cytokine secretion and expression of costimulatory molecules. It seems likely that IL-10 secreting regulatory T cells act to inhibit Th1 cell activation, and that IL-10 produced locally in the intestine acts on macrophages to prevent their activation and elaboration of proinflammatory molecules and chemokines, thus inhibiting T cell recruitment into the intestine. Consistent with this, mice in which macrophages and neutrophils are unable to respond to IL-10 as a result of a cell type-specific deletion of Stat-3 developed enterocolitis, suggesting that IL-10-mediated macrophage and neutrophil deactivation contributes to the immune suppressive properties of IL-10 in the intestine (33). Whether the immune suppressive functions of IL-10 and TGF-β are linked or act entirely separately remains to be elucidated. The finding that IL-10$^{-/-}$ mice have immune pathology primarily in the intestine while TGF-$\beta^{-/-}$ mice develop multiple organ disease suggests that IL-10 is not required for the production of

TGFβ1. However, TGF-β has been shown to induce IL-10 secretion by APC (121), making it a possibility that TGF-β alters antigen presentation in favour of the generation of IL-10-secreting regulatory T cells.

There is also evidence that NK cells may negatively regulate the development of intestinal inflammation (122). Transfer of CD4$^+$ T cells from IL-10$^{-/-}$ mice to RAG1$^{-/-}$ mice was more severe if the recipients were depleted of NK cells. The ability of NK cells to suppress colitis was, in part, due to a perforin-dependent mechanism, suggesting that the mechanism may involve NK cell-mediated lysis of pathogenic T cells.

6. GENETICS

Inheritance of IBD in humans is not a simple Mendelian trait and involves multiple genes (123,124). There is also strain-dependent susceptibility to development of IBD in a number of the animal models. For example, mice with the IL-10$^{-/-}$ mutation on the 129SvEv background were more susceptible to colitis than those with this mutation on the C57BL/6 background (30). Similarly, 129/Sv G$\alpha_{i2}^{-/-}$ mice had a higher incidence and severity of colitis than when crossed with the C57BL/6 strain (42), and 129/Sv TCR$\alpha^{-/-}$ mice had more severe disease than when crossed with the BALB/c strain (36). The most extensive studies have involved analysis of susceptibility to DSS-induced colitis amongst inbred strains of mice. Interestingly, the C3H/HeJBir substrain that spontaneously develops colitis was highly susceptible, as was the insulin-dependent diabetes (IDDM)-prone NOD/LtJ strain, whereas the C57BL/6 strain was more resistant (125). Analysis of an outcross between susceptible C3H/HeJ mice with partially resistant C57BL/6 mice revealed that multiple genes controlled susceptibility to DSS. Several loci linked to disease susceptibility were identified, and current studies are aimed at identifying the genes involved (126). The animal models of IBD make excellent tools with which to identify genetic regions, and ultimately genes, that confer resistance or susceptibility to IBD. As studies on the animal models as well as on the genetics of the human disease progress, it will be of interest to see whether similar genetic regions are identified.

7. THERAPEUTIC IMPLICATIONS AND RELEVANCE TO HUMAN IBD

Many of the concepts that have emerged from the study of animal models are relevant to the human disease. Thus, similar immunological alterations have been noted in patients with IBD as were noted in animal models. Both CD and UC have been consistently associated with increased pro-inflammatory cytokines, such as IL-1 (127–129), IL-6 (128,130) and TNF-α (128,131). Furthermore, CD appears to be a Th1-related disease associated with increased INF-γ, IL-2, but not IL-4 and IL-5 (132–135), as is the case in the majority of animal models (see Table 1). In contrast, it has been suggested that UC may be a Th2-related disease based on decreased levels of INF-γ and IL-2, and increased levels of IL-5 (134,136), although further data are needed to more definitively characterise the immune response in UC. Several animal models have also been shown to develop Th2 responses (Table 1).

On the basis of the immunological alterations uncovered in animal models of IBD, as well as data from human IBD, a number of therapeutics have been recently developed and evaluated. One of the most successful is the chimeric anti-TNF-α mAb, which has been shown to reduce clinical signs and symptoms in patients with CD, including the healing of fistulae (137–139). Treatment was supported by the consistent findings of increased TNF-α in both animal models and human IBD, and the success of anti-TNF antibody treatment in animal models. As discussed earlier, strategies to inhibit Th1 responses prevented, and in some cases ameliorated, IBD in animal models. Based on these studies, it would be anticipated that similar strategies may be efficacious in CD, which involves differential activation of Th1 cells. However, this approach has yet to be tested in human IBD. Administration of IL-11 was beneficial in the HLA-B27/human β_2 rat (140), and preliminary human data showed IL-11 had some beneficial activity in CD patients (141).

The indigenous microbiota drive intestinal inflammation in many models of IBD, and there is evidence of clonality of T cells in the intestine of mice with colitis, suggesting that bacterial antigens drive the inflammatory process. Data in humans suggests that IBD may also be related to inflammation driven by indigenous microbiota, since diversion of the faecal stream with resultant decreased exposure to luminal antigens prevented relapses in CD patients (142). Oligoclonal T cell expansions have also been described in patients with CD (143), and T cell clones isolated from the peripheral blood and colon of patients with IBD raised against anaerobes were able to cross react with different enterobacteria (144). These findings suggest that common luminal antigens may ultimately be identified, opening the door for the development of antigen-specific therapies for IBD.

Investigation of animals models has identified a population of regulatory T cells that control intestinal inflammation via secetion of the immune suppressive cytokines, IL-10 and TGF-β. Mucosal T cell unresponsiveness to enteric antigens in humans has also been shown to involve antigen-specific CD4$^+$ T cells and production of IL-10 and TGF-β (145). These findings suggest that strategies that enhance Treg cells may be beneficial in IBD. There is evidence that oral administration of antigen leads to the development of TGF-β-dependent regulatory T cells, termed Th3 cells (146). This has been exploited therapeutically in a model of IBD. It was shown that feeding mice with haptenated colonic proteins led to the development of TGF-β-dependent regulatory T cells that were able to prevent the development of TNBS-induced colitis. Recently, culture of both human and mouse T cells with antigen in the presence of IL-10 led to the development of regulatory T cell clones, termed Tr1 cells. These cells had similar functional properties to the Treg cells contained within the CD45RB^{low}CD4$^+$ population, and were able to inhibit colitis induced by transfer of CD45RB^{high}CD4$^+$T cells (147). Identification of conditions for the cloning of these cells will greatly facilitate their characterisation and may ultimately lead to development of strategies to enhance Treg cells *in vivo* The utility of enhancing immune suppressive cytokines to inhibit intestinal inflammation is illustrated by the fact that administration of IL-10 was able to inhibit colitis in several models (rev. in 148). The efficacy of IL-10 is currently being evaluated in humans with CD (149).

8. CONCLUDING REMARKS

Animal models of IBD have provided a large body of new information on the induction, perpetuation and regulation of chronic intestinal inflammation. Disruptions in the function of the epithelial cell barrier as well as in the homeostasis of the immune response result in a T cell-dependent inflammatory response in the intestine driven by resident bacteria. There are marked similarities between the pathogenic and regulatory mechanisms revealed in studies of animal models and those in the human disease. As a consequence, novel therapeutics have been developed that have already shown therapeutic utility for treatment of the human disease. Considering the rapid progress that has been made it seems likely that further study of animal models will provide additional insight into immune regulation in the intestine, and that more novel therapeutics will be forthcoming.

References

1. Sartor, R.B. 1995. Current concepts of the etiology and pathogenesis of ulcerative colitis and Crohn's disease. *Gastroenterol. Clin. North. Am.* 24:475–507.
2. Fiocchi, C. 1998. Inflammatory bowel disease: etiology and pathogenesis. *Gastroenterology* 115:182–205.
3. MacPherson, B.R., and C.J. Pfeiffer. 1978. Experimental production of diffuse colitis in rats. *Digestion* 17:135–150.
4. Okayasu, I., S. Hatakeyama, M. Yamada, T. Ohkusa, Y. Inagaki, and R. Nakaya. 1990. A novel method in the induction of reliable experimental acute and chronic ulcerative colitis in mice. *Gastroenterology* 98:694–702.
5. Dieleman, L.A., M.J. Palmen, H. Akol, E. Bloemena, A.S. Pena, S.G. Meuwissen, and E.P., Van Rees. 1998. Chronic experimental colitis induced by dextran sulphate sodium (DSS) is characterized by Th1 and Th2 cytokines. *Clin. Exp. Immunol.* 114:385–391.
6. Videla, S., and J. Vilaseca, F. Guarner, A. Salas, F. Treserra, E. Crespo, M. Antolin, and J.R. Malagelada. 1994. Role of intestinal microflora in chronic inflammation and ulceration of the rat colon. *Gut* 35:1090–1097.
7. Neurath, M.F., I. Fuss, B.L. Kelsall, D.H. Presky, W. Waegell, and W. Strober. 1996. Experimental granulomatous colitis in mice is abrogated by induction of TGF-β-mediated oral tolerance. *J. Exp. Med.* 183:2605–2616.
8. Dohi, T., K. Fujihashi, P.D. Rennert, K. Iwatani, H. Kiyono and J.R. McGhee. 1999. Hapten-induced colitis is associated with colonic patch hypertrophy and T helper cell 2-type responses. *J. Exp. Med.* 189:1169–1180.
9. Pantzar, N., G.M. Ekstrom, Q. Wang, and B.R. Westrom. 1994. Mechanisms of increased intestinal [51Cr]EDTA absorption during experimental colitis in the rat. *Dig. Dis. Sci.* 39:2327–2333.
10. Ekstrom, G.M. 1998. Oxazolone-induced colitis in rats: effects of budesonide, cyclosporin A, and 5-aminosalicylic acid. *Scand. J. Gastroenterol* 33:174–179.
11. Boirivant, M., I.J. Fuss, A. Chu, and W. Strober. 1998. Oxazolone colitis: A murine model of T helper cell type 2 colitis treatable with antibodies to interleukin 4. *J. Exp. Med.* 188:1929–1939.
12. Sartor, R.B., W.J. Cromartie, D.W. Powell, and J.H. Schwab. 1985. Granulomatous enterocolitis induced in rats by purified bacterial cell wall fragments. *Gastroenterology* 89:587–595.
13. Herfarth, H.H., S.P. Mohanty, H.C. Rath, S. Tonkonogy, and R.B. Sartor. 1996. Interleukin 10 suppresses experimental chronic, granulomatous inflammation induced by bacterial cell wall polymers. *Gut* 39:836–845.
14. Watt, J., R. Marcus. 1970. Hyperplastic mucosal changes in the rabbit colon produced by degraded carrageenin. *Gastroenterology* 59:760–768.
15. Onderdonk, A.B., M.L. Franklin, and R.L. Cisneros. 1981. Production of experimental ulcerative colitis in gnotobiotic guinea pigs with simplified microflora. *Infect Immun.* 32:225–231.
16. Powrie, F., M.W. Leach, S. Mauze, L.B. Caddle, and R.L. Coffman. 1993. Phenotypically distinct subsets of CD4$^+$ T cells induce or protect from chronic intestinal inflammation in C. B-17 scid mice. *Int. Immunol.* 5:1461–1471.
17. Powrie, F., M.W. Leach, S. Mauze, S. Menon, L. Barcomb Caddle, and R.L. Coffman. 1994. Inhibition of Th1 responses prevents inflammatory bowel disease in *scid* mice reconstituted with CD45RB^{hi}CD4$^+$T cells. *Immunity* 1:553–562.
18. Morrissey, P.J., and K. Charrier. 1994. Induction of wasting disease in SCID mice by the transfer of normal CD4+/CD45RBhi T cells and the regulation of this autoreactivity by CD4+/CD45RBlo T cells. *Res. Immunol.* 145:357–362.
19. Aranda, R., B.C. Sydora, P.L. McAllister, S.W. Binder, H.Y. Yang, S.R. Targan, and M. Kronenberg. 1997. Analysis of intestinal lymphocytes in mouse colitis mediated by transfer of CD4+, CD45RBhigh T cells to SCID recipients. *J. Immunol.* 158:3464–3473.
20. Claesson, M.H., A. Rudolphi, S. Kofoed, S.S. Poulsen, and J. Reimann. 1996. CD4+ T lymphocytes injected into severe combined immunodeficient (SCID) mice lead to an inflamma-

tory and lethal bowel disease. *Clin. Exp. Immunol.* 104:491–500.

21. Hollander, G.A., S.J. Simpson, E. Mizoguchi. A. Nichogiannopoulou, J. She, J.C. Gutierrez Ramos, A.K. Bhan, S.J. Burakoff, B. Wang, and C. Terhorst. 1995. Severe colitis in mice with aberrant thymic selection. *Immunity* 3:27–38.

22. Simpson, S.J., G.A. Hollander, E. Mizoguchi, D. Allen, A.K. Bhan, B. Wang, and C. Terhorst. 1997. Expression of pro-inflammatory cytokines by TCR alpha beta+ and TCR gamma delta+ T cells in an experimental model of colitis. *Eur. J. Immunol.* 27:17–25.

23. Steinhoff, U., V. Brinkmann, U. Klemm, P. Aichele, P. Seiler, U. Brandt, P.W. Bland, I. Prinz, U. Zugel, and S.H. Kaufmann. 1999. Autoimmune intestinal pathology induced by hsp60-specific CD8 T cells. *Immunity* 11:349–358.

24. Sadlack, B., H. Merz, H. Schorle, A. Schimpl, A.C. Feller, and I. Horak. 1993. Ulcerative colitis-like disease in mice with a disrupted interleukin-2 gene [see comments]. *Cell* 75:253–261.

25. Ehrhardt, R.O., B.R. Ludviksson, B. Gray, M. Neurath, and W. Strober. 1997. Induction and prevention of colonic inflammation in IL-2-deficient mice. *J. Immunol.* 158:566–573.

26. Schultz, M., S.L. Tonkonogy, R.K. Sellon C. Veltkamp, V.L. Godfrey, J. Kwon, W.B. Grenther, E. Balish, I. Horak, and R.B. Sartor. 1999. IL-2-deficient mice raised under germfree conditions develop delayed mild focal intestinal inflammation. *Am. J. Physiol.* 276:G1461–1472.

27. Willerford, D.M., J. Chen, J.A. Ferry, L. Davidson, A. Ma, and F.W. Alt. 1995. Interleukin-2 receptor alpha chain regulates the size and content of the peripheral lymphoid compartment. *Immunity* 3:521–530.

28. Watanabe, M., Y. Ueno, T. Yajima, S. Okamoto, T. Hayashi, M. Yamazaki, Y. Iwao, H. Ishii, S. Habu, M. Uehira, H. Nishimoto, H. Ishikawa, J. Hata, and T. Hibi. 1998. Interleukin 7 transgenic mice develop chronic colitis with decreased interleukin 7 protein accumulation in the colonic mucosa. *J. Exp. Med.* 187:389–402.

29. Kuhn, R., J. Lohler, D. Rennick, K. Rajewsky, and W. Muller. 1993. Interleukin-10-deficient mice develop chronic enterocolitis. *Cell* 75:263–274.

30. Berg, D.J., N. Davidson, R. Kuhn, W. Muller, S. Menon, G. Holland, L. Thompson Snipes, M.W. Leach, and D. Rennick. 1996. Enterocolitis and colon cancer in interleukin-10-deficient mice are associated with aberrant cytokine production and CD4(+) TH1-like responses. *J. Clin. Invest.* 98:1010–1020.

31. Davidson, N.J., M.W. Leach, M.M. Fort, L. Thompson-Snipes, R. Kuhn, W. Muller, D.J. Berg, and D.M. Rennick. 1996. T helper cell 1-type CD4+ T cells, but not B cells, mediate colitis in interleukin 10-deficient mice. *J. Exp. Med.* 184:241–251.

32. Sellon, R.K., S. Tonkonogy, M. Schultz, L.A. Dieleman, Grenther, E. Balish D.M. Rennick, and R.B. Sartor. 1998. Resident enteric bacteria are necessary for development of spontaneous colitis and immune system activation in interleukin-10-deficient mice. *Infect Immun.* 66:5224–5231.

33. Takeda, K., B.E. Clausen, T. Kaisho, T. Tsujimura, N. Terada, I. Forster, and S. Akira. 1999. Enhanced Th1 activity and development of chronic enterocolitis in mice devoid of Stat3 in macrophages and neutrophils. *Immunity* 10:39–49.

34. Spencer, S.D., F. Di Marco, J. Hooley, S. Pitts Meek, M. Bauer, A.M. Ryan, B. Sordat, V.C. Gibbs, and M. Aguet. 1998. The orphan receptor CRF2-4 is an essential subunit of the interleukin 10 receptor. *J. Exp. Med.* 187:571–578.

35. Kulkarni, A.B., J.M. Ward, L. Yaswen, C.L. Mackall, S.R. Bauer, C.G. Huh, R.E. Gress, and S. Karlsson. 1995. Transforming growth factor-beta 1 null mice. An animal model for inflammatory disorders. *Am. J. Pathol.* 146:264–275.

36. Mombaerts, P., E. Mizoguchi, M.J. Grusby, L.H. Glimcher, A.K. Bhan, and S. Tonegawa. 1993. Spontaneous development of inflammatory bowel disease in T cell receptor mutant mice. *Cell* 75:274–282.

37. Dianda, L., A.M. Hanby, N.A. Wright, A. Sebesteny, A.C. Hayday, and M.J. Owen. 1997. T cell receptor-alpha beta-deficient mice fail to develop colitis in the absence of a microbial environment. *Am. J. Pathol.* 150:91–97.

38. Iijima, H., I. Takahashi, D. Kishi, J.K. Kim, S. Kawano, M. Hori, and H. Kiyono. 1999. Alteration of interleukin 4 production results in the inhibition of T helper type 2 cell-dominated inflammatory bowel disease in T cell receptor alpha chain-deficient mice. *J. Exp. Med.* 190:607–615.

39. Mizoguchi, A., E. Mizoguchi, and A.K. Bhan. 1999. The critical role of interleukin 4 but not interferon gamma in the pathogenesis of colitis in T-cell receptor alpha mutant mice. *Gastroenterology* 116:320–326.

40. Kontoyiannis, D., M. Pasparakis, T.T. Pizarro, F. Cominelli, and G. Kollias. 1999. Impaired on/off regulation of TNF biosynthesis in mice lacking TNF AU-rich elements: implications for joint and gut-associated immunopathologies. *Immunity* 10:387–398.

41. Rudolph, U., M.J. Finegold, S.S. Rich, G.R. Harriman, Y. Srinivasan, P. Brabet, G. Boulay, A. Bradley, and L. Birnbaumer. 1995. Ulcerative colitis and adenocarcinoma of the colon in G alpha i2-deficient mice. *Nat. Genet.* 10:143–150.

42. Hornquist, C.E., X. Lu, P.M. Rogers-Fani, U. Rudolph, S. Shappell, L. Birnbaumer, and G.R. Harriman. 1997. Gα2-deficient mice with colitis exhibit a local increase in memory CD4+ T cells and proinflammatory Th1-type cytokines. *J. Immunol.* 158:1068–1077.

43. Wirtz, S. S. Finotto, S. Kanzler, A.W. Lohse, M. Blessing, H.A. Lehr, P.R. Galle, and M.F. Neurath. 1999. Chronic intestinal inflammation in STAT-4 transgenic mice: characterization of disease and adoptive transfer by TNF-plus IFN-gamma-producing CD4+ T cells that respond to bacterial antigens. *J. Immunol.* 162:1884–1888.

44. Hermiston, M.L., and J.I. Gordon. 1995. Inflammatory bowel disease and adenomas in mice expressing a dominant negative N-cadherin. *Science* 270:1203–1207.

45. Bush, T.G., T.C. Savidge, T.C. Freeman, H.J. Cox, E.A. Campbell, L. Mucke, M.H. Johnson, and M.V. Sofroniew. 1998. Fulminant jejuno-ileitis following ablation of enteric glia in adult transgenic mice. *Cell* 93:189–201.

46. Panwala, C.M., J.C. Jones, and J.L. Viney. 1998. A Novel model of inflammatory bowel disease: mice deficient for the multiple drug resistance gene, mdr1a, spontaneously develop colitis. *J. Immunol.* 161:5733–5744.

47. Snapper, S.B., F.S. Rosen, E. Mizoguchi, P. Cohen, W. Khan, C.H. Liu, T.L. Hagemann, S.P. Kwan, R. Ferrini, L. Davidson, A.K. Bhan, and F.W. Alt. 1998. Wiskott-Aldrich syndrome protein-deficient mice reveal a role for WASP in T but not B cell activation. *Immunity* 9:81–91.

48. Hammer, R.E., S.D. Maika, J.A. Richardson, J.P. Tang, and J.D. Taurog. 1990. Spontaneous inflammatory disease in transgenic rats expressing HLA-B27 and human beta 2m: an animal model of HLA-B27-associated human disorders. *Cell* 63:1099–1112.

49. Taurog, J.D., J.A. Richardson, J.T. Croft, W.A. Simmons, M. Zhou, J.L. Fernandez-Sueiro, E. Balish, and R.E. Hammer. 1994. The germfree state prevents development of gut and joint inflammatory disease in HLA-B27 transgenic rats. *J. Exp. Med.* 180:2359–2364.

50. Rath, H.C., H.H. Herfarth, J.S. Ikeda, W.B. Grenther, T.E. Hamm, Jr., E. Balish, J.D. Taurog, R.E. Hammer, K.H. Wilson, and R.B. Sartor. 1996. Normal luminal bacteria, especially Bacteroides species, mediate chronic colitis, gastritis, and arthritis in HLA-B27/human beta2 microglobulin transgenic rats. *J. Clin. Invest.* 98:945–953.

51. Ward, J.M., M.R. Anver, D.C. Haines, J.M. Melhorn, P. Gorelick, L. Yan, and J.G. Fox. 1996. Inflammatory large bowel disease in immunodeficient mice naturally infected with Helicobacter hepaticus. *Lab. Anim. Sci.* 46:15–20.

52. Kullberg, M.C., J.M. Ward, P.L. Gorelick, P. Caspar, S. Hieny, A. Cheever, D. Jankovic, and A. Sher. 1998. Helicobacter hepaticus triggers colitis in specific-pathogen-free interleukin-10 (IL-10)-deficient mice through an IL-12- and and gamma interferon-dependent mechanism. *Infect. Immun.* 66:5157–5166.

53. Haines, D.C., P.L. Gorelick, J.K. Battles, K.M. Pike, R.J. Anderson, J.G. Fox, N.S. Taylor, Z. Shen, F.E. Dewhirst, M.R. Anver, and J.M. Ward 1998. Inflammatory large bowel disease in immunodeficient rats naturally and experimentally infected with Helicobacter bilis. *Vet. Pathol.* 35:202–208.

54. Barthold, S.W., G.L. Coleman, R.O. Jacoby, E.M. Livestone, and A.M. Jonas. 1978. Transmissible murine colonic hyperplasia. *Vet. Pathol.* 15:223–236.

55. Higgins, L.M., G. Frankel, G. Douce, G. Dougan, and T.T. MacDonald. 1999. Citrobacter rodentium infection in mice elicits a mucosal Th1 cytokine response and lesions similar to those in murine inflammatory bowel disease. *Infect Immun.* 67:3031–3039.

56. Sundberg, J.P., C.O. Elson, H. Bedigian, and E.H. Birkenmeier. 1994. Spontaneous, heritable colitis in a new substrain of C3H/HeJ mice. *Gastroenterology* 107:1726–1735.

57. Cong, Y., S.L. Brandwein, R.P. McCabe, A. Lazeby, E.H. Birkenmeier, J.P. Sundberg, and C.O. Elson. 1998. CD4+ T cells reactive to enteric bacterial antigens in spontaneously colitic C3H/HeJBir mice: increased T helper cell type 1 response and ability to transfer disease. *J. Exp. Med.* 187:855–864.

58. Matsumoto, S., Y. Okabe, H. Setoyama, K. Takayama, J. Ohtsuka, H. Funahashi, A. Imaoka, Y. Okada, and Y. Umesaki. 1998. Inflammatory bowel disease-like enteritis and caecitis in a senescence accelerated mouse P1/Yit strain. *Gut* 43:71–78.

59. Madara, J.L., D.K. Podolsky, N.W. King, P.K. Sehgal, R. Moore, and H.S. Winter. 1985. Characterization of spontaneous colitis in cotton-top tamarins (Saguinus oedipus) and its response to sulfasalazine. *Gastroenterology* 88:13–19.

60. Okamoto, S., M. Watanabe, M. Yamazaki, T. Yajima, T. Hayashi, H. Ishii, H. Mukai, T. Yamada, N. Watanabe, B.A. Jameson, and T. Hibi. 1999. A synthetic mimetic of CD4 is able to suppress disease in a rodent model of immune colitis. *Eur. J. Immunol.* 29:355–366.

61. Morrissey, P.J., K. Charrier, S. Braddy, D. Liggitt, and J.D. Watson. 1993. CD4+ T cells that express high levels of CD45RB induce wasting disease when transferred into congenic severe combined immunodeficient mice. Disease development is prevented by contransfer of purified CD4+ T cells. *J. Exp. Med.* 178:237–244.

62. Takahashi, I., H. Kiyono, and S. Hamada. 1997. CD4+ T-cell population mediates development of inflammatory bowel disease in T-cell receptor alpha chain-deficient mice. *Gastroenterology* 112:1876–1886.

63. Simpson, S.J., E. Mizoguchi, D. Allen, A.K. Bhan, and C. Terhorst. 1995. Evidence that CD4+, but not CD8+ T cells are responsible for murine interleukin-2-deficient colitis. *Eur. J. Immunol.* 25:2618–2625.

64. Simpson, S.J., Y.P. De Jong, S.A. Shah, M. Comiskey, B. Wang, J.A. Spielman, E.R. Podack, E. Mizoguchi, A.K. Bhan, and C. Terhorst. 1998. Consequences of Fas-ligand and perforin expression by colon T cells in a mouse model of inflammatory bowel disease. *Gastroenterology* 115:849–855.

65. Shomer, N.H., C.A. Dangler, and M.D. Schrenzel, J.G. Fox. 1997. Helicobacter bilis-induced inflammatory bowel disease in scid mice with defined flora. *Infect. Immun.* 65:4858–4864.

66. Powrie, F., J. Carlino, M.W. Leach, S. Mauze, and R.L. Coffman. 1996. A critical role for transforming growth factor-β but not interleukin 4 in the suppression of T helper type 1-mediated colitis by CD45RB^low CD4^+ T cells. *J. Exp. Med.* 183:2669–2674.

67. Mizoguchi, A., E. Mizoguchi, R.N. Smith, F.I. Preffer, and A.K. Bhan. 1997. Suppressive role of B cells in chronic colitis of T cell receptor alpha mutant mice. *J. Exp. Med.* 186:1749–1756.

68. Ma, A., M. Datta, E. Margosian, J. Chen, and I. Horak. 1995. T cells, but not B cells, are required for bowel inflammation in interleukin 2-deficient mice. *J. Exp. Med.* 182:1567–1572.

69. Botto, M., C. Dell'Agnola, A.E. Bygrave, E.M. Thompson, H.T. Cook, F. Petry, M. Loos, P.P. Pandolfi, and M.J. Walport. 1998. Homozygous C1q deficiency causes glomerulonephritis associated with multiple apoptotic bodies. *Nat. Genet.* 19:56–59.

70. Mosmann, T.R., R.L. Coffman. 1989. TH1 and TH2 cells: different patterns of lymphokine secretion lead to different functional properties. *Annu. Rev. Immunol.* 7:145–173.

71. Romagnani, S. 1997. The Th1/Th2 paradigm. *Immunol. Today* 18:163–266.

72. Ito, H., C.G. Fathman. 1997. CD45RBhigh CD4+ T cells from IFN-gamma knockout mice do not induce wasting disease. *J. Autoimmun.* 10:455–459.

73. Simpson, S.J., S. Shah, M. Comiskey, Y.P. de Jong, B. Wang, E. Mizoguchi, A.K. Bhan, and C. Terhorst. 1998. T cell-mediated pathology in two models of experimental colitis depends predominantly on the interleukin 12/Signal transducer and activator of transcription (Stat)-4 pathway, but is not conditional on interferon gamma expression by T cells. *J. Exp. Med.* 187:1225–1234.

74. Kojouharoff, G., W. Hans, F. Obermeier, D.N. Mannel, T. Andus, J. Scholmerich, V. Gross, and W. Falk. 1997. Neutralization of tumour necrosis factor (TNF) but not of IL-1 reduces inflammation in chronic dextran sulphate sodium-induced colitis in mice. *Clin. Exp. Immunol.* 107:353–358.

75. Mackay, F., J.L. Browning, P. Lawton, S.A. Shah, M. Comiskey, A.K. Bhan, E. Mizoguchi, C. Terhorst, and S.J. Simpson. 1998. Both the lymphotoxin and tumor necrosis factor pathways are involved in experimental murine models of colitis. *Gastroenterology* 115:1464–1475.

76. Corazza, N., S. Eichenberger, H-P. Eugster, and C. Mueller. 1999. Nonlymphocyte-derived tumour necrosis factor is

required for induction of colitis in recombination activating gene (RAG)2-/- mice upon transfer of CD4+ CD45RBhi T cells. *J. Exp. Med.* 190:1479–1491.

77. Hsieh, C.S., S.E. Macatonia, C.S. Tripp, S.F. Wolf, A. O'Garra, and K.M. Murphy. 1993. Development of TH1 CD4+ T cells through IL-12 produced by Listeria-induced macrophages. *Science* 260:547–549.

78. O'Garra, A., N. Hosken, S. Macatonia, C.A. Wenner, and K. Murphy. 1995. The role of macrophage- and dendritic cell-derived IL12 in Th1 phenotype development. *Res. Immunol.* 146:466–472.

79. Claesson, M.H., S. Bregenholt, K. Bonhagen, S. Thoma, P. Moller, M.J. Grusby, F. Leithauser, M.H. Nissen, and J. Reimann. 1999. Colitis-inducing potency of CD4+ T cells in immunodeficient, adoptive hosts depends on their state of activation, IL-12 responsiveness, and CD45RB surface phenotype. *J. Immunol.* 162:3702–3710.

80. Davidson, N.J., S.A. Hudak, R.E. Lesley, S. Menon, M.W. Leach, and D.M. Rennick. 1998. IL-12, but not IFN-gamma, plays a major role in sustaining the chronic phase of colitis in IL-10-deficient mice. *J. Immunol.* 161:3143–3149.

81. Fuss, I.J., T. Marth, M.F. Neurath, G.R. Pearlstein, A. Jain, and W. Strober. 1999. Anti-interleukin 12 treatment regulates apoptosis of T helper 1 T cells in experimental colitis. *Gastroenterology* 117:1078–1088.

82. von Freeden Jeffry, U., N. Davidson, R. Wiler, M. Fort, S. Burdach, and R. Murray. 1998. IL-7 deficiency prevents development of a non-T cell non-B cell-mediated colitis. *J. Immunol.* 161:5673–5680.

83. Chambers, C.A., and J.P. Allison. 1999. Costimulatory regulation of T cell function. *Curr. Opin. Cell. Biol.* 11:203–210.

84. Grewal, I.S., and R.A. Flavell. 1998. CD40 and CD154 in cell-mediated immunity. *Annu. Rev. Immunol.* 16:111–135.

85. Stuber, E., W. Strober, and M. Neurath. 1996. Blocking the CD40L-CD40 interaction *in vivo* specifically prevents the priming of T helper 1 cells through the inhibition of interleukin 12 secretion. *J. Exp. Med.* 183:693–698.

86. Weinberg, A.D., A.T. Vella, and M. Croft. 1998. OX-40: life beyond the effector T cell stage. *Semin. Immunol.* 10:471–480.

87. Higgins, L.M., S.A. McDonald, N. Whittle, N. Crockett, J.G. Shields, and T.T. MacDonald. 1999. Regulation of T cell activation *in vitro* and *in vivo* by targeting the OX40–OX40 ligand interaction: amelioration of ongoing inflammatory bowel disease with an OX40-IgG fusion protein, but not with an OX40 ligand-IgG fusion protein. *J. Immunol.* 162:486–493.

88. Berlin, C., E.L. Berg, M.J. Briskin, D.P. Andrew, P.J. Kilshaw, B. Holzmann, I.L. Weissman, A. Hamann, and E.C. Butcher. 1993. Alpha 4 beta 7 integrin mediates lymphocyte binding to the mucosal vascular addressin MAdCAM-1. *Cell* 74:185–185.

89. Hesterberg, P.E., D. Winsor Hines, M.J. Briskin, D. Soler Ferran, C. Merrill, C.R. Mackay, W. Newman, and D.J. Ringler. 1996. Rapid resolution of chronic colitis in the cotton-top tamarin with an antibody to a gut-homing integrin alpha 4 beta 7. *Gastroenterology* 111:1373–1380.

90. Picarella, D., P. Hurlbut, J. Rottman, X. Shi, E. Butcher, and D.J. Ringler. 1997. Monoclonal antibodies specific for beta 7 integrin and mucosal addressin cell adhesion molecule-1 (MAdCAM-1) reduce inflammation in the colon of scid mice reconstituted with CD45RBhigh CD4+ T cells. *J. Immunol.* 158:2099–2106.

91. Cepek, K.L., S.K. Shaw, C.M. Parker, G.J. Russell, J.S. Morrow, D.L. Rimm, and M.B. Brenner. 1994. Adhesion between epithelial cells and T lymphocytes mediated by E-cadherin and the alpha E beta 7 integrin. *Nature* 372:190–193.

92. Karecla, P.I., S.J. Bowden, S.J. Green, and P.J. Kilshaw. 1995. Recoginition of E-cadherin on epithelial cells by the mucosal T cell integrin alpha M290 beta 7 (alpha E beta 7). *Eur. J. Immunol.* 25:852–856.

93. Ludviksson, B.R., W. Strober, R. Nishikomori, S.K. Hasan, and R.O. Ehrhardt. 1999. Administration of mAb against alpha E beta 7 prevents and ameliorates immunization-induced colitis in IL-2-/- mice. *J. Immunol.* 162:4975–4982.

94. Wittig, B., C. Schwarzler, N. Fohr, U. Gunthert, and M. Zoller. 1998. Curative treatment of an experimentally induced colitis by a CD44 variant V7-specific antibody. *J. Immunol.* 161:1069–1073.

95. Billiau, A. 1996. Interferon-gamma: biology and role in pathogenesis. *Adv. Immunol.* 62:61–130.

96. Guy Grand, D., J.P. DiSanto, P. Henchoz, M. Malassis Seris, and P. Vassalli. 1998. Small bowel enteropathy: role of intraepithelial lymphocytes and of cytokines (IL-12, IFN-gamma, TNF) in the induction of epithelial cell death and renewal. *Eur. J. Immunol.* 28:730–744.

97. Aiko, S., and M.B. Grisham. 1995. Spontaneous intestinal inflammation and nitric oxide metabolism in HLA-B27 transgenic rats. *Gastroenterology* 109:142–150.

98. Hogaboam, C.M., K. Jacobson, S.M. Collins, and M.G. Blennerhassett. 1995. The selective beneficial effects of nitric oxide inhibition in experimental colitis. *Am. J. Physiol.* 268:G673–684.

99. MacDonald, T.T., M. Bajaj Elliott, and S.L. Pender. 1999. T cells orchestrate intestinal mucosal shape and integrity. *Immunol. Today* 20:505–510.

100. Duchmann, R., I. Kaiser, E. Hermann, W. Mayet, K. Ewe, and K.H. Meyer zum Buschenfelde. 1995. Tolerance exists towards resident intestinal flora but is broken in active inflammatory bowel disease (IBD). *Clin. Exp. Immunol.* 102:448–455.

101. Duchmann, R., E. Schmitt, P. Knolle, K.H. Meyer zum Buschenfelde, and M. Neurath. 1996. Tolerance towards resident intestinal flora in mice is abrogated in experimental colitis and restored by treatment with interleukin-10 or antibodies to interleukin-12. *Eur. J. Immunol.* 26:934–938.

102. Powrie, F., S. Mauze, and R.L. Coffman. 1997. CD4+ T-cells in the regulation of inflammatory responses in the intestine. *Res. Immunol.* 148:576–581.

103. Fukushima, K., I. Sasaki, H. Ogawa, H. Naito, Y. Funayama, and S. Matsuno. 1999. Colonization of microflora in mice: mucosal defense against luminal bacteria. *J. Gastroenterol.* 34:54–60.

104. Mizoguchi, A., E. Mizoguchi, S. Tonegawa, and A.K. Bhan. 1996. Alteration of a polyclonal to an oligoclonal immune response to cecal aerobic bacterial antigens in TCR alpha mutant mice with inflammatory bowel disease. *Int. Immunol.* 8:1387–1394.

105. Takahashi, I., H. Iijima, R. Katashima, M. Itakura, and H. Kiyono. 1999. Clonal expansion of CD4+ TCRbetabeta+ T cells in TCR alpha-chain-deficient mice by gut-derived antigens. *J. Immunol.* 162:1843–1850.

106. Chester, J.F., J.S. Ross, R.A. Malt, and S.A. Weitzman. 1985. Acute colitis produced by chemotactic peptides in rats and mice. *Am. J. Pathol.* 121:284–290.

107. Higgins, L.M., G. Frankel, I. Connerton, N.S. Goncalves, G. Dougan, and T.T. MacDonald. 1999. Role of bacterial

intimin in colonic hyperplasia and inflammation. *Science* 285:588–591.

108. Madsen, K.L., J.S. Doyle, L.D. Jewell, M.M. Tavernini, and R.N. Fedorak. 1999. Lactobacillus species prevents colitis in interleukin 10 gene-deficient mice. *Gastroenterology* 116:1107–1114.

109. Fabia, R., A. Ar'Rajab, M.L. Johansson, R. Willen, R. Andersson, G. Molin, and S. Bengmark. 1993. The effect of exogenous administration of Lactobacillus reuteri R2LC and oat fiber on acetic acid-induced colitis in the rat. *Scand. J. Gastroenterol.* 28:155–162.

110. Campieri, M., and P. Gionchetti. 1999. Probiotics in inflammatory bowel disease: new insight to pathogenesis or a possible therapeutic alternative? *Gastroenterology* 116:1246–1249.

111. Russell, R.J., D.C. Haines, M.R. Anver, J.K. Battles, P.L. Gorelick, L.L. Blumenauer, M.A. Gonda, and J.M. Ward. 1995. Use of antibiotics to prevent hepatitis and typhlitis in male scid mice spontaneously infected with Helicobacter hepaticus. *Lab. Anim. Sci.* 45:373–378.

112. Fox, J.G., F.E. Dewhirst, J.G. Tully, B.J. Paster, L. Yan, N.S. Taylor, M.J. Collins, Jr., P.L. Gorelick, and J.M. Ward. 1994. Helicobacter hepaticus sp. nov., a microaerophilic bacterium isolated from livers and intestinal mucosal scrapings from mice. *J. Clin. Microbiol.* 32:1238–1245.

113. Groux, H., and F. Powrie. 1999. Regulatory T cells and inflammatory bowel disease. *Immunol Today* 20:442–445.

114. Hagenbaugh, A., S. Sharma, S.M. Dubinett, S.H. Wei, R. Aranda, H. Cheroutre, D.J. Fowell, S. Binder, B. Tsao, R.M. Locksley, K.W. Moore, and M. Kronenberg. 1997. Altered immune response in interleukine 10 transgenic mice. *J. Exp. Med.* 185:2101–2110.

115. Asseman, C., S. Mauze, M.W. Leach, R.L. Coffman, and F. Powrie. 1999. An essential role for interleukin 10 in the function of regulatory T cells that inhibit intestinal inflammation. *J. Exp. Med.* 190:995–1004.

116. Ludviksson, B.R., R.O. Ehrhardt, and W. Strober. 1997. TGF-beta production regulates the development of the 2,4,6-trintrophenol-conjugated keyhole limpet hemocyanin-induced colonic inflammation in IL-2-deficient mice. *J. Immunol.* 159:3622–3628.

117. Powrie, F., and M.W. Leach. 1995. Genetic and spontaneous models of inflammatory bowel disease in rodents: evidence for abnormalities in mucosal immune regulation. *Therapeutic Immunol.* 2:115–123.

118. Strober, W., B. Kelsall, I. Fuss, T. Marth, B. Ludviksson, R. Ehrhardt, and M. Neurath. 1997. Reciprocal IFN-gamma and TGF-beta responses regulate the occurrence of mucosal inflammation. *Immunol. Today* 18:61–64.

119. Moore, K.W., A. O'Garra, R. de Waal Malefyt, P. Vieira, and T.R. Mosmann. 1993. Interleukin-10. *Annu. Rev. Immunol.* 11:165–190.

120. Letterio, J.J., and A.B. Roberts. 1998. Regulation of immune responses by TGF-BETA. *Ann. Rev. Immunol.* 16:137–162.

121. Maeda, H., H. Kuwahara, Y. Ichimura, M. Ohtsuki, S. Kurakata, and A. Shiraishi. 1995. TGF-beta enhances macrophage ability to produce IL-10 in normal and tumor-bearing mice. *J. Immunol.* 155:4926–4932.

122. Fort, M.M., M.W. Leach, and D.M. Rennick. 1998. A role for NK cells as regulators of CD4+ T cells in a transfer model of colitis. *J. Immunol.* 161:3256–3261.

123. Hugot, J.P., P. Laurent Puig, C. Gower Rousseau, J.M. Olson, J.C. Lee, L. Beaugerie, I. Naom, J.L. Dupas, A. Van Gossum, M. Orholm, C. Bonaiti Pellie, J. Weissenbach, C.G. Mathew, J.E. Lennard Jones, A. Cortot, J.F. Colombel, and G. Thomas. 1996. Mapping of a susceptibility locus for Crohn's disease on chromosome 16. *Nature* 379:821–823.

124. Satsangi, J., M. Parkes, E. Louis, L. Hashimoto, N. Kato, K. Welsh, J.D. Terwilliger, G.M. Lathrop, J.I. Bell, and D.P. Jewell. 1996. Two stage genomewide search in inflammatory bowel disease provides evidence for susceptibility loci on chromosomes 3, 7 and 12. *Nat. Genet.* 14:199–202.

125. Mahler, M., I.J. Bristol, E.H. Leiter, A.E. Workman, E.H. Birkenmeier, C.O. Elson, J.P. and Sundberg. 1998. Differential susceptibility of inbred mouse strains to dextran sulfate sodium-induced colitis. *Am. J. Physiol.* 274:G544–551.

126. Mahler, M., I.J. Bristol, J.P. Sundberg, G.A. Churchill, E.H. Birkenmeier, C.O. Elson, and E.H. Leiter. 1999. Genetic analysis of susceptibility to dextran sulfate sodium-induced colitis in mice. *Genomics* 55:147–156.

127. Youngman, K.R., P.L. Simon, G.A. West, F. Cominelli, D. Rachmilewitz, J.S. Klein, and C. Fiocchi. 1993. Localization of intestinal interleukin 1 activity and protein and gene expression to lamina propria cells. *Gastroenterology* 104:749–758.

128. Reinecker, H.C., M. Steffen, T. Witthoeft, I. Pflueger, S. Schreiber, R.P. Mac Dermott, and, A. Raedler. 1993. Enhanced secretion of tumour necrosis factor-alpha, IL-6, and IL-1 beta by isolated lamina propria mononuclear cells from patients with ulcerative colitis and Crohn's disease. *Clin. Exp. Immunol.* 94:174–181.

129. Casini Raggi, V., L. Kam, Y.J. Chong, C. Fiocchi, T.T. Pizarro, and F. Cominelli. 1995. Mucosal imbalance of IL-1 and IL-1 receptor antagonist in inflammatory bowel disease. A novel mechanism of chronic intestinal inflammation. *J. Immunol.* 154:2434–2440.

130. Gross, V., T. Andus, I. Caesar, M. Roth, and J. Scholmerich. 1992. Evidence for continuous stimulation of interleukin-6 production in Crohn's disease [see comments]. *Gastroenterology* 102:514–519.

131. Breese, E.J., C.A. Michie, S.W. Nicholls, S.H. Murch, C.B. Williams, P. Domizio, J.A. Walker Smith, and T.T. MacDonald. 1994. Tumor necrosis factor alpha-producing cells in the intestinal mucosa of children with inflammatory bowel disease. *Gastroenterology* 106:1455–1466.

132. Breese, E., C.P. Braegger, C.J. Corrigan, J.A. Walker Smith, and T.T. MacDonald. 1993. Interleukin-2- and interferon-gamma-secreting T cells in normal and diseased human intestinal mucosa. *Immunology* 78:127–131.

133. Parronchi, P., P. Romagnani, F. Annunziato, S. Sampognaro, A. Becchio, L. Giannarini, E. Maggi, C. Pupilli, F. Tonelli, and S. Romagnani. 1997. Type 1 T-helper cell predominance and interleukin-12 expression in the gut of patients with Crohn's disease. *Am. J. Pathol.* 150:823–832.

134. Fuss, I.J., M. Neurath, M. Boirivant, J.S. Klein, C. de la Motte, S.A. Strong, C. Fiocchi, and W. Strober. 1996. Disparate CD4+ lamina propria (LP) lymphokine secretion profiles in inflammatory bowel disease. Crohn's disease LP cells manifest increased secretion of IFN-gamma, whereas ulcerative colitis LP cells manifest increased secretion of IL-5. *J. Immunol.* 157:1261–1270.

135. Monteleone, G., L. Biancone, R. Marasco, G. Morrone, O. Marasco, F. Luzza, and F. Pallone. 1997. Interleukin 12 is expressed and actively released by Crohn's disease intestinal lamina propria mononuclear cells. *Gastroenterology* 112:1169–1178.

136. Murata, Y., Y. Ishiguro, J. Itoh, A. Munakata, and Y. Yoshida. 1995. The role of proinflammatory and immunoregulatory cytokines in the pathogenesis of ulcerative colitis. *J. Gastroenterol.* 30:56–60.

137. Targan, S.R., S.B. Hanauer, S.J. van Deventer, L. Mayer, D.H. Present, T. Braakman, K.L. DeWoody, T.F. Schaible, and P.J. Rutgeerts. 1997. A short-term study of chimeric monoclonal antibody cA2 to tumor necrosis factor alpha for Crohn's disease. Crohn's Disease cA2 Study Group. *N. Engl. J. Med.* 337:1029–1035.

138. van Dullemen, H.M., S.J. van Deventer, D.W. Hommes, H.A. Bijl, J. Jansen, G.N. Tytgat, and J. Woody. 1995. Treatment of Crohn's disease with anti-tumor necrosis factor chimeric monoclonal antibody (cA2). *Gastroenterology* 109:129–135.

139. Present, D.H., P. Rutgeerts, S. Targan, S.B. Hanauer, L. Mayer, R.A. van Hogezand, D.K. Podolsky, B.E. Sands, T. Braakman, K.L. DeWoody, T.F. Schaible, and S.J. van Deventer. 1999. Infliximab for the treatment of fistulas in patients with Crohn's disease. *N. Engl. J. Med.* 340:1398–1405.

140. Peterson, R.L., L. Wang, L. Albert, J.C. Keith, Jr., and A.J. Dorner. 1998. Molecular effects of recombinant human interleukin-11 in the HLA-B27 rat model of inflammatory bowel disease. *Lab. Invest.* 78:1503–1512.

141. Sands, B.E., S. Bank, C.A. Sninsky, M. Robinson, S. Katz, J.W. Singleton, P.B. Miner, M.A. Safdi, S. Galandiuk, S.B. Hanauer, G.W. Varilek, A.L. Buchman, V.D. Rodgers, B. Salzberg, B. Cai, J. Loewy, M.F. DeBruin, H. Rogge, M. Shapiro, and U.S. Schwertschlag. 1999. Preliminary evaluation of safety and activity of recombinant human interleukin 11 in patients with active Crohn's disease. *Gastroenterology* 117:58–64.

142. Rutgeerts, P., K. Goboes, M. Peeters, M. Hiele, F. Penninckx, R. Aerts, R. Kerremans, and G. Vantrappen. 1991. Effect of faecal stream diversion on recurrence of Crohn's disease in the neoterminal ileum *Lancet* 338:771–774.

143. Probert, C.S., A. Chott, J.R. Turner, L.J. Saubermann, A.C. Stevens, K. Bodinaku, C.O. Elson, S.P. Balk, and R.S. Blumberg. 1996. Persistent clonal expansions of peripheral blood CD4+ lymphocytes in chronic inflammatory bowel disease. *J. Immunol.* 157:3183–3191.

144. Duchmann, R., E. May, M. Heike, P. Knolle, M. Neurath, and K.H. Meyer zum Buschenfelde. 1999. T cell specificity and cross reactivity towards enterobacteria, bacteroides, bifidobacterium, and antigens from resident intestinal flora in humans. *Gut* 44:812–818.

145. Khoo, U.Y., I.E. Proctor, and A.J. Macpherson. 1997. CD4+ T cell down-regulation in human intestinal mucosa: evidence for intestinal tolerance to luminal bacterial antigens. *J. Immunol.* 158:3626–3634.

146. Weiner, H.L. 1997. Oral tolerance: immune mechanisms and treatment of autoimmune diseases. *Immunol. Today* 18:335–343.

147. Groux, H., A. O'Garra, M. Bigler, M. Rouleau, S. Antonenko, J.E. de Vries, and M.G. Roncarolo. 1997. A CD4+ T-cell subset inhibits antigen-specific T-cell responses and prevents colitis. *Nature* 389:737–742.

148. Leach, M.W., N.J. Davidson, M.M. Fort, F. Powrie, and D.M. Rennick. 1999. The role of IL-10 in inflammatory bowel disease: "of mice and men". *Toxicol. Pathol.* 27:123–133.

149. van Deventer, S.J., C.O. Elson, and R.N. Fedorak. 1997. Multiple doses of intravenous interleukin 10 in steroid-refractory Crohn's disease. Crohn's Disease Study Group. *Gastroenterology* 113:383–389.

53 | Molecular Considerations and Immunopathology of Primary Biliary Cirrhosis

Akiyoshi Nishio, James Neuberger and M. Eric Gershwin

1. INTRODUCTION

The immune system has a tremendous diversity of mechanisms to protect the body from pathogens. Because the repertoire of specificity expressed by the T and B cell populations is generated randomly, it is bound to include many specific for self-components. Thus, the body must establish self-tolerance to distinguish between self and non-self determinants and avoid autoreactivity. However, all mechanisms have a risk of breakdown. As there are many diverse tolerance-inducing mechanisms, it is likely that there are multiple ways in which tolerance can be broken down, leading to autoimmunity; this may result in organ-specific or systemic autoimmune diseases. Organ-specific autoimmune diseases are characterized by the production of autoantibodies and the destruction of specific organs. Primary biliary cirrhosis (PBC) is an idiopathic hepatic disorder characterized by lymphoid infiltrates in the portal tracts of the liver, bile duct destruction, and the presence of disease-specific autoantibodies. The autoantigens recognized in PBC are predominantly components of normal mitochondria, although less common components of the nucleus and cytoplasm are also recognized. Primary biliary cirrhosis has been considered a paradigm for organ-specific autoimmune diseases such as Hashimoto's thyroiditis and type 1 diabetes mellitus. A great deal of attention has been paid to the disease because of the unique immunological abnormalities, particularly the association with high titers of anti-mitochondrial antibodies (AMA) How a mitochondrial antigen sequestered from the immune system by two membrane barriers can elicit an immune response or, once tolerance is broken, how autoimmunity causes tissue damage only in biliary epithelium, remains a mystery. However, the recent advent of

molecular biology has provided new approaches to address the pathogenesis of PBC. In this chapter, we will focus mainly on recent data from several areas that have provided a basis for defining the autoimmune response in PBC, and information suggesting an explanation for the selective immune damage.

2. CLINICAL ASPECTS

Primary biliary cirrhosis is a chronic cholestatic liver disease that predominantly involves middle-aged women and occurs in various ethnic populations, although rarely in people in Africa or the Indian sub-continent. Histologically, the disease is characterized by inflammatory and granulomatous destruction of interlobular and septal bile ducts, subsequent fibrosis, and finally liver failure. The first description of PBC was reported in 1851 (1). After 1900, associations between biliary cirrhosis and the presence of xanthomatous lesions in the bile ducts led to the designation of PBC as xanthomatous biliary cirrhosis (2). The term PBC became used to distinguish biliary cirrhosis in the absence of extrahepatic biliary obstruction from that associated with obstruction of the large bile ducts and other diseases of the biliary tree (3). In 1950, Ahrens and colleagues reported the first comprehensive description of the clinical features of PBC (4). In 1965, Rubin and associates described the characteristic histological process resulting in the destruction of intrahepatic bile ducts as "chronic non-suppurative destructive cholangitis" (5). In the same year, using an IF technique, Walker and colleagues identified AMA as the serological marker of the disease (6). The presence of AMA is recognized as the major serologic abnormality in PBC, since AMA is found in 90–95% of such patients and thus may be used as a diagnostic marker. The natural history and time course for the disease progression varies widely between patients.

PBC is more frequently diagnosed now than decades ago probably because of greater awareness of the disease, although a true increase in the prevalence of disease cannot be excluded. An increasing number of patients are found at the presymptomatic stage as a result of multiphasic laboratory screening tests. Biochemically, the most characteristic feature is an elevation of serum alkaline phosphatase and γ-glutamyltransferase (7). Elevation of serum aminotransferases are usually mild. In the later stages of the disease, patients with PBC present with a variety of symptoms such as general fatigue and pruitus (Table 1). Jaundice eventually develops. Later stages are associated with variceal bleeding, ascites and encephalopathy. The end-stage features of PBC include steatorrhea, osteopenia (rarely osteomalacia), melanosis, and liver failure. Histologically, PBC is characterized by "chronic nonsuppurative destructive cholangitis" that progresses to cirrhosis. The histological progression of PBC can be classified into four stages (8–10). Classification varies slightly in different systems, but broadly they are represented as: Stage 1) inflammatory destruction of interlobular and septal bile duct; stage 2) piecemeal necrosis and/or proliferation of bile ductules; stage 3) septal fibrosis and/or bridging necrosis; and stage 4) cirrhosis. The inflammation is sometimes accompanied by non-caseating granuloma formation that consists of histiocytes, lymphocytes, and occasionally giant cells. These granulomas are often seen in the early stage of the disease. Since the lesion develops at different rates in the different parts of the liver, several stages may be seen in one liver. In addition, there are no clear quantitative immunologic differences among these four stages. Most therapeutic efforts have been directed at altering the immune response. Ursodeoxycholic acid appears to be effective therapy in preventing or delaying the need for liver transplantation and improving survival (11).

Table 1 Clinical and pathological features of PBC

Predominantly involves middle-aged women
Progressive jaundice, recurrent pruritus, and hepatosplenomegaly
Elevation of serum alkaline phosphatase
Antimitochondrial antibody
Classified histologically into four stages:
1) Inflammatory destruction of interlobular bile ducts
2) Proliferation of ductules and/or piecemeal necrosis
3) Fibrosis and/or bridging necrosis
4) Cirrhosis
Non-caseating granuloma: sometimes but not always seen in the early stage

However, a number of patients develop end-stage disease requiring liver transplantation, which is an effective therapy at that stage.

3. MOLECULAR AND IMMUNOLOGICAL ASPECTS

3.1. Anti-mitochondrial Antibodies and 2-OADC

For more than two decades, PBC has been considered a model autoimmune disease, although the exact pathophysiology underlying its development has yet to be understood. The presence of AMA is a major serologic feature for the diagnosis of PBC. However, the titer of AMA does not correlate with the severity of the disease, its stage, or progression, and it is unknown whether AMA play an important role in the pathogenesis of the disease.

In 1958, a case report of PBC with high titers of complement-fixing antibodies to human tissue homogenates led to suggestions of an autoimmune process as the pathogenesis (12). In 1965, the application of indirect immunofluorescence with patient sera on frozen tissue sections revealed granular cytoplasmic fluorescence identifiable as antimitochondrial staining (6). This observation was later confirmed by the finding that PBC sera reacted *in vitro* with an isolated mitochondrial fraction (13). The mitochondrial autoantigen proved to be located on the inner mitochondrial membrane (14) and was trypsin-sensitive (15). During the 1970s, anti-mitochondrial reactions in conditions other than PBC were successively described. The first was the cardiolipin antigen used for serological tests for syphilis, which had a mitochondrial location (16). This was subsequently called M1 to distinguish it from the PBC-related antigen termed M2 (15). Subsequently, other putative mitochondrial antigens were reported and numbered M3 to M9 (17). Those associated with apparently variant forms of PBC were designated M4, M8, and M9, and those associated with other diseases as M3, M5, M6, and M7 (17). While the M1 and M2 antigens have been identified as cardiolipin and the pyruvate dehydrogenase complex, respectively, the identity of the other mitochondrial autoantigens is uncertain. It is claimed that the antigen for M4 is sulfite oxidase, M7 is sarcosine dehydrogenase, and M9 is glycogen phosphorylase (18). However, it seems appropriate to set aside the M3-M9 nomenclature since the data has not been reproduced.

In 1985, a new era for PBC began with the application of molecular biological techniques. Initially, immunoblotting allowed the recognition of discrete polypeptide antigens of defined molecular weight (19). Thereafter, immunoblotting studies demonstrated reactivity against one or more mitochondrial autoantigens, although the most common autoantigen recognized by patient sera was a 74 kD polypeptide found in mitochondria from many species. Other antigens of different molecular weights (56 kD, 48 kD, 42 kD, and 36 kD) were recognized by sera from small percentages of PBC patients.

This discovery was followed by the derivation by molecular cloning of cDNAs that encoded the major mitochondrial autoantigens in PBC. IN 1987, a cDNA for the major autoantigen recognized by AMA were cloned and sequenced by screening a rat liver cDNA library with sera from patients with PBC (20). The antigen was subsequently identified as dihydrolipoamide acetyltransferase, the E2 subunit of the pyruvate dehydrogenase complex (PDC) (21–24). Subsequently, a group of antigens associated with PDC and related enzymes were identified as the other antigens recognized in using patient sera by immunoblotting (Table 2) (25–29). Autoantibodies to the E3 component of the PDC have been reported (30). However, unlike the reactivity to the other components of the PDC, similar reactivity to the E3 is also found in control sera. Thus, the specificity of this reactivity and its significance is doubtful.

Pyruvate dehydrogenase complex is a multienzyme complex on the inner mitochondrial membrane and a component of the 2-oxo-acid dehydrogenase complexes (OADC). The 2-OADC are multienzyme complexes and essential in energy metabolism (31–33). This enzyme family comprises the pyruvate dehydrogenase complex (PDC), the oxo (keto) glutaric dehydrogenase complex (OGDC) and the branched-chain oxo-acid dehydrogenase complex (BCOADC). Each of the three enzyme complexes consists of three subunits, E1, E2 and E3. The E3 subunit is a common component of the three enzyme complexes (Table 3). All are nuclear-encoded proteins that are separately synthesized and imported into mitochondria for assembly into high molecular weight multimers on the inner membrane. Each of the 2-oxo acid dehydrogenase complexes occupies a central position in intermediary metabolism and the activity of each complex within mitochondria is under strict control of dietary factors and hormones. PDC links glycolysis to the Krebs cycle, OGDC is within the Krebs cycle itself, and BCOADC catalyses an irreversible step in the catabolism of several essential amino acids, including the branched-chain amino acid valine, leucine and isoleucine (Figure 1). The overall structure of each of the 2-OADC multienzyme complexes is similar in that each consists of multiple copies of three functionally-equivalent subunit enzymes (Table 3). For each multi-enzyme complex, the E2 component forms a symmetrical core around which the E1 and E3 components are arranged. For mammalian PDC, there are 60 E2 subunits and 6 E3 binding protein (E3BP) (protein X), arranged in icosahedral symmetry (32). The E2 components consist of several functional domains: An inner catalytic domain containing the active site, one or more lipoyl domains containing the lysine residue to which the essential cofactor lipoic acid is attached, and an E3-binding domain. In mammalian PDC, 20–30 copies of the $\alpha2\beta2$ tetramer that makes up the E1 component, and 6 copies of the E3 dimer, are associated with the complex. Each of the 2-OADC complexes is thus very large, with a molecular mass of several million, and a diameter of 510Å in solution (32).

In patients with PBC, the predominant AMA reactivity is directed to five autoantigens of the 2-OADC complexes, the E1α subunit of PDC, the E2 subunits of PDC, OGDC and BCOADC, and E3BP. Serum autoantibodies from more than 90% of patients with PBC react with PDC-E2 by immunoblotting, whereas the frequency of reactivity against the E2 subunits of OGDC and BCOADC is lower, around 50–70% (34). Antibodies to PDC-E1α are present in lower titers. Approximately 10% of patients react only with OGDC-E2 and/or BCOADC-E2. Thus, unless immunoblotting is performed, or an individual ELISA is set up to detect autoantibodies to each of the three 2-OADC enzyme autoantigens, these difficult autoantibodies will not be separately identified by the immunofluorescence methods used in routine diagnostic laboraties. Anti-mitochondrial reactivity is usually observed against two, or even all three of the 2-OADC, but serologic cross-reactivity is only found between PDC-E2 and E3BP (31). The absorption experiments have shown that there are two distinct populations among anti-PDC-E2 antibodies (35). One population, absorbed with E3BP, loses reactivity to PDC-E2 and the other still reacts with PDC-E2 after absorption with E3BP. It will be of interest to determine whether both populations react equally with the highly expressed protein found in the biliary epithelial cells of patients with PBC (36).

Table 2   Autoantigens recognized by PBC patient sera

Antigen	Prevalence (%)
Mitochondrial location	
PDC-E2	90–95
E3 binding protein	90–95
OGDC-E2	39–88
BCOADC-E2	53–55
PDC-E1α	41–66
PDC (OGDC, BCOADC)-E3	38
PDC-E1β	1–7
Nuclear location	
Histone	60–74
gp210	30–40
Sp100	20–30
Centromere	10
Lamin	<5

Table 3 Molecular weights and functions of the 2-oxo-acid dehydrogenase complex

Enzymes	MW (kD)	Function
Pyruvate dehydrogenase		
E1α decarboxylase	41	Decarboxylates pyruvate with thiamine pyrophosphate (TTP) as a co-factor
E1β decarboxylase	36	Decarboxylates pyruvate with TTP as a co-factor
E2 accetyltransferase	74	Transfers acetyl group from E1 to coenzyme A (CoA)
E3 lipoamide dehydrogenase	55	Regenerates disulfide of E2 by oxidation of lipoic acid
E3 binding protein (Protein X)	56	Anchoring E3 to the E2 core of pyruvate dehydrogenase complex
2-oxoglutarate dehydrogenase		
E1 oxoglutarate dehydrogenase	113	Decarboxylates α-ketoglutarate with TTP as a co-factor
E2 succinyl transferase	48	Transfers succinyl group from E2 to CoA
E3 lipoamide dehydrogenase	55	Regenerates disulfide of E2 by oxidation of lipoic acid
Branched chain 2-oxo-acid dehydrogenase		
E1α decarboxylase	46	Decarboxylates α-keto acids
E1β decarboxylase	38	Derived from leucine, isoleucine, and valine with TTP as a co-factor
E2 acyltransferase	52	Transfers acyl group from E1 to CoA
E3 lipoamide dehydrogenase	55	Regenerates disulfide of E2 by oxidation of lipoic acid

3.2. Immunodominant B Cell Epitopes

The advance of molecular biology has resulted in great progress in the molecular cloning of cDNAs that encode the major mitochondrial autoantigens for PBC, leading to the recognition of enzyme molecules of the 2-OADC complexes as the reactants revealed by immunoblotting. Among the 2-oxo acid dehydrogenase complexes, the antigenic determinants in PDC-E2 are best characterized. For fine epitope mapping of the human PDC-E2, truncated constructs of PDC-E2 were generated (37). Assays of multiple overlapping recombinant fusion proteins from human PDC-E2 cDNA indicated there were at least three autoepitopes present on human PDC-E2: The cross-reactive outer and inner lipoyl domains and a site surrounding the region of enzyme that binds the E1 and E3 subunits (Figure 2). The dominant epitope is localized to the inner lipoyl region. The outer lipoyl region has weaker reactivity, and only 1/26 PBC sera reacted with the E1/E3 binding region. Analysis of recombinant fusion proteins expressed from cDNA encoding the inner lipoyl domain revealed a minimal requirement of 76 amino acids (residues 146–221) for detectable autoantibody binding, and 94 amino acids (residues 128–221) for strong binding. This requirement for such a large peptide region for immunoreactivity is of interest and indicates that the autoepitope for the B cell response is conformational and is found with many other autoantigens (37,38). PDC-E1α differs from PDC-E2 in that it lacks covalently-bound lipoic acid. The autoepitope of PDC-E1α is located at the region that contains the enzyme functional sites, the phosphorylation site and TPP-binding sites (Table 4) (39).

The antigenic determinants of BCOADC-E2 (40) and OGDC-E2 (41) have been characterized using truncated constructs of BCOADC-E2 and OGDC-E2. Autoantibody reactivity to BCOADC-E2 mapped within residues 1–115 with strong binding, and residues 1–84 (lipoyl domain) with less strong binding. Only the full-length recombinant protein (residues 1–421) is sufficient to remove all detectable anti-BCOADC-E2 reactivity, indicating that proper folding of the protein is critical in the presentation of epitopes for recognition by anti-BCOADC-E2 antibodies. This suggests that the BCOADC-E2 epitopes are highly conformational. Similarly, a minimum of 81 amino acids (residues 67–147) corresponding to the lipoyl domain of OGDC-E2 are necessary for anti-OGDC-E2 reactivity (41). Recently the major epitope of E3BP recognized by AMA was also determined as the lipoic acid binding domain (35,42). There was little IgM response to the E3BP lipoyl domain, suggesting that this immune response is a secondary phenomenon probably resulting from antigen-determinant spreading.

Interestingly, the immunodominant epitopes of PDC-E2, and OGDC-E2 and BCOADC-E2 are all conformational lipoate binding sites and antibodies against them do not cross-react. Based on these epitope-mapping studies, triple expression hybrid clone consisting of three different lipoyl domains, PDC-E2 (residues 91–228), OGDC-E2 (67–147), and BCOADC-E2 (1–118), were constructed (Figure 2) and tested in immunoblotting and ELISA (43). Of 186 sera from patients with PBC, 152 (81.7%) reacted with recombinant fusion protein of PDC-E2, whereas 171 sera (91.9%) showed positive reactivities when probed by

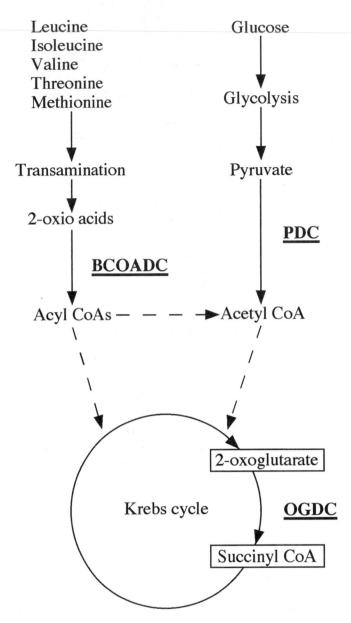

Figure 1.   Role of 2-OADC in metabolic pathway.

rich hinge regions. Although the immunodomainant epitope of each autoantigen has been shown to be located within the lipoate binding region, it is not clear whether the binding of lipoic acid to the epitope is necessary for autoantibody recognition. Lipoic acid has been shown to have adjuvant activities in an animal model system (44,45) and the observation that autoantibody epitopes are present in the lipoic acid binding domain is also consistent with the idea that it may have a direct role in inducing autoimmunity. Several experimental lines have attempted to characterize the degree of reactivity of autoantibodies to lipoic acid. In a study by Fussey et al. using native and recombinant *E. coli* PDC-E2, they suggested that the presence of lipoic acid was essential for the binding of PBC antibodies (46). In contrast, Leung et al. demonstrated by site-directed mutagenesis of lipoic acid binding amino acid, lysine with other amino acids, glutamine, histidine or tyrosine, that the removal of lipoic acid does not significantly alter antibody binding at the lipoyl domain (47). Koike et al. have shown that enzymatic delipoylation and relipoylation of the complexed and free PDC-E2 and OGDC-E2 components do not influence immunoreactivity with PBC sera (48). Together, these studies suggest that there is a population of antibodies directed against bound lipoic acid, but that it is a minor subpopulation among the antibodies directed to the entire inner lipoyl binding domain. It is supposed that the recognition of the lipoyl domain is a reflection of the surface-expressed and relatively mobile nature of this region of the autoantigen. However, of considerable interest is whether these antibodies are present as a dominant specificity early in the course of disease. If so, this may indicate that lipoic acid is relatively more important in inducing autoantibodies.

3.3.   Immunoglobulin Gene Usage

A great deal of interest has been focused on the issue of whether autoimmune responses arise from a restricted number of autoreactive clones of immune cells or whether the breakdown in tolerance is more general. In the experimental allergic encephalitis animal model, autoreactive T cells express a restricted set of V_β genes. However, clonal restriction is less commonly reported in human diseases. Although our understanding of the T cell autoimmune response in PBC is still rudimentary, there is more data available for the B cell response, which would help understanding the immunoglobulin gene usage of autoantibodies (49).

To dissect the population of polyclonal serum antibodies in PBC, it is necessary to obtain monoclonal antibodies (mAbs) that bear the same specificity and affinity as the antibodies in serum. Using hybridoma technology, five human PDC-E2-specific mAbs of both IgG and IgM isotypes were derived from a regional hepatic lymph node from a patient with PBC and their fine specificity was analyzed (50). As with polysera

immunoblotting against the recombinant triple hybrid protein. An ELISA or immunoblot using this recombinant protein seems to be a powerful and specific tool and will replace classical immunofluorescence for the detection of AMA.

The major autoantigens of PBC have been identified as the four closely-related mitochondrial enzymes, PDC-E2, OGDC-E2, BCOADC-E2 and E3BP. A major structural similarity of these enzymes is the presence of one or more lipoic acid binding domains that share an identical lysine residue, an E3 binding domain, and an inner core domain (containing the enzyme active site) connected by proline-

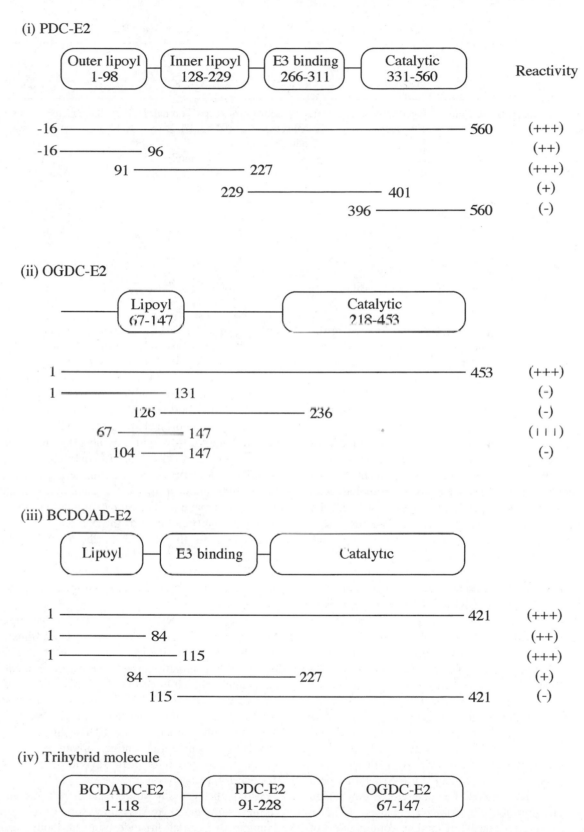

Figure 2. Reactivity of PBC sera to 2-OADC recombinant polypeptides. Recombinant polypeptide fragments were constructed from cDNAs of (i) PDC-E2, (ii) OGDC-E2 and (iii) BCOADC-E2. Reactivity of PBC sera to each fragment was tested by immunoblotting and/or ELISA. (iv) Trihybrid molecule was constructed from three different lipoyl domains, PDC-E2, OGDC-E2 and BCOADC-E2.

Table 4   Characteristic features of the antimitochondrial response in PBC

	Autoantibody				
	PDC-E2	OGDC-E2	BCOADC-E2	PDC-E1α	*E3BP*
Localization of the major B cell epitope	Inner lipoyl domain[a]	Lipoyl domain[b]	Lipoyl domain[c]	TTP binding site[d]	Lipoyl domain[e]
Amino acid residues	128–221	No distinct binding subregion	67–147	NA	1–84
Predominant Ig class and isotype	IgM and IgG3	NA	IgM and IgG2	NA	IgG
Inhibitory effect on the invitro enzyme activity[l]	+	+	+	+	NA

TTP: thiamine pyrophosphate. NA: not available
[a]Surh et al (1990)
[b]Moteki et al (1996a)
[c]Leung et al (1995)
[d]Iwayama et al (1991)
[e]Dubel et al. (1999); Palmer et al. (1999)
[f]Van de Water et al. (1988); Fregeau et al. (1989, 1990a, 1990b)

from PBC patients, these PDC-E2-specific human mAbs bind to the full length clone and the inner lipoyl domain of PDC-E2 and inhibit PDC enzyme activity. Nucleotide analysis of five human mAbs revealed that the four IgM human mAbs (B5, C10, D9, F6) were encoded by direct copies of V_H gene segments belonging to four different V_H families (51). B5 uses a V_H gene segment encoded by the germline V_H251 gene segment, one of the two functional members of the V_H5 family. C10 uses a V_H2 gene segment with 99.6% homology to a fetal V_H2 gene segment. The V_H gene segment expressed in D9 shares 99.6% homology with DP64, a recently reported germline V_H4 gene segment. F6 uses a direct copy of the germline V_H1.9III gene segment, a member of the V_H3 family. In contrast to the IgM mAbs, the only IgG mAb (C11) expressed a mutated V_H3 gene segment sharing 93% homology with the closest germline, V_H26. Moreover, mutations are scattered through the complementarity-determining region (CDR) and framework regions in C11. Thus, the anti-PDC-E2 human mAbs, mainly of the IgM isotype, were encoded by a vast array of V_H and V_L gene segments either as direct copies of germline or somatically-mutated variable region genes. The D segment expressed by the mAbs shows a series of characteristics similar to the above described V_H gene segment. This implies that the mAbs derive from a diverse array of germline genes without somatic mutations when expressed in the context of IgM molecules.

Using an alternative approach, Cha and colleagues isolated PDC-E2-specific Fab clones (LC1–5) against both recombinant and native PDC-E2 generated from human combinatorial libraries in lambda phages (52,53). Combinatorial libraries consist of random combinations of Ig H and Ig L chains and are potentially capable of producing a diverse array of antigen-specific Fab fragments (54). With respect to their antigen specificity, affinity and epitope, these PDC-E2 monoclonal Fab clones share striking similarities with autoantibodies found in the sera of patients with PBC. In contrast to the human mAbs, the recombinant Fab clones expressed clonally-related H chains with a high number of somatic mutations that are likely to be the results of antigen selection (51). The human combinatorial Fabs LC1, LC2 and LC4 were closest to the V_H9–14, with a number of common mutations shared by the Fabs. LC3 and LC5 are closest to germline V_H1.9III and DP50 H chain germline segments. Analysis of the light chain usage of the combinatorial antibodies revealed comparable data. Somatic mutations are evident in the light chain segments of the combinatorial antibodies, particularly in the CDRs. Furthermore, combinatorial pairing of clonally-related H-chains with highly homologous light chains suggests that these anti-PDC-E2 IgG Fabs are the result of clonal expression of a restricted set of autoimmune B cells.

Recently, an OGDC-E2-specific human IgG mAb, M37G037, was established from an EBV-transformed B cell clone (55). MAb M37G037 reacted with OGDC-E2, but not other mitochondrial proteins, on immunoblotting and completely inhibited the enzymatic activity of OGDC. The V_H gene of mAb M37G037 is derived from V_H3–7 of the V_H III gene family, and the V_H is derived from VkIV, with mutations primarily in the CDRs, suggesting that mAb M37G037 has undergone antigen-driven somatic hypermutation. The J_H and J_L of mAb M37G037 are derived from J_H4 and J_k4, respectively.

Although these data are obtained from the analysis of limited number of mAbs, they support the view that immune dysregulation underlying antibody response in patients with PBC results from the expansion and somatic diversification of a restricted set of self-reactive B cell clones.

3.4. Non-mitochondrial Autoantigens

Various autoantibodies against non-mitochondrial antigens have been reported. In particular, reactivity against nuclear antigens is quite striking and detected by immunofluorescence in 38 to 51% of sera from patients with PBC (56,57). Such reactivity is directed to histone (58), centromere (57,59), the SP-100 antigen, and nuclear envelope proteins (60,61).

Histones are small, basic nuclear proteins involved in the packing of DNA into the eukaryotic nucleus. Antibodies to histone proteins were exclusively found in sera from patients with PBC, but not other liver diseases (58). The predominant reactivity is to histone H1 and H2B, and amounts to 74% of patients for IgM antibodies and 60% for IgG antibodies.

There exists an overlap between PBC and scleroderma. The autoantibody to centromere has the same specificity as that seen in scleroderma and may coexist in PBC with the various clinical manifestations of the CREST (calcinosis cutis, Raynaud's phenomenon, esophageal dysmotility, sclerodactyly, telangiectasia) syndrome. In a study by Powell et al. (57,62), 11% of patients with PBC had centromere antibodies, and 3.9% had clinically-evident scleroderma. A cDNA for the major centromeric antigen, CENP-B, has been cloned and sequenced (63), but there are no structural homologies between the amino acid sequences of CENP-B and PDC-E2.

A third anti-nuclear reactivity is against nuclear antigens that give speckled patterns on immunofluorescence (56,64). One such antigen is called SP-100, and antibodies to this protein give a characteristic dot-like pattern on immunofluorescence, not associated with nucleoli or condensed chromatin (65). A cDNA has been sequenced for the SP-100 antigen that codes for an acidic protein with a molecular mass of 53 kD (66). This protein showed aberrant electrophoretic mobility with an apparent molecular mass of 95 kD. Antibodies to the recombinant protein occurred in 27% of patients with PBC by ELISA. The major immunoreactivity is directed to two antigenic regions consisting of 15 and 20 amino acids (67). The protein has not yet been identified, and its function is unknown, although the sequence showed striking similarities to both the $\alpha 1$ and $\alpha 2$ domains of MHC class I molecules that are part of the antigen groove, and also to several viral transcriptional regulatory proteins (66).

A fourth anti-nuclear reactivity of PBC sera is directed to antigens of the nuclear envelope that separate cytoplasm from the nuclear compartment in eukaryotic cells. The nuclear envelope is composed of a double nuclear membrane, pore complexes and the nuclear lamia (lamin), the latter consisting of three major proteins of 74, 68 and 60 kD (68). Autoantibodies to the nuclear envelope show a ring-like staining of the limiting membrane of the nucleus by immunofluoresence (60,61,62). Ring-like staining of the nucleus was initially associated with anti-DNA antibodies and is now also associated with autoantibodies to proteins of the nuclear lamin, and with a major glycoprotein of the nuclear pore complex (71). On immunoblotting, sera containing antibodies to lamins react with proteins of molecular weights of 74, 68 and 60 kD corresponding to lamins A, B and C, which can be regarded as members of the cytoskeletal proteins since there is a close sequence similarity between lamin proteins and intermediate filaments (72,73). It is, therefore, not surprising that autoantibodies to lamins appear as markers of autoimmune liver diseases. Anti-lamin antibodies may occur alone, but are usually associated with other antinuclear antibodies. Lassoued et al. (74) found that 11 of 50 sera with a ring-like nuclear fluorescence contained anti-lamin antibodies, and 7 of them were from cases of chronic liver disease. Wesierska-Gadek et al. (75) reported that 12 of 51 sera from patients with autoimmune hepatitis contained anti-lamin antibodies by immunofluorescence. However, antibodies to nuclear lamins occurred in only 3 of 37 cases of PBC.

A further pattern of reactivity on immunoblotting, associated with ring-like staining by immunofluorescence, was described by Lassoued et al. (61). In this study, 43 of 150 sera from patients with PBC displayed perinuclear fluorescence, and 40 of the 43 reacted with the 200 kD polypeptide of the nuclear envelope. In 1989, the primary structure of this protein was revealed (76). The 200 kD antigen was subsequently identified as gp210, an integral membrane protein believed to be involved in the formation of nuclear pores and the anchorage of pore complex constituents to the nuclear membrane (77). In contrast to autoantibodies to lamins, autoantibodies to the 200 kD nuclear pore glycoprotein appeared to be highly specific for PBC, and have not been detected in other autoimmune diseases (61). These autoantibodies recognize two epitopes within the carboxy-terminal tail and cytoplasmic domain of gp210 (78,79). In addition to reactivity to gp 210, reactivity to nucleoporin p62, a member of N-acetylglucosamine-modified proteins of nuclear pore complexes, has been reported to be specific for PBC (80).

The question arises as to whether the mitochondrial and nuclear autoantibodies in PBC represent independent populations. In the studies by Mackay et al. (81), sera that contained antibody to PDC-E2 and centromere, or to PDC-E2 and SP 100, were affinity purified on a column that contained PDC-E2. In each case, two antibody populations were identifiable. One population, eluted from the column, reacted only with PDC-E2, and the other unbound population reacted either with centromere or SP100. In the converse experiment using affinity purification on recombinant SP 100, Szostecki et al. (66) also showed no cross-reactivity between antibodies to mitochondria and SP100.

There are other autoantibodies in which there is an apparent increase in PBC, although the data are less clear than for nuclear antibodies. These include antibodies to thyroid and gastric antigens (82,83), and to liver-specific antigens (84,85). Moreover, certain of the non-mitochondrial autoantibodies appear to be increased in first-degree relatives of patients with PBC according to two family studies published in the 1970s (86).

3.5.   T Cell Response in PBC

The particular specificity of bile duct destruction, the presence of lymphoid infiltration in the portal tracts and the aberrant expression of MHC class II antigen on biliary epithelium in PBC have suggested that an intense autoimmune response is directed against biliary epithelial cells. It has been hypothesized that the destruction of biliary tract in PBC is mediated by autoreactive liver-infiltrating T cells through either cytotoxicity or lymphokine production. The evidence was initially derived from staining studies of histological samples and later by analyzing T cell lines that proliferate in the presence of putative mitochondrial autoantigens. In an early study by Meuer et al., phenotypic and functional analysis of isolated liver-infiltrating T cells in patients with PBC demonstrated a heterogeneous cell population. However, when compared to peripheral blood lymphocytes of the same patients, a marked enrichment of CD8[+] cytotoxic T cells was found to exist in the liver of these patients (87,88). Moebius et al. reported that infiltrating T cells in PBC livers are oligoclonal, in contrast to the remarkable clonal diversity of T cells from chronic hepatitis B livers (89). Immunohistochemical analysis of PBC livers also revealed a predominance of T cell infiltration in the liver, mainly CD3[+], CD4[+] T cells bearing the T cell receptor $\alpha\beta$, particularly around the portal tracts, suggesting that T cell mechanisms are involved in the pathological damage of the bile duct epithelial cells in PBC (90).

Despite accumulating data that T cell-mediated immune responses are involved in PBC, studies to determine the T cell epitopes of the mitochondrial antigens in PBC are limited. This is partly due to the technical difficulties in generating T cells from the cirrhotic PBC livers obtained at transplantation that reflect the *in vivo* condition. Antigen-specific T cell clones from peripheral blood are another good source of reactive clones.

Van de Water et al. have successfully obtained T cell lines from liver-biopsies of patients with PBC (91). Proliferation studies showed that cloned T cell lines specifically produced interleukin (IL)-2 when stimulated with PDC-E2 or BCOADC-E2, but not control proteins. Peripheral blood mononuclear cells (PBMC) from 16/19 of patients with PBC also responded to the entire PDC complex (Table 5) (92). Furthermore, PBMC from 10/19 patients, but none of the 12 control patients, responded to

the PDC-E2 subunit, and the response was directed to the inner and/or the outer lipoyl domains. This contrasts with the serologic observation that the autoantibody response is predominantly directed to the inner lipoyl domain. In addition, another series of T cell clones were derived from the liver explants of two patients. Lymphokine analysis of IL-2/IL-4/interferon (IFN)-γ production from individual liver-derived autoantigen-specific T cell clones suggests that both T helper (Th) 1- and Th2-like clones are present in the liver. The T cell receptor V_{β} repertoire utilized by PDC-reactive clones infiltrating the liver was remarkable in its heterogeneity, which was most apparent with subsets that responded to particular domains. For example, in both patients, T cell clones reactive to the inner lipoyl domain of PDC-E2 utilized multiple different V_{β} and only a few examples of repetitive V_{β} usage were noted. Of all cloned lines derived there were several CD4[+]/CD8[+] lines, but the vast majority of lines were CD4[+]. This is consistent with the lines being derived from patients late in the disease when CD4[+] T cells in the liver are predominant. Jones et al. have also characterized the peripheral blood T cell response

Table 5   Autoreactive T cell clones specific for the E2 component of the pyruvate dehydrogenase complex in PBC

Peripheral mononuclear cells from patients with PBC		
Positive/total		
Proliferative response to PDC		16/19[a]
PDC		1/19
PDC-E1		7/19[b]
PDC-E2		10/19
E2L1 (aal-98)		6/11 (54%)[c]
E2L2 (aa120–233)		4/11 (36%)
E2L1 + E2L2 (aa1–233)		4/11 (36%)
PDC-specific T cell clones isolated from the livers of PBC patients		
Positive/total		
Proliferative response to	Patient 1	patient 2
PDC-E1	17/28	0/23
PDC-E1 only	5/28	0/23
PDC-E2	6/28	8/23
E2L1 (aa1–98)	8/28	2/23
E2L2 (aa120–233)	11/28	19/23
E2L1 + E2L2 (aa1–233)	0/28	1/23
T cell phenotype	Predominance of CD4[+] T cell	
Lymphokine production	IL-2, Il-4 and IFN-γ	
T cell receptor phenotype α/β		
Vβ usage	Remarkable heterogeneity	
CD45RO[+]	*Positive/total*	
Patient 1	9/28 (32%)	
Patient 2	2/23 (8.6%)	

E2L1: outer lipoyl domain; E2L2: inner lipoyl domain.
[a]Positive to one or more mitochondrial antigens.
[b]Three of seven reactive patients responded only to PDC-E1.
[c]Two of 11 patients reacted to both E2L1 and E2L2.

(93). Using PBMC from 24 patients with PBC and 48 controls, the T cell responses to whole PDC (12 of 27 vs. 24 of 48) and PDC-E1 (15 of 24 vs. 25 of 48) were not significantly different between PBC and control patients. Responses to PDC-E3 were low in both PBC and control patients. However, a T cell response to PDC-E2/E3BP was detected in 14/24 PBC, but only 6/48 controls. This enhanced reactivity to PDC-E2/E3BP appeared to be more associated with precirrhotic stage than cirrhotic stage of PBC.

Recently, six T cell lines specific for PDC-E2 were established from PBMC from four patients with PBC using a panel of 33 different overlapping peptides spanning the entire PDC-E2 (94). The surface phenotypes of these T cell clones were CD4$^+$, CD45RO$^+$ and T cell receptor α/β. The minimal epitopes of these T cell clones were all mapped to amino acid residues 163–176 (GDLLAEIETDKATI) within the inner lipoyl domain of PDC-E2. Moreover, all T cell clones responded to residues 36–49 (GDLIAEVETD-KATV), which correspond to the outer lipoyl domain of PDC-E2. These reactive peptides all share common amino acids E, D, and K at positions 170, 172, and 173, respectively (EXDK sequence motif), whereas other peptides that share amino acid sequence homology with PDC-E2, but do not have the EXDK sequence, did not induce proliferation (Table 6). One particular T cell clone also cross-reacted with exogenous antigen peptide (EQSLITVEGDKASM) corresponding to the PDC-E2 from E. coli, which has an EXDK sequence. However, none of these PDC-E2-specific T cell clones cross-react with the HLA-DR α-chain peptide 82–95 (QGALANIAVDKANL) or the human glycogen phosphorylase β peptide 354–367 (KVDKERMDWD-JAWD), both of which have some amino acid homology

with PDC-E2 but lack the requisite EXDK motif. Furthermore, Shimoda et al. demonstrated that a disease-specific 100–150-fold increase in the precursor frequency of PDC-E2 163–176-specific T cells in the hilar lymph nodes and livers of patients compared with PBMC from PBC patients (95). Interestingly, autoreactive T cells and autoantibodies from PBC patients both recognize the same dominant epitope. In addition, they showed cross-reactivity of PDC-E2 peptide 163–176-specific T cell clone with OGDC-E2 peptide 100–113 (DEVVCEIETDKTSV), thereby identifying a common epitope "motif" EXETDK (Table 4).

T cell receptor analysis of these PDC-E2-specific T cells revealed diverse usage of V_β and J_β (96). In contrast, in the third complementarity-determining region (CDR3), G was frequently found and GXG or GXS sequence motif was found in all T cell clones, suggesting the preferential usage of limited motif in CDR3 region in PDC-E2-specific T cells. However, the results of T cell repertoire described here were obtained from the small number of clones and limit any conclusions.

The limited but detailed data available on the identification of the T cell epitope of PDC-E2 in PBC, as part of the lipoyl domain, show that there is some overlapping in the PDC-E2-specific T and B cell epitopes in PBC. This is similar to multiple sclerosis, where myelin basic protein (MBP)-specific autoantibodies react with an immunodominant MBP peptide, which is also the autoepitope of MBP-specific T cell clones (97). This may be a result not only of more efficient uptake and processing of the autoantigen by autoimmune B cells, but also could be due to the fact that the epitopes may be protected from degradation during antigen processing. It will be of interest

Table 6   Amino acid sequences of antigen peptides used for T cell stimulation

Antigen	Amino acid sequence		Proliferation
PDC-E2 (ILD) T I	163–176	G D L L A E I E T D K A	+
PDC-E2 (OLD) T V	36–49	G D L I A E V E T D K A	+
OGDC-E2 S V	100–113	D E V V C E I E T D K T	+
PDC-E2 (E.coli) A S M	31–44/134–147/235–248	E Q S L I T V E G D K	+
BCOADC-E2 (Pseudomonas putida) D I	104–117	D E L L A T I E T D K I	–
Human glycogen phosphorylase β W D	354–367	L V D L E R M D W D K A	–
HLA DR α chain N L	82–95	Q G A L A N I A V D K A	–

Underlines indicate identical or homologous amino acids.
ILD: inner lipoyl domain; OLD: outer lipoyl domain.

to find out whether T cell epitopes of BCOADC-E2 and OGDC-E2 are likewise similar to their reported respective B cell epitopes. If they are, it is likely that the immune response is triggered by a molecule that bears a common structural motif among these molecules and such finding would further support the hypothesis of molecular mimicry.

3.6.   Molecular Mimicry and PBC

PDC-E2 is evolutionarily highly conserved among various species, especially at the lipoic acid binding sites. The sera from patients with PBC have been shown to react with both human and E. coli PDC-E2 (46). Such reactivity of AMA to both human and bacterial mitochondria has stimulated speculation that PBC may result from occult chronic bacterial infection (98). Indeed chronic bacteriuria (99) and colonic infection with rough-colony (R) mutants of E. coli (100) have been accorded pathogenetic significance in patients with PBC (findings not confirmed by others). However, the immunodominant epitope of the inner lipoyl domain of the human PDC-E2 is conformational, and requires a polypeptide of at least 76 amino acids for reactivity. When conservative substitutions are included, the homology between human and E. coli PDC-E2 is 60% for the 10 amino acids surrounding the lysine residue to which lipoic acid is attached, whereas there is little homology (~30%) over the entire 76 amino acid epitope. In a study measuring the functional activity of serum autoantibodies by means of an enzyme inhibition assay against PDC from different sources, including mammalian, *Saccharomyces cerevisiae* and E. coli, Teoh et al. found the relatively weaker reactivity of PBC sera with bacterial or yeast PDC vis-a-vis mammalian PDC (101). Thus, with a functional assay that depends on epitope recognition of PBC sera, cross-reactivity between mammalian and bacterial PDC enzymes was low and the concept of "molecular mimicry", at least at the B cell level, was not supported.

Still, a mechanism of molecular mimicry has been proposed as a possible cause of PBC at the T cell level (102). According to this hypothesis, in response to an infection, the T cell would recognize the lipoyl domain of E. coli PDC-E2 and other enzymes containing AVDKA sequence from the HLA-DR α-chain or EDTDK sequence from human PDC-E2 by aberrantly expressed MHC class II molecules on bile duct cells. Finally, this reactivity would initiate the autoimmune cascade, leading to the destruction of the bile ducts in which AMA and/or T cells specific for mitochondrial antigens may play a pathogenic role. At this stage, the autoimmune process would take place in the absence of the exogenous antigens, such as E. coli PDC-E2, which initiate the original immune response. In some animal models of autoimmune diseases, such as experimental allergic encephalomyelitis and autoimmune oophoritis, the mechanism of molecular mimicry has been proven to

be operative *in vivo* (103,104). In the study by Shimoda et al., the demonstration of T cell clones specific for human PDC-E2 163–176 that cross-react with the lipoic acid binding site of E. coli PDC-E2 suggests that molecular mimicry between human and E. coli can be operative at the T cell clonal level in PBC. However, none of these PDC-E2-specific T cell clones cross-reacted with the HLA-DR α-chain peptide 82–95 or the human glycogen phosphorylase β peptide 354–367, both of which have some amino acid homology with PDC-E2, but lack EXDK sequence motif. Moreover, it is important to note that the T cell clone cross-reactive with E. coli PDC-E2 did not react to P. putida, which not only has EXDK sequence, but also has greater amino acid homology with human PDC-E2 than E. coli PDC-E2 (Table 6). Although of interest, there is no data as to whether this similarity *has pathogenetic experience*, and certainly the large difference between the prevalence of PBC and the frequency of bacterial infection in the population would suggest that other factors must be involved.

3.7.   Immunogenetics

The increase in prevalence of various autoantibodies among relatives of PBC patients (105,106) and data on the prevalence of PBC in different ethnic groups have emphasized the importance of genetic background. As PBC is believed to be an autoimmune disorder, studies of the genetic background have largely focused on the HLA complex (107) and, more recently on the molecular mechanism of the MHC class II-antigen peptide-T cell receptor interaction, because the recognition of peptide fragments by T cells in the context of MHC class II molecules is a central event in the development of immune responses.

HLA class II genes are highly polymorphic genes located at the HLA-D region and divided into three major subregions: HLA-DR, HLA-DQ and HLA-DP. Each HLA class II molecule consists of an α-chain and a β-chain. Specific association of HLA alleles with a number of autoimmune diseases has been described (108,109). A number of studies have been conducted to investigate whether there is a restricted or specific linkage between PBC and HLA class II antigens by serological HLA typing. The data are conflicting. Based on serological typing, some studies have shown that PBC is associated with HLA-DR3 (110), -DR8 (111–113) and possibly-DR4 (114) in Caucasians, while in Japanese an association with HLA-DR2 has been reported (115). In other studies, no association was found between PBC and the MHC class I or II antigens (116,117). Genes within the class III region have also been implicated (113,117).

Recently, techniques such as restriction fragment length polymorphism (RFLP), polymerase chain reaction (PCR) and sequence-specific oligonucleotides (SSOs) have been used for HLA typing. Morling et al. (118) used RFLP analysis to show that susceptibility of PBC in the Danish population was associated with the extended haplotype: HLA-B8, DR3, DQA1*0501, DQB1*0201. Begovich et al. (107), using PCR and SSOs, assessed the distribution of the polymorphic class II antigens. The study showed an increased frequency of DRB1*0801- DQA1*0401/0601- DQB1*4 haplotype and the decreased frequency of DRB1*1501-DQA1*0102-DQB1*0602 and DRB1*1302- DQA1*0102-DQB1*0604 in patients with PBC. In a study of German patients, HLA-DPB*0301 was clearly associated with PBC (119). Underhill et al. (United Kingdom) reported a weak association of PBC with DR8- DQB1*0402, which includes DPB1*0301 (120,121). In Japanese patients, an increased association with DR8 was shown to be due to linkage disequilibrium with DPB1*0501, suggesting that a single amino acid residue (leucine) at position 35 of the DPB polypeptide may be crucial for disease susceptibility (122). These varied results may reflect the difference in the sensitivity and methods of analysis, the differences in ethnic groups of the patients studied and possible errors in diagnosis. Some phenotypes may be associated with disease progression ratio non-susceptibility. However, accumulated data will be helpful for developing immune therapy in the future. Shimoda et al. (94) have shown that all T cell clones established from four patients with PBC recognize the inner lipoyl domain peptide residues 163–176 only in the context of HLA-DRB4*0101. This implies that PDC-E2 peptide 163–176 may play a central role in the PBC patients with HLA-DRB4*0101 allele. However, the frequency of HLA-DRB4*0101 allele among patients with PBC is about 77%, similar to a control Japanese population, indicating that possession of HLA-DRB4*0101 is not a genetic predisposition for PBC. Because polymorphic residues of DR alleles are scattered within the peptide-binding grooves, different DR molecules are capable of binding peptides with different structural motifs. This phenomenon contributes to the HLA-linked polymorphism of the immune system. In contrast to the highest degree of polymorphism of HLA-DRB1, polymorphism of HLA-DRB4*0101 has not been reported. Therefore, it is possible that the HLA-DRB4*0101-binding PDC-E2 peptide identified in Shimoda's study could provide important information for the development of peptide-specific immunotherapy in the majority of patients with PBC who have an HLA-DRB4*0101 allele. To date, motif for binding peptides have been reported in various DR molecules, indicating that two or three hydrophobic amino acids are necessary for the binding of the peptide to HLA hydrophobic grooves. Although the precise analysis of the peptide-binding motif of HLA-DRB4*0101 is now underway, a preliminary study indicates that LXXIXXD is an HLA-DRB4*0101 binding motif (94). If this is the case, E and K at positions 170 and 173 may be crucial TCR binding sites, and amino acid at position 169 may be important for the structure of the TCR binding site.

The mechanism for the participation of HLA in the pathogenesis of PBC is not fully understood. Membrane expression of MHC class I molecules increases on hepatocytes during cholestasis. Cytokines are known to induce MHC class I expression, particularly interferon-γ. It has also been reported that the increase of ursodeoxycholic acid in cholestasis activates protein kinase C and protein kinase A, resulting in enhanced MHC class I molecules on the human hepatocytes in primary culture (123). Recent understanding of the relationship between structure and function of HLA favor the hypothesis that the role of HLA in presenting peptide antigens to the autoreactive T cells is important in disease susceptibility. Although the association of a disease with a certain HLA allele may induce the enhanced antigen presentation to responsive T cells, direct experimental proof that such a mechanism is involved in presenting the mitochondrial antigens is lacking.

3.8. Apoptosis of Bile Duct Cells

Cell death has been described in two forms, necrosis and apoptosis. Apoptosis is a morphologically distinct form of programmed cells death that plays a major role during development, homeostasis, and in many diseases, including cancer, autoimmune disease, acquired immunodeficiency syndrome, and neurodegenerative disorders. It is uncertain whether apoptosis is involved in the pathogenesis of primary biliary cirrhosis. Published data addressing this hypothesis are contradictory. Using a in situ DNA nick-end labeling technique, Afford and colleagues (124) did not detect apoptosis in bile ducts in PBC, but only in inflammatory cells. In contrast, increased nuclear fragmentation, a hallmark of apoptosis, was observed in most cases of PBC (125,126). Supporting evidence of apoptosis came from an immunohistochemical study showing the increased expression of CD95 (Fas antigen) on the injured bile duct cells of PBC, which accompanied infiltration of CD95 ligand-expressing mononuclear cells (127). They also showed prominent granzyme B immunoreactivities on bile duct cells in PBC. Graham and colleagues showed that, compared to controls, bile duct cells in PBC displayed a "pro-apoptotic" phenotype expressing proteins involved in the execution of apoptosis such as CD95, Bax, and Bcl-x, while the anti-apoptotic Bcl-2 is only weakly expressed (128). This finding is in line with a recent study by Kuroki et al. showing expression of CD95 and lack of Bcl-2 in biliary epithelial cells. In addition, there is evidence that biliary cells retain the ability to proliferate, as shown by the positive Ki-67 staining seen in PBC liver. This supports the previous observation of increased proliferative activity of epithelial cells in affected ducts (129). When CD95 is already expressed on biliary epithelia cells in PBC, a question arises as to whether Fas/Fas ligand interaction is involved in the pathogenesis of disease. It is well established that portal tracts in PBC show infiltration of NK and activated T cells, the latter population dominating activated CD4$^+$ T cells (130) and cytotoxic T lymphocytes (CTL) (130). Most CD4$^+$ T cells in PBC exhibit a Th1 phenotype, as determined by their cytokine repertoire, with a predominance of IFN-γ (132,133). All of these T cell subsets found in PBC are known to have the capacity to express CD95 ligand. Moreover, IFN-γ up-regulates CD95 expression together with the adhesion molecule ICAM-1 and MHC class II antigens, and increases cellular susceptibility to CD95-induced apoptosis. Tumor necrosis factor-α and transforming growth factor-β have also been shown to cause apoptosis of hepatocytes, and mRNA for each has been detected in human PBC liver. In fact, bile duct cells in PBC exhibit aberrant expression of MHC class II antigens (134) (probably IFN-γ induced), increased ICAM-1 (135) and VCAM-1 (136). These adhesion molecules may serve as receptor/ligand in the recruitment of lymphocytes in inflamed areas. Lim et al. have reported increased levels of soluble E-selectin and VCAM-1 in PBC sera compared to non-PBC controls (137). The increased expression of the adhesion molecules important in the migration and adhesion of inflammatory cells and their fine interaction with lymphocyte trafficking in the portal tracts may partly account for the pathological destruction in PBC livers. It is, therefore, very likely that CD95/CD95 ligand interaction is involved in the bile duct injury in PBC, although the initial events leading to activation of CD95 ligand-bearing autoreactive T cells are only partially understood. In addition, different mechanisms may be involved in the apoptosis of hepatobiliary cells in PBC. Recently, toxic bile salts have been shown to induce apoptosis in cultured hepatocytes, suggesting that intracellular retention of the bile salts may contribute to apoptosis of hepatocytes, and possibly biliary epithelial cells, during cholestatsis in PBC (138).

3.9. Biliary Epithelium in PBC

In primary biliary cirrhosis, the immune responses are directed against intracellular mitochondrial antigens sheltered from the immune system by two membrane barriers. It is poorly understood how the immune response is confined to biliary ductular tissue of the liver despite ubiquitous distribution of mitochondrial antigens in the body. Several studies have suggested that autoantigens relevant to PBC may be present on the surface of hepatocytes or biliary epithelium. PBC sera react with the surface of teased-out hepatocytes and intact hepatoma cells, and such reactivity is abolished by prior absorption of sera with mitochondria (139). Van de Water et al. have examined antigen distribution in liver tissue from PBC patients by use of mouse mAbs that reacted with

the inner lipoyl domain of PDC-E2, as do PBC sera (90). They showed a distinct and intense reactivity with bile ducts of patients with PBC. In a similar study by Joplin et al. there was very strong staining with antibodies to PDC-E2 in the biliary epithelium and in a subset of macrophages in the portal lymph nodes (140). Cultured biliary epithelial cells from PBC patients also showed surface membrane-bound PDC-E2 (141). Using a panel of mouse mAbs and a human combinatorial antibody specific for PDC-E2 (52,142), Van de Water et al. examined, by indirect immunofluorescence and confocal microscopy, sections of liver from patients with PBC and other liver diseases (36). Murine mAbs map to four different regions of PDC-E2 when studies by ELISA with overlapping recombinant polypeptides (142). All produce intense immunofluorescence of mitochondria when used to stain Hep-2 cells. One of eight murine mAbs (C355.1) and the human combinatorial antibody (LC5) reacted with great intensity and specificity with the apical region of biliary epithelial cells from patients with PBC, but not controls. Such selective reactivity of some anti-PDC-E2 reagents may be interpreted as indicating that modified PDC-E2 or a molecule cross-reactive with PDC-E2 is expressed at high levels in the apical region of biliary epithelial cells in PBC. It seems unlikely that native PDC-E2 is accumulated in the bile duct cells because several of a panel of murine monoclonal antibodies were shown to map to regions very close to the binding site of a particular monoclonal antibody, C355.1, yet they were not reactive like C355.1. Furthermore, an *in situ* nucleic acid hybridization study has shown that PDC-E2 is not over-produced in bile duct cells (143). Staining could be due to a modified form of PDC-E2 that lacks all other epitopes except that recognized by C355.1. However, such grossly truncated products of a longer molecule are generally highly unstable and are usually rapidly removed from the cell by uniqutin tagging and subsequent proteolysis. An alternative explanation is that the serologically-reactive material is due to the presence of another, as yet unidentified, molecule that shares a single cross-reactive epitope with PDC-E2. It is unlikely that this molecule is a lipid, as the tissue sections are devoid of lipids following xylene and alcohol treatment during processing.

To determine the nature of the molecule cross-reactive with PDC-E2-specific mAb C355.1, a random phage-epitope library expressing random dodecapeptides was screened with C355 (144). Eight different mimotope sequences with three common amino acid motifs (W-SYS, TYVS and VRH) were identified in 36 phage clones. Interestingly, distinct peptides selectively inhibited C355.1 from unique apical staining of biliary epithelium of PBC, suggesting that such peptides might represent a mimotope of a cross-reactive molecule present in the bile ducts of patients with PBC. Moreover, rabbit sera raised against a chosen mimotope peptide-stained bile duct cells from PBC patients with higher intensity than controls. Similar data were obtained with electron microscopy, which identified the molecules targeted by rabbit sera as being located on the exterior of biliary epithelium in PBC, but not in normal controls. The mimotopes generated by this study represent powerful reagents in the pressing need to identify the target molecule responsible for the apical staining of bile duct cells in PBC.

More recently, in a study by Migliaccio et al., mAbs to the other mitochondrial antigens OGDC-E2 and BCOADC-E2 demonstrated disease-specific patterns of reactivity (145). Using a recombinant trihybrid protein containing the lipoyl domains of PDC-E2, OGDC-E2 and BCOADC-E2 (43), 35 mAbs specific for one or more of the above mitochondrial autoantigens were produced, and 7 of these mAbs uniquely stained the apical region of biliary epithelial cells in PBC. One mAb was reactive to PDC-E2, two recognized BCOADC-E2, three were reactive to OGDC-E2, and one recognized all three antigens. Similar to the previous study by Van de Water et al. regarding PDC-E2 (37), mAbs to OGDC-E2 and BCOADC-E2, or a mAb that cross-reacts with the inner lipoyl domains of all three enzymes, also show a uniquely intense staining of the apical region of bile duct cells in patients with PBC compared with diseased controls. Increased production of OGDC-E2 and BCOADC-E2 in bile duct cells in PBC seems unlikely (146). The abundance of such disease-specific determinants in the target cells of PBC raises interesting possibilities regarding the role of these autoantigens in the pathogenesis of this disease.

Importantly, the existence of this surface-expressed PDC-like molecule creates a compelling scenario in which the initial events in the pathogenesis of PBC would be a tissue-specific over-expression of a novel substance with consequent presentation by MHC proteins on the biliary ductular surface. Although biliary epithelium usually expresses only MHC class I molecules, the presence of class II molecules has been demonstrated in PBC (134). This would allow induction of helper CD4$^+$ cells, which provide help for induction of CD8$^+$ cytotoxic cells. It is also possible that this PDC-E2-like neo-antigen may be surface expressed. A high level of expression of the neo-antigen would lead to an association with MHC class I and class II molecules. Reactivity against the biliary specific neo-antigen, perhaps by CD4$^+$ and CD8$^+$ cytotoxic lymphocytes, would then lead to cell lysis and release of mitochondrial products. Presentation of released proteins to lymphocytes primed to the cross-reactive PDC-E2 could result in presentation of physically associated proteins such as other components of the PDC-complex and synthesis of a wider range of autoantibodies. It remains to be established to what degree the biliary epithelium is able to take over the function of a professional antigen-presenting cell in a way that allows provision of the essential second signal such as that provided by the co-stimulatory molecule B7.

The relationship between MHC class II and B1/B7, and the pathophysiology of PBC, was studied using direct immuno-histochemical analysis of biliary epithelial cells (147). Needle biopsy specimens from either stage I or stage II PBC patients were simultaneously stained with antibodies to PDC-E2, MHC class II and BB1/B7. MHC class II and BB1/B7 are two important ligands necessary for the activation of T cells, in which MHC class II delivers the first message through its interaction with T cell receptor and BB1/B7 present on antigen presenting cells that provide the second signal to T cells through CD28. Immunohistochemical staining shows that nearly all early stage PBC patients do not express the HLA-DR antigen or the BB1/B7 antigens. A previous report, however, has shown a high rate of HLA-DR antigen expression in PBC liver, but these observations did not distinguish between early and late stage disease. Taken together, these data can be interpreted as the appearance of PDC-E2 or a molecule cross-reactive with PDC-E2 that precedes the expression of HLA-DR and BB1/B7 co-stimulatory molecule.

It is of further interest that the bile from patients with PBC contains IgG and IgA anti-PDC-E2 antibodies. The human bile duct circulatory system is distinct from that in many species in that IgA is transported to the lumen via a secretory component found only within bile duct epithelial cells. In contrast, in rodents and many other mammals, IgA is transported to the lumen via hepatocytes. Thus, IgA is transported from the apex to the luminal surface of the biliary ductural epithelial cells. Moreover, in man, IgA is produced locally within the liver and extrahepatic biliary tissues, and its presence subserves local liver immunity. Whether autoreactive IgA is capable of reacting with proteins in the biliary cell cytoplasm during its transport is unknown. The PDC-E2-specific IgA may then, during its normal course from the apex to the lumen of the biliary epithelial cell, recognize the PDC-E2-like neo-antigen, or PDC-E2 itself, as it is being produced and/or transported in the cytosol of the cell. Thus, if the IgA anti-PDC-E2 antibodies bind to PDC-E2 either before it reaches the mitochondrial receptor or on the mitochondrion itself, it may prevent uptake of PDC-E2 into the inner membrane and the assembly of the enzyme complex. Eventually, the lack of PDC-E2 transported into the mitochondria would cause metabolic dysfunction, resulting in cell death. Our laboratory has shown that IgA derived from patients with PBC penetrated Madine-Darby canine kidney cells transected with the human IgA receptor and co-localized with PDC-E2 inside the cells when they were cultured with PBC-purified IgA. This suggests that these autoantibodies have a direct effect on the mitochondrial function of biliary epithelial cells (148). Furthermore, co-localization of IgA and PDC-E2 was demonstrated, both cytoplasmically and at the apical surface, by dual staining of liver sections from PBC patients. In a previous study, we demonstrated the presence of mitochondrial autoantigens and AMA in bile from patients with PBC and found a positive correlation between the two (149). Such mitochondrial antigens, especially coupled with IgA AMA, may be trapped or accumulated in bile duct cells during their normal transport to the bile duct lumen via the polyimmunoglobulin receptor found only on bile duct cells. Although the presence of AMA in bile may merely reflect the presence of these autoantibodies in sera, the simultaneous detection of mitochondrial autoantigens in bile suggests the presence of antigen-antibody complexes. Such immune complexes may serve to augment the local immune response and accelerate disease progression. Furthermore, the immune complex formation in bile duct cells may cause conformational change of mitochondrial antigens or a cross-reactive molecule, which bring about limited expression of antigen determinant, resulting in the recognition of antigens by only a subset of mAbs. Further work will be required to show whether such a molecule may be a normal cellular component of biliary tissue that is usually expressed only at very low levels, or alternatively, a PDC-E2-like host molecule aberrantly expressed in biliary cells, or even a neo-antigen of viral origin. Table 7 summarizes pathogenetic possibilities in PBC.

4. ANIMAL MODELS OF PBC

The low incidence of PBC makes the conduct of prospective studies extremely difficult, as a huge number of people would need to be followed. Furthermore, physicians are understand-

Table 7   Possible pathogenetic determinants in PBC

INCREASED EXPRESSION OF AUTOANTIGENS IN BILE DUCT CELLS
    Modified form of PDC-E2/Cross-reactive antigens?
    Other neo-antigen?
ABERRANT MOLECULAR TRAFFICKING
    MHC class I and class II association of autoantigen molecules
    Surface expression of autoantigens with MHC
    Up-regulation of adhesion molecules and co-stimulatory factor
    Expression of CD95 (Fas antigen)
STIMULATION OF CD4 CELLS
    Genetic background: HLA-Drw8 possible but not evident in all studies
    Induction of CD8 cells
    Expression of CD95 (Fas) ligand
    Autoantibody production
STIMULATION OF CD8 CELLS
    Genetic background?
    Cytokine release: inteferon γ and other cytokines
    Perforin/Granzyme B
DESTRUCTION OF BILE DUCT CELLS
    Apoptosis/necrosis

---

ably reluctant to perform liver biopsies on patients with few or no symptoms who would have no therapeutic application. To address some of unsolved issues of the pathogenesis of PBC, it would be helpful to analyze an appropriate animal model. Animal models of human autoimmune disease may occur spontaneously, or can be developed by immunization procedures. While there is one spontaneous model of PBC (150), "active" and "passive" immunization has been utilized. Various species of animals have been immunized with the PDC-E2 recombinant fusion protein and generally similar serological results have been obtained (151). Although these include typical AMA by immunofluorescence and antibodies to PDC-E2 by immunoblotting and ELISA, bile duct lesions have not been observed in immunized animals. However, in a recent study by Bassendine et al., immunization of mice with bovine PDC induced the development of a nonsuppurative destructive cholangitis in the livers of SJL mice, suggesting that the multi-subunit quaternary structure of intact PDC was critical for this immunostimuratory activity (152). Passive immunization has been also successful in inducing a valid animal model of PBC. Injection of peripheral blood lymphocytes from patients with PBC into servere combined immunodeficient (SCID) mice induced AMA production and an intense infiltration of mononuclear cells around the portal areas with subsequent destruction of bile duct (153). A mild hepatic inflammatory reaction was also observed in mice that received cells from normal donors, suggesting the presence of graft versus host (GvH)-like disease. It is hypothesized that PBC is similar to GvH disease, with reactivity ascribed to MHC antigens on host biliary epithelial cells. Bile duct lesions resembling chronic nonsuppurative destructive cholangitis seen in PBC were also reported in a murine model in which B6 spleen cells were injected into B6 X bm12 F_1 recipients (154). These mice developed epithelioid granuloma formation and inflammatory infiltrates, consisting of lymphocytes, plasma cells, and eosinophils around the small bile ducts. This model was postulated to be a model of PBC even though AMA were not observed. There have been similar speculations over the biliary ductular lesions seen in human GvH disease after bone marrow transplants. However, such GvH models seem to be of limited use because alloreactive T cells, but not autoreactive T cells, play an central role in disease induction.

An alternative potential approach is the construction of transgenic animals in which peripheral expression of autoantigens of interest is placed under the control of an appropriate promotor. Such transgenic animals could be constructed using PDC-E2, although this may not be the right molecule to overexpress. In addition, several technical problems need to be solved, such as the identification of a tissue-specific promotor for bile duct cells. Nevertheless, if these problems can be solved, such an animal model would help elucidate the difficult issue of studying the whole disease process.

5. CONCLUSIONS

During the past decade, our understanding of the immuno-biology of PBC has greatly increased with the advent of molecular biology. First, the mitochondrial autoantigens and their B cell epitopes have been defined at the molecular level. These results have made it possible to utilize cloned recombinant autoantigens for more reliable assays for AMA, replacing the traditional immunofluorescence techniques. Second, we now have available antigen-specific T cell lines and limited data on T cell epitopes for exploring the issue of whether specific immunotherapy is feasible. Third, the association of MHC class II in PBC has focused on possible relationships to clonally-derived T cells. Finally, there is increased evidence that mitochondrial antigens or cross-reactive molecules are accumulated on the surface of biliary epithelial cells, suggesting the fundamental role of bile duct cells in initiating the disease process. Utilization of animal models for PBC would help clarify the disease mechanism. We also believe that the study of patients at the early stages will provide the clues for understanding the pathogenesis of the disease.

References

1. Addison, T., and W. Gull. 1851. On a certain affection of the skin-vitiligoidea alpha plana, beta tuberosa. *Guys Hosp. Rep.* 7:265–276.
2. Thannhauser, S.J., and H. Magendantz. 1938. The different clinical groups of xanthomatous diseases; a clinical physiological study of 22 cases. *Ann. Intern. Med.* 11:1162.
3. Dauphinee, J.A., and J.C. Sinclair. 1949. Primary biliary cirrhosis. *Canad. Med. Assoc. J.* 61:1–6.
4. Ahrens, E.H., M.A. Payne, H.G. Kunkel, W.J. Eisenmenger, and S.H. Blondheim. 1950. Primary biliary cirrhosis. *Medicine (Baltimore)* 29:299–364.
5. Rubin, E., E. Schaffner, and H. Popper. 1965. Primary biliary cirrhosis. Chronic non-suppurative destructive cholangitis. *Am. J. Pathol.* 46:387–407.
6. Walker, J.G., D. Doniach, I.M. Roitt, and S. Sherlock. 1965. Serological tests in diagnosis of primary biliary cirrhosis. *Lancet* i:827–831.
7. Kaplan, M.M. 1987. Primary biliary cirrhosis. *Ann. Intern. Med.* 32:359–378.
8. Scheuer, P.J. 1967. Primary biliary cirrhosis. *Proc. R. Soc. Med.* 60:1257–1260.
9. Ludwig, J., E.R. Dickson, and G.S.A. McDonald. 1978. Staging of chronic nonsuppurative destructive cholangitis (syndrome of primary biliary cirrhosis). *Virchows Arch. A [Pathol. Anat.]* 379:103–112.
10. Portmann, B., H. Popper, J. Neuberger, and R. Williams. 1985. Sequential and diagnostic features in primary biliary cirrhosis based on serial histologic study in 209 patients. *Gastroenterology* 88:1777–1790.
11. Poupon, R.E., A.-M. Bonnand, Y. Chretien, R. Poupon, and the UDCA-PBC Study Group. 1999. Ten-year survival in ursodeoxycholic acid-treated patients with primary biliary cirrhosis. *Hepatology* 29:1668–1671.
12. Mackay, I.R. 1958. Primary biliary cirrhosis showing a high titer of autoantibody. *N. Eng. J. Med.* 258:185–188.
13. Berg, P.A., D. Doniach, and I.M. Roitt. 1967. Mitochondrial antibodies in primary biliary cirrhosis. I. Localization of the antigen to mitochondrial membranes. *J. Exp. Med.* 126:277–290.
14. Berg, P.A., U. Muscatello, R.W. Horne, I.M. Roitt, and D. Doniach. 1969. Mitochondrial antibodies in primary biliary cirrhosis. II. The complement fixing antigen as a component of mitochondrial inner membranes. *Br. J. Exp. Path.* 50:200–208.
15. Baum, H., and P.A. Berg. 1981. The complex nature of mitochondrial antibodies and their relation to primary biliary cirrhosis. *Sem. Liver. Dis.* 1:309–321.
16. Doniach, D., J. Delhanty, H.J. Lindqvist, and R.D. Catterall. 1970. Mitochondrial and other tissue autoantibodies in patients with biological false positive reactions for syphilis. *Clin. Exp. Immunol.* 6:871–884.
17. Berg, P.A., R. Klein, and J. Lindenborn-Fotinos. 1986. Antimitochondrial antibodies in primary biliary cirrhosis. *J. Hepatol.* 2:123–131.
18. Berg, P.A., and R. Klein. 1992. Antimitochondrial antibodies in primary biliary cirrhosis and other disorders: definition and clinical relevance. *Dig. Dis.* 10:85–101.
19. Frazer, I.H., I.R. Mackay, T.W. Jordan, S. Whittingham, and S. Marzuki. 1985. Reactivity of anti-mitochondrial autoantibodies in primary biliary cirrhosis: definition of two novel mitochondrial polypeptide autoantigens. *J. Immunol.* 135:1739–1745.
20. Gershwin, M.E., I.R. Mackay, A. Sturgess, and R.L. Coppel. 1987. Identification and specificity of a cDNA encoding the 70 kD mitochondrial antigen recognized in primary biliary cirrhosis. *J. Immunol.* 138:3525–3531.
21. Yeaman, S.J., S.P.M. Fussey, D.J. Danner, O.F.W. James, D.J. Mutimer, and M.F. Bassendine. 1988. Primary biliary cirrhosis: identification of two major M2 mitochondrial autoantigens. *Lancet* i:1067–1070.
22. Coppel, R.L., L.J. McNeilage, C.D. Surh, J. Van de Water, T.W. Spithill, S. Whittingham, and M.E. Gershwin. 1988. Primary structure of the human M2 mitochondrial autoantigen of primary biliary cirrhosis: dihydrolipoamide acetyltransferase. *Proc. Natl. Acad. Sci. USA* 85:7317–7321.
23. Van de Water, J., D. Fregeau, P. Davis, A. Ansari, D. Danner, P. Leung, R. Coppel, and M.E. Gershwin. 1988. Autoantibodies of primary biliary cirrhosis recognize dihydrolipoamide acetyltransferase and inhibit enzyme function. *J. Immunol.* 141:2321–2324.
24. Fussey, S.P.M., J.R. Guest, O.F.W. James, M.F. Bassendine, and S.J. Yeaman. 1988. Identification and analysis of the major M2 autoantigens in primary biliary cirrhosis. *Proc. Natl. Acad. Sci. USA* 85:8654–8658.
25. Fregeau, D.R., P.A. Davis, D.J. Danner, A. Ansari, R.L. Coppel, E.R. Dickson, and M.E. Gershwin. 1989. Antimitochondrial antibodies of primary biliary cirrhosis recognize dihydrolipoamide acyltransferase and inhibit enzyme function of the branched chain α-ketoacid dehydrogenase complex. *J. Immunol.* 142:3815–3820.
26. Fregeau, D.R., T. Prindiville, R.L. Coppel, M. Kaplan, E.R. Dickson, and M.E. Gershwin. 1990. Inhibition of α-ketoglutarate dehydrogenase activity by a distinct population of autoantibodies recognizing dihydrolipoamide succinyltransferase in primary biliary cirrhosis. *Hepatology* 11:975–981.
27. Fregeau, D.R., T.E. Roche, P.A. Davis, R. Coppel, and M.E. Gershwin. 1990. Primary biliary cirrhosis. Inhibition of pyruvate dehydrogenase complex activity by autoantibodies specific for E1α, a non-lipoic acid containing mitochondrial enzyme. *J. Immunol.* 144:1671–1676.

28. Surh, C.D., D.J. Danner, A. Ahmed, R.L. Coppel, I.R. Mackay, E.R. Dickson, and M.E. Gershwin. 1989. Reactivity of primary biliary cirrhosis sera with a human fetal liver cDNA clone of branched-chain α-keto acid dehydrogenase dihydrolipoamide acyltransferase, the 52 kD mitochondrial autoantigen. *Hepatology* 9:63–68.

29. Surh, C.D., T.E. Roche, D.J. Danner, A. Ansari, R.L. Coppel, T. Prindiville, E.R. Dickson, and M.E. Gershwin. 1989. Antimitochondrial autoantibodies in primary biliary cirrhosis recognized cross-reactive epitope(s) on protein X and dihydrolipoamide acetyltransferase of pyruvate dehydrogenase complex. *Hepatology* 10:127–133.

30. Maeda, T., B.E. Loveland, M.J. Rowley, and I.R. Mackay. 1991. Autoantibody against dihydrolipoamide dehydrogenase, the E3 subunit of the 2-oxoacid dehydrogenase complexes: significance for primary biliary cirrhosis. *Hepathology* 14:994–999.

31. Yeaman, S.J. 1986. The mammalian 2-oxoacid dehydrogenases: a complex family. *Trends in Biochem. Sci.* 11:293–296.

32. Patel, M.S., and T.E. Roche. 1990. Molecular biology and biochemistry of pyruvate dehydrogenase complexes. *FASEB J.* 4:3224–3233.

33. Reed, L.J., and M.L. Hackert. 1990. Structure-function relationships in dihydrolipoamide acyltransferases. *J. Biol. Chem.* 265:8971–8974.

34. Bassendine, M.F., S.P.M. Fussey, D.J. Mutimer, O.F.W. James, and S.J. Yeaman. 1989. Identification and characterization of four M2 mitochondrial autoantigens in primary biliary cirrhosis. *Sem. Liver Dis.* 9:124–131.

35. Dubel, L., A. Tanaka, P.S.C. Leung, J. Van de Water, R. Coppel, T. Roche, C. Johanet, Y. Motokawa, A. Ansari, and M.E. Gershwin. 1999. Autoepitope mapping and reactivity of autoantibodies to the dihydrolipoamide dehydrogenase-binding protein (E3BP) and the glycine cleavage proteins in primary biliary cirrhosis. *Hepatology* 29:1013–1018.

36. Van de Water, J., J. Turchany, P.S.C. Leung, J. Lake, S. Munoz, C.D. Surh, R. Coppel, A. Ansari, Y. Nakanuma, and M.E. Gershwin. 1993. Molecular mimicry in primary biliary cirrhosis. Evidence for biliary epithelial expression of a molecule cross-reactive with pyruvate dehydrogenase complex-E2. *J. Clin. Invest.* 91:2653–2664.

37. Surh, C.D., R. Coppel, and M.E. Gershwin. 1990. Structural requirement for autoreactivity on human pyruvate dehydrogenase-E2, the major autoantigen of primary biliary cirrhosis. Implication for a conformational autoepitope. *J. Immunol.* 144:3367–3374.

38. Rowley, M.J., L.J. McNeilage, J.M. Armstrong, and I.R. Mackay. 1991. Inhibitory autoantibody to a conformational epitope of the pyruvate dehydrogenase complex, the major autoantigen in primary biliary cirrhosis. *Clin. Immunol. Immunopathol.* 60:356–370.

39. Iwayama, T., P.S.C. Leung, R.L. Coppel, T.E. Roche, M.S. Patel, Y. Mizushima, T. Nakagawa, R. Dickson, and M.E. Gershwin. 1991. Specific reactivity of recombinant human PDC-E1α in primary biliary cirrhosis. *J. Autoimmun.* 4:769–778.

40. Leung, P.S.C., D.T. Chuang, R.M. Wynn, S. Cha, D.J. Danner, A. Ansari, R.L. Coppel, and M.E. Gershwin. 1995. Autoantibodies to BCOADC-E2 in patients with primary biliary cirrhosis recognize a conformational epitope. *Hepatology* 25:505–513.

41. Moteki, S., P.S.C. Leung, E.R. Dickson, D.H. Van Thiel, C. Galperin, T. Buch, D. Alarcon-Segovia, D. Kershenobich, K. Kawano, R.L. Coppel, S. Matsuda, and M.E. Gershwin. 1996. Epitope mapping and reactivity of autoantibodies to the E2 component of 2-oxoglutarate dehydrogenase complex in primary biliary cirrhosis using recombinant 2-oxoglutarate dehydrogenase complex. *Hepatology* 23:436–444.

42. Palmer, J.M., D.E. Jones, J. Quinn, A. McHugh, and S.J. Yeaman. 1999. Characterization of the autoantibody responses to recombinant E3 binding protein (protein X) of pyruvate dehydrogenase in primary biliary cirrhosis. *Hepatology* 30:21–26.

43. Moteki, S., P.S.C. Leung, R.L. Coppel, E.R. Dickson, M.M. Kaplan, S. Munoz, and M.E. Gershwin. 1996. Use of a designer triple expression hybrid clone for three different lipoyl domains for the detection of antimitochondrial autoantibodies. *Hepatology* 24:97–103.

44. Ohmori, H., I. Yamauchi, and I. Yamamoto. 1986. Augmentation of the antibody response by lipoic acid in mice. I. Analysis of the mode of action in an *in vitro* culture system. *Jpn. J. Pharmacol.* 42:135–140.

45. Ohmori, H., T. Yamauchi, and I. Yamamoto. 1986. Augmentation of the antibody response by lipoic acid in mice. II. Restoration of the antibody response in immunosuppressed mice. *Jpn. J. Pharmacol.* 42:275–280.

46. Fussey, S.P.M., S.T. Ali, J.R. Guest, O.F.W. James, M.F. Bassendine, and S.J. Yeaman. 1990. Reactivity of primary biliary cirrhosis sera with Escherichia coli dihydrolipoamide acetyltransferase (E2p): characterization of the main immunogenic region. *Proc. Natl. Acad. Sci. USA* 87:3987–3991.

47. Leung, P.S.C., T. Iwayama, R.L. Coppel, and M.E. Gershwin. 1990. Site-directed mutagenesis of lysine within the immunodominant autoepitope of PDC-E2. *Hepatology* 12:1321–1328.

48. Koike, K., H. Ishibashi, and M. Koike. 1998. Immunoreactivity of porcine heart dihydrolipoamide acetyl- and succinyl-transferase (PDC-E2, OGDC-E2) with primary biliary cirrhosis sera: characterization of the autoantigenic region and effects of enzymatic delipoylation and relipoylation. *Hepatology* 27:1467–1474.

49. Leung, P.S.C., and M.E. Gershwin. 1993. Immunoglobulin genes in autoimmunity. *Int. Arch. Allergy Immunol.* 101:113–118.

50. Leung, P.S., S. Krams, S. Munoz, C.D. Surh, A. Ansari, T. Kenny, D.L. Robbins, J. Fung, T.E. Starzl, W. Maddrey, R.L. Coppel, and M.E. Gershwin. 1992. Characterization and epitope mapping of human monoclonal antibodies to PDC-E2, the immunodominant autoantigen of primary biliary cirrhosis. *J. Autoimmun.* 5:703–718.

51. Pascual, V., S. Cha, M.E. Gershwin, J.D. Capra, and P.S.C. Leung. 1994. Nucleotide sequence analysis of natural and combinatorial anti-PDC-E2 antibodies in patients with primary biliary cirrhosis. Recapitulating immune selection with molecular biology. *J. Immunol.* 152:2577–2585.

52. Cha, S., P.S.C. Leung, M.E. Gershwin, M.P. Fletcher, A.A. Ansari, and R.L. Coppel. 1993. Combinatorial autoantibodies to dihydrolipoamide acetyltransferase, the major autoantigen of primary biliary cirrhosis. *Proc. Natl. Acad. Sci. USA* 90:2527–2531.

53. Cha, S., P.S.C. Leung, R.L. Coppel, J. Van de Water, A.A. Ansari, and M.E. Gershwin. 1994. Heterogeneity of combinatorial human autoantibodies against PDC-E2 and biliary epithelial cells in patients with primary biliary cirrhosis. *Hepatology* 20:574–583.

54. Huse, W.D., L. Sastry, S.A. Iverson, A.S. Kang, M. Alting-Mees, D.R. Burton, S.J. Benkovic, and R.A. Lerner. 1989. Generation of a large combinatorial library of the

immunoglobulin repertoire in phage lambda. *Science* 246:1275–1281.

55. Fukushima, N., M. Nakamura, M. Matsui, H. Ikematsu, K. Koike, H. Ishibashi, K. Hayashida, and Y. Niho. 1995. Establishment and structural analysis of human mAb to the E2 component of the 2-oxoglutarate dehydrogenase complex generated from a patient with primary biliary cirrhosis. *Int. Immunol.* 7:1047–1055.

56. Bernstein, R.M., J.M. Neuberger, C.C. Bunn, M.E. Callender, G.R.V. Hughes, and R. Williams. 1984. Diversity of autoantibodies in primary biliary cirrhosis and chronic active hepatitis. *Clin. Exp. Immunol.* 55:553–560.

57. Powell, F., A.L. Schroeter, and E.R. Dickson. 1984. Antinuclear antibodies in primary biliary cirrhosis. *Lancet* i:288–289.

58. Penner, E., S. Muller, D. Zimmermann, and M.H.V. Van Regenmortel. 1987. High prevalence of antibodies to histones among patients with primary biliary cirrhosis. *Clin. Exp. Immunol.* 70:47–52.

59. Bernstein, R.M., M.E. Callender, J.M. Neuberger, G.R.V. Hughes, and R. Williams. 1982. Anticentromere antibody in primary biliary cirrhosis. *Ann. Rheum. Dis.* 41:612–614.

60. Ruffatti, A., P. Arslan, A. Floreani, G. De Silvestro, A. Calligaro, R. Naccarato, and S. Todesco. 1985. Nuclear membrane-staining antinuclear antibody in patients with primary biliary cirrhosis. *J. Clin. Immunol.* 5:357–361.

61. Lassoued, K., R. Brenard, F. Degos, J.-C. Courvalin, C. Andre, F. Danon, J.-C. Brouet, Y. Zine-El-Abidine, C. Degott, S. Zafrani, D. Dhumeaux, and J.-P. Benhamou. 1990. Antinuclear antibodies directed to a 200-kilodalton polypeptide of the nuclear envelope in primary biliary cirrhosis. A clinical and immunological study of a series of 150 patients with primary biliary cirrhosis. *Gastroenterology* 99:181–186.

62. Powell, F.C., A.L. Schroeter, and E.R. Dickson. 1987. Primary biliary cirrhosis and the CREST syndrome: a report of 22 cases. *Q. J. Med.* 237:75–82.

63. Earnshaw, W.C., K.F. Sullivan, P.S. Machlin, C.A. Cooke, D.A. Kaiser, T.D. Pollard, N.F. Rothfield, and D.W. Cleveland. 1987. Molecular cloning of cDNA for CENP B, the major human centromere autoantigen. *J. Cell Biol.* 104:817–829.

64. Fusconi, M., F. Cassani, M. Govoni, F. Caselli, F. Farabegoli, M. Lenzi, G. Ballardini, D. Zauli, and F.B. Bianchi. 1991. Anti-nuclear antibodies of primary biliary cirrhosis recognize 78–92-kD and 96–100-kD proteins of nuclear bodies. *Clin. Exp. Immunol.* 83:291–297.

65. Szostecki, C., H. Krippner, E. Penner, and F.A. Bautz. 1987. Autoimmune sera recognize a 100 kD nuclear protein antigen (sp-100). *Clin. Exp. Immunol.* 68:108–116.

66. Szostecki, C., H.H. Guldner, H.J. Netter, and H. Will. 1990. Isolation and characterization of cDNA encoding a human nuclear antigen predominantly recognized by autoantibodies from patients with primary biliary cirrhosis. *J. Immunol.* 145:4338–4347.

67. Bluthner, M., C. Schafer, C. Schneider, and F.A. Bautz. 1999. Identification of major linear epitopes on the sp100 nuclear PBC autoantigen by the gene-fragment phage-display technology. *Autoimmunity* 29:33–42.

68. Newport, J.W., and D.J. Forbes. 1987. The nucleus: structure, function, and dynamics. *Ann. Rev. Biochem.* 56:535–565.

69. Lassoued, K., M.-N. Guilly, C. Andre, M. Paintrand, D. Dhumeaux, F. Danon, J.C. Brouet, and J.-C. Courvalin. 1988a. Autoantibodies to 200 kD polypeptide(s) of the nuclear envelope: a new serologic marker of primary biliary cirrhosis. *Clin. Exp. Immunol.* 74:283–288.

70. Lozano, F., A. Pares, L. Borche, M. Plana, T. Gallart, J. Rodes, and J. Vives. 1988. Autoantibodies against nuclear envelope-associated proteins in primary biliary cirrhosis. *Hepatology* 8:930–938.

71. Senecal, J.-L., and Y. Raymond. 1991. Autoantibodies to DNA, lamins, and pore complex proteins produce distinct peripheral fluorescent antinuclear antibody patterns on the HEp-2 substrate. *Arthritis Rheum.* 34:249–251.

72. Raymond, Y., and G. Gagnon. 1988. Lamin B shares a number of distinct epitopes with lamins A and C and with intermediate filament proteins. *Biochemistry* 27:2590–2597.

73. Tan, E.M. 1988. Autoantibodies to nuclear lamins. *Ann. Intern. Med.* 108:897–898.

74. Lassoued, K., M.-N. Guilly, F. Danon, C. Andre, D. Dhumeaux, J.-P. Clauvel, J.-C. Brouet, M. Seligmann, and J.-C. Courvalin. 1988. Antinuclear autoantibodies specific for lamins. Characterization and clinical significance. *Ann. Intern. Med.* 108:829–833.

75. Wesierska-Gadek, J., E. Penner, E. Hitchman, and G. Sauermann. 1988. Antibodies to nuclear lamins in autoimmune liver disease. *Clin. Immunol. Immunopathol.* 49:107–115.

76. Wozniak, R.W., E. Bartnik, and G. Blobel. 1989. Primary structure analysis of an integral membrane glycoprotein of the nuclear pore. *J. Cell Biol.* 108:2083–2092.

77. Courvalin, J.-C., K. Lassoued, E. Bartnik, G. Blobel, and R.W. Wozniak. 1990. The 210-kD nuclear envelope polypeptide recognized by human autoantibodies in primary biliary cirrhosis is the major glycoprotein of the nuclear pore. *J. Clin. Invest.* 86:279–285.

78. Nickowitz, R.E., and H.J. Worman. 1993. Autoantibodies from patients with primary biliary cirrhosis recognize a restricted region within the cytoplasmic tail of nuclear pore membrane glycoprotein gp210. *J. Exp. Med.* 178:2237–2242.

79. Wesierska-Gadek, J., H. Hohenauer, E. Hitchman, and E. Penner. 1995. Autoantibodies from patients with primary biliary cirrhosis preferentially react with the amino-terminal domain of nuclear pore complex glycoprotein gp210. *J. Exp. Med.* 182:1159–1162.

80. Wesierska-Gadek, J., H. Hohenauer, E. Hitchman, and E. Penner. 1996. Autoantibodies against nucleoporin p62 constitute a novel marker of primary biliary cirrhosis. *Gastroenterology* 110:840–847.

81. Mackay, I.R., M.J. Rowley, and S.F. Whittingham. 1993. Nuclear autoantibodies in primary biliary cirrhosis. *Immunology & Liver: Proceedings of Basel Liver Week, 1992*, Meyer, K., Manns, M., and Hoofnagel, J. eds. Kluwer Academic Publishers, Lancaster, U. K.

82. Doniach, D., I.M. Roitt, J.G. Walker, and S. Sherlock. 1966. Tissue antibodies in primary biliary cirrhosis, active chronic (lupoid) hepatitis, cryptogenic cirrhosis and other liver diseases and their clinical implications. *Clin. Exp. Immunol.* 1:237–262.

83. Christensen, E., J. Crowe, D. Doniach, H. Popper, L. Ranek, J. Rodes, N. Tygstrup, and R. Williams. 1980. Clinical pattern and course of disease in primary biliary cirrhosis based on an analysis of 236 patients. *Gastroenterology* 78:236–246.

84. Tsantoulas, D., A. Perperas, B. Portmann, A.L.W.F. Eddleston, and R. Williams. 1980. Antibodies to a human liver membrane lipoprotein (LSP) in primary biliary cirrhosis. *Gut* 21:557–560.

85. Bedlow, A.J., P.T. Donaldson, B.M. McFarlane, M. Lombard, I.G. McFarlane, and R. Williams. 1989. Autoreactivity to hepatocellular antigens in primary biliary cirrhosis and primary sclerosing cholangitis. *J. Clin. Lab. Immunol.* 30:103–109.

86. Mackay, I.R. 1984. Genetic aspects of immunologically mediated liver disease. *Sem. Liver Dis.* 4:13–25.

87. Meuer, S.C., U. Moebius, M.M. Manns, H.P. Dienes, G. Ramadori, G. Hess, T. Hercend, and K.H. Meyer zum Buschenfelde. 1988. Clonal analysis of human T lymphocytes infiltrating the liver in chronic active hepatitis B and primary biliary cirrhosis. *Eur. J. Immunol.* 18:1447–1452.

88. Hoffmann, R.M., G.R. Pape, U. Spengler, E.P. Rieber, J. Eisenburg, J. Dohrmann, G. Paumgartner, and G. Riethmuller. 1989. Clonal analysis of liver-derived T cells of patients with primary biliary cirrhosis. *Clin. Exp. Immunol.* 76:210–215.

89. Moebius, U., M. Manns, G. Hess, G. Kober, K.-H. Meyer zum Buschenfelde, and S.C. Meuer. 1990. T-cell receptor gene rearrangements of T lymphocytes infiltrating the liver in chronic active hepatitis B and primary biliary cirrhosis (PBC): oligoclonality of PBC-derived T cell clones. *Eur. J. Immunol.* 20:889–896.

90. Krams, S.M., J. Van de Water, R.L. Coppel, C. Esquivel, J. Roberts, A. Ansari, and M.E. Gershwin. 1990. Analysis of hepatic T lymphocyte and immunoglobulin deposits in patients with primary biliary cirrhosis. *Hepatology* 12:306–313.

91. Van de Water, J., A.A. Ansari, C.D. Surh, R. Coppel, T. Roche, H. Bonkovsky, M. Kaplan, and M.E. Gershwin. 1991. Evidence for the targeting by 2-oxo-dehydrogenase enzymes in the T cell response of primary biliary cirrhosis. *J. Immunol.* 146:89–94.

92. Van de Water, J., A. Ansari, T. Prindiville, R. Coppel, N. Ricalton, B.L. Kotzin, S. Liu, T.E. Roche, S.M. Krams, S. Munoz, and M.E. Gershwin, M.E. 1995. Heterogeneity of autoreactive T cell clones specific for the E2 component of the pyruvate dehydrogenase complex in primary biliary cirrhosis. *J. Exp. Med.* 181:723–733.

93. Jones, D.E.J., J.M. Palmer, O.F.W. James, S.J. Yeaman, M.F. Bassendine, and A.G. Diamond. 1995. T-cell responses to the components of pyruvate dehydrogenase complex in primary biliary cirrhosis. *Hepatology* 21:995–1002.

94. Shimoda, S., M. Nakamura, H. Ishibashi, K. Hayashida, and Y. Niho. 1995. HLA DRB4 0101-restricted immunodominant T cell autoepitope of pyruvate dehydrogenase complex in primary biliary cirrhosis: evidence of molecular mimicry in human autoimmune diseases. *J. Exp. Med.* 181:1835–1845.

95. Shimoda, S., J. Van de Water, A. Ansari, M. Nakamura, H. Ishibashi, R.L. Coppel, J. Lake, E.B. Keeffe, T.E. Roche, and M.E. Gershwin. 1998. Identification and precursor frequency analysis of a common T cell epitope motif in mitochondrial autoantigens in primary biliary cirrhosis. *J. Clin. Invest.* 102:1831–1840.

96. Ichiki, Y., S. Shimoda, H. Hara, H. Shigematsu, M. Nakamura, K. Hayashida, H. Ishibashi, and Y. Niho. 1997. Analysis of T-cell receptor β of the T-cell clones reactive to the human PDC-E2 163–176 peptide in the context of HLA-DR53 in patients with primary biliary cirrhosis. *Hepatology* 26:728–733.

97. Wucherpfenning, K.W., I. Catz, S. Hausmann, J.L. Strominger, L. Steinmann, and K.G. Warren. 1997. Recognition of the immunodominant myelin basic protein peptide by autoantibodies and HLA-DR-2restricted T cell

98. Stemerowicz, R., U. Hopf, B. Moller, C. Wittenbrink, A. Rodloff, R. Reinhardt, M. Freudenberg, and C. Galanos. 1988. Are antimitochondrial antibodies in primary biliary cirrhosis induced by R(rough)-mutants of enterobacteriaceae? *Lancet* ii:1166–1170.

99. Burroughs, A.K., I.J. Rosenstein, O. Epstein, J.M.T. Hamilton-Miller, W. Brumfitt, and S. Sherlock. 1989. Bacteriuria and primary biliary cirrhosis. *Gut* 25:133–137.

100. Hopf, U., B. Moller, R. Stemerowicz, H. Lobeck, A. Rodloff, M. Freudenberg, C. Galanos, and D. Huhn. 1989. Relation between Escherichia coli R(rough) -forms in gut, lipid A in liver, and primary biliary cirrhosis. *Lancet* ii:1419–1422.

101. Teoh, K.-L., I.R. Mackay, M.J. Rowley, and S.P.M. Fussey. 1994. Enzyme inhibitory autoantibodies to pyruvate dehydrogenase complex in primary biliary cirrhosis differ for mammalian, yeast and bacterial enzymes: implications for molecular mimicry. *Hepatology* 19:1029–1033.

102. Burroughs, A.K., P. Butler, M.J.E. Stemberg, and H. Baum. 1992. Molecular mimicry in liver disease. *Nature* 358:377–378.

103. Gautam, A.M., C.B. Lock, D.E. Smilek, C.I. Pearson, L. Steinman, and H.O. McDevitt. 1994. Minimum structural requirements for peptide presentation by major histocompatibility complex class II molecules: implications in induction of autoimmunity. *Proc. Natl. Acad. Sci. USA* 91:767–771.

104. Luo, A.M., K.M. Garza, D. Hunt, and K.S. Tung. 1993. Antigen mimicry in autoimmune disease sharing of amino acid residues critical for pathogenic T cell activation. *J. Clin. Invest.* 92:2117–2123.

105. Galbraith, R.M., M. Smith, R.M. Mackenzie, D.E. Tee, D. Doniach, and R. Williams. 1974. High prevalence of seroimmunologic abnormalities in relatives of patients with active chronic hepatitis or primary biliary cirrhosis. *N. Eng. J. Med.* 290:63–69.

106. Salaspuro, M.P., O.I. Laitinen, J. Lehtola, H. Makkonen, J.A. Rasanen, and P. Sipponen. 1976. Immunological parameters, viral antibodies and biochemical and histological findings in relatives of patients with chronic active hepatitis and primary biliary cirrhosis. *Scand. J. Gastroenterol.* 20:313–320.

107. Begovich, A.B., W. Klitz, P.V. Moonsamy, J. Van de Water, G. Peltz, and M.E. Gershwin. 1994. Genes within the HLA class II region confer both predisposition and resistance to primary biliary cirrhosis. *Tissue Antigens* 43:71–77.

108. Begovich, A.B., T.L. Bugawan, B.S. Nepom, W. Klitz, G.T. Nepom, and H.A. Erlich. 1989. A specific HLA-DPβ allele is associated with pauciarticular juvenile rheumatoid arthritis but not adult rheumatoid arthritis. *Proc. Natl. Acad. Sci. USA* 86:9489–9493.

109. Dong, R.-P., A. Kimura, R. Okubo, H. Shinagawa, H. Tamai, Y. Nishimura, and T. Sasazuki. 1992. HLA-A and DPB1 loci confer susceptibility to Graves' disease. *Human Immunol.* 35:165–172.

110. Arriaga, F., G. Ercilla, A. Pares, E. Bergada, M. Bruguera, R. Castillo, L. Revert, J. Rodes, and J. Vives. 1980. Association of HLA-DR3 antigen to disease with immunological components. *Sangre (Barc)* 25:430–437.

111. Gores, G.J., S.B. Moore, L.D. Fisher, F.C. Powell, and E.R. Dickson. 1987. Primary biliary cirrhosis: associations with class II major histocompatibility complex antigens. *Hepatology* 7:889–892.

112. Prochazka, E.J., P.I. Terasaki, M.S. Park, L.I. Goldstein, and R.W. Busuttil. 1990. Association of primary sclerosing cholangitis with HLA-DRw52a. *N. Eng. J. Med.* 322:1842–1844.

113. Manns, M.P., A. Bremm, P.M. Schneider, A. Notghi, G. Gerken, M. Prager-Eberle, B. Stradmann-Bellinghausen, K.-H. Meyer zum Buschenfelde, and C. Rittner. 1991. HLA DRw8 and complement C4 deficiency as risk factors in primary biliary cirrhosis. *Gastroenterology* 101:1367–1373.

114. Johnston, D.E., M.M. Kaplan, K.B. Miller, C.M. Connors, and E.L. Milford. 1987. Histocompatibility antigens in primary biliary cirrhosis. *Am. J. Gastroenterol.* 82:1127–1129.

115. Miyamori, H., Y. Kato, K. Kobayashi, and N. Hattori. 1983. HLA antigens in Japanese patients with primary biliary cirrhosis and autoimmune hepatitis. *Digestion* 26:213–217.

116. Bassendine, M.F., P.J. Dewar, and O.F.W. James. 1985. HLA-DR antigens in primary biliary cirrhosis: lack of association. *Gut* 26:625–628.

117. Briggs, D.C., P.T. Donaldson, P. Hayes, K.I. Welsh, R. Williams, and J.M. Neuberger. 1987. A major histocompatibility complex class III allotype (C4B 2) associated with primary biliary cirrhosis. *Hepatology* 29:141–145.

118. Morling, N., K. Dalhoff, L. Fugger, J. Georgsen, B. Jakobsen, L. Ranek, N. Otdum, and A. Svejgaard. 1992. DNA polymorphism of HLA class II genes in primary biliary cirrhosis. *Immunogenetics* 35:112–116.

119. Mella, J.G., E. Roschmann, K.P. Maier, and B.A. Volk. 1995. Association of primary biliary cirrhosis with the allele HLA DPB1*0301 in a German population. *Hepatology* 21:398–402.

120. Underhill, J., P. Donaldson, G. Bray, D. Doherty, B. Portmann, and R. Williams. 1992. Susceptibility to primary biliary cirrhosis is associated with the HLA-DR8-DQB1*0402 haplotype. *Hepatology* 16:1404–1408.

121. Underhill, J.A., P.T. Donaldson, D.G. Doherty, K. Manabe, and R. Williams. 1995. HLA DPB polymorphism in primary sclerosing cholangitis and primary biliary cirrhosis. *Hepatology* 21:959–962.

122. Seki, T., K. Kiyosawa, M. Ota, S. Furuta, H. Fukushima, E. Tanaka, K. Yoshizawa, T. Kumagai, N. Mizuki, A. Ando, and H. Inoko. 1993. Association of primary biliary cirrhosis with human leukocyte antigen DPB1*0501 in Japanese patients *Hepatology* 18:73–78.

123. Hillaire, S., E. Boucher, Y. Calmus, P. Gane, F. Ballet, D. Franco, M. Moukthar, and R. Poupon. 1994. Effects of bile acids and cholestasis on major histocompatibility complex class I in human and rat hepatocytes. *Gastroenterology* 107:781–788.

124. Afford, S.C., S. Hubscher, A.J. Strain, D.H. Adams, and J.M. Neuberger. 1995. Apoptosis in the human liver during allograft rejection and end-stage liver disease. *J. Hepatol.* 176:373–380.

125. Kuroki, T., S. Seki, N. Kawakita, K. Nakatani, T. Hisa, T. Kitada, and H. Sakaguchi. 1996. Expression of antigens related to apoptosis and cell proliferation in chronic nonsuppurative destructive cholangitis in primary biliary cirrhosis. *Virshows Arch.* 429:119–129.

126. Koga, H., S. Sakisaka, M. Ohishi, M. Sata, and K. Tanikawa. 1997. Nuclear DNA fragmentation and expression of Bcl-2 in primary biliary cirrhosis. *Hepatology* 25:1077–1084.

127. Harada, K., S. Ozaki, M.E. Gershwin, and Y. Nakanuama. 1997. Enchanced apoptosis relates to bile duct loss in primary biliary cirrhosis. *Hepatology* 26:1399–1405.

128. Graham, A.M., M.M. Dollinger, S.E.M. Howie, and D.J. Harrison. 1998. Bile duct cells in primary biliary cirrhosis are "primed" for apoptosis. *Eur. J. Hepatol. Gastroenterol.* 10:553–557.

129. Nakanuma, Y., and K. Harada. 1993. Florid duct lesion in primary biliary cirrhosis shows highly proliferative activities. *J. Hepatol.* 19:216–221.

130. Leon, M.P., G. Spickett, D.E.J. Jones, and M.F. Bassendine. 1995. CD4+ T cell subsets defined by isoforms of CD45 in primary biliary cirrhosis. *Clin. Exp. Immunol.* 99:233–239.

131. Bjorkland, A., R. Festin, I. Mendel-Hartvig, A. Nyberg, L. Loof, and T. Totterman. 1991. Blood and liver-infiltrating lymphocytes in primary biliary cirrhosis: increase in activated T and natural killer cells and recruitment of primed memory T cells. *Hepatology* 13:1106–1111.

132. Dienes, H.P., A.W. Lohse, G. Gerken, P. Schirmacher, H. Gallati, H.F. Lohr, and K.H. Meyer zum Buschenfelde. 1997. Bile duct epithelia as target cells in primary biliary cirrhosis and primary sclerosing cholangitis. *Virshows Arch.* 431:119–124.

133. Harada, K., J. Van de Water, P.S.C. Leung, R.L. Coppel, A. Ansari, Y. Nakanuma, and M.E. Gershwin. 1997. In situ nucleic acid hybridization of cytokines in primary biliary cirrhosis: predominance of the Th1 subset. *Hepatology* 25:791–796.

134. Ballardini, G., R. Mirakian, E.B. Bianchi, E. Pisi, D. Doniach, and G.F. Bottazzo. 1984. Aberrant expression of HLA-DR antigens on bileduct epithelium in primary biliary cirrhosis: relevance to pathogenesis. *Lancet* ii:1009–1013.

135. Adams, D.H., S.G. Hubscher, J. Shaw, G.D. Johnson, C. Babbs, R. Rothlein, and J.M. Neuberger. 1991. Increased expression of intercellular adhesion molecule 1 on bile ducts of primary biliary cirrhosis and primary sclerosing cholangitis. *Hepatology* 14:426–431.

136. Yasoshima, M., Y. Nakanuma, K. Tsuneyama, J. Van de Water, and M.E. Gershwin. 1995. Immunohistochemical analysis of adhesion molecules in the micro-environment of portal tracts in relation to aberrant expression of PDC-E2 and HLA-DR on the bile ducts in primary biliary cirrhosis. *J. Pathol.* 175:319–325.

137. Lim, A.G., R.P. Jazrawi, J.H. Levy, M.L. Petroni, A.C. Douds, J.D. Maxell, and T.C. Northfield. 1995. Soluble E-selection and vascular cell adhesion molecule-1 (VCAM-1) in primary biliary cirrhosis. *J. Hepatol.* 22:416–422.

138. Faubion, W.A., M.E. Guicciardi, H. Miyoshi, S.F. Bronk, P.J. Roberts, P.A. Svingen, S.H. Kaufmann, and G.J. Gores. 1999. Toxic bile salts induce rodent hepatocyte apoptosis via direct activation of Fas. *J. Clin. Invest.* 103:137–145.

139. Ghadiminejad, I., and H. Baum. 1987. Evidence for the cell-surface localization of antigens cross-reacting with the "mitochondrial antibodies" of primary biliary cirrhosis. *Hepatology* 7:743–749.

140. Joplin, R., J.G. Lindsay, S.G. Hubscher, G.D. Johnson, J.C. Shaw, A.J. Strain, and J.M. Neuberger. 1991. Distribution of dihydrolipoamide acetyltransferase (E2) in the liver and portal lymph nodes of patients with primary biliary cirrhosis: an immunohistochemical study. *Hepatology* 14:442–447.

141. Joplin, R., J.G. Lindsay, G.D. Johnson, A. Strain, and J. Neuberger. 1992. Membrane dihydrolipoamide acetyltransferase (E2) on human biliary epithelial cells in primary biliary cirrhosis. *Lancet* 339:93–94.

142. Surh, C.D., A. Ahmed-Ansari, and M.E. Gershwin. 1990. Comparative epitope mapping of murine monoclonal and human autoantibodies to human PDH-E2, the major mito-

chondrial autoantigen of primary biliary cirrhosis. *J. Immunol.* 144:2647–2652.

143. Harada, K., J. Van de Water, P.S.C. Leung, R.L. Coppel, Y. Nakanuma, and M.E. Gershwin. 1997. In situ nucleic acid hybridization of pyruvate dehydrogenase complex-E2 in primary biliary cirrhosis: pyruvate dehydrogenase complex-E2 messenger RNA is expressed in hepatocytes but not in biliary epithelium. *Hepatology* 25:27–32.

144. Cha, S., P.S.C. Leung, J. Van de Water, K. Tsuneyama, R.E. Joplin, A.A. Ansari, Y. Nakanuma, P.J. Schatz, S. Cwirla, L.E. Fabris, J.M. Neuberger, M.E. Gershwin, and R.L. Coppel. 1996. Random phage mimotopes recognized by monoclonal antibodies against the pyruvate dehydrogenase complex-E2 (PDC-E2). *Proc. Natl. Acad. Sci. USA* 93:10949–10954.

145. Migliaccio, C., A. Nishio, J. Van de Water, A.A. Ansari, P.S.C. Leung, Y. Nakanuma, R.L. Coppel, and M.E. Gershwin. 1998. Monoclonal antibodies to mitochondrial E2 components define autoepitopes in primary biliary cirrhosis. *J. Immunol.* 161:5157–5163.

146. Harada, K., Y. Sudo, N. Kono, S. Ozaki, K. Tsuneyama, M.E. Gershwin, and Y. Nakanuma. 1999. In situ nucleic acid detection of PDC-E2, BCOADC-E2, OGDC-E2, PDC-E1α, BCOADC-E1α, OGDC-E1, and the E3 binding protein (protein X) in primary biliary cirrhosis. *Hepatology* 30:36–45.

147. Tsuneyama, K., J. Van de Water, P.S.C. Leung, S. Cha, Y. Nakanuma, M. Kaplan, R. De Lellis, R. Coppel, A. Ansari, and M.E. Gershwin. 1995. Abnormal expression of the E2 component of the pyruvate dehydrogenase complex on the luminal surface of biliary epithelium occurs before major histocompatibility complex class II and BB1/B7 expression. *Hepatology* 21:1031–1037.

148. Malmborg, A.-C., D.B. Shultz, F. Luton, K. Mostov, E. Richly, P.S.C. Leung, G.D. Benson, A.A. Ansari, R.L. Coppel, M.E. Gershwin, and J. Van de Water. 1998. Penetration and co-localization in MDCK cell mitochondria of IgA derived from patients with primary biliary cirrhosis. *J. Autoimmun.* 11:573–580.

149. Nishio, A., J. Van de Water, P.S.C. Leung, R. Joplin, J.M. Neuberger, J. Lake, A. Bjorkland, T.H. Totterman, M. Peters, H.J. Worman, A.A. Ansari, R.L. Coppel, and M.E. Gershwin. 1997. Comparative studies of antimitochondrial autoantibodies in sera and bile in primary biliary cirrhosis. *Hepatology* 25:1085–1089.

150. Tison, V., F. Callea, C. Morisi, A.M. Mancini, and V.J. Desmet. 1982. Spontaneous "primary biliary cirrhosis" in rabbits. *Liver* 2:152–161.

151. Krams, S.M., C.D. Surh, R.L. Coppel, A. Ansari, B. Ruebner, and M.E. Gershwin. 1989. Immunization of experimental animals with dihydrolipoamide acetyltransferase, as a purified recombinant polypeptide, generates mitochondrial antibodies but not primary biliary cirrhosis. *Hepatology* 9:411–416.

152. Bassendine, M.F., J.M. Palmer, D. Decruz, G.W. McCaughan, D. Strickland, J.D. Sedgewick, and S.J. Yeaman. 1998. Approaches to a murine model of AMA positive non-suppurative destructive cholangitis (NSDC) [Abstract]. *J. Hepatol.* 28:59.

153. Krams, S.M., K. Dorshkind, and M.E. Gershwin. 1989. Generation of biliary lesions after transfer of human lymphocytes into severe combined immunodeficient mice. *J. Exp. Med.* 170:1919–1930.

154. Saito, T., M. Fujiwara, M. Nomoto, M. Makino, H. Watanabe, K. Ishihara, T. Kamimura, and F. Ichida. 1988. Hepatic lesions induced by graft-versus-host reaction across MHC class II antigens: an implication for animal model of primary biliary cirrhosis. *Clin. Immunol. Immunopathol.* 49:166–172.

54 | Autoimmune Hepatitis

Michael P. Manns, Christian P. Strassburg, and Petra Obermayer-Straub

1. INTRODUCTION

Autoimmune hepatitis (AIH) is a rare disease with an estimated prevalence of 170 cases per 1 million in the Northern European Caucasian population (1). The disease is characterized by a female predominance, hypergammaglobulinemia (>30 g/l), association with particular human leukocyte antigens (HLA) (DR3, DR4 in AIH type 1) and circulating antibodies against tissue antigens (2,3). At the time of diagnosis, the disease often has been present for several months and patients show high levels of serum transaminases. However, in 40% of patients, acute or even fulminant onset of hepatitis is found that may be mistaken for acute viral hepatitis (4). Histology usually reveals periportal and/or periseptal hepatitis (piecemeal necrosis) and lymphocytic infiltrates (5). The disease progresses further to bridging necrosis, panlobular and multilobular necrosis and active cirrhosis. Without treatment, up to 50% of patients with severe AIH will die within five years (6–10). The diagnosis of AIH is based on an array of clinical, serological and immunological features typical for hepatitis and exclusion of patients with other causes of liver disease (Table 1). These features are summarized in a provisional scoring system for the diagnosis of AIH (Table 1) (11). According to the autoantibodies found in AIH, it was proposed to classify AIH into subtypes 1–3 (AIH 1–3) (12) (Table 2). According to this nomenclature, AIH type 1 is characterized by antinuclear and/or anti-smooth muscle autoantibodies (13,14), AIH type 2 by antibodies specific for liver and kidney microsomes (LKM) (15) and/or liver cytosolic protein type 1 (anti-LC1) (16) and AIH type 3 by autoantibodies against anti-soluble liver antigen (anti-SLA) (17) or antibodies to liver—pancrease antigen (anti-LP) (18). Other autoantibodies associated with AIH are directed against the asialoglycoprotein receptor (ASGP-R) (19) and neutrophil cytoplasmic antigens (ANCA) (20).

As in other autoimmune diseases, genetic predisposition for autoimmune hepatitis is believed to be polygenic. However, even in the presence of a strong genetic predisposition, e.g. in identical twins with one twin affected, penetrance is not complete (21). Known risk factors for AIH type 1 are specific MHC class II antigens, namely DR3 and DR4, which increase the risk for AIH type 1 and also correlate with the clinical outcome (22,23). The incomplete penetrance of AIH indicates the necessity of a trigger to precipitate the disease. Many hepatotropic viruses have been suggested to trigger AIH type 1: Measles viruses (24), hepatitis A virus (25,26), hepatitis C virus (27), herpes simplex virus type 1 (28), human herpes simplex virus 6 (29) and Epstein Barr virus (EBV) (30). On the other hand, hepatitis caused by autoimmune processes may be triggered by chemicals or nutritional components. Several chemicals are known to induce immune-mediated drug-induced hepatitis (31,32). In these conditions, drug metabolizing enzymes of phase I, cytochromes P450 (CYPs) are often targets of autoimmunity (31,33) (Figure 1). Interestingly, a specific CYP, CYP2D6 and family 1 UDP-glucuronosyltransferases (UGTs), enzymes of the phase 2 of drug metabolism are targets of autoimmune processes in AIH type 2 and also in autoimmunity associated with chronic virus hepatitis C and D (Figure 1). So far, no particular drug has been identified as a trigger for AIH type 2. However, mechanisms leading to the induction of immune-mediated drug-induced hepatitis may provide insights that may inspire experiments to explore the mechanisms active in induction of AIH type 2 and about virus associated autoimmunity. Since we are not able to correlate AIH in general with exposure to a specific drug or to

Table 1   Scoring system for diagnosis of autoimmune hepatitis: minimum required parameters

Parameter	Score
Gender	
Female	+ 2
Male	0
Serum biochemistry	
Ratio of elevation of serum alkaline phosphatase vs aminotransferase	
< 1,5	+ 2
1,5–3.0	0
> 3,0	–2
Total serum globulin, γ-globulin or IgG (times upper normal limit)	
> 2.0	+ 3
1.5–2.0	+ 2
1.0–1.5	+ 1
< 1.0	0
Autoantibodies (titers by immunfluorescence on rodent tissues)	
ANA, SMA or LKM-1	
> 1:80	+ 3
1:80	+ 2
1:40	+ 1
< 1:40	0
Antimitochondrial antibody	
Positive	–4
Hepatitis viral markers	
Positive	–3
Negative	+ 3
Drug History	
Positive	–4
Negative	+ 1
Alcohol intake (average consumption)	
<25 g/d	+ 2
>60 g/d	–2
Other autoimmune diseases	+ 2
Liver histology	
Interface hepatitis	+ 3
Predominantly lymohocytic infiltrate	+ 1
Rosetting liver cells	+ 1
None of the above	–5
Biliary changes	–3
Other changes	–3
Optioal additional parameters:	
Seropositivity for other *defined* autoantibodies	+ 2
HLA DR3 or DR4	+ 1
Response to therapy:	
Complete	+ 2
Relapse	+3

Interpretation of aggregate scores:
Pre-treatment:

Definite AIH	>15
Probable AIH	10–15

Post-treatment:

Definite AIH	>17
Probable AIH	12–17

Abbreviations used: immunglobulin G (IgG), anti-nuclear antibodies (ANA), smooth muscle antibodies (SMA), liver microsomal autoantibodies type 1 (LKM-1), human leukocyte antigen (HLA).
The scoring system shown is according to Alvarez et al., 1999 (11).

Table 2 Classification of chronic hepatitis on the basis of etiology

Hepatitis type	HBsAg	anti-HBV HBV DNA	anti-HDV (HDV RNA)	anti-HCV (HCV RNA)	Autoantibodies
Virus Hepatitis					
B	+	+	–	–	–
D	+	+	+	–	10–20% LKM3
C	–	–	+		2–10% LKM
					2% anti-CYP2A6
Autoimmune hepatitis					
type 1	–	–	–	–	ANA/SMA, pANCA
type 2	–	–	–	–	LKMl, LC-1, LKM3
type 3	–	–	–	–	SLA/LP
Drug-induced hepatitis	–	–	–	–	Some: ANA, LKM, LH
					anti-CYPI A2, anti-CYP2C9
					anti-CYP3A, anti-CYP2EI
Cryptogenic hepatitis	–	–	–		–

Classification shown was modified according to Desmet et al. [12]

Abbreviations used: anti-nuclear antibodies (ANA), smooth muscle antibodies (SMA), soluble liver antigen antibody (SLA), liver-pancreas antigen antibody (LP), liver/kidney microsomal autoantibodies (LKM), cytochrome P450 (CYP), liver cytosolic protein type 1 (LC-1).

infection with a specific virus, multiple agents may exist that trigger autoimmune processes against hepatocytes.

2. AUTOANTIBODIES IN AUTOIMMUNE HEPATITIS

2.1. Antinuclear and Anti-Smooth Muscle Antibodies

Typical of AIH type 1 are significant titers of antinuclear (ANA) and smooth muscle antibodies (SMA) (12,34) (Figure 2A,B). ANAs are most frequently determined by indirect immunofluorescence on Hep-2 cells. Frequently used cut-off titers for autoantibody positivity are 1:40 for ANA and 1:80 for SMA. Antibody patterns detected in AIH type 1 are variable. In Caucasians, ANA alone are detected in 15% of sera, SMA alone in 35% and both antibodies in 49% of sera (14). ANA production was associated more commonly with HLA-DR4 and with a lower frequency of liver transplantation (35,36).

Not all ANA fluorescence patterns look alike. Dependent on the composition of antigens detected, different patterns of fluorescence are found. The majority of patient sera show either a speckled (38%) or a diffuse (34%) staining of the nuclei. Other patterns are relatively rare: Centromeric (6%), diffuse granular (4%), nucleolar (3%) or mixed (10%) (37). Clinically, speckled patterns are associated with a younger age of onset, with HLA A1-B8-DR3 and higher aminotransferase activity (37). Molecular targets recognized by ANAs are very heterogeneous, including single-stranded DNA (ss-DNA), double-stranded DNA (ds-DNA), snRNPs,

tRNA, lamins A and C, histones or even cyclin A (37–40). Centromere (42%) and the 52 K ribonucleoprotein complex (23%) were recognized with a high prevalence in AIH type 1 (37). ANA patterns could not be correlated with the detection of specific proteins and ANAs usually recognize more than one molecular target (37). Patterns of ANA in indirect immunofluorescence did not correlate with clinical outcome and lacked practical implications (37).

Autoantibody patterns in AIH type 1 are strongly fluctuating. During immunosuppressive treatment, disappearance of one or both autoantibodies are noted in about 76% of patients. Loss of autoantibodies is associated with improved laboratory tests, however, disappearance is not prognostic for a positive outcome of treatment (41). Neither autoantibody titers at presentation nor autoantibody behavior during therapy are accurate markers for severity or prognosis of disease. These results indicate that both ANA and SMA are not pathogenic.

Anti-histone autoantibodies. In American patients, anti-histone antibodies were detected in 35% of patients with AIH type 1. 52% of anti-histone-positive patients were reactive only to histones, while the others recognized multiple nuclear antigens. Presence of anti-histone autoantibodies did not correlate with a specific pattern in immunofluorescence. Furthermore anti-histone antibodies failed to identify a subgroup of patients with differences in age, gender, clinical or laboratory findings, HLA-antigens or prognosis (42).

In Japanese patients with AIH type 1, reactivity to individual histones was studied. Japanese patient sera also

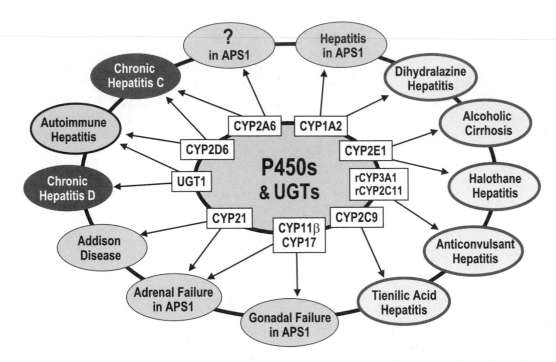

Figure 1.   Cytochromes P450s and UDP-glucuronosyltransferases are targets of autoimmunity in hepatic and nonhepatic immune-mediated diseases
  Abbreviations used in this figure are autoimmune polyglandular syndrome type 1 (APS1), cytochrome P450 (CYP) and UDP-glucuronosyltransferase (UGT).
  This figure was modified from (6).

recognized histones with a high prevalence (40%). Most anti-histone autoantibodies were directed against anti-H3 and were of the IgG type. These autoantibodies decreased significantly after immunosuppressive treatment and this decrease correlated with serum aminotransferase (43). In contrast to the American study, Japanese patients with anti-histone antibodies were characterized by higher ALT values, IgG and higher prevalence of the HLA A2-DR4 haplotype. However, in Japanese patients, ANAs or anti-histone antibodies failed to distinguish patients with special clinical features (43).

Anti-DNA autoantibodies.   Antibodies directed against ds-DNA are detected in autoimmune hepatitis with a prevalence of 34% or higher (44–46). When anti-ds-DNA antibodies were either detected by ELISA or by a *Crithidia luciliae* substrate-based immunofluorescence assay, anti-ds-DNA positive patients were characterized by higher IgG levels and higher prevalence of HLA DR4. Patients positive by the ELISA assay failed corticosteroid therapy more commonly, however no difference was noted in treatment failure when anti-ds DNA autoantibodies were determined by the *Crithidia* assay (46). Autoantibodies to single-stranded DNA were present in 85% of patients. No clinically relevant correlations were associated with antibodies to ss-DNA (46).

Anti-cyclin A autoantibodies.   Cyclins play a central role in the regulation of cell cycle, DNA transcription and cell proliferation. Anti-cyclin A autoantibodies were detected in 20% of patients with AIH type 1 and in 14% of patients with rheumatic diseases. Prevalence of anti-cyclin antibodies in AIH type 3 was similar to prevalence in healthy controls, with 7% and 9% anti-cyclin A-positive sera, respectively. Anti-cyclin A autoantibodies were not detected in any of 18 sera from patients with AIH type 2. This result demonstrates that there is little overlap between ANA and LKM1 autoantibodies (40).

Anti-actin autoantibodies.   SMA are present in 84% of American Caucasian patients with AIH type 1. Among SMA-positive patients in AIH type 1, 86% had antibodies directed against actin (47). Nonactin antibodies detected by SMA immunofluorescence are usually directed against tubulin and intermediate filaments (48). SMA antibodies in general are associated with the HLA A1-B8-DR3 haplotype and, possibly as a reflection of the HLA status, these patients are younger and had a poorer prognosis. Interestingly, association of actin-positive patients with the HLA A1-B8-DR3 haplotype was stronger than association of SMA with this HLA haplotype. Non-actin-positive patients with SMA showed a closer association with HLA-DR4 (47). Furthermore, a recent report described actin as a target antigen of c-ANCA autoantibodies (49).

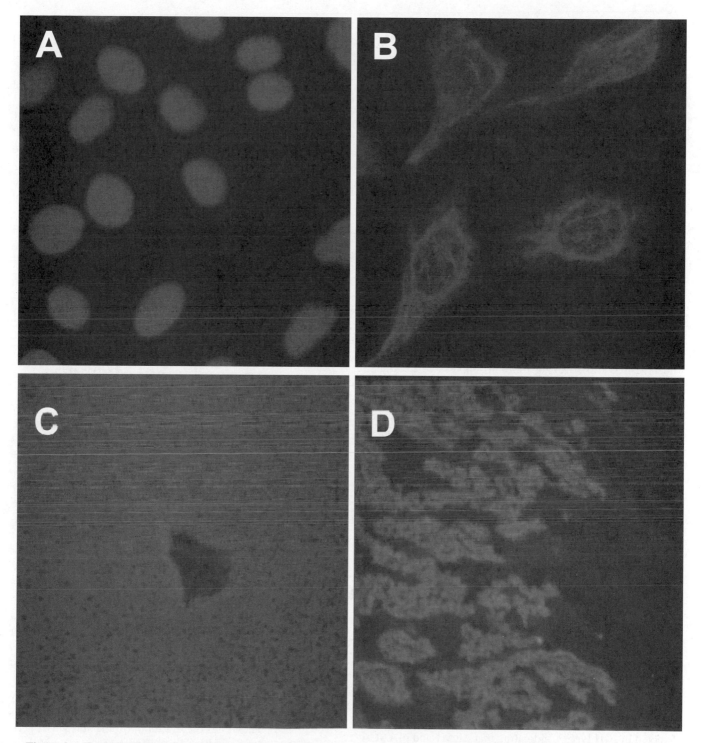

Figure 2.   Typical fluorescence patterns of autoantibodies detected in autoimmune hepatitis
A. antinuclear antibodies (ANA), diffuse, detected on HepG2 cells
B. anti-smooth muscle antibodies (SMA), detected on HepG2 cells
C. anti-liver/kidney microsomal (LKM1) fluorescence pattern, detected on cryostate sections of liver tissue
D. LKM1 fluorescence pattern, detected on cryostate sections of kidney tissue

2.2.   Anti-Neutrophil Cytoplasmic Antibodies

Antineutrophil cytoplasmic antibodies (ANCA) are directed against cytoplasmic components of neutrophilic granulocytes. By indirect immunofluorescence, two major subtypes can be distinguished: c-ANCAs, which show a cytoplasmic staining pattern and p-ANCAs, characterized by a perinuclear staining pattern. C-ANCAs were described in Wegener's granulomatosis and p-ANCAs in systemic vasculitis, systemic lupus erythematosus, ulcerative colitis, primary sclerosing cholangitis and autoimmune hepatitis (50). A main autoantigen associated with c-ANCA is proteinase-3, which may be detected in Wegener's granulomatosis or other forms of systemic vasculitis (51). Target antigens for p-ANCAs are myeloperoxidase (52,53) and elastase (54) in systemic vasculitis, and both cathepsin-G (55) and lactoferrin (56) in inflammatory bowel disease. P-ANCA are detected with a high prevalence in patients with autoimmune liver diseases (20,49,57–61). In an early report, p-ANCA were detected in 79% of patients with primary sclerosing cholangitis (PSC), in 88% of patients with AIH and in 28% of patients with primary biliary cirrhosis (PBC) (20). During immunosuppressive treatment, p-ANCA did not decrease significantly in titers and no correlation between p-ANCA titers and transaminase levels or duration of disease was detected. A second study confirmed the high prevalence of p-ANCA in AIH type 1. In spite of extremely high mean titers of p-ANCA (11410 ± 1875), again no correlation of p-ANCA with AST, γ-globulin concentration or IgG levels or presence and absence with concurrent autoimmune disorders was found (59). Furthermore, titers of p-ANCA and ANA did not correlate with each other, suggesting that they are independent phenomena. In contrast to the earlier study (20), no correlation with histological patterns of disease activity or cirrhosis was noted (59). P-ANCA may help in differential diagnosis, since p-ANCA in AIH type 1 differ from p-ANCA in PSC in two respects: 1.) Titers in AIH type 1 (11410 ± 1875) are significantly higher than in PSC (225 ± 50) (P<0.001) (59,62), and 2.) 80% of p-ANCA in AIH type 1 are only of the IgG1 subtype, while sera from patients with PSC are positive for IgG1 and IgG3 (59). Interestingly, p-ANCA seem to be more frequent in males than in females (P = 0.009) (60). While clearly p-ANCA were shown to be a frequent marker in AIH type 1 this antibody is not detected in AIH type 2, indicating that LKM1 and p-ANCA are mutually exclusive (60).

Molecular targets of P-ANCA and C-ANCA in autoimmune hepatitis. Recently, anti-actin autoantibodies were described as an important autoantigen of c-ANCA autoantibodies (49). In this study, ANCA were detected in 75% of patients with AIH type 1 with a prevalence of c-ANCA of 31% and a prevalence of p-ANCA of 44%. C-ANCA staining in indirect immunofluorescence associated with a 43 kDa band in Western Blots that was determined as actin by protein sequencing. Absorption experiments demonstrated that c-ANCA staining was indistinguisable with the staining pattern revealed by a specific anti-actin antibody. Furthermore, preabsorption of sera from patients with AIH type 1 resulted in a reduction of c-ANCA reactivity in all c-ANCA-positive sera and in 80% of p-ANCA-positive sera (49).

In AIH type 1, a major target of p-ANCA was identified with high mobility group (HMG) non-histone chromosomal proteins, HMG1 and HMG2 (61). HMG1 and HMG2 are transcription factors and detected in cytoplasm and nuclei of eucaryotic cells (63). Anti-HMG1/2 autoantibodies are detected in 89% of patients with AIH type 1, 70% of patients with PBC, 26% of patients with chronic hepatitis C and 9% of patients with chronic hepatitis B. HMG1/2 autoantibodies were detectable in 96% of AIH sera containing pANCA and anti-HMG1/2 titers correlated significantly with p-ANCA titers. Preincubation of sera with HMG1/2 proteins was able to abolish detection of p-ANCA in immunofluorescence indicating that HMG1/2 proteins are major target proteins of p-ANCA (61). In contrast, no correlation between anti-HMG1/2 autoantibodies and titers of ANA and SMA were detectable.

2.3.   Anti-Asialoglycoprotein Receptor Autoantibodies

The asialoglycoprotein receptor (ASGP-R) is a hetero-oligomeric glycoprotein located in the plasma membrane of hepatocytes. It permits the internalization of asialoglycoproteins by binding of a terminal galactose residue to coated pits, which pinch off as coated vesicles. Inside the cell, clathrin coats are quickly removed and the vesicles are believed to be delivered to an endosomal compartment (64). The asialoglycoprotein receptor is rescued and reshuttled to the cell surface. On the cell surface about 100,000–500,000 ASGP-R molecules have been determined that are exclusively expressed on the sinusoidal surface hepatocytes (65).

Autoantibodies directed against ASGP-R are frequently detected in sera from patients with chronic liver diseases (19,66–68). They bind preferentially to the surface of periportal hepatocytes, where liver damage is observed histologically. Therefore, it was proposed the anti-ASGP-R autoantibodies may play a pathogenic role in AIH (69). Prevalence of anti-ASGP-R autoantibodies was 76% in patients with AIH, 11% of patients with viral hepatitis, 19% in patients with PBC and 8% in patients with other liver diseases (67). In spite of differences in the clinical profile and in HLA associations in AIH type 1 between Caucasian and Japanese patients, prevalence of anti-ASGP-

R autoantibodies was similar in both ethnic groups (67). Titers of anti-ASGP-R autoantibodies were significantly higher in AIH than in PBC or chronic viral hepatitis (66,67,70). In autoimmune hepatitis, titers of anti-ASGP-R autoantibodies correlated with disease activity. Mild disease as determined by histological parameters associated with lower titers anti-ASGP-R autoantibodies than moderate or severe disease (66).

Biochemically, an association between low AST values and low anti-ASGP-R titers was found, however, this association did not extend to a direct quantitative relationship between both titers. This finding indicates that both parameters are independent correlates of disease activity (67). Similarly, during immunosuppressive treatment titers of anti-ASGP-R autoantibodies sharply decreased (67,70–72). An early study performed with a crude preparation of liver-specific protein. containing anti-ASGP-R as a major antigenic component, even suggested that anti-liver specific protein antibodies can predict outcome of chronic active hepatitis (71,73). Recently, a significant correlation between positive anti-ASGP-R titers and elevated soluble interleukin 2-receptor values was demonstrated (72). When epitope mapping was performed for anti-ASGP-R autoantibodies, autoantibodies were only able to recognize native glycosilated proteins, suggesting that the epitope is conformational and probably contains carbohydrate chains (74).

Cellular immune response to ASGP-R. In autoimmune hepatitis-activated T cells are found in the hepatic infiltrates, where hepatocytes were shown to express large amounts of HLA class I and II molecules (75–77). As age of onset and longterm prognosis in AIH correlate with HLA DR, it is believed that T-cell-mediated processes may play a major role in the pathogenesis of AIH (23). Next to a humoral response directed against ASGP-R in AIH also a cellular response directed against the ASGP-R was noted (78). Peripheral blood lymphocytes of 37% of patients with autoimmune hepatitis showed a proliferative response to purified human ASGP-R. In contrast, no proliferative was induced by human ASGP-R in peripheral blood lymphocytes from patients with chronic viral hepatitis and healthy controls. When liver biopsy samples or peripheral blood lymphocytes were used to establish T cell clones, between 2.8% and 14.3% of clones showed specific proliferative response to purified human ASGP-R (78,79). This reaction was restricted to autologous antigen-presenting cells and was blocked by antibodies to HLA DR (78).

2.4. LKM Autoantibodies

LKM1 autoantibodies are the serological markers of AIH type 2 (11,12,80). These autoantibodies were first described by Rizzetto et al. using indirect immunofluorescence on rodent liver and kidney sections (81). The characteristic feature of LKM1 autoantibodies is an even cytoplasmic staining of the whole liver lobule and a staining of proximal renal tubules (Figure 2 C,D). Western blots with hepatic and renal microsomes reveal a protein band at 50 kD.

Cytochrome P450 2D6 (CYP2D6) is the major antigen for LKM1 autoantibodies (82–84). Anti-CYP2D6 autoantibodies are found in 95–100% of patients with AIH type 2 (85–87). This reaction is highly specific for CYP2D6, since autoantibodies failed to cross-react with any of ten different closely-related members of the cytochrome P450 supergene family (Figure 3). CYP2D6 is an enzyme active in the phase 1 of detoxification. It is involved in the detoxification of at least 40 different drugs (88). CYP2D6 shows a significant polymorphism. The absence of functional CYP2D6 results in a low metabolizer phenotype for debrisoquine or sparteine, both substrates for CYP2D6 (89).

In vitro, the enzymatic activity of CYP2D6 is inhibited by LKM1 autoantibodies (83,90). Studies with sparteine as test substrate in patients permitted measurement of CYP2D6 activity *in vivo*. All patients with LKM1 antibodies tested were of the extensive metabolizer phenotype and expressed functionally-intact cytochrome P450 2D6 protein (90,91). Hence, adequate expression of CYP2D6 seems to be a prerequisite for the development of LKM1 autoantibodies. The epitopes recognized by LKM1 autoantibodies have been characterized. Autoimmune hepatitis sera recognize at least four different linear as well as additional conformational epitopes on CYP2D6, namely amino acids 257–260, 321–351, 373–389 and 419–429 (28,85,92,93) (figure 4). The major epitope recognized is a peptide of amino acids 257–269, which is recognized in about 70% of patients with AIH type 2, but only in 20% of patients with LKM1-positive chronic hepatitis C (85,94). Interestingly, this major recognition site of LKM1 binding shows a six amino acid sequence identity with an immediate early protein (IEP175) of herpes simplex virus (28) (Figure 4). A pair of identical twins was reported who showed discordant

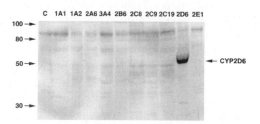

Figure 3. LKM1 autoantibodies detect CYP2D6 but do not recognize other closeley related human CYPs
Abbreviations: CYP 1A1 (1A1), CYP 1A2 (1A2), CYP 2A6 (2A6), CYP 3A4 (3A4), CYP 2B6 (2B6), CYP 2C8 (2C8), CYP 2C9 (2C9), CYP 2C19 (2C19), CYP 2D6 (2D6), CYP 2E1 (2E1).
Numbers of the left side of this Western Blot show the position and size of several molecular weight markers in kDa.

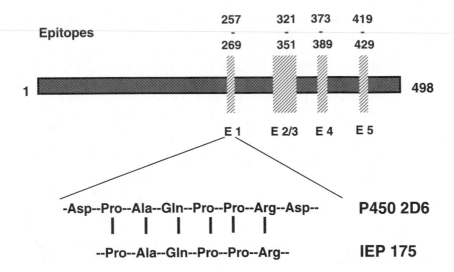

Figure 4.   Linear epitopes recognized by LKM1 autoantibodies
The main epitope E1, recognized by 70% of patients with AIH type 2 shows a six amino acid sequence identity with the immediate early protein 175 (IEP175), a transcription factor of herpes simplex virus type 1. This figure was modified from (6).

manifestation of LKM1-positive AIH, with only one sister affected by the disease. Interestingly, the healthy twin was negative for herpes simplex virus (HSV) infection, while the child affected by AIH type 2 had antibodies directed against herpes simplex virus (21,28). Anti-CYP2D6 specific antibodies were isolated from the girl with AIH and used to immunoprecipitate proteins from HSV-infected cells. By this experiment, it was demonstrated that LKM1 autoantibodies in this patient were able to precipitate HSV IEP175 (21,28). This, may indicate that viral infections by HSV may trigger autoimmune hepatitis by the mechanism of molecular mimikry (21,28). Once a self-perpetuating autoimmune process is induced, epitope spreading may occur, explaining the detection of four different linear epitopes on CYP2D6 and also the induction of autoantibodies directed against conformational epitopes on CYP2D6, which are responsible for the inhibitory activity of LKM1 antibodies (93). The identification of a major three-dimensional epitope was reported recently (85).

LKM1 autoantibodies are widely used as diagnostic markers for AIH type 2 or for autoimmunity associated with chronic hepatitis C (12,95). However, the role of these autoantibodies in liver cell injury is not known. Recently, surface expression of different CYPs was demonstrated by several labs (96–101). Therefore, it is possible that LKM1 autoantibodies may induce liver cell injury by direct binding of LKM autoantibodies to hepatocytes. This may result in lysis of hepatocytes either by complement or by antibody-directed cell-mediated cytotoxicity (ADCC).

Cellular immune response to CYP2D6   Liver cell injury may also be induced by infiltrating T-lymphocytes. Löhr et al. (102) investigated the T-cell response to CYP2D6. One hundred eighty nine T-cell clones were generated from four

liver biopsies of LKM1 autoantibody-positive patients. Approximately 85% of the T-cell clones were were CD4+CD8−. Five CD4+CD8− T cell clones specifically proliferated in the presence of purified recombinant CYP2D6. This reaction was dependent on autologous antigen- presenting cells and HLA class II molecules, indicating that CYP2D6-specific T_H-cells are part of the lymphocytic infiltrate of livers of patients with AIH type 2. Furthermore, CYP2D6 was also able to induce proliferation in peripheral blood lymphocytes (PBL). PBL from 3/4 patients proliferated in response to CYP2D6, while no proliferative response was detectable in LKM1-negative patients with chronic liver diseases and in healthy controls (102). Then peptides of the immunodominant epitope of CYP2D6 were used for stimulation. A peptide of aa 262–285 of CYP2D6 induced a proliferative response in 8/8 patients with AIH type 2, in 6/12 patients with LKM1 negative AIH type 1, and in 4/31 patients with chonic hepatitis C. After start of immunosuppressive treatment, T-cell responses to this epitope decreased. Sixteen CD4+-specific T-cell lines were generated and shown to release predominantly interferon-γ, but no or little interleukin-4. These results are in accordance with a Th1-phenotype of these CYP2D6 specific T-cell clones (103).

LKM3 autoantibodies.   In about 10% of patients with AIH type 2 next to CYP2D6 autoantibodies, LKM3 autoantibodies may be detected directed against the 1 UDP-glucuronosyltransferase family (UGT1(104–106). LKM3 antibodies were first described in 10–20% of patients with chronic hepatitis D (107). Epitope mapping of UDP-glucuronosyltransferases in autoimmune hepatitis revealed a large minimal epitope from amino acids 264–373, indicating that the autoantibody is binding to conformation dependent epitopes (108).

LM autoantibodies.   Antimicrosomal antibodies that react only with liver tissue and recognize CYP1A2 are called LM antibodies (39,109,110). LM antibodies are found in drug-induced hepatitis caused by dihydralazine and in autoimmune hepatitis as part of APS-1 (39,109,110). In Finish patients, there does not seem to be a serological overlap between LKM1 antibodies in autoimmune hepatitis type 2 and LM antibodies in APS-1 hepatitis (111).

2.5.   Antibodies to Liver Cytosolic Protein Type 1 (LC1)

In indirect immunofluorescence, anti-LC1 autoantibodies are characterized by a cytoplasmic staining of periportal but not perivenal compartments, indicating that the target is not evenly distributed (16,112) In gel filtration, the target protein appears at 240–290 kD. However, Western Blots under denaturing conditions with human LC1 autoantibodies reveal 58–62 kDa target protein. These results indicate polymer formation of the native target protein (112). The target protein of anti-LC1 autoantibodies was recently identified as the formiminotransferase cyclodeaminase (113).

Antibodies to liver cytosol type 1 (anti LC1) are closely associated with AIH type 2. Fifty percent of patients with anti-LKM autoantibodies also produce anti-LC1 autoantibodies as a second disease marker (114). However, both anti-LKM1 and anti-LC1 autoantibodies may be present as only disease markers in patients with AIH (16). Anti-LC1 autoantibodies are found in LKM1-positive sera with and without chronic hepatitis C (114,115). Anti-LC1 autoantibodies were absent in 100 anti-actin-positive patients with AIH type 1, 100 patients with primary biliary cirrhosis and 157 patients with drug-induced hepatitis and 100 healthy controls (16). It was reported that LC-1-positive patients with HCV-negative AIH tend to be younger at disease onset and may have higher AST values at presentation (16). Compared to LKM1 autoantibodies, titers of LC-1 autoantibodies seem to correlate better with disease activity. When titers of anti-LC1 were compared at onset of AIH and after induction of remission by immunosuppressive therapy, it showed that anti-LC1 titers had considerably decreased or were even lost in two patients. Anti-LC1-titers determined at onset of disease were considerably higher than anti-LC1 titers detected after remission. This result indicates that anti-LC1 autoantibodies may play a pathogenic role in AIH (116).

2.6.   Anti-SLA/LP Autoantibodies

AIH type 3 was proposed to comprise a group of patients with autoantibodies directed against the soluble liver antigen (SLA), which recognize one or two bands of about 50 kDa (17). SLA autoantibodies closely associated with autoimmune hepatitis (17,117,118). Patients with anti-SLA are young and predominatly female (17). Severity of disease and clinical markers of SLA patients are similar to those with AIH type 1 (34). Approximately 30% of AIH patients are SLA-positive and some overlap with SMA and ANA, however no overlap with LKM1 autoantibodies was found (118). Antibody overlap with ANA and SMA and similar-ities in the clinical profile of patients with AIH1- and SLA-positive hepatitis suggest that anti-SLA might be an important additional marker in AIH type 1, rather than a marker for a third type of AIH (34).

Conflicting results about the molecular target of SLA were reported. An early report suggested that cytokeratins 8 and 18 might be molecular targets of SLA autoantibodies (119). A recent report suggested members of the glutathione S-transferase family to be recognized by SLA autoantibodies, however the molecular weight of the target proteins was only 25kDa and 27 kDa (119,120). Recently, Wies et al. reported cloning of a novel cDNA that was specifically recognized by all SLA-sera tested and was shown to be able to absorb SLA immunoreactivity. The molecular target is 50 kDa protein with unknown function, present in the S100-supernatant of hepatocytes. The two protein targets detected by SLA proteins might be explained by the existence of two splice variants of this new target protein (121).

LP was described as a marker antigen for autoimmune hepatitis, reacting with liver and pancreas tissues. The target protein was identified by a band of 52 kDa and 48 kDa molecular weight (18). This antigen was predominantly detected in the S100 supernatant of liver and pankreas homogenates, indicating that the liver-pancreas (LP) antigen was a soluble protein. One hundred-eleven sera positive for anti-LP autoantibodies were detected, 86 recognized the 52 kDa protein, 33 with the 48 kDa determinant and 2 with both. Overlap with SMA and ANA was reported. These characteristics are similar to the SLA antigen. Also, anti-LP autoantibodies were proposed to identify a subgroup of patients with a different type of AIH (18). Recently, anti-LP sera were shown to react with a protein derived from a recombinant SLA cDNA (121).

3.   DIFFERENT SUBTYPES OF AUTOIMMUNE HEPATITIS

Different types of autoimmune hepatitis were distinguished by autoantibody patterns: AIH type 1 was characterized by ANA and SMA, AIH type 2 was characterized by LKM1, LKM3 and LC-1 antibodies and AIH type 3 were associated with SLA/LP autoantibodies (11,12) It was shown that patients with SLA antibodies have clinical profiles similar to patients positive for ANA/SMA and therefore a distinction only of AIH type 1 and AIH type 2 is preferred (34) (Table 3). Patients with AIH type 2 are characterized by LKM1 and/or anti-LC1 autoantibodies (15,16,112). At disease onset, patients with AIH type 2 are younger and had greater disease activity. AIH type 2 patients relapse more frequently under immunosuppressive therapy and progress to cirrhosis with a higher prevalence. Gregorio et al. (122) reported that no sustained remission without immuno-suppressive therapy was registered in a 20 year experience with patients with AIH type 2, while about 20% with AIH type 1 experienced sustained remission after weaning off immunosuppression (34). Similarly, even as only low numbers of patients with AIH type 2 were typed for HLA antigens, it becomes clear that different HLA-antigens are associated with AIH type 2 than with AIH type 1. These results may indicate a different ethiology of the two subtypes of AIH (34). Of special interest is the fact that LKM1 autoan-tibodies are a rare event in the United States and Australia and were not detectable in Swedish patients with AIH (123,124). These geographical differences indicate different abundance of risk factors or triggers for AIH type 2 in the United States, Australia and Sweden compared to Japan, Germany and the Mediterranian countries.

Table 3   AIH Type 1 and AIH Type 2 are distinguished by serological, clinical and genetic parameters

Distinction	AIH-1		AIH-2	
autoantibody pattern	heterogeneous and fluctuating autoantibody patterns		few well defined autoantibodies	
	ANA.	multiple and heterogeneous molecular targets	LKM-1:	CYP2D6
	SMA.	anti-actin, anti tubulin and anti-intermediate filaments	LKM-3:	UGT1A
	pANCA.	HMG1, HMG2, cathepsin G, minor target: lactoferrin	LC-1:	formiminotransferase cyclo-deaminase
	SLA/LP.	cloned protein of 50 kDa of unknown function glutathion-S-transferases ?		
age at onset	predominantly adult		predominantly pediatric	
	HLA DR3. correlation with early onset AIH (< 30 years)			
	HLA DR4. correlation with late onset AIH (> 30 years)			
serological profile at presentation	lower total bilirubin		higher total bilirubin	
	lower AST		higher AST	
	lower GGT		higher GGT	
	higher IgA		lower IgA	
	low C4		low C4	
response to immunosuppressive treatment	in general good		in general good	
	20% sustained remission after		no sustained remission after	
	discontinuation of treatment		discontinuation of treatment	
	HLA DR4: better response		frequent progression to cirrhosis HLA	
	DR3: more frequent relapses more frequent progression to cirrhosis than HLA DR4			
genetic predisposition	HLA Al-B8-DR3 and HLA DR4		HLA DR3, HLA DQ2	
	female gender		female gender	
association with extrahepatic auto-immune diseases	20–30%		20-30%	
	more common in adult patients with DR4 than in adult patients with DR3			

Modified from Czaja and Manns [34].
Abbreviations used: anti-nuclear antibodies (ANA), smooth muscle antibodies (SMA), anti-neutrophil cytoplasmic autoantibodies (ANCA), soluble liver antigen antibody (SLA), liver-pancreas antigen antibody (LP), liver/kidney microsomal antibody (LKM), cytochrome P450 (CYP), human leukocyte antigen (HLA), immunoglobulin A (IgA), complement factor 4 (C4).

Table 4   Clinical profiles of patients with autoimmune hepatitis and different HLA DR profiles.

Features at presentation	HLA DR3 n = 41	DR4 n = 44	HLA DR3 and DR4 negative, n = 16
Age (yr)	381 ± 3	51 ± 2	40 ± 4
Sex (FIM)	28:13 (68%)	39: 5 (89%)	6:10
Immunglobulin G	2732 ± 192	3300 ± 216	2722 ± 358
Concurrent immunologic diseases	27%	59%	13%
Autoimmune thyroiditis	3/41	9/44	2/16
Graves' disease	1/41	3/44	0/16
Ulcerative colitis	2/41	3/44	1/16
PerniCius anemia	2/41	0/44	0/16
Others	2/41	8/44	0/16
mMultiple	2/44	5/44	0/16
Outcome			
Remission	24/38 (63%)	34/40 (85%)	8/12 (75%)
Treatment failure	12/38 (31%)	4/40 (10%)	3/12 (25%)

Data included in this table were modified from [139].

4.   GENETIC PREDISPOSITION FOR AIH

4.1.   MHC Receptors Confer Increased Risk for Autoimmune Diseases

All autoimmune responses involve T cells, which only recognize specific peptides in the context of MHC antigens. MHC antigens do not only play a central role in immune defense against intracellular and extracellular parasites. Evidence exists that MHC molecules also play a role if misguided immune responses cause tissue damage in the absence of infection (22,23). Conditions in which the immune system attacks self-targets may result in autoimmune diseases and tend to be associated with the production of antibodies directed against various self-antigens (125). Even as the pathological role of MHC molecules in autoimmune diseases remains unclear, associations of specific HLA alleles have been reported consistently. Examples are HLA DR3 in Grave's disease, Myastenia gravis, systemic lupus erythematosus or HLA DR4 in rheumatoid arthritis and pemphigus vulgaris (126). These associations were found in Caucasian populations. Since HLA antigens are highly polymorphic and show extensive variability between humans of different ethnic groups, associations of HLA-alleles with autoimmunity reported in a Caucasian population may significantly differ from HLA associations found in Japanese, Mexicans or Argentinians (127–129).

HLA antigens in AIH type 1.

Clinical relevance of HLA antigens in autoimmune hepatitis type 1. HLA A1, B8 and DR3 were identified as susceptibility markers for AIH type 1 in Caucasians (130–136). Due to the localisation of these genes in close proximity to each other in the MHC complex, these alleles are subject to linkage disequilibrium and form the tightly-linked A1-B8-DR3 haplotype. This linkage disequilibrium makes it very difficult to decide whether the MHC alleles themselves are the true susceptibility markers or whether they are located close to one or several real risk factors.

In Caucasian AIH type 1 patients, the A1-B8-DR3 haplotype is strongly overrepresented (38% AIH type 1 vs 11% controls, p < 0.0005) (137). When all DR3-positive patients are eliminated from the evaluation, a secondary association of DR4 with AIH type 1 is detected (80% vs 39%, pc = 0,0013) (137). HLA DR alleles not only conferred increased risk for AIH type 1, but were also associated with different clinical profiles of the patients (137–139). Patients with A1-B8-DR3 were seen at significantly younger age, (39.75 yr vs. 48.21 yr, p > 0.025) (Figure 5), relapsed more frequently under immunosuppressive treatment (52% A1-B8-DR3 positive vs. 34% A1-B8-DR3 negative) and were more frequently referred to liver transplantation. (137). A large study was performed in 101 American Caucasians detected an association of HLA DR4 with the presence of concurrent autoimmune diseases. Autoimmune diseases

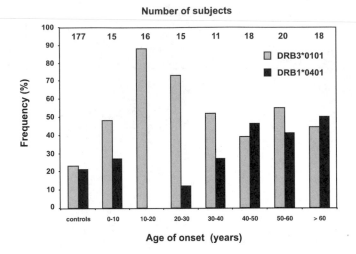

Figure 5.   Association of HLA DR3 and HLA DR4 with age of onset of autoimmune hepatitis.
This figure was modified from (35).

associated with the highest prevalences were autoimmune thyroiditis, Grave's disease, ulcerative collitis and pernicious anemia (139). Concurrent autoimmune diseases are also seen in chronic viral hepatitis (CVH) (HBV: n = 16; HCV: n = 46). Even as concurrent autoimmune diseases were more prevalent in patients with AIH type 1 (38% AIH vs 22% CVH, p = 0.04), the nature of autoimmune diseases was similar in both AIH and CVH. Interestingly, in CVH patients concurrent autoimmune diseases were strongly associated HLA DR4 (79% CVH vs 29% controls, p = 0.009) (140).

With the introduction of PCR-based methods, HLA typing at very high resolution became available. In British patients, it was shown that of DR3-linked alleles, DRB3*0101 predisposed patients most strongly to the disease (RR 4,2). In DR3-negative patients, the prevalence of DRB1*0401 was raised. Eighty-one percent of all patients with AIH were either DRB3*0101- or DRB4*0401-positive compared with 42% of controls (P < 0.0000001, RR 5,1). Heterozygosity for DR3 and DR4 was not associated with an additionally increased risk of hepatitis.

Furthermore, effects of homozygosity were different for DRB1*0401 and DRB3*0101. While no additional effect was recorded for homozygosity of DRB1*0401, homozygosity for DRB3*0101 increased the relative risk from 4.2 to 14.7 and was associated with a reduction of 10 yr survival (35). DRB3*0101 correlates with the clinically unfavorable outcome of AIH type 1 described previously for HLA DR3-positive patients (35,137,139). A large study in Northern American patients, however, associated the greatest risk with DRB1*0301 and not DRB3*0101, as reported in an earlier British study. The secondary association with DRB1*0401 was confirmed and, for the first time, the

DRB5*0101-DRB1*1501 was identified as protective haplotype (141).

Studies on HLA Cw and HLA DP.   The influence of the HLA DP locus was weak. None of 17 DB1 alleles was significantly associated with the susceptibility to AIH type 1. Only in young patients, first seen before the age of 16 years, was an increased risk associated with an extended seven-locus haplotype (A1-B8-DRB3*0101-DRB1*0301-DQA1*0501-DQB1*0201-DPB1*0401) compared to a six-locus haplotype without HLA DP (142). For HLA C alleles, an increased frequency of HLA Cw*0701 was noted in the AIH type 1 group. Currently it is not clear whether this effect is due to a linkage disequilibrium of Cw*0701 with both B8 and DRB1*0301 or whether this allele itself may contribute to disease susceptibility (143).

The Role of a Lysine at Position 71 of the HLA DR Receptor.   To evaluate whether clinical features can be correlated with the molecular structure of MHC class II molecules, the β-domain amino acids of HLA-DR and HLA-DQ proteins were deduced in patients and controls. The positions on these molecules, where specific residues conferred a relative risk (RR) of 6-fold or higher, were Glu9, Leu67, Lys71 in the DRβ polypeptide, which are encoded by DRB1*0301, DRB1*0401 and DRB3*0101. Amino acid 71 was identified as the primary determinant of susceptibility. A lysine at position 71 was present in 94.1% of the patients and in 68.3% of controls, resulting in a relative risk of 9.1. In contrast, alanine, glutamic acid or arginine at this position had a protective effect. The greatest risk for AIH type 1 correlated with specific amino acid substitutions at positions 67 to 72 on the DRβ-polypeptide (Figure 6). Interestingly, when the same analysis was performed with DQβ alleles, only protective amino acids were detected. Therefore, it was concluded that most likely DR rather than DQ-genes confer susceptibility (35). When the same analysis was performed in North American patients with AIH type 1, lysine at position 71 was found to confer the highest risk (RR = 8.6). Furthermore, both protective alleles, namely DRB1*1501 and DRB5*0101, encoded different amino acids at position 71. X-ray crystallography may explain the strong association of susceptibility with lysine 71 (Figure 6). Position 71 is situated on the edge of the α-helix of the DRβ-polypeptide chain, where it may make contact with both the incoming peptides and the T-cell receptor (144). It was hypothesized that MHC alleles associated with increased risk of autoimmunity show a weak binding of the respective self-peptide to class II molecules. In contrast, peptides that reduce the risk for autoimmunity may bind the autoantigenic peptide with high affinity, resulting in deletion or suppression of self-reactive T cells. This would suggest that lysine DRβ71 (DRB1*0301) results in low binding affinity for autoantigenic peptides and alanine DRβ71 (DRB1*1501) favors high affinity binding.

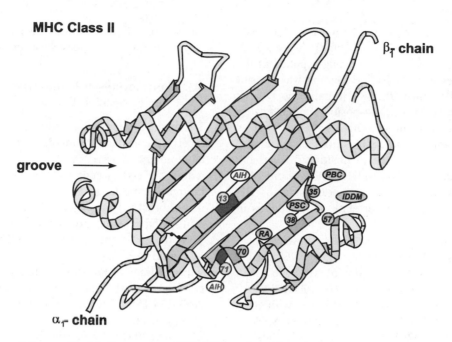

Figure 6. **Structure of the peptide binding groove of HLADR and position of residues that confer increased risk for different autoimmune diseases**
abbreviations: major histocombatibility complex (MHC), autoimmune hepatitis (AIH), rheumatoid arthritis (RA), primary sclerosing cholangitis (PSC), primary biliary cirrhosis (PBC), insulin dependent diabetes melitus (IDDM)

Studies of HLA-DRA/DRB1*0101 have shown that MHC class II molecules may form parallel dimers (144). Dimerisation of two HLA DR molecules may permit simultaneous engagement of two αβ-T cell receptors and facilitate crosslinking, a step necessary for intracellular signalling. By this mechanism, a greater density of lysine bearing DRβ71 molecules may increase chances for the formation of stable dimers and facilitate T cell activation. This hypothesis is in accordance with the finding that relative risk directly correlated with the number of lysine DRβ71 molecules (Table 5).

HLA Associations in the Japanese, Mexican and Argentinian populations. The studies summarized above were performed in Caucasian patients. When HLA typing was performed in Japanese patients, susceptibility was associated with DR4 (127,145–147). One reason for the lack of DR3 association may be a generally low prevalence of this HLA antigen among Japanese patients. Japanese patients were characterized by a strong female preponderance (female:male ratio 8·1) and a late age of onset (32 to 72 years) (127,147). HLA-Bw54, DR4, DR53 and DQ4 were significantly associated with AIH (Bw51: 39,6% vs. 11%; DR4: 88,7% vs 39%; DR 53: 88.7% vs 64,8%; DQ4: 30.2 vs. 10.2%). B54-DR4-DR53-DQ4 is one of the most common haplotypes in the Japanese population. When six DR4-negative patients were further investigated, all were positive for DR2.

Genotyping of HLA-DRB1 genes revealed that only DRB1*0405 was significantly associated with AIH type 1 in Japanese patients. However, the association of DRB1*0405 with AIH type 1 was weaker than of the serological DR4 antigen. DR2 alleles in patients with AIH were

Table 5   Dose-effect of lysine DRβ71 residues

Number of lysines DRβ71	Patients with AIH type 1	Controls	Relative Risk for AIH-1	P
0	7%	39%	0.12	0.0000003
1	27%	31%	0.80	NS
>2	66%	29%	4.72	0.0000004
>3	26%	10%	3.16	0.004 4
4	14%	1%	16.38	0.0000342

Data shown in this table were derived from [141].

the same as in healthy controls. Furthermore, an increased incidence for DQA1*301 and DQB1*0401 were noted. However, these alleles were not as strongly linked as the DR4 antigen, suggesting that DQ association was a result of linkage disequilibrium with DR4. Finally, no DR-BP allele was linked to hepatitis. DR4-positive patients had a lower age of onset than DR4-negative patients (48.9 vs 60.2 years). When amino acid compositions of DR alleles were compared, a basic residue at position 13 was the feature that distinguished DR4 and DR2 alleles from other DR molecules and was suggested to contribute to AIH (127,146). DRβLys71 did not correlate with hepatitis in Japanese patients (127,146).

In *white patients* from Argentinia, linkage of AIH type 1 to HLA alleles was different from Japanese or Caucasian patients (128). A very weak association of AIH type 1 with DR4 was present (44% patients vs 29% controls; RR 2.1; P = 0.02). The major association, however was with HLA A11 (31% in patients vs 6% in controls; RR6.8; Pc = 0.001). Extrahepatic autoimmunity was significantly increased in HLA A11-positive patients (90% HLA A11 vs. 60% total, RR 22.2, Pc = 0.00008). When HLA A11 and HLA DR4 occurred together, a synergistic effect was noted, yielding a relative risk of 357 (128). This effect was not noted for HLA A11 and HLA DR3. When AIH type 1 in childhood was investigated an association with HLA DR6 was noted (128).

In Mexican Mestizo patients, autoimmune hepatitis also associated with a HLA DR4 allele, namely DRB1*0404, which conferred a relative risk of 7.71. Furthermore, the DQB1*0301 allele was found to have a protective effect. As seen in Japanese and Caucasians, the DR4 allele was frequently associated with late onset of AIH. In Mexicans, early onset disease associated with DQA1*0501 (129).

4.2. Polymorphisms in the Human Complement C4 Genes

Early studies showed that AIH type 1 was frequently associated with persistently low serum complement levels (148). C4 phenotyping revealed association of null allotypes at the *C4A* or the *C4B* loci with AIH type 1 and that serum C4 levels are indeed lower in patients with C4A deletions (0.15 ± 0.05 g/l vs 0.23 ± 0.06g/l, P = 0.03) (149–151). Furthermore, *C4A* deletions are associated with increased mortality and a higher tendency to relapse during immunosuppressive treatment (150). Correlation with a younger age of onset in patients with C4A deletions is controversial (149,150). Homozygosity for C4A deletion correlated with a much stronger increase in the relative risk for hepatitis (RR = 18.1) than did a single deletion (RR = 3.3) (149). While the increased risk for AIH type 1 in patients with single gene deletion may be the result of a strong linkage disequilibrium with HLA B3*0101, the strong

increase in risk in homozygotous patients suggests an additional role for C4 in disease susceptibility (150).

4.3. Polymorphisms in the Gene Locus Coding for Tumor Necrosis Factor

The gene for tumor necrosis factor alpha (TNF-α) has received consideable attention as a candidate gene in autoimmune diseases (152–154). Several polymorphisms within the TNF-α gene have been determined. Four polymorphisms are located in the promoter region, one in the first exon (155). A polymorphism at position −308 of the TNF-alpha promoter consists in a biallelic variant, namely TNF308G, designated TNF1 and TNF308A (TNF2). An increased prevalence of the rare TNF2 allele associated with rheumatoid arthritis, systemic lupus erythematosus, coeliac disease and primary sclerosing cholangitis (152–154,156). Recently, a higher incidence of TNF2 was also associated with AIH type 1 (56% TNF2 in AIH versus 26% in controls; P < 0.001). Patients with the TNF2 allele exhibited a worse clinical course, with more patients not responding to immunosuppressive therapy and an increased number of patients progressing to cirrhosis. Although it is very tempting to discuss the TNF2 variant as a progression marker in AIH type 1, the increased prevalence of TNF2 in patients with poor prognoses has to be discussed in the light of a strong linkage disequilibrium of the TNF2 allele with HLA DRB1*0301, which also correlates with a poor clinical outcome of AIH. Therefore, studies in Japanese patients, where AIH-1 associes with HLA DR4, would be of importance to determine whether the TNF2 mutant is an independent disease factor or whether the effect on the clinical outcome of this allele is due to linkage disequilibrium with HLA DR3 (35,127,147).

5. OVERLAP SYNDROMES

Primary Biliary cirrhosis (PBC) is usually not a diagnostic problem due to the detection of antimitochrondrial antibodies against acyltransferases, particularly antibodies to the E2 subunit of pyruvate dehydrogenase. There is a debate whether AMA-negative PBC exists. Several groups have observed patients with clinical and morphological characteristics of primary biliary cirrhosis being negative for antimitochondrial, but positive for antinuclear, antibodies. Such studies do not include enough patients to draw final conclusions. These cases are called autoimmune cholangitis and seem to respond to immunosuppressive treatment. The condition of another group of patients became known as the CAH/PBC overlap syndrome. These are patients that share the characteristics of both autoimmune hepatitis as well as primary biliary cirrhosis. Histologically, these patients present with piecemeal necrosis and periductular infiltration of the portal tracts with bile duct destruction. These patients are positive for AMA and ANA and seem to profit from immunosuppressive treatment. A specific antimitochondrial antibody does not seem to be associated with this syndrome, as had been presumed. While the overlap syndrome between PBC and autoimmune hepatitis was described in the adult population, overlap between primary sclerosing cholangitis and autoimmune hepatitis was described in the pediatric population. These patients show typical bile duct strictures and dilatations on endoscopic retrograde cholangiography (ERCP), while typical histological lesions for chronic hepatitis are evident on liver biopsy. Again, high titer ANA are serological hallmarks and immunosuppression reliefs the chronic hepatitis part of this overlap syndrome.

6. TREATMENT AND PROGNOSIS

6.1. Standard Treatment with Predisone/Prednisolone Alone or in Combination with Azathioprine

Patients with severe autoimmune hepatitis are effectively treated by immunosuppression (7–10). Without treatment, prognosis of patients with severe autoimmune hepatitis is poor. Patients with biochemical or histological indicators of severe autoimmune hepatitis, such as high ALT values, hypergammaglobulinemia (>2-fold normal), bridging or multilobar necrosis show a 5-year mortality of 50% and a 10 year mortality of 90% if the condition is left untreated (7–10). Currently the most effective therapy for autoimmune hepatitis is immunosuppression. Prednisone in combination with azathioprine or a higher dose of prednisone alone are the treatments of choice (7–10) (Table 6). Both protocols are equally effective in the management of severe autoimmune hepatitis and confer 10-year survival rates of up to 90% (157). A major problem associated with prednisone monotherapy are severe side effects that occur frequently at high dosages (10). Therefore, the combination therapy, which is associated with lower frequency of adverse side effects, is the preferred initial treatment (10). Therapy is started with 40–60 mg/day of prednisone and 1–2 mg/kg azathioprine. A significant decrease of AST values is usually noted within two weeks (Table 6). Then prednisone is reduced weekly by 10 mg/day until 20 mg/day and later more slowly by 5 mg/day until a maintenance level of 5–10 mg/day prednisone is reached. Azathioprine is kept at a constant dosage. If complete biochemical remission is achieved, therapy is continued for a total of at least two years. Once remission is achieved, it is also possible to perform a maintenance therapy with azathioprine alone at a dosage of 2 mg/kg. While azathioprine is not able to achieve remission alone, an azathioprine monotherapy, is sufficient to keep remission in 80% of patients (158) (Table 6). The success rate of this therapy is high, with an overall rate of remission of 78% (159). The average treatment until remission is 22 ± 2 months, with treatment requirements ranging from 6 months to 3 years (159). With immunosuppressive treatment, patients with AIH experience 5- and 10-years life-expectancies of 94% and 90%, respectively, similar to age- and gender-matched healthy individuals (157,159). Importantly, the presence of cirrhosis at the beginning of therapy does not affect the response to treatment or reduce immediate life expectancy (157). Unfortunately, this permanent treatment cure, as defined by a complete disappearance of biochemical, serological and histological disease manifestations or sustained remission with residual manifestations of disease, are ulti-

Table 6   Treatment of autoimmune hepatitis

Therapeutic stages	Single drug regimen	Combination Regimen
Induction therapy		
week 1	40-60 mg prednisone / day	40–60 mg prednisone / day + 1–2 mg azathioprine / kg body weight
week 2	40 mg prednisone I day	15 mg prednisone / day + 1–2 mg azathioprine / kg body weight
weeks 3 and 4	30 mg prednisone / day	15 mg prednisone / day + 1–2 mg azathioprine / kg body weight
Maintenance therapy	taper 2.5 mg every week 20 mg prednisone or lower (5–15 mg/day are common dosages)	10 mg prednisone I day + 50–100 mg azathioprine / day **alternative option:** 2 mg/kg azathioprine as monotherapy
Upon relapse	like induction therapy	like induction therapy

Criteria included in this table were modified from [161].

mately achievable only in 31% of patients (160). Moreover, in the patients with complete responses and histological remission, relapse after drug withdrawal occurs with a frequency of 74% (157). In particular normal liver histology while under immunosuppression cannot guarantee lack of relapse after the end of treatment. After relapse, therapy has to be reinitiated at high dosages of immunosupressive treatment (161).

Poor prognosis is seen in 13% of patients, who experience an incomplete response after a 3-years treatment period and 9% of patients, who experience treatment failure (159). These patients are likely to deteriorate rapidly and are therefore candidates for liver transplantation, which is applied as a final rescue (162). Furthermore, both monotherapy and combination therapy are associated with side effects that lead to a discontinuation of treatment in about 13% of patients (159). Common side effects of prednisone treatment are usually mild. They include cosmetic changes, such as facial rounding, acne, hirsutism, dorsal hump formation and obesity and are reversible with reduction of drug dosage or drug withdrawal (10). Severe adverse side effects that necessitate withdrawal of treatment include osteoporosis, diabetes, cataracts and psychosis (159). These side effects are mostly due to prednisone. However, azathioprine has its own spectrum of side effects, including nausea, emesis, rash, hepatotoxicity and bone marrow suppression (159). Complications develop in less than 10% of patients receiving azathioprine at a dosage of 50 mg/day and improve with dosage reduction or drug withdrawal.

Prednisolone can be used instead of predisone. Although predisone is converted into prednisolone in the liver, the metabolism is unaffected by liver diseases and cirrhosis.

6.2. Response to Other Drugs

If standard treatment with prednisone/prednisolone alone or in combination with azathioprine fails to induce remission, other immunosuppressive drugs, including budesonide, cyclosporin A, tacrolimus, cyclophosphamide or mycophenolate may be tried.

Cyclosporine. Small numbers of adult patients who failed to respond to therapy with corticosteroids were treated by cyclosporine. Cyclosporine therapy was effective and induced remission in most AIH patients treated (163–168). Thirteen of 15 adult patients treated with cyclosporine (dosage: 2–5 mg/kg) showed complete or near to normalization of aminotransferase activities at a mean 2.4 months (range: 1–4 months) after the beginning of CsA treatment (167,169). However, even in patients with normal liver tests, relapse was experienced when the dose of CsA was lowered (169).

Recently, a multicenter study with short-term cyclosporine therapy in 32 children was reported, 28 with AIH type 1 and four with AIH type 2. This study was designed to induce remission by cyclosporine to protect children and adolescents from the side-effects of high doses of prednisone (170). As initiating therapy, cyclosporine was administered at 2.8–6.5 mg/kg per day for six months. Then combination treatment at low doses of prednisone and azathioprine for 1 month was added to cyclosporine. While patients were on triple treatment, cyclosporine was gradually decreased and and finally stopped (170). Treatment was continued with low dose combination treatment. This protocol successfully normalized ALT values in 25 patients by six months of treatment, and in all patients by 1 year of treatment. Side effects of cyclosporine treatment were well tolerated and disapeared after weaning off this medication. There was no difference in response between patients with AIH type 1 and AIH type 2 (170). The high efficiency of cyclosporine treatment was further demonstrated by a study in 15 children and adolescents with AIH type 2, who received cyclosprine due to risk factors for poor tolerance to steroids or failure to comply with steroid therapy. In both groups, ALT values returned to normal within six months, side effects were minimal and well tolerated and no relapse occurred in 10 patients after 1 to six years (171).

Tacrolimus (FK-506). Tacrolimus is a macrolide antibiotic with an immunosuppressive activity estimated to be 10–200 times greater than that of cyclosporine. Efficiency and toxicity of tacrolimus has been evaluated in an open-label preliminary trial with 21 patients. Mean dosage was 3 mg/kg twice daily. Treatment resulted in an improvement of AST, ALT and bilirubin values after 3 months (172). Controlled treatment trials are necessary to further evaluate this potentially promising new therapy.

Budesonide. Budesonide, a second generation glucocorticosteroid, may present an option to treat autoimmune hepatitis and to experience markedly reduced side effects compared to prednisone/prednisolone. Budesonide, which is taken orally, has a 90% first pass effect in the liver. The concept of budesonide treatment includes uptake via gastrointestinal tract and transport to the liver, where it may reach pathogenic lymphocytes at high pharmacological concentrations before it is detoxified. This mechanism may induce immunosuppression of hepatic lymphocytes while, due to detoxification, systemic levels of budesonide remain low and little systemic side effects may be experienced. A Swedish group has shown that budesonide normalized, elevated transaminase levels in autoimmune hepatitis (173). Our own studies in a limited number of patients have confirmed that budesonide is efficient in reducing ALT levels, while cortisol levels in the peripheral blood are hardly changed if the patient did not yet develop cirrhosis with portosystemic shunts (174). However, further studies are needed to establish efficacy of budesonide treatment. The use of budesonide in autoimmune hepatitis has to be evaluated in two ways. First, budesonide has to be analyzed for its ability to induce remission in comparison to prednisolone and in relation to side effects. Second, budesonide has to be evaluated for maintenance therapy once remission is achieved by standard immunosuppression therapy with prednisolone in combination with azathioprine.

6-Mercaptopurine. Promising results were seen with 6-mercaptopurine, which is the pharmacologically-active metabolite of azathioprine. Even if treatment with azathioprine was ineffective or caused adverse side effects, 6-mercaptopurine proved efficient in a study with three patients with AIH. This outcome may be a result of differences in the metabolic and toxicity profiles of 6-mercaptopurine and azathioprine (175). However, 6-mercaptopurine has not yet replaced azathioprine.

Cyclophosphamide. Successful long-term treatment of three patients with cyclophosphamide was reported. The patients had either severe side effects from, or were intolerant of standard therapy. Remission was induced with cyclophosphamide at a dose of 1–1.5 mg/kg bodyweight in combination with a tapering dose of corticosteroids beginning at 1 mg/kg bodyweight. Histology proven remission was achieved and could be maintained with 2.5–10 mg/day corticosteroids and 50 mg cyclophosphamide every other day (176).

Mycophenolate mofetil. Mycophenolate mofetil (MMF) is a new immunosuppressive agent that is now gaining wide-spread use in organ transplantation. MMF is believed to reduce lymphocyte proliferation by inhibition of purine nucleotide synthesis. When compared to azathioprine, MMF is more selective for lymphocytes. Another interesting difference is the fact that MMF suppresses immunglobulin synthesis to a greater extent. If immunoglobulins are playing a pathogenic role in AIH, MMF may be a promising new agent for the development of new treatment regimens in the future.

6.3. Liver Transplantation

Prognosis of liver transplantation in AIH. Liver transplantation was shown to be very effective in end-stage cirrhosis of autoimmune hepatitis, which may develop despite complete biochemical remission under long-term immunosuppressive treatment (162). Autoimmune hepatitis is among the best indications for liver transplantation and accounts for 4% of liver transplants in Europe (177). Five year survival rates of more than 90% were reported (162,178).

Recurrence of autoimmune hepatitis after liver transplantation. Recent reports stated recurrence of autoimmune hepatitis at frequencies of 0–80% after transplantation (162,178–182). Prados et al. (182) investigated 27 patients who were transplanted for autoimmune hepatitis. Nine (33%) developed chronic hepatitis with a mean time of recurrence of 2.6 ± 1.5 years. Estimated risk for recurrence was 8% over the first year and 68% five years after transplantation. Seven patients included in this study had AIH type 2. While autoantibodies prevailed after transplantation, none of these patients showed recurrence of hepatitis (182). This result is in accordance with a study performed by Ratiziu et al. (178), who investigated 15 patients transplanted for AIH. Five of 15 patients were affected by AIH type 2. While 3/10 (33%) patients with AIH type 1 showed recurrence of AIH, none of the patients with AIH type 2 did (178). HLADR3-positive patients are not only overrepresented among transplant patients, they also seem to show a higher incidence of recurrence of AIH (178,179,182). Response of recurrent disease to treatment was poor. This may be partially due to a negative selection of patients by transplantation, who had been treated unsuccessfully before transplantation and are characterized by more severe autoimmune hepatitis than patients without transplantation (182). High frequency of recurrence of autoimmune hepatitis and poor response to treatment did not decrease 5-year survival rates (178,182). However, as survival times increase, recurrent disease in patients transplanted for AIH may become an increasingly important issue. The best way to prevent recurrent disease is to maintain a rather high level of immunosuppression after transplantation and to do slow decrease.

De-novo autoimmune hepatitis after liver transplantation. One hundred eighty children (range 0.7–19.7 years) received liver transplantation for reasons other than autoimmune hepatitis. With a median of 24 months, 7 children (4%) developed a characteristic form of graft dysfunction, which is compatible with autoimmune hepatitis. Features include (183): 1. increases in aminotransferase values; 2. autoantibodies including p-ANCA, LKM1, ANA, SMA; 3. hypergammaglobulinemia; 4. liver biopsies with a dense lymphocytic portal infiltrate with plasma cells, periportal hepatitis and bridging collapse; and 5. response to the immunosuppressive schedule used for treatment of AIH Similarly, of 157 adult patients, 11% developed autoantibodies after liver transplantation and 14%, according to the scoring system of the International Autoimmune Hepatitis Group, developed possible or definite AIH (184). De novo autoimmune hepatitis after liver transplantation is a novel interesting area of clinical research in autoimmune hepatitis, which may give new insight in the ethiopathogenesis of autoimmune hepatitis.

7. HEPATITIS IN THE AUTOIMMUNE POLYGLANDULAR SYNDROME TYPE 1

The autoimmune polyglandular syndrome type 1 (APS1) is a rare autosomal recessive disorder encoded by mutations in a single gene (185). In contrast to other autoimmune diseases, APS1 is characterized by Mendelian inheritance, 100% penetrance, lack of HLA dependence and of female preponderance (185,186). APS1 is characterized by a variable combination of disease components: 1. mucocutaneous candidiasis, 2. autoimmune tissue destruction, and 3. ectodermal dystrophy (187). Typically, onset of APS1 occurs in childhood and multiple autoimmune manifestations evolve throughout the lifetime (188,189) (Table 7, Figure 7). The most frequent disease component is mucocutaneous candidiasis, which is found at least temporarily in every APS1 patient from Finland (188). The most frequent endocrine disease components are hypoparathyroidism (89%) and adrenal insufficiency (70%) (188) (Table 6). About 10–20% of APS1 patients are affected with hepatitis, which may occur without warning and may have a lethal outcome due to either fulminant or chronic hepatitis (188,190). In general, most APS1 patients are affected by 4–6 different disease components (Figure 7), which often are associated with organ-specific autoantibodies. Interestingly, autoantibodies may be detected one or several years before the onset of disease and provide useful markers for future risks (191,192).

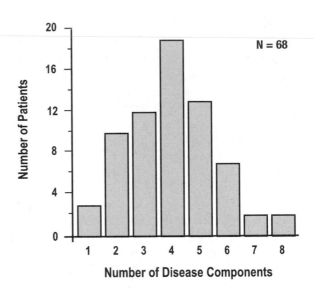

Figure 7. Patients with APS1 are affected by multiple disease components
This figure was modified from (188).

APS1 is caused by gene defects in the AIRE (autoimmune regulator) gene locus on the long arm of chromosome 21 (193,194). AIRE is one of the first gene loci known outside the HLA-locus. Several mutations in the AIRE gene are known to cause APS-1 (193–200) (Figure 8). AIRE encodes a protein expressed in the thymus medulla, in some cells in paracortex and medulla of lymph nodes, in

Table 7   Disease Components in APS1

Disease Components	Prevalence (%)		
	Neufeld 1981	Ahonen 1990	Betterle 1998
Endocrine Components			
Hypoparathyroidism	76	79	93
Adrenal failure	100	72	73
IDDM	4	12	2
Parietal cell atrophy	13	13	15
Autoimmune thyroid disease	11	2	10
Gonadal failure	17	50	43
Nonendocrine Components			
Mucocutaneous candidiasis	73	100	83
Alopecia	32	29	37
Vitiligo	8	13	15
Keratoconjunctivitis		35	12
Chronic hepatitis	13	12	20
Intestinal malabsorption	22	18	15
Enamel hypoplasia		77	
Nail dystrophy		52	

Data from this table are derived from [188, 205, 285].

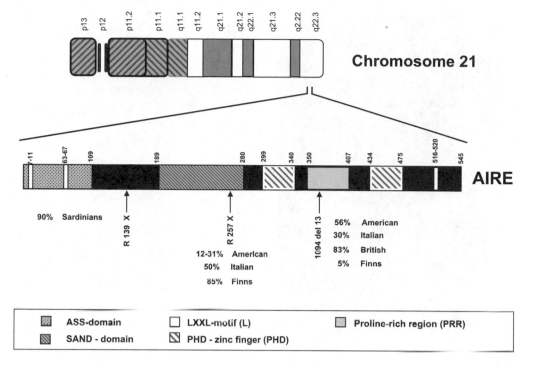

Figure 8. **Position of the *AIRE*** gene locus on chromosome 21, structure of the AIRE protein and localization of the most frequent mutations that cause APS-1

spleen and in fetal liver (201,202). On the subcellular level, the AIRE protein is localized in the nucleus, where it is distributed in speckled domains (202,203). Since it has two PHD-finger motives and an LXXXLL motif, the AIRE protein is believed to be a transcription factor (193,194) (Figure 8). Mutations causing APS1 usually destroy one or both PHD-finger motives (193,194,204).

Chronic hepatitis is a serious disease component present in 10–20% of patients with APS1 (187,188,205,206) and occasional deaths related to hepatitis are reported to occur in APS1 without warning (188,190). Recently, the first hepatic autoantigen in AIH related to APS1 was identified as cytochrome CYP1A2 (207). Anti-CYP1A2 autoantibodies may be detected by a predominant staining of the perivenous rat hepatocytes (Figure 9A). In Western Blots with human hepatic microsomes or recombinant CYP1A2, a band of 54 kDa is detected (Figure 9B) (110,207). Retrospectively, this finding identified "an unusual case of autoimmune hepatitis" that was reported before (39). This APS1 patient suffered from hepatitis, vitiligo, alopecia, nail dystrophy and had a brother who died from Addison is disease (39,110). In the serum, cytochrome CYP1A2 autoantibodies were detected that were able to inhibit the enzymatic activity of CYP1A2 (110). This patient was successfully treated with immunosuppressive therapy. Upon treatment, transaminase levels returned to normal and both

vitiligo and alopecia showed significant improvement, which confirms recent report by Gebre-Medhin of the association of CYP1A2 with hepatitis in APS1. Anti-CYP1A2 autoantibodies were detected in 3/8 patients with APS1, and all three patients with anti-CYP1A2 were affected by hepatitis (208). It is interesting to note that in 40 sera from patients with idiopathic AIH type 2, CYP1A2 could not be detected as hepatic autoantigen (209). In 68 sera from Finish patients with APS1 however, none of the patients was anti-CYP2D6 autoantibodies positive. Recently, however, a female patient with M. Addison, hypoparathyroidism, vitiligo, gastritis, pancreatitis and mild hepatitis was found to be anti-CYP2D6-positive (Obermayer-Straub, unpublished data).

A second hepatic autoantigen was identified by Rorsman with AADC (210). AADC is expressed in the hepatic cytosol and was originally described as a β-cell autoantigen (210). When 69 APS1 patients from Finland were screened for AADC autoantibodies, about 50% of patients expressed anti-AADC autoantibodies. The prevalence of AADC autoantibodies was significantly increased in APS1 patients with vitiligo and hepatitis, where 88% and 92% of patients were found to express AADC autoantibodies. So far, AADC antibodies have only been reported in APS1, and further work is needed to establish the role of IDDM in idiopathic AIH type 2 and in vitiligo.

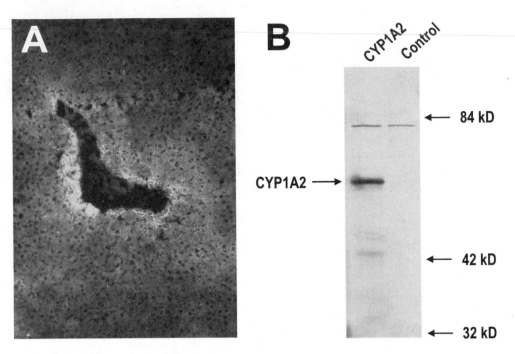

Figure 9. Liver microsomal (LM) fluorescence pattern typical for anti-CYP1A2 autoantibodies and Western Blot with recombinant CYP1A2, expressed in E. coli.
Figure A is derived from (207).

8. AUTOIMMUNITY ASSOCIATED WITH VIRAL INFECTIONS

8.1. Chronic Hepatitis C

Chronic infection with hepatitis C is known to induce autoimmune reactions (211,212). Hepatitis C is associated with many extrahepatic manifestations of immuno-pathogenetic relevance, including mixed cryoglobulinemia, membranoproliferative glomerulonephritis, polyarthritis, porphyria cutanea tarda, Sjögren's syndrome and autoimmune thyroid disease (213–219). Not surprisingly, many autoantibodies with hepatic and extrahepatic targets are detected in patients with chronic hepatitis C (60,70,95, 219–221). Similar to autoimmune hepatitis anti-tissue antibodies: ANA (95), SMA (95), p-ANCA (60,61), anti-asialoglycoprotein receptor (70) and anti-thyroid (212) autoantibodies are detected in chronic hepatitis C. However, autoantibody titers and prevalence of autoantibodies usually were lower in patients with chronic hepatitis C than in AIH (60,70,95). When patients with chronic hepatitis C were treated by α- interferon, the percentage of responding patients did not differ between patients with and without ANA or SMA (95). On the other hand, under α-interferon treatment, a small percentage of patients with chronic hepatitis C was repeatedly shown to exacerbate during α-interferon treatment (222,223). In a Spanish study (223),

144 patients (95 male, 49 female) with chronic hepatitis C were treated by α-interferon therapy. Seven patients experienced exacerbated disease under interferon treatment, and remission in these patients could be induced by corticosteroid therapy. At the beginning of interferon therapy, autoantibodies were negative in 4/7 patients with exacerbation and only low titers of ANA or SMA were detectable in three patients. All patients with exacerbation were female. After α-interferon treatment, all seven had developed significant titers of autoantibodies: 5 patients were ANA-and/or SMA-positive, 3 patients were LKM1-positive, among them a patient with LKM1 and SMA. Each of these patients was positive for at least one of the HLA DR antigens, which associate with an increased risk for auto-immune hepatitis, namely HLA DR3, HLA DR52 or HLA DR4. After initiating immunosuppressive therapy with corticosteroid and azathioprine, all seven patients responded during 8–16 months of therapy (223).

In a small percentage of patients, a serological overlap of autoantibody patterns with AIH type 2 is found (Table 8). Depending of the geographic origin LKM1 autoantibodies are detected in 0–7% of patients with HCV infection (211), and LC-1 autoantibodies are found in about 10% of patients with chronic hepatitis C (94,114). To search for distinguishing features between LKM1 autoantibodies in chronic hepatitis C and AIH type 2, epitope mapping experiments were performed. It was shown that the major epitope in AIH type 2, consisting of amino acids 257–269 of CYP2D6 (28), was detected in about 70% of sera with AIH type 2, but only in 20% of sera from patients with chronic hepatitis C (94). However, prevalence and titers of anti-CYP2D6 antibodies were 98–100% in LKM1-positive sera whether the sera were HCV-positive or not (86,87). Next to anti-CYP2D6 antibodies, antibodies directed against a 59 kDa and a 70 kDa protein were detected in sera from patients with chronic hepatitis C, but not in sera from HCV-negative patients (104). It was suggested that the reason for association of LKM1 autoantibodies with chronic hepatitis C may be the limited sequence homology between CYP2D6 and sequences of the HCV polyprotein (224). One of the best indicators of a causal role of HCV in the induction of LKM1 responses comes from a case report where in LKM1 autoantibodies developed after liver transplantation, with a donor liver carrying HCV virus to a HCV-negative recipient (225). On one hand, patients with chronic hepatitis C with and without LKM1 autoantibodies are similar in age, sex, HLA antigens and clinical symptoms (226,227), while on the other, low hepatitis C viremia levels were noted in LKM1-positive patients with hepatitis C (228). Exacerbation of LKM1-positive patients with chronic hepatitis C was reported to occur in about 10% (229–233). It remains to be investigated whether de novo induction of AIH type 2 by interferon therapy occured or whether preexisting undetected AIH type 2 with chronic

hepatitis C was exacerbated by treatment with α-interferon. Preexisting autoimmune hepatitis was indicated by a recent report, wherein 1/7 patients with LKM1-positive chronic hepatitis C were shown to be characterized by female gender, by the highest anti-CYP2D6 titers and by a pattern in epitope mapping, which is detected in about 20% of patients with chronic hepatitis C, but in 70% of patients with AIH type 2 (233). In any case, LKM1-positive patients, especially females with chronic hepatitis C, should be followed closely to detect exacerbation of disease.

Another humoral marker of autoimmunity associated with hepatitis C virus infection is the anti-GOR response, present in at least 80% of sera from patients with hepatitis C. The epitope recognized by anti-GOR is GRRGQKAK-SNPNRPL. Antibodies to the GOR antigen crossreact with the hepatitis C core region and a yet unidentified nuclear protein that is over-expressed in hepatocellular carcinoma (234). Anti-GOR antibodies are strictly associated with replicating hepatitis C virus infection. They do not occur in autoimmune liver diseases (235). A recent study has shown that anti-GOR-antibodies and their titers correlate with antibodies to HCV core antigen and necroinflammatory disease activity in liver biopsy (236). Therefore, an immunopathogenic role of the anti-GOR immune response in chronic hepatitis C seems possible. Interestingly, anti-GOR is not associated with autoimmune hepatitis, but is specific for hepatitis induced by HCV (234,235). Anti-GOR is not a pure autoantibody. It is a anti-HCV core anti-

Table 8   Clinical features of autoimmune hepatitis type 2 and hcv associated autoimmunity (LKM1 positive)

	Autoimmune hepatitis Type 2	Chronic hepatitis C associated with LKM 1 autoantibodies
Age	young	older
Sex	90% female	no prevalence
ALT	↑↑↑	↑
LKM-1 titer	↑↑↑	↑↑↑
LKM-1 antigens	CYP2D6	CYP2D6
	UGTIA	59kDa
	64 kDa	70 kDa
Autoepitopes:		
aa257-269	+++	+
aa321-350	+	(+)
Conformational	++++	++++
Immunosuppression effective	+++	–
Interferon effective	–	+
	exacerbation	5-10% risk for development of AIH
HLADR3	++	+
C4A-QO	+	+
anti-HCV RNA	–	+

This table was modified from [211].

body crossreacting with an unknown cellular self-antigen. Its potential role in immunopathogenesis of hepatitis C might deserve further evaluation.

8.2. Chronic Hepatitis D

In their original report, Crivelli et al. (107) observed in 11/81 patients with chronic hepatitis D autoantibodies, directed against hepatic microsomes and the proximal renal tubules. The reaction was strongest when human and primate tissues were used and progressively declined in ox, pig, rabbit and rat. In contrast to LKM1 and LKM2 auto-antibodies, which react with liver and kidney tissues only, additional fluorescence was detected with pancreas, adrenal gland, thyroid and stomach. The novel autoantibody was called LKM3. It is distinguished from LKM1 and LKM2 autoantibodies by the molecular weight of the target antigen, which is about 55 kDa (107). The molecular target of the LKM3 autoantibody was identified by screening of a cDNA library. Sequence analysis revealed that the molecu-lar target was a UDP-glucuronosyltransferase (UGT) (105). Therefore, Western blotting with recombinant rabbit UGT 1.6 was used to characterize the clinical significance of LKM3 autoantibodies. LKM3 autoantibodies were detected only in patients with hepatitis D and patients with autoim-mune hepatitis (105,106). They were not detected in sera from patients with hepatitis B, hepatitis C, primary biliary cirrhosis, primary sclerosing cholangitis and non-hepatic autoimmune diseases like systemic lupus erythematosus (SLE) (105).

UGTs are encoded by a large gene cluster. At least nine different exons 1 are connected to the constant exons 2–5 by differential splicing. Testing LKM3 autoantibodies with several family 1 UGTs revealed reactivity to UGT 1.1, UGT 1.4 and UGT 1.6 in all sera tested (105). As exons 1 differ between all these UGTs, this result indicates that the epitope is located in the C-terminal part of family 1 UGTs. Epitope mapping experiments confirmed that most sera rec-ognized the C-terminus. However, the signal obtained with the sera was very weak and disappeared if the C-terminus was further truncated. The signal was much higher if some N-terminal sequence was included in the clones, resulting in a minimal epitope of amino acids 264–373 (33). As truncation from the C-terminus, the N-terminus or central deletions resulted in a total loss of reactivity, it can be con-cluded that the epitope is conformation dependent (33). In addition to the major epitope on family 1 UGTs, a minor epitope was found on UGT 2B13 that was recognized by 2/8 LKM3-positive patient sera with HDV and the signal was much lower than signals detected with UGT family 1 (105). With the detection of UGTs, for the first time a drug metabolizing enzyme of phase 2 was identified as an autoantigen.

9. IMMUNE MEDIATED DRUG INDUCED HEPATITIS

9.1. Toxic Liver Injury Versus Drug-Induced Hepatitis

Hepatic cell injury, caused by direct toxicity, is dependent on the chemical characteristics of a specific drug. It is caused by a covalent modification of cellular components such as proteins, membranes or DNA by the drug itself or by reactive metabolites of the drug. Modification of cellular components may result in cell injury and necrosis or alternatively in cell death by the mechanism of apoptosis (Figure 10). Severity of symptoms caused by direct toxicity is directly related to the drug dose applied, and symptoms occur shortly after drug treatment. Furthermore, most test subjects are affected by this type of toxicity, and usually toxicity is also detected in relevant animal models. In contrast, in immune-mediated cytotoxicity, covalent modification of cellular proteins may result in very mild, often almost undetectable cytotoxic reaction. However, in a small number of genetically-predisposed persons, a typical prevalence is 1 in 10 000 people treated, covalent modification of cellular compounds results in an immune response (Figure 10). This immune response may be directed against the metabolite that was bound to specific proteins, against hapten protein domains or even against an unmodified native protein. Immune-mediated drug induced toxicity may be recognized by the following characteristics (31,237):

1. Disease does not occur after first drug exposure or directly after drug usage, but with a significant delay, in the range of a few weeks up to several months.
2. There is no dose-response relationship between drug intake and toxicity.
3. Symptoms disappear after the drug is withdrawn, and recur after resumption of drug treatment, usually with a shorter lag-period.
4. Often disease is accompanied by typical signs of an immune reaction, such as fever, eosinophily, or rash.
5. Usually autoantibodies directed against hepatic proteins are detected.
6. The disease is characterized by a female pre-dominance.

9.2. The Role of Cytochromes P450 and UDP-Glucuronoslytransferases in Detoxification and in the Induction of Drug-Induced Immune-Mediated Hepatitis, a Hypothesis

Detoxification of lipophilic substances usually occurs in two steps. The first is a hydroxylation reaction mediated by enzymes of the cytochrome P450 supergene family.

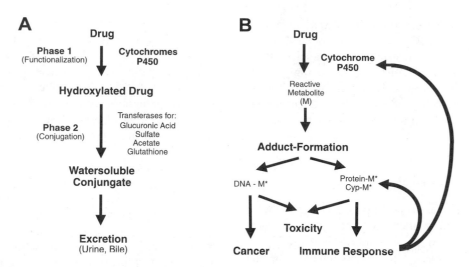

Figure 10.   Role of cytochromes P450 (CYPs) and UDP-glucuonosyltransferases (UGTs) in detoxification and bioactivation of xenobiotics. This figure was modified from (33).

In the second step, hydroxylated reaction products are conjugated with hydrophilic compounds e.g. glucuronic acid (238) (Figure 10A). Phase 2 reactions generate hydrophilic conjugates, which often have lost their biological activity and may be easily excreted via bile and kidney (239). One of the phase 2 reactions, conjugation with glucuronic acid is mediated by the multigene family of UDP-glucuronosyltransferases (240). These reactions usually are beneficial and lead to the detoxification of a wide variety of diet-borne components and environmental chemicals. Sometimes however, during a "detoxification" reaction, a reactive metabolite is generated (Figure 10B). Reactive metabolites often are produced during a hydroxylation reaction by CYPs. These reactive metabolites may bind directly to a nucleophilic group in the active center of CYP, which activated the drug. This reaction is called suicide inactivation since it results in a chemically-modified CYP that is enzymatically inactive. If the metabolite is more stable, it may leave the active center and bind to other cellular proteins. Upon repeated adduct formation the immune system may get activated against these modified-self proteins, resulting in an immune response that is self-perpetuating as long as protein-adducts are present. The majority of proteins are often expressed by hepatocytes, which become targets of immune-mediated cytotoxic processes, resulting in severe hepatitis. Immune-mediated attacks will last as long as the target cells express modified proteins. Therefore, withdrawal of the drug, resulting in a decline of protein modification, usually leads to disappearance of immune-mediated adverse reactions. However, since memory cells are generated during this process, drug-induced hepatitis will reappear upon rechallenge with the drug (31,32,241).

9.3.   Dihydralazine Hepatitis

Long-term treatment with dihydralazine, a vasodilatory substance, resulted in the induction of severe hepatitis in many patients. In the pathological institute in Berlin-Friedrichshain, Germany, 70 cases of severe dihydralazine hepatitis were registered during 1981–1985 (237). Seventy-five percent of patients were female, and most patients were slow acetylators (242). Hepatitis usually occured with a significant delay in the range of 2 weeks and 11 months (237). Complete recovery is usually noted after termination of dihydralazine treatment. About 70% of patients needed 1–4 weeks for full recovery, while 30% of patients needed 5–8 weeks (237). Rechallenge with dihydralazine leads to the recurrence of disease, but after a shorter lag-period than before (243).

Antibodies were detected in the blood of patients with dihydralazine hepatitis, which are directed against microsomal proteins of the liver (LM-autoantibodies). The pattern in immunofluorescence is the same as shown in Figure 9A. Renal tissue was not stained by these autoantibodies (244). The molecular target of these autoantibodies is CYP1A2, which is recognized with a high degree of specificity. Even as CYP1A1 is 80% homologous to CYP1A2, no cross reactivity was detected (109,245).

It is believed that in the active center of CYP1A2, a reactive metabolite is generated that may bind directly to CYP1A2 (246). Modified CYP1A2 is believed to be recognized as a neoantigen in susceptible patients and an immune reaction directed against CYP1A2 is induced (31,245–247). Hydroxylation by CYP1A2 is not the only pathway for dihydralazine metabolism. The second pathway is an acetylation reaction, which is believed to be beneficial. However, N-acetyltransferase, which mediates this second pathway for dihydralazine detoxification, is only active in 50% of Caucasians (88,248). Patients deficient in acetylation are restricted to metabolize dihydralazine exclusively via CYP1A2. Interestingly, slow acetylators are overrepresented in the patient population with dihydralazine hepatitis indicating that adduct formation via CYP1A2 is the ethiologic pathogenic pathway (242).

9.4.   Hepatitis Induced by Tienilic Acid

Tienilic acid was used as an anti-hypertensive drug but it was withdrawn from the market because of severe cases of hepatitis resulting from its use in (249) 0.1–0.7% of patients in equal number of male and female. The lag period between onset of treatment and beginning hepatitis was 2–35 weeks. Severity of hepatitis was independent of the doses of tienilic acid applied, and discontinuation of drug treatment resulted in recovery from hepatitis. Upon rechallenge with tienilic acid hepatitis recurred with a shorter lag period than before (249,250).

Livers of affected patients were infiltrated by lymphocytes and by neutrophilic and eosinophilic granulocytes. Antibodies directed against liver and kidney microsomes were detected and called anti-LKM2 autoantibodies (80). The molecular target of LKM2 autoantibodies is CYP2C9 (251). CYP2C9 is the most abundant hepatic CYP and the major tienilic acid-metabolizing enzyme (252). Auto-antibodies are directed against the unmodified enzyme as well as against the enzyme modified by tienilic acid (97). The major epitope recognized on CYP2C9 is three dimensional (253). CYP2C9 and tienilic acid modified adducts both were detected on the surface of the hepatocyte membrane, indicating that antibodies might be able to induce cell lysis via complement and that LKM2 autoantibodies may play a pathogenetic role in tienilic acid induced hepatitis (97).

9.5.   Hepatitis induced by Halothane and Chemically-Related Compounds

Halothane hepatitis.   Until recently, halothane was one of the most common anesthetics. Direct toxicity of halothane is found in about 20% of patients with mild tissue damage and only modest increases in transaminase values (131). In about 1 of 10,000 patients treated with halothane, severe hepatitis will develop, characterized by jaundice, fever, high transaminase values and severe centrilobular necrosis. Often hepatic encephalopathy occurs, resulting in a high rate of of mortality (254,255). Risk factors are female gender (male:female ratio 1:2), obesity and multiple exposures (256,257). Usually 1–2 weeks are noted between first exposure and disease.

Halothane was reported to be metabolized by several CYPs, resulting in two different pathways of metabolism, an oxydative and a reductive pathway (258,259). The reductive pathway may cause a mild form of liver cell injury, but the oxydative pathway is believed to generate a highly-reactive trifluoroacetylchloride (TFA) (258). If pharmacological doses of halothane are applied, the metabolism is predominantly mediated by CYP2E1 (98,260–263).

Initial adduct formation is detected with CYP2E1, which generates the reactive metabolite (98). The majority of reactive metabolites are believed to leave the active center of CYP2E1 and to modify ε-amino goups of lysins, which are present in many liver proteins (264). Therefore, many TFA-adducts are generated with molecular weights between 50 kDa–170 kDa (Table 9). ELISA experiments with purified proteins showed that autoantibodies not only detected TFA-modified domains, but also conformational epitopes on native proteins (265–268). A pathologic relevance of these autoantibodies is suggested by the presence of TFA-conjugates on the surface of hepatocytes, which were able to induce a cytotoxic reaction on isolated hepatocytes (98,269).

Hepatitis induced by enflurane, isoflurane and desflurane.   Further evidence that the mechanism proposed by Beaune et al. (31) may describe the etiology of halothan-induced hepatitis comes from a major effort to decrease toxicity of anesthetics. New components have been generated that show a much lower degree of metabolism and adduct formation (32). Metabolism decreased from halothane (20%) to enfluran (2,4%), isofluran (0.2%) and desfluran (0,01%) (32). Adducts generated by these new components are either cross reactive (enfluorane) or identical (desflurane and isoflurane) to those adducts generated by halothane. Metabolism correlates with the numbers of reported patients, who experienced toxicity: 900 patients with halothane hepatitis, 15–24 patients with enflurane hepatitis, 5 patients with isoflurane hepatitis and one case of desfluronane hepatitis (32). The patient with desflurane-induced hepatitis was previously exposed twice to halothane, indicating that the initial immunisation of this patient was due to previous halothane exposures (270).

Hepatitis induced by hydrofluorocarbons.   A mechanism parallel to the induction of halothane hepatitis was suggested to have caused an epidemic of liver disease induced by hydrofluorocarbons (HCFC) (271). Nine industrial workers had repeated accidental exposures to a mixture of 1,1-dichloro-2,2,2-trifluoroethane (HCFC 123) and 1-chloro-1,2,2,2-tetrafluoroethane (HCFC 124). All exposed workers were affected to some degree. Both compounds are metabolized to form TFA intermediates and

Table 9   Liver proteins recognized by autoantibodies in drug-induced hepatitis

Drug induced hepatitis	Protein	MW	References
Dihdralazine Hepatitis	CYP1A2	54 kDa	246
Tienilic acid induced hepatitis	CYP2C9	50 kDa	80
Anti-convulsant hepatitis	CYP3A1 -like domain in rat	50 kDa	275
Halothane hepatitis	UDP-glucose:glycoprotein glucosyltransferase	170 kDa	286
	Erp99	100 kDa	287
	BiPIGRP78	82 kDa	288
	ERp72	80 kDa	266
	Caireticulin	63 kDa	289
	Carboxylesterase	59 kDa	290
	Isomerase	58 kDa	291
	Protein disulfide isomerase	57 kDa	265
	CYP2E1	50 kDa	98, 260–263.
	Epoxidhydrolase	50 kDa	241

TFA-adducts were detected in the liver of a worker with severe hepatitis. In the serum of six workers, autoantibodies were detected directed against CYP2E1 and protein disulphide isomerase. These results demonstrate that repeated exposure of humans to HCFCs 123 and 124 may cause serious liver injury. The high prevalence of disease was probably due to repeated long term exposure to the agent, while halothane exposure was usually an acute short-term exposure (271). Furthermore, *in vitro* metabolic studies with human liver CYP2E1 suggest that HCFC 123 may result in higher concentrations of TFA-adducts in liver, and therefore show an increased risk for immune-mediated drug induced toxicity (272).

9.6.   Idiosyncratic Reactions to Anticonvulsants

With a low prevalence, patients (1:10,000) treated with aromatic anticonvulsants such as phenobarbital, phenytoin and carbamazepine may experience life-threatening systemic reactions. Symptoms at onset are fever and skin lesions. Secondary organ involvement may include varying combinations of hepatitis, hematologic abnormalities, nephritis, pneumonitis and lymphadenopathy and, in some cases, hepatitis or nephritis (273). The lag period between the beginning of drug treatment and the onset of symptoms ranges from 1 week for 3 months. No correlation between drug dosage and severity of symptoms was found. Patients who were exposed to two or all three anticonvulsants showed adverse reactions to each, and *in vitro* lymphocytes showed reactions to all three drugs upon rechallenge (274).

In 9/24 patients with anticonvulsant-induced idiosyncratic reactions, autoantibodies were detected that were directed against a 53 kDa hepatic protein. Protein targets that react with these antibodies are rat CYP3A1 and rat CYP2C11 (274). However, when an array of recombinant human CYPs was tested, neither human CYP1A1, CYP1A2, CYP2A6, CYP2B6, CYP2D6, CYP2E1 or CYP3A4 were recognized by these autoantibodies. Furthermore, in Western Blots with human liver proteins, only three weak signals were detected at 50–55 kDa. In contrast, a strong signal was recognized in liver proteins derived from a patient who died from hepatitis after phenytoin/phenobarbital therapy (274). The identity of this protein is still unknown. In an effort to elucidate the identity of the molecular target of anticonvulsant hepatits, a gene bank of fusion proteins with partial sequences of rat CYP3A1 was generated and screened for recognition sequences. Positive clones overlapped at a consensus sequence of amino acids 355–367 of rat CYP3A1 (275). This epitope was recognized by sera of all patients with idiosyncratic reactions to anticonvulsants. This epitope differs from the respective sequence of human CYP3A4 by a V361L substitution. If this substitution is introduced into the rat enzyme, binding activity is lost (275). Therefore, the molecular target of antibodies in anticonvulsant hepatitis remains still unknown. However, characteristic of anti-convulsant-induced idiosyncratic reactions fulfill all criteria for an immune-mediated disorder, which may be induced by adduct formation with a reactive intermediate.

9.7.   Alcoholic Liver Disease: Indications for an Involvement of the Immune System

Chronic consumption of large quantities of alcoholic beverages is a main cause of hepatitis and cirrhosis. Prevalence of hepatitis in severe cases of alcohol abuse, however, is only 10–20% of patients, indicating that host factors may be involved in the pathogenesis. If alcohol is withdrawn, liver injury will persist for a certain time period, although

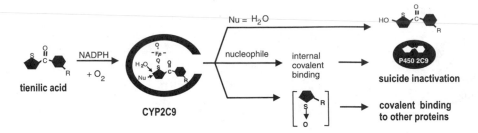

Figure 11.   Adduct formation of tienilic acid with CYP2C9.
This figure was modified from (31).

recovery frequently occurs that can be proved by a normal histology. Indicative of immune-mediated pathogenesis are the following characteristics of alcoholic liver disease (276,277):

1. In patients, with histologically complete recovery from alcoholic liver disease, consumption of alcohol induces a quick recurrence of hepatitis.
2. Hypergammaglobulinemia and circulating autoantibodies were detected in patients with alcoholic liver disease.
3. Lymphocytic infiltrates are detected in areas of hepatocellular degeneration.
4. A subgroup of patients was shown to improve upon corticosteroid therapy.

Several clinical studies and results obtained in animal models suggest that adduct formation and subsequent induction of an immune response may play an important role in the pathogenesis of alcoholic liver disease (277–282). In 1986, Israel and colleagues were able to demonstrate that chronic exposure to alcohol leads to the induction of antibodies directed against aldehyde-adducts of proteins (281). Reactivity to aldehyde-protein adducts was independent of the protein carrier used. Furthermore, immunization of ethanol-fed guinea pigs with acetaldehyde-modified hemoglobin was reported to reproduce several features of alcoholic hepatitis experimentally (283). However, acetaldehyde is not the only reactive intermediate during alcohol metabolism. Hepatic ethanol oxidation by CYP2E1 is also known to generate hydroxyethyl-radicals (278,284). In accordance with the hapten hypothesis of induction of alcoholic liver disease, antibodies directed against hydroxyethyl-adducts of proteins were detected by sera from patients with alcoholic liver disease (279). Since both hydroxyethyl-adducts of human serum albumin or bovine fibrinogen were detected by these antibodies, the major epitope detected was not a native protein, but the hydroxyethyl-determinant on the modified proteins (279). Autoantibodies directed against hydoxyethyl domains did not crossreact with autoantibodies directed against acetaldehyde modified proteins, indicat-

ing that two populations of autoantibodies exist (279). CYP2E1 is involved in the generation of hydroxyethyl radicals, and therefore this enzyme may play a critical role in the development of autoimmunity. *In vivo* experiments using inhibitors of expression and function of CYP2E1 showed that modifications induced in the CYP2E1 distribution correlated with changes in the pattern of hydroxyethyl radicals and subsequently with the pattern of degeneration and the distribution of fat (278,282). Incubation of liver microsomes with ethanol as a substrate under conditions supporting CYP activity, resulted in the formation of hydroxyethyl radical-derived liver antigens that were preferentially bound by sera of patients with alcoholic cirrhosis. Western Blots revealed that proteins of 78 kDa, 60 kDa, 52 kDa and 40 kDa were modified. Analysis of 23 randomly-chosen patient sera with alcoholic cirrhosis revealed binding of the 52 kDa protein in 86% of sera, reactions towards the other bands in 30–48% of sera. In contrast, unmodified hepatic proteins were not detected. The 52 kDa protein was identified as CYP2E1 (280). Clot et al. (280) concluded that CYP2E1 causes the formation of hydroxyethyl-adducts with CYP2E1 and three other proteins. These modified proteins may induce an immune response in susceptible patients and provoke the formation of autoantibodies. Therefore, these autoantibodies may play a role in the pathogenesis of alcoholic liver injury and they present useful markers for the severity of liver cirrhosis (280).

References

1. Boberg, K.M., E. Aadland, J. Jahnsen, N. Raknerud, M. Stiris, and H. Bell. 1998. Incidence and prevalence of primary biliary cirrhosis, primary sclerosing cholangitis, and auto-immune hepatitis in a Norwegian population. *Scand. J. Gastroenterol.* 33:99–103.
2. Waldenström, J., Leber, und Blutproteine und Nahrungseiweiss. 1950. *Dtsch. Gesellsch. Verd. Stoffw.* 15:113–119.
3. Johnson, P.J., I.G. McFarlane, and ALWF. Eddleston. 1991. The natural course and heterogeneity of autoimmune-type immune-type chronic active hepatitis. *Semin. Liver. Dis.* 11:187–196.

4. Nikias, G.A., K.P. Batts, and A.J. Czaja. 1994. The nature and prognostic implications of autoimmune hepatitis with acute presentation. *J. Hepatol.* 21:866–871.

5. Burgart, L.J., K.P. Batts, J. Ludwig, G.A. Nikias, and A.J. Czaja. 1995. Recent-onset autoimmune hepatitis. Biopsy findings and clinical correlations. *Am. J. Surg. Oathol.* 19:699–708.

6. Manns, M.P., and P. Obermayer-Straub. 1997. Cytochromes P450 and UDP-Glucuronosyltransferases: Model autoantigens to study drug-induced, virus-induced and autoimmune liver disease. *Hepatology* 26:1054–1066.

7. Cook, G.C., R. Mulligan, and S. Sherlock. 1971. Controlled protective trial of corticoid therapy in active chronic hepatitis. *Q.J. Med.* 40:159–185.

8. Soloway, R.D., W.H.J. Summerskill, A.H. Baggenstoss, M.G. Geall, G.L. Gitnick, E.R. Elveback, and L.J. Schoenfield. 1972. Clinical, biochemical and histologic remission of severe chronic active liver disease: A controlled study of treatments and early prognosis. *Gastroenterology* 63:820–833.

9. Murray-Lyon, I.M., R.B. Stern, and R. Williams. 1973. Controlled trial of prednisone and azathioprine in active chronic hepatitis. *Lancet* 1:735–737.

10. Summerskill, W.H.J., M.G. Korman, H.V. Ammon, and A.H. Baggenstoss. 1975. Prednisone for chronic active liver disease: dose titration, standard dose and combination with azathioprine compound. *Gut* 16:876–883.

11. Alvarez, F., P.A. Berg, F.B. Bianchi, L. Bianchi, A.K. Burroughs, E.L. Cancado, and R.W. Chapman. 1999. International Autoimmune Hepatitis Group Report: Review of criteria for diagnosis of autoimmune hepatitis. *J. Hepatol.* 31:929–938.

12. Desmet, V.J., M. Gerber, J.H. Hoofnagle, M. Manns, and P.J. Scheuer. 1994. Classification of chronic hepatitis: diagnosis, grading and staging. *Hepatology* 19:1513–1520.

13. Mackay, I.R. 1983. Immunological aspects of chronic active hepatitis. *Hepatology* 3:724–728.

14. Czaja, A.J., F. Cassani, M. Catala, P. Valentini, and F.B. Bianchi. 1997. Antinuclear antibodies and patterns of nuclear immunofluorescence type 1 autoimmune hepatitis. *Dig. Dis. Sci.* 8:1688–1696.

15. Homberg, J.C., N. Abuaf, O. Bernard, S. Islam, F. Alvarez, S.H. Khalil, R. Poupon, F. Darnis, V.G. Levy, and P. Grippon. 1987. Chronic active hepatitis associated with antiliver/kidney microsome type 1: a second type of "autoimmune" hepatitis. *Hepatology* 7:1333–1339.

16. Martini, E., N. Abuaf, F. Cavalli, V. Durand, C. Johanet, and J.C. Homberg. 1988. Antibody to liver cytosol anti-LC1 in patients with autoimmune chronic active hepatitis type 2. *Hepatology* 8:1662–1666.

17. Manns, M., G. Gerken, A. Kyriatsoulis, M. Staritz, and K.H. Meyer zum Büschenfelde. 1987. Characterization of a new subgroup of autoimmune chronic active hepatitis by autoantibodies against a soluble liver antigen. *Lancet* 1:292–294.

18. Stechemesser, E., R. Klein, and P.A. Berg. 1993. Characterization and clinical relevance of liver-pancreas antibodies in autoimmune hepatitis. *Hepatology* 18:1–9.

19. Treichel, U., T. Portalla, G. Hess, M. Manns, and K.H. Meyer zum Büschenfelde. 1990. Autoantibodies to human asialoglycoprotein receptor in autoimmune-type chronic active hepatitis. *Hepatology* 11:606–612.

20. Mulder, A.H.L., G. Horst, E.B. Haagsma, P.C. Limburg, J.H. Kleibeuker, and G.M. Kallenberg. 1993. Prevalence and characterization of neutrophil cytoplasmic antibodies in autoimmune liver diseases. *Hepatology* 17:411–417.

21. Manns, M.P., M. Jentzsch, K. Mergener, G. Gerken, V. Thiers, C. Brechot, K.H. Meyer zum Büschenfelde, and M. Eichelbaum. 1990. Discordant manifestation of LKM-1 antibody positive autoimmune hepatitis in identical twins. *Hepatology* 12(suppl.):840–843.

22. Manns, M.P., and M. Kruger. 1994. Genetics in liver diseases. *Gastroenterology* 106:1676–1697.

23. Donaldson, P., D. Doherty, J. Underhill, and R. Williams. 1994. The molecular genetics of autoimmune liver disease. *Hepatology* 20:225–229.

24. Vento, S., F. Cainelli, T. Ferraro, and E. Conicá. 1996. Autoimmune hepatitis type 1 after measles. *Am. J. Gastroenterol.* 91:2618–2620.

25. Vento, S., T. Garofano, G. Di Perri, L. Dolci, E. Conica, and D. Bassetti. 1991. Identification of hepatitis A virus as a trigger for autoimmune chronic hepatitis type 1 in susceptible individuals. *Lancet* 337:1183–1187.

26. Huppertz, H.I., U. Treichel, A.M. Gassel, R. Jeschke, and K.H. Meyer zum Büschenfelde. 1995. Autoimmune hepatitis following hepatitis A virus infection. *J. Hepatol.* 23:204–208.

27. Vento, S., F. Cainelli, C. Renzini, and E. Conica. 1997. Autoimmune hepatitis type 2 induced by HCV and persisting after viral clearance. *Lancet* 350:1298–1299.

28. Manns, M.P., K.J. Griffin, K.F. Sullivan, and E.F. Johnson. 1991. LKM-1 autoantibodies recognize a short linear sequence in P450IID6, a cytochrome P-450 monooxygenase. *J. Clin. Invest.* 88:1370–1378.

29. Schmitt, K., J. Deutsch, G. Tulzer, R. Meindi, and S. Aberle. 1996. Autoimmune hepatitis and adrenal insufficiency in an infant with human herpesvirus-6 infection. *Lancet* 348:966.

30. Vento, S., L. Guella, F. Mirandola, F. Cainelli, G. Di Perri, T. Ferraro, and E. Conica. 1995. Epstein-Barr virus as a trigger for autoimmune hepatitis in susceptible individuals. *Lancet* 346:608–609.

31. Beaune, P.H., D. Pessayre, P. Dansette, D. Mansuy, and M.P. Manns. 1994. Autoantibodies against cytochromes P450: Role in human diseases. *Adv. Pharamacol.* 30:199–245.

32. Pohl, L.R., N.R. Pumford, and J.L. Martin. 1996. Mechanisms, chemical structures and drug metabolism. *Eur. J. Haematol.* 57:98–104.

33. Obermayer-Straub, P., and M.P. Manns. 1996. Cytochromes P450 and UDP-glucuronosyltransferases as hepatocellular autoantigens. *Baillieres Clin. Gastroenterol.* 10:501–532.

34. Czaja, A.J., and M.P. Manns. 1995. The validity and importance of subtypes in autoimmune hepatitis: a point of view. *Am. J. Gastroenterol.* 90:1206–1211.

35. Doherty, D.G., P.T. Donaldson, J.A. Underhill, J.M. Farrant, A. Duthie, G. Mieli-Vergani, I.G. McFarlane, P.J. Johnson, A.L. Eddleston, and A.P. Mowat. et al. 1994. Allelic sequence variation in the HLA class II genes and proteins on patients with autoimmune hepatitis. *Hepatology* 19:609–615.

36. Czaja, A.J., M. Strettell, L.J. Thomson, P. Santrach, S.B. Moore, P.T. Donaldson, and R. Williams. 1997. Associations between alleles of the major histocompatibility complex and type 1 autoimmune hepatitis. *Hepatology* 25:317–323.

37. Czaja, A.J., M. Nishioka, S.A. Morhed, and T. Haciya. 1994. Patterns of nuclear immunofluorescence and reactivities to recombinant nuclear antigens in autoimmune hepatitis. *Gastroenterology* 107:200–207.

38. Nishioka, M. 1993. Nuclear antigens in autoimmune hepatitis. In: Meyer zum Büschenfelde, K-H, J. Hoofnagle, M.

Manns, eds. Immunology and the Liver. London, Dordrecht, Boston: Kluwer Academics 193–205.

39. Sacher, M., P. Blümel, H. Thaler, and M. Manns. 1990. Chronic active hepatitis associated with vitiligo, nail dystrophy, alopecia and a new variant of LKM antibodies. *J. Hepatol.* 10:364–369.

40. Strassburg, C., B. Alex, F. Zindy, G. Gerken, B. Lüttig, K.H. Meyer zum Büschenfelde, C. Bréchot, and M. Manns. 1996. Identification of cyclin A as a molecular target of antinuclear antibodies (ANA) in heptatic and non-hepatic diseases. *J. Hepatol.* 25:859–866.

41. Czaja, A.J. 1999. Behavior and significance of autoantibodies in type 1 autoimmune hepatitis. *J. Hepatol.* 30:394–401.

42. Czaja, A.J., C. Ming, M. Shirai, and M. Nishioka. 1995. Frequency and significance of antibodies to histones in autoimmune hepatitis. *J. Hepatol.* 23:32–38.

43. Chen, M., M. Shirai, A.J. Czaja, K. Kurokohchi, T. Arichi, K. Arima, T. Kodama, and M. Nishioka. 1998. Characterization of anti-histone antibodies in patients with type 1 autoimmune hepatitis. *J. Gastroenterol. Hepatol.* 13:483–489.

44. Wood, J.R., A.J. Czaja, S.J. Beaver, S. Hall, W.W. Ginsburg, D.K. Kaufman, and H. Markowitz. 1986. Frequency and significance of antibody to double-stranded DNA in chronic active hepatitis. *Hepatology* 6:976–980.

45. Tsuchiya, K., K. Kiyosawa, H. Imai, T. Sodeyama, and S. Furuta. 1994. Detection of anti-double and anti-singlr stranded DNA antibodies in chronic liver disease: significance of anti-double stranded DNA antibody in autoimmune hepatitis. *J. Gastroenterol.* 29:152–158.

46. Czaja, A.J., S.A. Morshed, S. Parveen, and M. Nishioka. 1997. Antibodies to single-stranded and double-stranded DNA in antinuclear antibody-positive type 1 autoimmune hepatitis. *Hepatology* 26:567–572.

47. Czaja, A.J., F. Cassani, M. Cataleta, P. Valenti, and F.B. Bianchi. 1996. Frequency and significance of antibodies to actin in type 1 autoimmune hepatitis. *Hepatology* 24:1068–1073.

48. Toh, B.H. 1979. Smooth muscle autoantibodies and autoantigens. *Clin. Exp. Immunol.* 38:621–628.

49. Orth, T., G. Gerken, R. Kellner, K-H. Meyer zum Büschenfelde, and W-J. Mayet. 1997. Actin is a target antigen of anti-neutrophil cytoplasmic antibodies (ANCA) in autoimmune hepatitis type-1. *J. Hepatol.* 26:37–47.

50. Hagen, E.C., BEPB. Ballieux, L.A. Van Es, M.R. Daha, and F.J. Van der Woulde. 1993. Antineutrophil cytoplasmic autoantibodies: A review of antigens involved, the assays, and the clinical and possible pathogenic consequences. *Blood* 81:1996–2002.

51. Hausschild, S., W.H. Schmitt, E. Csernok, B. Flesch, A. Rautman, and W.L. Gross. 1992. ANCA in Wegener's granulomatosias and related vasculitides. In: Advances in Experimental Medicine and Biology. Anstract Book of the Fouth International Workshop on ANCA. London, U.K: Plenum. 1992.

52. Falk, R.J., and J.C. Jennette. 1988. Anti-neutrophil cytoplasmic autoantibodies with specificity for myeloperoxidase in patients with systemic vasculitis and idiopathic necrotizing and crescentic glomeru-lonephritis. *N. Engl. J. Med.* 318:1651–1657.

53. Cohen Tervaert, J.W., R. Goldschmeding, J.D. Elema, P.C. Limburg, M. van der Giessen, M.G. Huitema, R.J. Hene, T.H. The, G.K. Van der Hem, and et al. 1990. Association of autoantibodies to myeloperoxidase with different forms of vasculitis. *Arthritis Rheum.* 33:1264–1272.

54. Ludermann, J., B. Utrecht, and W.L. Gross. 1990. Anti-neutrophil cytoplasm antibodies in Wegener's granulomatosis recognize an elastinolytic enzyme. *J. Exp. Med.* 171:357–362.

55. Halbwachs-Mecarelli, L., P. Nusbaum, L.H. Noel, D. Reumaux, S. Erlimger, J.P. Grunfeld, and P. Lesavre. 1992. Antineutrophil cytoplasmic antibodies (ANCA) directed against cathepsin G in ulcerative colitis, Crohn's disease and primary sclerosing cholangitis. *Clin. Exp. Immunol.* 90:79–84.

56. Peen, E., S. Almer, G. Bodemar, B.O. Ryden, C. Sjolin, K. Teje, and T. Skogh. 1993. Anti-lactoferrin antibodies and other types of ANCA in ulcerative colitis, primary sclerosing cholangitis, and Crohn's disease. *Gut* 34:56–62.

57. Warny, M., R. Brenard, C. Cornu, J.P. Tomasi, and A.P. Geubel. 1993. Anti-neutrophil antibodies in chronic hepatitis and the effect of alpha-interferon therapy. *J. Hepatol.* 17:294–300.

58. Hardarson, S., D.R. Labrecque, F.A. Mitros, G.A. Neil, and J.A. Goeken. 1993. Antineutrophil cytoplasmic antibody in inflammatory bowel and hepatobiliary diseases. High prevalence in ulcerative colitis, primary sclerosing colangitis, and autoimmune hepatitis. *Am. J. Clin. Pathol.* 99:277–281.

59. Targan, S.R., C. Landers, A. Vidrich, and A.J. Czaja. 1995. High titer antineutrophil cytoplasmic antibodies in type-1 autoimmune hepatitis. *Gastroenterology* 108:1159–1166.

60. Zauli, D., S. Ghetti, A. Grassi, C. Descovich, F. Cassani, G. Ballardini, L. Muratori, and F.B. Bianchi. Antineutrophil cytoplasmic antibodies in type 1 and 2 autoimmune hepatitis. *Hepatology* 25:1105–1107.

61. Sobajima, J., S. Ozaki, H. Uesugi, F. Osakada, M. Inoue, Y. Fukuda, H. Shirakawa, M. Yoshida, A. Rokuhura, H. Imai, K. Kiyosawa, and K. Nakao. 1999. High mobility group (HMG) non-histone chromosomal proteins HMG1 and HMG2 are significant target antigens of perinuclear antineutrophil cytoplasmic antibodies in autoimmune hepatitis. *Gut* 44:867–873.

62. Duerr, R.H., S.R. Targan, C.J. Landers, N.F. LaRusso, K.L. Lindsay, and R.H. Wiesner. 1991. Neutrophil cytoplasmatic antibodies: a link between primary sclerosing cholangitis and ulcerative colitis. *Gastroenterology* 100:1385–1391.

63. Einck, L., and M. Bustin. 1985. The intracellular distribution and function of the high mobility group chromosomal proteins. *Gut* 44:867–873.

64. Spiess, M. 1990. The asialoglycoprotein receptor: a model for endocytic transport receptors. *Biochemistry* 29:100009–100022.

65. Wali, D.A., G. Wilson, and A.L. Hubbard. 1980. The route of ligand internalization in rat hepatocytes. *Cell* 21:79–93.

66. McFarlane, B.M., C.G. McSorley, D. Vergani, I.G. McFarlane, and R. Williams. 1986. Serum antibodies reacting with the hepatic asialoglycoprotein receptor protein (hepatic lectin) in acute and chronic liver disorders. *J. Hepatol.* 3:196–205.

67. Treichel, U., B.M. McFarlane, T. Seki, E.L. Krawitt, N. Alessi, F. Sickel, I.G. McFarlane, K. Kiosawa, S. Furuta, M.A. Freni, G. Gerken, and K-H. Meyer zum Büschenfelde. 1994. Demographics of anti-asialoglycoprotein receptor autoantibodies in autoimmune hepatitis. *Hepatology* 107:799–804.

68. Czaja, A.J., K.D. Pfeifer, R.H. Decker, and A.S. Vallari. 1996. Frequency and significance of antibodies to asialogly-

AUTOIMMUNE HEPATITIS 943

coproteinreceptor in type 1 autoimmune hepatitis. *Dig. Dis. Sci.* 41:1733–1740.

69. McFarlane, B.M., J. Sipos, C.D. Gove, I.G. McFarlane, and R. Williams. 1990. Antibodies against the hepatic asialoglycoprotein receptor perfused in situ preferentially attach to periportal liver cells in the rat. *Hepatology* 11:408–415.

70. Treichel, U., G. Gerken, S. Rossol, H.W. Rotthaue, K-H. Meyer zum Büschenfelde, and T. Portalla. 1993. Autoantibodies against the human asialoglycoprotein receptor. Effects of therapy in autoimmune and virus-induced chronic active hepatitis. *J. Hepatol.* 19:55–63.

71. McFarlane, I.G., B. McFarlane, G.N. Major, P. Tolley, and R. Williams. 1984. Identification of the hepatic-receptor (hepatic lectin) as a component of liver specific membrane lipoprotein (LSP). *Clin. Exp. Immunol.* 55:347–354.

72. Dejica, D., U. Treiche, A. Par, O. Chira, and K.H. Meyer-zum-Buschenfelde. 1997. Anti asialoglycoprotein receptor antibodies and soluble interleukin-2 receptor levels as marker for inflammation in autoimmune hepatitis. *Z. Gastroenterol* 35:15–21.

73. McFarlane, I.G., J.E. Hegarty, and C.G. McSorley, et al. 1984. Antibodies to liver-specific protein predict outcome of treatment withdrawal in autoimmune chronic active hepatitis. *Lancet* 2:954–956.

74. Hajoui, O., S. Martin, and F. Alvarez. 1998. Study of antigenic sites on the asialoglycoprotein receptor recognized by autoantibodies. *Clin. Exp. Immunol.* 113:339–345.

75. Eggink, H.F., H.J. Houthoff, S. Hultema, C.H. Gips, and S. Poppema. 1982. Cellular and humoral immune reactions in chronic active liver disease. Lymphocyte subsets in liver biopsies of patients with untreated autoimmune hepatitis, chronic active hepatitis B and primary biliary cirrhosis. *Clin. Exp. Immunol.* 50.17–24.

76. Dienes, H.P., H. Popper, M. Manns, W. Baumann, W. Thoenes, and K-H. Meyer zum Büschenfelde. 1989. Histologic features in autoimmune hepatitis. *Z. Gastroenterol* 27:327–330.

77. Lobo-Yco, A., G. Senaldi, B. Portmann, A.P. Mowat, G. Mieli-Vergani, and D. Vergani. 1990. Class I and II major histocompatibility complex antigen expression on hepatocytes: a study in children with liver disease. *Hepatology* 12:224–232.

78. Löhr, H., U. Treichel, T. Poralla, M. Manns, K-H. Meyer zum Büschenfelde, and B. Fleischer. 1990. The human hepatic asialoglycoprotein receptor is a target antigen for liver infiltrating T cells in autoimmune chronic active hepatitis and primary biliary cirrhosis. *Hepatology* 12:1314–1320.

79. Löhr, H., U. Treichel, T. Portalla, M. Manns, and K.H. Meyer zum Büschenfelde. 1992. Liver-infiltrating T helper cells in autoimmune chronic active hepatitis stimulate the production of autoantibodies against the human asialoglycoprotein receptor *in vitro*. *Clin. Exp. Immunol.* 88:45–49.

80. Homberg, J.C., C. Andre, and N. Abuaf. 1984. A new anti-liver/kidney-microsome antibody (anti-LKM2) in tienilic induced hepatitis. *Clin. Exp. Immunol.* 55:561–570.

81. Rizzetto, M., G. Swana, and D. Doniach. 1973. Micorsomal antibodies in active chronic hepatitis and other disorders. *Clin. Exp. Immunol.* 15:331–344.

82. Manns, M.P., E.F. Johnson, K.J. Griffin, E.M. Tan, and K.F. Sullivan. 1989. Major antigen of liver kidney microsomal antibodies in idiopathic autoimmune hepatitis is cytochrome P450db1. *J. Clin. Invest.* 83:1066–1072.

83. Zanger, U.M., H.P. Hauri, J. Loeper, J.C. Homberg, and U.A. Meyer. 1988. Antibodies against human cytochrome P-450db1 in autoimmune hepatitis type 2. *Proc. Natl. Acad. Sci. USA* 85:8256–8260.

84. Guenguen, M., A.M. Yamamotoh, O. Bernard, and F. Alvarez. 1989. Anti-liver kidney microsome antibody type 1 recognizes human cytochrome P450 db1. *Biochem. Biophys. Res. Comm.* 159:542–547.

85. Obermayer-Straub, P., T. Sugimura, S. Braun, B. Lüttig, S. Loges, A. Kayser, M. Durazzo, C.P. Strassburg, and M.P. Manns. 1998. Definition of a novel CYP2D6 epitope in autoimmune hepatitis type 2 and in chronic hepatitis C. *J. Hepatol.* (Abstract) 28 (Suppl.1):139.

86. Ma, Y., G. Gregorio, J. Gäken, L. Muratori, F.B. Bianchi, G. Mieli-Vergani, and D. Vergani. 1997. Establishment of a novel radioligand assay using eukaryotically expressed cytochrome P4502D6 for the measurement of liver kidney microsomal type 1 antibody in patients with autoimmune hepatitis and hepatitis C virus infection. *J. Hepatol.* 26:1396–1402.

87. Yamamoto, A.M., C. Johanet, J.C. Duclos-Vallee, F.A. Bustarret, F. Alvarez, J.C. Homberg, and J.F. Bach. 1997. A new approach to cytochrome CYP2D6 antibody detection in autoimmune hepatitis type-2 (AIH-2) and chronic hepatitis C virus (HCV) infection: a sensitive and quantitative radioligand assay. *Clin. Exp. Immunol.* 108:396–400.

88. Meyer, U.A., and U.M. Zanger. 1997. Molecular mechanisms of genetic polymorphisms of drug metabolism. *Annu. Rev. Pharmacol. Toxicol.* 37:269–296.

89. Gonzalez, F.J., R.C. Skoda, S. Kimura, M. Umeno, U.M. Zanger, D.W. Nebert, H.V. Gelboin, J.P. Hardwick, and U.A. Meyer. 1988. Characterization of the common genetic defect in debrisoquine metabolism. *Nature* 331:442–449.

90. Manns, M., U. Zanger, G. Gerken, K.F. Sullivan, K.H. Meyer zum Büschenfelde, U.A. Meyer, and M. Eichelbaum. 1990. Patients with type II autoimmune hepatitis express functionally intact cytochrome P450 db1 that is inhibited by LKM1 autoantibodies *in vitro* but not *in vivo*. *Hepatology* 12:127–132.

91. Jacz-Aigain, E., J. Laurent, and F. Alvarez. 1990. Dextrometorphan phenotypes in paediatric patients with auoimmune hepatitis. *Br. J. Pharmacol.* 30:153–154.

92. Yamamoto, A.M., D. Cresteil, O. Boniface, F.F. Clerc, and F. Alvarez. 1993. Identification and analysis of cytochrome P450IID6 antigenic sites recognized by anti-liver-kidney microsome type-1 antibodies (LKM1). *Eur. J. Immunol.* 23:1105–1111.

93. Duclos-Vallee, J.C., O. Hajoui, A.M. Yamamoto, E. Jacqz-Aigrin, and F. Alvarez. 1995. Conformational epitopes on CYP2D6 are recognized by liver/kidney microsomal antibodies. *Gastroenterology* 108:470–476.

94. Muratori, L., M. Lenzi, Y. Ma, M. Cataleta, G. Mieli-Vergani, D. Vergani, and F.B. Bianchi. 1995. Heterogeneity of liver/kidney microsomal antibody type 1 in autoimmune hepatitis and hepatitis C virus related liver disease. *Gut* 37:406–412.

95. Clifford, B.D., D. Donahue, S. Smith, E. Cable, B. Luttig, M.P. Manns, and H.L. Bonkovsky. 1995. High prevalence of serological markers of autoimmunity in patients with chronic hepatitis C. *Hepatology* 231:613–619.

96. Robin, M.A., M. Maratrat, J. Loeper, A.M. Durand-Schneider, M. Tinel, F. Ballet, P. Beaune, G. Feldman, and D. Pessayre. 1995. Cytochrome P4502B follows a vesicular route to the plasma membrane in cultured rat hepatocytes. *Gastroenterology* 108:1110–1123.

97. Robin, M-A., M. Marat, M. Le Roy, F-P. Le Breton, E. Bonierbale, P. Dansette, F. Ballet, D. Mansuy, and D.

Pessayre. 1996. Antigenic targets in tienilic acid hepatitis. Both cytochrome P450 2C11 and 2C11-tienilic acid adducts are transported to the plasma membrane of rat hepatocytes and recognized by human sera. *J. Clin. Invest.* 98:1471–1480.

98. Eliasson, E., and G. Kenna. 1996. Cytochrome P450 2E1 is a cell surface autoantigen in halothane hepatitis. *Mol. Pharmacol.* 50:573–582.

99. Neve, E.P.A., E. Eliasson, M.A. Pronzato, E. Albano, U. Marinari, and M. Ingelman-Sundberg. 1996. Enzyme-specific transport of rat liver cytochrome P450 to the golgi apparatus. *Arch. Biochem. Biophys.* 333:459–465.

100. Loeper, J., V. Descartoire, M. Maurice, P. Beaune, J. Belghiti, D. Houssin, F. Ballet, G. Feldman, F.P. Guenguerich, and D. Pessayre. 1993. Cytochromes P450 in human hepatocyte plasma membrane: recognition by several autoantibodies. *Gastroenterology* 104:203–216.

101. Loeper, J., V. Descartoire, M. Maurice, P. Beaune, G. Feldmann, D. Larrey, and D. Pessayre. 1990. Presence of functional cytochrome P450 on isolated rat hepatocyte plasma membrane. *Hepatology* 11:850–858.

102. Löhr, H., M. Manns, A. Kyriatsoulis, A.W. Lohse, C. Trautwein, K.H. Meyer-zum-Büschenfelde, and B. Fleischer. 1991. Clonal analysis of liver infiltrating cells in patients with chronic active hepatitis (Al-CAH). *Clin. Exp. Immunol.* 84:297–302.

103. Löhr, H.F., J.F. Schlaak, A.W. Lohse, W.O. Böcher, M. Arenz, G. Gerken, and K-H. Meyer zum Büschenfelde. 1996. Autoreactive CD4+ LKM-specific and anticlonotypic T-cell responses in LKM-1 antibody-positive autoimmune hepatitis. *Hepatology* 24:1416–1421.

104. Durazzo, M., T. Philipp, F.N.A.M. Van Pelt, B. Lüttig, E. Borghesio, G. Michel, E. Schmidt, S. Loges, M. Rizzetto, and M.P. Manns. 1995. Heterogeneity of microsomal autoantibodies (LKM) in chronic hepatitis C and D virus infection. *Gastroenterology* 108:455–462.

105. Philipp, T., M. Durazzo, C. Trautwein, B. Alex, P. Straub, J.G. Lamb, E.F. Johnson, R.H. Tukey, and M.P. Manns. 1994. Recognition of uridine diphosphate glucuronosyl transferases by LKM-3 antibodies in chronic hepatitis D. *Lancet* 344:578–581.

106. Strassburg, C., P. Obermayer-Straub, B. Alex, M. Durazzo, M. Rizzetto, R.H. Tukey, and M.P. Manns. 1996. Autantibodies against glucuronosyltransferases differ between viral hepatitis and autoimmune hepatitis. *Gastroenterology* 11:1582–1592.

107. Crivelli, O., C. Lavarini, E. Chiaberge, A. Amoroso, P. Farci, F. Negro, and M. Rizzetto. 1983. Microsomal autoantibodies in chronic infection with HBsAg associated delta (delta) agent. *Clin. Exp. Immunol.* 54:232–238.

108. Obermayer-Straub, P., C. Strassburg, M-G. Clemente, T. Philipp, R.H. Tukey, and M.P. Manns. 1995. Recognition of three different epitopes on UDP-glucuronosyltransferases by LKM-3 antibodies in patients with autoimmune hepatitis and hepatitis D. *Gut* 37(2):A100.

109. Bourdi, M., D. Larrey, J. Nataf, J. Bernuau, D. Pessayre, M. Iwasaki, F.P. Guengerich, and P.H. Beaune. 1990. Anti-liver endoplasmic reticulum antibodies are directed against human cytochrome P-4501A2. A specific marker of dihydralazine hepatitis. *J. Clin. Invest.* 85:1967–1973.

110. Manns, M.P., K.J. Griffin, L. Quattrochi, M. Sacher, H. Thaler, R. Tukey, and E.F. Johnson. 1990. Identification of cytochrome P450 IA2 as a human autoantigen. *Arch. Biochem. Biophys.* 280:229–232.

111. Obermayer-Straub, P., S. Braun, S. Loges, M.G. Clemente, G. Dalekos, B. Lüttig, B. Grams, J. Perheentupa, and M.P. Manns. 1998. Cytochromes P4501A2 and P4502A6 are hepatic autoantigens in the autoimmune polyglandular syndrome type 1 (APS-1). 12th International Symposium on Microsomes and Drug Oxidations in Montpellier, France, 20–24 July 1998: Abstract 416.

112. Abuaf, N., C. Johanet, P. Chretien, E. Soulier, S. Laperche, and J.C. Homberg. 1992. Characterization of the liver cytosol antigen type 1 reacting with autoantibodies in chronic active hepatitis. *Hepatology* 16:892–898.

113. Lapierre, P., O. Hajoui, J-C. Homberg, and F. Alvrez. 1999. Fomiminotransferase cyclodeaminase is an organ specific autoantigen recognized by sera of patients with autoimmune hepatitis. *Gastroenterology* 116:643–649.

114. Lenzi, M., P. Manotti, L. Muratori, M. Cataleta, G. Ballardini, F. Cassani, and F.B. Bianchi. 1995. Liver cytosolic 1 antigen-antibody system in type 2 autoimmune hepatitis and hepatitis C virus infection. *Gut* 36:749–754.

115. Muratori, L., M. Cataleta, P. Muratori, P. Manotti, M. Lenzi, F. Cassani, and F.B. Bianchi. 1995. Detection of anti-liver cytosol antibody type 1 (anti-LC1) by immunodiffusion, counterimmunoelectrophoresis and immunoblotting: comparison of different techniques. *J. Immunol. Meth.* 187:259–264.

116. Muratori, L., M. Cataleta, P. Muratori, M. Lenzi, and F.B. Bianchi. 1998. Liver/kidney microsomal antibody type 1 and liver cytosol antibody type 1 concentrations in type 2 autoimmune hepatitis. *Gut* 42:721–726.

117. Czaja, A.J., H.A. Carpenter, and M.P. Manns. 1993. Antibodies to soluble liver antigen, P450IID6, and mitochondrial complexes in chronic hepatitis. *Gastroenterology* 105:1522–1528.

118. Lohse, A.W., G. Gerkem, H. Mohr, H.F. Löhr, U. Treichel, H.P. Dines, and K.H. Meyer zum Büschenfelde. 1995. Relation between autoimmune hepatitis and viral hepatitis: clinical and serological characteristics in 859 patients. *Z. Gastroenterol.* 33:527–533.

119. Wächter, B., A. Kyriasoulis, A.W. Lohse, G. Gerken, K.H. Meyer zum Büschenfelde, and M. Manns. 1990. Characterization of liver cytokeratin as a major antigen of anti-SLA antibodies. *J. Hepatol.* 11:232–239.

120. Wesierska-Gaedek, J., R. Frimm, E. Hitchman, and E. Penner. 1998. Members of the glutathione S-transferase gene family are antigens in autoimmune hepatitis. *Gastroenterology* 1998:329–335.

121. Wies, I., J. Henninger, S. Brunner, C. Waldman, U. Denzer, S. Kanzler, G. Gerken, M. Arand, M.P. Manns, K.H. Meyer zum Büschenfelde, and A.W. Lohse. 1998. Cloning of the target-antigen of antibodies to soluble liver antigen: identification of a novel 50 kDa protein with two splice-variants. *Hepatology* 28 (Suppl.4):A112.

122. Gregorio, G.V., B. Portman, F. Reid, P.T. Donaldson, D.G. Doherty, M. McCartney, D. Vergani, and G. Mieli–Vergani. 1997. Autoimmune hepatitis in childhood: a 20-year experience. *Hepatology* 25:541–547.

123. Czaja, A.J., M.P. Manns, and H.A. Homburger. 1992. Frequency and significance of antibodies to liver/kidney microsome type 1 in adults with chronic active hepatitis. *Gastroenterology* 103:1290–1295.

124. Lindgren, S., H.B. Braun, G. Michel, A. Nemeth, S. Nilsson, B. Thome-Kromer, and S. Eriksson. 1997. Absence of LKM-1 antibody reactivity in autoimmune and hepatitis-

C-related chronic liver disease in Sweden. Swedish Internal Medicine Liver club. *Scand-J-Gastroenterol.* 32:175–178.

125. Tan, Tan, E.M. 1991. Autoantibodies in pathology and cell biology. *Cell* 67:841–842.

126. Svejgaard, A., P. Platz, and L.P. Ryder. 1983. HLA and disease 1982—a survey. *Immunol. Rev.* 70:193–218.

127. Seki, T., M. Ota, S. Furuta, H. Fukushima, T. Kondo, K. Hino, and N. Mizuki. 1992. HLA class II molecules and autoimmune hepatitis susceptibility in Japanese patients. *Gastroenterology* 103:1041–1047.

128. Marcos, Y., H.A. Fainboim, M. Capucchio, J. Findor, J. Daruich, B. Reyes, M. Pando, G.C. Theiler, N. Mendez, and M.L. Satz, al. et. 1994. Two locus involvement in the association of human leukocyte antigen with the extrahepatic manifestations of autoimmune chronic active hepatitis. *Hepatology* 19:1371–1374.

129. Vazquez-Garcia, M.N., C. Alaez, A. Olivo, H. Debaz, E. Perez-Luque, A. Burguete, S. Cano, G. de la Rosa, N. Bautisa, A. Hernandez, J. Bandera, L.F. Torres, F. Alvarez, and C. Gorodezky. 1998. MHC class II sequences of susceptibility and protection in Mexicans with autoimmune hepatitis. *J. Hepatol.* 28:985–990.

130. Freudenberg, J., H. Baumann, W. Arnold, J. Berger, and K.H. Meyer zum Büschenfelde. 1977. HLA in different forms of chronic active hepatitis: a comparison between adult patients and children. *Digestion* 15:260–270.

131. Wright, R., O.F. Eade, M. Chisholm, M. Hawksley, B. Lloyd, T.M. Moles, J.C. Edwards, and M.J. Gardner. 1975. Controlled prospective study of the effect on liver function of multiple exposures to halothane. *Lancet* 1:817–820.

132. Mackay, I.R., and B.D. Tait. 1980. HLA association with autoimmune type chronic active hepatitis: identification of B8-DRw3 haplotype by family studies. *Gastroenterology* 79:95–98.

133. Mackay, I.R., and P.J. Morris. 1972. Association of autoimmune chronic hepatitis with HLA-A1-B8. *Lancet* 2:793–795.

134. Tait, B., I.R. Mackay, P. Board, M. Coggan, P. Emery, and G. Eckhardt. 1989. HLA-A1, -B8, -DR3 extended haplotypes in autoimmune chronic hepatitis. *Gastroenterology* 97:479–481.

135. Nouri Aria, K.T., P.T. Donaldson, J.E. Hegarty, A.L. Eddleston, and R. Williams. 1985. HLA A1-B8-DR3 and suppressor cell function in first degree relatives of patients with autoimmune chronic active hepatitis. *J. Hepatol.* 1:235–241.

136. Whittingham, S., J.D. Mathews, M.S. Schanfield, B.D. Tait, and I.R. Mackay. 1981. Interaction of HLA and Gm in autoimmune chronic active hepatitis. *Clin. Exp. Immunol.* 26:102–108.

137. Donaldson, P.T., D.G. Doherty, K.M. Hayllar, I.G. Mc Farlane, P.J. Johnson, and R. Williams. 1991. Susceptibility to autoimmune chronic active hepatitis: human leukocyte antigens DR 4 and A1-B8-DR-3 are independent risk factors. *Hepatology* 13:701–706.

138. Czaja, A.J., J. Rakela, J.E. Hay, and S.B. Moore. 1990. Clinical and prognostic implications of human leucocyte antigen B8 in corticosteroid-treated severe chronic active hepatitis. *Gastroenterology* 98:1587–1593.

139. Czaja, A.J., H.A. Carpenter, P.J. Santrach, and S.B. Moore. 1993. Significance of HLA DR4 in type 1 autoimmune hepatitis. *Gastroenterology* 105:1502–1507.

140. Czaja, A.J., H.A. Carpenter, P.J. Santrach, and S.B. Moore. 1993. Genetic predispositions for the immunological features of chronic active hepatitis. *Hepatology* 18:816–822.

141. Strettel, M.D.J., P.T. Donaldson, L.J. Thomson, P.J. Santrach, S. Brenndan Moore, A.J. Czaja, and R. Williams. 1997. Allelic basis for HLA-encoded susceptibility to type 1 autoimmune hepatitis. *Gastroenterology* 112:2028–2035.

142. Manabe, K., P.T. Donaldson, J.A. Underhill, D.G. Doherty, G. Mieli-Vergani, I.G. McFarlane, A.L.W.F. Eddlestin, and R. Williams. 1993. Human Leukocyte antigen A1-B8-DR3-DQ2-DPB1*0401 extended haplotype in autoimmune hepatitis. *Hepatology* 18:1334–1337.

143. Strettell, M.D., L.J. Thomson, P.T. Donaldson, M. Bunce, C.M. O'Neill, and R.W. Williams. 1997. HLA-C genes and susceptibility to type 1 autoimmune hepatitis. *Hepatology* 26:1023–1026.

144. Brown, J.H., T.S. Jardetski, J.C. Gorga, L.J. Stern, R.G. Urban, J.L. Strominger, and D.C. Wiley. 1993. Three-dimensional structure of the human class II histocompatibility antigen HLA-DR1. *Nature* 364:33–39.

145. Miyamori, H., Y. Kato, K. Kobayashi, and N. Hattori. 1983. HLA antigens in Japanese patients with primary biliary cirrhosis and autoimmune hepatitis. *Digestion* 26:213–217.

146. Ota, M., T. Seki, K. Kiyoshawa, S. Futura, K. Hino, T. Kondo, H. Fukushima, K. Tsuji, and H. Inoko. 1992. A possible association between basic amino acids of position 13 of DRB1 chains and autoimmune hepatitis. *Immunogenetics* 36:40–55.

147. Seki, T., K. Kiyosawa, H. Inoko, and M. Ota. 1990. Association of autoimmune hepatitis with HLA-Bw54 and DR4 in Japanese patients. *Hepatology* 12:1300–1304.

148. Munoz, L.E., D. DeVilliers, D. Markham, K. Whaley, and H.C. Thomas. 1982. Complement activation in chronic liver disease. *Clin. Exp. Immunol.* 47:548–554.

149. Scully, L.J., C. Toze, D.P.S. Sengar, and R. Goldstein. 1993. Early-onset autoimmune hepatitis is associated with a C4A gene deletion. *Gastroenterology* 104:1478–1484.

150. Doherty, D.G., J.A. Underhill, P.T. Donaldson, K. Manabe, G. Mieli-Vergani, A.L. Eddleston, D. Vergani A.G. Demaine, and R. Williams. 1994. Polymorphism in the human complement C4 genes and genetic susceptibility to autoimmune hepatitis. *Autoimmunity* 18:243–249.

151. Vergani, D., L. Wells, V.F. Larcher, B.A. Nasaruddin, E.T. Davies, G. Mieli-Vergani, and A.P. Mowat. 1985. Genetically determined low C4: a predisposing factor to autoimmune chronic active hepatitis. *Lancet* 2:294–298.

152. Wilson, A.G., F.S. di Giovine, and G.W. Duff. 1995. Genetics of tumor necrosis factor alpha in autoimmune, infectious and neoplastic diseases. *J. Inflamm.* 45:1–12.

153. Bernal, W., M. Moloney, J. Underhill, and P.T. Donaldson. 1999. Association of tumor necrosis factor polymorphism with primary sclerosing cholangitis. *J. Hepatol.* 30:237–241.

154. Jones, D.E., F.E. Watt, J. Grove, J.L. Newton, A.K. Daly, W.L. Gregory, C.P. Day, O.F. James, and M.F. Bassendine. 1999. Tumor necrosis factor-alpha promoter polymorphisms in primary biliary cirrhosis. *J. Hepatol.* 30:232–236.

155. Verweij, C.L., and T.W. Huizinga. 1998. Tumor necrosis factor alpha gene polymorphisms and rheumatic diseases. *Br. J. Rheumatol.* 37:923–926.

156. Bradham, C.A., J. Plumpe, M.P. Manns, D.A. Brenner, and C. Trautwein. 1998. Mechanisms of hepatic toxicity. 1. TNF-induced liver injury. *Am. J. Physiol.* 275:G385–392.

157. Roberts, S.K., T.M. Therneau, and A.J. Czaja. 1996. Prognosis of histological cirrhosis in type 1 autoimmune hepatitis. *Gastroenterology* 110:848–857.

158. Johnson, P.J., I.G. McFarlane, and R. Williams. 1995. Azathioprine for long-term maintenance of remission in autoimmune hepatitis. *N. Engl. J. Med.* 333:958–963.

159. Czaja, A.J. 1999. Drug therapy in the management of type 1 autoimmune hepatitis. *Drugs* 57:49–68.

160. Czaja, A.J., G.L. Davis, J. Ludwig, and H.F. Tashwell. 1984. Complete resolution of inflammatory activity following corticosteroid treatment of HBsAg-negative chronic active hepatitis. *Hepatology* 4:622–627.

161. Rambusch, E.G., and M.P. Manns. 1996. Therapie der Autoimmunhepatitis. *Dtsch. med. Wschr.* 121:1539–1542.

162. Sanchez-Urdazpal, L., A.J. Czaja, and B. Van Holk. 1991. Prognostic features and role of liver transplantation in severe corticoid-treated autoimmune chronic active hepatitis. *Hepatology* 15:215–221.

163. Mistilis, S.P., C.R. Vickers, M.H. Darroch, and S.W. McCarthy. 1985. Cyclosporin, a new treatment for autoimmune chronic active hepatitis. *Med. J. Aust.* 143:463–465.

164. Hyams, J.S., M. Ballow, and A.M. Leichtner. 1987. Cyclosporine treatment of autoimmune chronic active hepatitis. *Gastroenterology* 93:890–893.

165. Minuk, G.Y. 1989. A. Cyclosporin in nontransplant-related liver disease. *Am. J. Gastroenterol.* 84:1345–1350.

166. Person, J.L., J.G. McHutchison, T-L. Fong, and A.G. Redeker. 1993. A case of cyclosporine-sensitive, steroid resistant, autoimmune chronic active hepatitis. *J. Clin. Gastroenterol.* 17:317–320.

167. Sherman, K.E., M. Narkewicz, and P.C. Pinto. 1994. Cyclosporine in the management of corticosteroid-resistant type 1 autoimmune chronic active hepatitis. *J. Hepatol.* 21:1040–1047.

168. Jackson, L.D., and E. Song. 1995. Cyclosporn in the treatment of corticosteroid resistant autoimmune chronic active hepatitis. *Gut* 36:459–461.

169. Fernandez, N.F., A.G. Redeker, J.M. Vierling, F.G. Villamil, and T-L. Fong. 1999. Cyclosporine therapy in patients with steroid resistant autoimmune hepatitis. *Am. J. Gastroenterol.* 94:241–248.

170. Alvarez, F., M. Ciocca, C. Canero-Velasco, M. Ramonet, M.T.G. Davila d, M. Cuaterolo, T. Gonzalez, P. Jara-Vega, C. Camarena, P. Brochu, R. Drut, and E. Alvarez. 1999. Short-term cyclosporine induces a remission of autoimmune hepatitis in children. *J. Hepatol.* 30:222–227.

171. Debray, D., G. Maggiore, J.P. Giradet, E. Mallet, and O. Bernard. 1999. Efficacy of cyclosporin A in children with type 2 autoimmune hepatitis. *J. Pediatr.* 135:111–114.

172. Van Thiel, D.H., H. Wright, P. Carroll, K. Abu-Elmagd, H. Rodriguez-Rilo, J. McMichael, W. Irish, and T.E. Starzl. 1995. Tacrolimus: A potential new treatment for autoimmune chronic active hepatitis: results of an open-label preliminary trial. *Am. J. Gastroenterol.* 90:771–776.

173. Danielson, A., and H. Prytz. 1994. Oral budesonide for treatment of autoimmune chronic hepatitis. *Aliment. Pharmacol. Ther.* 8:585–590.

174. Schüler, A., and M.P. Manns. 1995. Treatment of autoimmune hepatitis. In: V. Arroyo, J. Bosch, J. Rodés, eds. Treatment in Hepatology: Masson, S.A., 375–383.

175. Pratt, D.S., D.P. Flavin, and M.M. Kaplan. 1996. The successful treatment of autoimmune hepatitis with 6-mercaptopurine after failure with azathioprine. *Gastroenterology* 110:271–274.

176. Kanzler, S., G. Gerken, H.P. Dienes, K-H. Meyer zum Büschenfelde, and A.W. Lohse. 1996. Cyclophosphamide as alternative immunosuppressive therapy for autoimmune hepatitis—report of three cases. *Z. Gastroenterol.* 35:571–578.

177. European Liver Transplant Registry, June 1996.

178. Ratziu, V., D. Samuel, M. Sebagh, O. Farges, F. Saliba, P. Ichai, H. Farahmand, M. Gigou, C. Feray, M. Reynes, and H. Bismuth. 1999. Long-term follow up after liver transplantation for autoimmune hepatitis: evidence of recurrence of primary disease. *J. Hepatol.* 30:121–141.

179. Wright, H.L., C.F. Bou-Abboud, T. Hassanenstein, G.D. Block, A.J. Demetris, T.E. Starzi, and D.H. Van Thiel. 1992. Disease recurrence and rejection following liver transplantation for autoimmune chronic active liver disease. *Transplatation* 53:136–139.

180. Devlin, J., P. Donaldson, and B. Portman. et al. 1995. Recurrence of autoimmune hepatitis following liver transplantation. *Liver Transpl. Surg.* 1:162–165.

181. Ahmed, M., D. Mutimer, M. Hathaway, S. Hubscher, P. McMaster, and E. Elias. 1997. Liver transplantation for autoimmune hepatitis: a 12-year experience. *Transplant. Proc.* 29:496.

182. Prados, E., V. Cuervas-Mons, M. De La Mata, E. Fraga, A. Rimola, M. Prieto, G. Clemente, E. Vicente, T. Casanovas, and E. Fabrega. 1998. Outcome of autoimmune hepatitis after liver transplantation. *Transplantation* 66:1645–1650.

183. Kerkar, N., N. Hadizic, E. Davies, B. Portman, P.T. Donaldson, M. Rela, N.D. Heaton, D. Vergani, G. Mieli-Vergani. 1998. De-novo autoimmune hepatitis after liver transplantation. *Lancet* 351:409–413.

184. Hernandez, Albujar, A., E. Alvarez, M. Salcedo, B. Piqueras, F.G. Duran, R. Banares, M. Rodriguez Mahou, E. Cos, and G. Clemente. 1997. Autoimmune mediated liver disease after liver transplantation (LT). *J. Hepatol.* 26:A152.

185. Aaltonen, J., P. Börses, L. Sandkuiji, J. Perheentupa, and L. Peltonen. 1994. An autosomal locus causing autoimmune disease: autoimmune polyglandular disease type 1 assigned to chromosome 21. *Nature Genet.* 8:83–87.

186. Ahonen, P., S. Koskimies, M.L. Lokki, A. Tiilikainen, and J. Perheentupa. 1988. The expression of autoimmune polyglandular disease type 1 appeared associated with several HLA-A antigens but not with HLA-DR. *J. Clin. Endocrinol. Metabol.* 66:1152–1157.

187. Perheentupa, J. 1996. Autoimmune polyendocrinopathy-candidiasis-ectodermal dystrophy (APECED). *Horm. Metab. Res.* 28:353–356.

188. Ahonen, P., S. Myllärniemi, I. Sipilä, and J. Perheentupa. 1990. Clinical variation of autoimmune polyendocrinopathy-candidiasis-ectodermal dystrophy (APECED) in a series of 68 patients. *N. Engl. J. Med.* 322:1829–1836.

189. Neufeld, M., N.K. Maclaren, and R.M. Blizzard. 1980. Two types of autoimmune Addison's disease associated with different Polyglandular Autoimmune (PGA) Syndromes. *Medicine* 60:355–362.

190. Michele, T.M., J. Fleckenstein, A.R. Sgrignoli, and P.J. Tuluvath. 1994. Chronic active hepatitis in the type 1 polyglandular autoimmune syndrome. *J. Postgrad. Med.* 70:128–131.

191. Ahonen, P., A. Miettinen, and J. Perheentupa. 1987. Adrenal and steroidal cell antibodies in patients with autoimmune polyglandular disease type 1 and risk of adrenocortical and ovarian failure. *J. Clin. Endocrinol. Metabol.* 64:494–500.

192. Betterle, C., A. Caretto, B. Pedini, F. Rigon, P. Bertoli, and A. Peserico. 1992. Complement-fixing activity to melanin-producing cells preceding the onset of vitiligo in a patient with type 1 polyglandular failure [letter]. *Arch. Dermatol.* 128:123–124.

193. The Finnish-German APECED Consortium. 1997. An autoimmune disease, APECED, caused by mutations in a novel gene featuring two. PHD-type zinc finger domains. *Nat. Genet.* 17:399–403.

194. Nagamine, K., P. Peterson, H.S. Scott, J. Kudoh, S. Minoshima, M. Heino, K.J.E. Krohn, M.D. Lalioti, P.E. Mullis, S.E. Antonarakis, K. Kawasaki, S. Asakawa, F. Ito, and N. Shimiziu. 1997. Positional cloning of the APECED gene. *Nature Genetics* 17:393–398.

195. Padeh, S., R. Theodor, A. Jonas, and J.H. Passwell. 1997. Severe malabsorption in autoimmune polyendo-crinopathy-candidosis-ectodermal dystrophy syndrome successfully treated with immunosuppression. *Arch. Dis. Child.* 76:532–534.

196. Wang, C.Y., H.H. Davppdi-Semiromi, E. Connor, J.-D. Shi, and J.-X. She. 1998. Characterization of mutations in patients with autoimmune polyglandular syndrome type 1 (APS1). *Hum Genet.* 103:681–685.

197. Pearce, S.H.S., T. Cheetman, H. Imrie, B. Vaidya, N.D. Barnes, R.W. Bilous, D. Carr, K. Meeran, N.J. Shaw C.S. Smith, A.D. Toft, W. Gareth, and P. Kendall-Taylor. 1998. A common and recurrent 13-bp deletion in the autoimmune regulator gene in British Kindreds with autoimmune polyendocrinopathy type 1. *Am. J. Hum. Genet.* 63:1675–1684.

198. Scott, H.S., M. Heino, P. Peterson, L. Mittaz, M.D. Lalioti, C. Betterle, A. Cohen, M. Seri, M. Lerone, P. Collin, M. Salo, R. Metcalfe, A. Weetman, M.P. Paoasavvas, C. Rossier, K. Nagamine, J. Kudoh, N. Shimizu, K. Krohn, and S. Antonarakis. 1998. Common mutations in autoimmune polyendocrinopathy-candidiasis ectodermal dystrophy patients of different origins. *Mol. Endocrinol.* 12:1112–1119.

199. Heino, M., H.S. Scott, Q. Chen, P. Peterson, U. Maebppaa, M.P. Papasavvas, L. Mittaz, C. Barras, C. Rossier, G.P. Chousos, C.A. Stratakis, K. Nagamine, J. Kudoh, N. Shimizu, N. Maclaren, S.E. Antonarakis, and K. Krohn. 1999. Mutation analyses of North American APS-1 patients. *Hum. Mutat.* 13:69–74.

200. Rosatelli, M.C., A. Meloni, A. Meloni, M. Devoto, A. Cao, H.S. Scott, P. Peterson, M. Heino, K.J.E. Krohn, K. Nagamine, J. Kudoh, N. Shimizu, and S. Antonakis. 1998. A common mutation in Sardinian autoimmune polyen-docrinopathy-candidiasis-ectodermal dystrophy patients. *Hum. Genet.* 103:428–434.

201. Blechschmidt, K., M. Schweiger, K. Wertz, R. Poulson, H.M. Christensen, A. Rosenthal, H. Lehrach, and M.L. Yaspo. 1999. The mouse Aire gene: comparative genomic sequencing, gene organization, and expression. *Genome Res.* 9:158–166.

202. Heino, M., P. Peterson, J. Kudoh, K. Nagamine, A. Lagerstedt, V. Ovod, A. Ranki, I. Rantala, M. Nieminen, J. Tuukkanen, H.S. Scott, S.E. Antonarakis, N. Shimizu, and K. Krohn. 1999. Autoimmune regulator is expressed in the cells regulating immune tolerance in thymus medulla. *Biochem. Biophys. Res. Comm.* 257:821–825.

203. Rinderle, C., H.M. Christensen, S. Schweiger, H. Lehrach, and M.L. Yaspo. 1999. AIRE encodes a nuclear protein co-localizing with cytoskeletal filaments: altered sub-cellular distribution of mutants lacking the PHD zinc fingers. *Hum. Mol. Genet.* 8:277–290.

204. Betterle, C., M. Volpato, B.R. Smith, J. Fumaraniak, S. Chen, N.A. Greggio, M. Sanzari, F. Tedesco, B. Pedini, M. Boscaro, and F.I. Presoto. 1997. Adrenal cortex and steroid 21-hydroxylase autoantibodies in adult patients with organ-specific autoimmune diseases: markers of low progression to clinical Addison's disease. *J. Clin. Endocrinol. Metabol.* 82:932–938.

205. Neufeld, M., N. Maclaren, and R. Blizzard. 1980. Autoimmune polyglandular syndromes. *Pediatr. Ann.* 9:154–162.

206. Riley, W.J. 1992. Autoimmune polyglandular syndromes. *Horm. Res.* 38:9–15.

207. Clemente, M.G., P. Obermayer-Straub, A. Meloni, C.P. Strassburg, V. Arangino, R.H. Tukey, S. De Virgiliis, and M.P. Manns. 1997. Cytochrome P4501A2 is a hepatic autoantigen in autoimmune polyglandular syndrome type 1. *J. Clin. Endocrinol. Metabol.* 1353–1361.

208. Gebre-Medhin, G., E.S. Husebye, J. Gustafsson, O. Winqvist, A. Goksoyr, F. Rorsman, and O. Kämpe. 1997. Cytochrome P4501A2 and aromatic L-amino acid decar-boxylase are hepatic autoantigens in autoimmune polyen-docrine syndrome type 1. *FEBS Lett.* 412:439–445.

209. Obermayer-Straub, P., S. Braun, B. Grams, S. Loges, M.G. Clemente, B. Lüttig, J. Perheentupa, and M.P. Manns. 1996. Different liver cytochromes P450s are autoantigens in patients with autoimmune hepatitis and with auto-immune polyglandular syndrome type 1 (APS1). *Hepatol.* 24:429.

210. Rorsman, F., E.S. Husebye, O. Winqvist, E. Björk, F.A. Karlsson, and O. Kämpe. 1995. Aromatic-L-amino acid decarboxylase, a pyridoxal phosphate-dependent enzyme, is a beta cell autoantigen. *Proc. Natl. Acad. Sci. USA* 92:8626–8629.

211. Strassburg, C.P., and M.P. Manns. 1995. Autoimmune hepatitis virsus viral hepatitis C. *Liver* 15:225–232.

212. Pawlotsky, J.M., M. Ben Yahia, C. Andre, M.-C. Voisin, L. Intrator, F. Roudot-Thoraval, L. Deforges, C. Duvoux, E.-L. Zafrani, J. Duval, and D. Dhumeaux. 1994. Immunological disorders in C virus chronic active hepatitis: a prospective case-control study. *Hepatology* 19:841–848.

213. Agnello, V., R.T. Chung, and L. Kaplan. 1992. A role of hepatitis C virus infection in type II cryoglobulinemia. *N. Engl. J. Med.* 327:1490–1495.

214. Cacoub, P., F. Lunel-Fabiani, and L.T. Huong Du. 1992. Polyarteritis nodosa and hepatitis C infection. *Ann. Int. Med.* 116:605–606.

215. Fargion, S., A. Peperno, M.D. Cappellini, P. Palla, A. Moretti, E. Marzo, and A. Mazzoni. 1992. Hepatitis C virus and porphyria cutanea tarda: evidence of a strong associ-ation. *Hepatology* 16:1322–1326.

216. Haddad, J., P. Deny, C. Munz-Gotheil, G. Pasero, S. Bombardieri, and P. Highfield. 1992. Lymphocytic sialadenitis of Sjörgen's syndrome associated with chronic hepatitis c virus liver disease. *Lancet* 339:321–323.

217. DeCastro, M., J. Sanchez, J.F. Herrera, A. Chaves, R. Duran, L. Garcia-Buey, and C. Garcia-Monzon. 1993. Hepatitis C virus antibodies and liver disease in patients with porphyria cutanea tarda. *Hepatology* 17:551–557.

218. Ferri, C., F. Greco, G. Longombardo, P. Palla, A. Moretti, E. Marzo, A.P. Mazzoni, G., S. Bombardieri, and P. Highfield, al. e. 1991. Association between hepatitis C virus and mixed cryoglobulinemia. *Clin. Exp. Rheumatol.* 9:621–624.

219. Tran, A., J.F. Quaranta, S. Benzaken, V. Thiers, H. Chan, P. Hastier, D. Regnier, and G. Dreyfus. 1993. High prevalence of thyroid antibodies in a prospective series of patients with chronic hepatitis C before interferon therapy. *Hepatology* 18:253–257.

220. Yamamoto, A.M., D. Cresteil, J.C. Homberg, and F. Alvarez. 1993. Characterization of anti-liver-kidney microsome antobody (anti-LKM1) from hepatitis C virus-positive and -negative sera. *Gastroenterology* 104:1762–1767.

221. Hadziyannis, S.J. 1997. The spectrum of exttrahepatic manifestations in hepatitis C infection. *J. Viral. Hepat.* 4:9–28.

222. Shindo, M., A.M. DiBisceglie, J.H. H. 1992. Acute exacerbation of liver disease during interferon alpha therapy for chronic hepatitis C. *Gastroenterology* 102:1406–1408.

223. Garcia-Buey, L., C. Garcia-Monzon, S. Rodriguez, M.J. Borque, A. Garcia-Sanchez, R. Iglesias, M. DeCastro, F.G. Mateos, J.L. Vicaro, A. Balas, and R. Moreno-Otero. 1995. Latent autoimmune hepatitis triggered during interferon therapy in patients with chronic hepatitis C. *Gastroenterology* 108:1770–1777.

224. Eddleston, A.L.W. 1996. Hepatitis C infection and autoimmunity. *J. Hepatol.* 24 (Suppl.2):55–60.

225. Mackie, F.D., M. Peakman, Y. Ma, R. Sallie, H. Smith, E.T. Davis, G. Mieli-Vergani, and G. Vergani. 1994. Primary and secondary liver/kidney microsomal response following infection with hepatitis C virus. *Gastroenterology* 106:1672–1675.

226. Miyakawa, H., E. Kitazawa, K. Abe, N. Kawaguchi, H. Fuzikawa, K. Kikuchi, M. Kako, T. Komastsu, N. Kayashi, and K. Kiyosawa. 1997. Chronic hepatitis C associated with anti-liver/kidney microsome-1 antibody is not a subgroup of autoimmune hepatitis. *J. Gastroenterol* 32:769–776.

227. Abuaf, N., F. Lunel, P. Giral, E. Borotto, S. Laperche, R. Poupon, P. Opolon, J.M. Huraux, and J.C. Homberg. 1993. Non-organ specific autoantibodies associated with chronic C virus hepatitis. *J. Hepatol.* 18:359–364.

228. Giostra, F., A. Manzin, M. Lenzi, R. Francesconi, L. Solforosi, P. Manotti, L. Muratori, D. Zauli, M. Clementi, and F.B. Bianchi. 1996. Low hepatitis C viremia levels in liver/kidney microsomal antibody type 1-positive chronic hepatitis. *J. Hepatol.* 25:433–438.

229. Todros, L., G. Touscoz, N. D'Urso, M. Durazzo, E. Albano, G. Poli, M. Baldi, and M. Rizzetto. 1991. Hepatitis C virus-related chronic liver disease with autoantibodies to liver-kidney microsomes (LKM). Clinical characterization from idiopathic LKM-positive disorders. *J. Hepatol.* 13:128–131.

230. Nishioka, M., S.A. Morshed, K. Kono, T. Himoto, S. Parveen, K. Arima, S. Watanabe, and M.P. Manns. 1997. Frequency and significance of antibodies to P450IID6 protein in Japanese patients with chronic hepatiis C. *J. Hepatol.* 26:992–1000.

231. Muratori, L., M. Lenzi, M. Cataleta, M. Giostra, and G. Ballardini. 1994. Interferon therapy in liver/kidney microsomal antibody type 1-positive patients with chronic hepatitis C. *J. Hepatology* 21:199–203.

232. Duclos-Vallee, J.-C., M. Nishioka, N. Hosomi, K. Arima, A. Leclerc, J.-F. Bach, and A.M. Yamamoto. 1998. Interferon therapy in LKM-1 positive patients with chronic hepatitis C: follow-up by a quantitative radioligand assay for CYP2D6 antibody detection. *J. Hepatol.* 28:965–970.

233. Dalekos, G.N., H. Wedemeyer, P. Obermayer-Straub, A. Kayser, A. Barut, H. Frank, and M.P. Manns. 1999. Epitope mapping of cytochrome P4502D6 autoantigen in patients with chronic hepatitis C during α-interferon treatment. *J. Hepatol.* 30:366–375.

234. Mishiro, S., Y. Hoshi, K. Takeda, A. Yoshihiro, T. Gotanda, K. Takahashi, Y. Akahane, H. Yoshizawa, H. Okamoto, F. Tsuda, D.A. Peterson, and E. Michmore. 1990. Non-A, non-B hepatitis specific antibodies directed at host-derived epitope: implication for an autoimmune process. *Lancet* 336:1400–1403.

235. Michel, G., A. Ritter, G. Gerken, K.H. Meyer zum Büschenfelde, R. Decker, and M.P. Manns. 1992. Anti-GOR and hepatitis C virus in autoimmune liver diseases. *Lancet* 339:267–269.

236. Quiroga, J.A., M. Pardo, S. Navas, J. Martin, and V. Carreno. 1996. Patterns of immune responses to host-encoded GOR and hepatitis C virus-core derived epitopes with relation to hepatitis C viremia, genotypes and liver disease severity. *J. Infect. Dis.* 173:300–305.

237. Roschlau, G., R. Baumgarten, and J.D. Fengler. 1990. Dihydralazine hepatitis. Morphologic and clinical criteria for diagnosis. *Zentralbl. Allg. Pathol.* 136:127–134.

238. Van Pelt, F., P. Straub, and M. Manns. 1995. Molecular basis of drug-induced immunological liver injury. *Semin. Liv. Dis.* 15:283–300.

239. Dutton, G.J. 1980. In: Dutton G.J. ed. Glucuronidation of drugs and other compounds. *Boca Raton,* FL: CRC Press, 69–78.

240. Mackenzie, O.I., I.S. Owens, B. Burchell, K.W. Bock, A. Bairoch, A. Belanger, S. Fournel-Gigleux, M. Green, D.W. Hum, T. Iyanagi, D. Lancet, P. Louisot, J. Magdalou, J. Roy Chowdhury, J.K. Ritter, H. Schachter, T.R. Tephly, K.F. Tipton, and D.W. Nebert. 1997. The UDP glucronosyltransferase gene superfamily: recommended nomenclature update based on evolutionary divergence. *Pharmaco-genetics* 7:255–269.

241. Kenna, J.G. 1997. Immunoallergic drug-induced hepatitis: lessons from halothane. *J. Hepatology* 26:5–12.

242. Siegmund, W., G. Franke, K.E. Biebler, I. Donner, R. Kawellis, M. Kairies, A. Scherber, and H. Huller. 1985. The influence of the acetylator phenotype on the clinical use of dihydralazine. *Int. J. Clin. Pharmacol. Ther. Toxicol.* 23:74–78.

243. Roschlau, G. 1983. Hepatitis mit konfluierenden Nekrosen durch Dihydralazin (Depressan). *Ztrlblt. Allg. Pathol.* 127:385–393.

244. Nataf, J., J. Berunau, and D. Larrey. 1986. A new anti-liver microsome antibody: a specific marker of dihydralazine hepatitis. *Gastroenterology* 90:A1751.

245. Bourdi, M., J.C. Gautier, J. Mircheva, D. Larrey, A. Guillonzo, C. Andre, C. Belloc, and P. Beaune. 1992. Anti-liver microsomes autoantibodies and dihydralazine induced hepatitis: Specificity of autoantibodies and inductive capacity of the drug. *Mol. Pharmacol.* 42:280–285.

246. Bourdi, M., M. Tinel, P.H. Beaune, and D. Pessayre. 1994. Interactions of dihydralazine with cytochromes P450 1A: a possible explanation for the appearance of anti-cytochrome P4501A2 autoantibodies. *Mol. Pharmacol.* 45:1287–1295.

247. Pessayre, D. 1993. Toxic and immune mechanisms leading to acute and subacute drug induced liver injury. In: Miguet J.P. D. Dhumeaux, eds. *Progr. Hepatol.* Paris: John Libbey Eurotext, 23–39.

248. Bock, K.W. 1992. Metabolic polymorpshisms affecting activation of toxic and mutagenic arylamines. *Trends Pharmacol.* 13:223–226.

249. Zimmerman, H.J., J.H. Lewis, K.G. Ishak, and W. Maddrey. 1984. Ticrynafen-associated hepatic injury: Analysis of 340 cases. *Hepatology* 4:315–323.

250. Bernuau, J., L. Mallet, and J.P. Benhamou. 1981. Hepatotoxicite due a l'acide tienilique. *Gastroenterologie Clinique et Biologique* 5:692–693.

251. Lecoeur, S., E. Bonierbale, D. Challin, J.C. Gatier, P. Valadon, P.M. Dansette, R. Catinot, F. Ballet, D. Mansuy, and P.H. Beaune. 1994. Specificity of *in vitro* covalent binding of tineilic acid metabolites to human liver microsomes in relationship with the type of hepatotoxicity: Comparison with two directly hepatototoxic drugs. *Chem. Res. Toxicol.* 7:434–442.

252. Lopez-Garcia, M.P., P.M. Dansette, P. Valadon, C. Amar, P.H. Beaune, F.P. Guenguerich, and D. Mansuy. 1993. Human liver P450s expressed in yeast as tools for reactive-metabolite formation studies. Oxidative activation of tienilic acid by cytochrome P4502C9 and P4502C10. *Eur. J. Biochem.* 213:223–232.

253. Lecoeur, S., C. Andre, and P.H. Beaune. 1996. Tienilic acid induced autoimmune hepatitis: anti-liver and-kidney microsomal type 2 autoantibodies recognize a three-site conformational epitope on cytochrome P4502C9. *Mol. Pharmacol.* 50:326–333.

254. Ray, D.C., and G.B. Drummond. 1991. Halothane hepatitis. *Br. J. Anaesth.* 67:84–99.

255. Cousins, M.J., J.L. Plummer, and P.M. Hall. 1989. Risk factors for halothane hepatitis. *Aust. NZ. J. Surg.* 59:5–14.

256. Farrell, G., D. Prendergast, and M. Murray. 1985. Halothane hepatitis. Detection of a constitutional susceptibility factor. *N. Engl. J. Med.* 313:1310–1314.

257. Zimmerman, H.J. 1978. The Adverse Effects of Drugs and Other Chemicals on the Liver. In: Zimmerman H.J. ed. Hepatotoxicity. New York: Appleton Century Crofts.

258. Sipes, I.G., J. Gandolfi, L.R. Pohl, G. Krishna, and B.R. Brown. 1980. Comparison of the biotransformation and hepatotoxicity of halothane and deuterated halothane. *J. Exp. Ther.* 214:716–720.

259. Spracklin, D.K., K.E. Thummel, and E.D. Kharasch. 1996. Human reductive halothane metabolism *in vitro* is catalyzed by cytochrome P4502A6 and 3A4. *Drug Metab. Dispos.* 24:976–983.

260. Brown, R.M., T. T, F.P. Guenguerich, M. Wood, and Waj J. 1995. Halothane microsomal oxidation is inhibited by diethyldthiocarbamate, an inhibitor of CYP2E1. *Anesth. Analg.* 80:60.

261. Madan, A., and A. Parkinson. 1996. Characterization of the NADPH-dependent covalent binding of [14C]halothane to human liver microsomes: a role for cytochrome P4502E1 at low substrate concentrations. *Drug Metab. Dispos.* 24:1307–1313.

262. Kharasch, E.D., D. Hankins, D. Mautz, and K.E. Thummel. 1996. Identification of the enzyme responsible for oxidative halothane metabolism: implications for prevention of halothane hepatitis. *Lancet* 347:1367–1371.

263. Spracklin, D.K., D.C. Hankins, J.M. Fisher, K.E. Thummel, and E.D. Kharasch. 1997. Cytochrome P450 2E1 is the principal catalyst of human oxidative halothane metabolism *in vitro*. *J. Pharmacol. Exp. Ther.* 281:400–411.

264. Kenna, J.G., J.L. Martin, H. Satoh, and L.R. Pohl. 1990. Purification of trifluoroacetylated protein antigens from livers of halothane-treated rats. *Eur. J. Pharmacol.* 183:1139–1140.

265. Martin, J.L., J.G. Kenna, B.M. Martin, D. Thomassen, G.F. Reed, and L.R. Pohl. 1993. Halothane hepatitis patients have serum antibodies that react with protein disulfide isomerase. *Hepatology* 18:858–863.

266. Pumford, N.R., B.M. Martin, D. Thomassen, J.A. Burris, J.G. Kenna, J.L. Martin, and L.R. Pohl. 1993. Serum antibodies from halothane hepatitis patients react with the rat endoplamic reticulum protein ERp72. *Chem. Res. Toxicol.* 6:609–615.

267. Smith, G.C.M., J.C. Kenna, D.J. Harrison, D. Tew, and C.R. Wolf. 1993. Antibodies to hepatic microsomal carboxylesterase in halothane hepatitis. *Lancet* 342:963–964.

268. Bourdi, M., W. Chen, R.M. Peter, J.L. Martin, J.T. Buters, S.D. Nelson, and L.R. Pohl. 1996. Human P4502E1 is a major autoantigen associated with halothane hepatitis and the epitopes are conformational. *Chem. Res. Toxicol.* 9:1159–1166.

269. Vergani, D., G. Mieli-Vergani, and A. Alberti. al. e. 1985. Antibodies to the surface of halothane-altered rabbit hepatocytes in patients with severe halothane-associated hepatitis. *N. Engl. J. Med.* 303:66–71.

270. Martin, J.L., D.J. Plevak, K.D. Flannery, M. Charlton, J.J. Poterucha, C.E. Humphreys, G. Derfus, and L.R. Pohl. 1995. Heopatotoxicity after desflurane anesthesia. *Anesthesiology* 83:1125–1129.

271. Hoet, P., M.L.M. Graf, M. Bourdi, L.R.D. Pohl, P.H., W. Chen, R.M. Peter, S.D. Nelson, N. Verlinden, and D. Lison. 1997. Epidemic of liver disease caused by hydrofluorocarbons used as ozone-sparing substitutes of chlorofluorocarbons. *Lancet* 350:556–559.

272. Yin, H., M.W. Anders, K.R. Korzekwa, L. Higgins, K.E. Thummel, F.D. Kharasch and J.P. Jones. 1995. Designing safer chemicals: predicting the rates of metabolism of halogenated alkanes. *Proc. Natl. Acad. Sci. USA* 92:11076–11080.

273. Shear, N.H., and S.P. Spielberg. 1988. Anticonvulsant hypersensitivity syndrome: *In vitro* assessment of risk. *J. Clin. Invest.* 82:1826–1832.

274. Leeder, J.S., R.J. Riley, V.A. Cook, and S.P. Spielberg 1992. Human anti-cytochrome P450 antibodies in aromatic anticonvulsant-induced hypersensitivity reactions. *J. Pharmacol. Ex. Ther.* 263:360–367.

275. Leeder, S., A. Gaedigk, X. Lu, and V.A. Cook. 1996. Epitope mapping studies with human anti-cytochrome P4503A autoantibodies. *Mol. Pharmacol.* 49:234–243.

276. Achord, J.L. 1993. Review of alcoholic hepatitis and its treatment. *Am. J. Gastroenterol.* 88:1822–1831.

277. Klassen, L.W., D. Tuma, and M.F. Sorrell. 1995. Immune mechanisms of alcohol-induced liver disease. *Hepatology* 22:355–357.

278. Albano, E., P. Clot, M. Morimoto, A. Tomasi, M. Ingelman-Sundberg, and S.W. French. 1996. Role of cytochrome P4502E1-dependent formation if hydroxyethyl free radical in the development of liver damage in rats intragastrically fed with ethanol. *Hepatology* 23:155–163.

279. Clot, P., G. Bellomo, M. Tabone, S. Arico, and E. Albano. 1995. Detection of antibodies against proteins modified by hydroxyethyl free radicals in patients with alcoholic cirrhosis. *Gastroenterology* 108:201–207.

280. Clot, P., E. Albano, E. Eliasson, M. Tabone, A. Sarino, Y. Israel, C. Moncada, and M. Ingelman-Sundberg. 1996. Cytochrome P450 2E1 hydroxyethyl radical adducts as the major antigen in autoantibody formation among alcoholics. *Gastroenterology* 111:206–216.

281. Israel, Y., E. Hurwitz, O. Niemela, and R. Arnon. 1996. Monoclonal and polyclonal antibodies against acetaldehyde-containing epitopes in acetaldehyde-protein adductus. *Proc. Natl. Acad. Sci. USA* 83:7923–7927.

282. Morimoto, M., A-L. Hagbjök, Y-J.Y. Wan, P.C. Fu, P. Clot, E. Albano, M. Ingelman-Sundberg, and S.W. French. 1995. Modulation of experimental alcohol-induced liver disease by cytochrome P4502E1 inhibitors. *Hepatology* 21:1610–1617.

283. Yokoyama, H., H. Ishii, S. Nagata, S. Kato, K. Kamegaya and M. Tsuchiya. 1993. Experimental hepatitis induced by ethanol after immunization with acetaldehyde adducts. *Hepatology* 17:14–19.

284. Albano, E., A. Tornasi, L. Goria-Gatti, J.O. Persson, Y. Terelius, M. Ingelman-Sundberg, and M.U. Dianzi. 1991. Role of ethanol-inducible cytochrome P450 (P450IIE1) in catalysing the free radical activation of alifatic alcohols. *Biochem. Pharmacol.* 41:1895–1902.

285. Betterle, C., N.A. Greggio, and M. Volpato. 1998. Autoimmune Polyglandular Syndrome Type 1. *J. Clin. Endocrinol. Metab.* 83:1049–1055.

286. Amouzadeh, H.R., M. Bourdi, J.L. Martin, B.M. Martin, and L.R. Pohl. 1997. UDP-glucose-glucoprotein gluco-syltransferase associates with endoplasmic reticulum chaperons and its activity is decreased *in vivo*. *Chem. Res. Toxicol.* 10:59–63.

287. Thomassen, D., B.M. Martin, J.L. Martin, N.R. Pumford, and L.R. Pohl. 1989. The role of a stress protein in the development of a drug-induced allergic response. *Eur. J. Pharmacol.* 183:1138–1139.

288. Davila, J.C., B.M. Martin, and L.R. Pohl. 1992. Patients with halothane hepatitis have serum antibodies directed against glucose regulated stress-protein GRP78/BiP. *Toxicologist* 12:255.

289. Butler, L.E., D. Thomassen, J.L. Martin, B.M. Martin, J.G. Kenna, and L.R. Pohl. 1992. The calcium binding protein calreticulin is covalently modified in rat liver by a reactive metabolite of the inhalation anaesthetic halothane. *Chem. Res. Toxicol.* 5:406–410.

290. Satoh, H., B.M. Martin, A.H. Schulick, D.D. Christ, J.G. Kenna, and L.R. Pohl. 1989. Human anti-endoplasmatic reticulum antibodies in sera from halothane induced hepatitis are directed against a trifluoroacetylated carboxylesterase. *Proc. Natl. Acad. Sci. USA* 86:322–326.

291. Martin, J.L., N.R. Pumford, A.C. LaRosa, B.M. Martin, H.M.S. Gonzaga, M.A. Beaven, and L.R. Pohl. 1991. A metabolite of halothane covalently binds to an endoplasmic reticulum protein that is highly homologous to phosphatidylinositol-specific phospholipase C-alpha but has no activity. *Biochem. Biophys. Res. Comm.* 178:679–685.

55 | Autoimmune Myocarditis

Susan L. Hill, Marina Afanasyeva, and Noel R. Rose

1. INTRODUCTION

Studies of immune-mediated heart disease represent, at the same time, one of the oldest and one of the most current areas of research in clinical immunology and immunopathology. The classical investigations of rheumatic heart disease and Chagas' disease go back many decades and set the stage for our contemporary concepts of infection-induced autoimmune disease. More recently, virus-induced myocarditis and its sequel, idiopathic dilated cardiomyopathy, have emerged as major topics of research. The availability of well-defined animal models, especially cardiac myosin-induced myocarditis in rodents, has permitted mechanistic studies of pathogenesis of immune-mediated heart disease.

This chapter summarizes the major trends in research into the molecular and genetic basis of autoimmune myocarditis.

2. CLINICAL ASPECTS

Myocarditis is an important cause of heart failure, especially among adolescents and young adults. It is an incompletely understood syndrome of multiple etiologies affecting at least five thousand individuals in the United States annually (1,2). A remarkable observation is the discrepancy between the limited overt evidence of myocyte injury and the global impairment of left ventricular function culminating in heart failure. This discrepancy has stimulated the suggestion that immunologic mechanisms contribute to cardiac damage. In the United States, infection by cardiotropic virus has been implicated in 18% to 42% of cases of known etiology and 40% to 60% of all patients have heart autoantibodies or other evidence of immune-

mediated pathogenesis (3–5). Human myocarditis represents a spectrum ranging from viral myocarditis to development of idiopathic dilated cardiomyopathy (IDCM), which presents in patients as congestive heart failure. Significantly, 30% to 50% of myocarditis and IDCM patients share similar immunological profiles; that is, the production of autoantibodies to cardiac myosin, adenine nucleotide translocator (ANT), and branch chained ketoacid dehydrogenase (BCKD) (6,7). A smaller percentage of patients exhibits T-cell proliferation to cardiac antigens (8). The presence of circulating autoantibody and antigen-specific T-cell proliferation suggests that an autoimmune component is triggered by a viral infection in myocarditis and IDCM.

In the United States, enteroviruses of the family Picornaviridae, particularly Coxsackieviruses, are the most common etiologic agents associated with myocarditis and IDCM (2). The Coxsackieviruses are divided into two groups, A and B, based on pathology observed in newborn mice. The former group produces flaccid paralysis, whereas the latter induces spastic paralysis. Coxsackieviruses of group A (CVA) show limited organ specificity replicating mainly in skeletal muscle. Group B Coxsackieviruses (CVB) infect and replicate in a large number of tissues, including the central nervous system, liver, exocrine pancreas, brown fat, and striated muscle depending on the viral strain. Of the enteroviruses, CVB type 3 is most often detected in heart muscle disease.

The Coxsackieviruses are small (28–30 nm diameter), non-enveloped, positive-strand RNA (approximately 7500 nucleotides) viruses (9). They contain an icosahedral capsid comprised of four peptides (VP1–4) with molecular weights of 29,000, 25,000, 21,000 and 5,500, respectively. They are ubiquitous in nature and are commonly associated with respiratory and gastrointestinal disease during the late summer and early fall in temperate climates. Approximately 5% of patients develop myocarditis.

3.   AUTOIMMUNE PHENOMENA

The number of cardiac diseases exhibiting an autoimmune component is mounting. Cytomegalovirus infection, rheumatic fever and Chagas' disease (American trypanosomiasis) represent well established models of post-infection autoimmune disease that show immunological findings similar to those of CVB3-associated myocarditis. In humans, patients with ischemic heart disease, myocardial infarction, cardiac transplant rejection, doxorubicin (Adriamycin) cardiotoxicity and Dressler's syndrome have evidence of circulating autoantibodies to heart tissue (10–12).

Our research team has developed two models of autoimmune myocarditis in mice. The first was produced by infecting mice with a cardiovirulent strain of CVB3 virus (13). The viral model was designed to simulate the broad spectrum of pathologic changes seen in humans with virus-induced myocarditis (14). The second model was produced by immunizing certain inbred strains of mice with murine cardiac myosin emulsified in complete Freund's adjuvant (CFA) (15–17). The latter model allows for the evaluation of autoimmune myocarditis without the complication of virus-induced disease.

3.1.   Coxsackievirus B3 Model

Coxsackievirus B3, the common etiological agent in human myocarditis, induces a histologically similar disease in mice (18). A detailed analysis of cardiac pathologic lesions allowed us to discern two distinct pathologic processes; an initial (viral) phase in which myocarditis is associated with the presence of infectious virus and focal myocardial necrosis with accompanying infiltrates of polymorphonuclear leukocytes and mononuclear inflammatory cells, and a later (autoimmune) phase in which infectious virus is absent and there is a more widely distributed cardiac interstitial infiltrate of mononuclear cells without accompanying myocyte necrosis. The biphasic nature of the virus-induced disease suggested that the later phase represents an autoimmune response instigated by the initial viral infection of the heart (19).

During the early (viral) phase, all of the mouse strains tested developed antibody to CVB3 had replicating virus within the heart and showed early infiltration with natural killer (NK) cells. By evaluating many multiple congenic mouse strains, we demonstrated that susceptibility to viral myocarditis is genetically determined and relies on both major histocompatibility complex (MHC) and non-MHC genes (20,21). During the late (autoimmune) phase,

susceptible mice progressed to chronic myocarditis and exhibited autoantibodies and splenocyte proliferation to the alpha isoform of murine cardiac myosin heavy chain. Susceptibility to the later phase was pronounced among inbred strains with an "A" background; that is, A/J or its congenics; A.SW, A.CA or A.BY. "B" background strains such as C57B1/6, C57B1/10 and their congenics; B10.A and B10.M, were relatively resistant to the development of autoimmune myocarditis.

In our model of virus-induced myocarditis, the early phase of disease began approximately three days after CVB3 inoculation (22). CVB3-neutralizing antibody was detected on day 2 or 3 and the kinetics of antibody production was genetically determined. "B" background strains showed extended production of neutralizing antibody compared to "A" background mouse strains (20). Cardiac viral replication began on day 3 after infection, reached its peak between days 5–7, and declined perceptibly between days 7 and 15. Natural autoantibody (IgM) to cardiac myosin was detectable during this phase; however, the lack of IgG autoantibody indicates failure to activate myosin-specific T cells.

The late (autoimmune) phase of disease began approximately on the ninth day after infection (16,22). Accompanying the interstitial cardiac infiltrates, myosin-specific IgG autoantibody levels started to increase. Circulating auto antibody levels peaked around day 14 and slowly declined after day 21. The mouse strains susceptible to post-infection autoimmune myocarditis produced high levels of circulating autoantibodies to murine cardiac myosin compared to resistant mice. They also developed a moderate humoral response to the other myosin isoforms (23). Absorption of serum with murine cardiac myosin completely abolished reactivity to murine cardiac myosin using enzyme-linked immunosorbent assay whereas absorption with skeletal or brain myosin only partially decreased anti-cardiac myosin activity. Thus, subpopulations of autoantibody reacted with epitopes that were shared between the myosin isoforms, while others were specific for cardiac myosin. Between days 8 and 14, T-cell proliferation to murine cardiac myosin was detected (24). The cardiac infiltrates during the late phase of disease included CD4[+] and CD8[+] T cells and macrophages. The results suggested persistent activation of the antigen-specific T-cell subpopulations and macrophages within the heart late in disease progression.

Prominent reactivity of circulating autoantibody to murine cardiac myosin during the late phase of disease prompted the direct evaluation of murine cardiac myosin as a potential autoantigen. Neu et al. (17) tested the hypothesis that the interstitial form of myocarditis could be elicited in susceptible strains of mice by immunizing the animals with murine cardiac myosin. Susceptible (A.CA and A.SW) and resistant (B10.PL and C57BL/6) mouse strains were immunized on days 0 and 7 with purified cardiac or skeletal myosin. Susceptible mouse strains immunized with cardiac myosin showed high titers of IgG antibody to cardiac myosin on day 21, hearts that were grossly enlarged and pale, and infiltrates of mononuclear cells (18).

3.2. Myosin-Induced Model

Based on the evidence presented in the last section cardiac myosin is an autoantigen capable of inducing autoimmune myocarditis. A molecule consisting of globular and rod regions, it is composed of two heavy chains (200 kDa) and two pairs of light chains (17 kDa). The two isoforms, alpha and beta cardiac myosin heavy chain, are expressed at different times during development. The former is produced in adult mice and the latter isoform is expressed in the fetus (25). Immunization with purified alpha heavy chain antigen emulsified in CFA resulted in the development of cardiac lesions in some mouse strains (26). Similar to the findings in CVB3-induced autoimmune myocarditis, "A" background mouse strains were susceptible to myosin-induced cardiac disease, whereas "B" background strains were relatively resistant (23). The CVB3-infected "A" background strains produced significant levels of circulating autoantibody to ANT and BCKD as well as cardiac myosin, whereas myosin-immunized "A" strains only demonstrated significant levels of circulating antibody to cardiac myosin antibody (27). Yet antibodies eluted from cardiac tissue of myosin-immunized and CVB3-inoculated animals revealed a more diverse population of autoantibodies with specificity for cardiac myosin, ANT and BCKD. The results indicated that susceptible mouse strains have a population of myosin-specific T cells capable of proliferation upon myosin immunization. Animals showing a greater range of cardiac-specific autoantibodies suggest that viral replication and immune-mediated cardiac damage lead to mobilization of additional intracellular autoantigens.

4. INDUCTIVE MECHANISMS

A number of mechanisms are probably responsible for the development of post-infection autoimmune myocarditis. Early (viral) myocarditis is attributed to direct cardiomyocyte damage due to the virus or to activated NK cells, macrophages and their soluble products. CVB3-specific T cells may damage virus-infected cardiomyocytes. In the late stage (autoimmune) disease observed in susceptible mouse strains, persistent cardiac infiltrates of T cells and macrophages and the intracardiac deposition of autoantibody suggest that T cells, antibodies and cytokines all may contribute to myocardial damage.

The major mechanisms of autoimmunity considered in the murine model of post-infection autoimmune myocarditis include: (i) molecular mimicry, (ii) viral induction of antigen release or of cryptic antigens, (iii) immune modulation,

(iv) alterations in target organ susceptibility, and (v) viral persistence. These possibilities are not mutually exclusive; more than one mechanism may be involved.

4.1. Molecular Mimicry

A number of viruses, such as measles virus (28,29) and herpes simplex virus (30), have shown sequence homology between viral antigens and host cell proteins. Additionally, antibodies against the streptococcus, the bacterium responsible for the production of rheumatic heart disease, have shown cross-reactivity with streptococcal M protein, myosin, and CVB3 virus (31). Bachmaier et al. (31a) has demonstrated an antigenic mimicry between cardiac-specific α myosin heavy chain and chlamydial peptides. Immunization of BALB/c mice with these peptides reproduced myocarditis induced by immunization with a cardiac myosin heavy chain peptide. Evaluation of CVB3 sequence and murine cardiac myosin failed to demonstrate significant sequence homology between the two molecules. This suggests that there are no shared T-cell epitopes. Other data preclude the presence of shared B cell epitopes; the latter are primarily dependent on conformational determinants and may be formed by non-linear segments of the molecule (32).

Gauntt et al. have shown that mice immunized with human murine cardiac myosin develop antibodies that cross-react with CVB3 virus (33). Pre-immunization of mice with human myosin protected against subsequent CVB3-inoculation indicating that myosin injection produced neutralizing antibody against the virus. The evaluation of murine cardiac myosin antibody produced in the myosin-immunized mice did not decrease viral replication in vitro (34). There are a number of studies evaluating cross reactivity between antibodies in various models of myocarditis. For example, studies of streptococcal-induced rheumatic heart disease have demonstrated cross-reactivity between pathogen-specific antibodies and cardiac autoantigens (31). The cross-reactivity was not attributed to significant sequence homology between pathogen proteins and cardiac myosin, but rather to a common helical structure.

In an examination of potential molecular mimicry in CVB3-infected mice, Schwimmbeck et al. (8) identified VP1 of CB3 as the predominant B- and T-cell epitope harboring viral coat protein. Using synthetic overlapping peptides, they found 6 peptides that induced significant T-cell proliferation (AA 1–15, AA 21–35, AA 79–93, AA 119–133, AA 129–143, and AA 199–213). The first two peptides were also B cell epitopes. Further, they discovered that preimmunization of peptide AA 1–15 followed by CVB3 injection reduced the titer of virus in the heart and resulted in decreased severity of cardiac interstitial infiltrates. Equally interesting was the finding that preimmunization with peptide AA 21–35 had no effect on cardiac viral titers and enhanced mortality and the severity of cardiac lesions compared to mice not previously immunized. The results do not necessarily support molecular mimicry as a mechanism in the progression to autoimmune myocarditis in this mouse model since sequence analysis failed to detect any significant homology between the molecules. However, the findings indicate that the response to different CVB3 T-cell epitopes can enhance or ameliorate cardiac disease. This finding is important in the design of potential vaccines to the virus. Further studies are needed to evaluate whether AA 21–35 is capable of inducing autoimmune myocarditis without the co-administration of CVB3 virus. Previously in our laboratory we tested VP2 viral capsid protein as a potential cross-reactive antigen exhibiting B cell epitopes shared with murine cardiac myosin (MCM). Antibody to VP2 in mice injected with CVB3 virus failed to cross-react with cardiac myosin. The study did not examine T-cell proliferation to VP2 and the antigen's ability to elicit myocarditis when emulsified with CFA and injected into mice, and therefore failed to rule out VP2 as a potential candidate for molecular mimicry. No experiments have examined viral proteins VP3 and VP4 for shared MCM B- or T-cell epitopes. Additionally, other potential cardiac antigens, such as ANT, BCKD and cardiac sarcoplasmic reticulum calcium ATPase, have not been evaluated for cross-reactivity with CVB3. Future studies evaluating shared epitopes between MCM and CVB3 are important not only for the assessment of the pathogenesis of autoimmune myocarditis, but are also critical in the development of CVB3 vaccines. Development of a vaccine containing a shared epitope could result in enhancement of autoimmune myocarditis in susceptible individuals.

4.2.   Cryptic or Novel Antigens

The second potential mechanism for the development of autoimmune myocarditis in postviral myocarditis is the production of novel antigens or exposure of cryptic intracellular antigens seen as foreign by the immune system. A number of enveloped viruses are known to alter cell surface protein expression during the process of budding. As CVB3 is a nonenveloped lytic virus, alteration of the plasma membrane seems less likely. However changes in the flux of ions across membranes have been noted pointing towards possible changes in ion channels and/or enzyme activities (34). Huber et al. (35,36) found a subpopulation of cytotoxic T cells which lysed only cardiomyocytes demonstrating altered metabolism. This population of CD4+ cytotoxic lymphocytes lysed cultured myocytes treated with adriamycin, actinomycin D or cells metabolically altered by infection with picornavirus. The cytotoxic T cells failed to damage cells exhibiting normal metabolism. Gauntt et al. (37) have found that fibroblasts infected *in vitro* with CVB3 produce novel surface structures.

Viral proteases have been shown to cleave host cell proteins involved in transcription, resulting in the shut-down of host cell protein synthesis (38). CB3 protease 2A has been shown to cleave dystrophin suggesting a mechanism of a direct damage by CB3 which may lead to dilated cardiomyopathy (38a). Viral proteases have also been implicated in changing host cell protein isoform expression. Hamrell et al. (39) discovered that CVB3-infection in mice resulted in a reduction in the unloaded sarcomere shortening velocity. This functional alteration in the cardiomyocytes was associated with a shift from the adult (V_1) myosin isoform (α-MHC) to the fetal (V_3) slower myosin (β-MHC) isoform and represented remodeling of the contractile apparatus. Viral proteases may have been responsible for the modification of the contractile subunit. Very few studies have addressed directly the cleavage and modification of host cell structural proteins and the potential development of novel antigens. Alternatively, oxidative stress may have been responsible for the switch to a slower myosin isoform in mice with virus-induced myocarditis. In a study conducted by Huber et al. (36), virus-infected mice developed autoantibody to a novel undefined antigen. Whether it was modified cardiac myosin or another cardiac antigen was not determined. Hamrell's findings dispute the development of a novel antigen and suggest a switch from the adult form of myosin (V_1) to the fetal myosin isoform (V_3). The group also concluded that the intracardiac infiltrate was crucial for the initiation of the myosin isoform switch. The fetal myosin isoform cannot be considered a novel antigen since it is presented to immune cells via the thymus early in development.

Another plausible mechanism for the development of autoimmune disease in lytic infections is the release of sequestered intracellular antigen. A number of diseases have been attributed to the release of proteins that were not originally presented in the thymus during the development of central tolerance. There is debate as to whether cardiac myosin qualified as a sequestered antigen. Some investigators have suggested that, in the normal heart, myosin is presented to T cells by interstitial dendritic cells and possibly by myocytes following cardiac myosin immun-ization (40).

When Macfarlane Burnet proposed the clonal selection hypothesis in the 1950s, he suggested that autoreactive lymphocytes are deleted during development. Subsequent studies have partially supported his hypothesis but have shown that many autoreactive cells escape deletion and are controlled in the periphery by various mechanisms that prevent the development of autoimmune disease. This conclusion is supported by the presence of natural autoantibodies in a large portion of the human population, particularly in aged individuals. The presence of autoantibodies to myosin without the development of autoimmune disease suggests that other mechanisms are required for the progression of the immune response to disease, such as increased target organ susceptibility, T-cell activation and the local production of cytokines or other mediators.

4.3. Immune Modulation

Comparison of susceptible and resistant mouse strains allows for the evaluation of immune modulation in the progression of CVB3-induced myocarditis. Immune modulation indicates an abnormal or exaggerated response to virus resulting in progression to autoimmune disease. Autoreactive T and B cells may become activated by the infectious agent. In the case of Coxsackievirus-induced autoimmune myocarditis, this represents the development of cardiac myosin specific autoantibody and T-cell proliferation with significant cardiac interstitial mononuclear cell infiltrates; all of these features are observed in susceptible mouse strains and are rare in resistant strains.

CVB3 virus has been shown to associate with, and replicate in, lymphoid cells (42,43). Replication of the virus results in lymphoid cell depletion in the spleen and lymph nodes. Although not yet proven, selective infection of particular lymphoid cell subpopulations in susceptible mouse strains may contribute to the progression to late stage disease by elimination of suppressor cells or enhancement of the activation state of CD4+ and/or CD8+ cytotoxic T cells. Humans with myocarditis and IDCM have exhibited imbalances in T-cell subpopulations (44–46). The strongest evidence concerning the contribution of T suppressor cells to disease progression comes from studies conducted in female BALB/c mice. Transfer of splenocytes from CVB3 inoculated female mice to susceptible CVB3-inoculated male BALB/c mice results in amelioration of cardiac infiltrates, autoantibodies and antigen-specific T-cell proliferation (47). The findings indicate that in "resistant" animals, a T-cell suppressor response prevents disease progression and may explain the differences in disease progression observed in the human population. Hormones may also modulate the autoimmune response. Unlike most other autoimmune diseases, autoimmune myocarditis may be more common in males, suggesting that steroid hormones are involved in disease progression (48). The susceptibility may be enhanced by masculinizing hormones and reflected in a number of functions, such as viral receptors, the immune response or intrinsic target organ vulnerability (49,50).

4.4. Target Organ Susceptibility

Recently, the role of the target organ in the progression to postinfectious autoimmune heart disease has been explored. In the myosin model of disease, early expression of endothelial ICAM-1, an adhesion molecule, and interstitial MHC Class II antigen were necessary for the induction of cardiac infiltrates (40). Blocking of ICAM-1 using monoclonal antibody resulted in decreased cardiac pathology by preventing entry of antigen-specific and/or activated T cells into the myocardium (25). Blocking of class II MHC in murine models of autoimmune myocarditis using competitive

peptides resulted in abrogation of cardiac lesions by impeding T-cell recognition of autoantigen in the context of class II MHC (51). Adoptive transfer studies have concluded that recipient mice required prior stimulation with lipopolysaccharide (LPS) or multiple injections of CFA (25,51) for successful transfer of disease. Priming with LPS resulted in a similar distribution of ICAM-1 and MHC Class II expression compared to mice immunized and boosted with murine cardiac myosin emulsified in CFA (25). Humans with myocarditis showed increased expression of class I MHC on cardiomyocytes and aberrant expression of class II MHC within the heart (52). Presumably, the class II expression was present on antigen presenting cells and may reflect an increase in cardiac dendritic cells.

A study conducted by Liao et al. (26) addressed allelic differences in alpha cardiac myosin heavy chain in predicting susceptibility to autoimmune myocarditis. BALB/c (susceptible) and C57B1/6 (resistant) cardiac myosin were compared for differences in immunogenicity and the ability to elicit cardiac lesions. Alpha cardiac myosin heavy chains differ by two amino acid residues, AA 838 and AA 955. The group found that BALB/c cardiac myosin induced myocarditis in both C57B1/6 and BALB/c mice, though increased prevalence and severity were noted in the latter susceptible strain. C57B1/6 myosin induced myocarditis in BALB/c mice at a lower prevalence compared to those immunized with BALB/c myosin. Both myosin preparations elicited anti-cardiac myosin antibodies in the strains examined. The results indicate that murine cardiac myosin preparations from both "susceptible" and "resistant" strains are immunogenic, but that the myosin from the former is a better autoantigen in that BALB/c myosin is capable of inducing cardiac lesions, whereas C57B1/6 myosin is more frequently associated with only autoantibody production. The investigators speculate that the allelic differences noted may be responsible for altered antigen processing that enhances the antigenicity of the molecule in susceptible mouse strains. This notion is supported by the fact that C57B1/6 myosin failed to induce significant myocarditis when injected into syngeneic mice.

4.5. Virus Persistence

Despite the lack of replicating virus approximately 9 days after infection, viral nucleic acid has been detected in the hearts of susceptible mouse strains (531). The significance of this finding is strengthened by the presence of viral genome in 50% of patients with IDCM (54), implicating a viral etiology in chronic cardiac disease and raising additional questions concerning the role of persistent virus in disease progression. Klingel and Kandolf (55) found that susceptible (A.CA, H–2f) mice inoculated with CVB3 exhibited persistent cardiac viral nucleic acid in association with cardiomyocyte damage and interstitial infiltrates. The investigators demonstrated the inflammatory infiltrates were dependent on the presence of persistent virus in this mouse strain. It is possible that failure to detect replicating virus late in disease is due to the significant reduction in the number of virally-infected cardiomyocytes at this time point. Isolated CD4[+] and CD8[+] T cells in the virus-induced murine model have shown cytolytic activity against both infected and uninfected cardiomyocytes *in vitro*, indicating that despite the presence of virus there is still an autoimmune component in susceptible individuals. The continued presence of viral nuclear acid and possible low level of replication in susceptible mouse strains may aid in the initiation or maintenance of the autoimmune response to cardiac antigens.

5. EFFECTOR MECHANISMS

One or more of the above mechanisms may play a role in the development of post-infection autoimmune myocarditis. Studies have been conducted in a number of mouse strains to analyze immune-mediated cardiac damage. The effector mechanisms utilized may vary according to mouse strain, age, sex and environmental factors.

5.1. T Cells—Human Disease

Autoimmune myocarditis has been described in the literature as a T cell-mediated disease (8). Endomyocardial biopsies from patients with myocarditis and IDCM show infiltration with activated lymphocytes. The predominant cell types included CD4[+] and CD8[+] T cells and macrophages. Some patients demonstrated autoreactant T cells *in vitro*. In approximately 50% of the biopsy specimens, primary lymphocyte cultures were established when the cells were co-cultured with irradiated autologous peripheral blood leukocytes, IL-2 and myocardial tissue. A small percentage (5/118) of the cultured cells showed significant proliferation to ANT. The studies described failed to test other potential autoantigens such as cardiac myosin, which may have yielded a larger proportion of autoreactive cultures. The investigators also examined the ability of peripheral blood mononuclear cells from IDCM patients devoid of cardiac enteroviral nucleic acid to induce disease in 8–10 week old male C.B-17 SCID mice. Animals examined 60 days after transfer had significant levels of circulating autoantibody against ANT, CD3-positive human T cells within the myocardium, and decreased left ventricular function. Depletion of CD4[+] T cells from peripheral blood mononuclear cells resulted in very low levels of human antibody and infiltrating cells within the myocardium, confirming that CD4[+] T cells are important in the initiation of cardiac disease as well as the production of autoantibodies in myocarditis and IDCM.

In another study, patients with active myocarditis showed an increase in intracardiac T helper cells and a higher peripheral blood helper to suppressor T cell ratio compared to subjects exhibiting healing myocarditis and aged-matched controls (46). Chronic disease resulted in a decrease in the T helper cell to suppressor cell ratio that may signify an increase in T suppressor cells or a decrease in T helper cell numbers. This decline in the peripheral blood T helper to suppressor ratio could have represented the resolution of cardiac lesions or an increase in trafficking of T helper cells to the heart. Consistently, investigators examining endomyocardial biopsies have shown a decrease in the T cell suppressor cell subpopulations in patients with chronic myocarditis or IDCM (56).

5.2. T Cells—Murine Myosin-Induced Model

Evaluation of human endomyocardial biopsy specimens and peripheral blood mononuclear cells provided evidence for the role of T cells in the development of autoimmune myocarditis. Animal models have allowed a more in-depth study of T cell effector functions in the progression of disease. Studies depleting CD4[+] and/or CD8[+] T cells have shown that CD4[+] T cells are required in the initiation of cardiac myosin-induced autoimmune myocarditis, whereas CD8[+] T cell are important in disease progression (57–59). Using the rat model of cardiac myosin-induced giant cell myocarditis, Hanawa et al. (60) found that long term blockage of the alpha beta T cell receptor ($\alpha\beta$TcR) prevented progression of autoimmune myocarditis in a dose-dependent manner. Penninger et al. (61) tested CD4[+] and CD8[+] knockout mice for induction of myocarditis using the myosin-induced model of disease. They found that CD8[+] knockout mice developed more severe disease, whereas CD4[+] knockout mice had cardiac infiltrates of similar severity compared to heterozygotes. Evaluation of T cell markers revealed that infiltrating cells were composed of alpha beta CD4[−] CD8[−] T cells in the former treatment group (59). The results underline the utilization of alternative pathways by knockout mice. In the absence of the normal T cell repertoire, other less differentiated T-cell subpopulations may be mobilized. Further data supporting the role of T cells in autoimmune myocarditis come from successful adoptive transfer of T cells from immunized C.B-17 mice with active myocarditis to SCID recipients (61a). T cell receptor signaling and antigen recognition in the context of MHC antigen were also important in this model of autoimmune myocarditis. Blockage of TcR co-receptors such as CD45, a tyrosine phosphatase involved in p56[lck] enzyme activity, prevented the development of autoimmune myocarditis (62). p56[lck] knockout mice, which lack a necessary src-family protein tyrosine kinase in TcR signaling, failed to develop myosin-induced myocarditis. The use of non-immunogenic competitive peptides that bound I-A[k] in susceptible A/J mice were capable of preventing the induction of myosin-induced myocarditis (62a). The findings indicate that antigen-specific T cell recognition of immunogenic peptides in the context of class II MHC, followed by T-cell activation and signal transduction are necessary for disease progression. T-cell activation requires a second signal that is provided by co-stimulatory molecules such as CD28. Without this signal, T cells become anergic. Studies have tested the requirement of CD28 signaling in the induction of myocarditis in this model. Using mice with the CD28 null mutation on a genetically-susceptible background, the investigators discovered that animals immunized with cardiac myosin heavy chain emulsified in CFA developed disease, though of lower severity and prevalence compared to their control littermates. Interestingly, heterozygotes exhibited a different cytokine profile compared with homozygotes based on the myosin-specific IgG isotypic response (63). The findings suggest that the level of T cell activation may dictate the downstream cytokine response.

5.3. T Cells – Murine CVB3-Induced Model

There has been considerable debate concerning the effector cells involved in viral myocarditis. It is generally accepted that NK cells, the first infiltrating cells, are important in control of viral replication, though the mechanisms have not been clearly identified. In addition to NK cells, cytolytic CD4+ and CD8+ T cells induced lysis of virus-infected cardiomyocytes. In the mouse strains studied, CD8+ T cells also lysed noninfected cardiomyocytes (64). Interestingly, animals depleted of T cells showed minimal to no cardiac lesions compared to immunocompetent mice, but the former exhibited similar to increased cardiac viral titers compared to the immunocompetent group. Reconstitution of T cells in T cell-deficient mice resulted in reestablishment of myocarditis susceptibility (65). Further supporting the necessity of T cells in CVB3-induced myocarditis, athymic nude mice failed to develop myocardial infiltrates compared to euthymic controls (66). Mouse strain differences in the utilization of various effector cell populations in cardiac damage become apparent in CD4+ and CD8+ T cell depletion studies. In A/J mice (68), anti-CD4+ monoclonal antibody had little effect on the development of myocarditis; however, anti-CD8+ monoclonal antibody treatment significantly enhanced cardiac lesion severity. The study suggested that CD8+ T cells were important in the destruction of virus-infected cells. CD4+ T cell depletion in DBA/2 and MRL +/+ mice significantly decreased cardiac damage but was less effective in BALB/c mice in which CD8+ T cell depletion significantly reduced cardiac lesions (69). Studies examining CD4+ and CD8+ knockout mice on a C57BL/6 background showed that CD4+ T cells had a protective role in the induction of myocarditis as animals bearing the null mutation exhibited severe cardiac infiltrates composed of CD8+ T cells. CD8+ knockout mice showed minimal cardiac pathology indicating that they were important mediators of cardiac damage (59).

CVB3-induced myocarditis utilizes a number of different cell populations in cardiac viral clearance. NK cells have been implicated as a first line of defense followed by a specific T-cell response to virus-infected cells (69). In susceptible mouse strains additional autoantigen-specific T cells were generated resulting in chronic cardiac inflammation and autoantibody generation. Autoreactive T cells appear to play a major role in the development of post-infection autoimmune myocarditis in humans and in the murine models examined. The preeminent cell populations may differ somewhat between murine cardiac myosin- and CVB3-induced myocarditis. NK cells are more important in the virus-induced model, but it is clear that both CD4+ and CD8+ T cells are also important in autodestruction of cardiomyocytes. The question arises whether or not the cells are responsible for direct or indirect destruction of cardiomyocytes by the production of cytokines and/or oxygen-free radicals. Release of soluble mediators may contribute to cardiomyocyte dysfunction and damage.

5.4. Antibody

A plethora of human cardiac diseases are associated with the presence of circulating autoantibodies to several cardiac antigens (70). Whether or not these antibodies played a role in disease initiation or progression has been debated. Studies attempting transfer of hyperimmune serum from myosin-immunized mice to syngeneic recipients have failed (71,72). In one study, however, serum transfer of disease was accomplished in DBA/2 mice (73). The same investigators failed to transfer disease to BALB/c mice and concluded that mouse strain differences existed in the production of pathogenic autoantibodies, with the former strain showing immunopathogenic autoantibodies and the latter lacking this subpopulation of antibody. The investigators went on to find that the presence of pathogenic autoantibody was associated with the deposition of myosin or a myosin-like molecule within the cardiac interstitium of DBA/2 mice. Further supporting this strain difference in the production of immunopathogenic antibodies, Huber and Lodge (74) reported that complement depletion using cobra venom factor abrogated cardiac inflammation and necrosis in DBA/2 mice, but failed to alter disease progression in BALB/c mice. Autoantibodies have been eluted from the hearts of susceptible mice infected with CVB3 (75). Immunohistochemistry has detected immunoglobulin and complement deposition in areas of cardiac damage. The results suggested that either (i) the eluted antibody was reacting to α-myosin released from damaged cardiomyocytes, or (ii) viral infection had modified the cardiac interstitium. Antibody can transfer disease in one mouse strain and not another under the same conditions; successful transfer correlates with the presence of a myosin-like molecule within the cardiac interstitium and implicates multiple effector mechanisms in the progression to autoimmune myocarditis.

6.   CYTOKINE PRODUCTION

Studies evaluating the roles of T cells in disease susceptibility have pointed towards cytokines as potential effector molecules responsible for cardiac dysfunction and cardiac damage. Endomyocardial biopsies have shown expression of IL-1 and TNF-α in patients with myocarditis and IDCM, although the role of proinflammatory cytokines in the latter group is less clearly defined (76). Henke et al. (77) examined monocyte release of cytokines post-CVB3 infection. They found that purified human monocytes upon exposure to the virus released IFN, IL-1β, and TNFα, which could account for increased levels of circulating proinflammatory cytokine detected in the examined patients. The role of proinflammatory cytokines in disease progression has been further evaluated using animal models.

6.1.   Myosin-Induced Model

Using experimental myocarditis model in rats, Okura et al. (78) assayed intracardiac mRNA expression of several cytokines. These investigators found significant production of IFN-γ and TNF-α as well as nitric oxide (NO) during peak myocarditis, followed by increasing levels of TGF-β and IL-10. The results suggested that there are two different phases of cytokine production in the myosin-induced rat model; first, a proinflammatory phase followed by production of regulatory cytokines that attempt to downregulate the proinflammatory cytokine cascade. Tumor necrosis factor alpha (TNF-α) is viewed as a critical cytokine in autoimmune myocarditis initiation and progression. The *in vitro* production of TNF-α correlates with the presence of disease upon cardiac myosin immunization (79). Neutralization of TNF-α using monoclonal antibody prevents the development of myocarditis (80). Bachmaier et al. (81) tested susceptible mice (A/J, H-2^a) lacking the TNF-receptor p55 (TNF-Rp55) using the myosin-induced model. Mice with the null mutation failed to develop myocardial infiltrates, but did produce autoantibodies. Heterozygotes had both severe cardiac lesions and circulating cardiac myosin antibody. Adoptive transfer of concanavalin A-stimulated splenocytes failed to induce disease in TNF-Rp55 knockout mice, but transfers were successful in the wild-type mice. The investigators additionally established that TNF-Rp55 modulated class II MHC expression in the heart. They concluded that TNF-Rp55 is important in target organ susceptibility.

IFN-γ, a prototypic Th1 cytokine, exhibits a protective role in experimental autoimmune myocarditis. Blocking IFN-γ with a monoclonal antibody leads to the development of a very severe myocarditis (80,82). A large proportion of mice treated with the antibody exhibit features of dilated cardiomyopathy, such as enlarged hearts with dilated ventricles and extensive fibrosis. Studies using IFN-γ knockout mice confirmed the protective role of this cytokine (our-unpublished observations). IFN-γ knockout mice also show exacerbation of disease compared to the wild-type controls. Additionally, the absence of IFN-γ increases mortality from heart failure in mice immunized with cardiac myosin. IFN-γ knockout mice immunized with cardiac myosin have increased splenocyte population with increased percentages of T cells (both CD4- and CD8-positive T cells), suggesting that IFN-γ may limit the progression of autoimmune disease by suppressing proliferation or survival of activated lymphocyte. Similarly, the expansion of CD4-positive T cells in IFN-γ knockout mice was observed in experimental autoimmune encephalomyelitis (82a). The role of IL-12, a proximal cytokine that directs the development of a Th1 response, has not been fully investigated. Okura et al. (83) have demonstrated that the addition of IL-12 to the cell culture of cardiac myosin-specific T cells increases their pathogenicity in a T cell transfer model of autoimmune myocarditis in rats. If this disease-promotimg role of IL-12 is confirmed by future studies, it would raise some intriguing questions regarding the mechanisms by which IL-12 could promote disease, since IFN-γ, which is induced by IL-12, has been proven to be protective.

IL-4, a prototypic Th2 cytokine, seems to play an important role in the development of cardiac myosin-induced myocarditis. Blocking IL-4 with a monoclonal antibody significantly reduces the severity of myocarditis (82). This reduction in severity is associated with an increased ability of splenocytes to produce IFN-γ in response to cardiac myosin, suggesting that IL-4 may contribute to the disease progression by suppressing a disease-limiting factor, IFN-γ.

6.2.   CVB3-Induced Model

Post-viral autoimmune myocarditis induced by injecting mice with encephalomyocarditis virus (EMC), a cardiotropic picornavirus, showed that IFN-γ, IL-1β, and TNF-α expression peaked at day 7 after inoculation and persisted throughout the course of the study. Although cardiomyocytes have been reported to produce proinflammatory cytokines (84), this study failed to demonstrate cardiomyocyte involvement and showed cytokine production only by infiltrating cells, fibroblasts and endothelial cells. Myocarditogenic CVB3 strains similarly induce proinflammatory cytokines in susceptible mouse strains (85). The studies described indicated the presence of circulating and intracardiac cytokine expression but failed to demonstrate their role in disease induction or progression.

Studies conducted in our laboratory showed that a susceptible mouse strain (A/J) exhibited increased circulating levels of TNF-α and intracardiac expression of TNF-α and IL-1β (86). More important, treatment of resistant B10. A mice with either TNF-α or LPS resulted in the development of autoimmune myocarditis (87,88). Additionally, blockage of IL-1β using IL-1 receptor antagonist (IL-1ra) in susceptible A/J mice ameliorated post-infectious autoimmune myocarditis (89). The studies indicated that these pro-inflammatory cytokines were critical in cardiac disease development.

Interferons, both type I (IFN-α and IFN-β) and type II (IFN-γ), exhibit potent antiviral activity. Oral treatment with type I IFNs suppressed the inflammatory response in cytomegalovirus (CMV)-induced myocarditis in mice (89a). Intranasal administration of IFN-γ suppressed viral replication and improved the prognosis of EMC-induced myocarditis in mice (89b). Transgenic expression of IFN-γ in the pancreas protected mice from CB3-induced myocarditis (89c). Treatment with IFN-α in addition to a conventional treatment regimen improved the outcomes of idiopathic myocarditis and IDCM patients (89d). These studies demonstrate that IFNs represent an effective therapy in murine viral myocarditis regardless of the viral agent (CMV, EMC, or CB3) and point to the effectiveness of such therapy in clinical settings.

Other models underline the importance of the cytokine milieu in the development of autoimmune myocarditis. Repeated injection with IL-2 resulted in the development of myocardial inflammatory infiltrates (90,91). Using the rat, an IL-2-induced myocarditis model was developed to determine the mechanism(s) of myocardial injury. A single intra-peritoneal injection of IL-2 resulted in myocardial infiltrates similar to those observed in human patients. Interestingly, the kinetics of the cellular infiltration also resembled those seen in CVB3-infected mice; that is, NK cell infiltration followed by infiltration by T cells. The investigators did not evaluate immunologic abnormalities in these animals, but it would be informative to examine autoantibody production and auto-reactive T-cell stimulation in this model. The results from the rat suggested that alteration of the cytokine milieu induced by the administration of exogenous IL-2 promoted the development of myocarditis (92). Kishimoto et al. (93) tested the effects of recombinant human IL-2 on the progression of CVB3-induced myocarditis. Reiterating the biphasic nature of disease in susceptible mouse strains, the investigators determined that early treatment on days 0–7 resulted in decreased mortality, lower cardiac viral titers and decreased cellular infiltration compared to untreated BALB/c mice. Late IL-2 treatment (on days 7–14) increased mortality and enhanced cardiac interstitial infiltrates. The protective role of IL-2 in the early phase of disease was attributed to increased NK cell activity, while the detrimental role in the late phase of disease was linked to enhanced T-cell activation.

In addition to the examination of the disease-modifying capabilities of proinflammatory cytokines on CVB3-induced myocarditis, other studies have evaluated the role of inhibitory or regulatory cytokines on the development of autoimmune disease. Kulkarni et al. (94) developed a transforming growth factor-beta 1 (TGF-β1) knockout mouse that spontaneously developed inflammatory heart disease in addition to inflammatory lesions in other organs. In collaboration with Dr. Kulkarni, we tested serum samples for the presence of cardiac myosin specific antibody. Mice homozygous for the mutation showed significant levels of circulating autoantibody compared to littermate controls, which did not exhibit cardiac infiltrates and did not produce autoantibody to heart. Therefore, cytokines are critical in the regulation of autoimmune myocarditis. They may function to enhance or diminish T cell activation or to alter target organ susceptibility. Other possible functions of the proinflammatory cytokines include the induction of oxygen-free radicals which can directly induce cardiac dysfunction and cardiomyocyte damage.

6.3. Nitric Oxide

Nitric oxide (NO) is an oxygen free-radical molecule that is important in cardiac physiology and pathology (95–97). It is synthesized by nitric oxide synthase (NOS) by the conversion of L-arginine to L-citrulline. The enzyme occurs in three recognized isoforms: n-NOS (neuronal NOS), i-NOS (inducible NOS), and e-NOS (endothelial NOS) n-NOS and e-NOS are constitutive calcium-dependent enzymes that produce nitric oxide (NO) at low levels. i-NOS, inducible NOS, is stimulated by LPS and proinflammatory cytokines, such as IL-1, IFN-γ and TNF-α. It is expressed by a variety of cells, including macrophages, polymorphonuclear leukocytes and cardiomyocytes (98–100). Inducible NOS activity is calcium-independent and the enzyme produces large quantities of NO.

Nitric oxide may be a double-edged sword in myocarditis. It provides beneficial functions because it is viricidal, but it is also detrimental as it has exhibited cytotoxic activity and has been shown to depress cardiac contractility in vitro (101). Endomyocardial biopsy specimens from patients with myocarditis and IDCM have demonstrated diffuse i-NOS expression not observed in age-matched controls (102,103). The development of several i-NOS inhibitors have been useful in examining the role of NO in autoimmune myocarditis.

The necessity of IFN-γ in the expression of i-NOS has been questioned (104). Investigators have utilized interferon regulatory factor-1 (IRF-1) knockout mice, animals deficient in an important IFN-γ-induced transcription factor, to examine the role of IFN-γ dependent NO production in myosin-induced myocarditis (105). The group showed that IRF-1 knockout mice developed cardiac disease of similar severity and prevalence compared to heterozygous littermate controls. They further examined the effects of blocking NOS using NOS inhibitor, N^G-nitro-L-arginine methyl ester (L-NAME), in myosin-immunized A/J mice. In this experiment, they also failed to alter the extent of cardiac lesions. The group concluded that IFN-γ signaling is critical for NO production in A/J mice in this experimental model of autoimmune myocarditis. However, the results suggest that NO does not play an important role in myosin-induced cardiac disease.

In a study comparing an outbred resistant mouse strain (CD-1) to an inbred susceptible strain (C3H/HeJ), Freeman et al. (106) found the CVB3 infection resulted in chronic myocarditis in the latter and resolution of cardiac lesions in the former animals on day 21. Between days 14 and 21 C3H/HeJ mice showed continued depression in cardiac functional performance and substantial expression of intracardiac IL-1β, TNF-α and i-NOS. The expression of proinflammatory cytokines and i-NOS appeared to parallel a reduction in cardiac contractility. Further evidence supporting the ability of CVB3 virus to induce i-NOS expression was provided by Mikami et al. (107) who showed intracardiac expression of α-NOS beginning on approximately day 4 post-inoculation in C3H/HeJ mice. The enzyme reached peak expression on days 8–9 and declined thereafter, approaching baseline levels on day 30. i-NOS was detected primarily in macrophages and polymorphonuclear leukocytes. *In vitro* studies exposing cardiomyocytes to proinflammatory cytokines supported the ability of cardiomyocytes to express i-NOS (108). Enzyme activity has been confirmed by the detection of nitrosylation of protein tyrosine residues using immunohistochemistry.

Several investigators have tested *in vivo* the role of NO in CVB3-induced myocarditis. Hiraoka et al. (109) examined the effects of blocking NOS using L-NAME in virus-inoculated C3H/HeJ mice. Both early (on days 0–14) and late (on days 14–35) treatment resulted in enhancement of cardiac lesions indicating that NO plays a protective role during both early and later stages of CVB3-induced myocarditis. L-NAME is a nonspecific inhibitor of NOS, suppressing both e-NOS and i-NOS activity. Therefore this study failed to distinguish between the contributions of the individual enzymes in disease induction and progression. In another study rats injected with L-NAME in the absence of virus inoculation or myosin immunization developed cardiac lesions, whereas animals injected with the inactive enantomer (D-NAME) failed to develop cardiac necrosis (110). The lesions were attributed to an increase in mean arterial blood pressure and a transient decrease in coronary blood flow. This study underlined the importance of NO in vascular relaxation and cardiac perfusion, functions primarily controlled by e-NOS under normal conditions. L-NAME is not the optimal NOS inhibitor to test in the CVB3-induced myocarditis model, because its effects through e-NOS on peripheral blood flow may contribute to disease development.

Other studies using nonspecific NOS inhibitors have concentrated on the effects of blocking NOS during the early viral phase of disease. Consistently investigators have found that early production of NO contributes to viral clearance from the heart (107,109,111). These findings are further supported by *in vitro* studies showing the anti-viral activity NO in CVB3-infected cultures of macrophages and HeLa cells (109,111).

7. APOPTOSIS

There are at least two possible mechanisms of cardiomyocyte destruction in myocarditis, necrosis or apoptosis. Both have come under scrutiny as potential means of cardiomyocyte death. To date, one mouse strain (BALB/c) tested has exhibited apoptosis of cardiomyocytes upon inoculation with CVB3 virus. Other mouse strains have shown minimal apoptosis of infiltrating mononuclear inflammatory cells in the heart (75). Proinflammatory cytokines such as TNF-α are known to cause apoptosis (112). Nitric oxide has also been linked with apoptosis of cardiomyocytes in ischemic heart disease (113,114). In other mouse strains tested, it appears that necrosis, nonprogrammed cell death, is the major mechanism of cardiomyocyte destruction. In a study examining C3H/HeJ mice, inoculation with CB3 virus resulted in an increase in an anti-apoptotic factor, Bcl-2, and very little cardiac apoptosis was observed (115). The results indicate that, in some mouse strains, the induction of protective molecules may prevent apoptosis. Further studies are needed to address the question of pro- and anti-apoptotic factors in CVB3 myocarditis.

8. CONCLUSIONS

Based on the evidence presented in this chapter, it is clear that susceptibility to autoimmune myocarditis depends on genetic factors and viral strain. We hypothesize that viral infection results in cardiomyocyte damage directly, or indirectly via the induction of an early inflammatory infiltrate that contributes to cardiomyocyte destruction. Cardiomyocyte lysis leads to the release of cardiac myosin into the interstitium, where it is taken up by resident dendritic cells and presented to cardiac myosin-specific T cells in a susceptible host. Release of the appropriate cardiac myosin epitopes, frequency of auto-reactive T and B cells, and target organ expression of class I and II MHC as well as adhesion molecules are critical in the induction of autoimmune myocarditis. Alterations in the cytokine milieu are capable of converting resistant animals to a susceptible phenotype. Such proinflammatory cytokines as IL-1β and TNF-α mediate inflammation in the heart as well as local tissue destruction. The balance between Th1 and Th2 immune responses is important for the progression of autoimmune myocarditis. We have demonstrated that a Th2 response, and IL-4 in particular, could be harmful and promote the autoimmune process. Remarkably, IFN-γ is protective in both models of autoimmune myocarditis exerting antiviral and anti-inflammatory effects. The exact molecular mechanisms of action of these cytokines as well as the contributions of other cytokines requires further investigations. Certain molecules, such as IL-2, may play a protective, antiviral, role in the viral phase of disease and a dentrimental, proinflammatory, role during the autoimmune phase. NO has been shown to have antiviral activity. However, its contribution to an autoimmune process is still unclear. The fact that the same mediators may play either a beneficial or detrimental role depending on the phase of the disease process complicates the development of therapeutic strategies in patients with autoimmune myocarditis and IDCM. The agents that are beneficial regardless of the stage of disease, such as IFNs, should be explored for the optimal therapeutic intervention.

ACKNOWLEDGMENTS

We are pleased to acknowledge the contributions to our studies of autoimmune myocarditis made by former colleagues in our laboratory. They include Gregory Allen, Floria Alvarez, Kirk Beisel, Ahvie Herskowitz, Ann Lafond-Walker, James Lane, Nicholas Neu, David Neumann, Luanne Wolfgram, and Stuart Wulff.

Our research was supported by NIH research grants HL33878 and KOI0095.

References

1. Smith, W.G. 1970. Coxsackie B myopericarditis in adults. *Am. Heart J.* 80:34 46.
2. Grist, N.R., and E.J. Bell. 1974. A six-year study of Coxsackievirus B infections in heart disease. *J. Hyg.* (London) 73:165–172.
3. Maisch, B.R. Trostel-Soeder, E. Steuhumoouer, P.A. Berg, and K. Kochsiek. 1982. Diagnostic relevance of humoral and cell-mediated immune reactions in patients with acute viral myocarditis. *Clin. Exp. Immunol.* 48:533–545.
4. Schultheis, H.P., and H.P. Bolte. 1985. Immunological analysis of auto-antibodies against the adenine nucleotide translocator in dilated cardiomyopathy. *J. Mol. Cell, Cardiol* 17:603–617.
5. Woodruff, J.F. 1980. Viral myocarditis: A review. *Am. J. Pathol.* 101:426.
6. Schultheiss, H.P., K. Schulze, U. Kühl, G. Ulrich, and M. Klingenberg. 1986. The ADP/ATP carrier as mitochondrial autoantigen—facts and perspectives. *Ann. NY Acad. Sci.* 488:44–64.
7. Ansari, A.A., A. Herskowitz, and D.J. Danner. 1988. Identification of mitochondrial proteins that serve as targets for autoimmunity in human dilated cardiomyopathy. *Circulation* 78(Suppl.):457.
8. Swimmbeck, P.L., C. Badorf, G. Rohn, K. Schulze, and P. Schultheiss. 1996. The role of sensitized T-cells in myocarditis and dilated cardiomyopathy. *Int. J. Cardiol.* 54:117–125.
9. Hyypia, T., T. Hovi, N.J. Knowles, and G. Stanway. 1997. Classification of enteroviruses based on molecular and biological properties. *J. Gen. Virol.* 78:1–11.
10. Maisch, B., H. Wilke, S. Marcin, C. Werner, and W. Gebhardt. 1984. Adriamycin cardiotoxicity: An echocardiographic and immunologic follow-up study. *Circulation* 70 (Suppl.II):149–157.
11. Lange, L.G. 1988. Immunologic mechanisms of cardiac disease. In: *Current concepts of* relations to autoimmunity in rheumatic fever, post-cardiotomy and post-infarction syndromes—Heart disease. W.B. Saunders, Philadelphia p. 1521.
12. Kaplan, M.H., and J.D. Frengley. 1969. Autoimmunity to heart in cardiac disease. In Current concepts of relations to autoimmunity in rheumatic fever, postcardiotomy and post-infarction syndromes. *Am. J. Cardiol.* 24:459–473.
13. Rose N.R., A. Herskowitz, D.A. Neumann, and N. Neu. 1988. Autoimmune myocarditis: a paradigm of post-infection autoimmune disease. *Immunology Today* 9:117–119.
14. Wolfgram, L.J., K.W. Beisel, A. Herskowitz, and N.R. Rose. 1986. Variations in the susceptibility to Coxsackievirus B3-induced myocarditis among different strains of mice. *J. Immunol.* 136:1846–1852.
15. Wolfgram, L.J., K.W. Beisel, and N.R. Rose. 1985. Heart-specific autoantibodies following murine Coxsackievirus B$_3$, myocarditis. *J. Exp. Med.* 161:1112–1121.
16. Alvarez F.L., N. Neu, N.R. Rose, S.W. Craig, and K.W. Beisel. 1987. Heart-specific autoantibodies induced by Coxsackievirus B$_3$: Identification of heart autoantigens. *Clin. Immunol. Immunopathol.* 43:129–139.
17. Neu, N., N.R. Rose, K.W. Beisel, A. Herskowitz, G. Gurri-Glass, and S.W. Craig. 1987. Cardiac myosin induces myocarditis in genetically predisposed mice. *J. Immunol.* 139:3630–3636.
18. Lerner, A.M., and F.M. Wilson. 1973. Virus myocardiopathy. *Prog. Med. Virol.* 15:63–91.
19. Rose, N.R., L.J. Wolfgram, A. Herskowitz, and K.W. Beisel. 1986. Postinfectious autoimmunity: two distinct phases of Coxsackievirus B$_3$-induced myocarditis. *Ann. NY Acad. Sci.* 475:146–156.
20. Herskowitz, A., L.J. Wolfgram, N.R. Rose, and K.W. Beisel. 1987. Coxsakievirus B$_3$ murine myocarditis – Histopathologic spectrum of myocarditis in genetically defined inbred strains. *J. Am. Coll. Cardiol.* 9:1311–1319.
21. Rose, N.R., D.A. Neumann, A. Herskowitz, M. Traystman, and K.W. Beisel. 1988. Genetics of susceptibility to viral myocarditis in mice. *Pathol. Immunopathol. Res.* 7:266–278.
22. Rose, N.R., N. Neu, D.A. Neumann, and A. Herskowitz. 1988. Myocarditis: A post-infectious autoimmune disease. In: *New concepts in viral heart disease—Virology, immunology and clinical management*, H.P. Schultheiss, editor. Springer—Verlag, Berlin. pp. 139–147.
23. Neu N., K.W. Beisel, M.D. Traystman, N.R. Rose, and S.W. Craig. 1987. Autoantibodies specific for the cardiac myosin isoform are found in mice susceptible to Coxsackievirus B$_3$-induced myocarditis. *J. Immunol.* 138:2488–2492.
24. Rose, N.R., and S.L. Hill. 1996. The pathogenesis of post-infectious myocarditis. *Clin. Immunol. Immunopathol.* 80:S92–S99.
25. Pummerer, C.L., L. Kerstin, G. Grassl, K. Bachmaier, F. Offner, S.K. Burrell, D.M. Lenz, T.J. Zamborelli, J.M. Penninger, and N. Neu. 1996. Cardiac myosin-induced myocarditis: Target recognition of autoreactive T cells requires prior activation of cardiac interstitial cells. *Lab. Invest.* 74:845–851.
26. Liao, L., R. Sindhwani, L. Leinwand, B. Diamond, and S. Factor. 1993. Cardiac alpha-myosin heavy chains differ in their induction of myocarditis: identification of pathogenic epitopes. *J. Clin. Invest.* 92:2877–2882.
27. Neumann, D.A., N.R. Rose, A.A. Ansari, and A. Herskowitz. 1994. Induction of multiple heart autoantibodies in mice with Coxsackievirus B3- and cardiac myosin-induced autoimmune myocarditis. *J. Immunol.* 152:343–350.
28. Sheshberadaraw, H., and E. Norrby. 1984. Three monoclonal antibodies against measles virus F protein cross-react with the cellular stress proteins. *J. Virol.* 52:995–999.
29. Norrby, E., H. Sheshberadaraw, and B. Rafner. 1985. Antigen mimicry involving the measles virus hemagglutinin and the

human respiratory syncytial virus nucleoprotein. *J. Virol.* 53:456–460.

30. Zhao, Z.S., F. Granucci, L. Yeh, P.A. Schaffer, and H. Cantor. 1998. Molecular mimicry by herpes simplex virus-type 1: Autoimmune disease after viral infection. *Science* 279:1344–1347.

31. Gauntt, C.J., A.L. Higdon, H.M. Arizde, M.R. Tamayo, R. Crawley, R.D. Henkel, M.E.N. Pereira, S.M. Tracy, and M.W. Cunningham. 1993. Epitopes shared between Coxsackievirus B3 (CVB3) and normal heart tissue contribute to CVB3-induced murine myocarditis. *Clin. Immunol. Immunopathol.* 68:129–134.

31A. Bachmaier, K., N. Neu, L.M. de la Maza, S. Pal, A Hessel, and J.M. Penninger. 1999. *Science* 283:1335–1339.

32. Neu, N., S.W. Craig, N.R. Rose, F. Alvarez, and K.W. Beisel. 1987. Coxsackievirus induced myocarditis in mice: cardiac myosin autoantibodies do not cross-react with the virus. *Clin. Exp. Immunol.* 69:566–574.

33. Gauntt, C.J., S.M. Tracy, N. Chapman, H.J. Wood, P.C. Kolbeck, A.G. Karaganis, C.J. Winfrey, and M.W. Cunningham. 1995. Coxsackievirus induced chronic myocarditis in murine model. *Eur. Heart. J.* 16 (Suppl.D):56–58.

34. Sharaf, A.R., J. Narula, P.D. Nicol, J.F. Southern, and B.A. Khaw. 1994. Cardiac sarcoplasmic reticulum calcium ATPase, an autoimmune antigen in experimental cardio-myopathy. *Circulation* 89:1217–1228.

35. Huber, S.A. 1992. Heat-shock protein induction in adriamycin and picornavirus-infected cardiocytes. *Lab. Invest.* 67:218–224.

36. Huber, S.A., N. Heintz, and R. Tracy. 1988. Coxsackievirus B3-induced myocarditis: virus and actinomycin D treatment of myocytes induces novel antigen recognized by cytotoxic T lymphocytes. *J. Immunol.* 141:3214–3219.

37. Gauntt, C.J., M.D. Trousdale, D.R.L. LaBadie, R.E. Paque, and T. Nealon. 1979. Properties of coxsackievirus B3 variants, which are amyocarditic or myocarditic for mice. *J. Med. Virol.* 3:207–220.

38. Ryan, M.D., and M. Flint. 1997. Virus encoded proteinases of picornavirus super-group. *J. Gen. Virol.* 78:699–723.

38A. Badorff, C., G.H. Lee, B.J. Lamphear, M.E. Martone, K.P. Campbell, R.E. Rhoads, and K.U. Knowlton. 1999. *Nat. Med.* 5:320–326.

39. Hamrell, B.B., S.A. Huber, and K.O. Leslie. 1994. Reduced unloaded sarcomere shortening velocity and a shift to a slower myosin isoform in acute murine coxsackievirus myocarditis. *Circulation Res.* 75:462–472.

40. Pummerer, C., P. Berger, M. Fruhwirth, C. Offner, and N. Neu. 1991. Cellular infiltrate, major histocompatibility antigen expression and immunopathogenic mechanisms in cardiac myosin-induced myocarditis. *Lab. Invest.* 65:538–547.

41. Kay, M.M., K. Sorensen, P. Wong, and P. Bolton. 1982. Antigenicity, storage, and aging: physiologic autoantibodies to cell membrane and serum proteins and the senescent cell antigen. *Mol. Cell. Biochem.* 49:65–85.

42. Anderson, D.R., J.E. Wilson, C.M. Carthy, D. Yang, R. Kandolf, and B.M. McManus. 1996. Direct interactions of Coxsackievirus B3 with immune cells in the splenic compartment of mice susceptible or resistant to myocarditis. *J. Virol.* 70:4632–4645.

43. Vuorinen, T., R. Vainioapaa, R. Vanharanta, and T. Hyypia. 1996. Susceptibility of human bone marrow cells and hematopoietic cell lines to Coxsackievirus B3 infection. *J. Virol.* 70:9018–9023.

44. Kanda, T., S. Ohshima, K. Yuasa, T. Watanabe, T. Suzuki, and K. Murata. 1990. Idiopathic myocarditis associated with T-cell subset changes and depressed natural killer activity. *Jap. Heart J.* 31:741–744.

45. Deguchi, H., Y. Kitaura, T. Hayashi, M. Kotaka, and K. Kawamura. 1989. Cell mediated immune cardiomyocyte injury in viral myocarditis of mice and patients. *Jap. Circ. J.* 53:61–77.

46. Saito, T., A. Shiokawa, and S. Inoue. 1989. Lymphatic sub-populations and their transition in myocardial tissue and peripheral blood of patients with biopsy-proven myocard-itis. *Jap. Circ. J.* 53:1–6.

47. Job, L.P., D.C. Lynden, and S.A. Huber. 1986. Demonstration of suppressor cells in coxsackievirus group B type 3 infected female Balb/c mice which prevent myocarditis. *Cell. Immunol.* 98:104–113.

48. Huber, S.A., L.P. Job, and K.R. Ault. 1982. Influence of sex hormones on coxsackievirus B3 infection in Balb/c mice. *Cell. Immunol.* 67:173–189.

49. Lynden, P.C., J. Olszewski, M. Feran, L.P. Job, and S.A. Huber. 1987. Coxsackievirus B3-induced myocarditis. Effects of sex steroids on viremia and infectivity of cardiocytes. *Am. J. Pathol.* 126:432–438.

50. Huber, S.A., and B. Pfaeffle. 1994. Differential Th1 and Th2 cell responses in male and female Balb/c mice infected with coxsackievirus group B type 3. *J. Virol.* 68:5126–5132.

51. Smith, S.C., and P.M. Allen. 1993. Expression of myosin-class II major histocompatibility complex in the normal myocardium occurs before inducing autoimmune myocard-itis. *Clin. Immunol. Immunopathol.* 68:100–106.

52. Ansari, A.A., W. Yi-Chong, D.J. Daner, M.B. Gravanis, A. Mayne, A. Neckelmann, K.W. Sell, and A. Herskowitz. 1991. Abnormal expression of histocompatibility and mitochondrial antigens by cardiac tissue from patients with myocarditis and dilated cardiomyopathy. *Clin. Immunol. Immunopathol.* 139:337–354.

53. Okada, I., A. Matsumori, and B. Kyu. 1991. Detection of viral RNA in experimental coxsackievirus B3 myocarditis of mice using polymerase chain reaction. *Int. J. Exp. Pathol.* 73:721–731.

54. Andreoletti, L., D. Hober, C. Decoene, M.C. Copin, P.E. Lobert, A. Dewilde, C. Stankowiac, and P. Wattre. 1996. Detection of enteroviral RNA by polymerase chain reaction in endomyocardial tissue of patients with chronic cardiac diseases. *J. Med. Virol.* 48:53–59.

55. Klingel, K., and R. Kandolf. 1993. The role of enterovirus replication in development of acute and chronic heart muscle disease in different immunocompetent mouse strains. *Scand. J. Infect. Dis.* 88(Suppl.):79–85.

56. Kipshidze, N.N., V.B. Chumburidze, L.M. Dzidziguri, and M.N. Detashidze. 1984. Characteristics of immuno-regulatory lymphocyte subpopulations in patients with congestive cardiomyopathy and nonrheumatic myocarditis studied by monoclonal antibodies. *Terapeuticheskii Arkhiv* 56:56–58.

57. Neu, N., C. Pummerer, T. Riekitz, and P. Berger. 1993. T cells in cardiac myosin-induced myocarditis. *Clin. Immunol. Immunopathol.* 68:107–110.

58. Wong, C.Y., J.J. Woodruff, and J.F. Woodruff. 1977. Generation of cytotoxic T lymphocytes during CB3 infection. II. Role of sex. *J. Immunol.* 118:1165–1169.

59. Henke, A., S. Huber, A. Stelzner, and L. Whitton. 1995. The role of CD8+ T lymphocytes in coxsackievirus B3-induced myocarditis. *J. Virol.* 69:6720–6728.

60. Hanawa, H., M. Kodama, T. Inomata, T. Izumi, A. Sihbata, M. Tuchida, Y. Matsumoto, and T. Abo. 1994. Anti-alpha beta T cell receptor antibody prevents the progression of experi-

mental autoimmune myocarditis. *Clin. Exp. Immunol.* 96:470–475.

61. Penninger, M., N. Neu, E. Timms, V.A. Wallace, D.R. Koh, K. Kishihara, C. Pummerer, and T.W. Mak. 1993. The induction of experimental autoimmune myocarditis in mice lacking CD4 or CD8 molecules. *J. Exp. Med.* 178:1837–1842.

61A. Smith, S.C., and P.M. Allen. 1991. Myosin-induced acute myocarditis is a T cell-mediated disease. *J. Immunol.* 147:2141–2147.

62. Bachmaier, K., C. Pummerer, A. Shahinian, J. Ionesco, N. Neu, T.W. Mak, and J.M. Penninger. 1996. Induction of autoimmunity in the absence of CD28 costimulation. *J. Immunol.* 157:1752–1757.

62A. Smith, S.C., and P.M. Allen. 1993. Prevention of myosin-induced autoimmune myocarditis with competitor peptides *Immunol. Ser.* 59:377–386.

63. Boise, L.H., A.J. Minn, P.J. Noel, C.H. June, M.A. Accavitti, T. Lindsten, and C.B. Thompson. 1995. CD28 costimulation can promote T cell survival by enhancing the expression of Bcl-XL. *Immunity* 3:87–98.

64. Estrin, M., and S.A. Huber. 1987. Coxsackievirus B3-induced myocarditis. Autoimmunity is L3T4+ T helper cell and IL-2 independent in Balb/c mice. *Am. J. Pathol.* 127:335–341.

65. Hashimoto, I., M. Iatsumi, and M. Nakagana. 1983. The role of T lymphocytes in the pathogenesis of Coxsackievirus B3 heart disease. *Brit. J. Exp. Pathol.* 64:497–504.

66. Kishimoto, C., and W.H. Abelmann. 1990. *In vivo* significance of T cells in the development of Coxsackievirus B3 myocarditis in mice. Immature but antigen specific T cells aggravate cardiac injury. *Circulation Res.* 67:589–598.

67. Lodge, P.A., M. Herzum, J. Olszewski, and S.A. Huber. 1987. Coxsackievirus B3 myocarditis: acute and chronic forms of disease caused by different immunopathogenic mechanisms. *Am. J. Pathol.* 128:455–463.

68. Huber, S.A. 1997. Coxsackievirus-induced myocarditis is dependent on distinct immunopathogenic responses in different strains of mice. *Lab. Invest.* 76:691–701.

69. Godney, E.K., and C.J. Gauntt. 1987. Murine natural killer cells limit coxsackievirus B3 replication. *J. Immunol.* 139:913–918.

70. Braunwald, E. 1988. Heart Disease. W.B. Saunders, Philadelphia, P.A., p. 1521.

71. Smith, S.C., and P.M. Allen. 1993. The role of T cells in myosin-induced autoimmune myocarditis. *Clin. Immunol. Immunopathol.* 68:100–106.

72. Neu, N., B. Ploier, and C. Offner. 1990. Cardiac myosin-induced myocarditis: Heart autoantibodies are not involved in the induction of disease. *J. Immunol.* 145:4094.

73. Liao, L., R. Sindhwani, M. Rojkino, S. Factor, L. Leinwand, and B. Diamond. 1995. Antibody mediated autoimmune myocarditis depends on genetically determined target organ sensitivity. *J. Exp. Med.* 181:1123–1131.

74. Huber, S.A., and P.A. Lodge. 1986. Coxsackievirus B3 myocarditis: Identification of different pathogenic mechanisms in DBA/2 and Balb/c mice. *Am. J. Pathol.* 122:284–291.

75. Neumann, D.A., J.R. Lane, A. LaFond-Walker, G.S. Allen, S.M. Wulff, A. Herskowitz, and N.R. Rose. 1991. Heart-specific autoantibodies can be eluted from the hearts of Coxsackievirus B3-infected mice. *Clin. Exp. Immunol.* 86:405–412.

76. Satoh, M., G. Tamura, I. Segawa, A. Tashiro, K. Hiramori, and R. Satodate. 1996. Expression of cytokine genes and presence of enteroviral genomic RNA in endomyocardial biopsy tissues of myocarditis and dilated cardiomyopathy. *Virchow's Archiv.* 427:503–509.

77. Henke, A., M. Nain, A. Stelzner, and D. Gemsa. 1991. Induction of cytokine release from human monocytes by coxsackievirus infection. *Eur. Heart. J.* 12(Suppl.D): 134–136.

78. Okura, Y., T. Yamamoto, S. Goto, T. Inomata, S. Hirono, H. Hanawa, L. Feng, C.B. Wilson, I. Kihara, T. Izumi, A. Shibata, Y. Aizawa, S. Seki, and T. Abo. 1997. Characterization of cytokine and i-NOS mRNA expression in situ during the course of experimental autoimmune myocarditis in rats. *J. Mol. Cell. Cardiol.* 29:491–502.

79. Wang, Y., M. Afanasyeva, S.L. Hill, and N.R. Rose. 1999. Characterization of murine autoimmune myocarditis induced by self and foreign cardiac myosin. *Autoimmunity* 31:151–162.

80. Smith, S.C., and P.M. Allen. 1992. Neutralization of endogenous tumor necrosis factor ameliorates the severity of myosin-induced myocarditis. *Circulation Res.* 70:856–863.

81. Bachmaier, K., C. Pummerer, I. Kozieradzki, K. Pfeffer, T.W. Mak, N. Neu, and J.M. Penninger. 1997. Low-molecular weight tumor necrosis factor receptor p55 controls induction of autoimmune heart disease. *Circulation* 95:655–661.

82. Afanasyeva, M., Y. Wang, Z. Kaya, S.L. Hill, and N.R. Rose. 2000. Experimental autoimmune myocarditis: an organ-specific autoimmune disease with a Th2 phenotype. *FASEB. J.* 14:A994.

82A. Chu, C.Q., S. Wittmer, and D.K. Dalton. 2000. Failure to suppress the expansion of the activated CD4 T cell population in interferon γ-deficient mice leads to exacerbation of experimental autoimmune encephalomyelitis. *J. Exp. Med.* 192:123–128.

83. Okura, Y., K. Takeda, S. Honda, H. Hanawa, H. Watanabe, M. Kodama, T. Izumi, Y. Aizawa, S. Seki, and T. Abo. 1998. Recombinant murine interleukin-12 facilitates induction of cardiac myosin-specific type 1 helper T cells in rats. *Circulation Res.* 82:1035–1042.

84. Shioi, T., A. Matsumori, and S. Sasayama. 1996. Persistent expression of cytokine in the chronic stage of viral myocarditis in mice. *Circulation* 94:2930–2937.

85. Seko, Y., N. Takahashi, H. Yagita, K. Okumura, and Y. Yazaki. 1997. Expression of cytokine mRNAs in murine hearts with acute myocarditis caused by coxsackievirus B3. *J. Pathol.* 183:105–108.

86. Lane, J.R., D.A. Neumann, A. Lafond-Walker, A. Herskowitz, and N.R. Rose. 1993. Role of IL-1 and tumor necrosis factor in Coxsackie virus-induced autoimmune myocarditis. *J. Immunol.* 151:1682–1690.

87. Lane, J.R., D.A. Neumann, A. Lafond-Walker, A. Herskowitz, and N.R. Rose. 1991. LPS promotes CB3-induced myocarditis in resistant B10.A mice. *Cell. Immunol.* 136:219–233.

88. Lane, J.R., D.A. Neumann, A. Lafond-Walker, A. Herskowitz, and N.R. Rose. 1992. Interleukin 1 or tumor necrosis factor can promote Coxsackie B3-induced myocarditis in resistant B10.A mice. *J. Exp. Med.* 175:1123–1129.

89A. Lawson, C.M. 1999. Low-dose oral use of interferon inhibits virally induced myocarditis. *J. Interferon Cytokine Res.* 19:863–867.

89B. Yamamoto, N., M. Shibamori, M. Ogura, Y. Seko, M. Kikuchi. 1998. Effects of intranasal administration of recombinant murine interferon-γ on murine acute myocarditis caused by encephalomyelitis virus. *Circulation* 97:1017–1023.

89C. Horwitz, M.S. 2000. Pancreatic expression of interferon-gamma protects mice from lethal coxsackievirus B3 infection and subsequent myocarditis. *Nat. Med.* 6:693–697.

89D. Miric, M., J. Vasiljevic, M. Bojic, Z. Popovic, N. Keserovic, and M. Pesic. 1996. Long-term follow up of patients with dilated heart muscle disease treated with human leucocytic interferon alpha or thymic hormones; initial results. *Heart* 75:596-601.

89. Neumann, D.A., J.R. Lane, G.S. Allen, A. Herskowitz, and N.R. Rose. 1993. Viral myocarditis leading to cardiomyopathy: do cytokines contribute to pathogenesis? *Clin. Immunol. Immunopathol.* 68:181–190.

90. Kragel, A.H., W.D. Travis, R.G. Steis, S.A. Rosenberg, and W.C. Roberts. 1990. Myocarditis or acute myocardial infarction associated with interleukin-2 therapy for cancer. *Cancer* 66:1513–1516.

91. Samlowski, W.E., J.H. Ward, C.M. Craven, and R.A. Freedman. 1989. Severe myocarditis following high-dose interleukin-2 administration. *Arch. Pathol. Lab. Med.* 113:838–841.

92. Yamamoto, A., R.J. Wenthold, J. Zhang, E.H. Herman, and V.J. Ferrans. 1995. Immunofluorescence techniques for the identification of immune effector cells in rat heart: application to the study of the myocarditis induced by interleukin-2. *J. Mol. Cell. Cardiol.* 27:307–319.

93. Kishimoto, C., Y. Kuroki, Y. Hiraoka, H. Ochiai, M. Kurakawa, and S. Sasayama. Cytokine and murine coxsackievirus B3 myocarditis: Interleukin-2 suppresses myocarditis in the acute stage but enhances the condition in the subsequent stage. *Circulation* 89:2836–2842.

94. Kulkarni, A.B., J.M. Ward, L. Yaswen, C.L. Mackall, S.R. Bauer, C.G. Huh, R.E. Gress, and S. Karlsson. 1995. Transforming growth factor beta-1 null mice. An animal model for inflammatory disorders. *Am. J. Pathol.* 146:264–275.

95. Moncada, S., and A. Higgs. 1993. The L-arginine-nitric oxide pathway. *N. Engl. J. Med.* 329:2002–2012.

96. Schulz, R., E. Nava, and S. Moncada. 1992. Induction and potential biological relevance of a Ca(2+)-independent nitric oxide synthase in the myocardium. *Brit. J. Pharmacol.* 105:575–580.

97. Balligand, J.L., R.A. Kelly, P.A. Marsden, T.W. Smith, and T. Michel. 1993. Control of cardiac muscle cell function by an endogenous nitric oxide signalling system. *Proc. Natl. Acad. Sci. USA* 90:347–351.

98. Xie, Q.W., H.J. Cho, J. Calaycay, R.A. Mumford, K.M. Swiderek, T.D. Lee, A. Ding, T. Troso, and C. Nathan. 1992. Cloning and characterization of inducible nitric oxide synthase from mouse macrophages. *Science* 256:225–228.

99. Tsujino, M., Y. Hirata, T. Imai, K. Kanno, S. Euguchi, H. Ito, and F. Marumo. 1994. Induction of nitric oxide synthase gene by interleukin-1 beta in cultured rat cardiocytes. *Circulation* 90:375–383.

100. Cifone, M.G., C. Festuccia, L. Cironi, G. Cavallo, M.A. Cessa, V. Pensa, E. Tubaro, and A. Santoni. 1994. Induction of the nitric oxide synthesizing pathway in fresh and interleukin 2-cultured rat natural killer cells. *Cell. Immunol.* 157:181–194.

101. Finkel, M.S., C.V. Oddis, T.D. Jacob, S.C. Watkins, B.G. Hattler, and R.L. Simmons. 1992. Negative inotropic effects of cytokines on the heart mediated by nitric oxide. *Science* 257:387–389.

102. de-Belder, A.J., M.W. Radomski, H.J. Why, P.J. Richardson, C.A. Bucknall, E. Salas, J.F. Martin, and S. Moncada. 1993. Nitric oxide synthase activities in human myocardium. *Lancet* 341:84–85.

103. de-Belder, A.J., M.W. Radomski, H.J. Why, P.J. Richardson, and J.F. Martin. 1995. Myocardial calcium-independent nitric oxide synthase activity is present in dilated cardiomyopathy, myocarditis, and postpartum cardiomyopathy but not in ischaemic or valvular heart disease. *Brit. Heart J.* 74:426–430.

104. Kamijo, R., D. Shapiro, J. Le, S. Huang, M. Aguet, and J. Vilsek. 1993. Generation of nitric oxide and induction of major histocompatibility complex class II antigen in mice lacking the interferon gamma receptor. *Proc. Natl. Acad. Sci. USA* 90:6626–6630.

105. Bachmaier, K., N. Neu, C. Pummerer, G.S. Duncan, T.W. Mak, T. Matsuyama, and J.F. Penninger. 1997. i-NOS expression and nitrotyrosine formation in the myocardium in response to inflammation is controlled by the interferon regulatory transcription factor 1. *Circulation* 96:585–591.

106. Freeman, G.L., J.T. Colston, M. Zabalgottia, and B. Chandrasekar. 1998. Contractile depression and expression of proinflammatory cytokines and i-NOS in viral myocarditis. *Am. J. Pathol.* 274:249–258.

107. Mikami, S., S. Kawashima, K. Kanazawa, K.-I. Hirata, Y. Katayama, H. Hotta, Y. Hayashi, H. Ito, and M. Yokoyama. 1996. Expression of nitric oxide synthase in a murine model of viral myocarditis induced by Coxsackievirus B3. *Biochem. Biophys. Res. Comm.* 220:983–989.

108. Kooy, N.W., S.J. Lewis, J.A. Royall, Y.Z. Ye, D.R. Kelly, and J.S. Beckman. 1997. Extensive nitration in human myocardial inflammation: presence of peroxynitrite. *Critical Cardiol. Med.* 25:812–819.

109. Hiraoka, Y., C. Kishimoto, H. Takada, M. Nakamura, M. Kurokawa, H. Ochiai, and K. Shiraki. 1996. Nitric oxide and murine coxsackievirus B3 myocarditis: Aggravation of myocarditis by inhibition of nitric oxide synthase. *J. Am. Coll. Cardiol.* 28:1610–1615.

110. Moreno, H., K. Metze, A.C. Bento, E. Antunes, R. Zatz, and G. de-Nucci. 1996. Chronic nitric oxide inhibition as a model of hypertensive heart muscle disease. *Basic Res. Cardiol.* 91:248–255.

111. Lowenstein, C.J., S.L. Hill, A. Lafond-Walker, J. Wu, G. Allen, M. Landavere, N.R. Rose, and A. Herskowitz. 1996. Nitric oxide inhibits viral replication in murine myocarditis. *J. Clin. Invest.* 97:1837–1843.

112. Geng, Y.-J., Q. Wu, M. Muszynski, G.K. Hansson, and P. Libby. 1996. Apoptosis of vascular smooth muscle cells induced by *in vitro* stimulation with IFN-gamma, tumor necrosis factor-alpha, and interleukin-1beta. *Arterioscl. Thromb. Vasc. Biol.* 16:19–27.

113. Gottlieb, R.A., K.O. Burleson, R.A. Kloner, B.M. Bablor, and R.L. Engler. 1994. Reperfusion injury induces apoptosis in rabbit cardiomyocytes. *J. Clin. Invest.* 94:1621.

114. Tanaka, M.H., H. Ito, S. Adachi, H. Akimoto, T. Nishikawa, T. Kasajima, F. Marumo, and M. Hiroe. 1994. Hypoxia induces apoptosis with enhanced expression of Fas antigen messenger RNA in cultured neonatal rat cardiomyocytes. *Circulation Res.* 75:426–433.

115. Colston, J.T., B.M. Chandrasekar, and G.L. Freeman. 1998. Expression of apoptosis-related proteins in experimental coxsackievirus myocarditis. *Cardiovasc. Res.* 38:158–168.

56 | Autoimmunity in Atherosclerosis

George S. Wick and Qingbo Xu

1. INTRODUCTION

The arterial wall consists of an innermost endothelial layer sitting on a basement membrane, followed by the intimal connective tissue and a few smooth muscle cells (SMCs), the media, which is a muscular layer, and an outer connective tissue layer, the adventitia. Atherosclerotic lesions develop in the intima and, according to general belief, the earliest lesions are whitish cushion-like changes, so-called "fatty streaks", that may later develop into advanced lesions, called atherosclerotic plaques (1).

In a strict sense, *arteriosclerosis* is a disease characterized by thickening of the arterial wall due to infiltration of blood-borne mononuclear cells (macrophages by classical dogma) and SMC proliferation as well as extracellular matrix (ECM, mostly collagen) deposition. In later stages, calcification of arteriosclerotic lesions may occur. *Atherosclerosis* is a special form of arteriosclerosis, the additional hallmark of which is the emergence of so-called foam cells, as well as extracellular cholesterol deposition, even in the form of cholesterol crystals (1). Foam cells are mainly formed by macrophages and SMC that both express the scavenger receptor to which biochemically-altered, e.g. oxidized, low density lipoproteins (oxLDL), are bound and taken up, resulting in the formation of abundant intracellular lipid droplets (2). In contrast to receptors that bind native LDL, the scavenger receptor is not saturable and, therefore, its expression is not downregulated by an abundance of biochemically-altered LDL (3). There is evidence that dendritic cells also possess the scavenger receptor, and may thus become foam cells under atherogenic conditions.

The *"response-to-injury"* hypothesis (2) postulates endothelial damage to be the initiating factor in atherogenesis. Endothelial damage has several consequences, such as platelet aggregation, the release of cytokines (interleukin-1, IL-1; tumor necrosis factor $\alpha \rightarrow$ TNFα) monocyte chemoattractant protein (MCP-1) and various other chemokines, granulocyte-monocyte colony stimulating factor (GM-CSF) and growth factors (platelet-derived growth factor, PDGF; transforming growth factor β, TGFβ). Native and biochemically-altered LDL then diffuse into the intima, where it acts as a chemoattractant for blood-borne monocytes that then become macrophages, as well as media-derived SMC. Both of the latter express the scavenger receptor and eventually become foam cells.

The *"altered lipoprotein"* hypothesis (3) considers biochemically-altered LDL, e.g. oxLDL, to be the initiating factor by first transgressing through an intact endothelium into the intima, where it acts as a chemoattractant and foam cell-forming agent for monocytes/macrophages and SMC. Recently, however, it became, clear that it is native LDL rather than altered LDL that transgresses into the intima, where it is then secondarily oxidized and exerts its atherogenic function. This latter modification of the altered-LDL theory is called the *"retention of altered LDL"* hypothesis (4).

Finally, although less well-supported by the available data, the possibility has also been advanced that atherosclerosis starts as a benign polifération of vascular SMC, a concept that has been termed the *"monoclonal SMC proliferation"* hypothesis.

In recent years, it has become increasingly clear from both clinical and experimental data that inflammatory-immunological processes seem to play an important role in atherogenesis. This *"immunological"* hypothesis (5) originally did not distinguish primary from secondary effects of humoral and cellular immune reactions during atherogenesis, nor were the possible autologous or exogenous antigens identified by which such immune reactions could be initiated. Indications for the participation of humoral and cellular immune reactions in the development of this disease came from the demonstration of immune complexes (6) and numerous T cells in advanced lesions (7). Neither the specificity of the immune complex-forming antibodies nor that of the infiltrating T cells has, however, been determined. Analyzing the T cell receptor (TCR) repertoire in atherosclerotic plaques, the group of Hansson showed it to be polyclonal (8).

2. IMMUNOLOGIC INDUCTION OF ATHEROSCLEROSIS IN EXPERIMENTAL ANIMALS

2.1. Rabbits

To determine the possible autoreactivity of immune complex-forming antibodies and infiltrating T cells in atherosclerotic lesions, an approach for the induction of experimentally-induced autoimmune diseases originally described by Rose and Witebsky (9) was chosen. Our reasoning was that the putative autoantigen(s) should be present in atherosclerotic lesions and, therefore, total plaque proteins were extracted from human atherosclerotic vessels as well as from those of Watanabe rabbits, which lack the LDL receptor and develop a hereditary form of atherosclerosis (10). The delipidated plaque proteins were emulsified with complete Freund's adjuvant (CFA) consisting of a suspension of heat-killed mycobacteria in mineral oil together with an emulsifier. Rabbits selected as being normocholesterolemic were immunized with the emulsified plaque proteins and ovalbumin plus CFA for control purposes. Immunizations were performed on day 0, and after 5 and 10 weeks. The animals were sacrificed after 16 weeks for pathohistological analyses of aortae and immunological investigations of sera and T cells. Surprisingly, all three groups developed arteriosclerotic lesions at the known predilection sites, while unimmunized controls remained unaffected. As CFA was the only known common denominator in these experiments, normocholesterolemic rabbits were immunized with CFA alone and the development of arteriosclerosis was again observed. Immunization with non-HSP 65 containing adjuvants, such as Ribi or lipopeptide, had no effect. Since heat shock proteins (HSP) of the 60 kD family are known to be a major component of mycobacteria and had previously been incriminated as potential autoantigens in a variety of autoimmune diseases, including rheumatoid arthritis and diabetes, experiments were repeated by immunizing rabbits with recombinant mycobacterial HSP 65 (mHSP 65) alone and the same results were achieved, i.e. macroscopically visible lesions at sites of the arterial tree known to be subjected to major haemodynamic stress. Histologically, these lesions consisted of intimal infiltrations by mononuclear cells, i.e. T cells, macrophages and SMCs, as well as increased ECM deposition, but no foam cells (10). Feeding rabbits a high cholesterol diet only has been routinely used as a classical method to induce atherosclerotic lesions, the hallmark of which, as mentioned above, is the presence of abundant numbers of foam cells derived from macrophages and SMC as well as extracellular cholesterol deposition. Studies in our laboratory have shown that immunization of rabbits with mHSP 65 plus supplementation of a cholesterol-rich diet (2% cholesterol) induces significantly more severe atherosclerotic lesions compared to those induced by a cholesterol-rich diet alone. Furthermore, we have demonstrated that the first inflammatory stage of atheroclerosis is still reversible in the rabbit model, while severe atherosclerotic changes induced by immunization and the additional feeding of a cholesterol-rich diet are not (11).

As expected, the peripheral blood of rabbits immunized with mHSP 65 contained antibodies and T cells, respectively, reacting with this antigen. Analyses of mHSP 65-specific T cell lines derived from either peripheral blood or the atherosclerotic lesions from a given rabbit showed an intralesional accumulation of such cells. Surprisingly, this was also found to be true in unimmunized rabbits that had developed atherosclerosis upon receiving a cholesterol-rich diet only: mHPS 65-reactive T cells were also enriched in the lesions compared to T cells derived from the peripheral blood from the same animal (12).

In conclusion, HSP 60 has been found to be an optimal candidate for an antigen that may be involved in the pathogenesis of the very early inflammatory stage of arteriosclerosis, and these changes may progress into atherosclerosis when high blood levels of LDL are present, leading to foam cell formation and extracellular lipid deposition.

2.2. Mice

Mice are notoriously resistant to the induction of atherosclerosis by conventional dietary measures. Among all strains, C57BL/6J mice are still the most susceptible, but lesions that develop under conditions of a high cholesterol diet are much milder than those observed in rabbits. The recent availability of mice with defects of various genes (knock-out—KO, –/–) coding for proteins of crucial importance in the lipid metabolism has opened new possibilities for experimental atherosclerosis research. Thus, LDL receptor-deficient or apoprotein E KO (ApoE–/–) mice develop severe atherosclerosis even when only given mildly cholesterol-enriched food, such as the so-called. "Western" diet (13). Therefore, these latter models lend themselves as valuable tools to the analysis of various factors, including immune reactions, involved in the development of atherosclerosis.

It has long been known that clinically healthy people and, more importantly, those with proven atheroclerosis, develop autoantibodies against chemically-modified LDL (oxLDL) (14). Therefore, oxLDL was also considered a possible candidate for the induction of the immunologic reactions that occur in the arterial intima in the first stages of the disease. To clarify this question, George et al. (15,16) immunized C57BL/6J mice and ApoE$^{-/-}$ with mHSP 65 or biochemically-altered LDL and conclusively demonstrated that immune responses occur against both of these antigens, but that the reaction against HSP 65 aggravated diet-induced atherosclerosis, while immunization with oxLDL had a protective effect.

Finally, Zou et al. (17) recently developed a carotid artery venous bypass model in the mouse that will allow for the elucidation of the mechanisms underlying bypass restenosis by delineating the possible contribution of HSP 60 expression in the venous graft, which is subjected to higher arterial blood pressure (Figure 1).

3.   STRESS PROTEINS (HEAT SHOCK PROTEINS)

Stress proteins are expressed by prokaryontic and eukaryontic cells constitutively or under mild stressful conditions,

such as heat (hence the synonym heat shock proteins). HSPs and cognates thereof are classified into various families according to their molecular weight, i.e. the 100, 90, 70, 60, 27 kD and smaller molecular weight families (18). In the present discussion, we will concentrate on the 60 kD family, members of which play an important role in physiological processes, such as protein folding and intracellular protein transport. Under stressful conditions, they act as chaperones protecting other proteins from denaturation and subsequent malfunction (19).

HSPs are phylogenetically highly conserved. Thus, there is an over 95% homology on the protein and DNA level of HSP 60 from various bacteria, e.g. mycobacterial HSP 60 (mHSP 65), HSP 65 from *E. coli* (called GroEL) and HSP 60 of *Chlamydia pneumoniae* (cHSP 60). Even between bacterial and human HSP 60 (hHSP 60), there is an approximately 50–55% sequence homology (20). HSP 60 can also be found in parasites and even in the envelope layers of viruses such as HIV, which acquires this molecule when budding off the surface of host cells (21).

Since HSP 60 is not only quantitatively a major microbial component, but also qualitatively an important and very immunogenic antigen, practically all humans and animals develop humoral and cellular immunity against this antigen as a consequence of infection or vaccination. This protective immune reaction may have to be "paid for" by

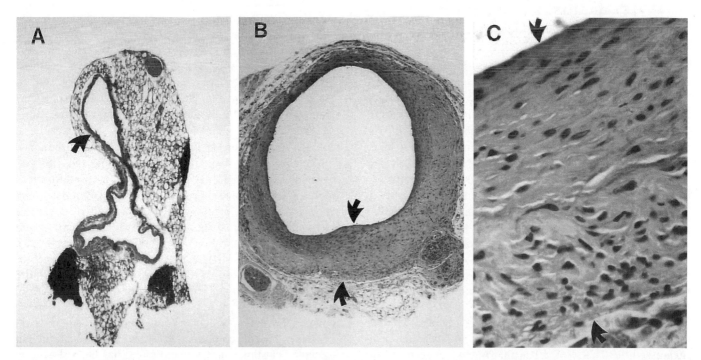

Figure 1.   Hematoxylin eosin-(HE) stained sections of a mouse control vein and vein bypass grafts. Mice underwent anesthesia and the autologous external jugular donor vein was grafted into the common carotid artery of a syngeneic recipient. Animals were sacrificed 0 (A; control) or 8 weeks (B and C) after surgery, and the grafted tissue fragments fixed in 4% phosphate-buffered (pH 7.2) formaldehyde, embedded in paraffin, sectioned, and stained with HE. Arrows indicate the control vessel wall (A) and thickened neointima (B and C). Original magnifications × 40 (A and B), × 250 (C).

the risk of cross-reactivity with autologous HSP 60 (22) that can, for example, be expressed by stressed or mal-treated vascular endothelial cells, as discussed later in this chapter.

In eukaryotic cells, e.g. human endothelial cells, HSP 60 expression can be induced by many different types of stressors, such as mechanical stress, temperature, oxygen radicals, infections, toxins, proinflammatory cytokines (e.g. TNFα), etc. (19). Importantly, the same stressors lead to the simultaneous expression of HSP 60 and adhesion molecules (intercellular adhesion molecule-1 = ICAM-1, E-selectin = ELAM-1, vascular cell adhesion molecule-1 = VCAM-1) (23), thus providing the prerequisites for an interaction of potentially bacterial/human HSP 60 cross-reactive T cells and antibodies with endothelial targets. At later stages of atherogenesis, intralesional cells (macrophages, SMC) also express HSP 60, and the anti-HSP 60 cellular immune reaction could, therefore, be perpetuated *in situ*.

The *in vivo* treatment of rats with LPS leads to the simul-taneous expression of HSP 60 and ICAM-1 by endothelial cells, which entails the attachment of leukocytes at exactly those sites known to be subjected to major haemodynamic stress, e.g. the branching points of intercostal arteries from the aorta (24).

4.   INVESTIGATIONS IN HUMANS

4.1.   Immunohistological Studies

In an extensive immunohistological investigation of arterial specimens from young (< 35 years) and elderly (> 65 years) patients, it was found that, contrary to the current dogma, the first cells to infiltrate the intima are T cells (25), while macrophages (developing from blood-derived mono-cytes) and SMC (immigrating from the media) arrive later. Scarce mast cells were also found, but B cells and K/NK cells were practically absent. In more advanced atheroscle-rotic lesions, macrophages and SMC prevail. The early T cells are predominantly CD4+ with markers of activation (interleukin-2 receptor positive—IL-2R+, HLA-DR+), and the majority carry the TCRα/β. However, an unexpectedly high proportion of these intimal T cells (10–15%) are TCRγ/δ+ (25,26), a frequency far above that of TCRγ/δ+ cells in the peripheral blood (1–2%). Further characteriza-tion of the latter cells revealed that TCR Vγ9δ2+ cells, i.e. a configuration characteristic for human peripheral blood, were present in low frequency (1–2%), while the majority expressed the TCR Vδ1 chain that is characteristic of the human mucosa-associated lymphoid tissue (MALT). It must be noted in the present context that TCRγ/δ+ T cells derived from sites of local immunity have previously been shown to exhibit an exceptionally strong reactivity with

HSP 60 without classical MHC restriction. Due to technical limitations, no functional data on the characteristics of early human intralesional T cells has yet been obtained, but promising work in this area is in progress.

4.2.   Humoral Antibodies

Serum antibodies to HSP 60   Within the framework of a large atherosclerosis prevention study in northern Italy (the Bruneck Study), titers of antibodies against mHSP 65 by ELISA (27) were determined and it was shown that all these symptomless subjects had serum antibodies against mHSP 65. Those with sonographically-demonstrable athero-sclerotic lesions in their *A. carotis*, however, exhibited significantly higher titers. It was then shown that these antibodies not only reacted with the standard antigen mHSP 65, but crossreacted with GroEL, cHSP and even hHSP 60 (28). The presence of high titers (over 320) of anti-HSP 60 antibodies emerged as a new diagnostic parameter for atherosclerosis independent of other classical risk factors. In a follow-up study performed five years later (29), the anti-HSP 60/65 titers in these clinically-healthy volunteers were shown to represent a very robust parameter, i.e. high and low titers remained relatively constant over this time period. In addition, persons who had died during the five year observation period were generally those with the highest titers. From these data, it was concluded that the determination of anti-HSP 60/65 antibodies is not only a predictive parameter for morbidity, but also mortality. Furthermore, these antibodies were shown to be able to lyse stressed [42°C, 30 min, H_2O_2, bacterial lipopolysaccharide (LPS), tumor necrosis factor (TNFα), etc.], but not unstressed, endothelial cells by complement-mediated or antibody-dependent cellular cytotoxicity (ADCC) (30). The same phenomenon has been observed on stressed *versus* unstressed macrophages (31).

To clarify whether the association of anti-HSP 60 antibodies was unique to carotid atherosclerosis or represented a more general phenomenon, patients with coronary atherosclerosis were investigated in a similar fashion, and statistically-significantly elevated titers compared to healthy controls were again observed (32). Interestingly, these titers dropped after the occurrence of myocardial infarction. A tentative explanation of this latter behaviour was that the release of autologous HSP 60 from the infarcted heart muscle cells could then lead to immune complex formation with preexisting antibodies and removal of the immune complexes by the reticuloendothelial system. That this concept may be true has been shown by experiments where *in vitro* perfused rat hearts were subjected to various times of ischemia and the coronary eluate from these preparations was assessed for rat HSP 60 (33). Indeed, it was shown that, depending on the length of ischemia, such treated hearts did release HSP 60 and this eluate could one hand react with anti-HSP 60/65 antibodies in Western blots, and on the other stimulate HSP 65-specific rat T cell lines and clones.

In the course of the ever-increasing discussion about the potentially infectious etiology of atherosclerosis and other cardiovascular diseases, different viruses and bacteria have been incriminated as possible causative agents. Among these, *Chlamydiae* are perhaps the most prominent example (34). Our laboratory, therefore, tested the antibodies of sera from the Bruneck Study volunteers for their reactivity with *C. pneumoniae*, and found a significant and linear correlation between the titers against mHSP 65 and those against total chlamydial antigens, i.e. including cHSP 60 plus non-HSP 60 components (33a). Using an ELISA for antibodies against cHSP 60 only, this correlation was shown to also exist with the latter purified recombinant antigenic preparation. Moreover, a linear correlation of reactivity of affinity chromatography-purified human anti-cHSP 60 antibodies was found with mHSP 65, GroEL and hHSP 60 and *vice versa* (28). Finally, we showed that purified human anti-cHSP 60 antibodies as well as anti-GroEL antibodies lysed stressed human endothelial cells as effectively as purified anti-mHSP 65 antibodies (33a). These data support the notion that immune reactions against *Chlamydiae* differ from those to other microorganisms that have been discussed in the context of atherogenesis in a quantitative rather than a qualitative fashion, because (a) cHSP 60 is an abundant constituent of *Chlamydiae*, (b) cHSP 60 is highly immunogenic, (c) *Chlamydiae* have a tropism for endothelial cells, and (d) chlamydial infection of endothelial cells not only leads to the presence of cHSP 60, but also the potent induction of hHSP 60.

Absorption studies of human sera with mHSP 65, cHSP 60 and GroEL proved the antigenic cross-reactivity, and also provided proof for antigenic differences by showing that absorption of the reactivity with mHSP 65 was more efficient with the nominal antigen compared to mHSP 65 or GroEL (28).

In a recent study, it was additionally shown that titers of sera from the clinically-healthy participants in the Bruneck-Study against mHSP 65 and human HSP 60 also correlated with those obtained in an ELISA against total *Helicobacter pylori* antigens. *H. pylori* HSP 60 was not available to us, but a correlation may also be found when the sera are tested against this antigen.

Periodontal disease and atherosclerosis   An interesting and still insufficiently elucidated aspect in this area of study concerns the possible involvement of anti-HSP 60/65 immune reactivity in patients with periodontal diseases in the development of atherosclerosis (35). A statistically-significant association of periodontal diseases and atherosclerosis has been described in several publications (36,37), but no

common denominator for this observation and an association with other infections that do not involve the oral cavity with atherogenesis has emerged. It has previously been shown that IgG and IgM class antibodies were predominant among the anti-HSP 60/65 in the sera of the participants in the Bruneck-Study (30), but the IgA fraction was significantly increased within this antibody population compared to the composition of normal human serum. There is, as yet, no explanation for this phenomenon, particularly with respect to its potential relevance as an indicator for the involvement of the local immune system. It may, however, be of relevance in this context that, in a study aimed at comparing the saliva of people with completely healthy dentures with that of patients with gingivitis and paradontitis, significantly increased anti-HP 60/65 secretory IgA antibody titers were found in the gingivitis group (30). Investigations to elucidate a possible link of this phenomenon with atherogenesis are underway.

HSP 60 in the circulation   Recently, reports from two independent groups have demonstrated that both chlamydial and human HSP60 can act as extracellular agonists of cytokines and induce TNF-α and matrix metalloproteinase (MMP-9) production by human and mouse macrophages (38,39). Interestingly, both chlamydial and human HSP60 induced E-selectin, ICAM-1, and VCAM-1 expression and IL-6 production by endothelial cells (40). These findings suggest that HSP60 directly stimulates vascular endothelial cells, leading to an inflammatory response that contributes to the pathophysiology of atherosclerosis.

The question then arose as to whether HSP60, normally an intracellular mitochondrial protein, exists in the circulation of healthy individuals, where it could directly contact the arterial wall and immune cells. To explore the possibility that HSP60 exists in circulation where it can exert its functions, a population-based investigation was performed within the framework of the Bruneck-Study (40a). A total of 826 subjects aged 40–79 years were recruited in 1990 and had a follow-up evaluation five years later (1995). All participants were subjected to determination of serum-soluble HSP60 (sHSP60), anti-E coli LPS, anti-C. pneumoniae, anti-H. pylori, anti-HSP 65/60 antibodies, and a variety of acute phase reactants (C-reactive protein, α1-antitrypsin, coeruloplasmin) as markers of systemic inflammation. The data show that serum levels of sHSP60 were significantly elevated in subjects with prevalent/incident carotid atherosclerosis and correlated to the common carotid artery intima-media (ITM) thickness. These associations are independent of age, sex and other established risk factors. The risk of atherosclerosis associated with high sHSP60 levels was amplified when subjects also had clinical (i.e., respiratory) and/or laboratory evidence of chronic infections. These findings provide solid evidence that sHSP60 contacts vascular and immune cells, and thus could significantly enhance our understanding the role of HSP60 in the pathobiology of atherosclerosis.

5.   HSP 60/65 EPITOPE MAPPING

The first step in our attempts to delineate the specificity of anti-HSP 60/65 antibodies concerned their reactivity with overlapping 15mer linear peptides with an 10 aminoacid (AA) overlap spanning the whole length of the mHSP 65 molecule (41). Using affinity chromatography-purified, anti-mHSP 65, pooled antibodies from the Bruneck project, three epitiopes (AA91–105, AA171–185, AA501–515) emerged. One of them (AA171–185) contains the toxic shock syndrome toxin (TSST) sequence that has no homology to human HSP 60, while the C- and N-terminal epitopes do exhibit such a homology (41). Humoral anti-HSP 60/65 antibodies will recognize conformational rather than linear epitopes, and our studies are presently attempting to identify this type of determinant. In this respect, it is of interest that, in a computerized model of hHSP 60 designed on the basis of the known three dimensional crystalline structure of GroEL, the two linear N- and C-terminal mHSP 65 epitopes identified seem to aggregate into one conformational epitope (Figure 2).

Using a similar approach, i.e. assessing the reactivity of overlapping peptides with saliva, the same HSP 60/65 epitopes emerged but, in addition, secretory IgA (sIgA) reactivity with a variety of other epitopes not recognized by serum antibodies was demonstrated, perhaps reflecting the special microbial flora of the oral cavity. Again, ongoing experiments are aimed at determining conformational cross-reactive HSP 60/65 epitopes that are recognized by sIgA antibodies in the saliva.

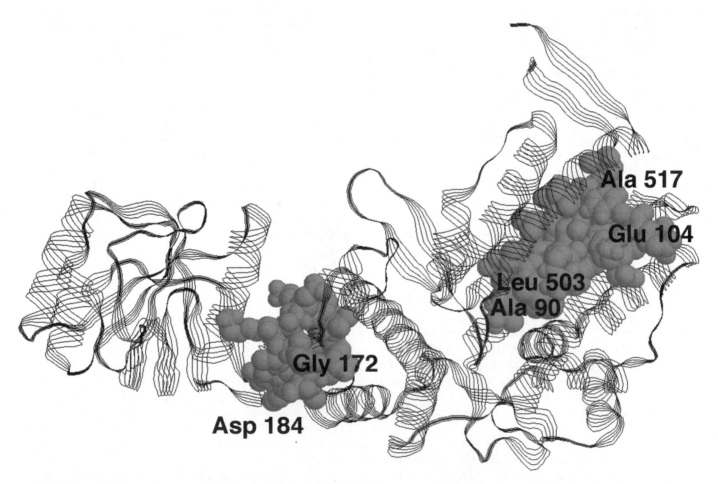

Figure 2.   The figure shows a calculated model of a human HSP60 monomer. The structure was obtained by homology-modelling according to the Swiss-Modell based on known X-ray-structures of four 60 KD bacterial chaperones. Three possible linear epitopes, identified on the homologous mycobacterial HSP65 (ref. 41), are shown with spacefilling balls. In the three-dimensional structure, two of the linear epitopes (Ala 90–Glu 104, Leu 503–Ala 517) seem to be assembled into a single conformational epitope. (Courtesy of Mag. Hannes Perschinka)

6. IMMUNOSUPPRESSION

Data on the effect of immunosuppression and non-specific anti-inflammatory treatment on the development of atherosclerosis are controversial. Feeding a cholesterol-rich diet to RAG-1- or RAG-2-deficient mice or a chow diet to crosses of RAG-1, RAG-2 –/– and ApoE –/– mice revealed that these immunodeficient animals did, nevertheless, develop atherosclerosis (42–44). A careful reassessment of these data, however, still shows a significant decrease in the incidence and severity of atherosclerosis, supporting the idea of an involvement of the immune system, especially in the earliest stages of the disease (Table 1). In our own experiments, we showed that arteriosclerosis induced by immunization of rabbits with recombinant mHSP 65 could be completely prevented by treatment with monoclonal anti-T-cell antibodies combined with prednisolone to avoid the formation of antibodies against murine immunoglobulins (45). Prednisolone alone had only a modest inhibitory effect. On the other hand, it is known that patients receiving the immunosuppressive drug cyclosporin A (CsA) are prone to develop severe atherosclerosis. In an in vitro study wherein human umbilical vein endothelial cells (HUVEC) were subjected to various forms of stress, e.g. TNFα combined with different immunosuppressive (CsA) and anti-inflammatory (aspirin, indomethacin) drugs, it was shown that CsA does not affect TNFα-induced adhesion molecule expression, but enhances TNFα-induced HSP 60 expression (46). As a matter of fact, CsA proved to be a potent inducer of HSP 60 by itself at concentrations equivalent to those that can be reached therapeutically in vivo. Aspirin also induces HSP 60 expression, but concurrently leads to a significant downregulation of the expression of adhesion molecules (ICAM-1, ELAM-1, VCAM-1), thus potentially preventing the interaction of HSP 60-reactive T cells with stressed endothelial cell (46). This may be a further protective principle of aspirin, in addition to its known anti-inflammatory effect based on blocking the arachidonic acid

Table 1 Comparative recalculated atherosclerotic lesion scores in cited references where various combinations of immunocompromised and ApoE $^{-/-}$ mice with high serum cholesterol levels were used.

Genotype	Immune defect	Cholesterol mg/dl	Lesion[a] score	Ref
E$^{-/-}$xRAG1$^{-/-}$	T & B	388 (chow)	57	(a)
E$^{-/-}$xRAG1$^{-/-}$	T & B	989 (WD)	88	(a)
E$^{-/-}$xRAG2$^{-/-}$	T & B	2263 (WD)	83	(b)
E$^{-/-}$xRAG2$^{-/-}$	TH1	843 (WD)	39	(c)

chow = standard diet
WD = Western diet (= cholesterol enriched diet)
a = percent of similarly fed wild-type control.

metabolism and its anti-coagulatory activity. CsA only affects HSP 60, not adhesion molecule expression.

From all these data, one may conclude that there is a whole spectrum of atherogenetic mechanisms: On one hand, atherosclerosis develops in experimental animals and human beings that have genetic defects, such as the LDL receptor deficiency in human familial hypercholesterolemia, Watanabe rabbits or LDL receptor KO mice and achieve excessively high serum levels of cholesterol that entail the development of severe atherosclerotic lesions with only minor contribution of the immune system. On the other side of this spectrum is the phenomenon of transplant arteriosclerosis, where it seems that immune reactivity is the prime driving force, perhaps in the form of a combination of alloreactivity and anti-HSP 60 reactivity. Between these rather artificial (KO mice and allotransplants) or genetically-determined (e.g. LDL receptor deficiency) situations, we have the vast majority of what we would like to call "conventional" atherosclerosis. We postulate an autoimmune inflammatory process as the initiating event in the latter situation followed by the contribution of cholesterol-dependent alterations in case that the risk factor of high serum cholesterol levels persists (47,48).

7. THE VASCULAR-ASSOCIATED LYMPHOID TISSUE (VALT)

During immunohistological investigations of early atherosclerotic lesions, specimens of carotid arteries from children and adolescents (8 months–15 years) that died from causes other than diseases involving the cardiovascular system (sudden infant death, accidents) were also used for control purposes (49). Surprisingly, accumulations of mononuclear cells were found in the intima of these arteries in areas subjected to major haemodynamic stress, i.e. at the branching points, known to be predilection sites for the possible later development of atherosclerosis. These newly-discovered cellular accumulations also consisted mainly of T cells, a smaller proportion of macrophages and SMC, and some mast cells. Importantly, dendritic cells were also found at these locations (Figure 3), while B cells, granulocytes and K/NK cells were practically absent. The intimal T cells in these infants and young children were already mostly activated, i.e. IL-2 receptor + and HLA-DR +, and most expressed, the TCR α/β, although a considerable number were TCR γ/δ +. These intimal mononuclear cell accumulations have been tentatively termed the "vascular-associated lymphoid tissue" (VALT), analogous to the mucosa-associated lymphoid tissue (MALT). Although there are, so far, no functional data available on the VALT, we hypothesize that it may play a similar role as the MALT, i.e. monitoring the surface, in this case of areas of the arterial tree subjected to major haemodynamic stress, and thus may represent a *locus minoris resistantiae* for potentially dangerous endogenous or exogenous factors. While arterial and venous endothelial cells in experimental animals and man are generally MHC class II-, they do express HLA-DR above the VALT, where they seem to be exposed to IFNγ released by the underlying activated T cells. For the presentation of autologous or exogenous antigenic peptides one would, therefore, not need to postulate this to occur only in local draining lymph nodes, but endothelial cells and intra-VALT dendritic cells may also exert this function.

8. STRESS-INDUCED HSP EXPRESSION IN ATHEROGENESIS

The signal transduction pathways leading to the expression of adhesion molecules upon exertion of various forms of stress on different types of cells, including endothelial cells, have been studied in great detail by many groups. In contrast, very little is known about signal transduction pathways leading to the expression of HSP 60 and the regulation of HSP 60 gene expression in general. In this context, it is important to remember that humans do develop arteriosclerosis, but not venosclerosis. Under special circumstances, such as the use of venous fragments

for bypass operations, the latter are also subject to intimal thickening and later progression into atherosclerosis-like lesions (51). Apparently, subjecting these venous bypass grafts to arterial flow conditions, specifically arterial blood pressure, predisposes these vascular grafts to the development of atherosclerosis in a similar fashion as arteries. In those instances where, for example, only a single obstructed coronary artery has to be bypassed, vascular surgeons have long used the *A. mammaria interna* and only resorted to venous grafts, such as the *V. saphena*, when more bypasses were required. Interestingly, the *A. mammaria* interna has been a very good choice, because it is known to be rather resistant against the development of atherosclerosis, even in persons with risk factors. The reason for the peculiar resistance of this vessel is not known, but seems to be due to very special flow conditions that are still insufficiently studied. According to our concept, the *A. mammaria interna* is not subjected to a high degree of turbulant mechanical stress that acts on the arterial wall in general, and endothelial cells in particular, at sites of the arterial tree (mainly branching points) known to be predisposed to the development of atherosclerotic lesions. So far, no investigations of the *A. mammaria interna in situ* with respect to the expression of HSP 60 and adhesion molecules have been conducted, although these vessels have been studied after developing intimal thickening and even

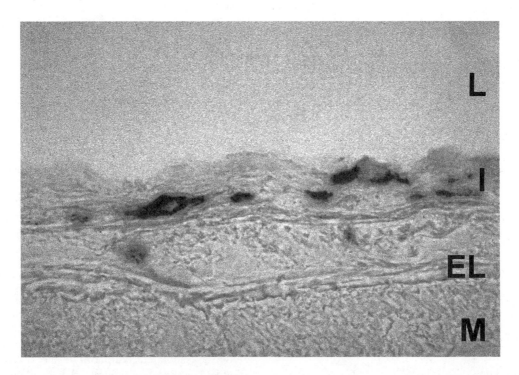

Figure 3. CD1a⁺ dendritic cells in the intima of the carotid artery of a 10 year-old boy stained in indirect immunohistochemistry (horseradish peroxidase-labeled secondary antibody; visualisation with DAB and metal enhancer) Original magnification 200 × .
L: vascular lumen; I: intima; EL: elastica interna; M: media.
(Courtesy of Dr. Gunda Millonig)

restenosis when used in bypass operations. In these instances, strong HSP 60 expression of the endothelium as well as still unidentified cells in the intima (probably macrophages and SMC) has been demonstrated (Fraedrich and Wick, unpublished). According to our concept, lifelong exposure of arterial endothelial cells and, by the same token, also of SMC in the media to the arterial blood pressure and pulsations lowers the threshold for the effects of other types of stressors, such as classical risk factors for atherosclerosis. Thus, while the effect of oxLDL as a risk factor for atherosclerosis cannot be denied, we assign a triple function to it: First, oxLDL is a stressor for endothelial cells, later it acts as a chemoattractant for mononuclear cells infiltrating the intima from the blood stream and SMC immigrating from the media, and finally leads to the formation of foam cells via binding to the non-saturable scavenger receptors on macrophages, SMC and dendritic cells. Thus, oxLDL leads to the expression of HSP 60 (unpublished results) and adhesion molecules in cultures of human arterial, but not venous endothelial cells (51).

9. STRESS-INDUCED SIGNALING

The passage of blood through the vascular system generates hemodynamic forces. Fluid flow across the cell surface results in shear stress, while strain stress, which produces elongational stretch, is caused by circumferential deformations resulting from transmural pressure gradients and vascular smooth muscle tone (52). In fact, altered hemodynamic stress indeed induces HSP60 expression in the external carotid arteries of rats, when the internal carotid artery was ligated (53). Increased HSP60 production in endothelial cells subjected to alterations in shear stress has also been observed, indicating that mechanical stress is an effective stressor responsible for HSP60 expression.

The HSP production is primarily mediated by heat shock transcription factors (HSFs) that interact with a specific regulatory element, the heat shock element (HSE), present in the HSP gene promoters (54–57). HSFs are present constitutively in the cell in a non-DNA binding state and are activated to a DNA binding form in response to various stressors. Although the activation process appears to involve HSF oligomerization from a monomeric to a trimeric state, stress-initiated signal transduction pathways leading to HSF activation are largely unknown (54–57). Very recently, Xu et al. (unpublished observations) demonstrated that HSF1 phosphorylation or activation in SMCs is induced by mechanical forces (Figure 4). Conditioned medium from mechanically-stressed SMCs did not result in HSF-DNA binding activation. Furthermore, mitogen-activated protein kinases (MAPKs), including extracellular signal-regulated kinases (ERK), c-Jun NH$_2$-terminal protein

kinases or stress-activated protein kinases and p38 MAPKs, were also highly activated in response to cyclic strain stress (60). Inhibition of ERK and p38 MAPK activation by their specific inhibitors did not influence HSF1 activation. Interestingly, SMC lines stably expressing dominant negative rac (rac N17) abolished HSF1 activation induced by cyclic strain stress, while a significant reduction of HSF1 activation was seen in ras N17-transfected-SMC lines. Thus, in SMC, cyclic strain stress-induced-HSF1 activation is regulated by rac/ras GTP-binding proteins. Similar investigations concerning mechanical stress-induced signal transduction mechanisms in endothelial cells are now underway in our laboratory.

10. CONCLUSION

This review presents experimental and clinical data that led to the conception and corroboration of a new immunological hypothesis for the development of atherosclerosis, as summarized in Figure 5. We postulate that atherosclerosis starts as an inflammatory-autoimmune process representing

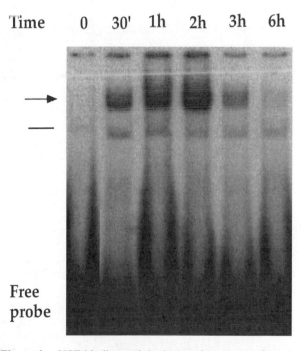

Figure 4. HSF-binding activity in protein extracts of mechanically-stressed smooth muscle cells. Smooth muscle cells were treated with cyclic strain stress (60 cycles/min, 15% elongation) at 37°C. The cells were harvested and nuclear proteins were prepared. Protein extracts (5 μg per lane) were incubated with a radiolabeled oligonucleotide encompassing a 24 bp HSE of the hsp70 promoter. Gel mobility shift assay was performed in a 4% gel. Arrows indicate specific HSF–HSE binding complexes; lines indicate nonspecific binding.

the price we pay for our protective cellular and humoral immunity against microbial HSP 60. As shown in Figure 5, bacteria (and also virus envelopes and parasites) express HSP 60 and non-HSP antigens. HSP 60 is highly immunogenic and shows over 95% sequence homology on the DNA and protein levels between different bacteria (e.g. *C. pneumoniae, H. pylori M. tuberculosis* and *E. coli*). Also, due to its high degree of phylogenetic conservation, bacterial HSP 60 still shows about 55% homology with eukaryontic, e.g. human, HSP 60. When arterial endothelial cells are maltreated by risk factors for atherosclerosis, the danger of immunological cross-reactivity arises and leads to the early inflammatory stage of the disease. Due to the lifelong exposure to the higher arterial blood pressure, arterial endothelial cells exhibit a lower threshold than venous endothelial cells for HSP 60 expression upon exposure to various stress factors. Unstressed endothelial cells (Fig. 5, left panel) do not represent targets for an anti-HSP 60 response. The occasional individual who does not develop atherosclerosis despite being exposed to classical risk factors seems to lack a cellular and/or humoral immune response against potentially dangerous cross-reactive HSP 60 epitopes. The early inflammatory stage of atherosclerosis can be identified by appropriate serological tests and is still reversible. However, if atherogenic risk factors, e.g. hypertension, elevated blood cholesterol levels etc. persist, advanced atherosclerotic lesions (plaques) develop to levels that are difficult, if not impossible to regress. Finally, under certain circumstances, a *bona fide* autoimmune reaction, e.g. directed against biochemically-altered autologous HSP 60 released during tissue necrosis, must also be taken into account as an initiating factor in atherogenesis.

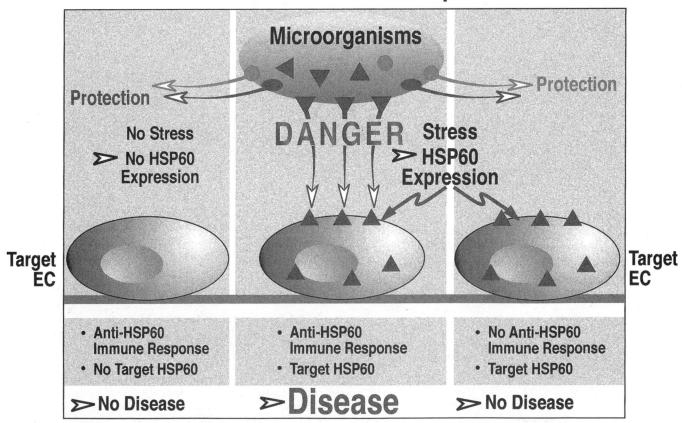

Figure 5. The "autoimmune" hypothesis for the pathogenesis of arterio-/atherosclerosis based on the concept of a humoral/cellular immunological crossreactivity against microbial HSP60 and HSP60 expressed on arterial endothelial cells that are stressed by atherogenic risk factors (hypertension, oxygen radicals, toxins, infections etc.). Different symbols for bacterial antigens represent non-HSP60 and crossreactive HSP60 epitopes, respectively.

ACKNOWLEDGEMENTS

The work of the authors has been supported by grants from the Austrian Science Fund (project No. 12213-MED, G.Wick; project No. 13099–BIO, Q. Xu).

We thank our collaborators, whose work is cited in this review.

We especially appreciate the preparation of Figures 2 and 3 by Mag. H. Perschinka and Dr. Gunda Millonig, respectively, and thank Miss Anita Ender for preparation of the manuscript.

References

1. Stary, H.C., A.B. Chandler, R.E. Dinsmore, V. Fuster, S. Glagov, W. Insull, Jr., M.E. Rosenfeld, C.J. Schwartz, W.D. Wanger, and R.W. Wissler. 1995. A definition of advanced types of atherosclerotic lesions and a histological classification of atherosclerosis. A report from the Committee on Vascular Lesions of the Council on Arteriosclerosis, American Heart Association. *Circulation* 92:1355–1374.
2. Ross, R. 1993. The pathogenesis of atherosclerosis: a perspective for the 1990s. *Nature* 362:801–809.
3. Steinberg, D., and J.L. Witztum. 1990. Lipoproteins and atherogenesis: current concepts. *J. Am. Med. Assoc.* 264:3047–3052.
4. Schwartz, S.M., and C.E. Murry. 1998. Proliferation and the monoclonal origins of atherosclerotic lesions. *Annu. Rev. Med.* 49:437–460.
5. Hansson, G.K., L. Jonasson, P.S. Seifert, and S. Stemme. 1989. Immune mechanisms in atherosclerosis. *Arteriosclerosis* 9:567–578.
6. Vlaicu, R., F. Niculescu, H.G. Rus, and A. Cristea. 1985. Immunohistochemical localization of the terminal C5b-9 complement complex in human aortic fibrous plaque. *Atherosclerosis* 57:163–177.
7   Jonasson, L., J. Holm, O. Skalli, G. Dondjers, and G.K. Hansson. 1986. Regional accumulations of T cells, macrophages and smooth muscle cells in the human atherosclerotic plaque. *Atherosclerosis* 6:131–1384.
8. Stemme, S., L. Rymo, and G.K. Hansson. 1991. Polyclonal origin of T lymphocytes in human atherosclerotic plaques. *Lab. Invest.* 65:654–660.
9. Rose, N.R., and E. Witebsky. 1956. Studies in organ specificity. V. Changes in the thyroid glands of rabbits following active immunization with rabbit thyroid extracts. *J. Immunol.* 76:417–427.
10. Xu, Q., H. Dietrich, H.J. Steiner, A.M. Gown, B. Schoel, G. Mikuz, S.H.E. Kaufmann, and G. Wick. 1992. Induction of arteriosclerosis in normocholesterolemic rabbits by immunization with heat shock protein 65. *Arterioscl. Throm. Vas.* 12:789–799.
11. Xu, Q., R. Kleindienst, G. Schett, W. Waitz, S.J. Jindal, R.S. Gupta, H. Dietrich, and G. Wick. 1996. Regression of arteriosclerotic lesions induced by immunization with heat shock protein 65-containing material in normocholesterolemic, but not hypercholesterolemic, rabbits. *Atherosclerosis* 123:145–155.
12. Xu, Q., R. Kleindienst, W. Waitz, H. Dietrich, and G. Wick. 1993. Increased expression of heat shock protein 65 coincides with a population of infiltrating T lymphocytes in athero-sclerotic lesions of rabbits specifically responding to heat shock protein 65. *J. Clin. Invest.* 91:2693–2702.
13. Zhang, S.H., R.L. Reddick, J.A. Piedrahita, and N. Maeda. 1992. Spontaneous hyper-cholesterolemia and arterial lesions in mice lacking apolipoprotein E. *Science* 258:468–471.
14. Salonen, J.T., S. Yla-Herttuala, R. Yamamoto, S. Butler, H. Korpela, R. Salonen, K. Nyyssonen, W. Palinski, and J.L. Witztum. 1992. Autoantibody against oxidised LDL and progression of carotid atherosclerosis. *Lancet* 339:883–887.
15. George, J., A. Afek, B. Gilburd, H. Levkovitz, A. Shaish, I. Goldberg, Y. Kopolovic, G. Wick, Y. Shoenfeld, and D. Harats. 1998. Hyperimmunization of apo-E-deficient mice with homologous malondialdehyde low-density lipoprotein suppresses early atherogenesis. *Atherosclerosis* 138:147–152.
16. George, J., Y. Shoenfeld, A. Afek, B. Gilburd, P. Keren, A. Shaish, J. Kopolovic, G. Wick, and D. Harats. 1999. Enhanced fatty streak formation in C57BL/6J mice by immunization with heat shock protein-65. *Arterioscler. Thromb. Vasc. Biol.* 19:505–510.
17. Zou, Y., H. Dietrich, Y. Hu, B. Metzler, G. Wick, and Q. Xu. 1998. Mouse model of venous bypass graft arteriosclerosis. *Am. J. Pathol.* 153:1301–1310.
18. Morimoto, R.I. 1998. Regulation of the heat shock transcriptional responses cross talk between a family of heat shock factos, molecular chaperones, and negative regulators. *Genes Dev.* 12:3788–2796.
19 Xu, Q., and G. Wick. 1996. The role of heat shock proteins in protection and pathophysiology of the arterial wall. *Mol. Med. Today* 2:372 379.
20. Jones, D.B., A.F. Coulson, and G.W. Duff. 1993. Sequence homologies between hsp60 and autoantigens. *Immunol. Today* 14:115–118.
21. Bartz, S.R., C.D. Pauza, J. Ivanyi, S. Jindal, W.J. Welch, and M. Malkovsky. 1994. An Hsp60 related protein is associated with purified HIV and SIV. *J. Med. Primatol.* 23:151–154.
22. Wick, G., G. Schett, A. Amberger, R. Kleindienst, and Q. Xu. 1995. Is atherosclerosis an immunologically mediated disease? *Immunol. Today* 16:27–33.
23. Amberger, A., C. Maczek, G. Jürgens, D. Michaelis, G. Schett, K. Trieb, S. Jindal, Q. Xu, and G. Wick. 1997. Co-expression of ICAM-1, VCAM-1, ELAM-1 and hsp 60 in human arterial and venous endothelial cells in response to cytokines and oxidized low density lipoproteins. *Cell Stress & Chaperones* 2:94–103.
24. Seitz, C.S., R. Kleindienst, Q. Xu, and G. Wick. 1996. Coexpression of intercellular adhesion molecule-1 and heat shock protein 60 is related to increased adherent monocytes and T cells on aortic endothelium of rats in response to endotoxin. *Lab. Invest.* 74:241–252.
25. Xu, Q., G. Oberhuber, M. Gruschwitz, and G. Wick. 1990. Immunology of atherosclerosis: Cellular composition and major histocompatibility complex class II antigen expression in aortic intima, fatty streaks and atherosclerotic plaques in young and aged human specimens. *Clin. Immunol. Immunopathol.* 56:344–359.
26. Kleindienst, R., Q. Xu, J. Willeit, F. Waldenberger, S. Weimann, and G. Wick. 1993. Immunology of atherosclerosis: Demonstration of heat shock protein 60 expression and T-lymphocytes bearing α/β or γ/δ receptor in human atherosclerotic lesions. *Am. J. Pathol.* 142:1927–1937.

27. Xu, Q., J. Willeit, M. Marosi, R. Kleindienst, F. Oberhollenzer, S. Kiechl, T. Stulnig, G. Luef, and G. Wick. 1993. Association of serum antibodies to protein 65 with carotid atherosclerosis. *Lancet* 341:255–259.

28. Mayr, M., B. Metzler, S. Kiechl, J. Willeit, G. Schett, Q. Xu, and G. Wick. 1999. Endothelial cytotoxicity mediated by serum antibodies to heat shock proteins of *E. coli* and *Chlamydia pneumoniae*: Immune reactions to heat shock proteins as a possible link between infections and atherosclerosis. *Circulation* 99:1560–1566.

29. Xu, Q., S. Kiechl, M. Mayr, B. Metzler, G. Egger, F. Oberhollenzer, J. Willeit, and G. Wick. 1999. Association of serum antibodies to heat shock-protein 65 with carotid atherosclerosis: Clinical significance determined in a follow-up study. *Circulation* 100:1169–1174.

30. Schett, G., Q. Xu, A. Amberger, R. van der Zee, H. Recheis, J. Willeit, and G. Wick. 1995. Autoantibodies against heat shock protein 60 mediate endothelial cytotoxicity. *J. Clin. Invest.* 96:2569–2577.

31. Schett, G., B. Metzler, M. Mayr, A. Amberger, R.S. Gupta, L. Mizzen, Q. Xu, and G. Wick. 1997. Macrophage-lysis mediated by autoantibodies to heat shock protein 65/60. *Atherosclerosis* 128:27–38.

32. Hoppichler, F., M. Lechleitner, C. Tragweger, G. Schett, A. Dzien, W. Sturm, and Q. Xu. 1996. Changes of serum antibodies to heat-shock protein 65 in coronary heart disease and acute myocardial infarction. *Atherosclerosis* 126:333–338.

33. Schett, G., B. Metzler, R. Kleindienst, A. Amberger, H. Recheis, Q. Xu, and G. Wick. 1999. Myocardial injury leads to a release of heat shock protein (hsp) 60 and a suppression of the anti-hsp65 immune response. *Cardiovasc. Res.* 42:685–695.

33a. Mayr, M., S. Kiechl, J. Willeit, G. Wick, and Q. Xu. 2000. Infections, immunity, and atherosclerosis. Associations of antibodies to Chlamydia pneumoniae, Helicobacter pylori, and cytomegalovirus with immune reactions to heat-shock protein 60 and carotid of femoral atherosclerosis. *Circulation* 102:833–839.

34. Saikku, P., K. Mattila, M. Nieminen, J.K. Huttunen, M. Leinonen, M.-R. Ekman, P.H. Mäkelä, and V. Valtonen. 1988. Serological evidence of an association of a novel chlamydia, TWAR, with chronic coronary heart disease and acute myocardial infarction. *Lancet* 2:983–985.

35. Schett, G., B. Metzler, R. Kleindienst, I. Moschèn, R. Hattmannsdorfer, H. Wolf, Q. Xu, and G. Wick. 1997. Salivary anti-hsp65 antibodies as a diagnostic marker for gingivitis and a possible link to atherosclerosis. *Int. Arch. Allergy Immunol.* 114:246–250.

36. Kinane, D.F. 1998. Periodontal diseases' contributions to cardiovascular disease: an overview of potential mechanisms. *Ann. Periodontol.* 3:142–150.

37. Beck, J.D., J. Pankow, H.A. Tyroler, and S. Offenbacher. 1999. Dental infections and atherosclerosis. *Am. Heart J.* 138:528–533.

38. Kol, A., G.K. Sukhova, A.H. Lichtman, and P. Libby. 1998. Chlamydial heat shock protein 60 localizes in human atheroma and regulates macrophage tumor necrosis factor-α and matrix metalloproteinase expression. *Circulation* 98:300–307.

39. Chen, W., U. Syldath, K. Bellmann, V. Burkart, and H. Kolb. 1999. Human 60-kDa heat-shock protein: a danger signal to the innate immune system. *J. Immunol.* 162:3212–3219.

40. Kol, A., T. Bourcier, A.H. Lichtman, and P. Libby. 1999. Chlamydial and human heat shock protein 60s activate human vascular endothelium, smooth muscle cells, and macrophages. *J. Clin. Invest.* 103:571–577.

40a. Xu, Q., G. Schett, H. Perschinka, M. Mayr, G. Egger, F. Oberhollenzer, J. Willeit, S. Kiechl and G. Wick. 2000. Serum soluble heat shock protein 60 is elevated in subjects with atherosclerosis in a general population. *Circulation* 101:16–22.

41. Metzler, B., G. Schett, R. Kleindienst, Q. Xu, R. van der Zee, T. Ottenhoff, A. Hajeer, R. Bernstein, and G. Wick. 1997. Epitope specificity of anti-heat shock protein 65/60 serum antibodies in atherosclerosis. *Arterioscler. Thromb. Vasc. Biol.* 17:536–541.

42. Dansky, H.M., S.A. Charlton, M.M. Harper, and J.D. Smith. 1997. T and B lymphocytes play a minor role in atherosclerotic plaque formation in the apolipoprotein E-deficient mouse. *Proc. Natl. Acad. Sci. USA* 94:4642–4646.

43. Daugherty, A., E. Pure, D. Delfel-Butteiger, S. Chen, J. Leferovich, S.E. Roselaar, and D.J. Rader. 1997. The effects of total lymphocyte deficiency on the extent of atherosclerosis in apolipoprotein E–/– mice. *J. Clin. Invest.* 100:1575–1580.

44. Gupta, S., A.M. Pablo, X.C. Jiang, N. Wang, A.R. Tall, and C. Schindler. 1997. IFN-gamma potentiates atherosclerosis in ApoE knock-out mice. *J. Clin. Invest.* 99:2752–2761.

45. Metzler, B., M. Mayr, H. Dietrich, E. Wiebe, Q. Xu, and G. Wick. 1999. Inhibition of arteriosclerosis by T-cell-depletion of normocholesterolemic rabbits immunized with heat shock protein 65. *Arterioscler. Thromb. Vasc. Biol.* 19:1905–1911.

46. Amberger, A., M. Hala, M. Saurwein-Teissl, B. Metzler, B. Grubeck-Loebenstein, Q. Xu, and G. Wick. 1999. Suppressive effects of anti-inflammatory agents on human endothelial cell activation and induction of heat shock proteins. *Mol. Med.* 5:117–128.

47. Wick, G., R. Kleindienst, H. Dietrich, and Q. Xu. 1992. Is atherosclerosis an autoimmune disease? *Trends Food Sci. Tech.* 3:114–119.

48. Wick, G., M. Romen, A. Amberger, B. Metzler, M. Mayr, G. Falkensamer, and Q. Xu. 1997. Atherosclerosis, autoimmunity and vascular-associated lymphoid tissue. *FASEB. J.* 11:1199–1207.

49. Waltner-Romen, M., G. Falkensammer, W. Rabl, and G. Wick. 1998. A previously unrecognized site of local accumulation of mononuclear cells. The vascular-associated lymphoid tissue. *J. Histochem. Cytochem.* 46(12):1347–1350.

50. Motwani J.G., and E.J. Topol. 1998. Aortocoronary saphenous vein graft disease: Pathogenesis, predisposition, and prevention. *Circulation* 97:916–931.

51. Frostegard J., B. Kjellman, M. Gidlund, B. Andersson, and R. Kiessling. 1996. Induction of heat shock protein in monocytic cells by oxidized low density lipoprotein. *Atherosclerosis* 121:93–103.

52. McIntire L.V. 1994. Bioengineering and vascular biology. *Ann. Biomed. Eng.* 22:2–13.

53. Hochleitner, B., E.-O. Hochleitner, P. Obrist T. Eberl, A. Amberger, Q. Xu, R. Margreiter, and G. Wick. Fluid shear stress induces heat shock protein 60 expression in endothelial cells *in vitro* and *in vivo. Arth. Throm. Vasc.*: in press.

54. Lis, J., and C. Wu. 1993. Protein traffic on the heat shock promoter: Parking, stalling, and trucking along. *Cell* 74:1–4.

55. Morimoto, R.I. 1998. Regulation of the heat shock transcriptional responses cross talk between a family of heat shock factors, molecular chapersones, and negative regulators. *Genes. Dev.* 12:3788–96.

56. Sorger, P.K. 1991. Heat shock factor and the heat shock response. *Cell* 65:363–366.

57. Welch, W.J. 1993. Heat shock proteins functioning as molecular chaperones: their roles in normal and stressed cells. *Phil. Trans. R. Soc. Land.* 339:327–333.

58. Kim, D., H. Ouyang, and G.C. Li. 1995. Heat shock protein hsp70 accelerates the recovery of heat-shocked mammalian cells through its modulation of heat shock transcription factor HSF1. *Proc. Natl. Acad. Sci. USA* 92:2126–2130.

59. Sistonen, L., K.D. Sarge, and R.I. Morimoto. 1994. Human heat shock factor 1 and 2 are differentially activated and can synergistically induce hsp70 gene transcription. *Mol. Cell. Biol.* 14:2087–2099.

60. Li, C., Y. Hu, M. Mayr, and Q. Xu. 1999. Cyclic strain stress-induced mitogen-activated protein kinase (MAPK) phosphatase-1 expression in vascular smooth muscle cells is regulated by ras/rac-MAPK pathways. *J. Biol. Chem.* 274:25273–25280.

57 | Autoimmune Diseases of the Eye

Igal Gery, Robert B. Nussenblatt, Chi-Chao Chan, and Rachel R. Caspi

1. INTRODUCTION

Research of immunological phenomena unique to the eye has substantially contributed to the study of autoimmunity and its potential pathogenicity. The eye's anatomy is unique and allows high levels of sequestration of certain tissues from the immune system. Consequently, certain ocular-specific molecules may function as autoimmunogens and provoke autoimmunity, a capacity that was noticed almost a century ago. In fact, the production of autoantibodies against lens proteins by Uhlenhuth in 1903 was the first successful demonstration that autoimmunity is feasible, in defiance of the "horror autotoxicus" concept of Paul Ehrlich (cited in Ref. 1,2). Shortly thereafter, in 1906, Hess and Romer demonstrated autoimmunogenicity of proteins from the eye's posterior segment (cited in Ref. 2) and provided conceptual basis for Elschnig's suggestion (3) that sympathetic ophthalmia is caused by an autoimmune process. Sympathetic ophthalmia, that offers the most conspicuous demonstration of a human disease with an autoimmune etiology, develops as a consequence to a penetrating injury to one eye and consists of severe inflammation in both the injured eye, as well as in the untraumatized fellow eye (see Section 8).

Similar to other organs, the concept of autoimmunity causing eye conditions has been further supported by the finding that inflammatory eye diseases can be induced in animals following immunization with ocular-specific antigens. In addition to revealing unique features of autoimmune processes in the eye, these studies have shed new light on several issues related to autoimmunity such as mechanisms of immunotolerance toward self antigens, genetics of susceptibility to disease and immunosuppressive mechanisms that modulate the pathogenic process.

2. UNIQUE ANATOMY AND IMMUNOLOGICAL FEATURES OF EYE TISSUES

The eye is composed of various tissues that differ profoundly in their blood supply and exposure to the immune system. Of particular interest to the present monograph are tissues with limited exposure, i.e., the lens and central portion of the cornea, which are a vascular, and the retina, which resides behind efficient barriers composed of tight junctions between endothelial cells of blood vessels and between cells of the retinal pigment epithelium. These tissues are considered, therefore, to be "immune privileged", along with three sites that are poorly exposed to the immune system, namely, the anterior chamber, vitreous cavity and subretinal space. In addition to poor exposure, the immune privilege of the eye is achieved by a multitude of factors that include lack of MHC molecules on many cells (rev. in 4,5) and the presence of immunosuppressive molecules, in particular Fas (6), inhibitory cytokines such as TGF-β, neuropeptides VIP and MSH, as well as molecules that inhibit complement activation (rev. in 4,5). Additionally, certain sites are protected from inflammatory-mediating immune reactions by an elaborate immunomodulatory mechanism termed "anterior chamber-associated immune deviation" (ACAID). This mechanism involves a plethora of different cells and mediators that selectively inhibit delayed type hypersensitivity responses and production of complement-fixing antibody (rev. in 4,5,7). It is noteworthy that a recent study by Sonoda et al. (8) has revealed the pivotal role that natural killer T-cells play in the ACAID process. It is assumed the immune privilege of the eye is essential for protection against destructive inflammation that could affect clarity of ocular tissues, damage non-regenerating structures and impair vision. Similar phenomena were noted in other tissues, such as the brain and testis, but it seems they are particularly efficient in the eye. Despite the aforementioned immunosuppressive mechanisms, inflammation does develop in the eye and in many cases the initiating process is presumed to be autoimmunity. Indeed, the barriers and sequestration of ocular tissues from the immune system may be responsible for ocular antigens being unable to induce effective immunotolerance and thus becoming autoimmunogenic and targets for pathogenic autoimmunity.

The notion that ocular antigens are autoimmunogenic is in line with the finding that inflammatory eye diseases may be induced by immunization of experimental animals with a growing number of ocular-specific antigens (see Section 4). These conditions are termed "autoimmune", but it should be noted that in the great majority of these studies, the disease-inducing antigens were purified from eyes of xenogenesis species. On the other hand, ocular antigens of the same species were found inferior in their immunopathogenic capacity: rat S-Ag was found to be poorly immunopathogenic in Lewis rats (9), and monkey interphotoreceptor retinoid-binding protein (IRBP) failed to induce uveitis or immune response in experimental monkeys (10). It is assumed, therefore, that xenogeneic ocular antigens are superior immunogens, while the lymphocytes they sensitize induce inflammation upon recognition of the cross-reactive ocular autoantigens. The relationship between xenogeneic and autologous ocular antigens has been further analyzed in recent years by testing peptides derived from the sequence of the immunopathogenic xenogeneic proteins. As detailed below (Section 4), strongly immunopathogenic peptides are usually highly conserved, with identity between the sequence of xenogeneic and autologous peptides, but non-identical peptides may also exhibit disease-inducing capacity (Table 1).

The non tolerogenicity of sequestered retinal antigens was recently demonstrated in an elegant study by Gregerson et al. (11). Using transgenic mice that express β-galactosidase under the control of rhodopsin promoter, these authors showed that no tolerance develops in these mice against the β-galactosidase, and lymphocytes sensitized against it initiate uveoretinitis. Furthermore, Gregerson's group, and that of Caspi have very recently shown that transgenic extraocular expression of S-antigen (12) or IRBP (13) induces specific tolerance and inhibition of EAU induction by these proteins.

The notion concerning the lack of exposure of ocular antigens to the immune system was challenged, however, by the finding that the T-cell compartment in mice and rabbits are tolerant toward the heavily encapsulated lens crystallins (14,15). This unexpected finding was formerly attributed to "low dose tolerance", induced by hypothetical leaking of minute amounts of crystallins (14). More recent studies have indicated, however, that the state of tolerance to lens crystallins is mainly due to expression of these proteins in the thymus (16) and the ensuing deletion of T-cells specific against them (17). Furthermore, a study by Charukamnoetkanok et al. (18) revealed that murine thymi

also express several retinal antigens, S-Ag, IRBP, opsin and recoverin. The biological significance of this observation was underscored by the finding of Egwuagu et al. (19), that an inverse correlation exists between thymic expression of immunopathogenic retinal antigens and susceptibility to experimental autoimmune uveoretinitis (EAU) induced by these antigens. No expression of IRBP or S-Ag was detected by these authors in thymi of Lewis rats, a strain highly susceptible to EAU induced by these two proteins. On the other hand, S-Ag was expressed in thymi of all tested mouse strains, all of which are resistant to EAU induced by this protein (20). Furthermore, IRBP was expressed in thymi of mouse strains BALB/c and AKR/J, which are resistant to IRBP-induced disease, but not in thymi of the susceptible strains B10.A and B10RIII (19,20). Thymic expression of ocular antigens was determined according to the presence of both the corresponding mRNA and the proteins themselves (19). In a more recent study, using immunohistochemical techniques, M.C. Kennedy (National Eye Institute, Bethesda) succeeded in identifying thymic cells that express S-antigen or IRBP (Fig. 1). The presence of these cells correlated well with detection of the retinal proteins by the aforementioned methods in thymic preparations from different animals (21). These cells exhibit dendritic morphology and are localized in the thymic medulla, at the medulla-cortex boundary (Fig. 1). The medulla is the area where autoreactive maturing lymphocytes undergo apoptosis (22). It is proposed, therefore, that thymic cells expressing ocular specific antigens are responsible for elimination of potentially pathogenic lymphocytes with specificity against these self antigens. Thymic expression of ocular antigens has been examined in only a small number of experimental animals and more investigation is needed to discern how common the phenomenon is. It is of interest that analysis of thymi from four monkeys for retinal antigen expression revealed differences among individual animals (19). This observation suggests that variability in thymic expression of ocular antigens also exists among humans and may be related to differences in susceptibility to autoimmune ocular diseases.

3. MECHANISMS THAT TRIGGER AUTOIMMUNE EYE DISEASES

Thymic deletion of autoreactive lymphocytes, when occurring, is incomplete. Lymphocytes that recognize ocular-specific antigens may escape clonal deletion and are normally present in healthy subjects. Thus, low numbers of lymphocytes specific toward S-Ag are found in the circulation of healthy individuals (23). It is assumed that such autoreactive lymphocyte populations are responsible for initiating pathogenic autoimmune processes in affected individuals, but the triggering mechanisms are not well understood. Data collected in experimental animals have

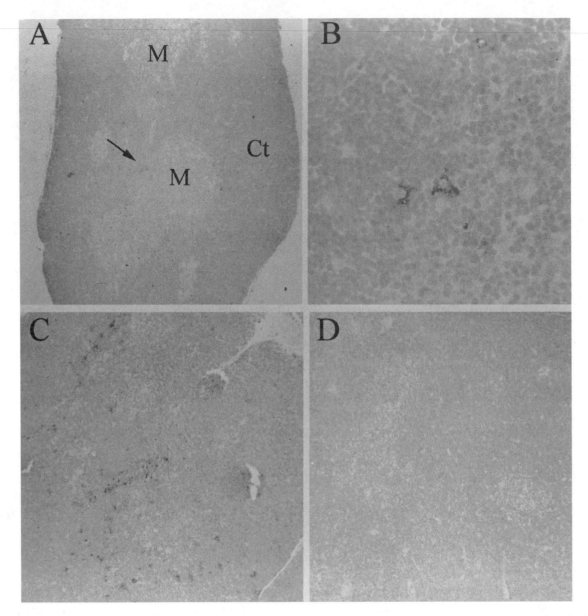

Figure 1.   Expression of retinal antigens by murine thymus cells. Sections of thymic tissue were stained with rabbit polyclonal antibodies against bovine S-antigen (A, B) or IRBP (C, D). **A,** thymus of a BALB/c mouse at a low magnification ($\times$ 50), showing stained cells localizing mainly in the medulla (M), along the boundary with the cortex (Ct). **B,** the area indicated by the arrow in frame A, at a higher magnification ($\times$ 400). Cells stained by the antibody are large and exhibit dendritic morphology. **C,** IRBP antibody stains many cells in a section of thymus from a FVB/N mouse, a strain resistant to IRBP-induced EAU (X200). **D,** no stained cells are seen in a thymus section of a B10.R III mouse, a strain susceptible to IRBP-induced EAU ($\times$ 200). (Courtesy of Dr. Michael C. Kennedy, National Eye Institute)

provided information concerning these mechanisms. Of particular interest are the following related findings:

(i) Only activated lymphocytes are capable of initiating ocular inflammation. Data accumulated in numerous experimental systems have established the pivotal role of lymphocyte activation in the process of inflammation induction; only activated

lymphocytes invade normal tissues and initiate the complex process of inflammation. This difference between activated and non-activated lymphocytes was repeatedly observed in systems in which EAU was adoptively transferred with lymphocytes. EAU was readily induced in recipient animals when the transferred cells were previously stimulated in culture with the uveitogenic antigen or with a polyclonal stimulant such as concanavalin A (Con A). In contrast, sensitized, but resting (unstimulated), cells had no effect even when injected in much higher numbers (24,25). The features that allow activated lymphocytes to enter normal tissues are not completely clear, but accumulating data attribute the lymphocyte-enhanced motility to the increased expression of surface molecules, such as integrin α4 (VLA-4) (26). Likewise, acquisition by lymphocytes of the capacity to induce ocular inflammation was found to be accompanied by increased expression of VLA-4, ICAM-1 and CD69 (27).

(ii) Lymphocytes are activated by microorganisms. Induction of most experimental autoimmune diseases depends on bacterial products, such as *Mycobacterium tuberculosis*, the major component of complete Freund's adjuvant. It is assumed, therefore, that many "spontaneous" autoimmune conditions are triggered by microbial infections that activate autoreactive lymphocytes. Activation of lymphocytes by microbial components could be direct (e.g., Staphylococcal enterotoxin (28), or indirect, via stimulation of macrophages or other cells that release lymphocyte-stimulating cytokines (e.g., endotoxin (29)).

(iii) Microorganisms trigger autoimmunity by mimicry. In addition to stimulating the immune system by non-specific mechanisms, microorganisms may activate clones of autoreactive lymphocytes by possessing peptide sequences similar or identical to those of the target epitopes of the lymphocytes. The phenomenon, termed "molecular mimicry", has been studied in several autoimmune systems (30,31) and its relevance to ocular autoimmune diseases has been extensively investigated by T. Shinohara's group. These investigators discovered homology between an immunopathogenic peptide of S-Ag designated "M" (sequence 303–320) and peptides from several microorganisms, including *E. coli* (32), *Saccharomyces cerevisiae* (33) and several viruses (34). Moreover, immunization of Lewis rats with the microbial peptides induced EAU. It is of interest that peptide M is only moderately uveitogenic in Lewis rats, whereas no information is available on mimicry with the highly uveitogenic peptide of S-Ag, i.e., sequence 343–362. In addition to their studies with retinal antigens, Shinohara and coworkers found homology between lens βA3-crystallin and three microbial proteins (35).

(iv) Spontaneous autoimmune-mediated inflammatory eye disease due to the absence of immunoregulatory cells. Nude mice transplanted with xenogeneic thymi develop spontaneous uveitis that is attributed to deficiency in a population of regulatory T-cells in the chimeric mice (36,37). These regulatory cells are normally generated in the thymus shortly after birth (38). No information is available on the existence of such cells in humans.

4. IMMUNOPATHOGENIC ANTIGENS OF THE EYE

The unique and diverse functions of eye tissues require a large number of molecules specific to these tissues. Several of these molecules were found to be immunopathogenic, i.e., capable of inducing ocular inflammation in experimental animals, and it is assumed that additional ocular-specific molecules will be identified as immunopathogenic once they become available for testing. Tissue-specific antigens were identified in the lens, cornea and retinal pigment epithelium, but the majority of studies have focused on retinal antigens. As detailed below, several retinal proteins have been found to be immunopathogenic. The majority of these antigens also localize in the pineal gland, a vision-associated organ, and animals immunized with these molecules usually develop pineal inflammation as well (see Section 6). It is further assumed that one or more of the ocular-specific molecules serve(s) as target for pathogenic autoimmune conditions in humans (39,40). Cellular and humoral immunity against ocular-specific antigens that are uveitogenic in animals have been detected in patients with uveitis, as well as in healthy subjects. The uveitogenicity of these molecules in experimental animals and the similarity of ocular changes support the notion that autoimmune processes are involved in uveitis in humans.

4.1. S-Antigen (S-Ag)

This molecule was the first immunopathogenic retinal protein to be purified (41). This protein's more recent designation, "arrestin", describes its pivotal role in the vision process, namely, binding to phosphorylated rhodopsin and inhibiting the activation of transducin molecules (42). This function was discovered by Kuhn et al., who designated the molecule "the 48K protein" and collaborated with Faure's group to identify it with S-antigen (43). This ~45kDa protein is highly conserved and is used by photoreceptor cells in eyes of a broad range of species, including invertebrates such as starfish or Drosophila (44,45). The biophysical properties of S-antigen are thoroughly reviewed by Shinohara et al. (46). A related protein designated "β-arrestin", was isolated by Lohse et al. (47) who identified its function in desensitizing β-adrenergic receptors.

S-Ag is a powerful immunopathogen, inducing severe EAU in a variety of species, including rats (9), guinea pigs (41) and primates (48,49). The potent uveitogenicity of this molecule is demonstrated by its capacity to induce EAU in Lewis rats at doses as low as 1 μg/rat (50). Significantly, however, S-Ag failed to induce EAU in all mouse strains tested so far, with the exception of mild disease in a small portion of AKR mice (20,51). This finding could be attributed, at least in part, to the expression of S-Ag in thymi of all mouse strains tested so far (19). The presence of antibodies and cellular immunity against S-Ag in uveitis patients (39,52–55) provides a basis for the notion that this retinal antigen is a target for autoimmune responses in these individuals. The finding of autoimmunity against S-Ag in healthy subjects (23) is interpreted to suggest that activation processes are necessary to initiate disease (see Section 3).

The immunopathogenic epitopes of S-Ag have been well investigated (Table 1). Early studies have focused on sequence 303–320, designated "peptide M", the first S-Ag peptide to demonstrate immunopathogenicity (46,56). In addition to Lewis rats, peptide M induced uveitis in primates, with changes similar to those induced by whole bovine S-Ag (57). Also of interest is the finding that peptide M exhibits sequence homology with several microbial products that are immunopathogenic, in support of the notion that molecular mimicry plays a role in initiating autoimmune conditions (see Section 3). The role that peptide M plays in the uveitogenic process initiated by whole S-Ag seems, however, to be minor if any. The minimal uveitogenic dose of peptide M is approximately 30 nmole (50 μg) per rat, that is 1500-fold higher than that of the whole protein (~20pmole). In addition, this peptide was found to be non-immunodominant, namely, lymphocytes sensitized against the whole protein did not recognize peptide M in culture. In more recent studies, Gregerson,

Donoso and their coworkers discovered two adjacent sequences of S-antigen that are highly uveitogenic. These sequences, at 333–352 and 352–364, induce EAU in Lewis rats at minimal doses of 0.5 and 5 μg, respectively (58,59 and Table 1). In addition, these two peptides are immunodominant in Lewis rats and are assumed, therefore, to be responsible for most of the immunopathogenic activity of the whole protein.

Two additional immunopathogenic peptides were identified by de Smet et al. (60) in the sequence of human S-antigen. These authors tested 40 overlapping peptides of 20 residues each spanning the whole length of the protein and detected significant uveitogenicity in peptides at sequences 181–200 and 281–300.

4.2. Interphotoreceptor Retinoid-Binding Protein (IRBP)

A glycoprotein of approximately 140kDa, IRBP is the major soluble protein of the interphotoreceptor matrix of vertebrate retinas. The main function of IRBP is assumed to be the transport of retinoids between the neural retina and the RPE (61). IRBP contains both carbohydrates and fatty acids and is remarkable by its fourfold repeat structure, with each repeat consisting of approximately 300 amino acid residues (61). Partial homology among the four repeats (30–40%) facilitates immunological cross-reactivity between corresponding sequences on different repeats (62) and was found crucial for the pathogenic activity of an immunodominant epitope (63). IRBP is highly conserved and remarkable levels of cross reactivity were shown by antibodies among various verte-

Table 1   Major uveitogenic epitopes of S-Ag and IRBP in the Lewis rat

Antigen	Epitope	Species	Sequence	Uveitogenicity* (μg/rat)
S-Ag	303–320	Bovine	D T N L A S S T I I K E G I D K T V	50
		Rat	: : : : : : : : : : : : : : R : :	N.D.[+]
S-Ag	333–352	Bovine	L T V S G L L G E L T S S E V A T E V P	0.5
		Rat	: : : : : : : : : : : : : : : : : : : :	:
	352–364	Bovine	P F R L M H P Q P E D P D	5
		Rat	: : : : : : : : : : : : :	N.D.
IRBP	521–540	Human	Y L L T S H R T A T A A A E F A F L M Q	0.1
		Rat	: : : : : : : : : : : : : : : : : : : :	:
	1177–1191	Bovine	A D G S S W E G V G V V P D V	0.01
		Rat	: : : : : : : : : : : T : N :	> 200‡
	1181–1191	Bovine	S W E G V G V V P D V	0.2
	273–283	Bovine	T W E G S G V L P C V	200
		Rat	: : : : : : : : : : :	:
	273–283[V277, D282]	(analog)	T W E G V G V L P D V	0.02

* Minimal dose to cause EAU in Lewis rats.
† Has not been determined.
‡ No disease detected at this maximum dose tested.

brates as well as cephalopods (64). A recent study by Gelderman et al. (65) showed that IRBP from a frog (*Xenopus laevis*) induced severe EAU in Lewis rats when given at relatively high doses (≥ 50 μg/rat).

IRBP is highly uveitogenic in rats (66,67), mice (51), primates (68) and rabbits (69), but is poorly active in guinea pigs (70). In Lewis rats, bovine IRBP induces EAU at the low dose of 0.3 μg, or 2 pmole per rat (50). Of particular importance is the immunopathogenic activity of IRBP in mice. IRBP is the only known molecule to cause EAU in mice and has been used extensively, therefore, to learn about the genetic control of the disease, the pathogenic process and approaches to its regulation (20,40, see section 7). Also of interest is the finding that IRBP is very likely the target of the pathogenic process of uveoretinitis that develops in nude mice grafted with rat thymus (36).

Identification of the immunopathogenic peptides of IRBP was carried out by two approaches. Donoso and associates tested 120 overlapping peptides corresponding to the entire span of 1262 residues of human IRBP (71). Nine of the peptides induced EAU in Lewis rats, with three sequences located at homologous domains on three of the four repeats. One of these peptides, at 521–540 (Table 1), exhibited the highest immunopathogenicity, inducing EAU in Lewis rats at doses, as low as 0.1/μg/rat (~50 pmole/rat) (71). The second approach selected bovine IRBP peptides according to their high amphipathicity (72). Of the ten selected peptides, three were found uveitogenic in Lewis rats, i.e., 1091–1115 (73), 1158–1180 (74) and 1169–1191 (75). Two of these peptides, 1158–1180 and 1169–1191, were also found to be uveitogenic in primates (76). Of particular interest is peptide 1169–1191 (*Table 1*). This peptide, or its truncated sequence 1177–1191, is uveitogenic in Lewis rats at the exceedingly low dose of 10 pmole/rat and was found to be a major immunodominant epitope in this strain of rats (75). The core sequence responsible for this activity is sequence 1182–1190, with residues 1182 and 1190 forming the binding sites with the MHC molecule and amino acids 1188 and 1189 interacting with the lymphocyte TCR (77). Surprisingly, bovine IRBP peptide 1181–1191 was found to differ from its rat IRBP homologue by two pivotal amino acids at positions 1188 and 1190 (63 and Table 1). Moreover, rat peptide 1181–1191 is immunologically-inactive in Lewis rats and cannot serve as target for lymphocytes sensitized against bovine 1181–1191 (63). Additional investigation revealed that, uniquely, lymphocytes sensitized against bovine IRBP peptide 1181–1191 recognized another IRBP peptide in the rat eye at sequence 273–283 (Table 1). Peptide 273–283, that is the homologue for 1181–1191 on repeat "1" of the IRBP molecule, is identical in bovine and rat, is recognized by lymphocytes sensitized against bovine 1181–1191, and is assumed to stimulate these cells to initiate the immunopathogenic process of

EAU in the rat eye, thus serving as a "surrogate epitope" (63). The potent immunogenicity of bovine IRBP peptide 1181–1191 in Lewis rats derives from its high affinity toward the MHC molecules on APC of these rats (63). This piece of information was used by Kozhich et al. (78) to design an analog of peptide 273–283 in which the residues responsible for MHC binding (277 and 282) were substituted with the corresponding amino acids of bovine IRBP 1181–1191 (Table 1). The substituted analog exhibited immunopathogenic capacity ~10 fold higher than that of bovine 1181–1191.

IRBP peptides that induce EAU in mice were also identified. Silver et al. (79) discovered that human IRBP peptide 161–180 is uveitogenic in mice of the H-2^r haplotype, whereas Avichezer et al. (80) showed that peptide 1–20 of this protein induces EAU in H-2^b mice. An IRBP peptide uveitogenic for H-2^k was found by Namba et al. at sequence 201–216 (81). IRBP peptides involved in spontaneous EAU in chimeric mice were identified by Takeuchi et al. (82). These authors revealed that bovine IRBP peptide 1182–1194 is uveitogenic in the BALB/c nude mice grafted with rat thymus. Interestingly, peptide 518–529 is also recognized by lymphocytes of these mice, but the lymphocytes it stimulates produce Th2 type cytokines and, therefore, the peptide suppresses the disease (82)

IRBP-induced EAU in primates closely resembles uveitic conditions such as Vogt-Koyanagi-Harada (VKH) disease or sympathetic ophthalmia (68). Surprisingly, however, cellular or humoral responses to IRBP are not common among patients with these eye diseases (unpublished data).

4.3. Rhodopsin

Rhodopsin, a glycoprotein with a molecular weight of ~40 KDa that plays crucial role in vision, is present in large amounts in the retina and meets all criteria for being an organ-specific antigen. Yet, rhodopsin is poorly uveitogenic in animals and its capacity to induce EAU has been in doubt for some time. Its uveitogenicity has been clearly established only in recent years, when highly purified rhodopsin became available. Rhodopsin induces EAU in Lewis rats when injected at the relatively high doses of 50 to 100 μg/rat (83); i.e., higher by two orders of magnitude than the minimal uveitogenic doses of S-antigen or IRBP (50). Schalken and associates later confirmed that rhodopsin is also uveitogenic in primates, producing severe chorioretinitis with concomitant anterior uveitis in these animals (84). These scientists also proved that rhodopsin is more pathogenic than opsin, both in the Lewis rat (83) and in monkeys (84). Adamus, Moticka, and their associates identified several peptide determinants of rhodopsin that are uveitogenic in Lewis rats (85).

4.4. Recoverin

This ~23KDa calcium-binding protein was found to induce uveoretinitis as well as pinealitis in Lewis rats at doses as low as 10 μg/rat (86,87). This observation is of interest in view of recoverin being identified as the major target for a putative autoimmune response that mediates retinal destruction in cases of cancer-associated retinopathy (CAR) (88). The actual pathogenic mechanism of CAR has been suggested to be via antibody effect and, indeed, recoverin antibodies are detected in the majority of cases. On the other hand, a high level of cellular immunity to recoverin was also detected in a CAR-like case (89) and it is conceivable cellular immunity could also participate in this process.

4.5. Phosducin

This ~33 KDa protein is found in both photoreceptor cells and pineal gland and is believed to participate in photo-transduction (90). Phosducin induces EAU in Lewis rats (91), and Satoh et al. (92) have identified five uveitogenic peptides from its sequence.

4.6. Melanin-Associated Proteins

Poorly defined preparations of melanin had been used in early studies for induction of experimental melanin protein-induced uveitis (EMIU) (93). Later, Broekhuyse et al. (94) treated melanin granules by chemical extraction and obtained several highly uveitogenic fractions, whereas an EMIU-inducing protein was also extracted from bovine melanin by Bora and her coworkers (95). More recently, Yamaki and his colleagues investigated the better defined tyrosinase family proteins and found them to be uveito-genic in pigmented rats, inducing eye disease with close similarity to VKH syndrome (96).

4.7. Other Uveitogenic Proteins

The unusual high susceptibility of Lewis rats to autoim-mune diseases made it possible to detect the capacity to Induce EAU by several other proteins. These include a gamma submit of cyclic guanosine monophosphate phos-phodiesterase (97) and the astrocyte-derived calcium-binding protein S100β (98).

5. IMMUNOPATHOGENIC MECHANISMS OF OCULAR AUTOIMMUNITY

Similar to immune-mediated inflammation in other tissues, uveitic processes are mediated by a variety of cells and their products, many of which are yet to be revealed. The topic has been thoroughly reviewed in a recent monograph (99).

5.1. Antibodies

Antibodies against ocular self-antigens are thought to be the main pathogenic agent in two eye diseases, lens-associated uveitis (LAU) and cancer-associated retinopathy (CAR). Animal models for LAU, described below (Section 6), were developed and analyzed, mostly by Marak and his colleagues (100), who showed that complement-fixing antibodies against lens crystallins mediate inflammatory processes in animal eyes in which the lens capsule is compromised. It is notewor-thy that while humoral immunity against autologous lens crystallins can be readily induced, cellular immunity to these antigens is selectively suppressed by an effective immuno-tolerance mechanism due to thymic expression of these lens proteins (see Section 2).

The notion that CAR is mediated by antibodies against recoverin or other retinal antigens is supported by the finding of such antibodies in most patients with this condition (88). In addition, antibodies against recoverin were found to induce apoptosis of certain retinal cells in vitro (101) and damage the retina in rats injected intravitreally (102).

In addition to their pathogenic role in LAU and CAR, antibodies were suggested to be involved in EAU by activating mast cells (103), while Kasp and associates (104) presented data proposing that circulating immune complexes could exert either regulatory or pathogenic effects in EAU.

5.2. Mast Cells

Several lines of evidence indicate that mast cells play an active role in the pathogenic process of EAU: (i) A correla-tion was found between susceptibility to EAU of rat inbred strains and the number of mast cells in the animals' choroid (105) and ciliary body (106); (ii) the number of choroidal mast cells increases in rat eyes 5 to 6 days post-immuniza-tion with S-antigen (107); (iii) a massive mast cell degranu-lation develops 1 or 2 days prior to the disease onset (107); (iv) EAU development can be suppressed by drugs that affect mast cells (107). It is assumed that mast cell products assist the disease process by affecting the blood-tissue barriers.

5.3. Adhesion Molecules

Adhesion molecules play a pivotal role in inflammatory processes by facilitating invasion of target tissues by both the initiating lymphocytes and the inflammatory cells they recruit. The role adhesion molecules play in autoimmune ocular disease was demonstrated by the finding that treat-ment with antibodies against adhesion molecules inhibited development of EAU in mice (108). The adhesion mole-cules that have been investigated in these studies included ICAM-1, LFA-1 and Mac-1, but it is assumed that addi-tional molecules are involved in this highly complex process. Expression of adhesion molecules was also noted in eye tissues of uveitis patients (109).

5.4. Lymphocytes

All EAU and EMIU models in rodents are cell-mediated, as demonstrated by the capacity of T-lymphocytes from affected donors to adoptively transfer the disease to naïve recipients (24,25,110) and the inability of athymic rats to develop EAU (111). The uveitogenic lymphocytes were identified to be CD4 (helper) T-cells (24,25) and more recent studies have provided evidence showing that disease induction is a feature of Th1 type cells (99). Of particular importance is the finding that only T-cell populations that produce interferon-γ (IFN-γ), the hallmark for Th1 cells, were uveitogenic (112,113). Moreover, non-uveitogenic lymphocyte populations became pathogenic following incubation with IL-12, a cytokine that induces polarization of naïve (Th0) lymphocytes toward Th1 type (112,113). Additional support for the notion that Th1 cells are the uveitogenic population was provided by Saoudi et al. (114), who showed that rats immunized with S-Ag and treated with HgCl$_2$ become resistant to EAU due to polarization of their T cell population toward Th2 type.

It is also of note that Caspi's group has recently shown that, under some conditions, IFN-γ producing T cells are not essential for EAU induction; severe disease developed in IFN-γ deficient mice following immunization with IRBP (115). Marked differences between the histological changes in affected eyes of IFN-γ deficient mice and wild-type animals suggest that Th-2-like cells mediated the disease in the former mice.

Uveitogenic lymphocytes were also analyzed for their usage of TCR genes. Studies of Egwuagu, Gregerson and their coworkers showed that all S antigen-specific uveitogenic cell lines express the Vβ8.2 gene product, whereas nonuveitogenic lines usually do not (116,117). Lymphocyte lines specific toward IRBP, on the other hand, express the Vβ8.3 gene product as well (118). The relationship between the expression of these TCR components and the immunopathogenicity of lymphocytes is not clear, but it is noteworthy that Vβ8.2 is also selectively expressed by lymphocytes capable of inducing another autoimmune disease, experimental allergic encephalomyelitis (EAE) (119). Moreover, in another study, Egwuagu and associates found that lymphocytes that express Vβ8.2 accumulate in rat eyes with S-Ag-induced EAU in proportions much higher than in the peripheral blood of these animals, whereas lymphocytes that express both Vβ8.2 and Vβ8.3 selectively accumulate in eyes with IRBP-induced EAU (120).

Limited information is available concerning the type(s) of T-cells and their TCR usage in patients with uveitis.

5.5. Cytokines and Chemokines

Cytokines initiate and modulate all inflammatory processes, and their involvement in autoimmune eye conditions has been well documented. Participation of cytokines in ocular pathogenic processes has been investigated by multiple approaches, including (i) detection and measurement of cytokines or their mRNA in eyes or lymphoid organs (121,122); (ii) administration of cytokines or antibodies against them in animals developing autoimmune eye diseases (123,124); (iii) testing disease development in knock-out mice deficient in individual cytokines (115,125,126). As mentioned above, and in line with observations on autoimmune diseases in other tissues, accumulating data indicate that Th1 type cytokines normally play a major role in mediating and promoting autoimmune ocular diseases, whereas Th2 cytokines mainly exert modulatory effects. Noteworthy observations include the following: (a) Both Th1 (IL-2, IFN-γ, IL-12) and Th2 type cytokines (IL-4, IL-10) are produced in inflamed eyes (121,122). (b) Susceptibility to EAU induction in rat and mouse strains correlates with polarization toward Th1 cytokines, whereas resistance to disease correlates with higher Th2 cytokine production (112-114). (c) IL-10 has a protective role in EAU: Treatment with this cytokine inhibited EAU development in normal mice, whereas systemic neutralization of IL-10 elevated disease scores (125). (d) Of particular interest are the complex effects of IFN-γ and IL-12 on EAU development. IFN-γ is produced by all uveitogenic cell lines but treatment with IFN-γ antibody exacerbates EAU (123) and IFN-γ deficient mice do develop severe disease (115). In addition, and in accord with these data, treatment with IL-12, a cytokine that stimulates IFN-γ production, was found to inhibit EAU development (124). These paradoxical observations were explained by the finding that IFN-γ stimulates synthesis of nitric oxide, a molecule that induces apoptosis of T-cells during the priming phase (124). It is of note, however, that IL-12 is essential for EAU development; no disease develops in IL-12-deficient mice (126), or in wild-type mice treated with antibodies against IL-12 (Silver et al., unpublished data). Moreover, even IFN-γ deficient mice require IL-12 for EAU development (Silver et al., unpublished data).

Similar to their effect in other inflammatory processes (127), chemokines are responsible for the actual recruitment process in ocular inflammation. Little information is available, however, concerning the identity of individual chemokines that participate in this process. Our attempt to inhibit various ocular inflammatory models by blocking chemokines such as MCP-1 or RANTES have been unsuccessful (128), presumably due to existence of multiple alternative shunts. Identification of individual chemokines in ocular and pineal inflammatory process should yield interesting data in view of the marked difference between the infiltrating cell populations in these two organs in the same animals (see Section 6).

5.6. Development of Autoimmune Ocular Disease: The Presumed Process

Data accumulated in various studies made it possible to construct an outline for the presumed process that brings about diseases such as EAU or EMIU. The early phase has been recently dissected by the elegant study of Prendergast et al. (129). These authors showed that activated lymphocytes enter eye tissues by a random, non-antigen specific mechanism. The wandering lymphocytes disappear rapidly from the ocular tissues, unless they are locally exposed to their target antigen, presented on APC. As shown by Prendergast et al. (129), a strikingly small number of lymphocytes sensitized against S-Ag are required to initiate the massive inflammatory response that follows the initial event. The identity of APC involved in this event has not been determined. McMenamin et al. (130) suggested that MHC class II-positive dendritic cells in the choroid are involved in this process, whereas Prendergast et al. (129) proposed that dendritic cells in the anterior segment are responsible for presenting the retinal antigens. Once the sensitized lymphocytes are stimulated by their target antigen, they initiate the inflammatory process by releasing a battery of cytokines that trigger a cascade of events involving other cells and their products. This recruitment, which is essential for the inflammatory reaction (131), is mediated by a variety of chemokines that have not been defined yet in ocular inflammation. The process of tissue invasion by inflammatory cells in eyes developing EAU has been thoroughly studied by McMenamin et al. (130,132) and Greenwood et al. (133). Using electron microscopy, these authors observed breakdown of blood barriers, changes in endothelial cells and actual infiltration of lymphoid cells. Ensuing tissue destruction is assumed to be carried out mainly by macrophages that release damaging products, such as nitric oxide or peroxide (99,134,135).

It is assumed that ocular inflammation in humans is initiated by a process similar to that in experimental animals. As discussed above (Section 3), lymphocytes sensitized against ocular antigens are normally present in humans and several different mechanisms can activate these lymphocytes and transform them into tissue invading and inflammation triggering cells.

6. ANIMAL MODELS FOR OCULAR AUTOIMMUNE DISEASES

Animal models of autoimmunity for different ocular tissues have been developed and investigated since the beginning of the century. After the autoimmunogenicity of lens was shown in 1903 by Uhlenhuth, and that of the retina by Hess and Romer (see Section 1), attention focused for several decades on the uveal pigment (melanin) as a potential target for autoimmune uveitis. "Experimental allergic uveitis" was first produced by Collins in 1949 by immunizing guinea pigs with uveal tissues (136). In the 1960s, Wacker and Lipton discovered, however, that the retina is much more effective than the uvea in inducing ocular inflammation (137). The disease induced by retinal antigens, named "experimental autoimmune uveoretinitis" (EAU), is elicited by different retinal antigens, in various species, including guinea pigs, rabbits, rats, monkeys, and mice (2). Thus, EAU has become a stereotype of animal model for ocular autoimmunity. In 1991, Broekhyse and associates described an ocular autoimmune model induced with posterior uveal melanin protein (93) and named it "experimental autoimmune anterior uveitis" (EAAU). This model was later termed "experimental melanin-protein induced uveitis" (EMIU) (138).

6.1. Experimental Autoimmune Uveoretinitis (EAU)

EAU can be induced either by active immunization with retinal antigens or adoptive transfer with lymphoid cells derived from immunized animals. Clinical and histopathological features of EAU vary according to the animal species or strain, the antigen used and its dose in actively induced EAU, or by the number of injected cells in adoptively transferred EAU. These factors also determine the onset time (9 days to more than a month in active EAU, or 2–8 days in adoptively transferred disease) and duration of the disease (a few days to more than a year). In the highly susceptible Lewis rats, EAU is a subacute panuveitis of a short duration, leading to loss of photoreceptors (Fig. 2A) and retinal gliosis (9,139) The disease is more moderate and less subacute in less susceptible rat strains and in mice, guinea pigs and primates. The most typical change observed in these species includes granulomatous inflammation. The various aspects of EAU therefore resemble a wide range of human ocular inflammatory diseases affecting the retina and uvea (Fig. 2D). Over the years, EAU has served as a model to elucidate basic mechanisms involved in pathogenesis and immune regulation of human uveitides and to develop new therapies for uveitic patients.

6.2. Rats

Most work on EAU has been carried out in Lewis rats immunized with bovine S-Ag. This strain is also susceptible to EAU by all other uveitogenic retinal antigens. A single immunization consisting of 30–50 μg of S-Ag emulsified in complete Freund's adjuvant produces severe ocular inflammation that develops 12–14 days later (9,139). Disease onset can be detected by flash-light examination of the eye as conjunctival congestion and dull red reflex, or by slit lamp examination of the eye with findings of flare and cells in the anterior chamber, iris vascular dilation, and anterior synechiae. When immunization is accompanied by injection of *B pertussis*, disease onset occurs more abruptly, in 9–12 days, and the inflammation is massively exudative and hemorrhagic, often leading within a few days to phthisis bulbi ("shrinking eye").

Histologically, the first abnormalities are subacute irido-cyclitis and perivascular mononuclear infiltration of the retina in Lewis rats with EAU. Various degrees of vitritis, choroiditis, retinitis, retinal edema and detachment follow (Fig. 2A). In the early stage the main inflammatory cells include polymorphonuclear neutrophils, macrophages, and T lymphocytes. Later, the infiltrating cells compose of mainly macrophages and T lymphocytes. After one to two weeks photoreceptor cells are mostly destroyed and the retina becomes gliotic. Chorioretinal adhesion is often noted. In other rat strains, the disease is less severe and of a longer duration.

6.3.   Primates

EAU has been induced in monkeys by immunization with rhodopsin (84), S-Ag (48) or IRBP (68). Clinically, EAU in monkeys resembles human recurrent chorioretinitis with retinal vascular sheathing, retinal edema, chroioretinal infiltrates, and even retinal detachment. Histologically, retinal periphlebitis and granulomatous inflammation in the choroid and retina are characteristic features of EAU in monkeys (Figs. 2E & 2F). Infiltrating giant cells surrounding the Bruch's membrane in the choroid are described. Dalen-Fuchs nodules, i.e., aggregation of macrophages at the level of the retinal pigment epithelium, a characteristic finding of non-infectious granulomatous uveitis in humans, are also observed in primate eyes. The disease may last up to more than a year, resulting in loss of photoreceptors and retinal gliosis.

6.4.   Mice

EAU is induced by immunization with the IRBP; the retinal S-Ag is poorly uveitogenic in mice (51). Most mouse strains are relatively resistant to induction of EAU, and most immunization protocols have required the use of per-tussis toxin to elicit disease (51,140). In addition to facilitating the breakage of ocular-blood barriers, the pertussis toxin is able to polarize the immune response toward the Th1 pathway (140).

Clinical presentations of EAU in mice show predominant inflammation involving the posterior segment of the eye. Fundoscopic examination demonstrates hazy vitreous, retinal vascular sheathing, retinal folds, focal retinal infiltrates and detachment, choroidal infiltrates and chorioretinal scar. The disease is usually a focal inflammation, with slow progression and recurrent course leading to a total loss of photoreceptors in 2–3 months. The pathology of murine EAU reveals vitritis, choroiditis, retinitis, retinal edema, vasculitis, granulomas and folds (141) (Fig. 2G). Focal loss of photoreceptors and chorioretinal scar are also characteristic. In addition, a mild iridocyclitis is often noted. The recurrent nature and focal pathology described in the B10.A mouse strain have made this animal disease an excellent model for the study of human uveitis (99).

6.5.   Experimental Autoimmune Pinealitis (EAP)

The pineal gland, the "third eye", functions as a vision organ in lower vertebrates, but in mammals it has evolved into a secretory organ and no longer functions in photoreception. Several retinal antigens have been phylogenetically conserved in pinealocytes, and an inflammatory response is elicited in the pineal gland following immunization with these antigens, consisting of patches of inflammation throughout the organ (Fig. 2H). EAP was induced by S-Ag in guinea pigs (41), rats (142), or monkeys (143), by IRBP in rats (66,67) and monkeys (68), as well as by recoverin in rats (86).

EAP develops in rats as early as day 9 post-immunization and, unlike EAU, was retained for at least two months post-immunization, long after the ocular inflammation subsided (144). EAP also differs from EAU by the population of infiltrating cells. In contrast to the mixed population of polymorphonuclear and mononuclear cells in eyes with EAU, the pineal gland of the same rats exhibit exclusive involvement of lymphocytes and a few macrophages (142) (Fig. 2I). No information is available concerning the pineal involvement in uveitis patients.

6.6.   Experimental Melanin-Protein-Induced Uveitis (EMIU)

EMIU is induced by melanin-protein associated insoluble antigen(s) isolated from bovine ocular pigmented tissues of the iris, ciliary body, choroid, and remnant retinal pigment epithelium (93,95). Like EAU, this model is a T cell-mediated ocular disease (110), but unlike EAU, the target tissue in EMIU is the uvea, not the retina of the eye. Therefore, ocular inflammation is mostly limited to the iris, ciliary body, and choroid. EMIU is characterized by bilateral, recurrent uveitis in the rats and monkeys.

Clinical manifestations of EMIU show anterior uveitis with an onset of 10–14 days after immunization. It presents with classical pictures of conjunctival congestion, dilation of iris vessels, cells and flare in the anterior chamber, irregular pupils, and synechiae. Hypopyon, hyphema and keratic precipitates are the findings in the animals with severe disease by slit lamp examination. Fundoscopic examination

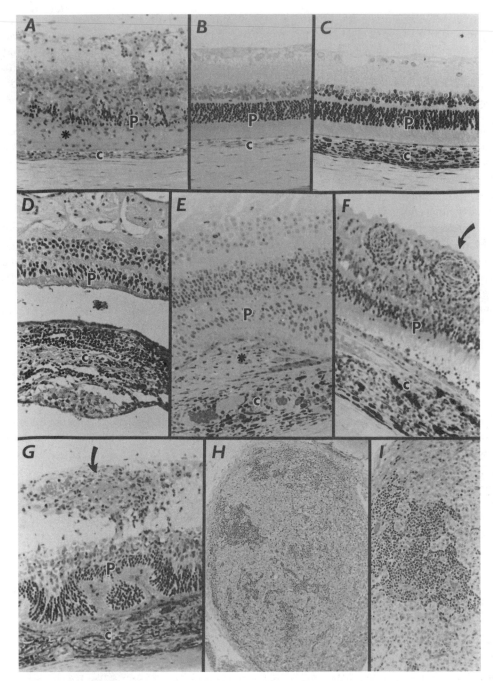

Figure 2. Autoimmunity-induced inflammatory changes in the eye and pineal gland. **A,** posterior segment of a Lewis rat eye with experimental autoimmune uveoretinitis (EAU) induced by IRBP. The retina is detached and infiltrated with inflammatory cells. The photoreceptor cell layer (P) is almost completely destroyed. Cellular and proteinaceous infiltrate in the subretinal space (asterisk). Significantly, the choroid (C) is only slightly inflamed. **B,** posterior segment of a normal Lewis rat eye, showing the characteristic multi-layer morphology of the retina and choroid. **C,** posterior segment of a Lewis rat eye with experimental melanin-induced uveitis (EMIU). The retina is intact, but the choroid is heavily infiltrated with inflammatory cells. **D,** posterior segment of a human eye with sympathetic ophthalmia. The choroid is infiltrated with inflammatory cells. The retina is thinned due mainly to damage of photoreceptor cells. **E,** posterior segment of a monkey eye with IRBP-induced EAU. The photoreceptor cell layer is thinned due to loss of cells in the outer layer. The choroid is inflamed with early subretinal fibrosis (asterisk). **F,** posterior segment of a monkey eye with early changes of IRBP-induced EAU. Perivascular inflammation in the retina (arrow). Many inflammatory cells throughout the retina and the subretinal space. The choroid is infiltrated with inflammatory cells. **G,** Posterior segment of a B10.A mouse eye with IRBP-induced EAU. Typical changes in the retina, including perivasculitis (arrow) and retinal folds. The choroid is infiltrated. **H,** pinealitis in a Lewis rat immunized with IRBP. The homogeneous pineal tissue is infiltrated with inflammatory cells. **I,** higher magnification of the affected pineal gland, showing that the infiltrate consists exclusively of mononuclear cells.

reveals choroidal infiltration and thickening. The acute disease usually lasts 1–2 weeks. Similar to EAU in mice, spontaneous recurrence can be observed in some EMIU rats.

The histopathology of EMIU in the Lewis rat demonstrates inflammatory cellular infiltration in the uvea. The pars plana and vitreous may be involved in severe cases. Macrophages and T lymphocytes are the main infiltrating cells in the eye. Similar to EAU in the Lewis rat, polymorphonuclear neutrophils, plasma cells, and occasional eosinophils are also frequently present. Proteinaceous exudates and fibrin are often observed in the anterior chamber and may spill over into the posterior chamber. The retinal pigment epithelium, retina, and optic nerve are seldom involved except in some severe recurrent cases (145). In the eyes with multiple recurrent EMIU, loss of stroma in the iris and ciliary body may be seen.

In the pigmented rats and primates, choroidal inflammation (Fig. 2C) with little iridocyclitis is unique. Dalen-Fuchs nodules and choroidal granuloma are prominent in the monkeys with EMIU (145). Additional B lymphocytes are also identified in monkeys, in contrast with mostly T lymphocytes in the rat (145). The pineal gland is not affected in animals with EMIU, in line with the absence of melanin in this organ.

Recently, Yamaki and associates described a model for VKH syndrome by immunization of pigmented rats with tyrosinase family protein, one of the melanocyte differentiation antigens (96). The rats developed anterior uveitis and "sunset glow fundus" with loss of fundus pigmentation. Histologically, retinal detachment, Dalen-Fuchs nodules, and granulomatous choroiditis with little retinal involvement are noted. Because the melanocyte differentiation antigens include tyrosinase family protein and are part of the melanin-protein components injected for the induction of the classical EMIU, this model may become a better defined model for these human uveitides.

6.7. Lens-Associated Uveitis (LAU)

Lens-induced uveitis is produced by disruption of the lens capsule in animals primed with xenogenic or allogeneic lens crystallins (100). The disease can be transferred by antibodies against lens antigens. LAU has been investigated mostly in rats, but it is also reported in other animal species including rabbits (146), or dogs (147). LAU is an acute granulomatous uveitis with characteristics of an immune complex-mediated inflammation surrounding the injured lens. This is a model for human phacoanaphylactic endophthalmitis, a rare ocular inflammation and can be cured by surgical removal of the lens material.

Recently, a lens-associated uveitis that is mediated by lymphocytes targeted to a lens antigen has been developed (148).

Transgenic mice expressing hen egg lysozyme (HEL) in their lens develop ocular inflammation after being injected with syngeneic wild type lymphocytes sensitized against HEL. The disease consists of intense intraocular inflammation characterized by severe limbitis, keratitis, iridocyclitis, vitritis, and retinitis. The cellular infiltration is of subacute type consisting of mononuclear and polymorphonuclear leukocytes. This new model is cell-mediated, and the mice do not produce antibody against HEL.

7. GENETIC CONTROL OF EAU

In the human, uveitis shows strong genetic influences (39,40). MHC genes appear to have a major impact. However, due to the complex nature of human heredity, it is difficult to perform sophisticated genetic analyses, especially in the case of a relatively rare and heterogeneous disease entity such as uveitis. We have, therefore, turned to the rodent models of EAU to study the genetic control of ocular autoimmunity.

EAU in rodents is genetically controlled. In mice and rats there are inbred strains of varying susceptibility, from highly susceptible to highly resistant (40,50,149,150). The observed patterns of susceptibility depend in part on the species and strain, and in part on the particular antigen being used. Some strains of rats develop EAU after immunization with IRBP more readily than when immunized with S-Ag, or vice versa (50,149) For example, Lewis rats develop high EAU scores after immunization with either S-Ag or IRBP, but F344 rats are resistant. BN rats are highly resistant to S-Ag, but are moderately susceptible to IRBP. IRBP, but not S-Ag, is uveitogenic in certain strains of mice; for example, B10.RIII and B10.A mice are susceptible, whereas BALB/c and AKR mice are resistant (20,40). In contrast to IRBP, S-Ag is a poor uveitogen in mice (20,40) while in guinea pigs, the opposite seems to be the case (70). The complex genetic influences that control these observed patterns of susceptibility will be examined in the following paragraphs.

Susceptibility to EAU is controlled by both MHC and non-MHC genes. Expression of disease in mice and in rats requires the presence of both a susceptible MHC haplotype and a "permissive" genetic background (40,150). We have mapped the MHC control of susceptibility to IRBP-induced EAU in mice to MHC class II genes (40). In H2^k haplotype mice, the I-A subregion (HLA-DQ equivalent) appeared to confer susceptibility. In contrast, expression of the I-E gene product (HLA-DR equivalent) appeared to have an ameliorating effect on disease (20,40). Because class II genes control which epitopes of the IRBP molecule will be presented to the T cells, this implicates epitope recognition as a major mechanism determining susceptibility. Currently ongoing studies with HLA-transgenic mice rendered deficient for mouse MHC genes will help to understand the mechanisms behind MHC control of susceptibility to uveitis in humans (151). However, it should be pointed out that the MHC cluster contains additional loci that are not directly involved in antigen presentation, such as the TNF gene. Although direct studies connecting polymorphisms in the TNF locus to EAU susceptibility have not been done, de Kozak et al. (152) have shown that retinal cells of susceptible Lewis rats produce more TNF in culture than those of resistant BN and LE rats, providing suggestive evidence linking local TNF production to susceptibility.

Importantly, in strains having a susceptible MHC type, the non-MHC genes can permit or prevent disease expression. Susceptible mouse strains exhibited highest disease scores if they had the B10 genetic background, and disease was reduced, or even absent, in strains having the same MHC on another (nonpermissive) background (20,40). Thus, even in individuals having a susceptible MHC, the final expression of disease is determined by the genetic background. Many factors may determine "permissiveness" or "nonpermissiveness" of a particular genetic background. These may include regulation of responses to lymphokines, hypothalamic-adrenal-pituitary axis (HPA) hormones, mast cell/vascular effects, and the thymic selection of the T cell repertoire.

The HPA axis in the autoimmunity-prone Lewis rat is defective, resulting in a blunted corticosteroid stress response in comparison to the resistant F344 (153). Corticosteroids are potent natural regulators of the immune response, and defective corticosteroid responses have been linked to a heightened propensity to the development of inflammatory and autoimmune responses (154). However, it is clear that the effects of HPA axis regulation are insufficient by themselves to account for the difference in EAU susceptibility between the two strains. This is indicated by the observation that although replacement doses of the synthetic corticosteroid dexamethasone given to Lewis rats did indeed confer resistance to EAU, corticosteroid depletion by adrenalectomy, or by chronic treatment with the corticosteroid receptor antagonist RU-486, did not abrogate resistance to EAU in F344 (Caspi et al., unpublished).

The genetically-predetermined tendency to generate a specific cytokine response profile in response to an antigen is an important determinant in susceptibility to autoimmune disease. Numerous studies show that EAU is mediated by cellular immunity and that the pathogenic effector T cells are of the Th1 subset (112). Conversely, Th2 cells can prevent EAU, and skewing the response in an animal given a uveitogenic immunization away from Th1 pathway and towards the Th2 pathway is protective (114,125). This raised the possibility that genotypes predisposed to generate predominantly Th1 responses might be more genetically prone to develop EAU. Indeed, a study of the responses in a series of EAU-susceptible and EAU-resistant rat and mouse strains revealed a consistent pattern. Highly susceptible strains such as the Lewis rat and B10.RIII mouse are dom-inant Th1 responders. Less susceptible strains mount responses progressively lower in IFN-γ, and the extent of polarization of the response towards Th1 correlates roughly with the extent of susceptibility of that strain (112,155). Interestingly, pertussis toxin, long used to enhance induction of autoimmunity in resistant strains and thought to act by breaking the blood-organ barrier, has been shown by us to promote the Th1 response when administered concurrently with immunization (112,140).

The Th1-low response pattern associated with resistance does not have to be synonymous with a dominant Th2 response to the uveitogen, such as in BALB/c mice. Genotypes with a "null" response that is low in both Th1-type and Th2-type cytokines, such as the F344 rat and the AKR mouse, are also highly resistant to EAU. Thus, while there seems to be one pathway to susceptibility, there is more than one pathway to resistance. Interestingly, in both the Th1-predisposed (B10.A) and the Th2-predisposed (BALB/c) mouse strains, the cytokine response to the uveitogenic antigen IRBP is initially a neutral, Th0-like response when measured during the first week after immunization. As the response evolves, it progressively polarizes towards Th1 in the susceptible, and towards Th2 in the resistant strain (156).

Degranulation of mast cells at the time of disease onset helps to break down the blood-retinal barrier. This facilitates the entry of cells into the eye, and the exit of tissue breakdown products and soluble mediators into the circulation, thus fueling the progression of the autoimmune process. An association between the number of ocular mast cells and susceptibility to EAU in rodents has been observed. Susceptible strains of rats were shown in several studies to have more numerous mast cells in the front as well as the back of the eye, suggesting that the blood-retinal barrier may be a genetically controlled trait that affects susceptibility (40,105,106).

Autoimmune disease is caused by the inappropriate activation of T cells capable of recognizing self-antigens. Therefore, presence of such cells in the repertoire can influence the predisposition of an individual to autoimmunity. It is now known that high affinity self-reactive clones are normally deleted in the thymus, as they encounter their cognate antigen during a critical stage in their maturation. This, of course, requires that the antigen in question is functionally expressed and presented in the thymic environment. Until recently, it has been assumed that this operates for soluble or ubiquitously expressed tissue antigens. Interestingly, Egwuagu et al. (19) have shown expression of S-Ag and IRBP in the thymus of several mouse and rat strains that roughly correlated with their susceptibility to uveitis induced by these two antigens. Although this is indirect evidence, and although the effectiveness of such thymic expression in terms of negative selection is not known (the negative selection process is not 100% efficient for antigens strongly represented in the thymus), this is nevertheless provocative evidence for another genetically controlled factor that might influence susceptibility.

The approaches above compared the immunological characteristics of susceptible and resistant strains in an attempt to draw genetic conclusions. Another approach is to identify genomic regions controlling susceptibility by using a direct genetic analysis. One such approach relies on generating an F2 cross between a susceptible and a resistant genotype, and analyzing the co-segregation of the susceptibility phenotype with known genetic markers. In a recent study, we performed a genome-wide analysis of quantitative trait loci (QTL) using simple sequence length polymorphisms (SSLP) in F2 progeny of the susceptible Lewis and the resistant F344 rat strains. Because these strains are MHC-compatible, any differences revealed by this type of analysis will be restricted to non-MHC loci. The analysis revealed 3 independent genetic regions on chromosomes 4, 10 and 12, that were associated with susceptibility to EAU (157). Interestingly, these particular regions are known to contain numerous immunologically-relevant genes (158). To mention just a few: on chromosome 4, this genetic interval contains genes for the T cell receptor β chain families, Neuropeptide Y, Substance P receptor, TGF-α, the Igκ light chain locus, the CD8 α and β chains, the CD4 antigen the IL-5 receptor α chain, and others. On chromosome 10, there are genes for the IL-12R beta chain, IL-4, IL-3, IL-5, NOS2 (= iNOS), IFN-γ and on chromosome 12, the gene for NOS1 and others. Although the genetic regions defined in this analysis are still relatively large, the ultimate goal of such studies is to progressively refine the genetic intervals to identify individual genes. In this manner, one of the susceptibility loci-controlling autoimmune diabetes and experimental autoimmune encephalomyelitis (EAE) in the NOD mouse was mapped to the IL-2 gene, with the disease-associated allele conferring a reduced half-life on the IL-2 molecule (159,160). This underscores the notion that diseaseassociated genes are allelic variants of "normal" genes, which may have evolved in response to selective biological pressures.

Comparative analysis of many independent genetic mapping studies shows that loci associated with different autoimmune diseases, including arthritis, diabetes, encephalitis, orchitis, uveitis and others, cluster in the same chromosomal regions. Furthermore, in various animal models and in humans, these autoimmune-related gene clusters are on homologous chromosomal segments (161,162). It thus appears that the same loci may be responsible for genetic control of autoimmune and inflammatory responses in different diseases and across multiple species.

This opens the possibility of therapeutically targeting common pathways to pathogenesis to treat different autoimmune diseases.

In summary, the experimental evidence indicates that genetic control of susceptibility to autoimmune disease, including uveitis, is complex. It is composed of MHC-controlled mechanisms and non-MHC controlled mechanisms. These include not only the ability to present and recognize immunodominant pathogenic epitopes, but also regulation of responses to lymphokines, HPA axis hormones, mast cell/vascular effects, and thymic selection of the T cell repertoire. Thus, it is the cumulative interaction of multiple genes inherited in various combinations, rather than any single trait alone, that will determine the susceptibility phenotype.

8. OCULAR DISEASES WITH PRESUMED RETINAL AUTOIMMUNITY

The concept that an autoimmune process plays a role in intraocular inflammatory disease or uveitis is not new, originating at the turn of the 20th Century (see Section 1). Some examples of uveitic autoimmune conditions are listed here to give the reader a sense of what has been observed and how the diseases present clinically.

8.1. Sympathetic Ophthalmia

This disease is the best known example of the uveitides associated with a potential autoimmune mechanisms. Sympathetic ophthalmia is rare and occurs after penetrating injury to one eye, either surgical or accidental trauma. A bilateral granulomatous uveitis may appear after a period of time from 10 days to decades after the traumatic event. It was theorized by researchers early in the 20th century that uveal pigment was the immunizing antigen, with antigen being released and then processed by the immune system at the time of ocular injury. Histologically, the choroid of the eye is filled with lymphocytes surrounding a granulomatous reaction. Immunohistochemical staining has demonstrated that in the recovery phase of the disease (163), there is an influx of OKT8+ cells, while in the earlier stages of the disease, there appears to be a greater number of OKT 4+ cells (164). Reports have demonstrated that many sympathetic ophthalmia patients' lymphocytes have positive cell-mediated immune responses to various antigens from the back of the eye. One mechanistic concept for this disease hypothesizes that with the penetrating injury there is the development of intraocular lymphatics (which are normally not present). With this event comes release of ocular antigens and presentation to the immune system in such a way that a CD4+ (Th1) immune response ensues (39).

8.2. Birdshot Retinochoroidopathy

This intraocular inflammation was described by Ryan and Maumenee in 1980 and by Gass in 1981 (165,166). This disease appears to affect individuals who are in their middle age and is more commonly seen in women. It is characterized as being a non-infectious bilateral posterior uveitis. Its most characteristic feature is round, cream colored lesions found to be strewn throughout the fundus. A retinal vasculitis is also present and is best noted on fluorescein angiography. This disorder has a clinical similarity with that induced with the retinal S-Ag or IRBP in nonhuman primates (49,68), where cream colored lesions similar to what is seen in the human situation can be noted. Further, the histologic appearance of S-Ag-induced uveitis bears striking similarity to the few cases of birdshot retinochoroidopathy that have been examined (167). Several HLA studies have been performed. The initial observation, which has been supported by subsequent studies, demonstrated a strong association between this disorder and HLA-A29 (167). Of interest has been the observation that the two isotypes of HLA-A29 are associated with this disease. Further evaluation of this relationship is now actively being pursued with the use of transgenic mice expressing the HLA-A29 antigen.

8.3. Behçet's disease

This systemic disorder bears the name of the Turkish dermatologist Hulusi Behçet. The disease is diagnosed by four major criteria: genital and mucosal ulcers, skin eruptions, and uveitis (39). The disease is closely associated with HLA-B51. The ocular changes characteristically present with either an hypopyon and/or a severe occlusive retinitis. Arguably the most devastating part of the disease's major criteria is the severe inflammatory process that affects the poster portion of the eye. The retinal vessels typically become occluded and markedly attenuated, with ultimately the retina becoming atrophic. The characteristic aspect of the ocular attacks in this disease is that they disappear but then repeatedly return. The repeated attacks can ultimately lead to severe visual impairment.

Though it has been theorized that there is an exogenous trigger to this disorder, it is clear that multiple endogenous immune changes can be noted, including circulating interferon. Evidence of ocular autoimmunity has been studied and reported by many. Studies by Nussenblatt et al. (168) suggested that positive cell-mediated responses to the retinal S-antigen by some patients with this disease could be found. Circulating soluble IL-2 receptor has been found in Behçet's disease patients more often than in controls (169). The histology of this disorder would suggest that a focused inflammatory response is occurring against a retinal antigen(s).

8.4. Vogt-Koyanagi-Harada (VKH) Disease

The ocular manifestation of this syndrome is a bilateral granulomatous uveitis, while other symptoms include whitening of hair and eyelashes, meningeal irritation, dysacusis and tinnitus. The etiology of the disease is not known, but the possible participation of immunopathogenic mechanisms is indicated by (i) the similarity between VKH disease and sympathetic ophthalmia; (ii) the recent development of an animal model induced by melanin-derived proteins (96) and (iii) cellular and humoral immune responses of VKH patients toward various ocular antigens. It is noteworthy that antigens active in these immune responses include crude uveal extracts or melanocytes or their antigens.

8.5. Therapies

Therapies for these noninfectious disorders generally center around the use of immunosuppressive agents. While corticosteroids are still one of the mainstays of such therapy, it is interesting to note that therapies that are more specifically directed against various arms of immune system (but particularly the T-cell arm) have also been successful. Initial studies with cyclosporine demonstrated the effectiveness of this therapy (170), and recently the use of monoclonal antibodies, such as the one directed against the Tac portion of the interleukin-2 receptor, have also been used (171).

References

1. Waksman, B.H. 1962. Autoimmunization and the lesions of autoimmunity. *Medicine* 41:93–141.
2. Faure, J.P. 1980. Autoimmunity and the retina. *Curr. Topics. Eye Res.* 2:215–301.
3. Elschnig, A. 1910. Studien sur sympathischen ophthalmis. Die antigene wirkung des augenpigmentes. *Albrecht von Graefe's Arch. Ophthalmol.* 76:509–546.
4. Streilein, J.W. 1993. Immune privilege as the result of local tissue barriers and immunosuppressive microenvironments. *Curr. Opin. Immunol.* 5:428–432.
5. Streilein, J.W. 1999. Immunologic privilege of the eye. *Springer Semin. Immunopathol.* 21:95–111.
6. Griffith, T.S., and T.A. Ferguson. 1997. The role of FasL-induced apoptosis in immune privilege. *Immunol. Today* 18:240–244.
7. Hong, S., and L.V. Kaer. 1999. Immune privilege: Keeping an eye on natural killer T cells. *J. Exp. Med.* 190:1197–1200.
8. Sonoda, K.H., M. Exley, S. Snapper, S.P. Balk, and J. Stein-Streilein. 1999. CD1-reactive natural killer T cells are required for development of systemic tolerance through an immune-privileged site [see comments]. *J. Exp. Med.* 190:1215–1226.
9. de Kozak, Y., J. Sakai, B. Thillaye, and J.P. Faure. 1981. S antigen-induced experimental autoimmune uveo-retinitis in rats. *Curr. Eye Res.* 1:327–337.

10. Hirose, S., B. Wiggert, T.M. Redmond, T. Kuwabara, R.B. Nussenblatt, G.J. Chader, and I. Gery. 1987. Uveitis induced in primates by IRBP: humoral and cellular immune responses. *Exp. Eye Res.* 45:695–702.

11. Gregerson, D.S., J.W. Torseth, S.W. McPherson, J.P. Roberts, D. Shinohara, and D.J. Zack. 1999. Retinal expression of a neo-self antigen, beta-galactosidase, is not tolerogenic and creates a target for autoimmune uveoretinitis. *J. Immunol.* 163:1073–1080.

12. McPherson, S.W., J.P. Roberts, and D.S. Gregerson. 1999. Systemic expression of rat soluble retinal antigen induces resistance to experimental autoimmune uveoretinitis. *J. Immunol.* 163:4269–4276.

13. Xu, H., E.F. Wawrousek, T.M. Redmond, J.M. Nickerson, B. Wiggert, C.C. Chan, and R.R. Caspi. 2000. Transgenic expression of an immunologically privileged retinal antigen extraocularly enhances self tolerance and abrogrates susceptibility to autoimmune uveitis. *Eur. J. Immunol.* 30:272–278.

14. Goldschmidt, L., M. Goldbaum, S.M. Walker, and W.O. Weigle. 1982. The immune response to homologous lens crystallin. I. Antibody production after lens injury. *J. Immunol.* 129:1652–1657.

15. Gery, I., R. Nussenblatt, and D. BenEzra. 1981. Dissociation between humoral and cellular immune responses to lens antigens. *Invest. Ophthalmol. Vis. Sci.* 20:32–39.

16. Srinivasan, A.N., C.N. Nagineni, and S.P. Bhat. 1992. alpha A-crystallin is expressed in non-ocular tissues. *J. Biol. Chem.* 267:23337–23341.

17. Lai, J.C., A. Fukushima, E.F. Wawrousek, M.C. Lobanoff, P. Charukamnoetkanok, S.J. Smith-Gill, B.P. Vistica, R.S. Lee, C.E. Egwuagu, S.M. Whitcup, and I. Gery. 1998. Immunotolerance against a foreign antigen transgenically expressed in the lens. *Invest. Ophthalmol. Vis. Sci.* 39:2049–2057.

18. Charukamnoetkanok, P., A. Fukushima, S.M. Whitcup, I. Gery, and C.E. Egwuagu. 1998. Expression of ocular autoantigens in the mouse thymus. *Curr. Eye Res.* 17:788–792.

19. Egwuagu, C.E., P. Charukamnoetkanok, and I. Gery. 1997. Thymic expression of autoantigens correlates with resistance to autoimmune disease. *J. Immunol.* 159:3109–3112.

20. Caspi, R.R., B.G. Grubbs, C.C. Chan, G.J. Chader, and B. Wiggert. 1992. Genetic control of susceptibility to experimental autoimmune uveoretinitis in the mouse model: Concomitant regulation by MHC and non-MHC genes. *J. Immunol.* 148:2384–2389.

21. Gery, I., M.P. Gelderman, and M.C. Kennedy. 1999. Thymic medullary dendritic cells express retinal and lens specific antigens. *Invest. Ophthalmol. Vis. Sci.* 40:S861.

22. Sprent, J., and S.R. Webb. 1995. Intrathymic and extrathymic clonal deletion of T Cells. *Curr. Opin. Immunol.* 7:196–205.

23. Hirose, S., T. Tanaka, R.B. Nussenblatt, A.G. Palestine, B. Wiggert, T.M. Redmond, G.J. Chader, and I. Gery. 1988. Lymphocyte responses to retinal-specific antigens in uveitis patients and healthy subjects. *Curr. Eye Res.* 7:393–402.

24. Mochizuki, M., T. Kuwabara, C. McAllister, R.B. Nussenblatt, and I. Gery. 1985. Adoptive transfer of experimental autoimmune uveoretinitis in rats. Immunopathogenic mechanisms and histologic features. *Invest. Ophthalmol. Vis. Sci.* 26:1–9.

25. Caspi, R.R., F.G. Roberge, C.G. McAllister, M. el Saied, T. Kuwabara, I. Gery, E. Hanna, and R.B. Nussenblatt. 1986. T cell lines mediating experimental autoimmune uveoretinitis (EAU) in the rat. *J. Immunol.* 136:928–933.

26. Baron, J.L., J.A. Madri, N.H. Ruddle, G. Hashim, and C.A. Janeway, Jr. 1993. Surface expression of alpha 4 integrin by CD4 T cells is required for their entry into brain parenchyma. *J. Exp. Med.* 177:57–68.

27. Fukushima, A., B.P. Vistica, R.R. Caspi, E.F. Wawrousek, A.T. Kozhich, S.M. Whitcup, and I. Gery. 1996. Lymphocyte stimulation and expression of cell adhesion molecules are critical for adoptive transfer of cellular immunity and disease. *Invest. Ophthalmol. Vis. Sci.* 37:S540.

28. Renno, T., and H. Acha-Orbea. 1996. Superantigens in autoimmune diseases: still more shades of gray. *Immunol. Rev.* 154:175–191.

29. Beutler, B., and A. Cerami. 1989. The biology of cachectin/TNF—a primary mediator of the host response. *Annu. Rev. Immunol.* 7:625–655.

30. Barnaba, V., and F. Sinigaglia. 1997. Molecular mimicry and T cell-mediated autoimmune disease. *J. Exp. Med.* 185:1529–1531.

31. Oldstone, M.B. 1998. Molecular mimicry and immune-mediated diseases. *FASEB J.* 12:1255–1265.

32. Singh, V.K., K. Yamaki, T. Abe, and T. Shinohara. 1989. Molecular mimicry between uveitopathogenic site of retinal S-antigen and Escherichia coli protein: induction of experimental autoimmune uveitis and lymphocyte cross-reaction. *Cell Immunol.* 122:262–273.

33. Singh, V.K., K. Yamaki, L.A. Donoso, and T. Shinohara. 1989. Molecular mimicry. Yeast histone H3-induced experimental autoimmune uveitis. *J. Immunol.* 142:1512–1517.

34. Singh, V.K., H.K. Kalra, K. Yamaki, T. Abe, L.A. Donoso, and T. Shinohara. 1990. Molecular mimicry between a uveitopathogenic site of S-antigen and viral peptides. Induction of experimental autoimmune uveitis in Lewis rats. *J. Immunol.* 144:1282–1287.

35. Singh, D.P., T. Sueno, T. Kikuchi, S.C. Guru, S. Yu, J. Horwitz, L.T. Chylack, Jr., and T. Shinohara. 1999. Antibodies to a microbial peptide sharing sequence homology with betaA3-crystallin damage lens epithelial cells in vitro and in vivo. *Autoimmunity* 29:311–322.

36. Ichikawa, T., O. Taguchi, T. Takahashi, H. Ikeda, M. Takeuchi, T. Tanaka, M. Usui, and Y. Nishizuka. 1991. Spontaneous development of autoimmune uveoretinitis in nude mice following reconstitution with embryonic rat thymus. *Clin. Exp. Immunol.* 86:112–117.

37. Nishigaki-Maki, K., T. Takahashi, K. Ohno, T. Morimoto, H. Ikeda, M. Takeuchi, M. Ueda, and O. Taguchi. 1999. Autoimmune diseases developed in athymic nude mice grafted with embryonic thymus of xenogeneic origin. *Eur. J. Immunol.* 29:3350–3359.

38. Sakaguchi, S., N. Sakaguchi, M. Asano, M. Itoh, and M. Toda. 1995. Immunologic self-tolerance maintained by activated T cells expressing IL-2 receptor alpha-chains (CD25). Breakdown of a single mechanism of self-tolerance causes various autoimmune diseases. *J. Immunol.* 155:1151–1164.

39. Nussenblatt, R.B., S.M. Whitcup, and A.G. Palestine. 1996. In: Uveitis: Fundamentals and Clinical Practice. Mosby-Year Book, Inc., St. Louis, MO.

40. Caspi, R.R. 1992. Immunogenetic aspects of clinical and experimental uveitis. *Reg. Immunol.* 4:321–330.

41. Wacker, W.B., L.A. Donoso, C.M. Kalsow, J.A. Yankeelov, Jr., and D.T. Organisciak. 1977. Experimental allergic uveitis. Isolation, characterization, and localization of a soluble uveitopathogenic antigen from bovine retina. *J. Immunol.* 119:1949–1958.

42. Wilden, U., S.W. Hall, and H. Kuhn. 1986. Phosphodiesterase activation by photoexcited rhodopsin is quenched when rhodopsin is phosphorylated and binds the intrinsic 48-kDa protein of rod outer segments. *Proc. Natl. Acad. Sci. USA* 83:1174–1178.

43. Pfister, C., M. Chabre, J. Plouet, V.V. Tuyen, Y. De Kozak, J.P. Faure, and H. Kuhn. 1985. Retinal S Antigen identified as the 48K protein regulating light-dependent phosphodiesterase in rods. *Science* 228:891–893.

44. Mirshahi, M., C. Boucheix, G. Collenot, B. Thillaye, and J.P. Faure. 1985. Retinal S-antigen epitopes in vertebrate and invertebrate photoreceptors. *Invest. Ophthalmol, Vis. Sci.* 26:1016–1021.

45. Lieb, W.E., L. Smith-Lang, H.S. Dua, A.C. Christensen, and L.A. Donoso. 1991. Identification of an S-antigen-like molecule in Drosophila melanogaster: an immunohistochemical study. *Exp. Eye Res.* 53:171–178.

46. Shinohara, T., L. Donoso, M. Tsuda, K. Yamaki, and V.K. Singh. 1988. S-antigen: structure, function, and experimental autoimmune uveitis (EAU). *Progress in Retinal Res.* 8:51–66.

47. Lohse, M.J., J.L. Benovic, J. Codina, M.G. Caron, and R.J. Lefkowitz. 1990. beta-Arrestin: a protein that regulates beta-adrenergic receptor function. *Science* 248:1547–1550.

48. Faure, J.P., Y. de Kozak, C. Dorey, and V.V. Tuyen. 1977. Activite de differentes preparations antigeniques de la retine dans l'induction de l'uveo-retinite autoimmune experimentale. *Arch. Ophthalmol.* 37:47–60.

49. Nussenblatt, R.B., T. Kuwabara, F.M. de Monasterio, and W.B. Wacker. 1981. S-antigen uveitis in primates. A new model for human disease. *Arch. Ophthalmol.* 99:1090–1092.

50. Fox, G.M., T., Kuwabara, B. Wiggert, T.M. Redmond, H.H. Hess, G.J. Chader, and I. Gery. 1987. Experimental autoimmune uveoretinitis (EAU) induced by retinal interphotoreceptor retinoid-binding protein (IRBP): differences between EAU induced by IRBP and by S-antigen. *Clin. Immunol. Immunopathol.* 43:256–264.

51. Caspi, R.R., F.G. Roberge, C.C. Chan, B. Wiggert, G.J. Chader, L.A. Rozenszajn, Z. Lando, and R.B. Nussenblatt. 1988. A new model of autoimmune disease. Experimental autoimmune uveoretinitis induced in mice with two different retinal antigens. *J. Immunol.* 140:1490–1495.

52. Nussenblatt, R.B., I. Gery, E.J. Ballintine, and W.B. Wacker. 1980. Cellular immune responsiveness of uveitis patients to retinal S-antigen. *Am. J. Ophthalmol.* 89:173–179.

53. de Smet, M.D., J.H. Yamamoto, M. Mochizuki, I. Gery, V.K. Singh, T. Shinohara, B. Wiggert, G.J. Chader, and R.B. Nussenblatt. 1990. Cellular immune responses of patients with uveitis to retinal antigens and their fragments. *Am. J. Ophthalmol.* 110:135–142.

54. Doekes, G., L. Luyendijk, M.J. Gerritsen, and A. Kijlstra. 1992. Anti-retinal S-antigen antibodies in human sera: a comparison of reactivity in ELISA with human or bovine S-antigen. *Int. Ophthalmol.* 16:147–152.

55. Gregerson, D.S., I.W. Abrahams, and C.E. Thirkill. 1981. Serum antibody levels of uveitis patients to bovine retinal antigens. *Invest. Ophthalmol. Vis. Sci.* 21:669–680.

56. Donoso, L.A., C.F. Merryman, T. Shinohara, B. Dietzschold, G. Wistow, C. Craft, W. Morley, and R.T. Henry. 1986. S-antigen: identification of the MAbA9-C6 monoclonal antibody binding site and the uveitopathogenic sites. *Curr. Eye Res.* 5:995–1004.

57. Hirose, S., V.K. Singh, L.A. Donoso, T. Shinohara, S. Kotake, T. Tanaka, T. Kuwabara, K. Yamaki, I. Gery, and R.B. Nussenblatt. 1989. An 18-mer peptide derived from the retinal S antigen induces uveitis and pinealitis in primates. *Clin. Exp. Immunol.* 77:106–111.

58. Gregerson, D.S., C.F. Merryman, W.F. Obritsch, and L.A. Donoso. 1990. Identification of a potent new pathogenic site in human retinal S-antigen which induces experimental autoimmune uveoretinitis in LEW rats. *Cell Immunol.* 128:209–219.

59. Merryman, C.F., L.A. Donoso, X.M. Zhang, E. Heber-Katz, and D.S. Gregerson. 1991. Characterization of a new, potent, immunopathogenic epitope in S-antigen that elicits T cells expressing V beta 8 and V alpha 2-like genes. *J. Immunol.* 146:75–80.

60. de Smet, M.D., G. Bitar, F.G. Roberge, I. Gery, and R.B. Nussenblatt. 1993. Human S-antigen: Presence of multiple immunogenic and immunopathogenic sites in the Lewis rat. *J. Autoimmun.* 6:587–599.

61. Chader, G.J. 1989. Interphotoreceptor retinoid-binding protein (IRBP): a model protein for molecular biological and clinically relevant studies. Friedenwald lecture. *Invest. Ophthalmol. Vis. Sci.* 30:7–22.

62. Kotake, S., B. Wiggert, T.M. Redmond, D.E. Borst, J.M. Nickerson, H. Margalit, J.A. Berzofsky, G.J. Chader, and I. Gery. 1990. Repeated determinants within the retinal interphotoreceptor retinoid-binding protein (IRBP): immunological properties of the repeats of an immunodominant determinant. *Cell. Immunol.* 126:331–342.

63. Kozhich, A.T., Y. Kawano, C.E. Egwuagu, R.R. Caspi, R.K. Maturi, J.A. Berzofsky, and I. Gery. 1994. A pathogenic autoimmune process targeted at a surrogate epitope. *J. Exp. Med.* 180:133–140.

64. Wiggert, B., L. Lee, M. Rodrigues, H. Hess, T.M. Redmond, and G.J. Chader. 1986. Immunochemical distribution of interphotoreceptor retinoid-binding protein in selected species. *Invest. Ophthalmol. Vis. Sci.* 27:1041–1049.

65. Gelderman, M.P., F. Gonzales-Fernandez, C.A. Baer, B. Wiggert, C.C. Chan, B.P. Vistica, and I. Gery. 2000. Xenopus IRBP, a phylogenetically remote protein, is uveitogenic in Lewis rats. *Exp. Eye Res.* 70:731–736.

66. Gery, I., B. Wigger, T.M. Redmond, T. Kuwabara, M.A. Crawford, B.P. Vistica, and G.J. Chader. 1986. Uveoretinitis and pinealitis induced by immunization with interphotoreceptor retinoid-binding protein. *Invest. Ophthalmol. Vis. Sci.* 27:1296–1300.

67. Broekhuyse, R.M., H.J. Winkens, and E.D. Kuhlmann. 1986. Induction of experimental autoimmune uveoretinitis and pinealities by IRBP. Comparison to uveoretinitis induced by S-antigen and opsin. *Curr. Eye Res.* 5:231–240.

68. Hirose, S., T. Kuwabara, R.B. Nussenblatt, B. Wiggert, T.M. Redmond, and I. Gery. 1986. Uveitis induced in primates by interphotoreceptor retinoid-binding protein. *Arch. Ophthalmol.* 104:1698–1702.

69. Eisenfeld, A.J., A.H. Bunt-Milam, and J.C. Saari. 1987. Uveoretinitis in rabbits following immunization with interphotoreceptor retinoid-binding protein. *Exp. Eye Res.* 44:425–438.

70. Vistica, B.P., M. Usui, T. Kuwabara, B. Wiggert, L. Lee, T.M. Redmond, G.J. Chader, and I. Gery. 1987. IRBP from bovine retina is poorly uveitogenic in guinea pigs and is identical to A-antigen. *Curr. Eye Res.* 6:409–417.

71. Donoso, L.A., C.F. Merryman, T. Sery, R. Sanders, T. Vrabec, and S.L. Fong. 1989. Human interstitial retinoid binding protein. A potent uveitopathogenic agent for the

induction of experimental autoimmune uveitis. *J. Immunol.* 143:79–83.

72. Margalit, H., J.L. Spouge, J.L. Cornette, K.B. Cease, C. Delisi, and J.A. Berzofsky. 1987. Prediction of immunodominant helper T cell antigenic sites from the primary sequence. *J. Immunol.* 138:2213–2229.

73. Kotake, S., T.M. Redmond, B. Wiggert, B. Vistica, H. Sanui, G.J. Chader, and I. Gery. 1991. Unusual immunologic properties of the uveitogenic interphotoreceptor retinoid-binding protein-derived peptide R23. *Invest. Ophthalmol. Vis. Sci.* 32:2058–2064.

74. Sanui, H., T.M. Redmond, L.H. Hu, T. Kuwabara, H. Margalit, J.L. Cornette, B. Wiggert, G.J. Chader, and I. Gery. 1988. Synthetic peptides derived from IRBP induce EAU and EAP in Lewis rats. *Curr. Eye Res.* 7:727–735.

75. Sanui, H., T.M. Redmond, S. Kotake, B. Wiggert, L.H. Hu, H. Margalit, J.A. Berzofsky, G.J. Chader, and I. Gery. 1989. Identification of an immunodominant and highly immunopathogenic determinant in the retinal interphotoreceptor retinoid-binding protein (IRBP). *J. Exp. Med.* 169:1947–1960.

76. Sanui, H., T.M. Redmond, S. Kotake, B. Wiggert, T. Tanaka, G.J. Chader, and I. Gery. 1990. Uveitis and immune responses in primates immunized with IRBP-derived synthetic peptides. *Curr. Eye Res.* 9:193–199.

77. Kotake, S., M.D. de Smet, B. Wiggert, T.M. Redmond, G.J. Chader, and I. Gery. 1991. Analysis of the pivotal residues of the immunodominant and highly uveitogenic determinant of interphotoreceptor retinoid-binding protein. *J. Immunol.* 146:2995–3001.

78. Kozhich, A.T., R.R. Caspi, J.A. Berzofsky, and I. Gery. 1997. Immunogenicity and immunopathogenicity of an autoimmune epitope are potentiated by increasing MHC binding through residue substitution. *J. Immunol.* 158:4145–4151.

79. Silver, P.B., L.V. Rizzo, C.C. Chan, L.A. Donoso, B. Wiggert, and R.R. Caspi. 1995. Identification of a major pathogenic epitope in the human IRBP molecule recognized by mice of the H-2r haplotype. *Invest. Ophthalmol. Vis. Sci.* 36:946–954.

80. Avichezer, D., P.B. Silver, C.C. Chan, B. Wiggert, and R.R. Caspi. 2000. Identification of a New Epitope of Human IRBP that Induces Autoimmune Uveoretinitis in mice of the H-2^b Haplotype. *Invest. Ophthalmol. Vis. Sci.* 41:127–131.

81. Namba, K., K. Ogasawara, N. Kitaichi, N. Matsuki, A. Takahashi, Y. Sasamoto, S. Kotake, H. Matsuda, K. Iwabuchi, S. Ohno, and K. Onoe. 1998. Identification of a peptide inducing experimental autoimmune uveoretinitis (EAU) in H-2Ak-carrying mice. *Clin. Exp. Immunol.* 111:442–449.

82. Takeuchi, M., T. Kezuka, H. Inoue, J. Sakai, M. Usui, T. Takahashi, and O. Taguchi. 1998. Suppression of spontaneous uveoretinitis development by non-immunopathogenic peptide immunization. *Eur. J. Immunol.* 28:1578–1586.

83. Schalken, J.J., A.H. van Vugt, H.J. Winkens, P.H. Bovee-Geurts, W.J. De Grip, and R.M. Broekhyse. 1988. Experimental autoimmune uveoretinitis in rats induced by rod visual pigment: rhodopsin is more pathogenic than opsin. *Graefes Arch. Clin. Exp. Ophthalmol.* 226:255–261.

84. Schalken, J.J., H.J. Winkens, A.H. Van Vugt, W.J. De Grip, and R.M. Broekhuyse. 1989. Rhodopsin-induced experimental autoimmune uveoretinitis in monkeys. *Br. J. Ophthalmol.* 73:168–172.

85. Adamus, G., J.L. Schmied, P.A. Hargrave, A. Arendt, and E.J. Moticka. 1992. Induction of experimental autoimmune

uveitis with rhodopsin synthetic peptides in Lewis rats. *Curr. Eye Res.* 11:657–667.

86. Gery, I., N.P. Chanaud, 3rd, and E. Anglade. 1994. Recoverin is highly uveitogenic in Lewis rats. *Invest. Ophthalmol. Vis. Sci.* 35:3342–3345.

87. Adamus, G., H. Ortega, D. Witkowska, and A. Polans. 1994. Recoverin: a potent uveitogen for the induction of photoreceptor degeneration in Lewis rats. *Exp. Eye Res.* 59:447–455.

88. Keltner, J.L., and C.E. Thirkill. 1998. Cancer-associated retinopathy vs recoverin-associated retinopathy. *Am. J. Ophthalmol.* 126:296–302.

89. Whitcup, S.M., B.P. Vistica, A.H. Milam, R.B. Nussenblatt, and I. Gery. 1998. Recoverin-associated retinopathy: a clinically and immunologically distinctive disease. *Am. J. Ophthalmol.* 126:230–237.

90. Lee, R.H., B.S. Lieberman, and R.N. Lolley. 1987. A novel complex from bovine visual cells of a 33,000-dalton phosphoprotein with beta- and gamma-transducin: purification and subunit structure. *Biochemistry* 26:3983–3990.

91. Dua, H.S., R.H. Lee, R.N. Lolley, J.A. Barrett, M. Abrams, J.V. Forrester, and L.A. Donoso. 1992. Induction of experimental autoimmune uveitis by the retinal photoreceptor cell protein, phosducin. *Curr. Eye Res.* 11:107–111.

92. Satoh, N., T. Abe, A. Nakajima, M. Ohkoshi, T. Koizumi, H. Tamada, and S. Sakuragi. 1998. Analysis of uveitogenic sites in phosducin molecule. *Curr. Eye Res.* 17:677–686.

93. Broekhuyse, R.M., E.D. Kuhlmann, H.J. Winkens, and A.H. Van Vugt. 1991. Experimental autoimmune anterior uveitis (EAAU), a new form of experimental uveitis. I. Induction by a detergent-insoluble, intrinsic protein fraction of the retinal pigment epithelium. *Exp. Eye Res.* 52:465–474.

94. Broekhuyse, R.M., E.D. Kuhlmann, and H.J. Winkens. 1996. Experimental melanin-protein induced uveitis (EMIU) is the sole type of uveitis evoked by a diversity of ocular melanin preparations and melanin-derived soluble polypeptides. *Jpn. J. Ophthalmol.* 40:459–468.

95. Simpson, S.C., H.J. Kaplan, and N.S. Bora. 1997. Uveitogenic proteins isolated from bovine iris and ciliary body. *Eye* 11:206–208.

96. Yamaki, K., K. Hayakawa, H. Nakamura, M. Miyano, I. Kondo, and K. Gocho. 1999. Experimental Vogt-Koyanagi-Harada disease induced in pigmented rats. *Invest. Ophthalmol. Vis. Sci.* 40:S400.

97. Ren, J., V.A. Bonderenko, A. Yamazaki, and H. Shichi. 1996. Experimental autoimmune uveoretinitis induced by the gamma-subunit of cyclic guanosine monophosphate phosphodiesterase in rats. *Invest. Ophthalmol. Vis. Sci.* 37:2527–2531.

98. Kojima, K., T. Berger, H. Lassmann, D. Hinze-Selch, Y. Zhang, J. Gehrmann, K. Reske, H. Wekerle, and C. Linington. 1994. Experimental autoimmune panencephalitis and uveoretinitis transferred to the Lewis rat by T lymphocytes specific for the S100 beta molecule, a calcium binding protein of astroglia. *J. Exp. Med.* 180:817–829.

99. Caspi, R.R. 1999. Immune mechanisms in uveitis. *Springer Semin. Immunopathol.* 21:113–124.

100. Marak, G.E., Jr. 1992. Phacoanaphylactic endophthalmitis. *Surv. Ophthalmol.* 36:325–339.

101. Adamus, G., M. Machnicki, H. Elerding, B. Sugden, Y.S. Blocker, and D.A. Fox. 1998. Antibodies to recoverin induce apoptosis of photoreceptor and bipolar cells *in vivo*. *J. Autoimmun.* 11:523–533.

102. Ohguro, H., K. Ogawa, T. Maeda, A. Maeda, and I. Maruyama. 1999. Cancer-associated retinopathy induced by

both anti-recoverin and anti- hsc70 antibodies *in vivo*. *Invest. Ophthalmol. Vis. Sci.* 40:3160–3167.

103. de Kozak, Y., J. Sainte-Laudy, J. Benveniste, and J.P. Faure. 1981. Evidence for immediate hypersensitivity phenomena in experimental autoimmune uveoretinitis. *Eur. J. Immunol.* 11:612–617.

104. Kasp, E., R. Whiston, D. Dumonde, E. Graham, M. Stanford, and M. Sanders. 1992. Antibody affinity to retinal S-antigen in patients with retinal vasculitis. *Am. J. Ophthalmol.* 113:697–701.

105. Mochizuki, M., T. Kuwabara, C.C. Chan, R.B. Nussenblatt, D.D. Metcalfe, and I. Gery. 1984. An association between susceptibility to experimental autoimmune uveitis and choroidal mast cell numbers. *J. Immunol.* 133:1699–1701.

106. Li, Q., Y. Fujino, R.R. Caspi, F. Najafian, R.B. Nussenblatt, and C.C. Chan. 1992. Association between mast cells and the development of experimental autoimmune uveitis in different rat strains. *Clin. Immunol. Immunopathol.* 65:294–299.

107. de Kozak, Y., J. Sakai, J. Sainte-Laudy, J.P. Faure, and J. Benveniste. 1983. Pharmacological modulation of IgE-dependent mast cell degranulation in experimental autoimmune uveoretinitis. *Jpn. J. Ophthalmol.* 27:598–608.

108. Whitcup, S.M., L.R. DeBarge, R.R. Caspi, R. Harning, R.B. Nussenblatt, and C.C. Chan. 1993. Monoclonal antibodies against ICAM-1 (CD54) and LFA-1 (CD11a/CD18) inhibit experimental autoimmune uveitis. *Clin. Immunol. Immunopathol.* 67:143–150.

109. Whitcup, S.M., C.C. Chan, Q. Li, and R.B. Nussenblatt. 1992. Expression of cell adhesion molecules in posterior uveitis. *Arch. Ophthalmol.* 110:662–666.

110. Broekhuyse, R.M., E.D. Kuhlmann, and H.J. Winkens. 1992. Experimental autoimmune anterior uveitis (EAAU). II. Dose-dependent induction and adoptive transfer using a melanin-bound antigen of the retinal pigment epithelium. *Exp. Eye Res.* 55:401–411.

111. Salinas-Carmona, M.C., R.B. Nussenblatt, and I. Gery. 1982. Experimental autoimmune uveitis in the athymic nude rat. *Eur. J. Immunol.* 12:480–484.

112. Caspi, R.R., P.B. Silver, C.C. Chan, B. Sun, R.K. Agarwal, J. Wells, S. Oddo, Y. Fujino, F. Najafian, and R.L. Wilder. 1996. Genetic susceptibility to experimental autoimmune uveoretinitis in the rat is associated with an elevated Th1 response. *J. Immunol.* 157:2668–2675.

113. Xu, H., L.V. Rizzo, P.B. Silver, and R.R. Caspi. 1997. Uveitogenicity is associated with a Th1-like lymphokine profile: cytokine-dependent modulation of primary and committed T cells in EAU. *Cell. Immunol.* 178:69–78.

114. Saoudi, A., J. Kuhn, K. Huygen, Y. de Kozak, T. Velu, M. Goldman, P. Druet, and B. Bellon. 1993. TH2 activated cells prevent experimental autoimmune uveoretinitis, a TH1-dependent autoimmune disease. *Eur. J. Immunol.* 23:3096–3103.

115. Jones, L.S., L.V. Rizzo, R.K. Agarwal, T.K. Tarrant, C.C. Chan, B. Wiggert, and R.R. Caspi. 1997. IFN-gamma-deficient mice develop experimental autoimmune uveitis in the context of a deviant effector response. *J. Immunol.* 158:5997–6005.

116. Egwuagu, C.E., C. Chow, E. Beraud, R.R. Caspi, R.M. Mahdi, A.P. Brezin, R.B. Nussenblatt, and I. Gery. 1991. T cell receptor beta-chain usage in experimental autoimmune uveoretinitis. *J. Autoimmun.* 4:315–324.

117. Gregerson, D.S., S.P. Fling, C.F. Merryman, X.M. Zhang, X.B. Li, and E. Heber-Katz. 1991. Conserved T cell receptor V gene usage by uveitogenic T cells. *Clin. Immunol. Immunopathol.* 58:154–161.

118. Egwuagu, C.E., S. Bahmanyar, R.M. Mahdi, R.B. Nussenblatt, I. Gery, and R.R. Caspi. 1992. Predominant usage of V beta 8.3 T cell receptor in a T cell line that induces experimental autoimmune uveoretinitis (EAU). *Clin. Immunol. Immunopathol.* 65:152–160.

119. Heber-Katz, E., and H. Acha-Orbea. 1989. The V-region disease hypothesis: evidence from autoimmune encephalomyelitis. *Immunol. Today* 10:164–169.

120. Egwuagu, C.E., R.M. Mahdi, R.B. Nussenblatt, I. Gery, and R.R. Caspi. 1993. Evidence for selective accumulation of V beta 8+ T lymphocytes in experimental autoimmune uveoretinitis induced with two different retinal antigens. *J. Immunol.* 151:1627–1636.

121. Charteris, D.G., and S.L. Lightman. 1993. *In vivo* lymphokine production in experimental autoimmune uveoretinitis. *Immunology* 78:387–392.

122. Li, Q., B. Sun, D.M. Matteson, T.P. O'Brien, and C.C. Chan. 1999. Cytokines and apoptotic molecules in experimental melanin-protein induced uveitis (EMIU) and experimental autoimmune uveoretinitis (EAU). *Autoimmunity* 30:171–182.

123. Caspi, R.R., C.C. Chan, B.G. Grubbs, P.B. Silver, B. Wiggert, C.F. Parsa, S. Bahmanyar, A. Billiau, and H. Heremans. 1994. Endogenous systemic IFN-gamma has a protective role against ocular autoimmunity in mice. *J. Immunol.* 152:890–899.

124. Tarrant, T.K., P.B. Silver, J.L. Wahlsten, L.V. Rizzo, C.C. Chan, B. Wiggert, and R.R. Caspi. 1999. Interleukin 12 protects from a T helper type 1-mediated autoimmune disease, experimental autoimmune uveitis, through a mechanism involving interferon gamma, nitric oxide, and apoptosis. *J. Exp. Med.* 189:219–230.

125. Rizzo, L.V., H. Xu, C.C. Chan, B. Wiggert, and R.R. Caspi. 1998. IL-10 has a protective role in experimental autoimmune uveoretinitis. *Int. Immunol.* 10:807–814.

126. Tarrant, T.K., P.B. Silver, C.C. Chan, B. Wiggert, and R.R. Caspi. 1998. Endogenous IL-12 is required for induction and expression of experimental autoimmune uveitis. *J. Immunol.* 161:122–127.

127. Mantovani, A. 1999. The chemokine system: redundancy for robust outputs. *Immunol. Today* 20:254–257.

128. Larkin, G., S.M. Whitcup, I. Gery, A. Proudfoot, and M.T. Magone. 1999. Chemokine antagonist in murine models of ocular inflammation. *Invest. Ophthalmol. Vis. Sci.* 40:S139.

129. Prendergast, R.A., C.E. Iliff, N.M. Coskuncan, R.R. Caspi, G. Sartani, T.K. Tarrant, G.A. Lutty, and D.S. McLeod. 1998. T cell traffic and the inflammatory response in experimental autoimmune uveoretinitis. *Invest. Ophthalmol. Vis. Sci.* 39:754–762.

130. McMenamin, P.G., R.M. Broekhuyser, and J.V. Forrester. 1993. Ultrastructural pathology of experimental autoimmune uveitis: a review. *Micron* 24:521–546.

131. Caspi, R.R., C.C. Chan, Y. Fujino, F. Najafian, S. Grover, C.T. Hansen, and R.L. Wilder. 1993. Recruitment of antigen-nonspecific cells plays a pivotal role in the pathogenesis of a T cell-mediated organ-specific autoimmune disease, experimental autoimmune uveoretinitis. *J. Neuroimmunol.* 47:177–188.

132. McMenamin, P.G., J.V. Forrester, R.J. Steptoe, and H.S. Dua. 1992. Ultrastructural pathology of experimental autoimmune uveitis. Quantitative evidence of activation and possible high endothelial venule-like changes in retinal vascular endothelium. *Lab. Invest.* 67:42–55.

133. Greenwood, J., R. Howes, and S. Lightman. 1994. The blood-retinal barrier in experimental autoimmune uveore-

tinitis. Leukocyte interactions and functional damage. *Lab. Invest.* 70:39–52.

134. Hoey, S., P.S. Grabowski, S.H. Ralston, J.V. Forrester, and J. Liversidge. 1997. Nitric oxide accelerates the onset and increases the severity of experimental autoimmune uveoretinitis through an IFN-gamma-dependent mechanism. *J. Immunol.* 159:5132–5142.

135. Wu, G.S., J. Zhang, and N.A. Rao. 1997. Peroxynitrite and oxidative damage in experimental autoimmune uveitis. *Invest. Ophthalmol. Vis. Sci.* 38:1333–1339.

136. Collins, R.C. 1949. Experimental studies in sympathetic ophthalmia. *Am. J. Ophthalmol.* 32:1687–1699.

137. Wacker, W.B., and M.M. Lipton. 1968. Experimental allergic uveitis. I. Production in the guinea pig and rabbit by immunization with retina in adjuvant. *J. Immunol.* 101:151–156.

138. Chan, C.C., N. Hikita, K. Dastgheib, S.M. Whitcup, I. Gery, and R.B. Nussenblatt. 1994. Experimental melanin-protein-induced uveitis in the Lewis rat. Immunopathologic processes. *Ophthalmology* 101:1275–1280.

139. Gery, I., M. Mochizuki, and R.B. Nussenblatt. 1986. Retinal specific antigens and immunopathogenic processes they provoke. In: Progress in Retinal Research, N. Osborne, and J. Chader, editors. Pergamon Press, Oxford, pp. 75–109.

140. Silver, P.B., C.C. Chan, B. Wiggert, and R.R. Caspi. 1999. The requirement for pertussis to induce EAU is strain-dependent: B10.RIII, but not B10. A mice, develop EAU and Th1 responses to IRBP without pertussis treatment. *Invest. Ophthalmol. Vis. Sci.* 40:2898–2905.

141. Chan, C.C., R.R. Caspi, M. Ni, W.C. Leake, B. Wiggert, G.J. Chader, and R.B. Nussenblatt. 1990. Pathology of experimental autoimmune uveoretinitis in mice. *J. Autoimmun.* 3:247–255.

142. Mochizuki, M., J. Charley, T. Kuwabara, R.B. Nussenblatt, and I. Gery. 1983. Involvement of the pineal gland in rats with experimental autoimmune uveitis. *Invest. Ophthalmol. Vis. Sci.* 24:1333–1338.

143. Kalsow, C.M., and W.B. Wacker. 1985. Pineal gland involvement in S-antigen-induced experimental allergic uveitis. In: *Pineal and retinal relationships*, P.J. O'Brien, and D.C. Klein, editors. Academic Press, Orlando. pp. 315–329.

144. Gery, I., S. Hirose, C. McAllister, G. Fox, B. Wiggert, T.M. Redmond, C. G.J., and T. Kuwabara. 1989. Differences between the inflammatory reactions in the retina and pineal gland in rats with EAU. Eds., A.G. Secchi, and I.A. Fregona). Masson, Milano, pp. 89–91.

145. Chan, C.C., K. Dastgheib, N. Hikita, R.C. Walton, R.B. Nussenblatt, and R.M. Broekhuyse. 1994. Experimental melanin-protein induced uveitis ("EMIU", formerly "EAAU"); Immunopathology, susceptibility and therapy. In: Advances in Ocular Immunology, R.B. Nussenblatt, S.M. Whitcup, R.R. Caspi, and I. Gery, editors. Elservier Science, Amsterdam. pp. 91–94.

146. Marak, G.E., Jr., N.A. Rao, G. Antonakou, and A. Sliwinski. 1982. Experimental lens-induced granulomatous endophthalmitis in common laboratory animals. *Ophthalmic Res.* 14:292–297.

147. Fischer, C.A. 1971. Lens-induced uveitis and secondary glaucoma in a dog. *J. Am. Vet. Med. Assoc.* 158:336–341.

148. Lai, J.C., M.C. Lobanoff, A. Fukushima, E.F. Wawrousek, C.C. Chan, S.M. Whitcup, and I. Gery. 1999. Uveitis induced by lymphocytes sensitized against a transgenically expressed lens protein. *Invest. Ophthalmol. Vis. Sci.* 40:2735–2739.

149. Gery, I., W.G. Robinson, Jr., H. Shichi, M. El-Saied, M. Mochizuki, R.B. Nussenblatt, and R.M. Williams. 1985. Differences in susceptibility to experimental autoimmune uveitis among rats of various strains. In: Advances in Immunology and Immunopathology of the Eye (Proceedings of the Third International Symposium on Immunology and Immunopathology of the Eye), J.W. Chandler, and G.R. O'Conner, editors. Masson Publishing, NY. pp. 242–245.

150. Hirose, S. 1990. [Genetic control of experimental autoimmune uveitis in rats]. *Hokkaido Igaku Zasshi.* 65:604–611.

151. Pennesi, G., S.H. Sun, C.S. David, P.A. Hargrave, H. McDowell, B. Wiggert, C.C. Chan, and R.R. Caspi. 2000. A humanized model of experimental autoimmune uveitis in HLA transgenic mice. *Invest. Ophthalmol. Vis. Sci.* 41:S106.

152. de Kozak, Y., M.C. Naud, J. Bellot, J.P. Faure, and D. Hicks. 1994. Differential tumor necrosis factor expression by resident retinal cells from experimental uveitis-susceptible and-resistant rat strains. *J. Neuroimmunol.* 55:1–9.

153. Sternberg, E.M., J.M. Hill, G.P. Chrousos, T. Kamilaris, S.J. Listwak, P.W. Gold, and R.L. Wilder. 1989. Inflammatory mediator-induced hypothalamic-pituitary- adrenal axis activation is defective in streptococcal cell wall arthritis-susceptible Lewis rats. *Proc. Natl. Acad. Sci. USA* 86:2374–2378.

154. Wilder, R.L. 1995. Neuroendocrine-immune system interactions and autoimmunity. *Annu. Rev. Immunol.* 13:307–338.

155. Sun, B., L.V. Rizzo, S.H. Sun, C.C. Chan, B. Wiggert, R.L. Wilder, and R.R. Caspi. 1997. Genetic susceptibility to experimental autoimmune uveitis involves more than a predisposition to generate a T helper-1-like or a T helper-2-like response. *J. Immunol.* 159:1004–1011.

156. Sun, B., S.H. Sun, C.C. Chan, B. Wiggert, and R.R. Caspi. 1999. Autoimmunity to a pathogenic retinal antigen begins as a balanced cytokine response that polarizes towards type 1 in a disease-susceptible and towards type 2 in a disease-resistant genotype. *Int. Immunol.* 11:1307–1312.

157. Sun, S.H., P.B. Silver, R.R. Caspi, Y. Du, C.C. Chan, R.L. Wilder, and E.F. Remmers. 1999. Identification of genomic regions controlling experimental autoimmune uveoretinitis in rats. *Int. Immunol.* 11:529–534.

158. Rat genetic databases. Internet. ARB Rat Genetic Database (http://www.nih.gov/niams/scientific/ratgbase/index.htm); Wellcome Trust/Oxford University Rat Genetic Database (http://www.well.ox.ac.uk/~bihoreau/key.html); Whitehead Institute/MIT Rat Genetic Database (http://www.genome.wi.mit.edu/rat/public).

159. Denny, P., C.J. Lord, N.J. Hill, J.V. Goy, E.R. Levy, P.L. Podolin, L.B. Peterson, L.S. Wicker, J.A. Todd, and P.A. Lyons. 1997. Mapping of the IDDM locus Idd3 to a 0.35-cM interval containing the interleukin-2 gene. *Diabetes* 46:695–700.

160. Encinas, J.A., L.S. Wicker, L.B. Peterson, A. Mukasa, C. Teuscher, R. Sobel, H.L. Weiner, C.E. Seidman, J.G. Seidman, and V.K. Kuchroo. 1999. QTL influencing autoimmune diabetes and encephalomyelitis map to a 0.15-cM region containing Il2. *Nat. Genet.* 21:158–160.

161. Becker, K.G., R.M. Simon, J.E. Bailey-Wilson, B. Freidlin, W.E. Biddison, H.F. McFarland, and J.M. Trent. 1998. Clustering of non-major histocompatibility complex susceptibility candidate loci in human autoimmune diseases. *Proc. Natl. Acad. Sci. USA* 95:9979–9984.

162. Kawahito, Y., G.W. Cannon, P.S. Gulko, E.F. Remmers, R.E. Longman, V.R. Reese, J. Wang, M.M. Griffiths, and R.L. Wilder. 1998. Localization of quantitative trait loci regulating adjuvant-induced arthritis in rats: evidence for genetic factors common to multiple autoimmune diseases. *J. Immunol.* 161:4411–4419.

163. Jakobiec, F.A., C.C. Marboe, D.M.d. Knowles, T. Iwamoto, W. Harrison, S. Chang, and D.J. Coleman. 1983. Human sympathetic ophthalmia. An analysis of the inflammatory infiltrate by hybridoma-monoclonal antibodies, immunochemistry, and correlative electron microscopy. *Ophthalmology* 90:76–95.

164. Chan, C.C., D. Benezra, M.M. Rodrigues, A.G. Palestine, S.M. Hsu, A.L. Murphree, and R.B. Nussenblatt. 1985. Immunohistochemistry and electron microscopy of choroidal infiltrates and Dalen-Fuchs nodules in sympathetic ophthalmia. *Ophthalmology* 92:580–590.

165. Gass, J.D. 1981. Vitiliginous chorioretinitis. *Arch. Ophthalmol.* 99:1778–1787.

166. Ryan, S.J., and A.E. Maumenee. 1980. Birdshot retinochoroidopathy. *Am. J. Ophthalmol.* 89:31–45.

167. Nussenblatt, R.B., K.K. Mittal, S. Ryan, W.R. Green, and A.E. Maumenee. 1982. Birdshot retinochoroidopathy associated with HLA-A29 antigen and immune responsiveness to retinal S-antigen. *Am. J. Ophthalmol.* 94:147–158.

168. Nussenblatt, R.B. 1991. Proctor Lecture. Experimental autoimmune uveitis: mechanisms of disease and clinical therapeutic indications. *Invest. Ophthalmol. Vis. Sci.* 32:3131–3141.

169. BenEzra, D., G. Maftzir, I. Kalichman, and V. Barak. 1993. Serum levels of interleukin-2 receptor in ocular Behcet's disease. *Am. J. Ophthalmol.* 115:26–30.

170. Nussenblatt, R.B., A.G. Palestine, and C.C. Chan. 1983. Cyclosporin A therapy in the treatment of intraocular inflammatory disease resistant to systemic corticosteroids and cytotoxic agents. *Am. J. Ophthalmol.* 96:275–282.

171. Nussenblatt, R.B., E. Fortin, R. Schiffman, L. Rizzo, J. Smith, P. Van Veldhuisen, P. Sran, A. Yaffe, C.K. Goldman, T.A. Waldmann, and S.M. Whitcup. 1999. Treatment of noninfectious intermediate and posterior uveitis with the humanized anti-Tac mAb: a phase I/II clinical trial. *Proc. Natl. Acad. Sci. USA* 96:7462–7466.

58 | Autoimmune Renal Disease Involving Basement Membrane Antigens

Curtis B. Wilson, Dorin-Bogdan Borza, and Billy G. Hudson

1. IMMUNE MECHANISMS OF RENAL INJURY

The kidney is a prominent target of immune and autoimmune injury, with glomerulonephritis (GN) and tubulointerstitial nephritis (TIN) major causes of primary renal injury leading to end-stage renal failure. Experimental GN and TIN have served not only to increase the understanding of the immunopathogenesis of renal disease, but also have been widely used in the study of the mechanism of immune-directed inflammation in general (1).

A number of different types of immune or autoimmune mechanisms of induction of renal injury have been defined over the years using experimental models (Table 1) (1–4). Observations in serum sickness leading to the concept of immune complex (IC)-induced injury flowed from the ideas of von Pirquet in 1911, catching the interest of Rich, Janway, Germuth, and others to reach full refinement by the group headed by Dixon (rev. in 5). This mechanism involves the dynamics of immune complex formation of antibody (Ab) and antigen (Ag) in the circulation, with continued rearrangement at the site of accumulation in the renal glomerulus or other vessels.

The concept of direct Ab interaction with kidney Ags was initiated with the work of Lindemann in 1900, using guinea pig anti-rabbit nephrotoxic Ab, and expanded by Masugi, Smadel, Kay, Steblay, Dixon's group, and others, leading to the identification of anti-glomerular basement membrane (GBM) Ab and anti-tubular basement membrane (TBM) disease (the subject of this chapter). The autologous phase of experimental anti-GBM Ab-induced GN, in which the host's anti-Ig response to the renal-bound heterologous anti-GBM Ab caused additional glomerular inflammation, demonstrated the nephritogenic potential of foreign antigenic material trapped in the glomerulus. This concept was later expanded to include materials trapped in the kidney by physicochemical or other interactions with glomerular cells, the glomerular capillary wall or other renal sites. More recently, experimental studies have identified mechanisms of selective glomerular or tubular cell injury in which cell surface Ag unique to a single cell type in the

Table 1 Immune mechanisms of renal disease

Ab reactions with renal Ags
   Basement membrane Ags—GBM, GBM/TBM, TBM, TBM-hapten complexes
   Glomerular cell Ags—epithelial cells (Heymann antigen complex), mesangial cells (Thy-1), endothelial cell Ags
   Tubular cell Ags—Tubular brush-border, Tamm-Horsfall protein, other tubular Ags?
Ab reactions with circulating or planted non-renal Ags
   IC mechanisms—circulating IC formation, deposition, and dynamic equilibration in the renal deposit
   Planted Ag mechanisms—infectious agents, lectins, charge-related binding, immune binding, mesangial uptake, Arthus reactions, etc.
Cellular immune reactions with glomerular Ags or planted non-renal Ags
Immune activation of mediator systems—ANCA??

multicellular kidney can be targeted by Ab (or cellular immune) injury, leaving surrounding cell types relatively free of direct involvement. Ab reactive with mediators of immune injury such as the anti-neutrophil cytoplasmic Ab (ANCA) may also contribute (see ahead).

2.   Abs REACTIVE WITH GBM Ags

2.1.   Background from Experimental Models.

One of the clearest examples of auto-Ab-induced tissue injury in humans is that associated with the reaction of anti-GBM Ab with its target Ag in the glomerulus (Fig. 1). This mechanism of Ab-induced renal injury was clearly defined by work in animal models beginning with the studies of Lindemann, and later Masugi, using administration of heterologous anti-kidney Abs, so-called "nephrotoxic serum." Subsequently, the glomeruli and their prominent GBMs were identified as the nephritogenic Ag site. The anti-GBM Abs used for study are absorbed with serum proteins and peripheral blood cells to partially remove extraneous Ab reactivity, but these Abs remain more polyclonal than their human counterparts. Anti-GBM Ab-induced injury occurs in two phases, and the amounts of Ab required for injury can be quantitated. About 75 μg of anti-GBM Ab/g of kidney is required to cause acute heterologous phase injury (abnormal proteinuria in the first 24 hours) in the rat (6). In addition to the quantities of Ab, the rate at which it binds is important, so that the "bolus" doses just mentioned do not cause heterologous phase injury in the rat when given in small divided doses over several hours (7). Sheep and rabbit are more sensitive, requiring only 5 and 15 μg of Ab bound/gm of kidney, respectively. Heterologous anti-GBM Abs that fix complement can induce injury with lower amounts of Ab bound to the GBM than fractions that don't

fix complement. Other factors also contribute to the severity of the anti-GBM Ab-induced lesions, including endotoxin and purposeful addition of cytokines, such as TNFα and IL-1β (8,9). Kidneys are the main target of heterologous anti-GBM Abs, but occasionally lungs are also affected. Pulmonary involvement can be stimulated by factors that alter the alveolar capillary wall, such as oxygen toxicity, intra-tracheal instillation of gasoline, or cytokines (10–13).

A second phase of injury in the anti-GBM Ab model occurs 7–10 days after the administration of the Ab when the recipient produces Abs reactive with the foreign Ig now bound to its GBM. Only small amounts of anti-GBM Ab are required as a planted Ag (6), and the model can be augmented by prior immunization to the heterologous Ig or by passive administration of Ab against it.

Active immunization of experimental animals with GBM preparations can induce active forms of anti-GBM Ab GN in sheep (14) (see ahead), rats (15–19), monkeys (20), rabbits (21), and guinea pigs (22), etc. The usual nephritogenic Ag consists of insoluble GBM, collagenase-solubilized GBM, or other GBM fractions. In rats, active immunization can induce both anti-GBM Ab and Ab specific for the TBM, depending on the immunogen and the strain (see ahead). Mild pulmonary hemorrhage may be associated (17).

The role of immune cells in active immunization forms of the anti-GBM model that employ heterologous GBM immunogens has been developed. In the bursectomized chicken, injury proceeds in the absence of Ab production, with the lesion transferable with immune cells, indicating an immune cellular component (23,24). The contribution of delayed type hypersentitivity in crescent formation is supported in a similar experiment in μ-chain-deficient mice, which lack a functional B cell response (25). We showed that anti-macrophage Ab modulated experimental glomeru-

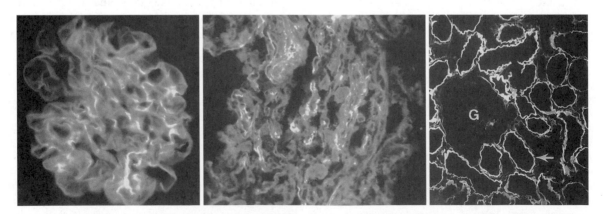

Figure 1.   The striking patterns of linear binding of anti-GBM Abs to the glomerular GBM (left panel), anti-ABM Abs to the lung ABM (middle panel), and anti-TBM Abs to the renal TBM (right panel) are shown. In the right panel, note the absence of reaction of the anti-TBM Abs with the glomerulus (G).

lar injury associated with macrophage infiltration (26). The role of cell-mediated immunity in glomerular disease has been the subject of recent study and review (27–31).

In rats, the Wistar-Kyoto (WKY) strain model is particularly susceptible to immunization with autologous or heterologous GBM Ag preparations or passive administration of anti-GBM Abs, with induction of severe crescentic GN (19,32–34). Isologous monoclonals of the IgG2a subclass can bind GBM and TBM and cause proteinuria (35). The crescentic GN in this strain has striking features of predominant cellular immune mediation in contrast to the humoral mediation in passive anti-GBM Ab disease in other rat strains. The WKY lesion can be modulated with anti-CD8$^+$ T cell Abs, anti-CD4 Abs, or Abs to factors involved in attraction of monocytes such as LFA-1 or MCP-1 (36–42). Crescentic lesions in the WKY as well as other rat strains and in mice may contain multinuclear giant cells also identified in some cases of human anti-GBM GN (43,44). The multinucleate giant cells relate to local macrophage proliferation in the experimental augmented autologous phase anti-GBM GN models, which are dependent on CD4 and CD8 T cells in some studies (39,45–50). Osteopontin expressed in glomerular epithelial cells is reported to promote macrophage and T cell infiltration and crescent formation with multinucleate giant cell formation (51). Upregulation of osteopontin expression is related to IL-1 (52). A Th-1 type response by the host fosters the delayed type hypersensitivity response, and can be attenuated by administration of TH-2 type cytokines, IL-4 and IL-10 (53–57). Chemokine expression of both RANTES and MCP-1 can be related to macrophage infiltration and crescent formation (58). The lack of the chemokine receptor CCR1, which binds RANTES and MIP-1α, enhances the Th-1 response and injury in anti-GBM Ab GN (59). Leukemia inhibitory factor infusion reduced macrophage influx in experiments in anti-GBM Ab GN (60). Anti-macrophage migration factor (MIF) Ab reduced crescent formation in the anti-GBM model (61). A role for Inf-γ and the Fc receptor in the glomerular response to injury is also suggested (62–66).

2.2. Anti-GBM Ab Disease in Humans

Beginning in 1964, deposits of Ig along the GBM and alveolar basement membrane (ABM) were reported in patients with GN and pulmonary hemorrhage (67–69), a condition termed "Goodpasture's syndrome" because of its clinical similarity to the description of an association of GN and hemorrhagic pneumonitis reported by Goodpasture during the 1918–19 influenza pandemic (70–72). We suggested in 1973 that the condition be termed anti-GBM Ab GN, or anti-GBM Ab GN with pulmonary hemorrhage as an etiologic diagnosis to exclude confusion with other causes of clinical "Goodpasture's syndrome" (73). Both terms continue to be used, with anti-GBM GN with or without pulmonary hemorrhage sometimes called Goodpasture disease rather than syndrome. The reactive GBM auto-Ag(s) have been referred to as Goodpasture (GP) Ag by some authors. In this chapter, the terms anti-GBM Ab disease and GBM auto-Ag will be used.

The induction of a severe anti-GBM Ab form of GN (without lung involvement) in sheep immunized with human GBM by Stebley in 1962 set the stage for recognition of the human anti-GBM Ab disease (14). The sheep had GBM bound and circulating anti-GBM Ab that could transfer anti-GBM GN to naive sheep (74,75). Much later, the reactivity of the renal-bound anti-GBM Ab was shown to be directed toward the collagenase-resistant globular region of type IV collagen, similar to the reaction of human anti-GBM Abs (76). Recurrences can also occur after sheep renal allografts (77). In a classic study in humans with GN in 1967, Lerner and collaborators (78) found that anti-GBM Abs were deposited in the glomeruli in a linear pattern along the GBM, and could be eluted from the kidneys for confirmation of the specificity of the reaction. Anti-GBM Abs were also found in the circulation of the patients and the anti-GBM Abs, either from the circulation or eluted from the kidneys, could transfer GN to nonhuman primates. The most convincing evidence of the pathogenic nature of the anti-GBM Abs came from the observation that the disease can recur in a renal transplant placed in a patient with circulating anti-GBM Ab (78), and the observation was later expanded (73). It is now recognized, as would be expected from the quantitative animal studies, that the severity of the recurrent disease will be influenced by the levels of circulating Ab at the time of transplant. The clinical recurrence of Ab production may be blunted by concomitant immunosuppression for graft acceptance, as demonstrated by the observation of recrudescence of a quiescent anti-GBM Ab response in a nonimmunosuppressed patient receiving an isograft or during withdrawal of immunosuppression (79,80).

Diagnosis. The diagnosis of anti-GBM Ab disease is based on identification of the typical smooth linear Ig deposits found along the GBM (and also TBM in 60–70% of cases) of the kidney and ABM of the lung (in patients with the Goodpasture's syndrome presentation), usually by immunofluorescence study of biopsy samples (Fig. 1). The TBM linear deposits may be confined to a limited population of tubules or, in some patients, be more diffuse. In the lung, it is thought that the linear Ig deposits are more focal than diffuse, since biopsies may be negative in patients with active pulmonary hemorrhage. Confirmation of the Ab specificity is done by elution study when sufficient tissue is available, i.e., the tissue is treated with an agent capable of dissociating anti-GBM Ab from the GBM for recovery and subsequent testing. Anti-GBM Abs were initially detected in serum by indirect immuno-flourescence assay using sections of "normal human kidney", usually obtained during nephrectomy for renal carcinoma. It was difficult to find true negative kidney specimens, since improper handling or other reasons often led to linear background staining possibly related to interruption of blood flow during biopsy, surgery, or autopsy (81,82).

The anti-GBM Ab-reactive Ag(s) was shown to be present in the non-collagenous proteins solubilized by diso-lution of human GBM by collagenase (83), leading to the development of a radioimmunoassay (RIA) for anti-GBM Ab in 1974 (84,85). The assay used a human GBM Ag purified from a collagenase extract of human GBM by immunoabsorption with human anti-GBM Abs. The RIA was the basis of our identification of over a thousand patients with anti-GBM Ab disease over the next 20 or so years. The complexity of the anti-GBM Ab response in terms of numbers of reactive epitopes as well as possible variations in anti-GBM Ab reactivity remains to be fully defined. Close similarity in epitope structure is suggested by cross-inhibition studies using a monoclonal Ab (termed P1) to the reactive Ag (86). Also, inhibition studies using human anti-GBM Abs suggest that the number of epitopes is limited (87). The major reactivity of anti-GBM Ab has been shown to reside in the noncollagenous globular region, termed NC1, of the α3(IV) collagen chain (see ahead).

The clinician must be critically aware of the specificity, sensitivity, and reproducibility of the anti-GBM assay in current use (88) to avoid both false-positives and-negatives. For example, HIV infection is said to cause false-positive reactions in the current anti-GBM Ab ELISA assays (89). Prompt and correct identification of the disease with rapid treatment and follow up are crucial for optimum outcome. Over the years, misdiagnoses related to problems with commercial anti-GBM assays have occurred. For example, a man with lung cancer who was initially thought to have anti-GBM Ab disease based on such an assay. When evalu-ated, the serum was negative in our anti-GBM RIA, and detailed study revealed the individual had Abs reactive with the α1(IV) collagen chain (see ahead), but not the GBM auto-Ag on the α3-(IV) NC1 (90). Other examples of anti-α1(IV) chain Abs have been described (91). The diagnosis of anti-GBM Ab disease can be suspected from either linear Ig deposits along the GBM, or from evidence of cir-culating anti-GBM Ab; the clinical diagnosis usually requires both. To confirm the diagnosis, anti-GBM Ab must be eluted from the kidney or lung and shown to be present in higher concentration than in the serum showing specific binding. The elution step was much easier years ago when nephrectomy specimens were available since nephrectomy was part of the treatment. Since anti-GBM Ab disease is only one of several conditions that can present with GN and

pulmonary hemorrhage, clinical Goodpasture's syndrome, the differential diagnosis, must also consider other possi-bilities, including several types of systemic vasculitis as well as systemic lupus erythematosis (see ahead).

The anti-GBM Abs are primarily IgG (73,92). Rarely, IgA or IgM anti-GBM Abs have been reported (93–95). Among IgG subclasses, IgG1 and IgG4 are particularly well-represented, both in the circulation and in renal eluates (96–98). Interestingly, these classes of IgG seem to differ in their pathogenicity. Recurring Abs of the IgG1 class were pathogenic, whereas IgG4 were benign (99). Complement deposits along the GBM are often irregular and less striking than the Ig deposits. The use of a control anti-albumin helps to separate specific IgG deposits from combined IgG and albumin, such as those that accumulate in diabetic kidneys (100).

Clinical presentation and course.   Anti-GBM Abs are associated with a spectrum of GNs, most often with rapidly progressive crescent forming histologic types of anti-GBM GN (Table 2) (1,94,101–107). A minority of patients have less aggressive proliferative forms of GN. About half have pulmonary hemorrhage due to Ab reactions with the ABM, producing a clinical picture termed "Goodpasture's syndrome" (108–112). Abs reactive with TBMs are also seen in the majority of patients with anti-GBM Ab disease. A different type of anti-TBM is found in patients with primary TIN or TIN associated with other forms of GN (Table II) (see ahead). Anti-GBM Abs can sometimes react with choroid plexus and possibly intestinal basement membranes. We have seen central nervous system involve-ment in a few patients in our large studies of patients with anti-GBM Abs, and a case report regarding this subject has recently appeared (113). Whether some form of reaction occurs in the joints is unknown; however, 5–10% of

Table 2   Clinical manifestations of anti-GBM and anti-TBM Abs

---

Anti-GBM Abs
  GN (often RPGN) and pulmonary hemorrhage (clinical
    Goodpasture's syndrome)
  RPGN and occasionally milder forms of GN
  Infrequently presents as pulmonary hemosiderosis
  ? Injury at other sites—choroid plexus, etc.
  Complicating renal transplantation—recurrent GN, or *de novo*
    anti-GBM Ab GN in Alport's Syndrome
Anti-TBM Abs
  TIN complicating anti-GBM Ab GN, other forms of GN
    (membranous nephropathy), drug toxicity
  ? Primary TIN
  Recurrent or *de novo* TIN after renal transplantation

RPGN—Rapidly progressive GN, usually crescentic in nature.

patients have some form of joint complaint, often at presentation. In experimental animals, anti-GBM Abs are reported to bind to the cochlear plexus with inflammation (114).

One to two percent of patients with anti-GBM Abs present with pulmonary hemorrhage alone and no overt clinical findings of GN are detected (1,115,116). In such patients a renal biopsy will reveal typical smooth linear deposits of IgG along the GBM, and circulating anti-GBM Abs can be detected in the circulation. These patients typically are identified during evaluations of suspected idiopathic pulmonary hemosiderosis. The severity of clinical involvement is possibly related to the amounts and temporal features of anti-GBM Ab production/binding in individual patients as well as possible variations in Ab reactivity and Ag epitope distribution. Anti-GBM disease can occur in the first decade through old age with bimodal peaks; the first peak in the 15–40 year old age group occurs more commonly in males than in females, with pulmonary hemorrhage more common than in older patients. A second peak occurs after 50–55, in which females are more often afflicted than males and GN alone is more common. Anti-GBM Ab disease is very infrequent in children. The youngest patient in our series was 2 years old, and other children have been reported (117–119). This may relate to developmental changes in the GBM based on observations that kidneys from very young children may not react with human anti-GBM Abs (120). Of interest, the racial distribution of the disease is very skewed toward the Caucasian race.

The clinical course often seems to have been determined before the patient seeks medical treatment, i.e., in our experience, the initial anti-GBM Ab measurement is almost always the highest when the patient is first seen. The Ab then falls at a rate unique to the particular patient, which can be modified by immunosuppressive agents and plasma exchange or other measures to reduce Ab in the circulation (94,121–126). Despite treatment, the degree of renal damage at its initiation is a major determining factor in salvage of renal function. Bouts of pulmonary hemorrhage can be life threatening, but are often manageable with high-dose steroids and the immunosuppression and plasma exchange mentioned above. The long-term outlook for pulmonary function is good (127). Events precipitating bouts of pulmonary hemorrhage may include pulmonary infection and for irritants, including smoking (128–130). Once circulating anti-GBM Ab has disappeared, which takes on average about 1.5 years (1–2 months to over 2–3 years), recurrence of the Ab response is unusual (80,131–135).

Anti-GBM Ab Disease and Vasculitis. There is an overlap of anti-GBM Abs and associated disease with some systemic vasculitis syndromes and pulmonary renal syndromes that are characterized by the presence of anti-neutrophil cytoplasmic Ab (ANCA) (136,137). Thus, in some patients with anti-GBM Ab disease, vasculitis is observed and the presence of ANCA can be detected in up to about 20% of patients with anti-GBM Abs (138–143). In some patients, ANCA appear or continue after anti-GBM Abs have disappeared (144). ANCA is a marker of pauci immune necrotizing and crescentic GN (minimial immune deposits in glomeruli) associated with microscopic polyangitis, as well as Wegener's granulomatosis and other types of systemic vasculitis, including Churg-Strauss syndrome (145–150). IgM ANCA have been associated with systemic vasculitis and pulmonary hemorrhage (151,152). ANCA can also react with monocytes (153). ANCA are divided into two major categories based on their immunofluorescent patterns on ethanol-fixed neutrophil targets. Cytoplasmic staining, called C-ANCA, usually results from a reaction with proteinase-3 (PR-3), and perinuclear staining, called P-ANCA, is usually due to a reaction with myeloperoxidase (MPO) (145,154,155). ANCA associated with anti-GBM Abs is usually, but not always, P-ANCA or anti-MPO. Other Ags have been identified in the ANCA reactions, including lactoferrin, lysozyme, elastase, and cathepsin G, although their clinical significance is unclear (156–159). For example, one of ANCA's targets is bacterial/permeability-increasing protein (BPI), which is found in a number of conditions, but is not unique to kidney disease (160–166). The immunoflourescent ANCA assays have been standardized with purified Ag assay (167–170). Ongoing studies are aimed at refining epitope identification and Ig subclass distribution, and include the use of recombinant proteins and Ab preparations (171–176). Attempts to correlate ANCA with genetic factors including mediation are underway (177–181). As with anti-GBM Ab disease, early diagnosis and treatment are important (182–184). Immunosuppression, plasma exchange, and immuno-absorption treatments are being evaluated (185–188). ANCA-associated glomerular injury can recur after renal transplantation as well as extrarenal small vessel vasculitis (189).

Additional studies are ongoing to determine if ANCA is involved in the immunopathogenesis of the vascular injury, perhaps via activating or enhancing neutrophil interactions with vascular endothelium. Activated neutrophils are found in the glomeruli of patients with ANCA, and cells expressing TNFα, IL-1β, and IL-2-receptors are present in periglomerular sites (190,191). Activated monocytes are also present in Wegner's granulomatosis (192). ANCA can induce degranulation as well as an oxidative burst in primed neutrophils *in vitro*, and can alter signal transduction pathways (193–199). Ig fractions of anti-PR3 ANCA are more active than anti-MPO ANCA in *in vitro* neutrophil activation (200). ANCA can cause neutrophil adherence and injury to endothelial cell cultures (201–203). These and other mechanisms, including anti-endothelial

Abs, leukocyte adhesion molecules, chemokines and their receptors, and other aspects of endothelial function, have been the subject of several reviews (204–209).

In experimental studies, MPO can bind to GBM *in vivo* and induce oxidative reactions (210). Perfusion of MPO (along with hydrogen peroxide or after renal ischemia) into the renal artery of a rat preimmunized with neutrophil extracts and hydrogen peroxide leads to MPO and Ig found along the GBM accompanied by inflammation (211,212). Others feel this is a form of immune complex injury and inconsistent with pauci immune necrotizing and crescentic GN (213). Experimental induction of auto Abs to MPO can cause a subclinical heterologous anti-GBM GN to become clinically evident in rats (214). Heterologous anti-rat MPO can augment glomerular injury in the rat anti-GBM model (215). Brown Norway rats develop anti-GBM Abs and other auto-Abs, including anti-MPO Ab after treatment with $HgCl_2$ (216,217). The gut, rather than the kidney, is the main focus of injury that can be manipulated with antibiotic treatment, and the injury is not transferable with immune serum (218). Vasculitis is common in acute serum sickness in rabbits and is found in some strains of lupus mice. A recombinant inbred strain derived from BXSB and MRL/l was reported to have crescentic GN and small vessel vasculitis with somewhat fewer glomerular deposits than the parental strains (219). The mice have anti-MPO ANCA. None of the models has yet established a convincing pathogenic role for ANCA in pauci-immune necrotizing and crescentic GN (220).

Molecular identity of the GBM auto-Ag. The linear pattern of deposition of anti-GBM Abs along the GBM indicates that the GBM auto-Ag(s) is an intrinsic component of the GBM, restricting the search for its identity to the normal basement membrane (BM) constituents. All BMs share the same building blocks (type IV collagen, laminin, entactin/nidogen, and heparan sulfate proteoglycans) and have the same overall organization (221,222). Type IV collagen and laminin self-assemble into two distinct networks joined by nidogen, thus forming the structural backbone of BMs, whereas proteoglycans impart an overall negative charge important for the selective ultrafiltration of charged molecules. Despite their overall structural similarity, BMs have subtle differences in composition among tissues and during development, which is thought to reflect adaptation to specific functions. These differences arise from two factors: a) type IV collagen and laminin exist as multiple genetically-distinct isoforms, and b) they are trimeric molecules, which allows for many possible combinations of isoforms.

The GBM is insoluble and hence difficult to study *in vitro*, but the GBM auto-Ag-reactive component of GBM could be solubilized by digestion with bacterial collagenase (223–228). In the collagenase-solubilized GBM extract,

anti-GBM Abs bind peptides with molecular masses of approximately 27 kDa and 54 kDa, which stand in a monomer-dimer relationship (223,225,229). These peptides were shown to be the globular, non-collagenous (NC1) domain of type IV collagen (230–233). Specifically, the GBM auto-Ag has been identified as the NC1 domain of $\alpha3(IV)$, a then novel chain of type IV collagen (Fig. 2) (234–236). The identity of GBM auto-Ag was confirmed by *in vitro* expression of the recombinant $\alpha3$ NC1 domain capable of binding anti-GBM Abs (237–240). Moreover, while the $\alpha3(IV)$ chain has a tissue-restricted distribution, it was found in the BMs of all tissues known to bind anti-GBM Abs, including lung and choroid plexus.

Most of these investigations were performed using anti-GBM Abs from the sera of patients. However, early studies showed that IgG eluted from kidney biopsies of GBM patients recognize the same Ags as circulating anti-GBM Abs. More recently, IgG eluted from the kidneys and lungs of patients with anti-GBM Ab disease was shown to bind specifically to $\alpha3$ NC1, like the circulating anti-GBM Abs (241,242). This would be expected, since both the kidney eluted and serum Abs can transfer disease (see above).

Besides the $\alpha3$ NC1 domain, anti-GBM Abs from some patients may react with other NC1 domains, but this reactivity is generally weak and much more variable among patients (87,239,240). At least in part, reactivity toward other NC1 domains may be explained by cross-reactivity, given the relatively high homology between NC1 domains (60–70%). Cross-reactivity of the anti-GBM Abs $\alpha1$ and/or $\alpha2$ NC1 domains has been demonstrated using inhibition ELISA (87,243,244). Additional Abs directed against the 7S domain of type IV collagen, laminin, and entactin occasionally have been reported. This broader specificity may, in part, be due to epitope spreading, a secondary phenomenon triggered by the inflammatory process, which results in the production of Abs to other molecules co-localized with the primary auto-Ag. In conclusion, anti-GBM Ab disease can be classified as anti-type IV collagen disease, with the primary GBM auto-Ag shared by all patients being the NC1 domain of the $\alpha3(IV)$ chain (245).

Structure and organization of normal GBM type IV collagen and Alport's syndrome. The $\alpha3(IV)$ chain is a member of a family of six homologous chains, designated $\alpha1$-$\alpha6$, that comprise type IV collagen (246). Each chain is characterized by a long collagenous domain of ~1400 residues of Gly-Xaa-Yaa repeats, interrupted by ~20 short non-collagenous sequences, and by a non-collagenous (NC1) domain of ~230 residues at the carboxyl terminus. A type IV collagen promoter is formed by assembly of three α chains into triple-helical molecules (Fig. 3, top). The triple-helix formation proceeds in a zipper-like fashion from the C-terminus to the N-terminus (247). The NC1 domain is thought to direct the lateral association and the

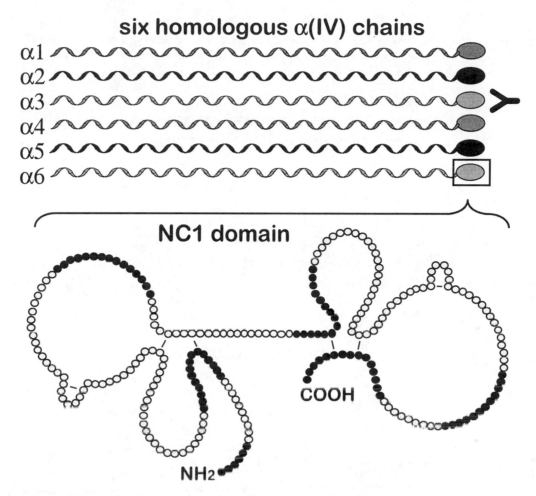

Figure 2. A protomer of type IV collagen (top) consists of three chains, selected from a repertoire of six genetically distinct, but homologous α chains (middle). The α3(IV) chain is the GBM auto-Ag. It binds anti-GBM Ab at its carboxyl terminal globular, non-collagenous (NC1) domain (bottom).

alignment of the three chains, a role analogous to that of the C-propeptides in fibrillar collagens (such as types I, II or III). However, unlike the propeptides, the NC1 domain of type IV collagen is not cleaved after association, but instead is involved in the inter-protomer interactions (see below).

Protomers of type IV collagen self-associate to form supramolecular networks with the aspect of an irregular polygonal lattice by electron microscopy (248). Three types of interactions are observed between protomers (Fig. 3, bottom). At the amino terminus, four protomers associate to form the 7S domain. At the carboxyl terminus, two promoters associate through NC1-to-NC1 interactions to form a NC1 hexamer. These interactions are further stabilized by inter-protomer disulfide bonds, which cross-link the type IV collagen network and render it insoluble. Lateral associations between the NC1 domains and the triple helix (249) further stabilize the network and increase its complexity. Selection of three α chains from a repetoire of six isoforms

allows for 56 distinct protomers. Further interactions among promoters can theoretically produce a very large number of combinations. Nevertheless, only a few networks with distinct chain compositions have been characterized to date, indicating the specificity in the selection of isoforms for protomer formation and selection of protomers for networks assembly. An α1.α2 network appears to exist in all BMs, and it is the major one in the lens BM (250). An α1-α6 and an α3-α6 network exist in the seminiferous BM along with the α1-α2 network (251). An α3.α4.α5 network, characterized by looping and supercoiling of the triple helices and by disulfide cross-links, exists in the GBM along with the α1.α2 network (252). This α3.α4.α5 network is the target of GBM Abs.

Interestingly, anti-GBM Abs do not bind to the GBM of patients with Alport syndrome (253–255), establishing an unexpected link between these glomerular diseases in which type IV collagen is affected. Alport syndrome is a hereditary disease caused by mutations in any of the genes

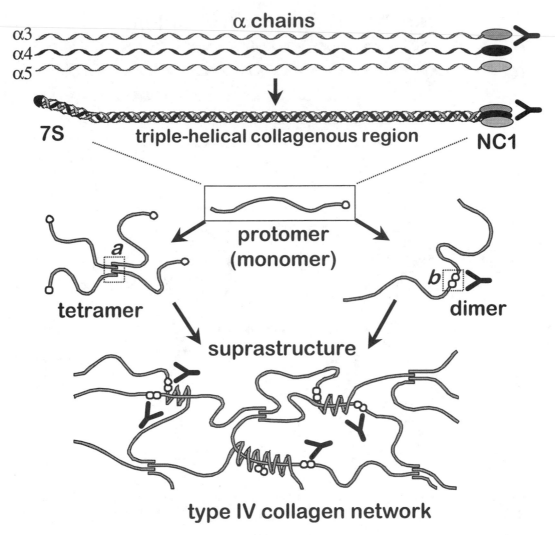

Figure 3. The organization of type IV collagen network. Three α chains assemble into a type IV collagen protomer (top). Protomers interact with each other through the 7S (a) and NC1 (b) domains, forming a network with the aspect of chicken wire (bottom). Anti-GBM Abs binding to the α3 NC1 domain are also shown (**Y**).

coding for the α3, α4, or α5 chains of type IV collagen. Although mutations affect only one chain, all three chains are absent from the GBM and other BMs in Alport patients. Anti-GBM Ab GN may develop in some Alport patients who receive a kidney transplant (256). These allo-Abs react with normal GBM, but not with Alport GBM (255,257) and have been shown to target the NCl domain of α3(IV) and α5(IV) chains (258–263).

Investigation of the molecular basis of Alport syndrome and its animal model established in dogs (264–266) and mice (267–268) revealed distinct roles for different networks of type IV collagen networks. In normal glomerular development, the α1.α2 network is assembled first in the embryonic glomerulus, but then it is replaced by the α3.α4.α5 network that forms the GBM of the mature glomerulus (269). In Alport syndrome, mutations in any of the three chains arrest this developmental switch, causing the GBM to be composed of the embryonic α1.α2 network only (270,271). The absence of the α3.α4.α5 network leads to the deterioration of the GBM and the progressive loss of renal function over a period of 10–20 years. This indicates that the α1.α2 network is important in glomerular development, whereas the α3.α4.α5 network is required for long-term stability, perhaps due to its higher stability against proteolytic degradation (270).

Identification of the location of GBM auto-Ag epitopes. The identification of the precise location of the GBM auto-Ag epitopes has long been sought because it can facilitate the design of more specific therapeutic strategies for treatment of anti-GBM Ab disease. Several lines of evidence suggested the existence of a shared immunodominant GBM auto-Ag epitope within the α3 NCl domain. Binding of anti-GBM Abs to immobilized

GBM auto-Ag is competitively inhibited by certain monoclonal Abs to α3(IV) NCl, such as P1 (86,87,272), and anti-GBM Abs from different patients also compete with each other (87,272). A polyclonal anti-idiotype directed against human anti-α3(IV) NCl auto-Ab from one patient with anti-GBM Ab disease demonstrated the presence of shared antigenic determinants in all other anti-GBM Abs tested (273).

Early studies attempted to map the epitope by using synthetic peptides derived from the α3(IV) NCl sequence (272,274–277) or by site-directed mutagenesis of the α3(IV) NCl expressed in *E. coli* (278). Although linear sequences that bind anti-GBM Abs were thus identified, the findings were at variance with each other. Moreover, these studies did not address whether the linear sequences constituted the major conformational, disulfide bond-dependent epitopes. The majority of anti-GBM Abs are probably directed against conformational epitopes, because reduction of the disulfide bonds of the α3 NCl domain drastically reduces its reactivity (226,274), and misfolded α3(IV) NCl produced in *E. coli* has only partial reactivity (~25%) with anti-GBM Abs (237).

Successful mapping of the conformational GBM auto-Ag epitopes was achieved using a novel strategy, homologue-scanning mutagenesis (279). For this approach, chimeric α1/α3 NCl domains were designed and produced, in which α3-specific sequences were substituted into an α1 NCl scaffolding (240,280,281). The non-immunogenic but homologous α1 NCl domain fulfills the role of an inert scaffolding that ensures the correct conformation of the guest α3 sequences. Using this strategy, the GBM auto-Ag epitope was localized roughly to the N-terminal third of α3(IV) NCl (280,281), and specifically to residues 17–31, designated **E$_A$** (240). A minor epitope region has been identified as the central portion of the α3 NCl domain (281), specifically as residues 127–141, designated **E$_B$** (240).

The **E$_A$** and **E$_B$** regions constitute distinct epitopes for anti-GBM Abs (244), as shown schematically in Fig. 4. Using affinity chromatography, at least four subpopulations of anti-GBM Abs have been isolated with specificity for distinct epitopes within α3 NCl. These subpopulations, designated **GP$_A$**, **GP$_B$**, **GP$_{AB}$** and **GP$_X$**, react with the E$_A$ epitope, the **E$_B$** epitope, both epitopes, or neither one, respectively (244). All patients' sera consistently show high reactivity with **E$_A$** (or the N-terminal third of α3 NCl), whereas reactivity with **E$_B$** (or the central portion of α3 NCl) is more variable and completely absent in some patients (240,281). The immunodominant Ab population **GP$_A$** accounts for approximately 60% of the total reactivity against α3 NCl (244).

Despite this heterogeneity, the epitopes of anti-GBM Abs must be relatively close in space, because they complete with the monoclonal Ab Mab17 for binding to α3 NCl

(87,244,272). In fact, the E$_A$ and E$_B$ regions jointly form the epitope of Mab17, which indicates their spatial proximity and explains the competition observed between anti-GBM Abs and Mab17 (244). Nevertheless, the GBM auto-Ab epitopes and the Mab17 epitope merely overlap without being identical, as is shown by their differential accessibility in the native form of the GBM auto-Ag (see below).

Cryptic properties of the GBM auto-Ag epitopes. *In vivo*, the NCl domains of type IV collagen, including the GBM auto-Ag, occur in the form of a hexamer complex, formed by the interaction at the carboxyl end between two protomers of type IV collagen. The NCl hexamer can be released from the insoluble component of BMs by digestion with bacterial collagenase. Further dissociation of the

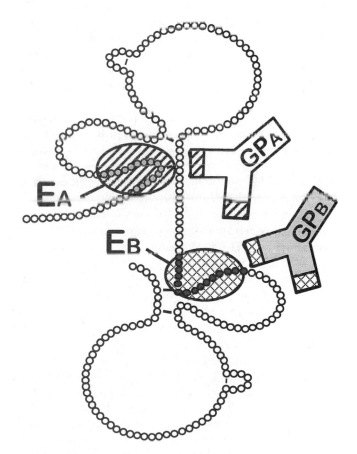

Figure 4. Localization of the GBM auto-Ag epitopes (GP) within the α3(IV) NC1 domain. By use of recombinant chimeric α1/α3 NC1 domains, two distinct GP epitopes designated **E$_A$** (residues 17–31) and **E$_B$** (residues 127–141) have been identified within the *α3(IV) NC1 domain. Abs to the E$_A$* epitope are immunodominant and account for 60–65% of the total reactivity against α3(IV) NC1. **E$_A$** and **E$_B$** must be located in close proximity in the folded α3 NC1 domain, because they jointly comprise the epitope of Mab17, a monoclonal Ab that competes with human anti-GBM Abs.

hexamers yields a mixture of NCl monomers (Mr ~24–28 kD by SDS-PAGE) and disulfide-linked dimers (Mr ~50–56 kD). The dimers, mostly homodimers (236), are generated by an exchange of disulfide bonds between two NCl monomers within a NCl hexamer (282). The ratio of dimers to monomers varies among tissues, being about 2:1 in GBM and ABM. Both α3 NCl monomers and dimers react with anti-GBM Abs.

The reactivity of anti-GBM Ab NCl hexamers isolated from bovine GBM was reversibly increased upon hexamer dissociation under denaturing conditions (232). This finding, later extended to the NCl hexamers from human GBM (283), indicates that the GBM auto-Ag epitopes are sequestered within the hexamer, but may be reversibly exposed upon dissociation. In immunofluorescence studies, only native GBM showed reactivity with anti-GBM Abs without denaturation, whereas after treatment of tissues with acid-urea, binding also occurred to hidden antigenic determinants in the basement membranes of lungs, skin, and placenta (284). Interestingly, all subpopulations of circulating anti-GBM Abs appeared to recognize that epitopes sequestered are cryptic in the hexamer complex (244) suggesting a possible etiologic mechanism for anti-GBM Ab disease. The sequestered GBM auto-Ag epitopes are immunologically privileged and do not induce tolerance. Anti-GBM Ab disease may be triggered by unidentified pathogenic factors that cause dissociation of NCl hexamers and exposure of the cryptic epitopes, which would then be perceived as foreign by the immune system. In contrast, Alport post-transplant allo-Abs can bind to epitopes that are exposed in the α3.α4.α5 NCl hexamers of the transplanted kidneys, but which are not present in the Alport GBM due to the absence of α3-α5(IV) chains.

If GBM auto-Ag epitopes are sequestered in the native form of the α3 NCl, then how do autoauto-Abs gain access to their binding sites *in vivo*? Whether certain GP auto-Abs can recognize exposed epitopes and thus trigger the nephritogenic process is not clear. One study showed that anti-GBM Abs eluted from the kidney of a patient with anti-GBM Ab disease bound to antigenic determinants exposed in the NCl hexamer, whereas circulating IgG recognized sequestered epitopes (285). However, another study found that kidney and lung-bound Abs bound cryptic epitopes (242). A different mechanism in which anti-GBM-Abs themselves can induce the dissociation of NCl hexamers and exposure of their epitopes for binding has been recently suggested (244). Upon reaction with native NCl hexamers from human GBM, a control monoclonal Ab (Mab17) and Alport allo-Abs to the α3 NCl domain precipitated NCl hexamers containing monomers and dimers of α3, α4, and α5(IV) chains, while anti-GBM Ab IgG precipitated α3 NCl monomers, but not α3 dimers nor α3-containing hexamers (Fig. 5). This suggests that auto-Abs are capable of extracting α3 NCl monomers from NCl

hexamers (presumably, hexamers containing α3 dimers are more stable), and a similar mechanism may be operative *in vivo*.

Induction of experimental anti-GBM Ab disease by active immunization with GBM components including the "GBM auto-Ag".   As noted above, the GBM is made up of a number of extracellular matrix proteins, including type IV collagen, laminin, heparan sulfate proteoglycan, and entactin/nidogen. A number of active or passive immunization models have been reported using some of these components without producing a true Goodpasture's syndrome picture (286–289), and Abs reactive with some of the components have been found as part of polyclonal B cell reactions after administration of $HgCl_2$. The same appears to be true in humans; for example, anti-entactin Abs have been found in some patients with glomerular immune deposits in a condition termed "non-Goodpasture anti-GBM nephritis" (290,291).

The nephritogenic fraction consists of the NCl domains of type IV collagen, which occur *in vivo* as a hexameric complex. Various methods have been used to identify the actual nephritogenic Ag of the NC1 hexamer fraction, similar to those used for identification of the GBM auto-Ag. Thus, animals have been immunized with NCl hexamers isolated from various tissues (which have distinct α chain compositions), with purified NCl domains isolated from native sources, with recombinant NCl domains, as well as with synthetic peptides.

Further fractionation of GBM, which contains α1-α5 chains, showed that the nephritogenic fraction contains α3, α4 and α5 NCl domains (292–294). Purified α3 NCl dimers from bovine GBM were nephritogenic in rabbits, whereas α1 and α2 NCl domains were not (295). More recently, purified bovine α3 NCl monomers and dimers were shown to induce experimental anti-GBM disease in rats (296). Comparison of the nephritogenic ability of the α1–α6 NCl domains, using recombinant human proteins expressed in mammalian cells for correct folding, showed that α3 and α4 produced anti-GBM disease, while α5 caused mild proteinuria (297).

To identify the nephritogenic epitope, the pathogenicity of various synthetic peptides was tested in animal models. A peptide consisting of the C-terminal 36 amino acids of the α3 NCl domain (then thought to encompass the GBM auto-Ag epitope based on epitope mapping studies with linear peptides) gave rise to Abs, but did not induce disease in rats (298). In another study, 25-mer C-terminal peptides from the NCl domain of the α1–α6(IV) chains were compared. Disease was induced in some rats by α3 (2/20), α4 (8/20), and α5 (1/20) peptides (299). Given the low rate of success with the α3 peptide compared with 100% induction of anti-GBM disease with purified α3 NCl or with whole NCl hexamers, it must be concluded that the nephritogenic

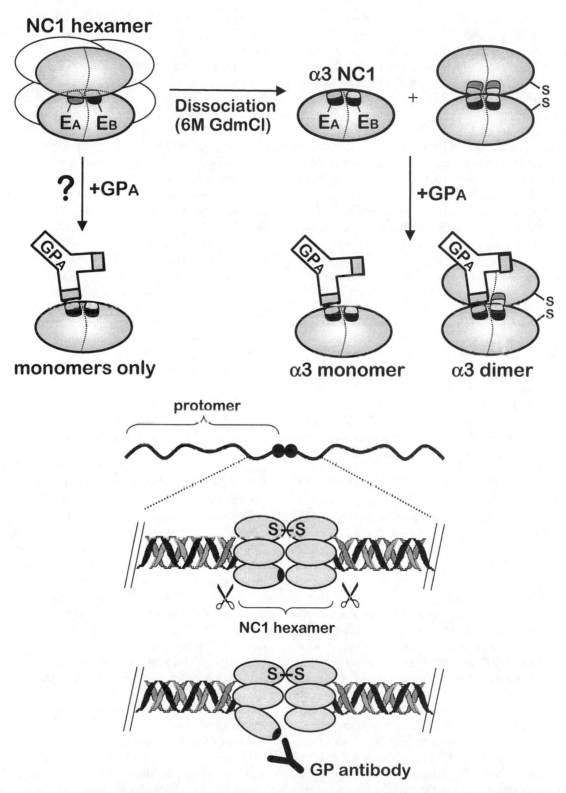

Figure 5. **(A)** Sequestering of GBM auto-Ag epitopes (GP) within the NC1 hexamer complex. Both E_A and E_B epitopes are partially buried. The solvent-exposed regions are fully accessible for unhindered binding of Mab17 monoclonal Ab (not shown) to intact NC1 hexamers. The binding sites of anti-GBM Abs are partially burried. Therefore, *in vitro* binding of anti-GBM Abs to the GBM is much enhanced after dissociation of the GBM hexamers. As yet, *in vivo* and *in vitro* anti-GBM Ab binding studies have not been compared. **(B)** A model depicting the hypothetical location of GP epitopes in the type IV collagen superstructure and the effect of anti-GBM Ab binding on the structure.

epitope of the GBM auto-Ag remains unidentified. The $\alpha3$ NCl sequences 17–31 and 127–141, encompassing the recently identified E_A and E_B epitopes, have not yet been tested in an animal model. Given the conformational nature of the GBN auto-Ag epitopes, it will be interesting to determine whether linear peptides spanning these sequences are nephritogenic.

HLA associations.   As with other autoimmune diseases (300,301), genetic factors appear to be important in anti-GBM Ab disease. Beginning in the late 1970s, a high frequency of HLA-DR2 was found in patients with the disease (302–304). Ig Gm allotype associations and Ig subclass restrictions were also noted (96,97,305), and the appearance of this unusual disease in identical twins also suggests a genetic component (306).

Subsequent studies have associated the HLA-DR2 specifity with the HLA-DRB1*1501, (DQB*0602) (307–311). In the study by Burns et al. (310), a sequence-specific oligonucleotide detected a six-amino-acid sequence shared by DRw15 and DRw4 that included four ploymorphic amino acids was found in 45/49 (91.8%) patients. The particular motif is in the floor of the Ag binding grove and had a stronger association with Goodpasture's disease than any individual allele. Studies using human B cells homozygous for HLA-DR15 demonstrated that $\alpha3(IV)$NCl is presented as at least two sets of 3–5 peptides centered on common core sequences (nested sets) (312). In subsequent studies, three nested sets of naturally presented $\alpha3(IV)$NCl peptides were detected bound to DR15 molecules (313). Of interest, peptides representative of each nested set bound to DR15, with the majority having equal or greater DR15 affinity than the naturally processed counterparts. This suggests factors during processing favor presentations rather than the ability of DR15 to bind many sequences derived from $\alpha3(IV)$NCl.

Data from 139 patients in the studies of Dunckley, Heuy, and Fisher (307,309,311), were reanalysed by Phelps and Rees (314) who confirmed the strong association with DRB1*1501 (and DQB1*0602), weak association with DRB1*04 and DRB1*03, and a neutral or slightly negative association with DRB1*01 to a strongly negative association with DRB1*07. DR3 and DR4 alleles were thought to have an influence on the presence of DR15. The strong increase in the odds ratio for disease with inheritance of the DRB1*1501 allele was modulated by the co-inherited DRB1 allele in the order 1501,1501 ~ 1501,4 ~ 1501,3 ~ 1501,14 > 1501,1 > 1501,7. The authors suggest that the differences in pockets 4 and 7 in the peptide binding grove distinguish DR2 alleles associated with Goodpasture's disease from those that are not. The largest differences in pocket 4 are found in DRB1*07, which was thought to have a protective effect. Additional definition of the nature of the T cell and $\alpha3(IV)$ epitope interactions would be expected to better define specific immune modulation of the anti-GBM Ab response (315). Perhaps studies using HLA transgenic mice could enhance understanding of the role of HLA in this disease (316–318).

Experimentally, in mice, all MHC haplotypes (H-2^a, k,s,b, and d) developed anti-$\alpha3(IV)$ NCl Abs when immunized with $\alpha3(IV)$NCl (319). In contrast, only a few of these strains developed anti-GBM-associated disease involving kidney or lung. Lymphocytes or Abs from nephritogenic strains could transfer disease to syngeneic recipients. Additional studies showed that disease correlated with the emergence of an IL-12/Th1-like T cell phenotype, suggesting that anti-GBM Abs in mice can cause injury only in MHC haplotypes that can generate nephritogenic lymphocytes with certain T cell characteristics. Human T cells react with GBM-auto-Ags (320). In humans, T cell clones specific for $\alpha3(IV)$ NCl have been recovered from two patients with anti-GBM Ab disease (321). One of the three CD8$^+$ clones was MHC class I-restricted (HLA-A11) and expressed the T-cell receptor Vβ5.1 chain. The clone recognized a motif at the N-terminal area of the $\alpha3(IV)$ NCl domain (AA 51–59, GSPATWTTR), a site different than the putative Ab-binding site.

Induction of the Anti-GBM Ab.   In thinking about the induction of the anti-GBM Ab autoimmune response, the observation that the response is generally short-lived, lasting on average over a year with little chance of recurrence (see above), suggests that the antigenic stimulus is relatively short-lived. It is also of interest that small clusters of anti-GBM Ab disease have been noted in our large series and other reports (322). It is unknown if the cases of GN and pulmonary hemorrhage described by Goodpasture in association with the 1918 influenza pandemic represented anti-GBM Ab disease or not, but it does raise the question of a possible association with influenza infection. Occasional patients have been described since in which a possible influenza association was noted (73,323). One patient had a rising influenza A2 titer and the other had influenza A2 virus isolated. In an unpublished study, we found no anti-GBM Abs in sera from a group of patients with rising influenza A2 titers. In terms of a possible molecular mimicry site between influenza A2 virus and the non-collagenous portion of $\alpha3(IV)$ collagen, a linear seven amino acid sequence homology was identified (324).

Isolated examples of anti-GBM Ab disease have been seen with other infections. Although post-streptococcal GN is not an anti-GBM Ab disease, streptococci may share cross-reactive Ags with glomeruli (325–327). In this situation, a monoclonal Ab raised against streptococci reacts with human GBM (328) and, conversely, a monoclonal Ab raised against human kidney plasma membrane is reactive with a 43 kD protein in glomeruli and also with streptococcal M protein (329). A common tetrapeptide amino acid sequence was found to occur between a portion of the M protein of certain streptococcal strains and the human glomerulus (330). Different M proteins were shown to share Ab reactions with different epitopes on vimentin present on glomerular mesangial cells (331,332), which may explain the older ideas of GBM and streptococcal cross-reactivity.

There is a report of IgM Ab against crude human GBM in 77% of patients with serologically-verified hantavirus infection (*Puumala* serotype), but no Ab to the NCl portion of type IV collagen (333). As noted earlier, HIV infection may be associated with "false-positive" anti-GBM Ab ELISA results (89). Another study found anti-GBM Abs with little relationship to disease in patients with HIV and *Pneumocystis carinii* pneumonia, and noted additional patients who may have had Goodpastures syndrome and anti-GBM Abs associated with *P. carinii* (334).

Studies have suggested that exposure to hydrocarbon solvents might be related to development of Goodpasture's or unspecified GN (335–345). In our own series of anti-GBM Ab disease patients, extensive hydrocarbon exposure was infrequent (less than 5 percent). Little good experimental evidence is available on this point; benzene and N,N'-diacetybenzidine have been reported to induce glomerular injury, but anti-GBM Abs were not identified (346). Carbon tetrachloride administration in rats and petroleum middle distillates given to mice caused renal injury but no report of anti-GBM Abs (347,348).

In studies in our laboratory, 6-week exposure to gasoline vapor in rats or rabbits did not lead to anti-GBM Ab production. Longer term exposure in mice and rats did not cause formation of anti-GBM Ab, even though renal tubular abnormalities and carcinogenesis were found (349). Case reports have coupled anti-GBM Ab disease with trichloroethane, carbon tetrachloride, herbicides, and insecticides; toulene sniffing is associated with renal damage, with improvement sometimes related to avoidance (350–352). Since the relationship between inhalants and induction of anti-GBM Abs is unclear, their role may be to alter the relationship of the ABM to the circulation so that anti-GBM Abs formed for other reasons may bind to the ABM with pulmonary hemorrhage, causing the patient to seek medical attention. For example, smoking has been stressed as a possible inducer pulmonary hemorrhage episodes in patients with anti-GBM Abs (129). Another possible example of an inhaled potential toxin is that anti-GBM Ab disease has been reported in two patients using cocaine (353,354).

Anti-GBM Abs have also been identified in the polyclonal Ab response generated in rats with HgCl2 (see above). As noted elsewhere in this chapter, differences in the GBM type IV collagen makeup of patients with Alport's syndrome can serve as allo-Ags in renal transplantation (253,255,355–357). Anti-GBM Ab-reactive Ag fractions, including type IV collagen, have been identified in human urine (358–361), and anti-GBM Abs could be induced in experimental animals by immunization with the antigenic urine fraction (362). This has led to speculation that urinary GBM Ags could be involved in anti-GBM Ab induction. Identification of alternative splicing of the $\alpha3$(IV) NCl where the GBM auto-Ag is located perhaps offer a possible antigenic source (324,363,364).

Physical or immunological injury to the GBM sometimes antedates the onset of anti-GBM Ab disease, as suggested by development of an anti-GBM Ab response with renal failure secondary to trauma induce cortical necrosis (365). Anti-GBM Abs may also occur in association with other forms of GN, such as membranous nephropathy or diabetic nephropathy (366–373). Anti-TBM Abs can complicate some instances of membranous nephritis, especially in children, and some drug reactions are also associated with the development of anti-TBM Abs (see ahead).

Anti-GBM Abs have been noted in three patients after lithotripsy (374–376). Of interest, all three patients were HLA DR2 or DR15, which is associated with most anti-GBM Ab disease (see above). Studies in nine randomly-selected patients demonstrated that circulating Abs against several GBM components, including type IV collagen, were negative, as were a number of auto-Abs, including anti-GBM Abs, in 59 consecutively studied patients in another study (377,378). Also, serum type IV collagen levels from 13 patients with renal stones and 15 with ureteral stones showed no changes after lithotripsy (379).

The report of the development of anti-GBM Ab diseases with pulmonary hemorrhage in a patient with multiple sclerosis points to the HLA haplotype in the individual (DRB1*1501-DQA1*0102-DQb1*06020) associated with both diseases (380).

3. ANTI-TBM AB DISEASE

Tubulointerstitial renal injury and TIN can occur as part of the renal injury associated with GN and its fibrotic sequelae, an autoimmune reaction, the aftermath of a toxic exposure, or associated with infections (Table 1). The latter may cause damage directly or possibly serve as a source of foreign Ag for immune mechanisms. The recent identification of Epstein-Barr virus infection in renal proximal tubules of patients with idiopathic chronic interstitial nephritis invites studies to determine the infection related mechanism of injury (381–383), as has the recent report of Epstein-Barr infection in a patient with the unusual uveitis and TIN syndrome, seen most often in young women (384).

Abs reactive with the TBM are present in 60–70% of patients with anti-GBM Ab disease and, when present, are associated with increased tubulointerstitial damage (385,386). The anti-GBM Ab reaction can involve only scattered tubules or be diffuse (as noted above), which might relate to differences in Ab reactivity or perhaps epitope distribution. Studies suggest that the anti-GBM Abs may recognize the same epitopes on GBM and TBM (387).

3.1.   Anti-TBM Ab Disease in Humans

In this section, we will consider the autoimmune reaction to TBM Ags in humans and experimental models (Table II) (3,388–390). This type of anti-TBM Ab reacts with the TBM, but not GBM. There are human examples of linear TBM Ig deposits and circulating anti-TBM Abs without glomerular involvement (391–395) and anti-TBM Abs can also be associated with non-anti-GBM Ab-induced glomerular disease, and may have evidence of tubular dysfunction, such as Fanconi syndrome. Anti-TBM Abs have also been reported after poststreptococcal GN (predominately membranous nephropathy) and in children with immune complex forms of GN as well as in SLE (396–401). We observed an infant with intractable diarrhea and severe membranous nephropathy who had anti-TBM Abs, and found both anti-jejunal BM reactivity and anti-TBM Abs in renal eluates (unpublished). Anti-TBM Abs have also been reported in patients with GN and diarrhea, nephrotic syndrome and celiac disease, possibly in villous atrophy of the small intestine, and with oxalosis and chronic TIN associated with intestinal bypass (398, 402–404). The association of anti-TBM Abs with membraneous nephropathy has now been related to Ab reactions with a specific TBM auto-Ag (405) (see ahead).

Anti-TBM Abs infrequently are observed in drug related forms of TIN (406,407, sometimes associated with methacillin (408,409), phenytoin (410), and possibly with allopurinol related TIN (411). Drug metabolites can bind and may serve as hapten-carrier immunogens, leading to the formation of anti-TBM Abs (409,410). The most frequent occurrence of anti-TBM Abs is in renal transplant recipients; however, the contribution of the Abs to transplant pathology is unclear (412–414). The anti-TBM Abs can form in response to allogenic differences in TBM Ags between donor and recipient (415). More commonly, no allogenic differences are detected, and the mechanisms responsible for induction of an anti-TBM Ab response in the immunosuppresed recipient are not defined. Anti-TBM Abs associated with the patient's primary disease can also sometimes be detected after transplantation (401,416,417).

3.2.   Experimental Models

Model systems of anti-TBM Ab-associated disease have been studied in several species (3,388–390,418), but only a brief mention of some of the findings in rat and mouse models will be covered here. Brown-Norway (BN) rats immunized with heterologous (bovine) TBM develop anti-TBM Abs with binding to the TBM, C fixation, and an intense neutrophil infiltration 7–8 days later, followed by a mononuclear infiltrate composed of T cells and increasing numbers of monocytes/macrophages (419,420). The nephritogenic TBM Ags are alloantigens and are not present in all rat strains in a reactive form (253,421). The Lewis (LEW) rat, for example, is TBM Ag-negative in terms of binding circulating anti-TBM Abs, and although making a good anti-TBM Ab response, the LEW rat escapes the nephritogenic effects of the Ab. Due to the allogenic difference, an anti-TBM Ab response can be induced in the TBM Ag-negative LEW rat by transplantation of TBM Ag-positive kidneys or other maneuvers that present TBM Ag (422,423). TBM Ag alloantigenic differences and induction of anti-TBM Ab by renal transplantation have been noted infrequently in humans (415).

As noted above, immunization of LEW rats with TBM Ag-positive BN rat renal BMs induces high levels of circulating anti-TBM Ab, since it does not bind to the LEW TBM. This Ab can be used to passively transfer anti-TBM Ab disease to BN rats (424). The kinetics of bindings are much different than the rapid binding of anti-GBM Ab. In the case of the passively administered LEW anti-TBM Ab kidney binding peaks at 5–6 days with about 170 μg IgG bound/gm of kidney, which is the amount needed to induce a lesion. The binding begins in focal areas in the cortex associated with complement fixation and a cellular inflammatory focus. Of interest, the LEW rat Ab donors develop focal granulomatous interstitial lesions which can be transferred to other LEW rats with immune cells (425).

In contrast to the rat, SJL mice develop a form of TIN several weeks after immunization with heterologous TBM in which cellular immune processes predominate, with little pathogenic role defined for the accompanying anti-TBM Ab. The SJL develops MHC class I restricted Lyt-2$^+$ effector cells, which can adoptively transfer the TIN (426–429). These effector cells express idiotypes shared with kidney-bound anti-TBM Ab (430). Susceptibility to the lesion is suggested to relate to modulation of the phenotypic selection of effector cells related to suppression and countersuppression, with the possible involvement of soluble factors (428,429,431,432). A soluble T helper cell factor that is Ag-specific and I-A-restricted can induce nephritogenic effector T cells (433). The factor has two chains, one related to the TCR/Id and the other to I-A$^+$. [The regulation of T cell function in experimental TIN is the subject of review (389,390,434).]

3.3.   Reactive TBM Ags

Trypsin digests of murine or human TBM contain a 30-kD Ag that can induce anti-TBM Ab and TIN without detected

GBM reactions in goats and BALB/c mice (435,436). As in anti-GBM Ab disease, the Ag(s) responsible for anti-TBM Ab-associated TIN are present among the collagenase-solubilized fractions of TBM (421,437). In the BN rat, TIN induced by immunization with BOV TBM, an antigen (42–45 kD) can be solubilized from BN TBM by collagenase, which accounts for the majority of the reactivity of the eluted anti-TBM Ab (421). Of interest, a small amount of the reactive material can be recovered from LEW TBM, which is Ag negative in terms of *in vivo* TBM reactions. Others have also identified TBM Ag in the LEW and have extended the study to show that genomic DNA for the gene segment of the Ag is present (438). This suggests that the LEW counterpart may be sequestered or modified in some way to mask its reactive epitopes. A monoclonal Ab reactive with BN, but not LEW, TBM was used to isolate a noncollagenous nephritogenic Ag from collagenase solubilized rabbit TBM fractions (439). The gene polymorphism involves both Ag and the susceptibility to infiltrative cellular lesions (440).

The Ag responsible for the SJL lesion is termed 3M-1 and was identified in a proximal tubular epithelial cell line which can support proliferation of MHC class II restricted T cell lines in culture (441). Ab reactive with 3M-1 can down-regulate MHC class II expression in this cell line (442). A nephritogenic CD8+ T cell line has been established which is cytotoxic to a 3M-1 expressing tubular epithelial line *in vitro* and is capable of transfer of TIN (443). Clonal analysis of these CD8+ cells indicates distinct functional phenotypes within the cell line. A 3M-1 common framework domain cDNA has been isolated from an expression library using a polyclonal anti-3M-1 Ab (444). This domain has an evolutionary relationship with a family of intermediate filament-associated proteins. A synthetic peptide derived from the predicted amino acid sequence reacts with monoclonal anti-3M-1 Ab. The peptide also stimulated growth of 3M-1 reactive T helper cells and can induce nephritogenic effector T cells. The monoclonal Ab used to isolate the rabbit 3m-1 Ag also recovered a 48-kD Ag from collagenase solubilized human renal basement membranes which reacted with two human anti-TBM Abs, but not with anti-GBM Abs (445). The reaction is primarily focused to a single epitope (446).

An Ag reactive with human anti-TBM Abs has been extracted from rabbit TBM using guanidine which is predominately a 58 kD molecule with small amounts of other components (447,448). Monoclonal Abs to the isolate recognize minor components up to 300 kD (449). The fractions can induce TIN in BM rats (450). The 58-kD Ag was found to be present in highest concentration in proximal renal tubule and small intestine which support epithelia with large absorptive capacity, suggesting a possible function (449,451). In other studies, 54 and 48 kD Ags with many of the same properties have been recovered from col-

lagenase solubilized BOV TBM and have been shown to be present along the interstitial side of the TBM in association with collagen fibers (387,452,453). The 58 kD Ag has been shown to interact with laminin and type IV collagen and to promote cell adhesion (454). In addition, the molecule can interfere with the ability of laminin to self-associate. The TBM auto-Ag is involved in adhesion interactions with proximal tubular epithelial cells (455). The 58 kD TBM auto-Ag can detect the anti-TBM Abs in patients with membanous nephropathy with anti-TBM Abs (401,405). The cDNA sequence of the rabbit, human, and murine TBM auto-Ag have about 85% homology (456–458). The TBM auto-Ag is regulated during development. In nephronophthisis, decreased reactivity of TBM with human anti-TBM Abs, but not monoclonal anti-TBM Abs, has been suggested to correlate with defective TBM structure (460).

ACKNOWLEDGEMENTS

This is Publication No. 13214-IMM from the Department of Immunology, The Scripps Research Institute, La Jolla, CA. This work was supported in part by United States Public Health Service grants No. DK20043, DK18381, and American Heart Association Grant No. 9920539Z.

References

1. Wilson, C.B. 1996. Renal response to immunologic glomerular injury. In: The Kidney, 5th Edition, Vol. 2. B.M. Brenner, editor. Saunders, Philadelphia. pp. 1253–1391.
2. Wilson, C.B. 1997. Immune models of glomerular injury. In: Immunologic Renal Diseases. E.G. Neilson and W.G. Couser, editors. Lippincott-Raven, Philadelphia. 729–773.
3. Wilson, C.B. 1997. Immune models of tubulointerstitial injury. In: Immunologic Renal Diseases. E.G. Neilson and W.G. Couser, editors. Lippincott-Raven, Philadelphia. pp. 775–799.
4. Jennette, J.C., and R.J. Falk. 1997. Diagnosis and management of glomerular diseases. *Med. Clin. No. Amer.* 81:653–677.
5. Dixon, F.J. 1963. The role of antigen-antibody complexes in disease. *Harvey Lect.* 58:21–52.
6. Unanue, E.R., and F.J. Dixon. 1967. Experimental glomerulonephritis: Immunological events and pathogenetic mechanisms. *Adv. Immunol.* 6:1–90.
7. Van Zyl Smit, R., A.J. Rees, and D.K. Peters, 1983. Factors affecting severity of injury during nephrotoxic nephritis in rabbits. *Clin. Exp. Immunol.* 54:366–372.
8. Karkar, A.M., Y. Koshino, S.J. Cashman, A.C. Dash, J. Bonnefoy, A. Meager, and A.J. Rees. 1992. Passive immunization against tumour necrosis factor-alpha (TNF-alpha) and IL-1 beta protects from LPS enhancing glomerular injury in nephrotoxic nephritis in rats. *Clin. Exp. Immunol.* 90:312–318.
9. Karkar, A.M., and A.J. Rees. 1997. Influence of endotoxin contamination on anti-GBM antibody induced glomerular injury in rats. *Kidney Int.* 52:1579–1583.

10. Jennings, L., O.A. Roholt, D. Pressman, M. Blau, G.A. Andres, and J.R. Brentjens. 1981. Experimental anti-alveolar basement membrane antibody-mediated pneumonitis. I. The role of increased permeability of the alveolar capillary wall induced by oxygen. *J. Immunol.* 127:129–134.

11. Downie, G.H., O.A. Roholt, L. Jennings, M. Blau, J.R. Brentjens, and G.A. Andres. 1982. Experimental anti-alveolar basement membrane antibody-mediated pneumonitis. II. Role of endothelial damage and repair, induction of autologous phase, and kinetics of antibody deposition in Lewis rats. *J. Immunol.* 129:2647–2652.

12. Yamamoto, T., and C.B., Wilson. 1987. Binding of anti-basement membrane antibody to alveolar basement membrane after intratracheal gasoline instillation in rabbits. *Am. J. Pathol.* 126:497–505.

13. Queluz, T.H., I. Pawlowski, M.J. Brunda, J.R. Brentjens, A.O. Vladutiu, and G. Andres. 1990. Pathogenesis of an experimental model of Goodpasture's hemorrhagic pneumonitis. *J. Clin. Invest.* 85:1507–1515.

14. Steblay, R.W. 1962. Glomerulonephritis induced in sheep by injections of heterologous glomerular basement membrane and Freund's complete adjuvant. *J. Exp. Med.* 116:253–271.

15. Robertson, J.L., G.S. Hill, and D.T. Rowlands, Jr. 1977. Tubulointerstitial nephritis and glomerulonephritis in Brown-Norway rats immunized with heterologous glomerular basement membrane. *Am. J. Pathol.* 88:53–64.

16. Stuffers-Heiman, M., E. Günther, and L.A. Van Es. 1979. Induction of autoimmunity to antigens of the glomerular basement membrane in in bred Brown-Norway rats. *Immunology* 36:759–767.

17. Sado, Y., T. Okigaki, H. Takamiya, and S. Seno. 1984. Experimental autoimmune glomerulonephritis with pulmonary hemorrhage in rats. The dose-effect relationship of the nephritogenic antigen from bovine glomerular basement membrane. *J. Clin. Lab. Immunol.* 15:199–204.

18. Pusey, C.D., M.J. Holland, S.J. Cashman, R.A. Sinico, J.-J. Lloveras, D.J. Evans, and C.M. Lockwood. 1991. Experimental autoimmune glomerulonephritis induced by homologous and isologous glomerular basement membrane in Brown Norway rat. *Nephrol. Dial. Transplant.* 6:457–465.

19. Bolton, W.K., W.J. May, and B.C. Sturgill. 1993. Proliferative autoimmune glomerulonephritis in rats: A model for autoimmune glomerulonephritis in humans. *Kidney Int.* 44:294–306.

20. Steblay, R.W. 1963. Glomerulonephritis induced in monkeys by injections of heterologous glomerular basement membrane and Freund's adjuvant. *Nature* 197:1173–1176.

21. Unanue, E.R., and F.J. Dixon. 1967. Experimental allergic glomerulonephritis induced in rabbits with heterologous renal antigens. *J. Exp. Med.* 125:149–162.

22. Couser, W.G., M. Stilmant, and E.J. Lewis. 1973. Experimental glomerulonephritis in the guinea pig. I. Glomerular lesions associated with antiglomerular basement membrane antibody deposits. *Lab. Invest.* 29:236–243.

23. Bolton, W.K., F.L. Tucker, and B.C. Sturgill. 1984. New avian model of experimental glomerulonephritis consistent with mediation by cellular immunity. Nonhumorally mediated glomerulonephritis in chickens. *J. Clin. Invest.* 73:1263–1276.

24. Bolton, W.K., M. Chandra, T.M. Tyson, P.R. Kirkpatrick, M.J. Sadovnic, and B.C. Sturgill. 1988. Transfer of experimental glomerulonephritis in chickens by mononuclear cells. *Kidney Int.* 34:598–610.

25. Li, S., S.R. Holdsworth, and P.G. Tipping. 1997. Antibody independent crescentic glomerulonephritis in μ chain deficient mice. *Kidney Int.* 51:672–678.

26. Holdsworth, S.R., T.J. Neale, and C.B. Wilson. 1981. Abrogation of macrophage-dependent injury in experimental glomerulonephritis in the rabbit. Use of an antimacrophage serum. *J. Clin. Invest.* 68:686–698.

27. Oliveira, D.B., and D.K. Peters. 1990. Autoimmunity and the pathogenesis of glomerulonephritis. *Pediatr. Nephrol.* 4:185–192.

28. Rovin, B.H., and G.F. Schreiner. 1991. Cell-mediated immunity in glomerular disease. *Annu. Rev. Med.* 42:25–33.

29. Lowe, M.G., S.R. Holdsworth, and P.G. Tipping. 1991. T lymphocyte participation in acute serum sickness glomerulonephritis in rabbits. *Immunol. Cell Biol.* 69:81–87.

30. Eldredge, C., S. Merritt, M. Goyal, H. Kulaga, T.J. Kindt, and R. Wiggins. 1991. Analysis of T cells and major histocompatibility complex class I and class II mRNA and protein content and distribution in antiglomerular basement membrane disease in the rabbit. *Am. J. Pathol.* 139:1021–1035.

31. Couser, W.G. 1999. Sensitized cells come of age: A new era in renal immunology with important therapeutic implications. *J. Am. Soc. Nephrol.* 10:664–665.

32. Sado, Y., I. Naito, M. Akita, and T. Okigaki. 1986. Strain specific responses of inbred rats on the severity of experimental autoimmune glomerulonephritis. *J. Clin. Lab. Immunol.* 19:193–199.

33. Granados, R., D.L. Mendrick, and H.G. Rennke. 1990. Antibody-induced crescent formation in WKY rats: potential role of antibody-dependent cell cytotoxicity (ADCC) *in vivo*. *Kidney Int.* 37:414.

34. Reynolds, J., K. Mavromatidis, S.J. Cashman, D.J. Evans, and C.D. Pusey. 1998. Experimental autoimmune glomerulonephritis (EAG) induced by homologous and heterologous gloumerular basement membrane in two substrains of Wistar-Kyoto rat. *Nephrol., Dial., Transplantation* 13:44–52.

35. Sado, Y., M. Kagawa, S. Rauf, I. Natio, C. Moritoh, and T. Okigaki. 1992. Isologous monoclonal antibodies can induce anti-GBM glomerulonephritis in rats. *J. Pathol.* 168:221–227.

36. Kawasaki, K., E. Yaoita, T. Yamamoto, and I. Kihara. 1992. Depletion of CD8 positive cells in nephrotoxic serum nephritis of WKY rats. *Kidney Int.* 41:1517–1526.

37. Kawasaki, K., E. Yaoita, T. Yamamoto, T. Tamatani, M. Miyasaka, and I. Kihara. 1993. Antibodies against intercellular adhesion molecule-I and lymphocyte function-associated antigen-1 prevent glomerular injury in rat experimental crescentic glomerulonephritis. *J. Immunol.* 150:1074–1083.

38. Nishikawa, K., P.S. Linsley, A.B. Collins, I. Stamenkovic, R.T. McCluskey, and G. Andres. 1993. Antibodies to intercellular adhesion molecule 1/lymphocyte function-associated antigen 1 prevent crescent formation in rat autoimmune glomerulonephritis. *J. Exp. Med.* 177:667–677.

39. Huang, X.R., S.R. Holdsworth, and P.G. Tipping. 1994. Evidence for delayed-type hypersensitivity mechanisms in glomerular crescent formation. *Kidney Int.* 46:69–78.

40. Yamamoto, T., H. Fujinaka, K. Kawasaki, E. Yaoita, L. Feng, C.B. Wilson, and I. Kihara. 1996. CD8+ lymphocytes play a central role in the development of anti-GBM nephritis through induction of ICAM-I and chemokines in WKY rats. *Contrib. Nephrol.* 118:109–112.

41. Fujinaka, H., T. Yamamoto, M. Takeya, L. Feng, K. Kawasaki, E. Yaoita, D. Kondo, C.B. Wilson, M. Uchiyama, and I Kihara. 1997. Suppression of anti-glomerular base-

ment membrane nephritis by administration of anti-monocyte chemoattractant protein-1 antibody in WKY rats. *J. Am. Soc. Nephrol.* 8:1174–1178.

42. Fujinaka, H., T. Yamamoto, L. Feng, K. Kawasaki, E. Yaoita, S. Hirose, S. Goto, C.B. Wilson, M. Uchiyama, and I. Kihara. 1997. Crucial role of CD8-positive lymphocytes in glomerular expression of ICAM-1 and cytokines in crescentic glomerulonephritis of WKY rats. *J. Immunol.* 158:4978–4983.

43. Kalowski, S., D.G. McKay, E.L. Howes, Jr., I. Csavossy, and M. Wolfson. 1976. Multinucleated giant cells in antiglomerular basement membrane antibody-induced glomerulonephritis. *Nephron* 16:415–426.

44. Yang, N., N.M. Isbel, D.J. Nikolic-Paterson, Y. Li, R. Ye, R.C. Atkins, and H.Y. Lan. 1998. Local macrophage proliferation in human glomerulonephritis. *Kidney Int.* 54:143–151.

45. Lan, H.Y., D.J. Nikolic-Paterson, W. Mu, and R.C. Atkins. 1995. Local macrophage proliferation in multinucleated giant cell and granuloma formation in experimental Goodpasture's syndrome. *Am. J. Pathol.* 147:1214–1220.

46. Lan, H.Y., D.J. Nikolic-Paterson, W. Mu, and R.C. Atkins. 1995. Local macrophage proliferation in the progression of glomerular and tubulointerstitial injury in rat anti-GBM glomerulonephritis. *Kidney Int.* 48:753–760.

47. Van Alderwegen, I.E., J.A. Bruijn, and E. de Heer. 1997. T cell subsets in immunologically-mediated glomerulonephritis. *Histol. Histopathol.* 12:241–250.

48. Lan, H.Y., D.J. Nikolic-Paterson, W. Mu, and R.C. Atkins. 1997. Local macrophage proliferation in the pathogenesis of glomerular crescent formation in rat anti-glomerular basement membrane (GBM) glomerulonephritis. *Clin. Exp. Immunol.* 110:233–240.

49. Huang, X.R., P.G. Tipping, J. Apostolopoulos, C. Oettinger, M. D'Souza, G. Milton, and S.R. Holdsworth. 1997. Mechanisms of T cell-induced glomerular injury in anti-glomerular basement membrane (GBM) glomerulonephritis in rats. *Clin. Exp. Immunol.* 109:134–142.

50. Tipping, P.G., X.R. Huang, M. Qi, G.Y. Van, and W.W. Tang. 1998. Crescentic glomerulonephritis in CD4- and CD8-deficient mice. *Am. J. Pathol.* 152:1541–1548.

51. Lan, H.Y., X.Q. Yu, N. Yang, D.J. Nikolic-Paterson, W. Mu, R. Pichler, R.J. Johnson, and R.C. Atkins. De novo glomerular osteopontin expression in rat crescentic glomerulonephritis. *Kidney Int.* 53:136–145.

52. Yu, X.Q., J.M. Fan, D.J. Nikolic-Paterson, N. Yang, W. Mu, R. Pichler, R.J. Johnson, R.C. Atkins, and H.Y. Lan. 1999. IL-1 up-regulates osteopontin expression in experimental crescentic glomerulonephritis in the rat. *Am. J. Pathol.* 154:833–841.

53. Tipping, P.G., A.R. Kitching, X.R. Huang, D.A. Mutch, and S.R. Holdsworth. 1997. Immune modulation with interleukin-4 and interleukin-10 prevents crescent formation and glomerular injury in experimental glomerulonephritis. *Eur. J. Immunol.* 27:530–537.

54. Tam, F.W.K., J. Smith, A.M. Karkar, C.D. Pusey, and A.J. Rees. 1997. Interleukin-4 ameliorates experimental glomerulonephritis and up-regulates glomerular gene expression of IL-1 decoy receptor. *Kidney Int.* 52:1224–1231.

55. Coelho, S.N., S. Saleem, B.T. Konieczny, K.R. Parekh, F.K. Baddoura, and F.G. Lakkis. 1997. Immunologic determinants of susceptibility to experimental glomerulonephritis: Role of cellular immunity. *Kidney Int.* 51:646–652.

56. Ring, G.H., and F.G. Lakkis. 1998. T lymphocyte-derived cytokines in experimental glomerulonephritis: testing the Th1/Th2 hypothesis. Editorial. *Nephrol. Dial. Transplantation* 13:1101–1103.

57. Schatzmann, U., C. Haas, and M. Le Hir. 1999. A Th1 response is essential for induction of crescentic glomerulonephritis in mice. *Kidney Blood Pressure Res.* 22:135–139.

58. Lloyd, C.M., A.W. Minto, M.E. Dorf, A. Proudfoot, T.N.C. Wells, D.J. Salant, and J.-C. Gutierrez-Ramon. 1997. RANTES and monocyte chemoattractant protein-1 (MCP-1) play an important role in the inflammatory phase of crescentic nephritis, but only MCP-1 is involved in crescent formation and interstitial fibrosis. *J. Exp. Med.* 185:1371–1380.

59. Topham, P.S., V. Csizmadia, D. Soler, D. Hines, C.J. Gerard, D.J. Salant, and W.W. Hancock. 1999. Lack of chemokine receptor CCR1 enhances Th1 responses and glomerular injury during nephrotoxic nephritis. *J. Clin. Invest.* 104:1549–1557.

60. Tang, W.W., M. Qi, G.Y. Van, G.P. Wariner, and B. Samal. 1996. Leukemia inhibitory factor ameliorates experimental anti-GBM Ab glomerulonephritis. *Kidney Int.* 50:1922–1927.

61. Lan, H.Y., M. Bacher, N. Yang, W. Mu, D.J. Nikolic-Paterson, C. Metz, A. Meinhardt, R. Bucala, and R.C. Atkins. 1997. The pathogenic role of macrophage migration inhibitory factor in immunologically induced kidney disease in the rat. *J. Exp. Med.* 185:1455–1465.

62. Haas, C., B. Ryffel, and M. LeHir. 1995. Crescentic glomerulonephritis in interferon-gamma receptor deficient mice. *J. Inflamm.* 96:206–213.

63. Balomenos, D., R. Rumold, and A.N. Theofilopoulos. 1998. Interferon-gamma is required for lupus-like disease and lymphoaccumulation in MRL-1pr mice. *J. Clin. Invest.* 101:364–371.

64. Park, S. YH., S. Ueda, H. Ohno, Y. Hamano, M. Tanaka, T. Shiratori, T. Yamazaki, H. Arase, N. Arase, A. Karasawa, S. Sato, B. Ledermann, Y. Kondo, K. Okumura, C. Ra, and T. Saito. 1998. Resistance of Fc receptor-deficient mice to fatal glomerulonephritis. *J. Clin. Invest.* 102:1229–1238.

65. Kitching, A.R., S.R. Holdsworth, and P.G. Tipping. 1999. IFN-gamma mediates crescent formation and cell-mediated immune injury in murine glomerulonephritis. *J. Am. Soc. Nephrol.* 10:752–759.

66. Ring, G.H., Z. Dai, S. Saleem, F.K. Baddoura, and F.G. Lakkis. 1999. Increased susceptibility to immunologically mediated glomerulonephritis in IFN-γ-deficient mice. *J. Immunol.* 163:2243–2248.

67. Scheer, R.L., and M.A. Grossman. 1964. Immune aspects of the glomerulonephritis associated with pulmonary hemorrhage. *Ann. Int. Med.* 60:1009–1021.

68. Sturgill, B.C., and F.B. Westervelt. 1965. Immunofluorescence studies in a case of Goodpasture's syndrome. *JAMA* 194:914–916.

69. Duncan, D.A., K.N. Drummond, A.F. Michael, and R.I. Vernier. 1965. Pulmonary hemorrhage and glomerulonephritis. Report of six cases and study of the renal lesion by the fluorescent antibody technique and electron microscopy. *Ann Intern. Med.* 62:920–938.

70. Goodpasture, E.W. 1919. The significance of certain pulmonary lesions in relation to the etiology of influenza. *Am. J. Med. Sci.* 158:863–870.

71. Parkin, T.W., I.E. Rusted, H.B. Burchell, and J.E. Edwards. 1955. Hemorrhagic and interstitial pneumonitis with nephritis. *Am. J. Med.* 18:220.

72. Benoit, F.L., D.B. Rulon, G.B. Theil, P.D. Doolan, and R.H. Watten. 1964. Goodpasture's syndroime: a clinicopathologic entity. *Am. J. Med.* 37:424–444.

73. Wilson, C.B., and F.J. Dixon. 1973. Anti-glomerular basement membrane antibody-induced glomerulonephritis. *Kidney Int.* 3:74–89.

74. Steblay, R.W. 1964. Transfer of nephritis from sheep with autoimmune nephritis to recipient sheep by artery-to-artery cross-circulation. *Fed. Proc.* 23:449.

75. Lerner, R.A., and F.J. Dixon. 1966. Transfer of ovine experimental allergic glomerulonephritis (EAG) with serum. *J. Exp. Med.* 124:431–442.

76. Bygren, P., J. Wieslander, and D. Heinegärd. 1987. Glomerulonephritis induced in sheep by immunization with human glomerular basement membrane. *Kidney Int.* 31:25–31.

77. James, M.P., P.B. Herdson, and J.B. Gavin. 1981. Recurrence of antiglomerular basement membrane glomerulonephritis in sheep renal allografts. *Pathology* 13:335–344.

78. Lerner, R.A., R.J. Glassock, and F.J. Dixon. 1967. The role of anti-glomerular basement membrane antibody in the pathogenesis of human glomerulonephritis. *J. Exp. Med.* 126:989–1004.

79. Almkuist, R.D., V.M. Buckalew, Jr., P. Hirszel, J.F. Maher, P.M. James, and C.B. Wilson. 1981. Recurrence of anti-glomerular basement membrane antibody mediated glomerulonephritis in an isograft. *Clin. Immunol. Immunopathol.* 18:54–60.

80. Fonck, C., G. Loute, J.P. Cosyns, and Y. Pirson. 1998. Recurrent fulminant anti-glomerular basement membrane nephritis at a 7-year interval. *Am. J. Kidney Dis.* 32:323–327.

81. Ryan, G.B., and M.J. Karnovsky. 1976. Distribution of endogenous albumin in the rat glomerulus: role of hemodynamic factors in glomerular barrier function. *Kidney Int.* 9:36–45.

82. Ryan, G.B., S.J. Hein, and M.J. Karnovsky. 1976. Glomerular permeability to proteins. Effects of hemodynamic factors on the distribution of endogenous immunoglobulin G and exogenous catalase in the rat glomerulus. *Lab. Invest.* 34:415–427.

83. Marquardt, H., C.B. Wilson, and F.J. Dixon. 1973. Isolation and immunological characterization of human glomerular basement membrane antigens. *Kidney Int.* 3:57–65.

84. Wilson, C.B., H. Marquardt, and F.J. Dixon. 1974. Radioimmunoassay (RIA) for circulating antiglomerular basement membrane (GBM) antibodies. *Kidney Int.* 6:144a (Abstr.).

85. Wilson, C.B. 1980. Radioimmunoassay for anti-glomerular basement membrane antibodies. In: Manual of Clinical Immunology, Second Edition. N.R. Rose, and H. Friedman, editors. American Society for Microbiology, Washington, DC, pp. 376–379.

86. Pusey, C.D., A. Dash, M.J. Kershaw, A. Morgan, A. Reilly, A.J. Rees, and C.M. Lockwood. 1987. A single autoantigen in Goodpasture's syndrome identified by a monoclonal antibody to human glomerular basement membrane. *Lab. Invest.* 56:23–31.

87. Hellmark, T., C. Johansson, and J. Wieslander. 1994. Characterization of anti-GBM antibodies involved in Goodpasture's syndrome. *Kidney Int.* 46:823–829.

88. Litwin, C.M., C.L. Mouritsen, P.A. Wilfahrt, M.C. Schroder, and H.R. Hill. 1996. Anti-glomerular basement membrane disease: role of enzyme-linked immunosorbent assays in diagnosis. *Biochem. Mol. Med.* 59:52–56.

89. Savige, J.A., L. Chang, S. Horn, and S.M. Crowe. 1994. Antinuclear, anti-neutrophil cytoplasmic and anti-glomerular basement membrane antibodies in HIV-infected individuals. *Autoimmunity* 18:205–211.

90. Kalluri, R., S. Petrides, C.B. Wilson, J.E. Tomaszewski, H.I. Palevsky, M.A. Grippi, M.P. Madaio, and E.G. Neilson. 1996. Anti-alpha (IV) collagen autoantibodies associated with lung adenocarcinoma presenting as the Goodpasture syndrome. *Ann. Int. Med.* 124:651–653.

91. Johansson, C., R. Butkowski, P. Swedenborg, P. Alm, and J. Weislander. 1993. Characterization of a non-Goodpasture autoantibody to type IV collagen. *Nephrol. Dial. Transplantation* 8:1205–1210.

92. McPhaul, J.J., Jr., and F.J. Dixon. 1971. Characterization of immunoglobulin G anti-glomerular basement membrane antibodies eluted from kidneys of patients with glomerulonephritis. II. IgG subtypes and *in vitro* complement fixation. *J. Immunol.* 107:678–684.

93. Border, W.A., R.W. Baehler, D. Bhathena, and R.J. Glassock. 1979. IgA antibasement membrane nephritis with pulmonary hemorrhage. *Ann. Intern. Med.* 91:21–25.

94. Savage, C.O., C.D. Pusey, C. Bowman, A.J. Rees, and C.M. Lockwood. 1986. Antiglomerular basement membrane antibody mediated disease in the British Isles 1980–4. *Br. Med. J.* 292:301–304.

95. de Caestecker, M.P., C.L. Hall, and A.G. MacIver. 1990. Atypical antiglomerular basement membrane disease associated with thin membrane nephropathy. *Nephrol. Dial. Transplant.* 5:909–913.

96. Noel, L.H., P. Aucouturier, R.C. Monteiro, J.L. Preud 'Homme, and P. Lesavre. 1988. Glomerular and serum immunoglobulin G subclasses in membranous nephropathy and anti-glomerular basement membrane nephritis. *Clin. Immunol. Immunopathol.* 46:186–194.

97. Weber, M., A.W. Lohse, M. Manns, K.H. Meyer zum Buschenfelde, and H. Kohler. 1988. IgG subclass distribution of autoantibodies to glomerular basement membrane in Goodpasture's syndrome compared to other autoantibodies. *Nephron.* 49:54–57.

98. Segelmark, M., R. Butkowski, and J. Wieslander. 1990. Antigen restriction and IgG subclasses among anti-GBM autoantibodies. *Nephrol. Dial. Transplant.* 5:991–996.

99. Bowman, C., K. Ambrus, and C.M. Lockwood. 1987. Restriction of human IgG subclass expression in the population of auto-antibodies to glomerular basement membrane. *Clin. Exp. Immunol.* 69:341–349.

100. Melvin, T., Y. Kim, and A.F. Michael. 1984. Selective binding of IgG4 and other negatively charged plasma proteins in normal and diabetic human kidneys. *Am. J. Pathol.* 115:443–446.

101. Salant, D.J. 1987. Immunopathogenesis of crescentic glomerulonephritis and lung purpura. *Kidney Int.* 32:408–425.

102. Couser, W.G. 1988. Rapidly progressive glomerulonephritis: Classifications, pathogenetic mechanisms, and therapy. *Am. J. Kidney. Dis.* 11:449–464.

103. Turner, A.N., and A.J. Rees. 1996. Goodpasture's disease and Alport's syndromes. *Ann. Rev. Med.* 47:377–386.

104. Daly, C., P.J. Conlon, W. Medwar, and J.J. Walshe. 1996. Characteristics and outcome of anti-glomerular basement disease: a single center experience. *Renal Failure* 18:105–112.

105. Kluth, D.C., and A.J. Rees. 1999. Anti-glomerular basement membrane disease. *J. Am. Soc. Nephrol.* 10:2446–2453.

106. Falk, R.J., J.C. Jennette, and P.H. Nachman. 2000. Primary glomerular disease. In: The Kidney, 6th Edition. B.M. Brenner, editor. Saunders, Philadelphia. pp. 1263–1249.

107. Appel, G.B., J. Radhakrishnan, and V. D'Agati. 2000. Secondary glomerular disease. In: The Kidney, 6th Edition. B.M. Brenner, editor. Saunders, Philadelphia. pp. 1350–1448.

108. Wilson, C.B. 1988. Immunologic diseases of the lung and kidney (Goodpasture's syndrome). In: Pulmonary Disease and Disorders, 2nd Edition. A.P. Fishman, editor.

109. Wilson, C.B. 1991. Goodpasture's syndrome. In: Therapy of Renal Diseases and Related Disorders, second edition. W.N. Suky and S.G. Massry, editors. Kluwer Academic Publishers, Boston. pp. 333–342.

110. Bolton, W.K. 1996. Goodpasture's syndrome. Kidney Int. 50:1753–1766.

111. Ball, J.A., and K.R. Young, Jr. 1998. Pulmonary manifestations of Goodpasture's syndrome. Antiglomerular basement membrane disease and related disorders. Clin. Chest. Med. 19:777–791.ix.

112. Bosch, X., and J. Font. 1999. The pulmonary-renal syndrome: a poorly understood clinicopathologic condition. Lupus 8:258–262.

113. Rydel, J.J., and R.A. Rodby. 1998. An 18-year-old man with Goodpasture's syndrome and ANCA-negative central nervous system vasculitis. Am. J. Kidney Dis. 31:345–349.

114. Satoh, H. 1997. Anti-glomerular basement membrane antibody-induced inflammation in rat cochlear plexus. Acta Otolaryngol. 117:80–86

115. Min, S.A., P. Rutherford, M.K. Ward, I Wheeler, H. Robertson, and T.H. Goodship. 1997. Goodpasture's syndrome with normal renal function. Nephrol., Dial., Transplantation 11:2302–2305.

116. Ang, C., J. Savige, J. Dawborn, P. Miach, W. Heale, B, Clarke, and R.S. Sinclair. 1998. Anti-glomerular basement membrane (GBM)-antibody-mediated disease with normal renal function. Nephrol., Dial., Transplantation 13:935–939.

117. McCarthy, L.J., J. Cotton, C. Danielson, V. Graves, and J. Bergstein. 1994. Goodpasture's syndrome in childhood: treatment with plasmapheresis and immunosuppression. J. Clin. Apheresis 9:116–119.

118. Bigler, S.A., W.M. Parry, D.S. Fitzwater, and R. Baliga. 1997. An 11-month-old with anti-glomerular basement membrane disease. Am. J. Kidney Dis. 30:710–712.

119. Trivedi, V.A., J.S. Malter, and E. Guillery. 1998. Pediatric anti-glomerular basement membrane glomerulonephritis. J. Clin. Apheresis. 13:69.

120. Jeraj, K., A.J. Fish, K. Yoshioka, and A.F. Michael. 1984. Development and heterogeneity of antigens in the immature nephron. Reactivity with human antiglomerular basement membrane autoantibodies. Am. J. Pathol. 117:180–183.

121. Lockwood, C.M., J.M. Boulton-Jones, R.M. Lowenthal, I.J. Simpson, D.K. Peters, and C.B. Wilson. 1975. Recovery from Goodpasture's syndrome after immunosuppressive treatment and plasmapheresis. Br. Med. J. 2:252–254.

122. Johnson, J.P., J. Moore, Jr., H.A. Austin, J.E. Balow, T.T. Antonovych, and C.B. Wilson. 1985. Therapy of anti-glomerular basement membrane antibody disease: Analysis of prognostic significance of clinical, pathologic and treatment factors. Medicine 64:219–227.

123. Madore, F., J.M. Lazarus, and H.R. Brady. 1996. Therapeutic plasma exchange in renal disease. J. Am. Soc. Nephrol. 7:367–386.

124. Levy, J.B., and C.D. Pusey. 1997. Still a role for plasma exchange in rapidly progressive glomerulonephritis? J. Nephrol. 10:7–13.

125. Merkel, F., K.O. Netzer, O. Gross, M. Marx, and M. Weber. 1998. Therapeutic options for critically ill patients suffering from progressive lupus nephritis or Goodpasture's syndrome. Kidney Int. 64:S31–S38.

126. Jindal, K.K. 1999. Management of idiopathic crescentic and diffuse proliferative glomerulonephritis: Evidence based recommendations. Kidney Int. 55:S33–S40.

127. Conlon, P.J., Jr., J.J. Walshe, C. Daly, M. Carmody, B. Keogh, J. Donohoe, and S. O'Neill. 1994. Antiglomerular basement mnembrane disease: The long-term pulmonary outcome. Am. J. Kidney. Dis. 23:794–796.

128. Rees, A.J., C.M. Lockwood, and D.K. Peters. 1977. Enhanced allergic tissue injury in Goodpasture's syndrome by concurrent bacterial infection. Br. Med. J. 2:723–726.

129. Donachy, M., and A.J. Rees. 1983. Cigarette smoking and lung haemorrhage in glomerulonephritis caused by autoantibodies to glomerular basement membrane. Lancet 2:1390–1392.

130. Keogh, A.M., L.S. Ibels, D.H. Allen, J.P. Isbister, and M.C. Kennedy. 1984. Exacerbation of Goodpasture's syndrome after inadvertent exposure to hydrocarbon fumes. Br. Med. J. 288:188.

131. Dahlberg, P.J., S.B. Kurtz, J.V. Donadio, K.E. Holley, J.A. Velosa, and C.B. Wilson. 1978. Recurrent Goodpasture's syndrome. Mayo Clin. Proc. 53:533.

132. Hind, C.R.K., C. Bowman, C.G. Winearls, and C.M. Lockwood. 1984. Recurrence of circulating anti-glomerular basement membrane antibody three years after immunosuppressive treatment and plasma exchange. Clin. Nephrol. 21:244–246.

133. Mehler, P.S., M.W. Brunvand, M.P. Hutt, and R.J. Anderson. 1987. Chronic recurrent Goodpasture's syndrome. Am. J. Med. 82:833–835.

134. Klasa, R.J., R.T. Abboud, H.S. Ballon, and L. Grossman. 1988. Goodpasture's syndrome. Recurrence after a five-year remission. Case report and review of the literature. Am. J. Med. 84:751–755.

135. Levy, J.B., R.H. Lachmann, and C.D. Pusey. 1996. Recurrent Goodpasture's disease. Am. J. Kidney Dis. 27:573–578.

136. Saxena, R., P. Bygren, N. Rasmussen, and J. Wieslander. 1991. Circulating autoantibodies in patients with extracapillary glomerulonephritis. Nephrol. Dial. Transplantation 6:389–397.

137. Saxena, R., P. Bygren, B. Arvastson, and J. Wieslander. 1995. Circulating autoantibodies as serological markers in the differential diagnosis of pulmonary renal syndromes. J. Int. Med. 238:143–152.

138. Jayne, D.R., P.D. Marshall, S.J. Jones, and C.M. Lockwood. 1990. Autoantibodies to GBM and neutrophil cytoplasm in rapidly progressive glomerulonephritis. Kidney Int. 37:965–970.

139. Bosch, X., E. Mirapeix, J. Font, X. Borrellas, R. Rodríguez, A. López-Soto, M. Ingelmo, and L. Revert. 1991. Prognostic implication of anti-neutrophil cytoplasmic autoantibodies with myeloperoxidase specificity in anti-glomerular basement membrane disease. Clin. Nephrol. 36:107–113.

140. McCance, D.R., A.P. Maxwell, C.M. Hill, and C.C. Doherty. 1992. Glomerulonephritis associated with antibodies to neutrophil cytoplasm and glomerular basement membrane. Postgrad. Med. J. 68:186–188.

141. Weber, M.F., K. Andrassy, O. Pulling, J. Koderisch, and K. Netzer. 1992. Antineutrophil-cytoplasmic antibodies and antiglomerular basement membrane antibodies in

Goodpasture's syndrome and in Wegener's granulomatosis. *J. Am. Soc. Nephrol.* 2:1227–1234.

142. Short, A.K., V.L. Esnault, and C.M. Lockwood. 1995. Anti-neutrophil cytoplasm antibodies and anti-glomerular basement membrane antibodies: two coexisting distinct autoreactivities detectable in patients with rapidly progressive glomerulonephritis. *Am. J. Kidney. Dis.* 26:439–445.

143. Kalluri, R., K. Meyers, A. Mogyorosi, M.P. Madaio, and E.G. Neilson. 1997. Goodpasture syndrome involving overlap with Wegener's granulomatosis and anti-glomerular basement membrane disease. *J. Am. Soc. Nephrol.* 8:1795–800.

144. Verburgh, C.A., J.A. Bruijn, M.R. Daha, and L.A. van Es. 1999. Sequential development of anti-GBM nephritis and ANCA-associated Pauci-immune glomerulonephritis. *Am. J. Kidney Dis.* 34:344–348.

145. Falk, R.J., and J.C. Jennette. 1988. Anti-neutrophil cytoplasmic autoantibodies with specificity for myeloperoxidase in patients with systemic vasculitis and idiopathic necrotizing and crescentic glomerulonephritis. *N. Engl. J. Med.* 318:1651–1657.

146. Jennette, J.C., and R.J. Falk. 1990. Antineutrophil cytoplasmic autoantibodies and associated diseases: a review. *Am. J. Kidney Dis.* 15:517–529.

147. Niles, J.L., G. Pan, A.B. Collins, T. Shannon, S. Skates, R. Fienberg, M.A. Arnaout, and R.T. McCluskey. 1991. Antigen-specific radioimmunoassays for anti-neutrophil cytoplasmic antibodies in the diagnosis of rapidly progressive glomerulonephritis. *J. Am. Soc. Nephrol.* 2:27–36.

148. Kallenberg, C.G.M., E. Brouwer, J.J. Weening, and J.W. Tervaert. 1994. Anti-neutrophil cytoplasmic antibodies: Current diagnostic and pathophysiological potential. *Kidney Int.* 46:1–15.

149. Jennette, J.C., and R.J. Falk. 1997. Medical progress: small-vessel vasculitis. *N. Engl. J. Med.* 337:1512–1523.

150. Savage, C.O.S., L. Harper, and D. Adu. 1997. Primary systemic vasculitis. *Lancet* 349:553–558.

151. Esnault, V.L., B. Soleimani, M.T. Keogan, A.A. Brownlee, D.R. Jayne, and C.M. Lockwood. 1992. Association of IgM with IgG ANCA in patients presenting with pulmonary hemorrhage. *Kidney Int.* 41:1304–1310.

152. Thomas, D.M., R. Moore, K. Donovan, D.C., Wheeler, V.L. Esnault, and C.M. Lockwood. 1992. Pulmonary-renal syndrome in association with anti-GBM and IgM ANCA *Lancet* 339:1304.

153. Charles, L.A., R.J. Falk, and J.C. Jennette. 1992. Reactivity of antineutrophil cytoplasmic autoantibodies with monoculear phagocytes. *J. Leukocyte. Biol.* 51:65–68.

154. Niles, J.L., R.T. McCluskey, M.F. Ahmed, and M.A. Arnaout. 1989. Wegener's granulomatosis autoantigen is a novel neutrophil serine proteinase. *Blood* 74:1888–1893.

155. Goldschmeding, R., C.E. van der Schoot, D. ten Bokkel Huinink, C.E. Hack, M.E. van den Ende, C.G.M. Kallenberg, and A.E. von dem Borne. 1989. Wegener's granulomatosis autoantibodies identify a novel diisopropylfluorophosphate-binding protein in the lysosomes of normal human neutrophils. *J. Clin. Invest.* 84:1577–1587.

156. Wiik, A., L. Stummann, L. Kjeldsen, N. Borregaard, S. Ullman, S. Jacobsen, and P. Halberg. 1995. The diversity of perinuclear antineutrophil cytoplasmic antibodies (pANCA) antigens. *Clin. Exp. Immunol.* 101:15–17.

157. Savige, J.A., B. Paspaliaris, R. Silvestrini, D. Davies, T. Nikoloutsopoulos, A. Sturgess, J. Neil, W. Pollock, K. Dunster, and M. Hendle. 1998. A review of immunofluorescent patterns associated with antineutrophil cytoplasmic antibodies (ANCA) and their differentiation from other antibodies. *J. Clin. Pathol.* 51:568–575.

158. Wong, R.C.W., R.A. Silvestrini, J.A. Savige, D.A. Fulcher, and E.M. Benson. 1999. Diagnostic value of classical and atypical antineutrophil cytoplasmic antibody (ANCA) immunofluorescence patterns. *J. Clin. Pathol.* 52:124–128.

159. Sobajima, J., S. Ozaki, H. Uesugi, F. Osakada, M. Inoue, Y. Fukuda, H. Shirakawa, M. Yoshida, A. Rokuhara, H. Imai, K. Kiyosawa, and K. Nakao. 1999. High mobility group (HMG) non-histone chromosomal proteins HMG1 and HMG2 are significant target antigens of perinuclear anti-neutrophil cytoplasmic antibodies in autoimmune hepatitis. *Gut* 44:867–873.

160. Zaho, M.H., S.J. Jones, and C.M. Lockwood. 1996. A comprehensive method of purify three major protein (BPI) is an important antigen for anti-neutrophil cytoplasmic autoantibodies (ANCA) in vasculitis. *Clin. Exp. Immunol.* 99:49–56.

161. Zhao, M.H., and C.M. Lockwood. 1996. A comprehensive method to purify three major ANCA antigens: proteinase 3, myeloperoxidase and bactericidal/permeability-increasing protein from human neutrophil granule acid extract. *J. Immunol. Methods* 197:121–130.

162. Yang, J.J., R. Tuttle, R.J. Falk, and J.C. Jennette. 1996. Frequency of anti-bactericidal/permeability-increasing protein (BPI) and anti-azurocidin in patients with renal disease. *Clin. Exp. Immunol.* 105:125–131.

163. Roozendaal, C., and C.G. Kallenberg. 1999. Are anti-neutrophil cytoplasmic antibodies (ANCA) clinically useful in inflammatory bowel disease (IBD)? *Clin. Exp. Immunol.* 116:206–213.

164. Dunn, A.C., R.S. Walmsley, R.L. Dedrick, A.J. Wakefield, and C.M. Lockwood. 1999. Anti-neutrophil cytoplasmic autoantibodies (ANCA) to bactericidal/permeability-increasing (BPI) protein recognize the carboxyl terminal domain. *J. Infect.* 39:81–87.

165. Elzouki, A.N., S. Eriksson, R. Lofberg, L. Nassberger, J. Wieslander, and S. Lindgren. 1999. The prevalence and clinical significance of alpha 1-antitrypsin deficiency (PiZ) and ANCA specificities (proteinase 3, BPI) in patients with ulcerative colitis. *Inflammatory Bowel Dis.* 5:246–252.

166. Mahadeva, R., A.C. Dunn, R.C. Westerbeek, L. Sharples, D.B. Whitehouse, N.R. Carroll, R.I. Ross-Russell, A.K. Webb, D. Bilton, D.A. Lomas, and C.M. Lockwood. 1999. Anti-neutrophil cytoplasmic antibodies (ANCA) against bactericidal/permeability-increasing protein (BPI) and cystic fibrosis lung disease. *Clin. Exp. Immunol.* 117:561–567.

167. Baslund, B., M. Segelmark, A. Wiik, W. Szpirt, J. Petersen, and J. Wieslander. 1995. Screening for anti-neutrophil cytoplasmic antibodies (ANCA): is indirect immunofluorescence the method of choice? *Clin. Exp. Immunol.* 99:486–492.

168. Hagen, E.C., K. Andrassy, E. Csernok, M.R. Daha, G. Gaskin, W.L. Gross, B. Hansen, Z. Heigl, J. Hermans, D. Jayne, C.G. Kallenberg, P. Lesavre, C.M. Lockwood, J. Ludemann, F. Mascart-Lemone, E. Mirapeix, C.D. Pusey, N. Rasmussen, R.A. Sinico, A. Tzioufas, J. Wieslander, A. Wiik, and F.J. Van der Woude. 1996. Development and standardization of solid phase assays for the detection of anti-neutrophil cytoplasmic antibodies (ANCA). A report on

the second phase of an international cooperative study on the standardization of ANCA assays. *J. Immunol. Methods* 196:1–15.

169. Lim, L.C., J.G. Taylor, 3rd, J.L. Schmitz, J.D. Folds, A.S. Wilkman, R.J. Falk, and J.C. Jennette. 1999. Diagnostic usefulness of antineutrophil cytoplasmic autoantibody serology. Comparative evaluation of commercial indirect fluorescent antibody kits and enzyme immunoassay kits. *Am. J. Clin. Pathol.* 111:363–369.

170. Savige, J., D. Gillis, E. Benson, D. Davies, V. Esnault, R.J. Falk, E.C. Hagen, D. Jayne, J.C. Jennette, B. Paspaliaris, W. Pollock, C. Pusey, C.O. Savage, R. Silvestrine, F. van der Woude, J. Wieslander, and A. Wiik. 1999. International consensus statement on testing and reporting of antineutrophil cytoplasmic antibodies (ANCA). *Am. J. Clin. Pathol.* 111:507–513.

171. Jayne, D.R., A.P. Weetman, and C.M. Lockwood. 1991. IgG subclass distribution of autoantibodies to neutrophil cytoplasmic antigens in systemic vasculitis. *Clin. Exp. Immunol.* 84:476–481.

172. Segelmark, M., and J. Wieslander. 1993. IgG subclasses of antineutrophil cytoplasm autoantibodies (ANCA). *Nephrol. Dial. Transplantation* 8:696–702.

173. Sommarin, Y., N. Rasmussen, and J. Wieslander. 1995. Characterization of monoclonal antibodies to proteinase-3 and application in the study of epitopes for classical antineutrophil cytoplasm antibodies. *Exp. Nephrol.* 3:249–256.

174. Finnern, R., E. Pedrollo, I. Fisch, J. Wieslander, J.D. Marks, C.M. Lockwood, and W.H. Ouwehand. 1997. Human autoimmune anti proteinase 3 scFv from a phage display library. *Clin. Exp. Immunol.* 107:269–281.

175. Audrain, M.A., T.A. Baranger, N. Moguilevski, S.J. Martin, A. Devys, C.M. Lockwood, J.Y. Muller, and V.L. Esnault. 1997. Anti native and recombinant myeloperoxidase monoclonals and human autoantibodies. *Clin. Exp. Immunol.* 107:127–134.

176. Sun, J., D.N. Fass, J.A. Hudson, M.A. Viss, J. Wieslander, H.A. Homburger, and U. Specks. 1998. Capture-ELISA based on recombinant PR3 is sensitive for PR3-ANCA testing and allows detection of PR3 and PR3-ANCA/PR3 immune complexes. *J. Immunol. Methods.* 211:111–123.

177. Segelmark, M., A.N. Elzouki, J. Wieslander, and S. Eriksson. 1995. The PiZ gene of alpha 1-antitrypsin as a determinant of outcome in PR3-ANCA-positive vasculitis. *Kidney Int.* 48:844–850.

178. Gencik, M., S. Borgmann, R. Zahn, E. Albert, T. Sitter, J.T. Epplen, and H. Fricke. 1999. Immunogenetic risk factors for anti-neutrophil cytoplasmic antibody (ANCA)-associated systemic vasculitis. *Clin. Exp. Immunol.* 117:412–417.

179. Persson, U., L. Truedsson, K.W. Westman, and M. Segelmark. 1999. C3 and C4 allotypes in anti-neutrophil cytoplasmic autoantibody (ANCA)-positive vasculitis. *Clin. Exp. Immunol.* 116:379–382.

180. Tse, W.Y., S. Abadeh, A. McTiernan, R. Jefferis, C.O. Savage, and D. Adu. 1999. No association between neutrophil Fcgamma RIIa allelic polymorphism and anti-neutrophil cytoplasmic antibody (ANCA)-positive systemic vasculitis. *Clin. Exp. Immunol.* 117:198–205.

181. Hansch, G.M., M. Radsak, C. Wagner, B. Reis, A. Koch, A. Breitbart, and K. Andrassy. 1999. Expression of major histocompatibility class II antigens on polymorphonuclear neutrophils in patients with Wegner's granulomatosis. *Kidney Int.* 55:1811–1818.

182. Westman, K.W., P.G. Bygren, I. Eilert, A. Wiik, and J. Wieslander. 1997. Rapid screening assay for anti-GBM antibody and ANCAs; an important tool for the differential diagnosis of pulmonary renal syndromes. *Nephrol. Dial. Transplantation.* 12:1863–1868.

183. Ara, J., E. Mirapeix, R. Rodriguez, A. Saurina, and A. Darnell. 1999. Relationship between ANCA and disease activity in small vessel vasculitis patients with anti-MPO ANCA. *Nephrol. Dial. Transplantation.* 14:1667–1672.

184. Kyndt, X., D. Reumaux, F. Bridoux, B. Tribout, P. Bataille, E. Hachulla, P.Y. Hatron, P. Duthilleul, and P. Vanhille. 1999. Serial measurements of antineutrophil cytoplasmic autoantibodies in patients with systemic vasculitis. *Am. J. Med.* 106:527–533.

185. Pusey, C.D., A.J. Rees, D.J. Evans, D.K. Peters, and C.M. Lockwood. 1991. Plasma exchange in focal necrotizing glomerulonephritis without anti-GBM antibodies. *Kidney Int.* 40:757–763.

186. Nachman, P.H., S.L. Hogan, J.C. Jennette, and R.J. Falk. 1996. Treatment response and relapse in antineutrophil cytoplasmic autoantibody-associated microscopic polyangiitis and glomerulonephritis. *J. Am. Soc. Nephrol.* 7:33–39.

187. Elliott, J.D., C.M. Lockwood, G. Hale, and H. Waldmann. 1998. Semi-specific immuno-absorption and monoclonal antibody therapy in ANCA positive vasculitis: experience in four cases. *Autoimmunity* 28:163–171.

188. Akimoto, T., Y. Ando, C. Ito, S. Muto, E. Kusano, and Y. Asano. 1999. Effect of plasmapheresis as initial monotherapy in a case of anti-neutrophil cytoplasmic autoantibody positive crescentic glomerulonephritis. *ASAIO J.* 45:509–513.

189. Nachman, P.H., M. Segelmark, K. Westman, S.L. Hogan, K.K. Satterly, J.C. Jennette, and R. Falk. 1999. Recurrent ANCA-associated small vessel vasculitis after transplantation: A pooled analysis. *Kidney Int.* 56:1544–1550.

190. Noronha, I.L., C. Kruger, K. Andrassy, E. Ritz, and R. Waldherr. 1993. In situ production of TNFα, IL-1β, and IL-2R in ANCA-positive glomerulonephritis. *Kidney Int.* 43:682–692.

191. Brouwer, E., M.G. Huitema, A.H. Leontine Mulder, P. van Goor, J.W. Cohen Tervaert, J.J. Weening, and C.G.M. Kallenberg. 1994. Neutrophil activation in vitro and in vivo in Wegner's granulomatosis. *Kidney Int.* 45:1120–1131.

192. Kobold, M., C. Anneke, C.G.M. Kallenberg, C. Tervaert, and J. Willem. 1999. Monocyte activation in patients with Wegner's granulomatosis. *Ann. Rheum. Dis.* 58:237–245.

193. Falk, R.J., R.S. Terrell, L.A. Charles, and J.C. Jennette. 1990. Anti-neutrophil cytoplasmic autoantibodies induce neutrophils to degranulate and produce oxygen radicals in vitro. *Proc. Natl. Acad. Sci. USA* 87:4115–4119.

194. Lia, K.N., and C.M. Lockwood. 1991. The effect of anti-neutrophil cytoplams autoantibodies on the signal transduction in human neutrophils. *Clin. Exp. Immunol.* 85:396–401.

195. Charles, L.A. M.L. Caldas, R.J. Falk, R.S. Terrell, and J.C. Jennette. 1991. Antibodies against granule proteins activate neutrophils in vitro. *J. Leukocyte Biol.* 50:539–546.

196. Keogan, M.T., V.L. Esnault, A.J. Green, C.M. Lockwood, and D.L. Brown. 1992. Activation of normal neutrophils by anti-neutrophil cytoplasm antibodies. *Clin. Exp. Immunol.* 90:228–234.

197. Li, K.N., J.C. Leung, I. Rifkin, and C.M. Lockwood. 1994. Effect of anti-neutrophil cytoplasm autoantibodies on the

intracellular calcium concentration of human neutrophils. *Lab. Invest.* 70:152–162.

198. Kettritz, R., J.C. Jennette, and R.J. Falk. 1997. Crosslinking of ANCA-antigens stimulates superoxide release by human neutrophils. *J. Am. Soc. Nephrol.* 8:386–394.

199. Radford, D.J., J.M. Lord, and C.O. Savage. 1999. The activation of the neutrophil respiratory burst by anti-neutrophil cytoplasm autoantibody (ANCA) from patients with systemic vasculitis requires tyrosine kinases and protein kinase C activation. *Clin. Exp. Immunol.* 118:171–179.

200. Franssen, C.F., M.G. Huitema, A.C. Muller Kobold, W.W. Oost-Kort, P.C. Limburg, A. Tiebosch, C.A. Stegeman, C.G. Kallenberg, and J.W. Tervaert. 1999. *In vitro* neutrophil activation by antibodies to proteinase 3 and myeloperoxidase from patients with crescentic glomerulonephritis. *J. Am. Soc. Nephrol.* 10:1506–1515.

201. Ewert, B.H., J.C. Jennette, and R.J. Falk. 1992. Anti-myeloperoxidase antibodies stimulate neutrophils to damage human endothelial cells. *Kidney Int.* 41:375–383.

202. Keogan, M.T., I. Rifkin, N. Ronda, C.M. Lockwood, and D.L. Brown. 1993. Anti-neutrophil cytoplasm antibodies (ANCA) increase neutrophil adhesion to cultured human endothelium. *Adv. Exp. Med. Biol.* 336:115–119.

203. Ewert, B.H., M.E. Becker, J.C. Jennette, and R.J. Falk. 1995. Antimyeloperoxidase antibodies induce neutrophil adherence to cultured human endothelial cells. *Renal Failure* 17:125–133.

204. Jennette, J.C. 1994. Pathogenic potential of anti-neutrophil cytoplasmic autoantibodies. *Lab Invest.* 70:135–137.

205. Kain, R., K. Matsui, M. Exner, S. Binder, G. Schaffner, E.M. Sommer, and D. Kerjaschki. 1995. A novel class of autoantigens of nati-neutrophil cytoplasmic antibodies in necrotizing and crescentic glomerulonephritis: the lysosomal membrane glycoprotein h-lamp-2 in neutrophil granulocytes and a related membrane protein in glomerular endothelial cells. *J. Exp. Med.* 181:585–597.

206. Jennette, J.C., and R.J. Falk. 1998. Pathogenesis of the vascular and glomerular damage in ANCA-positive vasculities. *Nephrol. Dial. Transplantation* 13 Suppl 1:16–20.

207. Harper, L., and C.O.S. Savage. 1999. Mechanisms of endothelial injury in systemic vasculitis. *Adv. Nephrol.* 29:1–15.

208. Langford, C.A., and G.S. Hoffman. 1999. Wegener's granulomatosis. *Thorax* 53:629–637.

209. Kallenberg, C.G., and J.W. Tervaert. 1999. What is new with anti-neutrophilic cytoplasmic antibodies: diagnostic, pathogenetic, and therapeutic implications. *Cur. Opin. Nephrol. Hypertension* 8:307–315.

210. Johnson, R.J., W.G. Couser, E.Y. Chi, S. Adler, and S.I. Klebanoff. 1987. New mechanism for glomerular injury. Myeloperoxidase-hydrogen peroxide-halide system. *J. Clin. Invest.* 79:1379–1387.

211. Brouwer, E., M.G. Huitema, P.A. Klok, H. de Weerd, J.W. Tervaert, J.J. Weening, and C.G.M. Kallenberg. 1993. Antimyeloperoxidase-associated proliferative glomerulonephritis: An animal model. *J. Exp. Med.* 177:905–914.

212. Brouwer, E., P.A. Klok, M.G. Huitema, J.J. Weening, and C.G. Kallenberg. 1995. Renal ischemia/reperfusion injury contributes to renal damage in experimental anti-myeloperoxidase-associated proliferative glomerulonephritis. *Kidney Int.* 47:1121–1129.

213. Yang, J.J., J.C. Jennette, and R.J. Falk. 1994. Immune complex glomerulonephritis is induced in rats immunized

with heterologous myeloperoxidase. *Clin. Exp. Immunol.* 97:466–473.

214. Heeringa, P.E. Brouwer, P.A. Klok, M.G. Huitema, J. van den Born, J.J. Weening, and C.G.M. Kallenberg. 1996. Autoantibodies to myeloperoxidase aggravate mild antiglomerular basement-membrane-mediated glomerular injury in the rat. *Am. J. Pathol.* 149:1695–1706.

215. Kobayashi, K., T. Shibata, and T. Sugisaki. 1993. Aggravation of rat Masugi nephritis by heterologous anti-rat myeloperoxidase (MPO) antibody. *Clin. Exp. Immunol.* 93:20.

216. Mathieson, P.W., S. Thiru, and D.B.G. Oliveria. 1992. Mercuric chloride-treated Brown Norway rats develop widespread tissue injury including necrotizing vasculitis. *Lab. Invest.* 67:121–129.

217. Esnault, V.L., P.W. Mathieson, S. Thiru, D.B.G. Oliveira, and C.M. Lockwood. 1992. Autoantibodies to myeloperoxidase in Brown Norway Rats treated with mercuric chloride. *Lab. Invest.* 1992;67:114–120.

218. Qasim, F.J., P.W. Mathieson, S. Thiru, D.B. Oliveira, and C.M. Lockwood. 1993. Further characterization of an animal model of systemic vasculitis. *Adv. Exp. Med. Biol.* 336:133–137.

219. Kinjoh, K., M. Kyogoku, and R.A. Good. 1993. Genetic selection for crescent formation yields mouse strain with rapidly progressive glomerulonephritis and small vessel vasculitis. *Proc. Natl. Acad. Sci. USA* 90:3413–3417.

220. Kettritz, R., J.J. Yang, K. Kinjoh, J.C. Jennette, and R.J. Falk. 1995. Animal models in ANCA-vasculitis. *Clin. Exp. Immunol.* 101:12–15.

221. Yurchenco, P.D., and J.J. O'Rear. 1994. Basal lamina assembly. *Curr. Opin. Cell Biol.* 6:674–81.

222. Timpl, R., and J.C. Brown. 1996. Supramolecular assembly of basement membranes. *BioEssays.* 18:123–32.

223. Holdsworth, S.R., S.M. Golbus, and C.B. Wilson. 1979. Characterization of collagenase solubilized human glomerular basement membrane antigens reacting with human antibodies. *Kidney Int.* 16:797.

224. Hunt, J.S., P.R. Macdonald, and A.R. McGiven. 1980. Isolation of human glomerular basement membrane antigens by affinity chromatography utilising Goodpasture's kidney antibody eluates. *Ren. Physiol.* 3:156–62.

225. Wilson, C.B., S.R. Holdsworth, and T.J. Neale. 1981. Anti-basement membrane antibodies in immunologic renal disease. *Aust. N. Z. J. Med.* 11:94–100.

226. Wieslander, J., P. Bygren, and D. Heinegard. 1984. Isolation of the specific glomerular basement membrane antigen involved in Goodpasture syndrome. *Proc. Natl. Acad. Sci. USA* 81:1544–1548.

227. Fish, A.J., M.C. Lockwood, M. Wong, and R.G. Price. 1984. Detection of Goodpasture antigen in fractions prepared from collagenase digests of human glomerular basement membrane. *Clin. Exp. Immunol.* 55:58–66.

228. Yoshioka, K., M. Kleppel, and A.J. Fish. 1985. Analysis of nephritogenic antigens in human glomerular basement membrane by two-dimensional gel electrophoresis. *J. Immunol.* 134:3831–3837.

229. Wilson, C.B., and S.R. Holdsworth. 1981. Anti-glomerular basement membrane (GBM) antibody-induced diseases. In: *Proceedings of the 8th International Congress of Nephrology*, W. Zurukzoglu, M. Papadimitriou, M. Pyrpasopoulos, M. Sion, and C. Zamboulis, editors. Karger, Basel. pp. 910–916.

230. Wieslander, J., J.F. Barr, R.J. Butkowski, S.J. Edwards, P. Bygren, D. Heinegard, and B.G. Hudson. 1984. Goodpasture

antigen of the glomerular basement membrane: localization to noncollagenous regions of type IV collagen. *Proc. Natl. Acad. Sci. USA* 81:3838–3842.

231. Butkowski, R.J., J. Wieslander, B.J. Wisdom, J.F. Barr, M.E. Noelken, and B.G. Hudson. 1985. Properties of the globular domain of type IV collagen and its relationship to the Goodpasture antigen. *J. Biol. Chem.* 260:3739–47.

232. Wieslander, J., J. Langeveld, R. Butkowski, M. Jodlowski, M. Noelken, and B.G. Hudson. 1985. Physical and immuno-chemical studies of the globular domain of type IV collagen. Cryptic properties of the Goodpasture antigen. *J. Biol. Chem.* 260:8564–8570.

233. Wieslander, J., M. Kataja, and B.G. Hudson. 1987. Characterization of the human Goodpasture antigen. *Clin. Exp. Immunol.* 69:332–40.

234. Butkowski, R.J., J.P. Langeveld, J. Wieslander, J. Hamilton, and B.G. Hudson. 1987. Localization of the Goodpasture epitope to a novel chain of basement membrane collagen. *J. Biol. Chem.* 262:7874–7877.

235. Saus, J., J. Wieslander, J.P. Langeveld, S. Quinones, and B.G. Hudson. 1988. Identification of the Goodpasture antigen as the alpha 3(IV) chain of collagen IV. *J. Biol. Chem.* 263:13374–80.

236. Gunwar, S., F. Ballester, R. Kalluri, J. Timoneda, A.M. Chonko, S.J. Edwards, M.E. Noelken, and B.G. Hudson. 1991. Glomerular basement membrane. Identification of dimeric subunits of the noncollagenous domain (hexamer) of collagen IV and the Goodpasture antigen. *J. Biol. Chem.* 266:15318–15324.

237. Neilson, E.G., R. Kalluri, M.J. Sun, S. Gunwar, T. Danoff, M. Mariyama, J.C. Myers, S.T. Reeders, and B.G. Hudson. 1993. Specificity of Goodpasture autoantibodies for the recombinant noncollagenous domains of human type IV collagen. *J. Biol. Chem.* 268:8402–8405.

238. Turner, N., J. Forstova, A. Rees, C.D. Pusey, and P.J. Mason. 1994. Production and characterization of recombinant Goodpasture antigen in insect cells. *J. Biol. Chem.* 269:17141–17145.

239. Dehan, P., M. Weber, X. Zhang, S.T. Reeders, J.M. Foidart, and K. Tryggvason. 1996. Sera from patients with anti-GBM nephritis including Goodpasture syndrome show heterogeneous reactivity to recombinant NC1 domain of type IV collagen alpha chains. *Nephrol. Dial. Transplant.* 11:2215–2222.

240. Netzer, K.O., A. Leinonen, A. Boutaud, D.B. Borza, P. Todd, S. Gunwar, J.P. Langeveld, and B.G. Hudson. 1999. The Goodpasture autoantigen. Mapping the major conformational epitope(s) of alpha3(IV) collagen to residues 17–31 and 127–141 of the NC1 domain. *J. Biol. Chem.* 274:11267–11274.

241. Saxena, R., P. Bygren, R. Butkowski, and J. Weislander. 1989. Specificity of kidney-bound antibodies in Goodpasture's syndrome. *Clin. Exp. Immunol.* 78:31–36.

242. Kalluri, R., E. Melendez, K.W. Rumpf, K. Sattler, G.A. Muller, F. Strutz, and E.G. Neilson. 1996. Specificity of circulating and tissue-bound autoantibodies in Goodpasture syndrome. *Proc. Assoc. Am. Physicians* 108:134–139.

243. Matsukura, H., R.J. Butkowski, and A.J. Fish. 1993. The Goodpasture antigen: common epitopes in the globular domains of collagen IV. *Nephron* 64:532–539.

244. Borza, D.B., K.O. Netzer, A. Leinonen, P. Todd, J. Cervera, J. Saus, and B.G. Hudson. 2000. The Goodpasture autoantigen: Identification of multiple cryptic epitopes on the NC1 domain of the alpha3(IV) collagen chain. *J. Biol. Chem.* 275:6030–6037.

245. Kalluri, R., C.B. Wilson, M. Weber, S. Gunwar, A.M. Chonko, E.G. Neilson, and B.G. Hudson. 1995. Identification of the alpha 3 chain of type IV collagen as the common autoantigen in antibasement membrane disease and Goodpasture syndrome. *J. Am. Soc. Nephrol.* 6:1178–1185.

246. Hudson, B.G., S.T. Reeders, and K. Tryggvason. 1993. Type IV collagen: structure, gene organization, and role in human diseases. Molecular basis of Goodpasture and Alport syndromes and diffuse leiomyomatosis. *J. Biol. Chem.* 268:26033–26036.

247. Dolz, R., J. Engel, and K. Kuhn. 1988. Folding of collagen IV. *Eur. J. Biochem.* 178:357–366.

248. Timpl, R., H. Wiedemann, V. van Delden, H. Furthmayr, and K. Kuhn. 1981. A network model for the organization of type IV collagen molecules in basement membranes. *Eur. J. Biochem.* 120:203–211.

249. Yurchenco, P.D., and G.C. Ruben. 1987. Basement membrane structure in situ: evidence for lateral associations in the type IV collagen network. *J. Cell Biol.* 105:2559–2568.

250. Gunwar, S., M.E. Noelken, and B.G. Hudson. 1991. Properties of the collagenous domain of the alpha 3(IV) chain, the Goodpasture antigen, of lens basement membrane collagen. Selective cleavage of alpha (IV) chains with retention of their triple helical structure and noncollagenous domain. *J. Biol. Chem.* 266:14088–14094.

251. Kahsai, T.Z., G.C. Enders, S. Gunwar, C. Brunmark, J. Wieslander, R. Kalluri, J. Zhou, M.E. Noelken, and B.G. Hudson. 1997. Seminiferous tubule basement membrane. Composition and organization of type IV collagen chains, and the linkage of alpha3(IV) and alpha5(IV) chains. *J. Biol. Chem.* 272:17023–17032.

252. Gunwar, S., F. Ballester, M.E. Noelken, Y. Sado, Y. Ninomiya, and B.G. Hudson. 1998. Glomerular basement membrane. Identification of a novel disulfide cross linked network of a3, a4 and a5 chains of type IV collagen and its implications for the pathogenesis of Alport syndrome. *J. Biol. Chem.* 273:8767–8775.

253. Wilson, C.B. 1980. Individual and strain differences in renal basement membrane antigens. *Transplant. Proc.* 12:69–73.

254. Olson, D.L., S.K. Anand, B.H. Landing, E. Heuser, C.M. Grushkin, and E. Lieberman. 1980. Diagnosis of hereditary nephritis by failure of glomeruli to bind anti-glomerular basement membrane antibodies. *J. Pediatr.* 96:697–699.

255. McCoy, R.C., H.K. Johnson, W.J. Stone, and C.B. Wilson. 1982. Absence of nephritogenic GBM antigen(s) in some patients with hereditary nephritis. *Kidney Int.* 21:642–52.

256. Kashtan, C.E., and A.F. Michael. 1993. Alport syndrome: from bedside to genome to bedside. *Am. J. Kidney Dis.* 22:627–640.

257. Kashtan, C.E., R.J. Butkowski, M.M. Kleppel, M.R. First, and A.F. Michael. 1990. Posttransplant anti-glomerular basement membrane nephritis in related males with Alport syndrome. *J. Lab. Clin. Med.* 116:508–515.

258. Kleppel, M.M., W.W. Fan, H.I. Cheong, C.E. Kashtan, and A.F. Michael. 1992. Immunochemical studies of the Alport antigen. *Kidney Int.* 41:1629–1637.

259. Hudson, B.G., R. Kalluri, S. Gunwar, M. Weber, F. Ballester, J.K. Hudson, M.E. Noelken, M. Sarras, W.R. Richardson, J. Saus, D.R. Abrahamson, A.D. Glick, M.A. Haralson, J.H. Helderman, W.J. Stone, and H.R. Jacobson. 1992. The pathogenesis of Alport syndrome involves type IV collagen molecules containing the alpha 3(IV) chain: evidence from anti-GBM nephritis after renal transplantation. *Kidney Int.* 42:179–187.

260. Kalluri, R., M. Weber, K.O. Netzer, M.J. Sun, E.G. Neilson, and B.G. Hudson. 1994. COL4A5 gene deletion and production of post-transplant anti-alpha 3(IV) collagen alloantibodies in Alport syndrome. *Kidney Int.* 45:721–726.

261. Ding, J., C.E. Kashtan, W.W. Fan, M.M. Kleppel, M.J. Sun, R. Kalluri, E.G. Neilson, and A.F. Michael. 1994. A monoclonal antibody marker for Alport syndrome identifies the Alport antigen as the alpha 5 chain of type IV collagen. *Kidney Int.* 45:1504–1506.

262. Dehan, P., L.P. Van den Heuvel, H.J. Smeets, K. Tryggvason, and J.M. Foidart. 1996. Identification of post-transplant anti-alpha 5(IV) collagen alloantibodies in X-linked Alport syndrome. *Nephrol. Dial. Transplant.* 11:1983–1988.

263. Brainwood, D., C. Kashtan, M.C. Gubler, and A.N. Turner. 1998. Targets of alloantibodies in Alport anti-glomerular basement membrane disease after renal transplantation. *Kidney Int.* 53:762–766.

264. Thorner, P., R. Baumal, V.E. Valli, D. Mahuran, R. McInnes, and P. Marrano. 1989. Abnormalities in the NC1 domain of collagen type IV in GBM in canine hereditary nephritis. *Kidney Int.* 35:843–850.

265. Zheng, K., P.S. Thorner, P. Marrano, R. Baumal, and R.R. McInnes. 1994. Canine X chromosome-linked hereditary nephritis: a genetic model for human X-linked hereditary nephritis resulting from a single base mutation in the gene encoding the alpha 5 chain of collagen type IV. *Proc. Natl. Acad. Sci. USA* 91:3989–3993.

266. Thorner, P.S., K. Zheng, R. Kalluri, R. Jacobs, and B.G. Hudson. 1996. Coordinate gene expression of the alpha3, alpha4, and alpha5 chains of collagen type IV. Evidence from a canine model of X-linked nephritis with a COL4A5 gene mutation. *J. Biol. Chem.* 271:13821–13828.

267. Miner, J.H., and J.R. Sanes. 1996. Molecular and functional defects in kidneys of mice lacking collagen alpha 3(IV): implications for Alport syndrome. *J. Cell Biol.* 135:1403–1413.

268. Cosgrove, D., D.T. Meehan, J.A. Grunkemeyer, J.M. Kornak, R. Sayers, W.J. Hunter, and G.C. Samuelson. 1996. Collagen COL4A3 knockout: a mouse model for autosomal Alport syndrome. *Genes Dev.* 10:2981–2992.

269. Miner, J.H., and J.R. Sanes. 1994. Collagen IV alpha 3, alpha 4, and alpha 5 chains in rodent basal laminae: sequence, distribution, association with laminins, and developmental switches. *J. Cell. Biol.* 127:879–891.

270. Kalluri, R., C.F. Shield, P. Todd, B.G. Hudson, and E.G. Neilson. 1997. Isoform switching of type IV collagen is developmentally arrested in X-linked Alport syndrome leading to increased susceptibility of renal basement membranes to endoproteolysis. *J. Clin. Invest.* 99:2470–2478.

271. Harvey, S.J., K. Zheng, Y. Sado, I. Naito, Y. Ninomiya, R.M. Jacobs, B.G. Hudson, and P.S. Thorner. 1998. Role of distinct type IV collagen networks in glomerular development and function. *Kidney Int.* 54:1857–1866.

272. Levy, J.B., A.N. Turner, A.J. George, and C.D. Pusey. 1996. Epitope analysis of the Goodpasture antigen using a resonant mirror biosensor. *Clin. Exp. Immunol.* 106:79–85.

273. Meyers, K.E., P.A. Kinniry, R. Kalluri, E.G. Neilson, and M.P. Madaio. 1998. Human Goodpasture anti-alpha3(IV)NC1 autoantibodies share structural determinants. *Kidney Int.* 53:402–407.

274. Kalluri, R., S. Gunwar, S.T. Reeders, K.C. Morrison, M. Mariyama, K.E. Ebner, M.E. Noelken, and B.G. Hudson. 1991. Goodpasture syndrome. Localization of the epitope for the autoantibodies to the carboxyl-terminal region of the alpha 3(IV) chain of basement membrane collagen. *J. Biol. Chem.* 266:24018–24024.

275. Kefalides, N.A., N. Ohno, C.B. Wilson, H. Fillit, J. Zabriski, and J. Rosenbloom. 1993. Identification of antigenic epitopes in type IV collagen by use of synthetic peptides. *Kidney Int.* 43:94–100.

276. Hellmark, T., C. Brunmark, J. Trojnar, and J. Wieslander. 1996. Epitope mapping of anti-glomerular basement membrane (GBM) antibodies with synthetic peptides. *Clin. Exp. Immunol.* 105:504–510.

277. Levy, J.B., A. Coulthart, and C.D. Pusey. 1997. Mapping B cell epitopes in Goodpasture's disease. *J. Am. Soc. Nephrol.* 8:1698–1705.

278. Kalluri, R., M.J. Sun, B.G. Hudson, and E.G. Neilson. 1996. The Goodpasture autoantigen. Structural delineation of two immunologically privileged epitopes on alpha3(IV) chain of type IV collagen. *J. Biol. Chem.* 271:9062–9068.

279. Cunningham, B.C., P. Jhurani, P. Ng, and J.A. Wells. 1989. Receptor and antibody epitopes in human growth hormone identified by homolog-scanning mutagenesis. *Science* 243:1330–1336.

280. Ryan, J.J., P.J. Mason, C.D. Pusey, and N. Turner. 1998. Recombinant alpha-chains of type IV collagen demonstrate that the amino terminal of the Goodpasture autoantigen is crucial for antibody recognition. *Clin. Exp. Immunol.* 113:17–27.

281. Hellmark, T., M. Segelmark, C. Unger, H. Burkhardt, J. Saus, and J. Wieslander. 1999. Identification of a clinically relevant immunodominant region of collagen IV in Goodpasture disease. *Kidney Int.* 55:936–944.

282. Weber, S., R. Dolz, R. Timpl, J.H. Fessler, and J. Engel. 1988. Reductive cleavage and reformation of the interchain and intrachain disulfide bonds in the globular hexameric domain NC1 involved in network assembly of basement membrane collagen (type IV). *Eur. J. Biochem.* 175:229–236.

283. Weber, M., K.H. Meyer zum Buschenfelde, and H. Kohler. 1988. Immunological properties of the human Goodpasture target antigen. *Clin. Exp. Immunol.* 74:289–294.

284. Yoshioka, K., A.F. Michael, J. Velosa, and A.J. Fish. 1985. Detection of hidden nephritogenic antigen determinants in human renal and nonrenal basement membranes. *Am. J. Pathol.* 121:156–65.

285. Thorner, P.S., R. Baumal, A. Eddy, and P. Marrano. 1989. Characterization of the NC1 domain of collagen type IV in glomerular basement membranes (GBM) and of antibodies to GBM in a patient with anti-GBM nephritis. *Clin. Nephrol.* 31:160–168.

286. Abrahamson, D.R., and J.P. Caulfield. 1982. Proteinuria and structural alterations in rat glomerular basement membranes induced by intravenously injected anti-laminin immunoglobulin G. *J. Exp. Med.* 156:128–145.

287. Yaar, M., J.M. Foidart, K.S. Brown, S.I. Rennard, G.R. Martin, and L. Liotta. 1982. The Goodpasture-like syndrome in mice induced by intravenous injections of anti-type IV collagen and anti-laminin antibody. *Am. J. Pathol.* 107:79–91.

288. Makino, H., B. Lelongt, and Y.S. Kanwar. 1988. Nephritogenicity of proteoglycans. III. Mechanism of immune deposit formation. *Kidney Int.* 34:209–219.

289. Makino, H., J.T. Gibbons, M.K. Reddy, and Y.S. Kanwar. 1986. Nephritogenicity of antibodies to proteoglycans of the

glomerular basement membrane-1. *J. Clin. Invest.* 77:142–156.

290. Saxena, R., P. Bygren, R. Butkowski, and J. Wieslander. 1990. Entactin: A possible auto-antigen in the pathogenesis of non-Goodpasture anti-GBM nephritis. *Kidney Int.* 38:263–272.

291. Saxena, R., P. Bygren, B. Cederholm, and J. Wieslander. 1991. Circulating anti-entactin antibodies in patients with glomerulonephritis. *Kidney Int.* 39:996–1004.

292. Sado, Y., M. Kagawa, I. Naito, and T. Okigaki. 1991. Properties of bovine nephritogenic antigen that induces anti-GBM nephritis in rats and its similarity to the Goodpasture antigen. *Virchows Arch. B Cell Pathol. Incl. Mol. Pathol.* 60:345–351.

293. Rauf, S., M. Kagawa, Y. Kishiro, S. Inoue, I. Naito, T. Oohashi, M. Sugimoto, Y. Ninomiya, and Y. Sado. 1996. Nephritogenicity and alpha-chain composition of NC1 fractions of type IV collagen from bovine renal basement membrane. *Virchows Arch.* 428:281–288.

294. Sado, Y., M. Kagawa, Y. Kishiro, I. Naito, K. Joh, and Y. Ninomiya. 1997. Purification and characterization of human nephritogenic antigen that induces anti-GBM nephritis in rats. *J. Pathol.* 182:225–232.

295. Kalluri, R., V.H. Gattone, 2nd, M.E. Neolken, and B.G. Hudson. 1994. The alpha 3 chain of type IV collagen induces autoimmune Goodpasture syndrome. *Proc. Natl. Acad. Sci. USA* 91:6201–6205.

296. Abbate, M., R. Kalluri, D. Corna, N. Yamaguchi, R.T. McCluskey, B.G. Hudson, G. Andres, C. Zoja, and G. Remuzzi. 1998. Experimental Goodpasture's syndrome in Wistar-Kyoto rats immunized with a3 chain of type IV collagen. *Kidney Int.* 54:1550–1561.

297. Sado, Y., A. Boutaud, M. Kagawa, I. Naito, Y. Ninomiya, and B.G. Hudson. 1998. Induction of anti-GBM nephritis in rats by recombinant a3(IV)NC1 and a4(IV)NC1 of type IV collagen. *Kidney Int.* 53:664–671.

298. Bolton, W.K., A.M. Luo, P. Fox, W. May, and J. Fox. 1996. Goodpasture's epitope in development of experimental autoimmune glomerulonephritis in rats. *Kidney Int.* 49:327–334.

299. Sugihara, K., Y. Sado, Y. Ninomiya, and H. Wada. 1996. Experimental anti-GBM glomerulonephritis induced in rats by immunization with synthetic peptides based on six alpha chains of human type IV collagen. *J. Pathol.* 178:352–358.

300. Theofilopoulos, A.N. 1995. The basis of autoimmunity: Part 1. Mechanisms of aberrant self-recognition. *Immunology Today* 16:90–98.

301. Theofilopoulos, A.N. 1995. The basis of autoimmunity: Part II. Genetic predisposition. *Immunology Today* 16:150–159.

302. Rees, A.J., D.K. Peters, D.A.S. Compston, and J.R. Batchelor. 1978. Strong association between HLA-DRw2 and antibody-mediated Goodpasture's syndrome. *Lancet* 1:966–968.

303. Perl, S.I., B.A. Pussell, J.A. Charlesworth, G.J. MacDonald, and M. Wolnizer. 1981. Goodpasture's (anti-GBM) disease and HLA-DRw2. *New Engl. J. Med.* 305:463–464.

304. Rees, A.J., D.K. Peters, N. Amos, K.I. Welsh, and J.R. Batchelor. 1984. The influence of HLA-linked genes on the severity of anti-GBM antibody-mediated nephritis. *Kidney Int.* 26:445–450.

305. Rees, A.J., A.G. Demaine, and K.I. Welsh. 1984. Association of immunoglobulin Gm allotypes with antiglomerular basement membrane antibodies and their titer. *Hum. Immunol.* 10:213–220.

306. D'Apice, A.J.F., P. Kincaid-Smith, G.J. Becker, M.G. Loughhead, J.W. Freeman, and J.M. Sands. 1978. Goodpasture's syndrome in identical twins. *Ann. Int. Med.* 88–61.

307. Dunckley, H., J.R. Chapman, J. Burke, J. Charlesworth, J. Hayes, E. Haywood, B. Hutchison, L. Ibels, S. Kalowski, P. Kincaid-Smith, S. Lawrence, D. Lewis, J. Moran, B. Pussell, A. Restifo, J. Stewart, G. Thatcher, R. Walker, D. Waugh, D. Wilson, and R. Wyndham. 1991. HLR-DR and -DQ genotyping in anti-GBM disease. *Disease Markers* 9:249–256.

308. Mercier, B., B. Bourbigot, O. Raguenes, E. Rondeau, P. Simon, G. Mourad, D. Legrand, P. Coville, and C. Ferec. 1992. HLA class II typing of Goodpasture's syndrome affected patients. *J. Am. Soc. Nephrol.* 3:658 (Abstr.)

309. Huey, B., K. McCormick, J. Capper, C. Ratliff, B.W. Colombe, M.R. Garovoy, and C.B. Wilson. 1993. Associations of HLA-DR and HLA-DQ types with anti-GBM nephritis by sequence-specific oligonucleotide probe hybridization. *Kidney Int.* 44:307–312.

310. Burns, A.P., M. Fisher, P. Li, C.D. Pusey, and A.J. Rees. 1995. Molecular analysis of HLA class II genes in Goodpasture's disease. *Quart. J. Med.* 88:93–100.

311. Fisher, M., C.D. Pusey, R.W. Vaughan, and A.J. Rees. 1997. Susceptibility to anti-glomerular basement membrane disease is strongly associated with HLA-DRB1 genes. *Kidney Int.* 51:222–229.

312. Phelps, R.G., A.N. Turner, and A.J. Rees. 1996. Direct identification of naturally processed autoantigen-derived peptides bound to HLA-DR15. *J. Biol. Chem.* 271.18549–18553.

313. Phelps, R.G., V.L. Jones, J.M. Coughlan, A.N. Turner, and A.J. Rees. 1998. Presentation of the Goodpasture autoantigen to CD4 T cells is influenced more by processing constraints than by HLA class II peptide binding preferences. *J. Biol. Chem.* 273:11440–11447.

314. Phelps, R.G., and A.J. Rees. 1999. The HLA complex in Goodpasture's disease: a model for analyzing susceptibility to autoimmunity. *Kidney International* 56:1638–1653.

315. Murphy, B., and A.M. Krensky. 1999. HLA-derived peptides as novel immunomodulary therapeutics. *J. Am. Soc. Nephrol.* 10:1346–1355.

316. Hammer, R.E., S.D. Maika, J.A. Richardson, J.P. Tang, and J.D. Taurog. 1990. Spontaneous inflammatory disease in transgenic rats expressing HLA-B27 and human beta 2m: an animal model of HLA-B27-associated human disorders. *Cell* 63:1099–1112.

317. Khare, S.D., J. Hansen, H.S. Luthra, and C.S. David. 1996. HLA-B27 heavy chains contribute to spontaneous inflammation disease in b27/human beta 2-microglobulin (β_2m) double transgenic mice with disrupted mouse β_2m. *J. Clin. Invest.* 98:2746–2755.

318. Taneja, V., and C.S. David. 1998. HLA Transgenic mice as humanized mouse models of disease and immunity. *J. Clin. Invest.* 101:921–926.

319. Kalluri, R., T.M. Danoff, H. Okada, and E.G. Neilson. 1997. Susceptibility to anti-glomerular basement membrane disease and Goodpasture Syndrome is linked to MHC Class II genes and the emergence of T cell-mediated immunity in mice. *J. Clin. Invest.* 100:2263–2275.

320. Derry, C.J., C.N. Ross, G. Lombardi, P.D. Mason, A.J. Rees, R.I. Lechler, and C.D. Pusey. 1995. Analysis of T cell responses to the autoantigen in Goodpasture's disease. *Clin. Exp. Immunol.* 100:262–268.

321. Merkel, F., R. Kalluri, M. Marx, U. Enders, S. Stevanovic, G. Giegerich, E.G. Neilson, H.-G. Rammensee, B.G. Hudson, and M. Weber. 1996. Autoreactive T-cells in Goodpasture's syndrome recognize the N-terminal NC1 domain on $\alpha3$ type IV collagen. *Kidney Int.* 49:1127–1133.

322. Williams, P.S., A. Davenport, I. McDicken, D. Ashby, H.J. Goldsmith, and J.M. Bone. 1988. Increased incidence of anti-glomerular basement membrane antibody (anti-GBM) nephritis in the Mersey Region, September 1984-October 1985. *Q. J. Med.* 68:727–733.

323. Wilson, C.B., and R.C. Smith. 1972. Goodpasture's syndrome associated with influenza A2 virus infection. *Ann. Intern. Med.* 76:91–94.

324. Feng, L., Y. Xia, and C.B. Wilson. 1994. Alternative splicing of the NC1 domain of the human $\alpha3(IV)$ collagen gene. Differential expression of mRNA transcripts that predict three protein variants with distinct carboxyl regions. *J. Biol. Chem.* 269:2342–2348.

325. Blue, W.T., and C.F. Lange. 1975. Increased immunologic reactivity between human glomerular basement membrane and group A type 12 streptococcal cell membrane after carbohydrase treatment. *J. Immunol.* 114:306–309.

326. Froude, J., A. Gibofsky, D.R., Buskirk, A. Khanna, and J.B. Zabriskie. 1989. Cross-reactivity between streptococcus and human tissue: A model of molecular mimicry and autoimmunity. *Cur. Top. Microbiol. Immunol.* 145:5–26.

327. Stollerman, G.H. 1991. Short Analytical Review. Rheumatogenic streptococci and autoimmunity. *Clin. Immunol. Immunopathol.* 61:131–142.

328. Fitzsimons, E.J., Jr., M. Weber, and C.F. Lange. 1987. The isolation of cross-reactive monoclonal antibodies: Hybridomas to streptococcal antigens cross-reactive with mammalian basement membrane. *Hybridoma* 6:61–69.

329. Goroncy-Bermes, P., J.B. Dale, E.H. Beachey, and W. Opferkuch. 1987. Monoclonal antibody to human renal glomeruli cross-reacts with streptococcal M protein. *Infect. Immun.* 55:2416–2419.

330. Kraus, W., and E.H. Beachey. 1988. Renal autoimmune epitope of group A streptocci specified by M protein tetrapeptide Ile-Arg-Leu-Arg. *Proc. Natl. Acad. Sci. USA* 85:4516–4520.

331. Kraus, W., J.M. Seyer, and E.H. Beachey. 1989. Vimentin-cross-reactive epitope of type 12 streptococcal M protein. *Infect. Immun.* 57:2457–2461.

332. Kraus, W., K. Ohyama, D.S. Snyder, and E.H. Beachey. 1989. Autoimmune sequence of streptococcal M protein shared with the intermediate filament protein, vimentin. *J. Exp. Med.* 169:481–492.

333. Billheden, J., J. Boman, B. Stegmayr, J. Wieslander, and B. Settergren. 1997. Glomerular basement membrane antibodies in hantavirus disease (hemorrhagic fever with renal syndrome). *Clin. Nephrol.* 48:137–140.

334. Elder, G., S. Perl, J.L. Yong, J. Fletcher, and J. Mackie. 1995. Progression from Goodpasture's disease to membranous glomerulonephritis. *Pathology* 27:233–236.

335. Beirne, G.J., and J.T. Brennan. 1972. Glomerulonephritis associated with hydrocarbon solvents: mediated by antiglomerular basement membrane antibody. *Arch. Environ. Health* 25:265–369.

336. Zimmerman, S.W., K. Groehler, and G.J. Beirne. 1975. Hydrocarbon exposure and chronic glomerulonephritis. *Lancet* 2:199–202.

337. Lagrue, G. 1976. Hydrocarbon exposure and chronic glomerulonephritis. *Lancet* 1:1191.

338. Case Records of the Massachusetts General Hospital (Case 17–1976) 1976. *N. Engl. J. Med.* 294:944.

339. Beirne, G.J., J.P. Wagnild, S.W. Zimmerman, P.D. Macken, and P.M. Burkholder. 1977. Idiopathic crescentic glomerulonephritis. *Medicine* 56:349–381.

340. Kleinknecht, D., L. Morel-Maroger, P. Callard, J.P. Adhémar, and P. Mahieu. 1979. Antiglomerular basement membrane (GBM) antibody-induced glomerulonephritis after solvent exposure. *Kidney Int.* 15:450.

341. Kleinknecht, D., L. Morel-Maroger, P. Callard, J.P. Adhemar, and P. Mahieu. 1980. Antiglomerular basement membrane nephritis after solvent exposure. *Arch. Intern. Med.* 140:230–232.

342. Ravnskov, U.L.S., and A. Norden. 1983. Hydrocarbon exposure and glomerulonephritis: Evidence from patients' occupations. *Lancet* 2:1214–1216.

343. Ravnskov, V. 1985. Possible mechanisms of hydrocarbon associated glomerulonephritis. *Clin. Nephrol.* 23:294.

344. Daniell, W.E., W.G. Couser, and L. Rosenstock. 1988. Occupational solvent exposure and glomerulonephritis. A case report and review of the literature. *JAMA* 259:2280–2283.

345. Bombassei, G.J., and A.A. Kaplan. 1992. The association between hydrocarbon exposure and anti-glomerular basement membrane antibody-mediated disease (Goodpasture's syndrome). *Am. J. Ind. Med.* 21:141–153.

346. Klavis, G., and W. Drommer. 1970. Goodpasture-syndrom und benzineinwirkung. *Arch. Toxikol.* 26:40–55.

347. Zimmerman, S.W., and D.H. Norbach. 1980. Nephrotoxic effects of long-term carbon tetrachloride administration in rats. *Arch. Pathol. Lab. Med.* 104:94–99.

348. Easley, J.R., J.M. Holland, L.C. Gibson, and M.J. Whitaker. 1982. Renal toxicity of middle distillates of shale oil and petroleum in mice. *Toxicol. Appl. Pharmacol.* 65:84–91.

349. Wilson, C. 1982. Drug and toxin-induced nephritides: Anti-kidney antibody and immune complex mediation. In: Nephrotoxic Mechanisms of Drugs and Environmental Toxins, G. Porter, editor. Plenum, New York. pp. 383–392.

350. Carlier, B., E. Schroeder, and P. Mahieu. 1980. A rapidly and spontaneously reversible Goodpasture's syndrome after carbon tetrachloride inhalation. *Acta. Clin. Belg.* 35:193–198.

351. Bernis, P., J. Hamels, A. Quoidbach, and P. Bouvy. 1985. Remission of Goodpasture's syndrome after withdrawal of an unusual toxic. *Clin. Nephrol.* 23:312–317.

352. Bonzel, K.-E., D.E. Müller-Wiefel, H. Ruder, A.M. Wingen, R. Waldherr, and M. Weber. 1987. Anti-glomerular basement membrane antibody-mediated glomerulonephritis due to glue sniffing. *Eur. J. Pediatr.* 146:296–300.

353. Garcia-Rostan Y Perez, G.M., F. Garcia Bragado, and A.M. Puras Gil. 1997. Pulmonary hemorrhage and antiglomerular basement membrane antibody-mediated glomerulonephritis after exposure to smoked cocaine (crack): a case report and review of the literature. *Pathol. Int.* 47:6923–6927.

354. Peces, R., R.A. Navascues, J. Baltar, M. Seco, and J. Alvarez. 1999. Antiglomerular basement membrane antibody-mediated glomerulonephritis after intranasal cocaine use. *Nephron* 81:434–438.

355. Milliner, D.S., A.M. Pierides, and K.E. Holley. 1982. Renal transplantation in Alport's syndrome. Anti-glomerular basement membrane glomerulonephritis in the allograft. *Mayo Clin. Proc.* 57:35–43.

356. Quérin, S., L.-H. Noëlm, H.-P. Grünfeld, D. Droz, P. Mahieu, J. Berger, and H. Kreis. 1986. Linear glomerular

IgG fixation in renal allografts: incidence and significance in Alport's syndrome. *Clin. Nephrol.* 25:134–140.

357. Teruel, J.L., F. Liaño, F. Mampaso, J. Moreno, A. Serrano, C. Quereda, and J. Ortuño. 1987. Allograft antiglomerular basement membrane glomerulonephritis in a patients with Alport's syndrome. *Nephron* 46:43–44.

358. McPhaul, J.J., Jr., and F.J. Dixon. 1969. Immunoreactive basement membrane antigens in normal human urine and serum. *J. Exp. Med.* 130:1395–1409.

359. Huttunen, N.P., M.W. Turner, and T.M. Barratt. 1979. Physico-chemical characteristics of glomerular basement membrane antigens in urine, *Kidney Int.* 16:322–328.

360. Lubec, G., and H. Coradello. 1979. Urinary excretion of glomerular basement membrane antigens in premature infants and the newborn. *Biol. Neonate* 36:277–281.

361. Bowman, B.H., L. Schneider, D.R. Barnett, A. Kurosky, and R.M. Goldblum. 1980. Novel urinary fragments from human basement membrane collagen. *J. Biol. Chem.* 255:9484–9489.

362. Lerner, R.A., and Dixon, F.J. 1968. The induction of acute glomerulonephritis in rabbits with soluble antigens isolated from normal homologous and autologous urine. *J. Immunol.* 100:1277–1287.

363. Bernal, D., S. Quinones, and J. Saus. 1993. The human mRNA encoding the Goodpasture antigen is alternatively spliced. *J. Biol. Chem.* 268:12090–12094.

364. Penadés, J.R., D. Bernal. F. Revert, C. Johansson, V.J. Fresquet, J. Cervera, J. Wieslander, S. Quinones, and J. Saus. 1995. Characterization and expression of multiple alternatively spliced transcripts of the Goodpasture antigen gene region. Goodpasture antibodies recognize recombinant proteins representing the autoantigens and one of its alternative forms. *Eur. J. Biochem.* 229:754–760.

365. Hume, D.M., W.A. Sterling, R.J. Weymouth, H.R. Siebel, G.E. Madage, and H.M. Lee. 1970. Glomerulonephritis in human renal homotransplants. *Transplant. Proc.* 2:361–412.

366. Kurki, P., T. Helve, M. von Bonsdorff, T. Törnroth, E. Pettersson, H. Riska, and A. Miettinen. 1984. Transformation of membranous glomerulonephraitis into crescentic glomerulonephritis with glomerular basement membrane antibodies. Serial determinations of anti-GBM before the transformation. *Nephron* 38:134–137.

367. Rajaraman, S., J.A. Pinto, and T. Cavallo. 1984. Glomerulonephritis with coexistent immune deposits and antibasement membrane activity. *J. Clin. Pathol.* 37:176–181.

368. Pettersson, E., T. Törnroth, and A. Miettinen. 1984. Simultaneous anti-glomerular basement membrane and membranous glomerulonephritis: case report and literature review. *Clin. Immunol. Immunopathol.* 31:171–180.

369. Zevin, D., M. Ben-Bassat, T. Weinstein, Z. Shapira, and J. Levi. 1985. Rejection-related nephrotic syndrome associated with massive antiglomerular and antitubular basement membrane deposits. *Isr. J. Med. Sci.* 21:915–918.

370. Thitiarchakul, S., S.M. Lal, A. Luger, and G. Ross. 1995. Goodpasture's syndrome superimposed on membranous nephropathy. A case report. *Int. J. Artif. Organs* 18:763–765.

371. Calderon, E.J., I. Wishmann, J.M. Varela, N. Respaldiza, C. Regordan, J. Fernandez-Alonso, F.J. Medrano, S. Cano, J.A. Cuello, and A. Nuñez-Roldan. 1997. Presence of glomerular basement membrane (GBM) antibodies in HIV⁻ patients with *Pneumocystis carinii* pneumonia. *Clin. Exp. Immunol.* 107:448–450.

372. Ahuja, T.S., A. Velasco, W. Deiss, Jr., A.J. Indrikovs, and S. Rajaraman. 1998. Diabetic nephropathy with anti-GBM nephritis. *Am. J. Kidney Dis.* 31:127–130.

373. Sanchez-Fructuoso, A.I., J. Blanco, P. Naranjo, D. Prats, and A. Barrientos. 1998. Antiglomerular basement membrane antibody-mediated nephritis in two patients with type II diabetes mellitus. *Nephrol. Dial. Transplant.* 13:2674–2678.

374. Guerin, V., C. Rabian, L.H. Noel, D. Droz, C. Bacon, F. Lallemand, and P. Jungers. 1990. Anti-glomerular basement membrane disease after lithotripsy. *Lancet* 335:856–857.

375. Iwamoto, I., S. Yonekawa, T. Takeda, M. Sakaguchi, T. Ohno, H. Tanaka, H. Hasegawa, A. Imada, A. Horiuchi, T. Umekawa, and T. Kurita. 1998. Anti-glomerular basement membrane nephritis after extracorporeal shock wave lithotripsy. *Am. J. Nephrol.* 18:534–537.

376. Xenocostas, A., S. Jothy, B. Collins, R. Loertscher, and M. Levy. 1999. Anti-glomerular basement membrane glomerulonephritis after extracorporeal shock wave lithotripsy. *Am. J. Kidney Dis.* 33:128–132.

377. Umekawa, T., K. Kohri, K. Yoshioka, M. Iguchi, and T. Kurita. 1994. Production of anti-glomerular basement membrane antibody after extracorporeal shock wave lithotripsy. *Uro. Int.* 52:106–108.

378. Westman, K.W., U.B. Ericsson, M. Hoier-Madsen, J. Wieslander, E. Lindstedt, P.G. Bygren, and E.M. Erfurth. 1997. Prevalence of autoantibodies associated with glomerulonephritis, unaffected after extracorporeal shock wave lithotripsy for renal calculi, in a three-year follow-up. *Scand. J. Urol. Nephrol.* 31:463–467.

379. Umekawa, T., T. Yamate, N. Amasaki, and T. Kurita. 1995. Clinical efficiency of type IV collagen in extracorporeal shock wave lithotripsy. *Uro. Int.* 54:154–156.

380. Henderson, R.D., D. Saltissi, and M.P. Pender. 1998. Goodpasture's syndrome associated with multiple sclerosis. *Acta Neurol. Scand.* 98:134–135.

381. Joh, K., Y. Kanetsuna, Y. Ishikawa, S. Aizawa, A. Imadachi, O. Tastusawa, and T. Ohishi. 1998. Epstein-Barr virus genome-positive tubulointerstitial nephritis associated with immune complex-mediated glomerulonephritis in chronic active EB virus infection. *Virchows Arch.* 432:567–573.

382. Becker, J.L., F. Miller, G.J. Nuovo, C. Josepovitz, W.H. Schubach, and E.P. Nord. 1999. Epstein-Barr virus infection of renal proximal tubule cells: possible role in chronic interstitial nephritis. *J. Clin. Invest.* 104:1673–1681.

383. Neilson, E.G. 1999. Interstitial nephritis: another kissing disease? *J. Clin. Invest.* 104:1671–1672.

384. Grefer, J., R. Santer, T. Ankermann, S. Faul, B. Nolle, and P. Eggert. 1999. Tubulointerstitial nephritis and uveitis in association with Epstein-Barr virus infection. *Pediatr. Nephrol.* 13:336–339.

385. Lehman, D.H., C.B. Wilson, and F.J. Dixon. 1975. Extraglomerular immunoglobulin deposits in human nephritis. *Am. J. Med.* 58:765–786.

386. Andres, G., J. Brentjens, R. Kohli, R. Anthone, S. Anthone, T. Baliah, M. Montes, B.K. Mookerjee, A. Presyna, M. Sepulveda, M. Venuto, and C. Elwood. 1978. Histology of human tubulo-interstitial nephritis associated with antibodies to renal basement membranes. *Kidney Int.* 13:480–491.

387. Yoshioka, K., Y. Morimoto, T. Iseki, and S. Maki. 1986. Characterization of tubular basement membrane antigens in human kidney. *J. Immunol.* 136:1654–1660.

388. Wilson, C.B. 1991. Nephritogenic tubulointerstitial antigens. *Kidney Int.* 39:501–517.

389. Kelly, C.J. 1999. Cellular immunity and the tubulointerstitium. *Semin. Nephrol.* 19:182–187.

390. Kelly, C.J., and E.G. Neilson. 2000. Tubulointerstitial diseases. In: The Kidney, Sixth Edition, B.M. Brenner, editor. Saunders, Philadelphia, pp. 1509–1536.

391. Bergstein, J., and N. Litman. 1975. Interstitial nephritis with anti-tubular-basement-membrane antibody. *N. Engl. J. Med.* 292:875–878.

392. Andres, G.A., and R.T. McCluskey. 1975. Tubular and interstitial renal disease due to immunologic mechanisms. *Kidney Int.* 7:271–289.

393. Rakotoarivony, J., C. Orfila, A. Segonds, P. Giraud, D. Durand, Ch. DuBois, Ph. Mahieu, and J.-M. Suc. 1981. Human and experimental nephropathies associated with antibodies to tubular basement membrane. *Adv. Nephrol.* 10:187–212.

394. Brentjens, J.R., S. Matsuo, A. Fukatsu, I. Min, R. Kohli, R. Anthone, S. Anthone, G. Biesecker, and G. Andres. 1989. Immunologic studies in two patients with antitubular basement membrane nephritis. *Am. J. Med.* 86:603–608.

395. Lindqvist, B., L. Lundberg, and J. Wieslander. 1994. The prevalence of circulating anti-tubular basement membrane-antibody in renal diseases, and clinical observations. *Clin. Nephrol.* 41:199–204.

396. Morel-Maroger, L., O. Kourilski, F. Mignon, and G. Richet. 1974. Antitubular basement membrane antibodies in rapidly progressive poststreptococcal glomerulonephritis. Report of a case. *Clin. Immunol. Immunopathol.* 2:185–194.

397. Tung, K.S.K., and W.C. Black. 1975. Association of renal glomerular and tubular immune complex disease and anti-tubular basement membrane antibody. *Lab. Invest.* 32:696–700.

398. Levy, M., P. Guesry, C. Loirat, J.P. Dommergues, H. Nivet, and R. Habib. 1979. Immunologically mediated tubulo-interstitial nephritis in children. *Contrib. Nephrol.* 16:132–140.

399. Makker, S.P. 1980. Tubular basement membrane antibody-induced interstitial nephritis in systemic lupus etythematosus. *Am. J. Med.* 69:949–952.

400. Wood, E.G., B.H. Brouhard, L.B. Travis, T. Cavallo, and R.E. Lynch. 1982. Membranous glomerulonephropathy with tubular dysfunction and linear tubular basement membrane IgG deposition. *J. Pediatr.* 101:414–417.

401. Katz, A., A.J. Fish, P. Santamaria, T.E. Nevins, Y. Kim, and R.J. Butkowski. 1992. Role of antibodies to tubulointerstitial nephritis antigen in human anti-tubular basement membrane nephritis associated with membranous nephropathy. *Am. J. Med.* 93:691–698.

402. Zawada. E/T., Jr., W.H. Johnston, and J. Bergstein. 1981. Chronic interstitial nephritis. Its occurrence with oxalosis and anti-tubular basement membrane antibodies after jejunoileal bypass. *Arch. Pathol. Lab. Med.* 105:379–383.

403. Ellis, D., S.E. Fisher, W.I. Smith, Jr., and R. Jaffe. 1982. Familial occurrence of renal and intestinal disease associated with tissue autoantibodies. *Am. J. Dis. Child.* 136:323–326.

404. Griswold, W.R., H.F. Krous, V. Reznik, J. Lemire, N.W. Wilson, J. Bastian, and H. Spiegelberg. 1997. The syndrome of autoimmune interstitial nephritis and membranous nephropathy. *Pediatr. Nephrol.* 11:699–702.

405. Iványi, B., I. Haszon, E. Endreffy, P. Szenohradszky, A.S. Charonis, and S. Türi. 1998. Childhood membranous nephropathy, circulating antibodies to the 58-kD TIN antigen, and anti-tubular basement membrane nephritis: An 11-year follow-up. *Am. J. Kidney Dis.* 32:1068–1074.

406. Appel, G.B. 1980. A decade of penicillin related acute interstitial nephritis-more questions than answers. *Clin. Nephrol.* 13:151–154.

407. Kleinknecht, D., Ph. Vanhille, L. Morel-Maroger, A. Kanfer, V. LeMaitre, J.P. Mery, J. Laederich, and P. Callard. 1983. Acute interstitial nephritis due to drug hypersensitivity. An up-to-date review with a report of 19 cases. *Adv. Nephrol.* 12:277–308.

408. Baldwin, D.S., B.B. Levine, R.T. McCluskey, and G.R. Gallo. 1968. Renal failure and interstitial nephritis due to pencillin and methicillin. *N. Engl. J. Med.* 279:1245–1252.

409. Border, W.A., D.H. Lehman, J.D. Egan, H.J. Sass, J.E. Glode, and C.B. Wilson. 1974. Anti-tbuluar basement-membrane antibodies in methicillin-associated interstitial nephritis. *N. Engl. J. Med.* 291:381–384.

410. Hyman, L.R., M. Ballow, and M.R. Knieser. 1978. Diphenylhydantoin interstitial nephritis. Roles of cellular and humoral immunologic injury. *J. Pediatr.* 92:915–920.

411. Grussendorf, M., K. Andrassy, R. Waldherr, and E. Ritz. 1981. Systemic hypersensitivity to allopurinol with acute interstitial nephritis. *Am. J. Nephrol.* 1:105–109.

412. Klassen, J., K. Kano, F. Milgrom, A.B. Menno, S. Anthone, R. Anthone, M. Sepulveda, C.M. Elwood, and G.A. Andres. 1973. Tubular lesions produced by autoantibodies to tubular basement membrane in human renal allografts. *Int. Arch. Allergy Appl. Immunol.* 45:675–689.

413. Rotellar, C., L.H. Noel, D. Droz, H. Kreis, and J. Berger. 1986. Role of antibodies directed against tubular basement membranes in human renal transplantation. *Am. J. Kidney Dis.* 7:157–161.

414. Orfila, C., D. Durand, C. Vega-Vidalle, and J.M. Suc. 1991. Immunofluorescent deposits on the tubular basement membrane in human renal transplant. *Nephron* 57:149–155.

415. Wilson, C.B., D.H. Lehman, R.C. McCoy, J.C. Gunnells, Jr., and D.L. Stickel. 1974. Antitubular basement membrane antibodies after renal transplantation. *Transplantation* 18:447–452.

416. Cattran, D.C. 1980. Circulating anti-tubular basement membrane antibody in a variety of human renal diseases. *Nephron* 26:13–19.

417. Jordan, S.C., S.C. Barkley, J.M. Lemire, R.S. Sakai, A. Cohen, and R.N. Fine. 1986. Spontaneous anti-tubular-basement-membrane antibody production by lymphocytes isolated from a rejected allograft. *Transplantation* 41:173–176.

418. Kelly, C.J., D.A. Roth, and C.M. Meyers. 1991. Immune recognition and response to the renal interstitium. *Kidney Int.* 39:518–530.

419. Lehman, D.H., C.B. Wilson, and F.J. Dixon. 1974. Interstitial nephritis in rats immunized with heterologous tubular basement membrane. *Kidney Int.* 5:187–195.

420. Mampaso, F.M., and C.B. Wilson. 1983. Characterization of inflammatory cells in autoimmune tubulointerstitial nephritis in rats. *Kidney Int.* 23:448–457.

421. Zanetti, M., and C.B. Wilson. 1983. Characterization of anti-tubular basement membrane antibodies in rats. *J. Immunol.* 130:2173–2179.

422. Lehman, D.H., S. Lee, C.B. Wilson, and F.J. Dixon. 1974. Induction of antitubular basement membrane antibodies in rats by renal transplantation. *Transplantation* 17:429–431.

423. Sugisaki, T., K. Kano, G. Andres, and F. Milgrom. 1982. Transplacental transfer of antibodies to tubular basement membrane. *J. Clin. Lab. Immunol.* 7:167–171.

424. Bannister, K.M., and C.B. Wilson. 1985. Transfer of tubulointerstitial nephritis in the Brown-Norway rat with anti-tubular basement membrane antibody: Quantitation and kinetics of binding and effect of decomplementation. *J. Immunol.* 135:3911–3917.

425. Bannister, K.M., T.R. Ulich, and C.B. Wilson. 1987. Induction, characterization, and cell transfer of autoimmune tubulointerstitial nephritis in the Lewis rat. *Kidney Int.* 32:642–651.

426. Neilson, E.G., and S.M. Phillips. 1982. Murine interstitial nephritis. I. Analysis of disease susceptibility and its relationship to pleiomorphic gene products defining both immune-response genes and a restrictive requirement for cytotoxic T cells at H-2K. *J. Exp. Med.* 155:1075–1085.

427. Zakheim, B., E. McCafferty, S.M. Phillips, M. Clayman, and E.G. Neilson. 1984. Murine interstitial nephritis II. The adoptive transfer of disease with immune T lymphocytes produces a phenotypically complex interstitial lesion. *J. Immunol.* 133:234–239.

428. Mann, R., C.J. Kelly, W.H. Hins, M.D. Clayman, N. Blanchard, M.J. Sun, and E.G. Neilson. 1987. Effector T cell differentiation in experimental interstitial nephritis. I. The development and modulation of effector lymphocyte maturation by I-J+ regulatory T cells. *J. Immunol.* 138:4200–4208.

429. Kelly, C.J., H. Mok, and E.G. Neilson. 1988. The selection of effector T cell phenotype by contrasuppression modulates susceptibility to autoimmune injury. *J. Immunol.* 141:3022–3028.

430. Neilson, E.G., E. McCafferty, R. Mann, L. Michaud, and M. Clayman. 1985. Murine interstitial nephritis. III. The selection of phenotypic (Lyt and L3T4) and idiotypic (Re-Id) T cell preferences by genes in Igh-1 and H-2K characterizes the cell-mediated potential for disease expression: Susceptible mice provide a unique effector T cell repertoire in response to tubular antigen. *J. Immunol.* 134:2375–2382.

431. Mann, R., and E.G. Neilson. 1986. Murine interstitial nephritis. V. The auto-induction of antigen-specific Lyt-2+ suppressor T cells diminishes the expression of interstitial nephritis in mice with antitubular basement membrane disease. *J. Immunol.* 136:908–912.

432. Neilson, E.G., C.J. Kelly, M.D. Clayman, W.H. Hines, T. Haverty, M.J. Sun, and N. Blanchard. 1987. Murine interstitial nephritis. VII. Suppression of renal injury after treatment with soluble suppressor factor TsF₁¹. *J. Immunol.* 139:1518–1524.

433. Hines, W.H., R.A. Mann, C.J. Kelly, and E.G. Neilson. 1990. Murine interstitial nephritis. IX. Induction of the nephritogenic effector T cell repertoire with an antigen-specific T cell cytokine. *J. Immunol.* 144:75–83.

434. Kelly, C.J. 1990. T cell regulation of autoimmune interstitial nephritis. *J. Am. Soc. Nephrol.* 1:140–149.

435. Wakashin, M., Y. Wakashin, S. Ueda, I. Takei, Y. Mori, T. Mori, K. Iesato, and K. Okuda. 1980. Murine autoimmune interstitial nephritis and associated antigen: purification of a soluble tubular basement membrane antigen from mice kidneys. *Renal Physiol.* 3:360–367.

436. Wakashin, Y., I. Takei, S. Ueda, Y. Mori, K. Iesato, M. Wakashin, and K. Okuda. 1981. Autoimmune interstitial disease of the kidney and associated antigen purification and characterization of a soluble tubular basement membrane antigen. *Clin. Immunol. Immunopathol.* 19:360–371.

437. Lehman, D.H., H. Marquardt, C.B. Wilson, and F.J. Dixon. 1974. Specificity of autoantibodies to tubular and glomerular basement membranes induced in guinea pigs. *J. Immunol.* 112:241–248.

438. Nelson, T.R., R.J. Butkowski, A.F. Michael, and A.S. Charonis. 1997. Detection of tubulointerstitial nephritis antigen (TIN-ag) in Lewis rat. *Connect. Tissue Res.* 36:223–229.

439. Clayman, M.D., A. Martinez-Hernandez, L. Michaud, R. Alper, R. Mann, N.A. Kefalides, and E.G. Neilson. 1985. Isolation and characterization of the nephritogenic antigen producing anti-tubular basement membrane disease. *J. Exp. Med.* 161:290–305.

440. Neilson, E.G., D.L. Gasser, E. McCafferty, B. Zakheim, and S.M. Phillips. 1983. Polymorphism of genes involved in anti-tubular basement membrane disease in rats. *Immunogenetics* 17:55–65.

441. Haverty, T.P., C.J. Kelly, W.H. Hines, P.S. Amenta, M. Watanabe, R.A. Harper, N.A. Kefalides, and E.G. Neilson. 1988. Characterization of a renal tubular epithelial cell line which secretes the autologous target antigen of autoimmune experimental interstitial nephritis. *J. Cell Biol.* 107:1359–1368.

442. Haverty, T.P., M. Watanabe, E.G. Neilson, and C.J. Kelly. 1989. Protective modulation of class II MHC gene expression in tubular epithelium by target antigen-specific antibodies. Cell-surface directed down-regulation of transcription can influence susceptibility to murine tubulointerstitial nephritis. *J. Immunol.* 143:1133–1141.

443. Meyers, C.M., and C.J. Kelly. 1991. Effector mechanisms in organ-specific autoimmunity. I. Characterization of a CD8+ T cell line that mediates murine interstitial nephritis. *J. Clin. Invest.* 88:408–416.

444. Neilson, E.G., M.J. Sun, C.J. Kelly, W.H. Hines, T.P. Haverty, M.D. Clayman, and N.E. Cooke. 1991. Molecular characterization of a major nephritogenic domain in the autoantigen of anti-tubular basement membrane disease. *Proc. Natl. Acad. Sci. USA* 88:2006–2010.

445. Clayman, M.D., L. Michaud, J. Brentjens, G.A. Andres, N.A. Kefalides, and E.G. Neilson. 1986. Isolation of the target antigen of human anti-tubular basement membrane antibody-associated interstitial nephritis. *J. Clin. Invest.* 77:1143–1147.

446. Clayman, M.D., M.J. Sun, L. Michaud, J. Brill-Dashoff, R. Riblet, and E.G. Neilson. 1988. Clonotypic heterogeneity in experimental interstitial nephritis. Restricted specificity of the anti-tubular basement membrane B cell repertoire is associated with a disease-modifying crossreactive idiotype. *J. Exp. Med.* 167:1296–1312.

447. Fliger, F.D., J. Wieslander, J.R. Brentjens, G.A. Andres, and R.J. Butkowski. 1987. Identification of a target antigen in human anti-tubular basement membrane nephritis. *Kidney Int.* 31:800–807.

448. Butkowski, R.J., J.P.M. Langeveld, J. Wieslander, J.R. Brentjens, and G.A. Andres. 1990. Characterization of a tubular basement membrane component reactive with autoantibodies associated with tubulointerstitial nephritis. *J. Biol. Chem.* 265:21091–21098.

449. Butkowski, R.J., M.M. Kleppel, A. Katz, A.F. Michael, and A.J. Fish. 1991. Distribution of tubulointerstitial nephritis antigen and evidence for multiple forms. *Kidney Int.* 40:838–846.

450. Crary, G.S., A. Katz, A.J. Fish, A.F. Michael, and R.J. Butkowski. 1993. Role of a basement membrane glycoprotein in anti-tubular basement membrane nephritis. *Kidney Int.* 43:140–146.

451. Nelson, T.R., Y. Kim, A.F. Michael, R.J. Butkowski, and A.S. Charonis. 1998. Tubulointerstitial nephritis antigen (TIN-ag) is expressed in distinct segments of the developing human nephron. *Connective Tissue Res.* 37:53–60.

452. Yoshioka, K., S. Hino, T. Takemura, H. Miyasato, E. Honda, and S. Maki. 1992. Isolation and characterization of the tubular basement membrane antigen associated with human tubulo-interstitial nephritis. *Clin. Exp. Immunol.* 90:319–325.

453. Miyazato, H., K. Yoshioka, S. Hino, N. Aya, S. Matsuo, N. Suzuki, Y. Suzuki, H. Sinohara, and S. Maki. 1994. The target antigen of anti-tubular basement antibody-mediated interstitial nephritis. *Autoimmunity* 18:259–265.

454. Kalfa, T.A., J.D. Thull, R.J. Butkowski, and A.S. Charonis. 1994. Tubulointerstitial nephritis antigen interacts with laminin and type IV collagen and promotes cell adhesion. *J. Biol. Chem.* 269:1654–1659.

455. Chen, Y., U. Krishnamurti, E.A. Wayner, A.F. Michael, and A.S. Charonis. 1996. Receptors in proximal tubular epithelial cells for tubulointerstitial nephritis antigen. *Kidney Int.* 49:153–157.

456. Nelson, T.R., A.S. Charonis, R.S. McIvor, and R.J. Butkowski. 1995. Identification of a cDNA encoding tubulointerstitial nephritis antigen. *J. Biol. Chem.* 270:16265–16270.

457. Kanwar, Y.S., A. Kumar, Q. Yang, Y. Tian, J. Wada, N. Kashihara, and E.I. Wallner. 1999. Tubulointerstitial nephritis antigen: An extracellular matrix protein that selectively regulates tubulogenesis vs. glomerulogenesis during mammalian renal development. *Proc. Natl. Acad. Sci. USA* 96:11323–11328.

458. Ikeda, M., T. Takemura, S. Hino, and K. Yoshioka. 2000. Molecular cloning, expression, and chromosomal localization of a human tubulointerstitial nephritis antigen. *Biochem. Biophys. Res. Commun.* 268:225–230.

459. Kumar, A., K. Ota, J. Wada, E.I. Wallner, A.S. Charonis, F.A. Carone, and Y.S. Kanwar. 1997. Developmental regulation and partial-length cloning of tubulointerstitial nephritis antigen of murine metanephros. *Kidney Int.* 52:620–627.

460. Cohen, A.H., and J.R. Hoyer. 1986. Nephrophthisis. A primary tubular basement membrane defect. *Lab. Invest.* 55:564–572.

ASSOCIATED AUTOIMMUNITY

59 | Autoimmune Disease of the Spermatozoa, Ovary and Testis

Kenneth S.K. Tung, Francesco Fusi, and Cory Teuscher

1. INTRODUCTION

Basic research on autoimmunity has generally outpaced our understanding of clinical autoimmune disease, and this is also true for autoimmune diseases of spermatozoa and the gonads. Infertility in men and women with serum- or sperm-bound antibody may represent immunological infertility; however, a causal relation between antibody and infertility has not been established. Similarly, human ovarian and testicular autoimmune diseases are incompletely defined with respect to prevalence, mechanism, diagnosis and treatment. In contrast, research based on the versatile experimental autoimmune models of the gonads has contributed significantly to the general concepts of autoimmune disease pathogenesis and prevention. In this chapter we will describe the clinical diseases, and then the basic principles derived from research on the experimental models in more depth.

2. CLINICAL AUTOIMMUNE DISEASE OF SPERM, TESTIS AND OVARY

2.1. Male Infertility Associated with Sperm Autoantibodies

Male infertility may result from binding of sperm antibodies to ejaculated sperm. In most patients, infertility occurs without associated diseases; however, patients with cystic fibrosis may have occluded vas deferens, and are reported to have a higher prevalence of sperm antibodies (1). Antibodies that result from surgical vasectomy may also be responsible for infertility in some patients following vaso-vasostomy (2). In addition, sperm antibodies have been reported in patients with testicular trauma (3), torsion (4), biopsy (5), tumor (6), and genital tract infections (7).

The mixed agglutination method and the immunobead method are frequently used to detect sperm antibodies in the clinical laboratories. Mixed antiglobulin reaction detects sperm surface immunoglobulin (Ig) (8). Rh-positive RBC coated with human anti-D IgG is mixed with a drop of the patient's motile ejaculated sperm, and rabbit anti human IgG is added to bring about mixed agglutination. Direct and indirect mixed agglutination assays detect antibodies on sperm and in serum, respectively. The mixed agglutination assay, popular in Europe, lacks sensitivity because of the large size of the RBC. With the more popular immunobead assay, the RBC is replaced by the small polyacrylamide beads, covalently coated with rabbit antibody to human IgG, IgA or IgM (9). After a brief incubation period, motile sperm are evaluated microscopically for the frequency of sperm with bound beads, for the Ig class of the sperm antibody and the site of antibody binding (sperm head, midpiece or tail). The test is positive when 20% or more motile sperm are coated with beads. The Ig class of the antibody appears to influence diagnostic accuracy. Altered semen quality is most frequently found when both IgM and IgG antibodies are detected, but not with IgG alone (10). In retrospective analysis of patients in the *in vitro* Fertilization programs, a low fertilization rate was associated with sperm decorated with combined IgA and IgG head-bound antibodies (11,12).

In addition to antibody detection, patients with immunological infertility are also evaluated for the effect of antibody action upon the different steps of fertilization. Sperm antibodies impede penetration of the cervical mucus and this is detected by the post-coital test in which the number and motility of sperm within the cervical mucus are evaluated 8–12 hr after sexual intercourse (13). For fertilization to occur, the ejaculated sperm must modify its plasma membrane (capacitation), penetrate the cumulus, bind to the zona pellucida (ZP), undergo the calcium dependent acrosome reaction, and fuse with the oolema (rev. in 14). Although routine laboratory test is not available, antibodies to sperm have been found to accelerate (15) or retard acrosome reaction (16–18), block ZP binding (15,19,20), inhibit oocyte fusion (16,21–24), and affect early cleavage of fertilized oocytes (25). Currently, treatment of infertility associated with sperm antibodies is not satisfactory. However, some successes are reported with the use of assisted reproduction techniques (26).

The identification of functional human sperm antigens will help clarify this group of human diseases, and has recently been assisted by active research on a contraceptive vaccine based on functional human sperm antigens. Many potentially important human and animal sperm molecules have been identified by means of serum antibody from infertile patients or monoclonal antibodies that affect fertilization events. These well-characterized antigens of human and animal sperm, summarized in Table 1, are fully described in an excellent review (45). Future contraceptive vaccine studies based on characterized sperm antigens should evaluate their relevance in fertilization events and in human autoimmune disease. In addition to the nature of sperm antigens, it is critical to learn more about the mechanism behind sperm autoantibody development, and how serum antibody accesses the ejaculated sperm in the apparently normal male reproductive tract.

2.2. Infertility Associated with Autoimmune Disease of the Testis and the Ovary

A second form of male immunological infertility results from immunopathology of the testis and its excurrent ducts. Autoimmune orchitis is well documented in spontaneously infertile animals, and it may cause a similar disease in human (Table 2). Human testicular diseases with a probable autoimmune basis have been described, and they resemble the two major changes in the autoimmune testicular disease found in the infertile dark mink: 1) granulomatous orchitis, 2) aspermatogenesis without orchitis in testis that may have tissue-bound immune complexes. In addition, patients with epididymal granulomas of noninfectious origin may also have an autoimmune basis (Rev. in 74).

In contrast to autoimmune orchitis, human autoimmune ovarian disease is better documented. It has been estimated that 10% of women with premature ovarian failure have an autoimmune basis (75). Ovarian inflammation (oophoritis)

Table 1   Characterized sperm antigens in fertilization events

Antigens	Species	Location or source	Effect on fertilization		References
			In vitro	*In vivo*	
Fertilin β	Nonhuman	Intrinsic	Yes	No	27
	Human		No	No	28
Fertilin α	Nonhuman	Intrinsic	Yes	Not done	27
Cyritestin	Nonhuman	Intrinsic	Yes	Not done	27
PH20	Nonhuman	Intrinsic	Yes	Yes	29
	Human		Not done	Not done	30
SP10	Nonhuman	Intrinsic	Yes	Not done	31
	Human		Not known	Not done	32
SP17	Nonhuman	Intrinsic	Not done	Yes	33
	Human		Not done	Not done	34
FA-1	Nonhuman	Intrinsic	Not done	Not done	35
	Human		Yes	Yes	35
P26H	Nonhuman	Intrinsic	Yes	Yes	36
NZ-1	Nonhuman	Intrinsic	Not done	Not done	37
SIAA	Human	Intrinsic	Yes	Not done	38
Zonadhesin	Nonhuman	Intrinsic	Not done	Not done	39
Equatorin	Nonhuman	Intrinsic	Yes	Not done	40
	Human		Not done	Not done	40
SOB2	Human	Intrinsic	Yes	Not done	41
LDHC4	Nonhuman	Intrinsic	Not done	Yes	42
Prot.DE/AEG	Nonhuman	Epididymal	Yes	Yes	43
Gp20	Human	Epididymal	Yes	Not done	44

Table 2 Autoimmune ovarian and testicular disease models

A) *Experimental autoimmune (allergic) orchitis (EAO) and experimental autoimmune oophoritis that result from immunization with tissue antigen:*

1) Classical EAO induced by immunization with testis antigen with adjuvant (46–48).
2) EAO induced by immunization with testis antigen without adjuvant (49,50).
3) Autoimmune oophoritis induced by immunization with a murine ZP3 peptide (51).

B) *Autoimmune diseases of ovary, testis (and/or other organs) that result from manipulations of the normal immune system:*

1) Thymectomize mice between days 1 to 4 (D3TX) after birth (52–55).
2) Transfer adult murine CD5[low] or CD25-T cells to athymic mice (56–59).
3) Treat normal mice with antibody to CD25 (60).
4) Treat neonatal mice with Cyclosporine A (61).
5) Engraft fetal rat thymus in athymic mice (62).
6) Engraft neonatal mouse thymus in athymic mice (63).
7) Transfer T cells from adult or neonatal thymus, neonatal spleen to athymic mice (58).
8) Mice with a transgenic Vα protein of the T cell receptor (64).
9) Transfer RT6-depleted rat spleen T cells to athymic rats (65).
10) Transfer OX22[high] (or CD45RC[high]) rat spleen T cells to athymic rats (66).

C) *Other models of autoimmune orchitis:*

1) Spontaneous autoimmune orchitis in dog (67), mink (68), rat (69) and man (70).
2) Postvasectomy autoimmune orchitis (71,72).
3) Orchitis in rats with the transgenic HLA/B27 molecule (73).

---

occurs early in the course of disease. Serum autoantibodies appear to react with a wide range of ovarian antigens. The common autoantibodies appear to react with antigens of steroid-producing cells common to ovaries, placenta, adrenal and testis (rev. in 76). A known antigenic molecule is the p450 side chain cleavage enzyme and the 17α OH enzyme (77). Antibodies that recognize oocyte cytoplasm and the ZP have also been described (78). In this regard, similar antibodies are detected in the serum of mice with experimental ovarian disease that follows neonatal thymectomy (Table 1); and ZP2 and ZP3, the major glycoproteins of the ZP, elicit autoantibodies that block fertility (79,80). Some have reported antibodies to follicular-stimulating hormone receptors and luteinizing hormone receptors (76). Immunological investigation of patients with premature ovarian failure is limited and progress is slow in part due to difficulty in identifying patients with early disease.

Another important example of clinical gonadal autoimmune diseases is illustrated by patients with the various forms of autoimmune multi-endocrinopathy syndromes (81).

3. EXPERIMENTAL TESTICULAR AND OVARIAN AUTOIMMUNE DISEASES AND NATURE OF THE SELF ANTIGENS

EAO, and autoimmune ovarian disease (AOD), can be elicited by two experimental approaches (Table I). Immunization with testis or ovarian antigen or peptide in adjuvant is the classical approach. Severe diseases also occur following deliberate perturbation of T cell compositions of common laboratory mice and rats, such as thymectomy at a narrow time window of days 1 to 4, but usually on day 3 (D3TX) after birth, or the transfer of defined T cell populations from normal laboratory inbred mice to syngeneic athymic nu/nu mice recipients (Table II). Since these treatments also lead to autoimmune diseases of the thyroid, prostate, salivary gland, eye and pancreatic islets, research on this model will likely elucidate the fundamental mechanisms of self tolerance or unresponsiveness, relevant to the physiological control against pathogenic autoimmune responses. A detailed description of the histopathology and of the protocols for testicular and ovarian autoimmune disease induction has recently been described (82,83). While autoimmune orchitis occurs spontaneously in the dark mink, the beagle dog, aging rats and horses, spontaneous autoimmune disease of the ovary has not been reported.

Pathogenic autoimmune responses to antigens in the male gonads are directed to haploid germ cells in the testis, including spermatozoa. Previous studies on EAO identified orchitogenic testicular proteins of unknown function (84–87). Recently, EAO and reversible male infertility were induced in guinea pigs with the testicular isoform of hyaluronidase (PH20) (Table 1) (88). PH20, expressed in the male haploid germ cells of species including human, is located on both the sperm plasma membrane and the inner acrosomal membrane of acrosome-reacted sperm. Beside its potent hyaluronidase activity, PH20 is required for sperm penetration of the ZP in fertilization (89). Recently, a mouse expressing transgenic ovalbumin in the haploid germ cells developed EAO when immunized with the oval-bumin T cell epitope (90).

Experimental AOD is induced by a peptide from ZP3 a major protein of the ZP (ZP) (91). This model will be referred to as ZP3-AOD. The ZP is the acellular matrix that surrounds developing and ovulated oocytes, and exists as degraded proteins within atretic follicles. The murine *zp3* gene encodes a polypeptide of 424 amino acids. A 13-mer murine ZP3 peptide, ZP3 (330–342), contains a native B cell epitope, ZP3 (336–342) (79), that overlaps with two nested T cell epitopes. The shortest oophoritogenic epitope is the 8-mer sequence, ZP3 (330–337). The T cell peptides in CFA or IFA elicit severe oophoritis in mice (51). These mice developed ZP3-specific T cell responses and produced antibodies to ZP3 detectable in serum and bound to the ovarian ZP. Because the 13 mer ZP3 peptide possesses all the functional attributes of a self-protein antigen, it helps to dissect the complex autoimmune process and its regulation. Whether infertile patients also develop T cell response to the ZP3 or any ovarian peptide has not been formally explored. D3TX mice produce antibody to several distinct oocyte cytoplasmic antigens (Thatte and Tung, unpublished), and one of these (MATER1) has recently been cloned and sequenced (92).

In the sections that follow, we will review systemic and regional mechanisms that normally prevent the occurrence of gonadal autoimmune diseases, events that might overcome these control mechanisms, and pathogenetic pathways that amplify the disease processes. When appropriate, findings in the testes and the ovaries will be compared. Finally, mapping of genes that regulate gonadal autoimmunity within and outside the major histocompatibility complex (MHC) locus will be reviewed.

4. MECHANISM OF GONADAL AUTOIMMUNE DISEASE PREVENTION AND INITIATION

4.1. Systemic Immunoregulation Depends on Regulatory T Cells

The existence of pathogenic T cells in the normal individuals, and their regulation by another subset of normal T cells, have been documented (58,93). This principle has been demonstrated for AOD as well as autoimmune disease of several organs, including the stomach. When CD4[+] thymocyte from normal female adult or neonatal BALB/c mice are transferred to athymic BALB/c mice, approximately 75% of the recipients develop significant oophoritis and/or autoimmune gastritis within two months. This is associated with detectable serum autoantibody to the oocytes and the gastric autoantigen, H[+]K[+]ATPase, respectively. Since the CD4[+] subset represents mature thymocytes beyond deletion of self-reactive T cells, pathogenic self-reactive T cells for self antigens of the gonad and other organs are not deleted in the normal thymus. This conclusion is supported by a study on mice with the transgenic expression of the gastric H[+]K[+]ATPase β-chain in the thymus; their thymocytes no longer elicit gastritis but transfer oophoritis in athymic recipients (94).

Oophoritis and gastritis develop in athymic syngeneic recipients that receive neonatal, but not adult, splenic T cells. Therefore, the neonatal repertoire is enriched in self-reactive T cells. Presumably, thymectomy soon after birth limits the T cell repertoire to that of the neonate, and skews it to one enriched in self-reactive T cells; this could explain, in part, autoimmune diseases in the D3TX mice (95).

Although adult spleen T cells do not transfer oophoritis and gastritis to athymic recipients, a subset of adult splenic and thymic T cells can do so (56,59,96–98). The phenotype of these pathogenic T cells bear the following markers: CD5[low], CD45RB[high], CD25–, and in rats, RT6–. These studies have established the existence of potentially pathogenic T cells within the peripheral T cells of normal adult mice and rats.

When adult splenic T cells and neonatal T cells are co-injected into athymic recipients, oophoritis and gastritis do not develop. The nature of the regulatory T cells in these models has phenotypes that are reciprocal to those of the pathogenic T cells. They are: CD5[high], CD45RB[low], CD25[high], CD44[+], CD69[+] and L selectin[low], and in rats, RT6[+] and CD45RC– (59,65,66,96–98). This series of experiments emphasizes the physiological role of regulatory T cells in rendering oophoritogenic T cells non-functional in normal rodents. While this discussion has focused on ovarian and gastric autoimmunity, similar data have also been obtained for orchitogenic T cells.

The mechanism of regulation is being clarified. Recent results based on *in vitro* regulation of CD25– T cells by CD25[+] T cells have shown that suppression requires the presence of antigen-presenting cells, of cell contact, the maintenance of the anergic state of the CD25[+] regulatory T cells, and an apparent independence of several cytokines (IL4, IL10, TGFβ) (97,98). However, the antigen specificity of disease suppression in this model remains controversial.

4.2. Regional Immunoregulation

Testis self-antigens of late ontogeny are largely sequestered, and ovarian self antigens are fully accessible to the immune system from birth. Testicular isoforms of somatic antigens, expressed in/on haploid germ cells, appear after puberty and are not available to interact with lymphocytes early in life. Because immunological tolerance might require the interaction between self-antigens and developing lymphocytes before puberty, it has been assumed that tolerance to male germ cell antigens might not exist. Instead, testis antigen might be protected by a *complete* immunological blood-testis barrier. The main structural barrier consists of the peritubular myoid cells and the junctional complexes between adjacent Sertoli cells that effectively separate circulating antibodies and lymphocytes from the intra-tubular haploid germ cells (99). It has been assumed that the barrier also limits the access of germ cell antigens to the antigen-presenting cells outside the seminiferous tubules.

Although tissue barrier and antigen sequestration can limit accessibility, the barrier is incomplete. Antibody can enter the rete testis to bind to spermatozoa (100). Immunogenic autoantigens are detected on the diploid, preleptotene spermatocytes located outside the blood-testis barrier (101). In cell transfer experiments, pathology occurs in unique locations, thus target peptides may be presented to activated T cells outside the blood-testis barrier along the straight tubules linking the seminiferous tubules to the rete testis, and at the vas deferens (102).

There is also evidence for active local immunoregulation against autoimmune response to testis antigens. The testicular interstitial space may be immunologically privileged. Accordingly, the testicular autoantigens are likely protected by three independent mechanisms: 1) Confinement of most of the germ cell antigens by a strong but regionally *incomplete* tissue barrier, 2) systemic tolerance mechanism of regulatory T cells, and 3) regional mechanism or immune privilege.

The major ovarian autoantigens are associated with the oocytes in growing, mature and atretic follicles. Normal ovaries are endowed with a fixed number of primordial oocytes. In cycling female mice, a cohort of 30–40 oocytes develop and mature in every ovarian cycle that lasts for 4–5 days. The majority of the oocytes degenerate and undergo a process of atresia wherein oocyte antigens are phagocytosed by MHC class II-positive macrophages, accompanied by occasional CD4+ T cells. Thus, a potential source of ovarian peptides are presented by activated macrophages located in ovarian interstitial space. Indeed, ZP3 peptide-specific T cell clones transfer, within 2 days, granulomatous inflammation that targets atretic follicles in normal recipients. Antigens of the ZP in normal ovarian follicles and within atretic follicles are accessible to circulating antibody of IgG class and to circulating immune complexes (103,104). Moreover, there is evidence to support the notion that oocyte antigens can leave normal ovaries to reach lymphoid tissues. Thus, compared with the testis, ovarian autoantigens that turn over continuously in cycling female mice are accessible to the immune system.

Testicular interstitial environment.   Organ and tissue allograft, including parathyroid and pancreatic islets, which would be rapidly rejected under the renal capsule, survive

for prolonged periods when engrafted inside the testis, particularly in cryptorchid testes (105,106).

The T cell response elicited by mitogen or antibody to the T cell receptor is suppressed by proteins in fluid obtained from the testicular interstitial space, and in the supernatant of cultured testicular interstitial cells. Suppression may involve multiple factors derived from resident macrophages (107–109) or Sertoli cells (110,111). Candidate molecules include interleukin 1, TGF-β, basic fibroblast growth factor, transglutaminase, and prostaglandin E2, and activin (108,112). Proteins that inhibit complement activation may reduce complement-mediated inflammation and complement-dependent cytolysis (113). Sertoli cells produce sulfated glycoprotein 2, which can inhibit cytolysis by the complement C56789 attack complex (114,115).

Altered ovarian environment in post-recovery resistance to ovarian autoimmune disease.   In both EAO and ZP3-AOD, the inflammatory infiltrates eventually regress (88,116). Recovery from ovarian pathology is associated with resistance of the animals to re-induction of the autoimmune oophoritis. In ZP3-AOD, oophoritis resistance is not explicable either by immunosuppressive effect of adjuvant priming, or by suppression of pathogenic T cells. Moreover, the recovered mice produce ZP3 antibodies of IgG class when challenged with the ZP3 peptide (116). The oophoritis resistance state has been ascribed to an altered target organ. Thus, the recovered mice immunized with the ZP3 peptide develop oophoritis in normal ovaries implanted under the renal capsule, while their endogenous ovaries are spared. Ovarian disease resistance is not due to limitation of accessible target antigens. When mated, the recovered mice produce normal litters. Moreover, pathogenic ZP3-specific T cells elicit oophoritis when transferred to recovered mice (116).

4.3. Initiation and Modulation of Gonadal Autoimmune Diseases

The studies in Section 4.1 on autoimmune disease induction by normal murine T cells indicate that pathogenic self-reactive T cells exist in normal mice, and that they are normally controlled by regulatory T cells. This is consistent with the concept of T cell balance (117). It follows that if the balance is tipped in favor of effector T cell activation, autoimmune diseases might occur. This scenario has been substantiated in AOD wherein disease occurs when: 1) The depletion of regulatory cells by D3TX or other manipulations of the normal immune system, and 2) oophoritogenic T cells are activated by non-ovarian peptides that mimic an oophoritogenic T cell peptide. Furthermore, activation of autoreactive T cells is rapidly followed by spontaneous production of autoantibodies driven by endogenous antigens, and together T cells and autoantibodies can cause autoimmune tissue injury.

Regulatory T cell depletion.   Since 1967, Nishizuka and his colleagues have developed and investigated several important models of autoimmune disease that include orchitis and oophoritis (52,53,61,62, 64,118) (Table 2).

(C57BL/6 × A/J) F[1] mice with D3TX develop autoimmune oophoritis spontaneously, and the disease is transferred to young recipients by CD4[+] (but not CD8[+]) T cells from the diseased individuals (96,119). Of particular significance is the observation that the autoimmune disease is prevented when by T cells from normal adult mice are given to D3TX recipients that are less than 10 days of age (93,96). Both adult thymocytes and adult spleen cells are effective and the regulatory T cells express CD4[+] and a high level of CD5 and CD25 (96).

Disease suppression by normal T cells in D3TX mice indicates that autoimmune diseases in D3TX mice are due to deprivation of physiologically-relevant regulatory T cells. This conclusion is corroborated by similar findings in all of the autoimmune models based on the perturbation of the normal immune system (Table 2). Therefore, two mechanisms are invoked for the pathogenesis of D3TX oophoritis and orchitis: 1) The maintenance and expansion of the self-reactive, neonatal T cell repertoire, and 2) the preferential deprivation of regulatory T cells that exit the thymus after the pathogenic T cells. A third probability is that the lymphopenia in the manipulated mice predisposes them to autoimmune T cell activation and autoimmune disease (120). Whether T cells that develop extrathymically participate in this process remains to be determined.

Endogenous or exogenous antigens can activate autoreactive pathogenic T cells.   When the ovaries of D3TX mice are removed at birth, they do not develop oocyte antibody or AOD (in ovarian grafts) at 6 weeks of age (121). Thus, the ablation of regulatory T cells (by D3TX) leads to spontaneous pathogenic T cell activation, driven by endogenous antigens. It has been shown that the endogenous ovarian antigenic stimulation occurs within 2–3 weeks after D3TX, with activation of T cells followed by B cells (122).

Autoimmune oophoritis also develops when the host is immunized with peptides that mimic the self T cell peptide, as in the ZP3-AOD model. A peptide from the δ-chain of murine acetylcholine receptor is recognized by ZP3-specific T cell clones, and induces autoimmune oophoritis (123). Of the 9 amino acids in the ZP3 and AcCRδ peptides, 4 are shared between the peptides, and 3 of these 4 residues are critical for induction of autoimmune ovarian disease. Direct evidence for molecular mimicry based on sharing of the critical residues is obtained by induction of oophoritis and ZP3-specific T cell response by nanomer polyalanine peptides into which selected residues of the ZP3 and/or the AChRδ peptides are inserted. The study provides one of the first evidences of pathogenic molecular mimicry at the level of T cell peptides.

The phenomenon of molecular mimic also applies to foreign peptides, and this can occur rather frequently. Of 16 randomly-selected, non-ovarian peptides that share partial sequence homology with the critical residue motif as ZP3 (330–338), 7 (44%) induced oophoritis and autoantibody responses (124).

Therefore, autoimmune response and autoimmune disease may be elicited by endogenous or exogenous antigens, the former from loss of immunoregulation, and the latter from stimulation by T cell peptide mimic that may follow infection.

Diversified autoantibody response driven by endogenous antigens and activated helper T cells. Studies on the ZP3-AOD model have uncovered a novel and important mechanism of autoantibody induction. Mice injected with the ZP3 T cell peptide that lacks the B cell epitope spontaneously produce autoantibodies to the native ZP3 protein (125,126). A similar antibody response is also induced in mice by non-ovarian peptides that crossreact with the ZP3 T cell peptide, and even by polyalanine peptides that contain the critical residues of the ZP3 peptide (123,124). This T-to-B epitope spreading phenomenon is confirmed by reaction of antibodies to ZP3 B cell epitopes outside the ZP3-immunizing T cell peptide (126). Endogenous ovarian antigens are responsible for induction of ZP autoantibodies since antibodies are not detected in mice ovariectomized before immunization. However, ovarian pathology is not required for antibody induction since ovariectomy 2 days after immunization fails to completely abrogate antibody response (126). Thus, this is not merely a secondary response to antigens released from diseased ovaries. Importantly, the antibody response occurs rapidly, detected 2 days after T cell response, concordant with the onset of ovarian inflammation.

The phenomenon of autoantibody induction by T cell peptides is explicable by the following events: In normal mice, ovarian antigenic macromolecules or macromolecular complexes that include ZP3 normally reach the regional lymph nodes where they encounter ZP3-specific B cells. ZP3 is internalized, processed, and its T cell peptides presented on the class II MHC molecules. In normal mice, the series of events ends here, since such B cells normally undergo apoptotic cell death. However, in mice immunized with T cell peptide, the ZP3-specific T cells can recognize and be activated by the peptide/MHC complexes on the ZP3-specific B cells. In turn, the T cells stimulate the B cells to produce antibodies. Importantly, the antibody specificity would match that of the antigen receptor on B cells that initially capture the ovarian antigen. An implication is that self-reactive B cells in normal female mice can respond to self- ovarian antigen and are therefore not intrinsically tolerized. In addition, the occurrence of diversified autoantibodies indicates that serum autoantibodies need not mirror the immunogens that initiate an autoimmune disease, therefore, investigation of molecular mimic in autoimmunity based on autoantibodies must be interpreted with caution.

The phenomenon of autoantibody diversification supports the concept that autoimmune T cell response and antibody response with specificity other than the immunogen or the immunogenic epitope can occur frequently. This has been documented in immune response to a peptide from murine myelin basic protein in experimental autoimmune encephalomyelitis (127), in antibody response to murine gastric parietal cell K$^+$H$^+$ dependent ATPase in autoimmune gastritis of the D3TX mice (94), in antibody response to the pancreatic islet β cell autoantigen, glutamic acid decarboxylase, in the diabetic NOD mice (128,129). Even more important, epitope spreading from one antigen to another within nuclear macromolecular complexes has been reported and provides a potential mechanism for autoantibody induction in systemic autoimmunity, such as systemic lupus erythematosus (130–133).

Loss of tissue barrier.   Vasectomy, a common male contraceptive approach, results in the production of autoantibody response to sperm antigens of all subjects and the guinea pigs, T cell response to testicular antigens (134). In addition, postvasectomy autoimmune orchitis has been documented in vasectomized rabbits, guinea pigs and monkeys (71,72,135).

4.4.   Pathogenetic Mechanisms

Proinflammatory CD4$^+$ T cell activation and unique tissue targets.   The importance of the T cell-mediated mechanism is established by experiments involving disease transfer by lymphocytes of known function and antigen specificity. Testis or sperm Ag-specificT cells that have been activated *in vitro* transfer severe orchitis and vasitis to syngeneic euthymic mice (136–140). Similarly,

T cells from mice immunized with oophoritogenic ZP3 peptides rapidly transfer oophoritis to syngeneic recipients (51). In both cases, CD4$^+$ T cells are responsible (51,136,141), and this is substantiated by studies using CD4$^+$ T cell lines and clones (51,139,142).

A study based on T cell clones has further defined the pathogenetic mechanism of EAO (142). Despite the use of crude testis antigens in this study, both T cell lines and all of 16 independent T cell clones transfer EAO to normal syngeneic mice with pathology that affects the testis, epididymis and/or the vas deferens. Thus, orchitogenic peptides are likely immunodominant among peptides in the

crude testis antigenic preparation. Although testis antigen-derived T cell clones respond preferentially to testis antigen, and sperm antigen-derived clones respond more to sperm antigens, each of the sixteen clones responded to both antigens. Thus, unique and shared orchitogenic antigens exist in both germ cell populations, and their quantity may differ in distribution. The finding of pathogenic sperm-specific T cell clones indicates that germ cells alone can elicit EAO. All orchitogenic T cell clones express CD4, produced IL2, interferon γ (IFNγ), but not IL4, typical of the Th1 CD4$^+$ T cells. Moreover, disease transfer is dependent on the cytokine TNF.

Based on the unique distribution of testicular pathology in recipients of activated orchitogenic T cell lines or T cell clones, a main location of accessible testis peptides to the CD4$^+$ T cells has been defined (102,142). The predominant testicular inflammation invariably begins around the straight tubules. CD4$^+$ T cells recognize peptides in association with class II MHC molecules, and in normal mouse testis, MHC II-positive macrophages or dendritic cells are sparse, but form a dense cuff around the straight tubules. These are the likely antigen-presenting cells that present the germ-cell peptides to orchitogenic T cells to initiate EAO (102). The inflammatory infiltrate then blocks the passage of tubular spermatozoa and fluids to cause dilatation of the proximal seminiferous tubules. Severe orchitis spread centripetally to involve peripheral seminiferous tubules, and eventually testicular atrophy and necrosis ensue.

CD4+ Th1 cell clones against the ZP3 peptide are also responsible for oophoritis, and disease transfer is also dependent on the cytokine TNF (143). The atretic follicles are the target antigenic structure in the ovaries is the atretic follicles. In contrast, the growing follicles, the antral follicles, and the corpus luteum are spared in AOD following adoptive transfer (143,144).

Naive T cell activation by self peptides requires the costimulation between ligands on the T cells with the corresponding receptors on the antigen presenting cells (APC) (145). In ZP3-AOD, blockage of the CD40 ligand pathway results in failure to induce AOD and autoantibodies. Inhibition of ligand binding to the B7.1 and B7.2 also results in failure to generate antibody to ZP and significantly reduces disease severity and prevalence. However, the frequencies of antigen-specific T cells in mice with inhibition of either of these costimulatory pathway were similar to those of mice given control reagents (limiting dilution analysis: 1:5000). When both pathways are blocked, the effect is additive, leading to inhibition of T cell activation as determined by *in vitro* proliferation and limiting dilution analysis (1:190,000), and the mice do not develop disease or antibody responses. Thus, oophoritis and ZP3 antibody production are inhibitable by blocking either of the costimulatory pathways, whereas inhibition of clonal expansion of the pathogenic T cell population requires blockade of both pathways. This *in vivo* study has effectively dissociated two sequential activation steps of autoreactive ZP3-specific T cells.

Autoantibody retargets T cell-mediated inflammation and alters end organ function.   In both AOD and EAO, autoantibodies can be detected on cells or structures of the target organs (antibody and immune complex involving preleptotene spermatocyte, sperm, immune complex, ZP). However, autoantibody response alone is not sufficient to cause tissue inflammation. As an example, adult female mice injected with a chimeric peptide of ZP3 that contains a foreign T cell peptide and a native B cell peptide of ZP3 develop antibody to the ZP without concomitant T cell response (80). Antibody bound to the ZP *in vivo*, but the mice are free of oophoritis.

Although antibodies do not cause inflammations in the gonads, their reaction with the target tissue can influence the distribution of pathology mediated by T cells, as illustrated in the ZP3-AOD model. As described earlier, ZP3-specific T cells transfer pathology confined to the atretic follicles and spares the growing and antral follicles; and the ovarian function is retained (143). When mice also received ZP3 antibodies, which bind to the ZP of growing and mature follicles, the T cell-mediated inflammation was deviated to these follicles (144). Because the growing and antral follicles are the functional unit of the ovary, their destruction led to ovarian atrophy and may result in loss of ovarian function (144). Thus autoantibody binding to a tissue antigen can cause a re-distribution of T cell-mediated inflammation and alter the clinical course of the autoimmune disease.

4.5   Disease-Associated Genetic Loci

In the past few years, an impressive number of chromosome loci have been found to confer susceptibility or resistance to testicular or ovarian autoimmune diseases. They are summarized in Table 3.

Independent regulation of susceptibility to orchitis, epididymitis and vasitis in EAO.   A study on the differential susceptibility to EAO of the testis, the epididymis and the vas deferens has shown that mice with the H-2^S haplotype developed mainly orchitis, whereas mice of the H-2^K haplotype developed pathology mainly in the epididymis and the vas deferens (epididymo-vasitis) (146,147). This difference in response has been substantiated in the BXH series of recombinant inbred lines derived from the EAO-susceptible C57BL/6J and the EAO-resistant C3H/HeJ strains (148). A comparison of the strain distribution pattern of autoimmune orchitis with that of the typed alleles segregating in the BXH RI lines has not demonstrated definitive linkage. A recent microsatellite analysis has mapped distinct chromosomal regions that encode susceptibility to orchitis (*Orch-6*, chromosoma 8), epididymitis (*Epd-1*, chromosome 16), and vasitis (*Vas-1*, chromosome 1) (149).

Differential susceptibility to EAO in BALB/c sublines.   The role of disease resistance genes is also evident in studies on substrains of BALB/c mice. Disease resistance, inherited as a recessive trait in the BALB/cJ mice, has been shown by segregation analysis to be associated with a single genotypic difference, *Orch-2* (148). So far, the *Orch-2* gene has been shown not to be

Table 3   Genetic loci controlling susceptibility and resistance to murine experimental allergic orchitis (EAO) And D3TX-induced autoimmune ovarian dysgenesis (D3TX-AOD)

Locus	Phenotype	Chromosome (cM)[a]	Co-localization[b]
Orch1	EAO/orchitis	17	*H2*
Orch2	EAO/orchitis (BALB/c substrains)	Unmapped	*eae*
Orch3	EAO/orchitis	11 (43)	*eae7, Idd4*
Orch4	EAO/orchitis	1 (110)	*Slelc*
Orch5	EAO/orchitis	1 (20)	
Orch6	EAO/orchitis	8 (34)	*eae 14*
Bphs	*B. pertussis*-induced histamine sensitivity	6 (57)	*eae*
Epd1	EAO/epididymitis	16 (4)	
Vas1	EAO/vasitis	1 (33)	
Aod1	D3TX AOD oophoritis	16 (28)	*eael 1*
Aod2	D3TX AOD atrophy	3 (30)	*eae3, Idd3*
Aod3	D3TX AOD ovarian autoantibody	11 (17)	
Aod4	D3TX AOD ovarian autoantibody	7 (9)	
Aod5	D3TX AOD antinuclear antibody	6 (67)	

[a] Centimorgan assignment based on the MGD map (*www.informatics.jax.org/*)
[b] Co-localization with other autoimmune disease susceptibility loci.

related to elevated levels of serum alpha-fetoprotein in adult BALB/cJ animals (150), or to the *Bphs* locus (151). In view of the evidence for suppressor T cells in the BALB/cJ mice, it will be important to consider the *Orch-2* gene in the context of regulatory T cell expression in auto-immunity. Although *Orch-2* has not been mapped, it has been shown that susceptibility to EAO and experimental allergic encephalomyelitis among the BALB/c substrains is controlled by a common immunoregulatory locus (152).

A class III MHC locus for EAO susceptibility. Studies based on intra-MHC recombinant inbred mice indicated that *Orch-1* is an orchitis-susceptibility gene mapped within the *H-2S/H-2D* interval (146,153), between the *IR* gene controlling antibody responsiveness to trinitrophenol-Ficoll (*TNP-Ficoll*) and the locus encoding *TNF-α* (153). This has established the gene order in the region as: *H-2S—TNP-Ficoll—Orch-1—Tnfα—H-2D*. Recent molecular analysis has further mapped *Orch-1* within a 50–60 kb segment from *Hsp 70.1* to a region proximal to *G7* that encompasses the *Hsp 70.1*, *Hsp70.3*, *Hsc70t*, *G7b*, and *G7a/Bat6* genes (154).

EAO resistance gene loci. Genes located outside the MHC strongly influence EAO susceptibility. DBA/2J (*H-2D*) mice are resistant to EAO, BALB/cByJ (*H-2D*) mice are susceptible to EAO, whereas BALB/cByJ X DBA/2J F₁ (CD2F₁) mice are resistant (146). Among a (BALB/cBy X DBA/2J)F₁ X BALB/cByJ (BC1) population (n = 172), 54% of the animals are resistant, while 46% are susceptible. Using DNA isolated from the phenotyped BC1 population and previously mapped microsatellite markers distinguishing DBA/2J and BALB/cByJ, three genetic loci showed strong linkage to EAO: 1) *Orch-3*, mapped to chromosome 11, 2) *Orch-4*, at the telemeric region of chromosome 1, and 3) *Orch-5*, at the centromeric region of chromosome 1 (155).

The ZP3-AOD-susceptible and-resistant MHC haplotypes. Studies on H-2 congenic mice clearly indicate that AOD induced by the ZP3 peptide is restricted by the MHC haplotypes. Thus, mice of the H-2^{A,K,U,S} are responders, whereas mice of the H-2^{D,B,Q} are non-responders (80).

Multiple genetic loci govern distinct pathological expression of AOD induced by D3TX. Recent studies have mapped two gene loci that control susceptibility to D3TX-AOD. Kojima and Prehn (53) have shown H-2-linked genes play little, if any, role in determining disease outcome in the D3TX mice. In addition, in a recent study on AOD and other polyendocrinopathy that developed in mice with the transgenic T cell receptor Vα protein, the organ distribution of pathology is strictly controlled by the genetic background of the mice that carry the transgene (64).

A/J are susceptible to D3TX-AOD (90%), B6 are non-responders (8%), and B6AF₁ are responsibles (100%). Among 144 D3TX (C57BL/6J X A/J) F₁ X C57BL/6J backcross (BC1) mice, 77 exhibit oophoritis and oocytic autoimmunity whereas 67 are resistant. Microsatellites have been employed to generate a genomic exclusion map that localizes the autoimmune oophoritis and autoantibody susceptibility gene (*Aod-1*) to chromosome 16. Several candidate genes of immunological relevance reside on chromosome 16, including *Ly-7, Ifgt, Ifrc, Ig1-1, Mls-3, VpreB* (156).

An independent genetic locus, *Aod-2*, strongly influences the development of ovarian atrophy of this disease (157). *Aod-2* gene has been mapped to chromosome 3. This interval contains the *il2* gene and also the basic fibroblast growth factor (*bg/b*) gene. The former co-maps with idd3 of the diabetes susceptibility locus in NOD mice, and has been shown to exhibit polymorphism in its coding region (158), and the latter is known to regulate cell growth and differentiated functions of ovarian cells including granulosa cells (159). More recently, by means of quantitative trait analysis, additional disease susceptibility loci of D3TX-AOD have been mapped to chromsome 11 (*Aod-3*, linked to Itk/Tsk/Em1, a T cell-specific, IL-2 inducible tyrosine kinase), chromosome 7 (*Aod-4*, linked to MATER, a gene encoding an autoantigen recognized by antibody of the D3TX mice), and chromosome 6 (*Aod-5*) (160).

5. CONCLUSIONS

In this chapter, we have reviewed current knowledge concerning clinical autoimmune diseases affecting the sperm, testis and ovary. We have considered the tolerance mechanisms that prevent gonadal autoimmunity and the potential events that can overcome such mechanism to trigger autoimmune diseases. In addition, we have summarized our current knowledge on the mechanisms responsible for the immunopathology of autoimmune diseases of the gonads.

The local testicular immunoregulatory environment, provided by a strong blood-testis barrier and less well-defined intratesticular humoral factors, partially impedes the induction of autoimmune responses to testis antigens of late ontogeny. How haploid germ cell antigens access the antigen-presenting cells located outside the Sertoli cell barrier is poorly understood. Although ovarian self antigens are fully accessible to the immune system, the ovaries become resistant to disease induction in animals recovered from AOD and it is based on mechanism other than loss of self antigens.

In addition, there are systemic tolerance mechanisms that regulate autoreactive T cells that recognize testicular and ovarian antigen. The studies described herein indicate that pathogenic T cells capable of eliciting autoimmune diseases in the gonads develop in both the neonatal and adult thymuses and persist in the normal peripheral immune system. However, the expression of the pathogenic T cell function in adult mice is normally negated by regulatory T cells with the capacity to maintain peripheral tolerance, and important phenotypic differences are being defined between these two CD4+ T cell subsets of the opposing effector/regulatory functions. When the clonal balance of these T cell subsets is tipped in favor of pathogenic T cells, autoimmune diseases of the gonads could ensue.

The loss of regulatory T cells may occur through aberrant T cell development, as documented in several novel models of autoimmune oophoritis and orchitis caused by deliberate perturbation of the normal immune system, including D3TX. On the other hand, oophoritogenic T cells can be activated by non-ovarian peptides that crossreact with self peptides. This form of T cell epitope mimicry can occur frequently, and may depend on the partial sharing, between unrelated peptides, critical amino acids required for activation of pathogenic T cells.

The inflammatory (Th1) CD4 T cell mechanism has been established as a critical pathogenetic pathway for autoimmune orchitis and autoimmune oophoritis. In both, the pathogenic T cell clones secrete IL2, IFNγ and TNF; and as in many autoimmune diseases, neutralizing TNF *in vivo* markedly reduced disease severity transferred by pathogenic T cell clones. The tissue locations wherein pathogenic T cells encounter testicular and ovarian target antigens have been identified by the unusually precise localization of histopathology that follows adoptive transfer of the pathogenic T cell clones. The activation of the pathogenic CD4+ oophoritogenic T cells depends on the integrity of CD28 and CD40 costimulation pathways. Inhibition of either pathway does not inhibit clonal proliferation, but the pathogenicity of the expanded T cells is blocked.

Circulating antibodies can access both testicular and ovarian target antigens during the development of autoimmune orchitis and autoimmune oophoritis, but this is not sufficient to cause disease. In case of oophoritis, it cannot be induced by a chimeric peptide that contains a foreign T cell epitope and a native ZP3 B cell epitope, which induces antibody to ZP3 without concomitant T cell response. However, antibody to ZP has a new action: It retargets the distribution of T cell-mediated pathology, leading to destruction of the ovarian functional unit.

Surprisingly, immunization of female mice with a pure T cell peptide from ZP3 can result in spontaneous antibody response against ZP3 domains outside the immunogenic ZP3 peptide. Evidently, endogenous antigens from normal and pathologic ovaries can reach peripheral lymphoid tissues, and trigger an autoantibody response – a finding that argues against instrinsic B cell tolerance to ZP3. This occurs concomitant with detection of ZP3-specific T cells activation and is therefore not a consequence of tissue injury. Importantly, the autoantibodies react with native antigenic determinants, and can retarget T cell-mediated ovarian inflammation.

Finally, impressive progress has been made on the genetic linkage analysis of inbred mice by mapping gene loci responsible for the susceptibility and resistance to autoimmune diseases of the testis and ovary. This approach will eventually elucidate new and unexpected mechanisms that underlie the complex gonadal and other autoimmune diseases.

In conclusion, contemporary research on experimental testicular and ovarian autoimmune diseases have led to rapid accumulation of new knowledge that pertains to self-tolerance and autoimmune disease pathogenesis. The findings have emphasized the similarities between these two organs as well as their similarities with autoimmune diseases affecting non-gonadal organs. Research based on the unique models of gonadal autoimmune diseases and the well-defined self antigens can be expected to further clarify human autoimmune diseases of the gonads, and the autoimmune process in general.

ACKNOWLEDGEMENTS

We wish to thank Dr. Cherri Mahi-Brown, Dr. Teresita Yule, Dr. Hedy Smith, Dr. An-Ming Luo, Dr. Yahuan Lou, Dr. Nathan Griggs, Dr. Kristina Garza, Harini Bagavant and Pascale Alard for their conceptual and experimental contributions to our studies described in this chapter. We also thank Marianne Volpe for her meticulous editorial assistance. The studies are supported by NIH grants AI41236 and U54 HD29099.

References

1. D'Cruz, O.J., G.G. Haas, R. de La Rocha, and H. Lambert. 1991. Occurrence of serum antisperm antibodies in patients with cystic fibrosis. *Fertil. Steril.* 56:519–527.
2. Alexander, N.J., and D.J. Anderson. 1979. Vasectomy: consequences of autoimmunity to sperm antigens. *Fertil. Steril.* 32:253.
3. Haensch, R. 1973. Spermatozoen-autoimmunphaenomehe bei genitaltraumen undverschlubbazoospermie. *Andrologia.* 5:147–151.
4. Mastrogiacomo, I., R. Zanchetta, P. Graziotti, C. Betterle, P. Scrufari, and A. Lembo. 1982. Immunological and clinical study in patients after spermatic cord torsion. *Andrologia.* 14:25–29.
5. Hjort, T., S. Husted, and P. Linnet-Jepsen. 1974. The effect of testis biopsy on autosensitization against spermatozoal antigens. *Clin. Exp. Immunol.* 18:201–209.

6. Guazzieri, S., A. Lembo, G. Ferro, W. Artibani, F. Merlo, R. Zanchetta, and F. Pagano. 1985. Sperm antibodies and infertility in patients with testicular cancer. *Urology* 26:139–144.

7. Witkin, S.S., and A. Toth. 1983. Relationship between genital tract infections, sperm antibodies in seminal fluid, and infertility. *Fertil. Steril.* 40:805–808.

8. Jager, S., J. Kremer, and T. Van Slochteren-Draaisma. 1978. A simple method for antisperm antibodies detection in the human male. Detection of spermatozoal surface IgG with the direct mixed antiglobulin reaction carried out on untreated fresh human semen. *J. Fertil.* 23:1–12.

9. Bronson, R.A., G.W. Cooper, and D.L. Rosenfeld. 1984. Sperm antibodies: their role in infertility. *Fertil. Steril.* 42:171–183.

10. Gonzales, G.F., G. Kortebani, and A.B. Mazzolli. 1992. Effect of isotypes of antisperm antibodies on semen quality. *Int. J. Androl.* 15:220–228.

11. Clarke, G.N., A. Lopata, J.C. McBain, H.W.G. Baker, and W.I.H. Johnston. 1985. Effect of sperm antibodies in males in human *in vitro* fertilization (IVF). *Am. J. Reprod. Immunol. Microbiol.* 8:62–66.

12. DeAlmeida, M., I. Gazagne, C. Jeulin, M. Herry, J. Belaisch-Allart, R. Frydman, P. Jouannet, and J. Testart. 1989. *In vitro* processing of sperm with autoantibodies and *in vitro* fertilization results. *Human Reprod.* 4:49–53.

13. Fjallbrant, B. 1969. Cervical mucus penetration by human spermatozoa treated with anti-spermatozoal antibodies from rabbit and man. *Acta Obstet. Gynecol. Scand.* 48:71–77.

14. Yanagimachi, R. 1989. Sperm capacitation and gamete interaction. *J. Reprod. Fertil.* 38:27–33.

15. Lansford, R., G.G. Haas, L.E. DeDault, and D.P. Wolf. 1988. Effect of antisperm antibodies on the acrosomal reactivity of human sperm. The American Fertility Society Program and Abstracts. 44th Annual Meeting, Atlanta, October 10–13. (Abstr.)

16. Tung, K.S.K., C. Teuscher, and A.L. Meng. 1981. Autoimmunity to spermatozoa and the testis. *Immunol. Rev.* 55:217–255.

17. Tasdemir, I., M. Tasdemir, J. Fukuda, H. Kodama, T. Matsui, and T. Tanaka. 1995. Effects of sperm-immobilizing antibodies on the spontaneous and calcium-ionophore (A23187)-induced acrosome reaction. *Intern. J. Fertil.* 40:192–195.

18. Menge, A.C., G.M. Christman, D.A. Ohl, and R.K. Naz. 1999. Fertilization antigen-1 removes antisperm autoantibodies from spermatozoa of infertile men and results in increased rates of acrosome reaction. *Fertil. Steril.* 71:256–260.

19. Bronson, R.A., G.W. Cooper, and D.L. Rosenfeld. 1982. Sperm-specific isoantibodies and autoantibodies inhibit the binding of human sperm to the human ZP. *Fertil. Steril.* 38:724–729.

20. Tsukui, S., Y. Noda, J. Yano, A. Fukuda, and T. Mori. 1986. Inhibition of sperm penetration through human ZP by antisperm antibodies. *Fertil. Steril.* 46:92–96.

21. Requeda, E., J. Charron, K.D. Roberts, A. Chapdelaine, and G. Bleau. 1983. Fertilizing capacity and sperm antibodies in vasostomized men. *Fertil. Steril.* 39:197–203.

22. Haas, G.G., M. Ausmanas, L. Culp, R.W. Tureck, and L. Blasco. 1985. The effect of immunoglobulin occurring on human sperm *in vivo* on the human sperm/hamster ova penetration assay. *Am. J. Reprod. Immunol. Microbiol.* 7:109–112.

23. Abdel-Latif, A., S. Mathur, P.F. Rust, C.M. Fredericks, H. Abdel-Aal, and H.O. Williamson. 1986. Cytotoxic sperm antibodies inhibit sperm penetration of zona-free hamster eggs. *Fertil. Steril.* 45:542–549.

24. Bronson, R.A., G.W. Cooper, and D.L. Rosenfeld. 1981. Ability of a antibody-bound sperm to penetrate zona-free hamster eggs. *Fertil. Steril.* 36:778–783.

25. Naz, R.K. 1992. Effects of antisperm antibodies on early cleavage of fertilized ova. *Biol. Reprod.* 46:130–139.

26. Francavilla, F., R. Romano, R. Santucci, G. La Verghetta, P. D'Abrizio, and S. Francavilla. 1999. Naturally occurring antisperm antibodies in men: interference with fertility and implications for treatment. *Front. Biosci.* 4:E9–E25.

27. Wolfsberg, T.G., P.D. Straight, R.L. Gerena, A.-P.J. Huovila, P. Primakoff, D.G. Myles, and J.M. White. 1995. ADAM, a widely distributed and developmentally regulated gene family encoding membrane proteins with a disintegrin and metalloprotease domain. *Dev. Biol.* 169:378–383.

28. Gupta, S.K., K. Alves, L. O'Niell Palladino, G.E. Mark, and G.F. Hollis. 1996. Molecular cloning of the human fertilin β subunit. *Biochem. Biophys. Res. Comm.* 224: 318–326.

29. Primakoff, P., W. Lathrop, L. Woolman, A. Cowan, and D. Myles. 1988. Fully effective contraception in male and female guinea pigs immunized with the sperm protein PH-20. *Nature* 335:543–546.

30. Lin, Y., K. Mahan, W.F. Lathrop, D.G. Myles, and P. Primakoff. 1994. A hyaluronidase activity of the sperm plasma membrane protein PH-20 enables sperm to penetrate the cumulus cell layer surrounding the egg. *J. Cell Biol.* 125:1157–1163.

31. Coonrod, S.A., J.C. Herr, and M.E. Westhusin. 1996. Inhibition of bovine fertilization *in vitro* by antibodies to SP-10. *J. Reprod. Fertil.* 107:287–297.

32. Herr, J.C., C.J. Flickinger, M. Homyk, K. Klotz, and E. John. 1990. Biochemical and morphological characterization of the intra-acrosomal antigen SP10 from human sperm. *Biol. Reprod.* 43:181–193.

33. Kong, M., R.T. Richardson, E.E. Widgren, and M.G. O'Rand. 1995. Sequence and localisation of the mouse sperm autoantigenic protein Sp17. *Biol. Reprod.* 53:579–590.

34. Lea, I.A., R.T. Richardson, E.E. Widgren, and M.G. O'Rand. 1996. Cloning and sequencing of cDNAs encoding the human sperm protein Sp17. *Biochem. Biophys. Acta* 1307:263–266.

35. Naz, R.K. 1996. Application of sperm antigens in immunocontraception. *Front. Biosci.* 1:87–95.

36. Berube, B., and R. Sullivan. 1994. Inhibition of *in vivo* fertilization of male hamsters against a 26-KDa sperm glycoprotein. *Biol. Reprod.* 51:1255–1263.

37. Naz, R., and X. Zhu. 1997. Molecular cloning and sequencing of cDNA encoding for a novel testis-specific antigen. *Mol. Reprod. Dev.* 48:449–457.

38. Jimenez, C., B. Sion, G. Grizard, C. Artonne, J.L. Kemeny, and D. Boucher. 1994. Characterisation of a monoclonal antibody to a human intra-acrosomal antigen that inhibits fertilisation. *Biol. Reprod.* 51:1117–1125.

39. Hardy, D.M. and D.L. Garbers. 1994. Species-specific binding of the sperm proteins to the extracellular matrix (zona pellucida) of the egg. *J. Biol. Chem.* 269:19000–19004.

40. Toshimori, K., D.K. Saxena, I. Tanii, and K. Yoshinaga. 1998. An MN9 antigenic molecule, Equatorin, is required for successful sperm oocyte fusion in mice. *Biol. Reprod.* 59:22–29.

41. Lefevre, A., C.M. Ruiz, S. Chokomian, C. Duquenne, and C. Finaz. 1997. Characterisation and isolation of SOB2, a sperm protein with a potential role in oocyte membrane binding. *Mol. Hum. Reprod.* 3:507–516.

42. O'Hern, P.A., C.S. Bambra, M. Isahakia, and E. Goldberg. 1995. Reversible contraception in female baboons immunized with a synthetic epitope of sperm-specific lactate dehydrogenase. *Biol. Reprod.* 51:331–339.

43. Cuasnicu, P.S., D. Conesa, and L. Rochwerger. 1990. Potential contraceptive use of an epididymal protein that participates in fertilisation. In: Gamete Interaction. Prospects for Immunocontraception. N.J. Alexander, Griffin, D., Spieler, J.M., Waites, G.M.H., III, eds., Wiley-Liss, New York, NY. pp. 143–153.

44. Focarelli, R., A. Giuffrida, S. Capparelli, M. Scibona, F.M. Fabris, F. Francavilla, S. Francavilla, C.D. Giovampaola, and F. Rosati. 1998. Specific localization in the equatorial region of gp20, a 20Kda sialyloglycoprotein of the capacitated human spermatozoon acquired during epididymal transit which is necessary to penetrate zona-free hamster eggs. *Mol. Hum. Reprod.* 4:119–125.

45. Frayne, J., and L. Hall. 1999. The potential use of sperm antigens as targets for immunocontraception; past, present and future. *J. Reprod. Immunol.* 43:1–33.

46. Voisin, G.A., A. Delaunay, and M. Barber. 1951. Lesions testiculaires provoquées chez le cobay par injection d'éxtrait de testicule homologue. *CR Hedb, Séances. Acad. Sci.* 232:48–63.

47. Freund, J., M.M. Lipton, and G.E. Thompson. 1953. Aspermatogenesis in the guinea pig induced by testicular tissue and adjuvant. *J. Exp. Med.* 97:711–725.

48. Kohno, S., J.A. Munoz, T.M. Williams, C. Teuscher, C.C.A. Bernard, and K.S.K. Tung. 1983. Immunopathology of murine experimental allergic orchitis. *J. Immunol.* 130:2675–2582.

49. Sakamoto, Y., K. Himeno, H. Sanui, S. Yoshida, and K. Nomoto. 1985. Experimental allergic orchitis in mice. I. A new model induced by immunization without adjuvants. *Clin. Immunol. Immunopathol.* 37:360–368.

50. Itoh, M., C. Hiramine, and K. Hojo. 1991. A new murine model of autoimmune orchitis induced by immunization with viable syngeneic germ cells alone. 1. Immunological and histological studies. *Clin. Exp. Immunol.* 83:137–142.

51. Rhim S.H., S.E. Millar, F. Robey, A.M. Luo, Y.H. Lou, T. Yule, P. Allen, J. Dean, and K.S.K. Tung. 1992. Autoimmune diseases of the ovary induced by a ZP3 peptide from the mouse ZP. *J. Clin. Invest.* 89:28–35.

52. Nishizuka, Y., and T. Sakakura. 1969. Thymus and reproduction: Sex linked dysgenesis of the gonad after neonatal thymectomy. *Science* 166:753–755.

53. Kojima, A., and R.T. Prehn. 1981. Genetic susceptibility to post-thymectomy autoimmune disease in mice. *Immunogenetics* 14:15–27.

54. Taguchi, O., and Y. Nishizuka. 1981. Experimental autoimmune orchitis after neonatal thymectomy in the mouse. *Clin. Exp. Immunol.* 46:425–434.

55. Tung, K.S.K., S. Smith, C. Teuscher, C. Cook, and R.E. Anderson. 1987. Murine autoimmune oophoritis, epididymo-orchitis and gastritis induced by day-3 thymectomy: Immunopathology. *Am. J. Pathol.* 126:293–302.

56. Sakaguchi, S.K., K. Fukuma, K. Kuribayashi, and T. Masuda. 1985. Organ-specific autoimmune diseases induced in mice by elimination of T cell subset. I. Evidence for the active participation of T cells in natural self-tolerance: deficit of a T cell subset as a possible cause of autoimmune disease. *J. Exp. Med.* 161:72–87.

57. Sugihara, S.Y., T. Izumi, H. Yoshioka, T. Yagi, Y. Tsujimura, Y. Kohno, S. Murakami, T. Hamaoka, and H. Fujiwara. 1988. Autoimmune thyroiditis induced in mice depleted of particular T cell subsets. I. Requirement of Lyt-1^{dull} L3T4^{bright} normal T cells for the induction of thyroiditis. *J. Immunol.* 141:105–113.

58. Smith, H., Y.H. Lou, P. Lacy, and K.S.K. Tung. 1992. Tolerance mechanism in ovarian and gastric autoimmune disease. *J. Immunol.* 149:2212–2218.

59. Sakaguchi, S., K. Fukuma, K. Kuribayashi, and T. Masuda. 1995. Organ-specific autoimmune disease induced in mice by elimination of T cell subset. I. Evidence of the active participation of T cells in natural self-tolerance; deficit of a T cell subset as a possible cause of autoimmune disease. *J. Exp. Med.* 161:72–87.

60. Taguchi, O., and T. Takahashi. 1996. Administration of anti-interleukin-2 receptor alpha antibody *in vivo* induces localized autoimmune disease. *Eur. J. Immunol.* 26:1608–1612.

61. Sakaguchi, S., and N. Sakaguchi. 1989. Organ-specific autoimmune disease induced in mice by elimination of T cell subsets. V. Neonatal administration of cyclosporin A causes autoimmune disease. *J. Immunol.* 142:471–480.

62. Taguchi, O., T. Takahashi, S. Masao, R. Namikawa, M. Matsuyama, and Y. Nishizuka. 1986. Development of multiple organ localized autoimmune disease in nude mice after reconstitution of T cell function by rat fetal thymus graft. *J. Exp. Med.* 164:60–71.

63. Sakaguchi, S., and N. Sakaguchi. 1990. Thymus and auto-immunity: capacity of the normal thymus to produce self-reactive T cells and conditions required for their induction of autoimmune disease. *J. Exp. Med.* 172:537–545.

64. Sakaguchi, S., T.H. Ermak, M. Toda, L.J. Berg, W. Ho, B.F. de St. Groth, P.A. Peterson, N. Sakaguchi, and M.M. Davis. 1994. Induction of autoimmune disease in mice by germline alteration of the T cell receptor gene expression. *J. Immunol.* 152:1471–1484.

65. McKeever, U., J.P. Mordes, D.L. Greiner, M.C. Apple, J.R. Rozing, E.S. Handler, and A.A. Rossini. 1990. Adoptive transfer of autoimmune diabetes and thyroiditis to athymic rats. *Proc. Natl. Acad. Sci. USA* 87:7618–7622.

66. Fowell, K., A.J. McKnight, F. Powrie, R. Dyke, and D. Mason. 1991. Subsets of CD4 T cells and their roles in the induction and prevention of autoimmunity. *Immunol. Rev.* 123:37–64.

67. Fritz, T.E., L.A. Lombard, S.A. Tyler, and W.P. Norris. 1976. Pathology and familial incidence or orchitis and its relation to thyroiditis in a closed beagle colony. *Exp. Mol. Pathol.* 24:142–158.

68. Tung, K.S.K., L. Ellis, C. Teuscher, A. Meng, J.C. Blaustein, S. Kohno, and R. Howell. 1981. The black mink (*mustela vison*): A natural model of immunologic male infertility. *J. Exp. Med.* 154:1016–1032.

69. Furbeth, C., G. Hubner, and G.H. Thoenes. 1989. Spontaneous immune complex orchitis in Brown Norway rats. *Vir. Arch. B Cell. Pathol.* 57:37–45.

70. Morgan, A.D. 1976. Inflammation and infestation of the testis and paratesticular structures. In Pathology of the testis, R.C.B. Pugh editor. Blackwell, Oxford. 79–138.

71. Bigazzi, P.E., L.L. Kosuda, K.C. Hsu, and G.A. Andres. 1976. Immune complex orchitis in vasectomized rabbits. *J. Exp. Med.* 143:382–404.

72. Tung, K.S.K. 1978. Allergic orchitis lesions are adoptively transferred from vasoligated guinea pigs to syngeneic recipients. *Science* 201:833–835.

73. Hammer, R.E., S.D. Maika, J.A. Richardson, J.Y.P. Tang, and J.D. Taurog. 1990. Transgenic rats expressing HLA-B27 and human b2 microglobulin with spontaneous inflammatory disease in multiple organ systems: An animal model of HLA-B27-associated disease. *Cell* 63:1099–1112.

74. Tung, K.S.K., and Lu, C.Y. 1991. Immunologic basis of reproductive failure. In: Pathology of Reproductive Failure, F.T. Kraus, I. Damjanov, and N. Kaufman eds., Williams and Wilkins, New York, NY. pp. 308–333.

75. LaBarbera, A.R., M.M. Miller, C. Ober, and R.W. Rebar. 1988. Autoimmune etiology in premature ovarian failure. *Am. J. Reprod. Immunol. Microbiol.* 16:115–122.

76. Moncago, R., and H.E. Moncago. 1995. Ovarian Autoimmunity: Clinical and Experimental Data. R.G. Landes Co., Austria.

77. Song, Y.-H., X. Li, and N.K. Maclaren. 1996. The nature of autoantigens targeted in autoimmune endocrine diseases. *Immun. Today* 17:232–238.

78. Nishimoto, T., T. Mori, I. Yamada, and T. Nishimura. 1980. Autoantibodies to zona pellucida in infertile and aged women. *Fertil. Steril.* 34:552–556.

79. Millar, S.E., S.M. Chamow, A.W. Baur, C. Oliver, F. Robey, and J. Dean. 1989. Vaccination with a synthetic ZP peptide produces long-term contraception in female mice. *Science* 246:935–938.

80. Lou, Y.H., J. Ang, H. Thai, F. McElveen, and K.S.K. Tung. 1995. A ZP3 peptide vaccine induces antibody and reversible infertility without ovarian pathology. *J. Immunol.* 155:2715–2720.

81. Neufeld, M., N.K. Maclaren, and R. Blizzard. 1980. Autoimmune Polyglandular syndromes *Pediatr. Ann.* 9:154–162.

82. Tung, K.S.K., S. Agersborg, H. Bagavant, K. Garza, and K. Wei. 1999. Autoimmune ovarian disease induced by immunization with ZP3 peptide (Unit 15.17). In: Current Protocols in Immunology, J.E. Coligan, A.M. Kruisbeek, D.H. Margulies, E.M. Shevach, and W. Strober, eds. John Wiley & Sons, Inc., New York, NY.

83. Suri-Payer, E., K. Wei, and K.S.K. Tung. 1999. The day 3 thymectomy model of organ-specific autoimmunity (Unit 15.16). In: Current Protocols in Immunology, J.E. Coligan, A.M. Kruisbeek, D.H. Margulies, E.M. Shevach, and W. Strober, eds. John Wiley & Sons, Inc., New York, NY.

84. Jackson, J.J., A. Hagopian, D.J. Carlo, G.A. Limjuco, and E.H. Eylar. 1975. Experimental allergic aspermatogenic orchitis. I. Isolation of a spermatozoal protein (AP1) which induces allergic aspermatogenic orchitis. *J. Biol. Chem.* 250:6141–6150.

85. Hagopian, A., J.J. Jackson, D.J. Carlo, G.A. Limjuco, and E.H. Eylar. 1975. Experimental allergic aspermatogenic orchitis. III. Isolation of spermatozoal gycoproteins and their role in allergic aspermatogenic orchitis. *J. Immunol.* 115:1731–1743.

86. Teuscher, C., G.C. Wild, and K.S.K. Tung. 1983. Acrosomal autoantigens of guinea pig sperm. I. The purification of an aspermatogenic protein, AP2. *J. Immunol.* 130:317–322.

87. Teuscher, C., G.C. Wild, and K.S.K. Tung. 1983. Experimental allergic orchitis: the isolation and partial characterization of an aspermatogenic polypeptide (AP3) with an apparent sequential disease-inducing determinant(s). *J. Immunol.* 130:2683–2688.

88. Tung, K.S.K., P. Primakoff, L. Woolman, and D. Myles. 1997. Mechanism of infertility in male guinea pigs immunized with sperm PH-20. *Bil. Reprod.* 56:1133–1141.

89. Primakoff, P., H. Hyatt, and D.G. Myles. 1985. A role for the migrating sperm surface antigen, PH-20, in guinea pig sperm binding to the egg ZP. *J. Cell Biol.* 101:2239–2244.

90. Grafer, C., W. Sun, and K.S.K. Tung. 1999. Experimental autoimmune orchitis based on chicken ovalbumin as a transgenic haploid germ cell antigen. FASEB Conference on Autoimmunity. *Saxton River, VT.* (Abstr. 63.)

91. Dean, J. 1992. Biology of mammalian fertilization: Role of the ZP. *J. Clin. Invest.* 89:1055–1059.

92. Tong, Z.B., and L.M. Nelson. 1998. A novel ooplasm protein identified by autoimmune serum from mice with oophoritis. 1st international conference on the genetic origins of premature ovarian failure. Bethesda, MD. (Abstr.)

93. Sakaguchi, S., T. Takahashi, and Y. Nishizuka. 1982. Study on cellular events in post-thymectomy autoimmune oophoritis in mice. II. Requirement of Lyt-1 cells in normal female mice for the prevention of oophoritis. *J. Exp. Med.* 156:1577–1586.

94. Alderuccio, F., B.H. Toh, S.S. Tan, P. Gleeson, and I. Driel. 1993. An autoimmune disease with multiple molecular targets abrogated by the transgenic expression of a single autoantigen in the thymus. *J. Exp. Med.* 178:419–426.

95. Smith, H., I.M. Chen, R. Kubo, and K.S.K. Tung. 1989. Neonatal thymectomy results in a repertoire enriched in T cells deleted in adult thymus. *Science* 245:749–752.

96. Smith, H., Y. Sakamoto, L. Kasai, and K.S.K. Tung. 1991. Effector and regulatory cells in autoimmune oophoritis elicited by neonatal thymectomy. *J. Immunol.* 147:2928–2933.

97. Takahashi, T., Y. Kuniyasu, M. Toda, N. Sakaguchi, M. Itoh, M. Iwata, J. Shimizu, and S. Sakaguchi. 1998. Immunologic self-tolerance maintained by CD25+CD4+ naturally anergic and suppressive T cells: induction of autoimmune disease by breaking their anergic/suppressive state. *Internat. Immunol.* 10:1969–80.

98. Thornton, A.M., and E.M. Shevach. 1998. CD4+CD25+ immunoregulatory T cells suppress polyclonal T cell activation in vitro by inhibiting interleukin 2 production. *J. Exp. Med.* 188:287–96.

99. Pelletier, R.M., and S.W. Byers. 1992. The blood-testis barrier and Sertoli cell junctions: structural considerations. *Micro. Res. Tech.* 20:3–33.

100. Tung, K.S.K., E.R. Unanue, and F.J. Dixon. 1971. Pathogenesis of experimental allergic orchitis. II. The role of antibody. *J. Immunol.* 106:1463–1472.

101. Yule, T.D., G.D. Montoya, L.D. Russell, T.M. Williams, and K.S.K. Tung. 1988. Autoantigenic germ cells exist outside the blood testis barrier. *J. Immunol.* 141:1161–1167.

102. Tung, K.S.K., T.D. Yule, C.A. Mahi-Brown, and M.B. Listrom. 1987. Distribution of histopathology and Ia positive cells in actively-induced and adoptively-transferred experimental autoimmune orchitis. *J. Immunol.* 138:752–759.

103. Accinni, L., B. Albini, G. Andres, and F.J. Dixon. 1980. Deposition of immune complexes in ovarian follicles of mice with lupus-like syndrome. *Am. J. Pathol.* 99:589–596.

104. Matsuo, B., P.R.B. Caldwell, J.R. Brentjens, and G. Andres. 1985. In vitro interaction of antibodies with cell surface antigens: A mechanism responsible for in situ formation of immune deposits in the ZP of rabbit oocytes. *J. Clin. Invest.* 75:1369–1380.

105. Head, J.R., W.B. Neaves, and R.E. Billingham. 1983. Immune privilege in the testis. I. Basic parameters of allograft survival. *Transplantation* 36:423–431.

106. Selawry, H., R. Fojaco, and K. Whittington. 1987. Extended survival of MHC-compatible islet grafts from diabetes-

resistant donors in spontaneously diabetic BB/W rat. *Diabetes* 36:1061–1067.

107. Kern, S., S.A. Robertson, V.J. Mau, and S. Maddocks. 1995. Cytokine secretion by macrophages in the rat testis. *Biol. Reprod.* 53:1407–1416.

108. Kern, S., and S. Maddocks. 1995. Indomethacin blocks the immunosuppressive activity of rat testicular macrophages cutured *in vitro*. *J. Reprod. Immun.* 28:189–201.

109. Pollanen, P., O. Soder, and J. Uksila. 1988. Testicular immunosuppressive protein. *J. Reprod. Immunol.* 14:125–128.

110. Wyatt, C.R., L. Law, J.A. Magnuson, M.D. Griswold, and N.S. Magnuson. 1988. Suppression of lymphocyte proliferation by proteins secreted by cultured Sertoli cells. *J. Reprod. Immunol.* 14:27–40.

111. De Cesaris, P., A. Filippini, C. Cervelli, A. Riccioli, S. Muci, G. Starace, M. Stefanini, and E. Ziparo. 1992. Immunosuppressive molecules produced by Sertoli cells cultured *in vitro*: Biological effects on lymphocytes. *Biochem. Biophys. Res. Commun.* 186:1639–1646.

112. Hedger, M.P., J.X. Qin, D.M. Robertson, and D.M. de Kretser. 1990. Intragonadal regulation of immune system functions. *Reprod. Fertil. Develop.* 2:263–280.

113. Tarter, T.H., and N.J. Alexander. 1984. Complement-inhibiting activity of seminal plasma. *Am. J. Reprod. Immunol.* 6:28–32.

114. Griswold, M.D., K. Robert, and P. Bishop. 1986. Purification and characterization of a sulfated glycoprotein secreted by Sertoli cells. *Biochem. J.* 25:7265–7270.

115. Jenne, D.E., and J. Tschopp. 1989. Molecular structure and functional characterization of a human complement cytolysis inhibitor found in blood and seminal plasma: Identity to sulfated glycoprotein 2, a constituent of rat testis fluid. *Proc. Natl. Acad. Sci. USA* 86:7123–7127.

116. Lou, Y.H., F. McKelveen, S. Adams, and K.S.K. Tung. 1995. Altered target organ: a mechanism of postrecovery resistance to murine autoimmune oophoritis. *J. Immunol.* 155:3667–3673.

117. Tung, K.S.K., Y.H. Lou, K.M. Garza, and C. Teuscher. 1997. Autoimmune ovarian disease: mechanism of disease induction and prevention. *Curr. Opin. Immunol.* 9:839–845.

118. Taguchi, O., and Y. Nishizuka. 1987. Self-tolerance and localized autoimmunity: Mouse models of autoimmune disease that suggest that tissue-specific suppressor T-cells are involved in tolerance. *J. Exp. Med.* 165:146–156.

119. Taguchi, O., and Y. Nishizuka. 1980. Autoimmune oophoritis in the thymectomized mice: T-cell requirement in the adoptive cell transfer. *Clin. Exp. Immunol.* 42:324–331.

120. Gleeson, P.A., B.H. Toh, and I.R. van Driel. 1996. Organ-specific autoimmunity induced by lymphopenia. *Immun. Rev.* 149:97–125.

121. Tung, K.S.K., Y.H. Lou, H. Bagavant, and P. Alard. 1999. Ovarian autoimmune disease: Its prevention and pathogenesis. Keystone symposium on immunogenetics of human disease MHC/TCR and peptide. Taos, NM. (Abst. 025).

122. Alard, P., and K.S.K. Tung. 1999. Key role for antigen presenting cells in the inhibition of T cell response by CD4+CD25+ regulatory cells. FASEB Conference on Autoimmunity. Saxton River, VT. (Abstr. 8).

123. Luo, A.M., K.M. Garza, D. Hunt, and K.S.K. Tung. 1993. Antigen mimicry in autoimmune disease: sharing of amino acid residues critical for pathogenic T cellactivation. *J. Clin. Invest.* 92:2117–2123.

124. Garza, K., and K.S.K. Tung. 1995. Frequency of T cell peptide crossreaction determined by experimental autoimmune disease and autoantibody induction. *J. Immunol.* 155:5444–5448.

125. Lou, Y.H., and K.S.K. Tung. 1993. T cell peptide of a self protein elicits autoantibody to the protein antigen: Implications for specificity and pathogenetic role of antibody in autoimmunity. *J. Immunol.* 151:5790–5799.

126. Lou, Y.H., F. McElveen, K. Garza, and K.S.K. Tung. 1996. Rapid induction of autoantibodies by andogenous ovarian antigens and activated T cells: Implication in autoimmunity pathogenesis and B cell tolerance. *J. Immunol.* 156:3535–3540.

127. Lehmann, P.V., T. Forsthuber, A. Miller, and E. Sercarz. 1992. Spreading of T-cell autoimmunity to cryptic determinants of an autoantigen. *Nature* 358:155–157.

128. Kaufman, D., M. Clare-Salzler, J. Tian, T. Forsthuber, G. Ting, P. Robinson, M. Atkinson, E. Sercarz, A. Tobin, and P. Lehmann. 1993. Spontaneous loss of T-cell tolerance to glutamic acid decarboxylase in murine insulin-dependent diabetes. *Nature* 66:69–72.

129. Tisch, R., X.-D. Yang, S. Singer, R. Liblau, L. Fugger, and H. McDevitt. 1993. Immune response to glutamic acid decarboxylase correlates with insulitis in non-obese diabetic mice. *Nature* 366:72–75.

130. Fatenejad, S., M. Mamula, and J. Craft. 1993. Role of intermolecular/intrastruc-tural B-and T-cell determinants in the diversification of autoantibodies to ribonucleoprotein particles. *Proc. Natl. Acad. Sci. USA* 90:12010–12014.

131. James, J.A., T. Gross, R.H. Scofield, and J.B. Harley. 1995. Immunoglobulin epitope spreading and autoimmune disease after peptide immunization: Sm B/B′-derived PPPGMRPP and PPPGIRGP induce spliceosome autoimmunity. *J. Exp. Med.* 181:453–461.

132. Topfer, F., T. Gordon, and J. McCluskey. 1995. Intra- and intermolecular spreading of autoimmunity involving the nuclear self-antigens La (SS-B) and Ro (SS-A). *Proc. Natl. Acad. Sci. USA* 92:875–879.

133. Deshmukh, U., J.E. Lewis, F. Gaskin, C.C. Kannapell, S.T. Waters, Y.H. Lou, K.S.K. Tung, and S.M. Fu. 1999. Immune responses to Ro60 and its peptides in mice: The nature of the immunogen and endogenous autoantigen determine the specificities of the induced autoantibodies. *J. Exp. Med.* 189:531–540.

134. Meng, A.L., and K.S.K. Tung. 1983. Cell-mediated and humoral immune responses to aspermatogenic antigen in experimental allergic orchitis in guinea pig. *J. Reprod. Fertil.* 69:279–288.

135. Tung, K.S.K., and N.J. Alexander. 1980. Monocytic orchitis and aspermatogenesis in normal and vasectomized rhesus macaques (*macaca mulatta*). *Am. J. Pathol.* 101:17–29.

136. Mahi-Brown, C.A., T.D. Yule, and K.S.K. Tung. 1987. Adoptive transfer of murine autoimmune orchitis to naive recipients with immune lymphocytes. *Cell. Immunol.* 106:408–419.

137. Mahi-Brown, C.A., and K.S.K. Tung. 1989. Activation requirements of donor T cells and host T cell recruitment in adoptive transfer of murine experimental autoimmune orchitis (EAO). *Cell. Immunol.* 124:368–379.

138. Feng, Z.Y., L.D. Ming, L.J. Louis, and W.Y. Fei. 1990. Adoptive transfer of murine autoimmune orchitis with spermspecific cases of male infertility. *Arch. Androl.* 24:51–59.

139. Itoh, M., C. Hiramine, A. Mukasa, Y. Tokunaga, Y. Fukui, Y. Takeuchi, and K. Hojo. 1992. Establishment of an experimental model of autoimmune epididymo-orchitis induced by the transfer of a T-cell line in mice. *Int. J. Androl.* 15:170–181.

140. Mukasa, A., M. Itoh, Y. Tokunaga, C. Hiramine, and K. Hojo. 1992. Inhibition of a novel model of murine experimental autoimmune orchitis by intravenous administration with a soluble testicular antigen: participation of CD8+ regulatory T cells. *Clin. Immunol. Immunopathol.* 62:210–219.

141. Itoh, M., A. Mukasa, Y. Tokunaga, C. Hiramine, and K. Hojo. 1991. New experimental model for adoptive transfer of murine autoimmune orchitis. *Andrologia* 23:415–420.

142. Yule, T.D., and K.S.K. Tung. 1993. Experimental autoimmune orchitis induced by testis and sperm antigen-specific T cell clones: An important pathogenic cytokine is tumor necrosis factor. *Endocrinology* 133:1098–1107.

143. Bagavant, H., S. Adams, P. Terranova, A. Chang, F.W. Kraemer, Y.H. Lou, A.M. Luo, and K.S.K. Tung. 1999. Autoimmune ovarian inflammation triggered by proinflammatory (Th1) T cells is compatible with normal ovarian function. *Biol. Reprod.* In press.

144. Lou, Y.H., K-K. Park, and K.S.K. Tung. 1996. Binding of autoantibody to the tissue antigen directs targeting of T cell-mediated autoimmune inflammation. *Keystone Symposia on Lymphocyte Activation.* Hilton Head, South Carolina. (Abstr. 2051)

145. Unanue, E.R. 1989. Macrophages, antigen-presenting cells, and the phenomenon of antigen handling and presentation. In: Fundamental Immunology, 2nd ed., W.E. Paul editor Raven Press, New York, NY. pp. 95–115.

146. Teuscher, C., S.M. Smith, E.H. Goldberg, G.M. Shearer, and K.S.K. Tung. 1985. Experimental allergic orchitis in mice, I: genetic control of susceptibility and resistance to induction of autoimmune orchitis. *Immunogenetics* 22:323–333.

147. Tung, K.S.K., C. Teuscher, S. Smith, L. Ellis, and M.L. Dufau. 1985. Factors that regulate the development of testicular autoimmune diseases. In: Hormone Action and Testicular Function, K.J. Catt and M.L. Dufau, eds. New York, NY. Vol. 438, pp. 171–188.

148. Person, P.L., M. Snoek, P. Demant, S.R. Woodward, and C. Teuscher. 1992. The immunogenetics of susceptibility and resistance to murine experimental allergic orchitis. *Reg. Immunol.* 4:284–297.

149. Roper, R.J., R.W. Doerge, S.B. Call, K.S.K. Tung, W.F. Hickey, and C. Teuscher. 1998. Autoimmune orchitis, epididymitis, and vasitis are immunogenetically distinct lesions. *Am. J. Path.* 152:1337–1345.

150. Olsson, M., G. Lindahl, and E. Ruoslahti. 1977. Genetic control of alpha-fetoprotein synthesis in the mouse. *J. Exp. Med.* 145:819–827.

151. Teuscher, C., E.P. Blankenhorn, and W.F. Hickey. 1987. Differential susceptibility to actively induced experimental allergic encephalomyelitis (EAE) and experimental allergic orchitis (EAO) among BALB/c substrains. *Cell. Immunol.* 110:294–304.

152. Teuscher, C., W.F. Hickey, C.M., Grafer, and K.S.K. Tung. 1998. A common immunoregulatory locus controls susceptibility to actively induced experimental allergic encephalomyelitis and experimental allergic orchitis in BALB/c mice. *J. Immunol.* 160:2751–2756.

153. Teuscher, C., D.L. Gasser, S.R. Woodward, and W.F. Hickey. 1990. Experimental allergic orchitis in mice, VI: recombinations within the H-2S/H-2D interval define the map position of the H-2 associated locus controlling disease susceptibility. *Immunogenetics* 32:237–344.

154. Snoek, M., M. Jansen, M.G. Olavessen, D.R. Campbell, C. Teuscher, and H. van Vugt. 1993. The Hsp7-genes are located in the Cr-H-2D region: possible candidates for the Orch-1 locus. *Genomics* 15:350–356.

155. Meeker, N.D., W.F. Hickey, R. Korngold, W.K. Hansen, J.D. Sudweeks, B.B. Wardell, J.S. Griffith, and C. Teuscher. 1995. Multiple loci govern the bone marrow-derived immunoregulatory mechanism controlling dominant resistance to autoimmune orchitis. *Proc. Natl. Acad. Sci. USA* 92:5684–5688.

156. Wardell, B.B., and S.D. Michael, K.S.K. Tung, J.A. Todd, E.P. Blankenhorn, K. McEntee, J.D. Sudweeks, K.W. Hansen, N.D. Meeker, J.S. Griffith, K.D. Livingstone, and C. Teuscher. 1995. Multiple genes regulate neonatal thymectomy-induced autoimmune ovarian dysgenesis (AOD) with linkage of ovarian atrophy to Idd 3. *Proc. Natl. Acad. Sci. USA* 92:4758–4762.

157. Teuscher, C., B.B. Wardell, J.K. Lunceford, S.D. Michael, and K.S.K. Tung. 1996. *Aod2*, the locus controlling development of atrophy in neonatal thymectomy induced autoimmune ovarian dysgenesis of atrophy, is linked to *IL2*, *Fgfb*, and *Idd3*. *J. Exp. Med.* 183:631–637.

158. Denny, P., C.J. Lord, N.J. Hill, J.V. Goy, E.R. Levy, P.L. Podlin, L.B. Peterson, L.S. Wicker, J.A. Todd, and P.A. Lyons. 1997. Mapping of the IDDM locus *idd3* to a 0.35-cM interval containing the *interleukin-2* gene. *Diabetes* 46:695–700.

159. Asakai, R., S. Song, I. Nobuyuki, T. Yamakuni, K. Tamura, and R. Okamoto. 1994. Differential gene expression of fibroblast growth factor receptor isoforms in rat ovary. *Mol. Cell. Endocrinol.* 104:75–80.

160. Roper, R.J., R.W. Doerge, K.S.K. Tung, and C. Teuscher. 1999. Identification of new QTL controlling day three thymectomy induced autoimmune ovarian dysgenesis using novel quantitative genetics approaches. In preparation.

60 | HIV Infection and Autoimmunity

Alberto Amadori and Fulvia Chieco-Bianchi

1. INTRODUCTION

The human immunodeficiency virus type 1 (HIV-1) is the causative agent of the acquired immune deficiency syndrome (AIDS) in man (1–2). HIV-1 belongs to the family of retroviruses, which by definition are RNA viruses that need transcription to DNA to insert into the host genome and replicate. All retroviruses derive from an unknown common ancestor and, according to their biological and pathogenic characteristics, are classified into the *oncovirinae*, *lentivirinae*, and *spumavirinae* sub-families. Based on its *in vivo* and *in vitro* behaviour, morphologic features, and nucleotide sequences, HIV-1 belongs to the lentiviruses, which include cytopathic retroviruses, such as equine infectious anemia virus, caprine arthritis-encephalitis virus, and Visna-Maedi virus (3). The sub-family of lentiviruses also includes HIV-2, which is also associated with AIDS albeit with different epidemiologic characteristics and clinical course (4,5). HIV-1 and HIV-2 appear to be distinct but related viruses; viral protein cross-reactivity and nucleotide sequence homology studies suggest that HIV-1 and HIV-2 recently diverged from a common progenitor (6). Henceforth, we will mainly discuss the features of HIV-1, which will be referred to here as "HIV".

The basic structure of HIV is shown in Figure 1. Like all retroviruses, it contains three structural genes (Fig. 2): The *gag* gene (coding for core proteins, mainly represented by p24), the *pol* gene (coding for the enzymes integrase, protease and reverse transcriptase), and the *env* gene (coding for gp120 and gp41, which are non-covalently associated in the viral envelope). HIV also contains six non-structural genes that serve important regulatory functions (7).

After HIV enters the cell, its cycle depends on the transcription of viral RNA to double-stranded DNA through the reverse transcriptase (RT) enzyme, and subsequent integration of the DNA copy into the host's cell genome (*provirus*) (8). Provirus insertion is random, and no preferential sites of integration have been described. The integrated provirus may successively become activated and express viral proteins up to the formation and budding of complete virions, or it may remain in a latent status for an indeterminate period. The mechanisms underlying virus latency are unknown. Viral characteristics, such as the degree of viral genome methylation (9), or the balance of the regulatory genes, favouring the production of proteins that down-regulate or potentiate provirus transcription (10), may be responsible for virus latency or expression in the infected cell; it is generally believed that cell activation is a pre-requisite for HIV activation and expression.

A body of data suggests that CD4 is the major receptor for viral entry (11–13). However, recent evidence revealed the existence of "second" HIV receptors that cooperate with CD4 molecules for virus entry into target cells (rev. in 14). Following gp120/CD4 interaction, these coreceptors bind a conserved region of gp120, allowing exposure of the fusion domain in viral gp41, and eventual penetration of the core particle. Several coreceptors were identified, all belonging to the chemokine receptor family (15–18). Among these coreceptors, CCR-5 (which serves as a receptor for the CC-chemokines MIP-1α, MIP-1β and RANTES) and CXCR-4 (the binding unit for the CXC-chemokine SDF-1) play a major role (19,20). Even though such a dichotomy may be too simplistic (21,22), CCR-5 serves as a coreceptor for monocyte/macrophage-tropic HIV strains, whereas T cell-tropic strains bind preferentially to CXCR-4. The study of HIV coreceptors and their usage by the different HIV strains is presently a very active

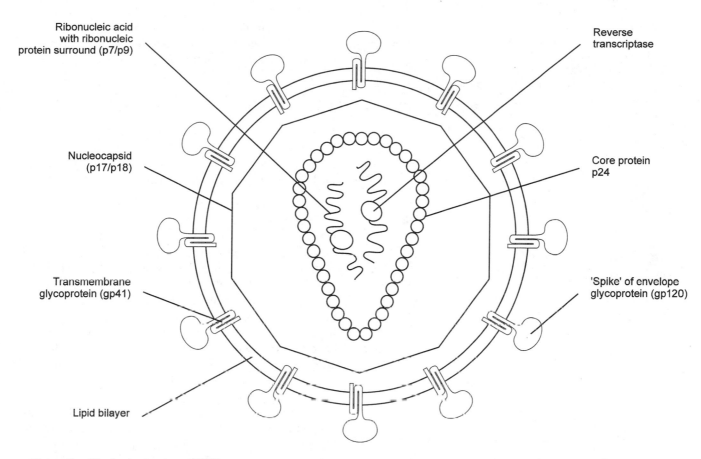

Figure 1. The basic structure of HIV

field of investigation due to implications in terms of AIDS prevention and immunotherapy.

The spectrum of target cells for HIV infection is large; in addition to CD4+ helper T cells, HIV is also able to infect a wide range of apparently CD4- cells (Table 1). The mechanisms leading to infection of these cells are presently undefined; it is possible that very low CD4 concentrations, which are undetectable by standard cytofluorographic analysis but sufficient to mediate virus binding, are present on some cell types. *In vivo*, HIV infection of CD4- cells might also occur via endocytosis of virion/antibody complexes by cells displaying the receptors for IgG or complement (23–25). This entry mechanism into CD4- cells, such as monocytes and astrocytes in the central nervous system (CNS), may play an important role in establishing viral reservoirs.

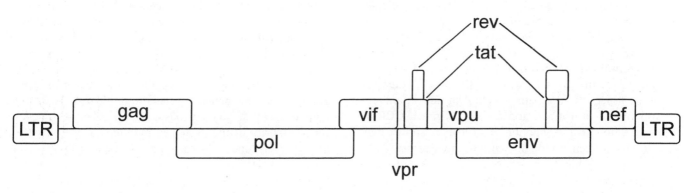

Figure 2. Genetic structure of HIV.

Table 1 *In vitro* and *in vivo* targets of HIV infection

Hematopoietic cells
- CD4⁺ lymphocytes
- B lymphocytes (*in vitro*, EBV-transformed)
- Monocytes/macrophages
- Dendritic cells
- Megakaryocytes
- Promyelocytes
- Stem cells

Skin
- Langerhans cells
- Fibroblasts

Brain
- Astrocytes
- Microglial cells
- Oligodendrocytes

Gastrointestinal
- Colon carcinoma cells
- Epithelial and columnar cells
- Enterochromaffin cells
- Kupffer's cells
- Liver sinusoidal cells

2. VIRUS-HOST INTERACTIONS

Virus spread in the host depends on a multitude of factors related to both the host and the features of the virus itself. Indeed, the interaction between HIV and the immune system has been recently defined as a "titanic struggle" between the virus that directly or indirectly destroys daily a huge number of CD4⁺ T cells, and the primary lymphoid organs, which attempt to replace the losses in the peripheral pool (26). This scenario involves several elements, only some of which are known.

2.1. Viral Factors

Several viral factors may influence HIV-host interactions. It is known that wild-type isolates are mostly tropic for mononuclear phagocytes, and do not exert and important cytopathic effect on CD4⁺ cells (27). Moreover, different HIV strains exhibit dissimilar biologic effects, and highly replicating cytopathic strains are associated with rapid transition to full-blown AIDS (28–29). In addition, the biological changes in the virus once inside the host are another important aspect of HIV pathogenicity; in this regard, co-receptor usage is critical, and a switch from a CCR-5-dependent to a T-tropic or dual-tropic strain may significantly influence viral spread into the organism (22). In fact, due to the highly error-prone transcription of molecular sequences by RT, with time the virus structure undergoes significant changes within the host; while this event

could result in the emergence of variants with increased pathogenicity and replication rates, it may also lead to virus escape from immune control (29–30).

2.2. Host Factors

Many host factors may influence the outcome of the infection (for review, see 31). These mainly include genetic factors, the humoral and cellular immune responses, and the behaviour of the cytokine network.

Genetic factors HIV-host interactions are obviously influenced by the host's genetic background. This is evidenced by paradigmatic situations where the infection by the same HIV isolate follows different clinical courses in different individuals (32–33). Among the many genetic factors that may participate in virus/host interactions, the most obvious candidate is the presence of particular HLA haplotypes, in view of the importance of histocompatibility genes in the immune response. Although in the '80s many workers attempted to establish a correlation between specific HLA haplotypes and disease evolution (34–37), the conclusions of these efforts are partly deceiving, and no conclusive HLA associations predictive of disease evolution could be identified. The very complexity of the factors participating in this scenario, including the variability of the different HIV strains, the racial factors, the nature of the risk group, and the type of immuno-deficiency manifestation, makes it difficult to define significant HLA markers; in addition, in view of the poor reliability of serological tissue typing techniques, molecular analysis of HLA specificities will be necessary to obtain sound results. In this regard, it is possible that other HLA-linked loci, such as the TAP gene involved in class I-restricted antigen presentation, could play a more significant role (38).

More recently, great attention was paid to understanding why several high-risk individuals escape HIV infection; a very intriguing observation was the discovery that natural resistance to HIV may derive from the inheritance of a defective allele of the CCR-5 gene (39–42). The reported 32-bp deletion prevents surface expression of the CCR-5 receptor, with eventual *in vitro* resistance of CD4⁺ cells to HIV infection. In addition, the finding of homozygous CCR-5-deficient patients within cohorts of seropositive subjects is extremely rare compared to the frequency of this genotype in the general population (43–45). However, inheritance of the homozygous defective genotype accounts for a minority of the protected individuals, and this trait may be only one of several genetic factors involved in escape from HIV infection.

We recently reported that the CD4/CD8 ratio in humans is under genetic control (46): in other words, some individuals preferentially produce CD4+ T cells from the thymus, whereas others make the opposite. This observation may be relevant to HIV natural history; in a setting of "tug-of-war" (Fig. 3) between HIV and the immune system (26), individuals genetically predisposed to a high CD4/CD8 ratio could withstand HIV-associated CD4+ cell losses better than those genetically predisposed to a low ratio (47).

Humoral responses   An antibody response to HIV is present in most individuals within two months following infection (for reviews, see 48–49). In a minority of subjects, however, the virus remains dormant for an unpredictable period of time, and no antibody can be detected in the serum, despite the presence of the viral genome in the cells (50). Disease progression is often associated with a decline in anti-HIV antibody titer, particularly antibodies against p24 antigen (51); however, this decrease is partly due to increased p24 shedding by the replicating virus, and complex formation with anti-p24 antibody (52).

Humoral responses in HIV-infected patients are profoundly deranged (for review, see 53). On one hand, seropositive subjects show poor *in vitro* responses to B cell mitogens and antigens. On the other, HIV infection is associated with an intense B cell activation; most patients are hypergammaglobulinemic, and show high numbers of activated B cells in circulation (54–55). B cell activation, however, does not involve memory B cells, and most activated B lymphocytes produce antibodies directed against HIV determinants (56–57).

In most viral infection models, antibodies serve a very important protective function (58). Several observations suggest that this might also be true for HIV infection (59–60), but the importance of the humoral anti-HIV response is unclear, and the role of antibodies in controlling virus spread remains debatable. On one hand, neutralizing antibody titers are usually low (49,60); on the other, long-term non-progressor patients who tolerate HIV infection without immune suppression for prolonged periods (61), show a higher and broader spectrum of serum-neutralizing

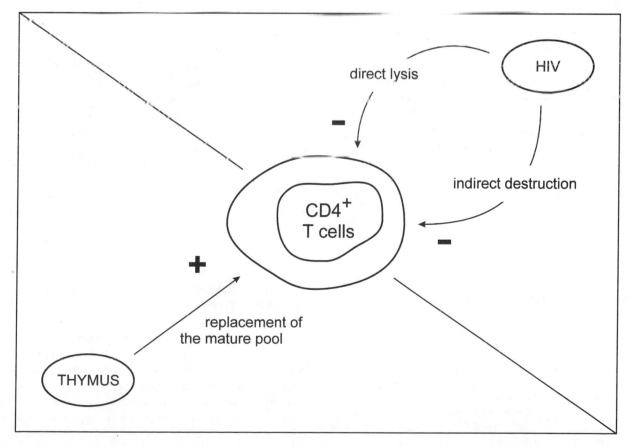

Figure 3.   The HIV "tug-of-the-war". Immune deficiency could result from the balance between HIV-related events that cause depletion of the circulating CD4+ cell pool, and the regenerating capacity of the primary lymphoid organs, that steadily attempt to replace peripheral CD4+ lymphocyte losses.

activity compared to seropositive patients progressing to AIDS (62). Compelling evidence suggests instead that anti-HIV antibodies might be involved in the amplification of virus-induced immune damage through several mechanisms (53), some of which are summarized in Table 2.

Cellular responses   The cellular response to HIV has been extensively studied (for reviews, see 63–64). Whereas the *in vitro* proliferative response to viral antigens is poor (65–66), several studies evidenced the presence of class I-restricted CD8+ cytotoxic T lymphocytes (CTL) directed against all HIV-coded proteins in peripheral blood, cerebrospinal fluid (CSF), and lymphoid organs; their frequency is very high in the early stages of infection, but CTL activity may be detected throughout the course of the disease, and only tends to decline in very advanced stages. In addition to this classical CTL response, virus replication may also be controlled by CD8+ lymphocytes through different mechanisms, as CD8+ cells were shown to down-regulate virus replication and expression in a non-MHC-restricted manner (67). However, although this inhibitory effect on HIV replication was consistently observed in many laboratories, the factors responsible for this phenomenon are still elusive; MIP-1α, MIP-1β and RANTES, although endowed with antiviral activity through competition for coreceptors (68), do not account for this phenomenon, as they are not produced exclusively by CD8+ T cells (69).

The significance of the cellular response to HIV is not fully understood. According to available evidence, a high frequency of specific CTL precursors is associated with limited disease progression, whereas increased disease severity is often preceded by a fall in their number (70–71). In addition, some investigators reported CTL activity in peripheral blood of high-risk, uninfected seronegative individuals (72). Nevertheless, it is not clear why these high numbers of cytotoxic precursors are not able to eradicate infection in most individuals. In this regard, the ability of HIV to escape immune responses may be of paramount importance; the mechanisms underlying this phenomenon

are complex, and include extremely high rate of sequence variation, altered antigen presentation (73–74), privileged sites of viral replication, and loss of effector cells by apoptosis or clonal exhaustion (75–76).

Alterations in the cytokine network   It is becoming increasingly clear that cytokines play a fundamental role in the immunobiology of HIV infection. In fact, cytokines may up-regulate or suppress viral expression and replication. Colony-stimulating factors (77–78), TNF-α and IL-6 (79–80) are able to potentiate viral replication in chronically-infected monocyte/macrophage cells. On the other hand, other cytokines, such as IFNs (81) and TGF-β (82), can down-regulate HIV expression and replication, but their activity may result in a down-regulating or enhancing effect, according to the different steps of the virus cell cycle (83).

More recently, the theory was advanced that a switch in cytokine production by T lymphocytes could account for progression of HIV infection to AIDS (84). Both murine (85) and human (86) CD4+ helper T lymphocytes may be functionally divided into two subsets that are not distinguishable by phenotypic properties, but are defined on the basis of the profile of cytokine production; these subpopulations are known as Th1 and Th2 cells. Th1 lymphocytes secrete IL-2, IFN-γ and TNF, and mostly promote cell-mediated immune mechanisms, whereas Th2 cells produce IL-4, IL-5, IL-6 and IL-10, which mostly drive humoral responses. Indeed, the hypothesis that a shift to preferential synthesis of Th2 cytokines might underlay HIV persistence and disease progression is supported by several data in both experimental models of infection and clinical situations (87). However, unlike the mouse, a strict dichotomy between Th1 and Th2 functions is too simplistic in humans (88). Moreover, this idea has found no practical applications so far, and the possibility of pharmacologically inducing a switch from a Th2 to a Th1 profile in HIV-infected patients remains elusive.

Table 2   Possible pathways of anti-HIV antibody-mediated damage

- Enhancement of infectivity
- Generation of suppressor T cells
- Complement and cell-mediated killing of:
  — infected CD4+ cells
  — non-infected CD4+ cells that processed gp 120
  — cells sharing homologous epitopes with HIV
- Immune suppression through anti-CD4 reactivity generation
  — anti-idiotypic mechanisms
  — binding of gp 120/anti-gp 120 complexes to CD4
- Lymphoma development

3.   AUTOIMMUNE PHENOMENA IN HIV INFECTIONS

The association between microbial agents and autoimmunity has long been known (89–90). Autoimmunity is also a common feature of retrovirus infection in animal models (rev. in 91). Although opportunistic infections, Kaposi's sarcoma, and lymphomas are major problems in seropositive patients, HIV infection is also associated with autoimmune phenomena. A non-exhaustive list of autoantibodies found in HIV-infected patients is shown in Table 3. The pathways leading to loss of tolerance and autoimmunity are fully discussed elsewhere in this book; here, we

Table 3   Autoantibodies in HIV-infected patients

— Antinuclear
— Anticardiolipin
— Lupus anticoagulant
— Anti-denatured collagen
— Anti-platelet
— Anti-lymphocyte
— Anti-red blood cells
— Anti-granulocyte
— Anti-sperm
— Anti-parietal cell
— Anti-myelin basic protein

will briefly recall some basic mechanisms through which retroviruses, and in particular HIV, may be involved in the generation of autoimmunity.

1. Autoimmunity may be generated through the well-known phenomenon of "molecular mimicry" (92), which is based on the relative structural or conformational homology of some microbial antigens to host proteins. Cross-reactivity between antigenic epitopes of infectious agents and host antigens may break tolerance to self determinants, resulting in the persistence of the immune response and eventual autoimmune disorders even after the infectious agent has been cleared (93). Indeed, structural homology of HIV proteins to a variety of cellular and soluble human antigens has been reported by several investigators (rev. in 94).

2. Autoimmunity may be mediated through the generation of idiotype/anti-idiotype circuits. Experimental studies in murine models of reovirus infection (95) evidenced that antibodies against viral proteins involved in cell binding could evoke anti-idiotypic antibodies that carry the internal image of the antigen, and behave as autoantibodies, binding to cell viral receptors and mimicking virus-cell interactions (96).

3. The impact of retrovirus transcription on host genome structure and function must also be considered. Ample evidence in murine models (97–98) indicates that the integration of proviral DNA near cellular genes or protooncogenes can profoundly modify gene expression through the phenomenon known as insertional mutagenesis. Although in man this event is less well documented (99–100), and HIV provirus insertion is random within the host genome, the possibility that HIV could modulate host genome function in *trans* cannot be overlooked (7).

4. Autoimmune responses could also be elicited by a perturbation of the normal antigen recognition

processes. It was recently shown that mice exposed to autologous apoptotic cells produce anti-nuclear antibodies; this phenomenon was also associated with renal immunopathology, as shown by IgG deposition in the glomeruli (101). Since apoptotic death of CD4$^+$ and CD8$^+$ T cells is a hallmark of seropositive patients (102), it is conceivable that aberrant exposure to autologous antigens may perturb antigen selection processes, thus breaking the tolerance and initiating an immune response against self antigens in AIDS patients.

On the other hand, the presence of circulating immune complexes (CIC) in many infections, as well as autoimmune and neoplastic disorders, is well documented (103). That CIC have been detected by several assays in the sera of a very high percentage of seropositive patients (104–106) is not surprising since CIC may originate from several sources. The presence of opportunistic infections in AIDS patients constitutes a formidable source of antigenic material. Indeed, hepatitis B virus (HBV), *Pneumocystis carinii*, and HIV antigens were detected in CIC from seropositive patients (105–106). Only HIV *gag* and *pol* products, but not gp120, were found. Given the high affinity of gp120 for CD4 (107), circulating gp120 molecules are not available for soluble complex formation (see below).

Autoimmune manifestations during HIV infection are not readily classified, as they are extremely polymorphic and often due to the intervention of multiple immunopathologic mechanisms, most of which are still unclear. We will discuss these manifestations according to the scheme shown in Table 4.

4. AUTOIMMUNE DISORDERS UNRELATED TO AIDS PATHOGENESIS

4.1. Rheumatologic Disorders

Musculo-skeletal involvement is a frequent feature in animal models of retrovirus infection (108). While musculo-skeletal complications are also being recognized with increasing frequency in HIV-infected patients (for review, see 109), their prevalence, compared to a population matched for age, sex, and life-style, is uncertain, and clinical pictures are often ill-defined and not readily classified. Moreover, the precise link between HIV infection and these syndromes is unclear, as are most of the underlying pathogenetic mechanisms. Nonetheless, further understanding in this sense might also shed some light on the immunopathologic mechanisms underlying these conditions in seronegative subjects. For instance, the frequent occurrence of Reiter's syndrome in association with AIDS seems to rule out a CD4$^+$ cell role in this condition, while

Table 4   Autoimmune manifestations during HIV infection

UNRELATED TO AIDS PATHOGENESIS
Rheumatologic disorders
- Arthritis
- Myopathies
- Vasculitis
- Sjogren's-like syndrome
- Lupus-like syndrome
Kidney disorders
Neurologic disorders
Hematological disorders
- Idiopathic thrombocytopenic purpura
- Thrombotic thrombocytopenic purpura
- Neutropenia
- Anemia
- Blood clotting alterations

POSSIBLY RELATED TO AIDS PATHOGENESIS
Cellular mechanisms
- Class II-restricted CD4$^+$ CTL
- Autoreactive lymphocytes
Humoral mechanisms
- Anti-lymphocyte antibodies
- Anti-IL-2 antibodies
- Anti-IFN-γ antibodies
Generation of anti-CD4 reactivity

---

the infrequent association of rheumatoid arthritis and systemic lupus erythematosus with HIV infection strongly indicates that helper lymphocytes play a pivotal role in the pathogenesis of these diseases.

Arthritis

Reiter's syndrome and reactive arthritis   Joint symptoms ranging from acute, intermittent arthralgia with no clinical evidence of arthritis to severe arthritis with synovitis are very common in HIV infections, and were reported in 12–45% of seropositive patients. In later stages of disease, a painful articular syndrome can be observed characterized by severe intermittent pain and oligoarticular involvement, but no clinical evidence of synovitis (110). Among the major clinical pictures, the most frequently reported types are Reiter's syndrome (RS) and reactive arthritis (ReA); the latter describes patients with arthritis who have evidence of infection distant from the joints preceding or concomitant with arthritis development, but do not fulfill extra-articular criteria for classical RS, nor show the increased prevalence of the HLA-B27 haplotype observed in RS patients (111).

Although the frequent association of HIV infection with RS and ReA strongly suggests a biologic connection between these conditions, the pathogenesis of these manifestations is not clear, and several non-mutually exclusive hypotheses were advanced.

1. The increased RS incidence among seropositive subjects could merely reflect an increased exposure to sexually-transmitted organisms known to trigger it, such as *Chlamydia trachomatis*. Since RS is often preceded by culture-negative diarrhea or urethritis, non-conventional pathogens could also be involved, especially in view of the concomitant immunodeficiency that predisposes seropositive individuals to infections by arthritogenic micro-organisms, such as *Salmonella typhimurium* and *enteritidis*, *Shigella flexneri*, *Campylobacter jejuni*, *Yersinia enterocolitica* etc. Indeed, the alteration of bacterial clearance mechanisms that accompany depletion of CD4$^+$ T cells may be of primary importance in conferring susceptibility to RS and ReA, because more than 30% of the cases are preceded by infection with enteric micro-organisms (111). Indeed, bacteria have an established role in the pathogenesis of septic arthritis, Lyme disease, and ReA. Unlike septic arthritis, Lyme disease and ReA were previously thought to be driven by an autoimmune process triggered from a distant site of infection, as the etiological bacteria could not be cultured from the inflamed joint. However, recent data indicate that *Chlamydia trachomatis* DNA and mRNA can be detected in the joints of patients with sexually-acquired ReA and non-differentiated oligoarthritis, and rarely in patients with other systemic rheumatic diseases, suggesting that intact organisms can reside in the synovium (112–113). Bacteria have been also implicated in the pathogenesis of arthritides of unknown etiology, such as rheumatoid arthritis (RA); recently, *Mycobacterium smegmatis* and *M. hodleri* (non-tubercolous mycobacteria) DNAs have been detected in the synovial membrane of three RA patients, although no evidence was obtained for a pathogenetic role (114). In view of the frequent infection by atypical mycobacteria in seropositive patients, their role in the pathogenesis of articular manifestations cannot be overlooked.

2. HIV might infect synovial cells, and thus directly cause immunopathology, or trigger a CTL response against infected synovial cells. HIV p24 antigen has been detected in cells of the synovial lining and in CD4$^+$ and CD8$^+$ T cells of the subsynovium of patients with arthritis, although viral particles were not demonstrated by immunohistochemical techniques and electron microscopy (115). HIV DNA has been evidenced by *in situ* hybridization within dendritic cells isolated from synovial tissues of infected individuals with arthritis. Furthermore, several HIV-related antigens, such as p24, gp41 and gp120, have been identified in articular tissues of seropositive individuals (116).

3. Finally, joint involvement might simply be mediated through CIC deposition in the synovial tissue, as in joint disorders associated with HBV, HCV or rubella infection (117), that share many characteristics with HIV-associated arthritis.

Other arthritides.   Since the original report by Winchester and colleagues (111) of Reiter's syndrome and ReA associated with HIV infection, a wider spectrum of rheumatic manifestations has been described, including psoriatic arthritis (PsA) and undifferentiated spondilo-arthritis (SpA). The prevalence of these forms varies between 1% to 32%, and the articular pattern is more frequently polyarticular and asymmetric, accompanied by dactilitis and enthesopathy. No association with HLA class I alleles was found among HIV+ individuals; furthermore, these patients exhibit a variety of cutaneous proliferative disorders, from psoriasis vulgaris to severe exfoliative erythrodermia (118). Unlike in developed countries, where accurate avoidance of bacterial persistence and decrease in viral load have been generally pursued, the relationship between SpA and HIV is striking in Africa, where HIV seroprevalence is about 30% among adults. As reported by Njobvu et al. (119) in 595 black Zambians, the prevalence of clinically-significant SpA over a 30-month period was twelve times greater in HIV+ individuals than in the general population (180/100,000, compared to 15/100,000). In addition, in a recent study conducted in Africa among 39 HIV+ patients, Bileckot and colleagues (120) identified 32 patients with HIV-associated arthritis, 2 ReA, 2 patients with staphylococcal septic arthritis and 3 with infectious discitis. These figures make HIV infection the leading reason for admission and the leading cause of arthritis in this country. In contrast, a study carried out at an infectious disease clinic in Argentina (121) described acute HIV-associated arthritis in only 8% of the population considered. The arthritis showed an acute onset, short duration, was self-limiting, with mono- or oligo-articular presentation, lack of recurrence, and no association with erosive changes in the involved joint. Moreover, it occurred more frequently in later stages of the disease. Septic arthritis due to *Staphylococcus aureus* or Streptococci has been reported in HIV+ hemophiliacs (122). A limited spectrum of bacteria is involved and there is no peripheral leukocytosis. Arthritis is localized exclusively to joints affected by hemophilic arthropathy so that the clinical picture mimics that of hemoarthrosis, often causing a delay in diagnosis.

The presence of CIC in synovial fluid of patients with HIV-associated arthritis has been shown, and their deposition in the synovial tissue is the most obvious pathogenic mechanism of joint involvement, as in HBV, HCV and rubella infection-associated joint involvement. On the other hand, two additional possibilities may explain why HIV increases the susceptibility to SpA (123). The first underlines the role of prolonged persistence of arthritogenic bacteria, especially of enteric or sexual origin, and the second mechanism could involve a role of HIV-infected macrophages or dendritic cells in dysruption of immuno-logical tolerance.

As far as classical rheumatoid arthritis (RA) is concerned, several reports of both amelioration and progression of rheumatic symptoms in HIV-infected patients exist. Lapadula and colleagues (124) described complete clinical remission of RA following HIV infection, with disappearance of rheumatoid factor and resolution of bone erosions. In contrast, in patients with SpA, HIV infection causes aggressive and progressive disease, refractory to conventional anti-rheumatic drugs (125). There are a number of reports in the literature concerning active RA and HIV infection coexistence, although all patients have a decreased number of CD4+ lymphocytes. It remains to be seen under what setting HIV may have a disease-modifying effect during RA, in view of the critical implications in the pathogenesis and therapy of RA (126).

Myopathies.   Myopathy is a common feature of viral infection (127), and several types of myopathy have been observed in HIV-infected patients (128). Polymyositis is the most commonly reported disorder, and shows numerous similarities with the experimental monkey model. Muscle weakness and electromyographic changes are usually associated with elevated serum creatine kinase levels and pathologic findings of muscle inflammatory infiltrate and fiber necrosis (129). In biopsy specimens, there is a variation in fibre size, vacuolar changes, fibre destruction and type II atrophy. However, the most striking feature is the inflammatory interstitial and perivascular cell infiltrate, mainly consisting of macrophages and CD8+ T lymphocytes. HIV transcriptional products have been detected (130) in the infiltrating macrophages, but not in cultured myotubules or within muscle fibres, suggesting that the myocyte damage could result from immunological effector mechanisms rather than the direct HIV action (131). Moreover, similar to their idiopathic counterparts, in HIV-related polymyositis and inflammatory neuropathies, T lymphocyte infiltration is not cleared by apoptosis, as assessed by nuclear morphological findings and *in situ* labeling techniques (132–133).

The myopathy may also be a result of polymyositis-like host response to the virus, opportunistic infections (such as toxoplasmosis), or zidovudine therapy (134). Both polymyositis and zidovudine-induced myopathy are accompanied by elevation of muscle enzymes and similar abnormalities on electromyogram. Muscle biopsy in zidovudine-associated myopathy demonstrates changes similar to those reported in polymyositis, with less inflammatory infiltrate and mitochondrial damage, but wider size variation, swelling and degeneration, as detected in ultrastructural studies (135).

Vasculitis. The association of vasculitis and viral infection is also a well-known phenomenon (136). Several types of vasculitis have been described in AIDS patients (137), although the association remains rare, being observed in about 1% of patients (138). Among these, a necrotizing vasculitis of the polyarteritis nodosa (PAN) type is best characterized (139). The vasculitis syndromes can involve vessels of any size, ranging from small vessels in the hypersensitivity group to medium-size vessels in PAN. Large vessel involvement has been reported, with aneurysm formation in the iliac, femoral and subclavian arteries and the aorta. The presenting symptoms are peripheral symmetric sensorimotor neuropathy, motoneuritis multiplex, muscle pain and digital ischemia. In some cases, the vasculitis is limited to CNS, whereas in others it involves several districts (skin, lung, liver etc.).

The most common cause of hypersensitivity angitis is exposure to commonly employed anti-microbial or anti-retroviral agents. However, several non-mutually exclusive pathogenetic possibilities must be considered. First, the vasculitic phenomena might be mediated through an inflammatory reaction triggered by CIC deposition in blood vessels. Second, direct HIV-mediated injury to endothelial cells is conceivable, in view of HIV's ability to infect these cells. HIV particles, mRNA and p24 antigen were identified within the vascular necrotizing lesions in patients with PAN-like syndrome (140–141), suggesting a possible direct viral role in disease pathogenesis. Finally, the possibility of an immunopathologic damage cannot be ruled out. Since endothelial cells express HLA-DR antigens and eventually process and present soluble viral proteins in association with HLA class II antigens, it is possible that they become targets of virus-specific CTL; generation of CD4+, class II-restricted CTL has been demonstrated (142–143).

Sjögren's-like syndrome. In HIV infection, 2–5% of patients develop a clinical picture closely resembling Sjögren's syndrome, named diffuse infiltrative lymphocytosis syndrome (DILS) (144–145). DILS is characterized by symptoms almost indistinguishable from those of primary Sjögren's syndrome (SS), with bilateral parotid gland enlargement in 90% of patients, together with xerostomia, xerophtalmia and keratoconjunctivitis sicca in 40% (146). The salivary glands show mucoid degeneration of the stroma (147), and areas of lymphoid cell infiltration. Opportunistic infections are rarely encountered and patients with DILS generally have less advanced clinical stage, although these patients are at higher risk of developing high-grade B cell salivary gland lymphomas (144). It was posited that lymphoplasmacytic infiltrates in salivary glands and other organs might represent, at least in some cases, a prodromal

stage of lymphoma development (148). Occasional lung lymphomas were, in fact, observed in seropositive subjects previously diagnosed as having SS.

The pathogenesis of DILS is unclear (for review, see 149). Despite the similarities in clinical presentation and pathologic findings in the salivary glands, several unusual features are observed in seropositive patients. In fact: 1) a male predominance exists, whereas classical SS is more common among women; this may simply reflect the special composition of groups at risk for AIDS in Western countries; 2) significant titres of autoantibodies, such as rheumatoid factor, antinuclear antibodies, anti-SSA/Ro, or anti-SSB/La are found in less than 20% of patients (144); 3) epithelial lesions usually associated with extensive infiltrates are not observed, and DILS salivary gland infiltrates are mainly composed of CD8+ T cells, while in SS only CD4+ T cells are found; 4) patients show high counts of circulating CD8+ T cells with a memory phenotype. In contrast to SS, CD8+ cells also infiltrate visceral organs, causing lymphocytic interstitial pneumonia (seen in about 50% of the cases), lymphocytic infiltration of gastric and intestinal mucosa, lymphocytic hepatitis, lymphocytic interstitial nephritis with progressive renal insufficiency, lymphocytic aseptic meningitis and symmetric sensorimotor neuropathies; 5) infiltrating cells have a restricted usage of T cell receptor gene segments (145), suggesting an antigen-driven response (150); 6) no predominance of the usual HLA markers of classical SS is found; in contrast, there is an excess of DR5 allele DRB1.1102 and DR6 allele BRB1.1301 in Blacks (86% versus 7% in controls) and DR6, DR7 in Caucasians. These HLA class II alleles encode structures that share almost all critical polymorphic aminoacids in the antigen-binding groove, and have identical β-chain CDR3 regions, suggesting that they can bind and present similar peptides (151–152). At the class I locus, there is increased frequency of B45, B49 and B50, all of which encode similar peptide-anchoring structures (152). Furthermore, there is also a reduced prevalence of HLA-B35, a specificity associated with accelerated progression to AIDS-related opportunistic infections (153). This association with both class I and class II alleles suggests that DILS is characterized by a host immune response against unidentified antigens that could depend on the interactions between HLA-peptide complexes with both CD4+ helper and CD8+ effector lymphocytes.

Lupus-like syndrome. Systemic lupus erythematosus (SLE)-like symptoms represent a special challenge for rheumatologists. In fact, SLE and HIV infection show many similar features: eye, skin, lung, joint, and kidney alterations, as well as hematologic changes, CNS and constitutional symptoms, and laboratory findings (for reviews, see 128, 154). Despite this resemblance, which makes HIV antibody testing in all patients showing SLE-like symptoms mandatory, classical SLE and HIV-associated lupus-like syndrome differ in some fundamental aspects (155). Compared to control groups, HIV-infected patients present a high frequency of low-titer autoantibodies, including antinuclear antibodies (ANA), anti-denatured DNA and anti-cardiolipin, while anti-native DNA antibodies (diagnostic of SLE) have never been reported (156).

Antibodies directed towards neutrophil cytoplasmic antigens were reported in up to 42% of HIV-infected individuals, but the association with cutaneous or systemic vasculitis is rare. Furthermore, anti-cardiolipin IgG antibodies were observed in high frequency in both symptomatic and asymptomatic patients (156). Weiss et al. (157) demonstrated that HIV-associated anti-phospholipid antibodies do not bind the β_2-glycoprotein I cofactor, unlike those developing in autoimmune diseases. Moreover, anti-phospholipid antibodies in HIV-infected patients are not associated with increased risk for thrombotic events, suggesting differences in epitope specificity. Whether these observations reflect different pathogenetic mechanisms in classical SLE and HIV-associated lupus-like syndromes is presently unclear, although it is likely that the presence of high CIC levels, together with still undefined host factors, might partly account for the immunopathologic findings in the latter.

HIV infection in individuals with previous diagnosis of classical SLE has been reported (154). In these cases, HIV infection seemed to exert a beneficial effect on the clinical course of SLE, as the progressive CD4+ lymphocyte depletion was associated with improvement in the SLE symptoms; zidovudine treatment was instead followed by an increase in circulating CD4+ cells and the reappearance of joint and pulmonary symptoms, as well as a rise in a serum anti-DNA antibody titers, thus emphasizing the central importance of CD4+ T cells in SLE pathogenesis.

On the other hand, antibodies reactive with retroviral *gag* proteins have been detected by ELISA screening in patients with SLE (158), as well as in about 30% of non-HIV-infected, otherwise seronegative individuals with classical SS (159). Western blot analysis revealed intermediate patterns of reactivity for HIV antigens, but none of the samples fulfilled the criteria for HIV infection. Most of the positive samples reacted with p18 antigen, but antibodies were found to other HIV proteins, including p24, although in a minority of patients (158). In any case, false-positive results by ELISA and indeterminate reactions by Western blotting are now a well-known phenomenon in SLE (160–161). Although this reactivity may be due to still undefined antigenic mimicries between self antigens and retroviruses, human antigen contaminants or artifacts in HIV preparations are also possible. Antibodies reactive mainly with p24 antigen were described in strict association to a small Sm-like ribonucleoprotein (162–163). Although

a structural homology between p24 gag and the Sm B/B' antigen was found, however, most p24-reactive antibodies are directed to epitopes other than the proline-rich sequences shared by p24 and Sm B/B' (161).

4.2. Kidney Disorders

Since the early years of the AIDS epidemic, many studies have suggested that HIV-infected patients are at risk of developing several types of acute and chronic renal diseases. Among them, HIV-associated nephropathy (HIVAN) has been extensively studied (for reviews, see 164–165), partly because is the most common cause of chronic renal disease in seropositive individuals (166–167), and partly because it is believed to be caused by the direct effect of HIV on renal cells. HIVAN presents clinically with massive proteinuria, enlarged echogenic kidneys and, in the absence of specific therapy, progression to end-stage renal failure within weeks to months. A combination of different pathologic lesions is usually described, including a collapsing form of focal segmental glomerulosclerosis, glomerular visceral epithelial cell hypertrophy and prominent tubulo-interstitial infiltration with edema, fibrosis and microcystic tubule dilatation (164). Host factors are very important in HIVAN pathogenesis. In fact, about 80% of the cases occur in young black patients, suggesting that genetic factors are involved in disease susceptibility (168). Moreover, Winston et al. (169) demonstrated that HIVAN is a late manifestation of HIV infection, with the overwhelming majority of patients having an AIDS-defining condition.

Initially, similarities between HIVAN and heroin nephropathy suggested that HIV could not be the only cause of the nephropathy. However, further reports of HIVAN in seropositive children, recipients of contaminated blood product, heterosexuals, and non-drug abusers confirmed that the virus may play a predominant role. It is currently known that HIVAN is a result of direct viral involvement of renal epithelium. Support for a role of HIV proteins in HIVAN pathogenesis comes from animal models. Focal and segmental glomerulosclerosis, a typical feature of HIVAN, was observed in monkeys infected with simian immunodeficiency virus and in cats infected with feline immunodeficiency virus. Furthermore, a transgenic murine model has been developed with a 3.1 kb deletion that disrupts the *gag* and *pol* genes of HIV, but leaves the *env* and regulatory genes in frame (165). All the heterozygous mice developed a progressive nephropathy that mimics HIVAN in humans. Transplantation of transgenic kidneys into normal mice leads to the development of HIVAN, suggesting that, at least in the mouse model, HIV expression in kidney is essential for HIVAN development. In addition, HIV DNA and mRNA was found in tubules from human renal biopsy specimens by *in situ* hybridization (166,170). Based on these studies, HIVAN is now considered a two-stage disease (171). The first stage is characterized by direct renal cell injury by HIV, which corresponds clinically to the development of proteinuria. The subsequent proliferative response, with kidney enlargement and rapid progression to renal insufficiency, seems a process driven by cytokines or growth factors. Cytokines such as IL-1, IL-6 and TNF-α released by HIV-infected lymphocytes may contribute to renal injury, while TGF-β and bFGF (all increased in HIVAN) induce sclerosis and proliferation, which are preeminent pathologic features of HIVAN. The role of cytokine dysregulation has not yet been fully clarified, but it could be induced and enhanced by viral gene products, and could promote both viral spread and renal disease progression.

4.3. Hematological Disorders

Hematological disorders constitute a large part of the autoimmune complications of HIV infection. The mechanisms underlying these manifestations might be operative at a central or peripheral level, or both. In reference to the former, it is known that HIV also infects bone marrow progenitors. In addition, it was shown that anti-gp120 antibodies can suppress *in vitro* colony formation by bone marrow progenitors from AIDS patients (172). On the other hand, peripheral events that mediate destruction of blood elements delivered into circulation at a normal or increased rate might also be involved. Obviously, central and peripheral alterations are not mutually exclusive, and may occur in different subjects or in different stages of the disease.

Idiopathic thrombocytopenic purpura. Since idiopathic thrombocytopenic purpura (ITP) is one of the most common autoimmune disorders in seropositive patients (for reviews, see 173–174), the search for HIV risk factors and HIV testing are now highly recommended in any patient presenting with a thrombocytopenia of unknown origin. The ITP incidence among HIV-infected persons ranges from 5% to 10% in asymptomatic carriers, and from 25% to 45% in full-blown AIDS patients (175), but disease severity in general does not correlate with the degree of immunodeficiency. Clinically, ITP shows decreased numbers of circulating platelets, ad increased megakaryocytes in the bone marrow, and thus is virtually indistinguishable from classic autoimmune thrombocytopenic purpura. Spleomegaly is mild, even though platelet destruction occurs mainly by peripheral clearance within the spleen (176). The clinical course of ITP ranges from asymptomatic thrombocytopenia to easy bruising and petechiae, up to moderate bleeding; severe hemorrhage has been occasionally reported (177–178), especially in seropositive hemophiliacs.

Three major mechanisms could be involved in ITP pathogenesis. First, as in other retroviral infections, HIV can directly infect megakaryocytic cells (179). Morphologic abnormalities have been described by electron microscopy in the megakaryocytes of HIV-infected individuals (180). Furthermore, HIV mRNA has been demonstrated in the megakaryocytes of 10 out of 10 patients with HIV-ITP (181). Additionally, human megakaryocytes and platelets can internalize HIV particles, as shown in co-cultivation experiments (182). Finally, Dominguez et al. (183) have established the presence of HIV p24 antigen in megakaryocytes by immunoistochemical studies.

Second, antibodies to platelet surface antigens could be involved. The nature of these platelet-associated Ig is a matter of debate. Anti-platelet autoantibodies directed against epitopes of the platelet membrane glycoprotein IIb/IIIa were reported (184). On the other hand, elution experiments have postulated that molecular mimicry between HIV gp160/gp120 and platelet antigens might explain some cases of HIV-related ITP (185). Finally, in the serum of most ITP patients, Stricker et al. (186) identified an anti-platelet antibody that recognized a 25-Kd antigen showing a high degree of homology with an antigen present on herpes virus-infected monkey kidney cells. It was theorized that herpes infection in immunodeficient patients could evoke anti-herpes antibodies that cross-react with host cells, including platelets. On the other hand, these antibodies could merely represent an epiphenomenon due to the recognition of antigens released following increased platelet destruction.

The third possible pathophysiologic mechanism involves immune-mediated destruction by CIC bound to the platelet membrane. Some workers (187) found no reactivity against normal platelets of serum and platelet eluates, and instead found that not only did CIC levels often correlate with platelet-bound IgG, but the complexes were also able to bind purified platelets. Immune complexes were identified in platelet eluates from seropositive patients (188–189); these complexes usually lack HIV antigens and proviral DNA, but were shown to contain anti-HIV antibodies (anti-gp120) and also anti-anti-HIV antibodies (anti-idiotypic antibodies) (190). The presence of these complexes could lead to increased platelet uptake and destruction within the reticulo-endothelial system.

It is not clear which of the proposed mechanisms is responsible for increased platelet destruction. These pathways are not mutually exclusive; some observations seem to indicate that they might be variably involved in different categories of HIV$^+$ patients, such as homosexuals, hemophiliacs, and i.v. drug abusers (191). A major problem is why only some HIV-infected patients show thrombocytopenia, and why only a minority have clinical symptoms, even though platelet-associated IgG is a widespread phenomenon among infected persons. The role of intercurrent opportunistic infections, and the relevance of host co-factors are not forthcoming. A defect in phagocyte function, and a block in the ability of the reticulo-endothelial system to clear CIC, as well as the very physical characteristics of the CIC, such as antigen-antibody ratio and complex solubility (192), may all contribute to increased platelet destruction in some individuals, but not in others.

Therapy in HIV-related ITP is a major concern. Thrombocytopenia improves after suppression of reticulo-endothelial function with steroid therapy (although only 10–20% of patients have a sustained response, with the potential risk of increased immunosuppression). Intravenous high doses of IgG give transient but effective responses, while splenectomy has proven useful, resulting in a sustained increased in platelet counts with no apparent risk of increased progression to AIDS. Consistent with the evidence of direct megakaryocyte infection by HIV, the beneficial effects of anti-retroviral drugs were verified in clinical trials; zidovudine is now recommended as the treatment of choice (193). A good response has also been reported with low-dose IFN-α, especially in patients who failed to respond to zidovudine, but the mechanisms underlying this effect remain elusive (194).

Thrombotic thrombocytopenic purpura. Thrombotic thrombocytopenic purpura (TTP), a rare hematologic disorder characterized by thrombocytopenia, microangiopathic hemolytic anemia, renal failure, fluctuating neurologic abnormalities, and fever (195–196), has been reported with increasing incidence among HIV-infected individuals (3.7 per 100,000, almost 40 times greater than TTP in the general population) (197). Its recognition is important in view of the severity of the clinical picture, and of the beneficial effects of appropriate, timely therapies.

In seronegative patients, TTP is associated with pregnancy, malignancies and infections (198), but the pathophysiologic mechanisms leading to increased platelet aggregation are unknown. Vascular injury due to CIC, endotoxins or drugs, as well as alterations in platelet aggregation, such as the presence of serum agglutinating factors (199) or the deficiency of a component that selectively inhibits platelet aggregation (200), have been suspected. In HIV infection as well, several factors might contribute to increased platelet destruction and aggregation. The pathophysiology may be related to the capacity of HIV to infect endothelial cells both *in vitro* and *in vivo*. Moreover, it has been demonstrated that plasma from HIV-TTP patients can induce apoptosis in primary microvascular endothelial cells, but not in cells of large vessel origin (201). Apoptosis was independent of TNF-α secretion or the presence of CD36 on endothelial cells, which was linked to the rapid induction of Fas in these cells. The nature of the primary plasma protein capable of inducing endothelial apoptosis remains to be elucidated. Apoptotic endothelial cells were

recently shown to exhibit some features of a procoagulant phenotype, including depressed production of prostaglandin I_2 (202).

Infectious agents have been indicated as an inciting cause of microangiopathy. Cytomegalovirus, a common opportunistic pathogen in patients with HIV infection, has been shown to have endothelial tropism and to increase the procoagulant activity of endothelial cells (for review, see 195). Several reports have also suggested a link between intestinal toxin-producing pathogens (including *Shigella dysenteriae* and *Escherichia coli*) and the development of thrombotic microangiopathy. It is possible that a direct effect of some bacterial toxins on endothelial cells may result in down-regulation of prostacyclin, a potent endothelial-derived inhibitor of platelet aggregation, thus providing the initial insult leading to the development of TTP (203).

Treatment of TTP includes plasmapheresis, antiplatelet agents, splenectomy and the use of anti-retroviral agents, mainly zidovudine; nonetheless, the prognosis is poor, with no seropositive patient presenting with TTP surviving longer than 24 months (204).

Other hematologic disorders. Various hematologic cytopenias were reported in HIV-infected patients.

Neutropenia. Neutropenia, which may contribute significantly to an inadequate cellular defense against opportunistic pathogens (205), as well as a limited efficacy of antibiotic treatments, is more common in advanced AIDS, often caused or exacerbated by concomitant myelosuppressive medications. The pathogenesis of neutropenia is multifactorial. Although viral infection of stem cells is possible, it is not thought to be a central event. The major mechanism appears to be stromal or accessory cell dysfunction, as HIV-infected stromal cells show an abnormal pattern of cytokine secretion, including TGF-β and TNF-α, which may play a pivotal role in suppressing hematopoiesis (206–207). On the other hand, antibodies against granulocytes were detected in a high number of seropositive individuals (208–209). This finding was not correlated with the extent of neutropenia, however, and its prevalence was also high in seronegative individuals at risk of AIDS. Some workers claim that the presence of IgG on granulocytes simply represents an epiphenomenon of high CIC levels. Finally, commonly used drugs, such as antiretroviral compounds, anti-neoplastic agents and antibiotics, may represent a major cause of neutropenia in AIDS patients. Low-dose myeloid colony-stimulating factors, such as GM-CSF and G-CSF, can increase neutrophil counts when concomitant myelosuppressive medication is needed (210).

Anemia. Anemia is often associated with neutropenia. Anemia in patients with AIDS is multifactorial, with several different mechanisms operating in the same patient (for review, see 211). Anemia may result from a direct effect of HIV infection of hematopoietic stem cells and bone marrow stromal cells or accessory cells such as monocytes and T cells, as for neutropenia (212). The red blood cells are normochromic and normocytic, and serum iron, iron-binding capacity and transferrin concentrations are low, whereas serum ferritin concentration is markedly increased. Elevated erythrocyte sedimentation rate is frequent, and suppression of the reticolocyte response is seen, as well as a decreased physiologic response to erythropoietin. This inappropriately low erytropoietin response in AIDS patients led to the use of recombinant erytropoietin, which reduces transfusion requirements.

Inflammatory cytokines may play a central role in the pathogenesis of anemia. TNF-α, IL-1β and TGF-β have been shown to suppress progenitor growth *in vitro* (213). The action of these cytokines may be synergistic, and may also be indirectly exerted through a cytokine cascade. Further studies are needed to clarify whether HIV itself or HIV-induced release of cytokines in the bone-marrow microenvironment leads to the suppression of hematopoietic progenitors. Anemia may also result from opportunistic infections, assuming the typical features of the anemia associated to chronic inflammatory conditions. The most common cause of this type of anemia in AIDS patients are atypical mycobacterial infections. Furthermore, many HIV-related infections cause non-specific suppression of erythropoiesis, such as parvovirus B19, that can directly infect erythroid precursors, causing a chronic pure red cell aplasia (214).

Drug therapy for HIV infection or its complications is another common cause of anemia, by inducing either bone marrow suppression or hemolysis. Myelosuppression is, actually, the most common side effect of zidovudine treatment (215), while primaquine and dapsone (used for *Pneumocystis carinii* infection therapy and prevention) can provoke oxidative hemolysis. Folic acid deficiency and low serum vitamin B$_{12}$ concentrations have been reported during HIV infection, but there is no response to vitamin B$_{12}$ supplementation (206).

Although a direct antiglobulin test (Coomb's test) is positive in about 20–40% of HIV-infected patients, and seropositive individuals show a broad panel of specific and non-specific anti-erythrocyte antibodies, overt autoimmune haemolytic anemia is rare. Cases have been reported as a result of hypersplenism associated with concomitant chronic HBV or HCV infection. AIDS-related lymphomas can also cause either bone marrow infiltration anemia, or chronic conditions.

Blood clotting disorders. While blood clotting disorders are infrequently clinically relevant, autoimmune phenomena involving the coagulative network are common in seropositive patients (216). Laboratory alterations include increased partial thromboplastin time (PTT), activated PTT, and Russell's viper venom clotting time due to the presence in serum of IgG and/or IgM directed against the phospholipid components of the prothrombin activator complex involved in the coagulation pathway. This inhibitory autoantibody is known as *lupus anticoagulant* in view of its high prevalence in SLE patients, and is often associated with the presence in serum of autoantibodies against other phospholipids, such as anti-cardiolipin antibodies (217). The prevalence of lupus anticoagulant in HIV-infected individuals ranges from 50 to 70% in some reports, but does not cause abnormal bleeding, unless accompanied by thrombocytopenia or clotting factor deficiency. Association with thrombotic events, typical of SLE, are rare in HIV-infected patients, as previously discussed.

Anti-cardiolipin antibodies have been reported in patients with AIDS, with no relationship with the presence of lupus anticoagulant (157). Rarely, they have been associated with thrombosis, a vascular necrosis and skin necrosis in HIV-infected individuals. The presence of anti-phospholipid antibodies is not strictly correlated with HIV infection nor progression to AIDS, however, and may also be demonstrated in seronegative hemophiliacs (218). In seropositive subjects, anti-phospholipid antibodies usually accompany opportunistic infections, and tend to disappear with their clinical resolution (216). The mechanisms leading to anti-phospholipid antibody generation are obscure; the observation that this phenomenon is present in autoimmune disorders as well as many infections suggests that cross-reactivity of viral and bacterial antigens could lead to breakdown of immunological tolerance to self phospholipidic components.

4.4. Neurologic Disorders

Neurologic dysfunction occurs frequently in AIDS. About 20% of seropositive adults and 50% of children with advanced disease have neurologic involvement, which may also constitute a presenting picture in subjects with no previously documented seropositivity (219–220); neuropathologic alterations are even more common at autopsy (221). Opportunistic infections and primary CNS lymphomas aside (140), neurologic symptoms indicate both central and peripheral abnormalities, including subacute encephalitis, aseptic meningitis, vacuolar myelopathy, and peripheral neuropathy. Among these, the subacute encephalitis embodies a particular entity, broadly termed AIDS-dementia complex (ADC), and characterized by poor memory, psychomotor retardation, and behavioral changes rapidly progressing to full-blown dementia. The pathologic picture is one of focal necrosis of the gray and white matter, perivascular inflammation, microglial nodule and multinucleated giant cell formation, and severe white matter demyelination (221–222). The neuropathologic manifestations also include hypertrophy and accumulation of astrocytes (astrocytosis), and infiltration and expansion of macrophages in brain tissues; neuronal death is apparently due to both apoptotic and necrotic phenomena (223). HIV can be isolated from brain tissues (224–225). However, unlike other viral encephalitis, where the infectious agent shows a selective tropism for neurons and macroglia (such as herpes symplex type 1 infection), HIV infection in CNS is limited to cells of the monocyte/macrophage lineage, such as microglia, resident macrophages and macrophage-derived multinucleated cells (226–227). Infection of neuron cells, oligodendrocytes, and astrocytes has not yet been convincingly demonstrated, even though neurons express significant amounts of surface CD4 (228).

A substantial body of evidence assigns HIV a direct role in this condition, but why CNS involvement is observed only in some seropositive patients is still an open question. In addition, although HIV enters CNS shortly after infection (229), the clinical neurologic manifestations occur relatively late. Both viral and host factors may play a critical role in the pathogenesis of ADC; in fact, the degree of CNS damage is not correlated with the systemic or local virus load (230). Several non-mutually exclusive hypotheses have been advanced to explain the neuropathologic involvement in HIV-infected patients.

1. First, HIV products released by infected cells within the CNS could directly affect neuron function. Although it was demonstrated that soluble gp120 is able to induce damage in in vitro cultures of fetal neuron cells (231–232), these findings have not yet been sufficiently corroborated in in vivo models. However, gp120 might also be able to induce damage by indirect mechanisms. A moderate sequence homology was described between a highly conserved region of HIV gp120 and neuroleukin, a 56-Kd neurotrophic factor (233–234). This homology might be biologically relevant, since soluble gp120 was shown to compete with neuroleukin for neuron growth and survival (235). On the other hand, gp120 is capable of inducing TNF-α and IL-1β in cultured glial cells (236), thus triggering the inflammatory circuits that constitute a primary mechanism of CNS damage.
2. Monocyte, B cell, and T cell activation is a prominent feature within the CNS compartment during HIV infection; antibodies against HIV are produced within the CNS itself (237). Activated T cells produce a variety of cytokines, including IFN-γ, while activated B cells from seropositive patients produce considerable amounts of TNF-α (238), and

monocytes high amounts of IL-6 (239). All these mediators enhance viral expression in chronically infected cell lines (80,82). In this setting, the role played by cells of the monocyte/macrophage lineage may be of paramount importance. A body of evidence demonstrates that HIV-infected or activated macrophages and microglial cells produce a variety of neurotoxic cytokines and mediators, including TNF-α, IL-1β, IFN-γ, different eicosanoids, platelet-aggregating factor, and nitric oxide (240–242). This mediator network creates an amplification loop in which HIV-associated T/B cell activation initiates the immunopathologic process, which culminates in monocyte/macrophage activation and eventual neuronal damage. Indeed, in vitro observations were validated by in vivo findings; the levels of the above mediators, in fact, are increased in CSF of ADC patients (243–246).

3. The possibility that classical autoimmune mechanisms might be responsible for CNS involvement in AIDS was raised by the finding of antimyelin basic protein antibodies in the serum and CSF of seropositive patients (247). Moreover, antibodies against peripheral neural tissues were found in serum from most HIV-infected patients with peripheral neuropathy, and plasmapheresis had beneficial effects in some cases (248). However, whether these findings truly reflect a pathogenetic event or simply represent an epiphenomenon of neural tissue damage due to different factors is uncertain.

5. AUTOIMMUNE PHENOMENA POSSIBLY RELATED TO AIDS PATHOLOGY

While the phenomena described above contribute, in many instances, to clinical disease and health deterioration, they do not seem to intervene substantially in AIDS progression. Here, we will address autoimmune phenomena that might be causally involved in the generation of immune deficiency. It is now widely accepted that the often rapid, progressive decrease in immune function cannot be explained by viral cytopathogenicity alone. Indeed, HIV replication causes cytopathic effects on cultured cells, balloon degeneration, multinucleated giant cell formation (syncytia), and eventually cell death (249). However, while in vitro evidence of the cytopathic effect is compelling, the same is not true for in vivo findings. Syncytial pictures are rarely found within lymphoid organs and, when present, are due more to monocyte/macrophage coalescence than lymphocyte fusion (250). In addition, low numbers of infected cells circulate in these individuals (251), even though lymphoid organs and CNS may act as important virus reservoirs (252). Moreover, functional defects in T cell activity, such as early loss of *in vivo* and *in vitro* responses to recall antigens, invariably precede depletion of the circulating CD4$^+$ cell pool (253). These observations prompted many investigators to look for factors that might amplify the very damage induced by the virus (254); autoimmune mechanisms are likely candidates.

5.1. Cellular Mechanisms

The autoimmune events that affect cellular effector mechanisms may involve both infected and non-infected cells. HIV-infected CD4$^+$ lymphocytes can be killed by several mechanisms, including antibody and complement, Fc receptor-bearing (CD16$^+$) cells armed with anti-HIV antibodies (255), and HLA class I-restricted CTL (256–257). This event, however, is a relatively physiological defense mechanism aimed at clearing virus from the organism, and infected cells are presumably destined to die due to the virus' cytopathic effect. It is of much more concern that non-infected CD4$^+$ lymphocytes may also be lysed, thereby amplifying the damage directly induced by the virus. This might occur through at least two different pathways. First, it was demonstrated *in vitro* that CD4$^+$ cells can bind, process, and present soluble gp120 in association with HLA class II products, and thus become targets for class II-restricted CD4$^+$ CTL (142–143). Although no proof of a sizable number of circulating CD4$^+$ CTL in infected patients has been provided so far, the possibility that uninfected CD4$^+$ T cells might be killed by virus-specific unconventional CTL cannot be overlooked. In addition, some studies suggest that most seropositive subjects have circulating autoreactive lymphocytes (258–259). Lymphocytes from these patients were able to lyse autologous as well as allogeneic uninfected CD4$^+$ cells in a non-MHC-restricted manner, an activity that was mediated by cells displaying a CD3$^+$CD8$^+$CD16$^-$ phenotype, thus indicating that NK activity was likely not involved. The mech-anisms underlying this phenomenon, as well as its role in CD4$^+$ cell depletion are still undetermined.

5.2. Humoral Mechanisms

Anti-lymphocyte antibodies. Numerous reports describe the presence of anti-lymphocyte antibodies in the serum of a high fraction of HIV-infected patients (for review, see 94). Such autoantibodies were found to be directed against CD4$^+$ lymphocytes (260–261), but also against CD8$^+$ cells (260,262) and B cells (263). Most reports were limited to an analysis of the phenomenon and its correlation with clinical and immunological parameters, and no attempts were made to identify the putative antigens recognized by the autoantibodies.

In a few cases, however, the antigen was identified, and at least partially characterized. Stricker et al. (264) demonstrated in serum of most AIDS patients an antibody that reacted with an 18-Kd antigen on the surface of HIV-infected and non-infected, activated CD4$^+$ lymphocytes. This autoantibody exerted a cytotoxic effect in the presence of complement, and was able to suppress *in vitro* CD4$^+$ cell proliferation. Although the biochemical nature of the target antigen was not identified, any relationship with HIV antigens could be excluded. Moreover, an autoantibody recognizing a 73-Kd antigen expressed on the surface of both CD4$^+$ and CD8$^+$ cells (and in most instances also on B cells) was also reported (265), and again the molecular nature of the antigen remains to be clarified.

Many reports documented antibodies recognizing HLA class II antigens in the serum of infected individuals due to molecular mimicry between class II products and HIV antigens. A strong homology between the sequence of HIV gp41 and DR antigens was found to elicit autoantibodies in about 1/3 of seropositive patients (266). These antibodies were able to inhibit T cell responses to antigens, and HLA class II$^+$ cells were killed *in vitro* in the presence of patients' serum and complement (267).

Antibodies reacting against CD4 in HIV infection could theoretically arise directly, or through the generation of anti-idiotypes to antibodies directed against the CD4-binding site of gp120. However, experimental findings do not support these assumptions. Anti-CD4 antibodies in serum of HIV-infected individuals were detected by ELISA, but they were not able to recognize native CD4 on the cell surface (268–269). Autoantibodies against native CD4 were demonstrated in serum from a few infected individuals (270), but they recognized an intracytoplasmic domain of CD4 instead of the extracellular portion and thus probably represent a mere epiphenomenon in which epitopes normally hidden to the immune system and released in body fluids following increased cell death, are recognized. A similar mechanism may explain the recent observation of anti-CD4 cellular reactivity, which correlates with poor prognosis and disease progression in a sizable portion of HIV-infected patients (271).

It was shown that sera from about 50% of seropositive subjects contained autoantibodies capable of recognizing the CD43 antigen in its partially sialylated form on normal thymocytes, but not on peripheral blood T cells (272). These antibodies could originate from an HIV-induced alteration in the glycosylation pattern of the molecule (273). In any case, in view of the importance of CD43 in T cell ontogeny, these antibodies could play a significant role by interfering in the normal intrathymic T cell maturation process. Indeed, lymphocytes from children with Wiskott-Aldrich syndrome (a congenital immunodeficiency disease sharing many features with AIDS) express reduced amounts of this antigen (274).

Finally, a molecular mimicry between gp120 and the Fas protein was demonstrated (275). This structural homology could be very important in view of two observations: 1) the role played by the Fas receptor in apoptotic cell death and 2) the extremely high rate of apoptosis in circulating lymphocytes from HIV-infected patients, a phenomenon that involves both CD4$^+$ and CD8$^+$ cells (102). It is conceivable that anti-gp120 antibodies directed against epitopes cross-reacting with Fas could also bind to the latter, triggering the apoptotic process into non-infected, innocent bystander T cells.

In any case, the biologic significance of anti-lymphocyte antibodies is still unclear, as no strict correlation between disease progression and autoantibody presence has been documented. Anti-lymphocyte antibodies could simply represent an epiphenomenon of the disease, and reflect an aberrant immune response to epitopes not recognized under normal circumstances.

Anti-cytokine antibodies.   Interleukin-2 (IL-2), a crucial molecule in T cell responses, shows a sequence homology with HIV gp41 (276); antibodies against this HIV peptide also bind to IL-2 (277). Antibodies reacting with human IL-2 were found in sera from infected individuals (278). These findings imply that HIV infection could trigger an autoimmune response against IL-2, but further data are needed to understand the immunopathologic significance of this phenomenon in the natural history of HIV infection.

High serum levels of antibodies against IFN-γ were observed in seropositive patients (279). Such antibodies most likely were not directed against the active site of the molecule, as they did not neutralize its antiviral effect in a biological assay; thus, their contribution to immune deficiency generation is presently unclear.

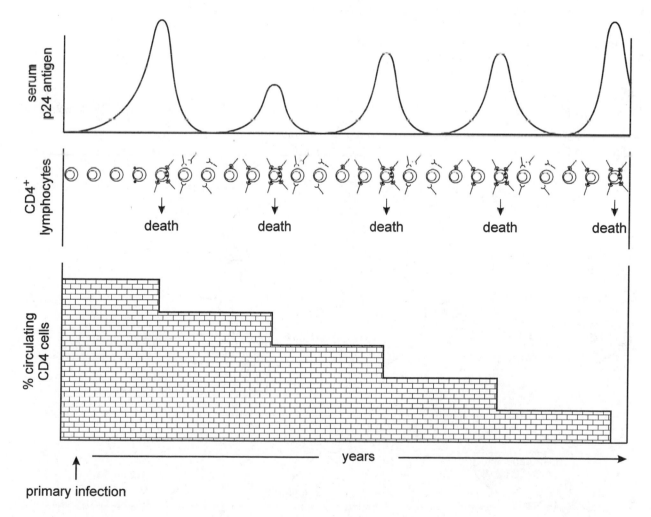

Figure 4.   Schematic representation of the events occurring during HIV infection. In correspondence to HIV antigen peaks, CD4$^+$ lymphocytes are covered by gp120/anti-gp120 antibody complexes, and could undergo death through the mechanisms described in the text.

5.3. Generation of Anti-CD4 Reactivity

That autoreactivity against lymphocytes may also be generated through indirect, non-conventional mechanisms during HIV infection is a thought-provoking possibility. This novel model of autoreactivity envisions the binding to CD4 of gp120/anti-gp120 antibody complexes, that mimic the activity of anti-CD4 antibodies.

Experimental evidence of the *in vitro* and *in vivo* activity of anti-CD4 antibodies is abundant. Both human and murine anti-CD4-treated cells *in vitro* become insensitive to subsequent stimulation via the CD3/TCR complex (280). The demonstration that anti-CD4-treated lymphocytes undergo endonuclease activation and cell death through apoptosis following stimulation via the CD3/TCR complex elucidated the mechanisms underlying this phenomenon (281). Anti-CD4 treatment *in vivo* is associated with unresponsiveness to co-administered antigens (282), and CD4 manipulation with monoclonal antibodies is now a very powerful means of interfering with immune function in several settings, particularly organ transplantation.

This model of anti-CD4-mediated immunosuppression might be relevant to AIDS pathogenesis. From a theoretical point of view, it is likely that gp120/anti-gp120 binding to the CD4⁺ cell surface occurs in seropositive individuals. Free gp120 in the serum of infected patients has not yet been demonstrated, but gp120 molecules are probably shed into body fluids by the replicating virus, analogous with the core p24 protein. Given the extremely high affinity of the CD4/gp120 interaction, it is also likely that most shed gp120 molecules bind rapidly to non-infected CD4⁺ lymphocytes. Subsequently, gp120-coated cells become the very target of anti-gp120 antibodies, which abound in the serum even in the final stages of full-blown AIDS.

In this setting, whether gp120/anti-gp120 complexes mimic anti-CD4 reactivity, and whether this event actually takes place in patients is a moot question. Ample evidence from our and other laboratories indicates that both occur. In reference to the former, *in vitro* incubation of normal CD4⁺ lymphocytes with gp120 and anti-gp120 antibodies was associated with profound down-regulation of CD4 expression, and gp120/anti-gp 120-coated cells were no longer able to proliferate in response, to mitogenic signals via CD3 (283). Regarding the second issue, several investigators obtained data consistent with the presence of gp120/anti-gp120 complexes on the surface of CD4⁺ lym-

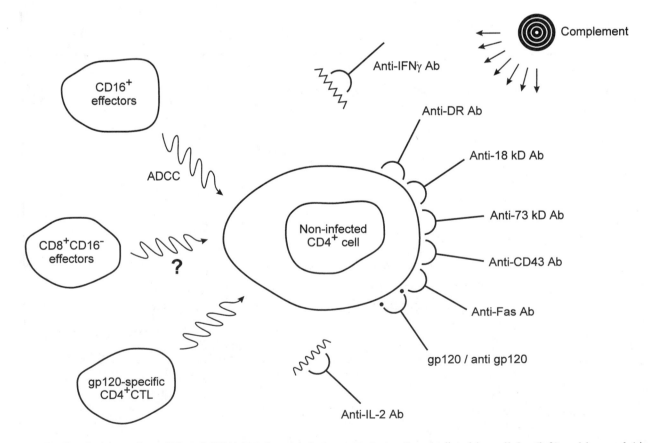

Figure 5. Involvement of non-infected CD4⁺ lymphocytes in immune destruction mediated by cellular (left) and humoral (right) autoimmune or immunopathologic mechanisms.

phocytes *in vivo* (284–285). Interestingly, the gp120/anti-gp120 coating of lymphocytes is down-regulated by anti-retroviral treatment (286).

gp120/anti-gp120 binding to CD4+ lymphocytes may be crucial in amplifying virus-induced immune damage (Fig. 4), and could explain the rapid loss of CD4+ cells that accompanies antigen peaks. In fact, gp120/anti-gp120-coated cells might undergo apoptosis and death following activation by antigens concomitant with HIV antigen peaks; it is known that highly replicating strains are associated with CD4+ cell depletion and clinical progression (28–29). Obviously, gp120/anti-gp120-coated cells also become suitable targets for the attack of complement or CD16+ killer cells. In addition, in view of the very powerful effects of CD4 manipulation, this event could also explain why T cell function alteration is often seen in patients in whom circulating CD4+ cells are not decreased (287). In any case, non-infected CD4+ T lymphocytes may be represented in the middle of a coordinate attack by several autoimmune mechanisms (summarized in Fig. 5) that may strongly influence their function or directly cause immune-mediated destruction, thus amplifying the immunological damage directly provoked by HIV infection.

References

1. Barré-Sinoussi, F., J.C. Chermann, F. Rey, M.T. Nugeyre, S. Chamaret, J. Gruest, C. Dauguet, C. Axler-Blin, F. Vezinet-Brun, C. Rouzioux, W. Rosenbaum, and L. Montagnier. 1983. Isolation of a T-lymphotropic retrovirus from a patient at risk for acquired immune deficiency syndrome (AIDS). *Science* 220:868–871.

2. Gallo, R.C., P.S. Sarin, E.P. Gelmann, M. Robert-Guroff, E. Richardson, V.S. Kalyanaraman, D. Mann, G.D. Sidhu, R.E. Stahl, S. Zolla-Pazner, J. Leibovitch, and M. Popovic. 1983. Isolation of human T-cell leukemia virus in acquired immune deficiency syndrome (AIDS). *Science* 220:865–867.

3. Ho, D.D., R.J. Pomerantz, and J.C. Kaplan. 1987. Pathogenesis of infection with human immunodeficiency virus. *N. Engl. J. Med.* 317:278–286.

4. Clavel, F., K. Mansinho, S. Chamaret, D. Guetard, V. Favier, J. Nina, M.O. Santos-Ferreira, J.L. Champalimaud, and L. Montagnier. 1987. Human immunodeficiency type 2 infection associated with AIDS in West Africa. *N. Engl. J. Med.* 316:1180–1185.

5. Brun-Vezinet, F., M.A. Rey, C. Katlama, P.M. Girard, D. Roulot, P. Yeni, L. Lenoble, F. Clavel, M. Alizon, and S. Gadelle. 1987. Lymphadenopathy-associated virus type 2 in AIDS and AIDS-related complex: clinical and virological features in four patients. *Lancet* 1:128–132.

6. Smith, T.F., A. Srinivasan, G. Schochetman, M. Marcus, and G. Myers. 1988. The phylogenetic history of immunodeficiency viruses. *Nature* 333:573–575.

7. Wong-Staal, F. 1989. Molecular biology of human immunodeficiency viruses. In: Current Topics in AIDS, M. Gottlieb, ed, J. Wiley and Sons.

8. Levy, J.A. 1988. The transmission of AIDS: the case of the infected cell. *J. Am. Med. Assoc.* 259:3037–3038.

9. Bednarik, D.P., J.D. Mosca, and N.B. Raj. 1987. Methylation as a modulator of expression of human immunodeficiency virus. *J. Virol.* 61:1253–1257.

10. Luciw, P.A., C. Cheng-Mayer, and J.A. Levy. 1987. Mutational analysis of the human immunodeficiency virus (HIV): the orf-B region down-regulates virus replication. *Proc. Natl. Acad. Sci. USA* 84:1434–1438.

11. Lifson, J.D., G.G. Reyes, M.S. McGrath, B.S. Stein, and E.G. Engleman. 1986. AIDS retrovirus-induced cytopathology: giant cell formation and involvement of CD4 antigen. *Science* 232:1123–1126.

12. Maddon, P.J., A.G. Dalgleish, J.S. McDougal, P.R. Clapham, R.A. Weiss, and R. Axel. 1986. The T4 gene encodes the AIDS virus receptor and is expressed in the immune system and the brain. *Cell* 47:333–346.

13. Fisher, R.A., J.M. Bertonis, W. Meier, V.A. Johnson, D.S. Costopoulos, T. Liu, R. Tizard, B.D. Walker, M.S. Hirsch, and R.T. Schooley. 1988. HIV infection is blocked *in vitro* by recombinant soluble CD4. *Nature* 331:76–78.

14. Moore, J.P., A. Trikola, and T. Dragic. 1997. Coreceptors for HIV-1 entry. *Curr. Opin. Immunol.* 9:551–562.

15. Alkhatib, G., C. Combadiere, C.C. Broder, Y. Feng, P.E. Kennedy, P.M. Murphy, and E.A. Berger. 1996. CC-CR5, RANTES, MIP-1α, MIP-1β receptor as a fusion cofactor for macrophage-tropic HIV-1, *Science* 272:1955–1958.

16. Choe, H., M. Farzan, Y. Sun, N. Sullivan, B. Rollins, P.D. Ponath, L. Wu, C.R. MacKay, G. LaRosa, W. Newman, N. Gerard, and J. Sodroski. 1996. The β chemokine receptors CCR3 and CCR5 facilitate infection by primary HIV-1 isolates. *Cell* 85:1135–1148.

17. Deng, H., R. Liu, W. Ellmeier, S. Choe, D. Unutmaz, M. Burkhart, P. Di Marzio, S. Marmon, R.E. Sutton, C.M. Hill, C.B. Davis, S.C. Peiper, T.J. Schall, D.R. Littman, and N.R. Landau. 1996. Identification of a major coreceptor for primary isolates of HIV-1. *Nature* 381:661–666.

18. Doranz, B.J., J. Rucker, Y. Yi, R.J. Smith, M. Samson, S.C. Peiper, M. Parmentier, R.G. Collman, and R.W. Doms. 1996. A dual-tropic primary HIV-1 isolate that uses fusin and the β-chemokine receptors CKR-5, CKR-3 and CKR-2b as fusion cofactors. *Cell* 85:1149–1158.

19. Bleul, C.C., M. Farzan, H. Choe, C. Parolin, I. Clark-Lewis, J. Sodroski, and T.A. Springer. 1996. The lymphocyte chemoattractant SDF-1 is a ligand for LESTR/fusin and blocks HIV-1 entry. *Nature* 382:829–833.

20. Dragic, T., V. Litwin, G.P. Allaway, S.R. Martin, Y. Huang, K.A. Nagashima, C. Cayanan, P.J. Maddon, R.A. Koup, J.P. Moore, and W.A. Paxton. 1996. HIV-1 entry into CD4+ cells is mediated by the chemokine receptor CC-CKR5. *Nature* 381:667–773.

21. Zhang, L., Y. Huang, T. He, Y. Cao, and D.D. Ho. 1996. HIV-1 subtype and second receptor use. *Nature* 383:768.

22. Dittmar, M.T., A. McKnight, G. Simmons, P.R. Clapham, R.A. Weiss, and P. Simmonds. 1997. HIV-1 tropism and coreceptor use. *Nature* 385:495–496.

23. Bolognesi, D.P. 1989. Do antibodies enhance the infection of cell by HIV? *Nature* 340:431–432.

24. Homsy, J., M. Meyer, M. Tateno, S. Clarkson, and J.A. Levy. 1989. The Fc and not CD4 receptor mediates antibody enhancement of HIV infection in human cells. *Science* 244:1357–1360.

25. Lund, O., J. Hansen, A.M. Soorensen, E. Mosekilde, J.O. Nielsen, and J.E. Hansen. 1995. Increased adhesion as a mechanism of antibody-dependent and antibody-decreased complement-mediated enhancement of human immunodeficiency virus infection. *J. Virol.* 69:2393–2400.

26. Ho, D.D., A.U. Neumann, A.S. Perelson, W. Chen, J.M. Leonard, and M. Markowitz. 1995. Rapid turnover of plasma virions and CD4 lymphocytes in HIV-1 infection. *Nature* 373:123–126.

27. Cheng-Mayer, C., and J.A. Levy. 1988. Distinct biologic and serologic properties of HIV isolated from the brain. *Ann. Neurol.* 23:583–615.

28. Asjo, B., L. Morfeldt-Manson, J. Albert, G. Biberfeld, A. Karlsson, K. Lidman, and E.M. Fenyo. 1986. Replicative capacity of human immunodeficiency virus from patients with varying severity of HIV infection. *Lancet* ii:660–662.

29. Cheng-Mayer, C., D. Seto, M. Tateno, and J.A. Levy. 1988. Biologic features of HIV-1 that correlate with virulence in the host. *Science* 240:80–82.

30. Tersmette, M., R.A. Gruters, F. de Wolf, R.E. de Goede, J.M. Lange, P.T. Schellekens, J. Goudsmit, H.G. Huisman, and F. Miedema. 1989. Evidence for a role of virulent HIV variants in the pathogenesis of AIDS obtained from studies on a panel of sequential HIV isolates. *J. Virol.* 63:2118–2125.

31. Fauci, A.S. 1996. Host factors and the pathogenesis of HIV-induced disease. *Nature* 384:529–533.

32. Steel, C.M., C.A. Ludlam, D. Beatson, J.F. Peutherer, R.J. Cuthbert, P. Simmonds, H. Morrison, and M. Jones. 1988. HLA haplotype A1 B8 DR3 as a risk factor for HIV-related disease. *Lancet* i:1185–1188.

33. Liu, S.L., T. Schacker, L. Musey, D. Shriner, M.J. McElrath, L. Corey, and J.I. Mullins. 1997. Divergent patterns of progression to AIDS after infection from the same source: human immunodeficiency virus type 1 evolution and antiviral responses. *J. Virol.* 71:4284–4295.

34. Pollack, M.S., J. Gold, C.E. Metroka, B. Safai, and B. Dupont. 1984. HLA-A, B, C and DR antigen frequencies in acquired immunodeficiency syndrome (AIDS) patients with opportunistic infections. *Hum. Immunol.* 11:99–103.

35. Raffoux, C., V. David, L.D. Couderc, C. Rabian, J.P. Clauvel, M. Seligmann, and J. Colombani. 1987. HLA-A, B and DR antigen frequencies in patients with AIDS-related persistent generalized lymphadenopathy (PGL) and thrombocytopenia. *Tissue Antigens* 29:60–62.

36. Mann, D.L., R.P. Garner, D.E. Dayhoff, K. Cao, M.A. Fernandez-Vina, C. Davis, N. Aronson, N. Ruiz, D.L. Birx, and N.L. Michael. 1998. Major histocompatibility complex genotype is associated with disease progression and virus load levels in a cohort of human immunodeficiency virus type 1-infected Caucasians and African Americans. *J. Infect. Dis.* 178:1799–1802.

37. Fabio, G., R.S. Smeraldi, A. Gringeri, M. Marchini, P. Bonara, and P.M. Mannucci. 1990. Susceptibility to HIV infection and AIDS in Italian hemophiliacs is HLA associated. *Br. J. Haematol.* 75:531–536.

38. Mann, D.L., C. Murray, R. Yarchoan, W.A. Blattner, and J.J. Goedert. 1988. HLA antigen frequencies in HIV-1 seropositive disease-free individuals and patients with AIDS. *J. AIDS* 1:13–17.

39. Dean, M., M. Carrington, C. Winkler, G.A. Huttley, M.W. Smith, R. Allikmets, J.J. Goedert, S.P. Buchbinder, E. Vittinghoff, E. Gomperts, S. Donfield, D. Vlahov, R. Kaslow, A. Saah, C. Rinaldo, R. Detels, and S.J. O'Brien. 1996. Genetic restriction of HIV-1 infection and progression to AIDS by a deletion allele of the CKR5 structural gene. *Science* 273:1856–1862.

40. Huang, Y., W.A. Paxton, S.M. Wolinsky, A.U. Neumann, L. Zhang, T. He, S. Kang, D. Ceradini, Z. Jin, K. Yazdanbakhsh, K. Kunstman, D. Erickson, E. Dragon, N.R. Landau, J. Phair, D.D. Ho, and R.A. Koup. 1996. The role of a mutant CCR5 allele in HIV-1 transmission and disease progression. *Nature Med.* 2:1240–1243.

41. Liu, R., W.A. Paxton, S. Choe, D. Ceradini, S.R. Martin, R. Horuk, M.E. MacDonald, H. Stuhlmann, R.A. Koup, and N.R. Landau. 1996. Homozygous defect in HIV-1 coreceptor accounts for resistance of some multiply-exposed individuals to HIV-1 infection. *Cell* 86:367–377.

42. Samson, M., F. Libert, B.J. Doranz, J. Rucker, C. Liesnard, C.M. Farber, S. Saragosti, C. Lapoumeroulie, J. Cognaux, C. Forceille, G. Muyldermans, C. Verhofstede, G. Burtonboy, M. Georges, T. Imai, S. Rana, Y. Yi, R.J. Smyth, R.G. Collman, R.W. Doms, G. Vassart, and M. Parmentier. 1996. Resistance to HIV-1 infection in Caucasian individuals bearing mutant alleles of the CCR-5 chemokine receptor gene. *Nature* 382:722–725.

43. Biti, R., R. French, J. Young, B. Bennetts, G. Stewart, and T. Liang. 1997. HIV-1 infection in an individual homozygous for the CCR5 deletion allele. *Nature Med.* 3:252–253.

44. O'Brien, T.R., C. Winkler, M. Dean, J.A. Nelson, M. Carrington, N.L. Michael, and G.C. White 2nd, 1997. HIV-1 infection in a man homozygous for CCR5. *Lancet* 349:1219.

45. Theodorou, I., L. Meyer, M. Magierowska, C. Katlama, and C. Rouzioux. 1997. HIV-1 infection in an individual homozygous for CCR5 δ 32. Seroco Study group. *Lancet* 349:1219–1220.

46. Amadori, A., R. Zamarchi, G. De Silvestro, G. Forza, G. Cavattoni, G.A. Danieli, M. Clementi, and L. Chieco-Bianchi. 1995. Genetic control of the CD4/CD8 T-cell ratio in humans. *Nature Med.* 1:1279–1283.

47. Amadori, A., R. Zamarchi, and L. Chieco-Bianchi. 1996. CD4:CD8 ratio and HIV infection: the "tap-and-drain" hypothesis. *Immunol. Today* 17:414–417.

48. Fenyö, E.M., J. Albert, and J. McKeating. 1996. The role of the humoral immune response in HIV infection. *AIDS* 10:S97–S106.

49. Burton, D.R., and D.C. Montefiori. 1997. The antibody response in HIV-1 infection. *AIDS* 11:S87–S98.

50. Jehuda-Cohen, T., B.A. Slade, J.D. Powell, F. Villinger, B. De, T.M. Folks, H.M. McClure, K.W. Sell, and A. Ahmed-Ansari. 1990. Polyclonal B-cell activation reveals antibodies against human immunodeficiency virus type 1 (HIV-1) in HIV-1 seronegative individuals. *Proc. Natl. Acad. Sci. USA* 87:3972–3976.

51. Lange, J.M.A., D.A. Paul, H.G. Huisman, F. de Wolf, H. van den Berg, R.A. Coutinho, S.A. Danner, J. van der Noordaa, and J. Goudsmit. 1986. Persistent HIV antigenaemia and decline of HIV core antibodies associated with transition to AIDS. *Br. Med. J.* 293:1459–1463.

52. Tsiquaye, K.N., M. Youle, and A.C. Chanas. 1988. Restriction of sensitivity of HIV-1 antigen ELISA by serum anti-core antibodies. *AIDS* 2:41–45.

53. Amadori, A., and L. Chieco-Bianchi. 1990. B-cell activation and HIV-1 infection: deeds and misdeeds. *Immunol. Today* 11:374–379.

54. Birx, D.L., R.R. Redfield, and G. Tosato. 1986. Defective regulation of Epstein-Barr virus infection in patients with acquired immunodeficiency syndrome (AIDS) or AIDS-related disorders. *N. Engl. J. Med.* 314:874–879.

55. Amadori, A., A. De Rossi, G.P. Faulkner-Valle, and L. Chieco-Bianchi. 1988. Spontaneous *in vitro* production of virus-specific antibody by lymphocytes from HIV-infected subjects. *Clin. Immunol. Immunopathol.* 46:342–351.

56. Amadori, A., R. Zamarchi, V. Ciminale, A. Del Mistro, S. Siervo, A. Alberti, M. Colombatti, and L. Chieco-Bianchi.

1989. HIV-1-specific B cell activation: a major constituent of spontaneous B cell activation during HIV-1 infection. *J. Immunol.* 143:2146–2152.

57. Pahwa, S., N. Chirmule, C. Leombruno, W. Lim, R. Harper, R. Bhalla, R. Pahwa, R.P. Nelson, and R.A. Good. 1989. *In vitro* synthesis of human immunodeficiency virus-specific antibodies in peripheral blood lymphocytes of infants. *Proc. Natl. Acad. Sci. USA* 86:7532–7536.

58. Arvin, A.M. 1996. Varicella-zoster virus. *Clin. Microbiol. Rev.* 9:361–381.

59. Gauduin, M-C., P.W. Parren, R. Weir, C.F. Barbas, D.R. Burton, and R.A. Koup. 1997. Passive immunization with a potent neutralizing human monoclonal antibody of HU-PBL-SCID mice against challenge by primary isolates of human immunodeficiency virus type 1. *Nature Med.* 3:1389–1393.

60. Haigwood, N.L., and S. Zolla-Pazner. 1998. Humoral immunity to HIV, SIV, and SHIV. *AIDS* 12:S121–S132.

61. Sheppard, H.W., W. Lang, M.S. Ascher, E. Vittinghoff, and W. Winkelstein. 1993. The characterization of non-progressors: long-term HIV-1 infection with stable CD4+ T-cell levels. *AIDS* 7:1159–1166.

62. Montefiori, D.C., G. Pantaleo, L.M. Fink, J.T. Zhou, J.Y. Zhou, M. Bilska, G.D. Miralles, and A.S. Fauci. 1996. Neutralzing and infection-enhancing antibody responses to human immunodeficiency virus type 1 in long term non-progressors. *J. Infect. Dis.* 173:60–67.

63. Gotch, F.M., R.A. Koup, and J.T. Safrit. 1997. New observations on cellular immune responses to HIV and T-cell epitopes. *AIDS* 11:S99–S107.

64. Johnson, R.P., R.F. Siliciano, and M.J. McElrath. 1998. Cellular immune responses to HIV-1. *AIDS* 12:S113–S120.

65. Wahren, B., L. Morfeldt-Manson, G. Biberfeld, L. Moberg, A. Sonneborg, P. Ljungman, A. Werner, R. Kurth, R. Gallo, and D. Bolognesi. 1987. Characteristics of the specific cell-mediated immune responses in human immunodeficiency virus infection. *J. Virol.* 61:2017–2023.

66. Krowka, J.F., D.P. Stites, S. Jain, K.S. Steimer, C. George-Nascimento, A. Gyenes, P.J. Barr, H. Hollander, A.R. Moss, and J.M. Homsy. 1989. Lymphocyte proliferative responses to human immunodeficiency virus antigens *in vitro*. *J. Clin. Invest.* 83:1198–1203.

67. Walker, C.M., D.J. Moody, D.P. Stites, and J.A. Levy. 1986. CD8+ lymphocytes can control HIV infection *in vitro* by suppressing viral replication. *Science* 234:1563–1566.

68. Cocchi, F., A.L. De Vico, A. Garzino-Demo, S.K. Arya, R.C. Gallo, and P. Lusso. 1995. Identification of RANTES, MIP-1α, and MIP-1β as the major HIV-immunosuppressive factors produced by CD8+ T cells. *Science* 270:1811–1815.

69. Furci, L., G. Scarlatti, S. Burastero, G. Tambussi, C. Colognesi, C. Quillent, R. Longhi, P. Loverro, B. Borgonovo, D. Gaffi, E. Carrow, M. Malnati, P. Lusso, A.G. Siccardi, A. Lazzarin, and A. Beretta. 1997. Antigen-driven C-C chemokine-mediated HIV-1 suppression by CD4+ T cells from exposed uninfected individuals expressing the wild-type CCR-5 allele. *J. Exp. Med.* 186;455–460.

70. Harrer, T., E. Harrer, S.A. Kalams, T. Elbeik, S.I. Staprans, M.B. Feinberg, Y. Cao, D.D. Ho, T. Yilma, A.M. Caliendo, R.P. Johnson, S.P. Buchbinde, and B.D. Walker. 1996. Strong cytotoxic T cell and weak neutralizing antibody responses in a subset of persons with stable nonprogressing HIV type 1 infection. *AIDS Res. Hum. Retroviruses* 12:585–592.

71. Greenough, T.C., D.B. Brettler, M. Somasundaran, D.L. Panicali, and J.L. Sullivan. 1997. Human immunodeficiency virus type 1-specific cytotoxic T lymphocytes (CTL), virus load, and CD4 T cell loss: evidence supporting a protective role for CTL *in vivo*. *J. Infect. Dis.* 176:118–125.

72. Rowland-Jones, S., J. Sutton, K. Aryoshi, T. Dong, F. Gotch, S. McAdam, D. Whitby, S. Sabally, A. Gallimore, and T. Corrah. 1995. HIV-specific cytotoxic T-cells in HIV-exposed but uninfected Gambian women. *Nature Med.* 1:59–64.

73. Schwartz, O., V. Marechal, S.S. Le Gall, F. Lemonnier, and J.M. Heard. 1996. Endocytosis of major histocompatibility complex class I molecules is induced by the HIV-1 Nef protein. *Nature Med.* 2:338–342.

74. Kerkau, T., I. Bacik, J.R. Bennink, J.W. Yewdell, T. Hunig, A. Schimpl, and U. Schubert. 1997. The human immunodeficiency virus type 1 (HIV-1) Vpu protein interferes with an early step in the biosynthesis of major histocompatibility complex (MHC) class I molecules. *J. Exp. Med.* 185:1295–1305.

75. Bevan, M.J., and T.J. Braciale. 1995. Why can't cytolytic T cells handle HIV? *Proc. Natl. Acad. Sci. USA* 92:5765–5767.

76. McMichael, A.J., and R.E. Phillips. 1997. Escape of human immunodeficiency virus from immune control. *Ann. Rev. Immunol.* 15:271–296.

77. Gendelman, H.E., J.M. Orenstein, M.A. Martin, C. Ferrua, R. Mitra, T. Phipps, L.A. Wahl, H.C. Lane, A.S. Fauci, and D.S. Burke. 1988. Efficient isolation and propagation of human immunodeficiency virus on recombinant colony-stimulating factor 1-treated monocytes. *J. Exp. Med.* 167:1428–1441.

78. Koyanagi, Y., W.A. O'Brien, J.Q. Zhao, D.W. Golde, J.C. Gasson, and I.S. Chen. 1988. Cytokines alter production of HIV-1 from primary mononuclear phagocytes. *Science* 241:1673–1675.

79. Poli, G., A. Kinter, J.S. Justement, J.H. Kehrl, P. Bressler, S. Stanley, and A.S. Fauci. 1990. Tumor necrosis factor alfa functions in an autocrine manner in the induction of human immunodeficiency virus expression. *Proc. Natl. Acad. Sci. USA* 87:782–785.

80. Poli, G., P. Bressler, A. Kinter, E. Duh, W.C. Timmer, A. Rabson, J.S. Justement, S. Stanley, and A.S. Fauci. 1990. Interleukin-6 induces human immunodeficiency virus expression in infected monocytic cells alone and in synergy with tumor necrosis factor alfa by transcriptional and post-transcriptional mechanisms. *J. Exp. Med.* 172:151–158.

81. Ho, D.D., K.L. Hartshorn, T.R. Rota, C.A. Andrews, J.C. Kaplan, R.T. Schooley, and M.S. Hirsch. 1985. Recombinant human interferon alfa suppresses HTLV-III replication *in vitro*. *Lancet* ii:602–604.

82. Poli, G., A.L. Kinter, J.S. Justement, P. Bressler, J.H. Kehrl, and A.S. Fauci. 1991. Transforming growth factor beta suppresses human immunodeficiency virus expression and replication in infected cells of the monocyte/macrophage lineage. *J. Exp. Med.* 173:589–597.

83. Peterson, P.K., G. Gekker, C.C. Chao, R. Schut, T.W. Molitor, and H.H. Balfour Jr. 1991. Cocaine potentiates HIV-1 replication in human peripheral blood mononuclear cell co-cultures. *J. Immunol.* 146:81–84.

84. Clerici, M., and G.M. Shearer. 1993. A Th1-Th2 switch is a critical step in the etiology of HIV infection. *Immunol. Today* 14:107–111.

85. Mosmann, T.R., and R.I. Coffman. 1987. Two types of mouse T helper cell clone: implication for immune regulation. *Immunol. Today* 8:223–226.

86. Romagnani, S. 1991. Th1 and Th2 subsets: doubt no more. *Immunol. Today* 12:256–257.

87. Clerici, M., and J.A. Berzofsky. 1994. Cellular immunity and cytokines in HIV infection. *AIDS* 8:S175–S182.

88. Kelso, A. 1995. Th1 and Th2 subsets: paradigms lost? *Immunol. Today* 16:374–377.

89. Phillips, P.E., and C.L. Christian. 1985. Infectious agents in chronic rheumatic disease. In: Arthritis and Allied Conditions, D.J. McCarthy Ed., 10th ed., Philadelphia, Lea and Febiger.

90. Saag, M.S., and J.C. Bennet. 1987. The infectious etiology of chronic rheumatic diseases. *Sem. Arthritis Rheum.* 17:1–23.

91. Krieg, A.M., and A.D. Steinberg. 1990. Retroviruses and autoimmunity. *J. Autoimmun.* 3:137–166.

92. Oldstone, M.B.A. 1987. Molecular mimicry and autoimmune disease. *Cell* 50:819–820.

93. Hausmann, S., and K.W. Wucherpfennig. 1997. Activation of autoreactive T cells by peptides from human pathogens. *Curr. Opin. Immunol.* 9:831–838.

94. Silvestris, F., R.C. Williams Jr., and F. Dammacco. 1995. Autoreactivity in HIV-1 infection: the role of molecular mimicry. *Clin. Immunol. Immunopathol.* 75:197–205.

95. Gaulton, G.N., and M.I. Greene. 1989. Inhibition of cellular DNA synthesis by reovirus occurs through a receptor-linked signaling pathway that is mimicked by anti-idiotypic, anti-receptor antibody. *J. Exp. Med.* 169:197–211.

96. Root-Bernstein, R.S. 1995. Preliminary evidence for idiotype-antiidiotype immune complexes cross-reactive with lymphocyte antigens in AIDS and lupus. *Med. Hypotheses* 44:20–27.

97. Breindl, M., K. Harbers, and R. Jaenisch. 1984. Retrovirus-induced lethal mutation in collagen I gene of mice is associated with altered chromatin structure. *Cell* 38:9–16.

98. Stoye, J.P., S. Fenner, G.E. Greenoak, C. Moran, and J.M. Coffin. 1988. Role of endogenous retroviruses as mutagens: the hairless mutation of mice. *Cell* 54:383–391.

99. Kazazian, H.H., C. Wong, H. Youssoufian, A.F. Scott, D.G. Phillips, and S.E. Antonarakis. 1988. Hemophylia A resulting from *de novo* insertion of L1 sequences represents a novel model for mutation in man. *Nature* 332:164–166.

100. Morse, B., P.G. Rothberg, V.J. South, J.M. Spandorfer, and S.M. Astrin. 1988. Insertional mutagenesis of the *myc* locus by a LINE-1 seuqence in a human breast carcinoma. *Nature* 333:87–90.

101. Mevorach, D., J.L. Zhou, X. Song, and K.B. Elkon. 1998. Systemic exposure to irradiated apoptotic cells induces autoantibody production. *J. Exp. Med.* 188:387–392.

102. Meyaard, L., S.A. Otto, R.R. Jonker, M.J. Mijnster, R.P. Keet, and F. Miedema. 1992. Programmed death of T cells in HIV-1 infection. *Science* 257:217–219.

103. Theophilopoulos, A.N., and F.J. Dixon. 1979. The biology and detection of immune complexes. *Adv. Immunol.* 28:89–220.

104. Euler, H.H., P. Kern, H. Loffler, and M. Dietrich. 1985. Precipitable immune complexes in healthy homosexual men, acquired immune deficiency syndrome and the related lymphadenopathy syndrome. *Clin. Exp. Immunol.* 40:515–524.

105. McDougal, J.S., M. Hubbard, J.K. Nicholson, B.M. Jones, R.C. Holman, J. Roberts, D.B. Fishbein, H.W. Jaffe, J.E. Kaplan, and T.J. Spira. 1985. Immune complexes in the acquired immunodeficiency syndrome (AIDS): relationship to disease manifestations, risk groups, and immunologic defect. *J. Clin. Immunol.* 5:130–138.

106. Morrow, W.J.W., M. Wharton, R.B. Stricker, and J.A. Levy. 1986. Circulating immune complexes in patients with acquired immune deficiency syndrome contain the AIDS-associated retrovirus. *Clin. Immunol. Immunopathol.* 40:515–524.

107. Lasky, L.A., G. Nakamura, D.H. Smith, C. Fennie, C. Shimasaki, E. Patzer, P. Berman, T. Gregory, and D.J. Capon. 1987. Delineation of a region of the human immunodeficiency virus type 1 ·gp 120 glycoprotein critical for interaction with the CD4 receptor. *Cell* 50:975–985.

108. Herrmann, M., M. Neidhart, S. Gay, M. Hagenhofer, and J.R. Kalden. 1998. Retrovirus-associated rheumatic syndromes. *Curr. Opin. Rheumatol.* 10:347–354.

109. Calabrese, L.H. 1998. Rheumatic aspects of human immunodeficiency virus infection and other immunodeficient states. In: Rheumatology, J.H. Klippel and P.A. Dieppe, eds., Mosby.

110. Berman, A., L.R. Espinoza, D.J. Diaz, J.L. Aguilar, T. Rolando, F.B. Vasey, B.F. Germain, and R.F. Lockey. 1988. Rheumatic manifestations of human immunodeficiency virus infection. *Am. J. Med.* 85:59–64.

111. Winchester, R., D.H. Bernstein, H.D. Fisher, R. Enlow, and G. Solomon. 1987. The co-occurrence of Reiter's syndrome and acquired immunodeficiency. *Ann. Intern. Med.* 106:19–26.

112. Wilkinson, N.Z., G.H. Kingsley, J. Sieper, J. Braun, and M.E. Ward. 1998. Lack of correlation between the detection of Chlamydia trachomatis DNA in synovial fluid from patients with a range of rheumatic diseases and the presence of an antichlamydial immune response. *Arthritis Rheum.* 41:845–854.

113. Wilkinson, N.Z., G.H. Kingsley, H.W. Jones, J. Sieper, J. Braun, and M.E. Ward. 1999. The detection of DNA from a range of bacterial species in the joints of patients with a variety of arthritides using a nested, broad-range polymerase chain reaction. *Rheumatology* 38:260–66.

114. van der Heijden, I.M., B. Wilbrink, L.M. Schouls, J.D.A. van Embden, F.C. Breedveld, and P.P. Tak. 1999. Detection of mycobacteria in joint samples from patients with arthritis using a genus-specific polymerase chain reaction and sequence analysis. *Rheumatology* 38:547–553.

115. Espinoza, L.R., J.L. Aguilar, C.G. Espinoza, A. Berman, F. Gutierrez, F.B. Vasey, and B.F. Germain. 1990. HIV-associated arthropathy. HIV antigen demonstration in the synovial membrane. *J. Rheumatol.* 17:1195–1201.

116. Withrington, R.H., P. Cornes, J.R. Harris, M.H. Seifert, E. Berrie, D. Taylor-Robinson, and D.J. Jeffries. 1987. Isolation of human immunodeficiency virus from synovial fluid of a patient with reactive arthritis. *Br. Med. J.* 294:484.

117. Hyer, F.H., and N.L. Gottlieb. 1978. Rheumatic disorders associated with viral infections. *Sem. Arthritis Rheum.* 8:17–31.

118. Cuellar, M.L. 1998. HIV infection-associated inflammatory disorders. *Rheum. Dis. Clin. North Am.* 24:403–421.

119. Njobvu, P., P. McGill, H. Kerr, J. Jellis, and J. Pobee. 1998. SpA and human immunodeficiency virus infection in Zambia. *J. Rheumatol.* 25:1553–1559.

120. Biłeckot, R., A. Mouaya, and M. Makuwa. 1998. Prevalence and clinical presentation of arthritis in HIV-positive patients seen at a rheumatology department in Congo-Brazzaville. *Rev. Rheum. Engl. Ed.* 65:549–554.

121. Berman, A., P. Cahn, H. Perez, A. Spindler, E. Lucero, S. Paz, and L.R. Espinoza. 1999. Human immunodeficiency virus infection-associated arthritis: clinical characteristics. *J. Rheumatol.* 26:1158–1162.

122. Barzilai, A., D. Varon, U. Martinowitz, M. Heim, and S. Schulman. 1999. Characteristics of septic arthritis in human immunodeficiency virus-infected haemophiliacs versus other risk groups. *Rheumatology* 38:139–142.

123. Winchester, R. 1999. Psoriatic arthritis and the spectrum of syndromes related to the SAPHO (synovitis, acne, pustulosis, hyperostosis, and osteitis) syndrome. *Curr. Opin. Rheumatol.* 11:251–256.

124. Lapadula, G., F. Iannone, C. Zuccaro, M. Covelli, and V. Pipitone. 1997. Recovery of erosive rheumatoid arthritis after human immunodeficiency virus-1 infection and hemiplegia. *J. Rheumatol.* 24:747–751.

125. Schewe, C.K., and H. Kellner. 1996. Rapidly progressive seronegative spondyloarthropathy with atlodental subluxation in a pateint with moderately advanced HIV infection. *Clin. Exp. Rheumatol.* 14:83–85.

126. Ornstein, M.H., L.D. Kerr, and H. Spiera. 1995. A reexamination of the relationship between active rheumatoid arthritis and the acquired immunodeficiency syndrome. *Arthritis Rheum.* 38:1701–1706.

127. Hays, A.P., and E.T. Gamboa. 1985. Acute viral myositis. In: Textbook of Rheumatology, W.N. Kelly, E.D. Harris, S. Ruddy, and C.B. Sledge, eds., 2nd ed., Philadelphia: WB Saunders Co.

128. Kaye, B.R. 1989. Rheumatologic manifestations of infection with human immunodeficiency virus (HIV). *Ann. Intern. Med.* 111:158–167.

129. Dalakas, M.C., and G.H. Pezeshkpour. 1988. Neuromuscular diseases associated with human immunodeficiency virus infection. *Ann. Neurol.* 23:S38–S48.

130. Leon-Monzon, M., I. Lamperth, and M.C. Dalakas. 1993. Search for HIV provirus DNA and amplified sequences in the muscle biopsies of patients with HIV polymyositis. *Muscle Nerve* 16:408–413.

131. Dalakas, M.C., M. Gravell, W.T. London, G. Cunningham, and J.L. Sever. 1987. Morphological changes of an inflammatory myopathy in Rhesus monkeys with simian immunodeficiency syndrome. *Proc. Soc. Exp. Biol. Med.* 185:368–376.

132. Gresh, J.P., J.L. Aguilar, and L.R. Espinoza. 1989. Human immunodeficiency virus infection-associated dermatomyositis. *J. Rheumatol.* 16:1397–1402.

133. Schneider, J.H., M.C. Dalakas, K.V. Toyka, G. Said, H.P. Hartung, and R. Gold. 1999. T-cell apoptosis in inflammatory neuromuscular disorders associated with human immunodeficiency virus infection. *Arch. Neurol.* 56:79–83.

134. Till, M., and K.B. McDonell. 1990. Myopathy with HIV-1 infection: HIV-1 or zidovudine? *Ann. Intern. Med.* 113:492–493.

135. Dalakas, M.C., I. Illa, G.H. Pezeshkpour, J.P. Laukaitis, B. Cohen, and J.L. Griffin. 1990. Mitochondrial myopathy caused by long-term zidovudine therapy. *N. Engl. J. Med.* 322:1098–1105.

136. Mandell, B.F., and L.H. Calabrese. 1998. Infection and systemic vasculitis. *Curr. Opin. Rheumatol.* 10:51–57.

137. Guillevin, L. 1999. Virus-associated vasculitides. *Rheumatology* 38:588–590.

138. Calabrese, L.H. 1991. Vasculitis and infection with the human immunodeficiency virus. *Rheum. Dis. Clin. North. Am.* 17:131–147.

139. Calabrese, L.H., M. Esyes, B. Yen-Liebermann, M.R. Proffitt, R. Tubbs, A.J. Fishleder, and K.H. Levin. 1989. Systemic vasculitis in association with the human immunodeficiency virus (HIV) infection. *Arthritis Rheum.* 32:569–576.

140. Anders, K.H., H. Latta, B.S. Chang, U. Tomiyasu, A.S. Quddusi, and V.H. Vinters. 1989. Lymphoid granulomatosis and malignant lymphoma of the central nervous system in acquired immunodeficiency syndrome. *Hum. Pathol.* 20:326–331.

141. Gherardi, R., L. Belec, C. Mhiri, F. Gray, M.C. Lescs, A. Sobel, L. Guillevin, and J. Wechsler. 1993. The spectrum of vasculitis in human immunodeficiency virus-infected patients. A clinicopathologic evaluation. *Arthritis Rheum.* 36:1164–1174.

142. Lanzavecchia, A., E. Roosneck, T. Gregory, P. Berman, and S. Abrignani. 1988. T cells can present antigens such as HIV gp 120 targeted to their own surface molecules. *Nature* 334:530–532.

143. Siliciano, R.F., T. Lawton, C. Knall, R.W. Karr, P. Berman, T. Gregory, and E.L. Reinherz. 1988. Analysis of host-virus interactions in AIDS with anti-gp 120 T cell clones: effect of HIV sequence variation and a mechanism for CD4+ cell depletion: *Cell* 54:561–575.

144. Itescu, S., and R. Winchester. 1992. Diffuse infiltrative lymphocytosis syndrome: disorder occurring in individuals infected with HIV-1 that may present as a sicca syndrome. *Rheum. Dis. Clin. North. Am.* 18:683–697.

145. Itescu, S., J. Dalton, H-Z. Zhang, and R. Winchester. 1993. Tissue infiltration in a CD8 lymphocytosis syndrome associated with HIV infection has a phenotype of an antigen-driven process. *J. Clin. Invest.* 91:2216–2225.

146. Williams, F.M., P.R. Cohen, J. Jumshyd, and J.D. Reveille. 1998. Prevalence of diffuse infiltrative lymphocytosis syndrome among human immunodeficiency virus type 1-positive outpatients. *Arthritis Rheum.* 41:863–868.

147. Kordossis, T., S. Paikos, K. Aroni, P. Kitsanta, A. Dimitrakopoulos, E. Kavouklis, V. Alevizou, P. Kyriaki, F.N. Skopouli, and H.M. Moutsopoulos. 1998. Prevalence of Sjögren's-like syndrome in a cohort of HIV-1-positive patients: descriptive pathology and immunopathology *Br. J. Rheumatol.* 37:691–695.

148. Harris, N.L. 1999. Lymphoid proliferations of the salivary glands. *Am. J. Clin. Pathol.* 111:S94–S103.

149. Venables, P.J.W., and S.P. Rigby. 1997. Viruses in the etiopathogenesis of Sjögren's syndrome. *J. Rheumatol.* 24:3–5.

150. Dwyer, E., S. Itescu, and R. Winchester. 1993. Characterization of the primary structure of the T cell receptor β chains in cells infiltrating the salivary glands in the sicca syndrome of HIV-1 infection. *J. Clin. Invest.* 92:496–501.

151. Itescu, S., S. Rose, E. Dweyer, and R. Winchester. 1994. Certain DR5 and DR6 MHC class II alleles are associated with a CD8 lymphocytic host response to HIV-1 characterized by low lymphocyte viral strain heterogeneity and slow disease progression. *Proc. Nat. Acad. Sci. USA* 91:11472–11476.

152. Itescu, S., S. Rose, E. Dweyer, and R. Winchester. 1995. Different HLA-B locus peptide anchoring pockets may determine outcome in HIV-1 infection. *Hum. Immunol.* 42:81–89.

153. Itescu, S., U. Mathur-Wagh, M.L. Skovron, L.J. Brancato, M. Marmor, A. Zeleniuch-Jaquotte, and R. Winchester. 1992. HLA-B35 is associated with accelerated progression to AIDS. *J. AIDS* 5:37–45.

154. Kopelman, R.H., and S. Zolla-Pazner. 1988. Association of human immunodeficiency virus infection and autoimmune phenomena. *Am. J. Med.* 84:82–88.

155. Fox, R.A., and D.A. Isenberg. 1997. Human immunodeficiency virus infection in systemic lupus erythematosus. *Arthritis Rheum.* 6:1168–1172.

156. Massabki, P.S., C. Accetturi, I.A. Nishie, N.P. da Silva, E.I. Sato, and L.E. Andrade. 1997. Clinical implications of autoantibodies in HIV infection. *AIDS* 11:1845–1850.

157. Weiss, L., G.-F. You, and P. Giral. 1995. Anti-cardiolipin antibodies are associated with anti-endothelial cell antibodies but not with anti-β_2-glycoprotein 1 antibodies in HIV infection. *Clin. Immunol. Immunopathol.* 77:69–74.

158. Gul, A., M. Inanc, G. Yilmaz, L. Ocal, M. Konice, O. Aral, S. Badur, and N. Dilsen. 1996. Antibodies reactive with HIV-1 antigens in systemic lupus erythematosus. *Lupus* 5:120–122.

159. Talal, N., M.J. Dauphinée, H. Dang, S.S. Alexander, D.J. Hart, and R.F. Garry. 1990. Detection of serum antibodies to retroviral proteins in patients with primary Sjögren's syndrome (autoimmune exocrinopathy). *Arthritis Rheum.* 33:774–781.

160. Font, J., J. Vidal, R. Cervera, A. Lopez-Soto, C. Miret, M.T. Jimenez de Anta, and M. Ingelmo. 1995. Lack of relationship between human immunodeficiency virus infection and systemic lupus erythematosus. *Lupus* 4:47–49.

161. Kammerer, R., P. Burgisser, and P.C. Frei. 1995. Anti-human immunodeficiency virus type 1 antibodies of non infected subjects are not related to autoantibodies occurring in systemic diseases. *Clin. Diagn. Lab. Immunol.* 4:458–461.

162. Talal, N., R.F. Garry, P.H. Schur, S. Alexander, M.J. Dauphinée, I.H. Livas, A. Ballester, M. Takei, and H. Dang. 1990. A conserved idiotype in antibodies to retroviral proteins in SLE. *J. Clin. Invest.* 85:1866–1871.

163. De Keyser, F., S.O. Hoch, M. Takei, H. Dang, H. De Keyser, L.A. Rokeach, and N. Talal. 1992. Cross-reactivity of the B/B' subunit of the Sm ribonucleoprotein autoantigen with proline-rich polypeptides. *Clin. Immunol. Immunopathol.* 62:69–74.

164. D'Agati, V., and G.B. Appel. 1998. Renal pathology of human immunodeficiency virus infection. *Sem. Nephrol.* 18:406–421.

165. Schwartz, E.J., and P.E. Klotman. 1998. Pathogenesis of human immunodeficiency virus (HIV)-associated nephropathy. *Sem. Nephrol.* 18:436–445.

166. Cohen, A.H. 1998. HIV-associated nephropathy: current concepts. *Nephrol. Dial. Transplant.* 13:540–542.

167. Winston, J.A., G.C. Burns, and P.E. Klotman. 1998. The human immunodeficiency virus (HIV) epidemic and HIV-associated nephropathy. *Sem. Nephrol.* 18:373–377.

168. Pardo, V., R. Meneses, L. Ossa, D.J. Jaffe, J. Strass, D. Roth, and J.J. Bourgoigne. 1987. AIDS-related glomerulopathy: occurrence in specific risk groups. *Kidney Int.* 31:1167–1173.

169. Winston, J.A., M.E. Klotman, and P.E. Klotman. 1999. HIV-associated nephropathy is a late, not early, manifestation of HIV-1 infection. *Kidney Int.* 55:1036–1040.

170. Cohen, A.H., N.C.Y. Sun, P. Shapshak, and D.T. Imagawa. 1989. Demonstration of human immunodeficiency virus in renal epithelium in HIV-associcated nephropathy. *Mod. Pathol.* 2:125–128.

171. Ray, P.E. 1999. Looking into the past and future of human immunodeficiency virus nephropathy. *Kidney Int.* 55:1123–1124.

172. Donahue, R.E., M.M. Johnson, L.I. Zon, S.C. Clark, and J.E. Groopman. 1987. Suppression of *in vitro* hematopoiesis following human immunodeficiency virus infection. *Nature* 326:200–203.

173. Coyle, T.E. 1997. Hematologic complications of human immunodeficiency virus infection and the acquired immunodeficiency syndrome. *Med. Clin. North. Am.* 81:449–470.

174. Gillis, S., and A. Eldor. 1998. Immune thrombocytopenic purpura in adults: clinical aspects. *Baillière's Clin. Haematol.* 11:361–372.

175. Zon, L.I., C. Arkin, and J.E. Groopman. 1987. Hematological manifestations of the human immunodeficiency virus (HIV). *Br. J. Haematol.* 66:251–256.

176. Bel-Ali, Z., V. Dufour, and Y. Najean. 1987. Platelet kinetics in human immunodeficiency virus-induced thrombocytopenia. *Am. J. Hematol.* 26:299–304.

177. Finazzi, G., P.M. Mannucci, A. Lazzarin, A. Gringieri, C. Arici, D. Ciaci, R. Terzi, and T. Barbui. 1990. Low incidence of bleeding from HIV-related thrombocytopenia in drug addicts and hemophiliacs: implications for therapeutic strategies. *Eur. J. Haematol.* 45:82–85.

178. Landonio, G., M. Galli, A. Nosari, A. Lazzarin, P. Crocchialo, L. Voltolin, F. Giannelli, L. Irato, and F. De Cataldo. 1990. HIV-related severe thrombocytopenia in intravenous drug users: prevalence, response to therapy in a medium-term follow-up, and pathogenetic evaluation. *AIDS* 4:29–34.

179. Grau, G.E., D. Morrow, S. Izui, and P.H. Lambert. 1986. Pathogenesis of the delayed phase of Rauscher virus-induced thrombocytopenia. *J. Immunol.* 136:686–691.

180. Zucker-Franklin, D., C.S. Termin, and M.C. Copper. 1989. Structural changes in the megakaryocytes of patients infected with the human immunodeficiency virus (HIV-1). *Am. J. Pathol.* 134:1295–303.

181. Zucker-Franklin, D., and Y.Z. Cao. 1989. Megakaryocytes of human immunodeficiency virus-infected individuals express viral RNA. *Proc. Natl. Acad. Sci. USA* 186:5595–5599.

182. Zucker-Franklin, D., S. Seremetis, and Z.Y. Zheng. 1990. Internalization of human immunodeficiency virus type 1 and other retroviruses by megakaryocytes and platelets. *Blood* 75:1920–1923.

183. Dominguez, A., G. Gamallo, R. Gaercia, A. Lopez-Pastor, J.M. Pena, and J.J. Vazquez. 1994. Pathophysiology of HIV-related thrombocytopenia: an analysis of 41 patients. *J. Clin. Pathol.* 47:999–1003.

184. Bettaieb, A., E. Oksenhendler, P. Fromont, N. Duedari, and P. Bierling. 1989. Immunochemical analysis of platelet autoantibodies in HIV-related thrombocytopenic purpura: a study of 68 patients. *Br. J. Haematol.* 73:241–247.

185. Bettaieb, A., P. Fromont, F. Louache, E. Oksenhendler, W. Vainchenker, N. Duedari, and P. Bierling. 1992. Presence of cross-reactive antibody between human immunodeficiency virus (HIV) and platelet glycoproteins in HIV-related immune thrombocytopenic purpura. *Blood* 80:162–169.

186. Stricker, R.B., D.I. Abrams, L. Corash, and M.A. Schuman. 1985. Target platelet antigen in homosexual men with immune thrombocytopenia. *N. Engl. J. Med.* 313:1375–1380.

187. Walsh, C.M., M.A. Nardi, and S. Karpatkin. 1984. On the mechanisms of thrombocytopenic purpura in sexually active homosexual men. *N. Engl. J. Med.* 311:635–639.

188. Yu, J-R., E.T. Lennette, and S. Karpatkin. 1986. Anti-F(ab)$_2$ antibodies in thrombocytopenic patients at risk for acquired immunodeficiency syndrome. *J. Clin. Invest.* 77:1756–1761.

189. Karpatkin, S., M. Nardi, E.T. Lennette, B. Byrne, and B. Poiesz. 1988. Anti-human immunodeficiency virus type 1 antibody complexes on platelets of seropositive thrombocytopenic homosexuals and narcotic addicts. *Proc. Natl. Acad. Sci. USA* 85:9763–9767.

190. Karpatkin, S., and M. Nardi. 1992. Autoimmune anti-HIV gp120 antibody with antiidiotype-like activity in sera and

immune complexes of HIV-1 related immunologic thrombocytopenia. *J. Clin. Invest.* 89:356–364.

191. Karpatkin, S. 1990. Autoimmune thrombocytopenia and AIDS-related thrombocytopenia. *Curr. Opin. Immunol.* 2:625–632.

192. Clark, W.F., G.J. Tevaarwerk, and B.D. Reid. 1982. Human platelet-immune complex interaction in plasma. *J. Lab. Clin. Med.* 100:917–931.

193. Oksenhendler, E., and M. Seligmann. 1990. HIV-related thrombocytopenia. *Immunodefic. Rev.* 2:221–31.

194. Northfelt, D.W., E.D. Charlebois, M.I. Mirda, C. Child, L.D. Kaplan, D.I. Abrams, and the Community. Consortium. 1995. Continuous low-dose interferon-α theraphy for HIV-related immune thrombocytopenic purpura. *J. Acquir. Immune Defic. Syndr. Hum. Retrovirol.* 8:45–50.

195. Hymes, K.B., and S. Karpatkin. 1997. Human immunodeficiency virus infection and thrombotic microangiopathy. *Sem. Hematol.* 34:117–125.

196. Eldor, A. 1998. Thrombotic thrombocytopenic purpura: diagnosis, pathogenesis and modern therapy. *Baillière's Clin. Haematol.* 11:475–495.

197. Torok, T.J., R.C. Holman, and T.L. Chorba. 1995. Increasing mortality from thrombotic thrombocytopenic purpura in the United States—Analysis of national mortality data, 1968–1991. *Am. J. Hematol.* 50:84–90.

198. Kwaan, A.C. 1987. Miscellaneous secondary thrombotic microangiopathy. *Sem. Hematol.* 24:141–147.

199. Kelton, J.G., J. Moore, A. Santos, and D. Sheridan. 1984. The detection of a platelet-agglutinating factor in thrombotic thrombocytopenic purpura. *Ann. Intern. Med.* 101:589–593.

200. Lian, E.C., P. Mui, F.A. Siddiqui, A.Y. Chiu, and L.L. Chiu. 1984. Inhibition of platelet aggregating activity in thrombotic thrombocytopenic purpura plasma by normal adult immunoglobulin G. *J. Clin. Invest.* 73:548–555.

201. Laurence, J., D. Mitra, M. Steiner, L. Staiano-Coico, and E. Jaffe. 1996. Plasma from patients with idiopathic and human immunodeficiency virus-associated thrombotic thrombocytopenic purpura induces apoptosis in microvascular endothelial cells. *Blood* 87:3245–3254.

202. Mitra, D., E.A. Jaffe, B. Weksler, K.A. Hajjar, C. Soderland, and J. Laurence. 1997. Thrombotic thrombocytopenic purpura and sporadic hemolytic-uremic syndrome plasmas induce apoptosis in restricted lineages of human microvasculare endothelial cells. *Blood* 84:1224–1234.

203. Farina, C., G. Gavazzeni, A. Caprioli, and G. Remuzzi. 1990. Hemolytic uremic sindrome associated with verocytotoxin-producing *Escherichia coli* infection in acquired immunodeficiency syndrome. *Blood* 75:2465–2468.

204. Gadallah, M.F., M.A. El-Shahawy, V.M. Campese, J.R. Todd, and J.W. King. 1996. Disparate prognosis of thrombotic microangiopathy in HIV-1 infected patients with and without AIDS. *Am. J. Nephrol.* 16:446–450.

205. Meyer, C.N., P. Skinhoj, and J. Prag. 1994. Bacteremia in HIV positive and AIDS patients: incidence, species distribution, risk factors, outcome and influence of long-term prophylactic antibiotic treatment. *Scand. J. Infect. Dis.* 26:635–642.

206. Bain, B.J. 1997. The haematological features of HIV infection. *Br. J. Haematol.* 99:1–8.

207. Moses, A., J. Nelson, and G.C. Bagby. 1998. The influence of human immunodeficiency virus-1 on hemopoiesis. *Blood* 91:1479–1495.

208. Abrams, D.I., E.K. Chinn, B.J. Lewis, P.A. Volberding, M.A. Conant, and R.M. Townsend. 1984. Hematologic manifestations in homosexual men with Kaposi's sarcoma. *Am. J. Clin. Pathol.* 81:13–18.

209. Van der Lelie, J., J.M. Lange, J.J. Vos, C.M. van Dalen, S.A. Danner, and A.E.G. von dem Borne. 1987. Autoimmunity against blood cells in human immunodeficiency virus (HIV) infection. *Br. J. Haematol.* 67:109–114.

210. Kimura, S., J. Matasuda, S. Ikematus, K. Miyazono, A. Ito, T. Nakahata, M. Minamitani, K. Shimada, Y. Shiokawa, and F. Takaku. 1990. Efficacy of recombinant human granulocyte colony-stimulating factor on neutropenia in patients with AIDS. *AIDS* 4:1251–1255.

211. Bain, B.J. 1999. Pathogenesis and pathophysiology of anemia in HIV infection. *Curr. Opin. Hematol.* 6:89–93.

212. Kreuzer, K.A., and J.K. Rockstroh. 1997. Pathogenesis and pathophysiology of anemia in HIV infection. *Ann. Hematol.* 75:179–187.

213. Davis, B.R., and G. Zauli. 1995. Effect of human immunodeficiency virus infection on hematopoiesis. *Baillère's Clin. Haematol.* 8:113–130.

214. Liu, W., M. Ittamann, J. Liu, R. Schoentag, P. Tierno, M.A. Greco, G. Sidhu, M. Nierodzik, and R. Wieczorek. 1997. Human parvovirus B19 in bone marrow from adults with acquired immunodeficiency syndrome: a comparative study using in situ hybridisation and immunohistochemistry. *Hum. Pathol.* 26:760–766.

215. Richman, D.D., M.A. Fischl, M.H. Grieco, M.S. Gottlieb, P.A. Volberding, O.L. Laskin, I.M. Leedom, G.E. Groopman, D. Mildvon, and M.S. Hirsch. 1987. The toxicity of azidotymidine (AZT) in the treatment of patients with AIDS and AIDS-related complex. A double-blind, placebo controlled trial. *N. Engl. J. Med.* 317:192–197.

216. Cohen, A.J., T.M. Philips, and C.M. Kessler. 1986. Circulating coagulation inhibitors in the acquired immunodeficiency syndrome. *Ann. Intern. Med.* 104:175–180.

217. Stimmler, M.M., F.P. Quismorio, W.G. McGehee, T. Boylen, and O.P. Sharma. 1989. Anticardiolipin antibodies in acquired immunodeficiency syndrome. *Arch. Intern. Med.* 149:1833–1835.

218. Panzer, S., C. Stain, H. Hartl, R. Dudczak, and K. Lechner. 1989. Anticardiolipin antibodies are elevated in HIV-1 infected hemophiliacs but do not predict for disease progression. *Thrombosis Hemostasis* 61:81–85.

219. Navia, B.A., B.D. Jordan, and R.W. Price. 1986. The AIDS dementia complex. I. Clinical features. *Ann. Neurol.* 19:517–524.

220. Gendelman, H.E., Y. Persidsky, A. Ghorpade, J. Limoges, M. Stins, M. Fiala, and R. Morrisett. 1997. The neuropathogenesis of AIDS dementia complex. *AIDS* 11:S35–S45.

221. Navia, B.A., E.S. Cho, C.K. Petito, and R.W. Price. 1986. The AIDS dementia complex. II. Neuropathology. *Ann. Neurol.* 19:525–535.

222. Anders, K.H., W.F. Guerra, U. Tomiyasu, M.A. Verity, and H.V. Vinters. 1986. The neuropathology of AIDS: UCLA experience and review. *Am. J. Pathol.* 124:537–558.

223. Gelbard, H.A., H.J. James, L.R. Sharer, S.W. Perry, Y. Saito, A.M. Kazee, B.M. Blumberg, and L.G. Epstein. 1995. Apoptotic neurons in brains from paediatric patients with HIV-1 encephalitis and progressive encephalopathy. *Neuropathol. Appl. Neurobiol.* 21:208–217.

224. Ho, D.D., T.R. Rota, R.T. Schooley, J.C. Kaplan, J.D. Allan, J.E. Groopman, L. Resnick, D. Felsenstein, C.A. Andrews, and M.S. Hirsch. 1985. Isolation of HTLV-III from cerebrospinal fluid and neural tissues of patients with neurologic

syndromes related to the acquired immunodeficiency syndrome. *N. Engl. J. Med.* 313:1493–1497.

225. Levy, J.A., J. Shimabukuro, H. Hollander, J. Mills, and L. Kaminsky. 1985. Isolation of AIDS-associated retroviruses from cerebrospinal fluid and brain of patients with neurological symptoms. *Lancet* ii:586–588.

226. Koenig, S., H.E. Gendelman, J.M. Orenstein, M.C. Dal Canto, G.H. Pezeshkpour, M. Yungbluth, F. Janotta, A. Aksamit, M.A. Martin, and A.S. Fauci. 1986. Detection of AIDS virus in macrophages in brain tissue from AIDS patients with encephalopathy. *Science* 233:1089–1093.

227. Gyorkey, F., J.L. Melnick, and P. Gyorkey. 1987. Human immunodeficiency virus in brain biopsies of patients with AIDS and progressive encephalopathy. *J. Infect. Dis.* 155:870–876.

228. Funke, I., A. Hahn, E.P. Rieber, E. Weiss, and G. Riethmuller. 1987. The cellular receptor (CD4) of the human immunodeficiency virus is expressed on neurons and glial cells in human brain. *J. Exp. Med.* 165:1230–1235.

229. Resnick, L., J.R. Berger, P. Shapshak, and W.W. Tourtellotte. 1988. Early penetration of the blood-brain barrier by HIV. *Neurology* 38:9–14.

230. Johnson, R.T., J.D. Glass, J.C. McArthur, and B.W. Chesebro. 1996. Quantitation of human immunodeficiency virus in brains of demented and nondemented patients with acquired immune deficiency syndrome. *Ann. Neurol.* 39:393–395.

231. Brenneman, D.E., G.L. Westbrook, S.P. Fitzgerald, D.L. Ennist, K.L. Elkins, M.R. Ruff, and C.B. Pert. 1988. Neuronal cell killing by the envelope protein of HIV and its prevention by vasoactive intestinal peptide. *Nature* 335:639–642.

232. Dreyer, E.B., P.K. Kaiser, J.T. Offermann, and S.A. Lipton. 1990. HIV-1 coat protein neurotoxicity prevented by calcium channel antagonists. *Science* 248:364–367.

233. Gurney, M.E., S.P. Heinrich, M.R. Lee, and H.S. Yin. 1986. Molecular cloning and expression of neuroleukin, a neurotrophic factor for spinal and sensory neurons. *Science* 234:566–574.

234. Gurney, M.E., B.R. Apatoff, G.T. Spear, M.J. Baumel, J.P. Antel, M.B. Bania, and A.T. Reder. 1986. Neuroleukin: a lymphokine product of lectin-stimulated T cells. *Science* 234:574–581.

235. Lee, M.R., D.D. Ho, and M.E. Gurney. 1987. Functional interaction and partial homology between human immunodeficiency virus and neuroleukin. *Science* 237:1047–1051.

236. Koka, P., K. He, J.A. Zack, S. Kitchen, W. Peacock, I. Fried, T. Tran, S.S. Yashar, and J.E. Merrill. 1995. Human immunodeficiency virus 1 envelope proteins induce interleukin-1, tumor necrosis factor-α, and nitric oxide in glial cultures derived from fetal, neonatal and adult human brain. *J. Exp. Med.* 182:941–952.

237. Amadori, A., A. De Rossi, P. Gallo, B. Tavolato, and L. Chieco-Bianchi. 1988. Cerebrospinal fluid lymphocytes from HIV-infected patients synthesize HIV-specific antibody *in vitro. J. Neuroimmunol.* 18:181–186.

238. Rieckman, P., G. Poli, J.H. Kehrl, and A.S. Fauci. 1991. Activated B lymphocytes from human immunodeficiency virus-infected individuals induce virus expression in infected T cells and a promonocytic cell line, U1. *J. Exp. Med.* 173:1–5.

239. Breen, E., A.R. Reezai, K. Nakajima, G.N. Beall, R.T. Mitsuyasu, T. Hirano, T. Kishimoto, and O. Martinez-Maza.

240. Genis, P., M. Jett, E.W. Bernton, T. Boyle, H.A. Gelbard, K. Dzenko, R.W. Keane, L. Resnick, Y. Mizrachi, and D.J. Volsky. 1992. Cytokines and arachidonic metabolites produced during human immunodeficiency virus (HIV)-infected macrophage-astroglia interactions: implications for the neuropathogenesis of HIV disease. *J. Exp. Med.* 172:1703–1718.

241. Nottet, H.S.L.M., and H.E. Gendelman. 1995. Unravelling the neuroimmune mechanisms for the HIV-1-associated cognitive/motor complex. *Immunol. Today* 16:441–448.

242. Nottet, H.S.L.M., M. Jett, C.R. Flanagan, Q.H. Zhai, Y. Persidsky, A. Rizzino, E.W. Bernton, PP. Genis, T. Baldwin, and J. Schwartz. 1995. A regulatory role for astrocytes in HIV-1 encephalitis. An overexpression of eicosanoids, platelet-activating factor, and tumor necrosis factor-α by activated HIV-1 infected monocytes is attenuated by primary human astrocytes. *J. Immunol.* 154:3567–3581.

243. Tyor, W.R., J.D. Glass, J.W. Griffin, P.S. Becker, J.C. McArthur, L. Bezman, and D.E. Griffin. 1992. Cytokine expression in the brain during the acquired immunodeficiency syndrome. *Ann. Neurol.* 31:349–360.

244. Wesselingh, S.L., C. Power, J.D. Glass, W.R. Tyor, J.C. McArthur, J.M. Farber, J.W. Griffin, and D.E. Griffin. 1993. Intracerebral cytokine messenger RNA expression in acquired immunodeficiency syndrome dementia. *Ann. Neurol.* 33:576–582.

245. Griffin, G.E., S.L. Wesselingh, and J.C. McArthur. 1994. Elevated central nervous system prostaglandins in human immunodeficiency virus-associated dementia. *Ann. Neurol.* 35:592–597.

246. Bukrinsky, M., H.S. Nottet, H. Schmidtmayerova, L. Dubrovsky, C.R. Flanagan, M.E. Mullins, S.A. Lipton, and H.E. Gendelman. 1995. Regulation of nitric oxide synthase activity in human immunodeficiency virus type 1 (HIV-1)-infected monocytes: implications for HIV-associated neurologic disease. *J. Exp. Med.* 181:735–745.

247. Mathiesen, T., A. Sonnerborg, and B. Wahren. 1989. Detection of antibodies against myelin basic protein and increased levels of HIV-IgG antibodies and HIV antigen after solubilization of immune complexes in sera and CSF of HIV infected patients. *Viral Immunol.* 2:1–8.

248. Kiprov, D., W. Pfaeffl, G. Parry, R. Lippert, W. Lang, and R. Miller. 1988. Antibody-mediated peripheral neuropathies associated with ARC and AIDS: successful treatment with plasmapheresis. *J. Clin. Apheresis* 4:3–7.

249. Levy, J.A. 1988. The mysteries of HIV: challenges for therapy and prevention. *Nature* 333:519–522.

250. Levy, J.A. 1989. The human immunodeficiency virus: detection and pathogenesis. In: AIDS Pathogenesis and Treatment, J.A. Levy Ed., New York, Marcel Dekker Inc.

251. Schnittman, S.M., M.C. Psallidopoulos, H.C. Lane, L. Thompson, M. Baseler, F. Massari, C.H. Fox, N.P. Salzman, and A.S. Fauci. 1989. The reservoir for HIV-1 in human peripheral blood is a T cell that maintains expression of CD4. *Science* 245:305–308.

252. Graziosi, C., G. Pantaleo, J.F. Demarest, O.J. Cohen, M. Vaccarezza, L. Butini, M. Montroni, and A.S. Fauci. 1993. HIV-1 infection in lymphoid organs. *AIDS* 7:S53–S58.

253. Clerici, M., N.I. Stocks, R.A. Zajac, R.N. Boswell, D.R. Lucey, C.S. Via, and G.M. Shearer. 1989. Detection of three distinct patterns of T helper cell dysfunction in asymptomatic, human immunodeficiency virus-seropositive patients. *J. Clin. Invest.* 84:1892–1899.

254. Zinkernagel, R.M., and H. Hengartner. 1994. T cell-mediated immunopathology versus direct cytolysis by virus: implications for HIV and AIDS. *Immunol. Today* 15:262–268.

255. Tyler, D.S., C.L. Nastala, S.D. Stanley, T.J. Matthews, H.K. Lyerly, D.P. Bolognesi, and K.J. Weinhold. 1989. gp120-specific cellular cytotoxicity in HIV-1 seropositive individuals. Evidence for circulating CD16+ effector cells armed *in vivo* with cytophilic antibody. *J. Immunol.* 142:1177–1182.

256. Plata, F., B. Autran, L.P. Martins, S. Wain-Hobson, M. Raphael, C. Mayaud, M. Denis, J.M. Guillon, and P. Debre. 1987. AIDS virus-specific cytotoxic T lymphocytes in lung disorders. *Nature* 328:348–351.

257. Langlade-Demoyen, P., F. Michel, A. Hoffenbach, E. Vilmer, G. Dadaglio, F. Garcia-Pons, C. Mayaud, B. Autran, S. Wain-Hobson, and F. Plata. 1988. Immune recognition of AIDS virus antigens by human and murine cytotoxic T lymphocytes. *J. Immunol.* 141:1949–1957.

258. Israel-Biet, D., A. Venet, K. Beldjord, J.M. Andrieu, and P. Even. 1990. Autoreactive cytotoxicity in HIV-infected individuals. *Clin. Exp. Immunol.* 81:18–24.

259. Zarling, J.M., J.A. Ledbetter, J. Sias, P. Fultz, J. Eichberg, G. Gjerset, and P.A. Moran. 1990. HIV-infected humans, but not chimpanzees, have circulating cytotoxic T lymphocytes that lyse uninfected CD4+ cells. *J. Immunol.* 144:2992–2998.

260. Ozturk, G.E., P.F. Kohler, C.R. Horsburgh, Jr., and C.H. Kirkpatrick. 1987. The significance of anti-lymphocyte antibodies in patients with acquired immune deficiency syndrome (AIDS) and their sexual partners. *J. Clin. Immunol.* 7.130–139.

261. Silvestris, F., B.S. Edwards, O.M. Sadeghi, M.A. Frassanito, R.C. Williams, Jr., and F. Dammacco. 1989. Isotype, distribution and target analysis of lymphocyte reactive antibodies in patients with human immunodeficiency virus infection. *Clin. Immunol. Immunopathol.* 53:329–340.

262. Williams, R.C., H. Masur, and T.J. Spira. 1984. Lymphocyte-reactive antibodies in acquired immune deficiency syndrome. *J. Clin. Immunol.* 4;118–123.

263. Tomar, R.H., P.A. John, A.K. Hennig, and B. Kloster. 1985. Cellular targets of antilymphocyte antibodies in AIDS and LAS. *Clin. Immunol. Immunopathol.* 37:37–47.

264. Stricker, R.B., T.M. McHugh, D.J. Moody, W.J. Morrow, D.P. Stites, M.A. Shuman, and J.A. Levy. 1987. An AIDS-related cytotoxic autoantibody reacts with a specific antigen on stimulated CD4+ T cells. *Nature* 327:710–713.

265. Warren, R.Q., E.A. Johnson, R.P. Donnelly, M.F. Lavia, and K.Y. Tsang. 1988. Specificity of anti-lymphocyte antibodies in sera from patients with AIDS-related complex (ARC) and healthy homosexuals. *Clin. Exp. Immunol.* 73:168–173.

266. Golding, H., F.A. Robey, F.T. Gates, 3rd, W. Linder, P.R. Biening, T. Hoffman, and B. Golding. 1988. Identification of homologous regions in human immunodeficiency virus 1 gp41 and human MHC Class II beta 1 domain. *J. Exp. Med.* 167:914–923.

267. Golding, H., G.M. Shearer, K. Hillman, P. Lucas, J. Manischewitz, R.A. Zajac, M. Clerici, R.E. Gress, R.N. Boswell, and B. Golding. 1989. Common epitope in human immunodeficiency virus (HIV) 1-gp41 and HLA Class II elicits immunosuppressive autoantibodies capable of contributing to immune dysfunction in HIV-1 infected individuals. *J. Clin. Invest.* 83:1430–1435.

268. Thiriart, C., J. Goudsmit, P. Schellekens, F. Barin, D. Zagury, M. De Wilde, and C. Bruck. 1988. Antibodies to

269. Moore, J.P., Q.J. Sattentau, and P.R. Clapham. 1990. Enhancement of soluble CD4-mediated HIV neutralization and gp120 binding by CD4 autoantibodies and monoclonal antibodies. *AIDS Res. Hum. Retroviruses* 6:1273–1279.

270. Kowalski, M., B. Ardman, L. Basiripour, Y.C. Lu, D. Blohm, W. Haseltine, and J. Sodroski. 1989. Antibodies to CD4 in individuals infected with human immunodeficiency virus type 1. *Proc. Natl. Acad. Sci. USA* 86:3346–3350.

271. Caporossi, A.P., G. Bruno, S. Salemi, C. Mastroianni, M. Falciano, A. Salotti, N. Bergami, I. Santilio, R. Nisini, and V. Barnaba. 1998. Autoimmune T-cell response to the CD4 molecule in HIV-infected patients. *Viral Immunol.* 11:9–17.

272. Ardman, B., M.A. Sikorski, M. Settles, and D.E. Staunton. 1990. Human immunodeficiency virus type 1-infected individuals make autoantibodies that bind to CD43 on normal thymic lymphocytes. *J. Exp. Med.* 172:1151–1158.

273. Lefebvre, J.C., V. Giordanengo, M. Limouse, A. Doglio, M. Cucchiarini, F. Monpoux, R. Mariani, and J.F. Peyron. 1994. Altered glycosylation of leukosialin, CD43, in HIV-1-infected cells of the CEM line. *J. Exp. Med.* 180:1609–1617.

274. Reisinger, D., and R. Parkman. 1987. Molecular heterogeneity of a lymphocyte glycoprotein in immunodeficient patients. *J. Clin. Invest.* 79:595–599.

275. Szawlowsky, P.W., T. Hanke, and R.E. Randall. 1993. Sequence homology between HIV-1 gp120 and the apoptosis-mediating protein Fas. *AIDS* 7:10–18.

276. Reiher, W.E., J.E. Blalock, and T.K. Brunck. 1986. Sequence homology between acquired immunodeficiency syndrome virus envelope protein and Interleukin-2. *Proc. Natl. Acad. Sci. USA* 83:9188–9192.

277. Bost, K.L., and D.W. Pascual. 1988. Antibodies against a peptide sequence within the HIV envelope protein cross-react with human Interleukin-2. *Immunol. Lett.* 17:577–586.

278. Bost, K.L., B.H. Hahn, M.S. Saag, D.A. Weigent, and J.E. Blalock. 1988. Individuals infected with HIV possess antibodies against IL-2. *Immunology* 65:611–615.

279. Caruso, A., C. Bonfanti, D. Colombrita, M. De Francesco, C. De Rango, I. Foresti, F. Gargiulo, R. Gonzales, G. Gribaudo, and S. Landolfo. 1990. Natural antibodies to IFN-gamma in man and their increase during viral infection. *J. Immunol.* 144:685–690.

280. Ledbetter, J.A., C.H. June, P.S. Rabinovitch, A. Grossmann, T.T. Tsu, and J.B. Imboden. 1988. Signal transduction through CD4 receptors: stimulator *vs* inhibitory activity is regulated by CD4 proximity to the CD3/T cell receptor. *Eur. J. Immunol.* 18:525–532.

281. Newell, M.K., L.J. Laughlan, C.R. Maroun, and M.H. Julius. 1990. Death of mature T cells by separate ligation of CD4 and the T-cell receptor for antigen. *Nature* 347:286–289.

282. Goronzy, J.J., and C.M. Weyand. 1989. Persistent suppression of virus-specific cytotoxic T cell responses after transient depletion of CD4+ T cells *in vivo*. *J. Immunol.* 142:4435–4440.

283. Mittler, R.S., and M.K. Hoffman. 1989. Synergism between HIV gp120 and gp120-specific antibody in blocking human cell activation. *Science* 245:1380–1382.

284. Amadori, A., G. De Silvestro, R. Zamarchi, M.L. Veronese, M.R. Mazza, G. Schiavo, M. Panozzo, A. De Rossi, L. Ometto, J. Mous, A. Barelli, A. Borri, L. Salmaso, and L. Chieco-Bianchi. 1992. CD4 epitope masking by gp120/anti-gp120 antibody complexes. A potential mechanism for CD4+ cell function down-regulation in AIDS patients. *J. Immunol.* 148:2709–2716.

soluble CD4 in HIV-1-infected individuals. *AIDS* 2:345–351.

285. Sunila, I., M. Vaccarezza, G. Pantaleo, A.S. Fauci, and J.M. Orenstein. 1997. gp120 is present on the plasma membrane of apoptotic CD4 cells prepared from lymph nodes of HIV-1-infected individuals: an immunoelectron microscopic study. *AIDS* 11:27–32.

286. Zamarchi, R., M. Panozzo, A. Del Mistro, A. Barelli, A. Borri, A. Amadori, and L. Chieco-Bianchi. 1994. B and T cell function parameters during zidovudine administration in human immunodeficiency virus-infected patients. *J. Infect. Dis.* 170:1148–1156.

287. Via, C.S., H.C. Morse, 3rd, and G.M. Shearer. 1990. Altered immunoregulation and autoimmune aspects of HIV infection: relevant murine models. *Immunol. Today* 11:250–255.

61 | Transplantation and Autoimmunity

Andrei Shustov and Charles S. Via

1. INTRODUCTION

Autoimmunity can be observed following transplantation of allogeneic cells into recipients incapable of rejecting the foreign cells. The autoimmune reaction is driven, in most cases, by immunocompetent T lymphocytes present in either the donor inoculum, resulting in graft-vs.-host disease (GVHD) or present in the host environment, resulting in host-vs graft disease (HVG). In this chapter, we will examine the mechanisms by which an initially allogeneic immune response results in autoimmunity.

2. AUTOIMMUNITY FOLLOWING BONE MARROW TRANSPLANTATION IN HUMANS

2.1. Acute GVHD

Bone marrow transplantation (BMT) in humans has been successfully used therapeutically for a variety of hematologic malignancies (rev. in 1). Donor bone marrow may be derived from genetically unrelated individuals (allogeneic BMT), genetically identical siblings (syngeneic BMT) or from the recipients themselves (autologous BMT). Successful engraftment of allogeneic donor bone marrow requires that the recipient be immunologically unable to reject the marrow. Typically, the recipient receives irradiation and possibly cytotoxic agents prior to BMT in order to not only eradicate the underlying malignancy, but also to prevent host rejection of donor marrow. A significant drawback to allogeneic BMT is the occurrence of acute GVHD, a potentially lethal syndrome consisting of dermatitis, hepatic damage (increased bilirubin) and intestinal involvement (diarrhea, ileus) (2). Acute GVHD occurs usually within two months after BMT and is graded on a five step scale from O to IV. Acute GVHD is mediated largely by mature T cells contained in the donor marrow. Unfortunately, efforts aimed at completely depleting contaminating donor T cells are associated with reduced marrow engraftment and/or a loss of graft-vs-leukemia (GVL) effect.

2.2. Chronic GVHD

BMT recipients may also develop an autoimmune syndrome termed chronic GVHD which is characterized by features of scleroderma, Sjögren's syndrome, autoimmune hepatitis and occasionally interstitial lung disease (3–8). Risk factors for the development of chronic GVHD are 1) prior acute GVHD, increasing patient age and treatment with viable donor buffy coat cells after transplantation (9–11). Additionally, cytomegalovirus infection may also contribute to chronic GVHD development (12).

2.3. Autoimmune Features of Chronic GVHD

Originally, chronic GVHD was defined as skin and organ changes occurring at greater than 100 days post BMT. The distinct autoimmune features of chronic GVHD are discussed below and have been noted to occur as early as 31 days after BMT (10).

Skin. The skin is the most common organ affected by chronic GVHD occurring in 80% of patients (1). Lesions consist of lichenoid papules, local erythema and hyper- or hypopigmentation. Importantly, skin lesions may resemble scleroderma with dermal fibrosis, induration, atrophy, joint contractures and ulceration.

Sjögren's syndrome.   Sicca syndrome (dry eyes, dry mouth) may accompany chronic GVHD and is due to involvement of the minor salivary glands and the ducts. Histologically the lesions are characterized by inflammation or fibrosis of acini as well as lymphocytic infiltration and necrosis of the ductal epithelium (13). These histologic changes are highly similar to that seen in primary or secondary Sjögren's syndrome. Eye involvement can be demonstrated as reduced tear production (Schirmer's test) and represents lacrimal gland involvement. Dryness of other mucosal surfaces can occur leading to dryness of the airway, the esophagus or the gynecological tract (14,15)

Interstitial lung disease.   Although uncommon, bronchiolitis obliterans is highly correlated with the presence of chronic GVHD. Symptoms include cough, dyspnea and signs of obstructive lung disease (16,17).

Other organs.   A chronic cholestatic liver disease resembling that seen in acute GVHD has been described (1). Additionally, patients with chronic GVHD may develop myasthenia gravis with anti-acetylcholine receptor antibodies (18), polymyositis (19) and autoimmune thyroid disease (20,21).

Serum autoantibodies.   A wide variety of serum autoantibodies have been described in patients with chronic GVHD including anti-nuclear, anti-dsDNA, anti-smooth muscle, anti-mitochondria, anti-microsomal and anti-epidermal (20,22,23).

2.4.   Mechanisms of Autoimmunity in Chronic GVHD

The immunopathogenisis of chronic GVHD is poorly understood at present. Nevertheless, it would appear that the mechanisms involved may be shared with other poorly understood autoimmune diseases, such as scleroderma and Sjögren's syndrome, given the clinical overlap between these three conditions. It has been reported that T cells from chronic GVHD mice secrete cytokines that promote fibroblast proliferation and collagen deposition (24), raising the possibility that the scleroderma-like features in chronic GVHD patients are a consequence of activated donor T cells. A second possible mechanism relates to the observation that patients with chronic GVHD exhibit severely damaged thymic architecture, as evidenced by involution of thymic epithelium, loss of Hassall's corpuscles and depletion of lymphocytes (1). Thymic damage is thought to occur as a result of acute GVHD and is supported by data in murine models demonstrating that acute GVHD can induce not only histological evidence of abnormal thymic architecture (25,26) but also a loss of the ability of the

thymus to educate new T cells (27) consistent with defective positive and/or negative thymic selection.

Despite the clinical similarities between chronic GVHD and scleroderma, it should be noted that the autoantibody profile typical of patients with scleroderma (anti-centromere, anti-nucleolar, anti-SCL-70) has not been reported to accompany chronic GVHD in a similar frequency. Moreover, collagen deposition patterns in chronic GVHD do not always resemble those of scleroderma. Taken together, it can be concluded from the above reports that a disease closely resembling scleroderma can be observed in humans with chronic GVHD. While the mechanism is not totally understood, the data supports the idea that chimerism or persistence of allogeneic T cells in an irradiated recipient can mediate autoimmune diseases such as scleroderma or Sjögren's syndrome.

3.   MATERNAL-FETAL CHIMERISM AND AUTOIMMUNITY

It has recently become evident that chimerism can be observed in individuals who have not had a preceding BMT (28). For example, it has been shown that during childbirth, fetal cord blood contains a substantial number of maternal cells, some of which are T cells (29). Maternal T cells are able to engraft in the fetal circulation and persist for years (30). Conversely, fetal cells can engraft and persist in the maternal circulation (28). Although the levels of chimerism in either instance are typically low (microchimerism), a given individual may not only have persistent maternal cells in their circulation but, in the case of parous females, they may also have circulating fetal cells. Regardless of the origin of the chimeric cells, the consequences of persistent maternal-fetal microchimerism are under investigation and have been linked to autoimmune diseases such as scleroderma (28,31,32). An interesting ancillary finding to these current studies is the relatively high frequency of microchimerism in normals who do not display clinical evidence of autoimmune disease (33). Thus it is possible that it is not microchimerism alone that induces autoimmunity, but perhaps an additional factor is required such as identity at one or more HLA loci. For example, it has been shown that transfusion-associated GVHD can occur an immunocompetent individual who is heterozygous at a particular HLA locus and then receives blood from a donor who is homozygous at the same locus (34). An extension of this analogy suggests that autoimmunity could be induced in immunocompetent hosts by the presence of homozygous alloreactive chimeric T cells, which share host HLA loci. Such a clinical situation is highly similar to the parent-into-F_1 model of induced autoimmunity. Possible mechanism(s) are described in detail in the next section.

4. AUTOIMMUNITY IN MURINE GVHD (PARENT-INTO-F_1 MODEL)

Transplantation of homozygous murine parental strain T lymphocytes into unirradiated F_1 recipients (P->F_1) results in a graft-vs.-host reaction in which donor strain (parental) T cells recognize the alloantigens of the opposite parent that are present on F_1 (host) cells. Typically, donor T cells are not rejected because the F_1 is tolerant to parental strain H-2 alloantigens. Nevertheless, F_1 anti-parent reactions have been described in this model and are thought to be mediated by host cytotoxic T lymphocyte (CTL) recognition of anti-F_1 specific TcR present on donor T cells (35). The host-anti-donor response is relatively weak and, as a result, is overshadowed by the ongoing donor-anti-host response.

4.1. Outcomes in P->F_1 GVHD

Recipient F_1 mice in this model develop either an acute suppressive GVHD or a lupus-like chronic (autoimmune) GVHD. Chronic GVHD is characterized by striking humoral autoimmunity consisting of high serum levels of IgG autoantibodies characteristic of human lupus (36,37). In addition, chronic GVHD mice exhibit other features of lupus, such as Ig deposition along the dermal basement membrane (38) and immune complex formation and deposition in the renal glomeruli resulting in eventual death due to renal failure (38,39). Acute GVHD is characterized by a predominantly cell mediated anti-host response manifested by demonstrable anti-host CTL activity that eliminate host lymphocytes and host hematopoietic elements. As a result, acute GVHD mice are profoundly immunodeficient. Interestingly, humoral autoimmunity is present in acute GVHD mice prior to the elimination of host B cells (40).

4.2. Mechanisms of Autoimmunity in Chronic GVHD

To understand the mechanisms involved in humoral autoimmunity occurring in chronic GVHD, it is instructive to first describe the cellular immunology of acute GVHD. Acute GVHD ensues following the injection of parental strain CD4+ and CD8+ T cells into an F_1 recipient differing from the donor at both MHC class I and II loci. A well-studied example of such a P->F_1 combination is the injection of C57Bl/6 parental cells (H-2^b) into B6D2F_1 (H-$2^{b/d}$) recipients. Initially, H-2^b donor CD4+ T cells specific for allogeneic H-2^d (host) MHC II become activated and produce several cytokines. Donor CD4+ T cell production of IL-2 promotes the subsequent activation, expansion and maturation of donor CD8+ precursor CTL specific for host H-2^d MHC class I (41). Mature CD8+ donor anti-host CTL

mediate many of the clinical features of acute GVHD by eliminating host lymphocytes.

In contrast, chronic GVHD is observed following the selective activation of donor CD4+ T cells in the absence of donor CD8+ T cell activation. Activation of donor CD8+ T cells can be prevented by either:

a) depletion of CD8+ T cells in the donor inoculum prior to injection (42);
b) injection of an undepleted donor inoculum into an F_1 recipient differing from the donor strain by MHC class II disparity only e.g., B6-> (B6 × bm12) F_1 (42);
c) injection of DBA/2 strain splenocytes into B6D2F_1 recipients (42,43); or
d) *in vivo* blockade of cytokines critical for CTL maturation e.g., IL-2 (44).

In the DBA->F_1 combination, the anti-F_1-specific pCTL frequency for DBA mice is approximately 10-fold less than that of B6 mice (43), resulting in suboptimal donor CD8+ T cell activation, making this combination analogous to (a) and (b).

4.3. Cellular Immunology of GVHD

The initiation of both acute and chronic GVHD follows a similar pattern (40) consisting of: 1) lymphoproliferative changes, such as an increase in to total spleen cells, host B cells and host T cells; and 2) B cell activation as measured by increased MHC class II expression and autoantibody production. These changes correlate with the engraftment and expansion of donor CD4+ T cells in both models (40).

At 1 week of disease, acute and chronic GVHD begin to diverge. In acute GVHD, significant expansion of donor CD8+ T cells has occurred and anti-host CTL becomes readily detectable. As a result, host lymphocytes, including autoantibody-secreting B cells are eliminated, serum autoantibody levels are reduced, and the characteristic profound immunodeficiency develops. Interestingly, the surviving B cells remain activated until their elimination. By contrast, in chronic GVHD, there is no significant expansion of donor CD8+ T cells, no development of anti host CTL (40) and, consequently, no reduction in the lymphoproliferation and B cell stimulation begun earlier. During the first 2 weeks of GVHD, both forms of GVHD exhibit an initial (days 1–3) increase in IL-2 production (41), followed by increased B cell stimulatory cytokines (IL-4 and IL-10) (40) beginning at day 4 of GVHD. Beginning at approximately day 5, increased IFN-γ production (>10-fold) is seen in acute, but not chronic, GVHD. Importantly, donor CD8+ T cells must be engrafted for both donor CD4+ and CD8+ T cells to secrete

IFN-γ in acute GVHD (40). These results indicate that: 1) lupus-like autoimmunity in this model is driven initially by Th2 cytokines, and 2) IFN-γ production and donor CD8$^+$ CTL activation distinguish acute GVHD from chronic GVHD.

4.4.   Defective Donor CD8$^+$ T Cell Activation Results in Chronic GVHD

In DBA->F$_1$ model of chronic GVHD, donor CD8$^+$ T cells are present in the donor inoculum and engraft in small numbers, but are defective in their ability to induce acute GVHD, in large part due to a reduced DBA-anti-F$_1$ precursor CTL frequency compared to that of B6 mice (43). Acute GVHD and IFN-γ production could be induced in DBA-> F$_1$ mice if purified (H-2^d) CD8$^+$ T cells were added to the donor inoculum, although DBA CD8$^+$ T cells were less effective numerically than B10.D2 CD8$^+$ T cells (40). Moreover, chronic GVHD in the DBA->B6D2F$_1$ can be converted into acute GVHD by *in vivo* treatment with IL-12, a potent Th1 cytokine and CTL promoter (45). Taken together, these data indicate an important role of CD8$^+$ T cells in controlling autoimmunity development in this system. These data indicate that chronic GVHD in DBA-> F$_1$ mice results from the combination of donor CD4$^+$ T cell activation AND suboptimal activation of donor CD8$^+$ T cells, suggesting that a similar combination in humans may result in lupus.

4.5.   Mechanisms of Autoantibody Production in Chronic GVHD

Polyclonal autoreactive B cell activation and autoantibody production are prominent features of chronic GVHD. From the foregoing, it is clear that B cell activation in this model is driven by alloreactive T cells, which provide help for all host B cells. In the absence of T cell help, autoreactive B cells that encounter autoantigens are unable to produce IgG autoantibodies. In the chronic GVHD model, however, alloreactive donor CD4$^+$ T cells provide the help required for autoreactive B cells that have encountered autoantigen to become mature autoantibody-secreting plasma cells.

Interestingly, autoantibody production in chronic GVHD is not randomly polyclonal (46). Only certain autoantigens elicit an autoantibody response. Gleichmann et al. have hypothesized (42) that donor CD4$^+$ Th cells specifically drive the expansion of autoreactive B cell clones that have their receptors crosslinked by antigen. Supporting this idea, autoantibodies found in GVHD mice are exclusively directed against antigens that are capable of crosslinking surface Ig on B cells, such as DNA. Autoantibodies to mouse thyroglobulin and insulin (self-antigens that do not posses the ability to crosslink B cell receptor) are absent (46).

In studies of the hybridomas derived from chronic GVHD mice, normal mice and mice injected with polyclonal B cell stimulators (bacterial lypopolysaccharide), it has been shown that, in general, the repertoire of autoreactive B cells in chronic GVHD does not differ from that of normal (nonautoimmune) mice (46,47). Autoantibodies produced in all three groups bind to the kidney cryosections, bind to DNA and bind to the cell surface of a variety of cells, including B cells, T cells, macrophages, and fibroblasts. Two specificities of autoantibodies were found only in GVHD mice, one of which was directed against 160 kD glycoprotein present in the microvilli of kidney brush border, and the second against 70 kD glycoprotein encoded by genes found in murine leukemia virus (48,49). The significance of these observations is not clear.

It should be noted that despite the fact that alloreactive donor CD4$^+$ T cells are thought to play a major role in the induction of chronic GVHD. alloreactive T cell clones specific for host MHC II have exhibited variable results when injected into appropriate recipients. In one report, MHC class II-reactive clones induced acute death of the recipient due to DTH-like syndrome (50). Interestingly, these T cell clones were able to trigger polyclonal B cell activation *in vitro*, confirming the T helper phenotype. However, other workers have reported B cell hyperactivity and autoantibody production following the injection of CD4$^+$ MHC class II restricted T cell clones (51). The reason for this discrepancy is not clear, but it is possible that T cell clones may exhibit altered organ homing due to defective adhesion molecule expression (52).

4.6.   Autoimmunity in Acute (P->F$_1$) Murine GVHD

Mice that do not succumb during the early stages of acute GVHD may eventually develop a chronic form of GVHD that resembles several human autoimmune diseases, including lupus, rheumatoid arthritis, Sjögren's syndrome, scleroderma and some other collagen vascular diseases. For example, the injection of 80–120 $\times$ 10^6 BALB/c spleen cells into unirradiated (BALB/c $\times$ A)F$_1$ mice leads to the development of chronic progressive polyarthritis with juxta-articular manifestations, including perivascular infiltrates, peritendinitis, myositis and inflammatory nodules (53). In addition, features of a variety of other connective tissue diseases are also present. For example, mice were noted to have Sjögren's-like salivary gland lesions, hepatic lesions resembling sclerosing cholangitis, scleroderma-like skin lesions and immune complex glomerulonephritis. These mice develop positive ANA (100%), anti-dsDNA (50%), anti-histone (25%) and low titer anti-snRNP (35%) antibodies (54,55). Importantly, the scleroderma-like changes require 9–12 months to develop (53).

5. AUTOIMMUNITY IN CYCLOSPORINE A-INDUCED SYNGENEIC GVHD

The fungal metabolite Cyclosporine A (CsA) is an immunosuppressive drug (56,57) used in the treatment of a variety of immune-mediated conditions (58–60), including the prevention of allograft rejection and as treatment for GVHD following BMT (61–63). It has been reported that CsA interferes with TCR signalling by inhibiting the calmodulin/calcineurin enzyme system and in so doing, disrupts the process of thymic positive and negative selection (64). Interestingly, CsA preferentially allows positive selection of self-reactive T cells that would normally be deleted but blocks the development of other α/β TCR-bearing cells (65). Thus, despite the fact that CsA suppresses T lymphocyte-dependent immune responses *in vivo* and *in vitro* (56,57,66,67), it paradoxically disrupts mechanisms controlling self-tolerance by inhibiting deletion of self reactive T cell clones in thymus (68) leading to systemic autoimmunity.

As first reported by Glazier et al. (69) and recently reviewed by Hess and Thoburn (70), administration of CsA after autologous or syngeneic bone marrow transplantation (BMT) in humans or rodents leads to the development of a T cell-dependent autoimmune (or autoaggressive) syndrome that resembles human GVHD occurring after allogeneic BMT. This syndrome, termed autologous or syngeneic GVHD (sGVHD), demonstrates that donor/host incompatibility across MHC I and MHC II barriers is not an absolute requirement for the induction of GVHD, as previously thought. In rats, syngeneic GVHD develops following discontinuation of a limited course of CsA treatment after irradiation and syngeneic or autologous BMT. The initial disease is characterized by erythroderma and dermatitis similar to that of acute GVHD seen in humans post BMT. Subsequently, there is a rapid progression to a chronic phase consisting of alopecia, fibrosis and scleroderma (69,71). The two stages of the sGVHD reflect different cellular elements infiltrating the target organs. At the onset of sGVHD, the majority of infiltrating lymphocytes are CD8+ cytotoxic T cells that likely mediate the epithelial destruction observed histologically. During the chronic phase, the majority of infiltrating cells are CD4+ T lymphocytes (69,72). Adoptive transfer studies indicated that both CD4+ and CD8+ T cell subsets are required for sGVHD induction and progression. CD8+ T cells from sGVHD animals transfer only acute self-limited disease into secondary recipients. Transfer of CD4+ T cells alone is ineffective in inducing disease (72). Thus, progression of acute disease to chronic disease requires CD4+ autoreactive T cell help. IL-2 appears to be an important, but insufficient, factor for CD4+ T cell help (72). Very likely, the production of cytokines and other factors contributes to the fibrosis observed in the chronic phase (73).

The induction and development of sGVHD requires two more conditions to be met in addition to CsA treatment. The first essential requirement is the presence of an intact thymus, since sGVHD cannot be induced in thymectomized animals (74). However, radiation damage to the thymus may accentuate the effect of CsA on preventing the clonal deletion of autoreactive cells. The mechanisms by which CsA alters thymic function are complex and are primarily related to changes in cellular architecture within the thymus. Treatment of both mice and rats with pharmacologic doses of CsA results in the rapid ablation of the thymic medulla with a loss of medullary epithelium and marked reduction in the expression of MHC class II antigens (75–77). These structural changes induced by CsA significantly alter the integrity of the thymic environment that governs T cell differentiation and the clonal deletion of autoreactive lymphocytes (78). In particular, the emergence of an autoreactive syndrome is associated with the ability of CsA to disrupt the positive selection of thymocytes and to retard the negative selection of thymocytes (65). It has been suggested that CsA may convert a normally negatively selecting TCR signal into a positive selection signal, thus distorting clonal selection in the thymus (65).

The second requirement for the induction of sGVHD is the ablation of peripheral regulatory mechanisms. The production of autoreactive clones in the thymus is not sufficient by itself for the induction of an autoimmune/ autoaggression syndrome. Autoreactive T cells can be detected in the periphery of normal (non transplanted) animals treated with CsA (65,68,79,80) or even in untreated mice without any evidence of autoimmune disease. This suggests that, under normal conditions, clonal deletion is incomplete (81) and that there are peripheral regulatory mechanisms that control the expression of autoreactive T cells. In fact, a permissive environment for the activation and expansion of autoreactive T cells in sGVHD develops only after disruption of this peripheral regulatory compartment (82–84). Thus, sGVHD can only be adaptively transferred into irradiated secondary recipients (69). Transfer of splenocytes from animals with active disease to normal animals is ineffective. Recent studies suggest that the host autoregulatory system undergoes upregulation in the presence of autoreactive T cells (85). For example, specific priming of normal animals with sGVHD effector cells results in marked increase in the ability of the peripheral regulatory T cells to prevent the adoptive transfer of the disease. Both CD4+ and CD8+ T cell compartments play a role in this peripheral autoregulatory activity. Of interest is the finding that, following priming, the primary autoregulatory activity resides within the CD4+ T-cell subset. Idiotype-anti-idiotype mechanism have been postulated as a possible explanation for the interaction between sGVHD effector cells and autoregulatory T cells (85).

Regarding effector mechanisms, recent studies have shown that the development of sGVHD is associated with the appearance of a highly restricted repertoire of CD8+ cytolytic T lymphocytes that promiscuously recognize major histocompatibility complex (MHC) class II determinants, including self-MHC (86). The vast majority of clones express Vβ8.3/8.5 and Vα11 TCR chains (86). Adoptive transfer studies confirmed that the CD8+ Vβ8.5 T cells are responsible for the initiation of this autoimmune syndrome. Depletion of CD8+ T lymphocytes expressing this TCR V region gene segment completely blocks the ability of effector spleen cells from animals with syngeneic GVHD to adaptively transfer the disease into secondary recipients. Amplification and progression of the disease also requires a Vβ8.5 CD4+ autoreactive T cells (86). Analysis of Vβ8.5 lymphocytes infiltrating target organs such as the tongue demonstrate restricted expression of Vα-chain. That is, Vβ8.5 CD4+ T cells expressed only Vα11, whereas Vα2, 11 and 23 were detected in Vβ8.5 CD8+T cells (87,88).

In sharp contrast to the limited diversity of the effector cell population in syngeneic GVHD, there is a promiscuous recognition of MHC class II determinants across histocompatibility barriers. Autoreactive CD8+ T cells recognize MHC class II determinants on target cells from several MHC-disparate strains of animals (78). Moreover, administration of anti-MHC class II, but not MHC class I, monoclonal antibodies *in vivo* delays or prevents the adoptive transfer of sGVHD (89,90). Recent studies have revealed that pathogenic CD8+ T cell clones recognize a peptide from the MHC class II invariant chain (CLIP) presented in the context of MHC class II (91,92). This peptide is highly conserved and contains a supermotif for binding to MHC class II molecules (93). Pretreatment of target cells with antibody to CLIP completely blocks lysis mediated by the CD8+Vβ8.5 CsA-induced autoreactive T lymphocytes (92,94). Moreover, target cells loaded with CLIP exhibit enhanced susceptibility to recognition and lysis. Thus, the recognition of this highly conserved element with its unique capacity to bind to virtually all MHC class II determinants may override the restriction of the T cell response to a specific MHC haplotype.

Clonal analysis of effector populations demonstrate a major pathogenic subset that requires the N-terminal flanking region of CLIP for activation and induction of Th1 cytokine (IL-2, IFN-γ) production. A second minor non-pathogenic subset of effector cells has been identified that requires the C-terminal flanking region for activation and induction of Th2 cytokine production (IL-4, IL-10) (91,95). It has been speculated that CLIP might have the same binding site on the T cell receptor as the superantigen staphylococcal enterotoxin B. Thus, in the setting of CsA-induced autoaggression, CLIP may be acting as a superantigen.

CsA treatment in humans who have received autologous BMT results in an autoaggressive/autoimmune disease similar to that in rodents and is termed autologous GVHD (96). Disease is self-limited, confined in most cases to skin without evidence of internal organ disease (97) and is mediated primarily by CD8+ CD4-α/β+ autocytotoxic T cells specific for the MHC II-CLIP complex (98,99). Double positive (CD4+, CD8+) lytic T cells have also been described in a smaller number of patients (98,99).

It has been suggested that the experimentally-induced CsA autoimmune syndrome could have useful anti-tumor activity. Therapeutic efforts aimed at reducing the occurrence of post-BMT acute GVHD in humans are associated with a high rate of tumor recurrence and an absent graft-versus-tumor activity. Conversely, high grade GVHD after allogeneic BMT is associated with significant mortality, but low tumor recurrence rates. Induction of sGVHD may be useful clinically in the elimination of MHC II-CLIP-expressing tumors. Expression of MHC class II-CLIP antigen in sGVHD can be upregulated by cytokines such as IFN-γ, thereby potentiating tumor targeting and effectiveness of tumor clearance (70,99,100). Furthermore, administration of IL-2 has been shown to enhance the expansion of autoreactive T cells, ensuring maximum tumor kill (100). Clinical trials are underway to evaluate the anti-tumor activity of sGVHD in humans. Ongoing analyses suggest that autologous GVHD has a beneficial effect in high-grade non-Hodgkin's lymphoma (70) and that the anti-tumor effect can be enhanced by rIFN-γ administration (100–102).

6. AUTOIMMUNITY AND NEONATAL TOLERANCE (F$_1$->P MODEL OF HVGD)

The injection of semiallogeneic (F$_1$) lymphocytes into homozygous parental strain mice during neonatal period induces a specific state of tolerance to the alloantigens of the opposite parent in the recipients, first described by Medavar et al. (103). The tolerance is characterized by the survival of skin grafts from the appropriate allogeneic donor later in life, long-lasting cellular chimerism in the host and specific cytolytic unresponsiveness of the host to corresponding donor alloantigens (104–108). Thus, the neonatal i.v. injection of (B10XA) F$_1$ (H-2$^{a/b}$) semiallogeneic spleen cells into A/J (H-2^a) recipients causes the state of partial tolerance to C57BL/10 (H-2^b) skin allografts (104).

The induction of this specific unresponsiveness to alloantigens is associated with the development of autoimmune syndrome termed Host-versus-Graft Disease (HVGD). HVGD is characterized by hypergammaglobulinemia, production of autoantibodies of multiple specificities, including antibodies against ssDNA, dsDNA, histones, cardiolipin and cytoskeleton as well as circulating immune complexes (CIC), increased serum IgE levels, lymphosplenomegaly, thrombocytopenia and membranoproliferative (lupus-like) glomerulonephritis (105,106, 109). Disease is further characterized by immunoglobulin deposits in both the chorioid plexus and dermal-epidermal junction and a positive direct Coomb's test (105).

The cellular immunology of HVGD appears to be complex and is believed to involve allogeneic cooperation between alloreactive host CD4+ Th2 cells and persistent donor semiallogeneic B cells (110–112). Athymic BALB/c nu/nu mice failed to develop neonatal tolerance after injection of (C57BL/6 × BALB.Ig^b) F$_1$ spleen cells, but reconstitution of the host with CD8$^-$, CD4$^+$ T cells from BALB/c mice rendered them toleragenic, suggesting that CD4$^+$ T cells from the tolerant mice are necessary for the activation of autoreactive donor B cells (111). Autoantibodies produced in HVGD carry almost exclusively the phenotype of F$_1$ donor B cells (110). It has been postulated that the development of autoimmunity in this model requires the persistence of F$_1$ hybrid donor lymphoid cells in the host and the effective induction of donor specific tolerance (105). Depletion studies have shown that donor CD4$^+$ T cells play only limited role in the pathogenesis of HVGD (113). Autoimmunity after induction of neonatal tolerance is self-limited and dependent on the presence of B cell chimerism. A decrease in B cell chimerism was accompanied by a progressive decline in autoantibody levels and the disappearance of membrane-proliferative glomerulonephritis. Re-injection of donor splenocytes lead to a flare in the serological and histopathological manifestations of autoimmune disease (106). These results demonstrate the central role of donor B cells in initiation and self-limitation of autoimmunity in HVGD.

MHC class II antigens have been thought to be the molecules recognized by alloreactive host Th2 cells on donor B cells leading to the production of autoantibodies by latter. Recent studies have shown that both I-A and I-E classes of MHC antigens can be involved (107,114,115).

Studies of the cytokines in HVGD reveal that unlike host CD8$^+$ T cells that undergo clonal deletion, host CD4$^+$ T cells are not deleted but differentiate into Th2 like cells and produce IL-4 and IL-6, although they deficient in IL-2 and IFN-γ production (108,116,117). Furthermore, *in vivo* treatment of HVGD mice with rIFN-γ significantly decreased the levels of hyper-IgE, anti-DNA and anti-laminin antibodies and prevented the occurrence of glomerulonephritis. This effect was not associated with a decrease in B cell chimerism, but appeared to restore intrinsic IL-2 and IFN-γ production by donor-specific T cells that could still produce high levels of IL-4 and IL-10 (118).

Another example of the severe immune dysregulation in HVGD is the high incidence of lymphoproliferative disorders (LPD) and tumors of the immune system. The majority of A/J (H2^a) mice neonatally tolerized by the injection of (A/J × C57BL/10) F$_1$ splenocytes develop severe LPD with a high mortality rate (104). Approximately 27% of these mice suffered from lymphoid malignancies. Histopathological changes were seen in the spleen, lymph nodes, liver and kidneys. The majority of the infiltrating cells were of immature and mature myeloid lineage and the proportion of CD4$^+$ cells was significantly decreased despite an increase in their absolute numbers. Lymphoproliferation observed was at the expense of host but not donor cells (104,119).

7. CONCLUSIONS

From the foregoing, it can be seen that the induction of autoimmunity in the P->F model (GVHD) or F$_1$->P model (HVGD) share several common features: a) Transfer of immunocompetent lymphocytes into a recipient that is immunocompetent but unable to reject the donor cells; b) homozygous T cells that recognize alloantigens on F$_1$B cells leading to polyclonal B cell activation and autoantibody production; c) defective CTL development either donor anti-host CTL in GVHD (43) or host-anti-donor in HVGD (105–108); and the proliferation and persistence of CD4$^+$ T cells secreting Th2 cytokines that drive B cell autoantibody production in GVHD (40) or HVGD (111,112,117,118).

The mechanisms of autoimmunity following human BMT are less well understood. Because the host is immunosuppressed prior to BMT, donor T cell activation of host B cells is not likely to occur. Instead, defective thymic deletion of autoreactive T cells (similar to that seen in CsA-induced sGVHD) may play a role. Future studies of long term survivors of P->F$_1$ model of acute GVHD may be instructive since these mice develop many of the features seen in humans with chronic GVHD following BMT.

References

1. Ferrara, J.L.M., and J.H. Deeg. 1991. Graft-versus-host disease. *N. Engl. J. Med.* 324:667–674.
2. Thomas, E., R. Storb, R.A. Clift, Fefer, F.L. Johnson, P.E. Neiman K.O. Lerner, H. Glucksberg, and C.D. Buckner. 1975. Bone-marrow transplantation. *N. Engl. J. Med.* 292:832–843.
3. Shulman, H.M., K.M. Sullivan, P.L. Weiden, G.B. McDonald, G.E. Striker, G.E. Sale, R. Hackman, M.S. Tsoi, R. Storb, and E.D. Thomas. 1980. Chronic graft-versus-host syndrome in man. A long-term clinicopathologic study of 20 Seattle patients. *Am. J. Med.* 69:204–217.
4. Van Vloten, W.A.E. Scheffer, and L.J. Dooren. 1977. Localized scleroderma-like lesions after bone marrow

transplantation in man. A chronic graft versus host reaction. *Br. J. Dermatol.* 96:337–341.

5. Lawley, T.J., G.L. Peck, and H.M. Moutsopoulos. 1977. Scleroderma. Sjögren's-like syndrome, and chronic graft-versus-host disease. *Ann. Intern. Med.* 87:707–709.

6. Shulman, H.M., K.M. Sullivan, P.L. Weiden, G.B. McDonald, G.E. Striker, G.E. Sale, R. Hackman, M.S. Tsoi, R. Storb, and E.D. Thomas. 1980. Chronic graft-versus-host syndrome in man: A long-term clinicopathologic study of 20 seattle patients. *Am. J. Med.* 69:204–217.

7. Graze, P.R., and R.P. Gale. 1979. Chronic graft versus host disease: A syndrome of disordered immunity. *Am. J. Med.* 66:611–620.

8. Sherer, Y., and Y. Shoenfeld. 1998. Autoimmune diseases and autoimmunity post-bone marrow transplantation. *Bone Marrow Transplant.* 22:873–881.

9. Storb, R., R.L. Prentice, K.M. Sullivan, H.M. Shulman, H.J. Deeg, K.C. Doney, C.D. Buckner, R.A. Clift, R.P. Witherspoon, F.A. Appelbaum, J.E. Sanders, P.S. Stewart, and E.D. Thomas. 1983. Predictive factors in chronic graft-versus-host disease in patients with aplastic anemia treated by marrow transplantation from HLA-identical siblings. *Ann. Intern. Med.* 98:461–466.

10. Ringden, O.T. Paulin, B. Lonnqvist, and B. Nilsson. 1985. An analysis of factors predisposing to chronic graft-versus-host disease. *Exp. Hematol.* 13:1062–1067.

11. Atkinson, K., M.M. Horowitz, R.P. Gale, D.W. van Bekkum, E. Gluckman, R.A. Good, N. Jacobsen, H.J. Kolb, A.A. Rimm, and O. Ringden. 1990. Risk factors for chronic graft-versus-host disease after HLA-identical sibling bone marrow transplantation. *Blood* 75:2459–2464.

12. Bostrom, L., O. Ringden, N. Jacobsen, F. Zwaan, and B. Nilsson. 1990. A European multicenter study of chronic graft-versus-host disease. The role of cytomegalovirus serology in recipients and donors-acute graft-versus-host disease, and splenectomy. *Transplantation* 49:1100–1105.

13. Schubert, M.M., K.M. Sullivan, T.H. Morton, K.T. Izutsu, D.E. Peterson, N. Flournoy, E.L. Truelove, G.E. Sale, C.D. Buckner, and R. Storb. 1984. Oral manifestations of chronic graft-v-host disease. *Arch. Intern. Med.* 144:1591–1595.

14. Corson, S.L., K. Sullivan, F. Batzer, C. August, R. Strob, and E.D. Thomas. 1982. Gynecologic manifestations of chronic graft-versus-host disease. *Obstet. Gynecol.* 60:488–492.

15. Ralph, D.D., S.C. Springmeyer, K.M. Sullivan, R.C. Hackman, R. Storb, and E.D. Thomas. 1984. Rapidly progressive air-flow obstruction in marrow transplant recipients. Possible association between obliterative bronchiolitis and chronic graft-versus-host disease. *Am. Rev. Respir. Dis.* 129:641–644.

16. Wyatt, S.E., P. Nunn, J.M. Hows, J. Yin, M.C. Hayes, D. Catovsky, E.C. Gordon-Smith, J.M. Hughes, J.M. Goldman, and D. Galton. 1984. Airways obstruction associated with graft versus host disease after bone marrow transplantation. *Thorax* 39:887–894.

17. Roca, J., A. Granena, R. Rodriguez-Roisin, P. Alvarez, A. Agusti-Vidal, and C. Rozman. 1982. Fatal airway disease in an adult with chronic graft-versus-host disease. *Thorax* 37:77–78.

18. Smith, C.I., J.A. Aarli, P. Biberfeld, P. Bolme, B. Christensson, G. Gahrton, L. Hammarstrom, A.K. Lefvert, B. Lonnqvist, and G. Matell. 1983. Myasthenia gravis after bone-marrow transplantation. Evidence for a donor origin. *N. Engl. J. Med.* 309:1565–1568.

19. Reyes, M.G., P. Noronha, W. Thomas, Jr., and R. Heredia. 1983. Myositis of chronic graft versus host disease. *Neurology* 33:1222–1224.

20. Carlson, K., G. Lonnerholm, B. Smedmyr, G. Oberg, and B. Simonsson. 1992. Thyroid function after autologous bone marrow transplantation. *Bone Marrow Transplant.* 10:123–127.

21. Ichihashi, T., H. Yoshida, H. Kiyoi, H. Fukutani, K. Kubo, T. Yamauchi, T. Naoe, and R. Ohno. 1992. Development of hyperthyroidism in donor and recipient after allogeneic bone marrow transplantation. *Bone Marrow Transplant.* 10:397–398.

22. Lister, J., H. Messner, E. Keystone, R. Miller, and M.J. Fritzler. 1987. Autoantibody analysis of patients with graft versus host disease. *J. Clin. Lab. Immunol.* 24:19–23.

23. Holmes, J.A., S.J. Livesey, A.E. Bedwell, N. Amos, and J.A. Whittaker. 1989. Autoantibody analysis in chronic graft-versus-host disease. *Bone Marrow Transplant.* 4:529–531.

24. DeClerck, Y., V. Draper, and R. Parkman. 1986. Clonal analysis of murine graft-vs-host disease: II. Leukokines that stimulates fibroblast proliferation and collagen synthesis in graft-vs. host disease. *J. Immunol.* 136:3549–3552.

25. Ghayur, T., T.A. Seemayer, A. Xenocostas, and W.S. Lapp. 1988. Complete sequential regeneration of graft-vs-host-induced severely dysplastic thymuses. Implications for the pathogenesis of chronic graft-vs-host disease. *Am. J. Pathol.* 133:39–46.

26. Onoe, Y., M. Harada, K. Tamada, K. Abe, T. Li, H. Tada, and K. Nomoto. 1998. Involvement of both donor cytotoxic T lymphocytes and host NK1.1+ T cells in the thymic atrophy of mice suffering from acute graft-versus-host disease. *Immunology* 95:248–256.

27. Fukuzawa, M., C.S. Via, and G.M. Shearer. 1988. Defective thymic education of L3T4+ T helper cells function in graft-vs-host mice. *J. Immunol.* 141:430–439.

28. Nelson, J.L. 1998. Microchimerism and the pathogenesis of systemic sclerosis. *Curr. Opin. Rheumatol.* 10:564–571.

29. Hall, J.M., P. Lingenfelter, S.L. Adams, D. Lasser, J.A. Hansen, and M.A. Bean. 1995. Detection of maternal cells in human umbilical cord blood using fluorescence in situ hybridization. *Blood* 86:2829–2832.

30. Boisset, M., and M.A. Fitzcharles. 1994. Alternative medicine use by rheumatology patients in a universal health care setting. *J. Rheumatol.* 21:148–152.

31. Nelson, J.L. 1998. Microchimerism and the causation of scleroderma. *Scand. J. Rheumatol. Suppl.* 107:10–13.

32. Nelson, J.L. 1998. Microchimerism and autoimmune disease. *N. Engl. J. Med* 338:1224–1225.

33. Evans, P.C., N. Lambert, S. Maloney, D.E. Furst, J.M. Moore, and J.L. Nelson. 1999. Long-term fetal microchimerism in peripheral blood mononuclear cell subsets in healthy women and women with scleroderma. *Blood* 93:2033–2037.

34. Orlin, J.B., and M.H. Ellis. 1997. Transfusion-associated graft-versus-host disease. *Curr. Opin. Hematol.* 4:442–448.

35. Kosmatopoulos, K., D.S. Algara, and S. Orbach-Arbouys. 1987. Anti-receptor anti-MHC cytotoxic T lymphocytes: their role in the resistance to graft vs host reaction. *J. Immunol.* 138:1038–1041.

36. Gleichmann, E., E.H. van Elven, and P.J.W. Van Der Veen. 1982. A systemic lupus erythematosus (SLE)-like disease in mice induced by abnormal T-B cell cooperation. Preferential formation of autoantibodies characteristic of SLE. *Eur. J. Immunol.* 12:152–159.

37. Portanova, J.P., H.N. Claman, and B.L. Kotzin. 1985. Autoimmunization in murine graft-vs-host disease: I. Selective production of antibodies to histones and DNA. *J. Immunol.* 135:3850–3856.

38. van Elven, E.H., J. Agterberg, S. Sadel, and E. Gleichmann. 1981. Diseases caused by reactions of T lymphocytes to incompatible structures of the major histocompatibility complex: II. Autoantibodies deposited along the basement membrane of skin and their relationship to immune-complex glomerulonepohritis. *J. Immunol.* 126:1684–1691.

39. Rolink, A.G., H. Gleichmann, and E. Gleichmann. 1983. Diseases caused by reaction of T lymphocytes to incompatible structures of the major histocompatibility complex: VII. Immune-complex glomerulonephritis. *J. Immunol.* 130:209–215.

40. Rus, V., A. Svetic, P. Nguyen, W.C. Gause, and C.S. Via. 1995. Kinetics of Th1 and Th2 cytokine production during the early course of acute and chronic murine graft-versus-host disease. Regulatory role of donor CD8⁺ T cells. *J. Immunol.* 155:2396–2406.

41. Via, C.S. 1991. Kinetics of T cell activation in acute and chronic forms of murine graft-versus-host disease. *J. Immunol.* 146:2603–2609.

42. Gleichmann, E., S.T. Pals, A.G. Rolink, T. Radaszkiewicz, and H. Gleichmann. 1984. Graft-versus-host reactions: clues to the etiopathology of a spectrum of immunological diseases. *Immunol. Today* 5:324–332.

43. Via, C.S., S.O. Sharrow, and G.M. Shearer. 1987. Role of cytotoxic T lymphocytes in the prevention of lupus-like disease occurring in a murine model of graft-vs-host disease. *J. Immunol.* 139:1840–1849.

44. Via, C.S., and F.D. Finkelman. 1993. Critical role of interleukin-2 in the development of acute graft versus host disease. *Int. Immunol.* 5:565–572.

45. Via, C.S., V. Rus, M.K. Gately, and F.D. Finkelman. 1994. IL-12 stimulates the development of acute graft-versus-host disease in mice that normally would develop chronic, autoimmune graft-versus-host disease. *J. Immunol.* 153:4040–4047.

46. Van Rappard-Van Der Veen, F.M., U. Kiesel, L. Poels, W. Schuler, C.J.M. Melief, J. Landegent, and E. Gleichmann. 1984. Further evidence against random polyclonal antibody formation in mice with lupus-like graft-vs-host disease. *J. Immunol.* 132:1814–1820.

47. Guilbert, B., G. Dighiero, and S. Avrameas. 1982. Naturally occurring antibodies against nine common antigens in human sera. I. Detection, isolation and characterization. *J. Immunol.* 128:2779–2787.

48. Rolink, A.G., T. Radaszkiewicz, and F. Melchers. 1988. Monoclonal autoantibodies specific for kidney proximal tubular brush border from mice with experimentally-induced chronic graft-versus-host disease. *Scand. J. Immunol.* 28:29–41.

49. Rolink, A.G., T. Radaszkiewicz, and F. Melchers. 1987. The autoantigen-binding B cell repertoires of normal and of chronically graft-versus-host-diseased mice. *J. Exp. Med.* 165:1675–1687.

50. Schreier, M.H., R. Tees, T. Radaszkiewicz, and A.G. Rolink. 1985. The *in vivo* effects of antigen-specific and I-A restricted T cell clones. In: T Cell Clones. H. von Boehmer and W. Haas, eds. Elsevier, pp. 173–182.

51. Tary-Lehmann, M., A.G. Rolink, P.V. Lehmann, Z.A. Nagy, and U. Hurtenbach. 1990. Induction of graft versus host-associated immunodeficiency by CD4+ T cell clones. *J. Immunol.* 145:2092–2098.

52. Dailey, M.O., C.G. Fathman, E.C. Butcher, E. Pillemer, and I. Weissman. 1982. Abnormal migration of T lymphocyte clones. *J. Immunol.* 128:2134–2136.

53. Pals, S.T., T. Radaszkiewicz, L. Rozendaal, and E. Gleichmann. 1985. Chronic progressive polyarthritis and other symptoms of collagen vascular disease induced by graft-vs-host reaction. *J. Immunol.* 134:1475–1482.

54. Schulick, R.D., S.C. Muluk, M. Clerici, B.L. Bermas, C.S. Via, M.R. Weir, and G.M. Shearer. 1994. Value of *in vitro* CD4+ T helper cell function test for predicting long-term loss of human renal allografts. *Transplantation* 57:480–482.

55. Gelpi, C., J.L. Rodriguez-Sanchez, M.A. Martinez, J. Craft, and J.A. Hardin. 1988. Murine graft vs host disease: A model for study of mechanisms that generate autoantibodies to ribonucleoproteins. *J. Immunol.* 140:4160–4166.

56. Kahan, B.D. 1989. Cylosporine. *N. Engl. J. Med.* 321:1725–1738.

57. Shevach, E.M. 1985. The effects of cyclosporin A on the immune system. *Ann. Rev. Immunol.* 3:397–423.

58. Nussenblatt, R.B., A.G. Palestine, A.H. Rook, I. Scher, W.B. Wacker, and I. Gery. 1983. Treatment of intraocular inflammatory disease with cyclosporin A. *Lancet* 2:235–238.

59. Stiller, C.R., J. Dupre, M. Gent, M.R. Jenner, P.A. Keown, A. Laupacis, R. Martell, N.W. Rodger, B. von Graffenried, and B.M. Wolfe. 1984. Effects of cyclosporine immunosuppression in insulin-dependent diabetes mellitus of recent onset. *Science* 223:1362–1367.

60. Caccavo, D., B. Lagana, A.P. Mitterhofer, G.M. Ferri, A. Afeltra, A. Amoroso, and L. Bonomo. 1997. Long-term treatment of systemic lupus erythematosus with cyclosporin A. *Arthritis Rheum.* 40:27–35.

61. Calne, R.Y. 1979. Immunosuppression for organ grafting—observations on cyclosporin A. *Immunol Rev.* 46:113–124.

62. White, D.J., and R.Y. Calne. 1982. The use of Cyclosporin A immunosuppression in organ grafting. *Immunol. Rev.* 65:115–131.

63. Green, C.J., and A.C. Allison. 1978. Extensive prolongation of rabbit kidney allograft survival after short-term cyclosporin-A treatment. *Lancet* 1:1182–1183.

64. Schreiber, S.L., and G.R. Crabtree. 1992. The mechanism of action of cyclosporin A and FK506. *Immunol. Today* 13:136–142.

65. Urdahl, K.B., D.M. Pardoll, and M.K. Jenkins. 1994. Cyclosporin A inhibits positive selection and delays negative selection in alpha beta TCR transgenic mice. *J. Immunol.* 152:2853–2859.

66. Fukuzawa, M., S.O. Sharrow, and G.M. Shearer. 1989. Effect of cyclosporin A on T cell immunity. II. Defective thymic education of CD4 T helper cell function in cyclosporin A-treated mice. *Eur. J. Immunol.* 19:1147–1152.

67. Fukuzawa, M., and G.M. Shearer. 1989. Effect of cyclosporin A on T cell immunity. I. Dose-dependent suppression of different murine T helper cell pathways. *Eur. J. Immunol.* 19:49–56.

68. Jenkins, M.K., R.H. Schwartz, and D.M. Pardoll. 1988. Effects of cyclosporine A on T cell development and clonal deletion. *Science* 241:1655–1658.

69. Glazier, A., P.J. Tutschka, E.R. Farmer, and G.W. Santos. 1983. Graft-versus-host disease in cyclosporin A-treated rats after syngeneic and autologous bone marrow reconstitution. *J. Exp. Med.* 158:1–8.

70. Hess, A.D., and C.J. Thoburn. 1997. Immunobiology and immunotherapeutic implications of syngeneic/autologous graft-versus-host disease. *Immunol. Rev.* 157:111–123.

71. Beschorner, W.E., C.A. Shinn, A.C. Fischer, G.W. Santos, and A.D. Hess. 1988. Cyclosporine-induced pseudo-graft-versus-host disease in the early post-cyclosporine period. *Transplantation* 46:112S–117S.

72. Hess, A.D., A.C. Fischer, and W.E. Beschorner. 1990. Effector mechanisms in cyclosporine A-induced syngeneic graft-versus-host disease. Role of CD4+ and CD8+ T lymphocyte subsets. *J. Immunol.* 145:526–533.

73. DeClerck, Y., V. Draper, and R. Parkman. 1986. Clonal analysis of murine graft-vs-host disease. II. Leukokines that stimulate fibroblast proliferation and collagen synthesis in graft-vs. host disease. *J. Immunol.* 136:3549–3552.

74. Sorokin, R.H. Kimura, K. Schroder, D.H. Wilson, and D.B. Wilson. 1986. Cyclosporine-induced autoimmunity. Conditions for expressign disease, requirement for intact thymus, and potency estimates of autoimmune lymphocytes in drug-treated rats. *J. Exp. Med.* 164:1615–1625.

75. Fabien, N.H., C. Auger, A. Moreira, and J.C. Monier. 1992. Effects of cyclosporin A on mouse thymus: immunochemical and ultrastructural studies. *Thymus* 20:153–162.

76. Beschorner, W.E., D.L. Suresch, T. Shinozawa, G.W. Santos, and A.D. Hess. 1988. Thymic immunopathology after cyclosporine: effect of irradiation and age on medullary involution and recovery. *Transplant. Proc.* 20:1072–1078.

77. Beschorner, W.E., J.D. Namnoum, A.D. Hess, C.A. Shinn, and G.W. Santos. 1987. Cyclosporin A and the thymus. Immunopathology. *Am. J. Pathol.* 126:487–496.

78. Hess, A.D., L. Horwitz, W.E. Beschorner, and G.W. Santos. 1985. Development of graft-vs-host disease-like syndrome in cyclosporine-treated rats after syngeneic bone marrow transplantation. I. Development of cytotoxic T lymphocytes with apparent polyclonal anti-Ia specificity, including autoreactivity. *J. Exp. Med.* 161:718–730.

79. Urdahl, K.B., D.M. Pardoll, and M.K. Jenkins. 1992. Self-reactive T cells are present in the peripheral lymphoid tissues of cyclosporin A-treated mice. *Int. Immunol.* 4:1341–1349.

80. Gao, E.K., D. Lo, R. Cheney, O. Kanagawa, and J. Sprent. 1988. Abnormal differentiation of thymocytes in mice treated with cyclosporin A. *Nature* 336:176–179.

81. Hammerling, G.J., G. Schonrich, F. Momburg, N. Auphan, M. Malissen, B. Malissen, A.M. Schmitt-Verhulst, and B. Arnold. 1991. Non-deletional mechanisms of peripheral and central tolerance: studies with transgenic mice with tissue-specific expression of a foreign MHC class I antigen. *Immunol. Rev.* 122:47–67.

82. Fischer, A.C., M.K. Laulis, L. Horwitz, W.E. Beschorner, and A. Hess. 1989. Host resistance to cyclosporine induced syngeneic graft-versus-host disease. Requirement for two distinct lymphocyte subsets. *J. Immunol.* 143:827–832.

83. Fischer, A.C., W.E. Beschorner, and A.D. Hess. 1989. Requirements for the induction and adoptive transfer of cyclosporine-induced syngeneic graft-versus-host disease. *J. Exp. Med.* 169:1031–1041.

84. Hess, A.D., A.C. Fischer, L.R. Horwitz, and M.K. Laulis. 1993. Cyclosporine-induced autoimmunity: critical role of autoregulation in the prevention of major histocompatibility class II-dependent autoaggression. *Transplant. Proc.* 25:2811–2813.

85. Hess, A.D., A.C. Fischer, L. Horwitz, E.C. Bright, and M.K. Laulis. 1994. Characterization of peripheral autoregulatory mechanisms that prevent development of cyclosporin-induced syngeneic graft-versus-host disease. *J. Immunol.* 153:400–411.

86. Fischer, A.C., P.P. Ruvolo, R. Burt, L.R. Horwitz, E.C. Bright, J.M. Hess, W.E. Beschorner, and A.D. Hess. 1995. Characterization of the autoreactive T cell repertoire in cyclosporin-induced syngeneic graft-versus-host disease. A highly conserved repertoire mediates autoaggression. *J. Immunol.* 154:3713–3725.

87. Ruvolo, P.P., A.C. Fischer, G.B. Vogelsang, R.J. Jones, and A.D. Hess. 1995. Analysis of the V beta T-cell receptor repertoire in autologous graft-versus-host disease. *Ann. NY. Acad. Sci.* 756:432–434.

88. Fischer, A.C., P.P. Ruvolo, L.R. Horwitz, and A.D. Hess. 1995. Analysis of V beta T-cell receptor repertoire of effector mechanisms in acute and chronic graft versus host disease. *Transplant. Proc.* 27:1366–1369.

89. Hess, A.D., L.R. Horwitz, M.K. Laulis, and E. Fuchs. 1993. Cyclosporine-induced syngeneic graft-vs-host disease: prevention of autoaggression by treatment with monoclonal antibodies to T lymphocyte cell surface determinants and to MHC class II antigens. *Clin. Immunol. Immunopathol.* 69:341–350.

90. Hess, A.D., L.R. Horwitz, and M.K. Laulis. 1993. Cyclosporine-induced syngeneic graft-versus-host disease: recognition of self MHC class II antigens *in vivo*. *Transplant. Proc.* 25:1218–1221.

91. Hess, A.D., C. Thoburn, and L. Horwitz. 1998. Promiscuous recognition of major histocompatibility complex class II determinants in cyclosporine-induced syngeneic graft-versus-host disease: specificity of cytolytic effector T cells. *Transplantation* 65:785–792.

92. Hess, A.D., E.C. Bright, C. Thoburn, G.B. Vogelsang, R.J. Jones, and M.J. Kennedy. 1997. Specificity of effector T lymphocytes in autologous graft-versus-host disease: role of the major histocompatibility complex class II invariant chain peptide. *Blood* 89:2203–2209.

93. Malcherek, G., V. Gnau, G. Jung, H.G. Rammensee, and A. Melms. 1995. Supermotifs enable natural invariant chain-derived peptides to interact with many major histocompatibility complex-class II molecules. *J. Exp. Med.* 181:527–536.

94. Chicz, R.M., R.G. Urban, J.C. Gorga, D.A. Vignali, W.S. Lane, and J.L. Strominger. 1993. Specificity and promiscuity among naturally processed peptides bound to HLA-DR alleles. *J. Exp. Med.* 178:27–47.

95. Chen, W., C. Thoburn, and A.D. Hess. 1998. Characterization of the pathogenic autoreactive T cells in cyclosporine-induced syngeneic graft-versus-host disease. *J. Immunol.* 161:7040–7046.

96. Jones, R.J., G.B. Vogelsang, A.D. Hess, E.R. Farmer, R.B. Mann, R.B. Geller, S. Piantadosi, and G.W. Santos. 1989. Induction of graft-versus-host disease after autologous bone marrow transplantation. *Lancet* 1:754–757.

97. Yeager, A.M., G.B. Vogelsang, R.J. Jones, E.R. Farmer, V. Altomonte, A.D. Hess, and G.W. Santos. 1992. Induction of cutaneous graft-versus-host disease by administration of cyclosporine to patients undergoing autologous bone marrow transplantation for acute myeloid leukemia. *Blood* 79:3031–3035.

98. Dale, B.M., K. Atkinson, D. Kotasek, J.C. Biggs, and R.E. Sage. 1989. Cyclosporine-induced graft vs host disease in two patients receiving syngeneic bone marrow transplants. *Transplant. Proc.* 21:3816–3817.

99. Ratanatharathorn, V., J. Uberti, C. Karanes, L.G. Lum, E. Abella, M.E. Dan, M. Hussein, and L.L. Sensenbrenner.

1994. Phase I study of alpha-interferon augmentation of cyclosporine-induced graft versus host disease in recipients of autologous bone marrow transplantation. *Bone Marrow Transplant.* 13:625–630.

100. Noga, S.J., L. Horwitz, H. Kim, M.K. Laulis, and A.D. Hess. 1992. Interferon-gamma potentiates the antitumor effect of cyclosporine-induced autoimmunity. *J. Hematother.* 1:75–84.

101. Hess, A.D., R.J. Jones, L.E. Morris, S.J. Noga, A.M. Yeager, G.B. Vogelsang, and G.W. Santos. 1992. Autologous graft-versus-host disease: a new frontier in immunotherapy. *Bone Marrow Transplant.* 1:16–21.

102. Geller, R.B., A.H. Esa, W.E. Beschorner, C.G. Frondoza, G.W. Santos, and A.D. Hess. 1989. Successful *in vitro* graft-versus-tumor effect against an Ia-bearing tumor using cyclosporine-induced syngeneic graft-versus-host disease in the rat. *Blood* 74:1165–1171.

103. Billingham, R.E., L. Brent, and P. Medawar. 1956. Quantitative studies of transplantation immunity. III. Actively acquired tolerance. *Phil. Trans. R. Soc. London Biol.* 239:357–361.

104. Vegh, P., L. Baranyi, and T. Janossy. 1990. Induction of transplantation tolerance and development of lymphomas in mice: lack of interdependence. *Cell Immunol.* 129:56–66.

105. Goldman, M., H.M. Feng, H. Engers, A. Hochman, J. Louis, and P.H. Lambert. 1983. Autoimmunity and immune complex disease after neonatal induction of transplantation tolerance in mice. *J. Immunol.* 131:251–258.

106. de la Hera, M., A. de la Hera, A. Ramos, L. Buelta, J.L. Alonso, V. Rodriguez Valverde, and J. Merino. 1992. Self-limited autoimmune disease related to transient donor B cell activation in mice neonatally injected with semi-allogeneic F1 cells. *Int. Immunol.* 4:67–74.

107. Schurmans, S., G. Brighouse, G. Kramer, L. Wen, S. Izui, J. Merino, and P.H. Lambert. 1991. Transient T and B cell activation after neonatal induction of tolerance to MHC class II or Mls alloantigens. *J. Immunol.* 146:2152–2160.

108. Abramowicz, D., P. Vandervorst, C. Bruyns, J.M. Doutrelepont, P. Vandenabeele, and M. Goldman. 1990. Persistence of anti-donor allohelper T cells after neonatal induction of allotolerance in mice. *Eur. J. Immunol.* 20:1647–1653.

109. Hard, R.C., and B. Kullgren. 1970. Etiology, pathogenesis and prevention of a fatal host-versus-graft syndrome in parent-F1 mouse chimeras. *Am. J. Path.* 59:203–206.

110. Luzuy, S., J. Merino, H. Engers, S. Izui, and P.H. Lambert. 1986. Autoimmunity after induction of neonatal tolerance to alloantigens: role of B cell chimerism and F1 donor B cell activation. *J. Immunol.* 136:4420–4426.

111. Merino, J., S. Schurmans, M.A. Duchosal, S. Izui, and P.H. Lambert. 1989. Autoimmune syndrome after induction of neonatal tolerance to alloantigens. CD4+ T cells from the tolerant host activate autoreactive F1 B cells. *J. Immunol.* 143:2202–2208.

112. Powell, T.J., Jr., and J.W. Streilein. 1990. Neonatal tolerance induction by class II alloantigens activates IL-4-secreting, tolerogen-responsive T cells. *J. Immunol.* 144:854–859.

113. Merino, J., S. Schurmans, L. Wen, G. Brighouse, S. Luzuy, and P.H. Lambert. 1990. Autoimmune syndrome after induction of neonatal tolerance to alloantigens: analysis of the role of donor T cells in the induction of autoimmunity. *Clin. Exp. Immunol.* 79:273–278.

114. White, J., A. Herman, A.M. Pullen, R. Kubo, J.W. Kappler, and P. Marrack. 1989. The V beta-specific superantigen staphylococcal enterotoxin B: stimulation of mature T cells and clonal deletion in neonatal mice. *Cell* 56:27–35.

115. Gonzalez, A.L., C. Conde, Revilla, A. Ramos, B. Renedo, and J. Merino. 1993. Autoimmune syndrome after induction of neonatal tolerance to 1-E antigens. *Eur. J. Immunol.* 23:2353–2357.

116. Feng, H.M., A.L. Glasebrook, H.D. Engers, and J.A. Louis. 1983. Clonal analysis of T cell unresponsiveness to alloantigens induced by neonatal injection of F1 spleen cells into parental mice. *J. Immunol.* 131:2165–2169.

117. Vandenabeele, P., D. Abramowicz, D. Berus, J. Van der Heyden, J. Grooten, V. Donckier, E.L. Hooghe-Peters, M. Goldman, and W. Fiers. 1993. Increased IL-6 production and IL-6-mediated Ig secretion in murine host-vs-graft disease. *J. Immunol.* 150:4179–4187.

118. Donckier, V., D. Abramowicz, C. Bruyns, S. Florquin, M.L. Vanderhaeghen, Z. Amraoui, C. Dubois, P. Vandenabeele, and M. Goldman. 1994. IFN-gamma prevents Th2 cell-mediated pathology after neonatal injection of semiallogenic spleen cells in mice. *J. Immunol.* 153:2361–2368.

119. Janossy, T., L. Baranyi, A.C. Knulst, C. Vizler, R. Benner, G. Kelenyi, and P. Vegh. 1993. Autoimmunity, hyporeactivity to T cell mitogens and lymphoproliferative disorders following neonatal induction of transplantation tolerance in mice. *Eur. J. Immunol.* 23:3011–3020.

62 | Self-Reactive Residual Repertoires after Tolerance Induction and Their Use in Anti-Tumor Immune Responses

Stephen P. Schoenberger and Eli E. Sercarz

1. INTRODUCTION

It is gradually becoming apparent, given the molecular identification of most tumor antigens as unmutated self proteins, that the most effective strategy for cancer immunotherapy may well be to mobilize the same self-directed immune repertoire thought to mediate pathogenic autoimmunity. Autoimmune damage often includes spreading of the response to additional, previously uninvolved determinants; harnessing a similar expanding response to tumors would be desirable (1). Although the concept of cancer immunotherapy has been around since the pioneering work of William Coley over 100 years ago, several important new considerations have raised the possibility that the immune system can be manipulated for tumor control. The first of these is the tremendous progress made in the molecular definition of cancer antigens, which has finally brought tumor immunology and immunotherapy into the mainstream of immunologic research. The second feature is the existence in normal individuals of an abundant self-directed T- and B-cell repertoire, which can be mobilized under appropriate conditions to mount an aggressive attack on the self. These potentially autoreactive lymphocytes have evaded negative selection because the determinants they recognize are poorly processed and presented from the native antigen. This hierarchy of determinant presentation and tolerance susceptibility in responding lymphocytes therefore represents another aspect of the recognition of self in cancer immunotherapy. Finally, the role of APC differentiation states in mediating immune activation versus tolerance induction has provided a theoretical framework for active regulation of the self-directed repertoire. In this chapter, we will consider each of these concepts as they relate to anti-self/anti-tumor immune responses.

2. THE RESIDUAL SELF-DIRECTED REPERTOIRE

The residual repertoire can serve as the source of both autoimmunity and tumor immunity. If it is recognized that potential autoimmune cells are effectively down-regulated in the healthy immune system, it may also be the case that this regulatory population prevents a group of tumor-reactive cells from becoming activated (2,3). This prediction was clearly verified in recent work involving a 5–10% regulatory subset of CD4+ T cells bearing CD25. Treatment of normal naïve mice with anti-CD25, with no deliberate immunization, produced autoimmune gastritis, thyroiditis and insulin-dependent diabetes. If the removed population was added back, autoimmunity could be prevented. Most interestingly, the CD25− CD4+ T cell subpopulation revealed by anti-CD25 treatment elicits potent tumor-specific immune responses that can eradicate a broad spectrum of tumors. They lead to the generation of a double-negative CD4−8− NK population, which acts as lymphokine-activated killer cells to abrogate tumor populations. Accordingly, within the normal and naïve populations of T lymphocytes are potentially reactive cells directed against self-determinants, kept in check by regulatory cells. Cells with other surface markers, such as CD45RB^{low}, CD38, and CTLA-4, are also involved in down-regulation (4,5). The principle we wish to emphasize is that of effector populations down-regulated by separate cell types whose removal reveals underlying autoimmune and tumor-reactive cells.

To take advantage of the residual repertoire, one strategy for tumor immunotherapy would be to seek determinants whose affinity for the MHC is medium to poor in order to avoid tolerance mechanisms. At that point, to heighten the chance that a protective Th1 response would be induced, the affinity of the interaction could then be increased by alteration of the crucial residues in the peptide determinant. This approach was used by Rosenberg and colleagues with great success in the melanoma system. With the powerful techniques of combinatorial peptide libraries, it should be possible to create variants of profoundly higher affinity. By altering a single residue from lysine to methionine, it has been possible to improve such affinity 10,000-fold and convert a weak to a strong Th1 response (6). It should be possible to initiate tumor-eradicating responses with any of several variants, one of which might have less dangerous cross-reactivity with self-molecules apt to raise an autoimmune response.

3. DOMINANCE AND RECESSIVENESS OF ANTIGENIC DETERMINANTS

Most responses to protein antigens, whether restricted by class I MHC or class II MHC molecules, are directed towards one or at most a small minority of the possible antigenic determinants that are able to bind to the presenting molecule. Consistently strong immune responses appear to such "dominant" determinants, which are readily expressed at the APC surface in an MHC context, available for reactivity with an ambient T cell. These responses arise even when the whole protein antigen is used for immunization. Determinants that are only poorly processed and presented by the APC, and are non-immunogenic within the context of the whole molecule, may be termed "recessive": These may be of two forms; "cryptic", when no proliferative response appears after protein immunization, or "subdominant", when only a small recall response is produced following *in vitro* challenge of protein-primed T cells with the peptide form of the determinant. The reasons for dominance and recessiveness (also referred to in the literature as "crypticity") have been outlined elsewhere (7) but in brief, they relate to the poor availability of the determinant, its low affinity for the corresponding MHC molecule, or the absence of a suitable T cell repertoire.

How is dominance achieved? With respect to class I MHC processing and presentation, dominance is achieved by a series of peptide selections. Peptides are first cleaved by a proteolytic enzyme system within the cytoplasmic proteasome, and then moved into the endoplasmic reticulum via the transporter associated with antigen processing (TAP). Different processing enzymes are engaged by particular amino acid residues positioned within the sequence that dictate the exact position of the peptide cleavage. A

further selective step occurs during the competition for binding to the class I groove, which is aided by special chaperones in the endoplasmic reticulum.

The whole strategy of binding in the MHC class II system is different. Whether a particular region of the protein molecule will become dominant is related to features of the protein antigen itself: How does it become unfolded and reduced, where are the sites of endopeptidic cleavage, and what are the affinities of the ambient MHC molecules for the newly disclosed sites on the unfolding protein? Here, if we consider a tightly-folded globular protein such as hen egg lysozyme (HEL), the initial steps in processing will involve an endopeptidase cleavage of the molecule followed by some opening up and unfolding of the tight structure that will be enhanced after reduction of the disulfide bonds. Since the class II molecule's binding groove is open at both ends, it is likely to be a long fragment or peptide derived from the unfolding molecule, or even the whole protein itself with a locally unfolded region, which makes an initial contact with the groove of the class II molecule. The most available determinant with reasonable affinity will eventually become the dominant determinant, and the dangling ends flanking the initially bound determinant will be trimmed by endopeptidases and exopeptidases. Concomitantly, the walls of the binding groove provide protection from enzymatic attack to the dominant determinant.

This scenario has some important consequences. First, there is competition for binding to the MHC among different determinants along the same protein molecule as well as from other antigens. Likewise, there is competition for newly available determinants between the ambient MHC class II molecules. These two types of competition will lead to an event termed determinant capture, in which the winner determinant (that binds first) to the winner MHC molecule will preempt the binding of other determinants to their MHC grooves. This struggle is a major influence in the phenomenon of immunodominance and will tend to make other determinants on the molecule subdominant and sometimes cryptic. In the latter two situations, the other, excluded determinants will be presented at much lower levels to the ambient T cell population. T cells generally exist in the complete repertoire at varying levels of avidity for each MHC-Ag complex and their involvement during the induction of immune responsiveness or at the other extreme, the process of negative selection in the thymus is avidity-linked. Usually, the T cells with the highest avidity will be engaged first in each of these situations, although it is possible, as levels of antigen increase, that T cells of lower avidity can be recruited.

A second consequence of this competition is that T cells directed against the "losing determinants" will be protected from any antigen-specific selective force, either towards induction of responsiveness or towards tolerization. Only dominant determinants and the most prevalent subdominant determinant on a molecule induce tolerance. In the tolerance scenario, T cells against the remaining determinants are saved from extinction and become part of the "residual repertoire". Although thymic tolerance suffices to purge the mature repertoire of overtly autoaggressive T cells, it can only do so for those self-determinants presented by thymic APC. Clearly, there are numerous antigens expressed outside the thymus or expressed after ontogeny that would be unable to induce thymic tolerance. For these self-directed cells, peripheral tolerance induction is likely the mechanism for avoiding pathology.

The final result of such competitive events in antigen-processing is the presentation to CD4 and CD8 T cells of a limited selection of the total number of determinants available on the protein, which will thereby become the immunodominant focus of the response. During the induction of tolerance in the thymus to cellular antigens, the identical features of this process occur, leading to efficient tolerance induction to a small set of determinants, with a correspondingly less efficient negative selection of T cells directed to less-efficiently displayed determinants. Accordingly, a rather substantial portion of the T cell repertoire exists directed against recessive determinants on self-proteins.

4. TUMOR ANTIGENS

Recent years have witnessed a revolution in our ability to identify tumor antigens capable of eliciting cellular immune responses (8,9). The most successful approaches have used MHC class I-restricted CTL as screening tools for genomic or cDNA libraries transfected into cells expressing the appropriate MHC molecule. Biochemical approaches have also been used in which peptides are first eluted from tumor cells or from MHC molecules isolated from tumor cells prior to their use in assays designed to detect fractions capable of stimulating tumor-specific T cells. Following identification of active fractions, their contents are subjected to triple quadrapole mass spectrometric analysis to obtain sequence information for the peptide species. Finally, the antigenic specificity of serological reactivities found in cancer patients has been successfully mapped using prokaryotically-expressed cDNA libraries prepared from tumors (10,11). This approach assumes a T helper-dependent response has resulted in antibody production.

These studies have revealed 3 main classes of tumor antigens: a) Unmutated tissue-specific differentiation antigens whose expression is shared by the tissue from which the cancer arose; b) "oncofetal" neoantigens expressed almost exclusively by the tumor, with some limited expression in other tissues, and c) mutated self-proteins. This latter class of antigens are most likely specific to a particular individual tumor and will not be considered here. The first two classes of tumor antigens display a wider pattern of expression among tumors and, as such, are better candidate target antigens for immunotherapy. The best-characterized examples of

shared tissue-specific differentiation antigens are the melanocyte-specific melanosomal proteins, such as tyrosinase, MART-1/Melan A, gp100, TRP-1, and TRP-2(9). Although these antigens were first identified as targets using CD8+ CTL and CD4+ T cells from melanoma patients, T cells from healthy individuals have also been shown to capable of recognizing determinants from each of these proteins (8). Other tissue- or organ-specific cancers have revealed similar T- and B- cell target antigens, such as PSA and PMSA in prostate cancer and CEA in colon cancer (12). Taken together, these studies demonstrate the existence in healthy individuals of a self-reactive repertoire that could potentially be directed against tumors. The fact that normal tissues expressing these antigens would also be subject to immune destruction may, upon initial consideration, appear to diminish their use in immunotherapy. As will be discussed, however, there may be ways to focus the attack on the tumor while sparing the normal cells that may antigenically resemble tumor cells. One excellent example of the relationship between tumor immunity and autoimmunity is seen in vitiligo, which often accompanies a successful response to melanoma antigens (13,14), Several groups have found that those anti-tumor treatments that are most effective against melanoma antigens, such as TRP1 and TRP2 and Pmel-17/gp100, concomitantly induce vitiligo. In certain cases, this can be shown to occur in the absence of CD4 T cells (15), whereas in some systems antibodies appear to be involved (16). When potent immunogenic regimens are used, such as recombinant vaccinia virus encoding TRP1, or after blocking CTLA-4, it is evident that deliberate induction of autoreactivity has a beneficial effect (1).

The other class of tumor antigens are exemplified by genes of the MAGE, BAGE, GAGE, and LAGE-1/NY-ESO-1 families, which are expressed in many solid tumor types and some myelomas, but not in normal tissues, except male germline cells lacking MHC class I expression. These antigens are, therefore, strictly tumor-specific and are not expected to have induced thymic tolerance in T cells capable of recognizing dominant determinants they encode. As will be discussed, however, this class of antigens may still induce peripheral extrathymic forms of tolerance.

5. MECHANISMS OF PERIPHERAL TOLERANCE INDUCTION

Induction of self-tolerance prevents autoimmune damage to self. Tolerance can be induced in newly-arising T cells through negative selection in the thymus, or can be induced in mature peripheral T cells. Thymic tolerance is currently understood in terms of affinity/avidity thresholds displayed by the TCR expressed by the T cells as they encounter self-peptides presented by thymic APC. Cells displaying a sufficiently high affinity for self-antigens are deleted, and those displaying too low an affinity fail to be positively selected. As discussed previously, the peripheral repertoire will, therefore, be comprised of cells that do not display strong reactivity to self, but that have nonetheless been sufficiently stimulated by self MHC plus self-peptide to allow positive selection. There are several possible reasons underlying the avoidance of negative selection in the thymus. For example, any factor that can diminish TCR signaling can potentially prevent negative selection. These may include low ligand density on selecting APC or on responding T cells. If one assumes a relatively uniform level of TCR and accessory molecule expression on developing T cells, then the APC emerges as an orchestrator of negative and positive selection events.

Peripheral tolerance mechanisms have, in general, been poorly understood, but recent insights have again implicated APC as playing a critical role (17). Data from transgenic mice expressing a model antigen under control of the rat insulin promoter support the idea that indirect or "cross-presentation" by bone marrow-derived APC results in the transient expansion and subsequent disappearance of antigen-specific CTL (18). Other experimental systems have shown that for antigens expressed outside of the lymphoid environment, cross-presentation by host APC can also result in T cell unresponsiveness (19). These results suggest that cross-presentation of antigens by APC may constitute an essential part of the tolerance mechanism used by peripheral organs to avoid self-reactive cells. Such a mechanism invokes yet another factor in determining whether a given determinant will induce tolerance; i.e., the efficiency with which it participates in cross-presentation. Recent evidence suggests that, in addition to otherwise healthy self-organs, tumors can use this APC-mediated pathway to escape from tumor-specific T cells (20). Interestingly, when APC are activated via CD40 signaling through the use of agonistic antibodies, the cross-tolerance was converted to cross-priming and significant anti-tumor efficacy. Thus, there may be antigens expressed by tumors for which specific determinants have been rendered invisible to the peripheral repertoire due to the induction of cross-tolerance by APC. By modification of their presentation context through APC activation, however, effective T cell responses may be induced. This approach may prove useful not only in identifying new tumor antigens, but also in mobilizing responses from T cells that have recently emerged from the thymus and have not yet encountered a tolerizing peripheral APC.

6. STRATEGIES FOR REGULATING THE SELF-DIRECTED RESIDUAL REPERTOIRE IN CANCER IMMUNOTHERAPY

The antigen presenting cell, by virtue of its ability to control the choice of determinants as well as the context of their presentation, is uniquely poised in T-cell activation and tolerization pathways to exert a profound regulatory

influence over the residual self-directed repertoire. It follows, therefore, that APC represent attractive targets on which to focus immunomodulatory strategies. Emerging data suggests that the activity of T cells can be regulated at the level of APC function, either by applying strategies aimed at activating APC to induce T cell responses or alternatively by blocking interactions which govern either APC activation such that T cells will either fail to be stimulated or will be actively tolerized.

APC activation strategies include the targeting of pathways known to increase the stimulatory capacity of APC, such as signaling through CD40, which results in the upregulation of key cytokines, costimulation and adhesion molecules (21–23). The longevity of APC, as well as their expression of critical inflammatory cytokines could also be increased by signaling through surface molecules, such as TRANCE (24,25). Conversely, APC could be induced to negatively regulate T cells by blockade of CD40 or TRANCE expressed on APC, or their ligands expressed on T cells. These approaches depend on modulating the presentation context of peripheral antigens that participate in the crosspresentation pathway. There is some evidence to suggest that the ability of an antigen to be cross-presented is partly dependent on its stability or level of expression. It is expected, however, that not all antigens will possess the capacity to be cross-presented, i.e., that some antigens, for unknown reasons, will not be represented among the repertoire of determinants on an APC. T cell responses against such antigens will, therefore, not benefit from solely APC-focused strategies. Such antigens will need to be introduced exogenously or endogenously to host APC through immunization before they can become part of any response induced via APC activation. Once this is done, however, it will be a matter of finding out which costimulatory molecules are most critical to inducing a strong and specific response against the desired determinants(s), and then insuring that these are part of the immunological "synapse" between the APC and T cell. In the case of poorly presented determinants, there are approaches to enhancing their presentation efficiency by APC. For CTL determinants, the MHC binding affinity can be enhanced by altering key residues not involved in TCR contacts. Alternatively, the addition of an endoplasmic reticulum (ER)-targeting sequence to the N-terminus of a peptide can result in the efficient presentation of an otherwise poorly-presented determinant. For MHC class II-restricted determinants, specific residues can be changed to create new protease cleavage sites leading to the enhanced generation of a subdominant determinant (26). Processing and presentation can also be enhanced through the addition of specific sequences which provide a target for endopeptidases or which target a given polypeptide for lysosomal degradation, leading to enhanced association with class II molecules. Each of these approaches can be envisioned to be used in powerful viral immunization systems, especially those vectors that can efficiently transduce APC directly.

7. CONCLUDING REMARKS

Within the mature T-cell repertoire of every individual lies a dormant population of potentially self-reactive effectors. Held in check by various regulatory mechanisms, these cells could be mobilized to mount a powerful response against tumors arising in specific tissues. Targets for this response include poorly-displayed determinants on self-proteins as well as determinants from the growing list of tissue-restricted differentiation antigens expressed by tumors of various histotypes (1). An important challenge for the future will be to elucidate the mechanisms through which these cells are regulated such that they can be utilized in a focused anti-self/anti-tumor attack. Promising avenues for this line of investigation will be to understand how regulatory T cells and APCs control the residual self-reactive repertoire, and how these mechanisms can be overcome for immunotherapeutic goals.

References

1. Schoenberger, S.P. and E.E. Sercarz. 1996. Harnessing self-reactivity in cancer immunotherapy. *Semin. Immunol.* 8:303–309.
2. Shimizu, J., S. Yamazaki, and S. Sakaguchi. 1999. Induction of tumor immunity by removing CD25+ CD4+ T cells: a common basis between tumor immunity and autoimmunity. *J. Immunol.* 163:5211–5218.
3. Thornton, A.M. and E.M. Shevach. 2000. Suppressor effector function of CD4+ CD25+ immunoregulatory T cells is antigen nonspecific. *J. Immunol.* 164:183–190.
4. Thompson, C.B. and J.P. Allison. 1997. The emerging role of CTLA-4 as an immune attenuator. *Immunity* 7:445–450.
5. Mason, D. and F. Powrie. 1998. Control of immune pathology by regulatory T cells. *Curr. Opin. Immunol.* 10:649–655.
6. Kumar, V. et al. 1995. Major histocompatibility complex binding affinity of an antigenic determinant is crucial for the differential secretion of interleukin 4/5 or interferon gamma by T cells. *Proc. Natl. Acad. Sci. USA* 92:9510–9514.
7. Sercarz, E.E. et al. 1993. Dominance and crypticity of T cell antigenic determinants. *Annu. Rev. Immunol.* 11:729–766.
8. Rosenberg, S.A. 1999. A new era for cancer immunotherapy based on the genes that encode cancer antigens. *Immunity* 10:281–287.
9. Boon, T. and L.J. Old. 1997. Cancer tumor antigens. *Curr. Opin. Immunol.* 9:681–683.
10. Sahin, U., O. Tureci, and M. Pfreundschuh. 1997. Serological identification of human tumor antigens. *Curr. Opin. Immunol.* 9:709–716.
11. Chen, Y.T. et al. 1997. A testicular antigen aberrantly expressed in human cancers detected by autologous antibody screening. *Proc. Natl. Acad. Sci. USA* 94:1914–1918.
12. Corman, J.M., E.E. Sercarz, and N.K. Nanda. 1998. Recognition of prostate-specific antigenic peptide determinants by human CD4 and CD8 T cells. *Clin. Exp. Immunol.* 114:166–172.
13. Okamoto, T. et al. 1998. Anti-tyrosinase-related protein-2 immune response in vitiligo patients and melanoma patients receiving active-specific immunotherapy. *J. Invest. Dermatol.* 111:1034–1039.
14. Ogg, G.S., P. Rod Dunbar, P. Romero, J.L. Chen, and V. Cerundolo. 1998. High frequency of skin-homing melanocyte-specific cytotoxic T lymphocytes in autoimmune vitiligo. *J. Exp. Med.* 188:1203–1208.
15. van Elsas, A., A.A. Hurwitz, and J.P. Allison. 1999. Combination immunotherapy of B16 melanoma using anti-cytotoxic T lymphocyte-associated antigen 4 (CTLA-4) and granulocyte/macrophage colony-stimulating factor (GM-CSF)-producing vaccines induces rejection of subcutaneous and metastatic tumors accompanied by autoimmune depigmentation. *J. Exp. Med.* 190:355–366.
16. Overwijk, W.W. et al. 1999. Vaccination with a recombinant vaccinia virus encoding a "self" antigen induces autoimmune vitiligo and tumor cell destruction in mice: requirement for CD4(+) T lymphocytes. *Proc. Natl. Acad. Sci. USA* 96:2982–2987.
17. Carbone, F.R., C. Kurts, S.R. Bennett, J.F. Miller, and W.R. Heath. 1998. Cross-presentation: a general mechanism for CTL immunity and tolerance. *Immunol Today* 19:368–373.
18. Heath, W.R., C. Kurts, J. Miller, and F.R. Carbone. 1998. Cross-tolerance: A pathway for inducing tolerance to peripheral tissue antigens. *J. Exp. Med.* 187:1549–1553.
19. Adler, A.J. et al. CD4+ T cell tolerance to parenchymal self-antigens requires presentation by bone marrow-derived antigen-presenting cells. *J. Exp. Med.* 187:1555–1564.
20. Sotomayor, E.M. et al. 1999. Conversion of tumor-specific CD4+ T-cell tolerance to T-cell priming through *in vivo* ligation of CD40. *Nat. Med.* 5:780–787.
21. Schoenberger, S.P., R.E. Toes, E.I. van der Voort, R. Offringa, and C.J. Melief. 1998. T-cell help for cytotoxic T lymphocytes is mediated by CD40-CD40L interactions. *Nature* 393:480–483.
22. Cella, M. et al. 1996. Ligation of CD40 on dendritic cells triggers production of high levels of interleukin-12 and enhances T cell stimulatory capacity: T-T help via APC activation. *Journal of Experimental Medicine* 184:747–752.
23. Koch, F. et al. 1996. High level IL-12 production by murine dendritic cells: upregulation via MHC class II and CD40 molecules and downregulation by IL-4 and IL-10. *Journal of Experimental Medicine* 184:741–746.
24. Wong, B.R. et al. 1997. TRANCE (tumor necrosis factor [TNF]-related activation-induced cytokine), a new TNF family member predominantly expressed in T cells, is a dendritic cell-specific survival factor. *J. Exp. Med.* 186:2075–2080.
25. Josien, R., B.R. Wong, H.L. Li, R.M. Steinman, and Y. Choi. 1998. TRANCE, a TNF family member, is differentially expressed on T cell subsets and induces cytokine production in dendritic cells. *J. Immunol.* 162:2562–2568.
26. Schneider, S.C., Ohmen, J., Fosdick, L., Gladstone, B., Guo, J., Ametani, A., Sercarz, E.E., and Deng, H. 2000. Cutting edge: introduction of an endopeptidase cleavage motif into a determinant flanking region of hen egg lysozyme results in enhanced T cell determinant display. *J Immunol.* 165:20–3.

63 | The Nature and Significance of Age-Associated Autoimmunity

Joong-Won Lee, Fang Jin, and Marc E. Weksler

1. INTRODUCTION

Our understanding of immunity in general and autoimmunity in particular has been derived almost exclusively from the study of young adult subjects. A fundamental observation concerning the operation of the immune system is that exposure of healthy young humans or experimental animals to molecules foreign to the host stimulates a specific immune response to the nominal antigen, although no immune response is normally made to the large number of molecules of the host. Ehrlich called this fundamental law of the immune system "horror autotoxicus" (1). When autoimmune reactions develop in young individuals, as the reader of this volume knows, autoimmune disease usually follows. In contrast, during aging, autoimmune reactions develop but rarely cause autoimmune diseases. In this chapter, the nature and significance of age-associated autoimmunity will be considered.

2. AN OVERVIEW OF IMMUNE SENESCENCE

Increased autoimmune reactions are just one of many changes in the immune system that occur during aging (Table 1). The involution of the thymus gland, the earliest age-associated change in the immune system, begins in childhood and is virtually complete by midlife (2). As a consequence of the age-associated involution of the thymus, there is a progressive decrease in the output of T cells from the thymus required to maintain a diverse and naive T cell repertoire (3,4). As a consequence of the reduced output of T cells from the thymus during aging, the number of peripheral T cells that express surface markers characterisitc of naive T cells and molecular markers char-

acteristic of recent T cell emigrants decrease (4,5). Despite the decreased output of T cells from the thymus during aging, the number of peripheral T cells changes little. This is partly due to the long life of peripheral T cells and their capacity for self-renewal, which compensates for the decreasing number of T cells exported from the thymus (6). However, when peripheral T cells are depleted following

Table 1 Age-associated changes in the lymphoid system

I. T cell Development and Function
  A. Thymic involution and decreased generation of T cells
  B. Decreased naive and increased memory T cells
  C. Decreased T cell diversity with increased frequency of T cell clonal expansions
  D. Decreased proliferative response to antigens and mitogens
  E. Decreased production of IL-2 and increased production IL-4 and IL-6
  F. Decreased number of germinal centers
  G. Decreased CD4 T cell helper and CD8 T cell cytotoxic activity
  H. Decreased CD8 T cell helper activity for bone marrow B cell development

II. B Cell Development and Function
  A. Decreased generation of conventional B2 cells within the bone marrow
  B. Increased number and activity of B1 cells
  C. Increased number of Ig-secreting B cells
  D. Increased concentration of serum immunoglobulins
  E. Appearance of B-cell clonal expansions and monoclonal immunoglobulins
  F. Decreased antibody response to foreign antigens
  G. Increased frequency of autoantibodies

chemotherapy or HIV infection, the regeneration rate of the peripheral T cell count is reduced in old compared to young individuals (7).

While the number of peripheral T cells changes little with age, there is an age-associated loss of cell-mediated immunity *in vitro* and *in vivo*. *In vitro*, T cells from elderly humans or old animals are impaired in their capacity to enter the cell cycle and undergo repeated divisions in culture with plant lectins, allogeneic leukocytes or mitogenic antibodies (8–10). The production of cytokines by T cells in culture also changes with age. Cultured T cells from elderly donors produce less IL-2, but more IL-4 and IL-6, than cultured T cells from young donors (11,12). The decreased proliferative capacity of T cells from old individuals in culture is due, in part, to the decreased secretion of IL-2, which can be reversed by adding exogenous IL-2 to the cultures. The reversal of the proliferative defect by exogenous IL-2 is only partial because half of the activated T cells from elderly donors cultured with mitogens do not develop high-affinity receptors for IL-2 (13,14). The decreased T cell proliferation probably explains the impaired development of cytotoxic T cells from old donors in culture (15).

The length of telomeres in cultured fibroblasts has been shown to be a critical factor in the proliferative limits of these cells in culture (16). Transfection of the telomerase gene into cultured fibroblasts extended their proliferative capacity in culture. The length of telomeres in T cells also decreases with the age of the T cell donor and with the number of T cell divisions in culture (17,18). It remains to be proven whether the transfection of the telomerase gene into T lymphocytes will extend their replicative capacity, as it does for fibroblasts.

In vivo, T cells mediate delayed cutaneous hypersensitivity reactivity and the antibody response to T-dependent antigens, both functions that are compromised during aging in humans (19–21). These functions depend on T cell clonal expansion, and the age-associated defect in T cell proliferation would be expected to be the mechanism underlying these functional defects. As observed *in vitro*, administration of IL-2 in association with influenza vaccine *in vivo* partially corrected the impaired antibody response to influenza vaccine (22). Finally, aging is associated with the appearance of T cell clonal expansions (23–25), and although their functional consequences are not totally clear, preliminary data suggests that they may inhibit the development of specific cytotoxic T cell responses.

The development and function of conventional B2 lymphocytes decline with age (26), as does the rate of generating conventional B cells from the bone marrow (26,27). This defect in B lymphocyte development reflects an impaired transition of Pro-B cells into Pre-B cells (28). The age-associated defect in the generation of Pre-B cells has been reported to be associated with impaired generation of

the Pre-B cell receptor and an increased rate of Pre-B cell apoptosis in bone marrow of old compared to young mice (28,29). The impaired production of Pre-B cells in old mice is associated with a decreased production of IL-16 by CD8 T cells (28). The decreased production of IL-16 appears to be functionally important, since the injection of rIL-16 into old mice decreases the rate of apoptosis of their bone marrow Pre-B cells and increases the number of Pre-B cells in their bone marrow.

The age-associated impairment in antibody production by B2 lymphocytes to most foreign antigens is also attributable to impaired T cell function (30), which leads not only to decreased production of antibody specific for the nominal antigen, but also a preferential loss of high-affinity IgG antibodies. High-affinity IgG antibodies develop within the germinal center during the processes of isotype switching and somatic mutation. The germinal center reaction, isotype switching, and somatic mutation are all highly T-cell dependent processes shown to be impaired in elderly mice (30–32). High-affinity antibodies are the most protective class of anti-bacterial antibodies (33) and their loss with age probably explains the observation that the same quantity of anti-pneumococcal antibody from old mice passively protected fewer young mice challenged with pneumococci than anti-pneumococcal antibody from young mice.

Despite the age-associated decrease in the antibody response to T-dependent foreign antigens by B2 lymphocytes, there is a marked increase in the number of Ig-secreting B cells in old compared to young mice (34) and an increased concentration of serum Ig in old compared to young mice (35,36). The polyclonal activation of B cells probably results from the increased secretion of IL-4 and IL-6 by T cells in old mice (12,37). The polyclonal B cell activation that develops during aging explains not only the increasing concentration of serum immunoglobulins, but also the appearance of serum autoantibodies. It is known that the repertoire of antibody specificities produced by B1 lymphocytes is skewed toward the production of autoantibodies (38). Overall, aging is associated with a decreased production by B2 lymphocytes of antibodies specific for foreign antigens, although the B1 lymphocyte production of autoantibodies and T-independent antibody responses to foreign antigens are undiminished with age (30,34).

3. AGE-ASSOCIATED AUTOIMMUNITY

3.1. Increased Frequency of Autoantibodies During Aging

The age-associated decline in the capacity of B cells to produce antibodies to virtually all foreign antigens would be expected to result in a decreased concentration of serum

Ig. In fact, serum Ig concentrations and the number of Ig-secreting B cells increase during aging (34–36). This paradox led to the suggestion that increased production of autoantibodies during aging compensates for the decreased production of antibodies to foreign antigens. More than 30 years ago, the serum concentration of an autoantibody and anti-bacterial antibody was compared (39). It was found that the percentage of individuals with serum anti-nuclear autoantibodies increased with age, while the serum concentration of anti-salmonella flagellin antibody titer decreased. These data have been redrawn in Figure 1 to demonstrate the shift in specificities of serum Ig from foreign to self antigens with age. Since this report, a number of studies have shown an increase in both organ-specific, e.g. anti-thyroglobulin, and non-organ-specific, e.g. anti-DNA, autoantibodies during aging (Table 2). In one study, 60% of healthy, elderly subjects were found to have at least one of the following autoantibodies: Rheumatoid factor, antinucleoprotein, or antithyroglobulin autoantibodies (40).

The health of elderly individuals influences their expression of autoantibodies. Chronically ill elderly humans have higher serum levels of non-organ-specific autoantibodies, such as rheumatoid factor and anti-DNA autoantibody, than do healthy elderly humans (41). Thus, co-morbidity may explain, at least in part, the reason why elderly persons with non-organ-specific autoantibodies are at higher risk of death than elderly without autoantibodies (42). In contrast to organ-non-specific autoantibodies, organ-specific autoantibodies, such as anti-thyroglobulin autoantibodies, increase with age in healthy individuals (43). Figure 2 shows the progressive increase in serum antithyroid peroxidase autoantibodies with age in healthy individuals less

than 100 years old (44). It should be noted that centenarians have a lower level of anti-thyroid autoantibodies than do other elderly subjects. This fact suggests that centenarians are an immunologically-privileged population of elderly subjects.

Although the specificity of autoantibodies detected in old persons, rheumatoid factor and anti-DNA antibodies, are the same as those detected in the serum of patients with rheumatoid arthritis and systemic lupus erythematosus, it is unlikely that elderly subjects with these autoantibodies have subclinical autoimmune disease. First of all, classical autoimmune diseases like rheumatoid arthritis and systemic lupus erythematosus rarely appear for the first time in persons over 65 years of age. Second, although the serum autoantibodies in elderly bind the same antigens, they differ

Table 2   Percentage of aged and young adults with non-organ-specific autoantibodies

Autoantibodies	Aged (>65 years)	Adult (18–45 years)
Antinuclear antibody[a]	27.0	7
Anticentromere	0.3	0
Anti-ssDNA	15.3	1
Anti-dsDNA	7.6	0
Anticardiolipin		
IgM	8.6	1
IgG	25.3	2
IgM rheumatoid factor	41.3	3

Adapted from Xavier *et al.* (1995) and Ruffati et al. (1990)

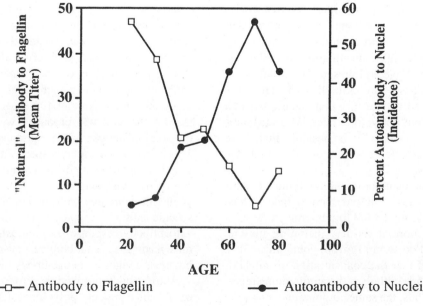

Figure 1.   Cross-wiring of the Humoral Immune Response with Age. Adapted from Rowley et al. (1968).

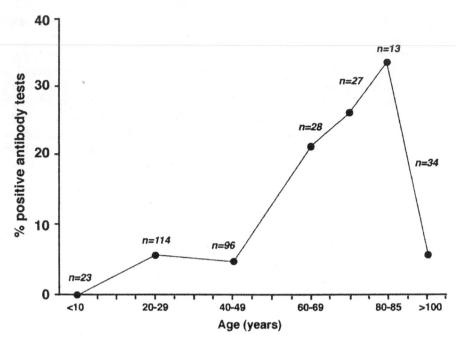

Figure 2. Change in Prevalence of Antithyroid Peroxidase (anti-TPO) in Health Individuals with Age. Adapted from Mariotti et al. (1992).

immunochemically from those present in young patients with autoimmune disease. Pathogenic autoantibodies in young persons with autoimmune disease are somatically-mutated, IgG antibodies with high affinity for the autoantigen. In contrast, serum autoantibodies in elderly individuals are usually in germ line, IgM antibodies of low affinity with polyreactivity (45).

A similar age-associated shift in the specificity of serum Igs from foreign to autologous antigens occurs in mice. Thus, the serum levels of autoantibodies that react with IgG, DNA, and thyroglobulin are higher in old compared to young mice (46). Among the most common serum autoantibodies found in mice are IgM, anti-thymocyte antibodies, first detected at 1 year of age and present in virtually all 2 year-old mice (47). These autoantibodies can enter the thymus as demonstrated by the fact that freshly-isolated thymocytes from old, but not young, mice have IgM antibodies bound to their surface. It is possible that these autoantibodies may interfere with normal thymocyte development.

The specificities of autoantibodies are not limited to antigens with which pathogenic autoantibodies react. Thus, more serum Ig from old mice bind to numerous molecules present in lysates of normal tissues, including muscle, spleen, and brain, than do serum Ig from young mice (48). These studies revealed that the concentration of autoantibodies increased with age, although the autoimmune repertoire did not change. Thus, the same autoimmune repertoire can be detected in sera from both young and old mice,

although at a very much lower concentrations in sera from young mice.

3.1. Cross-Wiring of the Immune Response

Figure 1 illustrates the age-associated shift in the specificities of serum antibodies with age. Serum from young adults have a high concentration of antibodies that react with foreign antigens and a low concentration of autoantibodies. In the elderly, the reverse is seen, with a low concentration of serum antibodies that react with foreign antigens and a high concentration of autoantibodies. This "cross-wiring" of the immune response is also observed after immunization. Thus, *in vivo* immunization of young mice stimulates a vigorous antibody response to the nominal antigen without stimulating many autoantibodies (34,49). In contrast old mice immunized in the same way produce a weak antibody response to the nominal antigen and a vigorous autoantibody response. This imbalance of antibody specificities observed in old mice is an inherent property of spleen cells, since the phenomenon can be demonstrated *in vitro* (49). Thus, spleen cells from young mice cultured with foreign antigens produce a highly specific antibody response to the nominal antigen with little autoantibody response. In contrast, spleen cells from old mice made a weak antibody response to the nominal antigen, but made a vigorous autoantibody response.

Autoantibody production before and after immunization of old mice reflects polyclonal activation of B1 B cells. First of all, old mice have two to three times the number of splenic Ig-secreting B cells prior to and after immunization *in vivo* and *in vitro* (37). Secondly, mice that express the Xid mutation and lack B1 B cells do not develop serum autoantibodies with age (46). It is likely that the increased secretion of IL-4 and IL-6 by T cells in old animals contribute importantly to the age-associated polyclonal B cell activation. In preliminary studies, we have shown that administration of exogenous IL-4 to young mice stimulates an increase in the number of Ig-secreting B cells, an increase in serum Ig and an increased level of serum autoantibodies.

Studies in humans also reveal that immunization stimulates an age-associated increase in autoimmune responses (50,51). Thus, immunization with tetanus toxoid stimulated an increased number of rheumatoid factor-specific precursor B cells in the peripheral blood of old compared to young adults. Furthermore, elderly humans immunized with influenza vaccine show an inverse correlation between the production of antibodies to the nominal influenza hemagglutin in (HA) and anti-DNA autoantibodies. Thus, elderly individuals with the highest anti-influenza HA antibody titers had the lowest anti-DNA antibody levels and vice versa.

3.2. Increased Auto-Anti-Idiotypic Autoantibodies During Aging

One type of autoantibody that increases with age after immunization is anti-idiotypic autoantibody (52). The production of autoanti-idiotypic antibody is regulated by long-lived peripheral T cells in old mice. The increase in anti-idiotypic autoantibodies not only reflects the age-associated change in antibody specificities from foreign to self-antigens, but also contributes to the impaired antibody response to foreign antigens in old mice. The contribution of anti-idiotypic autoantibodies to the age-associated decline in the antibody response to foreign antigens was demonstrated by showing that the antibody response to nominal antigen in old mice increased when the production of anti-idiotypic autoantibody was inhibited (53).

In humans, anti-idiotypic autoantibodies are present in the serum for prolonged periods of time and inhibit the antibody response to a booster immunization (54). Thus, the impaired anti-tetanus toxoid antibody response following boosting was associated with the presence of increased serum levels of anti-idiotypic autoantibodies specific for anti-tetanus toxoid antibodies at the time of boosting of elderly but not young adults.

3.3. Appearance of Monoclonal Immunoglobulins During Aging

The dysregulation of the immune response during aging is manifested not only by an increased production of autoanti-

bodies, but also by the increased frequency of serum monoclonal Ig (55). Age-associated monoclonal Ig were first discovered in elderly humans who had no evidence of multiple myeloma (56). However, it is now clear that although elderly individuals may have serum monoclonal immunoglobulins without multiple myeloma, these persons are at increased risk of developing multiple myeloma (57). The repertoire of monoclonal Ig, like all Ig specificities in the elderly, is skewed toward autoimmune reactions. Approximately 50% of age-associated monoclonal Ig react with autoantigens (58). Like the development of autoantibodies during aging, the incidence of monoclonal Ig is influenced by thymic function and chronic disease (43,59). Thus, the development of monoclonal immunoglobulines is increased by thymectomy and inhibited by the administration of young thymocytes.

The B lymphocytes that produce serum monoclonal Ig are not randomly selected from the B cell population. Age-associated monoclonal Ig that develop in transgeneic mice expressing an Ig heavy chain gene are selected from the small population of B cells that escape allelic exclusion and express endogenous Ig heavy chain genes (60). It is not certain which B cell subset gives rise to monoclonal Ig, but considerable circumstantial evidence suggests that B1 lymphocytes are responsible for the production of monoclonal Ig. Thus, Xid mice that do not have B1 lymphocytes do not develop monoclonal Ig during aging, and the clonal B cell expansions that appear with age are derived from the B1 population of lymphocytes (46,61).

It is also known that clonal expansions of B1 lymphocytes spontaneously develop in old mice (61). It has been shown that virtually all mice over 18 months of age have clonal B cell expansions demonstrable in the spleen and other peripheral lymphoid compartments. (61). The link between B1 lymphocyte clonal expansions, monoclonal Ig, and autoantibodies is supported by the observation that patients with chronic lymphocyte leukemia, a neoplasm of B1 lymphocytes, have monoclonal Ig in their serum and their CLL cells produce autoantibodies in culture (62,63).

3.4. Autoreactive T Cells

During aging, there is an increasing number and activity not only of autoreactive B cells, but also autoreactive T cells (64,65). It was demonstrated that injection of spleen cells from old, but not young, mice induced a syngeneic local graft-versus-host (GVH) in syngeneic young mice (64). T cells in the spleens of old mice mediate the local GVH reaction. In other studies, it has been shown that clonal expansions of an unusual CD4 T cell subset, which do not express CD28 on their surface, develop with age (66). These cells react with autoantigens *in vitro* and may contribute to the skewing of the immune compartment towards autoreactive responses and away from response to exogenous antigens. It has also been shown that thymectomized mice had an increased capacity to induce the syngeneic GVH reaction, suggesting that the age-associated thymic involution may be an important factor permitting the expression of autoreactivity among peripheral T cells (67).

In other studies, it was demonstrated that spleen cells from old mice not only stimulated a syngeneic GVH, but also a HVG reactions (65). The local, syngeneic HVG reaction is mediated by host T cells that react against irradiated spleen cells from old, but not young, mouse injected into the footpad. The targets of the HVG reaction are Ia-positive spleen cells. The nature of the age-associated changes in Ia-positive spleen cells that stimulate the HVG reaction is not known.

4. MECHANISMS OF AGE-ASSOCIATED AUTOIMMUNITY

Several factors contribute to age-associated autoimmunity: The appearance of neo-antigens during aging to which the immune system responds, a decline in regulatory forces that downregulate autoreactive lymphocyte activity, polyclonal B cell activation, which stimulates autoreactivity, and changes in the number and activity of autoreactive populations. As mentioned above, Ia+ spleen cells from old mice during aging develop the capacity to stimulate syngeneic T cells. Further support for the capacity of non-hematopoietic cells in old mice to stimulate autoreactivity was demonstrated in transfer experiments (68). Bone marrow from young or old mice reconstitutes, to a similar extent, the number and function of thymic and splenic lymphocytes in irradiated, syngeneic recipients. However, old recipients develop higher levels of serum autoantibodies and greater glomerular lesions after reconstitution with bone marrow from either young or old donors than do young mice reconstituted with bone marrow. These results indicate that peripheral tissues of old mice stimulate autoantibody formation and the deposition of immune complexes in the kidney.

Several reports suggest that non-specific suppressor activity declines with age (69). Non-specific suppressor activity decreased antibody activities by B cells. Other studies show that the increasing suppressor activity associated with aging act on B2, but not B1, lymphocytes (49). Thus, spleen cells from old mice suppress B2 lymphocyte responses to foreign antigens, but do not suppress, and may even enhance, autoantibody production by B1 lymphocytes.

Deletion of self-reactive lymphocytes plays an important role in self tolerance. We have reported that negative selection of T cells may be impaired in old mice (70). In these studies, young BALB/c mice, which delete the VB 11 TCR

family within the thymus, have a peripheral T cell repertoire almost completely lacking the VB 11 TCR. The negative selection of VB 11 T cells decreases with age so that there is a 3-to 4-fold increase in the number of T cells that express the VB 11 TCR in old BALB/c mice. Furthermore, these potentially autoreactive T cells in old mice are functionally competent, as judged by their capacity to proliferate when cultured with staphylococcal enterotoxin B, a superantigen that stimulates the proliferation of VB 11 TCR-bearing T cells. This superantigen stimulates a greater proliferative response in spleen cells from old compared to young BALB/c mice. In contrast, the proliferation of splenic T cells from old mice with other superantigens that stimulate T cells expressing other VB families not deleted during aging is actually lower in old than young mice.

In other strains of mice, there appears to be less age-associated escape of T cells normally deleted in the thymus (71). In these studies, the appearance of the normally-deleted T cells was seen in a smaller percentage of old mice and was usually limited to the very oldest mice studied. However, the T cells that escaped negative selection were competent, since antibodies to the normally deleted VB families stimulated greater T cell proliferation in old than young mice. The presence of potentially autoreactive T cells in old mice also contribute to the increased level of autoantibodies by stimulating polyclonal B cell activation. Polyclonal B cell activation follows the interaction of autoreactive T cells specific for MHC class II determinants on the surface of B cells and leads to the production of organ-specific autoantibodies in mice undergoing the syngeneic GVH rection (72). It is likely that the actions of these self-MHC class II-specific T cells are augmented by the effect of increased levels of IL-4 and IL-6 in old mice that also stimulate polyclonal B cell activation. Our preliminary results show that young mice given exogenous IL-4 develop greater numbers of Ig- and autoantibody-producing B cells. Thus, it is reasonable to propose that T cells with self-MHC reactivity, together with increased levels of IL-4 and IL-6 in old mice, contribute to both polyclonal B cell activation and autoantibody production.

Whatever the causes, there is no doubt that polyclonal B-cell activation is a characteristic of immune senescence. We have demonstrated that spleen cells from old mice have 5 to 10 times as many spontaneously Ig-secreting cells and twice the number of low density cycling B cells than young syngeneic mice (34,52). The link between polyclonal B cell activation and autoantibody production is supported by the capacity of lipopolysaccharide (LPS) to stimulate autoantibody production by murine lymphocytes *in vitro* or *in vivo* (73,74). Furthermore, old mice are more sensitive to LPS-stimulated autoantibody production. Thus, old mice produced more autoantibodies after injection of LPS than young mice. Similarly, spleen cells from old mice cultured *in vitro* with LPS produced more autoantibodies than spleen cells cultures from young mice. It is likely that age-associated thymic involution contributes to this phenomenon in old mice, since a greater autoantibody response following LPS injection was observed in thymectomized than in normal mice (75). In addition to gram-negative bacterial LPS stimulating autoantibodies, it is possible that latent infections with herpes zoster/varicella virus, EB virus, or mycobacteria tuberculosis stimulate polyclonal B cell activation and autoantibody production (76).

Polyclonal B cell activation would be expected to generate more autoantibodies in old than young mice, since the B cell repertoire becomes skewed toward autoantigen specificities during aging. The increased number of B cells with autoimmune specificities result from the increased number of precursors of autoantibody-forming B cells and the decreasing capacity of T cells in old mice to facilitate the light chain editing necessary to transform the preimmune repertoire, enriched for autoreactive B cells, to a postimmune repertoire enriched in foreign antigen specificities (77).

The increased production of autoantibodies during aging may also reflect the increased size and activity of the B1 lymphocyte population (78). It is known that the B1 lymphocyte population is enriched for autoimmune reactivity (38,77). Many of the autoantibodies that increase with age, including rheumatoid factor, anti-Br-MRBC, and antoanti-idiotypic antibodies, are produced to a great extent by B1 lymphocytes. The reaction of auto-anti-idiotypic antibodies with the antigen receptors on B2 lymphocytes inhibit their secretion of antibody specific for foreign antigens (79). Thus, the increase in anti-idiotypic autoantibodies not only reflects the dysregulated immune response associated with aging, but also contributes to the suppression of antibodies specific for foreign antigens. Thus, certain autoantibodies that arise with age contribute to the immune deficiency observed during immune senescence.

We have shown that the level of anti-idiotypic autoantibodies is regulated by peripheral T cells (52). Transfer experiments revealed that the injection of long-lived peripheral T cells from old, but not young, mice selected bone marrow B cells in syngeneic recipients that produce large amounts of autoanti-idiotypic antibodies. With age, long-lived memory peripheral T cells that stimulate the proliferation and activation of anti-idiotypic autoantibody-forming cells come to dominate the the T cell repertoire (80). It is possible that a vicious circle develops during aging with the production of antibody of limited diversity favoring the development of anti-idiotypic autoantibody that in turn, further limits the diversity of the antibody response.

We have used limiting dilution analysis of LPS-activated spleen cells from young and old mice to determine the frequency of precursors of autoantibody-forming cells (81).

Results of such analysis reveal that there is an increased frequency of autoantibody precursors in old compared to young mice. In contrast, the frequency of B cells specific for foreign antigen does not change with age (82). There are also links between the age-associated development of autoantibodies, of monoclonal expansions of B1 lymphocytes, and of monoclonal immunoglobulins. It is known that the antibody repertoire of B1 lymphocyte population is skewed toward autoantigens and that cultured CLL CD5+ B cells from elderly patients produce autoantibodies. The majority of these autoreactive immunoglobulins, like those typically found in the serum of elderly people, have characteristics of so-called natural antibodies: IgM isotype, low affinity for several autoantigens, and germline Vh region sequences (83,84). Furthermore, mice that express the Xid mutation and lack the B1 lymphocyte population do not develop serum autoantibodies or monoclonal Ig during aging (46). Thus, there is a considerable circumstantial evidence that changes in the B1 lymphocyte population play a large role in age-associated autoimmunity.

The development of autoreactive lymphocytes, by definition, indicates that self-tolerance declines with age. However, both B and T cell tolerance to foreign antigens are more difficult to induce in old compared to young mice (85,86). The degree of immune depression and the duration of immune tolerance is less in old than young mice. Furthermore, a higher dose of tolerogen is required to attain a comparable degree of tolerance.

There is, however, no evidence that autoantibodies or autoimmune diseases are more easily induced by foreign antigen in old than young mice. Cross-reacting xenoantigens induced no more autoantibodies in old than young mice (87). Furthermore, both collagen-induced arthritis and 16/6 anti-DNA antibody idiotype induced lupus were more difficult to induce in old than young mice (88,89). These observations probably can be explained by the impaired capacity of B2 lymphocyte in old mice to respond to foreign antigens in general and their preferential loss of high affinity, IgG antibodies in this response.

5. SIGNIFICANCE OF AGE-ASSOCIATED AUTOIMMUNITY

Autoantibodies can play an important role in the pathogenesis of disease. It is, therefore, important to consider whether the autoantibodies that appear during aging might contribute to the aging process and/or the diseases of aging. Thirty years ago, Walford proposed that neoantigens on senescent cells stimulate autoantibodies that, in turn, damaged cells and tissues, leading to the aging process and the diseases of aging (90). However, few neoantigens have been demonstrated on senescent cells. The best studied is an antigen on senescent erythrocytes with which a serum autoantibody reacts (91). This interaction has been proposed to play a role in the removal of effete erythocytes from the circulation. It is true that under special circumstances, elderly patients with lymphoma may develop pathogenic autoantibodies that cause hemolytic anemia, leukopenia or thrombopenia (92). However, these pathogenic autoantibodies are produced by neoplastic B cells and not by senescent lymphocytes.

Although the mechanism underlying the formation of age-associated autoantibodies remains uncertain and their consequences are debated, there is considerable epidemiological evidence that the presence of serum autoantibodies is a risk factor for death among the elderly (42, 44). Thus, elderly with serum autoantibodies had a survival significantly shorter than age-matched controls without serum autoantibodies. This may reflect that fact that chronically-ill, elderly individuals are more likely to have autoantibodies than age-matched controls (41,43). For this reason, it remains uncertain whether autoantibodies contribute to, or result from, the chronic diseases that cause increased morbidity and mortality among the elderly.

From the clinical point of view, autoimmune disease, with the exception of hypothyroidism, pernicious anemia, and polymyalgia rheumatica, generally affect young women between the ages of 30 and 40 and rarely have an onset late in life. (93). In fact, the elderly patient with arthritis and serum autoantibodies reactive with IgG and/or DNA, in the absence of previous autoimmune disease, is more likely to have osteoarthritis than rheumatoid arthritis or systemic lupus erythematosus. The pathological lesions seen in autoimmune diseases involve predominantly the hematopoietic, renal, and articular tissues in contrast to the pathology of aging, which involves the cardiovascular system, cancer, and the brain.

Pathogenic autoantibodies and antibodies induced by T-dependent foreign antigens share many molecular and genetic characteristics that differentiate them from autoantibodies found in the elderly. Pathogenic autoantibodies are usually highly specific, IgG antibodies with high affinity for the autoantigen and do not cross react with other antigens (94,95). The autoantibodies that occur with age, in contrast to pathogenic autoantibodies, are typically of the IgM isotype, have low to moderate affinity for autoantigen, and frequently are cross-reactive, polyspecific, and share idiotypes (79).

Genetic analysis of pathogenic antibodies reveals somatic mutations in regions coded for by variable region Ig genes (96). The presence of somatic mutations in pathogenic autoantibodies has been taken as evidence that the B lymphocytes that produce them have been selected for clonal expansions based on their high binding affinity for autoantigen or a foreign antigen that cross-reacts with an autoantigen. In contrast, the antibodies found in the serum of the elderly resemble natural autoantibodies that are coded for by gemline Ig variable region genes (77). This suggests that evolutionary pressure maintains the specificity of these autoantibodies.

Age-associated autoantibodies resemble so-called "natural" autoantibodies that are found in serum from newborns at high concentration and at lower concentrations in serum from young adults (97). These autoantibodies are members of the pre-immune repertoire and are coded for by unmutated germ-line genes. Similar antibodies are produced by antigen-deprived, germ-free animals and appear to be products of the "internal" activity of the immune system (98).

The fact that "natural" autoantibodies are coded for by germline Ig variable region genes suggests that evolutionary pressure maintains the specificity of these autoantibodies. Based on this assumption, it has been inferred that "natural" antibodies play physiological roles. One possible role of this class of antibodies is in the development of the immune system based on the dominance of antibodies with autoantigen specificities during the early development of the immune system (77). Thus, a high percentage of the hybridomas derived from neonatal B cells produce autoantibodies, have multiple idiotype-anti-idiotype interactions, and share immunochemical characteristics with natural antibodies. In addition, B1 lymphocytes that produce natural antibodies, represent a very significant percentage of the B cell repertoire during the neonatal period.

Natural antibodies are included in the so-called "internal activity" of the immune system that continues to be highly represented in adult germ-free mice deprived of antigenic stimulation (98). In contrast, in conventionally reared mice, the "external activity" of the immune system that develops in response to foreign antigens rapidly comes to dominate the immune system (99). However, the B cell repertoire that is represented within the "internal activity" of the immune system, although down-regulated in adulthood, survives and can be demonstrated as natural autoantibody-secreting B cells found within the low density, *in vitro*-activated B cell population in adults. Furthermore, polyclonal B cell activators stimulate the production of natural autoantibodies in normal adult mice. These autoantibodies share many of the characteristics of the pre-immune, natural antibodies, present in old animals.

The senescence of the immune system is associated with a return to the dominance of the "internal activity" of the immune system, including the increasing levels of natural autoantibodies and autoanti-idiotypic antibodies associated with a 50% increase in the number of activated, low density splenic lymphocytes, a doubling of the number of peritoneal CD5[+] B cell and a rising frequency of most B cell precursors for autoantibodies. In part, this may result from the greater production of autoanti-idiotypic antibodies with age that downregulates the "external activity" of the immune system.

It is also possible that age-associated autoantibodies are neither pathogenic nor natural autoantibodies, but pathological autoantibodies that do not cause disease, but are caused by disease. A number of pathological autoantibodies occur in chronic infectious diseases. For example, high serum concentrations of rheumatoid factor occur in approximately 50% of patients with subacute bacterial endocarditis (SBE) (100). The suggestion that rheumatoid factor in SBE is a pathological autoantibody is supported by the fact that these autoantibodies disappear following successful antibiotic treatment of SBE. Other examples of pathological autoantibodies include anti-cardiac muscle autoantibodies that develop after myocardial infarction and anti-erythrocyte autoantibodies that occur after viral infections, such as infectious mononucleosis. Subclinical infection has also been implicated as the cause of high rheumatoid factor levels in certain inbred mice strains (101). This conclusion was reached after it was shown that these mice, when raised under germ-free conditions, had little or no serum rheumatoid factor. Little is known about the molecular and genetic characteristics of these autoantibodies.

Grabar proposed yet another view of the autoantibodies that develop during aging (102). According to this view, autoantibodies that appear with age play a useful function in removing senescent cells and molecules from the body. This provocative hypothesis has remained without significant experimental support until recently. In the last few years, studies of patients with Alzheimer's Disease (AD) have suggested a defect in immunity to the pathogenic amyloid peptide that causes the characteristic cerebral lesions in this disease (103). In these studies, T cells from patients with AD were not stimulated to divide by amyloid peptides, although these peptides stimulated T cells from non-demented young and elderly individuals to divide in culture. These results raised the possibility that patients with AD have an impaired immune response to the pathogenic amyloid peptide that precipitates in their brain.

Additional support for a role of the immune system in preventing or reversing the deposition of the pathogenic amyloid peptide within the brain of transgenic mouse expressing a human AD gene coding for amyloid has recently been published (104). In these studies, immunization of young amyloid-transgenic mice with amyloid peptide led to high concentrations of anti-amyloid peptide antibodies in the serum of these mice and prevented the deposition of amyloid within the brain. Even more exciting was the observation that immunization of older animals, which already had amyloid deposits, led to a clearance of amyloid from the brain. It must be acknowledged that the mechanism by which immunization with amyloid peptide prevents or reduces the amyloid accumulation within the brain has not been directly linked to anti-amyloid peptide antibodies. However, if an immune response to amyloid peptide can clear pathogenic amyloid molecules that deposit in the brain late in life, it is possible that the autoimmunity to amyloid peptide, deficient in patients with AD, might play a protective role with respect to the development of AD very much as envisioned by Grabar.

5. CONCLUSION

In conclusion, autoantibodies found in elderly humans and experimental animals reflect the dysregulation of the immune response observed during aging. The level of some autoantibodies, particularly rheumatoid factor, anti-DNA autoantibody and other non-organ-specific autoantibodies, appears to be associated with the chronic diseases of aging. On the other hand, autoanti-idiotypic antibodies associated with aging may contribute to the impaired capacity of the elderly to mount a vigorous, diverse and protective immune response to foreign antigens. Thus, autoimmunity may not only be associated with immune deficiency of aging, but may also contribute to it. Recent studies in humans with AD and a mouse model of AD require us to reconsider the hypothesis proposed by Grabar some years ago that age-associated autoimmunity may be adaptive and even protect the host from disease.

ACKNOWLEDGEMENT

Supported in part by grants from the NIH AG-00541, AG 08707, AG 14669 and from the Gladys and Roland Harriman Foundation.

References

1. Langman, R.E., and M. Cohn. 1996. A short history of time and space in immune discrimination. *Scand. J. Immunol.* 44:544–548.
2. Boyd, E. 1932. The weight of the thymus gland in health and disease. *Am. J. Dis. Child.* 43:1162.
3. Scollay, R.G., E.C. Butcher, and I.L. Weissman. 1980. Thymus cell migration. Quantitative aspects of cellular traffic from the thymus to the periphery in mice. *Eur. J. Immunol.* 10:210–218.
4. Douek, D.C., R.D. McFarland, P.H. Keiser, E.A. Gage, J.M. Massey, B.F. Haynes, M.A. Polis, A.T. Haase, M.B. Feinberg, J.L. Sullivan, B.D. Jamieson, J.A. Zack, L.J. Picker, and R.A. Koup. 1998. Changes in thymic function with age and during the treatment of HIV infection. *Nature* 396:690–695.
5. De Paoli, P., S. Battistin, and G.F. Santini. 1998. Age-related changes in human lymphocyte subsets: progressive reduction of the CD4 CD45R (suppressor inducer) population. *Clin Immunol. Immunopathol.* 48:290–296.
6. Rocha, B., N. Dautigny, and P. Pereira. 1989. Peripheral T lymphocytes: expansion potential and homeostatic regulation of pool sizes and CD4/CD8 ratios *in vivo. Eur. J. Immunol.* 19:905–11.

7. Mackall, C.L., and R.E. Gress. 1997. Thymic aging and T-cell regeneration. *Immunol. Rev.* 160:91–102.

8. Inkeles, B., J.B. Innes, M.M. Kuntz, A.S. Kadish, and M.E. Weksler. 1977. Immunological studies of Aging. III. Cytokinetic basis for the impaired response of lymphocytes from aged humans to plant lectins. *J. Exp. Med.* 145:1176–1187.

9. Staiano-Coico, L., Z. Darzynkiewicz, M.R. Melamed, and M.E. Weksler. 1984. Immunological studies of aging. IX. Impaired proliferation of T lymphocytes detected in elderly humans by flow cytometry. *J. Immunol.* 132:1788–1792.

10. Miller, R.A. 1996. The aging immune system: primer and prospectus. *Science* 273:70–74.

11. Ernst, D.N., M.V. Hobbs, B.E. Torbett, A.L. Glasebrook, M.A. Rehse, K. Bottomly, K. Hayakawa, R.R. Hardy, and W.O. Weigle. 1990. Differences in the expression profiles of CD45RB, Pgp-1, and 3G11 membrane antigens and in the patterns of lymphokine secretion by splenic CD4+ T cells from young and aged mice. *J. Immunol.* 145:1295–1302.

12. Kirman, I., K. Zhao, I. Tschepen, P. Szabo, G. Richter, H. Nguyen, and M.E. Weksler. 1996. Treatment of old mice with IL-2 corrects dysregulated IL-2 and IL-4 production. *Int. Immunol.* 8:1009–1015.

13. Schwab, R., L.M. Pfeffer, P. Szabo, D. Gamble, C.M. Schnurr, and M.E. Weksler. 1990. Defective expression of high affinity IL-2 receptors on activated T cells from aged humans. *Int. Immunol.* 2:239–246.

14. Gillis, S., R. Kozak, M. Durante, and M.E. Weksler. 1981. Immunological studies of aging. Decreased production of and response to T cell growth factor by lymphocytes from aged humans. *J. Clin. Invest.* 67:937–942.

15. Effros, R.B., and R.L. Walford. 1983. The immune response of aged mice to influenza: diminished T-cell proliferation, interleukin 2 production and cytotoxicity. *Cell. Immunol.* 81:298–305.

16. Bodnar, A.G., M. Ouellette, M. Frolkis, S.E. Holt, C.P. Chiu, G.B. Morin, C.B. Harley, J.W. Shay, S. Lichtsteiner, and W.E. Wright. 1998. Extension of life-span by introduction of telomerase into normal human cells. *Science* 279:349–352.

17. Weng, N.P., B.L. Levine, C.H. June, and R.J. Hodes. 1995. Human naive and memory T lymphocytes differ in telomeric length and replicative potential. *Proc. Natl. Acad. Sci. USA.* 92:11091–11094.

18. Vaziri, H., F. Schachter, I. Uchida, L. Wei, X. Zhu, R. Efforos, D. Cohen, and C.B. Harley. 1993. Loss of telomeric DNA during aging of normal and trisomy 21 human lymphocytes. *Am. J. Hum. Genet.* 52:661–667.

19. Sabin, A.B., D.R. Ginder, M. Matumoto, et al. 1947. Serological response of Japanese children to Japanese B encephalitis mouse brain vaccine. *Proc. Soc. Exp. Biol. Med.* 65:135–140.

20. Waldorf, D.S., R.F. Wilken, and J.L. Decker. 1968. Impaired delayed hypersensitivity in an aging population: Association with antinuclear reactivity and rheumatoid factor. *JAMA* 203:831–834.

21. Ershler, W.B., A.L. Moore, and M.A. Socinski. 1984. Influenza and aging: Age-related changes and the effects of thymosin on the antibody response to influenza vaccine. *J. Clin. Immunol.* 4:445–454.

22. Provinciali, M., G. Di Stefano, M. Colombo, F. Della Croce, M.C. Gandolfi, L. Daghetta, M. Anichini, R. Della Bitta, and N. Fabris. 1994. Adjuvant effect of low-dose interleukin-2 on antibody response to influenza virus vaccination in healthy elderly subjects. *Mech. Ageing Dev.* 77:75–82.

23. Callahan, J.E., J.W. Kappler, and P. Marrack. 1993. Unexpected expansions of CD8-bearing cells in old mice. *J. Immunol.* 151:6657–6669.

24. Posnett, D.N., R. Sinha, S. Kabak, and C. Russo. 1994. Clonal populations of T cells in normal elderly humans: the T cell equivalent to "benign monoclonal gammapathy." *J. Exp. Med.* 179:609–618.

25. Schwab, R., P. Szabo, J.S. Manavalan, M.E. Weksler, D.N. Posnett, C. Pannetier, P. Kourilsky, and J. Even. 1997. Expanded CD4+ and CD8+ T cell clones in elderly humans. *J. Immunol.* 158:4493–4499.

26. LeMaoult, J., P. Szabo, and M.E. Weksler. 1997. Effect of age on humoral immunity, selection of the B-cell repertoire and B-cell development. *Immunol. Rev.* 160:115–126.

27. Zharhary, D. 1988. Age-related changes in the capability of the bone marrow to generate B cells. *J. Immunol.* 141:1863–1869.

28. Szabo, P., K. Zhao, I. Kirman, J. Le Maoult, R. Dyall, W. Cruikshank, and M.E. Weksler. 1998. Maturation of B cell precursors is impaired in thymic-deprived nude and old mice. *J. Immunol.* 161:2248–2253.

29. Sherwood, E.M., B.B. Blomberg, W. Xu, C.A. Warner, and R.L. Riley. 1998. Senescent BALB/c mice exhibit decreased expression of lambda5 surrogate light chains and reduced development within the pre-B cell compartment. *J. Immunol.* 161:4472–4475.

30. Goidl, E.A., J.B. Innes, and M.E. Weksler. 1976. Immunological studies of aging. II. Loss of IgG and high avidity plaque-forming cells and increased suppressor cell activity in aging mice. *J. Exp. Med.* 144:1037–1048.

31. Gonzalez-Fernandez, A., D. Gilmore, and C. Milstein. 1994. Age-related decrease in the proportion of germinal center B cells from mouse Peyer's patches is accompanied by an accumulation of somatic mutations in their immunoglobulin genes. *Eur. J. Immunol.* 24:2918–2921.

32. Yang, X., J. Stedra, and J. Cerny. 1996. Relative contribution of T and B cells to hypermutation and selection of the antibody repertoire in germinal centers of aged mice. *J. Exp. Med.* 183:959–970.

33. Nicoletti, C., X. Yang, and J. Cerny. 1993. Repertoire diversity of antibody response to bacterial antigens in aged mice. III. Phosphorylcholine antibody from young and aged mice differ in structure and protective activity against infection with Streptococcus pneumoniae. *J. Immunol.* 150:543–549.

34. Hu, A., D. Ehleiter, A. Ben-Yehuda, R. Schwab, C. Russo, P. Szabo, and M.E. Weksler. 1993. Effect of age on the expressed B cell repertoire: role of B cell subsets. *Int. Immunol.* 5:1035–1039.

35. Goidl, E.A., P.W. Stashak, S.J. Martin-McEvoy, and J.R. Hiernaux. 1989. Age-related changes in serum immunoglobulin isotypes and isotype sub-class levels among standard long-lived and autoimmune and immunodeficient strains of mice. *Aging: Immunol. Infect. Dis.* 1:227–237.

36. De Greef, G.E., M.J. Van Tol, J.W. Van Den Berg, G.J. Van Staalduinen, C.J. Janssen, J. Radl, and W. Hijmans. 1992. Serum immunoglobulin class and IgG subclass levels and the occurrence of homogeneous immunoglobulins during the course of ageing in humans. *Mech. Ageing Dev.* 66:29–44.

37. Zhao, K., I. Kirman, I. Tschepen, R. Schwab, and M.E. Weksler. 1997. Peritoneal lavage reduces lipopolysaccha-

ride-induced elevation of serum TNF-alpha and IL-6 mortality in mice. *Inflammation* 21:379–390.

38. Casali, P., and A.L. Notkins. 1989. CD5+ B lymphocytes, polyreactive antibodies and the human B-cell repertoire. *Immunol. Today* 10:364–368.

39. Rowley, M.J., H. Buchanan, and I.R. Mackay. 1968. Reciprocal change with age in antibody to extrinsic and intrinsic antigens. *Lancet* 2:24–26.

40. Hallgren, H.M., C.E. 3d. Buckley, V.A. Gilbertsen, and E.J. Yunis. 1973. Lymphocyte phytohemagglutinin responsiveness, immunoglobulins and autoantibodies in aging humans. *J. Immunol.* 111:1101–1107.

41. Litwin, S.D., and J.M. Singer. 1965. Studies of the incidence and significance of anti-gamma globulin factors in the aging. *Arthritis & Rheumatism* 8:538–550.

42. Mackay, I.R. 1972. Aging and immunological function in man. *Gerontologia* 18:285–304.

43. Ligthart, G.J., J.X. Corberand, H.G. Geertzen, A.E. Meinders, D.L. Knook, and W. Hijmans. 1990. Necessity of the assessment of health status in human immunogerontological studies: evaluation of the SENIEUR protocol. *Mech. Aging Dev.* 55:89–105.

44. Franceschi, C., D. Monti, P. Sansoni, and A. Cossarizza. 1995. The immunology of exceptional individuals; the lesson of centenarians. *Immunol. Today* 16:12–16.

45. Manavalan, J.S., I. Kirman, K. Zhao, and M.E. Weksler. 1998. Aging and autoimmunity. In: The Autoimmune Disease. Rose, N.R., and I.R. Mackay, eds), Academic Press, pp. 783–794.

46. Gueret, R., K.S. Zhao, and M.E. Weksler. 1996. Age-related changes in serum immunoglobulin levels, autoantibodies, and monoclonal immunoglobulins in Xid and C57BL/6 mice. *Aging: Immunol. Infect. Dis.* 6:177–187.

47. Adkins, B., and R.L. Riley. 1998. Autoantibodies to T lineage cells in aged mice. *Mech. Aging. Dev.* 103:147–164.

48. Nobrega, A., M. Haury, R. Gueret, A. Coutinho, and M.E. Weksler. 1996. The age-associated increase in autoreactive immunoglobulins reflects a quantitative increase in specificities detectable at lower concentrations in young mice. *Scand. J. Immunol.* 44:437–443.

49. Bovbjerg, D.H., Y.T. Kim, R. Schwab, K. Schmitt, T. DeBlasio, and M.E. Weksler. 1991. "Cross-wiring" of the immune response in old mice: increased autoantibody response despite reduced antibody response to nominal antigen. *Cell. Immunol.* 135:519–525.

50. Welch, M.J., S. Fong, J. Vaughan, and D. Carson. 1983. Increased frequency of rheumatoid factor precursor B lymphocytes after immunization of normal adults with tetanus toxoid. *Clin. Exp. Immunol.* 51:299–304.

51. Huang, Y.P., L. Gauthey, M. Michel, M. Loreto, M. Paccaud, J.C. Pechere, and J.P. Michel. 1992. The relationship between influenza vaccine-induced specific antibody responses and vaccine-induced nonspecific autoantibody responses in healthy older women. *J. Gerontol.* 47:M50–55.

52. Weksler, M.E., C. Russo, and G.W. Siskind. 1989. Peripheral T cells select the B-cell repertoire in old mice. *Immunol. Rev.* 110:173–185.

53. Tsuda, T., Y.T. Kim, G.W. Siskind, and M.E. Weksler. 1988. Old mice recover the ability to produce IgG and high-avidity antibody following irradiation with partial bone marrow shielding. *Proc. Natl. Acad. Sci. USA* 85:1169–1173.

54. Arreaza, E.E., J.J. Jr. Gibbons, G.W. Siskind, and M.E. Weksler. 1993. Lower antibody response to tetanus toxoid associated with higher auto-anti-idiotypic antibody in old compared with young humans. *Clin. Exp. Immunol.* 92:169–173.

55. Radl, J. 1990. Age-related monoclonal gammapathies: clinical lessons from the aging C57BL mouse. *Immunol. Today* 11:234–236.

56. Axelsson, U., R. Bachmann, and J. Hallen. 1966. Frequency of pathological proteins (M-components) in 6,995 sera from an adult population. *Acta Med. Scand.* 179:235–247.

57. Kyle, R.A. 1997. Monoclonal gammopathy of undetermined significance and solitary plasmacytoma. Implications for progression to overt multiple myeloma. *Hematol. Oncol. Clin. North Am.* 11:71–87.

58. Avrameas, S., G. Dighiero, P. Lymberi, and B. Guilbert. 1983. Studies on natural antibodies and autoantibodies *Ann. Immunol.* (*Inst. Pasture*) 134D:103–113.

59. van den Akker, T.W., A.P. Tio-Gillen, H.A. Solleveld, R. Benner, and J. Radl. 1988. The influence of T cells on homogenous immunoglobins in sera of athymic nude mice during aging. *Scand. J. Immunol.* 28:359–365.

60. Gueret, R., A. Grandien, J. Andersson, A. Coutinho, J. Radl, and M.E. Weksler. 1993. Evidence for selective pressure in the appearance of monoclonal immunoglobulins during aging: studies in M54 mu-transgenic mice. *Eur. J. Immunol.* 23:1735–1738.

61. LeMaoult, J., J.S. Manavalan, R. Dyall, P. Szabo, J. Nikolic-Zugic, and M.E. Weksler. 1999. Cellular basis of B cell clonal populations in old mice. *J. Immunol.* 162:6384–6391.

62. Sthoeger, Z.M., M. Wakai, D.B. Tse, V.P. Vinciguerra, S.L. Allen, D.R. Budman, S.M. Lichtman, P. Schulman, L.R. Weiselberg, and N. Chiorazzi. 1989. Production of autoantibodies by CD5-expressing B lymphocytes from patients with chronic lymphocytic leukemia. *J. Exp. Med.* 169:255–268.

63. Deegan, M.J., J.P. Abraham, M. Sawdyk, and E.J. Van Slyck. 1984. High incidence of monoclonal proteins in the serum and urine of chronic lymphocytic leukemia patients *Blood* 64:1207–1211.

64. Gozes, Y., T. Umiel, A. Meshorer, and N. Trainin. 1978. Syngeneic GvH induced in popliteal lymph nodes by spleen cells of old C57BL/6 mice. *J. Immunol.* 121:2199–2204.

65. Hosono, M., E. Toichi, M. Hosokawa, S. Imamura, J. Gyotoku, Y. Katsura, and T. Hosokawa. 1995. Development of autoreactivity and changes of T cell repertoire in different strains of aging mice. *Mech. Aging Dev.* 78:197–214.

66. Weyand, C.M., J.C. Brandes, D, Schmidt, J.W. Fulbright, and J.J. Goronzy. 1998. Functional properties of CD4+ CD28– T cells in the aging immune system. *Mech. Aging Dev.* 102:131–147.

67. Carnaud, C., J. Charreire, and J.F. Bach. 1977. Adult thymectomy promotes the manifestation of autoreactive lymphocytes. *Cell. Immunol.* 28:274–83.

68. Doria, G., C. Mancini, M. Utsuyama, D. Frasca, and K. Hirokawa 1997. Aging of the recipients but not of the bone marrow donors enhances autoimmunity in syngeneic radiation chimeras. *Mech. Ageing Dev.* 95:131–142.

69. Sawin, C.T., T. Herman, M.E. Molitch, M.H. London, and S.M. Kramer. 1983. Aging and the thyroid. Decreased requirement for thyroid hormone in older hypothyroid patients. *Am. J. Med.* 75:206–209.

70. Russo, C., R. Schwab, and M.E. Weksler. 1990. Immune Dysregulation associated with aging. *Aging: Immunol. Infect. Dis.* 2:111–116.

71. Crisi, G.M., V.K. Tsiagbe, C. Russo, R.S. Basch, and G.J. Thorbecke. 1996. Evaluation of presence and functional activity of potentially self-reactive T cells in aged mice. *Int. Immunol.* 8:387–395.

72. Sakaguchi, S., and N. Sakaguchi. 1990. Thymus and autoimmunity: capacity of the normal thymus to produce pathogenic self-reactive T cells and conditions required for their induction of autoimmune disease. *J. Exp. Med.* 172:537–545.

73. Izui, S., P.H. Lambert, G.J. Fournie, H. Turler, and P.A. Miescher. 1977. Features of systemic lupus erythematosus in mice injected with bacterial lipopolysaccharides: identification of circulating DNA and renal localization of DNA-anti-DNA complexes *J. Exp. Med.* 145:1115–1130.

74. Meredith, P.J., J.K. Kristie, and R.L. Walford. 1979. Aging increases expression of LPS-induced autoantibody-secreting B cells. *J. Immunol.* 123:87–91.

75. Tomer, Y., and Y. Shoenfeld. 1988. The significance of natural autoantibodies. *Immunol. Invest.* 17:389–424.

76. Ohashi, P.S., S. Oehen, K. Buerki, H. Pircher, C.T. Ohashi, B. Odermatt, B. Malissen, R.M. Zinkernagel, and H. Hengartner. 1991. Ablation of "tolerance" and induction of diabetes by virus infection in viral antigen transgenic mice. *Cell* 65:305–317.

77. Hardy, R.R., and K. Hayakawa. 1986. Development and physiology of Ly-1 B and its human homolog, Leu-1 B. *Immunol. Rev.* 93:53–79.

78. Ben-Yehuda, A., P. Szabo, J. LeMaoult, J.S. Manavalan, and M.E. Weksler. 1998. Increased VH 11 and VH Q52 gene use by splenic B cells in old mice associated with oligoclonal expansions of CD5+ B cells. *Mech. Aging. Dev.* 103:111–121.

79. Goidl, E.A., G.J. Thorbecke, M.E. Weksler, and G.W. Siskind. 1980. Production of auto-anti-idiotypic antibody during the normal immune response: changes in the auto-anti-idiotypic antibody response and the idiotype repertoire associated with aging. *Proc. Natl. Acad. Sci. USA* 77:6788–6792.

80. Kim, Y.T., E.A. Goidl, C. Samarut, M.E. Weksler, G.J. Thorbecke, and G.W. Siskind. 1985. Bone marrow function. I. Peripheral T cells are responsible for the increased auto-antiidiotype response of older mice. *J. Exp. Med.* 161:1237–1242.

81. Weksler, M.E., R. Schwab, F. Huetz, Y.T. Kim, and A. Coutinho. 1990. Cellular basis for the age-associated increase in autoimmune reactions. *Int. Immunol.* 2:329–335.

82. Callard, R.E., A. Basten, and L.K. Waters. 1997. Immune function in aged mice. II. B-cell function. *Cell. Immunol.* 31:26–36.

83. Baccala, R., T.V. Quang, M. Gilbert, T. Ternynck, and S. Avrameas. 1989. Two murine natural polyreactive autoantibodies are encoded by nonmutated germ-line genes. *Proc. Natl. Acad. Sci. USA* 86:4624–4628.

84. Kipps, T.J., E. Tomhave, P.P. Chen, and D.A. Carson. 1988. Autoantibody-associated kappa light chain variable region gene expressed in chronic lymphocytic leukemia with little or no somatic mutation. Implications for etiology and immunotherapy. *J. Exp. Med.* 167:840–852.

85. Dobken, J., M.E. Weksler, and G.W. Siskind. 1980. Effect of age on ease of B-cell tolerance induction *Cell. Immunol.* 55:66–73.

86. DeKruyff, R.H., E.A. Rinnooy Kan, M.E. Weksler, and G.W. Siskind. 1980. Effect of aging on T-cell tolerance induction. *Cell. Immunol.* 56:58–67.

87. Goidl, E.A., M.A. Michelis, G.W. Siskind, and M.E. Weksler. 1981. Effect of age on the induction of autoantibodies. *Clin. Exp. Immunol.* 44:24–30.

88. van Vollenhoven, R.P., C. Nagler-Anderson, V.J. Stecher, A. Soriano, K.M. Connolly, H. T. Nguyen, G.W. Siskind, and G.J. Thorbecke. 1988. Collegen induced arthritis and aging: influence of age on arthritis susceptibility and acute phase reponses. *Aging: Immunol. Infect. Dis.* 1:159–176.

89. Tomer, Y., S. Mendlovic, T. Kukulansky, E. Mozes, Y. Shoenfeld, and A. Globerson. 1991. Effects of aging on the induction of experimental systemic lupus erythematosus (SLE) in mice. *Mech. Aging Dev.* 58:233–244.

90. Walford, R.L. 1969. An Immunological Theory of Aging. Munksgaard, Copenhage: Williams & Wilkins, Baltimore. pp. 1–248.

91. Kay, M.M., J.J. Marchalonis, J. Hughes, K. Watanabe, and S.F. Schluter. 1990. Definition of a physiologic aging autoantigen by using synthetic peptides of membrane protein band 3: localization of the active antigenic sites. *Proc. Natl. Acad. Sci. USA* 87:5734–5738.

92. Shoenfeld, Y., and D.A. Isenberg (eds.) 1989. In: Aging and Autoimmunity. In: The mosaic of autoimmunity, New York, Elsevier Science Publishing, Inc.

93. Isselbacher, K.J., J.D. Wilson, J.B. Wartin, A.S. Fausi, and D.L. Kasper, eds. 1994. Part 11; Section 2: Disorders of Immune mediated injury. In: Harrison's Principles of Internal Medicine, 13th ed. McGraw-Hill, New York.

94. Winfield, J.B., D. Koffler, and H.G. Kunkel. 1975. Specific concentration of polynucleotide immune complexes in the cryoprecipitates of patients with systemic lupus erythematosus. *J. Clin. Invest.* 56:563–570.

95. Shlomchik, M., M. Mascelli, H. Shan, M.Z. Radic, D. Pisetsky, A. Marshak-Rothstein, and M. Weigert. 1990. Anti-DNA antibodies from autoimmune mice arise by clonal expansion and somatic mutation. *J. Exp. Med.* 171:265–292.

96. Manser, T. 1990. The efficiency of antibody affinity maturation: can the rate of B-cell division be limiting? *Immunol. Today* 11:305–308.

97. Baccala, R., T.V. Quang, M. Gilbert, T. Ternynck, and S. Avrameas. 1989. Two murine natural polyreactive autoantibodies are encoded by nonmutated germ-line genes. *Proc. Natl. Acad. Sci. USA* 86:4624–4628.

98. Pereira, P., L. Forni, E.L. Larsson, M. Cooper, C. Heusser, and A. Coutinho. 1986. Autonomous activation of B and T cells in antigen-free mice. *Eur. J. Immunol.* 16:685–688.

99. Coutinho, A., A. Andersson, A. Sundblad, et al. 1990. The dynamics of immune net works. In: Idiotype Networks in Biology and Medicine, (A.D.M.E. Osterhaus and F.G.C.M. UytdeHaag, eds), Elsevier Science Publishers B.V., p. 59.

100. Williams, R.C. 1977. In: Autoimmunity: Genetic, Immunologic and Clinical Aspects (N. Talal, ed), Academic Press, New York, Chapter 15, p. 457.

101. van Snick, J.L., and P.L. Masson. 1980. Incidence and specificities of IgA and IgM anti-AgG autoantibodies in various mouse strains and colonies. *J. Exp. Med.* 151:45–55.

102. Garbar, P. 1983. Autoantibodies and the physiological role of immunoglobulins. *Immunol. Today* 4:337–340.

103. Trieb, K., G. Ransmayr, R. Sgonc, H. Lassmann, and B. Grubeck-Loebenstein. 1996. APP peptides stimulate lymphocyte proliferation in normals, but not in patients with Alzheimer's disease. *Neurobiol. Aging* 17:541–547.

104. Schenk, D., R. Barbour, W. Dunn, G. Gordon, H. Grajeda, T. Guido, K. Hu, J. Huang, K. Johnson-Wood, K. Khan, D. Kholodenko, M. Lee, Z. Liao, I. Lieberburg, R. Motter, L. Mutter, F. Soriano, G. Shopp, N. Vasquez, C. Vandevert, S. Walker, M. Wogulis, T. Yednock, D. Games, and P. Seubert. 1999. Immunization with amyloid-beta attenuates Alzheimer-disease-like pathology in the PDAPP mouse. *Nature* 400:173–177.

SECTION V

IMMUNOTHERAPEUTIC APPROACHES

64 | Autoimmune Diseases as Stem Cell Disorders: Treatment By Allogeneic Bone Marrow Transplantation

Susumu Ikehara

1. INTRODUCTION

In the last decade, remarkable advances have been made in bone morrow transplantation (BMT), which is now becoming a powerful strategy in the treatment of life-threatening diseases such as leukemia, aplastic anemia, and congenital immunodeficiency.

Various mouse strains that spontaneously develop autoimmune diseases have contributed not only to better understanding of the fundamental nature of autoimmune diseases, but also to the analysis of their etiopathogenesis. The etiopathogenesis of systemic autoimmune diseases has previously been attributed to T cell deficiencies, polyclonal B cell activation, macrophage dysfunction, and environmental factors, such as hormonal disturbances (1). However, there has been an increase in information suggesting that autoimmune diseases originate from defects in hemopoietic stem cells (HSCs) (1–12).

These findings have recently been confirmed and extrapolated in humans. Autoimmune diseases such as rheumatoid arthritis (RA), systemic lupus erythematosus (SLE), multiple sclerosis (MS), and Crohn's disease were resolved after allogeneic BMT. However, there have recently been reports on the rapid recurrence or persistence of autoimmune diseases after autologous BMT. Conversely, the adoptive transfer of autoimmune diseases, such as myasthenia gravis (MG), insulin-dependent diabetes mellitus (IDDM) and Graves' disease by allogeneic BMT has been reported.

In this article, I provide evidence that autoimmune diseases are "stem cell disorders", and clarify the conditions for successful allogeneic BMT across major histocompatibility complex (MHC) barriers.

2. THYMIC ABORMALITIES IN AUTOIMMUNE DISEASES

The thymus contains more than 95% thymocytes and small numbers of macrophages, epithelial cells and nurse cells. Only very few plasma cells and B cells can be detected in the normal thymus (13). However, it is reported that lymphoid follicles have been detected in the thymuses of patients with autoimmune diseases such as MG and SLE (14,15). We have found that plasma cell infiltration into the thymus is a common feature in autoimmune-prone mice, and that the destruction of the blood-thymus barrier results in premature thymic involution in autoimmune-prone mice (16). However, it is not known why thymic abnormalities develop in autoimmune-prone mice.

To answer this question, we transplanted the thymus or bone marrow from normal into autoimmune-prone mice, and vice versa. The data are summarized in Fig. 1; thymic abnormalities originate from defects in the bone marrow of autoimmune-prone mice, and the transplantation of BMCs from normal to autoimmune-prone mice prevents both thymic abnormalities and autoimmune diseases (17). We have recently found that abnormal HSCs (but not the presence of extrinsic factors such as autoantibodies or intrinsic thymic abnormalities) induce thymic abnormalities (18); autoreactive T cells that have developed from abnormal HSCs destroy the blood thymus barrier, resulting in the infiltration of plasma cells and B cells in the thymus.

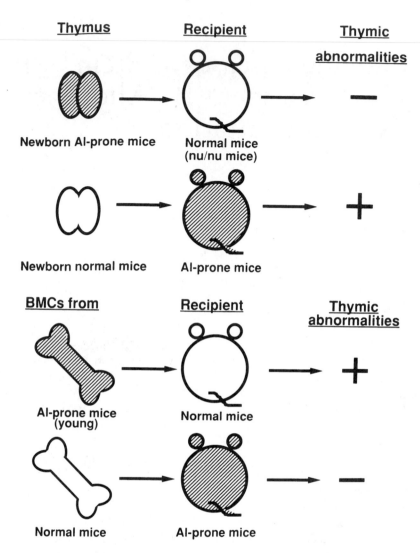

Figure 1.   The cause of thymic abnormalities in autoimmune (AI)-prone mice. When the newborn thymuses of AI-prone mice are engrafted under the renal capsule of normal mice (BALB/c nu/nu), the thymuses do not show abnormalities (premature involution or plasma cell infiltration, etc.), and the nu/nu mice do not show autoimmune diseases. In contrast, when BMCs from AI-prone mice are transferred into normal mice, the mice show thymic abnormalities.

3.   TREATMENT OF SYSTEMIC AUTOIMMUNE DISEASE BY BMT

When female (NZB × NZW) F_1 (B/W F_1) (> 6 months), female MRL/lpr (> 2 months), and male BXSB (> 6 months) mice that had already shown clear evidence of lupus nephritis were lethally irradiated and then reconstituted with either allogeneic BMCs of young (< 2 months) BALB/c nu/nu (H-2^d) mice or T cell-depleted BMCs of BALB/c mice, the recipients survived in good health for more than 3 months after BMT (7).

In BXSB and B/W F_1 mice, BMT had completely curative effects (11). Glomerular damage was ameliorated, and the levels of autoantibodies (anti-DNA and anti-Sm anti-

bodies [Abs]) and circulating immune complexes (CICs), particularly anti-gp-70 CICs, were reduced. The repair of glomerular damage was assessed by performing renal biopsies before and after BMT, as shown in Fig. 2. In addition, immunological functions were normalized; T cell functions, including IL-2 production, were restored, and hyperfunctions of macrophages and B cells decreased. Assays for both mixed-lymphocyte reaction (MLR) and the generation of cytotoxic T-lymphocytes (CTLs) revealed that newly developed T cells from BMT-treated mice were tolerant of both bone marrow donor-type and host-type MHC determinants, but responded vigorously to third-party cells. *In vitro* primary anti-sheep red blood cell (SRBC) plaque-forming cell (PFC) assay also showed that some

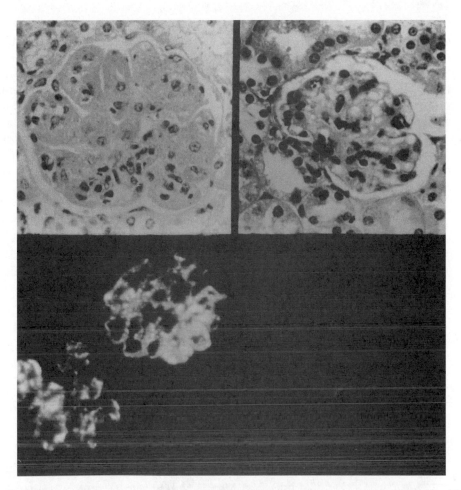

Figure 2. Histopathologic and immunofluorescent findings in the glomeruli of B/W F1 mice before and after BMT. Typical wire-loop lesions (top left) and IgG deposits (bottom left) are present in the glomeruli of the 8-month-old B/W F1 mouse before BMT. Five months after BMT, IgG deposits are markedly reduced (bottom right), and the glomeruli of the mouse exhibit a normal appearance on hematoxylin-eosin staining (top right).

degree of co-operation was achieved among APCs, helper T cells, and B cells.

In contrast to BXSB and B/W F_1 mice, MRL/lpr mice regularly suffered a relapse approximately 5 months after BMT (11). H-2 typing revealed that all the immunocompetent cells of the chimeras had been replaced by host (MRL/lpr)-derived cells by that time. The T cells of the chimeras showed responsiveness to donor BALB/c ($H-2^d$)-type, but not to recipient MRL/lpr ($H-2^K$)-type MHC determinants in both assays for MLR and the generation of CTLs. In addition, abnormal $B220^+$ $Thy-1^+$ cells reappeared in such mice. These results indicate that MRL/lpr mice possess abnormal radioresistant (9.5Gy) hemopoietic stem cells (HSCs), although the mice are radiosensitive (<8.5 Gy). This also provides additional evidence that the etiopathogenesis of autoimmune diseases resides in defects or characteristics located at the HSC level.

4. PREVENTION AND TREATMENT OF ORGAN-SPECIFIC AUTOIMMUNE DISEASE BY BMT PLUS ORGAN ALLOGRAFTS

We next examined whether organ-specific autoimmune diseases could be treated by BMT using an animal model for IDDM, the NOD mouse.

First, we attempted to prevent insulitis and overt diabetes by BMT. NOD mice (>4 months) were lethally irradiated and then reconstituted with T cell-depleted BALB/c BMCs. The mice were killed more than 3 months after BMT. No lymphocyte infiltration was observed in the islets of the BMT-treated NOD mice. Immunohistochemical studies revealed the presence of intact beta cells as well as alpha and delta cells. BMT-treated NOD mice showed a normal pattern in glucose tolerance tests (GTTs). Diabetic nephropathy was also corrected by BMT. Thus, BMT can prevent insulitis and overt diabetes (8). However, we could not treat overt diabetes in NOD mice by BMT, because mice with overt diabetes have no beta cells.

We next performed a combined transplantation of fetal or newborn pancreas plus allogeneic bone marrow, since we have shown that organ allografts are accepted if the organ is transplanted from the same donor as the bone marrow at the same time (19). NOD mice that had already developed overt diabetes were lethally irradiated and then reconstituted with allogeneic BALB/c bone marrow cells. The pancreatic tissues from fetal or newborn BALB/c mice were then engrafted under the renal capsules of NOD diabetic mice. Three months after transplantation, the mice exhibited a normal GTT pattern, and insulin levels in the sera were also normalized. Immunohistochemical studies revealed the presence of beta cells in the islets engrafted under the renal capsules of the NOD mice (Fig. 3). It should be noted that neither insulitis nor rejection occurred. Thus, we succeeded in treating diabetes by the combined transplantation of the pancreas and bone marrow (9).

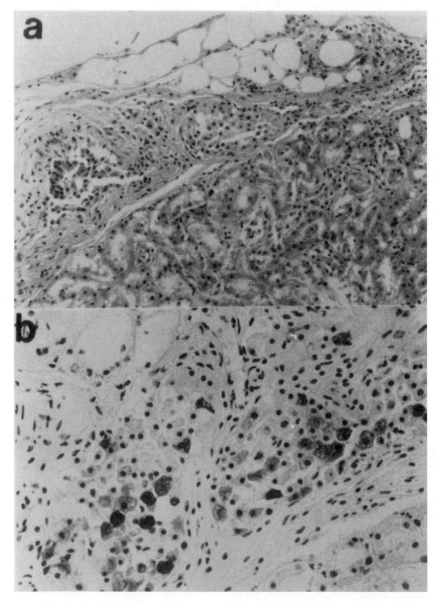

Figure 3.   Histology of engrafted pancreas. Clusters of islet cells are observed under the renal capsule by hematoxylin-eosin staining (a). These cells are shown to contain insulin by means of immunohistological staining (b).

5. PREVENTION AND TREATMENT OF BOTH ORGAN-SPECIFIC AND SYSTEMIC AUTOIMMUNE DISEASES BY BMT

We have recently found that (NZW × BXSB)F$_1$ (W/BF$_1$) mice, which develop lupus nephritis with myocardial infarction (20), show thrombocytopenia with age that is attributed to the presence of both platelet-associated and circulating anti-platelet antibodies (10). In addition, we found that myocardial infarction in W/BF$_1$ mice is associated with the presence of anti-cardiolipin Abs, and therefore, this mouse may be an animal model for anti-phospholipid Ab syndrome (21).

The transplantation of BMCs from normal to W/BF$_1$ mice was found to exert preventative and curative effects on lupus nephritis, thrombocytopenia and anti-phospholipid Ab syndrome; the platelet counts were normalized, and circulating anti-platelet Ab levels as well as anti-phospholipid Ab levels were reduced (10,21).

6. TRANSFER OF AUTOIMMUNE DISEASES INTO NORMAL MICE BY BMT

We attempted to transfer IDDM to normal mice by transplanting NOD BMCs to C3H/HeN mice. C3H/HeN mice express 1-Eα molecules and have an aspartic acid at residue 57 (Asp-57) of the 1-Aβ chain (22,23). We selected this strain because it has been postulated that the failure to express the Eα gene is the abnormality that permits NOD mice to develop insulitis, leading to diabetes (24,25). Also, it is thought that replacement of Asp-57 with Ser (non-Asp) in NOD mice (26) and with non-Asp in humans (27) may be the molecular anomaly responsible for the development of IDDM.

Female C3H/HeN (H-2^K) mice were lethally irradiated (9.5Gy) at the age of 8 weeks and then reconstituted with T cell-depleted BMCs of young (<8 weeks) female NOD (K^d, 1-Ag^7, D^b) mice. Two of four [NOD→C3H/HeN] chimeric mice developed both insulitis and overt diabetes more than 40 weeks after BMT (Fig. 4); beta cells were selectively destroyed by infiltration of T cells. These mice exhibited elevated glucose levels and abnormal glucose tolerance curves (12).

The next step was to investigate whether both systemic (SLE) and organ-specific (ITP) autoimmune diseases could be transferred to normal mice by BMT. Since the male W/BF1 mouse is an animal model of SLE and ITP, we used W/BF1 (H-2^Z/H-2^b) mice as donors and C3H/HeN (H-2^K) or C57BL/6J (H-2^b) mice as recipients. C3H/HeN or C57BL/6J mice were lethally irradiated (9.5Gy) and then reconstituted with T cell-depleted BMCs of young (<8

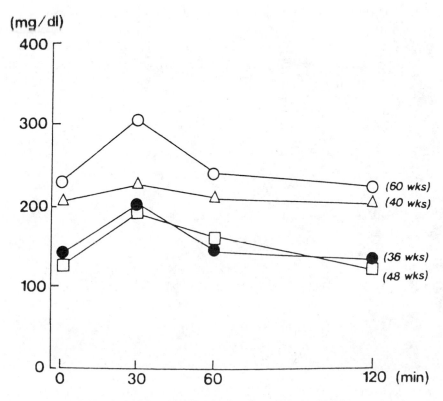

Figure 4.   Glucose tolerance tests (GTTs) in [NOD→C3H/HeN] mice. Two [NOD→C3H/HeN] mice show impaired GTTs (○ and △).

weeks) male W/BF$_1$ mice. [W/BF$_1$→C57BL/6J] mice showed thrombocytopenia (<10^5 platelets per mm^3; normal mice >10 × 10^5) in 5 of 11 mice (45%) 3 months after BMT, and in 5 of the same 11 mice (total 10/11:91%) by 5 months after BMT. [W/BF$_1$→C3H/HeN] mice also developed thrombocytopenia in 4 of 8 mice (50%) by 3 months after BMT and in 6 of 8 mice (75%) by 6 months after BMT. Cytofluorometric analyses demonstrated the presence of both platelet-associated antibodies (PAA) and anti-platelet bindable antibodies (PBA) in the thrombocytopenic mice. Immunohistopathological analyses revealed typical wire-loop lesions in the glomeruli of the [W/BF$_1$→C57BL/6J] or [W/BF$_1$→C3H/HeN] mice, as shown in Fig. 5 (12).

7. EVIDENCE FOR AUTOIMMUNE DISEASES AS STEM CELL DISORDERS IN MICE

To confirm that the defective HSCs were indeed the elements responsible for the development of the autoimmune diseases, we transferred the cells in a HSC-enriched fraction of W/BF$_1$ BMCs to C3H/HeN mice, since both Visser et al. (28), and we (29) have reported that spleen colony-forming units are enriched in wheat germ agglutinin (WGA)-binding (WGA$^+$) cells. We therefore attempted to transfer autoimmune diseases to normal mice by the transplantation of partially-purified HSCs (WGA$^+$ cells) from W/BF$_1$ mice. Although the C3H mice that had received 1 × 10^5 WGA$^+$ cells (without bone grafts) from W/BF$_1$ mice died due to graft failure, those that received WGA$^+$ cells plus bone grafts from W/BF$_1$ mice began to show proteinuria (++) and thrombocytopenia three months after the transplants (Fig. 6). H-2 typing revealed that the hematolymphoid cells were donor-derived (data not shown). All the mice died of renal failure due to lupus nephritis by 300 days after the transplants. The survival rates were similar to those in C3H mice that had received T cell-depleted (TCD) BMCs of W/BF$_1$ mice. PBAs were detected in the sera of these mice. We thus succeeded in inducing autoimmune diseases (SLE and ITP) in normal mice by transplanting partially purified HSCs with bone grafts from W/BF1 mice (30).

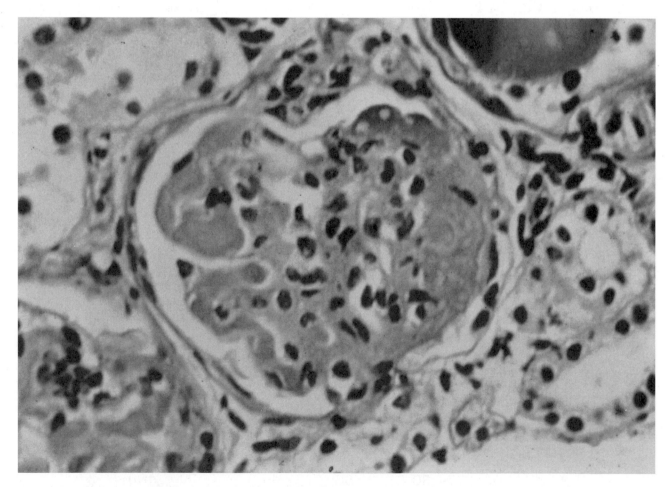

Figure 5.   Histology of a glomerulus of a [W/B F1 → C3H/HeN] mouse 5 months after BMT. Note the deposits of PAS-positive materials in both capillary and mesangial areas.

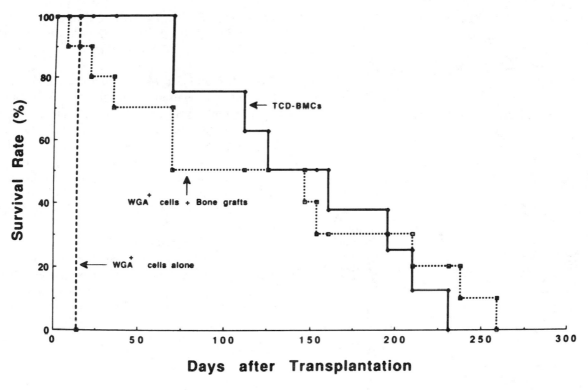

Figure 6. Survival rates in (W/BF1 → C3H) chimeric mice. C3H/HeN mice were irradiated (9.5 Gy) and then reconstituted with either 1–2 × 10⁷ TCD-BMCs (◆–◆) or 1–2 × 10⁵ WGA⁺ cells plus 40 Gy-irradiated bone grafts (□- - -□) or WGA⁺ cells alone from male W/BF1 mice (< 2 mo) (■- - -■).

8. EVIDENCE FOR AUTOIMMUNE DISEASES AS STEM CELL DISORDERS IN HUMANS

Seven RA patients received allogeneic BMT from HLA-identical siblings due to severe aplastic anemia supervening after gold and/or D-penicillamine therapy (31–33). Two patients are in complete remission after a follow-up of six years. It has also been reported that two cases of psoriasis vulgaris were resolved after BMT: One was associated with AML,(34) and the other with CML (35). Stable remission of ulcerative colitis has also been reported in a young woman who received BMT because of AML (35). Conversely, the adoptive transfer of autoimmune diseases after BMT has been reported. Grau et al. and others reported six cases of myasthenia gravis (MG) occurring after allogeneic BMT (36,37). Other adoptive, post-transplant autoimmune diseases include autoimmune thyroiditis (38,39), IDDM (40–43), and Graves' disease (44). Recently, data have accumulated that allogeneic BMT can be used to treat various autoimmune diseases (45,46). It should be noted that no autoimmune diseases have been seen to recur after allogeneic BMT in the 13 patients with autoimmune diseases plus leukemia or aplastic anemia during long-term observations (range 7 to 20 years) (47), although there have more recently been reports

on the rapid recurrence or persistence of autoimmune diseases after autologous BMT (48).

9. HYPOTHETICAL ETIOPATHOGENESIS OF SYSTEMIC AND ORGAN-SPECIFIC AUTOIMMUNE DISEASES

Regarding the etiopathogenesis of autoimmune diseases, our current hypothesis is as follows: As shown in Fig. 7, polygene abnormalities primarily exist at the gene level of P-HSCs. Endogenous or exogenous retroviruses are probably involved in the development of P-HSC abnormalities. Autoreactive immunocompetent cells, including T cells, differentiate from the abnormal pluripotent HSCs (P-HSCs). In systemic autoimmune diseases, autoreactive T cells attack various organs, as seen in chronic GVHR, resulting in systemic tissue injury (Fig. 7a). In contrast, regulatory suppressor T cells may play a crucial role in the development of organ-specific autoimmune diseases (Fig. 7b), since we have some data suggesting that there is a correlation between decreased CD8⁺ suppressor T cell counts and the development of organ-specific autoimmune diseases (unpublished data).

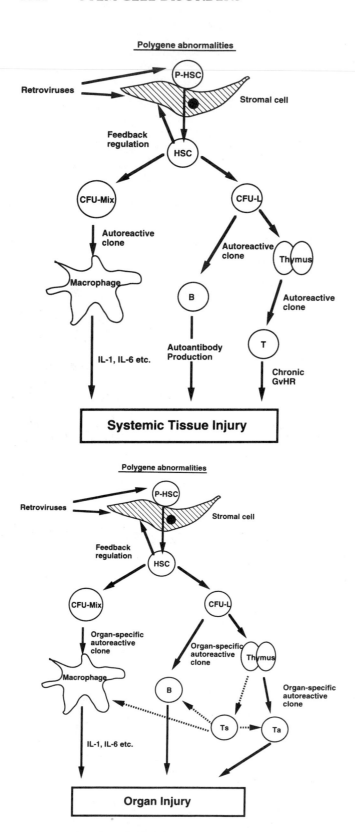

Figure 7. Hypothetical etiopathogenesis of systemic and organ-specific autoimmune diseases.

10. A NEW STRATEGY FOR TREATMENT OF AUTOIMMUNE DISEASES IN CHIMERIC RESISTANT MRL/IPR MICE

Since MRL/lpr mice possess abnormal radioresistant HSCs, they suffer a relapse 5 months after conventional BMT (11). We have very recently found that there is an MHC restriction between HSCs and stromal cells (49). To prevent the recurrence of autoimmune diseases in MRL/lpr mice, we therefore carried out BMT plus bone grafts to replace not only hemopoietic cells but also stromal cells by donor cells. As shown in Fig. 8, MRL/lpr mice that had been irradiated (8.5Gy) and then reconstituted with C57BL/6 BMCs plus bone grafts survived more than 48 wks after this treatment (50).

Although we have found that allogeneic BMT plus bone grafts (to recruit donor stomal cells) has completely preventative effects on autoimmune diseases in MRL/lpr mice (50), this strategy (8.5Gy/Bone/BMT) was found to have no effect on the treatment of autoimmune diseases in MRL/lpr mice after the onset of the disease (Fig. 9); MRL/lpr mice are radiosensitive (8.5 Gy is the maximum), become more sensitive to radiation due to renal failure after the onset of autoimmune diseases, and are resistant to allogeneic engraftment when irradiated at lower doses (11). Therefore, we have devised a new method that reduces the side effects of radiation and prevents graft rejection. MRL/lpr mice that had shown symptoms of autoimmune diseases (proteinuria and massive lymphadenopathy) were treated with CY, fractionated radiation (5 Gy × 2), followed by two transplantations of whole BMCs (WBMCs) plus bone grafts from normal B6 mice (CY/2X/Bone/2BMT). We carried out fractionated radiation (5Gy × 2) to reduce the acute radiation injury, and CY was used to eliminate host-derived activated T cells (51–53). MRL/lpr mice that received such treatments survived more than 40 wks after the treatment (1 mouse survived for >40 wks, 7 for >50 wks, and 4 for >60 wks) (Fig. 9). Furthermore, more than 70% of the recipients of CY/2X/Bone/BMT (without the first BMT) survived >25 wks. However, 67% of the mice treated with 2X/Bone/2BMT (without CY injection) died within 6 wks. Furthermore, all the recipients that had received T cell-depleted BMCs [CY/2X/Bone/2BMT (–T cells)] died within 10 wks without the reconstitution of donor-derived cells, indicating that transplantation with T cell-containing BMCs is essential to the engraftment and treatment. The mice treated with CY/2X/Bone/ 2BMT(–CD8) (CD8-depleted BMCs were used for both BMT) died within 15 wks due to graft rejection and lupus nephritis; cells recovered from these mice had H-2 of the recipient type when examined by a FACScan (H-2k[b+] cells were 97.5±0.5%). Furthermore, recipients treated with 2X/Bone/2BMT (without CY injection), CY/2X/Bone/

Survival Rate of MLR/lpr Mice with BMT Plus Bone Graft

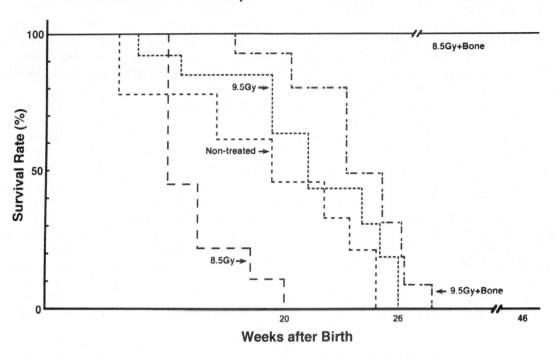

Figure 8. BMT plus bone grafts in [B6 → MRL/1pr] mice. The MRL/1pr mice were lethally (8.5Gy) irradiated and then reconstituted with T cell-depleted B6 BMCs. The mice were subcutaneously engrafted with B6 bone fragments. Stromal cells present in the B6 bone produce a HSC-chemotactic factor, which results in the migration of B6 HSCs into the engrafted bones. MHC-matched donor-derived stromal cells and HSCs stimulate each other and proliferate, and both migrate into the MRL/1pr bone marrow. The bone marrow in the MRL/1pr mice is thus replaced by B6-derived stromal cells and hemopoietic cells.

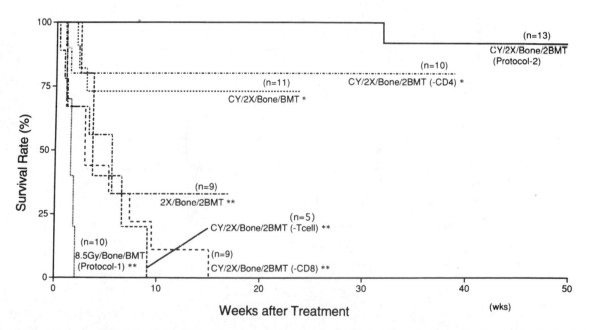

Figure 9. Survival rates in 6 groups. Numbers in parentheses are numbers of mice in each group. Treatment of mice is shown in the figure. Statistical analyses were performed by a logrank test, and asterisks (**) represent the *P* values of treated (CY/2X/Bone/2BMT) versus other groups; ** *P* < .01, * statistical insignificance.

2BMT(–T cells) or CY/2X/Bone/2BMT(–CD8) showed proteinuria; histological examinations revealed the presence of lupus nephritis. In contrast, 80% of the MRL/lpr mice treated with CY/2X/Bone/2BMT(–CD4) (CD4-depleted BMCs were used for both BMT) survived more than 40 wks (8 out of 10 mice), suggesting that CD8$^+$ cells in the BMCs are necessary for the engraftment of donor cells. As mentioned above, the presence of T cells in the transplanted BMCs is a critical factor in the long-term, disease-free survival of the recipient mice (mouse BMCs usually contain approximately 1% T cells). Therefore, various ratios (0.05, 0.1, 0.5, 1.0%) of mature T cells enriched from the peripheral blood were added to T cell-depleted BMCs (5×10^7) to provide the quantitative information as to the number of mature T cells necessary for the treatment. The recipients that received BMCs containing 0.05, and 0.1% mature T cells showed hematolymphoid cells with the recipient H-2 phenotype, and no donor-derived cells were detected in these recipients when tested 6 wks after the treatment, as observed in those treated with CY/2X/Bone/2BMT (–T

cells). However, the hematolymphoid cells of the recipients that had received BMCs containing >0.5% of T cells were almost all of donor origin (>98%). Thus, the presence of small numbers of T cells in BMCs (approximately 0.5% of BMCs, which contain 0.3% of CD8$^+$ T cells) seems to be necessary for the engraftment of donor BMCs and for long-term survival.

Recently, we have found that most donor HSCs are trapped and retained in the liver when they are injected either P.V. or even I.V. (54), and that the HSCs induce clonal anergy to host CD8$^+$ T cells (55). In addition, we have found that a strategy [P.V. (on day 0) plus I.V. (on day 5) injections of donor whole BMCs] can induce continuous tolerance in the skin allograft system (56). Based on these findings, we attempted to establish a new strategy for allogeneic BMT applicable to humans. MRL/lpr mice (4 to 5 months of age) that had developed the symptoms of autoimmune diseases, such as massive lymphadenopathy and proteinuria (>2.5+), were first treated with [5.5 Gy × 2+ P.V.]. As shown in Fig. 10, >70% of the mice thus treated survived more than one year, indicating that this

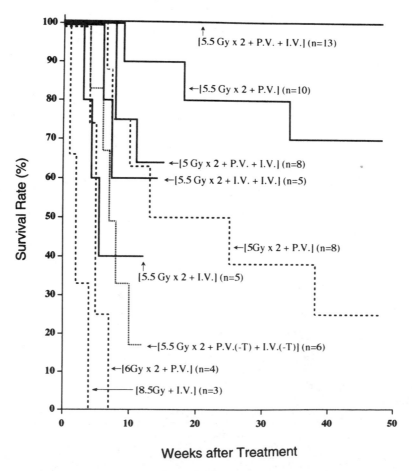

Figure 10. Survival rates after various treatments. Survival rates of the recipients treated with single or fractionated irradiation followed by P.V. and/or I.V. administrations of BMCs are shown. Numbers in parentheses represent the numbers of mice in each group. Statistical analyses were performed by a logrank test: P < 0.05, [5.5 Gy × 2 + P.V. + I.V.] vs. [5.5 Gy × 2 + P.V.].

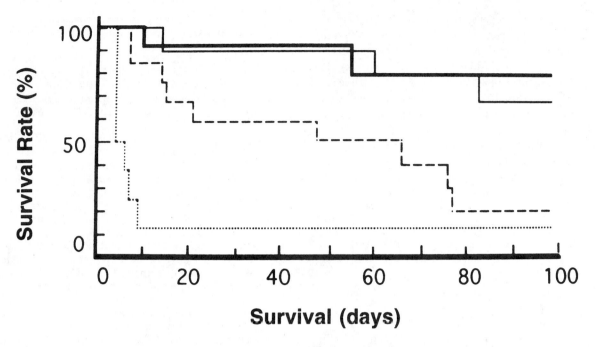

Figure 11.   Survival rates of mice in four groups. Lethally irradiated female MRL/+ mice (7 months) were injected i.v. with $3-4 \times 10^7$ T cell-depleted C57BL/6 BMCs and engrafted also with fragments of bone (group 1, . . .), or further engrafted with FT (group 2, ——). Irradiated MRL/+ mice were injected i.v. with $3-4 \times 10^7$ FHCs and engrafted with FB but no thymus graft (group 3, - - -), or irradiated female MRL/+ mice were injected i.v. with $3-4 \times 10^7$ FHCs, FB grafts and also engrafted with FT (group 4, —-).

treatment has some effects on the treatment of autoimmune diseases. We next treated autoimmune diseases in MRL/lpr mice with [5.5 Gy $\times$ 2 + P.V. + I.V.]. Treated MRL/lpr mice showed a 100% survival rate one year after the treatment, indicating that the supplemental I.V. injection is helpful for successful engraftment. In contrast, all recipients treated with [8.5 Gy + I.V.] died within 4 weeks due to the side effects of radiation, as previously reported (57). The MRL/lpr mice treated with either [5.5 Gy $\times$ 2+ I.V.] or [5.5 Gy $\times$ 2 + I.V. + I.V.] showed survival rates of 40% and 60%, respectively, 15 weeks after treatment. This appears to be due to graft rejection, since donor hematolymphoid cells cannot be detected in the recipients. These findings suggest that the P.V. injection of BMCs has much more effect on prolonging survival than the I.V. injection.

To investigate whether T cells in the BMCs are necessary for the engraftment, T cell-depleted BMCs (3×10^7) were used: [5.5 Gy $\times$ 2 + P.V. (T) + I.V. (–T)]. Forty percent of the mice that had received such treatment died by 20 weeks (Fig. 10), indicating that T cells (~1%) in BMCs are essential for the engraftment and treatment, as previously reported (57).

We next examined whether the radiation dose could be reduced when BMCs were P.V. injected. We treated MRL/lpr mice with [5 Gy $\times$ 2 + P.V.] or [5 Gy $\times$ 2 + P.V. + I.V.]. However, >70% of the recipients treated with [5 Gy $\times$ 2 + P.V.] died within 35 weeks due to a recurrence of the autoimmune diseases (renal failure), and the mice treated

with [5 Gy $\times$ 2 + P.V. + I.V.] showed a 65% survival rate 15 weeks after the treatment (Fig. 10), indicating that the 5 Gy $\times$ 2 irradiation (without recourse to CY administration plus bone grafts, as previously reported (57) was insufficient for depleting residual MRL/lpr HSCs. All recipients that had been treated with a higher dose of irradiation [6 Gy $\times$ 2 + P.V.] died due to intestinal infection by 7 weeks (Fig. 10). These findings suggest that [5.5 Gy $\times$ 2] is the most suitable irradiation dose. It should be noted that no immunosuppressants are necessary (58).

11. THYMUS TRANSPLANTATION, A CRITICAL FACTOR FOR SUCCESSFUL ALLOGENEIC BMT IN OLD MICE

MRL/+ mice develop pancreatitis and sialoadenitis after they reach 7 months of age. Conventional bone marrow transplantation has been found to be ineffective in the treatment of these forms of apparent autoimmune diseases. Old MRL/+ mice show a dramatic thymic involution with age. Hematolymphoid reconstitution is incomplete when fetal liver cells (as a source of hemopoietic stem cells) plus fetal bone (FB; which is used to recruit stromal cells) are transplanted from normal C57/BL/6 donor mice to MRL/+ female recipients. Fetal thymus from allogeneic C57BL/6 donors was therefore engrafted along with either bone marrow or fetal hematopoietic cells (FHCs) plus fragments of adult or fetal bone. More than seventy percent of old MRL/+ mice (>7 months) that had been given a fetal thymus (FT) transplant plus either bone marrow or FHCs and also bone fragments survived more than 100 days after treatment (Fig. 11). The mice that received FHCs, FB, plus FT from allogeneic donors developed normal T cell and B cell functions. Serum amylase levels decreased in these mice, whereas they increased in the mice that received FHCs and FB, but not FT. The pancreatitis and sialoadenitis already present at the time of transplantations were fully corrected according to histological analysis by transplants of allogeneic FHCs, FB and FT in the MRL/+ mice (59). These findings are taken as an experimental indication that stem cell transplants along with FT grafts might represent a useful strategy for the treatment of intractable diseases (including autoimmune diseases) in aged humans.

ACKNOWLEDGMENT

We would like to thank Ms K. Ando for her expert help in the preparation of the manuscript.This work was supported by a grant from the Japanese Ministry of Health and Welfare; a grant from "Haiteku Research Center" of the Ministry of Education, Science and Culture, Japan, the "Traditional Oriental Medical Science Program" of the Public Health Bureau of the Tokyo Metropolitan Government, and the Science Research Promotion Fund of the Japan Private School Promotion Foundation, Japan.

References

1. Theofilopoulos, A.R., and F.J. Dixon. 1985. Murine models of systemic lupus erythematosus. *Adv. Immunol.* 37:269–358.
2. Denman, A.M., A.S. Russell, and E.J. Denman. 1969. Adoptive transfer of the diseases of New Zealand Black mice to normal mouse strains. *Clin. Exp. Immunol.* 5:567–595.
3. Morton, J.I., and B.V. Siegel. 1974. Transplantation of autoimmune potential. I. Development of antinuclear antibodies in H-2 histocompatible recipients of bone marrow from New Zealand Black mice. *Proc. Natl. Acad. Sci. USA* 71:2162–2165.
4. Akizuki, M., J.P. Reeves, and A.D. Steinberg. 1978. Expression of autoimmunity by NZB/NZW marrow. *Clin. Immunol. Immunopathol.* 10:247–250.
5. Eisenberg, R.A., S. Izui, P.J. McConahey, L. Hang, C.J. Peters, A.N. Theofilopoulos, and F.J. Dixon. 1980. Male determined accelerated autoimmune disease in BXSB mice: Transfer by bone marrow and spleen cells. *J. Immunol.* 125:1032–1036.
6. Jyonouchi, H., P.W. Kincade, R.A. Good, and G.J. Fernandes. 1981. Reciprocal transfer of abnormalities in clonable B lymphocytes and myeloid progenitors between NZB and DBA/2 mice. *J. Immunol.* 127:1232–1238.
7. Ikehara, S., R.A. Good, T. Nakamura, K. Sekita, S. Inoue, O. Maung Maung, E. Muso, K. Ogawa, and Y. Hamashima. 1985. Rationale for bone marrow transplantation in the treatment of autoimmune diseases. *Proc. Natl. Acad. Sci. USA* 82:2483–2487.
8. Ikehara, S., H. Ohtsuki, R.A. Good, H. Asamoto, T. Nakamura, K. Sekita, E. Muso, Y. Tochino, T. Ida, H. Kuzuya, H. Imura, and Y. Hamashima. 1985. Prevention of type I diabetes in nonobese diabetic mice by allogeneic bone marrow transplantation. *Proc. Natl. Acad. Sci. USA* 82:7743–7747.
9. Yasumizu, R., K. Sugiura, H. Iwai, M. Inaba, S. Makino, T. Ida, H. Imura, Y. Hamashima, R.A. Good, and S. Ikehara. 1987. Treatment of type 1 diabetes mellitus in non-obese diabetic mice by transplantation of allogeneic bone marrow and pancreatic tissue. *Proc. Natl. Acad. Sci. USA* 84:6555–6557.
10. Oyaizu, N., R. Yasumizu, M. Inaba-Miyama, S. Nomura, H. Yoshida, S. Miyawaki, Y. Shibata, S. Mitsuoka, K. Yasunaga, S. Morii, R.A. Good, and S. Ikehara. 1988. (NZW × BXSB) F1 mouse, a new model of idiopathic thrombocytopenic purpura. *J. Exp. Med.* 167:2071–2022.
11. Ikehara, S., R. Yasumizu, M. Inaba, S. Izui, K. Hayakawa, K. Sekita, J. Toki, K. Sugiura, H. Iwai, T. Nakamura, E. Muso, Y. Hamashima, and R.A. Good. 1989. Long-term observations of autoimmune-prone mice treated for autoimmune disease by allogeneic bone marrow transplantation. *Proc. Natl. Acad. Sci. USA* 86:3306–3310.
12. Ikehara, S., M. Kawamura, F. Takao, M. Inaba, R. Yasumizu, Soe Than, H. Hisha, K. Sugiura, Y. Koide, T.O. Yoshida, T. Ida, H. Imura, and R.A. Good. 1990. Organ-specific and systemic autoimmune diseases originate from defects in hematopoietic stem cells. *Proc. Natl. Acad. Sci. USA* 87:8341–8344.
13. Miyama-Inaba, M., S. Kuma, K. Inaba, H. Ogata, H. Iwai, R. Yasumizu, S. Muramatsu, R.M. Steinman, and S. Ikehara. 1988. Unusual phenotype of B cells in the thymus of normal mice. *J. Exp. Med.* 168:811–816.
14. Good, R.A., C. Martinez, and A.E. Gabrielsen. 1964. Clinical considerations of the thymus in immunobiology. In: *The thymus in immunobiology* Good R.A. and Gabrielsen, A.E. (eds.). Hoeber-Harper, New York, pp. 3–48.
15. Mackay, I.R., and P. Gail. 1963. Thymic germinal centres and plasma cells in systemic lupus erythematosus. *Lancet* 2:667.
16. Ikehara, S., H. Tanaka, T. Nakamura, F. Furukawa, S. Inoue, K. Sekitta, J. Shimizu, Y. Hamashima, and R.A. Good. 1985. The influence of thymic abnormalities on the development of autoimmune diseases. *Thymus* 7:25–36.
17. Nakamura, T., S. Ikehara, R.A. Good, S. Inoue, K. Sekita, F. Furukawa, H. Tanaka, Oo. Maung Maung, and Y. Hamashima. 1985. Abnormal stem cells in autoimmune-prone mice are responsible for premature thymic involution. *Thymus* 7:151–160.
18. Adachi, Y., M. Inaba, K. Inaba, Soe Than, Y. Kobayashi, and S. Ikehara, 1993. Analyses of thymic abnormalities in autoimmune-prone (NZW × BXSB) F1 mice. *Immunobiol.* 188:340–354.
19. Nakamura, T., R.A. Good, S. Inoue, O. Maung Maung, Y. Hamashima, and S. Ikehara. 1986. Successful liver allografts in mice by combination with allogeneic bone marrow transplantation. *Proc. Natl. Acad. Sci. USA* 83:4529–4532.
20. Hang, L.M., S. Izui, and F.J. Dixo. 1981. (NZW × BXSB F1) hybrid A. model of acute lupus and coronary vascular disease with myocardial infarction. *J. Exp. Med.* 154:216–221.

21. Adachi, Y., M. Inaba, Y. Amoh, H. Yoshifusa, Y. Nakamura, H. Suzuki, S. Akamatu, S. Nakai, H. Haruna, M. Adachi, and S. Ikehara. 1995. Effect of bone marrow transplantation on anti-phospholipid antibody syndrome in murine lupus mice. *Immunobiol.* 192:218–230.

22. Estess, P., A.B. Begovich, M. Koo, P.P. Jones, and H.O. McDevitt. 1986. Sequence analysis and structure function of murine *q, k, u, s,* and *f* haplotype I-Aβ cDNA clones. *Proc. Natl. Acad. Sci. USA* 83:3594–3598.

23. Koide, Y., and T.O. Yoshida. 1989. The unique nucleotide sequence of the Ab gene in the NOD mouse is shared with its nondiabetic sister strains, the ILI and the CTS mouse. *Int. Immunol.* 2:189–192.

24. Nishimoto, H., H. Kikutani, K. Yamamura, and T. Kishimoto. 1989. Prevention of autoimmune insulitis by expression of I-E molecules in NOD mice. *Nature* 328:432–434.

25. Reich, E.P., R.S. Sherwin, O. Kanagawa, and C.A. Janeway, Jr. 1989. An explantation for the protective effect of the MHC class III-E molecule in murine diabetes. *Nature* 341:326–328.

26. Acha-Orbea, H., and H.O. McDevitt. 1987. The first external domain of the nonobese diabetic mouse class II I-Aβ chain is unique. *Proc. Natl. Acad. Sci. USA* 84:2435–2439.

27. Todd, J.A., J.I. Bell, and H.O. McDevill. 1987. HLA-DQβ gene contributes to susceptibility and resistance to insulin-dependent diabetes mellitus. *Nature* 329:599–604.

28. Visser, J.W.M., J.G.J. Bauman, A.H. Mulder, J.F. Eliason, and A.M. de Leeuw. 1984. Isolation of murine pluripotent hemopoietic stem cells. *J. Exp. Med.* 159:1576–1590.

29. Miyama-Inaba, M., H. Ogata, J. Toki, S. Kuma, K. Sugiura, R. Yasumizu, and S. Ikehara. 1987. Isolation of murine pluripotent hemopoietic stem cells in the Go phase. *Biochem. Biophys. Res. Commun.* 75:1809–1812.

30. Kawamura, M., H. Hisha, Y. Li, S. Fukuhara, and S. Ikehara. 1997. Distinct qualitative differences between normal and abnormal hemopoietic stem cells *in vivo* and *in vitro. Stem Cells* 15:56–62.

31. Baldwin, J.L., R. Storb, E.D. Thomas, and M. Mannik. 1977. Bone marrow transplantation in patients with gold-induced marrow aplasia. *Arthr. Rheum.* 20:1043–1048.

32. Jacobs, P., M.D. Vincent, and R.W. Martell. 1986. Prolonged remission of severe refractory rheumatoid arthritis following allogeneic bone marrow transplantation for drug-induced aplastic anemia. *Bone Marrow Transpl.* 1:237–239.

33. Lowenthal, R.M., M.L. Cohen, K. Atkinson, and J.C. Biggs. 1993. Apparent cure of rheumatoid arthritis by bone marrow transplantation. *J. Rheumatol.* 20:137–140.

34. Eedy, D.J., D. Burrows, J.M. Bridges, and F.G. Jones. 1990. Clearance of severe psoriasis after allogeneic bone marrow transplantation. *Br. Med. J.* 300:908–909.

35. Liu Yin, J.A., and S.N. Jowitt. 1992. Resolution of immune-mediated diseases following allogeneic bone marrow transplantation for leukaemia. *Bone Marrow Transpl.* 9:31–33.

36. Grau, J.M., J. Casademont, R. Monforte, P. Marin, A. Grañena, C. Rozman, and A. Urbano-Marquez. 1990. Myasthenia gravis after allogeneic bone marrow transplantation: report of a new case and pathogenetic consideratins. *Bone Marrow Transpl.* 5:435–437.

37. Melms, A., C. Faul, N. Sommer, H. Wietholter, C.A. Muller, and G. Ehninger. 1992. Myasthenia gravis after BMT: identification of patients at risk? *Bone Marrow Transpl.* 9:78–79.

38. Wyatt, D.T. L. Lum, J. Casper, J. Hunter, and B. Camitta. 1990. Autoimmune thyroiditis after bone marrow transplantation. *Bone Marrow Transpl.* 5:357–361.

39. Aldouri, M.A., R. Ruggier, O. Epstein, and H.G. Prentice. 1990. Adoptive transfer of hyperthyroidism and autoimmune thyroiditis following allogeneic bone marrow transplantation for chronic myeloid leukaemia. *Br. J. Haematol.* 74:118–119.

40. Hagopian, W., and A. Lernmark, 1992. Autoimmune diabetes mellitus. In: *The autoimmune diseases II.*, Rose, N.R., and Mackay, I.R. eds. Academic Press, San Diego-Toronto, pp. 235–278.

41. Lampeter, E.F., M. Homberg, K. Quabeck, U.W. Schaffer, P. Wernet, J. Bertrams, H. Grosse-Wilde, F.A. Gries, and H. Kolb. 1993. Transfer of insulin-dependent diabetes between HLA-identical siblings by bone marrow transplantation. *Lancet* 341:1243–1244.

42. Vialettes, B., D. Maraninchi, M.P. San Marco, F. Birg, A.M. Stoppa, C. Mattei-Zevaco, C. Thivolet, L. Hermitte, P. Vague, and P. Mercier. 1993. Autoimmune polyendocrine failure-Type I (insulin-dependent) diabetes mellitus and hypothyroidism-after allogeneic bone marrow transplantation in a patient with lymphoblastic leukaemia. *Diabetologia* 1993:541–546.

43. Lampeter, E.B. 1993. Discussion remark to Session 24: BMT in autoimmune diseases. *Exp. Hematol.* 21:1153–1156.

44. Holland, F.J., J.K. McConnon, R. Volpé, and E.F. Saunders. 1991. Concordant Graves' disease after bone marrow transplantation; implication for pathogenesis, *J. Clin. Endocrinol. Metab.* 72:837–840.

45. Mayumi, H., and R.A. Good, 1989, Long-lasting skin allograft tolerance in adult mice induced across fully allogeneic (multimajor H-2 plus multiminor histocompatibility) antigen barriers by a tolerance-inducing method using cyclophosphamide. *J. Exp. Med.* 169:213.

46. Role of hematopoietic stem cell transplantation for autoimmune diseases. Fred Hutchinson Cancer Research Center, Seattle, USA, October 27, 1995. In: Sullivan, K.M. and Furst, D.E. (eds.), *The Journal of Rheumatology*, Monthly International Journal, Canada, 1997. 24 (Suppl. 48), 102 pp.

47. Nelson, J.L., R. Torrez, F.M. Louie, O.S. Choe, R. Storb, and K.M. Sullivan. 1997. Pre-existing autoimmune diseases in patients with longterm survival after allogeneic bone marrow transplantation. *J. Rhematol.* 24:23–29.

48. Euler, H.H., A.M. Marmont, A. Bacigalupo, S. Fastenrath, P. Dreger, M. Hoffknecht, A.R. Zander, B. Schalke, U. Hahn, R. Haas, and N. Schmitz. 1996. Early recurrence or persistance of autoimmune diseases after unmanipulated autologous stem cell transplantation. *Blood* 88:3621–3625, 1996.

49. Hashimoto, F., K. Sugiura, K. Inoue, and S. Ikehara. 1997. Major histocompatibility complex restriction between hematopoietic stem cells and stromal cells *in vivo. Blood* 89:49–54.

50. Ishida, T., M. Inaba, H. Hisha, K. Sugiura, Y. Adachi, N. Nagata, R. Ogawa, R.A. Good, and S. Ikehara. 1994. Requirement of donor-derived stromal cells in the bone marrow for successful allogeneic bone marrow transplantation: complete prevention of recurrence of autoimmune diseases in MRL/MP-lpr/lpr mice by transplantation of bone marrow plus bone (stromal cells) from the same donor. *J. Immunol.* 152:3119–3127.

51. Mayumi, H., and R.A. Good. 1989. Long-lasting skin allograft tolerance in adult mice induced across fully allogeneic

(multimajor H-2 plus, multiminor histocompatibility) antigen barriers by a tolerance-inducing method using cyclophosphamide. *J. Exp. Med.* 169:213.

52. Shin, T., K. Himeno, H. Mayumi, and K. Nomoto, 1985. Drug-induced tolerance to allografts in mice. III. Prolongation of skin graft survival by tolerance induction in congenic mice disparate in the major H-2 antigens. *Transplantation* 39:333.

53. Mayumi, H., K. Kayashima, T. Shin, and K. Nomoto. 1985. Drug-induced tolerant to allografts in mice. V. Prolongation of skin graft survival in tolerance mice with combined immunosuppressive treatment. *Transplantation* 39:335.

54. Zhang, Y., R. Yasumizu, K. Sugiura, F. Hashimoto, Y. Amoh, Z. Lian, Cherry, N. Nishino, S. Ikehara. 1994. Fate of allogeneic or syngeneic cells in intravenous or portal vein injection: possible explanation by portal vein injection. *Eur. J. Immunol.* 24:1558–1565.

55. Sugiura, K., K. Kato, F. Hashimoto, T. Jin, Y. Amoh, Y. Tamamoto, H. Morita, T. Okumura, and S. Ikehara. 1997. Induction of donor-specific T cell anergy by portalvenous injection of allogeneic cells. *Immunobiol.* 197:460–477.

56. Morita, H., K. Sugiura, M. Inaba, T. Jin, J. Ishikawa, Z. Lian, Y. Adachi, S. Sogo, K. Yamanishi, H. Taki, M. Adachi, T. Noumi, Y. Kamiyama, R.A. Good, and S. Ikehara. 1998. A strategy for organ allografts without using immunosuppressants or irradiation. *Proc. Natl. Acad. Sci. USA* 95:6947–6952.

57. Takeuchi, K., M. Inaba, S. Miyashima, R. Ogawa, and S. Ikehara. 1998. A new strategy for treatment of autoimmune diseases in chimeric resistant MRL/lpr. mice. *Blood* 91:4616–4623.

58. Kushida, T., M. Inaba, K. Takeuchi, K. Sugiura, R. Ogawa, and S. Ikehara. 2000. Treatment of interactable autoimmune diseases in MRL/lpr mice using a new strategy for allogeneic bone marrow transplantation. *Blood* 95:1862–1868.

59. Hosaka, N., M. Nose, M. Kyogoku, N. Nagata, S. Miyashima, R.A. Good, and S. Ikehara. 1996. Thymus transplantation, a critical factor for correction of autoimmune disease in aging MRL/+ mice. *Proc. Natl. Acad. Sci. USA* 93:8558–8562.

65 | Immunointervention in Autoimmune Diseases

Luciano Adorini

1. INTRODUCTION

Activation of peripheral T cells by foreign and self antigens is under stringent control by different mechanisms, both thymic and peripheral (1). Control of T cell reactivity is accomplished by three major types of mechanisms: 1) deletion, the physical elimination of T cells specific for a given antigen, 2) anergy, the functional incapacity of T cells to respond to antigen, and 3) suppression, the inhibition of T cell function by a regulatory (suppressor) cell. The failure of these mechanisms may lead to autoimmune diseases.

Progress in understanding the mechanisms of T cell activation and inactivation, as well as modulation of T cell responses, is currently being translated into strategies able to induce selective immunosuppression to treat different pathological situations, notably autoimmune diseases, allergies, and allograft rejection. The medical need for selective immunosuppression is very high, as the available immunosuppressive drugs are substantially inadequate because of limited efficacy, modest selectivity, and considerable toxicity. Key attack points for selective immunointervention have been identified: Modulation of antigen recognition, costimulation blockade, induction of regulatory cells, deviation to non-pathogenic or protective responses, neutralization of proinflammatory cytokines, induction or administration of anti-inflammatory cytokines, and modulation of leukocyte trafficking (Table 1). Therefore, to selectively interfere with the activation of pathogenic T cells, immunosuppressive therapy can be directed to three cellular targets—antigen-presenting cells, autoreactive T cells and regulatory cells—with the common goal to selectively inhibit the activation of pathogenic class II-restricted CD4$^+$ T cells.

All these forms of immunointervention have been successfully used to prevent and sometimes treat experimental autoimmune diseases. Based on these results, expectations have been raised for exploiting the same strategies to inhibit the activation of human autoreactive T cells. In this review, we will examine recent advances towards immunointervention in autoimmune diseases highlighting their prospects for clinical applicability.

Table 1 Antigen and cytokine-based strategies for immunointervention in autoimmune diseases

---

Antigen-based
MHC blockade
Targeting the TCR complex
⇒ Parenteral antigen administration
⇒ Oral tolerance
⇒ Altered peptide ligands
⇒ Deletion of autoreactive T cells
Induction of regulatory cells/deviation from Th1 cells
⇒ Oral administration or inhalation of autoantigen
⇒ Presentation of autoantigen by non-IL-12-producing APC
⇒ Administration of peptide analogues of autoantigen
⇒ TCR peptide vaccination

Cytokine-based
Neutralization or inhibition of production of proinflammatory cytokines
⇒ IL-1, IL-2, IFN-γ, IL-12, IL-18, TNF-α
Administration or induction of anti-inflammatory cytokines
⇒ IL-4, IL-10, Type I interferons, TGF-β

---

2. TARGETING THE MHC-TCR COMPLEX

Cellular immune responses are primarily mediated by T cells. Binding of antigen/MHC ligand to the TCR complex initiates a cascade of signaling events, beginning with the activation of several cytoplasmic protein tyrosine kinases. Recruitment of the CD4 or the CD8 co-receptors and the associated protein kinase lck in the vicinity of the TCR complex induces phosphorylation of CD3 proteins, ultimately leading to downstream signal progression. Interruption of this signaling pathway at a number of points may lead to T cell unresponsiveness or tolerance. Different strategies targeting MHC, TCR and co-receptor molecules have been developed and some of them have already been applied clinically.

2.1. MHC Blockade

Experimental autoimmune diseases can be prevented by the administration of monoclonal antibodies (mAbs) directed to the MHC molecule presenting the autoantigen (2) or to complexes formed between class II molecules and autoantigenic peptides (3), indicating that blocking class II molecules can prevent disease.

Peptides binding to the same class II molecule can compete with each other, *in vitro* and *in vivo*, for presentation to T cells. This raises the possibility of inducing selective immunosuppression by blocking the binding site of class II molecules associated with autoimmune diseases, thus preventing their capacity to bind any antigen, including autoantigens (4). Direct evidence for MHC blockade upon administration of MHC antagonists has been obtained (5,6). Importantly, exogenous competitors can inhibit the presentation of endogenously synthesised antigens, likely the most relevant in the induction of autoreactive T cells leading to MHC class II-associated autoimmune diseases (7,8).

MHC blockade *in vivo* is capable of preventing autoimmune diseases. For example, co-administration of an encephalitogenic peptide together with an unrelated non-immunogenic peptide binding to the same MHC class II molecule prevents induction of EAE (9,10). More recently, *in vivo* treatment with a MHC class I-restricted blocking peptide has been shown to prevent virus-induced autoimmune diabetes (11). Administration of MHC antagonists may be an approach to preventing HLA-associated autoimmune diseases, assuming that the association between MHC molecules and the disease reflects their capacity to present autoantigens to autoreactive T cells. However, several points still need to be addressed to evaluate the practical feasibility of this form of immunointervention. Since MHC blockade is a passive type of treatment, a sustained plasma level of soluble MHC blockers necessary for effective inhibition of T cell activation must be achieved (12). Therefore, prevention of T cell activation by MHC blockade may not be of therapeutic value unless sustained high concentrations of MHC antagonist can be maintained in extracellular fluids. Peptidomimetic antagonists could exhibit adequate pharmacokinetic and pharmacodynamic characteristics (13,14).

In conclusion, MHC antagonists as immunosuppressive drugs in the treatment of autoimmune diseases have strengths, as information about the autoantigen/s is not required and a degree of selectivity is afforded by targeting MHC class II molecules associated to the disease. These are balanced by weaknesses: a sustained level of antagonist must be maintained for a long period of time and ongoing immune responses are not inhibited. The clinical development of pure MHC antagonists for the treatment of chronic autoimmune diseases will certainly not be easy, if possible at all.

However, a growing body of evidence indicates that synthetic peptides corresponding to linear sequences of MHC molecules have immunomodulatory effects *in vitro* and *in vivo* in animal models and in humans. Although they can act as MHC blockers, the mechanisms responsible for immunomodulation are more complex (15). Recent findings show that MHC-derived peptides can affect signal transduction, cell cycle progression and T cell apoptosis, down-regulating immune responses via a variety of mechanisms (16). Some soluble MHC molecules or synthetic peptides are capable of inducing and maintaining tolerance in animals (17). This suggests that MHC-derived peptides may represent a new class of immuno-therapeutics.

2.2. Targeting the TCR Complex

In most autoimmune disease situations, T cells recognizing self-antigens in the target organ are responsible, at least in part, for inducing tissue destruction. Using autoantigens to specifically inactivate autoreactive pathogenic T cells is thus an obvious approach to inhibiting autoimmune diseases. However, several issues render the clinical applicability of this approach problematic. First, a number of potential target antigens have been identified in several human autoimmune diseases and it is unclear which is the first autoantigen inciting the autoimmune process. Second, due to epitope spreading, it is likely that many combinations of MHC/peptide ligands, derived from different self-proteins, already contribute to autoimmune disease processes at the time of clinical presentation. Third, recent data indicate that the frequency of autoreactive T cells may be much higher than previously thought. Thus, approaches targeting a single TCR may be inadequate to modify the course of an autoimmune disease. Alternative approaches are, however, available.

Parenteral antigen administration. Systemic injection of antigen is one of the approaches that reproducibly induces effective antigen-specific hyporesponsiveness (18). This has been successfully exploited for inhibiting autoimmune disease in animal models and efforts are underway aimed at extending the success of antigen-based immunotherapeutics to human autoimmune diseases (19). This approach minimally requires two conditions: Knowledge of the autoantigen triggering the disease and the absolute certainty that antigen administration will never induce or recall a pathogenic response. Both conditions are demanding. A variety of mechanisms underlay systemic autoantigen-based immunotherapy, from T cell apoptosis to induction of regulatory cells, from TCR antagonism to immunodeviation. DNA vaccines encoding autoantigenic peptides, possibly useful tools in controlling autoimmune disease, provide an example. Vaccination with DNA encoding a dominant MBP epitope inhibited EAE by inducing T cell apoptosis (20) whereas administration of DNA encoding the insulin B chain reduced the incidence of IDDM by skewing the response to the Th2 phenotype (21). Irrespective of the mechanism induced, DNA immunization with plasmids expressing self-antigens might constitute a novel and attractive therapeutic approach to preventing autoimmune diseases.

Indeed, immunotherapies based on parenteral antigen administration are being evaluated clinically. A pilot trial of low-dose insulin to prevent diabetes in relatives of IDDM patients (22) has been followed by an expanded, still ongoing, multicenter trial. Parenteral antigen-based immunointervention can also be considered that afforded by the synthetic random amino acid copolymer glatiramer acetate (Copaxone®), also known as copolymer 1. This agent, which is immunologically cross-reactive with MBP, inhibits EAE by a variety of mechanisms, including MHC class II blockade and TCR antagonism (23). Significantly, it reduces MS relapses by over 30% with few side effects (24) and is now commercially available for MS treatment.

Oral tolerance. Oral administration of antigen can induce suppressor T cells that act by releasing inhibitory cytokines after antigen-specific stimulation, as demonstrated by adoptive transfer of suppression by CD8+ T cells, which produce transforming growth factor β (TGF-β) following antigen-specific triggering (25). In addition to CD8+ TGF-β secreting cells, CD4+ regulatory cells are also induced (26). These CD4+ cells are of Th2 type, secreting IL-4 and IL-10 in addition to TGF-β. The secretion of Th2-type cytokines induced by a myelin antigen such as MBP could also inhibit pathogenic responses driven by other myelin antigens, like PLP, by a mechanism of bystander suppression (27). Bystander suppression, which could be considered specular to epitope spreading (28), could be instrumental rendering

antigen-based immunointervention feasible. Accordingly, the induction of a Th2 response to a self-antigen present in the local microenvironment could convert the response to any autoantigen in the inflammatory site into a protective, or at least non-pathogenic, response. In different models of autoimmune diseases, transmucosal administration of antigen at low doses has been shown to favor induction of Th2 cells and to inhibit disease (29).

Feeding higher antigen doses, both Th1 and Th2 cells are deleted by apoptosis after their initial activation (30). The degree of peripheral T cell deletion was enhanced, via a Fas-mediated mechanism, when antigen feeding was combined with systemic administration of anti-IL-12 antibodies (31). These findings suggest that IL-12 negatively regulates T cell apoptosis, a major mechanism of peripheral tolerance. A combination of oral antigen feeding and administration of anti-IL-12 may thus be useful in the treatment of autoimmune diseases.

Suppression of EAE, either induced by oral tolerization or by natural recovery, could be related to the secretion of inhibitory cytokines, like TGF-β and IL-4, actively suppressing the inflammatory process in the target organ (32). TGF β can also enhance the antigen-dependent encephalitogenic function of T cells in EAE (33), suggesting different modes of action of this lymphokine depending on the cell types present in a particular microenvironment. Accordingly, it has been reported that an encephalitogenic peptide can inhibit EAE when administered by inhalation, but not orally (34).

Although extensive clinical trials have demonstrated the safety of oral tolerance, antigen feeding did induce cytolytic CD8+ T cells and autoimmune diabetes in a TCR transgenic mouse model (35). Initial human trials of oral tolerance showed positive findings in patients with multiple sclerosis (MS) and rheumatoid arthritis (RA). A double blind pilot trial of oral tolerization with myelin antigens in MS patients showed a lower incidence of disease exacerbations in the myelin-fed group (36). A randomized, double-blind trial involving 60 patients with severe active RA demonstrated a statistically significant decrease in the number of swollen joints and tender joints in subjects fed chicken type II collagen for 3 months (37), encouraging further research on clinical application of oral tolerance. However, additional phase II clinical trials investigating the oral administration of type II collagen in RA (38), S-antigen in uveitis (39), and insulin in type I diabetes have not yielded spectacular results in terms of efficacy. Most importantly, a pivotal double blind, placebo-controlled, phase III multicenter trial of oral myelin in 515 relapsing-remitting MS patients, concluded in 1998, failed to show efficacy over placebo.

Altered peptide ligands. The TCR can sense minor changes in the ligand and can transduce signals inducing the T cell to respond with a gradation of effector functions. In this respect, the TCR has a pharmacological behaviour similar to other receptor systems, in which agonists, partial agonists, and antagonists can be identified by modifying the ligand (40). Antigen processing can generate both agonist and antagonist peptides from a single immunogenic epitope (41).

Two separate signals are required for T cell activation, phosphorilation of CD3ζ and CD3$\gamma\delta\epsilon$ chains (42) and it has been hypothezised that partial agonists induce phosphorylation of the CD3$\gamma\delta\epsilon$ complex in the absence of phosphorylation of the ζ chain (40). Support for qualitative differences in the signal delivered by altered peptide ligands (APLs) is provided by the observation that phosphorylation of the ζ chain without activation of the ZAP-70 kinase is linked to the antagonist effect (43,44). This partial T cell signaling could take different forms, like energy induction, cytokine production without proliferation, or cytolysis without proliferation. Both Th1 (45) and Th2 (46) clones can be rendered anergic *in vitro* by APLs delivering an incomplete signal, allowing cytokine production in the absence of clonal expansion. Failure of T cells specific for self-epitopes to proliferate may be a highly efficient way to control autoimmune responses *in vivo*.

APLs behaving as TCR antagonists for either CD4+ (47) or CD8+ (48) T cells are about two orders of magnitude more efficient than MHC antagonists inhibiting T cell activation are. This would, in principle, render APLs attractive tools to induce selective immunosuppression. TCR antagonists have been shown to occur naturally in HIV (49) and hepatitis B (50) virus infections. Mutated viral epitopes behaving as TCR antagonists inhibit the capacity of CTLs to destroy infected target cells, thus effectively protecting the cell harbouring the mutant virus from elimination by CTLs specific for the wild-type epitope. Naturally occurring APLs can also induce a rapid shift from IFN-γ to IL-10 production (51) and narrow the repertoire of cellular immune response by interfering with T-cell priming (52). Modulation of the immune response by naturally occurring APLs may thus provide a general mechanism for immunodominance and for persistence by many polymorphic pathogens.

The potent bioactivity of TCR antagonists may have therapeutic potential. TCR antagonist peptides can inhibit EAE induced by different epitopes, their activity is determined by the capacity to modulate a diverse repertoire of autoreactive T cells (53,54). Interestingly, microbial peptides mimicking MBP87–99 can function as APLs and inhibit EAE (55). In addition, EAE can be reversed by administration of a soluble APL variant of an MBP epitope, which reduces IFN-γ and TNF-α production (56,57). Immunization with an autoantigenic peptide altered at the primary TCR contact point protects mice from the development of EAE by inducing T cells that are cross-reactive

with the native PLP peptide and secrete Th0 and Th2 cytokines (58). The observations that anti-IL-4 mAb treatment abrogates protection induced by the APL and that transfer of APL-induced T cell lines protects mice from EAE, further support the hypothesis that APLs mediate protection by inducing regulatory Th2 cells *in vivo* (57). Modulation of cytokine patterns of MBP-specific human autoreactive T cell clones by a single amino acid substitution of their peptide ligand has also been demonstrated (59). In this case, the APL induced selective secretion of TGF-β, raising the possibility that modulation of T cell cytokine secretion may be important in regulating the response of potentially pathogenic human T cells specific for self-antigens.

Theoretically, if the epitopes inducing pathogenic T cells could be identified, administration of TCR antagonists could prevent, or possibly even treat, autoimmune diseases. The development of this approach for clinical use would certainly be accelerated by the implementation of protocols able to induce some form of bystander immunosuppression, thereby converting a microenvironment dominated by proinflammatory Th1-type cytokines into one characterized by anti-inflammatory, Th2-type, cytokines. However, the feasibility of selective immunosuppression or immunodeviation by APLs acting as TCR antagonists requires detailed knowledge of the antigenic epitopes involved and homogeneity of these epitopes in different individuals. Obviously, the antagonist APL has to be designed in such a way that it will never, under any circumstance, become an agonist. These requirements may limit the clinical applicability of APLs.

Two clinical trials have already tested these concepts in MS patients using an APL of the immunodominant MBP peptide 83–99. Kappos et al. (59a) performed a double-blind placebo-controlled study involving 142 patients testing three different doses of the APL. No difference in the relapse rate between APL and placebo-treated patients was found, but the volume and number of gadolinium-enhancing lesions were reduced in patients receiving the lowest APL dosage. Interestingly, the APL administration induced a Th2 response to MBP in a subset of patients, which could explain the relatively high incidence of immediate-type hypersensitivity reactions observed in the treated group, an event that caused the termination of the trial. Bielekova et al. (59b) tested only the highest APL dose in eight patients, three of whom developed exacerbations of MS that was linked by immunological studies in two cases to the APL treatment, suggesting an encephalitogenic capacity of the peptide used. Both trials teach important lessons on the safety and efficacy of APL-based therapies.

Deletion of autoreactive T cells. Induction of apoptosis in autoreactive peripheral T cells is an intriguing approach to immunointervention in autoimmune diseases. Antigen-induced apoptosis is mediated by the death cytokines Fas ligand and TNF bound to their receptors that recruit and activate caspases, which initiate a chain of lethal proteolytic events (60).

The Fas receptor, also known as APO-1 or CD95, has emerged as a key initiator of apoptotic cell death in a variety of cell types. CD4+ T cells are unique in their ability to undergo apoptosis by stimulating their own Fas receptors with secreted or membrane-bound Fas ligand. Fas ligand-expressing APCs induce rapid and profound clonal deletion of antigen-specific peripheral T cells. T cells can also trigger apoptosis in B cells, macrophages, and other cell types through Fas ligand. These interactions negatively regulate the immune system but can also contribute to immunopathology, as occurs in Fas-mediated damage of target tissues in organ-specific autoimmune diseases (61). The dual role of Fas in the immune response may limit its potential as a therapeutic target. Despite the many roles of Fas in immunoregulation, analysis of genetic deficiencies in the Fas pathway have shown that the main result of disrupting this pathway in vivo is systemic autoimmunity.

The autoimmune lymphoproliferative syndrome (ALPS) affords novel insights into the mechanisms that regulate lymphocyte homeostasis and underlie the development of autoimmunity. In ALPS, defective lymphocyte apoptosis permits chronic, non-malignant adenopathy and splenomegaly, the survival of normally uncommon CD3+ CD4 CD8− T cells, and the development of autoimmune diseases, such as haemolytic anaemia and thrombocytopenia (62). This syndrome arises in early childhood in individuals with inherited mutations in genes that mediate apoptosis. Most cases of ALPS involve heterozygous mutations in Fas or FasL (63), analogous to the mutations observed in the autoimmune-prone strains lpr and gld (64), but a novel form involves a defect in caspase-10 resulting in defective apoptosis of dendritic cells (65). The sustained life-span of mature dendritic cells presenting self antigens can lead to increased autoreactivity (66), suggesting new avenues for immunointervention.

A significantly increased frequency of MBP-reactive T cells was found in patients with MS relative to healthy individuals only when anti-Fas-ligand mAb was used to block apoptosis (67). This indicates that a significant proportion of MBP-reactive T cells are sensitive to apoptosis and are not deleted in vivo in patients with MS, unlike in healthy individuals, suggesting a functional deficit in apoptotic deletional mechanism. Elimination of autoreactive T cells by selective induction of their apoptosis is, in principle, an ideal approach to the treatment of autoimmune diseases. Repeated intravenous administration of large amounts of soluble MBP could delete peripheral autoreactive T cells, and thereby improve the course of EAE (68). This implies

that the stronger the proliferative response to antigen, the more vulnerable T cells are to apoptosis by TCR re-engagement. Interestingly, soluble antigen therapy induces apoptosis of autoreactive T cells preferentially in the target organ (69). Feeding high antigen doses can also induce peripheral deletion of antigen-reactive cells (30).

Deletion of pathogenic autoreactive T cells by administration of high dose autoantigen would have obvious therapeutic potential for the treatment of autoimmune diseases, but possible side effects should not be underestimated. A critical danger intrinsic to this strategy is the activation of autoreactive T cells, before their activation-induced cell death. In addition, this approach faces three major challenges: i) the definition of pathogenic autoantigens at the population level, ii) the spreading of the response to different antigens, iii), the targeting of pathogenic T cell while sparing regulatory cells with protective activity. A fusion protein between MBP and PLP (MP4) was highly effective in suppressing EAE caused by multiple neuroantigen epitopes in mice (70). However, using the marmoset, a nonhuman primate, demyelination induced by priming with MBP and PLP was associated with spreading of the response to myelin oligodendrocyte glycoprotein (MOG) determinants that generated anti-MOG serum antibodies and Ig deposition in central nervous system white matter lesions (71). These data associate intermolecular "determinant spreading" (28) with clinical autoimmune disease in primates and raise important issues for the applicability of apoptosis-inducing regimens.

2.3. Non-Specific Targeting of the TCR Complex

The mouse anti-CD3 mAb OKT3 is a potent immunosuppressive agent inducing both T-cell depletion and antigenic modulation of the CD3/T-cell receptor complex, which is routinely used for prevention of acute allograft rejection. CD3+ lymphocytes are cleared from the circulation within 1 hour of intravenous administration of OKT3. OKT3 is also a potent human T-lymphocyte mitogen and an acute clinical syndrome is regularly observed upon the first OKT3 injection, related to massive, although transient, release of several cytokines (72). In addition, OKT3 triggers a potent humoral response that neutralizes the antibody activity during subsequent administrations. A "humanized" form of OKT3 has been developed to minimize immunogenicity and permit repetitive administrations (73). Its therapeutic potential in human autoimmune diseases will be evaluated shortly.

Administration of anti-CD3 mAb can also induce tolerance. A short anti-CD3 treatment of adult NOD mice significantly inhibits the spontaneous development of autoimmune diabetes, even when applied within 7 days of the onset of overt disease (74). One unique feature was that the anti-CD3-induced tolerance ensued only from treatment of overtly diabetic NOD mice. Durable protection was exclusively observed when treating mice with recent-onset IDDM. Treatment of young mice was without effect, and complete but transient protection followed the treatment of pre-diabetic NOD mice (75). The immunosuppression is specific for β cell-associated antigens, suggesting that self tolerance can be restored in adult mice by transient targeting of the CD3/TCR complex, albeit only when autoimmunity is fully established. Peritransplant administration of anti-CD3 immunotoxin was able to promote long-term acceptance of xenogenic islets in monkeys with spontaneous IDDM (76). The results suggest tolerance induction, at least in operational terms, and have potential implications for the therapy of IDDM.

2.4. Targeting TCR Co-Receptors

Administration of non-depleting (77) or depleting (78) anti-CD4 mAbs is effective in preventing and treating autoimmune diseases. Long-term administration of depleting anti-CD4 antibodies to autoimmune disease patients, however, is limited by the side effects of chronic antibody administration, such as immunogenicity, and by the scarce selectivity, because all CD4+ cells are affected. The use of depleting anti-CD4 mAb in RA has been disappointing as they do not penetrate the synovial joint in sufficient quantity to suppress disease without producing severe and protracted peripheral blood lymphopenia (79). In contrast, a short-course therapy with non-depleting anti-CD4 antibodies may achieve long-lasting effects, since tolerance can be induced to antigens co-administered with the antibody (80). Non-depleting anti-CD4 treatment is also capable of re-establishing tolerance to an unknown pancreatic β cell antigen in NOD mice, even when diabetogenic spleen cells already infiltrate the pancreas (81). Although the mechanism inducing tolerance is not clear, the maintenance of transplantation tolerance induced in adult mice by short treatment with antibodies to CD4 and CD8 molecules appears to involve active suppression, because CD4+ cells from tolerant mice can transfer tolerance (82). This does not exclude the concurrent activation of other immunosuppressive mechanisms (80). This approach may serve, in principle, to tolerize T cells specific for unknown autoantigens. Several open clinical trials have been performed to test this hypothesis in established human autoimmune diseases, and particularly in RA (79). Early clinical trials showed some efficacy, although long-lasting tolerance was not induced and withdrawal of anti-CD4 treatment was followed by disease relapse (83,84). A reduction of synovial inflammation after anti-CD4 mAb treatment in early RA was seen, but the patients did not experience clinical improvement (85). Anti-CD4 treatment was not very effective in Crohn's disease either. Only 2 out of 8 patients who were followed-up for 10 weeks after treatment with a chimeric anti-CD4 mAb were in remission and

there was only a minor effect on endoscopically evaluated disease activity (86).

3. DISRUPTION OF COSTIMULATION

Induction of T-cell responses requires T-cell receptor activation and costimulatory interactions between DCs and T cells, in the absence of costimulatory interactions T cells become anergic (87). The two major costimulatory pathways for T cell activation depend on engagement of CD28 and CD154 on T cells by CD80/CD86 and CD40 on DC, respectively (88,89). Once activated, T cells also express CD152, a CD28 homologue that binds to CD80/CD86 with higher affinity than CD28 and inhibits IL-2 production, IL-2 receptor expression, and cell cycle progression in activated T cells (90). Disruption of these costimulatory pathways by biological agents such as CD152-Ig and anti-CD154 mAb has been shown to be beneficial in autoimmune diseases and allograft rejection (reviewed in (88,89). Interestingly, a short treatment with CD152-Ig and anti-CD154 mAb can induce tolerance to allografts in mice (91) and, partially, also in non-human primates (92,93). The relative efficiency of APLs and costimulation blockade for the induction of CD8+ T cell unresponsiveness *in vivo* has recently been compared (94). The results demonstrate that inhibition of CD28-mediated costimulation in the presence of a strong TCR-mediated signal most efficiently induces T cell unresponsiveness. In contrast, APLs failed to interfere with T cell responsiveness *in vivo*. Thus, short-term blockade of CD28 during antigenic stimulation-rather than the use of APLs-appears to be most effective to down-modulate responsiveness of naive CD8+ T cells, at least in the TCR-transgenic mouse model analyzed.

CD40-CD154 interaction is crucial for the development of EAE, as indicated by its requirement for IL-12 secretion by microglia during antigen presentation to Th1 cells (95) and by suppression of EAE relapses upon its disruption (96). Anti-CD154 mAb treatment at either the peak of acute disease or during remission effectively blocked clinical disease progression and CNS inflammation. This treatment also impaired the expression of clinical disease in adoptive recipients of encephalitogenic T cells, suggesting that CD40-CD154 interactions may be involved in directing the CNS migration of these cells and/or in their effector ability to activate CNS macrophages/microglia (96). Thus, blockade of CD154-CD40 interactions is a promising immunotherapeutic strategy for treating ongoing T cell-mediated autoimmune diseases. However, the clinical trials with anti-CD154 mAb were suspended in November 1999 because of severe side effects, notably thrombo-embolic events, possibly due to the expression of CD154 by activated platelets.

CD152 can also regulate relapsing EAE. Anti-CD152 mAbs (or their F(ab) fragments) enhanced *in vitro* proliferation and proinflammatory cytokine production by PLP139–151-primed primed lymph node cells. *In vivo* administration of anti-CD152 mAb to recipients of PLP139–151-specific T cells resulted in accelerated and exacerbated EAE (97). DNA immunization with the myelin minigene for PLP also altered expression of CD80 and CD86 on APCs in the spleen, suggesting that suppressive immunization against DNA-encoded self may be exploited to disrupt costimulatory pathways (98).

Collectively, these results suggest that CD152 can down-regulate ongoing autoimmune diseases. This possibility has recently been confirmed in a phase I clinical trial demonstrating the immunosuppressive activity of CD152-Ig in psoriasis, a T cell-mediated autoimmune disease of the skin (99). Although efficacy results in phase I trials should be interpreted with caution, the treatment appeared safe and at least as effective as conventional therapy. This work will undoubtedly pave the way for the clinical testing of costimulation blockade in a variety of autoimmune diseases.

4. IMMUNE DEVIATION

Mouse and human CD4⁺ T cells can be distinguished, based on their pattern of cytokine production, into three major subsets: Th1, Th2, and Th0 (100,101). Cytokines associated primarily with Th1 responses include IL-2, IL-15 and TNF-β, in addition to IFN-γ. Th2 cytokines include IL-5, IL-6, IL-10 and IL-13, in addition to IL-4.

The development of Th1 and Th2 cells is influenced by several factors, but three are more important: ligand-TCR interactions, genetic polymorphism and cytokines. Decisive roles in the polarization of T cells are played by IL-12 and IL-4, guiding T cell responses towards the Th1 or Th2 phenotype, respectively (102,103). These, however, are not the only cytokines involved in the polarization of T cell responses, IL-18 and IL-13 for example, also play a role (104). Polarized Th1 and Th2 subsets can be generated from CD4⁺ cells *in vitro* (105), can also be recovered from primed animals (106) and are found in patients suffering from autoimmune or allergic diseases (107). However, polarized Th1 and Th2 cells represent extremes of a spectrum. Within this spectrum, discrete subsets of differentiated T cells secreting a mixture of Th1 and Th2 cytokines, for example IFN-γ and IL-10, have been identified (108).

Although firm evidence for therapeutic manipulation of the Th1/Th2 balance in experimental models has been obtained, its efficacy in autoimmune disease patients is not yet proven. The goal is to develop effective protocols capable of inhibiting established pathogenic Th1 responses

by converting the response in the lesion site from aggressive to protective, in short from Th1 to Th2. This should provide a suppressive environment for the response to any autoantigen and hopefully interfere with disease progression.

4.1. Th1 and Th2 Cells in Autoimmune Diseases

The relative role of Th1 and Th2 cells in autoimmune diseases has been very actively explored. Th1 cells appear to be critically involved in the pathogenesis of many organ-specific autoimmune disease models such as insulin-dependent diabetes mellitus (IDDM), experimental allergic encephalomyelitis (EAE), collagen induced arthritis (CIA), experimental autoimmune uveitis (EAU), thyroiditis and experimental autoimmune myasthenia gravis (EAMG) (reviewed in (109,110)). The role of Th2 cells, however, is still unclear, although indirect evidence for their protective capacity has been provided.

Experimental models. Th1 cells have been found to be involved in the induction of several experimental autoimmune diseases (109,111–113). Evidence for this is based on adoptive transfer experiments demonstrating that CD4⁺ cells producing Th1-type lymphokines can transfer disease, both in EAE (114) and in IDDM (115–117) models. However, cytokine regulation is complex, for example TNF-α and IL-10 have opposite effects on IDDM depending on the developmental stage of the immune system (118,119). This could also explain why, in some cases, β cell destruction in IDDM has been associated with Th2 rather than Th1 cells (120,121).

IL-12 deficiency consistently leads to decreased autoantigen-specific Th1 responses in induced autoimmune diseases such as CIA (122), EAMG (123), EAU (124), EAE (125) and thyroiditis (126). Conversely, spontaneous IDDM in NOD mice is unique among the autoimmune disease models so far examined because of its capacity to develop as efficiently in IL-12-deficient and in control mice (127). Interestingly, IL-12-deficient mice are only partially protected from CIA and EAMG, whereas they appear to be completely protected from EAE and EAU (Table 2). In these cases, complete protection from autoimmunity seems associated with an immunoregulatory circuit involving IL-10 (124,125). Thus, an impaired development of Th1 cells may not be sufficient for complete inhibition of an autoimmune disease, and the induction of an immunoregulatory pathway could be necessary. This regulation could depend more on IL-10 than IL-4, as indicated by the observation that IL-4 transgenic mice do develop EAE but IL-10 transgenic are completely protected (128) and by the capacity of IL-10-producing Tr1 cells to inhibit autoimmune colitis (129). In contrast to IL-12 antagonist-treated NOD mice which show a deviation of pancreas-infiltrating CD4⁺ T cells to the Th2

Table 2 Th1-mediated autoimmune diseases in IL-12-deficient mice

Absent
- Experimental allergic encephalomyelitis (Segal, 1998)
- Experimental autoimmune uveoretinitis (Tarrant, 1998)

Reduced
- Collagen-induced arthritis (McIntyre, 1996)
- Experimental autoimmune myasthenia gravis (Moiola, 1998)
- Autoimmune thyroiditis (Zaccone, 1999)

Unmodified
- Insulin-dependent diabetes mellitus (Trembleau, 1999)

phenotype (130), IL-12-deficient NOD mice show no reduction of IFN-γ producing cells and have few IL-4 or IL-10-producing cells (127). A defective IL-4 production by NOD CD4⁺ cells has been implicated in IDDM development (131), possibly through impairment of NK1.1⁺CD4⁺ cells which could be involved in early IL-4 production (132); it is possible that immunoregulatory pathways involving IL-10 are impaired as well. Consistent with this assumption, administration of a noncytolitic IL-10-fusion protein completely protects NOD mice from IDDM (133). In addition, IL-10- transduced islet-specific Th1 cells prevent IDDM transfer in NOD mice (134).

The reciprocal regulation between T cell subsets predicts a role for Th2 cells in the inhibition of autoimmune diseases. Regulatory T cells that suppress the development of EAE produce Th2-type cytokines (135) and recovery from EAE is associated with increased Th2 cytokines in the CNS (32). In addition, administration of IL-4 to mice with EAE ameliorates the disease (136). These results clearly suggest that activation of Th2 cells may prevent EAE. However, in immunodeficient hosts, MBP-specific Th2 cells cause EAE, suggesting that Th2 cells are not the final effectors of protection (137). Moreover, induction of a MOG-specific Th2 response exacerbates EAE via an antibody-dependent mechanism in a nonhuman primate, pointing out possible side effects of immunodeviation (138).

A protective role for Th2 cells has also been proposed for inhibition of IDDM development, based on the reduced IDDM incidence following IL-4 (139) or IL-10 (140) administration to NOD mice. However, at variance with the latter result, transgenic expression of IL-10 accelerates IDDM, possibly because B cells stimulated by IL-10 would activate T cells specific for cryptic determinants of self antigens (141). A role for Th2 cells regulating the onset of IDDM is also suggested by their capacity to inhibit the spontaneous onset of diabetes in rats (142) and by the correlation between protection from IDDM and IL-4 production in double-transgenic mice on BALB/c background (143). In contrast to their benign role in normal NOD mice, Th2 cells

have been shown to induce acute pancreatitis and IDDM in NOD-scid recipients, via production of IL-10 but not IL-4 (144). This suggests that lymphocyte-deficient recipients lack T cells able to regulate Th2 responses in normal mice. NOD mice that express IL-4 in their pancreatic β cells are protected from insulitis and IDDM (145). Protection in NOD-IL-4 mice appears to be mediated by the pancreatic tissue itself, which causes the activation of distinct, non-pathogenic T cell clonotypes. Regulatory T cells are not induced, as shown by the failure of spleen cells from NOD-IL-4 mice to inhibit IDDM transfer by diabetogenic T cells. These results are consistent with the observation that Th2 cells transgenic for a TCR derived from a clone able to transfer IDDM, when injected into neonatal NOD mice, invaded the islets but neither provoked disease nor provided substantial protection (117). Similar results were also obtained by adoptive transfer of non-transgenic Th1 and Th2 cell lines (146). Therefore, these data do not support the concept that Th2 cells afford protection from IDDM, at least in the effector phase of the disease.

Collectively, the available data point to a critical role of Th1 cells in the induction of autoimmune diseases, whereas the influence of Th2 cells is still unclear. Whether or not Th2 cells exert a direct protective role, diversion away from proinflammatory Th1 cells has mostly been found to reduce the chronic inflammatory response typical of autoimmune diseases.

Human diseases. Th1 cells also appear to be involved in human organ-specific autoimmune diseases (Table 2). CD4⁺ T cell clones isolated from lymphocytic infiltrates of Hashimoto's thyroiditis or Graves' disease exhibit a clear-cut type 1 phenotype (147). Also in Crohn's disease, the critical event in the initiation of bowel inflammatory lesions may involve up-regulation of IL-12 production, resulting in conditions that maximally promote type 1 T-helper immune responses (148). In addition, most T cell clones derived from peripheral blood or cerebrospinal fluid of MS patients show a Th1 lymphokine profile (149). Expression of IL-12p40 mRNA has been detected in acute MS lesions, particularly from early disease cases (150), suggesting that IL-12 up-regulation may be an important event in disease initiation. T cells from MS patients induce CD40L-dependent IL-12 secretion in the progressive but not in the relapsing-remitting form of the disease, suggesting a link to disease pathogenesis (151,152).

Involvement of Th1 cells has also been suggested in other human autoimmune diseases. Insulitis in IDDM patients has been shown to comprise a large number of IFN-γ producing lymphocytes (153). T cell clones derived from the synovial membrane of RA patients also display a Th1 phenotype as they produce, upon activation, large amounts of IFN-γ and no or very little IL-4 (154). Another study has shown that most CD4⁺ and CD8⁺ clones recovered from synovial fluid of RA patients display a Th1 phenotype (155). Interestingly, in situ hybridization for T cell cytokine expression demonstrates a Th1-like pattern in most synovial samples from RA patients, whereas samples from patients with reactive arthritis, a disorder with similar synovial pathology but driven by persisting exogenous antigen, express a Th0 phenotype (156). The situation is less clear in most systemic autoimmune disorders. In general, heterogeneous cytokine profiles are found in the serum or target organs of patients with systemic autoimmunity, such as systemic lupus, Sjögren's syndrome, and primary vasculitis (107).

4.2. Immunointervention in Autoimmune Diseases by Altering the Th1/Th2 Balance

The Th1/Th2 paradigm contains in itself the concept of immunodeviation (157). A good example of immunodeviation is provided by autoantigen-specific therapy of IDDM in NOD mice. Autoantigens in IDDM are fairly well known (158). Among them, glutamic acid decarboxylase (GAD) appears to be most important, because responses to GAD are detected before responses to other autoantigens, including insulin, heat shock protein, peripherin and carboxypeptidase H (159,160). Intravenous or intrathymic administration of GAD to 3-week-old NOD mice, which do not yet display islet infiltration, prevents T cell proliferation to GAD in 12-week-old mice, and also prevents the development of intra-islet infiltration and IDDM in adult NOD mice. The reduction in the number of IFN-γ secreting GAD specific T cells (159) associated with the continuing production of autoantibodies to GAD (160) suggests that parenteral GAD administration has indeed induced a biased switch towards a Th2 response. Evidence for induction of a protective Th2 response has been obtained by nasal administration of GAD peptides to NOD mice (161). These data indicate that nasal administration of GAD65 peptides induces a Th2 cell response that inhibits the spontaneous development of autoreactive Th1 responses and the progression of β cell autoimmunity in NOD mice. Induction of Th2 cells by intravenous administration of GAD65, but not by other β cell autoantigens, has also been found associated with inhibition of an ongoing diabetogenic response in NOD mice (162). GAD65-specific peptide immunotherapy, able to effectively suppress via IL-4 progression to overt IDDM, is dependent on the epitope targeted and the extent of pre-existing β cell autoimmunity in the recipient (163).

Additional approaches to immunodeviation are emerging, including DNA vaccination. The TCR variable chain Vβ8.2 is expressed on pathogenic T cells that induce EAE in H-2ᵘ mice immunized with myelin basic protein (MBP) (164). Vaccination of these mice with naked DNA encoding Vβ8.2 protects them from EAE, and protection has

been found associated to a shift from a Th1 to a Th2 response (165). In addition, administration of DNA encoding the insulin B chain has been shown to reduce the incidence of IDDM by skewing the response to the Th2 phenotype (21).

Collectively, these results suggest that immunodeviation towards the Th2 phenotype may be effective in treating Th1-mediated autoimmune diseases. A critical point for its clinical applicability is the possibility to modify the Th1/Th2 balance in primed individuals and thus treat established autoimmune diseases. The clinical verification of this possibility is eagerly awaited.

5. CYTOKINE-BASED IMMUNOINTERVENTION

Cytokines are essential components of the immune response, and an imbalance in the cytokine network plays an important role in the initiation and perpetuation of autoimmune diseases. Although preclinical models do not always accurately predict the efficacy of cytokine manipulation in human patients, considerable progress is being made in cytokine-based immunointervention. This can be implemented by administration of "anti-inflammatory" cytokines such as IFN-β, TGF-β, IL-4 and IL-10 or by neutralizing "proinflammatory" cytokines like IL-2, IFN-γ, IL-12, TNF-α and IL-1. Several approaches can be used to inhibit a given cytokine. The cytokine itself can be neutralized by specific monoclonal antibodies or by soluble cytokine receptors and the cytokine receptor can be inhibited by monoclonal antibodies or by receptor antagonists competing with the ligand for the receptor binding site (see Table 1).

Cytokine-based manipulation offers a unique possibility to interfere with autoimmune diseases. However, the potency of cytokines coupled to the complexity of the cytokine network can lead to severe side effects, which can still occur after careful preclinical evaluation. The use of cytokine inducers could permit to avoid systemic administration of cytokines and their severe side effects. An alternative approach relies on the local delivery of anti-inflammatory cytokines by gene therapy (166,167).

5.1. Neutralization of Proinflammatory Cytokines

IL-1. IL-1 is a highly proinflammatory cytokine and agents that reduce the production and/or activity of IL-1 are likely to have an impact in the immunotherapy of autoimmune diseases. The production and activity of IL-1, particularly IL-1β, are tightly regulated by three types of inhibitors: IL-1 receptor antagonist (IL-1Ra), soluble IL-1R and membrane type II R, a nonsignaling decoy (168). IL-1Ra showed both efficacy and safety in a large cohort of patients with active and severe RA. This was the first biological agent to demonstrate a beneficial effect on the rate of joint erosion (169). Local gene delivery of IL-1Ra or soluble IL-1R type I has therapeutic efficacy in animal models of RA (170) and has also been tested in a phase I clinical trial (171). In addition, caspase-1, the IL-1β-converting enzyme that processes IL-1β and IL-18 to their mature forms, could represent a target to prevent autoimmune diseases (172).

IL-2. Antibodies specific for the IL-2Rα chain are now in clinical use for the prevention of allograft rejection. These agents are effective and safe, suggesting the possibility to use the same approach in autoimmune diseases. Humanized anti-IL-2R mAb therapy, given intravenously with intervals of up to 4 weeks in lieu of standard immunosuppressive therapy, appeared to prevent the expression of uveitis in 8 of 10 patients treated over a 12-month period (173). These initial findings would suggest that anti-IL-2 receptor therapy may be an effective therapeutic approach for uveitis and, by implication, other disorders with a predominant Th1 profile.

IFN-γ. IFN-γ, produced by activated T and NK cells, is a potent activator of macrophages and monocytes and induces a variety of inflammatory mediators in these cells. It also up-regulates MHC class I and class II molecules, facilitating antigen presentation. Therefore, it is not surprising that a pilot trial of IFN-γ administration to MS patients resulted in a sharp increase in disease exacerbations (174). IFN-γ may be a reasonable target in itself, but most of the ongoing work is concentrated on its inducers, IL-12 in particular.

IL-12. IL-12 is a heterodimer composed of two covalently linked glycosylated chains, p35 and p40, encoded by distinct genes (102). This cytokine, produced predominantly by activated monocytes and dendritic cells but also by other cell types such as microglia (175), enhances proliferation and cytolytic activity of NK and T cells, and stimulates their IFN-γ production (102). Most importantly, IL-12 induces the development of Th1 cells *in vitro* (105,176) and *in vivo* (177). The important and non-redundant role of IL-12 in the induction of Th1 responses has been demonstrated in mice deficient for IL-12 (178), IL-12Rβ1 (179), or Stat4 (180).

IL-12-dependent Th1 responses have been implicated in a number of experimental autoimmune disorders, including IDDM (181), EAE (182), CIA (183), EAU (184), graulomatous colitis (185), EAMG (123), and thyroiditis (126). Neutralization of endogenous IL-12, by anti-IL-12 mAb or by IL-12R antagonists, has significantly contributed to clarify the important role of this cytokine in the pathogenesis of IDDM (130,186), granulomatous colitis (185) and EAE (182). Anti-IL-12 mAb treatment also prevents superantigen-

induced EAE and subsequent relapses (187). Increased IL-12 production by monocytes and increased IFN-γ production by T cells associated with disease activity has been observed in MS patients (152,188). These findings suggest that targeting IL-12 may prove beneficial in some forms of MS, and it is likely that IL-12 antagonists can be useful in other autoimmune conditions, such as inflammatory bowel disease (189). Given the critical role of IL-12 in the induction of Th1-mediated autoimmune diseases (181), IL-12 antagonists could be candidates for immunointervention (110). In addition, small molecular weight agents able to target IL-12 *in vivo* and active in autoimmune diseases are becoming attractive (190,191).

TNF-α. Consistent results from clinical trials have demonstrated the efficacy of TNF-α neutralizing therapies in different autoimmune diseases. Animal studies first clearly documented the important role of TNF-α in RA. Mice transgenic for the human TNF-α gene produce high levels of this cytokine and develop arthritis beginning at 4 weeks of age (192). In addition, in a model of type II collagen-induced arthritis, administration of anti-mouse TNF-α, even after disease onset, significantly reduced inflammation and tissue destruction (193).

Based on these results, chimeric anti-TNF-α mAb was administered to RA patients (194). Treatment with anti-TNF-α was safe and well tolerated, and led to significant clinical and laboratory improvements. After the first administration of anti-TNF-α mAb, remissions lasted, on the average, about three months. Reinjection of the mAb, however, induced a significant anti-globulin response in most patients, reducing considerably the efficacy of the treatment. Clinical improvement after anti-TNF-α mAb therapy was also seen in active Crohn's disease, accompanied by significant healing of endoscopic lesions and disappearance of the mucosal inflammatory infiltrate (195).

A pivotal clinical trial administrating multiple intravenous infusions of anti-TNF-α mAb combined with low-dose weekly methotrexate in RA patients displayed efficacy and a lack of major side effects (196). Longitudinal analysis demonstrated rapid down-regulation of a spectrum of cytokines, cytokine inhibitors, and acute-phase proteins (197). IL-6 reached normal levels within 24 h. Serum levels of cytokine inhibitors, such as soluble p75 and p55 TNFR, were reduced, as was IL-1 receptor antagonist. Reduction in acute-phase proteins was also observed. These results are consistent with the concept of a cytokine-dependent cytokine cascade. The degree of clinical benefit noted after anti-TNF-α therapy is probably due to the reduction of many proinflammatory mediators apart from TNF-α.

An alternative approach, using the soluble TNFR p55 chain fused to the constant region of human IgG1 heavy chain (sTNFR-IgG1), has been demonstrated to be about

10-fold more effective than anti-TNF-α mAb at neutralising the activity of endogenous TNF (198,199). This fusion protein appears to achieve the same clinical effects as anti-TNF-α mAb administration without strong induction of neutralizing antibodies. In a phase II randomized, double-blind, placebo-controlled trial, recombinant human TNFR (p75):Fc fusion protein safely produced rapid, significant and sustained dose-dependent improvement in RA patients (200).

The chimeric anti-TNF-α mAb (Infliximab, Remicade®) and recombinant human TNFR(p75):Fc fusion protein (Etanercept, Enbrel®), both approved by the FDA in 1998, are examples of a new class of disease-modifying anti-inflammatory drugs that interfere with the action of a proto-typical proinflammatory cytokine. At present, Infliximab is approved for Crohn's disease and Etanercept for RA. Both show promise in treating disease, although the long-term risks and benefits of these drugs are not yet known. In any case, their clear-cut efficacy and modest toxicity vividly demonstrate the power of appropriate immunointervention in autoimmune diseases.

Even if clinical results of anti-TNF-α therapy in RA and Crohn's disease patients are very promising, the role of TNF-α in other autoimmune diseases, like IDDM and EAE/MS is still puzzling. The fact that anti-TNF-α mAb treatment initiated before 3 weeks of age prevents insulitis and IDDM clearly suggests that TNF-α may be an essential mediator for the generation and/or activation of autoreactive lymphocytes (118). Intriguingly, administration of TNF-α to adult NOD mice could also prevent IDDM, but the mechanism is still unclear (118). More recently, TNF-α has been shown to partially protect β cells in syngeneic islet grafts from recurrent autoimmune destruction by reducing CD4+ and CD8+ T cells and down-regulating type 1 cytokines, both systemically and locally in the islet graft (201). TNF-α thus appears to have distinct effect on the diabetogenic process depending upon the developmental stage of the immune system and of the target organ, perhaps in a manner analogous to IL-10. These results stress the importance of the time window of cytokine or anti-cytokine treatment to obtain the desired effect. If this concept cannot be translated to clinical practice, conditions to recreate a situation favoring the protective effects of the anti-cytokine treatment should be optimized.

A complex situation exists also in EAE/MS. Although TNF-α has a demyelinating effect in vitro (202) and TNF-α administration enhances EAE (203), TNF-α deficient mice immunized with MOG develop severe neurological impairment with extensive inflammation and demyelination leading to high mortality (204). These results suggest that TNF-α may actually limit the extent and duration of severe CNS pathology. This view is consistent with increased gadolinium-enhancing lesions and lack of efficacy of anti-TNF-α mAb treatment in MS patients (205). This stresses the fact that immunotherapies effective in a given autoimmune condition cannot be automatically translated to any autoimmune disease.

5.2. Administration of Anti-Inflammatory Cytokines

IL-4. IL-4 is the cytokine that plays the most important role in Th2 cell development. Using TCR transgenic T cells, it has been shown that IL-4 drives the development of Th2 cells, and this effect is dominant over that of IL-12 (105). An important function of Th2 cells, and hence of IL-4, could be the control of the tissue-damaging effects of proinflammatory cytokines secreted or induced by Th1 cells.

Transgenic NOD mice that express IL-4 in their pancreatic β cells are protected from insulitis and IDDM, indicating the feasibility of a peripheral approach to the treatment of autoimmunity (145). IL-4 administration to adult NOD mice inhibits IDDM development, although insulitis is not blocked (131). This is consistent with the hypothesis that islet infiltration is not necessarily associated to IDDM, if Th2-type cells predominate (206). For example, in NOD male mice, which develop insulitis but only low incidence of IDDM, IL-4 mRNA expression is associated with non-destructive insulitis (207). In streptococcal-induced arthritis IL-4 administration suppresses the chronic destructive phase (208), and in EAE it inhibits considerably the clinical manifestations of the disease (136). These results look promising, but systemic administration of IL-4 in humans leads to severe side effects, which prevent its use in chronic diseases like autoimmune diseases. Conversely, local delivery of IL-4 by gene therapy has been shown to be beneficial without major side effects in EAE (209), although other anti-inflammatory cytokines such as TGF-β may be more effective (210).

IL-10. IL-10 is a potent suppressor of several effector functions of macrophages, T cells and NK cells. In addition, it contributes to regulate proliferation and differentiation of B cells, mast cells, and thymocytes (211). The most important property of IL-10, from an immuno-therapeutic perspective, is its capacity to inhibit Th1 cells. The inhibition of the Th1 cell pathway by IL-10 is mediated by several mechanisms, including inhibition of IL-12 production by APC (212) and blocking of IFN-γ synthesis by differentiated Th1 cells (211). In addition, IL-10 strongly inhibits production of proinflammatory monokines as IL-1, IL-6, IL-8, TNF-α and GM-CSF (213,214) as well as of reactive oxygen and nitrogen species (215) following activation of human or mouse macrophages. Thus, IL-10 has strong anti-inflammatory properties. In addition, IL-10 can induce regulatory cells (Tr1) able to inhibit autoimmune diseases (216).

IL-10 injection in adult NOD mice has been shown to decrease insulitis and IDDM (140) indicating that systemic administration of this cytokine may affect the course of autoimmune diseases. However, this may not be true in any situation, as demonstrated by the observation that transgenic IL-10 expression under the insulin promoter does not prevent islet destruction and may actually enhance the inflammatory response (119,134). The discrepancy between the protective effect of IL-10 administration on adult NOD mice versus the precipitating effect of transgenic IL-10 expression in pancreatic islets points to the importance of whether cytokines are produced locally or given systemically. Alternatively, the different effects of IL-10 on IDDM induction could be related to the development stage of the immune system or of the target cell at the time of cytokine delivery.

It is possible that these inhibitory functions of IL-10 can be exploited clinically to prevent graft rejection (217). Its activity in inhibiting macrophage activation and Th1 cytokine synthesis also suggests a possible use as a non antigen-specific suppressor factor in the treatment of autoimmune diseases. This possibility would be in accordance with the observation that mice failing to make IL-10 because of targeted disruption of its gene, develop a severe inflammatory bowel disease (218). The disease may reflect an overstimulation of Th1 cells, not controlled by Th2-derived IL-10, to gut antigenic stimulation. Supporting this hypothesis, severe colitis was abrogated in a model of inflammatory bowel disease by systemic administration of IL-10 but, interestingly, not of IL-4 (219). IL-10 administration is currently being tested in inflammatory bowel disease and in RA, but the results, so far, are inferior to the expectations.

Type I interferons. Interferons comprise a family of cytokines that interfere with viral replication. All IFNs increase expression of MHC class I molecules, but, unlike IFN-γ, IFN-α and β (type I IFNs) inhibit MHC class II expression (220). This could be important to explain their immunosuppressive activity and also, at least in part, the disease enhancing effect of IFN-γ (221). IFN-β is now an established therapeutic option for relapsing remitting MS. Attack frequency is reduced by 30% and major attacks by an even greater margin. Accumulating disease burden as measured by annual magnetic resonance imaging (MRI) is markedly lessened, and disease activity as measured by serial gadolinium-enhanced MRI scanning is reduced by over 80%. More recently, IFN-β has also been shown to slow down disease progression in secondary progressive MS (222). The clinical effect of IFN-β is reflected in MRI studies demonstrating a dramatic effect in reducing disease activity. The drug is generally well tolerated, but its efficacy can be compromised in some patients by the emergence of neutralizing antibodies (223). Although its mode of action is still unclear, there is ample evidence from

in vitro studies that IFN-β directly modulates the function of immune cells. IFN-β treatment was found to transiently increase plasma IL-10 levels whereas IL-12p40 was not affected. A significantly lower ratio of Th1 vs. Th2-type cells was observed in CD8$^+$ but not in the CD8$^-$ T-cell subset. Interestingly, an initial rise in the mean percentage of CD95$^+$ T cells and a gradual increase in the mean level of soluble CD95 in plasma was seen (224). Enhanced IL-10 secretion may have anti-inflammatory effects and the increased CD95 expression may directly interfere with T-cell survival.

IFN-β treatment was also evaluated in collagen type II-induced arthritis (CIA) in rhesus monkeys and in RA patients (225). Rapid clinical improvement during IFN-β therapy was observed in three of the four rhesus monkeys with CIA. There was also a marked decrease in serum C-reactive protein (CRP) levels with a subsequent increase after discontinuation of the treatment in all monkeys. The 10 RA patients who completed the study exhibited on average gradual clinical improvement suggesting that IFN-β treatment has a beneficial effect on arthritis.

TGF-β. TGF-β, a molecule known for its pleiotropic activities, can promote or inhibit cell growth and function. TGF-β1 is produced by every leukocyte lineage, including lymphocytes, macrophages, and dendritic cells. It can modulate expression of adhesion molecules, provide a chemotactic gradient for leukocytes and other cells participating in an inflammatory response, and inhibit them once they have become activated (226). Its beneficial role in autoimmune diseases is shown, for example, by the inhibition of EAE following TGF-β administration (227), and by enhancement of EAE upon its neutralization (228). In addition, TGF-β is considered a major mediator in oral tolerance (29). Although the disease limiting properties of TGF-β in autoimmune diseases seem attractive, disruption of the balance between its opposing activities can contribute to aberrant development, malignancy, or pathogenic immune and inflammatory responses characterized by widespread tissue fibrosis and deposition of extracellular matrix (229). The safety of TGF-β2 was tested in an open-label trial of 11 patients with secondary progressive MS. There was no change in expanded disability status scale or MRI lesions during treatment, but five patients experienced a reversible decline in the glomerular filtration rate (230). Systemic TGF-β2 treatment thus appears to be associated with reversible nephrotoxicity, and further investigation of its therapeutic potential should be performed with caution.

6.   INDUCTION OF REGULATORY CELLS

Regulatory T cells represent a potentially interesting target for induction of selective immunosuppression. As discussed above, they can be induced using different strategies, from oral tolerance to immune deviation, from anti-CD4 to IL-10 administration. In addition, the restricted TCR V gene repertoire expressed by autoreactive pathogenic T cells in some animal models of autoimmune diseases, for example in EAE, raises the possibility to control self reactivity at the network level (231). The aim is to boost specific anti-idiotypic T cell responses able to inhibit the pathogenic activity of T cells expressing the target TCR V region gene product (232).

6.1. T Cell Vaccination

Induction of regulatory T cells by T cell vaccination has been accomplished in several experimental models (233). T cell vaccination has also been tested in a limited number of autoimmune disease patients. Inoculation of MS patients with irradiated MBP-reactive T cells enhanced anti-clonotypic class I-restricted T cells able to deplete circulating MBP-reactive T cells, suggesting that clonotypic interactions regulating autoreactive T cells can be induced by T cell vaccination (234,235). The anti-clonotypic response induced by the T cell vaccine was characterized by class I-restricted cytotoxic CD8$^+$ cells recognizing the hypervariable regions of the TCR expressed by clones used for immunization, but not of other MPB-specific clones (236). However, comparable results were not obtained by T cell vaccination of RA patients (237). Although T cell vaccination may represent an interesting model of immune regulation, it could hardly be considered a feasible treatment for autoimmune diseases. DNA-based TCR vaccination techniques are also being developed. Administration of recombinant vaccinia virus (238) or of naked DNA (165) encoding Vβ8.2 is able to prevent EAE by inducing a Th2-type response. It remains to be seen whether these strategies may be applied clinically.

6.2. TCR Peptide Vaccination

A simplified version of T cell vaccination involves the administration of only a specific sequence of the T cell receptor expressed by pathogenic T cells, instead of entire, inactivated T cells. Administration of peptides corresponding to TCR sequences utilized by autoreactive T cells has indeed been reported to down-regulate EAE, presumably via induction of anti-idiotypic T cells with suppressive activity (239,240). The requirement for CD8$^+$ cells in TCR peptide-induced unresponsiveness suggests that such immunoregulatory T cells may participate in the normal course of EAE (241). Synthetic TCR peptides can also be used therapeutically in established EAE (242). Interestingly, the cytokine secretion pattern of TCR peptide-specific regulatory CD4 T cells can profoundly influence whether a type 1 or type 2 population predominates among MBP-specific CD4 effectors. The priming of type 1 regulatory T cells results in deviation of the antigen-specific effector T cell population in a type 2 direction and protection from EAE. In contrast, induction of type 2 regulatory T cells results in exacerbation of EAE, poor recovery, and an increased frequency of type 1 effectors. Thus, the encephalitogenic potential of the MBP-reactive effector population appears to be dominantly influenced by the cytokine secretion phenotype of regulatory CD4 T cells (243).

Based on results obtained in EAE, induction or enhancement of immunoregulatory T cells has been attempted, in MS patients, by injection of synthetic peptides derived from the TCR specific for a putative autoantigen, MBP. The synthetic TCR peptides injected were based on the over-utilization of Vβ5.2 and Vβ6.1 by MBP-specific T cells from MS patients (244). Preliminary results of this trial indicate that anti-TCR immunity can be safely up-regulated in MS patients by human TCR peptide administration. Subsequent studies from the same group concluded that, with the possible exception of Vβ5.2 in DR2/Dw2 patients, there is no single Vβ gene that is consistently used by MBP-reactive T cells in all MS patients (245,246). However, based on the increased frequency of TCR peptide-specific T cells in patients vaccinated with CDR2 peptides from Vβ5.2 and Vβ6.1 TCR (247), a double blind trial was carried out in MS patients (248). Patients who responded to the vaccine showed a reduced frequency of MBP-specific T cells and remained clinically stable. Interestingly, vaccine-specific T cells were Th2-type and inhibited MBP-specific Th1 cells via IL-4 and IL-10, but not TGF-β. Conversely, patients who did not respond to the vaccine had an increased anti-MBP T cell response and progressed clinically. The trial was not powered to show efficacy. Thus, it is not clear to what extent immunity to TCR peptides can regulate MBP-specific responses, nor whether this immunity may have any beneficial effect on the course of MS. Clinical efficacy could be expected only if common TCR V genes are utilized by pathogenic autoreactive T cells in different MS patients, as is the case in EAE (164).

The issue of restricted TCR usage by myelin basic protein-specific T cells from MS patients (249), as well as the role of human MBP as major autoantigen in MS (250), renders the TCR peptide approach more problematic than anticipated. The major challenges for its application are the complexity of T cell responses to encephalitogenic epitopes, the multiplicity of CNS autoantigens, the identification of V genes over-expressed by pathogenic T cells, and a clear-cut demonstration that TCR-specific regulation can control MS. A distinct possibility exists that a biased V gene expression in pathogenic T cells may turn out to be subject to individual regulation (251), thus requiring a patient-specific treatment with a personalized cocktail of preselected antigenic peptides corresponding to the relevant V genes expressed. From a pharmacological point of view, clinical applicability of this strategy is probably too laborious.

Nevertheless, this strategy could have some prospects. In RA a prominent T-cell infiltration in the synovial lining layer has been found, with TCR Vβ3, and Vβ14, and Vβ17 over-represented among IL-2R$^+$ T-cells. A phase II clinical trial in RA, using a combination of three peptides derived from Vβ3, Vβ14, and Vβ17, has yielded promising results. The therapeutic vaccine therapy was safe and well tolerated, immunogenic, and demonstrated clinical improvement in RA patients (252).

7.   TARGETING LEUKOCYTE TRAFFICKING

7.1.   Adhesion Molecules

Adhesion molecules participate in leukocyte circulation, transendothelial migration and homing. Selectins, integrins and immunoglobulin (Ig) gene superfamily adhesion receptors mediate the different steps of the migration of leukocytes from the blood stream towards sites of inflammation (253). Although differences between specific autoimmune diseases exist, key interactions in the development of autoimmune inflammation include L-selectin/ P-selectin/E-selectin, lymphocyte function-associated antigen-1 (LFA-1)/intercellular adhesion molecule-1 (ICAM-1), very late antigen-4 (VLA-4)/vascular cell adhesion molecule-1 (VCAM-1), and α4β7/MadCAM or VCAM-1 adhesion (254). The endothelium, the lining layer of the vasculature, controls the traffic of cells and molecules from the bloodstream into underlying tissues and is the primary target for circulating mediators. For example, inhibition of TNF-α activity results in deactivation of the endothelium, mani-fested as reduced expression of adhesion molecules and chemottractant cytokines, leading to diminished trafficking of inflammatory cells to synovial joints (255).

Blockade of the function or expression of cell adhesion molecules has emerged as an important new therapeutic target in inflammatory diseases. Different drugs are able to interfere with cell adhesion phenomena. In addition, new anti-adhesion therapeutic approaches including blocking mAbs, soluble receptors, synthetic peptides and peptido-mimetics are currently being developed (256).

7.2.   Chemokines and Chemokine Receptors

Directional movement of leukocytes in any aspect of their activities, such as development, homeostatic circulation and inflammatory response is controlled by chemotactic cytokines, the chemokines (257). The interest in the chemokine network also stems from its relevance in pathological states, from inflammation to autoimmunity and viral infection to tumorigenesis. The control of inflammation and other pathological processes with chemokine receptor antagonists has been achieved in preclinical models and will soon be tested in clinical trials (258). Indeed, the structure of chemokine receptors, which are all seven transmembrane domain G protein coupled receptors, makes the development of receptor antagonists chemically tractable and promising as potential drugs, provided that issues related to *in vivo* pleiotropy and redundancy of the chemokine system are handled.

Anti-chemokine mAbs and modified chemokines illustrate the potential of this form of immunointervention in autoimmune diseases. Met-RANTES, a functional antagonist of RANTES receptors, can prevent, although not treat, CIA (259). Blocking the function of MCP-1 or RANTES in the inflammatory phase of glomerulonephritis results in significant decreases in proteinuria as well as in numbers of infiltrating leukocytes and development of interstitial fibrosis (260).

During MS attacks, elevated CSF levels of three chemokines active on T cells and mononuclear phagocytes: IP-10, Mig, and RANTES, have been observed. In addition, CXCR3, an IP-10/Mig receptor, was expressed on lymphocytic cells in virtually every perivascular inflammatory infiltrate in active MS lesions. CCR5, a RANTES receptor, was detected on lymphocytic cells, macrophages, and microglia in actively demyelinating MS brain lesions. Compared with circulating T cells, CSF T cells were significantly enriched in cells expressing CXCR3 or CCR5, chemokine receptors selectively expressed on Th1 cells (261).

Chemokines receptors are differentially expressed on Th1 and Th2 cells. CCR4 and CCR5, have been shown to be preferentially expressed on Th2 and Th1 cells, respectively (262). Although both CCR5$^+$ and CCR4$^+$ CD4$^+$ T cell populations were observed in peripheral blood mononuclear cells from healthy controls and osteoarthritis patients, these cell populations were decreased in patients with active RA. In contrast, the vast majority of synovial fluid (SF) T cells from active RA patients expressed CCR5 but not CCR4. CCR5 ligands, MIP-1 alpha and RANTES, were found in RA SF at high levels. CCR5$^+$ CD4$^+$ T cells from SF mononuclear cells of RA patients produced IFN-γ, but not IL-4, in response to anti-CD3 stimulation *in vitro* (263).

These results indicate that differential expression of chemokine receptors plays a critical role in the selective recruitment, of proinflammatory T cells into the target organ in different autoimmune diseases, and also suggest additional molecular targets for therapeutic intervention.

8. PROSPECTS FOR IMMUNOTHERAPY OF HUMAN AUTOIMMUNE DISEASES

The efficacy of cytokine-specific treatments in chronic inflammatory disease states, in particular anti-TNF-α therapies in RA and Crohn's disease patients, and IFN-β in MS patients, documents the coming-of-age of immunointervention in autoimmune diseases. Targeted delivery by gene therapy of cytokines or cytokine antagonists should permit even more effective and less toxic treatments. Antigen-based immunointervention will continue to be tested clinically, irrespective of the failed attempts. We will certainly witness the application of more articulate strategies able to selectively target cytokine production by Th1 or Th2 cells or to modify the Th1/Th2 balance in clinical situations. The most effective manipulation of pathogenic and protective cells in autoimmunity may eventually rely on a combination of antigen- and cytokine-based approaches to selectively target autoreactive T cells and divert them from autoaggression. Most importantly, new avenues are opening, such as the use of chemokine receptor antagonists, and others approaches look very promising, such as costimulation blockade. Overall, the prospects look relatively bright.

References

1. Stockinger, B. 1999. T lymphocyte tolerance: from thymic deletion to peripheral control mechanisms. *Adv. Immunol.* 71:229–265.
2. Steinman, L., J.T. Rosenbaum, S. Sriram, and H.O. McDevitt. 1981. *In vivo* effects of antibodies to immune response gene products: prevention of experimental allergic encephalitis. *Proc. Natl. Acad. Sci. USA* 78:7111–7114.
3. Aharoni, R., D. Teitelbaum, R. Arnon, and J. Puri. 1991. Immunomodulation of experimental allergic encephalomyelitis by antibodies to the antigen-Ia complex. *Nature* 351:147–150.
4. Adorini, L. 1993. Selective suppression of T cell responses by antigenic proteins and MHC-binding peptides. *Curr. Opin. Invest. Drugs* 2:183–193.
5. Guéry, J.-C., A. Sette, J. Leighton, A. Dragomir, and L. Adorini. 1992. Selective immunosuppression by administration of MHC class II-binding peptides. I. Evidence for *in vivo* MHC blockade preventing T cell activation. *J. Exp. Med.* 175:1345–1352.
6. Guéry, J.-C., M. Neagu, G. Rodriguez-Tarduchy, and L. Adorini. 1993. Selective immunosuppression by administration of major histocompatibility complex class II-binding peptides. II. Preventive inhibition of primary and secondary antibody responses. *J. Exp. Med.* 177:1461–1468.
7. Adorini, L., J. Moreno, F. Momburg, G.J. Hämmerling, J.-C. Guéry, A. Valli, and S. Fuchs. 1991. Exogenous peptides compete for the presentation of endogenous antigens to major histocompatibility complex class II-restricted T cells. *J. Exp. Med.* 174:945–948.
8. Guéry, J.-C., and L. Adorini. 1993. Selective immunosuppression of class II-restricted T cells by MHC class II-binding peptides. *Crit. Rev. Immunol.* 13:195–206.
9. Lamont, A.G., A. Sette, R. Fujinami, S.M. Colon, G. Miles, and H.M. Grey. 1990. Inhibition of experimental autoimmune encephalomyelitis induction in SJL/J mice by using a peptide with high affinity for IA^s molecules. *J. Immunol.* 145:1687–1693.
10. Gautam, A.M., C.I. Pearson, A.A. Sinha, D.E. Smilek, L. Steinman, and H.O. McDevitt. 1992. Inhibition of experimental autoimmune encephalomyelitis by a nonimmunogenic non-self peptide that binds to I-A^u. *J. Immunol.* 148:3049–3054.
11. von Herrath, M.G., B. Coon, H. Lewicki, H. Mazarguil, J.E. Gairin, and M.B. Oldstone. 1998. *In vivo* treatment with a MHC class I-restricted blocking peptide can prevent virus-induced autoimmune diabetes. *J. Immunol.* 161:5087–5096.
12. Ishioka, G.Y., L. Adorini, J.-C. Guéry, F.C.A. Gaeta, R. LaFond, J. Alexander, M.F. Powell, A. Sette, and H.M. Grey. 1994. Failure to demonstrate long-lived MHC saturation both *in vitro* and *in vivo*: implication for therapeutic potential of MHC-blocking peptides. *J. Immunol.* 152:4310–4319.
13. Krebs, S., and D. Rognan. 1998. From peptides to peptidomimetics: design of nonpeptide ligands for major histocompatibility proteins. *Pharm. Acta Helv.* 73:173–181.
14. Falcioni, F., K. Ito, D. Vidovic, C. Belunis, R. Campbell, S.J. Berthel, D.R. Bolin, P. B. Gillespie, N. Huby, G.L. Olson, R. Sarabu, J. Guenot, V. Madison, J. Hammer, F. Sinigaglia, M. Steinmetz, and Z.A. Nagy. 1999. Peptidomimetic compounds that inhibit antigen presentation by autoimmune disease-associated class II major histocompatibility molecules. *Nat. Biotechnol.* 17:562–567.
15. Benichou, G., P.A. Takizawa, P.T. Ho, C.C. Killion, C.A. Olson, M. McMillan, and E.E. Sercarz. 1990. Immunogenicity and tolerogenicity of self-major histocompatibility complex peptides. *J. Exp. Med.* 172:1341–1346.
16. Murphy, B., C.C. Magee, S.I. Alexander, A.M. Waaga, H.W. Snoeck, J.P. Vella, C. B. Carpenter, and M.H. Sayegh. 1999. Inhibition of allorecognition by a human class II MHC-derived peptide through the induction of apoptosis. *J. Clin. Invest.* 103:859–867.
17. Murphy, B., and A.M. Krensky. 1999. HLA-derived peptides as novel immunomodulatory therapeutics. *J. Am. Soc. Nephrol.* 10:1346–1355.
18. Liblau, R., R. Tisch, N. Bercovici, and H.O. McDevitt. 1997. Systemic antigen in the treatment of T-cell-mediated autoimmune disease. *Immunol Today* 18:599–604.
19. Tian, J., A. Olcott, L. Hanssen, D. Zekzer, and D.L. Kaufman. 1999. Antigen-based immunotherapy for autoimmune disease: from animal models to humans? *Immunol. Today* 20:190–195.
20. Lobell, A., R. Weissert, M.K. Storch, C. Svanholm, K.L. de Graaf, H. Lassmann, R. Andersson, T. Olsson, and H. Wigzell. 1998. Vaccination with DNA encoding an immunodominant myelin basic protein peptide targeted to Fc of immunoglobulin G suppresses experimental autoimmune encephalomyelitis. *J. Exp. Med.* 187:1543–1548.
21. Coon, B., L.L. An, J.L. Whitton, and M.G. von Herrath. 1999. DNA immunization to prevent autoimmune diabetes. *J. Clin. Invest.* 104:189–194.
22. Keller, R.J., G.S. Eisenbarth, and R.A. Jackson. 1993. Insulin prophylaxis in individuals at high risk of type I diabetes [see comments]. *Lancet* 341:927–928.
23. Aharoni, R., D. Teitelbaum, R. Arnon, and M. Sela. 1999. Copolymer 1 acts against the immunodominant epitope 82–100 of myelin basic protein by T cell receptor antago-

nism in addition to major histocompatibility complex blocking. *Proc. Natl. Acad. Sci. USA* 96:634–639.

24. Johnson, K.P., B.R. Brooks, J.A. Cohen, C.C. Ford, J. Goldstein, R.P. Lisak, L.W. Myers, H.S. Panitch, J.W. Rose, R.B. Schiffer, T. Vollmer, L.P. Weiner, and J.S. Wolinsky. 1998. Extended use of glatiramer acetate (Copaxone) is well tolerated and maintains its clinical effect on multiple sclerosis relapse rate and degree of disability. Copolymer 1 Multiple Sclerosis Study Group. *Neurology* 50:701–708.

25. Miller, A., O. Lider, A.B. Roberts, M.B. Sporn, and H.L. Weiner. 1992. Suppressor T cells generated by oral tolerization to myelin basic protein suppress both *in vitro* and *in vivo* immune responses by the release of transforming growth factor beta after antigen-specific triggering. *Proc. Natl. Acad. Sci. USA* 89:421–425.

26. Chen, Y., V.K. Kuchroo, J.-I. Inobe, D.A. Hafler, and H.L. Weiner. 1994. Regulatory T cell clones induced by oral tolerance: suppression of autoimmune encephalomyelitis. *Science* 265:1237–1240.

27. Weiner, H.L., A. Friedman, A. Miller, S.J. Khoury, A. Al-Sabbagh, L. Santos, M. Sayegh, R.B. Nussenblatt, D.E. Trentham, and D.A. Hafler. 1994. Oral tolerance: immunologic mechanisms and treatment of animal and human organ-specific autoimmune diseases by oral administration of autoantigens. *Annu. Rev. Immunol.* 12:809–837.

28. Lehmann, P., E. Sercarz, T. Forsthuber, C. Dayan, and G. Gammon. 1993. Determinant spreading and the dynamics of the autoimmune T cell repertoire. *Immunol. Today* 14:203–208.

29. Garcia, G., and H.L. Weiner. 1999. Manipulation of Th responses by oral tolerance. *Curr. Top Microbiol. Immunol.* 238:123–145.

30. Chen, Y., J.-I. Inobe, R. Marks, P. Gonnella, V.K. Kuchroo, and H.L. Weiner. 1995. Peripheral deletion of antigen-reactive T cells in oral tolerance. *Nature* 376:177–180.

31. Marth, T., M. Zeitz, B.R. Ludviksson, W. Strober, and B.L. Kelsall. 1999. Extinction of IL-12 signaling promotes Fas-mediated apoptosis of antigen-specific T cells. *J. Immunol.* 162:7233–7240.

32. Khoury, S.J., W.W. Hancock, and H.L. Weiner. 1992. Oral tolerance to myelin basic protein and natural recovery from experimental autoimmune encephalomyelitis are associated with downregulation of inflammatory cytokines and differential upregulation of transforming growth factor β, interleukin 4, and prostaglandin E expression in the brain. *J. Exp. Med.* 176:1355–1364.

33. Weinberg, A.D., R. Whitham, S.L. Swain, W.J. Morrison, G. Wyrick, C. Hoy, A.A. Vandenbark, and H. Offner. 1992. Transforming growth factor beta enhances the *in vivo* effector function and memory phenotype of antigen-specific T helper cells in experimental autoimmune encephalomyelitis. *J. Immunol.* 148:2109–2117.

34. Metzler, B., and D.C. Wraith. 1992. Inhibition of experimental autoimmune encephalitis by inhalation but not by oral administration of encephalitogenic peptide: influence of MHC binding. *Int. Immunol.* 5:1159–1165.

35. Blanas, E., F.R. Carbone, J. Allison, J.F. Miller, and W.R. Heath. 1996. Induction of autoimmune diabetes by oral administration of autoantigen. *Science* 274:1707–1709.

36. Weiner, H.L., G.A. Mackin, M. Matsui, E.J. Orav, S.J. Khoury, D.M. Dawson, and D.A. Hafler. 1993. Double-blind pilot trial of oral tolerization with myelin antigens in multiple sclerosis. *Science* 259:1321–1324.

37. Trentham, D.E., R.A. Dynesius-Trentham, E.J. Orav, D. Combitchi, C. Lorenzo, K.L. Sewell, D.A. Hafler, and H.L.

Weiner. 1993. Effects of oral administration of type II collagen on rheumatoid arthritis. *Science* 261:1727–1730.

38. Barnett, M.L., J.M. Kremer, E.W. St. Clair, D.O. Clegg, D. Furst, M. Weisman, M.J. Fletcher, S. Chasan-Taber, E. Finger, A. Morales, C.H. Le, and D.E. Trentham. 1998. Treatment of rheumatoid arthritis with oral type II collagen. Results of a multicenter, double-blind, placebo-controlled trial. *Arthritis Rheum.* 41:290–297.

39. Nussenblatt, R.B., I. Gery, H.L. Weiner, F.L. Ferris, J. Shiloach, N. Remaley, C. Perry, R.R. Caspi, D.A. Hafler, C.S. Foster, and S.M. Whitcup. 1997. Treatment of uveitis by oral administration of retinal antigens: results of a phase I/II randomized masked trial. *Am. J. Ophthalmol.* 123:583–592.

40. Evavold, B.D., J. Sloan-Lancaster, and P.M. Allen. 1993. Tickling the TCR: selective T cell functions stimulated by altered peptide ligands. *Immunol. Today* 14:602–609.

41. Carson, R.T., D.D. Desai, K.M. Vignali, and D.A. Vignali. 1999. Immunoregulation of Th cells by naturally processed peptide antagonists. *J. Immunol.* 162:1–4.

42. Wegener, A.M., F. Letourner, A. Hoeveler, T. Broker, F. Luton, and B. Malissen. 1992. The T cell receptor/CD3 complex is composed of at least two autonomous transduction modules. *Cell* 68:83–95.

43. Sloan-Lancaster, J., A.S. Shaw, J.B. Rothbard, and P.M. Allen. 1994. Partial T cell signaling: altered phospho-ζ and lack of zap70 recruitment in APL-induced T cell anergy. *Cell* 79:913–922.

44. Madrenas, J., R.L. Wange, J.L. Wang, N. Isakov, L.E. Samelson, and R.N. Germain. 1995. ζ phosphorylation without zap-70 activation induced by TCR antagonists or partial agonists. *Science* 267:515–518.

45. Sloan-Lancaster, J., B.D. Evavold, and P.M. Allen. 1993. Induction of T cell anergy by altered T cell receptor ligand on live antigen-presenting cells. *Nature* 363:156–159.

46. Sloan-Lancaster, J., B. Evavold, and P.M. Allen. 1994. Th2 cell clonal anergy as a consequence of partial activation. *J. Exp. Med.* 180:1195–1205.

47. De Magistris, T.M., J. Alexander, M. Coggeshall, A. Altman, F.C.A. Gaeta, H.M. Grey, and A. Sette. 1992. Antigen analog-major histocompatibility complexes act as antagonists of the T cell receptor. *Cell* 68:625–634.

48. Jameson, S.C., F.R. Carbone, and M.J. Bevan. 1993. Clone-specific T cell receptor antagonists of major histocompatibility complex class I-restricted cytotoxic T cells. *J. Exp. Med.* 177:1541–1550.

49. Klenerman, P., S. Rowland-Jones, S. McAdam, J. Edwards, S. Daenke, D. Lalloo, B. Koppe, W. Rosenberg, D. Boyd, A. Edwards, and A.J. McMichael. 1994. Cytotoxic T-cell activity antagonized by naturally occurring HIV-1 Gag variants. *Nature* 369:403–407.

50. Bertoletti, A., A. Sette, F.V. Chisari, A. Penna, M. Levrero, M. De Carli, F. Fiaccadori, and C. Ferrari. 1994. Natural variants of cytotoxic epitopes are T-cell receptor antagonists for antiviral cytotoxic T cells. *Nature* 369:407–410.

51. Plebanski, M., K.L. Flanagan, E.A. Lee, W.H. Reece, K. Hart, C. Gelder, G. Gillespie, M. Pinder, and A.V. Hill. 1999. Interleukin 10-mediated immunosuppression by a variant CD4 T cells epitope of Plasmodium falciparum. *Immunity* 10:651–660.

52. Plebanski, M., E.A. Lee, C.M. Hannan, K.L. Flanagan, S.C. Gilbert, M.B. Gravenor, and A.V. Hill. 1999. Altered peptide ligands narrow the repertoire of cellular immune responses by interfering with T-cell priming. *Nat. Med.* 5:565–571.

53. Franco, A., S. Southwood, T. Arrhenius, V.K. Kucroo, H.M. Grey, A. Sette, and G.Y. Ishioka. 1994. T cell receptor antago-

nist peptides are effective inhibitors of experimental allergic encephalomyelitis. *Eur. J. Immunol.* 24:940–946.

54. Anderton, S.M., S. Kissler, A.G. Lamont, and D.C. Wraith. 1999. Therapeutic potential of TCR antagonists is determined by their ability to modulate a diverse repertoire of autoreactive T cells. *Eur. J. Immunol.* 29:1850–1857.

55. Ruiz, P.J., H. Garren, D.L. Hirschberg, A.M. Langer-Gould, M. Levite, M.V. Karpuj, S. Southwood, A. Sette, P. Conlon, and L. Steinman. 1999. Microbial epitopes act as altered peptide ligands to prevent experimental autoimmune encephalomyelitis. *J. Exp. Med.* 189:1275–1284.

56. Karin, N., D.J. Mitchell, S. Brocke, N. Ling, and L. Steinman. 1994. Reversal of Experimental Autoimmune Encephalomyelitis by a soluble peptide variant of a myelin basic protein epitope: T cell receptor antagonism and reduction of interferon γ and tumor necrosis factor α production. *J. Exp. Med.* 180:2227–2237.

57. Brocke, S., K. Gijbels, M. Allegretta, I. Ferber, C. Piercy, T. Blankenstein, R. Martin, U. Utz, N. Karin, D. Mitchell, T. Veromaa, A. Waisman, A. Gaur, D.C. Conlon, N. Ling, P.J. Fairchild, D.C. Wraith, A. O'Garra, C.G. Fathman, and L. Steinman. 1996. Treatment of experimental encephalomyelitis with a peptide analogue of myelin basic protein. *Nature* 379:343–346.

58. Nicholson, L.B., J.M. Greer, R.A. Sobel, M.B. Lees, and V.K. Kuchroo. 1995. An altered peptide ligand mediates immune deviation and prevents autoimmune encephalomyelitis. *Immunity* 3:397–405.

59. Windhagen, A., C. Scholz, P. Hollsberg, H. Fukaura, A. Sette, and D.A. Hafler. 1995. Modulation of cytokine patterns of human autoreactive T cell clones by a single amino acid substitution of their peptide ligand. *Immunity* 2:373–380.

59a. Kappos, L., G. Comi, H. Panitch, J. Oger, J. Antel, P. Conlon, L. Steinman, A. Rae-Grant, J. Castaldo, N. Eckert, J.B. Guarnaccia, P. Mills, G. Johnson, P.A. Calabresi, C. Pozzilli, S. Bastianello, E. Giugni, T. Witjas, P. Cozzone, J. Pelletier, E. Pohlau, H. Przuntek, V. Hoffman, C. Bever, Jr., E. Katz, M. Clanet, I. Berry, D. Brassat, I. Brunet, G. Edan, P. Duquette, E.W. Radue, D. Schott, C. Lienert, A. Taksaoui, M. Rodegher, M. Filippi, A. Evans, P. Bourgouin, A. Zijdenbos, S. Salem, N. Ling, D. Alleva, E. Johnson, A. Gaur, P. Crowe, X.J. Liu. 2000. Induction of a non-encephalitogenic type 2 T helper-cell autoimmune response in multiple sclerosis after administration of an altered peptide ligand in a placebo-controlled, randomized phase II trial. *Nat. Med.* 6:1176–1182.

59b. Bielekova, B., B. Goodwin, N. Richert, I. Cortese, T. Kondo, G. Afshar, B. Gran, J. Eaton, J. Antel, J.A. Frank, H.F. McFarland, R. Martin. 2000. Encephalitogenic potential of the myelin basic protein peptide (amino acids 83–99) in multiple sclerosis: Results of a phase II clinical trial with an altered peptide ligand. *Nat. Med.* 6:1167–1175.

60. Lenardo, M., K. Chan, F. Hornung, H. McFarland, R. Siegel, J. Wang, and L. Zheng. 1999. Mature T lymphocyte apoptosis-immune regulation in a dynamic and unpredictable antigenic environment. *Annu. Rev. Immunol.* 17:221–253.

61. Siegel, R.M., and T.A. Fleisher. 1999. The role of Fas and related death receptors in autoimmune and other disease states. *J. Allergy Clin. Immunol.* 103:729–738.

62. Canale, V.C., and C.H. Smith. 1967. Chronic lymphadenopathy simulating malignant lymphoma. *J. Pediatr* 70:891–899.

63. Fisher, G.H., F.J. Rosenberg, S.E. Straus, J.K. Dale, L.A. Middleton, A.Y. Lin, W. Strober, M.J. Lenardo, and J.M.

Puck. 1995. Dominant interfering Fas gene mutations impair apoptosis in a human autoimmune lymphoproliferative syndrome. *Cell* 81:935–946.

64. Nagata, S. 1998. Human autoimmune lymphoproliferative syndrome, a defect in the apoptosis-inducing Fas receptor: a lesson from the mouse model. *J. Hum. Genet.* 43:2–8.

65. Wang, J., L. Zheng, A. Lobito, F.K. Chan, J. Dale, M. Sneller, X. Yao, J.-M. Puck, S.E. Straus, and M.J. Lenardo. 1999. Inherited human Caspase 10 mutations underlie defective lymphocyte and dendritic cell apoptosis in autoimmune lymphoproliferative syndrome type II. *Cell* 98:47–58.

66. Ludewig, B., B. Odermatt, S. Landmann, H. Hengartner, and R.M. Zinkernagel. 1998. Dendritic cells induce autoimmune diabetes and maintain disease via de novo formation of local lymphoid tissue. *J. Exp. Med.* 188:1493–1501.

67. Zang, Y.C., M.M. Kozovska, J. Hong, S. Li, S. Mann, J.M. Killian, V.M. Rivera, and J.Z. Zhang. 1999. Impaired apoptotic deletion of myelin basic protein-reactive T cells in patients with multiple sclerosis. *Eur. J. Immunol.* 29:1692–1700.

68. Critchfield, J.M., M.K. Racke, J.C. Zuniga-Pflucker, B. Cannella, C.S. Raine, J. Goverman, and M.J. Lenardo. 1994. T cell deletion in high antigen dose therapy of autoimmune encephalomyelitis. *Science* 263:1139–1143.

69. Ishigami, T., C. White, and M. Pender. 1998. Soluble antigen therapy induces apoptosis of autoreactive T cells preferentially in the target organ rather than in the peripheral lymphoid organs. *Eur. J. Immunol.* 28:1623–1635.

70. Elliott, E., H. McFarland, S. Nye, R. Cofiell, T. Wilson, J. Wilkins, S. Squinto, L. Matis, and J. Mueller. 1996. Treatment of experimental encephalomyelitis with a novel chimeric fusion protein of myelin basic protein and proteolipid protein. *J. Clin. Invest.* 87:1602–1612.

71. McFarland, H., A. Lobito, M. Johnson, J. Nyswaner, J. Frank, G. Palardy, N. Tresser, C. Genain, J. Mueller, L. Matis, and M. Lenardo. 1999. Determinant spreading associated with demyelination in a nonhuman primate model of multiple sclerosis. *J. Immunol.* 162:2384–2390.

72. Bach, J.F., G.N. Fracchia, and L. Chatenoud. 1993. Safety and efficacy of therapeutic monoclonal antibodies in clinical therapy. *Immunol. Today* 14:421–425.

73. Richards, J., J. Auger, D. Peace, D. Gale, J. Michel, A. Koons, T. Haverty, R. Zivin, L. Jolliffe, and J.A. Bluestone. 1999. Phase I evaluation of humanized OKT3: toxicity and immunomodulatory effects of hOKT3gamma4. *Cancer Res.* 59:2096–2101.

74. Chatenoud, L., E. Thervet, J. Primo, and J.-F. Bach. 1994. Anti-CD3 antibody induces long-term remission of overt autoimmunity in nonobese diabetic mice. *Proc. Natl. Acad. Sci. USA* 91:123–127.

75. Chatenoud, L., J. Primo, and J.F. Bach. 1997. CD3 antibody-induced dominant self tolerance in overtly diabetic NOD mice. *J. Immunol.* 158:2947–2954.

76. Thomas, F.T., C. Ricordi, J.L. Contreras, W.J. Hubbard, X.L. Jiang, D.E. Eckhoff, S. Cartner, G. Bilbao, D.M. Neville, Jr., and J.M. Thomas. 1999. Reversal of naturally occuring diabetes in primates by unmodified islet xenografts without chronic immunosuppression. *Transplantation* 67:846–854.

77. Brostoff, S.W., and D.W. Mason. 1984. Experimental allergic encephalomyelitis: successful treatment *in vivo* with a monoclonal antibody that recognizes T helper cells. *J. Immunol.* 133:1938–1942.

78. Cobbold, S.P., A. Jayasuruya, A. Nash, T. Prospero, and H. Waldmann. 1985. Therapy with monoclonal antibodies

by elimination of T-cell subsets *in vivo*. *Nature* 312:348–351.

79. Choy, E.H., G.H. Kingsley, and G.S. Panayi. 1998. Anti-CD4 monoclonal antibodies in rheumatoid arthritis. *Springer Semin. Immunopathol.* 20:261–273.

80. Waldmann, H., and S. Cobbold. 1993. The use of monoclonal antibodies to achieve immunological tolerance. *Immunol. Today* 14:247–251.

81. Hutchings, P., L. O'Reilly, N.M. Parish, H. Waldmann, and A. Cooke. 1992. The use of a non-depleting anti-CD4 monoclonal antibody to re-establish tolerance to β cells in NOD mice. *Eur. J. Immunol.* 22:1913–1918.

82. Qin, S., S.P. Cobbold, H. Pope, J. Elliott, D. Kioussis, J. Davies, and H. Waldmann. 1993. "Infectious" transplantation tolerance. *Science* 259:974–977.

83. Horneff, G., G.R. Burmester, F. Emmrich, and J.R. Kalden. 1991. Treatment of rheumatoid arthritis with an anti-CD4 monoclonal antibody. *Arthritis Rheum.* 34:129–139.

84. Goldberg, D., P. Morel, L. Chatenoud, C. Boitard, C.J. Menkes, P.-H. Bertoye, J.-P. Revillard, and J.-F. Bach. 1991. Immunological effects of high dose administration of anti-CD4 antibody in rheumatoid arthritis patients. *J. Autoimmunity* 4:617–630.

85. Tak, P.P., P.A. van der Lubbe, A. Cauli, M.R. Daha, T.J. Smeets, P.M. Kluin, A.E. Meinders, G. Yanni, G.S. Panayi, and F.C. Breedveld. 1995. Reduction of synovial inflammation after anti-CD4 monoclonal antibody treatment in early rheumatoid arthritis. *Arthritis Rheum.* 38:1457–1465.

86. Stronkhorst, A., S. Radema, S.L. Yong, H. Bijl, I.J. ten Berge, G.N. Tytgat, and S.J. van Deventer. 1997. CD4 antibody treatment in patients with active Crohn's disease: a phase 1 dose finding study. *Gut* 40:320–327.

87. Schwartz, R.H. 1990. A cell culture model for T cell anergy. *Science* 248:1349–1356.

88. Lenschow, D.J., T.L. Walunas, and J.A. Bluestone. 1996. CD28/B7 system of T cell costimulation. *Annu. Rev. Immunol.* 14:233–258.

89. Grewal, I., and R. Flavell. 1998. CD40 and CD154 in cell-mediated immunity. *Annu. Rev. Immunol.* 16:111–135.

90. Walunas, T.L., C.Y. Bakker, and J.A. Blustone. 1996. CTLA-4 ligation blocks CD28-dependent T cell activation. *J. Exp. Med.* 183:2541–2550.

91. Larsen, C., E. Elwood, D. Alexander, S. Ritchie, R. Hendrix, C. Tucker-Burden, H. Cho, A. Aruffo, D. Hollenbaugh, P. Linsley, K. Winn, and T. Pearson. 1996. Long-term acceptance of skin and cardiac allografts after blocking CD40 and CD28 pathways. *Nature* 381:434–438.

92. Kenyon, N., M. Chatzipetrou, M. Masetti, A. Ranuncoli, M. Oliveira, J. Wagner, A. Kirk, D. Harlan, L. Burkly, and C. Ricordi. 1999. Long-term survival and function of intrahepatic islet allografts in rhesus monkeys treated with humanized anti-CD154. *Proc. Natl. Acad. Sci. USA* 96:8132–8137.

93. Kirk, A., L. Burkly, D. Batty, R. Baumgartner, J. Berning, K. Buchanan, J.J. Fechner, R. Germond, R. Kampen, N. Patterson, S. Swanson, D. Tadaki, C. TenHoor, L. White, S. Knechtle, and D. Harlan. 1999. Treatment with humanized monoclonal antibody against CD154 prevents acute renal allograft rejection innonhuman primates. *Nature Med.* 6:686–693.

94. Bachmann, M.F., D.E. Speiser, T.W. Mak, and P.S. Ohashi. 1999. Absence of co-stimulation and not the intensity of TCR signaling is critical for the induction of T cell unresponsiveness *in vivo*. *Eur. J. Immunol.* 29:2156–2166.

95. Aloisi, F., G. Penna, E. Polazzi, L. Minghetti, and L. Adorini. 1999. CD40-CD154 interaction and IFN-γ are required for IL-12 but not prostaglandin E_2 secretion by microglia during antigen presentation to Th1 Cells. *J. Immunol.* 162:1384–1391.

96. Howard, L.M., A.J. Miga, C.L. Vanderlugt, M.C. Dal Canto, J.D. Laman, R.J. Noelle, and S.D. Miller. 1999. Mechanisms of immunotherapeutic intervention by anti-CD40L (CD154) antibody in an animal model of multiple sclerosis. *J. Clin. Invest.* 103:281–290.

97. Karandikar, N.J., C.L. Vanderlugt, T.L. Walunas, S.D. Miller, and J.A. Bluestone. 1996. CTLA-4: a negative regulator of autoimmune disease. *J. Exp. Med.* 184:783–788.

98. Ruiz, P.J., H. Garren, I.U. Ruiz, D.L. Hirschberg, L.V. Nguyen, M.V. Karpuj, M.T. Cooper, D.J. Mitchell, C.G. Fathman, and L. Steinman. 1999. Suppressive immunization with DNA encoding a self-peptide prevents autoimmune disease: modulation of T cell costimulation. *J. Immunol.* 162:3336–3341.

99. Abrams, J.R., M.G. Lebwohl, C.A. Guzzo, B.V. Jegasothy, M.T. Goldfarb, B.S. Goffe, A. Menter, N.J. Lowe, G. Krueger, M.J. Brown, R.S. Weiner, M.J. Birkhofer, G.L. Warner, K.K. Berry, P.S. Linsley, J.G. Krueger, H.D. Ochs, S.L. Kelley, and S. Kang. 1999. CTLA-4Ig-mediated blockade of T-cell costimulation in patients with psoriasis vulgaris. *J. Clin. Invest.* 103:1243–1252.

100. Mosmann, T.R., H. Cherwinski, M.W. Bond, M.A. Giedlin, and R.L. Coffmann. 1986. Two types of murine helper T cell clone. I Definition according to profile of lymphokine activities and secreted proteins. *J. Immunol.* 136:2348–2357.

101. Del Prete, G., M. De Carli, C. Mastromauro, R. Biagiotti, D. Macchia, P. Falagiani, M. Ricci, and S. Romagnani. 1991. Purified protein derivative of mycobacterium tuberculosis and escretory-secretory antigen(s) of toxocara canis expand *in vitro* human T cells with stable and opposite (type 1 T helper or type 2 T helper) profile of cytokine production. *J. Clin. Invest.* 88:346–350.

102. Gately, M.K., L.M. Renzetti, J. Magram, A.S. Stern, L. Adorini, U. Gubler, and D.H. Presky. 1998. The interleukin-12/interleukin-12-receptor system: role in normal and pathologic immune responses. *Annu. Rev. Immunol.* 16:495–512.

103. Paul, W.E., and R.A. Seder. 1994. Lymphocytes responses and cytokines. *Cell* 76:241–251.

104. O'Garra, A. 1998. Cytokines induce the development of functionally heterogeneous T helper cell subsets. *Immunity* 8:275–283.

105. Hsieh, C.-S., S.E. Macatonia, C.S. Tripp, S.F. Wolf, A. O'Garra, and K.M. Murphy. 1993. Development of Th1 CD4⁺ T cells through IL-12 produced by *Listeria*-induced macrophages. *Science* 260:547–549.

106. Reiner, S.L., and R.M. Locksley. 1995. The regulation of immunity to Leishmania major. *Annu. Rev. Immunol.* 13:151–177.

107. Romagnani, S. 1994. Lymphokine production by human T cells in disease states. *Annu. Rev. Immunol.* 12:227–257.

108. Mosmann, T.R., and S. Sad. 1996. The expanding universe of T-cell subsets: Th1, Th2 and more. *Immunol. Today* 17:138–146.

109. Liblau, R.S., S.M. Singer, and H.O. McDevitt. 1995. Th1 and Th2 CD4⁺ T cells in the pathogenesis of organ-specific autoimmune diseases. *Immunol. Today* 16:34–38.

110. Adorini, L., F. Aloisi, F. Galbiati, M.K. Gately, S. Gregori, G. Penna, F. Ria, S. Smiroldo, and S. Trembleau. 1997. Targeting IL-12, the key cytokine driving Th1-mediated autoimmune disease. *Chem. Immunol.* 68:175–197.

111. Powrie, F., and R.L. Coffmann. 1993. Cytokine regulation of T cell function: potential for therapeutic intervention. *Immunol. Today* 14:270–274.

112. O'Garra, A., and K. Murphy. 1993. T-cell subsets in autoimmunity. *Curr. Op. Immunol.* 5:880–886.

113. Trembleau, S., T. Germann, M.K. Gately, and L. Adorini. 1995. The role of IL-12 in the induction of organ-specific autoimmune diseases. *Immunol. Today* 16:383–386.

114. Ando, D.G., J. Clayton, D. Kong, J.L. Urban, and E.E. Sercarz. 1989. Encephalitogenic T cells in the B10.PL model of experimental allergic encephalomyelitis (EAE) are of the Th1 lymphokine subtype. *Cell. Immunol.* 124:132–143.

115. Haskins, K., and M. McDuffie. 1990. Acceleration of diabetes in young NOD mice with a CD4+ islet specific T cell clone. *Science* 249:1433–1436.

116. Bergman, B., and K. Haskins. 1994. Islet-specific T-cell clones from the NOD mouse respond to beta-granule antigen. *Diabetes* 43:197–203.

117. Katz, J.D., C. Benoist, and D. Mathis. 1995. T helper cell subsets in insulin-dependent diabetes. *Science* 268:1185–1188.

118. Yang, X.-D., R. Tisch, S.M. Singer, Z.A. Cao, R.S. Liblau, R.D. Schreiber, and H.O. McDevitt. 1994. Effect of tumor necrosis factor α on insulin-dependent diabetes mellitus in NOD mice. I. The early development of autoimmunity and the diabetogenic process. *J. Exp. Med.* 180:995–1004.

119. Wogensen, L., M.-S. Lee, and N. Sarvetnick. 1994. Production of interleukin 10 by islet cells accelerates immune mediated destruction of β cells in nonobese diabetic mice. *J. Exp. Med.* 179:1379–1384.

120. Anderson, J.T., J.G. Cornelius, A.J. Jarpe, W.E. Winter, and A.B. Peck. 1993. Insulin-dependent diabetes in the NOD mouse model. II. Beta cell destruction in autoimmune diabetes is a Th2 and not a Th1 mediated event. *Autoimmunity* 15:113–122.

121. Akhtar, I., J.P. Gold, L.-Y. Pan, J.L. Ferrara, X.-D. Yang, J.I. Kim, and K.-N. Tan. 1995. CD4+ β islet cell-reactive T cell clones that suppress autoimmune diabetes in nonobese diabetic mice. *J. Exp. Med.* 182:87–97.

122. McIntyre, K.W., D.J. Shuster, K.M. Gillooly, R.R. Warrier, S.E. Connaughton, R.B. Hall, L.H. Arp, M.K. Gately, and J. Magram. 1996. Reduced incidence and severity of collagen-induced arthritis in interleukin-12-deficient mice. *Eur. J. Immunol.* 26:2933–2938.

123. Moiola, L., F. Galbiati, G. Martino, S. Amadio, E. Brambilla, G. Comi, L.M.E. Grimaldi, and L. Adorini. 1998. IL-12 is involved in the induction of experimental autoimmune myasthenia gravis, an antibody-mediated disease. *Eur. J. Immunol.* 28:2487–2497.

124. Tarrant, T.K., P.B. Silver, C.-C. Chan, B. Wiggert, and R.R. Caspi. 1998. Endogenous IL-12 is required for induction and expression of experimental autoimmune uveitis. *J. Immunol.* 161:122–127.

125. Segal, B.M., B.K. Dwyer, and E.M. Shevach. 1998. An interleukin (IL)-10/IL-12 immunoregulatory circuit controls susceptibility to autoimmune disease. *J. Exp. Med.* 187:537–546.

126. Zaccone, P., P. Hutchings, F. Nicoletti, G. Penna, L. Adorini, and A. Cooke. 1999. The involvement of IL-12 in experimentally induced autoimmune thyroid disease. *Eur. J. Immunol.* 29:1933–1942.

127. Trembleau, S., G. Penna, S. Gregori, H.D. Chapman, D.V. Serreze, J. Magram, and L. Adorini. 1999. Pancreas-infiltrating Th1 cells and diabetes develop in IL-12-deficient nonobese diabetic mice. *J. Immunol.* 163:2960–2968.

128. Bettelli, E., M. Prabhu Das, E.D. Howard, H.L. Weiner, R.A. Sobel, and V.K. Kuchroo. 1998. IL-10 is critical in the regulation of autoimmune encephalomyelitis as demonstrated by studies of IL-10- and IL-4-deficient and transgenic mice. *J. Immunol.* 161:3299–3306.

129. Groux, H., A. O'Garra, M. Bigler, M. Rouleau, S. Antonenko, J.E. De Vries, and M.G. Roncarolo. 1997. A CD4+ T-cell subset inhibits antigen-specific T-cell responses and prevents colitis. *Nature* 389:737–742.

130. Trembleau, S., G. Penna, S. Gregori, M.K. Gately, and L. Adorini. 1997. Deviation of pancreas-infiltrating cells to Th2 by interleukin-12 antagonist administration inhibits autoimmune diabetes. *Eur. J. Immunol.* 27:2230–2239.

131. Rapoport, M.J., A. Jaramillo, D. Zipris, A.H. Lazarus, D.V. Serreze, E.H. Leiter, P. Cyopick, J.S. Danska, and T.L. Delovitch. 1993. Interleukin 4 reverses T cell proliferative unresponsiveness and prevents the onset of diabetes in nonobese diabetic mice. *J. Exp. Med.* 178:87–99.

132. Lehuen, A., O. Lantz, L. Beaudoin, V. Laloux, C. Carnaud, A. Bendelac, J.-C. Bach, and R.C. Monteiro. 1998. Overexpression of natural killer T cells protects Vα14-Jα281 transgenic nonobese diabetic mice against diabetes. *J. Exp. Med.* 188:1–9.

133. Zheng, X.X., A.W. Steele, W.W. Hancock, A.C. Stevens, P.W. Nickerson, P.R. Chaudhury, Y. Tian, and T.B. Strom. 1997. A noncytolytic IL-10/Fc fusion protein prevents diabetes, blocks autoimmunity, and promotes suppressor phenomena in NOD mice. *J. Immunol.* 158:4507–4513.

134. Moritani, M., K. Yoshimoto, F. Tashiro, C. Hashimoto, J. Miyazaki, S. Ii, E. Kudo, H. Izahana, Y. Hayashi, T. Sano, and M. Itakura. 1994. Transgenic expression of IL-10 in pancreatic islet A cells accelerates autoimmune insulitis and diabetes in non-obese diabetic mice. *Int. Immunol.* 6:1927–1936.

135. van der Veen, R.C., and S.A. Stohlman. 1993. Encephalitogenic Th1 cells are inhibited by Th2 cells with related peptide specificity: relative roles of interleukine (IL)-4 and IL-10. *J. Neuroimmunol.* 48.213–220.

136. Racke, M.K., A. Bonomo, D.E. Scott, B. Cannella, A. Levine, C.S. Raine, E.M. Shevach, and M. Roecken. 1994. Cytokine-induced immune deviation as a therapy for inflammatory autoimmune disease. *J. Exp. Med.* 180:1961–1966.

137. Lafaille, J.J., F. Van-de-Keere, A.L. Hsu, J.L. Baron, W. Haas, C.S. Raine, and S. Tonegawa. 1997. Myelin basic protein-specific T helper 2 (Th2) cells cause experimental autoimmune encephalomyelitis in immunodeficent hosts rather than protect them from disease. *J. Exp. Med.* 186:307–312.

138. Genain, C.P., K. Abel, N. Belmar, F. Villinger, D.P. Rosenberg, C. linington, C.S. Raine, and S.L. Hauser. 1996. Late complications of immune deviation therapy in a nonhuman primate. *Science* 274:2054–2056.

139. Cameron, M.J., G.A. Arreaza, P. Zucker, S.W. Chensue, R.M. Strieter, S. Chakrabarti, and T.L. Delovitch. 1997. IL-4 prevents insulitis and insulin-dependent diabetes mellitus in nonobese diabetic mice by potentiation of regulatory T helper-2 cell function. *J. Immunol.* 159:4686–4692.

140. Pennline, K.J., E. Roquegaffney, and M. Monahan. 1994. Recombinant human IL-10 prevents the onset of diabetes in the nonobese diabetic mouse. *Clin. Immunol. Immunopathol.* 71:169–175.

141. Lee, M.-S., R. Mueller, L.S. Wicker, L.B. Peterson, and N. Sarvetnick. 1996. IL-10 is necessary and sufficient for autoimmune diabetes in conjunction with NOD MHC homozygosity. *J. Exp. Med.* 183:2663–2668.

142. Fowell, D., and D. Mason. 1993. Evidence that the T cell repertoire of normal rats contains cells with the potential to cause diabetes. Characterization of the CD4$^+$ T cell subset that inhibits this autoimmune potential. *J. Exp. Med.* 177:627–636.

143. Scott, B., R. Liblau, S. Degermann, L.A. Marconi, L. Ogata, A.J. Caton, H.O. McDevitt, and D. Lo. 1994. A role for non-MHC genetic polymorphism in susceptibility to spontaneous autoimmunity. *Immunity* 1:1–20.

144. Pakala, S.V., M.O. Kurrer, and J.D. Katz. 1997. T helper 2 (Th2) T cells induce acute pancreatitis and diabetes in immune-compromised nonobese diabetic (NOD) mice. *J. Exp. Med.* 186:299–306.

145. Mueller, R., T. Krahl, and N. Sarvetnick. 1996. Pancreatic expression of interleukin-4 abrogates insulitis and autoimmune diabetes in nonobese diabetic (NOD) mice. *J. Exp. Med.* 184:1093–1099.

146. Healey, D., P. Ozegbe, S. Arden, P. Chandler, J. Hutton, and A. Cooke. 1995. *In vivo* activity and *in vitro* specificity of CD4$^+$ Th1 and Th2 cells derived from the spleens of diabetic NOD mice. *J. Clin. Invest.* 95:2979–2985.

147. De Carli, M., M. D'Elios, S. Mariotti, C. Marcocci, A. Pinchera, M. Ricci, S. Romagnani, and G.F. Del Prete. 1993. Cytolytic T cells with Th1-like cytokine profile predominate in retroorbital lymphocytic infiltrates of Graves' ophthalmopathy. *J. Clin. Endocrinol. Metab.* 77:1120–1124.

148. Parronchi, P., P. Romagnani, F. Annunziato, S. Sampognaro, A. Becchio, L. Giannarini, E. Maggi, C. Pupilli, F. Tonelli, and S. Romagnani. 1997. Type 1 T-helper cell predominance and interleukin-12 expression in the gut of patients with Crohn's disease. *Am. J. Pathol.* 150:823–832.

149. Brod, S.A., D. Benjamin, and D.A. Hafler. 1991. Restricted T cell expression of IL-2, IFN-γ mRNA in human inflammatory disease. *J. Immunol.* 147:810–815.

150. Windhagen, A., J. Newcombe, F. Dangond, C. Strand, M.N. Woodroofe, M.L. Cuzner, and D.A. Hafler. 1995. Expression of costimulatory molecules B7-1 (CD80) and B7-2 (CD86), and interleukin 12 cytokine in multiple sclerosis lesions. *J. Exp. Med.* 182:1985–1996.

151. Balashov, K.E., D.R. Smith, S.J. Khoury, D.A. Hafler, and H.L. Weiner. 1997. Increased interleukin 12 production in progressive multiple sclerosis: induction by activated CD4$^+$ T cells via CD40 ligand. *Proc. Natl. Acad. Sci. USA* 94:599–603.

152. Comabella, M., K. Balashov, S. Issazadeh, D. Smith, H.L. Weiner, and S.J. Khoury. 1998. Elevated interleukin-12 in progressive multiple sclerosis correlates with disease activity and is normalized by pulse cyclophosphamide therapy. *J. Clin. Invest.* 102:671–678.

153. Foulis, A.K., M. McGill, and M.A. Farquahrson. 1991. Insulitis in type I (insulin-dependent) diabetes mellitus in man. Macrophages, lymphocytes and interferon-γ containing cells. *J. Pathol.* 165:97–103.

154. Miltenburg, A.M., J.M. van Laar, R. de Kuiper, M.R. Daha, and F.C. Breedveld. 1992. T cells cloned from human rheumatoid synovial membrane functionally represent the Th1 subset. *Scand. J. Immunol.* 35:603–610.

155. De Carli, M., M.M. D'Elios, G. Zancuoghi, S. Romagnani, and G. Del Prete. 1994. Human TH1 and TH2 cells: func-tional properties, regulation of development and role in autoimmunity. *Autoimmunity* 18:301–308.

156. Simon, A.K., E. Seipelt, and J. Sieper. 1994. Divergent T-cell cytokine patterns in inflammatory arthritis. *Proc. Natl. Acad. Sci. USA* 91:8562–8566.

157. Roecken, M., and E.M. Shevach. 1996. Immune deviation—the third dimension of nondeletional T cell tolerance. *Immunol. Rev.* 149:175–194.

158. Harrison, L.C. 1992. Islet cell autoantigens in insulin-dependent diabetes: Pandora's box re-visited. *Immunol. Today* 13:348–352.

159. Kaufman, D.L., M. Clare-Salzler, J. Tian, T. Forsthuber, G.S.P. Ting, P. Robinson, M.A. Atkinson, E.E. Sercarz, A.J. Tobin, and P.V. Lehmann. 1993. Spontaneous loss of T-cell tolerance to glutamic acid decarboxylase in murine insulin-dependent diabetes. *Nature* 366:69–72.

160. Tisch, R., X.-D. Yang, S.M. Singer, R.S. Liblau, L. Fugger, and H.O. McDevitt. 1993. Immune response to glutamic acid decarboxylase correlates with insulitis in non-obese diabetic mice. *Nature* 366:72–75.

161. Tian, J., M.A. Atkinson, M. Clare-Salzer, A. Herschenfeld, T. Forsthuber, P.V. Lehmann, and D.L. Kaufman. 1996. Nasal administration of glutamate decarboxylase (GAD65) peptides induces Th2 responses and prevents murine insulin-dependent diabetes. *J. Exp. Med.* 183:1561–1567.

162. Tisch, R., R.S. Liblau, X.D. Yang, P. Liblau, and H.O. McDevitt. 1998. Induction of GAD65-specific regulatory T-cells inhibits ongoing autoimmune diabetes in nonobese diabetic mice. *Diabetes* 47:894–899.

163. Tisch, R., B. Wang, and D.V. Serreze. 1999. Induction of glutamic acid decarboxylase 65-specific Th2 cells and suppression of autoimmune diabetes at late stages of disease is epitope dependent. *J. Immunol.* 163:1178–1187.

164. Acha-Orbea, H., D.J. Mitchell, L. Timmermann, D.C. Wraith, G.S. Tausch, M.K. Waldor, S.S. Zamvil, H.O. Mc Devitt, and L. Steinman. 1988. Limited heterogeneity of T cell receptors in experimental allergic encephalomyelitis. *Cell* 54:263–273.

165. Waisman, A., P.J. Ruiz, D.L. Hirschenberg, A. Gelman, J.R. Oskenberg, S. Brocke, F. Mor, I.R. Cohen, and L. Steinman. 1996. Suppressive vaccination with DNA encoding a variable region gene of the T-cell receptor prevents autoimmune encephalomyelitis and activates Th2 immunity. *Nature Med.* 2:899–905.

166. Mathisen, P.M., and V.K. Tuohy. 1998. Gene therapy in the treatment of autoimmune disease. *Immunol. Today* 19:103–105.

167. Martino, G., R. Furlan, F. Galbiati, P.L. Poliani, A. Bergami, L.M. Grimaldi, L. Adorini, and G. Comi. 1998. A gene therapy approach to treat demyelinating diseases using non-replicative herpetic vectors engineered to produce cytokines. *Mult. Scler.* 4:222–227.

168. Dinarello, C.A. 1998. Interleukin-1, interleukin-1 receptors and interleukin-1 receptor antagonist. *Int. Rev. Immunol.* 16:457–499.

169. Bresnihan, B., J.M. Alvaro-Gracia, M. Cobby, M. Doherty, Z. Domljan, P. Emery, G. Nuki, K. Pavelka, R. Rau, B. Rozman, I. Watt, B. Williams, R. Aitchison, D. McCabe, and P. Musikic. 1998. Treatment of rheumatoid arthritis with recombinant human interleukin-1 receptor antagonist. *Arthritis Rheum.* 41:2196–2204.

170. Evans, C.H., S.C. Ghivizzani, and P.D. Robbins. 1998. Blocking cytokines with genes. *J. Leukoc. Biol.* 64:55–61.

171. Evans, C.H., P.D. Robbins, S.C. Ghivizzani, J.H. Herndon, R. Kang. A.B. Bahnson, J.A. Barranger, E.M. Elders, S. Gay, M.M. Tomaino, M.C. Wasko, S.C. Watkins, T. L. Whiteside, J.C. Glorioso, M.T. Lotze, and T.M. Wright. 1996. Clinical trial to assess the safety, feasibility, and efficacy of transferring a potentially anti-arthritic cytokine gene to human joints with rheumatoid arthritis. *Hum. Gene. Ther.* 7:1261–1280.

172. Furlan, R., G. Martino, F. Galbiati, P.L. Poliani, S. Smiroldo, A. Bergami, G. Desina, G. Comi, R. Flavell, M.S. Su, and L. Adorini. 1999. Caspase-1 regulates the inflammatory process leading to autoimmune demyelination. *J. Immunol.* 163:2403–2409.

173. Nussenblatt, R.B., E. Fortin, R. Schiffman, L. Rizzo, J. Smith, P. Van Veldhuisen, P. Sran, A. Yaffe, C.K. Goldman, T.A. Waldmann, and S.M. Whitcup. 1999. Treatment of noninfectious intermediate and posterior uveitis with the humanized anti-Tac mAb: a phase I/II clinical trial. *Proc. Natl. Acad. Sci. USA* 96:7462–7466.

174. Panitch, H.S., R.L. Hirsch, J. Schindler, and K.P. Johnson. 1987. Treatment of multiple sclerosis with gamma interferon: exacerbations associated with activation of the immune system. *Neurology* 37:1097–1102.

175. Aloisi, F., G. Penna, J. Cerase, B. Menéndez Iglesias, and L. Adorini. 1997. IL-12 production by central nervous system microglia is inhibited by astrocytes. *J. Immunol.* 159:1604–1612.

176. Manetti, R., P. Parronchi, M.G. Giudizi, M.-P. Piccinni, E. Maggi, G. Trinchieri, and S. Romagnani. 1993. Natural killer cell stimulatory factor (interleukin 12, IL-12) induces T helper type 1 (Th1)-specific immune responses and inhibits the development of IL-4-producing Th cells. *J. Exp. Med.* 177:1199–1204.

177. Afonso, L.C.C., T.M. Scharton, L.Q. Vieira, M. Wysocka, G. Trinchieri, and P. Scott. 1994. The adjuvant effect of interleukin-12 in a vaccine against *Leishmania major*. *Science* 263:235–237.

178. Magram, J., S. Connaughton, R. Warrier, D. Carvajal, C. Wu, J. Ferrante, C. Stewart, U. Sarmiento, D. Faherty, and M.K. Gately. 1996. IL-12 deficient mice are defective in IFN-γ production and type 1 cytokine responses. *Immunity* 4:471–482.

179. Wu, C.-Y., J. Ferrante, M.K. Gately, and J. Magram. 1997. Characterization of IL-12 receptor β1 chain (IL-12Rβ1)-deficient mice: IL-12Rβ1 is an essential component of the functional mouse IL-12R. *J. Immunol.* 159:1658–1665.

180. Kaplan, M.H., Y.-L. Sun, T. Hoey, and M.J. Grusby. 1996. Impaired IL-12 responses and enhanced development of Th2 cells in Stat4-deficient mice. *Nature* 382:174–177.

181. Trembleau, S., G. Penna, E. Bosi, A. Mortara, M.K. Gately, and L. Adorini. 1995. IL-12 administration induces Th1 cells and accelerates autoimmune diabetes in NOD mice. *J. Exp. Med.* 181:817–821.

182. Leonard, J.P., K.E. Waldburger, and S.J. Goldman. 1995. Prevention of experimental autoimmune encephalomyelitis by antibodies against interleukin-12. *J. Exp. Med.* 181:381–386.

183. Germann, T., J. Szeliga, H. Hess, S. Stoerkel, F.J. Podlaski, M.K. Gately, E. Schmitt, and E. Ruede. 1995. Administration of IL-12 in combination with type II collagen induces severe arthritis in DBA/1 mice. *Proc. Natl. Acad. Sci. USA* 92:4823–4827.

184. Xu, H., L.V. Rizzo, P.B. Silver, and R.R. Caspi. 1997. Uveitogenicity is associated with a Th1-like lymphokine profile: cytokine-dependent modulation of early and committed effector T cells in experimental autoimmune uveitis. *Cell. Immunol.* 178:69–78.

185. Neurath, M.F., I. Fuss, B.L. Kelsall, E. Stueber, and W. Strober. 1995. Antibodies to interleukin 12 abrogate established experimental colitis in mice. *J. Exp. Med.* 182:1281–1290.

186. Rothe, H., R.M. O'Hara, S. Martin, and H. Kolb. 1997. Suppression of cyclophosphamide induced diabetes development and pancreatic Th1 reactivity in NOD mice treated with the interleukin (IL)-12 antagonist IL-12(p40)$_2$. *Diabetologia* 40:641–646.

187. Constantinescu, C.S., M. Wysocka, B. Hilliard, E. Ventura, E. Lavi, G. Trinchieri, and A. Rostami. 1998. Antibodies against interleukin-12 prevent superantigen-induced and spontaneous relapses of experimental allergic encephalomyelitis. *J. Immunol.* 161:5097–5104.

188. Ferrante, P., M. Fusi, M. Sarasella, D. Caputo, M. Biasin, D. Trabattoni, A. Salvaggio, E. Clerici, J. de Vries, G. Aversa, C. Cazzullo, and M. Clerici. 1998. Cytokine production and surface marker expression in acute and stable multiple sclerosis: altered IL-12 production augmented signaling lymphocytic activation molecule (SLAM)-expressing lymphocytes in acute multiple sclerosis. *J. Immunol.* 160:1514–1521.

189. Rogler, G., and T. Andus. 1998. Cytokines in inflammatory bowel disease. *World J. Surg.* 22:382–389.

190. Verstuyf, A., S. Segaert, L. Verlinden, K. Casteels, R. Bouillon, and C. Mathieu. 1998. Recent developments in the use of vitamin D analogues. *Curr. Opin. Nephrol. Hypertens.* 7:397–403.

191. Bright, J.J., C. Du, M. Coon, S. Sriram, and S.J. Klaus. 1998. Prevention of experimental allergic encephalomyelitis via inhibition of IL-12 signaling and IL-12-mediated Th1 differentiation: an effect of the novel anti-inflammatory drug lisofylline. *J. Immunol.* 161:7015–7022.

192. Keffer, J., L. Probert, H. Cazlaris, S. Georgopulos, E. Kaslaris, D. Kioussis, and G. Kollias. 1991. Transgenic mice expressing human tumor necrosis factor—a predictive genetic model of arthritis. *EMBO J.* 13:4025–4031.

193. Williams, R.O., M. Feldmann, and R.N. Maini. 1992. Anti-tumor necrosis factor ameliorates joint disease in murine collagen-induced arthritis. *Proc. Natal. Acad. Sci. USA* 89:9784–9788.

194. Elliott, M.J., R.N. Maini, M. Feldmann, A. Long-Fox, P. Charles, P. Katsikis, F.M. Brennan, J. Walker, H. Bijl, J. Ghrayeb, and J.N. Woody. 1993. Treatment of rheumatoid arthritis with chimeric monoclonal antibodies to tumor necrosis factor α. *Arthritis Rheum.* 36:1681–1690.

195. D'Haens, G., S. Van Deventer, R. Van Hogezand, D. Chalmers, C. Kothe, F. Baert, T. Braakman, T. Schaible, K. Geboes, and P. Rutgeerts. 1999. Endoscopic and histological healing with infliximab anti-tumor necrosis factor antibodies in Crohn's disease: A European multicenter trial. *Gastroenterology* 116:1029–1034.

196. Maini, R.N., F.C. Breedveld, J.R. Kalden, J.S. Smolen, D. Davis, J.D. Macfarlane, C. Antoni, B. Leeb, M.J. Elliott, J.N. Woody, T.F. Schaible, and M. Feldmann. 1998. Therapeutic efficacy of multiple intravenous infusions of anti-tumor necrosis factor alpha monoclonal antibody combined with low-dose weekly methotrexate in rheumatoid arthritis. *Arthritis Rheum.* 41:1552–1563.

197. Charles, P., M.J. Elliott, D. Davis, A. Potter, J.R. Kalden, C., Antoni, F.C. Breedveld, J.S. Smolen, G. Eberl, K. deWoody, M. Feldmann, and R.N. Maini. 1999. Regulation of cytokines, cytokine inhibitors, and acute-phase proteins

following anti-TNF-alpha therapy in rheumatoid arthritis. *J. Immunol.* 163:1521–1528.

198. Haak-Frendscho, M., S.A. Marsters, J. Mordenti, S. Brady, N.A. Gillett, S.A. Chen, and A. Ashkenazi. 1994. Inhibition of TNF by a TNF receptor immunoadhesin. Comparison to an anti-TNF monoclonal antibody. *J. Immunol.* 152:1347–1353.

199. Baker, D., D. Butler, B.J. Scallon, J.K. O'Neill, J.L. Turk, and M. Feldmann. 1994. Control of established experimental allergic encephalomyelitis by inhibition of tumor necrosis factor (TNF) activity within the central nervous system using monoclonal antibodies and TNF receptor-immunoglobulin fusion proteins. *Eur. J. Immunol.* 24:2040–2048.

200. Moreland, L.W., M.H. Schiff, S.W. Baumgartner, E.A. Tindall, R.M. Fleischmann, K.J. Bulpitt, A.L. Weaver, E.C. Keystone, D.E. Furst, P.J. Mease, E.M. Ruderman, D.A. Horwitz, D.G. Arkfeld, L. Garrison, D.J. Burge, C.M. Blosch, M.L. Lange, N. D. McDonnell, and M.E. Weinblatt. 1999. Etanercept therapy in rheumatoid arthritis. A randomized, controlled trial. *Ann. Intern. Med.* 130:478–486.

201. Rabinovitch, A., W.L. Suarez-Pinzon, O. Sorensen, R.V. Rajotte, and R.F. Power. 1997. TNF-alpha down-regulates type 1 cytokines and prolongs survival of syngeneic islet grafts in nonobese diabetic mice. *J. Immunol.* 159:6298–6303.

202. Selmaj, K., and C.S. Raine. 1988. Tumor necrosis factor mediates myelin damage in organotypic cultures of nervous tissue. *Ann. NY Acad. Sci.* 540:568–570.

203. Kuroda, Y., and Y. Shimamoto. 1991. Human tumor necrosis factor-alpha arguments experimental allergic encephalomyelitis in rats. *J. Neuroimmunol.* 34:159–164.

204. Liu, J., M.W. Marino, G. Wong, D. Grail, A. Dunn, J. Bettadapura, A.J. Slavin, L. Old, and C.C.A. Bernard. 1998. TNF is a potent anti-inflammatory cytokine in autoimmune-mediated demyelination. *Nature Med.* 4:78–83.

205. van Oosten, B.W., F. Barkhof, L. Truyen, J.B. Boringa, F.W. Bertelsmann, B.M. von Blomberg, J.N. Woody, H.P. Hartung, and C.H. Polman. 1996. Increased MRI activity and immune activation in two multiple sclerosis patients treated with the monoclonal anti-tumor necrosis factor antibody cA2. *Neurology* 47:1531–1534.

206. Shehadeh, N., F. Calcinaro, B.J. Bradley, I. Bruchlim, P. Vardi, and K.J. Lafferty. 1994. Effect of adjuvant therapy on development of diabetes in mouse and man. *Lancet* 343:706–707.

207. Rabinovitch, A., W.L. Suarez-Pinzon, O. Sorensen, R.C. Bleackley, and R.F. Power. 1995. IFN-γ gene expression in pancreatic islet-infiltrating mononuclear cells correlates with autoimmune diabetes in nonobese diabetic mice. *J. Immunol.* 154:4874–4882.

208. Allen, J.B., H.L. Wong, G.L. Costa, M.J. Bienkowski, and S.M. Wahl. 1993. Suppression of monocyte function and differential regulation of IL-1 and IL-1ra by IL-4 contribute to resolution of experimental arthritis. *J. Immunol.* 151:4344–4351.

209. Furlan, R., P.L. Poliani, F. Galbiati, A. Bergami, L. Grimaldi, G. Comi, L. Adorini, and G. Martino. 1998. Central nervous system delivery of interleukin 4 by a non-replicative herpes simplex type 1 viral vector ameliorates autoimmune demyelination. *Hum. Gene Ther.* 9:2605–2617.

210. Piccirillo, C.A., and G.J. Prud'homme. 1999. Prevention of experimental allergic encephalomyelitis by intramuscular gene transfer with cytokine-encoding plasmid vectors [In Process Citation]. *Hum. Gene Ther.* 10:1915–1922.

211. Moore, K., A. O'Garra, R. de Waal Malefyt, P. Vieira, and T.R. Mosmann. 1993. Interleukin-10. *Annu. Rev. Immunol.* 11:165–190.

212. D'Andrea, A., M. Aste-Amezaga, N.M. Valiante, X. Ma, M. Kubin, and G. Trinchieri. 1993. Interleukin 10 (IL-10) inhibits human lymphocyte interferon-γ production by suppressing natural killer cell stimulatory factor/IL-12 synthesis in accessory cells. *J. Exp. Med.* 178:1041–1048.

213. Fiorentino, D.F., A. Zlotnik, T.R. Mosmann, M.H. Howard, and A. O'Garra. 1991. IL-10 inhibits cytokine production by activated macrophages. *J. Immunol.* 147:3815–3622.

214. de Waal Malefyt, R., J. Abrams, B. Bennett, C. Figdor, and J. de Vries. 1991. IL-10 inhibits cytokine synthesis by human monocytes: an auto-regulatory role of IL-10 produced by monocytes. *J. Exp. Med.* 174:1209–1220.

215. Bogdan, C., Y. Vodovotz, and C. Nathan. 1991. Macrophage deactivation by interleukin 10. *J. Exp. Med.* 174:1549–1555.

216. Groux, H., M. Bigler, J.E. De Vries, and M.G. Roncarolo. 1996. Interleukin-10 induced a long-term antigen specific anergic state in human CD4+ T cells. *J. Exp. Med.* 184:19–29.

217. Bromberg, J.S. 1995. IL-10 immunosuppression in transplantation. *Curr. Opin. Immunol.* 7:639–643.

218. Kuehn, R., J. Loehler, D. Rennick, K. Rajewsky, and W. Mueller. 1993. Interleukin-10-deficient mice develop chronic enterocolitis. *Cell* 75:263–274.

219. Powrie, F., M.W. Leach, S. Mauze, S. Menon, L.B. Caddle, and R.L. Coffman. 1994. Inhibition of Th1 responses prevents inflammatory bowel disease in *scid* mice reconstituted with CD45RB^hi CD4+ T cells. *Immunity* 1:553–562.

220. Inaba, K., M. Kitaura, T. Kato, Y. Watanabe, Y. Kawade, and S. Muramatsu. 1986. Contrasting effect of alpha/beta- and gamma-interferons on expression of macrophage Ia antigens. *J. Exp. Med.* 163:1030–1035.

221. Panitch, H.S., R.L. Hirsch, A.S. Haley, and K.P. Johnson. 1987. Exacerbations of multiple sclerosis in patients treated with gamma interferon. *Lancet* 1:893–895.

222. Arnason, B.G. 1999. Immunologic therapy of multiple sclerosis. *Annu. Rev. Med.* 50:291–302.

223. Jacobs, L.D., D.L. Cookfair, R.A. Rudick, R.M. Herndon, J.R. Richert, A.M. Salazar, J.S. Fischer, D.E. Goodkin, C.V. Granger, J.H. Simon et al. 1995. A phase III trial of intramuscular recombinant interferon beta as treatment for exacerbating-remitting multiple sclerosis: design and conduct of study and baseline characteristics of patients. Multiple Sclerosis Collaborative Research Group (MSCRG). *Mult. Scler.* 1:118–135.

224. Rep, M.H., H.M. Schrijver, T. van Lopik, R.Q. Hintzen, M.T. Roos, H.J. Ader, C. H. Polman, and R.A. van Lier. 1999. Interferon (IFN)-beta treatment enhances CD95 and interleukin 10 expression but reduces interferon-gamma producing T cells in MS patients. *J. Neuroimmunol.* 96:92–100.

225. Tak, P.P., B.A. t Hart, M.C. Kraan, M. Jonker, T.J. Smeets, and F.C. Breedveld. 1999. The effects of interferon beta treatment on arthritis. *Rheumatology (Oxford)* 38:362–369.

226. Letterio, J.J., and A.B. Roberts. 1998. Regulation of immune responses by TGF-beta. *Annu. Rev. Immunol.* 16:137–161.

227. Schluesener, H.J., and O. Lider. 1989. Transforming growth factors beta 1 and beta 2: cytokines with identical immunosuppressive effects and a potential role in the regulation of autoimmune T cell function. *J. Neuroimmunol.* 24:249–258.

228. Johns, L.D., and S. Sriram. 1993. Experimental allergic encephalomyelitis: neutralizing antibody to TGF beta 1 enhances the clinical severity of the disease. *J. Neuroimmunol.* 47:1–7.

229. McCartney-Francis, N.L., M., Frazier-Jessen, and S.M. Wahl. 1998. TGF-beta: a balancing act. *Int. Rev. Immunol.* 16:553–580.

230. Calabresi, P.A., N.S. Fields, H.W. Maloni, A. Hanham, J. Carlino, J. Moore, M.C. Levin, S. Dhib-Jalbut, L.R. Tranquill, H. Austin, H.F. McFarland, and M.K. Racke. 1998. Phase 1 trial of transforming growth factor beta 2 in chronic progressive MS. *Neurology* 51:289–292.

231. Ben-Nun, A., H. Wekerle, and I.R. Cohen. 1981. Vaccination against autoimmune encephalomyelitis with T-lymphocyte line cells reactive against myelin basic protein. *Nature* 292:60–61.

232. Cohen, I.R. 1986. Regulation of autoimmune disease: physiological and therapeutic. *Immunol. Rev.* 94:5–21.

233. Lider, O., T. Reshef, E. Beraud, A. Ben-Nun, and I.R. Cohen. 1989. Anti-idiotypic network induced by T cell vaccination against EAE. *Science* 239:181–183.

234. Zhang, J., R. Medaer, P. Stinissen, D. Hafler, and J. Raus. 1993. MHC-restricted depletion of human myelin basic protein-reactive T cells by T cell vaccination. *Science* 261:1451–1454.

235. Medaer, R., P. Stinissen, L. Truyen, J. Raus, and J. Zhang. 1995. Depletion of myelin-basic-protein autoreactive T cells by T-cell vaccination: pilot trial in multiple sclerosis. *Lancet* 346:807–808.

236. Zhang, J., C. Vandevyver, P. Stinissen, and J. Raus. 1995. *In vivo* clonotypic regulation of human myelin basic protein-reactive T cells by T cell vaccination. *J. Immunol.* 155:5868–5877.

237. Van Laar, J.M., A.M.M. Miltenburg, M.J.A. Verdonk, A. Leow, B.G. Elferink, M. R. Daha, I.R. Cohen, R R P de Vries, and F.C. Breedveld. 1993. Effects of inoculation with attenuated autologous T cells in patients with rheumatoid arthritis. *J. Autoimmun.* 6:159–167.

238. Chunduru, S.K., R.M. Sutherland, G.A. Stewart, R.W. Doms, and Y. Paterson. 1996. Exploitation of the Vbeta8.2 T cell receptor in protection against experimental autoimmune encephalomyelitis using a live vaccinia virus vector. *J. Immunol.* 156:4940–4945.

239. Howell, M.D., S.T. Winters, T. Olee, H.C. Powell, D.J. Carlo, and S.W. Brostoff. 1989. Vaccination against experimental allergic encephalmyelitis with T cell receptor peptides. *Science* 246:668–670.

240. Vandenbark, A.A., G. Hashim, and H. Offner. 1989. Immunization with a synthetic T-cell receptor V-region peptide protects against experimental autoimmune encephalomyelitis. *Nature* 341:541–544.

241. Gaur, A., R. Haspel, J.P. Mayer, and C.G. Fathman. 1993. Requirement for CD8+ cells in T cell receptor peptide-induced clonal unresponsiveness. *Science* 259:91–94.

242. Offner, H., G.A. Hashim, and A.A. Vandenbark. 1991. T cell receptor peptide therapy triggers autoregulation of experimental encephalomyelitis. *Science* 251:430–432.

243. Kumar, V., and E. Sercarz. 1998. Induction or protection from experimental autoimmune encephalomyelitis depends on the cytokine secretion profile of TCR peptide-specific regulatory CD4 T cells. *J. Immunol.* 161:6585–6591.

244. Kotzin, B.L., S. Karuturi, Y.K. Chou, J. Lafferty, J.M. Forrester, M. Better, G.E. Nedwin, H. Offner, and A.A. Vandenbark. 1991. Preferential T cell receptor Vβ gene usage by myelin basic protein-specific T cell clones from patients with multiple sclerosis. *Proc. Natl. Acad. Sci. USA* 88:9161–9165.

245. Satyanarayana, K., Y.K. Chou, D. Bourdette, R. Whitham, G.A. Hashim, H. Offner, and A.A. Vandenbark. 1993. Epitope specificity and V gene expression of cerebrospinal fluid T cells specific for intact versus cryptic epitopes of myelin basic protein. *J. Neuroimmunol.* 44:57–68.

246. Chou, Y.D., A.C. Buenafe, R. Dedrick, W.J. Morrison, D.N. Bourdette, R. Whitham, J. Atherton, J. Lane, E. Spoor, G.A. Hashim, H. Offner, and A.A. Vandenbark. 1994. T cell receptor Vβ gene usage in the recognition of myelin basic protein by cerebrospinal fluid-and blood-derived T cells from patients with multiple sclerosis. *J. Neurosci. Res.* 37:169–181.

247. Bourdette, D.N., R.H. Whitham, Y.K. Chou, W.J. Morrison, J. Atherton, C. Kenny, D. Liefeld, G.A. Hashim, H. Offner, and A.A. Vandenbark. 1994. Immunity to TCR peptides in multiple sclerosis. I. Successful immunization of patients with synthetic V beta 5.2 and V beta 6.1 CDR2 peptides [published erratum appears in *J. Immunol.* 1994 Jul 15;153(2):910]. *J. Immunol.* 152:2510–2519.

248. Vandenbark, A.A., Y.K. Chou, R. Whitham, M. Mass, A. Buenafe, D. Liefeld, D. Kavanagh, S. Cooper, G.A. Hashim, and H. Offner. 1996. Treatment of multiple sclerosis with T-cell receptor peptides: results of a double-blind pilot trial. *Nat. Med.* 2:1109–1115.

249. Hafler, D.A., M.G. Saadeh, V.K. Kuchroo, E. Milford, and L. Steinman. 1996. TCR usage in human and experimental demyelinating disease. *Immunol Today* 17:152–159.

250. Valli, A., A. Sette, L. Kappos, C. Oseroff, J. Sidney, G. Miescher, M. Hochberger, E. D. Albert, and L. Adorini. 1993. Binding of myelin basic protein peptides to human histocompatibility leukocyte antigen class II molecules and their recognition by T cells from multiple sclerosis patients. *J. Clin. Invest.* 91:616–628.

251. Ben-Nun, A., R.S. Liblau, L. Cohen, D. Lehmann, E. Tournier-Lasserve, A. Rosenzweig, Z. Jingwu, J.C.M. Raus, and M.A. Bach. 1991. Restricted T-cell receptor Vβ usage by myelin basic protein-specific T cell clones in multiple sclerosis: Predominant genes vary in individuals. *Proc. Natl. Acad. Sci. USA* 88:2466–2470.

252. Moreland, L.W., E.E. Morgan, T.C. Adamson, 3rd, Z. Fronek, L.H. Calabrese, J.M. Cash, J.A. Markenson, A.K. Matsumoto, J. Bathon, E.L. Matteson, K.M. Uramoto, C.M. Weyand, W.J. Koopman, L.W. Heck, V. Strand, J.P. Diveley, D.J. Carlo, C.J. Nardo, S.P. Richieri, and S.W. Brostoff. 1998. T cell receptor peptide vaccination in rheumatoid arthritis: a placebo-controlled trial using a combination of Vbeta3, Vbeta14, and Vbeta17 peptides. *Arthritis Rheum.* 41:1919–1929.

253. Springer, T.A. 1994. Traffic signals for lymphocyte recirculation and leukocyte emigration: the multistep paradigm. *Cell* 76:301–314.

254. McMurray, R.W. 1996. Adhesion molecules in autoimmune disease. *Semin. Arthritis Rheum.* 25:215–233.

255. Feldmann, M., M.J. Elliott, J.N. Woody, and R.N. Maini. 1997. Anti-tumor necrosis factor-alpha therapy of rheumatoid arthritis. *Adv. Immunol.* 64:283–350.

256. Gonzalez-Amaro, R., F. Diaz-Gonzalez, and F. Sanchez-Madrid. 1998. Adhesion molecules in inflammatory diseases. *Drugs* 56:977–988.

257. Baggiolini, M. 1998. Chemokines and leukocyte traffic. *Nature* 392:565–568.

258. Wells, T.N., C.A. Power, and A.E. Proudfoot. 1998. Definition, function and pathophysiological significance of chemokine receptors. *Trends Pharmacol. Sci.* 19:376–380.

259. Plater-Zyberk, C., A.J. Hoogewerf, A.E. Proudfoot, C.A. Power, and T.N. Wells. 1997. Effect of a CC chemokine receptor antagonist on collagen induced arthritis in DBA/1 mice. *Immunol. Lett.* 57:117–120.

260. Lloyd, C.M., A.W. Minto, M.E. Dorf, A. Proudfoot, T.N. Wells, D.J. Salant, and J. C. Gutierrez-Ramos. 1997. RANTES and monocyte chemoattractant protein-1 (MCP-1) play an important role in the inflammatory phase of crescentic nephritis, but only MCP-1 is involved in crescent formation and interstitial fibrosis. *J. Exp. Med.* 185:1371–1380.

261. Sorensen, T.L., M. Tani, J. Jensen, V. Pierce, C. Lucchinetti, V.A. Folcik, S. Qin, J. Rottman, F. Sellebjerg, R.M. Strieter, J.L. Frederiksen, and R.M. Ransohoff. 1999. Expression of specific chemokines and chemokine receptors in the central nervous system of multiple sclerosis patients. *J. Clin. Invest.* 103:807–815.

262. Bonecchi, R., G. Bianchi, P.P. Bordignon, D. D'Ambrosio, R. Lang, A. Borsatti, S. Sozzani, P. Allavena, P.A. Gray, A. Mantovani, and F. Sinigaglia. 1998. Differential expression of chemokine receptors and chemotactic responsiveness of type 1 T helper cells (Th1s) and Th2s. *J. Exp. Med.* 187:129–134.

263. Suzuki, N., A. Nakajima, S. Yoshino, K. Matsushima, H. Yagita, and K. Okumura. 1999. Selective accumulation of CCR5+ T lymphocytes into inflamed joints of rheumatoid arthritis. *Int. Immunol.* 11:553–559.

INDEX